# The Princeton Companion to Applied Mathematics

# The Princeton Companion to Applied Mathematics

EDITOR

**Nicholas J. Higham**

*The University of Manchester*

ASSOCIATE EDITORS

**Mark R. Dennis**

*University of Bristol*

**Paul Glendinning**

*The University of Manchester*

**Paul A. Martin**

*Colorado School of Mines*

**Fadil Santosa**

*University of Minnesota*

**Jared Tanner**

*University of Oxford*

**Princeton University Press**

**Princeton and Oxford**

Published by Princeton University Press,
41 William Street, Princeton, New Jersey 08540

In the United Kingdom: Princeton University Press,
6 Oxford Street, Woodstock, Oxfordshire OX20 1TW

press.princeton.edu

Jacket image courtesy of iStock



*Library of Congress Cataloging-in-Publication Data*

The Princeton companion to applied mathematics / editor,
Nicholas J. Higham, The University of Manchester ;
associate editors, Mark R. Dennis, University of Bristol
[and four others].
pages cm
Includes bibliographical references and index.
ISBN 978-0-691-15039-0 (hardcover : alk. paper)
1. Algebra. 2. Mathematics. 3. Mathematical models.
I. Higham, Nicholas J., 1961- editor. II. Dennis, Mark R.,
editor. III. Title: Companion to applied mathematics.
IV. Title: Applied mathematics.
QA155.P75 2015
510—dc23 2015013024

*British Library Cataloging-in-Publication Data is available*

This book has been composed in LucidaBright

Project management, composition and copyediting
by T&T Productions Ltd, London

Printed on acid-free paper ∞
Printed in the United States of America

1 2 3 4 5 6 7 8 9 10

# Contents

## Part IV Areas of Applied Mathematics



## Part V Modeling



## Part VI Example Problems

## Part VII Application Areas



## Part VIII Final Perspectives

# Preface

## 1 What Is *The Companion*?

*The Princeton Companion to Applied Mathematics* describes what applied mathematics is about, why it is important, its connections with other disciplines, and some of the main areas of current research. It also explains what applied mathematicians do, which includes not only studying the subject itself but also writing about mathematics, teaching it, and influencing policy makers.

*The Companion* differs from an encyclopedia in that it is not an exhaustive treatment of the subject, and it differs from a handbook in that it does not cover all relevant methods and techniques. Instead, the aim is to offer a broad but selective coverage that conveys the excitement of modern applied mathematics while also giving an appreciation of its history and the outstanding challenges. *The Companion* focuses on topics felt by the editors to be of enduring interest, and so it should remain relevant for many years to come.

With online sources of information about mathematics growing ever more extensive, one might ask what role a printed volume such as this has. Certainly, one can use Google to search for almost any topic in the book and find relevant material, perhaps on Wikipedia. What distinguishes *The Companion* is that it is a self-contained, structured reference work giving a consistent treatment of the subject. The content has been curated by an editorial board of applied mathematicians with a wide range of interests and experience, the articles have been written by leading experts and have been rigorously edited and copyedited, and the whole volume is thoroughly cross-referenced and indexed.

Within each article, the authors and editors have tried hard to convey the motivation for each topic or concept and the basic ideas behind it, while avoiding unnecessary detail. It is hoped that *The Companion* will be seen as a friendly and inspiring reference, containing both standard material and more unusual, novel, or unexpected topics.

## 2 Scope

It is difficult to give a precise definition of applied mathematics, as discussed in WHAT IS APPLIED MATHEMATICS? [I.1] and, from a historical perspective, in THE HISTORY OF APPLIED MATHEMATICS [I.6]. *The Companion* treats applied mathematics in a broad sense, and it cannot cover all aspects in equal depth. Some parts of mathematical physics are included, though a full treatment of modern fundamental theories is not given. Statistics and probability are not explicitly included, although a number of articles make use of ideas from these subjects, and in particular the burgeoning area of UNCERTAINTY QUANTIFICATION [II.34] brings together many ideas from applied mathematics and statistics. Applied mathematics increasingly makes use of algorithms and computation, and a number of aspects at the interface with computer science are included. Some parts of discrete and combinatorial mathematics are also covered.

## 3 Audience

The target audience for *The Companion* is mathematicians at undergraduate level or above; students, researchers, and professionals in other subjects who use mathematics; and mathematically interested lay readers. Some articles will also be accessible to students studying mathematics at pre-university level.

Prospective research students might use the book to obtain some idea of the different areas of applied mathematics that they could work in. Researchers who regularly attend seminars in areas outside their own specialities should find that the articles provide a gentle introduction to some of these areas, making good pre- or post-seminar reading.

In soliciting and editing the articles the editors aimed to maximize accessibility by keeping discussions at the lowest practical level. A good question is how much of the book a reader should expect to understand. Of course "understanding" is an imprecisely defined concept. It is one thing to read along with an argument and find it plausible, or even convincing, but another to reproduce it on a blank piece of paper, as every undergraduate discovers at exam time. The very wide range of topics covered means that it would take a reader with an unusually broad knowledge to understand everything, but every reader from undergraduate level upward should find a substantial portion of the book accessible.

## 4 Organization

*The Companion* is organized in eight parts, which are designed to cut across applied mathematics in different ways.

Part I, "Introduction to Applied Mathematics," begins by discussing what applied mathematics is and giving examples of the use of applied mathematics in everyday life. THE LANGUAGE OF APPLIED MATHEMATICS [I.2] then presents basic definitions, notation, and concepts that are needed frequently in other parts of the book, essentially giving a brief overview of some key parts of undergraduate mathematics. This article is not meant to be a complete survey, and many later articles provide other introductory material themselves. METHODS OF SOLUTION [I.3] describes some general solution techniques used in applied mathematics. ALGORITHMS [I.4] explains the concept of an algorithm, giving some important examples and discussing complexity issues. The presence of this article in part I reflects the increasing importance of algorithms in all areas of applied mathematics. GOALS OF APPLIED MATHEMATICAL RESEARCH [I.5] describes the kinds of questions and issues that research in applied mathematics addresses and discusses some strategic aspects of carrying out research. Finally, THE HISTORY OF APPLIED MATHEMATICS [I.6] describes the history of the subject from ancient times up until the late twentieth century.

Part II, "Concepts," comprises short articles that explain specific concepts and their significance. These are mainly concepts that cut across different models and areas and provide connections to other parts of the book. This part is not meant to be comprehensive, and many other concepts are well described in later articles (and discoverable via the index).

Part III, "Equations, Laws, and Functions of Applied Mathematics," treats important examples of what its title describes. The choice of what to include was based on a mix of importance, accessibility, and interest. Many equations, laws, and functions not contained in this part are included in other articles.

Part IV, "Areas of Applied Mathematics," contains longer articles giving an overview of the whole subject and how it is organized, arranged by research area. The aim of this part is to convey the breadth, depth, and diversity of applied mathematics research. The coverage is not comprehensive, but areas that do not appear as or in article titles may nevertheless be present in other articles. For example, there is no article on geoscience, yet EARTH SYSTEM DYNAMICS [IV.30], INVERSE PROBLEMS [IV.15], and IMAGING THE EARTH USING GREEN'S THEOREM [VII.16] all cover specific aspects of this area. Nor is there a part IV article on numerical analysis, but this area is represented by APPROXIMATION THEORY [IV.9], NUMERICAL LINEAR ALGEBRA AND MATRIX ANALYSIS [IV.10], CONTINUOUS OPTIMIZATION (NONLINEAR AND LINEAR PROGRAMMING) [IV.11], NUMERICAL SOLUTION OF ORDINARY DIFFERENTIAL EQUATIONS [IV.12], and NUMERICAL SOLUTION OF PARTIAL DIFFERENTIAL EQUATIONS [IV.13].

Part V, "Modeling," gives a selection of mathematical models, explaining how the models are derived and how they are solved.

Part VI, "Example Problems," contains short articles covering a variety of interesting applied mathematics problems.

Part VII, "Application Areas," comprises articles on connections between applied mathematics and other disciplines, including such diverse topics as integrated circuit (chip) design, medical imaging, and the screening of luggage in airports.

Part VIII, "Final Perspectives," contains essays on broader aspects, including reading, writing, and typesetting mathematics; teaching applied mathematics; and how to influence government as a mathematician.

The articles within a given part vary significantly in length. This should not be taken as an indication of the importance of the corresponding topic, as it is partly due to the number of pages that could be allocated to each article, as well as to how authors responded to their given page limit.

The ordering of articles within a part is alphabetical for parts II and III. For part IV some attempt was made to place related articles together and to place one article before another if there is a natural order in which to read the two articles. The ordering is nevertheless somewhat arbitrary, and the reader should feel free to read the articles in any order. The articles within parts V–VIII are arranged only loosely by theme.

The editors made an effort to encourage illustrations and diagrams. Due to the cost of reproduction, color has been used only where necessary and is restricted to the color plates following page 364.

Despite the careful organization, the editors expect that many readers will flick through the book to find something interesting, start reading, and by following cross-references navigate the book in an unpredictable fashion. This approach is perfectly reasonable, as is starting from a page found via the table of contents or the index. Whatever the reading strategy, we hope the book will be hard to put down!

## 5 Relation to *The Princeton Companion to Mathematics*

This *Companion* follows the highly successful *Princeton Companion to Mathematics* (PCM), edited by Gowers, Barrow-Green, and Leader (2008), which focuses on modern pure mathematics. We have tried to build on the PCM and avoid overlap with it. Thus we do not cover many of the basic mathematical concepts treated in parts I and II of the PCM, but rather assume the reader is familiar with them.

Some crucial concepts that are already in the PCM are included here as well, but they are approached from a rather different viewpoint, typically with more discussion of applications and computational aspects.

Some articles in the PCM, listed in table 1, could equally well have appeared here, and the editors therefore made a decision not to solicit articles on these same general topics. However, particular aspects of several of these topics are included.

**Table 1** Relevant articles from *The Princeton Companion to Mathematics* whose topic is not duplicated here.

| Title | PCM article number |
|---|---|
| Mathematics and Chemistry | VII.1 |
| Mathematical Biology | VII.2 |
| Wavelets and Applications | VII.3 |
| The Mathematics of Traffic in Networks | VII.4 |
| Mathematics and Cryptography | VII.7 |
| Mathematical Statistics | VII.10 |
| Mathematics and Medical Statistics | VII.11 |
| Mathematics and Music | VII.13 |

## 6 How to Use This Book

Authors were asked to make their articles as self-contained as possible and to define specific notation and technical terms. You should familiarize yourself with the material in THE LANGUAGE OF APPLIED MATHEMATICS [I.2], as this background is assumed for many of the later articles. If you are unsure about notation consult table 3 in I.2 or, for a definition, see if it is in the index. The editors, with the help of a professional indexer, have tried very hard to produce a thorough and usable index, so there is a good chance that the index will lead you to a place in the book where a particular piece of notation or a definition is clarified.

The extensive cross-references provide links between articles. A phrase such as "this vector can be computed by the FAST FOURIER TRANSFORM [II.10]" indicates that article 10 in part II contains more information on the fast Fourier transform.

In the research literature it is normal to support statements by citations to items in a bibliography. It is a style decision of *The Companion* not to include citations in the articles. The articles present core knowledge that is generally accepted in the field, and omitting citations makes for a smoother reading experience as the reader does not constantly have to refer to a list of references a few pages away. Many authors found it quite difficult to adopt this style, being so used to liberally sprinkling `\cite` commands through their LaTeX documents! Most articles have a further reading section, which provides a small number of sources that provide an entrée into the literature on that topic.

## 7 *The Companion* Project

I was invited to lead the project in late 2009. After the editorial board was assembled and the format of the book and the outline of its contents were agreed at a meeting of the board in Manchester, invitations to authors began to go out in 2011. We aimed to invite authors who are both leaders in their field and excellent writers. We were delighted with the high acceptance rate.

Ludwig Boltzmann was a contributor to the six-volume *Encyclopedia of Mathematical Sciences* (1898–1933) edited by Felix Klein, Wilhelm Meyer, and others. In 1905, he wrote, apropos of the selected author of an article:

> He must first be persuaded to promise a contribution; then, he must be instructed and pressed with all means of persuasion to write a contribution which fits into the general framework; and last, but not least, he must be urged to fulfill his promise in a timely matter.

Over a hundred years later these comments remain true, and one of the reasons for the long gestation period of a book such as this is that it does take a long time to sign up authors and collect articles. Occasionally, we were unable to find an author willing and able to deliver an article on a particular topic, so a small number of topics that we would have liked to include had to be omitted.

Of the 165 authors, at least two had babies during the course of preparing their articles. Sadly, one author, David Broomhead, did not live to see the project completed.

If the project has gone well, one of the reasons is the thoroughly professional and ever-cheerful support provided by Sam Clark of T&T Productions Ltd in London. Sam acted as project manager, copy editor, and typesetter, and made the process as smooth and painless for the editors and contributors as it could be. Sam played the same role for the *The Princeton Companion to Mathematics*, and his experience of that project was invaluable.

## 8 Acknowledgments

I am grateful to the members of the editorial board for their advice and their work on the project, and for traveling to three editorial board meetings, in Manchester, Minneapolis, and London.

A number of people beyond the editorial board have given me valuable comments on my own draft articles, especially those in part I, at different stages of the project. I thank, in particular, Doug Arnold, Des Higham, David Nelson, Nick Trefethen, and my Manchester colleagues Mary Aprahamian, Edvin Deadman, Stefan Güttel, Sven Hammarling, Oliver Jensen, Ramaseshan Kannan, Lijing Lin, Martin Lotz, Jennifer Pestana, Sam Relton, Vedran Šego, Nataša Strabić, Leo Tasleman, and Françoise Tisseur. Nick Trefethen also provided many valuable comments on a late version of the whole book.

It has been a pleasure to work with Sam Clark, with our level of cooperation being demonstrated by the 3000-plus emails that have traveled between us. I was constantly impressed not only by Sam's ability to find improvements in authors' submitted text during the copyediting phase but also by his ability to turn author-supplied figures of a widely varying standard into high-quality figures with a consistent style.

The original Princeton University Press editor for the project was Anne Savarese, who had also worked on *The Princeton Companion to Mathematics*. Anne's experience and advice was invaluable in formulating the plans for *The Companion*, and she shepherded the project through to the late stages with skill and patience. Vickie Kearn took over from Anne for the last eighteen months of the project, after Anne moved to another editorial role at the Press. A smooth transition was aided by Vickie's extensive knowledge of the world of mathematics publishing and by my having known her since the early 1990s when, as an editor at the Society for Industrial and Applied Mathematics (SIAM), she published my first book.

I would also like to thank Terri O'Prey, who oversaw the production at the Press, and indexer Julie Shawvan.

Finally, I thank my wife Françoise Tisseur and my sons Freddie and Thomas for uncomplainingly allowing me to spend many evenings and weekends working on *The Companion*.

*Nicholas J. Higham*

# Contributors

**David Acheson**, *Emeritus Fellow, Jesus College, University of Oxford*
INVERTED PENDULUMS [VI.4], TEACHING APPLIED MATHEMATICS [VIII.7]

**Miguel A. Alonso**, *Associate Professor, The Institute of Optics at the University of Rochester*
MODERN OPTICS [V.14]

**Douglas N. Arnold**, *McKnight Presidential Professor of Mathematics, University of Minnesota*
THE FLIGHT OF A GOLF BALL [VI.6]

**Karl Johan Åström**, *Emeritus Professor, Department of Automatic Control, Lund Institute of Technology/University of Lund*
CONTROL THEORY [IV.34]

**David H. Bailey**, *Lawrence Berkeley National Laboratory (retired); Research Fellow, University of California, Davis*
EXPERIMENTAL APPLIED MATHEMATICS [VIII.6]

**June Barrow-Green**, *Senior Lecturer in the History of Mathematics, The Open University*
THE HISTORY OF APPLIED MATHEMATICS [I.6]

**Peter Benner**, *Director, Max Planck Institute for Dynamics of Complex Technical Systems*
MODEL REDUCTION [II.26]

**Andrew J. Bernoff**, *Kenneth and Diana Jonsson Professor of Mathematics, Harvey Mudd College*
THE THIN-FILM EQUATION [III.29]

**Michael V. Berry**, *Melville Wills Professor of Physics (Emeritus), University of Bristol*
DIVERGENT SERIES: TAMING THE TAILS [V.8]

**Michael W. Berry**, *Professor, Department of Electrical Engineering and Computer Science, University of Tennessee*
TEXT MINING [VII.24]

**Brett Borden**, *Professor of Physics, The Naval Postgraduate School, Monterey, California*
RADAR IMAGING [VII.17]

**Jeffrey T. Borggaard**, *Professor of Mathematics, Virginia Tech*
OPTIMAL SENSOR LOCATION IN THE CONTROL OF ENERGY-EFFICIENT BUILDINGS [VI.13]

**Jonathan M. Borwein**, *Laureate Professor, School of Mathematical and Physical Sciences, University of Newcastle, Australia*
EXPERIMENTAL APPLIED MATHEMATICS [VIII.6]

**Fred Brauer**, *Professor Emeritus of Mathematics, University of Wisconsin-Madison*
THE SPREAD OF INFECTIOUS DISEASES [V.16]

**Thomas J. Brennan**, *Professor of Law, Harvard Law School*
PORTFOLIO THEORY [V.10]

**David S. Broomhead**, *Professor of Applied Mathematics, The University of Manchester (deceased)*
APPLICATIONS OF MAX-PLUS ALGEBRA [VII.4]

**Kurt Bryan**, *Professor of Mathematics, Rose-Hulam Institute of Technology*
CLOAKING [VI.1]

**Dorothy Buck**, *Reader in BioMathematics, Imperial College London*
KNOTTING AND LINKING OF MACROMOLECULES [VI.8]

**Chris Budd**, *Professor of Applied Mathematics, University of Bath; Professor of Mathematics, Royal Institution of Great Britain*
SLIPPING, SLIDING, RATTLING, AND IMPACT: NONSMOOTH DYNAMICS AND ITS APPLICATIONS [VI.15]

**John A. Burns**, *Hatcher Professor of Mathematics and Technical Director for the Interdisciplinary Center for Applied Mathematics, Virginia Tech*
OPTIMAL SENSOR LOCATION IN THE CONTROL OF ENERGY-EFFICIENT BUILDINGS [VI.13]

**Daniela Calvetti**, *The James Wood Williamson Professor, Department of Mathematics, Applied Mathematics and Statistics, Case Western Reserve University*
DIMENSIONAL ANALYSIS AND SCALING [II.9]

**Eric Cancès**, *Professor of Analysis, Ecole des Ponts and INRIA*
ELECTRONIC STRUCTURE CALCULATIONS (SOLID STATE PHYSICS) [VII.14]

**René Carmona**, *Paul M. Wythes '55 Professor of Engineering and Finance, Bendheim Center for Finance, ORFE, Princeton University*
FINANCIAL MATHEMATICS [V.9]

**C. J. Chapman**, *Professor of Applied Mathematics, University of Keele*
SHOCK WAVES [V.20], AIRCRAFT NOISE [VII.1]

**S. Jonathan Chapman**, *Professor of Mathematics and Its Applications, University of Oxford*
THE GINZBURG–LANDAU EQUATION [III.14]

**Gui-Qiang G. Chen**, *Statutory Professor in the Analysis of Partial Differential Equations and Professorial Fellow of Keble College, University of Oxford*
THE TRICOMI EQUATION [III.30]

**Margaret Cheney**, *Professor of Mathematics and Albert C. Yates Endowment Chair, Colorado State University*
RADAR IMAGING [VII.17]

**Peter A. Clarkson**, *Professor of Mathematics, University of Kent*
THE PAINLEVÉ EQUATIONS [III.24]

**Eugene M. Cliff**, *Professor Emeritus, Interdisciplinary Center for Applied Mathematics, Virginia Tech*
OPTIMAL SENSOR LOCATION IN THE CONTROL OF ENERGY-EFFICIENT BUILDINGS [VI.13]

**Paul G. Constantine**, *Ben L. Fryrear Assistant Professor of Applied Mathematics and Statistics, Colorado School of Mines*
RANKING WEB PAGES [VI.9]

**William Cook**, *Professor of Combinatorics and Optimization, University of Waterloo*
THE TRAVELING SALESMAN PROBLEM [VI.18]

**Robert M. Corless**, *Distinguished University Professor, Department of Applied Mathematics, The University of Western Ontario*
THE LAMBERT $W$ FUNCTION [III.17]

**Darren Crowdy**, *Professor of Applied Mathematics, Imperial College London*
CONFORMAL MAPPING [II.5]

**James M. Crowley**, *Executive Director, Society for Industrial and Applied Mathematics*
MATHEMATICS AND POLICY [VIII.9]

**Annie Cuyt**, *Professor, Department of Mathematics & Computer Science, University of Antwerp*
APPROXIMATION THEORY [IV.9]

**E. Brian Davies**, *Emeritus Professor of Mathematics, King's College London*
SPECTRAL THEORY [IV.8]

**Timothy A. Davis**, *Professor, Department of Computer Science and Engineering, Texas A&M University*
GRAPH THEORY [II.16], SEARCHING A GRAPH [VI.10]

**Florent de Dinechin**, *Professor of Applied Sciences, INSA—Lyon*
EVALUATING ELEMENTARY FUNCTIONS [VI.11]

**Mark R. Dennis**, *Professor of Theoretical Physics, University of Bristol*
INVARIANTS AND CONSERVATION LAWS [II.21], TENSORS AND MANIFOLDS [II.33], THE DIRAC EQUATION [III.9], MAXWELL'S EQUATIONS [III.22], SCHRÖDINGER'S EQUATION [III.26]

**Jack Dongarra**, *Professor, University of Tennessee; Professor, Oak Ridge National Laboratory; Professor, The University of Manchester*
HIGH-PERFORMANCE COMPUTING [VII.12]

**David L. Donoho**, *Anne T. and Robert M. Bass Professor in the Humanities and Sciences, Stanford University*
REPRODUCIBLE RESEARCH IN THE MATHEMATICAL SCIENCES [VIII.5]

**Ivar Ekeland**, *Professor Emeritus, CEREMADE and Institut de Finance, Université Paris-Dauphine*
MATHEMATICAL ECONOMICS [VII.20]

**Yonina C. Eldar**, *Professor of Electrical Engineering, Technion—Israel Institute of Technology, Haifa*
COMPRESSED SENSING [VII.10]

**George F. R. Ellis**, *Professor Emeritus, Mathematics Department, University of Cape Town*
GENERAL RELATIVITY AND COSMOLOGY [IV.40]

**Charles L. Epstein**, *Thomas A. Scott Professor of Mathematics, University of Pennsylvania*
MEDICAL IMAGING [VII.9]

**Bard Ermentrout**, *Distinguished University Professor of Computational Biology and Professor of Mathematics, University of Pittsburgh*
MATHEMATICAL NEUROSCIENCE [VII.21]

**Maria Esteban**, *Director of Research, CNRS*
MATHEMATICS AND POLICY [VIII.9]

**Lawrence C. Evans**, *Professor, Department of Mathematics, University of California, Berkeley*
PARTIAL DIFFERENTIAL EQUATIONS [IV.3]

**Hans G. Feichtinger**, *Faculty of Mathematics, University of Vienna*
FUNCTION SPACES [II.15]

**Martin Feinberg**, *Morrow Professor of Chemical & Biomolecular Engineering and Professor of Mathematics, The Ohio State University*
CHEMICAL REACTIONS [V.7]

**Alistair D. Fitt**, *Vice-Chancellor, Oxford Brookes University*
MATHEMATICS AND POLICY [VIII.9]

**Irene Fonseca**, *Mellon College of Science University Professor of Mathematics and Director of Center for Nonlinear Analysis, Carnegie Mellon University*
CALCULUS OF VARIATIONS [IV.6]

**L. B. Freund**, *Adjunct Professor, Department of Materials Science and Engineering, University of Illinois at Urbana-Champaign*
MECHANICS OF SOLIDS [IV.32]

**David F. Gleich**, *Assistant Professor of Computer Science, Purdue University*
RANKING WEB PAGES [VI.9]

**Paul Glendinning**, *Professor of Applied Mathematics, The University of Manchester*
CHAOS AND ERGODICITY [II.3], COMPLEX SYSTEMS [II.4], HYBRID SYSTEMS [II.18], THE EULER–LAGRANGE EQUATIONS [III.12], THE LOGISTIC EQUATION [III.19], THE LORENZ EQUATIONS [III.20], BIFURCATION THEORY [IV.21]

**Joe D. Goddard**, *Professor of Applied Mechanics and Engineering Science, University of California, San Diego*
GRANULAR FLOWS [V.13]

**Kenneth M. Golden**, *Professor of Mathematics/Adjunct Professor of Bioengineering, University of Utah*
THE MATHEMATICS OF SEA ICE [V.17]

**Timothy Gowers**, *Royal Society Research Professor, Department of Pure Mathematics and Mathematical Statistics, University of Cambridge*
MATHEMATICAL WRITING [VIII.1]

**Thomas A. Grandine**, *Senior Technical Fellow, The Boeing Company*
A HYBRID SYMBOLIC–NUMERIC APPROACH TO GEOMETRY PROCESSING AND MODELING [VII.2]

**Andreas Griewank**, *Professor of Mathematics, Humboldt University of Berlin*
AUTOMATIC DIFFERENTIATION [VI.7]

**David Griffiths**, *Emeritus Professor of Physics, Reed College*
QUANTUM MECHANICS [IV.23]

**Peter Grindrod**, *Professor of Mathematics, University of Oxford*
EVOLVING SOCIAL NETWORKS, ATTITUDES, AND BELIEFS—AND COUNTERTERRORISM [VII.5]

**Julio C. Gutiérrez-Vega**, *Director of the Optics Center, Technológico de Monterrey*
MATHIEU FUNCTIONS [III.21]

**Ernst Hairer**, *Honorary Professor of Mathematics, University of Geneva*
NUMERICAL SOLUTION OF ORDINARY DIFFERENTIAL EQUATIONS [IV.12]

**Ian Hawke**, *Associate Professor, Mathematical Sciences, University of Southampton*
NUMERICAL RELATIVITY [V.15]

**Stephan Held**, *Professor, Research Institute for Discrete Mathematics, Bonn University*
CHIP DESIGN [VII.6]

**Didier Henrion**, *Professor, LAAS-CNRS, University of Toulouse; Professor, Faculty of Electrical Engineering, Czech Technical University in Prague*
CONVEXITY [II.8]

**Willy A. Hereman**, *Professor of Applied Mathematics, Colorado School of Mines*
THE KORTEWEG-DE VRIES EQUATION [III.16]

**Desmond J. Higham**, *1966 Professor of Numerical Analysis, University of Strathclyde*
BAYESIAN INFERENCE IN APPLIED MATHEMATICS [V.11]

**Nicholas J. Higham**, *Richardson Professor of Applied Mathematics, The University of Manchester*
WHAT IS APPLIED MATHEMATICS? [I.1], THE LANGUAGE OF APPLIED MATHEMATICS [I.2], METHODS OF SOLUTION [I.3], ALGORITHMS [I.4], GOALS OF APPLIED MATHEMATICAL RESEARCH [I.5], CONTROL [II.7], FINITE DIFFERENCES [II.11], THE FINITE-ELEMENT METHOD [II.12], FLOATING-POINT ARITHMETIC [II.13], FUNCTIONS OF MATRICES [II.14], INTEGRAL TRANSFORMS AND CONVOLUTION [II.19], THE JORDAN CANONICAL FORM [II.22], ORTHOGONAL POLYNOMIALS [II.29], THE SINGULAR VALUE DECOMPOSITION [II.32], VARIATIONAL PRINCIPLE [II.35], THE BLACK–SCHOLES EQUATION [III.3], THE SYLVESTER AND LYAPUNOV EQUATIONS [III.28], NUMERICAL LINEAR ALGEBRA AND MATRIX ANALYSIS [IV.10], COLOR SPACES AND DIGITAL IMAGING [VII.7], PROGRAMMING LANGUAGES: AN APPLIED MATHEMATICS VIEW [VII.11], HOW TO READ AND UNDERSTAND A PAPER [VIII.2], WORKFLOW [VIII.4]

**Theodore P. Hill**, *Professor Emeritus of Mathematics, Georgia Institute of Technology*
BENFORD'S LAW [III.1]

**Philip Holmes**, *Eugene Higgins Professor of Mechanical and Aerospace Engineering and Professor of Applied and Computational Mathematics, Princeton University*
DYNAMICAL SYSTEMS [IV.20]

**Stefan Hougardy**, *Professor of Mathematics, University of Bonn*
CHIP DESIGN [VII.6]

**Christopher J. Howls**, *Professor of Mathematics, University of Southampton*
DIVERGENT SERIES: TAMING THE TAILS [V.8]

**Yifan Hu**, *Principal Research Scientist, Yahoo Labs*
GRAPH THEORY [II.16]

**David W. Hughes**, *Professor of Applied Mathematics, University of Leeds*
MAGNETOHYDRODYNAMICS [IV.29]

**Julian C. R. Hunt**, *Emeritus Professor of Climate Modelling and Honorary Professor of Mathematics, University College London*
TURBULENCE [V.21]

**Stefan Hutzler**, *Associate Professor, School of Physics, Trinity College Dublin*
FOAMS [VI.3]

**Richard D. James**, *Professor, Department of Aerospace Engineering and Mechanics, University of Minnesota*
CONTINUUM MECHANICS [IV.26]

**David J. Jeffrey**, *Professor, Department of Applied Mathematics, The University of Western Ontario*
THE LAMBERT $W$ FUNCTION [III.17]

**Oliver E. Jensen**, *Sir Horace Lamb Professor, School of Mathematics, The University of Manchester*
MATHEMATICAL BIOMECHANICS [V.4]

**Chris R. Johnson**, *Director, Scientific Computing and Imaging Institute; Distinguished Professor, School of Computing, University of Utah*
VISUALIZATION [VII.13]

**Chandrika Kamath**, *Member of Technical Staff, Lawrence Livermore National Laboratory*
DATA MINING AND ANALYSIS [IV.17]

**Randall D. Kamien**, *Vicki and William Abrams Professor in the Natural Sciences, University of Pennsylvania*
SOFT MATTER [IV.33]

**Jonathan Peter Keating**, *Henry Overton Wills Professor of Mathematics, University of Bristol*
RANDOM-MATRIX THEORY [IV.24]

**David E. Keyes**, *Professor of Applied Mathematics and Computational Science and Director, Extreme Computing Research Center, King Abdullah University of Science and Technology; Professor of Applied Mathematics and Applied Physics, Columbia University*
COMPUTATIONAL SCIENCE [IV.16]

**Barbara Lee Keyfitz**, *Dr. Charles Saltzer Professor of Mathematics, The Ohio State University*
CONSERVATION LAWS [II.6], SHOCKS [II.30]

**David Krakauer**, *President and Professor, Santa Fe Institute*
THE MATHEMATICS OF ADAPTATION (OR THE TEN AVATARS OF VISHNU) [V.1]

**Rainer Kress**, *Professor Emeritus, Institut für Numerische und Angewandte Mathematik, University of Göttingen*
INTEGRAL EQUATIONS [IV.4]

**Alan J. Laub**, *Professor, Department of Electrical Engineering/ Mathematics, University of California, Los Angeles*
THE RICCATI EQUATION [III.25]

**Anita T. Layton**, *Robert R. and Katherine B. Penn Associate Professor, Department of Mathematics, Duke University*
MATHEMATICAL PHYSIOLOGY [V.5]

**Tanya Leise**, *Associate Professor of Mathematics, Amherst College*
CLOAKING [VI.1]

**Giovanni Leoni**, *Professor, Department of Mathematical Sciences, Carnegie Mellon University*
CALCULUS OF VARIATIONS [IV.6]

**Randall J. LeVeque**, *Professor, Department of Applied Mathematics, University of Washington*
TSUNAMI MODELING [V.19]

**Rachel Levy**, *Associate Professor of Mathematics, Harvey Mudd College*
TEACHING APPLIED MATHEMATICS [VIII.7]

**W. R. B. Lionheart**, *Professor of Applied Mathematics, The University of Manchester*
AIRPORT BAGGAGE SCREENING WITH X-RAY TOMOGRAPHY [VII.19]

**Andrew W. Lo**, *Charles E. and Susan T. Harris Professor, Massachusetts Institute of Technology*
PORTFOLIO THEORY [V.10]

**Christian Lubich**, *Professor, Mathematisches Institut, Universität Tübingen*
NUMERICAL SOLUTION OF ORDINARY DIFFERENTIAL EQUATIONS [IV.12]

**Peter Lynch**, *Emeritus Professor, School of Mathematical Sciences, University College, Dublin*
NUMERICAL WEATHER PREDICTION [V.18]

**Malcolm A. H. MacCallum**, *Emeritus Professor of Applied Mathematics, Queen Mary University of London*
EINSTEIN'S FIELD EQUATIONS [III.10]

**Dian I. Martin**, *CEO, Senior Consultant, Small Bear Technologies, Inc.*
TEXT MINING [VII.24]

**P. A. Martin**, *Professor of Applied Mathematics, Colorado School of Mines*
ASYMPTOTICS [II.1], BOUNDARY LAYER [II.2], INTEGRAL TRANSFORMS AND CONVOLUTION [II.19], SINGULARITIES [II.31], WAVE PHENOMENA [II.36], BESSEL FUNCTIONS [III.2], THE BURGERS EQUATION [III.4], THE CAUCHY-RIEMANN EQUATIONS [III.6], THE DIFFUSION EQUATION [III.8], THE EULER EQUATIONS [III.11], THE GAMMA FUNCTION [III.13], HOOKE'S LAW [III.15], LAPLACE'S EQUATION [III.18], THE SHALLOW-WATER EQUATIONS [III.27], THE WAVE EQUATION [III.31], COMPLEX ANALYSIS [IV.1]

**Youssef Marzouk**, *Associate Professor, Department of Aeronautics and Astronautics and Center for Computational Engineering, Massachusetts Institute of Technology*
UNCERTAINTY QUANTIFICATION [II.34]

**Moshe Matalon**, *Caterpillar Distinguished Professor, Mechanical Science and Engineering, University of Illinois at Urbana-Champaign*
FLAME PROPAGATION [VII.15]

**Sean McKee**, *Research Professor of Mathematics and Statistics, University of Strathclyde, Glasgow*
MODELING A PREGNANCY TESTING KIT [VII.18]

**Ross C. McPhedran**, *Emeritus Professor in Physics, CUDOS, University of Sydney*
EFFECTIVE MEDIUM THEORIES [IV.31]

**John G. McWhirter**, *Distinguished Research Professor, School of Engineering, Cardiff University*
SIGNAL PROCESSING [IV.35]

**Beatrice Meini**, *Professor of Numerical Analysis, University of Pisa*
MARKOV CHAINS [II.25]

**James D. Meiss**, *Professor, Department of Applied Mathematics, University of Colorado at Boulder*
ORDINARY DIFFERENTIAL EQUATIONS [IV.2]

**Heather Mendick**, *Reader in Education, Brunel University*
MEDIATED MATHEMATICS: REPRESENTATIONS OF MATHEMATICS IN POPULAR CULTURE AND WHY THESE MATTER [VIII.8]

**Peter D. Miller**, *Professor of Mathematics, The University of Michigan, Ann Arbor*
PERTURBATION THEORY AND ASYMPTOTICS [IV.5]

**H. K. Moffatt**, *Emeritus Professor of Mathematical Physics, University of Cambridge*
THE NAVIER-STOKES EQUATIONS [III.23], FLUID DYNAMICS [IV.28]

**Esteban Moro**, *Associate Professor, Department of Mathematics, Universidad Carlos III de Madrid*
NETWORK ANALYSIS [IV.18]

**Clément Mouhot**, *Professor of Mathematical Sciences, University of Cambridge*
KINETIC THEORY [IV.25]

**Jean-Michel Muller**, *Directeur de Recherche, CNRS*
EVALUATING ELEMENTARY FUNCTIONS [VI.11]

**Tri-Dung Nguyen**, *Associate Professor in Operational Research and Management Sciences, University of Southampton*
PORTFOLIO THEORY [V.10]

**Qing Nie**, *Professor, Department of Mathematics, Center for Mathematical and Computational Biology, Center for Complex Biological Systems, University of California, Irvine*
SYSTEMS BIOLOGY [VII.22]

**Harald Niederreiter**, *Senior Scientist, RICAM, Austrian Academy of Sciences, Linz*
RANDOM NUMBER GENERATION [VI.12]

**Amy Novick-Cohen**, *Professor, Department of Mathematics, Technion—Israel Institute of Technology, Haifa*
THE CAHN-HILLIARD EQUATION [III.5]

**Bernt Øksendal**, *Professor, Department of Mathematics, University of Oslo*
APPLICATIONS OF STOCHASTIC ANALYSIS [IV.14]

**Alexander V. Panfilov**, *Professor, Department of Physics and Astronomy, Gent University*
CARDIAC MODELING [V.6]

**Nicola Parolini**, *Associate Professor of Numerical Analysis, Dipartimento di Matematica, MOX Politecnico di Milano*
SPORT [V.2]

**Kristin Potter**, *Scientific Software Consultant, University of Oregon*
VISUALIZATION [VII.13]

**Andrea Prosperetti**, *C. A. Miller Jr. Professor of Mechanical Engineering, Johns Hopkins University; G. Berkhoff Professor of Applied Physics, University of Twente*
BUBBLES [VI.2]

**Ian Proudler**, *Professor of Signal Processing, Loughborough University*
SIGNAL PROCESSING [IV.35]

**Alfio Quarteroni**, *Professor and Director, Chair of Modelling and Scientific Computing, Ecole Polytechnique Fédérale de Lausanne*
SPORT [V.2]

**Anders Rantzer**, *Professor, Automatic Control, LTH Lund University*
CONTROL THEORY [IV.34]

**Marcos Raydan**, *Professor, Departamento de Cómputo Científico y Estadística, Universidad Simón Bolívar*
NONLINEAR EQUATIONS AND NEWTON'S METHOD [II.28]

**Daniel N. Rockmore**, *William H. Neukom 1964 Professor of Computational Science, Dartmouth College*
THE FAST FOURIER TRANSFORM [II.10], THE MATHEMATICS OF ADAPTATION (OR THE TEN AVATARS OF VISHNU) [V.1]

**Donald G. Saari**, *Distinguished Professor and Director, Institute for Mathematical Behavioral Sciences, University of California, Irvine*
FROM THE *N*-BODY PROBLEM TO ASTRONOMY AND DARK MATTER [VI.16], VOTING SYSTEMS [VII.25]

**Fadil Santosa**, *Professor, School of Mathematics, University of Minnesota; Director, Institute for Mathematics and its Applications*
HOMOGENIZATION [II.17], THE LEVEL SET METHOD [II.24], MULTISCALE MODELING [II.27], INVERSE PROBLEMS [IV.15]

**Guillermo Sapiro**, *Edmund T. Pratt, Jr. School Professor of Electrical and Computer Engineering, Duke University*
MATHEMATICAL IMAGE PROCESSING [VII.8]

**Arnd Scheel**, *Professor, School of Mathematics, University of Minnesota*
PATTERN FORMATION [IV.27]

**Emily Shuckburgh**, *Head of Open Oceans, British Antarctic Survey*
EARTH SYSTEM DYNAMICS [IV.30]

**Reinhard Siegmund-Schultze**, *Faculty of Engineering and Science, University of Agder*
THE HISTORY OF APPLIED MATHEMATICS [I.6]

**Valeria Simoncini**, *Professor of Numerical Analysis, Alma Mater Studiorum Università di Bologna*
KRYLOV SUBSPACES [II.23]

**Ronnie Sircar**, *Professor, Operations Research & Financial Engineering Department, Princeton University*
FINANCIAL MATHEMATICS [V.9]

**Malcolm C. Smith**, *Professor, Department of Engineering, University of Cambridge*
INERTERS [V.3]

**Roel Snieder**, *W. M. Keck Distinguished Professor of Basic Exploration Science, Colorado School of Mines*
IMAGING THE EARTH USING GREEN'S THEOREM [VII.16]

**Erkki Somersalo**, *Professor, Department of Mathematics, Applied Mathematics and Statistics, Case Western Reserve University*
DIMENSIONAL ANALYSIS AND SCALING [II.9]

**Frank Sottile**, *Professor of Mathematics, Texas A&M University*
ALGEBRAIC GEOMETRY [IV.39]

**Ian Stewart**, *Emeritus Professor of Mathematics, University of Warwick*
SYMMETRY IN APPLIED MATHEMATICS [IV.22], HOW TO WRITE A GENERAL INTEREST MATHEMATICS BOOK [VIII.3]

**Victoria Stodden**, *Associate Professor, Graduate School of Library and Information Science, University of Illinois at Urbana-Champaign*
REPRODUCIBLE RESEARCH IN THE MATHEMATICAL SCIENCES [VIII.5]

**Gilbert Strang**, *Professor of Mathematics, Massachusetts Institute of Technology*
A SYMMETRIC FRAMEWORK WITH MANY APPLICATIONS [V.12], TEACHING APPLIED MATHEMATICS [VIII.7]

**Agnès Sulem**, *Researcher, INRIA Paris-Rocquencourt*
APPLICATIONS OF STOCHASTIC ANALYSIS [IV.14]

**Endre Süli**, *Professor of Numerical Analysis, University of Oxford*
NUMERICAL SOLUTION OF PARTIAL DIFFERENTIAL EQUATIONS [IV.13]

**William W. Symes**, *Noah G. Harding Professor in Computational and Applied Mathematics and Professor of Earth Science, Rice University*
INVERSE PROBLEMS [IV.15]

**Nico M. Temme**, *Emeritus Researcher, Centrum Wiskunde & Informatica, Amsterdam*
SPECIAL FUNCTIONS [IV.7]

**David Tong**, *Professor of Theoretical Physics, University of Cambridge*
CLASSICAL MECHANICS [IV.19]

**Warwick Tucker**, *Professor of Mathematics, Uppsala University*
INTERVAL ANALYSIS [II.20], COMPUTER-AIDED PROOFS VIA INTERVAL ANALYSIS [VII.3]

**Peter R. Turner**, *Dean of Arts and Sciences and Professor of Mathematics, Clarkson University*
TEACHING APPLIED MATHEMATICS [VIII.7]

**P. J. Upton**, *Lecturer, Department of Mathematics and Statistics, The Open University*
THE DELTA FUNCTION AND GENERALIZED FUNCTIONS [III.7]

**P. van den Driessche**, *Professor Emeritus of Mathematics and Statistics, University of Victoria*
THE SPREAD OF INFECTIOUS DISEASES [V.16]

**Sergio Verdú**, *Eugene Higgins Professor of Electrical Engineering, Princeton University*
INFORMATION THEORY [IV.36]

**Cédric Villani**, *Professor of Mathematics, University Claude Bernard Lyon I; Director, Institut Henri Poincaré (CNRS/UPMC)*
KINETIC THEORY [IV.25]

**Jens Vygen**, *Professor, Research Institute for Discrete Mathematics, University of Bonn*
COMBINATORIAL OPTIMIZATION [IV.38], CHIP DESIGN [VII.6]

**Charles W. Wampler**, *Technical Fellow, General Motors Global Research and Development*
ROBOTICS [VI.14]

**Z. Jane Wang**, *Professor, Department of Physics, Cornell University*
INSECT FLIGHT [VI.5]

**Denis Weaire**, *Emeritus Professor, School of Physics, Trinity College Dublin*
FOAMS [VI.3]

**Karen Willcox**, *Professor of Aeronautics and Astronautics, Massachusetts Institute of Technology*
UNCERTAINTY QUANTIFICATION [II.34]

**Walter Willinger**, *Chief Scientist, NIKSUN, Inc.*
COMMUNICATION NETWORKS [VII.23]

**Peter Winkler**, *William Morrill Professor of Mathematics and Computer Science, Dartmouth College*
APPLIED COMBINATORICS AND GRAPH THEORY [IV.37]

**Stephen J. Wright**, *Professor, Department of Computer Sciences, University of Wisconsin-Madison*
CONTINUOUS OPTIMIZATION (NONLINEAR AND LINEAR PROGRAMMING) [IV.11]

**Lexing Ying**, *Professor of Mathematics, Stanford University*
THE *N*-BODY PROBLEM AND THE FAST MULTIPOLE METHOD [VI.17]

**Ya-xiang Yuan**, *Professor, Institute of Computational Mathematics and Scientific/Engineering Computing, Chinese Academy of Sciences*
MATHEMATICS AND POLICY [VIII.9]

# The Princeton Companion to Applied Mathematics

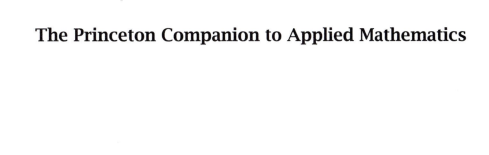

# Part I
# Introduction to Applied Mathematics

## I.1 What Is Applied Mathematics?
*Nicholas J. Higham*

### 1 The Big Picture

Applied mathematics is a large subject that interfaces with many other fields. Trying to define it is problematic, as noted by William Prager and Richard Courant, who set up two of the first centers of applied mathematics in the United States in the first half of the twentieth century, at Brown University and New York University, respectively. They explained that:

> Precisely to define applied mathematics is next to impossible. It cannot be done in terms of subject matter: the borderline between theory and application is highly subjective and shifts with time. Nor can it be done in terms of motivation: to study a mathematical problem for its own sake is surely not the exclusive privilege of pure mathematicians. Perhaps the best I can do within the framework of this talk is to describe applied mathematics as the bridge connecting pure mathematics with science and technology.
>
> Prager (1972)

> Applied mathematics is not a definable scientific field but a human attitude. The attitude of the applied scientist is directed towards finding clear cut answers which can stand the test of empirical observation. To obtain the answers to theoretically often insuperably difficult problems, he must be willing to make compromises regarding rigorous mathematical completeness; he must supplement theoretical reasoning by numerical work, plausibility considerations and so on.
>
> Courant (1965)

Garrett Birkhoff offered the following view in 1977, with reference to the mathematician and physicist Lord Rayleigh (John William Strutt, 1842–1919):

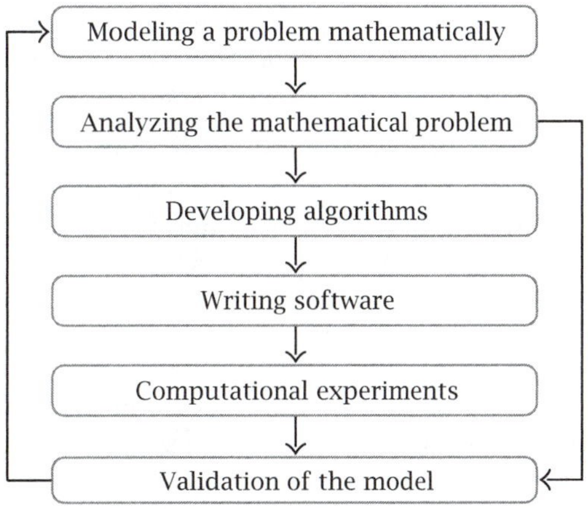


**Figure 1** The main steps in solving a problem in applied mathematics.

> Essentially, mathematics becomes "applied" when it is used to solve real-world problems "neither seeking nor avoiding mathematical difficulties" (Rayleigh).

Rather than define what applied mathematics is, one can describe the methods used in it. Peter Lax stated of these methods, in 1989, that:

> Some of them are organic parts of pure mathematics: rigorous proofs of precisely stated theorems. But for the greatest part the applied mathematician must rely on other weapons: special solutions, asymptotic description, simplified equations, experimentation both in the laboratory and on the computer.

Here, instead of attempting to give our own definition of applied mathematics we describe the various facets of the subject, as organized around solving a problem. The main steps are described in figure 1. Let us go through each of these steps in turn.

**Modeling a problem.** Modeling is about taking a physical problem and developing equations—differential, difference, integral, or algebraic—that capture the essential features of the problem and so can be used to obtain qualitative or quantitative understanding of its behavior. Here, "physical problem" might refer to a vibrating string, the spread of an infectious disease, or the influence of people participating in a social network. Modeling is necessarily imperfect and requires simplifying assumptions. One needs to retain enough aspects of the system being studied that the model reproduces the most important behavior but not so many that the model is too hard to analyze. Different types of models might be feasible (continuous, discrete, stochastic), and for a given type there can be many possibilities. Not all applied mathematicians carry out modeling; in fact, most join the process at the next step.

**Analyzing the mathematical problem.** The equations formulated in the previous step are now analyzed and, ideally, solved. In practice, an explicit, easily evaluated solution usually cannot be obtained, so approximations may have to be made, e.g., by discretizing a differential equation, producing a reduced problem. The techniques necessary for the analysis of the equations or reduced problem may not exist, so this step may involve developing appropriate new techniques. If analytic or perturbation methods have been used then the process may jump from here directly to validation of the model.

**Developing algorithms.** It may be possible to solve the reduced problem using an existing algorithm—a sequence of steps that can be followed mechanically without the need for ingenuity. Even if a suitable algorithm exists it may not be fast or accurate enough, may not exploit available structure or other problem features, or may not fully exploit the architecture of the computer on which it is to be run. It is therefore often necessary to develop new or improved algorithms.

**Writing software.** In order to use algorithms on a computer it is necessary to implement them in software. Writing reliable, efficient software is not easy, and depending on the computer environment being targeted it can be a highly specialized task. The necessary software may already be available, perhaps in a package or program library. If it is not, software is ideally developed and documented to a high standard and made available to others. In many cases the software stage consists simply of writing short programs, scripts, or notebooks that carry out the necessary computations and summarize the results, perhaps graphically.

**Computational experiments.** The software is now run on problem instances and solutions obtained. The computations could be numeric or symbolic, or a mixture of the two.

**Validation of the model.** The final step is to take the results from the experiments (or from the analysis, if the previous three steps were not needed), interpret them (which may be a nontrivial task), and see if they agree with the observed behavior of the original system. If the agreement is not sufficiently good then the model can be modified and the loop through the steps repeated. The validation step may be impossible, as the system in question may not yet have been built (e.g., a bridge or a building).

Other important tasks for some problems, which are not explicitly shown in our outline, are to calibrate parameters in a model, to quantify the uncertainty in these parameters, and to analyze the effect of that uncertainty on the solution of the problem. These steps fall under the heading of UNCERTAINTY QUANTIFICATION [II.34].

Once all the steps have been successfully completed the mathematical model can be used to make predictions, compare competing hypotheses, and so on. A key aim is that the mathematical analysis gives new insights into the physical problem, even though the mathematical model may be a simplification of it.

A particular applied mathematician is most likely to work on just some of the steps; indeed, except for relatively simple problems it is rare for one person to have the skills to carry out the whole process from modeling to computer solution and validation.

In some cases the original problem may have been communicated by a scientist in a different field. A significant effort can be required to understand what the mathematical problem is and, when it is eventually solved, to translate the findings back into the language of the relevant field. Being able to talk to people outside mathematics is therefore a valuable skill for the applied mathematician.

It would be wrong to give the impression that all applied mathematics is done in the context of modeling. Frequently, a mathematical problem will be tackled because of its inherent interest (see the quote from Prager above) with the hope or expectation that a relevant application will be found. Indeed some applied mathematicians spend their whole careers working in this way. There are many examples of mathematical results that provide the foundations for important practical applications but were developed without knowledge of those applications (sections 3.1 and 3.2 provide such examples).

Before the twentieth century, applied mathematics was driven by problems in astronomy and mechanics. In the twentieth century physics became the main driver, with other areas such as biology, chemistry, economics, engineering, and medicine also providing many challenging mathematical problems from the 1950s onward. With the massive and still-growing amounts of data available to us in today's digital society we can expect information, in its many guises, to be an increasingly important influence on applied mathematics in the twenty-first century.

For more on the definition and history of applied mathematics, including the development of the term "applied mathematics," see the article HISTORY OF APPLIED MATHEMATICS [I.6].

## 2 Applied Mathematics and Pure Mathematics

The question of how applied mathematics compares with pure mathematics is often raised and has been discussed by many authors, sometimes in controversial terms. We give a few highlights.

Paul Halmos wrote a 1981 paper provocatively titled "Applied mathematics is bad mathematics." However, much of what Halmos says would not be disputed by many applied mathematicians. For example:

> Pure mathematics can be practically useful and applied mathematics can be artistically elegant....
>
> Just as pure mathematics can be useful, applied mathematics can be more beautifully useless than is sometimes recognized....
>
> Applied mathematics is an intellectual discipline, not a part of industrial technology....
>
> Not only, as is universally admitted, does the applied need the pure, but, in order to keep from becoming inbred, sterile, meaningless, and dead, the pure needs the revitalization and the contact with reality that only the applied can provide.

G. H. Hardy's book *A Mathematician's Apology* (1940) is well known as a defense of mathematics as a subject that can be pursued for its own sake and beauty. As such it contains some criticism of applied mathematics:

> But is not the position of an ordinary applied mathematician in some ways a little pathetic? If he wants to be useful, he must work in a humdrum way, and he cannot give full play to his fancy even when he wishes to rise to the heights. "Imaginary" universes are so much more beautiful than this stupidly constructed "real" one; and most of the finest products of an applied mathematician's fancy must be rejected, as soon as they have been created, for the brutal but sufficient reason that they do not fit the facts.

Halmos and Hardy were pure mathematicians. Applied mathematicians C. C. Lin and L. A. Segel offer some insights in the introductory chapter of their classic 1974 book *Mathematics Applied to Deterministic Problems in the Natural Sciences*:

> The differences in motivation and objectives between pure and applied mathematics—and the consequent differences in emphasis and attitude—must be fully recognized. In pure mathematics, one is often dealing with such abstract concepts that logic remains the only tool permitting judgment of the correctness of a theory. In applied mathematics, empirical verification is a necessary and powerful judge. However... in some cases (e.g., celestial mechanics), rigorous theorems can be proved that are also valuable for practical purposes. On the other hand, there are many instances in which new mathematical ideas and new mathematical theories are stimulated by applied mathematicians or theoretical scientists.

They also opine that:

> Much second-rate pure mathematics is concealed beneath the trappings of applied mathematics (and vice versa). As always, knowledge and taste are needed if quality is to be assured.

The applied versus pure discussion is not always taken too seriously. Chandler Davis quotes the applied mathematician Joseph Keller as saying, "pure mathematics is a subfield of applied mathematics"!

The discussion can also focus on where in the spectrum a particular type of mathematics lies. An interesting story was told in 1988 by Clifford Truesdell of his cofounding in 1952 of the *Journal of Rational Mechanics and Analysis* (which later became *Archive for Rational Mechanics and Analysis*). He explained that

> In those days papers on the foundation of continuum mechanics were rejected by journals of mathematics as being applied, by journals of "applied" mathematics as being physics or pure mathematics, by journals of physics as being mathematics, and by all of them as too long, too expensive to print, and of interest to no one.

## 3 Applied Mathematics in Everyday Life

We now give three examples of applied mathematics in use in everyday life. These examples were chosen because they can be described without delving into too many technicalities and because they illustrate different characteristics. Some of the terms used in the descriptions are explained in THE LANGUAGE OF APPLIED MATHEMATICS [I.2].

### 3.1 Searching Web Pages

In the early to mid-1990s—the early days of the World Wide Web—search engines would find Web pages that matched a user's search query and would order the results by a simple criterion such as the number of times that the search query appears on a page. This approach became unsatisfactory as the Web grew in size and spammers learned how to influence the search results. From the late 1990s onward, more sophisticated criteria were developed, based on analysis of the links between Web pages. One of these is Google's PAGERANK ALGORITHM [VI.9]. Another is the hyperlink-induced topic search (HITS) algorithm of Kleinberg.

The HITS algorithm is based on the idea of determining hubs and authorities. *Authorities* are Web pages with many links to them and for which the linking pages point to many authorities. For example, the *New York Times* home page or a Wikipedia article on a popular topic might be an authority. *Hubs* are pages that point to many authorities. An example might be a page on a programming language that provides links to useful pages about that language but that does not necessarily contain much content itself. The authorities are the pages that we would like to rank higher among pages that match a search term. However, the definition of hubs and authorities is circular, as each depends on the other.

To resolve this circularity, associate an authority weight $x_i$ and a hub weight $y_i$ with page $i$, with both weights nonnegative. Let there be $n$ pages to be considered (in practice this is a much smaller number than the total number of pages that match the search term). Define an $n \times n$ matrix $A = (a_{ij})$ by $a_{ij} = 1$ if there is a hyperlink from page $i$ to page $j$ and by $a_{ij} = 0$ otherwise. Let us make initial guesses $x_i^{(0)} = 1$ and $y_i^{(0)} = 1$, for $i = 1, 2, \dots, n$. It is reasonable to update the authority weight $x_i$ for page $i$ by replacing it by the sum of the weights of the hubs that point to it. Similarly, the hub weight $y_i$ for page $i$ can be replaced by the sum of the weights of the authorities to which it points. In equations, these updates can be written as $x_i^{(1)} = \sum_{a_{ji} \neq 0} y_j^{(0)}$ and $y_i^{(1)} = \sum_{a_{ij} \neq 0} x_j^{(1)}$; note that in the latter equation we are using the updated authority weights, and the sums are over those $j$ for which $a_{ji}$ or $a_{ij}$ is nonzero, respectively. This process can be iterated:

$$\left.\begin{aligned} x_i^{(k+1)} &= \sum_{a_{ji} \neq 0} y_j^{(k)} \\ y_i^{(k+1)} &= \sum_{a_{ij} \neq 0} x_j^{(k+1)} \end{aligned}\right\} \quad k = 0, 1, 2, \dots.$$

The circular definition of hubs and authorities has been turned into an iteration. The iteration is best analyzed by rewriting it in matrix-vector form. Defining the $n$-vectors

$$x_k = [x_1^{(k)}, x_2^{(k)}, \dots, x_n^{(k)}]^{\mathrm{T}},$$
$$y_k = [y_1^{(k)}, y_2^{(k)}, \dots, y_n^{(k)}]^{\mathrm{T}}$$

and recalling that the elements of $A$ are 0 or 1, we can rewrite the iteration as

$$\left.\begin{aligned} x_{k+1} &= A^{\mathrm{T}} y_k \\ y_{k+1} &= A x_{k+1} \end{aligned}\right\} \quad k = 0, 1, 2, \dots,$$

where $A^{\mathrm{T}} = (a_{ji})$ is the transpose of $A$. Combining the two formulas into one gives $x_{k+1} = A^{\mathrm{T}} y_k = A^{\mathrm{T}}(A x_k) = (A^{\mathrm{T}} A) x_k$. Hence the $x_k$ are generated by repeatedly multiplying by the matrix $A^{\mathrm{T}}A$. Each element of $A^{\mathrm{T}}A$ is either zero or a positive integer, so the powers of $A^{\mathrm{T}}A$ will usually grow without bound. In practice we should therefore normalize the vectors $x_k$ and $y_k$ so that the largest element is 1; this avoids overflow and has no effect on the relative sizes of the components, which is all that matters. Our iteration is then

$$x_{k+1} = c_k^{-1} A^{\mathrm{T}} A x_k,$$

where $c_k$ is the largest element of $A^{\mathrm{T}} A x_k$. If the sequences $x_k$ and $c_k$ converge, say to $x_*$ and $c_*$, respectively, then $A^{\mathrm{T}} A x_* = c_* x_*$. This equation says that $x_*$ is an eigenvector of $A^{\mathrm{T}}A$ with corresponding eigenvalue $c_*$. A similar argument shows that, if the normalized sequence of vectors $y_k$ converges, then it must be to an eigenvector of $AA^{\mathrm{T}}$.

This process of repeated multiplication by a matrix is known as the POWER METHOD [IV.10 §5.5]. The PERRON–FROBENIUS THEOREM [IV.10 §11.1] can be used to show that, provided the matrix $A^{\mathrm{T}}A$ has a property called irreducibility, it has a unique eigenvalue of largest magnitude and this eigenvalue is real and positive, with an associated eigenvector $x$ having positive entries. Convergence theory for the power method then shows that

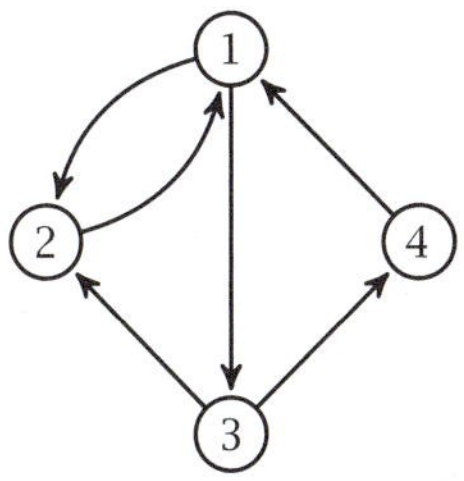


**Figure 2** Graph representing four Web pages and the links between them.

$x_k$ converges to a multiple of $x$. Therefore, the vectors $x_*$ and $y_*$ required for the HITS algorithm are the eigenvectors corresponding to the dominant eigenvalues of $A^TA$ and $AA^T$, respectively. Another interpretation of these vectors is that they are the right and left SINGULAR VECTORS [II.32] corresponding to the dominant singular value of $A$.

We give a simple example to illustrate the algorithm. Consider the network of four Web pages shown in figure 2 in a DIRECTED GRAPH [II.16] representation. The corresponding matrix is

$$A = \begin{bmatrix} 0 & 1 & 1 & 0 \\ 1 & 0 & 0 & 0 \\ 0 & 1 & 0 & 1 \\ 1 & 0 & 0 & 0 \end{bmatrix}.$$

The HITS algorithm produces

$$x = \begin{bmatrix} 0 \\ 1 \\ 0.5 \\ 0.5 \end{bmatrix}, \quad y = \begin{bmatrix} 1 \\ 0 \\ 1 \\ 0 \end{bmatrix}.$$

This tells us that page 2 is the highest-ranked authority and pages 1 and 3 are the jointly highest-ranked hubs, which is an intuitively reasonable ranking.

This Web search example illustrates the steps in figure 1. The first step, modeling, yields the notion of hubs and authorities. The production of the iteration for the weights is the analysis step, which for this problem overlaps with the third step of developing algorithms, because the iteration is a practical means for computing the hubs and authorities. However, other algorithms could be used for this purpose, so much more can be done in the third step. Implementing the iteration in software (the fourth step) is straightforward if we assume the matrix $A$ is given. The computational experiments in the fifth step should use real data from particular Web searches. Probably the most difficult step is the final one: validating the model. The reason for this is that there is no natural quantitative measure of the goodness of a ranking based on the computed authorities; this assessment requires human judgment.

A final point to note is that further insight into the HITS algorithm can be obtained using GRAPH THEORY [II.16]. In general, there may be multiple ways to analyze a problem and it may be necessary to employ more than one technique to obtain a complete understanding.

### 3.2 Digital Imaging

Image retouching refers to the process of changing a digital image to make it more visually pleasing by removing a color cast, creatively adjusting color and contrast, smoothing out wrinkles on a subject's face in a portrait, and so on. In the days of film cameras, image retouching was carried out by professionals on scanned images using expensive hardware and software. Since the advent of digital cameras retouching has become something that anyone can attempt, even on a smartphone with a suitable app. An operation called "cloning" is of particular interest. Cloning is the operation of copying one area of an image over another and is commonly used to remove defects or unwanted elements from an image, such as dust spots on the camera sensor, litter in a landscape, or someone's limbs intruding into the edge of an image.

Cloning tools in modern software use sophisticated mathematics to blend the image fragment being copied into the target area. Let us represent the image as a function $f$ of two variables, where $f(x,y)$ is an RGB (red, green, blue) triple corresponding to the point $(x,y)$. In practice, an image is a discrete grid of points and the RGB values are integers. Our goal is to replace an open target region $\Omega$ by a source region that has the same shape and size but is in a different location in the image, and that therefore corresponds to a translation $(\delta x, \delta y)$. If we simply copy the source region into the target, the result will not be convincing visually, as it neither preserves texture nor matches up well around the boundary $\partial\Omega$ (see figure 3). To alleviate these problems we can replace $f$ inside $\Omega$ by the function $g$ defined by the partial differential equation (PDE) problem

$$\begin{aligned} \Delta g(x,y) &= \Delta f(x+\delta x, y+\delta y), \quad && (x,y) \in \Omega, \\ g(x,y) &= f(x,y), && (x,y) \in \partial\Omega, \end{aligned}$$

where $\Delta = \partial^2/\partial x^2 + \partial^2/\partial y^2$ is the LAPLACE OPERATOR [III.18]. We are forcing $g$ to be identical to $f$ on the boundary of the target region, but inside the region the Laplacian of $g$ has to match that of $f$ in the source region. This is a Poisson equation with Dirichlet boundary condition. Since the Laplace operator is connected with diffusion effects in equilibrium, we can think of the pixels in the image diffusing to form a more convincing visual result. Another interpretation, which shows that the solution has optimal smoothness in a certain sense, is that we are minimizing $\int_{\Omega} \|\nabla g - \nabla f\|^2 \, \mathrm{d}\Omega$, where $\nabla = [\partial/\partial x \;\; \partial/\partial y]^{\mathrm{T}}$ is the gradient operator and $\|\cdot\|$ is the $L_2$-NORM [I.2 §19.3]. In practice, the PDE is solved by a NUMERICAL METHOD [IV.13].

Adobe Photoshop, the digital image manipulation software originally released in 1990, introduced a new feature called the healing brush in 2002. It carries out the sophisticated cloning just outlined but solves the biharmonic equation $\Delta^2 g(x,y) = \Delta^2 f(x + \delta x, y + \delta y)$, where $\Delta^2 = \partial^4/\partial x^4 + 2\partial^4/(\partial x^2 \partial y^2) + \partial^4/\partial y^4$, because this has been found to provide better matching of derivatives on the boundary and thereby produces a better blending of the source into the area containing the target. In this case we are minimizing $\int_{\Omega} (\Delta g - \Delta f)^2 \, \mathrm{d}\Omega$. See figure 3.

The idea of using the Poisson equation or the biharmonic equation to fill in gaps in two-dimensional data appears in other application areas too, such as in mapping and contouring of geophysical data, where it was proposed as early as the 1950s.

A related problem is how to detect when an image has been subject to cloning. Applications include checking the veracity of images appearing in the media or used as evidence in legal cases and checking the eligibility of entries to a photographic competition in which image manipulation is disallowed. The same ideas that make cloning possible also enable its use to be detected. For example, even though the values in the target and source areas are different, the Laplacian or biharmonic operators yield the same values, so a systematic search can be done to find different areas with this property. One complication is that if the image has been compressed after cloning, as is typically done with JPEG COMPRESSION [VII.7 §5], the uncompressed image contains noise that causes the derivatives in question no longer to match exactly. Having to deal with noise is a common requirement in applied mathematics.

### 3.3 Computer Arithmetic

In modern life we rely on computers to carry out arithmetic calculations for a wide range of tasks, including computation of financial indices, aircraft flight paths, and utility bills. We take it for granted that these calculations produce the "correct result," but they do not always do so, and there are infamous examples where inaccurate results have had disastrous consequences. Applied mathematicians and computer scientists have been responsible for ensuring that today's computer calculations are more reliable than ever.

Probably the most important advance in computer arithmetic in the last fifty years was the development of the 1985 IEEE standard for binary FLOATING-POINT ARITHMETIC [II.13]. Prior to the development of the standard, different computers used different implementations of floating-point arithmetic that were of varying quality. For example, on some computers of the 1980s a test such as

```
if x ≠ y
   f = f/(x − y)
end
```

could fail with division by zero because the difference $x - y$ evaluated to zero even though $x$ was not equal to $y$. This failure is not possible in IEEE-compliant arithmetic.

Some features of the IEEE standard are more unexpected. For example, it includes a special number inf that represents $\infty$ and satisfies the natural relations. For example, $1/0$ evaluates to inf, $1/\mathrm{inf}$ evaluates to 0, and $1 + \mathrm{inf}$ and $\mathrm{inf} + \mathrm{inf}$ produce inf. If a program does divide by zero then this does not halt the program and it can be tested for and appropriate action taken. But division by zero does not necessarily signify a problem. For example, if we evaluate the function

$$f(x) = \frac{1}{1 + \frac{1}{1-x}}$$

at $x = 1$, IEEE arithmetic correctly yields $f(1) = 1/(1 + \infty) = 0$. In 1997 the USS Yorktown, a guided missile cruiser, was paralyzed ("dead in the water" according to the commander) for $2\frac{3}{4}$ hours when a crewman entered a zero into a database and a division by zero was triggered that caused the program to crash. The incident could have been avoided if IEEE standard features such as inf had been fully utilized.

The development of the IEEE standard took many years and required much mathematical analysis of the various options. The benefits brought by the standard include increased portability of programs

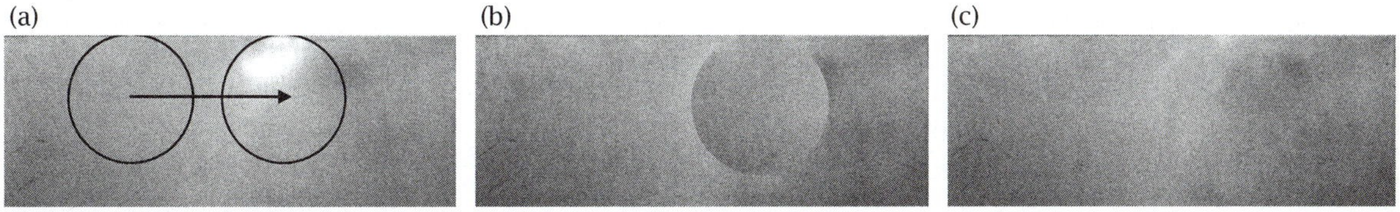


**Figure 3** Close-up of part of a metal sign with a hot spot from a reflection. (a) Original image showing source and target regions. (b) The result of copying source to target. (c) The result of one application of Photoshop healing brush with same target area. In practice, multiple applications of the healing brush would be used with smaller, overlapping target areas.

between different computers and an improved ability for mathematicians to understand the way algorithms will behave when implemented in floating-point arithmetic.

In IEEE double-precision arithmetic, numbers are represented to a precision equivalent to about sixteen significant decimal digits. In many situations in life, results are needed to far fewer figures and a final result must be *rounded*. For example, a conversion from euros to dollars producing an answer \$110.89613 might be rounded up to \$110.90: the nearest amount in whole cents. A bank paying the dollars into a customer's account might prefer to round down to \$110.89 and keep the remainder. However, deciding on the rules for rounding was not so simple when the euro was founded in 1997. A twenty-nine-page document was needed to specify precisely how conversions among the fifteen currencies of the member states and the euro should be done. Its pronouncements included how many significant figures each individual conversion rate should have (six was the number that was chosen), how rounding should be done (round to the nearest six-digit number), and how ties should be handled (always round up).

Even when rounding should be straightforward it is often carried out incorrectly. In 1982 the Vancouver Stock Exchange established an index with an initial value of 1000. After twenty-two months the index had been hitting lows in the 520s, despite the exchange apparently performing well. The index was recorded to three decimal places and it was discovered that the computer program calculating the index always rounded down, hence always underestimating the index. Upon recalculation (presumably with round to nearest) the index almost doubled.

In 2006 athlete Justin Gatlin was credited with a new world record of 9.76 seconds for the 100 meters. Almost a week after the race the time was changed to 9.77 seconds, meaning that he had merely equaled the existing record held by Asafa Powell. The reason for the change was that his recorded time of 9.766 had incorrectly been rounded down to the nearest hundredth of a second instead of up as the International Association of Athletics Federations rules require.

## 4 What Do Applied Mathematicians Do?

Applied mathematicians can work in academia, industry, or government research laboratories. Their work may involve research, teaching, and (especially for more senior mathematicians) administrative tasks such as managing teams of people. They usually spend only part of their time doing mathematics in the traditional sense of sitting with pen and paper scribbling formulas on paper and trying to solve equations or prove theorems. Under the general heading of research, a lot of time is spent writing papers, books, grant proposals, reports, lecture notes, and talks; attending seminars, conferences, and workshops; writing and running computer programs; reading papers in the research literature; refereeing papers submitted to journals and grant proposals submitted to funding bodies; and commenting on draft papers and theses written by Ph.D. students.

Mathematics can be a lonely endeavor: one may be working on different problems from one's colleagues or may be the only mathematician in a company. Although some applied mathematicians prefer to work alone, many collaborate with others, often in faraway countries. Collaborations are frequently initiated through discussions at conferences, though sometimes papers are coauthored by people who have never met, thanks to the ease of email communication.

Applied mathematics societies provide an important source of identity and connectivity, as well as opportunities for networking and professional development. They mostly focus on particular countries or regions, an exception being the Society for Industrial and Applied Mathematics (SIAM), based in Philadelphia. SIAM is the largest applied mathematics organization in the world and has a strong international outlook, with about one-third of its members residing outside the United States. A mathematician's activities are frequently connected with societies, whether it be through publishing in or editing their journals, attending their conferences, or keeping up with news through their magazines and newsletters. Most societies offer greatly reduced membership fees (sometimes free membership) for students.

Applied mathematicians can be part of multidisciplinary teams. Their skills in problem solving, thinking logically, modeling, and programming are sought after in other subjects, such as medical imaging, weather prediction, and financial engineering.

In the business world, applied mathematics can be invisible because it is called "analytics," "modeling," or simply generic "research." But whatever their job title, applied mathematicians play a crucial role in today's knowledge-based economy.

## 5 What Is the Impact of Applied Mathematics?

The impact of applied mathematics is illustrated in many articles in this volume, and in this section we provide just a brief overview, concentrating on the impact outside mathematics itself.

Applied mathematics provides the tools and algorithms to enable understanding and predictive modeling of many aspects of our planet, including WEATHER [V.18] (for which the accuracy of forecasts has improved greatly in recent decades), ATMOSPHERE AND THE OCEANS [IV.30], TSUNAMIS [V.19], and SEA ICE [V.17]. In many cases the models are used to inform policy makers.

At least two mathematical algorithms are used by most of us almost every day. The FAST FOURIER TRANSFORM [II.10] is found in any device that carries out signal processing, such as a smartphone. Photos that we take on our cameras or view on a computer screen are usually stored using JPEG COMPRESSION [VII.7 §5].

X-ray tomography devices, ranging from AIRPORT LUGGAGE SCANNERS [VII.19] to HUMAN BODY SCANNERS [VII.9], rely on the fast and accurate solution of INVERSE PROBLEMS [IV.15], which are problems in which we need to recover information about the internals of a system from (noisy) measurements taken outside the system.

Investments are routinely made on the basis of mathematical models, whether for individual options or collections of assets (portfolios): see FINANCIAL MATHEMATICS [V.9] and PORTFOLIO THEORY [V.10].

The clever use of mathematical modeling offers a competitive advantage in sports, such as YACHT RACING [V.2], SWIMMING [V.2], and FORMULA ONE RACING [V.3], where small improvements can be the difference between success and failure.

# I.2 The Language of Applied Mathematics

*Nicholas J. Higham*

This article provides background on the notation, terminology, and basic results and concepts of applied mathematics. It therefore serves as a foundation for the later articles, many of which cross-reference it.

In view of the limited space, the material has been restricted to that common to many areas of applied mathematics. A number of later articles provide their own careful introduction to the language of their particular topic.

## 1 Notation

Table 1 lists the Greek alphabet, which is widely used to denote mathematical variables. Note that almost always $\delta$ and $\varepsilon$ are used to denote small quantities, and $\pi$ is used as a variable as well as for $\pi = 3.14159\ldots$.

Mathematics has a wealth of notation to express commonly occurring concepts. But notation is both a blessing and a curse. Used carefully, it can make mathematical arguments easier to read and understand. If overused it can have the opposite effect, and often it is better to express a statement in words than in symbols (see MATHEMATICAL WRITING [VIII.1]). Table 2 gives some notation that is common in informal contexts such as lectures and is occasionally encountered in this book. Table 3 summarizes basic notation used throughout the book.

## 2 Complex Numbers

Most applied mathematics takes place in the set of complex numbers, $\mathbb{C}$, or the set of real numbers, $\mathbb{R}$. A complex number $z = x + \mathrm{i}y$ has real and imaginary parts $x = \operatorname{Re} z$ and $y = \operatorname{Im} z$ belonging to $\mathbb{R}$, and the *imaginary unit* i denotes $\sqrt{-1}$. The imaginary unit is sometimes written as j, e.g., in electrical engineering and in the programming language PYTHON [VII.11].

We can represent complex numbers geometrically in the *complex plane*, in which a complex number $a + \mathrm{i}b$ is represented by the point with coordinates $(a, b)$

**Table 1** The Greek alphabet. Where an uppercase Greek letter is the same as the Latin letter it is not shown.

| | | | |
|---|---|---|---|
| $\alpha$ | alpha | $\nu$ | nu |
| $\beta$ | beta | $\xi, \Xi$ | xi |
| $\gamma, \Gamma$ | gamma | $o$ | omicron |
| $\delta, \Delta$ | delta | $\pi, \varpi, \Pi$ | pi |
| $\epsilon, \varepsilon$ | epsilon | $\rho, \varrho$ | rho |
| $\zeta$ | zeta | $\sigma, \varsigma, \Sigma$ | sigma |
| $\eta$ | eta | $\tau$ | tau |
| $\theta, \vartheta, \Theta$ | theta | $\upsilon, \Upsilon$ | upsilon |
| $\iota$ | iota | $\phi, \varphi, \Phi$ | phi |
| $\kappa$ | kappa | $\chi$ | chi |
| $\lambda, \Lambda$ | lambda | $\psi, \Psi$ | psi |
| $\mu$ | mu | $\omega, \Omega$ | omega |

**Table 2** Other notation.

| | | | |
|---|---|---|---|
| $\Rightarrow$ | Implies | $\exists$ | There exists |
| $\Leftarrow$ | Implied by | $\nexists$ | There does not exist |
| $\Leftrightarrow$ | If and only if | $\forall$ | For all |

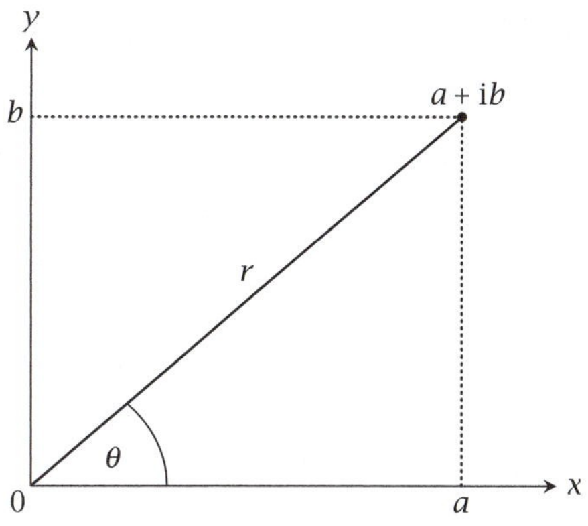


**Figure 1** Complex plane with $z = a + \mathrm{i}b = r\mathrm{e}^{\mathrm{i}\theta}$.

(see figure 1). The corresponding diagram is called the *Argand diagram*. Important roles are played by the *right half-plane* $\{z\colon \operatorname{Re} z \geqslant 0\}$ and the *left half-plane* $\{z\colon \operatorname{Re} z \leqslant 0\}$. If we exclude the pure imaginary numbers ($\operatorname{Im} z = 0$) from these sets we obtain the *open* half-planes. Euler's formula, $\mathrm{e}^{\mathrm{i}\theta} = \cos\theta + \mathrm{i}\sin\theta$, is fundamental.

The *polar form* of a complex number is $z = r\mathrm{e}^{\mathrm{i}\theta}$, where $r \geqslant 0$ and the *argument* $\arg z = \theta$ are real, and $\theta$ can be restricted to any interval of length $2\pi$, such as $[0, 2\pi)$ or $(-\pi, \pi]$. The *complex conjugate* of $z = x + \mathrm{i}y$ is $\bar{z} = x - \mathrm{i}y$, sometimes written $z^*$. The *modulus*, or *absolute value*, $|z| = (\bar{z}z)^{1/2} = (x^2 + y^2)^{1/2} = r$.

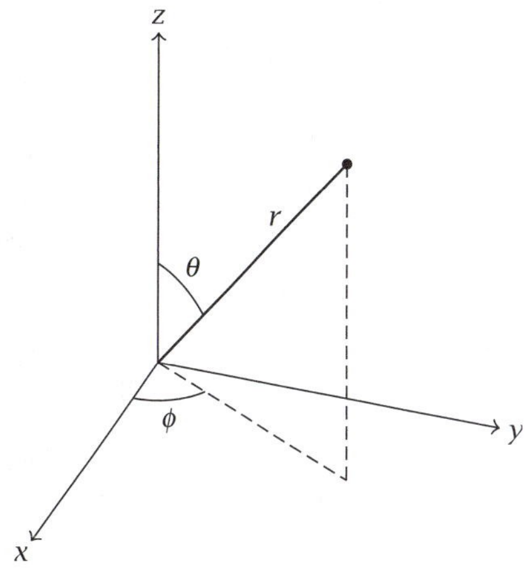


**Figure 2** Spherical coordinates.

Complex arithmetic is defined in terms of real arithmetic according to the following rules, for $z_1 = x_1 + \mathrm{i}y_1$ and $z_2 = x_2 + \mathrm{i}y_2$:

$$\begin{aligned} z_1 \pm z_2 &= x_1 \pm x_2 + \mathrm{i}(y_1 \pm y_2), \\ z_1 z_2 &= x_1 x_2 - y_1 y_2 + \mathrm{i}(x_1 y_2 + x_2 y_1), \\ \frac{z_1}{z_2} &= \frac{x_1 x_2 + y_1 y_2}{x_2^2 + y_2^2} + \mathrm{i}\frac{x_2 y_1 - x_1 y_2}{x_2^2 + y_2^2}. \end{aligned}$$

In polar form multiplication and division become notationally simpler: if $z_1 = r_1 \mathrm{e}^{\mathrm{i}\theta_1}$ and $z_2 = r_2 \mathrm{e}^{\mathrm{i}\theta_2}$ then $z_1 z_2 = r_1 r_2 \mathrm{e}^{\mathrm{i}(\theta_1 + \theta_2)}$ and $z_1/z_2 = (r_1/r_2)\mathrm{e}^{\mathrm{i}(\theta_1 - \theta_2)}$.

## 3 Coordinate Systems

We are used to specifying a point in two dimensions by its $x$- and $y$-coordinates, and a point in three dimensions by its $x$-, $y$-, and $z$-coordinates. These are called *Cartesian coordinates*. In two dimensions we can also use *polar coordinates*, which are as described in the previous section if we identify $(x, y)$ with $x + \mathrm{i}y$. *Spherical coordinates*, illustrated in figure 2, are an extension of polar coordinates to three dimensions. Here, $(x, y, z)$ is represented by $(r, \theta, \phi)$, where

$$x = r\sin\theta\cos\phi, \quad y = r\sin\theta\sin\phi, \quad z = r\cos\theta,$$

with nonnegative radius $r$ and angles $\theta$ and $\phi$ in the ranges $0 \leqslant \theta \leqslant \pi$ and $0 \leqslant \phi < 2\pi$.

*Cylindrical coordinates* provide another three-dimensional coordinate system. Here, polar coordinates are used in the $xy$-plane and $z$ is retained, so $(x, y, z)$ is represented by $(r, \theta, z)$.

**Table 3** Notation frequently used in this book.

| Notation | Meaning | Example |
|---|---|---|
| $\mathbb{R}$, $\mathbb{C}$ | The real numbers, the complex numbers | |
| $\mathbb{R}^n$, $\mathbb{R}^{m\times n}$ | The real $n$-vectors and real $m \times n$ matrices; similarly for $\mathbb{C}^n$ and $\mathbb{C}^{m\times n}$ | |
| $\operatorname{Re} z$, $\operatorname{Im} z$ | Real and imaginary parts of the complex number $z$ | |
| $\mathbb{Z}$, $\mathbb{N}$ | The integers, $\{0, \pm1, \pm2, \dots\}$, and the positive integers, $\{1, 2, \dots\}$ | |
| $i = 1, 2, \dots, n$ | The integer variable $i$ takes on the values 1, 2, 3, and so on, up to $n$; also written $1 \leqslant i \leqslant n$ and $i = 1\colon n$ | |
| $\approx$ | Approximately equal; also written $\simeq$ | $\pi \approx 3.14$ |
| $\in$ | Belongs to | $x \in \mathbb{R}$, $n \in \mathbb{Z}$ |
| $\equiv$ | Identically equal to $f \equiv 0$ means that $f$ is the zero function, that is, $f$ is zero for all values, not just some values, of its argument(s) | |
| $n!$ | Factorial, $n! = n(n-1)\cdots 1$ | |
| $\to$ | Tends to, or converges to | $n \to \infty$ |
| $\sum$ | Summation | $\sum_{i=1}^{3} x_i = x_1 + x_2 + x_3$ |
| $\prod$ | Product | $\prod_{i=1}^{3} x_i = x_1 x_2 x_3$ |
| $\ll$, $\gg$ | Much less than, much greater than | $n \gg 1$, $0 \leqslant \varepsilon \ll 1$ |
| $\delta_{ij}$ | Kronecker delta: $\delta_{ij} = 1$ if $i = j$ and $\delta_{ij} = 0$ if $i \neq j$ | |
| $[a,b]$, $(a,b)$, $[a,b)$ | The closed interval $\{x\colon a \leqslant x \leqslant b\}$, the open interval $\{x\colon a < x < b\}$, and the half-closed, half-open interval $\{x\colon a \leqslant x < b\}$ | |
| $f\colon P \to Q$ | The function $f$ maps the set $P$ to the set $Q$, that is, $x \in P$ implies $f(x) \in Q$ | |
| $f', f'', f''', f^{(k)}$ | First, second, third, and $k$th derivatives of the function $f$ | |
| $\dot{f}, \ddot{f}$ | First and second derivatives of the function $f$ | |
| $C[a,b]$ | Real-valued continuous functions on $[a,b]$ | $f \in C[a,b]$ |
| $C^k[a,b]$ | Real-valued functions with continuous derivatives of order $0, 1, \dots, k$ on $[a,b]$ | $f \in C^2[a,b]$ |
| $L^2[a,b]$ | The functions $f\colon \mathbb{R} \to \mathbb{R}$ such that the Lebesgue integral $\int_a^b f(x)^2\,dx$ exists | |
| $f \circ g$ | Composition of functions: $(f \circ g)(x) = f(g(x))$ | $e^{x^2} = e^x \circ x^2$ |
| $:=$, $=:$ | Definition of a variable or function, to distinguish from mathematical equality | $y' = 1 + y^4 =: f(y)$ |

## 4 Functions

A *function* $f$ is a rule that assigns for each value of $x$ a unique value $f(x)$. It can be thought of as a black box that takes an input $x$ and produces an output $y = f(x)$. A function is sometimes called a *mapping*. If we write $y = f(x)$ then $y$ is the *dependent variable* and $x$ is the *independent variable*, also called the *argument* of $f$.

For some functions there is not a unique value of $f(x)$ for a given $x$, and these *multivalued functions* are not true functions unless restrictions are imposed. For example, consider $y = \log x$, which in general denotes any solution of the equation $e^y = x$. There are infinitely many solutions, which can be written as $y = y_0 + 2\pi i k$ for $k \in \mathbb{Z}$, where $y_0$ is the *principal logarithm*, defined as the logarithm whose imaginary part lies in $(-\pi, \pi]$. The principal logarithm is often the one that is needed in practice and is usually the one computed by software. Multivalued functions of a complex variable can be elegantly handled using RIEMANN SURFACES [IV.1 §2] and BRANCH CUTS [IV.1 §2].

A function is *linear* if the independent variable appears only to the first power. Thus the function $f(x) = ax + b$, where $a$ and $b$ are constants, is linear in $x$. In some contexts, e.g., in convex optimization, $ax + b$ is called an *affine function* and the term linear is reserved for $f(x) = ax$, for which $f(tx) = tf(x)$ for all $t$.

A function $f$ is *odd* if $f(x) = -f(-x)$ for all $x$ and it is *even* if $f(x) = f(-x)$ for all $x$. For example, the sine function is odd, whereas $x^2$ and $|x|$ are even.

It is worth noting the distinction between the function $f$ and its value $f(x)$ at a particular point $x$. Sometimes this distinction is blurred; for example, one might write "the function $f(u,v)$," in order to emphasize the symbols being used for the independent variables.

Functions with more than one independent variable are called *multivariate functions*. For ease of notation the independent variables can be collected into a vector. For example, the multivariate function $f(u,v) = \cos u \sin v$ can be written $f(x) = \cos x_1 \sin x_2$, where $x = [x_1, x_2]^T$.

## 5 Limits and Continuity

The notion of a function converging to a limit as its argument approaches a certain value seems intuitively obvious. For example, the statement that $x^2 \to 4$ as $x \to 2$, where the symbol "$\to$" means tends to or converges to, is clearly true, as can be seen by considering the graph of $x^2$. However, we need to make the notion of convergence precise because a large number of definitions are built on it.

Let $f$ be a real function of a real variable. We say that $f(x) \to \ell$ as $x \to a$, and we write $\lim_{x \to a} f(x) = \ell$, if for every $\varepsilon > 0$ there is a $\delta > 0$ such that $0 < |x-a| < \delta$ implies $|f(x) - \ell| < \varepsilon$. In other words, by choosing $x$ close enough to $a$, $f(x)$ can be made as close as desired to $\ell$. Showing that the definition holds in a particular case boils down to determining $\delta$ as a function of $\varepsilon$.

It is implicit in this definition that $\ell$ is finite. We say that $f(x) \to \infty$ as $x \to a$ if for every $\rho > 0$ there is a $\delta > 0$ such that $|x - a| < \delta$ implies $f(x) > \rho$.

In practice, mathematicians rarely prove existence of a limit by exhibiting the appropriate $\delta = \delta(\varepsilon)$ in these definitions. For example, one would argue that $\tan x \to \infty$ as $x \to \pi/2$ because $\sin x \to 1$ and $\cos x \to 0$ as $x \to \pi/2$. However, the definition might be used if $f$ is an implicitly defined function whose behavior is not well understood.

We can also define one-sided limits, in which the limiting value of $x$ is approached from the right or the left. For the right-sided limit $\lim_{x \to a^+} f(x) = \ell$, the definition of limit is modified so that $0 < |x - a| < \delta$ is replaced by $a < x < a + \delta$, and the left-sided limit $\lim_{x \to a^-} f(x)$ is defined analogously. The standard limit exists if and only if the right- and left-sided limits exist and are equal.

The function $f$ is *continuous* at $x = a$ if $f(a)$ exists and $\lim_{x \to a} f(x) = f(a)$.

The definitions of limit and continuity apply equally well to functions of a complex variable. Here, the condition $|x - a| < \delta$ places $x$ in a disk of radius less than $\delta$ in the complex plane instead of an open interval on the real axis.

The function $f$ is continuous on $[a,b]$ if it is continuous at every point in that interval. A more restricted form of continuity is Lipschitz continuity. The function $f$ is *Lipschitz continuous* on $[a,b]$ if

$$|f(x) - f(y)| \leqslant L|x - y| \quad \text{for all } x, y \in [a,b]$$

for some constant $L$, which is called the *Lipschitz constant*. This definition, which is quantitative as opposed to the purely qualitative usual definition of continuity, is useful in many settings in applied mathematics. A function may, however, be continuous without being Lipschitz continuous, as $f(x) = x^{1/2}$ on $[0,1]$ illustrates.

A *sequence* $a_1, a_2, a_3, \dots$ of real or complex numbers, written $\{a_n\}$, has limit $c$ if for every $\varepsilon > 0$ there is a positive integer $N$ such that $|a_n - c| < \varepsilon$ for all $n \geqslant N$. We write $c = \lim_{n \to \infty} a_n$. An *infinite series* $\sum_{i=1}^{\infty} a_i$ converges if the sequence of *partial sums* $\sum_{i=1}^{n} a_i$ converges.

## 6 Bounds

In applied mathematics we are often concerned with deriving bounds for quantities of interest. For example, we might wish to find a constant $u$ such that $f(x) \leqslant u$ for all $x$ on a given interval. Such a $u$, if it exists, is called an *upper bound*. Similarly, a lower bound is a constant $\ell$ such that $f(x) \geqslant \ell$ for all $x$ on the interval. Of particular interest is the *least upper bound*, also called the *supremum* or sup, which is the smallest possible upper bound. The supremum might not actually be attained, as illustrated by the function $f(x) = x/(1+x)$ on $[0, \infty)$, which has supremum 1. The *infimum*, or inf, is the greatest possible lower bound.

A function that has an upper (or lower) bound is said to be *bounded above* (or *bounded below*). If the function is bounded both above and below it is said to be *bounded*. A function that is not bounded is *unbounded*.

Determining whether a certain function, perhaps a function of several variables or one defined in a FUNCTION SPACE [II.15], is bounded can be nontrivial and it is often a crucial step in proving the convergence of a process or determining the quality of an approximation.

Physical considerations sometimes imply that a function is bounded. For example, a function that represents energy must be nonnegative.

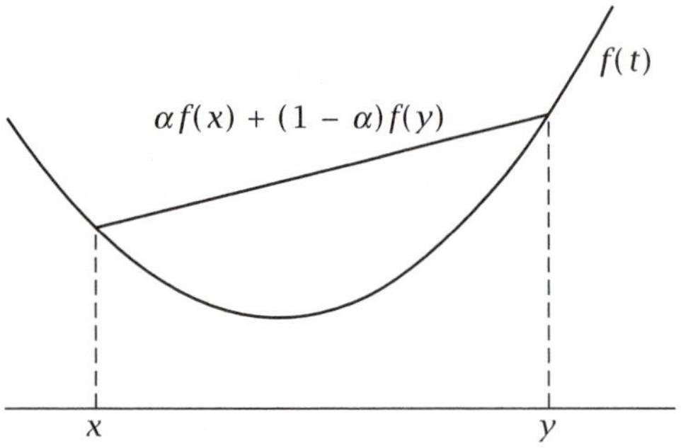


**Figure 3** A convex function, illustrating the inequality (1).

## 7 Sets and Convexity

Three types of sets in $\mathbb{R}$ or $\mathbb{C}$ are commonly used in applied mathematics.

An *open set* is a set such that for every point in the set there is an open disk around it lying entirely in the set. An *open disk* (or *open ball*) around a point $a$ in $\mathbb{R}$ or $\mathbb{C}$ is the set of all points $z$ satisfying $|z-a| < \varepsilon$ for some specified $\varepsilon > 0$. For example, $\{z \in \mathbb{C}\colon |z| < 1\}$ is an open disk. In $\mathbb{R}$, an open disk reduces to an open interval $(a-\varepsilon, a+\varepsilon)$. A *closed set* is a set that is the complement of an open set; that is, it comprises all the points that are not in some open set. For example, $\{z \in \mathbb{C}\colon |z| \leqslant 1\}$ is a closed disk, the complement of the open set $\{z \in \mathbb{C}\colon |z| > 1\}$.

A *bounded set* is one for which there is a constant $M$ such that $|x| \leqslant M$ for all $x$ in the set. A set is *compact* if it is closed and bounded.

A *convex set* is a set for which the line joining any two points $x$ and $y$ in the set lies in the set, that is, $\alpha x + (1-\alpha)y$ is in the set for all $\alpha \in [0,1]$. A related notion is that of a *convex function*. A real-valued function $f$ is *convex* on a convex set $S$ if

$$f(\alpha x + (1-\alpha)y) \leqslant \alpha f(x) + (1-\alpha)f(y) \qquad (1)$$

for all $x, y \in S$ and $\alpha \in [0,1]$. This inequality says that on the interval defined by $x$ and $y$ the function $f$ lies below the line joining $f(x)$ and $f(y)$ (see figure 3). An example of a convex function is $f(x) = x^2$ on the real line. A *concave function* is one satisfying (1) with the inequality reversed.

## 8 Order Notation

We write $x \approx y$ to mean that $x$ is approximately equal to $y$. The accuracy of the approximation may be implied by the context or the way $y$ is written. For example, the statement that $\pi \approx 3.14$ implies that the approximation is correct to two decimal places.

The big-oh and little-oh notations, $O(\cdot)$ and $o(\cdot)$, are used to give information about the relative behavior of two functions. We write

(i) $f(z) = O(g(z))$ as $z \to \infty$ (or $z \to 0$) if $|f(z)| \leqslant c|g(z)|$ for some constant $c$ for all sufficiently large $|z|$ (or all sufficiently small $|z|$);

(ii) $f(z) = o(g(z))$ as $z \to \infty$ (or $z \to 0$) if $f(z)/g(z) \to 0$ as $z \to \infty$ (or $z \to 0$).

In both cases, $g$ is usually a well-understood function and $f$ is a function whose behavior we are trying to understand.

To illustrate: $z^3 + z^2 + z + 1 = O(z^3)$ as $z \to \infty$ and $z^3 + z^2 + z = O(z)$ as $z \to 0$, while $z = o(\mathrm{e}^z)$ as $z \to \infty$.

Big-oh notation is frequently used when comparing the cost of algorithms measured as a function of problem size. For example, the cost of multiplying two $n \times n$ matrices by the usual formulas is $n^3 + q(n)$ additions and $n^3$ multiplications, where $q$ is a quadratic function. We can say that matrix multiplication costs $2n^3 + O(n^2)$ operations, or simply $O(n^3)$ operations.

We write $f(z) \sim g(z)$ (in words, "$f(z)$ twiddles $g(z)$") if $f(z)/g(z)$ tends to 1 as $z$ tends to some quantity $z_0$ (sometimes the ratio is required only to tend to a finite, nonzero limit). For example, $\sin z \sim z$ as $z \to 0$, $\sum_{i=1}^{n} i^2 \sim n^2/3$ as $n \to \infty$, and $n! \sim \sqrt{2\pi n}(n/\mathrm{e})^n$ as $n \to \infty$, the last approximation, called *Stirling's approximation*, being good even for small $n$.

## 9 Calculus

The rate of change of a quantity is a fundamental concept. The rate of change of the distance of a moving object from a given point is its speed, and the rate of change of speed is acceleration. In economics, inflation is the rate of change of a price index. The rate of change of a function is its derivative. Let $f$ be a real function of a real variable. Intuitively, the rate of change of $f$ at $x$ is obtained by making a small change in $x$ and taking the ratio of the corresponding change in $f$ to the change in $x$, that is, $(f(x+\varepsilon) - f(x))/\varepsilon$ for small $\varepsilon$. In order to get a unique quantity we take the limit as $\varepsilon \to 0$, which gives the derivative

$$\frac{\mathrm{d}f}{\mathrm{d}x} = f'(x) = \lim_{\varepsilon \to 0} \frac{f(x+\varepsilon) - f(x)}{\varepsilon}.$$

The derivative may or may not exist. For example, the absolute value function $f(x) = |x|$ is not differentiable at the origin because the left- and right-sided limits

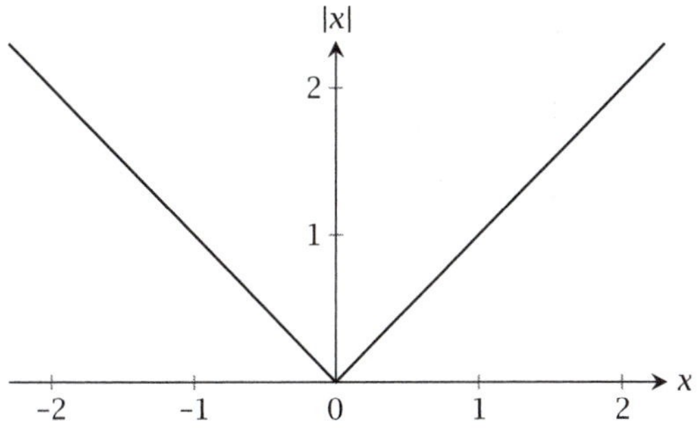


**Figure 4** The absolute value function, $|x|$.

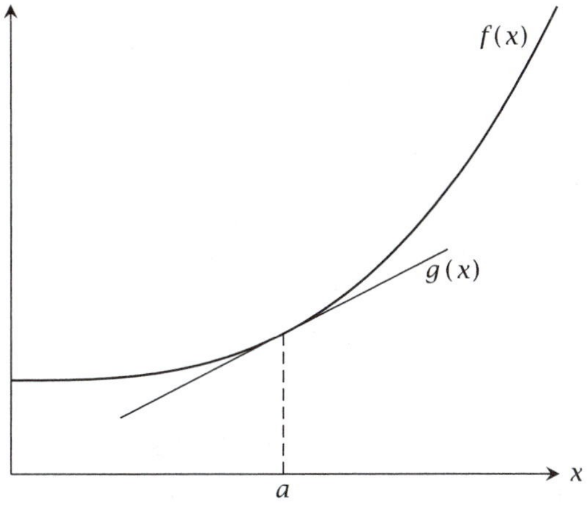


**Figure 5** The function $f(x)$ and the tangent $g$ to $f$ at $x = a$. The tangent is $g(x) = f'(a)(x - a) + f(a)$.

are different: $\lim_{\varepsilon \to 0^-}(f(x+\varepsilon) - f(x))/\varepsilon = -1$ and $\lim_{\varepsilon \to 0^+}(f(x+\varepsilon) - f(x))/\varepsilon = 1$ (see figure 4). Higher derivatives are defined by applying the definition recursively; thus $f''(x)$ is the derivative of $f'(x)$.

A graphical interpretation of the derivative is that it is the slope of the tangent to the curve $y = f(x)$ (see figure 5).

Another way to write the definition of derivative is as

$$f(x+\varepsilon) - f(x) - f'(x)\varepsilon = o(\varepsilon).$$

This definition has the benefit of generalizing naturally to FUNCTION SPACES [II.15], where it yields the *Fréchet derivative*.

A zero derivative identifies stationary points of a function, with the type of stationary point—maximum, minimum, or saddle point (also called a point of inflection)—being determined by the second and possibly higher derivatives. This can be seen with the aid

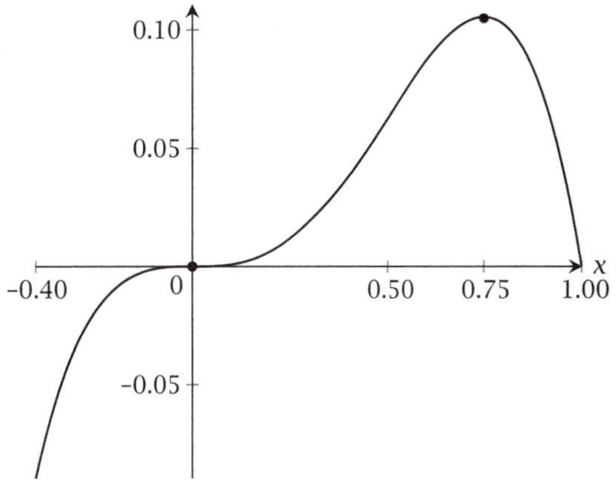


**Figure 6** The function $f(x) = x^3 - x^4$ with a saddle point at $x = 0$ and a maximum at $x = 3/4$.

of a *Taylor series* about the point $a$ of interest:

$$f(x) = f(a) + f'(a)(x-a) + f''(a)\frac{(x-a)^2}{2!} + f'''(a)\frac{(x-a)^3}{3!} + \cdots.$$

If $f'(a) = 0$ and $f''(a) \neq 0$ then, for $x$ sufficiently close to $a$, $f(x) \approx f(a) + f''(a)(x-a)^2/2$, and so $a$ is a maximum point if $f''(a) < 0$ and a minimum point if $f''(a) > 0$. If $f'(a) = f''(a) = 0$ then we need to look at the higher-order derivatives to determine the nature of the stationary point; in particular, if $f'''(a) \neq 0$ then $a$ is a saddle point (see figure 6).

The error in truncating a Taylor series is captured in the *Taylor series with remainder*:

$$f(x) = \sum_{k=0}^{n} f^{(k)}(a)\frac{(x-a)^k}{k!} + f^{(n+1)}(\xi)\frac{(x-a)^{n+1}}{(n+1)!},$$

where $\xi$ is an unknown point on the interval with endpoints $a$ and $x$. For $n = 0$ this reduces to $f(x) - f(a) = f'(\xi)(x-a)$, which is the *mean-value theorem*.

Rigorous statements of results must include assumptions about the smoothness of the functions involved, that is, how many derivatives are assumed to exist. For example, the Taylor series with remainder is valid if $f$ is $(n+1)$-times continuously differentiable on an interval containing $x$ and $a$. In applied mathematics we often avoid clutter by writing "for smooth functions $f$" to indicate that the existence of continuous derivatives up to some order is assumed. Underlying such a statement might be some known minimal assumption on $f$, or just the knowledge that the existence of continuous derivatives of all orders is sufficient and that less restrictive conditions can usually be derived if necessary. Sometimes, when deriving or using results, it is not possible to check smoothness conditions and one simply carries on anyway ("making compromises," as mentioned in the quote by Courant on page 1). It may be possible to verify by other means that an answer obtained in a nonrigorous way is valid.

For a function $f(x, y)$ of two variables, partial derivatives with respect to each of the two variables are defined by holding one variable constant and varying the other:

$$\frac{\partial f}{\partial x} = \lim_{\varepsilon \to 0} \frac{f(x+\varepsilon, y) - f(x, y)}{\varepsilon},$$
$$\frac{\partial f}{\partial y} = \lim_{\varepsilon \to 0} \frac{f(x, y+\varepsilon) - f(x, y)}{\varepsilon}.$$

Higher derivatives are defined recursively. For example,

$$\frac{\partial^2 f}{\partial x^2} = \lim_{\varepsilon \to 0} \frac{\frac{\partial f}{\partial x}(x+\varepsilon, y) - \frac{\partial f}{\partial x}(x, y)}{\varepsilon},$$
$$\frac{\partial^2 f}{\partial x \partial y} = \lim_{\varepsilon \to 0} \frac{\frac{\partial f}{\partial x}(x, y+\varepsilon) - \frac{\partial f}{\partial x}(x, y)}{\varepsilon},$$
$$\frac{\partial^2 f}{\partial y \partial x} = \lim_{\varepsilon \to 0} \frac{\frac{\partial f}{\partial y}(x+\varepsilon, y) - \frac{\partial f}{\partial y}(x, y)}{\varepsilon}.$$

Common abbreviations are $f_x = \partial f/\partial x$, $f_{xy} = \partial^2 f/(\partial x \partial y)$, $f_{yy} = \partial^2 f/\partial y^2$, and so on. As long as they are continuous the two mixed second-order partial derivatives are equal: $f_{xy} = f_{yx}$.

For a function of $n$ variables, $F\colon \mathbb{R}^n \to \mathbb{R}$, a Taylor series takes the form, for $x, a \in \mathbb{R}^n$,

$$F(x) = F(a) + \nabla F(a)^{\mathrm{T}}(x-a) + \tfrac{1}{2}(x-a)^{\mathrm{T}}\nabla^2 F(a)(x-a) + \cdots,$$

where $\nabla F(x) = (\partial F/\partial x_j) \in \mathbb{R}^n$ is the *gradient vector* and $\nabla^2 F(x) = (\partial^2 F/(\partial x_i \partial x_j)) \in \mathbb{R}^{n \times n}$ is the symmetric *Hessian matrix*, with $x_j$ denoting the $j$th component of the vector $x$. The symbol $\nabla$ is called nabla. Stationary points of $F$ are zeros of the gradient and their nature (maximum, minimum, or saddle point) is determined by the eigenvalues of the Hessian (see CONTINUOUS OPTIMIZATION [IV.11 §2]).

Now we return to functions of a single (real) variable. The *indefinite integral* of $f(x)$ is $\int f(x)\,\mathrm{d}x$, while integrating between limits $a$ and $b$ gives the *definite integral* $\int_a^b f(x)\,\mathrm{d}x$. The definite integral can be interpreted as the area under the curve $f(x)$ between $a$ and $b$. The inverse of differentiation is integration, as shown by the *fundamental theorem of calculus*, which states that, if $f$ is continuous on $[a, b]$, then the function $g(x) = \int_a^x f(t)\,\mathrm{d}t$ is differentiable on $(a, b)$ and $g'(x) = f(x)$. Generalizations of the fundamental theorem of calculus to functions of more than one variable are given in section 24.

For functions of two or more variables there are other kinds of integrals. When there are two variables, $x$ and $y$, we can integrate over regions in the $xy$-plane (double integrals) or along curves in the plane (line integrals). For functions of three variables, $x$, $y$, and $z$, there are more possibilities. We can integrate over volumes (triple integrals) or over surfaces or along curves within $xyz$-space. As the number of variables increases, so does the number of different kinds of integrals. Multidimensional calculus shows how these different integrals can be calculated, used, and related. The number of variables can be very large (e.g., in mathematical finance) and the CURSE OF DIMENSIONALITY [I.3 §2] poses major challenges for numerical evaluation. Numerical integration in more than one dimension is an active area of research, and Monte Carlo methods and quasi-Monte Carlo methods are among the methods in use.

The *product rule* gives a formula for the derivative of a product of two functions:

$$\frac{\mathrm{d}}{\mathrm{d}x} f(x)g(x) = f'(x)g(x) + f(x)g'(x).$$

Integrating this equation gives the rule for *integration by parts*:

$$\int f(x)g'(x)\,\mathrm{d}x = f(x)g(x) - \int f'(x)g(x)\,\mathrm{d}x.$$

In many problems functions are composed: the argument of a function is another function. Consider the example $f(x) = g(h(x))$. We would hope to be able to determine the derivative of $f$ in terms of the derivatives of $g$ and $h$. The *chain rule* provides the necessary formula: $f'(x) = h'(x)g'(h(x))$. An equivalent formulation is that, if $f$ is a function of $u$, which is itself a function of $x$, then

$$\frac{\mathrm{d}f}{\mathrm{d}x} = \frac{\mathrm{d}f}{\mathrm{d}u}\frac{\mathrm{d}u}{\mathrm{d}x}.$$

For example, if $f(x) = \sin x^2$ then with $f(x) = \sin u$ and $u = x^2$ we have $\mathrm{d}f/\mathrm{d}x = 2x\cos x^2$.

## 10 Ordinary Differential Equations

A differential equation is an equation containing one or more derivatives of an unknown function. It provides a relation among a function, its rate of change, and (possibly) higher-order rates of change. The independent variable usually represents a spatial coordinate ($x$) or time ($t$). The differential equation may be accompanied by additional information about the function, called *boundary conditions* or *initial conditions*, that serve to uniquely determine the solution. A solution to a differential equation is a function that satisfies the equation for all values of the independent variables (perhaps in some region) and also satisfies the required boundary conditions or initial conditions. A differential equation can express a law of motion, a conservation law, or concentrations of constituents of a chemical reaction, for example.

*Ordinary differential equations* (ODEs) contain just one independent variable. The simplest nontrivial ODE is $\mathrm{d}y/\mathrm{d}t = ay$, where $y = y(t)$ is a function of $t$. This equation is linear in $y$ and it is *first order* because only the first derivative of $y$ appears. The general solution is $y(t) = ce^{at}$, where $c$ is an arbitrary constant. To determine $c$, some value of $y$ must be supplied, say $y(0) = y_0$, whence $c = y_0$.

A general first-order ODE has the form $y' = f(t, y)$ for some function $f$ of two variables. The *initial-value problem* supplies an initial condition and asks for $y$ at later times:

$$y' = f(t, y), \quad a \leqslant t \leqslant b, \quad y(a) = y_a.$$

A specific example is the Riccati equation

$$y' = t^2 + y^2, \quad 0 \leqslant t \leqslant 1, \quad y(0) = 0,$$

which is nonlinear because of the appearance of $y^2$.

For an example of a second-order ODE initial-value problem, that is, one involving $y''$, consider a mass $m$ attached to a vertical spring and to a damper, as shown in figure 7. Let $y = y(t)$ denote how much the spring is stretched from its natural length at time $t$. Balancing forces using Newton's second law (force equals mass times acceleration) and HOOKE'S LAW [III.15] gives

$$my'' = mg - ky - cy',$$

where $k$ is the spring constant, $c$ is the damping constant, and $g$ is the gravitational constant. With prescribed values for $y(0)$ and $y'(0)$ this is an initial-value problem. More generally, the spring might also be subjected to an external force $f(t)$, in which case the equation of motion becomes

$$my'' + cy' + ky = mg + f(t).$$

Second-order ODEs also arise in electrical networks. Consider the flow of electric current $I(t)$ in a simple RLC circuit composed of an inductor with inductance $L$, a resistor with resistance $R$, a capacitor with capacitance $C$, and a source with voltage $v_S$, as illustrated in figure 8. The Kirchhoff voltage law states that the sum of the voltage drops around the circuit equals the input voltage, $v_S$. The voltage drops across the resistor, inductor, and capacitor are $RI$, $L\mathrm{d}I/\mathrm{d}t$, and $Q/C$, respectively, where $Q(t)$ is the charge on the capacitor, so

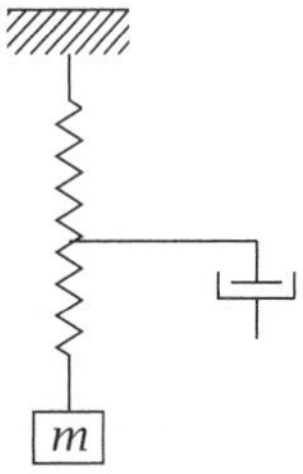


**Figure 7** A spring system with damping.

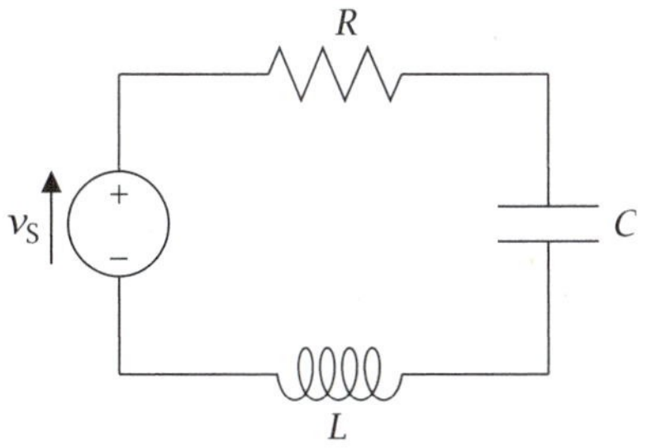


**Figure 8** A simple RLC electric circuit.

$$L\frac{\mathrm{d}I}{\mathrm{d}t} + RI + \frac{Q}{C} = v_S(t).$$

Since $I = \mathrm{d}Q/\mathrm{d}t$, this equation can be rewritten as the second-order ODE

$$L\frac{\mathrm{d}^2Q}{\mathrm{d}t^2} + R\frac{\mathrm{d}Q}{\mathrm{d}t} + \frac{1}{C}Q = v_S(t).$$

The unknown function $y$ may have more than one component, as illustrated by the predator–prey model derived by Lotka and Volterra in the 1920s. In a population of rabbits (the prey) and foxes (the predators) let $r(t)$ be the number of rabbits at time $t$ and $f(t)$ the number of foxes at time $t$. The model is

$$\frac{\mathrm{d}r}{\mathrm{d}t} = r - \alpha rf, \quad r(0) = r_0,$$
$$\frac{\mathrm{d}f}{\mathrm{d}t} = -f + \alpha rf, \quad f(0) = f_0.$$

The $rf$ term represents an interaction between the foxes and the rabbits (a fox eating a rabbit) and the parameter $\alpha \geqslant 0$ controls the amount of interaction. For $\alpha = 0$ there is no interaction and the solution is

$r(t) = r_0 e^t$, $f(t) = f_0 e^{-t}$: the foxes die from starvation and the rabbits go forth and multiply, unhindered. The aim is to investigate the behavior of the solutions for various parameters $\alpha$ and starting populations $r_0$ and $f_0$.

As we have described it, the predator–prey model has the apparent contradiction that $r$ and $f$ are integers by definition yet the solutions to the differential equation are real-valued. The way around this is to assume that $r$ and $f$ are large enough for the error in representing them by continuous variables to be small.

A *boundary-value problem* specifies the function at more than one value of the independent variable, as in the two-point boundary-value problem

$$y'' = f(t, y, y'), \quad a \leqslant t \leqslant b, \quad y(a) = y_a, \; y(b) = y_b.$$

An example is the Thomas–Fermi equation

$$y'' = t^{-1/2} y^{3/2}, \qquad y(0) = 1, \quad y(\infty) = 0,$$

which arises in a semiclassical description of the charge density in atoms of high atomic number. Another example, this time of third order, is the BLASIUS EQUATION [IV.28 §7.2]

$$2y''' + yy'' = 0, \qquad y(0) = y'(0) = 0, \quad y'(\infty) = 1,$$

which describes the boundary layer in a fluid flow.

A special type of ODE boundary-value problem is the *Sturm–Liouville problem*

$$-(p(x)y'(x))' + q(x)y(x) = \lambda r(x)y(x),$$
$$x \in [a, b], \quad y(a) = y(b) = 0.$$

This is an *eigenvalue problem*, meaning that the aim is to determine values of the parameter $\lambda$ for which the boundary-value problem has a solution that is not identically zero.

## 11 Partial Differential Equations

Many important physical processes are modeled by partial differential equations (PDEs): differential equations containing more than one independent variable. We summarize a few key equations and basic concepts. We write the equations in forms where the unknown $u$ has two space dimensions, $u = u(x, y)$, or one space dimension and one time dimension, $u = u(x, t)$. Where possible, the equations are given in parameter-free form, a form that is obtained by the process of NONDIMENSIONALIZATION [II.9]. Recall the abbreviations $u_t = \partial u/\partial t$, $u_{xx} = \partial^2 u/\partial x^2$, etc.

LAPLACE'S EQUATION [III.18] is

$$u_{xx} + u_{yy} = 0.$$

The left-hand side of the equation is the *Laplacian* of $u$, written $\Delta u$. This equation is encountered in electrostatics (for example), where $u$ is the potential function. The equation $\Delta u = f$, for a given function $f(x, y)$, is known as *Poisson's equation*.

To define a problem with a unique solution it is necessary to augment the PDE with conditions on the solution: either boundary conditions for static problems or, for time-dependent problems, initial conditions. In the former class there are three main types of boundary conditions, with the problem being to determine $u$ inside the boundary of a closed region.

- *Dirichlet conditions*, in which the function $u$ is specified on the boundary.
- *Neumann conditions*, where the inner product (see section 19.1) of the gradient
$$\nabla u = [\partial u/\partial x, \partial u/\partial y]^{\mathrm{T}}$$
with the normal to the boundary is specified.
- *Cauchy conditions*, which comprise a combination of Dirichlet and Neumann conditions.

For time-dependent problems, which are known as evolution problems and represent equations of motion, initial conditions at the starting time, usually taken to be $t = 0$, are needed, the number of initial conditions depending on the highest order of time derivative in the PDE.

The WAVE EQUATION [III.31] is

$$u_{tt} = u_{xx}.$$

It describes linear, nondispersive propagation of a wave, represented by the wave function $u$, e.g., a vibrating string. Two initial conditions, prescribing $u(x, 0)$ and $u_t(x, 0)$, for example, are needed to determine $u$.

The HEAT EQUATION [III.8] (*diffusion equation*) is

$$u_t = u_{xx}, \tag{2}$$

which describes the diffusion of heat in a solid or the spread of a disease in a population. An initial condition prescribing $u$ at $t = 0$ is usual. When a term $f(x, t, u)$ is added to the right-hand side of (2) the equation becomes a *reaction–diffusion equation*.

The *advection–diffusion equation* is

$$u_t + vu_x = u_{xx},$$

where $v$ is a given function of $x$ and $t$. Again, $u$ is usually given at $t = 0$. For $v = 0$ this is just the heat equation. This PDE models the convection (or transport) of a quantity such as a pollutant in the atmosphere.

The general linear second-order PDE

$$au_{xx} + 2bu_{xt} + cu_{tt} = f(x, t, u, u_x, u_t) \quad (3)$$

is classified into different types according to the (constant) coefficients of the second derivatives. Let $d = ac - b^2$, which is the determinant of the symmetric matrix $\left[\begin{smallmatrix} a & b \\ b & c \end{smallmatrix}\right]$.

- If $d > 0$ the PDE is *elliptic*. These PDEs, of which the Laplace equation is a particular case, are associated with equilibrium or steady-state processes. The independent variables are denoted by $x$ and $y$ instead of $x$ and $t$.
- If $d = 0$ the PDE is *parabolic*. This is an evolution problem governing a diffusion process. The heat equation is an example.
- If $d < 0$ the PDE is *hyperbolic*. This is an evolution problem, governing wave propagation. The wave equation is an example.

Some elliptic PDEs and parabolic PDEs have *maximum principles*, which say that the solution must take on its maximum value on the boundary of the domain over which it is defined.

In (3) we took $a$, $b$, and $c$ to be constants, but they may also be specified as functions of $x$ and $t$, in which case the nature of the PDE can change as $x$ and $t$ vary in the domain. For example, the TRICOMI EQUATION [III.30]

$$u_{xx} + xu_{yy} = 0$$

is hyperbolic for $x < 0$, elliptic for $x > 0$, and parabolic for $x = 0$.

The PDEs stated so far are all linear. Nonlinear PDEs, in which the unknown function appears nonlinearly, are of great practical importance. Examples are the KORTEWEG–DE VRIES EQUATION [III.16]

$$u_t + uu_x + u_{xxx} = 0,$$

the CAHN–HILLIARD EQUATION [III.5]

$$u_t = \Delta(-u + u^3 + \varepsilon^2 \Delta u),$$

and *Fisher's equation*

$$u_t = u_{xx} + u(1 - u),$$

a reaction–diffusion equation that describes PATTERN FORMATION [IV.27] and the propagation of genes in a population.

PDEs also occur in the form of eigenvalue problems. A famous example is the eigenvalue problem corresponding to the Laplace equation:

$$\Delta u + \lambda u = 0$$

on a membrane $\Omega$, with boundary conditions that $u$ vanishes on the boundary of $\Omega$. A nonzero solution $u$ is called an *eigenfunction* and $\lambda$ is the corresponding *eigenvalue*. In a 1966 paper titled "Can one hear the shape of a drum?" Mark Kac asked the question of whether one can determine $\Omega$ given all the eigenvalues. In other words, do the frequencies at which a drum vibrates uniquely determine its shape? It was shown in a 1992 paper by Gordon, Webb, and Wolpert that the answer is no in general.

Higher-order PDEs also arise. For example, fluid dynamics problems involving surface tension forces are generally modeled by PDEs in space and time with fourth-order derivatives in space. The same is true of the *Euler–Bernoulli equation* for a beam, which has the form

$$\rho A \frac{\partial^2 u}{\partial t^2} + EI \frac{\partial^4 u}{\partial x^4} = f(x, t),$$

where $u(x, t)$ is the vertical displacement of the beam at time $t$ and position $x$ along the beam, $\rho$ is the density of the beam, $A$ its cross-sectional area, $E$ is Young's modulus, $I$ is the second moment of inertia, and $f(x, t)$ is an applied force.

## 12 Other Types of Differential Equations

*Delay differential equations* are differential equations in which the derivative of the unknown function $y$ at time $t$ (in general, a vector function) depends on past values of $y$ and/or its derivatives. For example, $y'(t) = Ay(t - 1)$ is a delay differential equation analogue of the familiar $y'(t) = Ay(t)$. Looking for a solution of the form $y(t) = e^{wt}$ leads to the equation $we^w = A$, whose solutions are given by the LAMBERT $W$ FUNCTION [III.17].

INTEGRAL EQUATIONS [IV.4] contain the unknown function inside an integral. Examples are *Fredholm equations*, which are of the form either

$$\int_0^1 K(x, y) f(y)\, dy = g(x),$$

where $K$ and $g$ are given and the task is to find $f$, or

$$\lambda \int_0^1 K(x, y) f(y)\, dy + g(x) = f(x),$$

where $\lambda$ is an eigenvalue and again $f$ is unknown. These two types of equations are analogous to a matrix linear system $Kf = g$ and an eigenvalue problem $(I - \lambda K)f = g$, respectively. *Integro-differential equations* involve both integrals and derivatives (see, for example, MODELING A PREGNANCY TESTING KIT [VII.18 §2]).

*Fractional differential equations* contain fractional derivatives. For example, $(\mathrm{d}/\mathrm{d}x)^{1/2}$ is defined to be an operator such that applying $(\mathrm{d}/\mathrm{d}x)^{1/2}$ twice in succession to a function $f(x)$ is the same as differentiating it once (that is, applying $\mathrm{d}/\mathrm{d}x$).

*Differential-algebraic equations* (DAEs) are systems of equations that contain both differential and algebraic equations. For example, the DAE

$$x'' = -2\lambda x,$$
$$y'' = -2\lambda y - g,$$
$$x^2 + y^2 = L^2$$

describes the coordinates of an infinitesimal ball of mass 1 at the end of a pendulum of length $L$, where $g$ is the gravitational constant and $\lambda$ is the tension in the rod. DAEs often arise in the form $My' = f(t, y)$, where the matrix $M$ is singular.

## 13 Recurrence Relations

*Recurrence relations* are the discrete counterpart of differential equations. They define a sequence $x_0, x_1, x_2, \dots$ recursively, by specifying $x_n$ in terms of earlier terms in the sequence. Such equations are also called *difference equations*, as they arise when derivatives in differential equations are replaced by FINITE DIFFERENCES [II.11].

A famous recurrence is the three-term recurrence that defines the *Fibonacci numbers*:

$$f_n = f_{n-1} + f_{n-2}, \quad n \geqslant 2, \quad f_0 = f_1 = 1.$$

This recurrence has the explicit solution $f_n = (\phi^n - (-\phi)^{-n})/\sqrt{5}$, where $\phi = (1 + \sqrt{5})/2$ is the *golden ratio*. An example of a two-term recurrence is $f(n) = nf(n-1)$, with $f(0) = 1$, which defines the factorial function $f(n) = n!$. Both the examples so far are linear recurrences, but in some recurrences the earlier terms appear nonlinearly, as in the LOGISTIC RECURRENCE [III.19] $x_{n+1} = \mu x_n(1 - x_n)$.

Although one can evaluate the terms in a recurrence one often needs an explicit formula for the general solution of the recurrence. Recurrence relations have a theory analogous to that of differential equations, though it is much less frequently encountered in courses and textbooks than it was fifty years ago.

The elements in a recurrence can be functions as well as numbers. Most transcendental functions that carry subscripts satisfy a recurrence. For example, the BESSEL FUNCTION [III.2] $J_n(x)$ of order $n$ satisfies the three-term recurrence

$$J_{n+1}(x) = \frac{2n}{x} J_n(x) - J_{n-1}(x).$$

An important source of three-term recurrences is ORTHOGONAL POLYNOMIALS [II.29].

## 14 Polynomials

Polynomials are one of the simplest and most familiar classes of functions and they find wide use in applied mathematics. A degree-$n$ polynomial

$$p_n(x) = a_0 + a_1 x + \cdots + a_n x^n$$

is defined by its $n+1$ coefficients $a_0, \dots, a_n \in \mathbb{C}$ (with $a_n \neq 0$). Addition of two polynomials is carried out by adding the corresponding coefficients. Thus, if $q_n(x) = b_0 + b_1 x + \cdots + b_n x^n$ then $p_n(x) + q_n(x) = a_0 + b_0 + (a_1 + b_1)x + \cdots + (a_n + b_n)x^n$. Multiplication is carried out by expanding the product term by term and collecting like powers of $x$:

$$p_n(x)q_n(x) = a_0 b_0 + (a_0 b_1 + a_1 b_0)x + \cdots + (a_0 b_n + a_1 b_{n-1} + \cdots + a_n b_0)x^n.$$

The coefficient of $x^n$, $\sum_{i=0}^{n} a_i b_{n-i}$, is the *convolution* of the vectors $a = [a_0, a_1, \dots, a_n]^{\mathrm{T}}$ and $b = [b_0, b_1, \dots, b_n]^{\mathrm{T}}$. Polynomial division is also possible. Dividing $p_n$ by $q_m$ with $m \leqslant n$ results in

$$p_n(x) = q_m(x)g(x) + r(x), \tag{4}$$

where the quotient $g$ and remainder $r$ are polynomials and the degree of $r$ is less than that of $q_m$.

The *fundamental theorem of algebra* says that a degree-$n$ polynomial $p_n$ has a *root* in $\mathbb{C}$; that is, there exists $z_1 \in \mathbb{C}$ such that $p_n(z_1) = 0$. If we take $q_m(x) = x - z_1$ in (4) then we have $p_n(x) = (x - z_1)g(x) + r(x)$, where $\deg r < 1$, so $r$ is a constant. But setting $x = z_1$ we see that $0 = p_n(z_1) = r$, so $p_n(x) = (x - z_1)g(x)$ and $g$ clearly has degree $n - 1$. Repeating this argument inductively on $g$, we end up with a factorization $p_n(x) = (x - z_1)(x - z_2)\cdots(x - z_n)$, which shows that $p_n$ has $n$ roots in $\mathbb{C}$ (not necessarily distinct). If the coefficients of $p_n$ are real it does not follow that the roots are real, and indeed there may be no real roots at all, as the polynomial $x^2 + 1$ shows; however, nonreal roots must occur in complex conjugate pairs $x_j \pm \mathrm{i}y_j$.

Three basic problems associated with polynomials are as follows.

**Evaluation:** given the polynomial (specified by its coefficients), find its value at a given point. A standard way of doing this is HORNER'S METHOD [I.4 §6].

**Interpolation:** given the values of a degree-$n$ polynomial at a set of $n+1$ distinct points, find its coefficients. This can be done by various INTERPOLATION SCHEMES [I.3 §3.1].

**Root finding:** given the polynomial and the ability to evaluate it, find its roots. This is a classic problem with a vast literature, including methods specific to polynomials and specializations of general-purpose nonlinear equation solvers.

## 15 Rational Functions

A rational function is the ratio of two polynomials:

$$r_{mn}(x) = \frac{p_m(x)}{q_n(x)} = \frac{\sum_{i=0}^m a_i x^i}{\sum_{i=0}^n b_i x^i}, \quad a_m, b_n \neq 0.$$

Rational functions are more versatile than polynomials as a means of approximating other functions. As $x$ grows larger, a polynomial of degree 1 or higher necessarily blows up to infinity. In contrast, a rational function $r_{mm}$ with equal-degree numerator and denominator is asymptotic to $a_m/b_m$, as $x \to \infty$, while for $m < n$, $r_{mn}(x)$ converges to zero as $x \to \infty$. Moreover, a rational function has *poles*: certain finite values of $x$ for which it is infinite (the roots of the denominator polynomial $q_n$).

The representation of a rational function as a ratio of polynomials is just one of several possibilities. We can write $r_{mn}$ in *partial fraction* form, for example. If $m < n$ and $q_n$ has distinct roots $x_1, \dots, x_n$, then

$$r_{mn}(x) = \sum_{i=1}^n \frac{c_i}{x - x_i} \tag{5}$$

for some $c_1, \dots, c_n$. One reason to put a rational function in partial fraction form is in order to integrate it, since the integral of (5) is immediate: $\int r_{mn}(x)\,dx = \sum_{i=1}^n c_i \log|x - x_i| + C$, where $C$ is a constant.

An important class of rational functions is the *Padé approximants* to a given function $f$, which are defined by the property that $r_{mn}(x) - f(x) = O(x^k)$ with $k$ as large as possible. Since $r_{mn}$ has $m + n + 1$ degrees of freedom (one having been lost due to the division), generically $k = m + n + 1$, but $k$ can be smaller or larger than this value (see APPROXIMATION THEORY [IV.9 §2.4]). When $m = 0$, a Padé approximant reduces to a truncated Taylor series.

## 16 Special Functions

Applied mathematicians make much use of functions that are not polynomial or rational, though they may ultimately use polynomial or rational approximations to such functions. A larger class of functions is the *elementary functions*, which are made up of polynomials, rationals, the exponential, the logarithm, and all functions that can be obtained from these by addition, subtraction, multiplication, division, composition, and the taking of roots. Another important class is the *transcendental functions*: those that are not algebraic, that is, that are not the solution $f(x)$ of an equation $p(x, f(x)) = 0$, where $p(x, y)$ is a polynomial in $x$ and $y$ with integer coefficients. Examples of transcendental functions include the exponential, the logarithm, the trigonometric functions, and the hyperbolic functions.

In solving problems we talk about the ability to obtain the solution in *closed form*, which is an informal concept meaning that the solution is expressed in terms of elementary functions or functions that are "well understood," in that they have a significant literature and good algorithms exist for computing them.

The SPECIAL FUNCTIONS [IV.7] provide a large set of examples of well-understood functions. They arise in different areas, such as physics, number theory, and probability and statistics, often as the solution to a second-order ODE or as the integral of an elementary function. A general example is the HYPERGEOMETRIC FUNCTION [IV.7 §5]

$$\begin{aligned} F(a, b; c; x) &= 1 + \frac{ab}{c}x + \frac{a(a+1)b(b+1)}{2!c(c+1)}x^2 + \cdots \\ &= \sum_{i=0}^{\infty} \frac{(a)_i (b)_i}{i!(c)_i} x^i. \end{aligned}$$

Here, $a, b, c \in \mathbb{R}$, $c$ is not zero or a negative integer, and $(a)_i \equiv a(a+1)\cdots(a+i-1)$ for $i \geqslant 1$, with $(a)_0 = 1$. The hypergeometric function is a solution of the second-order differential equation

$$x(1-x)w''(x) + (c - (a+b+1)x)w'(x) - abw(x) = 0.$$

The hypergeometric functions contain many interesting special cases, such as $F(a, b; b; x) = (1-x)^{-a}$ and $F(1, 1; 2; x) = -x^{-1}\log(1-x)$.

Other special functions include the following.

- The *error function*

  $$\operatorname{erf}(x) = \frac{2}{\sqrt{\pi}} \int_0^x \mathrm{e}^{-t^2}\,dt,$$

  which is closely related to the standard normal distribution in probability and statistics.
- The GAMMA FUNCTION [III.13]

  $$\Gamma(z) = \int_0^\infty \mathrm{e}^{-t} t^{z-1}\,dt,$$

  which satisfies $\Gamma(n) = (n-1)!$ for positive integers $n$ and so generalizes the factorial function. Note that the argument $z$ is a complex number.

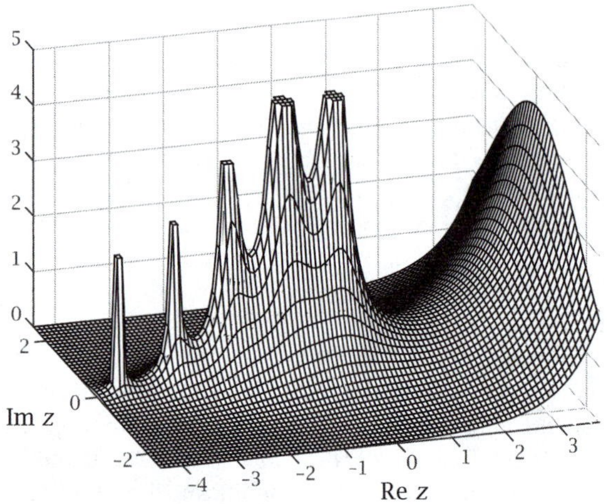


**Figure 9** The gamma function in the complex plane. The height of the surface is $|\Gamma(z)|$. The function has poles at the negative integers; in this plot the infinite peaks have been truncated at different heights.

Figure 9 is modeled on a classic, hand-drawn plot of the gamma function in the complex plane from the book *Tables of Functions with Formulas and Curves* by Eugene Jahnke and Fritz Emde, first published in 1909.

- BESSEL FUNCTIONS [III.2], the LAMBERT $W$ FUNCTION [III.17], elliptic functions, and the RIEMANN ZETA FUNCTION [IV.7 §4].

The class of special functions can be enlarged by identifying useful functions, giving them a name, studying their properties, and deriving algorithms and software for evaluating them. Of the examples mentioned above, the most recent is the Lambert $W$ function, whose significance was realized, and to which the name was given, only in the 1990s.

## 17 Power Series

A power series is an infinite expansion of the form

$$a_0 + a_1 z + a_2 z^2 + a_3 z^3 + \cdots,$$

where $z$ is a complex variable and the $a_i$ are complex constants. Results from COMPLEX ANALYSIS [IV.1 §5] tell us that such a series has a *radius of convergence* $R$ such that the series converges for $|z| < R$, diverges for $|z| > R$, and may either converge or diverge for $|z| = R$. For example, the power series $1 + z + z^2 + \cdots$ converges for $|z| < 1$, and inside this disk it agrees with the function $f(z) = (1-z)^{-1}$. More generally, a power-series expansion can be taken about an arbitrary point $z_0$: $a_0 + a_1(z - z_0) + a_2(z - z_0)^2 + a_3(z - z_0)^3 + \cdots$.

Some functions have power series with an infinite radius of convergence, $R = \infty$. Perhaps the most important example is the exponential:

$$\mathrm{e}^z = 1 + z + \frac{z^2}{2!} + \frac{z^3}{3!} + \cdots.$$

Suppose a function $f$ has a power-series expansion $f(z) = a_0 + a_1 z + a_2 z^2 + a_3 z^3 + \cdots$. Then $f(0) = a_0$ and differentiating gives $f'(z) = a_1 + 2a_2 z + 3a_3 z^2 + \cdots$ and, hence, on setting $z = 0$, $a_1 = f'(0)$. What we have just done is to differentiate this infinite series term by term, something that in general is of dubious validity but in this case is justified because a power series can always be differentiated term by term within its radius of convergence. Continuing in this way we find that all the $a_k$ are derivatives of $f$ evaluated at the origin and the expansion can be written as the Taylor series expansion

$$f(z) = f(0) + f'(0)z + f''(0)\frac{z^2}{2!} + f'''(0)\frac{z^3}{3!} + \cdots.$$

## 18 Matrices and Vectors

A matrix is an $m \times n$ (read as "$m$-by-$n$") array of real or complex numbers, written as

$$A = (a_{ij}) = \begin{bmatrix} a_{11} & a_{12} & \cdots & a_{1n} \\ a_{21} & a_{22} & \cdots & a_{2n} \\ \vdots & \vdots & \ddots & \vdots \\ a_{m1} & a_{m2} & \cdots & a_{mn} \end{bmatrix}.$$

The element at the intersection of row $i$ and column $j$ is $a_{ij}$. The matrix is *square* if $m = n$ and *rectangular* otherwise. A vector is a matrix with one row or column: an $m \times 1$ matrix is a column vector and a $1 \times n$ matrix is a row vector. A number is often referred to as a *scalar* in order to distinguish it from a vector or matrix.

The sets of $m \times n$ matrices and $n \times 1$ vectors over $\mathbb{R}$ are denoted by $\mathbb{R}^{m \times n}$ and $\mathbb{R}^n$, respectively, and similarly for $\mathbb{C}$.

A notation that is common, though not ubiquitous, in applied mathematics employs uppercase letters for matrices and lowercase letters for vectors or, when subscripted, matrix elements. Similarly, matrices or vectors are sometimes written in boldface.

What distinguishes a matrix from a mere array of numbers is the algebraic operations defined on it. For two matrices $A$, $B$ of the same dimensions, addition is defined element-wise: $C = A + B$ means that

$c_{ij} = a_{ij} + b_{ij}$ for all $i$ and $j$. Multiplication by a scalar is defined in the natural way, so $C = \alpha A$ means that $c_{ij} = \alpha a_{ij}$ for all $i$ and $j$. However, matrix multiplication is *not* defined element-wise. If $A$ is $m \times r$ and $B$ is $r \times n$ then the product $C = AB$ is $m \times n$ and is defined by

$$c_{ij} = \sum_{k=1}^{r} a_{ik} b_{kj}.$$

This formula can be obtained as follows. Write $B = [b^1, b^2, \dots, b^n]$, where $b^j$ is the $j$th column of $B$; this is a *partitioning* of $B$ into its columns. Then $AB = A[b^1, b^2, \dots, b^n] = [Ab^1, Ab^2, \dots, Ab^n]$, where each $Ab^j$ is a matrix–vector product. Matrix–vector products $Ax$ with $x$ an $r \times 1$ vector are in turn defined by

$$Ax = [a^1, a^2, \dots, a^r] \begin{bmatrix} x_1 \\ x_2 \\ \vdots \\ x_r \end{bmatrix} = x_1 a^1 + x_2 a^2 + \cdots + x_r a^r,$$

so that $Ax$ is a *linear combination* of the columns of $A$.

Matrix multiplication is not commutative: $AB \neq BA$ in general, as is easily checked for $2 \times 2$ matrices. In some contexts the *commutator* (or *Lie bracket*) $[A, B] = AB - BA$ plays a role.

A linear system $Ax = b$ expresses the vector $b$ as a linear combination of the columns of $A$. When $A$ is square and of dimension $n$, this system provides $n$ linear equations for the $n$ components of $x$. The system has a unique solution when $A$ is nonsingular, that is, when $A$ has an inverse. An *inverse* of a square matrix $A$ is a matrix $A^{-1}$ such that $AA^{-1} = A^{-1}A = I$, where $I$ is the *identity matrix*, which has ones on the diagonal and zeros everywhere else. We can write $I = (\delta_{ij})$, where $\delta_{ij}$ is the *Kronecker delta* defined in table 3. The inverse is unique when it exists. If $A$ is nonsingular then $x = A^{-1}b$ is the solution to $Ax = b$. While this formula is useful mathematically, in practice one almost never solves a linear system by inverting $A$ and then multiplying the right-hand side by the inverse. Instead, GAUSSIAN ELIMINATION [IV.10 §2] with some form of pivoting is used.

*Transposition* turns an $m \times n$ matrix into an $n \times m$ one by interchanging the rows and columns: $C = A^{\mathrm{T}} \iff c_{ij} = a_{ji}$ for all $i$ and $j$. *Conjugate transposition* also conjugates the elements: $C = A^* \iff c_{ij} = \overline{a_{ji}}$ for all $i$ and $j$. The conjugate transpose of a product satisfies a useful reverse-order law: $(AB)^* = B^*A^*$.

Matrices can have a variety of different structures that can be exploited both in theory and in computation. A matrix $A \in \mathbb{R}^{n\times n}$ is *upper triangular* if $a_{ij} = 0$ for $i > j$, *lower triangular* if $A^{\mathrm{T}}$ is upper triangular, and *diagonal* if $a_{ij} = 0$ for $i \neq j$. For $n = 3$, such matrices have the forms

$$\begin{bmatrix} \times & \times & \times \\ 0 & \times & \times \\ 0 & 0 & \times \end{bmatrix}, \quad \begin{bmatrix} \times & 0 & 0 \\ \times & \times & 0 \\ \times & \times & \times \end{bmatrix}, \quad \begin{bmatrix} d_1 & 0 & 0 \\ 0 & d_2 & 0 \\ 0 & 0 & d_3 \end{bmatrix},$$

respectively, where $\times$ denotes a possibly nonzero entry; the third matrix is abbreviated $\mathrm{diag}(d_1, d_2, d_3)$. The matrix $A \in \mathbb{R}^{n\times n}$ is *symmetric* if $A^{\mathrm{T}} = A$, while $A \in \mathbb{C}^{n\times n}$ is *Hermitian* if $A^* = A$. If in addition the quadratic form $x^{\mathrm{T}}Ax$ (or $x^*Ax$) is always positive for nonzero vectors in $\mathbb{R}^n$ (or $\mathbb{C}^n$), then $A$ is *positive-definite*. The term *self-adjoint* is sometimes used instead of symmetric or Hermitian. Also fundamental is the notion of orthogonality: $A \in \mathbb{R}^{n\times n}$ is *orthogonal* if $A^{\mathrm{T}}A = I$, and $A \in \mathbb{C}^{n\times n}$ is *unitary* if $A^*A = I$. These properties mean that the inverse of $A$ is its (conjugate) transpose, but deeper properties of unitary matrices such as preservation of angles, norms, etc., under multiplication are what make them so important.

Structures can correspond to the pattern of the elements. A *Toeplitz matrix* has constant diagonals, made up from $2n - 1$ parameters $a_i$, $i = -(n-1), \dots, n-1$. Thus a $5 \times 5$ Toeplitz matrix has the form

$$\begin{bmatrix} a_0 & a_1 & a_2 & a_3 & a_4 \\ a_{-1} & a_0 & a_1 & a_2 & a_3 \\ a_{-2} & a_{-1} & a_0 & a_1 & a_2 \\ a_{-3} & a_{-2} & a_{-1} & a_0 & a_1 \\ a_{-4} & a_{-3} & a_{-2} & a_{-1} & a_0 \end{bmatrix}.$$

Toeplitz matrices arise in SIGNAL PROCESSING [IV.35]. A *circulant matrix* is a special type of Toeplitz matrix in which each row is a cyclic permutation (one element to the right) of the row above. Circulant matrices have many special properties, including that an explicit formula exists for their inverses and their eigenvalues.

A *Hamiltonian matrix* is a $2n \times 2n$ matrix of the block form

$$\begin{bmatrix} A & F \\ G & -A^* \end{bmatrix},$$

where $A$, $F$, and $G$ are $n \times n$ matrices and $F$ and $G$ are Hermitian. Hamiltonian matrices play an important role in CONTROL THEORY [III.25].

The *determinant* of an $n \times n$ matrix $A$ is a scalar that can be defined inductively by

$$\det(A) = \sum_{j=1}^{n} (-1)^{i+j} a_{ij} \det(A_{ij})$$

for any $i \in \{1, 2, \dots, n\}$, where $A_{ij}$ denotes the $(n-1) \times (n-1)$ matrix obtained from $A$ by deleting row $i$ and column $j$, and $\det(a) = a$ for a scalar $a$. This formula is called the expansion by minors because $\det(A_{kj})$ is a *minor* of $A$. The determinant is sometimes written with vertical bars, as $|A|$. Although determinants came before matrices historically, determinants have only a minor role in applied mathematics.

The quantity obtained by modifying the definition of determinant to remove the $(-1)^{i+j}$ term is the *permanent*, which is the sum of all possible products of $n$ elements of $A$ in which exactly one is taken from each row and each column. The permanent arises in combinatorics and in quantum mechanics.

## 19 Vector Spaces and Norms

A *vector space* is a mathematical structure in which a linear combination of elements can be taken, with the result remaining in the vector space. A vector space $V$ has a binary operation, which we will write as addition, that is associative, is commutative, and has an identity (the "zero vector," written 0) and additive inverses. In other words, for any $a, b, c \in V$ we have $(a + b) + c = a + (b + c)$, $a + b = b + a$, $a + 0 = a$, and there is a $d$ such that $a + d = 0$. There is also an underlying set of scalars, $\mathbb{R}$ or $\mathbb{C}$, such that $V$ is closed under scalar multiplication. Moreover, for all $x, y \in V$ and scalars $\alpha$, $\beta$ we have $\alpha(x + y) = \alpha x + \alpha y$, $(\alpha + \beta)x = \alpha x + \beta x$, and $\alpha(\beta x) = (\alpha\beta)x$.

A vector space can take many possible forms. For example, the set of real-valued functions on an interval $[a, b]$ is a vector space over $\mathbb{R}$, and the set of polynomials of degree less than or equal to $n$ with complex coefficients is a vector space over $\mathbb{C}$. Most importantly, the sets of $n$-vectors with real or complex coefficients are vector spaces over $\mathbb{R}$ and $\mathbb{C}$, respectively.

An important concept is that of linear independence. Vectors $v_1, v_2, \dots, v_n$ in $V$ are *linearly independent* if no nontrivial linear combination of them is zero, that is, if the equation $\alpha_1 v_1 + \alpha_2 v_2 + \cdots + \alpha_n v_n = 0$ holds only when the scalars $\alpha_i$ are all zero. If a collection of vectors is not linearly independent then it is *linearly dependent*.

Given vectors $v_1, v_2, \dots, v_n$ in $V$ we can form their *span*, which is the set of all possible linear combinations of them. A linearly independent collection of vectors whose span is $V$ is a *basis* for $V$, and any vector in $V$ can be written uniquely as a linear combination of these vectors.

The number of vectors in a basis for $V$ is the *dimension* of $V$, written $\dim V$, and it can be finite or infinite. The vector space of functions mentioned above is infinite dimensional, while the vector space of polynomials of degree at most $n$ has dimension $n + 1$, with a basis being $1, x, x^2, \dots, x^n$ or any other sequence of polynomials of degrees $0, 1, 2, \dots, n$.

A *subspace* of a vector space $V$ is a subset of $V$ that is itself a vector space under the same operations of addition and scalar multiplication.

### 19.1 Inner Products

Some vector spaces can be equipped with an *inner product*, which is a function $\langle x, y \rangle$ of two arguments that satisfies the conditions (i) $\langle x, x \rangle \geqslant 0$ and $\langle x, x \rangle = 0$ if and only if $x = 0$, (ii) $\langle x + y, z \rangle = \langle x, z \rangle + \langle y, z \rangle$, (iii) $\langle \alpha x, y \rangle = \alpha \langle x, y \rangle$, and (iv) $\langle x, y \rangle = \overline{\langle y, x \rangle}$ for all $x, y, z \in V$ and scalars $\alpha$. The usual (Euclidean) inner product on $\mathbb{R}^n$ is $\langle x, y \rangle = x^{\mathrm{T}} y$; on $\mathbb{C}^n$ the conjugate transpose must be used: $\langle x, y \rangle = x^* y$. For the vector space $C[a, b]$ of real-valued continuous functions on $[a, b]$ an inner product is

$$\langle f, g \rangle = \int_a^b w(x) f(x) g(x) \, \mathrm{d}x, \tag{6}$$

where $w(x)$ is some given, positive weight function, while for the vector space of $n$-vectors of the form $[f(x_1), f(x_2), \dots, f(x_n)]^{\mathrm{T}}$ for fixed points $x_i \in [a, b]$ and real-valued functions $f$ an inner product is

$$\langle f, g \rangle = \sum_{i=1}^{n} w_i f(x_i) g(x_i), \tag{7}$$

where the $w_i$ are positive weights. Note that (7) is not an inner product on the space of real-valued continuous functions because $\langle f, f \rangle = 0$ implies only that $f(x_i) = 0$ for all $i$ and not that $f \equiv 0$.

The vector space $\mathbb{R}^n$ with the Euclidean inner product is known as *$n$-dimensional Euclidean space.*

### 19.2 Orthogonality

Two vectors $u$, $v$ in an inner product space are *orthogonal* if $\langle u, v \rangle = 0$. For $\mathbb{R}^n$ and $\mathbb{C}^n$ this is just the usual notion of orthogonality: $u^{\mathrm{T}} v = 0$ and $u^* v = 0$, respectively. A set of vectors $\{u_i\}$ forms an *orthonormal set* if $\langle u_i, u_j \rangle = \delta_{ij}$ for all $i$ and $j$.

For an inner product space with inner product (6) or (7), useful examples of orthogonal functions are ORTHOGONAL POLYNOMIALS [II.29], which have the important property that they satisfy a three-term recurrence relation. For example, the *Chebyshev polynomials*

$T_k$ satisfy $T_0(x) = 1$, $T_1(x) = x$, and

$$T_{k+1}(x) = 2xT_k(x) - T_{k-1}(x), \quad k \geqslant 1, \tag{8}$$

and they are orthogonal on $[-1, 1]$ with respect to the weight function $(1 - x^2)^{-1/2}$:

$$\int_{-1}^{1} \frac{T_i(x)T_j(x)}{(1 - x^2)^{1/2}}\,dx = 0, \quad i \neq j.$$

Another commonly occurring class of orthogonal polynomials is the *Legendre polynomials* $P_k$, which are orthogonal with respect to $w(x) \equiv 1$ on $[-1, 1]$ and satisfy the recurrence

$$P_{k+1}(x) = \frac{2k+1}{k+1}xP_k(x) - \frac{k}{k+1}P_{k-1}(x), \tag{9}$$

with $P_0(x) = 1$ and $P_1(x) = x$, when they are normalized so that $P_i(1) = 1$.

Figure 10 plots some Chebyshev polynomials and Legendre polynomials on $[-1, 1]$. Both sets of polynomials are odd for odd degrees and even for even degrees. The values of the Chebyshev polynomials oscillate between $-1$ and 1, which is explained by the fact that $T_k(x) = \cos(k\theta)$, where $\theta = \cos^{-1} x$.

A beautiful theory surrounds orthogonal polynomials and their relations to various other areas of mathematics, including Padé approximation, spectral theory, and matrix eigenvalue problems.

If $\phi_1, \phi_2, \ldots$ is an orthogonal system, that is, $\langle \phi_i, \phi_j \rangle = 0$ for $i \neq j$, then the $\phi_i$ are necessarily linearly independent. Moreover, in an expansion

$$f(x) = \sum_{i=1}^{\infty} a_i \phi_i(x) \tag{10}$$

there is an explicit formula for the $a_i$. To determine it, we take the inner product of this equation with $\phi_j$ and use the orthogonality:

$$\langle f, \phi_j \rangle = \sum_{i=1}^{\infty} a_i \langle \phi_i, \phi_j \rangle = a_j \langle \phi_j, \phi_j \rangle,$$

so that $a_j = \langle f, \phi_j \rangle / \langle \phi_j, \phi_j \rangle$.

An important example of an orthogonal system of functions that are not polynomials is 1, $\cos x$, $\sin x$, $\cos(2x)$, $\sin(2x)$, $\cos(3x)$, $\ldots$, which are orthogonal with respect to the weight function $w(x) \equiv 1$ on $[-\pi, \pi]$, and for this basis (10) is a *Fourier series expansion*.

### 19.3 Norms

A common task is to approximate an element of a vector space $V$ by the closest element in a subspace $S$. To define "closest" we need a way to measure the size of a vector. A norm provides such a measure.

A *norm* is a mapping $\|\cdot\|$ from $V$ to the nonnegative real numbers such that $\|x\| = 0$ precisely when $x = 0$, $\|\alpha x\| = |\alpha|\,\|x\|$ for all scalars $\alpha$ and $x \in V$, and the *triangle inequality* $\|x + y\| \leqslant \|x\| + \|y\|$ holds for all $x, y \in V$. There are many possible norms, and on a finite-dimensional vector space all are *equivalent* in the sense that for any two norms $\|\cdot\|$ and $\|\cdot\|'$ there are positive constants $c_1$ and $c_2$ such that $c_1\|x\|' \leqslant \|x\| \leqslant c_2\|x\|'$ for all $x \in V$.

An example of a norm on $C[a, b]$ is

$$\|f\|_\infty = \max_{x \in [a,b]} |f(x)|, \tag{11}$$

known as the $L_\infty$-norm, the supremum norm, the maximum norm, or the uniform norm. For $p \in [1, \infty)$,

$$\|f\|_p = \left( \int_a^b |f(x)|^p\,dx \right)^{1/p}$$

is the $L_p$-norm on the space $L^p[a, b]$ of functions for which the (Lebesgue) integral is finite. Important special cases are the $L_2$-norm and the $L_1$-norm.

In an inner product space the natural norm is $\|x\| = \langle x, x \rangle^{1/2}$, and indeed the $L_2$-norm corresponds to the inner product (6) with unit weight function. A very useful inequality involving this norm is the *Cauchy–Schwarz inequality*:

$$|\langle x, y \rangle|^2 \leqslant \langle x, x \rangle \langle y, y \rangle = \|x\|^2 \|y\|^2$$

for all $x, y \in V$. This inequality shows that we can define the *angle* $\theta$ between two vectors $x$ and $y$ by $\cos\theta = \langle x, y \rangle / (\|x\|\,\|y\|) \in [-1, 1]$. Thus orthogonality corresponds to an angle $\theta = \pm\pi/2$.

Several different norms are commonly used on the vector spaces $\mathbb{R}^n$ and $\mathbb{C}^n$. The vector $p$-norm is defined for real $p$ by

$$\|x\|_p = \left( \sum_{i=1}^{n} |x_i|^p \right)^{1/p}, \quad 1 \leqslant p < \infty.$$

It includes the important special cases

$$\|x\|_1 = \sum_{i=1}^{n} |x_i|,$$

$$\|x\|_2 = \left( \sum_{i=1}^{n} |x_i|^2 \right)^{1/2} = (x^* x)^{1/2},$$

$$\|x\|_\infty = \max_{1 \leqslant i \leqslant n} |x_i|.$$

The 2-norm is Euclidean length. The 1-norm is sometimes called the "Manhattan" or "taxi cab" norm, as when $x, y \in \mathbb{R}^2$ contain the coordinates of two locations in Manhattan (which has a regular grid of streets), $\|x - y\|_1$ measures the distance by taxi cab from $x$ to $y$. Figure 11 shows the boundaries of the unit balls

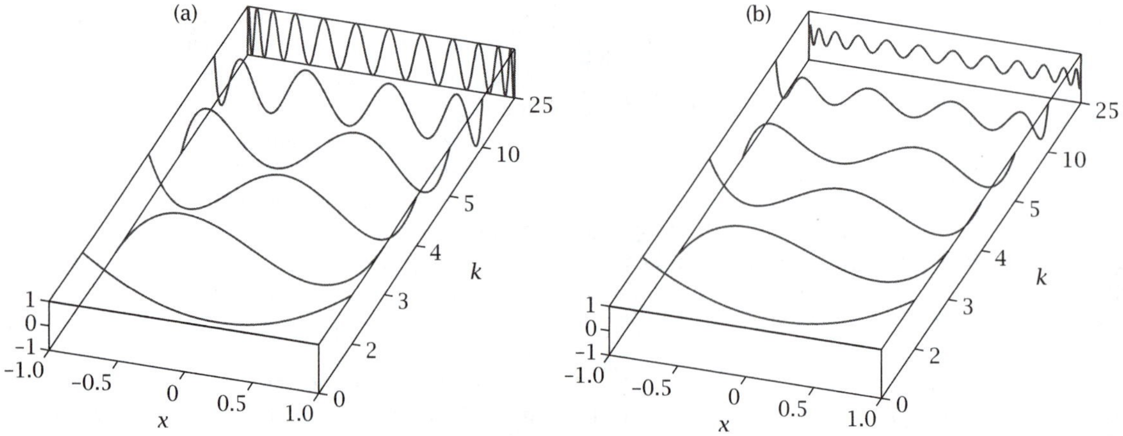


**Figure 10** Selected (a) Chebyshev polynomials $T_k(x)$ and (b) Legendre polynomials $P_k(x)$ on $[-1,1]$.

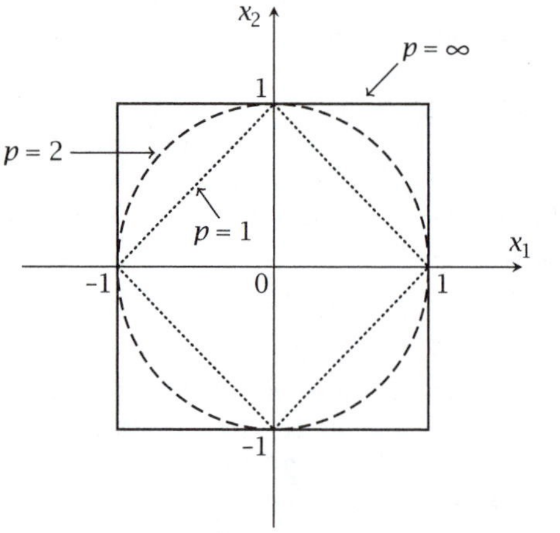


**Figure 11** The boundary of the unit ball in $\mathbb{R}^2$ for the 1-, 2-, and $\infty$-norms.

$\{x \in \mathbb{R}^n : \|x\| = 1\}$ for the latter three $p$-norms. The very different shapes of the unit balls suggest that the appropriate choice of norm will depend on the problem, as is the case, for example, in DATA FITTING [IV.9 §3.2].

Related to norms is the notion of a *metric*, defined on a set $M$ called a *metric space*. A metric on $M$ is a nonnegative function $d$ such that $d(x,y) = d(y,x)$ (symmetry), $d(x,z) \leqslant d(x,y) + d(y,z)$ (the *triangle inequality*), and for all $x,y,z \in M$, $d(x,y) = 0$ precisely when $x = y$. An example of a metric on the set of positive real numbers is $d(x,y) = |\log(x/y)|$. For a normed vector space, the function $d(x,y) = \|x - y\|$ is always a metric, so a normed vector space is always a metric space.

### 19.4 Convergence

We say that a sequence of points $x_1, x_2, \dots$, each belonging to a normed vector space $V$, *converges* to a limit $x_* \in V$, written $\lim_{i\to\infty} x_i = x_*$ (or $x_i \to x_*$ as $i \to \infty$), if for any $\varepsilon > 0$ there exists a positive integer $N$ such that $\|x_* - x_i\| < \varepsilon$ for all $i \geqslant N$.

The sequence is a *Cauchy sequence* if for any $\varepsilon > 0$ there exists a positive integer $N$ such that $\|x_i - x_j\| < \varepsilon$ for all $i,j \geqslant N$. A convergent sequence is a Cauchy sequence, but whether or not the converse is true depends on the space $V$.

A normed vector space is *complete* if every Cauchy sequence in $V$ has a limit in $V$. A complete normed vector space is called a *Banach space*. In a Banach space we can therefore prove convergence of a sequence without knowing its limit by showing that it is a Cauchy sequence.

A complete inner product space is called a *Hilbert space*. The spaces $\mathbb{R}^n$ and $\mathbb{C}^n$ with the Euclidean inner product are standard examples of Hilbert spaces.

## 20 Operators

An *operator* is a mapping from one vector space, $U$, to another, $V$ (possibly the same one). A *linear operator* (or *linear transformation*) $A$ is an operator such that $A(\alpha_1 x_1 + \alpha_2 x_2) = \alpha_1 A x_1 + \alpha_2 A x_2$ for all scalars $\alpha_1$, $\alpha_2$ and vectors $x_1, x_2 \in U$. For example, the differentiation operator is a linear operator that maps the vector space of polynomials of degree at most $n$ to the vector space of polynomials of degree at most $n-1$.

A natural measure of the size of a linear operator $A$ mapping $U$ to $V$ is the *induced norm* (also called the operator norm or subordinate norm),

$$\|A\| = \max\left\{\frac{\|Ax\|}{\|x\|} : x \in U,\ x \neq 0\right\},$$

where on the right-hand side $\|\cdot\|$ denotes both a norm on $U$ (in the denominator) and a norm on $V$ (in the numerator). For the rest of this section we assume that $U = V$ for simplicity. If $\|A\|$ is finite then $A$ is said to be a *bounded* linear operator. On a finite-dimensional vector space all linear operators are bounded.

The definition of an operator norm yields the inequalities $\|Ax\| \leqslant \|A\|\,\|x\|$ (immediate) and $\|AB\| \leqslant \|A\|\,\|B\|$ (using the previous inequality), both of which are indispensable.

The operator $A$ maps vectors in $U$ to other vectors in $U$, and it may change the norm by as much as $\|A\|$. For some vectors, called *eigenvectors*, it is only the norm, and not the direction, that changes. A nonzero vector $v$ is an eigenvector, with *eigenvalue* $\lambda$, if $Av = \lambda v$. Eigenvalues and eigenvectors play an important role in many areas of applied mathematics and appear in many places in this book. For example, SPECTRAL THEORY [IV.8] is about the eigenvalues and eigenvectors of linear operators on appropriate function spaces. The adjective *spectral* comes from *spectrum*, which is a set that contains the eigenvalues of an operator.

On taking norms in the relation $Av = \lambda v$ and using $\|v\| \neq 0$ we obtain $|\lambda| \leqslant \|A\|$. Thus all the eigenvalues of the operator $A$ lie in a disk of radius $\|A\|$ centered at the origin. This is an example of a localization result.

An *invariant subspace* of an operator $A$ that maps a vector space $U$ to itself is a subspace $X$ of $U$ such that $AX$ is a subset of $X$, so that $x \in X$ implies $Ax \in X$. An eigenvector is the special case of a one-dimensional invariant subspace.

For $n \times n$ matrices, the eigenvalue equation $Av = \lambda v$ says that $A - \lambda I$ is a singular matrix, which is equivalent to the condition $p(\lambda) = \det(\lambda I - A) = 0$. The polynomial $p$ is the *characteristic polynomial* of $A$, and since it has degree $n$ it follows from the fundamental theorem of algebra (section 14) that it has $n$ roots in the complex plane, which are the eigenvalues of $A$. Whether there are $n$ linearly independent eigenvectors associated with the eigenvalues depends on $A$ and can be elegantly answered in terms of the JORDAN CANONICAL FORM [II.22]. For real symmetric and complex Hermitian matrices, the eigenvalues are all real and there is a set of $n$ linearly independent eigenvectors, which can be taken to be orthonormal. If $A$ is in addition positive-definite, then the eigenvalues are all positive.

For matrices on $\mathbb{C}^{m\times n}$ the operator matrix norms corresponding to the 1, 2, and $\infty$ vector norms have explicit formulas:

$$\begin{aligned}
\|A\|_1 &= \max_{1\leqslant j\leqslant n} \sum_{i=1}^{m} |a_{ij}|, && \text{“max column sum,”}\\
\|A\|_\infty &= \max_{1\leqslant i\leqslant m} \sum_{j=1}^{n} |a_{ij}|, && \text{“max row sum,”}\\
\|A\|_2 &= (\rho(A^*A))^{1/2}, && \text{spectral norm,}
\end{aligned}$$

where the *spectral radius*

$$\rho(B) = \max\{|\lambda| : \lambda \text{ is an eigenvalue of } B\}.$$

Another useful formula is $\|A\|_2 = \sigma_{\max}(A)$, where $\sigma_{\max}(A)$ is the largest SINGULAR VALUE [II.32] of $A$. A further matrix norm that is commonly used is the Frobenius norm, given by

$$\|A\|_F = \left(\sum_{i=1}^{m}\sum_{j=1}^{n} |a_{ij}|^2\right)^{1/2} = (\operatorname{trace}(A^*A))^{1/2},$$

where the *trace* of a square matrix is the sum of its diagonal elements. Note that $\|A\|_F$ is just the 2-norm of the vector obtained by stringing the columns of $A$ out into one long vector. The Frobenius norm is not induced by any vector norm, as can be seen by taking $A$ as the identity matrix.

## 21 Linear Algebra

Associated with a matrix $A \in \mathbb{C}^{m\times n}$ are four important subspaces, two in $\mathbb{C}^m$ and two in $\mathbb{C}^n$: the ranges and the nullspaces of $A$ and $A^*$. The *range* of $A$ is the set of all linear combinations of the columns: $\operatorname{range}(A) = \{Ax : x \in \mathbb{C}^n\}$. The *null space* of $A$ is the set of vectors annihilated by $A$: $\operatorname{null}(A) = \{x \in \mathbb{C}^n : Ax = 0\}$.

The two most important laws of linear algebra are

$$\begin{aligned}
\dim\operatorname{range}(A) &= \dim\operatorname{range}(A^*),\\
\dim\operatorname{range}(A) + \dim\operatorname{null}(A) &= n,
\end{aligned}$$

where dim denotes dimension. These equalities can be proved in various ways, one of which is via the SINGULAR VALUE DECOMPOSITION [II.32].

Suppose $x \in \operatorname{null}(A)$. Then $x$ is orthogonal to every row of $A$ and hence is orthogonal to the subspace spanned by the rows of $A$. Since the rows of $A$ are the columns of $A^*$, it follows that $\operatorname{null}(A)$ is orthogonal to $\operatorname{range}(A^*)$, where two subspaces are said to be orthogonal if every vector in one of the subspaces is orthogonal to every vector in the other. In fact, it can be shown that $\operatorname{null}(A)$ and $\operatorname{range}(A^*)$ together span $\mathbb{C}^n$, and this implies that $\dim \operatorname{range}(A^*) + \dim \operatorname{null}(A) = n$, which can also be obtained by combining the two laws.

The *rank* of $A$ is the maximum number of linearly independent rows or columns of $A$. The rank plays an important role in linear equation problems. For example, a linear system $Ax = b$ has a solution if and only if $A$ and the augmented matrix $[A\ b]$ have the same rank.

The *Fredholm alternative* says that the equation $Ax = b$ has a solution if and only if $b^*v = 0$ for every vector $v$ satisfying $A^*v = 0$. This is a special case of more general versions of the alternative, e.g., in INTEGRAL EQUATIONS [IV.4 §3]. The "only if" part is easy, since if $A^*v = 0$ and $Ax = b$ then $b^*v = (v^*b)^* = (v^*Ax)^* = ((A^*v)^*x)^* = 0$. For the "if" part, suppose $b^*v = 0$ for every vector $v$ such that $A^*v = 0$. The latter equation says that $v \in \operatorname{null}(A^*)$, and from what we have just seen this means that $v$ is orthogonal to $\operatorname{range}(A)$. So every vector orthogonal to $\operatorname{range}(A)$ is orthogonal to $b$, which means that $b$ is in $\operatorname{range}(A)$ and so $Ax = b$ has a solution.

## 22 Condition Numbers

A condition number of a problem measures the sensitivity of the solution to perturbations in the data. For some problems there is not a unique solution and the problem can be regarded as infinitely sensitive; such problems fall into the class of ILL-POSED PROBLEMS [I.5 §1.2]. Consider a function $f$ mapping a vector space to itself such that $f(x)$ is defined in some neighborhood of $x$. A (relative) condition number for $f$ at $x$ is defined by

$$\operatorname{cond}(f,x) = \lim_{\varepsilon \to 0} \sup_{\|\delta x\| \leqslant \varepsilon \|x\|} \frac{\|f(x+\delta x) - f(x)\|}{\varepsilon \|f(x)\|}.$$

The condition number cond measures by how much, at most, small changes in the data can be magnified in the function value when both changes are measured in a relative sense. This definition implies that

$$\frac{\|f(x+\delta x) - f(x)\|}{\|f(x)\|} \leqslant \operatorname{cond}(f,x)\frac{\|\delta x\|}{\|x\|} + o(\|\delta x\|) \tag{12}$$

and so provides an approximate perturbation bound for small perturbations $\delta x$. In practice, $\delta x$ in the latter bound might represent inherent errors in the data from a physical experiment or rounding errors when the data is stored on a computer.

A problem is said to be *ill-conditioned* if its condition number is large and *well-conditioned* if its condition number is small, where the meaning of "large" and "small" depends on the context.

For many problems, explicit expressions can be obtained for the condition number. For a continuously differentiable function $f\colon \mathbb{R} \to \mathbb{R}$, $\operatorname{cond}(f,x) = |xf'(x)/f(x)|$. For the problem of matrix inversion, $f(A) = A^{-1}$, the condition number turns out to be $\kappa(A) = \|A\|\,\|A^{-1}\|$ for any matrix norm; this is known as the *condition number of $A$ with respect to inversion*. For the linear system $Ax = b$, with data the matrix $A$ and vector $b$, the condition number is also essentially $\kappa(A)$.

One role of the condition number is to provide a link between the residual of an approximate solution of an equation and the error of that approximation. This is most easily seen for a nonsingular linear system $Ax = b$. For any approximate solution $\hat{x}$ the residual satisfies $r = b - A\hat{x} = A(x - \hat{x})$, so the error is related to the residual by $x - \hat{x} = A^{-1}r$, which leads to the upper bound $\|x - \hat{x}\| \leqslant \kappa(A)\|r\|/\|A\|$.

## 23 Stability

The term "stability" is widely used in applied mathematics, with different meanings that depend on the context. A general meaning is that errors introduced in the initial stages of a process do not grow (or at least are bounded) as the process evolves. Here "process" could mean an iteration, a recurrence, or the evolution of a time-dependent differential equation. Stability is usually a necessary attribute and so a lot of effort is put into analyzing whether processes are stable or not. Discussions of stability can be found throughout this book.

Here we focus on *numerical stability*, in the context of evaluating a function $y = f(x)$ in floating-point arithmetic by some given algorithm, where $x$ and $y$ are scalars. If $\hat{y}$ is an approximation to $y$ then one measure of its quality is the *forward error* $\hat{y} - y$, which is often called, simply, "the error." The forward error is usually unknown and may be difficult to estimate. As an alternative we can ask whether we can perturb the data $x$ so that $\hat{y}$ is the exact solution to the perturbed problem; that is, can we find a $\delta x$ such that $\hat{y} = f(x + \delta x)$? In general, there may be many such $\delta x$; the smallest possible value of $|\delta x|$ is called the *backward error*. If the backward error is sufficiently small relative to the precision of the underlying arithmetic, then the algorithm is said to be *backward stable*.

It can be much easier to analyze the backward error than the forward error. Backward error analysis originates in NUMERICAL LINEAR ALGEBRA [IV.10 §8], where the underlying errors are rounding errors, but it has been used in various other contexts, including in the numerical solution of ordinary differential equations. Once the backward error is known, the forward error can be bounded by using the inequality (12), provided that an estimate of the relevant condition number is available.

### 24 Vector Calculus

While $n$-dimensional vector spaces, with $n$ possibly infinite, are the appropriate setting for much applied mathematics, the world we live in is three dimensional and so three coordinates are enough in many situations, such as in mechanics. Let $\boldsymbol{i}$, $\boldsymbol{j}$, and $\boldsymbol{k}$ denote unit vectors along the $x$-, $y$-, and $z$-axes, respectively. As this notation suggests, we will use boldface to denote vectors in this subsection. A vector $\boldsymbol{x}$ in $\mathbb{R}^3$ can then be expressed as $\boldsymbol{x} = x_1\boldsymbol{i} + x_2\boldsymbol{j} + x_3\boldsymbol{k}$. The *scalar product* or *dot product* of two vectors $\boldsymbol{x}$ and $\boldsymbol{y}$ is $\boldsymbol{x}\cdot\boldsymbol{y} = x_1y_1+x_2y_2+x_3y_3$, which is a special case of the Euclidean inner product of vectors in $\mathbb{R}^n$. The *cross product* or *vector product* does not have an $n$-dimensional analogue; it is the vector

$$\begin{aligned}\boldsymbol{x}\times\boldsymbol{y} &= \begin{vmatrix} \boldsymbol{i} & \boldsymbol{j} & \boldsymbol{k} \\ x_1 & x_2 & x_3 \\ y_1 & y_2 & y_3 \end{vmatrix} \\ &= (x_2y_3 - x_3y_2)\boldsymbol{i} + (x_3y_1 - x_1y_3)\boldsymbol{j} \\ &\quad + (x_1y_2 - x_2y_1)\boldsymbol{k},\end{aligned}$$

which is orthogonal to the plane in which $\boldsymbol{x}$ and $\boldsymbol{y}$ lie. Note that $\boldsymbol{x}\times\boldsymbol{y} = -\boldsymbol{y}\times\boldsymbol{x}$. The *vector triple product* of three vectors $\boldsymbol{x}$, $\boldsymbol{y}$, and $\boldsymbol{z}$ is the vector $\boldsymbol{x}\times(\boldsymbol{y}\times\boldsymbol{z})$, which can be expressed as

$$\boldsymbol{x}\times(\boldsymbol{y}\times\boldsymbol{z}) = (\boldsymbol{x}\cdot\boldsymbol{z})\boldsymbol{y} - (\boldsymbol{x}\cdot\boldsymbol{y})\boldsymbol{z}.$$

If $f$ is a scalar function of three variables, then its *gradient* is

$$\nabla f = \frac{\partial f}{\partial x}\boldsymbol{i} + \frac{\partial f}{\partial y}\boldsymbol{j} + \frac{\partial f}{\partial z}\boldsymbol{k}.$$

We can think of

$$\nabla = \frac{\partial}{\partial x}\boldsymbol{i} + \frac{\partial}{\partial y}\boldsymbol{j} + \frac{\partial}{\partial z}\boldsymbol{k}$$

as an operator: the *gradient operator*. There is nothing to stop us forming the dot product of the two vectors $\nabla$ and $\nabla f$:

$$\nabla\cdot\nabla f = \frac{\partial^2 f}{\partial x^2} + \frac{\partial^2 f}{\partial y^2} + \frac{\partial^2 f}{\partial z^2}.$$

This "del squared" operator is called the *Laplacian*, $\Delta = \nabla^2 \equiv \nabla\cdot\nabla$.

Now let $\boldsymbol{F} = F_1\boldsymbol{i} + F_2\boldsymbol{j} + F_3\boldsymbol{k}$ be a vector function mapping $\mathbb{R}^3$ to $\mathbb{R}^3$. The *divergence* of $\boldsymbol{F}$ is the dot product of $\nabla$ and $\boldsymbol{F}$:

$$\operatorname{div}\boldsymbol{F} = \nabla\cdot\boldsymbol{F} = \frac{\partial F_1}{\partial x} + \frac{\partial F_2}{\partial y} + \frac{\partial F_3}{\partial z}.$$

Another operator on vector functions that is commonly encountered is the *curl*: $\operatorname{curl}\boldsymbol{F} = \nabla\times\boldsymbol{F}$.

The *divergence theorem* says that, if $V$ is a vector field enclosed by a smooth surface $S$ oriented by an outward-pointing unit normal $\boldsymbol{n}$ and $F$ is a continuously differentiable vector field over $V$, then

$$\iiint_V \operatorname{div}\boldsymbol{F}\,\mathrm{d}V = \iint_S \boldsymbol{F}\cdot\boldsymbol{n}\,\mathrm{d}S.$$

In other words, the triple integral of the divergence of $\boldsymbol{F}$ over $V$ is equal to the surface integral of the normal component, $\boldsymbol{F}\cdot\boldsymbol{n}$. Many equations of physical interest can be derived using the divergence theorem.

Another important theorem is *Stokes's theorem*. It says that, for an oriented smooth surface $S$ with outward-pointing unit normal $\boldsymbol{n}$, bounded by a smooth simple closed curve $C$, if $F$ is a continuously differentiable vector field over $S$, then

$$\iint_S (\nabla\times\boldsymbol{F})\cdot\boldsymbol{n}\,\mathrm{d}S = \int_C \boldsymbol{F}\cdot\boldsymbol{t}\,\mathrm{d}s,$$

where $\boldsymbol{t} = \boldsymbol{t}(x,y,z)$ is a unit vector tangential to the curve $C$. Stokes's theorem says that the integral of the normal component of the curl of $\boldsymbol{F}$ over a surface $S$ is equal to the integral of the tangential component of $\boldsymbol{F}$ along the boundary $C$ of the surface.

## I.3 Methods of Solution

*Nicholas J. Higham*

Problems in applied mathematics come in many shapes and forms, and a wide variety of methods and techniques are used to solve them. In this article we outline some key ideas that underlie many different solution approaches.

### 1 Specifying the Problem

Before we can set about choosing a method to solve a problem we need to be clear about our assumptions. For example, if our problem is defined by a function (which could be the right-hand side of a differential equation), what can we assume about the smoothness of the function, that is, the number of continuous derivatives? If our problem is to find the eigenvalues of an $n\times n$ matrix $A$, are the elements of $A$ explicitly stored and accessible or is $A$ given only in the form of a "black box" that takes a vector $x$ as input and returns the product $Ax$?

The problem of finding a minimum or maximum of a scalar function $f$ of $n$ variables provides a good example of a wide range of possible scenarios. At one extreme, $f$ has derivatives of all orders and we can compute $f$ and its first and second derivatives at any point (most methods do not use derivatives of higher than second order). At another extreme, $f$ may be discontinuous and only function values may be available. It may even be that we are not able to evaluate $f$ but only to test whether, for a given $x$ and $y$, $f(x) < f(y)$ or vice versa. This is precisely the scenario for an optometrist formulating a prescription for a patient. The optometrist asks the patient to compare pairs of lenses and say which one gives the better vision. By suitably choosing the lenses the optometrist is able to home in on a prescription within a few minutes. In numerical optimization, DERIVATIVE-FREE METHODS [IV.11 §4.3] use only function values and many of them are based solely on comparisons of these values.

Another fundamental question is what is meant by a solution. If the solution is a function, would we accept its representation as an infinite series in some basis functions, or as an integral, or would we accept values of the function on a finite grid of points? If an inexact representation is allowed, how accurate must it be and what measure of error is appropriate?

## 2 Dimension Reduction

A common theme in many contexts is that of approximating a problem by one of smaller dimension and using the solution of the smaller problem to approximate the solution of the original problem. The motivation is that the large problem may be too expensive to solve, but of course this approach is viable only if the smaller problem can be constructed at low cost. In some situations the smaller problem is solved repeatedly, perhaps as some parameter varies, thereby amortizing the cost of producing it.

A ubiquitous example of this general approach concerns images displayed on Web pages. Modern digital cameras (even smartphones) produce images of 5 megapixels (million pixels) or more. Yet even a 27-inch monitor with a resolution of $2560 \times 1440$ pixels displays only about 3.7 megapixels. Since most images on Web pages are displayed at a small size within a page, it would be a great waste of storage and bandwidth to deal with them at their original size. They are therefore interpolated down to smaller dimensions appropriate for the intended usage (e.g., with longest side 400 pixels for an image on a news site). Here, dimension reduction is relatively straightforward and error is not an issue.

Often, though, an image is of intrinsic interest and we wish to keep it at its original dimensions and reduce the required storage, with minimal loss of quality. This is the more typical scenario for dimension reduction. The reason that dimension reduction is possible is that many images contain a high degree of redundancy. The SINGULAR VALUE DECOMPOSITION [II.32] (SVD) provides a way of capturing the important information in an image in a small number of vectors, at least for some images. A generally more effective reduction is produced by JPEG COMPRESSION [VII.7 §5], which uses two successive changes of basis in order to identify information that can be discarded.

A dynamical system may have many parameters but the behavior of interest may take place in a low-dimensional subspace. In this case we can try to identify that subspace and work within it, gaining a reduction in computation and storage. The general term for reducing dimension in a dynamical system is MODEL REDUCTION [II.26]. Model reduction has been an area of intensive research in the last thirty years, with applications ranging from the design of very large scale integration circuits to data assimilation in modeling the atmosphere.

Dimension reduction is fundamental to DATA ANALYSIS [IV.17 §4], where large data sets are transformed via a change of basis into lower-dimensional spaces that capture the behavior of the original data. Classic techniques are principal component analysis and application of the SVD. In the context of linear matrix equations such as the LYAPUNOV EQUATION [III.28], an approximation to a dominant invariant subspace of the solution (that is, an invariant subspace corresponding to the $k$ eigenvalues of largest magnitude, for some $k$) can be as useful as an approximation to the whole solution, and such an approximation can often be computed at much lower cost.

A term often used in the context of dimension reduction is *curse of dimensionality*, which refers to the fact that many problems become much harder in higher dimensions and, more informally, that intuition gained from two and three dimensions does not necessarily translate to higher dimensions. A simple illustration is given by an $n$-sphere, or hypersphere, of radius $r$ in $\mathbb{R}^n$, which comprises all $n$-vectors of 2-norm (Euclidean norm) $r$. A hypersphere has volume

$$S_n = \frac{\pi^{n/2} r^n}{\Gamma(n/2+1)},$$

where $\Gamma$ is the GAMMA FUNCTION [III.13]. Since $\Gamma(x) \sim \sqrt{2\pi/x}(x/e)^x$ (STIRLING'S APPROXIMATION [IV.7 §3]), for any fixed $r$ we find that $S_n$ tends to 0 as $n$ tends to $\infty$, that is, the volume of the hypersphere tends to zero, which is perhaps surprising. For a means of comparison, consider the $n$-cube, or hypercube, with sides of length $2r$. It has volume

$$H_n = (2r)^n$$

and therefore $S_n/H_n \to 0$ as $n \to \infty$. In other words, most of the volume of a hypercube lies away from the enclosed hypersphere, and hence "in the corners." For $n = 2$, the ratio $S_2/H_2 = 0.785$ (see figure 11 in THE LANGUAGE OF APPLIED MATHEMATICS [I.2 §19.3]), which is already substantially less than 1. The sequence $S_n/H_n$ continues 0.524, 0.308, 0.164, 0.081, .... This behavior is not too surprising when one realizes that any corner of the unit hypercube centered on the origin has coordinates $[\pm 1, \pm 1, \dots, \pm 1]^T$, and so is at distance $\sqrt{n}$ from the origin, whereas any point on the unit hypersphere centered on the origin is at distance 1 from the origin, so the latter distance divided by the former tends to 0 as $n \to \infty$. The term curse of dimensionality was introduced by Richard Bellman in 1961, with reference to the fact that sampling a function of $n$ variables on a grid with a fixed spacing requires a number of points that grows exponentially with $n$.

## 3 Approximation of Functions

We consider the problem of approximating a scalar function $f$, which may be given either as an explicit formula or implicitly, for example as the solution to an algebraic or differential equation. How the problem is solved depends on what is known about the function and what is required of the solution. We summarize some of the questions that must be answered before choosing a method.

- What form do we want the approximation to take: power series, polynomial, rational, Fourier series, ...?
- Do we want an approximation that has a desired accuracy within a certain region? If so, what measure of error should be used?
- Do we want an approximation that has certain qualitative features, such as convexity, monotonicity, or nonnegativity?

In this section we discuss a few examples of different types of approximation, touching on all the questions in this list. In the next three subsections $f$ is assumed to be real (its argument being written $x$), whereas in the fourth subsection it can be complex (so its argument is written $z$). We consider first approximations based on polynomials.

### 3.1 Polynomials

Perhaps the simplest class of approximating functions is the polynomials, $p_n(x) = a_0 + a_1 x + \cdots + a_n x^n$. Polynomials are easy to add, multiply, differentiate, and integrate, and their roots can be found by standard algorithms. Justification for the use of polynomials comes from *Weierstrass's theorem* of 1885, which states that for any $f \in C[a,b]$ and any $\varepsilon > 0$ there is a polynomial $p_n(x)$ such that $\|f - p_n\|_\infty < \varepsilon$, where the norm is the $L_\infty$-NORM [I.2 §19.3] given by $\|f\|_\infty = \max_{x\in[a,b]} |f(x)|$. Weierstrass's theorem assures us that any desired degree of accuracy in the maximum norm can be obtained by polynomials, though it does not bound the degree $n$, which may have to be high. Here are some of the ways in which polynomial approximations are constructed.

**Truncated Taylor series.** If $f$ is sufficiently smooth that it has a Taylor series expansion and its derivatives can be evaluated, then a polynomial approximation can be obtained simply by truncating the Taylor series. The Taylor series with remainder tells us that we can write $f(x) = p_n(x) + E_n(x)$, where $p_n(x) = f(0) + f'(0)x + \cdots + f^{(n)}(0)x^n/n!$ is a degree-$n$ polynomial and the remainder term has the form $E_n(x) = f^{(n+1)}(\xi)x^{n+1}/(n+1)!$ for some $\xi$ on the interval with endpoints 0 and $x$. The value of $n$ and the range of $x$ for which the approximation $f(x) \approx p_n(x)$ is applied will depend on $f$ and the desired accuracy. Figure 1 shows the degree-1, degree-3, and degree-5 Taylor approximants to the sine function.

**Interpolation.** We may require $p_n(x)$ to agree with $f(x)$ at certain specified points $x_i \in [a,b]$. Since $p_n$ contains $n+1$ coefficients and each condition $p_n(x_i) = f(x_i)$ provides one equation, we need $n+1$ points in order to specify $p_n$. It can be shown that the $n+1$ interpolation equations in $n+1$ unknowns have a unique solution provided that the interpolation points $\{x_i\}_{i=0}^n$ are distinct, in which case there is a unique *interpolating polynomial.* There is a variety of ways of representing $p_n$ (e.g., Lagrange form, barycentric form, and

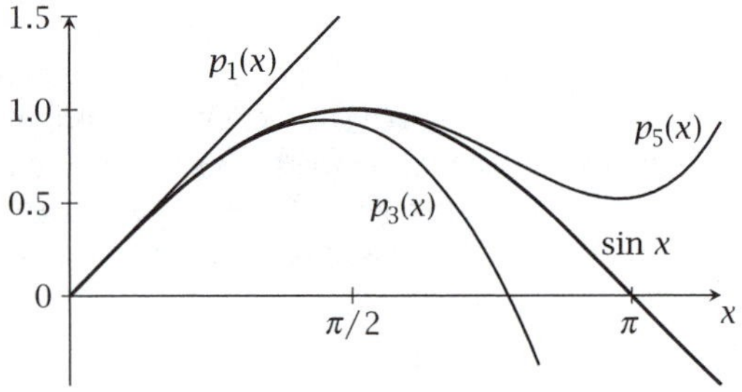


**Figure 1** $\sin x$ and its Taylor approximants $p_1(x) = x$, $p_3(x) = x - x^3/3!$, and $p_5(x) = x - x^3/3! + x^5/5!$.

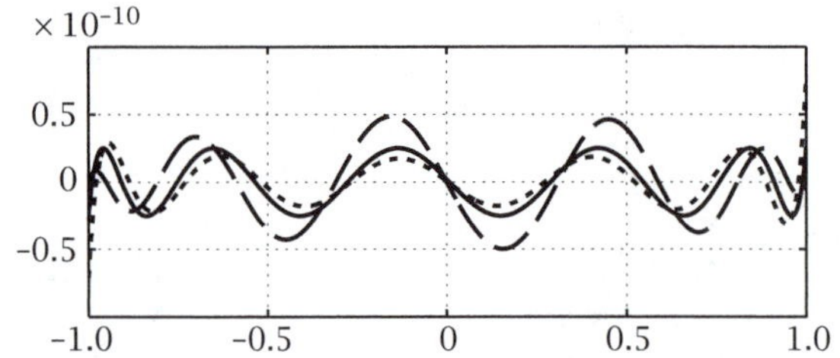


**Figure 2** Error in polynomial approximations to $e^x$ on $[-1, 1]$: solid line, $L_\infty$ approximation; dashed line, Chebyshev interpolant; dotted line, least squares ($L_2$ approximation).

divided difference form). An explicit formula is available for the error: if $f$ has $n+1$ continuous derivatives on $[a, b]$ then for any $x \in [a, b]$

$$f(x) - p_n(x) = \frac{f^{(n+1)}(\xi_x)}{(n+1)!} \prod_{i=0}^{n} (x - x_i),$$

where $\xi_x$ is some unknown point in the interval determined by $x_0, x_1, \dots, x_n$, and $x$. This error formula can be used to obtain insight into how to choose the $x_i$. It turns out that equally spaced points are poor, whereas points derived by rescaling to $[a, b]$ the ZEROS OR EXTREMA OF THE CHEBYSHEV POLYNOMIAL [IV.9 §2.2] of degree $n+1$ or $n$, respectively, are good.

**Least-squares approximation.** In least-squares approximation we fix the degree $n$ and then choose the polynomial $p_n$ to minimize the $L_2$-norm

$$\left(\int_a^b |f(x) - p_n(x)|^2 \, dx\right)^{1/2},$$

where $[a, b]$ is the interval of interest. It turns out that there is a unique $p_n$ minimizing the error, and its coefficients satisfy a linear system of equations called the *normal equations*. The normal equations tend to be ill-conditioned when $p_n$ is represented in the *monomial basis*, $\{1, x, x^2, \dots\}$, so in this context it is usual to write $p_n = \sum_{i=0}^{n} a_i \phi_i(x)$, where the $\phi_i$ are ORTHOGONAL POLYNOMIALS [II.29] on $[a, b]$. In this case the normal equations are diagonal and there is an explicit expression for the optimal coefficients: $a_i = \int_a^b \phi_i(x) f(x) \, dx / \int_a^b \phi_i(x)^2 \, dx$.

**$L_\infty$ approximation.** Instead of using the $L_2$-norm we can use the $L_\infty$-norm and so minimize $\|f - p_n\|_\infty$. A best $L_\infty$ approximation always exists and is unique, and there is a beautiful theory that characterizes the solution in terms of *equioscillation*, whereby the error achieves its maximum magnitude at a certain number of points with alternating sign. An algorithm called the Remez algorithm is available for computing the best $L_\infty$ approximation. One use of it is in EVALUATING ELEMENTARY FUNCTIONS [VI.11].

Figure 2 plots the absolute error $|f - p_n(x)|$ in three degree-10 polynomial approximations to $e^x$ on $[-1, 1]$: the least-squares approximation; the $L_\infty$ approximation; and a polynomial interpolant based on the Chebyshev points, $\cos(j\pi/n)$, $j = 0{:}n$. Note that the $L_\infty$ approximation has equioscillating error with maximum error strictly less than that for the other two approximations, and that the error of the Chebyshev interpolant is zero at the eleven points where it interpolates, which include the endpoints. It is also clear that the Chebyshev approximation is not much worse than the $L_\infty$ one—something that is true in general.

### 3.2 Piecewise Polynomials

High-degree polynomials have a tendency to wiggle. A degree-100 polynomial $p$ has up to 100 points at which it crosses the $x$-axis on a plot of $y = p(x)$: the distinct real zeros of $p$. This can make high-degree polynomials unsatisfactory as approximating functions. Instead of using one polynomial of large degree it can be better to use many polynomials of low degree. This can be done by breaking the interval of interest into pieces and using a different low-degree polynomial on each piece, with the polynomials joined together smoothly to make up the complete approximating function. Such *piecewise polynomials* can produce functions with high approximating power while avoiding the oscillations possible with high-degree polynomials.

A trivial example of a piecewise polynomial is the absolute value function $|x|$, which is equal to $-x$ for $x \leqslant 0$ and $x$ for $x \geqslant 0$ (see figure 4 on p. 13 in THE LANGUAGE OF APPLIED MATHEMATICS [I.2]). More generally, a piecewise polynomial $g$ defined on an interval

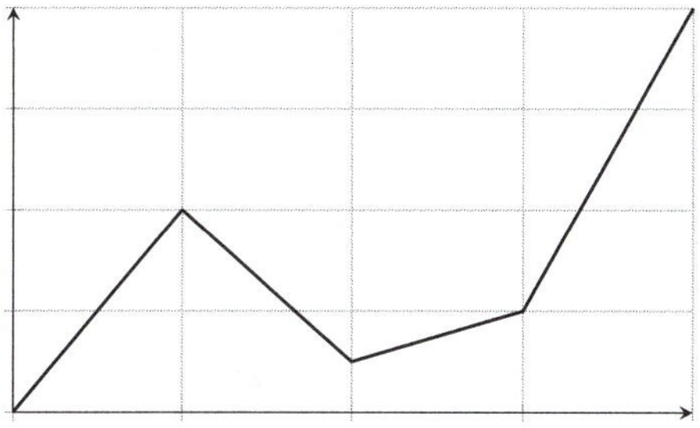

**Figure 3** A piecewise-linear function (spline).

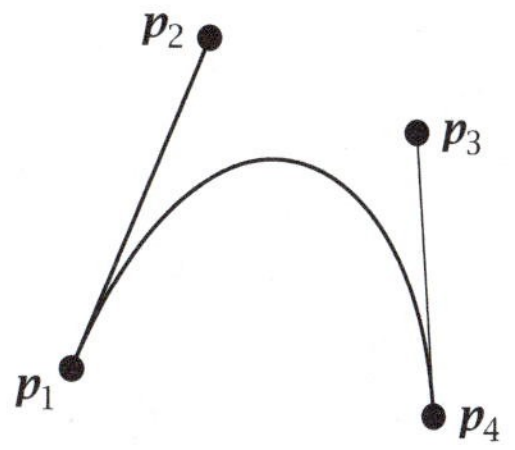


**Figure 4** A cubic Bézier curve with four control points $\boldsymbol{p}_1$, $\boldsymbol{p}_2$, $\boldsymbol{p}_3$, $\boldsymbol{p}_4$.

$[a, b] =: [x_0, x_n]$ that is the union of $n$ subintervals $[x_0, x_1], [x_1, x_2], \dots, [x_{n-1}, x_n]$ is defined by the property that $g(x) = p_i(x)$ for $x \in [x_i, x_{i+1}]$, where each $p_i$ is a polynomial. Thus on each interval $g$ is a polynomial, but each of these individual polynomials is in general different and possibly of different degree. Such a function is generally discontinuous, but we can ensure continuity by insisting that $p_{i-1}(x_i) = p_i(x_i)$, $i = 1 : n - 1$.

Important examples of piecewise polynomials are *splines*, which are piecewise polynomials $g$ for which each individual polynomial has degree $k$ or less and for which $g$ has $k - 1$ continuous derivatives on the interval. A spline therefore has the maximum possible smoothness. The most commonly used splines are linear splines and cubic splines, and an important application is in the FINITE-ELEMENT METHOD [II.12]. Figure 3 shows an example of a linear spline. Splines are commonly used in plotting data, where they provide a way of "joining up the dots," e.g., by straight lines in the case of a linear spline.

In computer-aided design the individual polynomials in a piecewise polynomial are often constructed as *Bézier curves*, which have the form

$$B_n(x) = \sum_{i=0}^{n} \binom{n}{i} \frac{(b-x)^{n-i}(x-a)^i}{(b-a)^n} \boldsymbol{p}_i$$

for an interval $[a, b]$. The $\boldsymbol{p}_i$ are control points in the plane that the user chooses via a graphical interface in order to achieve a desired form of curve. Figure 4 shows a cubic Bézier curve. The polynomials that multiply the $\boldsymbol{p}_i$ are called *Bernstein polynomials*, and they were originally introduced by Bernstein in 1912 in order to give a constructive proof of Weierstrass's theorem. The use of Bézier curves as a design tool to intuitively construct and manipulate complex shapes was initiated at the Citroën and Renault car companies in the 1960s. Today, cubic Bézier curves are widely used, e.g., in the design of fonts, in image manipulation programs such as Adobe Photoshop, and in the ISO standard for the Portable Document Format (PDF).

### 3.3 Wavelets

FOURIER ANALYSIS [I.2 §19.2] decomposes a function into a linear combination of trigonometric functions (sines and cosines) with different frequencies and so is a natural way to deal with periodic functions. Wavelet analysis, which was first developed in the 1980s, is designed to handle nonperiodic functions and does so by using basis functions that are rough and localized. Rather than varying the frequency as with the Fourier basis, a wavelet basis is constructed by translation $(f(x) \to f(x-1))$ and dilation $(f(x) \to f(2x))$. Given a mother wavelet $\psi(x)$, which has compact support (that is, it is zero outside a bounded interval), translations and dilations are created as $\psi(2^n x - k)$ with integer $n$ and $k$. This leads to many different resolutions, and hence the term *multiresolution analysis* is used in this context. Larger $n$ correspond to finer resolutions, and as $k$ varies the support moves around.

The localized nature of the wavelet basis functions makes wavelet representations of many functions and data relatively sparse, which makes wavelets particularly suitable for data compression, detection of features in images (such as edges and other discontinuities), and noise reduction. These are some of the reasons for the success of wavelets in (for example) imaging, where they are used in the JPEG2000 STANDARD [VII.7 §5].

### 3.4 Series Solution

We now turn to the development of explicit series representations of a function. As an example we take the Airy function $w(z)$, which satisfies the differential equation

$$w'' - zw = 0.$$

We can look for a solution $w(z) = \sum_{k=0}^{\infty} a_k z^k$, where $a_0 = w(0)$ and $a_1 = w'(0)$ can be regarded as given. For simplicity we will take $a_0 = 1$ and $a_1 = 0$. Differentiating twice gives $w''(z) = \sum_{k=2}^{\infty} k(k-1)a_k z^{k-2}$. Substituting the power series for $w$ and $w''$ into the differential equation we obtain $\sum_{k=2}^{\infty} k(k-1)a_k z^{k-2} - \sum_{k=0}^{\infty} a_k z^{k+1} = 0$. Since this equation must hold for all $z$ we can equate coefficients of $z^0, z^1, z^2, \dots$ on both sides to obtain a sequence of equations that provide recurrence relations for the $a_k$, specifically $(k+1)(k+2)a_{k+2} = a_{k-1}$ along with $a_2 = 0$. We find that

$$w(z) = 1 + \frac{z^3}{6} + \frac{z^6}{180} + \frac{z^9}{12\,960} + \cdots.$$

The modulus of the ratio of successive nonzero terms tends to zero as the index of the terms tends to infinity, which ensures that the series is convergent for all $z$. Since a power series can be differentiated term by term within its radius of convergence, it follows that our series does indeed satisfy the Airy equation.

Constructing a series expansion does not always lead to a convergent series. Consider the exponential integral

$$E_1(z) = \int_z^{\infty} \frac{e^{-t}}{t}\, dt.$$

Integrating by parts repeatedly gives

$$\begin{aligned} E_1(z) &= \frac{e^{-z}}{z} - \int_z^{\infty} \frac{e^{-t}}{t^2}\, dt \\ &= \frac{e^{-z}}{z} - \frac{e^{-z}}{z^2} + 2\int_z^{\infty} \frac{e^{-t}}{t^3}\, dt \\ &= \frac{e^{-z}}{z}\left(1 - \frac{1}{z} + \frac{2!}{z^2} + \cdots + (-1)^{k-1}\frac{(k-1)!}{z^{k-1}}\right) + R_k. \end{aligned}$$

The remainder term, $R_k = (-1)^k k! \int_z^{\infty} (e^{-t}/t^{k+1})\, dt$, does not tend to zero as $k \to \infty$ for fixed $z$, so the series is not convergent. Nevertheless, $|R_k|$ does decrease with $k$ before it increases, and a reasonable approximation to $E_1(z)$ can be obtained by choosing a suitable value of $k$. For example, with $z = 10$ the remainder starts increasing at $k = 11$, and taking $k = 10$ we obtain the approximation $E_1(10) \approx 4.156 \times 10^{-6}$, which is to be compared with $E_1(10) = 4.157 \times 10^{-6}$, where both results have been rounded to four significant figures. The series above is an example of an *asymptotic series*. In general, we say that the series $\sum_{k=0}^{\infty} a_k z^{-k}$ is an *asymptotic expansion* of $f$ as $z \to \infty$ if

$$\lim_{z\to\infty} z^n \left(f(z) - \sum_{k=0}^{n} a_k z^{-k}\right) = 0$$

for every $n$, and we write $f(z) \sim \sum_{k=0}^{\infty} a_k z^{-k}$, where the symbol "$\sim$" is read as "is asymptotic to." This condition can also be written as

$$f(z) = \sum_{k=0}^{n-1} a_k z^{-k} + O(z^{-n}).$$

For the series for $E_1$ we have

$$\begin{aligned} |z^k R_k| &= |z|^k k! \left| \int_z^{\infty} \frac{e^{-t}}{t^{k+1}}\, dt \right| \\ &\leqslant \frac{k!}{|z|} \left| \int_z^{\infty} e^{-t}\, dt \right| = \frac{k!}{|z|} |e^{-z}|, \end{aligned}$$

and the latter bound tends to zero as $|z| \to \infty$ if $\arg z \in (-\pi/2, \pi/2)$, so the series is asymptotic under this constraint on $z$.

By summing an appropriate number of terms, asymptotic series can deliver approximations of up to a certain, possibly good, accuracy, for large enough $|z|$, but beyond a certain point the accuracy worsens.

Suppose we have the quadratic $q_\varepsilon(x) = x^2 - x + \varepsilon = 0$, where $\varepsilon$ is a small parameter and we wish to obtain a series expansion for $x$ as a function of $\varepsilon$. This can be done by substituting $x(\varepsilon) = \sum_{k=0}^{\infty} a_k \varepsilon^k$ into the equation and setting the coefficients of each power of $\varepsilon$ to zero. This produces a system of equations that can be used to express $a_1, a_2, \dots$ in terms of $a_0$. The two solutions of $q_\varepsilon(x) = 0$ for $\varepsilon = 0$ are 0 and 1, so we take $a_0 = 0, 1$ and obtain the series

$$x(\varepsilon) = \begin{cases} \varepsilon + \varepsilon^2 + 2\varepsilon^3 + \cdots, & a_0 = 0, \\ 1 - \varepsilon - \varepsilon^2 - 2\varepsilon^3 + \cdots, & a_0 = 1, \end{cases} \tag{1}$$

which describe how the roots 0 and 1 of $q_0(x)$ behave for small $\varepsilon$. Suppose now that it is the leading term that is small and that we have the quadratic $\tilde{q}_\varepsilon(x) = \varepsilon x^2 - x + 1 = 0$. If we repeat the process of looking for an expansion of $x(\varepsilon)$, we obtain $x(\varepsilon) = 1 + \varepsilon + 2\varepsilon^2 + 5\varepsilon^3 + \cdots$ describing the behavior of the root 1 of $\tilde{q}_0(x)$. But $\tilde{q}$ is a quadratic and so has two roots. What has happened to the other one? There is a change of degree as we go from $\varepsilon = 0$ to $\varepsilon \neq 0$, and this takes us into SINGULAR PERTURBATION THEORY [IV.5 §3.2]. In this simple case we can use the transformation $y = 1/x$ to write $\tilde{q}_\varepsilon(x) = q_\varepsilon(y)/y^2$, and so we obtain expansions for $x(\varepsilon)$ by inverting those in (1). Indeed, inverting the second expression in (1) and expanding in a power series recovers the expansion we just derived.

## 4 Symbolic Solution

Sometimes a useful representation of a solution can be obtained using a computer symbolic manipulation package. Such packages are, for example, very good at determining closed forms for indefinite integrals that would be too tedious to derive by hand and for finding series expansions. We give four examples, each with a different package.

The MATLAB Symbolic Math Toolbox shows that the terms in $x^0, x^1, \dots, x^6$ of the Taylor series of $\tan(\sin x) - \sin(\tan x)$ are zero:

```
>> syms x
>> taylor(tan(sin(x)) - sin(tan(x)),...
          'order', 8)  % Terms up to x^7
ans =
x^7/30
```

The SymPy package in Python provides a concise form for a definite integral (here, the backslash \ denotes that a continuation line follows):

```
>>> from sympy import *
>>> x = Symbol('x')
>>> integrate((sin(x) + cos(x) + 1) / \
... (sin(x) + cos(x) + 1 + sin(x)*cos(x)))
2*log(tan(x/2) + 1)
```

Maple finds an explicit form for a definite integral:

```
> int(x^2/sin(x)^2, x = 0 .. Pi/2)
  Pi*ln(2)
```

Mathematica solves the first-order PDE initial-value problem $u_t + 9u_x = u^2$, $u(x, 0) = \sin x$, via the input

```
DSolve[{D[u[x, t], t] + 9 D[u[x, t], x]
        == u[x, t]^2,
        u[x, 0] == Sin[x]}, u[x, t], {x, t}]
```

which yields the output

```
{{u[x,t]->-1/(t+Csc[9t-x])}}
```

where Csc denotes the cosecant, $\csc x = 1/\sin x$.

One should never take such results at face value and should always run some kind of check, e.g., by substituting the claimed solution into the equation or comparing a symbolic solution with a numerically computed one.

## 5 Working from First Principles

Many mathematical problems can successfully be attacked from first principles, or with the use of heuristics, and applied mathematicians often take this approach rather than have recourse to general theory. Indeed, for unsolved research problems there may be no other way forward. For easier problems these approaches can provide useful insight and experience. The techniques in question include

- looking for a solution of a particular form, which may be constructed from experience or intuition or just by making an educated guess, and then showing that a solution of that form exists, and
- deducing what the general form of a solution must be and then determining the solution.

A standard example taught to undergraduate students is finding the general solution of the second-order ordinary differential equation (ODE)

$$y'' + ay' + by = 0 \tag{2}$$

for the unknown function $y = y(t)$. The starting point is to look for a solution of the form $y(t) = \mathrm{e}^{\lambda t}$. Substituting this purported solution into the equation gives $(\lambda^2 + a\lambda + b)\mathrm{e}^{\lambda t} = 0$, which implies $\lambda^2 + a\lambda + b = 0$. If this quadratic has distinct roots $\lambda_1$ and $\lambda_2$, then two linearly independent solutions of the differential equation, $\mathrm{e}^{\lambda_1 t}$ and $\mathrm{e}^{\lambda_2 t}$, have been determined. The general solution can be built up, in all cases, from this starting point.

A guess guided by intuition or experience is called an *ansatz*. The assumed form of solution $y(t) = \mathrm{e}^{\lambda t}$ to (2) is an ansatz. The corresponding ansatz for the difference equation $y_n + ay_{n-1} + by_n = 0$ is $y_n = \lambda^n$, which again yields the quadratic $\lambda^2 + a\lambda + b = 0$, and continuing with this line of investigation leads to the theory of difference equations.

An example of working from first principles is to determine a formula for the Vandermonde determinant

$$D(x_1, x_2, x_2) = \det\left(\begin{bmatrix} 1 & 1 & 1 \\ x_1 & x_2 & x_3 \\ x_1^2 & x_2^2 & x_3^2 \end{bmatrix}\right).$$

Instead of laboriously expanding the determinant one can observe that, since it is a sum of products of terms with one taken from each row, the result must be a multivariate polynomial of the form $\sum x_1^i x_2^j x_3^k$, with $i$, $j$, and $k$ distinct nonnegative integers summing to 3. If $x_1 = x_2$ then the first two columns are linearly dependent and the determinant is zero; hence $x_1 - x_2$ is a factor of the determinant. Continuing to argue in this fashion, $(x_1 - x_2)(x_2 - x_3)(x_3 - x_1)$ must be a factor. Since this product has degree 3, it must be that $D(x_1, x_2, x_3) = c(x_1 - x_2)(x_2 - x_3)(x_3 - x_1)$ for some constant $c$, and by considering the terms in $x_2 x_3^2$ on both sides of this equation it is clear that $c = 1$. The same argument extends straightforwardly to an $n \times n$ Vandermonde determinant. An advantage of this kind of reasoning is that it is adaptable: if the first row of the matrix is replaced by $[x_1^3\ x_2^3\ x_3^3]$, then the same form of argument can be used.

## 6 Iteration

Suppose we wish to solve the three equations in three unknowns

$$\begin{aligned} 4x\ -y\qquad &= a,\\ -x\ +4y\ -z &= b,\\ -y\ +4z &= c \end{aligned}$$

for some given $a$, $b$, and $c$. Because the coefficients on the diagonal (the 4s) are the largest coefficients in each row, it is reasonable to expect that $x \approx a/4$, $y \approx b/4$, $z \approx c/4$ are reasonable approximations. Note that this corresponds to rewriting the equations as

$$\begin{aligned} x &= \frac{a}{4} + \frac{y}{4},\\ y &= \frac{b}{4} + \frac{x+z}{4},\\ z &= \frac{c}{4} + \frac{y}{4} \end{aligned} \tag{3}$$

and setting $x = y = z = 0$ on the right-hand side. If we want to improve our approximation we can try plugging it into the right-hand side of (3) and reading off the new values from the left-hand side. By repeating this process we obtain the iteration

$$\left.\begin{aligned} x_{k+1} &= \frac{a}{4} + \frac{y_k}{4}\\ y_{k+1} &= \frac{b}{4} + \frac{x_k+z_k}{4}\\ z_{k+1} &= \frac{c}{4} + \frac{y_k}{4} \end{aligned}\right\} \quad k = 1, 2, \dots,$$

with $x_0 = y_0 = z_0 = 0$. This notation means that we are defining infinite sequences $\{x_k\}$, $\{y_k\}$, and $\{z_k\}$ by the given formulas.

If the iteration converges, that is, if $\lim_{k\to\infty} x_k$, $\lim_{k\to\infty} y_k$, and $\lim_{k\to\infty} z_k$ all exist, then the limits must satisfy (3), which is equivalent to the original system, so the only possible limits are the required solution components. This iteration is known as the *Jacobi iteration.* It is defined in an analogous way for any linear system of equations $Ax = b$ for which the diagonal elements of $A$ are nonzero. Convergence can be proved if the matrix $A$ has a large diagonal in the sense of being STRICTLY DIAGONALLY DOMINANT BY ROWS [IV.10 §1].

The Jacobi iteration is a special case of a powerful technique known as *fixed-point iteration* (or functional iteration) for solving a nonlinear system of equations. Consider an equation $x = f(x)$, where $f\colon \mathbb{R} \to \mathbb{R}$. Any scalar nonlinear equation can be put in this form. We can set up the iteration $x_{k+1} = f(x_k)$, with some choice of $x_0 \in \mathbb{R}$. The iteration may or may not converge, as can be seen by considering the case $f(x) = x^2$, for which we have convergence to 0 for $|x_0| < 1$ and divergence for $|x_0| > 1$. But if it does converge, to $\hat{x}$, say, then $\hat{x}$ must be a solution of the equation because taking limits in the iteration gives $\hat{x} = f(\hat{x})$.

To analyze the convergence of fixed-point iteration we note that if $x_*$ is a solution then $x_* - x_{k+1} = f(x_*) - f(x_k) = f'(\theta_k)(x_* - x_k)$ for some $\theta_k$ lying between $x_*$ and $x_k$, by the mean-value theorem. Hence $|x_* - x_{k+1}| < |x_* - x_k|$ if $|f'(\theta_k)| < 1$. This observation can be made into a proof that $x_k$ converges to $x_*$ if $x_0$ lies in a neighborhood of $x_*$ in which $|f'(x)|$ is less than 1.

The most widely used iteration for solving nonlinear equations is NEWTON'S METHOD [II.28] and its variants. Newton's method for $f(x) = 0$ is

$$x_{n+1} = x_n - \frac{f(x_n)}{f'(x_n)},$$

where $x_0$ is given. In order to apply the method we need to be able to compute the derivative $f'(x)$. Newton's method has a tendency to wander away from the root we are trying to compute if not started close enough to it. It is therefore common to start the method with a good approximation, possibly one computed by some other method, or to modify Newton's method in some way that encourages convergence.

In comparing different classes of iteration, the rate of convergence is an important concept. Let $\{x_k\}$ be a sequence of scalars converging to $x_*$ and denote the error in $x_n$ by $e_n = x_* - x_n$. If

$$\lim_{n\to\infty} \frac{|e_{n+1}|}{|e_n|^p} = C \neq 0,$$

where $C$ is a constant, the sequence is said to converge with *rate p* (or *order p*). This definition generalizes to vectors by replacing the absolute values with a vector norm. Fixed-point iteration has *linear convergence* in general, for which $p = 1$. Newton's method has (local) *quadratic convergence* ($p = 2$) to a simple root $x_*$, that is, one for which $f'(x_*) \neq 0$. Quadratic convergence is very desirable as it roughly doubles the number of correct figures on each step, while linear convergence merely reduces the error by a fixed percentage each time. Iterations of arbitrarily high order can be derived (e.g., by combining several Newton steps into one).

Fixed-point iteration can be used to iterate on functions as well as numbers. Consider the ODE initial-value problem

$$y'(x) = f(x, y), \qquad y(0) = y_0.$$

Integrating between 0 and $x$ leads to the equivalent problem

$$y(x) = y_0 + \int_0^x f(x, y(x))\,dx,$$

which is a type of equation known as an INTEGRAL EQUATION [IV.4] because the unknown function occurs within an integral. Applying the fixed-point iteration idea we can make a guess $\phi_0$ for $y$, plug it into the right-hand side of the integral equation, and call the result $\phi_1$. The process can be iterated to produce a sequence of functions $\phi_k$ defined by

$$\phi_{k+1}(x) = y_0 + \int_0^x f(x, \phi_k(x))\,dx, \quad k \geqslant 1.$$

In general, none of the $\phi_k$ will satisfy the differential equation, but we might hope that the sequence has a limit that does. Let us try out this idea on the problem

$$y' = 2x(1 + y), \qquad y(0) = 0$$

using first guess $\phi_0(x) = 0$. Then $\phi_1(x) = \int_0^x 2x\,dx = x^2$ and $\phi_2(x) = \int_0^x 2x(1 + x^2)\,dx = x^2 + x^4/2$. Continuing in this fashion yields $\phi_k(x) = x^2 + x^4/2! + x^6/3! + \cdots + x^{2k}/k!$. The limit as $k \to \infty$ exists and is $e^{x^2} - 1$, which is the required solution.

The procedure we have just carried out is known as *Picard iteration*, or the method of successive approximation. Of course, in most cases it will not be possible to evaluate the integrals in closed form, and so Picard iteration is not a practical means for computing a solution. However, Picard iteration is the basis of the proof of the standard result on existence and uniqueness of solutions for ODEs. The result says that, if $f(x, y)$ is continuous for $x \in [a, b]$ and for all $y$ and satisfies a *Lipschitz condition*

$$|f(x, u) - f(x, v)| \leqslant L|u - v| \quad \forall x \in [a, b],\ \forall u, v,$$

with Lipschitz constant $L$, then for any $y_0$ there is a unique continuously differentiable function $y(x)$ defined on $[a, b]$ that satisfies $y' = f(x, y)$ and $y(a) = y_0$.

## 7 Conversion to Another Problem

When we cannot solve a problem it can be useful to convert it to a different problem that is more amenable to attack. In this section we give several examples of such conversions.

We note first that it is not always obvious what is meant by a solution to a problem. Consider the ODE problem

$$\frac{dy}{dx} = 1 - 2xy, \qquad y(0) = 0.$$

The solution $y$ can be written as

$$y(x) = e^{-x^2} \int_0^x e^{t^2}\,dt,$$

which is known as *Dawson's integral* or *Dawson's function*. Which representation of $y$ is better? If we need to obtain higher derivatives $d^k y/dx^k$, the differential equation is more convenient. To evaluate $y(x)$ for a given $x$, numerical methods can be applied to either representation. Both representations therefore have their uses.

### 7.1 Uncoupling

When we are solving equations, of whatever type, a particularly favorable circumstance is when the first equation involves only one unknown and each successive equation introduces only one new unknown. We can then solve the equations from first to last. The simplest example is a triangular system of linear equations, such as

$$\begin{aligned} a_{11}x_1 \qquad\qquad\qquad &= b_1, \\ a_{21}x_1 + a_{22}x_2 \qquad\quad &= b_2, \\ a_{31}x_1 + a_{32}x_2 + a_{33}x_3 &= b_3, \end{aligned}$$

which can be solved by finding $x_1$ from the first equation, then $x_2$ from the second, and finally $x_3$ from the third. This is the process known as *substitution*.

Most linear equation problems do not have this triangular structure, but the process of GAUSSIAN ELIMINATION [IV.10 §2] converts an arbitrary linear system into triangular form.

More generally we might have $n$ nonlinear equations in $n$ unknowns, and a natural way to solve them is to try to manipulate them into an analogous triangular form. In computer algebra a way of doing this for polynomial equations is provided by Buchberger's algorithm for computing a GRÖBNER BASIS [IV.39 §2.1].

A triangular problem is partially uncoupled. In a fully uncoupled system each equation contains only one unknown. A linear system of ODEs $y' = Ay$ with an $n \times n$ coefficient matrix $A$ can be uncoupled if $A$ is diagonalizable. Indeed, if $A = XDX^{-1}$ with $X$ nonsingular and $D = \mathrm{diag}(\lambda_i)$, then the transformation $z = X^{-1}y$ gives $z' = Dz$, which represents $n$ uncoupled scalar equations $z_i' = \lambda_i z_i$, $i = 1:n$. The behavior of the vector $y$ can now be understood by looking at the behavior of the $n$ independent scalars $z_i$.

### 7.2 Polynomial Roots and Matrix Eigenvalues

Consider the problem of finding the roots (zeros) of a polynomial $p_n(x) = a_n x^n + a_{n-1}x^{n-1} + \cdots + a_0$ with $a_n \neq 0$, that is, the values of $x$ for which $p_n(x) = 0$. It is known from Galois theory that there is no explicit formula for the roots when $n \geqslant 5$. Many methods are available for computing polynomial roots, but not all are able to compute all $n$ roots reliably and software might not be readily available. Consider the $n \times n$ matrix

$$C = \begin{bmatrix} -a_{n-1}/a_n & -a_{n-2}/a_n & \cdots & \cdots & -a_0/a_n \\ 1 & 0 & \cdots & \cdots & 0 \\ 0 & 1 & \ddots & & 0 \\ \vdots & & \ddots & 0 & \vdots \\ 0 & \cdots & \cdots & 1 & 0 \end{bmatrix}.$$

Let $\lambda$ be a root of $p_n$. For the vector defined by $y = [\lambda^{n-1}\ \lambda^{n-2}\ \cdots\ 1]^{\mathrm{T}}$ we have $Cy = \lambda y$, so $\lambda$ is an eigenvalue of $C$ with eigenvector $y$. In fact, the set of roots of $p$ is the set of eigenvalues of $C$, so the polynomial root problem has been converted into an eigenvalue problem—albeit a specially structured one. The matrix $C$ is called a *companion matrix*. Of course, one can go in the opposite direction: to find the eigenvalues of $C$ one might look for solutions of $\det(C - \lambda I) = 0$, and the determinant is precisely $(-1)^n p_n(\lambda)/a_n$.

The eigenvector problem $Ax = \lambda x$ can be converted into a nonlinear system of equations $F(v) = 0$, where

$$F(v) = \begin{bmatrix} (A - \lambda I)x \\ e_s^{\mathrm{T}} x - 1 \end{bmatrix}, \qquad v = \begin{bmatrix} x \\ \lambda \end{bmatrix}.$$

The last component of $F$ serves to normalize the eigenvector and here $s$ is some fixed integer, with $e_s$ denoting the $s$th column of the identity matrix. By solving $F(v) = 0$ we obtain both an eigenvalue of $A$ and the corresponding eigenvector.

### 7.3 Dubious Conversions

Converting one problem to an apparently simpler one is not always a good idea. The problem of solving the scalar nonlinear equation $f(x) = 0$ can be converted to the problem of minimizing the function $g(x) = f(x)^2$. Since the latter problem has a global minimum attained when $f(x) = 0$, the conversion might look attractive. However, it has a pitfall: since $g'(x) = 2f'(x)f(x)$, the derivative of $g$ is zero whenever $f'(x) = 0$, and this means that methods for minimizing $g$ might converge to points that are stationary points of $g$ but not zeros of $f$.

For another example, consider the generalized eigenproblem in $n \times n$ matrices $A$ and $B$, $Ax = \lambda Bx$, which arises in problems in engineering and physics. It is natural to attempt to convert it to the standard eigenproblem $B^{-1}Ax = \lambda x$ and then apply standard theory and algorithms. However, if $B$ is singular this transformation is not possible, and when $B$ is nonsingular but ILL-CONDITIONED [I.2 §22] the transformation is inadvisable in floating-point arithmetic as it will be numerically unstable. A further drawback is that if $B$ is SPARSE [IV.10 §6] (has many zeros) then $B^{-1}A$ can have many more nonzeros than $A$ or $B$.

### 7.4 High-Order Differential Equations

Methods of solution of differential equations have been more extensively developed for first-order equations than for higher-order ones, where order refers to the highest derivative in the equation. Fortunately, higher-order equations can always be converted to first-order ones. Consider the $q$th-order ODE

$$y^{(q)} = f(t, y, y', \dots, y^{(q-1)})$$

with $y, y', \dots, y^{(q-1)}$ given at $t = t_0$. Define new variables

$$z_1 = y, \quad z_2 = y', \quad \dots, \quad z_q = y^{(q-1)}.$$

Then we have the first-order system of equations

$$\begin{aligned} z_1' &= z_2, \\ z_2' &= z_3, \\ &\vdots \\ z_{q-1}' &= z_q, \\ z_q' &= f(t, z_1, z_2, \dots, z_q), \end{aligned}$$

with $z_1, z_2, \dots, z_q$ given at $t = t_0$. We can write this system in vector form:

$$z' = f(t, z), \quad z = [z_1, z_2, \dots, z_q]^{\mathrm{T}}. \tag{4}$$

So we have traded high order for high dimension. Fortunately, the theory and the numerical methods developed for scalar first-order ODEs generally carry over straightforwardly to vector ODEs. We can go further and remove the explicit time dependence from (4) to put the system in *autonomous form*: with $w = [t, z^{\mathrm{T}}]^{\mathrm{T}}$, we have

$$w' = \begin{bmatrix} 1 \\ f(z) \end{bmatrix} = \begin{bmatrix} 1 \\ f(w_2, \dots, w_n) \end{bmatrix} =: g(w).$$

### 7.5 Continuation

Suppose we have a hard problem "solve $f(x) = 0$" and another problem "solve $g(x) = 0$" that is trivial to solve. Consider the parametrized problem "solve $h(x,t) = tf(x) + (1-t)g(x) = 0$." We know the solution for $t = 0$ and wish to find it for $t = 1$. The idea of *continuation* (also called *homotopy*, or *incremental loading* in elasticity) is to traverse the interval from 0 to 1 in several steps: $0 < t_1 < t_2 < \cdots < t_n = 1$. On the $k$th step we use the solution $x_{k-1}$ of the problem $h(x, t_{k-1}) = 0$ as the starting point for an iteration for solving $h(x, t_k) = 0$. We are therefore solving the original problem by approaching it gradually from a trivial problem. Continuation cannot be expected to work well in all cases. It is particularly well suited to cases where $f$ already depends on a parameter and the problem is simpler for some value of that parameter.

Continuation is a very general technique and has close connections with BIFURCATION THEORY [IV.21]. A special case of it is the idea of SHRINKING [V.10 §2.2], whereby a convex combination is taken of a given object with another having more desirable properties.

## 8 Linearization

A huge body of mathematics is concerned with problems that are linear in the variables of interest, such as a system $Ax = b$ of $n$ linear equations in $n$ unknowns or a system of ODEs $\mathrm{d}y/\mathrm{d}t = A(t)y$. For linear problems it is usually easy to analyze the existence of solutions, to obtain an explicit formula for a solution, and to derive practical methods of solution that exploit the linearity. Unfortunately, many real-world processes are inherently nonlinear. This means, first of all, that it may not be easy to determine whether or not there is a solution at all or, if a solution exists, whether it is unique. Secondly, finding a solution is in general difficult. A general technique for solving nonlinear problems is to transform them into linear ones, thereby converting a problem that we cannot solve into one that we can. The transformation can rarely be done exactly, so what is usually done is to *approximate* the nonlinear problem by a linear one—the process of *linearization*—and carry out some sort of iteration or refinement process.

To illustrate the idea of linear approximations we consider the quadratic equation

$$x^2 - 10x + 1 = 0. \tag{5}$$

Because the coefficient of the linear term, 10, is large compared with that of the quadratic term, 1, we can think of (5) as a linear equation with a small quadratic perturbation:

$$x = \frac{1}{10} + \frac{x^2}{10}. \tag{6}$$

Indeed, if we solve the linear part we obtain $x = 1/10$, which leaves a residual of just $1/100$ when substituted into the left-hand side of (5). We can therefore say that $x \approx 1/10$ is a reasonable approximation to a root (in fact, to the smallest root, since the product of the roots must be 1). Note that this approximation is obtained by putting $x = 0$ in the right-hand side of (6). To obtain a better approximation we might try putting $x = 1/10$ into the right-hand side. Repeating this process leads to the fixed-point iteration

$$x_{k+1} = \frac{1 + x_k^2}{10}, \qquad x_0 = 0,$$

which yields 0, 0.10, 0.101, .... After ten iterations we have $x_{10} = 0.101020514433644$, which is correct to the fifteen significant digits shown. Of course we could have obtained this solution as $x = 5 - \sqrt{24}$ using the quadratic formula, but the linearization approach gives an instant approximation and provides insight. For the equation $x^7 - 10x + 1 = 0$, for which there is no explicit formula for the roots, $1/10$ is an even better approximation to the smallest root and the analogue of the iteration above converges even more quickly.

Linearization is the key concept underlying NEWTON'S METHOD [II.28], which we discussed in section 6. Suppose we wish to solve a nonlinear system $f(x) = 0$, where $f \colon \mathbb{R}^n \to \mathbb{R}^n$, and let $x$ be an approximation to a solution $x_*$. Writing $x_* = x + h$, for sufficiently smooth $f$ we have $0 = f(x_*) = f(x) + J(x)h + O(\|h\|^2)$, where $J(x) = (\partial f_i/\partial x_j) \in \mathbb{R}^{n\times n}$ is the *Jacobian matrix* and the big-oh term includes the second- and higher-order terms from a multidimensional Taylor series. Newton's method approximates $f$ by the linear part of the series and so solves the linear system $J(x)h = -f(x)$ in order to produce a new approximation $x + h$. The process is iterated, yielding $x_{k+1} = x_k - J(x_k)^{-1}f(x_k)$. Theorems are available that guarantee when the linear approximations of the Newton method are good enough to ensure convergence to a solution. Indeed the Newton–Kantorovich theorem even uses Newton's method itself to prove the existence of a solution under certain conditions.

An equilibrium point (or critical point) of a nonlinear autonomous system of ODEs $y'(t) = f(y)$, where $f \colon \mathbb{R}^n \to \mathbb{R}^n$, is a vector $y_0$ such that $f(y_0) = 0$. For such a point, $y(t) = y_0$ is a constant solution to the differential equations. Linear stability analysis determines the effect of small perturbations away from the equilibrium point. Let $y(t) = y_0 + h(t)$ with $h(0) = h_0$ small. We wish to determine the behavior of $h(t)$ as $t \to \infty$. A linear approximation to $f$ at $y_0$ yields $h'(t) = y'(t) = f(y_0) + J(y_0)h = J(y_0)h$. The solution to this first-order system is $h(t) = \mathrm{e}^{J(y_0)t}h_0$, and so the behavior of $h$ depends on the behavior of the MATRIX EXPONENTIAL [II.14] $\mathrm{e}^{J(y_0)t}$. In particular, whether or not $h(t)$ grows or decays as $t \to \infty$ depends on the real parts of the eigenvalues of $J(y_0)$. For the case where $y$ has two components ($n = 2$), it is possible to give detailed classifications and plots (called phase-plane portraits) of the different qualitative behaviors that can occur. For more on the stability of ODEs see ORDINARY DIFFERENTIAL EQUATIONS [IV.2 §§8, 9].

An example of a nonlinear problem that can be linearized exactly, without any approximation, is the QUADRATIC EIGENVALUE PROBLEM [IV.10 §5.8].

Many other uses of linearization can be found throughout this book.

## 9 Recurrence Relations

A useful tactic for solving a problem whose solution is a number or function depending on a parameter is to try to derive a recurrence. For example, consider the integral

$$x_n = \int_0^1 \frac{t^n}{t+5}\,\mathrm{d}t.$$

It is easy to verify that $x_n$ satisfies the recurrence $x_n + 5x_{n-1} = 1/n$ and $x_0 = \log(6/5)$, so values of $x_n$ can easily be generated from the recurrence. However, when evaluating a recurrence numerically, one always needs to be aware of possible instability. Evaluating the recurrence in IEEE double-precision arithmetic (corresponding to about sixteen significant decimal digits) we find that $\hat{x}_{21} = -0.0159\ldots$, where the hat denotes the computed result. But

$$\frac{1}{6(n+1)} = \int_0^1 \frac{t^n}{6}\,\mathrm{d}t < x_n < \int_0^1 \frac{t^n}{5}\,\mathrm{d}t = \frac{1}{5(n+1)}$$

for all $n$, so this result is clearly not even of the right sign. The cause of the inaccuracy can be seen by considering the ideal case in which the only error, $\varepsilon$, say, occurs in evaluating $x_0$. That error is multiplied by $-5$ in computing $x_1$ and by a further factor of $-5$ on each step of the recurrence; overall, $x_n$ will be contaminated by an error of $(-5)^n\varepsilon$. This is an example of *numerical instability* and it is something that recurrences are prone to. We can obtain a more accurate result by using the recurrence in the backward direction, which will result in errors being divided by $-5$, so that they are damped out. From the inequalities above we see that for large $n$, $x_n \approx 1/(5(n+1))$. Let us simply set $y_{20} = 1/105$. Then, using the recurrence backward in the form $x_{n-1} = (1/n - x_n)/5$, we find that $x_0$ is computed with a relative error of order $10^{-16}$. For similar reasons, the recurrence relation in THE LANGUAGE OF APPLIED MATHEMATICS [I.2 §13] for the Bessel functions is also used in the backward direction for $x < n$.

## 10 Lagrange Multipliers

Optimization problems abound in applied mathematics because in many practical situations one wishes to maximize a desirable attribute (e.g., profit, or the strength of a structure) or minimize something that is desired to be small (such as cost or energy). More often than not, constraints impose limits on the variables and help to balance conflicting requirements. For example, in designing a tripod for cameras we may wish to minimize the weight of the tripod subject to it being able to support cameras up to a certain maximal weight, and a constraint might be a lower bound on the maximal height of the tripod.

Calculus enables us to characterize and find maxima and minima of functions. In the presence of constraints, though, the standard results are not so helpful. Consider the problem in three variables

$$\begin{aligned}&\text{minimize } f(x_1, x_2, x_3)\\&\text{subject to } c(x_1, x_2, x_3) = 0,\end{aligned} \tag{7}$$

where the objective function $f$ and constraint function $c$ are scalars. We know that any minimizer of the unconstrained problem $\min f(x_1, x_2, x_3)$ has to have a zero gradient; that is, $\nabla f(x) = [\partial f/\partial x_1, \partial f/\partial x_2, \partial f/\partial x_3]^{\mathrm{T}}$ must be the zero vector. How can we take account of the constraint $c(x_1, x_2, x_3) = 0$?

Let $x_* \in \mathbb{R}^3$ be a *feasible point*, that is, a point satisfying the constraint $c(x_*) = 0$. Consider a smooth curve $z(t)$ with $z(0) = x_*$ that remains on the constraint, that is, $c(z(t)) = 0$ for all sufficiently small $t$. Differentiating the latter equation and using the chain rule gives $(\mathrm{d}z(t)/\mathrm{d}t)^{\mathrm{T}}\nabla c(z(t)) = 0$. Setting $t = 0$ and putting $p_* = \mathrm{d}z/\mathrm{d}t|_{t=0}$ gives

$$p_*^{\mathrm{T}}\nabla c(x_*) = 0. \tag{8}$$

For $x_*$ to be optimal, the rate of change of $f$ along $z$ must be zero at $x_*$, so, using the chain rule again,

$$0 = \frac{\mathrm{d}}{\mathrm{d}t} f(z(t))\bigg|_{t=0} = \sum_{i=1}^{3} \frac{\partial f}{\partial z_i} \frac{\mathrm{d}z_i}{\mathrm{d}t}\bigg|_{t=0}$$
$$= \nabla f(x_*)^{\mathrm{T}} p_*. \qquad (9)$$

Now assume that $\nabla c(x_*) \neq 0$, which is known as a *constraint qualification*. This assumption ensures that every vector $p_*$ satisfying (8) is the tangent at $t = 0$ to some curve $z(t)$. It then follows that since (8) and (9) hold for all $p_*$,

$$\nabla f(x_*) = \lambda_* \nabla c(x_*) \qquad (10)$$

for some scalar $\lambda_*$. The scalar $\lambda_*$ is called a *Lagrange multiplier*. The constraint equation $c(x) = 0$ and (10) together constitute four equations in four unknowns, $x_1$, $x_2$, $x_3$, and $\lambda$. We have therefore reduced the original constrained minimization problem to a nonlinear system of equations. The latter system can be solved by any means at our disposal, though being nonlinear it is not necessarily an easy problem.

Another way to express our findings is in terms of the *Lagrangian function* $L(x, \lambda) = f(x) - \lambda c(x)$. Since $\nabla_x L(x, \lambda) = \nabla f(x) - \lambda \nabla c(x)$, the Lagrange multiplier condition (10) says that the solution $x_*$ is a stationary point of $L$ with respect to $x$ when $\lambda = \lambda_*$. Moreover, $\nabla_\lambda L(x, \lambda) = -c(x)$, so stationarity of $L$ with respect to $\lambda$ expresses the constraint $c(x) = 0$.

The development above was presented for a problem with three variables and one constraint, but it generalizes in a straightforward way to $n$ variables and $m$ constraints, with $\lambda$ becoming an $m$-vector of Lagrange multipliers.

Let us see how Lagrange multipliers help us to solve the problem

$$\text{maximize } 8xyz \text{ subject to } \frac{x^2}{a^2} + \frac{y^2}{b^2} + \frac{z^2}{c^2} = 1,$$

which defines the maximum rectangular block that fits inside the specified ellipsoid. Although our original problem (7) is a minimization problem, there is nothing in the development of (10) that is specific to minimization, and in fact the latter equation must be satisfied at any stationary point, so we can use it here. Setting $\tilde{x} = x/a$, $\tilde{y} = y/b$, $\tilde{z} = z/c$, the problem simplifies to

$$\text{maximize } 8abc\tilde{x}\tilde{y}\tilde{z} \text{ subject to } \tilde{x}^2 + \tilde{y}^2 + \tilde{z}^2 = 1.$$

The Lagrange multiplier condition is

$$8abc \begin{bmatrix} \tilde{y}\tilde{z} \\ \tilde{x}\tilde{z} \\ \tilde{x}\tilde{y} \end{bmatrix} = \lambda \begin{bmatrix} 2\tilde{x} \\ 2\tilde{y} \\ 2\tilde{z} \end{bmatrix}.$$

It is easily seen that these equations yield $\tilde{x} = \tilde{y} = \tilde{z} = 1/\sqrt{3}$ (and $\lambda_* = 4abc/\sqrt{3}$) and that the corresponding volume is $8abc/(3\sqrt{3})$. It is intuitively clear that this is a maximum, though in general checking for optimality requires further analysis involving inspection of second derivatives.

Lagrange multipliers and the Lagrangian function are widely used in applied mathematics in a variety of settings, including the CALCULUS OF VARIATIONS [IV.6] and LINEAR AND NONLINEAR OPTIMIZATION [IV.11]. One of the reasons for the importance of Lagrange multipliers is that they quantify the sensitivity of the optimal value to perturbations in the constraints. We can check this for our problem. If we perturb the constraint to $x^2/a^2 + y^2/b^2 + z^2/c^2 = 1 + \varepsilon$, then it is easy to see that the solution is $V(\varepsilon) = 8abc((1 + \varepsilon)/3)^{1/2}$, and hence $V'(0) = 4abc/\sqrt{3} = \lambda_*$.

## 11 Tricks and Techniques

As well as the general ideas and principles described in this article, applied mathematicians have at their disposal their own bags of tricks and techniques, which they bring into play when experience suggests they might be useful. Some will work only on very specific problems. Others might be nonrigorous but able to give useful insight. George Pólya is quoted as saying, "A trick used three times becomes a standard technique." Here are a few examples of tricks and techniques that prove useful on many different occasions, along with a very simple example in each case.

**Use symmetry.** When a problem has certain symmetries one can often argue that these must carry over into the solution. For example, the maximization problem at the end of the previous section is symmetric in $\tilde{x}$, $\tilde{y}$, and $\tilde{z}$, so one can argue that we must have $\tilde{x} = \tilde{y} = \tilde{z}$ at the solution.

**Add and subtract a term, or multiply and divide by a term.** As a very simple example, if $A$ and $B$ are $n \times n$ matrices with $A$ nonsingular, then $AB = AB \cdot AA^{-1} = A(BA)A^{-1}$, which shows that $AB$ and $BA$ are similar and that they therefore have the same eigenvalues. A common scenario is that $\hat{x}$ is an approximation to $x$ whose error cannot be directly estimated, but one can find another approximation $\tilde{x}$ whose relation to $x$ and $\hat{x}$ is understood. One then writes $x - \hat{x} = (x - \tilde{x}) + (\tilde{x} - \hat{x})$ and thereby obtains, using the triangle inequality, the bound $\|x - \hat{x}\| \leqslant \|x - \tilde{x}\| + \|\tilde{x} - \hat{x}\|$. For example, $x$

might be the solution to a PDE, $\tilde{x}$ the solution to a discretization of the PDE, and $\hat{x}$ an approximate solution to the discretized problem.

**Consider special cases.** Insight is often obtained by looking at special cases, such as a polynomial in place of a general function, or a diagonal matrix instead of a full matrix, or $n = 2$ or 3 in a problem in $n$ dimensions.

**Transform the problem.** It is always worth considering whether the problem is stated in the best way. As a simple example, suppose we are asked to find solutions of an equation of the form $1/f(x) = a$. If $f$ is a nearly linear function the problem is better written as $f(x) = 1/a$, and for computing a numerical solution Newton's method should work very well.

**Proof by contradiction.** A classic technique is to prove a result by assuming that the result is false and then obtaining a contradiction. Sometimes this approach can be used to show that an equation has no solution. For example, if the particular SYLVESTER EQUATION [III.28] in $n \times n$ matrices $AX - XA = I$ has a solution, then $0 = \operatorname{trace}(AX) - \operatorname{trace}(XA) = \operatorname{trace}(AX - XA) = \operatorname{trace}(I) = n$, which is a contradiction; we conclude that the equation has no solution.

**Going into the complex plane.** It is sometimes possible to solve a problem posed on the real line by making an excursion into the complex plane. This tactic is often used to evaluate real integrals by using the CAUCHY INTEGRAL FORMULA OR THE RESIDUE THEOREM [IV.1 §15]. For another example, let $f$ be an analytic function that is real for real arguments. Then, for real $x$,

$$f'(x) \approx \frac{\operatorname{Im} f(x + \mathrm{i}h)}{h},$$

where i is the imaginary unit and $h$ is a small real parameter. This *complex step approximation* has error $O(h^2)$. Here is an example with $f(x) = \cos x$ in MATLAB:

```
>> x = pi/6; h = 1e-8;
>> fdash_cs = imag( cos(x + i*h) )/h;
>> error = fdash_cs - (-sin(x))
error =
  -5.5511e-17
```

Finally, there are various basic techniques that are learned in elementary courses and are always useful, such as integration by parts, use of the Cauchy–Schwarz inequality, and interchange of the order of integration in a double integral or of summation in a double sum.

## I.4 Algorithms

*Nicholas J. Higham*

An algorithm is a procedure for accomplishing a certain task in a finite number of steps.

The terms "algorithm" and "method" are often used interchangeably, and there is no clear consensus on the distinction between them. However, many authors use "method" very generally and reserve "algorithm" for a procedure in which every step is precisely defined. For example, Newton's method solves a system of $n$ nonlinear equations in $n$ variables, $F(x) = 0$, using the iteration $x_{k+1} = x_k - J(x_k)^{-1}F(x_k)$, where $J$ is the $n \times n$ Jacobian matrix of $F$. An algorithm implementing Newton's method has to define how to compute the correction terms $J(x_k)^{-1}F(x_k)$ (e.g., by Gaussian elimination with partial pivoting), what to do if $J(x_k)$ is singular, and when to terminate the iteration (e.g., when $\|F(x_k\|$ is less than some convergence tolerance).

Developing algorithms has always been an important activity in applied mathematics. George Forsythe put it well when he said:

> A useful algorithm is a substantial contribution to knowledge. Its publication constitutes an important piece of scholarship.

This statement was made in the 1960s, when the emergence of standards for PROGRAMMING LANGUAGES [VII.11] had started to make it feasible to publish and share computer implementations of algorithms.

### 1 Summation

Let us begin with the problem of forming the sum of $n$ numbers, $x_1 + x_2 + \cdots + x_n$, on a computer. This problem is deceptively simple, partly because the problem statement is so suggestive of an algorithm. But note that the numbers can be added in any order and that parentheses can be inserted in many ways to break up the overall sum into smaller sums. So there are many ways in which the sum can be computed, and if we simply choose one of them without thinking, we may miss a better alternative. Thinking about the summation process leads to the notion of repeatedly adding two numbers from the set then putting the sum back into the set. This process can be expressed formally in the following algorithm.

**Algorithm 1.** *Given numbers $x_1, \dots, x_n$ this algorithm computes $S_n = \sum_{i=1}^{n} x_i$.*

```
1 Let X = {x_1, ..., x_n}.
2 while X contains more than one element
3         Remove two numbers y and z from X
          and put their sum y + z back in X.
4 end
5 Assign the remaining element of X to S_n.
```

This algorithm is written in *pseudocode*, which refers to any informal syntax that resembles a programming language. The purpose of pseudocode is to specify an algorithm in a concise, readable way without having to worry about the details of any particular programming language.

It is clear that after execution of line 3, the number of elements of $\mathcal{X}$ has decreased by one. The algorithm therefore terminates after exactly $n - 1$ iterations of the while loop, at which point it has carried out $n - 1$ additions.

Clearly, different ways of choosing $x$ and $y$ on line 3 lead to different summation algorithms. The most obvious choice is to take $y$ as the sum computed at the previous step and $z$ as the element with smallest index (and $y = x_1$ and $z = x_2$ at the first step). The resulting algorithm is known as *recursive summation* and can be expressed as

```
s = x_1
for i = 2:n   % i goes from 2 to n in steps of 1
    s = s + x_i
end
S_n = s
```

In this pseudocode everything from a % sign to the end of a line is a comment.

An alternative is to add the elements pairwise, then add their sums pairwise, and so on, repeatedly. For $n = 8$ this corresponds to the formula

$$S_8 = [(x_1 + x_2) + (x_3 + x_4)] + [(x_5 + x_6) + (x_7 + x_8)].$$

This summation algorithm is forming a BINARY TREE [II.16], as illustrated in figure 1. An attractive feature of pairwise summation is that on a parallel computer the summations on each level of the tree can be done in parallel, so the computation time is proportional to $\log_2 n$ instead of $n$.

A third summation algorithm, called the *insertion algorithm*, takes the summands $y$ and $z$ as those with the smallest absolute value at each stage.

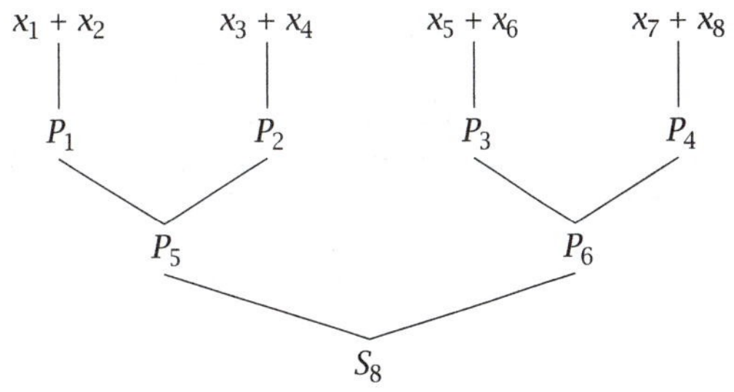


**Figure 1** A binary tree for pairwise summation, with partial sums $P_i$.

One reason for considering these different summation algorithms as special cases of algorithm 1 is that a common analysis can be done of the effects of rounding errors in floating-point arithmetic. Express the $i$th execution of the while loop in algorithm 1 as $P_i = y_i + z_i$. The standard model of floating-point arithmetic (equation (2) in FLOATING-POINT ARITHMETIC [II.13]) says that the computed $\hat{P}_i$ satisfies

$$\hat{P}_i = \frac{y_i + z_i}{1 + \delta_i}, \quad |\delta_i| \leqslant u, \ i = 1:n-1,$$

where $u$ is the unit roundoff. (The model actually says we should *multiply* by $1 + \delta_i$, but it can be shown to be equally valid to divide by $1 + \delta_i$, which is more convenient here.) The error introduced in forming $\hat{P}_i$, namely $y_i + z_i - \hat{P}_i$, is therefore $\delta_i \hat{P}_i$. By summing each of these individual errors we obtain the overall error,

$$E_n := S_n - \hat{S}_n = \sum_{i=1}^{n-1} \delta_i \hat{P}_i,$$

which is bounded by

$$|E_n| \leqslant u \sum_{i=1}^{n-1} |\hat{P}_i|.$$

This bound provides the insight that to maximize the accuracy of the sum it should be helpful to minimize the size of the intermediate partial sums. The insertion algorithm can be seen as a heuristic way of doing that.

When the $x_i$ are all nonnegative and we use recursive summation, $P_i = \sum_{j=1}^{i+1} x_j$, so the bound for $|E_n|$ is minimized if the $x_i$ are arranged in increasing order. Moreover, since $P_i \leqslant S_n$, the bound implies $|E_n| \leqslant (n-1)uS_n + O(u^2)$ for *any* ordering, which guarantees a relative error of order $nu$.

This summation example shows that taking a bird's eye view of a computational problem can be beneficial, in that it can reveal algorithmic possibilities that might have been missed and can help a single analysis to be developed that applies to a wide range of problems. Of course, algorithm 1 does not cover all the possibilities. Another way to compute the sum is as $S_n = \log \prod_{i=1}^{n} e^{x_i}$. This formula has little to recommend it, but it is not so different from the expression $\exp(n^{-1} \log \sum_{i=1}^{n} x_i)$, which is a log-Euclidean mean of the $x_i$ that has applications when the $x_i$ are structured matrices or operators.

## 2 Bisection

The summation problem is unusual in that there is no difficulty in seeing the correctness of algorithm 1 or its computational cost. A slightly trickier algorithm is the bisection algorithm for finding a zero of a continuous function $f(x)$. The bisection algorithm takes as input an interval $[a, b]$ such that $f(a)f(b) < 0$; the intermediate-value theorem tells us that there must be a zero of $f$ on this interval. The bisection algorithm repeatedly halves the interval and retains the half on which $f$ has different signs at the endpoints, that is, the interval on which we can be sure there is a zero. To make the algorithm finite we need a stopping criterion. The following algorithm terminates once the interval is of length at most tol, a given tolerance.

**Algorithm 2 (bisection algorithm).** *This algorithm finds a zero of a continuous function $f(x)$ given an interval $[a, b]$ such that $f(a)f(b) < 0$ and an error tolerance* tol.

```
1  while b - a > tol
2      c = (a + b)/2
3      if f(c) = 0, quit, end
4      if f(c)f(b) < 0
5          a = c
6      else
7          b = c
8      end
9  end
10 x = (a + b)/2
```

To show the correctness of this algorithm note first that at the end of the while loop $f(a)f(b) < 0$ still holds; in other words, this inequality is an invariant of the loop. Therefore we have a sequence of intervals each of length half the previous interval and all containing a zero. This means that after $k$ steps we have an interval of length $(b-a)/2^k$ containing a zero. The algorithm therefore terminates after $\lceil \log_2((|b - a|/\mathrm{tol})) \rceil$ steps. Here, we are using the *ceiling function* $\lceil x \rceil$, which is the smallest integer greater than or equal to $x$. In the next section we will also need the *floor function* $\lfloor x \rfloor$, which is the largest integer less than or equal to $x$.

The algorithm returns as the approximate zero the midpoint of the final interval, which has length at most tol; since a zero lies in this interval, the absolute error is at most tol/2.

Algorithm 2 needs a number of refinements to make it more reliable and efficient for practical use. First, testing whether $f(c)$ and $f(b)$ have opposite signs should not be done by multiplying them, as the product could overflow or underflow in floating-point arithmetic. Instead, the signs should be directly compared. Second, $f(c)$ should not be computed twice, on lines 3 and 4, but rather computed once and its value reused. Finally, the convergence test is an absolute one, so is scale dependent. A better alternative is $|b - a| > \mathrm{tol}(|a| + |b|)$, which is unaffected by scalings $a \leftarrow \theta a$, $b \leftarrow \theta b$.

Bisection is a widely applicable technique. For example, it can be used to search an ordered list to see if a given element is contained in the list; here it is known as *binary search*. It is also used for debugging. If the LaTeX source for this article fails to compile and I cannot spot the error, I will move the `\end{document}` command to the middle of the file and try again; I can thereby determine in which half of the file the error lies and can repeat the process to narrow the error down.

## 3 Divide and Conquer

The *divide and conquer* principle breaks a problem down into two (or more) equally sized subproblems and solves each subproblem recursively.

An example of how divide and conquer can be exploited is in the computation of a large integer power of a number. Computing $x^n$ in the obvious way takes $n - 1$ multiplications. But $x^{13}$, for example, can be written $x^8x^4x$, which can be evaluated in just five multiplications instead of twelve by first forming $x^2$, $x^4 = (x^2)^2$, and $x^8 = (x^4)^2$. Notice that $13 = (1101)_2$ in base 2, and in general the base 2 representation of $n$ tells us exactly how to break down the computation of $x^n$ into products of terms $x^{2^k}$. However, by expressing the computation using divide and conquer we can avoid the need to compute the binary representation of $n$. The idea is to write $x^n = (x^{n/2})^2$ if $n$ is even and $x^n = x(x^{\lfloor n/2 \rfloor})^2$ if $n$ is odd. In either case the problem is reduced to one of half the size. The resulting algorithm is most elegantly expressed in recursive form, as an algorithm that calls itself.

**Algorithm 3.** *This algorithm computes $x^n$ for a positive integer $n$.*

```
function y = power(x, n)
if n = 1, y = x, return
if n is odd
    y = x power(x, (n − 1)/2)^2     % Recursive call
else
    y = power(x, n/2)^2              % Recursive call
end
```

The number of multiplications required by algorithm 3 is bounded above by $2\lfloor \log_2 n \rfloor$.

Another example of how divide and conquer can be used is for computing the inverse of a nonsingular upper triangular matrix, $T \in \mathbb{C}^{n\times n}$. Write $T$ in partitioned form as

$$T = \begin{bmatrix} T_{11} & T_{12} \\ 0 & T_{22} \end{bmatrix}, \tag{1}$$

where $T_{11}$ has dimension $\lceil n/2 \rceil$. It is easy to check that

$$T^{-1} = \begin{bmatrix} T_{11}^{-1} & -T_{11}^{-1} T_{12} T_{22}^{-1} \\ 0 & T_{22}^{-1} \end{bmatrix}.$$

This formula reduces the problem to the computation of the inverses of two smaller matrices, namely, the diagonal blocks $T_{11}$ and $T_{22}$, and their inverses can be expressed in the same way. The process can be repeated until scalars are reached and the inversion is trivial.

**Algorithm 4.** *This algorithm computes the inverse of a nonsingular upper triangular matrix T by divide and conquer.*

```
function U = inv(T)
n = dimension of T
if n = 1, u_11 = t_11^{-1}, return
Partition T according to (1), where T_11 has
  dimension ⌈n/2⌉.
U_11 = inv(T_11)   % Recursive call
U_22 = inv(T_22)   % Recursive call
U_12 = −U_11 T_12 U_22.
```

Let us now work out the computational cost of this algorithm, in *flops*, where a flop is a multiplication, addition, subtraction, or division. Denote the cost of calling inv for an $n \times n$ matrix by $c_n$ and assume for simplicity that $n = 2^k$. We then have $c_n = 2c_{n/2} + 2(n/2)^3$, where the second term is the cost of forming the triangular–full–triangular product $U_{11}T_{12}U_{22}$ of matrices of dimension $n/2$. Solving this recurrence gives $c_n = n^3/3 + O(n^2)$, which is the same as the cost of inverting a triangular matrix by standard techniques such as solving $TX = I$ by substitution.

As these examples show, recursion is a powerful way to express algorithms. But it is not always the right tool. To illustrate, consider the Fibonacci numbers, 1, 1, 2, 3, 5, ..., which satisfy the recurrence $f_n = f_{n-1} + f_{n-2}$ for $n \geqslant 2$, with $f_0 = f_1 = 1$. The obvious way to express the computation of the $f_i$ is as a loop:

```
f_0 = 1, f_1 = 1
for i = 2:n
    f_i = f_{i-1} + f_{i-2}
end
```

If just $f_n$ is required then an alternative is the recursive function

```
function f = fib(n)
if n ⩽ 1
   f = 1
else
   f = fib(n − 1) + fib(n − 2)
end
```

The problem with this recursion is that it computes fib$(n-1)$ and fib$(n-2)$ independently instead of obtaining fib$(n-1)$ from fib$(n-2)$ with one addition as in the previous algorithm. In fact, the evaluation of fib$(n)$ requires $f_n \approx 1.6^n$ operations, so the recursive algorithm is exponential in cost versus the linear cost of the first algorithm. It is possible to compute $f_n$ with only logarithmic cost. The idea is to write

$$\begin{bmatrix} f_n \\ f_{n-1} \end{bmatrix} = \begin{bmatrix} 1 & 1 \\ 1 & 0 \end{bmatrix} \begin{bmatrix} f_{n-1} \\ f_{n-2} \end{bmatrix} = \begin{bmatrix} 1 & 1 \\ 1 & 0 \end{bmatrix}^2 \begin{bmatrix} f_{n-2} \\ f_{n-3} \end{bmatrix} = \cdots = \begin{bmatrix} 1 & 1 \\ 1 & 0 \end{bmatrix}^{n-1} \begin{bmatrix} f_1 \\ f_0 \end{bmatrix}.$$

The matrix $\left[\begin{smallmatrix} 1 & 1 \\ 1 & 0 \end{smallmatrix}\right]^{n-1}$ can be computed in $O(\log_2 n)$ operations using the analogue for matrices of algorithm 3.

A divide and conquer algorithm can break the problem into more than two subproblems. An example is the Karatsuba algorithm for multiplying two $n$-digit integers $x$ and $y$. Suppose $n$ is a power of 2 and write $x = x_1 10^{n/2} + x_2$, $y = y_1 10^{n/2} + y_2$, where $x_1$, $x_2$, $y_1$, and $y_2$ are $n/2$-digit integers. Then

$$xy = x_1 y_1 10^n + (x_1 y_2 + x_2 y_1) 10^{n/2} + x_2 y_2.$$

Computing $xy$ has been reduced to computing three half-sized products because $x_1y_2 + x_2y_1 = (x_1 + x_2)(y_1 + y_2) - x_1y_1 - x_2y_2$. This procedure can be applied recursively. Denoting by $C_n$ the number of arithmetic operations (on single-digit numbers) to form the product of two $n$-digit integers by this algorithm, we have $C_n = 3C_{n/2} + kn$ and $C_1 = 1$, where $kn$ is the cost of the additions. Then

$$\begin{aligned} C_n &= 3(3C_{n/4} + kn/2) + kn \\ &= 3(3(3C_{n/8} + kn/4) + kn/2) + kn \\ &= kn(1 + 3/2 + (3/2)^2 + \cdots + (3/2)^{\log_2 n}) \\ &\approx 3kn^{\log_2 3} \approx 3kn^{1.58}, \end{aligned}$$

where the approximation is obtained by assuming that $n$ is a power of 2. The cost is asymptotically less than the $O(n^2)$ cost of forming $xy$ by the usual long multiplication method taught in school.

## 4 Computational Complexity

The computational cost of an algorithm is usually defined as the total number of arithmetic operations it requires, though it can also be defined as the execution time, under some assumption on the time required for each arithmetic operation. The cost is usually a function of the problem size, $n$ say, and since the growth with $n$ is of particular interest, the cost is usually approximated by the highest-order term, with lower-order terms ignored.

The algorithms considered so far all have the property that their computational cost is straightforward to evaluate and essentially independent of the data. For many algorithms the cost can vary greatly with the data. For example, an algorithm to sort a list of numbers might run more quickly when the list is nearly sorted. In this case it is desirable to find a bound that applies in all cases (a worst-case bound)—preferably one that is attainable for some set of data. It is also useful to have estimates of cost under certain assumptions on the distribution of the data. In *average-case analysis*, a probability distribution is assumed for the data and the expected cost is determined. *Smoothed analysis*, developed since 2000, interpolates between worst-case analysis and average-case analysis by measuring the expected performance of algorithms under small random perturbations of worst-case inputs. A number of algorithms are known for which the worst-case cost is exponential in the problem dimension $n$ whereas the smoothed cost is polynomial in $n$, a prominent example being the SIMPLEX METHOD [IV.11 §3.1] for linear programming.

A good example of a problem for which different algorithms can have widely varying cost is the solution of a linear system $Ax = b$, where $A$ is an $n \times n$ matrix. Cramer's rule states that $x_i = \det(A_i(b))/\det(A)$, where $A_i(b)$ denotes $A$ with its $i$th column replaced by $b$. If the determinant is evaluated from the usual textbook formula involving EXPANSION BY MINORS [I.2 §18], the cost of computing $x$ is about $(n+1)!$ operations, making this method impractical unless $n$ is very small. By contrast, Gaussian elimination solves the system in $2n^3/3 + O(n^2)$ operations, with mere polynomial growth of the operation count with $n$. However, Gaussian elimination is by no means of optimal complexity, as we now explain.

The complexity of matrix inversion can be shown to be the same as that of matrix multiplication, so it suffices to consider the matrix multiplication problem $C = AB$ for $n \times n$ matrices $A$ and $B$. The usual formula for matrix multiplication yields $C$ in $2n^3$ operations. In a 1969 paper Volker Strassen showed that when $n = 2$ the product can be computed from the formulas

$$\begin{aligned} &p_1 = (a_{11} + a_{22})(b_{11} + b_{22}), \\ &p_2 = (a_{21} + a_{22})b_{11}, \qquad p_3 = a_{11}(b_{12} - b_{22}), \\ &p_4 = a_{22}(b_{21} - b_{11}), \qquad p_5 = (a_{11} + a_{12})b_{22}, \\ &p_6 = (a_{21} - a_{11})(b_{11} + b_{12}), \\ &p_7 = (a_{12} - a_{22})(b_{21} + b_{22}), \end{aligned}$$

$$C = \begin{bmatrix} p_1 + p_4 - p_5 + p_7 & p_3 + p_5 \\ p_2 + p_4 & p_1 + p_3 - p_2 + p_6 \end{bmatrix}.$$

The evaluation requires seven multiplications and eighteen additions instead of eight multiplications and eight additions for the usual formulas. At first sight, this does not appear to be an improvement. However, these formulas do not rely on commutativity so are valid when the $a_{ij}$ and $b_{ij}$ are matrices, in which case for large dimensions the saving of one multiplication greatly outweighs the extra ten additions. Assuming $n$ is a power of 2, we can partition $A$ and $B$ into four blocks of size $n/2$, apply Strassen's formulas for the multiplication, and then apply the same formulas recursively on the half-sized matrix products. The resulting algorithm requires $O(n^{\log_2 7}) = O(n^{2.81})$ operations. Strassen's work sparked interest in finding matrix multiplication algorithms of even lower complexity. Since there are $O(n^2)$ elements of data, which must each participate in at least one operation, the

**Table 1** The cost of solving an $n \times n$ linear system obtained by discretizing the two-dimensional Poisson equation.

| Year | Method | Cost | Type |
|---|---|---|---|
| 1948 | Banded Cholesky | $n^2$ | Direct |
| 1948 | Jacobi, Gauss–Seidel | $n^2$ | Iterative |
| 1950 | SOR (optimal parameter) | $n^{3/2}$ | Iterative |
| 1952 | Conjugate gradients | $n^{3/2}$ | Iterative |
| 1965 | Fast Fourier transform | $n \log n$ | Direct |
| 1965 | Block cyclic reduction | $n \log n$ | Direct |
| 1977 | Multigrid | $n$ | Iterative |

exponent of $n$ must be at least 2. The current world record upper bound on the exponent is 2.3728639, proved by François Le Gall in 2014. However, all existing algorithms with exponent less than that of Strassen's algorithm are extremely complicated and not of practical interest.

An area that has undergone many important algorithmic developments over the years is the solution of linear systems arising from the discretization of partial differential equations (PDEs). Consider the POISSON EQUATION [III.18] on a square with the unknown function specified on the boundary. When discretized on an $N \times N$ grid by centered differences, a system of $n = N^2$ equations in $n$ unknowns is obtained with a banded, symmetric positive-definite coefficient matrix containing $O(n)$ nonzeros. Table 1 gives the dominant term in the operation count (ignoring the multiplicative constant) for different methods, some of which are described in NUMERICAL LINEAR ALGEBRA AND MATRIX ANALYSIS [IV.10]. For the iterative algorithms it is assumed that the iteration is terminated when the error is of order $10^{-6}$. The year is the year of first publication, or, for the first two methods, the year that the first stored-program computer was operational. Since there are $n$ elements in the solution vector and at least one operation is required to compute each element, a lower bound on the cost is $O(n)$, and this is achieved by the multigrid method. The algorithmic speedups shown in the table are of a similar magnitude to the speedups in computer hardware over the same period.

### 4.1 Complexity Classes

The algorithms we have described so far all have a cost that is bounded by a polynomial in the problem dimension, $n$. For some problems the existence of algorithms with polynomial complexity is unclear. In analyzing this question mathematicians and computer scientists use a classification of problems that makes a distinction finer than whether there is or is not an algorithm of polynomial run time. This classification is phrased in terms of decision problems: ones that have a yes or no answer. The problem class *P* comprises those problems that can be solved in polynomial time in the problem dimension. The class *NP* comprises those problems for which a yes answer can be verified in polynomial time. An example of a problem in NP is a jigsaw puzzle: it is easy to check that a claimed solution is a correctly assembled puzzle, but solving the puzzle in the first place appears to be much harder.

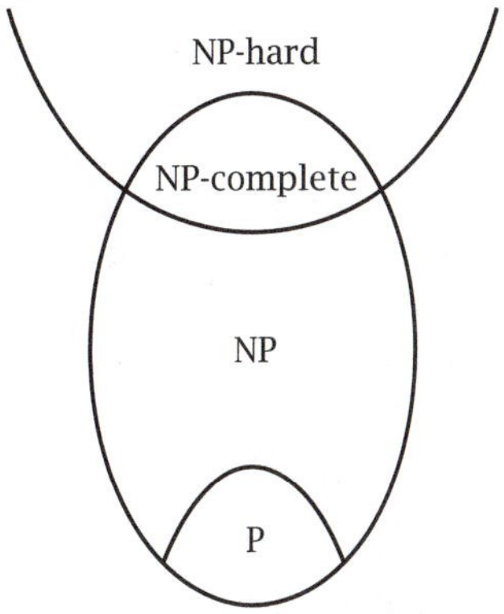


**Figure 2** Complexity classes. It is not known whether the classes P and NP are equal.

A problem is *NP-complete* if it is in NP and it is possible to reduce any other NP problem to it in polynomial time. Hence if a polynomial-time algorithm exists for an NP-complete problem then all NP problems can be solved in polynomial time. Many NP-complete problems are known, including Boolean satisfiability, graph coloring, choosing optimal page breaks in a document, and the Battleship game or puzzle.

A problem (not necessarily a decision problem) is *NP-hard* if it is at least as hard as any NP problem, in the sense that there is an NP-complete problem that is reducible to it in polynomial time. Thus the NP-hard problems are even harder than the NP-complete problems. Examples of NP-hard problems are the TRAVELING SALESMAN PROBLEM [VI.18], SPARSE APPROXIMATION [VII.10], and nonconvex QUADRATIC PROGRAMMING [IV.11 §1.3]. Figure 2 shows the relation among the classes.

An excellent example of the subtleties of computational complexity is provided by the determinant and the permanent of a matrix. The permanent of an $n \times n$

matrix $A$ is

$$\operatorname{perm}(A) = \sum_{\sigma} \prod_{i=1}^{n} a_{i,\sigma_i},$$

where the vector $\sigma$ ranges over all permutations of the set of integers $\{1, 2, \ldots, n\}$. The determinant has a similar expression differing only in that the product term is multiplied by the sign ($\pm 1$) of the permutation. Yet while the determinant can be computed in $O(n^3)$ operations, by Gaussian elimination, no polynomial-time algorithm has ever been discovered for computing the permanent. Leslie Valiant gave insight into this disparity when he showed in 1979 that the problem of computing the permanent is complete for a complexity class of counting problems called #P that extends NP.

The most famous open problem in computer science is "is P equal to NP?" It was posed by Stephen Cook in 1971 and is one of the seven Clay Institute Millennium Problems, for each of which a \$1 million prize is available for a solution. Informally, the question is whether the "easy to solve" problems are equal to the "easy to check" problems. It is known that $P \subseteq NP$, so the question is whether or not the inclusion is strict.

## 5 Trade-off between Speed and Accuracy

In designing algorithms that run in floating-point arithmetic it frequently happens that an increase in speed is accompanied by a decrease in accuracy. A classic example is the computation of the *sample variance* of $n$ numbers $x_1, \ldots, x_n$, which is defined as

$$s_n^2 = \frac{1}{n-1} \sum_{i=1}^{n} (x_i - \bar{x})^2, \tag{2}$$

where the sample mean

$$\bar{x} = \frac{1}{n} \sum_{i=1}^{n} x_i.$$

Computing $s_n^2$ from this formula requires two passes through the data, one to compute $\bar{x}$ and the other to accumulate the sum of squares. A two-pass computation is undesirable for large data sets or when the sample variance is to be computed as the data is generated. An alternative formula, found in statistics textbooks (and implemented on many pocket calculators and spreadsheets over the years), uses about the same number of operations but requires only one pass through the data:

$$s_n^2 = \frac{1}{n-1} \left( \sum_{i=1}^{n} x_i^2 - \frac{1}{n} \left( \sum_{i=1}^{n} x_i \right)^2 \right). \tag{3}$$

However, this formula behaves badly in floating-point arithmetic. For example, if $n = 3$ and $x_1 = 10\,000$, $x_2 = 10\,001$, and $x_3 = 10\,002$, then, in IEEE single-precision arithmetic (with unit roundoff $u \approx 6 \times 10^{-8}$), the sample variance is computed as 1.0 by the two-pass formula (relative error 0) but 0.0 by the one-pass formula (relative error 1). The reason for the poor accuracy of the one-pass formula is that there is massive SUBTRACTIVE CANCELATION [II.13] in (3). The original formula (2) always yields a computed result with error $O(nu)$. Is there a way of combining the speed of the one-pass formula with the accuracy of the two-pass one? Yes: the recurrence

$$M_1 = x_1, \qquad Q_1 = 0,$$

$$\left.\begin{aligned} M_k &= M_{k-1} + \frac{x_k - M_{k-1}}{k} \\ Q_k &= Q_{k-1} + \frac{(k-1)(x_k - M_{k-1})^2}{k} \end{aligned}\right\} \quad k = 2{:}n$$

calculates $Q_n$, which yields $s_n^2 = Q_n/(n-1)$ and produces an accurate result in floating-point arithmetic.

## 6 Choice of Algorithm

Much research in numerical analysis and scientific computing is about finding the best algorithm for solving a given problem, and for classic problems such as solving a PDE or finding the eigenvalues of a matrix there are many possibilities, with improvements continually being developed. However, even for some quite elementary problems there are several possible algorithms, some of which are far from obvious.

A first example is the evaluation of a polynomial $p(x) = a_0 + a_1 x + \cdots + a_n x^n$. The most obvious way to evaluate the polynomial is by directly forming the powers of $x$.

```
p = a0 + a1 x, w = x
for i = 2:n
    w = wx
    p = p + ai w
end
```

This algorithm requires $2n$ multiplications and $n$ additions (ignoring the constant term in the operation count).

An alternative method is *Horner's method* (nested multiplication). It is derived by writing the polynomial in nested form:

$$p(x) = (\cdots((a_n x + a_{n-1})x + a_{n-2})x + \cdots + a_1)x + a_0.$$

Horner's method is

```
p = a_n
for i = n - 1:-1:0   % i goes down in steps of -1
    p = px + a_i
end
```

Horner's method requires $n$ multiplications and $n$ additions, so is significantly less expensive than the first method.

However, even Horner's method is not optimal. For example, it requires eight multiplications for $p(x) = x^8$, but this polynomial can be evaluated in just three multiplications using algorithm 3 in section 3. Moreover, for general polynomials of degree $n > 4$ there exist evaluation schemes that require strictly less than the $2n$ total additions and multiplications required by Horner's rule; the catch is that the operation count excludes some precomputation of coefficients that, once computed, can be reused for every subsequent polynomial evaluation. This latter example emphasizes that if one wishes to investigate optimality of schemes for polynomial evaluation it is important to be precise about what is included in the operation count.

A second example concerns the continued fraction

$$r_n(x) = b_0 + \cfrac{a_1 x}{b_1 + \cfrac{a_2 x}{b_2 + \cfrac{a_3 x}{b_3 + \cdots + \cfrac{a_{n-1} x}{b_{n-1} + \cfrac{a_n x}{b_n}}}}}. \quad (4)$$

This continued fraction represents a rational function $r_n(x) = p_n(x)/q_n(x)$, where $p_n$ and $q_n$ are polynomials of degrees $\lceil n/2 \rceil$ and $\lfloor n/2 \rfloor$, respectively. Such continued fractions arise in the approximation of transcendental functions by PADÉ APPROXIMANTS [IV.9 §2.4].

Probably the most obvious way to evaluate the continued fraction is by the following bottom-up procedure.

**Algorithm 5 (continued fraction, bottom-up).** *This algorithm evaluates the continued fraction (4) in bottom-up fashion.*

```
y_n = (a_n/b_n)x
for j = n - 1:-1:1
    y_j = a_j x/(b_j + y_{j+1})
end
r_n = b_0 + y_1
```

**Cost.** The total number of operations is $n(D + M + A)$, where $A$, $D$, and $M$ denote an addition, a division, and a multiplication, respectively.

The bottom-up evaluation requires us to know the value of $n$ in advance and is not well suited to evaluating the sequence $r_1(x), r_2(x), \dots$, since it needs to start afresh each time. Top-down evaluation is better in this case. The following recurrence dates back to John Wallis in 1655.

**Algorithm 6 (continued fraction, top-down).** *This algorithm evaluates the continued fraction (4) in top-down fashion.*

```
p_{-1} = 1, q_{-1} = 0, p_0 = b_0, q_0 = 1
for j = 1:n
    p_j = b_j p_{j-1} + a_j x p_{j-2}
    q_j = b_j q_{j-1} + a_j x q_{j-2}
end
r_n = p_n q_n^{-1}
```

**Cost.** $5nM + 2nA + D$.

It is not obvious that algorithm 6 does actually evaluate $r_m(x)$, but this can be proved inductively. Note that top-down evaluation is substantially more expensive than bottom-up evaluation.

These are not the only possibilities. For example, when $n = 2m$ one can write $r_n$ in partial fraction form $r_n(x) = \sum_{j=1}^{m} \alpha_j x/(x - \beta_j)$, where the $\beta_j$ (assumed to be distinct) are the roots of $q_n$, which permits evaluation at a cost of $m(2M + A + D)$.

In practice, one needs to consider the effect of rounding errors on the evaluation, and this is in general dependent on the particular coefficients $a_i$ and $b_i$ and on the range of $x$ of interest.

When there are several algorithms for solving a problem one may need a *polyalgorithm*, which is a set of algorithms with rules for choosing between them. A good example of a polyalgorithm is the MATLAB backslash operator, which enables a solution to a linear system $Ax = b$ to be computed using the syntax $A\backslash b$. The matrix $A$ may be square or rectangular; diagonal, triangular, full, or sparse; and symmetric, symmetric positive-definite, or completely general. Underlying the backslash operator is an algorithm for identifying which of these properties hold and choosing the appropriate matrix factorization and solution process to apply.

## 7 Randomized Algorithms

All the algorithms described so far in this article are deterministic: if they are run repeatedly on the same data, they produce the same result every time. Some algorithms make random choices and so generally produce a different result every time they are run. For example, we might approximate

$$\int_0^1 f(x)\,\mathrm{d}x \approx \frac{1}{n}\sum_{i=1}^{n} f(x_i),$$

where the $x_i$ are independent random numbers uniformly distributed in $[0,1]$. The standard deviation of the error in this approximation is of order $n^{-1/2}$. This is an example of a *Monte Carlo algorithm*; such algorithms have a deterministic run time and produce an output that is correct (or has a given accuracy) with a certain probability. Of course, there are much more efficient ways to estimate a one-dimensional integral, but Monte Carlo algorithms come into their own for multidimensional integrals over complicated domains.

A *Las Vegas algorithm* always produces a correct result, but its run time is nondeterministic. A classic example is the quicksort algorithm for sorting a list of numbers, for which a randomized choice of the partition element makes the algorithm much faster on average than in the worst case ($O(n \log n)$ running time versus $O(n^2)$, for $n$ numbers).

Randomized algorithms can be much simpler than deterministic alternatives, they may be more able to exploit modern computing architectures, and they may be better suited to large data sets. There is a wide variety of randomized algorithms, and they are studied in mathematics, computer science, statistics, and other areas.

One active area of research is randomized algorithms for numerical linear algebra problems, based on random sampling and random projections. For example, fast algorithms exist for computing low-rank approximations to a given matrix. The general framework is that random sampling is used to identify a subspace that captures most of the action of the matrix, the matrix is then compressed to this subspace, and a low-rank factorization is computed from the reduced matrix.

Examples of randomized algorithms mentioned in this book are the Google PAGERANK ALGORITHM [VI.9], with its use of a random surfer, the $k$-MEANS ALGORITHM [IV.17 §5.3] for clustering, and MARKOV CHAIN MONTE CARLO ALGORITHMS [V.11 §3].

## 8 Some Key Algorithms in Applied Mathematics

Table 2 lists a selection of algorithms mentioned in this book. Very general methods such as preconditioning and the finite-element method, which require much more information to produce a particular algorithm, are omitted. The table illustrates the wide variety of important algorithms in applied mathematics, ranging from the old to the relatively new.

A notable feature of some of the algorithms is that they are iterative algorithms, which in principle take an infinite number of steps, for solving problems that can be solved directly, that is, in a finite number of operations. The conjugate gradient and multigrid methods are iterative methods for solving a linear system of equations, and for suitably structured systems they can provide a given level of accuracy much faster than Gaussian elimination, which is a direct method. Similarly, interior point methods are iterative methods for linear programming, competing with the simplex method, which is a direct method.

### Further Reading

A classic reference for algorithms and their analysis is Donald Knuth's *The Art of Computer Programming*. The first volume appeared in 1968 and the development is ongoing. Current volumes are *Fundamental Algorithms* (volume 1), *Seminumerical Algorithms* (volume 2), *Sorting and Searching* (volume 3), and *Combinatorial Algorithms* (volume 4), all published by Addison-Wesley (Reading, MA).

Bentley, J. L. 1986. *Programming Pearls*. Reading, MA: Addison-Wesley.

Brassard, G., and P. Bratley. 1996. *Fundamentals of Algorithmics*. Englewood Cliffs, NJ: Prentice-Hall.

Cormen, T. H., C. E. Leiserson, R. L. Rivest, and C. Stein. 2009. *Introduction to Algorithms*, 3rd edn. Cambridge, MA: MIT Press.

Higham, N. J. 2002. *Accuracy and Stability of Numerical Algorithms*, 2nd edn. Philadelphia, PA: SIAM.

# I.5 Goals of Applied Mathematical Research

*Nicholas J. Higham*

A large body of existing mathematical knowledge is encapsulated in theorems, methods, and algorithms, some of which have been known for centuries. But applied mathematics is not simply the application of

**Table 2** Some algorithms mentioned in this book.

| Algorithm | Reference | Key early figures |
|---|---|---|
| Gaussian elimination | IV.10 §2 | Ancient Chinese (ca. 1 C.E.), Gauss (1809); formulated as LU factorization by various authors from 1940s |
| Newton's method | II.28 | Newton (1669), Raphson (1690) |
| Fast Fourier transform | II.10 | Gauss (1805), Cooley and Tukey (1965) |
| Cholesky factorization | IV.10 §2 | Cholesky (1910) |
| Remez algorithm | IV.9 §3.5, VI.11 §2 | Remez (1934) |
| Simplex method (linear programming) | IV.11 §3.1 | Dantzig (1947) |
| Conjugate gradient and Lanczos methods | IV.10 §9 | Hestenes and Stiefel (1952), Lanczos (1952) |
| Ford–Fulkerson algorithm | IV.37 §7 | Ford and Fulkerson (1956) |
| $k$-means algorithm | IV.17 §5.3 | Lloyd (1957), Steinhaus (1957) |
| QR factorization | IV.10 §2 | Givens (1958), Householder (1958) |
| Dijkstra's algorithm | VI.10 | Dijkstra (1959) |
| Quasi-Newton methods | IV.11 §4.2 | Davidon (1959), Broyden, Fletcher, Goldfarb, Powell, Shanno (early 1960s) |
| QR algorithm | IV.10 §5.5 | Francis (1961), Kublanovskaya (1962) |
| QZ algorithm | IV.10 §5.8 | Moler and Stewart (1973) |
| Singular value decomposition | II.32 | Golub and Kahan (1965), Golub and Reinsch (1970) |
| Strassen's method | I.4 §4 | Strassen (1968) |
| Multigrid | IV.10 §9, IV.13 §3, IV.16 | Fedorenko (1964), Brandt (1973), Hackbusch (1977) |
| Interior point methods | IV.11 §3.2 | Karmarkar (1984) |
| Generalized minimal residual method | IV.10 §9 | Saad and Schulz (1986) |
| Fast multipole method | VI.17 | Greengard and Rokhlin (1987) |
| JPEG | VII.7 §5, VII.8 | Members of the Joint Photographic Experts Group (1992) |
| PageRank | VI.9 | Brin and Page (1998) |
| HITS | I.1 | Kleinberg (1999) |

existing mathematical ideas to practical problems: new results are continually being developed, usually building on old ones. Applied mathematicians are always innovating, and the constant arrival of new or modified problems provides direction and motivation for their research.

In this article we describe some goals of research in applied mathematics from the perspectives of the ancient problem of solving equations, the more contemporary theme of exploiting structure, and the practically important tasks of modeling and prediction. We also discuss the strategy behind research.

## 1 Solving Equations

A large proportion of applied mathematics research papers are about analyzing or solving equations. The equations may be algebraic, such as linear or nonlinear equations in one or more variables. They may be ordinary differential equations (ODEs), partial differential equations (PDEs), integral equations, or differential-algebraic equations.

The wide variety of equations reflects the many different ways in which one can attempt to capture the behavior of the system being modeled. Whatever the equation, an applied mathematician is interested in answering a number of questions.

### 1.1 Does the Equation Have a Solution?

We are interested in whether there is a unique solution and, if there is more than one solution, how many there are and how they are characterized. Existence of solutions may not be obvious, and one occasionally hears tales of mathematicians who have solved equations for which a proof is later given that no solution exists. Such a circumstance may sound puzzling: is it not easy to check that a putative solution actually is a solution? Unfortunately, checking satisfaction of the equation may not be easy, especially if one is working in a function space. Moreover, the problem specification may require the solution to have certain properties, such as existence of a certain number of derivatives, and the claimed solution might satisfy the equation but fail to have some of the required properties. Instead of analyzing the problem in the precise form in which it is given, it may be better to investigate what additional properties must be imposed for an equation to have a unique solution.

### 1.2 Is the Equation Well-Posed?

A problem is *well-posed* if it has a unique solution and the solution changes continuously with the data that define the problem. A problem that is not well-posed is *ill-posed*. For an ill-posed problem an arbitrarily small perturbation of the data can produce an arbitrarily large change in the solution, which is clearly an unsatisfactory situation.

An example of a well-posed problem is to determine the weight supported by each leg of a three-legged table. Assuming that the table and its legs are perfectly symmetric and the ground is flat, the answer is that each leg carries one-third of the total weight. For a table with four legs each leg supports one-quarter of the total weight, but if one leg is shortened by a tiny amount then it leaves the ground and the other three legs support the weight of the table (a phenomenon many of us have experienced in restaurants). For four-legged tables the problem is therefore ill-posed.

For finite-dimensional problems, uniqueness of the solution implies well-posedness. For example, a linear system $Ax = b$ of $n$ equations in $n$ unknowns with a nonsingular coefficient matrix $A$ is well-posed. Even so, if $A$ is nearly singular then a small perturbation of $A$ can produce a large change in the solution, albeit not arbitrarily large: the CONDITION NUMBER [I.2 §22] $\kappa(A) = \|A\| \, \|A^{-1}\|$ bounds the relative change. But for infinite-dimensional problems the existence of a unique solution does not imply that the problem is well-posed; examples are given in the article on INTEGRAL EQUATIONS [IV.4 §6].

The notion of well-posedness was introduced by Jacques Hadamard at the beginning of the twentieth century. He believed that physically meaningful problems should be well-posed. Today it is recognized that many problems are ill-posed, and they are routinely solved by reformulating them so that they are well-posed, typically by a process called REGULARIZATION [IV.15 §2.6] (see also INTEGRAL EQUATIONS [IV.4 §7]).

An important source of ill-posed problems is INVERSE PROBLEMS [IV.15]. Consider a mathematical model in which the inputs are physical variables that can be adjusted and the output variables are the result of an experiment. The *forward problem* is to predict the outputs from a given set of inputs. The *inverse problem* is to make deductions about the inputs that could have produced a given set of outputs. In practice, the measurements of the outputs may be subject to noise and the model may be imperfect, so UNCERTAINTY QUANTIFICATION [II.34] needs to be carried out in order to estimate the uncertainty in the predictions and deductions.

### 1.3 What Qualitative Properties Does a Solution Have?

It may be of more interest to know the behavior of a solution than to know the solution itself. One may be interested in whether the solution, $f(t)$ say, decays as $t \to \infty$, whether it is monotonic in $t$, or whether it oscillates and, if so, with what fixed or time-varying frequency. If the problem depends on parameters, it may be possible to answer these questions for a range of values of the parameters.

### 1.4 Does an Iteration Converge?

As we saw in METHODS OF SOLUTION [I.3], solutions are often computed from iterative processes, and we therefore need to understand these processes. Various facets of convergence may be of interest.

- Is the iteration always defined, or can it break down (e.g., because of division by zero)?
- For what starting values, and for what class of problems, does the iteration converge?
- To what does the iteration converge, and how does this depend on the starting value (if it does at all)?
- How fast does the iteration converge?
- How are errors (in the initial data, or rounding errors introduced during the iteration) propagated? In particular, are they bounded?

To illustrate some of these points we consider the iteration

$$x_{k+1} = \frac{1}{p}[(p-1)x_k + x_k^{1-p}a], \qquad (1)$$

with $p$ a positive integer and $a \in \mathbb{C}$, which is Newton's method for computing a $p$th root of $a$. We ask for which $a$ and which starting values $x_0$ the iteration converges and to what root it converges. The analysis is simplified by defining $y_k = \theta^{-1}x_k$, where $\theta$ is a $p$th root of $a$, as the iteration can then be rewritten

$$y_{k+1} = \frac{1}{p}[(p-1)y_k + y_k^{1-p}], \qquad y_0 = \theta^{-1}x_0, \quad (2)$$

which is Newton's method for computing a $p$th root of unity. The original parameters $a$ and $x_0$ have been combined into the starting value $y_0$.

Figure 1 illustrates the convergence of the iteration for $p = 2, 3, 5$. For $y_0$ ranging over a $400 \times 400$ grid with $\operatorname{Re} y_0, \operatorname{Im} y_0 \in [-2.5, 2.5]$, it plots the root to which $y_k$ from (2) converges, with each root denoted by a different grayscale from white (the principal root, 1) to black. Convergence is declared if after fifty iterations the iterate is within relative distance $10^{-13}$ of a root; the relatively small number of points for which convergence was not observed are plotted white. For $p = 2$ the figure suggests that the iteration converges to 1 if started in the open right half-plane and $-1$ if started in the open left half-plane, and this can be proved to be true. But for $p = 3, 5$ the regions of convergence have a much more complicated structure, involving sectors with petal-like boundaries.

The complexity of the convergence for $p \geqslant 3$ was first noticed by Arthur Cayley in 1879, and an analysis of convergence for all starting values requires the theory of Julia sets of rational maps. However, for practical purposes it is usually principal roots that need to be computed, so from a practical viewpoint the main implication to be drawn from the figure is that for $p = 3, 5$ Newton's method converges to 1 for $y_0$ sufficiently close to the positive real axis—and it can be proved that this is true.

We see from this example that the convergence analysis depends very much on the precise question that is being asked. The iteration (1) generalizes in a natural way to matrices and operators, for which the convergence results for the scalar case can be exploited.

## 2 Preserving Structure

Many mathematical problems have some kind of structure. An example with explicit structure is a linear system $Ax = b$ in which the $n \times n$ matrix $A$ is a TOEPLITZ MATRIX [I.2 §18]. This system has $n^2 + n$ numbers in $A$ and $b$ but only $3n - 1$ independent parameters. On the other hand, if for the vector ODE $y' = f(t, y)$ there is a vector $v$ such that $v^{\mathrm{T}}f(t, y) = 0$ for all $t$ and $y$, then $(\mathrm{d}/\mathrm{d}t)v^{\mathrm{T}}y(t) = v^{\mathrm{T}}f(t, y) = 0$, so $v^{\mathrm{T}}y(t)$ is constant for all $t$. This conservation or invariance property is a form of structure, though one more implicit than for the Toeplitz system.

An example of a nonlinear conservation property is provided by the system of ODEs

$$u'(t) = v(t),$$
$$v'(t) = -u(t).$$

For this system,

$$\frac{\mathrm{d}}{\mathrm{d}t}(u^2 + v^2) = 2(u'u + v'v) = 2(vu - uv) = 0,$$

so there is a quadratic invariant. In particular, for the initial values $u(0) = 1$ and $v(0) = 0$ the solution is $u(t) = \cos t$ and $v(t) = -\sin t$, which lies on the unit circle centered at the origin in the $uv$-plane. If we solve the system using a numerical method, we would like the numerical solution also to lie on the circle. In fact, one potential use of this differential equation is to provide a method for plotting circles that avoids the relatively expensive evaluation of sines and cosines. Consider the following four standard numerical methods applied to our ODE system. Here, $u_k \approx u(kh)$ and $v_k \approx v(kh)$, where $h$ is a given step size, and $u_0 = 1$ and $v_0 = 0$:

Forward Euler $\begin{cases} u_{k+1} = u_k + hv_k, \\ v_{k+1} = v_k - hu_k, \end{cases}$

Backward Euler $\begin{cases} u_{k+1} = u_k + hv_{k+1}, \\ v_{k+1} = v_k - hu_{k+1}, \end{cases}$

Trapezium method $\begin{cases} u_{k+1} = u_k + h(v_k + v_{k+1})/2, \\ v_{k+1} = v_k - h(u_k + u_{k+1})/2, \end{cases}$

Leapfrog method $\begin{cases} u_{k+1} = u_k + hv_k, \\ v_{k+1} = v_k - hu_{k+1}. \end{cases}$

Figure 2 plots the numerical solutions computed with 32 steps of length $h = 2\pi/32$. We see that the forward Euler solution spirals outward while the backward Euler solution spirals inward. The trapezium method solution stays nicely on the unit circle. The leapfrog method solution traces an ellipse. This behavior is easy to explain if we write each method in the form

$$z_{k+1} = Gz_k, \qquad z_k = \begin{bmatrix} u_k \\ v_k \end{bmatrix},$$

where $G = [\begin{smallmatrix} 1 & h \\ -h & 1 \end{smallmatrix}]$ for the Euler method, for example. Then the behavior of the sequence $z_k$ depends on the

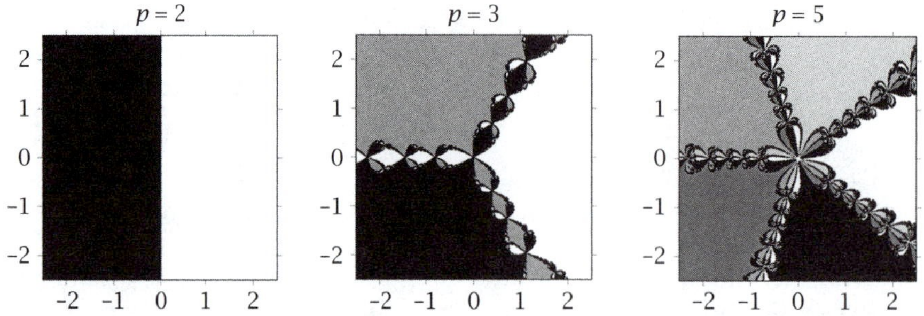


**Figure 1** Newton iteration for a $p$th root of unity. Each point $y_0$ in the region is shaded according to the root to which the iteration converges, with white denoting the principal root, 1.

eigenvalues of the matrix $G$. It turns out that the spectral radius of $G$ is greater than 1 for forward Euler and less than 1 for backward Euler, which explains the spiraling. For the trapezium rule $G$ is orthogonal, so $\|z_{k+1}\|_2 = \|z_k\|_2$ and the trapezium solutions stay exactly on the unit circle. For the leapfrog method the determinant of $G$ is 1, which means that areas are preserved, but $G$ is not orthogonal so the leapfrog solution drifts slightly off the circle.

The subject of GEOMETRIC INTEGRATION [IV.12 §5] is concerned more generally with methods for integrating nonlinear initial-value ODEs and PDEs in a way that preserves the invariants of the system, while also providing good accuracy in the usual sense. This includes, in particular, SYMPLECTIC INTEGRATORS [IV.12 §1.3] for Hamiltonian systems.

## 3 Modeling and Prediction

As WHAT IS APPLIED MATHEMATICS? [I.1 §1] explains, modeling is the first step in solving a physical problem. Models are necessarily simplifications because it is impractical to incorporate every detail. But simple models can still be useful as tools to explore the broad consequences of physical laws. Moreover, the more complex a model is the more parameters it has (all of which need estimating) and the harder it is to analyze.

In their 1987 book *Empirical Model-Building and Response Surfaces*, Box and Draper ask us to

> Remember that all models are wrong; the practical question is how wrong do they have to be to not be useful.

Road maps illustrate this statement. They are always a simplified representation of reality due to representing a three-dimensional world in two dimensions and displaying wiggly roads as straight lines. But road maps are very useful. Moreover, there is no single "correct" map but rather many possibilities depending on resolution and purpose. Another example is the approximation of $\pi$. The approximation $\pi \approx 3.14$ is a model for $\pi$ that is wrong in that it is not exact, but it is good enough for many purposes.

It is difficult to give examples of the modeling process because knowledge of the problem domain is usually required and derivations can be lengthy. We will use for illustration a very simple model of population growth, based on the *logistic equation*

$$\frac{\mathrm{d}N}{\mathrm{d}t} = rN\left(1 - \frac{N}{K}\right).$$

Here, $N(t)$ is a representation in a continuous variable of the number of individuals in a population at time $t$, $r > 0$ is the growth rate of the population, and $K > 0$ is the carrying capacity. For $K = \infty$, the model says that the rate of change of the population, $\mathrm{d}N/\mathrm{d}t$, is $rN$; that is, it is proportional to the size of the population through the constant $r$, so the population grows exponentially. For finite $K$, the model attenuates this rate of growth by a subtractive term $rN^2/K$, which can be interpreted as representing the increasing effects of competition for food as the population grows. The logistic equation can be solved exactly for $N(t)$ (see ORDINARY DIFFERENTIAL EQUATIONS [IV.2 §2]). Laboratory experiments have shown that the model can predict reasonably well the growth of protozoa feeding on bacteria. However, for some organisms the basic logistic equation is not a good model because it assumes instant responses to changes in population size and so does not account for gestation periods, the time taken for young to reach maturity, and other delays. A more realistic model may therefore be

$$\frac{\mathrm{d}N(t)}{\mathrm{d}t} = rN(t)\left(1 - \frac{N(t-\tau)}{K}\right),$$

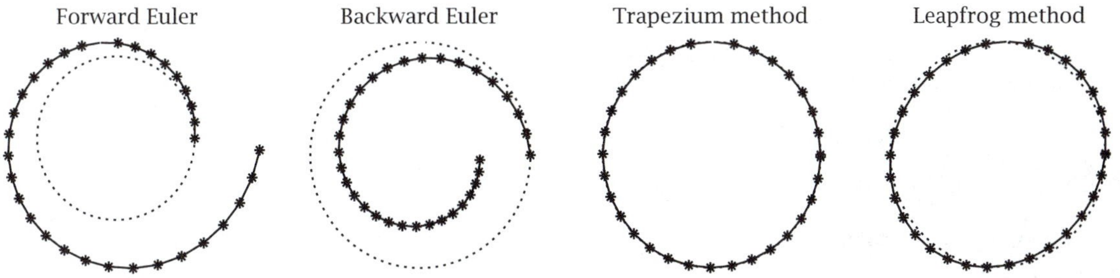


**Figure 2** Approximations to the unit circle computed by four different numerical integrators with step size $h = 2\pi/32$. The dotted line is the unit circle; asterisks denote numerical approximations.

where $\tau > 0$ is a delay parameter. At time $t$, part of the quadratic term is now evaluated at an earlier time, $t - \tau$. This delay differential equation has oscillatory solutions and has been found to model well the population of lemmings in the Arctic. Note that in contrast to the PREDATOR–PREY MODEL [I.2 §10], the delayed logistic model can produce oscillations in a population without the need for a second species acting as predator. There is no suggestion that either of these logistic models is perfect, but with appropriate fitting of parameters they can provide useful approximations to actual populations and can be used to predict future behavior.

### 3.1 Errors

A lot of research is devoted to understanding the errors that arise at the different stages of the modeling process. These can broadly be categorized as follows.

**Errors in the mathematical model.** Setting up the model introduces errors, since the model is never exact. These are the hardest errors to estimate.

**Approximation errors.** These are the errors incurred when infinite-dimensional equations are replaced by a finite-dimensional system (that is, a continuous problem is replaced by a discrete one: the process of discretization), or when simpler approximations to the equations are developed (e.g., by MODEL REDUCTION [II.26]). These errors include errors in replacing one approximating space by another (e.g., replacing continuous functions by polynomials), errors in FINITE-DIFFERENCE [II.11] approximations, and errors in truncating power series and other expansions.

**Rounding errors.** Once the problem has been put in a form that can be solved by an algorithm implemented in a computer program, the effects of the rounding errors introduced by working in finite-precision arithmetic need to be determined.

Analysis of errors may include looking at the effects of uncertainties in the model data, including in any parameters in the model that must be estimated. This might be tackled in a statistical sense using techniques from UNCERTAINTY QUANTIFICATION [II.34]—indeed, if the model has incompletely known data then probabilistic techniques may already be in use to estimate the missing data. Sensitivity of the solution of the model may also be analyzed by obtaining worst-case error bounds with the aid of CONDITION NUMBERS [I.2 §22].

### 3.2 Multiphysics and Multiscale Modeling

Scientists are increasingly tackling problems with one or both of the following characteristics: (a) the system has multiple components, each governed by its own physical principles; and (b) the relevant processes develop over widely different time and space scales. These are called *multiphysics* and *multiscale* problems, respectively. An example of both is the problem of modeling how *space weather* affects the Earth, and in particular modeling the interaction of the solar wind (the flow of charged particles emitted by the sun) with the Earth's magnetic field. Different physical models describe the statistical distribution of the plasma, which consists of charged particles, and the evolution of the electric and magnetic fields, and these form a coupled nonlinear system of PDEs. The length scales range from millions of kilometers (the Earth–sun distance) to hundreds of meters, and the timescales range from hours down to $10^{-5}$ seconds. Problems such as this pose challenges both for modeling and for computational solution of the models. The computations require HIGH-PERFORMANCE COMPUTERS [VII.12], and a particular task is to present the vast quantities of data generated in such a way that users, such as forecasters of space weather, can explore and interpret them. More on the issues of this section can be found in the articles on COMPUTATIONAL SCIENCE [IV.16] and VISUALIZATION [VII.13].

### 3.3 Computational Experiments

The step in the problem solution when computational experiments are carried out might seem to be the easiest, but it can in fact be one of the hardest and most time-consuming, for several reasons. It can be hard to decide what experiments to carry out, and it may be necessary to refine the experiments many times until useful or satisfactory results are obtained. The computations may have a long run time, even if executed on a high-performance computer.

Many pitfalls can be avoided by working to modern standards of REPRODUCIBLE RESEARCH [VIII.5], which require that programs, data, and results be recorded, documented, and made available in such a way that the results can be reproduced by an independent researcher and, just as importantly, by the original author.

### 3.4 Validation

The process of *validation* involves asking the question, "Have we solved the right equations?" This is not to be confused with *verification*, which asks whether the equations in the model have been solved correctly. Whereas verification is a purely mathematical question, validation intimately involves the underlying physical problem. A classic way to validate results from a model is to compare them with experimental results. However, this is not always feasible, as we may be modeling a device or structure that is still in the design phase or on which experiments are not possible (e.g., the Earth's climate).

Validation may not produce a yes or no answer but may instead indicate a range of parameters for which the model is a good predictor of actual behavior.

Validation may be the first step of an iterative refinement procedure in which the steps in figure 1 are repeated, with the second and subsequent invocations of the first step now comprising adjustments to the current model. Assuming it is feasible to carry out refinement, there is much to be said for starting with the simplest possible model and building gradually toward an effective model of minimal complexity.

## 4 Strategies for Research and Publishing

Analysis of the research literature in applied mathematics reveals some common features that can be built into a list of strategies for doing research.

(i) Solve an open problem or prove the truth or falsity of a conjecture that has previously been stated in the literature.

(ii) Derive a method for solving a problem that occurs in practice and has not been effectively solved previously. Problems of very large dimension, for which existing techniques might be impractical, are good hunting grounds.

(iii) Prove convergence of a method for which the existing convergence theory is incomplete.

(iv) Spot some previously unnoticed phenomenon and explain it.

(v) Generalize a result or algorithm to a wider class of problems, obtaining new insight in doing so.

(vi) Provide a new derivation of an existing result or algorithm that yields new insight.

(vii) Develop a new measure of cost or error for a problem and then derive a new algorithm that is better than existing algorithms with respect to that metric. For example, instead of measuring computational cost in the traditional way by the number of elementary arithmetic operations, also include the cost of data movement when the algorithm is implemented on a parallel computer.

(viii) Find hidden assumptions in an existing method and remove them. For example, it may seem obvious that multiplying two $2 \times 2$ matrices requires eight multiplications, but Strassen showed that only seven multiplications are needed (see [I.4 §4] for the relevant formulas), thereby deriving an asymptotically fast method for matrix multiplication.

(ix) Rehabilitate an out-of-favor method by showing that it can be made competitive again by exploiting new research results, problem requirements, or hardware developments.

(x) Use mathematical models to gain new insight into complex physical processes.

(xi) Use mathematical models to make quantitative predictions about physical phenomena that can lead to new procedures, standards, etc., in the target field. Here it will probably be necessary to work with researchers from other disciplines.

Nowadays, publishing the results of one's research is more important than ever. Funding bodies expect to see publications, as they provide evidence that the research has been successful and they help to disseminate the work. Assessment of researchers and their institutions increasingly makes use of metrics, some of which relate to publications, such as the number of citations a paper receives and the ranking of the journal in which it is published. There is therefore a tension between publishing prolifically (which, taken to the extreme, leads to breaking research up into "least publishable units") and publishing fewer, longer, more-considered papers.

In addition to the traditional journals and conference proceedings that publish (usually) refereed articles, nowadays there are many outlets for unrefereed manuscripts, including institutional eprint servers, the global arXiv eprint service, and personal Web pages. Blogs provide yet another venue for publishing research, usually in the form of shorter articles presented in a more accessible form than in a regular paper. The nonjournal outlets provide for instant publication but have varying degrees of visibility and permanence.

The balance between publication of journals and books in print only, in both print and electronic form, and purely in electronic form has been changing for the past decade or more, and the advent of handheld devices such as smartphones and tablets has accelerated developments. Equally disruptive is the movement toward open-access publishing. Traditionally, the publishers of mathematics journals did not charge an author to publish an article but did charge institutions to subscribe to the journal. In recent years a new model has been introduced in which the author pays to publish an article in a journal and the article is freely available to all.

While we can be sure that there will always be outlets for publishing research, it is difficult to predict how the forms that these outlets take will evolve in the future.

## I.6 The History of Applied Mathematics

*June Barrow-Green and Reinhard Siegmund-Schultze*

> And as for the Mixt Mathematikes I may only make this prediction, that there cannot faile to bee more kinds of them, as Nature growes furder disclosed.
>
> Francis Bacon: *Of the Proficience and Advancement of Learning* (1605)

**Figure 1** Detail from a pictorial representation of J. le R. d'Alembert's "Système des Connoissances Humaines" in the supplement to volume 2 of *l'Encyclopédie* (1769), mentioning "mixed mathematics," for which Francis Bacon had predicted a great future.

### 1 Introduction

What is applied mathematics? This is a difficult question—one to which there is no simple answer. The massive growth in applications of mathematics within and outside the sciences, especially since World War II, has made this question even more problematic, the increasing overlap with other disciplines and their methods adding further to the difficulties, creating problems that border on the philosophical.

Given the fact that almost every part of mathematics is potentially applicable, there are mathematicians and historians who consider the term "applied mathematics" primarily as a term of social distinction or a matter of attitude. One such was William Bonnor, the mathematician and gravitational physicist, who in 1962 in a lecture on "The future of applied mathematics" said the following:

> Applied mathematics, as I should like it to be understood, means the application of mathematics to *any* subject, physical or otherwise; with the proviso that the mathematics shall be interesting and the results nontrivial. An applied mathematician, on this view, is somebody who has been trained to make such applications, and who is always prepared to look for situations where fruitful application is possible. As such, he is not a physicist *manqué*. I therefore see applied mathematics as an activity, or attitude of mind, rather than as a body of knowledge.

Not everyone will agree with Bonnor. Some take a more methodological approach and almost equate applied mathematics with "mathematical modeling." Others are of a more concrete, mathematical mind and insist that there are parts of mathematics that are per se more or less applicable than others. We find Bonnor's definition appealing because it stresses the social dimension of the mathematical working process and allows a historical understanding of the notion of applied mathematics.

The importance of "attitudes" notwithstanding, by any definition applied mathematics has to be "genuine" mathematics in the sense that it aims at and/or uses general statements (theorems) even if the piece of mathematics in question has not yet been fully logically established. In fact, the applicability of mathematics is mainly based on its "generality," which in relation to fields of application often appears as "abstractness." This applies even to relatively elementary applications such as the use of positional number systems.

Applications of mathematics, even on a nonelementary level, have been possible because certain practices and properties, such as algorithms for approximations or geometrical constructions, have always existed within mathematics itself and have led to spontaneous or deliberate applications. While, as the universal mathematician John von Neumann observed in 1947, in pure mathematics many problems and methods are selected for aesthetic reasons, in applied mathematics, problems considered at the time as urgent have priority, and the choice of methods often has to be subordinated to the goals in question. However, attitudes and values, which often had and continue to have strong political and economic overtones, have always been instrumental in deciding exactly which parts of mathematics should be emphasized and developed. Since attitudes have to be promoted through education, this puts a great responsibility on teaching and training and makes developments in that area an important topic for a history of applied mathematics.

Of course, many modern and recent applications rely on older mathematical ideas in differential equations, topology, and discrete mathematics and on established notation and symbolism (matrices, quaternions, Laplace transforms, etc.), while important new developments in INTEGRAL EQUATIONS [IV.4], measure theory, vector and tensor analysis, etc., at the turn of the twentieth century have added to these ideas.

However, in the twentieth century, three major scientific and technical innovations both changed and enlarged the notion of applied mathematics. In rough chronological order, these are mathematical *modeling* in a broad, modern sense, *stochastics* (modern probability and statistics), and the *digital computer*. These three innovations have, through their interactions, restructured applied mathematics. They were principally established after World War II, and it was also only then that the term "mathematical modeling" came to be more frequently used for activities that had hitherto usually been expressed by less concise words such as "problem formulation and evaluation." In addition to these innovations, which are essentially concerned with methodology, several totally new fields of application, such as electrical engineering, economics, biology, meteorology, etc., emerged in the twentieth century.

While in 1914 one of the pioneers of modern applied mathematics, Carl Runge, still doubted whether "the name of 'applied mathematics' was chosen appropriately, because when applied to empirical sciences it still remains pure mathematics," the three major innovations listed above would radically alter and extend the notion of mathematics and, in particular, that of applied mathematics. Due to these innovations, the modern disciplines at the interface of mathematics and engineering, such as cybernetics, control theory, computer science, and optimization, were all able to emerge in the 1940s and 1950s in the United States (Wiener, Shannon, Dantzig) and the Soviet Union (Andronov, Kolmogorov, Pontryagin, Kantorovich) independently, and to a somewhat lesser degree in England (Turing, Southwell, Wilkinson), France (Couffignal), and elsewhere. These innovations also gradually changed "hybrid disciplines," such as electrical engineering and aerodynamics, that had originated at the turn of the century. In the case of aerodynamics, not only were statistical explanations of TURBULENCE [V.21] increasingly proposed after World War II, but also CONFORMAL MAPPINGS [II.5] gradually lost importance in favor of computational FLUID DYNAMICS [IV.28]. Within operations research, with its various approaches and techniques (linear programming, optimization methods, statistical quality control, inventory control, queuing analysis, network flow analysis), mathematical concepts, especially mathematical models, acquired an even stronger foothold than in the more traditional industrial engineering.

One typical modern mathematical discipline that intimately combines pure and applied aspects of the subject and that is intertwined with various other scientific (physical and biological) and engineering disciplines is the theory of DYNAMICAL SYSTEMS [IV.20]. After initial work in the field by Poincaré, Lyapunov, and Birkhoff, the theory fell into oblivion until the 1960s. This falling away can be explained by fashion (such as the trend toward the mathematics of Bourbaki), by new demands in applications connected to dissipative systems, and by the partial invisibility of the Russian school in the West. With the advent of modern computing devices, the shape of the discipline changed dramatically. Mathematicians were empowered computationally and graphically, the visualization of new objects such as fractals was made possible, and applications in fields such as CONTROL THEORY [IV.34] and meteorology—quantitative applications as well as qualitative ones—began to proliferate. The philosophical discussion about mathematics and applications has also been enriched by this discipline, with the public being confronted by catchwords such as CHAOS [II.3], catastrophe, and self-organization. However, the process whereby the various streams of problems converged and led to the subject's modern incarnation is complex:

> In the 1930s, for example, what could the socio-professional worlds of the mathematician Birkhoff (professor at Harvard), the "grand old man of radio" van der Pol (at the Philips Research Lab), and the Soviet "physico [engineer] mathematician" Andronov at Gorki have had in common? What, in the 1950s, had Kolmogorov's school in common with Lefschetz's? It is precisely this manifold character of social and epistemic landscapes that poses problem[s] in this history.
>
> Aubin and Dahan (2002)

The role played by the three major innovations continues largely unabated today, as is evident, for instance, from a 2012 report from the Society for Industrial and Applied Mathematics (SIAM) on industrial mathematics:

> Roughly half of all mathematical scientists hired into business and industry are statisticians. The second-largest group by academic specialty is applied mathematics. Compared to the 1996 survey, fewer graduates reported "modeling and simulation" as an important academic specialty for their jobs, and more reported "statistics." Programming and computer skills continue to be the most important technical skill that new hires bring to their jobs.

By separating statistics from applied mathematics, the SIAM report follows a certain tradition, caused in part by institutional boundaries, such as the existence of separate statistics departments in universities. This distinction is also partly followed in the present volume and in this article, although there is no doubt about the crucial role of probability and statistics in applications. For example, one need only consider the Monte Carlo method—notably developed at Los Alamos in the 1940s by Stanislaw Ulam and von Neumann, and continued by Nicholas Metropolis—which is now used in a wide variety of different contexts including numerical integration, optimization, and inverse problems. In addition, the combination of stochastics and modeling in biological and physical applications has had a philosophical dimension, contributing to the abandonment of rigid causality in science, e.g., through Karl Pearson's correlation coefficient and Werner Heisenberg's uncertainty principle. However, by the end of the 1960s the limitations of stochastics in helping us to understand the nature of disorder had become apparent, particularly in connection with the study of complex ("chaotic") dynamical systems. Nevertheless, stochastics continues to play an important role in the development of big theories with relevance for applications, including statistical models for weather forecasting.

### 1.1 Further Themes and Some Limitations

Putting stochastics on the sidelines is but one of several limitations of this article—limitations that are the result of a lack of space, a lack of distance, and more general methodological considerations. A further thematic restriction concerns industrial mathematics, which figures separately from applied mathematics in the very name of SIAM, although there are obvious connections between the two, in particular with respect to training and in developing attitudes toward applications. Knowing that industrial mathematics has changed, and above all expanded, from its origins in the early twentieth century to move beyond its purely industrial context, these connections become even clearer. Industrial mathematics is, today, an established subdiscipline, loosely described as the modeling of problems of direct and immediate interest to industry, performed partly in industrial surroundings and partly in academic ones.

The history of mathematical instruments, including both numerical and geometrical devices, and their underlying mathematical principles is another topic we have had to leave out almost completely. Some discussion of the history of mathematical table projects is the farthest we reach in this respect. This limitation also applies to the technological basis of modern computing and the development of software technology (covered by Cortada's excellent bibliography), which has provided, and continues to provide, an important stimulus for the development (and funding) of applied mathematics. In 2000, in the *Journal of Computational and Applied Mathematics*, it was estimated that of the increase in computational power, half should be attributed to improved algorithms and half should be attributed to the increase in computational hardware speeds. Computing technology has continued to advance rapidly, and companies are making more and more aggressive use of HIGH-PERFORMANCE COMPUTING [VII.12]

A detailed discussion of the fields of application of mathematics themselves—be it in (pure) mathematics, the sciences, engineering, economics and finance, industry, or the military—is absent from this article for a number of reasons, both practical and methodological, above all the huge variation of specific conditions in these fields.

This particularly affects the role of mathematics in the military, to which we will devote only scattered remarks and no systematic discussion. While there are still considerable lacunae in the literature on mathematics during World War I (although some of these have been filled by publications prepared for the centenary), there is more to be found in print about mathematics in World War II, not least because of the increased role of that discipline in it. (We recommend Booß-Bavnbek and Høyrup (2003) as a good place to start to find out more than is covered in our article.)

Another topic that deserves broader coverage than is possible here is the history of philosophical reflection about mathematical applications. This is particularly true for the notion of "mathematical modeling" taken in the sense of problem formulation. According to the *Oxford Encyclopedic Dictionary* (1996), the new notion of a mathematical model was used first in a statistical context in 1901. At about the same time, the French physicist and philosopher Pierre Duhem accused British physicists of still using the term "model" only in the older and narrower sense of material, mechanical, or visualizable models. Duhem therefore preferred the word "analogy" for expressing the relationship between a theory and some other set of statements. Particularly with the upswing of "mathematical modeling" since the 1980s, a broad literature, often with a philosophical bent, has discussed the specificity of mathematics as a language, as an abstract unifier and a source of concepts and principles for various scientific and societal domains of application. Another (though not unrelated) development in the philosophy of applied mathematics concerns the growing importance of algorithmic aspects within mathematics as a whole. It was no coincidence that in the 1980s, with the rise of scientific computing, several "maverick" philosophers of mathematics, such as Philip Kitcher and Thomas Tymosczko, entered the scene. They introduced the notion of "mathematical practice," by which they meant more than simply applications. One of the features of the maverick tradition was the polemic against the ambitions of mathematical logic to be a canon for the philosophy of mathematics, ambitions that have dominated much of the philosophy of mathematics in the twentieth century. The change was inspired by the work of both those mathematicians (such as Philip Davis and Reuben Hersh) and those philosophers (including Imre Lakatos and David Corfield) who were primarily interested in the actual working process of mathematicians, or what they sometimes called "real mathematics." Meanwhile, the philosophical discussion of mathematical practice has been professionalized and reconnected to the foundationalist tradition. It usually avoids premature discussion of "big questions" such as "Why is mathematics applicable?" or "Is the growth of mathematics rational?" restricting its efforts to themes of mathematical practice in a broader sense, like visualization, explanation, purity of methods, philosophical aspects of the uses of computer science in mathematics, and so on. An overview of the more recent developments in the philosophy of mathematical practice is given in the introduction to Mancosu (2008).

Unfortunately, there is also little space for biographical detail in this article, and thus no bow can be given to the great historical heroes of applied mathematics, such as Archimedes, Ptolemy, Newton, Euler, Laplace, and Gauss. Nor is there room to report on the conversions of pure mathematicians into applied mathematicians, such as those undergone by Alexander Ostrowski, John von Neumann, Solomon Lefschetz, Ralph Fowler, Garrett Birkhoff, and David Mumford, all personal trajectories that paralleled the global development of mathematics. In any case, any systematic inclusion of biographies could not be restricted to mathematicians, considering the term in its narrowest sense. In an influential report on industrial mathematics in the *American Mathematical Monthly* of

1941, Thornton Fry spoke about a "contrast between the ubiquity of mathematics and the fewness of the mathematicians." Indeed, historically, engineers such as Theodore von Kármán, Richard von Mises, Ludwig Prandtl, and Oliver Heaviside; physicists such as Walter Ritz, Aleksandr Andronov, Cornelius Lanczos, and Werner Romberg; and industrial mathematicians such as Balthasar van der Pol have, by any measure, made significant contributions to applied mathematics. In addition, several pioneers of applied mathematics, such as Gaspard Monge, Felix Klein, Mauro Picone, Vladimir Steklov, Vannevar Bush, and John von Neumann, actively used political connections. The actions of nonscientists, and particularly politicians, have also therefore played a part in the development of the subject. For a full history of applied mathematics the concrete interplay of the interests of mathematicians, physicists, engineers, the military, industrialists, politicians, and other appliers of mathematics would have to be analyzed, but this is a task that goes well beyond the scope of this article.

In general, the historical origin of individual notions or methods of applied mathematics, which often have a history spanning several centuries, will not be traced here; pertinent historical information is often included in the specialized articles elsewhere in this volume. By and large, then, this article will focus on the broader methodological trends and the institutional advances that have occurred in applied mathematics since the early nineteenth century.

### 1.2 Periodization

From the point of view of applications, the history of mathematics can be roughly divided into five main periods that reveal five qualitatively different levels of applied mathematics, the first two of which can be considered as belonging to the prehistory of the subject.

**(1) ca. 4000 B.C.E.–1400 C.E.** Emergence of mathematical thinking, and establishment of theoretical mathematics with spontaneous applications.

**(2) ca. 1400–1800.** Period of "mixed mathematics" centered on the Scientific Revolution of the seventeenth century and including "rational mechanics" of the eighteenth century (dominated by Euler).

**(3) 1800–1890.** Applied mathematics between the Industrial Revolution and the start of what is often called the second industrial (or scientific-technical) revolution. Gradual establishment of both the term and the notion of "applied mathematics." France and Britain dominate applied mathematics, while Germany focuses more on pure.

**(4) 1890–1945.** The so-called resurgence of applications and increasing internationalization of mathematics. The rise of new fields of application (electrical communication, aviation, economics, biology, psychology), and the development of new methods, particularly those related to mathematical modeling and statistics.

**(5) 1945–2000.** Modern internationalized applied mathematics after World War II, inextricably linked with industrial mathematics and high-speed digital computing, led largely by the United States and the Soviet Union, the new mathematical superpowers.

Arguably, one could single out at least two additional subperiods of applied mathematics: the eighteenth century, with Euler's "rational mechanics," and the technological revolution of the present age accompanied by the rise of computer science since the 1980s. However, in the first of these subperiods, which will be described in some detail below, mathematics as a discipline was still not yet fully established, either institutionally or with respect to its goals and values, so distinguishing between her pure and applied aspects is not straightforward. As to the second of the two subperiods, we believe that these events are so recent that they escape an adequate historical description. Moreover, World War II had such strong repercussions on mathematics as a whole—particularly on institutionalization (journals, institutes, professionalization), on material underpinning (state funding, computers, industry), and not least on the massive migration of mathematicians to the United States—that it can be considered a watershed in the worldwide development of both pure and applied mathematics. However, the dramatic prediction by James C. Frauenthal—in an editorial of *SIAM News* in 1980 on what he considered the "revolutionary" change in applied mathematics brought about by the invention of the computer—that by 2025 "in only a few places will there remain centers for research in pure mathematics as we know it today" seems premature.

## 2 Mathematics before the Industrial Revolution

Since the emergence of mathematical thinking around 4000 B.C.E., through antiquity and up to the start of the Renaissance (ca. 1400 C.E.), and embracing the cultures of Mesopotamia, Egypt, and ancient Greece, as well as those of China, India, the Americas, and the Islamic–Arabic world, applications arose as a result of various societal, technical, philosophical, and religious needs. Well before the emergence of the Greek notion of a mathematical proof around 500 B.C.E., areas of application of mathematics in various cultures included accountancy, agricultural surveying, teaching at scribal schools, religious ceremonies, and (somewhat later) astronomy. Among the methods used were practical arithmetic, basic geometry, elementary combinatorics, approximations (e.g., $\pi$), and solving quadratic equations. Instruments included simple measuring and calculation devices: measuring rods, compasses, scales, knotted ropes, counting rods, abaci, etc.

The six classical sciences—geometry, arithmetic, astronomy, music, statics, and optics—existed from the time of Greek antiquity and were based on mathematical theory, with the Greek word "mathemata" broadly referring to anything teachable and learnable. The first four of the classical sciences constituted the quadrivium within the Pythagorean–Platonic tradition. The theories of musical harmony (as applied arithmetic) and astronomy (as applied geometry) can thus be considered the two historically earliest branches of applied mathematics. The two outstanding applied mathematicians of Greek antiquity were Archimedes (statics, hydrostatics, mechanics) and Ptolemy (astronomy, optics, geography). Since the early Middle Ages, the Computus (Latin for computation)—the calculation of the date of Easter in terms of first the Julian calendar and later the Gregorian calendar—was considered to be the most important computation in Europe. In medieval times, particularly from the seventh century, the development of algebraic and calculative techniques and of trigonometry in the hands of Islamic and Indian mathematicians constituted considerable theoretical progress and a basis for further applications, with significant consequences for European mathematics. Particularly notable was the *Liber Abaci* (1202) of Leonardo of Pisa (Fibonacci), which heralded the gradual introduction of the decimal positional system into Europe, one of the broadest and most important applications of mathematics during the period. Chinese mathematics remained more isolated from other cultures at the time and is in need of further historical investigation, as are some developments within Christian scholastics. In spite of their relative fewness and their thematic restrictions, we consider the early applications to be a deep and historically important root for the emergence of theoretical mathematics and not as a mere follow-up of the latter.

From the beginning of the fifteenth century to the end of the eighteenth century, applications of mathematics were successively based on the dissemination of the decimal system, the rise of symbolic algebra, the theory of perspective, functional thinking (Descartes's coordinates), the calculus, and natural philosophy (physics). The teaching of practical arithmetic, including the decimal system, by professional "reckoning masters," such as the German Adam Ries in the sixteenth century, remained on the agenda for several centuries. Meanwhile, the first systematic discussion of decimal fractions appeared in a book by the Dutch engineer Simon Stevin in 1585.

During this period, and connected to the new demands of society, there emerged various hybrid disciplines combining elements of mathematics and engineering: architecture, ballistics, navigation, dynamics, hydraulics, and so on. Their origins can be traced back, at least in part, to medieval times. For example, partly as a result of fourteenth-century scholastic analysis, the subject of local motion was separated from the traditional philosophical problem of general qualitative change, thus becoming a subject of study in its own right.

The term "mixed mathematics" as a catch-all for the various hybrid disciplines seems to have been introduced by the Italian Marsilio Ficino during the fifteenth century in his commentary on Plato's *Republic*. It was first used in English by Francis Bacon in 1605. In his *Mathematicall Præface* to the first English translation of Euclid's *Elements* (1570), John Dee set out a "groundplat" or plan of the "sciences and artes mathematicall," which included astronomy and astrology. Due to the broad meaning of the original Greek word, the Latin name "mathematicus" was used for almost every European practitioner or artisan within one of these hybrid disciplines. As late as 1716, the loose use of "mathematicus" was deplored by the philosopher Christian Wolff (a follower of Leibniz) in his *Mathematisches Lexicon*, an influential dictionary of mathematics, because in his opinion it diminished the role of mathematics.

The emergence of the new Baconian sciences (magnetism, electricity, chemistry, etc.)—which went beyond mixed mathematics and were even partially opposed to the mathematical spirit of the classical sciences (Bacon's acknowledgment of the future of mixed mathematics, as expressed in the epigraph, was coupled with a certain distrust of pure mathematics)—signaled

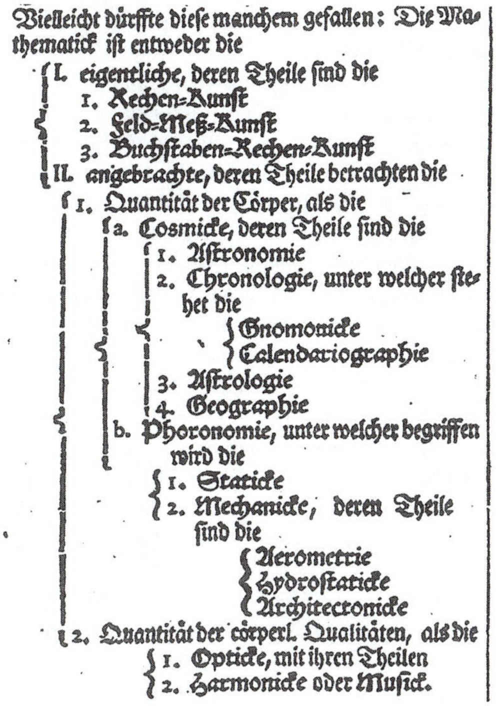
Vielleicht dürffte diese manchem gefallen: Die Mathematick ist entweder die

I. eigentliche, deren Theile sind die
1. Rechen-Kunst
2. Feld-Meß-Kunst
3. Buchstaben-Rechen-Kunst

II. angebrachte, deren Theile betrachten die
1. Quantität der Cörper, als die
a. Cosmicke, deren Theile sind die
1. Astronomie
2. Chronologie, unter welcher stehet die
Gnomonicke
Calendariographie
3. Astrologie
4. Geographie
b. Phoronomie, unter welcher begriffen wird die
1. Staticke
2. Mechanicke, deren Theile sind die
Aerometrie
Hydrostaticke
Architectonicke
2. Quantität der cörperl. Qualitäten, als die
1. Opticke, mit ihren Theilen
2. Harmonicke oder Musick.

**Figure 2** J. H. Zedler, *Universallexicon* (1731–1754).

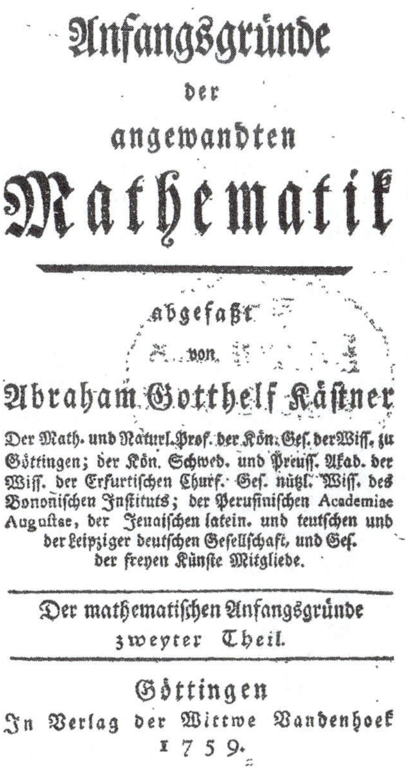
Anfangsgründe der angewandten Mathematik

abgefaßt von

Abraham Gotthelf Kästner

Der Math. und Naturl. Prof. der Kön. Ges. der Wiss. zu Göttingen; der Kön. Schwed. und Preuss. Akad. der Wiss. der Erfurtischen Churf. Ges. nützl. Wiss. des Bononischen Instituts; der Perusinischen Academiae Augustae, der Jenaischen latein. und teutschen und der Leipziger deutschen Gesellschaft, und Ges. der freyen Künste Mitgliede.

Der mathematischen Anfangsgründe zweyter Theil.

Göttingen

In Verlag der Wittwe Vandenhoek

1759.

**Figure 3** A. G. Kästner, *Anfangsgründe der Angewandten Mathematik* (1759).

the rise of systematic experimental methods. The latter reached a symbiosis with mixed mathematics in the hands of Galileo, Descartes, Huygens, Newton, and other pioneers of the Scientific Revolution. In their work, the guiding thought was of a fundamental connection between mathematical and mechanical exactness. However, a certain division between the tradition of mathematics and of the Baconian sciences remained palpable until the nineteenth century.

While talk about "applications of mathematics" was common in the English language from at least the seventeenth century (in, for example, the work of Isaac Barrow and others), the term "applied mathematics" (with "applied" as an adjective), as the successor to "mixed mathematics," was apparently not introduced prior to the eighteenth century. The name seems to have appeared first in German as "angebrachte Mathematik" ("angebracht" having roughly the meaning of "applied," if a bit more in the sense of "attached") in Wolff's dictionary of 1716. In Latin it appeared two years later, in Johann Friedrich Weidler's textbook *Institutiones mathematicae* (1718), as "Mathesis applicata quam nonnulli mixta appellant." In his *Universallexicon* (1731–54), Johann Heinrich Zedler followed Wolff and gave a detailed classification of the parts of "angebrachte Mathematik" (figure 2).

Finally, "applied mathematics" figures for the first time on the title page of a book as "Angewandte Mathematik" in the second volume of Abraham Gotthelf Kästner's mathematical textbook *Mathematische Anfangsgründe* ("mathematical elements") of 1759 (figure 3). The German philosopher Kant used the term "applied mathematics" in his Berlin *Preisschrift* of 1763/64. The latter was translated into English in 1798 and, according to the *Oxford English Dictionary*, it is in this translation that the term makes its first appearance in English. Meanwhile, "mixed mathematics" ("mathématiques mixtes") was still in use in French, appearing in the famous *Encyclopédie* of Diderot and d'Alembert (1750) (figure 1), and even in the second edition of J.-E. Montucla's *Histoire des Mathématiques* (1798–1802).

There is no doubt that the period of the Enlightenment of the eighteenth century deserves a special place in a history of applied mathematics, particularly as a bridge—via so-called rational mechanics, which uses the invention of the calculus (Newton/Leibniz)—between the natural philosophy of the seventeenth century and the mathematical engineering and physics of the nineteenth and twentieth centuries. There are six principal actors here: three from the famous Swiss Bernoulli family (the brothers Jakob (James) and Johann (John), and Daniel, the son of the latter), Alexis-Claude Clairaut, Jean d'Alembert, and, by far the most prolific of all, Leonhard Euler. The dominating role of applied mathematics at the time is perhaps most convincingly illustrated in the themes of the prize competitions run by the prominent academies of science, especially the Paris Academy. The topics set in Paris included the optimum arrangement of ship masts (1727), the motion of the moon (1764/68), the motion of the satellites of Jupiter (1766), the three-body problem (1770/72), the secular perturbations of the moon (1774), the perturbations of comet orbits (1776/78/80), the perturbation of the orbit of Pallas (in the beginning of the nineteenth century), the question of heat conduction (1810/12), and the propagation of sound waves in liquids (1816). Only occasionally, and mostly in a later period, would questions of pure mathematics be posed in these competitions, examples being questions on the theory of polyhedra (1810) and the Fermat problem (1816). Euler himself, who won the Paris prize twelve times, subscribed to the utilitarian mood of Enlightenment when in 1741 he wrote his essay "On the utility of higher mathematics," first published only in 1847. According to Goldstine (a pioneer and historian of numerical analysis):

> Euler did at least the ground work on virtually every topic in modern numerical analysis. This work included the basic notions for the numerical integration of differential equations. Moreover, his development of lunar theory made possible the accurate calculation of the moon's position and the founding of the Nautical Almanac in Great Britain.
>
> Goldstine (1977)

A typical example of Euler's influence can be found in Carl Runge's first two articles on the numerical solution of differential equations, which appeared in the *Mathematische Annalen* in 1894 and 1895. The second article connects explicitly to Euler's *Introductio in Analysin Infinitorum* (1748), in which one finds the first example of an approximation by a polygonal chain. Of Euler's more than 800 publications, two-thirds belong to mechanics of a varying degree of abstractness. The noted and controversially discussed continuum mechanist and historian Clifford Truesdell has traced and described the innovations that Euler, d'Alembert, and the Bernoullis brought into Newtonian mechanics, in terms of both modeling and mathematical methods, relying on experience but not undertaking systematic experiments. In an apparent allusion to the philosophical age of reason, he called these innovations "rational mechanics," a term that was occasionally used at the time in a broader sense but by which Truesdell meant more specifically mathematical mechanics (Truesdell 1960). Rational mechanics, as Newton had used the term in the preface to his *Principia* (1687), was one of two traditions of mechanics already known in Greek antiquity, namely the one that "proceeds accurately by demonstration" (the other being practical mechanics). For Newton, rational mechanics in this sense was the core of natural philosophy. Truesdell opposed the view (propounded, for instance, by the positivist philosopher Ernst Mach) that Euler and like-minded mathematicians simply systematically applied the new calculus of Newton and Leibniz to mechanic and thus did not contribute anything substantial to theoretical mechanics itself. Today, historians usually stress the conceptual progress both in mathematics (e.g., by removing certain geometric elements from the Leibnizian calculus) and physics (e.g., exploring in detail the relation between force and motion) accomplished in the work of Euler. It is probably this role of Euler as a "mathematical physicist," combined with the lack of immediate usefulness of Euler's rational mechanics, that caused Truesdell in 1960 to declare that the latter was not applied mathematics.

Among the new fields of mathematics was, for instance, partial differential equations (of which the first example appears in a work by Euler of 1734), which were successfully applied by Euler and d'Alembert in the second half of the 1740s in the analysis of the vibrating string. In his *Methodus inveniendi* of 1744, Euler picked up on the tradition of solving the PROBLEM OF THE BRACHISTOCHRONE [IV.6 §1] in the work of the Bernoulli brothers. He set standards in the calculus of variations developed later by Joseph Louis Lagrange and others. As an extension of Daniel Bernoulli's *Hydrodynamica* of 1738, and partly influenced by d'Alembert, Euler's equations of fluid dynamics (1755), which did not yet have a term for viscosity, remained a challenge for generations of mathematicians to come, not least due to their nonlinearity. Euler, like d'Alembert before him, was unable to produce from his equations of fluid dynamics a single new result fit for comparison with experiment.

In our opinion (deviating slightly from Truesdell), rational mechanics, in hindsight, bears almost all the characteristics of applied mathematics in the modern sense. At the time, however, in a predominantly utilitarian environment, it was the pinnacle of mathematics. It was then rarely counted as mixed mathematics, notwithstanding some occasional remarks by d'Alembert. The term mixed mathematics was more frequently used for the mathematically less advanced engineering mechanics (Bernard Forest de Bélidor and Charles-Augustin de Coulomb, etc.) of the time and for other fields of application.

Toward the end of the eighteenth century rational mechanics was somewhat narrowed down, both thematically and with respect to possible applications (although still including continuum mechanics), by further mathematical formalization, particularly at the hands of Lagrange, Euler's successor in Berlin, whose *Méchanique Analitique* first appeared in 1788. The towering figure of Pierre Simon Laplace in Paris—with his pioneering work since the late 1770s in celestial and terrestrial mechanics and in probability theory—foreshadowed much of the important French work in applied mathematics, such as that done by Poisson, Fourier, Cauchy, and others in the century that followed. To Laplace (generating functions, difference equations) and to his great younger contemporary Carl Friedrich Gauss in Göttingen (numerical integration, elimination, least-squares method) we owe much of the foundations of future numerical analysis. Parts of their work overlapped (interpolation), while parts were supplemented by Adrien-Marie Legendre (least-squares method), details of which can be traced from Goldstine's *A History of Numerical Analysis*.

## 3 Applied Mathematics in the Nineteenth Century

Around 1800, in the age of the Industrial Revolution and of continued nation building, state funding and political and ideological support (revolution in France, Neo-Humanism in Germany) led (mainly through teaching and journals) to a new level of recognition for mathematics as a discipline. The older bifurcation of pure/mixed mathematics was replaced in France and Germany (although not yet in England) by that of pure/applied. The difference was mainly that before 1800 only mixed mathematics together with rational mechanics had the support of patrons, while now, around 1800, the whole of mathematics was beginning to be supported and recognized. Somewhat paradoxically, then, in spite of the general importance of the Industrial Revolution as a historical background, it is pure mathematics that increasingly gets systematic public support for the first time. Indeed, for most of the nineteenth century, mathematics would not be strongly represented in either engineering or industrial environments.

The foundation of the École Polytechnique (EP) in Paris in 1794 is a good point of reference for the beginning of our third period. The EP, where military and civil engineers were trained, became the leading and "most mathematical" institution within a system of technical education. This included several "schools of applications," such as the École des Mines and the École Nationale des Ponts et Chaussées, to which the students of the EP proceeded. The EP became an example to be emulated by many technical colleges, particularly in German-speaking regions, throughout the nineteenth century. The most influential mathematician in the early history of the EP was Gaspard Monge, and it was in accordance with his ideas that mathematics became one of the bases of the EP curriculum. In 1795, in the introduction to his lectures on descriptive geometry, the theory that became the "language of the engineer" for more than a century, Monge wrote:

> In order to reduce the dependence of the French nation on foreign industry one has to direct public education to those subjects which require precision.

Monge's aspirations for a use of higher mathematics in industrial production remained largely unfulfilled at the time, except for the use of descriptive geometry. However, developments in industry and in educational systems led to a stronger focus on the criteria for precision and exactitude in the sciences (most notably in academic physics) and in engineering, preparing the ground for an increased use of mathematics in these fields of application at the beginning of the twentieth century. In fact, it could be argued that it required a logical consolidation and a more theoretical phase of the development of mathematical analysis before a new phase of more sophisticated applied mathematics could set in.

The first concrete institutional confirmation of the notion of "applied mathematics" was the appearance of the term in the names of journals. Again, the Germans were quicker than the French here. Two short-lived journals cofounded by the influential combinatorialist Carl Friedrich Hindenburg were the *Leipziger*

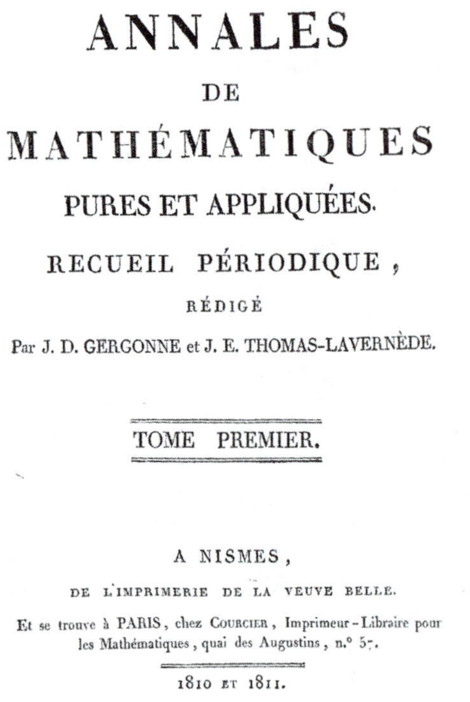
ANNALES
DE
MATHÉMATIQUES
PURES ET APPLIQUÉES.
RECUEIL PÉRIODIQUE,
RÉDIGÉ
Par J. D. GERGONNE et J. E. THOMAS-LAVERNÈDE.

TOME PREMIER.

A NISMES,
DE L'IMPRIMERIE DE LA VEUVE BELLE.
Et se trouve à PARIS, chez COURCIER, Imprimeur-Libraire pour les Mathématiques, quai des Augustins, n.° 57.
1810 ET 1811.

**Figure 4** Gergonne's *Annales de Mathématiques Pures et Appliquées* (1810–11).

*Magazin für reine und angewandte Mathematik* (1786–89) and the *Archiv für reine und angewandte Mathematik* (1795–99). A somewhat longer career was had by *Annales de Mathématiques Pures et Appliquées*, founded by Joseph Diaz Gergonne in 1810 (figure 4). While this journal survived only until 1832, the German *Journal für die reine und angewandte Mathematik* (which, according to the preface by its founder August Leopold Crelle in 1826, was largely modeled after Gergonne's journal) is still extant today. This is true too of the French *Journal de Mathématiques Pures et Appliquées*, founded by Joseph Liouville in 1836, and of the Italian *Annali di Matematica Pura ed Applicata*, launched by Francesco Brioschi and Barnaba Tortolini in Italy in 1858 as an immediate successor to the *Annali di Scienze Matematiche e Fisiche*. On the other hand, James Joseph Sylvester's *Quarterly Journal of Pure and Applied Mathematics*, which was founded in 1855, survived only until 1927.

The inclusion of "applied mathematics" in the names of these nineteenth-century journals did not necessarily guarantee a strong representation of applied topics, however, either in the journals themselves or in the mathematical culture at large. But neither were these journals the only outlets for articles on applied topics. Journals associated with national academies, such as the *Philosophical Transactions of the Royal Society*, carried articles on applied topics, while the *Philosophical Magazine* (launched in 1798) was the journal of choice for several leading nineteenth-century British applied mathematicians.

This was also the period in which positions explicitly devoted to applications were created at universities. In Norway, which had just introduced a constitution and was emancipating itself from Danish rule, Christopher Hansteen's position as "lecturer for applied mathematics" ("Lector i den anvendte Mathematik") at the newly founded university in Christiania was expressly justified in May 1814 by "the broad scope of applied mathematics and its importance for Norway." In 1815 Hansteen was promoted to "Professor Matheseos applicatae."

Throughout the nineteenth century, the mathematization of mechanics continued largely in the tradition of Lagrange's analytical mechanics, with a division of labor between physicists and mathematicians such as William Rowan Hamilton and Carl Jacobi, arguably neglecting some of the topics and insights of Euler's rational mechanics, particularly in continuum mechanics. However, from the 1820s, although the EP still gave preference to analytical mechanics in its courses, there were efforts among the professors there, and at the more practically oriented French engineering schools ("écoles d'application"), to develop a mechanics for the special needs of engineers, a discipline that would today be called technical mechanics. The latter drew strongly on traditions in mixed mathematics, such as the work of de Bélidor in hydraulics from the 1730s to the 1750s and that of de Coulomb in mechanics and electromagnetism from the 1780s onward. It found its first energetic proponents in Claude Navier, Jean Victor Poncelet, and Gaspard Gustave de Coriolis.

Around 1820, Poncelet separately developed his projective geometry, which became part of the mathematically rather sophisticated engineering education at several continental technical colleges. It led to methods such as graphical statics, founded by the German–Swiss Carl Culmann in the middle of the century, with applications in crystallography and civil engineering, the latter exemplified by the construction of the Eiffel Tower in 1889. Also in the 1820s, influenced by Euler's hydrodynamics and possibly by Navier's work in engineering,

Augustin-Louis Cauchy was the first to base the theory of elasticity on a general definition of internal stresses. The work of Cauchy, who was at the same time known for his efforts to introduce rigor into analysis, underscores the dominance of the French in both pure and applied mathematics during the early nineteenth century, with the singular work of Gauss in Göttingen being the only notable exception.

In England the development of both pure and applied mathematics during the nineteenth century showed marked differences from that in Continental Europe. One of the goals of the short-lived Analytical Society (1812–19), founded in Cambridge by Charles Babbage and others, was to promote Leibnizian calculus over Newtonian calculus, or, in Babbage's words, to promote "The Principles of pure D-ism in opposition to the Dotage of the University." The members of the Analytical Society were impressed by the new rigor in analysis achieved in France, especially in the work of Lagrange, and lobbied for a change in teaching and research in Cambridge mathematics, and in particular in the examinations of the Mathematical Tripos, which were very much based on traditional mixed and physical mathematics, as well as on Euclid's *Elements*. If anything, though, this aspect of the French influence led away from applications and toward a gradual purification of British mathematics.

Babbage was impressed by the French mathematical tables project directed by Gaspard de Prony at the end of the eighteenth century. In a similar vein to Monge before him, Babbage pointed to increased competition between nations in the age of industrialization, and he stressed the need for the development of calculating techniques. In *On the Economy of Machinery and Manufactures* (1832) he wrote:

> It is the science of *calculation*,—which becomes continually more necessary at each step of our progress, and which must ultimately govern the whole of the applications of science to the arts of life.

Another (at least indirect) impact of the Industrial Revolution on mathematics was the Russian Pafnuty Lvovich Chebyshev's study of James Watt's steam engine, in particular of the "governor," the theory of which proved to be a stimulus for the notion of feedback in control theory, and the modern theory of servomechanisms. Chebyshev's interest in the technical mechanics of links was also one of the stimuli for his studies concerning mathematical approximation theory in the 1850s. In addition, he was impressed with Poncelet's technical mechanics. As a result, and due to Chebyshev's great influence within Russian mathematics, applied mathematics remained much more part of mainstream mathematics in Russia during the latter half of the nineteenth century than it was in other parts of Europe, especially in Germany.

In the middle of the nineteenth century, the French engineering schools, in particular the EP, lost their predominant position in mathematics, due to slow industrial development in France and problems with the overcentralized and elitist educational system. The lead was taken by the German-speaking technical colleges ("Technische Hochschulen") in Prague, Vienna, Karlsruhe, and Zurich, in particular with respect to the mathematization of the engineering sciences. This was true for their emulation of the general axiomatic spirit of mathematics even more than for their concern for the actual mathematical details. Ferdinand Redtenbacher (in his analytical machine theory (1852)) and Franz Reuleaux (in his kinematics (1875)) aimed at "designing invention and construction deductively." So convinced of the important future role of mathematics were leading engineers at the Technische Hochschulen that they supported the appointment of academically trained mathematicians from the classical universities. In this way pure mathematicians, such as Richard Dedekind, Alfred Clebsch, and later Felix Klein, assumed positions at Technische Hochschulen in which they were responsible for the education of engineers.

In parallel, and also from the middle of the nineteenth century, mathematics at the leading German universities that did not have engineering departments increasingly developed into a pure science, detached from practical applications. Supported by the ideology of "Neo-Humanism" within a politically unmodernized environment, the discipline's educational goal (and its legitimation in society) was the training of high school teachers, who during their studies were often introduced to the frontiers of recent (pure) mathematical research. The result of this was that professors at the Technische Hochschulen who were hired from the traditional universities were not really prepared for training engineers. In the long run, the strategy of appointing university mathematicians backfired and this, together with general controversies about the social status of technical schools, led to the so-called anti-mathematical movement of engineers in Germany in the 1890s.

British and Irish applied mathematics, in the sense of mathematical physics, remained strong through the nineteenth century, with work by George Green, George Stokes, William Rowan Hamilton, James Clerk Maxwell, William Thomson (Lord Kelvin), Lord Rayleigh, William Rankine, Oliver Heaviside, Karl Pearson, and others, while in the works of James Joseph Sylvester and Arthur Cayley in the 1850s, the foundations of modern matrix theory were laid. That being said, there was no systematic, state-supported technical or engineering education in the British system until late in the nineteenth century. What was taught in this respect at schools and traditional universities was increasingly questioned by engineers such as John Perry (see below), particularly with respect to the mathematics involved. In England the term "mixed mathematics" was occasionally used interchangeably with "applied mathematics" up until the end of the century, a prominent example of this being the tribute by Richard Walker (then president of the London Mathematical Society) to Lord Rayleigh on winning the society's De Morgan Medal in 1890.

Applications of mathematics also featured among the activities of the British Association for the Advancement of Science, which, in 1871, formed a Mathematical Tables Committee for both cataloguing and producing numerical tables; the committee lasted, with varying levels of intensity, until 1948, when the Royal Society took over. A prime example of joint enterprise between pure and applied mathematicians—the original committee consisted of Cayley, Henry Smith, Stokes, and Thomson—the project catered for all tastes, its products including both factor tables and Bessel function tables, among others. As J. W. L. Glaisher, the project secretary, wrote in 1873: "one of the most valuable uses of numerical tables is that they connect mathematics and physics, and enable the extension of the former to bear fruit practically in aiding the advance of the latter." The project was finally dissolved in 1965, with some of the greatest British mathematicians, both pure and applied, having been active in its work.

With the upswing of electrical engineering, more sophisticated mathematics (operational calculus, complex numbers, vectors) finally entered industry in around 1890, e.g., through the work of Heaviside in England and of the German immigrant Charles P. Steinmetz in the United States. Mechanical engineering, on the other hand, e.g., in the construction of turbines, remained free of advanced mathematics until well into the twentieth century.

It was also not until the end of the nineteenth century that applied mathematics finally began to lose its almost exclusive bond to mathematical physics and mechanics; new fields of application, new methods such as statistics, and new professions such as actuarial and industrial mathematicians were largely matters for the twentieth century.

This change is nicely captured through the example of the English applied mathematician Karl Pearson. In his philosophical book *Grammar of Science* (1892), Pearson, who at the time was mainly known for his work on elasticity, defined as the "topic of Applied Mathematics... the process of analyzing inorganic phenomena by aid of ideal elementary motions." At the time Pearson was already working on biometrics, the subject that would lead him to found, together with Francis Galton, the journal *Biometrika* in 1901. Therefore, although Pearson was effectively extending the realm of applications of mathematics to the statistical analysis of biological (i.e., organic) phenomena, he apparently did not consider what he was doing to be applied mathematics.

## 4 The "Resurgence of Applications" and New Developments up to World War II

From the 1890s, the University of Göttingen (pure) mathematician Felix Klein saw the importance of taking the diverging interests of the engineering professors at technical colleges and those of German university mathematicians into account. Not only did different professions (teaching and engineering) require different education, but the gradual emergence of industrial mathematics had to be considered as well. Klein recognized the need for reform, including in teaching at high school level, and he developed Göttingen into a center of mathematics and the exact sciences (figure 5). Chairs for applied mathematics and applied mechanics were created there in 1904, with Carl Runge and Ludwig Prandtl being the first appointees. Meanwhile, from 1901, and under the editorship of Runge, the transformation of *Zeitschrift für Mathematik und Physik* into a journal exclusively for applied mathematics had begun. These events in Germany, contrasting with those of the period before, led to talk about a "resurgence of applications" ("Wiederhervorkommen der Anwendungen").

From 1898 and for several decades afterward, the famous German multivolume *Encyclopedia of the Mathematical Sciences including Their Applications* was edited by Klein together with Walther von Dyck and Arnold Sommerfeld, both from Munich, and others.

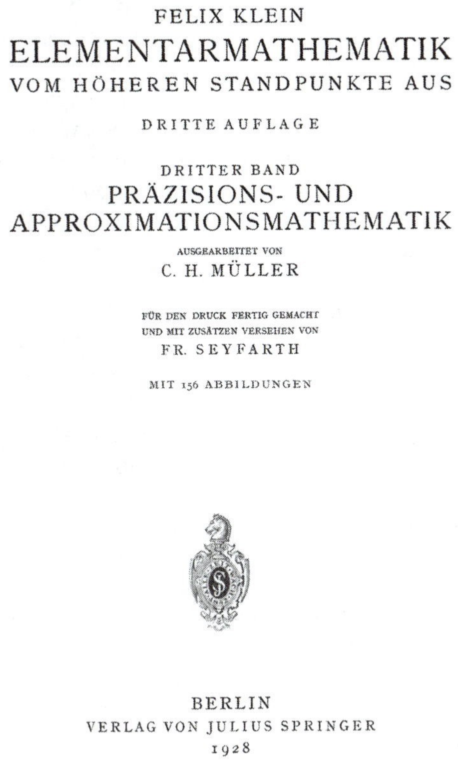

FELIX KLEIN

ELEMENTARMATHEMATIK

VOM HÖHEREN STANDPUNKTE AUS

DRITTE AUFLAGE

DRITTER BAND

PRÄZISIONS- UND APPROXIMATIONSMATHEMATIK

AUSGEARBEITET VON

C. H. MÜLLER

FÜR DEN DRUCK FERTIG GEMACHT
UND MIT ZUSÄTZEN VERSEHEN VON

FR. SEYFARTH

MIT 156 ABBILDUNGEN

BERLIN

VERLAG VON JULIUS SPRINGER

1928

**Figure 5** F. Klein, *Präzisions- und Approximationsmathematik* (1928). Posthumous publication of Klein's 1901 lectures in Göttingen where he tried to differentiate between a mathematics of precision and one of approximation.

The articles in it, all written in German but including authors from France and Britain, such as Paul Painlevé and Edmund Taylor Whittaker, contain valuable historical references that are still worth consulting today. "Applications" as emphasized in the title and in the program of the *Encyclopedia* meant areas of application, such as mechanics, electricity, optics, and geodesy. The articles were assigned to volumes IV–VI, which were in themselves divided into several voluminous books each. There were also articles on mechanical engineering, such as those by von Mises and von Kármán. However, topics that would today be classified as core subjects of applied mathematics—such as numerical calculation (Rudolph Mehmke), difference equations (Dmitri Seliwanoff), and interpolation and error compensation (both by Julius Bauschinger)—appeared as appendices within volume I, which was devoted to pure mathematics (arithmetic, algebra, and probability). Runge's contribution on "separation and approximation of roots" (1899) was subsumed under "algebra."

Klein also succeeded in introducing a state examination in applied mathematics for mathematics teachers, which focused on numerical methods, geodesy, statistics, and astronomy. In addition, he inspired educational reform of mathematics in high schools that he designed around the notions of "functional thinking" and "intuition," thereby trying to counteract the overly logical and arithmetical tendencies that had until then permeated mathematics education. Klein and his allies insisted on taking into account international developments in teaching and research, for instance by initiating a series of comparative international reports on mathematical education; these reports in turn led to the creation of what has now become the International Commission on Mathematical Instruction (ICMI). The *Encyclopedia* also provided evidence of the increasing significance of the international dimension. In his "introductory report" in 1904, von Dyck stressed the importance for the project of securing foreign authors in applied mathematics. Later, a French translation of the *Encyclopedia* began to appear in a considerably enlarged version, although the project was never completed due to the outbreak of World War I.

Around 1900, reform movements reacting to problems in mathematics education similar to those in Germany existed in almost all industrialized nations. In England, the engineer John Perry had initiated a reform of engineering education in the 1890s, and this reform played into the ongoing critical discussions of the antiquated Cambridge Mathematical Tripos examinations and their traditional reliance on Euclid. The "Perry Movement" was noticed in Germany and in the United States. On the pages of *Science* in 1903, the founding father of modern American mathematics, Eliakim Hastings Moore, himself very much a pure mathematician, declared himself to be in "agreement with Perry" and proposed a "laboratory method of instruction in mathematics and physics." At about the same time (1905), similar ideas "de créer de vrais laboratoires de Mathématiques" were proposed by Émile Borel in France. In Edinburgh, Whittaker instituted a "Mathematical Laboratory" in 1913 and later, together with George Robinson, published the influential *The Calculus of Observations: A Treatise on Numerical Mathematics* (1924), which derived from Whittaker's lectures given in the Mathematical Laboratory. In Germany, the term "Mathematisches Praktikum" was most often used to describe laboratory work in mathematics. Common to these efforts was the aim to foster among students practical abilities in calculation and drawing, extending beyond mere theoretical mathematical knowledge.

Along with the German *Encyclopedia* and Whittaker and Robinson's *Calculus of Observations*, several influential textbooks on applied mathematics, on both numerical and geometrical methods, appeared in various countries at the end of the nineteenth century and into the first decades of the new century, preparing the ground for future advances. Horace Lamb's *Hydrodynamics* (1895), an expanded version of his 1879 text deriving from his Cambridge course, and Augustus Love's *Elasticity* (1892) became staple fare for mathematicians in Britain, each running into several editions. In France Maurice d'Ocagne published his *Traité de Nomographie* (1899), which was initially developed for civil engineering (particularly railway and bridge construction) but was also used in the study of ballistics during Word War I and thereafter. Carl Runge's lectures at Columbia University in New York (1909/10) were published as *Graphical Methods* in the United States in 1912, and appeared in a German translation in 1914. Runge's joint book with Hermann König *Vorlesungen über numerisches Rechnen* (1924), part of Springer's *Grundlehren der mathematischen Wissenschaften* series, became a classic. Published that same year, and in the same series, *Differenzenrechnung* by the Dane Niels Erik Nörlund became "the standard treatise on new aspects of difference equations." This was stated in a report to the American National Research Council on *Numerical Integration of Differential Equations* (1933), authored by Albert Bennett, William Milne, and Harry Bateman, which contains many valuable references. Cyrus Colton MacDuffee's *The Theory of Matrices* (1933), also published by Springer in Germany, aimed, according to its preface, "to unify certain parts of the theory" because "many of the fundamental properties of matrices were first discovered in the notation of a particular application." Some influential Russian textbooks such as L. V. Kantorovich's and V. I. Krylov's *Approximate Methods of Higher Analysis* (1941) became known in the West only much later through translations.

While it was certainly more a chemists' and physicists' war than one of mathematicians, World War I involved mathematicians on all sides. But politicians were slow to recognize the specific contribution that mathematics could offer to the war effort. France did little to protect her most promising young mathematicians from the front, and about half of the science cohort studying at the École Normale Supérieure at the outbreak of the war was killed. Leading mathematicians such as Borel, however, gradually became involved as scientific advisors in warfare, while ballistics at the Gâvre proving ground was developed under Prosper Charbonnier, using graphical and numerical methods. In England Ralph Fowler, Edward Milne, and Herbert Richmond worked on the mathematics of antiaircraft gunnery, while J. E. Littlewood applied his talents to more general ballistics and Karl Pearson interrupted his statistical research in biomathematics to oversee the production of the corresponding range tables. In Italy the experiences of Mauro Picone as an artillery officer working on ballistics research were influential with respect to the formation of the institute that was later set up under his lead (see below). Meanwhile, a key event in Germany was the foundation of the Aerodynamische Versuchsanstalt ("Aerodynamic Proving Ground") in Göttingen in 1917, mainly financed by the military. This Proving Ground, together with the more theoretical Institute for Fluid Mechanics established in 1925 under the leadership of Ludwig Prandtl, would become the most advanced aerodynamic research installation in the world. In the United States systematic ballistics research was done in Washington under celestial mechanist Forest Ray Moulton and at the newly founded Aberdeen Proving Ground in Maryland under the geometer Oswald Veblen. In this context, finite-difference methods were introduced into ballistic computation by Moulton, as were higher methods of the calculus of variations by Gilbert Bliss in Aberdeen, which even in 1927 were described by Bliss as being "on the farther boundary of the explored mathematical domain of today." From 1917 the new National Advisory Committee for Aeronautics, the predecessor of NASA, had its Langley Laboratory in Virginia, where Prandtl's former student Max Munk was particularly influential in the 1920s when he explained the mathematics of aerodynamics to the laboratory's engineers and built a variable-density wind tunnel.

More important for applied mathematics than the war itself, though, were its consequences.

The war led to the gradual recognition of the importance of the fundamental sciences (including mathematics), beyond mere technical inventions, for industrial and military applications. Science systems that were under close political control and received their main funding from government, such as those in

Germany, Italy, and Russia, had the potential to quickly recognize and respond to the importance of applied mathematics. There is of course nothing guaranteed about this relationship, as the example of the rather slow development in France shows. In the United States, new modes of industrial mass production during the war included scientific management (Taylorism). The superior strength of material resources, particularly in the electrical industry and in the development of calculating devices—IBM was founded in 1925—pointed to the future role of the United States in applied mathematics. Nevertheless, the predominantly private, but nonindustrial, sponsorship of the American university system before World War II slowed down the arrival of academic applied mathematics in the United States.

The foundation of the institute for applied mathematics at the University of Berlin (in 1920, under von Mises) can be partly explained against this backdrop of war and international competition. From 1921 von Mises edited the *Zeitschrift für angewandte Mathematik und Mechanik* (ZAMM), which had a clear engineering context, unlike the older *Zeitschrift für Mathematik und Physik*. In his programmatic article "On the tasks and goals of applied mathematics," with which he started his new journal, von Mises gave his own definition of applied mathematics and added:

> As a matter of course we put ourselves on the basis of the present, particularly on the standpoint of the scientifically minded engineer.

The foundation of the ZAMM was significantly ahead of the first Russian journal that expressly (and exclusively) referred to applications in its title, *Prikladnaya Matematika i Mekhanika*, which began in 1933; it is today regularly translated into English as the *Journal of Applied Mathematics and Mechanics*. The first American journal with "applied mathematics" in its title was the *Quarterly of Applied Mathematics*, edited at Brown University from 1943, which was partly modeled on von Mises's ZAMM. However, the editors made a point in dropping "mechanics" from the title both in order to stress autonomy from the German example and to express the new level of independence of the field.

In spite of, or rather because of, increased economic competition after the war, internationalization remained the order of the day even in the applied sciences, as promoted by Felix Klein before the war. Indeed, the applied mathematician and fluid dynamics engineer von Kármán, who was closely connected to Göttingen, initiated the international congresses for applied mechanics in 1924 (after organizing a successful conference on hydrodynamics and aerodynamics in Innsbruck in 1922) because he felt that the critical mass for discussion was not big enough on a national level and because postwar policies on each side threatened communication. Being Hungarian by origin, he was less skeptical about restoring international collaboration than von Mises and others who had stronger nationalist feelings. As von Kármán had hoped, the congresses attracted participants from across Europe, and by 1930, when the congress was held in Stockholm, there was also a strong delegation from the United States. However, the story of Russian participation is not so good. Politics did get in the way, and Russian mathematicians barely made an appearance.

Applied mathematics exhibited particular challenges for internationalization, not least because of its engineering context, which often bore traits of national idiosyncrasies, as is clear from the international efforts to unify terminology in vector calculus in around 1900. As late as 1935, at the Volta congress "High Velocities in Aviation" in Rome, terminology in ballistics and fluid mechanics was considered to be far from internationally standardized.

On the French side, the foundation of the Institut Henri Poincaré (1926), on the initiative of Borel and funded with American Rockefeller money, was a clear token of internationalization. Although the institute was less instrumental in the development of applied mathematics than its founder, with his broad view of "Borelian mathematics" (which stressed links between pure mathematics and applications, mainly based on mathematical physics and the theory of probability), had anticipated and wished for, it did provide the seeds for the development of French applied mathematics after the war.

In other European countries, such as Italy and Soviet Russia, new efforts to further work on applications were launched in the 1920s, partly motivated by experiences during the war and by political revolutions. The foundation in 1921 of the Institute of Physics and Mathematics of the Soviet Academy of Sciences by Vladimir Steklov (a former student of Aleksandr Lyapunov), which was supported by Lenin, was crucial for the further development of applied mathematics in the Soviet Union; the history of the institute has yet to be written. In Italy, under Mussolini, the central event concerning applications of mathematics in industry and the military was the creation in 1927–31 of the Italian

Research Council's National Institute for the Application of the Calculus under Mauro Picone in Naples and later in Rome.

In contrast, Britain created no new institutes for mathematics. Even G. H. Hardy, who had left Cambridge for Oxford shortly after the war, was unable to persuade his new university to build one. Nevertheless, the war left a tangible legacy for applied mathematics. Imperial College received a substantial grant to finance its Department of Aeronautics, while Cambridge established a new chair in aeronautical engineering. As a result of increased funding after the war, establishments such as the Royal Aircraft Establishment and the National Physical Laboratory were able to retain a number of their wartime staff, several of whom were mathematicians. Notable inclusions were Hermann Glauert, who made a career in aerodynamics at the Royal Aircraft Establishment, and Robert Frazer, who worked on wing flutter at the National Physical Laboratory. In the 1930s Frazer and his colleagues W. J. Duncan and A. R. Collar were "the first to use matrices in applied mathematics." In addition, theoreticians and practitioners who were brought together because of the war worked together afterward. Sometimes, as in the case of the Cambridge mathematician Arthur Berry and the aeronautical engineer Leonard Bairstow, the end of hostilities meant only the end of working in the same location, it did not mean the end of collaboration.

In the interwar period the degree of industrialization in a particular country was without doubt one of the defining factors in that country's support of applied mathematics. This is well exemplified by the solid development of applied mathematics in industrialized Czechoslovakia compared with the strong tradition in pure mathematics in less industrialized Poland.

Indeed, it became increasingly obvious after the war that engineering mathematics and insurance mathematics, both of which corresponded to the developing needs of the new professions and industries, had become legitimate parts of applied mathematics. Not only were they the most promising areas of the subject, but they were economically the most rewarding. Students trained at von Mises's institute in Berlin and at Prandtl's institute in Göttingen found jobs in various aerodynamic laboratories and proving grounds, as well as in industry. Von Mises himself both undertook governmental assignments and acted as an advisor for industry. At Siemens, AEG, and Zeiss (all in Germany), the General Electric Company (in Britain), Philips (in the Netherlands), and General Electric and Bell Laboratories (in the United States) (Millman 1984), industrial laboratories (mainly in electrical engineering but also, for instance, in the optical and aviation industries) developed a demand for trained mathematicians. It was the study of the propagation of radio waves and of the electrical devices required to generate them that led in 1920 to the Dutchman van der Pol working out the equation that is to this day considered as the prototype of the nonlinear feedback oscillator. VAN DER POL'S EQUATION [IV.2 §10]) and his modeling approach have repeatedly been cited as exemplars for modern applied mathematics. Van der Pol's contribution, together with theoretical work by Henri Poincaré on limit cycles, strongly influenced Russian work on nonlinear mechanics. Its mathematical depth gained the approval (albeit somewhat reluctant approval) even of André Weil, a foremost member of the Bourbaki group of French mathematicians, who in 1950 called it "one of the few interesting problems which contemporary physics has suggested to mathematics." Van der Pol, who worked at the Philips Laboratories in Eindhoven from 1922, also contributed to the justification of the Heaviside operational calculus in electrical engineering. Around 1929 he used integral transformation methods similar to those developed before him by the English mathematician Thomas Bromwich and the American engineer John Carson at Bell Laboratories, who in 1926 wrote the influential book *Electric Circuit Theory and the Operational Calculus*. Somewhat later, the German Gustav Doetsch provided a more systematic justification of Heaviside's calculus based on the theory of the Laplace transform in his well-received book *Theorie und Anwendung der Laplace-Transformation* (1937). In another influential book, *Economic Control of Quality of Manufactured Product* (1931), the physicist Walter Shewart, a colleague of Carson's at Bell Laboratories, was one of the first to promote statistics for industrial quality control using so-called control charts.

However, many of these developments in applied and industrial mathematics, both in Europe and America, occurred outside their national academic institutions, notwithstanding the beginnings of systematic academic training in applied mathematics in new institutes such as the one led by von Mises. A number of academically trained mathematicians and physicists were impressed by the spectacular and revolutionary ideas of relativity theory and quantum theory, but they were slow to recognize the importance of those new applications, often in engineering, that relied on classical mechanics.

This aloofness of academic scientists prevailed in the United States. The undisputed leader of American mathematicians, George Birkhoff, was aware of that when he addressed the American Mathematical Society at its semicentennial in 1938 with the following words:

> The field of applied mathematics always will remain of the first order of importance inasmuch as it indicates those directions of mathematical effort to which nature herself has given approval.
>
> Unfortunately, American mathematicians have shown in the last fifty years a disregard for this most authentically justified field of all.

There were exceptions, such as Norbert Wiener at the Massachusetts Institute of Technology, who was based in the mathematics department but interacted with the electrical engineering department run by Vannevar Bush (the inventor of the "differential analyzer," an analogue computer) and through it with Bell Laboratories, and there were also the individual efforts of a number of mathematicians with a European background. One of the most successful of the latter was Harry Bateman, Professor of Aeronautical Research and Mathematical Physics at Caltech in Pasadena, who became a champion of special functions during the 1930s and 1940s, and who had earlier (immediately prior to his emigration from England in 1910) discussed the Laplace transform and applications to differential equations. However, American academia was late in recognizing applied mathematics, as exemplified by the abovementioned report on *Numerical Integration of Differential Equations* (1933), in which the authors write that the report was produced "without special grant for relief from teaching from any of the institutions represented." Mathematical physicist Warren Weaver (who later, in World War II, would lead the Applied Mathematics Panel within the American war effort) was surprised, as late as 1930, "at the emphasis given, in the discussion [on a planned journal for applied mathematics], to the field between mathematics and engineering." During the 1920s and 1930s, Rockefeller money had primarily been geared toward supporting pure academic mathematical and physical research, leaving applied research in the hands of industry. It was left to the clever negotiations of Richard Courant (Göttingen's adherent to applied mathematics) to win Rockefeller fellowships for the applied candidates under his tutelage, such as Wilhelm Cauer and Alwin Walther.

The 1920s and 1930s were also a time in which mathematical modeling came to the fore, although the term "mathematical modeling" was rarely used before World War II. In a 1993 article on the emergence of biomathematics in *Science in Context*, the author Giorgio Israel emphasizes the increasing role of mathematical modeling in the nonphysical sciences:

> Another important characteristic of the new trends of mathematical modeling and applied mathematics is interest in the mathematization of the nonphysical sciences. The 1920s offer in fact an extraordinary concentration of new research in these fields, which is developed from points of view more or less reflecting the modeling approach. So the systematic use of mathematics in economics (both in the context of microeconomics and game theory) is found in the work of K. Menger, J. Von Neumann, O. Morgenstern, and A. Wald, starting from 1928. The basic mathematical model of the spread of an epidemic (following the research of R. Ross on malaria) was published in 1927 [by W. O. Kermack and A. G. McKendrick]; the first papers by S. Wright, R. A. Fisher and J. B. S. Haldane on mathematical theory of population genetics appeared in the early twenties; the first contributions of Volterra and Lotka to population dynamics and the mathematical theory of the struggle for existence were published in 1925 and 1926; and many isolated contributions (such as van der Pol's model) also appeared in these years.

Moreover, during the twentieth century there was a certain tendency for mathematicians to be less inspired by physics and to resort instead to less rigorous or less complete models from other sciences, including engineering. In 1977 Garrett Birkhoff, George Birkhoff's son, wrote:

> Engineers and physicists create and adopt mathematical models for very different purposes. Physicists are looking for universal laws (of "natural philosophy"), and want their models to be exact, universally valid, and philosophically consistent. Engineers, whose complex artifacts are usually designed for a limited range of operating conditions, are satisfied if their models are reasonably realistic under those conditions. On the other hand, since their artifacts do not operate in sterilized laboratories, they must be "robust" with respect to changes in many variables. This tends to make engineering models somewhat fuzzy yet kaleidoscopic. In fluid mechanics, Prandtl's "mixing length" theory and von Kármán's theory of "vortex streets" are good examples; the "jet streams" and "fronts" of meteorologists are others.

The same author, himself a convert from abstract algebra to hydrodynamics, explains resistance to mathematical models in economics, pointing to the fact that they did not fit well into Bourbaki's "conventional framework of pure mathematics." The latter has often been described as in some respects being inimical to applications and (since the "New Math" of the 1960s) as being pedagogically disastrous. However, as detailed by Israel in the paper quoted above, the relationship between Bourbaki and the new practices in modeling has not necessarily been negative. Some mathematicians considered Bourbaki's notion of mathematics as an "abstract scheme of possible realities" to be the right way to liberate mathematics from the classical reductionist mechanistic approach that had often relied on linearization methods. There have even been efforts, for instance by logicians, to introduce "planned artificial language" into the sciences, as exemplified in J. H. Woodger's *The Axiomatic Method in Biology* (1937). However, these efforts seem to have had limited success. It took another step in the development of computers in the 1980s before necessarily simplified models of biological processes could be abandoned, and investigations of cellular automata, membrane computing, simulation of ecological systems, and similar tasks from modern mathematical biology could be undertaken.

During the 1920s and 1930s, many further results in different fields of application were obtained. Well-known examples include Alan Turing's work during the 1930s on the theory of algorithms and computability, and the Russian Leonid Kantorovich's work on linear programming within an economic context (1939), which escaped the attention of Western scholars for several decades.

This was also a period in which some of the foundations were laid for what would, from the late 1940s on, be called numerical analysis. In 1928 Courant and his students Kurt Friedrichs and Hans Lewy, all three of whom eventually emigrated to the United States, published "On the partial difference equations of mathematical physics" in *Mathematische Annalen.* The paper was translated in the *IBM Journal of Research and Development* as late as 1967 on the grounds that it was "one of the most prophetically stimulating developments in numerical analysis... before the appearance of electronic digital computers.... The ideas exposed still prevail." In the history of numerical analysis, the paper gained special importance because it contains the germ of the notion of numerical stability and involves the problem of well-posedness of partial differential equations (as proposed by Hadamard in 1902).

## 5 Applied Mathematics during and after World War II

World War II, like World War I, was not a mathematicians' war. Indeed, in early 1942 the chemist, and Harvard president, James Conant said: "The last was a war of chemistry but this one is a war of physics." This of course partially reflected the increasing role of mathematics in World War II, revealed by the use of ballistics, operations research, statistics, and cryptography throughout the conflict. In fact, the president of the American National Academy of Sciences, the physicist Frank B. Jewett, responded to Conant with the words: "It may be a war of physics but the physicists say it is a war of mathematics." However, at the time, due to lingering tradition, mathematics was not given the same high priority as the other sciences either in the preparation for warfare nor in war-related research. In the early 1940s within the leading research organizations in the United States, in Germany, and in other countries, mathematics was still subordinate to other fields, such as engineering and physics. In addition, the mathematicians themselves were not prepared for a new and broader social role, e.g., as professionals in industry, such as might be demanded by the war. When considering the future of their field during and after the war, many pure mathematicians were worried that mathematics would suffer from a too utilitarian point of view. This is exemplified by the well-known essay *A Mathematician's Apology* written by the leading English mathematician G. H. Hardy in 1940.

But not long after Hardy's essay was written, another Cambridge mathematician, Alan Turing, demonstrated the potential of sophisticated mathematics—a mix of logic, number theory, and Bayesian statistics—for warfare, when he and his collaborators at Bletchley Park broke the code of the German Enigma machine.

In Germany, the Diplommathematiker (mathematics degree with diploma), which was designed for careers in industry and the civil service, was officially introduced in 1942, and teaching as a career for mathematicians began to lose its monopoly.

The entry of the United States into World War II in December 1941 brought with it deep changes in the way mathematicians worked together with industry, the military, and government. In the *American Mathematical Monthly*, rich memoirs on the state of industrial mathematics and (academic) applied mathematics in the United States by Thornton Fry (1941) and Roland

Richardson (1943), respectively, were published. Probably the most spectacular development in communications mathematics took place in the 1940s at Bell Laboratories with the formulation of information theory by Claude Shannon.

Based on their European experiences, immigrants to the United States such as von Kármán, Jerzy Neyman, von Neumann, and Courant contributed substantially to a new kind of collaboration between mathematicians and users of mathematics. In the mathematical war work organized by the Applied Mathematics Panel, where the leading positions were occupied by Americans, with Warren Weaver at the head, the applied mathematicians cooperated with mathematicians of an originally purer persuasion, natives of the United States (Oswald Veblen, Marston Morse) and immigrants (von Neumann) alike.

As well as their political and administrative experience, the immigrants brought to their new environment European research traditions from engineering mathematics, classical analysis, and discrete mathematics. Ideas, such as those of von Neumann in theoretical computing, could gradually mature and materialize within the industrial infrastructure of the United States (Bell Laboratories, etc.), aided during the war by seemingly unlimited public money (Los Alamos, etc.). In March 1945, while the war was still on, von Neumann sent a famous memo on the "Use of variational methods in hydrodynamics" to Veblen. Von Neumann recommended the "great virtue of Ritz's method" and deplored that before, and even during, the war mathematical work had not been sufficiently centralized for a systematic attack on the nonlinear equations occurring in fluid mechanics and related fields. In the same memo von Neumann pointed to the "increasing availability of high-power computing devices," a development to which he had of course contributed substantially. As mentioned in the introduction to the SIAM "History of numerical analysis and scientific computing" Web pages:

> Modern numerical analysis can be credibly said to begin with the 1947 paper by John von Neumann and Herman Goldstine, "Numerical inverting of matrices of high order" (*Bulletin of the AMS*, Nov. 1947). It is one of the first papers to study rounding error and include discussion of what today is called scientific computing.

Von Neumann and Goldstine's results were soon followed up and critically discussed by English mathematicians (Leslie Fox and James Wilkinson, as well as Alan Turing) at the National Physical Laboratory at Teddington.

After the war, the increased level of U.S. federal funding for mathematics was maintained. Although partly fueled by the beginning of the Cold War, it was nevertheless no longer restricted to applications. Much of it was channeled through the department of defense (e.g., by the Office of Naval Research) and the new National Science Foundation (NSF), which was founded in 1950. The NSF was initiated by the electrical engineer Vannevar Bush, who had led the Office for Scientific Research and Development, the American war-research organization. The concerns about exaggerated utilitarianism that were harbored by pure mathematicians before the war therefore turned out to be groundless.

As George Dantzig, the creator of the SIMPLEX METHOD [IV.11 §3.1], observed, the outpouring of papers in linear programming between 1947 and 1950 coincided with the building of the first digital computers, which made applications in the field possible. Mathematical approaches to logistics, warehousing, and facility location were practiced from at least the 1950s, with early results in optimization by Dantzig, William Karush, Harold Kuhn, and Albert Tucker being enthusiastically received by (and utilized in the logistics programs of) the United States Air Force and the Office of Naval Research. These optimization techniques are still highly relevant to industry today.

The papers of the 1950 Symposium on Electromagnetic Waves, sponsored by the United States Air Force and published in the new journal *Communications on Pure and Applied Mathematics*, summarized the war effort in the field. Richard Courant, by then at New York University, pointed to the importance of a new approach to classical electromagnetism, where "a great number of new problems were suggested by engineers." This strengthened the feeling, already evident before the war, that the predominance of academic mathematical physics as the main source of inspiration for mathematical applications had begun to wane. Ironically, Courant's paper of 1943 on variational methods for the solution of problems of equilibrium and vibrations, which would later be widely considered to be one of the starting points for the FINITE-ELEMENT METHOD [II.12] (the name being coined by R. W. Clough in 1960), lay in obscurity for many years because Courant, not being an engineer, did not link the idea to networks of discrete elements. Another reason for the later breakthrough was a development of variational methods of approximation in the theory of partial differential equations, which would ultimately prove to be vital to the development of finite-element methods in the 1960s (see J. T. Oden's chapter in Nash (1990)). A thoughtful historical look at the different ways mathematicians and engineers use finite-element methods is given in Babuška (1994).

New institutions for mathematical research, both pure and applied, were created after the war. Among them the institute under Richard Courant at New York University developed the most strongly. In a 1954 government report it was stated that Courant's institute had an

> enrollment of over 400 graduate students in mathematics of which about half have a physics or engineering background.... The next largest figure, reported from Brown University, is a whole order of magnitude smaller! In the way of a rough estimate this means that New York University alone provides about one third of this country's annual output of applied mathematicians with graduate training.

The figures are based on a questionnaire prepared in connection with a conference organized in 1953 at Columbia University in New York by F. Joachim Weyl, the son of Hermann Weyl, as part of a Survey of Training and Research in Applied Mathematics sponsored by the American Mathematical Society and by the National Research Council under contract with the NSF. The conference proceedings and the report both included discussions on not only the training of applied mathematicians (particularly for industry, and including international comparisons) but also the increasing use of electronic computing; a summary was published in the *Bulletin of the American Mathematical Society* in 1954.

Between 1947 and 1954 the Institute for Numerical Analysis at the University of California, Los Angeles, sponsored by the National Bureau of Standards, played a special role in training university staff in numerical analysis and computer operations. The institute was closed in 1954, a victim of McCarthyism.

Brown University's summer school of applied mechanics, which were organized by Richardson from 1941 onward, had relied heavily on the contributions of immigrants. This is also partly true of the first American journal of applied mathematics, the *Quarterly of Applied Mathematics*, which began in 1943, and of *Mathematical Tables and Other Aids to Computation*, another Brown journal, which started the same year under Raymond Archibald.

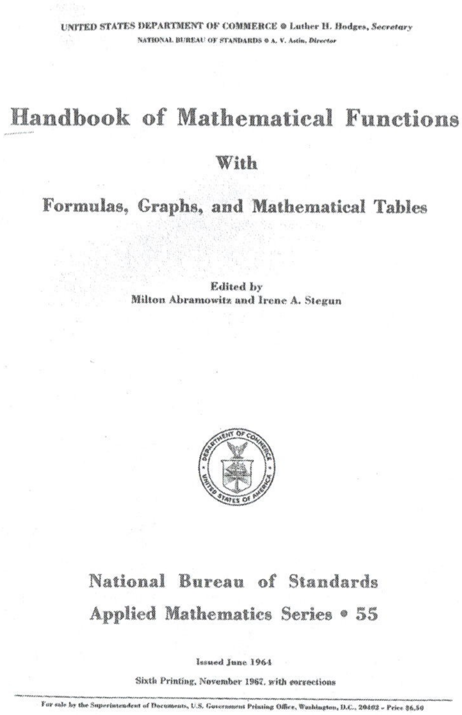

UNITED STATES DEPARTMENT OF COMMERCE • Luther H. Hodges, *Secretary*
NATIONAL BUREAU OF STANDARDS • A. V. Astin, *Director*

**Handbook of Mathematical Functions**

**With**

**Formulas, Graphs, and Mathematical Tables**

Edited by
Milton Abramowitz and Irene A. Stegun

**National Bureau of Standards**
**Applied Mathematics Series • 55**

Issued June 1964
Sixth Printing, November 1967, with corrections

For sale by the Superintendent of Documents, U.S. Government Printing Office, Washington, D.C., 20402 - Price $6.50

**Figure 6** M. Abramowitz and I. A. Stegun, eds, *Handbook of Mathematical Functions* (1964).

Various projects on mathematical tables and special functions that had their origins early in the twentieth century in various countries (the United Kingdom, Germany, and the United States) received a boost from the war. At Caltech, Arthur Erdélyi, with financial support from the Office of Naval Research, oversaw the Bateman Manuscript Project—the collation and publication of material collected by Harry Bateman, who had died in 1946—which led to the three-volume *Higher Transcendental Functions* (1953–55).

The Mathematical Tables Project, which had been initiated by the Works Progress Administration in New York in 1938, with Gertrude Blanch as its technical director, was disbanded after the war but many of its members moved to Washington in 1947 to become part of the new National Applied Mathematics Laboratories of the National Bureau of Standards. The latter's conference of 1952 resulted in one of the best-selling applied mathematics books of all time, *Handbook of Mathematical Functions with Formulas, Graphs and Mathematical Tables* (1964) by Milton Abramowitz and Irene

Stegun, both of whom had worked under Blanch (figure 6). Meanwhile, Blanch herself, rather than go to Washington, took a position at the Institute for Numerical Analysis at University of California, Los Angeles until its closure in 1954, where she worked with John Todd and Olga Taussky. Taussky, a rare example of a female pure mathematician who made contributions to applied mathematics, had worked during the war in Frazer's group at the National Physical Laboratory on flutter problems in supersonic aircraft and was an important figure in the development of matrix theory. Aware of certain resentments against applied mathematics from some mathematical quarters, in 1988 Taussky said:

> When people look down on matrices, remind them of great mathematicians such as Frobenius, Schur, C. L. Siegel, Ostrowski, Motzkin, Kac, etc., who made important contributions to the subject. I am proud to have been a torchbearer for matrix theory, and I am happy to see that there are many others to whom the torch can be passed.

Computer developments rapidly changed the character of the table-making aspect of applied mathematics. As a result, Archibald's journal had a rather short life in its original form, being renamed *Mathematics of Computation* in 1960 in order "to reflect the broadened scope of the journal," which, as the editor Harry Polachek described, had "expanded to meet the need in [the United States] for a publication devoted to numerical analysis and computation." Private and government institutes such as the Research and Development Corporation (RAND), founded within the Douglas Aircraft Company in October 1945, and professional organizations such as the Association for Computing Machinery (1947), the Operations Research Society of America (1952), the Society for Industrial and Applied Mathematics (1952), and the Institute of Management Sciences (1953) all testified to the broadening social base for applied mathematics.

The technological development of computers and software had ramifications for the development of mathematical algorithms, particularly with respect to their speed and reliability. In his treatment of ordinary differential equations around 1910, Carl Runge had no need of systematic error estimation—it was not until after World War II that John Couch Adams's multistep methods for the numerical solution of ordinary differential equations from the mid-nineteenth century were analyzed with respect to error estimation, after they had been brought into a more practical form by W. E. Milne in the United States ("Milne device") in 1926. However, postwar increases in computing speed led to greater concern about numerical stability and to the abandonment of methods like those of Milne. Europeans were also strongly involved in these theoretical developments. For example, in the 1930s the Englishman Richard Southwell had developed the so-called relaxation method, an iterative method connected to the numerical solution of partial differential equations in elasticity theory. The story goes that Garrett Birkhoff gave his doctoral student David M. Young the task of "automating relaxation," and Young introduced successive overrelaxation in his 1950 thesis (see Young's chapter in Nash (1990)). Young's deep analysis showed how to choose the relaxation parameters automatically in some important cases and thus provided a method suitable for programming on digital computers. This was an important advance on Southwell's method, which had been designed for computation by hand. The Norwegian–German Werner Romberg's elimination approach (1955) for improving the accuracy of the trapezoidal rule relied on Richardson extrapolation, a technique developed by the Englishman Lewis Fry Richardson early in the century. Richardson introduced finite-difference methods for the numerical solution of partial differential equations in 1910, of which extrapolation is just one part. And he was the first to apply mathematics, in particular the method of finite differences, to WEATHER PREDICTION [V.18]. His book *Weather Prediction by Numerical Process* (1922) was republished by Cambridge University Press in 2007 with a new foreword. The Richardson–Romberg procedure became widely known after it had been subjected to a rigorous error analysis by the German Friedrich L. Bauer and the Swiss Eduard L. Stiefel in the 1960s. In 1956 the Swede Germund Dahlquist introduced, within his theory of numerical instability, what Nick Trefethen (in *The Princeton Companion to Mathematics*) called "the fundamental theorem of numerical analysis: consistency + stability = convergence."

A major step in understanding the effects of rounding errors in linear algebra algorithms such as Gaussian elimination was the development, principally by the Englishman James Wilkinson, of BACKWARD ERROR ANALYSIS [I.2 §23], as described in Wilkinson's influential 1963 and 1965 research monographs. The all-important role of computer developments for the practical realization of previously existing theory is also evident in the history of the FAST FOURIER TRANSFORM

[II.10] (FFT), made famous by Cooley and Tukey in 1965: the FFT was originally discovered by Gauss but it could not be effectively applied before the advent of digital computers.

Many of the European mathematicians mentioned above worked in the United States for some of their careers. The "Brain Drain" gradually replaced the flight from Europe for political reasons, as observed in 1953 by the Dutch mathematician A. van Wijngaarden when attending a Columbia University symposium:

> The one thing that continues to worry us is that we lose many of our best people to that big sink of European scientific talent, the U.S.A.

However, there were many European mathematicians who did not (or could not) follow the Brain Drain and who worked on an equal level with the Americans, among them many Russians. Nevertheless, there is no doubt about the superior technological and industrial infrastructure, particularly with respect to computing facilities and software development, that existed in the United States. Although this superiority was sometimes met with resentment, including by Alan Turing, it was admitted by Russians such as A. P. Ershov and M. R. Shura-Bura (Metropolis et al. 1980) and by Western European applied mathematicians such as the Frenchmen Louis Couffignal, the cybernetics pioneer, and Jacques-Louis Lions, the numerical analyst. For this reason, the latter cooperated strongly with Russian applied mathematicians, and he sometimes felt that the lack of access to cutting-edge technology increased the theoretical depth of their collaborative work.

From the 1930s Russian mathematicians had begun publishing exclusively in Russian and no longer also published in French and German. This practice, which continued after the war, prompted the American Mathematical Society, with funding from the Office of Naval Research, to begin a Russian translation project in 1947. SIAM followed suit in 1956, with support from the NSF. The "Sputnik crisis" in 1957 caused American mathematicians to look even more closely at the work being done in the Soviet Union. The Russian-born American topologist Solomon Lefschetz, who was prompted out of retirement by the event, persuaded the Martin Aircraft Company to set up a mathematics research center, at which he directed a large group working on nonlinear differential equations.

And yet during the two decades following the war, despite the relatively favorable material conditions of the United States, applied mathematics lost some of its reputation in comparison with pure mathematics. In retrospect, of course, this period can be seen as one of consolidation of methods and of waiting for more powerful computational technology and for deeper theoretical foundations to arrive.

For example, one could argue that parts of the reform movement in teaching ("New Math") in the early 1960s, which was meant to correct obvious and perceived shortcomings in American secondary education, counteracted efforts to develop applied mathematics. This movement was strongly influenced by the "purist" ideology of Bourbaki and had worldwide ramifications, threatening to undermine the development of pupils' attitudes toward applications. In 1961, at a symposium organized by SIAM in Washington, DC, on research and education in applied mathematics, the chairman deemed that

> applied mathematics is something of a stepchild; I might even say an out-of-step child, whose creations are looked upon with equal disinterest by mathematicians, physicists and engineers.
>
> *SIAM Review* (1961)

On January 24 of the same year, the *New York Times* ran a column entitled, "Russia may be losing its traditional leadership in mathematics because of overemphasis on applied research," which prompted the applied mathematician Harvey Greenspan from MIT to make a critical response in *American Mathematical Monthly* saying that, to the contrary,

> the increased Russian emphasis in this area is really cause for some serious concern on our part.... The present system does not produce adequate numbers of applied mathematicians and remedial steps must soon be taken. The great majority of the senior faculty in applied mathematics are products of a European education.

The real breakthrough for applied mathematics in the United States came, however, with the rise of computer technology and computer science in the 1970s and 1980s, when, at the same time, the ideology of Bourbaki was in retreat. The so-called Davis Report of 1984 on "Renewing U.S. Mathematics" was an important event, leading to a near doubling of federal investment in mathematical research and including a special mathematics of computing initiative. In 1989 the Hungarian-born Peter Lax—who emigrated with his parents to the United States in 1941 and studied at New York University, and who in 2005 was a recipient of the Abel Prize in Mathematics "for his groundbreaking contributions to the theory and application of partial differential equations and to the computation of their solutions"—signaled a new level of acceptance for applied mathematics in his widely distributed paper in *SIAM Review* on "The flowering of applied mathematics in America":

> Whereas in the not so distant past a mathematician asserting "applied mathematics is bad mathematics" or "the best applied mathematics is pure mathematics" could count on a measure of assent and applause, today a person making such statements would be regarded as ignorant.

The publication of articles by working applied mathematicians in Metropolis et al. (1980) and Nash (1990), and the extensive seven-volume historical project undertaken by the *Journal of Computational and Applied Mathematics* (2000), which was republished in Brezinski and Wuytack (2001), seem to testify to a growing self-confidence of applied mathematics within mathematics more widely. Some efforts, such as the publication of the "Top ten algorithms" in *Computing in Science and Engineering* in 2000 (six of which are contained in [I.5, table 2]), provoked controversial discussions. Many practitioners of applied mathematics in these and other publications reveal awareness of problems regarding the rigor and reliability of their methods, showing that the links between pure and applied mathematics exist and continue to stimulate the field. However, the standard philosophical approaches to mathematics—circling repetitively around formalism, logicism, and intuitionism, with no consideration of applications and doing no justice to the ever-increasing range of mathematical practice—are no longer satisfying either to mathematicians or to the public.

The unabated loyalty to pure mathematics as the mother discipline sometimes leads to overcautious reflection on the part of the applied mathematician. A nice example is provided by Trefethen (again in *The Princeton Companion to Mathematics*) in the context of rounding errors and the problem of numerical stability:

> These men, including von Neumann, Wilkinson, Forsythe, and Henrici, took great pains to publicize the risks of careless reliance on machine arithmetic. These risks are very real, but the message was communicated all too successfully, leading to the current widespread impression that the main business of numerical analysis is coping with rounding errors. In fact, the main business of numerical analysis is designing algorithms that converge quickly; rounding-error analysis, while often a part of the discussion, is rarely the central issue. If rounding errors vanished, 90% of numerical analysis would remain.

But it is not only these methodological concerns that hold back applied mathematics. For example, Lax (1989) mentions persisting problems in education and the need to maintain training in classical analysis:

> The applied point of view is essential for the much-needed reform of the undergraduate curriculum, especially its sorest spot, calculus. The teaching of calculus has been in the doldrums ever since research mathematicians gave up responsibility for undergraduate courses.

The education and training of applied mathematicians remains a central concern, and it is not even clear whether the situation has changed significantly since the Columbia University Conference of 1953. At that time the applied mathematician and statistician John Wilder Tukey, best known for the development of the FFT algorithm and the box plot, declared with reference to what is now called modeling:

> Formulation is the most important part of applied mathematics, yet no one has started to work on the theory of formulation—if we had one, perhaps we could teach applied mathematics.

A 1998 report by the NSF states that:

> Careers in mathematics have become less attractive to U.S. students. [Several]... factors contribute to this change: (i) students mistakenly believe that the only jobs available are collegiate teaching jobs, a job market which is saturated (more than 1,100 new Ph.D.s compete for approximately 600 academic tenure-track openings each year); (ii) academic training in the mathematical sciences tends to be narrow and to leave students poorly prepared for careers outside academia; (iii) neither students nor faculty understand the kinds of positions available outside academia to those trained in the mathematical sciences.

The same report underscores the undiminished dependence of American pure and applied mathematics on immigration from Europe and (now) from Asia, South America, and elsewhere:

> Although the United States is the strongest national community in the mathematical sciences, this strength is somewhat fragile. If one took into account only

home-grown experts, the United States would be weaker than Western Europe. Interest by native-born Americans in the mathematical sciences has been steadily declining. Many of the strongest U.S. mathematicians were trained outside the United States and even more are not native born. A very large number of them emigrated from the former Soviet Union following its collapse. (Russia's strength in mathematics has been greatly weakened with the disappearance of research funding and the exodus of most of its leading mathematicians.) Western Europe is nearly as strong in mathematics as the United States, and leads in important areas. It has also benefited by the presence of émigré Soviet mathematical scientists.

The Fields Medals for Pierre-Louis Lions (son of Jacques-Louis Lions) (1994), Jean-Christophe Yoccoz (1994), Stanislav Smirnov (2010), and Cédric Villani (2010) testify to the growing strength of European applied mathematics and to the changed status of the field within mathematics. Likewise, the awarding of the Abel Prize of the Norwegian Academy of Science and Letters to Peter Lax (2005), Srinivasa Varadhan (2007), and Endre Szemerédi (2012) for predominantly applied topics is a further indication of this shift. In addition, prestigious prizes devoted specifically to applications, with particular emphasis on connections to technological developments, have been founded in recent decades. The fact that several of these prizes have been named for mathematicians of outstanding theoretical ability—the ACM A. M. Turing Award (starting in 1966), the IMU Rolf Nevanlinna Prize (1981), the DMV and IMU Carl Friedrich Gauss Prize (2006)—underscores the unity of mathematics in its pure and applied aspects.

Meanwhile, problems remain in the academic–industrial relationship and, connected to it, in the professional image of the applied mathematician, as described in the two most recent reports on "Mathematics in Industry" (1996 and 2012) published by SIAM. The report for 2012 summarizes the situation:

> Industrial mathematics is a specialty with a curious case of double invisibility. In the academic world, it is invisible because so few academic mathematicians actively engage in work on industrial problems. Research in industrial mathematics may not find its way into standard research journals, often because the companies where it is conducted do not want it to. (Some companies encourage publication and others do not; policies vary widely.) And advisors of graduates who go into industry may not keep track of them as closely as they keep track of their students who stay in academia.

However, most of the problems mentioned in this article with respect to academic applied mathematics (research funding, the lack of applications in mathematics education, the need for migration between national cultures) concern pure and applied mathematics alike. On the purely cognitive and theoretical level, the difference between the two aspects of mathematics—for all its interesting and important historical and sociological dimensions—hardly exists, as the above-quoted NSF report of 1998 underscores:

> Nowadays all mathematics is being applied, so the term applied mathematics should be viewed as a different cross cut of the discipline.

### Further Reading

Aspray, W. 1990. *John von Neumann and the Origins of Modern Computing*. Cambridge, MA: MIT Press.

Aubin, D., and A. Dahan Dalmedico. 2002. Writing the history of dynamical systems and chaos: longue durée and revolution, disciplines and cultures. *Historia Mathematica* 29:273–339.

Babuška, I. 1994. Courant element: before and after. In *Finite Element Methods: Fifty Years of the Courant Element*, edited by M. Křížek, P. Neittaanmäki, and R. Stenberg, pp. 37–51. New York: Marcel Dekker.

Bennett, S. 1979/1993. *A History of Control Engineering: Volume 1, 1800–1930; Volume 2, 1930–1950*. London: Peter Peregrinus.

Birkhoff, G. 1977. Applied mathematics and its future. In *Science and Technology in America: An Assessment*, edited by R. M. Thomson, pp. 82–103. Gaithersburg, MD: National Bureau of Standards.

Booß-Bavnbek, B., and J. Høyrup, eds. 2003. *Mathematics and War*. Basel: Birkhäuser.

Brezinski, C., and L. Wuytack, eds. 2001. *Numerical Analysis: Historical Developments in the 20th Century*. Amsterdam: Elsevier.

Campbell-Kelly, M., M. Croarken, R. Flood, and E. Robson, eds. 2003. *The History of Mathematical Tables: From Sumer to Spreadsheets*. Oxford: Oxford University Press.

Chabert, J.-L. 1999. *A History of Algorithms: From the Pebble to the Microchip*. Berlin: Springer.

Cortada, J. W. 1990/1996. *A Bibliographic Guide to the History of Computing, Computers, and the Information Processing Industry*, two volumes. New York: Greenwood Press. (Available online at www.cbi.umn.edu/.)

Darrigol, O. 2005. *Worlds of Flow: A History of Hydrodynamics from the Bernoullis to Prandtl*. Oxford: Oxford University Press.

Goldstine, H. H. 1977. *A History of Numerical Analysis from the 16th through the 19th Century*. New York: Springer.

Grattan-Guinness, I., ed. 1994. *Companion Encyclopedia of the History and Philosophy of the Mathematical Sciences*, two volumes. London: Routledge.

Grötschel, M., ed. 2012. Optimization Stories (21st International Symposium on Mathematical Programming, Berlin, August 19–24, 2012). *Documenta Mathematica* (extra volume).

Lax, P. 1989. The flowering of applied mathematics in America. *SIAM Review* 31:533–41.

Lenstra, J. K., A. Rinnooy Kan, and A. Schrijver, eds. 1991. *History of Mathematical Programming: A Collection of Personal Reminiscences.* Amsterdam: North-Holland.

Lucertini, M., A. Millán Gasca, and F. Nicolò, eds. 2004. *Technological Concepts and Mathematical Methods in the Evolution of Modern Engineering Systems: Controlling, Managing, Organizing.* Basel: Birkhäuser.

Mancosu, P., ed. 2008. *The Philosophy of Mathematical Practice.* Oxford: Oxford University Press.

Mehrtens, H., H. Bos, and I. Schneider, eds. 1981. *Social History of Nineteenth Century Mathematics.* Boston, MA: Birkhäuser.

Metropolis, N., J. Howlett, and G.-C. Rota. 1980. *A History of Computing in the Twentieth Century.* New York: Academic Press.

Millman, S., ed. 1984. *A History of Engineering and Science in the Bell System: Communication Sciences (1925–1980).* Murray Hill, NJ: AT&T Bell Laboratories.

Nash, S. G., ed. 1990. *A History of Scientific Computing.* New York: ACM Press.

Tournès, D. 1998. L'origine des méthodes multipas pour l'intégration numérique des équations différentielles ordinaires. *Revue d'Histoire des Mathématiques* 4:5–72.

Truesdell, C. 1960. A program toward rediscovering the rational mechanics of the age of reason. *Archive for History of Exact Sciences* 1:3–36.

# Part II
# Concepts

## II.1 Asymptotics
*P. A. Martin*

When sketching the graph of a function, $y = f(x)$, we may notice (or look for) lines that the graph approaches, often as $x \to \pm\infty$. For example, the graph of $y = x^2/(x^2+1)$ approaches the straight line $y = 1$ as $x \to \infty$ (and as $x \to -\infty$). This line is called an *asymptote*. Asymptotes need not be horizontal or straight, and they may be approached as $x \to x_0$ for some finite $x_0$. For example, $y = x^4/(x^2+1)$ approaches the parabola $y = x^2$ as $x \to \pm\infty$, and $y = \log x$ approaches the vertical line $x = 0$ as $x \to 0$ through positive values. Another example is that $\sinh x = \frac{1}{2}(e^x + e^{-x})$ approaches $\frac{1}{2}e^x$ as $x \to \infty$: we say that $\sinh x$ grows exponentially with $x$.

The qualitative notions exemplified above can be made much more quantitative. One feature that we want to retain when we say something like "$y = f(x)$ approaches $y = g(x)$ as $x \to \infty$" is that, to be useful, $g(x)$ should be simpler than $f(x)$, where "simpler" will depend on the context. This is a familiar idea; for example, we can approximate a smooth curve near a chosen point on the curve by the tangent line through that point.

When $\lim_{x\to x_0}[f(x)/g(x)] = 1$, we write $f(x) \sim g(x)$ as $x \to x_0$, and we say that $g(x)$ is an *asymptotic approximation* to $f(x)$ as $x \to x_0$. For example, $\sinh x \sim x$ as $x \to 0$ and $\tanh x \sim 1$ as $x \to \infty$. A famous asymptotic approximation of this kind is Stirling's formula from 1730: $n! \sim (2\pi n)^{1/2}(n/e)^n$ as $n \to \infty$.

According to our definition, we have $e^x \sim 1$, $e^x \sim 1 + x$, and $e^x \sim 1 + 2x$, all as $x \to 0$. On the other hand, we have the Maclaurin expansion, $e^x = 1 + x + \frac{1}{2}x^2 + \cdots$, which converges for all $x$; truncating this infinite series gives good approximations to $e^x$ near $x = 0$, and these approximations improve if we take more terms in the series. This suggests that we should select $1+x$ and not $1 + 2x$, so our definition of "$\sim$" is too crude. We want asymptotic approximations to be approximations, and we want to be able to improve them by taking more terms, if possible. With this in mind, suppose we have a sequence of functions, $\phi_n(x)$, $n = 0, 1, 2, \dots$, with the property that $\phi_{n+1}(x)/\phi_n(x) \to 0$ as $x \to x_0$. Standard examples are $\phi_n(x) = x^n$ as $x \to 0$ and $\phi_n(x) = x^{-n}$ as $x \to \infty$. Let $R_N(x) = \sum_{n=0}^{N} a_n\phi_n(x)$ for some coefficients $a_n$. We write

$$f(x) \sim \sum_{n=0}^{\infty} a_n\phi_n(x) \quad \text{as } x \to x_0,$$

and say that the series is an *asymptotic expansion* of $f(x)$ as $x \to x_0$ when, for each $N = 0, 1, 2, \dots,$

$$[f(x) - R_N(x)]/\phi_N(x) \to 0 \quad \text{as } x \to x_0. \qquad (1)$$

In words, the "error" $f - R_N$ is comparable to the first term omitted, the one with $n = N + 1$. Note that the definition does not require the infinite series to be convergent (so that $R_N(x)$ may not have a limit as $N \to \infty$ for fixed $x$). Instead, for each fixed $N$, we impose a requirement on the error as $x \to x_0$, namely (1).

Asymptotic approximations may be convergent. For example, we have $e^x \sim 1+x+\frac{1}{2}x^2+\cdots$ as $x \to 0$. However, many interesting and useful asymptotic expansions are divergent. As an example, the *complementary error function*

$$\begin{aligned}\operatorname{erfc}(x) &= \frac{2}{\sqrt{\pi}}\int_x^{\infty} e^{-t^2}\,dt \\ &\sim \frac{e^{-x^2}}{x\sqrt{\pi}}\left[1 + \sum_{n=1}^{\infty}(-1)^n\frac{1\cdot 3\cdots(2n-1)}{(2x^2)^n}\right]\end{aligned}$$

as $x \to \infty$, where the series is obtained by repeated integration by parts of the defining integral. The series is divergent, but taking a few terms gives a good approximation to $\operatorname{erfc} x$, an approximation that improves as $x$ becomes larger.

Many techniques have been devised for obtaining asymptotic expansions. Some are designed for functions defined by integrals (such as $\operatorname{erfc} x$), others for functions that solve differential equations. Asymptotic methods can also be used to estimate the complexity of an algorithm or to predict properties of large combinatorial structures and complex networks.

In many cases of interest, the function to be estimated depends on parameters. Their presence may be benign, but complications can arise. Suppose we are interested in $f(x, \lambda)$ when both $x$ and the parameter $\lambda$ are large. Using standard methods, we may be able to show that $f(x, \lambda) \sim g(x, \lambda)$ as $x \to \infty$ for fixed $\lambda$, and that $f(x, \lambda) \sim h(x, \lambda)$ as $\lambda \to \infty$ for fixed $x$. Then, the natural question to ask is: if we estimate $g(x, \lambda)$ for large $\lambda$ and $h(x, \lambda)$ for large $x$, do we get the same answer? Simpler questions of this kind arise when investigating the commutativity of limits: given that

$$L_1 = \lim_{x \to \infty} \lim_{y \to \infty} F(x, y), \qquad L_2 = \lim_{y \to \infty} \lim_{x \to \infty} F(x, y),$$

when does $L_1 = L_2$? In our context, with, for example, $f(x, \lambda) = (x + \lambda)/(x + 2\lambda)$, $f \sim 1$ as $x \to \infty$ but $f \sim \frac{1}{2}$ as $\lambda \to \infty$, so standard asymptotic approximations may fail when there are two or more variables. In these situations, *uniform asymptotic expansions* are needed. Inevitably, these are more complicated, but they are often needed to gain a full understanding of certain physical phenomena. For example, consider the shadow of an illuminated sphere on a flat screen. It is a dark circular disk. However, close inspection shows that the shadow boundary is not sharp: the image changes rapidly but smoothly as we cross the circular boundary. Standard asymptotic techniques explain what is happening in the illuminated and shadow regions (the wavelength of light is much smaller than the radius of the sphere, so their small ratio can be exploited in building asymptotic approximations), but uniform asymptotic approximations are needed to explain the transition between the two.

For more on asymptotic methods, see the articles on PERTURBATION THEORY AND ASYMPTOTICS [IV.5], SPECIAL FUNCTIONS [IV.7], and DIVERGENT SERIES: TAMING THE TAILS [V.8]. Another good source is chapter 2 of *NIST Handbook of Mathematical Functions* (Cambridge University Press, Cambridge, 2010), edited by F. W. J. Olver, D. W. Lozier, R. F. Boisvert, and C. W. Clark. (An electronic version of the book is available at http://dlmf.nist.gov.)

## II.2 Boundary Layer

The mathematical idea of a boundary layer emerged from an analysis of a physical problem arising in FLUID DYNAMICS [IV.28]; the flow of a viscous fluid past a rigid body. Suppose that the body has diameter $L$ and $U$ is the speed of the flow far from the body. The *Reynolds number* for the flow is $Re = UL/\nu$, where $\nu$ is the dynamic viscosity coefficient. Assume that the fluid is not very viscous (meaning that $Re \gg 1$) so that the flow can be well approximated by solving the governing equations for a nonviscous fluid, the EULER EQUATIONS [III.11]. This approximation is not good near the surface of the body. There, viscous effects are important, so the NAVIER–STOKES EQUATIONS [III.23] must be used. Nevertheless, good approximations can be constructed by exploiting the fact that $Re \gg 1$. This is done within a thin viscous *boundary layer*. The two solutions, one in the boundary layer and one in the outer region, are matched in a region where both are assumed to be valid, and we therefore obtain a good approximation everywhere in the fluid. This whole scheme was first described and implemented by Ludwig Prandtl in 1904.

The idea of joining two solutions together is part of an extensive suite of asymptotic techniques in which large (or small) dimensionless parameters are identified and exploited. There can be more than two regions and more than one large parameter. Sophisticated matching procedures have been developed and applied to many complicated problems, not just those arising in fluid dynamics. See the article on PERTURBATION THEORY AND ASYMPTOTICS [IV.5].

## II.3 Chaos and Ergodicity
*Paul Glendinning*

The term *chaos* is used with greater or lesser degrees of precision to describe deterministic dynamics that is nonperiodic and unpredictable. The word was first used in this context by Li and Yorke in the title of their article "Period three implies chaos" in 1975. Li and Yorke were careful not to define the term too narrowly and described their mathematical results using clearly defined properties of solutions. In one of the early textbooks on the subject (published in 1989), Devaney defined chaos for discrete maps $T$ on a metric space $X$ with $x_{n+1} = T(x_n)$, $x_n \in X$. Devaney's definition is that $T \colon X \to X$ is chaotic on $X$ if

(i) periodic orbits are dense in $X$,
(ii) the dynamics is transitive (so for any points $x, y \in X$ it is possible to find a point in $X$ that is arbitrarily close to $x$ whose orbit passes arbitrarily close to $y$), and
(iii) the system has sensitive dependence on initial conditions (*SDIC*) on $X$.

SDIC means that there exists a precision $\varepsilon > 0$ such that for every $x \in X$ and $\delta > 0$ there exists a point $y \in X$ within a distance $\delta$ of $x$ and a time $n$ such that solutions started at $x$ and $y$ are at least a distance $\varepsilon$ apart after time $n$. SDIC is often described as implying a loss of predictability, or a loss of memory of initial conditions, although it is a little more complicated than this suggests. Devaney's definition was intended to make it possible to ask sensible examination questions about chaos at undergraduate level, and by 1992 two groups had shown that (for appropriate systems) the first two conditions imply the third, suggesting it is not an ideal definition!

Another commonly used definition of chaos uses the idea of Lyapunov exponents. These measure the asymptotic local expansion properties of solutions, so if $(x_n)$ is an orbit of the one-dimensional map $x_{n+1} = T(x_n)$ and $T$ is differentiable, then the Lyapunov exponent of $x_0$ is

$$\lambda(x_0) = \lim_{n\to\infty} \frac{1}{n} \sum_{k=0}^{n-1} \log |T'(x_k)|$$

provided the limit exists. The dynamics is regarded as chaotic if it is not "simple" (a union of periodic points, for example) and typical points have positive Lyapunov exponents.

Issues around definition notwithstanding, the idea of chaos has radically changed the choices that applied mathematicians make when modeling systems with complicated temporal behavior. In particular, chaos theory shows that apparently random behavior may have a deterministic origin (see also DYNAMICAL SYSTEMS [IV.20] and THE LORENZ EQUATIONS [III.20]).

The analogy with probability theory is made mathematically precise using *ergodic theory*, which connects time averages of a quantity along an orbit with an expected value. This expected value is a spatial integral, and hence it is independent of the initial conditions.

The definition of ergodicity uses two ideas: invariant sets and invariant measures. An *invariant set* $X$ of a map $T$ satisfies $X = T^{-1}(X)$, and a map $T$ with an invariant set $X$ has an *invariant probability measure* if there is a measure $\mu$ such that $\mu(S) = \mu(T^{-1}(S))$ for all measurable sets $S$ in $X$ and $\mu(X) = 1$.

The map $T$ is *ergodic* with respect to an invariant measure $\mu$ if $\mu(S)$ is either zero or one for all invariant sets $S$. In other words, every invariant set is either very small (measure zero) or it is essentially (up to sets of measure zero) the whole set $X$. This means that the dynamics is not decomposable into smaller sets, a role played by the transitive property in Devaney's definition of chaos.

Ergodic maps have nice averaging properties. For all integrable functions $f$ on $X$ the time average of $f$ along typical orbits equals the spatial average obtained by integrating $X$ with respect to the invariant measure:

$$\lim_{n\to\infty} \frac{1}{n} \sum_{k=0}^{n-1} f(T^k(x)) = \int_X f \, d\mu$$

for almost all $x \in X$ (with respect to the invariant measure $\mu$). Thus $\mu$ can be interpreted as a probability measure of the dynamics, and if $d\mu = h(x)\,dx$, then $h$ can be interpreted as a sort of probability density function for the iterates of points under $T$. The "typical" Lyapunov exponent of a one-dimensional map can then be interpreted as the integral of $f = \log|T'|$ with respect to the invariant measure.

### Further Reading


Devaney, R. L. 1989. *An Introduction to Chaotic Dynamical Systems*. Reading, MA: Addison-Wesley.

Li, T.-Y., and J. A. Yorke. 1975. Period three implies chaos. *American Mathematical Monthly* 82:985–92.

Walters, P. 1982. *An Introduction to Ergodic Theory*. New York: Springer.


## II.4 Complex Systems

*Paul Glendinning*

Complex systems are dynamical models that involve the interaction of many components. The study of these systems is called complexity theory. Complex systems are characterized by having large dimension and complicated interactions.

There are many examples of complex systems across the sciences. The reactions between chemicals in a living cell can be considered as a complex system, where the variables are the concentrations of chemicals and the dynamics is defined by the chemical reactions that can take place. A chemical can react with only a small set of the chemicals present, so the connectivity is generally low.

Other examples include the Internet, where individual computers are linked together in complex ways, and electronic systems whose components are connected on a circuit board. Agent-based models of human behavior such as opinion formation within groups or the movement of crowds are also complex systems.

It is often hard to do much more than numerical simulations to determine the behavior of systems, leading to a somewhat limited phenomenological description of their behavior. There are two central methods in the study of complex systems that go further than this, though again with limited concrete predictive power. These are graph theory to characterize how components influence each other, and dimension reduction methods to capture (where applicable) any lower-dimensional approximations that determine the evolution of the system.

If the system has real variables $x_i$, $i = 1, \dots, N$, then each variable can be identified with the node of a graph labeled by $i$, with an edge from $i$ to $j$ if the dynamics of $x_j$ is directly influenced by $x_i$ (see GRAPH THEORY [II.16]). For example, if the evolution is determined by a differential equation then $\dot{x}_i = f_i(x_1, x_2, \dots, x_N)$, but not every variable need appear explicitly in the argument of $f_i$, so there is an edge from $x_j$ to $x_i$ only if $\partial f_i / \partial x_j$ is not identically zero. This graph can be represented by an adjacency matrix $(a_{ij})$ with $a_{ij} = 1$ if there is an edge from $i$ to $j$ and $a_{ij} = 0$ otherwise. The degree of a node is the number of edges at the node (this can be split into the in-degree (respectively, out-degree) if only edges ending (respectively, starting) at the node are counted). The proportion of nodes with degree $k$ is the degree distribution of the network. Properties of the degree distribution are often used to characterize the network. For example, if the degree distribution obeys a power law, the network is said to be scale free (the Internet is supposedly of this type; see NETWORK ANALYSIS [IV.18]).

By analyzing subgraphs of biological models it was found that some subgraphs appear in examples much more often than would be expected on the basis of a statistical analysis. This has led to the conjecture that these *motifs* may have associated functional properties.

In many complex systems the individual components of the system behave according to very simple, though often nonlinear, rules. For example, a bird in a flock may change its direction of flight as a function of the average direction of flight of nearby birds. Although this is a local rule, the effect across the entire flock of birds is to produce coherent movement of the flock as a whole. This effect, whereby simple local rules lead to interesting global results, is called *emergent behavior*. The emergent behavior resulting from given local rules is often unclear until the system is simulated numerically.

In some cases the dimension of the problem can be reduced, so fewer variables need to be considered, making the system easier to simulate and more amenable to analysis. The methods of dimension reduction often rely on SINGULAR VALUE DECOMPOSITION [II.32] techniques to identify the more dynamically active directions in phase space, and then an attempt is made to project the system onto these directions and analyze the resulting system.

In some systems the mean-field theory of theoretical physics can be used to understand collective behavior.

Since complexity theory encompasses so many different models, the range of possible dynamic phenomena is vast, even before further complications such as stochastic effects or network evolution are included. Complex systems describing neuron interactions in the brain can model pattern recognition and memory (see MATHEMATICAL NEUROSCIENCE [VII.21]). Numerical models of partial differential equations are complex systems, and the dynamical behavior can include synchronization, in which all components lock on to a similar pattern of behavior, and PATTERN FORMATION [IV.27]. Different parts of the system may behave in dynamically different ways, with regions of frustration (or fronts) separating them. Interactions may have different strengths, leading to different timescales in the problem. This is particularly true of many biological models and adds to the difficulty of modeling phenomena accurately.

### Further Reading

Ball, R., V. Kolokoltsov, and R. S. MacKay, eds. 2013. *Complexity Science*. Cambridge: Cambridge University Press.

Estrada, E. 2011. *The Structure of Complex Networks: Theory and Applications*. Oxford: Oxford University Press.

Watts, D. J. 1999. *Small Worlds: The Dynamics of Networks Between Order and Randomness*. Princeton, NJ: Princeton University Press.

## II.5 Conformal Mapping

*Darren Crowdy*

### 1 What Is a Conformal Mapping?

Conformal mapping is the name given to the idea of interpreting an analytic function of a complex variable in a geometric fashion. Let $z = x + iy$ and suppose that another complex variable $w$ is defined by

$$w = f(z) = \phi(x, y) + i\psi(x, y),$$

where $\phi$ and $\psi$ are, respectively, the real and imaginary parts of some function $f(z)$, an analytic function of $z$. One can think of this relation as assigning a correspondence between points in the complex $z$-plane and points in the complex $w$-plane. Under this function a designated region of the $z$-plane is transplanted, or "mapped," to some region in the $w$-plane, as illustrated in figure 1. The shape of the image will depend on $f$. The fact that $f$ is an *analytic* function implies certain special properties of this mapping of regions. If the mapping is to be one-to-one, then a necessary, but not sufficient, condition is that the derivative $f'(z) = \mathrm{d}f/\mathrm{d}z$ does not vanish in the $z$-region of interest.

A simple example is the *Cayley mapping*

$$w = f(z) = \frac{1+z}{1-z}.$$

This maps the interior of the unit disk $|z| < 1$ in the $z$-plane to the right half $w$-plane $\mathrm{Re}\, w > 0$. The point $z = 1$ maps to $w = \infty$, and $z = -1$ maps to $w = 0$. The unit circle $|z| = 1$ maps to the imaginary $w$-axis. Conformal mappings clearly preserve neither area nor perimeters; their principal geometrical feature is that they locally preserve angles. To see this, note that since $f(z)$ is analytic at a point $z_0$, it has a local Taylor expansion there:

$$w = f(z_0) + f'(z_0)(z - z_0) + \cdots .$$

If $\delta z = z - z_0$ is an infinitesimal line element through $z_0$ in the $z$-plane, its image $\delta w$ under the mapping defined as $\delta w = w - w_0$, where $w_0 = f(z_0)$, is, to leading order,

$$\delta w \approx f'(z_0)\delta z.$$

But $f'(z_0)$ is just a nonzero complex number so, under a conformal mapping, all infinitesimal line elements through $z_0$ are transplanted to line elements through $w_0$ in the $w$-plane that are simply rescaled by the modulus of $f'(z_0)$ and rotated by its argument. In particular, the angle between two given line elements through $z_0$ is preserved by the mapping.

## 2 The Riemann Mapping Theorem

The Riemann mapping theorem is considered by many to be the pinnacle of achievement of nineteenth-century mathematics. It is an existence theorem: it states that there exists a conformal mapping from the unit $z$-disk to *any* given simply connected region (no holes) in the $w$-plane, so long as it is not the entire plane.

## 3 Conformal Invariance

One reason why conformal mappings are an important tool in applied mathematics is the property of *conformal invariance* of certain boundary-value problems

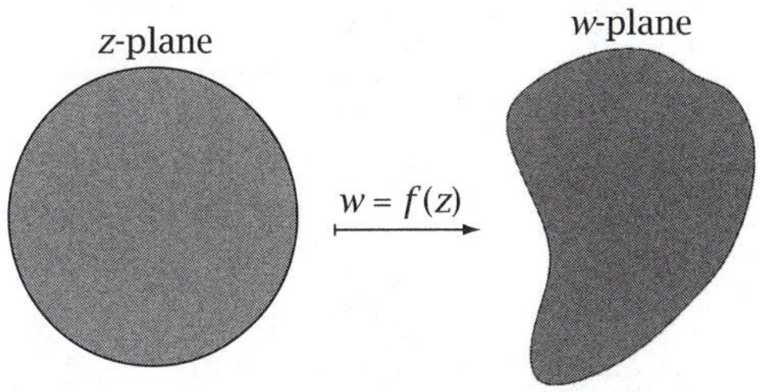


**Figure 1** A conformal mapping from a region in a complex $z$-plane to a region in a complex $w$-plane.

arising in applications. An example is the boundary-value problem determining Green's function $G(z; z_0)$ for the Laplace equation in a region $D$ in $\mathbb{R}^2$ with boundary $\partial D$, which can be written as

$$\nabla^2 G = \delta^{(2)}(z - z_0) \quad \text{in } D \text{ with } G = 0 \text{ on } \partial D,$$

where $z_0$ is some point inside $D$ and $\delta^{(2)}$ is the two-dimensional DIRAC DELTA FUNCTION [III.7]. The Green function for the unit disk $|z| < 1$ is known to be

$$G(z; z_0) = \mathrm{Im}\left[\frac{\mathrm{i}}{2\pi}\log\left(\frac{z - z_0}{|z_0|(z - 1/\overline{z_0})}\right)\right],$$

where $\overline{z_0}$ is the complex conjugate of $z_0$. Now if $D$ is any other simply connected region of a complex $w$-plane, the corresponding Green function in $D$ is nothing other than $G(f^{-1}(w); f^{-1}(w_0))$, where $f^{-1}(w)$ is the inverse function of the conformal mapping taking the unit $z$-disk to $D$. Geometrically, $f^{-1}(w)$ is just the inverse conformal mapping transplanting $D$ to the unit disk $|z| < 1$. The Green function in any simply connected region $D$ is therefore known immediately provided the conformal mapping between $D$ and the unit disk can be found.

## 4 Schwarz–Christoffel Mappings

The Riemann mapping theorem is nonconstructive and, while the existence of a conformal mapping between given simply connected regions is guaranteed, the practical matter of actually constructing it is another story. One of the few general constructions often used in applications is the *Schwarz–Christoffel mapping*. This is a conformal mapping from a standard region such as the unit $z$-disk $|z| < 1$ to the region interior or exterior to an $N$-sided polygon. At the preimage of any vertex of the polygon (a *prevertex*), the local argument outlined earlier demonstrating the preservation of angles between infinitesimal line elements must fail. Indeed, at any such prevertex it can be argued that the derivative $f'(z)$ of the conformal mapping must have a simple

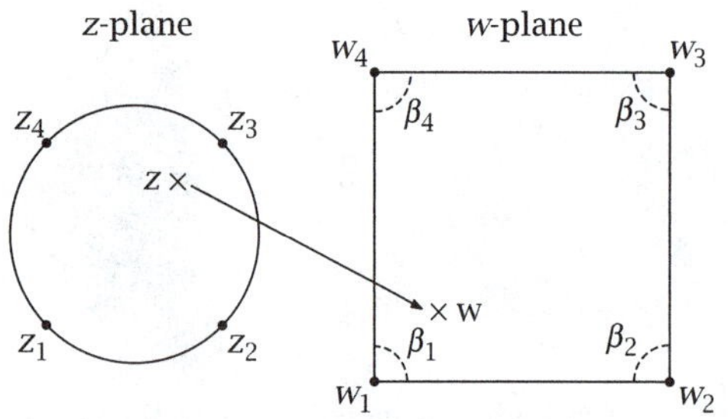


**Figure 2** A Schwarz–Christoffel mapping from the unit $z$-disk to the interior of a square in a $w$-plane. A function of the form (1) with $N = 4$ identifies a point $w$ with a point $z$. Here, $\beta_1 = \beta_2 = \beta_3 = \beta_4 = \pi/2$.

zero, a simple pole, or a branch point singularity. The general formula for a mapping from $|z| < 1$ to the interior of a bounded polygon in a $w$-plane is

$$w = f(z) = A + B\int^z \prod_{k=1}^{N}\left(1 - \frac{z'}{z_k}\right)^{(\beta_k/\pi - 1)} \mathrm{d}z', \quad (1)$$

while the formula for a mapping from $|z| < 1$ to the exterior of a bounded polygon in a $w$-plane, with $z = 0$ mapping to $w = \infty$, is

$$w = f(z) = A + B\int^z \prod_{k=1}^{N}\left(1 - \frac{z'}{z_k}\right)^{(\beta_k/\pi - 1)} \frac{\mathrm{d}z'}{z'^2}. \quad (2)$$

The parameters $\{\beta_k \mid k = 1, 2, \dots, N\}$ are the *turning angles* shown in figure 2; the points $\{z_k \mid k = 1, 2, \dots, N\}$ are the prevertices. $A$ and $B$ are complex constants. These so-called *accessory parameters* are usually computed numerically by fixing geometrical features such as ensuring that the sides of the polygon have the required length. A famous mapping of Schwarz–Christoffel type known for its use in aerodynamics is the *Joukowski mapping*,

$$w = f(z) = \frac{1}{2}\left(z + \frac{1}{z}\right),$$

which maps the unit disk $|z| < 1$ to the infinite region exterior to a flat plate, or airfoil, lying on the real $w$-axis between $w = -1$ and $w = 1$. It is a simple matter to derive it from (2) with the prevertices $z_1 = 1$, $z_2 = -1$ and turning angles $\beta_1 = \beta_2 = 2\pi$. Since it is natural, given any two-dimensional shape, to approximate it by taking a set of points on the boundary and joining them with straight line segments to form a polygon, the Schwarz–Christoffel formula has found many uses in applied mathematics. Versatile numerical software to compute the accessory parameters has also been developed.

### Further Reading

Courant, R. 1950. *Dirichlet's Principle, Conformal Mapping and Minimal Surfaces*. New York: Interscience.

Driscoll, T. A., and L. N. Trefethen. 2002. *Schwarz–Christoffel Mapping*. Cambridge: Cambridge University Press.

Nehari, Z. 1975. *Conformal Mapping*. New York: Dover.

## II.6 Conservation Laws

*Barbara Lee Keyfitz*

### 1 Quasilinear Hyperbolic Partial Differential Equations

A system of first-order partial differential equations (PDEs) in the form

$$\boldsymbol{u}_t + \sum_{i=1}^{d} A_i(x, t, \boldsymbol{u})\boldsymbol{u}_{x_i} + \boldsymbol{b}(x, t, \boldsymbol{u}) = 0, \quad (1)$$

where $\boldsymbol{u} \in \mathbb{R}^n$, $\boldsymbol{b} \in \mathbb{R}^n$, the $A_i$ are $n \times n$ matrices, and $\boldsymbol{u}_t \equiv \partial\boldsymbol{u}/\partial t$ and $\boldsymbol{u}_{x_i} \equiv \partial\boldsymbol{u}/\partial x_i$, is said to be *quasilinear*; the system is nonlinear as defined in the article PARTIAL DIFFERENTIAL EQUATIONS [IV.3], but the terms containing derivatives of $\boldsymbol{u}$ appear only in linear combination. Identifying $t$ as a time variable and $x = (x_1, \dots, x_d)$ as a space variable, the *Cauchy problem* asks for a solution to (1) for $t > 0$ with the initial condition

$$\boldsymbol{u}(x, 0) = \boldsymbol{u}_0(x). \quad (2)$$

By analogy with the theory of linear PDEs, one expects this problem to be well-posed only if the system is *hyperbolic*, which means that all the roots $\tau(\xi)$ (known as *characteristics*) of the polynomial equation

$$\det\left(\tau I + \sum_{i=1}^{d} A_i\xi_i\right) = 0 \quad (3)$$

are real for all $\xi \in \mathbb{R}^d$ and, as eigenvalues of the matrix $\sum_{i=1}^{d} A_i\xi_i$, each has equal ALGEBRAIC AND GEOMETRIC MULTIPLICITIES [II.22].

In 1974 Fritz John showed that if $d = 1$, and the system is *genuinely nonlinear* (meaning that $\nabla_\xi\tau_i \cdot \boldsymbol{r}_i \neq 0$ for each root $\tau_i$ of (3) and corresponding eigenvector $\boldsymbol{r}_i$), then for smooth Cauchy data at least one component of $\nabla\boldsymbol{u}$ tends to infinity in finite time, exactly as in the BURGERS EQUATION [III.4] (see also PARTIAL DIFFERENTIAL EQUATIONS [IV.3 §3.6]).

Characteristics in hyperbolic systems define the speed of propagation of signals in specific directions (normal to $\xi$), so genuine nonlinearity says that this speed is a nontrivial function of the state $\boldsymbol{u}$. This has physical significance as a description of the phenomena modeled by conservation laws, and it has mathematical implications for the existence of smooth solutions. Specifically, the behavior seen in solutions of the Burgers equation typifies solutions of genuinely nonlinear hyperbolic systems.

Furthermore, despite the fact that DISTRIBUTION SOLUTIONS [IV.3 §5.2] are well defined for linear hyperbolic equations, the concept fails for quasilinear systems since, in the first place, $A_i$ and $A_i\boldsymbol{u}_{x_i}$ are not defined if $\boldsymbol{u}$ lacks sufficient smoothness, and, in the second, the standard procedure of creating the weak form of an equation (multiply by a smooth test function and integrate by parts) does not usually succeed in eliminating $\nabla\boldsymbol{u}$ from the system when $A = A(\boldsymbol{u})$ depends in a nontrivial way on $\boldsymbol{u}$. The exception is when each $A_i\boldsymbol{u}_{x_i}$ is itself a derivative: $A_i\boldsymbol{u}_{x_i} = \partial_{x_i} f_i(\boldsymbol{u})$. This happens if each row of each $A_i$ is a gradient, and that happens only if the requisite mixed partial derivatives are equal. In this case, we have a *system of balance laws*:

$$\boldsymbol{u}_t + \sum_{i=1}^{d} (f_i(x,t,\boldsymbol{u}))_{x_i} + \boldsymbol{b}(x,t,\boldsymbol{u}) = 0. \tag{4}$$

In the important case in which $\boldsymbol{b} \equiv 0$, we have a system of *conservation laws*. The weak form of (4) is

$$\iint \Big[\boldsymbol{u}\varphi_t + \sum_{i=1}^{d} (f_i(x,t,\boldsymbol{u}))\varphi_{x_i} - \boldsymbol{b}(x,t,\boldsymbol{u})\varphi\Big]\,\mathrm{d}x\,\mathrm{d}t = 0. \tag{5}$$

Since this is the only case in which solutions to (1) can be unambiguously defined, the subject of quasilinear hyperbolic systems is often referred to as "conservation laws."

A mathematical challenge in conservation laws is to find spaces of functions that are inclusive enough to admit weak solutions for general classes of conservation laws but regular enough that solutions and their approximations can be analyzed. At this time, there is a satisfactory well-posedness theory only in a single space dimension.

## 2 How Conservation Laws Arise

Problems of importance in physics, engineering, and technology lead to systems of conservation laws; a sample selection of these problems follows.

### 2.1 Compressible Flow

The basic equations of compressible fluid flow, derived from the principles of conservation of mass, momentum, and energy, along with constitutive equations relating thermodynamic quantities, take the form

$$\left.\begin{aligned} \rho_t + \operatorname{div}(\rho\boldsymbol{u}) &= 0,\\ (\rho\boldsymbol{u})_t + \operatorname{div}(\rho\boldsymbol{u}\otimes\boldsymbol{u}) + \nabla p &= 0,\\ (\rho E)_t + \operatorname{div}(\rho\boldsymbol{u}H) &= 0, \end{aligned}\right\} \tag{6}$$

where $\rho$ represents density, $\boldsymbol{u}$ velocity, $p$ pressure, $E$ energy, and $H$ enthalpy, with

$$E = \tfrac{1}{2}|\boldsymbol{u}|^2 + \frac{1}{\gamma-1}\frac{p}{\rho}, \qquad H = \gamma E,$$

and $\gamma$ a constant that depends on the fluid ($\gamma = 1.4$ for air). To obtain the first equation in (6), one notes that the total amount of mass in an arbitrary control volume $D$ is the integral over $D$ of the density, and this changes in time if there is flux through the boundary $\Gamma$ of $D$. Furthermore, the flux is precisely the product of the density and the velocity normal to that boundary, from which we obtain

$$\frac{\mathrm{d}}{\mathrm{d}t}\iint_D \rho\,\mathrm{d}V = -\int_\Gamma \rho\boldsymbol{u}\cdot\nu\,\mathrm{d}A. \tag{7}$$

(The negative sign will remind the reader of the convention that $\nu$ is the outward normal, and flow out of $D$ will decrease the mass contained in $D$.) Interchanging differentiation and integration on the left in (7), along with an application of the DIVERGENCE THEOREM [I.2 §24] on the right, immediately yields

$$\iint_D (\rho_t + \operatorname{div}(\rho\boldsymbol{u}))\,\mathrm{d}V = 0. \tag{8}$$

Finally, the observation that $D$ is an arbitrary domain in the region allows one to pass to the infinitesimal version in (6). The integral version (8) also justifies the weak form (5), since if (8) holds on arbitrary domains then it is possible to form weighted averages with arbitrary differentiable functions $\varphi$ and to integrate by parts, which produces (5).

In compressible flow, the speed of sound is finite; in (6) it is one of the characteristics. Steady flow at speeds that exceed the speed of sound also gives a hyperbolic system of conservation laws (6) with the time derivatives absent. In this case, the hyperbolic direction (the time-like variable) is given by the flow direction.

Conservation principles also lead to equations for ELASTICITY [IV.26 §3.3] and MAGNETOHYDRODYNAMICS [IV.29]. Industrial applications include continuum models for multiphase flow (e.g., water mixed with steam in nuclear reactor cooling systems, or multicomponent flows in oil reservoirs).

The necessity of solving, or at least approximating, conservation laws for many of these applications has resulted in extensive techniques for numerical simulation of solutions, even when existence of solutions remains an open question.

### 2.2 Chromatography

Chromatography is a widely used industrial process for separating chemical components of a mixture by differential adsorption on a substrate. Modeling a chromatographic column leads to a system of conservation laws in a single space variable that takes the form

$$\boldsymbol{c}_x + (\boldsymbol{f}(\boldsymbol{c}))_t = 0,$$

where $\boldsymbol{c} = (c_1, \dots, c_n)$ is a vector of component concentrations and $\boldsymbol{f}$ is the equilibrium column isotherm. A common model for $\boldsymbol{f}$ uses the Langmuir isotherm and gives, with positive parameters $\alpha_i$ measuring the relative adsorption rates,

$$f_i = c_i + \frac{\alpha_i c_i}{1 + \sum c_j}, \quad 1 \leqslant i \leqslant n.$$

### 2.3 Other Models

Many other physical phenomena lead naturally to conservation laws. For example, a continuum model for vehicular traffic on a one-way road is the scalar equation

$$u_t + q(u)_x = 0,$$

where $u$ represents the linear density of traffic and $q(u) = uv(u)$ the flux, where $v$ is velocity. As in (7), this equation is a conservation law, the "law of conservation of cars." This model assumes that the velocity at which traffic moves depends only on the traffic density. Although this model is too simple to be of much practical use, it is appealing as a pedagogical tool. Adaptations of it are of interest in current research.

### Further Reading

Bellomo, N., and C. Dogbe. 2011. On the modeling of traffic and crowds: a survey of models, speculations, and perspectives. *SIAM Review* 53:409–63.

Courant, R., and K. O. Friedrichs. 1948. *Supersonic Flow and Shock Waves*. New York: Wiley-Interscience.

Dafermos, C. M. 2000. *Hyperbolic Conservation Laws in Continuum Physics*. Berlin: Springer.

Godlewski, E., and P.-A. Raviart. 1996. *Numerical Approximation of Hyperbolic Systems of Conservation Laws*. New York: Springer.

John, F. 1974. Formation of singularities in one-dimensional nonlinear wave propagation. *Communications on Pure and Applied Mathematics* 27:377–405.

Rhee, H.-K., R. Aris, and N. R. Amundson. 1989. *First-Order Partial Differential Equations: Volume II. Theory and Application of Hyperbolic Systems and Quasilinear Equations*. Englewood Cliffs, NJ: Prentice-Hall.

Whitham, G. B. 1974. *Linear and Nonlinear Waves*. New York: Wiley-Interscience.

## II.7 Control

A *system* is a collection of objects that interact and produce various outputs in response to different inputs. Systems arise in a wide variety of situations and include chemical plants, cars, the human body, and a country's economy. Control problems associated with these systems include the production of a chemical, control of self-driving cars, the regulation of bodily functions such as temperature and heartbeat, and the control of debt. In each case one wants to have a way of controlling these processes automatically without direct human intervention.

A general control system is depicted in figure 1. The state of the system is described by $n$ *state variables* $x_i$, and these span the *state space*. In general, the $x_i$ cannot be observed or measured directly, but $p$ *output variables* $y_i$, which depend on the $x_i$, are known. The system is controlled by manipulating $m$ *control variables* $u_i$.

The system might be expressed as a system of difference equations (discrete time) or differential equations (continuous time). In the latter case a linear, time-invariant control problem takes the form

$$\frac{\mathrm{d}x}{\mathrm{d}t} = Ax(t) + Bu(t),$$
$$y(t) = Cx(t) + Du(t),$$

where $A$, $B$, $C$, and $D$ are $n \times n$, $n \times m$, $p \times n$, and $p \times m$ matrices, respectively. This is known as a *state-space system*. In some cases an additional $n \times n$ matrix $E$, which is usually singular, premultiplies the $\mathrm{d}x/\mathrm{d}t$ term; these so-called *descriptor systems* or *generalized state-space systems* lead to DIFFERENTIAL-ALGEBRAIC EQUATIONS [I.2 §12].

A natural question is whether, given a starting value $x(0)$, the input $u$ can be chosen so that $x$ takes a given value at time $t$. Questions of this form are fundamental in classical control theory.

If feedback occurs from the outputs or state variables to the controller, then the system is called a *closed-loop system*. In *output feedback*, illustrated in figure 1, $u$ depends on $y$, while in *state feedback* $u$ depends on $x$.

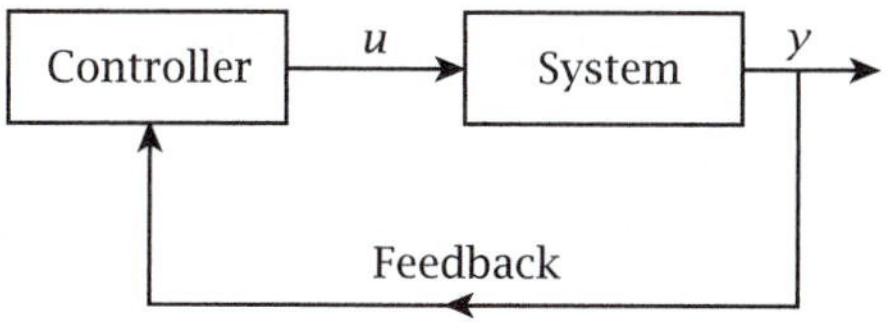


**Figure 1** Control system.

For example, in state feedback we may have $u = Fx$, for some $m \times n$ matrix $F$. Then $\mathrm{d}x/\mathrm{d}t = (A + BF)x$, which leads to questions about what properties the matrix $A + BF$ can be given by suitable choice of $F$.

For more details, see CONTROL THEORY [IV.34].

## II.8 Convexity
*Didier Henrion*

The notion of convexity is central in applied mathematics. It is also used in everyday life in connection with the curvature properties of a surface. For example, an optical lens is said to be convex if it is bulging outward.

Convexity appears in ancient Greek geometry, e.g., in the description of the five regular convex space polyhedra (the platonic solids). Archimedes (ca. 250 B.C.E.) seems to have been the first to give a rigorous definition of convexity, similar to the geometric definition we use today: a set is convex if it contains all line segments between each of its points.

In his study of singularities of real algebraic curves, Newton (ca. 1720) introduced a convex polygon in the plane built from the exponents of the monomials of the polynomial defining the curve; this is known as the Newton polygon. Cauchy (ca. 1840) studied convex curves and remarked, for example, that, if a closed convex curve is contained in a circle, then its perimeter is smaller than that of the circle. Convex polyhedra were studied by Fourier (ca. 1825) in connection with the problem of the solvability of linear inequalities.

A central figure in the modern development of convexity is Minkowski, who was motivated by problems from number theory. In 1891 Minkowski proved that, in Euclidean space $\mathbb{R}^n$, every compact convex set with center at the origin and volume greater than $2^n$ contains at least one point with integer coordinates different from the origin. From Minkowski's work follows the classical isoperimetric inequality, which states that among all convex sets with given volume, the ball is the one with minimal surface area. In 1896 Minkowski considered systems of the form $Ax \geqslant 0$, where $A$ is a real $m \times n$ matrix and $x \in \mathbb{R}^n$. Together with the above-mentioned contribution by Fourier, this laid the groundwork for LINEAR PROGRAMMING [IV.11 §3], which emerged in the late 1940s, with key contributions by Kantorovich (1912–86) and Dantzig (1914–2005). In the second half of the twentieth century, convexity was developed further by Fenchel (1905–88), Moreau (1923–), and Rockafellar (1935–), among many others. Convexity is now a key notion in many branches of applied mathematics: it is essential in mathematical programming (to ensure convergence of optimization algorithms), functional analysis (to ensure existence and uniqueness of solutions of problems of calculus of variations and optimal control), geometry (to classify sets and their invariants, or to relate geometrical quantities), and probability and statistics (to derive inequalities).

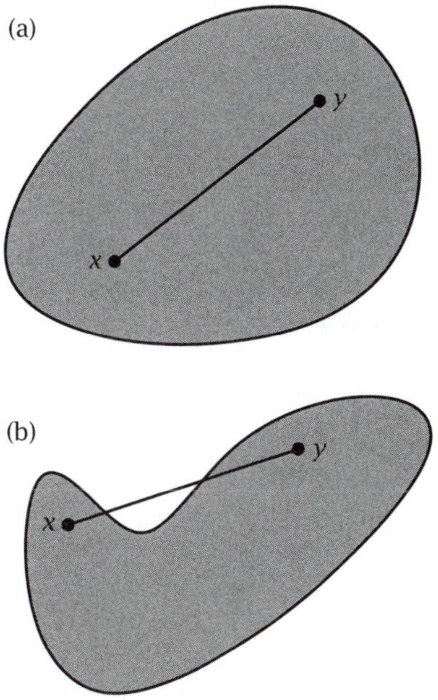


**Figure 1** (a) A convex set. (b) A nonconvex set.

Convex objects can be thought of as the opposite, geometrically speaking, to fractal objects. Indeed, fractal objects arise in maximization problems (sponges, lungs, batteries) and they have a rough boundary. In contrast, convex objects arise in minimization problems (isoperimetric problems, smallest energy) and they have a smoother boundary.

Mathematically, a set $X$ is *convex* if, for all $x, y \in X$ and for all $\lambda \in [0,1]$, $\lambda x + (1 - \lambda)y \in X$ (see figure 1). Geometrically, this means that the line segment between any two points of the set belongs to

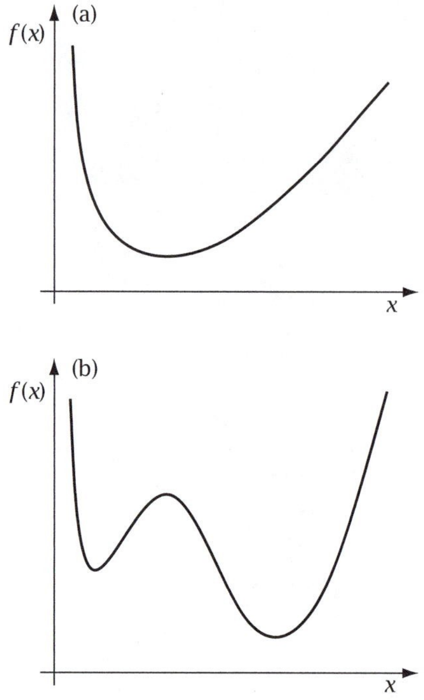


**Figure 2** (a) A convex function. (b) A nonconvex function.

the set. A real-valued function $f\colon X \to \mathbb{R}$ is *convex* if, for all $x, y \in X$ and for all $\lambda \in [0,1]$, it holds that $f(\lambda x + (1-\lambda)y) \leqslant \lambda f(x) + (1-\lambda)f(y)$. Geometrically, this means that the line segment between any two points on the graph of the function lies above the graph (see figure 2). This is the same as saying that the *epigraph* $\{(x,y)\colon x \in X,\ y \geqslant f(x)\}$ is a convex set. If a function is twice continuously differentiable, convexity of the function is equivalent to nonnegativity of the quadratic form of the matrix of second-order partial derivatives (the Hessian).

If the function $f$ is convex, then $f(\lambda_1 x_1 + \cdots + \lambda_m x_m) \leqslant \lambda_1 f(x_1) + \cdots + \lambda_m f(x_m)$ for all $x_1, \dots, x_m$ in $X$ and $\lambda$ in the $m$-dimensional unit simplex $\{\lambda \in \mathbb{R}^m\colon \lambda_1 + \cdots + \lambda_m = 1,\ \lambda_1 \geqslant 0, \dots, \lambda_m \geqslant 0\}$. This is called Jensen's inequality, and more generally it can be expressed as $f(\int x\mu(\mathrm{d}x)) \leqslant \int f(x)\mu(\mathrm{d}x)$ for every probability measure $\mu$ supported on $X$, or equivalently as $f(E[x]) \leqslant E[f(x)]$, where $E$ denotes the expectation of a random variable.

A function $f$ is *concave* whenever the function $-f$ is convex. If a function $f$ is both convex and concave, it is affine. For this reason, convexity can sometimes be interpreted as a one-sided linearity, and in some instances (e.g., in problems of calculus of variations and partial differential equations), nonlinear convex functions behave similarly to linear functions.

A set $X$ is a *cone* if $x \in X$ implies $\lambda x \in X$ for all $\lambda \geqslant 0$. A convex cone is therefore a set that is closed under addition and under multiplication by positive scalars. Convex cones are central in optimization, and conic programming is the minimization of a linear function over an affine section of a convex cone. Important examples of convex cones include the linear cone (also called the positive orthant), the quadratic cone (also called the Lorentz cone), and the semidefinite cone (which is the set of nonnegative quadratic forms, or, equivalently, the set of positive-semidefinite matrices).

The *convex hull* of a set $X$ is the smallest closed convex set containing $X$, which is sometimes denoted by $\operatorname{conv} X$. If $X$ is the union of a finite number of points, then $\operatorname{conv} X$ is the polytope with vertices among these points. A theorem by Carathéodory states that given a set $X \subset \mathbb{R}^{n-1}$, every point of $\operatorname{conv} X$ can be expressed as $\lambda_1 x_1 + \cdots + \lambda_n x_n$ for some choice of $x_1, \dots, x_n$ in $X$ and $\lambda$ in the $n$-dimensional unit simplex.

A theorem of Minkowski (generalized to infinite-dimensional spaces in 1940 by Krein and Milman) states that every compact convex set is the closure of the convex hull of its extreme points (a point $x \in X$ is *extreme* if $x = (x_1 + x_2)/2$ for some $x_1, x_2 \in X$ implies $x_1 = x_2$). Finally, we mention the Brunn–Minkowski theorem, which relates the volume of the sum of two compact convex sets (all points that can be obtained by adding a point of the first set to a point of the second set) to the respective volumes of the sets.

### Further Reading


Ben-Tal, A., and A. Nemirovski. 2001. *Lectures on Modern Convex Optimization.* Philadelphia, PA: SIAM.

Berger, M. 2010. *Geometry Revealed: A Jacob's Ladder to Modern Higher Geometry.* Berlin: Springer.

Evans, L. C. 2010. *Partial Differential Equations*, 2nd edn. Providence, RI: American Mathematical Society.

Gruber, P. M. 1993. History of convexity. In *Handbook of Convex Geometry*, edited by P. M. Gruber and J. M. Wills, volume A, pp. 1–15. Amsterdam: North-Holland.

Hiriart-Urruty, J. B., and C. Lemaréchal. 2004. *Fundamentals of Convex Analysis.* Berlin: Springer.


## II.9 Dimensional Analysis and Scaling

*Daniela Calvetti and Erkki Somersalo*

Dimensional analysis makes it possible to analyze in a systematic way dimensional relationships between physical quantities defining a model. In order to explain how this works, we need to introduce some definitions and establish the notation that will be used below. Consider a system that contains $n$ physical quantities, $q_1, q_2, \dots, q_n$, that we believe to be relevant for describing the system's behavior, the quantities being expressed using $r$ fundamental units, or dimensions, denoted by $d_1, d_2, \dots, d_r$. The generally accepted SI unit system consists of $r = 7$ basic dimensions and numerous derived dimensions. More precisely, $d_1$ is the meter (m), $d_2$ the second (s), $d_3$ the kilogram (kg), $d_4$ the ampere (A), $d_5$ the mole (mol), $d_6$ the kelvin (K), and $d_7$ the candela (cd). The number $r$ can be smaller, since not all units are always needed. The physical dimension of a quantity $q$ is denoted by $[q]$.

A meaningful mathematical relation between the quantities $q_j$ should obey the *principle of dimensional homogeneity*, which can be summarized as follows: *summing up quantities is meaningful only if all the terms have the same dimension.* Furthermore, any functional relation of the type

$$f(q_1, q_2, \dots, q_n) = 0 \tag{1}$$

should remain valid if expressed in different units. In other words, since *dimensional scaling must not change the equation*, it is natural to seek to express the relations in terms of dimensionless quantities. It is therefore not a surprise that dimensionless quantities, known as *Π-numbers*, have a central role in dimensional analysis. A canonical example of a Π-number is $\pi$, the invariant ratio of the circumference and the diameter of circles of all sizes.

Given a system described by the physical quantities $q_1, q_2, \dots, q_n$, we will define a Π-number, or a *dimensionless group*, to be any combination of those quantities of the form

$$R = q_1^{\mu_1} q_2^{\mu_2} \cdots q_n^{\mu_n}, \tag{2}$$

where the $\mu_j$ are *rational* numbers, not all equal to zero, and $R$ is dimensionless. If, in such a system, we are able to identify $k$ Π-numbers, $R_1, \dots, R_k$, that characterize it, we can describe it with a dimensionless version of (1) of the form

$$\varphi(R_1, R_2, \dots, R_k) = 0.$$

The advantage of the latter formulation is that it automatically satisfies the dimensional homogeneity; moreover, it does not change with any scaling of the model that leaves the values of the Π-numbers invariant. These points are best clarified by a classical example of dimensional analysis.

Consider steady fluid flow in a pipe of constant diameter $D$. The fluid is assumed to be incompressible, having density $\rho$ and viscosity $\mu$. By denoting the pressure drop across a distance $L$ by $\Delta p$ and the (average) velocity by $v$, we may assume that there is an algebraic relation between the quantities:

$$f(L, D, \rho, \mu, v, \Delta p) = 0. \tag{3}$$

In SI units, the dimensions of the variables involved are

$$\left.\begin{aligned} [L] = [D] = \mathrm{m}, \quad [\rho] = \frac{\mathrm{kg}}{\mathrm{m}^3}, \quad [\mu] = \frac{\mathrm{kg}}{\mathrm{m}\cdot\mathrm{s}}, \\ [v] = \frac{\mathrm{m}}{\mathrm{s}}, \quad [\Delta p] = \frac{\mathrm{kg}}{\mathrm{m}\cdot\mathrm{s}^2}. \end{aligned}\right\} \tag{4}$$

In his classic paper of 1883, Osborne Reynolds suggested a scaling law of the form

$$\Delta p = \rho v^2 \frac{L}{D} F\left(\frac{\rho v D}{\mu}\right), \tag{5}$$

where $F$ is some function; Reynolds himself considered the power law $F(R) = cR^{-n}$ with different values of $n$ and experimentally validated it. Equation (5) can be seen as a dimensionless version of (3),

$$\varphi(R_1, R_2, R_3) = 0,$$

where

$$R_1 = \frac{\rho v D}{\mu}, \qquad R_2 = \frac{\Delta p}{\rho v^2}, \qquad R_3 = \frac{L}{D}.$$

The quantities $R_1$ and $R_2$ are known as the *Reynolds number* and the *Euler number*, respectively, and it is a straightforward matter to check that $R_1$, $R_2$, and $R_3$ are dimensionless. The scaling law (5) has been experimentally validated in a range of geometric settings. An example of its use is the design of miniature models. If the dimensions are scaled by a factor $\alpha$, $L \to \alpha L$, $D \to \alpha D$, we may assume that the flow in the miniature model gives a good prediction for the actual system if we scale the velocity and pressure as $v \to v/\alpha$ and $\Delta p \to \Delta p/\alpha^2$, leaving the dimensionless quantities intact.

In view of the above example it is natural to ask how many Π-numbers characterize a given system and if there is a systematic way of finding them. To address these questions it is important to identify possible redundancy among the physical quantities, on the one hand, and the dimensions, on the other. With this in mind we introduce the concepts of *independency* and *relevance* of the dimensions.

The dimensions $d_1, \dots, d_r$ are *independent* if none can be expressed as a rational product of the others, that is,

$$d_1^{\alpha_1} d_2^{\alpha_2} \cdots d_r^{\alpha_r} = 1 \tag{6}$$

if and only if $\alpha_1 = \alpha_2 = \cdots = \alpha_r = 0$. The dimensions $d_j$ may be the fundamental dimensions of the SI unit system or derived dimensions, such as the newton (N = $\mathrm{kg/m \cdot s^2}$).

It is not a coincidence that this definition strongly resembles that of linear independency in linear algebra, as will become evident later.

Let a system be described by $n$ quantities, $q_1, \ldots, q_n$, and $r$ dimensions, $d_1, \ldots, d_r$, with the dimensional dependency

$$[q_j] = d_1^{\mu_{j1}} d_2^{\mu_{j2}} \cdots d_r^{\mu_{jr}}, \quad 1 \leqslant j \leqslant n. \tag{7}$$

We say that the dimensions $d_k$, $1 \leqslant k \leqslant r$, are *relevant* if for each $d_k$ there are rational coefficients $\alpha_{kj}$ such that

$$d_k = [q_1]^{\alpha_{k1}} [q_2]^{\alpha_{k2}} \cdots [q_n]^{\alpha_{kn}}. \tag{8}$$

In other words, the dimensions $d_k$ are relevant if they can be expressed in terms of the dimensions of the variables $q_k$. It follows immediately that, if the quantities $q_j$ can be measured, then there must exist an operational description of all units in terms of the measurements. Identifying relevant quantities may be more subtle than it seems.

For the sake of definiteness, assume that we adhere to the SI system, and denote the seven basic SI units by $e_1, e_2, \ldots, e_7$, the ordering being unimportant. We now proceed to define an associated dimension space: to each $e_i$ we associate a vector $\boldsymbol{e}_i \in \mathbb{R}^7$, where $\boldsymbol{e}_i$ is the $i$th unit coordinate vector. Further, we define a group homomorphism between the $\mathbb{Q}$-moduli of dimensions and vectors; since any dimension $d$ can be represented in the SI system in terms of the seven basic units $e_i$ as

$$d = e_1^{\nu_1} \cdots e_7^{\nu_7},$$

we associate $d$ with a vector $\boldsymbol{d}$, where

$$\boldsymbol{d} = \nu_1 \boldsymbol{e}_1 + \cdots + \nu_7 \boldsymbol{e}_7.$$

Along these lines, we associate with a quantity $q$ with dimensions

$$[q] = d_1^{\mu_1} \cdots d_r^{\mu_r}$$

the vector

$$\boldsymbol{q} = \mu_1 \boldsymbol{d}_1 + \cdots + \mu_r \boldsymbol{d}_r.$$

It is straightforward to verify that the representation of $\boldsymbol{q}$ in terms of the basis vectors $\boldsymbol{e}_j$ is unambiguous.

We are now ready to revisit independency of units in the light of the associated vectors. In linear algebraic terms, condition (6) is equivalent to saying that

$$\alpha_1 \boldsymbol{d}_1 + \cdots + \alpha_r \boldsymbol{d}_r = \boldsymbol{0},$$

and therefore *the independency of dimensions is equivalent to the linear independency of the corresponding dimension vectors.*

Next we look for a connection with linear algebra to help us reinterpret the concept of relevance. In the dimension space, condition (7) can be expressed as

$$\boldsymbol{q}_j = \mu_{j1} \boldsymbol{d}_1 + \cdots + \mu_{jr} \boldsymbol{d}_r = \sum_{k=1}^{r} \mu_{jk} \boldsymbol{d}_k,$$

which implies that every $\boldsymbol{q}_j$ is in the subspace spanned by the vectors $\boldsymbol{d}_k$, while the linear algebraic formulation of condition (8),

$$\boldsymbol{d}_k = \alpha_{k1} \boldsymbol{q}_1 + \cdots + \alpha_{kn} \boldsymbol{q}_n = \sum_{\ell=1}^{n} \alpha_{k\ell} \boldsymbol{q}_\ell,$$

states that the vectors $\boldsymbol{d}_k$ are in the subspace spanned by the vectors $\boldsymbol{q}_\ell$. We therefore conclude that the relevance of dimensions is equivalent to the condition that

$$\operatorname{span}\{\boldsymbol{q}_1, \ldots, \boldsymbol{q}_n\} = \operatorname{span}\{\boldsymbol{d}_1, \ldots, \boldsymbol{d}_r\}.$$

It is obvious that when $n > r$, there must be redundancy among the quantities because the subspace can be spanned by fewer than $n$ vectors. This redundancy is indeed the key to the theory of $\Pi$-numbers.

Let us take a second look at the definition of $\Pi$-number, (2). In order for a quantity to be dimensionless, the coefficients of the dimension vectors must all vanish, which, in the new formalism, is equivalent to the corresponding dimension vector being the zero vector. In other words, equation (2) is equivalent to

$$\mu_1 \boldsymbol{q}_1 + \mu_2 \boldsymbol{q}_2 + \cdots + \mu_n \boldsymbol{q}_n = \boldsymbol{R} = \boldsymbol{0}.$$

If we now define the *dimension matrix* of the quantities $q_1, \ldots, q_n$ to be

$$Q = \begin{pmatrix} \boldsymbol{q}_1 & \boldsymbol{q}_2 & \cdots & \boldsymbol{q}_n \end{pmatrix} \in \mathbb{R}^{r \times n},$$

we can immediately verify that the vector $\mu \in \mathbb{R}^n$ with entries $\mu_j$ must satisfy $Q\mu = \boldsymbol{0}$, so $\mu$ must be in the null space of $Q$, $\mathcal{N}(Q)$.

We can now restate the definition of $\Pi$-number in the language of linear algebra: $R = q_1^{\mu_1} \cdots q_n^{\mu_n}$ is a $\Pi$-number if and only if $\mu \in \mathcal{N}(Q)$.

It is a central question in dimensional analysis how many essentially different $\Pi$-numbers can be found that correspond to a given system. If $R_1$ and $R_2$ are two $\Pi$-numbers, their product and ratio are also $\Pi$-numbers, yet they are not independent. To find out how to determine which $\Pi$-numbers are independent, assume that $R_1$ and $R_2$ correspond to vectors $\mu$ and $\nu$ in the null space of $Q$. From the observation that

$$R_1 \times R_2 = q_1^{\mu_1} \cdots q_n^{\mu_n} \times q_1^{\nu_1} \cdots q_n^{\nu_n} = q_1^{\mu_1+\nu_1} \cdots q_n^{\mu_n+\nu_n},$$

it follows that multiplication of two $\Pi$-numbers corresponds to addition of the corresponding vectors in the null space of the dimension vector. This naturally leads to the definition that the $\Pi$-numbers $\{R_1, \dots, R_k\}$ are *essentially different* if the corresponding coefficient vectors in $\mathcal{N}(Q)$ are linearly independent. In particular, the number of essentially different $\Pi$-numbers is equal to the dimension of $\mathcal{N}(Q)$, and a maximal set of essentially different $\Pi$-numbers corresponds to a basis for $\mathcal{N}(Q)$.

It is now easy to state the following central theorem of dimensional analysis, which is a corollary of the theorem about the dimensions of the FOUR FUNDAMENTAL SUBSPACES [I.2 §21] of a dimension matrix.

**Buckingham's $\Pi$ theorem.** *If a physical problem is described by $n$ variables, with every variable expressed in terms of $r$ independent and relevant dimensions, the number of essentially different $\Pi$-numbers (dimensionless groups whose numerical values depend on the properties of the system) is at most $n - r$.*

It is important to stress that the number of essentially different $\Pi$-numbers is "at most" $n - r$ because the system may actually admit fewer. It is a nice corollary that the $\Pi$-numbers of a system can be found by computing a basis for the null space of the dimension matrix by Gaussian elimination, which results in rational coefficients.

Returning to our example from fluid dynamics, let a system be described by the five quantities length ($L$), a characteristic scalar velocity ($v_0$), density ($\rho$), viscosity ($\mu$), and pressure ($p$), the dimensions of which were given in (4). We characterize the system with three SI units, m, s, and kg. The dimension matrix in this case is

$$Q = \begin{pmatrix} 1 & 1 & -3 & -1 & -1 \\ 0 & -1 & 0 & -1 & -2 \\ 0 & 0 & 1 & 1 & 1 \end{pmatrix}.$$

To find a basis of the null space we reduce the matrix to its row echelon form by Gauss–Jordan elimination, which shows that its rank is three. This implies that the null space is two dimensional, with a basis consisting of the two vectors

$$\boldsymbol{u}_1 = \begin{pmatrix} 1 \\ 1 \\ 1 \\ -1 \\ 0 \end{pmatrix}, \qquad \boldsymbol{u}_2 = \begin{pmatrix} 0 \\ -2 \\ -1 \\ 0 \\ 1 \end{pmatrix},$$

corresponding to the Reynolds number and the Euler number, respectively:

$$R_1 = L^1 v_0^1 \rho^1 \mu^{-1} p^0, \quad R_2 = L^0 v_0^2 \rho^{-1} \mu^0 p^1.$$

To appreciate the usefulness of finding these $\Pi$-numbers, consider the nondimensionalization of the NAVIER–STOKES EQUATION [III.23],

$$\rho\left(\frac{\partial \boldsymbol{v}}{\partial t} + \boldsymbol{v} \cdot \nabla \boldsymbol{v}\right) = -\nabla p + \mu \Delta \boldsymbol{v},$$

where $\Delta = \nabla \cdot \nabla$. Assuming that a characteristic speed $v_0$ (e.g., an asymptotic value) and a characteristic length scale $L$ are given, first we nondimensionalize the velocity and the spatial variable, writing

$$\boldsymbol{v} = v_0 \boldsymbol{\vartheta}, \qquad \boldsymbol{x} = L\boldsymbol{\xi},$$

and then we define a dimensionless pressure field based on the nondimensionality of the Euler number $R_2$,

$$\pi(\boldsymbol{\xi}) = \frac{1}{\rho v_0^2} p(L\boldsymbol{\xi}),$$

arriving at the scaled version of the equation:

$$\frac{\rho v_0^2}{L}\left(\frac{L}{v_0}\frac{\partial \boldsymbol{\vartheta}}{\partial t} + \boldsymbol{\vartheta} \cdot \nabla' \boldsymbol{\vartheta}\right) = -\frac{\rho v_0^2}{L} \nabla' \pi + \frac{\mu v_0}{L^2} \Delta' \boldsymbol{\vartheta},$$

where $\nabla' = \nabla_{\boldsymbol{\xi}}$ and $\Delta' = \nabla' \cdot \nabla'$. By going further and defining the time in terms of the characteristic timescale $L/v_0$,

$$t = \frac{L}{v_0} \tau,$$

the nondimensional version of the Navier–Stokes equation ensues:

$$\frac{\partial \boldsymbol{\vartheta}}{\partial \tau} + \boldsymbol{\vartheta} \cdot \nabla' \boldsymbol{\vartheta} = -\nabla' \pi + \frac{1}{R_1} \Delta' \boldsymbol{\vartheta}.$$

This form provides a natural justification for the different approximations corresponding to, for example, nonviscous fluid flow ($R_1$ large) or nonturbulent flow ($R_1$ small).

## Further Reading

Barenblatt, G. I. 1987. *Dimensional Analysis.* New York: Gordon & Breach.

——. 2003. *Scaling.* Cambridge: Cambridge University Press.

Calvetti, D., and E. Somersalo. 2012. *Computational Mathematical Modeling: An Integrated Approach across Scales.* Philadelphia, PA: SIAM.

Mattheij, R. M. M., S. W. Rienstra, and J. H. M. ten Thije Boonkkamp. 2005. *Partial Differential Equations: Modeling, Analysis and Computation.* Philadelphia, PA: SIAM.

## II.10 The Fast Fourier Transform

*Daniel N. Rockmore*

In 1965 James Cooley and John Tukey wrote a brief article (a note, really) that laid out an efficient method for computing the various trigonometric sums necessary for computing or approximating the Fourier transform of a function on the real line. While theirs was not the first such article (it was later discovered that the algorithm's fundamental step was first sketched in papers of Gauss), what was very different was the context. Newly invented analog-to-digital converters had now enabled the accumulation of (for the time) extraordinarily large data sets of sampled time series, whose analysis required the computation of the underlying signal's Fourier transform. In this new world of 1960s "big data," a clever reduction in computational complexity (a term not yet widely in use) could make a tremendous difference.[1]

While the Cooley–Tukey approach is what is usually associated with the phrase "fast Fourier transform" (or "FFT"), this term more correctly refers to a family of algorithms designed to accomplish the efficient calculation of the FOURIER TRANSFORM [II.19] (or an approximation thereof) of a real-valued function $f$ sampled at points $x_j$ (on either the real line, the unit interval, or the unit circle): samples go in and Fourier coefficients are returned. The discrete sums of interest

$$\hat{f}(k) = \sum_{j=0}^{n-1} f(j)\omega_n^{jk} \tag{1}$$

computed for each $k = 0, \dots, n-1$, where $\omega_n = \mathrm{e}^{2\pi\mathrm{i}/n}$ is a primitive $n$th root of unity and $f(j) = f(x_j)$, make up what is usually called the "discrete Fourier transform" (DFT). This can be written succinctly as the outcome of the matrix–vector multiplication

$$\hat{f} = \Omega f, \tag{2}$$

where the $(j,k)$ element of $\Omega$ is $\omega_n^{jk}$.

### 1 The Cooley–Tukey FFT

If computed directly, the DFT requires $n^2$ multiplications and $n(n-1)$ additions, or $2n^2 - n$ arithmetic operations (assuming the $f(j)$ values and the powers of the root of unity have been precomputed and stored). Note that this is approximately $2n^2$ (and, asymptotically, $O(n^2)$) operations. The "classical" FFT (i.e., the Cooley–Tukey FFT) can be employed in the case in which $n$ can be factored, $n = pq$, whereupon we can take advantage of a concomitant factorization of the calculation (which, in turn, is a factorization of the matrix $\Omega$) that can be cast as a DIVIDE AND CONQUER ALGORITHM [I.4 §3], writing the DFT of order $n$ as $p$ DFTs of order $q$ (or $q$ DFTs of order $p$). More explicitly, in this case we can write

$$j = j(a,b) = aq + b, \quad 0 \leqslant a < p,\ 0 \leqslant b < q,$$
$$k = k(c,d) = cp + d, \quad 0 \leqslant c < q,\ 0 \leqslant d < p,$$

so that (1) can be rewritten as

$$\hat{f}(c,d) = \sum_{b=0}^{q-1} \omega_n^{b(cp+d)} \sum_{a=0}^{p-1} f(a,b)\omega_p^{ad} \tag{3}$$

using the fact that $\omega_n^{adq} = \omega_p^{ad}$.

Computation of $\hat{f}$ is now performed in two steps. First, compute for each $b$ the inner sums (for all $d$)

$$\tilde{f}(b,d) = \sum_{a=0}^{p-1} f(a,b)\omega_p^{ad}, \tag{4}$$

which have the form of DFTs of length $p$ equispaced among multiples of $q$. In engineering language, (4) would be called "a subsampled DFT of length $p$."

Direct calculation of all the $\tilde{f}(b,d)$ requires $pq[p + (p-1)]$ arithmetic operations. Step two is to then compute an additional $pq$ transforms of length $q$,

$$\hat{f}(c,d) = \sum_{b=0}^{q-1} \omega_n^{b(cp+d)} \tilde{f}(b,d),$$

requiring at most an additional $pq[q + (q-1)]$ operations to complete the calculation. Thus, instead of the approximately $2n^2 = 2(pq)^2$ operations required by direct computation, the above algorithm uses approximately $2(pq)(p+q)$ operations. If $n$ can be factored further, this approach works even better. When $n$ is a power of two, the successive splittings of the calculation give the well-known $O(n \log_2 n)$ complexity result (in comparison to $O(n^2)$).

Since $\Omega^*\Omega = nI$, from (2) we have $f = n^{-1}\Omega^*\hat{f}$, so the discretized function $f = (f(0), \dots, f(n-1))$ (sample values) can be recovered from its Fourier coefficients via

$$f(m) = \frac{1}{n}\sum_k \hat{f}(k)\omega_n^{-mk},$$

1. Many years later Cooley told me that he believed that the fast Fourier transform could be thought of as one of the inspirations for asymptotic algorithmic analysis and the study of computational complexity, as previous to the publication of his paper with Tukey very few people had considered data sets large enough to suggest the utility of an asymptotic analysis.

a so-called inverse transform. The inverse transform expresses $f$ as a superposition of (sampled) exponentials or, equivalently, sines and cosines of frequencies that are multiples of $2\pi/n$, so that if we think of $f$ as a function of time, the DFT is a change of basis from the "time domain" to the "frequency domain."

In the case in which $n = 2N - 1$ and the $f(x_j)$ represent equispaced samples of a bandlimited function on the circle (or, equivalently, on the unit interval), so that $x_j = j/n$, and of bandlimit $N$ (i.e., $\hat{f}(k) = 0$ for all $k > N$), then (up to a normalization) the sums exactly compute the Fourier coefficients of the function $f$ (suitably indexed). The form of the inverse transform can itself be restated as a DFT, so that an FFT enables the efficient change of basis between the time and frequency domains.

The utility of an efficient algorithm for computing these sums cannot be overstated—occupying as it does a central position in the world of SIGNAL PROCESSING [IV.35], IMAGE PROCESSING [VII.8], and INFORMATION PROCESSING [IV.36]—not only for the intrinsic interest in the Fourier coefficients (say, in various forms of spectral analysis, especially for time series) but also for their use in effecting an efficient convolution of data sequences via the relation (for two functions on $n$ points)

$$\widehat{(f \star g)}(k) = \hat{f}(k)\hat{g}(k),$$

where

$$(f \star g)(k) = \sum_{m=0}^{n-1} f(k-m)g(m). \tag{5}$$

If computed directly for all $k$, (5) requires $n[n + (n - 1)] = O(n^2)$ operations. An efficient FFT-based convolution is effected by first computing $\hat{f}$ and $\hat{g}$, then using $n$ operations for pointwise multiplication of the transformed sequences, and then using another FFT for the efficient inverse transform back to the time domain.

This relationship is the key to FFTs that work for data streams of prime length $p$. The best-known ideas make use of rewriting the DFT at nonzero frequencies in terms of a convolution of length $p-1$ and then computing the DFT at the zero frequency directly. One well-known example is Rader's prime FFT, which uses the fact that we can find a generator $g$ of $\mathbb{Z}/p\mathbb{Z}^{\times}$, a cyclic group (under multiplication) of order $p - 1$, to write $\hat{f}(g^{-b})$ as

$$\hat{f}(g^{-b}) = f(0) + \sum_{a=0}^{p-2} f(g^a)\mathrm{e}^{2\pi \mathrm{i} g^{a-b}/p}. \tag{6}$$

The summation in (6) has the form of a convolution of length $p - 1$ of the sequence $f'(a) = f(g^a)$ with the function $z(a) = \mathrm{e}^{2\pi \mathrm{i} g^a/p}$.

Through the use of these kinds of reductions—contributions by various members of the "FFT family"—we achieve a general $O(n \log_2 n)$ algorithm.

### Further Reading


Brigham, E. O. 1988. *The Fast Fourier Transform and Its Applications.* Englewood Cliffs, NJ: Prentice-Hall.

Cooley, J. W. 1987. The re-discovery of the fast Fourier transform algorithm. *Mikrochimica Acta* 3:33–45.

Heideman, M. T., D. H. Johnson, and C. S. Burrus. 1985. Gauss and the history of the fast Fourier transform. *Archive for History of Exact Sciences* 34(3):265–77.

Maslen, D. K., and D. N. Rockmore. 2001. The Cooley–Tukey FFT and group theory. *Notices of the American Mathematical Society* 48(10):1151–61.

van Loan, C. F. 1992. *Computational Frameworks for the Fast Fourier Transform.* Philadelphia, PA: SIAM.


## II.11 Finite Differences

In the definition of the derivative of a real function $f$ of a real variable, $f'(x) = \lim_{\varepsilon \to 0}(f(x+\varepsilon) - f(x))/\varepsilon$, we can take a small positive $\varepsilon = h > 0$ and form the approximation

$$f'(x) \approx \frac{f(x+h) - f(x)}{h}.$$

This process is called *discretization* and the approximation is called a *forward difference* because we evaluate $f$ at a point to the right of $x$. We could instead take a small negative $\varepsilon$, so that with $h = -\varepsilon$ we have

$$f'(x) = \frac{f(x-h) - f(x)}{-h} = \frac{f(x) - f(x-h)}{h}.$$

The latter approximation is a *backward difference.* Higher derivatives can be approximated in a similar fashion. An example is the *centered second difference* approximation

$$f''(x) \approx \frac{f(x+h) - 2f(x) + f(x-h)}{h^2}.$$

The term *finite differences* is used to describe such approximations to derivatives by linear combinations of function values. One way to derive finite-difference approximations, and also to analyze their accuracy, is by manipulating TAYLOR SERIES [I.2 §9] expansions. A more systematic approach is through the *calculus of finite differences,* which is based on operators such as the *forward difference operator* $\Delta f(x) = f(x+h) - f(x)$ and its powers: $\Delta^2 f(x) = \Delta(\Delta f(x)) = f(x +$

$2h) - 2f(x+h) + f(x)$ and so on. The term "calculus" is used because there are many analogies between these operators and the differentiation operator. Finite-difference calculus is thoroughly developed in classical numerical analysis texts of the last century, but it is less commonly encountered nowadays.

Finite differences can be used to approximate partial derivatives in an analogous way. For example, if $f(x,y)$ is a function of two variables, then

$$\frac{\partial f}{\partial x}(x,y) \approx \frac{f(x+\varepsilon,y) - f(x,y)}{\varepsilon}.$$

Finite differences are widely used in numerical methods for solving ORDINARY DIFFERENTIAL EQUATIONS [IV.12] and PARTIAL DIFFERENTIAL EQUATIONS [IV.13].

## II.12 The Finite-Element Method

The finite-element method is a method for approximating the solution of a partial differential equation (PDE) with boundary conditions over a given domain using piecewise polynomial approximations to the unknown function. The domain is partitioned into elements, typically triangles for a two-dimensional region or tetrahedrons in three dimensions, and on each element the solution is approximated by a low-degree polynomial. The approximations are obtained by solving a variational form of the PDE within the corresponding finite-dimensional subspace, which reduces to solving a sparse linear system of equations for the coefficients.

For more on the finite-element method, see NUMERICAL SOLUTION OF PARTIAL DIFFERENTIAL EQUATIONS [IV.13 §4] and MECHANICS OF SOLIDS [IV.32 §4.2].

## II.13 Floating-Point Arithmetic
*Nicholas J. Higham*

The real line, $\mathbb{R}$, contains infinitely many numbers, but in many practical situations we must work with a finite subset of the real line. For example, computers have a finite number of storage locations in their random access memory, so they can represent only a finite set of numbers, whereas in a bank savings account amounts of money are recorded to two decimal places (dollars and cents, for example) and may be limited to some maximum value. In the latter situation, assuming a representation with $n$ base-10 digits, all possible numbers can be expressed as $\pm d_1 d_2 \dots d_{n-2}.d_{n-1}d_n$, where the $d_i$ are integers between 0 and 9. This is called a *fixed-point number system* because the decimal point is in a fixed position, here just before the $(n-1)$st digit. A *floating-point number system* differs in that an extra multiplicative term that is a variable power of the base allows the decimal point (or its analogue for other bases) to move around.

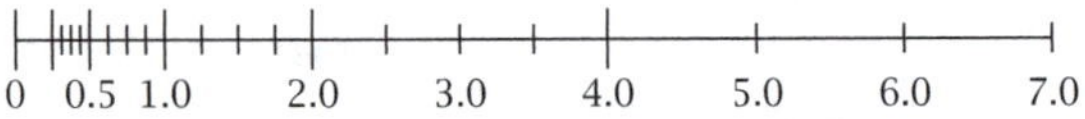


**Figure 1** The ticks denote the nonnegative numbers in the simple floating-point number system (1).

A simple example of a floating-point number system is the set of numbers, with base 2,

$$x = \pm 2^e \left(\frac{d_1}{2} + \frac{d_2}{4} + \frac{d_3}{8}\right) = \pm 2^e f, \qquad (1)$$

where the *exponent* $e \in \{-1, 0, 1, 2, 3\}$, and each binary digit $d_i$ is either 0 or 1. The number $f = (0.d_1d_2d_3)_2$ is called the *significand* (or mantissa). If we assume that $d_1 \neq 0$ (hence $d_1 = 1$) then each $x$ in the system has a unique representation (1) and the system is called *normalized*. The nonnegative numbers in this system are

0, 0.25, 0.3125, 0.3750, 0.4375, 0.5, 0.625, 0.750, 0.875, 1.0, 1.25, 1.50, 1.75, 2.0, 2.5, 3.0, 3.5, 4.0, 5.0, 6.0, 7.0,

and they are represented pictorially in figure 1.

Notice that the spacing of the numbers in this example increases by a factor 2 at every power of 2. It is a very important feature of all floating-point number systems that the spacing of the numbers is not constant (whereas for fixed-point systems the spacing *is* constant).

Other floating-point number systems are obtained by varying the range of values that the exponent $e$ can take, by varying the number of digits $d_i$ in the significand, and by varying the base. Historically, computers have mainly used base 16 or base 2. Pocket calculators instead use base 10, in order to avoid users being confused by the effects of the errors in converting from one base to another.

To use a floating-point number system, $F$, for practical computations, we need to put our data into it. Some real numbers can be exactly represented in $F$, while others can only be approximated. How should the conversion from $\mathbb{R}$ to $F$ be done, and how large an error is committed? The mapping from $\mathbb{R}$ to $F$ is called *rounding* and is denoted by "fl." The usual definition of $\mathrm{fl}(x)$, for $x \in \mathbb{R}$, is that it is the nearest number in $F$ to $x$. A rule is needed to break ties when $x$ lies midway between two members of $F$; the most common rule is to take whichever number has an even last digit in the significand. Following this rule in our toy system (1), we have $\mathrm{fl}(1.1) = 1.0$ and $\mathrm{fl}(5.5) = 6.0 = 2^3(0.110)_2$. If $x$ has magnitude greater than the magnitude of every number in $F$ then we say $\mathrm{fl}(x)$ overflows, while if $x \neq 0$ rounds to zero we say that $\mathrm{fl}(x)$ underflows.

Let $F$ be a floating-point system with base $\beta$ and $t$ digits $d_i$. It is possible to prove that, if $|x|$ lies between the smallest nonzero number in $F$ and the largest number in $F$, then $\mathrm{fl}(x) = x(1+\delta)$ with $|\delta| \leqslant u$, where $u = \frac{1}{2}\beta^{1-t}$ is called the *unit roundoff*. This means that the relative error in representing $x$ in $F$ is at most $u$. For example, in our toy system we can represent $\pi$ with relative error at most $\frac{1}{2}2^{1-3} = \frac{1}{8}$.

When we multiply two $t$-digit numbers $x_1, x_2 \in F$ the product has $2t-1$ or $2t$ significant digits, and thus in general is not itself in $F$. The best we can do is to round the result and take that as our approximation to the product: $y = \mathrm{fl}(x_1 x_2)$. Similarly, addition, subtraction, and division of numbers in $F$ also generally incur errors. If again we take the rounded version of the exact result then we can say that

$$\mathrm{fl}(x \text{ op } y) = (x \text{ op } y)(1+\delta), \quad |\delta| \leqslant u, \qquad (2)$$

for op $= +, -, *, /$.

Virtually all today's computers implement floating-point arithmetic in conformance with a 1985 IEEE standard. This standard defines two forms of base-2 floating-point arithmetic: one called single precision, with $t = 24$, and one called double precision, with $t = 53$. In IEEE arithmetic the elementary operations $+$, $-$, $*$, and $/$ are defined to be the rounded exact operations, and so (2) is satisfied. This makes (2) a very useful tool for analyzing the effects of rounding errors on numerical algorithms.

Historically, (2) has been just a model for floating-point arithmetic—an assumption on which analysis was based. Prior to the widespread adoption of the IEEE standard, different computer manufacturers used different forms of arithmetic, some of which did not satisfy (2) or lacked certain other desirable properties. This led to bizarre situations such as the expression $x/\sqrt{x^2+y^2}$ occasionally evaluating to a floating-point number greater than 1. Fortunately, the IEEE standard ensures that elementary floating-point computations retain many of the properties of arithmetic. However, special-purpose processors, such as graphics processing units (GPUs), do not necessarily (fully) comply with the IEEE standard.

It is often thought that subtraction of nearly equal floating-point numbers is dangerous because of cancellation. In fact, the subtraction is done exactly. Indeed, if $x$ and $y$ are floating-point numbers with $y/2 \leqslant x \leqslant 2y$ then $\mathrm{fl}(x-y) = x-y$ (as long as $x-y$ does not underflow). The danger of cancellation is that it causes a loss of significant digits and can thereby bring into prominence errors committed in earlier computations. A classic example is the usual formula for solving a quadratic equation such as $x^2 - 56x + 1 = 0$, which gives $x_1 = 28 + \sqrt{783}$ and $x_2 = 28 - \sqrt{783}$. Working to four significant decimal digits, $\sqrt{783} = 27.98$, and therefore the computed results are $\hat{x}_1 = 55.98 = 5.598 \times 10^1$ and $\hat{x}_2 = 0.02 = 2.000 \times 10^{-2}$. However, the correct results to four significant digits are $x_1 = 5.598 \times 10^1$ and $x_2 = 1.786 \times 10^{-2}$. So while the larger computed root has all four digits correct, the smaller one is very inaccurate due to cancellation. Fortunately, since $x^2 - 56x + 1 = (x - x_1)(x - x_2)$, the product of the roots is 1, and so we can obtain an accurate value for the smaller root $x_2$ from $x_2 = 1/x_1$.

### Further Reading

Higham, N. J. 2002. *Accuracy and Stability of Numerical Algorithms*, 2nd edn. Philadelphia, PA: SIAM.

Muller, J.-M., N. Brisebarre, F. de Dinechin, C.-P. Jeannerod, V. Lefèvre, G. Melquiond, N. Revol, D. Stehlé, and S. Torres. 2010. *Handbook of Floating-Point Arithmetic*. Boston, MA: Birkhäuser.

## II.14 Functions of Matrices

*Nicholas J. Higham*

Given the scalar function $f(x) = (x-1)/(1+x^2)$ we can define $f(A)$ for an $n \times n$ matrix $A$ by $f(A) = (A-I)(I+A^2)^{-1}$, as long as $A$ does not have $\pm \mathrm{i}$ as an eigenvalue, so that $I + A^2$ is nonsingular. This notion of "replacing $x$ by $A$" is very natural and produces useful results. For example, we can define the exponential of $A$ by

$$\mathrm{e}^A = I + A + \frac{A^2}{2!} + \frac{A^3}{3!} + \cdots.$$

The resulting function satisfies analogues of the properties of the scalar exponential, such as $\mathrm{e}^A \mathrm{e}^{-A} = I$, $\mathrm{e}^A = \lim_{s\to\infty}(I + A/s)^s$, and $(\mathrm{d}/\mathrm{d}t)\mathrm{e}^{At} = A\mathrm{e}^{At} = \mathrm{e}^{At}A$. Thanks to the latter relation, the matrix exponential plays a fundamental role in linear differential equations. In particular, the general solution to $\mathrm{d}y/\mathrm{d}t = Ay$ is $y(t) = \mathrm{e}^{At}c$, for a constant vector $c$. However, not every scalar relation generalizes. In particular, $\mathrm{e}^{A+B} =$

$e^A e^B$ holds if $A$ and $B$ commute, but it does not hold in general.

More generally, for any function with a Taylor series expansion the scalar argument can be replaced by a square matrix as long as the eigenvalues of the matrix are within the radius of convergence of the Taylor series. Thus we have

$$\cos(A) = I - \frac{A^2}{2!} + \frac{A^4}{4!} - \frac{A^6}{6!} + \cdots,$$
$$\sin(A) = A - \frac{A^3}{3!} + \frac{A^5}{5!} - \frac{A^7}{7!} + \cdots,$$
$$\log(I + A) = A - \frac{A^2}{2} + \frac{A^3}{3} - \frac{A^4}{4} + \cdots, \quad \rho(A) < 1,$$

where $\rho$ denotes the SPECTRAL RADIUS [I.2 §20]. The series for log raises two questions: does $X = \log(I + A)$ satisfy $e^X = I + A$ and, if so, which of the many matrix logarithms is produced (note that if $e^X = I + A$ then $e^{X+2k\pi iI} = e^X e^{2k\pi iI} = I + A$ for any integer $k$)? The answer to the first question is yes. The answer to the second question is that the logarithm produced is the *principal logarithm*, which for a matrix with no eigenvalues lying on the nonpositive real axis is the unique logarithm all of whose eigenvalues have imaginary parts lying in the interval $(-\pi, \pi)$.

Defining $f(A)$ via a power series may specify the function only for a certain range of $A$, as for the logarithm, and moreover, some functions do not have a (convenient) power series. For more general functions a different approach is needed.

If $f$ is analytic on and inside a closed contour $\Gamma$ that encloses the spectrum of $A$, then we can define

$$f(A) := \frac{1}{2\pi i} \int_\Gamma f(z)(zI - A)^{-1}\, dz,$$

which is a generalization to matrices of the CAUCHY INTEGRAL FORMULA [IV.1 §7]. Another definition can be given in terms of the JORDAN CANONICAL FORM [II.22] $Z^{-1}AZ = J = \operatorname{diag}(J_1, J_2, \dots, J_p)$, where $J_k$ is an $m_k \times m_k$ Jordan block with eigenvalue $\lambda_k$. The definition is

$$f(A) := Zf(J)Z^{-1} = Z \operatorname{diag}(f(J_k))Z^{-1},$$

where

$$f(J_k) := \begin{bmatrix} f(\lambda_k) & f'(\lambda_k) & \cdots & \dfrac{f^{(m_k-1)}(\lambda_k)}{(m_k-1)!} \\ & f(\lambda_k) & \ddots & \vdots \\ & & \ddots & f'(\lambda_k) \\ & & & f(\lambda_k) \end{bmatrix}.$$

This definition does not require $f$ to be analytic but merely requires the existence of the derivatives $f^{(j)}(\lambda_k)$ for $j$ up to one less than the size of the largest block in which $\lambda_k$ appears. Note that when $A$ is diagonalizable, that is, $A = ZDZ^{-1}$ for $D = \operatorname{diag}(\lambda_i)$, the definition is simply $f(A) = Zf(D)Z^{-1}$, where $f(D) = \operatorname{diag}(f(\lambda_i))$.

The Cauchy integral and Jordan canonical form definitions are equivalent when $f$ is analytic.

Some key properties that follow from the definitions are that $f(A)$ commutes with $A$, $f(X^{-1}AX) = X^{-1}f(A)X$ for any nonsingular $X$, and $f(A)$ is upper (lower) triangular if $A$ is. It can also be shown that certain forms of identity carry over from the scalar case to the matrix case, under assumptions that ensure that all the relevant matrices are defined. Examples are $\exp(iA) = \cos(A) + i\sin(A)$ and $\cos^2(A) + \sin^2(A) = I$. However, care is needed when dealing with multivalued functions; for example, for the principal logarithm, $\log(e^A)$ cannot be guaranteed to equal $A$ without restrictions on the spectrum of $A$.

Another important class of functions is the $p$th roots: the solutions of $X^p = A$, where $p$ is a positive integer. For nonsingular $A$ there are many $p$th roots. The one usually required in practice is the *principal $p$th root*, defined for $A$ with no eigenvalues lying on the nonpositive real axis as the unique $p$th root whose eigenvalues lie strictly within the wedge making an angle $\pi/p$ with the positive real axis, and denoted by $A^{1/p}$. Thus $A^{1/2}$ is the square root whose eigenvalues all lie in the open right half-plane.

The function $\operatorname{sign}(A) = A(A^2)^{-1/2}$, defined for any $A$ having no pure imaginary eigenvalues, is the *matrix sign function*. It has applications in control theory, in particular for solving ALGEBRAIC RICCATI EQUATIONS [III.25], and corresponds to the scalar function mapping complex numbers in the open left and right half-planes to $-1$ and 1, respectively.

Matrix functions provide one-line solutions to many problems. For example, the second-order ordinary differential equation initial-value problem

$$\frac{d^2y}{dt^2} + Ay = 0, \quad y(0) = y_0, \ y'(0) = y_0',$$

with $y$ an $n$-vector and $A$ an $n \times n$ matrix, has solution

$$y(t) = \cos(\sqrt{A}t)y_0 + (\sqrt{A})^{-1}\sin(\sqrt{A}t)y_0',$$

where $\sqrt{A}$ denotes *any* square root of $A$. Alternatively, by writing $z = \left[\begin{smallmatrix} y' \\ y \end{smallmatrix}\right]$ we can convert the problem into two first-order differential equations:

$$z' = \begin{bmatrix} y'' \\ y' \end{bmatrix} = \begin{bmatrix} 0 & -A \\ I & 0 \end{bmatrix}\begin{bmatrix} y' \\ y \end{bmatrix} = \begin{bmatrix} 0 & -A \\ I & 0 \end{bmatrix} z,$$

from which follows the formula

$$\begin{bmatrix} y'(t) \\ y(t) \end{bmatrix} = \exp\left(\begin{bmatrix} 0 & -tA \\ tI & 0 \end{bmatrix}\right)\begin{bmatrix} y_0' \\ y_0 \end{bmatrix}.$$

There is an explicit formula for a function of a $2 \times 2$ triangular matrix:

$$f\left(\begin{bmatrix} \lambda_1 & \alpha \\ 0 & \lambda_2 \end{bmatrix}\right) = \begin{bmatrix} f(\lambda_1) & \alpha f[\lambda_1,\lambda_2] \\ 0 & f(\lambda_2) \end{bmatrix},$$

where

$$f[\lambda_1,\lambda_2] = \begin{cases} \dfrac{f(\lambda_2) - f(\lambda_1)}{\lambda_2 - \lambda_1}, & \lambda_1 \neq \lambda_2, \\ f'(\lambda_2), & \lambda_1 = \lambda_2, \end{cases}$$

is a first-order divided difference. This formula extends to $n \times n$ triangular matrices $T$, although the formula for the $(i,j)$ element contains up to $2^n$ terms and so is not computationally useful unless $n$ is very small. It is nevertheless possible to compute $F = f(T)$ for an $n \times n$ triangular matrix in $n^3/3$ operations using the *Parlett recurrence*, which is obtained by equating elements in the equation $TF = FT$.

### Further Reading

Higham, N. J. 2008. *Functions of Matrices: Theory and Computation*. Philadelphia, PA: SIAM.

Higham, N. J., and A. H. Al-Mohy. 2010. Computing matrix functions. *Acta Numerica* 19:159–208.

## II.15 Function Spaces

*Hans G. Feichtinger*

While in the early days of mathematics each function was treated individually, it became appreciated that it was more appropriate to make collective statements for all continuous functions, all integrable functions, or all continuously differentiable ones. Fortunately, most of these collections of functions $(f_k)$ are closed under addition and allow the formation of linear combinations $\sum_{k=1}^{K} c_k f_k$ for real or complex coefficients $c_k$, $1 \leqslant k \leqslant K$. They are, therefore, vector spaces. In addition, most of these spaces are endowed with a suitable NORM [I.2 §19.3] $f \mapsto \|f\|$, allowing one to measure the size of their members and hence to introduce concepts of closeness by looking at the distance $d(f_1, f_2) := \|f_1 - f_2\|$. One can therefore say that a *function space* is a normed space consisting of (generalized) functions on some domain.

For the vector space $C_b(D)$ of bounded and continuous functions on some domain $D \subset \mathbb{R}^d$, the sup-norm $\|f\|_\infty := \sup_{z \in D} |f(z)|$ is the appropriate norm. With this norm, $C_b(D)$ is a Banach space (that is, a complete normed space), i.e., every CAUCHY SEQUENCE [I.2 §19.4] with respect to this norm is convergent to a unique limit element in the space. Hence, such Banach spaces of functions share many properties with the Euclidean spaces of vectors in $\mathbb{R}^d$, with the important distinction that they are not finite dimensional.

### 1 Lebesgue Spaces $L^p(\mathbb{R}^d)$

The completeness of the *Lebesgue space* $L^1(\mathbb{R}^d)$, consisting of all (measurable) functions with $\|f\|_1 := \int_{\mathbb{R}^d} |f(z)|\,dz < \infty$, is the reason why the Lebesgue integral is preferred over the Riemann integral. Note that in order to ensure the property that $\|f\|_1 = 0$ implies $f = 0$ (the null function), one has to regard two functions $f_1$ and $f_2$ as equal if they are equal up to a set of measure zero, i.e., if the set $\{z \mid f_1(z) \neq f_2(z)\}$ has Lebesgue measure zero.

Another norm that is important for many applications is the $L^2$-norm, $\|f\|_2 := (\int_{\mathbb{R}^d} |f(z)|^2)^{1/2}$. It is related to an inner product defined as $\langle f, g\rangle := \int_{\mathbb{R}^d} f(z)\overline{g(z)}\,dz$ via the formula $\|f\|_2 := \langle f, f\rangle^{1/2}$. $(L^2(\mathbb{R}^d), \|\cdot\|_2)$ is a Hilbert space, and one can talk about orthogonality and unitary linear mappings, comparable with the situation of the Euclidean space $\mathbb{R}^d$ with its standard inner product.

Having these three norms, namely $\|\cdot\|_1$, $\|\cdot\|_2$, and $\|\cdot\|_\infty$, it is natural to look for norms "in between." This leads to the $L^p$-spaces, defined by the finiteness of $\|f\|_p^p := \int_{\mathbb{R}^d} |f(z)|^p\,dz$ for $1 \leqslant p < \infty$. The limiting case for $p \to \infty$ is the space $L^\infty(\mathbb{R}^d)$ of essentially bounded functions.

Since these spaces are not finite dimensional, it is necessary to work with the set of all bounded linear functionals, the so-called dual space, which is often, but not always, a function space. For $1 \leqslant p < \infty$ the dual space to $L^p$ is $L^q$, with $1/p + 1/q = 1$, meaning that any continuous linear functional on $L^p(\mathbb{R}^d)$ has the form $f \mapsto \int_{\mathbb{R}^d} f(z)g(z)\,dz$ for a unique function $g \in L^q(\mathbb{R}^d)$.

$L^1(\mathbb{R}^d)$ also appears as the natural domain for the Fourier transform, given for $s \in \mathbb{R}^d$ by

$$\mathcal{F}\colon f \mapsto \hat{f}(s) = \int_{\mathbb{R}^d} f(t) \exp\left\{-2\pi i \sum_{j=1}^{d} s_j t_j\right\} dt,$$

while $(L^2(\mathbb{R}^d), \|\cdot\|_2)$ allows us to describe $\mathcal{F}$ as a *unitary* (and hence isometric) automorphism.

## 2 Related Function Spaces

The Lebesgue spaces are prototypical for a much larger class of *Banach function spaces* or *Banach lattices*, a class that also includes *Lorentz spaces* $L(p,q)$ or *Orlicz spaces* $L^{\Phi}$, which are all *rearrangement invariant*. This means that for any transformation $\alpha\colon \mathbb{R}^d \to \mathbb{R}^d$ that has the property that it preserves the (Lebesgue) measure $|M|$ of a set, i.e., with $|\alpha(M)| = |M|$, one has $\|f\| = \|\alpha^*(f)\|$, where $\alpha^*(f)(z) := f(\alpha(z))$.

In contrast, *weighted spaces* such as $L^p_w(\mathbb{R}^d)$, characterized by $\|f\|_{p,w} = \|fw\|_p < \infty$, allow us to capture the decay of $f$ at infinity using some strictly positive *weight function* $w$. For applications in the theory of partial differential equations (PDEs), polynomial weights such as $w_s(x) = (1+|x|^2)^{s/2}$ are important. For $s \geqslant 0$, *Sobolev spaces* $\mathcal{H}_s(\mathbb{R}^d)$ can be defined as inverse images of $L^2_{w_s}(\mathbb{R}^d)$ under the Fourier transform. For $s \in \mathbb{N}$ they consist of those functions that have (in a distributional sense) $s$ derivatives in $L^2(\mathbb{R}^d)$. *Mixed norm* $L^p$ spaces (using different $p$-norms in different directions) are also not invariant in this sense, but they are still very useful.

A large variety of function spaces arose out of the attempt to characterize smoothness, including fractional differentiability. Examples are *Besov spaces* $B^s_{p,q}(\mathbb{R}^d)$ and *Triebel-Lizorkin spaces* $F^s_{p,q}(\mathbb{R}^d)$; the classical Sobolev spaces are the only function spaces that belong to both families. The origin of this theory is in the theory of Lipschitz spaces $\mathrm{Lip}(\alpha)$, where the range $\alpha \in (0,1)$ allows us to express the degree of smoothness (differentiability corresponds intuitively to the case $\alpha = 1$).

## 3 Wavelets and Modulation Spaces

Many of the spaces mentioned above are highly relevant for PDEs, e.g., the description of elliptic PDEs. Their characterization using Paley-Littlewood (dyadic Fourier) decompositions has ignited wavelet theory. For $1 < p < \infty$ they can be characterized via (weighted) summability conditions of their wavelet coefficients with respect to (sufficiently "good") mother wavelets. In the limiting case, one obtains the *real Hardy space* $H^1(\mathbb{R}^d)$ and its dual, the BMO-space, which consists of functions of *bounded mean oscillation*. Both spaces are important for the study of Calderon-Zygmund operators or the Hardy-Littlewood maximal operator. Wavelets provide unconditional bases for these spaces, including *Besov* and *potential* spaces.

For the affine "$ax+b$"-group acting on the space $L^2(\mathbb{R}^d)$, function spaces are defined using the continuous wavelet transform, and atomic characterizations (involving *Banach frames*) of the above smoothness spaces are obtained. Alternatively, the Schrödinger representation of the Heisenberg group, again on $L^2(\mathbb{R}^d)$ via time-frequency shifts, gives rise to the family of *modulation spaces* $M^s_{p,q}$. They were introduced as *Wiener amalgam spaces* on the Fourier transform side, using uniform partitions of unity (instead of dyadic ones).

Using engineering terminology, the now-classical spaces $M^s_{p,q}(\mathbb{R}^d)$ are characterized by the behavior of the *short-time Fourier transform* of their members (replacing the continuous wavelet transform). They play an important role in time-frequency analysis, and their atomic characterizations use *Gabor expansions*.

A variety of Banach spaces of analytic or polyanalytic functions play an important role in complex analysis. Again, integrability conditions over their domain are typically used to define these spaces. The corresponding $L^2$-spaces are typically reproducing kernel Hilbert spaces, with good localization of these kernels allowing one to view them as continuous mappings on (weighted, mixed-norm) $L^p$-spaces as well. We mention some of the spaces that are important in the context of complex analysis or Toeplitz operators: Fock spaces, Bergman spaces, and Segal-Bargmann spaces.

## 4 Variations of the Theme

One of the first important examples of a Banach space of functions was the space *BV* of functions of *bounded variation*. One simple characterization of functions of this type (the so-called Jordan decomposition) is that they are the difference of two bounded and nondecreasing functions (the ascending part of the function minus the descending part of it). Via Fourier-Stieltjes integrals, F. Riesz showed that there is a one-to-one correspondence between the dual space of $(C[0,1], \|\cdot\|_\infty)$ and $BV[0,1]$ endowed with the variation norm. More recently, total variation in a two-dimensional setting has been fundamental to IMAGE RESTORATION ALGORITHMS [VII.8]. Another family of function spaces that captures variation at different scales are the Morrey-Campanato spaces.

In addition to Banach spaces of functions there are also topological vector spaces and Fréchet spaces of functions, among them the *spaces of test functions* that are used in distribution theory. *Generalized functions* consist of the continuous linear functionals on such spaces. The "theory of function spaces" as developed by Hans Triebel includes a large variety of Banach spaces of such generalized functions (or distributions).

### Further Reading


Ambrosio, L., N. Fusco, and D. Pallara. 2000. *Functions of Bounded Variation and Free Discontinuity Problems.* Oxford: Clarendon.

Gröchenig, K. 2001. *Foundations of Time–Frequency Analysis.* Boston, MA: Birkhäuser.

Leoni, G. 2009. *A First Course in Sobolev Spaces.* Providence, RI: American Mathematical Society.

Meyer, Y. 1992. *Wavelets and Operators*, translated by D. H. Salinger. Cambridge: Cambridge University Press.

Stein, E. M. 1970. *Singular Integrals and Differentiability Properties of Functions.* Princeton, NJ: Princeton University Press.

Triebel, H. 1983. *Theory of Function Spaces.* Basel: Birkhäuser.


## II.16 Graph Theory

*Timothy A. Davis and Yifan Hu*

At the back of most airline magazines you will find a map of airports and the airline routes that connect them. This is just one example of a *graph*, a widely used mathematical entity that represents relationships between discrete objects. More precisely, a graph $G = (V, E)$ consists of a set of *nodes* $V$ and a set of *edges* $E \subseteq \{(i, j) \mid i, j \in V\}$ that connect them. A graph is not a diagram but it can be drawn, as illustrated in figure 1.

Graphs arise in a vast array of applications, including social networks (a node is a person and an edge is a relationship between two people), computational fluid dynamics (a node is an unknown such as the pressure at a certain point and an edge is the physical connection between two unknowns), finding things on the web (a node is a web page and an edge is a link), circuit simulation (the wires are the edges), economics (a node is a financial entity and the edges represent trade between two entities), and many others.

In some problems, an edge connects in both directions, and in this case the graph is *undirected*. For example, friendship is mutual, so if Alice and Bob are friends, the edges (Alice, Bob) and (Bob, Alice) are the same. In other cases, the direction of the edge is important. If Alice follows Bob on Twitter, this does not mean that Bob follows Alice. In this *directed* graph, the edge (Alice, Bob) is not the same as the edge (Bob, Alice).

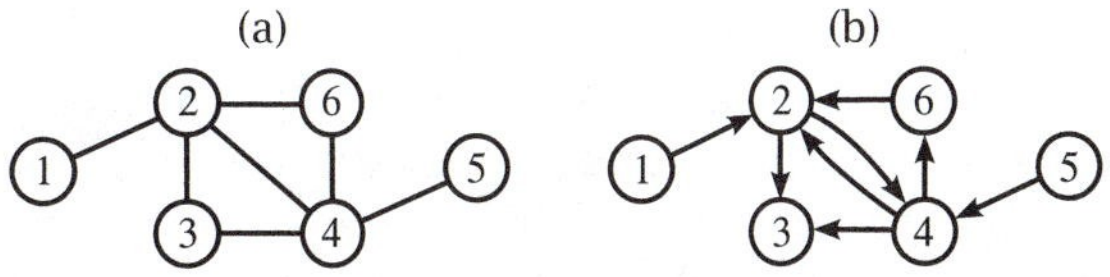


**Figure 1** Two example graphs: (a) undirected and (b) directed.

In a *simple* graph, an edge $(i, j)$ can appear just once, but in a *multigraph* it can appear multiple times ($E$ becomes a multiset). Simple graphs do not have *self-edges* $(i, i)$, but a *pseudograph* can have multiple edges and self-edges. The airline route map in the back of the magazine is an example of a simple undirected graph. Representing each flight for a whole airline would require a directed multigraph: the flight from Philadelphia to New York is not the same as the flight in the opposite direction, and there are many flights each day between the two airports. If sightseeing tours are added (self-edges), then a pseudograph would be needed.

The *adjacency set* of a node $i$, also called its *neighbors*, is the set of nodes $j$ where edge $(i, j)$ is in the graph. For a directed graph, this is the out-adjacency; the in-adjacency of node $i$ is the set $\{j \mid (j, i) \in E\}$. A graph can be represented as a binary *adjacency matrix*, with entries $a_{ij} = 1$ if $(i, j) \in E$, and $a_{ij} = 0$ otherwise. The *degree* of a node is the size of its adjacency set.

Graphs can contain infinite sets of nodes and edges. Consider the directed graph on the natural numbers $\mathbb{N}$ with the edges $(i, j)$, where $j$ is an integer multiple of $i$. A prime number $j > 1$ in this graph has in-adjacency $\{1, j\}$ and an in-degree of 2 (including the self-edge $(j, j)$); a composite number $j > 1$ has a larger in-degree.

Nodes $i$ and $j$ are *incident* on the edge $(i, j)$ and, likewise, the edge $(i, j)$ is incident on its two nodes. A *subgraph* of $G$ consists of a subset of its nodes and edges, $\bar{G} = (\bar{V}, \bar{E})$, where $\bar{V} \subseteq V$ and $\bar{E} \subseteq E$. If an edge $(i, j)$ appears in $\bar{E}$, then its two incident nodes must also appear in $\bar{V}$, but the opposite need not hold. Two special kinds of subgraphs are *node-induced* and *edge-induced* subgraphs. A node-induced subgraph starts with a subset of nodes $\bar{V}$; the edges $\bar{E}$ are all those edges whose two incident nodes are both in $\bar{V}$. An *edge-induced* subgraph starts with a subset of edges $\bar{E}$ and then $\bar{V}$ consists of all nodes incident on those edges. A graph is *completely connected* if it has an edge between every pair of nodes. A *clique* is a completely connected subgraph.

A *path* from $i$ to $j$ is a list of nodes $(i, \ldots, j)$ with edges between adjacent pairs of nodes. The path cannot traverse a directed edge backward. The *length* of the path is the number of nodes in the list minus one. In a *simple* path, a node can appear only once. If there is a path from $i$ to $j$, then node $j$ is *reachable* from node $i$. The set of all nodes reachable from $i$ is the *reach* of $i$. Among all paths from $i$ to $j$, one with the shortest length is a *shortest path*, its length the *(geodesic) distance* from $i$ to $j$. The *diameter* of a graph is the length of the longest possible shortest path. In a *small-world graph*, each node is a small distance (logarithmic in the number of nodes) away from any other node.

An undirected graph is *connected* if there is a path between each pair of nodes, but there are two kinds of connectivity in a directed graph. If a path exists between every pair of nodes, then a directed graph is *strongly connected*. A directed graph is *weakly connected* if its *underlying undirected graph* is connected; to obtain such a graph, all edge directions are dropped.

A *cycle* is a path that starts and ends at the same node $i$; the cycle is *simple* if no node is repeated (except for node $i$ itself). There are no cycles in an *acyclic* graph. The acronym DAG is often used for a directed acyclic graph.

The undirected graph in figure 1(a) is connected. Nodes $\{2,3,4\}$ form a clique, as do $\{2,4,6\}$. The path $(1,2,4,3,2,6)$ has length 5 and is not simple. A simple path from 1 to 6 is $(1,2,4,6)$ of length 3, but the shortest path is $(1,2,6)$ of length 2, which traverses the edges $(1,2)$ and $(2,6)$. The path $(2,3,4,2)$ is a cycle of length 3. Node 2 has degree 4, with neighbors $\{1,3,4,6\}$. The diameter of the graph is 3. Since the graph is connected, the reach of node 2 is the whole graph. This graph is the underlying undirected graph of the directed graph in part (b) of the figure.

The largest clique in this directed graph has only two nodes: $\{2,4\}$. The out-adjacency of node 2 is the set $\{3,4\}$ and its in-adjacency is $\{1,4,6\}$. The reach of node 2 is $\{2,3,4,6\}$. The graph is not strongly connected since there is no path from 1 to 5, but it is weakly connected since its underlying undirected graph is connected.

Figure 2 illustrates a *bipartite* graph. The nodes of a bipartite graph are partitioned into two sets, and no edge in the graph is incident on a pair of nodes in the same partition. Bipartite graphs arise naturally when modeling a relationship between two very different sets. For example, in term/document analysis, a bipartite graph of $m$ terms and $n$ documents has an edge $(i,j)$ if term $i$ appears in document $j$. No edge connects two terms, nor two documents. An undirected bipartite graph is often represented as a rectangular $m \times n$ adjacency matrix, where $a_{ij} = 1$ if the edge $(i,j)$ appears in the graph and $a_{ij} = 0$ otherwise.

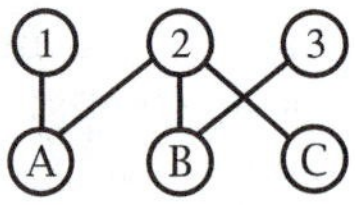


**Figure 2** An undirected bipartite graph.

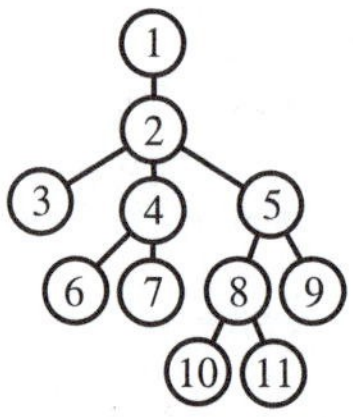


**Figure 3** A tree of height 4, with node 1 as the root.

An undirected acyclic graph is a *forest*. An important special case is a *tree*, which is a connected forest. In a tree, there is a unique simple path between each pair of nodes. In a *rooted* tree, one node is designated as the *root*. The *ancestors* of node $i$ are all the nodes on the path from $i$ to the root (excluding $i$ itself). The first node after $i$ in this path is the *parent* of $i$, and node $i$ is its *child*. The length of this path is the *level* of the node (the root has level zero). The *height* of a tree is the maximum level of its nodes.

In a tree, all nodes except the root have a single parent. Nodes can have any number of children, and a node with no children is a *leaf*. *Internal* nodes have at least one child. In a *binary* tree, nodes have at most two children, and in a *full binary* tree, all internal nodes have exactly two children. Node $i$ is a *descendant* of all nodes in the path from $i$ to the root (excluding $i$ itself). The *subtree* rooted at node $i$ is the subgraph induced by node $i$ and its descendants.

In the example in figure 3, the parent of node 5 is 2, its descendants are $\{8,9,10,11\}$, its ancestors are $\{1,2\}$, and its children are $\{8,9\}$. Since node 2 has three children, the tree is not binary.

Sometimes a graph with its nodes and edges is not enough to fully represent a problem. Edges in a graph do not have a length, but this is useful for the airline route map, and thus nodes and edges are often augmented with additional data. Attaching a single numerical value to each node and/or edge is common; this results in a *weighted* graph. In this graph, the length of a path is the sum of the weights of its edges.

Drawing a graph requires node positions, so they must be computed for graphs without natural node positions. In the *force-directed* method that created the graphs in plates 1–4, nodes are given an electrical charge and edges become springs. A low-energy state is found, which often leads to a visually pleasing layout that reveals the graph's large-scale structure.

Be aware that there are many minor variations in the terminology of graph theory. For example, nodes can be called *vertices*, edges are also called *arcs*, and self-edges are sometimes called *loops*. Directed graphs are also called *digraphs*. Sometimes the term *arc* is restricted to directed edges. In a common alternative terminology, a path becomes a *walk*, a simple path becomes a path (conflicting with the definition here), a *trail* is a walk with no repeated edges, and a cycle becomes a *closed walk*.

### Further Reading

Davis, T. A. 2006. *Direct Methods for Sparse Linear Systems.* Philadelphia, PA: SIAM.

Gross, J. L., and J. Yellen. 2005. *Graph Theory and Its Applications*, 2nd edn. Boca Raton, FL: CRC.

Rosen, K. H. 2012. *Discrete Mathematics and Its Applications*, 7th edn. Columbus, OH: McGraw-Hill.

## II.17 Homogenization

Homogenization is a method for obtaining an equation whose solution approximates the solution of a partial differential equation (PDE) with rapidly varying coefficients. In many cases the approximate equation is a PDE, but in some it may be an integro-differential equation. PDEs with rapidly varying coefficients arise in the study of the response of composite materials. The method is best illustrated by an example in heat conduction. Let $u(x_1, x_2)$ represent temperature in a composite medium occupying a domain $\Omega$. The composite is modeled by a conductivity $a > 0$ that is rapidly oscillating in both $x_1$ and $x_2$. To represent this mathematically, the conductivity is written as

$$a(x_1/\varepsilon, x_2/\varepsilon),$$

where the function $a(y_1, y_2)$ is unit periodic, i.e., $a(y_1 + 1, y_2) = a(y_1, y_2)$ and $a(y_1, y_2 + 1) = a(y_1, y_2)$. Therefore, one can see that $\varepsilon$ represents the periodicity of the medium and is small when the medium is rapidly oscillating. The governing equation is

$$\nabla \cdot a\nabla u = f,$$

where $f$ represents the heat source.

Homogenization, which is based on multiscale asymptotics, allows one to obtain an approximate equation for the heat conduction problem when $\varepsilon \ll 1$. The resulting equation,

$$\nabla \cdot A\nabla u = f,$$

is referred to as the homogenized equation and has a constant, albeit anisotropic ($A$ may be a matrix), conductivity. The conductivity in the homogenized equation in this particular case is easily identified as $A$ and is referred to as the effective property of the medium.

Periodicity is essential in deriving the homogenized equation. When the medium is not periodic, e.g., when it is random, the method of homogenization may still be applied, although its interpretation is different and more difficult. In this case there is no simple way to obtain the effective property of the random medium, but a connection to EFFECTIVE MEDIUM THEORIES [IV.31] can be made.

For periodic media, homogenization can be applied to PDEs that model other phenomena such as fluid flow in porous media, elasticity, wave propagation, vibration, electrostatics, and electromagnetics. Recent research in homogenization has explored its use in situations where the medium is almost periodic, that is, in PDEs where the coefficients are oscillatory with almost constant period or where the coefficients are periodic with the exception of a few "defect" regions.

## II.18 Hybrid Systems

*Paul Glendinning*

A hybrid system is a system that combines both continuous and discrete variables. A simple motivating example is the thermostat. A thermostat switches a heater to one of two discrete states, `on` or `off`, depending on the ambient temperature (a continuous variable), which evolves differently if the heater is in the `on` or `off` states. The specification of the thermostat includes transition rules. If the thermostat is in the `off` state and the temperature falls below a critical value, then the thermostat is switched to `on`, possibly with some (deterministic or random) delay in time. If the thermostat is in the `on` state and the temperature rises above a critical value, then the thermostat is switched to `off`, possibly with some (deterministic or random) delay in time.

This type of system—for which the value of a discrete variable determines the rules of evolution of continuous variables, and with jumps in the discrete variables according to another rule—provides a formal description of many systems in computer science, control theory, and dynamics. The bouncing ball is another simple example and the reader may like to think about how this fits into the formalism described below.

There are many different ways a hybrid system could be defined, and the choice of definition will depend on the application area or the level of generality that is required. However, the basic structure is as follows. A hybrid system's state is determined by two variables, a continuous variable $x$ on some manifold $M$ (often $\mathbb{R}^n$) and a discrete variable $d \in \{1, 2, \dots, D\}$. For each $d$ there is a domain $M_d$ in $M$ and an evolution equation that determines the time evolution of $x$. This may be a differential equation or a stochastic differential equation, for example. Therefore, given $d$ there is a well-defined rule for the evolution of $x$ in a subset of $M$.

The next part of the specification determines how the discrete variable can change. For each pair $e = (d_0, d_1)$ there is a set $G_e \subseteq M_{d_0}$, possibly empty, such that if $x \in G_e$ then there is a finite probability that the dynamics will be instantaneously reset so that the discrete variable becomes $d_1$ and the continuous variable becomes $R(x, e)$ for some reset map $R\colon M \times D \times D \to M$. After this, the dynamics continues in the new domain $M_{d_1}$ according to the evolution rule for $d_1$ until the next transition is made.

Of course, for this specification to make sense a number of consistency conditions must be satisfied, e.g., $R(x, e) \in M_{d_1}$, and it is possible that the continuous variable may leave the region $M_d$ on which its dynamics is defined before a transition is made, in which case the dynamics becomes undefined. There are therefore many checks that need to be made to ensure that, given an initial condition in some set $\mathrm{Init} \subseteq M \times D$, the solution is well defined. In the deterministic case uniqueness may also be an issue.

A solution with an infinite sequence of transitions in finite time is called a *Zeno solution*. In computer science, where the transitions are executions in some program, this is not useful. In other contexts, however, the infinite sequence may have physical significance. Chattering in mechanical impacts can be modeled as a Zeno phenomenon.

Hybrid systems are used to examine reachability and verification problems in computer science. The reachability problem is that of determining all the possible states that can be reached from a given set of initial conditions. This is also relevant to CONTROL THEORY [IV.34], where it might be important to know that a given control system keeps a car on the road or avoids collisions. The existence of transitions or jumps makes hybrid systems useful in models of mechanical impacts as well (see SLIPPING, SLIDING, RATTLING, AND IMPACT [VI.15]). Control systems with thresholds can also be modeled as hybrid systems.

The existence of transitions means that many classic results for smooth dynamical systems, such as stability theorems, are made considerably harder to prove, but it could be argued that the piecewise-smooth nature of these models is a much more generally applicable feature for modern applications in electronics and biology than the smooth dynamical systems approach that has dominated so much of dynamical systems theory.

### Further Reading

Hristu-Varsakelis, D., and W. S. Levine, eds. 2005. *Handbook of Networked and Embedded Control Systems*. Boston, MA: Birkhäuser.

Lygeros, J. 2004. *Lecture Notes on Hybrid Systems*. (Digital resource.)

## II.19 Integral Transforms and Convolution

An *integral transform* is an operator $J$ mapping a function $f$ of a real variable to another function $Jf$ according to

$$(Jf)(s) = \int_a^b K(s,t) f(t)\,\mathrm{d}t,$$

where $K$ is called the *kernel* of the transformation. There are many different integral transforms, depending on the choice of $a$, $b$, and $K$. They are used to transform one problem into another, the intent being that the new problem is simpler than the original one. Having solved the new problem, the solution $f$ is obtained by applying the appropriate inverse operator, $J^{-1}$; this is often another integral transform.

Important special cases include the *Fourier transform*

$$(\mathcal{F}f)(s) = \int_{-\infty}^{\infty} f(t) \mathrm{e}^{\mathrm{i}st}\,\mathrm{d}t$$

and the *Laplace transform*

$$(\mathcal{L}f)(s) = \int_0^\infty f(t)\mathrm{e}^{-st}\,\mathrm{d}t.$$

This definition of $\mathcal{L}$ is standard, but the definition of $\mathcal{F}$ is one of many; for example, some authors insert an extra factor $(2\pi)^{-1/2}$, and some use $\mathrm{e}^{-\mathrm{i}st}$. Always check the author's definition when reading a book or article in which Fourier transforms are used!

Many integral transforms have an associated *convolution*; given two functions of a real variable, $f$ and $g$, their convolution is another function of a real variable, denoted by $f * g$. It is defined so that $J(f * g) = (Jf)(Jg)$; the transform of the convolution is the product of the transforms. For the Fourier transform

$$(f * g)(t) = \int_{-\infty}^{\infty} f(t-s)g(s)\,\mathrm{d}s,$$

while for the Laplace transform

$$(f * g)(t) = \int_0^t f(t-s)g(s)\,\mathrm{d}s.$$

It is easy to see that, in both cases, $f * g = g * f$. Convolution is an important operation in SIGNAL PROCESSING [IV.35] and in many applications involving Fourier analysis and INTEGRAL EQUATIONS [IV.4].

There are also discrete versions of integral transforms in which the integral is replaced by a finite sum of terms. The *discrete Fourier transform* is especially important because it can be computed rapidly using the FAST FOURIER TRANSFORM [II.10] (FFT).

## II.20 Interval Analysis
*Warwick Tucker*

Interval analysis is a calculus based on set-valued mathematics. In its simplest (and by far most popular) form, it builds upon interval arithmetic, which is a natural extension of real-valued arithmetic. Despite its simplicity, this kind of set-valued mathematics has a very wide range of applications in computer-aided proofs for continuous problems. In a nutshell, interval arithmetic enables us to bound the range of a continuous function, i.e., it produces a set *enclosing* the range of a given function over a given domain. This, in turn, enables us to prove mathematical statements that use open conditions, such as strict inequalities, fixed-point theorems, etc.

### 1 Interval Arithmetic

In this section we will briefly describe the fundamentals of interval arithmetic. Let $\mathbb{IR}$ denote the set of closed intervals of the real line. For any element $\boldsymbol{a} \in \mathbb{IR}$, we use the notation $\boldsymbol{a} = [\underline{a}, \overline{a}]$. If $\star$ is one of the operators $+, -, \times, /$, we define arithmetic on elements of $\boldsymbol{a}, \boldsymbol{b} \in \mathbb{IR}$ by

$$\boldsymbol{a} \star \boldsymbol{b} = \{a \star b \colon a \in \boldsymbol{a},\ b \in \boldsymbol{b}\}, \tag{1}$$

except that $\boldsymbol{a}/\boldsymbol{b}$ is undefined if $0 \in \boldsymbol{b}$. Working exclusively with closed intervals, the resulting interval can be expressed in terms of the endpoints of the arguments. This makes the arithmetic very easy to implement in software.

Note that a generic element in $\mathbb{IR}$ has no additive or multiplicative inverse. For example, we have $[1,2] - [1,2] = [-1,1] \neq [0,0]$, and $[1,2]/[1,2] = [\frac{1}{2},2] \neq [1,1]$. This is known as the *dependency problem*, and it can cause large overestimations. In practice, however, the use of high-order (e.g., Taylor series) representations greatly mitigates this problem.

A key feature of interval arithmetic is that it is *inclusion monotonic*; i.e., if $\boldsymbol{a} \subseteq \boldsymbol{a}'$ and $\boldsymbol{b} \subseteq \boldsymbol{b}'$, then by (1) we have

$$\boldsymbol{a} \star \boldsymbol{b} \subseteq \boldsymbol{a}' \star \boldsymbol{b}'.$$

This is of fundamental importance: it says that, if we can enclose the arguments, we can enclose the result.

More generally, when we extend a real-valued function $f$ to an interval-valued one $F$, we demand that it satisfies the *inclusion principle*

$$\mathrm{range}(f; \boldsymbol{x}) = \{f(x) \colon x \in \boldsymbol{x}\} \subseteq F(\boldsymbol{x}). \tag{2}$$

If this can be arranged for a finite set of *standard* functions, then the inclusion principle will also hold for any *elementary* function constructed by arithmetic and composition applied to the set of standard functions.

Multivariate functions can be handled by working componentwise on interval vectors (boxes) $\boldsymbol{x} = (\boldsymbol{x}_1, \dots, \boldsymbol{x}_n)$.

When implementing interval arithmetic on a computer, the endpoints must be FLOATING-POINT NUMBERS [II.13]. This introduces rounding errors, which must be properly dealt with. As an example, interval addition becomes

$$\boldsymbol{a} + \boldsymbol{b} = [\triangledown(\underline{a} + \underline{b}), \triangle(\overline{a} + \overline{b})].$$

Here, $\triangledown(x)$ is the largest floating-point number no greater than $x$, and $\triangle(x) = -\triangledown(-x)$. The IEEE standard for floating-point computations guarantees that this type of *outward rounding* preserves the inclusion principle for $+, -, \times$, and $/$. For other operations (such as trigonometric functions) there are no such assurances; interval extensions of these functions must be built from scratch.

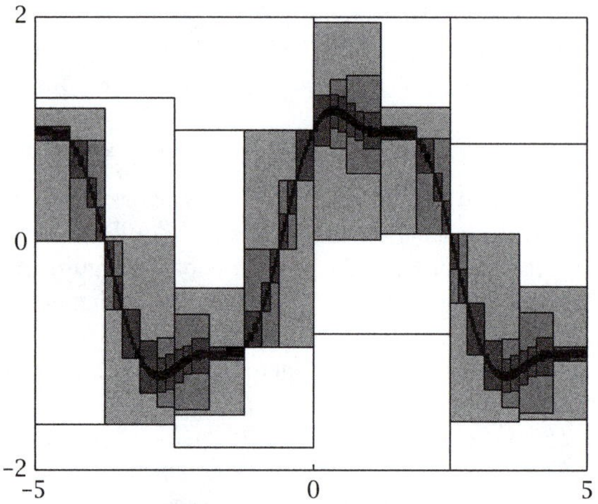


**Figure 1** Successively tighter enclosures of a graph.

## 2 Interval Analysis

The inclusion principle (2) enables us to capture continuous properties of a function, using only a finite number of operations. Its most important use is to explicitly bound discretization errors that naturally arise in numerical algorithms.

As an example, consider the function $f(x) = \cos^3 x + \sin x$ on the domain $\boldsymbol{x} = [-5, 5]$. For any decomposition of the domain $\boldsymbol{x}$ into a finite set of subintervals $\boldsymbol{x} = \bigcup_{i=1}^{n} \boldsymbol{x}_i$, we can form the set-valued graph consisting of the pairs $(\boldsymbol{x}_1, F(\boldsymbol{x}_1)), \dots, (\boldsymbol{x}_n, F(\boldsymbol{x}_n))$. As the partition is made finer (that is, as $\max_i \operatorname{diam}(\boldsymbol{x}_i)$ is made smaller), the set-valued graph tends to the graph of $f$ (see figure 1). And, most importantly, every such set-valued graph contains the graph of $f$.

This way of incorporating the discretization errors is extremely useful for quadrature, optimization, and equation solving. As one example, suppose we wish to compute the definite integral $I = \int_0^8 \sin(x + e^x)\,dx$.

A MATLAB function `simpson` that implements a simple textbook adaptive Simpson quadrature algorithm produces the following result.

```
% Compute integral I with tolerance 1e-6.
>> I = simpson(@(x) sin(x + exp(x)), 0, 8)
I =
   0.251102722027180
```

A (*very naive*) set-valued approach to quadrature is to enclose the integral $I$ via

$$I \in \sum_{i=1}^{n} F(\boldsymbol{x}_i) \operatorname{diam}(\boldsymbol{x}_i),$$

which, for a sufficiently fine partition, produces the integral enclosure

$$I \in 0.3474001726_{49}^{66}.$$

Thus, it turns out that the result from `simpson` was completely wrong! This is one example of the importance of rigorous computations.

## 3 Recent Developments

There is currently an ongoing effort within the IEEE community to standardize the implementation of interval arithmetic. The hope is that we will enable computer manufacturers to incorporate these types of computations at the hardware level. This would remove the large computational penalty incurred by repeatedly having to switch rounding modes—a task that central processing units were not designed to perform efficiently.

### Further Reading

Alefeld, G., and J. Herzberger. 1983. *Introduction to Interval Computations*. New York: Academic Press.

Kulisch, U. W., and W. L. Miranker. 1981. *Computer Arithmetic in Theory and Practice*. New York: Academic Press.

Lerch, M., G. Tischler, J. Wolff von Gudenberg, W. Hofschuster, and W. Krämer. 2006. filib++, a fast interval library supporting containment computations. *ACM Transactions on Mathematical Software* 32(2):299–324.

Moore, R. E., R. B. Kearfott, and M. J. Cloud. 2009. *Introduction to Interval Analysis*. Philadelphia, PA: SIAM.

Rump, S. M. 1999. INTLAB—INTerval LABoratory. In *Developments in Reliable Computing*, edited by T. Csendes, pp. 77–104. Dordrecht: Kluwer Academic.

Tucker, W. 2011. *Validated Numerics: A Short Introduction to Rigorous Computations*. Princeton, NJ: Princeton University Press.

## II.21 Invariants and Conservation Laws

*Mark R. Dennis*

As important as the study of *change* in the mathematical representation of physical phenomena is the study of *invariants*. Physical laws often depend only on the *relative* positions and times between phenomena, so certain physical quantities do not change; i.e., they are *invariant*, under continuous translation or rotation of the spatial axes. Furthermore, as the spatial configuration of a system evolves with time, quantities such as total energy may remain unchanged; that is, they are *conserved*. The study of invariants has been a remarkably successful approach to the mathematical formulation of physical laws, and the study of continuous symmetries and conservation laws—which are related by the result known as *Noether's theorem*—has become a systematic part of our description of physics over the last century, from the atomic scale to the cosmic scale.

As an example of so-called *Galilean invariance*, Newton's force law keeps the same form when the velocity of the frame of reference (i.e., the coordinate system specified by $x$-, $y$-, and $z$-axes) is changed by adding a constant; this is equivalent to adding the same constant velocity to all the particles in a mechanical system. Other quantities *do* change under such a velocity transformation, such as the kinetic energy $\frac{1}{2}m|\boldsymbol{v}|^2$ (for a particle of mass $m$ and velocity $\boldsymbol{v}$); however, for an evolving, nondissipative system such as a bouncing, perfectly elastic rubber ball, the total energy is constant in time—that is, energy is conserved.

The development of our understanding of fundamental laws of dynamics can be interpreted by progressively more sophisticated and general representations of space and time themselves: ancient Greek physical science assumed absolute space with a privileged spatial point (the center of the Earth), through static Euclidean space where all spatial points are equivalent, through CLASSICAL MECHANICS [IV.19] where all inertial frames, moving at uniform velocity with respect to each other, are equivalent according to Newton's first law, to the modern theories of special and general relativity. The theory of relativity (both general and special) is motivated by Einstein's principle of covariance, which is described below. In this theory, space and time in different frames of reference are treated as coordinate systems on a four-dimensional pseudo-Riemannian manifold (whose mathematical background is described in TENSORS AND MANIFOLDS [II.33]), which manifestly combines conservation laws and continuous geometric symmetries of space and time. In special relativity (described in some detail in this article), this manifold is flat *Minkowski space-time*, generalizing Euclidean space to include time in a physically natural way. In general relativity, described in detail in GENERAL RELATIVITY AND COSMOLOGY [IV.40], this manifold may be curved, depending in part on the distribution of matter and energy according to EINSTEIN'S FIELD EQUATIONS [III.10].

In quantum physics, the description of a system in terms of a complex vector in Hilbert space gives rise to new symmetries. An important example is the fact that physical phenomena do not depend on the overall phase (argument) of this vector. Extension of Noether's theorem here leads to the conservation of electric charge, and extension to Yang–Mills theories provides other conserved quantities associated with the nuclear forces studied in contemporary fundamental particle physics. Other phenomena, such as the Higgs mechanism (leading to the Higgs boson recently discovered in high-energy experiments), are a consequence of the breaking of certain quantum symmetries in certain low-energy regimes. Symmetry and symmetry breaking in quantum theory are discussed briefly at the end of this article.

Spatial vectors, such as $\boldsymbol{r} = (x, y, z)$, represent the spatial distance between a chosen point and the origin, and of course the vector between two such points $\boldsymbol{r}_2 - \boldsymbol{r}_1$ is independent of translations of this origin. Similarly, the scalar product $\boldsymbol{r}_2 \cdot \boldsymbol{r}_1$ is unchanged under rotation of the coordinate system by an orthogonal matrix $\boldsymbol{R}$, under which $\boldsymbol{r} \to \boldsymbol{R}\boldsymbol{r}$.

*Continuous groups* of transformations such as translation and rotation, and their matrix representations, are an important tool used in calculations of invariants. For example, the set of two-dimensional matrices $\left(\begin{smallmatrix}\cos\theta & -\sin\theta \\ \sin\theta & \cos\theta\end{smallmatrix}\right)$, representing rotations through angles $\theta$, may be considered as a continuous one-parameter Abelian group of matrices generated by the MATRIX EXPONENTIAL [II.14] $\mathrm{e}^{\theta A}$, where $\boldsymbol{A}$ is the generator $\left(\begin{smallmatrix}0 & -1 \\ 1 & 0\end{smallmatrix}\right)$. The generator itself is found as the derivative of the original matrix with respect to $\theta$, evaluated at $\theta = 0$. Translations are less obviously represented by matrices; one approach is to append an extra dimension to the position vector with unit entry, such as $(1, x)$ specifying one-dimensional position $x$; a translation by $X$ is thus represented by

$$\begin{pmatrix}1 & 0 \\ X & 1\end{pmatrix} = \exp\left(X\begin{pmatrix}0 & 0 \\ 1 & 0\end{pmatrix}\right). \tag{1}$$

When a physical system is invariant under a one-parameter group of transformations, the corresponding generator plays a role in determining the associated conservation law.

## 1 Mechanics in Euclidean Space

It is conventional in classical mechanics to define the positions of a set of interacting particles in a vector space. However, we do not observe any unique origin to the three-dimensional space we inhabit, which we therefore take to be the *Euclidean space* $\mathbb{E}^3$; only relative positions between different interacting subsystems (i.e., positions relative to the common center of mass) enter the equations of motion. The entire system may be translated in space without any effect on the phenomena.

Of course, *external forces* acting on the system may prevent this, such as a rubber ball in a linear gravity field. (In such situations the source of the force, such as the Earth as the source of gravity, is not considered part of the system.) The gravitational force may be represented by a potential $V = gz$ for height $z$ and gravitational acceleration $g$; the ball's mass $m$ times the negative gradient, $-m\nabla V$, gives the downward force acting on the ball. The contours of $V$, given by $z$ = const., nevertheless have a symmetry: they are invariant to translations of the horizontal coordinates $x$ and $y$. Since the gradient of the potential is proportional to the gravitational force—which, by virtue of Newton's law equals the rate of change of the particle's linear momentum—the horizontal component of momentum does not change and is therefore conserved even when the particle bounces due to an impulsive, upward force from the floor. The continuous, horizontal translational symmetry of the system therefore leads to conservation of linear momentum in the horizontal plane. In a similar argument employing Newton's laws in cylindrical polar coordinates, the invariance of the potential to rotations about the $z$-axis leads to the conservation of the vertical component a body's angular momentum, as observed for tops spinning frictionlessly.

## 2 Noether's Theorem

The Lagrangian framework for mechanics (CLASSICAL MECHANICS [IV.19 §2]), which describes systems acting under forces defined by gradients of potentials (as in the previous section), is a natural mathematical setting in which to explore the connection between a system's symmetries and its conservation laws. Here, a mechanical system evolving in time $t$ is described by $n$ generalized coordinates $q_j(t)$ and their time derivatives $\dot{q}_j$, for $j = 1, \dots, n$, where the initial values $q_j(t_0)$ at time $t_0$ and final values $q_j(t_1)$ at $t_1$ are fixed. The *action* of the system is the functional

$$S[\{q_j\}] = \int_{t_0}^{t_1} L(\{q_j\}, \{\dot{q}_j\}, t)\, dt,$$

where $L(\{q_j\}, \{\dot{q}_j\}, t)$ is the *Lagrangian*; this is a function of the coordinates, their corresponding velocities, and maybe time, specified here by the total kinetic energy minus the total potential energy of the system (thereby capturing the forces as gradients of the potential energy). Using the CALCULUS OF VARIATIONS [IV.6], the functions $q_j(t)$ that satisfy the laws of mechanics are those that make the action stationary, and these satisfy Lagrange's equations of motion

$$\frac{\partial L}{\partial q_j} - \frac{\mathrm{d}}{\mathrm{d}t}\frac{\partial L}{\partial \dot{q}_j} = 0, \quad j = 1, \dots, n. \tag{2}$$

The argument of the time derivative in this expression, $\partial L/\partial \dot{q}_j$, is called the *canonical momentum* $p_j$ for each $j$. The set of equations (2) involves the combination of partial derivatives of the Lagrangian with respect to the coordinates and velocities, together with the total derivative with respect to time. By the chain rule, this total derivative affects explicit time dependence in $L$ and the implicit time dependence in each $q_j$ and $\dot{q}_j$. Many of the conservation laws involving Lagrangians involve such an interplay of explicit and implicit time dependence.

Any transformation of the coordinates $q_j$ that does not change the Lagrangian is a symmetry of the system. If $L$ does not have explicit dependence on a coordinate $q_j$, then the first term in (2) vanishes: $\mathrm{d}p_j/\mathrm{d}t = 0$, i.e., the corresponding canonical momentum is conserved in time. In the example from the last section of a particle in a linear gravitational field, the coordinates can be chosen to be Cartesian $x, y, z$, or cylindrical polars $r, \phi, z$; $L$ is independent of $x$ and $y$, leading to conservation of horizontal momentum, and also $\phi$, leading to conservation of angular momentum about the $z$-axis. The theorem is proved for symmetries of this type in CLASSICAL MECHANICS [IV.19 §2.3]: if a system is homogeneous in space (translation invariant), then linear momentum is conserved, and if it is isotropic (independent of rotations, such as the Newtonian gravitational potential around a massive point particle exerting a central force), then angular momentum is conserved (equivalent to Kepler's second law of planetary motion for gravity).

Since it is the equations (2) that represent the physical laws rather than the form of $L$ or $S$, the system may admit a more general kind of symmetry whose transformation adds a time-dependent function to the Lagrangian $L$. If, under the transformation, the Lagrangian transforms $L \to L + \mathrm{d}\Lambda/\mathrm{d}t$ involving the total time derivative of some function $\Lambda$, the action transforms $S \to S + \Lambda(t_1) - \Lambda(t_0)$. Thus the transformed action is still made stationary by functions satisfying (2), so transformations of this kind are symmetries of the system, which are also continuous if $\Lambda$ also depends on a continuous parameter $s$ so that its time derivative is zero when $s = 0$. It is not then difficult to see that the quantity

$$\sum_{j=1}^{n} p_j \frac{\partial q_j}{\partial s}\bigg|_{s=0} - \frac{\partial \Lambda}{\partial s}\bigg|_{s=0}, \tag{3}$$

defined in terms of the generators of the transformation on each coordinate and the Lagrangian, is constant in time. This is Noether's theorem for classical mechanics.

An important example is when the Lagrangian has no explicit dependence on time, $\partial L/\partial t = 0$. In this case, under an infinitesimal time translation $t \to t + \delta t$, $L \to L + \delta t \mathrm{d}L/\mathrm{d}t$, so here $\Lambda$ is $L\delta t$, with $L$ evaluated at $t$, and $\delta t$ plays the role of $s$. Under the same infinitesimal transformation, $q_j \to q_j + \delta t \dot{q}_j$, so the relevant conserved quantity (3) is $\sum_j p_j \dot{q}_j - L$, which is the Hamiltonian of the system, which is equal to the total energy in many systems of interest. It is apparently a fundamental law of physics that the total energy in physical processes is conserved in time; energy can be in other forms such as electromagnetic, gravitational, or heat, as well as mechanical. Noether's theorem states that the law of conservation of energy is equivalent to the fact that the physical laws of the system, characterized by their Lagrangian, do not change with time.

The vanishing of the action functional's integrand (i.e., the Lagrangian $L$) is equivalent to the existence of a first integral for the system of Lagrange equations, which is interpreted in the mechanical setting as a constant of the motion of the system. In this sense, Noether's theorem may be applied more generally in other physical situations described by functionals whose physical laws are given by the corresponding Euler–Lagrange equations. In the case of the Lagrangian approach applied to fields (i.e., functions of space and time), Noether's theorem generalizes to give a continuous density $\rho$ (such as mass or charge density) and a flux vector $\boldsymbol{J}$ satisfying the continuity equation $\dot{\rho} + \nabla \cdot \boldsymbol{J} = 0$ at every point in space and time.

## 3 Galilean Relativity

Newton's first law of motion can be paraphrased as "all inertial frames, traveling at uniform linear velocity with respect to each other, are equivalent for the formulation of mechanics"—that is, without action of external forces, a system will behave in the same way regardless of the motion of its center of mass. The behavior of a mechanical system is therefore independent of its overall velocity; this is a consequence of Newton's second law, that force is proportional to acceleration. According to pre-Newtonian physics, forces were thought to be proportional to *velocity* (as the effect of friction was not fully appreciated), and it was not until Galileo's thought experiments in friction-free environments that the proportionality of force to *acceleration* was appreciated. In spite of Galilean invariance, problems involving circular motion do in fact seem to require a privileged frame of reference, called *absolute space*. One example due to Newton himself is the problem of explaining, without absolute space, the meniscus formed by the surface of water in a spinning bucket; such problems are properly overcome only in general relativity.

With Galilean relativity, absolute position is no longer defined: events occurring at the same position but at different times in one frame (such as a moving train carriage) occur at different positions in other frames (such as the frame of the train track). However, changes to the state of motion, i.e., accelerations, have physical consequences and are related to forces. This is an example of a *covariance principle*, whose importance for physical theories was emphasized by Einstein. According to this principle, from the statement of physical laws in one frame of reference (such as the laws of motion), one can derive their statement in a different frame of reference from the application of the appropriate transformation rule between reference frames. The statement in the new frame should have the same *mathematical form* as in the previous frame, although quantities may not take the same values in different frames.

Transformations between different inertial frames are represented mathematically in a similar way to the translations of (1); events are labeled by their positions in space and time, such as $(t, x)$ in one frame and $(t, x')$ in another moving at velocity $v$ with respect to the first. Since $x' = x - vt$, the transformation from $(t, x)$ to $(t, x')$ is represented by the matrix $\left(\begin{smallmatrix} 1 & 0 \\ -v & 1 \end{smallmatrix}\right)$. This Galilean transformation (or *Galilean boost*) differs from (1) in that time $t$ is here appended to the position vector, since the translation from the boost is time-dependent. Galilean boosts in three spatial dimensions, together with regular translations and rotations, define the *Galilean group*. It can be shown that the Lagrangian of a free particle follows directly from the covariance of the corresponding action under the Galilean group.

Infinitesimal velocity boosts generate a Noetherian symmetry on systems of particles interacting via forces that depend only on the positions of the others. Consider $N$ point particles of mass $m_k$ and position $\boldsymbol{r}_k$ such that $V$ depends only on $\boldsymbol{r}_k - \boldsymbol{r}_\ell$ for $k, \ell = 1, \dots, N$. Under an infinitesimal boost by $\delta\boldsymbol{v}$ for each $k$, $\boldsymbol{r}_k \to \boldsymbol{r}_k + t\delta\boldsymbol{v}$, $\dot{\boldsymbol{r}}_k \to \dot{\boldsymbol{r}}_k + \delta\boldsymbol{v}$, and

$$L \to L + \tfrac{1}{2}|\delta\boldsymbol{v}|^2 \sum_k m_k + \delta\boldsymbol{v} \cdot \sum_k m_k\dot{\boldsymbol{r}}_k.$$

The resulting Noetherian conserved quantity (3) is

$$\boldsymbol{C} = t\sum_k m_k\dot{\boldsymbol{r}}_k - \sum_k m_k\boldsymbol{r}_k. \tag{4}$$

This quantity is constant in time; its value is minus the product of the system's total mass with the position of its center of mass when $t = 0$. Thus, $\dot{\boldsymbol{C}} = t\sum_k m_k\ddot{\boldsymbol{r}}$, that is, $t$ times the sum of the forces on each of the $N$ particles, which is zero by assumption.

## 4 Special Relativity

Historically, the physics of fields, such as electromagnetic radiation described by MAXWELL'S EQUATIONS [III.22], was developed much later than the mechanics of Galileo and Newton. The simplest wave equations, in fact, suggest a different invariant space-time from the Galilean framework above, which was first investigated by Hendrik Lorentz and Albert Einstein at the turn of the twentieth century.

Consider the time-dependent wave equation $\nabla^2\varphi - c^{-2}\ddot{\varphi} = 0$ for some scalar function $\varphi(t,\boldsymbol{r})$ and $\nabla^2$ the LAPLACE OPERATOR [III.18]. If all observers agree on the same value of the wave speed $c$ (as proposed by Einstein for $c$ the speed of light, and verified by experiments), this suggests that the transformation between moving inertial frames should be in terms of *Lorentz transformations* (or *Lorentz boosts*) in the $x$-direction,

$$\begin{pmatrix} ct' \\ x' \end{pmatrix} = \gamma(v)\begin{pmatrix} 1 & -v/c \\ -v/c & 1 \end{pmatrix}\begin{pmatrix} ct \\ x \end{pmatrix}, \tag{5}$$

where $(ct', x')$ denotes position in space and time (multiplied by the invariant speed $c$) in a frame moving along the $x$-axis at velocity $v$ with respect to the frame with space-time coordinates $(ct, x)$, and

$$\gamma(v) = (1 - v^2/c^2)^{-1/2}.$$

Requiring $\gamma(v)$ to be finite, positive, and real-valued implies $|v| < c$. According to the Lorentz transformation with $x = 0$, $t' = \gamma(\boldsymbol{v})t$, suggesting that $\gamma(v)$ may be interpreted as the rate of flow of time in the transformed frame moving at speed $v$ with respect to time in the reference frame.

According to Einstein's theory of *special relativity*, *all* physical laws should be invariant to translations, rotations, and Lorentz boosts, and Galilean transformations and Newton's laws arise in the low-velocity limit. Rotations and Lorentz boosts, together with time reflection and spatial inversion, define the *Lorentz group*; the *Poincaré group* is the semidirect product group of the Lorentz group with space-time translations. Special relativity is therefore also based on the principle of covariance, but with a different set of transformations between frames of reference. All of the familiar results of CLASSICAL MECHANICS [IV.19] are recovered when the boost transformations are restricted to $v \ll c$.

In special relativity different observers measure different time intervals between events, as well as different spatial separations. For a relativistic particle, different observers will disagree on the particle's relativistic energy $E$ and momentum $\boldsymbol{p}$. Nevertheless, according to the principle of special relativity, all observers agree on the quantity

$$E^2 - |\boldsymbol{p}|^2c^2 = m^2c^4,$$

so $m$ is an invariant on which all observers agree, called the *rest mass* of the particle. In the frame in which the particle is at rest, $E = mc^2$ arises as a special case. Otherwise, $E = m\gamma(v)c^2$, so in special relativity the moving particle's energy is explicitly related to the flow of time in the particle's rest frame with respect to that in the reference frame.

The Lagrangian formalism can be generalized to special relativity, with all observers agreeing on the form of the Euler–Lagrange equations; instead of being equal to its kinetic energy, the Lagrangian of a free particle is $-mc^2/\gamma(v)$.

The set of space-time events $(ct, \boldsymbol{r})$ described by special relativity is known as *Minkowski space* or Minkowski space-time (discussed in TENSORS AND MANIFOLDS [II.33]). It is a flat four-dimensional manifold with one time coordinate and three space coordinates (i.e., a manifold with a 3 + 1 pseudo-Euclidean metric), and the inertial frames with perpendicular spatial axes are analogous to Cartesian coordinates in Euclidean space. Time intervals and (simultaneous) spatial distances between events are no longer separately invariant; only the *space-time interval*

$$s^2 = |\Delta\boldsymbol{r}|^2 - c^2\Delta t^2$$

between two events with spatial separation $\Delta\boldsymbol{r}$ and time separation $\Delta t$ is invariant to Lorentz transformations and hence takes the same value in all inertial frames. $s^2$ may be positive, zero, or negative: if positive, there is a frame in which the pair of events are simultaneous; if zero, the events lie on the trajectory of a light ray; if negative, there is a frame in which the events occur at the same position. Since nothing can travel faster than

$c$, only events with $s^2 \leqslant 0$ can be causally connected; the events that may be affected by a given space-time point are those within its *future light cone.*

Being a manifold, the symmetries of Minkowski space are easily characterized, and its continuous symmetries are generated by vector fields, called *Killing vector fields*: dynamics in the space-time manifold is not changed under infinitesimal pointwise translation along these fields. The symmetries are translations in four linearly independent space-time directions, rotation about three linearly independent spatial axes, and Lorentz boosts in three linearly independent spatial directions (which are equivalent to rotations in space-time that mix the $t$-direction with a spatial direction). There are therefore ten independent symmetry transformations in Minkowski space-time (generating the Poincaré group), such that, for special relativistic systems not experiencing external forces, relativistic energy, momentum, angular momentum, and the analogue of $\boldsymbol{C}$ in (4) are conserved.

Both electromagnetic fields described by MAXWELL'S EQUATIONS [III.22] and relativistic quantum matter waves (strictly for a single particle) described by the DIRAC EQUATION [III.9] can be expressed as Lagrangian field theories; the combination of these is called *classical field theory.* As described above, Noether's theorem relates continuous symmetries of these theories to continuity equations of relativistic 4-currents. For instance, space-time translation symmetry ensures that the relativistic rank-2 *stress-energy-momentum* tensor, which describes the space-time flux of energy (including matter), is divergence free.

## 5 Other Theories

Einstein's theory of *general relativity* is a yet more general approach, which admits transformations between any coordinate systems on space-time (not simply inertial frames). This general formulation now applies to an arbitrary coordinate system, such as rotating coordinates (resolving the paradoxes implicit in Newton's bucket problem). This is formulated on a possibly curved, pseudo-Riemannian manifold where local neighborhoods of space-time events are equivalent to Minkowski space and free particles follow geodesics on the manifold. The geometry of the manifold—expressed by the Einstein curvature tensor—is proportional, by EINSTEIN'S FIELD EQUATIONS [III.10], to the stress-energy-momentum tensor, and gravitational forces are fictitious forces as freely falling particles follow curved space-time geodesics. The symmetries of general relativistic space-time manifolds are much more complicated than those for Minkowski space but are still formulated in terms of Killing vector fields that generate transformations that keep equations invariant.

In quantum physics (in a Galilean or special relativistic setting), the ideas described in this article are very important in *quantum field theory,* required to describe the quantum nature of electromagnetic and other fields, and systems involving many quantum particles. In quantum field theory, all energies in the system are quantized, often around a minimum-energy configuration. However, many field theories involve potentials whose minimum energy is not the most symmetric choice of origin; for instance, a "Mexican hat potential" of the form $|\boldsymbol{v}|^4 - 2a|\boldsymbol{v}|^2$ for some field vector $\boldsymbol{v}$ and constant $a > 0$ is rotationally symmetric around $\boldsymbol{v} = 0$ but has a minimum for any $\boldsymbol{v}$ with $|\boldsymbol{v}| = \sqrt{a}$. A minimum-energy excitation of the system *breaks* this symmetry by choosing an appropriate minimum energy $\boldsymbol{v}$. Many phenomena in the quantum theory of condensed matter (such as the Meissner effect in superconductors) and fundamental particles (such as the Higgs boson) arise from this type of symmetry breaking.

In classical mechanics, the time direction is privileged (it is common to all observers), so its space-time structure cannot be phrased simply in terms of manifolds. Nevertheless, the Galilean group of transformations naturally has a fiber bundle structure, with the base space given by the one-dimensional Euclidean line $\mathbb{E}^1$ describing time and the fiber being the spatial coordinates given by $\mathbb{E}^3$.

Galilean transformations can naturally be built into this fiber bundle structure, known as *Newton–Cartan space-time,* by defining the paths of free inertial particles (i.e., straight lines) in the connections of the bundle. This provides a useful mathematical framework to compare the physical laws and symmetries of Newtonian and relativity theories, and interesting connections exist between general relativity and Newtonian gravity in the Newton–Cartan formalism.

### Further Reading

Lorentz, H. A., A. Einstein, H. Minkowski, and H. Weyl. 1952. *The Principle of Relativity.* New York: Dover.

Misner, C. W., K. S. Thorne, and J. A. Wheeler. 1973. *Gravitation.* New York: W. H. Freeman.

Neuenschwander, D. E. 2010. *Emmy Noether's Wonderful Theorem.* Baltimore, MD: Johns Hopkins University Press.

Penrose, R. 2004. *The Road to Reality* (chapters 17–20 are particularly relevant). London: Random House.

## II.22 The Jordan Canonical Form

*Nicholas J. Higham*

A canonical form for a class of matrices is a form of matrix—usually chosen to be as simple as possible—to which all members of the class can be reduced by transformations of a specified kind. The Jordan canonical form (JCF) is associated with similarity transformations on $n \times n$ matrices. A similarity transformation of a matrix $A$ is a transformation from $A$ to $X^{-1}AX$, where $X$ is nonsingular. The JCF is the simplest form that can be achieved by similarity transformations, in the sense that it is the closest to a diagonal matrix.

The JCF of a complex $n \times n$ matrix $A$ can be written $A = ZJZ^{-1}$, where $Z$ is nonsingular and the Jordan matrix $J$ is a block-diagonal matrix

$$\begin{bmatrix} J_1 & & & \\ & J_2 & & \\ & & \ddots & \\ & & & J_p \end{bmatrix}$$

with diagonal blocks of the form

$$J_k = J_k(\lambda_k) = \begin{bmatrix} \lambda_k & 1 & & \\ & \lambda_k & \ddots & \\ & & \ddots & 1 \\ & & & \lambda_k \end{bmatrix}.$$

Here, blanks denote zero blocks or zero entries. The matrix $J$ is unique up to permutation of the diagonal blocks, but $Z$ is not. Each $\lambda_k$ is an eigenvalue of $A$ and may appear in more than one Jordan block. All the EIGENVALUES [I.2 §20] of the Jordan block $J_k$ are equal to $\lambda_k$. By definition, an eigenvector of $J_k$ is a nonzero vector $x$ satisfying $J_k x = \lambda_k x$, and all such $x$ are nonzero multiples of the vector $x = [1\ 0\ \cdots\ 0]^{\mathrm{T}}$. Therefore $J_k$ has only one linearly independent eigenvector. Expand $x$ to a vector $\tilde{x}$ with $n$ components by padding it with zeros in positions corresponding to each of the other Jordan blocks $J_i$, $i \neq k$. The vector $\tilde{x}$ has a single 1, in the $r$th component, say. A corresponding eigenvector of $A$ is $Z\tilde{x}$, since $A(Z\tilde{x}) = ZJZ^{-1}(Z\tilde{x}) = ZJ\tilde{x} = \lambda_k Z\tilde{x}$; this eigenvector is the $r$th column of $Z$.

If every block $J_k$ is $1 \times 1$ then $J$ is diagonal and $A$ is similar to a diagonal matrix; such matrices $A$ are called *diagonalizable*. For example, real symmetric matrices are diagonalizable—and moreover the eigenvalues are real and the matrix $Z$ in the JCF can be taken to be orthogonal. A matrix that is not diagonalizable is *defective*; such matrices do not have a complete set of linearly independent eigenvectors or, equivalently, their Jordan form has at least one block of dimension 2 or greater.

To give a specific example, the matrix

$$A = \frac{1}{2}\begin{bmatrix} 3 & 1 & 1 \\ -1 & 1 & 0 \\ 0 & 0 & 1 \end{bmatrix} \tag{1}$$

has a JCF with

$$Z = \begin{bmatrix} 0 & \frac{1}{2} & 1 \\ -1 & -\frac{1}{2} & 0 \\ 1 & 0 & 0 \end{bmatrix}, \qquad J = \left[\begin{array}{c|cc} \frac{1}{2} & 0 & 0 \\ \hline 0 & 1 & 1 \\ 0 & 0 & 1 \end{array}\right].$$

As the partitioning of $J$ indicates, there are two Jordan blocks: a $1 \times 1$ block with eigenvalue $\frac{1}{2}$ and a $2 \times 2$ block with eigenvalue 1. The eigenvalue $\frac{1}{2}$ of $A$ has an associated eigenvector equal to the first column of $Z$. For the double eigenvalue 1 there is only one linearly independent eigenvector, namely the second column, $z_2$, of $Z$. The third column, $z_3$, of $Z$ is a generalized eigenvector: it satisfies $Az_3 = z_2 + z_3$.

The JCF provides complete information about the eigensystem. The *geometric multiplicity* of an eigenvalue, defined as the number of associated linearly independent eigenvectors, is the number of Jordan blocks in which that eigenvalue appears. The *algebraic multiplicity* of an eigenvalue, defined as its multiplicity as a zero of the characteristic polynomial $q(t) = \det(tI - A)$, is the number of copies of the eigenvalue among all the Jordan blocks. For the matrix (1) above, the geometric multiplicity of the eigenvalue 1 is 1 and the algebraic multiplicity is 2, while the eigenvalue $\frac{1}{2}$ has geometric and algebraic multiplicities both equal to 1.

The *minimal polynomial* of a matrix is the unique monic polynomial $\psi$ of lowest degree such that $\psi(A) = 0$. The degree of $\psi$ is certainly no larger than $n$ because the CAYLEY–HAMILTON THEOREM [IV.10 §5.3] states that $q(A) = 0$. The minimal polynomial of an $m \times m$ Jordan block $J_k(\lambda_k)$ is $(t - \lambda_k)^m$. The minimal polynomial of $A$ is therefore given by

$$\psi(t) = \prod_{i=1}^{s}(t - \lambda_i)^{m_i},$$

where $\lambda_1, \ldots, \lambda_s$ are the distinct eigenvalues of $A$ and $m_i$ is the dimension of the largest Jordan block in which $\lambda_i$ appears. An $n \times n$ matrix is *derogatory* if the minimal polynomial has degree less than $n$. This is equivalent to some eigenvalue appearing in more than one Jordan block. The matrix $A$ in (1) is defective but not derogatory. The $n \times n$ identity matrix is derogatory for $n > 1$: it has characteristic polynomial $(t-1)^n$ and minimal polynomial $t - 1$.

Two questions that arise in many situations are, "Do the powers of the matrix $A$ converge to zero?" and "Are the powers of $A$ bounded?" The answers to both questions are easily obtained using the JCF. If $A = ZJZ^{-1}$ then $A^2 = ZJZ^{-1} \cdot ZJZ^{-1} = ZJ^2Z^{-1}$ and, in general, $A^k = ZJ^kZ^{-1}$. Therefore the powers of $A$ converge to zero precisely when the powers of $J$ converge to zero, and this in turn holds when the powers of each individual Jordan block converge to zero. The powers of a $1 \times 1$ Jordan block $J_i = (\lambda_i)$ obviously converge to zero when $|\lambda_i| < 1$. In general, since $J_i(\lambda_i)^k$ has diagonal elements $\lambda_i^k$, for the powers of $J_k(\lambda_k)$ to converge to zero it is necessary that $|\lambda_k| < 1$, and this condition turns out to be sufficient. Therefore $A^k \to 0$ as $k \to \infty$ precisely when $\rho(A) < 1$, where $\rho$ is the *spectral radius*, defined as the largest absolute value of any eigenvalue of $A$.

Turning to the question of whether the powers of $A$ are bounded, by the argument in the previous paragraph it suffices to consider an individual Jordan block. The powers of $J_k(\lambda_k)$ are clearly bounded when $|\lambda_k| < 1$, as we have just seen, and unbounded when $|\lambda_k| > 1$. When $|\lambda_k| = 1$ the powers are bounded if the block is $1 \times 1$, but they are unbounded for larger blocks. For example, $\left[\begin{smallmatrix} 1 & 1 \\ 0 & 1 \end{smallmatrix}\right]^k = \left[\begin{smallmatrix} 1 & k \\ 0 & 1 \end{smallmatrix}\right]$, which is unbounded as $k \to \infty$. The conclusion is that the powers of $A$ are bounded as long as $\rho(A) \leqslant 1$ and any eigenvalues of modulus 1 are in Jordan blocks of size 1. Thus the powers of $A$ in (1) are not bounded.

In one sense, defective matrices—those with nontrivial Jordan structure—are very rare because the diagonalizable matrices are dense in the set of all matrices. Therefore if you generate matrices randomly you will be very unlikely to generate one that is not diagonalizable (this is true even if you generate matrices with random integer entries). But in another sense, defective matrices are quite common. Certain types of BIFURCATIONS [IV.21] in dynamical systems are characterized by the presence of nontrivial Jordan blocks in the Jacobian matrix, while in problems where some function of the eigenvalues of a matrix is optimized the optimum often occurs at a defective matrix.

While the JCF provides understanding of a variety of matrix problems, it is not suitable as a computational tool. The JCF is not a continuous function of the entries of the matrix and can be very sensitive to perturbations. For example, for $\varepsilon \neq 0$, $\left[\begin{smallmatrix} 1 & \varepsilon \\ 0 & 1 \end{smallmatrix}\right]$ (one Jordan block) and $\left[\begin{smallmatrix} 1 & 0 \\ 0 & 1 \end{smallmatrix}\right]$ (two Jordan blocks) have different Jordan structures, even though the matrices can be made arbitrarily close by taking $\varepsilon$ sufficiently small. In practice, it is very difficult to compute the JCF in floating-point arithmetic due to the unavoidable perturbations caused by rounding errors. As a general principle, the SCHUR DECOMPOSITION [IV.10 §5.5] is preferred for practical computations.

## II.23 Krylov Subspaces

*Valeria Simoncini*

### 1 Definition and Properties

The $m$th Krylov subspace of the matrix $A \in \mathbb{C}^{n \times n}$ and the vector $v \in \mathbb{C}^n$ is

$$\mathcal{K}_m(A, v) = \operatorname{span}\{v, Av, \ldots, A^{m-1}v\}.$$

The dimension of $\mathcal{K}_m(A, v)$ is at most $m$, and it is less if an invariant subspace of $A$ with respect to $v$ is obtained for some $m_* < m$. In general, $\mathcal{K}_m(A, v) \subseteq \mathcal{K}_{m+1}(A, v)$ (the spaces are nested); if $m_* = n$, then $\mathcal{K}_n(A, v)$ spans the whole of $\mathbb{C}^n$.

Let $v_1 = v/\|v\|_2$, with $\|v\|_2 = (v^*v)^{1/2}$ the 2-norm, and let $\{v_1, v_2, \ldots, v_m\}$ be an orthonormal basis of $\mathcal{K}_m(A, v)$. Setting $\mathcal{V}_m = [v_1, v_2, \ldots, v_m]$, from the nesting property it follows that the next basis vector $v_{m+1}$ can be computed by the following Arnoldi relation:

$$A\mathcal{V}_m = [\mathcal{V}_m, v_{m+1}]H_{m+1,m},$$

where $H_{m+1,m} \in \mathbb{C}^{(m+1)\times m}$ is an *upper Hessenberg* matrix (upper triangular plus nonzero entries immediately below the diagonal) whose columns contain the coefficients that make $v_{m+1}$ orthogonal to the already available basis vectors $v_1, \ldots, v_m$.

Suppose we wish to approximate a vector $y$ by a vector $x \in \mathcal{K}_m(A, v)$, measuring error in the 2-norm. Any such $x$ can be written as a polynomial in $A$ of degree at most $m - 1$ times $v$: $x = \sum_{i=0}^{m-1} \alpha_i A^i v$. If $A$ is Hermitian, then by using a spectral decomposition we can reduce $A$ to diagonal form by unitary transformations, which do not change the 2-norm, and it is then clear that the eigenvalues of $A$ and the decomposition of $v$

in terms of the eigenvectors of $A$ drive the approximation properties of the space. For non-Hermitian $A$ the approximation error is harder to analyze, especially for highly nonnormal or nondiagonalizable matrices.

By replacing $v$ by an $n \times s$ matrix $V$, with $s \geqslant 1$, spaces of dimension at most $ms$ can be obtained. An immediate matrix counterpart is

$$\mathbb{K}_m(A, V) = \left\{ \sum_{i=0}^{m-1} \gamma_i A^i V : \gamma_i \in \mathbb{C} \text{ for all } i \right\}.$$

A richer version is obtained by working with linear combinations of all the available vectors:

$$\mathbb{K}_m^{\square}(A, V) = \text{range}([V, AV, \dots, A^{m-1}V]).$$

Methods based on this latter space are called "block" methods, since all matrix structure properties are generalized to blocks (e.g., $H_{m+1,m}$ will be block upper Hessenberg, with $s \times s$ blocks). Block spaces are appropriate, for instance, in the presence of multiple eigenvalues or if the original application requires using the same $A$ and different vectors $v$.

## 2 Applications and Generalizations

Krylov subspaces are used in projection methods for solving large algebraic linear systems, eigenvalue problems, and matrix equations; for approximating a wide range of matrix functions (analytic functions, trace, determinant, transfer functions, etc.); and in model order reduction.

The general idea is to project the original problem of size $n$ onto the Krylov subspace of dimension $m \ll n$ and then solve the smaller $m \times m$ reduced problem with a more direct method (one that would be too computationally expensive if applied to the original $n \times n$ problem). If the Krylov subspace is good enough, then the projected problem retains sufficient information from the original problem that the sought after quantities are well approximated.

When equations are involved, Krylov subspaces usually play a role as approximation spaces, as well as test spaces. The actual test space used determines the resulting method and influences the convergence properties.

Generalized spaces have emerged as second-generation Krylov subspaces. In the eigenvalue context, the "shift-and-invert" Krylov subspace $\mathcal{K}_m((A - \sigma I)^{-1}, v)$ is able to efficiently approximate eigenvalues in a neighborhood of a fixed scalar $\sigma \in \mathbb{C}$; here $I$ is the identity matrix of size $n$. In matrix function evaluations and matrix equations, the extended space $\mathcal{K}_m(A, v) + \mathcal{K}_m(A^{-1}, A^{-1}v)$ has shown some advantages over the classical space, while for $\sigma_1, \dots, \sigma_m \in \mathbb{C}$, the use of the more general *rational* space

$$\text{span}\{(A - \sigma_1 I)^{-1}v, \dots, (A - \sigma_m I)^{-1}v\}$$

has recently received a lot of attention for its potential in a variety of advanced applications beyond eigenvalue problems, where it was first introduced in the 1980s. All these generalized spaces require solving systems with some shifted forms of $A$, so that they are in general more expensive to build than the classical one, depending on the computational cost involved in solving these systems. However, the computed space is usually richer in spectral information, so that a much smaller space dimension is required to satisfactorily approximate the requested quantities. The choice among these variants thus depends on the spectral and sparsity properties of the matrix $A$.

### Further Reading

Freund, R. W. 2003. Model reduction methods based on Krylov subspaces. *Acta Numerica* 12:126–32.

Liesen, J., and Z. Strakoš. 2013. *Krylov Subspace Methods: Principles and Analysis.* Oxford: Oxford University Press.

Saad, Y. 2003. *Iterative Methods for Sparse Linear Systems,* 2nd edn. Philadelphia, PA: SIAM.

Watkins, D. S. 2007. *The Matrix Eigenvalue Problem: GR and Krylov Subspace Methods.* Philadelphia, PA: SIAM.

# II.24 The Level Set Method

*Fadil Santosa*

## 1 The Basic Idea

The level set method is a numerical method for representing a closed curve or surface. It uses an implicit representation of the geometrical object in question. It has found widespread use in problems where the closed curve evolves or needs to be recomputed often. A main advantage of the method is that such a representation is very flexible and calculation can be done on a regular grid. In computations where surfaces evolve, changes in the topology of the surface are easily handled.

Consider an example in two dimensions in the $(x, y)$-plane. Suppose one is interested in the motion of a curve under external forcing terms. Let $C(t)$ denote the curve as a function of time $t$. One method for solving this problem is to track the curve, which can be done by choosing marker points, $(x_i(t), y_i(t)) \in C(t)$,

**Figure 1** The graphs of a level set function $z = \varphi(x,y,t)$ for three values of $t$ are shown at the bottom of the figure. The plane $z = 0$ intersects these functions. The domains $D(t) = \{(x,y)\colon \varphi(x,y,t) > 0\}$ are shown above. Note that, in this example, $D(t)$ has gone through a topological change as $t$ varies.

$i = 1,\dots,n$, whose motions are determined by the forcing. The curve itself may be recovered at any time $t$ by a prescribed spline interpolation.

The level set method takes a different approach; it represents the curve as the zero level set of a function $\varphi(x,y,t)$. That is, the curve is given by

$$C(t) = \{(x,y)\colon \varphi(x,y,t) = 0\}.$$

One can set up the function so that the interior of the curve $C(t)$ is the set

$$D(t) = \{(x,y)\colon \varphi(x,y,t) > 0\}.$$

In figure 1 the level set function $z = \varphi(x,y,t)$ can be seen to intersect the plane $z = 0$ at various times $t$. The sets of $D(t)$ (not up to the same scale) are shown above each three-dimensional figure.

An advantage of the level set method is demonstrated in the figure. One can see that a topological change in $D(t)$ has occurred as $t$ is varied. The level set method allows for such a change without the need to redefine the representation, as would be the case for the front-tracking method described previously.

## 2 Discretization

One of the attractive features of the level set method is that calculations are done on a regular Cartesian grid. Suppose we have discretized the computational domain and the nodes are at coordinates $(x_i, y_j)$ for $i = 1,\dots,m$ and $j = 1,\dots,n$. The values of the level set function $\varphi(x,y,t)$ are then stored at coordinate points $x = x_i$ and $y = y_j$. At any time, if one is interested in the curve $C(t)$, the zero level set, the set

$$C(t) = \{(x,y)\colon \varphi(x,y,t) = 0\}$$

needs to be approximated from the data $\varphi(x_i, y_j, t)$. One is typically interested in such quantities as the normal to the curve and the curvature at a point on the curve. These quantities are easily calculated by evaluating finite-difference approximations of

$$\nu = \frac{\nabla\varphi}{|\nabla\varphi|}$$

and

$$\kappa = \nabla\cdot\nu,$$

where the gradient operator $\nabla = [\partial/\partial x \;\; \partial/\partial y]^{\mathrm{T}}$.

In practice, it is not necessary to keep all values of the level set function on the nodes. Since one is often interested only in the motion of the curve $C(t)$, the zero level set, one needs only the values of the level set function in the neighborhood of the curve. Such approaches have been dubbed "narrow-band methods" and can potentially reduce the amount of computation in a problem involving complex evolution of surfaces.

It must be noted that, in the two-dimensional example here, $C(t)$ is a one-dimensional object, whereas the level set function $\varphi(x,y,t)$ is a two-dimensional function. Thus, one might say that the ability to track topological changes is made at the cost of increased computational complexity.

## 3 Applications

A simple problem one may pose is that of tracking the motion of a curve for which every point on the curve is moving in the direction normal to the curve with a given velocity. If the velocity is $v$, then the equation for the level sets is given by

$$\frac{\partial\varphi}{\partial t} = v|\nabla\varphi|.$$

If one is interested in tracking the motion of the zero level set $C(t)$, then one must specify an initial condition

$$\varphi(x,y,0) = \varphi_0(x,y),$$

where the initial zero level set is given by

$$C(0) = \{(x,y)\colon \varphi_0(x,y) = 0\}.$$

This evolution of such a curve may be very complicated and go through topological changes. The power of the level set method is demonstrated here because all one needs to do is solve the initial-value problem for $\varphi(x,y,t)$.

Another simple problem is the classical motion by mean curvature. In this "flow," one is interested in tracking the motion of a curve for which every point on the curve is moving normal to the curve at a velocity proportional to the curvature. The evolution equation is given by

$$\frac{\partial \varphi}{\partial t} = \nabla \cdot \frac{\nabla \varphi}{|\nabla \varphi|}.$$

A numerical solution of this evolution equation can be used to demonstrate the classical Grayson theorem, which asserts that, if the closed curve starts out without self-intersections, then it will never form self-intersections and it will become convex in finite time.

Significant problems arising from applications from diverse fields have benefited from the level set treatment. The following is an incomplete list meant to give a sense of the range of applications.

**Image processing.** The level set method can be used for segmentation of objects in a two-dimensional scene. It has also been demonstrated to be effective in modeling surfaces from point clouds.

**Fluid dynamics.** Two-phase flows, which involve interfaces separating the two phases, can be approached by the level set method. It is particularly effective for problems in which one of the phases is dispersed in bubbles.

**Inverse problems.** Inverse problems exist in which the unknown that one wishes to reconstruct from data is the boundary of an object. Examples include inverse scattering.

**Optimal shape design.** When the object is to design a shape that maximizes certain attributes (design objectives), it is often very convenient to represent the shape by a level set function.

**Computer animation.** The need for physically based simulations in the animation industry has been partially met by solving equations of physics using the level set method to represent surfaces that are involved in the simulation.

Current research areas include improved accuracy in the numerical schemes employed and in applying the method to ever more complex physics.

### Further Reading

Osher, S., and R. Fedkiw. 2003. *Level Set Methods and Dynamic Implicit Surfaces.* New York: Springer.

Osher, S., and J. A. Sethian. 1988. Fronts propagating with curvature-dependent speed: algorithms based on Hamilton–Jacobi formulations. *Journal of Computational Physics* 79:12–49.

Sethian, J. A. 1999. *Level Set Methods and Fast Marching Methods: Evolving Interfaces in Computational Geometry, Fluid Mechanics, Computer Vision, and Materials Science.* Cambridge: Cambridge University Press.

## II.25 Markov Chains

*Beatrice Meini*

A Markov chain is a type of random process whose behavior in the future is influenced only by its current state and not by what happened in the past. A simple example is a random walk on $\mathbb{Z}$, where a particle moves among the integers of the real line and is allowed to move one step forward with probability $0 < p < 1$ and one step backward with probability $q = 1 - p$ (see figure 1). The position of the particle at time $n + 1$ (in the future) depends on the position of the particle at time $n$ (at the present time), and what happened before time $n$ (in the past) is irrelevant.

To give a precise definition of a Markov chain we will need some notation. Let $E$ be a countable set representing the states, and let $\Omega$ be a set that represents the sample space. Let $X, Y \colon \Omega \to E$ be two random variables. We denote by $\mathbb{P}[X = j]$ the probability that $X$ takes the value $j$, and we denote by $\mathbb{P}[X = j \mid Y = i]$ the probability that $X$ takes the value $j$ given that the random variable $Y$ takes the value $i$. A discrete stochastic process is a family $(X_n)_{n\in\mathbb{N}}$ of random variables $X_n \colon \Omega \to E$.

A stochastic process $(X_n)_{n\in\mathbb{N}}$ is called a *Markov chain* if

$$\begin{aligned}\mathbb{P}[X_{n+1} = i_{n+1} \mid X_0 = i_0, \dots, X_n = i_n]\\ = \mathbb{P}[X_{n+1} = i_{n+1} \mid X_n = i_n]\end{aligned}$$

at any time $n \geqslant 0$ and for any states $i_0, \dots, i_{n+1} \in E$. This means that the state $X_n$ at time $n$ is sufficient to determine which state $X_{n+1}$ might be occupied at time $n + 1$, and we may forget the past history $X_0, \dots, X_{n-1}$.

It is often required that the laws that govern the evolution of the system be time invariant. The Markov chain is said to be *homogeneous* if the transitions from one state to another state are independent of the time $n$, i.e., if

$$\mathbb{P}[X_{n+1} = j \mid X_n = i] = p_{ij}$$

at any time $n \geqslant 0$ and for any states $i, j \in E$. The number $p_{ij}$ represents the probability of passing from state $i$ to state $j$ in one time step. The matrix $P = (p_{ij})_{i,j\in E}$ is called the *transition matrix* of the Markov chain. The matrix $P$ is a *stochastic matrix*: that is, it has nonnegative entries and unit row sums ($\sum_{j\in E} p_{ij} = 1$ for all $i \in E$). The dynamic behavior of the Markov chain is governed by the transition matrix $P$. In particular, the problem of computing the probability that, after $n$

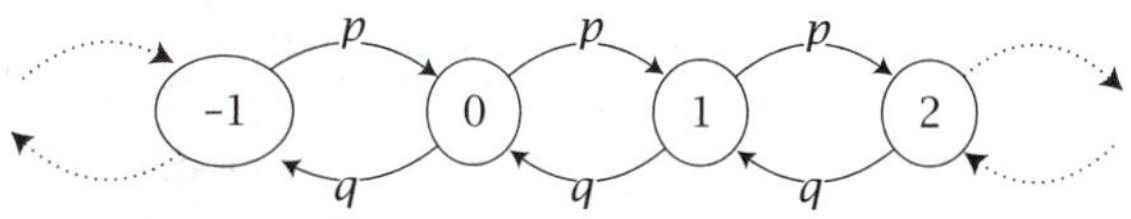


**Figure 1** A random walk on $\mathbb{Z}$.

steps, the Markov chain is in a given state reduces to calculating the entries of the $n$th power of $P$, since

$$\mathbb{P}[X_n = j \mid X_0 = i] = (P^n)_{ij},$$

where $(P^n)_{ij}$ denotes the $(i, j)$ entry of $P^n$.

In the random walk on $\mathbb{Z}$, the set of states is $E = \mathbb{Z}$, and the random variable $X_n$ is the position of the particle at time $n$. The stochastic process $(X_n)_{n\in\mathbb{N}}$ is a homogeneous Markov chain. The transition matrix $P$ is tridiagonal, with $p_{i,i+1} = p$, $p_{i,i-1} = q$, and $p_{ij} = 0$ for $j \neq i-1, i+1$.

In many applications, the interest is in the asymptotic behavior of the Markov chain. In particular, the question is to understand if $\lim_{n\to\infty} \mathbb{P}[X_n = j \mid X_0 = i]$ exists and if that limit is independent of the initial state $i$. If such a limit exists and is equal to $\pi_j > 0$ for any initial state $i$, then the vector $\pi = (\pi_j)_{j\in E}$ is such that $\sum_{j\in E} \pi_j = 1$ and $\pi^{\mathrm{T}} P = \pi^{\mathrm{T}}$. This vector $\pi$ is called the *steady state vector*, or the *probability invariant vector*. If $E$ is a finite set, then the steady state vector is a left eigenvector of $P$ corresponding to the eigenvalue 1; moreover, if in addition the matrix $P$ is irreducible, then there exists a unique steady state vector.

The matrix

$$P = \begin{bmatrix} \frac{1}{4} & \frac{1}{2} & \frac{1}{4} \\ 0 & \frac{2}{3} & \frac{1}{3} \\ \frac{1}{2} & 0 & \frac{1}{2} \end{bmatrix} \tag{1}$$

is the transition matrix of a Markov chain with space state $E = \{1, 2, 3\}$. The transitions among the states can be represented by the graph of figure 2. The powers of $P$ converge:

$$\lim_{n\to\infty} P^n = \begin{bmatrix} \frac{1}{4} & \frac{3}{8} & \frac{3}{8} \\ \frac{1}{4} & \frac{3}{8} & \frac{3}{8} \\ \frac{1}{4} & \frac{3}{8} & \frac{3}{8} \end{bmatrix}.$$

Hence the steady state vector is $\pi^{\mathrm{T}} = [\frac{1}{4}\ \frac{3}{8}\ \frac{3}{8}]$. A simple computation shows that $\pi^{\mathrm{T}} = \pi^{\mathrm{T}} P$.

Markov chains have applications across a wide range of topics in different areas, including mathematical biology, chemistry, queueing theory, information sciences, economics and finance, Internet applications, and more. According to the model, the transition matrix $P$ can be finite or infinite dimensional and can have

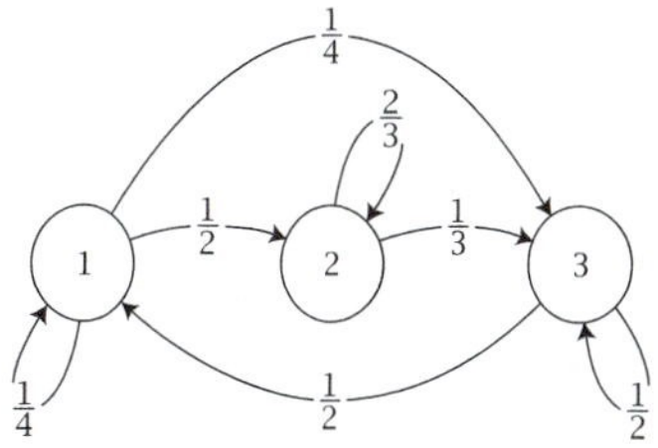


**Figure 2** Transitions of the Markov chain having the matrix $P$ of equation (1) as transition matrix.

specific structures, like sparsity or pattern structure. In most of the applications of Markov chains, the interest is in the computation of the steady state vector, assuming that it exists. In this regard, specific algorithms can be designed according to the properties of the matrix $P$.

### Further Reading

Stewart (1994) is an introduction to numerical methods for general Markov chains, while Bini et al. (2005) presents specific algorithms for Markov chains arising in queueing models. Norris (1999) gives a complete treatise on Markov chains.

Bini, D. A., G. Latouche, and B. Meini. 2005. *Numerical Methods for Structured Markov Chains*. New York: Oxford University Press.

Norris, J. R. 1999. *Markov Chains*, 3rd edn. Cambridge: Cambridge University Press.

Stewart, W. J. 1994. *Introduction to the Numerical Solution of Markov Chains*. Princeton, NJ: Princeton University Press.

## II.26 Model Reduction

*Peter Benner*

"Model reduction" is an ambiguous term; in this article it is understood to mean the reduction of the complexity of a mathematical model by (semi-) automatic mathematical algorithms. Often such techniques are also called "dimension reduction," "order reduction," or, inspired by system-theoretic terminology, "model order reduction." The concept is used in various application areas and in different contexts. Model order reduction has emerged in many disciplines, primarily in structural dynamics, systems and control theory, computational fluid mechanics, chemical process engineering, and, more recently, circuit simulation, microelectromechanical systems, and computational electromagnetics. It now also finds its way into numerous other application areas such as image processing, computational neuroscience, and computational/computer-aided engineering in general.

Here, we will focus on the reduction of the dimension of the state space of a dynamical system, for which we use the following model equation:

$$E(t,p)\dot{x}(t) = f(t,x(t),u(t),p), \tag{1a}$$

$$x(t_0) = x_0, \tag{1b}$$

$$y(t) = g(t,x(t),u(t),p). \tag{1c}$$

In this set of equations, $x(t) \in \mathbb{R}^n$ is the state of the system at time $t \in [t_0, T]$ ($t_0 < T \leqslant \infty$), $u(t) \in \mathbb{R}^m$ denotes inputs (control, time-varying parameters), $p \in \Omega \subset \mathbb{R}^d$ is a vector of stationary (material, geometry, design, etc.) parameters, $\Omega$ is usually a bounded domain, and $x_0 \in \mathbb{R}^n$ is the initial state of the system. The matrix $E(t,p) \in \mathbb{R}^{n\times n}$ determines the nature of the system. When it is uniformly nonsingular, (1a) represents a system of ordinary differential equations; otherwise (1) is a *descriptor system.* If $d > 0$, i.e., when parameters are present, one may also consider the case $E(t,p) \equiv 0$. Then (1a) becomes a system of (nonlinear) algebraic equations. Model reduction preserving the parameters as symbolic quantities will then accelerate the approximate solution of (1a) in the case of varying parameters (the "many-query context"). The equation (1c) describes an output vector $y(t) \in \mathbb{R}^q$. It may correspond to a practical setting where only a few measurements of the system or observables are available. This equation may also be used to identify quantities of interest when full state information is not required in the application. If full state information is needed, one simply sets $y(t) = x(t)$. The functions $f$ and $g$ are assumed to have sufficient smoothness properties, where for $f$, Lipschitz continuity is usually a minimum requirement for ensuring existence and uniqueness of local solutions of (1a).

The goal of model reduction is to replace (1) by a system of reduced state-space dimension $r \ll n$ of the same form,

$$\hat{E}(t,p)\dot{\hat{x}}(t) = \hat{f}(t,\hat{x}(t),u(t),p), \tag{2a}$$

$$\hat{x}(t_0) = \hat{x}_0, \tag{2b}$$

$$\hat{y}(t) = g(t,\hat{x}(t),u(t),p), \tag{2c}$$

with the inputs $u(t)$ and the parameter vector $p$ unchanged from (1), such that $\hat{y}$ matches $y$ as closely as possible for all admissible control inputs and parameters. Additionally, one may require the preservation of structural properties like stability, passivity, dissipativity, etc. It should be clear that it is difficult to fulfill all these demands at once, and various model order reduction methods for specific applications have therefore been developed. A common principle can be found in many of these, and the mathematical core of the methods is often very similar.

## 1 The Basic Concept

The basic principle behind most model order reduction methods is the projection of the state equation (1a) onto a low-dimensional subset $\mathcal{V} \subset \mathbb{R}^n$, possibly along a complementary subspace $\mathcal{W}$ of the same dimension. The projection onto nonlinear subsets like central or (approximate) inertial manifolds is the topic of the theory of dynamical systems and has been applied so far mainly in reaction kinetics, process engineering, and systems biology. Most other successful families of methods—such as *modal truncation, balanced truncation, Padé approximation/Krylov subspace methods/ moment matching* (which are all instances of rational interpolation), and *proper orthogonal decomposition* and *reduced basis methods*—use linear projection subspaces, and they can mostly be categorized as (Petrov-) Galerkin projection methods. Suppose we are given an orthogonal basis of $\mathcal{V}$, represented by the column space of $V \in \mathbb{R}^{n\times r}$, and a basis of $\mathcal{W}$ forming the column space of $W \in \mathbb{R}^{n\times r}$ (where $\mathcal{W} = \mathcal{V}$ and $W = V$ in the Galerkin case) such that $W^{\mathrm{T}}V = I_r$ (the $r \times r$ identity matrix). The reduced-order model is then obtained by setting $\hat{x} = W^{\mathrm{T}}x$, $\hat{x}_0 = W^{\mathrm{T}}x_0$, and making the residual of (1a) for $\tilde{x}$ orthogonal to $\mathcal{W}$:

$$\mathcal{W} \perp E(t,p)\dot{\tilde{x}} - f(t,\tilde{x},u,p) \quad \forall \text{ admissible } t,u,p$$
$$\Leftrightarrow \quad 0 = W^{\mathrm{T}}(E(t,p)\dot{\tilde{x}} - f(t,\tilde{x},u,p)).$$

The different methods now mainly differ in the way $V$ (and $W$) are computed. The full state $x$ of (1) is approximated by $\tilde{x} = V\hat{x}(t) = VW^{\mathrm{T}}x(t)$.

As an example, consider a linear time-invariant system without parameters:

$$E(t,p) \equiv E, \tag{3a}$$

$$f(t,x,u,p) = Ax + Bu, \tag{3b}$$

$$g(t,x,u,p) = Cx, \tag{3c}$$

with $A, E \in \mathbb{R}^{n\times n}$, $B \in \mathbb{R}^{n\times m}$, $C \in \mathbb{R}^{q\times n}$. For instance, interpolation methods (and also balanced truncation) utilize the transfer function

$$G(s) = C(sE - A)^{-1}B, \quad s \in \mathbb{C},$$

of (1), which in the linear time-invariant case is obtained by taking Laplace transforms and inserting the transformed equation (1a) into the transformed (1c). The transfer function represents the mapping of inputs $u$ to outputs $y$. As a rational matrix-valued function of a complex variable, it can be approximated in different ways. In rational interpolation methods, $V$, $W$ are computed so that

$$\frac{\mathrm{d}^j}{\mathrm{d}s^j}G(s_k) = \frac{\mathrm{d}^j}{\mathrm{d}s^j}\hat{G}(s_k), \quad k = 1,\dots,K,\ j = 0,\dots,J_k,$$

for $K$ interpolation points $s_k$ and derivatives up to order $J_k$ at each point. Here, $\hat{G}$ denotes the transfer function of (2) and (3), defined by $\hat{A} = W^{\mathrm{T}}AV$, $\hat{B} = W^{\mathrm{T}}B$, and $\hat{C} = CV$.

In the nonlinear case, a further question is how to obtain functions $\hat{f}$ and $\hat{g}$ allowing for fast evaluation. Simply setting $\hat{f}(t,x,u,p) = W^{\mathrm{T}}f(t,V\hat{x},u,p)$ obviously does not lead to faster simulation in general. Therefore, dedicated methods, such as (discrete) empirical interpolation, are needed to obtain a "reduced" $\hat{f}$ and $\hat{g}$.

## 2 An Example

As an example consider the mathematical model of a microgyroscope: a device used in stability control of vehicles. Finite-element discretization of this particular model leads to a linear time-invariant system of $n = 34\,722$ linear ordinary differential equations with $d = 4$ parameters, $m = 1$ input, and $q = 12$ outputs. Using a reduced-order model of size $r = 289$, a parameter study involving two parameters (defining $x$- and $y$-axes in figure 1) and the excitation frequency $\omega$ (i.e., the parametric transfer function $G(s,p)$ is evaluated for $s = \mathrm{i}\omega$ with varying $\omega$) could be accelerated by a factor of approximately 90 without significant loss of accuracy. The output $y$ was computed with an error of less than 0.01% in the whole frequency and parameter domain. Figure 1 shows the response surfaces of the full and reduced-order models at one frequency for variations of two parameters.

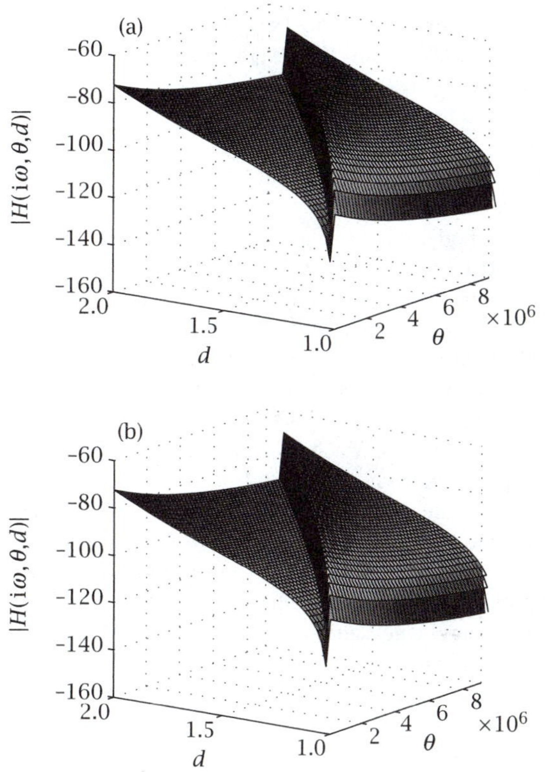


**Figure 1** The parametric transfer function of a microdevice (at $\omega = 0.025$): results from (a) the full model with dimension 34 722 and (b) the reduced-order model with dimension 289. (Computations and graphics by L. Feng and T. Breiten.)

### Further Reading

Antoulas, A. 2005. *Approximation of Large-Scale Dynamical Systems*. Philadelphia, PA: SIAM.

Benner, P., M. Hinze, and E. J. W. ter Maten, eds. 2011. *Model Reduction for Circuit Simulation*. Lecture Notes in Electrical Engineering, volume 74. Dordrecht: Springer.

Benner, P., V. Mehrmann, and D. C. Sorensen, eds. 2005. *Dimension Reduction of Large-Scale Systems*. Lecture Notes in Computational Science and Engineering, volume 45. Berlin: Springer.

Schilders, W. H. A., H. A. van der Vorst, and J. Rommes, eds. 2008. *Model Order Reduction: Theory, Research Aspects and Applications*. Mathematics in Industry, volume 13. Berlin: Springer.

# II.27 Multiscale Modeling

*Fadil Santosa*

To accurately model physical, biological, and other phenomena, one is often confronted with the need to capture complex interactions occurring at distinct temporal and spatial scales. In the language of multiscale modeling, temporal scales are usually differentiated by slow, medium, and fast timescales. Spatially, the phenomena are separated into micro-, meso-, and macroscales. In modeling the deformation of solids, for instance, the microscale phenomena could be atomistic interactions occurring on a femtosecond ($10^{-15}$ second) timescale. At the mesoscopic scale, one could be interested in the behavior of the constituent macromolecules, e.g., a tangled bundle of polymers. Finally, at the macroscopic scale, one might be interested in how a body, whose size could be in meters, deforms under an applied force. The challenge in multiscale modeling is that the interactions at one scale communicate with interactions at other scales. Thus, in the example given, the question we wish to answer is how the applied forces affect the atomistic interactions, and how those interactions impact the behavior of the macromolecules, which in turn affects how the overall shape of the body deforms.

Multiscale modeling is a rapidly developing field because of its enormous importance in applications. The range of applications is staggering. It has been applied in geophysics, biology, chemistry, meteorology, materials science, and physics.

We give another concrete example that arises in solid mechanics. Suppose we have a block of pure aluminum whose crystalline structure is known. How can we calculate its elastic properties, i.e., its Lamé modulus and Poisson ratio, *ab initio* from knowledge of its atomistic structure? Such a calculation would start by considering Schrödinger's equation for the multiparticle system. By solving for the ground states of the system, one can then extract the desired macroscopic properties of the bulk aluminum.

A classical example of multiscale modeling in applied mathematics is the HOMOGENIZATION METHOD [II.17], which allows for extraction of effective properties of composite materials. Consider the steady-state distribution of temperature in a rod of length $\ell$ made up of a material with rapidly oscillating conductivity. The conductivity is described by a periodic function $a(y) > 0$, such that $a(y+1) = a(y)$. A small-scale $\varepsilon$ is introduced to denote the actual period in the medium. The governing equation for temperature $u(x)$ is

$$\left(a\left(\frac{x}{\varepsilon}\right)u'\right)' = f, \quad 0 < x < \ell,$$

where a prime denotes differentiation with respect to $x$. Here, $f$ is the heat source distribution, with $x$ measuring distance along the rod. To solve the problem, the solution $u$ is developed in powers of $\varepsilon$. The macroscopic behavior of $u$ is identified with the zeroth order. This solution will be smooth as the small rapid oscillations are ignored.

Current research in multiscale modeling focuses on bridging the phenomena at the different scales and developing efficient numerical methods. There are efforts to develop rigorous multiscale models that agree with their continuum counterparts. CONTINUUM MODELS [IV.26] are macroscale models derived from first principles and where the material properties are usually measured. Other efforts concentrate more on developing accurate simulations, such as modeling the properties of Kevlar starting from the polymers in the resin and the carbon fibers used. All research in this area involves some numerical analysis and scientific computing.

## II.28 Nonlinear Equations and Newton's Method

*Marcos Raydan*

Nonlinear equations appear frequently in the mathematical modeling of real-world processes. They are usually written as a zero-finding problem: find $x_j \in \mathbb{R}$, for $j = 1, 2, \dots, n$, such that

$$f_i(x_1, \dots, x_n) = 0 \quad \text{for } i = 1, 2, \dots, n,$$

where the $f_i$ are given functions of $n$ variables. This system of equations is nonlinear if at least one of the functions $f_i$ depends nonlinearly on at least one of the variables. Using vector notation, the problem can also be written as find $x = [x_1, \dots, x_n]^{\mathrm{T}} \in \mathbb{R}^n$ such that

$$F(x) = [f_1(x), \dots, f_n(x)]^{\mathrm{T}} = 0.$$

If every function $f_i$ depends linearly on all the variables, then it is usually written as a linear system of equations $Ax = b$, where $b \in \mathbb{R}^n$ and $A$ is an $n \times n$ matrix.

The existence and uniqueness of solutions for nonlinear systems of equations is more complicated than for linear systems of equations. For solving $Ax = b$, the number of solutions must be either zero, infinity, or one (when $A$ is nonsingular), whereas $F(x) = 0$ can have zero, infinitely many, or any finite number of solutions. Fortunately, in practice, it is usually sufficient to find a solution of the nonlinear system for which a reasonable initial approximation is known.

Even in the simple one-dimensional case ($n = 1$), most nonlinear equations cannot be solved by a closed formula, i.e., using a finite number of operations. A well-known exception is the problem of finding the roots of polynomials of degree less than or equal to four, for which closed formulas have been known for

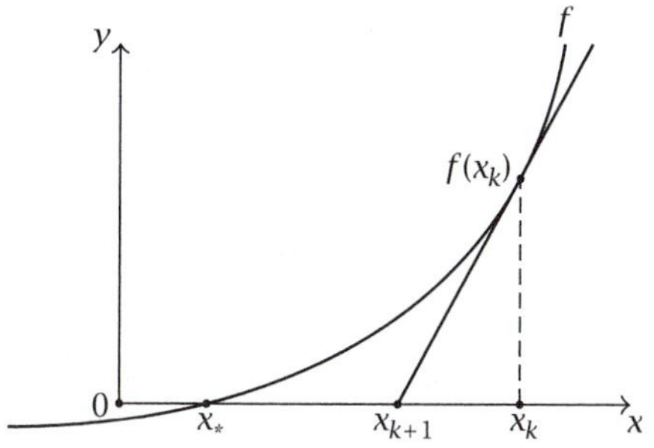


**Figure 1** One iteration of Newton's method in one dimension for $f(x) = 0$.

centuries. As a consequence, in general iterative methods must be used to produce increasingly accurate approximations to the solution. One of the oldest iterative schemes, which has played an important role in the numerical methods literature for solving $F(x) = 0$, is Newton's method.

## 1 Newton's Method

Newton's method for solving nonlinear equations was born in one dimension. In that case, the problem is find $x \in \mathbb{R}$ such that $f(x) = 0$, where $f: \mathbb{R} \to \mathbb{R}$ is differentiable in the neighborhood of a solution $x_*$. Starting from a given $x_0$, on the $k$th iteration Newton's method constructs the tangent line passing through the point $(x_k, f(x_k))$,

$$M_k(x) = f(x_k) + f'(x_k)(x - x_k),$$

and defines the next iterate, $x_{k+1}$, as the root of the equation $M_k(x) = 0$ (see figure 1). Hence, from a given $x_0 \in \mathbb{R}$, Newton's method generates the sequence $\{x_k\}$ of approximations to $x_*$ given by

$$x_{k+1} = x_k - f(x_k)/f'(x_k).$$

Notice that the tangent line or linear model $M_k(x)$ is equal to the first two terms of the Taylor series of $f$ around $x_k$.

Newton's idea in one dimension can be extended to $n$-dimensional problems. In $\mathbb{R}^n$ the method approximates the solution of a square nonlinear system of equations by solving a sequence of square linear systems. As in the one-dimensional case, on the $k$th iteration the idea is to define $x_{k+1}$ as a zero of the linear model given by

$$M_k(x) = F(x_k) + J(x_k)(x - x_k),$$

where the map $F: \mathbb{R}^n \to \mathbb{R}^n$ is assumed to be differentiable in a neighborhood of a solution $x_*$ and where $J(x_k)$ is the $n \times n$ Jacobian matrix with entries $J_{ij}(x_k) = \partial f_i / \partial x_j(x_k)$ for $1 \leqslant i, j \leqslant n$. Therefore, starting at a given $x_0 \in \mathbb{R}^n$, Newton's method carries out for $k = 1, 2, \dots$ the following two steps.

- Solve $J(x_k)s_k = -F(x_k)$ for $s_k$.
- Set $x_{k+1} = x_k + s_k$.

Notice that Newton's method is scale invariant: if the method is applied to the nonlinear system $AF(x) = 0$, for any nonsingular $n \times n$ matrix $A$, the sequence of iterates is identical to the ones obtained when it is applied to $F(x) = 0$. Another interesting theoretical feature is its impressively fast local convergence. Under some standard assumptions—namely that $J(x_*)$ is nonsingular, $J(x)$ is Lipschitz continuous in a neighborhood of $x_*$, and the initial guess $x_0$ is sufficiently close to $x_*$—the sequence $\{x_k\}$ generated by Newton's method converges *q-quadratically* to $x_*$; i.e., there exist $c > 0$ and $\hat{k} \geqslant 0$ such that for all $k \geqslant \hat{k}$,

$$\|x_{k+1} - x_*\| \leqslant c\|x_k - x_*\|^2.$$

Hence Newton's method is theoretically attractive, but it may be difficult to use in practice for various reasons, including the need to calculate the derivatives, the need to have a good initial guess to guarantee convergence, and the cost of solving an $n \times n$ linear system per iteration.

## 2 Practical Variants

If the derivatives are not available, or are too expensive to compute, they can be approximated by finite differences. A standard option is to approximate the $j$th column of $J(x_k)$ by a forward difference quotient: $(F(x_k + h_k e_j) - F(x_k))/h_k$, where $e_j$ denotes the $j$th unit vector and $h_k > 0$ is a suitable small number. Notice that, when using this finite-difference variant, the map $F$ needs to be evaluated $n + 1$ times per iteration, once for each column of the Jacobian and one for the vector $x_k$. Therefore, this variant is attractive when the evaluation of $F$ is not expensive.

Another option is to extend the well-known one-dimensional secant method to the $n$-dimensional problem $F(x) = 0$. The main idea, in these so-called *secant* or QUASI-NEWTON METHODS [IV.11 §4.2], is to generate not only a sequence of iterates $\{x_k\}$ but also a sequence of matrices $\{B_k\}$ that approximate $J(x_k)$ and satisfy the secant equation $B_k s_{k-1} = y_{k-1}$, where $s_{k-1} = x_k - x_{k-1}$

and $y_{k-1} = F(x_k) - F(x_{k-1})$. In this case, an initial matrix $B_0 \approx J(x_0)$ must be supplied. Clearly, infinitely many $n \times n$ matrices satisfy the secant equation. As a consequence, a wide variety of quasi-Newton methods (e.g., Broyden's method) with different properties have been developed.

When using Newton's method, or any of its derivative-free variants, a linear system needs to be solved at each iteration. This linear system can be solved by direct methods (e.g., LU or QR factorization), but if $n$ is large and the Jacobian matrix has a sparse structure, it may be preferable to use an iterative method (e.g., a KRYLOV SUBSPACE METHOD [IV.10 §9]). For that, note that $x_k$ can be used as the initial guess for the solution at iteration $k+1$. One of the important features of these so-called inexact variants of Newton's method is that modern iterative linear solvers do not require explicit knowledge of the Jacobian; instead, they require only the matrix–vector product $J(x_k)z$ for any given vector $z$. This product can be approximated using a forward finite difference:

$$J(x_k)z \approx (F(x_k + h_k z) - F(x_k))/h_k.$$

Hence, inexact variants of Newton's method are also suitable when derivatives are not available. In all the discussed variants, the local *q-quadratic convergence* is in general lost, but *q-superlinear convergence* can nevertheless be obtained, i.e., $\|x_{k+1} - x_*\|/\|x_k - x_*\| \to 0$.

Finally, in general Newton's method converges only locally, so it requires globalization strategies to be practically effective. The two most popular and best-studied options are line searches and trust regions. In any case, a merit function $\hat{f}: \mathbb{R}^n \to \mathbb{R}^+$ must be used to evaluate the quality of all possible iterates. When solving $F(x) = 0$, the natural choice is $\hat{f}(x) = F(x)^{\mathrm{T}}F(x)$.

### Further Reading

Dennis, J. E., and R. Schnabel. 1983. *Numerical Methods for Unconstrained Optimization and Nonlinear Equations.* Englewood Cliffs, NJ: Prentice Hall. (Republished by SIAM (Philadelphia, PA) in 1996.)

Kelley, C. T. 2003. *Solving Nonlinear Equations with Newton's Method.* Philadelphia, PA: SIAM.

Ypma, T. J. 1995. Historical development of the Newton–Raphson method. *SIAM Review* 37(4):531–51.

## II.29 Orthogonal Polynomials

Polynomials $p_0(x)$, $p_1(x)$, ..., where $p_i$ has degree $i$, are *orthogonal polynomials* on an interval $[a, b]$ with respect to a nonnegative weight function $w(x)$ if

$$\int_a^b w(x)p_i(x)p_j(x)\,\mathrm{d}x = 0, \quad i \neq j,$$

that is, if all distinct pairs of polynomials are orthogonal on $[a, b]$ with respect to $w$. For a given weight function and interval, the orthogonality conditions determine the polynomials $p_i$ uniquely up to a constant factor.

**Table 1** Parameters in the three-term recurrence (1) for some classical orthogonal polynomials.

| Polynomial | $[a,b]$ | $w(x)$ | $a_j$ | $b_j$ | $c_j$ |
|---|---|---|---|---|---|
| Chebyshev | $[-1,1]$ | $(1-x^2)^{-1/2}$ | 2 | 0 | 1 |
| Legendre | $[-1,1]$ | 1 | $\frac{2j+1}{j+1}$ | 0 | $\frac{j}{j+1}$ |
| Hermite | $(-\infty,\infty)$ | $\mathrm{e}^{-x^2}$ | 2 | 0 | $2j$ |
| Laguerre | $[0,\infty)$ | $\mathrm{e}^{-x}$ | $-\frac{1}{j+1}$ | $\frac{2j+1}{j+1}$ | $\frac{j}{j+1}$ |

An important property of orthogonal polynomials is that they satisfy a three-term recurrence relation

$$p_{j+1}(x) = (a_j x + b_j)p_j(x) - c_j p_{j-1}(x), \quad j \geqslant 1. \quad (1)$$

The weight functions, interval, and recurrence coefficients for some classical orthogonal polynomials are summarized in table 1, in which is assumed the normalization $p_0(x) = 1$, with $p_1(x) = x$ for the Chebyshev and Legendre polynomials, $p_1(x) = 2x$ for the Hermite polynomials, and $p_1(x) = 1 - x$ for the Laguerre polynomials.

Orthogonal polynomials have many interesting properties and find use in many different settings, e.g., in numerical integration, Krylov subspace methods, and the theory of continued fractions. In this volume they arise in LEAST-SQUARES APPROXIMATION [IV.9 §3.3], NUMERICAL SOLUTION OF PARTIAL DIFFERENTIAL EQUATIONS [IV.13 §6], RANDOM-MATRIX THEORY [IV.24], and as SPECIAL FUNCTIONS [IV.7 §7]. See SPECIAL FUNCTIONS [IV.7 §7] for more information.

## II.30 Shocks

*Barbara Lee Keyfitz*

### 1 What Are Shocks?

"Shocks" (or "shock waves") is another name for the field of quasilinear hyperbolic PDEs, or CONSERVATION LAWS [II.6]. When the mathematical theory of supersonic flow was in its infancy, the first text on the subject named it this way; and the first modern monograph to focus on the mathematical theory of quasilinear hyperbolic PDEs also used this terminology. Shocks are a dominant feature of the subject, for, as noted with reference to the BURGERS EQUATION [III.4] (see also PARTIAL DIFFERENTIAL EQUATIONS [IV.3 §3.6]), solutions to initial-value problems, even with smooth data, are not likely to remain smooth for all time. We return to the derivation to see what happens when solutions are not differentiable.

In the derivation of a system in one space dimension,

$$\boldsymbol{u}_t + \boldsymbol{f}(\boldsymbol{u})_x = 0, \tag{1}$$

one typically invokes the conservation of each component $u_i$ of $\boldsymbol{u}$. The rate of change of $u_i$ over a control length $[x, x+h]$ is the net flux across the endpoints:

$$\partial_t \int_x^{x+h} u_i(y,t)\,\mathrm{d}y = f_i(\boldsymbol{u}(x,t)) - f_i(\boldsymbol{u}(x+h,t)). \tag{2}$$

Under the assumption that $\boldsymbol{u}$ is differentiable, the mean value theorem of calculus yields (1) in the limit $h \to 0$. However, (2) is also useful in a different case. If $\boldsymbol{u}$ approaches two different limits, $\boldsymbol{u}_{\mathrm{L}}(X(t),t)$ and $\boldsymbol{u}_{\mathrm{R}}(X(t),t)$, on the left and right sides of a curve of discontinuity, $x = X(t)$, then taking the limit $h \to 0$ in (2) with $x$ and $x+h$ straddling the curve $X(t)$ yields a relationship among $\boldsymbol{u}_{\mathrm{L}}$, $\boldsymbol{u}_{\mathrm{R}}$, and the derivative of the curve:

$$\begin{aligned} X'(t)(\boldsymbol{u}_{\mathrm{R}}(X(t),t) &- \boldsymbol{u}_{\mathrm{L}}(X(t),t)) \\ &= \boldsymbol{f}(\boldsymbol{u}_{\mathrm{R}}(X(t),t)) - \boldsymbol{f}(\boldsymbol{u}_{\mathrm{L}}(X(t),t)). \end{aligned} \tag{3}$$

This is known as the (generalized) Rankine–Hugoniot relation. The quantity $X'(t)$ measures the speed of propagation of the discontinuity at $X(t)$.

Because solutions of conservation laws are not expected to be continuous for all time, even when the initial data are smooth, it is necessary to allow shocks in any formulation of what is meant by a "solution" of (1). Conservation law theory states that a solution of (1) may contain countably many shocks, the functions $X(t)$ may be no smoother than Lipschitz continuous, and there may be countably many points in physical $(x,t)$-space at which shock curves intersect. In the case of conservation laws in more than one space dimension, the notion of a "shock curve" can be generalized to that of a "shock surface" by supposing that the solution is piecewise differentiable on each side of such a surface. One obtains an equation similar to (3) that relates the states on either side of the surface to the normal to the surface at each point. However, as distinct from the case in a single space dimension, it is not known whether all solutions have this structure, or whether more singular behavior is possible.

## 2 Entropy, Admissibility, and Uniqueness

Although allowing for weak solutions, in the form of solutions containing shocks, is forced upon us by both mathematical considerations (they arise from almost all data) and physical considerations (they are seen in all the fluid systems modeled by conservation laws), a new difficulty arises: if shocks are admitted as solutions to a conservation law system, there may be *too many* solutions (this is also known, somewhat illogically, as "lack of uniqueness"). Here is a simple example, involving the Burgers equation. If at $t = 0$ we are given

$$u(x,0) = \begin{cases} 0, & x \leqslant 0, \\ 1, & x > 0, \end{cases}$$

then

$$u(x,t) = \begin{cases} 0, & x \leqslant \frac{1}{2}t, \\ 1, & x > \frac{1}{2}t, \end{cases}$$

is a shock solution in the sense of (3). But

$$u(x,t) = \begin{cases} 0, & x \leqslant 0, \\ x/t, & 0 < x \leqslant t, \\ 1, & x > t, \end{cases}$$

is also a solution, and in fact it is the latter, which is described as a "rarefaction wave," that is correct, while the former, known as a "RAREFACTION SHOCK [V.20 §2.2]," can be ruled out on both mathematical and physical grounds. A fluid that is rarefying (that is, one in which the force of pressure is decreasing), be it a gas or traffic, spreads out gradually and erases the initial discontinuity, while a fluid that is being compressed forms a shock.

Another mode of reasoning, which has both a mathematical and a physical basis, goes as follows. Suppose $\eta$ is a convex function of $\boldsymbol{u}$ for which another function $q(\boldsymbol{u})$ exists such that

$$\eta(\boldsymbol{u})_t + q(\boldsymbol{u})_x = 0 \tag{4}$$

whenever $\boldsymbol{u}$ is a smooth solution of (1). When this is the case, we say that (1) "admits a convex entropy." A calculation (easy for the Burgers equation and true in general) shows that we should not expect (4) to be satisfied (in the weak sense, as an additional Rankine–Hugoniot relation like (3)) in regions containing shocks. But since $\eta$ is convex, imposing the requirement that $\eta$ decrease in time when shocks are present forces a bound on solutions. For systems of conservation laws in a single space dimension, this condition, which admits some shocks and not others as weak solutions, is sufficient to guarantee uniqueness.

There are a number of other ways to formulate admissibility conditions for shocks, including modifying the system with so-called viscosity terms to make it parabolic,

$$\boldsymbol{u}_t + \boldsymbol{f}(\boldsymbol{u})_x = \varepsilon \boldsymbol{u}_{xx}, \tag{5}$$

and then admitting only shocks that are limits, as $\varepsilon \to 0$, of classical solutions of this semilinear parabolic system.

### 3 Shock Profiles

To a fluid dynamicist, the viscous equation (5) is more than an artifice to obtain uniqueness of solutions by winnowing out shocks that fail some test. The hyperbolic system (1) may be regarded as an approximation to a more realistic physical situation that takes into account viscosity, heat transfer effects, and even the mean free path of particles (for a gas). If, for example, viscous effects are included, the right-hand side of (5) takes the form $B\boldsymbol{u}_{xx}$, where $B = B(\boldsymbol{u})$ is a matrix that is typically diagonal, typically positive-semidefinite, and typically small when the system is measured on the length scale of interest. In particular, the hyperbolic system (1) gives a good description of a flow on that length scale. However, across a shock there is a rapid change in $\boldsymbol{u}$, or at least in some of its components, and the hyperbolic approximation is not adequate. One approach here is to use the hyperbolic system to uncover the macroscopic features of the shock—the speed $c = X'$ and the states on either side, using (3)—and then to formulate a traveling wave problem for the solution inside the shock. That is, one uses techniques from dynamical systems to study solutions $\boldsymbol{u}(\xi) = \boldsymbol{u}(x - ct)$ of

$$-c\boldsymbol{u}' + A(\boldsymbol{u})\boldsymbol{u}' = B(\boldsymbol{u})\boldsymbol{u}'',$$
$$\boldsymbol{u}(-\infty) = \boldsymbol{u}_{\mathrm{L}}, \qquad \boldsymbol{u}(\infty) = \boldsymbol{u}_{\mathrm{R}},$$

where $A$ is the Jacobian matrix of the flux function $\boldsymbol{f}$.

**Further Reading**

Bressan, A. 2000. *Hyperbolic Systems of Conservation Laws: The One-Dimensional Cauchy Problem.* Oxford: Oxford University Press.

Courant, R., and K. O. Friedrichs. 1948. *Supersonic Flow and Shock Waves.* New York: Wiley-Interscience.

Smoller, J. A. 1983. *Shock Waves and Reaction-Diffusion Equations.* New York: Springer.

## II.31 Singularities

*P. A. Martin*

The word "singularity" has a variety of meanings in mathematics but it usually means a place or point where something bad happens. For example, the function $f(x) = 1/x$ is not defined at $x = 0$; it has a singularity at $x = 0$: it is "singular" there. In this simple example, $f(x)$ is unbounded (infinite) at $x = 0$, but this need not be a defining property of a singularity. Thus, in COMPLEX ANALYSIS [IV.1], $f(z)$ is said to have a singularity at a point $z_0$ when $f$ is not differentiable at $z_0$. For example, $f_1(z) = 1/(z-1)^2$ has a singularity at $z = 1$, and $f_2(z) = z^{1/2}$ has a singularity at $z = 0$. Note that $f_1(z)$ is unbounded at $z = 1$, whereas $f_2(0) = 0$. In complex analysis, singularities of $f(z)$ are exploited to good effect, especially in the calculus of residues.

Before describing more benefits of singularities, let us consider some of the trouble that they may cause. Elementary examples occur in the evaluation of definite integrals when the integrand is unbounded at some point in the range of integration. For example, consider $I = \int_0^1 x^\alpha \,\mathrm{d}x$, where $\alpha$ is a parameter. Integrating, $I = 1/(\alpha + 1)$ provided $\alpha > -1$; the integral diverges for $\alpha \leqslant -1$. (When $\alpha = -1$, use $\int x^{-1}\,\mathrm{d}x = \log|x|$.) When $\alpha < 0$, the integrand $f(x) = x^\alpha$ is unbounded as $x \to 0$ through positive values. Nevertheless, even though $f(x)$ has a singularity at $x = 0$, the singularity is integrable when $-1 < \alpha < 0$, meaning that the area under the graph, $y = f(x)$, $0 < x < 1$, is finite; the area is infinite when $\alpha \leqslant -1$.

In the example just described we were able to evaluate the integral $I$ exactly, and this enabled us to examine the effect of the parameter $\alpha$. In practice, we may have to compute the value of an integral numerically using a quadrature rule (such as the trapezium rule or Simpson's rule). When the integrand has a singularity, we are often obliged to use a specialized rule tailored to that specific kind of singularity or to use a substitution designed to remove the singularity. Generally, a blend of analytical and numerical techniques is needed so as to mollify the effects of the singularity.

Similar difficulties can occur in many other problems, such as when solving boundary-value problems for a PARTIAL DIFFERENTIAL EQUATION [IV.3] (PDE). For a specific example, consider LAPLACE'S EQUATION [III.18], $\nabla^2 u = 0$, in the region $r > 0$, $0 < \theta < \beta$, where $r$ and $\theta$ are plane polar coordinates and the angle $\beta$ satisfies $0 < \beta < 2\pi$. We shall refer to this region as a *wedge.* If we have boundary conditions $u = 0$ on both sides of the wedge, $\theta = 0$ and $\theta = \beta$, one solution is $u(r,\theta) = r^{\nu}\sin(\nu\theta)$, where $\nu = \pi/\beta$. This solution is zero at the tip of the wedge (where $r = 0$), but derivatives of $u$ may be unbounded there; for example, if we take $\beta = \frac{4}{3}\pi$, then $\nu = \frac{3}{4}$ and the gradient of $u$ is unbounded at $r = 0$. This singular behavior is typical at corners and often degrades the performance of numerical methods for solving PDEs. In fact, it is not even necessary to have a corner, as a change in boundary condition can have the same effect. For example, take $\beta = \pi$, so that the wedge becomes the half-plane $y > 0$, and then require that $u = 0$ at $\theta = 0$ and $\partial u/\partial\theta = 0$ at $\theta = \pi$. Then one solution is $u = r^{1/2}\sin(\theta/2)$, and this has an unbounded gradient at $r = 0$, which is the point on the straight edge $y = 0$ where the boundary condition changes. This phenomenon was built into the *Motz problem*, devised in the 1940s for testing numerical methods; it is still in use today.

Returning to the wedge, take $\beta = 2\pi$ so that the wedge becomes the whole plane with a slit along the positive $x$-axis. The two sides of the slit are defined by $\theta = 0$ and $\theta = 2\pi$. Suppose that the boundary conditions are $\partial u/\partial\theta = 0$ at $\theta = 0$ and $\theta = 2\pi$. One solution of $\nabla^2 u = 0$ is then $u = r^{1/2}\cos(\theta/2)$, and this has an unbounded gradient at $r = 0$. This solution is of interest in the MECHANICS OF SOLIDS [IV.32], where the slit represents a crack, $u$ is a displacement, and the gradient of $u$ is related to the elastic stresses. The linear theory of elasticity predicts that the stresses are given approximately by $r^{-1/2}K(\theta)$ when $r$ (the distance from the crack tip) is small, where $K$ can be calculated. The form of the singular behavior, namely $r^{-1/2}$, is given by a local analysis near the crack tip, but the multiplier $K$, known as the *stress intensity factor*, requires knowledge of the geometry of the cracked body and the applied loads. Physically, the stresses cannot be infinite: the linear theory of elasticity breaks down at crack tips. A more accurate theory might involve plastic effects or consideration of atomic structures. However, useful predictions about when a cracked object will break as the applied loads are increased can be made by examining $K$. This is at the heart of engineering theories of *fracture mechanics.* It is another example where the singular behavior can be exploited.

Returning to mathematics, consider Laplace's equation in three dimensions, $\nabla^2 u = 0$. One solution is

$$G(x,y,z;x_0,y_0,z_0) = G(\mathrm{P},\mathrm{P}_0) = R^{-1},$$

where

$$R = \{(x-x_0)^2 + (y-y_0)^2 + (z-z_0)^2\}^{1/2}.$$

Here we can regard $(x,y,z)$ and $(x_0,y_0,z_0)$ as being the coordinates of points P and $\mathrm{P}_0$, respectively, and we have $\nabla^2 G = 0$ for fixed $\mathrm{P}_0$, provided $\mathrm{P} \neq \mathrm{P}_0$. As $R$ is the distance between P and $\mathrm{P}_0$, we see that $G$ is singular as $\mathrm{P} \to \mathrm{P}_0$. The function $G$ is an example of a *Green function*, named after George Green. (It is conventional to abandon the rules of English grammar and to speak of "a Green's function"; opposing this widespread misuse appears futile.) One use of $G$ comes when we want to solve Poisson's equation, $\nabla^2 u = f$, where $f$ is a given function. Thus

$$u(\mathrm{P}_0) = -\frac{1}{4\pi}\int G(\mathrm{P},\mathrm{P}_0)\,f(\mathrm{P})\,\mathrm{dP},$$

where the integration is over all P. Notice that $G$ could be replaced by $A/R + H(\mathrm{P},\mathrm{P}_0)$, where $A$ is a constant and $H$ is any solution of $\nabla^2 H = 0$. In principle, $H$ can be chosen so that $G$ satisfies additional conditions, such as boundary conditions; indeed, this was how Green conceived of his function, as the electrostatic field inside a conductor due to a point charge. However, in practice, it is usual to use a simple $G$ and then to impose boundary conditions on $u$ by solving an INTEGRAL EQUATION [IV.4]. We note that the alternative terminology *fundamental solution* is often used, meaning a simple singular solution of a governing PDE.

Singularities can occur in many other contexts. We mention two. Suppose we want to solve an initial-value problem for a nonlinear PDE, where the initial state at time $t = 0$ is specified and the goal is to calculate the solution as $t$ increases. It is then possible that the solution becomes unbounded as $t \to t_c$, where $t_c$ is some finite critical time. This is known as "blow-up": there is a finite-time singularity at $t = t_c$ (see PARTIAL DIFFERENTIAL EQUATIONS [IV.3 §3.6]). There is much interest within FLUID DYNAMICS [IV.28] in the existence or otherwise of finite-time singularities because it is thought that they may be relevant in understanding the nature of TURBULENCE [V.21].

Finally, we cannot end an article on singularities without mentioning cosmology. It is generally accepted that the Big Bang theory gives a model for how the universe evolves, starting from an initial singularity. Some cosmologists believe that the universe will end with a singularity; the trouble caused by this singularity (if it exists) is unlikely to bother readers of this article.

## II.32 The Singular Value Decomposition

*Nicholas J. Higham*

One of the most useful matrix factorizations is the *singular value decomposition* (SVD), which is defined for an arbitrary rectangular matrix $A \in \mathbb{C}^{m\times n}$. It takes the form

$$A = U\Sigma V^*, \quad \Sigma = \operatorname{diag}(\sigma_1, \sigma_2, \dots, \sigma_p) \in \mathbb{R}^{m\times n}, \quad (1)$$

where $p = \min(m, n)$, $\Sigma$ is a diagonal matrix with diagonal elements $\sigma_1 \geqslant \sigma_2 \geqslant \cdots \geqslant \sigma_p \geqslant 0$, and $U \in \mathbb{C}^{m\times m}$ and $V \in \mathbb{C}^{n\times n}$ are unitary. The $\sigma_i$ are the *singular values* of $A$, and they are the nonnegative square roots of the $p$ largest eigenvalues of $A^*A$. The columns of $U$ and $V$ are the left and right *singular vectors* of $A$, respectively.

Postmultiplying (1) by $V$ gives $AV = U\Sigma$ since $V^*V = I$, which shows that the $i$th columns of $U$ and $V$ are related by $Av_i = \sigma_i u_i$ for $i = 1:p$. Similarly, $A^*u_i = \sigma_i v_i$ for $i = 1:p$. A geometrical interpretation of the former equation is that the singular values of $A$ are the lengths of the semiaxes of the hyperellipsoid $\{Ax : \|x\|_2 = 1\}$.

Assuming that $m \geqslant n$ for notational simplicity, from (1) we have

$$A^*A = V(\Sigma^*\Sigma)V^*, \quad (2)$$

with $\Sigma^*\Sigma = \operatorname{diag}(\sigma_1^2, \sigma_2^2, \dots, \sigma_n^2)$, which shows that the columns of $V$ are eigenvectors of the matrix $A^*A$ with corresponding eigenvalues the squares of the singular values of $A$. Likewise, the columns of $U$ are eigenvectors of the matrix $AA^*$.

The SVD reveals a great deal about the matrix $A$ and the key subspaces associated with it. The rank, $r$, of $A$ is equal to the number of nonzero singular values, and the range and the null space of $A$ are spanned by the first $r$ columns of $U$ and the last $n - r$ columns of $V$, respectively.

The SVD reveals not only the rank but also how close $A$ is to a matrix of a given rank, as shown by a classic 1936 theorem of Eckart and Young.

**Theorem 1 (Eckart–Young).** *Let $A \in \mathbb{C}^{m\times n}$ have the SVD (1). If $k < r = \operatorname{rank}(A)$, then for the 2-norm and the Frobenius norm,*

$$\min_{\operatorname{rank}(B)=k} \|A - B\| = \|A - A_k\| = \begin{cases} \sigma_{k+1}, & \text{2-norm,} \\ \sqrt{\sum_{i=k+1}^{r} \sigma_i^2}, & \text{F-norm,} \end{cases}$$

*where*

$$A_k = UD_kV^*, \quad D_k = \operatorname{diag}(\sigma_1, \dots, \sigma_k, 0, \dots, 0).$$

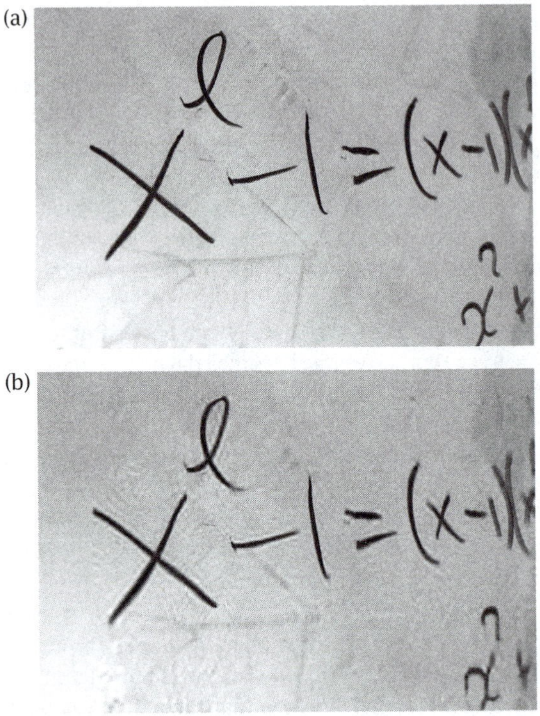

**Figure 1** Photo of a blackboard, inverted so that white and black are interchanged in order to show more clearly the texture of the board: (a) original 1067 × 1600 image; (b) image compressed using rank-40 approximation $A_{40}$ computed from SVD.

In many situations the matrices that arise are necessarily of low rank but errors in the underlying data make the matrices actually obtained of full rank. The Eckart–Young result tells us that in order to obtain a lower-rank matrix we are justified in discarding (i.e., setting to zero) singular values that are of the same order of magnitude as the errors in the data.

The SVD (1) can be written as an outer product expansion

$$A = \sum_{i=1}^{p} \sigma_i u_i v_i^*,$$

and $A_k$ in the Eckart–Young theorem is given by the same expression with $p$ replaced by $k$. If $k \ll p$ then $A_k$ requires much less storage than $A$ and so the SVD can provide *data compression* (or *data reduction*). As an example, consider the monochrome image in figure 1(a) represented by a 1067 × 1600 array of RGB values ($R = G = B$ since the image is monochrome). Let $A \in \mathbb{R}^{1067 \times 1600}$ contain the values from any one of the three channels. The singular values of $A$ range from $8.4 \times 10^4$ down to $1.3 \times 10^1$. If we retain only the singular values down to the 40th, $\sigma_{40} = 2.1 \times 10^3$ (a somewhat arbitrary cutoff since there is no pronounced gap in the singular values), we obtain the image in figure 1(b). The reduced SVD requires only 6% of the storage of the original matrix. Some degradation is visible in the compressed image (and more can be seen when it is viewed at 100% size on screen), but it retains all the key features of the original image. While this example illustrates the power of the SVD, image compression is in general done much more effectively by the JPEG SCHEME [VII.7 §5].

A pleasing feature of the SVD is that the singular values are not unduly affected by perturbations. Indeed, if $A$ is perturbed to $A + E$ then no singular value of $A$ changes by more than $\|E\|_2$.

The SVD is a valuable tool in applications where two-sided orthogonal transformations can be carried out without "changing the problem," as it allows the matrix of interest to be diagonalized. Foremost among such problems is the LINEAR LEAST-SQUARES PROBLEM [IV.10 §7.1] $\min_{x \in \mathbb{C}^n} \|b - Ax\|_2$.

The SVD was first derived by Beltrami in 1873. The first reliable method for computing it was published by Golub and Kahan in 1965; this method applies two-sided unitary transformations to $A$ and does not form and solve the equation (2), or its analogue for $AA^*$. Once software for computing the SVD became readily available, in the 1970s, the use of the SVD proliferated. Among the wide variety of uses of the SVD are for TEXT MINING [VII.24], deciphering encrypted messages, and image deblurring.

### Further Reading

Eldén, L. 2007. *Matrix Methods in Data Mining and Pattern Recognition.* Philadelphia, PA: SIAM.

Golub, G. H., and C. F. Van Loan. 2013. *Matrix Computations*, 4th edn. Baltimore, MD: Johns Hopkins University Press.

Moler, C. B., and D. Morrison. 1983. Singular value analysis of cryptograms. *American Mathematical Monthly* 90:78–87.

## II.33 Tensors and Manifolds

*Mark R. Dennis*

We know that the surface of the Earth is curved, despite the fact that it appears flat. This is easily understood from the fact that the Earth's radius of curvature is over 6000 km, vast on a human scale. This picture motivates the mathematical definition of a *manifold* (properly a *Riemannian manifold*): a space that appears to be Euclidean locally in a neighborhood of each point (or pseudo-Euclidean, as defined below) but globally may have curvature, such as the surface of a sphere.

Manifolds are most simply defined in terms of the coordinate systems on them, and of course there are uncountably many such systems. *Tensors* are mathematical objects defined on manifolds, such as vector fields, which are in a natural sense independent of the coordinate system used to define them and their components. The importance of tensors in physics stems from the fact that the description of physical phenomena ought to be independent of any coordinate system we choose to impose on space and hence should be tensorial.

Our description of manifolds and tensors will be rather informal. For instance, we will picture vector or tensor fields as defining a vector or tensor at each point of the manifold itself rather than more abstractly as a section of the appropriate tangent bundle. In applications, tensors are frequently used in the study of GENERAL RELATIVITY AND COSMOLOGY [IV.40], which involves describing the dynamics of matter and fields using any reference frame (coordinate system), assuming space-time is a four-dimensional pseudo-Riemannian curved manifold, as described below.

An $n$-dimensional manifold is a topological space such that a neighborhood around each point is equivalent (i.e., homeomorphic) to a neighborhood of a point in $n$-DIMENSIONAL EUCLIDEAN SPACE [I.2 §19.1]. More formally, it can be defined as the set of smooth coordinate systems that can be defined on the space, together with transformation rules between them. In a neighborhood around each point, a coordinate system can always be found that looks locally Cartesian, regardless of any global curvature (which can cause the system to fail to be Cartesian at other points).

In practice, each coordinate system on a Riemannian manifold has a *metric*, defined below, which is possibly position dependent. This enables inner products between pairs of vectors at each point in the space to be defined. The situation is complicated by the fact that, at each point, most coordinate systems are *oblique*, as in figure 1. The following description uses "index notation," which suggests the explicit choice of

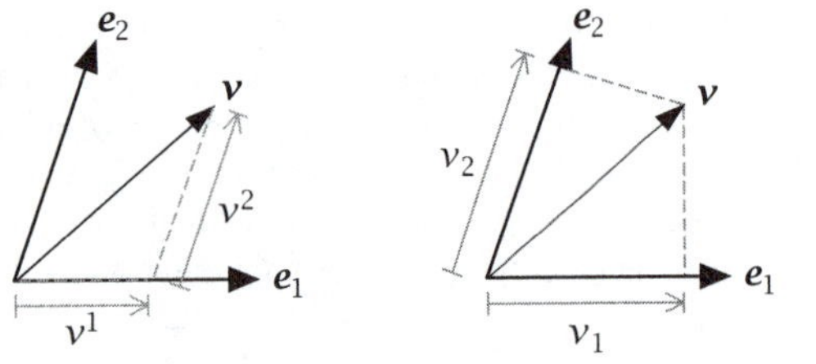


**Figure 1** The components of a vector $\boldsymbol{v}$ in an oblique basis of unit vectors $\{\boldsymbol{e}_1, \boldsymbol{e}_2\}$: contravariant components $v^1$, $v^2$ follow from the parallelogram rule, and covariant components $v_1$, $v_2$ are found by dropping perpendiculars onto the basis vectors. For basis vectors with arbitrary lengths, each covariant component is the length shown on the diagram multiplied by the norm of the corresponding basis vector.

a coordinate system. However, such expressions are always valid since all tensorial expressions should be the same in every coordinate system; that is, they are *covariant*. Explicitly coordinate-free formulations exist, although they require a greater familiarity with differential geometry than is assumed here and require the definition of much new notation.

Consider a vector $\boldsymbol{v}$ in $n$-dimensional Euclidean space with oblique coordinate axes locally defined by linearly independent but not necessarily orthonormal basis vectors $\boldsymbol{e}_j$, $j = 1, \dots, n$. The *contravariant* components of $\boldsymbol{v}$, represented with upper indices $v^j$ for $j = 1, \dots, n$, are defined by the parallelogram law; that is,

$$\boldsymbol{v} = \sum_{j=1}^{n} v^j \boldsymbol{e}_j \quad (v^j \text{ contravariant}).$$

Throughout this article, we will assume that the symbol $v^j$ represents the *vector itself*, and not simply the components. This will also apply to objects with multiple indices.

The *covariant* components $v_j$, represented with lower indices, are those defined by the scalar product with the basis vectors (i.e., formally, covariant components are the components of the vectors in the dual space to the tangent vector space, with "covariant" here not to be confused with the sense previously), i.e.,

$$v_j = \boldsymbol{v} \cdot \boldsymbol{e}_j, \quad j = 1, \dots, n \quad (v_j \text{ covariant}).$$

If the $\boldsymbol{e}_j$ are orthonormal, then the covariant components $v_j$ are the same as the contravariant $v^j$. The components of the *metric tensor* $g_{ij}$ for the coordinate system at this point are then given by $g_{ij} = \boldsymbol{e}_i \cdot \boldsymbol{e}_j$, with $g^{ij} \equiv (g_{ij})^{-1}$, the inverse of $g_{ij}$ considered as a matrix.

All of the geometry of the local basis is encoded in $g_{ij}$; for instance, covariant and contravariant components are found from each other using the metric tensor to "raise" and "lower" indices, such as $v^i = g^{ij} v_j$ and $v_i = g_{ij} v^j$. In these expressions, and for the remainder of the article, we adopt the *Einstein summation convention*; that is, when an index symbol is repeated, once each in an upper (contravariant) and a lower (covariant) position, we assume that the indices are summed from 1 to $n$. The summed index symbol $i, j, \dots$ is itself arbitrary, i.e., a "dummy index."

This procedure generalizes the inner product to oblique axes, defining *inner multiplication*, or *index contraction*. As usual in linear algebra, objects with multiple indices can be defined, e.g., $T^{ij} = u^i v^j + p^i q^j$ for vectors $u^j$, $v^j$, $p^j$, $q^j$. Forming a product of objects such as $u^i v^j$ without contracting the indices is referred to as *outer multiplication*, and in coordinate-free from, $u^i v^j$ is written as $\boldsymbol{u} \otimes \boldsymbol{v}$.

The components of the position vector, in terms of the chosen coordinate system, are denoted $x^i$, i.e., in Cartesian coordinates $x^1 = x$, $x^2 = y, \dots$. Differentiation with respect to a set of contravariant indices is in fact covariant, as can easily be verified from a Taylor expansion by a small displacement $\delta x^i$ of a scalar field $f(x^i)$ around a chosen point $x^i$:

$$f(x^i + \delta x^i) = f(x^i) + \delta x^i \partial_i f + O((\delta x)^2).$$

The term $\delta x^i \partial_i f$ is first order in $\delta x^i$, so it must be the same in all coordinate systems; since $\delta x^i$ is contravariant, $\partial_i \equiv \partial / \partial x^i$ must be covariant.

As coordinate systems on curved manifolds are not typically orthonormal everywhere, we must assume in general that the components of a vector are different from those of its dual (which is used to take inner products). Any vector object may be represented with upper or lower indices, which are related by the metric tensor. Objects with multiple indices, such as the metric tensor itself, can also have both indices upper, both lower, or a mixture. In fact, since $g^{ij}$ is the matrix inverse of $g_{ij}$, the matrix representation of the mixed metric tensor $g_i{}^k = g_{ij} g^{jk}$ is the identity matrix and so is often written $\delta_i^k$, to be understood as a generalization of the usual Kronecker symbol with mixed upper and lower indices.

With this formalism an alternative coordinate system has components $x^{j'}$, the different system being represented by the prime on the coordinate symbol; at each point, the linear transformation between the systems is given by the Jacobian $\partial x^i/\partial x^{j'} \equiv \Lambda^i_{j'}$; therefore, a vector's contravariant components are transformed to the new system by $v^{i'} = \Lambda^{i'}_j v^j$. The inverse transformation $\Lambda^j_{i'}$ not only transforms contravariant coordinates back from the primed to the unprimed system but also transforms unprimed covariant components to the primed system, i.e., $v_{i'} = \Lambda^j_{i'} v_j$. This equivalence guarantees that the squared length of the vector is independent of the coordinate system, i.e.,

$$v_{i'} v^{i'} = v_j \Lambda^j_{i'} \Lambda^{i'}_k v^k = v_j \delta^j_k v^k = v_j v^j,$$

where $\delta^j_k$ is the Kronecker symbol with one upper index and one lower index (both unprimed), which may be thought of as representing the identity transformation from the unprimed coordinate system to itself.

A general *tensor* is therefore a (possibly) multicomponent object, all of whose components transform between coordinate systems according to the local Jacobian transformation, namely,

$$T^{i'j'\cdots}{}_{k'\ell'\cdots} = \Lambda^{i'}_i \Lambda^{j'}_j \cdots \Lambda^k_{k'} \Lambda^\ell_{\ell'} \cdots T^{ij\cdots}{}_{k\ell\cdots}.$$

A *scalar* is a tensorial object with no indices, a *vector* has one index, and in general a tensor with $m$ distinct indices is said to have *rank* $m$. The ordering of the indices of a tensor is of course important, and this is maintained regardless of whether they appear in upper or lower positions. The *principle of covariance* (either special covariance or general covariance), due to Einstein, states that physical laws should be expressible in tensorial form, that is, they should be covariant under the class of coordinate transformations being considered.

Although it is tempting to identify tensors such as $g_{ij}$ as arrays of numbers in particular coordinate systems, a tensor on a manifold is in fact properly defined independently of any particular coordinate system, and instead its tensorality follows from its indices transforming in the appropriate way under coordinate transformations.

Coordinate derivatives $\partial_i v_j$ (also written $v_{j,i}$) should not be expected to be tensorial, as they follow the possibly curved coordinate lines of the arbitrarily chosen coordinate system. The *covariant derivative*

$$\nabla_i v_j = v_{i;j} \equiv \partial_i v_j - \Gamma^k_{ij} v_k \tag{1}$$

is tensorial, where the *connection coefficients* or *Christoffel symbols* $\Gamma^k_{ij}$ denote a nontensorial object defined in terms of coordinate derivatives of the metric tensor ($\Gamma^k_{ij} \equiv 2^{-1} g^{k\ell} \partial_j g_{i\ell} + \partial_i g_{\ell j} - \partial_\ell g_{ij}$), whose combination with the coordinate derivative in (1) does indeed yield a tensor. Covariant derivatives of general tensors pick up a $\Gamma$ term for each index (e.g., $T_{i\ell;j} = T_{i\ell,j} - \Gamma^k_{ij} T_{k\ell} - \Gamma^m_{\ell j} T_{im}$). The derivative of a scalar is therefore automatically tensorial.

No part of the discussion up to this point explicitly involves the manifold's curvature. Indeed, a complicated coordinate system may have nonzero connection coefficients and yet describe Euclidean space. The tensorality of the covariant derivative is an expression of the *parallel transport* of a vector along a curve. A *geodesic* on a manifold is a curve that is as straight as possible, given that the manifold may be curved. Such a curve can be constructed by parallel transporting a vector as a tangent vector along a curve; a geodesic curve $z^i(s)$, parametrized by $s$, therefore satisfies the geodesic equation

$$\frac{\mathrm{d}^2 z^k}{\mathrm{d}s^2} + \Gamma^k_{ij} \frac{\mathrm{d}z^i}{\mathrm{d}s} \frac{\mathrm{d}z^j}{\mathrm{d}s} = 0,$$

which generalizes the equation for a Euclidean straight line in Cartesian coordinates (for which the connection is zero), and is in general nonlinear.

*Curvature* exists around a point on a manifold when there exists a vector that, upon being parallel transported from the point around some infinitesimal closed path back to the point, does not return to its original direction. This is equivalent to *the failure of covariant derivatives to commute*. For any vector field $v_i$,

$$\nabla_j \nabla_k v_i - \nabla_k \nabla_j v_i = R^\ell{}_{ijk} v_\ell, \tag{2}$$

defining the *Riemann curvature tensor* $R^\ell{}_{ijk}$. This tensor has rank 4: the indices $j$ and $k$ are related to the plane defined by the derivatives (i.e., the closed path), and $i$ is related to the original direction of the vector, whose change in direction is related to $\ell$. The Riemann curvature tensor has many symmetries, such as $R_{\ell ijk} = -R_{i\ell jk} = R_{jk\ell i}$, with the result that, although a general rank-4 tensor has $n^4$ components, only $n^2(n^2-1)/12$ of the Riemann tensor's components are independent: one for $n = 2$, six for $n = 3$, twenty for $n = 4$, etc. A manifold where the curvature tensor vanishes everywhere is said to be *flat*.

The symmetric, rank-2 "trace" of the Riemann tensor, $R_{ik} \equiv R^j{}_{ijk}$, is known as the *Ricci tensor*. It plays a crucial role in the theory of general relativity, as the part of the curvature of the space-time manifold affected by energy-momentum and the cosmological constant. Its trace $R \equiv g^{ik} R_{ik}$ is called the *curvature scalar*.

In many applications it is mathematically easier to restrict the class of coordinate systems being considered in order to gain mathematical tractability at the cost of generality. For example, in problems of CLASSICAL MECHANICS [IV.19] in three-dimensional flat Euclidean space described by a position vector $\boldsymbol{r}$, it is conventional to work with *Cartesian tensors*; all coordinate systems are Cartesian, and the transformations between them are simply rotations and translations. In this case, covariant and contravariant indices transform in the same way, so the metric becomes the Kronecker symbol in all systems $g_{ij} = \delta_{ij}$ (the indices are both lower here, appropriate for a metric tensor).

Examples of Cartesian tensors include the inertia tensor of a solid body with density $\rho(\boldsymbol{r})$,

$$I_{ij} = \int_{\text{body}} \rho(\boldsymbol{r})(r^k r_k \delta_{ij} - r_i r_j)\,\mathrm{d}^3\boldsymbol{r},$$

which relates the rotating body's angular momentum $L_i = I_{ij}\omega^j$ to its angular velocity $\omega^j$. Another example that is important for continuum mechanics is a body's Cauchy stress tensor $\sigma_{ij}$; at each point, $f_j = \sigma_{ij}n^i$ is the force acting on a surface perpendicular to the unit vector $n^i$.

The tensorial framework for space-time was introduced into physics by Einstein, who applied it to the four-dimensional manifold of space-time events, in which all vectors become four-dimensional *4-vectors*. In this formalism, *time* becomes a spatial coordinate $ct = x^0$ (or, in older literature, $x^4$), where $c$ is the speed of light (approximately $3 \times 10^8$ meters per second), so that physical laws satisfy the *principle of special relativity* (or special covariance); they take the same form in all inertial frames, which are regarded effectively as orthonormal coordinate systems on flat space-time. This generalization of Newton's first law of mechanics was necessary to accommodate Maxwell's equations of electrodynamics, famously requiring all inertial observers to agree on the value of $c$. In Einstein's theory, all inertial observers agree on the value of $x^2 + y^2 + z^2 - c^2t^2$, written more compactly via the summation convention as $\eta_{ab}x^a x^b$, or $x^a x_a$; this is the "space-time separation" of the event $x^a = (ct, x, y, z)$ from the space-time origin, where $\eta_{ab}$ is the *Minkowski tensor* with the form $\operatorname{diag}(-1, +1, +1, +1)$. Conventionally, indices $a, b, \dots$ are used to denote four-dimensional space-time indices $0, \dots, 3$, whereas indices $i, j, \dots$ are used to denote spatial indices $1, 2, 3$. The different continuous symmetries of Euclidean, Newtonian, and Minkowski space-time are discussed in INVARIANTS AND CONSERVATION LAWS [II.21].

The four-dimensional flat manifold that admits the Minkowski tensor $\eta_{ab}$ at each point is called *Minkowski space*. Notably, $\eta_{ab}$ has some negative as well as positive entries (as such it is *pseudo-Euclidean*); Minkowski space has many similarities geometrically to Euclidean space. However, the crucial difference is that the squared length $x^a x_a$ of a space-time 4-vector may be positive, negative, or zero.

In special relativity, therefore, space-time is represented by Minkowski space, a flat pseudo-Riemannian manifold. Space-time coordinate transformations on Minkowski space are called *Lorentz transformations*. The choice of the overall sign of the tensor $\eta_{ab}$ (i.e., $\operatorname{diag}(-1, +1, +1, +1)$ or $\operatorname{diag}(+1, -1, -1, -1)$) is a convention, and it is common, particularly in the literature on relativistic quantum theory, to use the opposite sign to that chosen here (it is also common to represent the four space-time indices by Greek characters $\alpha, \mu, \dots$, reserving Roman characters for the space-like indices). The *signature* of a metric is the sum of signs of its entries in a coordinate system when it is diagonal: in a Euclidean space it is $n$, but in Minkowski space it is 2 (or $-2$, depending on the sign convention). The neighborhood of each point in a Riemannian manifold admits a coordinate system that is locally Euclidean. Manifolds in which the neighborhood of each point admits a coordinate system with metric signature not equal to $n$, such as Minkowski space, are said to be *pseudo-Riemannian*.

In *general relativity*, Einstein proposed the *principle of general covariance*, i.e., that physical laws be tensorial with respect to *all observers* (i.e., all frames on space-time), not just the inertial ones (for which tensors in special relativity become analogous to the restriction in Euclidean space to Cartesian tensors). In general relativity, therefore, space-time is no longer required to be flat (i.e., the neighborhood of each point locally looks like Minkowski space), allowing for the possibility that at larger length scales, space-time itself may be curved. The EINSTEIN FIELD EQUATIONS [III.10] relate the Ricci curvature of a manifold to the distribution of matter, energy, and momentum in space-time, thereby fundamentally tying the nature of space and time to the phenomena that occupy it.

### Further Reading

Bishop, R. L., and S. I. Goldberg. 1980. *Tensor Analysis on Manifolds*. New York: Dover.

Lorentz, H. A., A. Einstein, H. Minkowski, and H. Weyl. 1952. *The Principle of Relativity*. New York: Dover.

Schutz, B. 1980. *Geometrical Methods of Mathematical Physics*. Cambridge: Cambridge University Press.

## II.34 Uncertainty Quantification

*Youssef Marzouk and Karen Willcox*

Uncertainty quantification (UQ) involves the quantitative characterization and management of uncertainty in a broad range of applications. It employs both computational models and observational data, together with theoretical analysis. UQ encompasses many different tasks, including uncertainty propagation, sensitivity analysis, statistical inference and model calibration, decision making under uncertainty, experimental design, and model validation. UQ therefore draws upon many foundational ideas and techniques in applied mathematics and statistics (e.g., approximation theory, error estimation, stochastic modeling, and Monte Carlo methods) but focuses these techniques on complex models (e.g., of physical or sociotechnical systems) that are primarily accessible through computational simulation. UQ has become an essential aspect of the development and use of predictive computational simulation tools.

Modeling endeavors may contain multiple sources of uncertainty. A widely used classification contrasts *aleatory* or irreducible uncertainty, resulting from some inherent variability, with *epistemic* or reducible uncertainty that reflects a lack of knowledge. The more detailed classification of uncertainties below was proposed in the seminal work of Kennedy and O'Hagan, and it provides a useful foundation on which to establish mathematical approaches.

*Parameter uncertainty* refers to uncertain inputs to or parameters of a model. For example, parameters representing physical properties (permeability, porosity) of the Earth may be unknown in a computational model of the subsurface. *Parametric variability* captures the uncertainty due to uncontrolled or unspecified conditions in inputs or parameters. For example, aircraft design must account for uncertain operating conditions that arise from varying atmospheric conditions and gust encounters. *Residual variability* describes the uncertainty due to intrinsic random variation in the underlying physics being modeled or induced by behavior at physical scales that are not resolved by the model. Examples of residual variability include the uncertainty due to using models of turbulence that approximate the effects of the small scales that are not resolved. *Code uncertainty* refers to the uncertainty associated with not knowing the output of a computer model given any particular configuration until the code is run. For example, a Gaussian process emulator used as a surrogate for a higher-fidelity model has code uncertainty at parameter values away from those at which the emulator was calibrated. *Observation error* is the uncertainty associated with actual observations and measurements; it plays an important role in model calibration and inverse problems. *Model discrepancy* captures the uncertainty due to limitations of and assumptions in the model. This uncertainty is present in almost every model used in science and engineering.

This article takes a probabilistic view of all these sources of uncertainty. While other approaches, e.g., interval analysis, fuzzy set theory, and Dempster-Shafer theory, have also been employed to analyze particular UQ problems, the probabilistic approach to UQ offers a particularly rich and flexible structure, and has seen extensive development over the past two decades.

The challenge of *model validation* is closely related to UQ. A recent National Academy of Sciences report defines validation as "the process of determining the degree to which a model is an accurate representation of the real world from the perspective of the intended uses of the model." This process therefore involves assessing how the sources of uncertainty described above contribute to any prediction of interest. Quantifying model discrepancy is a particularly challenging aspect of the validation process; posterior predictive checks, cross validation, and other ways of assessing model error are important in this regard. In situations where data are few in number or where the model is intended for use in an extrapolatory setting, however, traditional techniques for model checking may not apply. This is important and somewhat uncharted territory, for which new mathematical and statistical approaches are being developed.

### 1 Characterizing Uncertainty

Probabilistic approaches characterize uncertain quantities using probability density or mass functions. These approaches require the specification of sufficient information to endow model inputs with probability distributions. This information can be drawn from various types of prior knowledge, including historical databases, physical constraints, previous computations, and the elicitation of expert opinion. The principle of maximum entropy is sometimes used to map from a few specified characteristics of the uncertainty (e.g., minimum and maximum values, or mean, variance, and other moments) to a probability distribution by determining the maximum entropy distribution that is compatible with the specified constraints. Inferential approaches, described below, can update probabilistic characterizations of uncertain quantities by conditioning these quantities on new observational data.

## 2 Forward Propagation of Uncertainty

Forward propagation of uncertainty addresses how uncertainty in model inputs translates into uncertainty in model outputs. The goal is often to provide distributional information (output means and variances, event probabilities) in support of uncertainty assessment or decision making. The most flexible method for estimating distributional information is Monte Carlo simulation, which draws random samples from the joint distribution of inputs and evaluates the output value corresponding to each input sample. Expectations over the output distribution are then estimated from these samples. Monte Carlo estimates converge slowly, typically requiring many samples to achieve acceptable levels of accuracy; the error in a simple Monte Carlo estimator converges as $O(N^{-1/2})$, where $N$ is the number of Monte Carlo samples. Variance-reduction techniques, such as importance sampling and the use of control variates, are therefore of great interest in UQ. Also, quasi-Monte Carlo approaches can offer faster convergence rates than random (or pseudorandom) sampling, while maintaining good scalability with respect to the number of uncertain parameters.

Approaches that rely on polynomial chaos and other spectral representations of random quantities, such as stochastic Galerkin and stochastic collocation methods, are important and widely used alternatives to Monte Carlo simulation. By exploiting the regularity of the input–output relationship induced by a model, these approaches can provide greater accuracy and efficiency than Monte Carlo simulation for moderate numbers of parameters. They can also provide more information about this input–output relationship, including sensitivity indices and approximations that are useful in inverse problems (see below). Stochastic spectral techniques have been developed for many classes of partial differential equations with random parameters or input data, as well as for more general black-box models. Methods for identifying and exploiting sparsity in polynomial representations, for evaluating the model on sparse grids in high-dimensional parameter spaces, and for performing dimension reduction have proven quite successful in expanding the range and size of problems to which spectral techniques can be successfully applied.

## 3 Sensitivity Analysis

Sensitivity analysis aims to elucidate how the uncertain inputs of a system contribute to system output uncertainty. Variance-based sensitivity analysis apportions the variance of an output quantity of interest among contributions from each of the system inputs and their interactions. This apportionment is based on the law of total variance, which for a given output quantity of interest $Q$ and a given factor $X_i$ is written as

$$\mathrm{Var}(Q) = \mathbb{E}[\mathrm{Var}(Q \mid X_i)] + \mathrm{Var}(\mathbb{E}[Q \mid X_i]).$$

From this, a main effect sensitivity index is defined as the expected fraction of the variance of $Q$ that would be removed if the variance of $X_i$ were reduced to zero:

$$S_i = \frac{\mathrm{Var}(\mathbb{E}[Q \mid X_i])}{\mathrm{Var}(Q)}.$$

The results of a global sensitivity analysis can be used for *factor prioritization* (identifying those inputs where future research is expected to bring the largest reduction in variance) and *factor fixing* (identifying those inputs that may be fixed to a deterministic value without substantially affecting probabilistic model outputs). Algorithms for computing main and total effect sensitivity indices may rely on Monte Carlo or quasi-Monte Carlo sampling, sparse quadrature, or post-processing of the stochastic spectral expansions described above.

## 4 Inverse Problems and Data Assimilation

INVERSE PROBLEMS [IV.15] arise from indirect observation of a quantity of interest. For example, one may wish to estimate certain parameters of a system given limited and noisy observations of the system's outputs. From the UQ perspective, one seeks not only a point estimate of the parameters but also a quantitative assessment of their uncertainty. UQ in inverse problems can therefore be addressed through the perspective of statistical inference. The statistical inference problems that arise in this context typically involve a likelihood function that contains a complex physical model (described by ordinary or partial differential equations) and inversion "parameters" that are in fact functions, and hence are infinite dimensional.

The Bayesian statistical approach provides a natural route to quantifying uncertainty in inverse problems by characterizing the posterior probability distribution. And yet the application of Bayesian inference to inverse problems raises a number of important computational challenges and foundational issues. From the computational perspective, important tasks include the design of Markov chain Monte Carlo sampling schemes for complicated and high-dimensional posterior distributions; the construction of controlled approximations to the likelihood function or forward model, whether through statistical emulation, model reduction, or function approximation; and the development of more efficient alternative algorithms, including variational approaches, for characterizing the posterior distribution. From the foundational perspective, important efforts center on incorporating model error or model discrepancy into the solution of the inverse problem and subsequent predictions, and on designing classes of prior distributions that are sufficiently rich or expressive to capture available information about the inversion parameters, while ensuring that the Bayesian formulation is well-posed and discretization invariant. We also note that not all methods for characterizing uncertainty in the inversion parameters and subsequent predictions are Bayesian; many frequentist methods for uncertainty assessment (using, for example, Tikhonov regularized estimators) have also been developed.

Data assimilation encompasses a related set of problems for which the goal is to estimate the time-evolving *state* of a dynamical system, given a sequence of observations. Applications include ocean modeling and NUMERICAL WEATHER PREDICTION [V.18]. Observations are typically available sequentially, and one therefore seeks algorithms that can be applied recursively—updating the state estimates as new data become available. When one's goal is to condition the state at time $t$ on observations received up to time $t$, the inference problem is known as *filtering*; when the state at time $t$ is conditioned on observations up to some time $T > t$, the problem becomes one of *smoothing*. For linear models and Gaussian error distributions, the recursive Kalman formulas for filtering and smoothing apply. For nonlinear models and non-Gaussian distributions, a host of other algorithms can be used to approximate the posterior distribution. Chief among these are ensemble methods, including ensemble Kalman filters and smoothers, and weighted particle methods, including many types of particle filters and smoothers. Practical applications of data assimilation involve high-dimensional states and chaotic dynamics, and the development of efficient and accurate probabilistic approaches for such problems is an active area of research.

## 5 Decision Making under Uncertainty

The results of uncertainty analysis and inference are often a prelude to *designing* a system, executing some *control* action, or otherwise making a decision, perhaps in an iterative fashion. Optimization techniques that account for uncertainty are therefore an important component of an end-to-end UQ approach. For example, robust optimization methods define an objective function that can incorporate both the mean of some performance metric and its variance, thus making a trade between absolute performance and variability. Multiobjective formulations allow this trade-off to be controlled and explored completely. One can also introduce chance constraints that specify an acceptable level of reliability for a system (e.g., by requiring the probability of failure to be less than a specified value); in the engineering literature, these are known as reliability-based design-optimization approaches. More general approaches to decision making under uncertainty use the framework of decision theory to incorporate the probabilistic background of any choice, e.g., choosing the action that maximizes some expected utility. Even questions of optimal *experimental design* can be cast in this framework: choosing where to place a sensor or how to interrogate a system, before the outcome of the experiment is known and while other uncertainties remain in the model, is yet another instance of decision making under uncertainty.

### Further Reading

Ghanem, R., and P. Spanos. 1991. *Stochastic Finite Elements: A Spectral Approach*. Berlin: Springer.

Kennedy, M. C., and A. O'Hagan. 2001. Bayesian calibration of computer models. *Journal of the Royal Statistical Society* B 63:425–64.

Le Maître, O. P., and O. M. Knio. 2010. *Spectral Methods for Uncertainty Quantification: With Applications to Computational Fluid Dynamics*. Berlin: Springer.

Saltelli, A., M. Ratto, T. Andres, F. Campolongo, J. Cariboni, D. Gatelli, M. Saisana, and S. Tarantola. 2008. *Global Sensitivity Analysis: The Primer*. New York: John Wiley.

Stuart, A. M. 2010. Inverse problems: a Bayesian perspective. *Acta Numerica* 19:451–559.

Xiu, D. 2010. *Numerical Methods for Stochastic Computation*. Princeton, NJ: Princeton University Press.

## II.35 Variational Principle

A variational principle is a method in the CALCULUS OF VARIATIONS [IV.6] for determining a function by identifying it as a minimum or maximum of a *functional*, which is a function that maps functions into scalars. An example of a functional on a HILBERT SPACE [I.2 §19.4] $H$ is the mapping from $f \in H$ to the inner product $\langle f, g \rangle$, where $g$ is any fixed element in $H$. In particular, $\int_0^1 f(x)\,dx$ is a functional on $C[0,1]$.

Many partial differential equations (PDEs) have the property that their solution is a minimum of a certain Lagrangian functional. The PDE is known as the EULER–LAGRANGE EQUATION [III.12] (see also PARTIAL DIFFERENTIAL EQUATIONS [IV.3 §4.3]) of the functional.

In matrix analysis an example of a variational principle is the expression $\lambda_1 = \max_{x \neq 0} x^*Ax/(x^*x)$ for the largest eigenvalue of a Hermitian matrix $A$, which is a particular case of the COURANT–FISCHER THEOREM [IV.10 §5.4]. The expression $x^*Ax/(x^*x)$ is called a *Rayleigh quotient*, and generalizations of it arise in the Rayleigh–Ritz approximation problem of finding optimal approximate eigenvectors of $A$ given an approximate invariant subspace.

## II.36 Wave Phenomena

Waves are everywhere. We immediately think of waves on the ocean, sound waves, electromagnetic waves such as light and radio, and seismic waves caused by earthquakes. Less obviously, there are waves of traffic on busy roads, ultrasonic waves used to image the insides of our bodies, and the Mexican wave around a football stadium.

From these examples, we all have a good idea of what a wave is, but it is not so easy to define a wave. Perhaps the key property is *propagation*: disturbances are propagated. Think of the Mexican wave; people in the crowd stand and sit in an organized manner so as to generate a disturbance that propagates around the stadium. Notice that the people do not propagate! Similarly, in a sound wave air particles move about their equilibrium positions. Electromagnetic waves do not require a medium in order to exist; they can propagate through empty space whereas sound waves cannot.

In addition to propagating disturbances, waves are often associated with the transfer of energy, and this is one reason they are useful. Waves can also interact with objects (giving rise to reflection, refraction, diffraction, or scattering), or even with other waves.

For a simple formula, suppose that the $x$-axis points to the right. A disturbance $u(x,t)$ at position $x$ and time $t$ will be a wave propagating to the right if it has the form

$$u(x,t) = f(x - ct),$$

where $f$ is a function of one variable and $c$ is a constant (the speed of propagation, or the *phase speed*). For an explanation, see the article on the WAVE EQUATION [III.31]. In this very simple one-dimensional example, the disturbance propagates without change of shape. In reality, the shape may change as the wave propagates (think of ocean waves approaching and then breaking on a beach).

Wave phenomena continue to fascinate and provoke the creation of new mathematics.

### Further Reading

Scales, J. A., and R. Snieder. 1999. What is a wave? *Nature* 401:739–40.

# Part III

# Equations, Laws, and Functions of Applied Mathematics

## III.1 Benford's Law

*Theodore P. Hill*

*Benford's law*, also known as the *first-digit law* or the *significant-digit law*, is the empirical observation from statistical folklore that in many naturally occurring tables of numerical data the leading significant digits are not equally likely. In particular, more than 30% of the leading significant (nonzero) digits are 1 and less than 5% are 9.

### 1 The First-Digit Law

Benford's law asserts that, instead of being uniformly distributed, as might be expected, the first significant decimal digit often tends to follow the logarithmic distribution

$$\text{Prob}(D_1 = d) = \log_{10}\left(\frac{d+1}{d}\right), \quad d = 1, 2, \dots, 9,$$

so

$$\begin{aligned} \text{Prob}(D_1 = 1) &= \log_{10}(2) = 0.3010\dots, \\ \text{Prob}(D_1 = 2) &= \log_{10}(3/2) = 0.1760\dots, \\ &\vdots \\ \text{Prob}(D_1 = 9) &= \log_{10}(10/9) = 0.04575\dots, \end{aligned}$$

where $D_1$ represents the first significant decimal digit (e.g., $D_1(0.0203) = D_1(203) = 2$).

### 2 History

The earliest known reference to this logarithmic distribution is a short article in 1881 by polymath Simon Newcomb in the *American Journal of Mathematics*, and the article contained not only the first-digit law above but also the second-digit law. This paper was forgotten, and in 1938 Frank A. Benford published an article containing the same first- and second-digit laws, as well as extensive empirical evidence of the law in tables ranging from baseball statistics to square-root tables and atomic weights. This article attracted much attention but Newcomb's note continued to be overlooked for decades and the name "Benford" came to be associated with the significant-digit law. Since then, over 700 articles have appeared on applications, statistical tests, and mathematical proofs of Benford's law.

### 3 Empirical Evidence

Many common tables of numerical data do not follow Benford's law. For example, the proportion of positive integers that begin with 1, i.e., $\{1, 10, 11, \dots, 19, 100, 101, \dots, 199, 1000, 1001, \dots\}$, oscillates between $\frac{1}{9}$ and $\frac{5}{9}$ as the sample size increases. The prime numbers also do not follow Benford's law. Similarly, many tables of real-world data, such as telephone numbers and lottery numbers, do not follow Benford's law.

On the other hand, in addition to Benford's original empirical data, an abundance of subsequent empirical evidence of Benford's law has appeared in a wide range of fields. This includes numbers gleaned from newspaper articles and almanacs, tables of physical constants and half-lives of radioactive substances, stock markets and financial data, demographic and geographical data, scientific calculations, eBay prices, and the collection of all numbers on the Internet, as reflected by magnitudes of Google hits on numbers.

### 4 Applications

One of the main applications of Benford's law has been for fraud detection. Since certain types of true tax data have been found to be a close fit to Benford's law (e.g., about 30% of the numbers begin with 1), chi-squared goodness-of-fit tests have been used successfully to detect fraud or error by checking conformance of the first digits of the data to Benford's law. Benford's law goodness-of-fit tests have also been used to identify anomalous signals in data. This has been employed, for example, to detect mild earthquakes, and to evaluate the detection efficiency of lightning location networks.

The output of numerical algorithms and other digital computer calculations often follows Benford's law, and based on a hypothesis of the output following Benford's law it is possible to obtain improved estimates of expected roundoff errors, and of likelihoods of overflow and underflow errors.

Benford's law has also been used as an effective teaching tool, to introduce students to basic concepts in statistics such as goodness-of-fit tests and basic data-collection methods, and to demonstrate tools in MATHEMATICA.

## 5 The General-Digit Law

The general form of Benford's law is a statement about the joint distribution of all decimal digits, namely:

$$\begin{aligned}\text{Prob}(D_1 = d_1, D_2 = d_2, \dots, D_m = d_m) \\ = \log_{10}\Big(1 + \Big(\sum_{j=1}^{m} 10^{m-j} d_j\Big)^{-1}\Big)\end{aligned}$$

holds for all $m$-tuples $(d_1, d_2, \dots, d_m)$, where $d_1$ is an integer in $\{1, 2, \dots, 9\}$ and, for $j \geqslant 2$, $d_j$ is an integer in $\{0, 1, \dots, 9\}$. Here $D_2$, $D_3$, $D_4$, etc., represent the second, third, fourth, etc., significant decimal digits, e.g., $D_2(0.0203) = 0$, $D_3(0.0203) = 3$.

Thus, for example, this general form of Benford's law implies that

$$\begin{aligned}\text{Prob}(D_1 = 3, D_2 = 1, D_3 = 4) &= \log_{10}(315/314) \\ &= 0.001380\dots.\end{aligned}$$

A corollary of the general form of Benford's law is that the significant digits are dependent and not independent, as one might expect.

Letting $S(x)$ denote the (floating-point) *significand* (see FLOATING-POINT ARITHMETIC [II.13]) of the positive number $x$, e.g., $S(0.0203) = S(2.03 \times 10^{-2}) = 2.03$, a more compact form of the general Benford's law is

$$\text{Prob}(S < t) = \log_{10} t \quad \text{for all } 1 \leqslant t < 10.$$

Analogs of Benford's law for nondecimal bases $b$ are obtained by simply replacing the decimal base by the new base, both in the significant digits or significand and in the logarithm.

## 6 The Mathematical Framework

One of the main tools used to study Benford's law is the fact that a data set $X$ (e.g., a sequence, function, or random variable) is Benford if and only if $\log X$ is uniformly distributed modulo 1, that is, if and only if the fractional part of $\log X$ is uniformly distributed between 0 and 1.

Two key characterizations of Benford's law are *scale invariance* and *base invariance*. The Benford distribution is the only distribution of significant digits that does not change under multiplicative changes of scale. For example, if a data set originally in euros or meters follows Benford's law, conversion of the data into dollars or feet will also follow Benford's law. Similarly, the Benford distribution is the only distribution of significant digits that is continuous and invariant under changes of base.

## 7 Sequences and Functions

Many common sequences follow Benford's law exactly. That is, the proportion of times that particular significant digits appear in the elements of the sequence converges to the exact Benford's law probabilities. For example, the sequence of powers of 2 (and of 3 or 5), the Fibonacci and Lucas numbers, and the sequence of factorials $1!, 2!, 3!, 4!, \dots = 1, 2, 6, 24, \dots$ all follow Benford's law exactly.

Sequences exhibiting exponential growth (or decay) generally obey Benford's law for almost all starting points and almost all bases. Similarly, many general classes of algorithms, including Newton's method, and multidimensional systems such as Markov chains can also be shown to obey Benford's law. Continuous functions with exponential or super-exponential growth or decay also typically exhibit Benford's law behavior, and thus wide classes of initial-value problems obey Benford's law exactly.

## 8 Random Variables and Probability Distributions

None of the classical probability distributions—uniform, exponential, normal, Pareto, etc.—is arbitrarily close to Benford's law for *any* values of the parameters, although the standard Cauchy distribution comes quite close. On the other hand, it is easy to construct distributions that satisfy Benford's law exactly, such as the continuous distribution on $[1, 10)$ with density proportional to $1/x$. Some of the basic probabilistic Benford's law results are given below.

- $X$ is Benford if and only if $1/X$ is Benford.
- If a random variable $X$ has a density, then the powers $X^n$ of $X$ converge in distribution to Benford's law (i.e., $P(S(X^n) < t) \to \log t$).
- If $X$ is Benford and $Y$ is positive and independent of $X$, then $XY$ is Benford.
- The product of independent and identically distributed continuous random variables converges in distribution to Benford's law.
- If random samples are taken from probability distributions chosen at random (in an unbiased way), the combined sample will converge to Benford's law.

### Further Reading

Benford Online Bibliography. 2012. An open-access database on Benford's law, which is available at www.benford online.net.

Berger, A., and T. P. Hill. 2015. *An Introduction to Benford's Law.* Princeton, NJ: Princeton University Press.

Nigrini, M. 2012. *Benford's Law: Applications for Forensic Accounting, Auditing, and Fraud Detection.* New York: John Wiley.

Raimi, R. A. 1976. The first digit problem. *American Mathematical Monthly* 83(7):521–38.

## III.2 Bessel Functions

*P. A. Martin*

F. W. Bessel (1784–1846) was a German astronomer. In an analysis of planetary motions, published in 1826, he investigated properties of the integral

$$J_n(z) = \frac{1}{2\pi}\int_0^{2\pi} \cos(n\theta - z\sin\theta)\,\mathrm{d}\theta.$$

(He denoted the integral by $I_z^n$.) Schlömilch (1856) regarded $J_n(z)$ as coefficients in the expansion

$$\exp\{(z/2)(t - 1/t)\} = \sum_{n=-\infty}^{\infty} t^n J_n(z); \qquad (1)$$

the left-hand side is called a *generating function.* Both approaches are convenient when $n$ is an integer. Another approach, which generalizes more easily, is via an ordinary differential equation,

$$z^2 w''(z) + zw'(z) + (z^2 - \nu^2)w(z) = 0, \qquad (2)$$

in which the independent variable, $z$, and the parameter $\nu$ can be complex. Solutions of *Bessel's equation* (2) can be constructed by the method of Frobenius. One of these is

$$J_\nu(z) = \sum_{m=0}^{\infty} \frac{(-1)^m (z/2)^{\nu+2m}}{m!\Gamma(\nu+m+1)},$$

which is known as the *Bessel function of the first kind.* Here, $\Gamma$ denotes the GAMMA FUNCTION [III.13]. If $\nu$ is not an integer, $J_{-\nu}(z)$ gives a second, linearly independent, solution of (2). However, when $\nu = n$, an integer, we have $J_n(z) = (-1)^n J_{-n}(z)$ (to see this, replace $t$ by $-1/t$ in (1)), so a new solution is required. It is

$$Y_\nu(z) = \frac{J_\nu(z)\cos\nu\pi - J_{-\nu}(z)}{\sin\nu\pi},$$

with $Y_n = \lim_{\nu\to n} Y_\nu$. This defines the *Bessel function of the second kind.*

The functions $J_\nu(z)$ and $Y_\nu(z)$ are examples of SPECIAL FUNCTIONS [IV.7 §9] of two variables, $\nu$ and $z$. In elementary applications, $\nu$ is an integer, $z$ is real, and both are nonnegative. For example, the vibrational modes of a circular membrane of radius $a$ (such as a drumhead) have the form

$$J_n(kr)\cos(n\theta)\cos(\omega t),$$

where $r$ and $\theta$ are plane polar coordinates, $t$ is time, $\omega$ is frequency, $k = \omega/c$, $c$ is the speed of sound in the membrane (it is a constant depending on what the membrane is made from), and $ka$ is chosen as any one of the positive zeros of $J_n(x)$ (there are infinitely many).

Much is known about the properties of Bessel functions. The standard reference is G. N. Watson's 800-page book *A Treatise on the Theory of Bessel Functions* (Cambridge University Press, Cambridge, 2nd edn, 1944). However, a good place to start is chapter 10 of *NIST Handbook of Mathematical Functions* (Cambridge University Press, Cambridge, 2010), edited by F. W. J. Olver, D. W. Lozier, R. F. Boisvert, and C. W. Clark. (An electronic version of the book is available at http://dlmf.nist.gov.)

## III.3 The Black–Scholes Equation

The Black–Scholes equation is a linear parabolic partial differential equation of the form

$$\frac{\partial V}{\partial t} + \tfrac{1}{2}\sigma^2 S^2 \frac{\partial^2 V}{\partial S^2} + rS\frac{\partial V}{\partial S} - rV = 0.$$

It is associated with the problem of pricing a financial option whose value is $V = V(S,t)$. The underlying asset has price $S \geqslant 0$ at time $t \in [0,T]$, where $T$ is the expiry time of the option. The equation also involves the volatility $\sigma$ and the interest rate $r$. Appropriate boundary conditions must be added in order to determine $V$ uniquely. Under appropriate changes of variable the Black–Scholes equation transforms into the DIFFUSION

(HEAT) EQUATION [III.8]. The Black–Scholes equation is used in the derivation of the BLACK–SCHOLES OPTION PRICING FORMULA [IV.14 §2.2].

## III.4 The Burgers Equation

The one-dimensional *Burgers equation* is a nonlinear PARTIAL DIFFERENTIAL EQUATION [IV.3] (PDE) for $u(x,t)$,

$$\frac{\partial u}{\partial t} + u\frac{\partial u}{\partial x} = \nu\frac{\partial^2 u}{\partial x^2}, \qquad (1)$$

where $\nu$ is a positive constant. It is named after J. M. Burgers (1895–1981), although the importance of (1) had been first recognized (1915) by Bateman in his study of the NAVIER–STOKES EQUATIONS [III.23] as the viscosity $\nu \to 0$.

When $\nu = 0$, (1) simplifies to

$$\frac{\partial u}{\partial t} + u\frac{\partial u}{\partial x} = 0. \qquad (2)$$

This is known as the *inviscid Burgers equation*, although the word "inviscid" is sometimes omitted.

There is a connection between (1) and the DIFFUSION EQUATION [III.8] known as the *Cole–Hopf transformation*. Thus, if $w(x,t)$ solves

$$\frac{\partial^2 w}{\partial x^2} = \frac{1}{\nu}\frac{\partial w}{\partial t},$$

then

$$u = -\frac{2\nu}{w}\frac{\partial w}{\partial x}$$

solves (1). This is an example of solving a nonlinear PDE using solutions of a related linear PDE.

The Cole–Hopf transformation can be used to show that, if (1) is solved with specified initial conditions at $t = 0$, then the solution is smooth for all $x$ and for all $t > 0$. On the other hand, the inviscid Burgers equation (2) can have discontinuous solutions ("shocks"). We can say that the presence of the diffusive term on the right-hand side of (1) "regularizes" the problem and prevents shocks from appearing. Energy is removed because $\nu > 0$.

If we change the sign of the diffusive term, energy is added. To compensate for this, another term can be appended. One PDE of this kind is the *Kuramoto–Sivashinsky equation*,

$$\frac{\partial u}{\partial t} + u\frac{\partial u}{\partial x} = -\frac{\partial^2 u}{\partial x^2} - \frac{\partial^4 u}{\partial x^4}.$$

It is encountered in the modeling of several physical phenomena, it has been studied extensively, and it has many interesting kinds of solutions.

## III.5 The Cahn–Hilliard Equation

*Amy Novick-Cohen*

### 1 The Equation

The partial differential equation for $u = u(x,t) \in \mathbb{R}$,

$$u_t = M\Delta(-u + u^3 + \varepsilon^2\Delta u), \quad (x,t) \in Q_T,$$

is known as the Cahn–Hilliard equation. Here, $\Delta := \sum_{i=1}^{N} \partial^2_{x_i}$ and $Q_T = \Omega \times (0,T)$, where $\Omega \subset \mathbb{R}^N$, $T > 0$. It was proposed in 1958 by John W. Cahn and John E. Hilliard to describe phase separation in binary alloys. In that context, $u$ represents the locally defined mass fraction of one of the two components of the binary alloy, $M$ is the "mobility," and $\varepsilon$ measures the effective length scale of the interatomic forces.

### 2 Structure

Equation (1) may by written in the coupled form

$$\left.\begin{aligned} u_t &= \nabla\cdot(M\nabla\mu), & (x,t) \in Q_T,\\ \mu &= f'(u) - \varepsilon^2\Delta u, & (x,t) \in Q_T,\end{aligned}\right\} \qquad (1)$$

where $\mu = \mu(x,t)$ is the "chemical potential," and $f(u) := \frac{1}{4}(1-u^2)^2$ has minima at $u = \pm 1$ and is referred to as a "double-well potential."

Typically, $\Omega \subset \mathbb{R}^N$ is a bounded domain, $N = 3$, and periodic or Neumann and no flux boundary conditions ($\hat{n}\cdot\nabla u = \hat{n}\cdot\nabla\mu = 0$ along $\partial\Omega$ for $\hat{n} \perp \partial\Omega$) are imposed. In both these cases, integrating (1) over $\Omega$, and then integrating by parts, yields that

$$\frac{\mathrm{d}}{\mathrm{d}t}\int_\Omega u(x,t)\,\mathrm{d}x = 0,$$

which may be interpreted as stating that mass is conserved in the system.

Multiplying the first equation in (1) by $\mu$ and then integrating over $\Omega$, we get

$$\frac{\mathrm{d}}{\mathrm{d}t}\int_\Omega (f(u) + \tfrac{1}{2}\varepsilon^2|\nabla u|^2)\,\mathrm{d}x = -\int_\Omega M|\nabla\mu|^2\,\mathrm{d}x,$$

which describes energy dissipation in the system.

### 3 Dynamics

It is often reasonable to assume that

$$u(x,0) = \bar{u} + \tilde{u}(x,0),$$

where $\bar{u} \in \mathbb{R}$ and $\tilde{u}(x,0)$ is a small disturbance or "perturbation" of $\bar{u}$ that satisfies $\int_\Omega \tilde{u}(x,0)\,\mathrm{d}x = 0$. The subsequent dynamics can be described by an early "spinodal" regime followed by a late "coarsening" regime. During the spinodal or "linear" regime, the

Cahn–Hilliard dynamics is dominated by its linearization about $\bar{u}$. Spatial perturbations of sufficiently long wavelength grow, while shorter-wavelength perturbations decay exponentially. Locally spatially uniform domains begin to form. The larger domains start to grow at the expense of the smaller domains; this behavior is known as "coarsening." The coarsening dynamics can be described via a free boundary problem for the motion of the interfaces, $\Gamma(t)$, which partition $\Omega$ into locally uniform subdomains. In this free boundary problem, known as the "Mullins–Sekerka" problem, the normal velocity of $\Gamma(t)$ is proportional to the jump in the normal derivative across $\Gamma(t)$ of $\mu$, the chemical potential. Moreover, $\mu = \kappa$ along $\Gamma(t)$, where $\kappa$ denotes the mean curvature of $\Gamma(t)$, and $\Delta\mu = 0$ away from the interfaces, in $\Omega \setminus \Gamma(t)$. The late-time dynamics predict that the average size of the uniform subdomains grows at a rate proportional to $t^{1/3}$.

### 4 Applications

Although the mass-conservative Cahn–Hilliard equation was conceived as a model for phase separation in binary alloys, the features of its dynamics, with an early linear regime followed by a late coarsening regime, make it an appropriate model in a wide range of settings. It has been used to describe pattern formation in populations, structure formation in biofilms, galaxy formation, as well as in image processing.

### 5 Generalizations

Many generalizations of the Cahn–Hilliard equation have been suggested. These include conserved phase field models (models that couple the Cahn–Hilliard equation with thermal effects), models for simultaneous phase separation and ordering (models coupling Cahn–Hilliard and Allen–Cahn equations), models coupling the Cahn–Hilliard equation with hydrodynamics effects, phase field crystal models (generalization that include crystalline anisotropy effects), and more.

#### Further Reading

Cahn, J. W. 1961. On spinodal decomposition. *Acta Metallurgica* 9:795–801.

Cahn, J. W., and J. E. Hilliard. 1958. Free energy of a nonuniform system. I. Interfacial free energy. *Journal of Chemical Physics* 28:258–67.

Novick-Cohen, A. Forthcoming. *The Cahn–Hilliard Equation: From Backwards Diffusion to Surface Diffusion.* Cambridge: Cambridge University Press.

## III.6 The Cauchy–Riemann Equations

Let $u(x,y)$ and $v(x,y)$ be two real functions of two real variables, $x$ and $y$. The *Cauchy–Riemann equations* are

$$\frac{\partial u}{\partial x} = \frac{\partial v}{\partial y} \quad \text{and} \quad \frac{\partial u}{\partial y} = -\frac{\partial v}{\partial x}. \tag{1}$$

These equations arise in COMPLEX ANALYSIS [IV.1]: if $f(z) = u(x,y) + \mathrm{i}v(x,y)$ is an analytic function of $z = x + \mathrm{i}y$, where $u$ and $v$ are real, then $u$ and $v$ satisfy the Cauchy–Riemann equations.

Eliminating $v$ from (1) shows that LAPLACE'S EQUATION [III.18] is satisfied by $u$:

$$\frac{\partial^2 u}{\partial x^2} + \frac{\partial^2 u}{\partial y^2} = 0;$$

$v$ satisfies the same equation.

In plane potential flow of a fluid, $u$ is identified as a *velocity potential* and $v$ as a *stream function*. If we define vectors $\boldsymbol{n} = (n_1, n_2)$ and $\boldsymbol{t} = (-n_2, n_1)$, so that $\boldsymbol{n} \cdot \boldsymbol{t} = 0$, from (1) we have

$$\boldsymbol{n} \cdot \operatorname{grad} u = \boldsymbol{t} \cdot \operatorname{grad} v.$$

If we take $\boldsymbol{n}$ as being a unit normal vector to a curve $C$, then $\boldsymbol{t}$ is a unit tangent vector. In fluid flow, a typical boundary condition is zero normal velocity on $C$, $\boldsymbol{n} \cdot \operatorname{grad} u = 0$. This derivative condition on $u$ can then be replaced by $v = \text{const.}$ on $C$, a condition that is often easier to enforce.

## III.7 The Delta Function and Generalized Functions

*P. J. Upton*

The Dirac $\delta$ function, denoted by $\delta(x)$ and named after its inventor, the twentieth-century theoretical physicist P. A. M. Dirac, may be thought of as a function that is so tightly peaked about the value $x = 0$ that it is zero everywhere except at $x = 0$ and yet its integral $\int_{-\infty}^{\infty} \delta(x)\,\mathrm{d}x = 1$. This means that, strictly speaking, $\delta(x)$ is not a function at all, since its value at $x = 0$ is ill defined. However, as will become clear later on, the "function" $\delta(x)$ can be used as a convenient shorthand notation for the limit of a sequence of well-defined functions, $g_n(x)$, as $n \to \infty$. The $\delta$ function is used widely across physical applied mathematics, to model, for example, impulses in mechanics and electric circuit theory, and point-like charge distributions in electromagnetism or gravity.

So why define such a strange object as $\delta(x)$? One answer is that many physical situations can be described well using the $\delta$ function. For example, in one-dimensional mechanics a particle with mass $m$ may be subject to a force $f(t)$ present for *only* a very short time interval $-t_0 < t < t_0$. Consider the total *impulse* defined by the integral

$$\int_{-t_0}^{t_0} f(t)\,\mathrm{d}t = \int_{-t_0}^{t_0} m\frac{\mathrm{d}v}{\mathrm{d}t}\,\mathrm{d}t = mv(t_0) - mv(-t_0),$$

where the first equality follows from Newton's second law, and $v(t)$ is the velocity of the particle at time $t$. Suppose that the force acts in such a way that it causes the particle, initially at rest, to move off with constant velocity and its momentum set to unity, so that $v(-t_0) = 0$ and $mv(t_0) = 1$. Furthermore, we require that the force acts over such a tiny time interval that $t_0$ is practically zero. In other words, the force acts *instantaneously*, as might be expected if $f(t)$ were the result of a sudden hammer blow or kick, and the precise form of $f(t)$ when $-t_0 < t < t_0$ does not significantly affect the outcome. In order to model this kick, $f(t)$ must be zero everywhere except at $t = 0$ but with unit total integral, corresponding to a unit impulse. We are therefore led to a $\delta$ function as a model for an *impulsive force*. The subsequent motion (or *response*) of a particle due to an impulsive force is an important function of $t$. It is an *initial-value Green function*, $G(t)$, named after the early nineteenth-century mathematical physicist George Green. Green functions are, more generally, the solutions of nonhomogeneous differential equations with a $\delta$ function as the source term.

## 1 Basic Properties

As discussed above, the $\delta$ function has the property that

$$\int_a^b \delta(x)\,\mathrm{d}x = \begin{cases} 1 & \text{if } a < 0 < b, \\ 0 & \text{if } 0 \notin [a,b]; \end{cases}$$

cases where $a = 0$ or $b = 0$ will not be considered. The $\delta$ function is not strictly a function at all, as $x = 0$ is in the domain of the "function" yet $\delta(0)$ cannot be assigned a value in the codomain. Strictly speaking, $\delta(x)$ has a meaning only when it appears under an integral sign, in which case the following important basic property is satisfied:

$$\int_{-\infty}^{\infty} f(x)\,\delta(x)\,\mathrm{d}x = f(0) \tag{1}$$

for a sufficiently well-behaved function $f(x)$. The identity $\int_{-\infty}^{\infty} \delta(x)\,\mathrm{d}x = 1$ is a special case of (1).

## 2 Green Functions

Linear initial-value problems, such as those in which we seek the position of a particle, $x(t)$, at time $t$, involve solving a differential equation $Lx = f(t)$, where $L$ is a linear differential operator and $f(t)$ some arbitrary force. The general solution is given by $x = x_c + x_p$. Here, $x_c$ is the complementary function and it solves $Lx_c = 0$, while $x_p$, the particular integral, can be expressed in terms of a Green function, $G(t)$, that is, the response to the impulsive force $\delta(t)$. Thus, with $LG = \delta(t)$, we have $x_p(t) = \int_0^\infty G(t-t')f(t')\,\mathrm{d}t'$.

The $\delta$ function defined on $\mathbb{R}^3$, where

$$\delta(\boldsymbol{r}) = \delta(x)\delta(y)\delta(z) \quad \text{for } \boldsymbol{r} = (x,y,z) \in \mathbb{R}^3,$$

can be used to model a point-like charge distribution. This then enables the determination of the potential, $\phi(\boldsymbol{r})$, $\boldsymbol{r} \in \mathbb{R}^3$, for general charge distributions, $\rho(\boldsymbol{r})$, which requires solving POISSON'S EQUATION [III.18] $-\Delta\phi = \rho(\boldsymbol{r})$ ($\Delta$ denotes the Laplace operator). Again, this can be done using a Green function, $G(\boldsymbol{r})$, which solves $-\Delta G = \delta(\boldsymbol{r})$, so that $G(\boldsymbol{r})$ is the potential for a point-like charge, from which we find that $G(\boldsymbol{r}) = 1/(4\pi|\boldsymbol{r}|)$, i.e., the Coulomb potential, and that $\phi(\boldsymbol{r}) = \int_{\mathbb{R}^3} G(\boldsymbol{r}-\boldsymbol{r}')\rho(\boldsymbol{r}')\,\mathrm{d}\boldsymbol{r}'$.

When treating problems involving both time evolution and spatial variation, such as in quantum theory and the theory of Brownian motion, the Green function depends on time as well as position. In these cases, $G(t,\boldsymbol{r})$ is often referred to as a *propagator*.

## 3 Delta-Convergent Sequences

The $\delta$ function can be constructed by taking a limit of a sequence of well-defined functions $g_n(x)$ as $n \to \infty$. If $g_n$ is *smooth* (that is, differentiable at all orders) for all $n$, then the sequence is said to define a *generalized function*. In addition, for all $n$, we insist on the condition $\int_{-\infty}^{\infty} g_n(x)\,\mathrm{d}x = 1$. An example of such a $\delta$-convergent sequence is given by the sequence of Gaussian functions

$$g_n(x) = \frac{n}{\sqrt{2\pi}}\mathrm{e}^{-(nx)^2/2}, \quad n \geqslant 1. \tag{2}$$

Examples of this $g_n(x)$ are plotted in figure 1(a). As $n$ increases, the functions $g_n(x)$ get more strongly peaked (both narrower and taller) about $x = 0$. Indeed, from the expression in (2) it follows that as $n \to \infty$, $g_n(x) \to 0$ for $x \neq 0$ but $g_n(0) \to \infty$, as expected if the limit is $\delta(x)$. Also, $\int_{-\infty}^{\infty} g_n(x)\,\mathrm{d}x = 1$, $n \geqslant 1$, must hold, in view of the expression $\int_{-\infty}^{\infty} \mathrm{e}^{-x^2}\,\mathrm{d}x = \sqrt{\pi}$ for the integral of a Gaussian function. Thus, taking into

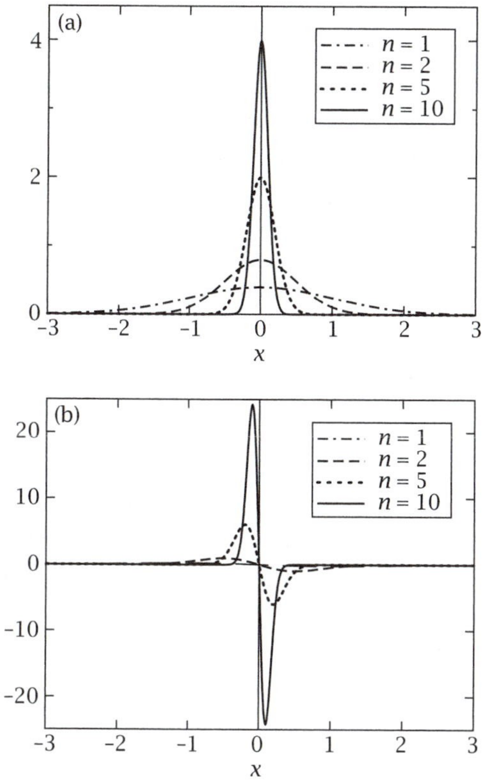


**Figure 1** (a) The functions $g_n(x)$ as defined in (2). (b) Their derivatives $g_n'(x)$ for $n = 1, 2, 5$, and 10. These sequences converge to $\delta(x)$ and $\delta'(x)$, respectively.

account these properties of $g_n(x)$ in (2), it begins to look plausible that $g_n(x) \to \delta(x)$ (in some sense) as $n \to \infty$. Indeed, one can prove that for sufficiently well-behaved functions $f(x)$ (such as Lipschitz-continuous functions), the limit

$$\lim_{n\to\infty} \int_{-\infty}^{\infty} g_n(x) f(x)\,\mathrm{d}x = f(0) \qquad (3)$$

holds. Hence, the limit of $g_n(x)$ as $n \to \infty$ satisfies the basic property of the $\delta$ function given by (1), at least for Lipschitz-continuous functions. But with a different $\delta$-convergent sequence, $g_n(x)$, $n \geqslant 1$, one can *prove* that (3) holds for *all* continuous functions $f(x)$. In this case, the functions $g_n(x)$ are smooth but *compactly supported*, that is, nonzero only on bounded intervals of $\mathbb{R}$.

## 4 Derivatives of the Delta Function

One can define a generalized function corresponding to the derivative of the delta function, $\delta'(x)$, through a ($\delta'$-convergent) sequence of functions, $g_n'(x)$, $n \geqslant 1$. An example of such a sequence is illustrated in figure 1(b), where $g_n'(x)$ is the derivative of $g_n(x)$ given by (2). Such a process has an appealing physical interpretation. Just as we can think of $\delta(x)$ as a model for a one-dimensional charge distribution of a point-like electric charge, we can regard $\delta'(x)$ as that of an electric dipole consisting of two oppositely signed point-like charges that are infinitesimally close together. From the limiting properties of $g_n'(x)$, one can show that the dipole moment of this charge distribution is given by $\int_{-\infty}^{\infty} x\delta'(x)\,\mathrm{d}x = -1$. Indeed, this is a special case of the more general result for derivatives of general order $k$,

$$\int_{-\infty}^{\infty} \delta^{(k)}(x) f(x)\,\mathrm{d}x = (-1)^k f^{(k)}(0),$$

for sufficiently well-behaved functions $f(x)$.

## 5 Toward a Rigorous Theory

A rigorous theory of generalized functions, also called *distributions*, can be constructed by defining them as *continuous linear functionals*, $T(\varphi)$, on the space of all smooth, compactly supported test functions $\varphi(x)$ on $\mathbb{R}$. The derivative of $T(\varphi)$ is defined by $\mathrm{d}T(\varphi)/\mathrm{d}x = -T(\varphi')$. If $T(\varphi)$ can be expressed in the form $T(\varphi) = \int_{-\infty}^{\infty} g(x)\varphi(x)\,\mathrm{d}x$, where $g(x)$ is a well-defined locally integrable function, then $T(\varphi)$ is said to be a *regular* generalized function; otherwise, it is called *singular*. The $\delta$ function is the singular generalized function defined by $T(\varphi) = \varphi(0)$. So, regular generalized functions assign a functional $T(\varphi)$ to each well-defined function $g(x)$. But, by extending the range of allowed $T(\varphi)$ to include singular generalized functions, one can think of $g(x)$ as being part of a much larger set, hence the term "generalized" function.

Another approach to the rigorous treatment of the $\delta$ function, and one that is particularly useful in probability theory, is to regard it as a *measure*. The Dirac $\delta$ measure centered at a point $a \in \mathbb{R}$, denoted by $\delta_a$, acts on sets $A \subset \mathbb{R}$ and is defined by the property that $\delta_a(A) = 1$ if $a \in A$ and $\delta_a(A) = 0$ otherwise. A measure-theoretic version of the basic property (1) then follows from $\int_{\mathbb{R}} f(x)\delta_a(\mathrm{d}x) = f(a)$.

### Further Reading

Kolmogorov, A. N., and S. V. Fomin. 1975. *Introductory Real Analysis*. New York: Dover.

Lighthill, M. J. 1958. *An Introduction to Fourier Analysis and Generalised Functions*. Cambridge: Cambridge University Press.

## III.8 The Diffusion Equation

The one-dimensional *diffusion* (or *heat*) *equation* is a PARTIAL DIFFERENTIAL EQUATION [IV.3] (PDE) for $u(x,t)$,

$$\frac{\partial^2 u}{\partial x^2} = \frac{1}{k}\frac{\partial u}{\partial t}, \tag{1}$$

where $k$ is a constant. It is classified as a linear second-order homogeneous parabolic equation.

The diffusion equation has solutions that are separated, $u(x,t) = \mathrm{e}^{\gamma x}\mathrm{e}^{\gamma^2 kt}$, where $\gamma$ is an arbitrary constant. There are also nonseparable solutions such as $u(x,t) = t^{-1/2}\mathrm{e}^{-\phi}$, where $\phi(x,t) = x^2/(4kt)$.

Equation (1) describes the conduction of heat along a bar of metal and many other diffusion processes. In financial mathematics, there is a famous PDE known as the BLACK–SCHOLES EQUATION [III.3]

$$\frac{\partial V}{\partial t} + \tfrac{1}{2}\sigma^2 S^2 \frac{\partial^2 V}{\partial S^2} + rS\frac{\partial V}{\partial S} - rV = 0, \tag{2}$$

where $\sigma$ and $r$ are constants and the independent variables are $S$ and $t$. Solutions of this PDE can be constructed from solutions of the diffusion equation. Thus, the substitutions $u(x,t) = \mathrm{e}^{\alpha x + \beta t}V$ and $\mathrm{e}^x = S$ in (1) show that $V$ solves (2) if we take $k = -\frac{1}{2}\sigma^2$, $\alpha = -\frac{1}{2}(1 + r/k)$, and $\beta = \frac{1}{4}k(1 - r/k)^2$.

THE BURGERS EQUATION [III.4] is another PDE that can be solved using solutions of (1).

## III.9 The Dirac Equation
*Mark R. Dennis*

In quantum theory, the Dirac equation is the relativistic counterpart to SCHRÖDINGER'S EQUATION [III.26], representing the space-time dependence of an electron wave packet. It is one of the most fundamental equations in physics, combining the formalism of quantum physics with special relativity, and its solutions naturally lead to the concept of antimatter.

In QUANTUM MECHANICS [IV.23], when considering the wave function $\Psi(\boldsymbol{r},t)$ of a quantum particle, quantities such as position in 3-space $\boldsymbol{r}$, time $t$, momentum $\boldsymbol{p}$, and energy $E > 0$ are related to multiplication operators or the product of differential operators multiplied by the imaginary unit i times the quantum of action, Planck's constant $\hbar$. These include the energy operator $\mathrm{i}\hbar\partial_t$ and the momentum operator with components $-\mathrm{i}\hbar\partial_j$, with $\partial_j$ for $j = 1,2,3$ denoting derivatives on the spatial coordinates of $\boldsymbol{r}$. The Schrödinger equation $\mathrm{i}\hbar\partial_t\Psi = \hat{H}\Psi$ thus expresses equality of the energy operator and a Hamiltonian operator $\hat{H}$ when each acts on the quantum wave function $\Psi(\boldsymbol{r},t)$. However, in special relativity, the energy–momentum relationship is

$$E^2 = |\boldsymbol{p}|^2c^2 + m^2c^4 \tag{1}$$

for a particle of rest mass $m$, with constant $c$ the speed of light. We will refer to the quantum particle of interest as an "electron" and denote its electric charge by $e$.

The *Klein–Gordon equation* is one possible quantum equation corresponding to (1),

$$c^{-2}\partial_t^2\varphi - \nabla^2\varphi + \frac{m^2c^2}{\hbar^2}\varphi = 0, \tag{2}$$

for a generally complex-valued wave function $\varphi(\boldsymbol{r},t)$. However, there are physical problems in interpreting a solution of (2) as the relativistic counterpart of a Schrödinger wave function: unlike the Schrödinger equation, (2) involves the second derivative of time, so two initial conditions at $t = 0$ are required to specify a solution (whereas only one is required for the Schrödinger equation), and there are problems with defining a continuity equation for $|\varphi|^2$ as a probability density. Furthermore, (2) does not include the two components of electron spin. Therefore, to write down a quantum mechanical partial differential equation (PDE) corresponding to (1) that is first order in time, one must find an appropriate square root of the overall operator acting on the left-hand side of (2).

Paul Dirac famously resolved this problem algebraically, proposing the Dirac equation

$$\mathrm{i}\hbar\partial_t\psi = \sum_{j=1}^{3}\alpha_j c\hat{p}_j\psi + \beta mc^2\psi. \tag{3}$$

Unlike the Schrödinger and Klein–Gordon equations, (3) is a vector-matrix equation, where $\psi \equiv \psi(\boldsymbol{r},t)$ is a wave function with four complex components, called a *Dirac spinor* (the components are not related to four-dimensional space-time), and $\alpha_j$, $\beta$ are the $4 \times 4$ *Dirac matrices*, most commonly written in block form:

$$\beta = \begin{pmatrix} 1 & 0 \\ 0 & -1 \end{pmatrix}, \quad \alpha_j = \begin{pmatrix} 0 & \sigma_j \\ \sigma_j & 0 \end{pmatrix}, \quad j = 1,2,3.$$

Here, 0 and 1 denote the $2 \times 2$ zero and identity matrices, and the $\sigma_j$ denote the three Pauli matrices:

$$\sigma_1 = \begin{pmatrix} 0 & 1 \\ 1 & 0 \end{pmatrix}, \quad \sigma_2 = \begin{pmatrix} 0 & -\mathrm{i} \\ \mathrm{i} & 0 \end{pmatrix}, \quad \sigma_3 = \begin{pmatrix} 1 & 0 \\ 0 & -1 \end{pmatrix}.$$

The square of each Dirac matrix so defined is the identity matrix, and otherwise the matrices anticommute; that is, if $i \neq j$ then $\alpha_j\alpha_k + \alpha_k\alpha_j = \alpha_j\beta + \beta\alpha_j = 0$. These properties imply that the square of the operators on the left- and right-hand sides of (3) gives the operator equivalent of (1). Thus, the $\alpha_j$, $\beta$ matrices themselves form a Clifford algebra, known as the *Dirac algebra.*

When considering solutions in the nonrelativistic, low-velocity limit $|\boldsymbol{p}|/m \ll c$, the *Dirac equation* asymptotically approaches the Schrödinger equation for a two-component spin (in the first two components of the Dirac spinor; the second two are vanishingly small), recovering the familiar nonrelativistic quantum behavior of the electron. Dirac therefore found a relativistic, quantum PDE that is first order in space and time, albeit requiring the quantum electron to be described by the four-component Dirac spinor $\psi$, each component of which, in fact, satisfies the Klein–Gordon equation (2). The fact that the Pauli spin matrices—previously included ad hoc in the Schrödinger equation to explain the behavior of the electron's quantum mechanical spin—appear naturally in the Dirac algebra has been seen as one of the great successes of the Dirac equation as the fundamental quantum equation for the electron.

Although a matrix-vector equation, the form (3) of the Dirac equation is similar to the Schrödinger equation, and indeed the operator $\sum_{j=1}^{3}\alpha_j c\hat{p}_j + \beta mc^2$ is referred to as the *Dirac Hamiltonian.* The Dirac equation may also be written in a relativistically covariant form, by defining the set of four *gamma matrices* $\gamma^0 = \beta$, $\gamma^j = \gamma^0\alpha_j$, enabling (3) to be rewritten as

$$\mathrm{i}\hbar\sum_{a=0}^{3}\gamma^a\partial_a\psi = mc\psi, \qquad (4)$$

with $\partial_0 = c^{-1}\partial_t$. Usually, this form of the Dirac equation is written in the Einstein summation convention (omitting explicit summation symbols for repeated 4-vector components, as in the article on TENSORS AND MANIFOLDS [II.33]).

It is natural to define a current 4-vector $j^a \equiv \psi^\dagger\gamma^0\gamma^a\psi$, where $a = 0,\dots,3$, and $\psi^\dagger$ is the adjoint (conjugate transpose) of $\psi$. The 0-component $\psi^\dagger\psi$ is the nonnegative-definite probability associated with the Dirac particle described by $\psi$, and the other components $\psi^\dagger\alpha_j\psi$ give a 3-velocity field, and hence the electric current on multiplication by $e$. The 4-divergence of this current vanishes: $\sum_a \partial_a j^a = 0$. This is interpreted as both the local conservation of probability and, after multiplication by $e$, the local conservation of the electric charge distribution determined by $\psi$. The 3-velocity so defined is different from the 3-momentum, $\psi^\dagger\hat{p}_j\psi$, $j = 1,2,3$. This distinction between velocity and momentum gives rise to some surprising aspects of Dirac particles, such as the difference between quantum numbers associated with the electron's magnetic moment (related to velocity) and its angular momentum (related to momentum). For the remainder of this article we will adopt the convention in relativistic quantum theory of working in units where the constants $c$ and $\hbar$ are unity.

The Dirac equation, being a PDE for a multicomponent wave field dependent on space and time compatible with special relativity, is comparable to the set of MAXWELL'S EQUATIONS [III.22], which has similar properties (particularly in the case when $m = 0$, which is sometimes referred to as *Weyl's equation*). In particular, it has four linearly independent plane-wave solutions, of the form $\exp(\mathrm{i}(-Et+\boldsymbol{p}\cdot\boldsymbol{r}))$ times the constant unnormalized Dirac spinors

$$\begin{aligned}
\psi^1_{\mathrm{pw}} &= (E+m, 0, p_3, p_1+\mathrm{i}p_2),\\
\psi^2_{\mathrm{pw}} &= (0, E+m, p_1-\mathrm{i}p_2, -p_3),\\
\psi^3_{\mathrm{pw}} &= (-p_3, -(p_1+\mathrm{i}p_2), -E+m, 0),\\
\psi^4_{\mathrm{pw}} &= (-(p_1-\mathrm{i}p_2), p_3, 0, -E+m).
\end{aligned}$$

The solutions involving $\psi^1_{\mathrm{pw}}$ and $\psi^2_{\mathrm{pw}}$ are interpreted as electron plane waves in spin up and spin down states, respectively, especially in the nonrelativistic regime $p \ll m$. However, $\psi^3_{\mathrm{pw}}$, $\psi^4_{\mathrm{pw}}$ appear to be *negative energy solutions*, reflecting the fact that the relativistic energy relation (1), involving only $E^2$, should mathematically admit negative energies as well as positive energies.

The existence of negative energy solutions is one of the most striking aspects of the Dirac equation (similar solutions also exist for the Klein–Gordon equation (2)), not least because physically they appear to preclude energetic ground states; it suggests that relativistic electrons may constantly decay to successively lower energies without bound, at odds with our physical experience. Dirac himself proposed the following resolution to this problem: the negative energies are already filled with a *Dirac sea* of electrons. Free electrons cannot then occupy these filled negative energy states and must have nonnegative energy. The existence of this postulated Dirac sea has the further consequences that a negative energy electron might, through some process, become excited into a positive energy state, leaving a positively charged "hole" in the sea whose plane-wave components act according to the negative energy solutions $\psi^3_{\mathrm{pw}}$ and $\psi^4_{\mathrm{pw}}$ (in a similar way to "holes" appearing in filled valence bands in semiconductors). Dirac originally identified these "positive charge electrons" as protons, despite the fact that a proton has a mass that is very different from the electron mass $m$. However, *positrons*, which appeared to have the same mass as electrons but opposite charge, were discovered experimentally by Anderson in 1932, just four years after Dirac's theory. This prediction of a new kind of particle earned Dirac the Nobel Prize in Physics in 1933, shared with Schrödinger.

To incorporate the interaction between an electron and the electromagnetic field, the momentum operator $\hat{p}_j$ must be replaced with the appropriate canonical momentum for a charged particle moving in a field from the Lorentz force equation (as in CLASSICAL MECHANICS [IV.19]). In a scalar potential $V$ and vector potential $\boldsymbol{A}$, giving a 4-potential $A_a$, this replacement in (3), (4) requires replacing the usual partial derivatives with the gauge-covariant derivative

$$\partial_a \to D_a \equiv \partial_a + ieA_a, \quad a = 0, 1, 2, 3.$$

Finding self-consistent solutions of Maxwell's equations and the Dirac equation with interactions is analytically difficult. Nevertheless, very good approximations are possible that agree well with experiment, particularly in quantum electrodynamics, which is a systematic quantum approach to solving systems with many electrons interacting quantum mechanically with the electromagnetic field. Quantum electrodynamics, like other quantum field theories, requires a more sophisticated mathematical approach based on the Dirac and Maxwell equations and relates the negative energy solutions of the Dirac and Klein–Gordon equations to positive energy states of *antimatter* (i.e., positrons are "anti-electrons"). Despite many mathematical complications, this theory successfully describes the evolution of many interacting quantum particles, including the possibility of their creation and annihilation in a fluctuating vacuum.

### Further Reading

Dirac, P. A. M. 1928a. The quantum theory of the electron. *Proceedings of the Royal Society of London* A 117:610–24.

——. 1928b. The quantum theory of the electron, part II. *Proceedings of the Royal Society of London* A 118:351–61.

Folland, G. B. 2008. *Quantum Field Theory: A Tourist Guide for Mathematicians*. Providence, RI: American Mathematical Society.

Lounesto, P. 2001. *Clifford Algebras and Spinors*, 2nd edn. Cambridge: Cambridge University Press.

Zee, A. 2010. *Quantum Field Theory in a Nutshell*, 2nd edn. Princeton, NJ: Princeton University Press.

## III.10 Einstein's Field Equations

*Malcolm A. H. MacCallum*

Einstein's general theory of relativity generalizes Newton's gravity theory to one compatible with special relativity. It models space and time points as a (pseudo-) Riemannian four-dimensional manifold (see TENSORS AND MANIFOLDS [II.33]) with a metric $g_{ab}$ of signature $\pm 2$ (the sign choice is conventional). Test particles move on space-time's geodesics. This formulation ensures the "weak equivalence principle" or "universality of free fall," which states that free fall under gravity depends only on a body's initial position and momentum. The other fundamental part of the theory is Einstein's field equations (EFEs), which relate the metric to the matter present. For more on the theory and its applications see GENERAL RELATIVITY AND COSMOLOGY [IV.40].

By considering a "gedanken" experiment in which bodies released at relative rest in a laboratory freely falling toward the Earth will appear to move toward one another, Einstein recognized that geodesics that are initially parallel meet due to gravity. This is described by the metric's curvature. To generalize Newton's theory, the curvature must be related to the space-time distribution of the energy-momentum tensor of the matter content, $T^{ab}$.

$T^{ab}$ is assumed to obey $T^{ab}{}_{;b} = 0$, the generalization to curved space of the Noetherian conservation laws obtained when the matter obeys a variational principle. Here, "$;b$" denotes a covariant derivative with respect to the $b$ index, while "$,b$" will denote a partial derivative below.

The formulas relating the metric, the connection $\Gamma^a{}_{bc}$, and the Riemannian curvature, in coordinate components, are

$$\Gamma^a{}_{bc} = \tfrac{1}{2} g^{ad}(g_{bd,c} + g_{dc,b} - g_{bc,d}),$$
$$R^a{}_{bcd} = \Gamma^a{}_{bd,c} - \Gamma^a{}_{bc,d} + \Gamma^e{}_{bd}\Gamma^a{}_{ec} - \Gamma^e{}_{bc}\Gamma^a{}_{ed},$$

where $g^{ad}$ is the inverse of $g_{bc}$.

Taking a weak-field, slow-motion limit, comparison of the geodesic equation with Newtonian free fall identifies corrections to an approximating flat (special-relativistic) metric with the Newtonian gravitational potential $\Phi$. One therefore wants to find equations relating the second derivative of the metric to $T_{ab}$. Defining the Ricci tensor $R_{ab}$ and the Ricci scalar $R$ by

$$R_{bd} := R^a{}_{bad}, \qquad R := g^{ab} R_{ab},$$

the EFEs

$$G_{ab} := R_{ab} - \tfrac{1}{2}Rg_{ab} = \kappa T_{ab} + \Lambda g_{ab} \qquad (1)$$

achieve this relation. Here, $\Lambda$ is a constant and $G_{ab}$ is called the Einstein tensor.

To agree with the Newtonian limit, the constant $\kappa$ has to be $8\pi G/c^4$, where $G$ is the Newtonian constant of gravitation and $c$ the speed of light.

The full curvature can be expressed, in four dimensions, as

$$R^{ab}{}_{cd} = C^{ab}{}_{cd} - \tfrac{1}{3}R\delta^{a}_{[c}\delta^{b}_{d]} + 2\delta^{[a}_{[c}R^{b]}_{d]},$$

where square brackets denote antisymmetrization, so that, for any tensor, $T_{[ab]} := \frac{1}{2}(T_{ab} - T_{ba})$. The Weyl tensor $C^{a}{}_{bcd}$ thus defined is conformally invariant and describes tidal gravitational forces and gravitational waves. Its value depends on distant matter and the boundary conditions: it is nonlocally determined by certain first-order differential equations, the Bianchi identities, with a source given by derivatives of $R_{ab}$.

There are a number of other ways to write (1). A first-order system arises by taking the metric and its connection as variables. Tetrad bases are widely used in place of coordinate bases. The EFEs themselves are given by a variational principle (assuming that $T_{ab}$ is), with the $G_{ab}$ part coming from an action $\int R\sqrt{-g}\,\mathrm{d}^4x$, where $g = \det(g_{ab})$.

Calculating in coordinates, the ten components of (1) each have on the order of $10^4$ terms in components of $g_{ab}$, they are quasilinear in second derivatives of $g_{ab}$, and they are nonlinear of degree 8 in $g_{ab}$ itself (after clearing denominators). The tensorial EFEs are therefore a set of coupled nonlinear inhomogeneous partial differential equations. Lovelock proved that in four dimensions, $G_{ab}$ is the only symmetric divergence-free tensorial concomitant of $g_{ab}$ that is linear in second derivatives, so the EFEs are the unique field equations of this character.

No general solution is known, except in the sense of the integral form given by Sciama, Waylen, and Gilman, but many specific solutions have been found. Their local geometry can be completely characterized by components of the curvature tensor and its covariant derivatives.

General relativity shares with other physical theories the property that the evolution is unique given initial values of the field and its first derivative. In this case these are the induced metric on an initial space-like surface, and its derivative off the surface (the first and second "fundamental forms"). Four of the EFEs constrain these initial values, the remaining equations giving six second-order evolution equations. The constraint equations are elliptic and the evolution equations hyperbolic; the characteristic speed is that of light.

Existence and uniqueness theorems have been obtained for the evolution equations in this form. Typically, the functions that appear lie in appropriate Sobolev spaces.

As well as such Cauchy problems, the EFEs can be studied in a "2 + 2" formalism, where data is given on a pair of intersecting two-dimensional characteristic surfaces.

It was recently recognized that the EFEs in standard form are only weakly hyperbolic, which explained problems that had arisen in numerical integrations, and strongly hyperbolic reformulations are now available that allow fully four-dimensional numerical computations (see NUMERICAL RELATIVITY [V.15]).

In both numerical and analytic studies, important problems such as the generation of gravitational waves require characterization of isolated bodies. This is achieved by defining *asymptotic flatness*, when at spatial or light-like (null) infinity the geometry approaches that of special relativity's empty Minkowski space. A conformal transformation by a factor $\Omega$ is used in the precise definition; this transformation preserves null directions and gives a single point, denoted by $i^0$, at spatial infinity. The definition, which is abbreviated as AEFANSI (for "asymptotically empty and flat at null and spatial infinity"), specifies the behavior of $\Omega$ and the Ricci tensor near $i^0$.

Space-times may have more than one asymptotic infinity (see, for example, the Kruskal diagram in GENERAL RELATIVITY AND COSMOLOGY [IV.40]). *Weakly asymptotically simple* (WAS) space-times are those for which only one asymptotic region need be considered. There is a large class of WAS space-times for which null infinity is smooth, for which fields such as the Weyl tensor can be expanded in inverse powers of a radial distance $r$, and for which there is an asymptotic symmetry, the Bondi-Metzner-Sachs group. These results have been generalized to cases where $\log r$ terms appear in expansions; it is conjectured that only a rather restricted set of cases avoids such terms.

There is no locally defined energy of the gravitational field in general relativity; such an energy could not be compatible with the local special relativistic limit required by the principle of equivalence. However, asymptotically flat spaces do possess globally defined energies, using integrals at infinity—the *ADM energy* for a spatial infinity (named after the formalism introduced by Arnowitt, Deser, and Misner) and the *Trautman–Bondi energy* for a future null infinity—thus enabling definition of the total gravitational energy of an isolated system.

The ADM energy has been proved to be nonnegative—assuming both that the matter that is present obeys the dominant energy condition that $T^{ab}v_a v_b \geqslant 0$ and that $T^{ab}$ is space-like for all time-like vectors $v^a$—and it is zero only in Minkowski space. The Trautman–Bondi energy is monotone decreasing with time, as radiation carries energy away, and agrees with ADM energy in its past limit and hence must also be nonnegative (as can also be proved directly). These results show in particular that gravitational waves carry energy. However, the waves cannot carry away more energy than an isolated system has initially, as total energy would then become negative. The positive energy results also imply that there is no analogue of the possibility, present in Newtonian gravity theory, that negative gravitational potential energy could be greater than the positive energy of the field's source.

### Further Reading


Griffiths, J., and J. Podolsky. 2009. *Exact Space-Times in Einstein's General Relativity*. Cambridge: Cambridge University Press.

Hawking, S., and G. Ellis. 1973. *The Large Scale Structure of Space-Time*. Cambridge: Cambridge University Press.

Stephani, H., D. Kramer, M. MacCallum, C. Hoenselaers, and E. Herlt. 2003. *Exact Solutions of Einstein's Field Equations*, 2nd edn. Cambridge: Cambridge University Press. (Corrected paperback reprint, 2009.)

Wald, R. 1984. *General Relativity*. Chicago, IL: University of Chicago Press.

See also http://relativity.livingreviews.org/ for many valuable articles.


## III.11 The Euler Equations

*P. A. Martin*

The motion of a fluid is well modeled by the NAVIER–STOKES EQUATIONS [III.23]. An underlying assumption is that the fluid is viscous (meaning that it is sticky, in the sense that there is some resistance to shearing motions). Assume further that the fluid is incompressible (meaning that the density is constant). Then, if the viscous effects are removed from the Navier–Stokes equations (even though real fluids are always viscous to some extent), the result is known as the *Euler equations*.

To state them, let $\boldsymbol{x} = (x_1, x_2, x_3)$ be the position vector of a point in the fluid and let $\boldsymbol{u}(\boldsymbol{x}, t) = (u_1, u_2, u_3)$ be the fluid velocity at $\boldsymbol{x}$ at time $t$. Then

$$\frac{\partial \boldsymbol{u}}{\partial t} + (\boldsymbol{u} \cdot \nabla)\boldsymbol{u} + \frac{1}{\rho}\nabla p = \boldsymbol{0}, \qquad (1)$$

where $p(\boldsymbol{x}, t)$ is the pressure and $\rho$ is the density. The vector equation (1) is to be solved together with the incompressibility constraint, which is

$$\nabla \cdot \boldsymbol{u} = 0. \qquad (2)$$

There are thus four partial differential equations (PDEs) for the four unknowns, $u_1$, $u_2$, $u_3$, and $p$. These form the incompressible Euler equations for inviscid (zero-viscosity) flows; there are also *compressible Euler equations* for flows in which the fluid density is not constant but has to be calculated. If body forces are acting (the most important of these is gravity), there would be an extra term on the right-hand side of (1).

To clarify the notation used in (1) and (2), we can write them in component form:

$$\frac{\partial u_i}{\partial t} + \sum_{j=1}^{3} u_j \frac{\partial u_i}{\partial x_j} + \frac{1}{\rho}\frac{\partial p}{\partial x_i} = 0, \qquad \sum_{j=1}^{3} \frac{\partial u_j}{\partial x_j} = 0,$$

where $i = 1, 2, 3$. Alternatively, if we use $\boldsymbol{x} = (x, y, z)$ and $\boldsymbol{u} = (u, v, w)$, we can write out Euler's equations explicitly:

$$\frac{\partial u}{\partial t} + u\frac{\partial u}{\partial x} + v\frac{\partial u}{\partial y} + w\frac{\partial u}{\partial z} + \frac{1}{\rho}\frac{\partial p}{\partial x} = 0,$$
$$\frac{\partial v}{\partial t} + u\frac{\partial v}{\partial x} + v\frac{\partial v}{\partial y} + w\frac{\partial v}{\partial z} + \frac{1}{\rho}\frac{\partial p}{\partial y} = 0,$$
$$\frac{\partial w}{\partial t} + u\frac{\partial w}{\partial x} + v\frac{\partial w}{\partial y} + w\frac{\partial w}{\partial z} + \frac{1}{\rho}\frac{\partial p}{\partial z} = 0,$$
$$\frac{\partial u}{\partial x} + \frac{\partial v}{\partial y} + \frac{\partial w}{\partial z} = 0.$$

An important special case arises when the motion is *irrotational*, which means that the vorticity $\nabla \times \boldsymbol{u} = \boldsymbol{0}$. In this case, $\boldsymbol{u} = \nabla\phi$ for some scalar (potential) function $\phi$. From (2), $\nabla^2\phi = 0$; that is, $\phi$ solves LAPLACE'S EQUATION [III.18]. Moreover, the nonlinear PDEs (1) reduce to a formula for $p$ that is known as *Bernoulli's equation*.

Irrotational flows of inviscid incompressible fluids have been studied extensively since the nineteenth century. However, it is also known that the underlying assumptions are too restrictive in some circumstances because they lead to some results that do not agree with our experience. Perhaps the most glaring example is the *d'Alembert paradox*, a mathematical theorem asserting that irrotational flow of an inviscid incompressible fluid about a rigid body generates no drag force on the body. The conventional way to overcome the paradox is to bring back viscosity but only inside a thin BOUNDARY LAYER [II.2] attached to the body (see also FLUID MECHANICS [IV.28 §7.2]).

## III.12 The Euler–Lagrange Equations
*Paul Glendinning*

The function $y(x)$ with derivative $y' = \mathrm{d}y/\mathrm{d}x$ that maximizes or minimizes the integral

$$\int F(y, y', x)\,\mathrm{d}x$$

with given endpoints satisfies the Euler–Lagrange equation

$$\frac{\mathrm{d}}{\mathrm{d}x}\left(\frac{\partial F}{\partial y'}\right) - \frac{\partial F}{\partial y} = 0. \tag{1}$$

There are many variants of this equation to deal with further complications, e.g., if $y$ or $x$ or both are vectors, and more details are given in CALCULUS OF VARIATIONS [IV.6], but this simple version is sufficient to demonstrate the power and ubiquity of variational problems of this form.

If $F = F(y, y')$ has no explicit $x$-dependence ($x$ is said to be *absent*), then the Euler–Lagrange equations can be simplified by finding a first integral. Using (1) it is straightforward to show that

$$\frac{\mathrm{d}}{\mathrm{d}x}\left(y'\frac{\partial F}{\partial y'} - F\right) = 0,$$

and hence that

$$y'\frac{\partial F}{\partial y'} - F = A \tag{2}$$

for some constant $A$.

### Application 1: Potential Forces

Classical mechanics can be formulated as a problem of minimizing the integral of a function called the *Lagrangian*, $\mathcal{L}$, which is the kinetic energy minus the potential energy. For a particle moving in one dimension with position $q$ (so the dependent variable $q$ plays the role of $y$ above and time $t$ plays the role of the independent variable $x$) in a potential $V(q)$, the Lagrangian is $\mathcal{L} = \frac{1}{2}m\dot{q}^2 - V(q)$ and the Euler–Lagrange equation (1) is simply Newton's law for the acceleration, $m\ddot{q} = -V'(q)$ (where the prime denotes differentiation with respect to $q$), while the autonomous version (2) shows that $\frac{1}{2}m\dot{q}^2 + V(q)$ is constant, which is the conservation of energy (see CLASSICAL MECHANICS [IV.19]).

The power of this approach (and a related version due to Hamilton) is such that much of modern theoretical physics revolves round a generalization called an *action*.

### Application 2: The Catenary

The problem of determining the curve describing the rest state of a heavy chain or cable with fixed endpoints can also be solved using the Euler–Lagrange formulation, although the original seventeenth-century solution uses simple mechanics. In the rest state the chain will assume a shape $y = y(x)$ that minimizes the potential energy $g\int y\,\mathrm{d}s$, where $g$ is the acceleration due to gravity and $s$ is the arc length along the chain. The length of the chain is $\int \mathrm{d}s$, and since this length is assumed to be constant, $L$ say, $\int \mathrm{d}s = L$. This acts as a constraint on the solutions of the energy-minimization problem and so the full problem can be approached by introducing a LAGRANGE MULTIPLIER [I.3 §10], $\lambda$. Scaling out the constant $g$ and noting that $\mathrm{d}s = \sqrt{1 + y'^2}\,\mathrm{d}x$, the shape of the curve minimizes

$$\int y\sqrt{1 + y'^2}\,\mathrm{d}x - \lambda\left(\int \sqrt{1 + y'^2}\,\mathrm{d}x - L\right). \tag{3}$$

(The second term represents the constraint and is zero when the constraint is satisfied.) The Euler–Lagrange equation with

$$F(y, y', \lambda) = \sqrt{1 + y'^2} - \lambda\sqrt{1 + y'^2}$$

can now be used since the $\lambda L$ term of (3) is constant with respect to variations in $y$. The Euler–Lagrange equation is supplemented by an additional equation obtained by extremizing with respect to the Lagrange multiplier, i.e., setting the derivative of (3) with respect to $\lambda$ to zero, but this is just the length constraint again. Since $x$ is absent, (2) implies that

$$(y - \lambda)\left(\frac{y'^2}{\sqrt{1 + y'^2}} - \sqrt{1 + y'^2}\right) = A.$$

Tidying up the left-hand side and rearranging gives $A^2(1 + y'^2) = (y - \lambda)^2$. Rewriting this as an expression for $y'$ gives a differential equation that can be solved by separation of variables to give

$$y - \lambda = A\cosh\left(\frac{x - B}{A}\right),$$

where $B$ is a further constant of integration. This is the catenary curve, and the constants are determined by the endpoints of the chain and the constraint that the total length is $L$.

## III.13 The Gamma Function

Euler's gamma function, $\Gamma$, is defined by

$$\Gamma(x) = \int_0^\infty t^{x-1}e^{-t}\,dt, \quad x > 0.$$

One integration by parts shows that

$$\Gamma(x+1) = x\Gamma(x), \quad x > 0.$$

A direct calculation gives $\Gamma(1) = 1$, and then an inductive argument gives $\Gamma(n) = (n-1)!$ when $n$ is any positive integer. For this reason, the alternative notation $x! = \Gamma(x+1)$ is also used.

Much is known about $\Gamma$ and its properties. It is classified as a SPECIAL FUNCTION [IV.7] of one variable. According to Davis (1959), of all special functions, $\Gamma$ "is undoubtedly the most fundamental." It is also ubiquitous, appearing in countless applications. In COMPLEX ANALYSIS [IV.1], $\Gamma(z)$ is defined as a function of a complex variable, $z$, for all $z \neq 0, -1, -2, \dots$.

An alternative but equivalent definition of $\Gamma$ is

$$\Gamma(x) = \frac{e^{-\gamma x}}{x}\prod_{n=1}^{\infty}\frac{n}{n+x}e^{x/n},$$

where $\gamma = 0.5772\dots$ is *Euler's constant*, defined by

$$\gamma = \lim_{n\to\infty}\left(1 + \frac{1}{2} + \frac{1}{3} + \cdots + \frac{1}{n} - \log n\right).$$

The definition of $\Gamma$ as an infinite product shows clearly that $\Gamma(x)$ is not defined when $x = 0, -1, -2, \dots$, and it reveals the singular nature at these points.

### Further Reading

Davis, P. J. 1959. Leonhard Euler's integral: a historical profile of the gamma function. *American Mathematical Monthly* 66:849–69.

## III.14 The Ginzburg–Landau Equation
*S. Jonathan Chapman*

The Ginzburg–Landau equations were written down in 1950 to describe the change of phase of a superconducting material in the presence of a magnetic field. They are partial differential equations for the complex-valued superconducting order parameter $\Psi$ and the (real-valued) magnetic vector potential $\boldsymbol{A}$. The parameter $\Psi$ can be thought of as a kind of macroscopic wave function and is such that $|\Psi|^2$ is the number density of superconducting electrons, while $\boldsymbol{A}$ is such that the magnetic field is curl $\boldsymbol{A}$.

In their normalized form the equations are

$$\left(\frac{1}{\kappa}\nabla - i\boldsymbol{A}\right)^2\Psi = (|\Psi|^2 - 1)\Psi,$$

$$\text{curl}^2\,\boldsymbol{A} = \frac{i}{2\kappa}(\Psi^*\nabla\Psi - \Psi\nabla\Psi^*) - |\Psi|^2\boldsymbol{A},$$

where $i = \sqrt{-1}$, $\kappa$ is a material constant known as the Ginzburg–Landau parameter, and the asterisk denotes complex conjugation.

### 1 The Ginzburg–Landau Free Energy

The Ginzburg–Landau equations arise from minimizing the Ginzburg–Landau free energy. For small applied magnetic fields the superconducting solution ($|\Psi| = 1$) has a lower energy than the nonsuperconducting (normal) solution ($\Psi = 0$), while for high applied magnetic fields the normal solution has the lower energy. At the critical magnetic field $H_c$, the two states have the same energy, and a normal-superconducting transition region is possible.

The main aim in developing the Ginzburg–Landau equations was to determine the energy of such a transition region (the so-called surface energy), since this determines the scale of the pattern of normal and superconducting domains when both are present. In particular, it was desirable to demonstrate that the surface energy was positive. In fact, it turns out that the surface energy is positive only if $\kappa < 1/\sqrt{2}$. Values of the Ginzburg–Landau parameter above this threshold were dismissed at the time as being unphysical.

### 2 Type-II Superconductors

A few years later, in 1957, Abrikosov published solutions of the equations for values of $\kappa > 1/\sqrt{2}$ that had a quite different structure. When surface energy is negative, a normal region shrinks until it is just a point. However, the scale of the solution is prevented from being infinitely fine by the complex nature of the order parameter; each zero of $\Psi$ has a winding number that is a topological invariant. Abrikosov's solutions are of the form $\Psi = f(r)e^{in\theta}$, $\boldsymbol{A} = A(r)\boldsymbol{e}_\theta$, where $r$ and $\theta$ are polar coordinates, the integer $n$ is the winding number, and $\boldsymbol{e}_\theta$ is the unit vector in the azimuthal direction. The electric current associated with such a solution is

$$\boldsymbol{j} = -f^2\left(A - \frac{n}{\kappa r}\right)\boldsymbol{e}_\theta,$$

which is why these solutions are known as superconducting vortices. Such solutions demonstrate the quantum nature of superconductivity on a macroscopic scale. Superconductors with $\kappa > 1/\sqrt{2}$ are now known as type-II superconductors, while those with $\kappa < 1/\sqrt{2}$ are known as type-I superconductors.

For type-II superconductors, the critical magnetic field $H_c$ splits into two critical fields, $H_{c1}$ and $H_{c2}$. Below $H_{c1}$ the superconducting state is energetically preferred, above $H_{c2}$ the normal state is preferred, but in between $H_{c1}$ and $H_{c2}$ a mixed state comprising a periodic array of superconducting vortices exists.

Following Abrikosov's work this lattice of vortices was demonstrated experimentally. It is an amazing triumph of the Ginzburg–Landau theory that it predicted the existence of such structures, not only before they had been observed experimentally but when the very idea of them was disturbing. Almost all technological applications of superconductivity involve type-II superconductors in the vortex state.

### Further Reading

Tinkham, M. 1996. *Introduction to Superconductivity*. New York: McGraw-Hill.

## III.15 Hooke's Law

*P. A. Martin*

Strictly, Hooke's law is not a law, as it is readily and frequently violated. Nevertheless, it can be a useful approximation, it can be generalized, and it leads to ideas of *elasticity* and *constitutive equations*.

In 1678, Robert Hooke (1635–1703) described his experiments in which he fixed one end of a long vertical wire to the ceiling and hung various weights to the other end. When there are no hanging weights, the wire has length $L$, say. When a weight of mass $m$ is added, the wire extends by an amount $\ell$, so that the new length is $L + \ell$. If the weight is removed, the wire's length returns to $L$: this is the signature of elastic behavior. Hooke also showed that doubling the mass (from $m$ to $2m$) gives double the extension. He inferred that the restoring force exerted by the wire on the mass, $F$, is proportional to the displacement of the mass from its equilibrium position,

$$F = k\ell,$$

where $k$ is the constant of proportionality, the *spring constant*. This is Hooke's law. Hooke also did experiments on the stretching and compression of springs and on the lateral deflections of wooden beams. He asserted that his law was applicable to "every springing body" and that it could be used to understand vibrations of such bodies.

To see this, return to the mass $m$ hanging on the wire. In equilibrium, its weight is balanced by the restoring force, $mg = k\ell$, where $g$ is the acceleration due to gravity. If the mass is pulled down further and then released, it will oscillate about its equilibrium position. In detail, if the mass is displaced by an amount $x$, the downward force on the mass is $mg - k(\ell + x) = -kx$, so, by Newton's second law, force = mass × acceleration, the equation of motion is $-kx = m(\mathrm{d}^2x/\mathrm{d}t^2)$. As $k = mg/\ell$, we obtain $\mathrm{d}^2x/\mathrm{d}t^2 = -\omega^2 x$, where $\omega = (g/\ell)^{1/2}$. This differential equation for $x(t)$ has the general solution $x = x_0 \cos(\omega t + \delta)$, where $x_0$ and $\delta$ are arbitrary constants. The oscillating mass exhibits *simple harmonic motion* with frequency $\omega$.

Hooke's law is an approximation. It models the mechanical behavior of his wire and many other elastic bodies: doubling the load doubles the extension. It is a linear approximation, where the constant of proportionality can be found by experiment; recall that $k = mg/\ell$. However, there will be limits to the validity of Hooke's law: if a very large load is applied, the wire will extend plastically (which means that the wire's length will not return to $L$ if the load is removed) and then it might break.

Returning to Hooke's experiments, suppose we hang a mass $m$ on a wire of length $2L$; the extension doubles to $2\ell$. This implies that $k$ must be proportional to $L^{-1}$ (the left-hand side of $mg = k \times$ extension has not changed). Similarly, if we suspend the mass by *two* wires in parallel, each of length $L$, we see half the extension; we can say that $k$ must be proportional to $A$, the cross-sectional area of the wire. Thus, we rewrite Hooke's law as

$$\frac{F}{A} = \frac{kL}{A}\frac{\ell}{L}.$$

The dimensionless quantity $\ell/L$ is the extension per unit length of wire; it is a measure of *strain* in the wire. The quantity $F/A$ is the force per unit area; it is a measure of *stress* in the wire. The quantity $kL/A$ does not depend on $L$ or $A$; it depends on what the wire is made from, so it is a *material constant*. Therefore, we write

$$\sigma = E\varepsilon, \tag{1}$$

stating that the stress, $\sigma$, is proportional to the strain, $\varepsilon$. The material constant $E$ is known as *Young's modulus*. The formula (1) is a basic assumption in the one-dimensional theory of the strength of materials. Suitably generalized, it is fundamental to the linear theory of elasticity.

Hooke's law is a *constitutive relation*: it is a model of how certain materials behave when they are subjected to forces. Constitutive relations are part of all continuum theories. Hooke's law is useful for elastic materials, but, as we saw for the oscillating mass, there is no damping; the oscillations do not decay with time. Incorporating damping leads to viscoelastic models. For fluids, it is common to replace Hooke's law with a relation between stresses and velocities; this leads to the Navier–Stokes equations.

Developing and selecting constitutive relations requires an interplay between good modeling of experimental observations, essential mathematical properties (such as causality and frame indifference), and simplicity.

## III.16 The Korteweg–de Vries Equation

*Willy A. Hereman*

### 1 Historical Perspective

In 1895 Diederik Korteweg (1848–1941) and Gustav de Vries (1866–1934) derived a partial differential equation (PDE) that models the "great wave of translation" that naval engineer John Scott Russell had observed in the Union Canal in 1834.

Assuming that the wave propagates in the $X$-direction, the evolution of the surface elevation $\eta(X,T)$ above the undisturbed water depth $h$ at time $T$ can be modeled by the Korteweg–de Vries (KdV) equation:

$$\frac{\partial \eta}{\partial T} + \sqrt{gh}\frac{\partial \eta}{\partial X} + \frac{3}{2}\frac{\sqrt{gh}}{h}\eta\frac{\partial \eta}{\partial X} + \tfrac{1}{2}h^2\sqrt{gh}\left(\frac{1}{3} - \frac{\mathcal{T}}{\rho g h^2}\right)\frac{\partial^3 \eta}{\partial X^3} = 0, \quad (1)$$

where $g$ is the gravitational acceleration, $\rho$ is the density, and $\mathcal{T}$ is the surface tension. The dimensionless parameter $\mathcal{T}/\rho g h^2$, called the *Bond number*, measures the relative strengths of surface tension and the gravitational force. Equation (1) is valid for long waves of relatively small amplitude, $|\eta|/h \ll 1$.

In dimensionless variables, (1) can be written as

$$u_t + \alpha u u_x + u_{xxx} = 0, \quad (2)$$

where subscripts denote partial derivatives. The term $\sqrt{gh}\eta_X$ in (1) has been removed by an elementary transformation. Conversely, a linear term in $u_x$ can be added to (2). The parameter $\alpha$ can be scaled to any real number. Commonly used values are $\alpha = \pm 1$ and $\alpha = \pm 6$.

The term $u_t$ describes the time evolution of the wave. Therefore, (2) is called an *evolution* equation. The nonlinear term $\alpha u u_x$ accounts for steepening of the wave. The linear dispersive term $u_{xxx}$ describes spreading of the wave.

It is worth noting that the KdV equation had already appeared in seminal work on water waves published by Joseph Boussinesq about twenty years earlier.

### 2 Solitary Waves and Periodic Solutions

The balance of the steepening and spreading effects gives rises to a stable solitary wave,

$$u(x,t) = \frac{\omega - 4k^3}{\alpha k} + \frac{12k^2}{\alpha}\operatorname{sech}^2(kx - \omega t + \delta), \quad (3)$$

where the wave number $k$, the angular frequency $\omega$, and the phase $\delta$ are arbitrary constants. Requiring that $\lim_{x\to\pm\infty} u(x,t) = 0$ for all $t$ leads to $\omega = 4k^3$, in which case (3) reduces to

$$u(x,t) = 12(k^2/\alpha)\operatorname{sech}^2(kx - 4k^3 t + \delta). \quad (4)$$

This hump-shaped solitary wave of finite amplitude $12k^2/\alpha$ travels to the right at constant phase speed $v = \omega/k = 4k^2$, and it models Scott Russell's "great wave of translation" that traveled without change of shape over a fairly long distance.

As shown by Korteweg and de Vries, (2) also has a periodic solution:

$$u(x,t) = \frac{\omega - 4k^3(2m-1)}{\alpha k} + 12(k^2/\alpha)m\,\operatorname{cn}^2(kx - \omega t + \delta; m). \quad (5)$$

They called this the *cnoidal wave* solution because it involves Jacobi's elliptic cosine function, cn, with modulus $m$, $0 < m < 1$. In the limit $m \to 1$, $\operatorname{cn}(\xi; m) \to \operatorname{sech}\xi$ and (5) reduces to (3).

### 3 Modern Developments

The solitary wave was, for many years, considered an unimportant curiosity in the field of nonlinear waves. That changed in 1965, when Zabusky and Kruskal realized that the KdV equation arises as the continuum limit of a one-dimensional anharmonic lattice used by Fermi, Pasta, and Ulam in 1955 to investigate how energy is distributed among the many possible oscillations in the lattice. Since taller solitary waves travel faster than shorter ones, Zabusky and Kruskal simulated the collision of two waves in a nonlinear crystal lattice and observed that each retains its shape and speed after collision. Interacting solitary waves merely experience a phase shift, advancing the faster wave and retarding the slower one. In analogy with colliding particles, they coined the word "solitons" to describe these elastically colliding waves.

To model water waves that are weakly nonlinear, weakly dispersive, and weakly two-dimensional, with all three effects being comparable, Kadomtsev and Petviashvili (KP) derived a two-dimensional version of (2) in 1970:

$$(u_t + 6uu_x + u_{xxx})_x + 3\sigma^2 u_{yy} = 0, \tag{6}$$

where $\sigma^2 = \pm 1$ and the $y$-axis is perpendicular to the direction of propagation of the wave (along the $x$-axis).

The KdV and KP equations, and the NONLINEAR SCHRÖDINGER EQUATION [III.26]

$$iu_t + u_{xx} + \kappa |u|^2 u = 0 \tag{7}$$

(where $\kappa$ is a constant and $u(x,t)$ is a complex-valued function), are famous examples of so-called completely integrable nonlinear PDEs. This means that they can be solved with the inverse scattering transform, a nonlinear analogue of the Fourier transform.

The inverse scattering transform is not applied to (2) directly but to an auxiliary system of linear PDEs,

$$\psi_{xx} + (\lambda + \tfrac{1}{6}\alpha u)\psi = 0, \tag{8}$$

$$\psi_t + \tfrac{1}{2}\alpha u_x \psi + \alpha u \psi_x + 4\psi_{xxx} = 0, \tag{9}$$

which is called the *Lax pair* for the KdV equation. Equation (8) is a linear Schrödinger equation for an eigenfunction $\psi$, a constant eigenvalue $\lambda$, and a potential $(-\alpha u)/6$. Equation (9) governs the time evolution of $\psi$. The two equations are compatible, i.e., $\psi_{xxt} = \psi_{txx}$, if and only if $u(x,t)$ satisfies (2). For given $u(x,0)$ decaying sufficiently fast as $|x| \to \infty$, the inverse scattering transform solves (8) and (9) and finally determines $u(x,t)$.

## 4 Properties and Applications

Scientists remain intrigued by the rich mathematical structure of completely integrable nonlinear PDEs. These PDEs can be written as infinite-dimensional bi-Hamiltonian systems and have additional, remarkable features. For example, they have an associated Lax pair, they can be written in Hirota's bilinear form, they admit Bäcklund transformations, and they have the Painlevé property. They have an infinite number of conserved quantities, infinitely many higher-order symmetries, and an infinite number of soliton solutions.

As well as being applicable to shallow-water waves, the KdV equation is ubiquitous in applied science. It describes, for example, ion-acoustic waves in a plasma, elastic waves in a rod, and internal waves in the atmosphere or ocean. The KP equation models, for example, water waves, acoustic waves, and magnetoelastic waves in anti-ferromagnetic materials. The nonlinear Schrödinger equation describes weakly nonlinear and dispersive wave packets in physical systems, e.g., light pulses in optical fibers, surface waves in deep water, Langmuir waves in a plasma, and high-frequency vibrations in a crystal lattice. Equation (7) with an extra linear term $V(x)u$ to account for the external potential $V(x)$ also arises in the study of Bose-Einstein condensates, where it is referred to as the time-dependent Gross-Pitaevskii equation.

### Further Reading

Ablowitz, M. J. 2011. *Nonlinear Dispersive Waves: Asymptotic Analysis and Solitons.* Cambridge: Cambridge University Press.

Ablowitz, M. J., and P. A. Clarkson. 1991. *Solitons, Nonlinear Evolution Equations and Inverse Scattering.* Cambridge: Cambridge University Press.

Kasman, A. 2010. *Glimpses of Soliton Theory.* Providence, RI: American Mathematical Society.

Osborne, A. R. 2010. *Nonlinear Ocean Waves and the Inverse Scattering Transform.* Burlington, MA: Academic Press.

# III.17 The Lambert *W* Function

*Robert M. Corless and David J. Jeffrey*

## 1 Definition and Basic Properties

For a given complex number $z$, the equation

$$w\mathrm{e}^w = z$$

has a countably infinite number of solutions, which are denoted by $W_k(z)$ for integers $k$. Each choice of $k$ specifies a *branch* of the Lambert $W$ function. By convention, only the branches $k = 0$ (called the *principal* branch) and $k = -1$ are real-valued for any $z$; the range of every other branch excludes the real axis, although the range of $W_1(z)$ includes $(-\infty, -1/\mathrm{e}]$ in its closure. Only $W_0(z)$ contains positive values in its range (see figure 1). When $z = -1/\mathrm{e}$ (the only nonzero branch point), there is a double root $w = -1$ of the basic equation $w\mathrm{e}^w = z$. The conventional choice of branches assigns

$$W_0(-1/\mathrm{e}) = W_{-1}(-1/\mathrm{e}) = -1$$

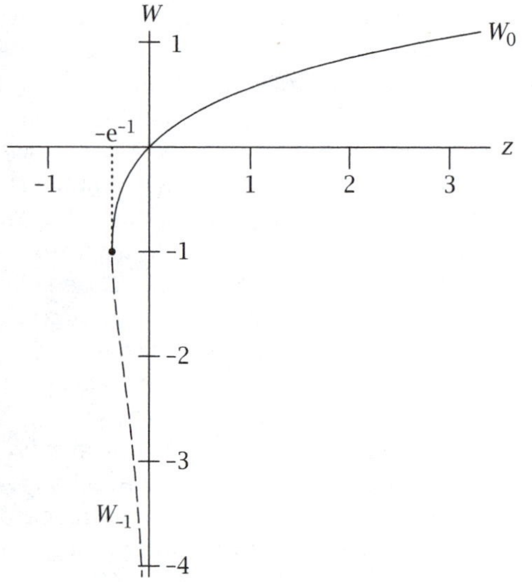


**Figure 1** Real branches of the Lambert $W$ function. The solid line is the principal branch $W_0$; the dashed line is $W_{-1}$, which is the only other branch that takes real values. The small filled circle at the branch point corresponds to the one in figure 2.

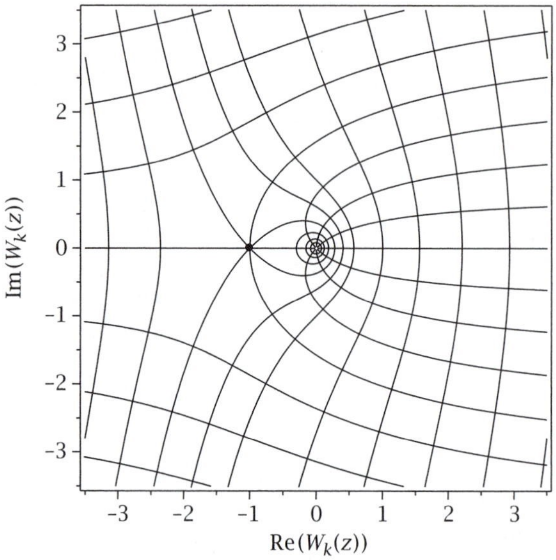


**Figure 2** Images of circles and rays in the $z$-plane under the maps $z \to W_k(z)$. The circle with radius $e^{-1}$ maps to a curve that goes through the branch point, as does the ray along the negative real axis. This graph was produced in Maple by numerical evaluation of $\omega(x+iy) = W_{\mathcal{K}(iy)}(e^{x+iy})$ first for a selection of fixed $x$ and varying $y$, and then for a selection of fixed $y$ and varying $x$. These two sets produce orthogonal curves as images of horizontal and vertical lines in $x$ and $y$ under $\omega$ or, equivalently, images of circles with constant $r = e^x$ and rays with constant $\theta = y$ under $W$.

and implies that $W_1(-1/e - i\varepsilon^2) = -1 + O(\varepsilon)$ is arbitrarily close to $-1$ because the conventional branch choice means that the point $-1$ is on the border between these three branches. Each branch is a single-valued complex function, analytic away from the branch point and branch cuts.

The set of all branches is often referred to, loosely, as the Lambert $W$ "function"; but of course $W$ is multivalued. Depending on context, the symbol $W(z)$ can refer to the principal branch ($k = 0$) or to some unspecified branch. Numerical computation of any branch of $W$ is typically carried out by Newton's method or a variant thereof. Images of $W_k(re^{i\theta})$ for various $k$, $r$, and $\theta$ are shown in figure 2.

In contrast to more commonly encountered multibranched functions, such as the inverse sine or cosine, the branches of $W$ are not linearly related. However, by rephrasing things slightly, in terms of the *unwinding number*

$$\mathcal{K}(z) := \frac{z - \ln(e^z)}{2\pi i}$$

and the related *single-valued* function

$$\omega(z) := W_{\mathcal{K}(z)}(e^z),$$

which is called the *Wright $\omega$ function*, we do have the somewhat simple relationship between branches that $W_k(z) = \omega(\ln_k z)$, where $\ln_k z$ denotes $\ln z + 2\pi i k$ and $\ln z$ is the principal branch of the logarithm, having $-\pi < \mathrm{Im}(\ln z) \leqslant \pi$.

The Wright $\omega$ function helps to solve the equation $y + \ln y = z$. We have that, if $z \neq t \pm i\pi$ for $t < -1$, then $y = \omega(z)$. If $z = t - i\pi$ for $t < -1$, then there is no solution to the equation; if $z = t + i\pi$ for $t < -1$, then there are two solutions: $\omega(z)$ and $\omega(z - 2\pi i)$.

### 1.1 Derivatives

Implicit differentiation yields

$$W'(z) = e^{-W(z)}/(1 + W(z))$$

as long as $W(z) \neq -1$. The derivative can be simplified to the *rational* differential equation

$$\frac{dW}{dz} = \frac{W}{z(1+W)}$$

if, in addition, $z \neq 0$. Higher derivatives follow naturally.

### 1.2 Integrals

Integrals containing $W(x)$ can often be performed analytically by the change of variable $w = W(x)$, used in an inverse fashion: $x = we^w$. Thus,

$$\int \sin W(x)\,dx = \int (1+w)e^w \sin w\,dw,$$

and integration using usual methods gives

$$\tfrac{1}{2}(1+w)e^w \sin w - \tfrac{1}{2}we^w \cos w,$$

which eventually gives

$$\int 2\sin W(x)\,dx = \left(x + \frac{x}{W(x)}\right)\sin W(x) - x\cos W(x) + C.$$

More interestingly, there are many *definite* integrals for $W(z)$, including one for the principal branch that is due to Poisson and is listed in the famous table of integrals by D. Bierens de Haan. The following integral, which is of relatively recent construction and which is valid for $z$ not in $(-\infty, -1/e]$, can be computed with spectral accuracy by the trapezoidal rule:

$$\frac{W(z)}{z} = \frac{1}{2\pi}\int_{-\pi}^{\pi} \frac{(1 - v\cot v)^2 + v^2}{z + v\csc v e^{-v\cot v}}\,dv.$$

### 1.3 Series and Generating Functions

Euler was the first to notice, using a series due to Lambert, that what we now call the Lambert $W$ function has a convergent series expansion around $z = 0$:

$$W(z) = \sum_{n\geqslant 1} \frac{(-n)^{n-1}}{n!} z^n.$$

Euler knew that this series converges for $-1/e \leqslant z \leqslant 1/e$. The nearest singularity is the branch point $z = -1/e$.

$W$ can also be expanded in series about the branch point. The series at the branch point can be expressed most cleanly using the *tree function* $T(z) = -W(-z)$ rather than $W$ or $\omega$, but keeping with $W$ we have

$$W_0(-e^{-1-z^2/2}) = -\sum_{n\geqslant 0} (-1)^n a_n z^n,$$

$$W_{-1}(-e^{-1-z^2/2}) = -\sum_{n\geqslant 0} a_n z^n,$$

where the $a_n$ are given by $a_0 = a_1 = 1$ and

$$a_n = \frac{1}{(n+1)a_1}\left(a_{n-1} - \sum_{k=2}^{n-1} k a_k a_{n+1-k}\right).$$

These give an interesting variation on STIRLING'S FORMULA [IV.7 §3] for the asymptotics of $n!$. Euler's integral

$$n! = \int_0^\infty t^n e^{-t}\,dt$$

is split at the maximum of the integrand ($t = n$), and each integral is transformed using the substitutions $t = -nW_k(-e^{-1-z^2/2})$, where $k = 0$ is used for $t \leqslant n$ and $k = -1$ otherwise. The integrands then simplify to $t^n e^{-t} = n^n e^{-n} e^{-nz^2/2}$ and the differentials $dt$ are obtained as series from the above expansions. Term-by-term integration leads to

$$n! \sim \frac{n^{n+1}}{e^n} \sum_{k\geqslant 0} (2k+1)a_{2k+1}\left(\frac{2}{n}\right)^{k+1/2} \Gamma(k+\tfrac{1}{2}),$$

where $\Gamma$ is the gamma function.

Asymptotic series for $z \to \infty$ have been known since de Bruijn's work in the 1960s. He also proved that the asymptotic series are actually convergent for large enough $z$. The series begin as follows: $W_k(z) \sim \ln_k(z) - \ln(\ln_k(z)) + o(\ln\ln_k z)$. Somewhat surprisingly, these series can be reversed to give a simple (though apparently useless) expansion for the logarithm in terms of compositions of $W$:

$$\ln z = W(z) + W(W(z)) + W(W(W(z))) + \cdots + W^{(N)}(z) + \ln W^{(N)}(z)$$

for a suitably restricted domain in $z$. The series obtained by omitting the term $\ln W^{(N)}(z)$ is not convergent as $N \to \infty$, but for fixed $N$ if we let $z \to \infty$ the approximation improves, although only tediously slowly.

## 2 Applications

Because $W$ is a so-called implicitly elementary function, meaning it is defined as an implicit solution of an equation containing only elementary functions, it can be considered an "answer" rather than a question. That it solves a simple rational differential equation means that it occurs in a wide range of mathematical models. Out of many applications, we mention just two favorites.

First, a serious application. $W$ occurs in a chemical kinetics model of how the human eye adapts to darkness after exposure to bright light: a phenomenon known as bleaching. The model differential equation is

$$\frac{d}{dt}O_p(t) = \frac{K_m O_p(t)}{\tau(K_m + O_p(t))},$$

and its solution in terms of $W$ is

$$O_p(t) = K_m W\left(\frac{B}{K_m} e^{B/K_m - t/\tau}\right),$$

where the constant $B$ is the initial value of $O_p(0)$, that is, the amount of initial bleaching. The constants $K_m$ and $\tau$ are determined by experiment. The solution in terms of $W$ enables more convenient analysis by allowing the use of known properties.

The second application we mention is nearly frivolous. $W$ can be used to explore solutions of the so-called astrologer's equation, $\dot{y}(t) = ay(t+1)$. In this equation, the rate of change of $y$ is supposed to be proportional to the value of $y$ one time unit into the *future*. Dependence on past times instead leads to DELAY DIFFERENTIAL EQUATIONS [I.2 §12], which of course are of serious interest in applications, and again $W$ is useful there in much the same way as for this frivolous problem.

Frivolity can be educational, however. Notice first that, if $e^{\lambda t}$ satisfies the equation, then $\lambda = ae^{\lambda}$, and therefore $\lambda = -W_k(-a)$. For the astrologer's equation, any function $y(t)$ that can be expressed as a *finite* linear combination $y(t) = \sum_{k \in M} c_k e^{-W_k(-a)t}$ for $0 \leqslant t \leqslant 1$ and some finite set $M$ of integers then solves the astrologer's equation for all time. Thus, perfect knowledge of $y(t)$ on the time interval $0 \leqslant t \leqslant 1$ is sufficient to predict $y(t)$ for all time. However, if the knowledge of $y(t)$ is imperfect, even by an infinitesimal amount (omitting a single term $\varepsilon e^{-W_K(-a)t}$, say, where $K$ is some large integer), then since the real parts of $-W_K(-a)$ go to infinity as $K \to \infty$ by the first two terms of the logarithmic series for $W_k$ given above, the "true" value of $y(t)$ can depart arbitrarily rapidly from the prediction. This seems completely in accord with our intuitions about horoscopes.

Returning to serious applications, we note that the tree function $T(z)$ has huge combinatorial significance for all kinds of enumeration. Many instances can be found in Knuth's selected papers, for example. Additionally, a key reference to the tree function is a note by Borel in *Comptes Rendus de l'Académie des Sciences* (volume 214, 1942; reprinted in his *Œuvres*). The generating function for probabilities of the time between periods when a queue is empty, given Poisson arrivals and service time $\sigma$, is $T(\sigma e^{-\sigma} z)/\sigma$.

## 3 Solution of Equations

Several equations containing algebraic quantities together with logarithms or exponentials can be manipulated into either the form $y + \ln y = z$ or $we^w = z$, and hence solved in terms of the Lambert $W$ function. However, it appears that not every exponential polynomial equation—or even most of them—can be solved in this way. We point out one equation, here, that starts with a nested exponential and can be solved in terms of branch *differences* of $W$: a solution of

$$z + v \csc v e^{-v \cot v} = 0$$

is $v = (W_k(z) - W_\ell(z))/(2i)$ for some pair of integers $k$ and $\ell$; moreover, every such pair generates a solution. This bi-infinite family of solutions has accumulation points of zeros near odd multiples of $\pi$, which in turn implies that the denominator in the above definite integral for $W(z)/z$ has essential singularities at $v = \pm\pi$. This example underlines the importance of the fact that the branches of $W$ are not trivially related.

Another equation of popular interest occurs in the analysis of the limit of the recurrence relation

$$a_{n+1} = z^{a_n}$$

starting with, say, $a_0 = 1$. This sequence has $a_1 = z$, $a_2 = z^z$, $a_3 = z^{z^z}$, and so on. If this limit converges, it does so to a solution of the equation $a = z^a$. By inspection, the limit that is of interest is $a = -W(-\ln z)/\ln z$. Somewhat surprisingly, this recurrence relation—which defines the so-called tower of exponentials—diverges for small enough $z$, even if $z$ is real. Specifically, the recurrence converges for $e^{-e} \leqslant z \leqslant e^{1/e}$ if $z$ is real and diverges if $z < e^{-e} = 0.0659880\ldots$. This fact was known to Euler. The detailed convergence properties for complex $z$ were settled only relatively recently. Describing the regions in the complex plane where the recurrence relation converges to an $n$-cycle is made possible by a transformation that is itself related to $W$: if $\zeta = -W(-\ln z)$, then the iteration converges if $|\zeta| < 1$, and also if $\zeta = e^{i\theta}$ for $\theta$ equal to some *rational* multiple of $\pi$, say $m\pi/k$. Regions where the iteration converges to a $k$-cycle may touch the unit circle at those points.

## 4 Retrospective

The Lambert $W$ function crept into the mathematics literature unobtrusively, and it now seems natural there. There is even a matrix version of it, although the solution of the matrix equation $Se^S = A$ is not always $W(A)$.

Hindsight can, as it so often does, identify the presence of $W$ in writings by Euler, Poisson, and Wright and in many applications. Its implementation in Maple in the early 1980s was a key step in its eventual popularity.

Indeed, its recognition and naming supports Alfred North Whitehead's opinion that:

> By relieving the brain of all unnecessary work, a good notation sets it free to concentrate on more advanced problems.

### Further Reading

Corless, R. M., G. H. Gonnet, D. E. G. Hare, D. J. Jeffrey, and D. E. Knuth. 1996. On the Lambert $W$ function. *Advances in Computational Mathematics* 5(1):329–59.

Knuth, D. E. 2000. *Selected Papers on the Analysis of Algorithms.* Palo Alto, CA: Stanford University Press.

——. 2003. *Selected Papers on Discrete Mathematics.* Stanford, CA: CSLI Publications.

Lamb, T., and E. Pugh. 2004. Dark adaptation and the retinoid cycle of vision. *Progress in Retinal and Eye Research* 23(3):307–80.

Olver, F. W. J., D. W. Lozier, R. F. Boisvert, and C. W. Clark, eds. 2010. *NIST Handbook of Mathematical Functions.* Cambridge: Cambridge University Press. (Electronic version available at http://dlmf.nist.gov.)

## III.18 Laplace's Equation

*P. A. Martin*

In 1789, Pierre-Simon Laplace (1749–1827) wrote down an equation,

$$\frac{\partial^2 V}{\partial x^2} + \frac{\partial^2 V}{\partial y^2} + \frac{\partial^2 V}{\partial z^2} = 0, \tag{1}$$

that now bears his name. Today, it is arguably the most important partial differential equation (PDE) in mathematics.

The left-hand side of (1) defines the *Laplacian* of $V$, denoted by $\nabla^2 V$ or $\Delta V$:

$$\nabla^2 V = \Delta V = \nabla \cdot \nabla V = \operatorname{div}\operatorname{grad} V.$$

Laplace's equation, $\nabla^2 V = 0$, is classified as a linear homogeneous second-order elliptic PDE for $V(x, y, z)$. The inhomogeneous version, $\nabla^2 V = f$, where $f$ is a given function, is known as *Poisson's equation.* The fact that there are three independent variables ($x$, $y$, and $z$) in (1) can be indicated by calling it the *three-dimensional Laplace equation.* The two-dimensional version,

$$\nabla^2 V \equiv \frac{\partial^2 V}{\partial x^2} + \frac{\partial^2 V}{\partial y^2} = 0, \tag{2}$$

also has important applications; it is a PDE for $V(x, y)$. There is a natural generalization to $n$ independent variables. Usually, the number of terms in $\nabla^2 V$ is determined by the context.

### 1 Harmonic Functions

Solutions of Laplace's equation are known as *harmonic functions.* It is easy to see (one of Laplace's favorite phrases) that there are infinitely many different harmonic functions.

A short list of solutions, $V(x, y)$, for (2) follows:

- there are polynomial solutions, such as 1, $x$, $y$, $xy$, and $x^2 - y^2$;
- there are solutions such as $e^{\alpha x}\cos\alpha y$ and $e^{\alpha x} \times \sin\alpha y$, where $\alpha$ is an arbitrary parameter; and
- $\log r$, $\theta$, $r^\alpha\cos\alpha\theta$, and $r^\alpha \sin\alpha\theta$ are solutions, where $x = r\cos\theta$ and $y = r\sin\theta$ define plane polar coordinates, $r$ and $\theta$.

Further solutions can be found by differentiating or integrating any solution with respect to $x$ or $y$; for example, $(\partial/\partial x)\log r = x/r^2$ is harmonic. One can also differentiate or integrate with respect to any parameter; for example,

$$\int_{\alpha_1}^{\alpha_2} g(\alpha)\, e^{\alpha x} \sin\alpha y \, d\alpha$$

is harmonic, where $g(\alpha)$ is an arbitrary (integrable) function of the parameter $\alpha$ and the integration could be over the real interval $\alpha_1 < \alpha < \alpha_2$ or along a contour in the complex $\alpha$-plane.

A list of solutions, $V(x, y, z)$, for (1) follows:

- any solution of (2) also solves (1);
- there are polynomial solutions, such as $xyz$ and $x^2 + y^2 - 2z^2$;
- there are solutions such as $e^{\gamma z}\cos\alpha x\cos\beta y$, where $\gamma^2 = \alpha^2 + \beta^2$ and $\alpha$ and $\beta$ are arbitrary parameters;
- $e^{\alpha z} J_n(\alpha r)\cos n\theta$ is a solution, where $x = r\cos\theta$, $y = r\sin\theta$, and $J_n$ is a Bessel function; and
- $R^{-1}$ is a solution, where $R = (x^2 + y^2 + z^2)^{-1/2}$ is a spherical polar coordinate.

Again, further solutions can be obtained by differentiating or integrating any solution with respect to $x$, $y$, $z$, or any parameter. For example, if $i$, $j$, and $k$ are any nonnegative integers,

$$V(x, y, z) = \frac{\partial^{i+j+k}}{\partial x^i \partial y^j \partial z^k} \frac{1}{R}$$

is harmonic; it is known as a *Maxwell multipole.*

As Laplace's equation is linear and homogeneous, more solutions can be constructed by superposition; if $V_1$ and $V_2$ satisfy $\nabla^2 V = 0$, then so does $AV_1 + BV_2$, where $A$ and $B$ are arbitrary constants.

### 2 Boundary-Value Problems

Although Laplace's equation has many solutions, it is usual to seek solutions that also satisfy boundary conditions. A basic problem is to solve $\nabla^2 V = 0$ inside a bounded region $D$ subject to $V = g$ (a given function) on the boundary of $D$, $\partial D$. This boundary-value problem (BVP) is known as the *interior Dirichlet problem*. Under certain mild conditions, this problem has exactly one solution. This solution can be constructed explicitly (by the method of separation of variables) when $D$ has a simple shape, such as a circular disk or a rectangle in two dimensions, and a ball or a cube in three dimensions. For more complicated geometries, progress can be made by reducing the BVP to a boundary integral equation around $\partial D$, or by solving the BVP numerically, using finite elements, for example.

Other BVPs can be formulated. For example, instead of specifying $V$ on $\partial D$, the normal derivative of $V$, $\partial V/\partial n$, could be given (this is the *interior Neumann problem*). One could specify a linear combination of $V$ and $\partial V/\partial n$ at each point on $\partial D$, or one could specify $V$ on part of $\partial D$ and $\partial V/\partial n$ on the rest of $\partial D$ (this is the *mixed problem*). One could specify both $V$ and $\partial V/\partial n$ on part of $\partial D$ but give no information on the rest of $\partial D$; this is called the *Cauchy problem*. There are also exterior versions of all these problems, where the goal is to solve $\nabla^2 V = 0$ in the unbounded region outside $\partial D$; for such problems, one also has to specify the behavior of $V$ "at infinity," far from $\partial D$.

## 3 Applications

Laplace's 1789 application of (1) was to gravitational attraction and the rings of Saturn. Gravitational forces can be written as $\boldsymbol{F} = \operatorname{grad} V$, where $\nabla^2 V = 0$ outside regions containing matter. In general, vector fields that can be written as the gradient of a scalar are called *conservative*; the scalar, $V$, is called a *potential*. Equivalently, conservative fields $\boldsymbol{F}$ satisfy $\operatorname{curl} \boldsymbol{F} = \mathbf{0}$: they are *irrotational*.

The velocity $\boldsymbol{v}$ for the motion of an incompressible (constant-density) fluid satisfies the *continuity equation*, $\operatorname{div} \boldsymbol{v} = 0$. If the motion is irrotational, then there exists a *velocity potential* $\phi$ such that $\boldsymbol{v} = \operatorname{grad} \phi$. The continuity equation then shows that $\nabla^2 \phi = 0$.

Similar equations are encountered in electrostatics and magnetostatics. For example, in empty space, the electric field, $\boldsymbol{E}$, satisfies $\operatorname{div} \boldsymbol{E} = 0$ and $\operatorname{curl} \boldsymbol{E} = \mathbf{0}$. Thus, we can write $\boldsymbol{E} = \operatorname{grad} \varphi$, where the potential $\varphi$ solves $\nabla^2 \varphi = 0$.

## 4 Analytic Function Theory

In this section we use $z = x + \mathrm{i}y$ to denote a complex variable. Let $f(z) = u(x,y) + \mathrm{i}v(x,y)$ be a function of $z$. If $f$ is analytic (that is, a differentiable function of $z$), $u$ and $v$ satisfy the *Cauchy–Riemann equations*:

$$\frac{\partial u}{\partial x} = \frac{\partial v}{\partial y} \quad \text{and} \quad \frac{\partial u}{\partial y} = -\frac{\partial v}{\partial x}.$$

Eliminating $v$ between these equations shows that $u(x,y)$ satisfies the two-dimensional Laplace equation, (2); $v(x,y)$ solves the same PDE. The real and imaginary parts of an analytic function are said to be harmonic.

This connection between functions of a complex variable and two-dimensional harmonic functions leads to powerful methods for solving BVPs for (2), especially when CONFORMAL MAPPINGS [II.5] are employed.

## 5 Generalizations

The Laplacian occurs in PDEs other than Laplace's equation. Here are a few examples:

| | |
|---|---|
| $\nabla^2 u = c^{-2}(\partial^2 u/\partial t^2)$, | the wave equation, |
| $\nabla^2 u = k^{-1}(\partial u/\partial t)$, | the heat/diffusion equation, |
| $\nabla^2 u + k^2 u = 0$, | the Helmholtz equation, |
| $\nabla^2 u + \mathrm{i}\ell(\partial u/\partial t) = Wu$, | the Schrödinger equation. |

In these equations, $c$, $k$, $\ell$, and $W$ are given (the first three are often constants) and $t$ is time. One reason that $\nabla^2$ occurs frequently is that space is often assumed to be isotropic, meaning that there is no preferred direction. Thus, in (1), $x$, $y$, and $z$ are Cartesian coordinates, but the value of $\nabla^2 V$ does not change if the coordinates are rotated or translated. The Laplacian is the simplest linear second-order operator with this property.

There is an anisotropic version of $\nabla^2 u$, namely, $\operatorname{div}(A \operatorname{grad} u)$, where $A$ is an $n \times n$ matrix with entries that could be constants or functions of the $n$ independent variables. If $A$ depends on $u$, a nonlinear operator is obtained. Another well-studied nonlinear variant is the $p$-Laplacian, defined by $\operatorname{div}(|\operatorname{grad} u|^{p-2} \operatorname{grad} u)$ with $1 < p < \infty$.

A useful fourth-order operator is $\nabla^2\nabla^2 = \nabla^4$. The *biharmonic equation*, $\nabla^4 u = 0$, arises in the theory of thin elastic plates, for example.

# III.19 The Logistic Equation

*Paul Glendinning*

The *logistic equation* is a simple differential equation or difference equation with quadratic nonlinearity. It arises naturally in POPULATION MODELS [I.5 §3] (for example) as a model of a process with population-limited growth, i.e., models that include the inhibitory effects of overcrowding.

The continuous-time version is a differential equation for a real variable $x$ that represents the size of a population. It is

$$\frac{dx}{dt} = rx\left(1 - \frac{x}{K}\right), \quad r, K > 0,$$

although there are higher-dimensional analogues and extensions to partial differential equations. The application to population dynamics means that it is usual to assume that $x(0) = x_0 > 0$. The two parameters are the reproduction rate $r > 0$ (the difference between the birth rate and the death rate), and the carrying capacity $K$, which, as shown below, is the equilibrium population level.

The logistic equation is separable and solutions may be calculated explicitly using partial fractions (see ORDINARY DIFFERENTIAL EQUATIONS [IV.2 §2]). The solution is

$$x(t) = \frac{x_0 K}{x_0 + (K - x_0)e^{-rt}},$$

so all solutions with $x_0 > 0$ tend to the carrying capacity, $K$, as $t$ tends to $\infty$.

The right-hand side of the logistic equation can be interpreted as the first two terms of the Taylor series expansion of a function $f$ with $f(0) = 0$, so the equation is a natural model in many other contexts.

The discrete-time version of the logistic equation is often called the *logistic map*. It is

$$x_{n+1} = \mu x_n(1 - x_n), \quad \mu > 0.$$

This difference equation is one of the paradigmatic examples of systems with chaotic attractors. One early interpretation is again from population biology: the equation may describe the successive population levels of an organism that has discrete generations, so $x_n$ is the normalized number of insects (for example) in the $n$th generation, and this depends on the population size of the previous generation.

The logistic map illustrates the different dynamics that can exist in general unimodal, or one-hump, maps. If $0 < \mu \leqslant 4$ then all solutions that start in $[0, 1]$ stay in that interval, so there is (at least) one bounded attractor of the system. The way this attractor varies with $\mu$ can be very complicated.

If $\mu \in (0, 1)$ then the origin is a stable fixed point (if $x_0 = 0$ then $x_n = 0$ for all $n \in \mathbb{N}$), and for all initial conditions $x_0 \in [0, 1]$, $x_n \to 0$ as $n \to \infty$. If $\mu \in (1, 3)$ then there is a stable nontrivial fixed point $(\mu - 1)/\mu$ that attracts all initial conditions in $(0, 1)$. At $\mu = 3$ this fixed point undergoes a PERIOD-DOUBLING BIFURCATION [IV.21]. The effect of this bifurcation is to create a stable period-two orbit as $\mu$ increases through 3 and the fixed point becomes unstable. As $\mu$ increases further there is a sequence of period-doubling bifurcations at parameter values $\mu_n$ at which orbits of period $2^n$ lose stability, and a stable periodic orbit of period $2^{n+1}$ is created.

These bifurcation values accumulate geometrically at a special value $\mu_\infty$, above which there are infinitely many periodic orbits in the system and the system is CHAOTIC [II.3], although the chaotic set may not be attracting. In this chaotic region of parameters there are "windows" (i.e., intervals of the parameter) for which the attracting behavior is again periodic. These orbits then lose stability, undergoing their own period-doubling sequences and having a similar bifurcation structure to the original map, but over a smaller parameter interval.

There are many interesting results about the dynamics in the chaotic region. For example, the set of parameters for which the map has a chaotic attractor has positive measure; and between any two parameters with chaotic attractors there are parameters with stable periodic behavior.

The order in which periodic orbits are created satisfies Sharkovskii's theorem (valid for any continuous map of the interval). Sharkovskii's theorem defines an order on the natural numbers that reflects the order in which different periods appear in families of maps. The Sharkovskii order, $\prec$, is a complete order on the positive integers, so for any positive integers $p$ and $q$ with $p \neq q$, either $p \prec q$ or $q \prec p$. The order is defined by the list below. To interpret this list, imagine you have two positive integers and think about where each one appears in the list (they must both appear somewhere!). If $p$ appears before $q$, then $p \prec q$. The list is

$$1 \prec 2 \prec 2^2 \prec 2^3 \prec \cdots$$

$$\cdots$$

$$\cdots \prec 2^{n+1} \times 9 \prec 2^{n+1} \times 7 \prec 2^{n+1} \times 5 \prec 2^{n+1} \times 3$$

$$\cdots \prec 2^n \times 9 \prec 2^n \times 7 \prec 2^n \times 5 \prec 2^n \times 3$$

$$\cdots$$

$$\cdots \prec 9 \prec 7 \prec 5 \prec 3,$$

i.e., powers of two ascending, then for each $n$, $2^n$ times the odds descending, with $n$ descending to zero. Sharkovskii's theorem states that, if a continuous map of the interval has an orbit of period $p$, then it also has orbits of period $q$ for all $q \prec p$ in this order.

The logistic equation has different orbits of the same period that can be described by labeling the periodic points $x_1 < x_2 < \cdots < x_p$ with $f(x_i) = x_{\pi(i)}$ for some permutation $\pi$ of $\{1,\dots,p\}$. Sharkovskii's theorem does not distinguish between orbits with the same period and different permutation type. It turns out that the period and associated permutation type define a complete order on the periodic orbits arising in the logistic equation.

### Further Reading

Collet, P., and J.-P. Eckmann. 1980. *Iterated Maps of the Interval as Dynamical Systems*. Basel: Birkhäuser.

de Melo, W., and S. van Strien. 1993. *One-Dimensional Dynamics*. Berlin: Springer.

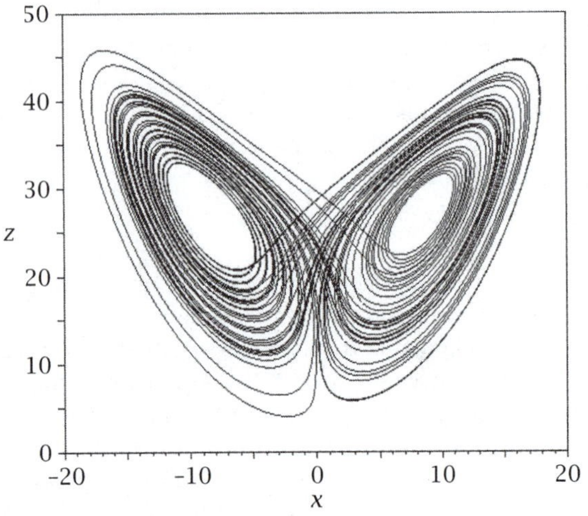


**Figure 1** The Lorenz attractor projected onto the $(x, z)$ coordinates.

## III.20 The Lorenz Equations

*Paul Glendinning*

The Lorenz equations provided one of the earliest examples of a nonlinear differential equation with chaotic behavior. They were derived in the early 1960s when the growing use and power of computers raised the exciting prospect of greatly improved weather forecasts, and particularly more reliable long-range forecasts. This generated work on improving the numerical techniques applied to meteorological models, and it also led to the study of simplified models and their properties. Saltzman used a crude truncation of the Fourier expansion of solutions to convert the partial differential equations describing convection in a horizontal layer into a finite set of coupled ordinary differential equations that are significantly easier to solve numerically. Saltzman concentrated on a fifty-two-mode truncation, but in an article published the next year (1963), Ed Lorenz took the model to its oversimplified extreme, retaining just three of the Galerkin modes.

Lorenz's model is normally written as a set of coupled differential equations in three variables, now known as the Lorenz equations:

$$\dot{x} = \sigma(y - x),$$
$$\dot{y} = rx - y - xz,$$
$$\dot{z} = -bz + xy,$$

where the dot denotes differentiation with respect to the independent variable $t$, and where $\sigma$ is a normalized Prandtl number, $r$ a normalized Rayleigh number, and $b$ the aspect ratio of the convecting cell. The parameters used by Lorenz were

$$\sigma = 10, \qquad r = 28, \qquad b = \tfrac{8}{3}.$$

The insight derived through the investigation of these equations has motivated a greater understanding of the mathematics of CHAOS [II.3] and has had a major impact on the way weather forecasts are created and reported. A sample trajectory of the Lorenz equations at the parameter values given above is shown in figure 1. The solution clearly settles down to a bounded attracting set (the *Lorenz attractor*), but it appears not to be periodic. Moreover, solutions that start close together eventually diverge and behave completely differently, a property now called *sensitive dependence on initial conditions*, which is one of the hallmarks of chaos. (A little care is needed here. Sensitive dependence on initial conditions as usually defined in the twentieth century is *not* a good definition of chaos; some exponential divergence in time is necessary in more modern definitions (see CHAOS AND ERGODICITY [II.3]).) The story goes that Lorenz discovered this phenomenon by a fortuitous mistyping of an initial condition when checking a modification of his computer code. Lorenz's genius was to recognize this as a significant property of the solutions.

The Lorenz attractor motivated several developments in bifurcation theory and chaos. In the late 1970s mathematicians developed a simple geometric model of the flow that could be reduced to the analysis of a one-dimensional map with a discontinuity that could be proved to be chaotic. At the same time, researchers started to describe the development of the attractor as a function of the parameter $r$, showing that the solutions in the strange attractor are created as an initially unstable set by a GLOBAL BIFURCATION [IV.21]. Sparrow brought all these results together, giving a description of the bifurcations of the Lorenz attractor as $r$ is changed, using a mixture of numerical simulations, mathematical proofs, and conjectured links between the two.

Sparrow's work provides strong evidence for the following description of the changes in the attractors of the Lorenz equations as $r$ increases. The origin, which is a globally attracting stationary point if $0 < r < 1$, loses stability as $r$ increases through 1 via a pitchfork bifurcation that creates a stable symmetric pair of stationary points corresponding to convective rolls (see BIFURCATION THEORY [IV.21 §2] for a description of different types of bifurcation). At $r = r_{\mathrm{H}} \approx 24.74$ there is a pair of subcritical Hopf bifurcations. The effect of this bifurcation is that the stationary points lose stability as $r$ increases through $rH$, and a pair of unstable periodic orbits are created in $r < r_{\mathrm{H}}$. Where did these orbits come from? Yorke and coworkers had shown in 1979 that the answer lies in a homoclinic bifurcation at $r \approx 13.926$. This creates an unstable chaotic set containing infinitely many unstable periodic orbits, many of which are destroyed in bifurcations before the chaotic set becomes attracting by a mechanism involving the two simple periodic orbits that are the orbits involved in the Hopf bifurcation at $r \approx 24.06$. There is therefore a brief interval of $r$ values for which a complicated attractor coexists with the stable stationary points, and if $r$ is a little greater than $r_{\mathrm{H}}$, the only attractor is the chaotic set. This is initially similar to the geometric model, but it develops contracting regions ("hooks") as $r$ increases further, which allows for the possibility of the creation of stable periodic orbits. At very large $r$ the only attractor is a simple symmetric periodic orbit, so a sequence of bifurcations destroying the orbits of the chaotic set needs to occur.

Despite a host of theoretical results on the geometric models and on the bifurcations in systems such as the Lorenz equations, a proof that the Lorenz equations at the standard parameter values really do have a strange attractor remained open until 2002, when Tucker used a combination of rigorous numerically computed bounds and mathematical analysis to show that the attractor has the chaotic properties required (see DYNAMICAL SYSTEMS [IV.20 §4.5] for more details).

While the mathematical issues were being resolved, questions about the physical relevance of the Lorenz equations continued to cause controversy. As more and more of the ignored modes of Saltzman's truncated model are added back into the equations, the chaotic region exists for larger and larger values of $r$ and is no longer present in the full partial differential equations. More imaginative physical situations, such as convection in a rotating hoop, have been devised, and for these the Lorenz equations *are* a good model.

Even though it is not an accurate model of convection in a fluid layer, the physical insight Lorenz brought to the problem of weather forecasting was hugely influential. As a result of the recognition that sensitive dependence on initial conditions can be a problem, forecasters now routinely run computer simulations of their models with a variety of initial conditions so that when it is appropriate they can comment on the probability of rain rather than issue a simple statement that it will or will not rain. The understanding of how chaos can be accommodated in nonlinear prediction has also influenced the way in which sophisticated models of the climate and other nonlinear phenomena are interpreted.

### Further Reading

Kaplan, J. L., and J. A. Yorke. 1979. Preturbulence: a regime observed in a fluid flow model of Lorenz. *Communications in Mathematical Physics* 67:93–108.

Lorenz, E. 1963. Deterministic nonperiodic flow. *Journal of the Atmospheric Sciences* 20:130–41.

Sparrow, C. T. 1982. *The Lorenz Equations: Bifurcations, Chaos, and Strange Attractors.* New York: Springer.

Tucker, W. 2002. A rigorous ODE solver and Smale's 14th problem. *Foundations of Computational Mathematics* 2: 53–117.

## III.21 Mathieu Functions

*Julio C. Gutiérrez-Vega*

Mathieu functions are solutions of the *ordinary Mathieu equation*

$$\frac{\mathrm{d}^2 y}{\mathrm{d}\theta^2} + (a - 2q\cos 2\theta)y = 0, \tag{1}$$

where $q$ is a free parameter and $a$ is the eigenvalue of the equation. Mathieu's equation was first studied by

Émile Mathieu in 1868 in the context of the vibrational modes of an elliptic membrane.

For arbitrary parameters $(a, q)$, (1) is a linear second-order differential equation with two independent solutions. In general, these solutions are not periodic functions and their behavior depends on the initial conditions $y(0)$ and $y'(0)$. Of particular interest are the periodic solutions with period $\pi$ or $2\pi$. In this case, according to Sturm–Liouville theory, there exists a countably infinite set of characteristic eigenvalues $a_m(q)$ that yield even periodic solutions of (1), and another set of characteristic eigenvalues $b_m(q)$ that yield odd periodic solutions of (1). The eigenfunctions associated with these sets of eigenvalues are known as the even and odd *Mathieu functions*:

$$y_m = \begin{cases} \mathrm{ce}_m(\theta; q), & m = 0, 1, 2, \dots, \\ \mathrm{se}_m(\theta; q), & m = 1, 2, 3, \dots, \end{cases}$$

where $m$ is the order. The notation ce and se comes from *cosine-elliptic* and *sine-elliptic*, and it is now a widely accepted notation for the periodic Mathieu functions.

Mathieu functions occur in two main categories of physical problems. First, they appear in applications involving the separation of the wave equation in elliptic coordinates, e.g., the vibrating modes in elliptic membranes, the propagating modes in elliptic pipes, or the oscillations of water in a lake of elliptic shape. Second, they occur in phenomena that involve periodic motion, e.g., the trajectory of an electron in a periodic array of atoms, the mechanics of the quantum pendulum, or the oscillations of floating vessels.

The behavior of Mathieu functions is rather complicated, and their analysis is difficult using standard methods, mainly due to their nontrivial dependence on the parameters $(a, q)$. In figure 1 we plot the functions $\mathrm{ce}_m(\theta; q)$ and $\mathrm{se}_m(\theta; q)$ for several values of $m$ over the plane $(\theta, q)$. Note that Mathieu's equation becomes the harmonic equation when $q \to 0$. Evidently, $\mathrm{ce}_m$ and $\mathrm{se}_m$ converge to the trigonometric functions $\cos(m\theta)$ and $\sin(m\theta)$ as $q$ tends to zero.

The parity, periodicity, and normalization of the periodic Mathieu functions are exactly the same as for their trigonometric counterparts. That is, $\mathrm{ce}_m$ is even and $\mathrm{se}_m$ is odd, and they have period $\pi$ when $m$ is even or period $2\pi$ when $m$ is odd. The Mathieu functions have $m$ real zeros in the open interval $\theta \in (0, \pi)$, and the zeros cluster around $\pi/2$ as $q$ increases. Because the Mathieu equation is of Sturm–Liouville type, the Mathieu functions form a complete family of orthogonal

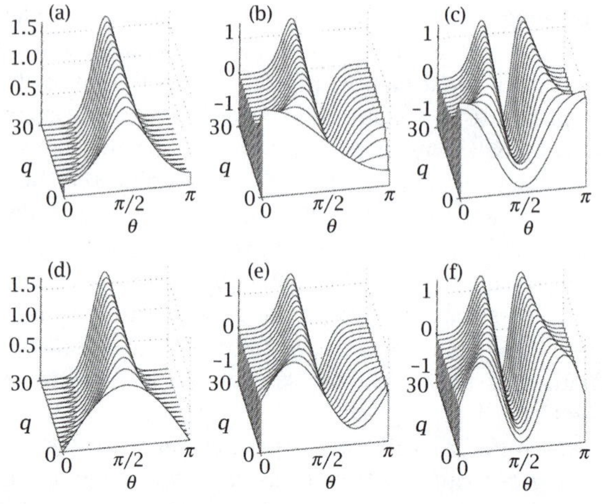


**Figure 1** The behavior of Mathieu functions over the plane $(\theta, q)$. The range of the plots has been limited to $[0, \pi]$, since their behavior over the entire range can be deduced from the parity and symmetry relations. (a) $\mathrm{ce}_0(\theta; q)$. (b) $\mathrm{ce}_1(\theta; q)$. (c) $\mathrm{ce}_2(\theta; q)$. (d) $\mathrm{se}_1(\theta; q)$. (e) $\mathrm{se}_2(\theta; q)$. (f) $\mathrm{se}_3(\theta; q)$.

functions whose normalization conditions are

$$\int_0^{2\pi} \mathrm{ce}_m \mathrm{ce}_n \, \mathrm{d}\theta = \int_0^{2\pi} \mathrm{se}_m \mathrm{se}_n \, \mathrm{d}\theta = \pi \delta_{m,n}. \tag{2}$$

If a function $f(\theta)$ is periodic with period $\pi$ or $2\pi$, then it can be expanded as a series of orthogonal Mathieu functions.

### Further Reading

Mathieu, T. 1868. Le mouvement vibratoire d'une membrane de forme elliptique. *Journal de Mathématiques: Pures et Appliquées* 13:137–203.

McLachlan, N. W. 1951. *Theory and Application of Mathieu Functions*. Oxford: Oxford University Press.

## III.22 Maxwell's Equations

*Mark R. Dennis*

In the early 1860s James Clerk Maxwell wrote down a set of equations summarizing the spatial and temporal behavior of electric and magnetic fields; these equations are the foundation of the physical theory of *electromagnetism*. In modern notation they are usually written as first-order vector differential equations depending on time $t$ and three-dimensional position $\boldsymbol{r}$. With $\nabla$ denoting the gradient operator and a dot above a quantity denoting its time derivative, Maxwell's equations are usually written

$$\nabla \cdot \boldsymbol{B} = 0, \tag{1}$$

$$\nabla \times \boldsymbol{E} + \dot{\boldsymbol{B}} = 0, \tag{2}$$

$$\nabla \cdot \boldsymbol{D} = \rho, \tag{3}$$

$$\nabla \times \boldsymbol{H} - \dot{\boldsymbol{D}} = \boldsymbol{J}. \tag{4}$$

The various electromagnetic fields are called the *electric field* $\boldsymbol{E}$, the *electric displacement* $\boldsymbol{D}$, the *magnetic induction* $\boldsymbol{B}$, and the *magnetic field* $\boldsymbol{H}$. $\boldsymbol{E}$ and $\boldsymbol{B}$ are usually viewed as more fundamental, and $\boldsymbol{D}$ and $\boldsymbol{H}$ can often be expressed as functions of them. Equations (3) and (4) are inhomogeneous and depend on sources determined by the scalar electric *charge density* $\rho$ and vector *current density* $\boldsymbol{J}$. Equations (1)–(4) may also be expressed in integral form, in terms of surface and volume integrals. Conventionally, they are referred to as Gauss's law (3), Gauss's law for magnetism (1) (also sometimes referred to as "no magnetic monopoles"), Faraday's law (2), and the Maxwell–Ampère law (4).

All electromagnetic phenomena are described by Maxwell's equations, including the behavior of electricity and electronic circuits, motors and dynamos, light and optics, wireless communication, microwaves, etc. The phenomena they describe are ubiquitous in the modern world, and much of the machinery of applied mathematics, such as partial differential equations, Green functions, DELTA FUNCTIONS [III.7], and vector calculus were introduced in part to describe electromagnetic situations. Maxwell's equations take on various special forms depending on the system being considered. For instance, when the fields are static (i.e., when time derivatives are zero), the electric field around charges resembles the gravitational field around masses determined by Newton's law.

In *free space*, all sources are zero, and $\boldsymbol{D} = \varepsilon_0 \boldsymbol{E}$, $\boldsymbol{H} = \mu_0^{-1} \boldsymbol{B}$, with constants the *permittivity of free space* $\varepsilon_0 = 8.85 \times 10^{-12}\ \mathrm{F\,m^{-1}}$ and the *permeability of free space* $\mu_0 = 4\pi \times 10^{-7}\ \mathrm{N\,A^{-2}}$, given in conventional SI units, originally determined experimentally using electrical currents, charges, and magnets. Written like this, Maxwell's equations are symmetric in form between $\boldsymbol{E}$ and $\boldsymbol{B}$ (with the exception of a minus sign), and can be combined to give the d'Alembert equation for $\boldsymbol{E}$ and $\boldsymbol{B}$, propagating at speed $c = (\varepsilon_0 \mu_0)^{-1/2} = 2.998 \times 10^8\ \mathrm{m\,s^{-1}}$. Maxwell himself originally noticed this fact, realizing that $c$ was close to the experimentally measured speed of light and therefore that light itself is an electromagnetic wave. *Plane-wave solutions* of Maxwell's equations in free space are $\boldsymbol{E}_0 \cos(\boldsymbol{k} \cdot \boldsymbol{r} - \omega t)$ $\boldsymbol{B}_0 = \cos(\boldsymbol{k} \cdot \boldsymbol{r} - \omega t)$, where $\boldsymbol{E}_0$, $\boldsymbol{B}_0$ are constant *polarization vectors* that form a right-handed orthogonal triple $(\boldsymbol{E}_0, \boldsymbol{B}_0, \boldsymbol{k})$ with the wave vector $\boldsymbol{k}$, and $c = \omega/|\boldsymbol{k}|$, with $\omega$ the angular frequency. Free-space solutions of Maxwell's equations are studied systematically in the field of OPTICS AND PHOTONICS [V.14].

Many naturally occurring materials without free charges (*dielectric materials*) are *linear*: $\boldsymbol{D}$ and $\boldsymbol{H}$ are linear functions of $\boldsymbol{E}$ and $\boldsymbol{B}$, but with different permittivities and permeabilities (possibly in different directions). The relative speed of electromagnetic wave propagation with respect to $c$ in these materials is called the *refractive index*. Laws of refraction, reflection, and transmission can all in principle be derived by applying Maxwell's equations to the interface region between materials. Other materials can have more complicated dependence between the various field quantities and sources. Some of the simplest nonlinear phenomena occur in materials where $\boldsymbol{D}$ depends quadratically on $\boldsymbol{E}$.

The homogeneity of (1) and (2) suggests that $\boldsymbol{E}$ and $\boldsymbol{B}$ may themselves be expressed as derivatives of the *scalar potential* $V$ and the *vector potential* $\boldsymbol{A}$, via

$$\boldsymbol{B} = \nabla \times \boldsymbol{A}, \qquad \boldsymbol{E} = -\nabla V - \dot{\boldsymbol{A}}. \tag{5}$$

In terms of these potentials, (1) and (2) are automatically satisfied, and (3) and (4) become second-order differential equations in $V$ and $\boldsymbol{A}$. As only their derivatives are defined by the electromagnetic fields, the absolute values of $V$ and $\boldsymbol{A}$ are somewhat arbitrary, and choices of potential field may be changed by a *gauge transformation* that preserves the derivative relations in (5), $\boldsymbol{A} \to \boldsymbol{A}' = \boldsymbol{A} + \nabla \chi$, $V \to V' = V - \dot{\chi}$, for some sufficiently differentiable scalar field $\chi$. It is often convenient to simplify (1) and (2) by reexpressing them in terms of the potentials and then choosing a restriction on the gauge transformations to be considered (known as *fixing the gauge*), by requiring the potential fields to satisfy some extra equation compatible with (5), such as $\nabla \cdot \boldsymbol{A} = 0$ (the Coulomb gauge) or $\nabla \cdot A + c^{-2} \dot{V} = 0$ (the Lorenz gauge).

Maxwell's equations describe how electric and magnetic fields depend spatially and temporally on electric charges and their motion. They do not, however, describe directly how the motion of charges depends on fields; the force $\boldsymbol{F}$ on a particle of electric charge $q$ at position $\boldsymbol{r}$ due to electromagnetic fields is given by the *Lorentz force law*

$$\boldsymbol{F} = q(\boldsymbol{E} + \dot{\boldsymbol{r}} \times \boldsymbol{B}). \tag{6}$$

The combined theory of electromagnetic fields, charged matter, and their mutual dependence is called *electrodynamics*.

A particularly interesting aspect of Maxwell's equations from the physical point of view is that they appear to apply equally well at many different length scales, from the astronomical (used to understand electromagnetic radiation in the cosmos) down to the microscopic (and even interactions at the scale of subatomic particles). Inside materials, at the atomic level, electrons and nuclei are separated by free space, and only the fundamental fields $\boldsymbol{E}$ and $\boldsymbol{B}$ play a role. In situations where length scales are longer, the electric displacement $\boldsymbol{D}$ and magnetic field $\boldsymbol{H}$ arise through HOMOGENIZATION [IV.6 §13.3] of length scales. Maxwell's equations can also be made compatible with the laws of QUANTUM MECHANICS [IV.23], both in treatments with the field as classical and in treatments in which the electromagnetic fields themselves are quantized, with their energy quanta $\hbar\omega$ referred to as *photons*, where $\hbar$ is Planck's constant. The combined quantum theory of charged matter and electromagnetic fields is *quantum electrodynamics*. Gauge transformations take on an important role and a new physical interpretation in quantum theory.

Unlike Newton's equations in CLASSICAL MECHANICS [IV.19], Maxwell's equations are not invariant with respect to Galilean coordinate transformations. Historically, this provided Einstein with the main motivation for the theory of special relativity, as providing the appropriate set of transformations keeping Maxwell's equations covariant. Maxwell's equations in free space can thus be expressed in 4-vector notation (which is explained in TENSORS AND MANIFOLDS [II.33]), in which the electromagnetic field is specified by an antisymmetric rank-2 tensor, the *Faraday tensor*:

$$F^{ab} = \begin{pmatrix} 0 & -E_x/c & -E_y/c & -E_z/c \\ E_x/c & 0 & -B_z & B_y \\ E_y/c & B_z & 0 & -B_x \\ E_z/c & -B_y & B_x & 0 \end{pmatrix}.$$

Reexpressed in tensor form, Maxwell's equations are

$$\partial_a F_{bc} + \partial_b F_{ca} + \partial_c F_{ab} = 0,$$
$$\partial_a F^{ab} = \mu_0 J^b,$$

with the first equation corresponding to (1) and (2), and the second to (3) and (4), with the current 4-vector $J^b = (c\rho, J_x, J_y, J_z)$. These equations transform covariantly under Lorentz transformations. The Lorentz law (6) can also be relativistically generalized to the equation for a force 4-vector $f^a = qF^{ab}u_b$, acting on a particle with velocity 4-vector $u^b$. These equations can be further generalized to fields in curved space-time, in which they play a role in GENERAL RELATIVITY AND COSMOLOGY [IV.40].

### Further Reading

Feynman, R. P. 2006. *QED: The Strange Theory of Light and Matter.* Princeton, NJ: Princeton University Press.

Harman, P. M. 2001. *The Natural Philosophy of James Clerk Maxwell.* Cambridge: Cambridge University Press.

Jackson, J. D. 1998. *Classical Electrodynamics*, 3rd edn. New York: John Wiley.

## III.23 The Navier–Stokes Equations

*H. K. Moffatt*

The Navier–Stokes equations are the partial differential equations that govern the flow of a fluid (a liquid or a gas) that is regarded as a continuum; that is to say, a medium whose density field $\rho(\boldsymbol{x}, t)$ and (vector) velocity field $\boldsymbol{u}(\boldsymbol{x}, t)$ may be considered to be smooth functions of position $\boldsymbol{x} = (x, y, z)$ and time $t$. These equations express in mathematical form the physical principles of conservation of mass and balance of momentum for each small element (or *parcel*) of fluid in the course of its motion.

### 1 The Mass-Conservation Equation

This equation relates $\rho(\boldsymbol{x}, t)$ and $\boldsymbol{u}(\boldsymbol{x}, t)$ in a very simple way:

$$\frac{\mathrm{D}\rho}{\mathrm{D}t} \equiv \frac{\partial\rho}{\partial t} + \boldsymbol{u} \cdot \nabla\rho = -\rho\nabla \cdot \boldsymbol{u}.$$

Here, the symbol $\nabla$ represents the vector differential operator $(\partial/\partial x, \partial/\partial y, \partial/\partial z)$. The operator $\mathrm{D}/\mathrm{D}t \equiv \partial/\partial t + \boldsymbol{u} \cdot \nabla$ is the *Lagrangian* derivative (or "derivative following the fluid"); this consists of two parts, the local time derivative $\partial/\partial t$ and the *convective* derivative $\boldsymbol{u} \cdot \nabla$. The equation indicates that the density of a fluid element decreases or increases according to whether the local divergence $\nabla \cdot \boldsymbol{u}$ is positive or negative, respectively.

An important subclass of flows is described as *incompressible*. For these, the density of any element of fluid is constant: $\mathrm{D}\rho/\mathrm{D}t = 0$, and so, from the above, $\nabla \cdot \boldsymbol{u} = 0$. Attention is focused on incompressible flows in the following section.

## 2 The Momentum Equation

This equation represents the fundamental Newtonian balance

$$\text{mass} \times \text{acceleration} = \text{force}$$

for each fluid element. Of course, density is just mass per unit volume, and the (Lagrangian) acceleration of a fluid element is $\mathrm{D}\boldsymbol{u}/\mathrm{D}t \equiv (\partial/\partial t + \boldsymbol{u}\cdot\nabla)\boldsymbol{u}$. The left-hand side of the following *momentum equation* should therefore come as no surprise:

$$\rho\left(\frac{\partial \boldsymbol{u}}{\partial t} + (\boldsymbol{u}\cdot\nabla)\boldsymbol{u}\right) = -\nabla p + \mu\nabla^2\boldsymbol{u} + \boldsymbol{f}.$$

The right-hand side contains three terms that represent the forces that may act on a fluid element. The term $-\nabla p$ represents (minus) the gradient of the local fluid pressure $p(\boldsymbol{x},t)$; the second term, $\mu\nabla^2\boldsymbol{u}$, represents the net force associated with *viscosity* $\mu$ ($>0$) (i.e., internal friction); and the third term, $\boldsymbol{f}(\boldsymbol{x},t)$, represents any *external* force per unit mass acting on the fluid. The most common force of this kind is that associated with gravity, $\boldsymbol{g}$, i.e., $\boldsymbol{f} = \rho\boldsymbol{g}$. This is by no means the only possibility; for example, in an electrically conducting fluid in which a current density $\boldsymbol{j}$ flows across a magnetic field $\boldsymbol{B}$, the force per unit mass is the Lorentz force given by the vector product $\boldsymbol{f} = \boldsymbol{j}\times\boldsymbol{B}$.

It might appear from the above that a further equation is needed to determine the pressure distribution $p(\boldsymbol{x},t)$. However, this equation is already implied by the above; it may be obtained by taking the divergence of the momentum equation to give a Poisson equation for $p$. In the simplest case of incompressible flow of constant density $\rho_0$, this Poisson equation takes the form

$$\nabla^2 p = -\rho_0\nabla\cdot(\boldsymbol{u}\cdot\nabla)\boldsymbol{u} = -\rho_0\frac{\partial u_i}{\partial x_j}\frac{\partial u_j}{\partial x_i},$$

using suffix notation and the summation convention (and, from incompressibility, $\partial u_i/\partial x_i = 0$).

In the inviscid limit ($\mu = 0$), the above equations were derived in 1758 by Euler, and they are known as the (incompressible) EULER EQUATIONS [III.11]. The viscous equations (with $\mu > 0$) are named after Claude-Louis Navier, who obtained them in 1822 assuming a particular "atomic" model to take account of internal friction, and George Gabriel Stokes, who in 1845 developed the more general continuum treatment that is still normally used today. A full derivation of these Navier–Stokes equations may be found in Batchelor's *An Introduction to Fluid Dynamics*.

### Further Reading

Batchelor, G. K. 1967. *An Introduction to Fluid Dynamics*. Cambridge: Cambridge University Press.

## III.24 The Painlevé Equations

*Peter A. Clarkson*

The six nonlinear second-order ordinary differential equations that follow are called the Painlevé equations:

$$w'' = 6w^2 + z,$$

$$w'' = 2w^3 + zw + \alpha,$$

$$w'' = \frac{(w')^2}{w} - \frac{w'}{z} + \frac{\alpha w^2 + \beta}{z} + \gamma w^3 + \frac{\delta}{w},$$

$$w'' = \frac{(w')^2}{2w} + \tfrac{3}{2}w^3 + 4zw^2 + 2(z^2 - \alpha)w + \frac{\beta}{w},$$

$$w'' = \left(\frac{1}{2w} + \frac{1}{w-1}\right)(w')^2 - \frac{w'}{z} + \frac{(w-1)^2}{z^2}\left(\alpha w + \frac{\beta}{w}\right) + \frac{\gamma w}{z} + \frac{\delta w(w+1)}{w-1},$$

$$\begin{aligned} w'' &= \frac{1}{2}\left(\frac{1}{w} + \frac{1}{w-1} + \frac{1}{w-z}\right)(w')^2 \\ &\quad - \left(\frac{1}{z} + \frac{1}{z-1} + \frac{1}{w-z}\right)w' \\ &\quad + \frac{w(w-1)(w-z)}{z^2(z-1)^2} \\ &\quad \times \left\{\alpha + \frac{\beta z}{w^2} + \frac{\gamma(z-1)}{(w-1)^2} + \frac{\delta z(z-1)}{(w-z)^2}\right\}, \end{aligned}$$

where $w = w(z)$; a prime denotes differentiation with respect to $z$; and $\alpha$, $\beta$, $\gamma$, and $\delta$ are arbitrary constants. The equations are commonly referred to as $\mathrm{P_I}$–$\mathrm{P_{VI}}$ in the literature, a convention that we will adhere to here.

These equations were discovered about a hundred years ago by Painlevé, Gambier, and their colleagues while studying second-order ordinary differential equations of the form

$$w'' = F(z; w, w'), \tag{1}$$

where $F$ is rational in $w'$ and $w$ and locally analytic in $z$. In general, the singularities of the solutions are movable in the sense that their location depends on the constants of integration associated with the initial or boundary conditions. An equation is said to have the *Painlevé property* if all its solutions are free from movable branch points, i.e., the locations of multivalued singularities of any of its solutions are independent of the particular solution chosen and so are dependent only on the equation; the solutions may have movable poles or movable isolated essential singularities.

Painlevé et al. showed that there are 50 canonical equations of the form (1) that have this property, up to a Möbius (bilinear rational) transformation. Of these 50 equations, 44 can be reduced to linear equations, solved in terms of elliptic functions, or are reducible to one of six new nonlinear ordinary differential equations that define new transcendental functions.

Although first discovered as a result of mathematical study, the Painlevé equations have arisen in a variety of applications including statistical mechanics (correlation functions of the $XY$ model and the Ising model), random-matrix theory, topological field theory, plasma physics, nonlinear waves (resonant oscillations in shallow water, convective flows with viscous dissipation, Görtler vortices in boundary layers, and Hele-Shaw problems), quantum gravity, quantum field theory, general relativity, nonlinear and fiber optics, polyelectrolytes, Bose–Einstein condensation, and stimulated Raman scattering. The Painlevé equations also arise as symmetry reductions of the soliton equations, such as the Korteweg–de Vries and nonlinear Schrödinger equations, which are solvable by inverse scattering.

The Painlevé equations may be thought of as nonlinear analogs of the classical special functions. They have a large number of interesting properties, some of which we summarize below.

(1) For arbitrary values of the parameters $\alpha$, $\beta$, $\gamma$, and $\delta$, the general solutions of $P_I$–$P_{VI}$ are *transcendental*, i.e., they cannot be expressed in terms of closed-form elementary functions.

(2) All Painlevé equations can be expressed as the compatibility condition of a linear system: the *isomonodromy problem* or *Lax pair*. Suppose that

$$\frac{\partial \Psi}{\partial \lambda} = A(z;\lambda)\Psi, \qquad \frac{\partial \Psi}{\partial z} = B(z;\lambda)\Psi$$

is a linear system in which $\Psi$ is a vector, $A$ and $B$ are matrices, and $\lambda$ is independent of $z$. Then the equation

$$\frac{\partial^2 \Psi}{\partial z \partial \lambda} = \frac{\partial^2 \Psi}{\partial \lambda \partial z}$$

is satisfied provided that

$$\frac{\partial A}{\partial z} - \frac{\partial B}{\partial \lambda} + AB - BA = 0,$$

which is the *compatibility condition*.

(3) Each of the Painlevé equations can be written as a *Hamiltonian system*,

$$q' = \frac{\partial \mathcal{H}_J}{\partial p}, \qquad p' = -\frac{\partial \mathcal{H}_J}{\partial q},$$

for suitable (nonautonomous) Hamiltonian functions $\mathcal{H}_J(q,p,z)$, $J = I, II, \dots, VI$. Furthermore, the function $\sigma = \mathcal{H}_J(q,p,z)$ satisfies a second-order, second-degree equation. For example, the Hamiltonian for $P_I$ is

$$\mathcal{H}_I(q,p,z) = \tfrac{1}{2}p^2 - 2q^3 - zq,$$

and so

$$q' = p, \qquad p' = 6q^2 + z,$$

and the function $\sigma = \mathcal{H}_I(q,p,z)$ satisfies

$$(\sigma'')^2 + 4(\sigma')^3 + 2z\sigma' - 2\sigma = 0.$$

(4) Equations $P_{II}$–$P_{VI}$ possess *Bäcklund transformations*, which relate one solution to another solution of the same equation, with different values of the parameters, or to another equation. For example, if $w = w(z;\alpha)$ is a solution of $P_{II}$, then so are

$$w(z;\alpha \pm 1) = -w - \frac{2\alpha \pm 1}{2w^2 \pm 2w' + z},$$

provided that $\alpha \neq \mp\frac{1}{2}$.

(5) For certain values of the parameters, $P_{II}$–$P_{VI}$ possess rational solutions, algebraic solutions, and solutions expressible in terms of classical special functions (Airy functions for $P_{II}$, Bessel functions for $P_{III}$, parabolic cylinder functions for $P_{IV}$, confluent hypergeometric functions for $P_V$, and hypergeometric functions for $P_{VI}$). These solutions, which are known as "classical solutions," can often be expressed in the form of determinants. For example, $P_{II}$ has rational solutions if $\alpha = n \in \mathbb{Z}$ and solutions in terms of Airy functions if $\alpha = n + \frac{1}{2}$, with $n \in \mathbb{Z}$.

(6) The asymptotic behavior of their solutions—together with the associated *connection formulas* that relate the asymptotic behaviors of the solutions as $|z| \to \infty$ in different regions of the complex plane—play an important role in the application of the Painlevé equations.

(7) The Painlevé equations possess a coalescence cascade, in that $P_I$–$P_V$ can be obtained from $P_{VI}$ by the cascade

$$\begin{array}{ccccc} P_{VI} & \longrightarrow & P_V & \longrightarrow & P_{IV} \\ & & \downarrow & & \downarrow \\ & & P_{III} & \longrightarrow & P_{II} & \longrightarrow & P_I \end{array}$$

For example, if we make the transformation

$$w(z;\alpha) = \varepsilon u(\zeta) + \varepsilon^{-5},$$
$$z = \varepsilon^2\zeta - 6\varepsilon^{-10}, \qquad \alpha = 4\varepsilon^{-15}$$

in $P_{II}$, then

$$\frac{d^2u}{d\zeta^2} = 6u^2 + \zeta + \varepsilon^6(2u^3 + \zeta u).$$

So in the limit as $\varepsilon \to 0$, $u(\zeta)$ satisfies $P_I$.

### Further Reading


Clarkson, P. A. 2006. Painlevé equations—non-linear special functions. In *Orthogonal Polynomials and Special Functions: Computation and Application*, edited by P. Marcellàn and W. van Assche. Lecture Notes in Mathematics, volume 1883, pp. 331–411. Berlin: Springer.

——. 2010. Painlevé transcendents. In *NIST Handbook of Mathematical Functions*, edited by F. W. J. Olver, D. W. Lozier, R. F. Boisvert, and C. W. Clark, pp. 723–40. Cambridge: Cambridge University Press.

Fokas, A. S., A. R. Its, A. A. Kapaev, and Y. Yu. Novokshenov. 2006. *Painlevé Transcendents: The Riemann–Hilbert Approach*. Mathematical Surveys and Monographs, volume 128. Providence, RI: American Mathematical Society.

Gromak, V. I., I. Laine, and S. Shimomura. 2002. *Painlevé Differential Equations in the Complex Plane*. Studies in Mathematics, volume 28. Berlin: Walter de Gruyter.


## III.25 The Riccati Equation

*Alan J. Laub*

### 1 History

The name Riccati equation is given to a wide variety of algebraic and differential (or difference) equations characterized by the quadratic appearance of the unknown $X$. Such equations are named after the Italian mathematician Count Jacopo Francesco Riccati (1676–1754). The variable $X$ was a scalar in the original writings of Riccati, but in modern applications $X$ is often a matrix, and that is the focus here.

### 2 The Algebraic Riccati Equation

The simplest form of the algebraic Riccati equation (ARE) arises in the so-called continuous-time linear-quadratic theory of control. Suppose we have a linear differential equation (with initial conditions)

$$\frac{d}{dt}x(t) = Ax(t) + Bu(t),$$

where $A \in \mathbb{R}^{n\times n}$ and $B \in \mathbb{R}^{n\times m}$ $(n \geqslant m)$ and the pair of matrices $(A, B)$ is *controllable*, that is, $\operatorname{rank}(A - \lambda I, B) = n$ for all $\lambda \in \Lambda(A)$, where $\Lambda(A)$ denotes the spectrum of $A$. Suppose, further, that only a subset of the *state variables* $x$ can be measured, namely the *outputs* $y = Cx$, where $C \in \mathbb{R}^{q\times n}$ $(q \leqslant n)$, and that the pair of matrices $(C, A)$ is *observable*, that is, $\operatorname{rank}(A - \lambda I, C) = n$ for all $\lambda \in \Lambda(A)$. The *control* $u(t)$ is to be chosen to minimize the quadratic functional

$$\int_0^\infty (y^T Qy + u^T Ru)\, dt,$$

where $Q$ and $R$ are given weighting matrices that are symmetric positive-semidefinite and symmetric positive-definite, respectively. It turns out that the optimal $u(t)$ is given by $u(t) = -R^{-1}B^T Xx(t)$, where the symmetric positive-definite matrix $X$ solves the ARE

$$A^T X + XA - XBR^{-1}B^T X + C^T QC = 0. \qquad (1)$$

The closed-loop matrix $A - BR^{-1}B^T X$ (formed by substituting the feedback control $u(t)$ above into the differential equation) is asymptotically stable; that is, its eigenvalues lie in the open left half-plane.

The assumptions made in the linear-quadratic problem are sufficient to guarantee that the $2n \times 2n$ system matrix associated with the problem, namely

$$H = \begin{bmatrix} A & -BR^{-1}B^T \\ -C^T QC & -A^T \end{bmatrix},$$

has no pure imaginary eigenvalues. There is a long history of the association between the system matrix $H$, its invariant subspaces, and solutions $X$ of the ARE.

It is easily shown that the matrix $H$ is Hamiltonian, that is, $JA$ is symmetric where $J = \left[\begin{smallmatrix} 0 & I_n \\ -I_n & 0 \end{smallmatrix}\right]$. The Hamiltonian structure has the consequence that if $\lambda$ is an eigenvalue of $H$ then so is $-\lambda$ with the same multiplicity. Thus, by our linear-quadratic assumptions, $H$ has precisely $n$ eigenvalues in the open left half-plane and $n$ eigenvalues in the open right half-plane. Hence we can find an orthogonal matrix $U \in \mathbb{R}^{2n\times 2n}$ with $n \times n$ blocks $U_{ij}$ (with $U_{11}$ nonsingular) that transforms $H$ to upper quasi-triangular REAL SCHUR FORM [IV.10 §5.5]

$$U^T HU = \begin{bmatrix} S_{11} & S_{12} \\ 0 & S_{22} \end{bmatrix},$$

where $\Lambda(S_{11})$ is contained in the open left half-plane. Setting $X = U_{21}U_{11}^{-1}$, it can then be verified (after a modest amount of matrix algebra) that $X$ solves (1) and is symmetric positive-definite and that the closed-loop eigenvalues are asymptotically stable, since $\Lambda(A - BR^{-1}B^T X) = \Lambda(S_{11})$. Note that the $n$ columns of $\left[\begin{smallmatrix} U_{11} \\ U_{21} \end{smallmatrix}\right]$ are the Schur vectors of $H$ that form a basis for the stable invariant subspace (corresponding to the stable left half-plane eigenvalues of $S_{11}$). Other orderings of the eigenvalues give rise to other solutions (even nonsymmetric ones) of the ARE. The ARE can have infinitely many solutions (think of the Riccati equation $X^2 = I$).

There are also results in terms of the SINGULAR VALUE DECOMPOSITION [II.32] of $U_{11}$ that explicitly exhibit the symmetry of the solution $X$.

The foregoing describes the case for the simplest linear–quadratic problem. Many of the assumptions can be weakened considerably. For example, controllability can be weakened to stabilizability, while observability can be weakened to detectability. In this case, $X$ is still symmetric and stabilizing but may only be positive-semidefinite.

The connection between the $2n \times 2n$ Hamiltonian matrix $H$ and the ARE has been known for well over a hundred years. The use of Schur vectors, as opposed to eigenvectors, as a basis for the stable invariant subspace is crucial for reliable numerical solution. This (and related topics) has been a fertile subject of research. Perhaps surprisingly, the Schur method was not published until the late 1970s, but since then methods based on Schur vectors have found numerous other applications in systems theory, control, and beyond.

In the discrete-time problem, the differential equation is replaced by the difference equation (with initial conditions)

$$x_{k+1} = Ax_k + Bu_k,$$

where $A \in \mathbb{R}^{n\times n}$ and $B \in \mathbb{R}^{n\times m}$. As before, $y_k = Cx_k$, but we will work with $C = I$ for convenience. The integral performance constraint is replaced by an appropriate summation, and (one form of) the resulting discrete-time ARE is given by

$$A^{\mathrm{T}}XA - X - A^{\mathrm{T}}XB(R + B^{\mathrm{T}}XB)^{-1}B^{\mathrm{T}}XA + Q = 0,$$

where the closed-loop matrix $A - B(R + B^{\mathrm{T}}XB)^{-1}B^{\mathrm{T}}XA$ is asymptotically stable (which now means that its eigenvalues lie inside the unit circle in the complex plane). Provided $A$ is nonsingular, the role of the $2n \times 2n$ Hamiltonian matrix $H$ is taken here by the $2n \times 2n$ matrix

$$S = \begin{bmatrix} A + BR^{-1}B^{\mathrm{T}}A^{-T}Q & -BR^{-1}B^{\mathrm{T}}A^{-T} \\ -A^{-T}Q & A^{-T} \end{bmatrix},$$

which is symplectic; that is, $S^{\mathrm{T}}JS = J$. The symplectic matrix $S$ has a $\lambda \leftrightarrow 1/\lambda$ symmetry to its eigenvalues, and appropriate assumptions guarantee that there are no eigenvalues on the unit circle. If $A$ is singular, it turns out to be much better to work with the $2n \times 2n$ symplectic pencil

$$\begin{bmatrix} A & 0 \\ -Q & I \end{bmatrix} - \lambda \begin{bmatrix} I & BR^{-1}B^{\mathrm{T}} \\ 0 & A^{\mathrm{T}} \end{bmatrix},$$

and this gives rise to a large number of new methods based on the Schur vectors of the corresponding generalized eigenproblem. This time the role of the invariant subspace is taken by a generalization for matrix pencils called the deflating subspace, but the essence of the method remains the same.

## 3 Extensions of the ARE

There are many generalizations of the linear–quadratic problem and the Riccati equation, such as

- extended (i.e., $(2n + m) \times (2n + m)$) pencils, in which even the $R$ matrix may be singular;
- versions with a cross-performance term in the integral (or a sum in the discrete-time case);
- versions for which the system constraint (the differential equation or the difference equation) is given in so-called descriptor form with a matrix $E$ (which may or may not be singular) multiplying the left-hand side;
- versions associated with the so-called $H_\infty$ problem that are Riccati equations but with different sets of assumptions on the coefficient matrices; and
- AREs with specific structure on the matrices that then gives rise to other features of the solution (such as componentwise nonnegativity).

Many other solution techniques have been developed, including

- structured methods that try to preserve the underlying Hamiltonian or symplectic nature of the matrices;
- methods based on the MATRIX SIGN FUNCTION [II.14] that exploit the connection with an appropriate invariant subspace;
- methods that are iterative in nature, such as NEWTON'S METHOD [II.28], which involves solving a sequence of much-easier-to-solve LYAPUNOV [III.28] (or Sylvester) equations at each step; and
- doubling methods, which constitute another large class of iterative methods.

## 4 The Riccati Differential Equation

The ARE can be thought of as a stationary point of the associated Riccati differential equation. For the algebraic equation above, its differential equation takes the form (with initial conditions)

$$\frac{\mathrm{d}}{\mathrm{d}t}X(t) = F(X(t)),$$

where $F(X(t))$ has the form of the left-hand side of the ARE (1) and where each of the coefficient matrices may also vary with time. Symmetry is not necessary in the sense that so-called nonsymmetric equations of the form

$$\frac{\mathrm{d}}{\mathrm{d}t}X(t) = A_1 + A_2X(t) + X(t)A_3 - X(t)A_4X(t)$$

can be studied, where $X(t) \in \mathbb{R}^{m\times n}$ and the coefficient matrices $A_i$ are general. In all cases, the defining property of equations of Riccati type is the characteristic appearance of the unknown matrix in quadratic form. Of course, there are also algebraic versions of these nonsymmetric equations.

### Further Reading

Bini, D. A., B. Iannazzo, and B. Meini. 2012. *Numerical Solution of Algebraic Riccati Equations*. Philadelphia, PA: SIAM.

Bittanti, S., A. J. Laub, and J. C. Willems, eds. 1991. *The Riccati Equation*. Berlin: Springer.

Lancaster, P., and L. Rodman. 1995. *Algebraic Riccati Equations*. Oxford: Oxford University Press.

Laub, A. J. 1979. A Schur method for solving algebraic Riccati equations. *IEEE Transactions on Automatic Control* 24(6):913–21.

## III.26 Schrödinger's Equation

In its full generality, Schrödinger's equation describes the quantum mechanical evolution under time $t$ of a vector (often called a *ket*) $|\Psi\rangle$ in a HILBERT SPACE [I.2 §19.4], according to a self-adjoint operator $\hat{H}$ called the *Hamiltonian operator*,

$$\mathrm{i}\hbar\frac{\partial}{\partial t}|\Psi\rangle = \hat{H}|\Psi\rangle, \tag{1}$$

where i is the imaginary unit and $\hbar$ is a constant, known as Planck's constant or the quantum of action. In SI units, $\hbar \approx 1.05 \times 10^{-34}$ J s.

$|\Psi\rangle$ represents the time-dependent total state of the quantum system, and $\hat{H}$ determines the energy of the system. If $\hat{H}$ is independent of $t$ and $|\Psi\rangle = \mathrm{e}^{-\mathrm{i}Et/\hbar}|E\rangle$ for $|E\rangle$ an eigenvector of $\hat{H}$ with energy eigenvalue $E$, then (1) can be written in time-independent form, $\hat{H}|E\rangle = E|E\rangle$.

When the quantum system consists of a single particle of mass $m$ (disregarding effects of quantum spin and electromagnetic interactions), the Hilbert space may be chosen to be the space of square-integrable complex-valued functions of $t$ and position $\boldsymbol{r}$ in Euclidean configuration space, and the quantum state is described by a *wave function* $\Psi(\boldsymbol{r},t)$. If the particle is subject to a potential $V(\boldsymbol{r},t)$, then Schrödinger's equation takes the form of the partial differential equation

$$\mathrm{i}\hbar\frac{\partial}{\partial t}\Psi(\boldsymbol{r},t) = -\frac{1}{2m}\nabla^2\Psi(\boldsymbol{r},t) + V(\boldsymbol{r},t)\Psi(\boldsymbol{r},t),$$

where $\nabla^2$ denotes the LAPLACE OPERATOR [III.18]. In the *nonlinear Schrödinger equation*, the potential is proportional to the modulus squared of the wave function: $V = \kappa|\Psi|^2$ for some positive or negative constant $\kappa$.

Study of the evolution of Schrödinger's equation is the subject of QUANTUM MECHANICS [IV.23], and it emerges naturally from Hamilton's approach to CLASSICAL MECHANICS [IV.19].

## III.27 The Shallow-Water Equations

*P. A. Martin*

As one might expect, the shallow-water equations are appropriate when considering the motion of a liquid occupying a layer. Major applications concern ocean waves, so we use appropriate terminology. We therefore identify the bottom of the layer with the sea floor at $z = -b(x,y)$, where $b(x,y)$ is given and positive, and $x$, $y$, $z$ are Cartesian coordinates with $z$ pointing upward. The top of the layer is the moving free surface, $z = \eta(x,y,t)$, where $t$ is time. The total depth of the water is $h(x,y,t) = b(x,y) + \eta(x,y,t)$. The water is assumed to be incompressible (with constant density $\rho$) and inviscid (viscous effects are ignored). Thus, the governing partial differential equations (PDEs) in the water are the EULER EQUATIONS [III.11],

$$\frac{\partial \boldsymbol{u}}{\partial t} + (\boldsymbol{u}\cdot\nabla)\boldsymbol{u} + \frac{1}{\rho}\nabla p = -g\hat{\boldsymbol{z}}, \tag{1}$$

and the continuity equation, $\nabla \cdot \boldsymbol{u} = 0$, where $\boldsymbol{u} = (u, v, w)$ is the fluid velocity, $p$ is the pressure, $g$ is the acceleration due to gravity, and $\hat{\boldsymbol{z}}$ is a unit vector in the $z$-direction. In addition, there are boundary conditions at the free surface and at the bottom. They are

$$p = 0 \quad\text{and}\quad \frac{\partial\eta}{\partial t} + u\frac{\partial\eta}{\partial x} + v\frac{\partial\eta}{\partial y} = w \quad \text{at } z = \eta$$

and

$$u\frac{\partial b}{\partial x} + v\frac{\partial b}{\partial y} + w = 0 \quad \text{at } z = -b.$$

Integrating $\nabla\cdot\boldsymbol{u} = 0$ with respect to $z$ gives

$$\frac{\partial\eta}{\partial t} + \frac{\partial}{\partial x}\int_{-b}^{\eta} u\,\mathrm{d}z + \frac{\partial}{\partial y}\int_{-b}^{\eta} v\,\mathrm{d}z = 0, \tag{2}$$

where the boundary conditions at $z = \eta$ and $z = -b$ have been used.

So far, we have not made any approximations. Next, we assume that the waves generated are much longer than the water depth: this is what is implied by the "shallow-water" terminology. One consequence is that we can neglect the vertical acceleration terms (involving $\boldsymbol{u}$) in the $z$-component of (1), giving $p(x,y,z,t) = \rho g\{\eta(x,y,t) - z\}$ after integration with respect to $z$.

Finally, we assume further that the fluid velocity is horizontal and that it does not vary with depth: $\boldsymbol{u} = (u(x,y,t), v(x,y,t), 0)$. Then, using the horizontal components of (1) together with (2), we obtain the *shallow-water equations*:

$$\frac{\partial u}{\partial t} + u\frac{\partial u}{\partial x} + v\frac{\partial u}{\partial y} + g\frac{\partial \eta}{\partial x} = 0,$$
$$\frac{\partial v}{\partial t} + u\frac{\partial v}{\partial x} + v\frac{\partial v}{\partial y} + g\frac{\partial \eta}{\partial y} = 0,$$
$$\frac{\partial \eta}{\partial t} + \frac{\partial}{\partial x}(hu) + \frac{\partial}{\partial y}(hv) = 0.$$

This is a nonlinear hyperbolic system of PDEs. It can be rewritten in other ways; see, for example, how it is presented in the article on TSUNAMI MODELING [V.19 §2].

If the motions are small, a linearized version of the shallow-water equations can be derived. The result is a two-dimensional WAVE EQUATION [III.31] for $\eta(x,y,t)$:

$$\frac{\partial^2 \eta}{\partial t^2} = \frac{\partial}{\partial x}\left(gb\frac{\partial \eta}{\partial x}\right) + \frac{\partial}{\partial y}\left(gb\frac{\partial \eta}{\partial y}\right).$$

The basic nonlinear shallow-water equations can be augmented to include other effects. For example, if we want to model global phenomena, we should take account of the rotation of the Earth.

### Further Reading

Vallis, G. K. 2006. *Atmospheric and Oceanic Fluid Dynamics*. Cambridge: Cambridge University Press.

## III.28 The Sylvester and Lyapunov Equations

*Nicholas J. Higham*

The *Sylvester equation* (named after James Joseph Sylvester, who introduced it in 1884) is the linear matrix equation

$$AX + XB = C,$$

where $A \in \mathbb{C}^{m\times m}$, $B \in \mathbb{C}^{n\times n}$, and $C \in \mathbb{C}^{m\times n}$ are given and $X \in \mathbb{C}^{m\times n}$ is to be determined. It has a unique solution provided that $A$ and $-B$ have no eigenvalues in common. One interesting property is that if the integral $\int_0^\infty e^{At}Ce^{Bt}\,dt$ exists, then minus that integral is a solution of the Sylvester equation. Another is that the block upper triangular matrix $\left[\begin{smallmatrix} A & C \\ 0 & -B \end{smallmatrix}\right]$ can be reduced by a similarity transformation to block-diagonal form $\left[\begin{smallmatrix} A & 0 \\ 0 & -B \end{smallmatrix}\right]$ if and only if the Sylvester equation has a solution.

The Sylvester equation can be generalized by including coefficient matrices on both sides of $X$ and increasing the number of terms:

$$\sum_{i=1}^{k} A_i X B_i = C, \quad A_i \in \mathbb{C}^{m\times m}, \ B_i \in \mathbb{C}^{n\times n}.$$

When $m = n$, numerical methods are available that solve the system with $k = 2$ terms in $O(n^3)$ operations, but for $k > 2$ the best available methods require $O(n^6)$ operations. In applications in which the coefficient matrices are large, sparse, and possibly highly structured, much recent research has focused on computing inexpensive approximations to the solution using iterative methods, with good low-rank approximations being possible in some cases.

The *Lyapunov equation* is the special case of the Sylvester equation with $B = A^* \in \mathbb{C}^{n\times n}$:

$$AX + XA^* = C.$$

It is common in control and systems theory. Usually, $C$ is Hermitian, in which case the solution $X$ is Hermitian when the equation has a unique solution, which is the case when $\lambda_i + \bar{\lambda}_j \neq 0$ for all eigenvalues $\lambda_i$ and $\lambda_j$ of $A$.

A classic theorem says that for any given Hermitian positive-definite $C$ the Lyapunov equation has a unique Hermitian negative-definite solution if and only if all the eigenvalues of $A$ lie in the open left half-plane. The latter condition is equivalent to the asymptotic stability of the linear system of differential equations $\dot{x}(t) = Ax(t)$, and the Lyapunov equation is correspondingly also known as the *continuous-time Lyapunov equation*.

The *discrete-time Lyapunov equation* (also known as the *Stein equation*) is

$$X - A^*XA = C.$$

Given any Hermitian positive-definite matrix $C$, there is a unique Hermitian positive-definite solution $X$ if and only if all the eigenvalues of $A$ lie within the unit circle, and the latter condition is equivalent to the asymptotic stability of the discrete system $x_{k+1} = Ax_k$.

The Sylvester and Lyapunov equations both arise when Newton's method is used to solve ALGEBRAIC RICCATI EQUATIONS [III.25].

### Further Reading


Bhatia, R., and P. Rosenthal. 1997. How and why to solve the operator equation $AX - XB = Y$. *Bulletin of the London Mathematical Society* 29:1–21.

Higham, N. J. 2014. Sylvester's influence on applied mathematics. *Mathematics Today* 50(4):202–6.

Lancaster, P., and M. Tismenetsky. 1985. *The Theory of Matrices*, 2nd edn. London: Academic Press.


## III.29 The Thin-Film Equation
*Andrew J. Bernoff*

The thin-film equation (TFE),

$$\frac{\partial h}{\partial t} = -\frac{\partial}{\partial x}\left(Q(h)\frac{\partial^3 h}{\partial x^3}\right),$$

is a nonlinear fourth-order partial differential equation describing the flow of thin viscous films with strong surface tension. It has received a great deal of attention over the last two decades, both as a tractable model of thin fluid films and as a paradigm of challenges in the study of partial differential equations. Here, $h(x,t)$ represents the nonnegative height of the film's free surface (as a function of position $x$ and time $t$ (see figure 1)), and $Q(h)$ is a mobility function that is degenerate ($Q(0) = 0$) and increasing ($Q'(h) > 0$ for $h > 0$). The degeneracy allows compactly supported solutions (with $h = 0$ except on some compact subset of the real line), and the increasing mobility, which is necessary for the problem to be well-posed, leads to solutions that are infinitely smooth within their support. The most common choice for the mobility is $Q(h) = h^n$ for some $n > 0$.

### 1 Physical Origins

The TFE is an example of a *lubrication theory*, whereby a problem is considered in the asymptotic limit where horizontal variation occurs on a scale much longer than the film thickness. Lubrication theory applied to a thin viscous film on a no-slip substrate yields the TFE with $Q(h) = h^3$; the introduction of slip on the boundary yields $Q(h) = \beta h^2 + h^3$ for some $\beta > 0$.

### 2 Mathematical Structure

A mathematical theory for the existence of compactly supported weak solutions emerged in the 1990s. The TFE conserves mass,

$$M \equiv \int h \, \mathrm{d}x \Rightarrow \frac{\mathrm{d}M}{\mathrm{d}t} = 0,$$

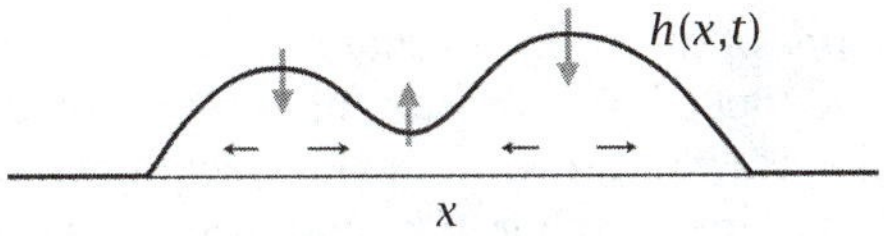


**Figure 1** A typical compactly supported configuration for a thin film of height $h(x,t)$. Variations in surface tension forces (proportional to $h_{xx}$ and indicated by the vertical gray arrow) create a pressure gradient that drives fluid motion (horizontal black arrows).

and dissipates surface energy (a lubrication approximation of the free surface's arc length),

$$E \equiv \frac{1}{2}\int (h_x)^2 \, \mathrm{d}x \Rightarrow \frac{\mathrm{d}E}{\mathrm{d}t} = -\int Q(h)(h_{xxx})^2 \, \mathrm{d}x \leqslant 0,$$

where integrals are over the support of $h(x,t)$.

It is believed that for $n \geqslant 3$ the support of the solution cannot increase; this means that the *contact line* (where the solution vanishes) cannot move for a no-slip boundary, which reflects the well-known contact line paradox in fluid mechanics (where we observe contact lines physically moving whenever a fluid wets a solid surface, but mathematically it is known that a contact line cannot move for the Navier–Stokes equation with a no-slip boundary condition). However, for $0 < n < 3$ one can find moving contact line solutions, which allow one to model spreading drops. Physically, this reflects the fact that the addition of slip to the boundary condition allows contact line motion, a modeling strategy that also resolves the contact line paradox for the full fluid equations.

The question of whether a film can rupture, leaving a dry spot (where $h = 0$), is only partially resolved; numerical results suggest that films may rupture for $n < \frac{3}{2}$ and will not rupture for greater $n$, but rigorous analytical results for this problem are lacking.

For $0 < n < 3$ one can find self-similar spreading drop solutions that are analogous to the Barenblatt solution of the porous media equation.

### 3 Higher Dimensions and Generalizations

The TFE generalizes naturally to $N$ dimensions; the film height $h(\boldsymbol{x},t)$, $\boldsymbol{x} \in \mathbb{R}^N$, satisfies

$$\frac{\partial h}{\partial t} = -\nabla \cdot [Q(h)\nabla(\Delta h)],$$

and the existence theory for weak solutions, energy dissipation, contact line motion, axisymmetric spreading droplets, and film rupture remains largely unchanged from the one-dimensional TFE.

Variants of the TFE incorporating gravity, inertia, substrate topography, variable surface tension due to heat or surfactants (known as Marangoni effects), van der Waals forces, evaporation, and many other physical effects have been highly successful in modeling thin films; the common theme in these models is that the additional terms are lower order, and the fourth-order thin-film term ensures well-posedness.

### Further Reading

Bertozzi, A. L. 1998. The mathematics of moving contact lines in thin liquid films. *Notices of the American Mathematical Society* 45(6):689–97.

Oron, A., S. H. Davis, and S. G. Bankoff. 1997. Long-scale evolution of thin liquid films. *Reviews of Modern Physics* 69:931–80.

## III.30 The Tricomi Equation

*Gui-Qiang G. Chen*

The *Tricomi equation* is a second-order partial differential equation of mixed elliptic-hyperbolic type. It takes the following form for an unknown function $u(x,y)$:

$$u_{xx} + xu_{yy} = 0.$$

The Tricomi equation was first analyzed in 1923 by Francesco Giacomo Tricomi when he was studying the well-posedness of a boundary-value problem. The equation is hyperbolic in the half-plane $x < 0$, elliptic in the half-plane $x > 0$, and degenerates on the line $x = 0$. Its characteristic equation is

$$\mathrm{d}y^2 + x\mathrm{d}x^2 = 0,$$

whose solutions are

$$y \pm \tfrac{2}{3}(-x)^{3/2} = C$$

for any constant $C$; the solutions are real for $x < 0$. The characteristics constitute two families of semicubical parabolas lying in the half-plane $x < 0$, with cusps on the line $x = 0$. This shows that the Tricomi equation is of hyperbolic degeneracy in the domain $x \leqslant 0$, for which the two characteristic families coincide, *perpendicularly* to the line $x = 0$.

For $\pm x > 0$, set $\tau = \frac{2}{3}(\pm x)^{3/2}$. The Tricomi equation then becomes the classical elliptic or hyperbolic *Euler–Poisson–Darboux equation*:

$$u_{\tau\tau} \pm u_{yy} + \frac{\beta}{\tau}u_\tau = 0.$$

The index $\beta = \frac{1}{3}$ determines the singularity of solutions near $\tau = 0$ (or, equivalently, near $x = 0$).

Many important problems in fluid mechanics and differential geometry can be reduced to corresponding problems for the Tricomi equation, particularly *transonic flow problems* and *isometric embedding problems*. The Tricomi equation is a prototype of the *generalized Tricomi equation*:

$$u_{xx} + K(x)u_{yy} = 0,$$

where $K(x)$ is a given function with $xK(x) > 0$ for $x \neq 0$. For a steady-state transonic flow in $\mathbb{R}^2$, $u(x,y)$ is the stream function of the flow, $x$ is a function of the velocity (which is positive at subsonic speeds and negative at supersonic speeds), and $y$ is the angle of inclination of the velocity. The solutions $u(x,y)$ also serve as entropy generators for entropy pairs of the potential flow system for the velocity. For the isometric embedding problem of two-dimensional Riemannian manifolds into $\mathbb{R}^3$, the function $K(x)$ has the same sign as the Gaussian curvature.

A closely related partial differential equation is the *Keldysh equation*:

$$xu_{xx} + u_{yy} = 0.$$

It is hyperbolic when $x < 0$, elliptic when $x > 0$, and degenerates on the line $x = 0$. Its characteristics are given by

$$y \pm \tfrac{1}{2}(-x)^{1/2} = C$$

for any constant $C$; the characteristics are real for $x < 0$. The two characteristic families are (quadratic) parabolas lying in the half-plane $x < 0$; they coincide *tangentially* to the degenerate line $x = 0$. This shows that the Keldysh equation is of parabolic degeneracy. For $\pm x > 0$, the Keldysh equation becomes the elliptic or hyperbolic *Euler–Poisson–Darboux equation* with index $\beta = -\frac{1}{4}$, by setting $\tau = \frac{1}{2}(\pm x)^{1/2}$. Many important problems in continuum mechanics can also be reduced to corresponding problems for the Keldysh equation, particularly shock reflection–diffraction problems in gas dynamics.

### Further Reading

Bitsadze, A. V. 1964. *Equations of the Mixed Type*. New York: Pergamon.

Chen, G.-Q., and M. Feldman. 2015. *Shock Reflection–Diffraction and von Neumann's Conjectures*. Annals of Mathematics Studies. Princeton, NJ: Princeton University Press.

Chen, G.-Q., M. Slemrod, and D. Wang. 2008. Vanishing viscosity method for transonic flow. *Archive for Rational Mechanics and Analysis* 189:159–88.

Chen, G.-Q., M. Slemrod, and D. Wang. 2010. Isometric immersions and compensated compactness. *Communications in Mathematical Physics* 294:411–37.

Han, Q., and J.-X. Hong. 2006. *Isometric Embedding of Riemannian Manifolds in Euclidean Spaces.* Providence, RI: American Mathematical Society.

Morawetz, C. M. 2004. Mixed equations and transonic flow. *Journal of Hyperbolic Differential Equations* 1:1–26.

## III.31 The Wave Equation

The one-dimensional *wave equation* is a PARTIAL DIFFERENTIAL EQUATION [IV.3] for $u(x,t)$,

$$\frac{\partial^2 u}{\partial x^2} = \frac{1}{c^2}\frac{\partial^2 u}{\partial t^2}, \tag{1}$$

where $c$ is a constant. It is classified as a linear second-order homogeneous hyperbolic equation. Unusually, its general solution (*d'Alembert's solution*) is known:

$$u(x,t) = f(x-ct) + g(x+ct). \tag{2}$$

Here, $f$ and $g$ are arbitrary twice-differentiable functions of one variable.

Small motions of a stretched string are governed by (1). Suppose that, in equilibrium, the string is along the $x$-axis. Then $u(x,t)$ gives the lateral displacement at position $x$ and time $t$. The solution $u(x,t) = f(x-ct)$ represents a wave traveling to the right ($x$ increasing) as $t$ increases: a photograph at $t = 0$ would show the string to have displacement $u(x,0) = f(x)$; a second photograph at time $1/c$ would show the same displacement but moved to the right by one unit, $u(x,1/c) = f(x-1)$. Equation (2) shows that the general solution consists of two waves, one moving to the right and one to the left, both moving at speed $c$.

# Part IV

# Areas of Applied Mathematics

## IV.1 Complex Analysis

*P. A. Martin*

### 1 Introduction

All calculus textbooks start with $f(x)$: $f$ is a function of one (real) variable $x$. Topics covered include limits, continuity, differentiation, and integration, with the associated notation, such as $\mathrm{d}f/\mathrm{d}x = f'(x)$ and $\int_a^b f(x)\,\mathrm{d}x$. It is also usual to include a discussion of infinite sequences and series. The rigorous treatment of all these topics constitutes *real analysis.*

Complex analysis starts with the following question. What happens if we replace $x$ by $z = x + \mathrm{i}y$, where $x$ and $y$ are two independent real variables and $\mathrm{i} = \sqrt{-1}$? Answering this question leads to rich new fields of mathematics; we shall be concerned with those parts that are used in applied mathematics.

Let us begin with basic terminology and concepts. We call $z = x + \mathrm{i}y$ a *complex variable.* The *imaginary unit* i should be treated as a symbol that obeys all the usual laws of algebra together with $\mathrm{i}^2 = -1$. We call $x = \operatorname{Re} z$ the *real part* of $z$ and $y = \operatorname{Im} z$ the *imaginary part* of $z$. We can identify $z = x+\mathrm{i}y$ with a point in the $xy$-plane (known as the *z-plane* or the *complex plane*).

The *complex conjugate* of $z$ is $\bar{z} = x - \mathrm{i}y$; complex conjugation is reflection in the $x$-axis. The *absolute value* (or *modulus* or *magnitude*) of $z$ is $|z| = +\sqrt{x^2 + y^2}$, the distance from $z$ to the origin. Given $w = u + \mathrm{i}v$, we define $z + w = (x + u) + \mathrm{i}(y + v)$. Addition of complex quantities is therefore equivalent to addition of two-dimensional vectors. For multiplication, $zw = xu - yv + \mathrm{i}(xv + yu)$. Putting $z = w$ shows that $\operatorname{Re} z^2 = x^2 - y^2 \neq x^2$ unless $z$ is real ($y = 0$). Also, $z\bar{z} = |z|^2$ and $z/w = z\bar{w}/|w|^2$ when $w \neq 0$.

Introducing plane polar coordinates, $r$ and $\theta$, we have $z = r\cos\theta + \mathrm{i}r\sin\theta = r\mathrm{e}^{\mathrm{i}\theta}$ by Euler's formula. Thus, $r = |z|$. The angle $\theta$ is called an *argument* of $z$, denoted by $\arg z$ or $\operatorname{ph} z$ (for *phase*). Notice that $\arg z$ is not unique, as we can always add any integer multiple of $2\pi$; this nonuniqueness is sometimes useful and sometimes a nuisance.

If we let $r \to \infty$, the point $z$ recedes to infinity. It is usual to state that there is a single "point at infinity," denoted by $z = \infty$, that is reached by letting $r \to \infty$ in any direction, $\theta$. Alternatively, we can state that the formula $z = 1/w$ takes the point $w = 0$ to the point $z = \infty$.

### 2 Functions

A function of a complex variable, $f(z)$, is a rule; given $z = x + \mathrm{i}y$ in some set (the *domain* of $f$), the rule provides a unique complex number denoted by $f(z) = u + \mathrm{i}v$, say, where $u = \operatorname{Re} f$ and $v = \operatorname{Im} f$ are real. We write $f(z) = u(x, y) + \mathrm{i}v(x, y)$ to emphasize the dependence on $x$ and $y$.

Simple examples of functions are $f(z) = z^2$ and $f(z) = \bar{z}$. Elementary functions are defined "naturally"; for example, $\mathrm{e}^z = \mathrm{e}^{x+\mathrm{i}y} = \mathrm{e}^x\mathrm{e}^{\mathrm{i}y}$ and $\cos z = \frac{1}{2}(\mathrm{e}^{\mathrm{i}z} + \mathrm{e}^{-\mathrm{i}z})$. For powers and logarithms, we have the formulas $z^\alpha = r^\alpha\mathrm{e}^{\mathrm{i}\alpha\theta}$ ($\alpha$ is real) and $\log z = \log(r\mathrm{e}^{\mathrm{i}\theta}) = \log r + \mathrm{i}\theta$. Strictly, these do not define functions because of the nonuniqueness of $\theta$; changing $\theta$ by $2\pi$ does not change $z$ but it does change the values of $z^\alpha$ (unless $\alpha$ is an integer) and $\log z$. One response to this phenomenon is to say that $\log z$, for example, is a multivalued "function"; increasing $\theta$ by $2\pi$ takes us onto another *branch* or *Riemann sheet* of $\log z$. However, in practice, it is usually better to introduce a *branch cut*, which, for $\log z$, is any line from $z = 0$ to $z = \infty$. This cut is regarded as an artificial barrier; we must not cross it. Its presence prevents us from increasing $\theta$ by $2\pi$. For example, we could restrict $\theta$ to satisfy $-\pi < \theta < \pi$ and put the cut on the negative $x$-axis. Once we have restricted $\theta$ to lie in some interval of length $2\pi$, $\log z$ and $z^\alpha$ become single-valued; they are now functions. We shall have more to say about branches in section 4.

There are many other ways to define functions. For example, $f(z) = \sum_{n=0}^{\infty} c_n z^n$ is a function provided the *power series* converges at $z$; characteristically, power series converge in disks, $|z| < R$, for some $R > 0$ ($R$ is the *radius of convergence*). The prototype power series is the *geometric series*; it converges inside the unit disk where its sum is known:

$$\sum_{n=0}^{\infty} z^n = \frac{1}{1-z}, \quad |z| < 1. \tag{1}$$

For another example, take

$$f(z) = \int_0^{\infty} g(t)\mathrm{e}^{-zt}\,\mathrm{d}t. \tag{2}$$

This defines the *Laplace transform* of $g$. Typically, such integrals converge for $\operatorname{Re} z > A$, where $A$ is a constant that depends on $g$. Another function defined by an integral is Euler's gamma function:

$$\Gamma(z) = \int_0^{\infty} t^{z-1}\mathrm{e}^{-t}\,\mathrm{d}t, \quad \operatorname{Re} z > 0. \tag{3}$$

Much is known about the properties of $\Gamma$. For example, $\Gamma(n) = (n-1)!$ when $n$ is any positive integer. There is more on this in section 13 below.

## 3 Analytic Functions

The notions of limit, continuity, and derivative are defined exactly as in real-variable calculus. In particular, the *derivative* of $f$ at $z$ is defined by

$$f'(z) = \frac{\mathrm{d}f}{\mathrm{d}z} = \lim_{h\to 0}\frac{f(z+h)-f(z)}{h}, \tag{4}$$

provided the limit exists. Here, $h$ is allowed to be complex; the point $z+h$ must be able to approach the point $z$ in any direction, and the limit must be the same. As a consequence, if $f(z) = u(x,y) + \mathrm{i}v(x,y)$ has a derivative, $f'(z)$, at $z$, then $u$ and $v$ satisfy the *Cauchy–Riemann equations*:

$$\frac{\partial u}{\partial x} = \frac{\partial v}{\partial y} \quad \text{and} \quad \frac{\partial u}{\partial y} = -\frac{\partial v}{\partial x}. \tag{5}$$

If these are not both satisfied, then $f'(z)$ does not exist. Two examples: $f(z) = \bar{z}$ is not differentiable for any $z$; and any real-valued function $f(z) = u(x,y)$ is not differentiable unless $u$ is a constant. If both Cauchy–Riemann equations (5) are satisfied and the partial derivatives in (5) are continuous functions, then $f'$ exists.

Using (5), if $f'$ exists, then

$$\begin{aligned} f'(z) &= \frac{\partial u}{\partial x} + \mathrm{i}\frac{\partial v}{\partial x} = \frac{\partial v}{\partial y} - \mathrm{i}\frac{\partial u}{\partial y} \\ &= \frac{\partial u}{\partial x} - \mathrm{i}\frac{\partial u}{\partial y} = \frac{\partial v}{\partial y} + \mathrm{i}\frac{\partial v}{\partial x}. \end{aligned}$$

The first equality follows by taking $h$ to be real in (4) and the second by taking $h$ to be purely imaginary. The four formulas for $f'$ show that we can calculate $f'$ from $\operatorname{Re} f$ or $\operatorname{Im} f$, or by differentiating with respect to $x$ or $y$.

Differentiability is a local property, defined at a point $z$. Usually, we are interested in functions that are differentiable at all points in their domains. Such functions are called *analytic* or *holomorphic*. Points at which a function is not differentiable are called *singularities*.

Derivatives of higher order (such as $f''(z)$) are defined in the natural way. One surprising fact is that a differentiable function can be differentiated any number of times; once differentiable implies infinitely differentiable (see (13) below for an indication of a proof). This result is certainly not true for real functions.

If we eliminate $v$ from (5), we obtain

$$\frac{\partial^2 u}{\partial x^2} + \frac{\partial^2 u}{\partial y^2} = 0. \tag{6}$$

Thus, the real part of an analytic function, $u(x,y)$, satisfies LAPLACE'S EQUATION [III.18] (6). The imaginary part, $v(x,y)$, satisfies the same partial differential equation (PDE). This reveals a close connection between analytic functions and solutions of one particular PDE. As Laplace's equation arises in the modeling of many physical phenomena, this connection has been exploited extensively.

## 4 More on Branches

Let us return to log, which we can define as a (single-valued) function by

$$\log z = \log r + \mathrm{i}\theta, \quad r > 0, \quad -\pi < \theta < \pi, \tag{7}$$

with $z = r\mathrm{e}^{\mathrm{i}\theta}$. There is a branch cut along the negative $x$-axis, with a branch point at $z = 0$ and a branch point at $z = \infty$. Thus, our domain of definition for $\log z$ is the cut plane, i.e., the whole complex plane with the cut removed. Then, $\log z$ is analytic; it is differentiable at all points in its domain of definition. Moreover, $(\mathrm{d}/\mathrm{d}z)\log z = z^{-1}$.

According to (7), $\log z$ is not defined *on* the negative $x$-axis. Some authors regard this as unacceptable, and so they replace the (open) interval $-\pi < \theta < \pi$ in (7) by $-\pi < \theta \leqslant \pi$ or $-\pi \leqslant \theta < \pi$. The first choice gives, for example, $\log(-1) = \mathrm{i}\pi$ and the second gives $\log(-1) = -\mathrm{i}\pi$. Either choice enlarges the domain of definition to the whole plane with $z = 0$ removed. However, we lose analyticity; $\log z$ is not differentiable on the line $\theta = \pi$ (first choice) because points on that line are not accessible in all directions (as they must be if one wants to compute limits, as in the definition of derivative) without leaving the domain of definition.

For some applications this may be acceptable, but, in practice, it is usual to simply move the cut. We can therefore replace $-\pi < \theta < \pi$ in (7) by another open interval, $\theta_0 < \theta < 2\pi + \theta_0$, implying a cut along the straight half-line $\theta = \theta_0$, $r \geqslant 0$. (In fact, the cut need not be straight; any line connecting the branch point at $z = 0$ to $z = \infty$ may be used.) Then $\log z$ is analytic in a new cut plane.

Once we define a function such as $\log z$ or $z^{1/2}$ with a specified range for $\theta$, we can say that we have defined a *principal value* of that function. Certain choices (such as $-\pi < \theta < \pi$ or $-\pi < \theta \leqslant \pi$) are common, but the reader should not overlook the option of moving cuts when it is convenient to do so.

There is another consequence of insisting on having (single-valued) functions: some standard identities, such as $\log(z^2) = 2\log z$, may no longer hold. For example, with $z = -1 + \mathrm{i}$ and the definition (7), we find $\log(z^2) = \log 2 - \frac{1}{2}\mathrm{i}\pi$ but $2\log z = \log 2 + \frac{3}{2}\mathrm{i}\pi$.

Summarizing, functions with branches are very common (for another example, see the article on THE LAMBERT $W$-FUNCTION [III.17]) but their presence often leads to complications, subtle difficulties, and calculational errors; care is always required.

## 5 Infinite Series

A power series about the point $z_0$ has the form

$$\sum_{n=0}^{\infty} c_n (z - z_0)^n, \tag{8}$$

where the coefficients $c_n$ are complex numbers. The series (8) converges for $|z - z_0| < R$ and diverges for $|z - z_0| > R$, where the radius of convergence, $R$, may be finite or infinite. (It may happen that (8) converges at $z = z_0$ only, with sum $c_0$.) When the series does converge, we denote its sum by $S(z)$. For an example, see the geometric series (1).

The sum $S(z)$ is analytic for $|z - z_0| < R$; power series define analytic functions.

Now we turn this around. We take an analytic function, $f(z)$, and we try to write it as a power series. Doing this is familiar from calculus, and the result is Taylor's theorem:

$$f(z) = \sum_{n=0}^{\infty} \frac{f^{(n)}(z_0)}{n!}(z - z_0)^n,$$

where $f^{(n)}$ is the $n$th derivative of $f$. The series is known as the *Taylor expansion* of $f(z)$ about $z_0$. It converges for $|z - z_0| < R$, where $R$ is the distance from $z_0$ to the nearest singularity of $f(z)$. A Taylor expansion about the origin ($z_0 = 0$) is known as a *Maclaurin expansion*. All these expansions are the same as those occurring in the calculus of functions of one real variable. For example, (1) gives the Maclaurin expansion of $1/(1 - z)$. Another familiar Maclaurin expansion is $\mathrm{e}^z = \sum_{n=0}^{\infty} z^n/n!$, which is convergent for all $z$.

A generalization of Taylor's theorem, Laurent's theorem, will be given in section 8.

Not all infinite series are power series. A famous series is the RIEMANN ZETA FUNCTION [IV.7 §4], which is defined by

$$\zeta(z) = \sum_{n=1}^{\infty} \frac{1}{n^z} \quad \text{for } \operatorname{Re} z > 1, \tag{9}$$

which is intimately connected with the distribution of the prime numbers.

It is possible to develop the theory of analytic functions by starting with power series; this approach, which goes back to Weierstrass, has a constructive flavor. We started with the notion of differentiability; this approach, which goes back to Riemann and Cauchy, is closer to real-variable calculus. The two approaches are equivalent; power series define analytic functions and analytic functions have power-series expansions.

## 6 Contour Integrals

In the calculus of functions of two real variables $x$ and $y$, double integrals over regions of the $xy$-plane and line integrals along curves in the $xy$-plane are defined. In complex analysis, we are mainly concerned with integrals along curves in the $z$-plane. They are defined similarly to line integrals. Thus, suppose that points on a curve $C$ are located by a parametrization,

$$C\colon z(t) = x(t) + \mathrm{i}y(t), \quad a \leqslant t \leqslant b,$$

where $a$ and $b$ are constants and $x(t)$ and $y(t)$ are real functions of the real variable $t$. As $t$ increases from $a$ to $b$, $z(t)$ moves from $z(a)$ to $z(b)$; the parametrization induces a direction or *orientation* on $C$. The curve $C$ is *smooth* if $x'(t)$ and $y'(t)$ exist and are continuous. Then, if $f(z)$ is defined for all points $z$ on a smooth curve $C$,

$$\int_C f(z)\,\mathrm{d}z = \int_a^b f(z(t))z'(t)\,\mathrm{d}t, \tag{10}$$

where $z'(t) = x'(t) + \mathrm{i}y'(t)$. In (10), the right-hand side defines the expression on the left-hand side as an integration with respect to the parameter $t$. More generally, suppose that $C$ is a *contour*, defined as a continuous curve made from smooth pieces joined at corners.

Then, to define $\int_C f\,dz$, we parametrize each smooth piece separately, and then sum the contributions from each piece, ensuring that the parametrizations are such that $z$ moves continuously along $C$.

If $C_o$ is the same curve as $C$ but traversed in the opposite direction (from $z(b)$ to $z(a)$), then $\int_{C_o} f(z)\,dz = -\int_C f(z)\,dz$; changing the direction changes the sign.

## 7 Cauchy's Theorem

A contour $C$ is *closed* if $z(a) = z(b)$, and it is *simple* if it has no self-intersections. Cauchy's theorem can be stated as follows. Suppose that $f(z)$ is analytic inside a simple closed contour $C$ and continuous on $C$. Then

$$\int_C f(z)\,dz = 0. \tag{11}$$

It is worth emphasizing the hypotheses. First, we do not need to know anything about $f(z)$ outside $C$; Cauchy assumed stronger conditions, but these were later weakened by Goursat. Second, $C$ is a contour, so corners are allowed. Third, by requiring that "$f$ is analytic inside $C$," we mean that $f(z)$ must be differentiable at all points $z$ inside $C$; singularities (including branch points) are not allowed (although they may be present outside $C$).

There are many consequences of Cauchy's theorem. One is known as *deforming the contour*. Suppose that $C_1$ and $C_2$ are simple closed contours, both traversed in the same direction, with $C_1$ enclosed by $C_2$. Suppose that $f(z)$ is analytic in the region between $C_1$ and $C_2$ and that it is continuous on $C_1$ and $C_2$. (Note that $f$ may have singularities inside the smaller contour $C_1$ or outside the larger contour $C_2$.) Then, $\int_{C_1} f(z)\,dz = \int_{C_2} f(z)\,dz$; one contour can be deformed into another without changing the value of the integral, provided the integrand is analytic between the contours. The same result is true when $C_1$ and $C_2$ are two contours with the same endpoints, provided $f$ is analytic between $C_1$ and $C_2$. These results are useful because they may allow us to deform a complicated contour into a simpler contour (such as a circle or a straight line).

Another consequence of Cauchy's theorem is the *Cauchy integral formula*. Under the same conditions, we have

$$f(z_0) = \frac{1}{2\pi i}\int_C \frac{f(z)}{z - z_0}\,dz, \tag{12}$$

where $z_0$ is an arbitrary point inside $C$, and $C$ is traversed counterclockwise. This shows that we can recover the values of an analytic function inside $C$ from its values on $C$.

More generally, and again under the same conditions, we have

$$f^{(n)}(z_0) = \frac{n!}{2\pi i}\int_C \frac{f(z)}{(z - z_0)^{n+1}}\,dz. \tag{13}$$

Formally, this can be seen as the $n$th derivative of the Cauchy integral formula (12), but it is deeper; it can be used to prove the existence of $f^{(n)}$, for $n = 2, 3, \dots$, assuming that $f'$ exists. This is done using an inductive argument. We have (compare with (4))

$$f^{(n+1)}(z_0) = \lim_{h\to 0} \frac{f^{(n)}(z_0 + h) - f^{(n)}(z_0)}{h},$$

provided the limit exists. Now, on the right-hand side, use (13) twice; the limit can then be taken.

Formula (13) with $n = 1$ can be used to prove *Liouville's theorem*. Suppose that $f(z)$ is analytic everywhere in the $z$-plane (that is, there are no singularities); such a function is called *entire*. Suppose further that $|f(z)| < M$ for some constant $M$ and for all $z$; we say that $f$ is *bounded*. Liouville's theorem states that a bounded entire function is necessarily constant. In other words, (nonconstant) entire functions must be large somewhere in the complex plane. For example,

$$\begin{aligned}|\cos z|^2 &= \cos z\,\overline{\cos z} = \tfrac{1}{4}(e^{iz} + e^{-iz})(e^{i\bar{z}} + e^{-i\bar{z}}) \\ &= \tfrac{1}{4}(e^{2ix} + e^{-2ix} + e^{2y} + e^{-2y}) \\ &= \tfrac{1}{2}(\cos 2x + \cosh 2y) = \cos^2 x + \cosh^2 y - 1\end{aligned}$$

using $z + \bar{z} = 2x$ and $z - \bar{z} = 2iy$. Thus, $|\cos z|$ grows rapidly as we move away from the real axis (where $y = 0$ and $\cosh 0 = 1$).

## 8 Laurent's Theorem

Suppose that $f(z)$ is analytic inside an annulus, $a < |z - z_0| < b$, centered at $z_0$. We say nothing about $f(z)$ when $z$ is in the "hole" of radius $a$ ($|z - z_0| < a$) or when $z$ is outside the annulus ($|z - z_0| > b$). We then have the Laurent expansion

$$f(z) = \sum_{n=-\infty}^{\infty} c_n (z - z_0)^n \tag{14}$$

for all $z$ in the annulus, where the coefficients are given by contour integrals,

$$c_n = \frac{1}{2\pi i}\int_C \frac{f(z)}{(z - z_0)^{n+1}}\,dz, \tag{15}$$

in which $C$ is a simple closed contour in the annulus that encircles the hole (once) in the counterclockwise direction. Note that the sum in (14) is over all $n$. It is often convenient to split the sum, giving, for all $z$ in the annulus,

$$f(z) = \sum_{n=0}^{\infty} a_n (z - z_0)^n + \sum_{n=1}^{\infty} \frac{b_n}{(z - z_0)^n}, \qquad (16)$$

where $a_n = c_n$ for $n = 0, 1, 2, \ldots$ and $b_n = c_{-n}$ for $n = 1, 2, \ldots$. In particular,

$$b_1 = \frac{1}{2\pi i} \int_C f(z)\,dz. \qquad (17)$$

Suppose that $f(z)$ is also analytic in the hole, so that $f(z)$ is analytic in the disk $|z - z_0| < b$. Then $b_n = 0$ $(n = 1, 2, \ldots)$ by Cauchy's theorem and $a_n = f^{(n)}(z_0)/n!$ by (13); Taylor's theorem is recovered. Note that, in general, when $f$ does have singularities in the hole, we cannot use (13) to evaluate the contour integrals defining $a_n$.

## 9 Singularities

A singularity is a point at which a function is not differentiable. There are several kinds of singularities. A point $z_0$ is called an *isolated singularity* if there is an annulus $0 < |z - z_0| < b$ (a "punctured disk") in which there are no other singularities. In this annulus, we have a Laurent expansion, (16). The first part (the sum over $a_n$) is a power series, and so it defines an analytic function on the whole disk. The singular behavior resides in the second sum (over $b_n$); it is called the *principal part*, $P(z)$. In practice, $P(z)$ often has a finite number of terms,

$$P(z) = \frac{b_1}{z - z_0} + \frac{b_2}{(z - z_0)^2} + \cdots + \frac{b_m}{(z - z_0)^m}, \qquad (18)$$

with $b_n = 0$ for all $n > m$ and $b_m \neq 0$. In this situation, we say that $f$ has a *pole of order* $m$ at $z_0$. A pole of order 1 is called a *simple pole* and a pole of order 2 is called a *double pole*. For example, all the following have simple poles at $z = 0$:

$$\frac{1}{z}, \quad \frac{1+z}{z}, \quad \frac{e^z}{z}, \quad \frac{\sin z}{z^2}, \quad \frac{\pi}{\sin \pi z}; \qquad (19)$$

the last in this list also has simple poles at $z = \pm 1, \pm 2, \ldots$. All the following have double poles at $z = 0$:

$$\frac{1}{z^2}, \quad \frac{1+z}{z^2}, \quad \frac{1}{z \sin z}, \quad \frac{\cos z}{z^2}, \quad \frac{1}{\sin^2 \pi z}. \qquad (20)$$

If the principal part of the Laurent expansion contains an infinite number of nonzero terms, $z_0$ is called an *isolated essential singularity*. For example, $e^{1/z}$ has such a singularity at $z = 0$.

The coefficient $b_1$ (given by (17)) will play a special role later; it is called the *residue* of $f$ at the isolated singularity, $z_0$, and it is denoted by $\operatorname{Res}[f; z_0]$.

There are also nonisolated singularities. The most common of these occur at branch points. For example, $f(z) = z^{1/2}$ has a branch-point singularity at $z = 0$. Note that any disk centered at $z = 0$ will include a piece of the branch cut emanating from the branch point; $f$ is discontinuous across the cut, so it is certainly not differentiable there, implying that $z = 0$ is not an isolated singularity.

## 10 Cauchy's Residue Theorem

If we want to evaluate $I = \int_C f(z)\,dz$, the basic method is to parametrize each smooth piece of $C$ and then use the definition (10). In principle, this works for any $f$ and for any $C$. However, in practice, $C$ is often closed and $f$ is analytic apart from some singularities. In these happy situations, we can calculate $I$ efficiently by using Cauchy's residue theorem. Thus, suppose that $f(z)$ is analytic inside the simple closed contour $C$ (and continuous on $C$) apart from isolated singularities at $z_j$, $j = 1, 2, \ldots, n$. (Note that $f$ may have other singularities, including branch points, outside $C$, but these are of no interest here.) Then

$$\int_C f(z)\,dz = 2\pi i \sum_{j=1}^{n} \operatorname{Res}[f; z_j], \qquad (21)$$

where $C$ is traversed counterclockwise. This important result is remembered as "$2\pi i$ times the sum of the residues at the isolated singularities inside the contour." If there are no singularities inside, we recover Cauchy's theorem (11).

To prove the theorem, we start with the case $n = 1$. There is a Laurent expansion about the sole singularity $z_1$, convergent in a punctured disk $0 < |z - z_1| < b$. We deform $C$ into a smaller contour (enclosing $z_1$) that is inside the disk. Then we use (17). In the general case, we deform $C$ and "pinch off," giving a sum of $n$ contour integrals, each one containing one singularity.

To understand the pinching-off process, suppose that $n = 2$ and deform $C$ into a dumbbell-shaped contour, with two circles joined by two parallel straight lines, $L_1$ and $L_2$, traversed in opposite directions. The contributions from $L_1$ and $L_2$ cancel in the limit as the lines go together, leaving the contributions from disjoint closed contours around each singularity. This process is readily extended to any (finite) number of isolated singularities.

In order to exploit the residue theorem, we need efficient methods for computing residues. Recall that $\operatorname{Res}[f; z_0]$ is the coefficient $b_1$ in the Laurent expansion about $z_0$ (see (18)). For simple poles, $b_1$ is the only nontrivial coefficient in the principal part; thus, at a simple pole $z_0$,

$$\operatorname{Res}[f; z_0] = \lim_{z\to z_0} \{(z - z_0)f(z)\}. \tag{22}$$

Often, simple poles are characterized by writing $f(z) = p(z)/q(z)$ with $q(z_0) = 0$, $p(z_0) \neq 0$, and $q'(z_0) \neq 0$. Then

$$\operatorname{Res}[f; z_0] = p(z_0)/q'(z_0). \tag{23}$$

For a pole of order $m$, we can use

$$\operatorname{Res}[f; z_0] = \frac{1}{m!} \lim_{z\to z_0} \frac{\mathrm{d}^{m-1}}{\mathrm{d}z^{m-1}}\{(z - z_0)^m f(z)\}.$$

However, it is sometimes quicker to construct the Laurent expansion directly and then to pick off $b_1$, the coefficient of $1/(z - z_0)$. Thus, almost by inspection, all the simple-pole examples in (19) have $\operatorname{Res}[f; 0] = 1$. The five double-pole examples in (20) have $\operatorname{Res}[f; 0] = 0$, 1, 0, 0, and 0, respectively.

## 11 Evaluation of Integrals

Cauchy's residue theorem gives a powerful method for evaluating integrals. We give a few examples.

Let $C$ be a circle of radius $a$ centered at the origin and traversed counterclockwise. The function $\mathrm{e}^z/z$ has a simple pole at $z = 0$ with residue $= 1$. Hence

$$\int_C \frac{\mathrm{e}^z}{z}\,\mathrm{d}z = 2\pi\mathrm{i}. \tag{24}$$

This result could have been obtained from the Cauchy integral formula (12) with $f(z) = \mathrm{e}^z$ and $z_0 = 0$. Note also that the value of the integral does not depend on $a$; this is not surprising because we know that we can deform $C$ into a concentric circular contour (for example) without changing the value of the integral.

Now, starting from the result (24), suppose we parametrize $C$ and then use (10); a suitable parametrization is $z(t) = a\mathrm{e}^{\mathrm{i}t}$, $-\pi \leqslant t \leqslant \pi$. As $z'(t) = \mathrm{i}a\mathrm{e}^{\mathrm{i}t} = \mathrm{i}z(t)$, we obtain

$$\int_{-\pi}^{\pi} \exp(a\mathrm{e}^{\mathrm{i}t})\,\mathrm{d}t = 2\pi.$$

By Euler's formula, $\mathrm{e}^{\mathrm{i}\theta} = \cos\theta + \mathrm{i}\sin\theta$, the integrand is $\mathrm{e}^{a\cos t}\cos(a\sin t) + \mathrm{i}\mathrm{e}^{a\cos t}\sin(a\sin t)$. The second term is an odd function of $t$ and so it integrates to zero, leaving

$$\int_0^{\pi} \mathrm{e}^{a\cos t}\cos(a\sin t)\,\mathrm{d}t = \pi. \tag{25}$$

Thus, from the known value of a fairly simple contour integral, (24), we obtained the value of a complicated real integral. Notice that the formula (25) was derived by assuming that the parameter $a$ is real and positive. In fact, it is valid for arbitrary complex $a$; this is an example of analytic continuation (see section 13).

We now consider doing the opposite: evaluating integrals by converting them into contour integrals, followed by use of the residue theorem.

For trigonometric integrals such as

$$I_1 = \int_0^{2\pi} \frac{\mathrm{d}\theta}{5 + 4\cos\theta},$$

the substitution $z = \mathrm{e}^{\mathrm{i}\theta}$ will convert $I_1$ into a contour integral around the unit circle, $|z| = 1$. Using $\mathrm{d}\theta/\mathrm{d}z = 1/(\mathrm{i}z)$ and $\cos\theta = \frac{1}{2}(z + z^{-1})$, we obtain

$$I_1 = \frac{1}{2\mathrm{i}} \int_{|z|=1} \frac{\mathrm{d}z}{(z + 2)(z + \frac{1}{2})}.$$

The integrand is analytic apart from simple poles at $z = -2$ and $z = -\frac{1}{2}$. The latter is inside the contour; its residue is $\frac{2}{3}$ (use (22)). Cauchy's residue theorem (21) therefore gives $I_1 = \frac{2}{3}\pi$.

The method just described requires that the range of integration for $\theta$ have length $2\pi$ and that the resulting integrand have only isolated singularities (not branch points) inside $|z| = 1$.

For a second example, consider

$$I_2 = \int_{-\infty}^{\infty} f(x)\,\mathrm{d}x \quad \text{with } f(x) = \frac{1}{x^4 + 1}.$$

In order to use the residue theorem, we need a closed contour $C$, so we try $\int_C f(z)\,\mathrm{d}z$ with $C$ consisting of a piece of the real axis from $z = -R$ to $z = R$ and a semicircle $C_R$ in the upper half-plane of radius $R$ and centered at $z = 0$. Then

$$\int_{-R}^{R} f(x)\,\mathrm{d}x + \int_{C_R} f(z)\,\mathrm{d}z = 2\pi\mathrm{i} \times \begin{cases}\text{residues} \\ \text{at poles} \\ \text{inside } C.\end{cases} \tag{26}$$

After calculation of the residues, we let $R \to \infty$, so that the first integral $\to I_2$. We will see in a moment that the second integral $\to 0$ as $R \to \infty$.

Now, $z^4 + 1 = 0$ at $z = z_n = \exp(\mathrm{i}(2n + 1)\pi/4)$, $n = 0, 1, 2, 3$. These are simple poles of $f(z)$ with residue $1/(4z_n^3)$ (use (23) with $p = 1$, $q = z^4 + 1$). The poles $z_0$ and $z_1$ are in the upper half-plane. Hence the right-hand side of (26) is $\pi/\sqrt{2}$, and this is $I_2$.

For $z \in C_R$ we parametrize using $z(t) = R\mathrm{e}^{\mathrm{i}t}$, $0 \leqslant t \leqslant \pi$. We see that $f(z)$ decays as $R^{-4}$, whereas the length of $C_R$, $\pi R$, increases; overall, $\int_{C_R} f\,\mathrm{d}z$ decays as $R^{-3}$. This rough argument can be made precise.

If we replace $f(z)$ by $\mathrm{e}^{\mathrm{i}kz}f(z)$, we can evaluate Fourier transforms such as

$$I_3 = \int_{-\infty}^{\infty} \mathrm{e}^{\mathrm{i}kx} f(x)\,\mathrm{d}x \quad \text{with } f(x) = \frac{1}{x^2 + 1},$$

where $k$ is a real parameter. However, some care is needed; as $e^{ikz} = e^{ikx}e^{-ky}$, we have exponential decay as $y \to \infty$ when $k > 0$ but exponential growth when $k < 0$. Therefore, we use $C_R$ when $k > 0$, but we close using a semicircle in the lower half-plane when $k < 0$. We find that $I_3 = \pi e^{-|k|}$.

Laplace transforms (2) can be inverted using

$$g(t) = \frac{1}{2\pi i}\int_{c-i\infty}^{c+i\infty} f(z)\, e^{zt}\, dz.$$

The contour (called the *Bromwich contour*) is parallel to the $y$-axis in the $z$-plane. The constant $c$ is chosen so that all the singularities of $f(z)$ are to the left of the contour. If $f(z)$ has poles only, the integral can be evaluated by closing the contour using a large semicircle on the left.

There are many other applications of contour-integral methods to the evaluation of integrals. They can also be used to find the sums of infinite series. Integrands containing branch points can also be considered. In all cases, one may need some ingenuity in selecting an appropriate closed contour and/or the function $f(z)$.

## 12 Conformal Mapping

Suppose that $f(z)$ is analytic for $z \in D$. We can regard $f$ as a mapping, taking points $z = x + iy$ to points $w = f(z) = u + iv$; denote the set of such points in the $uv$-plane for all points $z \in D$ by $R$. Given $f$ and $D$, we can determine $R$. More interestingly, given the regions $D$ and $R$, can we find an analytic function $f$ that maps $D$ onto $R$? The *Riemann mapping theorem* asserts that any simply connected region $D$ can be mapped to the unit disk $|w| < 1$. (A region bounded by a simple closed curve is simply connected if it does not contain any holes.) The analytic function $f$ effecting the mapping is called a CONFORMAL MAPPING [II.5]; two small lines meeting at a point $z_0 \in D$ will be mapped into two small lines meeting at a point $w_0 = f(z_0) \in R$, and the angles between the two pairs of lines will be equal. The conformality property holds for all $z_0 \in D$ except for *critical points* (where $f'(z_0) = 0$ or $\infty$). Many conformal mappings are known (there are dictionaries of them), but constructing them for regions $D$ with complicated shapes or holes remains a challenge. Once a conformal mapping is available, it can be used to solve boundary-value problems for Laplace's equation (6), for example.

## 13 Analytic Continuation

Return to the geometric series (1). Denote the infinite series on the left-hand side by $f(z)$, with domain $D$ ($|z| < 1$). Denote the sum on the right-hand side by $g(z) = 1/(1 - z)$, with domain $D'$ ($z \neq 1$). We observe that $f(z)$ is analytic in $D$ whereas $g(z)$ is analytic in the larger region $D'$. As $f(z) = g(z)$ for $z \in D$, we say that $g$ is the *analytic continuation* of $f$ into $D'$. In practice, we do not usually distinguish between $f$ and $g$, we just say that $g(z)$ is analytic for $z \in D'$ and that it can be defined for $z \in D \subset D'$ using $f(z)$. This point of view is surprisingly powerful.

There are several aspects to this, and it raises several questions. To begin with, suppose we are given $f$ and $D$ and we want to find $g$ outside $D$. There are analytical and numerical methods available for doing so. For example, we could use a chain of overlapping disks with a Taylor expansion about each center. The result will be locally unique (each step in the chain gives a unique result) but, if $g$ has a branch point, we could step onto another branch and thus lose global uniqueness.

Often, we do not know $D'$; typical analytic continuations will have singularities. For example, the gamma function, $\Gamma(z)$, is defined by the integral (3) for $\operatorname{Re} z > 0$; in this half-plane, $\Gamma$ is analytic. If we continue $\Gamma(z)$ into $\operatorname{Re} z \leqslant 0$, we find that there are simple poles at $z = -N$, $N = 0, 1, 2, \dots$ (so that $D'$ is the whole complex plane with the points $z = -N$ removed). Explicitly, we can use *Hankel's loop integral*:

$$\Gamma(z) = \frac{1}{2i\sin(\pi z)}\int_C t^{z-1}e^t\, dt.$$

This is a contour integral in the complex $t$-plane. There is a cut along the negative real-$t$ axis. The branch of $t^z$ is chosen so that $t^z = e^{z\log t}$ when $t$ is real and positive. The contour starts at $\operatorname{Re} t = -\infty$, below the cut, goes once around $t = 0$, and then returns to $\operatorname{Re} t = -\infty$ above the cut.

There are also loop integrals for the Riemann zeta function, $\zeta(z)$, defined initially for $\operatorname{Re} z > 1$ by the series (9). Thus, it turns out that $\zeta(z)$ can be analytically continued into the whole $z$-plane apart from a simple pole at $z = 1$.

## 14 Differential Equations

We usually think of a differential equation as being something to be solved for a real function of a real variable. However, it can be advantageous to "complexify" the problem. One good reason is that we may be able to construct solutions using a power-series expansion (8), and we know that the convergence of such a series is governed by singularity locations. (More generally, we could use the "method of Frobenius.") For example, one solution of *Airy's equation*, $w''(z) = zw(z)$, is

$$w(z) = 1 + \frac{1}{3!}z^3 + \frac{1 \cdot 4}{6!}z^6 + \frac{1 \cdot 4 \cdot 7}{9!}z^9 + \cdots,$$

which defines an entire function of $z$.

We may be able to write solutions as contour integrals, which then offers possibilities for further analysis. For example, solutions of Airy's equation can be written (or sought) in the form

$$w(z) = \int_C e^{-zt+t^3/3}\,dt,$$

where $C$ is a carefully chosen contour in the complex $t$-plane.

The study of linear differential equations is a well-established branch of complex analysis, especially in the context of the classical SPECIAL FUNCTIONS [IV.7] (e.g., Bessel functions and hypergeometric functions). Nonlinear differential equations and their associated special functions are also of interest. For example, there are the six PAINLEVÉ EQUATIONS [III.24], the simplest being $w''(z) = 6w^2 + z$; their solutions, known as *Painlevé transcendents*, have a variety of physical applications, but their properties are not well understood.

## 15 Cauchy Integrals

Let $C$ be a simple closed smooth contour. Denote the interior of $C$ by $D_+$ and the exterior by $D_-$. Define a function $F(z)$ by the *Cauchy integral*

$$F(z) = \frac{1}{2\pi i}\int_C \frac{g(\tau)}{\tau - z}\,d\tau, \quad z \notin C, \tag{27}$$

where $g(t)$ is defined for $t \in C$. For example, if $g(t) = 1$, $t \in C$, then

$$\frac{1}{2\pi i}\int_C \frac{d\tau}{\tau - z} = \begin{cases} 1, & z \in D_+, \\ 0, & z \in D_-. \end{cases} \tag{28}$$

The integral in (27) is similar to that which appears in Cauchy's integral formula (12), except we are not given any information about $g(t)$ when $t \notin C$. Nevertheless, under mild conditions on $g$, $F(z)$ is analytic for $z \in D_+ \cup D_-$, and $F(z) \to 0$ as $z \to \infty$. What are the values of $F$ on $C$? The example (28) suggests that we should expect $F(z)$ to be discontinuous as $z$ crosses $C$. Therefore, we consider the limits of $F(z)$ as $z$ approaches $C$ (if they exist), and write

$$F_\pm(t) = \lim_{z \to t} F(z) \quad \text{with } z \in D_\pm \text{ and } t \in C. \tag{29}$$

For the example (28), $F_+(t) = 1$ and $F_-(t) = 0$.

Notice that we cannot simply put $z = t \in C$ on the right-hand side of (27); the resulting integral diverges. However, if $g$ is differentiable at $t$ (in fact, Hölder continuity is sufficient), we can define the *Cauchy principal-value integral*

$$⨍_C \frac{g(\tau)}{\tau - t}\,d\tau = \lim_{\varepsilon \to 0}\int_{C_\varepsilon} \frac{g(\tau)}{\tau - t}\,d\tau, \quad t \in C,$$

where $C_\varepsilon$ is obtained from $C$ as follows: draw a little circle of radius $\varepsilon$, centered at $t \in C$, and then remove the piece of $C$ inside the circle.

Using this definition, define

$$F(t) = \frac{1}{2\pi i} ⨍_C \frac{g(\tau)}{\tau - t}\,d\tau, \quad t \in C.$$

This function is related to $F_\pm(t)$, defined by (29), by the *Sokhotski–Plemelj formula*:

$$F_\pm(t) = \pm\tfrac{1}{2}g(t) + F(t), \quad t \in C. \tag{30}$$

This describes the "jump behavior" of the Cauchy integral $F(z)$ as $z$ crosses $C$. In particular,

$$F_+(t) - F_-(t) = g(t), \quad t \in C. \tag{31}$$

One elegant consequence of (30) is that the solution, $w$, of the *singular integral equation*

$$\frac{1}{\pi i} ⨍_C \frac{w(\tau)}{\tau - t}\,d\tau = g(t), \quad t \in C,$$

is given by the formula

$$w(t) = \frac{1}{\pi i} ⨍_C \frac{g(\tau)}{\tau - t}\,d\tau, \quad t \in C.$$

## 16 The Riemann–Hilbert Problem

Let $D_\pm$ and $C$ be as in section 15. Suppose that two functions, $G(t)$ and $g(t)$, are given for $t \in C$. Then, the basic *Riemann–Hilbert problem* is to find two functions $\Phi_+(z)$ and $\Phi_-(z)$, with $\Phi_\pm$ analytic in $D_\pm$, that satisfy

$$\Phi_+(t) = G(t)\Phi_-(t) + g(t), \quad t \in C, \tag{32}$$

where $\Phi_+(t)$ and $\Phi_-(t)$ are defined as in (29); conditions on $\Phi_-(z)$ as $z \to \infty$ are usually imposed too. (There is a variant where $C$ is not closed; in this case, the behavior near the endpoints of $C$ plays a major role.)

When $G \equiv 1$, we can solve (32) using a Cauchy integral and (31). When $G \not\equiv 1$, we start with the following homogeneous problem ($g \equiv 0$).

Find functions $K_+$ and $K_-$, with $K_\pm$ analytic in $D_\pm$, that satisfy

$$K_+(t) = G(t)K_-(t), \quad t \in C. \tag{33}$$

Suppose we can find such functions and that they do not vanish. Then, eliminating $G$ from (32) gives

$$\frac{\Phi_+(t)}{K_+(t)} - \frac{\Phi_-(t)}{K_-(t)} = \frac{g(t)}{K_+(t)}, \quad t \in C,$$

which, again, we can solve using a Cauchy integral and (31).

The problem of finding $K_\pm$ is more delicate. At first sight, we could take the logarithm of (33), giving $\log K_+ - \log K_- = \log G$. This looks similar to (31), but it usually happens that $\log G(t)$ is not continuous for all $t \in C$, which means that we cannot use (30). However, this difficulty can be overcome.

The problem of finding $K_\pm$ such that (33) is satisfied is also the key step in the *Wiener–Hopf technique* (a method for solving linear PDEs with mixed boundary conditions and semi-infinite geometries). In that context, a typical problem would be: factor a given function $L(z)$ as $L(z) = L_+(z)L_-(z)$, where $L_+(z)$ is analytic in an upper half-plane, $\operatorname{Im} z > a$, $L_-(z)$ is analytic in a lower half-plane, $\operatorname{Im} z < b$, and $a < b$ so that the two half-planes overlap. There are also related problems where $L$ is a $2 \times 2$ or $3 \times 3$ matrix; it is not currently known how to solve such *matrix Wiener–Hopf problems* except in some special cases.

## 17 Closing Remarks

Complex analysis is a rich, deep, and broad subject with a history going back to Cauchy in the 1820s. Inevitably, we have omitted some important topics, such as approximation theory in the complex plane and analytic number theory. There are numerous fine textbooks, a few of which are listed below. However, do not get the impression that complex analysis is a dead subject; it is not. In this article we have tried to cover the basics, with some indications of where problems and opportunities remain.

### Further Reading

Ahlfors, L. V. 1966. *Complex Analysis*, 2nd edn. New York: McGraw-Hill.

Carrier, G. F., M. Krook, and C. E. Pearson. 1966. *Functions of a Complex Variable*. New York: McGraw-Hill.

Henrici, P. 1974, 1977, 1986. *Applied and Computational Complex Analysis*, volumes 1, 2, and 3. New York: Wiley.

Kristensson, G. 2010. *Second Order Differential Equations: Special Functions and Their Classification*. New York: Springer.

Needham, T. 1997. *Visual Complex Analysis*. Oxford: Oxford University Press.

Noble, B. 1988. *Methods Based on the Wiener–Hopf Technique*, 2nd edn. New York: Chelsea.

Smithies, F. 1997. *Cauchy and the Creation of Complex Function Theory*. Cambridge: Cambridge University Press.

Trefethen, L. N. 2013. *Approximation Theory and Approximation Practice*. Philadelphia, PA: SIAM.

Walsh, J. L. 1973. History of the Riemann mapping theorem. *American Mathematical Monthly* 80:270–76.

# IV.2 Ordinary Differential Equations

*James D. Meiss*

## 1 Introduction

Differential equations are near-universal models in applied mathematics. They encapsulate the idea that change occurs incrementally but at rates that may depend upon the state of the system. A system of *ordinary differential equations* (ODEs) prescribes the rate of change of a set of functions, $\boldsymbol{y}(t) = (y_1(t), y_2(t), \dots, y_k(t))$, that depend upon a single variable $t$, which may be real or complex. The functions $y_j$ are the *dependent* variables of the system, and $t$ is the *independent* variable. (If there is more than one independent variable, then the system becomes a PARTIAL DIFFERENTIAL EQUATION [IV.3] (PDE).) Perhaps the most famous ODE is Newton's second law of motion,

$$m\ddot{\boldsymbol{y}} = \boldsymbol{F}(\boldsymbol{y}, \dot{\boldsymbol{y}}, t),$$

which relates the acceleration of the center of mass $\boldsymbol{y} \in \mathbb{R}^3$ of a body of mass $m$ to an externally applied force $\boldsymbol{F}$. This force commonly depends upon the position of the body, $\boldsymbol{y}$; its velocity, $\dot{\boldsymbol{y}}$ (e.g., electromagnetic or damping forces); and perhaps upon time, $t$ (e.g., time-varying external control). The force may also depend upon positions of other bodies; a prominent example is the $n$-BODY PROBLEM [VI.16] of gravitation. We will follow the convention of denoting the first derivative by $\dot{\boldsymbol{y}}$ or $\boldsymbol{y}'$, the second by $\ddot{\boldsymbol{y}}$ or $\boldsymbol{y}''$, and, in general, the $k$th by $\boldsymbol{y}^{(k)}$.

Newton's law is a system of second-order ODEs. More generally, an ODE system is of $n$th order if it involves the first $n$ derivatives of a $k$-dimensional vector $\boldsymbol{y}$; formally, therefore, it is a relation of the form

$$G(\boldsymbol{y}, \boldsymbol{y}^{(1)}, \boldsymbol{y}^{(2)}, \dots, \boldsymbol{y}^{(n)}; t) = 0. \tag{1}$$

An example is Clairaut's differential equation for a scalar function $y(t)$:

$$-y + t\dot{y} + g(\dot{y}) = 0. \tag{2}$$

This is a first-order ODE since it involves only the first derivative of $y$. Equations like this are *implicit*, since they can be viewed as implicitly defining the highest derivative as a function of $\boldsymbol{y}$ and its lower derivatives. Clairaut's equation depends explicitly on the independent variable; such ODEs are said to be *nonautonomous*.

Clairaut showed in 1734 that (2) has a particularly simple family of solutions: $y(t) = ct + g(c)$ for any $c \in \mathbb{R}$. While it might not be completely obvious how to find this solution, it is easy to verify that it does solve (2) by simple substitution since, on the proposed solution $\dot{y} = c$, we have $-y+t\dot{y}+g(\dot{y}) = -(ct+g(c))+tc+g(c) \equiv 0$. More generally, a *solution* of the ODE (1) on an interval $(a, b)$ is a function $\boldsymbol{y}(t)$ that makes (1) identically zero for all $t \in (a, b)$. More specifically, a solution may be required to solve an *initial-value problem* (IVP) or a boundary-value problem (see section 5). For the first-order case, the former means finding a function with a given value, $y(t_0) = y_0$, at a given "initial" time $t_0$. For example, the family of solutions to (2) satisfies the initial condition $y(0) = y_0$ so long as there is a $c$ such that $y_0 = g(c)$, i.e., $y_0$ is in the range of $g$. A family $\boldsymbol{y}(t; \boldsymbol{c})$ that satisfies an IVP for a domain of initial values is known as a *general* solution.

Apart from these families of solutions, implicit ODEs can also have *singular* solutions. For example, (2) also has the solution defined parametrically by $(t(s), y(s)) = (-g'(s), g(s) - sg'(s))$. Again, it is easy to verify that this is a solution to (2) by substitution (and implicit differentiation), but it is perhaps not obvious how to find it. Lagrange showed that some singular solutions of an implicit ODE can be found as envelopes of the general solutions, but the general theory was developed later by Cayley and Darboux.

The classical theory of ODEs, originating with Newton in his *Method of Fluxions* in 1671, has as its goal the construction of the general and singular solutions of an ODE in terms of elementary functions. However, in most cases, ODEs do not have such explicit solutions. Indeed most of the well-known SPECIAL FUNCTIONS [IV.7] of mathematics are defined as solutions of differential equations. For example, the BESSEL FUNCTION [III.2] $J_n(x)$ is defined to be the unique solution of the second-order, explicit, nonautonomous, scalar IVP

$$\left.\begin{aligned} x^2y'' + xy' + (x^2 - n^2)y = 0,\\ y(0) = \delta_{n,0}, \qquad y'(0) = \tfrac{1}{2}\delta_{n,1},\end{aligned}\right\} \tag{3}$$

where $\delta_{i,j}$, the Kronecker delta, is nonzero only when $i = j$ and $\delta_{i,i} = 1$. This equation arises from a number of PDEs through separation of variables. Many of the properties of $J_n$ (e.g., its power-series expansion, asymptotic behavior, etc.) are obtained by direct manipulation of this ODE.

In most applications, the ODE (1) can be written in the explicit form

$$\frac{\mathrm{d}^n}{\mathrm{d}t^n}\boldsymbol{y} = H(\boldsymbol{y}, \dot{\boldsymbol{y}}, \ddot{\boldsymbol{y}}, \dots, \boldsymbol{y}^{(n-1)}; t). \tag{4}$$

Such systems can always be converted into a system of first-order ODEs. For example, if we let $x = (\boldsymbol{y}, \dot{\boldsymbol{y}}, \dots, \boldsymbol{y}^{(n-1)}, t)$ denote a list of $d = nk + 1$ variables, it is then easy to see that (4) can be rewritten as the autonomous first-order system

$$\dot{x} = f(x) \tag{5}$$

for a suitable $f$. Every coupled set of $k$, $n$th-order, explicit ODEs can be written in the form (5).[1] The ODE (5) is a common form in applications, e.g., in population models of ecology or in Hamiltonian dynamics. In general, $x \in M$, where $M$ is a $d$-dimensional manifold called the *phase space*. For example, the phase space of the planar pendulum is a cylinder with $x = (\theta, p_\theta)$, where $\theta$ and $p_\theta$ are the angle and angular momentum, respectively.

In general, the function $f$ in (5) gives a velocity vector (an element of the tangent space $TM$) for each point in the manifold $M$; thus $f\colon M \to TM$. Such a function is a *vector field*. A solution $\varphi\colon (a, b) \to M$ of (5) is a differentiable curve $x(t) = \varphi(t)$ in $M$ with velocity $f(\varphi(t))$; it is everywhere tangent to $f$. Given such a curve it is trivial to check to see if it solves (5); by contrast, the construction of solutions is a highly nontrivial task. A *general solution* of (5) has the form $x(t) = \varphi(t; c)$. Here, $c \in \mathbb{R}^d$ is a set of parameters such that, for each $t_0 \in (a, b)$ and each initial condition $x_0 \in M$, the equation $\varphi(t_0; c) = x_0$ can be solved for $c$.

A *general solution* of (5) is a solution $x(t) = \varphi(t; c)$ that depends upon $d$ parameters, $c$, such that for any IVP, $x(t_0) = x_0 \in M$ with $t_0 \in (a, b)$, there is a $c \in \mathbb{R}^d$ such that $\varphi(t_0; c) = x_0$. The search for explicit, general solutions of (5) is, in most cases, quixotic.

## 2 First-Order Differential Equations

Many techniques were developed through the first half of the eighteenth century for obtaining analytical solutions of first-order ODEs. The equations were often motivated by mechanical problems such as the

1. Though (5) is autonomous, the study of nonautonomous equations per se is not without merit. For example, stability of periodic orbits is most fruitfully studied as a nonautonomous linear problem.

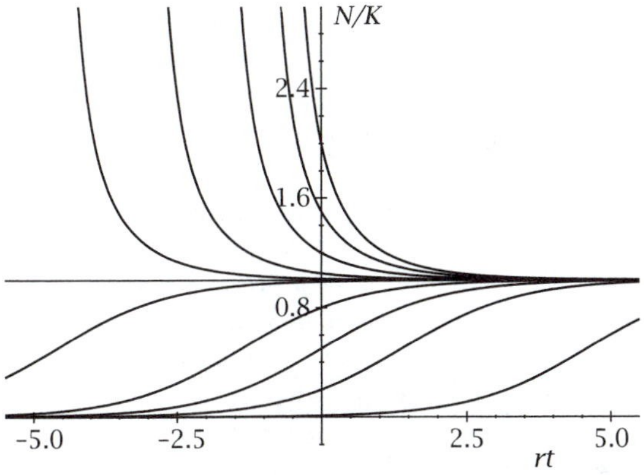


**Figure 1** Solutions of the logistic ODE (7). The equilibria $N = 0$ and $N = K$ are the horizontal lines.

isochrone (find a pendulum whose period is independent of amplitude), which was solved by James Bernoulli in 1690, and da Vinci's catenary (find the shape of a suspended cable), which was solved by John Bernoulli in 1691.

During this period a number of methods were devised that can be applied to general categories of systems. In 1691 Leibniz formulated the method of separation of variables: the formal solution of the ODE $dy/dx = g(x)h(y)$ has the implicit form

$$\int \frac{dy}{h(y)} = \int g(x)\,dx. \tag{6}$$

Any autonomous first-order ODE is separable. For example, for the LOGISTIC POPULATION MODEL [III.19]

$$\dot{N} = rN(1 - N/K), \tag{7}$$

the integrals can be performed and the result solved for $N$ to obtain

$$N(t) = \frac{N_0 K}{N_0 + (K - N_0)e^{-rt}}. \tag{8}$$

Representative solutions are sketched in figure 1. Note that, whenever $N_0 > 0$, this solution tends, as $t \to \infty$, to $K$, the "carrying capacity" of the environment. This value and $N = 0$ are the two *equilibria* of (7), since the vector field vanishes at these points.

The differential equation $N(x,y)y' + M(x,y) = 0$ can be formally rewritten as the vanishing of a differential one-form, $M(x,y)\,dx + N(x,y)\,dy = 0$. In 1739 Clairaut solved such equations when this one-form is exact, that is (for $\mathbb{R}^2$), when $\partial M/\partial y = \partial N/\partial x$. In 1734 Euler had already developed the more general method of integrating factors: if one can devise a function $F(x,y)$ such that the form $F(M\,dx + N\,dy)$ equals the total differential

$$dH \equiv \frac{\partial H}{\partial x}dx + \frac{\partial H}{\partial y}dy$$

of a function $H(x,y)$, then the solutions to $dH = 0$ lie on contours of $H(x,y)$. As an example, the integral curves of a Hamiltonian system with one degree of freedom,

$$\dot{x} = -\frac{\partial H}{\partial y}(x,y), \qquad \dot{y} = \frac{\partial H}{\partial x}(x,y), \tag{9}$$

are those curves that are everywhere tangent to the velocity; equivalently, they are orthogonal to the gradient vector $\nabla H \equiv (\partial H/\partial x, \partial H/\partial y)$. Denoting the infinitesimal tangent vector by $(dx, dy)$, this requirement becomes the exact one-form $(dx, dy) \cdot \nabla H = 0$. Its solutions lie on contours $H(x,y) = E$ with constant "energy." This method gives the phase curves or *trajectories* of the planar system but does not provide the time-dependent functions $x(t)$ and $y(t)$. However, using the constancy of $H$, the ODE for $x$, say, becomes $\dot{x} = \partial_y H(x, y(x;E))$, a separable first-order equation whose solution can be obtained *up to the quadrature* (6).

The technique of substitution was also used to solve many special cases (just as for integrating factors, there is no general prescription for finding an appropriate substitution). For example, James Bernoulli's nonautonomous, first-order ODE

$$y' = P(x)y + Q(x)y^n$$

is linearized by the change of variables $z = y^{1-n}$. Similarly, Leibniz showed that the degree-zero, homogeneous equation $y' = G(y/x)$ becomes separable with the substitution $y(x) = xv(x)$.

Discussion of these and other methods can be found in various classic texts, such as Ince (1956).

## 3 Linear ODEs

In 1743 Euler showed how to solve the $n$th-order linear constant-coefficient equation

$$\sum_{j=0}^{n} a_j \frac{d^j y}{dt^j} = 0 \tag{10}$$

by using the exponential ansatz, $y = e^{rt}$, to reduce the ODE to the $n$th-degree *characteristic equation* $p(r) = \sum_{j=0}^{n} a_j r^j = 0$. Each root, $r_k$, of $p$ provides a solution $y(t) = e^{r_k t}$. Linearity implies that a superposition of these solutions, $y(t) = \sum_{k=0}^{n} c_k e^{r_k t}$, is also a solution for any constant coefficients $c_k$. When $p(r)$ has a root $r^*$ of multiplicity $m > 1$, Euler's reduction-of-order method suggests the further ansatz $y(t) = e^{r^*t}u(t)$. This provides new solutions when $u$ satisfies $u^{(m)} = 0$, which has as its general solution a degree-$(m-1)$ polynomial in $t$. The general solution therefore becomes a superposition of $n$ linearly independent functions of the form $t^\ell e^{r_k t}$. Even when the ODE is real, the roots $r_k = \alpha_k + i\beta_k$ may be complex. In this case, the conjugate root can be used to construct real solutions of the form $t^\ell e^{\alpha_k t}\cos(\beta_k t)$ and $t^\ell e^{\alpha_k t}\sin(\beta_k t)$ for $\ell = 0,\dots,m-1$. A superposition of these real solutions has $n$ arbitrary real constants $c_k$, and since the functions are independent, there is a choice of these constants that solves the IVP $y^{(k)}(t_0) = b_k$, $k = 0,\dots,n-1$, for arbitrarily specified values $b_k$.

More generally, when (5) is linear, it reduces to

$$\dot{x} = Ax \qquad (11)$$

for a constant, $n \times n$ matrix $A$. Formally, the general solution of this system can be written as the matrix exponential: $x(t) = e^{tA}x(0)$. As for more general FUNCTIONS OF MATRICES [II.14], we can view this as defining the symbol $e^{tA}$ as the solution of the ODE. More explicitly, this exponential is defined by the same convergent MacLaurin series as $e^{at}$ for scalar $a$. If $A$ is semisimple (i.e., if it has a complete eigenvector basis), then $A$ is diagonalized by the matrix $P$ whose columns are eigenvectors: $A = P\Lambda P^{-1}$, where $\Lambda = \operatorname{diag}(\lambda_1,\dots,\lambda_n)$ is the diagonal matrix of eigenvalues. In this case,

$$e^{tA} = P\operatorname{diag}(e^{\lambda_1 t},\dots,e^{\lambda_n t})P^{-1}.$$

More generally, $e^{tA}$ also contains powers of $t$, generalizing the simpler, scalar situation.

The nonhomogeneous linear system

$$\dot{x} = Ax + g(t)$$

with forcing function $g \in \mathbb{R}^n$ can be solved by Lagrange's method of variation of parameters. The idea is to replace the parameters $x(0)$ in the homogeneous solution by functions $u(t)$. Substitution of $x(t) = e^{tA}u(t)$ into the ODE permits the unknown functions to be isolated and yields the integral form

$$x(t) = e^{tA}x(0) + \int_0^t e^{(t-s)A}g(s)\,ds.$$

Solution of linear, nonautonomous ODEs, $\dot{x} = A(t)x$, is much more difficult. The source of the difficulty is that $A(t)$ does not generally commute with $A(s)$ when $t \neq s$. Indeed, $e^A e^B \neq e^{A+B}$ unless the matrices do commute (the Baker–Campbell–Hausdorff theorem from Lie theory gives a series expansion for the product). The special case of a time-periodic family of matrices $A(t) = A(t+T)$ can be solved. Floquet showed that the general solution for this case takes the form $x(t) = P(t)e^{tB}x(0)$, where $B$ is a real, constant matrix and $P(t)$ is a periodic matrix with period $2T$. One much-studied example of this form is MATHIEU'S EQUATION [III.21].

More generally, finding a transformation to a set of coordinates in which the effective matrix is constant is called the *reducibility problem*; even for a quasiperiodic dependence on time, this is nontrivial.

## 4 Singular Points

Consider the linear ODE (10), now allowing the coefficients $a_j(t)$ to be analytic functions of $t \in \mathbb{C}$ so that it is nonautonomous. Cauchy showed that, if the coefficients are analytic in a neighborhood of $t_0$ and if $a_n(t_0) \neq 0$, this ODE has $n$ independent analytic solutions. The coefficients of the power series of $y$ can be determined from a recursion relation upon substitution of a series for $y$ into the ODE.

A point at which some of the ratios $a_j(t)/a_n(t)$ are singular is a (fixed) *singular point* of the ODE, and the solution need not be analytic at $t_0$. There are two distinct cases. A singular point is *regular* if $a_{n-j}(t)/a_n(t)$ has at most a $j$th-order pole for each $j = 1,\dots,n$. In this case, there is an $r \in \mathbb{C}$ such that there is at least one solution of the form $y(t) = (t-t_0)^r\phi(t)$ with $\phi$ analytic at $t_0$. Additional solutions may also have logarithmic singularities. An ODE for which all singular points are regular is called *Fuchsian*.

Most of the SPECIAL FUNCTIONS [IV.7] of mathematical physics are defined as solutions of second-order linear ODEs with regular singular points. Many are special cases of the hypergeometric equation

$$z(1-z)w'' + (\gamma - (\alpha+\beta+1)z)w' - \alpha\beta w = 0 \qquad (12)$$

for a complex-valued function $w(z)$. This ODE has regular singular points at $z = 0$, 1, and $\infty$ (the latter obtained upon transforming the independent variable to $u = 1/z$). For the singular point at 0, following Frobenius, we make the ansatz that the solution has the form of a series:

$$w(z) = z^r \sum_{j=0}^{\infty} c_j z^j.$$

Substitution into (12) yields $c_0 p(r) z^{r-1} + \mathcal{O}(z^r) = 0$, and if this is to vanish with $c_0 \neq 0$, then $r$ must satisfy the *indicial equation* $p(r) = r^2 + (\gamma-1)r = 0$, with roots $r_1 = 0$ and $r_2 = 1-\gamma$. A recursion relation for the $c_j$, $j > 0$, is obtained from the terms of order $z^{r+j-1}$.

For $r_1 = 0$, this yields the GAUSS HYPERGEOMETRIC FUNCTION [IV.7 §5]

$$F(\alpha, \beta, \gamma; z) = \frac{\Gamma(\gamma)}{\Gamma(\alpha)\Gamma(\beta)} \sum_{j=0}^{\infty} \frac{\Gamma(\alpha + j)\Gamma(\beta + j)}{j!\Gamma(\gamma + j)} z^j,$$

where the gamma function $\Gamma$ generalizes the factorial. When $\gamma \notin \mathbb{Z}$, the second solution turns out to be $x^{1-\gamma}F(\alpha-\gamma+1, \beta-\gamma+1, 2-\gamma; x)$, which is not analytic at $z = 0$.

When the indicial equation has roots that differ by an integer, a second solution can be found by the method of reduction of order. For example, for the second-order case, suppose $r_1 - r_2 \in \mathbb{N}$ and let $w_1(z) = e^{r_1 t}\phi_1(t)$ be the solution for $r_1$. Substitution of the ansatz $w(z) = w_1(z) \int v(z)\,dz$ shows that $v$ satisfies a first-order ODE with a regular singular point at 0 whose indicial equation has the negative integer root $r_2 - r_1 - 1$. If the power series for $v$ has $\mathcal{O}(z^{-1})$ terms, then $w(z)$ has logarithmic singularities; ultimately, the second solution has the form $w(z) = z^{r_2}\phi_2(z) + cw_1(z)\ln z$, where $\phi_2$ is analytic at 0 and $c$ might be zero. Thus, for example, the second hypergeometric solution for integral $\gamma$, where the roots of the indicial equation differ by an integer, has logarithmic singularities.

Near an *irregular* singular point, the solution may have essential singularities. For example, the first-order ODE $z^2 w' = w$ has an irregular singular point at 0; its solution, $w(z) = ce^{-1/z}$, has an essential singularity there. Similarly, Bessel's equation (3) has an irregular singular point at $\infty$.

Singular points of nonlinear ODEs can be fixed (i.e., determined by singularities of the vector field) or moveable. In the latter case, the position of the singularity depends upon initial conditions. The study of equations whose only movable singularities are poles leads to the theory of PAINLEVÉ TRANSCENDENTS [III.24].

## 5 Boundary-Value Problems

So far we have considered IVPs for systems of the form (5), that is, when the imposed values occur at one point, $t = t_0$. Another common formulation is that of a boundary-value problem (BVP), where properties of the solution are specified at two distinct points. Such problems commonly occur for ODEs that arise by separation of variables from PDEs. They also occur in control theory, where constraints may be applied at different times.

A classical BVP is the Sturm–Liouville equation:

$$\left.\begin{aligned} -(p(x)y')' + q(x)y &= \lambda r(x)y, \\ \alpha_1 y(a) + \alpha_2 y'(a) &= 0, \\ \beta_1 y(b) + \beta_2 y'(b) &= 0. \end{aligned}\right\} \quad (13)$$

Here, $\lambda$ is a parameter, e.g., the separation constant for the PDE case, $p \in C^1[a,b]$, $q, r \in C^0[a,b]$, $p$ and the weight function $r$ are assumed to be positive, and $\alpha_1\alpha_2, \beta_1\beta_2 \neq 0$. For example, Bessel's equation (3) takes this form, with $p(x) = -q(x) = x$ and $r(x) = 1/x$, if appropriate boundary conditions are imposed.

The Sturm–Liouville problem has (unique) solutions $y_n(x) \in C^2[a,b]$ only for a discrete set $\lambda_n$, $n \in \mathbb{N}$, of values of the separation constant. Moreover, these "eigenfunctions" and their corresponding "eigenvalues" have a number of remarkable properties.

**Ordering:** $\lambda_1 < \lambda_2 < \cdots < \lambda_n < \cdots$.
**Oscillation:** $y_n(x)$ has $n - 1$ simple zeros in $(a, b)$.
**Growth:** $\lambda_n \to \infty$ as $n \to \infty$.
**Orthogonality:** $\int_a^b r(x)y_n(x)y_m(x)\,dx = \delta_{m,n}$.
**Completeness:** the set $y_n$ is a basis for the space $L^2(a, b)$.

Perhaps the simplest such problem is $y'' = -\lambda y$ with $y(0) = y(1) = 0$. Here, the eigenvalues are $\lambda_n = (n\pi)^2$ and the eigenfunctions are $y_n = \sin(n\pi x)$. The completeness of these functions in $L^2(0, 1)$ is the expression of the convergence of the Fourier sine series. A more interesting problem is the quantum harmonic oscillator, which, when nondimensionalized, is governed by the Schrödinger equation

$$-\psi'' + x^2\psi = \lambda\psi. \quad (14)$$

Here, $\lambda$ is related to the energy $E = \frac{1}{2}\lambda\hbar\omega$ for classical frequency $\omega$. This is a Sturm–Liouville problem for $\psi \in L^2(-\infty, \infty)$. The solutions are most easily obtained by the substitution $\psi(x) = e^{-x^2/2}y(x)$ that transforms (14) to the Hermite equation $y'' - 2xy' + (\lambda - 1)y = 0$. This ODE has degree-$(n-1)$ polynomial solutions when $\lambda_n = 2n - 1$, $n \in \mathbb{N}$; otherwise, the wave function $\psi$ is not square integrable. The first five orthonormal eigenstates of (14) are shown in figure 2.

## 6 Equilibria and Stability

Apart from the solution of linear PDEs, linear systems of ODEs find their primary application in the study of the stability of equilibria of the nonlinear system (5). A point $x^*$ is an equilibrium if $f(x^*) = 0$. If $f \in C^1$, then

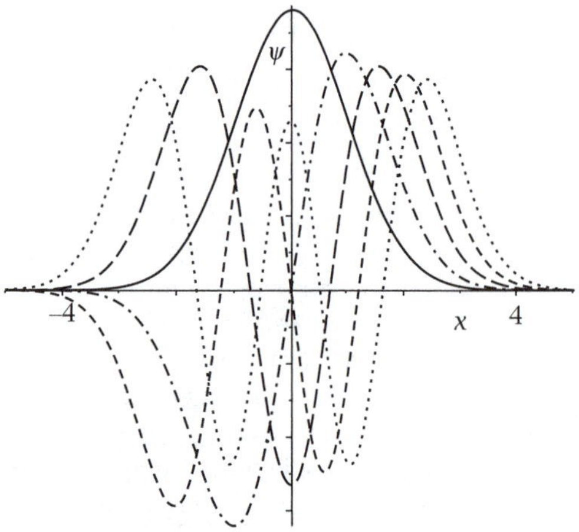


**Figure 2** The first five eigenstates of the Sturm–Liouville problem (14).

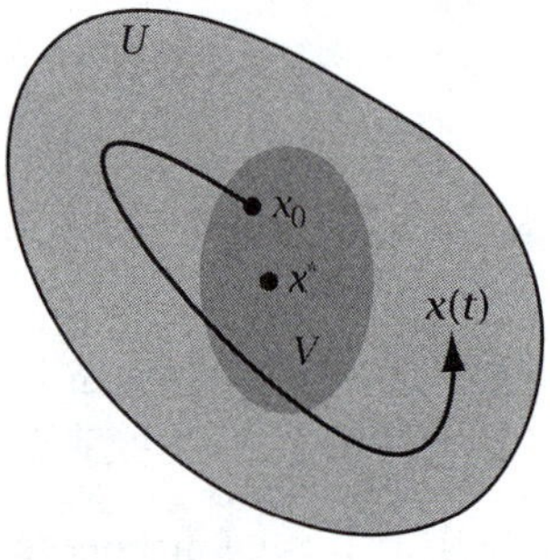


**Figure 3** A Lyapunov-stable equilibrium $x^*$.

the dynamics of a nearby point $x(t) = x^* + \delta x(t)$ may be approximated by $\delta\dot{x} = \mathrm{D}f(x^*)\delta x$, where $(\mathrm{D}f)_{ij} = \partial f_i/\partial x_j$ is the Jacobian matrix of the vector field.

An equilibrium is *Lyapunov stable* if, for *every* neighborhood $U$, there is a neighborhood $V \subset U$ such that if $x(0) \in V$ then $x(t) \in U$ for all $t > 0$ (see figure 3). For the case in which $A = \mathrm{D}f(x^*)$ is a hyperbolic matrix (its spectrum does not intersect the imaginary axis), the stability of $x^*$ can be decided by the eigenvalues of $A$. Indeed, the Hartman–Grobman theorem states that in this case there is a neighborhood $U$ of $x^*$ such that there is a coordinate change (a homeomorphism) that takes the dynamics of (5) in $U$ to that of (11). In this case we say that the two dynamical systems are topologically conjugate on $U$.

An equilibrium is stable if all of the eigenvalues of $A$ are in the left half of the complex plane, $\mathrm{Re}(\lambda) < 0$. Indeed, in this case it is *asymptotically* stable: there is a neighborhood $U$ such that every solution that starts in $U$ remains in $U$ and converges to $x^*$ as $t \to \infty$. In this case, $x^*$ is a stable *node*. When there are eigenvalues with both positive and negative real parts, then $x^*$ is a *saddle*. The case of complex eigenvalues deserves special mention, since the solution of the linear system then involves trigonometric functions and there are solutions in $\mathbb{R}^d$ that are infinite spirals. This is not necessarily the case when nonlinear terms are added, however: the homeomorphism that conjugates the system in $U$ may unwrap the spirals.

As an example consider the damped Duffing oscillator[2]

$$\dot{x} = y, \qquad \dot{y} = -\mu y + x(1 - x^2), \tag{15}$$

with the phase portrait shown in figure 4 when $\mu = \frac{1}{2}$. There are three equilibria: $(0,0)$ and $(\pm 1, 0)$. The Jacobian at the origin is

$$\mathrm{D}f(0,0) = \begin{pmatrix} 0 & 1 \\ 1 & -\frac{1}{2} \end{pmatrix},$$

with eigenvalues $\lambda_{1,2} = -\frac{1}{4}(1 \pm \sqrt{17})$. Since these are real and of opposite signs, the origin is a saddle. By contrast, the Jacobian of the other fixed points is

$$\mathrm{D}f(\pm 1,0) = \begin{pmatrix} 0 & 1 \\ -2 & -\frac{1}{2} \end{pmatrix},$$

with the complex eigenvalues $\lambda_{1,2} = -\frac{1}{4}(1 \pm \mathrm{i}\sqrt{31})$. Since the real parts are negative, these points are both attracting foci. They are still foci in the nonlinear system, as illustrated in figure 4, since trajectories that approach them cross the line $y = 0$ infinitely many times. Apart from the saddle and its *stable manifold* (the dotted curve in the figure), every other trajectory is asymptotic to one of the foci; these are *attractors* whose basins of attraction are separated by the stable manifold of the saddle.

The stability of a nonhyperbolic equilibrium (when $A$ has eigenvalues on the imaginary axis) is delicate and depends in detail on the nonlinear terms, i.e., the $\mathcal{O}(\delta x^2)$ terms in the expansion of $f$ about $x^*$. For example, the system

$$\dot{x} = -y + ax(x^2 + y^2), \qquad \dot{y} = x + ay(x^2 + y^2) \tag{16}$$

has only one equilibrium, $(0,0)$. The Jacobian at the origin has eigenvalues $\lambda = \pm\mathrm{i}$; its dynamics are that of a *center*. Nevertheless, the dynamics of (16) near $(0,0)$ depend upon the value of $a$. This can be easily seen by transforming to polar coordinates using $(x, y) =$

2. George Duffing studied the periodically forced version of (15) in 1918.

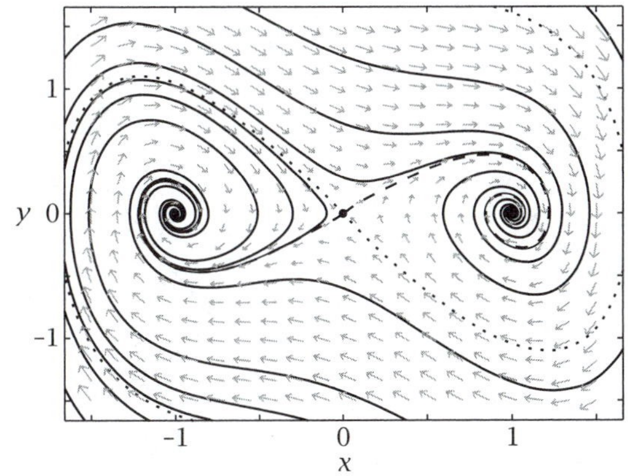


**Figure 4** The phase portrait of (15) for $\mu = \frac{1}{2}$. Arrows depict the vector field, and dots depict the three equilibria. The unstable (dashed) and stable (dotted) manifolds of the saddle are shown.

$(r\cos\theta, r\sin\theta)$:

$$\dot{r} = \frac{1}{r}(x\dot{x} + y\dot{y}) = ar^3,$$
$$\dot{\theta} = \frac{1}{r^2}(x\dot{y} - y\dot{x}) = 1.$$

Thus if $a < 0$, the origin is a global attractor: every trajectory limits to the origin as $t \to \infty$. If $a > 0$, the origin is a repellor. The study of nonhyperbolic equilibria is the first step in BIFURCATION THEORY [IV.21].

## 7 Existence and Uniqueness

Before one attempts to find solutions to an ODE, it is important to know whether solutions exist, and if they exist whether there is more than one solution to a given IVP. There are two types of problems that can occur.

The first is that the velocity $f$ may be unbounded on $M$; in this case, a solution might exist but only over a finite interval of time. For example, the system $\dot{x} = x^2$ for $x \in \mathbb{R}$ has the general solution $x(t) = x_0/(1 - tx_0)$. Note that $|x| \to \infty$ as $t \to 1/x_0$, even though $f$ is a "nice" function: it is smooth, and moreover, it is analytic. The problem is, however, that as $|x|$ increases, the velocity increases even more rapidly, leading to infinite speed in finite time. The existence theorem deals with this problem by being local; it guarantees existence only on a compact interval.

The second problem is that $f$ may not be smooth enough to guarantee a unique solution. One might expect that it is sufficient that $f$ be continuous. However, the simple system $\dot{x} = \sqrt{|x|}$ for $x \in \mathbb{R}$ has infinitely many solutions that satisfy the initial condition $x(0) = 0$. The obvious solution is $x(t) \equiv 0$, but $x(t) = \frac{1}{4}\operatorname{sgn}(t)t^2$ is also a solution. Moreover, any function $x(t)$ that is zero up to an arbitrary time $t_0 > 0$ and then connects to the parabola $\frac{1}{4}(t - t_0)^2$ also solves the IVP. Elimination of this problem requires assuming that $f$ is *more* than continuous; it must be at least Lipschitz. A function $f\colon M \to \mathbb{R}^d$ is *Lipschitz* on $M \subset \mathbb{R}^m$ if there is a constant $K$ such that for all $x, y \in M$, $\|f(x) - f(y)\| \leqslant K\|x - y\|$.

With this concept, we can state a theorem of existence and uniqueness. Let $B_r(x)$ denote the closed ball of radius $r$ about $x$.

**Theorem 1 (Picard–Lindelöf).** *Suppose that for $x_0 \in \mathbb{R}^d$ there exists $b > 0$ such that $f\colon B_b(x_0) \to \mathbb{R}^d$ is Lipschitz. Then the IVP (5) with $x(t_0) = x_0$ has a unique solution $x\colon [t_0 - a, t_0 + a] \to B_b(x_0)$ with $a = b/V$, where $V = \max_{x \in B_b(x_0)} \|f(x)\|$.*

This theorem can be proved iteratively (e.g., by Picard iteration), but the most elegant proof uses the contraction mapping theorem.

## 8 Flows

When the vector field of (5) satisfies the conditions of the Picard–Lindelöf theorem, the solution is necessarily a $C^1$ function of time. It is also a Lipschitz function of the initial condition. Suppose now that $f \in C^1(\mathbb{R}^d, \mathbb{R}^d)$ and is locally Lipschitz. Though the theorem guarantees the existence only on a (perhaps small) interval $t \in [t_0 - a, t_0 + a]$, this solution can be uniquely extended to a maximal open interval $J = (\alpha, \beta)$ such that the solution is unbounded as $t$ approaches $\alpha$ or $\beta$ when they are finite. As noted in section 7, unbounded solutions may arise even for "nice" vector fields; however, if $f$ is bounded or globally Lipschitz, then $J = \mathbb{R}$.

If $\varphi_t(x_0)$ denotes the maximally extended solution, then $\varphi\colon J \times \mathbb{R}^d \to \mathbb{R}^d$ satisfies a number of conditions:

- $\varphi \in C^1$,
- $\varphi_0(x) = x$, and
- $\varphi_t \circ \varphi_s = \varphi_{t+s}$ whenever $t$, $s$, and $t + s \in J$.

The last condition encapsulates the idea of autonomy: flowing from the point $\varphi_s(x)$ for a time $t$ is the same as flowing for time $t + s$ from $x$; the origin of time is a matter of convention.

For example, (8), the solution of the logistic ODE, gives such a function if it is rewritten as $\varphi_t(N_0) = N(t)$.

In this case (for $r$ and $K$ positive), $J = \mathbb{R}$ if $0 \leqslant N_0 \leqslant K$, and $J = (\alpha, \infty)$ with

$$\alpha = \frac{1}{r}\ln\left(1 - \frac{K}{N_0}\right) < 0$$

if $N_0 > K$. Indeed, it is apparent from figure 1 that solutions with initial conditions above the carrying capacity $K$ grow rapidly for decreasing $t$; the theory implies that $\varphi_t(N_0) \to \infty$ as $t \downarrow \alpha$.

More generally, any function satisfying these conditions is called a *flow*. The flow is *complete* if $J = \mathbb{R}$, and it is a *semiflow* if $\alpha$ is finite but $\beta$ is still $\infty$. It is not hard to see that every flow is the solution of a differential equation (5) for some $C^0$ vector field. Flows form one fundamental part of the theory of DYNAMICAL SYSTEMS [IV.20].

## 9 Phase-Plane Analysis

A system of two differential equations

$$\left.\begin{aligned}\dot{x} &= P(x, y),\\ \dot{y} &= Q(x, y)\end{aligned}\right\} \qquad (17)$$

can be qualitatively analyzed by considering a few simple properties of $P$ and $Q$. The goals of such an analysis include determining the asymptotic behavior for $t \to \pm\infty$ and the stability of any equilibria or periodic orbits.

The *nullclines* are the sets

$$N_{\mathrm{h}} = \{x, y\colon Q(x, y) = 0\},$$
$$N_{\mathrm{v}} = \{x, y\colon P(x, y) = 0\}.$$

Typically these are curves on which the instantaneous motion is horizontal or vertical, respectively. The set of equilibria is precisely the intersection of the nullclines, $E = N_{\mathrm{h}} \cap N_{\mathrm{v}}$. The web of nullclines divides the phase plane into sectors in which the velocity vector lies in one of the four quadrants.

For example, the Lotka–Volterra system

$$\left.\begin{aligned}\dot{x} &= bx(1 - x - 2y),\\ \dot{y} &= cy(1 - 2x - y)\end{aligned}\right\} \qquad (18)$$

can be thought of as a model of competition between two species with normalized populations $x \geqslant 0$ and $y \geqslant 0$. The species have per capita birth rates $b$ and $c$, respectively, when their populations are small, but these decrease if either or both of $x$ and $y$ grows because of competition for the same resource. In the absence of competition, the environment has a carrying capacity of one population unit. The nullclines are pairs of lines $N_{\mathrm{h}} = \{y = 0\} \cup \{y = 1 - 2x\}$ and

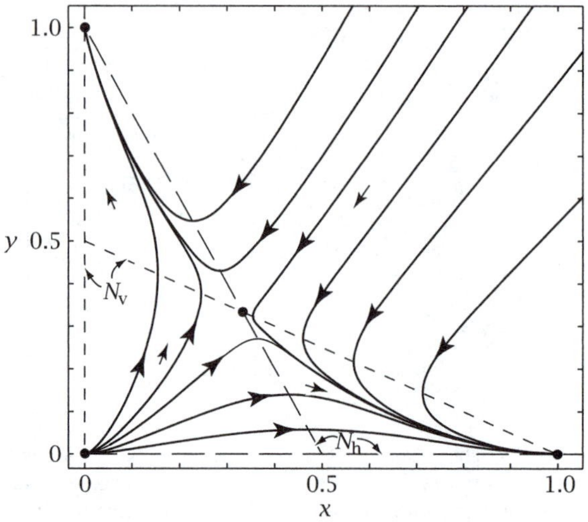


**Figure 5** The phase portrait for (18) for $2c = 3b > 0$. Representative velocity vectors are shown in each sector defined by the nullclines, as are several numerically generated trajectories.

$N_{\mathrm{v}} = \{x = 0\} \cup \{y = \frac{1}{2}(1 - x)\}$. Consequently, there are four equilibria: $(0, 0)$, $(1, 0)$, $(0, 1)$, and $(\frac{1}{3}, \frac{1}{3})$. The nullclines divide the biologically relevant domain into four regions within which the velocity lies in one of the four quadrants, as shown in figure 5. In particular, when both $x$ and $y$ are large (e.g., bigger than the carrying capacity), the velocity must be in the third quadrant since both $\dot{x} < 0$ and $\dot{y} < 0$. Since a component of the velocity can reverse only upon crossing a nullcline (and in this case does reverse), the remainder of the qualitative behavior is then determined.

From this simple observation one can conclude that the origin is a *source*, i.e., every nearby trajectory approaches the origin as $t \to -\infty$. By contrast, the two equilibria on the axes are *sinks* since all nearby trajectories approach them as $t \to +\infty$. The remaining equilibrium is a *saddle* since there are approaching and diverging solutions nearby. Moreover, every trajectory in the positive quadrant is bounded, and almost all trajectories asymptotically approach one of the two sinks. The only exceptions are a pair of trajectories that are on the stable manifold of the saddle. Details that are not determined by this analysis include the timescale over which this behavior occurs and the curvature of the solution curves, which depends upon the ratio $b/c$. This model demonstrates the ecological phenomenon of competitive exclusion; typically, only one species survives.

## 10 Limit Cycles

If the simplest solutions of ODE systems are equilibria, periodic orbits form the second class. A solution $\Gamma = \{x(t): 0 \leqslant t < T\}$ of an autonomous ODE is periodic with (minimal) period $T$ if $\Gamma$ is a simple closed loop in the phase space. Indeed, uniqueness of solutions implies that, if $x(T) = x(0)$, then $x(nT) = x(0)$ for all $n \in \mathbb{Z}$.

A one-dimensional autonomous ODE (5) cannot have any periodic solutions. Indeed, every solution of such a system is a monotone function of $t$. Periodic trajectories are common in two dimensions. For example, each planar Hamiltonian system (9) has periodic trajectories on every closed nondegenerate ($\nabla H \neq 0$) contour $H(x,y) = E$. These periodic trajectories are not isolated. An isolated periodic orbit is called a limit cycle. More generally, a *limit cycle* is a periodic orbit that is the forward ($\omega$) or backward ($\alpha$) limit of another trajectory.

The van der Pol oscillator,

$$\dot{x} = y, \qquad \dot{y} = -x + 2\mu y - x^2 y, \tag{19}$$

was introduced in 1922 as a model of a nonlinear circuit with a triode tube. Here, $x$ represents the current through the circuit, and $y$ represents the voltage drop across an inductor. The parameter $\mu$ corresponds to the "negative" resistance of the triode passing a small current. This system has a unique periodic solution when $\mu > 0$ (see figure 6). The creation of this limit cycle at $\mu = 0$ follows from the HOPF BIFURCATION [IV.21 §2] theorem. Its uniqueness is a consequence of a more general theorem due to Liénard.

Planar vector fields can therefore have equilibria and periodic orbits. Are there more complicated trajectories, e.g., quasiperiodic or chaotic orbits? The negation of this speculation is contained in the theorem proposed by Poincaré and proved later by Bendixson: the set of limit points of any bounded trajectory in the plane can contain only equilibria and periodic orbits. There is therefore *no* CHAOS *[II.3] in two dimensions*! From the point of view of finding periodic trajectories, this theorem implies the following.

**Theorem 2 (Poincaré–Bendixson).** *Suppose that $A \subset \mathbb{R}^2$ is bounded and positively invariant and that $\varphi$ is a complete semiflow in $A$. Then, if $A$ contains no equilibria, it must contain a periodic orbit.*

For example, consider the system

$$\dot{x} = y, \qquad \dot{y} = -x + yh(r),$$

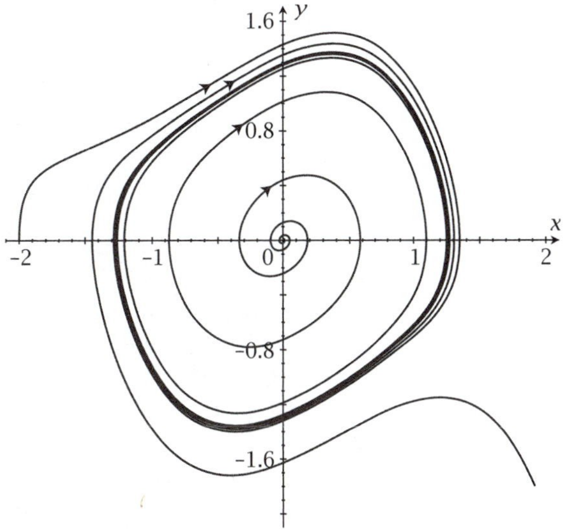


**Figure 6** The phase portrait of the van der Pol oscillator (19) for $\mu = 0.2$.

where $r = \sqrt{x^2 + y^2}$, and let $A$ be the annulus $\{(x,y): a < r < b\}$. Thus $A$ contains no equilibria for any $0 < a < b$. Converting to polar coordinates gives, for the radial equation,

$$\dot{r} = \frac{y^2}{r} h(r).$$

Now suppose that there exist $0 < a < b$ such that $h(b) < 0 < h(a)$. On the circle $r = a$ we then have $\dot{r} \geqslant 0$, implying that trajectories cannot leave $A$ through its inner boundary. Similarly, trajectories cannot leave through $r = b$ because $\dot{r} \leqslant 0$ on this circle. By the Poincaré–Bendixson theorem, therefore, there is a periodic orbit in $A$.

## 11 Heteroclinic Orbits

Suppose that $\varphi$ is a complete $C^{r+1}$ flow that has a saddle equilibrium at $x^*$. The $k$-generalized eigenvectors of $Df(x^*)$ corresponding to the stable eigenvalues, $\mathrm{Re}(\lambda_i) < 0$, define a $k$-dimensional tangent plane $E^{\mathrm{s}}$ at $x^*$. This linear plane can be extended to form a set of trajectories of the nonlinear flow whose forward evolution converges to $x^*$:

$$W^{\mathrm{s}}(x^*) = \left\{x \in M \setminus \{x^*\}: \lim_{t\to\infty} \varphi_t(x) = x^*\right\}.$$

The STABLE MANIFOLD THEOREM [IV.20] implies that this set is a $k$-dimensional, $C^r$, immersed manifold that is tangent to $E^{\mathrm{s}}$ at $x^*$. Similarly, a saddle has an unstable manifold

$$W^u(x^*) = \left\{x \in M \setminus \{x^*\} \colon \lim_{t \to -\infty} \varphi_t(x) = x^*\right\}$$

that is tangent to the $(n-k)$-dimensional plane spanned by the unstable eigenvectors of $x^*$. It is important to note that this set is defined by its backward asymptotic behavior and not by the idea that it escapes from $x^*$. These concepts can also be generalized to hyperbolic invariant sets.

Poincaré realized that intersections of stable and unstable manifolds can give rise to complicated orbits. He called an orbit $\Gamma$ *homoclinic* if $\Gamma \in W^u(x^*) \cap W^s(x^*)$. Similarly, an orbit is *heteroclinic* if $\Gamma \in W^u(a) \cap W^s(b)$ for distinct saddles $a$ and $b$.

Planar Hamiltonian systems often have homoclinic or heteroclinic orbits. For example, the conservative Duffing oscillator, (15) with $\mu = 0$, has Hamiltonian $H(x,y) = \frac{1}{2}(y^2 - x^2 + \frac{1}{2}x^4)$. This function has a critical level set $H = 0$ that is a figure-eight intersecting the saddle equilibrium at $(0,0)$. Since energy is conserved, trajectories remain on each level set; in particular, every trajectory on the figure-eight is biasymptotic to the origin (these are homoclinic trajectories). For this case, the stable and unstable manifolds coincide, and we say that there is a *homoclinic connection*. This set is also called a *separatrix* since it separates motion that encircles each center from that enclosing both centers. Such a homoclinic connection is fragile; for example, it is destroyed whenever $\mu \neq 0$ in (15). More generally, a homoclinic BIFURCATION [IV.21] corresponds to the creation/destruction of a homoclinic orbit from a periodic one.

If, however, the intersection of $W^u(a)$ with $W^s(b)$ is transverse, it cannot be destroyed by a small perturbation. A transversal intersection of two submanifolds is one for which the union of their tangent spaces at an intersection point spans $TM$:

$$T_x W^u(a) \oplus T_x W^u(b) = T_x M.$$

Note that for this to be the case, we must have $\dim(W^u) + \dim(W^s) \geqslant \dim(M)$. Every such intersection point lies on a heteroclinic orbit that is structurally stable. Poincaré realized that in certain cases the existence of such a transversal heteroclinic orbit implies infinite complexity. This idea was formalized by Steve Smale in his construction of the *Smale horseshoe*. The existence of a transversal heteroclinic orbit implies a chaotic invariant set.

## 12 Other Techniques and Concepts

Differential equations often have discrete or continuous SYMMETRIES [IV.22], and these are useful in constructing new solutions and reducing the order of the system. Given sufficiently many symmetries and invariants, a system of ODEs can be effectively solved, that is, it is integrable.

One often finds that no analytical method leads to explicit solutions of an ODE. In this case, NUMERICAL SOLUTION [IV.12] techniques are invaluable.

### Further Reading

Abramowitz, M., and I. Stegun. 1965. *Handbook of Mathematical Functions*. New York: Dover.

Ince, E. L. 1956. *Ordinary Differential Equations*. New York: Dover.

Kline, M. 1972. *Mathematical Thought from Ancient to Modern Times*. New York: Oxford University Press.

Meiss, J. D. 2007. *Differential Dynamical Systems*. Philadelphia, PA: SIAM.

Robinson, C. 1999. *Dynamical Systems: Stability, Symbolic Dynamics, and Chaos*. Boca Raton, FL: CRC Press.

# IV.3 Partial Differential Equations

*Lawrence C. Evans*

## 1 Overview

This article is an extremely rapid survey of the modern theory of partial differential equations (PDEs). Sources of PDEs are legion: mathematical physics, geometry, probability theory, continuum mechanics, optimization theory, etc. Indeed, *most of the fundamental laws of the physical sciences are partial differential equations* and most papers published in applied mathematics concern PDEs.

The following discussion is consequently very broad but also very shallow, and it will certainly be inadequate for any given PDE the reader may care about. The goal is rather to highlight some of the many key insights and unifying principles across the entire subject.

### 1.1 Confronting PDEs

Among the greatest accomplishments of the physical and other sciences are the discoveries of fundamental laws, which are usually PDEs. The great problems for mathematicians, both pure and applied, are then to understand the solutions of these equations, using theoretical analysis, numerical simulations, perturbation theory, and whatever other tools they can find.

But this very success in physics—that some fairly simple-looking PDEs, for example the Euler equations for fluid mechanics (see (11) below), model complicated and diverse physical phenomena—causes all sorts of mathematical difficulties. Whatever general assertion we try to show mathematically must apply to all sorts of solutions with extremely disparate behavior.

It is therefore a really major undertaking to understand solutions of partial differential equations, and there are at least three primary mathematical approaches for doing so:

- discovering analytical formulas for solutions, either exact or approximate,
- devising accurate and fast numerical methods, and
- developing rigorous theory.

In other words, we can aspire to actually solve the PDE more or less explicitly, to compute solutions, or to indirectly deduce properties of the solutions (without relying upon formulas or numerics). This article surveys these viewpoints, with particular emphasis on the last.

*Terminology*

A *partial differential equation* is an equation involving an unknown function $u$ of more than one variable and certain of its partial derivatives. The *order* of a PDE is the order of the highest-order partial derivative of the unknown appearing within it.

A *system* of PDEs comprises several equations involving an unknown vector-valued function $\boldsymbol{u}$ and its partial derivatives.

A PDE is *linear* if it corresponds to a linear operator acting on the unknown and its partial derivatives; otherwise, the PDE is *nonlinear*.

*Notation*

Hereafter, $u$ usually denotes the real-valued solution of a given PDE and is usually a function of points $x = (x_1, \dots, x_n) \in \mathbb{R}^n$, typically denoting a position in space, and sometimes also a function of $t \in \mathbb{R}$, denoting time. We write $u_{x_k} = \partial u / \partial x_k$ to denote the partial derivative of $u$ with respect to $x_k$, and $u_t = \partial u / \partial t$, $u_{x_k x_l} = \partial^2 u / \partial x_k \partial x_l$, etc., for higher partial derivatives. The *gradient* of $u$ in the variable $x$ is

$$\nabla u = (u_{x_1}, \dots, u_{x_n}).$$

(In this article, $\nabla u$ always denotes the gradient in the variables $x_1, \dots, x_n$, even if $u$ also depends on $t$.) We write the *divergence* of a vector field $\boldsymbol{F} = (F^1, \dots, F^n)$ as $\operatorname{div} \boldsymbol{F} = \sum_{i=1}^n F^i_{x_i}$.

The *Laplacian* of $u$ is the divergence of its gradient:

$$\Delta u = \nabla^2 u = \sum_{k=1}^n u_{x_k x_k}. \tag{1}$$

Let us also write $\boldsymbol{u} = (u^1, \dots, u^m)$ to display the components of a vector-valued function. We always use boldface for vector-valued mappings.

The solid $n$-dimensional ball with center $x$ and radius $r$ is denoted by $B(x, r)$, and $\partial B(x, r)$ is its boundary, a sphere. More generally, $\partial U$ means the boundary of a set $U \subset \mathbb{R}^n$; and we denote by

$$\int_{\partial U} f \, \mathrm{d}S$$

the integral of a function $f$ over the boundary, with respect to $(n-1)$-dimensional surface area.

### 1.2 Some Important PDEs

A list of some of the most commonly studied PDEs follows. To streamline and clarify the presentation, we have mostly set various physical parameters to unity in these equations.

*First-Order PDEs*

First-order PDEs appear in many physical theories—mostly in dynamics, continuum mechanics, and optics. For example, in the *scalar conservation law*

$$u_t + \operatorname{div} \boldsymbol{F}(u) = 0, \tag{2}$$

the unknown $u$ is the density of some physically interesting quantity and the vector field $\boldsymbol{F}(u)$, its flux, depends nonlinearly on $u$.

Another important first-order PDE, the *Hamilton–Jacobi equation*

$$u_t + H(\nabla u, x) = 0, \tag{3}$$

appears in classical mechanics and in optimal control theory. In these contexts, $H$ is called the *Hamiltonian*.

*Second-Order PDEs*

Second-order PDEs model a significantly wider variety of physical phenomena than do first-order equations. For example, among its many other interpretations, LAPLACE'S EQUATION [III.18]

$$\Delta u = 0 \tag{4}$$

records diffusion effects in equilibrium. Its time-dependent analogue is the HEAT EQUATION [III.8]

$$u_t - \Delta u = 0, \tag{5}$$

which is also known as the *diffusion equation.*

The WAVE EQUATION [III.31]

$$u_{tt} - c^2 \Delta u = 0 \tag{6}$$

superficially somewhat resembles the heat equation, but as the name suggests, it supports solutions with utterly different behavior.

SCHRÖDINGER'S EQUATION [III.26]

$$\mathrm{i}u_t + \Delta u = 0, \tag{7}$$

for which solutions $u$ are complex-valued, is the quantum mechanics analogue of the wave equation.

*Systems of PDEs*

In a SYSTEM OF CONSERVATION LAWS [II.6]

$$\boldsymbol{u}_t + \operatorname{div} \boldsymbol{F}(\boldsymbol{u}) = \boldsymbol{0}, \tag{8}$$

each component of $\boldsymbol{u} = (u^1, \dots, u^m)$ typically represents a mass, momentum, or energy density.

A *reaction–diffusion system* of PDEs has the form

$$\boldsymbol{u}_t - \Delta \boldsymbol{u} = \boldsymbol{f}(\boldsymbol{u}). \tag{9}$$

Here, the components of $\boldsymbol{u}$ typically represent densities of, say, different chemicals, whose interactions are modeled by the nonlinear term $\boldsymbol{f}$.

The simplest form of MAXWELL'S EQUATIONS [III.22] reads

$$\left.\begin{aligned} \boldsymbol{E}_t &= \operatorname{curl} \boldsymbol{B}, \\ \boldsymbol{B}_t &= -\operatorname{curl} \boldsymbol{E}, \\ \operatorname{div} \boldsymbol{E} &= \operatorname{div} \boldsymbol{B} = 0, \end{aligned}\right\} \tag{10}$$

in which $\boldsymbol{E}$ is the electric field and $\boldsymbol{B}$ the magnetic field.

Fluid mechanics provides some of the most complicated and fascinating systems of PDEs in applied mathematics. The most important are EULER'S EQUATIONS [III.11] for incompressible, inviscid fluid flow,

$$\left.\begin{aligned} \boldsymbol{u}_t + \boldsymbol{u} \cdot \nabla \boldsymbol{u} &= -\nabla p, \\ \operatorname{div} \boldsymbol{u} &= 0, \end{aligned}\right\} \tag{11}$$

and the NAVIER–STOKES EQUATIONS [III.23] for incompressible, viscous flow,

$$\left.\begin{aligned} \boldsymbol{u}_t + \boldsymbol{u} \cdot \nabla \boldsymbol{u} - \Delta \boldsymbol{u} &= -\nabla p, \\ \operatorname{div} \boldsymbol{u} &= 0. \end{aligned}\right\} \tag{12}$$

In these systems $\boldsymbol{u}$ denotes the fluid velocity and $p$ the pressure.

*Higher-Order PDEs*

Equations of order greater than two are much less common. Generally speaking, such higher-order PDEs do not represent fundamental physical laws but are rather derived from such.

For instance, we can sometimes rewrite a system of two second-order equations as a single fourth-order PDE. In this way, the *biharmonic equation*

$$\Delta^2 u = 0 \tag{13}$$

comes up in linear elasticity theory.

The KORTEWEG–DE VRIES (KDV) EQUATION [III.16]

$$u_t + auu_x + bu_{xxx} = 0, \tag{14}$$

a model of shallow-water waves, similarly appears when we combine a complicated system of lower-order equations appearing in appropriate asymptotic expansions.

### 1.3 Boundary and Initial Conditions

Partial differential equations very rarely appear alone; most problems require us to solve the PDEs subject to appropriate boundary and/or initial conditions. If, for instance, we are to study a solution $u = u(x)$, defined for points $x$ lying in some region $U \subset \mathbb{R}^n$, we usually also prescribe something about how $u$ behaves on the boundary $\partial U$. The most common prescriptions are *Dirichlet's boundary condition*

$$u = 0 \quad \text{on } \partial U \tag{15}$$

and *Neumann's boundary condition*

$$\frac{\partial u}{\partial \nu} = 0 \quad \text{on } \partial U, \tag{16}$$

where $\nu$ denotes the outward-pointing unit normal to the boundary and $\partial u / \partial \nu := \nabla u \cdot \nu$ is the *outer normal derivative.* If, say, $u$ represents a temperature, then (15) specifies that the temperature is held constant on the boundary and (16) specifies that the heat flux through the boundary is zero.

Imposing initial conditions is usually appropriate for time-dependent PDEs, for which we require of the solution $u = u(x, t)$ that

$$u(\cdot, 0) = g, \tag{17}$$

where $g = g(x)$ is a given function, comprising the *initial data.* For PDEs that are second order in time, such as the wave equation (6), it is usually appropriate to also specify

$$u_t(\cdot, 0) = h. \tag{18}$$

## 2 Understanding PDEs

In this section we explore several general procedures for understanding PDEs and their solutions.

### 2.1 Exact Solutions

The most effective approach is, of course, just to solve the PDE outright, if we can. For instance, the boundary-value problem

$$\Delta u = 0 \quad \text{in } B(0,1),$$
$$u = g \quad \text{on } \partial B(0,1)$$

is solved by *Poisson's formula*,

$$u(x) = \frac{1-|x|^2}{n\alpha(n)} \int_{\partial B(0,1)} \frac{g(y)}{|x-y|^n}\,\mathrm{d}S,$$

where $\alpha(n)$ denotes the volume of the unit ball in $\mathbb{R}^n$.

The solution of the initial-value problem for the wave equation in one space dimension,

$$u_{tt} - c^2 u_{xx} = 0 \quad \text{in } \mathbb{R}\times(0,\infty),$$
$$u = g, \quad u_t = h \quad \text{on } \mathbb{R}\times\{t=0\},$$

is provided by *d'Alembert's formula*:

$$u(x,t) = \frac{g(x+ct)+g(x-ct)}{2} + \frac{1}{2c}\int_{x-ct}^{x+ct} h(y)\,\mathrm{d}y. \tag{19}$$

The wave equation can also be solved in higher dimensions, but the formulas become increasingly complicated. For example, *Kirchhoff's formula*,

$$u(x,t) = \frac{1}{4\pi c^2 t}\int_{\partial B(x,ct)} h\,\mathrm{d}S + \frac{\partial}{\partial t}\left\{\frac{1}{4\pi c^2 t}\int_{\partial B(x,ct)} g\,\mathrm{d}S\right\}, \tag{20}$$

satisfies this initial-value problem for the wave equation in three space dimensions:

$$\left.\begin{aligned} u_{tt} - c^2\Delta u = 0 \quad &\text{in } \mathbb{R}^3\times(0,\infty), \\ u = g, \quad u_t = h \quad &\text{on } \mathbb{R}^3\times\{t=0\}. \end{aligned}\right\} \tag{21}$$

The initial-value problem for the heat equation,

$$\left.\begin{aligned} u_t - \Delta u = 0 \quad &\text{in } \mathbb{R}^n\times(0,\infty), \\ u = g \quad &\text{on } \mathbb{R}^n\times\{t=0\}, \end{aligned}\right\} \tag{22}$$

has for all dimensions the explicit solution

$$u(x,t) = \frac{1}{(4\pi t)^{n/2}}\int_{\mathbb{R}^n} \mathrm{e}^{-|x-y|^2/4t} g(y)\,\mathrm{d}y. \tag{23}$$

Certain nonlinear PDEs, including the KdV equation (14), are also exactly solvable; discovering these so-called *integrable partial differential equations* is a very important undertaking.

*It is however a fundamental truth that we cannot solve most PDEs*, if by to "solve" we mean to come up with a more or less explicit formula for the answer.

### 2.2 Approximate Solutions and Perturbation Methods

It is consequently important to realize that *we can often deduce properties of solutions without actually solving the PDE*, either explicitly or numerically.

One such approach develops systematic perturbation schemes to build small "corrections" to a known solution. There is a vast repertoire of such techniques. Given a PDE depending on a small parameter $\varepsilon$, the idea is to posit some form for the corrections and to plug this guess into the differential equation, trying then to fine-tune the form of the perturbations to make the error as small as possible. These procedures do not usually amount to proofs but rather construct *self-consistent* guesses.

#### *Multiple Scales*

HOMOGENIZATION [II.17] problems entail PDEs whose solutions act quantitatively differently on different spatial or temporal scales, say of respective orders 1 and $\varepsilon$. Often, a goal is to derive simpler *effective PDEs* that yield good approximations. We guess the form of the effective equations by supposing an asymptotic expansion of the form

$$u^\varepsilon(x) \sim \sum_{k=0}^{\infty} \varepsilon^k u^k(x, x/\varepsilon)$$

and showing that the leading term $u^0$ is a function of $x$ alone, solving some kind of simpler equation.

This example illustrates the insight that *simpler behavior often appears in asymptotic limits.*

#### *Asymptotic Matching*

Solutions of PDEs sometimes have quite different properties in different subregions. When this happens we can try to fashion an approximate solution by (a) constructing simpler approximate solutions in each subregion and then (b) appropriately *matching* these solutions across areas of overlap.

A common such application is to *boundary layers.* The *outer expansion* for the solution within some region often has a form like

$$u^\varepsilon(x) \sim \sum_{k=0}^{\infty} \varepsilon^k u^k(x). \tag{24}$$

Suppose we expect different behavior near the boundary, which we take for simplicity to be the plane $\{x_n = 0\}$. We can then introduce the *stretched variables* $y_n =$

$x_n/\varepsilon^\alpha$, $y_i = x_i$ $(i = 1, \dots, n-1)$ and define $\bar{u}^\varepsilon(y) = u^\varepsilon(x)$. We then look for an *inner expansion*:

$$\bar{u}^\varepsilon(y) \sim \sum_{k=0}^{\infty} \varepsilon^k \bar{u}^k(y). \qquad (25)$$

The idea now is to match terms in the outer expansion (24) in the limit $x_n \to 0$ with terms in the inner expansion (25) in the limit $y_n \to \infty$. Working this out determines, for instance, the value of $\alpha$ in the scaling.

### 2.3 Numerical Analysis of PDEs

Devising effective computer algorithms for PDEs is a vast enterprise—far beyond the scope of this article—and great ingenuity has gone into the design and implementation of such methods.

Among the most popular are the FINITE-DIFFERENCE [IV.13 §3] methods (which approximate functions by values at grid points), the *method of lines* (which discretizes all but the time variable), the FINITE-ELEMENT METHOD [II.12] and *spectral methods* (which represent functions using carefully designed basis functions), MULTIGRID METHODS [IV.13 §3] (which employ discretizations across different spatial scales), and the LEVEL SET METHOD [II.24] (which represents free boundaries as a level set of a function).

The design and analysis of such useful numerical methods, especially for nonlinear equations, depends on a good theoretical understanding of the underlying PDE.

### 2.4 Theory and the Importance of Estimates

The fully rigorous theory of PDEs focuses largely on the foundational issues of the existence, smoothness, and, where appropriate, uniqueness of solutions. Once these issues are resolved, at least provisionally, theorists turn their attention to understanding the behavior of solutions.

*A key point is availability, or not, of strong analytic estimates.* Many physically relevant PDEs predict that various quantities are conserved, but these identities are usually not strong enough to be useful, especially in three dimensions. For nonlinear PDEs the higher derivatives solve increasingly complicated, and thus intractable, equations. And so a major dynamic in modern theory is the interplay between (a) deriving "hard" analytic estimates for PDEs and (b) devising "soft" mathematical tools to exploit these estimates. In the remainder of this article we present for many important PDEs the key estimates upon which rigorous mathematical theory is built.

## 3 The Behavior of Solutions

Since PDEs model so vast a range of physical and other phenomena, their solutions display an even vaster range of behaviors. But some of these are more prevalent than others.

### 3.1 Waves

Many PDEs of interest in applied mathematics support at least some solutions displaying "wavelike" behavior.

*The Wave Equation*

The wave equation is, of course, an example, as is most easily seen in one space dimension from d'Alembert's formula (19). This dictates that the solution has the general form $u(x,t) = F(x+ct) + G(x-ct)$ and is consequently the sum of right- and left-moving waves with speed $c$. The wavelike behavior encoded within Kirchhoff's formula (20) in three space dimensions is somewhat less obvious.

*Traveling Waves*

A solution $u$ of a PDE involving time $t$ and the single space variable $x \in \mathbb{R}$ is a *traveling wave* if it has the form

$$u(x,t) = v(x - \sigma t) \qquad (26)$$

for some *speed* $\sigma$. More generally, a solution $u$ of a PDE in more space variables having the form

$$u(x,t) = v(y \cdot x - \sigma t)$$

is a *plane wave*. An extremely useful first step for studying a PDE is to look for solutions with these special structures.

*Dispersion*

It is often informative to look for plane-wave solutions of the complex form

$$u(x,t) = e^{i(y\cdot x - \sigma t)}, \qquad (27)$$

where $\sigma \in \mathbb{C}$ and $y \in \mathbb{R}^n$. We plug the guess (27) into some given linear PDE in order to discover the so-called *dispersion relationship* between $y$ and $\sigma = \sigma(y)$ forced by the algebraic structure.

For example, inserting (27) into the *Klein-Gordon equation*,

$$u_{tt} - \Delta u + m^2 u = 0, \qquad (28)$$

gives $\sigma = \pm(|y|^2 + m^2)^{1/2}$. Hence the speed $\sigma/|y|$ of propagation depends nonlinearly on the frequency of the initial data $e^{iy \cdot x}$. So *waves of different frequencies propagate at different speeds*; this is dispersion.

*Solitons*

As a nonlinear example, putting (26) into the KdV equation (14) with $a = 6$ and $b = 1$ leads to the ordinary differential equation

$$-\sigma v' + 6vv' + v''' = 0,$$

a solution of which is the explicit profile

$$v(s) = \tfrac{1}{2}\sigma \operatorname{sech}^2(\tfrac{1}{2}\sqrt{\sigma}s)$$

for each speed $\sigma$. The corresponding traveling wave, $u(x,t) = v(x - \sigma t)$, is called a *soliton*.

### 3.2 Diffusion and Smoothing

We can read off a lot of interesting quantitative information about the solution $u$ of the initial-value problem (22) for the heat equation from the explicit formula (23).

In particular, notice from (23) that, if the initial data function $g$ is merely integrable, the solution $u$ is infinitely differentiable in both the variables $x$ and $t$ at later times. So *the heat equation instantly smooths its initial data*; this observation makes sense as the PDE models diffusive effects.

### 3.3 Propagation Speeds

It is also easy to deduce from (23) that, if $u$ solves the heat equation, then values of the initial data $g(y)$ at all points $y \in \mathbb{R}^n$ contribute to determining the solution at $(x,t)$ for times $t > 0$. We can interpret this as an "infinite propagation speed" phenomenon.

By contrast, for many time-dependent PDEs we have instead "finite propagation speed." This means that some parts of the initial data do not affect the solution at a given point in space until enough time has passed. This is so for first-order PDEs in general, for the wave equation, and remarkably also for some nonlinear diffusion PDEs, such as the *porous medium equation*

$$u_t - \Delta(u^{\gamma}) = 0 \tag{29}$$

with $\gamma > 1$. The particular explicit solution

$$u(x,t) = \frac{1}{t^{\alpha}}\left(b - \frac{\gamma - 1}{2\gamma}\beta\frac{|x|^2}{t^{2\beta}}\right)_+^{1/(\gamma-1)} \tag{30}$$

for

$$\alpha = \frac{n}{n(\gamma-1)+2}, \qquad \beta = \frac{1}{n(\gamma-1)+2},$$

and $x_+ = \max\{x, 0\}$ shows clearly that the region of positivity moves outward at finite speed.

### 3.4 Pattern Formation

The interplay between diffusion and nonlinear terms can create interesting effects. For example, let $\Phi(z) = \frac{1}{4}(z^2-1)^2$ denote a "two-well" potential, having minima at $z = \pm 1$. Look now at this scalar reaction–diffusion problem in which $\varepsilon > 0$ is a small parameter:

$$\begin{aligned} u_t^{\varepsilon} - \Delta u^{\varepsilon} &= \frac{1}{\varepsilon^2}\Phi'(u^{\varepsilon}) \quad &&\text{in } \mathbb{R}^2 \times (0, \infty), \\ u^{\varepsilon} &= g^{\varepsilon} &&\text{on } \mathbb{R}^2 \times \{t = 0\}. \end{aligned}$$

For suitably designed, initial data functions $g^{\varepsilon}$, it turns out that

$$\lim_{\varepsilon \to 0} u^{\varepsilon}(x,t) = \pm 1;$$

so the solution asymptotically goes to one or the other of the two minima of $\Phi$. We can informally think of these regions as being colored black and white.

For each time $t \geqslant 0$, denote by $\Gamma(t)$ the curve between the regions $\{u^{\varepsilon}(\cdot,t) \to 1\}$ and $\{u^{\varepsilon}(\cdot,t) \to -1\}$. Asymptotic matching methods reveal that *the normal velocity of $\Gamma(t)$ equals its curvature.* This is a *geometric law of motion* for the evolving black/white patterns emerging in the asymptotic limit as $\varepsilon \to 0$.

Much more complex pattern formation effects can be modeled by systems of reaction–diffusion PDEs of the general form (9): see the article on PATTERN FORMATION [IV.27] elsewhere in this volume.

### 3.5 Blow-up

Solutions of time-dependent PDEs may or may not exist for all future times, even if their initial conditions at time $t = 0$ are well behaved. Note, for example, that among solutions of the nonlinear heat equation

$$u_t - \Delta u = u^2, \tag{31}$$

subject to Neumann boundary conditions (16), are those solutions $u = u(t)$ that do not depend on $x$ and consequently that solve the ordinary differential equation $u_t = u^2$. It is not hard to show that solutions of this equation go to infinity ("blow up") at a finite positive time, if $u(0) > 0$.

For more general initial data, there is an interesting competition between the diffusive, and therefore stabilizing, term $\Delta u$ and the destabilizing term $u^2$.

### 3.6 Shocks

As we have just seen, a solution of a time-dependent PDE can fail to exist for large times since its maximum may explode to infinity in finite time. But there are other mechanisms whereby a solution may cease to exist; it is possible, for example, that a solution remains bounded, but its gradient becomes singular in finite time.

This effect occurs for conservation laws (2). Consider, for example, the following initial-value problem for the BURGERS EQUATION [III.4]:

$$\left.\begin{aligned} u_t + \tfrac{1}{2}(u^2)_x &= 0 \quad \text{in } \mathbb{R} \times (0, \infty), \\ u &= g \quad \text{on } \mathbb{R} \times \{t = 0\}. \end{aligned}\right\} \tag{32}$$

Assume we have a smooth solution $u$ and define the *characteristic curve* $x(t)$ to solve the ordinary differential equation

$$\begin{aligned} \dot{x}(t) &= u(x(t), t) \quad (t \geqslant 0), \\ x(0) &= x_0. \end{aligned}$$

Then

$$\begin{aligned} \frac{\mathrm{d}}{\mathrm{d}t} u(x(t), t) &= u_x(x(t), t)\dot{x}(t) + u_t(x(t), t) \\ &= u_x(x(t), t) u(x(t), t) + u_t(x(t), t) \\ &= 0, \end{aligned}$$

according to the PDE (32). Consequently, $u(x(t), t) \equiv g(x_0)$, and also the characteristic "curve" $x(t)$ is in fact a straight line.

So far, so good; and yet the foregoing often implies that *the PDE does not in fact possess a smooth solution*, existing for all times. To see this, notice that we can easily build initial data $g$ for which the characteristic lines emanating from two distinct initial points cross at some later time, say at $(x, t)$. If we then use these two different characteristics to compute $u(x, t)$, we will get different answers. This seeming paradox is resolved once we understand that the Burgers equation with the initial data $g$ simply does not have a smooth solution existing until the time $t$.

A major task for the rigorous analysis of the Burgers equation and related conservation laws is characterizing surfaces of discontinuity (called SHOCKS [II.30]) for appropriately defined generalized solutions.

### 3.7 Free Boundaries

Some very difficult problems require more than just finding the solution of some PDE: one must also find the region within which that PDE holds. Consider, for example, the *Stefan problem*, which asks us to determine the temperature within some body of water surrounded by ice. The temperature distribution solves the heat equation inside a region whose shape changes in time as the ice melts and/or the water freezes. The unknowns are therefore both the temperature profile and the so-called *free boundary* of the water.

There are in general two sorts of such free boundary problems that occur in PDE theory: those for which the free boundary is explicit, such as the Stefan problem, and those for which it is implicit. An example of the latter is the *obstacle problem*:

$$\min\{u, -\Delta u - f\} = 0.$$

The free boundary is

$$\Gamma = \partial\{u > 0\},$$

along which the solution satisfies the *overdetermined boundary conditions* $u = 0, \partial u/\partial \nu = 0$. Many important physical and engineering free boundary problems can be cast as obstacle problems.

Much more complicated free boundary problems occur in fluid mechanics, in which the unknown velocity $\boldsymbol{u}$ satisfies differing sorts of PDE within the *sonic* and *subsonic* regions. We say that the equations *change type* across the free boundary.

## 4 Some Technical Methods

So vast is the field of PDEs that no small handful of procedures can possibly handle them all. Rather, mathematicians have discovered over the years, and continue to discover, all sorts of useful technical devices and tricks. This section provides a selection of some of the most important.

### 4.1 Transform Methods

A panoply of integral transforms is available to convert linear, constant-coefficient PDEs into algebraic equations. The most important is the FOURIER TRANSFORM [II.19]:

$$\hat{u}(y) := \frac{1}{(2\pi)^{n/2}} \int_{\mathbb{R}^n} \mathrm{e}^{-\mathrm{i}x \cdot y} u(x)\, \mathrm{d}x.$$

Consider, as an example, the equation

$$-\Delta u + u = f \quad \text{in } \mathbb{R}^n. \tag{33}$$

We apply the Fourier transform and learn that $(1 + |y|^2)\hat{u} = \hat{f}$. This algebraic equation lets us easily find $\hat{u}$, after which a somewhat tricky inversion yields the formula

$$u(x) = \frac{1}{(4\pi)^{n/2}} \int_0^\infty \int_{\mathbb{R}^n} \frac{\mathrm{e}^{-s-(|x-y|^2/4s)}}{s^{n/2}} f(y)\, \mathrm{d}y\, \mathrm{d}s.$$

Strongly related are *Fourier series* methods, which represent solutions of certain PDEs on bounded domains as infinite sums entailing sines and cosines. Another favorite is the LAPLACE TRANSFORM [II.19], which for PDEs is mostly useful as a transform in the time variable.

### 4.2 Energy Methods and the Functional Analytic Framework

For many PDEs, various sorts of "energy estimates" are valid, where we use this term loosely to mean integral expressions involving squared quantities.

*Integration by Parts*

Important for what follows is the *integration by parts formula*:

$$\int_U u_{x_i} v \,\mathrm{d}x = -\int_U u v_{x_i} \,\mathrm{d}x + \int_{\partial U} u v \nu^i \,\mathrm{d}S$$

for each $i = 1, \dots, n$. Here, $\nu$ denotes the outward-pointing unit normal to the boundary. This is a form of the DIVERGENCE THEOREM [I.2 §24] from multivariable calculus.

*Energy Estimates*

Assume that $u$ solves *Poisson's equation*:

$$-\Delta u = f \quad \text{in } \mathbb{R}^n. \tag{34}$$

Then, assuming that $u$ goes to zero as $|x| \to \infty$ fast enough to justify the integration by parts, we compute that

$$\begin{aligned}\int_{\mathbb{R}^n} f^2 \,\mathrm{d}x &= \int_{\mathbb{R}^n} \sum_{i,j=1}^n u_{x_i x_i} u_{x_j x_j} \,\mathrm{d}x \\ &= -\int_{\mathbb{R}^n} \sum_{i,j=1}^n u_{x_i x_i x_j} u_{x_j} \,\mathrm{d}x \\ &= \int_{\mathbb{R}^n} \sum_{i,j=1}^n (u_{x_i x_j})^2 \,\mathrm{d}x.\end{aligned}$$

This identity implies something remarkable: if the Laplacian $\Delta u$ (which is the sum of the pure second derivatives $u_{x_i x_i}$ for $i = 1, \dots, n$) is square integrable, then each individual second derivative $u_{x_i x_j}$ for $i, j = 1, \dots, n$ is square integrable, even those mixed second derivatives that do not even appear in (34).

This is an example of *regularity theory*, which aims to deduce the higher integrability and/or smoothness properties of solutions.

*Time-Dependent Energy Estimates*

As a next example suppose that $u = u(x,t)$ solves the wave equation (6), and define the *energy* at time $t$:

$$e(t) := \frac{1}{2} \int_{\mathbb{R}^n} (u_t^2 + c^2 |\nabla u|^2) \,\mathrm{d}x.$$

Then, assuming that $u$ goes to zero as $|x| \to \infty$ fast enough, we have

$$\begin{aligned}\dot{e}(t) &= \int_{\mathbb{R}^n} (u_t u_{tt} + c^2 \nabla u \cdot \nabla u_t) \,\mathrm{d}x \\ &= \int_{\mathbb{R}^n} u_t (u_{tt} - c^2 \Delta u) \,\mathrm{d}x = 0.\end{aligned}$$

This demonstrates *conservation of energy*.

For a *nonlinear wave equation* of the form

$$u_{tt} - \Delta u + f(u) = 0, \tag{35}$$

a similar calculation works for the modified energy

$$e(t) = \int_{\mathbb{R}^n} (\tfrac{1}{2} u_t^2 + \tfrac{1}{2} |\nabla u|^2 + F(u)) \,\mathrm{d}x,$$

where $f = F'$.

### 4.3 Variational Problems

By far the most successful of the nonlinear theories is THE CALCULUS OF VARIATIONS [IV.6]; indeed, a fundamental question to ask of any given PDE is whether or not it is *variational*, meaning that it appears as follows.

Given the *Lagrangian density function* $L = L(v, z, x)$, we introduce the functional

$$I[u] := \int_U L(\nabla u, u, x) \,\mathrm{d}x,$$

defined for functions $u \colon U \to \mathbb{R}$, subject to given boundary conditions that are not specified here. Suppose hereafter that $u$ is a *minimizer* of $I[\cdot]$.

We will show that $u$ *automatically solves an appropriate PDE*. To see this, put $i(\tau) := I[u + \tau v]$, where $v$ vanishes near $\partial U$. Since $i$ has a minimum at $\tau = 0$, we can use the chain rule to compute

$$0 = i'(0) = \int_U (\nabla_v L \cdot \nabla v + L_z v) \,\mathrm{d}x;$$

and so

$$0 = \int_U (-\operatorname{div}(\nabla_v L) + L_z) v \,\mathrm{d}x,$$

in which $L$ is evaluated at $(\nabla u, u, x)$. Here we write $\nabla_v L = (L_{v_1}, \dots, L_{v_n})$.

This integral identity is valid for all functions $v$ vanishing on $\partial U$, and from this the EULER-LAGRANGE EQUATION [III.12] follows:

$$-\operatorname{div}(\nabla_v L(\nabla u, u, x)) + L_z(\nabla u, u, x) = 0. \tag{36}$$

*The Nonlinear Poisson Equation*

For example, the Euler-Lagrange equation for

$$I[u] = \int_U \tfrac{1}{2} |\nabla u|^2 - F(u) \,\mathrm{d}x$$

is the *nonlinear Poisson equation*

$$-\Delta u = f(u), \tag{37}$$

where $f = F'$.

*Minimal Surfaces*

The surface area of the graph of a function $u$ is

$$I[u] = \int_U (1 + |\nabla u|^2)^{1/2}\, dx,$$

and the corresponding Euler–Lagrange equation is the *minimal surface equation*

$$\operatorname{div}\left(\frac{\nabla u}{(1 + |\nabla u|^2)^{1/2}}\right) = 0. \tag{38}$$

The expression on the left-hand side is ($n$ times) the *mean curvature* of the surface; and consequently, *a minimal surface has zero mean curvature.*

### 4.4 Maximum Principles

The integral energy methods just discussed can be augmented for certain PDEs with pointwise maximum principle techniques. These are predicated upon the elementary observation that, if the function $u$ attains its maximum at an interior point $x_0$, then

$$u_{x_k}(x_0) = 0, \quad k = 1, \dots, n, \tag{39}$$

and

$$\sum_{k,l=1}^{n} u_{x_k x_l}(x_0)\xi_k\xi_l \leqslant 0, \quad \xi \in \mathbb{R}^n. \tag{40}$$

*Linear Elliptic Equations*

Such insights are essential for understanding the general *second-order linear elliptic equation*

$$Lu = 0, \tag{41}$$

where

$$Lu = -\sum_{i,j=1}^{n} a^{ij}(x)u_{x_i x_j} + \sum_{i=1}^{n} b^i(x)u_{x_i} + c(x)u.$$

We say $L$ is *elliptic* provided the symmetric matrix $((a^{ij}(x)))$ is positive-definite. In usual applications $u$ represents the density of some quantity. The second-order term $\sum_{i,j=1}^{n} a^{ij}u_{x_i x_j}$ records *diffusion*, the first-order term $\sum_{i=1}^{n} b^i u_{x_i}$ represents *transport*, and the zeroth-order term $cu$ describes the local *increase* or *depletion*.

We use the maximum principle to show, for instance, that if $c > 0$ then $u$ *cannot attain a positive maximum at an interior point.* Indeed, if $u$ took on a positive maximum at some point $x_0$, then the first term of $Lu$ at $x_0$ would be nonnegative (according to (40)), the next term would be zero (according to (39)), and the last would be positive. But this is a contradiction, since $Lu(x_0) = 0$.

*Nonlinear Elliptic Equations*

Maximum principle techniques also apply to many highly nonlinear equations, such as the *Hamilton–Jacobi–Bellman* equation:

$$\max_{k=1,\dots,m} \{L^k u\} = 0. \tag{42}$$

This is an important equation in *stochastic optimization theory*, in which each elliptic operator $L^k$ is the infinitesimal generator of a different stochastic process. We leave it to the reader to use the maximum principle to show that a solution of (42) cannot attain an interior maximum or minimum.

Related, but much more sophisticated, maximum principle arguments can reveal many of the subtle properties of solutions to the linear elliptic equation (41) and the nonlinear equation (42).

### 4.5 Differential Inequalities

Since solutions of PDEs depend on many variables, another useful trick is to design appropriate integral expressions over all but one of these variables, in the hope that these expressions will satisfy interesting differential inequalities in the remaining variable.

*Dissipation Estimates and Gradient Flows*

For example, let $u = u(x,t)$ solve the *nonlinear gradient flow* equation

$$u_t - \operatorname{div}(\nabla L(\nabla u)) = 0 \tag{43}$$

in $\mathbb{R}^n \times (0, \infty)$. Put

$$e(t) := \frac{1}{2}\int_{\mathbb{R}^n} L(\nabla u)\, dx.$$

Then, assuming that $u$ goes to zero rapidly as $|x| \to \infty$, we have

$$\begin{aligned} \dot{e}(t) &= \int_{\mathbb{R}^n} \nabla L(\nabla u) \cdot \nabla u_t\, dx \\ &= -\int_{\mathbb{R}^n} (\operatorname{div} \nabla L(\nabla u))u_t\, dx \\ &= -\int_{\mathbb{R}^n} (u_t)^2\, dx \leqslant 0. \end{aligned}$$

This is a dynamic *dissipation* inequality.

*Entropy Estimates*

Related to dissipation inequalities are *entropy* estimates for conservation laws. To illustrate these, assume that $u^\varepsilon = u^\varepsilon(x,t)$ solves the *viscous conservation law*

$$u^\varepsilon_t + F(u^\varepsilon)_x = \varepsilon u^\varepsilon_{xx} \tag{44}$$

for $\varepsilon > 0$. Suppose $\Phi$ is a convex function and put

$$e(t) := \int_{\mathbb{R}} \Phi(u^{\varepsilon})\,\mathrm{d}x.$$

Then

$$\begin{aligned}\dot{e}(t) &= \int_{\mathbb{R}} \Phi' u_t^{\varepsilon}\,\mathrm{d}x = \int_{\mathbb{R}} \Phi'(-F_x + \varepsilon u_{xx}^{\varepsilon})\,\mathrm{d}x \\ &= -\int_{\mathbb{R}} (\Psi(u^{\varepsilon})_x + \varepsilon\Phi''(u_x^{\varepsilon})^2)\,\mathrm{d}x \\ &= -\int_{\mathbb{R}} \varepsilon\Phi''(u_x^{\varepsilon})^2\,\mathrm{d}x \leqslant 0,\end{aligned}$$

where $\Psi$ satisfies $\Psi' = \Phi' F'$. What is important is that we have found not just one but rather a large collection of dissipation inequalities, corresponding to each pair of *entropy/entropy flux* functions $(\Phi, \Psi)$.

Finding and utilizing entropy/entropy flux pairs for systems of conservation laws of the form (8) is a major challenge.

*Monotonicity Formulas*

For *monotonicity formulas* we try to find interesting expressions to integrate over balls $B(0, r)$, with center 0, say, and radius $r$. The hope is that these integral quantities will solve useful differential inequalities as functions of $r$.

As an example, consider the system

$$-\Delta \boldsymbol{u} = |D\boldsymbol{u}|^2 \boldsymbol{u}, \quad |\boldsymbol{u}|^2 = 1 \tag{45}$$

for the unknown $\boldsymbol{u} = (u^1, \dots, u^m)$, where we write $|D\boldsymbol{u}|^2 = \sum_{i=1}^{n} \sum_{j=1}^{m} (u_{x_i}^j)^2$. A solution $\boldsymbol{u}$ is called a *harmonic map* into the unit sphere. It is a challenging exercise to derive from (45) the differential inequality

$$\frac{\mathrm{d}}{\mathrm{d}r}\left(\frac{1}{r^{n-2}} \int_{B(0,r)} |D\boldsymbol{u}|^2\,\mathrm{d}x\right) = \frac{2}{r^n} \int_{\partial B(0,r)} \sum_{i,j,k} u_{x_i}^k x_i u_{x_j}^k x_j\,\mathrm{d}S \geqslant 0,$$

from which we deduce that

$$\frac{1}{r^{n-2}} \int_{B(0,r)} |D\boldsymbol{u}|^2\,\mathrm{d}x \leqslant \frac{1}{R^{n-2}} \int_{B(0,R)} |D\boldsymbol{u}|^2\,\mathrm{d}x$$

if $0 < r < R$. This inequality is often useful, as it lets us deduce fine information at small scales $r$ from that at larger scales $R$.

## 5 Theory and Application

The foregoing listing of mathematical viewpoints and technical tricks provides at best a glimpse into the immensity of modern PDE theory, both pure and applied.

### 5.1 Well-Posed Problems

A common goal of most of these procedures is to understand a given PDE (plus appropriate boundary and/or initial conditions) as a *well-posed problem*, meaning that (a) the solution exists, (b) it is unique, and (c) it depends continuously on the given data for the problem. This is usually the beginning of wisdom, as well-posed problems provide the starting point for further theoretical inquiry, for numerical analysis, and for construction of approximate solutions.

### 5.2 Generalized Solutions

*A central theoretical problem is therefore to fashion for any given PDE problem an appropriate notion of solution for which the problem is well-posed.* For linear PDEs the concept of "distributional solutions" is usually the best, but for nonlinear problems there are many, including "viscosity solutions," "entropy solutions," "renormalized solutions," etc.

For example, the unique *entropy solution* of the initial-value problem (2) for a scalar conservation law exists for all positive times, but it may support lines of discontinuities across so-called *shock waves*. Similarly, the unique *viscosity solution* of the initial-value problem for the Hamilton–Jacobi equation (3) generally supports surfaces of discontinuity for its gradient. The explicit solution (30) for the porous medium equation is, likewise, not smooth everywhere and so needs suitable interpretation as a valid generalized solution.

The research literature teems with many such notions, and some of the deepest insights in the field are uniqueness theorems for appropriate generalized solutions.

### 5.3 Learning More

As promised, this article is a wide-ranging survey that actually explains precious little in any detail.

To learn more, interested readers should definitely consult other articles in this volume as well as the following suggested reading. Markowich (2007) is a nice introduction to the subject, with lots of pictures, and Strauss (2008) is a very good undergraduate text, containing derivations of the various formulas cited here. The survey article by Klainerman (2008) is extensive and provides some different viewpoints. My graduate-level textbook (Evans 2010) carefully builds up much of the modern theory of PDEs, but it is aimed at mathematically advanced students.

### Further Reading

Brezis, H. 2012. *Functional Analysis, Sobolev Spaces and Partial Differential Equations*. New York: Springer.
DiBenedetto, E. 2009. *Partial Differential Equations*, 2nd edn. Boston, MA: Birkhäuser.
Evans, L. C. 2010. *Partial Differential Equations*, 2nd edn. Providence, RI: American Mathematical Society.
Garabedian, P. 1964. *Partial Differential Equations*. New York: John Wiley.
John, F. 1991. *Partial Differential Equations*, 4th edn. New York: Springer.
Klainerman, S. 2008. Partial differential equations. In *The Princeton Companion to Mathematics*, edited by T. Gowers. Princeton, NJ: Princeton University Press.
Markowich, P. A. 2007. *Applied Partial Differential Equations: A Visual Approach*. New York: Springer.
Rauch, J. 1991. *Partial Differential Equations*. New York: Springer.
Renardy, M., and R. Rogers. 2004. *An Introduction to Partial Differential Equations*, 2nd edn. New York: Springer.
Strauss, W. 2008. *Partial Differential Equations: An Introduction*, 2nd edn. New York: John Wiley.
Taylor, M. 2001. *Partial Differential Equations*, volumes I–III, 2nd edn. New York: Springer.

## IV.4 Integral Equations

*Rainer Kress*

### 1 Introduction

Some forty years ago when I was working on my thesis, I fell in love with integral equations, one of the most beautiful topics in both pure and applied analysis. This article is intended to stimulate the reader to share this love with me.

The term integral equation was first used by Paul du Bois-Reymond in 1888 for equations in which an unknown function occurs under an integral. Typical examples of such integral equations are

$$\int_0^1 K(x,y)\phi(y)\,\mathrm{d}y = f(x) \tag{1}$$

and

$$\phi(x) + \int_0^1 K(x,y)\phi(y)\,\mathrm{d}y = f(x). \tag{2}$$

In these equations the function $\phi$ is the unknown, and the *kernel* $K$ and the right-hand side $f$ are given functions. Solving one of these integral equations amounts to determining a function $\phi$ such that the equation is satisfied for all $x$ with $0 \leqslant x \leqslant 1$. Equations (1) and (2) carry the name of Ivar Fredholm and are called *Fredholm integral equations* of the *first* and *second kind*, respectively. In the first equation the unknown function only occurs under the integral, whereas in the second equation it also appears outside the integral. Later on we will show that this is more than just a formal difference between the two types of equations. A first impression of the difference can be obtained by considering the special case of a constant kernel $K(x,y) = c \neq 0$ for all $x, y \in [0,1]$. On the one hand, it is easily seen that the equation of the second kind (2) has a unique solution given by

$$\phi(x) = f(x) - \frac{c}{1+c}\int_0^1 f(y)\,\mathrm{d}y$$

provided $c \neq -1$. If $c = -1$ then (2) is solvable if and only if $\int_0^1 f(y)\,\mathrm{d}y = 0$, and the general solution is given by $\phi = f + \gamma$ with an arbitrary constant $\gamma$. On the other hand, the equation of the first kind (1) is solvable if and only if $f$ is a constant: $f(x) = \gamma$ for all $x$, say. In this case every function $\phi$ with mean value $\gamma$ is a solution.

The integration domains in (1) and (2) are not restricted to the interval $[0,1]$. In particular, the integration domain can be multidimensional, and for the integral equation of the first kind, the domain in which the equation is required to be satisfied need not coincide with the integration domain.

The first aim of this article is to guide the reader through part of the historical development of the theory and the applications of these equations. In particular, we discuss their close connection to partial differential equations and emphasize their fundamental role in the early years of the development of functional analysis as the appropriate abstract framework for studying integral (and differential) equations. Then, in the second part of the article, we will illustrate how integral equations play an important role in current mathematical research on *inverse* and *ill-posed problems* in areas such as medical imaging and nondestructive evaluation.

Two mathematical problems are said to be inverse to each other if the formulation of the first problem contains the solution of the second problem, and vice versa. According to this definition, at first glance it seems arbitrary to distinguish one of the two problems as an inverse problem. However, in general, one of the two problems is easier and more intensively studied, while the other is more difficult and less explored. The first problem is then denoted as the direct problem, and the second as the inverse problem.

A wealth of inverse problems arise in the mathematical modeling of noninvasive evaluation and imaging methods in science, medicine, and technology. For imaging devices such as conventional X-rays or X-ray tomography, the direct problem consists of determining the images, i.e., two-dimensional projections of the known density distribution on planar photographic films in conventional X-ray devices and projections along all lines through the object measured via intensity losses in X-ray tomography (for the latter see also section 8). Conversely, the inverse problem demands that we reconstruct the density from the images. More generally, inverse problems answer questions about the cause of a given effect, whereas in the corresponding direct problem the cause is known and the effect is to be determined. A common feature of such inverse problems is their ill-posedness, or instability, i.e., small changes in the measured effect may result in large changes in the estimated cause.

Equations (1) and (2) are linear equations since the unknown function $\phi$ appears in a linear fashion. Though nonlinear integral equations also constitute an important part of the mathematical theory and the applications of integral equations, we do not consider them here.

## 2 Abel's Integral Equation

As an appetizer we consider Abel's integral equation. It was one of the first integral equations in mathematical history. A tautochrone is a planar curve for which the time taken by an object sliding without friction in uniform gravity to reach its lowest point is independent of its starting point. The problem of identifying this curve was solved by Christiaan Huygens in 1659, who, using geometrical tools, established that the tautochrone is a cycloid.

In 1823 Niels Henrik Abel attacked the more general problem of determining a planar curve such that the time of descent for a given starting height $y$ coincides with the value $f(y)$ of a given function $f$. The tautochrone then reduces to the special case when $f$ is a constant. Following Abel we describe the curve by $x = \psi(y)$ (with $\psi(0) = 0$) and, using the principle of conservation of energy, obtain

$$f(y) = \int_0^y \frac{\phi(\eta)}{\sqrt{y-\eta}}\,\mathrm{d}\eta, \quad y > 0, \tag{3}$$

for the total time $f(y)$ required for the object to fall from $P = (\psi(y), y)$ to $P_0 = (0,0)$, where

$$\phi := \sqrt{\frac{1+(\psi')^2}{2g}}$$

and $g$ denotes the acceleration due to gravity. Equation (3) is known as *Abel's integral equation*. Given the shape $\phi$, the falling time $f$ is obtained by simply evaluating the integral on the right-hand side of (3). However, the solution of the generalized tautochrone problem requires the solution of the inverse problem; that is, given the function $f$, the solution $\phi$ of the integral equation (3) has to be found, which is certainly a more challenging task. This solution can be shown to be given by

$$\phi(y) = \frac{1}{\pi}\frac{\mathrm{d}}{\mathrm{d}y}\int_0^y \frac{f(\eta)}{\sqrt{y-\eta}}\,\mathrm{d}\eta, \quad y > 0. \tag{4}$$

For the special case of a constant $f = \pi\sqrt{a/2g}$ with $a > 0$ one obtains from (4), after some calculations, that $[\psi'(y)]^2 = (a/y) - 1$, and it can be seen that the solution of this equation is given by the cycloid with parametric representation

$$(x(t), y(t)) = \tfrac{1}{2}a(t + \sin t, 1 - \cos t), \quad 0 \leqslant t \leqslant \pi.$$

## 3 The Early Years

We proceed by giving a brief account of the close connections between the early development of integral equations and potential theory. For the sake of simplicity we confine the presentation to the two-dimensional case as a model for the practically relevant three-dimensional case. In what follows, $x = (x_1, x_2)$ and $y = (y_1, y_2)$ stand for points or vectors in the Euclidean space $\mathbb{R}^2$. Twice continuously differentiable solutions $u$ of Laplace's equation

$$\frac{\partial^2 u}{\partial x_1^2} + \frac{\partial^2 u}{\partial x_2^2} = 0$$

are called HARMONIC FUNCTIONS [III.18 §1]. They model time-independent temperature distributions, potentials of electrostatic and magnetostatic fields, and velocity potentials of incompressible irrotational fluid flows.

For a simply connected bounded domain $D$ in $\mathbb{R}^2$ with smooth boundary $\Gamma := \partial D$ the *Dirichlet problem* of potential theory consists of finding a harmonic function $u$ in $D$ that is continuous up to the boundary and assumes boundary values $u = f$ on $\Gamma$ for a given continuous function $f$ on $\Gamma$. A first approach to this problem, developed in the early nineteenth century, was to create a so-called single-layer potential by distributing point sources with a density $\phi$ on the boundary curve $\Gamma$, i.e., by looking for a solution in the form

$$u(x) = \int_\Gamma \phi(y) \ln|x-y|\,\mathrm{d}s(y), \quad x \in D. \tag{5}$$

Here, $|\cdot|$ denotes the Euclidean norm, i.e., $|x - y|$ represents the distance between the two points $x$ in $D$ and

$y$ on $\Gamma$. Since $\ln|x-y|$ satisfies Laplace's equation if $x \neq y$, the function $u$ given by (5) is harmonic and in order to satisfy the boundary condition it suffices to choose the unknown function $\phi$ as a solution of the integral equation

$$\int_{\Gamma} \phi(y) \ln|x-y| \, ds(y) = f(x), \quad x \in \Gamma, \tag{6}$$

which is known as *Symm's integral equation.* However, the analysis available at that time did not allow a successful treatment of this integral equation of the first kind. Actually, only in the second half of the twentieth century was a satisfying analysis of (6) achieved. Therefore, it represented a major breakthrough when, in 1856, August Beer proposed to place dipoles on the boundary curve, i.e., to look for a solution in the form of a double-layer potential:

$$u(x) = \int_{\Gamma} \phi(y) \frac{\partial \ln|x-y|}{\partial \nu(y)} \, ds(y), \quad x \in D, \tag{7}$$

where $\nu$ denotes the unit normal vector to the boundary curve $\Gamma$ directed into the exterior of $D$. Now the so-called jump relations from potential theory require that

$$\phi(x) + \frac{1}{\pi} \int_{\Gamma} \phi(y) \frac{\partial \ln|x-y|}{\partial \nu(y)} \, ds(y) = \frac{1}{\pi} f(x) \tag{8}$$

is satisfied for $x \in \Gamma$ in order to fulfill the boundary condition. This is an integral equation of the second kind and as such, in principle, was accessible to the method of successive approximations. However, in order to achieve convergence for the case of convex domains, in 1877 Carl Neumann had to modify the successive approximations into what he called the method of arithmetic means and what we would call a relaxation method in modern terms.

For the general case, establishing the existence of a solution to (8) had to wait until the pioneering results of Fredholm that were published in final form in 1903 in the journal *Acta Mathematica* with the title "Sur une classe d'équations fonctionelles." Fredholm considered equations of the form (2) with a general kernel $K$ and assumed all the functions involved to be continuous and real-valued. His approach was to consider the integral equation as the limiting case of a system of linear algebraic equations by approximating the integral by Riemannian sums. Using Cramer's rule for this linear system, Fredholm passes to the limit by using Koch's theory of infinite determinants from 1896 and Hadamard's inequality for determinants from 1893. The idea of viewing integral equations as the limiting case of linear systems had already been proposed by Volterra in 1896, but it was Fredholm who followed it through successfully.

In addition to equation (2), Fredholm's results also contain the adjoint integral equation that is obtained by interchanging the variables in the kernel function. They can be summarized in the following theorem, which is known as the Fredholm alternative. Note that all four of equations (9)–(12) in Theorem 1 are required to be satisfied for all $0 \leqslant x \leqslant 1$.

**Theorem 1.** Either *the homogeneous integral equations*

$$\phi(x) + \int_0^1 K(x,y)\phi(y) \, dy = 0 \tag{9}$$

*and*

$$\psi(x) + \int_0^1 K(y,x)\psi(y) \, dy = 0 \tag{10}$$

*only have the trivial solutions* $\phi = 0$ *and* $\psi = 0$, *and the inhomogeneous integral equations*

$$\phi(x) + \int_0^1 K(x,y)\phi(y) \, dy = f(x) \tag{11}$$

*and*

$$\psi(x) + \int_0^1 K(y,x)\psi(y) \, dy = g(x) \tag{12}$$

*have unique continuous solutions* $\phi$ *and* $\psi$ *for each continuous right-hand side* $f$ *and* $g$, *respectively,* or *the homogeneous equations (9) and (10) have the same finite number of linearly independent solutions and the inhomogeneous integral equations are solvable if and only if the right-hand sides satisfy* $\int_0^1 f(x)\psi(x) \, dx = 0$ *for all solutions* $\psi$ *to the homogeneous adjoint equation (10) and* $\int_0^1 \phi(x)g(x) \, dx = 0$ *for all solutions* $\phi$ *to the homogeneous equation (9).*

We explicitly note that this theorem implies that for the first of the two alternatives each one of the four properties implies the three others. Hence, in particular, uniqueness for the homogeneous equation (9) implies existence of a solution to the inhomogeneous equation (11) for each right-hand side. This is notable, since it is almost always much simpler to prove uniqueness for a linear problem than to prove existence.

Fredholm's existence results also clarify the existence of a solution to the boundary integral equation (8) for the Dirichlet problem for the Laplace equation. By inserting a parametrization of the boundary curve $\Gamma$ we can transform (8) into the form (2) with a continuous kernel for which the homogeneous equation only allows the trivial solution.

Over the last century this boundary integral equation approach via potentials of the form (5) and (7) has been successfully extended to almost all boundary and initial-boundary-value problems for second-order partial differential equations with constant coefficients, such as the time-dependent heat equation, the time-dependent and the time-harmonic wave equations, and Maxwell's equations, among many others. In addition to settling existence of solutions, boundary integral equations provide an excellent tool for obtaining approximate solutions of the boundary and initial-boundary-value problems by solving the integral equations numerically (see section 5). These so-called boundary element methods compete well with finite-element methods. It is an important part of current research on integral equations to develop, implement, and theoretically justify new efficient algorithms for boundary integral equations for very complex geometries in three dimensions that arise in real applications.

## 4 Impact on Functional Analysis

Fredholm's results on the integral equation (2) initiated the development of modern functional analysis in the 1920s. The almost literal agreement of the Fredholm alternative for linear integral equations as formulated in Theorem 1 with the corresponding alternative for linear systems soon gave rise to research into a broader and more abstract form of the Fredholm alternative. This in turn also allowed extensions of the integral equation theory under weaker regularity requirements on the kernel and solution functions. In addition, many years later it was found that more insight was achieved into the structure of Fredholm integral equations by abandoning the initially very fruitful analogy between integral equations and linear systems altogether.

Frigyes Riesz was the first to find an answer to the search for a general formulation of the Fredholm alternative. In his work from 1916 he interpreted the integral equation as a special case of an equation of the second kind,

$$\phi + A\phi = f,$$

with a compact linear operator $A\colon X \to X$ mapping a normed space $X$ into itself. The notion of a normed space that is common in today's mathematics was not yet available in 1916.

Riesz set his work up in the function space of continuous real-valued functions on the interval $[0,1]$—what we would call the space $C[0,1]$ in today's terminology. He called the maximum of the absolute value of a function $f$ on $[0,1]$ the *norm* of $f$ and confirmed its properties that we now know as the standard norm axioms. Riesz used only these axioms, not the special meaning as the maximum norm.

The concept of a compact operator was not yet available in 1916 either. However, using the notion of compact sets as introduced by Fréchet in 1906, Riesz proved that the integral operator $A$ defined by

$$(A\phi)(x) := \int_0^1 K(x,y)\phi(y)\,\mathrm{d}y, \quad x \in [0,1], \qquad (13)$$

on the space $C[0,1]$ maps bounded sets into relatively compact sets, i.e., in today's terminology, $A$ is a compact operator.

What is fascinating about the work of Riesz is that his proofs are still usable and can be transferred, almost unchanged, from the case of an integral operator in the space of continuous functions to the general case of a compact operator in a normed space. Riesz knew about the generality of his method, explicitly noting that the restriction to continuous functions was not relevant.

Summarizing the results of Riesz gives us the following theorem, in which $I$ denotes the identity operator.

**Theorem 2.** *For a compact linear operator $A\colon X \to X$ in a normed space $X$,* either *$I+A$ is injective and surjective and has a bounded inverse* or *the null space $N(I+A) := \{\phi\colon \phi + A\phi = 0\}$ has nonzero finite dimension and the image space $(I+A)(X)$ is a proper subspace of $X$.*

The central and most valuable part of Riesz's theory is again the equivalence of injectivity and surjectivity. Theorem 2 does not yet completely contain the alternative of Theorem 1 for Fredholm integral equations since a link with an adjoint equation and the characterization of the image space in the second case of the alternative are missing. This gap was closed by results of Schauder from 1929 and by more recent developments from the 1960s.

The following theorem is simply a consequence of the fact that the identity operator on a normed space is compact if and only if $X$ has finite dimension, which we refrain from discussing here. It explains why the difference between the two integral equations (1) and (2) is more than just formal.

**Theorem 3.** *Let $X$ and $Y$ be normed spaces and let $A\colon X \to Y$ be a compact linear operator. Then $A$ cannot have a bounded inverse if $X$ is infinite dimensional.*

## 5 Numerical Solution

The idea for the numerical solution of integral equations of the second kind that is conceptually the most straightforward dates back to Nyström in 1930 and consists of replacing the integral in (2) by numerical integration. Using a quadrature formula

$$\int_0^1 g(x)\,\mathrm{d}x \approx \sum_{k=1}^{n} a_k g(x_k)$$

with quadrature points $x_1,\dots,x_n \in [0,1]$ and quadrature weights $a_1,\dots,a_n \in \mathbb{R}$, such as the composite trapezoidal or composite Simpson rule, we approximate the integral operator (13) by the numerical integration operator

$$(A_n\phi)(x) := \sum_{k=1}^{n} a_k K(x,x_k)\phi(x_k) \qquad (14)$$

for $x \in [0,1]$, i.e., we apply the quadrature formula for $g = K(x,\cdot)\phi$. The solution to the integral equation of the second kind, $\phi + A\phi = f$, is then approximated by the solution of $\phi_n + A_n\phi_n = f$, which reduces to solving a finite-dimensional linear system as follows. If $\phi_n$ is a solution of

$$\phi_n(x) + \sum_{k=1}^{n} a_k K(x,x_k)\phi_n(x_k) = f(x) \qquad (15)$$

for $x \in [0,1]$, then clearly the values $\phi_{n,j} := \phi_n(x_j)$ at the quadrature points satisfy the linear system

$$\phi_{n,j} + \sum_{k=1}^{n} a_k K(x_j,x_k)\phi_{n,k} = f(x_j) \qquad (16)$$

for $j = 1,\dots,n$. Conversely, if $\phi_{n,j}$ is a solution of the system (16), the function $\phi_n$ defined by

$$\phi_n(x) := f(x) - \sum_{k=1}^{n} a_k K(x,x_k)\phi_{n,k} \qquad (17)$$

for $x \in [0,1]$ can be seen to solve equation (15). Under appropriate assumptions on the kernel $K$ and the right-hand side $f$ for a convergent sequence of quadrature rules, it can be shown that the corresponding sequence $(\phi_n)$ of approximate solutions converges uniformly to the solution $\phi$ of the integral equation as $n \to \infty$. Furthermore, it can be established that the error estimates for the quadrature rules carry over to error estimates for the Nyström approximations.

We conclude this short discussion of the numerical solution of integral equations by pointing out that in addition to the Nyström method many other methods are available, such as collocation and Galerkin methods.

## 6 Ill-Posed Problems

In 1923 Hadamard postulated three requirements for problems in mathematical physics: a solution should exist, the solution should be unique, and the solution should depend continuously on the data. The third postulate is motivated by the fact that the data will be measured quantities in applications and will therefore always be contaminated by errors. A problem satisfying all three requirements is called *well-posed*. Otherwise, it is called *ill-posed*. If $A\colon X \to Y$ is a bounded linear operator mapping a normed space $X$ into a normed space $Y$, then the equation $A\phi = f$ is well-posed if $A$ is bijective and the inverse operator $A^{-1}\colon Y \to X$ is bounded, i.e., continuous. Otherwise it is ill-posed. The main concern with ill-posed problems is *instability*, where the solution $\phi$ of $A\phi = f$ does not depend continuously on the data $f$.

As an example of an ill-posed problem we present backward heat conduction. Consider the forward heat equation

$$\frac{\partial u}{\partial t} = \frac{\partial^2 u}{\partial x^2}$$

for the time-dependent temperature $u$ in a rectangle $[0,1] \times [0,T]$ subject to the homogeneous boundary conditions

$$u(0,t) = u(1,t) = 0, \quad 0 \leqslant t \leqslant T,$$

and the initial condition

$$u(x,0) = \phi(x), \quad 0 \leqslant x \leqslant 1,$$

where $\phi$ is a given initial temperature. By separation of variables the solution can be obtained in the form

$$u(x,t) = \sum_{n=1}^{\infty} a_n \mathrm{e}^{-n^2\pi^2 t} \sin n\pi x, \qquad (18)$$

with the Fourier coefficients

$$a_n = 2\int_0^1 \phi(y) \sin n\pi y\,\mathrm{d}y \qquad (19)$$

of the given initial values. This initial-value problem is well-posed: the final temperature $f := u(\cdot\,,T)$ clearly depends continuously on the initial temperature $\phi$ because of the exponentially decreasing factors in the series

$$f(x) = \sum_{n=1}^{\infty} a_n \mathrm{e}^{-n^2\pi^2 T} \sin n\pi x. \qquad (20)$$

However, the corresponding inverse problem, i.e., determination of the initial temperature $\phi$ from knowledge of the final temperature $f$, is ill-posed. From (20) we deduce

$$\phi(x) = \sum_{n=1}^{\infty} b_n \mathrm{e}^{n^2\pi^2 T} \sin n\pi x, \qquad (21)$$

with the Fourier coefficients $b_n$ of the final temperature $f$. Changes in the final temperature will be drastically amplified by the exponentially increasing factors in the series (21).

Inserting (19) into (18), we see that this example can be put in the form of an integral equation of the first kind (1) with the kernel given by

$$K(x,y) = 2\sum_{n=1}^{\infty} e^{-n^2\pi^2 T} \sin n\pi x \sin n\pi y.$$

In general, integral equations of the first kind that have continuous kernels provide typical examples of ill-posed problems as a consequence of Theorem 3.

Of course, the ill-posed nature of an equation has consequences for its numerical solution. The fact that an operator does not have a bounded inverse means that the condition numbers of its finite-dimensional approximations grow with the quality of the approximation. Hence, a careless discretization of ill-posed problems leads to a numerical behavior that at first glance seems to be paradoxical: increasing the degree of discretization (that is, increasing the accuracy of the approximation for the operator) will cause the approximate solution to the equation to become more unreliable.

## 7 Regularization

Methods for obtaining a stable approximate solution of an ill-posed problem are called *regularization methods*. It is our aim to describe a few ideas about regularization concepts for equations of the first kind with a compact linear operator $A\colon X \to Y$ between two normed spaces, $X$ and $Y$. We wish to approximate the solution $\phi$ to the equation $A\phi = f$ from a perturbed right-hand side $f^\delta$ with a known error level $\|f^\delta - f\| \leqslant \delta$. Using the erroneous data $f^\delta$, we want to construct a reasonable approximation $\phi^\delta$ to the exact solution $\phi$ of the unperturbed equation $A\phi = f$. Of course, we want this approximation to be stable, i.e., we want $\phi^\delta$ to depend continuously on the actual data $f^\delta$. Therefore, assuming without major loss of generality that $A$ is injective, our task is to find an approximation of the unbounded inverse operator $A^{-1}\colon A(X) \to X$ by a bounded linear operator $R\colon Y \to X$. With this in mind, a family of bounded linear operators $R_\alpha\colon Y \to X$, $\alpha > 0$, with the property of pointwise convergence

$$\lim_{\alpha\to 0} R_\alpha A\phi = \phi \tag{22}$$

for all $\phi \in X$, is called a *regularization scheme* for the operator $A$. The parameter $\alpha$ is called the regularization parameter.

The regularization scheme approximates the solution $\phi$ of $A\phi = f$ by the regularized solution $\phi_\alpha^\delta := R_\alpha f^\delta$. For the total approximation error by the triangle inequality we then have the estimate

$$\|\phi_\alpha^\delta - \phi\| \leqslant \delta\|R_\alpha\| + \|R_\alpha A\phi - \phi\|.$$

This decomposition shows that the error consists of two parts: the first term reflects the influence of the incorrect data, and the second term is due to the approximation error between $R_\alpha$ and $A^{-1}$. Assuming that $X$ is infinite dimensional, $R_\alpha$ cannot be uniformly bounded, since otherwise $A$ would have a bounded inverse. Consequently, the first term will be increasing as $\alpha \to 0$, whereas the second term will be decreasing as $\alpha \to 0$ according to (22).

Every regularization scheme requires a strategy for choosing the parameter $\alpha$, depending on the error level $\delta$ and the data $f^\delta$, so as to achieve an acceptable total error for the regularized solution. On the one hand, the accuracy of the approximation requires a small error $\|R_\alpha A\phi - \phi\|$; this implies a small parameter $\alpha$. On the other hand, stability requires a small value of $\|R_\alpha\|$; this implies a large parameter $\alpha$. A popular strategy is given by the *discrepancy principle*. Its motivation is based on the consideration that, in general, for erroneous data the residual $\|A\phi_\alpha^\delta - f^\delta\|$ should not be smaller than the accuracy of the measurements of $f$, i.e., the regularization parameter $\alpha$ should be chosen such that $\|AR_\alpha f^\delta - f^\delta\| \approx \delta$.

We now assume that $X$ and $Y$ are Hilbert spaces and denote their inner products by $(\cdot,\cdot)$, with the space $L^2[0,1]$ of Lebesgue square-integrable complex-valued functions on $[0,1]$ as a typical example. Each bounded linear operator $A\colon X \to Y$ has a unique adjoint operator $A^*\colon Y \to X$ with the property $(A\phi, g) = (\phi, A^*g)$ for all $\phi \in X$ and $g \in Y$. If $A$ is compact, then $A^*$ is also compact. The adjoint of the compact integral operator $A\colon L^2[0,1] \to L^2[0,1]$ defined by (13) is given by the integral operator with the kernel $\overline{K(y,x)}$, where the bar indicates the complex conjugate.

Extending the singular value decomposition (SVD) for matrices from linear algebra, for each compact linear operator $A\colon X \to Y$ there exists a singular system consisting of a monotonically decreasing null sequence $(\mu_n)$ of positive numbers and two orthonormal sequences $(\phi_n)$ in $X$ and $(g_n)$ in $Y$ such that

$$A\phi_n = \mu_n g_n, \quad A^*g_n = \mu_n\phi_n, \quad n \in \mathbb{N}.$$

For each $\phi \in X$ we have the SVD

$$\phi = \sum_{n=1}^{\infty} (\phi, \phi_n)\phi_n + P\phi,$$

where $P: X \to N(A)$ is the orthogonal projection operator onto the null space of $A$ and

$$A\phi = \sum_{n=1}^{\infty} \mu_n (\phi, \phi_n) g_n.$$

From the SVD it can be readily deduced that the equation of the first kind, $A\phi = f$, is solvable if and only if $f$ is orthogonal to the null space of $A^*$ and satisfies

$$\sum_{n=1}^{\infty} \frac{1}{\mu_n^2} |(f, g_n)|^2 < \infty. \tag{23}$$

If (23) is fulfilled, a solution is given by

$$\phi = \sum_{n=1}^{\infty} \frac{1}{\mu_n} (f, g_n) \phi_n. \tag{24}$$

The solution (24) clearly demonstrates the ill-posed nature of the equation $A\phi = f$. If we perturb the right-hand side $f$ to $f^\delta = f + \delta g_n$, we obtain the solution $\phi^\delta = \phi + \delta \mu_n^{-1} \phi_n$. Hence, the ratio

$$\frac{\|\phi^\delta - \phi\|}{\|f^\delta - f\|} = \frac{1}{\mu_n}$$

can be made arbitrarily large due to the fact that the singular values tend to zero. This observation suggests that we regularize by damping the influence of the factor $1/\mu_n$ in the solution formula (24). In the *Tikhonov regularization* this is achieved by choosing

$$R_\alpha f := \sum_{n=1}^{\infty} \frac{\mu_n}{\alpha + \mu_n^2} (f, g_n) \phi_n. \tag{25}$$

Computing $R_\alpha f$ does not require the singular system to be used since for injective $A$ it can be shown that

$$R_\alpha = (\alpha I + A^* A)^{-1} A^*.$$

Hence $\phi_\alpha := R_\alpha f$ can be obtained as the unique solution of the well-posed equation of the second kind:

$$\alpha \phi_\alpha + A^* A \phi_\alpha = A^* f.$$

## 8 Computerized Tomography

In transmission COMPUTERIZED TOMOGRAPHY [VII.19] a cross section of an object is scanned by a thin X-ray beam whose intensity loss is recorded by a detector and processed to produce an image. Denote by $f$ the space-dependent attenuation coefficient within a two-dimensional medium. The relative intensity loss of an X-ray along a straight line $L$ is given by $\mathrm{d}I = -If\,\mathrm{d}s$, and by integration, it follows that

$$I_{\text{detector}} = I_{\text{source}} \exp\left(-\int_L f\,\mathrm{d}s\right),$$

i.e., in principle, the scanning process provides the line integrals over all lines traversing the scanned cross section. The transform that maps a function in $\mathbb{R}^2$ onto its line integrals is called the *Radon transform*, and the inverse problem of computerized tomography requires its inversion. Radon had already given an explicit inversion formula in 1917, but it is not immediately applicable for practical computations.

For the formal description of the Radon transform it is convenient to parametrize the line $L$ by its unit normal vector $\theta$ and its signed distance $s$ from the origin in the form $L = \{s\theta + t\theta^\perp : t \in \mathbb{R}\}$, where $\theta^\perp$ is obtained by rotating $\theta$ counterclockwise by 90°. Now, the two-dimensional Radon transform $R$ is defined by

$$(Rf)(\theta, s) := \int_{-\infty}^{\infty} f(s\theta + t\theta^\perp)\,\mathrm{d}t, \quad \theta \in S^1,\ s \in \mathbb{R},$$

and it maps $L^1(\mathbb{R}^2)$ into $L^1(S^1 \times \mathbb{R})$, where $S^1$ is the unit circle. Given the measured line integrals $g$, the inverse problem of computerized tomography consists of solving

$$Rf = g \tag{26}$$

for $f$. Although it is not of the conventional form (1) or (2), equation (26) can clearly be viewed as an integral equation. Its solution can be obtained using Radon's inversion formula

$$f = \frac{1}{4\pi} R^* H \frac{\partial}{\partial s} Rf \tag{27}$$

with the Hilbert transform

$$(Hg)(s) := \frac{1}{\pi} \int_{-\infty}^{\infty} \frac{g(t)}{s - t}\,\mathrm{d}t, \quad s \in \mathbb{R},$$

applied with respect to the second variable in $Rf$. The operator $R^*$ is the adjoint of $R$ with respect to the $L^2$ inner products on $\mathbb{R}^2$ and $S^1 \times \mathbb{R}$, which is given by

$$(R^* g)(x) = \int_{S^1} g(\theta, x \cdot \theta)\,\mathrm{d}\theta, \quad x \in \mathbb{R}^2,$$

i.e., it can be considered as an integration over all lines through $x$ and is therefore called the back-projection operator. Because of the occurrence of the Hilbert transform in (27), inverting the Radon transform is not local; i.e., the line integrals through a neighborhood of the point $x$ do not suffice for the reconstruction of $f(x)$. Due to the derivative appearing in (27), the inverse problem of reconstructing the function $f$ from its line integrals is ill-posed.

In practice, the integrals can be measured only for a finite number of lines and, correspondingly, a discrete version of the Radon transform has to be inverted. The most widely used inversion algorithm is the filtered back-projection algorithm, which may be considered as an implementation of Radon's inversion formula with the middle part $H(\partial/\partial s)$ replaced by a convolution, i.e., a filter in the terminology of image processing. However, so-called algebraic reconstruction techniques are also used where the function $f$ is decomposed into pixels, i.e., where it is approximated by piecewise constants on a grid of little squares. The resulting sparse linear system for the pixel values is then solved iteratively, by Kaczmarz's method, for example.

For the case of a radially symmetric function $f$ (that is, when $f(x) = f_0(|x|)$), $Rf$ clearly does not depend on $\theta$ (that is, $(Rf)(\theta, s) = g_0(s)$), where

$$g_0(s) = 2\int_0^\infty f_0(\sqrt{s^2+t^2})\,\mathrm{d}t, \quad s \geqslant 0.$$

Substituting $t = \sqrt{r^2 - s^2}$, this transforms into

$$g_0(s) = 2\int_s^\infty f_0(r)\frac{r}{\sqrt{r^2-s^2}}\,\mathrm{d}r, \quad s \geqslant 0,$$

which is an Abel-type integral equation again. Its solution is given by

$$f_0(r) = -\frac{1}{\pi}\int_r^\infty g_0'(s)\frac{1}{\sqrt{s^2-r^2}}\,\mathrm{d}s, \quad r \geqslant 0.$$

This approach can be extended to a full inversion formula by expanding both $f$ and $g = Rf$ as Fourier series with respect to the polar angle. The Fourier coefficients of $f$ and $g$ are then related by Abel-type integral equations involving Chebyshev polynomials.

X-ray tomography was first suggested and studied by the physicist Allan Cormack in 1963, and due to the efforts of the electrical engineer Godfrey Hounsfield, it was introduced into medical practice in the 1970s. For their contributions to X-ray tomography Cormack and Hounsfield were awarded the 1979 Nobel Prize for Medicine.

## 9 Inverse Scattering

Scattering theory is concerned with the effects that obstacles and inhomogeneities have on the propagation of waves and particularly time-harmonic waves. Inverse scattering provides the mathematical tools for fields such as radar, sonar, medical imaging, and nondestructive testing.

For time-harmonic waves the time dependence is represented in the form $U(x,t) = \mathrm{Re}\{u(x)\mathrm{e}^{-\mathrm{i}\omega t}\}$ with a positive frequency $\omega$; i.e, the complex-valued space-dependent part $u$ represents the real-valued amplitude and phase of the wave and satisfies the Helmholtz equation $\Delta u + k^2 u = 0$ with a positive wave number $k$. For a unit vector $d \in \mathbb{R}^3$, the function $\mathrm{e}^{\mathrm{i}kx\cdot d}$ satisfies the Helmholtz equation for all $x \in \mathbb{R}^3$. It is called a plane wave, since $\mathrm{e}^{\mathrm{i}(kx\cdot d-\omega t)}$ is constant on the planes $kx \cdot d - \omega t = \text{const}$. Assume that an incident field is given by $u^{\text{in}}(x) = \mathrm{e}^{\mathrm{i}kx\cdot d}$. Then the simplest obstacle scattering problem is to find the scattered field $u^{\text{sc}}$ as a solution to the Helmholtz equation in the exterior of a bounded scatterer $D \subset \mathbb{R}^3$ such that the total field $u = u^{\text{in}} + u^{\text{sc}}$ satisfies the Dirichlet boundary condition $u = 0$ on $\partial D$ modeling a sound-soft obstacle or a perfect conductor. In addition, to ensure that the scattered wave is outgoing, it has to satisfy the Sommerfeld radiation condition

$$\lim_{r\to\infty} r\left(\frac{\partial u^{\text{sc}}}{\partial r} - \mathrm{i}ku^{\text{sc}}\right) = 0, \qquad (28)$$

where $r = |x|$ and the limit holds uniformly in all directions $x/|x|$. This ensures uniqueness of the solution to the exterior Dirichlet problem for the Helmholtz equation. Existence of the solution was established in the 1950s by Vekua, Weyl, and Müller via boundary integral equations in the spirit of section 3.

The radiation condition (28) can be shown to be equivalent to the asymptotic behavior

$$u^{\text{sc}}(x) = \frac{\mathrm{e}^{\mathrm{i}k|x|}}{|x|}\left\{u_\infty(\hat{x}) + O\left(\frac{1}{|x|}\right)\right\}, \quad |x| \to \infty,$$

uniformly for all directions $\hat{x} = x/|x|$, where the function $u_\infty$ defined on the unit sphere $S^2$ is known as the *far-field pattern* of the scattered wave. We indicate its dependence on the incident direction $d$ and the observation direction $\hat{x}$ by writing $u_\infty = u_\infty(\hat{x}, d)$. The inverse scattering problem now consists of determining the scattering obstacle $D$ from a knowledge of $u_\infty$. As an example of an application, we could think of the problem of determining from the shape of the water waves arriving at the shore whether a ball or a cube was thrown into the water in the middle of a lake. We note that this inverse problem is nonlinear since the scattered wave depends nonlinearly on the scatterer $D$, and it is ill-posed since the far-field pattern $u_\infty$ is an analytic function on $S^2$ with respect to $\hat{x}$.

Roughly speaking, one can distinguish between three groups of methods for solving the inverse obstacle scattering problem: iterative methods, decomposition methods, and sampling methods. Iterative methods interpret the inverse problem as a nonlinear ill-posed operator equation that is solved by methods such as regularized Newton-type iterations. The main idea of decomposition methods is to break up the inverse scattering problem into two parts: the first part deals with the ill-posedness by constructing the scattered wave $u^{\text{sc}}$ from its far-field pattern $u_\infty$, and the second part deals with the nonlinearity by determining the unknown boundary $\partial D$ of the scatterer as the set of points where the boundary condition for the total field is satisfied. Since boundary integral equations play an essential role in the existence analysis and numerical solution of the direct scattering problem, it is not surprising that they are also an efficient tool within these two groups of methods for solving the inverse problem.

Sampling methods are based on choosing an appropriate indicator function $f$ on $\mathbb{R}^3$ such that its value $f(z)$ indicates whether $z$ lies inside or outside the scatterer $D$. In contrast to iterative and decomposition methods, sampling methods do not need any a priori information on the geometry of the obstacle. However, they do require knowledge of the far-field pattern for a large number of incident waves, whereas the iterative and decomposition methods, in principle, work with just one incident field.

For two of the sampling methods—the so-called linear sampling method proposed by Colton and Kirsch and the factorization method proposed by Kirsch—the indicator functions are defined using ill-posed linear integral equations of the first kind involving the integral operator $F\colon L^2(S^2) \to L^2(S^2)$ with kernel $u_\infty(\hat{x}, d)$ given by

$$(Fg) := \int_{S^2} u_\infty(\hat{x}, d) g(d)\, \mathrm{d}s(d), \quad \hat{x} \in S^2.$$

With the far-field pattern

$$\Phi_\infty(\hat{x}, z) = (4\pi)^{-1} \mathrm{e}^{-\mathrm{i}k\hat{x}\cdot z}$$

of the fundamental solution

$$\Phi(x, z) := \frac{\mathrm{e}^{\mathrm{i}k|x-z|}}{4\pi|x-z|}, \quad x \neq z,$$

of the Helmholtz equation with source point at $z \in \mathbb{R}^3$, the linear sampling method is based on the ill-posed equation

$$Fg_z = \Phi_\infty(\cdot, z), \qquad (29)$$

whereas the factorization method is based on

$$(F^* F g_z)^{1/4} = \Phi_\infty(\cdot, z). \qquad (30)$$

An essential tool in the linear sampling method is the Herglotz wave function with kernel $g$, defined as the superposition of plane waves given by

$$v_g(x) := \int_{S^2} \mathrm{e}^{\mathrm{i}kx\cdot d} g(d)\, \mathrm{d}s(d), \quad x \in \mathbb{R}^3.$$

It can be shown that, if $z \in D$, then the value of the Herglotz wave function $v_{g_{z,\alpha}}(z)$ with the kernel $g_{z,\alpha}$ given by the solution of (29) obtained by Tikhonov regularization with parameter $\alpha$ remains bounded as $\alpha \to 0$, whereas it is unbounded if $z \notin D$. Evaluating $v_{g_{z,\alpha}}(z)$ on a sufficiently fine grid of points $z$, the scatterer $D$ can be visualized as the set of those points where $v_{g_{z,\alpha}}(z)$ is small. The main feature of the factorization method is the fact that equation (30) is solvable in $L^2(S^2)$ if and only if $z \in D$. With the aid of the solubility condition (23) in terms of a singular system of $F$, this can be utilized to visualize the scatterer $D$ as the set of points $z$ from a grid where the series (23), applied to the equation (30), converges, that is, where its approximation by a finite sum remains small.

Three of the items in the further reading list below are evidence of my enduring love of integral equations.

### Further Reading

Colton, D., and R. Kress. 2013. *Inverse Acoustic and Electromagnetic Scattering Theory*, 3rd edn. New York: Springer.

Ivanyshyn, O., R. Kress, and P. Serranho. 2010. Huygens' principle and iterative methods in inverse obstacle scattering. *Advances in Computational Mathematics* 33:413–29.

Kirsch, A., and N. Grinberg. 2008. *The Factorization Method for Inverse Problems*. Oxford: Oxford University Press.

Kress, R. 2014. *Linear Integral Equations*, 3rd edn. New York: Springer.

Natterer, F. 2001. *The Mathematics of Computerized Tomography*. Philadelphia, PA: SIAM.

## IV.5 Perturbation Theory and Asymptotics

*Peter D. Miller*

### 1 Introduction

Perturbation theory is a tool for dealing with certain kinds of physical or mathematical problems involving parameters. For example, the behavior of many problems of turbulent fluid mechanics is influenced by the value of the Reynolds number, a dimensionless parameter that measures the relative strength of forces applied to the fluid compared with viscous damping forces. Likewise, in quantum theory the Planck constant $\hbar$ is a parameter in the Schrödinger equation that governs dynamics.

The key idea of perturbation theory is to try to take advantage of special values of parameters for which the problem of interest can be solved easily to get information about the solution for nearby values of the parameters. As the parameters are *perturbed* from their original simplifying values, one expects that the solution will also be correspondingly perturbed. Perturbation methods allow one to compute the way in which the solution changes under perturbation, and perturbation theory explains how the resulting computation is to be properly understood.

Clearly the potential for perturbation theory success is tied to the possibility that a parameter can be regarded as being tunable. Tunability is sometimes rather obvious; for instance, the Reynolds number is tuned in fluid experiments either by changing the applied forces (this changes the numerator) or by using different fluids with different viscosities (this changes the denominator). In quantum mechanics, however, it is not reasonable to tune Planck's constant, as it takes a fixed value: $\hbar \approx 1.05 \times 10^{-34}$ kg m$^2$ s$^{-1}$. Here, tunability can be recovered by nondimensionalizing the problem; one should introduce units $M$ of mass, $L$ of length, and $T$ of time that are characteristic of the problem at hand and then consider the dimensionless ratio $H := T\hbar/ML^2$ instead of $\hbar$. The dimensionless parameter $H$ then becomes tunable via $M$, $L$, and $T$.

In this introduction we have followed the overwhelming majority of the literature and used the terms "perturbation theory" and "perturbation methods" almost interchangeably. However, we will try to be more precise from now on, referring to perturbation methods when describing the mechanical construction of approximate solutions of perturbed problems, while describing the mathematical analysis of the approximations obtained and their convergence properties as perturbation theory.

### 1.1 A Basic Example

As a first example of perturbation methods, suppose we want to find real solutions $x$ of the polynomial equation

$$x^5 + a_4x^4 + a_3x^3 + a_2x^2 + a_1x + a_0 = 0. \quad (1)$$

Here, the real coefficients $a_j$ are the parameters of the problem. There is no explicit formula for the roots $x$ of a quintic polynomial in general, but for certain values of the parameters the situation is obviously much better. For example, if $a_4 = a_3 = a_2 = a_1 = 0$ and $a_0 = -1$, then (1) reduces to the problem $x^5 = 1$, which clearly has a unique real solution, $x = 1$. To perturb from this exactly solvable situation, make the replacements $a_j \to \varepsilon a_j$ for $j = 1, 2, 3, 4$ and $a_0 \to -1+\varepsilon(a_0+1)$. Then, when $\varepsilon = 0$ we have the simple exactly solvable case, and when $\varepsilon = 1$ we recover the general case. It is traditional for the Greek letter $\varepsilon$ to denote a small quantity in perturbation problems. Our problem can therefore be written in the form

$$P_0(x) + \varepsilon P_1(x) = 0, \quad (2)$$

where $\varepsilon \in [0, 1]$ is our tunable parameter, and

$$P_0(x) := x^5 - 1,$$
$$P_1(x) := a_4x^4 + a_3x^3 + a_2x^2 + a_1x + a_0 + 1.$$

At this point, we have redefined the original problem somewhat, as the goal is now to understand how the known roots of (2) for $\varepsilon = 0$ begin to change for small nonzero $\varepsilon$. (It will be clear that while perturbation methods suffice to solve this redefined problem, they may not be sufficiently powerful to describe the solutions of the original problem (1) as it may not be possible to allow $\varepsilon$ to be as large as $\varepsilon = 1$.) In this case, perturbation *theory* amounts to the invocation of the implicit function theorem, which guarantees that, since $P_0'(1) = 5 \neq 0$, there is a unique solution $x = x(\varepsilon)$ of (2) that satisfies $x(0) = 1$ and that can be expanded in a convergent (for $|\varepsilon|$ sufficiently small) power series in powers of $\varepsilon$. On the other hand, perturbation *methods* are concerned with the effective construction of the series itself. We do not attempt to find a closed-form expression for the general term in the series; rather, we find the terms iteratively by the following simple procedure amounting to an algorithm to compute the first $N + 1$ terms. We write

$$x(\varepsilon) = \sum_{n=0}^{N} x_n\varepsilon^n + R_N(\varepsilon),$$

where the remainder term in the Taylor expansion satisfies $\varepsilon^{-N}R_N(\varepsilon) \to 0$ as $\varepsilon \to 0$. By substituting this expression into (2) and expanding out the multinomials $x(\varepsilon)^p$ that occur there, one rewrites (2) in the form

$$\sum_{n=0}^{N} p_n\varepsilon^n + Q_N(\varepsilon) = 0, \quad (3)$$

where the $p_n$ are certain well-defined expressions in terms of $\{x_0, \dots, x_N\}$ and $Q_N(\varepsilon)$ is a remainder that, like $R_N(\varepsilon)$, satisfies $\varepsilon^{-N}Q_N(\varepsilon) \to 0$ as $\varepsilon \to 0$. Since (3) should hold for all sufficiently small $\varepsilon$, it is easy to show that we must have $p_n = 0$ for all $n = 0, \dots, N$. This is a system of equations for the unknown coefficients $\{x_0, \dots, x_N\}$. The first few values of $p_n$ are

$$p_0 := P_0(x_0),$$
$$p_1 := P_0'(x_0)x_1 + P_1(x_0),$$
$$p_2 := 2P_0'(x_0)x_2 + P_0''(x_0)x_1^2 + 2P_1'(x_0)x_1.$$

These display a useful triangular structure, in that $p_n$ depends only on $\{x_0, \dots, x_n\}$, and we also note that $p_n$ is linear in $x_n$. These features actually hold for all

$n$, and they allow the construction of the series coefficients in $x(\varepsilon)$ in a completely systematic fashion once the unperturbed solution $x_0 = 1$ is specified. Indeed, considering the equations $p_1 = 0$, $p_2 = 0$, and so on in order, we see that

$$x_1 = -\frac{P_1(x_0)}{P_0'(x_0)},$$

$$x_2 = -\frac{P_1'(x_0)x_1}{P_0'(x_0)} - \frac{P_0''(x_0)x_1^2}{2P_0'(x_0)}$$
$$= \frac{P_1(x_0)P_1'(x_0)}{P_0'(x_0)^2} - \frac{P_1(x_0)^2 P_0''(x_0)}{2P_0'(x_0)^3},$$

and so on. Note that the denominators are nonzero under exactly the same condition that the implicit function theorem applies. In this way, the perturbation series coefficients $x_n$ are systematically determined one after the other. Note also that if we were interested in complex roots $x$ of the quintic, we could equally well have started developing the perturbation series from any of the five complex roots of unity $x_0 = e^{2\pi ik/5}$ for $k = 0, 1, 2, 3, 4$.

This example shows several of the most elementary features of perturbation methods:

(i) The effect of perturbation of the parameter $\varepsilon$ from the special value $\varepsilon = 0$ is to introduce corrections to the unperturbed solution $x_0$ in the form of an infinite perturbation series of corrections of higher and higher order.

(ii) Once the leading term of the series has been obtained by solving the reduced problem with $\varepsilon = 0$, the subsequent terms of the perturbation series are all obtained by solving inhomogeneous linear equations of the form $P_0'(x_0)u = f$, where $f$ is given in terms of previously calculated terms and $P_0'(x_0)$ denotes the linearization of the unperturbed problem about its exact solution $x_0$.

The latter feature makes perturbation methods an attractive way to attack nonlinear problems, as the procedure for calculating corrections always involves solving only linear problems.

## 2 Asymptotic Expansions

### 2.1 Motivation

Consider the second-order differential equation for $y(x)$:

$$-\varepsilon x^3 y''(x) + y(x) = x^2. \tag{4}$$

Let us try to solve (4) for $x > 0$. This equation generally has no elementary solutions, but we may notice that when $\varepsilon = 0$ it is obvious that $y(x) = x^2$. Taking a perturbative approach to include the effect of the neglected term, we may seek a solution in the form of a power series in $\varepsilon$:

$$y(x) \sim \sum_{n=0}^{\infty} y_n(x)\varepsilon^n, \quad y_0(x) := x^2. \tag{5}$$

The notation "$\sim$" will be properly explained below in section 2.2; for now the reader may think of it as "=". Substituting this series into (4) and equating the terms with corresponding powers of $\varepsilon$ gives the recurrence relation

$$y_n(x) = x^3 y_{n-1}''(x), \quad n > 0. \tag{6}$$

This recurrence is easily solved, and one finds that $y_n(x) = (n+1)!\,n!\,x^{n+2}$ for all $n \geqslant 0$, and therefore the series (5) becomes

$$y(x) \sim \sum_{n=0}^{\infty} (n+1)!\,n!\,x^{n+2}\varepsilon^n. \tag{7}$$

Now let us consider carefully the meaning of the power series in $\varepsilon$ on the right-hand side of (7). The absolute value of the ratio of successive terms in the series is

$$\left|\frac{y_{n+1}(x)\varepsilon^{n+1}}{y_n(x)\varepsilon^n}\right| = (n+2)(n+1)x|\varepsilon|,$$

and this ratio blows up as $n \to \infty$ regardless of the value of $x > 0$, unless of course $\varepsilon = 0$. Therefore, by the ratio test, the series on the right-hand side of (7) diverges (has no finite sum) for every value of $x > 0$ unless $\varepsilon = 0$. Another way of saying the same thing is that the error or remainder $R_N(x, \varepsilon)$ upon truncating the series after the term proportional to $\varepsilon^N$,

$$R_N(x, \varepsilon) := y(x) - \sum_{n=0}^{N} (n+1)!\,n!\,x^{n+2}\varepsilon^n,$$

does not tend to zero (or, for that matter, any finite limit) as $N \to \infty$, no matter what values we choose for $x$ and $\varepsilon$. It is simply not possible to use partial sums of the series (7) to get better and better approximations to $y(x)$ by including more and more terms in the partial sum.

On the other hand, there does indeed exist a particular solution $y(x)$ of (4) for which the partial sums are good approximations. The key idea is the following: instead of trying to choose the parameter $N$ (the number of terms) to make the remainder $R_N(x, \varepsilon)$ small for fixed $\varepsilon > 0$, try to choose the parameter $\varepsilon$ small enough that $R_N(x, \varepsilon)$ is less than some tolerance for $N$ fixed. It turns out that there is a positive constant $K_N(x)$ independent of $\varepsilon$ such that

$$|R_N(x, \varepsilon)| \leqslant K_N(x)\varepsilon^{N+1}. \tag{8}$$

This fact shows that for each given $N$, the error in approximating $y(x)$ by the $N$th partial sum of the series (7) tends to zero with $\varepsilon$, and it does so at a rate depending on the number of retained terms in the sum. This is the property that makes the series on the right-hand side of (7) an *asymptotic expansion* of $y(x)$. We must understand that it makes no sense to add up all of the terms in the series, but the partial sums are good approximations of $y(x)$ when $\varepsilon$ is small, and the error of approximation goes to zero faster with $\varepsilon$ the more terms are kept in the partial sum.

### 2.2 Definitions and Notation

We begin to formalize some of this notation by introducing some simple standard notation due to Edmund Landau for estimates involving functions of $\varepsilon$. Let $f(\varepsilon, p)$ be a function of $\varepsilon$ for $\varepsilon$ sufficiently small, depending on some auxiliary parameter $p$. We write

$$f(\varepsilon, p) = O(g(\varepsilon)), \quad \varepsilon \to 0,$$

and say that "$f$ is *big-oh* of $g$" if there exists some $K(p) > 0$ such that, for each $p$,

$$|f(\varepsilon, p)| \leqslant K(p)|g(\varepsilon)|, \quad \forall p,$$

and for all $\varepsilon$ small enough. If $K(p)$ can be chosen to be independent of $p$, then $f$ is big-oh of $g$ *uniformly* with respect to $p$. We write

$$f(\varepsilon, p) = o(g(\varepsilon)), \quad \varepsilon \to 0,$$

and say that "$f$ is *little-oh* of $g$" if for *every* $K > 0$ there is some $\delta(p, K) > 0$ such that

$$|\varepsilon| \leqslant \delta(p, K) \implies |f(\varepsilon, p)| \leqslant K|g(\varepsilon)|, \quad \forall p.$$

As with big-oh, if $\delta$ is independent of $p$, then $f$ is little-oh of $g$ uniformly with respect to $p$. If $g$ is a function that is nonzero for all sufficiently small $\varepsilon \neq 0$, then $f = o(g)$ is the same thing as asserting that $f/g \to 0$ as $\varepsilon \to 0$. (This is often used in the special case when $g(\varepsilon) \equiv 1$.) Heuristically, $f = O(g)$ means that $f$ "is no bigger than" $g$ in a neighborhood of $\varepsilon = 0$, while $f = o(g)$ means that $f$ "is much smaller than" $g$ in the limit $\varepsilon \to 0$. The convenience of Landau's notation is that it avoids reference to various constants that always occur in estimates. For example, (8) could easily be written without reference to the constant $K_N(x)$ in the form $R_N(x, \varepsilon) = O(\varepsilon^{N+1})$ as $\varepsilon \to 0$.

A sequence of functions $\{\phi_n(\varepsilon)\} = \{\phi_n(\varepsilon)\}_{n=0}^{\infty}$ is called an *asymptotic sequence* in the limit $\varepsilon \to 0$ if, for each $n$, $\phi_{n+1}(\varepsilon) = o(\phi_n(\varepsilon))$ as $\varepsilon \to 0$. Given an asymptotic sequence $\{\phi_n(\varepsilon)\}$ and an arbitrary numerical sequence $\{a_n\} = \{a_n\}_{n=0}^{\infty}$, the purely formal series

$$\sum_{n=0}^{\infty} a_n \phi_n(\varepsilon)$$

is called an *asymptotic series*. Such a series is said to be an *asymptotic expansion* of a function $f(\varepsilon)$, written in the form

$$f(\varepsilon) \sim \sum_{n=0}^{\infty} a_n \phi_n(\varepsilon), \quad \varepsilon \to 0, \tag{9}$$

if, for each $N = 0, 1, 2, \dots,$

$$f(\varepsilon) - \sum_{n=0}^{N} a_n \phi_n(\varepsilon) = o(\phi_N(\varepsilon)), \quad \varepsilon \to 0. \tag{10}$$

From this relation it follows that, if $f(\varepsilon)$ has an asymptotic expansion with respect to $\{\phi_n(\varepsilon)\}$, then the coefficients $\{a_n\}$ are uniquely determined by the recursive sequence of limits

$$a_n := \lim_{\varepsilon \to 0} \frac{1}{\phi_n(\varepsilon)} \left[ f(\varepsilon) - \sum_{k=0}^{n-1} a_k \phi_k(\varepsilon) \right].$$

Indeed, the existence of each of these limits in turn is equivalent to the assertion that $f$ has an asymptotic expansion with respect to the sequence $\{\phi_n(\varepsilon)\}$. On the other hand, the function $f$ is most certainly not determined uniquely given the asymptotic sequence $\{\phi_n(\varepsilon)\}$ and the coefficients $\{a_n\}$; given $f(\varepsilon)$ satisfying (10), $f(\varepsilon) + g(\varepsilon)$ will also satisfy (10) if $g(\varepsilon) = o(\phi_n(\varepsilon))$ as $\varepsilon \to 0$ for all $n$. This condition by no means forces $g(\varepsilon) = 0$; such a function $g$ that is too small in the limit as $\varepsilon \to 0$ to have any effect on the coefficients $\{a_n\}$ is said to be *beyond all orders* with respect to $\{\phi_n(\varepsilon)\}$.

The simplest example of an asymptotic sequence, and one that occurs frequently in applications, is the sequence of integer powers $\phi_n(\varepsilon) := \varepsilon^n$. In this context, a function $g$ that is beyond all orders is sometimes called *transcendentally small* or *exponentially small*; indeed, a particular example of such a function is $g(\varepsilon) = \exp(-|\varepsilon|^{-1})$.

Now, the notation used in (9) should be strongly contrasted with the standard notation used for convergent series:

$$f(\varepsilon) = \sum_{n=0}^{\infty} a_n \phi_n(\varepsilon). \tag{11}$$

The use of "=" here implies in particular that the expressions on both sides are the same kind of object: functions of $\varepsilon$ with well-defined values. Furthermore, the only way to unambiguously assign a numerical value to the infinite series on the right-hand side given a value of $\varepsilon$ is to sum the series, that is, to compute the limit of the sequence of partial sums. By contrast, the meaning (10) given to the expression (9) in no way implies that the series on the right-hand side can be summed for any $\varepsilon$ at all. Therefore, we view the relation (9) as defining an infinite hierarchy of approximations to the function $f(\varepsilon)$ given by (well-defined) partial sums of the formal series; each subsequent partial sum is a better approximation than the preceding one when the error is made small by letting $\varepsilon$ tend to zero, precisely because (10) holds and the functions $\{\phi_n(\varepsilon)\}$ form an asymptotic sequence. However, it need not be the case that the error in approximating $f(\varepsilon)$ by the $N$th partial sum can be made small by fixing some $\varepsilon$ and letting $N$ increase. Only in the latter case can we use the convergent series notation (11).

The subject of perturbation *theory* is largely concerned with determining the nature of a series obtained via formal perturbation methods. Some perturbation series are both convergent (for sufficiently small $\varepsilon$) and asymptotic as $\varepsilon \to 0$; this was the case in the example of the perturbation of the unperturbed root $x_0 = 1$ in the root-finding problem in section 1.1. On the other hand, most perturbation series are divergent, as in the case of the expansion considered in section 2.1, and in such cases proving the validity of the perturbation series requires establishing the existence of a true, $\varepsilon$-dependent solution of the problem at hand to which the perturbation series is asymptotic in the sense of the definition (9), (10). This in turn usually amounts to formulating a mathematical problem (e.g., a differential equation with side conditions) satisfied by the remainder and applying an appropriate fixed-point or iteration argument.

Some related notation used in papers on the subject includes the following. The notation $f \ll g$ is frequently used in place of $f = o(g)$. Also, one sometimes sees the notation $f \lesssim g$ for $f = O(g)$. It should also be remarked that the symbol "$\sim$" often appears as a relation between functions in the following two senses:

(i) $f(\varepsilon) \sim g(\varepsilon)$ may indicate that both $f(\varepsilon) = O(g(\varepsilon))$ and also $g(\varepsilon) = O(f(\varepsilon))$ (that is, $f$ is bounded both above and below by multiples of $g$) as $\varepsilon \to 0$.

(ii) $f(\varepsilon) \sim g(\varepsilon)$ may indicate that $f(\varepsilon)/g(\varepsilon) \to 1$ as $\varepsilon \to 0$, a special case of the above notation.

To avoid any confusion, we will use the symbol "$\sim$" only in the sense defined by (9), (10).

The theory of asymptotic expansions applies in a number of contexts beyond its application to perturbation problems. As just one very important example, it is the basis for a collection of very well-developed methods for approximating certain types of integrals. Key methods include *Laplace's method* for the asymptotic expansion of real integrals with exponential integrands, Kelvin's *method of stationary phase* for the asymptotic expansion of oscillatory integrals, and the *method of steepest descent* (or saddle-point method) applying to integrals with analytic integrands. Readers can find detailed information about these useful methods in the books of Bleistein and Handelsman (1986), Wong (2001), and Miller (2006).

## 3 Types of Perturbation Problems

Perturbation problems are frequently categorized as being either regular or singular. The distinction is not a precise one, so there is not much point in giving careful definitions. However, the two kinds of problems often require different methods, so it is worth considering which type a given problem most resembles.

### 3.1 Regular Perturbation Problems

A regular perturbation problem is one in which the perturbed problem ($\varepsilon \neq 0$) is of the same general "type" as the unperturbed problem ($\varepsilon = 0$) that can be solved easily. Regular perturbation problems often lead to series that are both asymptotic as $\varepsilon \to 0$ and convergent for sufficiently small $\varepsilon$.

One example of a regular perturbation problem is that of finding the energy levels of a perturbed quantum mechanical system. Consider a particle moving in one space dimension subject to a force $F = -V'(x)$, where $x$ is the position of the particle. The problem is to find nontrivial square-integrable "bound states" $\psi(x)$ and corresponding energy levels $E \in \mathbb{R}$ such that Schrödinger's equation $\mathcal{H}\psi = E\psi$ holds, where $\mathcal{H} = \mathcal{H}_0 + \varepsilon\mathcal{H}_1$ and

$$\mathcal{H}_0\psi(x) := -\psi''(x) + V_0(x)\psi(x),$$
$$\mathcal{H}_1\psi(x) := V_1(x)\psi(x).$$

Here we have artificially separated the potential energy function $V$ into two parts, $V = V_0 + \varepsilon V_1$; the idea is to choose $V_0$ such that when $\varepsilon = 0$ it is easy to solve the problem by finding a nonzero function $\psi_0 \in L^2(\mathbb{R})$ and a number $E_0$ that satisfy $\mathcal{H}_0\psi_0 = E_0\psi_0$. In this one-dimensional setting it turns out that, given $E_0$, all solutions of this equation are proportional to $\psi_0$ (which makes the energy level $E_0$ "nondegenerate" in the language of quantum mechanics). By choosing an appropriate scaling factor, we may assume that $\psi_0$ is "normalized" to satisfy

$$\int_{\mathbb{R}} \psi_0(x)^2 \, dx = 1. \tag{12}$$

Now, to calculate the effect of the perturbation, we may suppose that both $\psi$ and $E$ are expandable in asymptotic power series in $\varepsilon$:

$$\psi \sim \sum_{n=0}^{\infty} \varepsilon^n \psi_n, \quad E \sim \sum_{n=0}^{\infty} \varepsilon^n E_n, \quad \varepsilon \to 0. \tag{13}$$

The coefficients then necessarily satisfy the hierarchy of equations:

$$\mathcal{H}_0 \psi_n - E_0 \psi_n = \sum_{j=1}^{n} E_j \psi_{n-j} - \mathcal{H}_1 \psi_{n-1}. \tag{14}$$

Denote the right-hand side of this equation by $f_n(x)$. Then (14) has a solution $\psi_n(x)$ if and only if $f_n(x)$ satisfies a *solvability condition* (stemming from the *Fredholm alternative*):

$$\int_{\mathbb{R}} \psi_0(x) f_n(x) \, dx = 0.$$

This solvability condition is in fact a recursive formula for $E_n$ in disguise:

$$\begin{aligned} E_n = {} & \int_{\mathbb{R}} \psi_0(x) \mathcal{H}_1 \psi_{n-1}(x) \, dx \\ & - \sum_{j=1}^{n-1} E_j \int_{\mathbb{R}} \psi_0(x) \psi_{n-j}(x) \, dx, \end{aligned} \tag{15}$$

where we have used (12). Once $E_n$ is determined from this relation, the equation (14) can be solved for $\psi_n$, but the latter is only determined modulo multiples of $\psi_0$; one typically chooses the correct multiple of $\psi_0$ to add in order that $\psi_n$ be orthogonal to $\psi_0$ in the sense that

$$\int_{\mathbb{R}} \psi_0(x) \psi_n(x) \, dx = 0, \quad n > 0. \tag{16}$$

Subject to (15), (14) has a *unique* solution determined by the auxiliary condition (16). Note that this condition actually ensures that the sum on the right-hand side of (15) equals zero.

The perturbation expansions (13) of the pair $(\psi, E)$ are known as *Rayleigh–Schrödinger series*. Under suitable conditions on the operators $\mathcal{H}_0$ and $\mathcal{H}_1$ it can be shown (by the method of Lyapunov–Schmidt reduction to eliminate the eigenfunction) that the power series (13) are actually convergent series, and hence "$\sim$" and "$=$" can be used interchangeably in this context.

### 3.2 Singular Perturbation Problems

In a singular perturbation problem, the perturbed and unperturbed problems are different in some essential way. The most elementary examples involve root finding. Consider the problem of finding the roots of the polynomial $P(x) := \varepsilon x^3 - x + 1 = 0$ when $\varepsilon$ is small and positive. The unperturbed problem (with $\varepsilon = 0$) is to solve the linear equation $-x + 1 = 0$, which of course has the unique solution $x = 1$. However, the perturbed problem for $\varepsilon \neq 0$ is to find the roots of a cubic, and by the fundamental theorem of algebra there are three such roots. The perturbed and unperturbed problems are of different types because setting $\varepsilon = 0$ changes the degree of the equation.

Somehow, two of the roots disappear altogether from the complex plane as $\varepsilon \to 0$. Where can they go? Some intuition is obtained in this case simply by looking at a graph of $P(x)$ when $\varepsilon$ is very small; while one of the roots looks to be close to $x = 1$ (the unique root of the unperturbed problem), the other two are very large in magnitude and of opposite signs. So the answer is that the two "extra" roots go to infinity as $\varepsilon \to 0$.

To completely solve this problem using perturbation methods, we need to capture all three roots. The root near $x = 1$ when $\varepsilon$ is small can be expanded in a power series in $\varepsilon$ whose coefficients can be found recursively using exactly the same methodology as in section 1.1. Finding the remaining two roots requires another idea.

The key idea is to try to pull the two escaping roots back to the finite complex plane by rescaling them by an appropriate power of $\varepsilon$. Let $p > 0$ be given, and write $x = y\varepsilon^{-p}$. This is a change of variables in our problem that takes a large value of $x$, proportional to $\varepsilon^{-p}$, and produces a value of $y$ that does not grow to infinity as $\varepsilon \to 0$. In terms of $y$, the root-finding problem at hand takes the form

$$\varepsilon^{1-3p} y^3 - \varepsilon^{-p} y + 1 = 0. \tag{17}$$

Now, $p > 0$ is undetermined so far, but we will choose it (and hence determine the rate at which the two roots are escaping to infinity) using the *principle of dominant balance.*

By a *balance* we simply mean a pair of terms on the left-hand side of (17) having the same power of $\varepsilon$ (by choice of $p > 0$). A balance is called *dominant* if all other terms on the left-hand side are big-oh of the terms involved in the balance. The principle of dominant balance asserts that *only dominant balances lead to possible perturbation expansions.* There are three pairs of terms to choose from, and hence three possible balances to consider:

(i) Balancing $\varepsilon^{1-3p}y^3$ with 1 requires choosing $p = \frac{1}{3}$. The terms involved in the balance are then both proportional to $\varepsilon^0$, while the remaining term is proportional to $\varepsilon^{-1/3}$, so this balance is not dominant.

(ii) Balancing $\varepsilon^{-p}y$ with 1 requires choosing $p = 0$. The terms involved in the balance are then proportional to $\varepsilon^0$, making the balance dominant over the remaining term as $\varepsilon \to 0$. Since $p = 0$ this rescaling has had no effect ($y = x$), and in fact setting the sum of the dominant balance terms to zero recovers the original unperturbed problem. No new information is gained.

(iii) Balancing $\varepsilon^{1-3p}y^3$ with $\varepsilon^{-p}y$ requires choosing $p = \frac{1}{2}$. The terms involved in the balance are then both proportional to $\varepsilon^{-1/2}$, making the balance dominant over the remaining term as $\varepsilon \to 0$. This is a new dominant balance.

The new dominant balance will lead to perturbation expansions of the two large roots of the original problem. Indeed, with $p = \frac{1}{2}$, for $\varepsilon \neq 0$ our problem takes the form

$$y^3 - y + \varepsilon^{1/2} = 0, \tag{18}$$

which now appears as a perturbation of the equation $y^3 - y = 0$. The latter has three roots: $y = y_0$ with $y_0 = 0$ or $y_0 = \pm 1$. Obviously, if (18) has solutions $y$ that are close to $y_0 = \pm 1$ when $\varepsilon$ is small, then the original problem will have two corresponding solutions that are roughly proportional to $\varepsilon^{-1/2}$. These will then be our two missing roots. The expansion procedure for (18) with $y = y_0 = \pm 1$ for $\varepsilon = 0$ is similar to that described in section 1.1, except that $y(\varepsilon)$ will be a power series in $\varepsilon^{1/2}$. The implicit function theorem applies to (18), showing that the perturbation series for $y(\varepsilon)$ will be convergent for $\varepsilon$ sufficiently small, in addition to being asymptotic in the limit $\varepsilon \to 0$. Scaling $y$ by $\varepsilon^{-1/2}$ then yields series representations for the two large roots $x$ of the original problem in the form

$$x \sim \sum_{n=0}^{\infty} y_n \varepsilon^{(n-1)/2}, \quad \varepsilon \to 0,$$

with $y_0 = \pm 1$. As already mentioned, in this case "$\sim$" could be replaced by "$=$" if $\varepsilon$ is small enough.

Another major category of singular perturbation problems are those involving differential equations in which the small parameter $\varepsilon$ multiplies the highest-order derivatives. In such a case, setting $\varepsilon = 0$ replaces the differential equation by another one of strictly lower order. This is clearly analogous to the algebraic degeneracy described above in that the number of solutions (this time counted in terms of the dimension of some space of integration constants) is different for the perturbed and unperturbed problems. We now discuss some common perturbation methods that are generally applicable to singular perturbation problems of this latter sort.

### 3.2.1 *WKB Methods and Generalizations*

WKB methods (named after Wentzel, Kramers, and Brillouin) concern differential equations that are so singularly perturbed that upon setting $\varepsilon$ to zero there are no derivatives left at all. A sufficiently rich example equation with this property is the Sturm–Liouville, or stationary Schrödinger, equation

$$\varepsilon^2 \psi''(x) + f(x)\psi(x) = 0, \tag{19}$$

where $f(x)$ is a given coefficient and solutions $\psi = \psi(x;\varepsilon)$ are desired for sufficiently small $\varepsilon > 0$. Obviously, for the two terms on the left-hand side to sum to zero when $f$ is independent of $\varepsilon$, derivatives of $\psi$ must be very large compared with $\psi$ itself. The essence of the WKB method is to take this into account by making a substitution of the form

$$\psi(x;\varepsilon) = \exp\left(\frac{1}{\varepsilon}\int_{x_0}^{x} u(\xi;\varepsilon)\,\mathrm{d}\xi\right), \tag{20}$$

and in terms of the new unknown $u$, (19) becomes a first-order nonlinear equation of Riccati type:

$$\varepsilon u'(x;\varepsilon) + u(x;\varepsilon)^2 + f(x) = 0. \tag{21}$$

Unlike (19), this equation admits two nontrivial solutions when $\varepsilon = 0$, namely $u(x;0) = \pm(-f(x))^{1/2}$. One can now develop a perturbation series for $u$ in powers of $\varepsilon$ in the usual way, starting from each of these two solutions. These series can be shown to be asymptotic to certain true solutions of (21) in the sense of (9), (10), but they are nearly always divergent. When partial sums of these two distinct series are substituted into (20), one obtains approximations to two linearly independent solutions of (19).

The WKB method as described above always fails near points $x$ where the coefficient $f$ vanishes. Such $x$ are called *turning points*. What we mean when we say that the method fails is that the little-oh relation (10) does not hold uniformly on any open interval of $x$ with a turning point as an endpoint and, moreover, there is no solution that is accurately approximated by partial sums of the WKB expansion over a neighborhood of $x$ containing a turning point. An important problem in the theory of equation (19) is the *connection problem*, in which a solution described accurately by a WKB expansion on one side of a turning point is to be approximated on the other side of the turning point by an appropriate linear combination of WKB expansions. This problem can sometimes be solved within the context of the WKB method by analytically continuing a solution $\psi$ into the complex $x$-plane and around a turning point. Such an approach always requires analyticity of the coefficient function $f$ and in any case fails to describe the solution $\psi$ in any neighborhood of the turning point.

A more satisfactory way of dealing with turning points and solving connection problems is to generalize the WKB method (that is, generalize the ansatz (20)). The following approach is due to Langer.

By a simultaneous change of independent and dependent variables of the form $y = g(x)$ and $\psi(x;\varepsilon) = a(x)\phi(y;\varepsilon)$, respectively, one tries to choose $g$ and $a$ (as smooth functions independent of $\varepsilon$) such that (19) becomes a perturbation of a model equation:

(i) $\varepsilon^2\phi''(y) \pm \phi(y) = \varepsilon^2\alpha_\pm(y)\phi$ in intervals of $x$ where $f(x) \gtrless 0$;
(ii) $\varepsilon^2\phi''(y) - y^n\phi(y) = \varepsilon^2\beta_n(y)\phi$ in intervals of $x$ where $f$ has a turning point (zero) of order $n$.

Here, $\alpha_\pm(y)$ and $\beta_n(y)$ are smooth functions of $y$ explicitly written in terms of $f$ and its derivatives, and avoiding terms proportional to $\phi'(y)$ requires relating $a$ and $g$ by $a(x) = |g'(x)|^{-1/2}$. In each case, the model equation is obtained by neglecting the terms on the right-hand side (which turn out to be "more" negligible than the singular perturbation term $\varepsilon^2\phi''(y)$).

To solve (19) on an interval without turning points, one may arrive at case (i) via the (Liouville–Green) transformation

$$y = g(x) = \int_{x_0}^{x} \sqrt{\pm f(\xi)}\,d\xi.$$

This transformation is smooth and invertible in the absence of turning points. It is easy to confirm that the use of the elementary exponential solutions of the model equation $\varepsilon^2\phi''(y) \pm \phi(y) = 0$ alone immediately reproduces the first two terms of the standard WKB expansion (the coefficient $a(x)$ corresponds to the term in $u$ that is proportional to $\varepsilon$). Treating the error term $\varepsilon^2\alpha_\pm(y)\phi(y)$ perturbatively reproduces the rest of the WKB series.

To solve (19) on an interval containing a simple turning point $x_0$, one should take $n = 1$ in case (ii). Arriving at this target requires choosing the (Langer) transformation

$$y = g(x) = -\operatorname{sgn}(f(x))\left|\frac{3}{2}\int_{x_0}^{x}\sqrt{|f(\xi)|}\,d\xi\right|^{2/3}.$$

This transformation is smooth and invertible near $x_0$ precisely because $x_0$ is a simple zero of $f$. In this case, the model equation $\varepsilon^2\phi''(y) - y\phi(y) = 0$ is not solvable by elementary functions, but it *is* solvable in terms of special functions known as *Airy functions*. Airy functions can be expressed as certain contour integrals, and this is enough information to allow the solution of a number of connection problems for simple turning points without detouring into the complex $x$-plane. For double turning points ($n = 2$) one needs a different transformation $g(x)$ that is again smooth, and instead of Airy functions one has *Weber functions*, but again these can be written as integrals, allowing connection problems to be solved. For $n \geqslant 3$ the solvability of the model equation (in terms of useful special functions) becomes a more serious issue. Nonetheless, when Langer's generalization applies it not only allows an alternative approach to connection problems but also provides accurate asymptotic formulas for solutions of (19) in full neighborhoods of turning points where standard WKB methodology fails.

One famous formula that can be obtained with the use of the WKB method and connection formulas for simple turning points is the *Bohr–Sommerfeld quantization rule* of quantum mechanics. Consider the case in which $f(x)$ takes the form $f(x) = E - V(x)$, where $V$ is a potential energy function with $V''(x) > 0$, and therefore its graph has the shape of a "potential well" whose minimum value we take to be $V = 0$. Then (19) with $\varepsilon^2 = \hbar^2/(2m)$ is the equation satisfied by stationary quantum states for a particle of mass $m$ and total energy $E$, and $E$ should take on a discrete spectrum of values such that there exists a solution $\psi$ that is square integrable on $\mathbb{R}$. If $E > 0$ then $f(x)$ will have exactly two simple turning points, $x_-(E) < x_+(E)$, and one can use WKB methods and connection formulas to obtain two nonzero solutions $\psi_\pm(x)$ that exhibit rapid exponential decay as $x \to \pm\infty$. $E$ is the energy of a stationary state if and only if $\psi_-$ is proportional to $\psi_+$. By computing a Wronskian of these two solutions in the region between the two turning points and equating the result to zero, one obtains the Bohr–Sommerfeld quantization rule as a condition for $E > 0$ to be an energy eigenvalue:

$$\int_{x_-(E)}^{x_+(E)} \sqrt{E - V(x)}\,\mathrm{d}x = \pi\varepsilon(n + \tfrac{1}{2}) + O(\varepsilon^2).$$

### 3.2.2 Multiple-Scale Methods

Another class of perturbation methods deals with singular perturbation problems of a different type in which what makes the problem singular is that an accurate solution of some differential equation is required over a very large range of values of the independent variables. Such problems often masquerade as regular perturbation problems until the need for accuracy over long time intervals or large distances is revealed and understood. A typical example is the weakly anharmonic oscillator, whose displacement $u = u(t;\varepsilon)$ as a function of time $t$ is modeled by the initial-value problem for

$$u'' + \omega_0^2 u = \varepsilon u^3, \tag{22}$$

subject to the initial conditions

$$u(0;\varepsilon) = A \quad \text{and} \quad u'(0;\varepsilon) = \omega_0 B.$$

When $\varepsilon = 0$, $u$ undergoes simple harmonic motion with frequency $\omega_0$: $u(t;0) = A\cos(\omega_0 t) + B\sin(\omega_0 t)$. The presence of the cubic perturbation term on the right-hand side does not modify the order of the differential equation, so this appears to be a regular perturbation problem (and indeed it is so for bounded $t$).

To begin to see the difficulty, we should first observe that (22) conserves the energy:

$$E = \tfrac{1}{2}u'(t;\varepsilon)^2 + \tfrac{1}{2}\omega_0^2 u(t;\varepsilon)^2 - \tfrac{1}{4}\varepsilon u(t;\varepsilon)^4.$$

As the energy function has a local minimum in the phase plane at $(u, u') = (0, 0)$, and as this is the global minimum when $\varepsilon = 0$, it follows that for each initial condition pair $(A, B)$, if $|\varepsilon|$ is sufficiently small the corresponding solution is time-periodic, following a closed orbit in the phase plane. Now consider solving (22) using a perturbation (power) series in $\varepsilon$. If the assumed series takes the form

$$u(t;\varepsilon) \sim \sum_{n=0}^{\infty} \varepsilon^n u_n(t), \quad \varepsilon \to 0, \tag{23}$$

with $u_0(t) = u(t;0)$, then by substitution and collection of coefficients of like powers of $\varepsilon$ it follows in particular that the first correction $u_1(t)$ solves

$$u_1'' + \omega_0^2 u_1 = u_0(t)^3, \quad u_1(0) = u_1'(0) = 0. \tag{24}$$

The forcing function $u_0^3 = (A\cos(\omega_0 t) + B\sin(\omega_0 t))^3$ is known, and it contains terms proportional to the first and third harmonics. Solving for $u_1(t)$ gives

$$\begin{aligned} u_1(t) = &-\frac{A}{32\omega_0^2}(A^2 - 3B^2)\cos(3\omega_0 t) \\ &-\frac{B}{32\omega_0^2}(3A^2 - B^2)\sin(3\omega_0 t) \\ &+\frac{3}{8\omega_0}(A^2 + B^2)[At\sin(\omega_0 t) - Bt\cos(\omega_0 t)] \\ &+\frac{A}{32\omega_0^2}(A^2 - 3B^2)\cos(\omega_0 t) \\ &+\frac{3B}{32\omega_0^2}(7A^2 + 3B^2)\sin(\omega_0 t). \end{aligned}$$

The terms on the first two lines here are the response to the third harmonic forcing terms in $u_0(t)^3$, the terms on the third line are the response to the first harmonic forcing terms in $u_0(t)^3$, and the last two lines constitute a homogeneous solution necessary to satisfy the initial conditions. This procedure can be easily continued, and all of the functions $u_n(t)$ are therefore systematically determined. The trouble with this procedure is that, while we know that the true solution $u(t;\varepsilon)$ is a periodic function of $t$, we have already found terms in $u_1(t)$ that are linearly growing in $t$, produced as a resonant response to forcing at the fundamental frequency.

Terms growing in $t$ are not troublesome only because they are nonperiodic; they also introduce nonuniformity with respect to $t$ into the asymptotic condition (10). Indeed, if $t$ becomes as large as $\varepsilon^{-1}$, then the term $\varepsilon u_1(t)$ becomes comparable to the leading term $u_0(t)$, and condition (10) is violated. Such growing terms in an asymptotic expansion are called *secular terms*, from the French word *siècle*, meaning century, because they were first recognized in perturbation problems of celestial mechanics where they would lead to difficulty over long time intervals of about 100 years.

It is not hard to understand what has gone wrong in the particular problem at hand. The point is that, while the exact solution is indeed periodic, its fundamental frequency is not exactly $\omega_0$ but rather is slightly dependent on $\varepsilon$. If we could know what the frequency is in advance, then we might expect the perturbation series to turn out to be a Fourier series in harmonics of the fundamental $\varepsilon$-dependent frequency.

The method of multiple scales is a systematic method of removing secular terms from asymptotic expansions, and in the problem at hand it leads automatically to a power-series expansion of the correct frequency as a function of $\varepsilon$. The "scales" in the name of the method are multiples of the independent variable of the problem; we introduce a number of variables of the form $T_k := \varepsilon^k t$ for $k = 0, 1, 2, \dots$. Given some finite $K$, we seek $u(t, \varepsilon)$ in the form $u(t, \varepsilon) = U(T_0, \dots, T_K; \varepsilon)$ so that, by the chain rule, the ordinary differential equation (22) becomes a partial differential equation:

$$\sum_{j=0}^{K}\sum_{k=0}^{K} \varepsilon^{j+k} \frac{\partial^2 U}{\partial T_j \partial T_k} + \omega_0^2 U = \varepsilon U^3.$$

Into this equation we now introduce an expansion analogous to (23):

$$U(T_0, \dots, T_K; \varepsilon) \sim \sum_{n=0}^{\infty} \varepsilon^n U_n(T_0, \dots, T_K), \quad \varepsilon \to 0.$$

The terms proportional to $\varepsilon^0$ are

$$\frac{\partial^2 U_0}{\partial T_0^2} + \omega_0^2 U_0 = 0,$$

the general solution of which is $U_0 = A\cos(\omega_0 T_0) + B\sin(\omega_0 T_0)$, where $A$ and $B$ are undetermined functions of the "slow times" $T_1, \dots, T_K$. The terms proportional to $\varepsilon^1$ are then

$$\frac{\partial^2 U_1}{\partial T_0^2} + \omega_0^2 U_1 = U_0^3 - 2\frac{\partial^2 U_0}{\partial T_0 \partial T_1}. \tag{25}$$

Comparing with (24), the final term on the right-hand side is the new contribution from the method of multiple scales. It contains only first harmonics:

$$\frac{\partial^2 U_0}{\partial T_0 \partial T_1} = -\omega_0 \frac{\partial A}{\partial T_1}\sin(\omega_0 T_0) + \omega_0 \frac{\partial B}{\partial T_1}\cos(\omega_0 T_0).$$

Therefore, if the dependence on $T_1$ of the coefficients $A$ and $B$ is selected so that

$$\left.\begin{aligned} 2\omega_0 \frac{\partial A}{\partial T_1} + \frac{3B}{4}(A^2 + B^2) &= 0, \\ -2\omega_0 \frac{\partial B}{\partial T_1} + \frac{3A}{4}(A^2 + B^2) &= 0, \end{aligned}\right\} \tag{26}$$

then the resonant forcing terms that are proportional to $\cos(\omega_0 T_0)$ and $\sin(\omega_0 T_0)$ and that are responsible for the secular response in $U_1$ will be removed from the right-hand side of (25), and consequently $U_1$ will now be a periodic function of $T_0$. The dependence of $A$ and $B$ on longer timescales $T_2, T_3, \dots, T_K$ is determined similarly, order by order, so that $U_2, U_3, \dots, U_K$ are periodic functions of $T_0$. Once this series has been constructed, the dependence on $t$ may be restored by the substitution $T_k = \varepsilon^k t$. This procedure produces an asymptotic expansion that is uniformly consistent with the ordering relation (10) for times $t$ that satisfy $t = O(\varepsilon^{-K})$ as $\varepsilon \to 0$.

The interpretation of the system (26) is that, as expected, the frequency of oscillation depends on the amplitude in nonlinear systems. Indeed, from (26) it is easy to check that the squared amplitude $A^2 + B^2$ is independent of $T_1$, and hence $A$ and $B$ undergo simple harmonic motion with respect to $T_1$ of amplitude-dependent frequency $\omega_1 := -3(A^2 + B^2)/(8\omega_0)$. By standard trigonometric identities it then follows that $U_0$ is a sinusoidal oscillation of frequency $\omega_0 + \varepsilon\omega_1 + O(\varepsilon^2)$, where $\omega_1$ depends on the amplitude (determined from initial conditions).

### 3.2.3 Matching of Asymptotic Expansions

Another set of methods for dealing with singularly perturbed differential equations involves using the principle of dominant balance to isolate different domains of the independent variables in which different asymptotic expansions apply and then "matching" the expansions together to produce an approximate solution that is uniformly accurate with respect to the independent variable. A typical context is a singularly perturbed boundary-value problem such as

$$\varepsilon u'' + u' + fu = 0, \quad u(0; \varepsilon) = u(1; \varepsilon) = 1. \tag{27}$$

Here $f = f(x)$ is given, and $u(x; \varepsilon)$ is desired for $\varepsilon$ sufficiently small (in which case one can prove that this problem has a unique solution). For the differential equation itself, one dominant balance is that between $u'$ and $fu$. The expansion based on this balance is just the power-series expansion

$$u(x; \varepsilon) \sim \sum_{n=0}^{\infty} \varepsilon^n u_n(x), \quad \varepsilon \to 0, \tag{28}$$

and the leading term $u_0(x)$ satisfies the limiting equation $u_0'(x) + f(x)u_0(x) = 0$, which has the general solution

$$u_0(x) = C_0 \exp\left(\int_x^1 f(\xi)\,\mathrm{d}\xi\right). \tag{29}$$

Given $u_0(x)$, the procedure can be continued in the usual way to obtain successively the coefficients $u_n(x)$. At each order, one additional arbitrary constant $C_n$ is generated.

The expansion (28) is insufficient to solve the boundary-value problem (27) because at each order there are two boundary conditions imposed on each of the functions $u_n(x)$ but there is only one constant available to satisfy them. To seek additional expansions, we can introduce scalings of the independent variable to suggest other dominant balances. For example, if we set $x = \varepsilon^p y$ and write $v(y; \varepsilon) = u(x; \varepsilon)$, then the differential equation becomes

$$\varepsilon^{1-2p} v''(y; \varepsilon) + \varepsilon^{-p} v(y; \varepsilon) + f(\varepsilon^p y) v(y; \varepsilon) = 0.$$

The first two terms balance if $p = 1$, and this is a dominant balance if $f$ is continuous on $[0,1]$ and hence bounded. Moreover, if $f$ is smooth, $f(\varepsilon y)$ can be expanded in Taylor series for small $\varepsilon$, and it becomes clear that this dominant balance leads to an expansion of $v$ of the form

$$v(y;\varepsilon) \sim \sum_{n=0}^{\infty} \varepsilon^n v_n(y), \quad \varepsilon \to 0. \tag{30}$$

Here the leading term $v_0(y)$ will satisfy the limiting equation $v_0''(y) + v_0'(y) = 0$, which has the general solution

$$v_0(y) = A_0 e^{-y} + B_0, \tag{31}$$

and the standard perturbation procedure allows one to calculate $v_n(y)$ order by order, introducing two new integration constants each time.

The expansion (28) turns out to be a good approximation for $u(x;\varepsilon)$ for $x$ in intervals of the form $[\delta, 1]$ for $\delta > 0$, as long as the constants are chosen to satisfy the boundary condition at $x = 1$: $u_0(1) = 1$ and $u_n(1) = 0$ for $n \geqslant 1$. This expansion therefore holds in "most" of the domain; in fluid dynamics problems this corresponds to flow in regions away from a problematic boundary, the so-called outer flow. Hence (28) is called an *outer expansion*.

On the other hand, the expansion (30) provides a good approximation for $u(x;\varepsilon) = v(x/\varepsilon;\varepsilon)$ in the "boundary layer" near $x = 0$ of thickness proportional to $\varepsilon$, provided that

(i) the integration constants are chosen to satisfy the boundary condition at $x = y = 0$, which forces

$$v_0(0) = 1 \quad \text{and} \quad v_n(0) = 0 \quad \text{for } n \geqslant 1;$$

and

(ii) the remaining constant at each order is determined so that the *inner expansion* (30) is compatible with the outer expansion (28) in some common "overlap domain" of $x$-values.

Choosing the constants to satisfy compatibility of the expansions is called *matching of asymptotic expansions*. In general, matching involves choosing some intermediate scale on which $y$ is large while $x$ is small as $\varepsilon \to 0$; for example, fixing $z > 0$, we could set $x = \varepsilon^{1/2} z$ and then $y = x/\varepsilon = \varepsilon^{-1/2} z$. With this substitution one writes both inner and outer expansions in terms of the common independent variable $z$ and re-expands both with respect to a suitable asymptotic sequence of functions of $\varepsilon$. Equating these expansions term-by-term then yields relations among the constants in the two expansions. The common expansion with $z$ fixed is sometimes called an *intermediate expansion*.

In the problem at hand, to satisfy the boundary condition at $x = 1$ we should choose $C_0 = 1$ in (29), while to satisfy $v_0(0) = 1$ we require $A_0 + B_0 = 1$ in (31). The matching condition at this leading order reads $u_0(x = 0) = v_0(y = +\infty)$, which implies that $B_0 = \exp \int_0^1 f(\xi)\,d\xi$. The constants $A_0$, $B_0$, and $C_0$ have clearly thus been determined by a combination of the two imposed boundary conditions and an asymptotic matching condition. The procedure can be continued to arbitrarily high order in $\varepsilon$.

Successful matching of asymptotic expansions for boundary-layer problems yields two different expansions that are valid in different parts of the physical domain. For some purposes it is useful to have a single approximation that is uniformly valid over the whole domain. Matching again plays a role here, as the correct formula for the uniformly valid approximation is the sum of corresponding partial sums of the inner and outer expansions, minus the corresponding terms of the intermediate expansion (which would otherwise be counted twice, it turns out).

Another application of matched asymptotic expansions is to problems involving periodic behavior that is alternately dominated by "fast" and "slow" dynamics, so-called *relaxation oscillations*. The slow parts of the cycle correspond to outer asymptotic expansions, and the rapid parts of the cycle are analyzed by rescaling and dominant balance arguments and resemble the inner expansions from boundary-layer problems. Matching is required to enforce the periodicity of the solution.

### Further Reading

Bleistein, N., and R. A. Handelsman. 1986. *Asymptotic Expansions of Integrals*, 2nd edn. New York: Dover.

Miller, P. D. 2006. *Applied Asymptotic Analysis*. Providence, RI: American Mathematical Society.

Wong, R. 2001. *Asymptotic Approximations of Integrals*. Philadelphia, PA: SIAM.

## IV.6 Calculus of Variations

*Irene Fonseca and Giovanni Leoni*

### 1 History

The calculus of variations is a branch of mathematical analysis that studies extrema and critical points of functionals (or energies). Here, by *functional* we mean a mapping from a function space to the real numbers.

One of the first questions that may be framed within this theory is *Dido's isoperimetric problem* (see section 2.3), finding the shape of a curve of prescribed perimeter that maximizes the area enclosed. Dido was a Phoenician princess who emigrated to North Africa and upon arrival obtained from the native chief as much territory as she could enclose with an ox hide. She cut the hide into a long strip and used it to delineate the territory later known as Carthage, bounded by a straight coastal line and a semicircle.

It is commonly accepted that the systematic development of the theory of the calculus of variations began with the *brachistochrone curve problem* proposed by Johann Bernoulli in 1696. Consider two points A and B on the same vertical plane but on different vertical lines. Assume that A is higher than B and that a particle M is moving from A to B along a curve and under the action of gravity. The curve that minimizes the time it takes M to travel between A and B is called the *brachistochrone*. The solution to this problem required the use of infinitesimal calculus and was later found by Jacob Bernoulli, Newton, Leibniz, and de l'Hôpital. The arguments thus developed led to the development of the foundations of the calculus of variations by Euler. Important contributions to the subject are attributed to Dirichlet, Hilbert, Lebesgue, Riemann, Tonelli, and Weierstrass, among many others.

The common feature underlying Dido's isoperimetric problem and the brachistochrone curve problem is that one seeks to maximize or minimize a functional over a class of competitors satisfying given constraints. In both cases the functional is given by an integral of a density that depends on an underlying field and some of its derivatives, and this will be the prototype we will adopt in what follows. To be precise, we consider a functional

$$u \in X \mapsto F(u) := \int_{\Omega} f(x, u(x), \nabla u(x))\,\mathrm{d}x, \tag{1}$$

where $X$ is a function space (usually an $L^p$ space or a Sobolev-type space), $u\colon \Omega \to \mathbb{R}^d$, with $\Omega \subset \mathbb{R}^N$ an open set, $N$ and $d$ are positive integers, and the density is a function $f(x, u, \xi)$, with $(x, u, \xi) \in \Omega \times \mathbb{R}^d \times \mathbb{R}^{d\times N}$. Here and in what follows, $\nabla u$ stands for the $d \times N$ matrix-valued distributional derivative of $u$.

The calculus of variations is a vast theory, so here we choose to highlight only some contemporary aspects of the field. We conclude the article by mentioning a few areas that are at the forefront of application and that are driving current research.

## 2 Extrema

In this section we address fundamental minimization problems and relevant techniques in the calculus of variations. In geometry, the simplest example is the problem of finding the curve of shortest length connecting two points: a *geodesic*. A (continuous) curve joining two points $\mathrm{A}, \mathrm{B} \in \mathbb{R}^d$ is represented by a (continuous) function $y\colon [0,1] \to \mathbb{R}^d$ such that $y(0) = \mathrm{A}$, $y(1) = \mathrm{B}$, and its length is given by

$$L(y) := \sup\left\{\sum_{i=1}^{n} |y(t_i) - y(t_{i-1})|\right\},$$

where the supremum is taken over all partitions $0 = t_0 < t_1 < \cdots < t_n = 1$, $n \in \mathbb{N}$, of the interval $[0, 1]$. If $y$ is smooth, then $L(y) = \int_0^1 |y'(t)|\,\mathrm{d}t$. In the absence of constraints, the geodesic is the straight segment with endpoints A and B, and so $L(y) = |A - B|$, where $|A - B|$ stands for the magnitude (or length) of the vector 0A–0B with 0 being the origin. In applications the curves are often restricted to lie on a given manifold, e.g., a sphere (in this case, the geodesic is the shortest great circle joining A and B).

### 2.1 Minimal Surfaces

A *minimal surface* is a surface of least area among all those bounded by a given closed curve. The problem of finding minimal surfaces, called the *Plateau problem*, was first solved in three dimensions in the 1930s by Douglas and, separately, by Rado, and then in the 1960s several authors, including Almgren, De Giorgi, Fleming, and Federer, addressed it using geometric measure-theoretical tools. This approach gives existence of solutions in a "weak sense," and establishing their regularity is significantly more involved. De Giorgi proved that minimal surfaces are analytic except on a singular set of dimension at most $N - 1$. Later, Federer, based on earlier results by Almgren and Simons, improved the dimension of the singular set to $N - 8$. The sharpness of this estimate was confirmed with an example by Bombieri, De Giorgi, and Giusti.

Important minimal surfaces are the *nonparametric* minimal surfaces, which are given as graphs of real-valued functions. To be precise, given an open set $\Omega \subset \mathbb{R}^N$ and a smooth function $u\colon \Omega \to \mathbb{R}$, the area of the graph of $u$, $\{(x, u(x))\colon x \in \Omega\}$, is given by

$$F(u) := \int_{\Omega} \sqrt{1 + |\nabla u|^2}\,\mathrm{d}x. \tag{2}$$

It can be shown that $u$ minimizes the area of its graph subject to prescribed values on the boundary of $\Omega$ if

$$\operatorname{div}\left(\frac{\nabla u}{\sqrt{1+|\nabla u|^2}}\right) = 0 \quad \text{in } \Omega.$$

### 2.2 The Willmore Functional

Many smooth surfaces, including tori, have recently been obtained as minima or critical points of certain geometrical functionals in the calculus of variations. An important example is the Willmore (or bending) energy of a compact surface $S$ embedded in $\mathbb{R}^3$, namely the surface integral $\mathcal{W}(S) := \int_S H^2\,d\sigma$, where $H := \frac{1}{2}(k_1 + k_2)$ and $k_1$ and $k_2$ are the principal curvatures of $S$. This energy has a wide scope of applications, ranging from materials science (e.g., elastic shells, bending energy) to mathematical biology (e.g., cell membranes) to image segmentation in computer vision (e.g., staircasing).

Critical points of $\mathcal{W}$ are called *Willmore surfaces* and satisfy the Euler–Lagrange equation

$$\Delta_S H + 2H(H^2 - K) = 0,$$

where $K := k_1 k_2$ is the Gaussian curvature and $\Delta_S$ is the Laplace–Beltrami operator.

In the 1920s it was shown by Blaschke and, separately, by Thomsen that the Willmore energy is invariant under conformal transformations of $\mathbb{R}^3$. Also, the Willmore energy is minimized by spheres, with resulting energy value $4\pi$. Therefore, $\mathcal{W}(S) - 4\pi$ describes how much $S$ differs from a sphere in terms of its bending. The problem of minimizing the Willmore energy among the class of embedded tori $T$ was proposed by Willmore, who conjectured in 1965 that $\mathcal{W}(T) \geqslant 2\pi^2$. This conjecture was proved by Marques and Neves in 2012.

### 2.3 Isoperimetric Problems and the Wulff Shape

The understanding of the surface structure of crystals plays a central role in many fields of physics, chemistry, and materials science. If the dimension of the crystals is sufficiently small, then the leading morphological mechanism is driven by the minimization of surface energy. Since the work of Herring in the 1950s, a classical problem in this field is to determine the crystalline shape that has the smallest surface energy for a given volume. To be precise, we seek to minimize the surface integral

$$\int_{\partial E} \psi(\nu(x))\,d\sigma \tag{3}$$

over all smooth sets $E \subset \mathbb{R}^N$ with prescribed volume and where $\nu(x)$ is the outward unit normal to $\partial E$ at $x$. The right variational framework for this problem is within the class of sets of finite perimeter. The solution, which exists and is unique up to translations, is called the Wulff shape. A key ingredient in the proof is the Brunn–Minkowski inequality

$$(\mathcal{L}^N(A))^{1/N} + (\mathcal{L}^N(B))^{1/N} \leqslant (\mathcal{L}^N(A+B))^{1/N}, \tag{4}$$

which holds for all Lebesgue measurable sets $A, B \subset \mathbb{R}^N$ such that $A + B$ is also Lebesgue measurable. Here, $\mathcal{L}^N$ stands for the $N$-dimensional Lebesgue measure.

## 3 The Euler–Lagrange Equation

Consider the functional (1), in the scalar case $d = 1$ and where $f$ is of class $C^1$ and $X$ is the *Sobolev space* $X = W^{1,p}(\Omega)$, $1 \leqslant p \leqslant +\infty$, of all functions $u \in L^p(\Omega)$ whose distributional gradient $\nabla u$ belongs to $L^p(\Omega; \mathbb{R}^N)$. Let $u \in X$ be a *local minimizer* of the functional $F$; that is,

$$\int_U f(x, u(x), \nabla u(x))\,dx \leqslant \int_U f(x, v(x), \nabla v(x))\,dx$$

for every open subset $U$ compactly contained in $\Omega$, and for all $v$ such that $u - v \in W_0^{1,p}(U)$, where $W_0^{1,p}(U)$ is the space of all functions in $W^{1,p}(U)$ that "vanish" on the boundary of $\partial U$. Note that $v$ will then coincide with $u$ outside the set $U$. If $\varphi \in C_c^1(\Omega)$, then $u + t\varphi$, $t \in \mathbb{R}$, are admissible, and thus

$$t \in \mathbb{R} \mapsto g(t) := F(u + t\varphi)$$

has a minimum at $t = 0$. Therefore, under appropriate growth conditions on $f$, we have that $g'(0) = 0$, i.e.,

$$\int_\Omega \left(\sum_{i=1}^N \frac{\partial f}{\partial \xi_i}(x, u, \nabla u)\frac{\partial \varphi}{\partial x_i} + \frac{\partial f}{\partial u}(x, u, \nabla u)\varphi\right) dx = 0. \tag{5}$$

A function $u \in X$ satisfying (5) is said to be a *weak solution of the Euler–Lagrange equation* associated to (1).

Under suitable regularity conditions on $f$ and $u$, (5) can be written in the strong form

$$\operatorname{div}(\nabla_\xi f(x, u, \nabla u)) = \frac{\partial f}{\partial u}(x, u, \nabla u), \tag{6}$$

where $\nabla_\xi f(x, u, \xi)$ is the gradient of the function $f(x, u, \cdot)$.

In the vectorial case $d > 1$, the same argument leads to a system of partial differential equations (PDEs) in place of (5).

## 4 Variational Inequalities and Free-Boundary and Free-Discontinuity Problems

We now add a constraint to the minimization problem considered in the previous section. To be precise, let $d = 1$ and let $\phi$ be a function in $\Omega$. If $u$ is a local minimizer of (1) among all functions $v \in W^{1,p}(\Omega)$ subject to the constraint $v \geqslant \phi$ in $\Omega$, then the variation $u + t\varphi$ is admissible if $\varphi \geqslant 0$ and $t \geqslant 0$. Therefore, the function $g$ satisfies $g'(0) \geqslant 0$, and the Euler–Lagrange equation (5) becomes the variational inequality

$$\int_\Omega \Big( \sum_{i=1}^N \frac{\partial f}{\partial \xi_i}(x, u, \nabla u) \frac{\partial \varphi}{\partial x_i} + \frac{\partial f}{\partial u}(x, u, \nabla u)\varphi \Big)\, \mathrm{d}x \geqslant 0$$

for all nonnegative $\varphi \in C_c^1(\Omega)$. This is called the *obstacle problem*, and the coincidence set $\{u = \phi\}$ is not known a priori and is called the *free boundary*. This is an example of a broad class of variational inequalities and free boundary problems that have applications in a variety of contexts, including the modeling of the melting of ice (the *Stefan problem*), lubrication, and the filtration of a liquid through a porous medium.

A related class of minimization problems in which the unknowns are both an underlying field $u$ and a subset $E$ of $\Omega$ is the class of *free discontinuity problems* that are characterized by the competition between a volume energy of the type (1) and a surface energy, e.g., as in (3). Important examples are in the study of liquid crystals, the optimal design of composite materials in continuum mechanics (see section 13.3), and image segmentation in computer vision (see section 13.4).

## 5 Lagrange Multipliers

The method of *Lagrange multipliers in Banach spaces* is used to find extrema of functionals $G \colon X \to \mathbb{R}$ subject to a constraint

$$\{x \in X \colon \Psi(x) = 0\}, \tag{7}$$

where $\Psi \colon X \to Y$ is another functional and $X$ and $Y$ are Banach spaces. It can be shown that if $G$ and $\Psi$ are of class $C^1$ and $u \in X$ is an extremum of $G$ subject to (7), and if the derivative $\mathrm{D}\Psi(u) \colon X \to Y$ is surjective, then there exists a continuous, linear functional $\lambda \colon Y \to \mathbb{R}$ such that

$$\mathrm{D}G(u) + \lambda \circ \mathrm{D}\Psi(u) = 0, \tag{8}$$

where $\circ$ stands for the composition operator between functions. The functional $\lambda$ is called a *Lagrange multiplier.*

In the special case in which $Y = \mathbb{R}$, $\lambda$ may be identified with a scalar, still denoted by $\lambda$, and (8) takes the familiar form

$$\mathrm{D}G(u) + \lambda \mathrm{D}\Psi(u) = 0.$$

Therefore, candidates for extrema may be found among all critical points of the family of functionals $G + \lambda\Psi$, $\lambda \in \mathbb{R}$.

If $G$ has the form (1) and $X = W^{1,p}(\Omega; \mathbb{R}^d)$, $1 \leqslant p \leqslant +\infty$, then typical examples of $\Psi$ are

$$\Psi(u) := \int_\Omega |u|^s \, \mathrm{d}x - c_1 \quad \text{or} \quad \Psi(u) := \int_\Omega u \, \mathrm{d}x - c_2$$

for some constants $c_1 \in \mathbb{R}$, $c_2 \in \mathbb{R}^d$, and $1 \leqslant s < +\infty$.

## 6 Minimax Methods

Minimax methods are used to establish the existence of *saddle points* of the functional (1), i.e., critical points that are not extrema. More generally, for $C^1$ functionals $G \colon X \to \mathbb{R}$, where $X$ is an infinite-dimensional Banach space, as introduced in section 5, the *Palais–Smale compactness condition* (hereafter simply referred to as the PS condition) plays the role of compactness in the finite-dimensional case. To be precise, $G$ satisfies the PS condition if whenever $\{u_n\} \subset X$ is such that $\{G(u_n)\}$ is a bounded sequence in $\mathbb{R}$ and $\mathrm{D}G(u_n) \to 0$ in the dual of $X$, $X'$, then $\{u_n\}$ admits a convergent subsequence.

An important result for the existence of saddle points that uses the PS condition is the *mountain pass lemma* of Ambrosetti and Rabinowitz, which states that, if $G$ satisfies the PS condition, if $G(0) = 0$, and if there are $r > 0$ and $u_0 \in X \setminus \overline{B(0,r)}$ such that

$$\inf_{\partial B(0,r)} G > 0 \quad \text{and} \quad G(u_0) \leqslant 0,$$

then

$$\inf_{\gamma \in C} \sup_{u \in \gamma} G(u)$$

is a critical value, where $C$ is the set of all continuous curves from $[0,1]$ into $X$ joining 0 to $u_0$.

In addition, minimax methods can be used to prove the existence of multiple critical points of functionals $G$ that satisfy certain symmetry properties, for example, the generalization of the result by Ljusternik and Schnirelmann for symmetric functions to the infinite-dimensional case.

## 7 Lower Semicontinuity

### 7.1 The Direct Method

The *direct method* in the calculus of variations provides conditions on the function space $X$ and on a functional

$G$, as introduced in section 5, that guarantee the existence of minimizers of $G$. The method consists of the following steps.

**Step 1.** Consider a *minimizing sequence* $\{u_n\} \subset X$, i.e., $\lim_{n\to\infty} G(u_n) = \inf_{u\in X} G(u)$.

**Step 2.** Prove that $\{u_n\}$ admits a subsequence $\{u_{n_k}\}$ converging to some $u_0 \in X$ with respect to some (weak) topology $\tau$ in $X$. When $G$ has an integral representation of the form (1), this is usually a consequence of a priori coercivity conditions on the integrand $f$.

**Step 3.** Establish the *sequential lower semicontinuity of $G$ with respect to $\tau$*, i.e., $\liminf_{n\to\infty} G(v_n) \geqslant G(v)$ whenever the sequence $\{v_n\} \subset X$ converges weakly to $v \in X$ with respect to $\tau$.

**Step 4.** Conclude that $u_0$ minimizes $G$. Indeed,

$$\begin{aligned}\inf_{u\in X} G(u) &= \lim_{n\to\infty} G(u_n) = \lim_{k\to\infty} G(u_{n_k})\\ &\geqslant G(u_0) \geqslant \inf_{u\in X} G(u).\end{aligned}$$

### 7.2 Integrands: Convex, Polyconvex, Quasiconvex, and Rank-One Convex

In view of step 3 above, it is important to characterize the class of integrands $f$ in (1) for which the corresponding functional $F$ is sequentially lower semicontinuous with respect to $\tau$. In the case in which $X$ is the Sobolev space $W^{1,p}(\Omega;\mathbb{R}^d)$, $1 \leqslant p \leqslant +\infty$, and $\tau$ is the weak topology (or weak-$\star$ topology $p = +\infty$), this is related to convexity-type properties of $f(x,u,\cdot)$. If $\min\{d,N\} = 1$, then under appropriate growth and regularity conditions it can be shown that convexity of $f(x,u,\cdot)$ is necessary and sufficient. More generally, if $\min\{d,N\} > 1$, then the corresponding condition is called *quasiconvexity*; to be precise, $f(x,u,\cdot)$ is said to be quasiconvex if

$$f(x,u,\xi) \leqslant \int_{(0,1)^N} f(x,u,\xi + \nabla\varphi(y))\,\mathrm{d}y$$

for all $\xi \in \mathbb{R}^{d\times N}$ and all $\varphi \in W_0^{1,\infty}((0,1)^N;\mathbb{R}^d)$, whenever the right-hand side in this inequality is well defined. Since this condition is nonlocal, in applications in mechanics one often studies related classes of integrands, such as polyconvex and rank-one convex functions, for which there are algebraic criteria.

## 8 Relaxation

In most applications, step 3 in section 7.1 fails, and this leads to an important topic at the core of the calculus of variations, namely, the introduction of a *relaxed*, or *effective*, energy $\mathcal{G}$ that is related to $G$, as introduced in section 5, as follows.

(a) $\mathcal{G}$ is sequentially lower semicontinuous with respect to $\tau$.
(b) $\mathcal{G} \leqslant G$ and $\mathcal{G}$ inherits coercivity properties from $G$.
(c) $\min_{u\in X} \mathcal{G} = \inf_{u\in X} G$.

When $G$ is of the type (1), a central problem is to understand if $\mathcal{G}$ has an integral form of the type (1) for some new integrand $h$ and then, if it has, to understand what the relation between $h$ and the original integrand $f$ is.

If $X = W^{1,p}(\Omega)$, $p \geqslant 1$, and $\tau$ is the weak topology, then under appropriate growth and regularity conditions it can be shown that $h(x,u,\cdot)$ is the *convex envelope* of $f(x,u,\cdot)$, i.e., the greatest convex function less than $f(x,u,\cdot)$. In the vectorial case $d > 1$, the convex envelope is replaced by a similar notion of quasiconvex envelope (see section 7.2).

## 9 $\Gamma$-Convergence

In physical problems the behavior of a system is often described in terms of a sequence $\{G_n\}$, $n \in \mathbb{N}$, of energy functionals $G_n\colon X \to [-\infty,+\infty]$, where $X$ is a metric space with a metric $d$. Is it possible to identify a limiting energy $G_\infty$ that sheds light on the qualitative properties of this family and that has the property that minimizers of $G_n$ converge to minimizers of $G_\infty$?

The notion of *$\Gamma$-convergence*, which was introduced by De Giorgi, provides a tool for answering these questions. To motivate this concept with an example, consider a fluid confined in a container $\Omega \subset \mathbb{R}^N$. Assume that the total mass of the fluid is $m$, so that admissible density distributions $u\colon \Omega \to \mathbb{R}$ satisfy the constraint $\int_\Omega u(x)\,\mathrm{d}x = m$. The total energy is given by the functional $u \mapsto \int_\Omega W(u(x))\,\mathrm{d}x$, where $W\colon \mathbb{R} \to [0,\infty)$ is the energy per unit volume. Assume that $W$ supports two phases $a < b$, that is, $W$ is a *double-well potential*, with $\{u \in \mathbb{R}\colon W(u) = 0\} = \{a,b\}$. Then any density distribution $u$ that renders the body stable in the sense of Gibbs is a minimizer of the following problem:

$$\min\left\{\int_\Omega W(u(x))\,\mathrm{d}x\colon \int_\Omega u(x)\,\mathrm{d}x = m\right\}. \qquad (\mathcal{P}_0)$$

If $\mathcal{L}^N(\Omega) = 1$ and $a < m < b$, then given any measurable set $E \subset \Omega$ with $\mathcal{L}^N(E) = (b-m)/(b-a)$, the function $u = a\chi_E + b\chi_{\Omega\setminus E}$ is a solution of problem $(\mathcal{P}_0)$. This lack of uniqueness is due to the fact that interfaces between the two phases $a$ and $b$ can be formed without increasing the total energy. The physically preferred solutions should be the ones that arise as limiting cases of a theory that penalizes interfacial energy, so it is expected that these solutions should minimize the surface area of $\partial E \cap \Omega$.

In the van der Waals–Cahn–Hilliard theory of phase transitions, the energy depends not only on the density $u$ but also on its gradient. To be precise,

$$\int_\Omega W(u(x))\,dx + \varepsilon^2 \int_\Omega |\nabla u(x)|^2\,dx.$$

Note that the gradient term penalizes rapid changes in the density $u$, and thus it plays the role of an interfacial energy. Stable density distributions $u$ are now solutions of the minimization problem

$$\min\left\{\int_\Omega W(u(x))\,dx + \varepsilon^2 \int_\Omega |\nabla u(x)|^2\,dx\right\}, \qquad (\mathcal{P}_\varepsilon)$$

where the minimum is taken over all smooth functions $u$ satisfying $\int_\Omega u(x)\,dx = m$. In 1983 Gurtin conjectured that the limits, as $\varepsilon \to 0$, of solutions of $(\mathcal{P}_\varepsilon)$ are solutions of $(\mathcal{P}_0)$ with minimal surface area. Using results of Modica and Mortola, this conjecture was proved independently by Modica and by Sternberg in the setting of $\Gamma$-convergence.

The *$\Gamma$-limit* $G_\infty\colon X \to [-\infty, +\infty]$ of $\{G_n\}$ with respect to a metric $d$, when it exists, is defined uniquely by the following properties.

**(i) The lim inf inequality.** For every sequence $\{u_n\} \subset X$ converging to $u \in X$ with respect to $d$,

$$G_\infty(u) \leqslant \liminf_{n\to\infty} G_n(u_n).$$

**(ii) The lim sup inequality.** For every $u \in X$ there exists a sequence $\{u_n\} \subset X$ converging to $u \in X$ with respect to $d$ such that

$$G_\infty(u) \geqslant \limsup_{n\to\infty} G_n(u_n).$$

This notion may be extended to the case in which the convergence of the sequences is taken with respect to some weak topology rather than the topology induced by the metric $d$. In this context, we remark that, when the sequence $\{G_n\}$ reduces to a single energy functional $\{G\}$, under appropriate growth and coercivity assumptions, $G_\infty$ coincides with the relaxed energy $\mathcal{G}$, as discussed in section 8.

Other important applications of $\Gamma$-convergence include the Ginzburg–Landau theory for superconductivity (see section 13.5), homogenization of variational problems (see section 13.3), dimension reduction problems in elasticity (see section 13.2), and free-discontinuity problems in image segmentation in computer vision (see section 13.4) and in fracture mechanics.

## 10 Regularity

Optimal regularity of minimizers and local minimizers of the energy (1) in the vectorial case $d \geqslant 2$, and when $X = W^{1,p}(\Omega; \mathbb{R}^d)$, $1 \leqslant p \leqslant +\infty$, is mostly an open question. In the scalar case $d = 1$ there is an extensive body of literature on the regularity of weak solutions of the Euler–Lagrange equation (5), stemming from a fundamental result of De Giorgi in the late 1950s that was independently obtained by Nash. For $d \geqslant 2$, (local) minimizers of (1) are not generally everywhere smooth. On the other hand, and under suitable hypotheses on the integrand $f$, it can be shown that *partial regularity* holds, i.e., if $u$ is a local minimizer, then there exists an open subset of $\Omega$, $\Omega_0$, of full measure such that $u \in C^{1,\alpha}(\Omega_0; \mathbb{R}^d)$ for some $\alpha \in (0, 1)$. Sharp estimates of $\alpha$ and of the Hausdorff dimension of the *singular set* $\Sigma_u := \Omega \setminus \Omega_0$ are still unknown.

## 11 Symmetrization

Rearrangements of sets preserve their measure while modifying their geometry to achieve specific symmetries. In turn, rearrangements of a function $u$ yield new functions that have desired symmetry properties and that are obtained via suitable rearrangements of the *$t$-superlevel sets* of $u$, $\Omega_t := \{x \in \Omega\colon u(x) > t\}$. These tools are used in a variety of contexts, from harmonic analysis and PDEs to the spectral theory of differential operators. In the calculus of variations, they are often found in the study of extrema of functionals of type (1). Among the most common rearrangements we mention the directional monotone decreasing rearrangement, the star-shaped rearrangement, the directional Steiner symmetrization, the Schwarz symmetrization, the circular and spherical symmetrization, and the radial symmetrization.

Of these, we highlight the Schwarz symmetrization, which is the most frequently used in the calculus of variations. If $u$ is a nonnegative measurable function with compact support in $\mathbb{R}^N$, then its *Schwartz symmetric rearrangement* is the (unique) spherically symmetric and decreasing function $u^\star$ such that for all $t > 0$ the $t$-superlevel sets of $u$ and $u^\star$ have the same measure.

When $\Omega = \mathbb{R}^N$, it can be shown that $u^\star$ preserves the $L^p$-norm of $u$ and the regularity of $u$ up to first order; that is, if $u$ belongs to $W^{1,p}(\mathbb{R}^N)$, then so does $u^\star$, $1 \leqslant p \leqslant +\infty$. Moreover, by the Pólya–Szegö inequality, $\|\nabla u^\star\|_p \leqslant \|\nabla u\|_p$, and we remark that for $p = \infty$

this is obtained using the Brunn–Minkowski inequality discussed in section 2.3.

Another important inequality relating $u$ and $u^\star$ is the Riesz inequality, and the Faber–Krahn inequality compares eigenvalues of the Dirichlet problems in $\Omega$ and in $\Omega^\star$. Classical applications of rearrangements include the derivation of the sharp constant in the Sobolev–Gagliardo–Nirenberg inequality in $W^{1,p}(\mathbb{R}^N)$, $1 < p < N$, as well as in the Young inequality and the Hardy–Littlewood–Sobolev inequality.

Finally, we remark that the first and most important application of Steiner symmetrization is the isoperimetric property of balls (see Dido's problem in section 1).

## 12 Duality Theory

Duality theory associates with a minimization problem $(\mathcal{P})$ a maximization problem $(\mathcal{P}^*)$, called the *dual problem*, and studies the relation between these two. It has important applications in several disciplines, including economics and mechanics, and different areas of mathematics, such as the calculus of variations, convex analysis, and numerical analysis.

The theory of dual problems is inspired by the notion of duality in convex analysis and by the *Fenchel transform* $f^\star$ of a function $f\colon \mathbb{R}^N \to [-\infty, +\infty]$, defined as

$$f^\star(\eta) := \sup\{\eta \cdot \xi - f(\xi) \colon \xi \in \mathbb{R}^N\} \quad \text{for } \eta \in \mathbb{R}^N.$$

As an example, consider the minimization problem

$$\inf\left\{\int_\Omega f(\nabla u)\,dx \colon u \in W_0^{1,p}(\Omega)\right\}, \qquad (\mathcal{P})$$

with $f\colon \mathbb{R}^N \to \mathbb{R}$. If $f$ satisfies appropriate growth and convexity conditions, then the dual problem $(\mathcal{P}^\star)$ is given by

$$\sup\left\{-\int_\Omega f^\star(v(x))\,dx \colon v \in L^q(\Omega;\mathbb{R}^N),\ \operatorname{div} v = 0 \text{ in } \Omega\right\},$$

where $1/p + 1/q = 1$. The latter problem may be simpler to handle in specific situations, e.g., for nonparametric minimal surfaces and with $f$ given as in (2), where, due to lack of coercivity, $(\mathcal{P})$ may not admit a solution in $X$.

## 13 Some Contemporary Applications

There is a plethora of applications of the calculus of variations. Classical ones include Hamiltonians and Lagrangians, the Hamilton–Jacobi equation, conservation laws, Noether's theorem, and optimal control. Below we focus on a few contemporary applications that are pushing the frontiers of the theory in novel directions.

### 13.1 Elasticity

Consider an elastic body that occupies a domain $\Omega \subset \mathbb{R}^3$ in a given reference configuration. The deformations of the body can be described by maps $u\colon \Omega \to \mathbb{R}^3$. If the body is homogeneous, then the total elastic energy corresponding to $u$ is given by the functional

$$F(u) := \int_\Omega f(\nabla u(x))\,dx, \qquad (9)$$

where $f$ is the *stored-energy density* of the material. In order to prevent interpenetration of matter, the deformations should be invertible and it should require an infinite amount of energy to violate this property, i.e.,

$$f(\xi) \to +\infty \quad \text{as } \det \xi \to 0^+. \qquad (10)$$

Also, $f$ needs to be *frame indifferent*, i.e.,

$$f(R\xi) = f(\xi) \qquad (11)$$

for all rotations $R$ and all $\xi \in \mathbb{R}^{3\times 3}$.

Under appropriate coercivity and convex-type conditions on $f$ (see section 7.2), and under suitable boundary conditions, it can be shown that $F$ admits a global minimizer $u_0$. However, the regularity of $u_0$ is still an open problem, and the Euler–Lagrange equation cannot therefore be derived (see section 3). In addition, the existence of local minimizers remains unsolved.

### 13.2 Dimension Reduction

An important problem in elasticity is the derivation of models for thin structures, such as membranes, shells, plates, rods, and beams, from three-dimensional elasticity theory. The mathematically rigorous analysis was initiated by Acerbi, Buttazzo, and Percivale in the 1990s for rods, and this was followed by the work of Le Dret and Raoult for membranes. Recent contributions by Friesecke, James, and Müller have allowed us to handle the physical requirements (10) and (11). The main tool underlying these works is $\Gamma$-convergence (see section 9).

To illustrate the deduction in the case of membranes, consider a thin cylindrical elastic body of thickness $2\varepsilon > 0$ occupying the reference configuration $\Omega_\varepsilon := \omega \times (-\varepsilon, \varepsilon)$, with $\omega \subset \mathbb{R}^2$. Using the typical rescaling

$$(x_1, x_2, x_3) \mapsto (y_1, y_2, y_3) := (x_1, x_2, x_3/\varepsilon),$$

the deformations $u$ of $\Omega_\varepsilon$ now correspond to deformation $v$ of the fixed domain $\Omega_1$, through the formula

$v(y_1, y_2, y_3) = u(x_1, x_2, x_3)$. Therefore,

$$\frac{1}{\varepsilon}\int_{\Omega_\varepsilon} f(\nabla u)\,\mathrm{d}x = \int_{\Omega_1} f\left(\frac{\partial v}{\partial y_1}, \frac{\partial v}{\partial y_2}, \frac{1}{\varepsilon}\frac{\partial v}{\partial y_3}\right)\mathrm{d}y.$$

The right-hand side of the previous equality yields a family of functionals to which the theory of $\Gamma$-convergence is applied.

### 13.3 Homogenization

HOMOGENIZATION THEORY [II.17] is used to describe the macroscopic behavior of heterogeneous composite materials, which are characterized by having two or more finely mixed material components. Composite materials have important technological and industrial applications as their effective properties are often better than the corresponding properties of the individual constituents. The study of these materials falls within the so-called *multiscale problems*, with the two relevant scales here being the *microscopic scale* at the level of the heterogeneities and the *macroscopic scale* describing the resulting "homogeneous" material. Mathematically, the properties of composite materials can be described in terms of PDEs with fast oscillating coefficients or in terms of energy functionals that depend on a small parameter $\varepsilon$. As an example, consider a material matrix $A$ with corresponding stored-energy density $f_A$, with periodically distributed inclusions of another material $B$ with stored-energy density $f_B$, whose periodicity cell has side-length $\varepsilon$. The total energy of the composite is then given by

$$\int_\Omega \left[\left(1 - \chi\left(\frac{x}{\varepsilon}\right)\right) f_A(\nabla u) + \chi\left(\frac{x}{\varepsilon}\right) f_B(\nabla u)\right]\mathrm{d}x,$$

where $\chi$ is the characteristic function of the locus of material $B$ contained in the unit cube $Q$ of material $A$, extended periodically to $\mathbb{R}^3$ with period $Q$. The goal here is to characterize the "homogenized" energy when $\varepsilon \to 0^+$ using $\Gamma$-convergence (see section 9).

### 13.4 Computer Vision

Several problems in computer vision can be treated variationally, including image segmentation (e.g., the Mumford–Shah and Blake–Zisserman models), image morphing, image denoising (e.g., the Perona–Malik scheme and the Rudin–Osher–Fatemi total variation model), and inpainting (e.g., recolorization).

The Mumford–Shah model provides a good example of the use of the calculus of variations to treat free discontinuity problems. Let $\Omega$ be a rectangle in the plane, representing the locus of the image, with gray levels given by a function $g\colon \Omega \to [0,1]$. We want to find an approximation of $g$ that is smooth outside a set $K$ of sharp contours related to the set of discontinuities of $g$. This leads to the minimization of the functional

$$\int_{\Omega\setminus K} (|\nabla u|^2 + \alpha(u-g)^2)\,\mathrm{d}x + \beta\,\mathrm{length}(K \cap \Omega)$$

over all contour curves $K$ and functions $u \in C^1(\Omega \setminus K)$. The first term in this energy functional is minimized when $u$ is constant outside $K$, and it therefore forces $u$ not to vary much outside $K$. The second term is minimized when $u = g$ outside $K$, and hence $u$ is required to stay close to the original gray level $g$. The last term is minimized when $K$ has length as short as possible. The existence of a minimizing pair $(u, K)$ was established by De Giorgi, Carriero, and Leaci, with $u$ in a class of functions larger than $C^1(\Omega \setminus K)$, to be precise, the space of functions of special bounded variation. The full regularity of these solutions $u$ and the structure of $K$ remain an open problem.

### 13.5 The Ginzburg–Landau Theory for Superconductivity

In the 1950s Ginzburg and Landau proposed a mathematical theory to study phase transition problems in superconductivity; there are similar formulations to address problems in superfluids, e.g., helium II, and in $XY$ magnetism. In its simplest form, the Ginzburg–Landau functional reduces to

$$F_\varepsilon(u) := \frac{1}{2}\int_\Omega |\nabla u|^2\,\mathrm{d}x + \frac{1}{4\varepsilon^2}\int_\Omega (|u|^2 - 1)^2\,\mathrm{d}x,$$

where $\Omega \subset \mathbb{R}^2$ is a star-shaped domain, the *condensate wave function* $u \in W^{1,2}(\Omega; \mathbb{R}^2)$ is an order parameter with two degrees of freedom, and the parameter $\varepsilon$ is a (small) characteristic length. Given $g\colon \partial\Omega \to \mathbb{S}^1$, with $\mathbb{S}^1$ the unit circle in $\mathbb{R}^2$ centered at the origin, we are interested in characterizing the limits of minimizers $u_\varepsilon$ of $F_\varepsilon$ subject to the boundary condition $u_\varepsilon = g$ on $\partial\Omega$. Under suitable geometric conditions on $g$ (related to the winding number), Bethuel, Brezis, and Hélein have shown that there are no limiting functions $u$ in $C(\bar{\Omega}, \mathbb{S}^1)$ that satisfy the boundary condition. Rather, the limiting functions are smooth outside a finite set of singularities, called *vortices*. $\Gamma$-convergence techniques may be used to study this family of functionals (see section 9).

### 13.6 Mass Transport

Mass transportation was introduced by Monge in 1781, studied by the Nobel Prize winner Kantorovich in the 1940s, and revived by Brenier in 1987. Since then, it has surfaced in a variety of areas, from economics to optimization. Given a pile of sand of mass 1 and a hole of volume 1, we want to fill the hole with the sand while minimizing the cost of transportation. This problem is formulated using probability theory as follows. The pile and the hole are represented by probability measures $\mu$ and $\nu$, with supports in measurable spaces $X$ and $Y$, respectively. If $A \subset X$ and $B \subset Y$ are measurable sets, then $\mu(A)$ measures the amount of sand in $A$ and $\nu(B)$ measures the amount of sand that can fill $B$. The cost of transportation is modeled by a measurable cost function $c\colon X \times Y \to \mathbb{R} \cup +\infty$. Kantorovich's optimal transportation problem consists of minimizing

$$\int_{X\times Y} c(x,y)\,\mathrm{d}\pi(x,y)$$

over all probability measures on $X \times Y$ such that $\pi(A \times Y) = \mu(A)$ and $\pi(X \times B) = \nu(B)$, for all measurable sets $A \subset X$ and $B \subset Y$. The main problem is to establish the existence of minimizers and to obtain their characterization. This depends strongly on the cost function $c$ and on the regularity of the measures $\mu$ and $\nu$. There is a multitude of applications of this theory, and here we mention only that it can be used to give a simple proof of the Brunn–Minkowski inequality (see (4)).

### 13.7 Gradient Flows

Given a function $h\colon \mathbb{R}^m \to \mathbb{R}$ of class $C^2$, the *gradient flow of* $h$ is the family of maps $S_t\colon \mathbb{R}^m \to \mathbb{R}^m$, $t \geqslant 0$, satisfying the following property. For every $w_0 \in \mathbb{R}^m$, $S_0(w_0) := w_0$ and the curve $w_t := S_t(w_0)$, $t > 0$, is the unique $C^1$ solution of the Cauchy problem

$$\frac{\mathrm{d}}{\mathrm{d}t} w_t = -\nabla h(w_t) \quad \text{for } t > 0, \quad \lim_{t \to 0^+} w_t = w_0, \tag{12}$$

if it exists.

If $\mathrm{D}^2 h \geqslant \alpha\mathbb{I}$ for some $\alpha \in \mathbb{R}$, then it can be shown that the gradient flow exists, that it is unique, and that it satisfies a semigroup property, i.e.,

$$S_{t+s}(w_0) = S_t(S_s(w_0)), \quad \lim_{t \to 0^+} S_t(w_0) = w_0,$$

for every $w_0 \in \mathbb{R}^m$.

A common way of approximating discretely the solution of (12) is via the *implicit Euler scheme*, as follows. Given a time step $\tau > 0$, consider the partition of $[0, +\infty)$

$$\{0 = t_\tau^0 < t_\tau^1 < \cdots < t_\tau^n < \cdots\},$$

where $t_\tau^n := n\tau$. Define recursively a discrete sequence $\{W_\tau^n\}$ as follows: assuming that $W_\tau^{n-1}$ has already been defined, let $W_\tau^n$ be the unique minimizer of the function

$$w \mapsto \frac{1}{2\tau}|w - W_\tau^{n-1}|^2 + h(w). \tag{13}$$

Introduce the piecewise-linear function $W_\tau\colon [0, +\infty) \to \mathbb{R}^m$ given by

$$W_\tau(t) = \frac{t - t_\tau^{n-1}}{\tau} W_\tau^{n-1} + \frac{t_\tau^n - t}{\tau} W_\tau^n$$

for $t \in [t_\tau^{n-1}, t_\tau^n]$. If $W_\tau^0 \to w_0$ as $\tau \to 0^+$, then it can be shown that $\{W_\tau\}_{\tau>0}$ converges to the solution of (12) as $\tau \to 0^+$.

This approximation scheme, here described for the finite-dimensional vector space $\mathbb{R}^d$, may be extended to the case in which $\mathbb{R}^d$ is replaced by an infinite-dimensional metric space $X$, the function $h$ is replaced by a functional $G\colon X \to \mathbb{R}$, and the minimization procedure in (13) is now a variational minimization problem of the type addressed in section 7. This method is known as *De Giorgi's minimizing movements*. Important applications include the study of a large class of parabolic PDEs.

### Further Reading

Ambrosio, L., N. Fusco, and D. Pallara. 2000. *Functions of Bounded Variation and Free Discontinuity Problems.* Oxford: Clarendon.

Ambrosio, L., N. Gigli, and G. Savaré. 2008. *Gradient Flows in Metric Spaces and in the Space of Probability Measures.* Basel: Birkhäuser.

Bethuel, F., H. Brezis, and F. Hélein. 1994. *Ginzburg–Landau Vortices.* Boston, MA: Birkhäuser.

Cioranescu, D., and P. Donato. 1999. *An Introduction to Homogenization.* New York: Oxford University Press.

Dacorogna, B. 2008. *Direct Methods in the Calculus of Variations.* New York: Springer.

Dal Maso, G. 1993. *An Introduction to Γ-Convergence.* Boston, MA: Birkhäuser.

Ekeland, I., and R. Téman. 1999. *Convex Analysis and Variational Problems.* Philadelphia, PA: SIAM.

Fonseca, I., and G. Leoni. 2007. *Modern Methods in the Calculus of Variations: $L^p$ Spaces.* New York: Springer.

Giaquinta, M., and L. Martinazzi. 2012. *An Introduction to the Regularity Theory for Elliptic Systems, Harmonic Maps and Minimal Graphs*, 2nd edn. Pisa: Edizioni della Normale.

Giusti, E. 1984. *Minimal Surfaces and Functions of Bounded Variation.* Boston, MA: Birkhäuser.

Kawohl, B. 1985. *Rearrangements and Convexity of Level Sets in PDE.* Berlin: Springer.

Kinderlehrer, D., and G. Stampacchia. 2000. *An Introduction to Variational Inequalities and Their Applications.* Philadelphia, PA: SIAM.

Struwe, M. 2008. *Variational Methods: Applications to Nonlinear Partial Differential Equations and Hamiltonian Systems.* Berlin: Springer.

Villani, C. 2003. *Topics in Optimal Transportation.* Providence, RI: American Mathematical Society.

# IV.7 Special Functions

*Nico M. Temme*

## 1 Introduction

Usually we call a function "special" when, like the logarithm, the exponential function, and the trigonometric functions (the elementary transcendental functions), it belongs to the toolbox of the applied mathematician, the physicist, the engineer, or the statistician. Each function has particular notation associated with it and has a great number of known properties.

The study of special functions is a branch of mathematics with a distinguished history involving great names such as Euler, Gauss, Fourier, Legendre, Bessel, and Riemann. Much of their work was inspired by problems from physics and by the resulting differential equations. This activity culminated in the publication in 1927 of the standard, and greatly influential, work *A Course of Modern Analysis* by Whittaker and Watson.

Many other monographs are now available, some that contain the formulas without explanation and others that explain how the special functions arise in problems from physics and statistics. A major project called the Digital Library of Mathematical Functions recently culminated in the *NIST Handbook of Mathematical Functions* (a successor to Abramowitz and Stegun's famed *Handbook of Mathematical Functions*), which is also readily accessible online.

In physics, special functions arise as solutions of the linear second-order differential equations that result from separating the variables in a partial differential equation in some special coordinate system (such as spherical or cylindrical). In this way solutions of the wave equation, the diffusion equation, and so on are written in the form of series or integrals.

In statistics, special functions arise as cumulative distribution functions (gamma and beta distributions, for example). In number theory, zeta functions, Dirichlet series, and modular forms are used.

Some topics that fall outside the scope of this article are mentioned in the final section.

## 2 Bernoulli Numbers, Euler Numbers, and Stirling Numbers

The Bernoulli numbers $B_n$ are defined by the generating function

$$\frac{z}{e^z - 1} = \sum_{n=0}^{\infty} \frac{B_n}{n!} z^n, \quad |z| < 2\pi.$$

Because the function $z/(e^z - 1) - 1 + \frac{1}{2}z$ is even, all $B_n$ with odd index $n \geqslant 3$ vanish: $B_{2n+1} = 0$, $n = 1, 2, 3, \dots$. The first nonvanishing numbers are

$$B_0 = 1, \quad B_1 = -\tfrac{1}{2}, \quad B_2 = \tfrac{1}{6}, \quad B_4 = -\tfrac{1}{30}, \quad B_6 = \tfrac{1}{42}.$$

The Bernoulli numbers are named after Jakob Bernoulli, who mentioned them in his posthumous *Ars Conjectandi* of 1713. He discussed *summae potestatum*, sums of equal powers of the first $n$ integers; for a nonnegative integer $p$,

$$\sum_{m=0}^{n-1} m^p = \frac{1}{p+1} \sum_{k=0}^{p} \binom{p+1}{k} B_k n^{p+1-k},$$

where $\binom{n}{k} = n!/[k!(n-k)!]$. The Bernoulli numbers occur in practically every field of mathematics and particularly in combinatorial theory, finite-difference calculus, numerical analysis, analytic number theory, and probability theory.

The Bernoulli polynomials are defined by the generating function

$$\frac{z e^{xz}}{e^z - 1} = \sum_{n=0}^{\infty} \frac{B_n(x)}{n!} z^n, \quad |z| < 2\pi.$$

The first few polynomials are

$$B_0(x) = 1, \quad B_1(x) = x - \tfrac{1}{2}, \quad B_2(x) = x^2 - x + \tfrac{1}{6}.$$

For the Euler numbers $E_n$, we have the generating function

$$\frac{1}{\cosh z} = \frac{2e^z}{e^{2z} + 1} = \sum_{n=0}^{\infty} \frac{E_n}{n!} z^n, \quad |z| < \tfrac{1}{2}\pi.$$

In contrast to the Bernoulli numbers, the Euler numbers are integers. The first few are

$$E_0 = 1, \quad E_2 = -1, \quad E_4 = 5, \quad E_6 = -61,$$

while those with odd index are zero.

The numbers $s(n,k)$ in the generating function

$$\sum_{n=0}^{\infty} s(n,k) \frac{x^n}{n!} = \frac{(\ln(1+x))^k}{k!}, \quad |x| < 1,$$

and the numbers $S(n,k)$ in the generating function

$$\sum_{n=0}^{\infty} S(n,k) \frac{x^n}{n!} = \frac{(e^x - 1)^k}{k!}$$

are called Stirling numbers of the first and second kind, respectively. They are defined for $0 \leqslant k \leqslant n$ and are named after James Stirling. The Stirling numbers of the second kind have the following combinatorial interpretation: $S(n,k)$ is the number of partitions of $\{1, 2, \ldots, n\}$ into exactly $k$ nonempty subsets. For example, $S(4, 2) = 7$, since

$$\begin{aligned}\{1,2,3,4\} &= \{1\} \cup \{2,3,4\} = \{2\} \cup \{1,3,4\} \\ &= \{3\} \cup \{1,2,4\} = \{4\} \cup \{1,2,3\} \\ &= \{1,2\} \cup \{3,4\} = \{1,3\} \cup \{2,4\} \\ &= \{1,4\} \cup \{2,3\}.\end{aligned}$$

## 3 The Gamma Function and Related Functions

The triangular numbers $T_n = 1 + 2 + \cdots + n$ can be written as $\frac{1}{2}n(n+1)$. Euler thought that it must be possible to express $n! = 1 \cdot 2 \cdot \cdots \cdot (n-1) \cdot n$ as a simple formula (the factorial notation was not used by Euler). In 1729 he proved that for $n!$ such a simple formula does not exist, but he did come up with the formula $n! = \int_0^1 (-\ln x)^n \, dx$. Nowadays, this integral is written as

$$\Gamma(z) = \int_0^\infty t^{z-1} e^{-t} \, dt, \quad \operatorname{Re} z > 0, \tag{1}$$

which is obtained when we use the values $t = -\ln x$ and $n = z - 1$. This notation was formulated by Legendre in 1809, and he also coined the term "gamma function" for $\Gamma$.

The fundamental property $\Gamma(z+1) = z\Gamma(z)$ follows easily by integrating by parts in (1), and this relation shows that the singularities at $z = 0, -1 - 2, \ldots$ are poles of first order. From the Maclaurin expansion of $e^{-t}$ and the Prym decomposition

$$\Gamma(z) = \int_0^1 t^{z-1} e^{-t} \, dt + \int_1^\infty t^{z-1} e^{-t} \, dt,$$

we obtain the expansion due to Mittag-Leffler

$$\Gamma(z) = \sum_{n=0}^\infty \frac{(-1)^n}{n!(n+z)} + \int_1^\infty t^{z-1} e^{-t} \, dt,$$

where $z \neq 0, -1, -2, \ldots$. As the integral is an entire function of $z$, we see that the pole at $z = -n$ has a residue $(-1)^n/n!$.

There is another definition as an infinite product,

$$\frac{1}{\Gamma(z)} = z e^{\gamma z} \prod_{n=0}^\infty \left(\left(1 + \frac{z}{n}\right) e^{-z/n}\right),$$

where $\gamma$ is Euler's constant, defined by the limit

$$\gamma = \lim_{N \to \infty} \left(\sum_{n=1}^N \frac{1}{n} - \ln N\right) = 0.5772\ldots. \tag{2}$$

From the infinite product it follows that

$$\Gamma(z) = \lim_{n \to \infty} \frac{n! \, n^z}{z(z+1)\cdots(z+n)}.$$

Important properties are the reflection formula (due to Euler (1771))

$$\Gamma(z)\Gamma(1-z) = \frac{\pi}{\sin(\pi z)}, \quad z \neq \text{an integer}, \tag{3}$$

and the duplication formula

$$\Gamma(\tfrac{1}{2})\Gamma(2z) = 2^{2z-1}\Gamma(z)\Gamma(z + \tfrac{1}{2}),$$

where $z \neq 0, -1, -2, \ldots$ and $\Gamma(\frac{1}{2}) = \sqrt{\pi}$.

A closely related special function is the beta integral, defined for $\operatorname{Re} p > 0$ and $\operatorname{Re} q > 0$ by

$$B(p, q) = \int_0^1 t^{p-1}(1-t)^{q-1} \, dt. \tag{4}$$

The relationship between the beta integral and the gamma function is given by

$$B(p, q) = \frac{\Gamma(p)\Gamma(q)}{\Gamma(p+q)}.$$

The derivative of the gamma function itself does not play an important role in the theory and applications of special functions. It is not a very manageable function. Much more interesting is the logarithmic derivative:

$$\psi(z) = \frac{d}{dz} \ln \Gamma(z) = \frac{\Gamma'(z)}{\Gamma(z)}.$$

It satisfies the recursion relation $\psi(z+1) = \psi(z) + 1/z$ and it has an interesting series expansion:

$$\psi(z) = -\gamma + \sum_{n=0}^\infty \left(\frac{1}{n+1} - \frac{1}{z+n}\right),$$

for $z \neq 0, -1, -2, \ldots$.

Stirling's asymptotic formula for factorials

$$n! \sim \sqrt{2\pi n}\, n^n e^{-n}, \quad n \to \infty, \tag{5}$$

has several refinements, first in the form

$$\Gamma(z) = \sqrt{2\pi/z}\, z^z e^{-z+\mu(z)}, \quad \mu(z) = \theta/z, \tag{6}$$

with $0 < \theta < 1$, if $z > 0$, and second in the form

$$\Gamma(z) \sim \sqrt{2\pi/z}\, z^z e^{-z} \sum_{n=0}^\infty \frac{a_n}{z^n}, \tag{7}$$

which is valid for large $z$ inside the sector $-\pi < \operatorname{ph} z < \pi$ (where $\operatorname{ph} z = \arg z$ is the phase or argument of $z$). The first few coefficients are

$$a_0 = 1, \quad a_1 = \tfrac{1}{12}, \quad a_2 = \tfrac{1}{288}, \quad a_3 = -\tfrac{139}{51\,840}.$$

For the asymptotic expansion of the logarithm of the gamma function the coefficients are explicitly known in terms of Bernoulli numbers:

$$\ln \Gamma(z) \sim (z - \tfrac{1}{2}) - z + \tfrac{1}{2}\ln(2\pi) + \sum_{n=1}^\infty \frac{B_{2n}}{2n(2n-1)z^{2n-1}}, \tag{8}$$

which is again valid for large $z$ inside the sector $-\pi < \operatorname{ph} z < \pi$. This expansion is more efficient than the one in (7) because it is in powers of $z^{-2}$. Moreover, estimates of the remainder in the expansion in (8) are available.

## 4 The Riemann Zeta Function

The Riemann zeta function is defined by the series

$$\zeta(s) = \sum_{n=1}^{\infty} \frac{1}{n^s}, \quad \operatorname{Re} s > 1. \tag{9}$$

This function was known to Euler, but its main properties were discovered by Riemann. A well-known result in analysis is the divergence of the harmonic series $\sum_{n=1}^{\infty}(1/n)$ (see (2)), and indeed, the Riemann zeta function has a singular point at $s = 1$, a pole of order 1. The limit $\lim_{s\to 1}(s-1)\zeta(s)$ exists and equals 1. Apart from this pole, $\zeta(s)$ is analytic throughout the complex $s$-plane.

The relationship between the Riemann zeta function and the gamma function is shown in the reflection formula:

$$\zeta(1-s) = 2(2\pi)^{-s}\Gamma(s)\zeta(s)\cos(\tfrac{1}{2}\pi s), \quad s \neq 0. \tag{10}$$

The most remarkable thing about $\zeta(s)$ is its relationship with the theory of prime numbers. We demonstrate one aspect of this relationship here. Assume that $\operatorname{Re} s > 1$. Subtract the series for $2^{-s}\zeta(s)$ from the one in (9). We then obtain

$$(1-2^{-s})\zeta(s) = \frac{1}{1^s} + \frac{1}{3^s} + \frac{1}{5^s} + \frac{1}{7^s} + \cdots.$$

Similarly, we obtain

$$(1-2^{-s})(1-3^{-s})\zeta(s) = \sum \frac{1}{n^s},$$

where the summation now runs over $n \geqslant 1$, except for multiples of 2 and 3. Now, let $p_n$ denote the $n$th prime number, starting with $p_1 = 2$. By repeating the above procedure we obtain

$$\zeta(s)\prod_{n=1}^{m}(1-p_n^{-s}) = 1 + \sum \frac{1}{n^s},$$

where the summation runs over integers $n > 1$, except for multiples of the primes $p_1, p_2, \ldots, p_m$. The sum of this series vanishes as $m \to \infty$ (since $p_m \to \infty$). From this we obtain the result:

$$\frac{1}{\zeta(s)} = \prod_{n=1}^{\infty}(1-p_n^{-s}), \quad \operatorname{Re} s > 1. \tag{11}$$

This formula is of fundamental importance to the relationship between the Riemann zeta function and the theory of prime numbers.

An immediate consequence is that $\zeta(s)$ does not have zeros in the half-plane $\operatorname{Re} s > 1$. The reflection formula (10) makes it clear that the only zeros in the half-plane $\operatorname{Re} s < 0$ occur at the points $-2, -4, -6, \ldots$. These are called the trivial zeros of the zeta function.

Riemann conjectured that all the zeros in the strip $0 \leqslant \operatorname{Re} s \leqslant 1$ (it is known that there are infinitely many of them) are located on the line $\operatorname{Re} s = \frac{1}{2}$. This conjecture, the Riemann hypothesis, has not yet been proved or disproved. An important part of number theory is based on this conjecture. Much time has been spent on attempting to verify or disprove the Riemann hypothesis, both analytically and numerically. It is one of the seven Millennium Prize Problems of the Clay Mathematics Institute, with a prize of \$1 000 000 on offer for its resolution.

## 5 Gauss Hypergeometric Functions

The Gauss hypergeometric function,

$$F(a,b;c;z) = 1 + \frac{ab}{c}z + \frac{a(a+1)b(b+1)}{c(c+1)\,2!}z^2 + \cdots,$$

plays a central role in the theory of special functions. (We always assume that $c \neq 0, -1, -2, \ldots$.) With more compact notation,

$$F(a,b;c;z) = \sum_{n=0}^{\infty} \frac{(a)_n(b)_n}{(c)_n\, n!} z^n, \quad |z| < 1, \tag{12}$$

where Pochhammer's symbol $(a)_n$ is defined by

$$(a)_n = \frac{\Gamma(a+n)}{\Gamma(a)} = \frac{(-1)^n\Gamma(1-a)}{\Gamma(1-a-n)},$$

using (3). Many special cases arise in the area of orthogonal polynomials (see section 7), in probability theory (as distribution functions: see section 6), and in physics (as Legendre functions: see section 10).

When $a = 0, -1, -2, \ldots$, the power series in (12) terminates and $F(a,b;c;z)$ becomes a polynomial of degree $-a$. The same holds for $b$; note the symmetry $F(a,b;c;z) = F(b,a;c;z)$.

Euler knew of the hypergeometric function, but Gauss made a more systematic study of it. The name "hypergeometric series" was introduced by John Wallis in 1655.

The geometric series $1/(1-z) = 1 + z + z^2 + \cdots$ is the simplest example, and more generally we have

$$(1-z)^{-a} = \sum_{n=0}^{\infty} \frac{(a)_n}{n!} z^n = F(a,b;b;z), \quad |z| < 1, \tag{13}$$

for any $b$. Other elementary examples are

$$F(\tfrac{1}{2},1;\tfrac{3}{2};z^2) = \tfrac{1}{2}z^{-1}\ln([1+z]/[1-z]),$$
$$F(\tfrac{1}{2},1;\tfrac{3}{2};-z^2) = z^{-1}\arctan z,$$
$$F(\tfrac{1}{2},\tfrac{1}{2};\tfrac{3}{2};z^2) = z^{-1}\arcsin z,$$
$$F(\tfrac{1}{2},\tfrac{1}{2};\tfrac{3}{2};-z^2) = z^{-1}\ln(z+\sqrt{1+z^2}),$$

as well as complete elliptic integrals. We have

$$K(k) = \int_0^{\pi/2} \frac{\mathrm{d}\theta}{\sqrt{1-k^2\sin^2\theta}} = \frac{\pi}{2}F(\tfrac{1}{2},\tfrac{1}{2};1;k^2)$$

for $0 \leqslant k^2 < 1$. Similarly, for $0 \leqslant k^2 \leqslant 1$,

$$E(k) = \int_0^{\pi/2} \sqrt{1-k^2\sin^2\theta}\,\mathrm{d}\theta = \frac{\pi}{2}F(-\tfrac{1}{2},\tfrac{1}{2};1;k^2).$$

The integral representation

$$F(a,b;c;z) = \frac{\Gamma(c)}{\Gamma(b)\Gamma(c-b)} \times \int_0^1 t^{b-1}(1-t)^{c-b-1}(1-tz)^{-a}\,\mathrm{d}t \tag{14}$$

can be used when $\operatorname{Re} c > \operatorname{Re} b > 0$ and for $|\mathrm{ph}(1-z)| < \pi$. This representation extends the $z$-domain of the function defined by the power series in (12) considerably. The relationship between the integral in (14) and the hypergeometric function follows if we expand $(1-tz)^{-a}$ as in (13) and using the beta integral (4). Several other integral representations along contours in the complex plane that have fewer or no restrictions on the parameters $a$, $b$, and $c$ can be obtained from (14).

The function $F(a,b;c;z)$ satisfies the differential equation (given by Gauss)

$$z(1-z)F'' + (c-(a+b+1)z)F' - abF = 0. \tag{15}$$

From the theory of differential equations it follows that (15) has three regular singular points, at $z = 0$, $z = 1$, and $z = \infty$.

A function may be expressed in terms of hypergeometric functions in several different ways. The example $(1-z)^{-1} = F(1,b;b;z)$ demonstrates the basic idea. We can write

$$\frac{1}{1-z} = \frac{-1}{z(1-1/z)} = -\frac{1}{z}F(1,b;b;1/z). \tag{16}$$

We therefore have a second representation of $(1-z)^{-1}$ but with a different domain of convergence of the power series, $|z| > 1$. In the general case, $F(a,b;c;z)$ can be written as a linear combination of other $F$-functions, with different $a$, $b$, and $c$, and power series in $z$, $1/z$, $1-z$, $z/(z-1)$, $1/(1-z)$, or $1-1/z$.

The simplest general set of such relations is

$$\begin{aligned} F(a,b;c;z) &= (1-z)^{-a}F(a,c-b;c;z/(z-1)) \\ &= (1-z)^{-b}F(c-a,b;c;z/(z-1)) \\ &= (1-z)^{-a-b}F(c-a,c-b;c;z). \end{aligned}$$

The first of these follows most easily from changing the variable of integration in (14), $t \mapsto (1-t)$, the second from the symmetry in $F(a,b;c;z)$ with respect to $a$ and $b$, and the third from using the first or second relation twice. This gives another way of extending the $z$-domain of the function defined by the power series in (12).

The Gauss hypergeometric function has many generalizations, of which we mention the most natural one, which takes the form

$${}_pF_q\left(\begin{matrix} a_1,a_2,\dots,a_p \\ b_1,b_2,\dots,b_q \end{matrix};z\right) = \sum_{n=0}^{\infty} \frac{(a_1)_n(a_2)_n\cdots(a_p)_n}{(b_1)_n(b_2)_n\cdots(b_q)_n}\frac{z^n}{n!}, \tag{17}$$

which is convergent for $|z| < 1$ if $p = q+1$ or for all $z$ when $p \leqslant q$. All $b_j$ should be different from $0, -1, -2, \dots$. Note that the Gauss hypergeometric function $F = {}_2F_1$.

## 6 Probability Functions

An essential concept in probability theory is the cumulative distribution function $F(x)$. It is defined on the real line, and it is nondecreasing with $F(-\infty) = 0$ and $F(\infty) = 1$. Other intervals are also used. The normal or Gaussian distribution

$$P(x) = \frac{1}{\sqrt{2\pi}} \int_{-\infty}^{x} \mathrm{e}^{-t^2/2}\,\mathrm{d}t$$

is a major example. The error functions

$$\operatorname{erf} x = \frac{2}{\sqrt{\pi}} \int_0^x \mathrm{e}^{-t^2}\,\mathrm{d}t, \qquad \operatorname{erfc} x = \frac{2}{\sqrt{\pi}} \int_x^{\infty} \mathrm{e}^{-t^2}\,\mathrm{d}t$$

are also used, and the following relationship holds: $\operatorname{erf} x + \operatorname{erfc} x = 1$. $P(x)$ and the complementary function $Q(x) = 1 - P(x)$ are related to erfc in the following ways:

$$P(x) = \tfrac{1}{2}\operatorname{erfc}(-x/\sqrt{2}), \qquad Q(x) = \tfrac{1}{2}\operatorname{erfc}(x/\sqrt{2}).$$

Error functions occur in other branches of applied mathematics, heat conduction, for example. Extra parameters such as mean and variance can be included in the basic forms. An extensive introduction to probability functions can be found in Johnson et al. (1994, 1995).

More degrees of freedom are available in the incomplete beta function ratio $I_x(p,q)$, which is based on the beta integral in (4) and is given by

$$I_x(p,q) = \frac{1}{B(p,q)} \int_0^x t^{p-1}(1-t)^{q-1}\,\mathrm{d}t,$$

with $x \in [0,1]$ and $p,q > 0$. Special cases are the $F$-(variance-ratio) distribution and Student's $t$-distribution. The function $I_x(p,q)$ can be written as a Gauss hypergeometric function.

For the gamma distribution we split up the integral in (1) that defines the gamma function. This gives the incomplete gamma function ratios

$$P(a,x) = \frac{1}{\Gamma(a)} \int_0^x t^{a-1}\mathrm{e}^{-t}\,\mathrm{d}t,$$

$$Q(a,x) = \frac{1}{\Gamma(a)} \int_x^\infty t^{a-1}\mathrm{e}^{-t}\,\mathrm{d}t,$$

where $a > 0$. The functions $P(a,x)$ and $Q(a,x)$ can be written in terms of confluent hypergeometric functions (see section 8).

A further generalization is the noncentral $\chi^2$ distribution, which can be defined by the integral

$$P_\mu(x,y) = \int_0^y \left(\frac{z}{x}\right)^{(\mu-1)/2} \mathrm{e}^{-x-z} I_{\mu-1}(2\sqrt{xz})\,\mathrm{d}z, \quad (18)$$

where $I_\mu$ is a modified Bessel function (see section 9). The complementary function is $Q_\mu = 1 - P_\mu$, and this function also plays a role in physics, for instance in problems on radar communications, where it is called the generalized Marcum $Q$-function.

In terms of the incomplete gamma function ratios we have the expansions

$$P_\mu(x,y) = \mathrm{e}^{-x} \sum_{n=0}^{\infty} \frac{x^n}{n!} P(\mu+n, y),$$

$$Q_\mu(x,y) = \mathrm{e}^{-x} \sum_{n=0}^{\infty} \frac{x^n}{n!} Q(\mu+n, y).$$

In statistics and probability theory one is more familiar with the definition through the $\chi^2$ probability functions, which are defined by

$$P(\chi^2 \mid \nu) = P(a,x), \qquad Q(\chi^2 \mid \nu) = Q(a,x),$$

where $\nu = 2a$ and $\chi^2 = 2x$. The noncentral $\chi^2$ distribution functions are then defined by

$$P(\chi^2 \mid \nu, \lambda) = \sum_{n=0}^{\infty} \mathrm{e}^{-\lambda/2} \frac{(\frac{1}{2}\lambda)^n}{n!} P(\chi^2 \mid \nu + 2n),$$

$$Q(\chi^2 \mid \nu, \lambda) = \sum_{n=0}^{\infty} \mathrm{e}^{-\lambda/2} \frac{(\frac{1}{2}\lambda)^n}{n!} Q(\chi^2 \mid \nu + 2n),$$

where $\lambda \geqslant 0$ is called the noncentrality parameter.

## 7 Orthogonal Polynomials

These special functions arise in many branches of pure and applied mathematics. For example, the Hermite polynomials $H_n$ play a role in the form $\mathrm{e}^{-x^2/2}H_n(x)$ as eigenfunctions of the Schrödinger equation for a linear harmonic oscillator.

Let $p_n$ be a polynomial of degree $n$ defined on $(a,b)$, a real interval, where $a = -\infty$ and/or $b = +\infty$ are allowed. Let $w$ be a nonnegative weight function defined on $(a,b)$. Suppose that

$$\int_a^b p_n(x)p_m(x)w(x)\,\mathrm{d}x = 0 \quad (19)$$

if and only if $n \neq m$. Then the family $\{p_n\}$ constitutes a system of orthogonal polynomials on $(a,b)$ with respect to weight $w$.

Orthogonal polynomials can also be defined with Lebesgue measures, on curves in the complex plane (such as the unit circle), and with respect to discrete weight functions or measures. In the last case, the integral in (19) becomes a sum.

The families of polynomials associated with the names of Jacobi, Gegenbauer, Chebyshev, Legendre, Laguerre, and Hermite are called the *classical orthogonal polynomials*. They share many features, and they have the following characteristics:

(i) the family $\{p_n'\}$ is also an orthogonal system;
(ii) the polynomial $p_n$ satisfies a second-order linear differential equation $A(x)y'' + B(x)y' + \lambda_n y = 0$, where $A$ and $B$ do not depend on $n$ and $\lambda_n$ does not depend on $x$; and
(iii) there is a *Rodrigues formula* of the form

$$p_n(x) = \frac{1}{K_n w(x)} \frac{\mathrm{d}^n}{\mathrm{d}x^n}(w(x)X^n(x)),$$

where $X(x)$ is a polynomial with coefficients not depending on $n$, and $K_n$ does not depend on $x$.

These three properties are so characteristic that any system of orthogonal polynomials that has them can be reduced to classical orthogonal polynomials.

Classical orthogonal polynomials satisfy recurrence relations of the form

$$p_{n+1}(x) = (a_n x + b_n)p_n(x) - c_n p_{n-1}(x) \quad (20)$$

for $n = 1, 2, \dots$.

In 1815, Gauss introduced the use of orthogonal polynomials in numerical quadrature for the Legendre case $a = -1$, $b = 1$, $w(x) = 1$. The general formula for a family $\{p_n\}$ of polynomials that are orthogonal on the interval $(a,b)$ with weight $w(x)$ is

$$\int_a^b w(x)f(x)\,\mathrm{d}x = \sum_{k=1}^{n} \lambda_{k,n} f(x_{k,n}) + R_n,$$

where $x_{k,n}$ (the nodes) are the zeros of $p_n$. Several forms are available for the weights $\lambda_{k,n}$; if the $p_n$ are orthonormal ($\int_a^b p_n^2(x)w(x)\,\mathrm{d}x = 1$),

$$\frac{1}{\lambda_{k,n}} = \sum_{j=1}^{n-1} p_j^2(x_{k,n}).$$

The remainder $R_n = 0$ for an arbitrary polynomial $f(x) = q_m(x)$ with degree $m$ not exceeding $2n-1$.

As an example, consider the Hermite polynomials $H_n(x)$. They have weight function $\mathrm{e}^{-x^2}$ on $(-\infty,\infty)$; they follow from the Rodrigues formula

$$H_n(x) = (-1)^n \mathrm{e}^{x^2} \frac{\mathrm{d}^n}{\mathrm{d}x^n} \mathrm{e}^{-x^2};$$

they satisfy the differential equation $y'' - 2xy' + 2ny = 0$; they satisfy the recursion (20) with $a_n = 2$, $b_n = 0$, $c_n = 2n$ and initial values $H_0(x) = 1$, $H_1(x) = 2x$; and they have the generating function

$$\mathrm{e}^{2xz-z^2} = \sum_{n=0}^{\infty} \frac{H_n(x)}{n!} z^n.$$

Jacobi, Gegenbauer, Legendre, and Chebyshev polynomials are special cases of the Gauss hypergeometric function. Laguerre and Hermite polynomials are special cases of the confluent hypergeometric function (see section 8).

## 8 Confluent Hypergeometric or Kummer Functions

When we take $b \to \infty$ in $F(a,b;c;z/b)$, using the series (12) and $\lim_{n\to\infty}(b)_n/b^n = 1$, the result is ${}_1F_1(a;c;z)$ (put $p = q = 1$ in (17)):

$${}_1F_1(a;c;z) = \sum_{n=0}^{\infty} \frac{(a)_n}{(c)_n} \frac{z^n}{n!}. \tag{21}$$

This series converges for all complex $z$, with the usual exception $c = 0,-1,-2,\dots$. When $a = c$, ${}_1F_1(a;a;z) = \mathrm{e}^z$.

The function $F(a,b;c;z/b)$ satisfies a differential equation with three regular singularities: at $z = 0$, $z = b$, and $z = \infty$. In the limit as $b \to \infty$, two singularities merge. This limiting process is called a *confluence*, and so ${}_1F_1(a;c;z)$ is called the confluent hypergeometric function. It satisfies the differential equation

$$zF'' + (c-z)F' - aF = 0, \tag{22}$$

which has a regular singularity at $z = 0$ and an irregular singularity at $z = \infty$.

We have the integral representation

$${}_1F_1(a;c;z) = \frac{\Gamma(c)}{\Gamma(a)\Gamma(c-a)} \int_0^1 t^{a-1}(1-t)^{c-a-1}\mathrm{e}^{zt}\,\mathrm{d}t,$$

which is valid when $\operatorname{Re} c > \operatorname{Re} a > 0$. When we expand the exponential function, we obtain the series in (21).

For a second solution of (22), we have

$$U(a,c,z) = \frac{1}{\Gamma(a)} \int_0^{\infty} t^{a-1}(1+t)^{c-a-1}\mathrm{e}^{-zt}\,\mathrm{d}t,$$

where we assume $\operatorname{Re} z > 0$ and $\operatorname{Re} a > 0$.

The functions ${}_1F_1(a;c;z)$ and $U(a,c,z)$ are named after Kummer. Special cases are Coulomb functions, Laguerre polynomials, Bessel functions, parabolic cylinder functions, incomplete gamma functions, Fresnel integrals, error functions, and exponential integrals.

The Whittaker functions are an alternative pair of Kummer functions and they have the following definitions:

$$M_{\kappa,\mu}(z) = \mathrm{e}^{-z/2} z^{1/2+\mu} {}_1F_1(\tfrac{1}{2}+\mu-\kappa; 1+2\mu; z),$$
$$W_{\kappa,\mu}(z) = \mathrm{e}^{-z/2} z^{1/2+\mu} U(\tfrac{1}{2}+\mu-\kappa, 1+2\mu, z).$$

These functions satisfy the differential equation

$$w'' + \left(-\frac{1}{4} + \frac{\kappa}{z} + \frac{\frac{1}{4}-\mu^2}{z^2}\right)w = 0.$$

Solutions of the differential equation

$$w'' - (\tfrac{1}{4}z^2 + a)w = 0 \tag{23}$$

can be expressed in terms of Kummer functions, but they are often called Weber parabolic cylinder functions, after Heinrich Weber, because they arise when solving the Laplace equation $\Delta V = 0$ by separating the variables into parabolic cylindrical coordinates, $(\xi,\eta,z)$. These parabolic cylindrical coordinates are related to rectangular coordinates $(x,y,z)$ through the equations $x = \frac{1}{2}c(\xi^2-\eta^2)$ and $y = c\xi\eta$, where $c$ is a scale factor. When $\xi$ or $\eta$ are kept constant, say $\xi = \xi_0$ or $\eta = \eta_0$, then we have

$$y^2 = -c\xi_0^2(2x - c\xi_0^2), \qquad y^2 = c\eta_0^2(2x - c\eta_0^2),$$

which are parabolas with foci at the origin.

Solutions of (23) can first be written in terms of Kummer functions, and then linear combinations give the Weber functions, $U(a,\pm z)$ and $V(a,\pm z)$. One can also use the notation $D_\nu(z) = U(-\frac{1}{2}-\nu, z)$; this function can be written in terms of a Hermite polynomial when $\nu = 0,1,2,\dots$.

## 9 Bessel Functions

Bessel functions show up in many physics and engineering problems, in Fourier theory and abstract harmonic analysis, and in statistics and probability theory. Most frequently they occur in connection with differential equations. The earliest systematic study was undertaken by Bessel in 1824 for a problem connected with planetary motion. For historical notes and an extensive treatment, see Watson (1944).

We have seen the modified Bessel function $I_\mu(z)$ in the noncentral $\chi^2$ distribution, (18). In mathematical physics Bessel functions are most commonly associated with the partial differential equations of the potential problem, wave motion, or diffusion, in cylindrical or spherical coordinates. By separating the variables with respect to these coordinates in the time-independent wave equation (the Helmholtz equation) $\Delta v + k^2 v = 0$, several differential equations are obtained and one of them can be put in the form of the *Bessel differential equation*:

$$z^2 w'' + zw' + (z^2 - \nu^2)w = 0. \tag{24}$$

Proper normalizations and combinations of solutions of (24) give the ordinary Bessel functions

$$J_\nu(z), \quad Y_\nu(z), \quad H_\nu^{(1)}(z), \quad H_\nu^{(2)}(z), \tag{25}$$

also called *cylinder functions*, where $\nu$ is the order of these functions. The modified Bessel functions $I_\nu(z)$ and $K_\nu(z)$ follow from these with $z$ replaced by $\pm \mathrm{i}z$.

In physical problems with circular or cylindrical symmetry Bessel functions with order $\nu = n$, an integer, are used, while in spherical coordinates the Bessel functions arise with $\nu = n + \frac{1}{2}$.

Bessel functions of order $\pm\frac{1}{3}$ with argument $\frac{2}{3}z^{3/2}$ are named Airy functions after George Biddell Airy, a British astronomer, who used them when studying rainbow phenomena. Airy functions are solutions of the differential equation $w'' = zw$. The real solutions are oscillatory for $z < 0$ and exponential for $z > 0$. Airy's equation is the simplest second-order linear differential equation that shows such a turning point (at $z = 0$).

Other special functions with turning-point behavior can be approximated in terms of Airy functions. The function $w(z) = \sqrt{z}C_\nu(\nu z)$, where $C_\nu(z)$ is any cylinder function that appears in (25), satisfies the differential equation

$$w'' + \left(\nu^2 \frac{1-z^2}{z^2} - \frac{1}{4z^2}\right)w = 0.$$

For large values of $\nu$ this equation has turning points at $z = \pm 1$. In fact, the cylinder functions $C_\nu(z)$ of large order $\nu$ show turning-point behavior at $z = \nu$. Airy functions can be used to give powerful asymptotic approximations (Olver 1997).

## 10 Legendre Functions

The associated Legendre functions $P_\nu^\mu(z)$ and $Q_\nu^\mu(z)$ satisfy Legendre's differential equation

$$(1-z^2)w'' - 2zw' + \left(\nu(\nu+1) - \frac{\mu^2}{1-z^2}\right)w = 0. \tag{26}$$

This equation has regular singularities at $\pm 1$ and $\infty$, and it can therefore be transformed into the equation of the Gauss hypergeometric functions (15). In fact,

$$P_\nu^\mu(z) = \frac{\zeta^{\mu/2}}{\Gamma(1-\mu)} F(-\nu, \nu+1; 1-\mu; \tfrac{1}{2} - \tfrac{1}{2}z), \tag{27}$$

where $\zeta = (z+1)/(z-1)$ and the branch cut is such that $\operatorname{ph}\zeta = 0$ if $z \in (1, \infty)$. When $z = x \in (-1, 1)$, a real solution is defined by replacing $\zeta$ with $(1+x)/(1-x)$ in (27).

We describe a few special cases that are relevant in boundary-value problems for special choices of Legendre functions and for specific domains: spheres, cones, and tori. Many problems in other domains, however, such as in a spheroid or a hyperboloid of revolution, can be solved using Legendre functions.

### 10.1 Spherical Harmonics

These functions arise in a variety of applications, in particular in investigating gravitational wave signals from a pulsar and in tomographic reconstruction of signals from outer space.

First we consider the subclass of associated Legendre functions defined by the Rodrigues-type formula

$$P_n^m(x) = (-1)^m \frac{(1-x^2)^{m/2}}{2^n\, n!} \frac{\mathrm{d}^{n+m}}{\mathrm{d}x^{n+m}}(x^2-1)^n,$$

where $n$, $m$ are nonnegative integers and $x \in (-1, 1)$. For fixed order $m$ they are orthogonal with respect to the degree $n$, with weight function $w(x) = 1$ on $(-1, 1)$. For $m = 0$ they become the well-known Legendre polynomials, $P_n(x)$.

The $P_n^m(x)$ are used in representations of spherical harmonics, which are given by

$$Y_n^m(\theta, \phi) = N_n^m P_n^m(\cos\theta)\mathrm{e}^{\mathrm{i}m\phi}, \tag{28}$$

where $N_n^m$ is used for normalization and does not depend on $\theta$ or $\phi$. These variables represent colatitude and longitude, respectively, on the sphere in the intervals $\theta \in [0, \pi]$ (from north pole to south pole) and $\phi \in [0, 2\pi)$ (also called the azimuth).

The general bounded solution to Laplace's equation, $\Delta f = 0$, inside the unit sphere centered at the origin, $r = 0$, is a linear combination

$$f(r, \theta, \phi) = \sum_{n=0}^{\infty} \sum_{m=-n}^{n} f_n^m r^n Y_n^m(\theta, \phi). \qquad (29)$$

The constants $f_n^m$ can be computed when boundary values on the sphere are given using orthogonality relationships for $P_n^m(x)$ and the trigonometric elements. When the boundary values are given as a real square-integrable function, the expansion (29) converges inside the unit sphere.

### 10.2 Conical Functions

Conical functions $P^{\mu}_{-1/2+\mathrm{i}\tau}(x)$ and $Q^{\mu}_{-1/2+\mathrm{i}\tau}(x)$ appear in a large number of applications in engineering, applied physics, cosmology, and quantum physics. They occur in boundary-value problems involving configurations of a conical shape. They are also the kernel of the Mehler–Fock transform, which has numerous applications. The functions were introduced by Gustav Ferdinand Mehler in 1868, when he was working with series that express the distance of a point on the axis of a cone to a point located on the surface of the cone.

The integral representation

$$P^{\mu}_{-1/2+\mathrm{i}\tau}(\cos\theta) = \sqrt{\frac{2}{\pi}} \frac{(\sin\theta)^{\mu}}{\Gamma(\frac{1}{2}-\mu)} \int_0^{\theta} \frac{\cosh(\tau t)\,\mathrm{d}t}{(\cos t - \cos\theta)^{\mu+1/2}},$$

with real $\tau$, $0 < \theta < \pi$, and $\operatorname{Re}\mu < \frac{1}{2}$, shows that this function is real for these values of the parameters.

The specific focus in physics is on the case when $\mu = m$, an integer. When this is the case, we can use

$$P^{m}_{-1/2+\mathrm{i}\tau}(x) = (\tfrac{1}{2}+\mathrm{i}\tau)_m(\tfrac{1}{2}-\mathrm{i}\tau)_m P^{-m}_{-1/2+\mathrm{i}\tau}(x)$$

to obtain representations for all $m$.

### 10.3 Toroidal Harmonics

Toroidal harmonics are used to solve the potential problem in a region bounded by a torus. The toroidal coordinates $(\xi, \eta, \phi)$, with $0 \leqslant \xi < \infty$, $-\pi < \eta, \phi \leqslant \pi$, are related to the rectangular coordinates $x$, $y$, $z$ (or $r$, $\phi$, $z$ in cylindrical coordinates) through $x = r\cos\phi$, $y = r\sin\phi$,

$$r = \frac{c\sinh\xi}{\cosh\xi - \cos\eta}, \qquad z = r\frac{\sin\eta}{\sinh\xi},$$

where $c$ is a scale factor. The coordinates are chosen so that $\xi = \xi_0$ represents the toroidal surface.

The general solution of $\Delta\Psi = 0$ in toroidal coordinates can be written in the form

$$\begin{aligned}\Psi(\xi, \eta, \phi) &= \sqrt{\cosh\xi - \cos\eta} \\ &\quad \times \sum_{n,m=0}^{\infty} \cos n(\eta - \eta_{mn}) \cos m(\phi - \phi_{mn}) \\ &\quad \times (A_{mn}P^m_{n-1/2}(\cosh\xi) + B_{mn}Q^m_{n-1/2}(\cosh\xi)),\end{aligned}$$

where $A_{mn}$, $B_{mn}$, $\eta_{mn}$, and $\phi_{mn}$ have to be determined from the boundary condition.

## 11 Functions from Other Boundary-Value Problems

The special functions mentioned in sections 5–10 are all of hypergeometric type. All these functions were known by 1850, usually as a result of the interaction between mathematicians and physicists. In 1868, Mathieu used a different curvilinear coordinate system from those used up to that point when he considered elliptic cylinder coordinates. Other, more general systems (such as oblate and prolate spheroidal and ellipsoidal systems) were introduced soon after.

Actually, there are eleven three-dimensional coordinate systems in which the time-independent wave equation, $\Delta v + k^2 v = 0$, is separable. The Laplace equation, $\Delta v = 0$, is separable in two more systems (the bipolar and bispherical systems).

The systems introduced since the time of Mathieu can be solved in terms of special functions, but these functions are not of hypergeometric type.

MATHIEU'S DIFFERENTIAL EQUATION [III.21] is

$$w'' + (\lambda - 2h^2\cos 2x)w = 0.$$

The substitution $t = \cos x$ transforms this equation into an algebraic form that has two regular singularities at $t = \pm 1$ and one irregular singularity at infinity. This implies that, in general, the solutions of Mathieu's equation cannot be expressed in terms of hypergeometric functions.

Another special feature is that, in general, no explicit representations of solutions of Mathieu's equation (in the form of integrals, say) are known. Nor can explicit power-series expansions or Fourier expansions be derived: the coefficients are solutions of three-term recurrence relations, and with a given value of $h$, convergent expansions can be obtained for an infinite number of eigenvalues $\lambda$. These features also apply for other differential equations that follow from the boundary-value problems with solutions that are beyond the class of hypergeometric functions.

## 12 Painlevé Transcendents

For linear second-order differential equations the location and nature of the singularities of the equation can be investigated easily; we need only know the coefficients in the differential equations.

For nonlinear equations this is not the case, as we see from the simple example $y' = 1 + y^2$. No singularities occur in the coefficients of this equation. The solution $y = \tan(x - x_0)$, however, has poles galore. In the example $y' = -y^2$, we have the solution $y = 1/(x - x_0)$, where $x_0$ is again a free constant.

The properties of these nonlinear equations are also found in PAINLEVÉ EQUATIONS [III.24], a topic that has attracted many researchers in recent decades. The solutions of the simple equations just mentioned are analytic except for poles. In the context of Painlevé equations, the poles of $y = \tan(x - x_0)$ and $y = 1/(x - x_0)$ are called movable poles: their locations change according to the initial values. Types of singularities other than poles, such as branch points and logarithmic singularities, may occur in other examples.

In 1900, Painlevé found the six equations now named after him by classifying second-order ordinary differential equations in a certain class. Painlevé equations have the property that all movable singularities in the complex plane of all solutions are poles, and this property is called the Painlevé property. The first three equations are

$$y'' = 6y^2 + x, \qquad y'' = 2y^3 + xy + \alpha,$$

$$y'' = \frac{(y')^2}{y} - \frac{y'}{x} + \frac{1}{x}(\alpha y^2 + \beta) + \gamma y^3 + \frac{\delta}{y},$$

where $\alpha$, $\beta$, $\gamma$, and $\delta$ are constants.

We have already discussed the important role that special functions (Bessel, Kummer, and so on) play in mathematical physics as solutions of linear differential equations. The Painlevé transcendents play an analogous role for nonlinear ordinary differential equations, and applications can be found in many areas of physics including nonlinear waves, plasma physics, statistical mechanics, nonlinear optics, and fibre optics.

## 13 Concluding Remarks

We have given an overview of a selection of the classical special functions, but there are many other functions we could have chosen to look at. We mention a few other important topics.

The theory of Lie groups, and in particular their representation theory, has shown how special functions can be interpreted from a completely different point of view. In the setting of $q$-functions, difference equations become a source for special functions. In recent decades we have seen a boom in $q$-hypergeometric function research.

Other areas of active research are the study of Jacobi elliptic functions, the study of theta functions, and the study of Weierstrass elliptic and related functions. Since the turn of the century the relationship between the theory of elliptic functions and the theory of elliptic curves has been extensively explored, and this relationship was used by Andrew Wiles to prove Fermat's last theorem. A class of elliptic curves is used in some cryptographic applications as well as for integer factorization. Applications of theta functions are found in physics, where they are used as solutions of diffusion equations.

In the further reading list below we mention only a few key works. In the *NIST Handbook of Mathematical Functions*, nearly all of the formulas we have discussed are given with extensive references to the relevant literature, including where to find a proof. Graphs of the functions are also shown.

### Further Reading

Andrews, G. E., R. Askey, and R. Roy. 1999. *Special Functions*. Cambridge: Cambridge University Press.

Johnson, N. L., S. Kotz, and N. Balakrishnan. 1994, 1995. *Continuous Univariate Distributions*, 2nd edn, volumes I and II. New York: John Wiley.

Olver, F. W. J. 1997. *Asymptotics and Special Functions*. Wellesley, MA: A. K. Peters. (Reprint of the 1974 original, published by Academic Press, New York.)

Olver, F. W. J., D. W. Lozier, R. F. Boisvert, and C. W. Clark, eds. 2010. *NIST Handbook of Mathematical Functions*. Cambridge: Cambridge University Press. (Electronic version available at http://dlmf.nist.gov.)

Temme, N. M. 1996. *Special Functions: An Introduction to the Classical Functions of Mathematical Physics*. New York: John Wiley.

Watson, G. N. 1944. *A Treatise on the Theory of Bessel Functions*, 2nd edn. Cambridge: Cambridge University Press.

Whittaker, E. T., and G. N. Watson. 1927. *A Course of Modern Analysis*. Cambridge: Cambridge University Press.

# IV.8 Spectral Theory

*E. Brian Davies*

## 1 Introduction

Applications of spectral theory arise in a wide range of areas in applied mathematics, physics, and engineering, but there are also applications to geometry, probability, and many other fields. The origins of spectral theory can be traced to Laplace at the start of the nineteenth century and, if one includes its connections with musical harmonies, back to Pythagoras 2500 years ago. Its present-day manifestations cover everything from the design and testing of mechanical structures to the algorithms that are used in the Google search engine. There is an extensive mathematical theory behind the subject, and there are computer packages (some highly specialized but others, such as MATLAB, of a more general character) that enable scientists and engineers to determine the spectral features of the problems that they are studying. Spectral theory is often important for its own sake, but it can also be used as a way of writing the solution of a problem as a linear combination of the solutions of a sequence of eigenvalue problems.

The word *eigenvalue* is a combination of the German root "eigen," meaning characteristic or distinctive, and the English word "value," which suggests that one is seeking a numerical quantity $\lambda$ that is naturally associated with the application being considered. This number is often interpreted as the energy or frequency of some nonzero solution $f$ of an equation of the form

$$Lf = \lambda f. \tag{1}$$

The symbol $L$ refers to the linear operator that describes the particular problem being considered. The solution $f$ of the equation is called the *eigenfunction* (or *eigenvector*, depending on the context) corresponding to the eigenvalue $\lambda$, and it may be a function of one or more variables. Although much of the literature concentrates on determining the eigenvalues of some model, the eigenfunctions carry much more information.

Depending on the particular evolution equation involved, if an eigenvalue $\lambda = u + iv$ associated with some problem is complex, then $u$ is interpreted as the *frequency of oscillation* or vibration of the system being studied. If $v < 0$ then the vibration being considered is stable and $v$ is interpreted as its *rate of decay*. If $v > 0$ then the vibration is unstable and the size of $v$ determines how unstable; unstable vibrations in engineering structures can lead to catastrophic failure.

We conclude the introduction by mentioning INVERSE PROBLEMS [IV.15], in which one seeks to determine important features of some problem from measurements of its associated spectrum. This field has a wide variety of important applications, ranging from engineering to seismology.

## 2 Some Applications

In this section we describe a few applications of spectral theory. These applications raise issues of great importance in their respective subjects as presently practised. Further applications are mentioned at the end of the section, but these by no means exhaust the possibilities.

Many physical structures, from violin strings and drums to turbines and skyscrapers, vibrate under suitable circumstances, and in general the larger the structure, the more slowly it vibrates. It is often important to know what the precise frequency of vibration is going to be before manufacturing a structure because adjustments after construction may be very expensive or even impossible. Calculating the frequencies of the important modes of vibration is now a well-developed branch of engineering. Nevertheless, mistakes are occasionally made.

As an example we mention the Millennium Bridge, which crosses the Thames river in London and was opened on June 10, 2000. Two days later it was closed, for two years, because of an unexpected flaw in its design. The problem was that when a large number of pedestrians were on the bridge, they experienced a lateral wobble, and this wobble caused them to adjust their steps in order to stay upright. Unfortunately, this caused them to start walking in synchrony, with the effect that the lateral vibrations were enhanced until they became potentially dangerous—to the people on the bridge rather than to the bridge itself. The modeling of the vibrations of the bridge had clearly not been adequate. The problem was eventually resolved by inserting a number of specially designed viscous dampers underneath the structure. In spectral terms the effect was to move the eigenvalues further away from the real axis so that there was still a damping effect even when the oscillations were driven by pedestrian effects. This solved the problem, but at a cost of about £5 million.

The design of aircraft involves similar problems but in a much more serious form. Since the cost of a modern airliner can be over 100 million dollars, it is imperative that unstable vibrations in the wings and control surfaces be anticipated and eliminated before manufacturing starts. Aircraft have to operate over a wide range of speeds and altitudes and with different loadings. Because an airliner uses so much fuel, its landing weight can be a third less than it was when taking off. The relevant eigenvalue calculations are time-consuming and expensive, and one cannot repeat them for every possible set of parameters that might be encountered in operation. The best that one can do is to try to ensure that all of the relevant eigenvalues have large enough negative imaginary parts for a range of plausible values of the operational parameters. For commercial airliners, going too far in this direction reduces the efficiency of the design and hence the profitability.

We turn to applications of spectral theory in chemistry. In the nineteenth century, Joseph von Fraunhofer and others discovered that if one heated a chemical element until it glowed and then passed the light emitted through a spectroscope, one could see a series of sharp lines. The frequencies of these lines could be measured precisely, and they could be used to identify the elements involved.

The *spectroscopic lines* for hydrogen have frequencies of the form

$$E_{m,n} = c\left(\frac{1}{m^2} - \frac{1}{n^2}\right),$$

where $m$ and $n$ are positive integers, but no reason was known for the validity of this formula until quantum theory was invented in 1925/26; we shall refer below to energies, which are proportional to frequencies in this context. It was found that the element hydrogen was described by a differential operator whose (discrete, or bound state) eigenvalues were given by $\lambda_n = -c/n^2$, where $c$ is the same constant as above. The spectroscopic energies are not the eigenvalues of the operator but differences of eigenvalues, the reason being that they measure the energy of the photon emitted by an atom when it makes a transition between one energy level and another. By the law of conservation of energy, this equals the difference between the two relevant energy levels of the hydrogen atom.

The new QUANTUM THEORY [IV.23] was fully accepted when the energy levels of helium were calculated and found to be in agreement with spectroscopic observations. Since that time, tens of thousands of spectral lines of elements and chemical compounds have been calculated and observed. If relativistic effects are included in the models, the calculations agree with observation in great detail.

The fact that such calculations are now possible depends on the astonishing growth of computer power over the last fifty years. In 1970 theoretical chemists were often dismissed by "real" chemists, who knew that their problems would never be solved by purely theoretical methods. Some people persisted in spite of this discouragement, and in the end two theoretical chemists, Walter Kohn and John Pople, were awarded a joint Nobel Prize in 1998 for their work developing computational quantum chemistry over three decades. The hard grind of these pioneers has now placed them at the center of chemistry and molecular biology.

One of the stranger "applications" of spectral theory is the apparent connection between the distribution of the zeros of the Riemann zeta function and the distribution of the eigenvalues of a large random self-adjoint matrix. There is no known rigorous argument relating the zeros of the Riemann zeta function to the eigenvalues of any self-adjoint matrix or operator, but the numerical similarities observed in the two fields have led to a number of deep conjectures, some of which have been proved rigorously. There is no basis for assuming that this line of investigation will lead to a proof of the Riemann hypothesis, but anything that prompts worthwhile conjectures must be taken seriously.

Finally, we mention that this volume contains a separate article on RANDOM-MATRIX THEORY [IV.24], in which spectral theory is only one of several techniques involved. We have also avoided any discussion of spectral issues in FLUID MECHANICS [IV.28] and SOLID MECHANICS [IV.32].

## 3 The Mathematical Context

Spectral theory involves finding the spectra of linear operators that approximate the behavior of a variety of systems; we mention some nonlinear eigenvalue problems in the final section. The subject has a body of general theory, some of which is quite difficult, but also a large range of techniques that are applicable only in particular cases. These are constantly being refined, and completely new ways of approaching problems also appear at irregular intervals.[1]

1. A variety of recent developments in the subject are described in Davies (2007).

Applications of spectral theory require the introduction of appropriate mathematical models. The choice of model always involves a compromise between simplicity and accuracy. Pure mathematicians are more interested in obtaining insights that can be applied to a range of similar operators, and they tend to consider simple generic models. Obtaining insights has traditionally been associated with the proof of theorems, but experimental mathematicians also obtain insights by testing a wide range of examples using appropriate software. The ultimate goal of numerical analysts is to produce software that can be used by a wide variety of people who have no interest in how it works and who may use it for purposes that were never envisaged when it was written. On the other hand, applied mathematicians and physicists may be willing to spend months or even years studying particular problems that are of crucial importance for their group of researchers.

### 3.1 Finite Matrices

We start by considering the simplest case, in which one wishes to solve $Af = \lambda f$, where $A$ is a general $n \times n$ matrix and $f$ is a column vector with length $n$. The existence of a nonzero solution $f$ is equivalent to $\lambda$ being a solution of the equation

$$\det(A - \lambda I) = 0.$$

This is a polynomial equation of degree $n$, so it must have $n$ solutions $\lambda_1, \dots, \lambda_n$ by the fundamental theorem of algebra. The simplest case arises when these solutions are all different. If this happens, let $v_r$ be an eigenvector associated with $\lambda_r$ for each $r$. Then $v_1, \dots, v_n$ is a basis for $\mathbb{C}^n$ and $A$ has a diagonal matrix with respect to this basis.

In general, no such basis exists. At the other extreme, the *Jordan block*

$$\begin{pmatrix} c & 1 & 0 & 0 \\ 0 & c & 1 & 0 \\ 0 & 0 & c & 1 \\ 0 & 0 & 0 & c \end{pmatrix}$$

has only one eigenvalue, namely $c$, and only one eigenvector, up to scalar multiples. Every finite matrix is similar to a sum of such Jordan blocks (see JORDAN CANONICAL FORM [II.22]).

It used to be argued that such pathologies are irrelevant in the real world; generically, all roots of a polynomial are different, and hence all eigenvalues of an $n \times n$ matrix are generically distinct. Unfortunately, highly non-self-adjoint matrices of even moderate size are often ill-conditioned, in the sense that the attempt to diagonalize them leads to highly unstable computations, in spite of the fact that the eigenvalues are all distinct. Numerical analysts have to take ill-conditioning into account for all algorithms that might be applied to non-self-adjoint matrices. One of the ways of doing this involves *pseudospectral theory*, a new branch of the subject that investigates the implications of, and connections between, different types of ill-conditioning (see Trefethen and Embree 2005).

Problems of the above type do not occur for *self-adjoint* or *normal* matrices, i.e., those that satisfy $A = A^*$ or $AA^* = A^*A$, respectively. In both cases one has

$$\|(\lambda I - A)^{-1}\| = \operatorname{dist}(\lambda, \operatorname{Spec}(A))^{-1}$$

for all $\lambda$ that do not lie in the spectrum $\operatorname{Spec}(A)$ of $A$, provided $\|\cdot\|$ is the operator norm, defined below. This is in sharp contrast to the general case, in which the left-hand side may be vastly bigger than the right-hand side, even for fairly small and "reasonable" matrices. In such cases, it has been argued that the value of determining the spectrum is considerably reduced.

### 3.2 Formalism

Most of the operators considered in spectral theory act in infinite-dimensional vector spaces over the complex number field. The detailed study of such operators is rather technical, not because analysts like technicalities but because spectra can have properties that do not conform to one's naive intuitions. One starts with a complex HILBERT OR BANACH SPACE [I.2 §19.4] $\mathcal{B}$, with norm $\|\cdot\|$. A linear operator $A \colon \mathcal{B} \to \mathcal{B}$ is said to be *bounded* if its *operator norm*

$$\|A\| = \sup\{\|Af\| \colon \|f\| \leqslant 1\}$$

is finite. One says that $\lambda \in \mathbb{C}$ lies in the *spectrum* $\operatorname{Spec}(A)$ of $A$ if $\lambda I - A$ does not have a two-sided inverse in the set of all bounded operators on $\mathcal{B}$. The spectrum of a bounded linear operator is always closed, bounded, and nonempty. It contains any eigenvalues that $A$ might have, but can be larger than the set of all eigenvalues. The *spectral radius* of $A$ is defined by

$$\rho(A) = \max\{|\lambda| \colon \lambda \in \operatorname{Spec}(A)\},$$

and it satisfies

$$\rho(A) = \lim_{n \to \infty} \|A^n\|^{1/n} \leqslant \|A\|.$$

The above definitions are not directly applicable to differential operators or other unbounded linear operators. In such cases one considers a linear operator

$A\colon \mathcal{D} \to \mathcal{B}$, where the *domain* $\mathcal{D}$ of $A$ is a norm-dense linear subspace of $\mathcal{B}$. One needs to assume or prove that $A$ is *closed* in the sense that, if $\lim_{n\to\infty} \|f_n - f\| = 0$ and $\lim_{n\to\infty} \|Af_n - g\| = 0$, then $f \in \mathcal{D}$ and $Af = g$. The definition of the spectrum of an unbounded operator is similar to that in the bounded case but also involves reference to its domain. The spectrum of an unbounded linear operator may be empty and it may equal $\mathbb{C}$, but it is always a closed set.

The spectrum of a linear operator is often hard to determine. The fact that the spectrum of an arbitrarily small perturbation of $A$ may differ radically from the spectrum of $A$ implies that one has to be extremely cautious about using numerical methods to determine spectra unless one knows that this extreme sensitivity does not arise for the operator of interest. A simple example of this phenomenon is obtained as follows. Given $s \in \mathbb{R}$, one defines the bounded linear operator $A_s$ acting on the space $\ell^2(\mathbb{Z})$ by

$$(A_s f)_n = \begin{cases} s f_{n+1} & \text{if } n = 0, \\ f_{n+1} & \text{otherwise.} \end{cases}$$

Some routine calculations establish that $\mathrm{Spec}(A_s) = \{z\colon |z| = 1\}$ if $s \neq 0$ but $\mathrm{Spec}(A_0) = \{z\colon |z| \leqslant 1\}$. Indeed, every $z$ such that $|z| < 1$ is an eigenvalue of $A_0$, in the sense that the corresponding eigenvector lies in $\ell^2(\mathbb{Z})$.

### 3.3 Self-adjoint Operators

A bounded linear operator $L$ acting on a Hilbert space $\mathcal{H}$ with inner product $\langle \cdot, \cdot \rangle$ is said to be *self-adjoint* if

$$\langle Av, w \rangle = \langle v, Aw \rangle \qquad (2)$$

for all $v, w \in \mathcal{H}$. The definition of self-adjointness for unbounded operators is more technical. It implies that (2) holds for all $v$, $w$ in the domain of $A$, but it is not implied by that condition, which is called *symmetry* by specialists. The difference between self-adjointness and symmetry was emphasized by John von Neumann in the early days of quantum mechanics. It is often evident that an operator of interest is symmetric, but proving that it is self-adjoint can be very hard; indeed, much of the literature before 1970 was devoted to this question. The spectral theory of self-adjoint operators is much more detailed and well understood than that of non-self-adjoint operators.

The center point of the self-adjoint theory is the *spectral theorem*, which was proved around 1930 by von Neumann. There are various statements of this, some easier to understand than others. Perhaps the simplest is the statement that, if $H$ is a self-adjoint operator on the abstract Hilbert space $\mathcal{H}$, then there exists a unitary operator $U$ from $L^2(X, \mathrm{d}x)$ to $\mathcal{H}$ for some measure space $(X, \mathrm{d}x)$ and a function $f\colon X \to \mathbb{R}$ such that

$$(U^{-1}HU\phi)(x) = f(x)\phi(x)$$

for all $\phi \in L^2(X, \mathrm{d}x)$ and almost every $x \in X$. Expressed more simply, $H$ is unitarily equivalent to the *multiplication operator* associated with $f$. This is an analogue of the spectral theorem for self-adjoint matrices because multiplication operators are the infinite-dimensional analogue of diagonal matrices. The spectral theorem implies that the spectrum of $H$ is real; in particular, every eigenvalue of $H$ is real. The implications of this theorem are wide-ranging: it reduces proofs of many of the properties of self-adjoint operators to the status of obvious trivialities, and its absence makes the spectral theory of non-self-adjoint operators far less transparent.

Another version of the spectral theorem introduces the *functional calculus* formula

$$H = \int_{\mathbb{R}} \lambda \, \mathrm{d}P(\lambda),$$

where $P(\lambda)$ is a *spectral projection* of $H$ for every choice of $\lambda \in \mathbb{R}$. This leads to the formula

$$f(H) = \int_{\mathbb{R}} f(\lambda) \, \mathrm{d}P(\lambda)$$

for a wide variety of bounded and unbounded functions $f$ provided one masters the technicalities involved. This version is more attractive, at least to mathematicians, than the previous one, but it is somewhat less easy to use.

If the spectrum of the operator $H$ consists of a countable set of eigenvalues $\lambda_n$, where $n \in \mathbb{N}$, then the spectral theorem may also be written in the form

$$H = \sum_{n=1}^{\infty} \lambda_n P_n.$$

In this formula each $P_n$ is an orthogonal projection whose range is the subspace consisting of all eigenvectors associated with the eigenvalue $\lambda_n$. The rank of $P_n$ equals the multiplicity of the eigenvalue $\lambda_n$. The projections are orthogonal in the sense that $P_m P_n = P_n P_m = 0$ if $m \neq n$, and $\sum_{n=1}^{\infty} P_n = I$.

A more general formulation of the spectral theorem applies to several commuting self-adjoint operators simultaneously. However, there is nothing analogous for even two self-adjoint operators that do not commute. Von Neumann's theory of operator algebras was an attempt to understand the complexities involved in the study of noncommuting families of operators.

### 3.4 Classification of the Spectrum

In infinite dimensions the spectrum of a self-adjoint operator may be divided into parts that have different qualitative features. These include the absolutely continuous spectrum, the singular continuous spectrum, the essential spectrum, the point spectrum, and the discrete spectrum. The definitions are technical and will be avoided here except for the following. One says that $\lambda$ lies in the *discrete spectrum* of a self-adjoint operator $H$ if it is an isolated eigenvalue with finite multiplicity. The rest of the spectrum is called the *essential spectrum.* For many operators the essential spectrum coincides with the continuous spectrum, but this is not true for the Anderson model, which is discussed below.

For non-self-adjoint operators the situation is even more complicated. Spectral theorists are aware that there are several distinct definitions of the essential spectrum, but many have contented themselves with using only one. This suffices for many purposes, but in some situations the others are of real importance.

Let $H$ be a self-adjoint differential operator acting in $L^2(\mathbb{R}^N)$. For many *but not all* such operators, one can divide classical solutions of the differential equation $Hg = \lambda g$ into three types. For some $\lambda$, $g$ may decay rapidly at infinity and $\lambda$ then lies in the discrete spectrum of $H$. For other $\lambda$ this may be false, but $g$ is bounded at infinity and $\lambda$ lies in the essential spectrum of $H$. If neither case pertains for any solution $g$ of $Hg = \lambda g$, then $\lambda$ does not lie in the spectrum of $H$. The discrete spectrum of $H$ is a finite or countable set, while the essential spectrum is typically an infinite interval or a union of several disjoint intervals. In many cases the essential spectrum can be determined in closed form by using perturbation arguments, but the analysis of the discrete spectrum is almost always harder, whether one approaches this task from the theoretical end or the computational one.

A further issue is that the spectrum of an operator can depend on the Banach space in which it operates. For example, the spectrum of the Laplace–Beltrami operator on the hyperbolic space $H_n$ of dimension $n \geqslant 2$ depends in an essential way on $p$ when considered as acting in $L^p(H_n)$. This phenomenon is much less common for differential operators acting in $L^2(\mathbb{R}^N)$.

### 3.5 The Variational Approach

One can obtain valuable upper and lower bounds on the eigenvalues of many self-adjoint operators by using *variational methods,* which go back to Rayleigh and Ritz in the early years of the twentieth century (see VARIATIONAL PRINCIPLE [II.35]). The simplest context assumes that $H$ has a complete orthonormal sequence of eigenvectors $\phi_n$, where $n \in \mathbb{N}$, that the corresponding eigenvalues satisfy $\lambda_n \leqslant \lambda_{n+1}$ for all $n$, and that $\lim_{n\to\infty} \lambda_n = +\infty$.

The variational formula involves the set $\mathcal{L}$ of all finite-dimensional subspaces $L$ of the domain of $H$. It is not directly useful, but it leads to rigorous spectral inequalities that have great value. Given $L \in \mathcal{L}$, one first defines

$$\lambda(L) = \sup\{\langle Hf, f\rangle : f \in L \text{ and } \|f\| = 1\}.$$

The variational formula is then

$$\lambda_n = \inf\{\lambda(L) : \dim(L) = n\}.$$

Applications of the variational formula depend on comparing the eigenvalues of two self-adjoint operators. If $H_1$ and $H_2$ have the same domain, then one writes $H_1 \leqslant H_2$ if $\langle H_1 f, f\rangle \leqslant \langle H_2 f, f\rangle$ for all $f$ in the common domain. Many important applications require one to define the notion $H_1 \leqslant H_2$ even when they do not have the same domain, and this can be done by using the theory of quadratic forms.

It follows immediately from the variational formula for the eigenvalues that $H_1 \leqslant H_2$ implies that $\lambda_n^{(1)} \leqslant \lambda_n^{(2)}$ for all $n$, using an obvious notation. If the eigenvalues of $H_1$ or $H_2$ are already known, this leads to rigorous upper or lower bounds on the eigenvalues of the other operator.

This idea can be recast in several ways, and it allows one to obtain rigorous numerical upper and lower bounds on the eigenvalues of some types of self-adjoint operator. It also has the following theoretical consequence. Let $\Omega_r$, $r = 1, 2$, be two bounded regions on $\mathbb{R}^N$ such that $\Omega_2 \subseteq \Omega_1$, and let $H_r = -\Delta$ act in $L^2(\Omega_r)$ subject to Dirichlet boundary conditions. The eigenvalues of the two operators then satisfy $\lambda_n^{(1)} \leqslant \lambda_n^{(2)}$ for all $n$, as above. This is called *domain monotonicity* and is one ingredient in the proof of Weyl's law, which dates from 1913 and states the following.

Let $H$ be the Laplacian acting in $L^2(\Omega)$ subject to Dirichlet boundary conditions, where $\Omega$ is any bounded region in $\mathbb{R}^N$. Let

$$N_H(s) = \#\{\lambda \in \mathrm{Spec}(H) : \lambda \leqslant s\},$$

where each eigenvalue is counted according to its multiplicity. Then

$$N_H(s) = c_N s^{N/2}|\Omega| + o(s^{N/2})$$

as $s \to \infty$, where

$$c_N = (4\pi)^{-N/2}\Gamma(N/2+1)^{-1},$$

and $\Gamma$ is the Gamma function.

A rigorous analysis of the next term in the asymptotic expansion of $N_H(s)$ was given by Victor Ivrii and Richard Melrose in the above context, but the general solution of this problem for a much wider class of operators was obtained only in the 1980s by Safarov and Vassiliev (1996).

Unfortunately, the domain monotonicity mentioned above holds only for Dirichlet boundary conditions, and the Weyl law can be false for $K = -\Delta$ acting in $L^2(\Omega)$ subject to Neumann or other boundary conditions unless one assumes that the boundary $\partial\Omega$ has some regularity properties. Assuming always that $\Omega$ is bounded, if $\partial\Omega$ is Lipschitz continuous, then $K$ has discrete spectrum and the asymptotic eigenvalue distribution of $K$ follows the same Weyl law as it does for Dirichlet boundary conditions. If $\partial\Omega$ is Hölder continuous, then $K$ has discrete spectrum, but its spectral asymptotics may be non-Weyl. If one makes no assumptions on $\partial\Omega$, then $K$ need not have discrete spectrum; indeed, the spectrum of $K$ may equal $[0,\infty)$.

We finally comment that there is no obvious analogue of the variational method for non-self-adjoint operators. When self-adjoint theorems have non-self-adjoint analogues, this is often because one can use analytic continuation arguments or some other aspect of analytic function theory. However, the most interesting aspects of non-self-adjoint spectral theory are those that have no self-adjoint analogues.

### 3.6 Evolution Equations

Applied mathematics and physics yield many examples of systems that evolve over time according to one of the following equations.

- The wave equation: $\mathrm{d}^2 f/\mathrm{d}t^2 = Lf$.
- The evolution equation: $\mathrm{d}f/\mathrm{d}t = Lf$, which is also called the heat equation when appropriate.
- The Schrödinger equation: $\mathrm{d}f/\mathrm{d}t = -\mathrm{i}Lf$, where i is the square root of $-1$.

In these equations $L$ is commonly a self-adjoint operator, but non-self-adjoint applications are of increasing interest and involve radically new ideas. In many applications $f$ is a function on some region $U$ in $\mathbb{R}^N$, or possibly on a Riemannian manifold, and $L$ is a partial differential operator. Solving these evolution equations starts with specifying the precise class of functions that is to be considered. If $U$ has a boundary, then all admissible $f$ must satisfy certain boundary conditions, which are intrinsic to the model. The same evolution equation has entirely different solutions if one changes the boundary conditions. Once one has specified the problem in sufficient technical detail, one may look for solutions that vary very simply in time. For example, solutions of the wave equation that are of the form $f(t) = \mathrm{e}^{\mathrm{i}kt}g$, where $g = f(0)$, correspond to solutions of the eigenvalue problem $Lg = \lambda g$, where $\lambda = -k^2$.

We finally mention that the abstract study of evolution equations has led to a well-developed theory of one-parameter semigroups.

## 4 Schrödinger Operators

Spectral theory can help one to understand differential operators of any order, whether or not they are self-adjoint. These operators can act on $L^2(U)$, where $U$ is a region in Euclidean space or in a manifold, if the manifold is provided with a measure. This section focuses on one class of differential operators, which have been studied with great intensity because of their applications to quantum theory.

### 4.1 Spectral Theory of Schrödinger Operators

Schrödinger operators are self-adjoint second-order partial differential operators. They play a fundamental role in describing the properties of elementary particles, atoms, and molecules as well as in describing the collisions between particles that occur continually inside fluids. There is a separate article in this volume on the underlying physics (see QUANTUM THEORY [IV.23]), and we restrict attention here to the qualitative properties of a few simple examples, in which Planck's constant $\hbar$ and the masses, charges, and spins of the particles do not appear. If relativistic effects are important, one needs to use the Dirac operator, whose spectral properties are quite different.

Quantum theory has a special vocabulary because of its history. For example, eigenvectors in the relevant Hilbert space are called *bound states* and eigenvalues are often called *energy levels*, but this is not a substantial problem, and we shall use both languages.

At the above level of simplification a *Schrödinger operator* is a differential operator of the form

$$(Hf)(x) = -\tfrac{1}{2}\Delta f(x) + V(x)f(x)$$

acting on functions $f \in L^2(\mathbb{R}^N)$, where $\Delta$ is the Laplace operator. The function $V$ is called the *potential*. For quantum particles moving in three dimensions, one puts $N = 3$, while the case $N = 2$ describes particles moving on a flat interface between two media. Both of these are subjects of great current interest in connection with electronic components and computers.

The spectral properties of $H$ depend heavily on the choice of the potential $V$. Much, but by no means all, of the analysis to date has focused on one of the following classes of potential. (All of the results below, and many other that we cannot state here, have rigorous proofs under suitable technical assumptions.)

If $V(x) \to +\infty$ as $|x| \to \infty$, then $H$ has discrete spectrum. In other words, there exists a complete orthonormal sequence of eigenfunctions $\phi_n$ of $H$ whose corresponding eigenvalues $\lambda_n$ are monotonically increasing with $\lim_{n\to\infty} \lambda_n = +\infty$. The smallest eigenvalue $\lambda_1$ has multiplicity 1 and the corresponding eigenfunction $\phi_1$ is strictly positive, after multiplying by a suitable constant.

If $V(x) \to 0$ as $|x| \to \infty$, then the spectrum is the union of $[0, \infty)$ and a (possibly empty) sequence of negative eigenvalues $\lambda_n$ that can be written in increasing order. If there are infinitely many eigenvalues, then they must converge to 0 as $n \to \infty$. If there exist positive constants $c$ and $\varepsilon$ such that $|V(x)| \leqslant c/|x|^{2+\varepsilon}$ for all large enough $x$, then there can only be a finite number of negative eigenvalues.

One says that $H$ is *periodic* if there exists a discrete group $G$ of translations acting on $\mathbb{R}^N$ such that $V(x + y) = V(x)$ for all $x \in \mathbb{R}^N$ and all $y \in G$. It may be proved that $\mathrm{Spec}(H)$ is the union of a finite or infinite sequence of intervals $[a_n, b_n]$ called *bands* that are separated by *gaps* $(b_n, a_{n+1})$. One may label the bands and gaps so that $a_1 < b_1 < a_2 < b_2 < a_3 < \cdots$. If $N = 1$ then generically there are infinitely many gaps, and the conditions on $V$ under which there are only finitely many have been studied in great detail. In higher dimensions it is known that there are only finitely many gaps.

In the *Anderson model* one studies Schrödinger operators for which the potential $V$ is random in the sense that it is any potential chosen from a precisely defined class, which is provided with a probability measure. The class and the probability measure are assumed to be invariant under a discrete group of translations of $\mathbb{R}^N$, but individual potentials are not. An ergodicity assumption implies that the spectrum of $H$ almost surely does not depend on the choice of $V$ within the class. The spectrum almost surely does not contain any isolated eigenvalues. However, its detailed structure depends on the dimension $N$ and is not fully understood at a rigorous level, in spite of very substantial progress for $N = 1$.

A simple *waveguide* is obtained by considering a Schrödinger operator on $\mathbb{R} \times U$, where $U$ is a bounded set in $\mathbb{R}^{N-1}$; most of the publications on this problem assume that $N = 2$ or $N = 3$. This model is called a quantum waveguide if one imposes Dirichlet boundary conditions on $\mathbb{R} \times (\partial U)$; there is also a substantial literature on similar operators subject to Neumann boundary conditions because of their applications in fluid mechanics. The two ends of the waveguide may point in different asymptotic directions and the shape of the waveguide may vary substantially in a bounded region within $\mathbb{R}^N$. However, it is usually assumed that the potential and the cross section of the waveguide are asymptotically constant far enough away from the origin. Standard methods in spectral and scattering theory allow one to determine the continuous or essential spectrum of such operators, but there may also be eigenvalues. The dependence of the eigenvalues on the geometry of the waveguide has been studied in some detail because of its potential applications to quantum devices.

All of the above problems have discrete analogues, but the case $N = 1$ has had the most attention. One replaces $L^2(\mathbb{R})$ by $\ell^2(\mathbb{Z})$ and considers discrete Schrödinger operators of the form

$$(Hf)_n = f_{n-1} + v_n f_n + f_{n+1}.$$

These operators are simpler to analyze than the usual type of Schrödinger operators because they are usually bounded and one may often base proofs on inductive arguments that are not possible in the continuous context or in higher dimensions.

### 4.2 Scattering Theory

Scattering theory is a difficult subject to explain at an elementary level, but it has important spectral implications. The subject may be studied at an abstract operator-theoretic level, using trace class operators or other technology, but this section will describe the theory only for the time-dependent Schrödinger equation under very standard conditions. The idea is to compare the evolution of a system with respect to two different Schrödinger operators, $H_0 = -\frac{1}{2}\Delta$ and $H = -\frac{1}{2}\Delta + V$, assuming that the potential $V(x)$ decreases rapidly enough as $|x| \to \infty$.[2] If $V$ is central (rotationally invariant), it is possible to obtain a very explicit analysis by decomposing the Hilbert space $L^2(\mathbb{R}^N)$ with respect to the irreducible representations of the rotation group. Current research focuses on more general potentials and multibody scattering. There is also a well-developed scattering theory for the wave equation.

The above assumptions imply that the spectrum of $H_0$ equals $[0, \infty)$ while the spectrum of $H$ equals $[0, \infty)$ together with a finite or infinite set of negative eigenvalues. Solving the Schrödinger equation $\mathrm{i}f'(t) = Hf(t)$ leads to the unitary operators

$$f(t) = \mathrm{e}^{-\mathrm{i}Ht} f(0).$$

If there exist positive constants $c, \varepsilon$ such that $|V(x)| \leqslant c/|x|^{1+\varepsilon}$ for all large enough $x$, then it may be shown that the *wave operators* $W_{\pm}$, defined by

$$W_{\pm} f = \lim_{t \to \pm\infty} \mathrm{e}^{-\mathrm{i}Ht} \mathrm{e}^{\mathrm{i}H_0 t} f,$$

exist for all $f \in L^2(\mathbb{R}^N)$. The wave operators are isometries mapping $L^2(\mathbb{R}^N)$ one-one onto the linear subspace $\mathcal{L}$ of $L^2(\mathbb{R}^N)$ that consists of all $f$ that are orthogonal to every eigenvector of $H$. The wave operators $W_{\pm}$ intertwine $H$ and $H_0$ in the sense that $HW_{\pm} = W_{\pm}H_0$, and this implies that the nonnegative spectra of $H$ and $H_0$ are identical in a very strong sense. It may be shown that the *scattering operator* $S$ defined by

$$S = W_-^* W_+ = \lim_{t \to +\infty} \mathrm{e}^{\mathrm{i}H_0 t} \mathrm{e}^{-2\mathrm{i}Ht} \mathrm{e}^{\mathrm{i}H_0 t}$$

is a unitary operator that commutes with $H_0$. By using Fourier transforms, one may now analyze $S$ in great detail, and in simple cases this yields a complete analysis of $H$ in the sense of the spectral theorem for self-adjoint operators.

This section has not considered the complete analysis of short-range $N$-body scattering by Sigal and Soffer, with important simplifications by Graf and others. This outstanding result was achieved in 1987, but even describing the result informally would take far more space than is available here.

### 4.3 Resonances

In this section we consider only Schrödinger operators $H$, although it will be clear that some aspects of the theory can be developed at a greater level of generality. The theory is based on the discovery that certain classes of Schrödinger operators have pseudo-eigenvalues, usually called *resonances*, with nonzero imaginary parts; resonances are supposed to describe unstable states and their imaginary parts determine the decay rates of the corresponding states. However, every Schrödinger operator is self-adjoint, so its spectrum is necessarily real and one seems to have a contradiction. In fact, a resonance $\lambda$ is associated with a "resonance eigenfunction" according to the equation $Hf = \lambda f$, but the function $f$ satisfies a condition at infinity that implies that it does not lie in the Hilbert space $\mathcal{H}$. Because of this, a number of mathematicians still feel dissatisfied with the foundational aspects of resonance theory. Nevertheless, the calculated resonances correspond to quantities that can be measured experimentally, and some comments are called for. The most important is that any definition of resonance must depend not only on the operator of interest but also on some further background data.

Let $H$ be a Schrödinger operator acting on $L^2(\mathbb{R}^N)$ and suppose that the potential $V$ decays at infinity rapidly enough. The resolvent operators $(zI - H)^{-1}$ are then integral operators for all $z \notin \mathbb{R}$. In other words,

$$((zI - H)^{-1} f)(x) = \int_{\mathbb{R}^N} G_z(x, y) f(y) \, \mathrm{d}^N y$$

for all $z \notin \mathbb{R}$ and all sufficiently well-behaved $f \in L^2(\mathbb{R}^N)$; $G$ is called the *Green function* for the resolvent operator. The resolvent operators are norm-analytic functions of $z$ and the Green functions $G_z(x, y)$ are pointwise analytic functions of $z$ for every $x, y$ (excluding the case $x = y$, for which the Green function is infinite unless $N = 1$).

Under certain reasonable conditions one can prove that the Green function (but not the associated resolvent operator) can be analytically continued through the real axis into a region in the lower half of the complex plane called the "unphysical sheet." In this region it may have isolated poles, and by definition these are the resonances of the operator. It can be proved that the positions of the poles do not depend on the choice of $x$, $y$ and that they coincide with the poles of analytic functions that are associated with the scattering operator for $H$.

## 5 Calculating Eigenvalues

### 5.1 Exactly Soluble Problems

Much of the nineteenth-century research in spectral theory was devoted to finding problems that could be

2. For a much more systematic account, see Yafaev (2010).

solved in closed form, often by using a variety of special functions. In many cases we can now see that the ultimate explanation for this was the presence of some group symmetry. Some spectral problems for ordinary differential equations are soluble using orthogonal polynomials, but the theory of orthogonal polynomials has now grown far beyond the confines of spectral theory or differential equations.

A typical exactly soluble example is provided by *circulant matrices.* These are $n \times n$ matrices $A$ whose entries obey

$$A_{r,s} = \begin{cases} a_{r-s} & \text{if } r \geqslant s, \\ a_{r-s+n} & \text{if } r \leqslant s, \end{cases}$$

for some coefficients $a_r$. The eigenvalues of $A$ may be computed by using the finite Fourier transform to diagonalize $A$. They are the numbers

$$\lambda_r = \sum_{s=1}^{n} a_s e^{2\pi i r s/n},$$

where $1 \leqslant r \leqslant n$. The obvious infinite generalization of this is the operator $Af = a * f$ acting on $L^2(\mathbb{R})$, where $*$ denotes *convolution*, defined by

$$(Af)(x) = \int_{-\infty}^{\infty} a(x-y)f(y)\,dy.$$

Every convolution operator commutes with the group of all translations $(T_s f)(x) = f(x+s)$. $A$ may be diagonalized by using the Fourier transform, the spectrum of $A$ being 0 together with the set of all

$$\lambda_k = \int_{\mathbb{R}} a(x)e^{-ixk}\,dx,$$

where $k \in \mathbb{R}$.

Convolution operators lie at the heart of problems involving digital signal processing and image enhancement. The FAST FOURIER TRANSFORM [II.10] algorithm of Cooley and Tukey, which was actually first described by Gauss in 1805, was seminal in enabling computations to be carried out.

Any constant-coefficient partial differential operator commutes with the group of all translations on $\mathbb{R}^N$ and may similarly be diagonalized, i.e., represented as a multiplication operator, by means of Fourier transforms.

### 5.2 Perturbation Techniques

Before the advent of computers, calculating the eigenvalues of problems that are not exactly soluble was extremely laborious, and much effort was devoted to reducing the work involved. Inevitably, the results obtained were only applicable in special situations. Perturbation theory allows one to calculate the eigenvalues of an operator $A + tB$ for small $t \in \mathbb{C}$ when the eigenvalues and eigenvectors of $A$ are already known. There are two approaches to the theory, applicable in different situations.

*Analytic perturbation theory* provides theorems that prove, under suitable assumptions, that every eigenvalue of $A + tB$ may be written as a convergent series of the form

$$\lambda(t) = a_0 + a_1 t + a_2 t^2 + \cdots,$$

where $a_0$ is an eigenvalue of $A$ and $a_n$ may be calculated directly from $A$ and $B$. In practice, one often needs to calculate only a few terms of the series to obtain a reasonable approximation (see Kato 1995).

If the eigenvalue of $A$ that is being studied has multiplicity greater than 1, the above method needs modification. The power series above has to be replaced by a series involving fractional powers of $t$, as one may see by calculating the eigenvalues of the matrix

$$A + tB = \begin{pmatrix} 0 & t \\ 1 & 0 \end{pmatrix}$$

in closed form. The technology for dealing with such problems was well established by the 1960s.

The use of perturbation expansions of the above type is easiest to justify in finite dimensions. If the Hilbert or Banach space $\mathcal{H}$ is infinite dimensional, then some condition is needed and the easiest is the assumption that $B$ is bounded or relatively bounded with respect to $A$. The latter condition requires the existence of a constant $a \in (0,1)$ and a constant $b > 0$ such that

$$\|Bf\| \leqslant a\|Af\| + b\|f\|$$

for all $f \in \mathrm{Dom}(A)$.

*Asymptotic perturbation theory* allows one to treat certain situations in which convergent perturbation expansions do not exist. An alternative approach to some such problems depends on what is called semiclassical analysis (see Zworski 2012). This involves much more geometrical input than most theorems in analytic perturbation theory. Much of the literature in this field involves the theory of pseudodifferential operators, but an impression of the issues involved can be conveyed by the following example, in which the operator involved acts in $L^2(\mathbb{R}^N)$.

One assumes that the operator $H_h$ is self-adjoint and that it depends on a small parameter in a rather special manner; the parameter is normally denoted by $h$ because of the origins of the subject in quantum mechanics, where $h$ is Planck's constant. One assumes that $H_h$ is obtained by the "quantization" of a classical Hamiltonian $H_{\text{cl}}(p, q)$, where $p, q \in \mathbb{R}^N$. There are, in fact, several types of quantization, and the difference between these needs to be taken into account. A typical example is the Schrödinger operator

$$(H_h f)(x) = -h^2(\Delta f)(x) + V(x)f(x)$$

obtained by quantizing $H_{\text{cl}}(p, q) = p^2 + V(q)$. The idea is that one can understand the $h \to 0$ asymptotics of various features of $H_h$ by utilizing intuitions that come from differential geometry; for example, by an analysis of neighborhoods of the critical points of the potential. The nature of the results obtained precludes our delving more deeply into this important field. However, it should be stated that, when perturbation expansions in powers of $h$ exist, they are usually asymptotic series rather than convergent series.

### 5.3 Numerical Analysis

This is not the place to discuss methods for computing the eigenvalues of partial differential operators in detail, but some general remarks may be in order. We assume that one wishes to determine several eigenvalues of a self-adjoint differential operator $A$ that is known to have an infinite sequence of eigenvalues $\lambda_n$ that can be listed in increasing order, each eigenvalue being repeated according to its multiplicity, with $\lambda_n$ diverging as $n \to \infty$ at a known rate.

There are special methods that enable one to calculate large numbers of eigenvalues of an ordinary differential operator in one dimension with great accuracy. Most general-purpose programs concentrate on two-dimensional problems. In three dimensions all methods are computationally very demanding, because of the large size of the matrix approximations needed. As a result these programs are often optimized for a specific problem that is of commercial, military, or scientific importance.

There are two general approaches to the calculation of eigenvalues, which we call a priori and a posteriori, and each of them depends on calculating the eigenvalues associated with the truncation of $A$ to a finite-dimensional subspace $L$ of high dimension.

The a priori method is far older and depends on theorems that state that, if one carries out the computations for a properly selected increasing sequence of subspaces $L_h$, where $h > 0$ is a small parameter, then the computed eigenvalues converge to the true eigenvalues of $A$ as $h \to 0$. Such theorems also provide an estimate of the convergence rate. The weakness of this approach is that the apparent convergence is usually far faster than that yielded by the theorems.

Assuming that the differential operator acts in a bounded region $U$ in $\mathbb{R}^2$, the FINITE-ELEMENT METHOD [II.12] is a prescription for generating a large finite-dimensional space of functions on $U$ starting from a mesh, i.e., a subdivision of $U$ into a collection of small triangles, whose sizes are or order $h$ (see Boffi 2010). This is carried out by the program, as is the next stage, which is the construction of a large sparse matrix, whose eigenvalues approximate a substantial number of the smaller eigenvalues of the differential operator. Under favorable circumstances one can make a priori estimates of the difference between the computed eigenvalues and the true eigenvalues. However, programs of this type often try to determine which part of the mesh is responsible for most of the error and then refine the mesh locally. It is known that one has to pay particular attention to corners of the region, particularly reentrant corners, so the mesh is usually made much finer there from the start.

A posteriori methods obtain rigorous upper and lower bounds on the eigenvalues of $A$ from numerical computations; these bounds might be based on the variational method described earlier (see Rump 2010, part 3). This method has three major differences from the a priori method. The first is that implementing the method has a considerably higher computational cost than that of the a priori method. If it is implemented using interval arithmetic, then one needs to avoid various computational traps that are by now well known to experts. The final feature is that one does not need to prove that the upper and lower bounds converge to the eigenvalue of interest as $h \to 0$. One simply performs the calculation for smaller and smaller $h$ until either one obtains a result for which the error bounds are sufficiently small, or one accepts that the method is not useful with the available computational resources.

Which of the two methods one uses must depend on the circumstances. A lot of ready-made code exists for the a priori method. Results obtained by using the a priori method may be the starting point even if one intends to use the a posteriori method eventually, because it is always useful to have a good idea about the approximate location of the solution before carrying out further rigorous calculations.

## 6 Other Applications of Spectral Theory

In this section we gather together a miscellaneous collection of applications. Although it is far from complete, it may serve to illustrate the vigor of the field.

### 6.1 Fredholm Operators

The results below can be extended to unbounded operators and to Banach spaces, but it is easiest to explain them in a more restricted context.

A bounded operator $A$ acting on a Hilbert space $\mathcal{H}$ is said to be *Fredholm* if its kernel $\mathrm{Ker}(A)$ is finite dimensional, its range is closed, and the orthogonal complement $\mathrm{Coker}(A)$ of its range is also finite dimensional. The integer

$$\mathrm{Ind}(A) = \mathrm{Dim}(\mathrm{Ker}(A)) - \mathrm{Dim}(\mathrm{Coker}(A))$$

is called the *index* of $A$. The index is a very useful invariant of an operator because of the following fact. If $A$ is Fredholm then any small enough perturbation of $A$ is also Fredholm and the perturbation has the same index as $A$. Hence, if $A_s$ is a one-parameter family of operators depending norm continuously on $s$, and each $A_s$ is Fredholm, then $\mathrm{Ind}(A_s)$ does not depend on $s$. This may be used to prove that the index of an elliptic differential operator acting on the space of sections of a vector bundle does not depend on the detailed values of its coefficients. This fact is a key ingredient in the Atiyah–Singer index theorem, which has ramifications throughout global analysis.

Under the same hypotheses, if $\mathrm{Ind}(A_0) \neq 0$ then $\mathrm{Ind}(A_s) \neq 0$, so 0 is an eigenvalue of $A_s$ or of $A_s^*$. In either case, $0 \in \mathrm{Spec}(A_s)$. This type of argument can enable one to determine more about the spectra of non-self-adjoint operators than one could by more elementary methods.

A particular example of this is afforded by Toeplitz operators, whose applications spread far and wide (see Böttcher and Silbermann 2006). Given a sequence $a \in \ell^1(\mathbb{Z}_+)$, one can define the associated *Toeplitz operator* $A$ on $\ell^2(\mathbb{Z}_+)$ by

$$(Af)_n = \sum_{m=0}^{\infty} a_{n-m} f_m$$

for all $f \in \ell^2(\mathbb{Z}_+)$. It may be shown that $\|A\| \leqslant \|a\|_1$. The function

$$\sigma(\theta) = \sum_{n=0}^{\infty} a_n \mathrm{e}^{-\mathrm{i}n\theta}$$

is called the *symbol* of $A$, and under the above assumptions it is a bounded continuous function that is periodic with period $2\pi$. One may in fact define a Toeplitz operator associated with any bounded measurable periodic symbol, but the technicalities become much greater, and some of the key results are different.

Under the present assumptions one has

$$\{\sigma(\theta) \colon 0 \leqslant \theta \leqslant 2\pi\} \subseteq \mathrm{Spec}(A),$$

but more is actually true. If a complex number $\lambda$ does not lie in the range of $\sigma$, then $A$ is Fredholm and its index equals the winding number of the closed curve $\sigma$ around $\lambda$. A theorem of Gohberg states that the spectrum of $A$ consists of $S = \{\sigma(\theta) \colon 0 \leqslant \theta \leqslant 2\pi\}$ together with every $\lambda \notin S$ for which the winding number is nonzero.

All of these results can be generalized to vector-valued Toeplitz operators—a fact that extends their possible applications substantially.

### 6.2 Spectral Geometry

In this subject one investigates the relationship between the geometry of a Riemannian manifold, or a bounded region in Euclidean space, and the eigenvalues $\lambda_n$ of the associated Laplace–Beltrami operator $H = -\Delta$ (see Gilkey 2005). The Weyl law, already described, establishes that one may determine the volume of a region in $\mathbb{R}^N$ from the eigenvalues of a related Laplace operator, but far more than this is possible. In this context it turns out to be easier to investigate the small $t$ asymptotics of

$$\mathrm{tr}(\mathrm{e}^{-Ht}) = \sum_{n=1}^{\infty} \mathrm{e}^{-\lambda_n t}$$

than the large $s$ asymptotics of the spectral counting function $N_H(s)$ defined earlier. In 1966 Mark Kac asked whether one can "hear the shape of a drum"; more specifically he asked whether the size and shape of a bounded Euclidean region $U$ are uniquely determined by the eigenvalues of the associated Laplace operator, assuming Dirichlet boundary conditions. Many positive results of great interest were obtained in the course of studying this problem, but in 1992, Gordon, Webb, and Wolpert constructed a very simple counterexample in which the two plane regions concerned are fairly simple polygons.

### 6.3 Graph Laplacians

The spectral analysis of graphs has many aspects, but the first problem is that the word "graph" has more than one meaning. The following two interpretations lead to different mathematical results.

One may define a GRAPH [II.16] to be a finite or countable set $X$ of vertices together with a set $\mathcal{E}$ of undirected edges. In this case the *adjacency matrix* $A$ is defined by putting $A_{x,y} = 1$ if $(x, y) \in \mathcal{E}$, and $A_{x,y} = 0$ otherwise. It may be seen that $A$ is associated with a bounded operator on $\ell^2(X)$ provided the *degrees* of the vertices are uniformly bounded, where the degree of a vertex $x$ is defined to be $\#\{y\colon (x, y) \in \mathcal{E}\}$. The spectrum of $(X, \mathcal{E})$ is by definition the spectrum of $A$, which is real because $A$ is self-adjoint. The spectrum provides a set of invariants of a graph. The spectral theory of finite graphs is now a subject in its own right (see Brouwer and Haemers 2012).

Physicists are also interested in *quantum graphs*, in which the edges are continuous intervals of given lengths (see Berkolaiko and Kuchment 2013). The Hilbert space of interest is then $L^2(E)$, where $E$ is the union of the edges in $\mathcal{E}$. One considers the second-order differential operator on $L^2(E)$ given by $(Af)(x) = -f''(x)$ on each edge. In order to obtain a self-adjoint operator one has to impose appropriate boundary conditions at each of the vertices. The simplest way of doing this is to require that if $f$ lies in the domain of $A$, then $f$ is continuous at each vertex and the sum of the outward-pointing first derivatives at each vertex vanishes. These are called free or *Kirchhoff* boundary conditions.

There are two reasons for studying quantum graphs. The first is that they are mild generalizations of the one-dimensional theory on $\mathbb{R}$, and there is some interest in determining how the geometry of the graph affects the spectral theory of the operator $A$. The second is that quantum graphs arise as limits of quantum waveguides as the width of the waveguide decreases to zero. People are interested in quantum waveguides because of their potential applications, and quantum graphs provide a useful first approximation to their properties.

### 6.4 Nonlinear Eigenvalue Problems

This article has concentrated on linear spectral theory, which is by far the best-developed part of the subject. There are two types of generalization of this theory to a nonlinear context. In the first, one attempts to solve $Af = \lambda f$ when $f$ lies in some Banach or Hilbert space of functions and $A$ is a nonlinear operator. The earliest and best-known problem of this type is called the Korteweg–de Vries equation. It is exactly soluble, subject to the solution of an associated linear inverse scattering problem. There are many papers that generalize specified spectral properties of linear second-order elliptic eigenvalue problems to the nonlinear context, under suitable hypotheses. However, nonlinear problems are much harder, whether one approaches them analytically or numerically, and current methods of solution depend heavily on the equation of interest.

Another type of nonlinear eigenvalue problem looks for solutions $f$ of equations such as

$$\lambda^2 A_2 f + \lambda A_1 f + A_0 f = 0,$$

where $\lambda \in \mathbb{C}$ and $A_0$, $A_1$, $A_2$ are all linear operators. This is called a QUADRATIC EIGENVALUE PROBLEM [IV.10 §5.8], and it is the most important case of the more general polynomial eigenvalue problem. It has applications to engineering, the wave equation, control theory, nonlinear boundary-value problems, oceanography, and other fields.

If $A_2$ is invertible then the above problem is equivalent to a standard eigenvalue problem for the block matrix

$$\begin{pmatrix} -A_1 A_2^{-1} & I \\ -A_0 A_2^{-1} & 0 \end{pmatrix}. \tag{3}$$

This transformation does not reduce the theory to a triviality because qualitative properties of the spectrum may be less easy to deduce from (3) than from the original problem. For example, even if $A_0, A_1, A_2$ are all self-adjoint, the matrix (3) is non-self-adjoint, with no obvious special structure. There is a rich theory of finite-dimensional quadratic pencils, which classifies them into many different types, each with its own special features.

### Further Reading

Berkolaiko, G., and P. Kuchment. 2013. *Introduction to Quantum Graphs*. Providence, RI: American Mathematical Society.

Boffi, D. 2010. Finite element approximation of eigenvalue problems. *Acta Numerica* 19:1–120.

Böttcher, A., and B. Silbermann. 2006. *Analysis of Toeplitz Operators*, 2nd edn. Berlin: Springer.

Brouwer, A. E., and W. H. Haemers. 2012. *Spectra of Graphs*. New York: Springer.

Davies, E. B. 2007. *Linear Operators and Their Spectra*. Cambridge: Cambridge University Press.

Gilkey, P. B. 2004. *Asymptotic Formulae in Spectral Geometry*. Zug, Switzerland: CRC Press.

Kato, T. 1995. *Perturbation Theory of Linear Operators*, 2nd edn. New York: Springer.

Rump, S. M. 2010. Verification methods: rigorous methods using floating point arithmetic. *Acta Numerica* 19:287–451.

Safarov, Yu., and D. Vassiliev. 1996. *The Asymptotic Distribution of Eigenvalues of Partial Differential Operators.* Providence, RI: American Mathematical Society.

Trefethen, L. N., and M. Embree. 2005. *Spectra and Pseudospectra: The Behavior of Nonnormal Matrices and Operators.* Princeton, NJ: Princeton University Press.

Yafaev, D. R. 2010. *Mathematical Scattering Theory: Analytic Theory.* Providence, RI: American Mathematical Society.

Zworski, M. 2012. *Semiclassical Analysis.* Graduate Studies in Mathematics. Providence, RI: American Mathematical Society.

# IV.9 Approximation Theory

*Annie Cuyt*

## 1 Introduction

Approximation theory is an area of mathematics that has become indispensable to the computational sciences. The approximation of magnitudes and functions describing some physical behavior is an integral part of scientific computing, queueing problems, neural networks, graphics, robotics, network traffic, financial trading, antenna design, floating-point arithmetic, image processing, speech analysis, and video signal filtering, to name just a few areas.

The idea of seeking a simple mathematical function that describes some behavior approximately has two sources of motivation. The exact behavior that one is studying may not be able to be expressed in a closed mathematical formula. But even if an exact description is available it may be far too complicated for practical use. In both cases a best and simple approximation is required. What is meant by best and simple depends on the application at hand.

The approximation is often used in a computer implementation, and therefore its evaluation needs to be efficient. The simplest and fastest functions for implementation are polynomials because they use only the fast hardware operations of addition and multiplication. Next come rational functions, which also need the hardware operation of division, one or more depending on their representation as a quotient of polynomials or as a continued fraction. Rational functions offer the clear advantage that they can reproduce asymptotic behavior (vertical, horizontal, slant), which is something polynomials are incapable of doing. For periodic phenomena, linear combinations of trigonometric functions make good candidates. For growth models or decaying magnitudes, linear combinations of exponentials can be used.

In approximation theory one distinguishes between interpolation and best approximation problems. In the former one wants the approximate model to take exactly the same values as prescribed by data given at precise argument values. In the latter a set of data (not necessarily discrete) is regarded as a trend and approximated by a simple model in one or other best sense. The difference is formalized in the following sections.

Besides constructing a good and efficient mathematical model, one should also take the following two issues into account.

- What can be said about the convergence of the selected mathematical model? In more practical terms: does the model improve when one adds more data?
- How sensitive is the mathematical model to perturbations in the input data? Data errors are usually unavoidable, and one wishes to know how much they can be magnified in the approximation process.

In the following sections we comment on both issues where appropriate. We do not aim to discuss convergence or undertake a sensitivity analysis for every technique.

Despite the need for and interest in multidimensional models and simulations, we restrict ourselves here mostly to one-dimensional approximation problems. In the penultimate section we include some brief remarks on multivariate interpolation and approximation and its additional complexity.

## 2 Numerical Interpolation

Let data $f_i$ be given at points $x_i \in [a,b]$, where $i = 0,\dots,n$. We assume that if some of the points $x_i$ are repeated, then it is not only the value of some underlying function $f(x)$ that is given (or measured) at $x_i$ but also as many higher derivatives $f^{(j)}(x_i)$ as there are copies of the point $x_i$. The interpolation problem is to find a function of a specified form that matches all the data at the points. In this section we deal with the two extreme cases: the one in which all the points $x_i$ are mutually distinct and no derivative information is available, and the one in which the value of the function and that of the first $n$ derivatives are all given at one single point $x_0$. Of course, intermediate situations can also be dealt with. The approximating functions that we consider are polynomials, piecewise polynomials (splines), and rational functions, each of which has particular advantages. Finally, we present the connection between exponential models and sparse interpolation, on the one hand, and exponential models and Padé approximation, on the other.

## 2.1 Polynomial Interpolation

For $n+1$ given values $f_i = f(x_i)$ at mutually distinct points $x_i$, the polynomial interpolation problem of degree $n$,

$$p_n(x) = \sum_{j=0}^{n} a_j x^j, \qquad p_n(x_i) = f_i, \quad i = 0,\dots,n, \quad (1)$$

has a unique solution for the coefficients $a_j$. Now let us turn to the computation of $p_n(x)$. Essentially, two approaches can be used, depending on the intended subsequent use of the polynomial interpolant. If one is interested in easily updating the polynomial interpolant by adding an extra data point and consequently increasing the degree of $p_n(x)$, then Newton's formula for the interpolating polynomial is very suitable. If one wants to use the interpolant for several sets of values $f_i$ while keeping the points $x_i$ fixed, then Lagrange's formula is most appropriate. A simple rearrangement of the Lagrange form as in (2) below results in the barycentric form, which combines the advantages of both approaches.

In the Newton form one writes the interpolating polynomial $p_n(x)$ as

$$\begin{aligned} p_n(x) &= b_0 + b_1(x - x_0) + b_2(x - x_0)(x - x_1) \\ &\quad + \cdots + b_n(x - x_0)\cdots(x - x_{n-1}). \end{aligned}$$

The coefficients $b_j$ then equal the *divided differences* $b_j = f[0,\dots,j]$ obtained from the recursive scheme

$$\begin{aligned} f[j] &= f_j, \quad j = 0,\dots,n, \\ f[0,j] &= \frac{f_j - f_0}{x_j - x_0}, \quad j = 1,\dots,n, \\ f[0,1,\dots,&k-1,k,j] \\ &= \frac{f[0,1,\dots,k-1,j] - f[0,1,\dots,k-1,k]}{x_j - x_k}, \\ &\qquad\qquad k,j = 2,\dots,n. \end{aligned}$$

Newton's form for the interpolating polynomial is very handy when one wants to update the interpolation with an additional point $(x_{n+1}, f_{n+1})$. It suffices to add the term

$$b_{n+1}(x - x_0)\cdots(x - x_n)$$

to $p_n(x)$ (which does not destroy the previous interpolation conditions since it evaluates to zero at all the previous $x_i$) and to complement the recursive scheme for the computation of the divided differences with the computation of the

$$f[0,1,\dots,k,n+1], \quad k = 0,\dots,n.$$

In the Lagrange form, which is especially suitable if the interpolation needs to be repeated for different sets of $f_i$ at the same points $x_i$, another form for $p_n(x)$ is used. We write

$$p_n(x) = \sum_{j=0}^{n} c_j \beta_j(x), \quad \beta_j(x) = \prod_{\substack{k=0 \\ k\neq j}}^{n} \frac{(x - x_k)}{(x_j - x_k)}.$$

The basis functions $\beta_j(x)$ satisfy a simple interpolation condition themselves, namely,

$$\beta_j(x_i) = \begin{cases} 0 & \text{for } j \neq i, \\ 1 & \text{for } j = i. \end{cases}$$

The choice $c_j = f_j$ for the coefficients therefore solves the interpolation problem. So when altering the $f_i$, without touching the $x_i$ that make up the basis functions $\beta_j(x)$, it takes no computation at all to get the new coefficients $c_j$.

The *barycentric form* of the interpolation polynomial,

$$\left.\begin{aligned} p_n(x) &= (x - x_0)\cdots(x - x_n) \sum_{j=0}^{n} \frac{w_j}{x - x_j} f_j, \\ w_j &= \Big(\prod_{\substack{k=0 \\ k\neq j}}^{n} (x_j - x_k)\Big)^{-1}, \end{aligned}\right\} \quad (2)$$

is easy to update and is BACKWARD STABLE [I.2 §23] for evaluation of $p_n(x)$.

The sensitivity of the polynomial interpolant expressed in the Lagrange form is measured by the value

$$L_n = \max_{a \leqslant x \leqslant b} \sum_{j=0}^{n} |\beta_j(x)|, \quad (3)$$

which is also known as the *Lebesgue constant*. The growth rate of $L_n$ with $n$ is only logarithmic when the interpolation points are as in (5) below. This is the slowest possible growth for polynomial interpolation.

Despite the simplicity and elegance of polynomial interpolation, the technique has a significant drawback, as we discuss next: it may not converge for data $f_i = f(x_i)$ given at arbitrary points $x_i$, even if $f(x)$ is continuous on $[a, b]$.

### 2.2 The Runge Phenomenon

What happens if we continue updating the interpolation problem with new data? In other words, what happens if we let the degree $n$ of the interpolating polynomial $p_n(x)$ increase? Will the interpolating polynomial of degree $n$ become better and better? The answer is no, at least not for freely chosen points $x_i$. To see what can go wrong, consider

$$f(x) = \frac{1}{1+25x^2}, \quad -1 \leqslant x \leqslant 1, \tag{4}$$

and take equidistant interpolation points $x_i = -1 + 2i/n$, $i = 0, \dots, n$. The error $(f-p)(x)$ toward the endpoints of the interval then increases dramatically with $n$. Take a look at the bell-shaped $f(x)$ and the interpolating polynomial $p_n(x)$ for $n = 10$ and $n = 20$ in figure 1.

This phenomenon is called *Runge's phenomenon*, after Carl Runge, who described this behavior for real-valued interpolation in 1901. An explanation for it can be found in the fundamental theorem of algebra, which states that a polynomial has as many zeros as its degree. Each of these zeros can be real or complex. So if $n$ is large and the zeros are all real, the polynomial under consideration displays rather oscillatory behavior.

On the other hand, under certain simple conditions for $f(x)$ besides continuity in $[a, b]$, it can be proved that if the interpolation points $x_i$ equal

$$x_i = \frac{a+b}{2} + \frac{b-a}{2}\cos\left(\frac{(2i+1)\pi}{2(n+1)}\right), \quad i = 0, \dots, n, \tag{5}$$

where the values $\cos((2i+1)/(n+1)(\pi/2))$ are the zeros of the Chebyshev polynomial of the first kind of degree $n+1$ (defined in section 3.3), then

$$\lim_{n\to\infty} \|f - p_n\|_\infty = \lim_{n\to\infty} \max_{x\in[-1,1]} |(f-p_n)(x)| = 0.$$

The effect of this choice of interpolation points is illustrated in figure 2(a). A similar result holds if the zeros of the Chebyshev polynomial of degree $n+1$ are replaced by the extrema $\cos(i\pi/n)$, $i = 0, \dots, n$, of the Chebyshev polynomial of degree $n$.

In order to make use of this result in real-world applications, where the interpolation points $x_i$ cannot usually be chosen arbitrarily, interpolation at the Chebyshev zeros is mimicked, for instance by selecting a proper subset of interpolation points $\tilde{x}_i$ from a fine equidistant grid, with $\tilde{x}_i \approx x_i$ from (5). The grid is considered to be sufficiently fine when the distance between the points ensures that a grid point nearest to a Chebyshev zero $x_i$ is never repeated. In a coarse grid, the same grid point may be the closest one to more than one Chebyshev zero, especially toward the ends of the interval $[-1, 1]$.

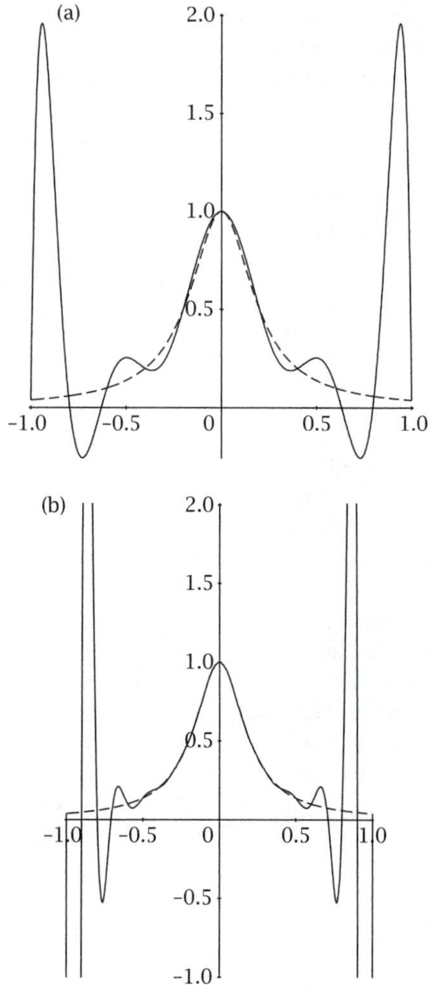


**Figure 1** (a) Degree-10 and (b) degree-20 equidistant interpolation (solid lines) for function $f$ in (4) (dashed lines).

This technique is called *mock-Chebyshev interpolation*. For comparison, in figure 2(b) we display the degree-20 mock-Chebyshev interpolant with the interpolation points selected from an equispaced grid with gap 1/155.

If a lot of accurate data points have to be used in an interpolation scheme, then splines, which are discussed in the next section, offer a better alternative than a monolithic high-degree polynomial interpolant.

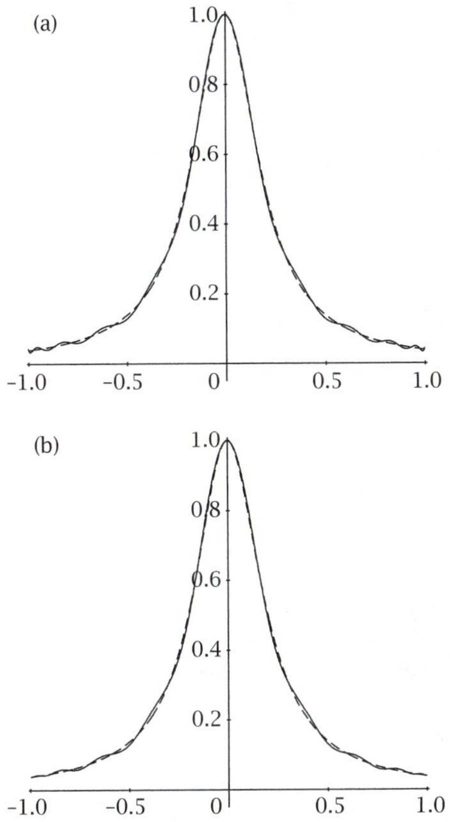


**Figure 2** Degree-20 (a) Chebyshev and (b) mock-Chebyshev interpolation (solid lines) for function $f$ in (4) (dashed lines).

## 2.3 Spline Interpolation

In order to avoid the Runge phenomenon when interpolating large data sets, piecewise polynomials, also called splines, can be used. To this end we divide the data set of $n + 1$ points into smaller sets, each containing two data points. Rather than interpolating the full data set by one polynomial of degree $n$, we interpolate each of the smaller data sets by a low-degree polynomial. These separate polynomial functions are then pieced together in such a way that the resulting function is as continuously differentiable as possible.

Take, for instance, the data set $(x_i, f_i)$ and consider linear polynomials interpolating every two consecutive $(x_i, f_i)$ and $(x_{i+1}, f_{i+1})$. These linear polynomial pieces can be joined together at the data points $(x_i, f_i)$ to produce a piecewise-linear continuous function or polygonal curve. Note that this function is continuous but not differentiable at the interpolation points since it is polygonal.

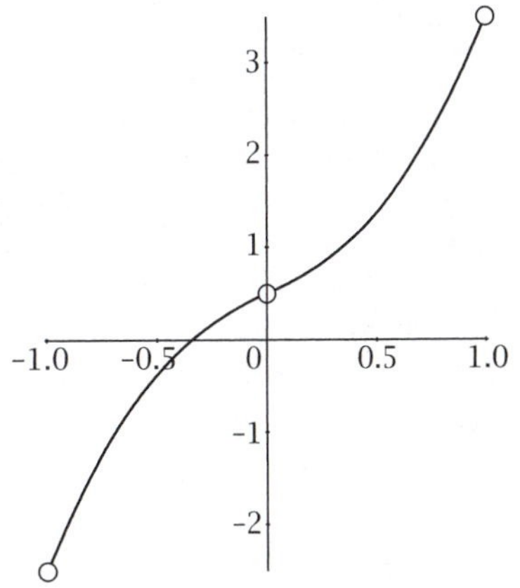


**Figure 3** A piecewise-cubic function that is not twice continuously differentiable.

If we introduce two parameters, $\Delta$ and $D$, to respectively denote the degree of the polynomial pieces and the differentiability of the overall function, where obviously $D \leqslant \Delta$ (even $D < \Delta$ to avoid an overdetermined system of defining equations, as explained below), then, for the polygonal curve, $\Delta = 1$ and $D = 0$. With $\Delta = 2$ and $D = 1$, a piecewise-quadratic and smooth (meaning continuously differentiable in the entire interval $[x_0, \dots, x_n]$) function is constructed. The slope of a smooth function is a continuous quantity. With $\Delta = 3$ and $D = 2$, a piecewise-cubic and twice continuously differentiable function is obtained. Twice continuously differentiable functions also enjoy continuous curvature. Can the naked eye distinguish between continuous and discontinuous curvature in a function? The untrained eye certainly cannot! As an example we take the cubic polynomial pieces

$$c_1(x) = x^3 - x^2 + x + 0.5, \quad x \in [-1, 0],$$
$$c_2(x) = x^3 + x^2 + x + 0.5, \quad x \in [0, 1],$$

and join these together at $x = 0$ to obtain a new piecewise-cubic function $c(x)$ on $[-1, 1]$. The result is a function that is continuous and differentiable at the origin, but for the second derivatives at the origin, we have $\lim_{x \to 0-} c^{(2)}(0) = -2$ and $\lim_{x \to 0+} c^{(2)}(0) = 2$. Nevertheless, the result of the gluing procedure shown in figure 3 is a very pleasing function that at first sight looks fine. But while $\Delta$ equals 3, $D$ is only 1.

Since a trained eye can spot these discontinuities, the most popular choice for piecewise-polynomial interpolation in industrial applications is $\Delta = 3$ and $D = 2$. Indeed, for manufacturing the continuity of the curvature is important.

Let us take a look at the general situation where $\Delta = m$ and $D = m-1$, for which the resulting piecewise polynomial is called a *spline*. Assume we are given the interpolation points $x_0, \dots, x_n$. With these $n+1$ points we can construct $n$ intervals $[x_i, x_{i+1}]$. The points $x_0$ and $x_n$ are the endpoints, and the other $n-1$ interpolation points are called the internal points. If $\Delta = m$, then for every interval $[x_i, x_{i+1}]$ we have to determine $m+1$ coefficients because the explicit formula for the spline on $[x_i, x_{i+1}]$ is a polynomial of degree $m$:

$$S(x) = s_i(x), \quad x \in [x_i, x_{i+1}], \ i = 0, \dots, n-1,$$
$$s_i(x) = \sum_{j=0}^{m} a_j^{(i)} x^j.$$

So, in total, $n(m+1)$ unknown coefficients $a_j^{(i)}$ have to be computed. From which conditions? There are the $n+1$ interpolation conditions $S(x_i) = f_i$, and we have the smoothness or continuity requirements at the internal points, meaning that a number of derivatives of $s_{i-1}(x)$ evaluated at the right endpoint of the domain $[x_{i-1}, x_i]$ should coincide with the derivatives of $s_i(x)$ when evaluated at the left endpoint of the domain $[x_i, x_{i+1}]$:

$$s_{i-1}^{(k)}(x_i) = s_i^{(k)}(x_i), \quad i = 1, \dots, n-1, \ k = 0, \dots, m-1.$$

The latter requirements add another $(n-1)m$ continuity conditions. This brings us to a total of $n+1+(n-1)m = n(m+1) - m + 1$ conditions for $n(m+1)$ unknowns. In other words, we lack $m-1$ conditions to determine the degree-$m$ piecewise-polynomial interpolant with overall smoothness of order $m-1$. When $m = 1$, which is the case for the piecewise-linear spline or the polygonal curve, no conditions are lacking. When $m = 2$, a value for $s_0'(x_0)$ is usually given as an additional piece of information. When $m = 3$, which is the case for the widely used cubic spline, values for $s_0''(x_0)$ and $s_{n-1}''(x_n)$ are often provided (the cubic spline with clamped end conditions) or they are set to zero (the natural cubic spline).

The natural cubic spline interpolant has a very elegant property, namely, that it avoids oscillatory behavior between interpolation points. More precisely, for every twice continuously differentiable function $f(x)$ defined on $[a, b]$ and satisfying $f(x_i) = f_i$ for all $i$, we have

$$\int_a^b S''(x)^2 \, dx \leqslant \int_a^b f''(x)^2 \, dx.$$

A simple illustration is given in figure 4 for $n = 6$ with $x_i = i$, $i = 1, \dots, 6$.

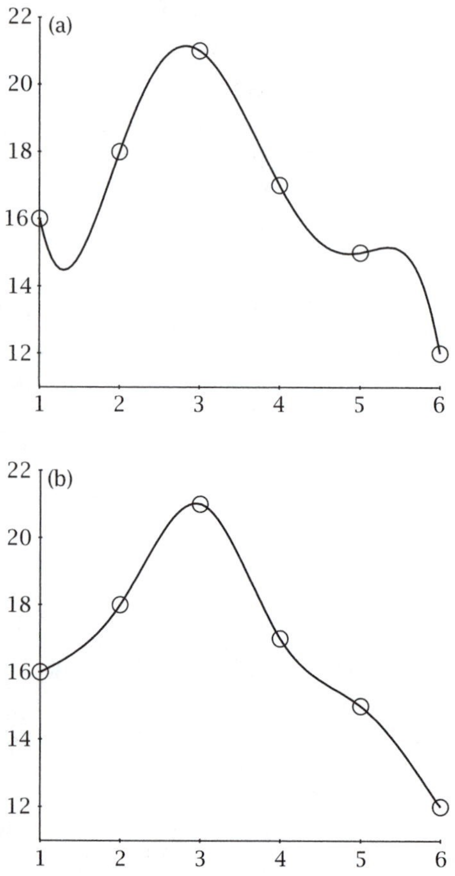


**Figure 4** (a) The polynomial interpolant and (b) the natural cubic spline.

### 2.4 Padé Approximation

The rational equivalent of the Taylor series partial sum is the irreducible rational function $r_{k,\ell}(x) = p_{k,\ell}(x)/q_{k,\ell}(x)$ with numerator of degree at most $k$ and denominator of degree at most $\ell$ that satisfies

$$r_{k,\ell}^{(i)}(x_0) = f^{(i)}(x_0), \quad i = 0, 1, \dots, n, \tag{6}$$

with $n$ as large as possible. It is also called the $[k/l]$ *Padé approximant*. The aim is to have $n = k + \ell$. Note that we are imposing one fewer condition than the total number $k + \ell + 2$ of coefficients in $r_{k,\ell}$. The reason is that one degree of freedom is lost because multiplying $p_{k,\ell}$ and $q_{k,\ell}$ by a scalar does not change $r_{k,\ell}$.

A key question is whether $n$ can be less than $k + \ell$. The answer to this question requires some analysis. Computing the numerator and denominator coefficients of $r_{k,\ell}(x)$ from (6) gives rise to a nonlinear system of equations. So let us explore whether the Padé approximant can also be obtained from the linearized approximation conditions

$$(fq_{k,\ell} - p_{k,\ell})^{(i)}(x_0) = 0, \quad i = 0, 1, \dots, k + \ell. \tag{7}$$

We denote $f^{(j)}(x_0)/j!$ by $d_j$, where $d_j = 0$ for $j < 0$. The linearized conditions (7) always have at least one nontrivial solution for the numerator coefficients $a_0, \dots, a_k$ and the denominator coefficients $b_0, \dots, b_\ell$ because they form a homogeneous linear system of $k + \ell + 1$ conditions in $k + \ell + 2$ unknowns:

$$\begin{aligned} d_0 b_0 &= a_0, \\ d_1 b_0 + d_0 b_1 &= a_1, \\ &\vdots \\ d_k b_0 + \cdots + d_{k-\ell} b_\ell &= a_k, \\ d_{k+1} b_0 + \cdots + d_{k-\ell+1} b_\ell &= 0, \\ &\vdots \\ d_{k+\ell} b_0 + \cdots + d_k b_\ell &= 0. \end{aligned}$$

Moreover, all solutions $p_{k,\ell}(x)$ and $q_{k,\ell}(x)$ of (7) are equivalent in the sense that they have the same irreducible form. Every solution of (6) with $n = k + \ell$ therefore also satisfies (7), but not vice versa. From $p_{k,\ell}(x)$ and $q_{k,\ell}(x)$ satisfying (7) we find, for the unique irreducible form $p^*_{k,\ell}(x)/q^*_{k,\ell}(x)$, that

$$(f - r_{k,\ell})^{(i)}(x_0) = 0, \quad i = 0, \dots, k' + \ell' + r,$$
$$k' = \partial p^*_{k,\ell}, \ \ell' = \partial q^*_{k,\ell}, \ r \geqslant 0,$$

where $\partial p$ denotes the degree of the polynomial $p$. In some textbooks, the $[k/l]$ Padé approximation problem is said to have no solution if $k' + \ell' + r < k + \ell$; in others, the Padé approximant $r_{k,\ell}$ is identified with $r_{k',\ell'} = p^*_{k,\ell}/q^*_{k,\ell}$ if that is the case (this is the convention we adopt here). Let us illustrate the situation with a simple example. Take $x_0 = 0$ with $d_0 = 1$, $d_1 = 0$, $d_2 = 1$, and $k = 1 = \ell$. The linearized conditions (7) are then

$$b_0 = a_0, \quad b_1 = a_1, \quad b_0 = 0.$$

A solution is given by $p_{1,1}(x) = x$ and $q_{1,1}(x) = x$. We therefore find $r_{1,1}(x) = 1$, $k' = 0$, $\ell' = 0$, and

$$(f - r_{1,1})^{(2)}(x_0) = 2 \neq 0.$$

Since $r = 1$, we have $k' + \ell' + r = 1 < k + \ell = 2$.

This kind of complication does not occur when $\ell = 0$. The Padé approximant $r_{k,0}(x)$ is then merely the Taylor series partial sum of degree $k$. But when asymptotic behavior needs to be reproduced, a polynomial function is not very useful. In figure 5 one can compare the Taylor series partial sum of degree 9 with the [5/4] Padé approximant for the function $f(x) = \arctan(x)$.

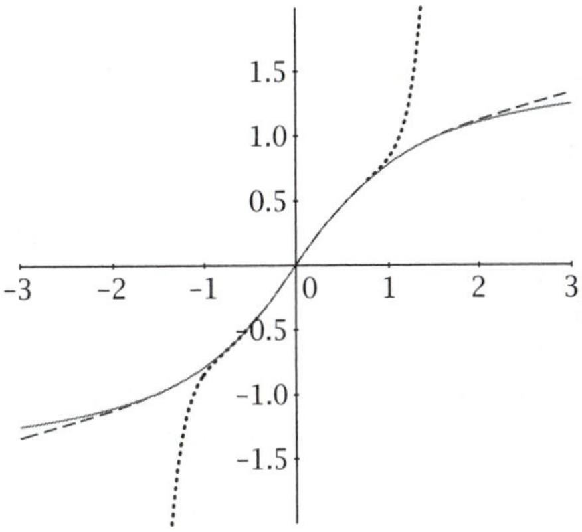


**Figure 5** Padé approximants $r_{9,0}(x)$ (dotted line) and $r_{5,4}(x)$ (dashed line) for $\arctan(x)$ (solid line).

**Table 1** The Padé table for $\sin(x)$.

| | 0 | 1 | 2 | |
|---|---|---|---|---|
| 1 | $x$ | $x$ | $\dfrac{x}{1+\frac{1}{6}x^2}$ | $\cdots$ |
| 2 | $x$ | $x$ | $\dfrac{x}{1+\frac{1}{6}x^2}$ | $\cdots$ |
| 3 | $x-\frac{1}{6}x^3$ | $x-\frac{1}{6}x^3$ | $\dfrac{(-\frac{7}{60}x^3+x)}{(1+\frac{1}{20}x^2)}$ | $\cdots$ |
| | $\vdots$ | $\vdots$ | $\vdots$ | $\ddots$ |

Padé approximants can be organized in a table, where the numerator degree indicates the row and the denominator degree the column. To illustrate this we give part of the Padé table for $f(x) = \sin(x)$ in table 1. A sequence of Padé approximants in the Padé table can converge uniformly or in measure only to a function $f(x)$ that is meromorphic in a substantial part of its domain.

### 2.5 Rational Interpolation

The rational equivalent of polynomial interpolation at mutually distinct interpolation points $x_i$ consists of finding an irreducible rational function $r_{k,\ell}(x)$, of numerator degree at most $k$ and denominator degree at most $\ell$, that satisfies

$$r_{k,\ell}(x_i) = f_i, \quad i = 0, \dots, k + \ell, \tag{8}$$

where $f_i = f(x_i)$. Instead of solving (8) one considers the linearized equations

$$(fq_{k,\ell} - p_{k,\ell})(x_i) = 0, \quad i = 0, \dots, k + \ell, \tag{9}$$

where $p_{k,\ell}(x)$ and $q_{k,\ell}(x)$ are polynomials of respective degree $k$ and $\ell$. Condition (9) is a homogeneous linear system of $k+\ell+1$ equations in $k+\ell+2$ unknowns, and it therefore always has a nontrivial solution. Moreover, as in the Padé approximation case, all solutions of (9) are equivalent in the sense that they deliver the same unique irreducible rational function.

In the construction of the irreducible form $r_{k,\ell}(x)$ of $p_{k,\ell}(x)/q_{k,\ell}(x)$, common factors in numerator and denominator are canceled, and it may well be that $r_{k,\ell}$ does not satisfy the interpolation conditions (8) anymore, despite $p_{k,\ell}$ and $q_{k,\ell}$ being solutions of (9), because one or more of the canceled factors may be of the form $x - x_i$ with $x_i$ an interpolation point. A simple example illustrates this. Let $x_0 = 0$, $x_1 = 1$, $x_2 = 2$ with $f_0 = 0$, $f_1 = 3$, $f_2 = 3$, and take $k = 1 = \ell$. The homogeneous linear system of interpolation conditions is then

$$\begin{aligned} a_0 &= 0, \\ 3(b_0 + b_1) - (a_0 + a_1) &= 0, \\ 3(b_0 + 2b_1) - (a_0 + 2a_1) &= 0. \end{aligned}$$

A solution is given by $p_{1,1}(x) = 3x$ and $q_{1,1}(x) = x$. Hence, $r_{1,1}(x) = 3$ and clearly $r_{1,1}(x_0) \neq f_0$. The interpolation point $x_0$ is then called unattainable. This problem can be fixed only by increasing the degrees $k$ and/or $\ell$ until the interpolation point is attainable. Note that unattainable interpolation points do not occur in polynomial interpolation ($\ell = 0$).

A well-known problem with rational interpolation and Padé approximation is the occurrence of undesirable poles in the interpolant $r_{k,\ell}(x)$. One way to avoid this is to work with preassigned poles, either explicitly or implicitly, by determining the denominator polynomial $q_{k,\ell}(x)$ a priori. All $k + \ell + 1$ interpolation conditions are then imposed on the coefficients of the numerator polynomial, and consequently the degree of the numerator is raised to $k + \ell$.

Let the interpolation points $x_i$ be ordered such that $x_0 < x_1 < \cdots < x_n$ with $k + \ell = n$. A popular choice for the denominator polynomial that guarantees a pole-free real axis, unless the location of the poles needs to be controlled by other considerations, is

$$q_{n,n}(x) = \sum_{j=0}^{n} (-1)^j \prod_{\substack{k=0 \\ k\neq j}}^{n} (x - x_k).$$

With this choice, the rational interpolant can be written in a barycentric form similar to that in (2):

$$r_{n,n}(x) = \frac{\sum_{j=0}^{n} f_j (-1)^j/(x - x_j)}{\sum_{j=0}^{n} (-1)^j/(x - x_j)}.$$

Again, this form is very stable for interpolation. Its numerical sensitivity is measured by

$$M_n = \max_{a\leqslant x\leqslant b} \sum_{i=0}^{n} \frac{|q_{n,n}(x_i)\beta_i(x)|}{|q_{n,n}(x)|}.$$

And there is more good news now: in the case of equidistant interpolation points, $M_n$ grows as slowly with $n$ as the Lebesgue constant $L_n$ in (3) for polynomial interpolation in the Chebyshev zeros. The latter makes the technique very useful in practice.

More practical choices for the denominator polynomial $q_{n,n}(x)$ are possible, guaranteeing other features, such as rapid convergence, comonotonicity, or coconvexity (coconcavity).

### 2.6 Sparse Interpolation

When interpolating

$$f(x) = \alpha_1 + \alpha_2 x^{100}$$

by a polynomial, the previous techniques require 101 samples of $f(x)$ to determine that $f(x)$ is itself a polynomial, while only four values need to be computed from the data points, namely the two exponents 0 and 100 and the two coefficients $\alpha_1$ and $\alpha_2$. So it would be nice if we could solve this polynomial reconstruction problem from only four samples.

The above is a special case of the more general *sparse interpolation*, which was studied as long ago as 1795, by Gaspard de Prony, in which the complex values $\phi_j$ and $\alpha_j$ in the interpolant

$$\phi(x) = \sum_{j=1}^{n} \alpha_j \mathrm{e}^{\phi_j x}, \quad \alpha_j, \phi_j \in \mathbb{C}, \tag{10}$$

are to be determined from only $2n$ samples of $\phi(x)$.

While the nonlinear interpolation problems of Padé approximation and rational interpolation are solved by linearizing the conditions as in (7) and (9), the nonlinear problem of sparse interpolation is solved by separating the computation of the $\phi_j$ and the $\alpha_j$ into two linear algebra subproblems. Let $\phi(x)$ be sampled at the equidistant points $x_i = i\Delta$, $i = 0, \dots, 2n-1$, and let us denote $\phi(x_i)$ by $f_i$. We introduce the $n \times n$ Hankel matrices

$$H_n^{(r)} := \begin{pmatrix} f_r & \cdots & f_{r+n-1} \\ \vdots & \iddots & \vdots \\ f_{r+n-1} & \cdots & f_{r+2n-2} \end{pmatrix}$$

and $\lambda_j = \mathrm{e}^{\phi_j \Delta}$, $j = 1, \dots, n$.

The $\lambda_j$ are then retrieved as the generalized eigenvalues of the problem

$$H_n^{(1)} v_j = \lambda_j H_n^{(0)} v_j, \quad j = 1, \dots, n,$$

where the $v_j$ are the generalized right eigenvectors. From the values $\lambda_j$, the complex numbers $\phi_j$ can be retrieved uniquely subject to the restriction that $|\mathrm{Im}(\phi_j \Delta)| < \pi$. In order to satisfy this restriction, the sampling interval $\Delta$ is usually adapted to the range of the values $\mathrm{Im}(\phi_j)$.

The $\alpha_j$ are computed from the interpolation conditions

$$\sum_{j=1}^{n} \alpha_j \mathrm{e}^{\phi_j x_i} = f_i, \quad i = 0, \dots, 2n-1, \tag{11}$$

either by solving the system in the least-squares sense or by solving a subset of $n$ consecutive interpolation conditions. Note that

$$\mathrm{e}^{\phi_j x_i} = \lambda_j^i$$

and that the coefficient matrix of (11) is therefore a Vandermonde matrix.

With $f_i = \phi(x_i)$ we now define

$$f(x) = \sum_{j=0}^{\infty} f_j x^j,$$

where $x_i = i\Delta$, $i \geqslant 0$. Since

$$f_i = \sum_{j=1}^{n} \alpha_j \mathrm{e}^{\phi_j x_i} = \sum_{j=1}^{n} \alpha_j \lambda_j^i,$$

we can rewrite $f(x)$ as follows:

$$f(x) = \sum_{j=1}^{n} \frac{\alpha_j}{1 - x\lambda_j}. \tag{12}$$

So we see that $f(x)$ is itself a rational function of degree $n-1$ in the numerator and $n$ in the denominator, with poles $1/\lambda_j$. Hence, from Padé approximation theory we know (as is to be expected) that $r_{n-1,n}(x)$ reconstructs $f(x)$; in other words,

$$r_{n-1,n}(x) = f(x)$$

with

$$q(x) = \prod_{j=1}^{n} (1 - x\lambda_j).$$

The partial fraction decomposition (12) is the Laplace transform of the exponential model (10), which explains why this approach is known as the *Padé–Laplace method.*

This connection between approximation theory and harmonic analysis is clearly not accidental. More constructions from harmonic analysis, including wavelets and Fourier series, also provide important insights into central problems in approximation theory. Other mathematical models in which the major features of a data set are represented using only a few terms are considered in the theory of COMPRESSED SENSING [VII.10].

## 3 Best Approximation

When the quality of the data does not justify the imposition of an exact match on the approximating function, or when the quantity of the data is simply overwhelming and depicts a trend rather than very precise measurements, interpolation techniques are of no use. It is better to find a linear combination of suitable basis functions that approximates the data in some best sense. We first discuss the existence and uniqueness of a best approximant and the discrete linear least-squares problem. How the bestness or nearness of the approximation is measured is then explained. Different measures lead to different approximants and are to be used in different contexts. We discuss the importance of orthogonal basis functions and describe the continuous linear least-squares problem and the minimax approximation. A discussion of a connection with Fourier series and the interpolation and approximation of periodic data concludes the section.

### 3.1 Existence and Uniqueness

First and foremost we discuss the existence and uniqueness of a best approximant $p^*$ from a finite-dimensional subspace $P$ to an element $f$ from a normed linear space $V$. More specifically, we ask for which of the $\ell_1$-, $\ell_2$-, or $\ell_\infty$-norms can we guarantee that either at least one or exactly one solution exists to the approximation problem of finding $p^* \in P$ such that

$$\|f - p^*\| \leqslant \|f - p\|, \quad p \in P.$$

The answer to the existence problem is affirmative for all three mentioned norms. To guarantee uniqueness of $p^*$, either the norm or the subspace $P$ under consideration must satisfy additional conditions. And we must distinguish between discrete and continuous approximation and norms.

When $V$ is *strictly convex*, in other words, when a sphere in $V$ does not contain line segments, so that

$$\|x_1 - c\| = r = \|x_2 - c\| \quad \Rightarrow \quad \|\lambda x_1 + (1-\lambda) x_2 - c\| < r, \quad 0 < \lambda < 1,$$

then the best approximant $p^*$ to $f$ is unique. This applies, for instance, to the $\ell_2$- or Euclidean norm, in both the discrete and continuous cases.

In the discussion of the role of $P$ with respect to the uniqueness of $p^*$, we deal with the continuous case first. When a basis $\{b_0(x), \dots, b_n(x)\}$ for $P$ satisfies the *Haar condition*, meaning that every linear combination

$$q_n(x) = \lambda_0 b_0(x) + \cdots + \lambda_n b_n(x)$$

has at most $n$ zeros, then the continuous best $\ell_1$ and best $\ell_\infty$ approximation problems also have a unique solution.

Let us now look at the discrete best approximation problem in somewhat more detail. We consider a large data set of values $f_i$ that we want to approximate by a linear combination of some linearly independent basis functions $b_j(x)$:

$$\lambda_0 b_0(x_i) + \cdots + \lambda_n b_n(x_i) = f_i, \quad i = 0, \dots, m > n. \tag{13}$$

This $(m+1) \times (n+1)$ linear system can be written compactly as

$$\left.\begin{aligned} &A\lambda = f, \quad \lambda = \begin{bmatrix} \lambda_0 \\ \vdots \\ \lambda_n \end{bmatrix}, \; f = \begin{bmatrix} f_0 \\ \vdots \\ f_m \end{bmatrix}, \\ &A = (b_{j-1}(x_{i-1})) \in \mathbb{R}^{(m+1)\times(n+1)}. \end{aligned}\right\} \tag{14}$$

Unless the right-hand side $f$ lies in the column space of $A$, the system cannot be solved exactly. The residual vector is given by

$$r = f - A\lambda \in \mathbb{R}^{m+1},$$

and the solution $\lambda$ we are looking for is the one that solves the system best, in other words, the system that makes the magnitude (or norm) of the residual vector minimal. The least-squares problem corresponds to using the Euclidean norm or the $\ell_2$-norm $\|r\|_2 = (r_1^2 + \cdots + r_m^2)^{1/2}$ to measure the residual vector, and the optimization problem translates to

$$(A^{\mathrm{T}}A)\lambda = A^{\mathrm{T}}f,$$

which is a square linear system of equations called the NORMAL EQUATIONS [IV.10 §7.1]. If the matrix $A$ of the overdetermined linear system has maximal column rank, then the matrix $A^{\mathrm{T}}A$ is nonsingular and the solution is unique.

When every $(n+1) \times (n+1)$ submatrix of the matrix $A$ in (14) is nonsingular, then the discrete best $\ell_\infty$ approximation problem has a unique solution as well. An example showing the lack of uniqueness of the best $\ell_1$ approximation under the same condition is easy to find. Take

$$A = \begin{pmatrix} 1 \\ -1 \end{pmatrix}, \qquad f = \begin{pmatrix} 1 \\ 2 \end{pmatrix}$$

in (14). Then the minimum of $\|A\lambda - f\|_1$ with $n = 0$ and $b_0(x) = 1$ is the same for all $-2 \leqslant \lambda_0 \leqslant 1$.

In practice, instead of solving the normal equations, more numerically stable techniques based on ORTHOGONAL TRANSFORMATIONS [IV.10 §7.1] are applied directly to the overdetermined system (14). These transformations do not alter the Euclidean norm of the residual vector $r$ and hence have no impact on the optimization criterion.

Let us now see whether the Euclidean norm is the correct norm to use.

### 3.2 Choice of Norm

If the optimal solution to the overdetermined linear system is the one that makes the norm $\|r\|$ of the residual minimal, then we must decide which norm to use to measure $r$. Although norms are in a sense equivalent, because they differ only by a scalar multiple depending only on the dimension, it makes quite a difference whether we minimize $\|r\|_1$, $\|r\|_2$, or $\|r\|_\infty$. Let us perform the following experiment.

Using a Gaussian random number generator with mean $\mu$ and standard deviation $\sigma$, we generate $m+1$ numbers $f_i$. The approximation problem we consider is the computation of an estimate for $\mu$ from the data points $f_i$, where $\sigma$ expresses a tight or loose spread around $\mu$. Compare this with a real-life situation where the data $f_i$ are collected by performing some measurements of a magnitude $\mu$, and $\sigma$ represents the accuracy of the measuring tool used to obtain the $f_i$.

In the notation of (13), we want to fit the $f_i$ by a multiple of the basis function $b_0(x) = 1$ because we are looking for the constant $\mu$. The overdetermined linear system takes the form

$$\lambda_0 \cdot 1 = f_i, \quad i = 0, \dots, m.$$

It is clear that this linear system does not have an exact solution. The residual vector is definitely nonzero. We shall see that different criteria or norms can be used to express the closeness of the estimate $\lambda_0$ for $\mu$ to the data points $f_i$ or, in other words, the magnitude of the residual vector $r$ with components $f_i - \lambda_0$, and that the standard deviation $\sigma$ will also play a role.

If the Euclidean norm is used, then the optimal estimate $\lambda_0^{(2)}$ is the mean of the $m$ measurements $f_i$:

$$\lambda_0^{(2)} = \frac{1}{m+1}\sum_{i=0}^{m} f_i.$$

If we choose the $\ell_1$-norm $\|r\|_1 = \sum_{i=1}^{m}|r_i|$ as a way to measure distances, then the value $\lambda_0^{(1)}$ that renders the $\ell_1$-norm of the residual vector minimal is the median of the values $f_i$. Any change that makes the larger values extremely large or the smaller values extremely small therefore has no impact on $\lambda_0^{(1)}$, which is rather insensitive to outliers.

When choosing as distance function the $\ell_\infty$-norm $\|r\|_\infty = \max_{i=1,\dots,m}|r_i|$, the optimal solution $\lambda_0^{(\infty)}$ to the problem is given by

$$\lambda_0^{(\infty)} = \tfrac{1}{2}\Big(\min_{i=0,\dots,m} f_i + \max_{i=0,\dots,m} f_i\Big).$$

This can also be understood intuitively. The value for $\lambda_1$ that makes $\|r\|_\infty$ minimal is the one that makes the largest deviation minimal, so it should be right in the middle between the extremes.

So the $\ell_\infty$-norm criterion performs particularly well in the context of rather accurate data (in this experiment meaning small standard deviation $\sigma$) that suffer relatively small input errors (such as roundoff errors). When outliers or additional errors (such as from manual data input) are suspected, use of the $\ell_1$-norm is recommended. If the measurement errors are believed to be normally distributed with mean zero, then the $\ell_2$-norm is the usual choice. Approximation problems of this type are therefore called least-squares problems.

### 3.3 Orthogonal Basis Functions

In the same way that we prefer to draw a graph using an orthogonal set of axes (the smaller the angle between the axes, the more difficult it becomes to make a clear drawing), it is preferred to use a so-called orthogonal set of basis functions $b_j(x)$ in (13). Orthogonal basis functions $b_j(x)$ can tremendously improve the conditioning or sensitivity of the problem (14). They are also useful in continuous least-squares problems.

The notion of orthogonality in a function space parallels that of orthogonality in the vector space $\mathbb{R}^k$: for a positive weight function $w(x)$ defined on the interval $[a,b]$, we say that the functions $f$ and $g$ are $w$-orthogonal if

$$\langle f,g\rangle_w = \int_a^b f(x)g(x)w(x)\,\mathrm{d}x = 0.$$

The function $w(x)$ can assign a larger weight to certain parts of the interval $[a,b]$. For instance, the function $w(x) = 1/\sqrt{1-x^2}$ on $[-1,1]$ assigns more weight toward the endpoints of the interval.

For $w(x) = 1$ and $[a,b] = [-1,1]$, a sequence of orthogonal polynomials $L_i(x)$ satisfying

$$\int_{-1}^{1} L_j(x)L_k(x)\,\mathrm{d}x = 0, \quad j \neq k,$$

is given by

$$L_0(x) = 1, \qquad L_1(x) = x,$$
$$L_{i+1}(x) = \frac{2i+1}{i+1}xL_i(x) - \frac{i}{i+1}L_{i-1}(x), \quad i \geqslant 1.$$

The polynomials $L_i(x)$ are called the *Legendre polynomials*. For $w(x) = 1/\sqrt{1-x^2}$ and $[a,b] = [-1,1]$, a sequence of orthogonal polynomials $T_i(x)$ satisfying

$$\int_{-1}^{1} T_j(x)T_k(x)\frac{1}{\sqrt{1-x^2}}\,\mathrm{d}x = 0, \quad j \neq k,$$

is given by

$$T_0(x) = 1, \qquad T_1(x) = x,$$
$$T_{i+1}(x) = 2xT_i(x) - T_{i-1}(x), \quad i \geqslant 1.$$

The polynomials $T_i(x)$ are called the *Chebyshev polynomials* (of the first kind). They are also very useful in (continuous as well as discrete) least-squares problems, as discussed below.

When the polynomials are to be used on an interval $[a,b]$ different from $[-1,1]$, the simple change of variable

$$x \to \frac{2}{b-a}\Big(x - \frac{a+b}{2}\Big)$$

transforms the interval $[a,b]$ to the interval $[-1,1]$, on which the orthogonal polynomials are defined.

Orthogonal polynomials also satisfy the Haar condition, so every linear combination

$$q_n(x) = a_0p_0(x) + \cdots + a_np_n(x)$$

of the orthogonal polynomials $p_i(x)$ of degree $i = 0,\dots,n$ has at most $n$ zeros. Therefore, orthogonal polynomials are also a suitable basis in which to express an interpolating function; the system of interpolation conditions

$$\sum_{j=0}^{n} a_jp_j(x_i) = f_i, \quad i = 0,\dots,n,$$

has a coefficient matrix that is guaranteed to be nonsingular for mutually distinct points $x_i$.

The importance of orthogonal basis functions in interpolation and approximation cannot be overstated. Problems become numerically better conditioned and formulas simplify. For instance, the Chebyshev polynomials $T_i(x)$ also satisfy the discrete orthogonality

$$\sum_{i=0}^{n} T_j(x_i)T_k(x_i) = (1+\delta_{k0})\frac{n+1}{2}\delta_{jk}, \qquad j,k = 0,1,\dots,$$

where $\delta_{ij}$ is the KRONECKER DELTA [I.2 §2, table 3] and

$$x_i = \cos\left(\frac{(2i+1)\pi}{2(n+1)}\right)$$

are the zeros of the Chebyshev polynomial $T_{n+1}$. When expressing the polynomial interpolant $p_n(x)$ in (1) of degree $n$ in the Chebyshev basis,

$$p_n(x) = \sum_{j=0}^{n} a_j T_j(x),$$

an easy explicit formula for $p_n(x)$ interpolating the values $f_i$ at the points $x_i$ can be given:

$$p_n(x) = \frac{\sum_{i=0}^{n} f_i}{n+1} + \sum_{j=1}^{n}\left(\frac{2\sum_{i=0}^{n} f_i T_j(x_i)}{n+1}\right)T_j(x).$$

Another elegant explicit formula, based on the continuous orthogonality property of the Chebyshev polynomials, is given in (16).

The practical utility of Chebyshev polynomials is illustrated by the open source Chebfun software system (see www.chebfun.org) for numerical computation with functions, which is built on piecewise-polynomial interpolation at the extrema of Chebyshev polynomials, or what is equivalent (via the FAST FOURIER TRANSFORM [II.10]), expansions in Chebyshev polynomials.

### 3.4 Chebyshev Series

Let us now choose the basis functions $b_j(x) = T_j(x)$ and look for the coefficients $\lambda_j$ that make the $\ell_2$-norm of

$$f(x) - \sum_{j=0}^{n} \lambda_j T_j(x), \quad -1 \leqslant x \leqslant 1,$$

minimal for $f(x)$ defined on $[-1,1]$, for simplicity. We are looking for the polynomial $p_n(x)$ of degree $n$ that is closest to $f(x)$, where we measure the distance between the functions using the inner product

$$\begin{aligned}\|f-p_n\|_2^2 &= \langle f-p_n, f-p_n\rangle \\ &= \int_{-1}^{1} \frac{(f-p_n)^2(x)}{\sqrt{1-x^2}}\,dx. \qquad (15)\end{aligned}$$

This is a continuous least-squares problem because the norm of a function is minimized instead of the norm of a finite-dimensional vector. Since

$$\begin{aligned}\left\|f - \sum_{j=0}^{n}\lambda_j T_j\right\|_2^2 &= \left\langle f - \sum_{j=0}^{n}\lambda_j T_j, f - \sum_{j=0}^{n}\lambda_j T_j\right\rangle \\ &= \|f\|_2^2 - \sum_{j=0}^{n}\langle f, T_j/\|T_j\|_2\rangle^2 \\ &\quad + \sum_{j=0}^{n}(\langle f, T_j/\|T_j\|_2\rangle - \lambda_j\|T_j\|_2)^2,\end{aligned}$$

in which only the last sum of squares depends on $\lambda_j$, the minimum is attained for the so-called Chebyshev coefficients

$$\lambda_j = \langle f, T_j\rangle / \langle T_j, T_j\rangle. \qquad (16)$$

The partial sum of degree $n$ of the Chebyshev series development of a function,

$$f(x) = \sum_{j=0}^{\infty} \frac{\langle f, T_j\rangle}{\langle T_j, T_j\rangle} T_j(x),$$

is therefore the best polynomial approximation of degree $n$ to $f(x)$ in the $\ell_2$ sense. Since

$$\left| f(x) - \sum_{j=0}^{n}\frac{\langle f, T_j\rangle}{\langle T_j, T_j\rangle}T_j(x)\right| \leqslant \sum_{j=n+1}^{\infty}\left|\frac{\langle f, T_j\rangle}{\langle T_j, T_j\rangle}\right|,$$

this error can be made arbitrarily small when the series of Chebyshev coefficients converges absolutely, a condition that is automatically satisfied for functions that are continuously differentiable in $[-1,1]$.

The above technique can be generalized to any weight function and its associated family of orthogonal polynomials: when switching the weight function, the norm criterion (15) changes and the orthogonal basis is changed.

The Chebyshev series partial sums are good overall approximations to a function $f(x)$ defined on the interval $[-1,1]$ (or $[a,b]$ after a suitable change of variable). To illustrate this, in figure 6 we compare, for the function $f(x) = \arctan(x)$, the error plots of the Chebyshev series partial sum of degree 9 with the Taylor series partial sum of the same degree. Its Chebyshev series and Taylor series developments are, respectively, given by

$$\arctan(x) = \sum_{i=0}^{\infty}(-1)^i \frac{2(\sqrt{2}-1)^{2i+1}}{2i+1}T_{2i+1}(x),$$

$$\arctan(x) = \sum_{i=0}^{\infty}(-1)^i \frac{1}{2i+1}x^{2i+1}.$$

Although explicit formulas for the Taylor series expansion of most elementary and special functions are known, the same is not true for Chebyshev series expansions. For most functions, the coefficients (16)

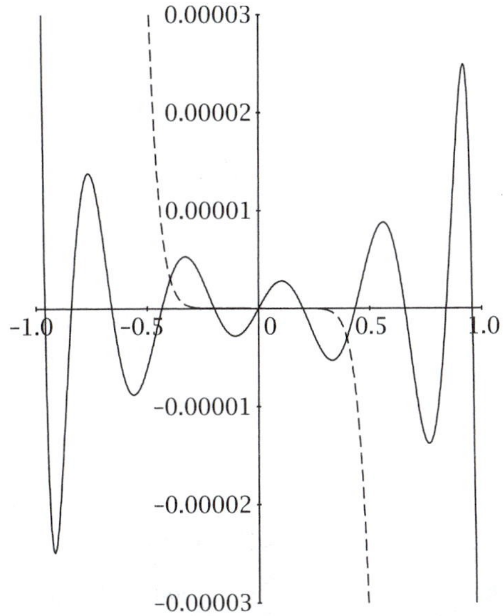


**Figure 6** Error plots of Chebyshev (solid line) and Taylor (dashed line) partial sums of degree 9 for arctan($x$).

have to be computed numerically because no analytic expression for (16) can be given.

### 3.5 The Minimax Approximation

Instead of minimizing the $\ell_2$-distance (15) between a function $f(x) \in C([a,b])$ and a polynomial model for $f(x)$, we can consider the problem of minimizing the $\ell_\infty$-distance. Every continuous function $f(x)$ defined on a closed interval $[a,b]$ has a unique so-called *minimax polynomial approximant* of degree $n$. This means that there exists a unique polynomial $p_n = p_n^*$ of degree $\partial p_n \leqslant n$ that minimizes

$$\|f - p_n\|_\infty = \max_{x\in[a,b]} \left| f(x) - \sum_{j=0}^{n} \lambda_j x^j \right|. \tag{17}$$

More generally, if the set of basis functions $\{b_0(x), \dots, b_n(x)\}$ satisfies the Haar condition, then there exists a unique approximant

$$q_n^*(x) = \lambda_0^* b_0(x) + \cdots + \lambda_n^* b_n(x)$$

that minimizes

$$\|f - q_n\|_\infty = \max_{x\in[a,b]} \left| f(x) - \sum_{j=0}^{n} \lambda_j b_j(x) \right|.$$

The minimum is attained and is not an infimum. When $b_j(x) = x^j$, the polynomial $p_n^*(x)$ is computed using the Remez algorithm, which is based on its characterization given by the alternation property of the function $(f - p_n^*)(x)$; $p_n^*$ is the best polynomial approximant of degree $n$ if the error $\|f - p_n^*\|_\infty$ is attained by the function $f - p_n^*$ in at least $n+2$ points $y_0, \dots, y_{n+1}$ in the interval $[a,b]$ and this with alternating sign, meaning that

$$\exists y_0 > y_1 > \cdots > y_{n+1} \in [a,b]:$$
$$(f - p_n^*)(y_i) = s(-1)^i \|f - p_n^*\|_\infty,$$
$$s = \pm 1,\ i = 0, \dots, n+1.$$

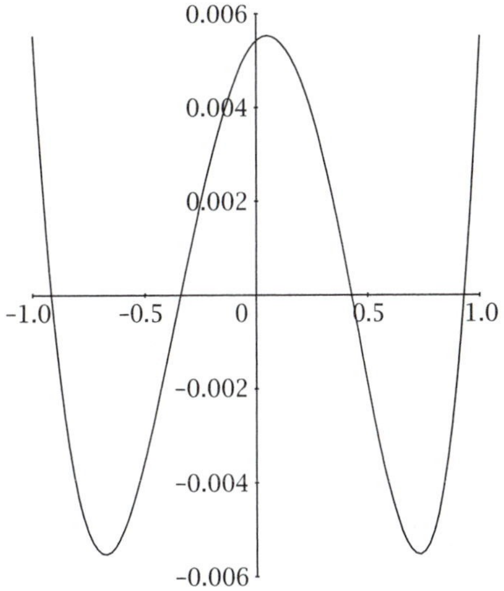


**Figure 7** Error plot of $e^x - p_3^*(x)$.

The Remez algorithm is an iterative procedure, and the polynomial $p_n^*(x)$ is obtained as the limit. The above characterization is also called the *equioscillation property*. We illustrate it in figure 7, where we plot the error $e^x - p_3^*(x)$ on $[-1,1]$. Compare this figure with figures 8 and 9, in which the error oscillates but does not equioscillate.

How much better the (nonlinear) minimax approximation is, compared with a linear approximation procedure of degree $n$ such as polynomial interpolation, Chebyshev approximation, and the like, is expressed by the norm $\|P_n\|_\infty = \sup_{\|f\|_\infty \leqslant 1} \|P_n(f)\|_\infty$ of the linear operator $P_n$ that associates with a function its particular linear approximant. Since $P_n(p_n^*) = p_n^*$, we have

$$\begin{aligned} \|f - P_n(f)\|_\infty &= \|f - p_n^* + p_n^* - P_n(f)\|_\infty \\ &= \|f - p_n^* + P_n(p_n^* - f)\|_\infty \\ &\leqslant (1 + \|P_n\|_\infty)\|f - p_n^*\|_\infty. \end{aligned}$$

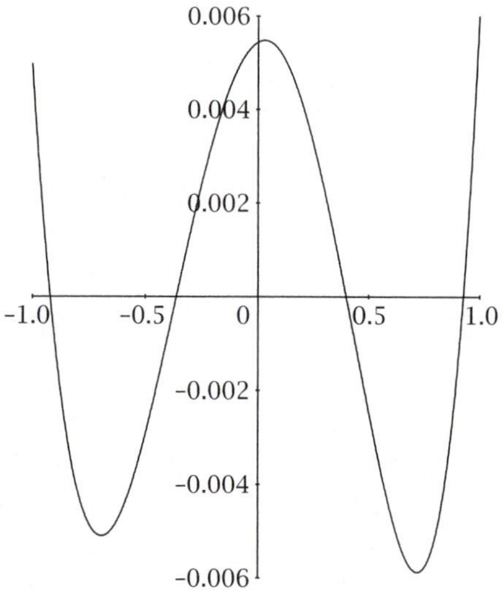


**Figure 8** Error plot of the Chebyshev partial sum of degree 3 for $e^x$.

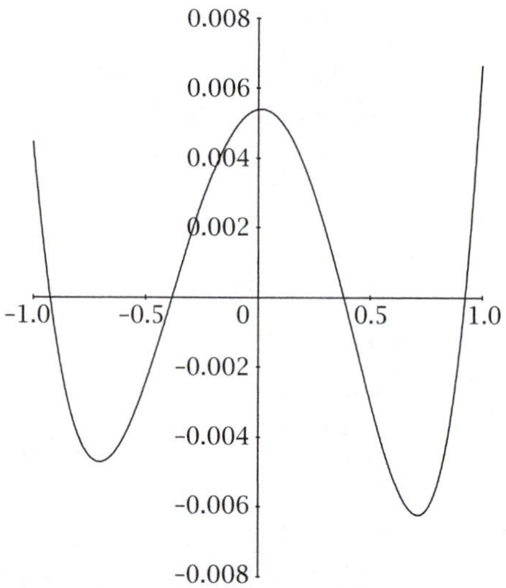


**Figure 9** Error plot of the polynomial interpolant of degree 3 in the Chebyshev zeros for $e^x$.

The value $\|P_n\|_\infty$ is called the *Lebesgue constant.* When dealing with polynomial interpolation, $\|P_n\|_\infty = L_n$, with $L_n$ given by (3).

The quality of the continuous best $\ell_2$ approximant (such as that in figure 8) is expressed by

$$\|P_n\|_\infty = \frac{1}{\pi}\int_0^\pi \left|\frac{\sin((n+\frac{1}{2})\theta)}{\sin(\frac{1}{2}\theta)}\right| \mathrm{d}\theta,$$

which again grows only logarithmically with $n$. Continuous $\ell_2$ polynomial approximation and Lagrange interpolation in the Chebyshev zeros (such as in figure 9) can therefore be considered near-best polynomial approximants.

### 3.6 Fourier Series

Let us return to a discrete approximation problem. Our interest is now in data exhibiting some periodic behavior, such as the description of rotation-invariant geometric figures or the sampling of a sound waveform. A suitable set of orthogonal basis functions is the set

$$\left.\begin{array}{c} 1,\ \cos(x),\ \cos(2x),\ \dots,\ \cos(nx), \\ \sin(x),\ \sin(2x),\ \dots,\ \sin(nx) \end{array}\right\} \tag{18}$$

as long as the distinct data points $x_i$ with $i = 0,\dots,m$ are evenly spaced on an interval of length $2\pi$ because then, for any two basis functions $b_j(x)$ and $b_k(x)$ from (18), we have

$$\langle b_j, b_k\rangle = \sum_{i=0}^m b_j(x_i)b_k(x_i) = \frac{m+1}{2}\delta_{jk}, \quad j \neq 0,$$

$$\langle b_0, b_k\rangle = \sum_{i=0}^m b_0(x_i)b_k(x_i) = (m+1)\delta_{0k},$$

where $\delta_{ij}$ is the Kronecker delta. For simplicity we assume that the real data $f_0,\dots,f_m$ are given on $[0, 2\pi)$ at

$$x_0 = 0, \quad x_1 = \frac{2\pi}{m}, \quad x_2 = \frac{4\pi}{m}, \quad \dots, \quad x_m = \frac{2m\pi}{m+1}.$$

Because of the periodicity, the value at $x_{m+1} = 2\pi$ equals the value at $x_0$ and therefore it is not repeated. Let $m \geqslant 2n$ and consider the approximation

$$\frac{\lambda_0}{2} + \sum_{j=1}^n \lambda_{2j}\cos(jx) + \sum_{j=1}^n \lambda_{2j-1}\sin(jx).$$

The values

$$\lambda_{2j} = \frac{2}{m+1}\sum_{i=0}^m f_i\cos(jx_i), \quad j = 0,\dots,n,$$

$$\lambda_{2j-1} = \frac{2}{m+1}\sum_{i=0}^m f_i\sin(jx_i), \quad j = 1,\dots,n,$$

minimize the $\ell_2$-norm

$$\sum_{i=1}^m \left(\frac{\lambda_0}{2} + \sum_{j=1}^n \lambda_{2j}\cos(jx_i) + \sum_{j=1}^n \lambda_{2j-1}\sin(jx_i) - f_i\right)^2.$$

When $m = 2n$, the minimum is zero because the least-squares approximant becomes a trigonometric interpolant. Note that we have replaced $\lambda_0$ by $\frac{1}{2}\lambda_0$ because $\langle b_j, b_j\rangle$ is smaller by a factor of 2 when $j = 0$.

If we form a single complex quantity $\Lambda_j = \lambda_{2j} - \mathrm{i}\lambda_{2j-1}$ for $j = 1, \dots, n$, where $\mathrm{i} = \sqrt{-1}$, these summations can be computed using a discrete Fourier transform that maps the data $f_i$ at the points $x_i$ to the $\Lambda_j$:

$$\Lambda_j = \frac{1}{m+1}\sum_{i=0}^{m} f_i \mathrm{e}^{-\mathrm{i}2\pi ij/(m+1)}, \quad j = 0, \dots, n.$$

The functions in (18) also satisfy a continuous orthogonality property:

$$\int_0^{2\pi} \cos(jx)\cos(kx)\,\mathrm{d}x = \pi\delta_{jk}, \quad j \neq 0,$$
$$\int_0^{2\pi} \cos(jx)\cos(kx)\,\mathrm{d}x = \frac{\pi}{2}\delta_{jk}, \quad j = 0,$$
$$\int_0^{2\pi} \cos(jx)\sin(kx)\,\mathrm{d}x = 0,$$
$$\int_0^{2\pi} \sin(jx)\sin(kx)\,\mathrm{d}x = \pi\delta_{jk}.$$

They can therefore be used for a Fourier series representation of a function $f(x)$:

$$f(x) = \frac{\lambda_0}{2} + \sum_{j=1}^{\infty} \lambda_{2j}\cos(jx) + \sum_{j=1}^{\infty} \lambda_{2j-1}\sin(jx),$$
$$\lambda_{2j} = \frac{\langle f(x), \cos(jx)\rangle}{\pi}, \quad j = 0, 1, \dots,$$
$$\lambda_{2j-1} = \frac{\langle f(x), \sin(jx)\rangle}{\pi}, \quad j = 1, 2, \dots.$$

The partial sum of trigonometric degree $n$ of this series minimizes the $\ell_2$-norm

$$\int_0^{2\pi}\biggl(f(x) - \frac{\lambda_0}{2} - \sum_{j=1}^{n}\lambda_{2j}\cos(jx) - \sum_{j=1}^{n}\lambda_{2j-1}\sin(jx)\biggr)^2 \mathrm{d}x.$$

## 4 Multivariate Interpolation and Approximation

The approximation of multivariate functions—continuous ones as well as discontinuous ones—is an active field of research due to its large variety of applications in the computational sciences and engineering. A wide range of multivariate generalizations of the above interpolation and approximation problems to functions of several variables $x, y, z, \dots$ have therefore been developed: polynomial and rational ones, discrete and continuous ones.

A fundamental issue in multivariate interpolation and approximation is the so-called CURSE OF DIMENSIONALITY [I.3 §2], meaning that, when the dimensionality increases, the number of different combinations of variables grows exponentially in the dimensionality. A polynomial of degree 3 in eight variables already has 165 terms! Another problem is that the polynomial basis of the multinomials does not satisfy the Haar condition and there is no easy generalization of this property to the multivariate case.

In order to counter both problems, the theory of radial basis functions has been developed.

Let us consider data $f_i$ given at corresponding multidimensional vectors $x_i$, $i = 0, \dots, n$. The data vectors $x_i$ do not have to form a grid but can be scattered. A *radial basis function* is a function whose value depends only on the distance from the origin or from another point, so its variable is $r = \|x\|_2$ or $r = \|x - c\|_2$. When centering a radial basis function $B(r)$ at each data point, there is a basis function $B(\|x - x_j\|_2)$ for each $j = 0, \dots, n$. The coefficients $a_j$ in a radial basis function interpolant $\sum_{j=0}^{n} a_j B(\|x - x_j\|_2)$ are then computed from the linear system

$$\sum_{j=0}^{n} a_j B(\|x_i - x_j\|_2) = f_i, \quad i = 0, \dots, n.$$

Several commonly used types of radial basis functions $B(r)$ guarantee nonsingular systems of interpolation conditions, in other words, nonsingular matrices

$$\bigl(B(\|x_i - x_j\|_2)\bigr)_{i,j=0,\dots,n}.$$

We mention as examples the Gaussian, multiquadric, inverse multiquadric, and a member of the Matern family, respectively, given by

$$B(r) = \mathrm{e}^{-(sr)^2},$$
$$B(r) = \sqrt{1 + (sr)^2},$$
$$B(r) = 1/\sqrt{1 + (sr)^2},$$
$$B(r) = (1 + sr)\mathrm{e}^{-sr}.$$

The real parameter $s$ is called a shape parameter. As can be seen in figure 10, different choices for $s$ greatly influence the shape of $B(r)$. Smaller shape parameters correspond to a flatter or wider basis function. The choice

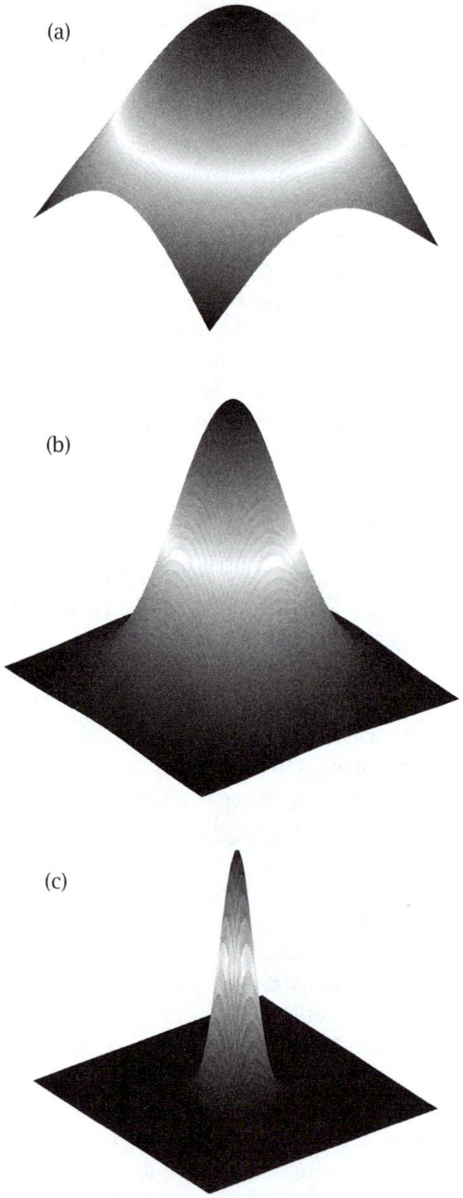


**Figure 10** Gaussian basis function with (a) $s = 0.4$, (b) $s = 1$, and (c) $s = 3$.

of $s$ has a significant impact on the accuracy of the approximation, and finding an optimal shape parameter is not an easy problem. Another concern is the numerical conditioning of the radial basis interpolation problem, especially when the shape parameter is small. The user often has to find the right trade-off between accuracy and conditioning.

The concept of radial basis function also allows one to work mesh free. In a mesh or grid of data points each point has a fixed number of neighbors, and this connectivity between neighbors is used to define mathematical operators such as the derivative and the divided difference. Multivariate mesh-free methods allow one to generalize these concepts and are especially useful when the mesh is difficult to maintain (e.g., in high-dimensional problems, when there is nonlinear behavior, discontinuities, singularities, etc.).

## 5 Future Research

Especially in multivariate approximation theory, many research questions remain unsolved: theory for the multivariate case is not nearly as well developed as it is for the univariate case. But researchers continue to push the boundaries in the one-variable case as well: what is the largest function class or the most general domain for which a result holds, for example? Many papers can be found on Jackson-type inequalities (approximation error bounds in terms of the function's smoothness), Bernstein-type inequalities (bounds on derivatives of polynomials), and convergence properties of particular approximations (polynomial, spline, rational, trigonometric), to name just a few fundamental topics. The development of orthogonal basis functions, on disconnected regions or in more variables, also deserves (and is getting) a lot of attention.

### Further Reading

Baker, G. Jr., and P. Graves-Morris. 1996. *Padé Approximants*, 2nd edn. Cambridge: Cambridge University Press.

Braess, D. 1986. *Nonlinear Approximation Theory*. Berlin: Springer.

Buhmann, M. D. 2003. *Radial Basis Functions: Theory and Implementations*. Cambridge: Cambridge University Press.

Cheney, E. W. 1982. *Introduction to Approximation Theory*. New York: McGraw-Hill.

Chihara, T. 2011. *An Introduction to Orthogonal Polynomials*. New York: Dover.

de Boor, C. 2001. *A Practical Guide to Splines*, revised edn. New York: Springer.

DeVore, R. A. 1998. Nonlinear approximation. *Acta Numerica* 7:51–150.

Powell, M. J. D. 1981. *Approximation Theory and Methods*. Cambridge: Cambridge University Press.

Szabados, J., and P. Vértesi. 1990. *Interpolation of Functions*. Teaneck, NJ: World Scientific.

Trefethen, L. N. 2013. *Approximation Theory and Approximation Practice*. Philadelphia, PA: SIAM.

# IV.10 Numerical Linear Algebra and Matrix Analysis

*Nicholas J. Higham*

Matrices are ubiquitous in applied mathematics. Ordinary differential equations (ODEs) and partial differential equations (PDEs) are solved numerically by finite-difference or finite-element methods, which lead to systems of linear equations or matrix eigenvalue problems. Nonlinear equations and optimization problems are typically solved using linear or quadratic models, which again lead to linear systems.

Solving linear systems of equations is an ancient task, undertaken by the Chinese around 1 C.E., but the study of matrices per se is relatively recent, originating with Arthur Cayley's 1858 "A memoir on the theory of matrices." Early research on matrices was largely theoretical, with much attention focused on the development of canonical forms, but in the twentieth century the practical value of matrices started to be appreciated. Heisenberg used matrix theory as a tool in the development of quantum mechanics in the 1920s. Early proponents of the systematic use of matrices in applied mathematics included Frazer, Duncan, and Collar, whose 1938 book *Elementary Matrices and Some Applications to Dynamics and Differential Equations* emphasized the important role of matrices in differential equations and mechanics. The continued growth of matrices in applications, together with the advent of mechanical and then digital computing devices, allowing ever larger problems to be solved, created the need for greater understanding of all aspects of matrices from theory to computation.

This article treats two closely related topics: matrix analysis, which is the theory of matrices with a focus on aspects relevant to other areas of mathematics, and numerical linear algebra (also called matrix computations), which is concerned with the construction and analysis of algorithms for solving matrix problems as well as related topics such as problem sensitivity and rounding error analysis.

Important themes that are discussed in this article include the matrix factorization paradigm, the use of unitary transformations for their numerical stability, exploitation of matrix structure (such as sparsity, symmetry, and definiteness), and the design of algorithms to exploit evolving computer architectures.

Throughout the article, uppercase letters are used for matrices and lowercase letters for vectors and scalars. Matrices and vectors are assumed to be complex, unless otherwise stated, and $A^* = (\overline{a_{ji}})$ denotes the conjugate transpose of $A = (a_{ij})$. An unsubscripted norm $\|\cdot\|$ denotes a general vector norm and the corresponding subordinate matrix norm. Particular norms used here are the 2-norm $\|\cdot\|_2$ and the Frobenius norm $\|\cdot\|_F$. All of these norms are defined in THE LANGUAGE OF APPLIED MATHEMATICS [I.2 §§19.3, 20]. The notation "$i = 1\colon n$" means that the integer variable $i$ takes on the values $1, 2, \dots, n$.

## 1 Nonsingularity and Conditioning

Nonsingularity of a matrix is a key requirement in many problems, such as in the solution of $n$ linear equations in $n$ unknowns. For some classes of matrices, nonsingularity is guaranteed. A good example is the diagonally dominant matrices. The matrix $A \in \mathbb{C}^{n\times n}$ is *strictly diagonally dominant by rows* if

$$\sum_{j\neq i} |a_{ij}| < |a_{ii}|, \quad i = 1\colon n,$$

and *strictly diagonally dominant by columns* if $A^*$ is strictly diagonally dominant by rows. Any matrix that is strictly diagonally dominant by rows or columns is nonsingular (a proof can be obtained by applying Gershgorin's theorem in section 5.1).

Since data is often subject to uncertainty, we wish to gauge the sensitivity of problems to perturbations, which is done using CONDITION NUMBERS [I.2 §22]. An appropriate condition number for the matrix inverse is

$$\lim_{\varepsilon\to 0} \sup_{\|\Delta A\|\leqslant\varepsilon\|A\|} \frac{\|(A+\Delta A)^{-1} - A^{-1}\|}{\varepsilon\|A^{-1}\|}.$$

This expression turns out to equal $\kappa(A) = \|A\|\,\|A^{-1}\|$, which is called *the condition number of A with respect to inversion*. This condition number occurs in many contexts. For example, suppose $A$ is contaminated by errors and we perform a similarity transformation $X^{-1}(A+E)X = X^{-1}AX + F$. Then $\|F\| = \|X^{-1}EX\| \leqslant \kappa(X)\|E\|$ and this bound is attainable for some $E$. Hence the errors can be multiplied by a factor as large as $\kappa(X)$. We therefore prefer to carry out similarity and other transformations with matrices that are *well-conditioned*, that is, ones for which $\kappa(X)$ is close to its lower bound of 1. By contrast, a matrix for which $\kappa$ is large is called *ill-conditioned*. For any unitary matrix $X$, $\kappa_2(X) = 1$, so in numerical linear algebra transformations by unitary or orthogonal matrices are preferred and usually lead to numerically stable algorithms.

In practice we often need an estimate of the matrix condition number $\kappa(A)$ but do not wish to go to the expense of computing $A^{-1}$ in order to obtain it. Fortunately, there are algorithms that can cheaply produce a reliable estimate of $\kappa(A)$ once a factorization of $A$ has been computed.

Note that the determinant, $\det(A)$, is rarely computed in numerical linear algebra. Its magnitude gives no useful information about the conditioning of $A$, not least because of its extreme behavior under scaling: $\det(\alpha A) = \alpha^n \det(A)$.

## 2 Matrix Factorizations

The method of *Gaussian elimination* (GE) for solving a nonsingular linear system $Ax = b$ of $n$ equations in $n$ unknowns reduces the matrix $A$ to upper triangular form and then solves for $x$ by SUBSTITUTION [I.3 §7.1]. GE is typically described by writing down the equations $a_{ij}^{(k+1)} = a_{ij}^{(k)} - a_{ik}^{(k)} a_{kj}^{(k)} / a_{kk}^{(k)}$ (and similarly for $b$) that describe how the starting matrix $A = A^{(1)} = (a_{ij}^{(1)})$ changes on each of the $n-1$ steps of the elimination in its progress toward upper triangular form $U$. Working at the element level in this way leads to a profusion of symbols, superscripts, and subscripts that tend to obscure the mathematical structure and hinder insights being drawn into the underlying process. One of the key developments in the last century was the recognition that it is much more profitable to work at the matrix level. Thus the basic equation above is written as $A^{(k+1)} = M_k A^{(k)}$, where $M_k$ agrees with the identity matrix except below the diagonal in the $k$th column, where its $(i,k)$ element is $m_{ik} = -a_{ik}^{(k)}/a_{kk}^{(k)}$, $i = k+1:n$. Recurring the matrix equation gives $U := A^{(n)} = M_{n-1} \cdots M_1 A$. Taking the $M_k$ matrices over to the left-hand side leads, after some calculations, to the equation $A = LU$, where $L$ is unit lower triangular, with $(i,k)$ element $m_{ik}$. The prefix "unit" means that $L$ has ones on the diagonal.

GE is therefore equivalent to factorizing the matrix $A$ as the product of a lower triangular matrix and an upper triangular matrix—something that is not at all obvious from the element-level equations. Solving the linear system $Ax = b$ now reduces to the task of solving the two triangular systems $Ly = b$ and $Ux = y$.

Interpreting GE as LU factorization separates the computation of the factors from the solution of the triangular systems. It is then clear how to solve efficiently several systems $Ax_i = b_i$, $i = 1:r$, with different right-hand sides but the same coefficient matrix $A$: compute the LU factors once and then reuse them to solve for each $x_i$ in turn.

This matrix factorization[1] viewpoint dates from around the 1940s and has been extremely successful in matrix computations. In general, a factorization is a representation of a matrix as a product of "simpler" matrices. Factorization is a tool that can be used to solve a variety of problems, as we will see below.

Two particular benefits of factorizations are unity and modularity. GE, for example, can be organized in several different ways, corresponding to different orderings of the three nested loops that it comprises, as well as the use of different blockings of the matrix elements. Yet all of them compute the same LU factorization, carrying out the same mathematical operations in a different order. Without the unifying concept of a factorization, reasoning about these GE variants would be difficult.

Modularity refers to the way that a factorization breaks a problem down into separate tasks that can be analyzed or programmed independently. To carry out a rounding error analysis of GE, we can analyze the LU factorization and the solution of the triangular systems by substitution separately and then put the analyses together. The rounding error analysis of substitution can be reused in the many other contexts in which triangular systems arise.

An important example of the use of LU factorization is in *iterative refinement*. Suppose we have used GE to obtain a computed solution $\hat{x}$ to $Ax = b$ in floating-point arithmetic. If we form $r = b - A\hat{x}$ and solve $Ae = r$, then in exact arithmetic $y = \hat{x} + e$ is the true solution. In computing $e$ we can reuse the LU factors of $A$, so obtaining $y$ from $\hat{x}$ is inexpensive. In practice, the computation of $r$, $e$, and $y$ is subject to rounding errors so the computed $\hat{y}$ is not equal to $x$. But under suitable assumptions $\hat{y}$ will be an improved approximation and we can iterate this refinement process. Iterative refinement is particularly effective if $r$ can be computed using extra precision.

Two other key factorizations are the following.

- *Cholesky factorization*: for Hermitian positive-definite $A \in \mathbb{C}^{n\times n}$, $A = R^*R$, where $R$ is upper triangular with positive diagonal elements, and this factorization is unique.
- *QR factorization*: for $A \in \mathbb{C}^{m\times n}$ with $m \geqslant n$, $A = QR$, where $Q \in \mathbb{C}^{m\times m}$ is unitary ($Q^*Q = I_m$) and $R \in \mathbb{C}^{m\times n}$ is upper trapezoidal, that is, $R = \left[\begin{smallmatrix} R_1 \\ 0 \end{smallmatrix}\right]$ with $R_1 \in \mathbb{C}^{n\times n}$ upper triangular.

1. Or decomposition—the two terms are essentially synonymous.

These two factorizations are related: if $A \in \mathbb{C}^{m \times n}$ with $m \geqslant n$ has full rank and $A = QR$ is a QR factorization, in which without loss of generality we can assume that $R$ has positive diagonal, then $A^*A = R^*R$, so $R$ is the Cholesky factor of $A^*A$.

The Cholesky factorization can be computed by what is essentially a symmetric and scaled version of GE. The QR factorization can be computed in three main ways, one of which is the classical Gram–Schmidt orthogonalization. The most widely used method constructs $Q$ as a product of *Householder reflectors*, which are unitary matrices of the form $H = I - 2vv^*/(v^*v)$, where $v$ is a nonzero vector. Note that $H$ is a rank-1 perturbation of the identity, and since it is Hermitian and unitary, it is its own inverse; that is, it is *involutory*. The third approach builds $Q$ as a product of *Givens rotations*, each of which is a $2 \times 2$ matrix $\left[\begin{smallmatrix} c & s \\ -s & c \end{smallmatrix}\right]$ embedded into two rows and columns of an $m \times m$ identity matrix, where (in the real case) $c^2 + s^2 = 1$.

The Cholesky factorization helps us to make the most of the very desirable property of positive-definiteness. For example, suppose $A$ is Hermitian positive-definite and we wish to evaluate the scalar $\alpha = x^*A^{-1}x$. We can rewrite it as $x^*(R^*R)^{-1}x = (x^*R^{-1})(R^{-*}x) = z^*z$, where $z = R^{-*}x$. So once the Cholesky factorization has been computed we need just one triangular solve to compute $\alpha$, and of course there is no need to explicitly invert the matrix $A$.

A matrix factorization might involve a larger number of factors: $A = N_1N_2 \cdots N_k$, say. It is immediate that $A^{\mathrm{T}} = N_k^{\mathrm{T}}N_{k-1}^{\mathrm{T}} \cdots N_1^{\mathrm{T}}$. This factorization of the transpose may have deep consequences in a particular application. For example, the discrete Fourier transform is the matrix–vector product $y = F_nx$, where the $n \times n$ matrix $F_n$ has $(p,q)$ element $\exp(-2\pi \mathrm{i}(p-1)(q-1)/n)$; $F_n$ is a complex, symmetric matrix. The FAST FOURIER TRANSFORM [II.10] (FFT) is a way of evaluating $y$ in $O(n \log_2 n)$ operations, as opposed to the $O(n^2)$ operations that are required by a standard matrix–vector multiplication. Many variants of the FFT have been proposed since the original 1965 paper by Cooley and Tukey. It turns out that different FFT variants correspond to different factorizations of $F_n$ with $k = \log_2 n$ sparse factors. Some of these methods correspond simply to transposing the factorization in another method (recall that $F_n^{\mathrm{T}} = F_n$), though this was not realized when the methods were developed. Transposition also plays an important role in AUTOMATIC DIFFERENTIATION [VI.7]; the so-called reverse or adjoint mode can be obtained by transposing a matrix factorization representation of the forward mode.

The factorizations described in this section are in "plain vanilla" form, but all have variants that incorporate pivoting. Pivoting refers to row or column interchanges carried out at each step of the factorization as it is computed, introduced either to ensure that the factorization succeeds and is numerically stable or to produce a factorization with certain desirable properties usually associated with rank deficiency. For GE, *partial pivoting* is normally used: at the start of the $k$th stage of the elimination, an element $a_{rk}^{(k)}$ of largest modulus in the $k$th column below the diagonal is brought into the $(k,k)$ (pivot) position by interchanging rows $k$ and $r$. Partial pivoting avoids dividing by zero (if $a_{kk}^{(k)} = 0$ after the interchange then the pivot column is zero below the diagonal and the elimination step can be skipped). More importantly, partial pivoting ensures numerical stability (see section 8). The overall effect of GE with partial pivoting is to produce an LU factorization $PA = LU$, where $P$ is a permutation matrix.

Pivoted variants of Cholesky factorization and QR factorization take the form $P^{\mathrm{T}}AP = R^*R$ and $AP = Q\left[\begin{smallmatrix} R \\ 0 \end{smallmatrix}\right]$, where $P$ is a permutation matrix and $R$ satisfies the inequalities

$$|r_{kk}|^2 \geqslant \sum_{i=k}^{j} |r_{ij}|^2, \quad j = k+1:n, \ k = 1:n.$$

If $A$ is rank deficient then $R$ has the form $R = \left[\begin{smallmatrix} R_{11} & R_{12} \\ 0 & 0 \end{smallmatrix}\right]$ with $R_{11}$ nonsingular, and the rank of $A$ is the dimension of $R_{11}$. Equally importantly, when $A$ is nearly rank deficient this tends to be revealed by a small trailing diagonal block of $R$.

A factorization of great importance in a wide variety of applications is the SINGULAR VALUE DECOMPOSITION [II.32] (SVD) of $A \in \mathbb{C}^{m \times n}$:

$$A = U\Sigma V^*, \quad \Sigma = \operatorname{diag}(\sigma_1, \sigma_2, \dots, \sigma_p) \in \mathbb{R}^{m \times n}, \quad (1)$$

where $p = \min(m,n)$, $U \in \mathbb{C}^{m \times m}$ and $V \in \mathbb{C}^{n \times n}$ are unitary, and the *singular values* $\sigma_i$ satisfy $\sigma_1 \geqslant \sigma_2 \geqslant \cdots \geqslant \sigma_p \geqslant 0$. For a square $A$ $(m = n)$, the 2-norm condition number is given by $\kappa_2(A) = \sigma_1/\sigma_n$.

The *polar decomposition* of $A \in \mathbb{C}^{m \times n}$ with $m \geqslant n$ is a factorization $A = UH$ in which $U \in \mathbb{C}^{m \times n}$ has orthonormal columns and $H \in \mathbb{C}^{n \times n}$ is Hermitian positive-semidefinite. The matrix $H$ is unique and is given by $(A^*A)^{1/2}$, where the exponent $1/2$ denotes the PRINCIPAL SQUARE ROOT [II.14], while $U$ is unique if $A$ has full rank. The polar decomposition generalizes to matrices the polar representation $z = re^{\mathrm{i}\theta}$ of a complex number. The Hermitian polar factor $H$ is also known as the *matrix absolute value*, $|A|$, and is much studied in matrix analysis and functional analysis.

One reason for the importance of the polar decomposition is that it provides an optimal way to orthogonalize a matrix; a result of Fan and Hoffman (1955) says that $U$ is the nearest matrix with orthonormal columns to $A$ in any unitarily invariant norm (a unitarily invariant norm is one with the property that $\|UAV\| = \|A\|$ for any unitary $U$ and $V$; the 2-norm and the Frobenius norm are particular examples). In various applications a matrix $A \in \mathbb{R}^{n\times n}$ that should be orthogonal drifts from orthogonality because of rounding or other errors; replacing it by the orthogonal polar factor $U$ is then a good strategy.

The polar decomposition also solves the *orthogonal Procrustes problem*, for $A, B \in \mathbb{C}^{m\times n}$,

$$\min\{\|A - BQ\|_F \colon Q \in \mathbb{C}^{n\times n},\ Q^*Q = I\},$$

for which any solution $Q$ is a unitary polar factor of $B^*A$. This problem comes from factor analysis and multidimensional scaling in statistics, where the aim is to see whether two data sets $A$ and $B$ are the same up to an orthogonal transformation.

Either of the SVD and the polar decomposition can be derived, or computed, from the other. Historically, the SVD came first (Beltrami, in 1873), with the polar decomposition three decades behind (Autonne, in 1902).

## 3 Distance to Singularity and Low-Rank Perturbations

The question commonly arises of whether a given perturbation of a nonsingular matrix $A$ preserves nonsingularity. In a sense, this question is trivial. Recalling that a square matrix is nonsingular when all its eigenvalues are nonzero, and that the product of two matrices is nonsingular unless one of them is singular, from $A + \Delta A = A(I + A^{-1}\Delta A)$ we see that $A + \Delta A$ is nonsingular as long as $A^{-1}\Delta A$ has no eigenvalue equal to $-1$. However, this is not an easy condition to check, and in practice we may not know $\Delta A$ but only a bound for its norm. Since any norm of a matrix exceeds the modulus of every eigenvalue, a sufficient condition for $A + \Delta A$ to be nonsingular is that $\|A^{-1}\Delta A\| < 1$, which is certainly true if $\|A^{-1}\|\,\|\Delta A\| < 1$. This condition can be rewritten as the inequality $\|\Delta A\|/\|A\| < \kappa(A)^{-1}$, where $\kappa(A) = \|A\|\,\|A^{-1}\| \geqslant 1$ is the condition number introduced in section 1. It turns out that we can always find a perturbation $\Delta A$ such that $A + \Delta A$ is singular and $\|\Delta A\|/\|A\| = \kappa(A)^{-1}$. It follows that the *relative distance to singularity*

$$d(A) = \min\{\|\Delta A\|/\|A\| \colon A + \Delta A \text{ is singular}\} \qquad (2)$$

is given by $d(A) = \kappa(A)^{-1}$. This reciprocal relation between problem conditioning and the distance to a singular problem (one with an infinite condition number) is common to a variety of problems in linear algebra and control theory, as shown by James Demmel in the 1980s.

We may want a more refined test for whether $A + \Delta A$ is nonsingular. To obtain one we will need to make some assumptions about the perturbation. Suppose that $\Delta A$ has rank 1: $\Delta A = xy^*$ for some vectors $x$ and $y$. From the analysis above we know that $A + \Delta A$ will be nonsingular if $A^{-1}\Delta A = A^{-1}xy^*$ has no eigenvalue equal to $-1$. Using the fact that the nonzero eigenvalues of $AB$ are the same as those of $BA$ for any conformable matrices $A$ and $B$, we see that the nonzero eigenvalues of $(A^{-1}x)y^*$ are the same as those of $y^*A^{-1}x$. Hence $A + xy^*$ is nonsingular as long as $y^*A^{-1}x \neq -1$.

Now that we know when $A + xy^*$ is nonsingular, we might ask if there is an explicit formula for the inverse. Since $A + xy^* = A(I + A^{-1}xy^*)$, we can take $A = I$ without loss of generality. So we are looking for the inverse of $B = I + xy^*$. One way to find it is to guess that $B^{-1} = I + \theta xy^*$ for some scalar $\theta$ and equate the product with $B$ to $I$, to obtain $\theta(1 + y^*x) + 1 = 0$. Thus $(I + xy^*)^{-1} = I - xy^*/(1 + y^*x)$. The corresponding formula for $(A + xy^*)^{-1}$ is

$$(A + xy^*)^{-1} = A^{-1} - A^{-1}xy^*A^{-1}/(1 + y^*A^{-1}x),$$

which is known as the *Sherman–Morrison formula*. This formula and its generalizations originated in the 1940s and have been rediscovered many times. The corresponding formula for a rank-$p$ perturbation is the *Sherman–Morrison–Woodbury formula*: for $U, V \in \mathbb{C}^{n\times p}$,

$$(A + UV^*)^{-1} = A^{-1} - A^{-1}U(I + V^*A^{-1}U)^{-1}V^*A^{-1}.$$

Important applications of these formulas are in optimization, where rank-1 or rank-2 updates are made to Hessian approximations in QUASI-NEWTON METHODS [IV.11 §4.2] and to basis matrices in the SIMPLEX METHOD [IV.11 §3.1]. More generally, the task of updating the solution to a problem after a coefficient matrix has undergone a low-rank change, or has had a row or column added or removed, arises in many applications, including SIGNAL PROCESSING [IV.35], where new data is continually being received and old data is discarded.

The minimal distance in the definition (2) of the distance to singularity $d(A)$ can be shown to be attained for a rank-1 matrix $\Delta A$. Rank-1 matrices often feature in the solutions of matrix optimization problems.

## 4 Computational Cost

In order to compare competing methods and predict their practical efficiency, we need to know their computational cost. Traditionally, computational cost has been measured by counting the number of scalar arithmetic operations and retaining only the highest-order terms in the total. For example, using GE we can solve a system of $n$ linear equations in $n$ unknowns with $n^3/3 + O(n^2)$ additions, $n^3/3 + O(n^2)$ multiplications, and $O(n)$ divisions. This is typically summarized as $2n^3/3$ flops, where a *flop* denotes any of the scalar operations $+, -, *, /$. Most standard problems involving $n \times n$ matrices can be solved with a cost of order $n^3$ flops or less, so the interest is in the exponent (1, 2, or 3) and the constant of the dominant term. However, the costs of moving data around a computer's hierarchical memory and the costs of communicating between different processors on a multiprocessor system can be equally important. Simply counting flops does not therefore necessarily give a good guide to performance in practice.

Seemingly trivial problems can offer interesting challenges as regards minimizing arithmetic costs. For matrices $A$, $B$, and $C$ of any dimensions such that the product $ABC$ is defined, how should we compute the product? The associative law for matrix multiplication tells us that $(AB)C = A(BC)$, but this mathematical equivalence is not a computational one. To see why, note that for three vectors $a, b, c \in \mathbb{R}^n$ we can write

$$\underbrace{(ab^*)}_{n\times n}c = a\underbrace{(b^*c)}_{1\times 1}.$$

Evaluation of the left-hand side requires $O(n^2)$ flops, as there is an outer product $ab^*$ and then a matrix–vector product to evaluate, while evaluation of the right-hand side requires just $O(n)$ flops, as it involves only vector operations: an inner product and a vector scaling. One should always be alert for opportunities to use the associative law to save computational effort.

## 5 Eigenvalue Problems

The eigenvalue problem $Ax = \lambda x$ for a square matrix $A \in \mathbb{C}^{n\times n}$, which seeks an eigenvalue $\lambda \in \mathbb{C}$ and an eigenvector $x \neq 0$, arises in many forms. Depending on the application we may want all the eigenvalues or just a subset, such as the ten that have the largest real part, and eigenvectors may or may not be required as well. Whether the problem is Hermitian or non-Hermitian changes its character greatly. In particular, while a Hermitian matrix has real eigenvalues and a linearly independent set of $n$ eigenvectors that can be taken to be orthonormal, the eigenvalues of a non-Hermitian matrix can be anywhere in the complex plane and there may not be a set of eigenvectors that spans $\mathbb{C}^n$.

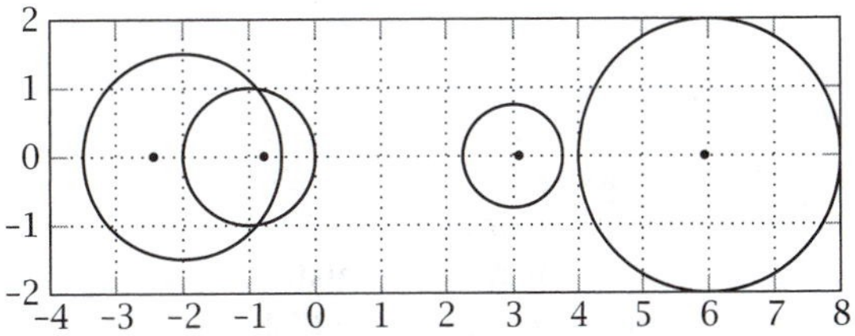

**Figure 1** Gershgorin disks for the matrix in (3); the eigenvalues are marked as solid dots.

### 5.1 Bounds and Localization

One of the first questions to ask is whether we can find a finite region containing the eigenvalues. The answer is yes because $Ax = \lambda x$ implies $|\lambda|\,\|x\| = \|Ax\| \leqslant \|A\|\,\|x\|$, and hence $|\lambda| \leqslant \|A\|$. So all the eigenvalues lie in a disk of radius $\|A\|$ about the origin. More refined bounds are provided by Gershgorin's theorem.

**Theorem 1 (Gershgorin's theorem, 1931).** *The eigenvalues of $A \in \mathbb{C}^{n\times n}$ lie in the union of the $n$ disks in the complex plane*

$$D_i = \Big\{z \in \mathbb{C} : |z - a_{ii}| \leqslant \sum_{j\neq i} |a_{ij}|\Big\}, \quad i = 1:n.$$

An extension of the theorem says that if $k$ disks form a connected region that is isolated from the other disks then there are precisely $k$ eigenvalues in this region. The Gershgorin disks for the matrix

$$\begin{bmatrix} -1 & 1/3 & 1/3 & 1/3 \\ 3/2 & -2 & 0 & 0 \\ 1/2 & 0 & 3 & 1/4 \\ 1 & 0 & -1 & 6 \end{bmatrix} \qquad (3)$$

are shown in figure 1. We can conclude that there is one eigenvalue in the disk centered at 3, one in the disk centered at 6, and two in the union of the other two disks.

Gershgorin's theorem is most useful for matrices that are close to diagonal, such as those eventually produced by the Jacobi iterative method for eigenvalues of Hermitian matrices. Improved estimates can be sought by applying Gershgorin's theorem to a matrix $D^{-1}AD$ similar to $A$, with the diagonal matrix $D$ chosen in an attempt to isolate and shrink the disks. Many variants of Gershgorin's theorem exist with disks replaced by other shapes.

The spectral radius $\rho(A)$ (the largest absolute value of any eigenvalue of $A$) satisfies $\rho(A) \leqslant \|A\|$, as shown above, but this inequality can be arbitrarily weak, as the matrix $\left[\begin{smallmatrix}1 & \theta \\ 0 & 1\end{smallmatrix}\right]$ shows for $|\theta| \gg 1$. It is natural to ask whether there are any sharper relations between the spectral radius and norms. One answer is the equality

$$\rho(A) = \lim_{k\to\infty} \|A^k\|^{1/k}. \tag{4}$$

Another is the result that given any $\varepsilon > 0$ there is a norm such that $\|A\| \leqslant \rho(A) + \varepsilon$; however, the norm depends on $A$. This result can be used to give a proof of the fact, discussed in the article on the JORDAN CANONICAL FORM [II.22], that the powers of $A$ converge to zero if $\rho(A) < 1$.

The *field of values*, also known as the *numerical range*, is a tool that can be used for localization and many other purposes. It is defined for $A \in \mathbb{C}^{n\times n}$ by

$$F(A) = \left\{\frac{z^*Az}{z^*z} : 0 \neq z \in \mathbb{C}^n\right\}.$$

The set $F(A)$ is compact and convex (a nontrivial property proved by Toeplitz and Hausdorff), and it contains all the eigenvalues of $A$. For normal matrices it is the CONVEX HULL [II.8] of the eigenvalues. The *normal matrices* $A$ are those for which $AA^* = A^*A$, and they include the Hermitian, the skew-Hermitian, and the unitary matrices. For a Hermitian matrix, $F(A)$ is a segment of the real axis, while for a skew-Hermitian matrix it is a segment of the imaginary axis. Figure 2 illustrates two fields of values, the second of which is the convex hull of the eigenvalues because a CIRCULANT MATRIX [I.2 §18] is normal.

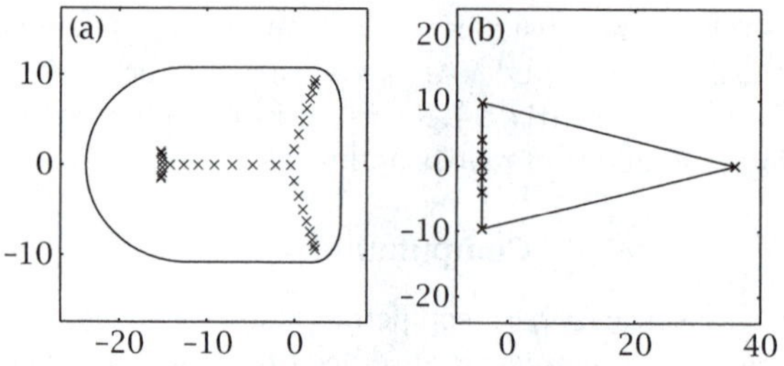


**Figure 2** Fields of values for (a) a pentadiagonal Toeplitz matrix and (b) a circulant matrix, both of dimension 32. The eigenvalues are denoted by crosses.

### 5.2 Eigenvalue Sensitivity

If $A$ is perturbed, how much do its eigenvalues change? This question is easy to answer for a simple eigenvalue $\lambda$—one that has ALGEBRAIC MULTIPLICITY [II.22] 1. We need the notion of a *left eigenvector* of $A$ corresponding to $\lambda$, which is a nonzero vector $y$ such that $y^*A = \lambda y^*$. If $\lambda$ is simple with right and left eigenvectors $x$ and $y$, respectively, then there is an eigenvalue $\lambda + \Delta\lambda$ of $A+\Delta A$ such that $\Delta\lambda = y^*\Delta Ax/(y^*x) + O(\|\Delta A\|^2)$ and so

$$|\Delta\lambda| \leqslant \frac{\|y\|_2\|x\|_2}{|y^*x|}\|\Delta A\| + O(\|\Delta A\|^2).$$

The term $\|y\|_2\|x\|_2/|y^*x|$ can be shown to be an (absolute) condition number for $\lambda$. It is at least 1 and tends to infinity as $y$ and $x$ approach orthogonality (which can never exactly be achieved for simple $\lambda$), so $\lambda$ can be very ill-conditioned. However, if $A$ is Hermitian then we can take $y = x$ and the bound simplifies to $|\Delta\lambda| \leqslant \|\Delta A\| + O(\|\Delta A\|^2)$, so all the eigenvalues of a Hermitian matrix are perfectly conditioned.

Much research has been done to obtain eigenvalue perturbation bounds under both weaker and stronger assumptions about the problem. Suppose we drop the requirement that $\lambda$ is simple. Consider the matrix and perturbation

$$A = \begin{bmatrix} 0 & 1 & 0 \\ 0 & 0 & 1 \\ 0 & 0 & 0 \end{bmatrix}, \qquad \Delta A = \begin{bmatrix} 0 & 0 & 0 \\ 0 & 0 & 0 \\ \varepsilon & 0 & 0 \end{bmatrix}.$$

The eigenvalues of $A$ are all zero, and those of $A + \Delta A$ are the third roots of $\varepsilon$. The change in the eigenvalue is proportional not to $\varepsilon$ but to a fractional power of $\varepsilon$. In general, the sensitivity of an eigenvalue depends on the JORDAN STRUCTURE [II.22] for that eigenvalue.

### 5.3 Companion Matrices and the Characteristic Polynomial

The eigenvalues of a matrix $A$ are the roots of its CHARACTERISTIC POLYNOMIAL [I.2 §20], $\det(\lambda I - A)$. Conversely, associated with the polynomial

$$p(\lambda) = \lambda^n - a_{n-1}\lambda^{n-1} - \cdots - a_0$$

is the companion matrix

$$C = \begin{bmatrix} a_{n-1} & a_{n-2} & \cdots & \cdots & a_0 \\ 1 & 0 & \cdots & \cdots & 0 \\ 0 & 1 & \ddots & & 0 \\ \vdots & & \ddots & 0 & \vdots \\ 0 & \cdots & \cdots & 1 & 0 \end{bmatrix},$$

and the eigenvalues of $C$ are the roots of $p$, as noted in METHODS OF SOLUTION [I.3 §7.2].

This relation means that the roots of a polynomial can be found by computing the eigenvalues of an $n \times n$ matrix, and this approach is used by some computer codes, for example the `roots` function of MATLAB. While standard eigenvalue algorithms do not exploit the structure of $C$, this approach has proved competitive with specialist polynomial root-finding algorithms. Another use for the relation is to obtain bounds for roots of polynomials from bounds for matrix eigenvalues, and vice versa.

Companion matrices have many interesting properties. For example, any NONDEROGATORY [II.22] $n \times n$ matrix is similar to a companion matrix. Companion matrices have therefore featured strongly in matrix analysis and also in control theory. However, similarity transformations to companion form are little used in practice because of problems with ill-conditioning and numerical instability.

Returning to the characteristic polynomial, $p(\lambda) = \det(\lambda I - A) = \lambda^n - a_{n-1}\lambda^{n-1} - \cdots - a_0$, we know that $p(\lambda_i) = 0$ for every eigenvalue $\lambda_i$ of $A$. The *Cayley–Hamilton theorem* says that $p(A) = A^n - a_{n-1}A^{n-1} - \cdots - a_0 I = 0$ (which cannot be obtained simply by putting "$\lambda = A$" in the previous expression!). Hence the $n$th power of $A$ and inductively all higher powers are expressible as a linear combination of $I$, $A$, ..., $A^{n-1}$. Moreover, if $A$ is nonsingular then from $A^{-1}p(A) = 0$ it follows that $A^{-1}$ can also be written as a polynomial in $A$ of degree at most $n - 1$. These relations are not useful for practical computation because the coefficients $a_i$ can vary tremendously in magnitude, and it is not possible to compute them to high relative accuracy.

### 5.4 Eigenvalue Inequalities for Hermitian Matrices

The eigenvalues of Hermitian matrices $A \in \mathbb{C}^{n\times n}$, which in this section we order $\lambda_n \leqslant \cdots \leqslant \lambda_1$, satisfy many beautiful inequalities. Among the most important are those in the *Courant–Fischer theorem* (1905), which states that every eigenvalue is the solution of a min-max problem over a suitable subspace $S$ of $\mathbb{C}^n$:

$$\lambda_i = \min_{\dim(S)=n-i+1} \max_{0\neq x\in S} \frac{x^*Ax}{x^*x}.$$

Special cases are $\lambda_n = \min_{x\neq 0} x^*Ax/(x^*x)$ and $\lambda_1 = \max_{x\neq 0} x^*Ax/(x^*x)$.

Taking $x$ to be a unit vector $e_i$ in the previous formula for $\lambda_1$ gives $\lambda_1 \geqslant a_{ii}$ for all $i$. This inequality is just the first in a sequence of inequalities relating sums of eigenvalues to sums of diagonal elements, obtained by Schur in 1923:

$$\sum_{i=1}^{k} \lambda_i \geqslant \sum_{i=1}^{k} \tilde{a}_{ii}, \quad k = 1:n, \tag{5}$$

where $\{\tilde{a}_{ii}\}$ is the set of diagonal elements of $A$ arranged in decreasing order: $\tilde{a}_{11} \geqslant \cdots \geqslant \tilde{a}_{nn}$. There is equality for $k = n$, since both sides equal trace$(A)$. These inequalities say that the vector $[\lambda_1, \dots, \lambda_n]$ of eigenvalues *majorizes* the vector $[\tilde{a}_{11}, \dots, \tilde{a}_{nn}]$ of diagonal elements.

In general there is no useful formula for the eigenvalues of a sum $A + B$ of Hermitian matrices. However, the Courant–Fischer theorem yields the upper and lower bounds

$$\lambda_k(A) + \lambda_n(B) \leqslant \lambda_k(A + B) \leqslant \lambda_k(A) + \lambda_1(B),$$

from which it follows that $|\lambda_k(A + B) - \lambda_k(A)| \leqslant \max(|\lambda_n(B)|, |\lambda_1(B)|) = \|B\|_2$. The latter inequality again shows that the eigenvalues of a Hermitian matrix are well-conditioned under perturbation.

The *Cauchy interlace theorem* has a different flavor. It relates the eigenvalues of successive leading principal submatrices $A_k = A(1:k, 1:k)$ by

$$\begin{aligned} \lambda_{k+1}(A_{k+1}) &\leqslant \lambda_k(A_k) \leqslant \lambda_k(A_{k+1}) \\ &\leqslant \cdots \leqslant \lambda_2(A_{k+1}) \leqslant \lambda_1(A_k) \leqslant \lambda_1(A_{k+1}) \end{aligned}$$

for $k = 1:n - 1$, showing that the eigenvalues of $A_k$ interlace those of $A_{k+1}$.

In 1962 Alfred Horn made a conjecture that a certain set of linear inequalities involving real numbers $\alpha_i$, $\beta_i$, and $\gamma_i$, $i = 1:n$, is necessary and sufficient for the existence of $n \times n$ Hermitian matrices $A$, $B$, and $C$ with eigenvalues the $\alpha_i$, $\beta_i$, and $\gamma_i$, respectively, such that $C = A + B$. The conjecture was open for many years but was finally proved to be true in papers published by Klyachko in 1998 and by Knutson and Tao in 1999, which exploited deep connections with algebraic geometry, representations of Lie groups, and quantum cohomology.

### 5.5 Solving the Non-Hermitian Eigenproblem

The simplest method for computing eigenvalues, the *power method*, computes just one: the largest in modulus. It comprises repeated multiplication of a starting vector $x$ by $A$. Since the resulting sequence is liable to overflow or underflow in floating-point arithmetic, one normalizes the vector after each iteration. Therefore one step of the power method has the form $x \leftarrow Ax$, $x \leftarrow \nu^{-1}x$, where $\nu = x_j$ with $|x_j| = \max_i |x_i|$. If $A$ has a unique eigenvalue $\lambda$ of largest modulus and the starting vector has a component in the direction of the corresponding eigenvector, then $\nu$ converges to $\lambda$ and $x$ converges to the corresponding eigenvector. The power method is most often applied to $(A - \mu I)^{-1}$, where $\mu$ is an approximation to an eigenvalue of interest. In this form it is known as *inverse iteration* and convergence is to the eigenvalue closest to $\mu$. We now turn to methods that compute all the eigenvalues.

Since similarities $X^{-1}AX$ preserve the eigenvalues and change the eigenvectors in a controlled way, carrying out a sequence of similarity transformations to reduce $A$ to a simpler form is a natural way to tackle the eigenproblem. Some early methods used nonunitary $X$, but such transformations are now avoided because of numerical instability when $X$ is ill-conditioned. Since the 1960s the focus has been on using unitary similarities to compute the *Schur decomposition* $A = QTQ^*$, where $Q$ is unitary and $T$ is upper triangular. The diagonal entries of $T$ are the eigenvalues of $A$, and they can be made to appear in any order by appropriate choice of $Q$. The first $k$ columns of $Q$ span an INVARIANT SUBSPACE [I.2 §20] corresponding to the eigenvalues $t_{11}, \dots, t_{kk}$. Eigenvectors can be obtained by solving triangular systems involving $T$.

For some matrices the Schur factor $T$ is diagonal; these are precisely the normal matrices defined in section 5.1. The *real Schur decomposition* contains only real matrices when $A$ is real: $A = QRQ^{\mathrm{T}}$, where $Q$ is orthogonal and $R$ is real upper quasitriangular, which means that $R$ is upper triangular except for $2 \times 2$ blocks on the diagonal corresponding to complex conjugate eigenvalues.

The standard algorithm for solving the non-Hermitian eigenproblem is the *QR algorithm*, which was proposed independently by John Francis and Vera Kublanovskaya in 1961. The matrix $A \in \mathbb{C}^{n \times n}$ is first unitarily reduced to upper Hessenberg form $H = U^*AU$ ($h_{ij} = 0$ for $i > j + 1$), with $U$ a product of Householder matrices. The QR iteration constructs a sequence of upper Hessenberg matrices beginning with $H_1 = H$ defined by $H_k - \mu_k I =: Q_k R_k$ (QR factorization, computed using Givens rotations), $H_{k+1} := R_k Q_k + \mu_k I$, where the $\mu_k$ are shifts chosen to accelerate the convergence of $H_k$ to upper triangular form. It is easy to check that $H_{k+1} = Q_k^* H_k Q_k$, so the QR iteration carries out a sequence of unitary similarity transformations.

Why the QR iteration works is not obvious but can be elegantly explained by analyzing the subspaces spanned by the columns of $Q_k$. To produce a practical and efficient algorithm various refinements of the iteration are needed, which include

- deflation, whereby when an element on the first subdiagonal of $H_k$ becomes small, that element is set to zero and the problem is split into two smaller problems that are solved independently,
- a double shift technique for real $A$ that allows two QR steps with complex conjugate shifts to be carried out entirely in real arithmetic and gives convergence to the real Schur form, and
- a multishift technique for including $m$ different shifts in a single QR iteration.

A proof of convergence is lacking for all current shift strategies. Implementations introduce a random shift when convergence appears to be stagnating. The QR algorithm works very well in practice and continues to be the method of choice for the non-Hermitian eigenproblem.

### 5.6 Solving the Hermitian Eigenproblem

The eigenvalue problem for Hermitian matrices is easier to solve than that for non-Hermitian matrices, and the range of available numerical methods is much wider.

To solve the complete Hermitian eigenproblem we need to compute the *spectral decomposition* $A = QDQ^*$, where $D = \mathrm{diag}(\lambda_i)$ contains the eigenvalues and the columns of the unitary matrix $Q$ are the corresponding eigenvectors. Many methods begin by unitary reduction to tridiagonal form $T = U^*AU$, where $t_{ij} = 0$ for $|i - j| > 1$ and the unitary matrix $U$ is constructed as a product of Householder matrices. The eigenvalue problem for $T$ is much simpler, though still nontrivial. The most widely used method is the QR algorithm, which has the same form as in the non-Hermitian case but with the upper Hessenberg $H_k$ replaced by the Hermitian tridiagonal $T_k$ and the shifts chosen to accelerate the convergence of $T_k$ to diagonal form. The

Hermitian QR algorithm with appropriate shifts has been proved to converge at a cubic rate.

Another method for solving the Hermitian tridiagonal eigenproblem is the *divide and conquer method*. This method decouples $T$ in the form

$$T = \begin{bmatrix} T_{11} & 0 \\ 0 & T_{22} \end{bmatrix} + \alpha v v^*,$$

where only the trailing diagonal element of $T_{11}$ and the leading diagonal element of $T_{22}$ differ from the corresponding elements of $T$ and hence the vector $v$ has only two nonzero elements. The eigensystems of $T_{11}$ and $T_{22}$ are found by applying the method recursively, yielding $T_{11} = Q_1 \Lambda_1 Q_1^*$ and $T_{22} = Q_2 \Lambda_2 Q_2^*$. Then

$$\begin{aligned} T &= \begin{bmatrix} Q_1\Lambda_1 Q_1^* & 0 \\ 0 & Q_2\Lambda_2 Q_2^* \end{bmatrix} + \alpha v v^* \\ &= \operatorname{diag}(Q_1, Q_2)(\operatorname{diag}(\Lambda_1, \Lambda_2) + \alpha \tilde{v}\tilde{v}^*)\operatorname{diag}(Q_1, Q_2)^*, \end{aligned}$$

where $\tilde{v} = \operatorname{diag}(Q_1, Q_2)^* v$. The eigensystem of a rank-1 perturbed diagonal matrix $D + \rho z z^*$ can be found by solving the *secular equation* obtained by equating the characteristic polynomial to zero:

$$f(\lambda) = 1 + \rho \sum_{j=1}^{n} \frac{|z_j|^2}{d_{jj} - \lambda} = 0.$$

Putting the pieces together yields the overall eigendecomposition.

Other methods are suitable for computing just a portion of the spectrum. Suppose we want to compute the $k$th smallest eigenvalue of $T$ and that we can somehow compute the integer $N(x)$ equal to the number of eigenvalues of $T$ that are less than or equal to $x$. Then we can apply the BISECTION METHOD [I.4 §2] to $N(x)$ to find the point where $N(x)$ jumps from $k-1$ to $k$. We can compute $N(x)$ by making use of the following result about the *inertia* of a Hermitian matrix, defined by $\operatorname{inertia}(A) = (\nu, \zeta, \pi)$, where $\nu$ is the number of negative eigenvalues, $\zeta$ is the number of zero eigenvalues, and $\pi$ is the number of positive eigenvalues.

**Theorem 2 (Sylvester's inertia theorem).** *If $A$ is Hermitian and $M$ is nonsingular, then* $\operatorname{inertia}(A) = \operatorname{inertia}(M^*AM)$.

Sylvester's inertia theorem says that the number of negative, zero, and positive eigenvalues does not change under *congruence transformations*. By using GE we can factorize[2] $T - xI = LDL^*$, where $D$ is diagonal and $L$ is unit lower bidiagonal (a bidiagonal matrix is one that is both triangular and tridiagonal). Then $\operatorname{inertia}(T - xI) = \operatorname{inertia}(D)$, so the number of negative or zero diagonal elements of $D$ equals the number of eigenvalues of $T - xI$ less than or equal to 0, which is the number of eigenvalues of $T$ less than or equal to $x$, that is, $N(x)$. The LDL* factors of a tridiagonal matrix can be computed in $O(n)$ flops, so this bisection process is efficient. An alternative approach can be built by using properties of *Sturm sequences*, which are sequences comprising the characteristic polynomials of leading principal submatrices of $T - \lambda I$.

2. The factorization may not exist, but if it does not we can simply perturb $T$ slightly and try again without any loss of numerical stability.

### 5.7 Computing the SVD

For a rectangular matrix $A \in \mathbb{C}^{m\times n}$ the eigenvalues of the Hermitian matrix $\left[\begin{smallmatrix} 0 & A \\ A^* & 0 \end{smallmatrix}\right]$ of dimension $m+n$ are plus and minus the nonzero singular values of $A$ along with $m + n - 2\min(m, n)$ zeros. Hence the SVD can be computed via the eigendecomposition of this larger matrix. However, this would be inefficient, and instead one uses algorithms that work directly on $A$ and are analogues of the algorithms for Hermitian matrices. The standard approach is to reduce $A$ to bidiagonal form $B$ by Householder transformations applied on the left and the right and then to apply an adaptation of the QR algorithm that works on the bidiagonal factor (and implicitly applies the QR algorithm to the tridiagonal matrix $B^*B$).

### 5.8 Generalized Eigenproblems

The *generalized eigenvalue problem* (GEP) $Ax = \lambda Bx$, with $A, B \in \mathbb{C}^{n\times n}$, can be converted into a standard eigenvalue problem if $B$ (say) is nonsingular: $B^{-1}Ax = \lambda x$. However, such a transformation is inadvisable numerically unless $B$ is very well-conditioned. If $A$ and $B$ have a common null vector $z$, the problem takes on a different character because then $(A - \lambda B)z = 0$ for *any* $\lambda$; such a problem is called *singular*. We will assume that the problem is *regular*, so that $\det(A - \lambda B) \not\equiv 0$. The linear polynomial $A - \lambda B$ is sometimes called a *pencil*.

It is convenient to write $\lambda = \alpha/\beta$, where $\alpha$ and $\beta$ are not both zero, and rephrase the problem in the more symmetric form $\beta Ax = \alpha Bx$. If $x$ is a nonzero vector such that $Bx = 0$, then, since the problem is assumed to be regular, $Ax \neq 0$ and so $\beta = 0$. This means that $\lambda = \infty$ is an eigenvalue. Infinite eigenvalues may seem a strange concept, but in fact they are no different in most respects to finite eigenvalues.

An important special case is the *definite generalized eigenvalue problem*, in which $A$ and $B$ are Hermitian and $B$ (say) is positive-definite. If $B = R^*R$ is a Cholesky factorization, then $Ax = \lambda Bx$ can be rewritten as $R^{-*}AR^{-1} \cdot Rx = \lambda Rx$, which is a standard eigenproblem for the Hermitian matrix $C = R^{-*}AR^{-1}$. This argument shows that the eigenvalues of a definite problem are all real. Definite generalized eigenvalue problems arise in many physical situations where an energy-minimization principle is at work, such as in problems in engineering and physics.

A generalization of the QR algorithm called the *QZ algorithm* computes a generalization to two matrices of the Schur decomposition: $Q^*AZ = T$, $Q^*BZ = S$, where $Q$ and $Z$ are unitary and $T$ and $S$ are upper triangular. The generalized Schur decomposition yields the eigenvalues as the ratios $t_{ii}/s_{ii}$ and enables eigenvectors to be computed by substitution.

The *quadratic eigenvalue problem* (QEP) $Q(\lambda)x = (\lambda^2 A_2 + \lambda A_1 + A_0)x = 0$, where $A_i \in \mathbb{C}^{n\times n}$, $i = 0{:}2$, arises most commonly in the dynamic analysis of structures when the finite-element method is used to discretize the original PDE into a system of second-order ODEs $A_2\ddot{q}(t) + A_1\dot{q}(t) + A_0 q(t) = f(t)$. Here, the $A_i$ are usually Hermitian (though $A_1$ is skew-Hermitian in gyroscopic systems) and positive (semi)definite. Analogously to the GEP, the QEP is said to be regular if $\det(Q(\lambda)) \not\equiv 0$. The quadratic problem differs fundamentally from the linear GEP because a regular problem has $2n$ eigenvalues, which are the roots of $\det(Q(\lambda)) = 0$, but at most $n$ linearly independent eigenvectors, and a vector may be an eigenvector for two different eigenvalues. For example, the QEP with

$$Q(\lambda) = \lambda^2 I + \lambda \begin{bmatrix} -1 & -6 \\ 2 & -9 \end{bmatrix} + \begin{bmatrix} 0 & 12 \\ -2 & 14 \end{bmatrix}$$

has eigenvalues 1, 2, 3, and 4, with eigenvectors $\left[\begin{smallmatrix}1\\0\end{smallmatrix}\right]$, $\left[\begin{smallmatrix}0\\1\end{smallmatrix}\right]$, $\left[\begin{smallmatrix}1\\1\end{smallmatrix}\right]$, and $\left[\begin{smallmatrix}1\\1\end{smallmatrix}\right]$, respectively. Moreover, there is no Schur form for three or more matrices; that is, we cannot in general find unitary matrices $U$ and $V$ such that $U^*A_iV$ is triangular for $i = 0{:}2$.

Associated with the QEP is the matrix $Q(X) = A_2X^2 + A_1X + A_0$, with $X \in \mathbb{C}^{n\times n}$. From the relation

$$\begin{aligned} Q(\lambda) - Q(X) &= A_2(\lambda^2 I - X^2) + A_1(\lambda I - X) \\ &= (\lambda A_2 + A_2X + A_1)(\lambda I - X), \end{aligned}$$

it is clear that if we can find a matrix $X$ such that $Q(X) = 0$, known as a *solvent*, then we have reduced the QEP to finding the eigenvalues of $X$ and solving one $n \times n$ GEP. For the $2 \times 2$ $Q$ above there are five solvents, one of which is $\left[\begin{smallmatrix}3&0\\1&2\end{smallmatrix}\right]$. The existence and enumeration of solvents is nontrivial and leads into the theory of *matrix polynomials*. In general, matrix polynomials are matrices of the form $\sum_{i=0}^{k} \lambda^i A_i$ whose elements are polynomials in a complex variable; an older term for such matrices is $\lambda$*-matrices*.

The standard approach for numerical solution of the QEP mimics the conversion of the scalar polynomial root problem into a matrix eigenproblem described in section 5.3. From the relation

$$\begin{aligned} L(\lambda)z \equiv \left(\begin{bmatrix} A_1 & A_0 \\ I & 0 \end{bmatrix} + \lambda \begin{bmatrix} A_2 & 0 \\ 0 & -I \end{bmatrix}\right)\begin{bmatrix} \lambda x \\ x \end{bmatrix} \\ = \begin{bmatrix} Q(\lambda)x \\ 0 \end{bmatrix}, \end{aligned}$$

we see that the eigenvalues of the quadratic $Q$ are the eigenvalues of the $2n\times 2n$ linear polynomial $L(\lambda)$. This is an example of an exact linearization process—thanks to the hidden $\lambda$ in the eigenvector! The eigenvalues of $L$ can be found using the QZ algorithm. The eigenvectors of $L$ have the form $z = \left[\begin{smallmatrix}\lambda x\\x\end{smallmatrix}\right]$, where $x$ is an eigenvector of $Q$, and so $x$ can be obtained from either the first $n$ (if $\lambda \neq 0$) or the last $n$ components of $z$.

## 6 Sparse Linear Systems

For linear systems coming from discretization of differential equations it is common that $A$ is *banded*, that is, the nonzero elements lie in a band about the main diagonal. An extreme case is a *tridiagonal matrix*, of which the classic example is the second-difference matrix, illustrated for $n = 4$ by

$$A = \begin{bmatrix} -2 & 1 & 0 & 0 \\ 1 & -2 & 1 & 0 \\ 0 & 1 & -2 & 1 \\ 0 & 0 & 1 & -2 \end{bmatrix}, \quad A^{-1} = -\frac{1}{5}\begin{bmatrix} 4 & 3 & 2 & 1 \\ 3 & 6 & 4 & 2 \\ 2 & 4 & 6 & 3 \\ 1 & 2 & 3 & 4 \end{bmatrix}.$$

This matrix corresponds to a centered FINITE-DIFFERENCE APPROXIMATION [II.11] to a second derivative: $f''(x) \approx (f(x+h) - 2f(x) + f(x-h))/h^2$. Note that $A^{-1}$ is a full matrix. For banded matrices, GE produces banded $LU$ factors and its computational cost is proportional to $n$ times the square of the bandwidth.

A matrix is *sparse* if advantage can be taken of the zero entries because of either their number or their distribution. A banded matrix is a special case of a sparse matrix. Sparse matrices are stored on a computer not as a square array but in a special format that records only the nonzeros and their location in the matrix. This can be done with three vectors: one to store the nonzero entries and the other two to define the row and column indices of the elements in the first vector.

Sparse matrices help to explain the tenet: *never solve a linear system $Ax = b$ by computing $x = A^{-1} \times b$*. The reasons for eschewing $A^{-1}$ are threefold.

- Computing $A^{-1}$ requires three times as many flops as solving $Ax = b$ by GE with partial pivoting.
- GE with partial pivoting is backward stable for solving $Ax = b$ (see section 8) but solution via $A^{-1}$ is not.
- If $A$ is sparse, $A^{-1}$ is generally dense and so requires much more storage than GE with partial pivoting.

When GE is applied to a sparse matrix, *fill-in* occurs when the row operations cause a zero entry to become nonzero during the elimination. To minimize the storage and the computational cost, fill-in must be avoided as much as possible. This can be done by employing row and column interchanges to choose a suitable pivot from the active submatrix. The first such strategy was introduced by Markowitz in 1957. At the $k$th stage, with $c_j^{(k)}$ denoting the number of nonzeros in rows $k$ to $n$ of column $j$ and $r_i^{(k)}$ the number of nonzeros in columns $k$ to $n$ of row $i$, the Markowitz strategy finds the pair $(r, s)$ that minimizes $(r_i^{(k)} - 1)(c_j^{(k)} - 1)$ over all nonzero potential pivots $a_{ij}^{(k)}$ and then takes $a_{rs}^{(k)}$ as the pivot. The quantity being minimized is a bound on the fill-in. In practice, the potential pivots must be restricted to those not too much smaller in magnitude than the partial pivot in order to preserve numerical stability. The result of GE with Markowitz pivoting is a factorization $PAQ = LU$, where $P$ and $Q$ are permutation matrices.

The analogue of the Markowitz strategy for Hermitian positive-definite matrices chooses a diagonal entry $a_{ii}^{(k)}$ as the pivot, where $r_i^{(k)}$ is minimal. This is the *minimum-degree algorithm*, which has been very successful in practice. Figure 3 shows in the first row a sparse and banded symmetric positive-definite matrix $A$ of dimension 225 followed to the right by its Cholesky factor. The Cholesky factor has many more nonzeros than $A$. The second row shows the matrix $PAP^{\mathrm{T}}$ produced by an approximate minimum-degree ordering (produced by the MATLAB `symamd` function) and its Cholesky factor. We can see that the permutations have destroyed the band structure but have greatly reduced the fill-in, producing a much sparser Cholesky factor.

As an alternative to GE for solving sparse linear systems one can apply iterative methods, described in section 9; for sufficiently large problems these are the only feasible methods.

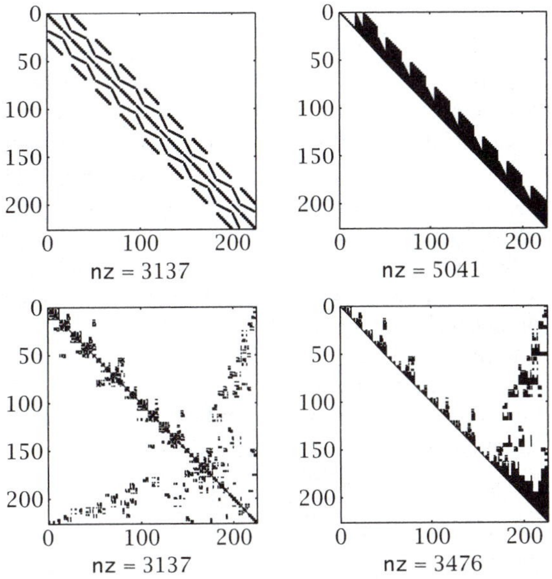


**Figure 3** Sparsity plots of a symmetric positive-definite matrix (left) and its Cholesky factor (right) for original matrix (first row) and reordered matrix (second row). `nz` is the number of nonzeros.

## 7 Overdetermined and Underdetermined Systems

Linear systems $Ax = b$ with a rectangular matrix $A \in \mathbb{C}^{m \times n}$ are very common. They break into two categories: *overdetermined systems*, with more equations than unknowns ($m > n$), and *underdetermined systems*, with fewer equations than unknowns ($m < n$). Since in general there is no solution when $m > n$ and there are many solutions when $m < n$, extra conditions must be imposed for the problems to be well defined. These usually involve norms, and different choices of norms are possible. We will restrict our discussion mainly to the 2-norm, which is the most important case, but other choices are also of practical interest.

### 7.1 The Linear Least-Squares Problem

When $m > n$ the residual $r = b - Ax$ cannot in general be made zero, so we try to minimize its norm. The most common choice of norm is the 2-norm, which gives the *linear least-squares problem*

$$\min_{x \in \mathbb{C}^n} \|b - Ax\|_2. \tag{6}$$

This choice can be motivated by statistical considerations (the Gauss–Markov theorem) or by the fact that the square of the 2-norm is differentiable, which makes the problem explicitly solvable. Indeed, by setting the gradient of $\|b - Ax\|_2^2$ to zero we obtain the *normal equations* $A^*Ax = A^*b$, which any solution of the least-squares problem must satisfy. If $A$ has full rank then $A^*A$ is positive-definite and so there is a unique solution, which can be computed by solving the normal equations using Cholesky factorization. For reasons of numerical stability it is preferable to use a QR factorization: if $A = Q\left[\begin{smallmatrix} R_1 \\ 0 \end{smallmatrix}\right]$ then the normal equations reduce to the triangular system $R_1 x = c$, where $c$ is the first $n$ components of $Q^*b$.

When $A$ is rank deficient there are many least-squares solutions, which vary widely in norm. A natural choice is one of minimal 2-norm, and in fact there is a unique minimal 2-norm solution, $x_{\text{LS}}$, given by

$$x_{\text{LS}} = \sum_{i=1}^{r} (u_i^* b/\sigma_i) v_i,$$

where

$$A = U\Sigma V^*, \quad U = [u_1, \dots, u_m], \ V = [v_1, \dots, v_n], \quad (7)$$

is an SVD and $r = \text{rank}(A)$. The use of this formula in practice is not straightforward because a matrix stored in floating-point arithmetic will rarely have any zero singular values. Therefore $r$ must be chosen by designating which singular values can be regarded as negligible, and this choice should take account of the accuracy with which the elements of $A$ are known.

Another choice of least-squares solution in the rank-deficient case is a *basic solution*: one with at most $r$ nonzeros. Such a solution can be computed via the QR factorization with column pivoting.

### 7.2 Underdetermined Systems

When $m < n$ and $A$ has full rank, there are infinitely many solutions to $Ax = b$ and again it is natural to seek one of minimal 2-norm. There is a unique such solution $x_{\text{LS}} = A^*(AA^*)^{-1}b$, and it is best computed via a QR factorization, this time of $A^*$. A basic solution, with $m$ nonzeros, can alternatively be computed. As a simple example, consider the problem "find two numbers whose sum is 5," that is, solve $[1\ 1]\left[\begin{smallmatrix} x_1 \\ x_2 \end{smallmatrix}\right] = 5$. A basic solution is $[5\ 0]^{\mathrm{T}}$, while the minimal 2-norm solution is $[5/2\ 5/2]^{\mathrm{T}}$. Minimal 1-norm solutions to underdetermined systems are important in COMPRESSED SENSING [VII.10].

### 7.3 Pseudoinverse

The analysis in the previous two subsections can be unified in a very elegant way by making use of the *Moore–Penrose pseudoinverse* $A^+$ of $A \in \mathbb{C}^{m\times n}$, which is defined as the unique $X \in \mathbb{C}^{n\times m}$ satisfying the Moore–Penrose conditions

$$AXA = A, \qquad XAX = X,$$
$$(AX)^* = AX, \qquad (XA)^* = XA.$$

(It is certainly not obvious that these equations have a unique solution.) In the case where $A$ is square and nonsingular, it is easily seen that $A^+$ is just $A^{-1}$. Moreover, if $\text{rank}(A) = n$ then $A^+ = (A^*A)^{-1}A^*$, while if $\text{rank}(A) = m$ then $A^+ = A^*(AA^*)^{-1}$. In terms of the SVD (7),

$$A^+ = V \operatorname{diag}(\sigma_1^{-1}, \dots, \sigma_r^{-1}, 0, \dots, 0)U^*,$$

where $r = \text{rank}(A)$. The formula $x_{\text{LS}} = A^+b$ holds for all $m$ and $n$, so the pseudoinverse yields the minimal 2-norm solution to both the least-squares (overdetermined) problem $Ax = b$ and an underdetermined system $Ax = b$. The pseudoinverse has many interesting properties, including $(A^+)^+ = A$, but it is not always true that $(AB)^+ = B^+A^+$.

Although the pseudoinverse is a very useful theoretical tool, it is rarely necessary to compute it explicitly (just as for its special case the matrix inverse).

The pseudoinverse is just one of many ways of generalizing the notion of inverse to rectangular matrices, but it is the right one for minimum 2-norm solutions to linear systems. Other generalized inverses can be obtained by requiring only a subset of the four Moore–Penrose conditions to hold.

## 8 Numerical Considerations

Prior to the introduction of the first digital computers in the 1940s, numerical computations were carried out by humans, sometimes with the aid of mechanical calculators. The human involvement in a sequence of calculations meant that potentially dangerous events such as dividing by a tiny number or subtracting two numbers that agree to almost all their significant digits could be observed, their effect monitored, and possible corrective action taken—such as temporarily increasing the precision of the calculations. On the very early computers intermediate results were observed on a cathode ray tube monitor, but this became impossible as problem sizes increased (along with available computing power). Fears were raised in the 1940s that algorithms such as GE would suffer exponential growth of errors as the problem dimension increased, due to the rapidly increasing number of arithmetic operations, each having its associated ROUNDING ERROR [II.13].

These fears were particularly concerning given that the error growth might be unseen and unsuspected.

The subject of *rounding error analysis* grew out of the need to understand the effect of rounding errors on algorithms. The person who did the most to develop the subject was James Wilkinson, whose influential papers and 1961 and 1965 books showed how BACKWARD ERROR ANALYSIS [I.2 §23] can be used to obtain deep insights into numerical stability. We will discuss just two particular examples.

Wilkinson showed that when a nonsingular linear system $Ax = b$ is solved by GE in floating-point arithmetic the computed solution $\hat{x}$ satisfies

$$(A + \Delta A)\hat{x} = b, \qquad \|\Delta A\|_\infty \leqslant p(n)\rho_n u\|A\|_\infty.$$

Here, $p(n)$ is a cubic polynomial, the *growth factor*

$$\rho_n = \frac{\max_{i,j,k} |a_{ij}^{(k)}|}{\max_{i,j} |a_{ij}|} \geqslant 1$$

measures the growth of elements during the elimination, and $u$ is the UNIT ROUNDOFF [II.13]. This is a *backward stability result*: it says that the computed solution $\hat{x}$ is the exact solution of a perturbed system. Ideally, we would like $\|\Delta A\|_\infty \leqslant u\|A\|_\infty$, which reflects the uncertainty caused by converting the elements of $A$ to floating-point numbers. The polynomial term $p(n)$ is pessimistic and might be more realistically replaced by its square root. The danger term is the growth factor $\rho_n$, and the conclusion from Wilkinson's analysis is that a pivoting strategy should aim to keep $\rho_n$ small. If no pivoting is done, $\rho_n$ can be arbitrarily large (e.g., for $A = \left[\begin{smallmatrix} \varepsilon & 1 \\ 1 & 1 \end{smallmatrix}\right]$ with $0 < \varepsilon \ll 1$, $\rho_n \approx 1/\varepsilon$). For partial pivoting, however, it can be shown that $\rho_n \leqslant 2^{n-1}$ and that this bound is attainable. In practice, $\rho_n$ is almost always of modest size for partial pivoting ($\rho_n \leqslant 50$, say); why this should be so remains one of the great mysteries of numerical analysis!

One of the benefits of Wilkinson's backward error analysis is that it enables us to identify classes of matrices for which pivoting is not necessary, that is, for which the LU factorization $A = LU$ exists and $\rho_n$ is nicely bounded. One such class is the matrices that are diagonally dominant by either rows or columns, for which $\rho_n \leqslant 2$.

The potential instability of GE can be attributed to the fact that $A$ is premultiplied by a sequence of nonunitary transformations, any of which can be ill-conditioned. Many algorithms, including Householder QR factorization and the QR algorithm for eigenvalues, use exclusively unitary transformations. Such algorithms are usually (but not always) backward stable, essentially because unitary transformations do not magnify errors: $\|UAV\| = \|A\|$ for any unitary $U$ and $V$ for the 2-norm and the Frobenius norm. As an example, the QR algorithm applied to $A \in \mathbb{C}^{n\times n}$ produces a computed upper triangular matrix $\hat{T}$ such that

$$\tilde{Q}^*(A + \Delta A)\tilde{Q} = \hat{T}, \qquad \|\Delta A\|_F \leqslant p(n)u\|A\|_F,$$

where $\tilde{Q}$ is some exactly unitary matrix and $p(n)$ is a cubic polynomial. The computed Schur factor $\hat{Q}$ is not necessarily close to $\tilde{Q}$—which in turn is not necessarily close to the exact $Q$!—but it is close to being orthogonal: $\|\hat{Q}^*\hat{Q} - I\|_F \leqslant p(n)u$. This distinction between the different $Q$ matrices is an indication of the subtleties of backward error analysis. For some problems it is not clear exactly what form of backward error result it is possible to prove while obtaining useful bounds. However, the purpose of a backward error analysis is always the same: either to show that an algorithm behaves in a numerically stable way or to shed light on how it might fail to do so and to indicate what quantities should be monitored in order to identify potential instability.

## 9 Iterative Methods

In numerical linear algebra, methods can broadly be divided into two classes: direct and iterative. Direct methods, such as GE, solve a problem in a fixed number of arithmetic operations or a variable number that in practice is fairly constant, as for the QR algorithm for eigenvalues. Iterative methods are infinite processes that must be truncated at some point when the approximation they provide is "good enough." Usually, iterative methods do not transform the matrix in question and access it only through matrix-vector products; this makes them particularly attractive for large, sparse matrices, where applying a direct method may not be practical.

We have already seen in section 5.5 a simple iterative method for the eigenvalue problem: the power method. The *stationary iterative methods* are an important class of iterative methods for solving a nonsingular linear system $Ax = b$. These methods are best described in terms of a *splitting*

$$A = M - N,$$

with $M$ nonsingular. The system $Ax = b$ can be rewritten $Mx = Nx + b$, which suggests constructing a sequence $\{x^{(k)}\}$ from a given starting vector $x^{(0)}$ via

$$Mx^{(k+1)} = Nx^{(k)} + b. \qquad (8)$$

Different choices of $M$ and $N$ yield different methods. The aim is to choose $M$ in such a way that it is inexpensive to solve (8) while $M$ is a good enough approximation to $A$ that convergence is fast. It is easy to analyze convergence. Denote by $e^{(k)} = x^{(k)} - x$ the error in the $k$th iterate. Subtracting $Mx = Nx + b$ from (8) gives $M(x^{(k+1)} - x) = N(x^{(k)} - x)$, so

$$e^{(k+1)} = M^{-1}Ne^{(k)} = \cdots = (M^{-1}N)^{k+1}e^{(0)}. \quad (9)$$

If $\rho(M^{-1}N) < 1$ then $(M^{-1}N)^k \to 0$ as $k \to \infty$ (see JORDAN CANONICAL FORM [II.22]) and so $x^{(k)}$ converges to $x$ at a linear rate. In practice, for convergence in a reasonable number of iterations we need $\rho(M^{-1}N)$ to be sufficiently less than 1 and the powers of $M^{-1}N$ should not grow too large initially before eventually decaying; in other words, $M^{-1}N$ must not be too nonnormal.

Three standard choices of splitting are as follows, where $D = \mathrm{diag}(A)$ and $L$ and $U$ denote the strictly lower and strictly upper triangular parts of $A$, respectively.

- $M = D$, $N = -(L + U)$: *Jacobi iteration* (illustrated in METHODS OF SOLUTION [I.3 §6]).
- $M = D + L$, $N = -U$: *Gauss-Seidel iteration.*
- $M = (1/\omega)D + L$, $N = ((1 - \omega)/\omega)D - U$, where $\omega \in (0, 2)$ is a relaxation parameter: *successive overrelaxation (SOR) iteration.*

Sufficient conditions for convergence are that $A$ is strictly diagonally dominant by rows for the Jacobi iteration and that $A$ is symmetric positive-definite for the Gauss-Seidel iteration. How to choose $\omega$ so that $\rho(M^{-1}N|_\omega)$ is minimized for the SOR iteration was elucidated in the landmark 1950 Ph.D. thesis of David Young.

The Google PAGERANK ALGORITHM [VI.9], which underlies Google's ordering of search results, can be interpreted as an application of the Jacobi iteration to a certain linear system involving the ADJACENCY MATRIX [II.16] of the graph corresponding to the whole World Wide Web. However, the most common use of stationary iterative methods is as preconditioners within other iterative methods.

The aim of *preconditioning* is to convert a given linear system $Ax = b$ into one that can be solved more cheaply by a particular iterative method. The basic idea is to use a nonsingular matrix $W$ to transform the system to $(W^{-1}A)x = W^{-1}b$ in such a way that (a) the preconditioned system can be solved in fewer iterations than the original system and (b) matrix-vector multiplications with $W^{-1}A$ (which require the solution of a linear system with coefficient matrix $W$) are not significantly more expensive than matrix-vector multiplications with $A$. In general, this is a difficult or impossible task, but in many applications the matrix $A$ has structure that can be exploited. For example, many elliptic PDE problems lead to a positive-definite matrix $A$ of the form

$$A = \begin{bmatrix} M_1 & F \\ F^T & M_2 \end{bmatrix},$$

where $M_1z = d_1$ and $M_2z = d_2$ are easy to solve. In this case it is natural to take $W = \mathrm{diag}(M_1, M_2)$ as the preconditioner. When $A$ is Hermitian positive-definite the preconditioned system is written in a way that preserves the structure. For example, for the Jacobi preconditioner, $D = \mathrm{diag}(A)$, the preconditioned system would be written $D^{-1/2}AD^{-1/2}\tilde{x} = \tilde{b}$, where $\tilde{x} = D^{1/2}x$ and $\tilde{b} = D^{-1/2}b$. Here, the matrix $D^{-1/2}AD^{-1/2}$ has unit diagonal and off-diagonal elements lying between $-1$ and 1.

The most powerful iterative methods for linear systems $Ax = b$ are the Krylov methods. In these methods each iterate $x^{(k)}$ is chosen from the shifted subspace $x^{(0)} + \mathcal{K}_k(A, r^{(0)})$, where

$$\mathcal{K}_k(A, r^{(0)}) = \mathrm{span}\{r^{(0)}, Ar^{(0)}, \dots, A^{k-1}r^{(0)}\}$$

is a KRYLOV SUBSPACE [II.23] of dimension $k$, with $r^{(k)} = b - Ax^{(k)}$. Different strategies for choosing approximations from within the Krylov subspaces yield different methods. For example, the *conjugate gradient method* (CG, for Hermitian positive-definite $A$) and the *full orthogonalization method* (FOM, for general $A$) make the residual $r^{(k)}$ orthogonal to the Krylov subspace $\mathcal{K}_k(A, r^{(0)})$, while the *minimal residual method* (MINRES, for Hermitian $A$) and the *generalized minimal residual method* (GMRES, for general $A$) minimize the 2-norm of the residual over all vectors in the Krylov subspace. How to compute the vectors defined in these ways is nontrivial. It turns out that CG can be implemented with a recurrence requiring just one matrix-vector multiplication and three inner products per iteration, and MINRES is just a little more expensive. GMRES, being applicable to non-Hermitian matrices, is significantly more expensive, and it is also much harder to analyze its convergence behavior. For general matrices there are alternatives to GMRES that employ short recurrences. We mention just *BiCGSTAB*, which has the distinction that the 1992 paper by Henk van der Vorst that introduced it was the most-cited paper in mathematics of the 1990s.

Theoretically, Krylov methods converge in at most $n$ iterations for a system of dimension $n$. However, in practical computation rounding errors intervene and the methods behave as truly iterative methods not having finite termination. Since $n$ is potentially huge, a Krylov method would not be used unless a good approximate solution was obtained in many fewer than $n$ iterations, and preconditioning plays a crucial role here. Available error bounds for a method help to guide the choice of preconditioner, but care is needed in interpreting the bounds. To illustrate this, consider the CG method for $Ax = b$, where $A$ is Hermitian positive-definite. In the $A$-norm, $\|z\|_A = (z^*Az)^{1/2}$, the error on the $k$th step satisfies

$$\|x - x^{(k)}\|_A \leqslant 2\|x - x^{(0)}\|_A \left(\frac{\kappa_2(A)^{1/2} - 1}{\kappa_2(A)^{1/2} + 1}\right)^k,$$

where $\kappa_2(A) = \|A\|_2\|A^{-1}\|_2$. If we can precondition $A$ so that its 2-norm condition number is very close to 1, then fast convergence is guaranteed. However, another result says that if $A$ has $k$ distinct eigenvalues then CG converges in at most $k$ iterations. A better approach might therefore be to choose the preconditioner so that the eigenvalues of the preconditioned matrix are clustered into a small number of groups.

Another important class of iterative methods is MULTIGRID METHODS [IV.13 §3], which work on a hierarchy of grids that come from a discretization of an underlying PDE (geometric multigrid) or are constructed artificially from a given matrix (algebraic multigrid).

An important practical issue is how to terminate an iteration. Popular approaches are to stop when the residual $r^{(k)} = b - Ax^{(k)}$ (suitably scaled) is small or when an estimate of the error $x - x^{(k)}$ is small. Complicating factors include the fact that the preconditioner can change the norm and a possible desire to match the error in the iterations with the discretization error in the PDE from which the linear system might have come (as there is no point solving the system to greater accuracy than the data warrants). Research in recent years has led to good understanding of these issues.

The ideas of Krylov methods and preconditioners can be applied to problems other than linear systems. A popular Krylov method for solving the least-squares problem (6) is *LSQR*, which is mathematically equivalent to applying CG to the normal equations. In large-scale eigenvalue problems only a few eigenpairs are usually required. A number of methods project the original matrix onto a Krylov subspace and then solve a smaller eigenvalue problem. These include the *Lanczos method* for Hermitian matrices and the *Arnoldi method* for general matrices. Also of much current research interest are *rational Krylov methods* based on RATIONAL GENERALIZATIONS OF KRYLOV SUBSPACES [II.23].

## 10 Nonnormality and Pseudospectra

Normal matrices $A \in \mathbb{C}^{n\times n}$ (defined in section 5.1) have the property that they are unitarily diagonalizable: $A = QDQ^*$ for some unitary $Q$ and diagonal $D = \operatorname{diag}(\lambda_i)$ containing the eigenvalues on its diagonal. In many respects normal matrices have very predictable behavior. For example, $\|A^k\|_2 = \rho(A)^k$ and $\|e^{tA}\|_2 = e^{\alpha(tA)}$, where the *spectral abscissa* $\alpha(tA)$ is the largest real part of any eigenvalue of $tA$. However, matrices that arise in practice are often very nonnormal. The adjective "very" can be quantified in various ways, of which one is the Frobenius norm of the strictly upper triangular part of the upper triangular matrix $T$ in the Schur decomposition $A = QTQ^*$. For example, the matrix $\left[\begin{smallmatrix} t_{11} & \theta \\ 0 & t_{22}\end{smallmatrix}\right]$ is nonnormal for $\theta \neq 0$ and grows increasingly nonnormal as $|\theta|$ increases.

Consider the moderately nonnormal matrix

$$A = \begin{bmatrix} -0.97 & 25 \\ 0 & -0.3 \end{bmatrix}. \tag{10}$$

While the powers of $A$ ultimately decay to zero, since $\rho(A) = 0.97 < 1$, we see from figure 4 that initially they increase in norm. Likewise, since $\alpha(A) = -0.3 < 0$ the norm $\|e^{tA}\|_2$ tends to zero as $t \to \infty$, but figure 4 shows that there is an initial hump in the plot. In stationary iterations the hump caused by a nonnormal iteration matrix $M^{-1}N$ can delay convergence, as is clear from (9). In finite-precision arithmetic it can even happen that, for a sufficiently large hump, rounding errors cause the norms of the powers to plateau at the hump level and never actually converge to zero.

How can we predict the shape of the curves in figure 4? Let us concentrate on $\|A^k\|_2$. Initially it grows like $\|A\|_2^k$ and ultimately it decays like $\rho(A)^k$, the decay rate following from (4). The height of the hump is related to pseudospectra, which have been popularized by Nick Trefethen.

The $\varepsilon$-*pseudospectrum* of $A \in \mathbb{C}^{n\times n}$ is defined, for a given $\varepsilon > 0$, to be the set

$$\Lambda_\varepsilon(A) = \{z \in \mathbb{C}\colon z \text{ is an eigenvalue of } A + E \text{ for some } E \text{ with } \|E\|_2 < \varepsilon\}, \tag{11}$$

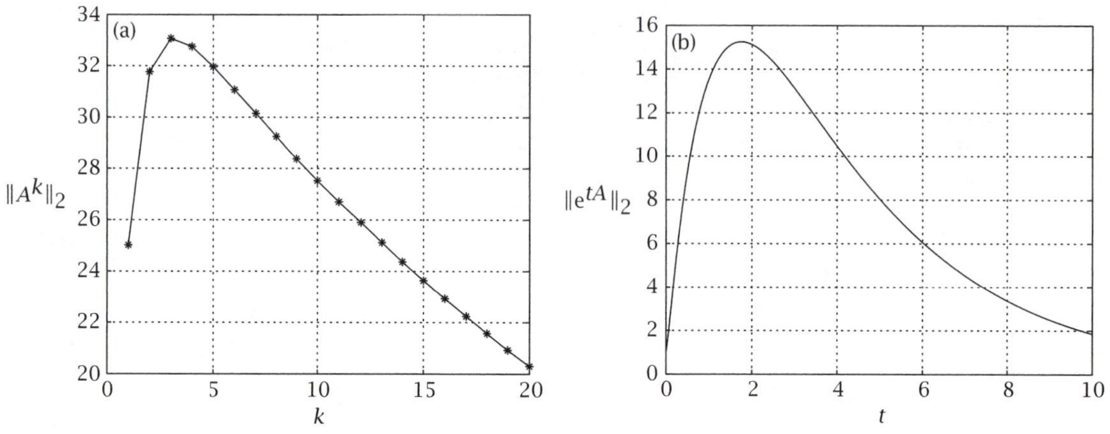


**Figure 4** 2-norms of (a) powers and (b) exponentials of $2 \times 2$ matrix $A$ in (10).

and it can also be represented, in terms of the *resolvent* $(zI - A)^{-1}$, as

$$\Lambda_\varepsilon(A) = \{z \in \mathbb{C} \colon \|(zI - A)^{-1}\|_2 > \varepsilon^{-1}\}.$$

The 0.001-pseudospectrum, for example, tells us the uncertainty in the eigenvalues of $A$ if the elements are known only to three decimal places. Pseudospectra provide much insight into the effects of nonnormality of matrices and (with an appropriate extension of the definition) linear operators. For nonnormal matrices the pseudospectra are much bigger than a perturbation of the spectrum by $\varepsilon$. It can be shown that for any $\varepsilon > 0$,

$$\sup_{k \geqslant 0} \|A^k\| \geqslant \frac{\rho_\varepsilon(A) - 1}{\varepsilon}, \qquad \|A^k\| \leqslant \frac{\rho_\varepsilon(A)^{k+1}}{\varepsilon},$$

where the *pseudospectral radius* $\rho_\varepsilon(A) = \max\{|\lambda| \colon \lambda \in \Lambda_\varepsilon(A)\}$. For $A$ in (10) and $\varepsilon = 10^{-2}$, these inequalities give an upper bound of 230 for $\|A^3\|$ and a lower bound of 23 for $\sup_{k\geqslant 0} \|A^k\|$, and figure 5 plots the corresponding $\varepsilon$-pseudospectrum.

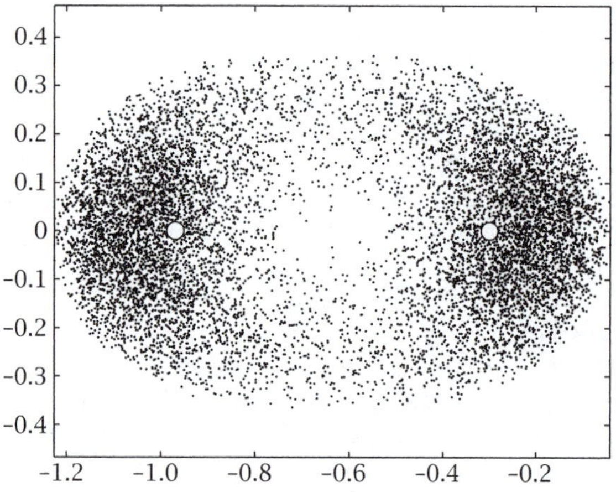


**Figure 5** Approximation to $10^{-2}$-pseudospectrum of $A$ in (10) comprising eigenvalues of 5000 randomly perturbed matrices $A + E$ in (11). The eigenvalues of $A$ are marked by white circles.

## 11 Structured Matrices

In a wide variety of applications the matrices have a special structure. The matrix elements might form a pattern, as for TOEPLITZ OR HAMILTONIAN MATRICES [I.2 §18], the matrix may satisfy a nonlinear equation such as $A^*\Sigma A = \Sigma$, where $\Sigma = \operatorname{diag}(\pm 1)$, which yields the *pseudo-unitary matrices* $A$, or the submatrices may satisfy certain rank conditions (as for *quasiseparable matrices*). We discuss here two of the oldest and most-studied classes of structured matrices, both of which were historically important in the analysis of iterative methods for linear systems arising from the discretization of differential equations.

### 11.1 Nonnegative Matrices

A *nonnegative matrix* is a real matrix all of whose entries are nonnegative. A number of important classes of matrices are subsets of the nonnegative matrices. These include adjacency matrices, STOCHASTIC MATRICES [II.25], and Leslie matrices (used in population modeling). Nonnegative matrices have a large body of theory, which originates with Perron in 1907 and Frobenius in 1908.

To state the celebrated Perron–Frobenius theorem we need the definition that $A \in \mathbb{R}^{n\times n}$ with $n \geqslant 2$ is *reducible* if there is a permutation matrix $P$ such that

$$P^{\mathrm{T}}AP = \begin{bmatrix} A_{11} & A_{12} \\ 0 & A_{22} \end{bmatrix},$$

where $A_{11}$ and $A_{22}$ are square, nonempty submatrices, and it is *irreducible* if it is not reducible. A matrix with positive entries is trivially irreducible. A useful characterization is that $A$ is irreducible if and only if the directed graph associated with $A$ (which has $n$ vertices, with an edge connecting the $i$th vertex to the $j$th vertex if $a_{ij} \neq 0$) is STRONGLY CONNECTED [II.16].

**Theorem 3 (Perron–Frobenius).** *If $A \in \mathbb{R}^{n\times n}$ is nonnegative and irreducible then*

(1) $\rho(A) > 0$,
(2) $\rho(A)$ *is an eigenvalue of* $A$,
(3) *there is a positive vector* $x$ *such that* $Ax = \rho(A)x$,
(4) $\rho(A)$ *is an eigenvalue of algebraic multiplicity* 1.

To illustrate the theorem consider the following two irreducible matrices and their eigenvalues:

$$A = \begin{bmatrix} 8 & 1 & 6 \\ 3 & 5 & 7 \\ 4 & 9 & 2 \end{bmatrix}, \quad \Lambda(A) = \{15, \pm 2\sqrt{6}\},$$

$$B = \begin{bmatrix} 0 & 0 & 6 \\ \frac{1}{2} & 0 & 0 \\ 0 & \frac{1}{3} & 0 \end{bmatrix}, \quad \Lambda(B) = \{1, \tfrac{1}{2}(-1 \pm \sqrt{3}\mathrm{i})\}.$$

The Perron–Frobenius theorem correctly tells us that $\rho(A) = 15$ is a distinct eigenvalue of $A$ and that it has a corresponding positive eigenvector, which is known as the *Perron vector*. The Perron vector of $A$ is the vector of all ones, as $A$ forms a magic square and $\rho(A)$ is the magic sum! The Perron vector of $B$, which is both a Leslie matrix and a companion matrix, is $[6\ 3\ 1]^{\mathrm{T}}$. There is one notable difference between $A$ and $B$: for $A$, $\rho(A)$ exceeds the other eigenvalues in modulus, but all three eigenvalues of $B$ have modulus 1. In fact, Perron's original version of Theorem 3 says that if $A$ has all positive elements then $\rho(A)$ is not only an eigenvalue of $A$ but is larger in modulus than every other eigenvalue. Note that $B^3 = I$, which provides another way to see that the eigenvalues of $B$ all have modulus 1.

We saw in section 9 that the spectral radius plays an important role in the convergence of stationary iterative methods, through $\rho(M^{-1}N)$, where $A = M - N$ is a splitting. In comparing different splittings we can use the result that for $A, B \in \mathbb{R}^{n\times n}$, with $|A|$ denoting the matrix $(|a_{ij}|)$,

$$|a_{ij}| \leqslant b_{ij} \quad \forall i,j \quad \Rightarrow \quad \rho(A) \leqslant \rho(|A|) \leqslant \rho(B).$$

### 11.2 *M*-Matrices

$A \in \mathbb{R}^{n\times n}$ is an *M-matrix* if it can be written in the form $A = sI - B$, where $B$ is nonnegative and $s > \rho(B)$. *M*-matrices arise in many applications, a classic one being Leontief's input–output models in economics.

The special sign pattern of an *M*-matrix—positive diagonal elements and nonpositive off-diagonal elements—combines with the spectral radius condition to give many interesting characterizations and properties. For example, a nonsingular matrix $A$ with nonpositive off-diagonal elements is an *M*-matrix if and only if $A^{-1}$ is nonnegative. Another characterization, which makes connections with section 1, is that $A$ is an *M*-matrix if and only if $A$ has positive diagonal entries and $AD$ is diagonally dominant by rows for some nonsingular diagonal matrix $D$.

An important source of *M*-matrices is discretizations of differential equations, and the archetypal example is the second-difference matrix, described at the start of section 6, which is an *M*-matrix multiplied by $-1$. For this application it is an important result that when $A$ is an *M*-matrix the Jacobi and Gauss–Seidel iterations for $Ax = b$ both converge for any starting vector—a result that is part of the more general theory of regular splittings.

Another important property of *M*-matrices is immediate from the definition: the eigenvalues all lie in the open right half-plane. This means that *M*-matrices are special cases of positive stable matrices, which in turn are of great interest due to the fact that the stability of various mathematical processes is equivalent to positive (or negative) stability of an associated matrix.

The class of matrices whose inverses are *M*-matrices is also much studied. To indicate why, we state a result about matrix roots. It is known that if $A$ is an *M*-matrix then $A^{1/2}$ is also an *M*-matrix. But if $A$ is stochastic (that is, it is nonnegative and has unit row sums), $A^{1/2}$ may not be stochastic. However, if $A$ is both stochastic *and* the inverse of an *M*-matrix, then $A^{1/p}$ is stochastic for all positive integers $p$.

## 12 Matrix Inequalities

There is a large body of work on matrix inequalities, ranging from classical nineteenth-century and early twentieth-century inequalities (some of which are described in section 5.4) to more recent contributions, which are often motivated by applications, notably in statistics, physics, and control theory. In this section we describe just a few examples, chosen for their interest or practical usefulness.

An important class of inequalities on Hermitian matrices is expressed using the *Löwner (partial) ordering* in which, for Hermitian $X$ and $Y$, $X \geqslant Y$ denotes that $X - Y$ is positive-semidefinite while $X > Y$ denotes that $X - Y$ is positive-definite. Many inequalities between real numbers generalize to Hermitian matrices in this ordering. For example, if $A$, $B$, $C$ are Hermitian and $A$ commutes with $B$ and $C$, then

$$A \geqslant 0, \quad B \leqslant C \quad \Rightarrow \quad AB \leqslant AC.$$

A function $f$ is *matrix monotone* if it preserves the order, that is, $A \leqslant B$ implies $f(A) \leqslant f(B)$, where $f(A)$ denotes a FUNCTION OF A MATRIX [II.14]. Much is known about this class of functions, including that $t^{1/2}$ and $\log t$ are matrix monotone but $t^2$ is not.

Many matrix inequalities involve norms. One example is

$$\| \, |A| - |B| \, \|_F \leqslant \sqrt{2} \|A - B\|_F,$$

where $A, B \in \mathbb{C}^{m \times n}$ and $| \cdot |$ is the matrix absolute value defined in section 2. This inequality can be regarded as a perturbation result that shows the matrix absolute value to be very well-conditioned.

An example of an inequality that finds use in the analysis of convergence of methods in nonlinear optimization is the *Kantorovich inequality*, which for Hermitian positive-definite $A$ with eigenvalues $\lambda_n \leqslant \cdots \leqslant \lambda_1$ and $x \neq 0$ is

$$\frac{(x^* A x)(x^* A^{-1} x)}{(x^* x)^2} \leqslant \frac{(\lambda_1 + \lambda_n)^2}{4 \lambda_1 \lambda_n}.$$

This inequality is attained for some $x$, and the left-hand side is always at least 1.

Many inequalities are available that generalize scalar inequalities for means. For example, the arithmetic–geometric mean inequality $(ab)^{1/2} \leqslant \frac{1}{2}(a + b)$ for positive scalars has an analogue for Hermitian positive-definite $A$ and $B$ in the inequality $A \# B \leqslant \frac{1}{2}(A + B)$, where $A \# B$ is the geometric mean defined as the unique Hermitian positive-definite solution to $XA^{-1}X = B$. The geometric mean also satisfies the extremal property

$$A \# B = \max \left\{ X \colon X = X^*, \begin{bmatrix} A & X \\ X & B \end{bmatrix} \geqslant 0 \right\},$$

which hints at *matrix completion problems*, in which the aim is to choose missing elements of a matrix in order to achieve some goal, which could be to satisfy a particular matrix property or, as here, to maximize an objective function. Another mean for Hermitian positive-definite matrices (and applicable more generally) is the *log-Euclidean mean*, $\exp(\frac{1}{2}(\log A + \log B))$, where log is the PRINCIPAL LOGARITHM [II.14], which is used in image registration, for example.

Finally, we mention an inequality for the matrix exponential. Although there is no simple relation between $e^{A+B}$ and $e^A e^B$ in general, for Hermitian $A$ and $B$ the inequality $\operatorname{trace}(e^{A+B}) \leqslant \operatorname{trace}(e^A e^B)$ was proved independently by S. Golden and J. Thompson in 1965. Originally of interest in statistical mechanics, the Golden–Thompson inequality has more recently found use in RANDOM-MATRIX THEORY [IV.24]. Again for Hermitian $A$ and $B$, the related inequalities $\|e^{A+B}\| \leqslant \|e^{A/2} e^B e^{A/2}\| \leqslant \|e^A e^B\|$ hold for any unitarily invariant norm.

## 13 Library Software

From the early days of digital computing the benefits of providing library subroutines for carrying out basic operations such as the addition of vectors and the formation of vector inner products was recognized. Over the ensuing years many matrix computation research codes were published, including in the linear algebra volume of the *Handbook for Automatic Computation* (1971) and in the *Collected Algorithms of the ACM*. Starting in the 1970s the concept of standardized subprograms was developed in the form of the Basic Linear Algebra Subprograms (BLAS), which are specifications for vector (level 1), matrix–vector (level 2), and matrix–matrix (level 3) operations. The BLAS have been widely adopted, and highly optimized implementations are available for most machines. The freely available LAPACK library of Fortran codes represents the current state of the art for solving dense linear equations, least-squares problems, and eigenvalue and singular value problems. Many modern programming packages and environments build on LAPACK.

It is interesting to note that the TOP500 list (www.top500.org) ranks the world's fastest computers by their speed (measured in flops per second) in solving a random linear system $Ax = b$ by GE. This benchmark has its origins in the 1970s LINPACK project, a precursor to LAPACK, in which the performance of contemporary machines was compared by running the LINPACK GE code on a $100 \times 100$ system.

## 14 Outlook

Matrix analysis and numerical linear algebra remain very active areas of research. Many problems in applied mathematics and scientific computing require the solution of a matrix problem at some stage, so there is always a demand for better understanding of matrix problems and faster and more accurate algorithms for their solution. As the overarching applications evolve, new problem variants are generated, often involving new assumptions on the data, different requirements on the solution, or new metrics for measuring the success of an algorithm. A further driver of research is computer hardware. With the advent of processors with many cores, the use of accelerators such as graphics processing units, and the harnessing of vast numbers of processors for parallel computing, the standard algorithms in numerical linear algebra are having to be reorganized and possibly even replaced, so we are likely to see significant changes in the coming years.

### Further Reading

Three must-haves for researchers are the influential treatment of numerical linear algebra by Golub and Van Loan and the two volumes by Horn and Johnson, which contain a comprehensive treatment of matrix analysis.


Bhatia, R. 1997. *Matrix Analysis.* New York: Springer.

——. 2001. Linear algebra to quantum cohomology: the story of Alfred Horn's inequalities. *American Mathematical Monthly* 108(4):289–318.

——. 2007. *Positive Definite Matrices.* Princeton, NJ: Princeton University Press.

Golub, G. H., and C. F. Van Loan. 2013. *Matrix Computations*, 4th edn. Baltimore, MD: Johns Hopkins University Press.

Higham, N. J. 2002. *Accuracy and Stability of Numerical Algorithms*, 2nd edn. Philadelphia, PA: SIAM.

Horn, R. A., and C. R. Johnson. 1991. *Topics in Matrix Analysis.* Cambridge: Cambridge University Press.

——. 2013. *Matrix Analysis*, 2nd edn. Cambridge: Cambridge University Press.

Parlett, B. N. 1998. *The Symmetric Eigenvalue Problem.* Philadelphia, PA: SIAM. (Unabridged, amended version of book first published by Prentice-Hall in 1980.)

Saad, Y. 2003. *Iterative Methods for Sparse Linear Systems*, 2nd edn. Philadelphia, PA: SIAM.

Stewart, G. W., and J. Sun. 1990. *Matrix Perturbation Theory.* London: Academic Press.

Tisseur, F., and K. Meerbergen. 2001. The quadratic eigenvalue problem. *SIAM Review* 43(2):235–86.


# IV.11 Continuous Optimization (Nonlinear and Linear Programming)

*Stephen J. Wright*

## 1 Overview

At the core of any optimization problem is a mathematical model of a system. This model could be constructed according to physical, economic, behavioral, or statistical principles, and it describes relationships between variables that define the state of the system; it may also place restrictions on the states, in the form of constraints on the variables. The model also includes an objective function that measures the desirability of a given set of variables. The optimization problem is to find the set of variables that achieves the best possible value of the objective function, among all those values that satisfy the given constraints.

### 1.1 Examples

Optimization problems are ubiquitous, as we illustrate with some examples.

(1) A firm wishes to maximize its profit, given constraints on availability of resources (equipment, labor, raw materials), production costs, and forecast demand.

(2) In order to forecast weather, we first need to solve a problem to identify the state of the atmosphere a few hours ago. This is done by finding the state that is most consistent with recent meteorological observations (temperature, wind speed, humidity, etc.) taken at a variety of locations and times. The model contains differential equations that describe evolution of the atmosphere, statistical elements that describe prior knowledge of the atmospheric state, and an objective function that measures the consistency between the atmospheric state and the observations.

(3) Computer systems for recognizing handwritten digits contain models that read the written character, in the form of a pixelated image, and output their best guess as to the digit that is represented in the image. These models can be "trained" by presenting them with a (typically large) set of images containing known digits. An optimization problem is solved to adjust the parameters in the model so that the error count on the training set is minimized. If the training set is representative of the images that the system will see in future,

this optimized model can be trusted to perform reliable digit recognition.

(4) Given a product that is produced in a number of cities and consumed in other cities, we wish to find the least expensive way to transport the product from supply locations to demand locations. Here, the model consists of a graph that describes the transportation network, capacity constraints, and the cost of transporting one unit of the product between two adjacent locations in the network.

(5) Given a set of possible investments along with the means, variances, and correlations of their expected returns, an investor wishes to allocate his or her funds in a way that balances the expected mean return of the portfolio with its variance, in a way that fits his or her appetite for risk.

These examples capture some of the wide variety of applications currently seen in the field. As they suggest, the mathematical models that underlie optimization problems vary widely in size, complexity, and structure. They may contain simple algebraic relationships, systems of ordinary or partial differential equations, models derived from Bayesian statistics, and "black-box" models whose internal details are not accessible and can be accessed only by supplying inputs and observing outputs.

### 1.2 Continuous Optimization

In continuous optimization, the variables in the model are nominally allowed to take on a continuous range of values, usually real numbers. This feature distinguishes continuous optimization from discrete or combinatorial optimization, in which the variables may be binary (restricted to the values 0 and 1), integer (for which only integer values are allowed), or more abstract objects drawn from sets with finitely many elements. (DISCRETE OPTIMIZATION [IV.38] is the subject of another article in this volume.)

The algorithms used to solve continuous optimization problems typically generate a sequence of values of the variables, known as *iterates*, that converge to a solution of the problem. In deciding how to step from one iterate to the next, the algorithm makes use of knowledge gained at previous iterates, and information about the model at the current iterate, possibly including information about its sensitivity to perturbations in the variables. The continuous nature of the problem allows sensitivities to be defined in terms of first and second derivatives of the functions that define the models.

### 1.3 Standard Paradigms

Research in continuous optimization tends to be organized into several paradigms, each of which makes certain assumptions about the properties of the objective function, variables, and constraints. To define these paradigms, we group the variables into a real vector $x$ with $n$ components (that is, $x \in \mathbb{R}^n$) and define the general continuous optimization problem as follows:

$$\min_{x\in\mathbb{R}^n} f(x) \tag{1 a}$$

$$\text{subject to } c_i(x) = 0, \quad i \in \mathcal{E}, \tag{1 b}$$

$$c_i(x) \geqslant 0, \quad i \in \mathcal{I}, \tag{1 c}$$

where the objective $f$ and the constraints $c_i$, $i \in \mathcal{E} \cup \mathcal{I}$, are real-valued functions on $\mathbb{R}^n$. To this formulation is sometimes added a geometric constraint

$$x \in \Omega, \tag{2}$$

where $\Omega \subset \mathbb{R}^n$ is a closed convex set. (Any nonconvexity in the feasible set is conventionally captured by the algebraic constraints (1 $b$) and (1 $c$) rather than the geometric constraint (2).) All functions in (1) are assumed to at least be continuous. A point $x$ that satisfies all the constraints is said to be *feasible*.

There is considerable flexibility in the way that a given optimization problem can be formulated; the choice of formulation has a strong bearing on the effectiveness with which the problem can be solved. One common reformulation technique is to replace an inequality constraint by an equality constraint plus a bound by introducing a new "slack" variable:

$$c_i(x) \leqslant 0 \Leftrightarrow c_i(x) + s_i = 0, \quad s_i \geqslant 0.$$

Referring to the general form (1), we distinguish several popular paradigms.

- In *linear programming*, the objective function and the constraints are affine functions of $x$; that is, they have the form $a^{\mathrm{T}}x + b$ for some $a \in \mathbb{R}^n$ and $b \in \mathbb{R}$.
- In *quadratic programming*, we have

  $$f(x) = \tfrac{1}{2}x^{\mathrm{T}}Qx + c^{\mathrm{T}}x + d$$

  for some $n \times n$ symmetric matrix $Q$, vector $c \in \mathbb{R}^n$, and scalar $d \in \mathbb{R}$, while all constraints $c_i$ are linear. When $Q$ is positive-semidefinite, we have a *convex quadratic program*.

- In *convex programming*, the objective $f$ and the negated inequality constraint functions $-c_i$, $i \in \mathcal{I}$, are convex functions, while the equality constraints $c_i$, $i \in \mathcal{E}$, are affine functions. (These assumptions, along with the convexity and closedness of $\Omega$ in the case when (2) is included in the formulation, imply that the set of feasible points is closed and convex.)
- *Conic optimization* problems have the form (1), (2), where the set $\Omega$ is assumed to be a closed, convex cone that is pointed (that is, it contains no straight line), while the objective $f$ and equality constraints $c_i$, $i \in \mathcal{E}$, are assumed to be affine. There are no inequalities; that is, $\mathcal{I} = \varnothing$.
- In *unconstrained optimization*, there are not any constraints (1 $b$), (1 $c$), and (2), while the objective $f$ is usually assumed to be smooth, with at least continuous first derivatives. *Nonsmooth optimization* allows $f$ to have discontinuous first derivatives, but it is often assumed that $f$ has some other structure that can be exploited by the algorithms.
- In *nonlinear programming*, the functions $f$ and $c_i$, $i \in \mathcal{E} \cup \mathcal{I}$, are generally nonlinear but smooth, at least having continuous first partial derivatives on the region of interest.

An important special class of conic optimization problems is *semidefinite programming*, in which the vector $x$ of unknowns contains the elements of a symmetric $m \times m$ matrix $X$ that is required to be positive-semidefinite. It is natural and useful to write this problem in terms of the matrix $X$ as follows:

$$\min_{X \in \mathbb{SR}^{m \times m}} \; C \bullet X \tag{3 a}$$

$$\text{subject to } A_i \bullet X = b_i, \quad i = 1, 2, \dots, p, \tag{3 b}$$

$$X \succeq 0. \tag{3 c}$$

Here, $\mathbb{SR}^{m \times m}$ denotes the set of symmetric $m \times m$ matrices, the matrices $C$ and $A_i$, $i = 1, 2, \dots, p$, all belong to $\mathbb{SR}^{m \times m}$, and the operator $\bullet$ is defined on pairs of matrices in $\mathbb{SR}^{m \times m}$ as follows:

$$X \bullet Z = \sum_{i=1}^{m} \sum_{j=1}^{m} x_{ij} z_{ij} = \operatorname{trace}(XZ).$$

The constraint (3 $c$) instantiates the geometric constraint (2).

### Terminology

"Mathematical programming" is a historical term that encompasses optimization and closely related areas such as complementarity problems. Its origins date to the 1940s, with the development of the simplex method of George Dantzig, the first effective method for linear programming (7). The term "programming" originally referred to the formalized, systematic mathematical procedure by which problems can be solved. Only later did "programming" become roughly synonymous with "computer programming," causing some confusion that optimization researchers have often been called on to explain. The more modern term "optimization" is generally preferred, although the term "programming" is still attached (probably forever) to such problems as linear programming and integer programming.

## 1.4 Scope of Research

Research in optimization encompasses

- study of the mathematical properties of the problems themselves;
- development, testing, and analysis of algorithms for solving particular classes of problems (such as one of the paradigms described above); and
- development of models and algorithms for specific application areas.

We give a brief description of each of these aspects.

One topic of fundamental interest is the characterization of solution sets: are there verifiable conditions that we can check to determine whether a given point is a solution to the optimization problem? Given the uncertainty that is present in many practical settings, we may also be interested in the sensitivity of the solution to perturbations in the data or in the objective and constraint functions. Ill-conditioned problems are those in which the solution can change significantly when the data or functions change slightly. Another important fundamental concept is *duality*. Often, the data and functions that define an optimization problem can be rearranged to produce a new "dual" problem that is related to the original problem in interesting ways. The concept of duality can also be of great practical importance in designing more efficient formulations and algorithms.

The study of algorithms for optimization problems blends theory and practice. The design of algorithms that work well on practical problems requires a good deal of intuition and testing. Most algorithms in use today have a solid theoretical basis, but the theory often allows wide latitude in the choice of certain parameters, and algorithms are often "engineered" to find suitable values for these parameters and to incorporate other heuristics. Analysis of algorithms tackles such issues as whether the iterates can be guaranteed to converge to a solution (or some other point of interest); whether there is an upper bound on the number of iterations needed, as a function of the size or complexity of the problem; and the rate of convergence, particularly after the iterates enter a certain neighborhood of the solution. Algorithmic analysis is typically worst-case in nature. It gives important indications about how the algorithm will behave in practice, but it does not tell the whole story. Famously, the simplex method (described in section 3.1) is known to perform badly in the worst case—its running time may be exponential in the problem size—yet its performance on most practical problems is impressively good.

Development of software that implements efficient algorithms is another important activity. High-quality codes are available both commercially and in the public domain. Modeling tools—high-level languages that serve as a front end to algorithmic software packages—have become more popular in recent years. They relieve users of much of the burden of transforming their practical problem to a set of functions (the objective and constraint functions in (1)), allowing the model to be expressed in intuitive terms, related more directly to the application.

With the growth in the size and complexity of practical optimization problems, issues of modeling, formulation, and customized algorithm design have become more prominent. A particular application can be formulated as an optimization problem in many different ways, and different formulations can lead to very different solver performance. Experience and testing is often required to identify the most effective formulation.

Many modern applications cannot be solved effectively with packaged software for one of the standard paradigms of section 1.3. It is necessary to assemble a customized algorithm, drawing on a variety of algorithmic elements from the optimization toolbox and also on tools from other disciplines in scientific computing. This approach allows the particular structure or context of the problem to be exploited. Examples of special context include the following.

- Low-accuracy solutions may suffice for some problems.
- Algorithms that require less data movement—or sampling from a large data set, or the ability to handle streaming data—may be essential in other settings.
- Algorithms that produce (possibly suboptimal) solutions in real time may be essential in such contexts as industrial control.

### 1.5 Connections

Continuous optimization is a highly interconnected discipline, having close relationships with other areas of mathematics, with scientific computing, and with numerous application areas. It also has close connections to discrete optimization, which often requires continuous optimization problems to be solved as subproblems or relaxations.

In mathematics, continuous optimization relies heavily on various forms of mathematical analysis, especially real analysis and functional analysis. Certain types of analysis have been developed in close association with the discipline of optimization, including convex analysis, nonsmooth analysis, and variational analysis. The theory of computational complexity also plays a role in the study of algorithms. Game theory is particularly relevant when we examine duality and optimality conditions for optimization problems. Control theory is also relevant: for framing problems involving dynamical models and as an important source of applications for optimization. Statistics provides vital tools for stochastic optimization and for optimization in machine learning, in which the model is available only through sampling from a data set.

Continuous optimization also intersects with many areas in numerical analysis and scientific computing. Numerical linear algebra is vitally important, since many optimization algorithms generate a sequence of linear approximations, and these must be solved with linear algebra tools. Differential equation solvers are important counterparts to optimization in applications such as data assimilation and distributed parameter identification, which involve optimization of ordinary differential equation and partial differential equation models. The ubiquity of multicore architectures and the wide availability of cluster computing have given new prominence to parallel algorithms in some areas (such as machine learning), requiring the use of software tools for parallel computing.

Finally, we mention some of the many connections between optimization and the application areas within which it has become deeply embedded. Machine learning uses optimization algorithms extensively to perform classification and learning tasks. The challenges posed by machine learning applications (e.g., large data sets) have driven recent developments in stochastic optimization and large-scale unconstrained optimization. Compressed sensing, in which sparse signals are recovered from randomized encodings, also relies heavily on optimization formulations and specialized algorithms. Engineering control is a rich source of challenging optimization problems at many scales, frequently involving dynamic models of plant processes. In these and many other areas, practitioners have made important contributions to all aspects of continuous optimization.

## 2 Basic Principles

We mention here some basic theory that underpins continuous optimization and that serves as a starting point for the algorithms outlined in later sections.

Possibly the most fundamental issues are how we define a solution to a problem and how we recognize a given point as a solution. The answers become more complicated as we expand the classes of functions allowed in the formulation. The type of solution most amenable to analysis is a *local solution*. The point $x^*$ is a local solution of (1) if $x^*$ is feasible and if there is an open neighborhood $\mathcal{N}$ of $x^*$ such that $f(x) \geqslant f(x^*)$ for all feasible points $x \in \mathcal{N}$. Furthermore, $x^*$ is a *strict local solution* if $f(x) > f(x^*)$ for all feasible $x \in \mathcal{N}$ with $x \neq x^*$. A *global solution* is a point $x^*$ such that $f(x) \geqslant f(x^*)$ for *all* feasible $x$.

As we see below, we can use the derivatives of the objective and constraint functions to construct testable conditions that verify that $x^*$ is a local solution under certain assumptions. It is difficult to verify global optimality, even when the objective and constraint functions are smooth, because of the difficulty of gaining a global perspective on these functions. However, in *convex optimization*, where the objective $f$ is a convex function and the set of feasible points is also convex, all local solutions are global solutions. (Convex optimization includes linear programming and conic optimization as special cases.)

Global optimization techniques have also been devised for certain classes of nonconvex problems. It is possible to prove results about the performance of such methods when the function $f$ satisfies additional properties (such as Lipschitz continuity, with known Lipschitz constant) and the feasible region is bounded. One class of methods for solving the global optimization problem uses a process of subdividing the feasible region and using information about $f$ to obtain a lower bound on the objective in that region, leading to a branch-and-bound algorithm akin to methods used in integer programming.

We now turn to characterizations of local solutions for problems defined by smooth functions, assuming for simplicity that $f$ and $c_i$, $i \in \mathcal{E} \cup \mathcal{I}$, have continuous second partial derivatives. We use $\nabla f(x)$ to denote the gradient of $f$ (the vector $[\partial f/\partial x_i]_{i=1}^n$ of first partial derivatives) and $\nabla^2 f(x)$ to denote the Hessian of $f$ (the $n \times n$ matrix $[\partial^2 f/\partial x_i \partial x_j]_{i,j=1}^n$ of second partial derivatives). An important tool, both in the characterization of solutions for smooth problems and in the design of algorithms, is *Taylor's theorem*. This result can be used to estimate the value of $f$ by using its derivative information at a nearby point. For example, we have

$$f(x+p) = f(x) + \nabla f(x)^{\mathrm{T}} p + o(\|p\|), \tag{4a}$$

$$f(x+p) = f(x) + \nabla f(x)^{\mathrm{T}} p + \tfrac{1}{2} p^{\mathrm{T}} \nabla^2 f(x) p + o(\|p\|^2), \tag{4b}$$

where the notation $o(t)$ indicates a quantity that goes to zero faster than $t$. These formulas can be used to construct low-order approximations to (1) that are valid in the neighborhood of a current iterate $x$ and can thus be used to identify a possibly improved iterate $x + p$.

For unconstrained optimization of a smooth function $f$, we have the following necessary condition.

> If $x^*$ is a local solution of $\min_x f(x)$, then $\nabla f(x^*) = 0$.

Note that this is only a *necessary* condition; it is possible to have $\nabla f(x) = 0$ without $x$ being a minimizer. (An example is the scalar function $f(x) = x^3$, which has no minimizer but which has $\nabla f(0) = 0$.) To complement this result, we have the following sufficient condition.

> If $x^*$ is a point such that $\nabla f(x^*) = 0$ with $\nabla^2 f(x^*)$ positive-definite, then $x^*$ is a strict local solution of $\min_x f(x)$.

Turning to constrained optimization—the general form (1), with smooth functions—characterization of local solutions becomes somewhat more complex. A central role is played by the *Lagrangian* function, defined as follows:

$$\mathcal{L}(x, \lambda) = f(x) - \sum_{i \in \mathcal{E} \cup \mathcal{I}} \lambda_i c_i(x). \tag{5}$$

This is a linear combination of objective and constraint functions, where the weights $\lambda_i$ are called *Lagrange multipliers*. The following set of conditions, known as the *Karush–Kuhn–Tucker conditions* or *KKT conditions* after their inventors, are closely related to local optimality of $x^*$ for the problem (1): there exist $\lambda_i^*$, $i \in \mathcal{E} \cup \mathcal{I}$, such that

$$\nabla_x \mathcal{L}(x^*, \lambda^*) = 0, \tag{6 a}$$

$$c_i(x^*) = 0, \quad i \in \mathcal{E}, \tag{6 b}$$

$$c_i(x^*) \geqslant 0, \quad i \in \mathcal{I}, \tag{6 c}$$

$$\lambda_i^* \geqslant 0, \quad i \in \mathcal{I}, \tag{6 d}$$

$$\lambda_i^* c_i(x^*) = 0, \quad i \in \mathcal{I}. \tag{6 e}$$

Condition (6 *e*) is a *complementarity condition* that indicates complementarity between each inequality constraint value $c_i(x^*)$ and its Lagrange multiplier $\lambda_i^*$; for each $i$, at least one of these two quantities must be zero. Roughly speaking, the Lagrange multipliers measure the sensitivity of the optimal objective value $f(x^*)$ to perturbations in the constraints $c_i$. When $x^*$ is a local solution of (1), the KKT conditions will hold, provided an additional condition called a *constraint qualification* is satisfied. The constraint qualification requires the true geometry of the feasible set near $x^*$ to be captured by linear approximations to the constraint functions around $x^*$.

When the functions in (1) are nonsmooth, it becomes harder to define optimality conditions, as even the concept of derivative becomes more complicated. We consider the simplest problem of this type: the unconstrained problem $\min_x f(x)$, where $f$ is a convex (possibly nonsmooth) function. The *subdifferential* of $f$ at a point $x$ is defined from the collection of supporting hyperplanes to $f$ at $x$:

$$\partial f(x) := \{v : f(z) \geqslant f(x) + v^{\mathrm{T}}(z - x) \text{ for all } z \text{ in the domain of } f\}.$$

For example, the function $f(x) = \|x\|_1 = \sum_{i=1}^n |x_i|$ is nonsmooth, with subdifferential consisting of the vectors $v$ such that

$$\begin{aligned} v_i &= +1 && \text{if } x_i > 0, \\ v_i &\in [-1, 1] && \text{if } x_i = 0, \\ v_i &= -1 && \text{if } x_i < 0. \end{aligned}$$

When $f$ is smooth at $x$ in addition to being convex, we have $\partial f(x) = \{\nabla f(x)\}$. A necessary and sufficient condition for $x^*$ to be a solution of $\min_x f(x)$ is that $0 \in \partial f(x^*)$.

## 3 Linear Programming

Consider the problem

$$\min_x c^{\mathrm{T}} x \quad \text{subject to } Ax = b, \ x \geqslant 0, \tag{7}$$

where $x \in \mathbb{R}^n$ as before, $b \in \mathbb{R}^m$ is the right-hand side, and $A \in \mathbb{R}^{m \times n}$ is the constraint matrix. Any optimization problem with an affine objective function and affine constraints can, after some elementary transformations, be written in this standard form. As illustrated by the example in figure 1, the feasible region for the problem (7) is polyhedral, and the contours of the objective function are lines.

There are three possible outcomes for a linear program.

(a) The problem is *infeasible*; that is, there is no point $x$ that satisfies $Ax = b$ and $x \geqslant 0$.
(b) The problem is *unbounded*; that is, there is a sequence of feasible points $x^k$ such that $c^{\mathrm{T}} x^k \downarrow -\infty$.
(c) The problem has a solution; that is, there is a feasible point $x^*$ such that $c^{\mathrm{T}} x^* \leqslant c^{\mathrm{T}} x$ for all feasible $x$.

When a solution exists (case (c)), it may not be uniquely defined. However, we can note that the set of solutions itself forms a polyhedron and that at least one solution lies at a *vertex* of this polyhedron, that is, at a point that does not lie in the interior of a line joining any other two points in the polyhedron.

By rearranging the data in (7), we obtain another linear program called the *dual*:

$$\max_{\lambda, s} b^{\mathrm{T}} \lambda \quad \text{subject to } A^{\mathrm{T}} \lambda + s = c, \ s \geqslant 0. \tag{8}$$

(In discussions of duality, the original problem (7) is called the *primal* problem.) The primal and dual problems are related by a powerful *duality theory* that has important practical implications. *Weak duality* states that, if $x$ is a feasible point for (7) and $(\lambda, s)$ is a feasible point for (8), then the primal objective is greater than or equal to the dual objective. This statement is easily proved in a single line:

$$c^{\mathrm{T}} x = (A^{\mathrm{T}} \lambda + s)^{\mathrm{T}} x \geqslant \lambda^{\mathrm{T}} A x = \lambda^{\mathrm{T}} b.$$

The other fundamental duality result—*strong duality*—states that there are three possible outcomes for the pair of problems (7) and (8):

(a) one of the two problems is infeasible and the other is unbounded;
(b) both problems are infeasible; or
(c) (7) has a solution $x^*$ and (8) has a solution $(\lambda^*, s^*)$ with objective functions equal (that is, $c^{\mathrm{T}} x^* = b^{\mathrm{T}} \lambda^*$).

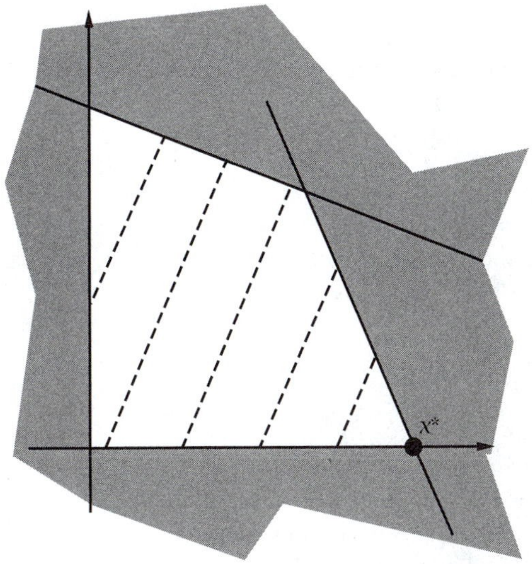


**Figure 1** Feasible region (unshaded), objective function contours (dashed lines), and optimal vertex $x^*$ for a linear program in two variables.

Specializing (6) to the case of linear programming, we see that the primal and dual problems share a common set of KKT conditions:

$$Ax = b, \qquad A^T\lambda + s = c, \tag{9a}$$

$$x \geqslant 0, \quad s \geqslant 0, \tag{9b}$$

$$x_i s_i = 0, \quad i = 1, 2, \dots, n. \tag{9c}$$

If $(x^*, \lambda^*, s^*)$ is any vector triple that satisfies these conditions, $x^*$ is a solution of (7) and $(\lambda^*, s^*)$ is a solution of (8).

We now discuss the two most important classes of algorithms for linear programming.

### 3.1 The Simplex Method

The simplex method, devised by George Dantzig in the 1940s, remains a fundamental approach of practical and theoretical importance in linear programming. Geometrically speaking, the simplex method moves from vertex to neighboring vertex of the feasible set, decreasing the objective function with each move and terminating when it cannot find a neighboring vertex with a lower objective value. The method is implemented by maintaining a *basis*: a subset of $m$ out of the $n$ components of $x$ that are allowed to be nonzero at the current iteration. The values of these basic components of $x$ are determined uniquely by the $m$ linear constraints $Ax = b$. Each step of the simplex method starts by choosing a nonbasic variable to enter the basis. This variable is allowed to increase away from zero, a process that, because of the requirement to maintain feasibility of the linear constraints $Ax = b$, causes the values of the existing basic variables to change. The entering variable is allowed to increase to the point where one of the basic variables reaches zero, upon which it leaves the basis, and the iteration is complete.

Efficient implementation of the simplex method depends both on good "pricing" strategies (to choose which nonbasic variable should enter the basis) and on efficient linear algebra (to update the values of the basic variables as the entering variable increases away from zero). Both topics have seen continued development over the years, and highly effective software is available, both commercially and in the public domain. Specialized, highly efficient versions of the simplex method exist for some special cases of linear programming, such as those arising from transportation or routing over networks.

The simplex method is an example of an active-set method: a subset of the inequality constraints (which are the bounds $x \geqslant 0$) is held to be *active* at each iteration. In the simplex method, the active set consists of the nonbasic variables, those indices $i$ for which $x_i = 0$. This active set changes only slightly from one iteration to the next; in fact, the simplex method changes just a single component of the active (nonbasic) set at each iteration.

The theoretical properties of the simplex method remain a source of fascination because, despite its practical efficiency, its worst-case behavior is poor. In an article from 1972, Klee and Minty famously showed that the number of steps may be exponential in the dimension of the problem. There have been various attempts to understand the "average-case" behavior, in which the number of iterations required is roughly linear in the problem dimensions, that is, the numbers of variables and constraints. The "smoothed analysis" of Spielman and Teng shows that any linear program for which the simplex method behaves badly can be modified, with small perturbations, to become a problem that requires only polynomially many iterations.

An algorithm with polynomial complexity (in the worst case) was announced in 1979: Khachiyan's ellipsoid algorithm. Though of great theoretical interest, it was not a practical alternative to simplex. The interior-point revolution began with Karmarkar's algorithm (in 1984); this is also a polynomial-time approach, but it has much better computational properties than the ellipsoid approach. It motivated a new class of algorithms—primal–dual interior-point methods—that not only had attractive theoretical properties but were also truly competitive with the simplex method on practical problems. We describe these next.

### 3.2 Interior-Point Methods

As their name suggests, primal–dual interior-point methods generate a sequence of iterates $(x^k, \lambda^k, s^k)$, $k = 1, 2, \dots$, in both primal and dual variables, in which $x^k$ and $s^k$ contain all positive numbers (that is, they are strictly feasible with respect to the constraints $x \geqslant 0$ and $s \geqslant 0$ in (9 *b*)). Steps between iterates are obtained by applying Newton's method to a perturbed form of the condition (9 *c*) in which the right-hand side 0 is replaced by a positive quantity $\mu_k > 0$, which is gradually decreased to zero as $k \to \infty$. The Newton equations for each step $(\Delta x^k, \Delta\lambda^k, \Delta s^k)$ are obtained from a linearization of these perturbed KKT conditions: specifically,

$$\begin{bmatrix} 0 & A^{\mathrm{T}} & I \\ A & 0 & 0 \\ S & 0 & X \end{bmatrix} \begin{bmatrix} \Delta x^k \\ \Delta\lambda^k \\ \Delta s^k \end{bmatrix} = - \begin{bmatrix} A^{\mathrm{T}}\lambda^k + s^k - c \\ Ax^k - b \\ X^k S^k e - \mu_k \mathbf{1} \end{bmatrix},$$

where $X^k$ is the diagonal matrix whose diagonal elements come from $x^k$, $S^k$ is defined similarly, and $\mathbf{1}$ is the vector of length $n$ whose elements are all 1. The new iteration is obtained by setting

$$(x^{k+1}, \lambda^{k+1}, s^{k+1}) \\ = (x^k + \alpha_k \Delta x^k, \lambda^k + \beta_k \Delta\lambda^k, s^k + \beta_k \Delta s^k),$$

where $\alpha_k$ and $\beta_k$ are step lengths in the range $[0,1]$ chosen so as to ensure that $x^{k+1} > 0$ and $s^{k+1} > 0$, among other goals. Convergence, with polynomial complexity, can be demonstrated under appropriate schemes for choosing $\mu_k$ and the step lengths $\alpha_k$ and $\beta_k$. Clever schemes for choosing these parameters and for enhancing the search directions using "second-order corrections" lead to good practical behavior.

Primal–dual interior-point methods have the additional virtue that they are easily extendible to convex quadratic programming and monotone linear complementarity problems, with only minor changes to the algorithm and the convergence theory. As we see in section 6.3, they can also be extended to general nonlinear programming, though the modifications in this case are more substantial and the convergence guarantees are weaker.

## 4 Unconstrained Optimization

Consider the problem of simply minimizing a function without constraints:

$$\min_x f(x),$$

where $f$ has at least continuous first derivatives. This problem is important in its own right. It also appears as a subproblem in many methods for constrained optimization, and it serves to illustrate several algorithmic techniques that can also be applied to the constrained case.

### 4.1 First-Order Methods

The Taylor approximation (4 *a*) shows that $f$ decreases most rapidly in the direction of the negative gradient vector $-\nabla f(x)$. Steepest-descent methods move in this direction, each iteration having the form

$$x^{k+1} = x^k - \alpha_k \nabla f(x^k),$$

for some positive step length $\alpha_k$. A suitable value of $\alpha_k$ can be found by performing (approximately) a one-dimensional search along the direction $-\nabla f(x^k)$, thus guaranteeing a decrease in $f$ at every iteration. When further information about $f$ is available, it may be possible to choose $\alpha_k$ to guarantee descent in $f$ without doing a line search. A nonstandard approach, which first appeared in a 1988 paper by Barzilai and Borwein, chooses $\alpha_k$ according to a formula that allows $f$ to increase (sometimes dramatically) on some iterations, while often achieving better long-term behavior than standard steepest-descent approaches.

For the case of convex $f$, there has been renewed focus on accelerated first-order methods that still require only the calculation of a gradient $\nabla f$ at each step but that have more attractive convergence rates than steepest descent, both in theory and in practice. The common aspect of these methods is a "momentum" device, in which the step from $x^k$ to $x^{k+1}$ is based not just on the latest gradient $\nabla f(x^k)$ but also on the step from the previous iterate $x^{k-1}$ to the current iterate $x^k$. In *heavy-ball* and *conjugate gradient* methods, the steps have the form

$$x^{k+1} = x^k - \alpha_k \nabla f(x^k) + \beta_k (x^k - x^{k-1})$$

for positive parameters $\alpha_k$ and $\beta_k$, which are chosen in a variety of ways. Accelerated methods that have been proposed more recently also make use of momentum together with the latest gradient $\nabla f(x^k)$, but combine these factors in different ways. Some methods separate the steepest-descent steps from the momentum steps, alternating between these two types of steps to produce two interleaved sequences of iterates, rather than the single sequence $\{x^k\}$.

For convex $f$, first-order methods are characterized in some cases by linear convergence (with the error in $x^k$ decreasing to zero in a geometric sequence) or sublinear convergence (with the error decreasing to zero but not geometrically, typically at a rate of $1/k$ or $1/k^2$, where $k$ is the iteration number).

### 4.2 Superlinear Methods

When second derivatives of $f$ are available, we can use the second-order Taylor approximation (4 *b*) to motivate *Newton's method*, a fundamental algorithm in both optimization and nonlinear equations. When $\nabla^2 f$ is positive-definite at the current iterate $x^k$, we can define the step $p^k$ to be the minimizer of the right-hand side of (4 *b*) (with the $o(\|p\|^2)$ term omitted), yielding the formula

$$p^k = -[\nabla^2 f(x^k)]^{-1} \nabla f(x^k). \tag{10}$$

The next iterate is defined by choosing a step length $\alpha_k > 0$ and setting $x^{k+1} = x^k + \alpha_k p^k$. This method is characterized by *quadratic* convergence, in which the error in $x^{k+1}$ is bounded by a constant multiple of the *square* of the error in $x^k$, for all $x^k$ sufficiently close to the solution $x^*$. (The number of digits of agreement between $x^k$ and $x^*$ doubles at each of the last few iterations.)

Enhancements of the basic approach based on (10) yield more robust and general implementations. For example, the Hessian matrix $\nabla^2 f(x^k)$ may be modified during the computation of $p^k$ to ensure that it is a descent direction for $f$. Another important class of methods known as *quasi-Newton* methods avoids the calculation of second derivatives altogether, instead replacing $\nabla^2 f(x^k)$ in (10) by an approximation $B_k$ that is constructed using first-derivative information. The possibility of such an approximation is a consequence of another form of Taylor's theorem, which posits the following relationship between two successive gradients:

$$\nabla f(x^{k+1}) - \nabla f(x^k) \approx \nabla^2 f(x^k)(x^{k+1} - x^k).$$

In updating the Hessian approximation to $B_{k+1}$ after the step to $x^{k+1}$ is taken, we ensure that $B_{k+1}$ mimics this property of the true Hessian; that is, we enforce the condition

$$\nabla f(x^{k+1}) - \nabla f(x^k) \approx B_{k+1}(x^{k+1} - x^k).$$

We obtain a variety of quasi-Newton methods by imposing various other conditions on $B_{k+1}$: closeness to $B_k$ in some metric, for example, and positive-semidefiniteness. *Limited-memory* quasi-Newton methods store $B_k$ implicitly by means of the difference vectors between successive iterates and successive gradients at a limited number of prior iterations (typically between five and twenty).

### 4.3 Derivative-Free Methods

Methods that require the user to supply only function values $f$ (and not gradients or Hessians) have been enormously popular for many years. More recently, they have attracted the attention of optimization researchers, who have tried to improve their performance, equip them with a convergence theory, and customize them to certain specific classes of problems, such as problems in which $f$ is obtained from a simulation.

In the absence of gradient or Hessian values, it is sometimes feasible to use finite differencing to construct approximations to these higher-order quantities and then apply the methods described above. Another possible option is to use algorithmic differentiation to obtain derivatives directly from computer code and, once again, use them in the algorithms described above.

Methods that use only function values are usually best suited to problems of modest dimension $n$. *Model-based methods* use interpolation among function values at recently visited points to construct a model of the function $f$. This model is used to generate a new candidate, which is accepted as the next iterate if it yields a sufficient improvement in the function value over the best point found so far. The model is updated by changing the set of points on which the interpolation is based, replacing older points with higher values of $f$ by newer points with lower function values. *Pattern-search methods* take candidate steps along a fixed set of directions, shrinking step lengths as needed to evaluate a new iterate with a lower function value. After a successful step, the step length may be increased, to speed future progress. Appropriate maintenance of the set of search directions is crucial to efficient implementation and valid convergence theory. Another derivative-free method is the enormously popular *simplex method* of Nelder and Mead from 1965. This method—unrelated to the method of the same name for linear programming—maintains a set of $n+1$ points that form the vertex of a simplex in $\mathbb{R}^n$. At each iteration, it replaces one of these points with a new one, by expanding or contracting the simplex along promising directions or reflecting one of the vertices through its opposite face. Various attempts have been made in the years since to improve the performance of this method and to develop a convergence theory.

### 4.4 Stochastic Gradient Methods

Important problems have recently been identified for which evaluation of $\nabla f$ or even $f$ is computationally expensive but where it is possible to obtain an unbiased estimate of $\nabla f$ cheaply. Such problems are common in data analysis, where $f$ typically has the form

$$f(x) = \frac{1}{N} \sum_{i=1}^{N} f_i(x)$$

for large $N$, where each $f_i$ depends on a single item in the data set. If $i$ is selected at random from $\{1, 2, \dots, N\}$, the vector $g^k = \nabla f_i(x^k)$ is an unbiased estimate of $\nabla f(x^k)$. For convex $f$, methods that use this approximate gradient information have been a focus of work in the optimization and machine learning communities for some years, and efforts have recently intensified as their wide applicability has become evident. The basic iteration has the form $x^{k+1} = x^k - \alpha_k g^k$, where the choice of $g^k$ may be based on additional information about $f$, such as lower and upper bounds on its curvature. (Line searches are not practical in this setting, as evaluation of $f$ is assumed to be too expensive.) Additional devices such as averaging of the iterates $x^k$ or the gradient estimates $g^k$ enhance the properties of the method in some settings, such as when $f$ is only weakly convex. Typical convergence analysis shows that the *expected* value of the error in $x^k$, or of the difference between the function value after $k$ iterations and its optimal value, approaches zero at a sublinear rate, such as $O(1/k)$ or $O(1/\sqrt{k})$.

## 5 Conic Optimization

Conic optimization problems have the form

$$\min c^{\mathrm{T}} x \quad \text{subject to } Ax = b, \ x \in \Omega,$$

where $\Omega$ is a closed, convex, pointed cone. They include linear programming (7) and semidefinite programming (3) as special cases. It is possible to design generic algorithms with good complexity properties for this problem class provided that we can identify a certain type of *barrier function* for $\Omega$. A barrier function $\varphi$ is convex with domain the interior of $\Omega$, with $\varphi(x) \to \infty$ as $x$ approaches the boundary of $\Omega$. The additional property required for an efficient algorithm is *self-concordancy*, which is the property that for any $x$ in the domain of $\varphi$, and any $v \in \mathbb{R}^n$, we have that

$$|t'''(0)| \leqslant 2|t''(0)|^{3/2},$$

where $t(\alpha) := \varphi(x + \alpha v)$. Because the third derivatives are bounded in terms of the second derivatives, the function $\varphi$ is well approximated (locally at least) by a quadratic, so we can derive complexity bounds on Newton's method applied to $\varphi$, with a suitable step-length scheme. We can use this barrier function to define an interior-point method in which each iterate $x^k$ is obtained by finding an approximate minimizer of the following equality-constrained optimization problem:

$$\min_x c^{\mathrm{T}} x + \mu_k \varphi(x) \quad \text{subject to } Ax = b,$$

where the positive parameter $\mu_k$ can be decreased gradually to zero as $k$ increases, as in interior-point methods for linear programming. One or more steps of Newton's method can be used to find the approximate solution to this subproblem, starting from the previous iterate.

For linear programming, the cone $\Omega = \{x \mid x \geqslant 0\}$ admits a self-concordant barrier function $\varphi(x) = -\sum_{i=1}^{n} \log x_i$. In semidefinite programming, where $\Omega$ is the cone of positive-semidefinite matrices, we have $\varphi(X) = -\log \det X$.

In practice, the most successful interior-point methods for semidefinite programming are primal–dual methods rather than primal methods. These methods are (nontrivial) extensions of the linear programming approaches of section 3.2.

## 6 Nonlinear Programming

Next we turn to methods for nonlinear programming, in which the functions $f$ and $c_i$ in (1) are smooth nonlinear functions. A basic principle used in constructing algorithms for this problem is successive approximation of the nonlinear program by simpler problems, such as quadratic programming or unconstrained optimization, to which methods from the previous sections can be applied. Taylor's theorem is instrumental in constructing these approximations, using first- or second-order expansions of functions around the current iterate $x^k$ and possibly also the current estimates of the Lagrange multipliers for the constraints (1 $b$) and (1 $c$). The optimality conditions described in section 2 also play a central role in algorithm design.

### 6.1 Gradient Projection

Gradient projection is an extension of the steepest-descent approach for unconstrained optimization in which steps are taken along the negative gradient direction but projected onto the feasible set. Considering the formulation

$$\min f(x) \quad \text{subject to } x \in \Omega,$$

the basic gradient projection step is

$$x^{k+1} = P_\Omega(x^k - \alpha_k \nabla f(x^k)),$$

where $P_\Omega(\cdot)$ denotes projection onto the closed convex constraint set $\Omega$. This approach may be practical if the projection can be computed inexpensively, as is the case when $\Omega$ is a "box" defined by bounds on the variables. It is possible to enhance the gradient method by using second-order information in a selective way (simple projection of the Newton step does not work).

### 6.2 Sequential Quadratic Programming

In sequential quadratic programming we use Taylor's theorem to form the following approximation of (1) around the current point $x^k$:

$$\min_{d \in \mathbb{R}^n} \nabla f(x^k)^{\mathrm{T}} d + \tfrac{1}{2} d^{\mathrm{T}} H_k d \qquad (11\,a)$$

$$\text{subject to } c_i(x^k) + \nabla c_i(x^k)^{\mathrm{T}} d = 0, \quad i \in \mathcal{E}, \qquad (11\,b)$$

$$c_i(x^k) + \nabla c_i(x^k)^{\mathrm{T}} d \geqslant 0, \quad i \in \mathcal{I}, \qquad (11\,c)$$

where $H_k$ is a symmetric matrix. Denoting the solution of (11) by $d^k$, the next iterate is obtained by setting

$$x^{k+1} = x^k + \alpha_k d^k$$

for some step length $\alpha_k > 0$. The subproblem (11) is a quadratic program; it can be solved with methods of active-set or interior-point type. The matrix $H_k$ may contain second-order information from both objective and constraint functions; an "ideal" value is the Hessian of the Lagrangian function defined in (5), that is, $H_k = \nabla^2_{xx}\mathcal{L}(x^k, \lambda^k)$, where $\lambda^k$ are estimates of the Lagrange multipliers, obtained for example from the solution of the subproblem (11) at the previous iteration. When second derivatives are not readily available, $H_k$ could be a quasi-Newton approximation to the Lagrangian Hessian, updated by formulas similar to those used in unconstrained optimization.

A line search can be performed to find a suitable value of $\alpha_k$ in the update step. An alternative approach to stabilizing sequential quadratic programming is to add a "trust region" to the subproblem (11), in the form of a constraint $\|d\|_\infty \leqslant \Delta_k$, for some $\Delta_k > 0$.

### 6.3 Interior-Point Methods

The interior-point methods for linear programming described in section 3.2 can be extended to nonlinear programming, and software based on such extensions has been highly successful. To avoid notational clutter, we consider a formulation of nonlinear programming containing nonnegativity constraints on $x$ along with equality constraints:

$$\min f(x) \quad \text{subject to } c_j(x) = 0,\ j \in \mathcal{E}; \quad x \geqslant 0. \quad (12)$$

(This problem is no less general than (1); simple transformations can be used to express (1) in the form (12).) Following (6), and introducing an additional vector $s$ in the style of (9), we can write the optimality conditions for this problem as

$$\nabla f(x) - \sum_{j \in \mathcal{E}} \lambda \nabla c_j(x) - s = 0, \qquad (13\,a)$$

$$c_j(x) = 0, \quad j \in \mathcal{E}, \qquad (13\,b)$$

$$x \geqslant 0, \quad s \geqslant 0, \qquad (13\,c)$$

$$x_i s_i = 0, \quad i = 1, 2, \dots, n. \qquad (13\,d)$$

As in linear programming, interior-point methods generate a sequence of iterates $(x^k, \lambda^k, s^k)$ in which all components of $x^k$ and $s^k$ are strictly positive. The basic primal–dual step is obtained by applying Newton's method at $(x^k, \lambda^k, s^k)$ to the nonlinear equations defined by (13 *a*), (13 *b*), and (13 *d*), with the right-hand side in (13 *d*) replaced by a positive parameter $\mu_k$, which is reduced to zero gradually as the iterations progress. The basic approach can be enhanced in various ways: quasi-Newton approximations, line searches or trust regions, second-order corrections to the search direction, and so on.

### 6.4 Augmented Lagrangian Methods

An approach for solving (12) that was first proposed in the early 1970s is enjoying renewed popularity because of its successful use in new application areas. Originally known as the "method of multipliers," it is founded on the augmented Lagrangian function

$$\mathcal{L}_A(x, \lambda; \mu) := f(x) + \sum_{j \in \mathcal{E}} \lambda_j c_j(x) + \frac{1}{2\mu} \sum_{j \in \mathcal{E}} c_j^2(x)$$

for some positive parameter $\mu$. The method defines a sequence of primal–dual iterates $(x^k, \lambda^k)$ for a given sequence of parameters $\{\mu_k\}$, where each iteration is defined as follows:

- obtain $x^{k+1}$ by solving (approximately) the problem

$$\min_x \mathcal{L}_A(x, \lambda^k; \mu_k) \quad \text{subject to } x \geqslant 0; \qquad (14)$$

- update Lagrange multipliers

$$\lambda_j^{k+1} = \lambda_j^k + c_j(x^k)/\mu_k, \quad j \in \mathcal{E};$$

- choose $\mu_{k+1} \in (0, \mu_k]$ by some heuristic.

The original problem with nonlinear constraints is replaced by a sequence of bound-constrained problems (14). Unlike in interior-point methods, it is not necessary to drive the parameters $\mu_k$ to zero to obtain satisfactory convergence. Although the motivation for this approach is perhaps not as clear as for other algorithms, it can be seen that, if the Lagrange multipliers $\lambda^k$ happen to be optimal in (14), then the solution of the original nonlinear program (12) would also be optimal for this subproblem. Under favorable assumptions, and provided that the sequence $\{\mu_k\}$ is chosen judiciously, we find that the sequence $(x^k, \lambda^k)$ converges to a point satisfying the optimality conditions (13).

Augmented Lagrangian methods were first proposed by Hestenes (in 1969) and Powell (in 1969). A 1982 book by Bertsekas was influential in later developments. The approach has proved particularly useful in "splitting" schemes, where the objective $f$ is decomposed naturally into a sum of functions, each of which is assigned its own copy of the variable vector $x$. Equality of the different copies is enforced via equality constraints, and the augmented Lagrangian method is applied to the resulting equality-constrained problem. The appeal of this approach is that minimization with respect to each copy of $x$ can be performed independently and these individual minimizations may be simpler to perform than minimization of the original function $f$. Moreover, the possibility arises of performing these minimizations simultaneously on a parallel computer.

### 6.5 Penalty Functions and Filters

Penalty functions combine the objective and constraint functions for a nonlinear program (1) into a single function, yielding an alternative problem whose solution is an approximate solution to the original constrained problem. The augmented Lagrangian function of section 6.4 can be viewed as a penalty function. Another important case is the $\ell_1$ penalty function, which is defined as follows:

$$f(x) + \nu \sum_{i \in \mathcal{E}} |c_i(x)| + \nu \sum_{i \in \mathcal{I}} \max(-c_i(x), 0), \qquad (15)$$

where $\nu > 0$ is a chosen *penalty parameter*. Note that each term in the summations is positive if and only if the corresponding constraint in (1) is violated. Under certain conditions, we have for $\nu$ sufficiently large that a local solution of (1) is an *exact* minimizer of (15). In other words, we can replace the constrained problem (1) by the unconstrained, but nonsmooth, problem of minimizing (15). One possible way to make use of this observation is to choose $\nu$ and minimize (15) directly, increasing $\nu$ as needed to ensure that the solutions of (15) and (1) coincide. More commonly, (15) is used as a *merit* function to evaluate the quality of proposed steps $d^k$ generated by some other algorithm, such as sequential quadratic programming or an interior-point method. Such steps are accepted only if they produce a sufficient reduction in (15) or some other merit function.

An alternative device to decide whether proposed steps are acceptable is a *filter*, which dispenses with the penalty parameter $\nu$ in (15) and considers the objective function $f$ and the constraint violations separately. Defining the violation measure by

$$h(x) = \sum_{i \in \mathcal{E}} |c_i(x)| + \sum_{i \in \mathcal{I}} \max(-c_i(x), 0),$$

the filter consists of a set of pairs $\{(f_\ell, h_\ell) : \ell \in F\}$ of objective and constraint values such that no pair dominates another: that is, we do not have $f_\ell \leqslant f_j$ and $h_\ell \leqslant h_j$ for any $\ell \in F$ and $j \in F$. An iterate $x^k$ is acceptable provided that $(f(x^k), h(x^k))$ is not dominated by any point in the filter $F$. When accepted, the new pair is added to the filter, and any pairs that are dominated by it are removed. (This basic strategy is amended in several ways to improve its practical performance and to facilitate convergence analysis.)

## 7 Final Remarks

Our brief description of major problem classes in continuous optimization, and algorithms for solving them, has necessarily omitted several important topics. We mention several of these before closing.

*Stochastic* and *robust* optimization deal with problems in which there is uncertainty in the objective functions or constraints but where the uncertainty can be quantified and modeled. In these problems we may seek solutions that minimize the *expected value* of the uncertain objective or solutions that are guaranteed to satisfy the constraints with a certain specified probability. The stochastic gradient method of section 4.4 provides one tool for solving these problems, but there are many other relevant techniques from optimization and statistics that can be brought to bear.

*Equilibrium problems* are not optimization problems, in that there is no objective to be minimized, but they use a range of algorithmic techniques that are closely related to optimization techniques. The basic formulation is as follows: given a function $F\colon \mathbb{R}^n \to \mathbb{R}^n$, find a vector $x \in \mathbb{R}^n$ such that

$$x \geqslant 0, \qquad F(x) \geqslant 0, \qquad x_i F_i(x) = 0, \quad i = 1, 2, \dots, n.$$

(Note that the KKT conditions in (6) have a similar form.) Equilibrium problems arise in economic applications and game theory. More recently, applications have been identified in contact problems in mechanical simulations.

*Nonlinear equations*, in which we seek a vector $x \in \mathbb{R}^n$ such that $F(x) = 0$ for some smooth function $F\colon \mathbb{R}^n \to \mathbb{R}^n$, arise throughout scientific computing. Newton's method, so fundamental in continuous optimization, is also key here. The Newton step is obtained by solving

$$\nabla F(x^k) d^k = -F(x^k)$$

(compare this with (10)), where $\nabla F(x) = [\partial F_i / \partial x_j]_{i,j=1}^n$ is the $n \times n$ Jacobian matrix.

### Further Reading


Bertsekas, D. P. 1999. *Nonlinear Programming*, 2nd edn. Nashua, NH: Athena Scientific.

Boyd, S., and L. Vandenberghe. 2003. *Convex Optimization.* Cambridge: Cambridge University Press.

Nesterov, Y., and A. S. Nemirovskii. 1994. *Interior-Point Polynomial Methods in Convex Programming.* Philadelphia, PA: SIAM.

Nocedal, J., and S. J. Wright. 2006. *Numerical Optimization*, 2nd edn. New York: Springer.

Todd, M. J. 2001. Semidefinite optimization. *Acta Numerica* 10:515–60.

Wright, S. J. 1997. *Primal-Dual Interior-Point Methods.* Philadelphia, PA: SIAM.


# IV.12 Numerical Solution of Ordinary Differential Equations

*Ernst Hairer and Christian Lubich*

## 1 Introduction: Euler Methods

Ordinary differential equations are ubiquitous in science and engineering: in geometry and mechanics from the first examples onward (Newton, Leibniz, Euler, Lagrange), in chemical reaction kinetics, molecular dynamics, electronic circuits, population dynamics, and in many more application areas. They also arise, after semidiscretization in space, in the numerical treatment of time-dependent partial differential equations, which are even more impressively omnipresent in our technologically developed and financially controlled world.

The standard initial-value problem is to determine a vector-valued function $y\colon [t_0, T] \to \mathbb{R}^d$ with a given initial value $y(t_0) = y_0 \in \mathbb{R}^d$ such that the derivative $y'(t)$ depends on the current solution value $y(t)$ at every $t \in [t_0, T]$ in a prescribed way:

$$y'(t) = f(t, y(t)) \quad \text{for } t_0 \leqslant t \leqslant T, \qquad y(t_0) = y_0.$$

Here, the given function $f$ is defined on an open subset of $\mathbb{R} \times \mathbb{R}^d$ containing $(t_0, y_0)$ and takes values in $\mathbb{R}^d$. If $f$ is continuously differentiable, then there exists a unique solution at least locally on some open interval containing $t_0$. In many applications, $t$ represents time, and it will be convenient to refer to $t$ as time in what follows.

In spite of the ingenious efforts of mathematicians throughout the eighteenth and nineteenth centuries, in most cases the solution of a differential equation cannot be given in closed form by functions that can be evaluated directly on a computer. This even applies to linear differential equations $y' = Ay$ with a square matrix $A$, for which $y(t) = e^{(t-t_0)A} y_0$, as computing the MATRIX EXPONENTIAL [II.14] is a notoriously tricky problem. One must therefore rely on numerical methods that are able to approximate the solution of a differential equation to any desired accuracy.

### 1.1 The Explicit Euler Method

The ancestor of all the advanced numerical methods in use today was proposed by Leonhard Euler in 1768. On writing down the first terms in the Taylor expansion of the solution at $t_0$ and using the prescribed initial value and the differential equation at $t = t_0$, it is noted that $y(t_0 + h) = y(t_0) + hy'(t_0) + \cdots = y_0 + hf(t_0, y_0) + \cdots$. Choosing a small step size $h > 0$ and neglecting the higher-order terms represented by the dots, an approximation $y_1$ to $y(t_1)$ at the later time $t_1 = t_0 + h$ is obtained by setting

$$y_1 = y_0 + hf(t_0, y_0).$$

The next idea is to take $y_1$ as the starting value for a further step, which then yields an approximation to the solution at $t_2 = t_1 + h$ as $y_2 = y_1 + hf(t_1, y_1)$. Continuing in this way, at the $(n+1)$st step we take $y_n \approx y(t_n)$

as the starting value for computing an approximation at $t_{n+1} = t_n + h$ as

$$y_{n+1} = y_n + hf(t_n, y_n),$$

and after a sufficient number of steps, we reach the final time $T$. The computational cost of the method lies in the evaluations of the function $f$. The step size need not be the same in each step and could be replaced by $h_n$ in the formula, so that $t_{n+1} = t_n + h_n$.

It is immediate that the quality of the approximation $y_n$ depends on two aspects: the error made by truncating the Taylor expansion and the error introduced by continuing from approximate solution values. These two aspects are captured in the notions of *consistency* and *stability*, respectively, and are fundamental to all numerical methods for ordinary differential equations.

### 1.2 The Implicit Euler Method and Stiff Differential Equations

A minor-looking change in the method, already considered by Euler in 1768, makes a big difference; taking as the argument of $f$ the new value instead of the previous one yields

$$y_{n+1} = y_n + hf(t_{n+1}, y_{n+1}),$$

from which $y_{n+1}$ is now determined implicitly. In general, the new solution approximation needs to be computed iteratively, typically by a modified Newton method such as $y_{n+1}^{(k+1)} = y_{n+1}^{(k)} + \Delta y_{n+1}^{(k)}$, where the increment is computed by solving a linear system of equations

$$(I - hJ_n)\Delta y_{n+1}^{(k)} = -r_{n+1}^{(k)}$$

with an approximation $J_n$ to the Jacobian matrix $\partial_y f(t_n, y_n)$ and the residual

$$r_{n+1}^{(k)} = y_{n+1}^{(k)} - y_n - hf(t_{n+1}, y_{n+1}^{(k)}).$$

The computational cost per step has increased dramatically; whereas the explicit Euler method requires a single function evaluation, we now need to compute the Jacobian and then solve a linear system and evaluate $f$ on each Newton iteration.

Why it may nevertheless be preferable to perform the computation using the implicit rather than the explicit Euler method is evident for the scalar linear example, made famous by Germund Dahlquist in 1963,

$$y' = \lambda y,$$

where the coefficient $\lambda$ is large and negative (or complex with large negative real part). Here the exact solution $y(t) = e^{(t-t_0)\lambda} y_0$ decays to zero as time increases, and

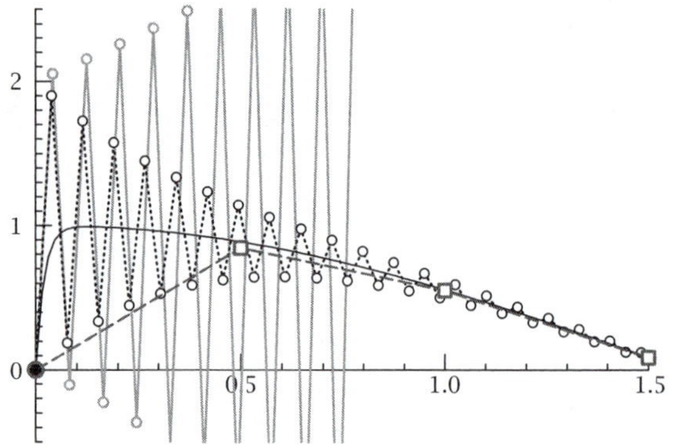


**Figure 1** The exact solution (solid line), the implicit Euler solution ($h = 0.5$, dashed line), and two explicit Euler solutions ($h = 0.038$, dotted line; $h = 0.041$, gray line) for the problem $y' = -50(y - \cos t)$, $y(0) = 0$.

so does the numerical solution given by the implicit Euler method *for every step size* $h > 0$:

$$y_n^{\text{impl}} = (1 - h\lambda)^{-n} y_0.$$

In contrast, the explicit Euler method yields

$$y_n^{\text{expl}} = (1 + h\lambda)^n y_0,$$

which decays to zero for growing $n$ only when $h$ is so small that $|1 + h\lambda| < 1$. This imposes a severe step-size restriction when $\lambda$ is a negative number of large absolute value (just think of $\lambda = -10^{10}$). For larger step sizes the numerical solution suffers an instability that is manifested in wild oscillations of increasing amplitude. The problem is that the explicit Euler method and the differential equation have completely different stability behaviors unless the step size is chosen extremely small (see figure 1).

Such behavior is not restricted to the simple scalar example considered above but extends to linear systems of differential equations in which the matrix has some eigenvalues with large negative real part and to classes of nonlinear differential equations with a Jacobian matrix $\partial_y f$ having this property. The explicit and implicit Euler methods also give rise to very different behaviors for nonlinear differential equations in which the function $f(t, y)$ has a large (local) Lipschitz constant $L$ with respect to $y$, while for some inner product the inequality

$$\langle f(t,y) - f(t,z), y - z \rangle \leqslant \ell \|y - z\|^2$$

holds for all $t$ and $y, z$ with a moderate constant $\ell \ll L$ (called a one-sided Lipschitz constant).

Differential equations for which the numerical solution using the implicit Euler method is more efficient than that using the explicit Euler method are called *stiff* differential equations. They include important applications in the description of processes with multiple timescales (e.g., fast and slow chemical reactions) and in spatial semidiscretizations of time-dependent partial differential equations. For example, for the heat equation, stable numerical solutions are obtained with the explicit Euler method only when temporal step sizes are bounded by the square of the spatial grid size, whereas the implicit Euler method is unconditionally stable.

### 1.3 The Symplectic Euler Method and Hamiltonian Systems

An important class of differential equations for which neither the explicit nor the implicit Euler method is appropriate is *Hamiltonian differential equations*,

$$p' = -\nabla_q H(p,q), \qquad q' = +\nabla_p H(p,q),$$

which are fundamental to many branches of physics. Here, the real-valued Hamilton function $H$, defined on a domain of $\mathbb{R}^{d+d}$, represents the total energy, and $q(t) \in \mathbb{R}^d$ and $p(t) \in \mathbb{R}^d$ represent the positions and momenta, respectively, of a conservative system at time $t$. The total energy is conserved:

$$H(p(t),q(t)) = \text{const.}$$

along any solution $(p(t),q(t))$ of the Hamiltonian system. It turns out that a partitioned method obtained by applying the explicit Euler method to the position variables and the implicit Euler method to the momentum variables (or vice versa) behaves much better than either Euler method applied to the system as a whole. The *symplectic Euler method* reads

$$p_{n+1} = p_n - h\nabla_q H(p_{n+1},q_n),$$
$$q_{n+1} = q_n + h\nabla_p H(p_{n+1},q_n).$$

For a separable Hamiltonian $H(p,q) = T(p) + V(q)$ the method is explicit.

Figure 2 illustrates the qualitative behavior of the three Euler methods applied to the differential equations of the mathematical pendulum,

$$p' = -\sin q, \qquad q' = p,$$

which are Hamiltonian with $H(p,q) = \frac{1}{2}p^2 - \cos q$. The energy of the implicit Euler solution decreases, while that of the explicit Euler solution increases. The symplectic Euler method nearly conserves the energy over extremely long times.

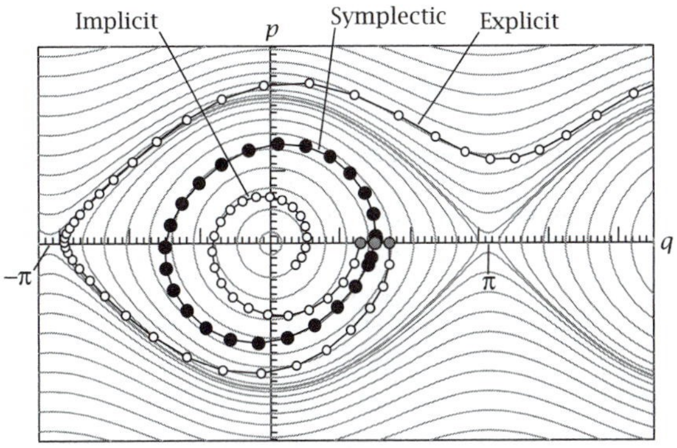


**Figure 2** The pendulum equation: Euler polygons with step size $h = 0.3$; initial value $p(0) = 0$ and $q(0) = 1.7$ for the explicit Euler method, $q(0) = 1.5$ for the symplectic Euler method, and $q(0) = 1.3$ for the implicit Euler method. The solid lines are solution curves for the differential equations.

## 2 Basic Notions

In this section we describe some of the mechanisms that lead to the different behaviors of the various methods.

### 2.1 Local Error

For the explicit Euler method, the error after one step of the method starting from the exact solution, called the *local error*, is given as

$$d_{n+1} = (y(t_n) + hf(t_n, y(t_n))) - y(t_n + h).$$

By estimating the remainder term in the Taylor expansion of $y(t_n + h)$ at $t_n$, we can bound $d_{n+1}$ by

$$\|d_{n+1}\| \leqslant Ch^2 \quad \text{with } C = \tfrac{1}{2} \max_{t_0 \leqslant t \leqslant T} \|y''(t)\|,$$

provided that the solution is twice continuously differentiable, which is the case if $f$ is continuously differentiable.

### 2.2 Error Propagation

Since the method advances in each step with the computed values $y_n$ instead of the exact solution values $y(t_n)$, it is important to know how errors, once introduced, are propagated by the method. Consider explicit Euler steps starting from different starting values:

$$u_{n+1} = u_n + hf(t_n, u_n),$$
$$v_{n+1} = v_n + hf(t_n, v_n).$$

When $f$ is (locally) Lipschitz continuous with Lipschitz constant $L$, the difference is controlled by the *stability estimate*

$$\|u_{n+1} - v_{n+1}\| \leqslant (1 + hL)\|u_n - v_n\|.$$

### 2.3 Lady Windermere's Fan

The above estimates can be combined to study the error accumulation, as illustrated by the fan of figure 3 (named by Gerhard Wanner in the 1980s after a play by Oscar Wilde). Each arrow from left to right represents a step of the numerical method, with different starting values. The fat vertical bars represent the local errors, whose propagation by the numerical method is controlled using the stability estimate repeatedly from step to step; the *global error*

$$e_n = y_n - y(t_n)$$

is the sum of the propagated local errors (represented as the distances between two adjacent arrowheads ending at $t_n$ in figure 3). The contribution of the first local error $d_1$ to the global error $e_n$ is bounded by $(1 + hL)^{n-1}\|d_1\|$, as is seen by applying the stability estimate $n - 1$ times following the numerical solutions starting from $y_1$ and $y(t_1)$. The contribution of the second local error $d_2$ is bounded by $(1 + hL)^{n-2}\|d_2\|$, and so on. Since the local errors are bounded by $Ch^2$, the global error is thus bounded by

$$\|e_n\| \leqslant \sum_{j=0}^{n-1}(1 + hL)^j Ch^2$$
$$= \frac{(1 + hL)^n - 1}{1 + hL - 1}Ch^2$$
$$\leqslant \frac{e^{nhL} - 1}{L}Ch.$$

With $M = (e^{(T-t_0)L} - 1)C/L$, the global error satisfies

$$\|e_n\| \leqslant Mh \quad \text{for } t_n \leqslant T.$$

The numerical method thus converges to the exact solution as $h \to 0$ with $nh$ fixed, but only at first order, that is, with an error bound proportional to $h$. We will later turn to higher-order numerical methods, with an error bound proportional to $h^p$ with $p > 1$.

### 2.4 Stiff Differential Equations

The above error bound becomes meaningless for stiff problems, where $L$ is large. The implicit Euler method admits an analogous error analysis in which only the one-sided Lipschitz constant $\ell$ appears in the stability estimate; provided that $h\ell < 1$,

$$\|u_{n+1} - v_{n+1}\| \leqslant \frac{1}{1 - h\ell}\|u_n - v_n\|$$

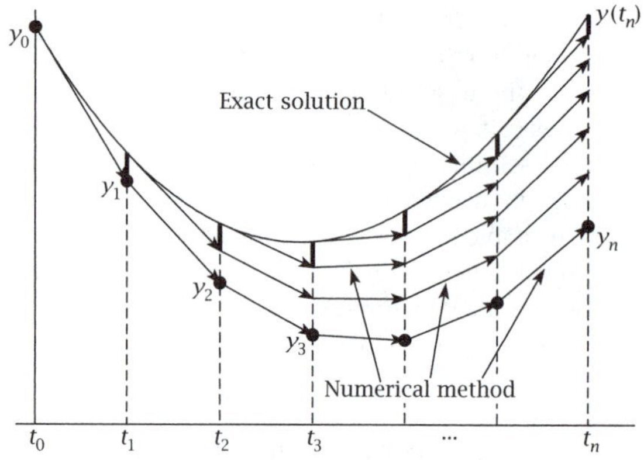


**Figure 3** Lady Windermere's fan.

holds for the results of two Euler steps starting from $u_n$ and $v_n$. For stiff problems with $\ell \ll L$ this is much more favorable than the stability estimate of the explicit Euler method in terms of $L$. It leads to an error bound $\|e_n\| \leqslant mh$, in which $m$ is essentially of the same form as $M$ above but with the Lipschitz constant $L$ replaced by the one-sided Lipschitz constant $\ell$.

The above arguments explain the convergence behavior of the explicit and implicit Euler methods and their fundamentally different behavior for large classes of stiff differential equations. They do not explain the favorable behavior of the symplectic Euler method for Hamiltonian systems. This requires another concept, backward analysis, which is treated next.

### 2.5 Backward Analysis

Much insight into numerical methods is obtained by interpreting the numerical result after a step as the (almost) exact solution of a *modified differential equation*. Properties of the numerical method can then be inferred from properties of a differential equation. For each of the Euler methods applied to $y' = f(y)$ an asymptotic expansion

$$\tilde{f}(\tilde{y}) = f(\tilde{y}) + hf_2(\tilde{y}) + h^2 f_3(\tilde{y}) + \cdots$$

can be uniquely constructed recursively such that, up to arbitrarily high powers of $h$,

$$y_1 = \tilde{y}(t_1),$$

where $\tilde{y}(t)$ is the solution of the modified differential equation $\tilde{y}' = \tilde{f}(\tilde{y})$ with initial value $y_0$. The remarkable feature is that, when the symplectic Euler method is applied to a Hamiltonian system, the modified differential equation is again Hamiltonian. The modified

Hamilton function has an asymptotic expansion

$$\tilde{H} = H + hH_2 + h^2H_3 + \cdots .$$

The symplectic Euler method therefore conserves the modified energy $\tilde{H}$ (up to arbitrarily high powers of $h$), which is close to the exact energy $H$. This conservation of the modified energy prevents the linearly growing drift in the energy that is present along numerical solutions of the explicit and implicit Euler methods. For these two methods the modified differential equation is no longer Hamiltonian.

## 3 Nonstiff Problems

### 3.1 Higher-Order Methods

A method is said to have *order* $p$ if the local error (recall that this is the error after one step of the method starting from the exact solution) is bounded by $Ch^{p+1}$, where $h$ is the step size and $C$ depends only on bounds of derivatives of the solution $y(t)$ and of the function $f$. As for the Euler method in section 2.1, the order is determined by comparing the Taylor expansions of the exact solution and the numerical solution, which for a method of order $p$ should agree up to and including the $h^p$ term.

A drawback of the Euler methods is that they are only of order 1. There are different ways to increase the order: using additional, auxiliary function evaluations in passing from $y_n$ to $y_{n+1}$ (one-step methods); using previously computed solution values $y_{n-1}, y_{n-2}, \dots$ and/or their function values (multistep methods); or using both (general linear methods). For nonstiff initial-value problems the most widely used methods are explicit Runge–Kutta methods of orders up to 8, in the class of one-step methods, and Adams-type multistep methods up to order 12. For very stringent accuracy requirements of 10 or 100 digits, high-order extrapolation methods or high-order Taylor series expansions of the solution (when higher derivatives of $f$ are available with automatic differentiation software) are sometimes used.

### 3.2 Explicit Runge–Kutta Methods

Two ideas underlie Runge–Kutta methods. First, the integral in

$$y(t_0 + h) = y(t_0) + h \int_0^1 y'(t_0 + \theta h)\,\mathrm{d}\theta,$$

with $y'(t) = f(t, y(t))$, is approximated by a quadrature formula with weights $b_i$ and nodes $c_i$:

$$y_1 = y_0 + h \sum_{i=1}^{s} b_i Y_i', \quad Y_i' = f(t_0 + c_i h, Y_i).$$

Second, the internal stage values $Y_i \approx y(t_0 + c_i h)$ are determined by another quadrature formula for the integral from 0 to $c_i$:

$$Y_i = y_0 + h \sum_{j=1}^{s} a_{ij} Y_j', \quad i = 1, \dots, s,$$

with the *same* function values $Y_j'$ as for $y_1$. If the coefficients satisfy $a_{ij} = 0$ for $j \geqslant i$, then the above sum actually extends only from $j = 1$ to $i - 1$, and hence $Y_1, Y_1', Y_2, Y_2', \dots, Y_s, Y_s'$ can be computed explicitly one after the other. The methods are named after Carl Runge, who in 1895 proposed two- and three-stage methods of this type, and Wilhelm Kutta, who in 1901 proposed what is now known as the *classical Runge–Kutta method* of order 4, which extends the Simpson quadrature rule from integrals to differential equations:

$$\begin{array}{c|c} c_i & a_{ij} \\ \hline & b_j \end{array} \equiv \begin{array}{c|cccc} 0 & & & & \\ \frac{1}{2} & \frac{1}{2} & & & \\ \frac{1}{2} & 0 & \frac{1}{2} & & \\ 1 & 0 & 0 & 1 & \\ \hline & \frac{1}{6} & \frac{2}{6} & \frac{2}{6} & \frac{1}{6} \end{array}$$

Using Lady Windermere's fan as in section 2, one finds that the global error $y_n - y(t_n)$ of a $p$th-order Runge–Kutta method over a bounded time interval is $\mathcal{O}(h^p)$.

The order conditions of general Runge–Kutta methods were elegantly derived by John Butcher in 1964 using a tree model for the derivatives of $f$ and their concatenations by the chain rule, as they appear in the Taylor expansions of the exact solution and the numerical solution. This enabled the construction of methods of even higher order, among which excellently constructed methods of orders 5 and 8 by Dormand and Prince (from 1980) have found widespread use. These methods are equipped with lower-order error indicators from embedded formulas that use the same function evaluations. These error indicators are used for an adaptive selection of the step size that is intended to keep the local error close to a given error tolerance in each time step.

### 3.3 Extrapolation Methods

A systematic, if suboptimal, construction of explicit Runge–Kutta methods of arbitrarily high order is provided by *Richardson extrapolation* of the results of the explicit Euler method obtained with different step sizes. This technique makes use of an asymptotic expansion of the error,

$$y(t,h) - y(t) = e_1(t)h + e_2(t)h^2 + \cdots,$$

where $y(t,h)$ is the explicit Euler approximation at $t$ obtained with step size $h$. At $t = t_0 + H$, the error expansion coefficients up to order $p$ can be eliminated by evaluating at $h = 0$ (*extrapolating*) the interpolation polynomial through the Euler values $y(t_0 + H, h_j)$ for $j = 1, \dots, p$ corresponding to different step sizes $h_j = H/j$. This gives a method of order $p$, which formally falls into the general class of Runge–Kutta methods. Instead of using the explicit Euler method as the basic method, it is preferable to use *Gragg's method* (from 1964), which uses the explicit midpoint rule

$$y_{n+1} = y_{n-1} + 2hf(t_n, y_n), \quad n \geqslant 1,$$

and an explicit Euler starting step to compute $y_1$. This method has an error expansion in powers of $h^2$ (instead of $h$, above) at even $n$, and with the elimination of each error coefficient one therefore gains a power of $h^2$. Extrapolation methods have built-in error indicators that can be used for order and step-size control.

### 3.4 Adams Methods

The methods introduced by astronomer John Couch Adams in 1855 were the first of high order that used only function evaluations, and in their current variable-order, variable-step-size implementations they are among the most efficient methods for general nonstiff initial-value problems.

When $k$ function values $f_{n-j} = f(t_{n-j}, y_{n-j})$ $(j = 0, \dots, k-1)$ have already been computed, the integrand in

$$y(t_{n+1}) = y(t_n) + h \int_0^1 f(t_n + \theta h, y(t_n + \theta h))\,\mathrm{d}\theta$$

is approximated by the interpolation polynomial $P(t)$ through $(t_{n-j}, f_{n-j})$, $j = 0, \dots, k-1$, yielding the *explicit Adams method* of order $k$,

$$y_{n+1} = y_n + h \int_0^1 P(t_n + \theta h)\,\mathrm{d}\theta,$$

which, upon inserting the Newton interpolation formula, becomes (for constant step size $h$)

$$y_{n+1} = y_n + hf_n + h \sum_{i=1}^{k-1} \gamma_i \nabla^i f_n,$$

with the backward differences $\nabla f_n = f_n - f_{n-1}$, $\nabla^i f_n = \nabla^{i-1} f_n - \nabla^{i-1} f_{n-1}$, and with coefficients $(\gamma_i) = (\frac{1}{2}, \frac{5}{12}, \frac{3}{8}, \frac{251}{720}, \dots)$. The method thus corrects the explicit Euler method by adding differences of previous function values.

Especially for higher orders, the accuracy of the approximation suffers from the fact that the interpolation polynomial is used outside the interval of interpolation. This is avoided if the (as yet) unknown value $(t_{n+1}, f_{n+1})$ is added to the interpolation points. Let $P^*(t)$ denote the corresponding interpolation polynomial, which is now used to replace the integrand $f(t, y(t))$. This yields the *implicit Adams method* of order $k + 1$, which takes the form

$$y_{n+1} = y_n + hf_{n+1} + h \sum_{i=1}^{k} \gamma_i^* \nabla^i f_{n+1},$$

with $(\gamma_i^*) = (-\frac{1}{2}, -\frac{1}{12}, -\frac{1}{24}, -\frac{19}{720}, \dots)$. The equation for $y_{n+1}$ is solved approximately by one or at most two fixed-point iterations, taking the result from the explicit Adams method as the starting iterate (the *predictor*) and inserting its function value on the right-hand side of the implicit Adams method (the *corrector*).

In a variable-order, variable-step-size implementation, the required starting values are built up by starting with the methods of increasing order $1, 2, 3, \dots$, one after the other. Strategies for increasing or lowering the order are based on monitoring the backward difference terms. Changing the step size is computationally more expensive, since it requires a recalculation of all method coefficients. It is facilitated by passing information from one step to the next by the *Nordsieck vector*, which collects the values of the interpolation polynomial and all its derivatives at $t_n$ scaled by powers of the step size.

### 3.5 Linear Multistep Methods

Both explicit and implicit Adams methods (with constant step size) belong to the class of *linear multistep methods*

$$\sum_{j=0}^{k} \alpha_j y_{n+j} = h \sum_{j=0}^{k} \beta_j f_{n+j},$$

with $f_i = f(t_i, y_i)$ and $\alpha_k \neq 0$. This class also includes important methods for stiff problems, in particular the backward differentiation formulas to be described in the next section. The theoretical study of linear multistep methods was initiated by Dahlquist in 1956. He showed that for such methods

$$\text{consistency} + \text{stability} \iff \text{convergence},$$

which, together with the contemporaneous Lax equivalence theorem for discretizations of partial differential equations, forms a basic principle of numerical analysis. What this means here is described in more detail below.

In contrast to one-step methods, having high order does not by itself guarantee that a multistep method converges as $h \to 0$. In fact, choosing the method coefficients in such a way that the order is maximized for a given $k$ leads to a method that produces wild oscillations, which increase in magnitude with decreasing step size. One requires in addition a stability condition, which can be phrased as saying that all solutions to the linear difference equation $\sum_{j=0}^{k} \alpha_j y_{n+j} = 0$ stay bounded as $n \to \infty$, or equivalently:

> All roots of the polynomial $\sum_{j=0}^{k} \alpha_j \zeta^j$ are in the complex unit disk, and those on the unit circle are simple.

If this stability condition is satisfied and the method is of order $p$, then the error satisfies $y_n - y(t_n) = \mathcal{O}(h^p)$ on bounded time intervals, provided that the error in the starting values is $\mathcal{O}(h^p)$.

Dahlquist also proved *order barriers*: the order of a stable $k$-step method cannot exceed $k+2$ if $k$ is even, $k+1$ if $k$ is odd, and $k$ if the method is explicit ($\beta_k = 0$).

### 3.6 General Linear Methods

Predictor-corrector Adams methods fall neither into the class of multistep methods, since they use the predictor as an internal stage, nor into the class of Runge–Kutta methods, since they use previous function values. Linear multistep methods and Runge–Kutta methods are extreme cases of a more general class of methods

$$u_{n+1} = S u_n + h\Phi(t_n, u_n, h),$$

where $u_n$ is a vector (usually of dimension a multiple of the dimension of the differential equation) from which the solution approximation $y_n \approx y(t_n)$ can be obtained by a linear mapping, $S$ is a square matrix, and $\Phi$ depends on function values of $f$. (For example, for predictor-corrector methods we would have $u_n = (y_n, y_n^{\text{pred}}, y_{n-1}, \dots, y_{n-k+1})$ in this framework.)

More general methods like these have been studied since the mid-1960s with the objective of looking for the "greatest good as a mean between extremes" (in the words of Aristotle and John Butcher). They include a number of methods of potential interest for both nonstiff and stiff problems, such as two-step Runge–Kutta methods or general linear methods with inherent Runge–Kutta stability, but as of now do not appear to have found their way into applications via competitive software.

## 4 Stiff Problems

We saw in the introduction that for important classes of differential equations, called stiff equations, the implicit Euler method yields a drastic improvement over the explicit Euler method. Are there higher-order methods with similarly good properties?

### 4.1 Backward Differentiation Formula Methods

The $k$-step implicit Adams methods, though naturally extending the implicit Euler method, perform disappointingly on stiff problems for $k > 1$. Multistep methods from another extension of the implicit Euler method, which is based on numerical differentiation rather than integration, turn out to be better for stiff problems. Suppose that $k$ solution approximations $y_{n-k+1}, \dots, y_n$ have already been computed, and consider the interpolation polynomial $u(t)$ passing through $y_{n+1-j}$ at $t_{n+1-j}$ for $j = 0, \dots, k$, including the as yet unknown approximation $y_{n+1}$. We then require the *collocation* condition

$$u'(t) = f(t, u(t)) \quad \text{at } t = t_{n+1},$$

or equivalently, in the case of a constant step size $h$,

$$\sum_{j=1}^{k} \frac{1}{j} \nabla^j y_{n+1} = h f_{n+1}.$$

This *backward differentiation formula* (BDF) is an implicit linear multistep method of order $k$, which is found to be unstable for $k > 6$. Methods for smaller $k$, however, up to $k \leqslant 5$, are currently the most widely used methods for stiff problems, which are implemented in numerous computer codes. The usefulness of these methods was first observed by Curtiss and Hirschfelder in 1952, who also coined the notion of stiff differential equations. Bill Gear's BDF code "DIFSUB" from 1971 was the first widely used code for stiff problems. It brought BDF methods ("Gear's method") to the attention of practitioners in many fields.

### 4.2 A-Stability and Related Notions

Which properties make BDF methods successful for stiff problems? In 1963, Dahlquist systematically studied the behavior of multistep methods on the scalar linear differential equation

$$y' = \lambda y \quad \text{with } \lambda \in \mathbb{C}, \ \operatorname{Re}\lambda \leqslant 0,$$

whose use as a test equation can be justified by linearization of the differential equation and diagonalization of the Jacobian matrix. The behavior of a numerical method on this deceivingly simple scalar linear differential equation gives much insight into its usefulness for more general stiff problems, as is shown by both numerical experience and theory.

Clearly, the exact solution $y(t) = e^{t\lambda}y_0$ remains bounded for $t \to +\infty$ when $\operatorname{Re}\lambda \leqslant 0$. Following Dahlquist, a method is called *A-stable* if for every $\lambda \in \mathbb{C}$ with $\operatorname{Re}\lambda \leqslant 0$, the numerical solution $y_n$ stays bounded as $n \to \infty$ for every step size $h > 0$ and every choice of starting values. The implicit Euler method and the second-order BDF method are A-stable, but the BDF methods of higher order are not. *Dahlquist's second-order barrier* states that the order of an A-stable linear multistep method cannot exceed 2. This fundamental, if negative, theoretical result has led to much work aimed at circumventing the barrier by using other methods or weaker notions of stability.

The *stability region S* is the set of all complex $z = h\lambda$, such that every numerical solution of the method applied to $y' = \lambda y$ with step size $h$ stays bounded. The stability regions of explicit and implicit $k$-step Adams methods with $k > 1$ are bounded, which leads to step-size restrictions when $\lambda$ has large absolute value. The BDF methods up to order 6 are $A(\alpha)$-stable; that is, the stability region contains an unbounded sector $|\arg(-z)| \leqslant \alpha$ with $\alpha = 90°, 90°, 86°, 73°, 51°, 17°$ for $k = 1, \dots, 6$, respectively. The higher-order BDF methods therefore perform well for differential equations where the Jacobian has large eigenvalues near the negative real half-axis, but they behave poorly when there are large eigenvalues near the imaginary axis.

### 4.3 Implicit Runge–Kutta Methods

It turns out that there is no order barrier for A-stable Runge–Kutta methods.

Explicit Runge–Kutta methods cannot be A-stable because application of such a method to the linear test equation yields $y_{n+1} = P(h\lambda)y_n$, where $P$ is a polynomial of degree $s$, the number of stages. The stability region of such a method is necessarily bounded, since $|P(z)| \to \infty$ as $|z| \to \infty$.

On the other hand, an *implicit* Runge–Kutta method has

$$y_{n+1} = R(h\lambda)y_n$$

with a rational function $R(z)$, called the *stability function* of the method, which is an approximation to the exponential at the origin, $R(z) = e^z + \mathcal{O}(z^{p+1})$ as $z \to 0$. The method is A-stable if $|R(z)| \leqslant 1$ for $\operatorname{Re} z \leqslant 0$. The subtle interplay between order and stability is clarified by the theory of *order stars*, developed by Wanner, Hairer, and Nørsett in 1978. In particular, this theory shows that among the PADÉ APPROXIMANTS [IV.9 §2.4] $R_{k,j}(z)$ to the exponential (the rational approximations of numerator degree $k$ and denominator degree $j$ of highest possible order $p = j + k$), precisely those with $k \leqslant j \leqslant k + 2$ are A-stable. Optimal-order implicit Runge–Kutta methods having the diagonal Padé approximants $R_{s,s}$ as stability function are the collocation methods based on the Gauss quadrature nodes, while those having the subdiagonal Padé approximants $R_{s-1,s}$ are the collocation methods based on the right-hand Radau quadrature nodes. We turn to these important implicit Runge–Kutta methods next.

### 4.4 Gauss and Radau Methods

A *collocation method* based on the nodes $0 \leqslant c_1 < \cdots < c_s \leqslant 1$ determines a polynomial $u(t)$ of degree at most $s$ such that $u(t_0) = y_0$ and the differential equation is satisfied at the $s$ points $t_0 + c_i h$:

$$u'(t) = f(t, u(t)) \quad \text{at } t = t_0 + c_i h,\ i = 1, \dots, s.$$

The solution approximation at the endpoint is then

$$y_1 = u(t_0 + h),$$

which is taken as the starting value for the next step. As was shown by Ken Wright (in 1970), such a collocation method is equivalent to an implicit Runge–Kutta method, the order of which is equal to the order of the underlying interpolatory quadrature with nodes $c_i$. The highest order $p = 2s$ is thus obtained with Gauss nodes. Nevertheless, Gauss methods have found little use in stiff initial-value problems (as opposed to boundary-value problems; see section 6). The reason for this is that the stability function here satisfies $|R(z)| \to 1$ as $z \to -\infty$, whereas $e^z \to 0$ as $z \to -\infty$.

The desired damping property at infinity is obtained for the subdiagonal Padé approximant. This is the stability function for the collocation method at Radau points, which are the nodes of the quadrature formula of order $p = 2s - 1$ with $c_s = 1$. Let us collect the basic properties: the $s$-stage Radau method is an implicit Runge–Kutta method of order $p = 2s - 1$; it is A-stable and has $R(\infty) = 0$.

The Radau methods have some more remarkable features: they are nonlinearly stable with the so-called *algebraic stability* property; their last internal stage equals the starting value for the next step (this property is useful for very stiff and for differential-algebraic equations); and their internal stages all have order $s$.

The last property does indeed hold for every collocation method with $s$ nodes. It is important because of the phenomenon of *order reduction*; in the application of an implicit Runge–Kutta method to stiff problems, the method may have only the stage order, or stage order + 1, with stiffness-independent error constants, instead of the full classical order $p$ that is obtained with nonstiff problems.

The implementation of Radau methods by Hairer and Wanner (from 1991) is known for its robustness in dealing with stiff problems and differential-algebraic equations of the type $My' = f(t, y)$ with a singular matrix $M$.

### 4.5 Linearly Implicit Methods

BDF and implicit Runge–Kutta methods are fully implicit, and the resulting systems of nonlinear equations need to be solved by variants of Newton's method. To reduce the computational cost while retaining favorable linear stability properties, linearly implicit methods have been proposed, such as the *linearly implicit Euler method*, in which only a single iteration of Newton's method is done in each step:

$$(I - hJ_n)(y_{n+1} - y_n) = hf_n,$$

where $J_n \approx \partial_y f(t_n, y_n)$. Thus just one linear system of equations is solved in each time step. The method is identical to the implicit Euler method for linear problems and therefore inherits its A-stability. Higher-order linearly implicit methods can be obtained by Richardson extrapolation of the linearly implicit Euler method, or they are specially constructed *Rosenbrock methods*. Like explicit Runge–Kutta methods, these methods determine the solution approximation as

$$y_1 = y_0 + h \sum_{i=1}^{s} b_i Y_i', \quad Y_i = y_0 + h \sum_{j=1}^{i-1} a_{ij} Y_j',$$

but compute the derivative stages consecutively by solving $s$ linear systems of equations (written here for an autonomous problem, $f(t, y) = f(y)$ and $J = \partial_y f(y_0)$):

$$(I - \gamma hJ)Y_i' = f(Y_i) + hJ \sum_{j=1}^{i-1} \gamma_{ij} Y_j'.$$

Such methods are easy to implement, and they have gained popularity in the numerical integration of spatial semidiscretizations of partial differential equations. For large problems, the dominating numerical cost is in the solution of the systems of linear equations, using either direct sparse solvers or iterative methods such as preconditioned Krylov subspace methods.

### 4.6 Exponential Integrators

While it appears an obvious idea to use the exponential of the Jacobian in a numerical method, this was for a long time considered impractical, and particularly so for large problems. This attitude changed, however, in the mid-1990s when it was realized that Krylov subspace methods for approximating a matrix exponential times a vector, $e^{\gamma hJ} v$, show superlinear convergence, whereas there is generally only linear convergence for solving linear systems $(I - \gamma hJ)x = v$. Unless a good preconditioner for the linear system is available, computing the action of the matrix exponential is therefore computationally less expensive than solving a corresponding linear system. This fact led to a revival of methods using the exponential or related functions like $\varphi(z) = (e^z - 1)/z$, such as the *exponential Euler method*

$$y_{n+1} = y_n + h\varphi(hJ_n) f_n.$$

The method is exact for linear $f(y) = Jy + c$. It differs from the linearly implicit Euler method in that the entire function $\varphi(z)$ replaces the rational function $1/(1 - z)$. Higher-order exponential methods of one-step and multistep type have also been constructed. Exponential integrators have proven useful for large-scale problems in physics and for nonlinear parabolic equations, as well as for highly oscillatory problems like those considered in section 5.6.

### 4.7 Chebyshev Methods

For moderately stiff problems one can avoid numerical linear algebra altogether by using *explicit* Runge–Kutta methods of low order (2 or 4) and high stage number, which are constructed to have a large stability domain covering a strip near the negative real semi-axis. The stability function of such methods is a high-degree polynomial related to Chebyshev polynomials. The stage number is chosen adaptively to include the product of the step size with the dominating eigenvalues of the Jacobian in the stability domain. With $s$ stages, one can cover intervals on the negative real axis of a length proportional to $s^2$. The quadratic growth of the stability interval with the stage number makes these methods suitable for problems with large negative real eigenvalues of the Jacobian, such as spatial semidiscretizations of parabolic partial differential equations.

## 5 Structure-Preserving Methods

The methods discussed so far are designed for general differential equations, and a distinction was drawn only between nonstiff and stiff problems. There are, however, important classes of differential equations with a special, often geometric, structure, whose preservation in the numerical discretization leads to substantially better methods, especially when integrating over long times. The most prominent of these are Hamiltonian systems, which are all-important in physics. Their flow has the geometric property of being symplectic.

In respecting the phase space geometry under discretization and analyzing its effect on the long-time behavior of a numerical method, there is a shift of viewpoint from concentrating on the approximation of a single solution trajectory to considering the numerical method as a discrete dynamical system that approximates the flow of a differential equation.

While many differential equations have interesting structures to preserve under discretization—and much work has been done in devising and analyzing appropriate numerical methods for doing so—we will restrict our attention to Hamiltonian systems here. Their numerical treatment has been an active research area for the past two decades.

### 5.1 Symplectic Methods

The time-$t$ *flow* of a differential equation $y' = f(y)$ is the map $\varphi_t$ that associates with an initial value $y_0$ at time 0 the solution value at time $t$: $\varphi_t(y_0) = y(t)$. Consider a Hamiltonian system

$$p' = -\nabla_q H(p,q), \qquad q' = \nabla_p H(p,q),$$

or equivalently, for $y = (p,q)$,

$$y' = J^{-1}\nabla H(y) \quad \text{with } J = \begin{pmatrix} 0 & I \\ -I & 0 \end{pmatrix}.$$

The flow $\varphi_t$ of a Hamiltonian system is *symplectic*; that is, the derivative $\mathrm{D}\varphi_t$ with respect to the initial value satisfies

$$\mathrm{D}\varphi_t(y)^{\mathrm{T}} J\, \mathrm{D}\varphi_t(y) = J$$

for all $y$ and $t$ for which $\varphi_t(y)$ exists. This is a quadratic relation formally similar to orthogonality, with $J$ in place of the identity matrix $I$, but it is related to the preservation of areas rather than lengths in phase space.

A numerical one-step method $y_{n+1} = \Phi_h(y_n)$ is called symplectic if the numerical flow $\Phi_h$ is a symplectic map:

$$\mathrm{D}\Phi_h(y)^{\mathrm{T}} J\, \mathrm{D}\Phi_h(y) = J.$$

Such methods exist; the "symplectic Euler method" of section 1.3 is indeed symplectic. Great interest in symplectic methods was spurred when, in 1988, Lasagni, Sanz-Serna, and Suris independently characterized symplectic Runge–Kutta methods as those whose coefficients satisfy the condition

$$b_i a_{ij} + b_j a_{ji} - b_i b_j = 0.$$

Gauss methods (see section 4.4) were already known to satisfy this condition and were thus found to be symplectic. Soon after those discoveries it was realized that a numerical method is symplectic if and only if the modified differential equation of backward analysis (section 2.5) is again Hamiltonian. This made it possible to prove rigorously the almost-conservation of energy over times that are exponentially long in the inverse step size, as well as further favorable long-time properties such as the almost-preservation of KAM (Kolmogorov–Arnold–Moser) tori of perturbed integrable systems over exponentially long times.

### 5.2 The Störmer–Verlet Method

By the time symplecticity was entering the field of numerical analysis, scientists in molecular simulation had been doing symplectic computations for more than 20 years without knowing it; the standard integrator of molecular dynamics, the method used successfully ever since Luc Verlet introduced it to the field in 1967, is symplectic. For a Hamiltonian $H(p,q) = \frac{1}{2}p^{\mathrm{T}}M^{-1}p + V(q)$ with a symmetric positive-definite mass matrix $M$, the method is explicit and is given by the formulas

$$\begin{aligned} p_{n+1/2} &= p_n - \tfrac{1}{2}h\nabla V(q_n), \\ q_{n+1} &= q_n + hM^{-1}p_{n+1/2}, \\ p_{n+1} &= p_{n+1/2} - \tfrac{1}{2}h\nabla V(q_{n+1}). \end{aligned}$$

Such a method was also formulated by the astronomer Störmer in 1907, and in fact it can even be traced back to Newton's *Principia* from 1687, where it was used as a theoretical tool in the proof of the preservation of angular momentum in the two-body problem (Kepler's second law), which is indeed preserved by this method. Given that there are already sufficiently many Newton methods in numerical analysis, it is fair to refer to the method as the *Störmer–Verlet method* (the Verlet method and the leapfrog method are also often-used names). As will be discussed in the next three subsections, the symplecticity of this method can be understood in various ways by relating the method to different classes of methods that have proven useful in a variety of applications.

### 5.3 Composition Methods

Let us denote the one-step map $(p_n, q_n) \mapsto (p_{n+1}, q_{n+1})$ of the Störmer–Verlet method by $\Phi_h^{\mathrm{SV}}$, that of the symplectic Euler method of section 1.3 by $\Phi_h^{\mathrm{SE}}$, and that of the adjoint symplectic Euler method by $\Phi_h^{\mathrm{SE}*}$, where instead of the argument $(p_{n+1}, q_n)$ one uses $(p_n, q_{n+1})$. The second-order Störmer–Verlet method can then be interpreted as the *composition* of the first-order symplectic Euler methods with halved step size:

$$\Phi_h^{\mathrm{SV}} = \Phi_{h/2}^{\mathrm{SE}*} \circ \Phi_{h/2}^{\mathrm{SE}}.$$

Since the composition of symplectic maps is symplectic, this shows the symplecticity of the Störmer–Verlet method. We further note that the method is *time reversible* (or *symmetric*): $\Phi_{-h} \circ \Phi_h = \mathrm{id}$, or equivalently $\Phi_h = \Phi_h^*$ with the adjoint method $\Phi_h^* := \Phi_{-h}^{-1}$. This is known to be another favorable property for conservative systems.

Moreover, from first-order methods we have obtained a second-order method. More generally, starting from a low-order method $\Phi_h$, methods of arbitrary order can be constructed by suitable compositions

$$\Phi_{c_s h} \circ \cdots \circ \Phi_{c_1 h} \quad \text{or} \quad \Phi_{b_s h}^* \circ \Phi_{a_s h} \circ \cdots \circ \Phi_{b_1 h}^* \circ \Phi_{a_1 h}.$$

The first to give systematic approaches to high-order compositions were Suzuki and Yoshida in 1990. Whenever the base method is symplectic, so is the composed method. The coefficients can be chosen such that the resulting method is symmetric.

### 5.4 Splitting Methods

Splitting the Hamiltonian $H(p,q) = T(p) + V(q)$ into its kinetic energy $T(p) = \frac{1}{2} p^{\mathrm{T}} M^{-1} p$ and its potential energy $V(q)$, we have that the flows $\varphi_t^T$ and $\varphi_t^V$ of the systems with Hamiltonians $T$ and $V$, respectively, are obtained by solving the trivial differential equations

$$\varphi_t^T : \begin{cases} p' = 0, \\ q' = M^{-1} p, \end{cases} \qquad \varphi_t^V : \begin{cases} p' = -\nabla V(q), \\ q' = 0. \end{cases}$$

We then note that the Störmer–Verlet method can be interpreted as a composition of the exact flows of the split differential equations:

$$\Phi_h^{\mathrm{SV}} = \varphi_{h/2}^T \circ \varphi_h^V \circ \varphi_{h/2}^T.$$

Since the flows $\varphi_{h/2}^T$ and $\varphi_h^V$ are symplectic, so is their composition.

Splitting the vector field of a differential equation and composing the flows of the subsystems is a structure-preserving approach that yields methods of arbitrary order and is useful in the time integration of a variety of ordinary and partial differential equations, such as linear and nonlinear Schrödinger equations.

### 5.5 Variational Integrators

For the Hamiltonian $H(p,q) = \frac{1}{2} p^{\mathrm{T}} M^{-1} p + V(q)$, the Hamilton equations of motion $\dot{p} = -\nabla V(q)$, $\dot{q} = M^{-1} p$ can be combined to give the second-order differential equation

$$M\ddot{q} = -\nabla V(q),$$

which can be interpreted as the Euler–Lagrange equations for minimizing the action integral

$$\int_{t_0}^{t_N} L(q(t), \dot{q}(t))\, \mathrm{d}t \quad \text{with } L(q, \dot{q}) = \tfrac{1}{2} \dot{q}^{\mathrm{T}} M \dot{q} - V(q)$$

over all paths $q(t)$ with fixed endpoints. In the Störmer–Verlet method, eliminating the momenta yields the second-order difference equations

$$M(q_{n+1} - 2q_n + q_{n-1}) = -h^2 \nabla V(q_n),$$

which are the discrete Euler–Lagrange equations for minimizing the discretized action integral

$$\sum_{n=0}^{N-1} \tfrac{1}{2} h \Big( L\Big(q_n, \frac{q_{n+1} - q_n}{h}\Big) + L\Big(q_{n+1}, \frac{q_{n+1} - q_n}{h}\Big) \Big),$$

which results from a trapezoidal rule approximation to the action integral and piecewise-linear approximation to $q(t)$. The Störmer–Verlet method can thus be interpreted as resulting from the direct discretization of the Hamilton variational principle. Such an interpretation can in fact be given for every symplectic method. Conversely, symplectic methods can be *constructed* by minimizing a discrete action integral. In particular, approximating the action integral by a quadrature formula and the positions $q(t)$ by a piecewise polynomial leads to a symplectic partitioned Runge–Kutta method, which in general uses different Runge–Kutta formulas for positions and momenta. With Gauss quadrature one reinterprets in this way the Gauss methods of section 4.4, and with higher-order Lobatto quadrature formulas one obtains higher-order relatives of the Störmer–Verlet method.

### 5.6 Oscillatory Problems

Highly oscillatory solution behavior in Hamiltonian systems typically arises when the potential is a multiscale sum $V = V^{[\text{slow}]} + V^{[\text{fast}]}$, where the Hessian of $V^{[\text{fast}]}$ has positive eigenvalues that are large compared with those of $V^{[\text{slow}]}$. (Here we assume that $M = I$ for simplicity.) With standard methods such as the Störmer–Verlet method, very small time steps would be required, for reasons of both accuracy and stability. Various numerical methods have been devised with the aim of overcoming this limitation. We describe just one such method here: a *multiple-time-stepping method* that reduces the computational work significantly when the slow force $f^{[\text{slow}]} = -\nabla V^{[\text{slow}]}$ is far more expensive to evaluate than the fast force $f^{[\text{fast}]} = -\nabla V^{[\text{fast}]}$. A basic principle is to rely on averages instead of pointwise force evaluations. In the *averaged-force method*, the force $f_n = -\nabla V(q_n)$ in the Störmer–Verlet method is replaced by an averaged force $\bar{f}_n$ as follows. We freeze the slow force at $q_n$ and consider the auxiliary differential equation

$$\ddot{u} = f^{[\text{slow}]}(q_n) + f^{[\text{fast}]}(u)$$

with initial values $u(0) = q_n$, $\dot{u}(0) = 0$. We then define the averaged force as

$$\bar{f}_n = \int_{-1}^{1} (1 - |\theta|)(f^{[\text{slow}]}(q_n) + f^{[\text{fast}]}(u(\theta h)))\, \mathrm{d}\theta,$$

which equals

$$\bar{f}_n = \frac{1}{h^2}(u(h) - 2u(0) + u(-h)).$$

The value $u(h)$ is computed approximately using smaller time steps, noting that $u(h) = u(-h)$.

The argument of $f^{[\text{slow}]}$ might preferably be replaced with an averaged value $\bar{q}_n$ in order to mitigate the adverse effect of possible step-size resonances that appear when the product of $h$ with an eigenfrequency of the Hessian is close to an integral multiple of $\pi$.

If the fast potential is quadratic, $V^{[\text{fast}]}(q) = \frac{1}{2}q^{\mathrm{T}}Aq$, the auxiliary differential equation can be solved exactly in terms of trigonometric functions of the matrix $h^2A$. The resulting method can then be viewed as an exponential integrator, as considered in section 4.6.

## 6 Boundary-Value Problems

In a two-point boundary-value problem, the differential equation is coupled with boundary conditions of the same dimension:

$$y'(t) = f(t, y(t)), \quad a \leqslant t \leqslant b,$$
$$r(y(a), y(b)) = 0.$$

As an important class of examples, such problems arise as the Euler–Lagrange equations of variational problems, typically with separated boundary conditions $r_a(y(a)) = 0$, $r_b(y(b)) = 0$.

### 6.1 The Sensitivity Matrix

The problem of existence and uniqueness of a solution is more subtle than for initial-value problems. For a linear boundary-value problem

$$y'(t) = C(t)y(t) + g(t), \quad a \leqslant t \leqslant b,$$
$$Ay(a) + By(b) = q,$$

a unique solution exists if and only if the *sensitivity matrix* $E = A + BU(b, a)$ is invertible, where $U(t, s)$ is the propagation matrix yielding $v(t) = U(t, s)v(s)$ for every solution of the linear differential equation $v'(t) = C(t)v(t)$.

A solution of a nonlinear boundary-value problem is locally unique if the linearization along this solution has an invertible sensitivity matrix.

### 6.2 Shooting

Just as Newton's method replaces a nonlinear system of equations with a sequence of linear systems, the shooting method replaces a boundary-value problem with a sequence of initial-value problems. The objective is to find an initial value $x$ such that the solution of the differential equation with this initial value, denoted $y(t; x)$, satisfies the boundary conditions

$$F(x) := r(x, y(b; x)) = 0.$$

Newton's method is now applied to this nonlinear system of equations; starting from an initial guess $x^0$, one iterates

$$x^{k+1} = x^k + \Delta x^k \quad \text{with } DF(x^k)\Delta x^k = -F(x^k).$$

Here, the derivative matrix $DF(x^k)$ turns out to be the sensitivity matrix $E^k$ of the linearization of the boundary-value problem along $y(t; x^k)$. In the $k$th iteration, one solves numerically the initial-value problem with initial value $x^k$ together with its linearization

$$(Y^k)'(t) = \partial_y f(t, y(t; x^k))Y^k(t), \quad Y^k(a) = I.$$

### 6.3 Multiple Shooting

The conceptual elegance of the shooting method—that it reduces everything to the solution of initial-value problems over the whole interval—can easily turn into its computational obstruction. Newton's method may be very sensitive to the choice of the initial value $x^0$. The norms of the matrices $E^{-1}$ and $U(b,a)$, which determine the effect of perturbations in the boundary-value problem and the initial-value problem, respectively, are unrelated and may differ widely.

The problem can be avoided by subdividing the interval $a = t_0 < t_1 < \cdots < t_N = b$, shooting on every subinterval, and requiring continuity of the solution at the nodes $t_n$. With $y(t; t_n, x_n)$ denoting the solution of the differential equation that starts at $t_n$ with initial value $x_n$, this approach leads to a larger nonlinear system with the continuity conditions

$$F_n(x_{n-1}, x_n) = y(t_n; t_{n-1}, x_{n-1}) - x_n = 0$$

for $n = 1, \dots, N$ together with the boundary conditions

$$F_0(x_0, x_N) := r(x_0, x_N) = 0.$$

Newton's method is now applied to this system of equations. In each iteration one solves initial-value problems on the subintervals together with their linearization, and then a linear system with a large sparse matrix is solved for the increments in $(x_0, \dots, x_N)$.

### 6.4 Collocation

In the collocation approach to the boundary-value problem, one determines an approximation $u(t)$ that is a continuous, piecewise polynomial of degree at most $s$ and that satisfies the boundary conditions and the differential equation at a finite number of collocation points $t_{n,i} = t_{n-1} + c_i(t_n - t_{n-1})$ (for $n = 1, \dots, N$ and $i = 1, \dots, s$):

$$u'(t) = f(t, u(t)) \quad \text{at } t = t_{n,i},$$
$$r(u(a), u(b)) = 0.$$

The method can be interpreted, and implemented, as a multiple-shooting method in which a single step with a collocation method for initial-value problems, as considered in section 4.4, is made to approximate the solution in each subinterval. The most common choice, as first implemented by Ascher, Christiansen, and Russell in 1979, is collocation at Gauss nodes, which has good stability properties in the forward and backward directions. The order of approximation at the grid points $t_n$ is $p = 2s$. Moreover, if the boundary-value problem results from a variational problem, then Gauss collocation can be interpreted as a direct discretization of the variational problem (see section 5.5).

## 7 Summary

The numerical solution of ordinary differential equations is an area driven by both applications and theory, with efficient computer codes alongside beautiful theorems, both relying on the insight and knowledge of the researchers that are active in this field. It is an area that interacts with neighboring fields in computational mathematics (NUMERICAL LINEAR ALGEBRA [IV.10], the NUMERICAL SOLUTION OF PARTIAL DIFFERENTIAL EQUATIONS [IV.13], and OPTIMIZATION [IV.11]), with the theory of DIFFERENTIAL EQUATIONS [IV.2] and DYNAMICAL SYSTEMS [IV.20], and time and again with the application areas in science and engineering in which numerical methods for differential equations are used.

### Further Reading


Ascher, U. M., R. M. M. Mattheij, and R. D. Russell. 1995. *Numerical Solution of Boundary Value Problems for Ordinary Differential Equations.* Philadelphia, PA: SIAM.

Ascher, U. M., and L. R. Petzold. 1998. *Computer Methods for Ordinary Differential Equations and Differential-Algebraic Equations.* Philadelphia, PA: SIAM.

Butcher, J. C. 2008. *Numerical Methods for Ordinary Differential Equations*, 2nd revised edn. Chichester, UK: John Wiley.

Crouzeix, M., and A. L. Mignot. 1989. *Analyse Numérique des Équations Différentielles*, 2nd revised and expanded edn. Paris: Masson.

Deuflhard, P., and F. Bornemann. 2002. *Scientific Computing with Ordinary Differential Equations.* New York: Springer.

Gear, C. W. 1971. *Numerical Initial Value Problems in Ordinary Differential Equations.* Englewood Cliffs, NJ: Prentice-Hall.

Hairer, E., C. Lubich, and G. Wanner. 2006. *Geometric Numerical Integration: Structure-Preserving Algorithms for Ordinary Differential Equations*, 2nd revised edn. Berlin: Springer.

Hairer, E., S. P. Nørsett, and G. Wanner. 1993. *Solving Ordinary Differential Equations. I: Nonstiff Problems*, 2nd revised edn. Berlin: Springer.

Hairer, E., and G. Wanner. 1996. *Solving Ordinary Differential Equations. II: Stiff and Differential-Algebraic Problems*, 2nd revised edn. Springer, Berlin.

Henrici, P. 1962. *Discrete Variable Methods in Ordinary Differential Equations.* New York: John Wiley.

Iserles, A. 2009. *A First Course in the Numerical Analysis of Differential Equations*, 2nd edn. Cambridge: Cambridge University Press.

Leimkuhler, B., and S. Reich. 2004. *Simulating Hamiltonian Dynamics.* Cambridge: Cambridge University Press.

# IV.13 Numerical Solution of Partial Differential Equations

*Endre Süli*

## 1 Introduction

Numerical solution of partial differential equations (PDEs) is a rich and active field of modern applied mathematics. The subject's steady growth is stimulated by ever-increasing demands from the natural sciences, from engineering, and from economics to provide accurate and reliable approximations to mathematical models involving PDEs whose exact solutions are either too complicated to determine in closed form or (as is the case for many) are not known to exist. While the history of numerical solution of ordinary differential equations is firmly rooted in eighteenth- and nineteenth-century mathematics, the mathematical foundations of the field of numerical solution of PDEs are much more recent; they were first formulated in the landmark paper "Über die partiellen Differenzengleichungen der mathematischen Physik" ("On the partial difference equations of mathematical physics") by Richard Courant, Karl Friedrichs, and Hans Lewy, which was published in 1928. Today, there is a vast array of powerful numerical techniques for specific PDEs, including the following:

- level set and fast-marching methods for front-tracking and interface problems;
- numerical methods for PDEs on (possibly evolving) manifolds;
- immersed boundary methods;
- mesh-free methods;
- particle methods;
- vortex methods;
- various numerical homogenization methods and specialized numerical techniques for multiscale problems;
- wavelet-based multiresolution methods;
- sparse finite-difference and finite-element methods, greedy algorithms, and tensorial methods for high-dimensional PDEs;
- domain-decomposition methods for geometrically complex problems; and
- numerical methods for PDEs with stochastic coefficients that arise in a variety of applications, including UNCERTAINTY QUANTIFICATION [II.34] problems.

Our brief review cannot do justice to this huge and rapidly evolving subject. We shall therefore confine ourselves to the most standard and well-established techniques for the numerical solution of PDEs: finite-difference methods, finite-element methods, finite-volume methods, and spectral methods. Before embarking on our survey, it is appropriate to take a brief excursion into the theory of PDEs in order to fix the relevant notational conventions and to describe some typical model problems.

## 2 Model Partial Differential Equations

A linear partial differential operator $L$ of *order* $m$ with real-valued coefficients $a_\alpha = a_\alpha(x)$, $|\alpha| \leqslant m$, on a domain $\Omega \subset \mathbb{R}^d$, defined by

$$L := \sum_{|\alpha|\leqslant m} a_\alpha(x)\partial^\alpha, \quad x \in \Omega,$$

is called *elliptic* if, for every $x := (x_1, \dots, x_d) \in \Omega$ and every nonzero $\xi := (\xi_1, \dots, \xi_d) \in \mathbb{R}^d$,

$$Q_m(x,\xi) := \sum_{|\alpha|=m} a_\alpha(x)\xi^\alpha \neq 0.$$

Here, $\alpha := (\alpha_1, \dots, \alpha_d)$ is a $d$-component vector with nonnegative integer entries, called a *multi-index*, $|\alpha| := \alpha_1 + \cdots + \alpha_d$ is the *length* of the multi-index $\alpha$, $\partial^\alpha := \partial_{x_1}^{\alpha_1} \cdots \partial_{x_d}^{\alpha_d}$, with $\partial_{x_j} := \partial/\partial x_j$, and $\xi^\alpha := \xi_1^{\alpha_1} \cdots \xi_d^{\alpha_d}$. In the case of complex-valued coefficients $a_\alpha$, the definition above is modified by demanding that $|Q_m(x,\xi)| \neq 0$ for all $x \in \Omega$ and all nonzero $\xi \in \mathbb{R}^d$. A typical example of a first-order elliptic operator with complex coefficients is the *Cauchy–Riemann operator* $\partial_{\bar{z}} := \frac{1}{2}(\partial_x + \mathrm{i}\partial_y)$, where $\mathrm{i} := \sqrt{-1}$. With this general definition of ellipticity, even-order operators can exhibit some rather disturbing properties. For example, the Bitsadze equation $\partial_{xx}u + 2\mathrm{i}\partial_{xy}u - \partial_{yy}u = 0$ admits infinitely many solutions on the unit disk $\Omega$ in $\mathbb{R}^2$ centered at the origin, all of which vanish on the boundary $\partial\Omega$ of $\Omega$. Indeed, with $z = x + \mathrm{i}y$, $u(x,y) = (1 - |z|^2)f(z)$ is a solution that vanishes on $\partial\Omega$ for any complex analytic function $f$. A stronger requirement, referred to as *uniform ellipticity*, is therefore frequently imposed: for real-valued coefficients $a_\alpha$, $|\alpha| \leqslant m$, and $m = 2k$, where $k$ is a positive integer, uniform ellipticity demands the existence of a constant $C > 0$ such that $(-1)^k Q_{2k}(x,\xi) \geqslant C|\xi|^{2k}$ for all $x \in \Omega$ and all nonzero $\xi \in \mathbb{R}^d$.

The archetypal linear second-order uniformly elliptic PDE is $-\Delta u + c(x)u = f(x)$, $x \in \Omega$. Here, $c$ and $f$ are real-valued functions defined on $\Omega$, and $\Delta := \sum_{i=1}^d \partial_{x_i}^2$ is the *Laplace operator*. When $c < 0$ the equation is called the *Helmholtz equation*. In the special case when $c(x) \equiv 0$ the equation is referred to as *Poisson's equation*, and when $c(x) \equiv 0$ and $f(x) \equiv 0$ it is referred to as *Laplace's equation*. Elliptic PDEs arise in a range of mathematical models in continuum mechanics, physics, chemistry, biology, economics, and finance. For example, in a two-dimensional flow of an incompressible fluid with flow velocity $u = (u_1, u_2, 0)$, the stream function $\psi$, related to $u$ by $u = \nabla \times (0, 0, \psi)$, satisfies Laplace's equation. The potential $\Phi$ of a gravitational field, due to an attracting massive object of density $\rho$, satisfies Poisson's equation $\Delta\Phi = 4\pi G\rho$, where $G$ is the universal gravitational constant.

More generally, one can consider fully nonlinear second-order PDEs:

$$F(x, u, \nabla u, D^2 u) = 0,$$

where $F$ is a real-valued function defined on the set $\Upsilon := \Omega \times \mathbb{R} \times \mathbb{R}^d \times \mathbb{R}^{d\times d}_{\mathrm{symm}}$, with a typical element $\upsilon := (x, z, p, R)$, where $x \in \Omega$, $z \in \mathbb{R}$, $p \in \mathbb{R}^d$, and $R \in \mathbb{R}^{d\times d}_{\mathrm{symm}}$; $\Omega$ is an open set in $\mathbb{R}^d$; $D^2 u$ denotes the Hessian matrix of $u$; and $\mathbb{R}^{d\times d}_{\mathrm{symm}}$ is the $d(d+1)/2$-dimensional linear space of real symmetric $d \times d$ matrices, $d \geqslant 2$. An equation of this form is said to be elliptic on $\Upsilon$ if the $d \times d$ matrix whose entries are $\partial F/\partial R_{ij}$, $i, j = 1, \dots, d$, is positive-definite at each $\upsilon \in \Upsilon$. An important example, encountered in connection with optimal transportation problems, is the Monge–Ampère equation: $\det D^2 u = f(x)$ with $x \in \Omega$. For this equation to be elliptic it is necessary to demand that the twice continuously differentiable function $u$ be uniformly convex at each point of $\Omega$, and for such a solution to exist we must also have $f$ positive.

Parabolic and hyperbolic PDEs typically arise in mathematical models in which one of the independent physical variables is time, denoted by $t$. For example, the PDEs

$$\partial_t u + Lu = f \quad \text{and} \quad \partial_{tt} u + Lu = f,$$

where $L$ is a uniformly elliptic partial differential operator of order $2m$ and $u$ and $f$ are functions of $(t, x_1, \dots, x_d)$, are *uniformly parabolic* and *uniformly hyperbolic*, respectively. The simplest examples are the (uniformly parabolic) unsteady heat equation and the (uniformly hyperbolic) second-order wave equation, where

$$Lu := -\sum_{i,j=1}^{d} \partial_{x_j}(a_{ij}(t, x)\partial_{x_i} u),$$

and where $a_{ij}(t, x) = a_{ij}(t, x_1, \dots, x_d)$, $i, j = 1, \dots, d$, are the entries of a $d \times d$ matrix, which is positive-definite, uniformly with respect to $(t, x_1, \dots, x_d)$.

Not all PDEs are of a certain fixed type. For example, the following PDEs are *mixed elliptic-hyperbolic*; they are elliptic for $x > 0$ and hyperbolic for $x < 0$:

$$\partial_{xx} u + \operatorname{sign}(x)\partial_{yy} u = 0 \quad \text{(Lavrentiev equation)},$$
$$\partial_{xx} u + x\partial_{yy} u = 0 \quad \text{(Tricomi equation)},$$
$$x\partial_{xx} u + \partial_{yy} u = 0 \quad \text{(Keldysh equation)}.$$

Stochastic analysis is a fertile source of PDEs of *nonnegative characteristic form*, such as

$$\partial_t u - \sum_{i,j=1}^{d} \partial_{x_j}(a_{ij}\partial_{x_i} u) + \sum_{i=1}^{d} b_i \partial_{x_i} u + cu = f,$$

where $b_i$, $c$, and $f$ are real-valued functions of $(t, x_1, \dots, x_d)$, and $a_{ij} = a_{ij}(t, x_1, \dots, x_d)$, $i, j = 1, \dots, d$, are the entries of a *positive-semidefinite* matrix; since the $a_{ij}$ are dependent on the temporal variable $t$, the equation is, potentially, of *changing type*. An important special case is when the $a_{ij}$ are all identically equal to zero, resulting in the following first-order hyperbolic equation, which is also referred to as the *advection* (or *transport*) *equation*:

$$\partial_t u + \sum_{i=1}^{d} b_i(t, x)\partial_{x_i} u + c(t, x)u = f(t, x).$$

The nonlinear counterpart of this equation,

$$\partial_t u + \sum_{i=1}^{d} \partial_{x_i}[f(t, x, u)] = 0,$$

plays an important role in compressible fluid dynamics, traffic flow models, and flow in porous media. Special cases include the Burgers equation $\partial_t u + \partial_x(\frac{1}{2}u^2) = 0$ and the Buckley–Leverett equation $\partial_t u + \partial_x(u^2/(u^2 + \frac{1}{4}(1-u)^2)) = 0$.

PDEs are rarely considered in isolation; additional information is typically supplied in the form of boundary conditions (imposed on the boundary $\partial\Omega$ of the domain $\Omega \subset \mathbb{R}^d$ in which the PDE is studied) or, in the case of parabolic and hyperbolic equations, as initial conditions at $t = 0$. The PDE in tandem with the boundary/initial conditions is referred to as a *boundary-value problem/initial-value problem* or, when both boundary and initial data are supplied, as an *initial–boundary-value problem*.

## 3 Finite-Difference Methods

We begin by considering finite-difference methods for elliptic boundary-value problems. The basic idea behind the construction of finite-difference methods is to *discretize* the closure, $\bar{\Omega}$, of the (bounded) domain of definition $\Omega \subset \mathbb{R}^d$ of the solution (the so-called *analytical solution*) to the PDE by approximating it with a finite set of points in $\mathbb{R}^d$, called the *mesh points* or *grid points*, and replacing the partial derivatives of the analytical solution appearing in the equation by *divided differences* (difference quotients) of a *grid function*, i.e., a function that is defined at all points in the finite-difference grid. The process results in a finite set of equations with a finite number of unknowns, the values of the grid function representing the finite-difference approximation to the analytical solution over the finite-difference grid. We illustrate the construction by considering a simple second-order uniformly elliptic PDE subject to a *homogeneous Dirichlet boundary condition*:

$$-\Delta u + c(x,y)u = f(x,y) \quad \text{in } \Omega, \tag{1}$$

$$u = 0 \quad \text{on } \partial\Omega \tag{2}$$

on the unit square $\Omega := (0,1)^2$; here, $c$ and $f$ are real-valued functions that are defined and continuous on $\Omega$, and $c \geqslant 0$ on $\Omega$. Let us suppose for simplicity that the grid points are equally spaced. Thus we take $h := 1/N$, where $N \geqslant 2$ is an integer. The corresponding finite-difference grid is then $\bar{\Omega}_h := \{(x_i, y_j) : i, j = 0, \dots, N\}$, where $x_i := ih$ and $y_j := jh$, $i, j = 0, \dots, N$. We also define $\Omega_h := \bar{\Omega}_h \cap \Omega$ and $\partial\Omega_h := \bar{\Omega}_h \setminus \Omega_h$.

It is helpful to introduce the following notation for *first-order divided differences*:

$$D_x^+ u(x_i, y_j) := \frac{u(x_{i+1}, y_j) - u(x_i, y_j)}{h}$$

and

$$D_x^- u(x_i, y_j) := \frac{u(x_i, y_j) - u(x_{i-1}, y_j)}{h},$$

with $D_y^+ u(x_i, y_j)$ and $D_y^-(x_i, y_j)$ defined analogously. Then,

$$D_x^2 u(x_i, y_j) := D_x^- D_x^+ u(x_i, y_j),$$
$$D_y^2 u(x_i, y_j) := D_y^- D_y^+ u(x_i, y_j)$$

are referred to as the *second-order divided difference* of $u$ in the $x$-direction and the $y$-direction, respectively, at $(x_i, y_j) \in \Omega_h$.

Assuming that $u \in C^4(\bar{\Omega})$ (i.e., that $u$ and all of its partial derivatives up to and including those of fourth order are defined and continuous on $\bar{\Omega}$), we have that, at any $(x_i, y_j) \in \Omega_h$,

$$D_x^2 u(x_i, y_j) = \frac{\partial^2 u}{\partial x^2}(x_i, y_j) + \mathcal{O}(h^2) \tag{3}$$

and

$$D_y^2 u(x_i, y_j) = \frac{\partial^2 u}{\partial y^2}(x_i, y_j) + \mathcal{O}(h^2) \tag{4}$$

as $h \to 0$. Omission of the $\mathcal{O}(h^2)$ terms in (3) and (4) above yields that

$$D_x^2 u(x_i, y_j) \approx \frac{\partial^2 u}{\partial x^2}(x_i, y_j),$$
$$D_y^2 u(x_i, y_j) \approx \frac{\partial^2 u}{\partial y^2}(x_i, y_j),$$

where the symbol "$\approx$" signifies approximate equality in the sense that as $h \to 0$ the expression on the left of the symbol converges to the expression on its right. Hence,

$$-(D_x^2 u(x_i, y_j) + D_y^2 u(x_i, y_j)) + c(x_i, y_j)u(x_i, y_j) \approx f(x_i, y_j) \quad \text{for all } (x_i, y_j) \in \Omega_h, \tag{5}$$

$$u(x_i, y_j) = 0 \quad \text{for all } (x_i, y_j) \in \partial\Omega_h. \tag{6}$$

It is instructive to note the similarity between (1) and (5), and between (2) and (6). Motivated by the form of (5) and (6), we seek a grid function $U$ whose value at the grid point $(x_i, y_j) \in \bar{\Omega}_h$, denoted by $U_{ij}$, approximates $u(x_i, y_j)$, the unknown exact solution to the boundary-value problem (1), (2) evaluated at $(x_i, y_j)$, $i, j = 0, \dots, N$. We define $U$ to be the solution to the following system of linear algebraic equations:

$$-(D_x^2 U_{ij} + D_y^2 U_{ij}) + c(x_i, y_j)U_{ij} = f(x_i, y_j) \quad \text{for all } (x_i, y_j) \in \Omega_h, \tag{7}$$

$$U_{ij} = 0 \quad \text{for all } (x_i, y_j) \in \partial\Omega_h. \tag{8}$$

As each equation in (7) involves five values of the grid function $U$ (namely, $U_{ij}$, $U_{i-1,j}$, $U_{i+1,j}$, $U_{i,j-1}$, $U_{i,j+1}$), the finite-difference method (7) is called the *five-point difference scheme*. The matrix of the linear system (7), (8) is sparse, symmetric, and positive-definite, and for given functions $c$ and $f$ it can be efficiently solved by iterative techniques from NUMERICAL LINEAR ALGEBRA [IV.10], including KRYLOV SUBSPACE [II.23] type methods (e.g., the conjugate gradient method) and *multigrid methods*. Multigrid methods were developed in the 1970s and 1980s and are widely used as the iterative solver of choice for large systems of linear algebraic equations that arise from finite-difference and finite-element approximations in many industrial applications. The key objective of a multigrid method is to accelerate the convergence of standard iterative methods (such as Jacobi iteration and successive over-relaxation) by using a hierarchy of coarser-to-finer grids. A multigrid method with an intentionally reduced convergence tolerance can also be used as an efficient *preconditioner* for a Krylov subspace iteration. The preconditioner $P$ for a nonsingular matrix $A$ is an approximation of $A^{-1}$ whose purpose is to ensure that $PA$ is a good approximation of the identity matrix, thereby ensuring that iterative algorithms for the solution of the preconditioned version, $PAx = Pb$, of the system of linear algebraic equations $Ax = b$ exhibit rapid convergence.

One of the central questions in the numerical analysis of PDEs is the mathematical study of the approximation properties of numerical methods. We will illustrate this by considering the finite-difference method (7), (8). The grid function $T$ defined on $\Omega_h$ by

$$T_{ij} := -(D_x^2 u(x_i, y_j) + D_y^2 u(x_i, y_j)) + c(x_i, y_j)u(x_i, y_j) - f(x_i, y_j) \quad (9)$$

is called the *truncation error* of the finite-difference method (7), (8). Assuming that $u \in C^4(\bar{\Omega})$, it follows from (3)–(5) that, at each grid point $(x_i, y_j) \in \Omega_h$, $T_{ij} = \mathcal{O}(h^2)$ as $h \to 0$. The exponent of $h$ in the statement $T_{ij} = \mathcal{O}(h^2)$ (which, in this case, is equal to 2) is called the *order of accuracy* (or *order of consistency*) of the method.

It can be shown that there exists a positive constant $c_0$, independent of $h$, $U$, and $f$, such that

$$\left(h^2 \sum_{i=1}^{N} \sum_{j=1}^{N-1} |D_x^- U_{ij}|^2 + h^2 \sum_{i=1}^{N-1} \sum_{j=1}^{N} |D_y^- U_{ij}|^2 + h^2 \sum_{i=1}^{N-1} \sum_{j=1}^{N-1} |U_{ij}|^2\right)^{1/2} \leqslant c_0 \left(h^2 \sum_{i=1}^{N-1} \sum_{j=1}^{N-1} |f(x_i, y_j)|^2\right)^{1/2}. \quad (10)$$

Such an inequality (which expresses the fact that the numerical solution $U \in S_{h,0}$ is bounded by the data (in this case $f \in S_h$) uniformly with respect to the grid size $h$, where $S_{h,0}$ denotes the linear space of all grid functions defined on $\bar{\Omega}_h$ that vanish on $\partial\Omega_h$ and where $S_h$ is the linear space of all grid functions defined on $\Omega_h$) is called a *stability inequality*. The smallest real number $c_0 > 0$ for which (10) holds is called the *stability constant* of the method. It follows in particular from (10) that if $f_{ij} = 0$ for all $i, j = 1, \dots, N-1$, then $U_{ij} = 0$ for all $i, j = 0, \dots, N$. Therefore, the matrix of the system of linear equations (7), (8) is nonsingular, which then implies the existence of a unique solution $U$ to (7), (8) for any $h = 1/N$, $N \geqslant 2$. Consider the difference operator $L_h \colon U \in S_{h,0} \mapsto f = L_h U \in S_h$ defined by (7), (8). The left-hand side of (10) is sometimes denoted by $\|U\|_{1,h}$ and the right-hand side by $\|f\|_{0,h}$; hence, the stability inequality (10) can be rewritten as

$$\|U\|_{1,h} \leqslant c_0 \|f\|_{0,h}$$

with $f = L_h U$, and stability can then be seen to be demanding the existence of the inverse to the linear finite-difference operator $L_h \colon S_{h,0} \to S_h$, and its boundedness, uniformly with respect to the discretization parameter $h$. The mapping $U \in S_{h,0} \mapsto \|U\|_{1,h} \in \mathbb{R}$ is a norm on $S_{h,0}$, called the *discrete (Sobolev) $H^1(\Omega)$-norm*, and the mapping $f \in S_h \mapsto \|f\|_{0,h} \in \mathbb{R}$ is a norm on $S_h$, called the *discrete $L^2(\Omega)$-norm*. It should be noted that the stability properties of finite-difference methods depend on the choice of norm for the data and for the associated solution.

In order to quantify the closeness of the approximate solution $U$ to the analytical solution $u$ at the grid points, we define the *global error* $e$ of the method (7), (8) by $e_{ij} := u(x_i, y_j) - U_{ij}$. Clearly, the grid function $e = u - U$ satisfies (7), (8) if $f(x_i, y_j)$ on the right-hand side of (7) is replaced by $T_{ij}$. Hence, by the stability inequality, $\|u - U\|_{1,h} = \|e\|_{1,h} \leqslant c_0 \|T\|_{0,h}$. Under the assumption that $u \in C^4(\bar{\Omega})$ we thus deduce that $\|u - U\|_{1,h} \leqslant c_1 h^2$, where $c_1$ is a positive constant, independent of $h$. The exponent of $h$ on the right-hand side (which is 2 in this case) is referred to as the *order of convergence* of the finite-difference method and is equal to the order of accuracy. Indeed, the fundamental idea that stability and consistency together imply convergence is a recurring theme in the analysis of numerical methods for differential equations.

The five-point difference scheme can be generalized in various ways. For example, instead of using the same grid size $h$ in both coordinate directions, one could have used a grid size $\Delta x = 1/M$ in the $x$-direction and a possibly different grid size $\Delta y = 1/N$ in the $y$-direction, where $M, N \geqslant 2$ are integers. One can also consider boundary-value problems on more complicated polygonal domains $\Omega$ in $\mathbb{R}^2$ such that each edge of $\Omega$ is parallel with one of the coordinate axes, for example, the L-shaped domain $(-1, 1)^2 \setminus [0, 1]^2$. The construction above can be extended to domains with curved boundaries in any number of dimensions; at grid points that are on (or next to) the boundary, divided differences with unequally spaced grid points are then used.

In the case of nonlinear elliptic boundary-value problems, such as the Monge–Ampère equation on a bounded open set $\Omega \subset \mathbb{R}^d$, subject to the nonhomogeneous Dirichlet boundary condition $u = g$ on $\partial\Omega$, a finite-difference approximation is easily constructed by replacing at each grid point $(x_i, y_j) \in \Omega$ the value $u(x_i, y_j)$ of the analytical solution $u$ (and its partial derivatives) in the PDE with the numerical solution $U_{ij}$

(and its divided differences) and by then imposing the numerical boundary condition $U_{ij} = g(x_i, y_j)$ for all $(x_i, y_j) \in \partial\Omega_h$. Unfortunately, such a simple-minded method does not explicitly demand the convexity of $U$ in any sense, and this can lead to instabilities. In fact, there is no reason why the sequence of finite-difference solutions should converge to the (convex) analytical solution of the Monge–Ampère equation as $h \to 0$. Even in two space dimensions the resulting method may have multiple solutions, and special iterative solvers need to be used to select the convex solution. Enforcing convexity of the finite-difference solution in higher dimensions is much more difficult. A recent development in this field has been the construction of so-called *wide-stencil finite-difference methods*, which are monotone, and the convergence theory of Barles and Souganidis therefore ensures convergence of the sequence of numerical solutions, as $h \to 0$, to the unique viscosity solution of the Monge–Ampère equation.

We close this section on finite-difference methods with a brief discussion about their application to time-dependent problems. A key result is the *Lax equivalence theorem*, which states that, for a finite-difference method that is consistent with a well-posed initial-value problem for a linear PDE, stability of the method implies convergence of the sequence of grid functions defined by the method on the grid to the analytical solution as the grid size converges to zero, and vice versa. Consider the unsteady heat equation $u_t - \Delta u + u = 0$ for $t \in (0, T]$, with $T > 0$ given, and $(x, y)$ in the unit square $\Omega = (0,1)^2$, subject to the homogeneous Dirichlet boundary condition $u = 0$ on $(0, T] \times \partial\Omega$ and the initial condition $u(0, x, y) = u_0(x, y)$, $(x, y) \in \Omega$, where $u_0$ and $f$ are given real-valued continuous functions. The computational domain $[0, T] \times \bar{\Omega}$ is discretized by the grid $\{t^m = m\Delta t \colon m = 0, \dots, M\} \times \bar{\Omega}_h$, where $\Delta t = T/M$, $M \geqslant 1$, and $h = 1/N$, $N \geqslant 2$. We consider the $\theta$-method

$$\frac{U_{ij}^{m+1} - U_{ij}^{m}}{\Delta t} - (D_x^2 U_{ij}^{m+\theta} + D_y^2 U_{ij}^{m+\theta}) + U_{ij}^{m+\theta} = 0$$

for all $i, j = 1, \dots, N-1$ and $m = 0, \dots, M-1$, supplemented with the initial condition $U_{ij}^0 = u_0(x_i, y_j)$, $i, j = 0, \dots, N$, and the boundary condition $U_{ij}^{m+1} = 0$, $m = 0, \dots, M-1$, for all $(i, j)$ such that $(x_i, y_j) \in \partial\Omega_h$. Here, $\theta \in [0, 1]$ and

$$U_{ij}^{m+\theta} := (1-\theta) U_{ij}^{m} + \theta U_{ij}^{m+1},$$

with $U_{ij}^m$ and $U_{ij}^{m+1}$ representing the approximations to $u(t^m, x_i, y_j)$ and $u(t^{m+1}, x_i, y_j)$, respectively. The values $\theta = 0, \frac{1}{2}, 1$ are particularly relevant; the corresponding finite-difference methods are called the *forward* (or *explicit*) *Euler method*, the *Crank–Nicolson method*, and the *backward* (or *implicit*) *Euler method*, respectively; their truncation errors are defined by

$$\begin{aligned} T_{ij}^{m+1} := {} & \frac{u(t^{m+1}, x_i, y_j) - u(t^m, x_i, y_j)}{\Delta t} \\ & - (1-\theta)(D_x^2 u(t^m, x_i, y_j) + D_y^2 u(t^m, x_i, y_j)) \\ & - \theta(D_x^2 u(t^{m+1}, x_i, y_j) + D_y^2 u(t^{m+1}, x_i, y_j)) \\ & + (1-\theta) u(t^m, x_i, y_j) + \theta u(t^{m+1}, x_i, y_j) \end{aligned}$$

for $i, j = 1, \dots, N-1$, $m = 0, \dots, M-1$. Assuming that $u$ is sufficiently smooth, Taylor series expansion yields that $T_{ij} = \mathcal{O}(\Delta t + h^2)$ for $\theta \neq \frac{1}{2}$ and $T_{ij} = \mathcal{O}((\Delta t)^2 + h^2)$ for $\theta = \frac{1}{2}$. Thus, in particular, the forward and backward Euler methods are first-order accurate with respect to the temporal variable $t$ and second-order accurate with respect to the spatial variables $x$ and $y$, whereas the Crank–Nicolson method is second-order accurate with respect to both the temporal variable and the spatial variables. The stability properties of the $\theta$-method are also influenced by the choice of $\theta \in [0, 1]$; we have that

$$\max_{1 \leqslant m \leqslant M} \|U^m\|_{0,h}^2 + \Delta t \sum_{m=0}^{M-1} \|U^{m+\theta}\|_{1,h}^2 \leqslant \|U^0\|_{0,h}^2$$

for $\theta \in [0, \frac{1}{2})$, provided that $2d(1-2\theta)\Delta t \leqslant h^2$, with $d = 2$ (space dimensions) in our case; and for $\theta \in [\frac{1}{2}, 1]$, irrespective of the choice of $\Delta t$ and $h$. Thus, in particular, the forward (explicit) Euler method is *conditionally stable*, the condition being that $2d\Delta t \leqslant h^2$, with $d = 2$ here, while the Crank–Nicolson and backward (implicit) Euler methods are *unconditionally stable*.

A finite-difference method approximates the analytical solution using a grid function that is defined over a finite-difference grid contained in the computational domain. Next we will consider finite-element methods, which involve piecewise polynomial approximations of the analytical solution, defined over the computational domain.

## 4 Finite-Element Methods

Finite-element methods (FEMs) are a powerful and general class of techniques for the numerical solution of PDEs. Their historical roots can be traced back to a paper by Richard Courant, published in 1943, that proposed the use of continuous piecewise affine approximations for the numerical solution of variational problems. This represented a significant advance from a practical point of view over earlier techniques by Ritz and Galerkin from the early 1900s, which were based on the use of linear combinations of smooth functions (e.g., eigenfunctions of the differential operator under consideration). The importance of Courant's contribution was, unfortunately, not recognized at the time and the idea was forgotten until the early 1950s, when it was rediscovered by engineers. FEMs have since been developed into an effective and flexible computational tool with a firm mathematical foundation.

### 4.1 FEMs for Elliptic PDEs

Suppose that $\Omega \subset \mathbb{R}^d$ is a bounded open set in $\mathbb{R}^d$ with a Lipschitz-continuous boundary $\partial\Omega$. We will denote by $L^2(\Omega)$ the space of square-integrable functions (in the sense of Lebesgue) equipped with the norm $\|v\|_0 := (\int_\Omega |v|^2\,dx)^{1/2}$. Let $H^m(\Omega)$ denote the *Sobolev space* consisting of all functions $v \in L^2(\Omega)$ whose (weak) partial derivatives $\partial^\alpha v$ belong to $L^2(\Omega)$ for all $\alpha$ such that $|\alpha| \leqslant m$. $H^m(\Omega)$ is equipped with the norm $\|v\|_m := (\sum_{|\alpha|\leqslant m} \|\partial^\alpha v\|_0^2)^{1/2}$. We denote by $H_0^1(\Omega)$ the set of all functions $v \in H^1(\Omega)$ that vanish on $\partial\Omega$.

Let $a$ and $c$ be real-valued functions, defined and continuous on $\bar{\Omega}$, and suppose that there exists a positive constant $c_0$ such that $a(x) \geqslant c_0$ for all $x \in \bar{\Omega}$. Assume further that $b_i$, $i = 1,\dots,d$, are continuously differentiable real-valued functions defined on $\bar{\Omega}$ such that $c - \frac{1}{2}\nabla \cdot b \geqslant c_0$ on $\bar{\Omega}$, where $b := (b_1,\dots,b_d)$, and let $f \in L^2(\Omega)$. Consider the boundary-value problem

$$-\nabla \cdot (a(x)\nabla u) + b(x) \cdot \nabla u + c(x)u = f(x)$$

for $x \in \Omega$, with $u|_{\partial\Omega} = 0$. The construction of the finite-element approximation of this boundary-value problem commences by considering the following *weak formulation* of the problem: find $u \in H_0^1(\Omega)$ such that

$$B(u,v) = \ell(v) \quad \forall v \in H_0^1(\Omega), \tag{11}$$

where the bilinear form $B(\cdot,\cdot)$ is defined by

$$B(w,v) := \int_\Omega [a(x)\nabla w \cdot \nabla v + b(x) \cdot \nabla w v + c(x) w v]\,dx$$

and $\ell(v) := \int_\Omega f v\,dx$, with $w, v \in H_0^1(\Omega)$. If $u$ is sufficiently smooth (for example, if $u \in H^2(\Omega) \cap H_0^1(\Omega)$), then integration by parts in (11) implies that $u$ is a *strong solution* of the boundary-value problem, i.e., $-\nabla \cdot (a(x)\nabla u) + b(x) \cdot \nabla u + c(x)u = f(x)$ almost everywhere in $\Omega$, and $u|_{\partial\Omega} = 0$. More generally, in the absence of such an additional assumption about smoothness, the function $u \in H_0^1(\Omega)$ satisfying (11) is called a *weak solution* of this elliptic boundary-value problem. Under our assumptions on $a$, $b$, $c$, and $f$, the existence of a unique weak solution follows from the Lax–Milgram theorem.

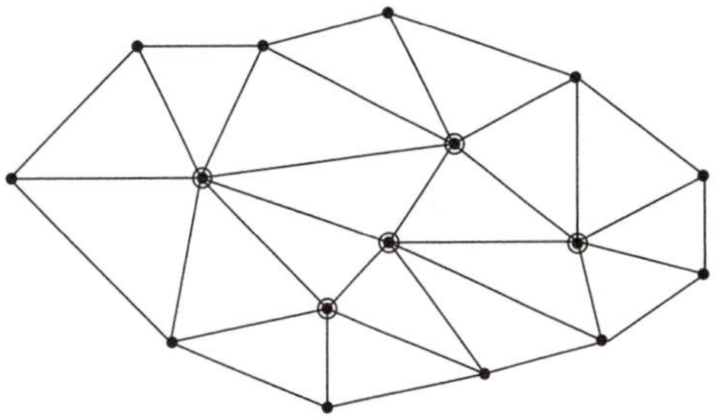

**Figure 1** Finite-element triangulation of the computational domain $\bar{\Omega}$, a polygonal region of $\mathbb{R}^2$. Vertices on $\partial\Omega$ are denoted by solids dots, and vertices internal to $\Omega$ by circled solid dots.

We will consider the finite-element approximation of (11) in the special case when $\Omega$ is a bounded open polygonal domain in $\mathbb{R}^2$. The first step in the construction of the FEM is to define a *triangulation* of $\bar{\Omega}$. A triangulation of $\bar{\Omega}$ is a tessellation of $\bar{\Omega}$ into a finite number of closed triangles $T_i$, $i = 1,\dots,M$, whose interiors are pairwise disjoint, and for each $i, j \in \{1,\dots,M\}$, $i \neq j$, for which $T_i \cap T_j$ is nonempty, $T_i \cap T_j$ is either a common vertex or a common edge of $T_i$ and $T_j$ (see figure 1). The vertices in the triangulation are also referred to as *nodes*.

Let $h_T$ denote the longest edge of a triangle $T$ in the triangulation, and let $h$ be the largest among the $h_T$. Furthermore, let $S_h$ denote the linear space of all real-valued continuous functions $v_h$ defined on $\bar{\Omega}$ such that the restriction of $v_h$ to any triangle in the triangulation is an affine function, and define $S_{h,0} := S_h \cap H_0^1(\Omega)$. The finite-element approximation of the problem (11) is as follows: find $u_h$ in the *finite-element space* $S_{h,0}$ such that

$$B(u_h, v_h) = \ell(v_h) \quad \forall v_h \in S_{h,0}. \tag{12}$$

Let us denote by $x_i$, $i = 1,\dots,L$, the set of all vertices (nodes) in the triangulation (see figure 1), and let $N = N(h)$ denote the dimension of the finite-element space $S_{h,0}$. We will assume that the vertices $x_i$, $i = 1,\dots,L$, are numbered so that $x_i$, $i = 1,\dots,N$, are within $\Omega$ and the remaining $L - N$ vertices are on $\partial\Omega$. Furthermore, let $\{\varphi_j\colon j = 1,\dots,N\} \subset S_{h,0}$ denote the so-called *nodal basis* for $S_{h,0}$, where the basis functions are defined by $\varphi_j(x_i) = \delta_{ij}$, $i = 1,\dots,L$, $j = 1,\dots,N$. A typical piecewise-linear nodal basis function is shown in figure 2. Thus, there exists a vector $U = (U_1,\dots,U_N)^{\mathrm{T}} \in$

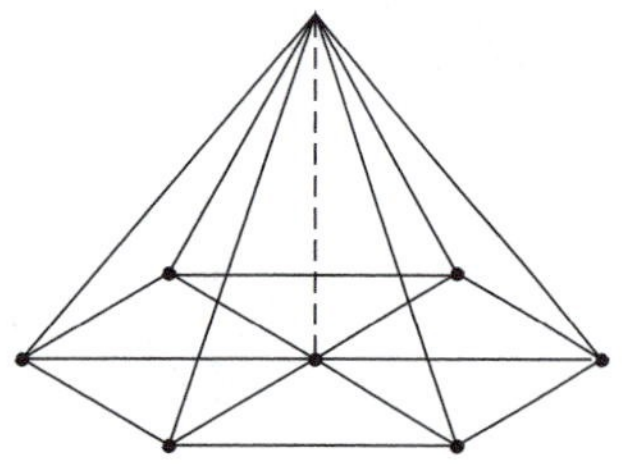

**Figure 2** A typical piecewise-linear nodal basis function. The basis function is identically zero outside the patch of triangles surrounding the central node, at which the height of the function is equal to 1.

$\mathbb{R}^N$ such that

$$u_h(x) = \sum_{j=1}^{N} U_j \varphi_j(x). \tag{13}$$

Substitution of this expansion into (12) and taking $v_h = \varphi_k$, $k = 1,\dots,N$, yields the following system of $N$ linear algebraic equations in the $N$ unknowns, $U_1,\dots,U_N$:

$$\sum_{j=1}^{N} B(\varphi_j, \varphi_k)U_j = \ell(\varphi_k), \quad k = 1,\dots,N. \tag{14}$$

By recalling the definition of $B(\cdot,\cdot)$, we see that the matrix $A := ([B(\varphi_j,\varphi_k)]_{j,k=1}^{N})^{\mathrm{T}}$ of this system of linear equations is sparse and positive-definite (and, if $b$ is identically zero, then the matrix is symmetric as well). The unique solution $U = (U_1,\dots,U_N)^{\mathrm{T}} \in \mathbb{R}^N$ of the linear system, upon substitution into (13), yields the computed approximation $u_h$ to the analytical solution $u$ on the given triangulation of the computational domain $\bar{\Omega}$, using numerical algorithms for SPARSE LINEAR SYSTEMS [IV.10 §6], including Krylov subspace type methods and multigrid methods.

As $S_{h,0}$ is a (finite-dimensional) linear subspace of $H_0^1(\Omega)$, $v = v_h$ is a legitimate choice in (11). By subtracting (12) from (11), with $v = v_h$, we deduce that

$$B(u - u_h, v_h) = 0 \quad \forall v_h \in S_{h,0}, \tag{15}$$

which is referred to as the *Galerkin orthogonality property* of the FEM. Hence, for any $v_h \in S_{h,0}$,

$$\begin{aligned} c_0\|u - u_h\|_1^2 &\leqslant B(u - u_h, u - u_h) \\ &= B(u - u_h, u - v_h) \\ &\leqslant c_1\|u - u_h\|_1\,\|u - v_h\|_1, \end{aligned}$$

where

$$c_1 := (M_a^2 + M_b^2 + M_c^2)^{1/2},$$

with $M_v := \max_{x\in\bar{\Omega}} |v(x)|$, $v \in \{a, b, c\}$. We therefore have that

$$\|u - u_h\|_1 \leqslant \frac{c_1}{c_0} \min_{v_h\in S_{h,0}} \|u - v_h\|_1. \tag{16}$$

This result is known as *Céa's lemma*, and it is an important tool in the analysis of FEMs. Suppose, for example, that $u \in H^2(\Omega) \cap H_0^1(\Omega)$, and denote by $I_h$ the *finite-element interpolant* of $u$ defined by

$$I_h u(x) := \sum_{j=1}^{N} u(x_j)\varphi_j(x).$$

It follows from (16) that $\|u - u_h\|_1 \leqslant (c_1/c_0)\|u - I_h u\|_1$. Assuming further that the triangulation is *shape regular* in the sense that there exists a positive constant $c_*$, independent of $h$, such that for each triangle in the triangulation the ratio of the longest edge to the radius of the inscribed circle is bounded below by $c_*$, arguments from approximation theory imply the existence of a positive constant $\hat{c}$, independent of $h$, such that $\|u - I_h u\|_1 \leqslant \hat{c}h\|u\|_2$. Hence, the following *a priori error bound* holds in the $H^1$-norm:

$$\|u - u_h\|_1 \leqslant (c_1/c_0)\hat{c}h\|u\|_2.$$

We deduce from this inequality that, as the triangulation is refined by letting $h \to 0$, the sequence of finite-element approximations $u_h$ computed on successively refined triangulations converges to the analytical solution $u$ in the $H^1$-norm. It is also possible to derive a priori error bounds in other norms, such as the $L^2$-norm.

The inequality (16) of Céa's lemma can be seen to express the fact that the approximation $u_h \in S_{h,0}$ to the solution $u \in H_0^1(\Omega)$ of (11) delivered by the FEM (12) is a *near-best approximation* to $u$ from the linear subspace $S_{h,0}$ of $H_0^1(\Omega)$. Clearly, $c_1/c_0 \geqslant 1$. When the constant $c_1/c_0 \gg 1$, the numerical solution $u_h$ supplied by the FEM is typically a poor approximation to $u$ in the $\|\cdot\|_1$-norm, unless $h$ is very small; for example, if $a(x) = c(x) \equiv \varepsilon$ and $b(x) = (1,1)^{\mathrm{T}}$, then $c_1/c_0 = \sqrt{2}(1+\varepsilon^2)^{1/2}/\varepsilon \gg 1$ if $0 < \varepsilon \ll 1$. Such non-self-adjoint elliptic boundary-value problems arise in mathematical models of diffusion–advection–reaction, where advection dominates diffusion and reaction in the sense that $|b(x)| \gg a(x) > 0$ and $|b(x)| \gg c(x) > 0$ for all $x \in \bar{\Omega}$. The stability and approximation properties of the classical FEM (12) for such advection-dominated problems can be improved by modifying, in a consistent manner, the definitions of $B(\cdot,\cdot)$ and $\ell(\cdot)$ through the addition of "stabilization terms" or by enriching the finite-element space with special basis functions that are designed so as to capture sharp boundary and interior layers exhibited by typical solutions of advection-dominated problems. The resulting FEMs are generally referred to as *stabilized finite-element methods*. A typical example is the *streamline-diffusion finite-element method*, in which the bilinear form of the standard FEM is supplemented with an additional numerical diffusion term, which acts in the streamwise direction only, i.e., in the direction of the vector $b$, in which classical FEMs tend to exhibit undesirable numerical oscillations.

If, on the other hand, $b$ is identically zero on $\bar{\Omega}$, then $B(\cdot,\cdot)$ is a *symmetric* bilinear form, in the sense that $B(w,v) = B(v,w)$ for all $w,v \in H_0^1(\Omega)$. The norm $\|\cdot\|_B$ defined by $\|v\|_B := [B(v,v)]^{1/2}$ is called the *energy norm* on $H_0^1(\Omega)$ associated with the elliptic boundary-value problem (11). In fact, (11) can then be restated as the following (equivalent) variational problem: find $u \in H_0^1(\Omega)$ such that

$$J(u) \leqslant J(v) \quad \forall v \in H_0^1(\Omega),$$

where

$$J(v) := \tfrac{1}{2}B(v,v) - \ell(v).$$

Analogously, the FEM (12) can then be restated equivalently as follows: find $u_h \in S_{h,0}$ such that $J(u_h) \leqslant J(v_h)$ for all $v_h \in S_{h,0}$. Furthermore, Céa's lemma, in terms of the energy norm, $\|\cdot\|_B$, becomes $\|u - u_h\|_B = \min_{v_h \in S_{h,0}} \|u - v_h\|_B$. Thus, in the case when the function $b$ is identically zero the numerical solution $u_h \in S_{h,0}$ delivered by the FEM is the *best approximation* to the analytical solution $u \in H_0^1(\Omega)$ in the energy norm $\|\cdot\|_B$.

We illustrate the extension of these ideas to nonlinear elliptic PDEs through a simple model problem. For a real number $p \in (1,\infty)$, let $L^p(\Omega) := \{v\colon \int_\Omega |v|^p \,dx < \infty\}$ and $W^{1,p}(\Omega) := \{v \in L^p(\Omega)\colon |\nabla v| \in L^p(\Omega)\}$. Furthermore, let $W_0^{1,p}(\Omega)$ denote the set of all $v \in W^{1,p}(\Omega)$ such that $v|_{\partial\Omega} = 0$. For $f \in L^q(\Omega)$, where $1/p + 1/q = 1$, $p \in (1,\infty)$, consider the problem of finding the minimizer $u \in W_0^{1,p}(\Omega)$ of the functional

$$J(v) := \frac{1}{p}\int_\Omega |\nabla v|^p \,dx - \int_\Omega fv \,dx, \quad v \in W_0^{1,p}(\Omega).$$

With $S_{h,0}$ as above, the finite-element approximation of the problem then consists of finding $u_h \in S_{h,0}$ that minimizes $J(v_h)$ over all $v_h \in S_{h,0}$. The existence and uniqueness of the minimizers $u \in W_0^{1,p}(\Omega)$ and $u_h \in S_{h,0}$ in the respective problems is a direct consequence of the convexity of the functional $J$. Moreover, as $h \to 0$, $u_h$ converges to $u$ in the norm of the Sobolev space $W^{1,p}(\Omega)$.

Problems in electromagnetism and continuum mechanics are typically modeled by systems of PDEs involving several dependent variables, which may need to be approximated from different finite-element spaces because of the disparate physical nature of the variables and the different boundary conditions that they may be required to satisfy. The resulting finite-element methods are called *mixed finite-element methods*. In order for a mixed FEM to possess a unique solution and for the method to be stable, the finite-element spaces from which the approximations to the various components of the vector of unknowns are sought cannot be chosen arbitrarily; they need to satisfy a certain compatibility condition, usually referred to as the *inf-sup condition*.

FEMs of the kind described in this section—where the finite-element space containing the approximate solution is a subset of the function space in which the weak solution to the problem is sought—are called *conforming finite-element methods*. Otherwise, the FEM is called *nonconforming*. Nonconforming FEMs are necessary in some application areas because in certain problems (such as div-curl problems from electromagnetism and variational problems exhibiting a Lavrentiev phenomenon, for example), conforming FEMs may converge to spurious solutions. *Discontinuous Galerkin finite-element methods* (DGFEMs) are extreme instances of nonconforming FEMs, in the sense that pointwise interelement continuity requirements in the piecewise polynomial approximation are completely abandoned, and the analytical solution is approximated by discontinuous piecewise polynomial functions. FEMs have several advantages over finite-difference methods: the concept of higher-order discretization is inherent to FEMs; it is, in addition, particularly convenient from the point of view of adaptivity that FEMs can easily accommodate very general tessellations of the computational domain, with local polynomial degrees in the approximation that may vary from element to element. Indeed, the notion of *adaptivity* is a powerful and important idea in the field of numerical approximation of PDEs, and it is this that we will further elaborate on in the context of finite-element methods below.

### 4.2 A Posteriori Error Analysis and Adaptivity

Provided that the analytical solution is sufficiently smooth, a priori error bounds guarantee that, as the grid size $h$ tends to 0, the corresponding sequence of numerical approximations converges to the exact solution of the boundary-value problem. In practice, one may unfortunately only be able to afford to compute on a small number of grids/triangulations, the minimum grid size attainable being limited by the computational resources available. A further practical consideration is that the regularity of the analytical solution may exhibit large variations over the computational domain, with singularities localized at particular points (e.g., at corners and edges of the domain) or low-dimensional manifolds in the interior of the domain (e.g., shocks and contact discontinuities in nonlinear conservation laws, or steep internal layers in advection-dominated diffusion equations). The error between the unknown analytical solution and numerical solutions computed on locally refined grids, which are best suited for such problems, cannot be accurately quantified by typical a priori error bounds and asymptotic convergence results that presuppose uniform refinement of the computational grid as the grid size tends to 0. The alternative is to perform a computation on a chosen computational grid/triangulation and use the computed approximation to the exact solution to quantify the approximation error a posteriori, and also to identify parts of the computational domain in which the grid size was inadequately chosen, necessitating local, so-called *adaptive*, refinement or coarsening of the computational grid/triangulation (*$h$-adaptivity*). In FEMs it is also possible to locally vary the degree of the piecewise polynomial function in the finite-element space (*$p$-adaptivity*). Finally, one may also make adjustments to the computational grid/triangulation by moving/relocating the grid points (*$r$-adaptivity*). The adaptive loop for an $h$-adaptive FEM has the following form:

SOLVE → ESTIMATE → MARK → REFINE.

Thus, a finite-element approximation is first computed on a certain fixed, typically coarse, triangulation of the computational domain. Then, in the second step, an a posteriori error bound is used to estimate the error in the computed solution; a typical a posteriori error bound for an elliptic boundary-value problem $Lu = f$ (where $L$ is a second-order uniformly elliptic operator and $f$ is a given right-hand side) is of the form $\|u - u_h\|_1 \leqslant C_* \|R(u_h)\|_*$, where $C_*$ is a (computable) constant; $\|\cdot\|_*$ is a certain norm, depending on the problem; and $R(u_h) = f - Lu_h$ is the (computable) *residual*, which measures the extent to which the computed numerical solution $u_h$ fails to satisfy the PDE $Lu = f$. In the third step, on the basis of the a posteriori error bound, selected triangles in the triangulation are marked as having an inadequate size (i.e., too large or too small, relative to a fixed local tolerance, which is usually chosen as a suitable fraction of the prescribed overall tolerance, TOL). Finally, in the fourth step, the marked triangles are refined or coarsened, as the case may be. This four-step adaptive loop is repeated either until a certain termination criterion is reached (e.g., $C_* \|R(u_h)\|_* <$ TOL) or until the computational resources are exhausted. A similar adaptive loop can be used in $p$-adaptive FEMs, except that the step REFINE is then interpreted as adjustment (i.e., an increase or decrease) of the local polynomial degree, which may then vary from triangle to triangle instead of being a fixed integer over the entire triangulation. It is also possible to combine different adaptive strategies; for example, simultaneous $h$- and $p$-adaptivity is referred to as *$hp$-adaptivity*; thanks to the simple communication at the boundaries of adjacent elements in the subdivision of the computational domain, $hp$-adaptivity is particularly easy to incorporate into DGFEMs (see figure 3).

## 5 Finite-Volume Methods

Finite-volume methods have been developed for the numerical solution of PDEs in divergence form, such as CONSERVATION LAWS [II.6] that arise in continuum mechanics. Consider, for example, the following system of nonlinear PDEs:

$$\frac{\partial u}{\partial t} + \nabla \cdot f(u) = 0, \tag{17}$$

where $u := (u_1, \dots, u_n)^{\mathrm{T}}$ is an $n$-component vector function of the variables $t \geqslant 0$ and $x := (x_1, \dots, x_d)$; the vector function $f(u) := (f_1(u), \dots, f_d(u))^{\mathrm{T}}$ is the corresponding *flux function*. The PDE (17) is supplemented with the initial condition $u(0, x) = u_0(x)$, $x \in \mathbb{R}^d$. Suppose that $\mathbb{R}^d$ has been tessellated into disjoint closed simplexes $\kappa$ (intervals if $d = 1$, triangles if $d = 2$, and tetrahedrons if $d = 3$), whose union is the whole of $\mathbb{R}^d$ and such that each pair of distinct simplexes from the tessellation is either disjoint or has only closed simplexes of dimension less than or equal to $d - 1$ in common. In the theory of finite-volume methods the simplexes $\kappa$ are usually referred to as *cells* (rather than elements). For each particular cell $\kappa$ in the tessellation of $\mathbb{R}^d$ the PDE (17) is integrated over $\kappa$, which gives

$$\int_\kappa \frac{\partial u}{\partial t}\,\mathrm{d}x + \int_\kappa \nabla \cdot f(u)\,\mathrm{d}x = 0. \tag{18}$$

By defining the *volume average*

$$\bar{u}_\kappa(t) := \frac{1}{|\kappa|} \int_\kappa u(t, x)\,\mathrm{d}x, \quad t \geqslant 0,$$

where $|\kappa|$ is the measure of $\kappa$, and applying the divergence theorem, we deduce that

$$\frac{\mathrm{d}\bar{u}_\kappa}{\mathrm{d}t} + \frac{1}{|\kappa|} \oint_{\partial\kappa} f(u) \cdot \nu\,\mathrm{d}S = 0,$$

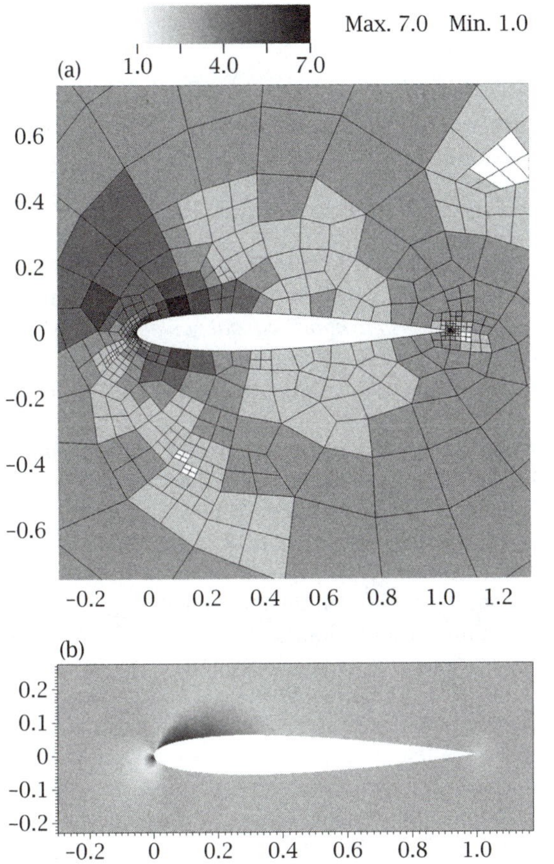


**Figure 3** An $hp$-adaptive finite-element grid: (a) in a discontinuous Galerkin finite-element approximation of the compressible Euler equations of gas dynamics with local polynomial degrees ranging from 1 to 7; and (b) the approximate density on the grid. (Figure courtesy of Paul Houston.)

where $\partial\kappa$ is the boundary of $\kappa$ and $\nu$ is the unit outward normal vector to $\partial\kappa$.

In the present construction the constant volume average is assigned to the barycenter of a cell, and the resulting finite-volume method is therefore referred to as a *cell-center finite-volume method*. In the theory of finite-volume methods the local region $\kappa$ over which the PDE is integrated is called a *control volume*. In the case of cell-center finite-volume methods, the control volumes therefore coincide with the cells in the tessellation. An alternative choice, resulting in *vertex-centered finite-volume methods*, is that for each vertex in the computational grid one considers the patch of cells surrounding the vertex and assigns to the vertex a control volume contained in the patch of elements (e.g., in the case of $d = 2$, the polygonal domain defined by connecting the barycenters of cells that surround a vertex).

Thus far, no approximation has taken place. In order to construct a practical numerical method, the integral over $\partial\kappa$ is rewritten as a sum of integrals over all $(d-1)$-dimensional open faces contained in $\partial\kappa$, and the integral over each face is approximated by replacing the normal flux $f(u)\cdot\nu$ over the face, appearing as the integrand, by interpolation or extrapolation of control volume averages. This procedure can be seen as a replacement of the exact normal flux over a face of a control volume with a *numerical flux function*. Thus, for example, denoting by $e_{\kappa\lambda}$ the $(d-1)$-dimensional face of the control volume $\kappa$ that is shared with a neighboring control volume $\lambda$, we have that

$$\oint_{\partial\kappa} f(u)\cdot\nu\,\mathrm{d}S \approx \sum_{\lambda:\, e_{\kappa\lambda}\subset\partial\kappa} g_{\kappa\lambda}(\bar{u}_\kappa, \bar{u}_\lambda),$$

where the numerical flux function $g_{\kappa\lambda}$ is required to possess the following two crucial properties.

**Conservation** ensures that fluxes from adjacent control volumes that share a mutual interface exactly cancel when summed. This is achieved by demanding that the numerical flux satisfies the identity

$$g_{\kappa\lambda}(u,v) = -g_{\lambda\kappa}(v,u)$$

for each pair of neighboring control volumes $\kappa$ and $\lambda$.

**Consistency** ensures that, for each face of each control volume, the numerical flux with identical state arguments reduces to the true total flux of that same state passing through the face, i.e.,

$$g_{\kappa\lambda}(u,u) = \int_{e_{\kappa\lambda}} f(u)\cdot\nu\,\mathrm{d}S$$

for each pair of neighboring control volumes $\kappa$ and $\lambda$ with common face $e_{\kappa\lambda} := \kappa\cap\lambda$.

The resulting spatial discretization of the nonlinear conservation law is then further discretized with respect to the temporal variable $t$ by time stepping, in steps of $\Delta t$, starting from the given initial datum $u_0$, the simplest choice being to use the explicit Euler method.

The historical roots of this construction date back to the work of Sergei Godunov in 1959 on the gas dynamics equations; Godunov used piecewise-constant solution representations in each control volume with value equal to the average over the control volume and calculated a single numerical flux from the local solution of the Riemann problem posed at the interfaces. A Riemann problem (named after Bernhard Riemann) consists of a conservation law together with piecewise-constant data having a single discontinuity. Additional resolution beyond the first-order accuracy of the Godunov scheme can be attained by *reconstruction/recovery* from the computed cell averages (as in the MUSCL (monotonic upstream-centered scheme for conservation laws) scheme of Van Leer, which is based on piecewise-linear reconstruction, or by piecewise quadratic reconstruction, as in the piecewise parabolic method of Colella and Woodward) by exactly evolving discontinuous piecewise-linear states instead of piecewise-constant states, or by completely avoiding the use of Riemann solvers (as in the Nessyahu–Tadmor and Kurganov–Tadmor central difference methods).

Thanks to their built-in conservation properties, finite-volume methods have been widely and successfully used for the numerical solution of both scalar nonlinear conservation laws and systems of nonlinear conservation laws, including the compressible Euler equations of gas dynamics. There is a satisfactory convergence theory of finite-volume methods for scalar multidimensional conservation laws; efforts to develop a similar body of theory for multidimensional systems of nonlinear conservation laws are, however, hampered by the incompleteness of the theory of well-posedness for such PDE systems.

## 6 Spectral Methods

While finite-difference methods provide approximate solutions to PDEs at the points of the chosen computational grid, and finite-element and finite-volume methods supply continuous or discontinuous piecewise polynomial approximations on tessellations of the computational domain, spectral methods deliver approximate solutions in the form of polynomials of a certain fixed degree, which are, by definition, smooth functions over the entire computational domain. If the solution to the underlying PDE is a smooth function, a spectral method will provide a highly accurate numerical approximation to it.

Spectral approximations are typically sought as linear combinations of ORTHOGONAL POLYNOMIALS [II.29] over the computational domain. Consider a nonempty open interval $(a, b)$ of the real line and a nonnegative *weight function* $w$, which is positive on $(a, b)$, except perhaps at countably many points in $(a, b)$, and such that

$$\int_a^b w(x)|x|^k \,dx < \infty \quad \forall k \in \{0, 1, 2, \dots\}.$$

Furthermore, let $L^2_w(a, b)$ denote the set of all real-valued functions $v$ defined on $(a, b)$ such that

$$\|v\|_w := \left(\int_a^b w(x)|v(x)|^2 \,dx\right)^{1/2} < \infty.$$

Then $\|\cdot\|_w$ is a norm on $L^2_w(a, b)$, induced by the inner product $(u, v)_w := \int_a^b w(x)u(x)v(x)\,dx$. We say that $\{P_k\}_{k=0}^{\infty}$ is a *system of orthogonal polynomials* on $(a, b)$ if $P_k$ is a polynomial of exact degree $k$ and $(P_m, P_n)_w = 0$ when $m \neq n$. For example, if $(a, b) = (-1, 1)$ and $w(x) = (1-x)^{\alpha}(1+x)^{\beta}$, with $\alpha, \beta \in (-1, 1)$ fixed, then the resulting system of orthogonal polynomials are the Jacobi polynomials, special cases of which are the Gegenbauer (or ultraspherical) polynomials ($\alpha = \beta \in (-1, 1)$), Chebyshev polynomials of the first kind ($\alpha = \beta = -\frac{1}{2}$), Chebyshev polynomials of the second kind ($\alpha = \beta = \frac{1}{2}$), and Legendre polynomials ($\alpha = \beta = 0$). On a multidimensional domain $\Omega \subset \mathbb{R}^d$, $d \geqslant 2$, that is the Cartesian product of nonempty open intervals $(a_k, b_k)$, $k = 1, \dots, d$, of the real line and a multivariate weight function $w$ of the form $w(x) = w_1(x_1) \cdots w_d(x_d)$, where $x = (x_1, \dots, x_d)$ and $w_k$ is a univariate weight function of the variable $x_k \in (a_k, b_k)$, $k = 1, \dots, d$, orthogonal polynomials with respect to the inner product $(\cdot, \cdot)_w$ defined by $(u, v)_w = \int_{\Omega} w(x)u(x)v(x)\,dx$ are simply products of univariate orthogonal polynomials with respect to the weights $w_k$, defined on the intervals $(a_k, b_k)$, $k = 1, \dots, d$, respectively.

*Spectral Galerkin methods* for PDEs are based on transforming the PDE problem under consideration into a suitable weak form by multiplication with a *test function*, integration of the resulting expression over the computational domain $\Omega$, and integration by parts, if necessary, in order to incorporate boundary conditions. As in the case of finite-element methods, an approximate solution $u_N$ to the analytical solution $u$ is sought from a finite-dimensional linear space $S_N \subset L^2_w(\Omega)$, which is now, however, spanned by the first $(N + 1)^d$ elements of a certain system of orthogonal polynomials with respect to the weight function $w$. The function $u_N$ is required to satisfy the same weak formulation as the analytical solution, except that the test functions are confined to the finite-dimensional linear space $S_N$. In order to exploit the orthogonality properties of the chosen system of orthogonal polynomials, the weight function $w$ has to be incorporated into the weak formulation of the problem, which is not always easy, unless of course the weight function $w$ already appears as a coefficient in the differential equation, or if the orthogonal polynomials in question are the Legendre polynomials (since then $w(x) \equiv 1$). We describe the construction for a uniformly elliptic PDE subject to a *homogeneous Neumann* boundary condition:

$$-\Delta u + u = f(x) \quad x \in \Omega := (-1,1)^d,$$
$$\frac{\partial u}{\partial \nu} = 0 \qquad \text{on } \partial\Omega,$$

where $f \in L^2(\Omega)$ and $\nu$ denotes the unit outward normal vector to $\partial\Omega$ (or, more precisely, to the $(d-1)$-dimensional open faces contained in $\partial\Omega$). Let us consider the finite-dimensional linear space

$$S_N := \operatorname{span}\{L_\alpha := L_{\alpha_1} \cdots L_{\alpha_d} : 0 \leqslant \alpha_k \leqslant N,\ k = 1,\dots,d\},$$

where $L_{\alpha_k}$ is the univariate Legendre polynomial of degree $\alpha_k$ of the variable $x_k \in (-1,1)$, $k = 1,\dots,d$. The *Legendre–Galerkin spectral approximation* of the boundary-value problem is defined as follows: find $u_N \in S_N$ such that

$$B(u_N, v_N) = \ell(v_N) \quad \forall v_N \in S_N, \tag{19}$$

where the linear functional $\ell(\cdot)$ and the bilinear form $B(\cdot,\cdot)$ are defined by

$$\ell(v) := \int_\Omega f v \,\mathrm{d}x$$

and

$$B(w,v) := \int_\Omega (\nabla w \cdot \nabla v + wv)\,\mathrm{d}x,$$

respectively, with $w, v \in H^1(\Omega)$. As $B(\cdot,\cdot)$ is a symmetric bilinear form and $S_N$ is a finite-dimensional linear space, the task of determining $u_N$ is equivalent to solving a system of linear algebraic equations with a symmetric square matrix $A \in \mathbb{R}^{K\times K}$ with $K := \dim(S_N) = (N+1)^d$. Since $B(V,V) = \|V\|_1^2 > 0$ for all $V \in S_N \setminus \{0\}$, where, as before, $\|\cdot\|_1$ denotes the $H^1(\Omega)$-norm, the matrix $A$ is positive-definite, and therefore invertible. Thus we deduce the existence and uniqueness of a solution to (19). Céa's lemma (see (16)) for (19) takes the form

$$\|u - u_N\|_1 = \min_{v_N \in S_N} \|u - v_N\|_1. \tag{20}$$

If we assume that $u \in H^s(\Omega)$, $s > 1$, results from approximation theory imply that the right-hand side of (20) is bounded by a constant multiple of $N^{1-s}\|u\|_s$, and we thus deduce the error bound

$$\|u - u_N\|_1 \leqslant C N^{1-s}\|u\|_s, \quad s > 1.$$

Furthermore, if $u \in C^\infty(\bar{\Omega})$ (i.e., all partial derivatives of $u$ of any order are continuous on $\bar{\Omega}$), then $\|u - u_N\|_1$ will converge to zero at a rate that is faster than any algebraic rate of convergence; such a superalgebraic convergence rate is usually referred to as *spectral convergence* and is the hallmark of spectral methods.

Since $u_N \in S_N$, there exist $U_\alpha \in \mathbb{R}$, with multi-indices $\alpha = (\alpha_1,\dots,\alpha_d) \in \{0,\dots,N\}^d$, such that

$$u_N(x) = \sum_{\alpha \in \{0,\dots,N\}^d} U_\alpha L_\alpha(x).$$

Substituting this expansion into (19) and taking $v_N = L_\beta$, with $\beta = (\beta_1,\dots,\beta_d) \in \{0,\dots,N\}^d$, we obtain the system of linear algebraic equations

$$\sum_{\alpha \in \{0,\dots,N\}^d} B(L_\alpha, L_\beta) U_\alpha = \ell(L_\beta), \quad \beta \in \{0,\dots,N\}^d, \tag{21}$$

for the unknowns $U_\alpha$, $\alpha \in \{0,\dots,N\}^d$, which is reminiscent of the system of linear equations (14) encountered in connection with finite-element methods. There is, however, a fundamental difference: whereas the matrix of the linear system (14) was symmetric, positive-definite, and *sparse*, the one appearing in (21) is symmetric, positive-definite, and *full*. It has to be noted that because

$$B(L_\alpha, L_\beta) = \int_\Omega \nabla L_\alpha \cdot \nabla L_\beta \,\mathrm{d}x + \int_\Omega L_\alpha L_\beta \,\mathrm{d}x,$$

in order for the matrix of the system to become diagonal, instead of Legendre polynomials one would need to use a system of polynomials that are orthogonal in the *energy inner product* $(u,v)_B := B(u,v)$ induced by $B$.

If the homogeneous Neumann boundary condition considered above is replaced with a 1-periodic boundary condition in each of the $d$ coordinate directions and the function $f$ appearing on the right-hand side of the PDE $-\Delta u + u = f(x)$ on $\Omega = (0,1)^d$ is a 1-periodic function in each coordinate direction, then one can use trigonometric polynomials instead of Legendre polynomials in the expansion of the numerical solution. This will then result in what is known as a *Fourier–Galerkin spectral method*. Because trigonometric polynomials are orthogonal in both the $L^2(\Omega)$ and the $H^1(\Omega)$ inner product, the matrix of the resulting system of linear equations will be diagonal, which greatly simplifies the solution process. Having said this, the presence of (periodic) nonconstant coefficients in the PDE will still destroy orthogonality in the associated energy inner product $(\cdot,\cdot)_B$, and the matrix of the resulting system of linear equations will then, again, become full. Nevertheless, significant savings can be made in spectral computations through the use of fast transform methods, such as the fast Fourier transform or the fast Chebyshev transform, and this has contributed to the popularity of Fourier and Chebyshev spectral methods.

*Spectral collocation methods* seek a numerical solution $u_N$ from a certain finite-dimensional space $S_N$, spanned by orthogonal polynomials, just as spectral Galerkin methods do, except that after expressing $u_N$ as a finite linear combination of orthogonal polynomials and substituting this linear combination into the differential equation $Lu = f$ under consideration, one demands that $Lu_N(x_k) = f(x_k)$ at certain carefully chosen points $x_k$, $k = 1, \dots, K$, called the *collocation points*. Boundary and initial conditions are enforced analogously. A trivial requirement in selecting the collocation points is that one ends up with as many equations as the number of unknowns, which is, in turn, equal to the dimension of the linear space $S_N$.

We illustrate the procedure by considering the parabolic equation

$$\partial_t u - \partial^2_{xx} u = 0, \quad (t,x) \in (0,\infty) \times (-1,1),$$

subject to the initial condition $u(0,x) = u_0(x)$ with $x \in [-1,1]$ and the homogeneous Dirichlet boundary conditions $u(t,-1) = 0$, $u(t,1) = 0$, $t \in (0,\infty)$. A numerical approximation $u_N$ is sought in the form of the finite linear combination

$$u_N(t,x) = \sum_{k=0}^{N} a_k(t) T_k(x)$$

with $(t,x) \in [0,\infty) \times [-1,1]$, where

$$T_k(x) := \cos(k \arccos(x)), \quad x \in [-1,1],$$

is the *Chebyshev polynomial* (of the first kind) of degree $k \geqslant 0$. Note that there are $N+1$ unknowns, the coefficients $a_k(t)$, $k = 0, 1, \dots, N$. We thus require the same number of equations. The function $u_N$ is substituted into the PDE and it is demanded that, for $t \in (0,\infty)$ and $k = 1, \dots, N-1$,

$$\partial_t u_N(t,x_k) - \partial^2_{xx} u_N(t,x_k) = 0.$$

It is further demanded that $u_N(t,-1) = 0$ and $u_N(t,1) = 0$ for $t \in (0,\infty)$ and that $u_N(0,x_k) = u_0(x_k)$ for $k = 0, \dots, N$, where the $(N+1)$ collocation points are defined by $x_k := \cos(k\pi/N)$, $k = 0, \dots, N$; these are the $(N+1)$ points of extrema of $T_N$ on the interval $[-1,1]$. By writing $u^k(t) := u_N(t,x_k)$, after some calculation based on properties of Chebyshev polynomials one arrives at the following set of ordinary differential equations:

$$\frac{\mathrm{d}u^k(t)}{\mathrm{d}t} = \sum_{l=1}^{N-1} (D_N^2)_{kl} u^l(t), \quad k = 1, \dots, N-1,$$

where $D_N^2$ is the *spectral differentiation matrix of second order*, whose entries $(D_N^2)_{kl}$ can be explicitly calculated. One can then use any standard numerical method for a system of ordinary differential equations to evolve the values $u^k(t) = u_N(t,x_k)$ of the approximate solution $u_N$ at the collocations points $x_k$, $k = 1, \dots, N-1$, contained in $(-1,1)$, from the values of the initial datum $u_0$ at the same points.

## 7 Concluding Remarks

We have concentrated on four general and widely applicable families of numerical methods: finite-difference, finite-element, finite-volume, and spectral methods. For additional details the reader is referred to the books listed below and to the rich literature on numerical methods for PDEs for the construction and analysis of other important techniques for specialized PDE problems.

### Further Reading

Bangerth, W., and R. Rannacher. 2003. *Adaptive Finite Element Methods for Differential Equations.* Lectures in Mathematics. Basel: Birkhäuser.

Brenner, S. C., and L. R. Scott. 2008. *The Mathematical Theory of Finite Element Methods*, 3rd edn. New York: Springer.

Brezzi, F., and M. Fortin. 1991. *Mixed and Hybrid Finite Element Methods.* New York: Springer.

Canuto, C., M. Hussaini, A. Quarteroni, and T. Zhang. 2006. *Spectral Methods: Fundamentals in Single Domains.* Berlin: Springer.

Ciarlet, P. G. 2002. *The Finite Element Method for Elliptic Problems.* Philadelphia, PA: SIAM.

Di Pietro, D. A., and A. Ern. 2012. *Mathematical Aspects of Discontinuous Galerkin Methods.* Heidelberg: Springer.

Eymard R., T. Galluoët, and R. Herbin. 2000. Finite volume methods. In *Handbook of Numerical Analysis*, volume 7, pp. 713–1020. Amsterdam: North-Holland.

Gustafsson, B., H.-O. Kreiss, and J. Oliger. 1995. *Time Dependent Problems and Difference Methods.* New York: Wiley-Interscience.

Hackbusch, W. 1985. *Multigrid Methods and Applications.* Berlin: Springer.

Johnson, C. 2009. *Numerical Solution of Partial Differential Equations by the Finite Element Method*, reprint of the 1987 edn. New York: Dover.

LeVeque, R. J. 2002. *Finite Volume Methods for Hyperbolic Problems*. Cambridge: Cambridge University Press.

Richtmyer, R. D., and K. W. Morton. 1994. *Difference Methods for Initial-Value Problems*, reprint of the 2nd edn. Malabar, FL: Robert E. Krieger.

Schwab, Ch. 1998. *p- and hp-Finite Element Methods*. New York: Oxford University Press.

Trefethen, L. N. 2000. *Spectral Methods in MATLAB*. Philadelphia, PA: SIAM.

Verfürth, R. 2013. *A Posteriori Error Estimation Techniques for Finite Element Methods*. Oxford: Clarendon/Oxford University Press.

# IV.14 Applications of Stochastic Analysis

*Bernt Øksendal and Agnès Sulem*

## 1 Introduction

Stochastic analysis is a relatively young mathematical discipline, characterized by a unification of *probability theory* (stochastics) and *mathematical analysis*. Key elements of stochastic analysis are, therefore, integration and differentiation of random/stochastic functions. Stochastic analysis is currently applied to a wide variety of applications in a number of different areas, including finance, physics, engineering, biology, and also within other fields of mathematics itself. We can mention only some of these applications here, and we refer the reader to the list of further reading at the end of the article for more information.

Probability theory and mathematical analysis were traditionally disjoint mathematical disciplines that had little or no interaction. In 1923 Norbert Wiener gave a mathematically rigorous description of the erratic motion of pollen grains in water, a phenomenon that was observed by Robert Brown in 1828. This motion, called Brownian motion, was modeled by Wiener as a *stochastic process* called the *Wiener process*, denoted by $W(t) = W(t,\omega)$ or $B(t) = B_t = B_t(\omega) = B(t,\omega)$. Heuristically, one could say that the position of a pollen grain at time $t$ is represented by $B(t,\omega)$. Here, $\omega$ is a "scenario" parameter that could represent, for example, a pollen grain or an experiment, depending on the model setup.

Wiener showed that $t \to B(t,\omega)$ is continuous but nowhere differentiable for almost all $\omega \in \Omega$ (i.e., for all $\omega$ except on a set of $P$-measure zero, where $P$ is the probability law of $B(\cdot)$). Nevertheless, he showed that it is possible to define what was later called the *Wiener integral*, namely,

$$I(f)(\omega) := \int_0^T f(t)\,\mathrm{d}B(t,\omega) \quad (T > 0 \text{ fixed}), \qquad (1)$$

as a square-integrable element with respect to $P$ for all deterministic square-integrable functions $f$ with respect to the Lebesgue measure on $[0,T]$. This construction represents the first combination of probability and analysis, and as such marks the birth of stochastic analysis.

Subsequently, in 1942 Kiyoshi Itô extended Wiener's construction to include stochastic integrands $\varphi(t,\omega)$ that are *adapted* (to $\mathcal{F}_t$), in the sense that for each $t$ the value of $\varphi(t,\omega)$ can be expressed in terms of the *history* $\mathcal{F}_t$ of $B(t,\omega)$, i.e., in terms of previous values of $B(s,\omega)$, $s \leqslant t$. He showed that in this case the integral can in some sense be represented as a limit of Riemann sums:

$$\int_0^T \varphi(t,\omega)\,\mathrm{d}B(t,\omega) = \lim_{N\to\infty} \sum_{i=1}^{N} \varphi(t_{i+1})(B_{t_{i+1}} - B_{t_i}) \quad (2)$$

($0 = t_0 < t_1 < \cdots < t_N = T$ being a partition of $[0,T]$) if $\varphi$ satisfies

$$\int_0^T \varphi^2(t,\omega)\,\mathrm{d}t < \infty \text{ a.s.},$$

and then the limit in (2) exists in a weak sense (in probability with respect to $P$; "a.s." is an abbreviation for "almost surely," i.e., with probability 1).

Itô proceeded to study such integrals and their properties in a series of papers in the 1950s. He proved a useful chain rule, now called the Itô formula, for processes that are sums of Itô integrals and integrals with respect to Lebesgue measure. Such processes are today known as *Itô processes*. Moreover, he studied corresponding *stochastic differential equations* of the form

$$\left.\begin{aligned} \mathrm{d}X(t) &= b(t,X(t))\,\mathrm{d}t + \sigma(t,X(t))\,\mathrm{d}B_t, \quad 0 \leqslant t \leqslant T, \\ X(0) &= x \in \mathbb{R}, \end{aligned}\right\} \qquad (3)$$

where $b\colon [0,T] \times \mathbb{R} \to \mathbb{R}$ and $\sigma\colon [0,T] \times \mathbb{R} \to \mathbb{R}$ are given functions satisfying certain conditions, and he proved the existence and uniqueness of solutions $X(\cdot)$ of such equations, later known as (Itô) stochastic differential equations (SDEs). Note that (3) is just a shorthand notation for the stochastic *integral* equation

$$X(t) = x + \int_0^t b(s,X(s))\,\mathrm{d}s + \int_0^t \sigma(s,X(s))\,\mathrm{d}B_s, \quad 0 \leqslant t \leqslant T. \qquad (4)$$

For several years the work of Itô was considered an interesting theoretical construction, but one without any applications. But this changed in the late 1960s and early 1970s when Henry McKean published a book with a useful introduction to the stochastic analysis of Itô and Samuelson proposed modeling the price of a stock in a financial market by the solution $X(t)$ of an SDE of the form

$$\left.\begin{aligned} \mathrm{d}X(t) &= \alpha X(t)\,\mathrm{d}t + \beta X(t)\,\mathrm{d}B(t), \quad 0 \leqslant t \leqslant T, \\ X(0) &= x > 0, \end{aligned}\right\} \tag{5}$$

with $\beta > 0$ and $\alpha$ constants. This was the beginning of a wide range of applications of stochastic analysis. The amazing subsequent development of the subject can be seen as the result of fruitful interplay between applications and theory. We think this strong two-way communication between problems and concepts in applications and the development of corresponding mathematical theory is unique to stochastic analysis.

In the following sections we will briefly review some examples of this interaction.

## 2 Applications to Finance

The application of stochastic analysis to finance is undoubtedly one of the most spectacular examples of the application of mathematics in our society. Before stochastic analysis entered the field, finance was, with a few exceptions, an area almost devoid of mathematics. Today, though, finance is strongly influenced by the theory and methods of stochastic analysis. Here we give just a brief introduction to this area of application, while we refer to the articles on FINANCIAL MATHEMATICS [V.9] and PORTFOLIO THEORY [V.10] elsewhere in this volume for more information on this huge and active research topic.

The breakthrough came in the late 1960s/early 1970s, when

- Jan Mossin studied a discrete-time financial model and solved the problem of finding the portfolio that maximizes the expected utility of the terminal wealth, using dynamic programming,
- Robert Merton used the (continuous-time) financial market model of Samuelson (see above) and solved the optimal-portfolio problem by using the Hamilton–Jacobi–Bellman equation, and
- Fischer Black and Myron Scholes developed their famous Black–Scholes option-pricing formula.

Merton and Scholes were awarded the Nobel Memorial Prize in Economic Sciences for these achievements in 1997 (Black died in 1995).

To see why stochastic analysis is such a natural tool for finance, let us consider a simple financial market model with two investment possibilities.

(1) A risk-free investment with constant unit price $S_0(t) = 1$.

(2) A risky investment, with unit price $S_1(t)$ given by

$$\begin{aligned} \mathrm{d}S_1(t) &= S_1(t)[\alpha\,\mathrm{d}t + \beta\,\mathrm{d}B(t)], \quad t \geqslant 0, \\ S_1(0) &> 0, \end{aligned}$$

as in Samuelson's model.

If $\varphi(t) = (\varphi_0(t), \varphi_1(t))$ is a *portfolio*, giving the number of units held at time $t$ of the risk-free and risky investments, respectively, then

$$X^{\varphi}(t) := \varphi_0(t)S_0(t) + \varphi_1(t)S_1(t)$$

is the total value of the portfolio at time $t$. We say that the portfolio is *self-financing* if the infinitesimal change in the wealth at a given time $t$, $\mathrm{d}X^{\varphi}(t)$, comes from the change in the market prices only, i.e., if

$$\mathrm{d}X^{\varphi}(t) = \varphi_0(t)\,\mathrm{d}S_0(t) + \varphi_1(t)\,\mathrm{d}S_1(t).$$

We assume from now on that all portfolios are self-financing. Since $\mathrm{d}S_0(t) = 0$, this gives the integral representation

$$\begin{aligned} X_x^{\varphi}(t) &= x + \int_0^t \varphi_1(s)\,\mathrm{d}S_1(s) \\ &= x + \int_0^t \varphi_1(s)\alpha S_1(s)\,\mathrm{d}s + \int_0^t \varphi_1(s)\beta S_1(s)\,\mathrm{d}B_s, \end{aligned} \tag{6}$$

where $x$ is the initial value and the last integral is the Itô integral.

Note that this integral representation is natural because of the interpretation (2) of the Itô integral as a limit of Riemann sums, where the portfolio choice $\varphi_1(t_i)$ is taken as the *left-hand side* of the partitioning intervals $[t_i, t_{i+1})$ of $[0, T]$. Heuristically, first we decide on the portfolio, then comes the price change. With this mathematical setup the breakthroughs of Mossin and Merton can be formulated more precisely, as follows.

### 2.1 The Portfolio-Optimization Problem (Mossin/Merton)

Given a *utility* function (i.e., a continuous, increasing, concave function) $U\colon [0,\infty) \to \mathbb{R}$, find a portfolio $\varphi^* \in \mathcal{A}$ (the family of admissible portfolios) such that

$$\sup_{\varphi\in\mathcal{A}} E[U(X^{\varphi}(T))] = E[U(X^{\varphi^*}(T))], \tag{7}$$

where $T > 0$ is a given terminal time and $E[\cdot]$ denotes expectation with respect to $P$. This may be regarded as a stochastic control problem. With specific choices of utility functions, e.g., $U(x) = \ln x$ or $U(x) = (1/\gamma)x^\gamma$ for some constant $\gamma \in (-\infty, 1)\backslash\{0\}$, this problem can be solved explicitly by dynamic programming, as Merton did (see section 4).

### 2.2 The Option-Pricing Problem (Black–Scholes)

Suppose that at time $t = 0$ you are offered a contract that pays $(S(T) - K)^+$ $(= \max\{S(T) - K, 0\})$ at a specified future time $T$. Here, $K > 0$ is a constant, also specified in the contract. Such a contract is called a *European call option*, and $K$ is called the exercise price. This contract is equivalent to a contract that gives the owner the right, but not the obligation, to buy one unit of the risky asset at time $T$ at the price $K$. To see this equivalence, we argue as follows. If the market price $S(T)$ turns out to be greater than $K$, then according to the second contract the owner can buy one unit at the price $K$ and then immediately sell it at $S(T)$, giving a profit of $S(T) - K$. If, however, $S(T) \leqslant K$, the contract is worthless and the payoff is 0. This leads to the contract payoff $(S(T)-K)^+$ in general, which is in agreement with the first contract. The question is, then, how much would you be willing to pay at time $t = 0$ for such a contract? If you are a careful person, you might argue as follows.

> If I pay an amount $z$ for the contract, I start with an initial wealth $-z$, and with this initial wealth it should be possible for me to find a portfolio $\varphi \in \mathcal{A}$ (in particular, a self-financing portfolio) such that the sum of the corresponding terminal wealth $X^{\varphi}_{-z}(T)$ and the payoff of the contract is, almost surely, nonnegative.

This gives the following expression for the buyer's price of a call option:

$$p_{\text{buyer}} := \sup\{z;\ \text{there exists } \varphi \in \mathcal{A} \text{ such that } X^{\varphi}_{-z}(T) + (S(T) - K)^+ \geqslant 0 \text{ a.s.}\}.$$

This problem of finding the option price $p_{\text{buyer}}$ is of a different type from the Merton problem in (7). The *Black–Scholes option-pricing formula* states that

$$p_{\text{buyer}} = E_Q[(S(T) - K)^+], \tag{8}$$

where $E_Q$ means expectation with respect to the measure $Q$, defined to be the unique probability measure $Q$ equivalent to $P$ such that $S_1(t)$ is a *martingale* under $Q$, i.e.,

$$E_Q[S_1(t) \mid \mathcal{F}_s] = S_1(s)$$

for all $0 \leqslant s \leqslant t \leqslant T$.

Such a measure is called an *equivalent martingale measure*. In this financial market there is only one equivalent martingale measure, and it is given by

$$\mathrm{d}Q(\omega) = \exp\Big(-\frac{\alpha}{\beta}B(T) - \frac{1}{2}\Big(\frac{\alpha}{\beta}\Big)^2 T\Big)\,\mathrm{d}P(\omega) \text{ on } \mathcal{F}_T.$$

Substituting this into (8) and using the probabilistic properties of Brownian motion we get an explicit formula for the call option price $p_{\text{buyer}}$ in terms of $K$, $\beta$, and $T$. A surprising feature of this formula is that it does not contain $\alpha$, and we are therefore led to the important conclusion that the option price does not depend on the drift coefficient $\alpha$.

The ground-breaking results in sections 2.1 and 2.2 motivated a lot of research activity. Other models were introduced and studied, and at the same time other areas of applications were found (see section 7).

We now proceed to present some applications that are not necessarily connected to finance.

## 3 Backward Stochastic Differential Equations

Returning to the equation (6) for the wealth $X^{\varphi}_x(t)$ corresponding to a given (self-financing) portfolio $\varphi(t) = (\varphi_0(t), \varphi_1(t))$, one may ask the following question. Given a random variable $F(\omega)$, assumed to depend only on the history $\mathcal{F}_T$ of $B(t,\omega)$ up to time $t$, does there exist a portfolio $\varphi$ such that $X^{\varphi}_x(T) = F$ a.s.?

If we substitute

$$Z(t) := \varphi_1(t)S(t)$$

into (6), we see that this question can be formulated as follows. Given $F$, find $X(t)$ and $Z(t)$ such that

$$\mathrm{d}X(t) = \frac{\alpha}{\beta}Z(t)\,\mathrm{d}t + Z(t)\,\mathrm{d}B(t), \quad 0 \leqslant t \leqslant T,$$
$$X(T) = F \text{ a.s.}$$

This is called a *backward* stochastic differential equation (BSDE) in the two unknown (adapted) processes $X$ and $Z$. In contrast to (3), it is the terminal value $F$ of $X$ that is given, not the initial value. More generally, given a function $g(t, y, z, \omega)\colon [0,T] \times \mathbb{R} \times \mathbb{R} \times \Omega \to \mathbb{R}$ and a random variable $F$, as above, the corresponding BSDE in the two unknown (adapted) processes $Y(t)$, $Z(t)$ has the form

$$\mathrm{d}Y(t) = -g(t, Y(t), Z(t), \omega)\,\mathrm{d}t + Z(t)\,\mathrm{d}B_t, \quad 0 \leqslant t \leqslant T,$$
$$Y(T) = F \text{ a.s.}$$

The process $g$ is sometimes called the *driver*.

The original motivation for studying BSDEs came from stochastic control theory (see section 4), but a number of other developments have since been found, such as the probabilistic representations of solutions of nonlinear parabolic partial differential equations developed by Pardoux and Peng in 1990. In the 1990s, Duffie and Epstein used BSDEs to introduce the concept of the *recursive utility* of a consumption process, and around ten to fifteen years ago BSDEs were used to define *convex risk measures* as a model for the risk of a financial position.

## 4 Optimal Stochastic Control

The portfolio-optimization problem presented in section 2 may be regarded as a special case of a general *optimal stochastic control problem* of the following type. Suppose that the state $X(t) = X_{s,x}(t)$ of a system at time $t$ is described by a stochastic differential equation of the form

$$\mathrm{d}X(t) = b(t, X(t), u(t), \omega)\,\mathrm{d}t + \sigma(t, X(t), u(t), \omega)\,\mathrm{d}B_t \tag{9}$$

for $s \leqslant t \leqslant T$, with $X(s) = x$, ($x \in \mathbb{R}$ and $0 \leqslant s \leqslant T$ given), where $b\colon [0,T]\times\mathbb{R}\times\mathbb{V}\times\Omega \to \mathbb{R}$ and $\sigma\colon [0,T]\times\mathbb{R}\times\mathbb{V}\times\Omega \to \mathbb{R}$ are given functions such that $b(\cdot, x, v, \cdot)$ and $\sigma(\cdot, x, v, \cdot)$ are adapted processes for each given $x \in \mathbb{R}$ and $v \in \mathbb{V}$, a given set of control values. The process $u(t) = u(t,\omega)$ is our *control* process. To be admissible it is required that $u$ be adapted and that $u(t,\omega) \in \mathbb{V}$ for all $t \in [0,T]$ and almost all $\omega \in \Omega$. The set of admissible control processes is denoted by $\mathcal{A}$. With each choice of $u \in \mathcal{A}$, $(s,x) \in [0,T] \times \mathbb{R}$, we associate a *performance* $J_u(s,x)$ given by

$$J_u(s,x) = E\biggl[\int_0^T f(t, X_{s,x}(t), u(t), \omega)\,\mathrm{d}t + g(X_{s,x}(t), \omega)\biggr], \tag{10}$$

where $f\colon [0,T] \times \mathbb{R} \times \mathbb{V} \times \Omega \to \mathbb{R}$ and $g\colon \mathbb{R} \times \Omega \to \mathbb{R}$ are given functions. Sometimes $f$ is called the *profit rate* and $g$ is called the *bequest* or *salvage* function. We assume that $f(\cdot, x, v, \cdot)$ is adapted and that $g(x,\cdot)$ depends only on the history $\mathcal{F}_T$ of the underlying Brownian motion $B(t,\omega)$, $s \leqslant t \leqslant T$, for each $x$ and $v$.

The problem is to find a control $u^* \in \mathcal{A}$ and a function $\Phi(s,x)$ such that

$$\Phi(s,x) = \sup_{u\in\mathcal{A}} J_u(s,x) = J_{u^*}(s,x), \quad (s,x) \in [0,T]\times\mathbb{R}. \tag{11}$$

Such a process $u^*$ (if it exists) is called an *optimal control*, and $\Phi$ is called the *value function*.

For example, if we represent the portfolio in (6) by the fraction $\pi(t)$ of the total wealth $X(t)$ invested in the risky asset at time $t$, then

$$\pi(t) = \frac{\psi_1(t)S_1(t)}{X(t)}$$

and (6) can be written

$$\left.\begin{aligned} \mathrm{d}X(t) &= \pi(t)X(t)[\alpha\,\mathrm{d}t + \beta\,\mathrm{d}B(t)], \quad 0 \leqslant t \leqslant T, \\ X(0) &= x > 0. \end{aligned}\right\} \tag{12}$$

Comparing (12) and (7) with (9)–(11), we see that the optimal-portfolio problem is an example of an optimal stochastic control problem, as claimed.

There are two main solution methods for stochastic control problems: the dynamic programming principle and the maximum principle.

### 4.1 The Dynamic Programming Principle

The *dynamic programming principle*, introduced by R. Bellman in the 1950s, applies only to *Markovian* problems. These are problems where the coefficients $b(s,x,v)$, $\sigma(s,x,v)$, $f(s,x,v)$, and $g(x)$, for fixed values of $s$, $x$, and $v$, do not depend on $\omega$. Moreover, the control process $u(t)$ must be Markovian, in the sense that $u(t) = u_0(t, X(t))$ for some deterministic function $u_0\colon [0,T]\times\mathbb{R} \to \mathbb{V}$. (Such a control is called a feedback control.) In this case, the dynamic programming principle leads to the *Hamilton–Jacobi–Bellman equation*, which (under some conditions) states the following.

Suppose that $\varphi$ is a smooth function such that

$$f(s,x,v) + A^v\varphi(s,x) \leqslant 0, \quad 0 \leqslant s \leqslant T,\ x \in \mathbb{R},\ v \in \mathbb{V},$$

and that

$$\varphi(T,x) = g(x).$$

Then

$$\varphi(s,x) \geqslant \Phi(s,x), \quad (s,x) \in [s,T] \times \mathbb{R}.$$

Moreover, assume that, in addition, for all $(s,x) \in [0,T] \times \mathbb{R}$ there exists $u_0(s,x) \in \mathbb{V}$ such that

$$f(s,x,u_0(s,x)) + A^{u_0(s,x)}\varphi(s,x) = 0. \tag{13}$$

Then

$$\varphi(s,x) = \Phi(s,x), \quad (s,x) \in [s,T] \times \mathbb{R}$$

and

$$u^*(t) := u^0(t, X(t))$$

is an optimal (Markovian) control.

Here, $A^v$ denotes the *generator* of the Markov process $X^v_{s,x}(t)$ obtained by using the constant control

$u(t) := v \in \mathbb{V}$. It is a parabolic second-order partial differential operator given by

$$A^v \varphi(s,x) := \frac{\partial \varphi}{\partial s}(s,x) + b(s,x,v)\frac{\partial \varphi}{\partial x}(s,x) + \tfrac{1}{2}\sigma^2(s,x,v)\frac{\partial^2 \varphi}{\partial x^2}(s,x).$$

For example, for the problem (7) with (12) we get

$$A^v \varphi(s,x) := \frac{\partial \varphi}{\partial s}(s,x) + \alpha v x\frac{\partial \varphi}{\partial x}(s,x) + \tfrac{1}{2}\beta^2 v^2 x^2(s,x,v)\frac{\partial^2 \varphi}{\partial x^2}(s,x),$$

which is maximal when

$$v = v^* = -\frac{\alpha\varphi'(s,x)}{\beta^2 x\varphi''(s,x)},$$

where

$$\varphi'(s,x) = \frac{\partial \varphi}{\partial x}(s,x) \quad \text{and} \quad \varphi''(s,x) = \frac{\partial^2 \varphi}{\partial x^2}(s,x).$$

Substituting into (13) we get

$$\frac{\partial \varphi}{\partial s}(s,x) - \frac{\alpha^2(\varphi')^2(s,x)}{2\beta^2\varphi''(s,x)} = 0.$$

In particular, if

$$U(x) = g(x) = \frac{1}{\gamma}x^\gamma, \quad x > 0,$$

for some parameter $\gamma \in (-\infty,1)\backslash 0$, then we see that the above equation holds for

$$\varphi(s,x) = h(s)x^\gamma,$$

where

$$h'(s) - \frac{\alpha^2\gamma}{2\beta^2(\gamma-1)}h(s) = 0, \quad 0 \leqslant s \leqslant T,$$
$$h(T) = 1/\gamma.$$

This gives an optimal control (portfolio)

$$u^*(t) = \pi^*(t) = -\frac{\alpha h(t)\gamma x^{\gamma-1}}{\beta^2 x\gamma(\gamma-1)h(t)x^{\gamma-2}} = \frac{\alpha}{\beta^2(1-\gamma)},$$

which is one of the classical results of Mossin and Merton.

## 4.2 The Maximum Principle

The *maximum principle* goes back to L. Pontryagin and his group, who developed this method for deterministic control problems in the 1950s. It was adapted to stochastic control by J.-M. Bismut and subsequently extended further by A. Bensoussan, E. Pardoux, S. Peng, and others throughout the 1970s, 1980s, and 1990s. Basically, the maximum principle approach to the stochastic control problem (11) is as follows.

First, we define the associated *Hamiltonian* $H\colon \mathbb{R}^5 \to \mathbb{R}$ by

$$J(t,x,u,p,q) = f(t,x,u) + b(t,x,u)p + \sigma(t,x,u)q. \tag{14}$$

Here, $p$, $q$ are adjoint variables, somehow related to Lagrange multipliers.

Next, associated with the Hamiltonian we consider, for given $u \in \mathcal{A}$, the BSDE

$$\left.\begin{aligned} \mathrm{d}p(t) &= -\frac{\partial H}{\partial x}(t,X(t),u(t),p(t),q(t))\,\mathrm{d}t + q(t)\,\mathrm{d}B_t, \\ &\qquad\qquad 0 \leqslant t \leqslant T, \\ p(T) &= g'(X(T)) \end{aligned}\right\} \tag{15}$$

in the two unknown adjoint processes $p(t) = p^{(u)}(t)$, $q(t) = q^{(u)}(t)$, where $X(t) = X^{(u)}(t)$ is the solution of (9) corresponding to the control $u$.

The *maximum principle* relates the maximization of $J_u(s,x)$ in (10) to the maximization of the Hamiltonian. For example, under some concavity assumptions we have the following result.

**The sufficient maximum principle (assumes concavity).** *Suppose that* $\hat{u} \in \mathcal{A}$, *with associated solution* $\hat{X}(t) = X^{(\hat{u})}(t)$, $\hat{p}(t) = p^{(\hat{u})}(t)$, *and* $\hat{q}(t) = q^{(\hat{u})}(t)$ *of the forward-backward SDE system (9), (15). Suppose that for each* $t$, $u = \hat{u}(t)$ *maximizes the Hamiltonian, in the sense that*

$$u \to H(t,\hat{X}(t),u,\hat{p}(t),\hat{q}(t))$$

*is maximal for* $u = \hat{u}(t)$. *Then,* $\hat{u}(\cdot)$ *is an optimal control for the problem (11).*

To illustrate how this result works, let us apply it to once again solve the problem (12), (7); in this case, the Hamiltonian becomes, with $u = \pi$,

$$H(t,x,\pi,p,q) = \pi x\alpha p + \pi x\beta q \tag{16}$$

and

$$\left.\begin{aligned} \mathrm{d}\hat{p}(t) &= -(\hat{\pi}(t)\alpha\hat{p}(t) + \hat{\pi}(t)\beta\hat{q}(t))\,\mathrm{d}t + \hat{q}(t)\,\mathrm{d}B(t), \\ \hat{p}(t) &= U'(\hat{X}(T)). \end{aligned}\right\} \tag{17}$$

Since $H$ is linear in $\pi$, the only possibility for the existence of a maximum of

$$\pi \to H(t,\hat{X}(t),\pi,\hat{p}(t),\hat{q}(t))$$

is that

$$\alpha\hat{p}(t) + \beta\hat{q}(t) = 0,$$

i.e.,

$$\hat{q}(t) = -\frac{\alpha}{\beta}\hat{p}(t). \tag{18}$$

Substituting this into (17) we get

$$d\hat{p}(t) = -\frac{\alpha}{\beta}\hat{p}(t)\,dB(t),$$

which has the solution

$$\hat{p}(t) = \hat{p}(0)\exp\left(-\frac{\alpha}{\beta}B(t) - \frac{1}{2}\left(\frac{\alpha}{\beta}\right)^2 t\right). \tag{19}$$

Now assume, as above, that

$$U(x) = \frac{1}{\gamma}x^{\gamma}. \tag{20}$$

The requirement

$$\hat{p}(T) = U'(\hat{X}(T)) \tag{21}$$

then gives the equation

$$\begin{aligned}\hat{p}(0)\exp&\left(-\frac{\alpha}{\beta}B(T) - \frac{1}{2}\left(\frac{\alpha}{\beta}\right)^2 T\right)\\ &= x^{\gamma-1}\exp\left((\gamma-1)\int_0^T \hat{\pi}(t)\beta\,dB(t)\right.\\ &\qquad\left.+(\gamma-1)\int_0^T(\hat{\pi}(t)\alpha - \tfrac{1}{2}\hat{\pi}^2(t)\beta^2)\,dt\right).\end{aligned} \tag{22}$$

From (22), it is natural to try $\hat{\pi}$ such that

$$(\gamma-1)\hat{\pi}(t)\beta = -\frac{\alpha}{\beta},$$

i.e.,

$$\hat{\pi}(t) = \frac{\alpha}{\beta^2(1-\gamma)}. \tag{23}$$

Since, by (19), $\hat{p}(t)$ is a martingale, from (21) we obtain $\hat{p}(0) = E[U'(\hat{X}(T))]$, and substituting this into (22) we verify that (22) holds with $\hat{\pi}$ as in (23).

This confirms the result we found in section 4.1.

## 5 Optimal Stopping

To describe the problem of optimal stopping of a stochastic process $S(t)$, we first need to explain the crucial concept of a *stopping time.*

A random time $\tau\colon \Omega \to [0,\infty]$ is called a *stopping time* (with respect to the history of $S(\cdot)$) if, for any time $t$, the decision of whether or not to stop at time $t$ is based only on the history of $S(s)$ for $s \leqslant t$.

For example, if $S(t)$ is the position at time $t$ of a car driving along a road in a city, then

$\tau_1$ := the *first* time we come to a traffic light

will be a stopping time because this instant can be decided simply by recording the history of $S$ up to that time. On the other hand, if we define

$\tau_2$ := the *last* time we come to a traffic light,

then $\tau_2$ will *not* be a stopping time because we would need to know the future in order to decide whether or not a given traffic light is the last one.

Thus, since we cannot assume knowledge about the future, we see that stopping times are the natural random times to consider in applications: the optimal time to sell (or buy) an object with a stochastic price process, the optimal time to start a new business, the optimal time to close down a factory, and so on.

If the state $Y(t) \in \mathbb{R}^k$ at time $t$ is described by an SDE of the form

$$\left.\begin{aligned} dY(t) &= b(Y(t))\,dt + \sigma(Y(t))\,dB(t), \quad t \geqslant 0,\\ y(0) &= y \in \mathbb{R}^k,\end{aligned}\right\} \tag{24}$$

then the associated *optimal stopping problem* is to find a stopping time $\tau^*$ (with respect to $Y(\cdot)$), called an *optimal* stopping time, such that

$$\begin{aligned}\Phi(y) &:= \sup_{\tau\in\mathcal{T}} E^y\left[\int_0^{\tau} f(Y(t))\,dt + g(Y(\tau))\right]\\ &= E^y\left[\int_0^{\tau^*} f(Y(t))\,dt + g(Y(\tau^*))\right],\end{aligned} \tag{25}$$

where $\mathcal{T}$ is the set of all $Y(\cdot)$ stopping times and we interpret $g(Y(\tau))$ as 0 if $\tau = \infty$. Here, $f$ and $g$ are given functions, and $E^y$ denotes the expectation assuming that $Y(0) = y$. The function $\Phi$ is called the *value function* of the optimal stopping problem.

As in stochastic control problems, it often turns out that in order to find an optimal stopping time $\tau^*$, it helps to simultaneously try to find the value function $\Phi$. In fact, under some technical conditions one can prove the following.

**The optimal stopping theorem.** *Let $A$ be the generator of $Y(\cdot)$ (see section 4).*

(a) *Then*

  (a) $A\Phi(y) + f(y) \leqslant 0$ *for all* $y$ *and*

  (b) $\Phi(y) \geqslant g(y)$ *for all* $y$.

  *Moreover, at all points $y$, at least one of the two inequalities holds with equality. We can therefore combine (i) and (ii) into the equation*

$$\max\{A\Phi(y) + f(y), g(y) - \Phi(y)\} = 0 \quad \textit{for all } y. \tag{26}$$

(b) *Let $D$ be the continuation region, defined by $D = \{y;\ g(y) < \Phi(y)\}$. Then it is optimal to stop the first time that $Y(t)$ exits from $D$, i.e.,*

$$\tau^* = \inf\{t > 0;\ Y(t) \notin D\}. \tag{27}$$

(c) *Moreover, $\Phi$ and $g$ meet smoothly at $\partial D$, i.e.,*

$$\nabla\Phi(y) = \nabla g(y) \quad \textit{for } y \in \partial D \tag{28}$$

*(this is called the high contact or smooth fit principle).*

This theorem shows that there is a close connection between optimal stopping problems (which are stochastic) and *variational inequality problems* and *free boundary problems*, which are nonstochastic classical analysis problems.

Regarding *variational inequality problems*, the equation (26) is an example of a classical *variational inequality*. The operator $A$ and the functions $f$ and $g$ are given, and the problem is to find $\Phi$ such that (26) holds. A classical example in which this problem appears is the problem of finding the wet region in a porous sand wall of a dam.

A *free boundary problem* is a problem of the following type. Given a differential operator $A$ and functions $f$ and $g$, find a function $\Phi$ *and a domain* $D$ such that

(i) $A\Phi(y) + f(y) = 0$ for $y \in D$,
(ii) $\Phi(y) = g(y)$ for $y \in \partial D$, and
(iii) $\nabla\Phi(y) = \nabla g(y)$ for $y \in \partial D$,

where $\nabla$ denotes the gradient operator. Applications of this type of problem include the problem of finding the boundary of a melting ice cube.

The link between the optimal stopping problem and the free boundary problem is provided by the high contact principle (28).

It can also be shown that optimal stopping problems are related to *reflected* BSDEs. In a reflected BSDE we are given a driver $g$ as in section 3 as well as a *lower barrier process* $L(t)$, and we want to find processes $Y(t)$, $Z(t)$ and a nondecreasing process $K(t)$ (all of them adapted) with $K(0) = 0$ such that the following hold:

- $\mathrm{d}Y(t) = -g(Y(t), Z(t), \omega)\,\mathrm{d}t + Z(t)\,\mathrm{d}B(t) - \mathrm{d}K(t)$, and
- $Y(t) \geqslant L(t)$ for all $t$.

## 6 Filtering Theory

Suppose we model the size of a population, e.g., the fish population in a lake, by a stochastic logistic differential equation of the form

$$\mathrm{d}X(t) = X(t)[K - X(t)](\alpha\,\mathrm{d}t + \beta\,\mathrm{d}B(t)), \quad t \geqslant 0, \tag{29}$$

where $\alpha$, $\beta$, and $K > 0$ are constants. $K$ is called the carrying capacity of the lake; it represents the maximal population size that the lake can sustain. Heuristically, the "noisy" factor $\alpha + \beta(\mathrm{d}B(t)/\mathrm{d}t)$ representing the nutritional quality of the lake, subject to random fluctuations (the quantity $\mathrm{d}B(t)/\mathrm{d}t$, called *white noise*), can be rigorously defined as a *stochastic distribution*. Even if there are good theoretical reasons for choosing such a model, the problem of how to estimate the coefficients $\alpha$, $\beta$, and $K$, and also $X(t)$ itself, remains. To simplify matters, let us assume that $\alpha$, $\beta$, and $K$ are known. How do we find $X(t)$?

The problem is that we do not know the precise value of $X(t_0)$ for any $t_0$. To compensate for this, we make (necessarily imprecise, or "noisy") observations $Y(t)$ on $X(t)$, i.e., we observe

$$Y(t) = X(t) + \gamma\frac{\mathrm{d}B_1(t)}{\mathrm{d}t}$$

or

$$\mathrm{d}Y(t) = X(t)\,\mathrm{d}t + \gamma\,\mathrm{d}B_1(t), \quad t \geqslant t_0,$$

where $\gamma > 0$ is a known constant and $B_1(\cdot)$ is another Brownian motion, usually assumed to be independent of $B(\cdot)$.

The *filtering problem* is as follows: what is the *best* estimate, $\hat{X}(t)$, of $X(t)$ *based on* the observations $Y(s)$, $s \leqslant t$?

By saying that an estimate $\tilde{X}(t)$ is *based on* $Y(s)$, $s \leqslant t$, we mean that it should be possible to express $\tilde{X}(t)$ by means of the values $Y(s)$, $s \leqslant t$, only. In other words, $\tilde{X}(\cdot)$ should be *adapted* to the history (filtration) $\{\mathcal{F}_t^Y\}_{t\leqslant 0}$ generated by $Y(t)$, $t \geqslant 0$. By saying that $\hat{X}(t)$ is the *best* estimate based on $Y$, we mean best in the sense of *minimal mean square error*, i.e.,

$$E[(X(t) - \hat{X}(t))^2] = \inf_{\tilde{X}\in\mathcal{Y}} E[(X(t) - \tilde{X}(t))^2], \tag{30}$$

where $\mathcal{Y}$ is the set of all estimates based on $Y$. The solution of the problem (30) can be expressed as

$$\hat{X}(t) = E[X(t) \mid \mathcal{F}_t^Y], \tag{31}$$

where the right-hand side denotes the *conditional expectation* of $X(t)$ with respect to $\mathcal{F}_t^Y$. The filtering problem is therefore the problem of finding this conditional expectation of $X(t)$.

The filtering problem is difficult in general, and an explicit solution is known only in special cases. The most famous solvable case is the *linear* case, in which we have

$$\mathrm{d}X(t) = F(t)X(t)\,\mathrm{d}t + C(t)\,\mathrm{d}B(t) \quad \text{(signal process)},$$
$$\mathrm{d}Y(t) = G(t)X(t)\,\mathrm{d}t + D(t)\,\mathrm{d}B_1(t) \quad \text{(observations)},$$

where $F(t)$, $C(t)$, $G(t)$, and $D(t) > 0$ are known, bounded deterministic functions, $D(t)$ being bounded away from 0.

The *Kalman–Bucy filtering theorem* then states that the best estimate $\hat{X}(t)$ solves the SDE

$$\left.\begin{aligned} \mathrm{d}\hat{X}(t) &= \left(F(t) - \frac{G^2(t)S(t)}{D^2(t)}\right)\hat{X}(t)\,\mathrm{d}t \\ &\qquad\qquad + \frac{G(t)S(t)}{D^2(t)}\,\mathrm{d}Y(t), \\ \hat{X}(0) &= E[X(0)], \end{aligned}\right\} \tag{32}$$

where $S(t) := E[(X(t) - \hat{X}(t))^2]$ (the minimal mean square error) satisfies the (deterministic) *Riccati equation*

$$\left.\begin{aligned} \frac{\mathrm{d}S(t)}{\mathrm{d}t} &= 2F(t)S(t) - \frac{G^2(t)}{D^2(t)}S^2(t) + C^2(t), \quad t \geqslant 0, \\ S(0) &= E[(X(0) - E[X(0)])^2]. \end{aligned}\right\} \tag{33}$$

Note that in (32), $\hat{X}(t)$ is indeed expressed in terms of the observations $Y(s)$, $s \leqslant t$, only, assuming that the initial distribution of $X(0)$ is Gaussian, with given mean $E[X(0)]$ and variance $S(0)$.

## 7 Outlook

In this short article we have been able to discuss only the classical stochastic calculus based on Brownian motion and some of the well-known (and spectacular) applications of this theory. But we should point out that in recent years there has been rapid research development both in the mathematical foundations of stochastic analysis and in further applications of the subject.

- For example, the theory of stochastic integration has been extended to other integrator processes, such as Lévy processes, Poisson random measures, and, more generally, semimartingales.
- Stochastic differential equations and more general *stochastic integral* or stochastic *functional* differential equations, and even stochastic *partial* differential equations driven by such processes, are being studied. In particular, *mean-field* SDEs are used in connection with the modeling of *systemic risk* in finance.
- Stochastic optimization methods (stochastic control, singular control, impulse control, optimal stopping) for such systems are being developed accordingly, with associated generalized BSDEs.
- Various ways of representing *model uncertainty* in applications have been introduced and studied, including *nonlinear expectation* theory and the theory of *G-Brownian motion*.
- An axiomatic approach to the concept of *convex risk measures* has been developed. Such risk measures can be represented either by a BSDE or by a dual approach, using a family of probability measures that are absolutely continuous with respect to $P$.
- The relationship between performance and available information is being studied. In particular, what is the optimal performance of a controller for which only *partial (e.g., delayed) information* is available? This topic may also combine optimal control with filtering.
- In the opposite direction, *anticipative* stochastic calculus, i.e., stochastic calculus with nonadapted integrands, has been developed. Combined with the recently developed *stochastic calculus of variations (Malliavin calculus)* and the *Hida white noise calculus*, this gives an efficient mathematical tool for investigating the actions of insiders in financial markets.
- *Stochastic differential games* involving players with *asymmetric information* is another hot topic with obvious applications in several areas, including biology, engineering, investment theory, and finance. This is a challenging area of research in which the full force of the mathematical machinery discussed above is needed.

### Further Reading

Black, F., and M. Scholes. 1973. The pricing of options and corporate liabilities. *Journal of Political Economy* 81:637–754.

Crisan, D., and B. Rozovskii. 2011. *The Oxford Handbook of Nonlinear Filtering*. Oxford: Oxford University Press.

Dixit, A. K., and R. S. Pindyck. 1994. *Investment under Uncertainty*. Princeton, NJ: Princeton University Press.

El Karoui, N., S. Peng, and M.-C. Quenez. 1997. Backward stochastic differential equations in finance. *Mathematical Finance* 7:1–71.

Kinderlehrer, D., and G. Stampacchia. 2000. *An Introduction to Variational Inequalities and Their Applications*. Philadelphia, PA: SIAM.

McKean, H. P. 1969. *Stochastic Integrals*. New York: Academic Press.

Merton, R. 1992. *Continuous-Time Finance*. Oxford: Blackwell.

Murray, J. D. 1993. *Mathematical Biology*, 2nd edn. Berlin: Springer.

Øksendal, B. 2013. *Stochastic Differential Equations*, 6th edn. Berlin: Springer.

Øksendal, B., and A. Sulem. 2007. *Applied Stochastic Control of Jump Diffusions*, 2nd edn. Berlin: Springer.

Øksendal, B., and A. Sulem. 2014. Risk minimization in financial markets modeled by Itô-Lévy processes. *Afrika Matematika*, DOI:10.1007/s13370-014-0248-9.

# IV.15 Inverse Problems

*Fadil Santosa and William W. Symes*

## 1 What Is an Inverse Problem?

In inverse problems one is interested in determining parameters of a system from measurements. Much of applied mathematics is about modeling and understanding phenomena that occur in the world. The models are often based on first principles: physical laws, empirical laws, etc. A common use of models is for prediction; if we know all the parameters of a model, we can predict how the model will respond to a given excitation. In inverse problems we perform experiments in which the response of a model to prescribed excitations is measured. The goal is to determine the unknown parameters of the model from the measured data.

### 1.1 Background and History

Perhaps the oldest formally studied inverse problem—dating back to the work of Herglotz (1907) and Wiechert (1910)—is that of determining properties of the Earth's subsurface from measurement of seismological data. This problem falls into the category of geophysical inverse problems, which includes techniques for imaging the near subsurface for deposits of oil and other minerals.

In physics the classical inverse problem involves finding the potential in the Schrödinger equation from scattering experiments. Still in the same family are scattering problems involving geometrical scatterers where one is interested in determining the geometrical shape from measurements of the acoustic or electromagnetic response of the object. In military applications such problems fall under the study of sonar and RADAR [VII.17].

Medical imaging has been a driving force in the development of inverse problems. The ability to visualize and characterize the internal structures of a patient is of great value in diagnostics. A very important imaging modality, X-RAY COMPUTED TOMOGRAPHY [VII.9, VII.19] (abbreviated X-ray CT and also known as computer-assisted tomography), can be viewed as an inverse problem. The conductivities of biological tissues can also be a target of medical imaging and could potentially provide further diagnostic capabilities that complement other imaging modalities.

## 2 Language and Concepts

One of the most influential researchers in the area of inverse problems was Pierre Sabatier, who is responsible for formalizing its study. He provided much of the vocabulary of the subject, which we describe in some detail below.

### 2.1 The Forward Map

In inverse problems, we are interested in determining the properties of a system that we call a *model*, $m$. The model is related to observations that we call *data*, $d$. To be mathematically precise, we call $M$ the set of all possible models under consideration. We would normally attach mathematical properties to this set that are consistent with the physics. The space of data $D$ also needs to be characterized. The *forward map* is the mathematical relationship between a model and its associated measurement. Let us indicate this by $F(m)$. In an inverse problem we are given data $d$ from which we wish to find, to the extent possible, the unknown model $m$. Thus, we wish to "solve" $F(m) \approx d$ for a given $d$. We use the approximation symbol "$\approx$" above because noise is inherent in any measurement. In most inverse problems it is unlikely that the data $d$ is in the range of the forward map for all $m$ in $M$.

To illustrate the above concept let us consider the inverse problem of X-ray CT. The model in question will be X-ray attenuation, described by an attenuation coefficient $\rho(x)$ that is a function of position $x$. A possible space of functions for the model is a set of nonnegative functions with an upper bound, namely $L^\infty(\Omega)$, where $\Omega$ is the domain being scanned. The data are input and output X-ray intensities, $I_0$ and $I$, respectively, parametrized by X-ray trajectory $L$ (see figure 1), usually given as the logarithm of their ratio: $\log(I(L)/I_0(L))$. The forward map, based on Beer's law, is

$$R(\rho; I_0, L) = -\int_L \rho \, \mathrm{d}\ell,$$

that is, $R(\rho; I_0, L)$ is the line integral of $\rho(x)$ along line $L$. The inverse problem of a computer-assisted tomography scan is to determine $\rho(x)$ from the data $\log(I(L)/I_0(L))$. The found $\rho(x)$ is often displayed as a set of images, aiding in diagnostics.

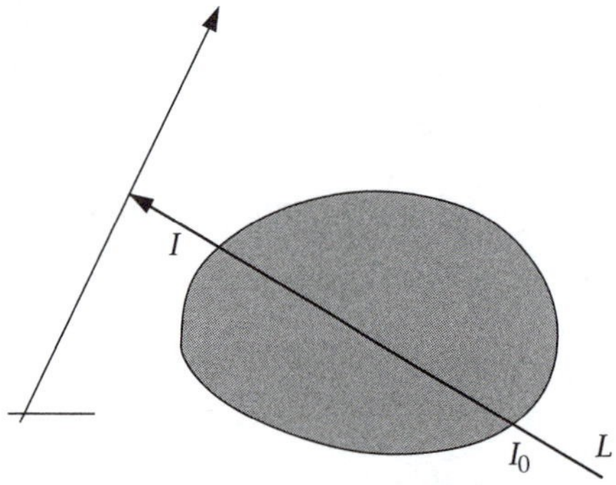


**Figure 1** Schematic of a generic X-ray CT setup. The intensity of the X-ray entering the body is $I_0$; the intensity at exit is $I$. The data value is $\log(I/I_0)$ for the ray $L$. The forward map is the line integral of the attenuation coefficient $\rho(x)$ along the line $L$. The inversion of $\rho(x)$ depends on the measurement setup and is particularly straightforward in two dimensions, as described in section 3.1.

### 2.2 Inversion and Data Fitting

There are instances where the solution to the equation $F(m) = d$ can be written down explicitly. That is, we have a formula for $m$ in terms of the given data $d$. An example is when $F$ is a linear operator whose inverse can be calculated. Another popular approach is to view the inverse problem as a data-fitting problem. In this case we seek a model $m$ that minimizes the misfit $\|F(m) - d\|$. The choice of norm is critical and depends on what additional information is available about the map $F$ and the data $d$. The most important information is the regularity of the map and the statistical properties of the noise embedded in the data.

### 2.3 Linear and Nonlinear Inverse Problems

In a linear inverse problem, the forward map satisfies the property

$$F(m_1 + m_2) = F(m_1) + F(m_2).$$

In practical terms, linearity often means that the problem is easy to solve. If both $m$ and $d$ are finite dimensional, then a linear forward map can be represented as a matrix. In this instance, the linear inverse problem amounts to a linear system of equations.

For nonlinear inverse problems, the above relationship does not hold. Nonlinear inverse problems are usually more difficult to analyze and to solve, and iterative techniques are often employed for solving them.

One can linearize a nonlinear inverse problem by assuming that the unknown model can be written as

$$m = m_0 + \Delta m,$$

where $m_0$ is known and $\Delta m$ is small. The concept of smallness can be made precise mathematically. Then, if the forward map $F(\cdot)$ associated with the problem is differentiable, we approximate it by a two-term Taylor series:

$$F(m) \approx F(m_0) + G\Delta m.$$

Here, $G$ is a linear operator representing the derivative (Jacobian) of the forward map $F(\cdot)$. Thus, given data $d$, the linearized inverse problem seeks to determine $\Delta m$ from the equation

$$G\Delta m = d - F(m_0).$$

In some situations we can solve a nonlinear inverse problem by successively seeking solutions to a sequence of linearized inverse problems. Such an approach, when applied to a least-squares formulation, may be viewed as a Gauss–Newton method.

### 2.4 Ill-Posedness and Ill-Conditioning

Hadamard defined a mathematical problem as *well-posed* if it has a unique solution that depends continuously on the data. Another way to put this is to say that a problem is well-posed if (i) a solution to the problem exists, (ii) the solution is unique, and (iii) the solution depends continuously on the data. A mathematical problem that does not possess these three attributes is called *ill-posed*. There are instances in which a problem is well-posed but is unstable to perturbations in the data. That is, the continuity in (iii) is weak. Such problems are often called ill-conditioned or unstable. It turns out that most inverse problems of interest are ill-posed, or at least ill-conditioned.

To put these concepts in the context of an inverse problem, violating (i) is equivalent to saying that, given data $d$ and a forward map $F(\cdot)$, the equation $F(m) = d$ does not have a solution. Violating condition (ii) is equivalent to the existence of two models, $m_1$ and $m_2$, distinct from each other, such that

$$F(m_1) = F(m_2).$$

Finally, violating condition (iii) is equivalent to saying that there are $m_1$ and $m_2$ such that

$$\|m_1 - m_2\| \text{ is arbitrarily large}$$

even if

$$\|F(m_1) - F(m_2)\| \text{ is finite.}$$

This notion is best understood in terms of limits of sequences.

These ideas are very easy to see in the case of a finite-dimensional linear inverse problem where the mapping $F(\cdot)$ is a matrix operation: $F(m) = Gm$ for some matrix $G$. Consider an inverse problem of finding $m$ in $Gm = d$. Nonexistence is equivalent to having data $d$ that are not in the range of $G$. Nonuniqueness amounts to $G$ having a null space, i.e., nontrivial solutions to $Gx = 0$. If $G$ has a null space with $x$ being a null vector, then if $m_1 - m_2 = ax$, $Gm_1 - Gm_2 = 0$ while $a \to \infty$, thus violating condition (iii). Finally, ill-conditioning means that the inverse of $F$ has a large norm, a circumstance that can be succinctly described by the SINGULAR VALUE DECOMPOSITION [II.32].

The study of inverse problems is often about mathematically overcoming ill-posedness or ill-conditioning. Various techniques can help ameliorate its disastrous effects, as we now explain.

### 2.5 A Priori Information and Preferences

We often have prior information about the unknown in an inverse problem, and such information can be valuable in solving the problem. We have previously alluded to situations where the model is of the form

$$m = m_0 + \Delta m.$$

Under such circumstance, we can view $m$ as a priori information. There are also situations in which we know some characteristics of the unknown. An example is in imaging, where the model is an image. Suppose we know that the image is piecewise constant or that the image is sparse. We can then formulate the problem such that the solution has the desired properties. The technique to be described next is often employed to enforce a preference.

### 2.6 Regularization

Tikhonov regularization is an approach for solving an ill-posed or ill-conditioned problem. It involves introducing auxiliary terms that make the problem well-posed. The classical inverse problem considered by Tikhonov is a linear inverse problem. Here, the forward map is a linear operator acting on the model parameters, given by $Gm$. We are given data $d$ and wish to find $m$. Instead of posing an equation for $m$, we seek the minimizer to

$$\|Gm - d\|^2 + \lambda\|Bm\|^2. \qquad (1)$$

Here, we have used the $L_2$-norm. The second term in (1) is often referred to as the penalty term. The linear operator $B$ is called the regularizing operator. Some mathematical properties are imposed on $B$. Examples include $B = I$ (the identity) when enforcing smallness in $m$, and $B$ equal to the equivalent of the spatial derivative with $m$ when it represents parameters distributed over space when enforcing smoothness of $B$.

The point of Tikhonov theory is a notion of "consistency." To get a sense of its mathematical significance, suppose that $G$ is not invertible. Let $m_0$ be the true model; thus, $d_0 = Gm_0$ is noiseless data for the inverse problem. We use $e = d - d_0$, to represent the noise in the data for a given experiment. Denote by $m$ the minimizer of (1) for a given $d$. Tikhonov theory states that under the right mathematical assumptions on $G$ and $B$, there exists a sequence of values for $\lambda$ such that if there is a sequence of data $d \to d_0$, then the minimizer $m \to m_0$.

While such a theory is of limited practical value, the use of regularization is very powerful for solving ill-conditioned inverse problems. The main challenge in its use is choosing the penalty parameter $\lambda$. There are two main methods used for setting $\lambda$: the L-curve method and the method of cross-validation. The former is intuitive, whereas the latter is based on statistical assumptions on the noise in the data.

It should be pointed out that different choices of the regularizing operator $B$ and the norms used can give very different answers. For example, when $B$ is the derivative operator and the $L_1$-norm is used, we get the well-known total variation regularization. When $B = I$ and the $L_1$-norm is again used, we get what is now known as COMPRESSED SENSING [VII.10].

### 2.7 A Statistical Approach

Bayesian statistics offer a different point of view when solving inverse problems. In this statistical approach we view the inverse problem as one where we have a prior distribution on the unknown model $m$. The data $d$ is used to update the prior.

To be more precise, instead of a model $m$ and an observation $d$, we consider random variables $M$ and $D$ representing models and data. By having a prior for the model, we mean that we have a probability distribution $\pi(m)$. For example, $m$ could be finite dimensional and $\pi(m)$ could be a multivariate normal distribution. In the statistical framework, data $d$ is observed and we wish to know the conditional probability distribution

$\pi(m|d)$. The question becomes that of updating the prior probability distribution for $m$ given that we have observed $d$.

The update is done using BAYES'S RULE [V.11 §1]. This requires a model for the joint probability distribution $\pi(m, d)$, which may be as simple as

$$\pi(m, d) = f(Gm - d),$$

depending on the assumptions.

Computationally, statistical methods can be very costly. However, the payoff can be large because one is often able to get an answer to an inverse problem with confidence intervals. The method of applying the Bayesian framework to inverse problems is often referred to as inversion with uncertainty quantification. Aside from the computational cost, a critique sometimes leveled against this approach is that the prior probability distribution for the model $\pi(m)$ is very difficult to obtain in practice. Nevertheless, Bayesian approaches have become a major tool for solving inverse problems.

## 3 Selected Examples of Inverse Problems

### 3.1 X-Ray Computed Tomography

X-ray CT is a method by which the X-ray absorption coefficients of a body are estimated. We have already briefly mentioned the mathematical problem associated with CT in section 2.1. We provide more detail here.

We consider the problem in two dimensions. In figure 2 the path of the X-ray is indicated by the dashed line, denoted by $L_{\theta,s}$. The angle $\theta$ is one of the parameters of the ray, the other being the displacement $s$. The ray path $L_{\theta,s}$ is given by

$$x\cos\theta + y\sin\theta - s = 0.$$

Letting $\rho(x, y)$ be the attenuation and letting $d\ell$ be the length element along the path, the forward map for the inverse problem of CT is

$$R_\theta[\rho](s) = \int_{L_{\theta,s}} \rho(x, y)\,\mathrm{d}\ell.$$

In the inverse problem, we are given data $P_\theta(s)$ for $0 \leqslant \theta < 2\pi$ and $a \leqslant s \leqslant b$, and we solve

$$R_\theta[\rho](s) = P_\theta(s)$$

for $\rho(x, y)$. We may assume that $\rho(x, y)$ is compactly supported. Modern CT can be attributed to the work of Cormack and Hounsfield, who shared the 1979 Nobel Prize in Physiology or Medicine for their work. However, the mathematics that makes CT a reality dates back to the work of Radon, who in 1907 showed that a function could be reconstructed from its projections. One can use Radon's theory to obtain an inversion formula for $\rho(x, y)$ given $P_\theta(s)$, which is given by

$$\rho(x, y) = \frac{1}{2\pi^2}\int_0^\pi \int_{-\infty}^{\infty} \frac{P_\theta'(s)}{x\cos\theta + y\sin\theta - s}\,\mathrm{d}s\,\mathrm{d}\theta,$$

although the formula is of limited practical use. Here, $P_\theta'(s)$ is the derivative of $P_\theta(s)$ with respect to $s$.

Several practical computational approaches are available for solving the inverse problem of X-ray CT. Some are based on Fourier transforms, while others are based on solving a system of linear equations. Modern CT scanners are designed for high-speed data acquisition and low X-ray dosage. Algorithms are designed for each machine in order to optimize computational efficiency and accuracy.

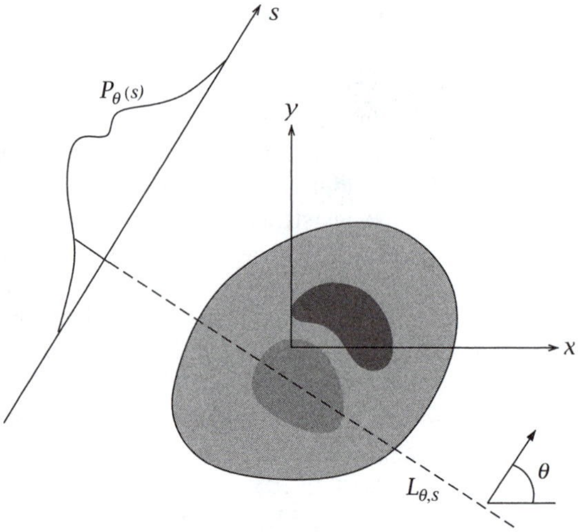


**Figure 2** Schematic of X-ray CT in two dimensions. The dashed line represents the path of the X-ray. The projection is parametrized by the angle $\theta$. Data at this angle are the function $P_\theta(s)$, where $s$ parametrizes the ray displacement as indicated.

### 3.2 Seismic Travel-Time Tomography

In travel-time tomography, the data consist of records of the time taken for a wave packet to leave a source, travel through the Earth's interior, and arrive at a point on the surface of the Earth. Such data are called travel-time data and can be extracted from seismic records. In the simplest model, the Earth is flat and two dimensional. The sound speed depends only on depth, and the source is located on the surface. Using geometrical

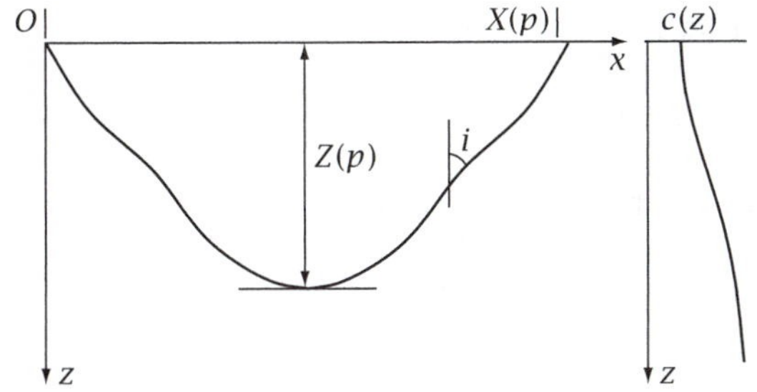


**Figure 3** Schematic of a travel-time inverse problem. A wave source is located at the origin. A ray emanates from the source, and its trajectory is determined by the take-off angle and the sound speed profile $c(z)$. It returns to the surface at $x = X(p)$, where $p$ is the ray parameter. The inverse problem is to find $c(z)$ from $X(p)$.

optics we follow a ray emanating from the source and trace its path back to the surface (see figure 3). A ray is parametrized by the angle it makes with the surface at the source.

Suppose that the unknown sound speed profile as a function of depth is $c(z)$. Let us locate the source at the origin. A ray leaves the origin, making an angle $i_0$ with respect to the vertical. According to geometrical optics, the quantity

$$p = \frac{\sin i}{c(z)}$$

is constant along the ray, where $i$ is the angle the ray makes at depth $z$, measured clockwise from the vertical. This quantity is called the ray parameter. Fermat's principle governs the ray trajectory, and the point at which a ray with parameter $p$ arrives on the surface, $X(p)$, is given by

$$X(p) = \int_0^{Z(p)} \frac{p\,\mathrm{d}z}{\sqrt{c(z)^{-2} - p^2}}.$$

The function $Z(p)$ is the maximum depth the ray reaches in its trajectory, and it is the solution to

$$c(Z(p)) = 1/p,$$

since the ray's angle $i$ is $\pi/2$ at this point. In the inverse problem, we are given travel-time data $X(p)$ for a range of values of $p$, and the goal is to recover $c(z)$ to a maximum depth possible.

This problem was considered by Herglotz (1907) and Wiechert (1910), who provided a formula for the solution. It is related to the well-known Abel problem in which one attempts to find the shape of a hill given the return times of a particle that goes up the hill at fixed velocities. Bôcher (1909) studied the Abel problem and arrived at the mathematical conditions under which the hill can be constructed uniquely. The techniques of Bôcher can be applied to the geophysical travel-time problem. A solution that gives $z$ as a function of $c$ is available for this problem:

$$z(c) = -\frac{1}{\pi}\int_{c_0^{-1}}^{c^{-1}} \frac{X(p)}{\sqrt{p^2 - c^{-2}}}\,\mathrm{d}p.$$

The formula is valid when $X(p)$ has continuous derivatives. In particular, it is not valid when $X(p)$ is multivalued, which occurs when $c(z)$ has a "low-velocity zone," i.e., an interval in which $c(z)$ dips below an otherwise-increasing trend.

### 3.3 Geophysical Inversion for Near-Surface Properties

In an attempt to determine the near-surface (several kilometers in depth) properties of the Earth, geophysicists perform experiments on the surface to gather data from which they hope to infer the properties of the subsurface. One such experiment is seismic exploration, wherein elastic waves are used to probe the Earth. As these waves propagate into the Earth, the inhomogeneities in the Earth diffract and reflect the waves in a process called scattering. The scattered waves are measured on the surface using geophones. The inverse problem is to determine the material properties of the Earth from the scattered data.

In a typical measurement, a localized source, in the form of heavy equipment that "thumps" the surface, is introduced. An array of geophones collects the response of the Earth to the source. Data collection is over a given time window. Such a measurement is called a "shot." The measurement is then repeated at another source location. The totality of the data consist of geophone readings from a number of shots (see figure 4).

Geophysicists have developed a number of approximation methods to interpret such data. These methods have been successful in situations where the wave phenomena are well modeled by the approximations. Current research focuses on more accurate modeling of the wave phenomena, the development of problem formulations that are computationally feasible, and computational methods that exploit the power of modern computers.

In the simplest useful approximation, the Earth is modeled as an acoustic fluid occupying a two- or three-dimensional half-space, with spatially variable bulk

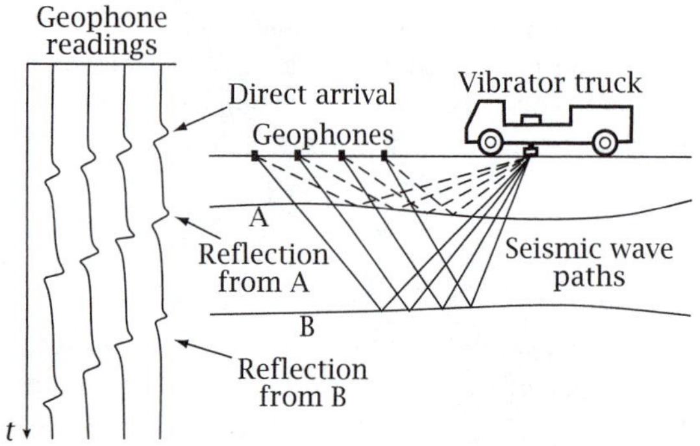


**Figure 4** Data collection in seismic exploration. The vibrator truck is capable of "thumping" the surface and produces waves that travel into the interior of the Earth. These waves are reflected by the inhomogeneities in the Earth and return to the surface. Geophones record the reflected waves. This "shot" is repeated after the source and the receiver array have been moved to a new location.

modulus $\kappa(x)$ and constant density $\rho$. Despite the neglect of significant physics, this model underlies the vast bulk of contemporary seismic imaging and inversion technology.

The examples that follow are two dimensional. The state of the acoustic Earth is captured by the pressure field $u(x,t)$: $x = (x_1, x_2)$ are spatial coordinates with $x_2 > 0$, and $t$ is time. The acoustic wave equation satisfied by $u(x,t)$ is

$$\frac{\partial^2 u}{\partial t^2} = \frac{\kappa}{\rho}\Delta u + f^{(i)}(x,t),$$

where $f^{(i)}(x,t)$ is a time-varying and spatially localized source of energy and $f^{(i)}(x,t) = 0$ for $t < 0$. The index $i$ refers to the location of the source. On the boundary $\{x\colon x_2 = 0\}$, representing the Earth's surface, we assume the Dirichlet boundary condition $u = 0$. This is a reasonable approximation, as pressures in the air are orders of magnitude smaller than pressures in the water or rock.

Since the seismic experiment consists of several shots, it produces several pressure fields $u^{(i)}(x,t)$, caused by several source fields $f^{(i)}(x,t)$ representing varying placements of the "thumping" machinery, $i = 1, 2, \dots, n$. The data recorded by the geophones is the pressure at a number of geophone locations $\{x_j^{(i)}\colon j = 1, 2, \dots, m\}$ near the surface (but not on it: there, $u = 0$!). Note that the geophone locations may depend on the shot index $i$; the entire apparatus of the survey may move from shot to shot, not just the energy source equipment.

The seismic inverse problem can be posed as a data-fitting optimization problem, as explained in section 2.2. In the terminology of that section, the model $m$ is the bulk modulus $\kappa(x)$ and the data $d$ are the vectors of functions of time, one for each shot and receiver location. The value of the forward map $F$ is the sampling of the pressure field predicted via solution of the wave equation:

$$F[\kappa] = \{u^{(i)}(x_j^{(i)}, t)\colon i = 1, \dots, n,\\ j = 1, \dots, m,\ 0 < t < T\}.$$

The inverse problem is to determine $\kappa(x)$ (insofar as possible) from the data $\{d_j^{(i)}(t)\colon i = 1, \dots, n,\ j = 1, \dots, m,\ 0 < t < T\}$:

$$F[\kappa] \simeq d.$$

The most commonly used optimization formulation of this problem seeks to choose $\kappa$ to minimize the mean-square residual

$$\|F[\kappa] - d\|^2 = \sum_{i=1}^{n}\sum_{j=1}^{m}\int_0^T |u^{(i)}(x_j^{(i)}, t) - d_j^{(i)}(t)|^2\,\mathrm{d}t.$$

In the current seismic literature, estimation of $\kappa$ (or other mechanical parameter fields) by minimization of the mean-square residual is known as *full waveform inversion* (FWI). This approach to extracting information about the Earth from seismic data was first studied in the 1980s. Early implementations were generally unsuccessful, partly because at that time only two-dimensional simulation was feasible. Since the turn of the twenty-first century, though, advances in algorithms and hardware performance have enabled three-dimensional simulation and iterative minimization of the mean-square residual for three-dimensional distributions of $\kappa$ and similar parameters. While the technology is still in the early stages of development, it has already become clear that FWI can produce information about the geometry of subsurface rock that is far better in terms of both quality and resolution than that obtainable with older methods based on more drastic approximation.

While plenty of FWI success stories can be found, we see failures too: failures driven by a fundamental *mathematical* difficulty. This issue was discovered in the 1980s, the first period of active research on FWI, and remains the main impediment to its widespread use.

The problem is easy to illustrate with a two-dimensional example. Figure 5(a) displays an example two-dimensional bulk modulus $\kappa$ that is widely used in

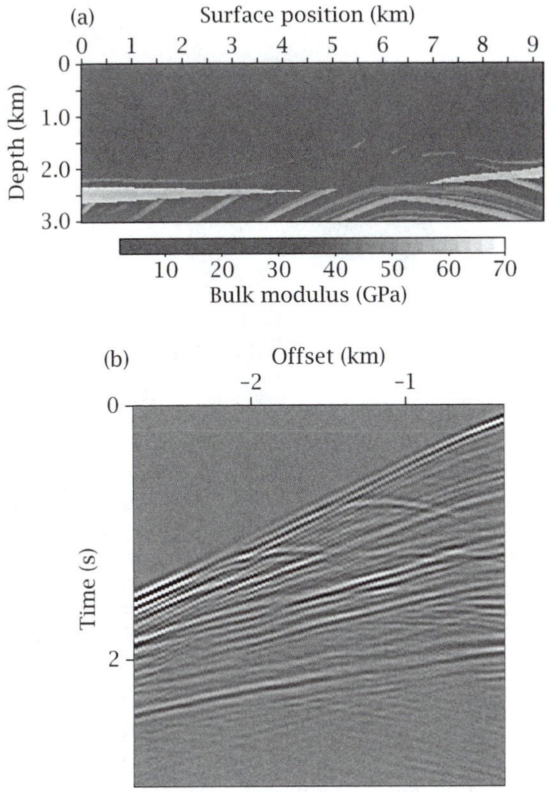


**Figure 5** (a) The Marmousi model for the bulk modulus. (b) The simulated seismic data associated with the model.

research on FWI: the Marmousi model. We choose one source location and solve the wave equation numerically to produce the simulated data depicted in figure 5(b). The simulated data consist of pressure readings at a set of geophones located equidistant from each other, starting some distance from the source. The figure displays the recordings for a single shot, at horizontal position 6 km from the left edge of the model in figure 5(a), at a depth of 6 m. The horizontal axis in figure 5(b) is receiver index ($j = 1, \dots, m$, $m = 96$) and the vertical axis is time. The pressure reading is given a gray level value.

To demonstrate the sensitivity of the cost function to a small change in the bulk modulus, we made a convex combination of the bulk modulus in figure 5(a) (95%) with a constant background of $\kappa_0 = 2.25$ GPa (5%). Let $\kappa_1$ be the bulk modulus in figure 5(a), and let $\kappa_2 = 0.95\kappa_1 + 0.05\kappa_0$. We will measure the difference in the forward maps for these two bulk moduli. The forward map at $\kappa_2$ is the simulated pressure field shown in

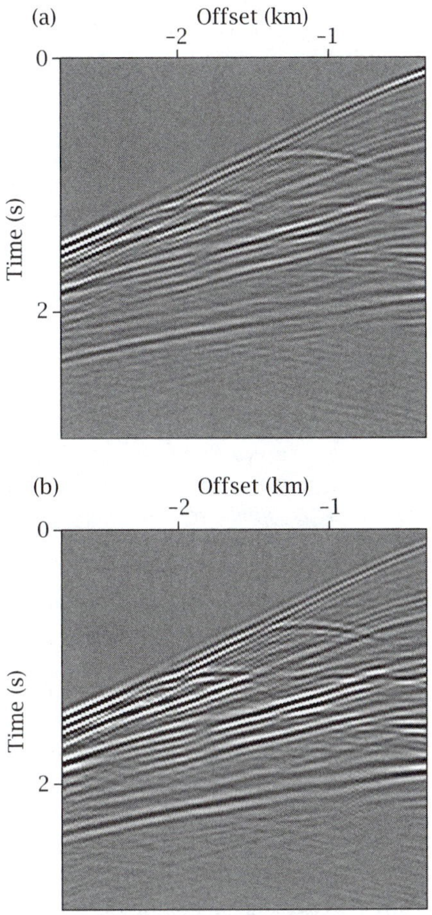


**Figure 6** (a) The simulated seismic data with a different model, consisting of 0.95 times the model in figure 5(a) added to 0.05 times a constant bulk modulus of 2.25 GPa; the net difference is well under 5% root mean square. (b) The residual, which is the difference between figure 5(a) and part (a) of this figure.

figure 6(a), and the residual, the difference between the two maps, $F(\kappa_2) - F(\kappa_1)$, is shown in figure 6(b). The difference $F(\kappa_2) - F(\kappa_1)$ is very visible. In fact, the $L_2$-norm of the difference is almost twice as large (184%) as the norm of the simulated data $F(\kappa_1)$.

From this example we can already see that the predicted data may change very rapidly as a function of the model. The reason for this rapid change is that the oscillatory signal has shifted in time by a large fraction of a wavelength, which yields a large mean-square ($L_2$) change. This "cycle skipping" phenomenon arises from the influence of the bulk modulus on the *speed* of the waves embedded in the solution of the wave equation.

The *size* of the predicted data (figure 6(a)) has not changed much, however. For a variety of reasons, principally conservation of acoustic energy, the overall size of the data (measured with the $L^2$-norm, for example) depends only weakly on the model ($\kappa$). This combination of very rapid change while staying the same size suggests that the predicted data cannot continue to "run away" from the "observed" data; the distance between them must oscillate.

In another numerical experiment we made a convex combination of 90% Marmousi bulk modulus and 10% constant bulk modulus, i.e., $\kappa_3 = 0.9\kappa_1 + 0.1\kappa_0$. We found $\|F(\kappa_3) - F(\kappa_1)\|/\|F(\kappa_1)\|$ to be 144%. We note that this is smaller than $\|F(\kappa_2) - F(\kappa_1)\|/\|F(\kappa_1)\|$. That is, we have observed that *the least-squares objective function (the mean-square residual) has local minima other than its global minimum*, at least when restricted to the line segment through the target model in figure 5(a) and the homogeneous $\kappa = \kappa_0$.

This is a very serious obstacle because the computational size of these problems is so large that only variants of local search algorithms (mainly, Newton's method) are (barely) feasible, and these find only local extrema.

The problem of local minima turns up with considerable regularity in research into, and field applications of, FWI. Since, in the field, we do not know the answer a priori, a frequent result is "zombie inversion": a failure to accurately estimate the structure of the Earth without any effective indication of failure. Quality control of FWI is a current research topic of great interest. Acquisition of unusually low-frequency data can lead to an escape from the local-minimum problem, and this is another topic of great current interest. At the time of writing, there is still no mathematical justification of the effectiveness of low-frequency inversion. Finally, considerable effort is being put into alternative objective functions with better convexity properties than the least-squares function studied here.

### 3.4 Electrical Impedance Tomography

In electrical impedance tomography we are given an object whose spatially dependent conductivity is unknown (see figure 7). Electrostatic measurements are taken on the boundary of the object. The objective is to determine the unknown conductivity from these boundary measurements. This problem is often referred to as the "Calderon problem" because Alberto Calderon was the first to study it mathematically, in 1980.

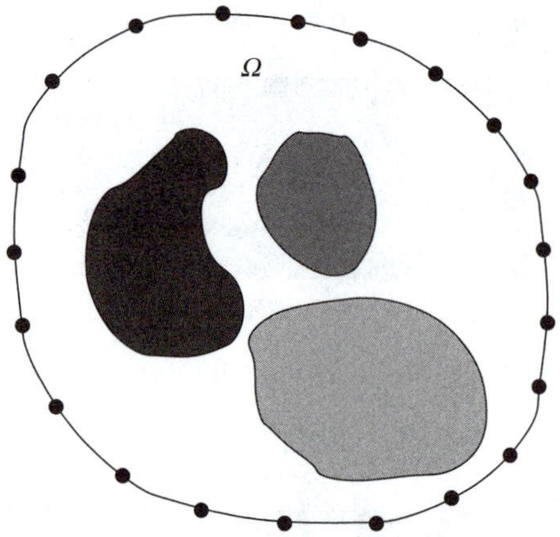


**Figure 7** A schematic of electrical impedance tomography. The body $\Omega$ has variable conductivity in its interior. The inverse problem is to determine the conductivity distribution from boundary measurements. In practice, the measurement is done by attaching electrodes to the boundary of $\Omega$, indicated in the figure by black circles. Current is assigned to the electrodes, while voltages are measured at all the electrodes. The data so collected represent a sampling of the Dirichlet-to-Neumann map.

Let $\sigma(x)$, where $x$ is the spatial variable in two or three dimensions, represent the conductivity of the object $\Omega$. Potential theory, which follows from Ohm's law, states that the electrical potential in $\Omega$ satisfies

$$\nabla \cdot \sigma(x)\nabla u = 0$$

if there are no sources in the interior of $\Omega$. In the idealized mathematical problem, we are allowed to prescribe any boundary value $g$ to $u(x)$, i.e.,

$$u|_{\partial\Omega} = g.$$

For each $g$ we measure the normal derivative of the potential $u(x)$ at the boundary:

$$\left.\frac{\partial u}{\partial \nu}\right|_{\partial\Omega} = f.$$

We therefore have an $f$ for every $g$. The totality of these pairs of functions is our data. The mathematical name for such data is the Dirichlet-to-Neumann map. It is clear that it depends on $\sigma(x)$. Denoting this map by $\Gamma_\sigma$, we have $f = \Gamma_\sigma g$. The inverse problem is to determine $\sigma(x)$ given $\Gamma_\sigma$.

In practice, we are not given the Dirichlet-to-Neumann map in its entirety. Rather, we are given a finite collection of pairs

$$\{f^{(i)}, g^{(i)}\}, \quad i = 1, 2, \ldots, n.$$

A practical approach to measurement is to attach a number of electrodes to the surface of the object.

Current is made to flow between a pair of electrodes, while voltage is measured on all the electrodes. Therefore, in practice, we do not have $(f^{(i)}, g^{(i)})$ but rather samples of their values at the electrodes.

The Calderon problem has received a lot of attention in the past thirty years due to its practical importance. Barber and Brown developed a practical medical imaging device using this principle in 1984. The same problem crops up in other applications such as nondestructive testing and geophysical prospecting.

The mathematical question of unique determination of $\sigma(x)$ from $\Gamma_\sigma$ is a well-developed area. The results depend on the number of dimensions. In two dimensions, the earliest uniqueness result (from 1984) is due to Kohn and Vogelius, who showed that if $\sigma(x)$ is analytic then it is uniquely determined by $\Gamma_\sigma$. The next seminal result on this subject was due to Sylvester and Uhlmann in 1987. They proved global uniqueness results in dimensions three and higher. Their approach is based on the powerful method of complex geometrical optics. Global uniqueness was proved a few years later by Nachman in two dimensions using a method called $\partial$-bar.

Recent work has focused on how rough $\sigma(x)$ can be while retaining uniqueness and on the case where $\sigma(x)$ is a matrix function. The latter study has led to the discovery of TRANSFORMATION OPTICS [VI.1], which has provided a strategy for cloaking and invisibility.

While uniqueness can be established, it is very difficult to reconstruct $\sigma(x)$ from actual measurements in practice. This is due to the fact that the problem of determining $\sigma(x)$ from $\Gamma_\sigma$ is very ill-conditioned. There have been several successful approaches. They are based on linearization of the relationship between $\Gamma_\sigma$ and $\sigma$, on least-squares fitting, and on a direct reconstruction method called the $\partial$-bar method. The last approach involves synthesizing data for a scattering problem from the measured data, which in itself is ill-conditioned.

## 4 Related Problems

There are problems that could be viewed as inverse problems but do not go by that name. One set arises in the study of control theory for distributed parameter systems. Statisticians view an inverse problem as a problem of parameter estimation from data. This point of view precedes the more recent statistical approaches to inverse problems.

## 5 Outlook

Inverse problems is an active research field with several devoted journals and a community of researchers coming from many disciplines. Many of the problems arise in engineering and scientific applications but the core language and tools are mathematical in nature. Progress in new imaging technologies, including functional magnetic resonance imaging and photoacoustic tomography, has been made possible by advances in inverse problems. Inverse problems techniques have also made important contributions to engineering fields such as nondestructive evaluation of materials and structures.

### Further Reading


Aki, K., and P. Richards. 1980. *Quantitative Seismology*, volume 2. San Francisco, CA: W. H. Freeman.

Hansen, P. C. 2010. *Discrete Inverse Problems: Insights and Algorithms*. Philadelphia, PA: SIAM.

Kaipio, J., and E. Somersalo. 2004 *Statistical and Computational Inverse Problems*. Berlin: Springer.

Natterer, F. 1985. *The Mathematics of Computerized Tomography*. New York: John Wiley.

Sabatier, P. 1978. Introduction to applied inverse problems. In *Applied Inverse Problems*, pp. 1–26. Lecture Notes in Physics, volume 85. Berlin: Springer.

Sylvester, J., and G. Uhlmann. 1987. A global uniqueness theorem for an inverse boundary value problem. *Annals of Mathematics* 125:153–69.

Tikhonov, A., and V. Arsenin. 1979. *Methods of Solution of Ill-Posed Problems*. Moscow: Nauka.

Virieux, J., and S. Operto. 2009. An overview of full waveform inversion in exploration geophysics. *Geophysics* 74: WCC1–WCC26.


# IV.16 Computational Science

*David E. Keyes*

## 1 Definitions

Computational science—the systematic study of the structure and behavior of natural and human-engineered systems accomplished by computational means —embraces the domains of mathematical modeling, mathematical analysis, numerical analysis, and computing. In order to reach the desired degrees of predictive power, fidelity, and speed of turnaround, computational science often stretches the state of the art of computing in terms of both hardware and software architecture. Because each decision made along the way—such as the choice between a partial differential equation (PDE) and a particle representation for the model, between a structured and an unstructured grid for the spatial discretization, between an implicit and an explicit time integrator, or between a direct and an iterative method for the linear solver—can greatly affect computational performance, computational science is a highly interdisciplinary field of endeavor.

Definitions of computational science and the related discipline of scientific computing are not universal, but we may usefully distinguish between them by considering computational science to be the vertically integrated *union* of models and data, mathematical technique, and computational technique, while scientific computing is the study of techniques in the *intersection* of many discipline-specific fields of computational science. Computational chemistry, computational physics, computational biology, computational finance, and all of the flavors of computational engineering (chemical, civil, electrical, mechanical, etc.) depend on a common set of techniques connecting the conceptualization of a system to its realization on a computational platform. These techniques—such as representing continuous governing equations with a discrete set of basis functions, encoding this representation for digital hardware, integrating or solving the discrete system, estimating the error in the result, adapting the computational approximation, visualizing functionals of the results, performing sensitivity analyses, or performing optimization or control—span diverse applications from science and engineering and are the elements of scientific computing. Chaining together such techniques to form a simulation to address a specific application constitutes computational science. Practitioners often use the more inclusive terms "computational science and engineering" and "scientific and engineering computing," but in this article the "engineering" applications are understood to be subsumed in the "scientific."

Practitioners also often distinguish between computational science executed by means of simulation and that executed by "mining" data, the latter without necessarily possessing a model. Simulation is often referred to as the "third paradigm" of science and data analytics as the "fourth paradigm." These are in apposition to the "first paradigm" of theory, which is millennia old, and the "second paradigm" of controlled experiment or observation, which is centuries old. The interplay of theoretical hypothesis and controlled experimentation defines the modern scientific method, in the era since, say, Galileo. Recently, the same standards of REPRODUCIBILITY [VIII.5] and reporting have been applied to simulation, and the interplay among the first three paradigms has become highly productive. Until recently, science generally tended to be "data poor," but now most scientific campaigns are data rich, with many "drowning" in data, so a contemporary challenge of computational science is to integrate the third and fourth paradigms. When a model is given, its simulation is called the "forward problem." The availability of data allows aspects of the underlying model to be inferred or improved upon; this is the domain of INVERSE PROBLEMS [IV.15]. Whereas forward problems are by design generally well-posed, inverse problems are often unavoidably ill-posed, due to ill-conditioning or nonuniqueness. There are many elements of scientific computing in this intersecting domain of the third and fourth paradigms, including data assimilation, parameter inversion, optimization and control, and UNCERTAINTY QUANTIFICATION [II.34]. These "post-forward problem" techniques are today applied throughout the computational sciences and they drive a considerable amount of research in applied and computational mathematics. Simulation and data analytics have attained peer status with theory and experiment in many areas of science.

Computer simulation and data analytics enhance or leapfrog theoretical and experimental progress in many areas of science critical to society, such as advanced energy systems (e.g., fuel cells, fusion), biotechnology (e.g., genomics, drug design), nanotechnology (e.g., sensors, storage devices), and environmental modeling (e.g., climate prediction, pollution remediation). Simulation and analytics also offer promising near-term hope for progress in answering a number of scientific questions in such areas as the fundamental structure of matter, the origin of the universe, and the functions of proteins and enzymes.

## 2 Historical Trends

The strains on theory were apparent to John von Neumann (1903–57) and drove him to pioneer computational fluid dynamics and computational radiation transport, and to contribute to supporting fields, especially numerical analysis and digital computer architecture. Models of fluid and transport phenomena, when expressed as mathematical equations, are inevitably nonlinear, while the bulk of mathematical theory (for algebraic, differential, and integral equations) is linear. Computation was to von Neumann, and remains today, the best means of making systematic progress in transport phenomena and many other scientific arenas. Breakthroughs in the theory of nonlinear systems come only occasionally, but computational understanding gains steadily with increases in speed and resolution.

The strains on experimentation—the gold standard of scientific truth—have grown along with expectations for it. Unfortunately, many systems and many questions are all but inaccessible to experiment. (We subsume under "experiment" both experiments designed and conducted by scientists and observations of natural phenomena, such as celestial events, that are out of the direct control of scientists.) The experiments that scientists need to perform to answer pressing questions are sometimes deemed unethical (e.g., because of their impact on living beings), hazardous (e.g., because of their impact on our planetary environment), untenable (e.g., because they are prohibited by treaties), inaccessible (e.g., those in astrophysics or core geodynamics), difficult (e.g., those requiring measurements that are too rapid or numerous to be instrumentable), too time-consuming (in that a decision would need to be made before the experimental results could be attained, e.g., certification of a system or a material lifetime), or simply too expensive (including detectors, the Large Hadron Collider facility cost over $10bn and the International Thermonuclear Experimental Reactor is projected to cost well over $20bn). There is thus a strong incentive to narrow the regimes that must be investigated experimentally using predictive computational simulation and data analytics.

As experimental means approach practical limits, for more than two decades computational performance on real applications has improved at a rate of more than three orders of magnitude per decade (see HIGH-PERFORMANCE COMPUTING [VII.12]). Over the same period, the acquisition cost of a high-performance computer system designed for scientific applications has fallen by almost three orders of magnitude per decade per unit of performance, together with the electrical power required per unit of performance, so that after a decade a system a thousand times more powerful costs about the same to own and operate. These trends in performance and cost are well documented in the history of the Gordon Bell Prizes and of the annual Top 500 lists (see www.top500.org). Increased computational power can be invested in fine resolution of multiscale phenomena, high fidelity, full dimensionality, integration of multiple interacting models in complex systems, and running large ensembles of forward problems to gain scientific understanding. Unfortunately, however, after more than two decades of riding Moore's law, physical barriers to extrapolating the favorable energy trends in computing are approaching, and these barriers are now significant drivers in computational science research.

Contemporary computational science stands at the confluence of four independently and fruitfully developing quests: the mathematization of nature, epitomized by Newton; numerical analysis, epitomized by von Neumann; high-performance computer performance, epitomized by Cray; and scientific software engineering, represented by numerous contemporaries. A predigital computer vision of computational science was proposed by L. F. Richardson in his 1922 monograph on numerical weather prediction. Richardson's parallel computer was an army of human calculators arrayed at latitudinal and longitudinal increments along the interior surface of a mammoth globe. Richardson's monograph appeared prior to the existence of not only any digital computer but also to the stability analysis of finite-difference methods for PDEs, which was addressed by Courant, Friedrichs, and Lewy in 1928. Since computers now commit $O(10^{15})$ floating-point rounding errors per second on simulations that execute for days, numerical analysis has a central role.

Until von Neumann and Goldstein's landmark 1947 work on the stability of Gaussian elimination, it was not clear how large a system (in this case, of linear equations) could be a candidate for computational solution using floating-point arithmetic, due to the accumulation of rounding errors. We can now make use of digital computers to solve systems of equations billions of times larger than the authors considered. Thanks to continued advances in numerical analysis, and particularly in error analysis and optimal algorithms, the size of simulations in many fields continues to grow to take advantage of all of the resolution that computers can provide. Of course, for systems that are fundamentally "chaotic" in continuous form—meaning that solution trajectories that started infinitesimally far apart can diverge exponentially in finite time—the scientific benefit of greater resolution must be carefully considered in relation to the exponent of divergence. In 1950, von Neumann (with Charney and Fjørtoft) introduced a form of PDE stability analysis that was complementary to that of Courant, Friedrichs, and Lewy. This analysis clarified a fundamental bound on the size of the time step of a simulation relative to the fastest signal speed of the phenomenon being modeled and the granularity of the spatial sampling. The 1950s, 1960s, and 1970s witnessed advances in discretization and equation solving, the development of high-level PROGRAMMING LANGUAGES [VII.11], and improvements in hardware that allowed ambition for computational simulation to soar. By 1979, computational fluid dynamicists were able to claim dramatic reductions in wind tunnel and flight testing of the Boeing 767 thanks to simulations performed to narrow the parameter spaces of design and testing. These simulations, heroic at the time though extremely modest by contemporary standards, gave practical and economic embodiment to the dream of computational science.

Two landmarks defining the ambitions and culture of computational science were the articles by Peter Lax in 1986 championing simulation as the "third leg" of the scientific platform and by Ken Wilson in 1989 on "grand challenges" in computational science. These articles were influential in unlocking funding for computational science campaigns in government, academia, and industry and in democratizing access to high-performance computers, which had previously been the province of relatively few pioneers at national and industrial laboratories.

The formation of the U.S. federal Networking and Information Technology Research and Development organization in 1992 coordinated investments in research, training, and infrastructure from ten federal agencies in high-performance computing and communications and provided recognition for the interdependent ecology of applications, algorithms, software, and architecture. The U.S. Department of Energy's Accelerated Strategic Computing Initiative of 1997 enshrined predictive simulation as a substitute for weapon detonations under a nuclear test ban in the United States and dramatically expanded investment in training computational scientists and engineers. The Department of Energy's Scientific Discovery through Advanced Computing (SciDAC) program, established in 2001, extended the culture of the Accelerated Strategic Computing Initiative program throughout mission space, from astrophysics and geophysics to quantum chemistry and molecular biology. The document that founded the SciDAC program declared the computer, after tuning for accuracy and performance, to be a reliable scientific instrument like any microscope, telescope, beamline, or spectrometer and of more general purpose.

As the Internet expanded to universities, and as the World Wide Web (developed at CERN for the sharing of experimental data sets among theoretical physicists) connected scientists and engineers globally, computational science and engineering forged a global identity. Today, scientific professional societies such as the Society for Industrial and Applied Mathematics, the Institute of Electrical and Electronics Engineers, the Association for Computing Machinery, and the American Physical Society sponsor activities and publications that promote the core enabling technologies of scientific computing. The expectations of large-scale simulation and data analytics to guide scientific discovery, engineering design, and corporate and public policy have never been greater. Nor have the cost-effectiveness and power of computing hardware, which are now driven by commercial market forces far beyond the scientific yearnings that gave birth to computing. Computational science and scientific computing span the gap between ever more demanding applications and ever more complex architectures. These fields provide enormous opportunities for applied and computational mathematicians.

## 3 Synergies and Hurdles

Simulation has aspects in common with both theory and experiment. First, it is fundamentally theoretical, in that it starts with a model, typically a set of equations. A powerful simulation capability breathes new life into theory by creating a demand for improvements in mathematical models. Simulation is also fundamentally experimental, in that upon constructing and implementing a model, one may systematically observe the transformation of inputs (or controls) into outputs (or observables). Simulation effectively bridges theory and experiment by allowing the execution of "theoretical experiments" on systems, including those that could never exist in the physical world, such as a fluid without viscosity or a perfectly two-dimensional form of turbulence. Computation also bridges theory and experiment by virtue of the computer, which serves as a universal and versatile data host. Once data have been digitized, they can be compared side by side with simulated results in visualization systems built for the simulations. They may also be reliably transmitted and retrievably archived. Moreover, simulation and experiment can complement each other by allowing a more complete picture of a system than either can provide on its own. Some data may be unmeasurable even with the best experimental techniques available, and some mathematical models may be too sensitive to unknown parameters to invoke with confidence. Simulation can be used to "fill in" the missing experimental fields, using experimentally measured fields as input data. Data assimilation can also be systematically employed throughout a simulation to keep it from "drifting" from measurements, thus overcoming the effect of modeling uncertainty. Many data-mining techniques in effect build statistical models from data that may be as predictive as executing the dynamics of a traditional model.

Some of the ingredients that are required for success in computational science include insights and models from scientists; theory, methods, and algorithms from mathematicians; and software and hardware infrastructure from computer scientists. There are numerous ways to translate a physical model into mathematical algorithms and to implement a computational program on a given computer. Decisions made without considering upstream and downstream stages may cut off highly productive options. A bidirectional dialogue, up and down the hierarchy at each stage, can ensure that the best resources and technologies are employed. To bridge the algorithmic gap between the application and the architecture, mathematicians must consider implications of both. For example, in representing a field that is anticipated to be smooth, one may select a high-order discretization. In anticipating a port to a hybrid architecture emphasizing SIMDization (SIMD stands for single-instruction-multiple-data), one may produce uniformly structured data aggregates even if this over-resolves the modeled phenomena somewhere relative to the required accuracy. If one requires the adjoint of an operator to perform optimization in the presence of PDE constraints, one may pay an extra price in the forward problem to employ an implicit Newton method, which creates a Jacobian matrix that is then available for adjoint use. However, this Jacobian matrix may be the largest single working set of data in the problem, and its layout may dominate implementation decisions.

While the promise of computational science is profound, so are its limitations. The limitations often come down to a question of resolution. Though of vast size, computers are triply finite: they represent individual quantities only to a finite precision, they keep track of only a finite number of such quantities, and they operate at a finite rate. Although all matter is, in fact, composed of a finite number of atoms, the number of atoms in a macroscopic sample of matter, on the scale of Avogadro's number, places simulations at macroscopic scales from "first principles" (i.e., from the quantum theory of electronic structure) well beyond any conceivable digital computational capability. Similar problems arise when timescales are considered. For example, the range of timescales in protein folding is at least twelve orders of magnitude, since a process that takes milliseconds to complete occurs in molecular dance steps (bond vibrations) that occur in femtoseconds (a trillion times shorter); again, this is far too wide a range to routinely simulate using first principles.

Some simulations of systems adequately described by PDEs of the macroscopic continuum (fluids such as air or water) can be daunting even for the most powerful computers foreseeably available. Today's computers are capable of execution rates in the tens of petaflop/s (1 petaflop/s is $10^{15}$ arithmetic operations per second) and can cost tens to hundreds of millions of dollars to purchase and millions of dollars per year to operate. However, to simulate fluid mechanical turbulence in the boundary and wake regions of a typical vehicle using "first principles" of continuum modeling (Navier–Stokes) would tie up such a computer for months, which makes this level of simulation too expensive and too slow for routine use. To attempt first-principles modeling based on Boltzmann kinetics would be far worse. Practitioners must decide whether upscaled continuum models such as Reynolds-averaged Navier–Stokes or large-eddy simulation will be adequate or whether discrete particle methods such as lattice Boltzmann or discrete simulation Monte Carlo would be more efficient to resolve certain scientific questions on certain architectures.

The "curse of dimensionality" leads to the phenomenon whereby increasing the resolution of a simulation in each relevant dimension quickly eats up any increases in processor power. For explicitly time-discretized problems in three dimensions, the computational complexity of a simulation grows like the fourth power of the resolution in one spatial dimension. Therefore, an increase in computer performance by a factor of 100 provides an increase in resolution in each spatial and temporal dimension by a factor of only slightly more than 3. The "blessing of dimensionality" refers to an entirely different observation: complex dynamics of real systems can often be represented in a relatively small number of principal components. If one represents the dynamics in terms of this efficient basis, rather than the nodal basis that is implicit in tabulating solution values on a mesh, one's predictive ability may effectively advance by the equivalent of several computer generations with a few strokes of a pen.

A standard means of bringing more computational power to bear on a problem is to divide the work to be done over many processors by means of domain decomposition that partitions mesh cells or particles and employs message-passing between the computational nodes. Typically, the same program runs on each node, but it may execute different instruction streams on different nodes due to encountering different data. This programming model is variously known as "single-program/multiple-data," "bulk synchronous processing," or "communicating sequential processes." Parallelism in computer architecture can provide additional factors of hundreds of thousands, or more, to the aggregate performance available for the solution of a given problem, beyond the factor available from Moore's law (a doubling of transistor density about every eighteen months) applied to individual processors alone. However, the desire for increased resolution and, hence, accuracy is seemingly insatiable. The CURSE OF DIMENSIONALITY [I.3 §2] means that "business as usual" in scaling up today's simulations by riding the computer performance curve alone, without attention to algorithms that squeeze more scientific information out of each byte stored or flop processed, will not be cost-effective.

The "curse" of knowledge explosion—namely, that no one computational scientist can hope to track advances in all of the facets of the mathematical theories, computational models and algorithms, applications and computing systems software, and computing hardware (computers, data stores, and networks) that may be needed in a successful simulation effort—is another substantial hurdle to the progress of simulation. Effective collaboration is a means by which it could be overcome.

## 4 Cross-cutting Themes

Computational science embraces diverse scientific disciplines; however, themes that cut across different disciplines, and even themes that appear to be universal, do emerge. We discuss some of these in this section.

### 4.1 Balance of Errors

Computational science proceeds by stages of successive transformation and approximation, including continuous modeling, numerical discretization, digital solution, and analysis and interpretation of the results, at each of which errors are made. The overall error can be estimated by recursive application of the triangle inequality. The error estimates for individual stages should be made commensurate, since it adds no value to produce an exact algebraic solution to a discrete system that is only an approximate representation, for instance, or to overly refine a discretization within which there are coefficients that are only imperfectly known, or where the underlying physical "laws" are merely approximate correlations. One of the primary and most difficult tasks in computational science is to estimate and balance the errors at each stage, to avoid overinvesting in work that does not ultimately contribute to the accuracy of the desired output. "Validation" and "verification" are two generalized stages of a comprehensive error analysis of a computational science campaign, the former measuring the degree to which the model represents the salient features under investigation and the latter measuring the quality of the solution to the model. The ultimate goal is to get the greatest scientific insight at the lowest "cost." Error estimation in an individual execution is an essential aspect of insight, but it must be accompanied by judicious selection of cases to run and by performance tuning to run them efficiently.

### 4.2 Uncertainty Quantification

To be accepted as a means of scientific discovery, engineering design, and decision support, computational science must embrace standards of predictivity and reproducibility. Uncertainty can enter computations in a number of ways, but fundamentally, there are two types: epistemic and aleatoric. Epistemic uncertainties are those that could, in principle, be reduced to zero, leading to accurate deterministic simulations, but for which the modeler simply lacks sufficient knowledge. Aleatoric uncertainties are those that are regarded as unknowable, except in a statistical sense, because of fundamental randomness. Uncertainty in the output of a simulation can enter via the structure of the model itself, via parameters, or via approximations committed by the algorithm. Classical approaches to uncertainty quantification are Monte Carlo in nature: one executes the model for a cloud of inputs distributed according to various assumptions and then performs statistics on the output. This is direct, but it can be wasteful of simulation resources. A more progressive approach is to derive models for the statistical properties of the distribution and propagate them along with the uncertain solution. Uncertainty quantification has justifiably become an interdisciplinary research area dependent upon mathematical, statistical, and domain knowledge. It is also a driver within computational science that justifies investment in scaling the hardware–software environment, since it greatly increases complexity relative to that of the execution of an individual forward model.

### 4.3 The Ideal Basis

The quest for the ideal basis in which to represent a problem is a recurring theme in scientific computing, since solving the same problem in an ideal basis can bring a critical improvement in computational complexity, that is, in the number of operations required to reach a solution of given accuracy. Rank, sparsity, and boundary conditions are among the features that allow intelligent choice of basis. Problems arrive at the doors of computational scientists with a choice of basis implicit; they are described by objects and operators that are convenient and natural to the scientific community. However, the best formulation for computation may be quite different. Some algorithms derive their own discrete bases that adapt to the operator, such as harmonic interpolants. Reduced-order models are important at the application level, where physical insight may guide a low-dimensional representation of an apparently high-dimensional system. They are also critical at the algorithmic kernel level, where a more automatically adaptive process may incrementally produce a reduced basis, such as a KRYLOV SUBSPACE [II.23] in linear algebra. Possessing multiple models for the same phenomena can increase algorithmic creativity. For instance, in linear problems, an inexpensive model may be used to construct a preconditioner for an expensive one. In nonlinear problems, the solution to a crude model may provide a more robust starting estimate than is otherwise available.

### 4.4 A Canonical "Algorithmic Basis"

Though the applications of computational science are very diverse, there are a number of algorithmic paradigms that recur throughout them. Seven of these floating-point algorithms were identified by Phillip Colella in 2004, and he called them the "seven dwarfs." This set was expanded by six integer algorithms in 2006 by the Berkeley scientific computing group, bringing us up to "13 dwarves": dense direct solvers, sparse direct solvers, fast transforms, $N$-body methods, structured (grid) iterations, unstructured (grid) iterations, Monte Carlo, combinatorial logic, graph traversal, graphical models, finite-state machines, dynamic programming, and backtrack and branch-and-bound. These kernels have been characterized by their amenability to scaling on what are known as "hybrid" or "hierarchical" architectures. In such architectures, cores are proliferated within a node of fixed memory and memory bandwidth resources in a shared-memory manner. The resulting nodes are then proliferated in a distributed-memory manner, being connected by a network that scales with the number of nodes. The fact that so many problems in computational science can be reduced to a relatively small set of similar tasks has profound implications for the scientific software environment and for research and development investment. It is possible for the work of a relatively small number of experts in computational mathematics and computer science to serve a relatively large number of scientists and engineers who are expert in something else, namely their fields of application. Vendors can invest in software libraries with well-defined interfaces (such as ones based on fast Fourier transforms or the basic linear algebra subroutines) that can access their hardware in custom ways. Researchers in numerical analysis can be confident, for instance, that work to improve a dense symmetric eigensolver has the potential to be adopted by thousands of chemists solving the Schrödinger equation or that work to improve a sparse symmetric eigensolver could be used by thousands of mechanical engineers analyzing vibrational modes. The list of "dwarfs/dwarves" is binned into fairly broad categories above and many subcategorizations are required in practice to arrive at a problem specification that leads directly to the selection of software. Many of the dwarfs do not yet have optimal implementations or even weakly scalable implementations. These dwarfs make good targets for ongoing mathematical research.

### 4.5 Polyalgorithms

While it may be possible to describe many computational science applications with reference to a relatively small number of kernels, each such specification can typically be approached by a large number of algorithms. A polyalgorithm is a set of algorithms that can accomplish the same task(s), together with a set of rules for selecting among the members of the set, based on the characteristics of the input, the requirements of the output, and the properties of the execution environment. Sometimes, one algorithm offers superior performance in terms of the number of operations, the pattern of memory references, running time, or other metrics over a wide range of inputs, output requirements, and environment, which renders the concept trivial. More typically, each algorithm has a niche in which it is superior to all others, and there is no transitivity with respect to performance metrics across all possible uses. Problem size (discrete dimension) is a fundamental input characteristic. A method that requires, asymptotically, $O(N^2)$ operations may beat a method that requires only $O(N)$ or $O(N \log N)$ operations if $N$ is small and the constant hidden by the order symbol is small compared with the constant for an asymptotically superior method. Between two such algorithms there is therefore a crossover point in $N$ that dictates selection between them. Other input characteristics that are often important include sparsity, symmetry, definiteness, heterogeneity, and isotropy. Output requirements may vary dramatically from one application to another. For instance, in an eigenanalysis scenario, are all eigenvalues required, or just the largest, the smallest, or those closest to an interior value? Are eigenvectors required? Environmental characteristics may include the number of processing elements, the availability of built-in floating-point precisions, the relative cost of communication versus computation, and the size of various memories, caches, or buffers. It is therefore beneficial to have access to a variety of well-characterized approaches to accomplish each kernel task.

### 4.6 Space-Time Trade-Offs

Often, scientific computing algorithms are parametrizable in ways that allow one to make space-time complexity trade-offs. If memory is limited, one may store less and compute more. If extra memory is available, one may store more and compute less. This principle has many manifestations in practical implementations and algorithmic tunings. Classical examples are the size of windows of past vectors to retain in a Krylov-style iterative method or a quasi-Newton iterative method. Contemporary examples emerge from the different latencies that different levels of computer memory possess from fast registers (which are essentially one cycle away), to caches of various levels, to main memory (thousands of cycles away), to disk files (millions of cycles away), to data distributed over the Internet. It may be cheaper to recompute some data than to store and retrieve them. On the other hand, as in modified Newton iterative methods, it may be cheaper and/or faster to reuse old data (in this case a Jacobian) than to constantly refresh them. In performance-oriented computing it may even be advantageous, on average, to assume values (on the basis of experience or rational assumption) for data that are not available when they appear on the critical path as an input for one process and to "roll back" the computation when they arrive if they are too far from the assumed values. Under the heading of space-time trade-offs, one may also employ multiple precisions of floating-point arithmetic, as in classical ITERATIVE REFINEMENT [IV.10 §2] for linear systems. In this algorithm, one stores simultaneously and manipulates both high- and low-precision forms of some objects, doing the bulk of the arithmetic on the low-precision objects in order to refine the high-precision ones.

### 4.7 Continuum–Discrete Duality

In computational science it is frequently convenient to have multiple views of the same object. Nature is fundamentally discrete but at a scale typically too fine for digital computers. We therefore work through idealized continuous models, such as Navier–Stokes PDEs to calculate momentum transfer in fluids with small mean free paths between molecular collisions. The continuous fields are then integrated over control volumes, parametrized by finite elements, or represented by pointwise values in finite differences, returning the model to finite cardinality, typically a much smaller cardinality than that of the original molecular model. However, adaptive error analyses for refining the spatial resolution or approximation order of the discretization reemploy the underlying continuous model. The optimal complexity solver known as the multigrid method makes simultaneous use of *several* different discrete approximations to the underlying continuum through recursive coarsening. Ordering multidimensional and possibly irregularly spaced coordinate data into the linear address space of computer memory may take advantage of a discrete analogue to Hilbert's continuous space-filling curves. Thus, computational mathematics moves fluidly back and forth between the continuous and the discrete. However, continuous properties such as conservation or zero divergence are not necessarily preserved in the dual discrete representation. The computational scientist needs to be aware of operations, such as variational differentiation or application of boundary conditions, that may not commute with the switch of representations. Thus, for example, "discretize then optimize" may yield a different result from "optimize then discretize" (and famously often does).

### 4.8 Numerical Conditioning

Numerical conditioning is a primary concern of the computational scientist; it affects the accuracy that can be guaranteed by theory or expected in practice and the convergence rate of many algorithms. It is therefore a key factor in the scientist's choice of how precise a floating-point arithmetic to employ. In computational science applications, the condition number of a discrete operator is often related to the quality of the discretization in resolving multiscale features in the solution. The continuous operator may have an infinite condition number, so the greater the resolution, the worse the numerical condition number. The progressively worsening conditioning of elliptic operators, in particular, as they better resolve fine-wavelength components in the output, is a bane of iterative methods whose iteration count grows with ill-conditioning. The computational work required to solve the problem to a given accuracy grows superlinearly overall, linearly in the cost per iteration times a factor that depends on the condition number. Optimal algorithms must generally possess a hierarchical structure that handles each component of the error on its own natural scale.

### 4.9 The Nested-Loop Co-design Process

The classic computational science development paradigm is reflected in figure 1, which is adapted from a figure in a report to the U.S. Department of Energy in 2000. On the left is what could be called a "validation and verification" loop. On the right is a performance tuning loop. The left loop is the province of the computational mathematician, who may make convenient assumptions, such as that memory is flat, in deriving algorithms that deliver bounded error while minimizing floating-point operations or memory capacity. The right loop is the province of computer scientists (and ultimately architects), who inherit a mathematical specification of an algorithm and optimize its implementation for power and runtime. At the time this scheme was introduced in aid of launching the unprecedentedly interdisciplinary SciDAC program, these loops were primarily envisioned to be sequential. First the validation and verification loop would converge and then the performance tuning loop would, resulting in a computational tool for scientific discovery. "Codesign" is a classic concept from embedded systems that is increasingly being invoked in the general-purpose scientific computing context; it puts an outer loop around these algorithmic and performance loops based on the recognition that isolated design keeps significant performance and efficiency gains off the table. An unstructured grid PDE may not be a good discretization for exploitation of hybrid hardware with significant SIMDization, for instance. Such feedback could lead to consideration of lattice Boltzmann methods in some circumstances, for example.

### 4.10 From Computing to Understanding

As hardware improves exponentially in performance, computational modeling has the luxury of becoming a science by means of which a simulated system is "poked" intelligently and repeatedly to reveal generalizable insight into behavior. Historically, computational science research was driven by the gap between the capabilities of the hardware and the complexity of the systems intended for modeling. Effort was concentrated on isolating a component of the overall system and improving computational capability thanks to improvements in algorithms, software, and hardware. This inevitably involves invoking assumptions for ignored coupling; for instance, an ocean model may be executed with simple assumptions for atmospheric forcing. Today, highly capable components are being reunified into complex systems that make fewer decoupling compromises. For instance, ocean and atmospheric models drive each other across the interface between them by passing fluxes of mass, momentum, and energy in various forms. With increased computational power and memory, these complex systems can be partially quantified with respect to uncertainty and then subjected to true experiments by being embedded into ensemble runs. "What if" questions can be posed and executed by controlling inputs. This progression is powered at every stage by new mathematics, and sometimes, in turn, it generates new mathematical questions. In building up capability, the focus is on reducing computational complexity. In coupling components, the focus is on balancing error and stability analysis, with a continued premium on complexity reduction. Finally, the focus is on uncertainty quantification and the formulation of hypotheses from observation—

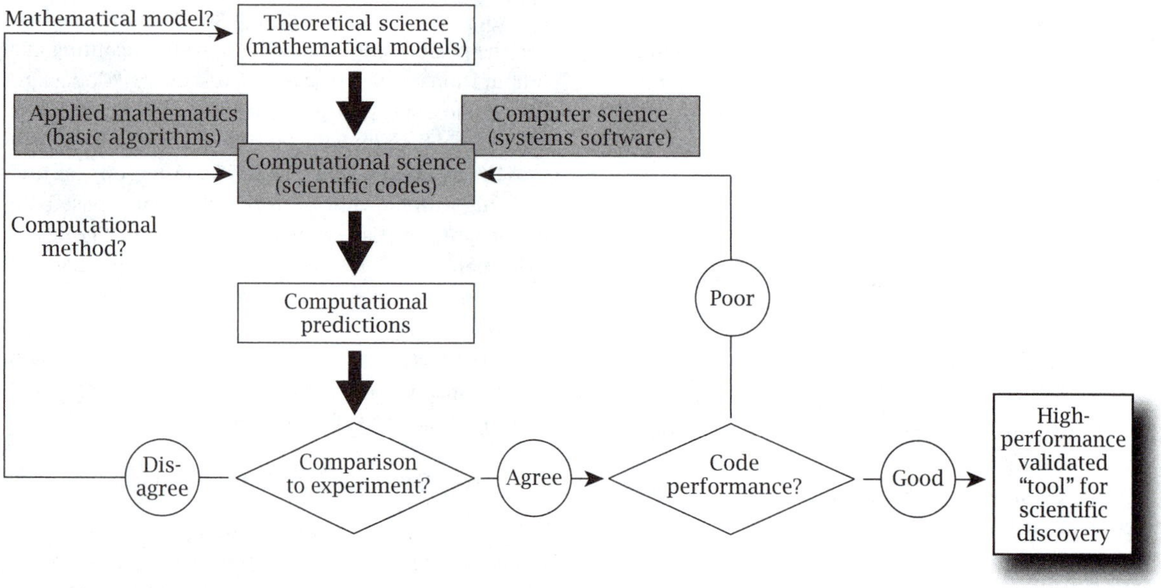


**Figure 1** Diagram that launched the SciDAC program in 2000, courtesy of Thom Dunning Jr.

hypotheses that may eventually be provable theoretically or be subject to controlled experimental testing in the real world.

### 4.11 The "Multis"

The discussion of the cross-cutting themes of computational science can be conveniently and mnemonically encapsulated in the list of the "multis." Scientific and engineering applications are typically multiphysics, multiscale, and multidimensional in nature. Algorithms to tackle them are typically hierarchical, using interacting multimodels, on multilevels of refinement, in multiprecisions. These then have to be implemented on multinode distributed systems, with multicore nodes, and multiprotocol programming styles, all of which requires a multidisciplinary approach in which the mathematician plays a critical bridging role. We elaborate on multiphysics in the next section.

### 4.12 Encapsulation in Software

There have been many efforts to respond to the targets presented by one or more of the "dwarfs" and the complexity of the applications and algorithms of computational science with well-engineered software packages. The Portable Extensible Toolkit for Scientific Computing (PETSc), a freely downloadable suite of data structures and routines from Argonne National Laboratory that has been in continuous development since 1992, is an example that incorporates generic parallel programming primitives, matrix and vector interfaces, and solvers for linear, nonlinear, and transient problems. PETSc emphasizes ease of experimentation. Its strong encapsulation design allows users to hierarchically compose methods at runtime for coupled problems. Several relatively recent components have substantially improved composability for multiphysics and multilevel methods. For example, `DMComposite` manages distributed-memory objects for coupled problems by handling the algebraic aspects of gluing together function spaces, decomposing them for residual evaluation, and setting up linear operators to act on the coupled function spaces. A matrix assembly interface is available that makes it possible for individual physics modules to assemble parts of global matrices without needing global knowledge or committing in advance to matrix format. The `FieldSplit` preconditioner solves linear block systems using either block relaxation or approximate block factorization (as in "physics-based" preconditioners for stiff waves or block preconditioners for incompressible flow). `FieldSplit` can be nested inside other preconditioners, including geometric or algebraic multigrid ones, with construction of the hierarchy and other algorithmic choices exposed as runtime options. Recent enhancements to the `TS` component support implicit–explicit time integration schemes. Multiphysics applications that have employed these capabilities include lithosphere dynamics, subduction and mantle convection, ice sheet dynamics, subsurface reactive flow, tokamak fusion, mesoscale materials modeling, and electrical power networks.

## 5 Multiphysics Modeling

A multiphysics system consists of multiple coupled components, each component governed by its own principle(s) for evolution or equilibrium, typically conservation or constitutive laws. Multiphysics simulation is an area of computational science possessing mathematical richness. Coupling individual simulations may introduce stability, accuracy, or robustness limitations that are more severe than the limitations imposed by the individual components. A major classification in such systems is whether the coupling occurs in the bulk (e.g., through source terms or constitutive relations that are active in the overlapping domains of the individual components) or whether it occurs over an idealized interface that is lower dimensional or over a narrow buffer zone (e.g., through boundary conditions that transmit fluxes, pressures, or displacements). Typical examples of bulk-coupled multiphysics systems that have their own extensively developed literature include radiation with hydrodynamics in astrophysics (radiation–hydrodynamics, or "rad-hydro"), electricity and magnetism with hydrodynamics in plasma physics (magnetohydrodynamics), and chemical reaction with transport in combustion or subsurface flows (reactive transport). Typical examples of interface-coupled multiphysics systems are ocean–atmosphere dynamics in geophysics, fluid-structure dynamics in aeroelasticity, and core–edge coupling in tokamaks. Beyond these classic multiphysics systems are many others that share important structural features.

The two simplest systems that exhibit the crux of a multiphysics problem are the coupled equilibrium problem

$$F_1(u_1, u_2) = 0, \qquad (1\,a)$$

$$F_2(u_1, u_2) = 0 \qquad (1\,b)$$

and the coupled evolution problem

$$\partial_t u_1 = f_1(u_1, u_2), \qquad (2\,a)$$

$$\partial_t u_2 = f_2(u_1, u_2). \qquad (2\,b)$$

When (2 *a*)–(2 *b*) is semidiscretized in time, the evolution problem leads to a set of problems that take the form (1 *a*)–(1 *b*) and that are solved sequentially to obtain values of the solution $u(t_n)$ at a set of discrete times. Here $u$ refers generically to a multiphysics solution, which has multiple components indicated by subscripts $u = (u_1, \dots, u_{N_c})$; the simplest case of $N_c = 2$ components is indicated here.

Initially, we assume for convenience that the Jacobian $J = \partial(F_1, F_2)/\partial(u_1, u_2)$ is diagonally dominant in some sense and that $\partial F_1/\partial u_1$ and $\partial F_2/\partial u_2$ are nonsingular. These assumptions are natural in the case where the system arises from the coupling of two individually well-posed systems that have historically been solved separately. In the equilibrium problem, we refer to $F_1$ and $F_2$ as the component *residuals*; in the evolution problem, we refer to $f_1$ and $f_2$ as the component *tendencies*.

The choice of solution approach for these coupled systems relies on a number of considerations. From a practical standpoint, existing codes for component solutions often motivate operator splitting as an expeditious route to a first multiphysics simulation, making use of the separate components. This approach, however, may ignore strong couplings between components. Solution approaches ensuring a tight coupling between components require smoothness, or continuity, of the nonlinear, problem-defining functions, $F_i$, and their derivatives.

Classic multiphysics algorithms preserve the integrity of the two uniphysics problems, namely, solving the first equation for the first unknown, given the second unknown, and solving the second equation for the second unknown, given the first. Multiphysics coupling is taken into account by iteration over the pair of problems, typically in a Gauss–Seidel manner (see algorithm 1), linearly or nonlinearly, according to context. Here we employ superscripts to denote iterates.

**Algorithm 1 (Gauss–Seidel multiphysics coupling).**

Given initial iterate $\{u_1^0, u_2^0\}$
**for** $k = 1, 2, \dots$ until convergence **do**
  Solve for $v$ in $F_1(v, u_2^{k-1}) = 0$; set $u_1^k = v$
  Solve for $w$ in $F_2(u_1^k, w) = 0$; set $u_2^k = w$
**end for**

Likewise, the simplest approach to the evolutionary problem employs a field-by-field approach in a way that leaves a first-order-in-time splitting error in the solution. Algorithm 2 gives a high-level description of this process that produces solution values at time nodes $t_0 < t_1 < \cdots < t_N$. Here, we use the notation $u(t_0), \dots, u(t_N)$ to denote discrete time steps. An alternative that staggers solution values in time is also possible.

**Algorithm 2 (multiphysics operator splitting).**

Given initial values $\{u_1(t_0), u_2(t_0)\}$
**for** $n = 1, 2, \ldots, N$ **do**
  Evolve one time step in $\partial_t u_1 + f_1(u_1, u_2(t_{n-1})) = 0$ to obtain $u_1(t_n)$
  Evolve one time step in $\partial_t u_2 + f_2(u_1(t_n), u_2) = 0$ to obtain $u_2(t_n)$
**end for**

If the residuals or tendencies and their derivatives are sufficiently smooth and if one is willing to write a small amount of solver code that goes beyond the legacy component codes, a good algorithm for both the equilibrium problem and the implicitly time-discretized evolution problem is the Jacobian-free Newton–Krylov method (see below). Here, the problem is formulated in terms of a single residual that includes all components in the problems,

$$F(u) \equiv \begin{bmatrix} F_1(u_1, u_2) \\ F_2(u_1, u_2) \end{bmatrix} = 0, \tag{3}$$

where $u = (u_1, u_2)$. The basic form of NEWTON'S METHOD [II.28] to solve (3), for either equilibrium or transient problems, is given by algorithm 3. Because of the inclusion of the off-diagonal blocks in the Jacobian, for example,

$$J = \begin{bmatrix} \dfrac{\partial F_1}{\partial u_1} & \dfrac{\partial F_1}{\partial u_2} \\[2ex] \dfrac{\partial F_2}{\partial u_1} & \dfrac{\partial F_2}{\partial u_2} \end{bmatrix},$$

Newton's method is regarded as being "tightly coupled."

**Algorithm 3 (Newton's method).**

Given initial iterate $u^0$
**for** $k = 1, 2, \ldots$ until convergence **do**
  Solve $J(u^{k-1})\delta u = -F(u^{k-1})$
  Update $u^k = u^{k-1} + \delta u$
**end for**

The operator and algebraic framework described here is relevant to many divide and conquer strategies in that it does not "care" (except in the critical matter of devising preconditioners and nonlinear component solvers for good convergence) whether the coupled subproblems are from different equations defined over a common domain, the same equations over different subdomains, or different equations over different subdomains. The general approach involves iterative corrections within subspaces of the global problem. All the methods have in common an amenability to exploiting a "black-box" solver philosophy that amortizes existing software for individual physics components. The differences are primarily in the nesting and ordering of loops and the introduction of certain low-cost auxiliary operations that transcend the subspaces.

Not all multiphysics problems can be easily or reliably cast into these equilibrium or evolution frameworks, which are primarily useful for deterministic problems with smooth operators for linearization. In formulating multiphysics problems, modelers first apply asymptotics to triangularize or even diagonalize the underlying Jacobian as much as possible, pruning provably insignificant dependences but bearing in mind the conservative rule: "coupled until proven uncoupled." One then applies multiscale analyses to simplify further, eliminating stiffness from mechanisms that are dynamically irrelevant to the goals of the simulation.

Perhaps the simplest approach for solving systems of nonlinear equations (3) is the fixed-point iteration, also known as the Picard or nonlinear Richardson iteration. The root-finding problem (3) is reformulated into a fixed-point problem $u = G(u)$ by defining a fixed-point iteration function, for example,

$$G(u) := u - \alpha F(u),$$

where $\alpha > 0$ is a fixed-point damping parameter that is typically chosen to be less than 1. Fixed-point methods then proceed through the iteration

$$u^{k+1} = G(u^k),$$

with the goal that $\|u^{k+1} - u^k\| < \varepsilon$. If the iteration function is a contraction, that is, if there exists some $\gamma \in (0, 1)$ such that

$$\|G(u) - G(v)\| \leqslant \gamma \|u - v\|$$

for all vectors $u$ and $v$ in a closed set containing the fixed-point solution $u^\star$, then the fixed-point iteration is guaranteed to converge. This convergence is typically linear, however, and can be slow even from a good initial guess.

Newton's method (algorithm 3) offers faster convergence, up to quadratic. However, direct computation of $\delta u$ in algorithm 3 may be expensive for large-scale problems. *Inexact* Newton methods generalize algorithm 3 by allowing computation of $\delta u$ with an iterative method, requiring only that $\|J(u^{k-1})\delta u + F(u^{k-1})\| < \varepsilon_k$ for some set of tolerances, $\varepsilon_k$. Newton–Krylov methods are variants of inexact Newton methods in which $\delta u$ is computed with a Krylov subspace method. This choice is advantageous because the only information required by the Krylov subspace method is a method for computing Jacobian-vector products $J(u^k)v$. A consequence of this reliance on only matrix-vector products is that these directional derivatives may be approximated and the Jacobian matrix $J(u^k)$ is never itself needed. The Jacobian-free Newton–Krylov method exploits this approach, using a finite-difference approximation to these products:

$$J(u^k)v \approx \frac{F(u^k + \sigma v) - F(u^k)}{\sigma},$$

where $\sigma$ is a carefully chosen differencing parameter and $F$ is sufficiently smooth. This specialization facilitates use of an inexact Newton method by eliminating the need to identify and implement the Jacobian. Of course, the efficiency of the Jacobian-free Newton–Krylov method depends on preconditioning the inner Krylov subspace method, and the changes in the Jacobian as nonlinear iterations progress place a premium on preconditioners with low setup cost. Implementations of inexact Newton methods are available in several high-performance software libraries, including PETSc, which was mentioned above.

Fixed-point and inexact Newton methods can be used to solve multiphysics problems in a fully coupled manner, but they can also be used to implement coupling strategies such as algorithms 1 and 2. With an implicit method for one of the components, one can directly eliminate it in a linear or nonlinear Schur complement formulation. The direct elimination process for $u_1$ in the first equation of the equilibrium system (1 $a$), given $u_2$, can be symbolically denoted

$$u_1 = G(u_2),$$

with which the second equation (1 $b$) is well defined in the form

$$F_2(G(u_2), u_2) = 0.$$

Each iteration thereof requires subiterations to solve (in principle to a high tolerance) the first equation. Unless the first system is much smaller or easier than the second, this is not likely to be an efficient algorithm, but it may have robustness advantages.

If the problem is linear,

$$\left.\begin{aligned} F_1(u_1, u_2) &= f_1 - A_{11}u_1 - A_{12}u_2,\\ F_2(u_1, u_2) &= f_2 - A_{21}u_1 - A_{22}u_2,\end{aligned}\right\} \qquad (4)$$

then $F_2(G(u_2), u_2) = 0$ involves the traditional Schur complement

$$S = A_{22} - A_{21}A_{11}^{-1}A_{12}.$$

If the problem is nonlinear and if Newton's method is used in the outer iteration, the Jacobian

$$\frac{dF_2}{du_2} = \frac{\partial F_2}{\partial u_1}\frac{\partial G}{\partial u_2} + \frac{\partial F_2}{\partial u_2}$$

is, to within a sign, the same Schur complement.

Similar procedures can be defined for the evolution problem, which, when each phase is implicit, becomes a modified root-finding problem on each time step with the Jacobian augmented with an identity or mass matrix.

If the types of nonlinearities in the two components are different, a better method may be the nonlinear Schwarz method or the additive Schwarz preconditioned inexact Newton (ASPIN) method. In nonlinear Schwarz, one solves component subproblems (by Newton or any other means) for componentwise corrections,

$$\begin{aligned} F_1(u_1^{k-1} + \delta u_1, u_2^{k-1}) &= 0,\\ F_2(u_1^{k-1}, u_2^{k-1} + \delta u_2) &= 0,\end{aligned}$$

and uses these legacy componentwise procedures implicitly to define modified residual functions of the two fields:

$$\begin{aligned} G_1(u_1, u_2) &\equiv \delta u_1,\\ G_2(u_1, u_2) &\equiv \delta u_2.\end{aligned}$$

One then solves the modified root-finding problem

$$\begin{aligned} G_1(u_1, u_2) &= 0,\\ G_2(u_1, u_2) &= 0.\end{aligned}$$

If one uses algorithm 3 to solve the modified problem, the Jacobian is

$$\begin{bmatrix} \dfrac{\partial G_1}{\partial u_1} & \dfrac{\partial G_1}{\partial u_2}\\[2ex] \dfrac{\partial G_2}{\partial u_1} & \dfrac{\partial G_2}{\partial u_2}\end{bmatrix} \approx \begin{bmatrix} I & \left(\dfrac{\partial F_1}{\partial u_1}\right)^{-1}\dfrac{\partial F_1}{\partial u_2}\\[2ex] \left(\dfrac{\partial F_2}{\partial u_2}\right)^{-1}\dfrac{\partial F_2}{\partial u_1} & I\end{bmatrix},$$

which clearly shows the impact of the cross-coupling in the off-diagonals. In practice, the outer Newton method must converge in a few steps for ASPIN to be worthwhile, since the inner iterations can be expensive. In nonlinear Schwarz, the partitioning of the global unknowns into $u_1$ and $u_2$ need not be along purely physical lines, as it would be if they came from a pair of legacy codes. The global variables can be partitioned into overlapping subsets, and the decomposition can be applied recursively to obtain a large number of small

Newton problems. The key to the convenience of nonlinear Schwarz is that the outer iteration is Jacobian-free and the inner iterations may involve legacy solvers. (This is also true for the Jacobian-free Newton–Krylov method, where the inner preconditioner may involve legacy solvers.)

Error bounds on linear and nonlinear block Gauss–Seidel solutions of coupled multiphysics problems show how the off-diagonal blocks of the Jacobian enter the analysis. Consider the linear case (4) for which the Jacobian of the physics-blocked preconditioned system is

$$\begin{bmatrix} I & A_{11}^{-1}A_{12} \\ A_{22}^{-1}A_{21} & I \end{bmatrix}.$$

Let the product of the coupling blocks be defined by the square matrix $C \equiv A_{11}^{-1}A_{12}A_{22}^{-1}A_{21}$, the norm of which may be bounded as $\|C\| \leqslant \|A_{11}^{-1}\|\,\|A_{12}\|\,\|A_{22}^{-1}\|\,\|A_{21}\|$. Provided $\|C\| \leqslant 1$, any linear functional of the solution $(u_1, u_2)$ of (4) solved by the physics-blocked Gauss–Seidel method (algorithm 1) satisfies conveniently computable bounds in terms of residuals of the individual physics blocks of (4). The cost of evaluating the bound involves the action of the inverse uniphysics operators $A_{11}^{-1}$ and $A_{22}^{-1}$ on vectors coming from the uniphysics residuals and the dual vectors defining the linear functionals of interest. The bounds provide confidence that the block Gauss–Seidel method has been iterated enough to produce sufficiently accurate outputs of interest, be they point values, averages, fluxes, or similar. The required actions of the inverse uniphysics operators are a by-product of an implicit method for each phase. The nonlinear case is similar, except that the Jacobian matrices from each physics phase that comprise $C$ may change on each block Gauss–Seidel iteration.

Preconditioning is essential for efficiency whenever a Krylov subspace method is used. While there has been considerable success in developing black-box algebraic strategies for preconditioning linear systems, preconditioners for multiphysics problems generally need to be designed by hand.

A widely used approach is to use a block-diagonal approximation of the Jacobian of the system. Improved performance can generally be achieved by capturing the strong couplings in the preconditioner and leaving the weak couplings to be resolved by the outer Krylov/Newton iterations. Such approaches generally lead to identification of a Schur complement that embodies the important coupling, and their success relies on judicious approximation of the Schur complement.

## 6 Anatomy of a Large-Scale Simulation

We illustrate many issues in computational science by discussing the multiphysics application of turbulent reacting flows. Fittingly, the first ACM–SIAM Prize in Computational Science and Engineering was awarded in 2003 to two scientists at Lawrence Berkeley National Laboratory who addressed this difficult problem by integrating analytical and numerical techniques. Apart from the scientific richness of the combination of fluid turbulence and chemical reaction, turbulent flames are at the heart of the design of equipment such as reciprocating engines, turbines, furnaces, and incinerators for efficiency and minimal environmental impact. Effective properties of turbulent flame dynamics are also required as model inputs in a broad range of larger-scale simulation challenges, including fire spread in buildings or wildfires, stellar dynamics, and chemical processing.

The first step in such simulations is to specify the models for the reacting flows. The essential feature of reacting flows is the set of chemical reactions at a priori unknown locations in the fluid. As well as chemical products, these reactions produce both temperature and pressure changes, which couple to the dynamics of the flow. Thus, an accurate description of the reactions is critical to predicting the shape and properties of the flame. Simultaneously, it is the fluid flow that transports the reacting chemical species to the reaction zone and transports the products of the reaction and the released energy away from the reaction zone. The location and shape of the reaction zone are determined by a delicate balance of species, energy, and momentum fluxes and are highly sensitive to how these fluxes are specified at the boundaries of the computational domain. Turbulence can wrinkle the reaction zone, giving it a much greater area than it would have in its laminar state, without turbulence. Hence, incorrect prediction of turbulence intensity may under- or over-represent the extent of reaction.

From first principles, the reactions of molecules are described by the SCHRÖDINGER EQUATION [III.26] and fluid flow is described by the NAVIER–STOKES EQUATIONS [III.23]. However, each of these equation sets is too difficult to solve directly, so we must rely on approximations. These approximations define the computational models used to describe the flame. For the combustion of methane, a selected set of eighty-four reactions involves twenty-one chemical species (methane, oxygen, water, carbon dioxide, and many trace products and reaction intermediates). A particularly useful approximation of the compressible Navier–Stokes model allows detailed consideration of the convection, diffusion, and expansion effects that shape the reaction, but it is filtered to remove sound waves, which pose complications that do not measurably affect combustion in the laboratory regime of the flame. The selection of models such as these requires context-specific expert judgment. For instance, in a jet turbine simulation, sound waves might represent a nonnegligible portion of the energy budget of the problem and would need to be modeled. However, in a jet, it might be valid (and very cost-effective) to use a less intricate chemical reaction mechanism to answer the questions concerning flame stability, thrust, noise, and the like. If atmospheric pollution were the main subject of the investigation, on the other hand, an even more detailed chemical model involving more photosensitive trace species might be required. (Development of the chemistry model and validation of the predictions of the simulation make an interesting scientific story, but one that is too long to be told here.)

The computations that led to the first ACM–SIAM Prize in Computational Science and Engineering were performed on an IBM SP supercomputer named "Seaborg" at the National Energy Research Scientific Computing Center, using as many as 2048 processors. At the time, they were among the most demanding combustion simulations ever performed, though they have since been eclipsed in terms of both complexity and computational resources by orders of magnitude. The ability to perform them was not, however, merely (or even mainly) a result of improvements in computer technology. Improvements in algorithm technology were even more instrumental in making the computations feasible. As mentioned above, mathematical analysis was used to reformulate the equations describing the fluid flow so that high-speed acoustic transients were removed analytically while compressibility effects due to chemical reactions were retained. The resulting model of the fluid flow was discretized using high-resolution finite-difference methods, combined with local adaptive mesh refinement by which regions of the finite-difference grid were automatically refined or coarsened in a structured manner to maximize overall computational efficiency. The implementation used an object-oriented, message-passing software

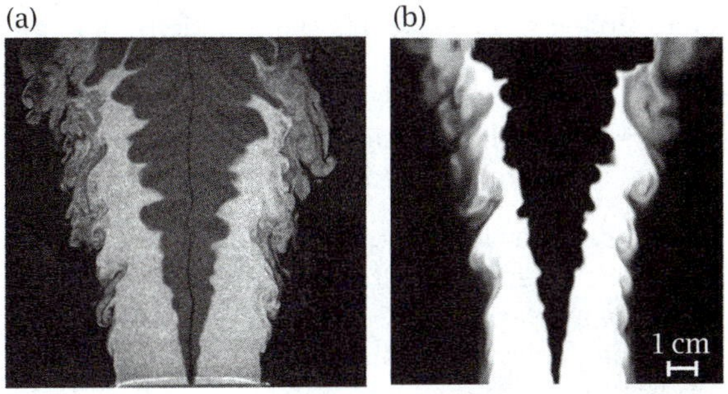


**Figure 2** A comparison of (a) the experimental particle image velocimetry crossview of the turbulent V flame and (b) a simulation. Reproduced from "Numerical simulation of a laboratory-scale turbulent V-flame" by J. B. Bell et al. (*Proceedings of the National Academy of Sciences of the USA* 102:10,009 (2005)).

framework that handled the complex data distribution and dynamic load balancing needed to effectively exploit thousandfold parallelism. The data analysis framework used to explore the results of the simulation and create visual images that lead to understanding is based on recent developments in "scripting languages" from computer science. The combination of these algorithmic innovations reduced the computational cost by a factor of 10 000 for the same effective resolution compared with a standard uniform-grid approach. (See figure 2 for a side-by-side comparison of a slotted V-flame experiment and a realization of a three-dimensional time-dependent simulation prepared for the same conditions.)

As the reaction zone shifts, the refinement automatically tracks it, adding and removing resolution dynamically. Unfortunately, the adaptivity and mathematical filtering employed to save memory and reduce the number of operations complicate the software and throw the execution of the code into a regime of computation that processes fewer operations per second and uses thousands of processors less uniformly than a "business as usual" algorithm would. As a result, the scientifically effective simulation runs at a small percentage of the theoretical peak rate of the hardware. In this case, a simplistic efficiency metric like "percentage of theoretical peak" is misleading.

Software of this complexity and versatility could never be assembled in the traditional mode of development whereby individual researchers with separate concerns asynchronously toss software written to a priori specifications "over the transom." Behind any simulation as complex as this stands a tightly coordinated, vertically integrated multidisciplinary team.

## 7 The Grand Challenge of Computational Science

Many "grand challenges" of computational science, in simulating complex multiscale multiphysics systems, await the increasing power of high-performance computing and insight and innovation in mathematics. It can be argued that any system that can deterministically and reproducibly simulate another system must be at least as "complex" as the system being simulated. Therefore, the high-performance computational environment itself is the most complex system to model. The grandest challenge of computational science is making this complex simulation system sufficiently manageable to use that scientists who are expert in something else (e.g., combustion, reservoir modeling, magnetically confined fusion) can employ it effectively. In over six decades of computational science, humans have so far bridged the gap between the complexity of the system being modeled and machine complexity, with von Neumann as an exemplar of an individual who made contributions at all levels, from the modeling of the physics to the construction of computer hardware. After all of the components of the chain are delivered, the remaining challenge is the vertical integration of models and technology, leading to a hardware-software instrument that can be manipulated to perform the theoretical experiments of computational science.

## Glossary

**Computational science (and engineering):** the vertically integrated multidisciplinary process of exploring scientific hypotheses using computers.

**Computational "X"** (where "X" is a particular natural or engineering science, such as physics, chemistry, biology, geophysics, fluid dynamics, structural mechanics, or electromagnetodynamics): a specialized subset of computational science concentrating on models, problems, techniques, and practices particular to problems from "X."

**Scientific computing:** an intersection of tools and techniques in the pursuit of different types of computational "X."

**Strong/weak coupling of physical models:** strong (respectively, weak) coupling refers to strong (respectively, weak) interactions between different physics models that are intrinsic in a natural process. Mathematically, the off-diagonal blocks of the Jacobian matrix of a strongly coupled multiphysics model may be full or sparse but contain relatively large entries. In contrast, a weakly coupled multiphysics model contains relatively small off-diagonal entries.

**Tight/loose coupling of numerical models:** tight (respectively, loose) coupling refers to a high (respectively, low) degree of synchronization of the state variables across different physical models. A tightly coupled scheme (sometimes referred to as a strongly coupled scheme) keeps all the state variables as synchronized as possible across different models at all times, whereas a loosely coupled scheme might allow the state variables to be shifted by one time step or be staggered by a fraction of the time step.

### Further Reading


Asanovic, K., R. Bodik, B. C. Catanzaro, J. J. Gebis, P. Husbands, K. Keutzer, D. A. Patterson, W. L. Plishker, J. Shalf, S. W. Williams, and K. A. Yelick. 2006. The landscape of parallel computing research: a view from Berkeley. Technical Report UCB/EECS-2006-183, Electrical Engineering and Computer Sciences Department, University of California, Berkeley.

Keyes, D. E., ed. 2003. A science-based case for large-scale simulation: volume 1. Report, Office of Science, United States Department of Energy.

Keyes, D. E., et al. 2013. Multiphysics simulations: challenges and opportunities. *International Journal of High Performance Computing Applications* 27:4–83.

Lax, P. D. 1986. Mathematics and computing. *Journal of Statistical Physics* 43:749–56.

Wilson, K. G. 1989. Grand challenges to computational science. *Future Generation Computer Systems* 5:171–89.


# IV.17 Data Mining and Analysis

*Chandrika Kamath*

## 1 The Need for Data Analysis

Technological advances are enabling us to acquire ever-increasing amounts of data. The amount of data, now routinely measured in petabytes, is matched only by its complexity, with the data available in the form of images, sequences of images, multivariate time series, unstructured text documents, graphs, sensor streams, and mesh data, sometimes all in the context of a single problem. Mathematical techniques from the fields of machine learning, optimization, pattern recognition, and statistics play an important role as we analyze

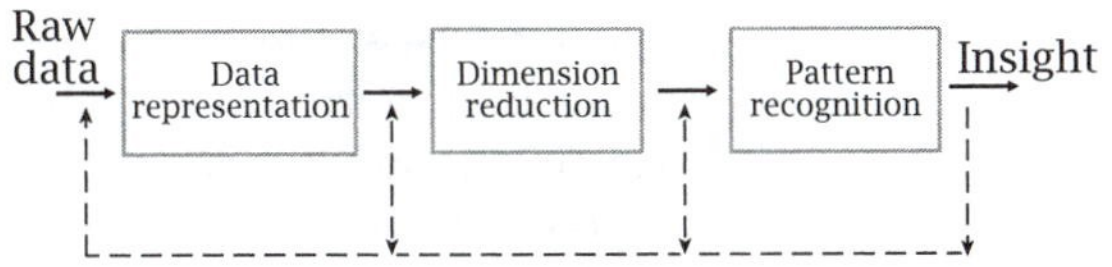


**Figure 1** A schematic showing the overall process of data analysis.

these data to gain insight and exploit the information present in them.

The types of analysis that can be done are often as varied as the data themselves. One of the best-known examples occurs in the task of searching, or information retrieval, where we want to identify documents containing a given phrase, retrieve images similar to a query image, or find the answer to a question from a repository of facts. In other analysis tasks, the focus is prediction, where, given historical data that relate input variables to output variables, we want to determine the outputs that result from a given set of inputs. For example, given examples of malignant and benign tumors in medical images, we may want to build a model that will identify the type of tumor in an image. Or we could analyze credit card transactions to build models that identify fraudulent use of a card. In other problems the focus is on building descriptive models. For example, we want to group chemicals with similar behavior to understand what might cause that behavior, or we may cluster users who like similar movies so we can recommend other movies to them.

*Data mining* is the semiautomatic process of discovering associations, anomalies, patterns, and statistically significant structures in data. At the risk of oversimplification, we can consider data mining, or data analysis, to have three phases, as shown in figure 1: the representation of the data, dimension reduction, and the identification of patterns.

First, the raw data are processed to bring them into a form more suitable for analysis. This is especially true when the data are in the form of images, text documents, or other formats that require us to identify the objects in the raw data and find appropriate representations for them. A common way of representing data is in the form of a table (see table 1) in which each row represents a data item (which could be a document, an image, a galaxy, or a chemical compound) and in which the columns are the features, or descriptors, that characterize the data item. These features are chosen to reflect the analysis task and could include the frequency of different words in a document, the texture of an image, the shape of the objects in an image, or the properties of the atomic species that constitute a chemical compound. Sometimes a data item may be associated with an output variable, which could be discrete (e.g., indicating if a transaction is valid or fraudulent) or continuous (such as the formation enthalpy of a chemical compound). This table, viewed as a matrix, could be sparse if not all features are available for a given data item, for example, words that do not occur in a document.

**Table 1** Table data, with each data item $I_i$ characterized by $d$ features, and potentially an output $O_i$.

| | $f_1$ | $f_2$ | $\cdots$ | $f_d$ | $O$ |
|---|---|---|---|---|---|
| $I_1$ | $f_{11}$ | $f_{12}$ | $\cdots$ | $f_{1d}$ | $O_1$ |
| $I_2$ | $f_{21}$ | $f_{22}$ | $\cdots$ | $f_{2d}$ | $O_2$ |
| $\vdots$ | $\cdots$ | $\cdots$ | $\cdots$ | $\cdots$ | $\cdots$ |
| $I_n$ | $f_{n1}$ | $f_{n2}$ | $\cdots$ | $f_{nd}$ | $O_n$ |

Each data item in the table can be considered as a point in a space spanned by the features; if the number of features is high, that is, the point is in a high-dimensional space, it can make the analysis challenging. This requires dimension reduction techniques to identify the key features in the data.

The final step of pattern recognition is often seen as the focus of analysis. However, its success is strongly dependent on how well the previous steps have been performed. Not identifying the data items correctly or representing them using inappropriate features can result in inaccurate insights into the data.

The process of data analysis is iterative: the representation of the data is refined and the dimension reduction and pattern recognition steps repeated until the desired accuracy is obtained. Data analysis is also interactive, with the domain experts providing input at each step to ensure the relevance of the process to the problem being addressed. Depending on the type of data being analyzed and the problem being addressed, these three phases of data representation, dimension reduction, and pattern recognition can be implemented using a host of mathematical techniques. In the remainder of this article we review some of these techniques, identify their relationship with methods in other domains, and discuss future trends.

## 2 History

Data analysis has a long history. We could say that it started with the ancient civilizations, which, observing the motions of celestial objects, identified patterns that led to the laws of celestial mechanics. Some of the early ideas in statistics arose when governments collected demographic and economic data to help them manage their populations better, while probability played an important role in the study of games of chance and the analysis of random phenomena. The field of statistics, as we know it today, goes back at least a couple of hundred years. By modern standards, the data sets being analyzed then were tiny, composed of a few hundred numbers. However, given that this analysis was done before the advent of computers, it was a remarkable achievement. The field has evolved considerably since then, prompted by new sources of data, a broader range of questions being addressed through the analysis, and different fields contributing different insights into the data.

The recent surge in both the size and the complexity of data started in the 1970s as technology enabled us to generate, store, and process vast amounts of data. Sensors provided some of the richest sources of data, ranging from those used in medicine, such as POSITRON EMISSION TOMOGRAPHY [VII.9] (PET) and MAGNETIC RESONANCE IMAGING [VII.10 §4.1] (MRI) scanning, to sensors monitoring equipment, such as cars and scientific experiments, and sensors observing the world around us through telescopes and satellites. Computers not only enabled the analysis of these data, they contributed to this data deluge as well. Numerical simulations are increasingly being used to model complex phenomena in problems where experiments are infeasible or expensive. The resulting spatiotemporal data, often in the form of variables at grid points, are analyzed to gain insight into the phenomena. The Internet, connecting computers worldwide, is one of the newest sources of data, providing insight into how the world is connected. These data are generated by sensors that observe network traffic, by web services that monitor the online browsing patterns of users, and by the network itself, which changes dynamically as connections are made or broken.

The tasks in data analysis have also evolved to handle the new types of data, the new questions being asked of the data, and the new ways in which users interact with the data. For example, network data, images, mesh data from simulations, and text documents are not in a form that can be directly used for pattern identification; they all have to be converted to an appropriate representation first. The high dimensionality of many data sets has led to the development of dimension reduction techniques, which shed light on the important features of the objects in a data set. The desire to improve the accuracy of predictions by analyzing many different modalities of data simultaneously has led to data fusion techniques. More recently, a recognition that the analysis of the data should be closely coupled with their generation has resulted in the reemergence of ideas from the field of design of experiments. Further, as data analysis gains acceptance and the results are used in making decisions, techniques to address missing and uncertain features have become important, and tools to reason under uncertainty are playing a greater role, especially when the risks and rewards associated with the decisions can be so high.

Along with these new sources of data and the novel questions that are being asked of the data, different fields have started to contribute to the broad area of data analysis. In addition to statistics, advances in the field of pattern recognition have focused on patterns in signals and images, while work in artificial intelligence and machine learning have tried to mimic human reasoning, and developments in rule-based systems have been used to extract meaningful rules from data. The more recent field of data mining was originally motivated by the need to make better use of databases, though it too is evolving into specialized topics such as web mining, scientific data mining, text mining, and graph mining. Domain scientists have also contributed their ideas for analyzing data. As a result, data analysis is replete with examples where the same algorithm is known by different names in different domains, and a technique proposed in one domain is found to be a special case of a technique from a different domain. The process of inference from data has benefited immensely from the contributions of these different fields, with each providing its own perspective on the data and, in the process, making data analysis a field rich in the diversity of mathematical techniques employed.

## 3 Data Representation

Many pattern recognition algorithms require the data to be in a tabular form, as shown in table 1. However, this is rarely the form in which the data are given to a data analyst, especially when the raw data are available as images, time series, documents, or network data. Converting the data into a representation suitable for analysis is a crucial first step in any analysis endeavor. However, this task is difficult as the representation is very dependent on the problem, the type of data, and the quality of the data. Domain- and problem-dependent methods are often used, and the representation is iteratively refined until the analysis results are acceptable.

Consider the problem of improving the quality of images, which are often corrupted by noise, making it difficult to identify the objects in the data. A simple solution is to apply a *mean filter*. In the case of a $3 \times 3$ filter, this would imply convolving an image with

$$\frac{1}{9}\begin{pmatrix} 1 & 1 & 1 \\ 1 & 1 & 1 \\ 1 & 1 & 1 \end{pmatrix},$$

thus replacing each image pixel by the weighted average of the pixel and its eight neighbors. While a mean filter assigns equal weight to all pixels within its support, a two-dimensional *Gaussian filter* with standard deviation $\sigma_{\mathrm{d}}$,

$$w_{\mathrm{d}}(p_i, p_j) = \frac{1}{2\pi\sigma_{\mathrm{d}}^2}\exp\biggl(-\frac{\|p_i - p_j\|^2}{2\sigma_{\mathrm{d}}^2}\biggr),$$

assigns a lower weight when the pixel $p_j$ is farther from the pixel $p_i$. A more recent development, the *bilateral filter*, applies this idea in both the domain and the range of the neighborhood around each pixel. Each weight in the filter is the normalized product of the corresponding weights $w_{\mathrm{d}}$ and $w_{\mathrm{r}}$:

$$w_{\mathrm{r}}(p_i, p_j) = \frac{1}{2\pi\sigma_{\mathrm{r}}^2}\exp\biggl(-\frac{\delta(I(p_i) - I(p_j))^2}{2\sigma_{\mathrm{r}}^2}\biggr),$$

where $\sigma_{\mathrm{r}}$ is the standard deviation of the range filter and $\delta(I(p_i) - I(p_j))$ is a measure of the distance between the intensities at pixels $p_i$ and $p_j$. When the intensity difference is high—near an edge of an object, for example—less smoothing is done, making it easier to identify the edge in the next step of the analysis.

In 1987 Perona and Malik noted that convolving an image with the Gaussian filter is equivalent to solving the diffusion equation

$$\frac{\partial}{\partial t}I(x,y,t) = \frac{\partial^2}{\partial x^2}I(x,y,t) + \frac{\partial^2}{\partial y^2}I(x,y,t),$$

where $I(x,y,t)$ is the two-dimensional image $I(x,y)$ at time $t = 0.5\sigma_{\mathrm{d}}^2$ and $I(x,y,0)$ is the original image. This allowed the adoption of ideas from partial differential equations (PDEs) to address the ill-posedness of the original formulation and to enhance de-noising by making suitable choices for the coefficients of the PDEs in order to restrict smoothing near the edges of objects in an image.

There has been similar evolution in the realm of segmentation techniques for identifying objects in images. Starting with the idea that there is a sharp gradient in the intensity at the boundary of an object, we can use a simple filter, such as the Sobel operators in the $x$- and $y$-dimensions,

$$L_x = \begin{pmatrix} -1 & 0 & 1 \\ -2 & 0 & 2 \\ -1 & 0 & 1 \end{pmatrix}, \qquad L_y = \begin{pmatrix} 1 & 2 & 1 \\ 0 & 0 & 0 \\ -1 & -2 & -1 \end{pmatrix},$$

to obtain the magnitude and orientation of the gradient. This basic idea is used in the *Canny edge detector*, which, despite its simplicity, performs quite well. It starts by smoothing the image using a Gaussian filter and then applies the Sobel operators. The edges found are thinned to a single pixel by suppressing values that are not maximal in the gradient direction. Finally, a two-parameter hysteresis thresholding retains both strong edge pixels and moderate-intensity edge pixels that are connected to the strong edge pixels. This results in a robust and well-localized detection of edges. However, in regions of low contrast, the gradient detectors may fail to identify any edge pixels, leading to edges that are not closed contours. To address this drawback, we need to use techniques such as snakes and implicit active contours, which again blend ideas from image processing and PDEs. In turn, this allows us to identify objects in mesh data from simulations by using the PDE version of a segmentation algorithm.

Other segmentation techniques, referred to as *region-growing methods*, identify objects in images by exploiting the fact that the pixels in the interior of an object are similar to each other. Starting with the highest intensity pixel, a similarity metric is used to grow a region around it. The process is repeated with the highest intensity pixel among the remaining ones, until all pixels are assigned to a region. Cleanup is often required to merge similar regions that are adjacent to each other or to remove very small regions. This idea of grouping data items based on their similarity is referred to as clustering and is described further in section 5.3. These clustering techniques can be applied to images and mesh data by defining a similarity metric that takes into account both the spatial locations of and the values at the pixels or mesh points.

Once we have identified objects in images, or sequences of images, we can represent them using features such as shape, texture, size, and various moments. There are different ways in which many of these features can be defined. For example, for objects in images, the shape feature can be represented as a linear combination of two-dimensional basis functions such as shapelets or the angular radial transform described in the MPEG-7 multimedia standard. The coefficients of the linear combination would be the features representing the object. Alternatively, we can focus on the boundary of an object and define the shape features as the coefficients of a Fourier series expansion of the curve that describes the boundary.

The features used to represent the objects in the data are very dependent on the problem and the type of data. It is often challenging to identify and extract features for some types of data, such as images or mesh data. Not only must the features characterize the objects in these data, they must also be invariant to rotation, translation, and scaling, as the patterns in the objects are often invariant to such transformations of the data. Further, if we are analyzing many images of varying quality, each with a variety of objects, it can be difficult to find a single set of algorithms, with a single set of parameters, that can be used to extract the objects and their features.

In contrast, the task of feature extraction is relatively easy for other types of data. For example, in documents it is clear that the features are related to the words or terms in the collection of documents being analyzed. As discussed in TEXT MINING [VII.24], a document can be represented using the term frequency of each term in the document, suitably weighted to reflect its importance in the document or collection of documents. However, representing documents is not without its own challenges, such as the need for word-sense disambiguation for words with multiple meanings, the representation of tables or diagrams in documents, and analysis across documents in different languages.

## 4 Dimension Reduction

The number of features used to represent the data items in a data set can vary widely, ranging from a handful to thousands. These features can be numeric or categorical. Often, not all the features are relevant to the task at hand, or some features may be correlated, resulting in a duplication of information. Since extracting and storing features can be expensive, and irrelevant features can adversely affect the accuracy of any models built from the data, it is often useful to identify the important features in a data set. This task of dimension reduction maps a data item described in the high-dimensional feature space into a lower-dimensional feature space.

There are two broad categories of dimension reduction methods. In the first category, the features are transformed, linearly or nonlinearly, into a lower-dimensional space. A popular method for linear transformation is *principal component analysis* (PCA), which has been rediscovered in many domains, ranging from fluid dynamics, climate, signal processing, and linear algebra to text mining, with each contributing different insights into what such techniques reveal about the data. PCA is an orthogonal transform that converts the data into a set of uncorrelated variables called the principal components. The first principal component captures the largest variability in the data; each of the remaining principal components has the highest variance subject to the constraint that it is orthogonal to all the previous principal components. PCA can be calculated by an eigendecomposition of the data covariance or correlation matrix or by the application of the SINGULAR VALUE DECOMPOSITION [II.32] to the data matrix. Usually, the data are first centered by subtracting the mean.

A challenging issue in the practical application of transform-based methods is the selection of the dimension of the lower-dimensional space. Often, a decision on the number of principal components to keep is made using a threshold on the percentage variance explained by the principal components that are retained. This essentially implies that the components that are discarded are considered as noise in the data. Thus, PCA can also be used to reduce the noise in data.

The second category of dimension reduction techniques selects a subset of the original features. These *feature selection techniques* are applicable to prediction tasks, such as classification and regression, where each data item is also associated with an output variable. By identifying important features, such techniques enable us to make judicious use of limited resources available for measuring the features. They can also be invaluable in problems where focusing on just the important variables makes it easier to gain insight into a complex phenomenon.

There are two types of feature selection algorithms: filters, which are not coupled to the prediction task, and wrappers, which are coupled to the prediction task.

As an example of a *filter method*, consider a simple problem where we have data items of two categories, or classes, and we need to find the features that are most useful in discriminating between these classes. For each feature we can build a histogram of the values that the feature takes for each of the two classes. A large distance between the histograms for the two

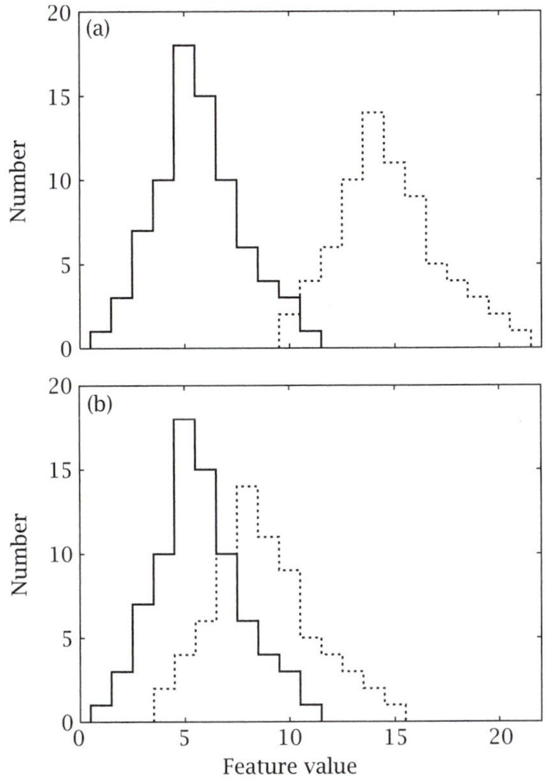


**Figure 2** The feature in (a) (with a larger distance between the histograms of feature values for two different classes (solid and dotted lines)) is considered more important than the feature in (b). The large overlap in (b) indicates that there is a larger range of values of the feature for which an object could belong to either of the two classes, making the feature less discriminating.

classes indicates that the feature is likely to be important, but if the histograms overlap, then the feature is unlikely to be helpful in differentiating between the classes (figure 2). The features can be ordered based on the distances between the histograms and the reduced dimension can be determined by placing an appropriate threshold on the distance. Any suitable measure for distances between histograms can be used. For example, we can first convert the histogram for each class into a probability distribution by normalizing it so that the area under the curve is 1. Then, we can either use the Kullback–Leibler divergence

$$d_{\mathrm{KL}}(P,Q) = \sum_{i=1}^{b} P_i \ln\left(\frac{P_i}{Q_i}\right)$$

as a measure of the difference between two distributions $P$ and $Q$ defined over $b$ bins, or we can create a symmetric version

$$d(P,Q) = d_{\mathrm{KL}}(P,Q) + d_{\mathrm{KL}}(Q,P),$$

which, unlike the Kullback–Leibler divergence, is a distance metric, as $d(P,Q) = d(Q,P)$. Other distances can also be used, such as the Wasserstein metric, or, as it is referred to in computer science, the earth mover's distance. This intuitively named metric considers the two distributions as two ways of piling dirt on a region, with the distance being defined as the minimum cost of turning one pile into the other. The cost is the amount of dirt moved times the distance over which it is moved.

The *wrapper approach* to feature selection evaluates each subset of features based on the accuracy of prediction using that subset. Suppose we use a decision tree classifier (section 5.2) to build a predictive model for the two-class problem. The forward selection wrapper method starts by selecting the single feature that gives the highest accuracy with the decision tree. It then chooses the next feature to add such that the subset of two features has the highest accuracy. Additional features are included to create larger subsets until the addition of any new feature does not result in an improvement in prediction accuracy. A backward selection approach starts with all the features and progressively removes features. The wrapper approach is more computationally expensive than the filter approach as it involves evaluating the accuracy of a classifier. However, it may identify subsets that are more appropriate when used in conjunction with a specific classifier.

## 5 Pattern Recognition

In pattern recognition tasks, we use the features that describe each data item to identify patterns among the data items. For example, if we process astronomical images to identify the galaxies in them and extract suitable features for each galaxy, we are then able to use the features in several different ways, as discussed next.

### 5.1 Information Retrieval

In *information retrieval*, given a query data item, described by its set of features, the task is to retrieve other items that are similar in some sense to the query. Usually, the retrieved items are returned in an order based on the similarity to the query. The similarity metric chosen depends on the problem and the representation of the data item. While Euclidean distance between feature vectors is a common metric, the cosine of the angle between the two vectors is used for document similarity, and more complex distances could be used if the data items are represented as graphs.

Data sets used in information retrieval are often massive, and a brute-force search by comparing each item with the query item is expensive and will not result in the real-time turnaround often required in such analysis. Ideas from computer science, especially the use of sophisticated data structures that allow for fast nearest-neighbor searches, provide solutions to this problem. As these data structures perform better than a brute-force search only when the data items have a limited number of features, dimension reduction techniques are often applied to the data before storing them in these efficient data structures.

### 5.2 Classification and Regression

In some pattern recognition tasks we are given examples of patterns, referred to as a training set, and the goal is to identify the pattern associated with a new data item. This is done by using the training set to create a predictive model that is then used to assign the pattern to the new data item. The items in the training set all have an output variable associated with them. If this variable is discrete, the problem is one of *classification*; if it is continuous, the problem is one of *regression*, where, instead of predicting a pattern, we need to predict the value of the continuous variable for the new data item.

Techniques for regression and classification tend to be quite similar. The simplest such technique is the *nearest-neighbor method*, where, given the data item to which we need to assign an output, we identify the nearest neighbors to this item in the feature space and use the outputs of these neighbors to calculate the output for the query. In classification problems we can use the majority class, while in regression problems we can use a weighted average of the outputs of the neighbors. The neighbors are usually identified either by specifying a fixed number of neighbors $k$, which leads to the $k$-nearest-neighbor method, or by specifying the neighbors within a radius $\varepsilon$, which leads to the $\varepsilon$-nearest-neighbor method. While nearest-neighbor techniques are intuitively appealing, it is a challenge to set a value for $k$ or $\varepsilon$, and the techniques may not work well if the query data item does not have neighbors in close proximity.

More complex predictive models can be built from a training set using other classification algorithms. *Decision trees* divide the data set into regions using hyperplanes parallel to the axis, such that each region has data items with the same output value (or similar ones). The tree is a data structure that is either a leaf, which indicates the output, or a decision node that specifies some test to be carried out on a feature, with a branch and subtree for each possible outcome of the test. For example, in figure 3 the open and closed circles are separated by first making the decision F1 < A1; data items that satisfy this constraint are next split using F2 < B2, while data items that do not satisfy the constraint are split using F2 < B1. Then, given a data item with certain values of features (F1, F2), we can follow its path down the decision tree and assign it the majority class of the leaf node at which the path ends.

A key task in building a decision tree is the choice of decision at each node of the tree. This is obtained by considering each feature in turn, sorting the values of the feature, and evaluating a quality metric at the midpoints between consecutive values of the feature. This metric is an indication of the suitability of a split using that feature and midpoint value. The feature–midpoint value combination that optimizes the quality

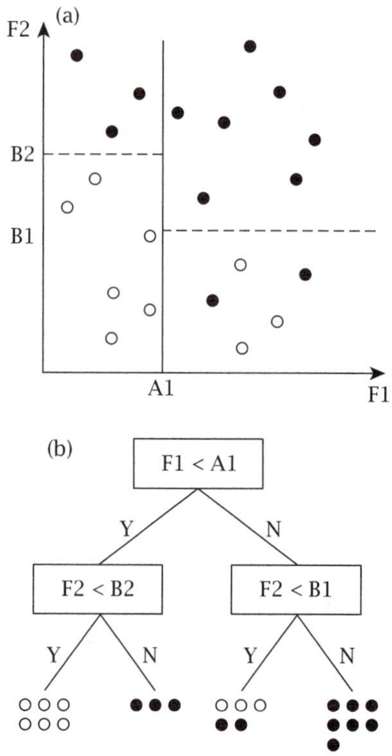


**Figure 3** A schematic of a decision tree for a data set with two features: (a) the decision boundaries between open and closed circles; (b) the corresponding decision tree.

metric across all features is chosen as the decision at the node. For example, in a two-class problem, a split that divides the data items at the node into two groups, each of which has a majority of items of one class, is preferable to a split where each of the two groups has an equal mix of the two classes. A commonly used metric is the Gini index, which finds the split that most reduces the node impurity, where the impurity for a $c$ class problem with $n$ data items is defined as follows:

$$L_{\text{Gini}} = 1.0 - \sum_{i=1}^{c} \left(\frac{|L_i|}{|T_L|}\right)^2,$$

$$R_{\text{Gini}} = 1.0 - \sum_{i=1}^{c} \left(\frac{|R_i|}{|T_R|}\right)^2,$$

$$\text{Impurity} = (|T_L|L_{\text{Gini}} + |T_R|R_{\text{Gini}})/n,$$

where $|T_L|$ and $|T_R|$ are the number of data items, $|L_i|$ and $|R_i|$ are the number of data items of class $i$, and $L_{\text{Gini}}$ and $R_{\text{Gini}}$ are the Gini indices on the left- and right-hand side of the split, respectively.

This simple idea of dividing the data items in feature space using decisions parallel to the axes appeared at nearly the same time in the literature relating to both machine learning and statistics, where it was referred to as the decision tree method and the classification tree method, respectively. Both communities have contributed to the advancement of the method. Ideas from machine learning have been used to create more complex decision boundaries by considering a linear combination of the features in the decision at each node. The process of making the decisions at each node of the tree has been related to probabilistic techniques, providing insight into the behavior of decision trees and indicating ways in which they can be improved. In particular, a statistical approach to decision tree modeling takes advantage of the trade-off between bias, which arises when the classifier underfits the data and cannot represent the true function being predicted, and variance, which arises when the classifier overfits the data.

Decision trees are not only simple to create and apply, they are also easy to interpret, providing useful insight into how a pattern is assigned to a data item. Further, the process used in making the decision at each node of the tree indicates which features are important in the data set. This makes the decision tree method a popular first choice for use in classification and regression problems.

Other, more complex, algorithms in common use are neural networks and support vector machines. The simplest *neural network*, called a perceptron, takes a weighted combination of the input features and outputs a 1 if the combination is greater than a threshold and 0 otherwise, thus acting as a two-class classifier. The task is to use the training set to learn the weights, an adaptive process that starts with random weights, applies the classifier, and modifies the weights whenever the classification is in error. When the two classes are not linearly separable, that is, they cannot be separated using a single line, we need to find the weights using OPTIMIZATION TECHNIQUES [IV.11] such as the gradient descent, conjugate gradient, quasi-Newton, or Levenberg–Marquardt methods.

More complex neural networks include multiple layers (in addition to the first layer, which represents the features) and, instead of a simple thresholding, use a continuously differentiable sigmoid function. This allows neural networks to model complex functions to differentiate among classes. However, they can be difficult to interpret, and designing the architecture of the network—which includes the number of layers, the number of nodes in a layer, and the connectivity among the nodes—can be a challenge.

Another classification algorithm that uses optimization techniques is the *support vector machine*. This uses a nonlinear function to map the input into a higher-dimensional space in which the data are linearly separable. There are many hyperplanes that linearly separate the data. The optimal one, as indicated by statistical learning theory, is the maximum margin solution, where the margin is the distance between the hyperplane and the closest example of each class. These examples are referred to as the support vectors. This solution is obtained by solving a QUADRATIC PROGRAMMING PROBLEM [IV.11] subject to a set of linear inequality constraints. Support vector machines provide some insight into the process of classification through the identification of the support vectors, though it is a challenge to find the nonlinear transformation that linearly separates the data.

A probabilistic approach to inference is provided by Bayesian reasoning, which assumes that quantities of interest are governed by probability distributions. This enables us to make decisions by reasoning about these probabilities in the presence of observed data. *Bayes's theorem*,

$$P(h \mid D) = P(D \mid h)P(h)/P(D),$$

is a way of calculating the posterior probability $P(h \mid D)$, or the probability of hypothesis $h$ given data $D$, from the prior probability of the hypothesis before we have seen the data, $P(h)$; the probability $P(D \mid h)$ of seeing the data $D$ in a world where the hypothesis $h$ holds; and the prior probability $p(D)$ of seeing the data $D$ without any knowledge of which hypothesis holds. Bayes's theorem is essentially derived from the two different ways of representing the probability of observing both $h$ and $D$ together by writing it conditional either on $D$ or $h$.

Given the probabilistic insight provided by Bayes's theorem, it is invaluable in tasks requiring reasoning under uncertainty. The theorem also plays a role in the *naive Bayes classifier*. The new data item is assigned the most probable output value given the values of its features. This most probable output value is obtained by applying Bayes's theorem using probabilities derived from the training data, combined with the added simplifying assumption that the feature values are conditionally independent, given the output.

A recent development in classification techniques is that of *ensemble learning*, where more than one classifier is created from the training data. This is done by introducing randomization into the process, by creating new training sets using sampling with replacement from the original training set, for example, or by selecting a random subset of features or samples to use in making the decision at each node. The class label assigned to a new data item is obtained by using a voting scheme on the labels assigned by each classifier in the ensemble. While ensembles are computationally more expensive as multiple classifiers have to be created, the ensemble prediction is often more accurate.

### 5.3 Clustering

*Clustering* is a descriptive technique used when the data items are not associated with an output value. It can be seen as complementary to classification. Instead of using the output value to identify boundaries between items of different classes near each other in feature space, we use the features to identify groups of items in feature space that are similar to, or close to, each other (figure 4). Intuitively, if the features are chosen carefully, two data items with similar features are likely to be similar. Thus, if we know that some chemical compounds, with desirable properties, occur in a certain part of an appropriately defined feature space, other compounds near them might also have the same property. Or, if a customer is interested in a specific movie or book, others that are nearby in feature space could be recommended to them.

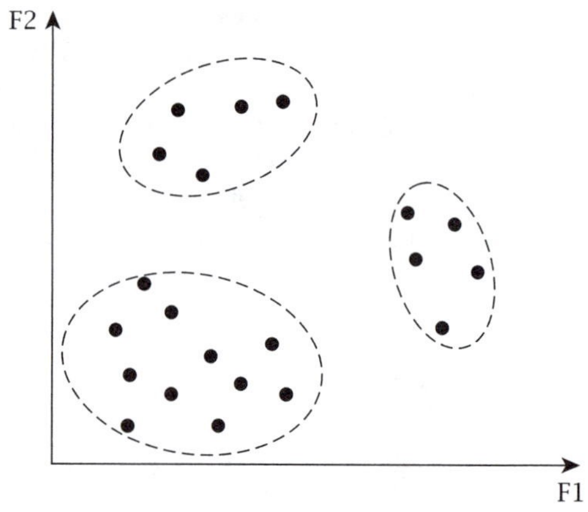


**Figure 4** A schematic showing three clusters in a two-dimensional data set.

Clustering techniques broadly fall into one of two categories. In *hierarchical methods* we have two options: start from the bottom, with each item belonging to a singleton cluster, and merge the clusters, two at a time, based on a similarity metric; or start at the top, with all items forming a single cluster, and split that into two, followed by a split of each of the two clusters, and so on. In both methods, we need to determine how many clusters to select in the hierarchy.

In contrast, *partitional methods* start with a predetermined number of clusters and divide the data items into these clusters such that the items within a cluster are more similar to each other than they are to the items in other clusters. The best-known partitional algorithm is the *k-means algorithm*, which divides the data items into $k$ clusters by first randomly selecting $k$ cluster centers. It then assigns each data item to the nearest cluster center. The items in each cluster are then used to update the cluster centers and the process continues until convergence. The $k$-means algorithm is very popular and, despite its simplicity, works well in practice. The main challenge is to determine the number of clusters; this can be done by evaluating a clustering criterion such as the sum of squared errors of each data item from its cluster center and then selecting the $k$ that minimizes this error.

The $k$-means algorithm is related to algorithms in other domains; examples include Voronoi diagrams from computational geometry, the expectation–maximization algorithm from statistics, and vector quantization in signal processing.

There is a third category of clustering algorithms that shares properties of both hierarchical and partitional methods. These *graph-based algorithms* are related to the reordering methods used in direct solvers and the domain decomposition techniques used in PDEs. Essentially, they represent the data items as nodes in a graph, with the edges weighted according to the similarity between the data items. Clustering is then done either by using graph partitioning algorithms, where clusters are identified by minimizing the weights of the edges that are cut by the partitioning, or by using spectral graph theory and calculating the second eigenvector of the GRAPH LAPLACIAN [V.12 §5]. The nodes of the graph are split into two using the eigenvector components, and the process is then repeated on the two parts.

## 6 The Future

As data-mining algorithms have gained a foothold in application domains such as astronomy and remote sensing, other domains such as medical imaging, with similar types of data and problems, have started considering data mining as a potential solution to their analysis problems. These new domains have in turn introduced new challenges to the analysis process and posed new questions, prompting the development of new algorithms. And then, when these algorithms are adopted by other communities, the cycle is continued.

There are several key developments from the last decade that will drive the future of data mining and analysis. These developments have been motivated both by the types of data being analyzed and by the requirements of the analysis.

There has been an explosion in the types of data being analyzed. While raw data in a tabular form is still the norm in some problem domains, other types of data—ranging from images and sequences of images to text documents, web pages, links between web pages, chemical compounds, deoxyribonucleic acid (DNA) sequences, mesh data from simulations, and graphs representing social networks—have prompted a new look at algorithms for finding accurate and appropriate representations of such data. This has also prompted greater interest in data fusion, where we analyze multiple modalities of data. For example, we may consider PET scans and MRI scans, as well as clinical data, to treat a patient. Or we may consider not just the links between web pages but also the text, figures, and images on those pages.

Adding to this complexity is the distributed nature of some of the data sets. For example, the use of sensor networks is becoming common in the surveillance and monitoring of experiments and complex systems. The networks may be autonomous and may reorganize, changing their positions in response to changes in their environment. In some problems, such as when monitoring climate, the sensors may be stationary but geographically scattered over the Earth. In all these cases, the data set to be analyzed is in several pieces that may not be collocated and the size of the data may make it difficult to collocate all the pieces in one place. There is a need for algorithms that analyze distributed data and build a model, or models, to represent the whole data set. In autonomous sensor networks there is an additional requirement that the amount of information exchanged between sensors be small, implying that it is the models and not the data that are exchanged. Such ideas are also relevant in telemedicine, as well as space exploration, where the distances and limited connectivity imply caution in determining what data and information are transferred.

This brings up the need for effective compression techniques: both lossy, for problems which can tolerate the loss of some information, and lossless, where it is important that the data be preserved in their original state. Compression will also play a role as we move to exascale computation, where it is becoming clear that the amount of data generated by simulations run on massively parallel machines will outpace the technology of the input–output system.

The ever-increasing volume of data will increase the need for algorithms that exploit parallel computers; otherwise we run the risk of the data not being analyzed at all. In addition, such algorithms will be invaluable in problems where a real-time or near-real-time response is required.

This short response time is required in the analysis of streaming data, where we analyze data as they are collected to identify untoward incidents, interesting events, or concept drift, where the statistical properties of the data change from one normal state to another. This is prompting the development of new algorithms that yield approximate results, as well as the creation of incremental versions of existing algorithms, where the models being built are constantly being updated to incorporate new data and discard old data. A particular challenge in such analysis is the need to minimize false positives while ensuring that we do not miss any positives, especially in problems where the analysis is used in decision support.

As data analysis comes to be viewed as one step in a closed system in which the data are analyzed and decisions are made based on the analysis, without a human in the loop, there is an increasing need to understand the uncertainty in the analysis results. Uncertainty can arise as a result of several factors, including the quality of the raw data, the suitability of the chosen analysis algorithms for the data being studied, the sensitivity of the results to the parameters used in the algorithms, a lack of complete understanding of the process or system being analyzed, and so on. Thus, ideas from uncertainty quantification and reasoning under uncertainty become important, especially when the risks associated with the decisions are high.

This brings up the interesting issue of using data analysis techniques to influence the data we collect so that they better meet our needs. So far we have viewed data analysis as a process that starts once the data have been collected, but sometimes we have control over what data to generate, such as which data items to label to create a training set or which input parameters to use for an experiment or a simulation. We can borrow ideas from the field of design of experiments, both physical and computational, to closely couple the generation of the data with their analysis, hopefully improving the quality of that analysis.

This borrowing of ideas from other domains will continue to increase as data analysis evolves to meet new demands. We have already seen how ideas from the traditional data analysis disciplines such as statistics, pattern recognition, and machine learning are being combined with ideas from fields such as image and video processing, mathematical optimization, natural language processing, linear algebra, and PDEs to address challenging problems in data analysis. This cycle involving the development of novel algorithms followed by their application to different problem domains, which in turn generates new analysis requirements, will continue, ensuring that data mining and analysis remain very exciting areas for the foreseeable future. Applied mathematics will remain a cornerstone of the field, not just by contributing to the extraction of insight into the data but also by playing a critical role in convincing the application experts that the insight obtained is based on sound mathematical principles.

### Further Reading

Duda, R. O., P. E. Hart, and D. G. Stork. 2000. *Pattern Classification*, 2nd edn. New York: Wiley-Interscience.

Hastie, T., R. Tibshirani, and J. Friedman. 2009. *The Elements of Statistical Learning: Data Mining, Inference, and Prediction*, 2nd edn. New York: Springer.

Kamath, C. 2009. *Scientific Data Mining: A Practical Perspective*. Philadelphia, PA: SIAM.

Mitchell, T. 1997. *Machine Learning*. Columbus, OH: McGraw-Hill.

Sonka, M., V. Hlavac, and R. Boyle. 2007. *Image Processing, Analysis, and Machine Vision*, 3rd edn. Toronto: Thomson Engineering.

Umbaugh, S. E. 2011. *Digital Image Processing and Analysis*, 2nd edn. Boca Raton, FL: CRC Press.

Witten, I., E. Frank, and M. Hall. 2011. *Data Mining: Practical Machine Learning Tools and Techniques*. Burlington, MA: Morgan Kaufmann.

## IV.18 Network Analysis
*Esteban Moro*

### 1 Introduction

Almost 300 years ago Euler posed the problem of finding a walk through the seven bridges of Königsberg and laid the foundations of graph theory. Euler's approach was probably one of the first examples of how to use network analysis to solve a real-world problem. Since then, network analysis has been used in many contexts, from biology to economics and the social sciences. The general approach is to map the constituent units of the system and its interdependencies onto a network and analyze that network in order to understand and predict a given process. For example, buyers and items form a network of purchases that is used in recommendation engines; protein–protein interaction networks are used to unveil functionally coherent families of proteins; social relationships might reveal potential adoption of products and services by social contagion.

The analysis of networks is an old subject in mathematics, and it has its roots in many other disciplines, such as engineering, the social sciences, and computer science. However, in recent times the digital revolution has brought with it easier access to detailed information about phenomena such as biological reactions, economic transactions, social interactions, and human movements. This has allowed us to study networks with an unprecedented level of detail. While this data revolution has produced an enormous boost in the models and applications of network theory, reaching unusual areas such as politics, crime, cooking, and so on, it has also challenged the available analytical methods because of the large size of real networks, which are typically made up of millions or even billions of nodes.

As a consequence, network analysis is a rapidly changing field that attracts many researchers from diverse disciplines. This article discusses the mathematical concepts behind network analysis and also some of the main applications to real-world problems.

## 2 Definitions

The main mathematical object in network analysis is the network itself. A network (or a graph) is a pair $G = (V, E)$ of a set $V$ of vertices (nodes) together with a set $E \in V \times V$ of edges (links). The numbers of nodes and edges are denoted by $n = |V|$ and $m = |E|$. Any edge is a relation between two vertices of $V$. Mathematically, a network can be represented by the *adjacency matrix* $A$, where $a_{ij}$ has the value 1 if there is an edge between vertices $i$ and $j$ and 0 otherwise (see plate 5). If the edge is undirected then $a_{ij} = a_{ji}$, while if the direction of the edge matters then the network is directed and $A$ is in general nonsymmetric. Another variant is a *weighted network*, in which the nonzero elements of $A$ represent the strength of the relationship: $a_{ij} = w_{ij} \in \mathbb{R}$.

Although a complete description of the network is given by the adjacency matrix, we can obtain valuable insights by measuring local (node-centric) and/or global properties of the graph. For example, the *degree* or *connectivity*, $k_i$, of node $i$ in an undirected network is the number of connections or edges that the node has. The *graph density*, $D = 2m/(n(n-1))$, measures how sparse or dense a graph is by the ratio of the number of edges to the maximum possible number of edges for an $n$-node graph. In a *complete graph* every node is connected to every other node, so $D = 1$. A *path* in the network is a sequence of edges that connects a distinct sequence of nodes. The graph is *connected* if one can get from any node to any other node by following a path. The *distance* $\delta(i, j)$ between two nodes $i$ and $j$ is the length of the shortest path (also known as a *geodesic*) between them; if no such path exists, we set $\delta(i, j) = \infty$. Distance can be also defined taking into account the weights of the edges. Finally, the *diameter* $\ell$ of a graph $G$ is the maximum value of $\delta(i, j)$ over all pairs of $i, j \in V$. Table 1 summarizes some key definitions and notation for easy reference.

One of the most common interests in network analysis is in the "substructures" that may be present in the network. The neighborhood of a node (its ego network), which comprises itself and its $k_i$ neighbors, can be thought of as a substructure. More generally, any subset of the graph is called a subgraph $G' = (E', V')$, where $E' \subseteq E$ and $V' \subseteq V$. The connected components of a graph are the subgraphs in which any two vertices are connected to each other by paths. We say that there is a *giant component* if there is a component that comprises a large fraction of the nodes. A *clique* in an undirected graph $G$ is a subgraph $G'$ that is complete. A maximal clique in $G$ is a clique of the largest possible size in $G$. Another measure of the "core" of the network is the $k$-*core* of $G$, which is defined as the maximal connected subgraph of $G$ in which all the vertices have degree at least $k$. Equivalently, it is one of the connected components of the subgraph of $G$ formed by repeatedly deleting all vertices of degree less than $k$. Finally, *network motifs* are recurrent and statistically significant subgraphs or patterns within a graph. More informal definitions of *clusters* in the network are used in the community-finding problem, in which a graph $G$ is divided into a partition of *dense* subgraphs (see section 3.5).

In some applications the static structure of networks or graphs is not enough to incorporate the dynamical nature of nodes and the interactions between them. For example, if we want to study the phone calls made between the customers of a mobile phone company, we will need to incorporate the fact that calls take place at different times. To this end, we can define *temporal graphs*, $G_t = (V_t, E_t)$, in which we have a different set of nodes and edges for each $t \geqslant 0$. Time-varying graphs are affected by the temporal aspects of interactions, like causality, which add a new perspective to network analysis. Also, when multiple types of edges are present (multiplexity), we must consider a description in which nodes have a different set of neighbors in each layer (type). For example, one can consider each layer as the different types of social ties among the same set of individuals. Or one could picture each layer as representing a particular time in a temporal graph, or the layers could be the mathematical setup for the bipartite network of users and items in recommendation algorithms.

## 3 Properties

The German mathematician Dietrich Braess noted that adding extra roads to a traffic network can lead to greater congestion. This paradox shows one of the characteristics of network analysis: namely, that networks have emergent properties that cannot be explained simply by the sum of their components. Here, we review some of the key properties of networks that are found

**Table 1** The key definitions and notation for a graph $G = (V, E)$ with $n$ nodes and $m$ edges.

| | | |
|---|---|---|
| Degree or connectivity | $k_i$ | Number of connections or edges that node $i$ has |
| Diameter | $\ell$ | $\max_{i,j \in V} \delta(i,j)$ |
| Degree distribution | $p_k$ | Fraction of nodes in $G$ that have degree $k$ |
| Average degree | $\bar{k}$ | $\sum_{i=1}^{n} k_i/n$ or $\sum_k k p_k$ |
| Average degree of neighbors | $\overline{k_{\text{nn}}}$ | Average degree of (next-nearest) neighbors of a node in $G$ |
| Graph density | $D$ | $2m/(n(n-1))$ |
| Distance | $\delta(i,j)$ | Length of the shortest path between nodes $i$ and $j$ |
| Clustering coefficient | $C$ | Relative frequency of triangles to triplets in $G$ |
| Individual clustering coefficient | $C_i$ | Relative frequency of triangles to triplets involving node $i$ |

in real examples and incorporated in the most representative models. The properties cover most scales of network, from local features like connectivity or clustering to the study of a network's modular structure (motifs, communities).

### 3.1 Heterogeneity

Many unique properties of networks are due to their heterogeneity. In the simplest approximation, networks are heterogeneous in the connectivity sense: a homogeneous network is one in which each node $i$ has the same degree $k_i \approx \bar{k}$, where $\bar{k}$ is the average degree in the network. However, in real-world networks the distribution of connectivities is highly heterogeneous. In fact, one is very likely to find nodes with $k_i \gg \bar{k}$. A simple way to characterize heterogeneity is by using the degree distribution of the network, $p_k$, the fraction of nodes in the network that have degree $k$. In a homogeneous network, $p_k$ peaks around the average value $\bar{k}$. However, real-world networks are usually heavily skewed, with a long tail of nodes having $k_i \gg \bar{k}$. These high-degree nodes or *hubs* have an important role in many properties of the network. Conversely, the network contains many nodes that are poorly connected. The connectivity description of the network is thus that of "few giants, many dwarves."

Although measuring the tail of heavily skewed distributions is statistically tricky, recent work has found that some real-world networks have power-law degree distributions $p_k \sim k^{-\alpha}$, where the scaling exponent typically lies in the range $2 < \alpha < 3$. Some instances of this observation are the network structure of the Internet, the network of links between Web pages, the network of citations between papers, phone communication networks, metabolic networks, and financial networks. Networks with power-law degree distributions are usually referred to as *scale-free networks* because $p_k$ lacks a characteristic degree. They have attracted wide attention in the literature due to their ubiquity in many complex systems and also because of their possible modeling by simple growth models. In other situations, degree distributions seem to be better described by exponentials or power-law distributions with exponential cutoffs. Regardless of which statistical model is best for describing $p_k$, the large heterogeneity found in networks implies that there is no such thing as typical connectivity in the network. Node degree is not the only network property that shows heterogeneity. For example, the weights (intensity) of edges, the frequency of motifs, and the distribution of community sizes are all described by broad distributions, showing that heterogeneity appears at different scales of the network and in various network descriptions.

### 3.2 Clustering

As well as being unequally shared among the nodes, edges tend to be clustered in a network. For example, the neighbors of a given node are very likely to themselves be linked by an edge. In the language of social networks, the probability that a friend of your friend is also your friend is very large. A way of measuring network clustering is to calculate the *transitivity* or *clustering coefficient*, $0 \leqslant C \leqslant 1$, which measures the relative frequency of triangles (cliques of size 3) in the network with respect to the total number of triplets (three nodes connected by at least two edges). In a fully connected graph, $C = 1$. Clustering can also be measured locally: $C_i$ measures the fraction of neighbors of a node $i$ that are also neighbors of each other. Social and biological networks display large clustering coefficients when compared with random network models, while technological networks like the World Wide Web, the Internet, or power grids have much less clustering. The origin of this difference lies in the potential mechanisms behind clustering or its absence: in social networks people who spend time with a common third

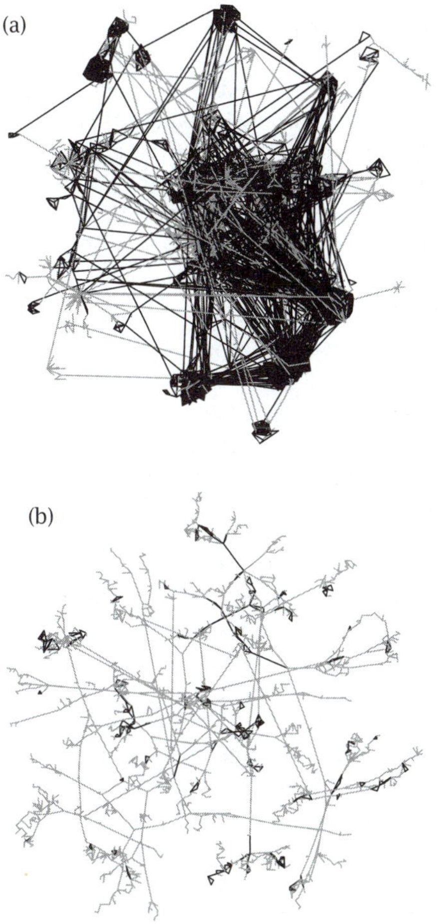


**Figure 1** (a) Yeast protein–protein interactions (data from von Mering, C., et al., 2002, *Nature*). (b) The power grid of the western states of the United States (data from Watts, D. J., and S. H. Strogatz, 1998, *Nature*). Links that belong to a triangle are shown in black, with the rest in gray. For reasons of clarity, nodes are removed in both cases. The clustering coefficients are (a) $C = 0.44$ and (b) $C = 0.09$.

person are likely to encounter each other, and thus *triadic closure* is favored by social interactions. Actually, triadic closure is exploited by many online social networks for friend-recommendation algorithms. In biological networks large clustering might be due to concurrent interaction of proteins in biological processes. However, efficiency in technological networks discourages the formation of redundant edges between nodes that are already close in the network, and $C$ is therefore very small (see figure 1).

In some networks $C_i$ decreases with the degree $k_i$. This means that low-degree nodes are densely clustered while hubs' neighborhoods are sparsely connected, a result which, together with the existence of densely connected groups of nodes (communities; see section 3.5), reflects the *hierarchical organization* of some networks: low-degree nodes are situated in dense communities, while hubs link different communities.

Triangles are not the only cliques or motifs that are over-represented in networks. For example, some three- or four-node motifs occur in large numbers even in small networks. However, most represented motifs in food webs, for example, are distinct from those found in transcriptional regulatory networks and from those in the World Wide Web, a finding that some authors ascribe to the function of the network. Networks may therefore be classified into distinct functional families based on their typical motifs. At a higher level, the relative size of the maximal clique or the $k$-core of the graph are also measures of the degree of clustering in the network.

### 3.3 Small World

Many naturally occurring networks exhibit the *small-world phenomenon*; that is, they have a small graph diameter. This was famously illustrated by the psychologist Stanley Milgram in 1967, who discovered the famous "six degrees of separation" (on average) between two persons in the worldwide social network. The experiment was later reproduced using email and measuring the distance in massive social graphs. The fact that networks are densely connected creates a wealth of short paths between nodes, and the typical distance between nodes is therefore very small. Illustrations of the small-world phenomenon are actors' "Bacon numbers," or mathematicians' Erdös numbers: the distance from Paul Erdös in the graph whose edges represent coauthorship of papers. The average Erdös number is only around 5.

Mathematically speaking, a small-world network refers to a network model in which the diameter increases sufficiently slowly with the number of nodes $n$ in the network—typically, as $\log n$ or slower. However, this property is trivially satisfied in a tree-like network if we assume that the number of nodes at distance $d$ from a given node scales as $d^{\bar{k}}$, where $\bar{k}$ is the average number of neighbors. The small-world property is therefore sometimes accompanied by the condition that the clustering coefficient is bounded away from zero as $n$ increases. In real-world networks, where $n$ is fixed, a small-world network is defined as one in which $\ell$ is smaller and $C$ is much bigger than the values found in statistically equivalent random network models.

### 3.4 Centrality

The concept of the centrality of a network tries to answer the question of which is its most important node. Depending on the nature of the relationships and the process under study, centrality means different things. For example, *degree centrality* approximates the centrality of a node by its degree in the network $k_i$. In most situations, this is a highly effective measure of centrality. These highly connected nodes, usually called *hubs*, might have a major impact on the robustness of the network if they are removed or damaged.

However, hubs might be located in the periphery of the network. More global and advanced measures of centrality are *eigenvector*, *betweenness*, and *closeness centrality*, which take into account the relative position of one node with respect to the rest.

Eigenvector centrality relies on the idea of centrality propagation: the centrality of a node $x_i$ is a linear function of the centrality of its neighbors $x_i = (1/\lambda)\sum_j a_{ij}x_j$. Writing $\boldsymbol{x} = (x_1, \dots, x_N)$, we obtain an eigenvector equation $A\boldsymbol{x} = \lambda\boldsymbol{x}$. Among all the possible solutions of this equation, the one with the greatest eigenvalue has all entries positive (by the PERRON–FROBENIUS THEOREM [IV.10 §11.1]). The components of the related eigenvector are taken to be the centrality of each node. Several generalizations of this method are possible; for example, a variant of eigenvector centrality is Google's PAGERANK ALGORITHM [VI.9], used to rank the importance of Web pages, which in turn is also used in network analysis to measure the centrality of nodes. Some other centrality measures, such as Katz centrality, the Bonacich power, and the Estrada index, can be obtained as solutions of eigenvalue problems of FUNCTIONS OF THE ADJACENCY MATRIX [II.14].

Another well-known centrality measure is betweenness, introduced by Freeman in 1979, which is defined as the number of times node $i$ appears in the shortest paths between any pair of nodes in the network. Specifically, if $g_{jk}$ is the number of geodesic paths from $i$ to $j$, and if $g_{jik}$ is the number of these geodesics that pass through node $i$, then the *betweenness centrality* of node $i$ is given by $b_i = \sum_{j \neq k \neq i} g_{ikj}/g_{ij}$. The idea behind betweenness is that it measures the volume of flow through a node of a process that is happening in the network. Finally, closeness centrality (also introduced by Freeman) is defined as the average distance from one node to the rest of the network, i.e., $c_i = (1/n)\sum_j \delta(i,j)$.

Centrality is widely used to determine the key nodes in the network. For example, it is used to find influential individuals in online social networks, to identify leaders in organizations, as a method for selecting individuals to target in viral marketing campaigns, or to identify central airports in the air transportation network.

### 3.5 Communities

In most networks (see plate 5 and figure 2) there are groups of nodes that are more densely connected internally than with the rest of the network. This feature of a network is called its *community structure*. Community structure is common in many networks, and its determination yields a mesoscopic description of the network. But primarily it is interesting because it can reveal groups of vertices that share common properties or subgraphs that play different functional roles in the system. In fact, communities in social networks are found to be related to social, economic, or geographical groups; communities in metabolic networks might reflect functional groups; and communities in citation networks are related to research topics.

The hypothesis underneath these findings is that the network itself contains the information needed to reveal the groups and that the communities can be obtained using a graph-partitioning technique that assigns vertices to each group (see figure 2). Mathematically speaking, the problem of identifying graph communities is not well defined. To start with, there is no clear definition of what a community is in a graph. This ambiguity is the reason behind the wealth of algorithms in the literature, each of which implicitly assumes its own mathematical and/or statistical definition of communities and thus produces different partitions in the graph. On top of that, graph-partitioning problems are typically NP-HARD [I.4 §4.1] and their solutions are generally achieved using heuristics and/or approximation algorithms that might not deliver the exact solution or even the same approximate solution. Most of the time, therefore, we need further external information to validate the partition obtained.

Many different methods have been developed and employed for finding communities in graphs. Some of them rely on graph-partitioning techniques (such as the minimum-cut method) or data-clustering analysis (such as hierarchical clustering) borrowed from computer science. Some problems with these methods are

**Plate 1 (II.16).** Force-directed graph visualizations. A sample of forty-nine graphs from the University of Florida Sparse Matrix Collection. The color is determined by the length of the edges: short ones are red, medium-length edges are green, and long edges are blue.

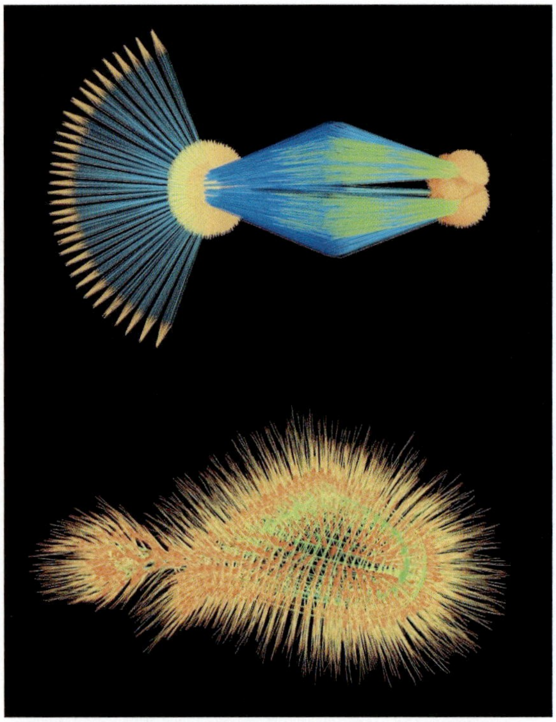

**Plate 2 (II.16).** Looking like Dr. Seuss's "red fish, blue fish," the top image is the graph from a constraint matrix for a linear programming problem, while the bottom image is the graph from a frequency-domain circuit simulation.

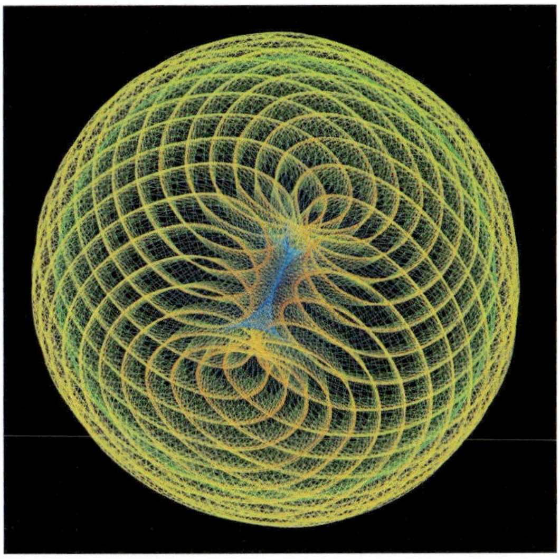

**Plate 3 (II.16).** The graph of a Hessian matrix from a convex quadratic programming problem.

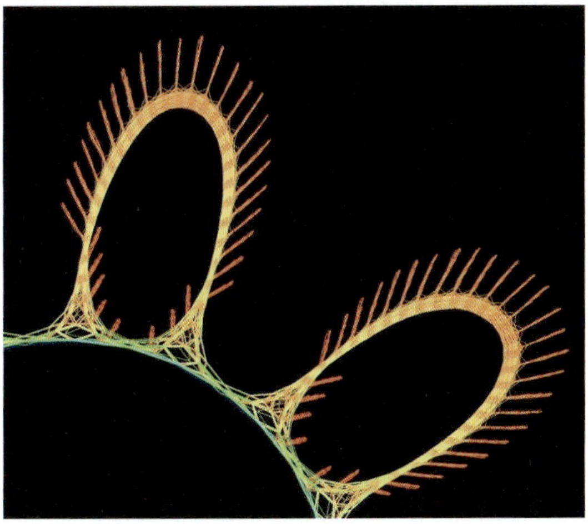

**Plate 4 (II.16).** A close-up of a graph from a financial portfolio optimization problem.

(a)

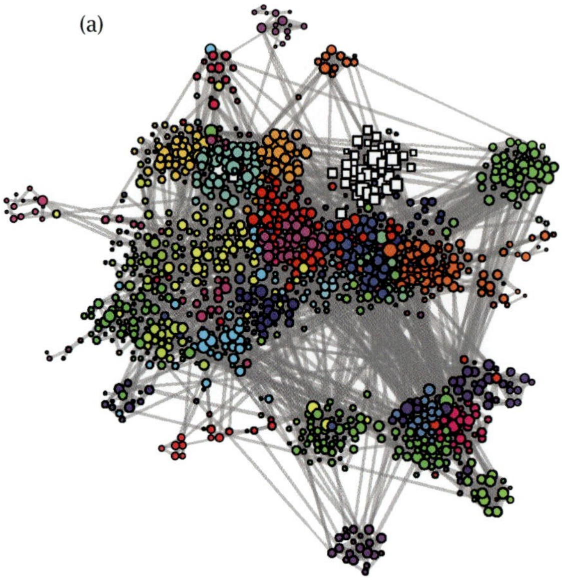

(b)

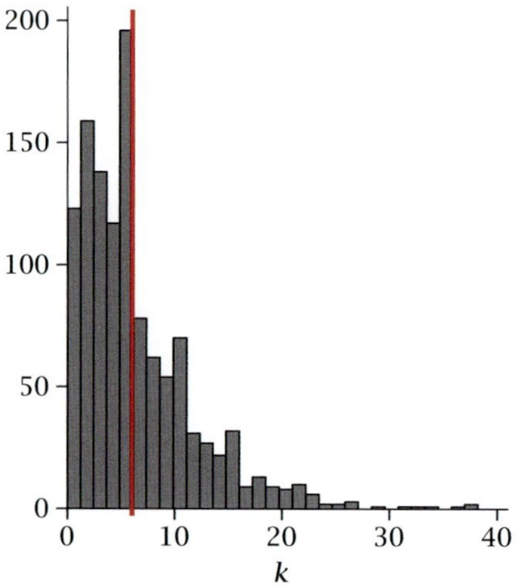


**Plate 5 (IV.18).** Network of email exchanges between academic staff at Universidad Carlos III de Madrid. The graph has 1178 nodes and 3830 links. Each link indicates that at least two emails were exchanged between those nodes, and the node colors correspond to the different departments within the institution. Node size and link width are log-proportional to their degree and weight (the number of emails exchanged), respectively. Square white nodes correspond to the mathematics department, which forms a dense community within the network. The two-dimensional layout was obtained using a force-directed graph-drawing algorithm. The network has average degree $\bar{k} = 6.5$, clustering coefficient $C = 0.21$, and diameter $\ell = 12$. (b) The distribution of connectivity $p_k$ for the network. The red vertical line corresponds to the value $k = \bar{k}$.

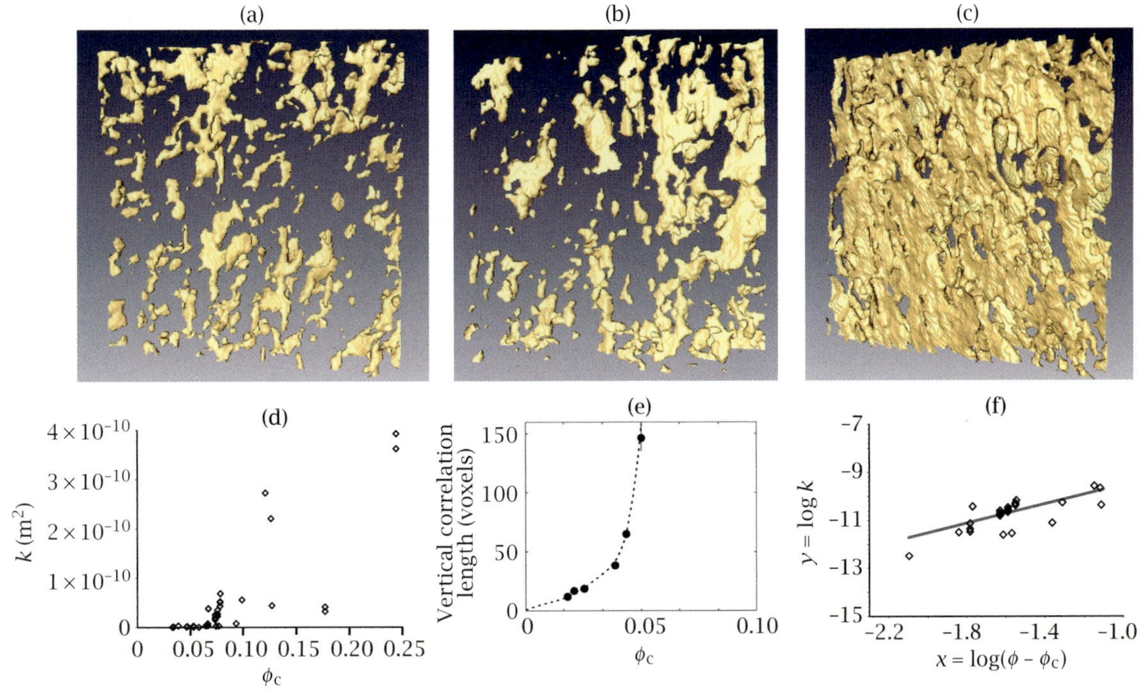


**Plate 6 (V.17).** An X-ray computed tomography volume rendering of brine layers within a lab-grown sea ice single crystal with $S = 9.3$ ppt. The (noncollocated) 8 mm × 8 mm × 2 mm subvolumes (a)–(c) illustrate a pronounced change in the microscale morphology and connectivity of the brine inclusions during warming ((a) $T = -15$ °C, $\phi = 0.033$; (b) $T = -6$ °C, $\phi = 0.075$; (c) $T = -3$ °C, $\phi = 0.143$). (d) Data for the vertical fluid permeability $k$ taken in situ on Arctic sea ice, displayed on a linear scale. (e) Divergence of the brine correlation length in the vertical direction as the percolation threshold is approached from below. (f) Comparison of Arctic permeability data in the critical regime (twenty-five data points) with percolation theory in (7). In logarithmic variables, the predicted line has the equation $y = 2x - 7.5$, while a best fit of the data yields $y = 2.07x - 7.45$, assuming $\phi_c = 0.05$. (Parts (a)–(d) are adapted from Golden, K. M., H. Eicken, A. L. Heaton, J. Miner, D. Pringle, and J. Zhu. 2007. Thermal evolution of permeability and microstructure in sea ice. *Geophysical Research Letters* 34:L16501. Copyright 2005 American Geophysical Union. Reprinted by permission of John Wiley & Sons, Inc.)

**Plate 7 (V.17).** Ocean swells propagating through a vast field of pancake ice in the Southern Ocean off the coast of East Antarctica (photo by K. M. Golden). These long waves do not "see" the individual floes, whose diameters are on the order of tens of centimeters. The bulk wave propagation characteristics are largely determined by the homogenized or effective rheological properties of the pancake/frazil conglomerate on the surface.

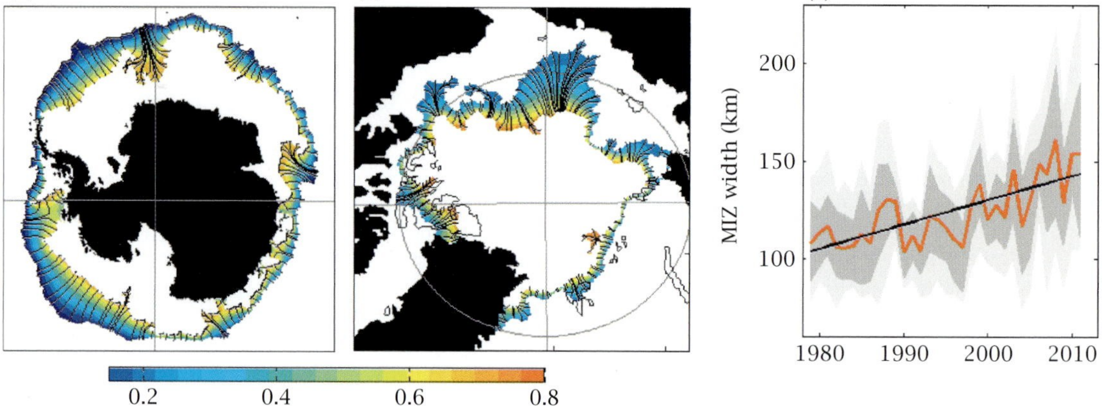


**Plate 8 (V.17).** (a) Shading shows the solution to Laplace's equation within the Antarctic MIZ ($\psi$) on August 26, 2010, and the black curves show MIZ width measurements following the gradient of $\psi$ (only a subset is shown for the sake of clarity) (courtesy of Courtenay Strong). (b) Same as (a) but for the Arctic MIZ on August 29, 2010. (c) Width of the July–September MIZ for 1979–2011 (red curve). Percentiles of daily MIZ widths are shaded dark gray (25th to 75th) and light gray (10th to 90th). Results are based on analysis of satellite-derived sea ice concentrations from the National Snow and Ice Data Center. (Parts (b) and (c) are adapted from Strong, C., and I. G. Rigor. 2013. Arctic marginal ice zone trending wider in summer and narrower in winter. *Geophysical Research Letters* 40(18):4864–68.)

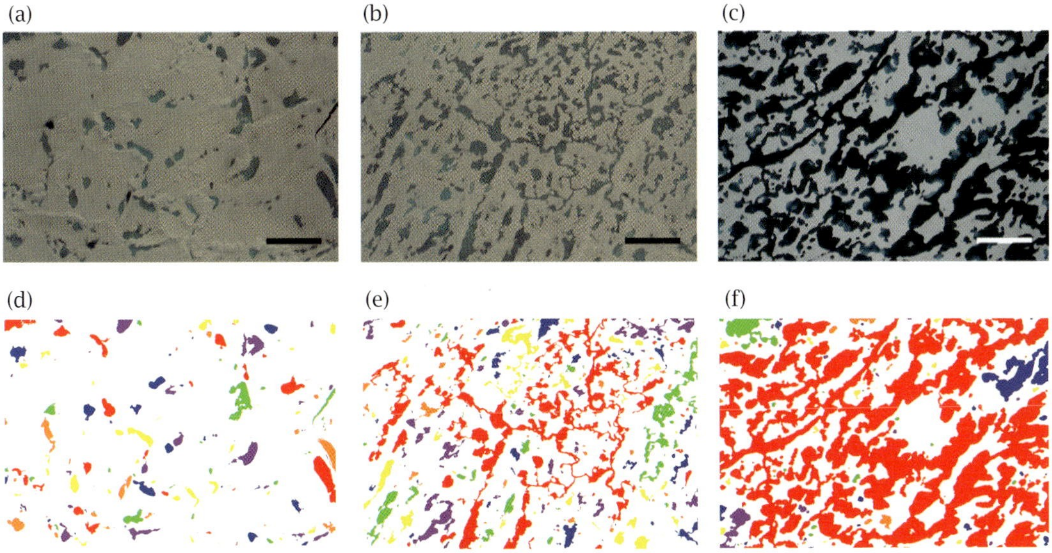


**Plate 9 (V.17).** The evolution of melt pond connectivity and color-coded connected components: (a) disconnected ponds, (b) transitional ponds, (c) fully connected melt ponds. The bottom row of figures shows the color-coded connected components for the corresponding image above: (d) no single color spans the image, (e) the red phase just spans the image, (f) the connected red phase dominates the image. The scale bars represent 200 m for (a) and (b), and 35 m for (c). (Adapted from Hohenegger, C., B. Alali, K. R. Steffen, D. K. Perovich, and K. M. Golden. 2012. Transition in the fractal geometry of Arctic melt ponds. *The Cryosphere* 6:1157–62 (doi:10.5194/tc-6-1157-2012).)

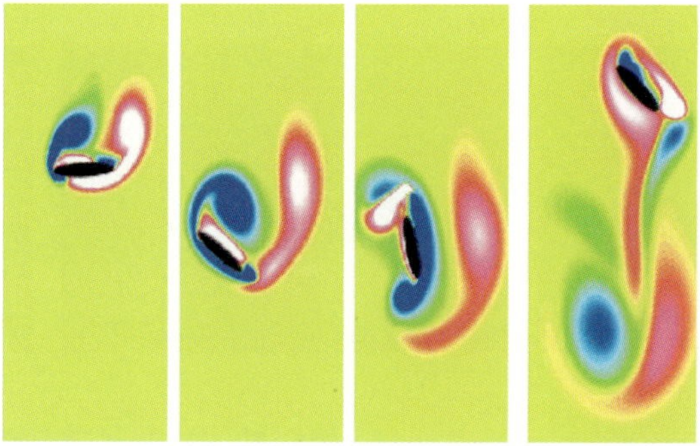

**Plate 10 (VI.5).** The flow generated by a two-dimensional flapping wing mimicking dragonfly wing motion. The colors indicate the vorticity field, with red and blue representing positive and negative vorticity, respectively. The wing motion creates a downward jet composed of counterrotating vortices. Each vortex pair can be viewed as the cross section of a donut-shaped vortex ring in three dimensions. From Z. J. Wang (2010), Two dimensional mechanism for insect hovering, *Physical Review Letters* 85(10):2216–19.

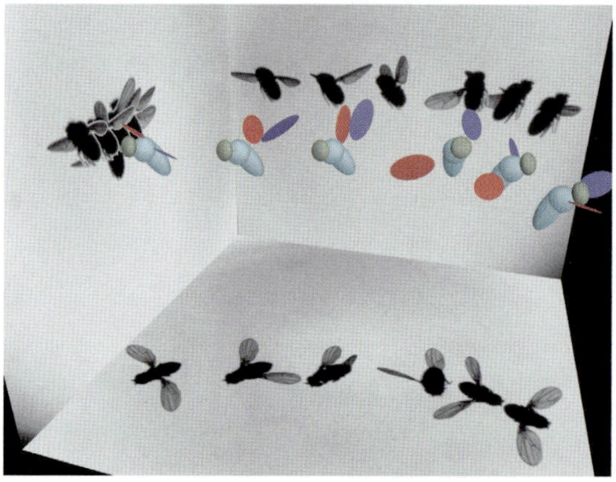

**Plate 11 (VI.5).** A fruit fly making a sharp yaw turn of 120° in about 20 wing beats, or 80 ms. The wing hinge acts as if it is a torsional spring. To adjust its wing motion, the wing hinge shifts the equilibrium position of the effective torsional spring, and this leads to a slight shift of the angle of attack of that wing. The asymmetry in the left and right wings creates a drag imbalance that causes the insect to turn. To turn 120°, the asymmetry in the wing angle of attack is only about 5° or so. From A. J. Bergou, L. Ristroph, J. Guckenheimer, I. Cohen, and Z. J. Wang (2010), Fruit flies modulate passive wing pitching to generate in-flight turns, *Physical Review Letters* 104:148101.

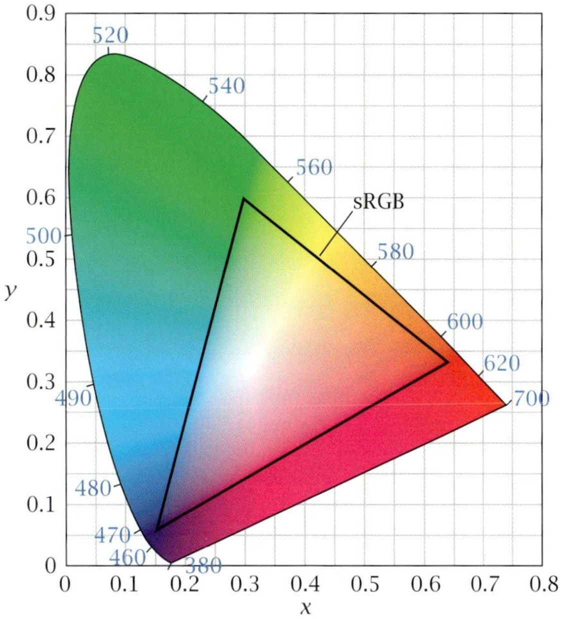


**Plate 12 (VII.7).** CIE 1931 color space chromaticity diagram, with the gamut of sRGB shown. (File adapted from an original on Wikimedia Commons.)

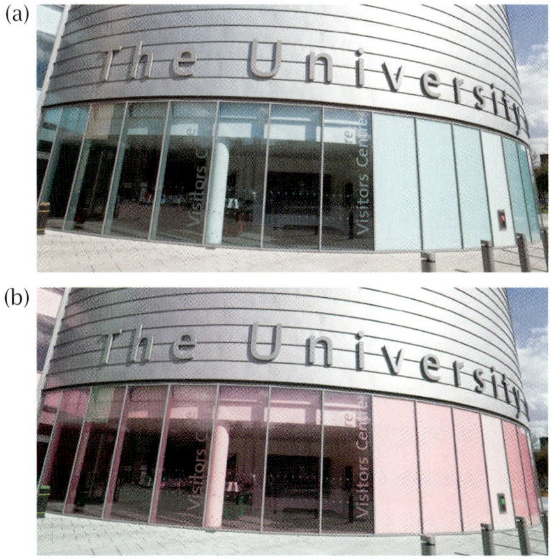


**Plate 13 (VII.7).** (a) Original image. (b) Image converted to LAB space and A channel negated $((L, A, B, ) \leftarrow (L, -A, B))$.

**Plate 14 (VII.8).** An inpainted image. (Courtesy of Bugeau, Bertalmio, Caselles, and Sapiro.)

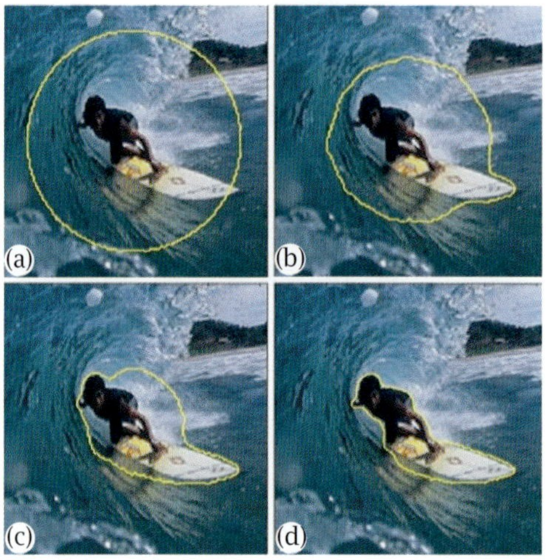


**Plate 15 (VII.8).** A contour (in green) evolving from the initial position in part (a) to the segmentation in part (d). The deformation (the two stages shown in parts (b) and (c)) is governed by a geometric partial differential equation. (Courtesy of Michailovich, Rathi, and Tannenbaum.)

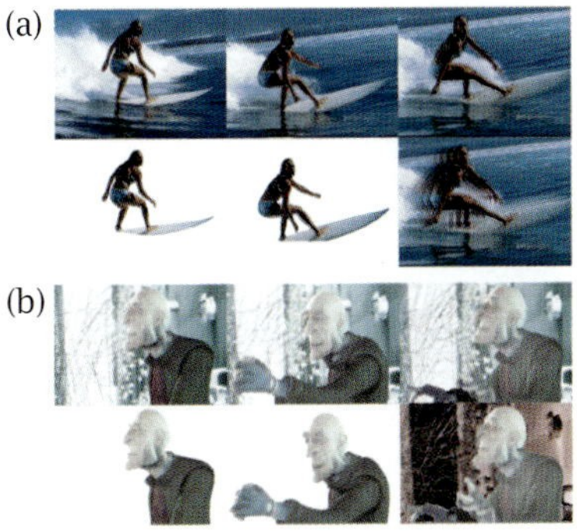


**Plate 16 (VII.8).** For each of the two examples, the subfigures in the top row correspond to three original frames, while those in the bottom row are (from left to right) two corresponding white-background composites and a (different) special effect. The special effects are (a) a delayed-fading foreground and (b) an inverted background. (Original videos courtesy of Artbeats (www.artbeats.com) and Getty Images (www.gettyimages.com).)

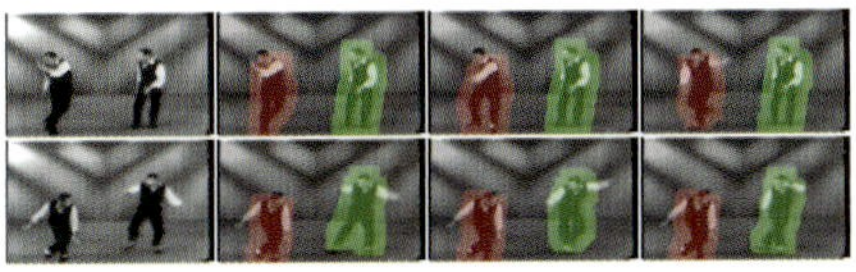

**Plate 17 (VII.8).** The first column shows two frames from a mimicking video, while the tracking/segmentation masks are displayed in different colors in the remaining columns. The two dancers are correctly detected as performing different actions. (Courtesy of Tang, Castrodad, Tepper, and Sapiro.)

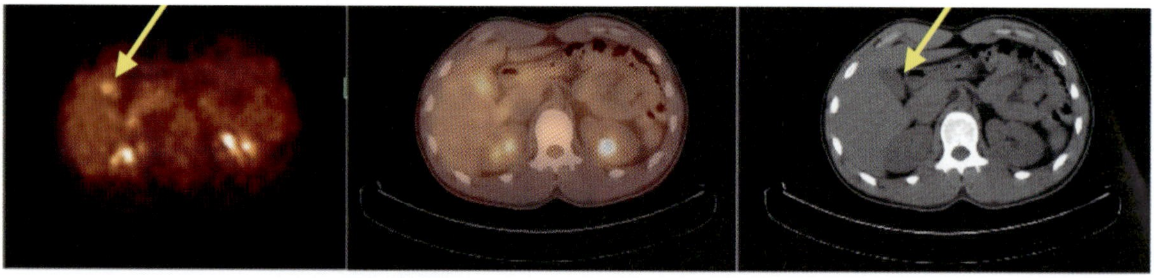

**Plate 18 (VII.9).** (a) A PET "heat map" image; (b) the image in (a) fused with the CT scan of the same section shown in (c). From the fused image it is apparent that the increased uptake of fluorodeoxyglucose, indicated by the yellow arrow, is in the gall bladder and is not the result of bowel activity. (Images courtesy of Dr. Joel Karp, Hospital of the University of Pennsylvania.)

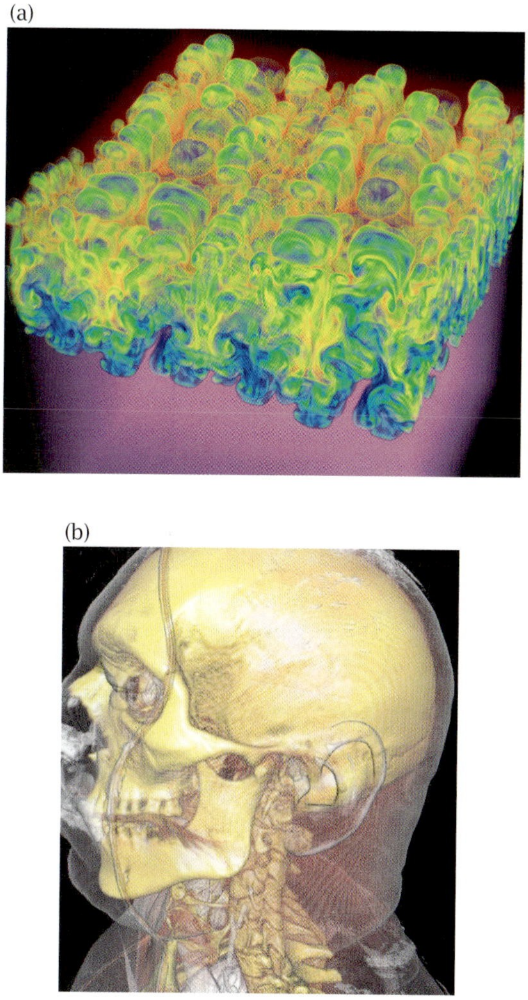

**Plate 19 (VII.13).** (a) Direct volume rendering of the instability of an interface between two fluids of different densities, termed the Rayleigh–Taylor instability. (b) Isosurfacing used to visualize the visible human data set (www.imagevis3D.org).

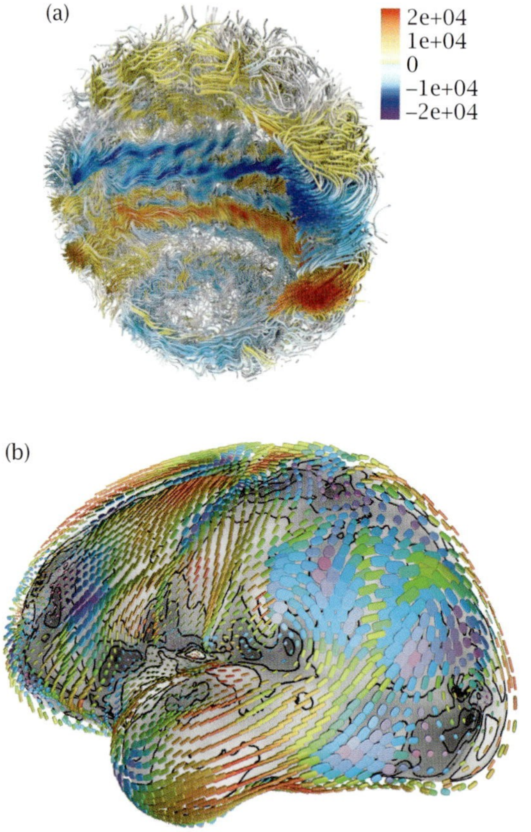


**Plate 20 (VII.13).** (a) Vector visualization of a stellar magnetic field using streamlines (Schott et al. 2012). (b) Glyph-based tensor visualization of anatomic covariance tensor fields (Kindlmann et al. 2004).

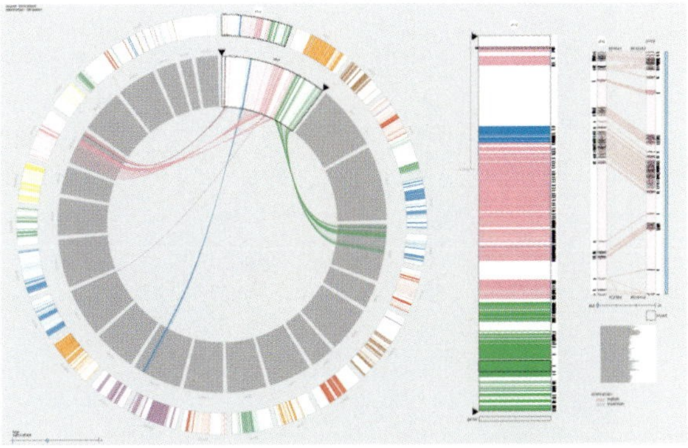

**Plate 21 (VII.13).** MizBee: a multiscale synteny browser for exploring comparative genomics data.

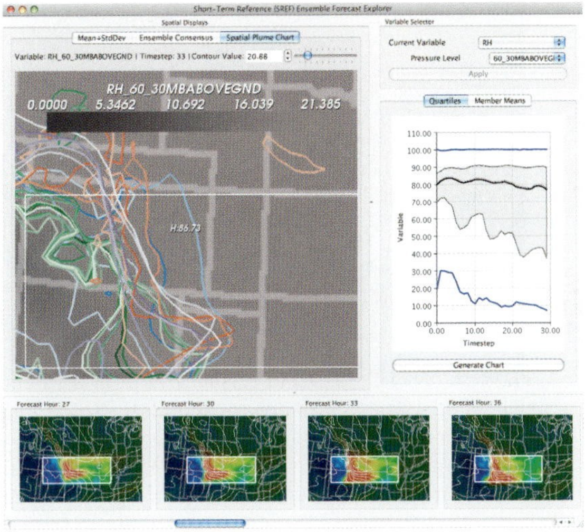


**Plate 22 (VII.13).** EnsembleVis: a framework designed for exploring short-term weather forecasts (Potter et al. 2009).

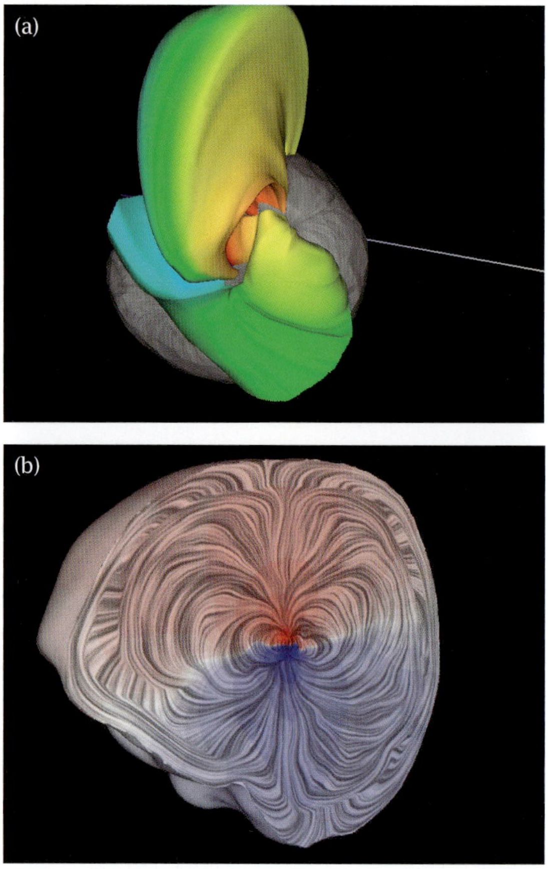


**Plate 23 (VII.13).** Visualizations of bioelectric fields in the heart and brain (Tricoche and Scheuermann 2003). (a) Stream surfaces show the bioelectric field in the direct vicinity of epicardium, or outer layer of the heart. (b) Textures applied across a cutting plane reveal details of a source of electric current in the brain and the interaction of the current with the surrounding tissue.

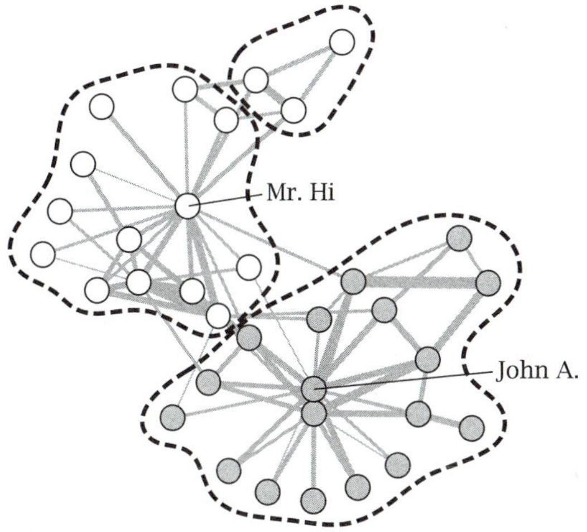


**Figure 2** A social network of friendships between thirty-four members of a karate club at a U.S. university in the 1970s. The club was led by president John A. and karate instructor Mr. Hi (pseudonyms). Edge width is proportional to the number of common activities the club members took part in. In 1997 Zachary studied how the club was split into two separate clubs (white and gray symbols) after long disputes between two factions within the club. A community-finding algorithm (label propagation) applied to the network before the split finds three communities (dashed lines) that have a large overlap with the factions after the split.

that the number of communities has to be given by the algorithm and/or there is no way to discriminate between the many possible partitions obtained by the procedure. To circumvent these issues a quality function of the partition has to be given. In a seminal paper, Newman and Girvan introduced the concept of the *modularity* of a partition, which measures the difference between the actual density of edges within communities and the fraction that would be expected if edges were distributed at random:

$$Q = \frac{1}{2m}\sum_{i,j}(a_{ij} - p_{ij})\gamma(c_i, c_j),$$

where the sum runs over all pairs of vertices, $\gamma(c_i, c_j) = 1$ if $i$ and $j$ are in the same community (zero otherwise), and $p_{ij}$ represents the expected number of edges between $i$ and $j$ in a null model, i.e., a model of a random graph with the same number of nodes and edges (see section 4). Given the large heterogeneity of degrees in the network, the most used model is the configuration model in which $p_{ij} = k_i k_j/(2m)$. Despite its limitations, modularity has had a great impact in community-finding algorithms. It gives a quantitative measure of the partition found, and it can therefore be employed to get the best (maximum-modularity) partition in divisive algorithms. In addition, modularity optimization is itself a popular method for community detection.

Other popular methods include clique-based methods such as the clique-percolation algorithm (which defines a community as percolation clusters of $k$-cliques), random-walk methods (where a community is a region of the graph in which a random walker spends a long time), and consensus algorithms (where a community is defined as a group of nodes that share common outcomes in a dynamical coordination process). Examples of consensus algorithms are label-propagation algorithms, in which nodes are given unique labels that are then updated by majority voting in the neighborhood of nodes. Labels reached asymptotically by this consensus process are taken as communities in the graph.

Community finding in networks is a computationally complex task. Typically, algorithm times scale with the number of nodes $n$ and edges $m$, and some methods are therefore not suitable for finding communities in very large networks. In recent years, however, much progress has been made in accelerating the algorithms, and it is possible to efficiently apply algorithms such as the Louvain method, Infomap, or the fast greedy algorithm of Clauset, Newman, and Moore to networks with millions of nodes.

## 4 Models

Building on the observations of real-world networks and their common properties found in many systems, one can create mathematical models of networks. Models are important for two reasons: the first is that simple null network models can be used to test the statistical significance of the results found in real-world applications. Of course, the very definition of null network models depends on the context and the process to be considered. On the other hand, modeling networks allows us to obtain good mathematical representations of the observed systems, which can then be used for further testing or even to make predictions about the future behavior of a process.

### 4.1 The Erdös–Rényi Model

The random graph model is the most used model type. In this technique, graphs are generated according to a probability distribution or a random process.

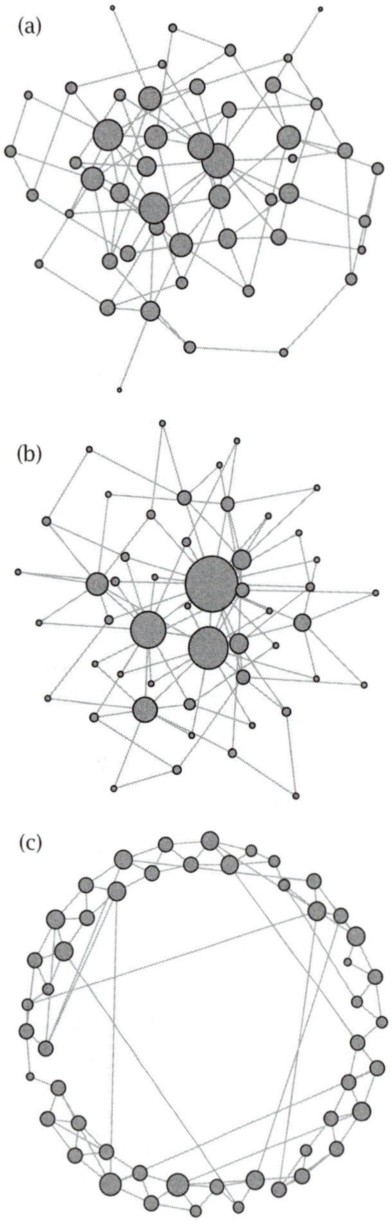


**Figure 3** Examples of networks generated by (a) the Erdös–Rényi model, (b) the preferential-attachment model, and (c) the small-world model. All networks have the same number of nodes $n = 50$ and edges $m = 100$. The size of a node is proportional to its connectivity $k_i$.

Here, we review some of the most popular examples in mathematical studies and applications.

In 1959 Paul Erdös, Alfréd Rényi, and (independently) Edgar Gilbert proposed a simple binomial random graph model, $G(n, p)$, in which the graph of $n$ nodes is constructed by connecting each pair of nodes with probability $p$ (see figure 3). The distribution of degrees in $G(n, p)$ is given by the binomial distribution, which becomes a Poisson distribution in the limit of large $n$:

$$p_k = \binom{n-1}{k} p^k (1-p)^{n-1-k} \approx \frac{\bar{k}^k \mathrm{e}^{-\bar{k}}}{k!},$$

where $\bar{k} = (n-1)p$ is the mean degree. Thus, $p_k$ does not have heavy tails. Despite this difference from real networks, many properties of $G(n, p)$ are exactly solvable in the limit of large $n$, and the random graph model has therefore attracted a lot of attention in the mathematics community. For example, the diameter is approximately given by $\ell \sim \log n / \log \bar{k}$, and thus $G(n, p)$ has the small-world property. On the other hand, the clustering coefficient is $C = p = \bar{k}/(n-1)$, which tends to zero in the limit of a large system (for finite $\bar{k}$), unlike real-world networks where $C$ is finite even for large $n$. Obviously, $G(n, p)$ does not have any community structure either. Nonetheless, $G(n, p)$ can be used as a zero-information model, providing a benchmark for comparison with other models and data.

### 4.2 The Configuration Model

To allow for non-Poisson degree distributions, we can generalize the Erdös–Rényi model to the configuration model, which was first given in its simple explicit form by Bollobás. In this model the degree distribution $p_k$ or degree sequence $k_1, \dots, k_n$ is given and a random graph is formed with that particular distribution or degree sequence. A simple algorithm to generate such random graphs is to give each vertex a number of "stubs" of edges, either according to the distribution or from the degree sequence, and then pick stubs in pairs uniformly at random and connect them to form edges. The ensemble of graphs produced in this way is called the *configuration model*. Many properties of the configuration model are known. Molloy and Reed showed that there is almost surely a giant component if $\overline{k^2} - 2\bar{k} > 0$ (where $\overline{k^2} = \sum_k k^2 p_k$), and that the probability of finding a loop on the graph decays like $n^{-1}$, i.e., the graph has a local tree-like structure for large $n$. This was used by Newman to show in a simple way that the clustering coefficient decays asymptotically like $C \sim n^{-1}$.

One of the most important properties of the configuration model is that the distribution $q_k$ of degrees of nodes obtained following a randomly chosen edge is not given by $p_k$. Instead, since in choosing a random edge there are $k$ ways to get to a node with connectivity $k$, we obtain that $q_k \sim kp_k$. The average degree of (next-nearest) neighbors of a node is therefore

$$\overline{k_{\mathrm{nn}}} = \sum_k kq_k \sim \frac{\overline{k^2}}{\bar{k}}.$$

Since real-world networks are very heterogeneous, we usually find that $\overline{k^2} > \bar{k}$, and the neighbors of a randomly chosen node therefore have more connectivity than the node itself. This fact has a key role in many processes that happen in networks (see section 6). In social networks it is known as the *friendship paradox*, posed by the sociologist Feld in 1991: on average, your friends have more friends than you do. It also has other direct consequences; for example, in scale-free networks with $2 < \alpha < 3$, we find that the diameter $\ell \sim \log\log N$, and, thus, scale-free networks are ultrasmall. For $\alpha > 3$ we recover the small-world behavior $\ell \sim \log N$. Finally, since edges are distributed randomly among nodes, the probability that two nodes with degrees $k_i$ and $k_j$ are connected is given by $k_i k_j/(2m)$, a result used in the definition of modularity $Q$ as a null model approximation for $p_{ij}$.

## 5 Other Random Models

Configuration models can be generalized to *exponential random graphs*, also called $p^*$ *models*. These are ensembles of graphs in which the probability of observing a graph $G$ with a given set of properties $\{x_i\}$ is $P(G) = \mathrm{e}^{H(G)}/Z$, where $Z = \sum_G \mathrm{e}^{H(G)}$ and the network *Hamiltonian* is given by $H(G) = \sum_i \theta_i x_i(G)$, with $\{\theta_i\}$ the ensemble parameters. For example, we could take $x_1$ to be the number of edges, $x_2$ to be the number of vertices with a given degree, $x_3$ to be the number of triangles in the network, and so on. Much of the progress made in this field has come through the use of Monte Carlo simulations of the ensemble and/or by using real network data to estimate the parameters of the model.

Another random graph model is the family of *stochastic block models*, in which nodes are assigned to $s$ different blocks and vertices are placed randomly between nodes of different blocks with a probability that depends only on the blocks of the nodes. Specifically, if $z_i$ denotes the block that the vertex $i$ belongs to, then we can define an $s \times s$ stochastic block matrix $M$, where $m_{ij}$ gives the probability that a vertex of type $z_i$ is connected to a node of group $z_j$. Blocks can be groups of nodes that have similar structural equivalence, have similar demographics, or belong to the same community.

### 5.1 The Small-World Model

The small-world model was introduced by Watts and Strogatz as a random graph that has two independent structural properties of real networks: a finite clustering coefficient, and the small-world property $\ell \sim \log n$ when $n$ increases. The basic idea of this model is to build a graph embedded in a one-dimensional lattice in which nodes are connected to a local neighborhood of size $d$ and then the edges are "rewired" with probability $p$ (see figure 3). When $p = 0$ we recover a one-dimensional regular graph, while when $p = 1$ we recover a random graph. The local tight neighborhood provides the finite clustering property, while the long-range rewired links are responsible for the low diameter of the network. Specifically, as $n \to \infty$ the clustering coefficient is $C \sim 3(d-1)/[2(2d-1)](1-p)^3$ and the diameter scales as $n$ and $\log n$ for $p = 0$ and $p = 1$, respectively. One interesting feature of the Watts–Strogatz model is that it interpolates between a regular graph and a completely random graph. A major limitation of the model is that it produces unrealistic degree distributions ($p_k$ decays exponentially). Many variations of the small-world model have been proposed and studied.

### 5.2 The Preferential-Attachment Model

While the models above incorporate the observed macrofeatures of networks, generative models try to explain those features as an emergent property that is due to microscopic mechanisms. A particularly popular class of models is the network growth model, in which the dynamics of node and edge creation create the observed network properties. Probably the best-known example is the so-called *preferential-attachment model* of Barabási and Albert, which aims to explain the scale-free property of the degree distribution. This is done using the rich-get-richer mechanism in edge creation: nodes are added to the network with a certain number $m$ of edges emerging from them, and those edges are connected to preexisting nodes with connectivity $k$ with probability $\pi_k$. This preferential-attachment mechanism has been observed in many growing networks, from citation and collaboration networks to the Internet and online social networks. Simple rate equations can be written for the evolution of the system, and we find that the degree of each node grows according to a power law $k_i(t) \sim t^{\beta}$ and that for all $t$ the degree distribution is scale free $p_k \sim k^{-\alpha}$, where the exponents $\alpha$, $\beta$ depend on the preferential-attachment probability. In the linear case, where $\pi_k = k$, we obtain that $\alpha = 3$, a value that is similar to those found in real-world networks such as the World Wide Web.

The preferential-attachment model has the small-world property with a double logarithmic correction $\ell \sim \log n / \log\log n$, and numerical results suggest that the cluster coefficient $C$ behaves as $C \sim n^{-0.75}$. Although $C$ vanishes in the limit $n \to \infty$, it can have large values for finite $n$. One of the main criticisms of the preferential-attachment mechanism is that it requires global information about the network, which is clearly an unrealistic assumption in many situations. This and other criticisms have been addressed by modifying the original model using local network formation rules, triadic closure, or finite memory of nodes.

## 6 Processes

Beyond the metrics and models presented in the previous sections, much of our knowledge about networks comes from our ability to explain processes happening on the network by analyzing its structure. For example, we would like to understand whether the node with the largest centrality is also the fastest to become aware of information spreading in the network. Or perhaps we want to know what the relationship is between an Internet network's degree of heterogeneity and its resilience to targeted attacks. Progress in this direction comes from simulating simple models of these processes on real or synthetic models of networks and from the availability of data that is empowering (and sometimes challenging) our understanding of how networks work.

### 6.1 Spreading

Probably the most-studied process in networks is how things spread around them. In particular, a wealth of work has been done on understanding how network structure (and dynamics) impacts the spreading of information, viruses, diseases, rumors, etc. For example, people have studied how the structure of networks of sexual contacts (or the network of flows of passengers between airports and cities) influences the spread of diseases; others have studied the best way of choosing the people who are initially targeted in a viral marketing campaign on a social network in order to optimize the reach and velocity of the campaign.

Spreading models are typically borrowed from epidemiology: diseases spread to susceptible (noninfected) (S) nodes when they are exposed to infected (I) nodes, and then they can decay into the recovered (R) state. This is the so-called SIR MODEL [V.16], which has a long history in mathematical epidemiology. Although the model is dynamical in nature, Grassberger found that it could be mapped exactly onto bond percolation in the network: outbreaks in the spreading process correspond to clusters in the percolation problem. Percolation is a well-known problem in mathematics; in its bond version it describes the behavior of connected clusters in a random graph in which edges are occupied with probability $\lambda$. The question is then whether there is a cluster that "percolates" the whole network, i.e., a connected cluster of a fraction of the network, the so-called giant component. This would correspond to a large disease outbreak in the spreading problem. In most systems there exists a critical $\lambda_c$ at which the percolation transition, or *epidemic breakout* in epidemiology, happens for $\lambda > \lambda_c$. In viral marketing we would like our campaigns to operate above $\lambda_c$, so the spreading "goes viral," while in disease spreading, vaccination and health policies are designed to maintain $\lambda < \lambda_c$. Although $\lambda_c$ depends on the virulence of the disease, it is also affected by the structure of the network in which disease propagates. For example, in the configuration model with degree distribution $p_k$, starting from one infected initial seed we have, on average, that $R_0 = \lambda\bar{k}$ of its neighbors are infected. Each of these $R_0$ infected neighbors goes on to infect an average of $R_1 = \lambda(\overline{k_{\mathrm{nn}}} - 1)$ new next-nearest neighbors. The same happens in the following steps. The size of the outbreak is therefore given by

$$\bar{s} = 1 + R_0 + R_0 R_1 + R_0 R_1^2 + \cdots = 1 + \frac{R_0}{1 - R_1}. \quad (1)$$

Thus, outbreak size diverges as $R_1 \to 1$, that is, when $\lambda \to \lambda_c = 1/(\overline{k_{\mathrm{nn}}} - 1) = \bar{k}/(\overline{k^2} - \bar{k})$. This is actually the Molloy and Reed criterion for the existence of a giant component applied to the corresponding configuration model in which edges are occupied with probability $\lambda$. Since heterogeneity in networks usually implies that $\overline{k^2} \gg \bar{k}$, the epidemic threshold is always very small. Moreover, in networks that are scale free with exponent $\alpha \leqslant 3$, we have that $\overline{k^2}$ diverges. This implies that in highly heterogeneous networks, the critical point vanishes and information or disease spreads all over the network, an interesting result of Pastor-Satorras and Vespignani that has been suggested as an explanation for the prevalence of computer viruses. Although this result was obtained for a very particular epidemic model in the configurational model, the general understanding is that network heterogeneity favors spreading. Indeed, $R_0$ and $R_1$ are widely used in epidemiology to assess epidemic outbreak, where they are known as the *basic reproduction numbers.* Away from the configuration model, epidemic spreading in real networks is controlled by the first eigenvalue $\sigma_1$ of the adjacency matrix, so that $\lambda_c = 1/\sigma_1$. Furthermore, network clustering and its community structure tends to reduce the spread of infection, since spreading gets trapped in densely connected areas of the network.

Two related questions are finding the nodes in the network that spread most efficiently and, conversely, finding the ones that are affected by the spreading process earliest. Both of these questions are related to centrality in the network: the more central a node is, the larger the outbreak cluster that can generate from it, and also the faster it becomes aware of the spreading process. This result allows us to strategically target initial spreaders in information spreading or to vaccinate more central people in social networks to stop epidemics. But it also provides a way of choosing a set of central nodes (*sensors*) in the network that can help us detect epidemic outbreaks or information spreading as quickly as possible. For example, Fowler and Christakis and collaborators have found that a set of sensors is highly effective in detecting outbreaks in information and disease spreading, even in massive networks.

### 6.2 Contagion

Closely related to information diffusion, yet mechanistically different, is behavior contagion, or how exposure to certain individual behavioral characteristics can drive the propagation of those characteristics from individual to individual in a network. The importance of social-network structure in promoting positive or arresting damaging behavior has recently been studied in many contexts, ranging from spreading of health behaviors, to product/service adoption, to political opinion, to participation in time-critical events, to the diffusion of innovations, and so on. Taken together, these studies hint at some generalities regarding behavior contagion: it critically depends on the structural diversity of the network (i.e., from how many different social communities the behavior is exposed) and on social reinforcement (i.e., how additional exposures change the probability of adoption and how clustering in networks promotes it). However, recent work has highlighted the problem that most of the observed causal influence in social networks could be chiefly due to latent homophily or other assortative confounder variables. For example, an individual might buy a particular product not because she is influenced by the network surrounding her but because she belongs to a group to which that product is appealing. A more refined statistical network would be needed to establish causality in behavior contagion.

### 6.3 Robustness

In many applications, networks should be robust to small topological or dynamical perturbations. For example, communication networks, power grids, or organizational business processes should be resilient to intentional attacks, random failures in stations, or organizational changes, respectively. In the most simple approximation network, robustness is studied under a random or intentional removal of nodes and/or edges. The question is at which point in this removal process does the network stop operating as intended, e.g., when a significant fraction of nodes in communication networks can no longer communicate with each other. This obviously happens when there is no path between those nodes, and in that case the network becomes disconnected with a number of small components. In this form, the robustness problem is then very similar to percolation, where nodes or edges are unoccupied when they are attacked or they fail. Depending on the removal strategy there will be a critical fraction $q_c$ such that when $q$ nodes are removed with $q > q_c$ the network becomes disconnected, while the removal of $q < q_c$ nodes leaves a large fraction of nodes in a connected component (the *giant component*). For example, if a fraction $q$ of randomly chosen nodes are removed from the configuration model with degree distribution $p_k$, then each node has $(1-q)\bar{k}$ neighbors on average and each of the neighbors will itself have $(1-q)\overline{k_{\text{nn}}}$ neighbors. Thus, starting from a node that has not been removed we can form a connected component that has average size given by equation (1), where $R_0 = (1-q)\bar{k}$ and $R_1 = (1-q)(\overline{k_{\text{nn}}} - 1)$, recovering the Molloy–Reed criterion that $q_c \approx 1 - \bar{k}/\overline{k^2}$. This implies that highly heterogeneous networks (where $\overline{k^2} \gg \bar{k}$) are very robust against random attacks. Conversely, given the structural importance (centrality) of hubs or bridges, an intentional attack to remove those nodes could disconnect the network very easily. Thus, although the heterogeneous and modular structure of networks makes them resilient against random attacks, the central role of hubs and bridges between communities makes networks very fragile when it comes to targeted attacks against them.

In some situations failures can propagate in the network like avalanches. This is the case in power transmission grids or the Internet, where the removal or failure of nodes changes the balances of flows and could lead to high loads in other nodes and their possible failure. Another example is cascades of bank failures in financial networks. In this case, the robustness of the network depends both on its topological structure and on how the capacities/tolerance of nodes are distributed over it. Finally, in some technological problems networks are interconnected; for example, in Italy's major 2003 blackout it was found that the coupling between the power grid network and the communication network caused further breakdown of power stations. The effect of heterogeneity in random attacks is the opposite in interconnected networks from the effect in simple networks: a broader distribution of degree increases the vulnerability of interconnected networks.

### 6.4 Consensus

Repeated information sharing and behavior contagion can lead to the formation of consensus or synchronization in the network. Examples of consensus formation in networks are the problems of opinion formation in society, synchronization of biological neural networks, or protocols in distributed computing. Simple models of consensus formation are the well-known mathematical problems of the voter model or the Kuramoto model of coupled oscillators. Important questions in consensus processes are whether the network structure favors the appearance of consensus and its impact on the time to reach it. The answer to the first question depends on the way in which models or experiments are engineered. For example, in the voter model, consensus is not reached in general in heterogeneous networks, although finite-size fluctuations induce consensus after a time $\tau \sim n$. Since most consensus problems can be written in a diffusion-like framework, the time to consensus can be obtained from the spectral properties of the graph. For example, consider the simple local average consensus problem

$$\dot{x}_i = \sum_{j \in \mathcal{N}_i} (x_j(t) - x_i(t)),$$

where the $\mathcal{N}_i$ are the neighbors of node $i$, and $x_i(t)$ is the state of node $i$. The collective dynamics of the agents can be written as $\dot{\boldsymbol{x}} = -L\boldsymbol{x}$, where $L$ is the Laplacian matrix $[L]_{ij} = k_i\delta_{ij} - a_{ij}$. Since $L$ always has a zero eigenvalue $\sigma_1 = 0$, the timescale in the consensus problem is given by the second smallest eigenvalue $\sigma_2$, which is usually called the *algebraic connectivity*. The time to achieve consensus is then given by $\tau \sim 1/\sigma_2$. In general, heterogeneity has a large influence on $\tau$: while in the voter model larger heterogeneity (more hubs) favors consensus, synchronization of oscillators is more difficult in highly heterogeneous networks. On the other hand, it appears that community structure has a large influence in reaching consensus. In fact, this is the main idea behind some community-finding algorithms that are based on consensus formation, such as the label-propagation algorithm (see section 3.5)

### 6.5 Network Prediction/Inference

Networks contain explicit information about the nodes they contain and how they interact, but they might also contain implicit information about interactions that might happen in the future or about the possibility that a given node, edge, or subgraph disappears from the network after some time. For example, in a social context, the fact that two people share a large number of friends indicates that those two people might be friends as well. This kind of analysis is not only useful in predicting how a network might evolve, but it also helps us to make inferences about links that are unobserved or are missing from the data. Most of the processes of link formation and decay can be predicted by looking at the local neighborhood or community of the link: nonconnected nodes that have structural similarity tend to become connected. The simplest similarity measure is the number of shared neighbors in the network (or the embeddedness of an edge), but other dyadic or neighborhood measures have been proposed. Conversely, studying four years' worth of banker relationships in a large organization, Burt found that edges between nodes that are very similar do not decay easily, and therefore network evolution happens mostly at the bridge positions between communities, where nodes are structurally different.

## 7 Applications

It would be impossible to cover all the applications of network analysis here. Although it has long been used in the social sciences, the last few decades have seen many technological, engineering, economic, and biological problems being studied using network analysis. Here we review some of those applications.

### 7.1 Social Networks

The analysis of social networks has a long tradition, and in fact many of the techniques and ideas in network analysis are derived from sociology. Social networks are defined by a set of actors (persons, organizations, etc.) and edges that represent relationships between them (friendships, business relationships, etc.). Traditional social network analysis was devoted to explaining the role of network structures, dyadic properties, or network positions in social processes. At the level of the network, for example, the connectedness or cohesion of a group or of the network itself could be an explanatory variable in consensus formation, shared norms, etc. Dyadic properties of social networks refer to concepts like distance between nodes, structural equivalence (sharing the same relationships to all other nodes), reciprocity (the tendency of nodes to form mutual connections), etc., that can be used to find structural equivalence classes, that is, types of actors in the network. Regarding network positions, the most studied concept is that of the use of centrality to determine the most important actor in a social network with respect to information sharing, economic opportunities, prestige, and so on.

Different social positions also lead to different network roles; for example, bridges are nodes that appear between communities and thus have a large centrality (see figure 4). Bridges are surrounded by structural holes (i.e., the lack of a connection between two nodes in communities connected by a bridge), and thus only bridges can access information from different sources and communities and benefit from their position in the network. The potential benefits that their position in a network could yield to a group or individual are known as social capital theory (introduced by Burt in the 1990s). Intimately related to the concept of bridges and structural holes is Granovetter's theory of weak ties: if strong ties are associated with intimate and intense relationships, Granovetter's theory is that weak ties are associated with bridges, that is, that our strongest edges happen within communities (see figure 4). Weak ties therefore enable people to reach information that is not accessible via strong ties. Granovetter used a survey of job seekers to prove this; he asked people who had found a job through contacts

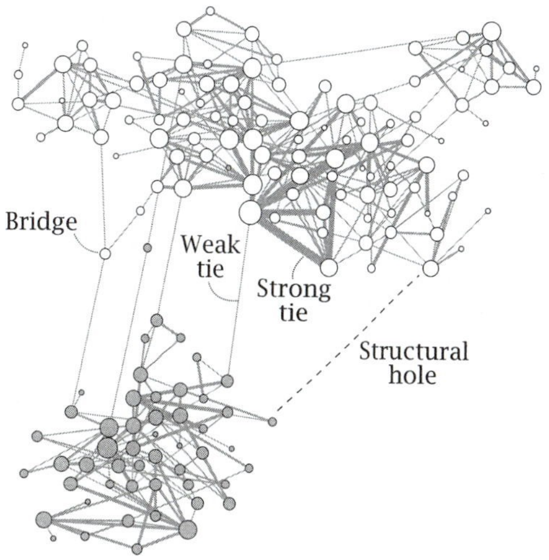


**Figure 4** The network of email communication between two departments (white and gray symbols) from plate 5. Most communication happens within departments, and some individuals have special roles as bridges between them. Also, there are a large number of structural holes in the network; weak ties happen between the departments while strong ties occur between people in the same department (the Granovetter hypothesis).

how often they saw the person who had helped them get their job. The majority were acquaintances rather than close friends. In more recent work, Onnela and collaborators measured these social theories quantitatively; using a large graph of mobile phone communications, they found that people with weak ties (those with a smaller number of calls) tend to have larger betweenness than those with strong ties.

Much work has been devoted to understanding when and why information propagates in social networks. Understanding how fast and how far information spreads is important in viral marketing, social mobilizations, innovation spreading, and computer viruses, for example. Although information spreading is affected by many factors, including the value of the content transmitted and exogenous campaigns, the very network structure shapes the speed and reach of information spreading. As we show in section 6, it is possible to study how the structural properties of the network—such as heterogeneity, clustering, and communities—influence the spreading, and to study what structural properties (centrality, degree, $k$-coreness) of an initial set of nodes are needed to maximize the spreading in the network.

Although behavior contagion is difficult to assess in many contexts, network analysis is widely used in marketing and social media to detect and predict changes of behavior. For example, it is used in the telecommunication industry to predict churn or product/service adoption using the social network of a company's clients obtained through the calls placed between them. The basic idea is that behavior like churning or adoption has a viral component and propagates on the network. Thus, the probability to churn or buy a particular product or service depends on the number of neighbors in the social network that have already done so. The analysis of the social network of clients permits one to identify and target those individuals most likely to make a purchase.

Disease spreading is also influenced by an infected person's social network. Here, the approach is different depending on the scale at which spreading is being studied. At small scale, the study of networks of sexual contacts can help to understand and control the spread of sexually transmitted diseases. These networks show high heterogeneity, modular structure, and small-world phenomena—properties that promote the spread of the disease across the network. In turn, since highly heterogeneous networks are susceptible to intentional attacks, targeted vaccination of super-spreaders might be enough to prevent an epidemic, a result that reinforces standard public-health guidelines. At a large scale, we can consider metapopulation network models, in which nodes are populations (groups, patches, cities) and edges account for the probability of transmission between populations. In these examples, human mobility networks play a major role. For example, the worldwide airport transportation network is highly connected; it is a small world, and its structure therefore favors epidemics on a global scale (see figure 5). In 2003 the SARS outbreak took just a few weeks to spread from Hong Kong to thirty-seven countries. Analysis of transportation networks allows us to predict which airports or hubs will be most likely to promote aggressive spatial spreading.

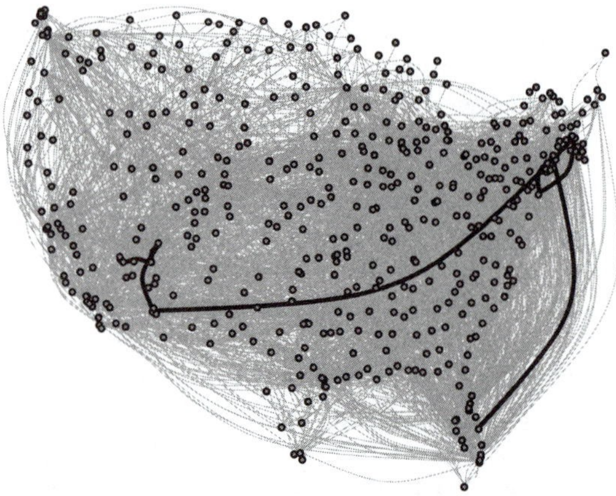

**Figure 5** A network of passenger flows among U.S. airports. The network has 489 airports and 5609 different routes between them. The diameter of the network is only 8. The black line shows one of the largest geodesics in the network: that between the airports of Grand Canyon West, AZ, and Fort Pierce, FL.

### 7.2 Economics and Finance

The complexity of interdependencies between different agents, financial instruments, traders, etc., can be also studied using network analysis. In these studies, the main aim is to understand how the network structure impacts the performance of institutions or economies and also the role it plays in economic risks and their possible mitigation. For example, the 2008 financial crisis has shown that systemic risk can propagate rapidly between interconnected financial structures. A potentially vulnerable market for contagion of financial shocks is the interbank loan market, where banks exchange large amounts of capital for short durations to accommodate temporal fluctuations. Network analysis has shown the important role of the topology of that market in the systemic risk of the system. Specifically, it has been found that contagion of bank failures can be promoted by the heterogeneity and density of the network, and the fragility of the system to external shocks has also been demonstrated. The impact of the structure of financial networks and global markets on their stability has attracted the interest of regulators and central bankers in using network analysis to evaluate systemic risk.

Network analysis is also used to map how organizational environments affect an organization's performance, that is, how market transactions, contracts, mergers, interlocking board directorates, or strategic alliances shape organizations, create innovation, or define the future performance of companies in a particular sector. For example, in 1994 Saxenian hypothesized that the dramatic growth of Silicon Valley in the previous decades could be explained, in part, by the cooperative and informal exchange of information among the organizations in the area. At the world level, the study of the network of international trade allows us to analyze indirect trade interactions between countries, the effect of globalization on the world economy, or the cascade effect of financial crises in different countries.

### 7.3 Biology

Many biological problems are intrinsically complex; they include the interaction of a large number of molecules, cells, individuals, or species. Mapping all this information into a network allows us to analyze the common structure of those problems and to develop tools that may prove useful for applications in a wide variety of biological problems.

For example, new noninvasive techniques for measuring brain structure and activity, such as neuroimaging (e.g., MAGNETIC RESONANCE IMAGING [VII.10 §4.1] (MRI), functional MRI, and diffusion tensor imaging) and neurophysiological recordings (e.g., electroencephalography, magnetoencephalography), have allowed us to collect large amounts of spatiotemporal data about brain structure and activity. Analyzing the anatomical and/or functional connectivity of different areas of the brain, researchers have constructed network models of the brain. Brain networks (*connectome*) seem to be organized in communities or modules, have the small-world property, and also present highly influential (hubs) nodes; they even seem to be scale free. Those communities appear to coincide with known cognitive networks or function subdivisions of the human brain, while the small diameter of the network seems to allow for efficient information processing.

The recent development of high-throughput techniques in molecular biology has led to an unprecedented amount of data about the molecular interactions that occur in biological organisms, e.g., the metabolic networks of biochemical reactions between metabolic substrates, the interaction networks between proteins (*interactome*), and the regulatory networks that represent the interactions between genes. A knowledge of the topologies of complex biological networks and their impact on biological processes is needed not only to understand those processes but also to develop more effective treatment strategies for complex diseases. Most of the available analyses are concerned with the application of the concept of centrality, which allows us to determine the *essential* protein or gene (or groups of them) in the organism and then apply the results to drug target identification.

### 7.4 Other Applications

Network analysis has expanded into many other areas. For example, it is used in sports to understand the style of play and the performance of teams. In football, by analyzing the network of passes between players it was found that a large clustering coefficient, or the diversity of the distribution of passes (entropy), correlates with the performance of the Spanish team that won the 2010 FIFA World Cup. Similar analysis has been done for basketball in the 2010 NBA Playoffs. Another interesting and recent application has been to map and understand the networks of cuisines, recipes, and ingredients in cooking. The availability of online recipe repositories has allowed researchers to find similarities among different regional cuisines in China, and to unveil the flavor network in culinary ingredients.

Since networks shape the information we receive and influence our behavior, a major concern is how much of our private life is encoded in the network structures and dynamics around us. For example, the information we leave behind in social networks can be used for identity disclosure or to unveil private personal traits. Network analysis can be used in this context to understand how to perform privacy-preserving network analysis, typically by graph-modification algorithms in which some edges or nodes are changed and/or removed. The idea is that these modifications conserve enough of the structure to perform analysis globally while hindering the identification of individuals and/or personal traits.

### 7.5 Software Tools

Much of the progress in network analysis in recent years is down to the availability of not only data but also the software tools needed to analyze it and visualize it. The most well-known tools are those that include a graphical user interface, such as UCINet, Pajek, Cytoscape, NetMiner, and Gephi. More powerful analyses of large networks can be done using packages such as the igraph package (ported to R, Python, and C) or the NetworkX library for Python, tools that are also widely used in producing high-quality graphics (the figures in this article were produced using igraph).

In the era of large data sets, efficient storage of networks can be achieved using the graph structure of nodes, edges, and the relationships between them. A number of *graph databases* have been developed in which the network structure is used internally to store and query network data. Typically, graph databases scale more naturally to large graphs and are more powerful for graph-like queries; for example, computing the neighborhood of a node or the shortest path between two nodes in the graph is much easier and faster than in relational databases. Prominent examples of these databases are Neo4j and Sparksee.

## 8 Outlook

In this article we have introduced the main mathematical tools needed to analyze networks, and we have also illustrated how a large variety of complex systems can be studied by mapping the interdependencies between their constituent units into a network. But network analysis is not just a powerful methodology for analyzing those graphs; it is also a different way of conceiving those systems as collective structures with emergent behaviors that cannot be explained by the sum of their individual parts. For example, the world economy, the biological processes that occur in cells or ecosystems, mass mobilization, and the performance of organizations all depend on the structure and dynamics of the whole network rather than on the sum of individual behaviors. Recent and future technologies will allow us to collect more data more quickly from those systems, allowing us to detect and promote more interdependencies between the units of the system. For example, in the near future it will be possible to completely map the connectome in the brain. Or perhaps we will be able to monitor the activity within cities on an unprecedented scale or reveal the interdependencies between financial and economic activities to help us prevent future economic and societal crises. We might therefore expect that the observation of systems will deliver more and more *networked data*. Network analysis will be the required tool for validating, modeling, and predicting the behavior of those systems.

### Further Reading

Albert, R., and A. L. Barabási. 2002. Statistical mechanics of complex networks. *Reviews of Modern Physics* 74:47–97.
Bollobás, B. 1998. *Modern Graph Theory*. Berlin: Springer.
Bollobás, B., R. Kozma, and D. Miklós. 2009. *Handbook of Large-Scale Random Networks*. Berlin: Springer.
Cross, R., R. J. Thomas, J. Singer, S. Colella, and Y. Silverstone. 2010. *The Organizational Network Fieldbook*. San Francisco, CA: Jossey-Bass.
Huisman, M., and M. A. van Duijn. 2011. A reader's guide to SNA software. In *SAGE Handbook of Social Network Analysis*, edited by J. Scott and P. J. Carrington, pp. 578–600. London: SAGE Publications.
Mason, O., and Verwoerd, M. 2007. Graph theory and networks in biology. *IET Systems Biology* 1:89–119.
Newman, M. E. J. 2010. *Networks: An Introduction*. Oxford: Oxford University Press.
Pinheiro, C. A. R. 2011. *Social Network Analysis in Telecommunications*. New York: John Wiley.
Scott, J. 2013. *Social Network Analysis*, 3rd edn. London: SAGE.
Sporns, O. 2011. *Networks of the Brain*. Cambridge, MA: MIT Press.
van den Bulte, C., and S. Wuyts. 2007. *Social Networks and Marketing*. Cambridge, MA: Marketing Science Institute.
Wasserman, S., and K. Faust. 1994. *Social Network Analysis*. Cambridge: Cambridge University Press.

# IV.19 Classical Mechanics

*David Tong*

Classical mechanics is an ambitious subject. Its purpose is to predict the future and reconstruct the past, to determine the history of every particle in the universe.

The fundamental principles of classical mechanics were laid down by Galileo and Newton in the sixteenth and seventeenth centuries. They provided a framework to explain vast swathes of the natural world, from planets to tides to falling cats. In later centuries the framework of classical mechanics was reformulated, most notably by Lagrange and Hamilton. While this new way of viewing classical mechanics often makes it simpler to solve problems, its main advantage lies in the new mathematical perspective it offers on the subject. In particular, it reveals the door to the quantum world that lies beyond the classical.

This article begins by reviewing the Newtonian framework, providing examples of important physical systems that can be solved. It then goes on to describe the Lagrangian and Hamiltonian frameworks of classical mechanics. Throughout, the emphasis of the article is more on the role that classical mechanics plays in the fundamental laws of physics than on practical engineering applications of the theory.

## 1 Newtonian Mechanics

Newtonian mechanics describes the motion of *particles*, which are defined to be objects of insignificant size. This means that if we want to say what a particle looks like at a given time $t$, the only information that we have to give is its position in $\mathbb{R}^3$, specified by a 3-vector $\boldsymbol{r} \equiv (x, y, z)$. The goal of classical dynamics is to determine the vector function $\boldsymbol{r}(t)$ in any given situation. This, in turn, tells us the velocity $\boldsymbol{v} = \mathrm{d}\boldsymbol{r}/\mathrm{d}t \equiv \dot{\boldsymbol{r}}$ of the particle.

### 1.1 Newton's Laws of Motion

The motion of a particle of mass $m$ at the position $\boldsymbol{r}$ is governed by a second-order differential equation known as *Newton's second law*:

$$\frac{\mathrm{d}\boldsymbol{p}}{\mathrm{d}t} = \boldsymbol{F}(\boldsymbol{r}, \dot{\boldsymbol{r}}; t), \tag{1}$$

where $\boldsymbol{p} = m\dot{\boldsymbol{r}}$ is the momentum of the particle and $\boldsymbol{F}$ is the force, an input that must be specified. Examples will be given below. In general, the force can depend on both the position $\boldsymbol{r}$ and the velocity $\dot{\boldsymbol{r}}$. It can also depend explicitly on time.

In most situations the mass is independent of time. In this case, Newton's second law reduces to the familiar form $\boldsymbol{F} = m\boldsymbol{a}$, where $\boldsymbol{a} \equiv \ddot{\boldsymbol{r}}$ is the acceleration. We will not consider phenomena with time-dependent mass in the following.

Because (1) is a *second-order* differential equation, a unique solution exists only if we specify two initial conditions. This has a consequence: if we are given a snapshot of some situation and asked what happens next, then there is no way of knowing the answer. It is not enough to be told only the positions of the particles at some point in time; we need to know their velocities too.

#### 1.1.1 Principle of Relativity

Equation (1) is not quite correct as stated; we must add the caveat that it holds in an *inertial frame*. This is a choice of coordinates appropriate for an observer who sees a free particle travel in a straight line, also known as uniform motion.

The statement that, in the absence of a force, a particle travels in a straight line is sometimes called *Newton's first law*. However, setting $\boldsymbol{F} = 0$ in (1) already tells us that free particles travel in straight lines. So is the first law nothing more than a special case of the second? In fact, a better formulation of the first law is the statement that inertial frames exist. This then sets the stage for the second law.

Inertial frames are not unique. Indeed, there are infinitely many of them. Let $S$ be an inertial frame in which the position of a particle is measured as $\boldsymbol{r}$. There are then $10 = 3 + 3 + 3 + 1$ independent transformations $S \to S'$ such that $S'$ is also an inertial frame. These transformations are the following.

**Spatial translation:** $\boldsymbol{r}' = \boldsymbol{r} + \boldsymbol{c}$ for any constant $\boldsymbol{c}$.
**Rotations:** $\boldsymbol{r}' = O\boldsymbol{r}$, where $O$ is a $3 \times 3$ orthogonal matrix with $O^{\mathrm{T}}O = I$.
**Boosts:** $\boldsymbol{r}' = \boldsymbol{r} + \boldsymbol{u}t$ for constant velocity $\boldsymbol{u}$.
**Time translation:** $t' = t + d$ for constant $d$.

If the motion of a particle is uniform in $S$, then it will also be uniform in $S'$. These transformations make up the *Galilean group* under which Newton's laws are invariant. The physical meaning of these transformations is that position, direction, and velocity are relative. But acceleration is not. One does not have to accelerate relative to something else. It makes perfect sense to simply say that one is, or is not, accelerating.

#### 1.1.2 Systems of Particles

The discussion above is restricted to the motion of a single particle. It is simple to generalize to many particles; we just add an index to everything in sight. Let particle $i$ have mass $m_i$ and position $\boldsymbol{r}_i$, where $i = 1, \dots, N$, with $N$ the number of particles. Newton's law now reads

$$\frac{\mathrm{d}\boldsymbol{p}_i}{\mathrm{d}t} = \boldsymbol{F}_i, \tag{2}$$

where $\boldsymbol{F}_i$ is the force on the $i$th particle. The novelty is that forces can now be working between particles. In general, we can decompose the force as

$$\boldsymbol{F}_i = \sum_{j \neq i} \boldsymbol{F}_{ij} + \boldsymbol{F}_i^{\mathrm{ext}},$$

where $\boldsymbol{F}_{ij}$ is the force acting on the $i$th particle due to the $j$th particle, while $\boldsymbol{F}_i^{\mathrm{ext}}$ is the external force on the $i$th particle.

The total mass of the system is $M = \sum_i m_i$. We define the *center of mass* as $\boldsymbol{R} = \sum_i m_i \boldsymbol{r}_i / M$ and the total momentum as $\boldsymbol{P} = \sum_i \boldsymbol{p}_i = M\dot{\boldsymbol{R}}$.

The center of mass motion is particularly simple. From (1),

$$\begin{aligned} M\ddot{\boldsymbol{R}} &= \sum_i \boldsymbol{F}_i = \sum_i \sum_{j \neq i} \boldsymbol{F}_{ij} + \sum_i \boldsymbol{F}_i^{\mathrm{ext}} \\ &= \sum_{i<j} (\boldsymbol{F}_{ij} + \boldsymbol{F}_{ji}) + \sum_i \boldsymbol{F}_i^{\mathrm{ext}}, \end{aligned}$$

where, in the second line, we have rewritten the sum to be over all pairs $i < j$. At this stage we invoke *Newton's third law of motion*: every action has an equal and opposite reaction. Or, in equation form, $\boldsymbol{F}_{ij} = -\boldsymbol{F}_{ji}$. We see that the first term vanishes and we are left with

$$M\ddot{\boldsymbol{R}} = \sum_i \boldsymbol{F}_i^{\mathrm{ext}}.$$

This is identical in form to Newton's second law (1) for a single particle. This is an important formula. It says that the center of mass of a system of particles acts as if all the mass were concentrated there. In other words, it does not matter if you throw a tennis ball or a very lively cat; the center of mass of each traces the same path.

## 1.2 Forces

Newton's second law is not useful until someone else tells us what the force $\boldsymbol{F}$ is in any given situation. Here we provide several examples of forces.

### 1.2.1 Conservative Forces and Energy

We start by considering time-independent forces that are a function only of the position of the particle, $\boldsymbol{F} = \boldsymbol{F}(\boldsymbol{r})$. Within this class there is a special class known as *conservative forces*. These can be expressed as

$$\boldsymbol{F} = -\nabla V(\boldsymbol{r})$$

for some scalar function $V(\boldsymbol{r})$. Systems that admit a potential of this form include gravitational, electrostatic, and interatomic forces. The importance of conservative forces lies in the fact that there is a *conserved quantity* called *energy*:

$$E = \tfrac{1}{2}m\dot{\boldsymbol{r}} \cdot \dot{\boldsymbol{r}} + V(\boldsymbol{r}) \equiv T + V.$$

(Recall that the scalar dot product is defined as $\dot{\boldsymbol{r}} \cdot \dot{\boldsymbol{r}} = \dot{x}^2 + \dot{y}^2 + \dot{z}^2$.) The function $T$ is known as the *kinetic energy*; $V$ is known as the *potential energy*. It is simple to show that *if* the equation of motion (1) is obeyed, then $E$ does not change over time:

$$\frac{\mathrm{d}E}{\mathrm{d}t} = (m\ddot{\boldsymbol{r}} + \nabla V) \cdot \dot{\boldsymbol{r}} = 0.$$

In section 2.3 we prove a result called Noether's theorem, which offers a deep explanation of why a conserved quantity called energy exists.

**Example (harmonic oscillator).** Perhaps the simplest example of a conservative force is provided by the *harmonic oscillator*, which describes a particle attached to a spring. The particle moves in one dimension, with position $x(t)$ and has potential energy $V = \frac{1}{2}kx^2$, where $k > 0$ is known as the spring constant. The resulting force, $F = -kx$, is known as HOOKE'S LAW [III.15]. The equation of motion $m\ddot{x} = -kx$ has the general solution $x = A\cos\omega_0 t + B\sin\omega_0 t$, where $A$ and $B$ are integration constants. This describes a particle oscillating around the origin with angular frequency $\omega_0 = \sqrt{k/m}$.

The harmonic oscillator is by far the most important system in all of theoretical physics. For any system described by a potential energy $V$, the stable equilibrium points are the minima of $V$. This means that if the particle is placed at an equilibrium point then, by construction, $\mathrm{d}V/\mathrm{d}x = 0$, ensuring that it remains at the equilibrium point for all time. Moreover, Taylor expanding the potential tells us that small deviations from equilibrium are generically governed by the harmonic oscillator.

### 1.2.2 Central Forces

Central forces form a subclass of conservative forces in which the potential depends only on the distance to the origin:

$$V(\boldsymbol{r}) = V(r),$$

where $r = |\boldsymbol{r}|$. The resulting force always points in the direction of the origin:

$$\boldsymbol{F}(r) = -\nabla V = -\frac{\mathrm{d}V}{\mathrm{d}r}\hat{\boldsymbol{r}}, \tag{3}$$

where $\hat{\boldsymbol{r}} \equiv \boldsymbol{r}/r$ is the unit radial vector. In addition to energy, central forces enjoy another conserved quantity, known as *angular momentum*:

$$\boldsymbol{L} = m\boldsymbol{r} \times \dot{\boldsymbol{r}},$$

where $\times$ denotes the CROSS PRODUCT [I.2 §24]. Notice that, in contrast to the linear momentum $\boldsymbol{p} = m\dot{\boldsymbol{r}}$, the angular momentum $\boldsymbol{L}$ depends on the choice of origin. It is perpendicular to both the position and the momentum.

When we take the time derivative of $\boldsymbol{L}$ we get two terms. But one of these contains $\dot{\boldsymbol{r}} \times \dot{\boldsymbol{r}} = \boldsymbol{0}$, and we are left with

$$\frac{\mathrm{d}\boldsymbol{L}}{\mathrm{d}t} = m\boldsymbol{r} \times \ddot{\boldsymbol{r}} = \boldsymbol{r} \times \boldsymbol{F}.$$

The quantity $\boldsymbol{\tau} = \boldsymbol{r} \times \boldsymbol{F}$ is called the *torque*. For a general force $\boldsymbol{F}$, we find an equation that is very similar to Newton's law: $\mathrm{d}\boldsymbol{L}/\mathrm{d}t = \boldsymbol{\tau}$. However, for central forces (3), $\boldsymbol{F}$ is parallel to the position $\boldsymbol{r}$ and the torque vanishes. We find that angular momentum is conserved:

$$\frac{\mathrm{d}\boldsymbol{L}}{\mathrm{d}t} = 0.$$

As with energy, we will gain a better understanding of why $\boldsymbol{L}$ is conserved when we prove Noether's theorem in section 2.3. For now, note that $\boldsymbol{L}$ is a constant vector and, by construction, $\boldsymbol{L} \cdot \boldsymbol{r} = 0$, which means that motion governed by a central potential takes place in a plane perpendicular to the vector $\boldsymbol{L}$.

### 1.2.3 Gravity

To the best of our knowledge, there are four fundamental forces in nature. They are gravity, electromagnetism, the strong nuclear force, and the weak nuclear force.

The two nuclear forces operate only on small scales, comparable with the size of the nucleus of an atom, and it makes little physical sense to discuss them without quantum mechanics. We will discuss the remaining two, starting with Newtonian gravity.

Gravity is a conservative force. Consider a particle of mass $M$ fixed at the origin. A particle of mass $m$ moving in its presence experiences a central potential energy

$$V(r) = -\frac{GMm}{r}. \tag{4}$$

Here $G$ is Newton's constant; it determines the strength of the gravitational force and is given by $G \approx 6.67 \times 10^{-11}\ \mathrm{m}^3\ \mathrm{kg}^{-1}\ \mathrm{s}^{-2}$. The force on the particle, $\boldsymbol{F} = -\nabla V$, points toward the origin and is proportional to $1/r^2$. For this reason it is called the *inverse-square law*.

Motion governed by the potential (4) describes the orbits of planets around the sun and is known as the Kepler problem. It is not difficult to solve Newton's equation for this potential, and solutions can be found in all of the references in the further reading section below. Instead, here we present a very slick, but indirect, method to find the orbits of the planets. We have already seen that the conservation of angular momentum ensures that the motion takes place in a plane. However, the potential (4) is special since it admits yet another conserved quantity known as the *Laplace–Runge–Lenz* vector:

$$\boldsymbol{e} = \frac{\dot{\boldsymbol{r}} \times \boldsymbol{L}}{GMm} - \frac{\boldsymbol{r}}{r}.$$

With a little algebra, one can show that $\mathrm{d}\boldsymbol{e}/\mathrm{d}t = 0$. Its magnitude $e = |\boldsymbol{e}|$ satisfies

$$er\cos\theta = L^2/GMm^2 - r,$$

where $\theta$ is the angle between $\boldsymbol{L}$ and $\boldsymbol{r}$. This is the equation for a conic section. The solution with $e < 1$ describes an ellipse, with the origin at one of the foci. For the special case of $e = 0$, the orbit is a circle. Note that these orbits are closed: the particle periodically returns to the same position. This is not generally true for orbits in central potentials other than (4). The solutions with $e > 1$ are not closed orbits; these describe hyperbolas.

The elliptical solutions with $e < 1$ describe the planetary orbits, with the sun sitting at the focus. Nearly all planets in the solar system have $e < 0.1$, which means that their orbits are approximately circular. The only exception is Mercury, the closest planet to the sun, which has $e \approx 0.2$. In contrast, some comets have very eccentric orbits. The most famous, Halley's Comet, has $e \approx 0.97$.

We could try to extend our analysis to the problem of three objects all moving under their mutual gravity, but here things are dramatically harder. No general solution to the three-body problem is known, and to answer any practical questions one must resort to numerical methods. Historically, though, the study of the three-body problem has led to a number of new mathematical developments, including chaos.

### 1.2.4 Electromagnetism

Throughout the universe, at each point in space there exist two vectors, $\boldsymbol{E}$ and $\boldsymbol{B}$. These are known as the *electric* and *magnetic* fields. The laws governing $\boldsymbol{E}$ and $\boldsymbol{B}$ are called MAXWELL'S EQUATIONS [III.22]. An important application of these equations is described in MAGNETOHYDRODYNAMICS [IV.29].

For the purposes of this article, the role of $\boldsymbol{E}$ and $\boldsymbol{B}$ is to guide any particle that carries electric charge. The force experienced by a particle with electric charge $q$ is called the *Lorentz force*:

$$\boldsymbol{F} = q(\boldsymbol{E}(\boldsymbol{r}) + \dot{\boldsymbol{r}} \times \boldsymbol{B}(\boldsymbol{r})). \tag{5}$$

Here, the notation $\boldsymbol{E}(\boldsymbol{r})$ and $\boldsymbol{B}(\boldsymbol{r})$ emphasizes that the electric and magnetic fields are functions of position. The term $\dot{\boldsymbol{r}} \times \boldsymbol{B}$ involves the vector cross product.

In principle, both $\boldsymbol{E}$ and $\boldsymbol{B}$ can change in time. However, here we will consider only situations in which they are static. In this case, the electric field is always of the form

$$\boldsymbol{E} = -\nabla\phi$$

for some function $\phi(\boldsymbol{r})$ called the *electric potential* or the *scalar potential*. This means that a static electric field gives rise to a conservative force. The electric potential is related to the potential energy by $V = q\phi$.

As an example, consider the electric field due to a particle of charge $Q$ fixed at the origin. This is given by

$$\boldsymbol{E} = -\nabla\left(\frac{Q}{4\pi\varepsilon_0 r}\right) = \frac{Q}{4\pi\varepsilon_0}\frac{\hat{\boldsymbol{r}}}{r^2}. \tag{6}$$

The quantity $\varepsilon_0$ has the grand name the *permittivity of free space* and is a constant given by $\varepsilon_0 \approx 8.85 \times 10^{-12}\ \mathrm{m}^{-3}\ \mathrm{kg}^{-1}\ \mathrm{s}^2\ \mathrm{C}^2$, where C stands for Coulomb, the unit of electric charge. The quantity $\varepsilon_0$ should be thought of as characterizing the strength of the electric interaction.

The force between two particles with charges $Q$ and $q$ is $\boldsymbol{F} = q\boldsymbol{E}$, with $\boldsymbol{E}$ given by (6). This is known as the *Coulomb force*. It is a remarkable fact that, mathematically, the force is identical to the Newtonian gravitational force arising from (4): both forces have the characteristic inverse-square form. This means that the classical solutions describing an electron orbiting a proton, say, are identical to those describing the planets orbiting the sun.

We turn now to magnetic fields. These give rise to a velocity-dependent force (5) with magnitude proportional to the speed of the particle, but with direction perpendicular to that of the particle. The magnetic force does not contribute to the energy of the particle.

For a constant magnetic field $\boldsymbol{B} = (0, 0, B)$, the equations of motion read

$$m\ddot{x} = qB\dot{y}, \qquad m\ddot{y} = -qB\dot{x},$$

together with $\ddot{z} = 0$. This last equation tells us that the particle travels at constant velocity in the $z$-direction parallel to $\boldsymbol{B}$. The equations in the $xy$-plane, perpendicular to $\boldsymbol{B}$, are also easily solved to reveal that a magnetic field causes particles to move in circles:

$$x = \frac{v}{\omega}(\cos\omega t - 1) \quad \text{and} \quad y = -\frac{v}{\omega}\sin\omega t,$$

where $v$ is the speed and $\omega = qB/m$ is the *cyclotron frequency*. The time to undergo a full circle is fixed: $T = 2\pi/\omega$, independent of $v$. The bigger the circle, the faster the particle moves.

### 1.2.5 Friction

At a fundamental level, energy is always conserved. However, in many everyday processes this does not appear to be the case. At a microscopic level, the kinetic energy of an object is transferred to the motion of many atoms, where it manifests itself as an increase in temperature. If we do not care about what all these atoms are doing, it is useful to package our ignorance into a single macroscopic force that we call *friction*. In practice, friction forces are important in nearly all applications, not just in questions in fundamental physics.

There are a number of different kinds of friction forces. When two solid objects are in contact they experience dry friction. Experimentally, one finds that the complicated dynamics involved in friction can often be summarized by a single force that opposes the motion. This force has magnitude $F = \mu R$, where $R$ is the component of the reaction force normal to the floor and $\mu$ is a constant called the *coefficient of friction*. For steel rubbing against steel, $\mu \approx 0.6$. With a layer of grease added between the metals, it drops to $\mu \approx 0.1$. For steel rubbing against ice, it is as low as $\mu \approx 0.02$.

A somewhat different form of friction, known as *drag*, occurs when an object moves through a fluid, either a liquid or a gas. The resistive force is opposite to the direction of the velocity and, typically, falls into one of two categories.

Linear drag is described by $\boldsymbol{F} = -\gamma\boldsymbol{v}$, where the coefficient of friction, $\gamma$, is a constant. This form of drag holds for objects moving slowly through very viscous fluids. For a spherical object of radius $L$, there is a formula due to Stokes that gives $\gamma = 6\pi\eta L$, where $\eta$ is the dynamic viscosity of the fluid.

In contrast, quadratic drag is described by $\boldsymbol{F} = -\gamma|\boldsymbol{v}|\boldsymbol{v}$, where, again, $\gamma$ is a constant. For quadratic friction, $\gamma$ is usually proportional to the surface area of the object, i.e., $\gamma \propto L^2$. Quadratic drag holds for fast-moving objects in less viscous fluids. This includes objects falling in air.

The kind of drag experienced by an object is determined by the *Reynolds number* $R \equiv \rho v L^2/\eta$, where $\rho$ is the density of the fluid and $\eta$ is the viscosity. For $R \ll 1$, linear drag dominates; for $R \gg 1$, quadratic friction dominates.

Systems that suffer any kind of friction are *dissipative*. They do not have a conserved energy.

To illustrate the effect of friction we return to the harmonic oscillator that we met in section 1.2.1. Adding a linear drag term to the equation of motion gives the damped harmonic oscillator:

$$m\ddot{x} = -\gamma\dot{x} - kx.$$

We can look for solutions of the form $x = e^{i\beta t}$. The results fall into one of the following three categories depending on the relative values of the natural frequency $\omega_0^2 = k/m$ and the magnitude of friction $\alpha = \gamma/2m$.

**Underdamped:** $\omega_0^2 > \alpha^2$**.** The solution takes the form $x = e^{-\alpha t}(Ae^{i\Omega t} + Be^{-i\Omega t})$, where $\Omega = \sqrt{\omega_0^2 - \alpha^2}$. Here, the system oscillates with a frequency $\Omega < \omega_0$, while the amplitude of the oscillations decays exponentially.

**Overdamped:** $\omega_0^2 < \alpha^2$**.** The general solution is now $x = e^{-\alpha t}(Ae^{\Omega t} + Be^{-\Omega t})$, where $\Omega = \sqrt{\alpha^2 - \omega_0^2}$. There are no oscillations.

**Critical damping:** $\omega_0^2 = \alpha^2$**.** For this special case, the general solution is $x = (A + Bt)e^{-\alpha t}$. Again, there are no oscillations, but the system does achieve some mild linear growth for times $t < 1/\alpha$, after which it decays away.

In each of these cases, the solutions tend asymptotically to $x = 0$ at large times. Energy is not conserved.

## 2 The Lagrangian Formulation

There are two reformulations of Newtonian mechanics: the first is due to Lagrange and the second to Hamilton. These new approaches are not particularly useful in modeling phenomena in which friction plays a dominant role. However, when energy is conserved these reformulations provide a new perspective on classical mechanics, one that is more elegant, is more powerful, and, ultimately, provides the link to more sophisticated theories of physics, such as quantum mechanics. The starting point is one of the most profound ideas in theoretical physics: the *principle of least action.*

### 2.1 The Principle of Least Action

First, let us get our notation right. Part of the power of the Lagrangian formulation over the Newtonian approach is that it does away with vectors in favor of more general coordinates. We start by doing this trivially. Let us rewrite the positions of $N$ particles with coordinates $\boldsymbol{r}_i$ as $x^a$, where $a = 1, \dots, 3N \equiv n$. Newton's equations (2) then read

$$\dot{p}_a = -\frac{\partial V}{\partial x^a}, \tag{7}$$

where $p_a = m_a \dot{x}^a$. The number of *degrees of freedom* of the system is said to be $n$. These parametrize an $n$-dimensional space known as the *configuration space* $C$. Each point in $C$ specifies a configuration of the system (i.e., the positions of all $N$ particles). Time evolution gives rise to a curve in $C$.

Define the *Lagrangian* to be a function of the positions $x^a$ and the velocities $\dot{x}^a$ of all the particles. It is formulated as follows:

$$L(x^a, \dot{x}^a) = T(\dot{x}^a) - V(x^a), \tag{8}$$

where $T = \frac{1}{2}\sum_a m_a (\dot{x}^a)^2$ is the kinetic energy and $V(x^a)$ is the potential energy. Note the minus sign between $T$ and $V$. To describe the principle of least action, we consider all smooth paths $x^a(t)$ in $C$ with fixed endpoints such that

$$x^a(t_{\mathrm{i}}) = x^a_{\text{initial}} \quad \text{and} \quad x^a(t_{\mathrm{f}}) = x^a_{\text{final}}.$$

Of all these possible paths, only one is the true path taken by the system. Which one?

To each path let us assign a number called the *action* $S$, defined as

$$S[x^a(t)] = \int_{t_{\mathrm{i}}}^{t_{\mathrm{f}}} L(x^a(t), \dot{x}^a(t))\, \mathrm{d}t.$$

The action is a functional (i.e., a function of the path that is itself a function). The principle of least action is the following result.

**Theorem (principle of least action).** *The actual path taken by the system is an extremum of $S$.*

*Proof.* Consider varying a given path slightly:

$$x^a(t) \to x^a(t) + \delta x^a(t),$$

where we fix the endpoints of the path by demanding that $\delta x^a(t_{\mathrm{i}}) = \delta x^a(t_{\mathrm{f}}) = 0$. Then the change in the action is

$$\begin{aligned} \delta S &= \delta \int_{t_{\mathrm{i}}}^{t_{\mathrm{f}}} L\, \mathrm{d}t \\ &= \int_{t_{\mathrm{i}}}^{t_{\mathrm{f}}} \delta L\, \mathrm{d}t \\ &= \int_{t_{\mathrm{i}}}^{t_{\mathrm{f}}} \left( \frac{\partial L}{\partial x^a}\delta x^a + \frac{\partial L}{\partial \dot{x}^a}\delta \dot{x}^a \right) \mathrm{d}t. \end{aligned}$$

In this last equation we are using the *summation convention*, according to which any term with a repeated $a$ or $b$ index is summed. This means that this term, and similar terms in subsequent equations, should be thought of as including an implicit $\sum_{a=1}^{3N}$. The second term above includes $\delta \dot{x}^a \equiv \mathrm{d}(\delta x^a)/\mathrm{d}t$ and can be integrated by parts to get

$$\delta S = \int_{t_{\mathrm{i}}}^{t_{\mathrm{f}}} \left( \frac{\partial L}{\partial x^a} - \frac{\mathrm{d}}{\mathrm{d}t}\left(\frac{\partial L}{\partial \dot{x}^a}\right)\right) \delta x^a\, \mathrm{d}t + \left[ \frac{\partial L}{\partial \dot{x}^a}\delta x^a \right]_{t_{\mathrm{i}}}^{t_{\mathrm{f}}}.$$

But the final term vanishes since we have fixed the endpoints of the path so that $\delta x^a(t_{\mathrm{i}}) = \delta x^a(t_{\mathrm{f}}) = 0$. The requirement that the action is an extremum says that $\delta S = 0$ for all changes in the path $\delta x^a(t)$. We see that this holds if and only if

$$\frac{\partial L}{\partial x^a} - \frac{\mathrm{d}}{\mathrm{d}t}\left(\frac{\partial L}{\partial \dot{x}^a}\right) = 0 \quad \forall a. \tag{9}$$

These are known as the EULER–LAGRANGE EQUATIONS [III.12]. To finish the proof we need only show that these equations are equivalent to Newton's. From the definition of the Lagrangian (8), we have $\partial L/\partial x^a = -\partial V/\partial x^a$, while $\partial L/\partial \dot{x}^a = p_a$. It is then easy to see that equations (9) are indeed equivalent to (7). □

The principle of least action is an example of a variational principle of the type discussed further in CALCULUS OF VARIATIONS [IV.6]. The path of the particle is viewed globally, as a whole rather than at any instance in time. At first this may appear somewhat teleological. Yet, as we have shown above, this perspective is entirely equivalent to the more local Newtonian methodology.

The principle of *least* action is a slight misnomer. The proof requires only that $\delta S = 0$; it does not specify whether it is a maximum or a minimum of $S$. Since $L = T - V$, we can always increase $S$ by taking a very fast, wiggly path with $T \gg 0$, so the true path is never a maximum. However, it may be either a minimum or a saddle point. "Principle of stationary action" would be a more accurate, but less catchy, name. It is sometimes called "Hamilton's principle."

Somewhat astonishingly, all the fundamental laws of physics can be written in terms of an action principle. This includes electromagnetism, general relativity, the standard model of particle physics, and attempts to go beyond the known laws of physics such as string theory. There is also a beautiful generalization of the action principle to quantum mechanics that is due to Richard Feynman. It is known as the theory of *path integrals*, and in it the particle takes *all paths* with some weight determined by $S$.

Returning to classical mechanics, there are two very important reasons for working with the Lagrangian formulation. The first is that the Euler–Lagrange equations hold in any coordinate system, while Newton's laws are restricted to an inertial frame. The second is the ease with which we can deal with certain types of constraints in the Lagrangian system. We discuss the former of these reasons below, but first let us look at an example.

**Example (the Lorentz force).** A particle with charge $q$ moving in a background electric and magnetic field experiences the Lorentz force law (5). A static electric field $\boldsymbol{E} = -\nabla\phi$ gives rise to a conservative force and fits naturally into the Lagrangian formulation. But in the presence of a magnetic field $\boldsymbol{B}$, it is less obvious that the equation of motion can be written using a Lagrangian. To do so, we first need to introduce the vector potential $\boldsymbol{A}$ and write (possibly time-dependent) magnetic and electric fields as

$$\boldsymbol{B} = \nabla \times \boldsymbol{A}, \qquad \boldsymbol{E} = -\nabla\phi - \frac{1}{c}\frac{\partial \boldsymbol{A}}{\partial t},$$

where $c$ is the speed of light. One can then check that the Euler–Lagrange equations arising from the Lagrangian

$$L = \tfrac{1}{2}m\dot{\boldsymbol{r}}\cdot\dot{\boldsymbol{r}} - q\Big(\phi - \frac{1}{c}\dot{\boldsymbol{r}}\cdot\boldsymbol{A}\Big) \qquad (10)$$

coincide with Newton's equations for the Lorentz force law (5).

## 2.2 Changing Coordinate Systems

We stressed in section 1 that Newton's equation of motion (2) holds only in inertial frames. In contrast, Lagrange's equations hold in any coordinate system. This means that we could choose to work with any coordinates

$$q^a = q^a(x_1,\dots,x_{3N};t),$$

where we have included the possibility of using a coordinate system that changes with time $t$. The Lagrangian can then be written as a function of $L = L(q^a,\dot{q}^a;t)$, and the equations of motion (9) are equivalent to

$$\frac{\mathrm{d}}{\mathrm{d}t}\left(\frac{\partial L}{\partial \dot{q}^a}\right) - \frac{\partial L}{\partial q^a} = 0. \qquad (11)$$

One can prove the equivalence of (9) and (11) through application of the chain rule. Alternatively, one can note that the principle of least action is a statement about the path taken and makes no mention of the coordinate used; it must therefore be true in all coordinate systems.

The general variables $q^a$ are called *generalized coordinates*; the variables $p_a = \partial L/\partial\dot{q}^a$ are called *generalized momenta*. These coincide with what we usually call "momenta" only in Cartesian coordinates.

### 2.2.1 Rotating Coordinate Systems

We can illustrate the flexibility of the Lagrangian approach by deriving the fictitious forces at play in a noninertial, rotating coordinate system. Consider a free particle with Lagrangian

$$L = \tfrac{1}{2}m\dot{\boldsymbol{r}}\cdot\dot{\boldsymbol{r}}.$$

Now measure the motion of the particle with respect to a coordinate system that is rotating with constant angular velocity $\boldsymbol{\omega} = (0,0,\omega)$ about the $z$-axis. Denote the coordinates in the rotating frame as $\boldsymbol{r}' = (x',y',z')$. We have the relationships $z' = z$ and

$$x' = x\cos\omega t + y\sin\omega t,$$
$$y' = y\cos\omega t - x\sin\omega t.$$

These expressions can be substituted directly into the Lagrangian to find $L$ in terms of the rotating coordinates:

$$L = \tfrac{1}{2}m[(\dot{x}' - \omega y')^2 + (\dot{y}' + \omega x')^2 + \dot{z}^2]$$
$$\equiv \tfrac{1}{2}m(\dot{\boldsymbol{r}}' + \boldsymbol{\omega}\times\boldsymbol{r}')\cdot(\dot{\boldsymbol{r}}' + \boldsymbol{\omega}\times\boldsymbol{r}').$$

We can now derive the Euler–Lagrange equations in the rotating frame, differentiating $L$ with respect to $r^{a'}$ and $\dot{r}^{a'}$. We find that

$$m(\ddot{\boldsymbol{r}}' + \boldsymbol{\omega}\times(\boldsymbol{\omega}\times\boldsymbol{r}') + 2\boldsymbol{\omega}\times\dot{\boldsymbol{r}}') = 0.$$

We learn that in the rotating frame the particle does not follow a straight line with $\ddot{\boldsymbol{r}}' = 0$. Instead, we find the appearance of two extra terms in the equation of motion.

The term $\boldsymbol{\omega}\times(\boldsymbol{\omega}\times\boldsymbol{r}')$ is called the *centrifugal force*. It points outward in the plane perpendicular to $\boldsymbol{\omega}$ with magnitude $m\omega^2|\boldsymbol{r}'_\perp| = m|\boldsymbol{v}_\perp|^2/|\boldsymbol{r}'_\perp|$, where the subscript $\perp$ denotes the projection perpendicular to $\boldsymbol{\omega}$. This is the force you feel pinning you to the side of a car when you take a corner too fast.

The term $-2\boldsymbol{\omega}\times\dot{\boldsymbol{r}}'$ is called the *Coriolis force*. Notice that it is mathematically identical to the Lorentz force (5) for a particle moving in a constant magnetic field. In myth the Coriolis force determines the direction in which the water draining from a sink rotates, but in practice it is too small on domestic scales to have a noticeable effect. It is, however, significant on large length scales, where it dictates the circulation of oceans and the atmosphere.

The centrifugal and Coriolis forces are referred to as *fictitious forces* because they are a result of the reference frame rather than any interaction. However, we should not underestimate their importance just because they are "fictitious." According to Einstein's theory of general relativity, gravity is also a fictitious force, on the same footing as the Coriolis and centrifugal forces.

### 2.3 Noether's Theorem

A *symmetry* of a physical system is an invariance under a transformation. We have already met a number of symmetries in our discussion of inertial frames; the laws of (nonrelativistic) physics are invariant under the Galilean group, composed of rotations and translations in space and time. A beautiful theorem due to Emmy Noether relates these symmetries to conservation laws.

Consider a one-parameter family of maps between generalized coordinates

$$q^a(t) \to Q^a(s,t), \quad s \in \mathbb{R},$$

such that $Q^a(0,t) = q^a(t)$. This transformation is said to be a continuous *symmetry* of the Lagrangian $L$ if

$$\frac{\partial}{\partial s}L(Q^a(s,t),\dot{Q}^a(s,t);t) = 0.$$

Noether's theorem states that for each such symmetry there exists a conserved quantity. The proof is straightforward. We compute

$$\frac{\partial L}{\partial s} = \frac{\partial L}{\partial Q^a}\frac{\partial Q^a}{\partial s} + \frac{\partial L}{\partial \dot{Q}^a}\frac{\partial \dot{Q}^a}{\partial s}.$$

Evaluated at $s = 0$, we have

$$\begin{aligned}\left.\frac{\partial L}{\partial s}\right|_{s=0} &= \left.\frac{\partial L}{\partial q^a}\frac{\partial Q^a}{\partial s}\right|_{s=0} + \left.\frac{\partial L}{\partial \dot{q}^a}\frac{\partial \dot{Q}^a}{\partial s}\right|_{s=0} \\ &= \frac{\mathrm{d}}{\mathrm{d}t}\left(\frac{\partial L}{\partial \dot{q}^a}\right)\left.\frac{\partial Q^a}{\partial s}\right|_{s=0} + \left.\frac{\partial L}{\partial \dot{q}^a}\frac{\partial \dot{Q}^a}{\partial s}\right|_{s=0},\end{aligned}$$

where we have used the Euler–Lagrange equations. The result is a total derivative:

$$\left.\frac{\partial L}{\partial s}\right|_{s=0} = \frac{\mathrm{d}}{\mathrm{d}t}\left(\left.\frac{\partial L}{\partial \dot{q}^a}\frac{\partial Q^a}{\partial s}\right|_{s=0}\right) = 0.$$

We learn that the quantity $\sum_a(\partial L/\partial\dot{q}^a)(\partial Q^a/\partial s)$, evaluated at $s = 0$, is constant for all time whenever the equations of motion are obeyed. Notice that the proof of Noether's theorem is constructive: it does not just tell us about the existence of a conserved quantity, it also tells us what that quantity is.

#### *Homogeneity of Space*

Consider a closed system of $N$ particles interacting through a potential $V(|\boldsymbol{r}_i - \boldsymbol{r}_j|)$ that, as the notation suggests, depends only on the relative distances between the particles $i,j = 1,\dots,N$. The Lagrangian

$$L = \frac{1}{2}\sum_i m_i\dot{\boldsymbol{r}}_i\cdot\dot{\boldsymbol{r}}_i - V(|\boldsymbol{r}_i - \boldsymbol{r}_j|) \qquad (12)$$

has the symmetry of translation: $\boldsymbol{r}_i \to \boldsymbol{r}_i + s\boldsymbol{n}$ for any vector $\boldsymbol{n}$ and for any real number $s$, so $L(\boldsymbol{r}_i,\dot{\boldsymbol{r}}_i,t) = L(\boldsymbol{r}_i + s\boldsymbol{n},\dot{\boldsymbol{r}}_i,t)$. This is the statement that space is homogeneous. From Noether's theorem we can compute the conserved quantity associated with translations. It is $\sum_i \boldsymbol{p}_i\cdot\boldsymbol{n}$, which we recognize as the total linear momentum in the direction $\boldsymbol{n}$. Since this holds for all $\boldsymbol{n}$, we conclude that $\sum_i \boldsymbol{p}_i$ is conserved. The familiar fact that the total linear momentum is conserved is due to the homogeneity of space.

#### *Homogeneity of Time*

The laws of physics are the same today as they were yesterday. This invariance under time translations also gives rise to a conserved quantity. Mathematically, this means that $L$ is invariant under $t \to t + s$, or, in other words, $\partial L/\partial t = 0$. It is straightforward to check that this condition ensures that

$$H = \sum_a \dot{q}^a(\partial L/\partial\dot{q}^a) - L$$

is conserved. This is the *energy* of the system. We learn that time is to energy what space is to momentum, a lesson that resonates into the relativistic world of Einstein. We will meet the quantity $H$ again in the next section, where, viewed from a slightly different perspective, it will be rebranded the Hamiltonian.

One can also show that the isotropy of space, meaning invariance under rotations, gives rise to the conservation of angular momentum. In fact, suitably generalized, it turns out that *all* conservation laws in nature are related to symmetries through Noether's theorem. This includes the conservation of electric charge and the conservation of particles such as protons and neutrons (known as baryons).

## 3 The Hamiltonian Formulation

The next step in the formulation of classical mechanics is due to Hamilton. The basic idea is to place the generalized coordinates $q^a$ and the generalized momenta $p_a = \partial L/\partial \dot{q}^a$ on a more symmetric footing.

We can start by thinking pictorially. Recall that the coordinates $\{q^a\}$ define a point in $n$-dimensional *configuration space* $C$. Time evolution is a path in $C$. However, the *state* of the system is defined by $\{q^a\}$ and $\{p_a\}$ in the sense that this information will allow us to determine the state at all times in the future. The pair $\{q^a, p_a\}$ defines a point in $2n$-dimensional *phase space*. Since a point in phase space is sufficient to determine the future evolution of the system, paths in phase space can never cross. We say that evolution is governed by a *flow* in phase space.

### 3.1 Hamilton's Equations

The Lagrangian $L(q^a, \dot{q}^a; t)$ is a function of the coordinates $q^a$, their time derivatives $\dot{q}^a$, and (possibly) time. We define the *Hamiltonian* to be the Legendre transform of the Lagrangian with respect to the $\dot{q}^a$ variables:

$$H(q^a, p_a; t) = \sum_{a=1}^{n} p_a \dot{q}^a - L(q^a, \dot{q}^a, t),$$

where $\dot{q}^a$ is eliminated from the right-hand side in favor of $p_a$ by using $p_a = \partial L/\partial \dot{q}^a = p_a(q^a, \dot{q}^a; t)$ and inverting to get $\dot{q}^a = \dot{q}^a(q^a, p_a; t)$. Now we look at the variation of $H$. Once again employing the summation convention, we have

$$\begin{aligned} \delta H &= (\delta p_a \dot{q}^a + p_a \delta \dot{q}^a) \\ &\quad - \left(\frac{\partial L}{\partial q^a}\delta q^a + \frac{\partial L}{\partial \dot{q}^a}\delta \dot{q}^a + \frac{\partial L}{\partial t}\delta t\right) \\ &= \delta p_a \dot{q}^a - \frac{\partial L}{\partial q^a}\delta q^a - \frac{\partial L}{\partial t}\delta t. \end{aligned}$$

But we know that this can be rewritten as

$$\delta H = \frac{\partial H}{\partial q^a}\delta q^a + \frac{\partial H}{\partial p_a}\delta p_a + \frac{\partial H}{\partial t}\delta t.$$

We now equate terms. We also make use of the Euler–Lagrange equations, which can be written as $\dot{p}_a = \partial L/\partial q^a$. The end result is

$$\dot{p}_a = -\frac{\partial H}{\partial q^a}, \qquad \dot{q}^a = \frac{\partial H}{\partial p_a}.$$

For Lagrangians with explicit time dependence these are supplemented with $-\partial L/\partial t = \partial H/\partial t$. These are *Hamilton's equations.* We have replaced $n$ second-order differential equations by $2n$ first-order differential equations for $q_a$ and $p_a$. Recast in this manner, Hamilton's equations are ideally suited for dealing with initial-value problems rather than the boundary-value problems that are more natural in the Lagrangian formulation.

#### *A Particle in a Potential*

The simplest example is a single particle moving in a potential. The Lagrangian is $L = \frac{1}{2}m\dot{\boldsymbol{r}}\cdot\dot{\boldsymbol{r}} - V(\boldsymbol{r})$ and the momentum $\boldsymbol{p} = m\dot{\boldsymbol{r}}$. The steps above give us the Hamiltonian

$$H = \boldsymbol{p}\cdot\dot{\boldsymbol{r}} - L = \frac{1}{2m}\boldsymbol{p}\cdot\boldsymbol{p} + V(\boldsymbol{r}).$$

Hamilton's equations are simply $\dot{\boldsymbol{r}} = \boldsymbol{p}/m$ and $\dot{\boldsymbol{p}} = -\nabla V$, both of which are familiar: the first is the definition of momentum in terms of velocity; the second is Newton's equation for this system.

#### *The Lorentz Force*

A charged particle moving in an electric and magnetic field is described by the Lagrangian (10). From this we can compute the momentum, $\boldsymbol{p} = m\dot{\boldsymbol{r}} + (q/c)\boldsymbol{A}$, which now differs from what we usually call momentum by the addition of the vector potential $\boldsymbol{A}$. The Hamiltonian is

$$H(\boldsymbol{p}, \boldsymbol{r}) = \frac{1}{2m}\left(\boldsymbol{p} - \frac{q}{c}\boldsymbol{A}\right)\left(\boldsymbol{p} - \frac{q}{c}\boldsymbol{A}\right) + q\phi.$$

One can check that Hamilton's equations reduce to Newton's equations with the Lorentz force law (5).

### 3.2 Looking Forward

The advantage of the Hamiltonian formulation over the Lagrangian is not really a practical one. Instead, the true value of the formulation lies in what it tells us about the structure of classical mechanics. It is, at heart, a geometric formulation of classical mechanics and can be expressed in more abstract form using the language of symplectic geometry. Moreover, the Hamiltonian framework provides a springboard for later developments, including chaos theory and integrability. (See the article on CHAOS [II.3] in this book.) Perhaps most importantly, the Hamiltonian offers the most direct link to more fundamental theories of physics and particularly to quantum mechanics.

### Further Reading

This article is based on two lecture courses given by the author to undergraduates at the University of Cambridge. Full lecture notes for both courses can be downloaded from the author's Web page: www.damtp.cam.ac.uk/user/tong/teaching.html.

Lev Landau and Evgeny Lifshitz's *Mechanics* (Butterworth-Heinemann, Oxford, 3rd edn, 1982) is one of the most concise, elegant, and conceptually gorgeous textbooks ever written.

A more modern, pedagogical approach to Newtonian mechanics can be found in Mary Lunn's *A First Course in Mechanics* (Oxford Science Publications, Oxford, 1991).

A good introduction to the Lagrangian and Hamiltonian formulation is Louis Hand and Janet Finch's *Analytic Mechanics* (Cambridge University Press, Cambridge, 1998).

# IV.20 Dynamical Systems

*Philip Holmes*

## 1 Introduction

The theory of dynamical systems describes the construction and analysis of models for things that move or evolve over time and space. It combines analytical, geometrical, topological, and numerical methods for studying differential equations and iterated mappings. These methods stem from the work of Newton and his successors, the great natural philosophers of the eighteenth and nineteenth centuries, and in particular from Henri Poincaré. As such, the study of dynamical systems is normal mathematics rather than the paradigm shift that some popular accounts have claimed for CHAOS THEORY [II.3]. Nonetheless, problems from the applied sciences have continued to strongly influence and motivate it, especially over the past half century (see Aubin and Dahan Dalmedico (2002) for a sociohistorical discussion of developments in the turbulent decade around 1970).

It is a capacious field, implying different things to different people, including deterministic and stochastic systems of finite or infinite dimensions, ERGODIC THEORY [II.3], and holomorphic dynamics (the study of iterated functions on the complex plane). I shall focus on ordinary differential equations (ODEs) and iterated maps defined on Euclidean space $\mathbb{R}^n$, but note that the theory generalizes to manifolds, much of it generalizes to infinite dimensions, and some of it generalizes to stochastic systems. In contrast to classical ODE theory, which focuses on specific initial or boundary-value problems, dynamical systems theory brings a qualitative and geometrical approach to the analysis of nonlinear ODEs, addressing the existence, stability, and global behavior of sets of solutions, rather than seeking exact or approximate expressions for individual solutions (see ORDINARY DIFFERENTIAL EQUATIONS [IV.2]).

We consider systems of ODEs (1) and discrete mappings (2):

$$\dot{x}_j = f_j(x_1, x_2, \dots, x_n; \mu), \quad (1)$$

$$x_j(l+1) = F_j(x_1(l), \dots, x_n(l); \mu), \quad (2)$$

where $j = 1, \dots, n$; $\dot{x}_j$ denotes the time derivative; $f_j$ and $F_j$ are smooth, real-valued functions; the $x_j$ are *state variables*; and $\mu$ is a *control parameter*. In solving (1) or (2) with given initial conditions $\boldsymbol{x}(0)$ to obtain *orbits* $\boldsymbol{x}(t) = (x_1(t), \dots, x_n(t))$ or $\{\boldsymbol{x}(l)\}_{l=0}^{\infty}$, $\mu$ is kept fixed.

In studying the ODE (1), one seeks to describe the behavior of the *flow map*

$$\boldsymbol{x}(t, \boldsymbol{x}(0)) = \boldsymbol{\phi}_t(\boldsymbol{x}(0)) \quad \text{or} \quad \boldsymbol{\phi}_t : U \to \mathbb{R}^n, \quad (3)$$

generated by the *vector field* $\boldsymbol{f}(\boldsymbol{x}) = (f_1(\boldsymbol{x}), \dots, f_n(\boldsymbol{x}))$, which transports initial points $\boldsymbol{x}(0) \in U \subseteq \mathbb{R}^n$ to their images at time $t$: $\boldsymbol{\phi}_t(\boldsymbol{x}(0))$. If $\boldsymbol{\phi}_t$ can be found, then, fixing a time interval $t = T$, (1) reduces to (2), but explicit formulas can be derived only in exceptional cases, and in any case the study of iterated maps is no less complicated than that of ODEs. Of course, numerical algorithms for ODEs can provide excellent approximations of $\boldsymbol{\phi}_T$ (see NUMERICAL SOLUTION OF ORDINARY DIFFERENTIAL EQUATIONS [IV.12]).

Before describing the theory, I sketch some historical threads through the work of A. A. Andronov, V. I. Arnold, G. D. Birkhoff, A. N. Kolmogorov, S. Smale, and other key figures. Two motivating examples are then introduced: one physical and one mathematical.

## 2 A Brief History

Dynamical systems theory began with Poincaré's work on differential equations and celestial mechanics from 1879 to 1912. In addition to special methods for two-dimensional ODEs, many other central concepts first appeared in Poincaré's work, including *invariant manifolds* (smooth hypersurfaces composed of families of orbits), *first-return (Poincaré) maps* for the study of periodic motions, *bifurcations*, coordinate changes to *normal forms* that simplify analyses, and *perturbation methods*. Notably, Poincaré realized that, due to the presence of "doubly asymptotic" points or *homoclinic and heteroclinic orbits*, certain differential equations describing mechanical systems with $N \geqslant 2$ degrees of freedom were not integrable. More precisely, they do not possess enough independent, analytic functions of $\boldsymbol{x}$ that remain constant on solutions, which would imply that the geometry of invariant manifolds is relatively simple. During this period A. M. Lyapunov also made important contributions to stability theory.

Birkhoff extended Poincaré's work, in particular proving that maps from an annulus to itself having two periodic orbits with different periods also contain complicated limit sets that separate the domains of attraction of those orbits. This prepared the way for Cartwright and Littlewood's proof that the van der Pol equation—a periodically forced nonlinear oscillator—possesses infinitely many periodic orbits and a set of nonperiodic orbits "of the power of the continuum." N. Levinson subsequently drew this to Smale's attention, prompting his construction of the *horseshoe map*: a prototypical example with an unstable chaotic set, which is nonetheless *structurally stable*, implying that the flows of the original and perturbed systems are topologically equivalent (homeomorphic). The qualitative behavior of structurally stable vector fields and maps survives small perturbations.

Structural stability had been introduced by Andronov and L. S. Pontryagin in 1937 under the name "systèmes grossières" (coarse systems). From this perspective, a bifurcation occurs when a system becomes structurally unstable as a parameter varies; its behavior changes as one passes through the bifurcation point. Andronov's group in Gorky (now Nizhni Novgorod) also did important work in bifurcation theory. M. M. Peixoto's characterization of structurally stable flows on two-dimensional manifolds required that they possess only finitely many fixed points and periodic orbits, leading to Smale's conjecture that the same should hold in higher dimensions. The horseshoe map, with its infinite set of periodic points, can be seen as a return map for a three-dimensional flow and thus provided a counterexample to this conjecture. Moreover, it showed that the chaos glimpsed by Poincaré was prevalent in ODEs of dimensions $n \geqslant 3$ as well as in maps of dimensions $n \geqslant 2$. Smale's influential work in the 1960s introduced mathematicians to the field, but these ideas did not reach the applied mathematical mainstream for some time.

As was common in the Soviet Union, Andronov's group maintained strong connections between abstract theory and applications, in its case, nonlinear oscillators and waves, electronic circuits, and control theory. In Moscow, Kolmogorov and his students (including D. V. Anosov, Ya. G. Sinai, and Arnold) did foundational work on ergodic theory, billiards, and geodesic flows, with links to mathematical physics. Smale's visits in 1961 and during the International Congress of Mathematicians in 1966 helped introduce their work to the wider mathematical world.

Lorenz's paper on a three-dimensional ODE modeling Rayleigh–Bénard convection was done almost independently of the work described above, although in presenting his discovery of *sensitive dependence on initial conditions* in 1963, Lorenz appealed to Birkhoff's work. An earlier, extra-mathematical, discovery had taken place in 1961 when Ueda, a graduate student in electrical engineering at Kyoto University, observed irregular motions in analogue computer simulations of a periodically forced van der Pol–Duffing equation.

## 3 Two Dynamical Systems

To motivate the theory described below, we first introduce a problem from classical mechanics and then a mathematical toy that is, perhaps surprisingly, related to it.

### 3.1 The Double Pendulum

Consider a pendulum comprising two rigid links, the first rotating about a fixed pivot, the second pivoting about the end of the first (see figure 1(a)). Under Newtonian mechanics, the four angles and angular velocities $\theta_1$, $\theta_2$, $\dot{\theta}_1$, $\dot{\theta}_2$ describe the pendulum's *state space*, and neglecting air resistance and friction, conservation of energy implies that all motions started with given potential and kinetic energy lie on a three-dimensional subset of state space (typically, a smooth manifold). However, unlike the one-link pendulum, its motions are not generally periodic, and small changes in initial conditions yield orbits that rapidly diverge. Part (b) of figure 1 illustrates this sensitive dependence.

### 3.2 The Doubling Machine

Next we describe a piecewise-linear mapping defined on the interval $[0, 1]$ by the rule

$$h(x) = \begin{cases} 2x & \text{if } 0 \leqslant x < \frac{1}{2}, \\ 2x - 1 & \text{if } \frac{1}{2} < x \leqslant 1 \end{cases} \tag{4}$$

(see figure 2). An orbit of $h$ is the sequence $\{x_n\}_{n=0}^{\infty}$ obtained by repeatedly doubling the initial value $x_0$ and

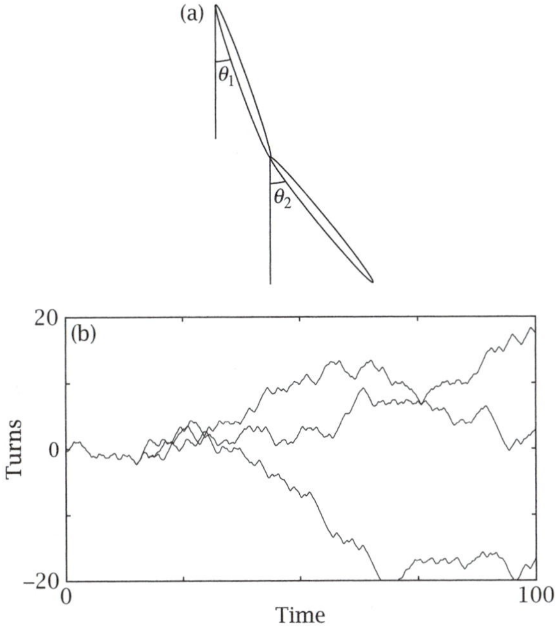


**Figure 1** (a) The double pendulum. (b) The numbers of full turns, $\theta_2$, executed by three different orbits released from rest with $\theta_1(0) = 90°$, $\theta_2(0) = -90°$, and $\theta_2(0) = -(90 \pm 10^{-6})°$.

subtracting the integer part at each step:

$$0.2753 \mapsto 0.5506 \mapsto 1.1012 = 0.1012 \mapsto 0.2024 \mapsto \cdots .$$

To understand the sensitive dependence on initial conditions and its consequences, we represent the numbers between 0 and 1 in binary form:

$$x_0 = \frac{a_1}{2} + \frac{a_2}{2^2} + \frac{a_3}{2^3} + \cdots + \frac{a_j}{2^j} + \cdots ,$$

where each coefficient $a_j$ is either 0 or 1. Thus,

$$x_1 = h(x_0) = 2\sum_{j=1}^{\infty} \frac{a_j}{2^j} = a_1 + \frac{a_2}{2} + \frac{a_3}{2^2} + \cdots ,$$

but since the integer part is removed on every iteration, we have

$$x_1 = \frac{a_2}{2} + \frac{a_3}{2^2} + \frac{a_4}{2^3} + \cdots ,$$

and in general,

$$x_k = \frac{a_{k+1}}{2} + \frac{a_{k+2}}{2^2} + \cdots = \sum_{j=1}^{\infty} \frac{a_{j+k}}{2^j}.$$

Applying $h$ is equivalent to shifting the "binary point" and dropping the leading coefficient,

$$(a_1a_2a_3a_4\cdots) \mapsto (a_2a_3a_4\cdots),$$

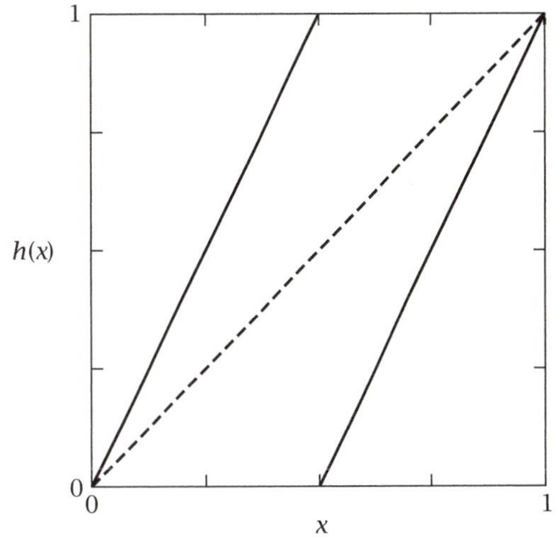


**Figure 2** The doubling map, $h$.

just as multiplication by 10 shifts a decimal point. Were $x_0$ known to infinite accuracy, with *all* the $a_k$ specified, the current state, $x_k$, at each step would also be known exactly. But given only the first $N$ coefficients $(a_1, a_2, \dots, a_N)$, after $N$ iterations, one cannot even determine whether $x_{N+1}$ lies above or below $\frac{1}{2}$. Moreover, if two points differ only at the $N$th binary place and thus lie within $(\frac{1}{2})^{N-1}$, after $N$ iterations they lie on opposite sides of $\frac{1}{2}$ and thereafter behave essentially independently. Repeated doubling amplifies small differences.

The binary representation exemplifies symbolic dynamics. To *every* infinite sequence of zeros and ones there corresponds a point in $[0, 1]$, and vice versa. Hence, any random sequence corresponds to a state $x_0 \in [0, 1]$ whose orbit $h^k(x_0)$ realizes that sequence; the map $h$ has infinitely many orbits whose itineraries are indistinguishable from random sequences. It also has infinitely many periodic orbits, corresponding to periodic sequences. These can be enumerated by listing all distinct sequences of lengths $1, 2, 3, \dots$ that contain no subsequences of smaller period, showing that there are approximately $2^N/N$ orbits of period $N$ and a countable infinity in all. Nonetheless, since numbers picked at random are almost always irrational (they form a set of full measure), almost all orbits are nonperiodic (see section 4.5).

## 4 Dynamical Systems Theory

As noted above, dynamical systems theory emphasizes the study of the global orbit structure or *phase portrait*, its dependence on parameters, and the description of qualitative properties such as the existence of closed orbits (periodic solutions). We start by describing some important tools and concepts.

### 4.1 Judicious Linearization

While few nonlinear systems can be "solved" completely, much can be deduced from linear analysis. The linearization of the ODE (1) near a *fixed point* or *equilibrium* $\boldsymbol{x}^{\mathrm{e}}$, where $\boldsymbol{f}(\boldsymbol{x}^{\mathrm{e}}) = 0$, is obtained by substituting $\boldsymbol{x} = \boldsymbol{x}^{\mathrm{e}} + \boldsymbol{\xi}$ into (1), expanding in a Taylor series, and neglecting quadratic and higher-order terms to obtain

$$\dot{\boldsymbol{\xi}} = \boldsymbol{Df}(\boldsymbol{x}^{\mathrm{e}})\boldsymbol{\xi} \quad \text{for } |\boldsymbol{\xi}| \ll 1. \tag{5}$$

Here, $\boldsymbol{Df}(\boldsymbol{x}^{\mathrm{e}})$ is the $n \times n$ Jacobian matrix of first partial derivatives $\partial f_i / \partial x_j$ evaluated at $\boldsymbol{x}^{\mathrm{e}}$.

Linear ODEs with constant coefficients, like (5), are completely soluble in terms of elementary functions. One assumes the exponential form

$$\boldsymbol{\xi} = \sum_{j=1}^{n} \boldsymbol{v}_j \exp(\lambda_j t)$$

and computes the eigenvalues $\lambda_j$ and eigenvectors $\boldsymbol{v}_j$ of $\boldsymbol{Df}(\boldsymbol{x}^{\mathrm{e}})$. If $\boldsymbol{Df}(\boldsymbol{x}^{\mathrm{e}})$ has $n$ linearly independent eigenvectors, then any solution may be expressed as a linear combination $\boldsymbol{x}(t) = \sum_{j=1}^{n} c_j \boldsymbol{v}_j \exp(\lambda_j t)$, and the $c_j$ can be uniquely determined by initial conditions. Eigenvalues and eigenvectors can, of course, be complex numbers and vectors, but using Euler's formula $\exp(\pm \mathrm{i}t) = \cos t \pm \mathrm{i} \sin t$, and allowing complex $c_j$, solutions to real-valued ODEs can be written as combinations of exponential and trigonometrical functions. (If $\boldsymbol{Df}(\boldsymbol{x}^{\mathrm{e}})$ has fewer than $n$ linearly independent eigenvectors, generalized eigenvectors are required and terms of the form $t^k \exp(\lambda_j t)$ appear.)

If every eigenvalue of $\boldsymbol{Df}(\boldsymbol{x}^{\mathrm{e}})$ has nonzero real part, $\boldsymbol{x}^{\mathrm{e}}$ is called a *hyperbolic* or *nondegenerate* fixed point. Excepting special cases, like the energy-conserving pendulum, fixed points are typically hyperbolic, and their stability in the original nonlinear system (1) can be deduced from the linearized system.

A fixed point $\boldsymbol{x}^{\mathrm{e}}$ of (1) is *Lyapunov* or *neutrally stable* if for every neighborhood $U \ni \boldsymbol{x}^{\mathrm{e}}$ there is a neighborhood $V \subseteq U$, also containing $\boldsymbol{x}^{\mathrm{e}}$, such that every solution $\boldsymbol{x}(t)$ of (1) starting in $V$ remains in $U$ for all $t \geqslant 0$. If $\boldsymbol{x}(t) \to \boldsymbol{x}^{\mathrm{e}}$ as $t \to \infty$ for all $\boldsymbol{x}(0) \in V$, then $\boldsymbol{x}^{\mathrm{e}}$ is *asymptotically stable*. If $\boldsymbol{x}^{\mathrm{e}}$ is not stable, it is *unstable*. More descriptively, if all nearby orbits approach $\boldsymbol{x}^{\mathrm{e}}$, it is called a *sink*; if all recede from it, it is a *source*; and

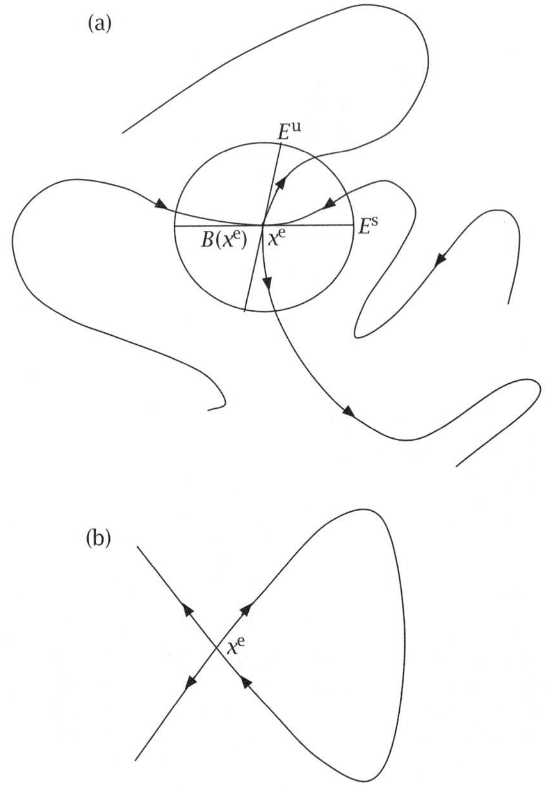


**Figure 3** Stable and unstable manifolds. (a) Near $\boldsymbol{x}^{\mathrm{e}}$ the local manifolds can be expressed as graphs, as in (7). (b) Globally, stable and unstable manifolds may intersect, forming a homoclinic orbit.

if some approach and some recede, it is a *saddle point*. Sinks are the simplest *attractors*: see section 4.5.

Choosing the neighborhood $V$ small enough that the linear part $\boldsymbol{Df}(\boldsymbol{x}^{\mathrm{e}})\boldsymbol{\xi}$ dominates the higher-order terms that were ignored in (5), it follows that, if all the eigenvalues of $\boldsymbol{Df}(\boldsymbol{x}^{\mathrm{e}})$ have strictly negative real parts, then $\boldsymbol{x}^{\mathrm{e}}$ is asymptotically stable but, if at least one eigenvalue has strictly positive real part, then $\boldsymbol{x}^{\mathrm{e}}$ is unstable. When one or more eigenvalues have zero real part, the local behavior is determined by the leading nonlinear terms, as described in section 4.3, but if $\boldsymbol{x}^{\mathrm{e}}$ is hyperbolic, then the entire orbit structure nearby is topologically equivalent to that of the nonlinear ODE (1). Similar results hold for the discrete mapping (2), and systems can also be linearized near periodic and other orbits.

### 4.2 Invariant Manifolds

Remarkably, the decomposition of the state space into invariant subspaces for the linearized system (5) also holds for the nonlinear system near $\boldsymbol{x}^{\mathrm{e}}$. Suppose that $\boldsymbol{Df}(\boldsymbol{x}^{\mathrm{e}})$ has $s \leqslant n$ and $u = n - s$ eigenvalues $\lambda_1, \dots, \lambda_s$ and $\lambda_{s+1}, \dots, \lambda_{s+u}$ with strictly negative and strictly positive real parts, respectively, and define the linear subspaces $E^{\mathrm{s}} = \mathrm{span}\{\boldsymbol{v}_1, \dots, \boldsymbol{v}_s\}$ and $E^{\mathrm{u}} = \mathrm{span}\{\boldsymbol{v}_{s+1}, \dots, \boldsymbol{v}_{s+u}\}$ spanned by their (generalized) eigenvectors. As $t$ increases, orbits of (5) in the stable subspace $E^{\mathrm{s}}$ decay exponentially, while those in the unstable subspace $E^{\mathrm{u}}$ grow.

The *stable manifold theorem* states that, near a hyperbolic fixed point $\boldsymbol{x}^{\mathrm{e}}$, equations (1) and (2) possess *local stable and unstable manifolds* $W^{\mathrm{s}}_{\mathrm{loc}}(\boldsymbol{x}^{\mathrm{e}})$, $W^{\mathrm{u}}_{\mathrm{loc}}(\boldsymbol{x}^{\mathrm{e}})$ of dimensions $s$ and $u$, respectively, tangent at $\boldsymbol{x}^{\mathrm{e}}$ to $E^{\mathrm{s}}$ and $E^{\mathrm{u}}$. The former consists of all orbits that start and remain near $\boldsymbol{x}^{\mathrm{e}}$ for all future time and approach it as $t \to \infty$:

$$W^{\mathrm{s}}_{\mathrm{loc}}(\boldsymbol{x}^{\mathrm{e}}) = \{\boldsymbol{x} \in V \mid \boldsymbol{\phi}_t(\boldsymbol{x}) \to \boldsymbol{x}^{\mathrm{e}} \text{ as } t \to \infty \text{ and } \boldsymbol{\phi}_t(\boldsymbol{x}) \in V \text{ for all } t \geqslant 0\}; \quad (6)$$

the unstable manifold $W^{\mathrm{u}}_{\mathrm{loc}}(\boldsymbol{x}^{\mathrm{e}})$ is defined similarly, with the substitutions "past time" and "$t \to -\infty$." These smooth, curved subspaces locally resemble their linear counterparts $E^{\mathrm{s}}$ and $E^{\mathrm{u}}$ (see figure 3(a)).

Near $\boldsymbol{x}^{\mathrm{e}}$ the local stable and unstable manifolds can be expressed as graphs over $E^{\mathrm{s}}$ and $E^{\mathrm{u}}$, respectively. Letting $E^{\mathrm{s}\perp}$ denote the $(n - s)$-dimensional orthogonal complement to $E^{\mathrm{s}}$ and letting $\boldsymbol{y} \in E^{\mathrm{s}}$ and $\boldsymbol{z} \in E^{\mathrm{s}\perp}$ be local coordinates, we can write

$$W^{\mathrm{s}}_{\mathrm{loc}}(\boldsymbol{x}^{\mathrm{e}}) = \{(\boldsymbol{y}, \boldsymbol{z}) \mid (\boldsymbol{y}, \boldsymbol{z}) \in B(\boldsymbol{0}),\ \boldsymbol{z} = \boldsymbol{g}(\boldsymbol{y})\} \quad (7)$$

for a smooth function $\boldsymbol{g}\colon E^{\mathrm{s}} \to E^{\mathrm{s}\perp}$. We cannot generally compute $\boldsymbol{g}$, but it can be approximated as described in section 4.3.

The *global* stable and unstable manifolds are defined as the unions of backward and forward images of the local manifolds under the flow map: $W^{\mathrm{s}}(\boldsymbol{x}^{\mathrm{e}})$ is the set of *all* points whose orbits approach $\boldsymbol{x}^{\mathrm{e}}$ as $t \to +\infty$, even if they leave $B(\boldsymbol{x}^{\mathrm{e}})$ for a while, and $W^{\mathrm{u}}(\boldsymbol{x}^{\mathrm{e}})$ is defined analogously for $t \to -\infty$. Stable manifolds can intersect neither themselves nor the stable manifolds of other fixed points, since this would violate uniqueness of solutions (the intersection point would lead to more than one future). The same is true of unstable manifolds, but intersections of stable and unstable manifolds can and do occur; they lie on solutions that lead from one fixed point to another. Intersection points of manifolds that belong to the same fixed point are called *homoclinic*, while those for manifolds that belong to different fixed points are called *heteroclinic* (see figure 3(b)).

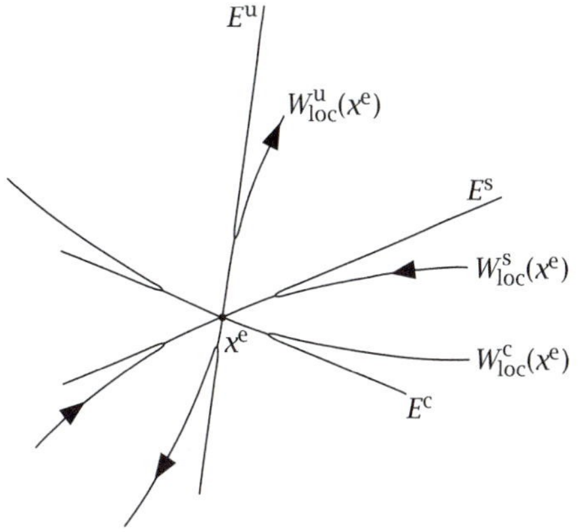


**Figure 4** The stable, center, and unstable manifolds.

### 4.3 Center Manifolds, Local Bifurcations, and Normal Forms

As parameters change, so do phase portraits and the resulting dynamics. New fixed points can appear, the stability of existing ones can change, and homoclinic and heteroclinic orbits can form and vanish. BIFURCATION THEORY [IV.21] addresses such questions, and it relies on three further ideas: structural instability, dimension reduction, and nonlinear coordinate changes.

As noted in section 2, a structurally stable system survives small perturbations of its defining vector field $\boldsymbol{f}$ or map $\boldsymbol{F}$ in the following sense. The phase portraits of the original and perturbed systems are *topologically equivalent*; they can be transformed into each other by a continuous coordinate change that preserves the sense of time, so that sinks remain sinks and sources, sources. Since eigenvalues of the Jacobian matrix determine stability, both the values and the derivatives of the perturbing functions must be small.

There are many ways in which structural stability can be lost, but the simplest is when a single (simple) eigenvalue passes through zero or when the real parts of a complex conjugate pair do the same. More generally, suppose that, in addition to $s$ and $u$ eigenvalues with negative and positive real parts, the Jacobian $\boldsymbol{Df}(\boldsymbol{x}^{\mathrm{e}})$ also has $c$ eigenvalues with zero real part ($s + c + u = n$).

The *center manifold theorem* asserts that, along with the $s$- and $u$-dimensional stable and unstable manifolds, a smooth $c$-dimensional local center manifold $W^{\mathrm{c}}_{\mathrm{loc}}$ exists, tangent to the subspace $E^{\mathrm{c}}$ spanned by the eigenvectors belonging to the eigenvalues with zero real part. As figure 4 suggests, this allows one to separate the locally stable and unstable directions from the structurally unstable ones and thus to reduce the analysis to that of a $c$-dimensional system restricted to the center manifold, a considerable simplification when $c \ll n$.

To describe the reduction process we assume that coordinates have been chosen with the degenerate equilibrium $\boldsymbol{x}^{\mathrm{e}}$ at the origin and such that the matrix $\boldsymbol{Df}(\boldsymbol{x}^{\mathrm{e}})$ is block diagonalized. Then (1) can be written in the form

$$\dot{\boldsymbol{x}} = \boldsymbol{Ax} + \boldsymbol{f}(\boldsymbol{x}, \boldsymbol{y}), \qquad \dot{\boldsymbol{y}} = \boldsymbol{By} + \boldsymbol{g}(\boldsymbol{x}, \boldsymbol{y}), \tag{8}$$

where $\boldsymbol{x} \in E^{\mathrm{c}}$ and $\boldsymbol{y} \in E^{\mathrm{s}} \oplus E^{\mathrm{u}}$, the span of the stable and unstable subspaces (explicit reference to the parameter $\mu$ is dropped). All eigenvalues of the $c \times c$ matrix $\boldsymbol{A}$ have zero real parts and the $(s+u) \times (s+u)$ matrix $\boldsymbol{B}$ has only eigenvalues with nonzero real parts. The center manifold can be expressed as a graph $\boldsymbol{y} = \boldsymbol{h}(\boldsymbol{x})$ over $E^{\mathrm{c}}$, and hence, as long as solutions remain on $W^{\mathrm{c}}_{\mathrm{loc}}$, the state $(\boldsymbol{x}, \boldsymbol{y})$ of the system is specified by the $\boldsymbol{x}$ variables alone. The *reduced system* is the projection of the vector field on $W^{\mathrm{c}}_{\mathrm{loc}}$ onto the linear subspace $E^{\mathrm{c}}$:

$$\dot{\boldsymbol{x}} = \boldsymbol{Ax} + \boldsymbol{f}(\boldsymbol{x}, \boldsymbol{h}(\boldsymbol{x})). \tag{9}$$

The graph $\boldsymbol{h}$ is found by substituting $\boldsymbol{y} = \boldsymbol{h}(\boldsymbol{x})$ into the second component of (8) and using the chain rule and the first component of (8) to obtain a (partial) differential equation:

$$\boldsymbol{Dh}(\boldsymbol{x})[\boldsymbol{Ax} + \boldsymbol{f}(\boldsymbol{x}, \boldsymbol{h}(\boldsymbol{x}))] = \boldsymbol{Bh}(\boldsymbol{x}) + \boldsymbol{g}(\boldsymbol{x}, \boldsymbol{h}(\boldsymbol{x})) \quad \text{with } \boldsymbol{h}(\boldsymbol{0}) = \boldsymbol{0},\ \boldsymbol{Dh}(\boldsymbol{0}) = \boldsymbol{0}. \tag{10}$$

The latter conditions are due to the tangency of $W^{\mathrm{c}}_{\mathrm{loc}}$ to $E^{\mathrm{c}}$ at $\boldsymbol{x}^{\mathrm{e}} = \boldsymbol{0}$. Solutions of (10) can be approximated as a Taylor series in $\boldsymbol{x}$, and stability and bifurcation behavior near the nonhyperbolic fixed point can be deduced from the resulting approximation to the reduced system (9).

The third idea—to choose a coordinate system that simplifies the nonlinear terms of the reduced system—effectively extends the use of similarity transformations to diagonalize and decouple components in linear systems. *Normal form theory* simplifies the Taylor series by iteratively performing nonlinear coordinate changes that successively remove, at each order, *nonresonant* terms that do not influence the qualitative behavior. Lie algebra provides bookkeeping methods, and the computations can be (semi-) automated using computer algebra.

To illustrate the resulting simplification, consider a two-dimensional ODE of the form

$$\dot{\boldsymbol{x}} = \boldsymbol{Ax} + \boldsymbol{f}(\boldsymbol{x}), \quad \boldsymbol{A} = \begin{bmatrix} 0 & -1 \\ 1 & 0 \end{bmatrix}, \tag{11}$$

the linear part of which is a harmonic oscillator whose phase plane is filled with periodic orbits surrounding a neutrally stable fixed point (a center). Asymptotic stability or instability of $\boldsymbol{x}^{\mathrm{e}} = \boldsymbol{0}$ depends on the nonlinear function $\boldsymbol{f}(\boldsymbol{x})$, and in particular on the leading terms in its Taylor series. There are six quadratic terms, eight cubic terms, and in general $2(k+1)$ terms of order $k$, but normal form transformations can successively remove all the even terms and all but two terms of each odd order, at the expense of modifying those terms. After such a transformation and written in polar coordinates ($x_1 = r\cos\theta$, $x_2 = r\sin\theta$), equation (11) becomes

$$\left.\begin{aligned} \dot{r} &= \alpha_3 r^3 + \alpha_5 r^5 + \mathcal{O}(r^7), \\ \dot{\theta} &= 1 + \beta_3 r^2 + \beta_5 r^4 + \mathcal{O}(r^6). \end{aligned}\right\} \tag{12}$$

Not only is the number of coefficients greatly reduced, but the circular symmetry implicit in the localized linear system is extended to higher order, uncoupling the azimuthal $\theta$ dynamics from the radial dynamics. Since the latter alone govern decay or growth of orbits, stability is determined by the first nonzero $\alpha_k$, explicit formulas for which emerge from the transformation.

Normal forms not only simplify the functions defining degenerate vector fields, they also allow the systematic introduction of parameters that *unfold* the bifurcation point to reveal the variety of structurally stable systems in its neighborhood, much as one can perturb a matrix to split a real eigenvalue of multiplicity 2 into a pair of distinct ones by adding a parameter. Equation (12), for example, is unfolded by the addition of a linear term $\mu r$ to the first component; as $\mu$ passes through zero, a Hopf bifurcation occurs, giving rise to a limit cycle (see BIFURCATION THEORY [IV.21]).

### 4.4 Limit Cycles and Poincaré Maps

The van der Pol equation—mentioned in section 2 as a motivating example in the study of chaotic orbits—also illustrates a simpler and more pervasive tool: the *first return* or *Poincaré map*. Without external forcing, this self-excited oscillator possesses a stable limit cycle, an isolated periodic orbit that attracts all nearby orbits. The ODE is

$$\dot{x}_1 = x_2 + \mu x_1 - \tfrac{1}{3}x_1^3, \qquad \dot{x}_2 = -x_1, \tag{13}$$

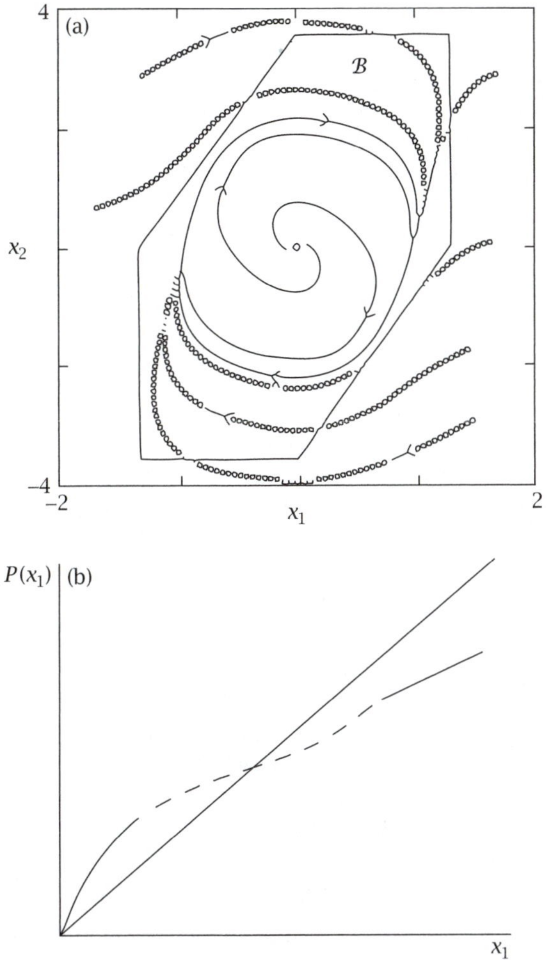


**Figure 5** (a) An annular trapping region $\mathcal{B}$ and the limit cycle of the van der Pol equation (13) with $\mu = 1$. (b) The Poincaré map for equation (13) expands lengths for $x_1$ small and contracts them for $x_1$ large, implying the existence of at least one fixed point corresponding to a periodic orbit.

and linearization reveals that, for $\mu > 0$, this planar system has a source at $(x_1, x_2) = (0, 0)$. It can, moreover, be shown that all orbits eventually enter and remain within an annular *trapping region* $\mathcal{B}$ surrounding the source. Since no fixed points lie in $\mathcal{B}$, the Poincaré–Bendixson theorem implies that it contains at least one periodic orbit (see figure 5(a)). As indicated at the end of section 4.3, one can also show that a stable periodic orbit appears in a Hopf bifurcation as $\mu$ passes through zero.

More generally, if $\gamma$ is a periodic orbit in an ODE of dimension $n \geqslant 2$, we may construct a *cross section* $\Sigma$, a subset of an $(n-1)$-dimensional manifold pierced by $\gamma$ at a point $\boldsymbol{p}$ and *transverse to the flow* in that all orbits cross $\Sigma$ with nonzero speed. Thus, by continuity, orbits starting at $\boldsymbol{q} \in \Sigma$ sufficiently close to $\boldsymbol{p}$ remain near $\gamma$ and next intersect $\Sigma$ again at a point $\boldsymbol{q}' \in \Sigma$, defining a Poincaré map

$$\boldsymbol{P}\colon \Sigma \to \Sigma \quad \text{or} \quad \boldsymbol{q} \mapsto \boldsymbol{q}' = \boldsymbol{P}(\boldsymbol{q}). \tag{14}$$

Evidently, $\boldsymbol{p}$ is a fixed point for the map $\boldsymbol{P}$.

The positive $x_1$-axis is a suitable cross section for equation (13), and the instability of the fixed point $(0, 0)$ and the attractivity of the trapping region $\mathcal{B}$ from outside imply that $P(x_1)$ takes the form sketched in figure 5(b). Since $P$ is continuous, it must intersect the diagonal at least once in a fixed point $p > 0$, corresponding to the limit cycle. In fact, $p$ is unique and the linearized map satisfies $0 < (\mathrm{d}P/\mathrm{d}x_1)|_{x_1=p} < 1$, implying asymptotic stability of both $p$ and the limit cycle. In general, if all the eigenvalues of the linearized map $\boldsymbol{DP}(\boldsymbol{p})$ have moduli strictly less than 1, then $\boldsymbol{p}$ and its associated periodic orbit are asymptotically stable; if at least one eigenvalue has modulus greater than 1, they are unstable.

This parallels the eigenvalue criteria for flows described in section 4.1, and the Poincaré map provides a second connection between ODEs and iterated maps (cf. the time $T$ flow map $\boldsymbol{\phi}_T$ of section 1). Conversely, a mapping $F\colon U \subset \mathbb{R}^n \to \mathbb{R}^n$ can be suspended to produce a $T$-periodic vector field on the $(n+1)$-dimensional space $\mathbb{R}^n \times S^1$. Analyses of the ODE (1) and the mapping (2) are therefore closely related, and analogs of the stable, unstable, and center manifold theorems hold for iterated maps.

### 4.5 Chaos and Strange Attractors

Several definitions of chaos have been advanced, but the following captures its key properties. A set of differential equations or an iterated map is called *chaotic*, or is said to have *chaotic solutions*, if it possesses a set $S$ of orbits such that (1) almost all pairs of orbits in $S$ display sensitive dependence on initial conditions; (2) there is an infinite set of periodic orbits that is dense in $S$; and (3) there is a dense orbit in $S$ (see CHAOS [II.3]).

*Sensitive dependence* means that, for any preassigned number $\beta < |S|$, any point $x_0$ in $S$, and any neighborhood $U$ of $x_0$, no matter how small, there exists a point $y_0 \in U$ and a time $T$ such that the orbits $x(T)$ and $y(T)$ starting at $x_0$ and $y_0$ are separated by at least $\beta$; almost all solutions diverge locally. An orbit $x(t)$ is *dense* in $S$ if, given any point $z$ in $S$ and any neighborhood $U$ of $z$, no matter how small, there exists a time $t_z$ such that $x(t_z)$ lies inside $U$: $x(t)$ passes arbitrarily close to all points in $S$.

For the doubling map, a binary sequence $\boldsymbol{a}^*$ corresponding to a dense orbit $x^*$ can be built by concatenating *all subsequences* of lengths 1, 2, 3, etc.: $\boldsymbol{a}^* = 0\ 1\ 00\ 01\ 10\ 11\ 000\ 001 \ldots.$ As one iterates and drops leading symbols, *every* subsequence appears at its head, implying that the orbit of $x^*$ contains points that lie arbitrarily close to every point in $[0,1]$. Hence $\{h^k(x^*)\}_{k=0}^{\infty}$ is dense and $h$ satisfies the definition, where $S = [0,1]$ is the entire state space.

*Positive topological entropy* also implies that typical orbits explore $S$, and there are examples of maps with this property that have *no* periodic orbits, so this definition is sometimes preferred. However, topological entropy (roughly speaking, the growth rate of distinct orbits as one progressively refines a discretization of $S$) is technically complicated, and the above definition will suffice here.

### 4.5.1 *Smale's Horseshoe*

The doubling map may seem to be a purely mathematical construct, but as promised in sections 2–3, it is related to the double pendulum and to forced nonlinear oscillators. To understand this, we now describe Smale's construction. Consider a piecewise-linear mapping $\boldsymbol{G}$ defined on the unit square $Q = [0,1] \times [0,1]$ by means of its action on two horizontal strips $H_0 = [0,1] \times [0,1/\mu]$ and $H_1 = [1-1/\mu, 1] \times [0,1]$ with images $V_0 = \boldsymbol{G}(H_0) = [0,\lambda] \times [0,1]$ and $V_1 = \boldsymbol{G}(H_1) = [1-\lambda, 1] \times [0,1]$, and having Jacobians

$$\boldsymbol{DG}|_{H_0} = \begin{bmatrix} \lambda & 0 \\ 0 & \mu \end{bmatrix}, \qquad \boldsymbol{DG}|_{H_1} = \begin{bmatrix} -\lambda & 0 \\ 0 & -\mu \end{bmatrix} \tag{15}$$

for $0 < \lambda < \frac{1}{2}$ and $\mu > 2$ (so that the images fit within $Q$ as defined). To make $\boldsymbol{G}$ continuous, the image of the central strip between $H_0$ and $H_1$ is taken as a semicircular arch joining $V_0$ to $V_1$. $\boldsymbol{G}$ can be thought of as the Poincaré map of a flow that compresses $Q$ horizontally, then stretches it vertically, and finally bends its middle to form the eponymous horseshoe (see figure 6).

The invariant set $\Lambda$ of $\boldsymbol{G}$ consists of points that remain in $Q$ under all forward and backward iterates: $\Lambda = \bigcap_{j=-\infty}^{+\infty} \boldsymbol{G}^j(Q)$. We construct $\Lambda$ step by step. Points that remain in $Q$ for one *backward* iteration occupy two vertical strips $V_0$ and $V_1$, the images of $H_0$ and $H_1$. $\boldsymbol{G}^{-2}(Q) \cap Q$ is obtained by considering the second iterates $\boldsymbol{G}(V_0)$ and $\boldsymbol{G}(V_1)$ of $H_0$ and $H_1$, which form nested arches standing on pillars, each of width $\lambda^2$, lying within $V_0$ and $V_1$. At each step, vertical distances are expanded by $\mu$ and horizontal distances shrunk by $\lambda$, and the middle $(1-2\lambda)$ fraction of the strips is removed. Continuing, $\Lambda_{-n} = \bigcap_{j=-n}^{0} \boldsymbol{G}^j(Q)$ is $2^n$ vertical strips, each of width $\lambda^n$, and passing to the limit, $\Lambda_{-\infty} = \bigcap_{j=-\infty}^{0} \boldsymbol{G}^j(Q)$ is a Cantor set of vertical line intervals. Similarly, $\Lambda_{+\infty} = \bigcap_{j=0}^{+\infty} \boldsymbol{G}^j(Q)$ is a Cantor set of horizontal intervals, and since any vertical and any horizontal segment intersect at a point, $\Lambda = \Lambda_{-\infty} \cap \Lambda_{+\infty}$ is a Cantor set of points.

A Cantor set is uncountable, closed, and contains no interior or isolated points; every point is an accumulation point. Cantor sets are examples of *fractals*, sets having fractional dimension.

Orbits of $\boldsymbol{G}$ can be described by *symbolic dynamics*, which codes points $\boldsymbol{x} \in \Lambda$ as binary sequences based on their visits to $H_0$ and $H_1$. A mapping $\boldsymbol{a}\colon \Lambda \to \{0,1\}^Z$, where $\boldsymbol{a} = \{a_j\}_{-\infty}^{+\infty}$ and $\{0,1\}^Z$ denotes the space of bi-infinite sequences with entries 0 or 1, is defined as follows:

$$a_j(\boldsymbol{x}) = \begin{cases} 0 & \text{if } \boldsymbol{G}(\boldsymbol{x}) \in H_0, \\ 1 & \text{if } \boldsymbol{G}(\boldsymbol{x}) \in H_1. \end{cases} \tag{16}$$

Given a suitable metric on $\{0,1\}^Z$, it can be shown that $\boldsymbol{a}$ is a one-to-one, continuous, invertible map, a homeomorphism. Every point in $\Lambda$ is faithfully coded by a sequence in $\{0,1\}^Z$, and, via equation (16), the action of $\boldsymbol{G}$ on $\Lambda$ becomes the shift map

$$\sigma\colon \{0,1\}^Z \to \{0,1\}^Z, \quad \text{with } a_j = \sigma(a_{j+1}). \tag{17}$$

This generalizes the binary representation of section 3.2 because $\boldsymbol{G}$ is invertible and points have past orbits as well as future orbits. The countably infinite set of periodic orbits is coded as before (for such an orbit, future and past are the same), but homoclinic and heteroclinic orbits to any periodic orbit or pair of orbits can now be formed by connecting semi-infinite periodic tails with an arbitrary central sequence. A dense orbit can be built by growing the sequence $0\ 1\ 00\ 01\ 10\ 11\ 000\ 001 \ldots$ forward and backward, and since all possible finite sequences appear in the set of all periodic orbits, this latter set is also dense in $\Lambda$.

$\boldsymbol{G}$ is linear on $H_0 \cup H_1$ with eigenvalues $\pm\lambda$ and $\pm\mu$ and $\lambda < 1 < \mu$, so $\Lambda$ is a hyperbolic set in which almost all pairs of orbits separate exponentially quickly and $\boldsymbol{G}$ is chaotic in the above sense. Moreover, it is structurally stable, so the chaos survives small perturbations, and Smale also proved that horseshoes appear in *any* smooth map that possesses a transverse homoclinic point.

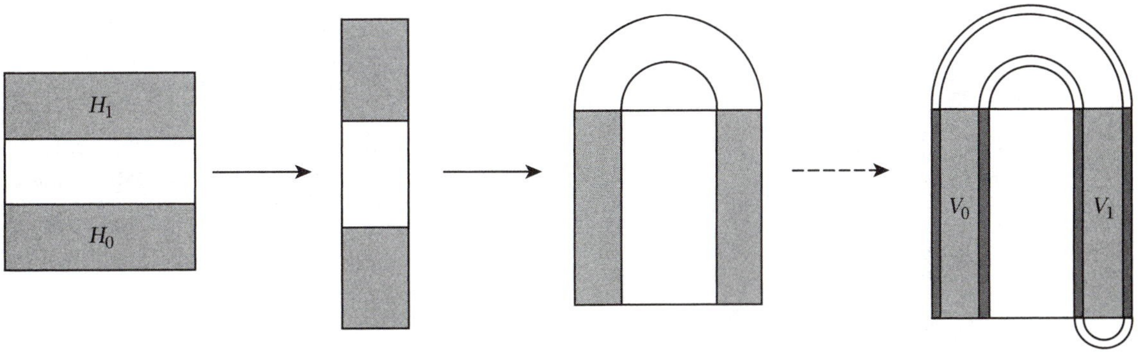


**Figure 6** Smale's horseshoe, showing the square $Q$ (left) and its images $\boldsymbol{G}(Q)$ (center right) and $\boldsymbol{G}^2(Q)$ (far right), with the strips $H_j$ and their images $V_j = \boldsymbol{G}(H_j)$ shaded.

There are computable conditions that detect homoclinic points for rather general classes of weakly perturbed systems. Applying these to the ODEs describing a double pendulum, like that of figure 1 but with a heavy upper link and a light lower one, allows one to prove that, given any alternating sequence of positive and negative integers $s_1, -s_2, s_3, \dots$, an orbit exists on which the lower link first turns $s_1$ times clockwise, then $s_2$ times counterclockwise, then $s_3$ times clockwise, and so on. Since any sequence chosen from a suitable probability distribution corresponds to an orbit, one cannot tell whether the motion is deterministic or stochastic.

The horseshoe $\Lambda$ is a set of saddle type; almost all orbits that approach it eventually leave, so its presence does not necessarily imply physically observable chaos. However, its stable manifold can form a *fractal boundary* separating orbits that have different fates, as in the forced van der Pol equation. To observe persistent chaos, one requires strange attractors.

### 4.5.2 *Strange Attractors*

Again there are subtle issues and competing definitions, but to convey the main ideas we define an *attracting set* $\mathcal{A}$ as the intersection of all forward images of a trapping region $\mathcal{B}$,

$$\mathcal{A} = \bigcap_{t \geqslant 0} \phi_t(\mathcal{B}), \qquad (18)$$

and an *attractor* as an attracting set that contains a dense orbit. Sinks and asymptotically stable limit cycles provide simple examples, as do stable invariant tori that carry quasiperiodic, and hence densely winding, motions. A *strange attractor* additionally exhibits sensitive dependence and chaos. The following example further emphasizes the geometrical viewpoint of dynamical systems theory.

The LORENZ EQUATIONS [III.20] are a (very) low-dimensional projection of the coupled Navier–Stokes and heat equations modeling convection in a fluid layer:

$$\left.\begin{aligned} \dot{x}_1 &= \sigma(x_2 - x_1), \\ \dot{x}_2 &= \rho x_1 - x_2 - x_1 x_3, \\ \dot{x}_3 &= -\beta x_3 + x_1 x_2. \end{aligned}\right\} \qquad (19)$$

The parameters $\sigma$ and $\beta$ are fixed and $\rho$ is proportional to the temperature difference across the layer. The origin is always a fixed point, representing stationary fluid and heat transport by conduction. For $\rho > 1$, $\boldsymbol{x} = \boldsymbol{0}$ is a saddle point with one positive and two negative real eigenvalues; two further fixed points $\boldsymbol{q}^{\pm}$, corresponding to steadily rotating convection cells, also exist. For $\sigma = 10$, $\rho = 28$, $\beta = \frac{8}{3}$ (the values used by Lorenz), these are also saddles, but a trapping region exists and (19) has an attracting set $\mathcal{A}$ containing the unstable fixed points.

Lorenz, who had studied with Birkhoff, realized that $\mathcal{A}$ has infinitely many "sheets." He also generated a one-dimensional return map related to the doubling map of section 3.2 and gave a symbolic description of its orbits. We adopt the geometric model of Guckenheimer and Williams here, defining a cross section $\Sigma$ lying above the origin. In suitable nonlinear coordinates, $\Sigma$ is a square $[-1,1] \times [-1,1]$ whose boundaries $\pm 1 \times [-1,1]$ contain $\boldsymbol{q}^{\pm}$ and their local stable manifolds, and whose centerline $0 \times [-1,1]$ lies in the stable manifold of $\boldsymbol{0}$. Assuming that all orbits leaving $\Sigma$ circulate around $\pm 1 \times [-1,1]$ (except those in $0 \times [-1,1]$, which flow into $\boldsymbol{x} = \boldsymbol{0}$), a Poincaré map $\boldsymbol{F}$ can be defined (see figure 7(a)).

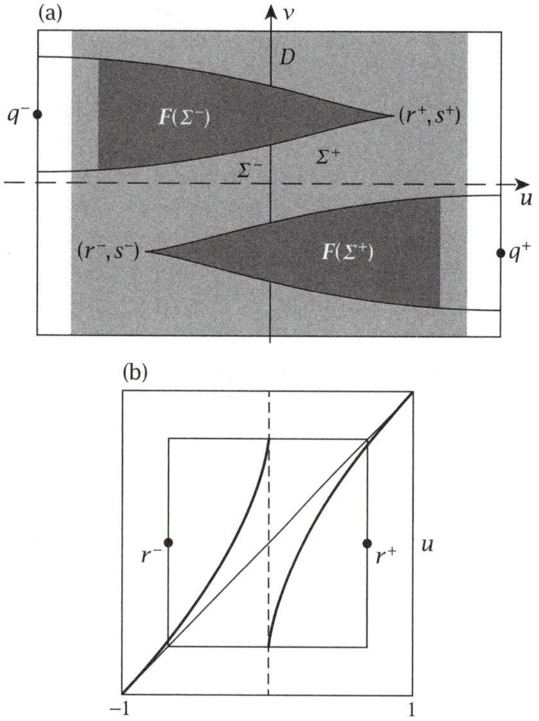


**Figure 7** The Poincaré map of the geometric Lorenz attractor. (a) A cross section $\Sigma$ showing $\Sigma^\pm$ and their images $\boldsymbol{F}(\Sigma^\pm)$, and the subset $V$ (light gray) and its image $\boldsymbol{F}(V)$ (dark gray). (b) The one-dimensional map $f(u)$.

More precisely, Guckenheimer and Williams assert that coordinates $(u, v)$ can be chosen in which the horizontal ($u$) dynamics decouples and $\boldsymbol{F}$ takes the form

$$\boldsymbol{F}(u, v) = (f(u), g(u, v)), \qquad (20)$$

where $g$ contracts in the $v$-direction, $f$ expands by a factor greater than or equal to $\sqrt{2}$ in the $u$-direction, and $\boldsymbol{F}(-u, -v) = -\boldsymbol{F}(u, v)$ respects the symmetry $(x_1, x_2, x_3) \to (-x_1, -x_2, x_3)$ of (19). Thus, $\boldsymbol{F}$ maps the open rectangles $\Sigma^- = (-1, 0) \times (-1, 1)$ and $\Sigma^+ = (0, 1) \times (-1, 1)$ into the interior of $\Sigma$, and since orbits starting near $u = 0$ pass close to $\boldsymbol{0}$, the strong stable eigenvalue of $\boldsymbol{0}$ pinches the images of $\Sigma^\pm$ at their endpoints $(r^\pm, s^\pm) = \lim_{u \to 0} \boldsymbol{F}(u, v)$. Continuing to iterate, a complicated attracting set appears: at the second step, $\boldsymbol{F}(\Sigma^\pm)$ comprises 4 thinner strips, 2 each inside $\boldsymbol{F}(\Sigma^-)$ and $\boldsymbol{F}(\Sigma^+)$, then 8, 16, etc., as in the Cantor set of Smale's horseshoe, but pinched together at their ends. Defining a subset $V \subset \Sigma$ (shaded in figure 7), it can be shown that $\mathcal{A} = \bigcap_{n \geqslant 0} \boldsymbol{F}^n(V)$ contains a dense orbit and so $\mathcal{A}$ is an attractor. Due to the expansive nature of $f$ (figure 7(b)), which resembles the doubling map of section 3.2, $\boldsymbol{F}$ has sensitive dependence and $\mathcal{A}$ is therefore strange.

The geometric picture has been verified by W. Tucker for equation (19) using a computer-assisted proof showing that a *stable foliation* exists, a continuous family of curves $\mathcal{F}$ such that, if $C \in \mathcal{F}$, then $\boldsymbol{F}(C) \in \mathcal{F}$ (here, $\mathcal{F}$ is composed of vertical line segments $u = \text{const.}$).

The key properties necessary for strange attractors are (1) stretching in some state-space directions and (2) contraction in others, so that volumes decrease, coupled with (3) bending (the horseshoe) or discontinuous cutting (the Lorenz example) to place forward images under the flow map $\boldsymbol{\phi}_t$ into a trapping region. There are now many examples, and smooth maps like that of the horseshoe, which often arise in applications, can produce very complicated dynamics, including infinite sequences of homoclinic bifurcations.

## 5 Conclusion

I have sketched some central ideas and themes in dynamical systems theory but have necessarily omitted much. In closing, here are some important topics and connections to other concepts, areas, and problems described in this volume.

Chaotic orbits admit statistical descriptions, and there is a flourishing *ergodic theory* of dynamical systems, describing, for example, invariant measures supported on strange attractors and decay of correlations along orbits. Stochastic ODEs are also treated probabilistically (see APPLICATIONS OF STOCHASTIC ANALYSIS [IV.14]). Interest in HYBRID SYSTEMS [II.18], with nonsmooth vector fields and discontinuous jumps, is growing, and the classification of their bifurcations proceeds (see SLIPPING, SLIDING, RATTLING, AND IMPACT: NONSMOOTH DYNAMICS AND ITS APPLICATIONS [VI.15]).

Many equations inherit symmetries from the phenomena they model, and *equivariant dynamical systems* is an important area. Symmetries can both constrain behaviors and stabilize objects that typically lack structural stability in nonsymmetric systems (e.g., heteroclinic cycles). Symmetries also profoundly affect normal forms and their unfoldings (see SYMMETRY IN APPLIED MATHEMATICS [IV.22]).

The Lorenz and other examples show that steady three-dimensional velocity fields of fluids or other continuums can exhibit chaotic mixing, which dramatically enhances transport of heat, pollutants, and reactants in such flows (see CONTINUUM MECHANICS [IV.26]).

### Further Reading


Aubin, D., and A. Dahan Dalmedico. 2002. Writing the history of dynamical systems and chaos: *longue durée* and revolution, disciplines and cultures. *Historia Mathematica* 29:273–339.

Barrow-Green, J. 1997. *Poincaré and the Three Body Problem.* Providence, RI: American Mathematical Society.

Constantin, P., C. Foias, B. Nicolaenko, and R. Temam. 1989. *Integral Manifolds and Inertial Manifolds for Dissipative Partial Differential Equations.* New York: Springer.

Guckenheimer, J., and P. Holmes. 1983. *Nonlinear Oscillations, Dynamical Systems and Bifurcations of Vector Fields.* New York: Springer.

Hirsch, M. W., S. Smale, and R. L. Devaney. 2004. *Differential Equations, Dynamical Systems and an Introduction to Chaos.* San Diego, CA: Academic Press/Elsevier.

Lorenz, E. N. 1963. Deterministic non-periodic flow. *Journal of Atmospheric Science* 20:130–41.

Poincaré, H. J. 1892, 1893, 1899. *Les Méthodes Nouvelles de la Mécanique Celeste*, volumes 1–3. Paris: Gauthiers-Villars. (English translation, edited by D. Goroff, published by the American Institute of Physics (New York, 1993).)

Smale, S. 1967. Differentiable dynamical systems. *Bulletin of the American Mathematical Society* 73:747–817.

Temam, R. 1997. *Infinite-Dimensional Dynamical Systems in Mechanics and Physics*, 2nd edn. New York: Springer.

Ueda, Y. 2001 *The Road to Chaos*, volume 2. Santa Cruz, CA: Aerial Press.


## IV.21 Bifurcation Theory

*Paul Glendinning*

### 1 Introduction

Bifurcation theory describes the qualitative changes in the dynamics of systems caused by small changes of parameters in the model. It uses DYNAMICAL SYSTEMS THEORY [IV.20] to determine the different conditions that lead to changes, and to describe the dynamics as a function of the parameter. This leads to a theoretical list of typical changes that can be used to understand the existence and stability of different solutions in a problem. Parameters are present in most applications. They are quantities that are constant in any realization of the system but that may take different values in different realizations depending on the details of the situation being modeled. Examples include the Raleigh and Reynolds numbers in fluid dynamics, virulence in disease models, and reaction rates in chemistry.

The "tipping points" that generate so much interest in economics, say, or climate change are good examples of the idea of a bifurcation: if the parameters satisfy some condition then the associated dynamics is of one type, but if they stray above a threshold, even by a small amount, then the resultant dynamics can change drastically.

There are two types of bifurcation, though the two can be mixed in more complicated problems. One involves changes in the spectrum of the linearization of a system about some solution; this type is an extension of the classical linear stability analysis that dominated so much of applied mathematics in the 1950s and 1960s. This leads to *local bifurcation theory*. The second type of bifurcation builds on Poincaré's qualitative analysis of dynamical systems to look at global features of the system, and in particular homoclinic and heteroclinic orbits (a homoclinic orbit approaches the same solution in forward and backward time, while a heteroclinic orbit is a solution that tends to one solution in forward time and another in backward time). The analysis of perturbations of these solutions is the basis of *global bifurcation theory*.

Although bifurcation theory can be applied to more general systems, we will restrict ourselves to the cases of autonomous ordinary differential equations in continuous time, which generate flows, and maps in discrete time, which generate sequences, i.e.,

$$\begin{aligned} \dot{x} &= f(x,\mu) \quad \text{(flow)}, \\ x_{n+1} &= f(x_n,\mu) \quad \text{(map)}, \end{aligned} \tag{1}$$

where $x \in \mathbb{R}^p$ are the dependent variables and $\mu \in \mathbb{R}^m$ are the parameters. The simplest solutions are those that are constant. For flows, this implies that $\dot{x} = 0$, so $f(x,\mu) = 0$. These solutions are called stationary points. For maps, a constant solution is called a fixed point. Fixed points satisfy $x_{n+1} = x_n$, or $x = f(x,\mu)$. Since periodic orbits of flows can be analyzed using a RETURN MAP [IV.20], the discrete-time case can describe behavior near periodic orbits of flows.

In some sense, then, bifurcation theory for the simplest dynamical objects is about the variation in the number of solutions of equations such as $f(x,\mu) = 0$ as the parameters vary. This leads to an approach through singularity theory. The linear stability of solutions is determined by the eigenvalues of the linearization (the Jacobian matrix) of the flow or map. In continuous time, simple eigenvalues of the Jacobian matrix $\lambda$ lead to solutions that are proportional to $e^{\lambda t}$ and eigenvalues with negative real parts are therefore stable while those with positive real parts are unstable. In the discrete-time case, solutions are proportional to $\lambda^n$, so the role of the imaginary axis is replaced by the unit circle since $\lambda^n \to 0$ as $n \to \infty$ if $|\lambda| < 1$. The power of local bifurcation theory lies in its ability to go beyond this linear approach.

Linearization determines the local behavior near stationary points and fixed points provided the eigenvalues of the Jacobian matrix are not on the imaginary axis (for continuous-time flows) or the unit circle (for discrete-time maps). Such solutions are called *hyperbolic*, and no local bifurcations can occur at hyperbolic solutions. Local bifurcation theory therefore considers behavior near nonhyperbolic solutions; local in both phase space and parameter space.

One further term that is used in bifurcation theory needs explanation before we move on to examples. The *codimension* of a bifurcation is essentially the number of parameters required to be able to observe the bifurcation in typical systems. Codimension-one bifurcations are therefore observed in one-parameter families of systems, while higher-codimension bifurcations are observed by varying more parameters. These can often act as organizing centers for bifurcations with lower codimension.

## 2 Five Canonical Examples

The "typical" local bifurcations can be classified using their Taylor series expansions. In section 3 we discuss why the changes in dynamics described here are actually more general, but first we simply describe the different possibilities using examples.

The first is the *saddle-node* or *tangent* bifurcation. Consider the following scalar equations, with $p = m = 1$ in (1):

$$\dot{x} = \mu - x^2 \qquad \text{(flow)},$$
$$x_{n+1} = x_n + \mu - x_n^2 \quad \text{(map)}.$$

Looking for stationary points of the continuous-time system ($\dot{x} = 0$) and fixed points of the discrete-time system ($x_{n+1} = x_n$) gives $x^2 = \mu$, so if $\mu < 0$ there are no such points, while if $\mu > 0$ there are two, $x_\pm = \pm\sqrt{\mu}$. Stability is determined by the $1 \times 1$ Jacobian matrix of derivatives. For the flow this is $-2x_\pm$, so $x_+$ is stable (negative eigenvalue) and $x_-$ is unstable. For the map the Jacobian is $1 - 2x_\pm$, which has modulus less than one at $x_+$ (for small $\mu$), indicating stability, while $x_-$ is unstable. The existence and stability of stationary points or fixed points are represented in the *bifurcation diagram* of figure 1(a), which shows the locus of stationary points and periodic orbits on the vertical axis as a function of the parameter (on the horizontal axis). By convention, the stability of solutions is indicated by plotting stable solutions with solid lines and unstable solutions with dotted lines.

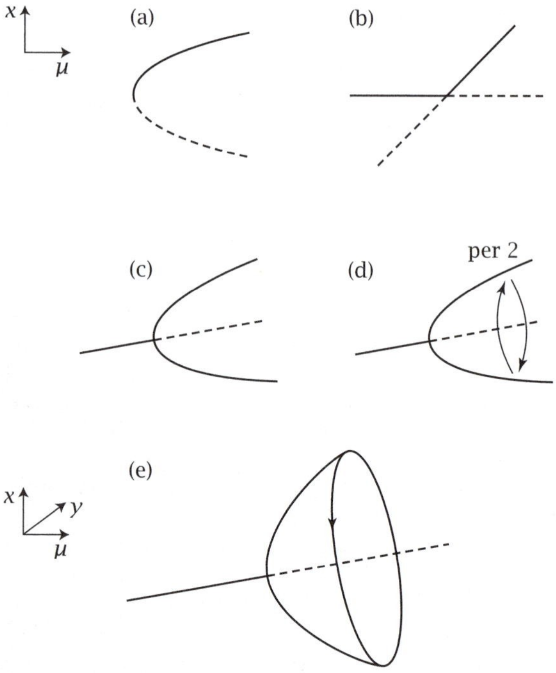

**Figure 1** The simple bifurcations.

If the origin is constrained to be a solution for all $\mu$, or if the linear term in the parameter of the Taylor series of the function $f(x, \mu)$ of (1) vanishes at the bifurcation point, then a simple scalar model is

$$\dot{x} = \mu x - x^2 \qquad \text{(flow)},$$
$$x_{n+1} = (1 + \mu)x_n - x_n^2 \quad \text{(map)}.$$

In these cases the stationary points are at $x = 0$ and $x = \mu$, and for small $|\mu|$ the origin is stable if $\mu < 0$ and unstable if $\mu > 0$, while the nontrivial solution has the opposite stability properties. This is called a *transcritical* bifurcation or *exchange of stability*, since the two branches of solutions cross and exchange their stability properties, as shown in figure 1(b).

If there is symmetry in the problem or if additional coefficients in the Taylor expansion of the map are zero, then the quadratic part may vanish identically, leaving

$$\dot{x} = \mu x - x^3 \quad \text{(flow)},$$
$$x_{n+1} = (1+\mu)x_n - x_n^3 \quad \text{(map)}.$$

This leads to a *pitchfork* bifurcation. If $|\mu|$ is sufficiently small, then the origin is stable if $\mu < 0$ and unstable if $\mu > 0$. If $\mu > 0$, a pair of new stable stationary solutions with $x^2 = \mu$ also exists. This is a *supercritical* pitchfork bifurcation: the pair of bifurcating solutions are stable (see figure 1(c)). If these solutions are unstable, then the bifurcation is said to be *subcritical*.

For maps there is another way in which a fixed point can lose stability in one-dimensional systems: the eigenvalue of the linearization can pass through $-1$. In this case the canonical example is

$$x_{n+1} = -(1+\mu)x_n + x_n^3,$$

which has a unique local fixed point at $x = 0$ that is stable if $\mu < 0$ and unstable if $\mu > 0$ (with $|\mu|$ sufficiently small). Due to the symmetry of the equations it is easy to see that there are two nontrivial solutions to the equation $x_{n+1} = -x_n$ if $\mu > 0$. These lie on a stable orbit of period two ($x \to -x \to x$), and this is therefore called a *period-doubling* bifurcation (see figure 1(d)). To indicate that the bifurcating period-two orbit is stable, it is called a supercritical period-doubling bifurcation; if the new orbit is unstable, coexisting with the stable fixed point, it is called subcritical. These bifurcations do not have a direct analogue for stationary points of flows, but they can occur as bifurcations of periodic orbits of flows, where the map represents a Poincaré map of the flow.

Another local bifurcation, shown in figure 1(e), occurs only in maps or flows with a phase space of dimension greater than one: a pair of complex conjugate eigenvalues pass through the imaginary axis for flows, or the unit circle for maps. Using polar coordinates $(r, \theta)$ for the flow and complex coordinates $z = x + \mathrm{i}y$ for the map, the canonical examples are

$$\dot{r} = \mu r - r^3, \qquad \dot{\theta} = \omega \quad \text{(flow)},$$
$$z_{n+1} = (1+\mu)\mathrm{e}^{2\pi\mathrm{i}\omega} z_n - |z_n|^2 z_n \quad \text{(map)}.$$

The origin is stable once again if $\mu < 0$ and $|\mu|$ is sufficiently small, but if $\mu > 0$ there is an attracting circle with radius $\sqrt{\mu}$. In the continuous case, this is a stable periodic orbit, and the bifurcation is called a supercritical *Hopf* bifurcation. In the discrete case, the dynamics on the invariant circle is more complicated: solutions are periodic if $\omega$ is rational, and nonperiodic and dense on the circle if $\omega$ is irrational. With more general nonlinear terms, this Hopf (or *Neimark–Sacker*) bifurcation of maps has more cases. The rational case usually breaks up into a finite collection of stable and unstable periodic orbits, and these exist over intervals of parameters in a phenomenon called *mode locking*.

## 3 Dimension Reduction

The examples above are relevant more generally because there is a (local) dimension reduction possible near nonhyperbolic fixed points. Intuitively, this can be described by noting that, in eigenspaces corresponding to eigenvalues that are not on the imaginary axis (for flows) or the unit circle (for maps), the dynamics is determined by the linearization and decays to zero in forward time for stable directions or backward time for unstable directions. The only interesting directions in which changes can occur are, therefore, the "central" directions associated with zero or purely imaginary eigenvalues (flows) or eigenvalues with modulus one (maps).

It turns out that this observation can be made rigorous in at least two ways. These are the CENTER MANIFOLD THEOREM [IV.20 §4.3] and Lyapunov–Schmidt reduction. These describe how to construct projections onto the nonhyperbolic directions (valid for parameters close to the values at which the nonhyperbolic solution exists) such that any change in dynamics occurs in this projection. In particular, if there is a simple nonhyperbolic eigenvalue (the "typical" case), then the analysis reduces to a system with the same dimension as the corresponding eigenspace, i.e., dimension one for a real eigenvalue and dimension two for a complex conjugate pair of eigenvalues.

On these projections the leading terms of the Taylor series expansion of the projected equations are essentially the examples of the previous section. In the Hopf bifurcation for maps there are some extra subtleties that are admirably described in Arrowsmith and Place's book *An Introduction to Dynamical Systems*.

The mathematical analysis of the bifurcations described above then follows using the dimension reduction and further analysis, e.g., by using the implicit function theorem to describe all possible local fixed points or periodic orbits. This produces a set of *genericity (or nondegeneracy) conditions* in terms of higher-order derivatives of the function $f$ of (1) at the bifurcation point that need to be satisfied in addition to the existence of a neutral direction if the corresponding bifurcation is to be observed.

Although the center manifold theorem and Lyapunov–Schmidt reduction lead to the same conclusions about dimension reduction, they use quite different methods. The center manifold theorem is established using invariant manifold theory for dynamical systems, e.g., by following the invariant manifold from the trivial linear case into the nonlinear regime. Lyapunov–Schmidt reduction relies on projection techniques onto linear eigendirections.

## 4 Global Bifurcations

Many global bifurcations involve homoclinic orbits. A homoclinic orbit to a stationary point of a flow is an orbit that approaches the stationary point in forward and backward time. This is a codimension-one phenomenon in typical families of differential equations, and so there is a straightforward bifurcation in which a given homoclinic orbit exists at an isolated value of the parameter. For periodic orbits the situation is rather different because homoclinic orbits typically persist over a range of parameter values, and the interest is therefore in how homoclinic orbits are created or destroyed as the parameter varies.

### 4.1 Homoclinic Bifurcations to Stationary Points: Planar Flows

Suppose an autonomous flow in the plane is defined by a differential equation that can be written (after a change of coordinates) as

$$\dot{x} = \lambda_1 x + f_1(x, y; \mu), \qquad \dot{y} = -\lambda_2 y + f_2(x, y; \mu), \quad (2)$$

where the functions $f_i$ vanish at the origin for all $\mu$ and denote nonlinear terms. The direction of time has been chosen such that $\lambda_2 > \lambda_1 > 0$, so locally the stationary point at the origin has a saddle structure as shown in figure 2. There is a one-dimensional stable manifold of solutions that tend to the origin as $t \to \infty$ that is tangential to the $y$-axis at the origin, and there is a one-dimensional unstable manifold of solutions that tend to the origin as $t \to -\infty$ that is tangential to the $x$-axis at the origin. Can these stable and unstable manifolds intersect?

Typically, one-dimensional sets can intersect transversely in two dimensions, so it might be expected that it is fairly easy for such intersections to occur and that these would be persistent under small changes of the parameters. However, if the two manifolds intersect at a point, then the trajectory through the intersection point must also be in the intersection of the manifolds. In other words, the intersection cannot be transverse, and a fairly straightforward argument shows that

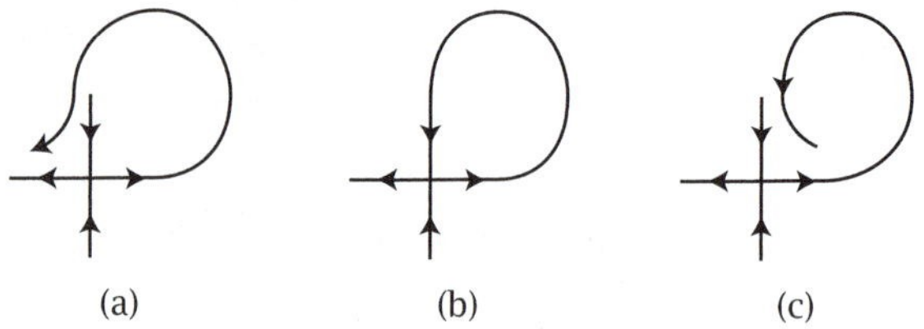


**Figure 2** The unstable manifold of a planar homoclinic bifurcation: (a) $\mu < 0$; (b) $\mu = 0$, showing the return planes used in the construction of the return map; and (c) $\mu > 0$.

such intersections typically occur on codimension-one manifolds in parameter space.

Now suppose that if $\mu = 0$ there is a homoclinic orbit, as shown in figure 2(b). Then as $\mu$ passes through zero we can expect that the arrangements of the stable and unstable manifolds vary as shown in parts (a) and (c) of figure 2. The local change in behavior can be determined using a return map approach. Close to the stationary point, the flow is approximated by the linear part of (2), as nonlinear terms are much smaller, so approximate solutions are

$$x = x_0 e^{\lambda_1 t}, \qquad y = y_0 e^{-\lambda_2 t}.$$

Thus if $h > 0$ is a small constant, solutions that start at $(x_0, h)$ with $x_0 > 0$ intersect $x = h$ after time $T$, where $h \approx x_0 e^{\lambda_1 T}$ or $T \approx (1/\lambda_1)\log(h/x_0)$. At this intersection, $y = y_1 \approx h e^{-\lambda_2 T} \approx h(x_0/h)^{\lambda_2/\lambda_1}$. In other words, a solution starting at $(x_0, h)$ with $x_0 > 0$ intersects $x = h$ at $(h, y_1)$ and then evolves close to the unstable manifold away from the stationary point until it returns to $y = h$. If we choose the parametrization so that the unstable manifold of the stationary point first strikes $y = h$ at $(\mu, h)$, then the solution starting at $(h, y_1)$ with $y_1 > 0$ small enough will intersect $y = h$ at $(x_1, h)$ close to $(\mu, h)$. Since this flow is bounded away from stationary points, the return map is smooth and can be approximated by a Taylor series: $x_1 = \mu + Ay_1 + \cdots$. Composing these two maps, the linear approximation near the origin and the return map close to the homoclinic orbit outside a neighborhood of the origin, we find that the full flow close to the homoclinic orbit for $|\mu|$ sufficiently small is modeled by the approximate return map

$$x_{n+1} \approx \mu + a x_n^{\delta}, \quad x_n > 0, \quad \delta = \lambda_2/\lambda_1,$$

with $a$ constant, and the map is undefined if $x_n \leqslant 0$ (if $x_n = 0$ the solution returns on the stable manifold and tends to the stationary point, while if $x_n < 0$ we have no information about the behavior of the unstable manifold). Since $\delta > 1$, there is a simple fixed point if

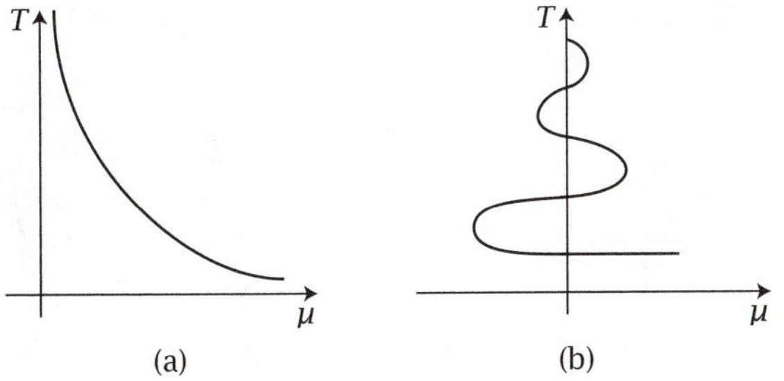


**Figure 3** Parameter ($\mu$) versus period ($T$) plots of the simple periodic orbit as it approaches the homoclinic orbit at $\mu = 0$ with $T \to \infty$. (a) Real eigenvalues, or a saddle-focus, with $\rho/\lambda > 1$. (b) A saddle-focus with $\rho/\lambda < 1$ (the Shilnikov case).

$\mu > 0$ at $x^* \approx \mu$ that is stable since the slope of the map is small, of order $x^{\delta-1}$. Most of the time is spent in the neighborhood of the stationary point, and so as $\mu$ tends to zero from above, the period scales like

$$T \sim -\frac{1}{\lambda_1}\log\mu,$$

i.e., it tends to infinity and approaches the homoclinic orbit itself. If $\mu < 0$ then there is no periodic solution locally. The effect of the bifurcation is to destroy or create a stable periodic orbit. The period of this orbit as a function of the parameter is sketched in figure 3(a).

### 4.2 Homoclinic Bifurcations to Stationary Points: Lorenz-Like Flows

In $\mathbb{R}^3$, the bifurcation of a homoclinic orbit to a stationary point with a Jacobian matrix that has real eigenvalues is typically similar to the planar case. A periodic orbit is created or destroyed by the bifurcation. An interesting exception occurs if the system has the symmetry $(x, y, z) \to (-x, -y, z)$ with linearization

$$\begin{aligned}\dot{x} &= \lambda_1 x + f_1(x, y, z; \mu),\\ \dot{y} &= -\lambda_2 y + f_2(x, y, z; \mu),\\ \dot{z} &= -\lambda_3 z + f_3(x, y, z; \mu),\end{aligned}$$

with $\lambda_2 > \lambda_3 > \lambda_1 > 0$, and again the functions $f_i$ vanish at the origin for all $\mu$ and denote nonlinear terms. In this case the second branch of the unstable manifold (the branch in $x < 0$) is the symmetric image of the branch in $x > 0$, and analysis similar to that of the previous subsection implies that the return map is defined in both $x > 0$ and $x < 0$ by symmetry:

$$x_{n+1} \approx \begin{cases} \mu + a x_n^{\delta}, & x_n > 0,\\ -\mu - a|x_n|^{\delta}, & x_n < 0,\end{cases}$$

with $\delta = \lambda_3/\lambda_1 < 1$. If $a > 0$ then the slope of the map tends to infinity as $x$ tends to 0 from above or below, so there are no stable periodic orbits. In fact, if $\mu > 0$ all solutions leave a neighborhood of $x = 0$. However, if $\mu < 0$ then there is an unstable chaotic set of solutions similar to the chaotic set for the map $2x$ (mod 1) in the article on DYNAMICAL SYSTEMS [IV.20 §3.2]. It is this chaotic set, or more accurately a subset of this set, that stabilizes at higher parameter values to become the LORENZ ATTRACTOR [III.20].

### 4.3 Homoclinic Bifurcations to Stationary Points: Shilnikov Flows

In generic systems a homoclinic orbit will approach the stationary point tangential to the eigenspace associated with the eigenvalue of the Jacobian matrix with smallest positive real part as $t \to -\infty$ and the eigenvalue with negative real part having the smallest modulus as $t \to \infty$. These are called the leading eigenvalues of the stationary point. In higher dimensions it is only the leading eigenvalues that determine the behavior of generic systems. The results can be phrased for general flows in $\mathbb{R}^p$, but for simplicity we consider only $p = 3$ and $p = 4$.

This means that there are two extra cases: the *saddle-focus*,

$$\begin{aligned}\dot{x} &= \lambda x + f_1(x, y, z; \mu),\\ \dot{y} &= -\rho y + \omega z + f_2(x, y, z; \mu),\\ \dot{z} &= -\omega y - \rho z + f_3(x, y, z; \mu),\end{aligned}$$

where $\rho$ and $\lambda$ are positive; and the *focus-focus*

$$\begin{aligned}\dot{x} &= \rho x + \omega y + f_1(x, y, z, w; \mu),\\ \dot{y} &= -\omega x + \rho y + f_2(x, y, z, w; \mu),\\ \dot{z} &= -Rz + \Omega w + f_3(x, y, z, w; \mu),\\ \dot{w} &= -\Omega z - Rw + f_4(x, y, z, w; \mu),\end{aligned}$$

with $R, \rho > 0$, and $\Omega$ and $\omega$ nonzero. In both sets of equations the functions $f_i$ are nonlinear and vanish at the origin for all $\mu$.

The dynamics associated with the existence of homoclinic orbits to stationary points described locally by the saddle-focus or focus-focus was analyzed by Leonid Shilnikov in the mid-to-late 1960s. For the saddle-focus with $\rho/\lambda > 1$, the situation is similar to the planar case: a periodic orbit can be continued with changing parameter and approaches the homoclinic orbit monotonically as in figure 3(a). In the remaining cases the situation is more exotic. Shilnikov proved that at the parameter value for which the homoclinic orbit exists there are a countable number of (unstable) periodic orbits and the dynamics contains horseshoes (and therefore unstable chaotic sets: see dynamical systems). Concentrating on the simplest periodic orbits—those that visit a neighborhood of the origin once in each period—it is possible to show that in the saddle-focus case with $\rho/\lambda < 1$ these exist with period $T$ at parameters $\mu$, where (to lowest order)

$$\mu \sim A\mathrm{e}^{-\rho T}\cos(\omega T + \phi),$$

which is an oscillatory approach to $\mu = 0$ as $T \to \infty$, as shown in figure 3(b). The turning points here correspond to saddle-node bifurcations, and on every branch involving stable orbits in the saddle-node bifurcations the orbit loses stability via period doubling before it reaches $\mu = 0$. Moreover, there are infinite sequences of more complicated homoclinic orbits accumulating on $\mu = 0$; these homoclinic orbits pass through a neighborhood of the stationary point more than once before tending to that point, and they are called *multipulse* homoclinic orbits. The focus-focus case is similar, but the accumulations can be more complicated.

### 4.4 Homoclinic Tangles

Whereas the intersection of stable and unstable manifolds is generally a codimension-one phenomenon for flows, it is persistent for maps. This is because the intersection must be at least one dimensional for flows, whereas for maps the existence of a point of intersection implies the existence of only a countable set of intersections. If a point $p$ is on the intersection of the stable and unstable manifolds of a fixed point of a smooth invertible map $f$, then so are its preimages $f^{-n}(p)$, $n = 1, 2, 3, \ldots$, which accumulate on the fixed point along its unstable manifold, and its images $f^n(x)$, which accumulate on the fixed point along its stable manifold. This gives rise to *homoclinic tangles*, as illustrated in figure 4(a). Transverse intersections of stable and unstable manifolds imply that chaotic sets exist, although the chaotic dynamics is not stable. This phenomenon also occurs in periodically forced systems, which can be analyzed via stroboscopic (or Poincaré) maps and provides one of the few ways of proving that a system has chaotic solutions. A range of techniques, known as Melnikov methods, have been developed to prove the existence of intersections between stable and unstable manifolds.

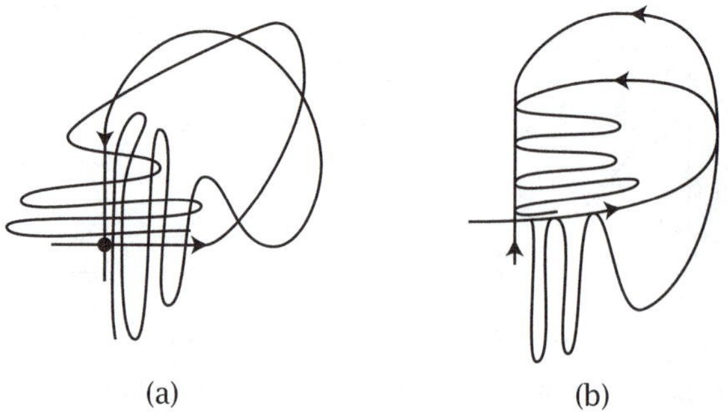

**Figure 4** (a) Homoclinic tangles for maps. (b) Homoclinic tangency.

### 4.5 Homoclinic Tangencies

The creation of persistent transverse intersections of stable and unstable manifolds as a parameter varies involves the existence of homoclinic tangencies at a critical parameter value as illustrated in figure 4(b). On one side of this value of the parameter there is no intersection locally, and on the other there is a transverse intersection. The existence of a homoclinic tangency implies all sorts of exotic dynamics at nearby parameters. In the 1970s Sheldon Newhouse proved that there are parameters close to the bifurcation point at which the system has a countably infinite number of *stable* periodic orbits, "infinitely many sinks," and by the end of the 1990s it had been established that there are nearby parameter values (a positive measure of them) at which the system has a strange attractor. Moreover, the existence of a homoclinic tangency implies the existence of homoclinic tangencies for higher iterates of the map at nearby parameter values at which the same analysis holds, so the whole picture repeats for higher periods.

## 5 Cascades of Bifurcations

By the mid-1970s numerical simulations had shown that simple families of maps such as the quadratic (or logistic) map

$$x_{n+1} = \mu x_n(1 - x_n)$$

can have many complicated bifurcations (see THE LOGISTIC EQUATION [III.19]). Mitchell Feigenbaum showed that a much stronger and more precise statement is possible. For smooth families of maps there are infinite sequences (cascades) of period-doubling bifurcations. Thus there are parameter values $\mu_n$ at which an orbit of period $2^n$ has a period-doubling bifurcation, creating an orbit of period $2^{n+1}$, $n = 0, 1, 2, \ldots$, and these accumulate on a limit $\mu_*$ at a rate

$$\lim_{n\to\infty} \frac{\mu_{n-1} - \mu_n}{\mu_n - \mu_{n+1}} = \delta$$

or $|\mu_n - \mu_*| \sim C\delta^{-n}$. Most remarkably, $\delta$ is a constant independent of the details of the map, a property referred to as quantitative universality. In fact, this "Feigenbaum constant" $\delta$ should really be thought of as a function of the order of the turning point of the map; if it behaves like $|x|^\alpha$ with $\alpha > 1$, then the rate is a function of $\alpha$. For the typical quadratic turning point,

$$\delta \approx 4.669201609\ldots.$$

This quantitative universality also manifests itself in the phase space as a sort of self-similarity under scaling. Feigenbaum was able to understand this by appealing to arguments based on renormalization analysis from statistical physics. At the accumulation point $\mu_*$, the second iterate of the map, restricted to a smaller interval and suitably rescaled, behaves in almost exactly the same way as the map itself: they both have periodic orbits of period $2^n$ for all $n = 0, 1, 2, 3, \ldots$. This idea can be exploited by defining a map $\mathcal{T}$ on unimodal functions $f\colon [-1,1] \to [-1,1]$ with a maximum at $x = 0$ and with $f(0) = 1$:

$$\mathcal{T} f(x) = -\frac{1}{\alpha} f \circ f(-\alpha x),$$

where $\alpha = -f(1)$. Seen as a map on an appropriately defined function space, it turns out that $\mathcal{T}$ has a fixed point $f_*$ (so $\mathcal{T} f_* = f_*$) with a codimension-one stable manifold consisting of maps corresponding to the special accumulation values of parameters $\mu_*$ for some family, and a one-dimensional unstable manifold with eigenvalue $\delta > 1$ (it is actually a little more complicated than this, but this statement contains the essential structure). The universal accumulation rate $\delta$ is therefore the unstable eigenvalue of a map in function space, and the universal scaling is due to the fact that under renormalization, maps on the stable manifold of $f_*$ accumulate on $f_*$ and so the spatial scaling $\alpha$ tends to $-f_*(1)$.

These period-doubling cascades are not limited to the doubling of fixed points. They can also be associated with periodic orbits of any period, creating cascades of period $2^n p$ for some fixed $p$. Period-doubling cascades also occur in higher dimensions and have been observed numerically in simulations of partial differential equations and experimentally in, for example, the changing convection patterns of liquid helium. The Lorenz maps of section 4.2 can also display cascades of homoclinic bifurcations if the saddle index $\lambda_3/\lambda_1$ is greater than one due to a simple correspondence between the Lorenz maps and unimodal maps based on the symmetry of the system.

Many codimension-two bifurcations involve the existence of infinitely many bifurcation curves, and these again can lead to cascades of bifurcations along appropriately chosen paths in parameter space. The special role played by period-doubling of orbits of period $2^n$ is that it is a natural route to chaos: on one side of the accumulation point of the period-doubling sequence associated with a fixed point of the map the map is simple, in the sense that it has only a finite number of periodic orbits, while on the other side it is chaotic with infinitely many periodic orbits of different periods (this is not true of cascades associated with other periods in general).

## 6 Codimension-Two Bifurcations

Codimension-two bifurcations are associated with special properties of a flow that can be observed in two-parameter systems but not (typically) in one-parameter families. Since there is often scope to vary more than one parameter in systems of interest, and because they act as useful organizing centers for describing the dynamics of one-parameter families and are the obvious next step after the codimension-one cases, many codimension-two bifurcations have been catalogued. Here we give two representative examples. One, the Takens–Bogdanov bifurcation, involves two zero eigenvalues at a stationary point and therefore generalizes the idea of local bifurcations to two-parameter systems. The other, a gluing bifurcation, occurs when there is a special value in parameter space at which two homoclinic orbits exist without a symmetry. This provides an example with an infinite set of bifurcations but no chaos.

### 6.1 Takens–Bogdanov Bifurcations

If a stationary point of a flow has two zero eigenvalues, then the linear part of the differential equation on the center manifold in appropriate coordinates will typically have the form

$$\begin{pmatrix} 0 & 1 \\ 0 & 0 \end{pmatrix} \qquad (3)$$

after a linear change of coordinates (there is obviously another possibility with every coefficient zero but this is more complicated and nongeneric). A natural way to unfold this singularity is to introduce two small parameters, $\mu$ and $\nu$, and consider the linear part

$$\begin{pmatrix} 0 & 1 \\ \mu & \nu \end{pmatrix}$$

with characteristic equation $s^2 - \nu s - \mu = 0$. If $\mu = 0$ there is a simple root (so a bifurcation such as a saddle node or pitchfork), and if $\nu = 0$ and $\mu < 0$ there is a purely imaginary pair of eigenvalues, suggesting a Hopf bifurcation (see section 2). By considering the nonlinear terms of lowest order after successive near-identity changes of coordinates, it is possible to show that the local behavior is typically modeled by the normal form

$$\dot{x} = y, \qquad \dot{y} = \mu + \nu y + x^2 + bxy,$$

where we treat $\mu$ and $\nu$ as small parameters and, after scaling, $b = \pm 1$. In what follows we treat the case $b = 1$ as an example. If $\mu = \nu = 0$ then the origin is a stationary point and the Jacobian is (3). The bifurcation at $\mu = 0$ is a saddle node (except in the degenerate case $\nu = 0$), and if $\mu = -\nu^2$ ($\mu < 0$), there is a Hopf bifurcation creating a periodic orbit that exists in $\nu^2 < -\mu$ immediately after the bifurcation. Thinking about a circular path in parameter space enclosing the origin, it is clear that there must be another bifurcation (the periodic orbit is created but never destroyed), and in fact there is a homoclinic bifurcation on a curve starting at the origin given to leading order by $\mu = -\frac{49}{25}\nu^2$ that creates/destroys the periodic orbit as expected.

### 6.2 A Gluing Bifurcation

Suppose that at some parameter a system has a pair of homoclinic orbits in the same configuration as the Lorenz equations but without the symmetry and with saddle index $\lambda_3/\lambda_1 = \delta > 1$ (see section 4.2). There is then a natural two-parameter unfolding locally such that if one parameter is zero, one of the homoclinic orbits persists, and if the other parameter is zero, then the other homoclinic orbit exists, with both therefore existing at the intersection of the two axes. In the same way that an approximate return map was derived for the Lorenz-type flows of section 4.2, the analysis of this more general configuration leads to the approximate map

$$x_{n+1} \approx \begin{cases} -\mu + a x_n^\delta, & x_n > 0, \\ \nu - a|x_n|^\delta, & x_n < 0, \end{cases}$$

with $\delta > 1$ and $a > 0$. There are now two small parameters, $\mu$ and $\nu$, and since $\delta > 1$ the derivative of the map is small if $|x|$ is small and nonzero. These maps have many similarities with maps of the circle that can be exploited to show that as the parameter varies in a path from $(\mu, \nu) = (\varepsilon, 0)$ to $(\varepsilon, 0)$ for some small $\varepsilon > 0$, the proportion of points of the attractor varies continuously from zero to one, and if it is rational then there is a stable periodic orbit with that proportion of points in $x > 0$. These stable periodic orbits exist on small intervals of the parameters in a form of mode locking.

## 7 Bits and Pieces

There are many other areas that could be covered in a review such as this, and in this final section we summarize just four of them.

### 7.1 Saddle–Node on Invariant Circle

Suppose that a saddle-node bifurcation of a stationary point occurs on a periodic orbit for a flow. At first sight this might appear to be a codimension-two bifurcation: one parameter for the saddle node and the other to locate it on a periodic orbit. However, if the periodic orbit is stable, then at the point of saddle-node bifurcation the connection between the weakly unstable direction and the stable manifold is persistent, so it is in fact codimension one. These bifurcations, called SNICs (saddle-node on invariant circle), create a stable periodic orbit from a stable-saddle pair of stationary points. The period of the orbit tends to infinity as the bifurcation point is approached like the inverse of the square root of the parameter (cf., the logarithmic divergence for homoclinic bifurcations in section 4).

The SNIC in discrete-time systems is again of codimension one, but this time chaotic solutions exist in a neighborhood of the bifurcation and there are complex foldings of the unstable manifold as it approaches the fixed point. This was described in detail by Newhouse, Palis, and Takens in the early 1980s.

### 7.2 Intermittency

Intermittency describes motion that is close to being periodic for a long time, called the laminar phase, and that then has a chaotic burst before returning to the laminar, almost periodic, phase. It can also be seen as a prototype for the more general mixed-mode oscillations observed in neuroscience and elsewhere. Although not strictly a bifurcation, the phenomenon of intermittency is seen as parameters vary and this makes it worth mentioning here. There are several different types of intermittency. The simplest case can be observed in the quadratic (or logistic) map $x_{n+1} = \mu x_n(1 - x_n)$. There are parameter values with stable periodic orbits that can lose stability via saddle-node bifurcations as the parameter is decreased through a critical value $\mu_c$. Just before the saddle-node bifurcation creates the stable periodic orbit, solutions that pass close to the locus of the periodic orbit will spend a long time, proportional to $|\mu - \mu_c|^{-1/2}$, close to the points where the periodic orbit will eventually be created, before moving away. If there is some reinjection mechanism due to global properties of the map, then after moving away the solution may find its way back into the laminar region and the process repeats. The SNIC is a special case in which the reinjection is always at the same point. This sequence of long laminar regions interspersed by bursts is generalized by the idea of mixed-mode oscillations, where the laminar region may be replaced by much more complicated but localized behavior.

### 7.3 Hamiltonian Flows

If the defining differential equation can be derived from a Hamiltonian, so the phase space can be divided into two sets of variables $(q, p) \in \mathbb{R}^n \times \mathbb{R}^n$ with a function $H(q, p)$ such that

$$\dot{q}_i = \frac{\partial H}{\partial p_i}, \quad \dot{p}_i = -\frac{\partial H}{\partial q_i}, \quad i = 1, \dots, n,$$

then this special structure (and generalizations of it) implies that the dynamics has very special features. These equations are important because they include many Newtonian models where friction is ignored. The equations of motion imply that $H$ is constant on solutions, so solutions lie on level sets of the function $H$. This in turn implies that features that are special (that is, have codimension one) in generic systems are robust in Hamiltonian systems. A good example is the existence of homoclinic orbits, so the global bifurcations of Hamiltonian systems are very different from those of the more general systems described in section 4. (Note that the global bifurcation theorems given above assumed that the homoclinic orbits exist on codimension-one surfaces in parameter space.) Even local bifurcations are special: a stationary point of a two-dimensional Hamiltonian system is typically either a saddle or a center, and this means that the saddle-node bifurcation involves the creation of a center and a saddle, with a set of periodic motions around the center, bounded by a robust homoclinic orbit! The differential equation with Hamiltonian

$$H(q, p) = \tfrac{1}{2}p^2 + \mu q - \tfrac{1}{3}q^3$$

has just such a transition as $\mu$ passes through zero.

### 7.4 Homoclinic Snaking

Since the existence of homoclinic orbits can be persistent with changing parameters in Hamiltonian systems, homoclinic orbits may be continued in parameter space in the same way that periodic orbits can be continued in more general systems. One interesting feature that is observed is *homoclinic snaking*, where the curve of homoclinic orbits oscillates as some measure (such as a norm) increases with parameters. Models of pattern formation such as the SWIFT–HOHENBERG EQUATIONS [IV.27 §2] exhibit this snaking if the time-dependent solutions are considered in one spatial dimension. Localized solutions (solitary waves) can be found that tend to zero at large spatial amplitudes but that have more and more complicated oscillations between these limits. In this case, the continuations of these solutions oscillate in parameter space, gaining an extra spatial maximum or minimum on each monotonic branch, in a pattern that is reminiscent of the oscillation of the periodic solution near the Shilnikov homoclinic orbits of figure 3, except that the oscillations have a bounded amplitude in parameter space rather than decreasing amplitude as in the figure.

### 7.5 Piecewise-Smooth Systems

Many applications in mechanics (friction and impacts, for example), biology (gene switching), and control (threshold control) are modeled by piecewise-smooth systems, where the dynamics evolves continuously until some switching surface is reached, and then there is a transition to a different (but still smooth) dynamics, possibly with a reset to another part of phase space. These are examples of HYBRID SYSTEMS [II.18]. The evolution is therefore a sequence of behaviors determined by smooth dynamical systems that are stitched together across the switching surfaces. New bifurcations occur when a periodic orbit intersects a switching surface tangentially, introducing a new segment to the dynamics, or if a fixed point or periodic orbit of a map intersects the switching surface. In the example of impacting systems, a standard model return map has a square root singularity:

$$x_{n+1} = \begin{cases} \mu - \sqrt{x_n}, & x_n > 0, \\ \mu + a x_n, & x_n < 0. \end{cases} \tag{4}$$

The maps obtained in these situations have much in common with the Lorenz maps discussed earlier with $\delta < 1$ as the derivative is unbounded as $x$ tends to zero from above, but they are also continuous unimodal maps and so general theorems such as Sharkovskii's theorem (which describes how the existence of a periodic orbit of a given period implies the existence of other periodic orbits in continuous maps) hold (see LOGISTIC EQUATION [III.19]). When applied to the special case of unimodal maps this restricts the order in which periodic orbits are created. However, while the logistic map has many windows of stable periodic orbits, each with its own period-doubling cascade and associated chaotic motion, the *stable* periodic orbits of the square root map (4) are much more constrained. For example, if $0 < a < \frac{1}{4}$, the stable periodic orbits form a period-adding sequence as $\mu$ decreases through zero. There is a parameter interval on which a period-$n$ orbit is the only attractor, followed by a bifurcation creating a stable period-$(n+1)$ orbit so that the two stable orbits of period $n$ and period $(n+1)$ coexist. Then there is a further bifurcation at which the period-$n$ orbit loses stability and the stable period-$(n+1)$ orbit is the only attractor. This sequence of behavior is repeated with $n$ tending to infinity as $\mu$ tends to zero.

## 8 Afterview

Bifurcation theory provides insights into why certain types of dynamics occur and how they arise. In cases such as period-doubling cascades it provides a framework in which to understand the changes in complexity of dynamics even if the behavior at a single parameter value might appear nonrepeatable. The number of different cases that may need to be considered can proliferate, and there is currently a nonuniformity of nomenclature that means that it is hard to tell whether a particular case has been studied previously in the literature. Given that the techniques are useful in any discipline that uses dynamic modeling, this aspect is unfortunate and leads to many reinventions of the same result. However, this only underlines the central role played by bifurcation theory in understanding the dynamics of mathematical models wherever they occur.

### Further Reading


Arrowsmith, D. K., and C. P. Place. 1996. *An Introduction to Dynamical Systems*. Cambridge: Cambridge University Press.

Glendinning, P. 1994. *Stability, Instability and Chaos*. Cambridge: Cambridge University Press.

Guckenheimer, J., and P. Holmes. 1983. *Nonlinear Oscillations, Dynamical Systems, and Bifurcations of Vector Fields*. New York: Springer.

Kusnetsov, Y. A. 1995. *Elements of Applied Bifurcation Theory*. New York: Springer.


# IV.22 Symmetry in Applied Mathematics

*Ian Stewart*

We tend to think of the applied mathematical toolkit as a collection of specific techniques for precise calculations, each intended for a particular kind of problem, solving a system of algebraic or differential equations numerically, for example. But some of the most powerful ideas in mathematics are so broad that at first sight they seem too vague and nebulous to have practical implications. Among them are probability, continuity, and symmetry.

Probability started out as a way to capture uncertainty in gambling games, but it quickly developed into a vitally important collection of mathematical techniques used throughout applied science, economics, sociology—even for formulating government policy.

Continuity proved such an elusive concept that it was used intuitively for centuries before it could be defined rigorously; now it underpins calculus, perhaps the most widely used mathematical tool of them all, especially in the form of ordinary and partial differential equations. But continuity is also fundamental to topology, a relative newcomer from pure mathematics that is starting to demonstrate its worth in the design of efficient trajectories for spacecraft, improved methods for forecasting the weather, frontier investigations in quantum mechanics, and the structure of biologically important molecules, especially deoxyribonucleic acid (DNA).

Symmetry, the topic of this article, was initially a rather ill-defined feeling that certain parts of shapes or structures were much the same as other parts of those shapes or structures. It has since become a vital method for understanding pattern formation throughout the scientific world, with applications that range from architecture to zoology. Symmetry, it turns out, underlies many of the deepest aspects of the natural world. Our universe behaves the way it does because of its symmetries—of space, time, and matter. Both relativity and particle physics are based on symmetry principles. Symmetry methods shed light on difficult problems by revealing general principles that can help us find solutions. Symmetry can be static, dynamic, even

**Figure 1** Which shapes are symmetric?

chaotic. It is a concept of great generality and deep abstraction, where beauty and power go hand in hand.

## 1 What Is Symmetry?

Symmetry is most easily understood in a geometric setting. Figure 1 shows a variety of plane figures, some symmetric, some not. Which are which?

The short answer is that the top row consists of symmetric shapes and the bottom row consists of asymmetric shapes. But there is more. Not only can shapes be symmetric, or not—they can have different *kinds* of symmetry. The heart (the middle of the top row) has the most familiar symmetry, one we encounter every time we look at ourselves in a mirror: *bilateral* symmetry. The left-hand side of the figure is an exact copy of the right-hand side, but flipped over. Three other shapes in the figure are bilaterally symmetric: the circle, the pentagon, and the circle with a square hole. The pentagon with a tilted square hole is not; if you flip it left–right, the pentagonal outline does not change but the square hole does because of the way it is tilted.

However, the circle has more than just bilateral symmetry. It would look the same if it were reflected in any mirror that runs through its center. The pentagon would look the same if it were reflected in any mirror that runs through its center and passes through a vertex.

What about the fourth shape in the top row, the flowery thing? If you reflect it in a mirror, it looks different, no matter where the mirror is placed. However, if you rotate it through a right angle about its center, it looks exactly the same as it did to start with. So this shape has *rotational* symmetry for a right-angle rotation. Thus primed, we notice that the pentagon also has rotational symmetry, for a rotation of 72°, and the circle has rotational symmetry for any angle whatsoever.

Most of the shapes on the bottom row look completely asymmetric. No significant part of any of them looks much like some other part of the same shape. The possible exception is the pentagon with a square hole. The pentagon is symmetric, as we have just seen, and so is the square. Surely combining symmetric shapes should lead to a symmetric shape? On the other hand, it does look a bit lopsided, which is not what we would expect from symmetry. With the current definition of symmetry, this shape is asymmetric, even though some pieces of it are symmetric.

In the middle of the nineteenth century mathematicians finally managed to define symmetry, for geometric shapes, by abstracting the common idea that unifies all of the above discussion. The background involves the idea of a *transformation* (or function or map). Some of the most common transformations, in this context, are rigid motions of the plane. These are rules for moving the entire plane so that the distances between points do not change. There are three basic types.

**Translations.** Slide the plane in a fixed direction so that every point moves the same distance.

**Rotations.** Choose a point, keep this fixed, and spin the entire plane around it through some angle.

**Reflections.** Choose a line, think of it as a mirror, and reflect every point in it.

These transformations do not exhaust the rigid motions of the plane, but every rigid motion can be obtained by combining them. One of the new transformations produced in this way is the *glide reflection*; reflect the plane in a line and then translate it along the direction of that line.

Having defined rigid motions, we can now define what a symmetry is. Given some shape in the plane, a symmetry of that shape is a rigid motion of the plane that leaves the shape *as a whole* unchanged. Individual points in the shape may move, but the end result looks exactly the same as it did to start with—not just in terms of shape, but also location.

For example, if we reflect the pentagon about a line through its center and a vertex, then points not on the line flip over to the other side. But they swap places in pairs, each landing where the other one started from, so the final position of the pentagon fits precisely on top of the initial position. This is false for the pentagon with a square hole, and this is the reason why that shape is not considered to possess symmetry.

The shapes in figure 1 are all of finite size, so they cannot possess translational symmetry. Translational symmetries require infinite patterns. A typical example is a square tiling of the entire plane, like bathroom tiles but continued indefinitely. If this pattern is translated sideways by the width of one tile, it remains unchanged. The same is true if it is translated upward by the width of one tile. It follows that, if the tiling is translated an integer number of widths in either of these two directions, it again remains unchanged. The symmetry group of translations is a *lattice*, consisting of integer combinations of two basic translations. The square tiling also has rotational symmetries, through multiples of 90° about the center of a tile or a corner where four tiles meet. Another type of rotational symmetry is rotation through 180° about the center of the edge of a tile. The square tiling pattern also has various reflectional symmetries.

Lattices in the plane can be interpreted as wallpaper patterns. In 1891 the pioneer of mathematical crystallography Yevgraf Fyodorov proved that there are exactly 17 different symmetry classes of wallpaper patterns. George Pólya obtained the same result independently in 1924. Lattice symmetries, often in combination with rotations and reflections, are of crucial importance in crystallography, but now the "tiling" is the regular atomic structure of the crystal, and it repeats in three-dimensional space along integer combinations of three independent directions. Again there may also be rotations and reflections. The physics of a crystal is strongly influenced by the symmetries of its atomic lattice. In the 1890s Fyodorov, Arthur Shönflies, and William Barlow proved that there are 230 symmetry types of lattice, or 219 if certain mirror-image pairs are considered to be the same.

Of course, no physical crystal can be of infinite extent. However, the idealization to infinite lattices can be a very accurate model for a real crystal because the size of the crystal is typically much larger than the lattice spacing. Real crystals differ from this ideal model in many ways: dislocations, where the lattice fails to repeat exactly; grain boundaries, where local lattices pointing in different directions meet; and so on. All applied mathematics involves a modeling step, representing the physical system by a simplified and idealized mathematical model. What matters is the extent to which the model provides useful insights. Its failure to include certain features of reality is not, of itself, a valid criticism. In fact, such a failure can be a virtue if it makes the analysis simpler without losing anything important.

Symmetries need not be rigid motions. Another geometric symmetry is dilation—change of scale. A logarithmic spiral, found in nature as a nautilus shell, remains unchanged if it is dilated by some fixed amount and also rotated through an appropriate angle.

Symmetry does not apply only to shapes; it is equally evident in mathematical formulas. For example, $x + y + z$ treats the three variables $x$, $y$, and $z$ in exactly the same manner. But the expression $x^3 + y - 2z^2$ does not. The first formula is symmetric in $x$, $y$, and $z$; the second is not. This time the transformations concerned are *permutations* of the three symbols—ways to swap them around. However we permute them, $x + y + z$ stays the same. For instance, if we swap $x$ and $y$ but leave $z$ the same, the expression becomes $y + x + z$. By the laws of algebra, this equals $x + y + z$. But the same permutation applied to the other expression yields $y^3 + x - 2z^2$, which is clearly different.

Symmetry thereby becomes a very general concept. Given any mathematical structure, and some class of transformations that can act on the structure, we define a symmetry to be any transformation that preserves the structure—that is, leaves it unchanged.

If a physical system has symmetry, then most sensible mathematical models of that system will have corresponding symmetries. (I say "most" because, for example, numerical methods cannot always incorporate all symmetries exactly. No computer model of a circle can be unchanged by all rotations. This inability of numerical methods to capture all symmetries can sometimes cause trouble.) The precise formulation of symmetry for a given model depends on the kind of model being used and its relation to reality.

## 2 Symmetry Groups

The above definition tells us that a symmetry of some structure (shape, equation, process) is not a thing but a transformation. However, it also tells us something deeper: structures may have several different symmetries. Indeed, some structures, such as the circle, have infinitely many symmetries. So there is a shift of emphasis from symmetry to symmetries, not symmetry as an abstract property of a structure but the *set of all symmetries* of the structure.

This set of transformations has a simple but vital feature. Transformations can be combined by performing them in turn. If two symmetries of some structure are combined in this way, the result is always a symmetry of that structure. It is not hard to see why: if you do not change something, and then you *again* do not change it ... you do not change it. This feature is known as the "group property," and the set of all symmetry transformations, together with this operation of composition, is the *symmetry group* of the structure.

The shapes in figure 1 illustrate several common kinds of symmetry group. The rotations of the circle form the *circle group* or *special orthogonal group in the plane*, denoted by $S^1$ or $SO(2)$. The reflections and rotations of the circle form the *orthogonal group in the plane*, denoted by $O(2)$. The rotations of the pentagon form the *cyclic group* $Z_5$ of order 5 (the order of a group is the number of transformations that it contains). The rotations of the flower shape form the cyclic group $Z_4$ of order 4. There is an analogous cyclic group $Z_n$ whose order is any positive integer $n$; it can be defined as the group of rotational symmetries of a regular $n$-sided polygon. If we also include the five reflectional symmetries of the pentagon, we obtain the *dihedral group* $D_5$, which has order 10. There is an analogous dihedral group $D_n$ of order $2n$ for any positive integer $n$: the rotations and reflections of a regular $n$-sided polygon. Finally, we mention the *symmetric group* $S_n$ of all permutations of a set with $n$ elements, such as $\{1, 2, 3, \dots, n\}$. This has order $n!$, that is, $n(n-1)(n-2)\cdots 3 \cdot 2 \cdot 1$.

Even asymmetric shapes have some symmetry; the identity transformation "leave everything as it is." The symmetry group contains only this trivial but useful transformation, and it is symbolized by $\mathbf{1}$.

We mention one useful piece of terminology. Often one group sits inside a bigger one. For example, $SO(2)$ is contained in $O(2)$, and $Z_n$ is contained in $D_n$. In such cases, we say that the smaller group is a *subgroup* of the bigger one. (They may also be equal, a trivial but sensible convention.)

The study of groups has led to a huge area of mathematics known as *group theory*. Some of it is part of abstract algebra, especially when the group is finite—that is, contains a finite number of transformations. Examples are the cyclic, dihedral, and symmetric groups. Another area, involving analysis and topology, is the theory of *Lie groups*, such as the circle group, the orthogonal group, and their analogues in spaces of any dimension. Here the main emphasis is on continuous families of symmetry transformations, which correspond to all choices of some real number. For example, a circle can be rotated through any real angle. Yet another important area is *representation theory*, which studies all the possible ways to construct a given group using matrices, linear transformations of some vector space.

One of the early triumphs of group theory in applied mathematics was Noether's theorem, proved by Emmy Noether in 1918. This applies to a special type of differential equation known as a Hamiltonian system, which arises in models of classical mechanics in the absence of frictional forces. Celestial mechanics—the motion of the planets—is a significant example. The theorem states that whenever a Hamiltonian system has a continuous family of symmetries, there is an associated conserved quantity. "Conserved" means that this quantity remains unchanged as the system moves.

The laws of nature are the same at all times: if you translate time from $t$ to $t + \theta$, the laws do not look any different. These transformations form a continuous family of symmetries, and the corresponding conserved quantity is energy. Translational symmetry in space (the laws are the same at every location) corresponds to conservation of momentum. Rotations about some axis in three-dimensional space provide another continuous family of symmetries; here the conserved quantity is angular momentum about that axis. All of the conservation laws of classical mechanics are consequences of symmetry.

## 3 Pattern Formation

Symmetry methods come into their own, and nowadays are almost mandatory, in problems about pattern formation. Often the most striking feature of some natural or experimental system is the occurrence of patterns. Rainbows are colored circular arcs of light. Ripples caused by a stone thrown into a pond are expanding circles. Sand dunes, ocean waves, and the stripes on a tiger or an angelfish are all patterns that can be modeled using repeating parallel features. Crystal lattices are repeating patterns of atoms. Galaxies form vast spirals, which rotate without (significantly) changing shape—a group of symmetries combining time translation with spatial rotation.

Many of these patterns arise through a general mechanism called "symmetry breaking." This is applicable whenever the equations that model a physical system have symmetry. I say "equations" here, even though I have already insisted that the symmetries of the system should appear in the equations, because it is not unusual for the model equations to have *more* symmetry than the pattern under consideration. Our theories of symmetry breaking and pattern formation rest on the structure of the symmetry group and its implications for mathematical models of symmetric systems.

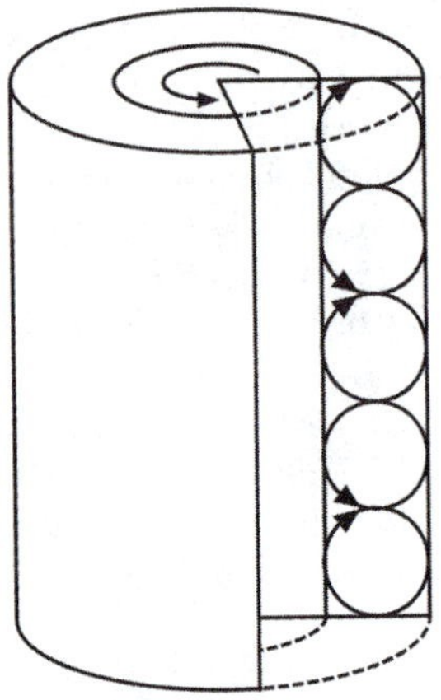

**Figure 2** Taylor–Couette apparatus, showing flow pattern for Taylor vortices.

A classic example is Taylor–Couette flow, in which fluid is confined between two rotating cylinders (figure 2). In experiments this system exhibits a bewildering variety of patterns, depending on the angular velocities of the cylinders and the relative size of the gap between them. The experiment is named after Maurice Couette, who used a fixed outer cylinder and a rotating inner one to measure the viscosity of fluids. At the low velocities he employed, the flow is featureless, as in figure 3(a). In 1923 Geoffrey Ingram Taylor noted that when the angular velocity of the inner cylinder exceeds a critical threshold, the uniform pattern of Couette flow becomes unstable and instead a stack of vortices appears; see figures 2 and 3(b). The vortices spiral round the cylinder, and alternate vortices spin the opposite way in cross section (small circles with arrows). Taylor calculated this critical velocity and used it to test the NAVIER–STOKES EQUATIONS [III.23] for fluid flow.

Further experimental and theoretical work followed, and the apparatus was modified to allow the outer cylinder to rotate as well. This can make a difference because, in a rotating frame of reference, the fluid is subject to additional centrifugal forces. In these more general experiments, many other patterns were observed. Figure 3 shows a selection of them.

The most obvious symmetries of the Taylor–Couette system are rotations about the common axis of the cylinders. These preserve the structure of the apparatus. But notice that not all patterns have full rotational symmetry. In figure 3 only the first two—Couette flow and Taylor vortices—are symmetric under all rotations. Another family of symmetries arises if the system is

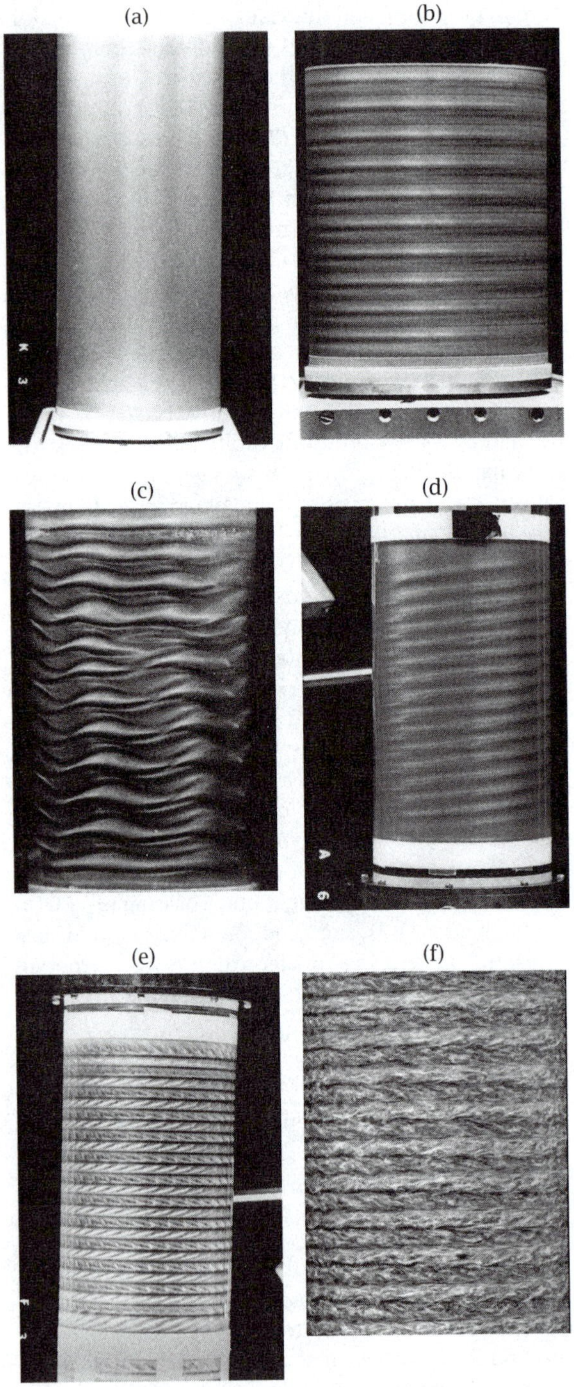


**Figure 3** Some of the numerous flow patterns in the Taylor–Couette system: (a) Couette flow; (b) Taylor vortices; (c) wavy vortices; (d) spiral vortices; (e) twisted vortices; and (f) turbulent vortices. Source: Andereck et al. (1986).

modeled (as is common for some purposes) by two infinitely long cylinders but restricted to patterns that repeat periodically along their lengths. In effect, this wraps the top of the cylinder round and identifies it with the bottom. Mathematically, this trick employs a modeling assumption: "periodic boundary conditions" that require the flow near the top to join smoothly to the flow near the bottom. With periodic boundary conditions, the equations are symmetric under all vertical translations. But again, not all patterns have full translational symmetry. In fact, the only one that does is Couette flow. So all patterns except Couette flow break at least some of the symmetries of the system.

On the other hand, most of the patterns retain some of the symmetries of the system. Taylor vortices are unchanged by vertical translations through distances equal to the width of a vortex pair (see figure 2). The same is true of wavy vortices, spirals, and twisted vortices. We will see in section 7 how each pattern in the figure can be characterized by its symmetry group.

There are at least two different ways to try to understand pattern formation in the Taylor–Couette system. One is to solve the Navier–Stokes equations numerically. The computations are difficult and become infeasible for more complex patterns. They also provide little insight into the patterns beyond showing that they are (rather mysterious) consequences of the Navier–Stokes equations. The other is to seek theoretical understanding, and here the symmetry of the apparatus is of vital importance, explaining most of the observed patterns.

## 4 Symmetry of Equations

To understand the patterns that arise in the Taylor–Couette system, we begin with a simpler example and abstract its general features. We then explain what these features imply for the dynamics of the system. In a later section we return to the Taylor–Couette system and show how the general theory of dynamics with symmetry classifies the patterns and shows how they arise. This theory is based on a mathematical consequence of the symmetry of the system being modeled; the differential equations used to set up the model have the same symmetries as the system. A symmetry of an equation is defined to be a transformation of the variables that sends solutions of the equation to (usually different) solutions.

The appropriate formulation of symmetry for an ordinary differential equation is called equivariance.

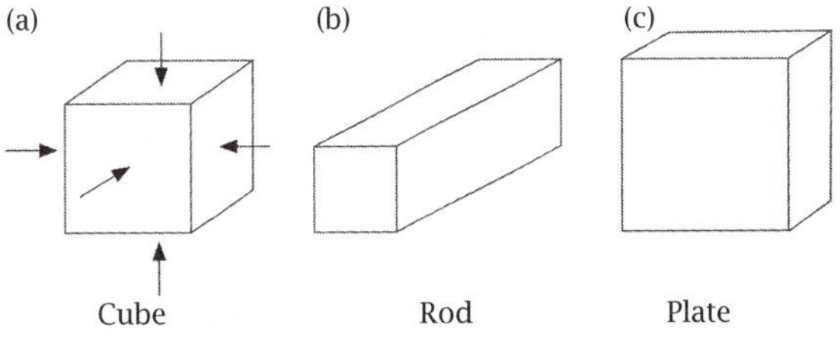


**Figure 4** (a) Cube under compression (force at back not shown). (b) Rod solution. (c) Plate solution.

Suppose that $\Gamma$ is a group of symmetries acting on the variables $x = (x_1, \dots, x_m)$ in the equation

$$\frac{dx}{dt} = F(x).$$

Then $F$ is *equivariant* if

$$F(\gamma x) = \gamma F(x) \quad \text{for all symmetries } \gamma \text{ in } \Gamma.$$

It follows immediately that, if $x(t)$ is any solution of the system, so is $\gamma x(t)$ for any symmetry $\gamma$. In fact, this condition is logically equivalent to equivariance. In simple terms, solutions always occur in symmetrically related sets. We will see an example of this phenomenon below.

A central concept in symmetric dynamics is BIFURCATION [IV.21], for which the context is a family of differential equations—an equation that contains one or more parameters. These are variables that are assumed to remain constant when solving the equation but can take arbitrary values. In a problem about the motion of a planet, for example, the mass of the planet may appear as a parameter. Such a family undergoes a *bifurcation* at certain parameter values if the solutions change in a qualitative manner near those values. For example, the number of equilibria might change, or an equilibrium might become a time-periodic oscillation. Bifurcations have a stronger effect than, say, moving an equilibrium continuously or slightly changing the shape and period of an oscillation. They provide a technique for proving the existence of interesting solutions by working out when simpler solutions become unstable and what happens when they do.

To show how symmetries behave at a bifurcation point, we consider a simple model of the deformation of an elastic cube when it is compressed by six equal forces acting at right angles to its faces (figure 4(a)). We consider deformations into any cuboid shape with sides $(a, b, c)$. When the forces are zero, the shape of the body is a cube, with $a = b = c$. The symmetry group consists of the permutations of the coordinate

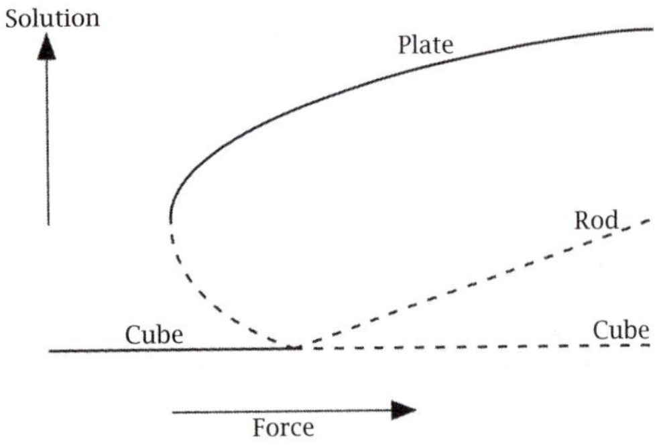


**Figure 5** Bifurcation diagram (solid lines, stable; dashed lines, unstable).

axes, changing $(a, b, c)$ to $(a, c, b)$, $(b, c, a)$, and so on. Physically, these rotate or reflect the shape, and they constitute the symmetric group $S_3$.

As the forces increase, always remaining equal, the cube becomes smaller. Initially, the sides remain equal, but analysis of a simplified but reasonable model shows that, when the forces become sufficiently large, the fully symmetric "cube" state becomes unstable. Two alternative shapes then arise: a *rod* shape $(a, b, c)$ with two sides equal and smaller than the third, and a *plate* shape $(a, b, c)$ with two sides equal and larger than the third (figure 4). The *bifurcation diagram* in figure 5 is a schematic plot of the shape of the deformed cube, plotted vertically, against the forces, plotted horizontally. The diagram shows how the existence and stability of the deformed states relate to the force. In this model only the plate solutions are stable.

In principle there might be a third shape, in which all three sides are of different lengths; however, such a solution does not occur (even unstably) in the model concerned.

Earlier, I remarked that solutions of symmetric equations occur in symmetrically related sets. Here, rod solutions occur in three symmetrically related forms, with the longer side of the rod pointing in any of the three coordinate directions. Algebraically, these solutions satisfy the symmetrically related conditions $a = b < c$, $a = c < b$, and $b = c < a$. The same goes for plate solutions.

## 5 Symmetry Breaking

The symmetry group for the buckling cube model contains six transformations:

$$I\colon (a, b, c) \mapsto (a, b, c),$$
$$X\colon (a, b, c) \mapsto (a, c, b),$$
$$Y\colon (a, b, c) \mapsto (c, b, a),$$
$$Z\colon (a, b, c) \mapsto (b, a, c),$$
$$R\colon (a, b, c) \mapsto (b, c, a),$$
$$S\colon (a, b, c) \mapsto (c, a, b).$$

We can consider the symmetries of the possible states, that is, the transformations that leave the shape of the buckled cube unchanged. Rods and plates have square cross section, and the two lengths in those directions are equal. If we interchange those two axes, the shape remains the same. Only the cube state has all six symmetries $S_3$. The rod with $a = b < c$ is symmetric under the permutations that leave $z$ fixed, namely $\{I, Z\}$. The rod with $a = c < b$ is symmetric under the permutations that leave $y$ fixed, namely $\{I, Y\}$. The rod with $b = c < a$ is symmetric under the permutations that leave $x$ fixed, namely $\{I, X\}$. The same holds for the plates. If a solution existed in which all three sides had different lengths, its symmetry group would consist of just $\{I\}$. All of these groups are subgroups of $S_3$.

Notice that the subgroups $\{I, X\}$, $\{I, Y\}$, $\{I, Z\}$ are themselves related by symmetry. For example, a solution $(a, b, c)$ with $a = b < c$ becomes $(b, c, a)$ with $b = c < a$ when the coordinate axes are permuted using $R$. In the terminology of group theory, these three subgroups are *conjugate* in the symmetry group.

The buckling cube is typical of many symmetric systems. Solutions need not have all of the symmetries of the system itself. Instead, some solutions may have smaller groups of symmetries—subgroups of the full symmetry group. Such solutions are said to *break* symmetry. For an equivariant system of ordinary differential equations, fully symmetric solutions may exist, but these may be unstable. If they are unstable, the system will find a stable solution (if it can), and this typically breaks the symmetry.

In general, the symmetry group of a solution is a subgroup of the symmetry group of the differential equation. Some subgroups may not occur here. Those that do are known as *isotropy subgroups*. For the buckling cube, the group $S_3$ has the five isotropy subgroups listed above. One subgroup is missing from that list, namely $\{I, R, S\}$. This is not an isotropy subgroup because a shape with these symmetries must satisfy the condition $(a, b, c) = (b, c, a)$, which forces $a = b = c$. This shape is the cube, which has additional symmetries $X$, $Y$, and $Z$. This situation is typical of subgroups that are not isotropy subgroups; the symmetries of such a subgroup force additional symmetries that are not in the subgroup.

A first step toward classifying the possible symmetry-breaking solutions of a differential equation with a known symmetry group is to use group theory to list the isotropy subgroups. Techniques then exist to find solutions with a given isotropy subgroup. Only one isotropy subgroup from each conjugacy class need be considered because solutions come in symmetrically related sets.

There are general theorems that guarantee the existence of solutions with certain types of isotropy subgroup, but they are too technical to state here. Roughly speaking, symmetry-breaking solutions often occur for large isotropy subgroups (but ones smaller than the full symmetry group). The theorems make this statement precise. In particular, they explain why rod and plate solutions occur for the buckling cube. Other general theorems help to determine whether a solution is stable or unstable.

## 6 Time-Periodic Solutions

A famous example of pattern formation is the Belousov–Zhabotinskii (or BZ) chemical reaction. This involves three chemicals together with an indicator that changes color from red to blue depending on whether the reaction is oxidizing or reducing. The chemicals are mixed together and placed in a shallow circular dish. They turn blue and then red. For a few minutes nothing seems to be happening; then tiny blue spots appear, which expand and turn into rings. As each ring grows, a new blue dot appears at its center. Soon the dish contains several expanding "target patterns" of rings. Unlike water waves, the patterns do not overlap and superpose. Instead, they meet to form angular junctions (figure 6(a)). Different target patterns expand at different rates, giving rings with differing thicknesses. However, each set of rings has a specific uniform speed of expansion, and the rings in that set all have the same width.

Another pattern can be created by breaking up a ring—by dragging a paperclip across it, for example. This new pattern curls up into a spiral (close to an Archimedean spiral with equally spaced turns), and the spiral slowly rotates about its center, winding more and more turns as it does so.

Neither pattern is an equilibrium, so the theorems alluded to above are not applicable. Instead, we can employ a different but related series of techniques. These apply to *time-periodic* solutions, which repeat their form at fixed intervals of time.

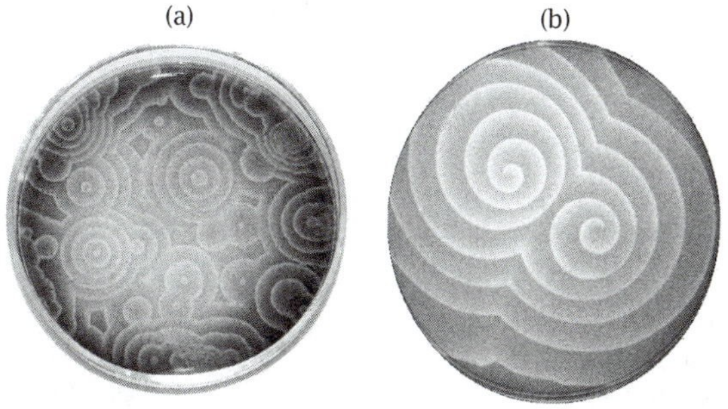


**Figure 6** Patterns in the BZ reaction: (a) snapshot of several coexisting target patterns; (b) snapshot of two coexisting spirals.

There are two main types of bifurcation in which the fully symmetric steady state loses stability as a parameter is varied, with new solutions appearing nearby. One is a steady-state bifurcation (which we met above), for which these new solutions are equilibria. The other is HOPF BIFURCATION [IV.21 §2], for which the new solutions are time-periodic oscillations. Hopf bifurcation can occur for equations without symmetry, but there is a generalization to symmetric systems. The main new ingredient is that symmetries can now occur not just in space (the shape of the pattern) and time (integer multiples of the period make no difference), but in a combination of both.

A single target pattern in the BZ reaction has a purely spatial symmetry: at any instant in time it is unchanged under all rotations about its center. It also has a purely temporal symmetry: in an ideal version where the pattern fills the whole plane, it looks identical after a time that is any integer multiple of the period.

A single spiral, occupying the entire plane, has no nontrivial spatial symmetry, but it has the same purely temporal symmetry as a target pattern. However, it has a further *spatiotemporal symmetry* that combines both. As time passes, the spiral slowly rotates without changing form. That is, an arbitrary translation of time, combined with a rotation through an appropriate angle, leaves the spiral pattern (and how it develops over time) unchanged.

For symmetric equations there is a version of the Hopf bifurcation theorem that applies to spatiotemporal symmetries when there is a suitable symmetry-breaking bifurcation. Again, its statement is too technical to give here, but it helps to explain the BZ patterns. In conjunction with the theory for steady-state bifurcation, it can be used to understand pattern formation in many physical systems.

## 7 Taylor–Couette Revisited

The Taylor–Couette system is a good example of the symmetry-breaking approach. A standard model, derived from the Navier–Stokes equations for fluid flow, involves three types of symmetry.

**Spatial:** rotation about the common axis of the cylinders through any angle.

**Temporal:** shift time by an integer multiple of the period.

**Model:** translation along the common axis of the cylinders through any distance (a consequence of the assumption of periodic boundary conditions).

Using bifurcation theory and a few properties of the Navier–Stokes equations, the dynamics of this model for the Taylor–Couette system can be reduced to an ordinary differential equation (the so-called center manifold reduction) in six variables. Two of these variables correspond to a steady-state bifurcation from Couette flow to Taylor vortices. This is steady in the sense that the fluid *velocity* at any point remains constant. The other four variables correspond to a Hopf bifurcation from Couette flow to spirals. These are the two basic "modes" of the system, and their combination is called a mode interaction.

Symmetric bifurcation theory's general theorems now prove the existence of numerous flow patterns, each with a specific isotropy subgroup. For example, if the rotational symmetry remains unbroken, but the group of translational symmetries breaks, a typical solution will have a specific translational symmetry, through some fraction of the length of the cylinder. All integer multiples of this translation are also symmetries of that solution. This combination of symmetries corresponds precisely to Taylor vortices; it leads to a discrete, repetitive pattern vertically, with no change in the horizontal direction.

A further breaking of the rotational symmetry produces a discrete set of rotational symmetries; these characterize wavy vortices. Symmetry under any rotation, if combined with a vertical translation through a corresponding distance, characterizes spiral vortices, and so on.

The only pattern in figure 3 that is not explained in this manner is the final one: turbulent vortices. This pattern turns out to be an example of symmetric chaos, chaotic dynamics that possesses "symmetry on average." At each instant the turbulent vortex state has no symmetry. But if the fluid velocity at each point is averaged over time, the result has the same symmetry as Taylor vortices. This is why the picture looks like Taylor vortices with random disturbances. A general theory of symmetric chaos also exists.

## 8 Conclusion

The symmetries of physical systems appear in the equations that model them. The symmetry affects the solutions of the equations, but it also provides systematic ways to solve them. One general area of application is pattern formation, and here the methods of symmetric bifurcation theory have been widely used. Topics include animal locomotion, speciation, hallucination patterns, the balance-sensing abilities of the inner ear, astrophysics, liquid crystals, fluid flow, coupled oscillators, elastic buckling, and convection.

In addition to a large number of specific applications, there are many other ways to exploit symmetry in applied mathematics. This article has barely scratched the surface.

### Further Reading

Andereck, C. D., S. S. Liu, and H. L. Swinney. 1986. Flow regimes in a circular Couette system with independently rotating cylinders. *Journal of Fluid Mechanics* 164:155–83.

Chossat, P., and G. Iooss. 1985. Primary and secondary bifurcations in the Couette–Taylor problem. *Japan Journal of Applied Mathematics* 2:37–68.

Field, M., and M. Golubitsky. 1992. *Symmetry in Chaos*. Oxford: Oxford University Press.

Golubitsky, M., and D. G. Schaeffer. 1985. *Singularities and Groups in Bifurcation Theory*, volume 1. New York: Springer.

Golubitsky, M., and I. Stewart. 1986. Symmetry and stability in Taylor–Couette flow. *SIAM Journal on Mathematical Analysis* 17:249–88.

——. 2002. *The Symmetry Perspective*. Basel: Birkhäuser.

Golubitsky, M., I. Stewart, and D. G. Schaeffer. 1988. *Singularities and Groups in Bifurcation Theory*, volume 2. New York: Springer.

Golubitsky, M., E. Knobloch, and I. Stewart. 2000. Target patterns and spirals in planar reaction–diffusion systems. *Journal of Nonlinear Science* 10:333–54.

Sattinger, D. H., and O. L. Weaver. 1986. *Lie Groups and Algebras with Applications to Physics, Geometry, and Mechanics*. New York: Springer.

## IV.23 Quantum Mechanics
*David Griffiths*

Quantum mechanics makes use of a wide variety of tools and techniques from applied mathematics—linear algebra, complex variables, differential equations, Fourier analysis, group representations—but the mathematical core of the subject is the theory of eigenvectors and eigenvalues of self-adjoint operators in Hilbert space.

### 1 A Single Particle in One Dimension

Let us begin with the simplest mechanical system: a particle (which is physicists' jargon for an object so small, compared with other relevant distances, that it can be considered to reside at a single point) of mass $m$, constrained to move in one dimension (say, along the $x$-axis) under the influence of a specified force $F$ (which may depend on position and time). The program of CLASSICAL MECHANICS [IV.19] is to determine the location of the particle at time $t$: $x(t)$. How do you calculate it? By applying Newton's second law of motion, $F = ma$ (or $\mathrm{d}^2x/\mathrm{d}t^2 = F/m$). If the force is conservative (nonconservative forces, such as friction, do not occur at a fundamental level), it can be expressed as the negative gradient of the *potential energy*, $V(x,t)$, and Newton's law becomes $\mathrm{d}^2x/\mathrm{d}t^2 = -m^{-1}\partial V/\partial x$. Together with appropriate initial conditions (typically, $x$ and $\mathrm{d}x/\mathrm{d}t$ at $t = 0$), this determines $x(t)$. From there it is easy to obtain the velocity ($v(t) \equiv \mathrm{d}x/\mathrm{d}t$), momentum ($p \equiv mv$), kinetic energy ($T \equiv \frac{1}{2}mv^2$), or any other dynamical quantity.

The program of quantum mechanics is quite different. Instead of $x(t)$, we seek the *wave function*, $\Psi(x,t)$, which we obtain by solving SCHRÖDINGER'S EQUATION [III.26],

$$\mathrm{i}\hbar\frac{\partial\Psi}{\partial t} = -\frac{\hbar^2}{2m}\frac{\partial^2\Psi}{\partial x^2} + V(x,t)\Psi, \qquad (1)$$

where $\hbar \equiv 1.05457 \times 10^{-34}$ J s is *Planck's constant*, the fingerprint of all quantum phenomena. Together with appropriate initial conditions (typically, $\Psi(x,0)$) this determines $\Psi(x,t)$.

But what *is* this wave function, and how can it be said to describe the state of the particle? After all, as its name suggests, $\Psi(x,t)$ is spread out in space, whereas a particle, by its nature, is localized at a point. The answer is provided by Born's statistical interpretation of the wave function: $\Psi(x,t)$ tells you the *probability* of finding the particle at the point $x$ if you conduct a (competent) measurement at time $t$. More precisely, $\int_a^b |\Psi(x,t)|^2\,\mathrm{d}x$ is the probability of finding the particle between point $a$ and point $b$. Evidently, the wave function must be *normalized*: $\int_{-\infty}^{\infty} |\Psi(x,t)|^2\,\mathrm{d}x = 1$ (the particle has got to be *somewhere*). Once normalized, at $t = 0$, $\Psi$ remains normalized for all time.

The statistical interpretation introduces a kind of *indeterminacy* into the theory, in the sense that even if you know everything quantum mechanics has to tell you about the particle (to wit: its wave function), you still cannot predict with certainty the outcome of a measurement to determine its position—all quantum mechanics has to offer is *statistical* information about the results from an ensemble of identically prepared systems (each in the state $\Psi$).

The average, or *expectation value*, of $x$ can be written in the form

$$\langle x\rangle = \int_{-\infty}^{\infty} \Psi(x,t)^* x \Psi(x,t)\,\mathrm{d}x;$$

similarly, the expectation value of *momentum* is

$$\langle p\rangle = \int_{-\infty}^{\infty} \Psi(x,t)^*\Big[-\mathrm{i}\hbar\frac{\partial}{\partial x}\Big]\Psi(x,t)\,\mathrm{d}x.$$

This is a consequence of de Broglie's formula relating momentum to the wavelength ($\lambda$) of $\Psi$: $\lambda = 2\pi\hbar/p$. We say that the observables $x$ and $p$ are "represented" by the operators

$$\hat{x} \equiv x, \qquad \hat{p} \equiv -\mathrm{i}\hbar\frac{\partial}{\partial x}. \qquad (2)$$

In general, a classical dynamical quantity $Q(x,p)$ is represented by the quantum operator

$$\hat{Q} \equiv Q(x, -\mathrm{i}\hbar(\partial/\partial x))$$

(simply replace every $p$ by $\hat{p}$), and its expectation value (in the state $\Psi$) is obtained by "sandwiching" $\hat{Q}$ between $\Psi$ and its conjugate $\Psi^*$, and integrating:

$$\langle Q\rangle = \int_{-\infty}^{\infty} \Psi(x,t)^*\hat{Q}\Psi(x,t)\,\mathrm{d}x.$$

Thus, for example, kinetic energy $T = \frac{1}{2}mv^2 = p^2/2m$ is represented by the operator

$$\hat{T} = -\frac{\hbar^2}{2m}\frac{\partial^2}{\partial x^2},$$

and the Schrödinger equation can be written

$$\mathrm{i}\hbar\frac{\partial\Psi}{\partial t} = \hat{H}\Psi,$$

where $\hat{H} = \hat{T} + \hat{V}$ is the *Hamiltonian operator*, representing the total energy.

If the potential energy is independent of time, the Schrödinger equation can be solved by separation of variables: $\Psi(x,t) = \mathrm{e}^{-\mathrm{i}Et/\hbar}\psi(x)$, where $E$ is the separation constant and $\psi(x)$ satisfies the *time-independent Schrödinger equation*:

$$\hat{H}\psi = -\frac{\hbar^2}{2m}\frac{\mathrm{d}^2\psi}{\mathrm{d}x^2} + V\psi = E\psi. \quad (3)$$

The expectation value of the total energy is

$$\begin{aligned}\langle H\rangle &= \int_{\infty}^{\infty} \Psi(x,t)^*\hat{H}\Psi(x,t)\,\mathrm{d}x \\ &= \int_{-\infty}^{\infty} \Psi(x,t)^*E\Psi(x,t)\,\mathrm{d}x \\ &= E,\end{aligned}$$

hence the choice of the letter $E$.

Suppose, for example, the particle is attached to a spring of force constant $k$. Classically, this would be a simple harmonic oscillator, with potential energy $V(x) = \frac{1}{2}kx^2$, and the (time-independent) Schrödinger equation reads

$$-\frac{\hbar^2}{2m}\frac{\mathrm{d}^2\psi}{\mathrm{d}x^2} + \tfrac{1}{2}kx^2\psi = E\psi. \quad (4)$$

The normalized solutions have energies $E_n = (n + \frac{1}{2})\hbar\omega$ $(n = 0, 1, 2, \dots)$ and are given by

$$\psi_n(x) = \frac{1}{\sqrt{2^n n! a\sqrt{\pi}}}\mathrm{e}^{-x^2/2a^2}H_n(x/a),$$

where $\omega \equiv \sqrt{k/m}$ is the classical (angular) frequency of oscillation, $a \equiv \sqrt{\hbar/m\omega}$, and $H_n$ is the $n$th HERMITE POLYNOMIAL [II.29]. (As a second-order ordinary differential equation, (4) admits *two* linearly independent solutions for *every* value of $E$, but all the other solutions are not normalizable, so they do not represent possible physical states.) The separable solutions of the time-dependent Schrödinger equation in this case are $\Psi_n(x,t) = \mathrm{e}^{-\mathrm{i}E_nt/\hbar}\psi_n(x)$. For instance, if the particle happens to be in the state $n = 0$, the probability of finding it outside the classical range $(\pm\sqrt{2E/k})$ is

$$2\int_a^{\infty} |\Psi_0(x,t)|^2\,\mathrm{d}x = 1 - \mathrm{erf}(1) = 0.1573,$$

where $\mathrm{erf}(x) \equiv (2/\sqrt{\pi})\int_0^x \mathrm{e}^{-t^2}\,\mathrm{d}t$ is the standard error function.

Of course, not all solutions to the Schrödinger equation are separable, but the separable solutions do constitute a complete set, in the sense that the most general (normalizable) solution can be expressed as a linear combination of them, $\Psi(x,t) = \sum_{n=0}^{\infty} c_n\Psi_n(x,t)$. Furthermore, they are orthonormal,

$$\int_{-\infty}^{\infty} \Psi_n(x,t)^*\Psi_m(x,t)\,\mathrm{d}x = \delta_{nm},$$

where $\delta_{nm}$ is the KRONECKER DELTA [I.2 §2, table 3], so the expansion coefficients can be determined from the initial wave function in the usual way:

$$c_n = \int_{-\infty}^{\infty} \psi_n(x)^*\Psi(x,0)\,\mathrm{d}x.$$

## 2 States and Observables

Formally, the state of a system is represented, in quantum mechanics, by a vector in HILBERT SPACE [I.2 §19.4]. In *Dirac notation*, vectors are denoted by *kets* $|s\rangle$ and their duals by *bras* $\langle s|$; the inner product of $|s_a\rangle$ and $|s_b\rangle$ is written as a "bra(c)ket" $\langle s_a|s_b\rangle = \langle s_b|s_a\rangle^*$. For instance, the wave function $\Psi(x,t)$ resides in the (infinite-dimensional) space $L^2$ of square-integrable functions $f(x)$ on $-\infty < x < \infty$, with the inner product

$$\langle f_a|f_b\rangle \equiv \int_{-\infty}^{\infty} f_a(x)^*f_b(x)\,\mathrm{d}x.$$

But there exist much simpler quantum systems in which the vector space is finite dimensional, and it pays to explore the general theory in this context first.

In an $n$-dimensional space a vector is conveniently represented by the column of its components (with respect to a specified orthonormal basis): $|s\rangle = (c_1, c_2, \dots, c_n)$ (as a column), $\langle s|$ is its conjugate transpose (a row), and $\langle s_a|s_b\rangle = \sum_{i=1}^n (c_a)_i^*(c_b)_i$ is their matrix product. *Observables* (measurable dynamical quantities) are represented by linear operators; for the system considered in section 1 they involve multiplication and differentiation, but in the finite-dimensional case they are matrices, $\hat{Q} \in \mathbb{C}^{n\times n}$ (with respect to a specified basis). In its most general form, the statistical interpretation reads as follows.

*If you measure observable $Q$, on a system in the state $|s\rangle$, you will get one of the eigenvalues of $\hat{Q}$; the probability of getting the particular eigenvalue $\lambda_i$ is $|\langle v_i|s\rangle|^2$, where $|v_i\rangle$ is the corresponding normalized eigenvector, $\hat{Q}|v_i\rangle = \lambda_i|v_i\rangle$, with $\langle v_i|v_i\rangle = 1$. In the act of measurement the state "collapses" to the eigenvector, $|s\rangle \to |v_i\rangle$. (This ensures that an immediately repeated measurement—on the same particle—will return the same value.)*

Of course, the outcome of a measurement must be a real number, and the probabilities must add up to 1. This is guaranteed if we stipulate that the operators representing observables be self-adjoint, i.e., $\langle s_a|\hat{Q}s_b\rangle = \langle\hat{Q}s_a|s_b\rangle$ for all vectors $|s_a\rangle$ and $|s_b\rangle$, which is to say that the matrix $\hat{Q}$ is *Hermitian* (equal to its transpose conjugate). (Physicists' notation is sloppy but harmless: technically, "$s$" is just the *name* of the vector, and you cannot multiply a matrix by a name. But $|\hat{Q}s\rangle$ means $\hat{Q}|s\rangle$, of course, and $\langle\hat{Q}s|$ is its dual.)

The expectation value of $Q$, for a system in the state $|s\rangle$, is $\langle Q\rangle = \langle s|\hat{Q}|s\rangle$, and the *standard deviation* $\sigma_Q$—known informally as the *uncertainty* in $Q$, and sometimes written $\Delta Q$—is given by

$$\sigma_Q^2 \equiv \langle(\hat{Q} - \langle Q\rangle)^2\rangle = \langle\hat{Q}^2\rangle - \langle Q\rangle^2.$$

It is zero if and only if $|s\rangle$ is an eigenvector of $\hat{Q}$ (in which case a measurement is certain to return the corresponding eigenvalue). Because noncommuting operators do not admit common eigenvectors, it is not possible, in general, to construct states with definite values of two different observables (say, $A$ and $B$). This is expressed quantitatively in the *uncertainty principle*:

$$\sigma_A\sigma_B \geqslant \tfrac{1}{2}|\langle[\hat{A},\hat{B}]\rangle|, \qquad (5)$$

where square brackets denote the *commutator*,

$$[\hat{A},\hat{B}] \equiv \hat{A}\hat{B} - \hat{B}\hat{A}.$$

Consider, for example, a two-dimensional space in which the state is represented by a (normalized) *spinor*, $|s\rangle = (c_1, c_2)$, such that $|c_1|^2 + |c_2|^2 = 1$. The most general $2\times 2$ Hermitian matrix has the form

$$\hat{Q} = \begin{pmatrix} a_0 + a_3 & a_1 - \mathrm{i}a_2 \\ a_1 + \mathrm{i}a_2 & a_0 - a_3 \end{pmatrix} = \sum_{i=0}^{3} a_i\sigma_i,$$

where the $a_j$s are real numbers, $\sigma_0$ is the unit matrix, and the other three $\sigma_j$s are the *Pauli spin matrices*:

$$\sigma_1 \equiv \begin{pmatrix} 0 & 1 \\ 1 & 0 \end{pmatrix}, \qquad \sigma_2 \equiv \begin{pmatrix} 0 & -\mathrm{i} \\ \mathrm{i} & 0 \end{pmatrix}, \qquad \sigma_3 \equiv \begin{pmatrix} 1 & 0 \\ 0 & -1 \end{pmatrix}. \qquad (6)$$

The Pauli matrices constitute a basis for the Lie algebra of SU(2), the group of "special" (unimodular) unitary $2\times 2$ matrices.

Suppose we measure the observable represented by $\sigma_3$. According to the statistical interpretation, the outcome could be $+1$ with probability $P_+ = |c_1|^2$ or $-1$ with probability $P_- = |c_2|^2$, since the eigenvalues and eigenvectors in this case are obviously $\lambda_+ = +1$, with $|s_+\rangle = (1,0)$, and $\lambda_- = -1$, with $|s_-\rangle = (0,1)$. But what if we chose instead to measure $\sigma_1$? The eigenvalues are again $\pm 1$, but the (normalized) eigenvectors are now $|s_+\rangle = \frac{1}{\sqrt{2}}(1,1)$ and $|s_-\rangle = \frac{1}{\sqrt{2}}(1,-1)$. The *possible* outcomes are the same as for $\sigma_3$, but the *probabilities* are quite different: $P_+ = \frac{1}{2}|c_1 + c_2|^2$ and $P_- = \frac{1}{2}|c_1 - c_2|^2$. Clearly, a system cannot simultaneously be in an eigenstate of $\sigma_3$ and of $\sigma_1$, and hence it cannot have definite values of both observables at the same time. What if I measure $\sigma_3$ and get (say) $+1$, and then you measure $\sigma_1$ and get (say) $-1$. Does this not mean that the system has definite values of both? No, because each measurement altered (collapsed) the state: $(c_1, c_2) \to (1,0) \to \frac{1}{\sqrt{2}}(1,-1)$. If I now repeat my measurement of $\sigma_3$, I am just as likely to get $-1$ as $+1$. It is not that I am *ignorant*—I know the state of the system precisely—but it simply does not *have* a definite value of $\sigma_3$ if it is in an eigenstate of $\sigma_1$. Indeed, since $[\sigma_3, \sigma_1] = 2\mathrm{i}\sigma_2$, the uncertainty principle says that $\Delta_1\Delta_3 \geqslant |\langle\sigma_2\rangle|$.

## 3 Continuous Spectra

In a *finite*-dimensional vector space every Hermitian matrix has a complete set of orthonormal eigenvectors, and implementation of the statistical interpretation is straightforward. But in an *infinite*-dimensional Hilbert space some or all of the eigenvectors may reside outside the space. For example, the eigenvectors of the position operator (2) are DIRAC DELTA FUNCTIONS [III.7]

$$xf_\lambda(x) = \lambda f_\lambda(x) \Rightarrow f_\lambda(x) = A\delta(x - \lambda), \qquad (7)$$

which are not square integrable, so they do not live in $L^2$ and cannot represent physically realizable states. The same goes for momentum,

$$-\mathrm{i}\hbar\frac{\partial f_\lambda}{\partial x} = \lambda f_\lambda \Rightarrow f_\lambda = A'\mathrm{e}^{\mathrm{i}\lambda x/\hbar}. \qquad (8)$$

Evidently, a particle simply cannot have a definite position or momentum in quantum mechanics. Moreover, since $\hat{x}$ and $\hat{p}$ do not commute ($[\hat{x},\hat{p}] = \mathrm{i}\hbar$), (5) says that $\sigma_x\sigma_p \geqslant \frac{1}{2}\hbar$, which is the original *Heisenberg uncertainty principle*.

Even though the eigenfunctions of $\hat{x}$ and $\hat{p}$ are not possible physical states, they are complete and orthogonal; the wave function can be expanded as a linear combination of them, and the (absolute) square of the coefficient represents the probability density for a measurement outcome. Note that nonnormalizable eigenfunctions are associated with a continuous spectrum of eigenvalues; probabilities become probability densities, and discrete sums are replaced by integrals. Adopting the convenient Dirac convention $\langle f_\lambda|f'_\lambda\rangle = \delta(\lambda - \lambda')$ (so that $A = 1$ in (7) and $A' = 1/\sqrt{2\pi\hbar}$ in (8)), we have

$$\Psi(x,t) = \int c_\lambda f_\lambda(x)\,\mathrm{d}\lambda$$

$$\Leftrightarrow \quad c_\lambda = \int f_\lambda(x)^*\Psi(x,t)\,\mathrm{d}x.$$

For eigenfunctions of position we get

$$c_\lambda = \int_{-\infty}^{\infty} \delta(x - \lambda)\Psi(x,t)\,\mathrm{d}x = \Psi(\lambda,t),$$

which is to say that the expansion coefficient is precisely the wave function itself (and we recover Born's original statistical interpretation). For eigenfunctions of momentum,

$$c_\lambda = \frac{1}{\sqrt{2\pi\hbar}} \int_{-\infty}^{\infty} e^{-i\lambda x/\hbar} \Psi(x,t)\, dx \equiv \Phi(\lambda, t),$$

which is the so-called *momentum space* wave function (mathematically, the Fourier transform of the position space wave function $\Psi$). Of course, $\Psi$ and $\Phi$ describe the same state vector, referred to two different bases.

In general, $E \geqslant \min\{V(x)\}$; in the case of a free particle ($V = 0$) of energy $E \geqslant 0$, we have, from (3),

$$-\frac{\hbar^2}{2m}\frac{d^2\psi}{dx^2} = E\psi \quad \Rightarrow \quad \psi(x) = Ae^{ikx},$$

so $\Psi(x,t) = Ae^{i(kx - Et/\hbar)}$, where $k^2 \equiv 2mE/\hbar^2$. If $k$ is positive, $\Psi$ is a wave propagating in the $+x$ direction (if $k$ is negative, it travels in the $-x$ direction). However, $\Psi$ is not normalizable, so it does not represent a possible physical state—there is *no such thing* as a free particle with precisely determined energy. But we can construct a normalizable *linear combination* of these states:

$$\Psi(x,t) = \int_{-\infty}^{\infty} c(k) e^{i(kx - k^2\hbar t/2m)}\, dk.$$

For instance, we might start with a *Gaussian wave packet* of width $\sigma_0$:

$$\Psi(x,0) = \frac{1}{(2\pi\sigma_0^2)^{1/4}} e^{-(x/2\sigma_0)^2} e^{ilx}.$$

Then $c(k) = (8\pi\sigma_0^2)^{1/4} e^{-[\sigma_0(k-l)]^2}$, so

$$\Psi(x,t) = \frac{e^{-l^2\sigma_0^2}}{(2\pi\sigma_0^2)^{1/4}\sqrt{1 + i\hbar t/2m\sigma_0^2}} \times \exp\left\{\frac{[(ix + 2l\sigma_0^2)/2\sigma_0]^2}{1 + i\hbar t/2m\sigma_0^2}\right\},$$

and hence

$$|\Psi(x,t)|^2 = (2\pi\sigma^2)^{-1/2} \exp\{-2[(x - l\hbar t/m)/2\sigma]^2\},$$

where $\sigma(t) \equiv \sigma_0\sqrt{1 + (\hbar t/2m\sigma_0^2)^2}$. The wave packet travels at speed $l\hbar/m$, spreading out as it goes. This is the quantum description of a free particle in motion.

Some Hamiltonians (e.g., the harmonic oscillator) have a discrete spectrum, with normalizable eigenstates; some (e.g., the free particle, $V = 0$) have a continuous spectrum and nonnormalizable eigenstates. Many systems have some of each. For example, the delta-function well, $V(x) = -\alpha\delta(x)$ (for some positive constant $\alpha$), admits a single normalizable state $\psi(x) = (\sqrt{m\alpha}/\hbar)e^{-m\alpha|x|/\hbar^2}$ with energy $E = -m\alpha^2/2\hbar^2$. This represents a *bound state*: the particle is "stuck" in the well. If (as here) $V(x) \to 0$ as $|x| \to \infty$, a bound state is typically characterized by a negative energy. But the delta-function well also admits *scattering states*, which have the form

$$\psi(x) = \begin{cases} Ae^{ikx} + Be^{-ikx}, & x \leqslant 0, \\ Fe^{ikx} + Ge^{-ikx}, & x \geqslant 0, \end{cases}$$

where $k \equiv \sqrt{2mE}/\hbar$ with $E > 0$. Here $A$ is the amplitude of an incoming wave from the left, and $G$ is the amplitude of an incoming wave from the right; $B$ is an outgoing wave to the left, and $F$ an outgoing wave to the right (remember, $\psi$ is to be combined with the time-dependent factor $\exp(-iEt/\hbar)$). The outgoing amplitudes are determined from the incoming amplitudes by the boundary conditions at $x = 0$:

$$\Delta\psi = 0, \quad \Delta\left(\frac{d\psi}{dx}\right) = -\frac{2m\alpha}{\hbar^2}\psi(0).$$

In the typical case of a particle incident from the left, $G = 0$, and the *reflection coefficient* (the probability of reflection back to the left) is

$$R = \frac{|B|^2}{|A|^2} = \frac{1}{1 + (2\hbar^2 E/m\alpha^2)},$$

while the *transmission coefficient* (the probability of transmission through to the right) is

$$T = \frac{|F|^2}{|A|^2} = \frac{1}{1 + (m\alpha^2/2\hbar^2 E)}.$$

Naturally, $R + T = 1$. Of course, these scattering states are not normalizable, and to represent an actual particle we should, in principle, form normalizable linear combinations of them.

## 4 Three Dimensions

Quantum mechanics extends to three dimensions in the obvious way: $\partial^2/\partial x^2 \to \nabla^2$ (the three-dimensional Laplacian), and the time-independent Schrödinger equation becomes

$$-\frac{\hbar^2}{2m}\nabla^2\psi + V\psi = E\psi.$$

We will use spherical coordinates $(r, \theta, \phi)$, where $\theta$ is the polar angle, measured down from the $z$-axis ($0 \leqslant \theta \leqslant \pi$), and $\phi$ is the azimuthal angle around from the $x$-axis ($0 \leqslant \phi < 2\pi$). In the typical case where $V$ is a function only of $r$, we solve by separation of variables; $\psi(r, \theta, \phi) = (1/r)u(r)Y_l^m(\theta, \phi)$, where

$$Y_l^m(\theta, \phi) = (-1)^m \sqrt{\frac{(2l+1)}{4\pi}\frac{(l-m)!}{(l+m)!}} e^{im\phi} P_l^m(\cos\theta) \tag{9}$$

is a *spherical harmonic* ($l = 0, 1, 2, \dots,\ m = -l, \dots, l$), and $P_l^m$ is an associated Legendre function. (Equation (9) applies for $m \geqslant 0$; for negative values, $Y_l^{-m} =$

$(-1)^m(Y_l^m)^*$.) Meanwhile, $u(r)$ satisfies the *radial equation*

$$-\frac{\hbar^2}{2m}\frac{d^2u}{dr^2} + \left[V + \frac{\hbar^2}{2m}\frac{l(l+1)}{r^2}\right]u = Eu, \qquad (10)$$

which is identical to the one-dimensional Schrödinger equation, except that the effective potential energy carries an extra centrifugal term.

The spherical harmonics are eigenfunctions of the angular momentum operator, $\hat{\boldsymbol{L}} \equiv \hat{\boldsymbol{r}} \times \hat{\boldsymbol{p}}$ ($\hat{L}_z = -i\hbar(x\partial/\partial y - y\partial/\partial x)$, and cyclic permutations thereof). The components of angular momentum do not commute,

$$[\hat{L}_i, \hat{L}_j] = i\hbar\hat{L}_k \quad \text{with } (ijk) = (xyz), \text{ etc.,} \qquad (11)$$

so it would be futile to look for simultaneous eigenstates of all three. However, each component does commute with the square of the total angular momentum, $\hat{L}^2 \equiv \hat{L}_x^2 + \hat{L}_y^2 + \hat{L}_z^2$, so it is possible to construct simultaneous eigenfunctions of $\hat{L}^2$ and (say) $\hat{L}_z$. In spherical coordinates, $\hat{L}^2$ is $-\hbar^2$ times the Laplacian restricted to the unit sphere ($r = 1$), $\hat{L}_z = -i\hbar\partial/\partial\phi$, and spherical harmonics are their eigenstates:

$$\hat{L}^2 Y_l^m = \hbar^2 l(l+1) Y_l^m, \qquad \hat{L}_z Y_l^m = \hbar m Y_l^m. \qquad (12)$$

In addition to its orbital angular momentum, $\boldsymbol{L}$, every particle carries an intrinsic *spin* angular momentum, $\boldsymbol{S}$ (distantly analogous to the daily rotation of the Earth on its axis, while the orbital angular momentum corresponds to its annual revolution about the sun). The components of $\hat{\boldsymbol{S}}$ satisfy the same fundamental commutation relations as $\hat{\boldsymbol{L}}$, namely, $[\hat{S}_x, \hat{S}_y] = i\hbar\hat{S}_z$ (and its cyclic permutations), and admit the same eigenvalues (with $s$ and $m_s$ in place of $l$ and $m$), $\hat{S}^2|sm_s\rangle = \hbar^2 s(s+1)|sm_s\rangle$, and $\hat{S}_z|sm_s\rangle = \hbar m_s|sm_s\rangle$. However, whereas $l$ can only be an integer, $s$ can be an integer or a half-integer; it is a fixed number, for a given type of particle (0 for $\pi$ mesons, $\frac{1}{2}$ for protons and electrons, 1 for photons, $\frac{3}{2}$ for the $\Omega^-$, and so on); we call it the *spin* of the particle. In the case of spin $\frac{1}{2}$, there are just two possible values for $m_s$: $\frac{1}{2}$ ("spin up") and $-\frac{1}{2}$ ("spin down"), and the resulting two-dimensional state space is precisely the one we studied in section 2, with the association $\hat{S}_x = \frac{1}{2}\hbar\sigma_1$, $\hat{S}_y = \frac{1}{2}\hbar\sigma_2$, $\hat{S}_z = \frac{1}{2}\hbar\sigma_3$.

Hydrogen is the simplest atom, consisting of a single electron bound to a single proton by the electrical attraction of opposite charges. The proton is almost 2000 times heavier than the electron, so it remains essentially at rest, and the potential energy of the electron is given by Coulomb's law, $V(r) = -e^2/(4\pi\varepsilon_0 r)$, where $e = 1.602 \times 10^{-19}$ C is the charge of the proton ($-e$ is the charge of the electron). The radial equation (10) reads

$$-\frac{\hbar^2}{2m}\frac{d^2u}{dr^2} + \left[-\frac{1}{4\pi\varepsilon_0}\frac{e^2}{r} + \frac{\hbar^2}{2m}\frac{l(l+1)}{r^2}\right]u = Eu.$$

It admits a discrete set of bound states, with energies

$$E_n = \frac{E_1}{n^2}, \qquad E_1 = -\frac{me^4}{2(4\pi\varepsilon_0)^2\hbar^2} = -13.606 \text{ eV} \qquad (13)$$

for $n = 1, 2, 3, \dots$ (as well as a continuum of scattering states with $E > 0$). The normalized wave functions are

$$\psi_{nlm} = \sqrt{\left(\frac{2}{na}\right)^3 \frac{(n-l-1)!}{2n[(n+l)!]^3}} \times e^{-r/na}\left(\frac{2r}{na}\right)^l \left[L_{n-l-1}^{2l+1}\left(\frac{2r}{na}\right)\right] Y_l^m, \qquad (14)$$

where $L_q^p(x)$ is an associated LAGUERRE POLYNOMIAL [II.29], and $a \equiv 4\pi\varepsilon_0\hbar^2/me^2 = 0.5292 \times 10^{-10}$ m is the *Bohr radius*.

The *principal quantum number*, $n$, tells you the energy of the state (13); the (misnamed) *azimuthal quantum number* $l$, which ranges from 0 to $n-1$, specifies the magnitude of the orbital angular momentum, and $m$ gives its $z$ component (12). Since $s = \frac{1}{2}$ for the electron, there are two linearly independent orientations for its spin. All told, the *degeneracy* of $E_n$ (that is, the number of distinct states sharing this energy) is $2n^2$.

If the atom makes a transition, or "quantum jump," to a state with lower energy, a photon (quantum of light) is emitted, carrying the energy released. (With the absorption of a photon it could make a transition in the other direction.) The energy of a photon is related to its frequency $\nu$ by the Planck equation $E = 2\pi\hbar\nu$, so the *spectrum* of hydrogen is given by

$$\lambda^{-1} = R(n_{\text{final}}^{-2} - n_{\text{initial}}^{-2}), \qquad (15)$$

with

$$R \equiv \frac{m}{4\pi c\hbar^3}\left(\frac{e^2}{4\pi\varepsilon_0}\right)^2 = 1.097 \times 10^7 \text{ m}^{-1}, \qquad (16)$$

where $\lambda = c/\nu$ is the wavelength (color) of the light. The *Rydberg formula* (15), with $R$ as an empirical constant, was discovered experimentally in the nineteenth century; Bohr derived it (and obtained the expression for $R$) in 1913, using a serendipitous mixture of inapplicable classical physics and primitive quantum theory. Schrödinger put it on a rigorous theoretical footing in 1924.

## 5 Composite Structures

The theory generalizes easily to systems with more than one particle. For example, two particles in one dimension would be described by the wave function $\Psi(x_1, x_2, t)$, where

$$\int_{x_1=a_1}^{b_1} \int_{x_2=a_2}^{b_2} |\Psi(x_1, x_2, t)|^2 \, dx_1 \, dx_2$$

is the probability of finding particle 1 between $a_1$ and $b_1$ and particle 2 between $a_2$ and $b_2$, if a measurement is made at time $t$.

But here quantum mechanics introduces a new twist: suppose the two particles are absolutely identical, so there is no way of knowing which is #1 and which #2. In classical physics such indistinguishability is unthinkable—you could always stamp a serial number on each particle. But you cannot put labels on electrons; they simply do not possess independent identities—the theory must treat the two particles on an equal footing. There are essentially two possibilities (others occur in special geometries): under interchange, $\Psi$ can be symmetric, $\Psi(x_2, x_1) = \Psi(x_1, x_2)$ (*bosons*), or antisymmetric, $\Psi(x_2, x_1) = -\Psi(x_1, x_2)$ (*fermions*). Some elementary particles (electrons and quarks, for example) are fermions; others (photons and pions, for instance) are bosons. The distinction is related to the spin of the particle: particles of integer spin are bosons, whereas particles of half-integer spin are fermions. (This "connection between spin and statistics" can be proved using relativistic quantum mechanics, but in the nonrelativistic theory it is simply an empirical fact.)

Suppose we have two particles, one in state $\psi_a(x)$ and the other in state $\psi_b(x)$. If the particles are distinguishable (an electron and a proton, say) and the first is in state $\psi_a$, then the composite wave function is $\psi(x_1, x_2) = \psi_a(x_1)\psi_b(x_2)$. But if they are identical bosons (two pions, say), we must use the symmetric combination,

$$\psi(x_1, x_2) = \frac{1}{\sqrt{2}}[\psi_a(x_1)\psi_b(x_2) + \psi_a(x_2)\psi_b(x_1)],$$

and for identical fermions (two electrons, say),

$$\psi(x_1, x_2) = \frac{1}{\sqrt{2}}[\psi_a(x_1)\psi_b(x_2) - \psi_a(x_2)\psi_b(x_1)].$$

(The normalization factor $1/\sqrt{2}$ assumes that $\psi_a$ and $\psi_b$ are orthonormal.)

For example, we might calculate the expectation value of the square of the separation distance between the two particles. If they are distinguishable, then

$$\langle (x_1 - x_2)^2 \rangle = \langle x^2 \rangle_a + \langle x^2 \rangle_b - 2\langle x \rangle_a \langle x \rangle_b \equiv \Delta^2,$$

where $\langle x \rangle_a$ is the expectation value of $x$ for a particle in the state $\psi_a$, and so on. If, however, the particles are identical bosons, then there is an extra term $\langle (x_1 - x_2)^2 \rangle = \Delta^2 - 2|\langle x \rangle_{ab}|^2$, and if they are identical fermions, then $\langle (x_1 - x_2)^2 \rangle = \Delta^2 + 2|\langle x \rangle_{ab}|^2$, where

$$\langle x \rangle_{ab} \equiv \int x\psi_a(x)^* \psi_b(x) \, dx.$$

Thus bosons tend to be closer together, and fermions farther apart, than distinguishable particles in the same two states. It is as if there were an attractive force between identical bosons and a repulsive force between identical fermions. We call these *exchange forces*, though there is no actual force involved—it is just an artifact of the (anti)symmetrization requirement. Exchange forces are responsible for ferromagnetism, and they contribute to *covalent bonding*. (If you include spin, it is the *total* wave function that must be antisymmetrized; if the spin part is antisymmetric, then the position wave function must actually be symmetric (for fermions), and the exchange force is attractive. Covalent bonding occurs when shared electrons cluster in the region between two nuclei, tending to pull them together.)

The (anti)symmetrization requirement gives rise to very different statistical mechanics for bosons and fermions. In particular, two identical fermions cannot occupy the same state, since if $\psi_a = \psi_b$ the composite wave function $\psi(x_1, x_2)$ vanishes. This is the famous *Pauli exclusion principle*. There is no such rule for bosons, and at extremely low temperatures identical bosons tend to congregate in the ground state, forming a *Bose-Einstein condensate*. In a large sample at (absolute) temperature $T$, the most probable number of particles in a (one-particle) state with energy $E$ is

$$n(E) = \begin{cases} e^{-(E-\mu)/kT}, & \text{Maxwell-Boltzmann,} \\ \dfrac{1}{e^{(E-\mu)/kT} + 1}, & \text{Fermi-Dirac,} \\ \dfrac{1}{e^{(E-\mu)/kT} - 1}, & \text{Bose-Einstein.} \end{cases}$$

Here, $k = 1.3807 \times 10^{-23}$ J K$^{-1}$ is Boltzmann's constant, and $\mu$ is the *chemical potential*—it is a function of temperature and depends on the nature of the particles. The Maxwell-Boltzmann distribution (the classical result) applies to distinguishable particles, the Fermi-Dirac distribution is for identical fermions, and the Bose-Einstein distribution is for identical bosons.

The Pauli principle is crucial in accounting for the periodic table of the elements. Atoms are labeled by their atomic number $Z$, the number of protons in the nucleus, which is also the number of electrons in orbit (the number of neutrons distinguishes different *isotopes*). Since the nucleus remains essentially at rest, only the behavior of the electrons is at issue in atomic physics. Hydrogen has one electron, helium has two (and two protons in the nucleus), lithium three, and so on. Ordinarily, each electron will settle into the lowest-energy accessible state. But the Pauli principle forbids any given state from being occupied by more than one electron, so they fill up the allowed states in succession, and this means that the outermost (valence) electrons (which are responsible for chemical behavior) are differently situated for different elements.

The Hamiltonian for an atom with atomic number $Z$ is

$$\hat{H} = \sum_{j=1}^{Z}\left\{-\frac{\hbar^2}{2m}\nabla_j^2 - \frac{1}{4\pi\varepsilon_0}\frac{Ze^2}{r_j}\right\} + \frac{e^2}{8\pi\varepsilon_0}\sum_{k\neq j}^{Z}\frac{1}{|\boldsymbol{r}_j - \boldsymbol{r}_k|},$$

where $\boldsymbol{r}_j$ is the position of the $j$th electron, $r_j = |\boldsymbol{r}_j|$, and $\nabla_j^2$ is the Laplacian with respect to $\boldsymbol{r}_j$. The term in curly brackets represents the kinetic and potential energy of the $j$th electron in the electric field of the nucleus, and the other term is the potential energy associated with the mutual repulsion of the electrons. (A tiny relativistic correction and magnetic coupling between the spin of the electron and the orbital motion account for *fine structure*, while an even smaller magnetic coupling between the spin of the electron and the spin of the nucleus leads to *hyperfine structure*.)

If we ignore the electron–electron repulsion altogether, we simply have $Z$ independent electrons, each in the field of a nucleus of charge $Ze$; we simply copy the results for hydrogen, replacing $e^2$ by $Ze^2$. The first two electrons would occupy the *orbital*, $n = 1$, $l = 0$, $m = 0$, the next eight fill out the $n = 2$ orbitals, and so on. This simple scheme explains the first two rows of the periodic table (up through neon). One would anticipate a third row of eighteen, but after argon the electron–electron repulsion finally catches up, and potassium skips to $n = 4$, $l = 0$ for the next electron, in preference to $n = 3$, $l = 2$. The details get complicated and are handled by sophisticated approximation schemes (hydrogen is the only atom for which the Schrödinger equation can be solved analytically).

The addition of angular momenta is an interesting problem in quantum mechanics. Suppose I want to combine $|l_1 m_1\rangle$ with $|l_2 m_2\rangle$; what is the resulting combined state, $|lm\rangle$? For instance, the electron in a hydrogen atom has orbital angular momentum and spin angular momentum; what is its total angular momentum? (I will use $l$ as the generic letter for the angular momentum quantum number—it could be orbital, or spin, or total, as the case may be.) Because $\boldsymbol{L} = \boldsymbol{L}_1 + \boldsymbol{L}_2$, the $z$ components add: $m = m_1 + m_2$. But the total angular momentum quantum number $l$ depends on the relative orientation of $\boldsymbol{L}_1$ and $\boldsymbol{L}_2$, and it can range from $|l_1 - l_2|$ (roughly speaking, when they are antiparallel) to $l_1 + l_2$ (parallel), in integer steps:

$$l = -|l_1 - l_2|, -|l_1 - l_2| + 1, \dots, l_1 + l_2 - 1, l_1 + l_2.$$

Specifically,

$$|l_1 m_1\rangle|l_2 m_2\rangle = \sum_{l=|l_1-l_2|}^{l_1+l_2} C_{m_1 m_2 m}^{l_1 l_2 l}|lm\rangle,$$

where $C_{m_1 m_2 m}^{l_1 l_2 l}$ are the so-called *Clebsch-Gordan coefficients*, which are tabulated in many handbooks. Mathematically, what we are doing is decomposing the direct product of two irreducible representations of SU(2) (the covering group for SO(3), the rotation group in three dimensions) into a direct sum of irreducible representations. For instance,

$$|2\,1\rangle|\tfrac{1}{2}\,\tfrac{1}{2}\rangle = \sqrt{\tfrac{4}{5}}|\tfrac{5}{2}\,\tfrac{3}{2}\rangle - \sqrt{\tfrac{1}{5}}|\tfrac{3}{2}\,\tfrac{3}{2}\rangle.$$

If you had a hydrogen atom in the state $Y_1^2$, and the electron had spin up ($m_s = \frac{1}{2}$), the probability is $\frac{4}{5}$ that a measurement of the total $l$ would return the value $\frac{5}{2}$ and $\frac{1}{5}$ that it would yield $\frac{3}{2}$ (a measurement of $m$ would be certain to give $\frac{3}{2}$). The Clebsch-Gordan coefficients work the other way, too. If I have a system in the angular momentum state $|lm\rangle$, composed of two particles with $l_1$ and $l_2$, and I want to know the possible values of $m_1$ and $m_2$, I would use

$$|l\,m\rangle = \sum_{m_1=-l_1}^{l_1}\sum_{m_2=-l_2}^{l_2} C_{m_1 m_2 m}^{l_1 l_2 l}|l_1\,m_1\rangle|l_2\,m_2\rangle,$$

where the sum is over all combinations of $m_1$ and $m_2$ such that $m_1 + m_2 = m$.

## 6 Implications, Applications, Extensions

In quantum mechanics the state of a system is represented by a (normalized) vector ($|s\rangle$) in Hilbert space; observables are represented by (self-adjoint) operators acting on vectors in this space. The theory rests on three pillars: the Schrödinger equation, which in its most general form tells you how the state vector $|s\rangle$ evolves in time,

$$\mathrm{i}\hbar\frac{\partial}{\partial t}|s\rangle = \hat{H}|s\rangle; \tag{17}$$

the statistical interpretation, which tells you how $|s\rangle$ determines the outcome of a measurement; and the (anti)symmetrization requirement, which tells you how to construct $|s\rangle$ for a collection of identical particles. Although the calculational procedures are unambiguous and the results are in spectacular agreement with experiment, the physical interpretation of quantum mechanics (the story we tell ourselves about what we are doing) has always been controversial. Some of the major issues are listed below.

**Indeterminacy.** Einstein argued—most persuasively in the famous *EPR* (Einstein-Podolsky-Rosen) *paradox*—that quantum indeterminacy means the theory is *incomplete*; other information (so-called *hidden variables*) must supplement the state vector to provide a full description of physical reality. By contrast, Bohr's *Copenhagen interpretation* holds that systems simply do not have definite properties (such as position or angular momentum) prior to measurement.

**Nonlocality.** The EPR argument is predicated on an assumption of *locality*: no influence can propagate faster than light. But in quantum mechanics, two particles can be *entangled* such that a measurement on one determines the state of the other and, if they are far apart, then that "influence" (the collapse of the wave function) is instantaneous. In 1964 John Bell proved that no local deterministic (hidden-variable) theory can be compatible with quantum mechanics. In subsequent experiments the quantum mechanical predictions were sustained. Evidently, nonlocality is a fact of nature, though the "influences" in question are not strictly causal—they produce no effects that can be discerned by examining the second particle alone. (A *causal* nonlocal influence would be incompatible with special relativity.)

**Measurement.** It has never been clear what constitutes a "measurement," as the word is used in the statistical interpretation. Why does a measurement force the system (which previously did not possess a determinate value of the observable in question) to "take a stand," and how does it collapse the wave function? Does "measurement" mean something people in white coats do in the laboratory, or can it occur in a forest, when no one is looking? Does measurement require the leaving of a permanent record or the interaction of a microscopic system with a macroscopic device? Does it involve the intervention of human consciousness? Or does it perhaps split reality into many worlds? After nearly a century of debate, there is no consensus on these questions.

**Decoherence.** If an electron can have an indeterminate position, what about a baseball? Could a baseball be in a linear combination of Seattle and San Francisco (until, I guess, the batters swing, and there is a home run in Seattle and nothing at all in San Francisco)? There is something absurd about the very idea of a macroscopic object being in two places at once (or an animal—in Schrödinger's famous *cat paradox*—being both alive and dead). Why do macroscopic objects obey the familiar classical laws, while microscopic objects are subject to the bizarre rules of quantum mechanics? Presumably if you *could* put a baseball into a state that stretched from Seattle to San Francisco, its wave function would very rapidly *decohere* into a state localized at some specific place. By what mechanism? Maybe multiple interactions with random impinging particles (photons left over from the Big Bang, perhaps) constitute a succession of "measurements" and collapse the wave function. But the details remain frustratingly elusive.

I have mentioned two kinds of solutions to the Schrödinger equation, for time-independent potentials: (normalizable) bound states, at discrete negative energies, and (nonnormalizable) scattering states, at a continuum of positive energies. There is a third important case, which occurs for periodic potentials, $V(x+a) = V(x)$, as in a crystal lattice. For these the spectrum forms *bands*, with continua of allowed energies separated by forbidden gaps. This band structure is crucial in accounting for the behavior of solids and underlies most of modern electronics.

Very few quantum problems can be solved exactly, so a number of powerful approximation methods have been developed. Here are some examples.

**The variational principle.** The lowest eigenvalue of a Hamiltonian $\hat{H}$ is less than or equal to the expectation value of $\hat{H}$ in *any* normalized state $|s\rangle$:

$$E_0 \leqslant \langle s|\hat{H}|s\rangle. \tag{18}$$

To get an upper bound on the ground-state energy of a system, then, you simply pick any (normalized) vector $|s\rangle$ and calculate $\langle s|\hat{H}|s\rangle$ in that state. Ordinarily, you will get a tighter bound if your "trial state" bears some resemblance to the actual ground state. In a typical application the trial state carries a number of adjustable parameters, which are then chosen so as to minimize $\langle\hat{H}\rangle$. The binding energy of helium, for example, has been calculated in this way and is consistent with the measured value to better than eight significant digits.

**Time-independent perturbation theory.** Suppose you know the eigenstates $\{|n\rangle\}$ and eigenvalues $\{E_n\}$ of some (time-independent) Hamiltonian $H$. Now you perturb that Hamiltonian slightly, $H \to H + H'$. The resulting change in the $n$th eigenvalue is approximately equal to the expectation value of $\hat{H}'$ in the unperturbed state $|n\rangle$:

$$\Delta(E_n) \approx \langle n|\hat{H}'|n\rangle. \tag{19}$$

For example, the fine structure of hydrogen can be calculated very accurately in this way, starting with the wave functions in (14).

**Time-dependent perturbation theory.** Again, suppose you know the eigenstates and eigenvalues of some time-independent Hamiltonian, and now you (briefly) turn on a small time-dependent perturbation, $H'(t)$. What is the transition rate (probability per unit time) to go from state $|n\rangle$ to state $|m\rangle$? The answer is given by

$$R_{n\to m} = \frac{2\pi}{\hbar}|H'_{mn}|^2\rho(E_m),$$

where $H'_{mn} \equiv \langle m|\hat{H}'|n\rangle$ is the so-called *matrix element* for the transition, and $\rho$ is the *density of states*—the number of states per unit energy. Fermi called this the *golden rule*. It is used, for example, to calculate the lifetime of an excited state.

In two important respects quantum mechanics is obviously not the end of the story. First of all, the Schrödinger equation as it stands is inconsistent with special relativity—as a differential equation it is second order in $x$ but first order in $t$, whereas relativity requires that they be treated on an equal footing. Dirac introduced the eponymous relativistic equation for particles of spin $\frac{1}{2}$ (the DIRAC EQUATION [III.9]) and others followed (Klein–Gordon for spin 0, Proca for spin 1, and so on). Second, while the *particles* have been treated quantum mechanically, the *fields* have not. In the hydrogen atom, for example, the electric potential energy was taken from classical electrostatics. In a fully consistent theory the fields, too, must be quantized. The quantum of the electromagnetic field is the photon, but although I have used this word once or twice, it has no place in quantum mechanics; it belongs instead to quantum field theory (specifically *quantum electrodynamics*). In the *standard model* of elementary particles, all known interactions save gravity are successfully handled by relativistic quantum field theory. But a fully consistent union of quantum mechanics and GENERAL RELATIVITY [IV.40] (Einstein's theory of gravity) still does not exist.

### Further Reading

Cohen-Tannoudji, C., B. Diu, and F. Laloe. 1977. *Quantum Mechanics*. New York: John Wiley.

Griffiths, D. J. 2005. *Introduction to Quantum Mechanics*, 2nd edn. Upper Saddle River, NJ: Pearson.

Merzbacher, E. 1998. *Quantum Mechanics*, 3rd edn. New York: John Wiley.

Townsend, J. S. 2012. *A Modern Approach to Quantum Mechanics*, 2nd edn. Herndon, VA: University Science Books.

## IV.24 Random-Matrix Theory

*Jonathan Peter Keating*

### 1 Introduction

Linear algebra and the analysis of systems of linear equations play a central role in applied mathematics; for example, QUANTUM MECHANICS [IV.23] is a linear wave theory in which observables are represented by linear operators; linear models are fundamental in electromagnetism, acoustics, and water waves; and linearization is important in stability analysis. When a linear system is intrinsically complex, either by virtue of external stochastic forcing or because of an underlying, persistent self-generated instability, it is natural to model it statistically by assuming that the elements of the matrices that appear in its mathematical description are in some sense random. This is similar, philosophically, to the way in which statistical features of long trajectories in complex dynamical systems are modeled statistically via notions of ergodicity and mixing. An example is when classical dynamics is modeled by statistical mechanics, that is, where one deduces statistical properties of the solutions of systems of equations, in this case Newton's equations of motion, by analyzing ensembles of similar trajectories and invoking notions of ergodicity (that time averages equal ensemble averages in the appropriate limit).

This is one significant motivation for exploring the properties of random matrices. There are, however, many others: linear algebra and probabilistic models both play a foundational role in mathematics, and it is therefore not surprising that they combine in a wide range of applications, including mathematical biology, financial mathematics, high-energy physics, condensed matter physics, numerical analysis, neuroscience, statistics, and wireless communications.

In many of the examples listed above, one has a system of linear equations that can be written in matrix form. One might then be interested in the eigenvalues or eigenvectors of the matrix in question, if it is square. For example, one might want to know the range of the spectrum, its density within that range, and the nature of fluctuations/correlations in the positions of the eigenvalues. Alternatively, one might want an estimate of the condition number of the matrix or the values taken by its characteristic polynomial.

If the matrix possesses no special features—that is, if its entries do not obey simple rules—it is natural to consider whether it behaves like one that is a "typical" member of some appropriate class of matrices. This again motivates the study of random matrices, that is, matrices whose entries are random variables.

The basic formulation of random-matrix theory is that one has a space of matrices $X$ endowed with a probability measure $P(X)$. This is termed a matrix *ensemble*. We shall here focus on square matrices of dimension $N$. One can then seek to determine the probability distribution of the eigenvalues and eigenvectors, and of other related quantities of interest. The motivation is that in many cases one can prove that spectral averages for a given matrix coincide, with probability 1, with ensemble averages, when $N \to \infty$.

The following are examples of such matrix ensembles.

- Let $X$ be an $N \times N$ matrix with real or complex-valued entries $X_{mn}$ satisfying $X_{mn} = \bar{X}_{nm}$ (so the matrix is either real symmetric or complex Hermitian). Take the entries $X_{nm}$, $n < m$, to be independent zero-mean real or complex-valued random variables, and the entries $X_{nn}$ to be independent, identically distributed, centered real-valued random variables, so $P(X)$ factorizes as $P(X) = \prod_{m \leqslant n} P_{mn}(X_{mn})$. This defines the ensemble of *Wigner random matrices.*
- Let $X$ be an $N \times N$ real symmetric matrix. When the probability measure is invariant under all orthogonal transformations of $X$, i.e., when $P(OXO^{\mathrm{T}}) = P(X)$ for all orthogonal ($OO^{\mathrm{T}} = I$) matrices $O$, then the ensemble is called *orthogonal invariant.*
- Let $X$ be an $N \times N$ complex Hermitian matrix. When the probability measure is invariant under all unitary transformations of $X$, i.e., when $P(UXU^{\dagger}) = P(X)$ for all unitary ($UU^{\dagger} = I$) matrices $U$, then the ensemble is called *unitary invariant.*
- The *Gaussian orthogonal ensemble* (GOE) is the (unique) orthogonal invariant ensemble of Wigner random matrices. It has $P(X) \propto \exp(-\frac{1}{4}\operatorname{Tr} X^2)$. Similarly, the *Gaussian unitary ensemble* (GUE) is the (unique) unitary invariant ensemble of Wigner random matrices. It has $P(X) \propto \exp(-\frac{1}{2}\operatorname{Tr} X^2)$.
- The most general class of random $N \times N$ matrices corresponds to taking the matrix elements $X_{mn}$ to be real or complex-valued independent identically distributed random variables with, for example, zero mean and unit variance. When these random variables have a Gaussian distribution, the matrices form what is known as the *Ginibre ensemble.*
- The *circular ensembles* correspond to $N \times N$ unitary matrices with probability measures that are invariant under all orthogonal (COE) or unitary (CUE) transformations. Matrices in the COE are unitary and symmetric. Alternatively, one can consider ensembles corresponding to the Haar measure on the classical compact groups (i.e., the measure that is invariant under the group action): for example, the orthogonal group $\mathrm{O}(N)$, comprising $N \times N$ orthogonal matrices, or the unitary group $\mathrm{U}(N)$, which coincides with the CUE, or the symplectic group $\mathrm{Sp}(2N)$.
- Let $X$ be an $N \times k$ matrix, each row of which is drawn independently from a $k$-variate normal distribution with zero mean. The *Wishart ensemble* corresponds to the matrices $X^{\mathrm{T}}X$.

These are far from being the only examples that are important—one can also define random matrices whose elements are quaternionic, or which have additional structure (i.e., invariance under additional symmetries)—but they will suffice to illustrate a number of general questions and themes.

For any of the above ensembles of random matrices, one can first ask where the eigenvalues typically lie, and what their mean density is. In ensembles where the eigenvalues are real, for example, how many can one expect to find in a given interval on average? One can, in particular, ask whether there is a limiting density when the matrix size tends to infinity.

Second, one can seek to understand fluctuations about the mean in the eigenvalue distribution. For example, in an interval of length such that the expected number of eigenvalues is $k$, what is the distribution of the actual numbers in the interval for different matrices in the ensemble? Again, is there a limiting distribution when the matrix size tends to infinity? Or, similarly, how do the gaps between adjacent eigenvalues fluctuate around their mean? How, if at all, are eigenvalues

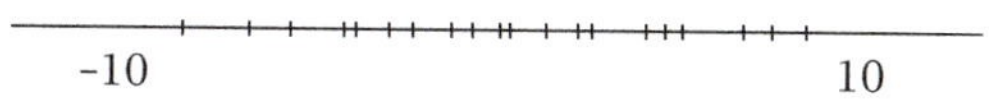


**Figure 1** The spectrum of a randomly chosen complex Hermitian (GUE) matrix of dimension $N = 20$.

correlated on the scale of their mean spacing? How are other functions over the ensemble distributed? For example, the values of the characteristic polynomials of the matrices, their traces, their condition numbers, the sizes of the largest and smallest eigenvalues?

Third, how do the answers to these questions depend on the particular choice of probability measure intrinsic to the definition of a given ensemble, and on the symmetries (if any) of the matrices involved? If they depend on the choice of the probability measure, then how do they? Is there a class of probability measures for which they can be said to be *universal* in an appropriate limit? Are there choices of probability measure that are in some sense exactly solvable?

Fourth, given answers to questions such as these, how can they be used to shed light on applications?

Our aim here is to give an introductory overview of some of these issues.

In order to motivate and illustrate the theory, it may be helpful to anticipate it with the results of some numerical computations. Figure 1 shows the spectrum of a complex Hermitian matrix of dimension $N = 20$, picked at random from the GUE by generating normal random variables to fill the independent matrix entries. One sees that the spectrum is denser at the center than at the edges and that the eigenvalues do not often lie close to each other.

To illustrate these points further, figure 2 shows the spectral measure (i.e., the local eigenvalue density; see equation (1) below) of a GUE matrix of dimension $N = 500\,000$, with the eigenvalues divided by $\sqrt{N}$. Figure 3 shows a similar plot with data averaged over the spectra of $2 \times 10^6$ matrices with $N = 500$. One sees that after dividing by $\sqrt{N}$ these spectra appear to lie between $-2$ and $+2$, that the picture for an individual matrix is very similar (indeed, on the scale used, figure 2 is indistinguishable from figure 3) to that of an average over a large number of matrices, and that the eigenvalue density appears to have a simple form that is well described by the result of an analytical random-matrix calculation (represented by curves) to be explained later. Similarly, figures 4 and 5 show the probability densities of the spacings between adjacent eigenvalues, further scaled to have unit mean separation (see below), for individual GOE and GUE matrices of dimension $N = 500\,000$

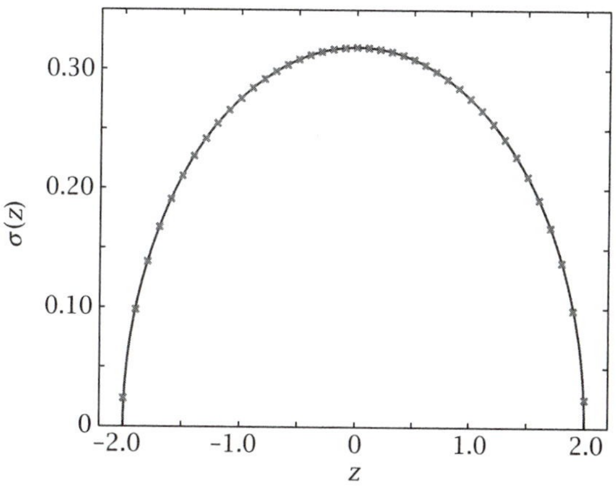


**Figure 2** The spectral measure (i.e., eigenvalue density (see (1))) of a GUE matrix of dimension $N = 500\,000$, with the eigenvalues divided by $\sqrt{N}$ (crosses). The curve is a prediction of random-matrix theory, which is described later in the text.

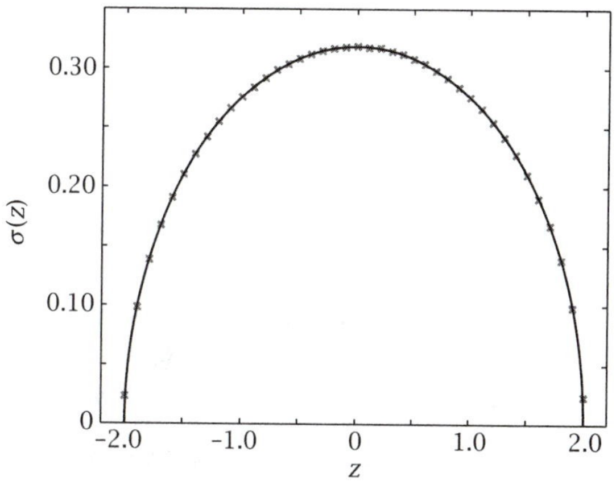


**Figure 3** The average of the spectral measures of $2 \times 10^6$ GUE matrices with dimension $N = 500$, with the eigenvalues divided by $\sqrt{N}$ (crosses). The curve is a prediction of random-matrix theory, which is described later in the text. Note that, on the scale shown, these data look identical to those in figure 2 despite the fact that they are in fact different (see section 7).

and separately averaged over GOE and GUE matrices of dimension $N = 500$, respectively. One again sees a marked similarity between the behavior of individual matrices and that of ensemble averages. In both the GOE and GUE examples, the probability density vanishes in the limit of small spacings; that is, the eigenvalues behave as if they repel each other. Note that the degree of the repulsion differs between the

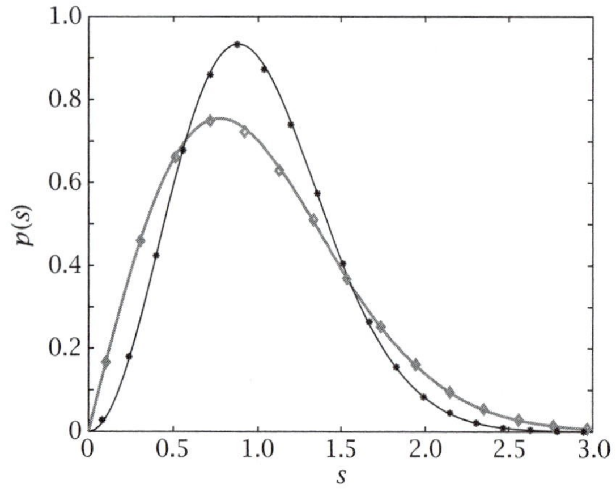


**Figure 4** The probability density of the spacings between adjacent eigenvalues for the GUE spectrum represented in figure 2 (asterisks, black line) (when these eigenvalues are scaled to have unit mean spacing), and the corresponding distribution for a GOE matrix of the same dimension (diamonds, gray line). The two curves represent the corresponding random-matrix formulas, which are described later in the text.

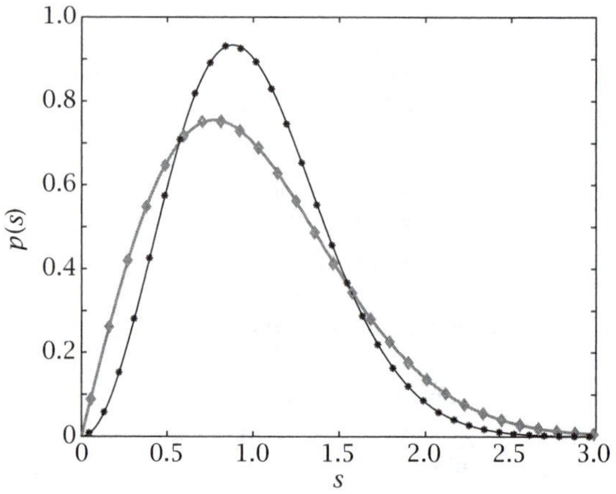


**Figure 5** The probability density of the spacings between adjacent eigenvalues, averaged over the GUE spectra represented in figure 3 (asterisks, black line) (when these eigenvalues are scaled to have unit mean spacing), and the corresponding distribution for $2 \times 10^6$ GOE matrices with dimension $N = 500$ (diamonds, gray line). The two curves represent the corresponding random-matrix formulas, which are described later in the text. Note that, on the scale shown, these data look identical to those in figure 4, despite the fact that they are in fact different (again see section 7).

two ensembles. Once more, these figures also show the results of analytical random-matrix calculations to be explained later.

## 2 History

The origins of random-matrix theory go back to work on multivariate statistics by Wishart, who introduced the Wishart ensemble in 1928 in his calculation of the maximum likelihood estimator for the covariance matrix of the multivariate normal distribution. Wishart worked at the Rothamsted Experimental Station, an agricultural research facility.

Many of the most significant developments in the field were stimulated by ideas of Wigner in the 1950s that related to nuclear physics. Wigner was interested in modeling the statistical distribution of the energy levels of heavy nuclei. He introduced the Wigner ensemble and the invariant Gaussian ensembles; calculated the mean eigenvalue density in these cases; and started to investigate spectral correlations, particularly the distribution of spacings between adjacent eigenvalues. In focusing on heavy nuclei, Wigner had in mind applications to many-body quantum systems, that is, systems with a large number of degrees of freedom.

In the 1960s Wigner's program was developed into a systematic area of mathematical physics by Dyson, Gaudin, and Mehta, who developed techniques for calculating eigenvalue statistics in the Gaussian and circular ensembles, and Marčenko and Pastur, who extended the analysis to the Wishart ensemble.

The idea of applying random-matrix theory to systems with a small number of degrees of freedom, in which complexity arises due to the internal dynamics, arose in the field of quantum chaos in the late 1970s and early 1980s in the work of Berry, Bohigas, and their coworkers. Here, the philosophy is that, when the classical dynamics of a system is chaotic, the quantum dynamics may manifest that in the semiclassical limit (i.e., in the limit as the de Broglie wavelength tends to zero) by the corresponding matrix elements behaving like random variables. This same philosophy immediately generalizes, *mutatis mutandis*, to other wave theories such as optics, acoustics, and so on.

The years since 1990 have seen a rapid development of interest in connections between random-matrix theory and, among other subjects, quantum field theory, high-energy physics, condensed matter physics, lasers, biology, finance, growth models, wireless communication, and number theory. There has also been considerable progress in proving universality and establishing links with the theory of integrable systems.

## 3 Eigenvalue Density

In the case of matrix ensembles in which the eigenvalues are real (e.g., Hermitian or real symmetric matrices), one can ask how many lie in a given interval. Let us consider first the case of $N \times N$ Wigner random matrices $X$. It turns out that when the matrix elements have finite second moment, the eigenvalues $\lambda_i^N(X)$ are, with extremely high probability, of the order of $\sqrt{N}$. This fact is illustrated in figures 2 and 3. Hence, defining the spectral measure by

$$\Lambda_N(z;X) = \frac{1}{N}\sum_{n=1}^{N} \delta\left(z - \frac{\lambda_i^N(X)}{\sqrt{N}}\right), \qquad (1)$$

where $\delta(x)$ is the Dirac delta function, the proportion of eigenvalues (normalized by $\sqrt{N}$) lying in the interval $[a,b]$ is

$$\frac{1}{N}\#\{n\colon \lambda_i^N(X) \in \sqrt{N}[a,b]\} = \int_a^b \Lambda_N(z;X)\,\mathrm{d}z.$$

As $N \to \infty$, $\Lambda_N$ converges, in the distributional sense, to a limiting function of $z$. Specifically, $\Lambda_N$ converges weakly to the *semicircle law* $\sigma$ given by

$$\mathrm{d}\sigma(z) = \frac{1}{2\pi}\sqrt{4-z^2}\,\mathrm{d}z$$

when $-2 \leqslant z \leqslant 2$, and by 0 when $|z| > 2$.

This was first established for convergence in probability by Wigner in 1955 under the condition that all the moments of the matrix elements are finite. However, it has subsequently been refined so that convergence holds almost surely and under the condition that the matrix elements have finite second moment. The semicircle law is the curve plotted in figures 2 and 3.

In the case of general Wigner matrices, one can establish the semicircle law by considering moments of traces of powers of $X$. In the special cases of the GOE and the GUE, one can also use techniques that rely on the invariance properties of the measure (see below) to obtain precise estimates for the rate of convergence. In the case of the other invariant ensembles, one can also establish a limiting density, but typically this is not semicircular; it may be calculated from the specific form of the probability.

For the Wishart ensemble and its generalizations, which play a major role in many applications, the analogue of the Wigner semicircle law is known as the *Marčenko–Pastur law*.

For the circular ensembles and matrices from the classical compact groups, the analogue of the semicircle law is that in the large matrix limit, the eigenvalues become uniformly dense on the unit circle (on which they are constrained to lie by unitarity).

In the case of non-Hermitian matrices, one can similarly calculate the density of eigenvalues in the complex plane. For example, in the Ginibre ensemble, the eigenvalues $\lambda_i^N(X)$, when divided by $\sqrt{N}$, have a limiting density that is given by the uniform measure on the unit disk.

## 4 Joint Eigenvalue Distribution

A more refined question about the eigenvalues $\lambda_i^N(X)$ of a random matrix $X$ than that of their limiting density relates to the probability that they all lie in some given set $S$ in the complex plane or, when they are constrained to be real, on the real line. For definiteness, we will focus here on the latter case. Essentially, the issue is then to calculate the probability that the first eigenvalue lies in the interval $(x_1, x_1 + \mathrm{d}x_1)$, the second in the interval $(x_2, x_2 + \mathrm{d}x_2)$, the third in the interval $(x_3, x_3 + \mathrm{d}x_3)$, etc. In general, this probability is hard to derive in a useful form. For the invariant ensembles, however, it can be computed. The idea is that the probability measure defined in terms of the matrix elements can be reexpressed in terms of the eigenvalues and the eigenvectors. In the invariant ensembles, the dependence on the eigenvectors may be straightforwardly integrated out, leaving just the dependence on the eigenvalues. The result is proportional to the product of a Vandermonde determinant factor $|\det(V)|^\beta$, with $V_{ij} = x_i^{j-1}$ and therefore

$$|\det(V)|^\beta = \prod_{1\leqslant i<j\leqslant N} |x_j - x_i|^\beta,$$

and a factor $w(x_1,\dots,x_N)$ coming from the precise form of the probability measure associated with the matrices. Here, $\beta = 1$ for orthogonal invariant ensembles (e.g., the GOE) and $\beta = 2$ for unitary invariant ensembles (e.g., the GUE). ($\beta = 4$ for quaternionic matrices.) The Vandermonde factor comes from the Jacobian associated with the change of variables from the matrix elements to the eigenvalues. For the Gaussian ensembles the weight factor is $w(x_1,\dots,x_N) = \exp(-\frac{1}{4}\beta\sum_{n=1}^{N} x_n^2)$. In the case of the Ginibre ensemble one has a similar result, but the eigenvalues are generally complex. For the circular ensembles, in which the eigenvalues lie on the unit circle, the Vandermonde factor takes the form $\prod_{1\leqslant i<j\leqslant N}|\mathrm{e}^{\mathrm{i}\theta_j} - \mathrm{e}^{\mathrm{i}\theta_i}|^\beta$ and the

weight factor $w$ is a constant. The joint eigenvalue distribution for the Wishart ensemble has a similar structure, but it also reflects the fact that the eigenvalues are nonnegative.

Perhaps the most significant consequence of the structure of the joint eigenvalue distribution functions outlined in the previous paragraph is that the eigenvalues of random matrices behave as if they repel each other. This follows from the Vandermonde factors, which vanish as any two eigenvalues approach each other, so the probability of having a pair of eigenvalues in close vicinity vanishes as their separation is reduced. Importantly, the rate of vanishing depends only on the symmetry of the matrix ensemble (via the value of $\beta$) and not on the detailed form of the probability measure (which determines the multiplying weight factor). By way of contrast, independent random numbers, for which $\beta = 0$, do not share this behavior.

A key feature of the formulas for the joint eigenvalue distribution functions for the invariant ensembles is that they have a determinantal structure. This is the starting point for the analysis of the statistical distribution of the eigenvalues. The essential idea is that, by forming linear combinations of the rows and columns of $V$, one can generate polynomials in the variables $x_n$ that are ORTHOGONAL [II.29] with respect to the (ensemble-dependent) weight $w$. This often allows integrals involving the joint eigenvalue distribution functions to be computed, either exactly or asymptotically. In the case of the GUE, this goes through straightforwardly for any matrix size $N$ (the polynomials in this case being the classical Hermite polynomials). The GOE requires more sophisticated analysis, but again the theory can be developed for any matrix size. In both cases the semicircle law can be obtained via this approach by integrating over all but one of the variables in the joint eigenvalue distribution function. The CUE and COE can also be analyzed in this way. In the case of more general invariant ensembles, the orthogonal polynomials that arise are nonclassical, and exact calculations for a finite matrix size are difficult. However, the large-$N$ asymptotics can be evaluated by an application of the steepest-descent analysis for the *Riemann-Hilbert problem*.

## 5 Eigenvalue Statistics

One theme that has been central to random-matrix theory has been to seek statistical measures that, unlike the full joint eigenvalue distribution, have a well-defined limit as $N \to \infty$. In order to compare different ensembles, and different parts of the spectra for a given ensemble, it is natural to rescale (or *unfold*) the eigenvalues to have unit-mean nearest-neighbor spacing. If we denote the mean eigenvalue density by $\rho(\lambda)$ (e.g., in the case of the Gaussian ensembles this would be the semicircle law for the range $-2\sqrt{N} \leqslant \lambda \leqslant 2\sqrt{N}$), then the scaled eigenvalues $\tilde{\lambda}_i^N(X)$ are given by

$$\tilde{\lambda}_i^N(X) = N \int_{-\infty}^{\lambda_i^N(X)} \rho(\lambda)\,\mathrm{d}\lambda.$$

One can then seek to calculate the probability distribution of the spacings between adjacent scaled eigenvalues, correlations between pairs (or more generally $n$-tuples) of eigenvalues, and more exotic measures of the eigenvalue distribution.

For the invariant ensembles, one can compute the eigenvalue statistics from the joint eigenvalue distribution function using the method of orthogonal polynomials. For example, the correlation function for pairs of scaled eigenvalues in the bulk of the spectrum (i.e., far from edges, such as $\pm 2$ in the case of the GOE and GUE) may be computed by integrating over all but two of the variables. The result encapsulates the eigenvalue repulsion discussed above, i.e., that the probability of finding pairs of eigenvalues close together vanishes with their separation, in a way determined by the symmetry of the ensemble. Higher correlation functions, involving $n$-tuples of scaled eigenvalues, may be computed in the same way.

Significantly, all correlation functions may be expressed in determinantal form. From the correlation functions one can then deduce, via a combinatorial calculation, the probability that nearest-neighbor eigenvalues in the bulk of the spectrum have a given spacing. The result may be expressed either as a Fredholm determinant or as a solution of a PAINLEVÉ EQUATION [III.24]. The probability of finding nearest-neighbor scaled eigenvalues a distance $s$ apart vanishes like $s^{\beta}$, where $\beta = 1$ for orthogonal invariant ensembles (e.g., the GOE) and $\beta = 2$ for unitary invariant ensembles (e.g., the GUE). This is in contrast to the spacings between uncorrelated random numbers (i.e., generated by a Poisson process and unfolded like the eigenvalues), for which the corresponding probability is $\mathrm{e}^{-s}$, and so increases as $s \to 0$. The GOE and GUE spacing distributions are plotted in figures 4 and 5.

From the correlation functions one can also characterize spectral fluctuations over longer ranges. For example, in a given interval of length $L$ one expects to find, on average, $L$ scaled (i.e., unfolded) eigenvalues. For each matrix the actual number will fluctuate around $L$. For random matrices the variance of these fluctuations is proportional to $\log L$ as $L \to \infty$, while for uncorrelated random numbers the variance is $L$. The much smaller variance in the random-matrix case demonstrates a distinctive *rigidity* in the spectrum.

In the case of the Gaussian ensembles, the spectral statistics can be computed explicitly for any size of matrix. The formulas simplify considerably in the large-matrix limit. The expressions that emerge in that limit are universal for invariant ensembles of the same symmetry type; that is, they do not depend on the specific probability weighting (unlike for the mean density, the dependence on which is counterbalanced by unfolding the spectrum).

This universality extends beyond the invariant ensembles to Wigner random matrices, which, in the large-matrix limit, have the same spectral statistics as the invariant ensembles, independently of the specific choice of distribution of the matrix elements. In this case the method of orthogonal polynomials is not available, and to establish the result one proceeds indirectly, by showing that the spectral statistics of a Wigner matrix are sufficiently close to those of some GOE/GUE matrix.

Proving universality for the invariant and Wigner ensembles has been one of the most significant developments of the past two decades. Another important development is the determination of the statistics of eigenvalues at a spectral edge. For example, one can determine the distribution of the largest eigenvalue of a GOE/GUE matrix. Typically, this will lie close to the upper limit of the range in which the semicircle law applies. The issue then is to establish the scale and nature of the fluctuations around this point. This problem was solved by Tracey and Widom, who showed that the distribution in question is also given in terms of the solution of a Painlevé equation.

## 6 Characteristic Polynomials

An alternative way of representing spectral statistics is via the value distribution of the characteristic polynomials of random matrices. For example, if $X$ denotes an $N \times N$ unitary matrix, its characteristic polynomial may be denoted $p(s;X) = \det(Is - X)$. The eigenvalues of $X$ are the zeros of this polynomial and so, by unitarity, they lie on the unit circle in the complex $s$-plane. On the unit circle one can determine the moments of, for example, $|p(s;X)|$, with respect to an average over $X$ in the CUE, by representing $p(s;X)$ as a product over the eigenvalues and then integrating over these using the joint eigenvalue distribution. Remarkably, the multiple integrals in question can be evaluated exactly by relating them to an integral computed by Selberg. From these moments one can prove that the values of the real and imaginary parts of $\log p(s;X)$, when divided by $(\frac{1}{2}\log N)^{1/2}$, are independent and normally distributed (with mean 0 and variance 1) in the limit as $N \to \infty$. Similar results hold for other ensembles of random matrices. The analysis of the value distribution of characteristic polynomials has been central to applications of random-matrix theory to number theory.

## 7 Ergodicity

The primary strategy in random-matrix calculations is to fix a point in the (scaled/unfolded) spectrum and average with respect to an ensemble. That is, one calculates average properties for a class of matrices. Crucially, however, in many cases of interest it can be proved that this averaging procedure is *ergodic* in the limit of large matrix size. That is, for a typical large matrix, the difference between the fluctuation statistics obtained by averaging over its spectrum and those calculated from the ensemble average vanishes in the limit as $N \to \infty$. To be more specific, if one considers sequences of matrices of increasing size, the probability of encountering a sequence for which spectral averages do not converge to the ensemble average vanishes as $N \to \infty$. In this sense, the ensemble averages one computes describe the properties of typical individual large matrices.

## 8 Connections

There are many significant connections between random-matrix theory and other areas of applied mathematics. I briefly note the following examples.

First, the fact that some ensembles are exactly solvable using orthogonal polynomials, and that the results can be expressed in determinantal form and are related to Fredholm theory and Painlevé transcendents, is indicative of deep underlying connections with the theory of integrable systems. Much of the theory for this has been developed in the last twenty years. The powerful techniques associated with integrable systems, such as the inverse scattering method and Riemann–Hilbert methods, have played a central role in the modern development of the subject.

Second, writing the Vandermonde factor in the joint eigenvalue distribution as

$$\prod_{1 \leqslant i<j \leqslant N} |x_j - x_i|^\beta = \exp\biggl(\beta \sum_{1 \leqslant i<j \leqslant N} \log |x_j - x_i|\biggr),$$

one sees that it takes the form of a Boltzmann weight associated with a one-dimensional gas of particles with positions $x_i$, interacting via the Coulomb potential $\sum_{1 \leqslant i<j \leqslant N} \log |x_j - x_i|$, at a temperature proportional to $\beta^{-1}$. This *Coulomb gas* analogy allows one to calculate many spectral properties using methods from equilibrium statistical mechanics. The connection is reinforced by an important idea due to Dyson: one can interpret the $x_i$ as positions of particles undergoing Brownian motion. They therefore satisfy stochastic partial differential equations, the precise forms of which (in particular, the conditioning on the solutions) depend on the ensemble in question. This picture has been further refined by relating the motion of the eigenvalues to nonintersecting random walks, allowing many detailed properties of the spectrum to be determined using techniques from probability theory.

Other important connections—to *free probability theory*, between products of random matrices and *Anderson localization* of waves in random media, to *random growth models*, to *enumerative geometry and topology*, and to *random permutations*, for example—lie outside the scope of this article.

## 9 Applications

As mentioned in the introduction, random-matrix theory has applications in virtually every area where matrices play a role, and in many others too. I outline below a few representative examples.

### 9.1 Condition Numbers

CONDITION NUMBERS [IV.10 §1] are of fundamental significance in characterizing the stability of numerical computations involving linear algebra. The question of determining the condition number of a random matrix was raised by von Neumann and Goldstine, and refined by Smale, in the context of characterizing the typical size of the condition number. For example, consider an $N \times N$ matrix $X$ with elements from a standard normal distribution. The 2-norm condition number $\kappa_X$ is the square root of the ratio of the largest eigenvalue of the (Wishart) matrix $X^{\mathrm{T}}X$ to its smallest eigenvalue. In this context the question was resolved by Edelman in 1988, who determined the distribution of the largest and smallest eigenvalues of $X^{\mathrm{T}}X$ from the joint eigenvalue distribution for the Wishart ensemble. The result is that the expected value of $\log \kappa_X$ increases like $\log N$ when $N \to \infty$ and that the distribution of $\kappa_X/N$ has a limit when $N \to \infty$ that can be written explicitly. The analysis extends to rectangular matrices.

### 9.2 Analysis of Large Data Sets

One of the canonical problems of the analysis of time series is that of determining correlations from a finite data set (see, for example, PORTFOLIO THEORY [V.10 §2.2]). Consider the situation of $N$ fluctuating quantities sampled at $T$ points in time. Let us denote these quantities by $q_n^t$, with $1 \leqslant n \leqslant N$ and $1 \leqslant t \leqslant T$. For example, these might be $N$ stocks sampled on each of $T$ days, or $N$ physiological variables measured at $T$ separate times. A key goal is to approximate any correlations underlying the dynamics from the finite sample of data available. Defining $X_{tn} = q_n^t/\sqrt{T}$, so that

$$E_{ij} = \frac{1}{T} \sum_{t=1}^{T} q_i^t q_j^t = (X^{\mathrm{T}}X)_{ij},$$

the question is how well the empirical correlation matrix $E_{ij}$ represents any genuine underlying correlations. One might anticipate that the empirical correlations will be representative when $N$ is fixed and as $T$ grows. However, in many applications, $N$ may not be small compared with $T$. This is particularly the case in financial data, in data from social science experiments, and in biological data. In this case it is natural to use random Wishart matrices to establish the typical difference between the limit $r = N/T \to 0$ and when $r \approx 1$. The Marčenko–Pastur law is a key tool as it describes the density of eigenvalues as a function of $r$. An important feature of this law is that, like the semicircle law, it is characterized by having edges. Thus the (Tracey–Widom) distribution of the largest eigenvalue near to the (soft) edge at the upper end of the spectrum and the corresponding distribution for the smallest eigenvalue (near to the hard edge at 0, associated with the nonnegativity of the eigenvalues) play central roles.

Applications to financial data often require one to consider random matrices where the distribution of the matrix elements is highly non-Gaussian, e.g., those possessing heavy tails, and where the matrix elements may by correlated. These remain major challenges.

### 9.3 Quantum Chaos

Wigner's contributions in the 1950s were motivated by the need to understand the quantum spectra of heavy atomic nuclei. This is a many-body problem in which the nucleons interact strongly with each other. The quantum Hamiltonian (i.e., the matrix whose eigenvalues are the energy levels) may thus be expected to be a complex object. Wigner's insight was to realize that statistical properties of the spectrum could be modeled by those of random matrices and that these could be determined by averaging over ensembles of matrices. This resembles the methodology of classical statistical mechanics, where one models statistical properties of solutions of the equations of motion for many interacting particles by averaging over all possible states or trajectories, subject to conservation of energy, etc., with each state/trajectory given its Boltzmann weight.

The same philosophy also underpins the applications of random-matrix theory to condensed matter physics. There, one often encounters the problem of an electron moving in a disordered medium, that is, in a medium where the forces behave like a random function of position. In this case, too, it is natural to model the quantum Hamiltonian by a random matrix.

One of the key ideas of chaos theory is that one does not need many interacting particles or random forces to generate complex dynamics. Instead, this can be (and typically is) found in systems with a small number of degrees of freedom where the dynamics is not maximally constrained by the symmetries but where the forces acting may, nevertheless, be simple. In such systems, typical long trajectories may also be ergodic, in that their statistical properties coincide with averages of ensembles of appropriately weighted trajectories and consequently with phase-space averages. Examples include the THREE-BODY PROBLEM [VI.16] and billiards (i.e., a freely moving particle making specular reflections at boundaries) in domains that are neither rectangular nor elliptical (more generally, in which the dynamics is not integrable). In the 1970s and 1980s it was suggested that the quantum eigenvalue statistics in such systems should, on the scale of the mean level spacing, be modeled by random-matrix theory. Specifically, this should hold in quantum systems whose classical limit is chaotic, in the *semiclassical limit* as Planck's constant $\hbar \to 0$, or, physically, in the limit of vanishing de Broglie wavelength. For example, in the case of billiard systems it should hold for the high-lying eigenvalues of the Laplacian with appropriate boundary conditions.

The idea that random-matrix theory should, in the semiclassical limit, describe quantum spectral statistics in classically chaotic systems has its origins in work of Berry and Tabor in 1977 (whose primary focus was on classically integrable systems, for which the energy levels are generically believed to have Poisson statistics, but who also speculated on chaotic systems) and was developed into a precise conjecture by Bohigas, Giannoni, and Schmit in 1984. It has been verified numerically in a very wide range of systems and is believed to hold generically (it can be subverted by the presence of symmetries). A related conjecture, due to Berry in 1977, is that the eigenfunctions behave like linear superpositions of randomly directed plane waves in the semiclassical limit and so exhibit the statistical features of Gaussian random functions. Taken together, these conjectures underpin the statistical modeling of quantum chaotic systems. However, neither has been proved mathematically in any system, and achieving this remains one of the outstanding open problems in the field.

While we may not have a proof of the random-matrix conjecture for quantum chaotic systems, it is supported by a highly sophisticated and subtle heuristic semiclassical analysis based on a relationship between quantum spectra and classical periodic orbits. This relationship emerges from a saddle-point evaluation, valid in the semiclassical limit, of the Feynman path integral representation of the energy-dependent Green function in terms of a sum over all possible paths weighted by $e^{iS(\text{path})/\hbar}$, where $S$ denotes the action. The saddle-point condition picks out those paths for which the action is stationary, that is, the classical trajectories. As first shown by Gutzwiller, a further application of the saddle-point method in evaluating the trace of the Green function, which determines the spectrum, selects out the periodic (i.e., self-retracing) orbits. Statistical properties of the quantum spectrum are thus linked semiclassically to statistical properties of the classical periodic orbits and hence to the chaotic nature of the classical dynamics. It was first demonstrated in this way by Hannay and Ozorio de Almeida, and later in more generality by Berry, that ergodicity of the classical dynamics governs some key features of the pair correlation function of the quantum spectrum and that these coincide with the corresponding features of the random-matrix results. This approach has subsequently been refined by a number of researchers, particularly Bogomolny and Keating, who extended it to all the main key features, and Sieber and Richter, who introduced the seminal idea of including pairs of orbits that experience a close encounter, which turn out to make a significant contribution.

To gain complete agreement with random-matrix expressions using semiclassical periodic orbit formulas requires either the use of sophisticated resummation techniques introduced by Berry and Keating, which relate long orbits to shorter ones, or a knowledge of the correlations between the actions of distinct pairs of long orbits. These correlations are not known a priori, but they can be shown to be equivalent to the random-matrix conjecture. In this sense, as shown by Argaman et al., the random-matrix conjecture for quantum spectra leads, remarkably, to an important prediction relating to the classical dynamics in chaotic systems. It is a major open problem to verify this prediction using classical mechanics.

Another key theme in quantum chaos has been the analysis of open chaotic systems; that is, scattering systems in which the interactions generate exponential instability in the dynamics. If the system is weakly open, in the sense that trajectories are trapped for a long time before escaping, quantum statistical properties (e.g., of the scattering resonances) can be modeled by random matrices that are close to being Hermitian. In chaotic systems that are strongly open, the semiclassical analysis is much less developed and remains a major challenge. The semiclassical theory of quantum chaotic scattering was developed by Smilansky and coworkers.

It is worth emphasizing that the ideas concerning quantum chaos outlined here apply in general to all complex wave problems and so have applications to the statistical analysis of optical, acoustic, and vibrational systems, as well as to essentially quantum phenomena (e.g., to lasers, superconductors, the motion of electrons in atomic, molecular, and solid-state systems, nuclei, etc.).

### Further Reading

Among the suggestions for further reading below, Mehta's book is the classic text on random-matrix theory; the book edited by Akemann, Baik, and Di Francesco contains excellent review articles covering much of the material included in this article (and more) and is an ideal first point of contact with the subject; Porter's book contains reprints of many of the important early physics papers in the subject; Haake's book is a useful introduction to quantum chaos; and the book edited by Wright and Weaver contains review articles aimed at applications to linear acoustics and vibration.


Akemann, G., J. Baik, and P. Di Francesco, eds. 2011. *The Oxford Handbook of Random Matrix Theory*. Oxford: Oxford University Press.

Anderson, G. W., A. Guionnet, and O. Zeitouni. 2010. *An Introduction to Random Matrices*. Cambridge: Cambridge University Press.

Argaman, N., F. M. Dittes, E. Doron, J. P. Keating, A. Kitaev, M. Sieber, and U. Smilansky. 1993. Correlations in the actions of periodic orbits derived from quantum chaos. *Physical Review Letters* 71:4326–29.

Deift, P. 1999. *Orthogonal Polynomials and Random Matrices: A Riemann–Hilbert Approach*. Providence, RI: American Mathematical Society.

Edelman, A., and N. R. Rao. 2005. Random matrix theory. *Acta Numerica* 14:233–97.

Forrester, P. J. 2010. *Log-Gases and Random Matrices*. Princeton, NJ: Princeton University Press.

Haake, F. 2010. *Quantum Signatures of Chaos*, 3rd edn. Heidelberg: Springer.

Mehta, M. L. 2004. *Random Matrices*, 3rd edn. Amsterdam: Elsevier.

Mezzadri, F., and N. C. Snaith, eds. 2005. *Recent Perspectives in Random Matrix Theory and Number Theory*. Cambridge: Cambridge University Press.

Porter, C. E., ed. 1965, *Statistical Theories of Spectra: Fluctuations—a Collection of Reprints and Original Papers*. New York: Academic Press.

Tulino, A. M., and S. Verdú. 2004. *Random Matrix Theory and Wireless Communication*. Boston, MA: Now Publishers.

Wright, M., and R. Weaver, eds. 2010. *New Directions in Linear Acoustics and Vibration: Quantum Chaos, Random Matrix Theory and Complexity*. Cambridge: Cambridge University Press.


## IV.25 Kinetic Theory

*Cédric Villani and Clément Mouhot*

### 1 The Birth of Kinetic Theory

Modern physics can be traced back to Newton and the advent of differential equations to substantiate the laws of classical mechanics. In the following centuries this was followed by more comprehensive theories of physical phenomena in the surrounding world: electric and magnetic forces were captured by the theory of electromagnetism (Ampère, Faraday, Maxwell); large velocities were handled by the theory of relativity (Lorentz, Poincaré, Minkowski, Einstein); small-scale particle physics could be taken care of by quantum mechanics (Planck, Einstein, Bohr, Heisenberg, Born, Jordan, Pauli, Fermi, Schrödinger, Dirac, de Broglie, Bose); and so on.

However, all of these theories are classically devised to study one physical system (a planet, a ship, a motor, a battery, an electron, a spaceship, etc.) or a small number of systems (planets in the solar system, electrons in a molecule, etc.). In many situations, though, one needs to deal with an assembly made up of elements so numerous that their individual tracking is neither useful nor possible: galaxies made up of hundreds of billions of stars, fluids made of more than $10^{20}$ molecules, crowds made of thousands of individuals, etc. Taking such large numbers into account leads to new effective laws of physics, requiring different models and concepts. This passage from microscopic rules to macroscopic laws is the founding principle of statistical physics. All branches of physics (classical, quantum, relativistic, etc.) can be studied from the point of view of statistical physics, in both stationary and dynamical perspectives. Classical mechanics was, naturally, one of the first laboratories for statistical physics, and thus in the nineteenth century kinetic theory was born.

Before we describe the key concepts of kinetic theory, let us recall the basic notion of *phase space*, which should be thought of as the space of all possible states occurring in a mathematical model of some physical system. If one studies a deterministic system obeying an evolution equation, then the phase space is, in principle, the "smallest" space in which the equation determines a unique "well-behaved" solution. For instance, the evolution of a classical point particle is governed by a second-order differential equation (Newton's law); so the position of the particle is not sufficient to predict its future positions, but the pair (position, velocity) is sufficient to predict future positions and velocities. The phase space of a classical particle is therefore made up of positions and velocities. On the other hand, if the physical system is, for instance, a rigid body with a certain shape, then the phase space should also include extra parameters related to the orientation of the body.

The main idea in kinetic theory is to replace a huge number of objects, whose physical states are completely described by points in a certain phase space and whose properties are otherwise identical, by a *statistical distribution* over that phase space. In particular, a large crowd of classical point particles will be described by a statistical distribution on the space of positions and velocities.

In retrospect, the conceptual leap from Newtonian mechanics to kinetic theory was quite significant: the new formalism involved a set of invisible variables, namely, the velocities of particles, that are inaccessible to observation. It was even counterintuitive; for instance, kinetic theory replaces the model of a fluid at rest (zero velocity) by a huge number of particles moving in all directions with great speed. This increase in complexity was not easy to justify, since at the time there was no way to measure any of these velocities—it is still barely possible today. This fundamental role of velocities accounts for the name *kinetic* theory.

With kinetic theory came the distinction between three scales: the macroscopic scale of phenomena that are accessible to observation, the microscopic scale of molecules and infinitesimal constituents, and an intermediate scale that is loosely defined and is often called *mesoscopic*. This is the scale of phenomena that are not accessible to macroscopic observation but already involve a large number of particles, so that statistical effects are significant.

With only a little stretch of the imagination, one can liken the principles of kinetic theory to those of certain contemporary models of theoretical physics, such as string theory, in which a set of hypothetical hidden variables is also taken into account (here we put aside any debates about the value of, and the possibility of validating, string theory).

The basic scheme of kinetic theory leaves room for obvious variations. If there are several species, one can consider several statistical distributions. (Think of air, which is mainly made up of a mixture of two gases; the two species have different properties, but within each species the molecules can be considered as identical.) If the position and velocity are not sufficient to describe the state of one object, one can enlarge the phase space. (In the case of air, one might wish to keep track of the orientation of a molecule of nitrogen or oxygen.)

Kinetic theory was first discussed by Daniel Bernoulli in the eighteenth century. The notions of *mean free path* and *mean free time*—which are the typical distance and typical time, respectively, that a particle can travel without hitting another particle—were studied by various authors (Herapath, Waterston, Joule, König, Clausius) between 1820 and 1860. At the same time, the very important notion of *cross section* emerged; this measures the likelihood of interaction between two particles, and it can be interpreted as an effective collision surface. The field as we know it, though, was really founded by Maxwell in a celebrated paper of 1867.

This theory was strongly influenced by two major earlier scientific developments. The first was the rise of thermodynamics throughout the eighteenth and nineteenth centuries. The laws governing exchanges of energy and variations of heat, density, pressure, and temperature did not seem to rely on fundamental equations and were discovered through a slow and confusing process; it was therefore desirable to grasp some more fundamental laws that would underlie thermodynamics. The second influence was the development of statistics, especially in the field of social sciences, with the empirical discovery by Galton and Quetelet of the omnipresence of simple statistical laws derived from probability theory—one prime example being the recognition that fluctuations in the size of individuals was essentially governed by Gaussian distributions. Both of these influences guided Bernoulli (whose father was one of the founders of probability theory), but by the time of Maxwell they had become much more mature.

In those days, the atomistic nature of matter was still largely hypothetical, and kinetic theory could be considered a thought experiment. Maxwell discussed the problem of the derivation of macroscopic laws from microscopic physics. He worked in a dilute regime to neglect collisions involving more than two bodies, and he assumed a clear separation between the inhomogeneity scale (a mesoscopic concept) and the interaction scale (at microscopic level). He then computed the effect of collisions on the distribution function via the solution of a classical scattering problem. In this way he came up with an evolution equation, equivalent to what we now call the *Boltzmann equation*. The unknown is a density function $f(t,x,v)$, standing for the density of particles at time $t$ in the phase space $(x,v)$ (equipped with the reference Liouville measure $\mathrm{d}x\,\mathrm{d}v$); and the equation, in modern notation, is

$$\frac{\partial f}{\partial t} + v \cdot \nabla_x f + F(t,x) \cdot \nabla_v f = Q(f,f). \quad (1)$$

Here, the left-hand side describes the evolution of $f$ under the action of the force $F(t,x)$, while the action of elastic collisions is described by the nonlinear operator $Q$ on the right-hand side:

$$Q(f,f) = \int_{\mathbb{R}^3}\int_{\mathbb{S}^2} \tilde{B}(v - v_*, \omega) \times (f(t,x,v')f(t,x,v'_*) - f(t,x,v)f(t,x,v_*))\,\mathrm{d}v_*\,\mathrm{d}\omega. \quad (2)$$

Note that this operator is localized in $t$ and $x$, it is quadratic, and it has the structure of a tensor product with respect to $f(t,x,\cdot)$. The velocities $v'$ and $v'_*$ should be thought of as the velocities of a pair of particles before collision, while $v$ and $v_*$ are the velocities after that collision; the formulas are

$$v' = v - \langle v - v_*, \omega\rangle\omega, \qquad v'_* = v_* + \langle v - v_*, \omega\rangle\omega.$$

When one computes $(v, v_*)$ from $(v', v'_*)$ (or does the reverse), conservation laws are not enough to yield the result, with only four scalar conservation laws for six degrees of freedom. The unit vector $\omega \in \mathbb{S}^2$ removes this ambiguity; in the case of colliding hard spheres, it can be thought of as the direction of the line joining the centers of the two particles. The kernel $\tilde{B}(v - v_*, \omega)$ describes the relative frequency of vectors $\omega$, depending on the relative impact velocity $v - v_*$; it depends on only the modulus $|v - v_*|$ and the deflection angle $\theta$ between $v - v_*$ and $v' - v'_*$. Maxwell computed it for hard spheres ($\tilde{B} \sim |v - v_*|\sin\theta$) and for inverse power forces. In the latter case, the kernel factorizes as the product of $|v - v_*|^\gamma$ with a function $\tilde{b}(\theta)$; Maxwell showed that, if the force is repulsive, proportional to $r^{-s}$ ($r$ being the interparticle distance), then $\gamma = (s-5)/(s-1)$ and $\tilde{b}(\theta) \simeq \theta^{-(1+\nu)}$ as $\theta \to 0$, where $\nu = 2/(s-1)$. In particular, the kernel is usually *nonintegrable* as a function of the angular variable: this is a general feature of long-range interactions and is nowadays called the "noncutoff property." Maxwell further noticed that the inverse power $s = 5$ leads to simplified formulas, which could lend themselves to more explicit computations.

Maxwell went on to discuss possible boundary conditions. Particles arriving at a point $x$ in the boundary with velocity $v$ may be assumed to acquire a new velocity $R_x v$, determined either by the model of specular reflection ($R_x v = v - 2v \cdot n_x n_x$, where $n_x$ is the unit ingoing normal vector at $x$) or by the model of bounce-back reflection ($R_x v = -v$). Either way, the boundary condition reads $f(x, R_x v) = f(x,v)$. In more sophisticated models, particles are assumed to be absorbed by the boundary and reemitted at a given rate, given, say, by a Gaussian distribution whose dispersion is dictated by the temperature of the wall, say:

$$f(x,v) = \rho_-(x) M_w(v), \quad v \cdot n_x > 0,$$

where

$$\rho_-(x) = \int_{v \cdot n_x < 0} f(x,v)|v \cdot n_x|\,\mathrm{d}v,$$

$$M_w(v) = \frac{\mathrm{e}^{-|v|^2/(2T_w)}}{2\pi T_w^2}.$$

Maxwell, understanding that the boundary behavior of a gas was a very complex matter, also considered combinations of the above models; these are nowadays called *Maxwell conditions*. In order to find the stationary solutions, that is, time-independent solutions of (2), he identified certain particular *hydrodynamic* solutions, which make the collision contribution vanish. These are *Gaussian distributions with a scalar covariance*:

$$f(v) = \frac{\rho e^{-|v-u|^2/2T}}{(2\pi T)^{3/2}},$$

where the parameters $\rho > 0$, $u \in \mathbb{R}^3$, and $T > 0$ can be identified, respectively, as the density, mean velocity, and temperature of the fluid. These parameters can be fixed throughout the whole domain (providing, in this case, an *equilibrium distribution*) or they can depend on the position $x$ and time $t$; in both cases, collisions will have no effect. It was remarkable that Maxwell could recover in this way the Gaussian distributions that already played a central role in probability theory; in the context of kinetic theory, these distributions are thus called *Maxwellian*.

Maxwell went further and made the connection with classical FLUID MECHANICS [IV.28], in which the equations are expressed in terms of $\rho$, $u$, and $T$. He suggested that one could go from the kinetic equations to hydrodynamic equations in certain regimes and therefore make some predictions about hydrodynamic behavior from kinetic theory. Let us give, for instance, two counterintuitive effects that Maxwell guessed through kinetic formalism. One is that the viscosity of a low-density fluid hardly depends on density. Another is the paradoxical "thermal creep effect": a gas that has a temperature gradient parallel to a fixed wall will have a tendency to flow from cold to hot near the wall.

A few years after Maxwell's masterpiece was published, Boltzmann rewrote and deepened the theory, completing the foundations of modern physical kinetic theory.

## 2 Boltzmann's Entropy and Collisional Relaxation

The word "entropy" was coined by Clausius to designate a certain quantity associated with the tendency to relax or achieve equilibrium. The properties of entropy in relation to exchanges of heat and energy were established empirically; in particular, the formula for infinitesimal variation of entropy was determined: $dS = \delta Q/T$ (variation in entropy is proportional to the exchanged heat divided by the temperature). In this vein came the well-known *second law of thermodynamics*, which states that entropy can never decrease in an isolated system. Even though there were rules to compute the entropy of an equilibrium system, the interpretation of that quantity remained somewhat elusive, and the second law was considered more or less as an axiom.

That changed radically with Boltzmann's contribution to the field (1872–77). In one of the most dramatic events in the history of statistical physics, Boltzmann introduced the following breakthroughs.

- A general *mathematical definition* of entropy: it is the logarithm of the volume of microscopic states that are compatible with the (observable) macrostate. This is the celebrated Boltzmann formula:
$$S = k \log W, \tag{3}$$
where $k$ is Boltzmann's constant (notation introduced by Planck, who was the first to estimate its value) and $W$ stands for the volume of microscopic states. Here, the volume may be computed with some natural measure on the phase space, which may be discrete or continuous, depending on the situation.
- A *practical formula* for computing the entropy of a kinetic system: if $f(x,v)$ is the distribution function, then $S = -\iint f \log f \, dx \, dv$. This is derived from Boltzmann's formula (3) through a discretization procedure; it can also be seen as an infinite-dimensional analogue of Liouville's volume measure.
- A *theorem* showing that the entropy of a gas that obeys the equations discovered by Maxwell can never decrease.

More precisely, Boltzmann's $H$ theorem states that for a rarefied gas modeled by a kinetic distribution $f(t,x,v)$, governed by the Boltzmann equation with appropriate boundary conditions, the functional $H = -S$ satisfies (i) $dH/dt \leqslant 0$ and (ii) $dH/dt = 0$ if and only if $f(t,x,v)$ is a Maxwellian distribution with possibly variable parameters $\rho$, $u$, $T$. Such a distribution can be called *hydrodynamic*, since it depends on only hydrodynamical quantities.

In fact, an exact formula can be given for the entropy production: for, say, specular reflection,

$$\left.\begin{aligned} \frac{dS}{dt} &= \int D(f(t,x,\cdot))\, dx, \\ D(f) &= \frac{1}{4}\iint_{\mathbb{R}^3\times\mathbb{R}^3}\int_{S^2} \tilde{B}(v - v_*, \omega) \\ &\qquad \times (f(v')f(v'_*) - f(v)f(v_*)) \\ &\qquad \times \log\frac{f(v')f(v'_*)}{f(v)f(v_*)}\, dv\, dv_*\, d\omega. \end{aligned}\right\} \tag{4}$$

Establishing this expression was one of Boltzmann's motivations for rewriting Maxwell's kinetic equation (expressed in some sort of weak formulation) in the modern form (1).

The $H$ theorem implies, in particular, that stationary states have to be hydrodynamic at all times; Boltzmann showed that in the absence of special symmetries this forces the distribution to be spatially homogeneous. This homogeneous Maxwellian is the only equilibrium state, and in this case the entropy coincides with Clausius's entropy. Boltzmann's beautiful proof was a giant conceptual leap. First, it provided a general definition of entropy, covering nonequilibrium situations: entropy in the Boltzmann theory can be considered as the typical uncertainty that remains in the state of a random particle taken in the system. Boltzmann then showed that the second law of thermodynamics could be considered as a logical consequence of fundamental postulates instead of simply being accepted as a God-given fact. He also showed that equilibrium thermodynamics could in principle follow from nonequilibrium dynamics, and he identified entropy increase, together with a scale separation, as the major factors behind the emergence of hydrodynamics.

Still, the most dramatic consequence of Boltzmann's work was the discovery of the *irreversibility* contained in (1), even though that model was derived from Newton's reversible equations of motion. This emergence of irreversibility in the many-particle limit would trigger a heated controversy involving preeminent scientists such as Poincaré, Loschmidt, and Zermelo; it is still considered as the classical explanation of the irreversibility of time at the macroscopic level of description, in spite of the reversibility of the full-scale evolution.

Entropy increase in the Boltzmann equation shows that particle configurations, loosely speaking, always evolve from unlikely to likely, from exceptional to typical. Information is therefore continuously lost (the gas may have started in a very interesting, exceptional configuration, but it soon becomes quite uninteresting). Actually, the information is gradually transferred from the macroscopic observable degrees of freedom to the microscopic invisible ones. This loss of information can be related to the separation of scales inherent in the derivation of the Boltzmann equation: at each encounter between particles, the parameters of the collision (say, the orientation of the colliding pair) are invisible because they occur on a scale much finer than the spatial scale, so collisions are treated as perfectly localized and the impact parameter is treated probabilistically. The inexorable increase in entropy can also be attributed to the huge numbers involved in the computation of probability: $N = 10^{20}$ is a large number, but when one enumerates possible configurations, this number appears in a combinatorial way, leading to numbers such as $2^{10^{20}}$, which are so large that they defy any human attempts to grasp them.

After being imported into mathematics, entropy was extraordinarily successful in helping to solve problems, both related and unrelated to kinetic theory. It was rediscovered by Shannon when he was building the theory of communication, and it still plays a central role in INFORMATION THEORY [IV.36]. It is the basis of Sanov's formula in the theory of large deviations for empirical measures. It was a key concept behind Nash's proof of the celebrated de Giorgi–Nash theorem of continuity of solutions of nonsmooth divergence parabolic equations. It lies at the core of the theory of logarithmic Sobolev inequalities, introduced by Nelson and Gross as an infinite-dimensional replacement for Sobolev inequalities. It was one of the key technical tools in the theory of probabilistic hydrodynamic limits that was initiated by Varadhan and his colleagues and students; the role of entropy in hydrodynamical limits was further reinforced with Yau's relative entropy method. It was an important tool in the DiPerna–Lions theory of weak solutions of the Boltzmann equation. Much further away from physics, entropy was adapted by Voiculescu in the context of free probability to help solve elusive problems from the theory of von Neumann algebras.

One of the reasons for the ubiquity of entropy is its extensivity property,

$$S(f \otimes g) = S(f) + S(g),$$

which is natural from the physical point of view (information associated with two independent variables should add up) and implies an additive dependence on dimension, eventually leading to dimension-independent inequalities.

Nowadays, entropy is still actively being used to develop new tools and techniques. A few recent works in which it has played a crucial role are the infinite-dimensional interpolation inequalities of Otto and Villani, the amazing solution of the Poincaré conjecture by Perelman, the theory of synthetic Ricci curvature bounds by Lott, Sturm, and Villani, and the solution of Kac's problem of propagation of chaos for the spatially homogeneous Boltzmann equation by Mischler and Mouhot (after preliminary work by Carlen, Villani, and others).

## 3 Landau Damping and Collisionless Relaxation

For the first forty years after its birth, kinetic theory mostly focused on the effect of collisions, which are brutal encounters between particles. A notable development, starting with Lorentz in 1905, was the introduction of transport equations to describe the motion of particles wandering in an array of scattering obstacles, such as beams of electrons or neutrons in metals. The resulting linear collisional equations would later be found to have considerable importance in nuclear physics.

However, around that time it was also realized that, in various cases, the collective effect of particles on each other is more important than collisions and leads to a rich variety of behaviors. This was the beginning of mean-field (noncollisional) theory.

Mean-field theory was first introduced in galactic dynamics. In 1915 Jeans discussed the use of the Boltzmann equation to model the evolution of galaxies over millions or billions of years, with each star considered as a particle. He came to the conclusion that, as a first approximation, collisions can be dropped and one can model the interaction by letting each particle feel a force field that is the resultant of all other particles.

The story was then repeated in plasma physics, which is primarily governed by the Coulomb interactions between electrons. For such interactions the collision kernel can be computed explicitly but leads to a diverging collision operator. In 1936 Landau remedied this situation by replacing the Boltzmann operator by an integro-differential collision operator:

$$Q_L(f,f) = \frac{\log \Lambda}{2\pi\Lambda}\nabla_v \times \left(\int_{\mathbb{R}^3} a(v-v_*)[f(v_*)\nabla f(v) - f(v)\nabla f(v_*)]\,dv_*\right), \quad (5)$$

where $\Lambda$ (the plasma parameter) is a large constant and where $a_{ij}(z) = (\delta_{ij} - z_i z_j/|z|^2)/|z|$. But in 1938 Vlasov pointed out that the effect of collisions can also be disregarded (except in long-time analysis), so the distribution of electrons is mainly subject to the force generated by electrons, coupled with the electrostatic equations of Maxwell.

Be it for galaxies or for plasmas, in both situations the basic evolution equation is

$$\left.\begin{aligned} \frac{\partial f}{\partial t} + v\cdot\nabla_x f + F[f]\cdot\nabla_v f &= 0, \\ F[f](t,x) = -\iint_{\Omega\times\mathbb{R}^3} f(t,y,w)&\nabla W(x-y)\,dy\,dw, \end{aligned}\right\} \quad (6)$$

where $\Omega$ is the position domain and $W$ is the interaction potential, which is assumed to be even. This equation, which comes with various admissible boundary conditions, is usually called the *Vlasov equation*, although the collisionless Boltzmann equation might be a historically more appropriate name. The most important cases are when $W$ is the fundamental solution of the POISSON EQUATION [III.18] $-\Delta W = \delta$ (Coulomb interactions, positive type) or $\Delta W = \delta$ (Newton interaction, negative type). Here, the notion of "type," coming from harmonic analysis, refers to the sign of the Fourier transform of the fundamental solution. These cases give rise to the so-called *Vlasov–Poisson equation*.

With collisions absent, the most striking features of the Boltzmann equation disappear; thus, (6) does not possess any meaningful Lyapunov functional, except that its solutions satisfy the conservation of energy,

$$E = \frac{1}{2}\iiiint f(x,v)f(y,w)W(x-y)\,dx\,dy\,dv\,dw + \iint f(x,v)\frac{|v|^2}{2}\,dx\,dv,$$

and conservation of all nonlinear functions of the density $\iint C(f)\,dx\,dv$. In particular, the entropy is constant.

Another way in which (6) contrasts with (1) is that it has a surprisingly large collection of steady states; in the absence of an external field or boundaries, this includes, in particular, homogeneous distributions $f^0(v)$ but also many inhomogeneous periodic stationary solutions.

All of this seems to oppose the idea that solutions of (6) should display definite long-time behavior. It therefore came as a huge surprise when, in 1946, Landau showed that the linearized analysis of (6) for Coulomb interactions led to the exponential decay of perturbations for a large class of equilibria and perturbations (e.g., if both the equilibrium and the perturbation are analytic distributions, and if the equilibrium has only one maximum in dimension 1 or depends only on $|v|$ in dimension 3). This effect has been dubbed *Landau damping*. Because it seemed to find irreversibility where everything looked reversible, it was probably as striking to contemporary physicists as Boltzmann's discovery of the collisional increase of kinetic entropy, even though its physical impact was much more restricted than that of the $H$ theorem.

In contrast with Boltzmann's $H$ theorem, which is genuinely nonlinear, Landau damping was based on a linearized computation. Landau's results were refined and extended by a number of physicists, including O'Neil, Penrose, Backus, Maslov, Fedoryuk, and others. When experiments became accessible, Landau's computations were verified with a good degree of accuracy, and Landau damping became one of the cornerstones of modern classical plasma physics. It was later exported to galactic dynamics by Lynden-Bell.

Having been discovered by mathematical computation, Landau damping has led to considerable speculation about its driving mechanism, and a number of misleading ideas have been generated. The most convincing interpretation is that collisionless transport phenomena involve a *mixing* of the distribution function via very fast kinetic oscillations, which in the stable case have a tendency to wipe out inhomogeneities.

Even though the collisional and collisionless analyses are both idealizations, they constitute the basis of most of our current understanding of kinetic theory. They can also interact with each other: for example, the tendency to homogenize faster than expected can enhance the impact of diffusion or collision on relaxation phenomena.

At the same time as all this analysis was being developed, mean-field analysis started to be applied to a number of situations outside kinetic theory, in both equilibrium and nonequilibrium systems. In particular, in a famous discussion of turbulence, Onsager studied the incompressible two-dimensional Euler equation in vorticity form as a mean-field system of "vortices," presenting many similarities with collisionless kinetic theory.

## 4 Driving Problems

The development of kinetic theory, and Boltzmann's very influential book, quickly attracted the attention of mathematicians, starting with Hilbert, who formulated his sixth problem (related to items (I) and (IV) in the list below) under the inspiration of Boltzmann. Hilbert himself did some early mathematical study of the Boltzmann equation, as did Carleman in the 1930s and then Grad and Kac in the 1950s. These works focused on Boltzmann's collision operator.

As the theory of partial differential equations (PDEs) was making progress, the effects of the tricky operator $v \cdot \nabla_x$ in the equation started to be analyzed, both in the case when the operator stands on its own and when it is coupled with other typical operators appearing in statistical evolution equations. This can be traced back to Kolmogorov's work on the fundamental solution of the kinetic Fokker–Planck equation.

It took longer for the Vlasov (collisionless) theory to make its way into mathematics; this task was undertaken only at the end of the 1970s in Russia with the work of Arsen'ev and Dobrushin. Soon after, Braun, Hepp, and Neunzert followed in the Western world.

A number of problems emerged from these works, at the interface between mathematics and physics; they have been driving the field for decades and they triggered far-reaching developments. For the most part, these problems fall into five general themes, all of which are related to each other.

**(I) Derivation from first-law principles.** Starting from fundamental equations such as Newton's laws or certain simple diffusive microscopic models, derive kinetic statistical equations. To derive collisional equations one is often led to justify, directly or indirectly, Boltzmann's *chaos assumption*: that precollisional configurations are uncorrelated. Chaos assumptions also play an important role in noncollisional models and more generally in the derivation of any deterministic equation on the distribution function.

**(II) The Cauchy problem and qualitative analysis.** Starting from an initial datum that satisfies certain assumptions about smoothness and decay at large velocities, prove that the solution is well behaved, make precise the way in which it solves the kinetic equation, and establish whether bounds of smoothness, large-velocity decay, and strict positivity are preserved in time. Is there regularization, or at least decay of the amplitude of singularities? In many situations, a lack of understanding of the Cauchy problem precludes progress in the derivation problem.

**(III) Long-time behavior.** Starting close to some equilibrium, does the solution remain close to the equilibrium for all times (orbital stability)? Does it converge to the equilibrium or to some other equilibrium (dynamical stability)? Starting far from equilibrium, does it converge to some equilibrium, and can one identify that equilibrium? Are there mixing properties (e.g., oscillations developing in time and leading to some weak convergence mechanism)?

**(IV) Relationships with other models.** Can one replace, in suitable asymptotic regimes, the kinetic equations with reduced models, such as compressible or incompressible hydrodynamic equations (in the hydrodynamic limit, that is when the mean free path becomes negligible with respect to the spatial scale), or boundary-layer equations (when one looks very close to an interface)? Can one couple kinetic models with other models? Or reduce the description by using a multiscale analysis? Can one use the kinetic equations to retrieve observable properties of fluids, such as thermodynamic laws of pressure, viscosity dependencies, or phase-transition diagrams? An important limitation of Boltzmann's classical theory is that it covers only perfect fluids, that is, those with a pressure law that is proportional to the product of the density and the temperature.

**(V) Numerical simulation.** Can one devise numerical methods that are fast and accurate? Ones that are particulary suitable to predict the value of a given quantity? Ones that satisfy given constraints? Can one prove that these schemes converge to the solutions of the corresponding kinetic equations? The latter problem may be strongly related to the derivation problem because a number of schemes are based on particle simulations. It is also obviously related to the analysis of the Cauchy problem.

In view of their archetypical nature and the role they play in fundamental issues such as the arrow of time, the basic equations of kinetic theory have aroused interest among theoretical physicists, going far beyond the range of application of these models.

We will describe some of these problems in more detail after discussing the models more precisely.

## 5 The Many Models of Kinetic Theory

Initially, kinetic theory was devised to model rarefied gas dynamics (the Boltzmann equation), galactic dynamics (the mean-field model with Newton interaction), and ideal plasma dynamics (the mean-field model with Coulomb interactions). All three domains of application are important; for instance, the Boltzmann equation is crucial in high-altitude aerodynamics, since the upper atmosphere is not dense enough for the laws of hydrodynamics to apply satisfactorily. More recently, Boltzmann equations have been found to be useful in the modeling of nanofluids.

The kinetic formalism is versatile, and its range of application has been widened considerably beyond these situations. The many resulting variants of the basic equations can be grouped into several categories.

(1) Classical models with an interaction kernel derived from various molecular interactions or modified from those that come from the laws of classical physics. In particular, since Grad's work on the properties of the linearized Boltzmann operator, one often truncates small deviation angles to ensure the angular integrability of the collision kernel. Under this assumption of *angular cutoff*, the collision operator can be split into two parts:

$$\begin{aligned} Q(f,f) &= Q^+(f,f) - Q^-(f,f) \\ &= \iint \tilde{B}(v - v_*, \omega) f(t,x,v') f(t,x,v'_*) \, \mathrm{d}v_* \, \mathrm{d}\omega \\ &\quad - \iint \tilde{B}(v - v_*, \omega) f(t,x,v) f(t,x,v_*) \, \mathrm{d}v_* \, \mathrm{d}\omega, \end{aligned}$$

which are called the gain and loss parts of the operator, respectively. By contrast, a kernel that is nonintegrable in the angular variable is called "noncutoff." This kernel corresponds to long-range interactions. Moreover, the interaction is called *hard* if the corresponding collision kernel is proportional to a positive power of the relative velocity, and it is called *soft* if the kernel is proportional to a negative power of the relative velocity. In between these extremes lies the *Maxwellian* case, where the kernel does not depend on the relative velocity. Hard, Maxwellian, and soft potentials often enjoy distinctive properties. A particular case is that of *hard spheres*, in which the kernel is simply proportional to $|\langle v - v_*, \omega\rangle|$.

(2) Models obtained from large particle systems by putting emphasis on various interactions according to physical conditions (density, strength of interaction, etc.). Popular and versatile models in this category are the Fokker–Planck equations, which date back to the 1930s and describe the evolution of a crowd of particles undergoing stochastic diffusion and deterministic drift. Systematic derivation of statistical models for particle systems goes back to Bogolyubov. It is especially in the field of plasma physics that this approach has led to a large number of variants. The Balescu–Lenard and Vlasov–Fokker–Planck–Landau equations are among the best known of these models; they incorporate both mean-field and collisional interactions,

with collision operators that behave like nonlinear diffusions in velocity space while bearing a resemblance to the integral and bilinear structure of the Boltzmann operator (recall (5)). These models try to reduce the great variety of processes that go on in plasmas to tractable equations.

(3) Linear models describing the interaction of a particle system with a given (deterministic or random) environment. Important examples that fall into this category include the linear Boltzmann equation, which describes the scattering of particles by a cloud of randomly located obstacles; the archetypal kinetic equation of Fokker–Planck type (studied by Kolmogorov),

$$\partial_t f + v \cdot \nabla_x f - \Delta_v f = 0; \tag{7}$$

and the equations of electron transport, which are useful in neutronics and in semiconductor theory. More generally, we can also include a variety of equations describing combinations of transport, scattering, diffusion, and so on.

(4) Spatially homogeneous models, in which one studies solutions that do not depend on the position variable, only on the kinetic variable. The most important of the resulting models is the spatially homogeneous Boltzmann equation $\partial_t f = Q(f, f)$, the study of which is very well developed; this equation allows the understanding of fine properties of the collision operator. Additional structure can be achieved by restricting the setting even further, e.g., by considering only Maxwellian interactions, in which the collision kernel $\tilde{B}(v - v_*, \omega)$ depends on only the deflection angle $\theta$. The dimension can also be reduced, leading for instance to Kac's one-dimensional caricature of a Boltzmann gas, in which velocities are one dimensional and the conservation of energy has been kept but the conservation of momentum has been dropped.

(5) Linearized equations, obtained by looking at first-order perturbations. For the Boltzmann equation near a homogeneous Maxwellian $M = M(v)$, the linearization is

$$\partial_t h + v \cdot \nabla_x h = Q(h, M) + Q(M, h),$$

which is often further transformed by conjugation with a multiplication operator. For the Vlasov equation near a homogeneous equilibrium $f^0 = f^0(v)$, this is

$$\partial_t h + v \cdot \nabla_x h + F[h] \cdot \nabla_v f^0 = 0.$$

In both cases, spectral properties depend strongly on the interaction potential and have been the object of numerous studies. Many variants of these archetypal models are available.

(6) Delocalized models, in which particles are allowed to have a nonnegligible interaction range. While this procedure is logically inconsistent with the many-particle limit, it does produce some useful equations, such as the Povzner equation and, especially, the Enskog equation, which is used in the description of granular matter.

(7) Models incorporating different physical laws: inelasticity (replacing the energy conservation by a dissipation law; this approach is especially important in the modeling of granular matter), quantum physics (either by modeling quantum phenomena in the interaction terms, thus leading to Boltzmann-Bose or Boltzmann–Fermi models for bosons or fermions collisions, or by keeping a classical description of collisions but incorporating quantum effects in the computation of the cross section), relativity (either by incorporating the geometry of special relativity into the laws of interaction, or by coupling a kinetic equation to the constitutive equations of general relativity), and so on. In relation to relativity it should be noted that the Einstein equations of general relativity cannot "stand on their own" unless one studies the vacuum: these equations need to be coupled to an evolution equation for matter, satisfying certain conditions. Since the pioneering works of Choquet-Bruhat on the Cauchy problem in general relativity, the Vlasov equation has been studied in this context, giving rise to the so-called Vlasov–Einstein model.

(8) Coagulation–fragmentation models incorporating crude modeling of chemical reactions, drop formation from molecules via larger and larger gatherings, gelation problems, etc. The Smoluchowski equation is one of the most popular models in this respect.

(9) Discrete velocity models and lattice models, devised to simplify the geometry of collisions and the phase space, e.g., for numerical simulations.

(10) Models appearing in various other physical contexts, such as interactions between waves in models of weak turbulence. Another example here is the collisionless kinetic equation that is obtained by application of the Wigner transform to the Schrödinger equation.

(11) Kinetic equations in interaction with other physical phenomena: coupling of radiative transfer and hydrodynamics (in astrophysics or nuclear physics), of particles and hydrodynamic fluids (e.g., in sprays), and so on.

(12) Phenomenological models for various interaction phenomena that are difficult to classify or evade precise physical modeling: crowds, traffic, disease transmission, sexual reproduction, etc.

## 6 The Many Mathematical Faces of Kinetic Theory

Modern kinetic theory enjoys an enviable position in the mathematical landscape: standing on top of it, the curious observer can view most of the regions of analysis, as well as significant territories of probability and geometry. In the past thirty years this theory has interacted with many other fields and displayed a number of sophisticated developments. In the same period, it has moved from being a rather minority field to being one that is center stage.

Some of the particular features of kinetic theory that differentiate it from other areas of mathematical physics are

- the presence of two variables (position and velocity);
- the omnipresence of large velocities, which cannot be truncated in the model;
- the degeneracy of most equations in the spatial variable;
- the intricate geometry of collisions;
- the fact that kinetic theory is at a cross-point between several areas of modeling; and
- the interplay of deterministic and chaotic behavior.

With these things in mind, here are some of the main mathematical tools and trends in kinetic theory.

**Spectral theory.** The linearized Boltzmann equation was one of the first model cases of study for integro-differential operators. Some of the important notions here are spectral gap estimates, the Fredholm alternative, self-adjointness, the localization of the essential spectrum, compact perturbations, compactness of the resolvent, accretivity, etc.

**Nonlinear analysis of the Cauchy problem.** Tools involve a priori estimates (starting with mass, energy, and entropy controls), the Cauchy–Kowalevskaya theorem (especially in the short-time derivation of the Boltzmann equation, or in early theories of the Vlasov–Poisson equation), Kolmogorov–Nash–Moser perturbation techniques, the Moser scheme (especially for the spatially homogeneous Boltzmann equation for long-range interactions, which has dissipative features), weak compactness theorems (particularly in the DiPerna–Lions theory of weak solutions), Sobolev trace theorems, and weighted functional spaces of Lebesgue, Sobolev, analytic, and Gevrey type. Bilinear and trilinear estimates with a strong input from harmonic analysis have recently been developed for the study of long-range interactions, in which the collision operator behaves more or less like a nonlinear fractional derivation. This list must also include both the nonlinear changes of variables used by DiPerna and Lions in their notion of "renormalized" solutions, and the "gliding" regularity analysis (regularity obtained after composing the function with a transport equation) used for the study of fast oscillations of the Vlasov equation in large time, etc.

**Harmonic analysis.** Fourier analysis, either in the position variable or the velocity variable, has played a crucial role in various parts of kinetic theory, most notably in the analysis of the spatially homogeneous Boltzmann equation with Maxwellian interactions (for which the Fourier transform of the collision operator is particularly simple and tractable), in the long-time perturbative analysis of the nonlinear Vlasov equation (Landau damping being analyzed mode by mode), in the regularity of the gain part of the Boltzmann operator (which is more regular, by a fractional amount, than the density function), and in velocity-averaging estimates. The latter are intended to answer the following type of question. Given an equation like $v \cdot \nabla_x f = g$, with certain regularity information on $f$ and $g$, show that, if a smooth test function $\varphi$ is given, then $\int f(x,v)\varphi(v)\,dv$ enjoys more regularity than can be predicted solely from the regularity of $f$. This point of view has been extremely fruitful in modern studies of the Cauchy problem and is based mainly on Fourier or X-ray transforms.

**Entropic inequalities.** The analysis of the long-time behavior of collisional kinetic equations naturally leads to the study of inequalities relating Boltzmann's entropy to its rate of production. Kac and McKean were the first to address these issues from a mathematical point of view, and they made the connection with information-theoretical inequalities, involving Fisher information, for example. A central topic in the field came to be known as *Cercignani's conjecture*: is it true that, under certain conditions of normalization or regularity, the entropy production (4) satisfies the functional inequality $D(f) \geqslant$

$K[S(M) - S(f)]$, where $M$ is a Maxwellian distribution? This problem, which is the entropic variant of a spectral gap inequality, has led to unexpected and rich developments related to logarithmic Sobolev inequalities and to the Shannon–Stam and Blachman–Stam inequalities.

**Semigroup arguments.** At the end of the 1990s, semigroup arguments made their way into kinetic theory, either through the use of auxiliary diffusion equations or via the second variation method introduced by Bakry and Émery in their study of logarithmic Sobolev inequalities. The ideas of Bakry and Émery were adapted in the context of PDEs by Toscani, Arnold, Markowich, and others; a large body of works followed on the entropic analysis of the convergence to equilibrium after a long time, for both linear and nonlinear models in kinetic theory, especially those of Fokker–Planck type. Some of the key concepts are the $\Gamma_2$ calculus, curvature-dimension inequalities, and dissipation of entropy production (i.e., the second time-derivative of the entropy).

**Specific techniques for degenerate operators.** Kolmogorov in the 1950s and Hörmander in the 1960s founded the theory of *hypoellipticity*, according to which certain degenerate operators, like $-v \cdot \nabla_x + \Delta_v$, generate a regularizing semigroup. This situation, which occurs frequently in linear-dissipation kinetic models, is often treated by commutator estimates. The more recent theory of *hypocoercivity* deals with the time decay of semigroups generated by degenerate operators, typically of the form $T + \Lambda$, where $\Lambda$ is coercive in some appropriate subspace and $T$ is skew-symmetric. Paradigmatic examples are $-v \cdot \nabla_x + \Delta_v - v \cdot \nabla_v$ in $L^2(M\,\mathrm{d}x\,\mathrm{d}v)$ and $-v \cdot \nabla_x + \Pi_M - \mathrm{Id}$ in $L^2(M\,\mathrm{d}x\,\mathrm{d}v)$, where $M$ is a Gaussian and $\Pi_M$ is the orthogonal projection on constant functions.

**Qualitative studies of solutions.** The Vlasov equation is of hyperbolic type, but the Boltzmann equation is of mixed hyperbolic/parabolic type in some sense; a number of works and techniques have been devoted to the study of the qualitative behavior of solutions, including regularization, propagation, or decay of singularities (often studied in Sobolev spaces), wave patterns (in connection with systems of conservation laws and compressible Navier–Stokes equations), harmonic analysis, pseudodifferential operators, Littlewood–Paley analysis, the Radon transform, quantitative uncertainty principles, the self-similar ansatz, concentration analysis, and so on.

**Singular limits.** These limits, in which a term of the equation is enhanced by a diverging coefficient, are studied via ansatzes, expansions, spectral theory, ergodic theory, etc. They appear in particular in connection with (a) inviscid or viscous hydrodynamic limits, in which the Knudsen number (the ratio of the mean free path to the typical length) goes to 0, typically leading to an enhanced collision operator, $\varepsilon^{-1}Q(f,f)$; (b) the homogenization of transport models, typically leading to an enhanced transport term, $\varepsilon^{-1}v \cdot \nabla_x$ or $\varepsilon^{-1}F \cdot \nabla_v$; (c) high-frequency semiclassical limits of Schrödinger equations, via the Wigner transform; (d) small mass ratio limits, e.g., in plasmas, where electrons are much lighter than nuclei.

**Differential geometry.** Curved phase spaces appear naturally in relativistic kinetic theory, either through the rules of collisions between particles or because the system is considered in a Lorentzian ambient space.

**Calculus of variations.** When stability is not ensured by a Lyapunov functional such as the entropy, stability issues can be very tricky. Convexity properties can then be crucial in studying the dynamic stability of particular equilibria that are energy minimizers. This approach, introduced into hydrodynamics in the 1960s, was systematically used from the 1980s on in the study of the Vlasov–Poisson equation, with the help of notions of concentration-compactness, rearrangement, etc.

**Many-particle techniques.** The quest for a rigorous foundation for the Boltzmann and Vlasov equations from particle systems has led to the analysis of many-particle systems, obeying the fundamental laws of classical or quantum mechanics, in the limit where the number $N$ of particles diverges to infinity. The microscopic equations then depend on all positions and velocities, say $(x_1, v_1), \dots, (x_N, v_N)$. The problem can be formulated in terms of the likely behavior of the empirical distribution, say, $\hat{\mu}^N = N^{-1}\sum_{i=1}^N \delta_{(x_i,v_i)}$, or in terms of the limit behavior of the first-particle marginal of an $N$-particle distribution $f^N$, satisfying the $N$-particle Liouville equation

$$\partial_t f^N + \sum v_i \cdot \nabla_{x_i} f^N - \sum_{i \neq j} \nabla W(x_i - x_j) \cdot \nabla_{v_i} f^N = 0,$$

in various asymptotic regimes where time, space, mass, and the strength of interaction may be rescaled. Popular scalings are the Boltzmann–Grad limit, pioneered by Grad, Cercignani, and Lanford, which led to the nonlinear Boltzmann equation, and the mean-field limit, established by Braun, Hepp, Dobrushin, and Neunzert for smooth interactions, which led to the Vlasov equation. Famous variants are the probabilistic approach of Kac, in which the deterministic Newton equations are replaced by a phenomenological, stochastic microscopic model, and the derivation of the linear Boltzmann equation (for a so-called Knudsen gas) pioneered by Gallavotti. Key concepts in this field are notions of molecular chaos (asymptotic independence of particles, i.e., low correlations), perturbative series, particle histories, functional inequalities in infinite dimension, quantitative laws of large numbers, central limit theorems, and orthogonal polynomials. For quantum particle systems, density matrices and Wigner transforms play a crucial role.

**Numerical analysis.** The simultaneous presence of very diverse terms in the equations, the high-dimensional phase space, the complexity of collisions, the presence of large velocities and small densities, and the difficulty of accurate experiments have all made numerical simulation of kinetic models a challenging area. Transport phenomena are most often simulated with the help of the methods of characteristics, that is, following particles in phase space; but the reconstruction of the density from one time step to the next leads to many subtleties, since the number of particles used in the simulation is always much smaller than the actual number of particles and since particle trajectories do not preserve grids or other discretizations of the phase space. Collisional phenomena are tricky to compute and were initially handled by stochastic methods, based on particle systems obeying more or less realistic interaction rules. These schemes were founded by Bird in the 1960s and remained dominant for more than forty years. It is only in the last decade that the progress of algorithms and computer power have made more accurate deterministic methods competitive in cost, at least in certain situations. Keywords in this area are the splitting method, Monte Carlo simulation, consistency analysis, Lagrangian and semi-Lagrangian methods, spectral analysis, the Fourier transform, the fast Fourier transform algorithm, finite elements, lattice simulation, conservative schemes, adaptive grids, etc. Specific methods were developed by Cheng and Knorr, Sone, Aoki, Babovsky, Neunzert, Wagner, Degond, Bobylev, Rjasanow, Sonnendrücker, Pareschi, Filbet, and many others in aeronautics, in astrophysics, in plasma physics—there is an enormous amount of literature and it is barely touched upon in this article.

Let us conclude our list with two subjects that were partly motivated by kinetic theory but where the main impact was made in other parts of mathematics.

**Ordinary differential equations (ODEs) with rough coefficients.** The classical theory of ODEs, say $\dot{x} = \xi(t,x)$, requires continuity of $\xi$ for the local existence of a flow, and Lipschitz regularity of $\xi$ for local uniqueness and continuous dependence. This requirement of Lipschitz regularity is often a strong restriction in applications to PDEs, especially when $\xi$ depends on the solution and its regularity is a priori unknown. As a by-product of their studies of the Cauchy problem in kinetic theory, DiPerna and Lions came up with a theory of ODEs that provides local existence and uniqueness for *almost every* initial data, under a more lenient assumption of Sobolev regularity (e.g., $\xi \in W^{1,1}_{\text{loc}}$, $\operatorname{div}\xi \in L^\infty$, and some growth condition at infinity). The original proof was based on the analysis of the transport equation $\partial_t f + \xi \cdot \nabla f = 0$ and the renormalization technique; more recently, the theory was refined by Ambrosio, de Lellis, and others to include the limit case of bounded variation regularity and to provide alternative, trajectorial proofs. This theory has been used in various types of PDEs, such as hyperbolic systems of conservation laws.

**Optimal transport and metric geometry.** In the 1970s it was shown by Tanaka that the spatially homogeneous Boltzmann equation with Maxwellian interactions is contracting for the Wasserstein (optimal transport) distance

$$W_2(\mu,\nu) = \left( \inf_{\pi \in \Pi(\mu,\nu)} \int_{\mathbb{R}^d \times \mathbb{R}^d} |v - v_*|^2 \, \mathrm{d}\pi(v,v_*) \right)^{1/2},$$

where $\mu$ and $\nu$ are two probability measures on $\mathbb{R}^d$, and the infimum is over all joint probability measures $\pi(\mathrm{d}v\,\mathrm{d}v_*)$ with marginals $\mu$ and $\nu$. After a period in which these connections sank more or less into oblivion, the links between optimal transport and kinetic equations were renewed in the 1990s and led to various uniqueness and stability results. Later, the interplay between optimal transport and Boltzmann's entropy played a key role in the theory of nonsmooth Ricci curvature.

## 7 Landmarks

A list of fifty striking works, arranged chronologically, appears below. These works have punctuated progress in kinetic theory and have become classical references. Some of these works have opened up a new field of research, while others have closed an existing one; some are gathered together if they are very close in terms of subject matter. The dates that are given are those of publication, which sometimes came a few years after the actual work was carried out.

This list is not intended to convey overall importance, and the selection is partly a personal matter. It is biased in favor of theoretical issues and does not do justice to some subjects that are of great importance in industrial applications, such as neutron transport or coagulation–fragmentation. Similarly, it does not even touch on the enormous and inventive body of work that has been undertaken in the field of numerical simulations.

The list also partly reflects the history of mathematical kinetic theory: at first, it was quite a minority subject, with rare contributions; but a few centers of wider study emerged after World War II (New York (at the Courant Institute), Osaka/Kyôtô, Göteborg, Rome, Moscow, and Zürich). In the 1980s and 1990s, the study of nonlinear problems started to flourish, with the French and Italian schools taking the lead, and the community became quite organized; at the same time, new research groups were emerging, most notably in Germany, the United States, Austria, Spain, Taiwan, and Canada. The past twenty years have been characterized by a growing and fruitful interplay between kinetic theory and other mathematical fields, by an emphasis on quantitative results and constructive methods, and by renewed study of the linearized and perturbative settings.

(1) *Hilbert (1912).* The first study of the linearized Boltzmann operator and the first formal expansion of the solution of the Boltzmann equation in powers of the (small) Knudsen number near the hydrodynamic regime.

(2) *Chapman and Enskog (1917).* Systematic formulas for deriving macroscopic transport coefficients from microscopic interactions and for perturbative expansion near the hydrodynamic regime (as an alternative to Hilbert's work).

(3) *Carleman (1932).* The first solution of the nonlinear Cauchy problem for the spatially homogeneous Boltzmann equation, for hard spheres; this included a qualitative study of the lower bound for the density and a study of the convergence to equilibrium.

(4) *Kolmogorov (1934).* Computation of the fundamental solution of the kinetic Fokker–Planck equation $\partial_t f + v \cdot \nabla_x f = \Delta_v f$, displaying (hypoelliptic) regularity properties.

(5) *Grad (1949).* A thirteen-moment system describing a high-order approximation to the hydrodynamic limit of the Boltzmann equation.

(6) *Kac (1954).* The probabilistic foundation of kinetic theory through a phenomenological stochastic many-particle model of the spatially homogeneous Boltzmann equation; this led to conjectures on quantitative relaxation rates.

(7) *Backus (1960), Penrose (1960).* A mathematical treatment of linear Landau damping, discovered by Landau in 1946, with sharp criteria for stability; statement of the nonlinear damping problem.

(8) *Carleman (1949, 1957), Grad (1963–65).* The modern spectral theory of the linearized Boltzmann equation with cutoff, for hard interactions, in the homogeneous and inhomogeneous settings.

(9) *McKean (1965).* Probabilistic study of Kac's caricature of the Boltzmann equation through molecular chaos, Fisher information estimates, and the quantitative central limit theorem for Maxwell interactions.

(10) *Bird (1966).* The first numerical scheme for the stochastic simulation of the Boltzmann equation; an alternative scheme was later introduced by Nanbu (1983).

(11) *Hörmander (1967).* General criteria for the hypoellipticity of degenerate diffusion equations, including a precursor to velocity-averaging lemmas.

(12) *Gallavotti (1969).* Derivation of the linear Boltzmann equation from the Lorentz gas through averaging over a random environment. Many developments, by Pulvirenti, Desvillettes, and others, followed from this work.

(13) *Arkeryd (1972).* The Cauchy problem for the spatially homogeneous Boltzmann equation with hard potentials, in weighted $L^1$ spaces, including weak compactness properties in $L^1$.

(14) *Tanaka (1973).* Contraction properties of the spatially homogeneous Boltzmann equation with Maxwell kernel in the Wasserstein $W_2$ distance.

(15) *Lanford (1974).* Short-time derivation of the Boltzmann equation from deterministic Newton laws.

(16) *Ukai (1974).* Perturbative solutions of the full inhomogeneous Boltzmann equation, based on the spectral theory of the linearized equation.

(17) *Bobylev (1976–88).* Systematic study of the spatially homogeneous Boltzmann equation with Maxwell interactions via the Fourier transform.

(18) *Braun and Hepp (1977), Dobrushin (1979), Neunzert (1984).* A rigorous mean-field limit for the Vlasov equation with smooth interactions.

(19) *Sznitman (1984).* Propagation of chaos and a probabilistic derivation of the spatially homogeneous Boltzmann equation with hard spheres.

(20) *Golse, Perthame, and Sentis (1985).* The start of the systematic study of velocity-averaging lemmas in Sobolev spaces, which had been independently introduced by Agoshkov (1984) shortly before.

(21) *Glassey and Strauss (1986).* The Cauchy problem for the relativistic Vlasov–Maxwell equation, conditional to a conjectured property of compact support.

(22) *Bony (1987).* A new Lyapunov functional for the discrete-velocity inhomogeneous Boltzmann equation in one space dimension; this was the starting point for various Lyapunov functionals for the Boltzmann equation in one space dimension.

(23) *DiPerna and Lions (1989).* The existence and stability of weak solutions ("renormalized solutions") in the large, for the nonlinear inhomogeneous Boltzmann equation.

(24) *Bardos, Golse, and Levermore (1991).* A systematic program for the proof of hydrodynamic limits of weak solutions, in particular in the incompressible regime; this program would take twenty years to complete.

(25) *Lions and Perthame (1991), Pfaffelmoser (1992).* The first proofs of existence and uniqueness of classical solutions for the three-dimensional Vlasov–Poisson equation, by two different approaches.

(26) *Desvillettes (1989), Carlen and Carvalho (1992, 1994).* The first lower bounds on the instantaneous rate of entropy production in the Boltzmann equation through the quantitative $H$ theorem and information theory.

(27) *Desvillettes (1993).* Refined moment estimates for the spatially homogeneous Boltzmann equation; in particular, their immediate appearance in the case of hard potentials.

(28) *Lions (1994).* The regularity of the gain term of the Boltzmann collision operator, which is shown to have the structure of a singular integral operator, gaining up to one derivative in three dimensions for smooth kernels.

(29) *Desvillettes (1995).* The first evidence of regularization due to long-range interactions in the Boltzmann equation on a spatially homogeneous caricature; the start of a long series of works on such regularization effects.

(30) *Gérard, Markowich, Mauser, and Poupaud (1997), Lions and Paul (1993).* The systematic study of high-frequency limits through the Wigner transform, with applications to quantum kinetic theory.

(31) *Erdös and Yau (1998).* Derivation of the linear quantum Boltzmann equation, in the weak coupling limit, for the Wigner distribution of a quantum particle in a random environment.

(32) *Mischler and Wennberg (1999), Lu (1999).* Optimal conditions for the well-posedness of the Cauchy problem of the spatially homogeneous Boltzmann equation with hard interaction and cutoff.

(33) *Carlen, Gabetta, and Toscani (1999).* Optimal rates of convergence to equilibrium for the spatially homogeneous Boltzmann equation with Maxwell interaction and angular cutoff (the removal of the cutoff was later obtained in subsequent works involving Wennberg, Dolera, and Regazzini).

(34) *Toscani and Villani (1999), Villani (2003).* Sharp entropy production bounds for the Boltzmann equation, solving, or nearly solving (depending on assumptions), Cercignani's conjecture; this work was based on semigroup methods, information theory, and the Landau equation.

(35) *Guo (2002).* The first of a series of works using energy methods to work out robust perturbative theories of the Boltzmann equation and other kinetic models.

(36) *Carlen, Carvalho, and Loss (2003), Maslen (2003).* Determination of the $L^2$ spectral gap for Kac's random walk in arbitrarily large dimension, after a uniform lower bound was established by Janvresse.

(37) *Carlen and Lu (2003).* Examples of arbitrarily slow convergence to equilibrium for the Boltzmann equation with Maxwell interactions.

(38) *Bobylev, Gamba, and Panferov (2004), Gamba, Panferov, and Villani (2004).* Moment estimates and the Cauchy problem for the inelastic spatially homogeneous Boltzmann equation with hard interactions.

(39) *Alexandre and Villani (2004).* Weak solutions for the spatially inhomogeneous Boltzmann equation without cutoff, and the asymptotic regime of predominantly grazing collisions in the spatially inhomogeneous case. This came after much progress in the understanding of grazing collisions by Alexandre, Desvillettes, Villani, and Wennberg.

(40) *Liu and Yu (2004).* The first works on pointwise stability and the Green function in Boltzmann theory, and in kinetic shock wave analysis (motivated by an earlier work of Caflisch), by analogy with systems of conservation laws; the start of a long series of works.

(41) *Golse and Saint-Raymond (2004).* A rigorous proof of the incompressible hydrodynamic limit for weak solutions of the Boltzmann equation; this came after works by Golse, Levermore, Masmoudi, and others.

(42) *Desvillettes and Villani (2005).* Quantitative convergence to equilibrium for the Boltzmann equation, far from equilibrium, by entropy methods, under a conjectural condition of regularity bounds.

(43) *Baranger and Mouhot (2005), Mouhot (2006), Gualdani, Mischler, and Mouhot (2013).* Optimal rates of convergence for the Boltzmann equation (homogeneous and inhomogeneous), coupling quantitative spectral analysis to entropy methods, conditional on regularity.

(44) *Mischler and Mouhot (2006).* A proof of Haff's law of decay of temperature and self-similar stability in the theory of granular (inelastic) gases.

(45) *Villani (2009).* General criteria for hypocoercivity, in both linear and nonlinear situations.

(46) *Gressman and Strain (2011), Alexandre, Morimoto, Ukai, Xu, and Yang (2011).* Construction of smooth solutions for the noncutoff spatially homogeneous Boltzmann equation, for potentials that are hard or not too soft.

(47) *Lemou, Méhats, and Raphaël (2011).* Orbital stability of spherical monotone equilibria of the gravitational Vlasov-Poisson equation; the culmination of a long series of works on the stability of the Vlasov–Poisson equation by Antonov, Wolansky, Strauss, Guo, Rein, and others.

(48) *Mouhot and Villani (2011).* A proof of Landau damping for the nonlinear Vlasov equation, near stable homogeneous equilibria, in analytic or Gevrey regularity, via phase mixing and gliding regularity; this was later adapted by Bedrossian and Masmoudi to inviscid damping near Couette flow.

(49) *Mischler and Mouhot (2013).* Significant progress on Kac's program: relaxation estimates for particle systems, quantitative and uniform in the number of particles, in the limit of the spatially homogeneous Boltzmann equation, using quantitative chaos properties and entropic estimates.

(50) *Escobedo and Velázquez (2013).* A rigorous proof of blow-up (Bose–Einstein condensation) for the quantum spatially homogeneous Boltzmann–Bose equation.

## 8 Challenges

While kinetic theory has come a tremendous distance, the field is still driven, among other motivations, by the ambition to understand certain famous, monstrously difficult problems. A list of some of these problem follows below, gathered together under a few main themes.

### 8.1 The Cauchy Problem, Regularity, Singularities, and Finite-Time Qualitative Behavior

The most important and annoying open problems in this list are certainly the related questions of the regularity and well-posedness of the Boltzmann equation when no perturbative or spatial homogeneity assumptions are imposed. Parallels could be drawn between this and the Millennium Prize Problem on the incompressible Navier–Stokes equation in three dimensions. For collisionless kinetic equations, even if the Cauchy problem for the Vlasov–Poisson equations has been tamed, other Herculean tasks concerning more intricate models remain. The Cauchy problems for the Vlasov–Maxwell and Vlasov–Einstein equations are of particular interest. Actually, for these two equations even the perturbative theory is far from well understood. In a completely different direction, the stability of homogeneous solutions remains almost untouched in the theory of the inhomogeneous inelastic Boltzmann equation; the annoying issue here is that nobody has been able to prove that clustering is possible, which is well accepted in physics. Finally, much remains to be understood about very soft interactions (when the collision kernel behaves like a large negative power of the relative velocity).

### 8.2 Long-Time Behavior

The entropic relaxation for the Boltzmann equation now seems quite well understood without boundaries, with robust estimates and recipes applying far from equilibrium as well as optimal decay from quantitative linearized arguments. Collisionless relaxation in Vlasov theory, on the other hand, is understood only near a stable homogeneous equilibrium, and the stability of inhomogeneous stationary solutions, such as so-called BGK waves (named after Bernstein, Greene, and Kruskal), remains a famous open problem. The mathematical theory of instability phenomena is also wide open. The study of the long-time behavior of "typical" data, e.g., via a statistical approach, is untouched.

A long-term goal is the combination of entropic and mixing effects in the study of convergence, e.g., for the so-called Vlasov–Poisson–Fokker–Planck equations, which combine mean-field mixing and hypoelliptic diffusion.

For dissipative equations with *nonreversible* stationary states, corresponding to nonlocal cancelations in the equation, long-time study is in its infancy. For nonlinear models, even when the existence of equilibria is proven, there is generally no Lyapunov-type approach to the long-time behavior.

### 8.3 Meso–Macro Limits: Hydrodynamic Limits

Huge progress has been made in understanding the hydrodynamic limit of Boltzmann equations since this problem was expressed by Hilbert more than a century ago. And yet many questions remain unanswered. The incompressible limit is now rather well understood, but this limit is quite specific, and one would like to understand the more natural *compressible* limit better. Examples of problems in this area include the long-time stability of Boltzmann solutions near a smooth solution of compressible Navier–Stokes equations, and the handling of shocks in the large.

As already mentioned, the hydrodynamic limit of the Boltzmann equation leads only to perfect fluids, unless the equations are modified in a phenomenological way. To retrieve alternative pressure laws from basic principles of classical mechanics, the most natural plan is to go directly from the equations of microscopic many-particle systems to hydrodynamic models, without passing through the mesoscopic scale. This strategy was first made precise in a program sketched by Morrey in the 1950s that proved to be extraordinarily difficult and is still largely open in spite of substantial progress by Varadhan, Yau, Olla, and others.

### 8.4 Microscopic Derivation

The derivation of kinetic models from the laws of atomistic matter is also part of Hilbert's sixth problem, and it is an emblematic issue in both kinetic theory and statistical physics.

In the collisional case, the most important open problem is certainly the validity of the Boltzmann–Grad limit for hard spheres in large time (i.e., in time significantly larger than the mean free time) and without any assumption of very small mass. This problem probably includes an understanding of the regularity of the inhomogeneous Boltzmann equation, so it can be considered a Holy Grail in the field. Another open problem is the low-density limit in the case of long-range collisional interactions; in this case, not even a short-time result has been established.

In the collisionless case, the main open problem is the rigorous justification of the mean-field limit in the case of Coulomb and Newton interactions. The best results so far were obtained by Hauray and Jabin around 2007, but they still require smoothing or cutoff of the interaction at small scales. A further goal is the understanding of the microscopic derivation of the many involved models that appear in plasma physics, one instance being the Balescu–Lenard equation (for which even the short-time well-posedness is still unclear).

Finally, in the case of diffusive kinetic equations, one of the most appealing open problems is the derivation of the heat equation from a set of interacting oscillators, as studied, for instance, by Rey-Bellet. While preliminary works have established, among other things, the existence of relevant equilibria, the derivation of the heat equation has been understood only in particular cases, with the help of hypoelliptic and hypocoercive tools.

### 8.5 The Challenge of Boundary Conditions

Questions about the interaction of gases with boundaries or external forces were raised in the early days of kinetic theory by both Maxwell and Boltzmann. In the real world, most phenomena involving many-particle systems include nontrivial geometries, boundary effects, or external fields. But the Boltzmann and Vlasov equations are still poorly understood in this respect, in particular concerning the geometry-driven asymptotic behavior. Even for the hypoelliptic kinetic Fokker–Planck operators in a domain, there is no equivalent to the huge body of work on the eigenvalue problem of the Laplace equation in a domain. The majority of the results and challenges discussed above lend themselves to boundary-driven formulations, which are mostly open.

Specific boundary-related problems raise beautiful challenges: propagation of singularities according to the shape of boundaries, ergodicity, relaxation to equilibrium, and so on.

An even more ambitious goal is the understanding of *self-induced nontrivial geometry*, as observed in particular in galactic dynamics, where the geometry of the confinement is influenced by the gravitational mean field of the system itself.

**Acknowledgments.** We warmly thank Claude Bardos, Damon Civin, Laurent Desvillettes, and Alexi Hoeft for their careful reading of our draft and for their constructive suggestions.

## Further Reading

In this last section we list some survey books and synthesis articles for further reading. We start with a *very short selection* of books and survey articles that can be used by a reader who wants to delve more deeply into the subject.

Binney, J., and S. Tremaine. 1987. *Galactic Dynamics*. Princeton, NJ: Princeton University Press.

Cercignani, C. 2000. *Rarefied Gas Dynamics: From Basic Concepts to Actual Calculations*. Cambridge: Cambridge University Press.

Cercignani, C., R. Illner, and M. Pulvirenti. 1994. *The Mathematical Theory of Dilute Gases*. New York: Springer.

Glassey, R. 1996. *The Cauchy Problem in Kinetic Theory*. Philadelphia, PA: SIAM.

Krall, N. A., and A. W. Trivelpiece. 1986. *Principles of Plasma Physics*. San Francisco, CA: San Francisco Press.

Lifshitz, E. M., and L. P. Pitaevski. 1979. *Teoreticheskaya fizika* ("Landau–Lifshitz"), Tom 10 (in Russian). Fizicheskaya Kinetika ("Physical Kinetics"). Moscow: Nauka.

Markowich, P. A., C. A. Ringhofer, and C. Schmeiser. 1990. *Semiconductor Equations*. Vienna: Springer.

Villani, C. 2002. *A Review of Mathematical Topics in Collisional Kinetic Theory*. Handbook of Mathematical Fluid Dynamics, volume I, pp. 71–305. Amsterdam: North-Holland.

——. 2006. Mathematics of granular materials. *Journal of Statistical Physics* 124(2–4):781–822.

——. 2013. (Ir)reversibility and entropy (French and English versions). In *Time: Poincaré Seminar 2010*, edited by B. Duplantier. Basel: Birkhaüser.

We now provide a longer list of articles and surveys (in addition to the ones mentioned just above) that could be useful to readers. We shall separate them into several categories.

We start with some of the *founding works* of kinetic theory.

Boltzmann, L. 1872. Weitere Studien über das Wärmegleichgewicht unter Gasmolekülen. *Wiener Berichte, Sitzungsberichte der Akademie der Wissenschaften* 66:275–370.

——. 1896, 1898. *Vorlesungen ueber Gastheorie*, volumes 1 and 2. Leipzig: J. A. Barth.

Jeans, J. H. 1915. On the theory of star-streaming and the structure of the universe. *Monthly Notices of the Royal Astronomical Society* 76:70–84.

Lorentz, H. 1905. Le mouvement des électrons dans les métaux. *Archives Neerlandaises* 10:336–71.

Maxwell, J. C. 1867. On the dynamical theory of gases. *Philosophical Transactions of the Royal Society of London* 157: 49–88.

Vlasov, A. A. 1938. On vibrational properties of an electron gas. *Journal of Experimental and Theoretical Physics* 8(3): 291–318. (In Russian.)

The *history of the subject* is described in many references: a selection follow.

Brush, S. 1965. *Kinetic Theory of Gases: An Anthology of Classic Papers with Historical Commentary*. London: Imperial College Press. (This book includes translations of some of the above-mentioned founding works, edited and with commentary.)

Cercignani, C. 1998. *Ludwig Boltzmann: The Man Who Trusted Atoms*. Oxford University Press.

Hénon, M. 1982. Vlasov equation? *Astronomy and Astrophysics* 114:211–12. (A short historical comment on the birth of the Vlasov equation.)

Levermore, C. D. Ongoing. History of kinetic theory. An unfinished Web project on kinetic theory containing incomplete but still relevant historical information: www.terpconnect.umd.edu/~lvrmr/History/index.shtml.

Lindley, D. 2001. *Boltzmann's Atom: The Great Debate That Launched a Revolution in Physics*. Glencoe, IL: Free Press.

For *general presentations* and treatises of kinetic theory, from the point of view of physics and modeling, one can consult the following works.

Akhiezer, A., I. Akhiezer, R. Polovin, A. Sitenko, and K. Stepanov. 1975. *Plasma Electrodynamics*, volume I (*Linear Theory*) and volume II (*Non-linear Theory and Fluctuations*), translated into English by D. ter Haar. Oxford: Pergamon Press.

Balescu, R. 1975. *Equilibrium and Nonequilibrium Statistical Mechanics*. New York: Wiley-Interscience.

Cercignani, C. 1975. *Theory and Application of the Boltzmann Equation*. New York: Elsevier.

——. 1995. Recent developments in the mechanics of granular materials. In *Fisica Matematica e Ingeneria delle Strutture: Rapporti e Compatibilità*, edited by G. Ferrarese, pp. 119–32. Bologna: Pitagora Editrice.

Chapman, S., and T. G. Cowling. 1970. *The Mathematical Theory of Non-uniform Gases: An Account of the Kinetic Theory of Viscosity, Thermal Conduction and Diffusion in Gases*, 3rd edn (prepared with the cooperation of D. Burnett). Cambridge: Cambridge University Press.

Decoster, A. 1998. Survey of the collisional kinetic theory of plasmas. Part I. In *Modeling of Collisions*, edited by A. Decoster, P. A. Markowich, and B. Perthame. Paris: Gauthier-Villars/Elsevier.

Delcroix, J. L., and A. Bers. 1994. *Physique des Plasmas*, volumes 1 and 2. Paris: InterEditions/CNRS. (In French.)

Lebowitz, J. 1994. Microscopic reversibility and macroscopic behavior: physical explanatoins [sic] and mathematical derivations. In *Twenty-Five Years of Non-equilibrium Statistical Mechanics: Proceedings of the XIII Sitges Conference*, edited by J. Brey, J. Marro, J. Rubi, and M. S. Miguel, pp. 1–20. Lecture Notes in Physics. New York: Springer.

Risken, H. 1989. *The Fokker–Planck Equation: Methods of Solution and Applications*. New York: Springer.

Ryutov, D. D. 1999. Landau damping: half a century with the great discovery. *Plasma Physics and Controlled Fusion* 41:A1–A12.

Some *more mathematically oriented surveys*, varying in size, timing, and scope, follow below.

Alexandre, R. 2009. A review of Boltzmann equation with singular kernels. *Kinetic and Related Models* 2(4):551–646.

Andréasson, H. 2011. The Einstein–Vlasov system/kinetic theory. *Living Reviews in Relativity* 14:4–55.

Bird, G. A. 1994. *Molecular Gas Dynamics and the Direct Simulation of Gas Flows*, 2nd edn. Oxford: Oxford University Press.

Bobylev, A. V. 1988. The theory of the nonlinear spatially uniform Boltzmann equation for Maxwell molecules. In *Mathematical Physics Reviews*. Soviet Scientific Reviews, Section C, volume 7, pp. 111–233. Amsterdam: Harwood Academic.

Carleman, T. 1957. *Problèmes Mathématiques dans la Théorie Cinétique des Gaz*, completed by L. Carleson. Publications Scientifiques de l'Institut Mittag-Leffler, volume 2. Uppsala: Almqvist and Wiksells Boktryckeri.

Case, K. M., and P. F. Zweifel. 1967. *Linear Transport Theory*. Reading, MA: Addison-Wesley.

Dautray, R., and J. L. Lions. 2000. *Mathematical Analysis and Numerical Methods for Science and Technology*, volume 6: *Evolution Problems II*, chapter 21, pp. 209–416. Contributions by C. Bardos, M. Cessenat, A. Kavenoky, B. Mercier, and R. Sentis. New York: Springer. (Translated from the 1985 French edition by A. Craig.)

Desvillettes, L., C. Mouhot, and C. Villani. 2011. Celebrating Cercignani's conjecture for the Boltzmann equation. *Kinetic and Related Models* 4(1):277–94.

Drake, R. L. 1972. A general mathematical survey of the coagulation equation. In *Topics in Current Aerosol Research*, edited by G. M. Hidy and J. R. Brock, part 2, pp. 201–376. Oxford: Pergamon.

Ehrenfest, P., and T. Ehrenfest. 1990. *The Conceptual Foundations of the Statistical Approach in Mechanics*. New York: Dover. (Translated from 1959 German original by M. J. Moravcsik; foreword by M. Kac and G. E. Uhlenbeck.)

Gallagher, I., L. Saint-Raymond, and B. Texier. 2014. *From Newton to Boltzmann: Hard Spheres and Short-Range Potentials*. Zürich: European Mathematical Society.

Illner, R., and H. Neunzert. 1987. On simulation methods for the Boltzmann equation. *Transport Theory and Statistical Physics* 16:141–54.

Kac, M. 1959. *Probability and Related Topics in Physical Sciences*. London: Interscience.

Laurençot, Ph., and S. Mischler. 2004. On coalescence equations and related models. In *Modeling and Computational Methods for Kinetic Equations*, edited by P. Degond and L. Pareschi, pp. 321–56. Boston, MA: Birkhäuser.

Liu, T.-P., and S.-H. Yu. 2009. Boltzmann equation, boundary effects. *Discrete and Continuous Dynamical Systems* 24(1):145–57.

Mouhot, C. 2007. Quantitative linearized study of the Boltzmann collision operator and applications. *Communications in Mathematical Science* 5:73–86.

Naranyan, A., and A. Klöckner. 2007. Deterministic numerical schemes for the Boltzmann equation. Technical Report, Brown University (arxiv: 0911.3589).

Rendall, A. D. 2008. *Partial Differential Equations in General Relativity*. Oxford: Oxford University Press.

Saint-Raymond, L. 2008. *Hydrodynamic Limits of the Boltzmann Equation: Lectures at SISSA, Trieste*. Lecture Notes in Mathematics. New York: Springer.

Sone, Y. 2002. *Kinetic Theory and Fluid Dynamics*. Modeling and Simulation in Science, Engineering and Technology. Boston, MA: Birkhäuser.

——. 2007. *Molecular Gas Dynamics*. Basel: Birkhaüser.

Spohn, H. 1991. *Large Scale Dynamics of Interacting Particles*. Heidelberg: Springer.

Truesdell, C., and R. G. Muncaster. 1980. *Fundamentals of Maxwell's Kinetic Theory of a Simple Monatomic Gas: Treated as a Branch of Rational Mechanics*. New York: Academic Press.

Uhlenbeck, G. E., and G. W. Ford. 1963. *Lectures in Statistical Mechanics*, chapter 3: *The Boltzmann Equation*. Providence, RI: American Mathematical Society.

Villani, C. 2006. Hypocoercive diffusion operators. In *International Congress of Mathematicians, Madrid, 2006*, volume III, pp. 473–98. Zürich: European Mathematical Society.

——. 2008. Entropy production and convergence to equilibrium (expanded version of notes from a series of lectures at the Institut Henri Poincaré, Paris, Fall 2001). In *Entropy Methods for the Boltzmann Equation*, pp. 1–70. Lecture Notes in Mathematics, volume 1916. Berlin: Springer.

——. 2008. $H$-theorem and beyond: Boltzmann's entropy in today's mathematics. In *Boltzmann's Legacy*, edited by G. Gallavotti, W. L. Reiter, and J. Yngvason, pp. 129–43. Zürich: European Mathematical Society.

Villani, C. 2009. *Hypocoercivity*. Memoirs of the American Mathematical Society. Providence, RI: American Mathematical Society.

——. Forthcoming. Landau damping. In *Numerical Models of Fusion*. Panoramas and Synthèses. Paris: Société Mathématique de France.

——. 2014. Particle systems and nonlinear Landau damping. *Physics of Plasmas* 21(3):030901.

Weinberg, A. W., and E. P. Wigner. 1958. *The Physical Theory of Neutron Chain Reactors*. Chicago, IL: University of Chicago Press.

Yu, S.-H. 2006. The development of the Green's function for the Boltzmann equation. *Journal of Statistical Physics* 124(2–4):301–20.

In closing we mention the *Porto Ercole Lecture Notes* series. This is a series of high-level lecture notes from the biennial "Methods and Models of Kinetic Theory" summer schools held in Porto Ercole in Italy. They have been published in the journal *Rivista di Matematica della Università di Parma* since 2002.

## IV.26 Continuum Mechanics

*Richard D. James*

### 1 What Is Continuum Mechanics?

Matter is composed of atoms. Atoms are composed of protons, neutrons, and electrons. Protons and neutrons are composed of elementary particles. Everything is discrete and, at the finest level, indivisible. It is quite surprising, then, that continuum mechanics, possibly the most successful theory of general use in applied mathematics, does not explicitly recognize the existence of atoms.

We will discuss the relationship between atoms and continuum mechanics later, but the subject is indeed used a lot; well over half the articles in this volume directly use one or another special case of the basic equations of continuum mechanics. Of course, these cases exist as subjects in their own domains, and the focus of these special cases may be on the beautiful phenomena they describe, as one can see. However, there is a structure that is common to all of them. From the perspective of continuum mechanics, theories that appear to be completely different become quite similar when viewed from within this structure.

As such, continuum mechanics has the same advantage as any other unifying theory of mathematics: by knowing the structure, one can understand many special cases by remembering only a few key concepts. We can in fact go further: by knowing the structure, one can more easily discover new special cases. This activity usually takes the form of the discovery of new mathematical theories for emerging materials.

Continuum mechanics usually gives rise to partial differential equations (PDEs). In modern research there is a healthy interaction between continuum mechanics and PDEs. Although linearization of these equations is possible and useful, the equations that arise are almost always nonlinear, and continuum mechanics has perhaps been the primary driving force behind the development of methods for solving nonlinear PDEs. Other subjects also give rise to PDEs, Maxwell's equations in electromagnetism, for example. But electromagnetism describes the electric and magnetic fields between the atoms, as well as macroscopic fields, and it is not usually considered a branch of continuum mechanics. On the other hand, micromagnetics—the theory of magnetism that describes magnetic domains, the magnetization of an iron bar by an applied field, and the writing of a bit in magnetic recording—is a continuum theory.

The three pillars of continuum mechanics are kinematics, balance laws, and constitutive equations. The central philosophy behind the subject is to separate as much as possible the hypotheses that are satisfied by all materials (or, realistically, large classes of materials) from those hypotheses that pertain to special materials, like elastic solids, viscous fluids, or magnetic materials.

Part of the reason continuum mechanics does not explicitly recognize the presence of atoms is that its main structure was described before the existence of atoms was accepted. The founder of continuum mechanics was Euler, and he conceived its main assumptions in the period 1740–60, half a century before Dalton's vague inferences about atoms. Other early contributors to continuum mechanics were Cauchy and Kirchhoff, and also Hooke, Navier, Poisson, Stokes, Maxwell, Saint-Venant, Kelvin, Gibbs, Duhem, and others. There was a resurgence of interest in the subject in the late 1940s and 1950s, coincident with the rise of materials science and polymer chemistry, as new solids and fluids emerged that were clearly not described at all well by the then-known equations of mechanics. This resurgence was led by Coleman, Ericksen, Noll, Oldroyd, Markovitz, Reiner, Rivlin, Serrin, Toupin, and Truesdell (along with many others) and was also synergistic with the emergence of numerical analysis, scientific computation, and, as noted above, materials science and PDEs. Today, there is a second resurgence, and this is described below.

## 2 Tensor Analysis

The conventional language of continuum mechanics is tensor analysis. To begin, this language concerns vectors. A vector in two or three dimensions is an arrow. Picture an arrow pointing at the sun, based at the center of Stonehenge, whose length is the intensity of light at noon on January 1, 2000. Its physical significance is clear, irrespective of how we describe it mathematically. If you ask four people to describe this arrow in mathematical terms (without communicating with each other), one person might give a list of three numbers $(a, b, c)$. Someone else might write numbers $(d, e, f)$. Each will likely have chosen a different basis. Inspired by the shape of Stonehenge, maybe someone will have used a cylindrical polar coordinate system, with a list $(r, \theta, z)$ denoting the standard polar coordinates of the tip of the arrow with respect to that person's choice of basis. Someone else might simply say $\boldsymbol{v}$. The purpose of tensor analysis is to give precise rules that relate $(a, b, c)$ and its basis to $(d, e, f)$ and its basis. The underlying principle is that everybody describes the same arrow.

In continuum mechanics one often deals with vector fields. These are familiar from weather maps. In fact, continuum mechanics has a lot to say about the patterns of vectors on those maps. To describe them quantitatively one needs a basis erected at each point on the map. Of course, each of these bases could be the same (up to their choice of origin); this case is called the *natural basis of a Cartesian coordinate system*. There are ways of constructing "natural" bases associated with other coordinate systems, i.e., systems of linearly independent vectors, erected at each point, that are parallel to the coordinate curves. Tensor analysis deals in an automatic (though somewhat laborious) way with this case too, giving laws for transforming lists $(a, b, c)$ that now depend on each point in $\mathbb{R}^3$ and a choice of basis at each point to new lists for another field of bases on $\mathbb{R}^3$. Common vector fields in continuum mechanics are position, velocity, acceleration, vorticity, and traction.

Typically, the vector field on the weather map (the velocity field) satisfies some equations of continuum mechanics. In general, the set of equations satisfied by those arrows is exceedingly complicated and actually not fully known. That is because, to construct those vector fields from measurements, there is a tremendous amount of averaging going on, the winds are decidedly turbulent, and the modeling of evaporation and condensation as occurs in clouds is not fully understood. Nevertheless, *if one did know these equations precisely*, no matter how complicated their forms or the methods needed to solve them, they would have the following property that is shared by all equations of continuum mechanics: if the components of the vector field with respect to one coordinate system satisfy these equations, and one changes the basis field, then the form of the equations has to change in just the right way to ensure that the components in the new basis field automatically satisfy the new equations.

As a simple example, the arrows on the weather map may approximately satisfy $\operatorname{div} \boldsymbol{v} = 0$. This is $\partial v_1/\partial x_1 + \partial v_2/\partial x_2 = 0$ expressed in the natural basis of a rectangular Cartesian system. The same vector field (the same arrows!) expressed in the natural (orthonormal) basis of a polar coordinate system, $(v_r(r, \theta), v_\theta(r, \theta))$, then automatically satisfies "the equation $\operatorname{div} \boldsymbol{v} = 0$ in polar coordinates," namely, $\partial(r v_r)/\partial r + \partial v_\theta/\partial\theta = 0$. These two PDEs describe the same property of the arrows on the map.

The same ideas apply to linear transformations. The analogue of an "arrow" for a linear transformation consists of two pictures: a unit *cube* and a *parallelepiped*, together with a *rule* that says which corner of the cube goes to which corner of the parallelepiped. In short, a linear transformation is a *cube-parallelepiped rule*. In cases in which the linear transformation is not invertible, the parallelepiped might degenerate; it might lie in a plane, for example.

We can describe any linear transformation quantitatively by introducing a basis, an orthonormal basis aligned with the cube we have chosen, say. Let us say that our rule says that the origin goes to itself (i.e., there is no translation). The vector associated with the $(1, 0, 0)$ corner of the cube is transformed by our rule to a vector with components, say, $(a, b, c)$. Similarly, say $(0, 1, 0) \to (d, e, f)$ and $(0, 0, 1) \to (g, h, i)$. The linear transformation is then represented by a matrix

$$\boldsymbol{F} = \begin{pmatrix} a & d & g \\ b & e & h \\ c & f & i \end{pmatrix}.$$

Matrix multiplication of $\boldsymbol{F}$ on any vector (expressed in the same basis) gives the components of the vector in the parallelepiped to which it is deformed (in the same basis). Just as for vectors, the important point is that many people will find many different matrices by choosing different bases, aligned with the cube or not aligned with it, but they all have to describe the same cube-parallelepiped rule.

In continuum mechanics, linear transformations are often called tensors. Continuum mechanics has many tensors and, like vector fields, they can depend on position ("tensor fields"). Typical examples are the deformation gradient and stress tensor.

The bottom line? All these components and bases and covariant components and Christoffel symbols at the beginning of a continuum mechanics course can be pretty daunting. Do they matter? Not really. The equations must already have the property of invariance under a change of coordinate system that was illustrated above through the example $\operatorname{div} \boldsymbol{v} = 0$. A vector is an arrow and a tensor is a cube-parallelepiped rule. You can always just use one particular rectangular Cartesian coordinate system, say, and its natural basis field. In that particular system vectors are represented by lists and tensors by matrices. A simpler and more elegant approach is to think of a vector as an arrow and write $\boldsymbol{v}$. Vectors are combined with each other, and with scalars, according to the rules of a vector space. If $\boldsymbol{v}(\boldsymbol{x})$, $\boldsymbol{x} \in \Omega \subset \mathbb{R}^n$, is a vector field, its gradient $\nabla \boldsymbol{v}$ is a tensor field. Every calculation, every computation, can be done using only the abstract rules for derivatives and vector spaces. Some researchers in continuum mechanics (myself, for example) do everyday calculations in this way, never writing a component. In this article we will write the equations of continuum mechanics in both an abstract way and in a rectangular Cartesian coordinate (RCC) system for convenience.

## 3 The Essential Structure of Continuum Mechanics

### 3.1 Kinematics

Kinematics is the geometry of motion. In continuum mechanics, which does not explicitly recognize the presence of atoms, motions of bodies are represented by functions. These are often assumed to be smooth, though some of the most important branches of continuum mechanics involve the study of the singularities of motions. Motions are fundamental to continuum mechanics because they can be studied independently of the material from which the body is made.

There are two ways of describing motions: *Eulerian* and *Lagrangian*. The terminology is standard but inaccurate: Euler introduced the Lagrangian description, while d'Alembert and Daniel Bernoulli introduced that called Eulerian! The Lagrangian description of motion is a natural generalization of the description of the

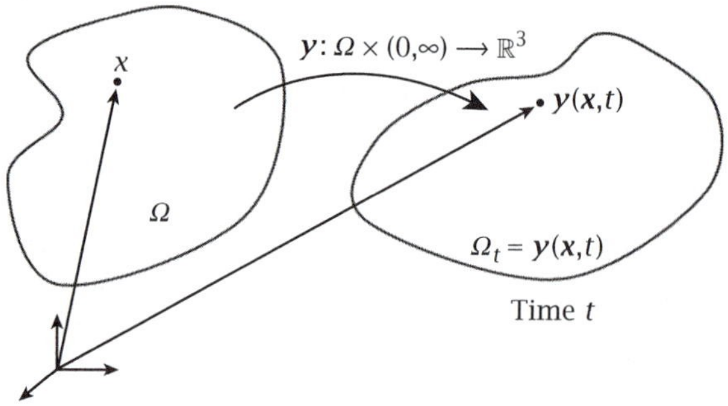


**Figure 1** The Lagrangian description of motion.

motions of individual particles. We name the particles $1, \dots, N$ and describe the motion of each particle as a vector-valued function of time: $\boldsymbol{y}_1(t), \dots, \boldsymbol{y}_N(t)$, $t > 0$, say. The vector $\boldsymbol{y}_1(t^*)$ is the position vector of particle 1 at time $t^*$. Inching closer to continuum mechanics, we could equally well use the notation $\boldsymbol{y}(1,t), \dots, \boldsymbol{y}(N,t)$, $t > 0$, for the same thing, where $\boldsymbol{y}\colon \{1,\dots,N\} \times (0,\infty) \to \mathbb{R}^3$. The Lagrangian description of motion in continuum mechanics allows the set of "particles" to belong to an open subset of $\mathbb{R}^3$, so that the motion is described by $\boldsymbol{y}\colon \Omega \times (0,\infty) \to \mathbb{R}^3$, where $\Omega \subset \mathbb{R}^3$. In this form, $\boldsymbol{y}(\boldsymbol{x},t)$ is the position of the particle $\boldsymbol{x}$ at time $t$. One can think of "the particle $\boldsymbol{x}$" as a small lump of solid or fluid, but more about that later. The choice of $\Omega$ is essentially arbitrary—it just serves as a way to label particles—but people often choose it to be the shape of the body at $t = 0$, i.e., $\boldsymbol{y}(\Omega, 0) = \Omega$. If a motion does not depend on $t$, it is called a *deformation*.

A picture of the Lagrangian description of motion is shown in figure 1. This description is particularly used in solid mechanics, as boundary conditions are usually idealized as the pushing or pulling of certain particles on the boundary.

Assuming that the motion is sufficiently smooth, the velocity is $\dot{\boldsymbol{y}} = \partial \boldsymbol{y}/\partial t$. It has exactly the same interpretation as in particle mechanics: $\dot{\boldsymbol{y}}(\boldsymbol{x},t)$ is the velocity of the particle $\boldsymbol{x}$ at time $t$.

Motions in the Lagrangian description are always assumed to be invertible. That is, the mapping $\boldsymbol{y}(\cdot,t)\colon \Omega \to \mathbb{R}^3$ is invertible at each fixed $t$. The inverse, $\boldsymbol{y}^{-1}(\boldsymbol{y},t)$, $\boldsymbol{y} \in \Omega_t$, is defined on the moving domain $\Omega_t = \boldsymbol{y}(\Omega,t)$. The failure of invertibility, i.e., the possibility that $\boldsymbol{y}(\boldsymbol{x}_1,t) = \boldsymbol{y}(\boldsymbol{x}_2,t)$ for $\boldsymbol{x}_1 \neq \boldsymbol{x}_2$, would be interpreted as the interpenetration of matter.

Now, using invertibility, construct the function

$$\boldsymbol{v}(\boldsymbol{y},t) = \dot{\boldsymbol{y}}(\boldsymbol{y}^{-1}(\boldsymbol{y},t),t).$$

In words, this is the velocity of the particle $\boldsymbol{x} = \boldsymbol{y}^{-1}(\boldsymbol{y}, t)$ at time $t$; that is, it is the velocity of the particular particle that happens to be at the location $\boldsymbol{y}$ at the same time $t$. This is the Eulerian description of motion, and $\boldsymbol{v}(\boldsymbol{y}, t)$ is the Eulerian velocity field. An example of an Eulerian velocity field is the arrows on the weather map, so long as we can reasonably associate a particular time with the map, i.e., so long as the velocities represented by the arrows were measured at the same time. Sometimes one sees weather videos in which the arrows change with time during a day, showing how the winds are changing; this is a direct visualization of the Eulerian velocity field. Above, we calculated the Eulerian description from the Lagrangian. The reverse can be done by noting that, by the definition of $\boldsymbol{v}(\boldsymbol{y}, t)$ given above, the Lagrangian description $\boldsymbol{y}(\boldsymbol{x}, t)$ satisfies the ordinary differential equation

$$\left.\begin{aligned} \dot{\boldsymbol{y}}(\boldsymbol{x}, t) &= \boldsymbol{v}(\boldsymbol{y}(\boldsymbol{x}, t), t), \\ \boldsymbol{y}(\boldsymbol{x}, 0) &= \boldsymbol{x} \in \Omega, \end{aligned}\right\} \tag{1}$$

where we have conveniently chosen $\Omega$ to be the shape of the body at $t = 0$. This is an ordinary differential equation in standard form, with $\boldsymbol{x}$ playing the role of a parameter. An interesting feature of this ordinary differential equation is that, even if the pattern of arrows in three dimensions is quite simple—if, for example, $\boldsymbol{v}$ is a low-order polynomial and has no explicit dependence on $t$—the solution $\boldsymbol{y}(\boldsymbol{x}, t)$ can be exceedingly complicated with highly intertwined orbits. (See the article on DYNAMICAL SYSTEMS [IV.20] for more on this.) People have developed this idea into a theory of the mixing of substances, like fluids or granular materials.

Many critically important kinematical quantities are calculated from the Eulerian or Lagrangian descriptions. We mention a few that appear later in this article. $\boldsymbol{F} = \nabla \boldsymbol{y}$ (or, in RCC, $F_{ij} = \partial y_i / \partial x_j$) is the *deformation gradient*. We can see directly from the abstract formula for the gradient,

$$\begin{aligned} \boldsymbol{y}(\boldsymbol{x} + \varepsilon \boldsymbol{z}, t) - \boldsymbol{y}(\boldsymbol{x}, t) &= \varepsilon \nabla \boldsymbol{y}(\boldsymbol{x}, t)\boldsymbol{z} + o(\varepsilon) \\ &= \varepsilon \boldsymbol{F}\boldsymbol{z} + o(\varepsilon), \end{aligned}$$

that $\boldsymbol{y}(\cdot, t)$ deforms a tiny cube centered at $\boldsymbol{x}$ to a tiny parallelepiped centered at $\boldsymbol{y}(\boldsymbol{x})$, these being scaled versions of the cube-parallelepiped rule associated with the tensor $\boldsymbol{F}$ (scaled by $\varepsilon$). As a local statement of invertibility it is always assumed in continuum mechanics that the parallelepiped is oriented and has positive volume: $\det \boldsymbol{F} > 0$. In short, $\boldsymbol{F}$ describes local deformation. $\boldsymbol{F}$ has the POLAR DECOMPOSITIONS [IV.10 §2] $\boldsymbol{F} = \boldsymbol{R}\boldsymbol{U} = \boldsymbol{V}\boldsymbol{R}$, where $\boldsymbol{R}$ is the *rotation tensor* ($\boldsymbol{R}^{\mathrm{T}}\boldsymbol{R} = \boldsymbol{I}$, $\det \boldsymbol{R} = 1$) and the positive-definite symmetric tensors $\boldsymbol{U}$ and $\boldsymbol{V}$ are the *right* and *left stretch tensors*, respectively. Again using the formula for the gradient above, we can think of the motion (at fixed time) as locally involving first stretching of the tiny cube by $\boldsymbol{U}$ and then rigid rotation by $\boldsymbol{R}$ to achieve the tiny parallelepiped associated with $\boldsymbol{F}$. If the edges of the cube happen to be oriented along the eigenvectors of $\boldsymbol{U}$, then this initial stretching by $\boldsymbol{U}$ produces a rectangular solid rather then a general parallelepiped, which is subsequently rotated by $\boldsymbol{R}$. In continuum mechanics people often associate with $\boldsymbol{F}$ a sphere–ellipsoid rule rather than a cube-parallelepiped rule. The resulting ellipsoid is called a *strain ellipsoid*. Its principal axes are eigenvectors of the left stretch tensor $\boldsymbol{V}$.

From the Eulerian description we get other kinematical quantities more typically used in fluid mechanics. $\boldsymbol{G} = \nabla \boldsymbol{v}$ (or, in RCC, $G_{ij} = \partial v_i / \partial y_j$) is the velocity gradient. Its symmetric part $\boldsymbol{D} = \frac{1}{2}(\boldsymbol{G} + \boldsymbol{G}^{\mathrm{T}})$ (or, in RCC, $D_{ij} = \frac{1}{2}(G_{ij} + G_{ji}) = \frac{1}{2}(\partial v_i / \partial y_j + \partial v_j / \partial y_i))$ is called the *stretching tensor* or the *strain-rate tensor*. It plays a central role in fluid mechanics. These tensors can be given physical interpretations in terms of instantaneous stretching of a small cube in space, as above. The key word here is "instantaneous" because the Eulerian description describes the velocity $\boldsymbol{v}(\boldsymbol{y}, t)$ of a particle located at $\boldsymbol{y}$ at time $t$. A short time later, $t + \delta$, it is a *different particle* at $\boldsymbol{y}$ whose velocity is given by $\boldsymbol{v}(\boldsymbol{y}, t + \delta)$.

### 3.2 Balance Laws

The balance laws of continuum mechanics express the fundamental conservation laws of mass, momentum, and energy. The reason these are so central to continuum mechanics is that they can be stated in ways that are independent of the constitution of the body, just as Newton's law $\boldsymbol{f}_i = m\ddot{\boldsymbol{y}}_i$ holds for a particle $i$ of mass $m$, regardless of whether it models a steel ball or a droplet, or whether the force $\boldsymbol{f}_i$ is produced by water resistance or air resistance.

The balance of mass is straightforward. We introduce a positive mass density $\rho_0 \colon \Omega \to \mathbb{R}$ on $\Omega$ with the interpretation that

$$\int_{\mathcal{D}} \rho_0(\boldsymbol{x})\,\mathrm{d}\boldsymbol{x},$$

$$\text{or, in RCC,} \quad \iiint_{\mathcal{D}} \rho_0(x_1, x_2, x_3)\,\mathrm{d}x_1\,\mathrm{d}x_2\,\mathrm{d}x_3, \tag{2}$$

is the mass of $\mathcal{D}$. Moving something around, or deforming it, even severely, does not change its mass, at least in classical mechanics. Therefore, since $\boldsymbol{y}(\mathcal{D},t)$ consists of the same particles as $\mathcal{D}$, its mass at every time $t$ must also be given by (2). To state this law it is also useful to introduce a mass density $\rho(\boldsymbol{y},t)$, $\boldsymbol{y} \in \Omega_t$, on the deformed region $\Omega_t = \boldsymbol{y}(\Omega,t)$. This mass density changes with time because the material is generally compressed or expanded as it deforms. The balance of mass is the statement that the mass of $\mathcal{D}$ never changes:

$$\int_{\mathcal{D}} \rho_0(\boldsymbol{x})\,\mathrm{d}\boldsymbol{x} = \int_{\boldsymbol{y}(\mathcal{D},t)} \rho(\boldsymbol{y},t)\,\mathrm{d}\boldsymbol{y} \quad \text{for all } \mathcal{D}\subset\Omega,\ t>0.$$

We see that the right-hand side is in exactly the right form to use the motion itself as a change of variables, $\boldsymbol{y}\to\boldsymbol{x}$:

$$\begin{aligned}\int_{\mathcal{D}} \rho_0(\boldsymbol{x})\,\mathrm{d}\boldsymbol{x} &= \int_{\boldsymbol{y}(\mathcal{D},t)} \rho(\boldsymbol{y},t)\,\mathrm{d}\boldsymbol{y}\\ &= \int_{\mathcal{D}} \rho(\boldsymbol{y}(\boldsymbol{x},t),t)J\,\mathrm{d}\boldsymbol{x},\end{aligned} \tag{3}$$

where $J$ is the Jacobian of the transformation $\boldsymbol{y}\to\boldsymbol{x}$; that is, $J = |\det \boldsymbol{F}| = \det \boldsymbol{F}$, where $\boldsymbol{F} = \nabla\boldsymbol{y}$ is the deformation gradient introduced above. Combining the left and right of (3) we get

$$\int_{\mathcal{D}} (\rho(\boldsymbol{y}(\boldsymbol{x},t),t)J - \rho_0(\boldsymbol{x}))\,\mathrm{d}\boldsymbol{x} = 0, \tag{4}$$

which must hold for all domains $\mathcal{D}\subset\Omega$ and all $t>0$, say. Suppose that for each fixed $t$ the integrand of (4) is continuous on the open region $\Omega$. Arguing by contradiction, suppose that the integrand is nonzero, say positive, when evaluated at some particular $\boldsymbol{x}_0\in\Omega$ and $t_0>0$. We fix $t=t_0$ and choose $\mathcal{D}$ to be a ball of radius $r$ centered at $\boldsymbol{x}_0$, with $r$ sufficiently small that this ball is contained in $\Omega$ and that the integrand is positive on this ball. Then, of course, the integral must be positive, contradicting (4). The conclusion is that the integrand must be zero at all $\boldsymbol{x}_0\in\Omega$ and all $t>0$. This is the *local form of the balance of mass*:

$$\rho(\boldsymbol{y}(\boldsymbol{x},t),t)J(\boldsymbol{x},t) = \rho_0(\boldsymbol{x}), \quad \text{or, briefly, } \rho J = \rho_0. \tag{5}$$

The line of argument just presented, which is called *localization*, is common in continuum mechanics. It permits the passage from statements summarizing laws satisfied "by all subregions" to differential equations. The assumption of continuity of the integrand can be considerably relaxed by using Lebesgue's differentiation theorem.

Careful differentiation of the statement in (5) with respect to $t$, using the chain rule and the formula for the differentiation of a determinant, yields

$$\begin{aligned} J\frac{\partial\rho}{\partial t} + J\nabla_{\boldsymbol{y}}\rho\cdot\dot{\boldsymbol{y}} + \rho\frac{\partial \det\nabla\boldsymbol{y}}{\partial t} &= J\frac{\partial\rho}{\partial t} + J\nabla\rho\cdot\dot{\boldsymbol{y}} + \rho J\nabla\boldsymbol{y}^{-\mathrm{T}}\cdot\nabla\dot{\boldsymbol{y}}\\ &= 0.\end{aligned} \tag{6}$$

If we now differentiate the fundamental relation (1) between the Eulerian and Lagrangian descriptions with respect to $\boldsymbol{x}$, we see that the term $\rho J\nabla\boldsymbol{y}^{-\mathrm{T}}\cdot\nabla\dot{\boldsymbol{y}}$ in (6) can be simplified to $\rho J \operatorname{div}_{\boldsymbol{y}}\boldsymbol{v}$. Dividing the result by $J>0$ and replacing $\boldsymbol{x}$ by $\boldsymbol{y}^{-1}(\boldsymbol{y},t)$ everywhere, we get the *Eulerian form of the balance of mass*:

$$\frac{\partial\rho}{\partial t} + \nabla\rho\cdot\boldsymbol{v} + \rho\operatorname{div}\boldsymbol{v} = \frac{\partial\rho}{\partial t} + \operatorname{div}(\rho\boldsymbol{v}) = 0,$$
$$\text{or, in RCC, } \frac{\partial\rho}{\partial t} + \sum_{i=1}^{3}\frac{\partial(\rho v_i)}{\partial y_i} = 0. \tag{7}$$

The independent variables in (7) are $\boldsymbol{y}$ and $t$ and (7) holds on $\Omega_t\times(0,\infty)$. One can see that it is critically important in continuum mechanics to keep track of independent and dependent variables and domains of functions.

The condition $\operatorname{div}\boldsymbol{v} = 0$ mentioned earlier in the context of weather maps can then be seen as a consequence of the balance of mass in the Eulerian description together with the assumption that the density $\rho$ is a constant. Materials for which all motions have constant density are called *incompressible materials*.

We will not repeat the arguments above for other laws, but they follow a similar pattern of introducing "densities," using the motion as a change of variables, localization, and passing between Eulerian and Lagrangian descriptions. For example, in the Lagrangian and Eulerian forms the *balance of linear momentum* is respectively given by

$$\begin{aligned}\frac{\mathrm{d}}{\mathrm{d}t}\int_{\mathcal{P}} \rho_0\frac{\partial\boldsymbol{y}}{\partial t}\,\mathrm{d}\boldsymbol{x} &= \text{force on } \boldsymbol{y}(\mathcal{P},t),\\ \frac{\mathrm{d}}{\mathrm{d}t}\int_{\boldsymbol{y}(\mathcal{P},t)} \rho\frac{\partial\boldsymbol{v}}{\partial t}\,\mathrm{d}\boldsymbol{y} &= \text{force on } \boldsymbol{y}(\mathcal{P},t).\end{aligned} \tag{8}$$

In these two cases there are various useful expressions for the "force on $\boldsymbol{y}(\mathcal{P},t)$." We will focus on the form typically used in the Eulerian description, and we will omit so-called body forces:

$$\text{force on } \boldsymbol{y}(\mathcal{P},t) = \int_{\partial\boldsymbol{y}(\mathcal{P},t)} \boldsymbol{t}\,\mathrm{d}A. \tag{9}$$

The integrand $\boldsymbol{t}$ is called the *traction*. From this formula one can see that it represents the force per unit area on the boundary of the region $\boldsymbol{y}(\mathcal{P},t)$.

In principle, the traction could depend on lots of things. Certainly, it depends on the material and the motion. It is likely to be generally different at different points in $\Omega_t$. One could also imagine that it depends on the surface $S = \partial \boldsymbol{y}(\mathcal{P}, t)$ in some complicated way. Fix $t$ and fix a point $\boldsymbol{y} \in \Omega_t$. Imagine lots of surfaces passing through $\boldsymbol{y}$. How are the tractions at $\boldsymbol{y}$ on these different surfaces related?

It was a brilliant insight of Cauchy to see that for a broad range of materials, both fluids and solids, the traction at $\boldsymbol{y}$ has a particularly simple dependence on the surface passing through $\boldsymbol{y}$. Cauchy's starting point was a plausible expression for $\boldsymbol{t}$ with a rather general dependence on $S$, from which he deduced, using directly the balance of linear momentum in Eulerian form, that the dependence of the traction on $S$ is through the unit normal only and that this dependence is linear: $\boldsymbol{t} = \boldsymbol{T}\boldsymbol{n}$ (or, in RCC, $t_i = \sum_j T_{ij} n_j$). The tensor $\boldsymbol{T}$ is called the *stress tensor*. This part of continuum mechanics is called the *theory of stress*. (See MECHANICS OF SOLIDS [IV.32 §2.3] for a helpful physical description of stress.) People do sometimes worry about Cauchy's starting point, i.e., whether some kind of exotic material might transmit forces across a surface in a more general way than by having just a linear dependence of the traction on the normal to the surface. But, generally, Cauchy's conclusion has been found to be widely applicable.

If we insert $\boldsymbol{t} = \boldsymbol{T}\boldsymbol{n}$ into (9), then use the divergence theorem in Eulerian form in (8), and finally localize, we get the local form of the balance of linear momentum:

$$\rho\left(\frac{\partial \boldsymbol{v}}{\partial t} + (\nabla \boldsymbol{v})\boldsymbol{v}\right) = \operatorname{div} \boldsymbol{T},$$

$$\text{or, in RCC, } \rho\left(\frac{\partial v_i}{\partial t} + \sum_{j=1}^{3} \frac{\partial v_i}{\partial y_j} v_j\right) = \sum_{j=1}^{3} \frac{\partial T_{ij}}{\partial y_j}. \quad (10)$$

There are two more conservation laws: the balance of rotational momentum and the balance of energy. The former leads to the symmetry of the stress tensor, $\boldsymbol{T} = \boldsymbol{T}^{\mathrm{T}}$ (or, in RCC, $T_{ij} = T_{ji}$), and the latter is, in its simplest local form,

$$\rho\left(\frac{\partial e}{\partial t} + \nabla e \cdot \boldsymbol{v}\right) = -\operatorname{div} \boldsymbol{q} + \operatorname{tr}(\boldsymbol{T}\boldsymbol{D}),$$

where $e$ is the *internal energy* and $\boldsymbol{q}$ is the heat flux.

Finally, there is a formulation of the second law of thermodynamics in continuum mechanics. It is most commonly represented by the *Clausius–Duhem inequality*. This inequality embodies in some way the fundamental statements of the second law, such as *it is impossible for a body to undergo a cyclic process that does work but emits no heat*, which goes back to Carnot. In continuum mechanics the Clausius–Duhem inequality plays two important roles. The first is restricting constitutive relations so that they do not allow behavior as indicated by the italicized statement above; for instance, the restricted relations do not permit the existence of a cyclic energy conversion device that produces electricity while completely immersed in a container of hot water. Needless to say, the precise form of these restrictions is exceedingly important these days. The other is as a restriction on processes for given constitutive relations. The latter is best represented by the theory of shock waves, where the Clausius–Duhem inequality declares some shock wave solutions to be "inadmissible."

### 3.3 Constitutive Equations

Everything we have said so far holds for broad classes of materials, both solids and fluids. Since each new equation we have introduced has introduced at least one new unknown function, none of the equations we have written could be used to predict anything. There have to be special equations that quantify the behavior of special classes of materials. These are called *constitutive equations*. Discovering a simple constitutive equation that, when combined with the balance laws, provides a good description of a material in a broad class of motions is a big success in continuum mechanics.

Often, constitutive equations take the form of a formula that relates the stress to the motion. Usually, this formula contains some constants that actually specify the material. These *material constants* have to be measured experimentally for each particular material described by the constitutive equation, and continuum mechanics has a lot to say about the design of these experiments. An example of a constitutive relation is that for the *Navier–Stokes fluid*, which is defined by

$$\boldsymbol{T} = -p\boldsymbol{I} + 2\mu(\boldsymbol{D} - \tfrac{1}{3}(\operatorname{tr}\boldsymbol{D})\boldsymbol{I}),$$

$$\text{or, in RCC, } T_{ij} = -p\delta_{ij} + 2\mu\left(D_{ij} - \frac{1}{3}\left(\sum_k D_{kk}\right)\delta_{ij}\right), \quad (11)$$

where $\boldsymbol{D}$ is the stretching tensor introduced above, $p$ is the pressure, and $\mu > 0$ is the viscosity. An important related constitutive relation is the one for the *incompressible Navier–Stokes fluid*, which is defined by

$$\boldsymbol{T} = -p\boldsymbol{I} + 2\mu\boldsymbol{D}. \quad (12)$$

The absence of the last term of (11) in (12) is due to the condition of incompressibility, which, as noted above, gives $\operatorname{div} \boldsymbol{v} = \operatorname{tr} \boldsymbol{D} = 0$. But also, $p$ means something different in (11) and (12). In the former it is usually specified as a function of the density—given in the simplest case by the ideal gas law—while in the latter it is treated as an independent function, unrelated to the motion. In equations of motion for an incompressible Navier-Stokes fluid, $p(\boldsymbol{y})$ becomes one of the unknowns, to be determined as part of the solution of a problem. This treatment of $p$ is a consequence of a general theory of *constrained materials* in continuum mechanics that includes incompressible materials but also treats other kinds of constraints, such as a material being inextensible in a certain direction.

If we substitute (12) into the balance of linear momentum (10), we get the famous Navier-Stokes equations: in RCC,

$$\rho\left(\frac{\partial v_i}{\partial t} + \sum_{j=1}^{3} \frac{\partial v_i}{\partial y_j} v_j\right) = -\frac{\partial p}{\partial y_i} + \mu \sum_{j=1} \frac{\partial^2 v_i}{\partial y_j \partial y_j},$$

$$\sum_j \frac{\partial v_j}{\partial y_j} = 0.$$

These equations represent one of the profound successes of continuum mechanics. Typically, one nondimensionalizes the first equation by scaling space by $\ell$ and time by $\tau$ (these being a length and a time arising in a problem of interest) and dividing by the (constant) density $\rho$, so that the resulting equation contains a single nondimensional material constant, $Re = \rho(\ell/T)\ell/\mu$, the Reynolds number. With only one material constant, $Re$, an enormous body of quantitative observation on the behavior of liquids (and even gases under some circumstances) can be understood with remarkable precision.

This does not mean that any fluid behaves exactly as predicted by the Navier-Stokes theory in all circumstances. Any fluid, if compressed enough, will become a solid or, if subjected to a sufficiently strong electric field, a plasma. Electrons can be ripped off nuclei and nuclei can be split, none of which is described by the Navier-Stokes theory. What is prized in continuum mechanics is not the reductionist's quest for "truth" but the elegance that comes with discovering an underlying simplicity that reveals many phenomena.

Alas, solids are not so simple, but much behavior can be understood from a constitutive equation of the form

$$\boldsymbol{T} = \hat{\boldsymbol{T}}(\boldsymbol{F}) = \boldsymbol{R}\hat{\boldsymbol{T}}(\boldsymbol{U})\boldsymbol{R}^{\mathrm{T}}, \tag{13}$$

where $\boldsymbol{F} = \boldsymbol{RU}$ is the polar decomposition discussed above. This constitutive equation describes a (nonlinear) elastic material. $\hat{\boldsymbol{T}}(\boldsymbol{U})$, a symmetric tensor-valued function of a symmetric tensor, can be pretty complicated; in RCC, $\hat{\boldsymbol{T}}(\boldsymbol{U})$ is six functions, each of six variables. On the other hand, this constitutive relation covers an enormous range of behavior that would intuitively be considered "elastic"—as well as, incidentally, some interesting behavior that would not be considered elastic.

There are lots of ideas that are used to simplify the constitutive relation (13) of elasticity. One of the most powerful is symmetry. If one deforms a ball $\mathcal{B}$ of rubber by a deformation $\boldsymbol{y}\colon \mathcal{B} \to \mathbb{R}^3$, one gets a certain stress field. If we know the constitutive relation, we get this stress field by using the formula $\boldsymbol{T}(\boldsymbol{x}) = \boldsymbol{R}(\boldsymbol{x})\tilde{\boldsymbol{T}}(\boldsymbol{U}(\boldsymbol{x}))\boldsymbol{R}(\boldsymbol{x})^{\mathrm{T}}$, where $\nabla\boldsymbol{y} = \boldsymbol{R}(\boldsymbol{x})\boldsymbol{U}(\boldsymbol{x})$ is the polar decomposition at each $\boldsymbol{x} \in \mathcal{B}$. If we take this ball of rubber, rotate it rigidly in any way, and, after doing so, place it exactly in the region $\mathcal{B}$, and deform it again using $\boldsymbol{y}\colon \mathcal{B} \to \mathbb{R}^3$, we will in general get a different stress field. For example, if $\boldsymbol{y}\colon \mathcal{B} \to \mathbb{R}^3$ primarily describes an extension in a certain direction, say "up," and we happen to rotate it so that a stiff direction of the rubber ball is oriented up, then we expect to have to exert larger forces to give it exactly this same deformation. In fact, rubber and many other materials are often well described by the assumption of *isotropy*. This means that if one undertakes the experiment described here, one gets exactly the same stress field, and this condition holds regardless of the rotation or subsequent deformation. The assumption of isotropy is exploited by phrasing all the steps in this paragraph in mathematical terms. The result is as follows: a nonlinear elastic material is isotropic if there are three functions $\varphi_1(\mathrm{I}, \mathrm{II}, \mathrm{III})$, $\varphi_2(\mathrm{I}, \mathrm{II}, \mathrm{III})$, $\varphi_3(\mathrm{I}, \mathrm{II}, \mathrm{III})$ such that

$$\boldsymbol{T} = \varphi_1\boldsymbol{I} + \varphi_2\boldsymbol{B} + \varphi_3\boldsymbol{B}^2, \tag{14}$$

where $\boldsymbol{B} = \boldsymbol{V}^2 = \boldsymbol{FF}^{\mathrm{T}}$, and where $\mathrm{I} = \operatorname{tr}\boldsymbol{B}$, $\mathrm{II} = \frac{1}{2}(\operatorname{tr}\boldsymbol{B}^2 - (\operatorname{tr}\boldsymbol{B})^2)$, and $\mathrm{III} = \det\boldsymbol{B}$ are the principal invariants of $\boldsymbol{B}$. At first, it may look like (14) is not a special case of (13), but it is; notice that by using the polar decomposition, $\boldsymbol{B} = \boldsymbol{FF}^{\mathrm{T}} = \boldsymbol{RU}^2\boldsymbol{R}^{\mathrm{T}}$. The form of (14) might suggest that some kind of Taylor expansion is involved, but this is not the case. The constitutive equation (14) holds for arbitrarily large deformations of isotropic elastic materials.

There is a set of general principles in continuum mechanics that are used to simplify constitutive relations. Perhaps the most powerful, and controversial, of these is the principle of material frame indifference (PMFI). We will not explain this principle in detail, but we will note that the particular kinematics used in the constitutive equations given here (notice the appearance of $\boldsymbol{D}$ in the Navier–Stokes relation, rather than simply $\nabla \boldsymbol{v}$, and the unexpected explicit dependence on $\boldsymbol{R}$ on the right-hand side of the constitutive relation (13) of elasticity) are a direct consequence of the PMFI. Even Stokes, when deriving the Navier–Stokes equations (building on work of Navier, Cauchy, Poisson, and Saint-Venant), did not begin with (12). Rather, he began with the hypothesis that the stress is affine in the full velocity gradient $\nabla \boldsymbol{v}$ rather than in its symmetrization $\boldsymbol{D}$. He then argued in words (using what can be recognized as a verbal, and less precise, form of the PMFI than is now accepted) that the stress should actually be affine in $\boldsymbol{D} = \frac{1}{2}(\nabla \boldsymbol{v} + \nabla \boldsymbol{v}^{\mathrm{T}})$. The ongoing controversy surrounding the PMFI is not so much concerned with its usefulness in continuum mechanics, which is nearly universally accepted, but rather with the absence of a direct experimental test of its validity (in the general case) and its obscure relation with atomistic and relativistic theories.

## 4 Phenomena

Even when it is restricted to the two constitutive relations given above, continuum mechanics explains a diverse collection of phenomena. It often seems to have applicability to materials on length or timescales that is quite unexpected based on our current understanding.

It is hopeless to try to give a representative glimpse of phenomena predicted by continuum mechanics. I will instead give two examples of phenomena that have been predicted as a direct result of research in the fundamentals. They are old examples, having emerged in the 1950s during the resurgence of interest in continuum mechanics that occurred in that period, but they retain their vibrancy today.

Under ordinary conditions water is described well as an incompressible Navier–Stokes fluid. Take a cup of water and vertically insert a rotating rod (of, say, 1 cm in diameter) spinning at a modest rotation rate (a few revolutions per second, say). By making some symmetry assumptions and undertaking some modest simplifications, this problem can be solved. Even without symmetry assumptions this problem can be solved to a quantifiable level of accuracy by any of several numerical methods that have been developed for the Navier–Stokes equations. The answer is the expected one. As

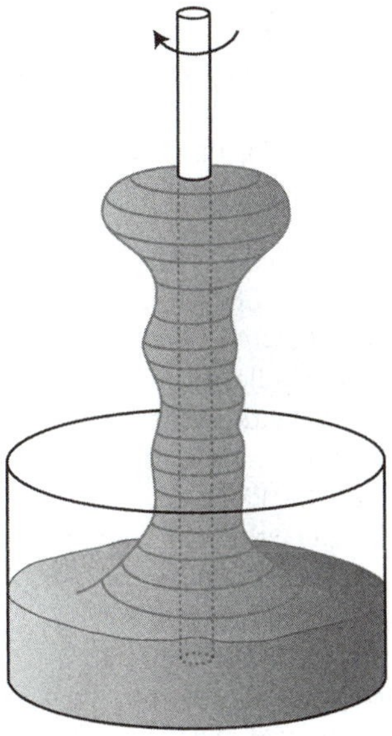

**Figure 2** The rod climbing of a viscoelastic fluid.

the water near the rod rotates, it is thrown outward. This causes the surface of the water to distort, resisted by the force of gravity, which tends to make the surface flat, and by viscosity, which tends to smooth the velocity field. At steady state the water level is slightly depressed near the spinning rod.

Entirely different behavior is observed when the fluid is *viscoelastic*. This subset of fluids includes many that have long-chain polymer molecules in solution, from paints to pancake batter. When one does the same experiment described above with these fluids, the fluid in fact climbs up the rod, as first demonstrated by Karl Weissenberg during what must have been a memorable meeting of the British Rheologists' Club in 1946; see figure 2, which is a sketch based on one of Weissenberg's early demonstrations. With a small cup of liquid and a strongly viscoelastic liquid, more than half the cup of liquid can climb up the rod after a short time, defying the force of gravity.[1]

There are many related examples; a solid cylinder falling in a container of Navier–Stokes fluid will turn so its axis becomes horizontal. In a viscoelastic fluid it turns so its axis is vertical!

What force pushes the fluid up the rod, against gravity? This question puzzled continuum mechanicians, especially Markus Reiner, the founder of the science

1. Commenting on a draft of this article, Oliver Penrose asked, "Is that why paints are so messy?"

of rheology, and Ronald Rivlin. Their first hypothesis was quite natural; guided by the developing principles of continuum mechanics, particularly the then-available form of PMFI, they theorized that a natural generalization of the Navier–Stokes fluid,

$$T = -pI + \alpha_1 D + \alpha_2 D^2 \tag{15}$$

with $\alpha_1$, $\alpha_2$ functions of the principal invariants of the stretching tensor $D$, would work for viscoelastic fluids. In fact, the formal similarity between (14) and (15) is no accident: the same mathematics is used, but the physical principles are different. For the former it is isotropy; for the latter it is the PMFI.

It cannot be overestimated how natural (15) is. Not only is it an obvious generalization of the Navier–Stokes fluid, by including nonlinear terms in $D$, but it is in fact the most general form of the relation $T = f(D)$ that is compatible with PMFI. Its only deficiency is that it turned out to be wrong! The Reiner–Rivlin relation does not describe Weissenberg's observations well. It turned out that the stress at time $t$ in a viscoelastic fluid is sensitive to its deformation at past times, longer ago than is captured by the first time derivative in $D$. The mathematical formulation of this idea was developed by many researchers, and there were many twists and turns along the way, including the observation by Pipkin and Tanner in 1969 that the standard method for interpreting measurements of traction on the boundary of the fluid was flawed, polluted by just those forces that drive the viscoelastic fluid up the rod. The result was that lots of measurements before that point were incorrectly interpreted. Now, of course, there are better constitutive relations, and the description of the "normal stresses" that drive the fluid up the rod are pretty well understood, but the accurate description of the behavior of viscoelastic fluids remains an active area of research today.

Another simple but influential example concerns the constitutive equation of isotropic elasticity (14). Consider the following deformation in RCC:

$$\left.\begin{aligned} y_1(x_1,x_2,x_3) &= x_1 + \kappa x_2,\\ y_2(x_1,x_2,x_3) &= x_2,\\ y_3(x_1,x_2,x_3) &= x_3. \end{aligned}\right\} \tag{16}$$

This deformation is known as simple shear, and it is represented in figure 3 using two different reference configurations. One can think of the shearing of a rubber block $\Omega$ aligned with this RCC basis, as shown in figure 3(a). The constant $\kappa$ is called the *amount of shear*. The components of the tensor $B$ in this same RCC basis

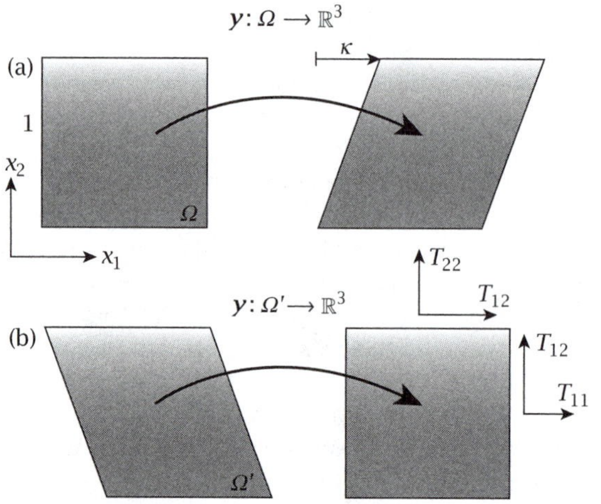


**Figure 3** Simple shear of a nonlinear elastic solid. The same deformation is applied to two different reference configurations, $\Omega$ and $\Omega'$, both of which represent relaxed configurations of a certain solid. The deformed configurations have been translated so as to be easily visible. Typically, $T_{11} < 0$ and $T_{22} < 0$ in real materials, in which case these components of traction are compressive.

are

$$B = FF^{\mathrm{T}} = \begin{pmatrix} 1 & \kappa & 0\\ 0 & 1 & 0\\ 0 & 0 & 1 \end{pmatrix} \begin{pmatrix} 1 & 0 & 0\\ \kappa & 1 & 0\\ 0 & 0 & 1 \end{pmatrix} = \begin{pmatrix} 1+\kappa^2 & \kappa & 0\\ \kappa & 1 & 0\\ 0 & 0 & 1 \end{pmatrix},$$

and the principal invariants I, II, III are particular functions of the amount of shear that are easily worked out. The coefficients $\varphi_1$, $\varphi_2$, $\varphi_3$ are then functions of $\kappa$ whose form depends on the material. Using the constitutive equation (14), we can calculate the stress:

$$T = \begin{pmatrix} T_{11} & T_{12} & 0\\ T_{21} & T_{22} & 0\\ 0 & 0 & T_{33} \end{pmatrix}, \tag{17}$$

where

$$\left.\begin{aligned} T_{11} &= \varphi_1 + (1+\kappa^2)\varphi_2 + (\kappa^2 + (1+\kappa^2)^2)\varphi_3,\\ T_{12} = T_{21} &= \kappa\varphi_2 + (2\kappa + \kappa^3)\varphi_3,\\ T_{22} &= \varphi_1 + \varphi_2 + (1+\kappa^2)\varphi_3,\\ T_{33} &= \varphi_1 + \varphi_2 + \varphi_3. \end{aligned}\right\} \tag{18}$$

The balance of linear momentum (10) is satisfied for the motion (16) because the velocity is zero and, since the stress is independent of position, $\operatorname{div} T = 0$.

A little study of (18) reveals that the stress components $T_{11}$, $T_{22}$, and $T_{12}$ are not independent. In fact, by direct observation of (18),

$$T_{11} - T_{22} = \kappa T_{12}, \qquad (19)$$

which is a *universal relation* discovered by Rivlin (1948). It is called universal because it holds for all isotropic elastic materials. It contains no material constants. It is not obvious.

It is illuminating to interpret the relation $T_{11} - T_{22} = \kappa T_{12}$. For this purpose we use figure 3(b) and focus on the choice of reference configuration $\Omega'$, arranged so that the deformed configuration is a rectangular solid. (The stress in the deformed configurations $\boldsymbol{y}(\Omega)$ and $\boldsymbol{y}(\Omega')$ is the same, but it is easier to understand this stress by calculating tractions on $\boldsymbol{y}(\Omega')$.) The components of traction (force per unit area) needed to hold the rubber block in the shape $\boldsymbol{y}(\Omega')$ are as shown. These are obtained from the general formula discussed above: $\boldsymbol{t} = \boldsymbol{T}\boldsymbol{n}$, with $\boldsymbol{n}$ chosen to be $(1,0,0)$ and $(0,1,0)$. On the top surface there is a shear component of traction $T_{12}$, as expected. Perhaps not so obvious is the fact that one generally needs a normal traction $T_{22}$. $T_{22}$ is usually negative, and, at large shears, it can be quite significant. Shearing a rectangular block of rubber horizontally typically causes it to expand in the vertical direction; only by applying an appropriate compressive traction does the height remain the same. These results would have been surprising at the time they were derived, since only the linear theory of elasticity was widely used at that time, and the linear theory predicts that $T_{22} = 0$ for an isotropic material. Rivlin's relation (19) says, quite unexpectedly, that the difference between the normal tractions on the right and top faces is determined by the shear traction and the amount of shear, *and it is independent of the (isotropic elastic) material being sheared.* It also shows that at least one of these normal stresses, $T_{11}$ or $T_{22}$, is quite large, of the order of the shear stress $T_{12}$ at $\kappa \approx 1$.

## 5 Current Research

Continuum mechanics is entering a new period of activity revolving around the main feature of matter that it suppresses: atoms. This development is due to the convergence of several factors.

One is that, following the successes of viscoelastic fluids, nonlinear elastic materials, the theory of liquid crystals, and several other similarly important examples, the discovery of broadly useful new constitutive relations has slowed. It is not that continuum mechanics has in any way lost its validity or applicability (the workhorse constitutive equations of continuum mechanics continue to find new and exciting applications) but the underlying theory has been less suggestive of truly new directions.

It is certainly true that materials science is producing a dizzying array of new materials, with properties that are not even named in any treatment of continuum mechanics. Biology, too, is identifying new, often highly heterogeneous materials, the critical properties of which are not as yet described within continuum mechanics. Increasingly, what matters in these subjects is the presence of certain atoms, arranged on a lattice in a particular way, or a specific biological molecule. Alternatively, it could be a specific kind of defect in an otherwise regular structure that produces the interesting behavior. The biologist says, "I do not care so much how a generic lump of soft matter deforms; I want to know how matter containing this particular molecule, critical to life itself, behaves." This attitude certainly turns the Navier–Stokes paradigm of "one Reynolds number needed to predict all behavior in all motions" on its head!

In 1929, following the spectacular discovery of quantum mechanics, Dirac wrote:

> The underlying physical laws necessary for the mathematical theory of a large part of physics and the whole of chemistry are thus completely known, and the difficulty is only that the exact application of these laws leads to equations much too complicated to be soluble.

Perhaps this was a bit optimistic, but it remains true today that the laws needed at atomic scale to entirely predict macroscopic behavior that should be under the purview of continuum mechanics are known, and it is essentially a mathematical problem to figure out what are the macroscopic implications of atomic theory, with all its wonderful specificity. This problem is called the *multiscale problem.* Today, it is widely theorized that the solution of this problem will involve the identification of a certain number of length or timescales, with separate theories on each scale and input to each theory coming from the output of the one at the next lowest scale. The author is skeptical.

The mathematical difficulties are easy to explain with examples. Consider just a single atom of carbon, atomic number 6, held at absolute zero temperature. To compute its electronic structure with quantum mechanics, one needs to solve an equation for the wave function $\psi((\boldsymbol{x}_1,s_1),\dots,(\boldsymbol{x}_6,s_6))$, $s_i = \pm\frac{1}{2}$ (the spins), describing the positions of the electrons probabilistically, and depending on $3\times 6 = 18$ independent spatial variables. If we modestly discretize each such variable by ten grid points, we obtain a mesh with $10^{18}$ grid points! Of course, a single atom of carbon is open to the methods of a simplified version of quantum mechanics called density functional theory (DFT) as well as other methods, and DFT would be considered accurate in this case for some purposes. But DFT has its own problems. Its status as an approximation of quantum mechanics is not understood beyond heuristics. And even for twenty carbon atoms, disturbingly, DFT does not get the most favorable geometries correct.

These observations may make it seem hopeless to try to pass from full quantum mechanics to some kind of useable version of quantum mechanics for many atoms, never mind passing from quantum mechanics to continuum mechanics. But there are other suggestions of deep underlying simplicities in quantum mechanics that are neither exploited nor understood. One example is the so-called electron-to-atom (e/a) ratio. The importance of e/a was recognized long ago by Hume-Rothery, and it was exploited in a particularly effective way in magnetism by Slater and Pauling. For an alloy such as $Cu_xMn_ySn_z$ consisting of $x$ atoms of copper, $y$ atoms of manganese, and $z$ atoms of tin, the e/a ratio is simply the valences of Cu, Mn, and Sn weighted by their atomic fractions:

$$\frac{e}{a} = \frac{x\,V_{Cu} + y\,V_{Mn} + z\,V_{Sn}}{x+y+z},$$

where $V_{Cu}$, $V_{Mn}$, $V_{Sn}$ are, respectively, the numbers of valance electrons of Cu, Mn, and Sn. When macroscopic properties of materials are plotted against e/a, there is often a remarkable collapse of data. Of course, e/a is just one of the parameters that enters a quantum mechanics calculation under the *Born–Oppenheimer approximation*. Under this assumption the inputs to the quantum mechanics calculation are the positions and atomic numbers of the nuclei. Correlation with e/a means that somehow the positions hardly matter! The e/a ratio is often most successful in cases where the underlying lattices are somewhat similar as the concentrations $x$, $y$, and $z$ are varied. And it must be admitted that the definition of valence itself is the result of a (single-atom) quantum mechanics calculation. Nevertheless, changing $x$, $y$, and $z$ is a change of order 1: some neighbors of some atoms change from one element to another. And the correlation often persists if new elements are introduced and the chemical formula gets very long. The properties that correlate with e/a are some of the most difficult positive-temperature properties to predict, like magnetization, or free energy difference between two phases. As a modern example, the Heusler family of alloys is currently perhaps the most fertile area in materials science for discovery of new alloys with applications in diverse areas, such as microelectronics (especially "spintronics"), information storage, biomedicine, actuation, refrigeration, energy conversion, and energy storage. Most of the discovery of new alloys for these applications is guided simply by e/a; it is the main theoretical tool. Why is e/a so important? What is the e/a dependence of a constitutive equation of continuum mechanics?

A fundamental conceptual problem for multiscale methods concerns time dependence. A standard approach at atomic level is to use the equations of molecular dynamics, based on Newton's laws of motion for the nuclei, at positions $\boldsymbol{y}_1(t),\dots,\boldsymbol{y}_n(t)$: that is,

$$\left.\begin{aligned} m_i\frac{\mathrm{d}^2\boldsymbol{y}_i}{\mathrm{d}t^2} &= \boldsymbol{f}_i(\boldsymbol{y}_1,\dots,\boldsymbol{y}_n),\\ \boldsymbol{y}_i(0) &= \boldsymbol{y}_i^0,\\ \frac{\mathrm{d}\boldsymbol{y}_i}{\mathrm{d}t}(0) &= \boldsymbol{v}_i^0,\quad i=1,\dots,n. \end{aligned}\right\} \tag{20}$$

The constants $m_1, m_2, \dots, m_n$ are the corresponding masses. The force $\boldsymbol{f}_i$ on nucleus $i$ depends on the positions of all the other nuclei, as reflected by the notation. This force could be given by quantum mechanics for all the electrons as described above, parametrized using the instantaneous nuclear positions. This is again the Born–Oppenheimer approximation. Of course, we would then have to do the very difficult quantum mechanical calculations described above at each time step, so in this case it would be essential to find simplifications. One such simplification would be to find accurate but simpler models of atomic forces. In any case, molecular dynamics is considered a rather general framework underlying continuum theories of many materials.

A fundamental dilemma is that, regardless of the atomic forces, the equations of molecular dynamics are time reversible, while every accepted sufficiently general model in continuum mechanics is time irreversible. That is, if we define $\bar{\boldsymbol{y}}_i(t) = \boldsymbol{y}_i(-t)$ and change the sign of $\boldsymbol{v}_0$, we see that $\bar{\boldsymbol{y}}_1(t),\dots,\bar{\boldsymbol{y}}_n(t)$ solves (20). If we do the analogous change in, say, the Navier–Stokes equations, i.e., if we begin with a solution $\boldsymbol{v}(\boldsymbol{y},t), p(\boldsymbol{y},t)$

and define $\bar{v}(y,t) = -v(y,-t)$, $\bar{p}(y,t) = p(y,-t)$, we see that $\bar{v}(y,t), \bar{p}(y,t)$ satisfies the Navier–Stokes equations if and only if the term multiplying the viscosity is zero. That is, the form of solution is unaffected by the viscosity of the material, a degenerate case indeed.

There is no widely accepted solution to this dilemma, though many ideas have been suggested. From a mathematical viewpoint, it is certainly difficult to imagine how any kind of rigorous averaging of the equations of molecular dynamics would somehow deliver a time-irreversible equation from a time-reversible one. And this is not just a quirk of molecular dynamics: every fundamental atomic-level equation of physics is dissipation free and time reversible. A suggestion made by Boltzmann that might in fact be consistent with a mathematical treatment has been elaborated in a recent article by Oliver Penrose. He says that flows seen in nature do not correspond to solutions of the general initial-value problem (20). Rather, any real flow corresponds to special "prepared" initial data of the equations of molecular dynamics. Who does this preparation in nature? Penrose suggests that merely running a dynamical system for some time could do this kind of preparation. (Admittedly, time does not begin at $t = 0$ in (20).) He discusses a method of weighted averaging, involving the initial conditions, that does introduce irreversibility. Penrose's suggestion is appealing and will probably resonate with anyone who has done molecular dynamics simulations. With almost any way of choosing initial data short of running it through the dynamical system, solutions of the equations of molecular dynamics invariably begin with a transient that would be regarded as unphysical from a macroscopic point of view.

Balancing these fundamental difficulties, there seems to be the possibility of tremendous simplification in some cases. Let us illustrate how easy it is to connect molecular dynamics with continuum theory by presenting a very simple way to average the equations (20) of molecular dynamics. We use a method of R. J. Hardy, which has been recently analyzed (and extended), together with related approaches by Admal and Tadmor. Let $\varphi\colon \mathbb{R}^3 \to \mathbb{R}$ be a simple averaging function; $\varphi$ is smooth, nonnegative, has compact support containing the origin, and has total integral equal to 1. Consider a solution $y_1(t), \dots, y_n(t)$, $t > 0$, of the equations of molecular dynamics (20). Recenter $\varphi$ on each instantaneous atomic position, multiply the equations of molecular dynamics by $\varphi$, and sum over the atoms:

$$\sum_{i=1}^{n} m_i \ddot{y}_i(t)\varphi(y - y_i(t)) = \sum_{i=1}^{n} f_i(y_1(t), \dots, y_n(t))\varphi(y - y_i(t)).$$

With this kind of spatial averaging it is natural to define the density as $\rho(y,t) = \sum m_i \varphi(y - y_i(t)) \geqslant 0$. We can also define the linear momentum $p(y,t)$ by averaging the linear momenta of the particles:

$$p(y,t) = \sum_{i=1}^{n} m_i \dot{y}_i(t)\varphi(y - y_i(t)). \tag{21}$$

The Eulerian velocity is then defined wherever $\rho > 0$ by $v(y,t) = p(y,t)/\rho(y,t)$. Notice that we have already avoided the tricky problem of averaging a product by simply defining it away. That is, for the purposes of averaging, the fundamental quantity is the momentum. If, instead of the above, we had defined the density and velocity first, in the obvious ways, then we would have had the nasty problem of trying to express the average of a product in terms of the product of averages in order to get the momentum. Based on this method of averaging, the velocity of continuum mechanics is not the average velocity of the particles (as is suggested above and in most continuum mechanics books) but rather it is the average momentum divided by the average density, which can be something quite different! On the other hand, we do get a balance of mass exactly as in continuum mechanics for free because by these definitions

$$\begin{aligned}\frac{\partial \rho}{\partial t} &= \frac{\partial}{\partial t}\sum_{i=1}^{n} m_i \varphi(y - y_i(t)) \\ &= -\sum_{i=1}^{n} m_i \nabla\varphi(y - y_i(t)) \cdot \dot{y}_i(t) \\ &= -\operatorname{div} p = -\operatorname{div}(\rho v).\end{aligned}$$

Continuing with our averaging, we bring the time derivative out of the right-hand side of (21) and introduce the definitions of velocity and density. We then get

$$\frac{\partial}{\partial t}(\rho v) + \operatorname{div}_y \sum_{i=1}^{n} m_i(\dot{y}_i(t) \otimes \dot{y}_i(t))\varphi(y - y_i(t)) = \sum_{i=1}^{n} f_i \varphi(y - y_i(t)).$$

We rewrite the second term by replacing $\dot{y}$ by $\dot{y} - v$ (which, incidentally, makes it insensitive to a Galilean transformation), and then we compensate for this insertion. After a few elementary manipulations we obtain

$$\rho\left(\frac{\partial \boldsymbol{v}}{\partial t} + \nabla \boldsymbol{v}\boldsymbol{v}\right) = \sum_{i=1}^{n} \boldsymbol{f}_i \varphi - \operatorname{div} \sum_{i=1}^{n} m_i(\dot{\boldsymbol{y}}_i - \boldsymbol{v}) \otimes (\dot{\boldsymbol{y}}_i - \boldsymbol{v})\varphi. \quad (22)$$

This takes exactly the same form as the balance of linear momentum (10) given above.

It is of course compelling to define div $\boldsymbol{T}$ to be equal to the right-hand side of (22) and to solve for $\boldsymbol{T}$ to get an atomistic definition of stress. But specification of the divergence of a tensor is a rather weak restriction on the tensor; in RCC, we can add the curl of a vector field to each row of $\boldsymbol{T}$, and the only restriction on these three vector fields comes from the symmetry of the stress. Various authors have implicitly made different choices of these curls, but there is no general agreement on which, if any, of the corresponding stresses ought to be the stress of continuum mechanics. And, of course, this may not be the best way to average. Averaging molecular dynamics to get anything that vaguely resembles a constitutive equation is much less clear.

There are additional hopeful directions of research. First, there is general mathematical experience with asymptotics. Many examples show that when doing asymptotics it is only certain quantities in the underlying theory that actually affect the asymptotic result. Identification of these quantities can be the beginning of a solution of the multiscale problem or the start of a new branch of continuum mechanics. The circle of ideas surrounding the Cauchy–Born rule, the quasicontinuum method of Tadmor, Ortiz, and Phillips, and the asymptotic methods of Blanc, Cances, Le Bris, and Lions are successes in this direction but mainly so far only in the static case. In light of these developments, one can almost imagine a finite-element method in which the subroutine that appeals to the constitutive relation is replaced by an efficient atomistic calculation.

Probabilistic approaches are also promising. These recognize that the equations of molecular dynamics are sufficiently irregular that they might be amenable to a probabilistic treatment, as successfully undertaken by statistical mechanics in the case of macroscopic equilibrium. Probability theory consists of an arsenal of highly developed techniques once a probability measure is found, but it does not say much about where to get the probability measure in the first place. As a starting point, perhaps it is time to revisit the kinetic theory of gases, the only truly nonequilibrium statistical mechanics we have.

### Further Reading

Admal, N. C., and E. B. Tadmor. 2010. A unified interpretation of stress in molecular systems. *Journal of Elasticity* 100:63–143.

Ball, J. M., and R. D. James. 2002. The scientific life and influence of Clifford Ambrose Truesdell III. *Archive for Rational Mechanics and Analysis* 161:1–26.

Blanc, X., C. Le Bris, and P.-L. Lions. 2002. From molecular models to continuum mechanics. *Archive for Rational Mechanics and Analysis* 164:341–81.

Knowles, J. K. 1998. *Linear Vector Spaces and Cartesian Tensors.* Oxford: Oxford University Press.

Reiner, M. 1960. *Lectures on Theoretical Rheology.* Amsterdam: North-Holland.

Truesdell, C., and W. Noll. 1965. *The Nonlinear Field Theories of Mechanics.* New York: Springer.

## IV.27 Pattern Formation

*Arnd Scheel*

### 1 Introduction

Patterns in nature fascinate observers and challenge scientists. We are particularly intrigued when simple systems generate complex patterns or when simple, highly organized patterns emerge in complex systems. Similarities in patterns across many fields suggest underlying mechanisms that dictate universal rules for pattern formation. We notice stripe and spot patterns on animal coats, but also in convection patterns. Rotating spiral waves organize collective behavior in bacterial motion, in chemical reaction, and on heart muscle tissue. Beyond simple observation, regular patterns are created in experiments, and a tremendous amount of theory has helped to predict phenomena. In this article we will discuss some of those phenomena and stress universality across the sciences. We focus on dissipative, or damped driven, systems, where a free energy is dissipated yet complex spatiotemporal behavior is sustained far from thermodynamic equilibrium. Specific applications arise in, but are not limited to, biology, chemistry, the social sciences, fluids, optics, and material science.

Historically, much of the research on pattern formation was motivated by fluid experiments, such as those on Rayleigh–Bénard convection. When a stationary fluid is heated from below, heat conduction is replaced by convective heat transport above a certain critical temperature gradient. Convective transport can occur through convection cells, often arranged in typical hexagonal arrays, or through convection rolls that form stripe patterns. Similar dichotomies were known from animal coats, between zebras and leopards, or, quite strikingly, in mutations of zebrafish (see figure 1).

Turing noticed in 1954 that the interplay of very simple mechanisms—diffusion and chemical reaction—can be responsible for the emergence of patterns. His predictions were experimentally realized only in 1990 by de Kepper and collaborators, who produced spot- and stripe-like patterns similar to figure 1 in an open-flow chemical reactor. Time-periodic rather than such spatially periodic behavior was observed much earlier, by Belousov, in the 1950s, and then rediscovered and popularized later by Zhabotinsky. They observed sustained temporal oscillations in a chemical reaction. In addition, these temporal oscillations in the reaction could cause complex spatial patterns. Yet much earlier than this, in 1896, Liesegang observed regular ring patterns when studying how electrolytes diffuse and precipitate in a gel. It is worth mentioning that Liesegang's patterns and their characteristic scaling laws are poorly understood even today, particularly when compared with Belousov's reaction or Turing-pattern formation.

In this discussion we have avoided the most basic question: *what is a pattern*? In a system that is invariant under translations in time and space, we naturally expect unpatterned solutions, that is, solutions that do not depend on time or space. We refer to such solutions as spatiotemporally uniform states. In fact, such uniform states "should be" the thermodynamic equilibrium and hence should be observed in experiments after initial temporal transients. A pattern is the opposite: a solution that is not spatiotemporally uniform. Narrowing this definition of a pattern further, one might also want to rule out solutions that are nonconstant in space and time only for an initial temporal transient, or near boundaries of the spatial domain. Note that in this characterization we started with a system that does not depend explicitly on time or space. This excludes patterns forced by external influences, such as masking or printing textures into substrates, and narrows our view to what one might refer to as self-organized patterns. Basic examples of spatial, temporal, and spatiotemporal states are shown in figure 3.

Explanations of patterns therefore often start with a partial differential equation

$$\partial_t u = F(\partial_x^m u, \dots, \partial_x u, u),$$

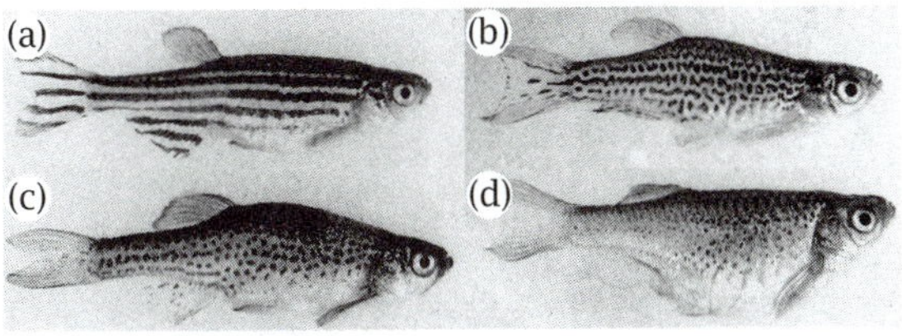


**Figure 1** Four different pigment patterns for the homozygous zebrafish corresponding to different alleles of the leopard gene.

for the dependent variables $u \in \mathbb{R}^N$, on an idealized unbounded, translation-invariant, spatial domain $x \in \mathbb{R}^n$, $n = 1, 2, 3$. Since $F$ does not depend on time $t$ or space $x$ explicitly, such systems typically support spatiotemporally uniform states $u(t,x) \equiv \bar{u} \in \mathbb{R}^N$ for all times $t \in \mathbb{R}$ and in the entire spatial domain $x \in \mathbb{R}^n$, satisfying $F(0, \dots, 0, \bar{u}) = 0$. They may, however, also accommodate solutions that depend on $x$ and $t$, even in the limit as $t \to \infty$.

Prototypical examples are reaction–diffusion systems,

$$\partial_t u = D\Delta u + f(u),$$

where $\Delta$ is the Laplacian, $D$ a positive diffusion matrix $D + D^{\mathrm{T}} > 0$, and $f(u)$ denote the reaction kinetics. They have been extensively studied as a prototype of pattern formation, motivated largely by Turing's observation that simple reaction and diffusion may explain many of the complex chemical and biological patterns that we see, possibly even patterns such as those in figure 1. Somewhat simpler (because it is scalar) is the Swift–Hohenberg equation,

$$\partial_t u = -(\Delta + 1)^2 u + \mu u - u^3,$$

mimicking instabilities in Rayleigh–Bénard convection or Turing-pattern formation. When considered on $x \in \mathbb{R}^n$, there always exists a trivial translation-invariant solution $u(t,x) \equiv 0$. For $\mu < 0$, most initial conditions $u(0,x)$ converge to 0, while for $\mu > 0$ there are stationary, spatially periodic solutions $u(t,x) \sim \sqrt{\mu}\cos(kx_1)$ with wave number $k \sim 1$.

A major theme of research in pattern formation is to describe, in this example and more generally, the longtime behavior of solutions based on simple coherent building blocks, such as spatially periodic patterns, defects, or fronts.

## 2 Linear Predictions

A much-studied scenario for pattern formation lets the spatiotemporally uniform state $\bar{u}$ destabilize while a

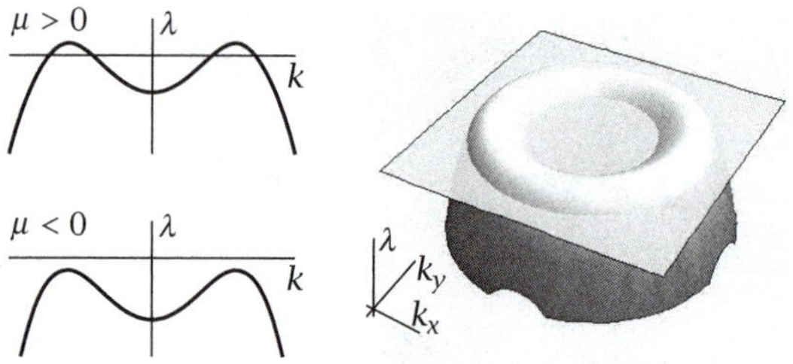


**Figure 2** The dispersion relation for Turing instabilities (Swift–Hohenberg).

parameter $\mu$ is increased. We illustrate this scenario in the Swift–Hohenberg equation, where we can analyze the linearization at $\bar{u} = 0$, $\partial_t u = \mathcal{L}u = [-(\Delta+1)^2+\mu]u$, using the Laplace–Fourier transform: solutions of the form $e^{\lambda t - ik\cdot x}$ exist when

$$d(\lambda, ik) = -(1-|k|^2)^2 + \mu - \lambda = 0,$$

an equation usually referred to as the dispersion relation.

Solving for $\lambda = \lambda(k)$, one finds $\lambda \in \mathbb{R}$ and the following regimes (see also figure 2):

$$\begin{aligned} &\mu < 0, \quad \lambda(k) < 0 \text{ for all } k &&\text{(stable)},\\ &\mu = 0, \quad \lambda(k) = 0 \text{ for } |k| = 1 &&\text{(critical)},\\ &\mu > 0, \quad \lambda(k) > 0 \text{ for } |k| \sim 1 &&\text{(unstable)}. \end{aligned}$$

Now consider evolving an initial condition consisting of a superposition of exponentials $\int_{\mathbb{R}^2} \hat{u}(k)e^{ik\cdot x}\,dk$ under the linear equation $\partial_t u = \mathcal{L}u$ for $\mu = 0$. Given the exponential decay $e^{\lambda t}$, $\lambda < 0$, for wave numbers $|k| \neq 1$, we expect that the solution is dominated by wave numbers $|k| = 1$ for large $t$, $u(t,x) \sim \int_{|k|=1} \hat{u}(k)e^{ik\cdot x}\,dk$. For $\mu > 0$ one still expects such wave numbers to dominate the solution given the relative faster growth. In fact, one often postulates that the linearly *fastest-growing modes* ($|k| = 1$ for Swift–Hohenberg) will also be dominantly observed in nonlinear systems.

Beyond the Swift–Hohenberg equation we are interested in dissipative systems, where most modes decay, $\operatorname{Re}\lambda(k) < 0$ for $|k|$ large. We also focus on isotropic systems, where $u(t,x)$ is a solution precisely when $u(t, g\cdot x)$ is a solution for any $g \in E(n)$, the Euclidean group of translations, rotations, and reflections in $\mathbb{R}^n$. At criticality, we *typically* expect $\operatorname{Re}\lambda(k) = 0$ at $|k| = k_*$ for some unique $k_*$ and $\lambda(k) = i\omega_*$. One can classify such instabilities by focusing on a simple Fourier mode, so that at criticality, $u(t,x) \sim e^{i(\omega_* t - k_* x_1)}$ (see figure 3 for the resulting basic patterns).

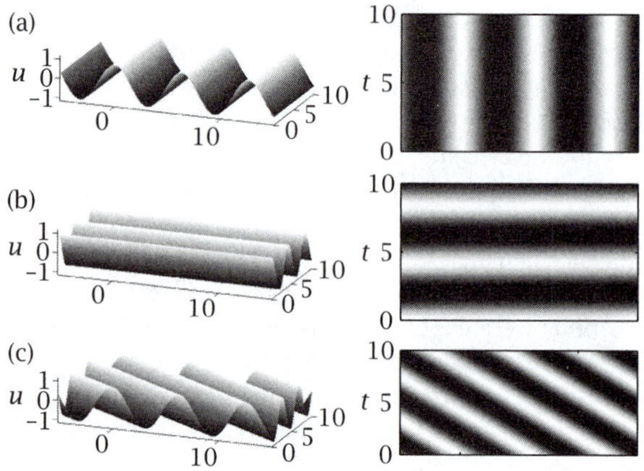


**Figure 3** Linear patterns with (a) $\omega_* = 0$, $k_* \neq 0$ (spatial pattern), (b) $\omega_* \neq 0$, $k_* = 0$ (temporal pattern), and (c) $\omega_*, k_* \neq 0$ (spatiotemporally periodic).

Linear predictions are, however, notoriously ambiguous. At $\mu = 0$ there are a plethora of bounded solutions formed by arbitrary superposition of critical modes. For instance, when $\omega_*, k_* \neq 0$, we find traveling waves $e^{i(\omega t - k\cdot x)}$ and standing waves $e^{i(\omega t - k\cdot x)} + e^{i(\omega t + k\cdot x)}$. In the two-dimensional Turing case, summing modes $k_j$ on an equilateral triangle gives hexagonal patterns, and choosing four $k_j$s on a square gives squares. Averaging uniformly over all critical modes $|k| = k_*$ gives Bessel functions, reminiscent of target patterns with maxima on concentric circles with radii $r_j \sim jk_*$ (see figure 5(c)). For $\mu > 0$ there is also ambiguity in the selected wave number. For instance, in the case $k_* = 0$, $\omega_* \neq 0$, we may find wave trains and standing waves of long wavelength $|k| \neq 0$ in the linear prediction. When we try to determine which patterns and wave numbers will actually be observed for most initial conditions, we therefore need to take nonlinearities into consideration.

## 3 Symmetry

Given the ambiguity in the linear predictions, it is quite surprising to find how many systems settle into simple spatiotemporally periodic states, involving but a few of the critical linearly unstable wave vectors $k$. Without explaining why spatial periodicity is favored, one can analyze systems that are invariant under the Euclidean group by a priori restricting to spatially periodic functions. In other words, we may restrict to functions $u(x) = u(x + p_j)$, where the vectors $p_j$ generate a lattice in $\mathbb{R}^n$. Going back to the striking dichotomy between spots and stripes in figure 1, consider the hexagonal lattice of width $L > 0$ generated by $p_1 = L(1,0)^{\mathrm{T}}$ and $p_2 = L(\sqrt{3}/2, 1/2)^{\mathrm{T}}$, which allows for both hexagons and stripes. The linear analysis of, say, the Swift–Hohenberg equation then needs to be restricted to wave vectors in the dual lattice, $k \cdot p_j = 2\pi$. For $k_* \sim 1$, the choice $L = 2\pi/k_*$ allows for six critical ($\operatorname{Re}\lambda = 0$) wave vectors.

The dynamics of systems near equilibria with finitely many critical eigenvalues in the linearization can in fact be reduced to a set of ordinary differential equations (ODEs). Such a reduction is exact in the vicinity of the trivial equilibrium, where all solutions converge to a finite-dimensional *center manifold* or escape the small neighborhood. Both the center manifold and the vector field on the center manifold can be computed to any order in the amplitude of the solutions and the parameter $\mu$. Typical equations on the center manifold are

$$\dot{A}_1 = \mu A_1 + \delta \bar{A}_2 \bar{A}_3 - A_1(|A_1|^2 + \kappa(|A_2|^2 + |A_3|^2) + \cdots),$$
$$\dot{A}_2 = \mu A_2 + \delta \bar{A}_3 \bar{A}_1 - A_2(|A_2|^2 + \kappa(|A_3|^2 + |A_1|^2) + \cdots),$$
$$\dot{A}_3 = \mu A_3 + \delta \bar{A}_1 \bar{A}_2 - A_3(|A_3|^2 + \kappa(|A_1|^2 + |A_2|^2) + \cdots)$$

with real parameters $\kappa$, $\delta$. The invariance under complex rotation $A_1 \mapsto \mathrm{e}^{\mathrm{i}\tau} A_1$, reflections $A_1 \mapsto \bar{A}_1$, $A_2 \mapsto -\bar{A}_3$, $A_3 \mapsto -\bar{A}_2$, and cyclic permutation $A_j \mapsto A_{j+1}$ is enforced by the Euclidean symmetry of the full system. More precisely, translations and reflection in $x_1$ as well as rotations by $2\pi/6$ leave the lattice invariant; they generate the isotropy group of the lattice $\mathbb{T} \rtimes D_6$. The action of this group on the critical wave vectors gives precisely the invariances of complex rotation, reflection, and permutation.

We always find nontrivial equilibria with maximal isotropy, that is, roughly speaking, equilibria that are invariant under a symmetry group that is a maximal subgroup of $\mathbb{T} \rtimes D_6$. In this case, these are stripes, $A_2 = A_3 = 0$, $A_1 \in \mathbb{R}$, or hexagons, $A_i = A_j \in \mathbb{R}$. Within the reduced differential equation one can study the stability of these equilibria and predict whether stripes or hexagons should be observed for parameter values near an instability. Depending on the parameters $\kappa$, $\delta$, one calculates equilibria and determines their stability within this reduced ODE. The fact that, typically, either hexagon equilibria or stripe equilibria will be stable reflects the universal ubiquity of stripes and spot patterns, as shown in figure 1.

As a second example, consider spatiotemporal instabilities, $\omega_*, k_* \neq 0$, $x \in \mathbb{R}$. Restriction to periodic functions yields reduced coupled-amplitude equations for left- and right-traveling waves,

$$\dot{A}_+ = (\mu + \mathrm{i}\omega_*)A_+ + A_+(|A_+|^2 + \kappa|A_-|^2) + \cdots,$$
$$\dot{A}_- = (\mu + \mathrm{i}\omega_*)A_- + A_-(|A_-|^2 + \kappa|A_+|^2) + \cdots,$$

with complex parameter $\kappa$. Spatial translations act as complex rotations $A_\pm \mapsto \mathrm{e}^{\pm\mathrm{i}\tau} A_\pm$, reflections act as $A_\pm \mapsto \bar{A}_\mp$. Maximal isotropy roughly corresponds to traveling waves $A_- = 0$, $A_+ \sim \mathrm{e}^{\mathrm{i}\omega t}$ or standing waves $A_+ = A_-$. Again, reduction combined with a bifurcation and symmetry analysis gives universal predictions for the competition between standing and traveling waves.

The reduced equations on the center manifold are often referred to as Landau equations, which describe dynamics of dominant modes. Fourier modes that are compatible with the lattice but decay exponentially for the linearized problem can be shown to follow the temporal evolution of neutral modes precisely. More precisely, all small solutions shadow solutions on the center manifold with exponentially decaying error. In geometric terms, this follows from the fact that the phase space is foliated by a strong stable fibration of the center manifold.

While the approach outlined here can discriminate between systems that favor hexagons over stripes, or traveling waves over standing ones, it cannot predict wave numbers since the analysis is a priori restricted to a set of functions with prescribed period. On the other hand, a center manifold analysis cannot be performed directly for the system posed on an unbounded (or very large) domain since neutral linear Fourier modes are not (well) separated from decaying modes.

We will look at three pattern-selection mechanisms below. First, periodic patterns such as hexagons may be unstable with respect to perturbations of the initial conditions. Our analysis in this section considered perturbations with the same period. However, a pattern might well be stable with respect to such coperiodic perturbations but unstable with respect to other ones, e.g., localized perturbations, a phenomenon usually referred to as a sideband instability. Such sideband-unstable patterns will typically not be observed in large systems, thereby restricting the set of wave numbers present in large systems. Second, initial conditions may evolve into periodic patterns outside of small areas in the domain where defects form. Such defects can have a significant influence on the wave numbers observed in the system. And finally, patterns are often created through spatial growth processes. In mathematical terms, patterns in the wake of invasion fronts often show distinguished, selected wavelengths.

Before we address these different mechanisms, we will briefly review amplitude equations, which, beyond the Landau equations, approximately describe, in a simplified and universal fashion, the evolution of systems near the onset of instability.

## 4 Modulation Equations

Center-manifold equations describe the long-term behavior of small-amplitude solutions exactly, for spatially periodic patterns. In most cases, leading-order approximations can be derived using scalings of amplitude and time. In the one-dimensional Swift–Hohenberg equation, the complex amplitude $a(t)$ of the Fourier mode $e^{ix}$ solves an equation of the form $\partial_t a = \mu a - 3a|a|^2 + \cdots$. Scaling $a = \mu^{1/2}A$, $T = \mu t$ for $\mu > 0$, we find $\partial_T A = A - 3A|A|^2 + O(\mu)$. Equivalently, we could substitute an ansatz $\mu^{1/2}A(\mu t)e^{ix}$ together with the complex conjugate (c.c.) into the Swift–Hohenberg equation and find $\partial_T A = A - 3A|A|^2$ as a compatibility condition at order $\mu^{3/2}$ in the expansion. This latter method can be generalized, allowing slow spatial variations of the amplitude, $u(t,x) = \mu^{1/2}A(\mu t, \mu^{1/2}x)e^{ix}$ + c.c. The scaling in $x$ is induced by the quadratic tangency in the linear relation, where $\lambda = -4(k-1)^2 + O((k-1)^3)$. An expansion now gives the compatibility condition

$$\partial_T A = 4\partial_{XX}A + A - 3A|A|^2.$$

This equation is known as the GINZBURG–LANDAU EQUATION [III.14]. It is a *modulation equation*, as it describes spatial modulations of the amplitude of critical modes. While in the case of periodic boundary conditions amplitude equations are the leading-order approximation to an ODE that describes the long-term dynamics of small solutions *exactly*, modulation equations give *approximations* to long-term dynamics, at best. Such approximation properties typically rely on some type of stability in the problem, following the mantra of "consistency + stability ⇒ convergence." It is not known if there exists an exact reduced description of the long-term dynamics in terms of a single partial differential equation, which would coincide with the modulation equation at leading order.

For Hopf bifurcations, $\omega_* \neq 0$, $k_* = 0$, one substitutes $u(t,x) = \mu^{1/2}A(\mu t, \mu^{1/2}x)e^{i\omega_* t}$ + c.c. and finds the complex Ginzburg–Landau equation

$$\partial_T A = (1 + i\alpha)\Delta_X A + A - (1 + i\beta)A|A|^2,$$

where the coefficient $\beta$ is responsible for frequency detuning of oscillations depending on the amplitude (nonlinear dispersion), and the coefficient $\alpha$ measures frequency dependence on wave number (linear dispersion).

When $k_* \neq 0$, in space dimension $n > 1$, this approach is limited by the fact that there is a continuum of critical modes $|k| = k_*$, while amplitudes $A_j$ can capture only bands near distinct wave numbers $k_j$. One can include modes with neighboring orientations, $u(t,x) = A(\mu t, \mu^{1/2}x, \mu^{1/4}y)e^{ik_* x}$ + c.c., but the resulting Newell–Whitehead–Segel equation

$$\partial_T A = -(\partial_X - i\partial_Y^2)^2 A + A - A|A|^2$$

poses several analytical challenges.

Both Landau equations on the center manifold and modulation equations can also be interpreted as universal normal forms near the onset of instability, thus explaining universality of patterns across the sciences to some extent. In both cases one eliminates fast spatiotemporal dependence via some effective averaging procedure. For the Landau equations, averaging can be more systematically understood in terms of normal-form transformations that simplify the equations. In particular, temporal oscillations can be exploited to eliminate coefficients in the Taylor jet of the vector field through polynomial coordinate changes, effectively averaging the vector field over the fast oscillations. For modulation equations, this procedure is less systematic, reminiscent of homogenization and effective medium theories. In that regard, modulation equations not only simplify the analysis but also provide approximations that allow for effective simulations.

## 5 Stability

Most of our discussion so far has been motivated by the presence of a trivial, spatiotemporally uniform state that loses stability as a parameter $\mu$ is increased. While this state still exists for $\mu > 0$, it would not be observed experimentally since small perturbations would grow exponentially and drive the system toward a different state. Restricting ourselves to periodic functions, we found ODEs that show that small periodic perturbations will result in spatially periodic, stable patterns. In this analysis, however, stability is understood only with respect to spatially coperiodic perturbations. More realistically, we should ask for stability against spatially random or, at least, spatially localized perturbations. In the Swift–Hohenberg equation we would study initial conditions $u_0(x) = u_{\text{per}}(x) + v(x)$, with $v(x)$ small,

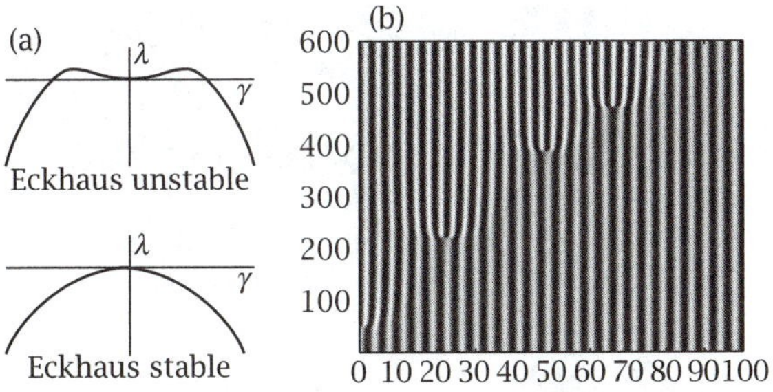


**Figure 4** (a) The Bloch dispersion relation before and after sideband instability. (b) The space-time diagram of Eckhaus coarsening. Here and throughout, time is plotted vertically.

localized, and $u_{\text{per}}(x)$ a stationary, spatially periodic pattern. To leading order, $v(x)$ satisfies the linearized equation at $u_{\text{per}}$:

$$\partial_t v = -(\partial_x^2 + 1)^2 v + \mu v - 3u_{\text{per}}^2(x)v.$$

The operator on the right-hand side possesses spatially periodic coefficients, and its properties can be expressed in terms of Fourier–Bloch eigenfunctions, $e^{i\gamma x}v_{\text{per}}(x)$, replacing the Fourier analysis near the spatially homogeneous steady state. One finds eigenvalues $\lambda_j(\gamma)$ that measure the temporal evolution of quasiperiodic perturbations to the periodic state. Translation of the pattern $u_{\text{per}}$ induces a neutral response; technically, $v_{\text{per}}(x) = \partial_x u_{\text{per}}(x)$ and $\gamma = 0$ correspond to a zero eigenvalue, sometimes referred to as a Goldstone mode. Varying $\gamma$, we can study modulations of this translational mode and possible instabilities.

The calculations are conceptually simpler in the modulation equation, where periodic patterns can be reduced to spatially constant states, such as $A(X) \equiv 1$ in the Ginzburg–Landau equation, so that the Bloch wave analysis reduces to Fourier analysis. One finds, for fixed $\mu$, a band of wave numbers $k$ where periodic patterns are stable. Patterns outside of this band are sideband unstable and typically cannot be observed in large domains. The simplest type of instability is the one-dimensional Eckhaus instability of stationary patterns that arise in Turing instabilities $k_* \neq 0$, $\omega_* = 0$ (see figure 4).

Phenomenologically, perturbations of unstable patterns grow and change the wave number by temporarily introducing defects into the pattern. In two space dimensions, rotational modes induce zigzag and skew-varicose instabilities. Generally, in the parameter plane spanned by wave number $k$ and system parameter $\mu$, stable patterns often occupy a bounded region commonly referred to as the Busse balloon. More dramatic instabilities occur in Hopf bifurcations, when $\alpha\beta < -1$: the band of wave numbers corresponding to stable spatiotemporal patterns vanishes, and dynamics in extended systems appear to sustain complex dynamics.

## 6 Defects

We care about imperfections or defects in periodic patterns not only because they naturally arise in experiments and simulations but also because they can play a crucial role in selecting wave numbers and wave vectors. Prominent examples are spiral waves in oscillatory media, which act as effective wave sources and select wave numbers and wave vectors in large parts of the domain. Also, interfaces between patches of stripes with different orientations tend to select wave numbers. In a similar vein, boundary conditions can select the orientation of convection rolls in Rayleigh–Bénard convection: typically, rolls orient themselves perpendicularly to the boundary, but heated boundaries allow for a parallel alignment.

When referring to defects we usually imply that the deviation from perfect spatiotemporally periodic structures is in some sense localized, and the temporal behavior is coherent, e.g., periodic in an appropriate coordinate frame. Such solutions can sometimes be found explicitly in amplitude equations; the Nozaki–Bekki holes in the complex Ginzburg–Landau equations are a prominent example. On the other hand, one would like to approach existence, stability, or even interaction of defects in a mathematically rigorous yet systematic fashion, similar to the treatment of spatially periodic patterns. We have already mentioned that center-manifold reductions are not available once we give up the restriction to spatial periodicity. We can, however, restrict to temporal periodicity, or even stationary solutions, possibly propagating or rotating with a fixed speed. Such solutions are amenable to an approach as systematic and rigorous as for spatially periodic structures: one interchanges the roles of space and time and studies *spatial dynamics*. We can, for instance, find small stationary solutions to a reaction–diffusion system

$$\partial_t u = D\partial_{xx}u + f(u;\mu), \quad u \in \mathbb{R}^N,$$

by looking for small bounded solutions to the ODE

$$u_x = v,$$
$$v_x = -D^{-1}f(u;\mu).$$

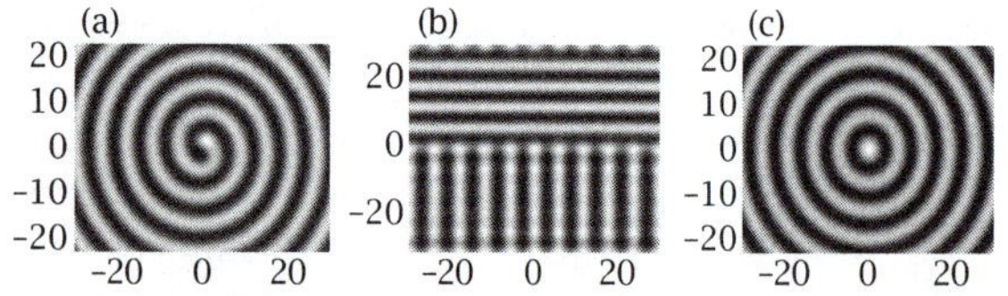


**Figure 5** (a) Spirals, (b) grain boundaries, and (c) a target pattern in the Swift–Hohenberg equation.

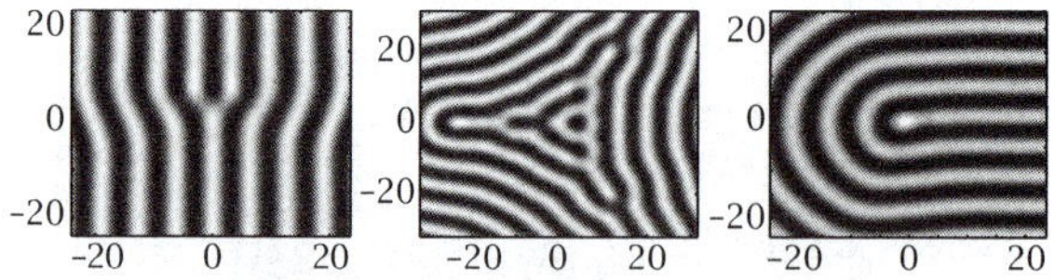


**Figure 6** Dislocations and concave and convex disclinations in the Swift–Hohenberg equation.

An $x$-independent equilibrium of the reaction–diffusion system corresponds to an equilibrium of this ODE; a spatially periodic pattern corresponds to a periodic orbit. Close to a Turing instability, center-manifold analysis and normal-form transformations reveal a universal description of stationary patterns:

$$A_x = \mathrm{i}kA + B + \cdots,$$
$$B_x = \mathrm{i}kB - \mu A + A|A|^2 + \cdots,$$

which at leading order recovers stationary solutions to the Ginzburg–Landau equation after passage to a corotating frame $(A, B) \mapsto \mathrm{e}^{\mathrm{i}kx}(A, B)$ and scaling. Similar constructions can find traveling waves $u(x - ct)$ and time-periodic traveling waves $u(x - ct, \omega t)$, $u(\xi, \tau) = u(\xi, \tau + 2\pi)$. In the latter case, the equation

$$u_\xi = v,$$
$$v_\xi = -D^{-1}(f(u; \mu) + cv - \omega\partial_\tau u)$$

is an ill-posed degenerate elliptic partial differential equation. Nevertheless, center-manifold reduction and normal-form techniques can be used here, too, to derive universal ODEs for the shape of coherent defects. Even more generally, we can use these methods to study coherent structures for $x = (x_1, \dots, x_n) \in \mathbb{R}^n$, provided that we impose periodic boundary conditions on $x_2, \dots, x_n$ while studying spatial dynamics in the $x_1$-coordinate. In this category one finds grain boundaries and certain types of dislocations. Somewhat more subtle reductions based on dynamics in the radial coordinate have been used to study point defects such as radially symmetric target patterns or spiral waves (see figure 5).

While these methods have been very successful in establishing existence and studying linearized stability through the eigenvalue problem in many examples, they fail to give a complete description in higher space dimensions: properties of dislocations and disclinations are not understood in this detailed fashion (see figure 6).

The key analogy in this spatial-dynamics setting relates defects and coherent structures to homoclinic and heteroclinic orbits, connecting equilibria or periodic orbits. Pushing this analogy further, we can reinterpret dynamical properties of heteroclinic orbits as properties of defects. One could now systematically reinterpret results in the literature and thereby effectively construct a dictionary that translates between dynamical-systems terminology in the spatial-dynamics description and properties of defects.

In oscillatory media ($\omega \neq 0$), defects can be classified according to dimensions of unstable and stable manifolds that intersect along a heteroclinic orbit in the ill-posed spatial-dynamics description. The codimension of the intersection can be translated into group velocities in the far field. Defects where group velocities point away from the defect (sources) correspond to codimension-two heteroclinic orbits. Sinks, where group velocities point toward the defect, correspond to transverse intersections.

An illustrative example is a small localized inhomogeneity $\varepsilon g(x)$ in a system that otherwise sustains a spatially homogeneous temporal oscillation. One finds that for $\varepsilon > 0$, say, the inhomogeneity typically acts as a wave source from which phase waves propagate into the medium, while for $\varepsilon < 0$ waves travel toward the inhomogeneity with speed $1/|x|$ for large distances $x$ from the inhomogeneity (see figure 7).

## 7 Fronts

The ability to describe in detail the fate of small random perturbations of an unstable homogeneous equilibrium in general pattern-forming systems such as the Swift–Hohenberg equation seems elusive. Wave numbers can vary continuously across the physical domain, while embedded defects move and undergo slow coarsening dynamics.

A more tractable situation arises when initial perturbations are spatially localized and patterns emerge in the wake of fronts as the instability spreads spatially. Similar in spirit to the fastest-growing-mode analysis that we described above, but quite different in the details, one tries to predict patterns that arise in the wake of an invasion front using the linearization first. For a linear system, the location of the leading edge of an invasion front can be determined by testing for *pointwise stability* in comoving frames: in a steady coordinate frame, one typically sees exponential growth in a finite window of observation, whereas in a frame moving with large speed, perturbations decay within such a finite window. The smallest of all speeds $c$ that outrun perturbations in this sense gives us the linear spreading speed. An observer traveling with the linear spreading speed will see a *marginally stable* state in a finite window of observation.

We can determine pointwise stability, that is, decay in a finite window of observation, using the Laplace–Fourier transform in a refined way as follows. The Laplace transform reduces the evolution problem $\partial_t u = \mathcal{L}u$ into a study of the resolvent $(\lambda - \mathcal{L})^{-1}$, which in turn can be analyzed using the Fourier transform. The key difference between pointwise and overall (say, $L^2$) stability analysis is that the resolvent is applied to spatially localized initial conditions. In fact, the resolvent can be represented as a convolution with a Green function $G_\lambda(x)$ that is readily computable via the Fourier transform. Boundedness of the resolvent as an operator typically requires integrability of the convolution kernel. Convolving with localized initial conditions, we can relax this condition, however, and merely require *pointwise* (fixed-$x$) analyticity of $G_\lambda(x)$ in $\lambda$. Equivalently, we are allowed to shift the Fourier contour that is used to calculate the convolution kernel off the real $k$-axis into the complex plane. Obstructions to shifting this contour—or, equivalently, pointwise singularities of the convolution kernel—typically occur at branch poles. These can be found by analyzing the complex dispersion relation $d(\lambda, \nu) = 0$, which is obtained from the ansatz $u \sim e^{\lambda t + \nu x}$. Rather than finding simple roots for given Fourier mode $\nu = ik$, one allows $\nu$ to be complex and looks for *double roots*:

$$d(\lambda, \nu) = 0, \quad \partial_\nu d(\lambda, \nu) = 0.$$

Such double roots (with an additional pinching condition) typically give singularities of the pointwise Green function and determine pointwise stability; we have pointwise instability if and only if such a double root is located in $\operatorname{Re} \lambda > 0$.

In summary, pointwise stability is determined by pinched double roots of the dispersion relation. Marginal stability in a frame moving with the linear spreading speed $c$ implies that such a double root is located on the imaginary axis, $\lambda = i\omega$. This linear oscillation

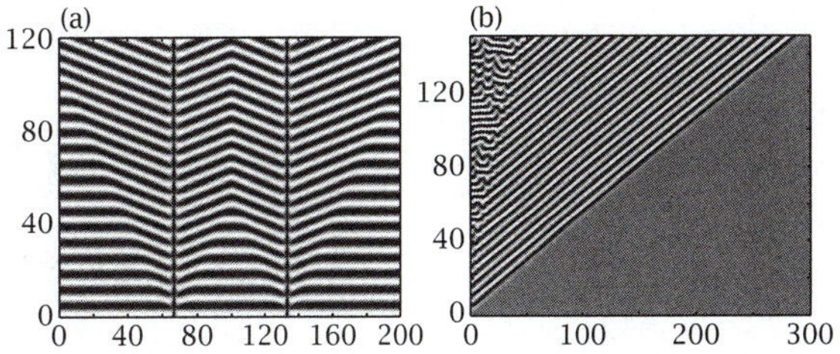


**Figure 7** (a) Two sources and a sink and (b) an invasion front creating an unstable pattern, followed by a turbulent state, both in an oscillatory system $\omega_* \neq 0$, $k_* = 0$.

with frequency $\omega$ in a frame moving with speed $c$ typically selects a resonant nonlinear pattern $u(\omega_0 t - kx)$ such that $\omega = \omega_0 - kc$.

For nonlinear systems there is ample experimental and numerical evidence that the linear predictions are often accurate, although convergence is usually slow, $O(t^{-1})$. Such convergence results have been established mathematically only for order-preserving systems and in particular for scalar reaction–diffusion systems. For pattern-forming systems, which intrinsically violate order preservation, existence and stability of invasion fronts have been established near Turing-like small-amplitude instabilities, in particular in the Swift–Hohenberg equation and the Couette–Taylor problem. In the complex Ginzburg–Landau equation, pattern-forming fronts can be found as traveling waves, finding connecting orbits in a three-dimensional ODE. Interestingly, the linear wave number predicted by the double-root criterion is typically nonzero, so invasion fronts create traveling waves in their wake (see figure 7). It is worth contrasting this with the linear fastest-growing-mode analysis based on the Fourier transform, which predicts spatially homogeneous oscillations. Since such wave trains can be sideband unstable, the state in the wake of the primary invasion front is subject to a secondary invasion, where a turbulent state typically takes over.

Fronts also play an important role in biological growth and crystal growth, as well as in phase-separation processes. Model problems include the CAHN–HILLIARD EQUATION [III.5], phase-field systems, and the Keller–Segel model. Fronts are known to exist in only a few cases. As for the situation in oscillatory media (figure 7), primary invasion fronts create unstable patterns that are invaded by secondary fronts (see figure 8).

Invasion fronts also provide key building blocks in spatial-growth or deposition processes. We can often model pattern growth via systems in which an

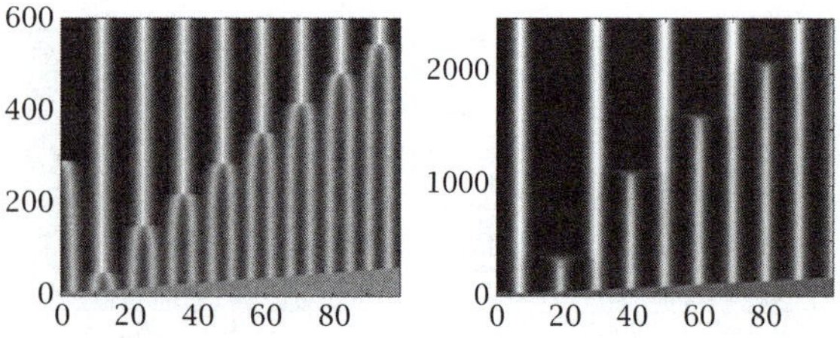


**Figure 8** Space-time plots of two-stage front invasion in the Keller–Segel model, with merging versus ripening in the secondary front.

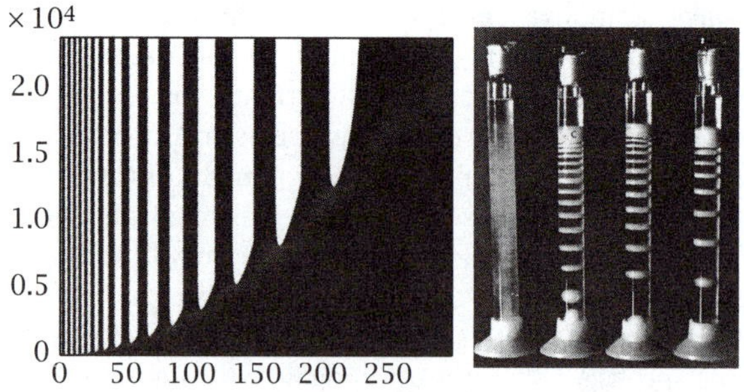


**Figure 9** Space-time plots in numerical simulations and experimental Liesegang patterns.

instability parameter $\mu$ is increased spatiotemporally in the form $\mu \sim -\tanh(x - st)$, so that instability is triggered externally in the region $x < st$. Liesegang's recurrent precipitation experiment falls into this category, as we will now explain. An outer electrolyte $A$ diffuses into a gel that is saturated with an inner electrolyte $B$ (see figure 9). Both react, and the reaction product accumulates until a supersaturation threshold is reached and precipitation is initiated. The concentration of $AB$ acts as an effective parameter that increases in the wake of the reaction–diffusion front. The main pattern-forming mechanism is the precipitation in the wake of the front. Intuitively speaking, precipitation is initiated only once the concentration locally exceeds a supersaturation threshold. Since the solute, providing the input to the precipitation process, can diffuse, it will be depleted in a neighborhood of the region where the process has been initiated, so that the supersaturation threshold will next be reached at a finite distance from where the first precipitate nucleated. More mathematically, precipitation can be understood as a simple conversion of solute $s$ into precipitate $p$,

$$\partial_t s = \Delta s - f(s, p),$$
$$\partial_t p = \kappa \Delta p + f(s, p),$$

with conversion rate $f$ and a small diffusion rate $\kappa \ll 1$ of the precipitate. The product $AB$ feeds as a source term into the equation for $s$, effectively increasing the value of $s$ until instability is triggered. Such instabilities turn out to be pattern-forming sideband instabilities, which in turn generate rhythmic oscillations in the wake of the $A$–$B$ reaction front. Wave numbers in the wake of such triggered fronts are therefore a central ingredient in the prediction of wave numbers in Liesegang patterns.

To summarize, our excursion into pattern formation in the wake of fronts points toward a promising route to a more systematic understanding of wave number and pattern selection, but many mathematical challenges lie ahead of us.

## 8 More Patterns

The point of view taken here emphasizes universality, mostly from the point of view of local bifurcations. Many exciting phenomena can be observed in reaction–diffusion systems in somewhat opposite parameter regimes when chemical species react and diffuse on disparate scales. Prototypical examples are the Gray–Scott and Gierer–Meinhardt equations, but one could also consider the FitzHugh–Nagumo, Hodgkin–Huxley, or Field–Noyes models for excitable and oscillatory media. Singular perturbation methods—both geometric methods inspired by dynamical systems and matched asymptotic methods—have helped us to gain tremendous insight into the complexity of spike and front dynamics in such systems. The basic building blocks here are scalar reaction–diffusion equations, where a variety of tools allows for quite explicit characterization of solutions and eigenvalue problems. The coupling between fast and slow components changes the interaction between spikes and fronts so that complex arrangements of fronts and spikes that would be unstable in scalar equations may form stable patterns here.

Beyond the pattern-forming instabilities that we have discussed in this article, more complex phenomena arise when conserved quantities interact with pattern-forming mechanisms. Examples include closed reaction–diffusion systems, fluid instabilities with neutral mean flow modes, and phase separation problems modeled by Cahn–Hilliard or phase-field equations. Conserved quantities can also be generated by symmetry through Goldstone modes, so descriptions of flame front instabilities or sideband instabilities usually involve coupling to conservation laws.

In a different direction, even simple systems such as the Swift–Hohenberg equation or the complex Ginzburg–Landau equation support patterns that escape our simplistic scheme based on periodic structures and embedded defects. Examples are quasicrystal patterns involving nonresonant spatial wave vectors and turbulent states, in which coherence in patterns is visible only after taking temporal averages.

Last but not least, patterns and coherent structures arise in spatially extended Hamiltonian systems, such as water-wave problems or nonlinear Schrödinger equations. Universal phenomena such as solitons, plane waves, and sideband instabilities abound, and partial descriptions of small-amplitude dynamics via universal modulation equations are possible.

### Further Reading

Ball, P. 2009. *Nature's Patterns: A Tapestry in Three Parts.* Oxford: Oxford University Press.

Haragus, M., and G. Iooss. 2011. *Local Bifurcations, Center Manifolds, and Normal Forms in Infinite-Dimensional Dynamical Systems.* Berlin: Springer.

Mielke, A. 2002. The Ginzburg–Landau equation in its role as a modulation equation. In *Handbook of Dynamical Systems*, volume 2, pp. 759–834. Amsterdam: North-Holland.

Sandstede, B., and A. Scheel. 2004. Defects in oscillatory media: toward a classification. *SIAM Journal on Applied Dynamical Systems* 3:1–68.

Scheel, A. 2003. *Radially Symmetric Patterns of Reaction–Diffusion Systems.* Providence, RI: American Mathematical Society.

van Saarloos, W. 2003. Front propagation into unstable states. *Physics Reports* 386:29–222.

## IV.28 Fluid Dynamics

*H. K. Moffatt*

### 1 Introduction

Fluid dynamics is a subject that engages the attention of mathematicians, physicists, engineers, meteorologists, oceanographers, geophysicists, astrophysicists, and, increasingly, biologists in almost equal measure. For mathematicians, the subject is the source of a wide range of problems involving both linear and nonlinear partial differential equations (PDEs). These equations arise from real-world natural phenomena and therefore provide serious motivation for exploring questions of existence, uniqueness, and stability of solutions. The nonlinear equations frequently involve one or more small parameters, allowing the development of perturbation techniques in their solution; thus, for example, *singular perturbation theory* is a branch of mathematics that finds its origin in the *boundary-layer theory* of fluid dynamics developed in the first half of the twentieth century, the modern study of *chaos* (since the 1960s) finds its origin in the (Lorenz) equations describing thermal convection in the atmosphere, and so on.

The word *fluid* covers liquids (such as water, oil, honey, blood, and liquid metals, to name but a few), gases (such as air, hydrogen, helium, carbon dioxide, and methane), plasma (i.e., fully ionized gas at extremely high temperatures), and *exotic fluids* such as liquid helium or Bose–Einstein condensates, which exist only at temperatures near absolute zero. Even conventional "solids" can behave like fluids if observed over a very long timescale; for example, the rock-like medium of the Earth's mantle flows slowly on a timescale of millions of years, and it is this that is responsible for the movement of the tectonic plates that gives rise to volcanic activity, earthquakes, and continental drift.

Fluid dynamics starts with the continuum approximation, whereby the fluid is regarded as a medium whose state can be expressed in terms of properties that are continuous functions of position $\boldsymbol{x} = (x, y, z)$ and time $t$. Chief among these properties are the density field $\rho(\boldsymbol{x}, t)$ and the velocity field $\boldsymbol{u}(\boldsymbol{x}, t)$ within the fluid domain. Thus the motion of individual molecules is ignored, and only properties that are averaged over at least millions of molecules are considered.

Fluid dynamics covers a vast range of phenomena on all length scales, from microns ($\sim 10^{-6}$ m) in biology and *nanoscale fluid dynamics* to kiloparsecs ($\sim 10^{21}$ m) in the fluid dynamics of the interstellar medium. At all these scales, the motion of the fluid medium is governed by the NAVIER–STOKES EQUATIONS [III.23], which are essentially derived from the principles of mass conservation and (Newtonian) momentum balance. In general, these equations must be coupled with thermodynamic equations of state and, when plasmas are considered, with Maxwell's equations for the electromagnetic field. In this article we shall, for simplicity, focus on the idealization of a nonconducting, incompressible fluid of constant density $\rho$, for which the Navier–Stokes (NS) equations become

$$\rho\left(\frac{\partial \boldsymbol{u}}{\partial t} + (\boldsymbol{u} \cdot \nabla)\boldsymbol{u}\right) = -\nabla p + \mu \nabla^2 \boldsymbol{u} + \boldsymbol{f}$$

and

$$\nabla \cdot \boldsymbol{u} = 0,$$

where $p(\boldsymbol{x}, t)$ is the pressure field, $\mu$ is the viscosity of the fluid, and $\boldsymbol{f}(\boldsymbol{x}, t)$ is any external force per unit mass (e.g., gravity) that acts upon the fluid.

The incompressibility condition calls for special comment. In adopting this idealization, sound waves are filtered from the more general equations of fluid dynamics that take density variations into account. The approximation is generally valid provided the fluid velocities considered are small compared with the speed of sound in the fluid ($\sim$340 m s$^{-1}$ in air and $\sim$1480 m s$^{-1}$ in water, at normal temperatures and pressures). Obviously, therefore, the incompressibility condition cannot be adopted if problems of transonic or supersonic flight are considered. For such problems, the coupling of fluid motion with thermodynamic effects cannot be ignored.

In the rest of the article we first consider the *kinematics* of flow governed by this incompressibility condition alone before turning to the dynamics of flow, with particular reference to the limits of large and small viscosity.

## 2 Kinematics of Flow

### 2.1 Streamlines and Particle Paths

Consider first the case of two-dimensional flows for which $\boldsymbol{u} = (u(x, y, t), v(x, y, t), 0)$. The incompressibility condition becomes $\partial u/\partial x + \partial v/\partial y = 0$, and it follows that there exists a *stream function* $\psi(x, y, t)$ such that

$$u = \partial\psi/\partial y, \qquad v = -\partial\psi/\partial x.$$

Since $\boldsymbol{u} \cdot \nabla\psi = 0$, it follows that $\boldsymbol{u}$ is everywhere parallel to the curves $\psi$ = const., which are therefore appropriately described as the *streamlines* of the flow. Note that $\psi$ has physical dimension $L^2/T$ (length$^2$/time).

A *steady* flow is one for which the velocity field is independent of $t$, i.e., $\boldsymbol{u} = \boldsymbol{u}(\boldsymbol{x})$ or, in the two-dimensional case, $\psi = \psi(x, y)$. In a steady flow, fluid particles follow the streamlines, which do not change with time. In a two-dimensional steady flow, initially adjacent particles on neighboring streamlines generally separate linearly in time due to the velocity gradient normal to the streamlines. If we consider a small patch of dye carried with the flow, then every pair of particles within the patch separates linearly in time, so the whole patch is similarly stretched by the flow.

The situation is very different in an unsteady flow. Now $\psi = \psi(x, y, t)$ with $\partial\psi/\partial t \neq 0$, and the streamline pattern changes with time. The path $\boldsymbol{x}(t)$ of a fluid particle released from a point $\boldsymbol{x}_0$ at time $t = 0$ is now determined by the dynamical system

$$\frac{\mathrm{d}x}{\mathrm{d}t} = \frac{\partial\psi}{\partial y}, \qquad \frac{\mathrm{d}y}{\mathrm{d}t} = -\frac{\partial\psi}{\partial x},$$

with initial condition $\boldsymbol{x}(0) = \boldsymbol{x}_0$. This is a second-order Hamiltonian system, in which the Hamiltonian is just the stream function $\psi$. Initially adjacent particles can now separate exponentially in time, a symptom of chaotic behavior. This obviously has important implications for the rate of stirring of any dynamically passive scalar contaminant in the flow.

In a fully three-dimensional flow, even in the steady case the streamlines (and therefore the particle paths) can diverge exponentially, a behavior conducive to the rapid dispersion of any passive contaminant.

The flow $\boldsymbol{u}(\boldsymbol{x}, t)$ (which is assumed to be smooth) determines a time-dependent mapping $\boldsymbol{X} \to \boldsymbol{x} = \boldsymbol{x}(\boldsymbol{X}, t)$ of the flow domain onto itself. Under the incompressibility condition, this mapping is volume-preserving. Since the volume element is given by $\mathrm{d}^3\boldsymbol{x} = |J|\,\mathrm{d}^3\boldsymbol{X}$, where $J$ is the Jacobian,

$$J = \frac{\partial(x_1, x_2, x_3)}{\partial(X_1, X_2, X_3)},$$

it follows that in this case $|J| = 1$. At time $t = 0$, the mapping is the identity $\boldsymbol{x} = \boldsymbol{X}$, with $J = 1$, so by continuity, $J = +1$ for all $t > 0$. The mapping is a volume-preserving diffeomorphism for all finite $t$.

### 2.2 Rate of Strain and Vorticity

In the neighborhood of any point (which may be chosen to be the origin $\boldsymbol{x} = 0$), the velocity field may be expanded, provided it is sufficiently smooth, as a Taylor series:

$$u_i(\boldsymbol{x}) = u_{0i} + c_{ij}x_j + O(|\boldsymbol{x}|^2),$$

where $c_{ij} = \partial u_i/\partial x_j|_{\boldsymbol{x}=0}$. This tensor may be split into a sum of symmetric and antisymmetric parts. The symmetric part,

$$e_{ij} = \frac{1}{2}\left(\frac{\partial u_i}{\partial x_j} + \frac{\partial u_i}{\partial x_j}\right),$$

is the *rate-of-strain tensor*. Note that $e_{ii} = 0$, by virtue of incompressibility. Referred to its principal axes, this symmetric tensor is diagonalized so that $e_{ij}x_j = (\alpha x, \beta y, \gamma z)$, where $(\alpha, \beta, \gamma)$ (with $\alpha + \beta + \gamma = 0$) are the *principal rates of strain*.

The antisymmetric part of $c_{ij}$ is

$$\omega_{ij} = \frac{1}{2}\left(\frac{\partial u_i}{\partial x_j} - \frac{\partial u_j}{\partial x_i}\right) = -\tfrac{1}{2}\varepsilon_{ijk}\omega_k,$$

where $\boldsymbol{\omega} = \nabla \times \boldsymbol{u}$ and $\omega_{ij}x_j = \frac{1}{2}\boldsymbol{\omega} \times \boldsymbol{x}$, a "rigid-body" rotation with angular velocity $\frac{1}{2}\boldsymbol{\omega}$. The pseudovector $\boldsymbol{\omega} = \nabla \times \boldsymbol{u}$, which can of course be defined at any $\boldsymbol{x}$, is the *vorticity* of the flow and is of the greatest importance in the dynamical theory that follows.

In two dimensions, $\boldsymbol{\omega} = (0, 0, \omega)$ with $\omega = -\nabla^2\psi$. In this case, the linear flow $c_{ij}x_j$ may be expressed in the form $\alpha(x, -y) + \frac{1}{2}\omega(-y, x)$, with corresponding stream function $\psi = \alpha xy - \frac{1}{4}\omega(x^2 + y^2)$. The streamlines $\psi = \text{const.}$ are elliptic or hyperbolic according to whether $|2\alpha/\omega| > 1$ or $|2\alpha/\omega| < 1$. In the special cases $2\alpha = \pm\omega$, the flow is a *pure shear flow*, with rectilinear streamlines parallel to $x = \pm y$, respectively.

## 3 The Navier–Stokes Equations

### 3.1 Stress Tensor and Pressure

The momentum balance equation may be written in a very fundamental form due to Cauchy:

$$\rho\frac{\mathrm{D}u_i}{\mathrm{D}t} = \frac{\partial}{\partial x_j}\sigma_{ij},$$

where $\sigma_{ij}$ is the *stress tensor* within the fluid, which may include the stress associated with any conservative force field of the form $\boldsymbol{f} = -\nabla V$ that acts on the fluid (e.g., gravity with $V = -\rho\boldsymbol{g} \cdot \boldsymbol{x}$).

A *Newtonian fluid*, which is here assumed to be incompressible, is one in which $\sigma_{ij}$ is related to the rate-of-strain tensor $e_{ij}$ through the linear isotropic equation $\sigma_{ij} = -p\delta_{ij} + 2\mu e_{ij}$, where $p = -\frac{1}{3}\sigma_{ii}$, the pressure in the fluid. Substitution into the Cauchy equation above leads immediately to the Navier–Stokes equation, now in the form

$$\frac{\mathrm{D}\boldsymbol{u}}{\mathrm{D}t} \equiv \frac{\partial \boldsymbol{u}}{\partial t} + \boldsymbol{u} \cdot \nabla\boldsymbol{u} = -\frac{1}{\rho}\nabla p + \nu\nabla^2\boldsymbol{u},$$

where $\nu = \mu/\rho$ is the *kinematic viscosity* of the fluid. Note that $\nu$ has physical dimension $L^2/T$ (like $\psi$ in section 2.1).

### 3.2 The Reynolds Number

Suppose that a flow is characterized by a length scale $L$ (usually associated with the boundary geometry) and a velocity scale $U$ (e.g., the maximum velocity in the fluid or on its boundary). Then, in order of magnitude, $|\mathrm{D}\boldsymbol{u}/\mathrm{D}t| \sim U^2/L$ and $|\nabla^2\boldsymbol{u}| \sim U/L^2$. Hence,

$$\frac{|\mathrm{D}\boldsymbol{u}/\mathrm{D}t|}{\nu|\nabla^2\boldsymbol{u}|} \sim \frac{UL}{\nu}.$$

The dimensionless number $Re = UL/\nu$ is the *Reynolds number* of the flow. If $Re \ll 1$, then viscous forces dominate over inertia forces, which are negligible in a first approximation. If $Re \gg 1$, then inertia forces are dominant; however, as we shall see later, viscous effects always remain important near fluid boundaries, no matter how large $Re$ may be; this is where *boundary-layer theory* must be invoked.

### 3.3 The Vorticity Equation

It is obviously possible to eliminate the pressure field by taking the curl of the NS equation. Using the vector identity $\boldsymbol{u} \cdot \nabla\boldsymbol{u} = \nabla(\boldsymbol{u}^2/2) - \boldsymbol{u} \times \boldsymbol{\omega}$, this yields the *vorticity equation*

$$\frac{\partial \boldsymbol{\omega}}{\partial t} = \nabla \times (\boldsymbol{u} \times \boldsymbol{\omega}) + \nu\nabla^2\boldsymbol{\omega}.$$

The first term on the right-hand side describes transport of the vorticity field by the velocity, while the second describes diffusion of vorticity relative to the fluid. This interpretation means that it is often helpful to focus on vorticity rather than velocity in the analysis of particular problems.

For two-dimensional flow we have seen that $\boldsymbol{\omega} = (0, 0, -\nabla^2\psi)$. The vorticity equation in this case reduces to a nonlinear equation for $\psi$:

$$\frac{\partial}{\partial t}(\nabla^2\psi) - \frac{\partial(\psi, \nabla^2\psi)}{\partial(x, y)} = \nu\nabla^4\psi,$$

where $\nabla^4 = (\nabla^2)^2$, the biharmonic operator.

### 3.4 Difficulties with the NS Equation

The nonlinearity of the NS equation represented by the term $\boldsymbol{u} \cdot \nabla\boldsymbol{u}$ presents a major difficulty for fundamental theory. This is not the only difficulty, however. The viscous term $\nu\nabla^2\boldsymbol{u}$ implies viscous dissipation of kinetic energy (to heat), and associated irreversibility in time. Furthermore, the influence of the pressure-gradient term is nonlocal. To see this, take the divergence of the equation, giving $\nabla^2 p = -s(\boldsymbol{x}, t)$, where $s = \rho\nabla \cdot [(\boldsymbol{u} \cdot \nabla)\boldsymbol{u}]$. The solution of this Poisson equation (in three dimensions) is

$$p(\boldsymbol{x}, t) = \frac{1}{4\pi}\int \frac{s(\boldsymbol{x}', t)}{|\boldsymbol{x} - \boldsymbol{x}'|}\,\mathrm{d}V',$$

plus possible boundary contributions. Thus, $p(\boldsymbol{x}, t)$ is influenced by values of $s(\boldsymbol{x}', t)$, and therefore of $\boldsymbol{u}(\boldsymbol{x}', t)$, at all points $\boldsymbol{x}'$ in the fluid, and the convergence of the integral for large $|\boldsymbol{x} - \boldsymbol{x}'|$ is slow.

This combination of difficulties in relation to the NS equations presents an enormous challenge; even the basic problem of proving regularity for all $t > 0$ of solutions of the NS equations evolving from smooth initial conditions remains unsolved at the present time. This is one of the Clay Mathematics Institute's Millennium Prize Problems.

## 4 Some Exact NS Solutions

Exact solutions of the NS equations fall into two categories: *trivial flows*, for which the inertia forces either vanish identically (as for steady rectilinear flow) or are exactly compensated by a pressure gradient, and *self-similar flows*, whose structure is to some extent dictated by dimensional considerations.

### 4.1 Trivial Flows

Chief among these are

**Poiseuille flow,** driven by a constant pressure gradient $\boldsymbol{G} = -(\partial p/\partial x)\boldsymbol{i}$, where $\boldsymbol{i} = (1, 0, 0)$, in a two-dimensional channel between rigid walls $y = \pm b$, for which the velocity field is $\boldsymbol{u} = (G/2\mu)(b^2 - y^2)\boldsymbol{i}$; and

**Couette flow,** driven by motion of the boundaries $y = \pm b$ with velocities $\pm V\boldsymbol{i}$, for which the velocity field is $\boldsymbol{u} = (Vy/b)\boldsymbol{i}$.

These flows serve as prototypes when questions of stability arise.

Similar flows exist for fluid contained in the annular region between two concentric cylinders $a < r < b$. The Poiseuille flow driven by pressure gradient $G$ parallel to the axis has velocity profile

$$u(r) = \frac{G}{4\mu}\left\{(b^2 - r^2) - (b^2 - a^2)\frac{\log(b/r)}{\log(b/a)}\right\}.$$

The Couette flow corresponding to rotation of the cylinders with angular velocities $\Omega_1$, $\Omega_2$ has circulating velocity (around the axis) $v(r) = Ar + B/r$, where $A$ and $B$ are determined from the boundary conditions $v(a) = \Omega_1 a$, $v(b) = \Omega_2 b$. These two flows may be linearly superposed, giving a flow with helical streamlines.

Slightly less trivial is the *Burgers vortex*, for which the vorticity distribution is in the $z$-direction and has the Gaussian form

$$\omega(r) = \frac{\Gamma\gamma}{4\pi\nu}\exp\left\{-\frac{\gamma r^2}{4\nu}\right\}.$$

Here, $\Gamma$ $(= 2\pi\int\omega(r)r\,\mathrm{d}r)$ is the total strength of the vortex, and $\gamma$ $(>0)$ is the rate-of-strain that must be imposed to keep the vortex steady against the erosive effect of viscous diffusion.

### 4.2 Self-Similar Flows

It will be sufficient to illustrate this type of flow by a simple example: *Jeffery–Hamel flow*. Let $(r, \theta)$ be plane polar coordinates, and suppose that fluid is contained between two planes $\theta = \pm\alpha$ and is extracted by a *line sink* $Q$ at the origin; then, if $u$ is the radial velocity, $\int_{-\alpha}^{\alpha} ur\,\mathrm{d}\theta = -Q$ for all $r$. Here, like $\nu$, $Q$ has dimension $L^2/T$, and we may define a Reynolds number $Re = Q/\nu$. Furthermore, on dimensional grounds, the stream function for the resulting flow must take the form $\psi = Qf(\theta)$, where $f$ is dimensionless. (The velocity $u = r^{-1}\partial\psi/\partial\theta$ is then purely radial.) Substitution into the equation for the stream function in plane polar coordinates,

$$\frac{\partial}{\partial t}(\nabla^2\psi) - \frac{1}{r}\frac{\partial(\psi, \nabla^2\psi)}{\partial(r, \theta)} = \nu\nabla^4\psi,$$

gives an ordinary differential equation (ODE) for $f(\theta)$,

$$f'''' + 4f'' + Re\, f'f'' = 0,$$

which may be integrated three times to give the velocity profile $r^{-1}f'(\theta)$. What is important here is that, from dimensional considerations alone, as described above, the nonlinear PDE for $\psi$ has been reduced to an ODE, still nonlinear but nevertheless relatively easy to solve.

A second example is that of two-dimensional flow toward a *stagnation point* on a plane boundary $y = 0$. The flow far from the boundary is the uniform strain flow for which $\psi \sim \alpha xy$, and conditions of impermeability and *no-slip*, $\psi = \partial\psi/\partial y = 0$, are imposed on $y = 0$. Here, dimensional analysis implies that $\psi = x(\nu\alpha)^{1/2}f(\eta)$, where $\eta = y(\alpha/\nu)^{1/2}$, and again the PDE for $\psi$ reduces to an ODE for $f(\eta)$, which can be easily solved numerically.

Further examples are

- the axisymmetric flow due to a concentrated point force $\boldsymbol{F}$ applied at the origin, for which the velocity field is everywhere inversely proportional to distance from the origin (the *Squire–Landau jet*); and
- the *von Karman flow* due to the differential rotation of two parallel discs about their axes.

In the latter case, the flow between the discs has helical streamlines. This flow has recently provided the basis for the *VKS* (von Karman sodium) experiment demonstrating dynamo action due to flow driven by counterrotating propellers.

## 5 Stokes Flow

When $Re \ll 1$, inertial forces are negligible, and the NS equations simplify to the quasistatic *Stokes equations*

$$\frac{\partial}{\partial x_i}\sigma_{ij} = 0, \qquad \mu\nabla^2 \boldsymbol{u} = \nabla p, \qquad \nabla \cdot \boldsymbol{u} = 0.$$

### 5.1 Some General Results

Certain results concerning *Stokes flows* as governed by these equations can be proved in a straightforward way. First, the equations are now linear, so that if any two solutions are known, then any linear combination of them is also a solution. Second, if a function $\boldsymbol{U}(\boldsymbol{x})$ is defined on a closed surface $S$ bounding a volume $V$ of fluid, satisfying the condition $\int_S \boldsymbol{U}(\boldsymbol{x}) \cdot \boldsymbol{n}\,\mathrm{d}S = 0$, then the solution of the Stokes equations in $V$ corresponding to the boundary condition $\boldsymbol{u}(\boldsymbol{x}) = \boldsymbol{U}(\boldsymbol{x})$ on $S$ exists and is unique. Moreover, the solution $\boldsymbol{v}(\boldsymbol{x})$ corresponding to the boundary condition $\boldsymbol{v}(\boldsymbol{x}) = -\boldsymbol{U}(\boldsymbol{x})$ is simply $\boldsymbol{v}(\boldsymbol{x}) = -\boldsymbol{u}(\boldsymbol{x})$, i.e., reversal of the boundary condition reverses the flow everywhere. Thus, for example, a small spherical patch of dye that is distorted by a flow into a long thin filament can be restored to its original spherical shape (apart from the effect of molecular diffusion) by time-reversal of the motion of the fluid boundary.

Next, there is a *minimum dissipation theorem* relating to the class $C$ of *kinematically possible* flows in a closed volume $V$ (i.e., flows satisfying merely $\nabla \cdot \boldsymbol{u} = 0$ in $V$ and the boundary condition $\boldsymbol{u}(\boldsymbol{x}) = \boldsymbol{U}(\boldsymbol{x})$ on $S$). Let $\Phi = \int_V \sigma_{ij}e_{ij}\,\mathrm{d}V$ be the rate of dissipation of kinetic energy of any such flow. Then the unique Stokes flow $\boldsymbol{u}(\boldsymbol{x})$ satisfying this boundary condition minimizes $\Phi$ within the class $C$. This result clearly does not extend to solutions of the NS equations (since these are within the class $C$ and have greater $\Phi$ than that of the unique Stokes flow).

Finally, there is a *reciprocity theorem* that finds important application in the following subsection: let $(u_i^{(1)}, \sigma_{ij}^{(1)})$ and $(u_i^{(2)}, \sigma_{ij}^{(2)})$ be the velocity and stress fields corresponding to two Stokes flows with different boundary conditions $u_i^{(1)} = U_i^{(1)}$, $u_i^{(2)} = U_i^{(2)}$ on the surface $S$; then

$$\int_S \sigma_{ij}^{(1)} U_j^{(2)} n_i\,\mathrm{d}S = \int_S \sigma_{ij}^{(2)} U_j^{(1)} n_i\,\mathrm{d}S.$$

### 5.2 Flow Due to the Motion of a Particle

From a historic viewpoint, the most important problem in this low-Reynolds-number regime is that solved by Stokes himself in 1851: the flow due to the motion of a rigid sphere through a viscous fluid. Here, we briefly consider the principles governing the flow due to the motion (translation plus rotation) of a rigid particle of arbitrary geometry, the fluid being assumed to be at rest *at infinity*. Let $a = (3V/4\pi)^{1/3}$, where $V$ is the volume of the particle. Attention is naturally focused on the force $\boldsymbol{F}$ and torque $\boldsymbol{G}$ acting on the particle. The instantaneous motion of the particle is determined by the velocity $\boldsymbol{U}$ of its center of volume and its angular velocity $\boldsymbol{\Omega}$, and the linearity of the Stokes equations implies linear relations between $\{\boldsymbol{F}, \boldsymbol{G}\}$ and $\{\boldsymbol{U}, \boldsymbol{\Omega}\}$ of the form

$$F_i = -\mu(aA_{ij}U_j + a^2 B_{ij}\Omega_j),$$
$$G_i = -\mu(a^2 C_{ij}U_j + a^3 D_{ij}\Omega_j),$$

where, by virtue of the above reciprocity theorem,

$$A_{ij} = A_{ji}, \qquad D_{ij} = D_{ji}, \qquad C_{ij} = B_{ji}.$$

These dimensionless tensor coefficients are determined solely by the shape of the particle. Any symmetry imposes further constraints. If the particle is *mirror symmetric* (invariant under reflection with respect to its center of volume), then the (pseudo)tensor $C_{ij}$ must vanish. If the particle has at least the symmetry of a cube (i.e., is invariant under the group of rotations of a cube), then $A_{ij}$ and $D_{ij}$ must be isotropic: $A_{ij} = \alpha\delta_{ij}$, $B_{ij} = \beta\delta_{ij}$. The case of a sphere is classic; in this case, as shown by Stokes, $\alpha = 6\pi$, $\beta = 8\pi$.

Particles of helical shape are obviously not mirror symmetric, and for these, $C_{ij} \neq 0$. This means that such a particle freely sedimenting through a fluid must experience a torque and will therefore also rotate. Equally, if such a particle rotates (through some internal mechanism), then it will experience a force causing it also to translate. Microscopic organisms can propel themselves through a viscous environment by adopting a swimming strategy that exploits this phenomenon.

## 6 Inviscid Vorticity Dynamics

In the formal *Euler limit* $Re = \infty$, it is tempting to simply set $\nu = 0$ in the NS equation and in the vorticity equation (section 3.3). Note, however, that, since $\nu$ multiplies the highest space derivative in these equations, the order of the equations is reduced in so doing, and it is not possible to satisfy the no-slip condition at rigid fluid boundaries in this Euler limit. We shall consider boundary effects in section 7; for the moment, we consider the evolution of an isolated blob of vorticity far from any fluid boundary. The *vortex ring* as visualized by convected smoke is a well-known prototype. Vortex rings occur in a wide variety of circumstances, e.g., in the gas emitted impulsively in volcanic eruptions, or in the ocean, created by dolphins in the course of their underwater antics.

### 6.1 Helmholtz's Laws of Vortex Motion

The study of vorticity goes back to Hermann von Helmholtz (1858) and the subsequent work of P. G. Tait and Lord Kelvin that this stimulated. Helmholtz recognized that vortex lines are transported with the fluid and that vortex tubes have constant strength in the course of this motion. This means that, if the flow is such as to stretch a vortex tube, then its cross section decreases but the flux of vorticity along the tube (equivalent to the *circulation* around it) is constant. Helmholtz also maintained that vortex lines must either be closed curves or end on fluid boundaries, but it is now known that this is incorrect; even for a localized blob of vorticity without any obvious symmetry, a vortex line will in general wind indefinitely, rather like a ball of wool, but in a chaotic manner.

### 6.2 Conservation of Helicity

The fact that vortex lines move like material lines within the fluid implies that any topological structure of the vorticity field is conserved for all time. For example, as recognized by Kelvin in 1867, if a vortex tube is knotted on itself, then this knot persists for all time. Let $S$ be any closed surface moving with the fluid (a *Lagrangian surface*) on which $\boldsymbol{\omega} \cdot \boldsymbol{n} = 0$, a condition that persists for all $t$. Then for each such surface containing a volume $V$, we may define

$$\mathcal{H} = \int_V \boldsymbol{u} \cdot \boldsymbol{\omega} \, \mathrm{d}V,$$

the *helicity* within $V$. This helicity is constant, as may be proved directly from the Euler equations, and it is known that this integral provides a measure of the linkage of vortex lines within $V$. For example, if the vorticity distribution consists of two linked vortex tubes with (constant) circulations $\kappa_1$ and $\kappa_2$, and if $V$ contains just these two linked tubes, then $\mathcal{H} = \pm 2n\kappa_1\kappa_2$, where $n$ is the Gauss linking number of the linkage, and the sign is chosen according to whether the linkage is right- or left-handed. This result provides a bridge between the Euler equations of inviscid fluid mechanics and a fundamental concept of topology.

### 6.3 The Biot–Savart Law

The relation inverse to $\boldsymbol{\omega} = \nabla \times \boldsymbol{u}$, $\nabla \cdot \boldsymbol{u} = 0$, is given by the *Biot–Savart law*

$$\boldsymbol{u}(\boldsymbol{x}, t) = \frac{1}{4\pi} \int \frac{\boldsymbol{\omega}(\boldsymbol{x}', t) \times (\boldsymbol{x} - \boldsymbol{x}')}{|\boldsymbol{x} - \boldsymbol{x}'|^3} \, \mathrm{d}V',$$

which shows how the velocity at any point may be instantaneously obtained from the vorticity distribution, a purely kinematic result. Knowing $\boldsymbol{u}(\boldsymbol{x}, t)$, Eulerian fluid dynamics is now completely contained in the statement that the vorticity field $\boldsymbol{\omega}(\boldsymbol{x}, t)$ is convected by the velocity field $\boldsymbol{u}(\boldsymbol{x}, t)$ that is *induced* in this way.

### 6.4 Vortex Ring Propagation

This consideration enabled Kelvin to calculate the velocity of propagation of a vortex ring of circulation $\Gamma$ in the form

$$V \sim \frac{\Gamma}{4\pi R} \left\{ \log\left(\frac{8R}{\sigma}\right) - \frac{1}{4} \right\},$$

where $R$ is the radius of the ring and $\sigma$ ($\ll R$) is the radius of its cross section.

More generally, the velocity of a thin curved vortex filament with parametric equation $\boldsymbol{x} = \boldsymbol{X}(s, t)$ is frequently assumed to be given by the *local induction approximation* in the compact normalized form $\boldsymbol{X}_t = \boldsymbol{X}_s \times \boldsymbol{X}_{ss}$.

## 7 The Aerodynamics of Flight

Mastery of the aerodynamics of flight was arguably the greatest engineering accomplishment of the twentieth century. This mastery required an understanding of the role played by viscosity in the immediate neighborhood of an aircraft wing—in other words, of boundary-layer theory, or what is known at a more sophisticated level as the theory of *matched asymptotic expansions.*

### 7.1 The External Irrotational Flow

It is convenient to adopt a frame of reference fixed in the aircraft; in this frame, the fluid velocity *at infinity* is uniform, $(-U, 0, 0)$, where $U$ is the speed of flight. In a steady two-dimensional flow, the vorticity equation reduces when $\nu = 0$ to $\mathrm{D}\omega/\mathrm{D}t = 0$, so that vorticity is constant on streamlines. The vorticity is zero on every streamline that comes from far upstream; thus $\nabla \times \boldsymbol{u} = 0$; i.e., the flow is *irrotational* in the Euler limit. The pressure in a steady irrotational flow is found by integrating Euler's equation, giving *Bernoulli's theorem* $p = p_0 - \frac{1}{2}\rho \boldsymbol{u}^2$, where $p_0$ is a constant reference pressure. Thus, where $|\boldsymbol{u}|$ is high, $p$ is low, and vice versa. Aerodynamic lift on the wings results from lower air speed (and thus higher pressure) on the lower surface of the wings, a matter of some comfort for jet-age travelers.

The classical theory of irrotational flow past a circular cylinder allows for a circulation $\kappa$, the flow due to an apparent point vortex at the center of the circle. Given an airfoil cross section with a sharp trailing edge, the exterior region can be conformally mapped to the exterior of a circle. This provides a solution for the irrotational flow past the airfoil, still including the arbitrary circulation $\kappa$. There are two stagnation points (where $\boldsymbol{u} = 0$) on the airfoil surface: one on the lower side and one on the upper side. The position of these stagnation points varies with $\kappa$, and there is a unique value of $\kappa = \kappa_c$ that moves the upper stagnation point exactly to the trailing edge; for this value, the airflow leaves the trailing edge smoothly. The *Kutta-Joukowski hypothesis* asserts that the critical value $\kappa_c$ is actually realized in practice. The lift per unit span of aircraft wing is then evaluated as $L = \rho U \kappa_c$. In order of magnitude, $\kappa_c \sim Ua \sin\alpha$, where $a$ is the streamwise extent of the wing and $\alpha$ is the *angle of attack* (see figure 1); so $L \sim \rho a U^2 \sin\alpha$, independent of the value of $\nu$. The lift on the whole wing is therefore $\rho U^2$ times the area of the wing projected on the flight direction.

If any lower value of $\kappa$ is chosen, then air flows round the sharp trailing edge T, at which point there is a singularity of air velocity. The vorticity generated by viscosity between T and the stagnation point S on the upper side of the wing is convected downstream into the wake, with a compensating increase in $\kappa$; this process persists until the Kutta-Joukowski condition is indeed satisfied (the stagnation point then suppressing the singularity).

It remains to explain why, if the lift force on which flight depends is independent of the very small value of $\nu$ in air, the fact that $\nu \neq 0$ is nevertheless essential for this lift to exist at all. This paradoxical role of weak viscosity forces us to focus attention on the immediate vicinity of the wing surface, i.e., the boundary layer.

### 7.2 The Boundary Layer

Let O be any point on the wing surface, and let $Oxy$ be a locally Cartesian coordinate system, with $Ox$ tangential to the boundary and $Oy$ normal to it. The tangential velocity increases from zero on the boundary to $U(x)$ (determined from irrotational theory) in the external stream. In this thin layer, $\nabla^2\psi \approx \partial^2\psi/\partial y^2 = \psi_{yy}$, so the steady vorticity equation simplifies to

$$-\frac{\partial(\psi, \psi_{yy})}{\partial(x, y)} = \nu\psi_{yyyy}.$$

This integrates with respect to $y$ to give

$$\psi_y\psi_{xy} - \psi_x\psi_{yy} = G(x) + \nu\psi_y yy,$$

or, in terms of the tangential and normal velocity components $u = \psi_y$, $v = -\psi_x$,

$$uu_x + vu_y = G(x) + \nu u_{yy}.$$

Here, $G(x)$ is the "constant of integration," which may be identified with the "$-\partial p/\partial x$" of the NS equation; it turns out, therefore, that the pressure is independent of the normal coordinate $y$; in fact, since $u \sim U(x)$ for large $y$, we must have $G(x) = U(\mathrm{d}U/\mathrm{d}x)$. The external pressure is *impressed on the boundary layer.* If $G(x) > 0$, the pressure gradient tends to accelerate the flow and is described as *favorable,* while if $G(x) < 0$, it tends to retard the flow and is described as *adverse.*

At this stage, dimensional analysis may be used to great effect. First note that $\nu$ may be scaled out of the above equation through defining $\hat{y} = y/\nu^{1/2}$, $\hat{\psi} = \psi/\nu^{1/2}$, and note that $y = O(\nu^{1/2})$ when $\hat{y} = O(1)$. We now ask under what circumstances a similarity solution of the form

$$\psi = (\nu x U(x))^{1/2} f(\eta), \quad \text{with } \eta = y\left(\frac{U(x)}{\nu x}\right)^{1/2}$$

is possible. It turns out that this requires that $U(x) = Ax^m$ (so $G(x) = mA^2x^{2m-1}$), for some constants $A$, $m$, and that $f(\eta)$ then satisfies the equation

$$f''' + \tfrac{1}{2}(m+1)ff'' + m(1 - f'^2) = 0,$$

with boundary conditions $f(0) = f'(0) = 0$, $f'(\infty) = 1$, the last of these coming from the required matching to the external flow. For the particular value $m = 0$, the equation is known as the *Blasius equation,* and it describes the boundary layer on a flat plate aligned with the stream, with zero pressure gradient. More generally, it is known as the *Falkner-Skan equation.*

A *well-behaved* solution of the Falkner-Skan equation is one for which $0 < f' < 1$ and $f'' > 0$ for all positive $\eta$, i.e., one for which the tangential velocity rises smoothly from zero on the boundary $y = 0$ to its asymptotic value in the external stream. It is known that such a solution exists for all positive $m$, and even for mildly negative $m$ in the range $m > -0.09$. However, no well-behaved solution exists for $m < -0.09$, and the only solutions in this range of adverse pressure gradient exhibit reversed flow ($f' < 0$) near the wall. This suggests that a boundary layer cannot survive against a strong adverse pressure gradient.

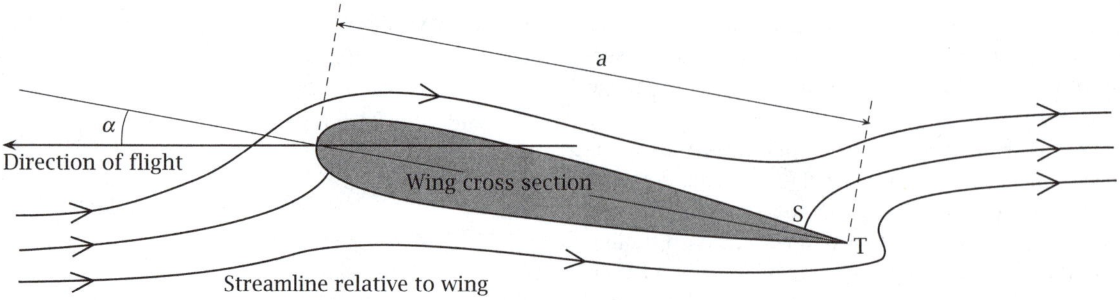


**Figure 1** Cross section of an aircraft wing (airfoil), and streamline of the irrotational flow relative to the wing in steady flight at angle of incidence $\alpha$. Additional circulation associated with viscous shedding of vorticity moves the stagnation point S into coincidence with the trailing edge T, causing the flow to leave the trailing edge smoothly.

This conclusion is supported both by numerical solution of the full PDE for $\psi$ and by observation of flow past *bluff* (rather than *streamlined*) bodies, for which the boundary layer is seen to *separate* from the boundary, creating a substantial *wake* region in which the flow recirculates with nonzero vorticity.

### 7.3 A Comment on Flow Separation

Boundary-layer separation at high Reynolds number is one of the most difficult aspects of fluid mechanics, and this has led to theoretical investigations of great sophistication. Separation is responsible for the aerodynamic phenomenon of *stall* with consequential loss of lift and increase of drag when an aircraft's angle of attack (the inclination of the wings to the oncoming stream) increases beyond a critical value. Control of separation is therefore crucial to maintenance of stable flight.

We may note, however, that separation occurs even in low-Reynolds-number situations. Consider, for example, an oncoming shear flow $(\alpha y, 0, 0)$ over a rigid boundary $y = 0$ that takes a sudden Gaussian dip in the neighborhood of $x = 0$:

$$y = -y_0 \mathrm{e}^{-(x/\delta)^2},$$

where $y_0 \gg \delta$. The Reynolds number can be taken to be $Re = \alpha\delta^2/\nu$. When $Re \ll 1$, this flow separates at some point $x = x_c = O(-\delta)$, in the sense that a streamline detaches from the boundary at this point, with reversed flow near the boundary immediately beyond it. This type of low-Reynolds-number separation is well documented. Now consider what happens as the Reynolds number is continuously increased by reducing $\nu$ while keeping the geometry and the upstream velocity constant. The separation will undoubtedly persist, although the precise location of the separation point $x_c$ may be expected to change slightly with increasing $Re$.

An entirely different consideration, namely that of flow instability, also arises with an increase in $Re$. We now turn to this important branch of fluid dynamics.

## 8 Flow Instability

In consideration of the instabilities to which steady flows may be subject, one may distinguish between *fast instabilities*, i.e., those that are of purely inertial origin and have growth rates that do not depend on viscosity, and *slow instabilities*, which are essentially of viscous origin and have growth rates that therefore tend to zero as $\nu \to 0$, or equivalently as $Re = UL/\nu \to \infty$. Examples of fast instabilities are the *Rayleigh–Taylor instability* that occurs when a heavy layer of fluid lies over a lighter layer, the *centrifugal instability* (leading to *Taylor vortices*) that occurs in a fluid undergoing differential rotation when the circulation about the axis of rotation decreases with radius, and the *Kelvin–Helmholtz instability* that occurs in any region of rapid shearing of a fluid. The best-known example of a slow instability is the instability of pressure-driven Poiseuille flow between parallel planes, which is associated with subtle effects of viscosity in *critical layers* near the boundaries. The *dynamo instability* of magnetic fields in electrically conducting fluids is usually also diffusive in origin (through magnetic diffusivity rather than viscosity) and may therefore also be classed as a slow instability (although exotic examples of fast dynamo instability have also been identified).

### 8.1 The Kelvin–Helmholtz Instability

This may be idealized as the instability of a *vortex sheet* located on the plane $y = 0$ in an inviscid fluid of infinite extent. The basic velocity field is then taken to be $\boldsymbol{U} = (\mp U, 0, 0)$ for $y > 0$ or $y < 0$, respectively. We suppose that the vortex sheet is slightly deformed to $y = \eta(x,t) = \hat{\eta}(t)e^{ikx}$, where the real part is understood. The associated perturbation velocity field is irrotational, and it is given by $\boldsymbol{u} = \nabla\phi_{1,2}$ for $y > \eta$ and $y < \eta$, respectively. Incompressibility implies that $\nabla^2\phi_{1,2} = 0$, and the requirement that the disturbance decays at $y = \pm\infty$ then leads to $\phi_{1,2} = \Phi_{1,2}(t)\exp(\mp ky + ikx)$. The sheet *moves with the fluid*, according to Helmholtz; moreover, pressure is continuous across it. These two conditions, when linearized in the small perturbation quantities, lead to the amplitude equation $d^2\hat{\eta}/dt^2 = \frac{1}{4}k^2U^2\hat{\eta}$. The vortex sheet is therefore subject to an instability $\eta \sim \exp e^{\sigma t}$, where $\sigma$ is given by the *dispersion relation* $\sigma = \frac{1}{2}|kU|$. The physical interpretation of this instability is that the surface vorticity on the sheet tends to concentrate at the inflection points where $\eta_x > 0$, $\eta_{xx} = 0$, and the induced velocity then tends to amplify the perturbation of the sheet.

A notable feature of this dispersion relation is that as the wavelength $\lambda = 2\pi/k$ of the disturbance decreases to zero, its growth rate $\sigma$ increases without limit. This instability is therefore quite dramatic. As the disturbance grows, nonlinear effects inevitably become important. In the nonlinear regime, the vorticity continues to concentrate at the upward-sloping inflection points, the distribution of vorticity on the sheet becoming cusp-shaped at a finite time, beyond which it is believed that *spiral windup* of the vortex sheet around each such cuspidal singularity occurs.

The same type of Kelvin–Helmholtz instability occurs for any parallel flow having an *inflectional velocity profile*; it is ubiquitous in nature and is one of the main mechanisms of transition from laminar to turbulent flow.

### 8.2 Transient Instability

This is a very different type of instability, but it is also a key ingredient in the process of transition to turbulence for many flows. It is well illustrated with reference to plane Couette flow $\boldsymbol{U} = (\alpha y, 0, 0)$ for $|y| < b$, a flow that is known to be stable to conventional disturbances of normal-mode type. Here, we consider the central region of such a flow and neglect the influence of the boundaries. In this region the flow is subject to disturbances of the form

$$\boldsymbol{u} = \boldsymbol{A}(t)e^{i\boldsymbol{k}(t)\cdot\boldsymbol{x}}, \qquad \boldsymbol{A}(t)\cdot\boldsymbol{k}(t) = 0,$$

where $\boldsymbol{k}(t) = (k_1, k_2 - \alpha k_1 t, k_3)$. This type of disturbance, whose wave vector $\boldsymbol{k}(t)$ is itself sheared by the flow, is called a *Kelvin mode*. Substitution into the NS equation and elimination of the pressure leads to an exact solution for the amplitude $\boldsymbol{A}(t)$. This solution reveals that, when $k_1^2 \ll k_2^2 + k_3^2$ and viscous effects are negligible, the component $A_1(t)$ increases linearly (rather than exponentially) for a long time. This instability may be attributed to the term $\boldsymbol{u}\cdot\nabla\boldsymbol{U} = u_2(d\boldsymbol{U}/dy)$ in the NS equation, which corresponds to persistent transport in the $y$-direction of the $x$-component of mean momentum. This is the essence of *transient*, as opposed to normal-mode, instability. Transient instability is responsible for the emergence of *streamwise vortices* in shear flows. These vortices can themselves become prone to *secondary instabilities*, almost inevitably leading to transition to turbulence.

### 8.3 Centrifugal Instability

Consider now the *circular Couette flow* between rotating cylinders, for which the velocity field takes the form $\boldsymbol{u} = (0, v(r), 0)$ in cylindrical polar coordinates $(r, \theta, z)$. The radial pressure gradient required to balance the centripetal force is $dp/dr = \rho v^2/r$. For the moment, neglect viscous effects. Suppose that a ring of fluid of radius $r$ expands to radius $r_1 = r + \delta r$, its angular momentum remaining constant; its velocity then becomes $v_1 = rv(r)/r_1$. The centripetal force acting on the ring will now be $\rho(rv/r_1)^2$, and if this is greater than the local restoring pressure gradient $\rho v_1^2/r_1$, the ring will continue to expand. The condition for instability is then $|rv| > |r_1 v(r_1)|$ or, equivalently,

$$\frac{d}{dr}|rv(r)| < 0.$$

Thus, if the angular momentum (or, equivalently, the circulation) decreases outward, then the flow is prone to centrifugal instability (the *Rayleigh criterion*).

Such a flow is realized between coaxial cylinders, when the inner cylinder (of radius $a$) rotates about its axis with angular velocity $\Omega$ while the outer cylinder is stationary. If the gap $d$ between the cylinders is small compared with $a$, then, as shown by G. I. Taylor in 1923, when due account is taken of viscous effects, the criterion for instability to axisymmetric disturbances of the above kind becomes, to good approximation,

$$\frac{ad^3\Omega^2}{\nu^2} > 1708.$$

The dimensionless number $Ta \equiv ad^3\Omega^2/\nu^2$ is known as the Taylor number. The instability when $Ta > 1708$

manifests itself as a sequence of axisymmetric *Taylor vortices* of nearly square cross section and alternating sign of azimuthal vorticity.

There is a close analogy between centrifugal instability and the convective instability of a horizontal layer of fluid of depth $d$, heated from below (*Rayleigh-Bénard convection*). Here, the analogue of the Taylor number is the *Rayleigh number* $Ra = g\beta d^3 \Delta T / \nu\kappa$, where $\kappa$ is the thermal diffusivity of the fluid, $\beta$ its coefficient of thermal expansion, and $\Delta T$ the temperature difference between the bottom of the layer and the top. The analogy is so close that the critical Rayleigh number for instability is also $Ra = 1708$. For $Ra$ just a little greater than 1708, a steady pattern of convection is established: either rolls with horizontal axes or convective cells with hexagonal planform, the choice being influenced by weak nonlinearity and other subtle effects.

### Further Reading

Aref, H. 1984. Stirring by chaotic advection. *Journal of Fluid Mechanics* 143: 1–21.

Batchelor, G. K. 1967. *An Introduction to Fluid Dynamics.* Cambridge: Cambridge University Press.

Drazin, P. G., and W. H. Reid. 2004. *Hydrodynamics Stability*, 2nd edn. Cambridge: Cambridge University Press.

Fefferman, C. L. 2000. Existence and smoothness of the Navier–Stokes equations. In *The Millennium Prize Problems*, pp. 57–67. Providence, RI: Clay Mathematics Institute.

Happel, J., and H. Brenner. 1965. *Low Reynolds Number Hydrodynamics.* Englewood Cliffs, NJ: Prentice Hall.

Pinter, A., B. Dubrulle, and F. Daviaud. 2010. Kinematic dynamo simulations of von Karman flows: application to the VKS experiment. *European Physical Journal* B 74: 165–76.

Saffman, P. G. 1992. *Vortex Dynamics.* Cambridge: Cambridge University Press.

Taylor, G. I. 1971. *Scientific Papers, Volume 4: Mechanics of Fluids*, edited by G. K. Batchelor. Cambridge: Cambridge University Press.

## IV.29 Magnetohydrodynamics

*David W. Hughes*

### 1 Introduction

Magnetohydrodynamics (MHD) is the study of the motion of an electrically conducting fluid or plasma in the presence of a magnetic field. It is employed principally in relation to astrophysical and geophysical magnetic fields, but there are also several important terrestrial and industrial applications. On the grand scale, magnetic fields are both ubiquitous and dynamically significant; determining a theoretical understanding of their behavior is therefore crucial to understanding the dynamics of star formation, the interstellar medium, accretion disks, stellar atmospheres, and stellar and planetary interiors. Figure 1 shows a high-resolution image taken from space of the solar atmosphere, revealing the scale and complexity of the magnetic field. The most ambitious terrestrial application of MHD lies in the quest to harness energy from nuclear fusion; controlling the confinement of fusion plasma in tokamaks through the imposition of strong magnetic fields poses a formidable theoretical and experimental challenge. In a rather different industrial direction, MHD seeks to explain the complex turbulent motion that arises via the regulation, by magnetic fields, of the flow of liquid metals in casting and refining operations.

The origins of MHD might be traced to Larmor's short, but highly influential, 1919 paper entitled "How could a rotating body such as the sun become a magnet?" in which it was postulated that the swirling motions inside stars could maintain a magnetic field. This was followed by the early theoretical development of the subject in the 1930s and 1940s by pioneers such as Alfvén, Cowling, Elsasser, and Hartmann. A century earlier, in what might be regarded as a precursor to MHD, Faraday had investigated the role of a moving fluid conductor by attempting to measure the potential difference induced by the Thames flowing in the Earth's magnetic field.

The governing equations of MHD are derived from combining the ideas of fluid dynamics with those of electromagnetism. In spirit, though, it is closer to the former. The "fluid" under consideration in MHD may be truly a fluid, such as the liquid iron in the Earth's outer core, or, as in astrophysical contexts, it may be an ionized gas or *plasma*. To a very good approximation, collision-dominated plasmas, as found in stellar interiors, can be treated as a single fluid, and their behavior is therefore well described within MHD. However, in other contexts, e.g., stellar atmospheres, a more complex (multifluid) plasma description may sometimes be required.

Magnetohydrodynamics extends fluid dynamics in two significant ways: by the addition of a new equation, the *magnetic induction equation*, and by the incorporation of the magnetic force, the *Lorentz force*, into the MOMENTUM (NAVIER–STOKES) EQUATION [III.23 §2]. The

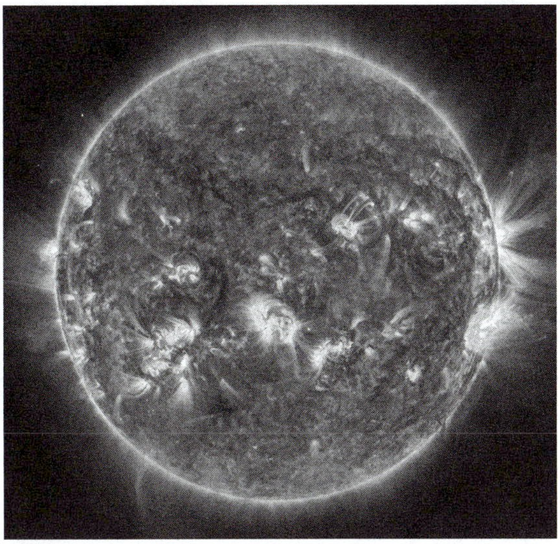

**Figure 1** An extreme ultraviolet image of the sun's atmosphere, taken by the Solar Dynamics Observatory. Bright regions are locations of intense magnetic activity. The magnetic field emerges from the solar interior, forming sunspots on the surface and arched structures (*coronal loops*) in the solar atmosphere.

induction equation describes the influence of the velocity on the magnetic field; conversely, the Lorentz force describes the back reaction of the magnetic field on the velocity. Thus, the field and flow are linked in a complex, nonlinear fashion. MHD therefore contains all of the complexities and subtleties of (nonmagnetic) fluid dynamics together with some unambiguously magnetic novel phenomena.

## 2 The Magnetic Induction Equation

Magnetohydrodynamics is a nonrelativistic theory, valid for fluid velocities much less than the speed of light; its starting point is, therefore, the set of "pre-Maxwell" equations of electromagnetism, in which the displacement current is neglected. The electric field $\boldsymbol{E}$, magnetic field $\boldsymbol{B}$, and electric current $\boldsymbol{J}$ (real three-dimensional vectors) then satisfy Ampère's law, Faraday's law, and the divergence-free condition on the magnetic field. In MKS units, these are expressed as

$$\nabla \times \boldsymbol{B} = \mu_0 \boldsymbol{J}, \qquad \frac{\partial \boldsymbol{B}}{\partial t} = -\nabla \times \boldsymbol{E}, \qquad \nabla \cdot \boldsymbol{B} = 0,$$

where $\mu_0$ is the magnetic permeability. To obtain a closed system of equations, one more relation (Ohm's law) is needed, linking $\boldsymbol{E}$, $\boldsymbol{B}$, and $\boldsymbol{J}$. In MHD it is customary to adopt the simplest form of Ohm's law, namely, $\boldsymbol{J}' = \sigma \boldsymbol{E}'$, where the fields $\boldsymbol{J}'$ and $\boldsymbol{E}'$ are measured in a frame of reference moving with a fluid element and where $\sigma$ is the electrical conductivity. Relative to a fixed reference frame, this becomes

$$\boldsymbol{J} = \sigma(\boldsymbol{E} + \boldsymbol{u} \times \boldsymbol{B}),$$

where $\boldsymbol{u}$ is the fluid velocity. (Additional plasma processes—such as an electron pressure gradient or the Hall effect, which are not included in classical MHD—can be incorporated via a generalized version of Ohm's law.) Eliminating $\boldsymbol{E}$ and $\boldsymbol{J}$ and, for simplicity, assuming that the conductivity is uniform leads to the magnetic induction equation:

$$\frac{\partial \boldsymbol{B}}{\partial t} = \nabla \times (\boldsymbol{u} \times \boldsymbol{B}) + \eta \nabla^2 \boldsymbol{B}, \tag{1}$$

where $\eta = 1/(\sigma \mu_0)$ is the magnetic diffusivity. In MHD primacy is afforded to the magnetic field; the electric field and current are secondary, but, if needed, they can be evaluated from a knowledge of $\boldsymbol{B}$. Similarly, Gauss's law relating the electric field to the electric charge plays no explicit role in the above arguments; it simply provides a means of determining the charge once $\boldsymbol{E}$ has been determined.

The induction equation describes the evolution of the magnetic field subject to advection and diffusion. A measure of the relative magnitude of these two terms is given by the *magnetic Reynolds number* $Rm = UL/\eta$, where $U$ and $L$ are representative velocity and length scales. In astrophysical bodies, characterized by vast length scales and high conductivities, $Rm$ is invariably large and often extremely so (in the solar convection zone; for example, $Rm \approx 10^9$ at depth and $Rm \approx 10^6$ close to the surface; in the interstellar medium, $Rm \approx 10^{17}$). On the other hand, for most industrial flows of liquid metals, $Rm$ is small ($\lesssim 10^{-2}$).

On expanding the curl of the vector product, and using $\nabla \cdot \boldsymbol{B} = 0$, equation (1) can alternatively be written as

$$\frac{\mathrm{D}\boldsymbol{B}}{\mathrm{D}t} = \boldsymbol{B} \cdot \nabla \boldsymbol{u} - (\nabla \cdot \boldsymbol{u})\boldsymbol{B} + \eta \nabla^2 \boldsymbol{B},$$

where $\mathrm{D}/\mathrm{D}t \equiv \partial/\partial t + \boldsymbol{u} \cdot \nabla$ denotes the Lagrangian or material derivative "following the fluid." The three terms on the right-hand side denote, respectively, contributions to magnetic field evolution due to field line stretching, the compressibility of the fluid, and magnetic diffusion.

It is often helpful to decompose the (solenoidal) magnetic field into poloidal and toroidal components:

$$\boldsymbol{B} = \boldsymbol{B}_{\mathrm{P}} + \boldsymbol{B}_{\mathrm{T}} = \nabla \times \nabla \times (P\boldsymbol{r}) + \nabla \times (T\boldsymbol{r}),$$

where $P$ and $T$ are scalar functions of position and $\boldsymbol{r}$ is the position vector. For axisymmetric fields, $\boldsymbol{B}_{\mathrm{T}}$ is azimuthal and $\boldsymbol{B}_{\mathrm{P}}$ meridional.

## 3 Perfectly Conducting Fluids

Motivated by the fact that $Rm \gg 1$ in astrophysics, it is instructive to consider the idealized dynamics of a perfectly conducting fluid (often referred to as *ideal MHD*), for which $Rm$ is formally infinite and, from (1), $\boldsymbol{B}$ is described simply by

$$\frac{\partial \boldsymbol{B}}{\partial t} = \nabla \times (\boldsymbol{u} \times \boldsymbol{B}) \tag{2}$$

(though of course one always needs to exercise caution when dropping the term with the highest derivative). Here, we consider two important properties of the magnetic field in a perfectly conducting fluid.

### 3.1 Flux Freezing

Equation (2) for the magnetic field $B$ is formally identical to the vorticity equation of an inviscid fluid and, as such, we may seek analogs of the HELMHOLTZ VORTEX THEOREMS [IV.28 §6.1]. The proofs go through unaltered, since they do not rely on the additional constraint that the vorticity is the curl of the velocity. Thus, for any material surface $S_{\mathrm{m}}$ moving with the fluid it can be shown that

$$\frac{\mathrm{D}}{\mathrm{D}t}\int_{S_{\mathrm{m}}} \boldsymbol{B}\cdot\boldsymbol{n}\,\mathrm{d}S = 0,$$

where $\boldsymbol{n}$ is the normal to the surface. In other words, the flux through a surface moving with the fluid is conserved. It is then straightforward to show that two fluid elements initially on the same magnetic field line will remain on that line for all subsequent times. This is Alfvén's famous result that the magnetic field lines move with the fluid; this is often referred to as the field lines being *frozen* into the fluid. One of the earliest results of MHD—Ferraro's 1937 law of isorotation—is a particular case of flux freezing. It states that if the angular velocity $\Omega$ of a flow is constant on lines of poloidal magnetic field (i.e., $\boldsymbol{B}_{\mathrm{P}}\cdot\nabla\Omega = 0$), then there is no tendency to generate toroidal field through the pulling out of poloidal field. Conversely, if $\boldsymbol{B}_{\mathrm{P}}\cdot\nabla\Omega \neq 0$, then toroidal field is "wound up" from a poloidal component in a differentially rotating flow, an important consideration given that most astrophysical bodies support large-scale shearing motions.

A related result comes from combining the induction equation with the equation for the conservation of mass to give

$$\frac{\mathrm{D}}{\mathrm{D}t}\left(\frac{\boldsymbol{B}}{\rho}\right) = \frac{\boldsymbol{B}}{\rho}\cdot\nabla\boldsymbol{u}.$$

Thus $\boldsymbol{B}/\rho$ satisfies the same equation as a material line element and, consequently, will increase or decrease in magnitude in direct proportion to the stretching or compression of such an element. One important consequence is the amplification of the magnetic field during the gravitational collapse of astronomical bodies. When this is particularly dramatic, as in the formation of a neutron star, in which the radius can decrease by a factor of $O(10^5)$, it leads to an immense ($O(10^{10})$) increase in the magnetic field strength.

### 3.2 Magnetic Helicity

Since the magnetic field $\boldsymbol{B}$ is solenoidal, it can be expressed as $\boldsymbol{B} = \nabla\times\boldsymbol{A}$, where $\boldsymbol{A}$ is the magnetic vector potential. Then, by analogy with the HELICITY OF A FLUID FLOW [IV.28 §6.2], the *magnetic helicity* of a volume $V$ bounded by a magnetic surface (i.e., one on which $\boldsymbol{B}\cdot\boldsymbol{n} = 0$) is defined as

$$\mathcal{H} = \int_V \boldsymbol{A}\cdot\boldsymbol{B}\,\mathrm{d}V;$$

$\mathcal{H}$ is invariant to gauge transformations $\boldsymbol{A}\to\boldsymbol{A}+\nabla\phi$.

The magnetic helicity provides a measure of the linkage of magnetic flux tubes. In a perfectly conducting fluid, in which field lines are frozen to the fluid, we would therefore expect this topological property to be preserved for all times. This can indeed be proved straightforwardly from a combination of the induction equation and the equation for the conservation of mass. With finite diffusivity, $\mathcal{H}$ is no longer conserved. It has, though, been conjectured that $\mathcal{H}$ is, nonetheless, "approximately conserved" in high-$Rm$ fluids, and that in decaying MHD turbulence, magnetic energy will decay, whereas magnetic helicity will, to a first approximation, remain constant.

## 4 Kinematic Considerations

The interaction between the two processes of advection and diffusion, encapsulated by the induction equation, is brought out in the following simple important examples, in which we shall suppose that the flow is prescribed.

### 4.1 Stagnation Point Flow

The evolution of a unidirectional magnetic field, $\boldsymbol{B} = (0, B(x,t), 0)$, say, in the two-dimensional stagnation point flow $\boldsymbol{u} = \lambda(-x, y, 0)$ is governed by the equation

$$\frac{\partial B}{\partial t} - \lambda x \frac{\partial B}{\partial x} = \lambda B + \eta \frac{\partial^2 B}{\partial x^2}.$$

Magnetic field is brought in toward $x = 0$ by the flow, but in a region of strong field gradients, it is then subject to diffusion. A steady state is attained in which these processes balance. Two solutions are of particular interest, providing prototypical examples of fundamental astrophysical processes.

For a magnetic field of one sign,

$$B(x) = B_{\max} \exp(-\lambda x^2/2\eta),$$

affording the simplest model of the concentration of field into a flux rope. If the total magnetic flux is finite, and equal to $B_0 L$, say, then it accumulates into a rope of width $O(Rm^{-1/2}L)$ and maximum field strength $O(Rm^{1/2}B_0)$.

For a field that changes sign, vanishing at $x = 0$,

$$B(x) = \frac{E}{\sqrt{\eta\lambda}} \exp(-x^2\lambda/2\eta) \int_0^{x\sqrt{\lambda/\eta}} \exp(s^2/2)\,\mathrm{d}s,$$

where $E$ is the (constant) electric field in the $z$-direction. This provides a simple description of a current sheet formed by the annihilation of magnetic fields of opposite sign. It is an important ingredient in the more complex process of *magnetic reconnection*, in which magnetic energy is released rapidly through a change in the topology of the field.

### 4.2 Flux Expulsion

In a swirling flow, a magnetic field is distorted and amplified; its transverse scale of variation decreases, leading to diffusion becoming important and the field being annihilated, or *expelled* from the flow. An extremely complicated version of this process takes place, for example, in the turbulent convection cells observed at the solar surface. The underlying physics, and in particular the timescale of expulsion, can, though, be understood via the idealized example of a single laminar eddy, with a field in the plane of the flow. Consider the flow $\boldsymbol{u} = r\Omega(r)\boldsymbol{e}_\theta$ in plane polar coordinates, with a field initially uniform of strength $B_0$ and parallel to the direction $\theta = 0$. Expressing the magnetic field as $\boldsymbol{B} = \nabla \times A\boldsymbol{e}_z$ and using the Fourier decomposition $A = B_0 \operatorname{Im}(f(r,t)\mathrm{e}^{\mathrm{i}\theta})$ gives the following equation for $f$:

$$\frac{\partial f}{\partial t} + \mathrm{i}\Omega(r) f = \eta\left(\frac{\partial^2}{\partial r^2} + \frac{1}{r}\frac{\partial}{\partial r} - \frac{1}{r^2}\right) f,$$

with $f(r, 0) = r$. This has the asymptotic solution

$$f \sim r \exp(-\mathrm{i}\Omega t - \eta\Omega'^2 t^3/3);$$

significantly, flux expulsion occurs on an $O(Rm^{1/3})$ timescale, as first proposed by Weiss in 1966, much faster than the $O(Rm)$ diffusive (or Ohmic) timescale.

## 5 The Lorentz Force

The induction equation (1) describes the evolution of the magnetic field $\boldsymbol{B}$ under the influence of a fluid velocity $\boldsymbol{u}$. However, except when very weak, the magnetic field is not passive, but itself exerts a force on the velocity, to be incorporated into the Navier–Stokes equation. This is the Lorentz force, defined by

$$\boldsymbol{J} \times \boldsymbol{B} = \frac{1}{\mu_0}(\nabla \times \boldsymbol{B}) \times \boldsymbol{B}, \tag{3}$$

after substituting for the current from Ampère's law. (Under the assumptions of MHD, the electrostatic force is negligible.) MHD is thus described by a coupled nonlinear system of partial differential equations. Note that the Lorentz force has no influence on the motions along magnetic field lines; these must result from other forces such as gravity or pressure gradients.

### 5.1 Magnetic Pressure and Tension

The Lorentz force (3) may be decomposed as

$$\boldsymbol{J} \times \boldsymbol{B} = -\nabla\left(\frac{B^2}{2\mu_0}\right) + \frac{1}{\mu_0}\boldsymbol{B} \cdot \nabla \boldsymbol{B}, \tag{4}$$

where $B$ denotes the magnitude of the magnetic field $\boldsymbol{B}$. The first term represents the gradient of *magnetic pressure*, $p_{\mathrm{m}} = B^2/2\mu_0$. The total pressure is then the sum of the gas and magnetic pressures; the ratio of these, $p/p_{\mathrm{m}}$, is known as the *plasma-$\beta$*. In gases, the ramifications of the magnetic pressure can be extremely important. An isolated tube of magnetic flux, in total pressure balance with a nonmagnetic atmosphere, will necessarily have a reduced gas pressure. If the tube is also at the same temperature as its surroundings, then, from the perfect gas law, it will have a lower density. It will therefore rise, a phenomenon known as *magnetic buoyancy*. As first proposed by Parker, this mechanism is believed to be instrumental in bringing up magnetic flux tubes from the interior of the sun to the solar surface and on into the solar atmosphere (see figure 1).

Writing $\boldsymbol{B} = B\hat{\boldsymbol{s}}$, in terms of the unit vector $\hat{\boldsymbol{s}}$ along the field, the second term in (4) can be decomposed as

$$\begin{aligned}\frac{B}{\mu_0}\frac{\mathrm{d}}{\mathrm{d}s}(B\hat{\boldsymbol{s}}) &= \frac{B}{\mu_0}\frac{\mathrm{d}B}{\mathrm{d}s}\hat{\boldsymbol{s}} + \frac{B^2}{\mu_0}\frac{\mathrm{d}\hat{\boldsymbol{s}}}{\mathrm{d}s} \\ &= \frac{\mathrm{d}}{\mathrm{d}s}\left(\frac{B^2}{2\mu_0}\right)\hat{\boldsymbol{s}} + \frac{B^2}{\mu_0}\frac{\hat{\boldsymbol{n}}}{R_{\mathrm{c}}},\end{aligned}$$

where $\hat{\boldsymbol{n}}$ is the principal normal to the magnetic field line and $R_c$ is its radius of curvature. The first term exactly opposes the component of the magnetic pressure along field lines. The second represents the *magnetic tension*, showing that field lines, when bent, possess an elastic restoring force.

### 5.2 Force-Free Fields

In low-$\beta$ plasmas, such as that of the solar corona, the magnetic field is dominant. Static or, more accurately, quasistatic magnetic structures can then be modeled as possessing magnetic fields that are force free, i.e., the Lorentz force vanishes. From (3) this implies that $\boldsymbol{B}$ and $\boldsymbol{J}$ are parallel and hence that

$$\nabla \times \boldsymbol{B} = \lambda \boldsymbol{B}, \qquad (5)$$

where $\lambda$ is a scalar, which may depend on position. Vector fields satisfying (5) are known as *Beltrami* fields. It follows that $\boldsymbol{B}\cdot\nabla\lambda = 0$, namely, that $\lambda$ cannot vary along an individual field line. The simplest case is when $\lambda$ is a constant; in a cylindrically symmetric system, for example, this leads to a magnetic field with the following components, expressed as Bessel functions:

$$B_r = 0, \qquad B_\theta = B_0 J_1(\lambda r), \qquad B_z = B_0 J_0(\lambda r).$$

### 5.3 Hartmann Flow

The influence of the magnetic tension is illustrated by the flow of an electrically conducting fluid along a channel with a transverse magnetic field. Flows such as this, which were first studied theoretically and experimentally by Hartmann in the 1930s, have received considerable attention owing to their importance in liquid-metal MHD. Suppose that an incompressible viscous fluid, of density $\rho$ and kinematic viscosity $\nu$, is driven by a uniform pressure gradient in the $x$-direction between two parallel planes at $y = \pm d$ with an imposed uniform magnetic field $B_0\hat{\boldsymbol{y}}$. The flow $u(y)\hat{\boldsymbol{x}}$ vanishes on the bounding planes and is fastest in the center of the channel; it thus pulls out the imposed field to generate a component of field $b(y)\hat{\boldsymbol{x}}$. The field lines then possess a magnetic tension, which acts to resist the motion. A steady state is possible in which the stretching of the field lines and their "slippage" due to diffusion are in balance. The nondimensional steady-state momentum and induction equations can be written concisely as

$$Ha\frac{\mathrm{d}b}{\mathrm{d}y} + \frac{\mathrm{d}^2u}{\mathrm{d}y^2} = -1, \qquad Ha\frac{\mathrm{d}u}{\mathrm{d}y} + \frac{\mathrm{d}^2b}{\mathrm{d}y^2} = 0,$$

where the Hartmann number $Ha = dB_0\sqrt{\sigma/\rho\nu}$. These are to be solved subject to the no-slip boundary condition ($u = 0$ at $y = \pm 1$) and a magnetic boundary condition

$$\pm\frac{\mathrm{d}b}{\mathrm{d}y} + \frac{1}{c}b = 0 \quad \text{at } y = \pm 1,$$

where $c = 0$ corresponds to electrically insulating walls and $c \to \infty$ to perfectly conducting boundaries. The resulting velocity is given by

$$u = \frac{1}{Ha}\left(\frac{1+c}{c\,Ha + \tanh Ha}\right)\left(1 - \frac{\cosh(Ha\,y)}{\cosh Ha}\right).$$

For small $Ha$, the classical parabolic hydrodynamic profile is recovered. For $Ha \gg 1$, the velocity is strongly reduced; furthermore, the profile becomes flat across the bulk of the channel, with exponential boundary layers of width $O(Ha^{-1})$.

## 6 Magnetohydrodynamic Waves

The interplay in MHD between the velocity and the magnetic field is brought out most vividly through the support of a variety of wave motions. Such waves, with an extended range of spatial and temporal scales, are revealed in spectacular movies of the magnetic field in the solar atmosphere.

Most striking is the occurrence of what are known as *Alfvén waves*. These can be analyzed in the simplest system of ideal incompressible MHD. Consider linear plane-wave perturbations, varying as $\exp(\mathrm{i}(\boldsymbol{k}\cdot\boldsymbol{x} - \omega t))$, to a homogeneous equilibrium state with a uniform magnetic field $\boldsymbol{B}_0$. Combining the momentum and induction equations yields the dispersion relation

$$\omega^2 = k_\parallel^2 V_{\mathrm{A}}^2,$$

where the Alfvén velocity $V_{\mathrm{A}} = B_0/\sqrt{\mu_0\rho}$, and $k_\parallel$ is the component of the wave vector parallel to the imposed field. Alfvén waves, which are transverse, are driven solely by the magnetic tension and are therefore, in some sense, analogous to waves on a stretched string.

Compressible atmospheres support two additional modes, known as *magnetoacoustic waves*, with dispersion relation

$$\omega^4 - \omega^2k^2(V_{\mathrm{S}}^2 + V_{\mathrm{A}}^2) + k_\parallel^2k^2V_{\mathrm{S}}^2V_{\mathrm{A}}^2 = 0,$$

where the speed of sound $V_{\mathrm{S}} = \sqrt{\gamma p/\rho}$. The two possible solutions for the modulus of the phase speed $|\omega/k|$ are designated as fast and slow modes. The properties of the phase and group velocities for the various MHD waves are demonstrated most clearly by a polar diagram, as shown in figure 2. The most significant differences are revealed by the group velocities: whereas for the fast mode, energy is propagated in all directions, for the slow mode, energy propagates in only a small

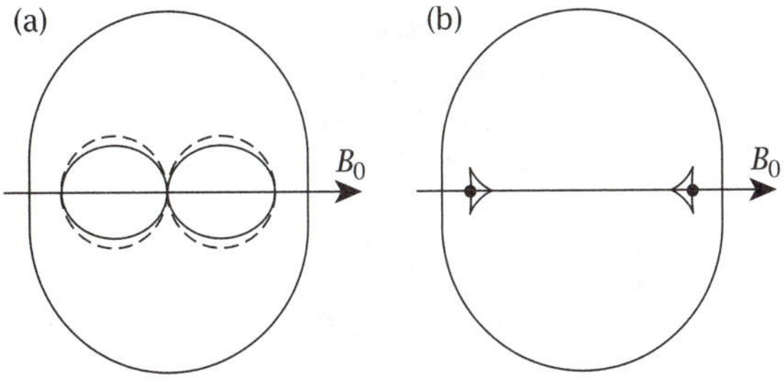


**Figure 2** (a) Phase velocities and (b) group velocities for the various MHD waves with an imposed magnetic field as shown. Solid lines denote the fast and slow waves; the Alfvén waves are denoted by the dashed lines in (a) and the solid circles in (b). The ratio of Alfvén to sound speeds (or the inverse ratio; the picture is unaltered) is 0.8.

range of angles around the direction of the imposed field and for the Alfvén waves, energy is conveyed, at speed $V_A$, only in the direction of $\boldsymbol{B}_0$.

## 7 Dynamos

Dynamo theory—the study of the maintenance of magnetic fields—constitutes one of the most mathematically intriguing and physically significant areas of MHD. Put starkly, there are just two possibilities for the occurrence of any magnetic field observed in an astrophysical body: either it is a slowly decaying "fossil field," trapped in the body at its formation, or, alternatively, it is maintained by the inductive motions of the plasma within the body. The latter process is known as a *magnetohydrodynamic dynamo*. For the Earth the issue is clear-cut. Paleomagnetism reveals that the Earth's field has persisted for $O(10^9)$ years, whereas its decay time ($L^2/\eta$, where $L$ is the length scale of the conducting outer core and $\eta$ its magnetic diffusivity) is several orders of magnitude shorter: $O(10^5)$ years. The geomagnetic field cannot therefore be simply a decaying fossil field; it must be maintained by some sort of dynamo resulting from the flows of the liquid metal in the outer core. The case for dynamo action in the sun (and, by extension, similar stars) is rather different. The Ohmic decay time for the solar field is long, $O(10^9)$ years, comparable with the lifetime of the sun itself, and so on these grounds alone one cannot rule out a primordial field explanation. However, the solar magnetic field exhibits variations on very *short* timescales, with the entire field reversing every eleven years or so, and it is extremely difficult to reconcile this behavior with that of a slowly decaying relic field.

A full description of any natural MHD dynamo can be obtained only through a solution of *all* the governing equations. That said, many of the most important aspects of dynamo theory, and many of its subtleties, can be captured via consideration of the induction equation alone.

### 7.1 The Kinematic Dynamo Problem

The kinematic dynamo problem asks whether it is possible to find a velocity $\boldsymbol{u}(\boldsymbol{x},t)$ such that the magnetic field $\boldsymbol{B}$, governed solely by the induction equation, does not decay. More formally, a flow $\boldsymbol{u}$ is said to act as a kinematic dynamo if the magnetic energy,

$$M(t) = \frac{1}{2\mu_0}\int_{\text{all space}} B^2\,\mathrm{d}V,$$

does not tend to zero as $t \to \infty$. For the simplest case of steady flows $\boldsymbol{u}(\boldsymbol{x})$, the magnetic field varies as $\mathrm{e}^{pt}$; the induction equation reduces to an eigenvalue problem for $p$, with dynamo action if there exists an eigenvalue with $\mathrm{Re}(p) > 0$.

Although easily stated, the kinematic dynamo problem is rather more demanding to solve. Indeed, whereas it is possible to prove rigorously a number of *anti*-dynamo theorems, proving the positive result is less straightforward.

### 7.2 Antidynamo Theorems

Antidynamo theorems address either the magnetic field or the velocity. Those that address the former, which are geometric in nature, reveal how certain types of field cannot be generated by dynamo action. Those that address the latter, which are concerned with either the flow geometry or, alternatively, some property of the flow, such as its amplitude or stretching capacity, demonstrate that specific classes of velocity cannot succeed as dynamos.

The most important result concerning the magnetic field is Cowling's theorem of 1933, which states that a steady axisymmetric field cannot be maintained by dynamo action. Cowling argued that such a magnetic field must have a closed contour on which the poloidal field $\boldsymbol{B}_P$ vanishes and around which the $\boldsymbol{B}_P$-lines are closed (an "O-type" neutral point); consideration of Ohm's law then shows that the induction effect cannot overcome diffusion in the neighborhood of the neutral point. It is important, however, to point out that *non-axisymmetric* magnetic fields *can* be generated from axisymmetric flows.

In terms of the geometrical properties of the velocity, the most revealing antidynamo theorem is due to Zeldovich, who proved that a planar two-dimensional flow, $\boldsymbol{u} = (u(x,y,z,t), v(x,y,z,t), 0)$, say, is incapable of dynamo action. The proof proceeds in two identical steps. First, the $z$-component of the induction equation,

$$\frac{\mathrm{D}B_z}{\mathrm{D}t} = \eta\nabla^2 B_z, \tag{6}$$

reveals that there is no source term for $B_z$; it therefore follows that $B_z \to 0$. Hence, only the field in the $xy$-plane need be considered. On writing $\boldsymbol{B} = \nabla \times (A\hat{\boldsymbol{z}})$, $A$ then also satisfies equation (6), from which it follows that $\boldsymbol{B}$ tends to zero. Thus, stretching of the field is not in itself sufficient for dynamo action; lifting the field out of the plane and folding it, constructively, is an essential factor.

Bounds on flow properties necessary for dynamo action may be obtained through consideration of the evolution of the magnetic energy. Suppose that an incompressible conducting fluid is contained in a volume $V$, external to which is an insulator. The scalar product of equation (1) with $\boldsymbol{B}/\mu_0$, after using the divergence theorem and the boundary conditions, gives

$$\frac{\mathrm{d}}{\mathrm{d}t}\int_V \frac{B^2}{2\mu_0}\,\mathrm{d}V = \int_V \boldsymbol{J}\cdot(\boldsymbol{u}\times\boldsymbol{B})\,\mathrm{d}V - \frac{1}{\sigma}\int_V J^2\,\mathrm{d}V.$$

The second term on the right-hand side can be bounded in terms of the magnetic energy using calculus of variations. The first can be bounded in terms of either the maximum velocity or the maximum of the rate-of-strain tensor. If $V$ is a sphere, then the former method reveals that dynamo action requires that $Rm > \pi$, where $Rm$ is based on the maximum fluid velocity; the latter gives the bound $\widetilde{Rm} > \pi^2$, where the magnetic Reynolds number $\widetilde{Rm}$ is now defined (unconventionally) using the maximum of the rate-of-strain tensor.

Putting everything together, antidynamo theorems reveal the essence of at least some of what *is* needed for a successful dynamo; complexity (in the field and flow) and sufficiently high $Rm$ are vital ingredients. Consequently, most demonstrations of dynamo action are the result of numerical (i.e., computational) solution of the induction equation for prescribed velocities.

### 7.3 Mean-Field MHD

One of the great advances in dynamo theory, pioneered by Steenbeck, Krause, and Rädler in the 1960s, was to seek a solution not for the magnetic field itself but for its mean value $\langle\boldsymbol{B}\rangle$, where the averaging operator $\langle\cdot\rangle$ obeys the Reynolds rules; for example, this could be an ensemble average or, more appropriately for isolated bodies such as the Earth, an azimuthal average.

Averaging equation (1), and supposing for the moment that there is no mean flow, gives

$$\frac{\partial\langle\boldsymbol{B}\rangle}{\partial t} = \nabla\times\langle\boldsymbol{u}\times\boldsymbol{b}\rangle + \eta\nabla^2\langle\boldsymbol{B}\rangle,$$

where $\boldsymbol{b}$ is the fluctuating magnetic field. The term $\boldsymbol{\mathcal{E}} = \langle\boldsymbol{u}\times\boldsymbol{b}\rangle$, which lies at the very heart of the theory and represents the essential difference between the averaged and unaveraged induction equations, denotes a *mean electromotive force* resulting from correlations between the fluctuating velocity and the fluctuating field. The system is closed through an argument relating $\boldsymbol{b}$, and hence $\boldsymbol{\mathcal{E}}$, linearly to $\langle\boldsymbol{B}\rangle$, and then postulating an expansion

$$\mathcal{E}_i = \alpha_{ij}\langle B\rangle_j + \beta_{ijk}\frac{\partial\langle B\rangle_j}{\partial x_k} + \cdots. \tag{7}$$

In a kinematic theory, the tensors $\alpha_{ij}$ and $\beta_{ijk}$ depend only on the statistical properties of the flow and on $Rm$. The first term (the "$\alpha$-effect" of mean-field MHD) provides a possible source of dynamo action; the second generally represents an additional, turbulent diffusion. From (7) it is immediately possible to make a very strong statement. Since $\boldsymbol{\mathcal{E}}$ is a polar vector, whereas $\langle\boldsymbol{B}\rangle$ is axial, $\alpha_{ij}$ must be a pseudotensor, i.e., a tensor that changes sign under a transformation from right-handed to left-handed coordinates. It therefore vanishes if the flow, on average, is reflectionally symmetric. Thus, for mean-field dynamo action through the $\alpha$-effect, the flow must possess "handedness," the simplest measure of this being the flow helicity $\langle\boldsymbol{u}\cdot\nabla\times\boldsymbol{u}\rangle$. Astrophysical flows, influenced by rotation, will typically be helical.

A mean-field dynamo may be thought of as a means of maintaining the "dynamo loop" between poloidal and toroidal fields. Large-scale differential rotation is a generic feature of astrophysical bodies and provides a natural means of generating toroidal magnetic field by stretching the poloidal component (often referred to as the $\Omega$-effect). Closing the loop by the regeneration of poloidal field from toroidal, which is the less physically transparent step, is then accomplished by the $\alpha$-effect. The combination of these processes is known as an $\alpha\Omega$-dynamo. Alternatively, both elements of the cycle can be achieved through the $\alpha$-effect, without the need for differential rotation (an $\alpha^2$-dynamo).

A model flow that has proved to be extremely influential in the development of dynamo theory is that introduced by G. O. Roberts in 1972, defined by

$$u = \left(\frac{\partial \psi}{\partial y}, -\frac{\partial \psi}{\partial x}, \psi\right), \quad \text{with } \psi = \cos x + \cos y. \quad (8)$$

The flow is maximally helical ($\boldsymbol{u}$ parallel to $\nabla \times \boldsymbol{u}$). For $z$-independent flows, such as (8), the induction equation supports solutions of the form $\boldsymbol{B}(x, y, t)\mathrm{e}^{\mathrm{i}kz}$; for any given wave number $k$, the problem then becomes simpler, involving only two, rather than three, spatial dimensions. In this geometry, a mean-field dynamo is one in which the field has a long $z$ wavelength (small $k$), with averages taken over the $xy$-plane.

An exciting experimental challenge, taken up by a number of groups worldwide, is to construct a laboratory MHD dynamo. One of the successes has been the Karlsruhe dynamo, in which liquid sodium is pumped through an array of helical pipes, thereby producing a flow similar to that conceived by Roberts.

### 7.4 Fast Dynamos

As discussed earlier, the magnetic diffusion timescale for astrophysical bodies is extremely long; it is thus of interest to ask whether dynamo action can proceed on a shorter, diffusion-independent timescale. The mathematical abstraction of the astrophysical problem, in which $Rm$ is large but finite, is to investigate kinematic dynamo action in the limit as $Rm \to \infty$. To fix ideas, consider a steady flow $\boldsymbol{u}(\boldsymbol{x})$ for which the magnetic field assumes the form $\boldsymbol{B}(\boldsymbol{x})\mathrm{e}^{pt}$. A flow will act as a dynamo if $\mathrm{Re}(p) > 0$ for some value of $Rm$; it is said to act as a *fast* dynamo if

$$\lim_{Rm \to \infty} \mathrm{Re}(p) > 0;$$

otherwise, it is said to be *slow*.

An extremely important result is the derivation of an upper bound for fast-dynamo growth, the growth rate for steady or time-periodic flows being bounded above by the *topological entropy* of the flow, $h_{\mathrm{top}}$. This is a measure of the complexity of the flow and is closely related to $h_{\mathrm{line}}$, the rate of stretching of material lines (for two-dimensional flows, they are identical; for three-dimensional flows, $h_{\mathrm{line}} \leqslant h_{\mathrm{top}}$). Since nonchaotic flows have zero topological entropy, a simple consequence is the following powerful anti-fast-dynamo theorem: *fast-dynamo action is not possible in an integrable flow*. Steady, two-dimensional flows, such as the Roberts flow (8), are therefore guaranteed to act (at best) only as slow dynamos. That said, it is of interest to see to what extent the Roberts dynamo fails to be fast; the answer turns out to be "not by much." In a powerful asymptotic analysis, Soward proved that the growth rate as $Rm \to \infty$ is given by

$$p \sim \frac{\ln(\ln Rm)}{\ln Rm},$$

with its "slowness" attributed to the long time spent by fluid elements in the neighborhood of the stagnation point.

Given the difficulties, in general, in providing rigorous demonstrations of dynamo action, it is perhaps not surprising that these are exacerbated for the more restrictive class of fast dynamos. The numerical approach is to consider dynamos at increasing values of $Rm$, in the hope of reaching an asymptotic regime in which the growth rate is positive and ceases to vary with $Rm$. For three-dimensional flows, this has so far proved inconclusive; the requisite computational resources increase with $Rm$ (in order to resolve finer-scale structures), and no plausible $Rm$-independent regime has yet been attained, even with today's computing facilities. Instead, the most convincing examples of fast-dynamo action have come from time-dependent modifications of the Roberts flow, allowing exploration of the dynamo up to $Rm = O(10^6)$, with the time dependence circumventing the anti-fast-dynamo theorem. Although clearly not a proof, the numerical evidence is strong that such flows can act as fast dynamos.

## 8 Instabilities

Magnetic fields can play a significant role in modifying classical hydrodynamic instabilities driven, for example, by shear flows or by convection. They can, though, also be the agent of instability. Here we consider two such examples, each extremely important in an astrophysical context.

### 8.1 Magnetic Buoyancy

In section 5.1 we discussed how magnetic pressure can cause the rise of isolated tubes of magnetic flux, a manifestation of a lack of equilibrium. But magnetic buoyancy can also act as an instability mechanism, one that is believed to be responsible for facilitating the breakup of a large-scale field in the solar interior into the flux tubes that subsequently rise to the surface.

Consider a static equilibrium atmosphere with a homogenous horizontal magnetic field whose strength varies with height $z$. For motions that do not bend the magnetic field lines (interchange modes), the criterion for instability can be derived from a fluid parcel argument, assuming that a displaced parcel conserves its mass, magnetic flux, and specific entropy. It is then straightforward to show that a horizontal field $B(z)\hat{\boldsymbol{x}}$ is unstable to interchange disturbances if

$$-\frac{gV_A^2}{V_S^2}\frac{d}{dz}\ln\left(\frac{B}{\rho}\right) > \frac{g}{\gamma}\frac{d}{dz}\ln\left(\frac{p}{\rho^\gamma}\right) = N^2,$$

where $N$ is the buoyancy frequency. Significantly, instability can occur even in convectively stable atmospheres ($N^2 > 0$) provided that $B/\rho$ falls off sufficiently rapidly with height. Somewhat surprisingly, three-dimensional modes, despite having to do work against magnetic tension, are more readily destabilized, requiring a sufficiently negative gradient only of $B$ (rather than $B/\rho$).

### 8.2 Magnetorotational Instability

A fundamental and long-standing problem in astrophysics concerns the mechanism by which matter accreting onto a massive central object (such as a neutron star or black hole) loses its angular momentum. Some sort of turbulent process is needed in order to transport angular momentum outward as the mass moves inward. From a purely hydrodynamic viewpoint, a Keplerian flow, with angular velocity depending on radius as $\Omega \sim r^{-3/2}$, is stable according to RAYLEIGH'S CRITERION [IV.28 §8.3], and thus there is no obvious route to hydrodynamic turbulence. But the picture is changed dramatically by the effect of a magnetic field, the significant astrophysical consequences of which were first realized by Balbus and Hawley in the 1990s. A new instability—the *magnetorotational instability*—can then ensue provided that the *angular velocity* decreases with radius, a less stringent criterion than that for hydrodynamic flows, for which a decrease of *angular momentum* is necessary. The nonlinear development of this instability may thus provide the turbulence required for angular momentum transport.

It is of great physical interest to ask why a swirling flow that is hydrodynamically stable can be destabilized by the inclusion of a magnetic field. The crucial factor is that the field can provide a means of relaxing the angular momentum constraint that underpins the hydrodynamic result. As such, it turns out that the underlying mechanism can then be explained by mechanical, rather than fluid mechanical, arguments. Suppose, as an analogy for fluid elements connected by a magnetic field line, one considers two spacecraft ($m_1$ and $m_2$) orbiting a central body, at different radii ($r_1 < r_2$) and joined by a weak elastic tether. Since the angular velocity is a decreasing function of radius,

**Figure 3** Computational simulation of the breakup of a magnetic layer, resulting from the nonlinear development of magnetic buoyancy instabilities; the magnetic field is pointing into the page. Such a process is of importance in triggering the escape of magnetic field from the solar interior.

the spacecraft at radius $r_1$ will move ahead, stretching the tether. In so doing, angular momentum is removed from $m_1$ and transferred to $m_2$. The spacecraft then readjust to orbits compatible with their new angular momenta; $m_1$ moves inward, $m_2$ outward. Since $d\Omega/dr < 0$, the difference in the angular velocity is increased; a fortiori the process is repeated, leading to an exponential separation of the spacecraft in time. Note that this argument fails if the tether is sufficiently strong, since in this case it acts to keep the spacecraft together. Instability therefore occurs for *weak* elastic coupling or, in the MHD context, for weak magnetic fields.

## 9 Current State of Play

Fundamental questions remain unanswered in all aspects of MHD; it thus remains an extremely active and exciting research area. The major theoretical difficulties, as with (nonmagnetic) hydrodynamics, arise from two directions: the extreme values of the parameters of interest and the inherent nonlinearity of the coupled equations (though it would be a mistake to think that everything is understood in, for example, the (kinematic) fast-dynamo problem).

Ever-increasing computing power has allowed investigation of the full MHD equations at moderately high values ($\lesssim 10^4$) of the fluid and magnetic Reynolds numbers (as shown, for example, in figure 3), which has in turn stimulated new theoretical understanding, particularly of nonlinear phenomena. A *direct* computational solution, i.e., of astrophysical or industrial flows at the true parameter values, is not feasible; nor, even if the current rate of increase in computing power is maintained, will it be so in the foreseeable future. Our understanding of MHD will therefore progress through the interaction of theoretical, computational, and experimental approaches.

**Further Reading**

Childress, S., and A. D. Gilbert. 1995. *Stretch, Twist, Fold: The Fast Dynamo.* New York: Springer.

Davidson, P. A. 2001. *An Introduction to Magnetohydrodynamics.* Cambridge: Cambridge University Press.

Mestel, L. 2012. *Stellar Magnetism.* Oxford: Oxford University Press.

Moffatt, H. K. 1978. *The Generation of Magnetic Fields in Electrically Conducting Fluids.* Cambridge: Cambridge University Press.

Molokov, S., R. Moreau, and H. K. Moffatt. 2007. *Magnetohydrodynamics: Historical Evolution and Trends.* New York: Springer.

Priest, E. R. 1982. *Solar Magnetohydrodynamics.* Dordrecht: Reidel.

Roberts, P. H. 1967. *An Introduction to Magnetohydrodynamics.* London: Longman.

# IV.30 Earth System Dynamics

*Emily Shuckburgh*

## 1 Introduction

Mathematics is fundamental to our understanding of the Earth's weather and climate. Over the last 200 years or so, a mathematical description of the evolution of the Earth system has been developed that allows predictions to be made concerning the weather and climate. This takes account of the fact that the atmosphere and oceans are thin films of fluid on the spherical Earth under the influence of (i) heating by solar radiation, (ii) gravity, and (iii) the Earth's rotation. The mathematician Lewis Fry Richardson first proposed a numerical scheme for forecasting the weather in 1922, and this paved the way for modern numerical models of the weather and climate.

The atmosphere varies on a range of timescales, from less than an hour for an individual cloud to a week or so for a weather system. The timescales of variability in the ocean are typically longer: the ocean surface layer, which is directly influenced by the atmosphere, exhibits variability on diurnal, seasonal, and interannual timescales, but the ocean interior varies significantly only on decadal to centennial and longer timescales. The different components of the Earth system (the atmosphere, oceans, land, and ice) are closely coupled. For example, roughly half of the carbon dioxide that is released into the atmosphere by human activities each year is taken up by the oceans and the land. Changes in atmospheric temperature impact the ice, which in turn influences the evolution of both the atmosphere and the ocean. Furthermore, changes in sea-surface temperature can directly affect the atmosphere and its weather systems. Each of these interactions and feedbacks is more or less relevant on different timescales. The *climate* is usually taken to mean the state of the Earth system over years to decades or longer. It is sometimes defined more precisely as the probability distribution of the variable weather, traditionally taken over a thirty-year period. While aspects of the weather are sensitive to initial conditions, as famously demonstrated by Edward Lorenz in his seminal work on chaos theory (see THE LORENZ EQUATIONS [III.20]), the statistics of the weather that define the climate do not exhibit such sensitivity.

## 2 The Temperature of the Earth

Understanding what determines the temperature of the surface of the Earth has been something that has long fascinated mathematicians. In 1827 the mathematician Joseph Fourier wrote an article on the subject, noting that he considered it to be one of the most important and most difficult questions of all of natural philosophy.

### 2.1 Solar Forcing

The atmosphere is continually bombarded by solar photons at infrared, visible, and ultraviolet wavelengths. It is necessary to consider the passage through the atmosphere of this incoming solar radiation to determine the temperature of the surface of the Earth. The air in the atmosphere is a mixture of different gases: nitrogen ($N_2$) and oxygen ($O_2$) are the largest by volume, but other gases including carbon dioxide ($CO_2$), water vapor ($H_2O$), methane ($CH_4$), nitrous oxide ($N_2O$), and ozone ($O_3$) play significant roles in influencing the passage of photons. Atmospheric water vapor is particularly important in this context. The amount of water vapor is variable (typically about 0.5% by volume), being strongly dependent on the temperature, and it is primarily a consequence of evaporation from the ocean (which covers some 70% of the surface of the Earth).

Some incoming solar photons are scattered back to space by atmospheric gases or reflected back to space by clouds or the Earth's surface (with snow and ice reflecting considerably more than darker surfaces); some are absorbed by atmospheric gases or clouds, leading to heating of parts of the atmosphere; and some reach the Earth's surface and heat it. Atmospheric gases, clouds, and the Earth's surface also emit and absorb infrared photons, leading to further heat transfer between one region and another, or loss of heat to space.

The amount of energy that the Earth receives from the sun has varied over geological history, but at present the incident solar flux, or power per unit area, of solar energy is $F = 1370\ \mathrm{W\ m^{-2}}$.

Given that the cross-sectional area of the Earth intercepting the solar energy flux is $\pi a^2$, where $a$ is the Earth's radius (mean value $a = 6370$ km), the total solar energy received per unit time is $F\pi a^2 = 1.74 \times 10^{17}$ W. As noted above, not all of this radiation is absorbed by the Earth; a significant fraction is directly reflected. The ratio of the reflected to incident solar energy is the *albedo*, $\alpha$. Under present conditions of cloudiness, snow, and ice cover, an average of about 30% of the incoming solar radiation is reflected back to space without being absorbed, i.e., $\alpha = 0.3$. The surface of the Earth has an area $4\pi a^2$. That means that per unit area,

$$\text{final incoming power} = \tfrac{1}{4}(1-\alpha)F,$$

which is approximately 240 W m$^{-2}$.

In physics, a *black body* is an idealized object that absorbs all radiation that falls on it. Because no light is reflected or transmitted, the object appears black when it is cold. However, a black body emits a temperature-dependent spectrum of light. As noted above, the Earth reflects much of the radiation that is incident upon it, but for simplicity let us initially assume that it emits radiation in the same temperature-dependent way as a black body. In this case the emitted radiation is given by the Stefan–Boltzmann law, which states that the power emitted per unit area of a black body at absolute temperature $T$ is $\sigma T^4$, where $\sigma$ is the Stefan–Boltzmann constant ($\sigma = 5.67 \times 10^{-8}\ \mathrm{W\ m^{-2}\ K^{-4}}$). This power is emitted in all directions from the surface of the Earth, so that per unit area,

$$\text{final outgoing power} = \sigma T_{\mathrm{bb}}^4$$

if the Earth has a uniform surface temperature $T_{\mathrm{bb}}$. This defines the *emission temperature* under the assumption that the Earth acts as a black body, which is the temperature one would infer by looking back at the Earth from space if a black body curve was fitted to the measured spectrum of outgoing radiation.

Assuming that the Earth is in thermal equilibrium, the incoming and outgoing power must balance. Therefore, by equating the equations for the final incoming power and the final outgoing power,

$$T_{\mathrm{bb}} = \left(\frac{(1-\alpha)F}{4\sigma}\right)^{1/4}.$$

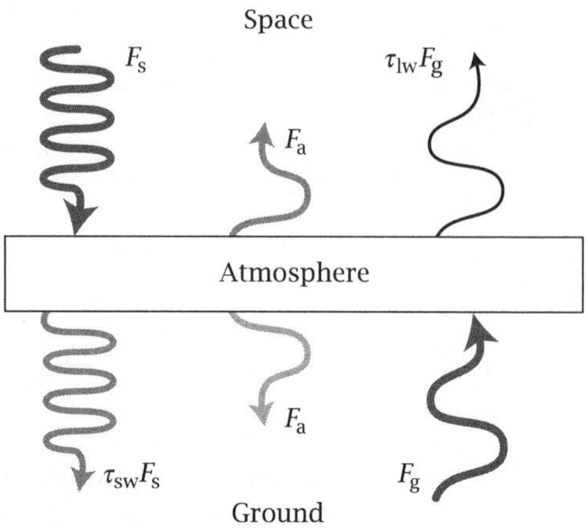


**Figure 1** A simple model of the greenhouse effect. The atmosphere is taken to be a shallow layer at temperature $T_a$ and the ground a black body at temperature $T_g$. The various solar and thermal fluxes are shown (see text for details).

Substituting standard values for $\alpha$, $F$, and $\sigma$ gives $T_{\mathrm{bb}} \approx 255$ K. This value is of the correct order of magnitude but is more than 30 K lower than the observed mean surface temperature of $T_{\mathrm{E}} \approx 288$ K. The simplest possible model of the climate system has therefore captured some of the key aspects, but it must have some important missing ingredients.

### 2.2 The Greenhouse Effect

To refine the calculation of the temperature of the surface of the Earth, the influence of atmospheric constituents in emitting, absorbing, and scattering radiation needs to be taken into consideration. This can be achieved by assuming that the system has a layer of atmosphere of uniform temperature $T_a$ that transmits a fraction $\tau_{sw}$ of incident solar (shortwave) radiation and a fraction $\tau_{lw}$ of any incident terrestrial (longwave) radiation while absorbing the remainder (see figure 1).

As noted above, the final incoming solar power per unit area at the top of the atmosphere is $F_s = \frac{1}{4}(1-\alpha)F$. Under the revised model, a proportion $\tau_{sw}F_s$ reaches the ground and the remainder is absorbed by the atmosphere. Let us assume that the ground has a temperature $T_g$ and that it emits as a black body. This gives an upward flux of $F_g = \sigma T_g^4$, of which a proportion $\tau_{lw}F_g$ reaches the top of the atmosphere, with the remainder being absorbed by the atmosphere. The atmosphere is not a black body, and instead, it emits radiation following Kirchhoff's law, such that the emitted flux is

$F_a = (1 - \tau_{lw})\sigma T_a^4$ both upward and downward, where $T_a$ is the temperature of the atmosphere.

Assuming that the system is in equilibrium, these fluxes must balance. At the top of the atmosphere we have $F_s = F_a + \tau_{lw}F_g$, and at the ground we have $F_g = F_a + \tau_{sw}F_s$. By eliminating $F_a$, and using the Stefan–Boltzmann law for $T_g$, we find that

$$T_g = \left(\frac{1+\tau_{sw}}{1+\tau_{lw}}\frac{(1-\alpha)F}{4\sigma}\right)^{1/4} = \left(\frac{1+\tau_{sw}}{1+\tau_{lw}}\right)^{1/4} T_{bb}.$$

Substituting reasonable values for the Earth's present-day atmosphere ($\tau_{sw} = 0.9$ and $\tau_{lw} = 0.2$) gives a surface temperature of $T_g \approx 286$ K, which is much closer to the observed value of $T_E \approx 288$ K. Including an atmosphere that allows greater transmission of shortwave radiation than longwave radiation has had the influence of increasing the surface temperature. This is known as the *greenhouse effect*. The temperature of the atmosphere $T_a$ under this model is

$$T_a = \left(\frac{1-\tau_{lw}\tau_{sw}}{1-\tau_{lw}^2}\right)^{1/4} T_{bb},$$

which gives $T_a \approx 245$ K.

The expression for $T_g$ indicates that the temperature of the surface of the Earth can change as a result of changes to the solar flux ($F$), changes to the albedo ($\alpha$), and/or changes to the transmission of radiation through the atmosphere ($\tau$). This provides a basic explanation for the cycle of ice ages that has been observed to occur on Earth every hundred thousand years or so. Over long timescales, changes to the orbit of the Earth around the sun occur that are associated with the Milankovitch cycles. Resulting changes to the annual global mean $F$ are small, but changes to the seasonal/latitudinal pattern of solar radiation reaching the Earth lead to the ice sheets in the Northern Hemisphere shrinking or growing, changing $\alpha$. This then modifies the temperature $T_g$, which further changes the ice sheets. Additionally, changes in $T_g$ can result in changes that alter the balance of exchange of carbon dioxide between the atmosphere and the land/ocean, which changes the transmission value $\tau$, modifying the temperature still further.

## 3 Atmospheric Properties

We now turn our attention to the vertical variation in temperature in the atmosphere. This is influenced by radiative processes and by convection.

### 3.1 Radiation

To refine the model of the transfer of radiation through the atmosphere further it is necessary to consider the atmospheric properties in more detail. To a good approximation, the atmosphere as a whole behaves as a simple ideal gas, with each mole of gas obeying the law $pV = R_gT$, where $p$ is the pressure, $V$ is the volume of one mole, $R_g$ is the universal gas constant, and $T$ is the absolute temperature. If $M$ is the mass of one mole, the density is $\rho = M/V$, and the ideal gas law may be written as $p/\rho = RT$, where $R = R_g/M$ is the gas constant per unit mass. The value of $R$ depends on the composition of the sample of air. For dry air it is $R = 287$ J kg$^{-1}$ K$^{-1}$.

Each portion of the atmosphere is approximately in what is known as *hydrostatic balance* (usually valid on scales greater than a few kilometers). This means that the weight of the portion of atmosphere is supported by the difference in pressure between the lower and upper surfaces, and that the following relationship between density $\rho$ and pressure $p$ holds to a good approximation:

$$g\rho = -\frac{\partial p}{\partial z}, \qquad (1)$$

where $g = 9.81$ m s$^{-2}$ is the gravitational acceleration and $z$ is the height above the ground. The ideal gas law can be used to replace $\rho$ in this equation by $p/RT$.

The temperature in the lowest section of the atmosphere, known as the *troposphere*, decreases with altitude, from the surface value of about 288 K to about 217 K at its upper limit (about 10–15 km altitude). In general, the temperature does not vary greatly in the atmosphere. The mass-weighted mean temperature is approximately 255 K, and over the first 100 km in altitude the temperature varies by no more than about ±15%. Approximating the atmospheric temperature by $T \approx T_0 =$ const. and using the ideal gas law, (1) can be integrated to obtain $p = p_0 e^{-gz/RT_0}$, where $p_0$ is the pressure at $z = 0$, which is approximately 1000 hPa. Therefore, at 5 km altitude the pressure is about 500 hPa and at 10 km it is about 250 hPa. A similar expression can be derived for the density. Therefore, gravity tends to produce a density stratification in the atmosphere, which means that the atmosphere can be considered from a dynamical perspective as a stratified fluid on a rotating sphere.

As discussed above, certain gases in the atmosphere, known as *greenhouse gases*, act to absorb radiation of certain wavelengths. At the surface, the relatively high temperature and pressure mean that these gases absorb radiation in broad bands around specific wavelengths. These bands are made up of many individual spectral lines. The individual lines are broadened by collisions (pressure broadening), and this broadening reduces in width with altitude as the pressure decreases. If we consider the atmosphere to be made up of many thin vertical layers, as radiation emitted from the Earth's surface moves up layer by layer through the troposphere, some is stopped in each layer. Each layer then emits radiation back toward the ground and up to higher layers. However, due to the reduction in width of the absorption lines with altitude, a level can be reached at which the radiation is able to escape to space. In addition, because the amount of gas between a given altitude and space decreases with increasing altitude, even the line centers are more able to emit directly to space, with increasing altitude. Adding more greenhouse gas molecules means that the upper layers will absorb more radiation and the altitude of the layer at which the radiation escapes to space increases, and hence its temperature decreases. Since colder layers do not radiate heat as well, all the layers from this height to the surface must warm to restore the incoming/outgoing radiation balance.

It has become standard to assess the importance of a factor (such as an increase in a greenhouse gas) as a potential climate change mechanism in terms of its *radiative forcing*, $\Delta F$. This is a measure of the influence the factor has in altering the balance of incoming and outgoing energy, and it is defined as the change in net irradiance (i.e., the difference between incoming and outgoing radiation) measured at the tropopause (the upper boundary of the troposphere). For carbon dioxide the radiative forcing is given to a good approximation by the simple algebraic expression $\Delta F = 5.35 \times \ln C/C_0$ W m$^{-2}$, where $C$ is the concentration of carbon dioxide and $C_0$ is a reference preindustrial value, taken to be 278 parts per million. For a doubling of carbon dioxide values above preindustrial values, this gives a radiative forcing of approximately 3.7 W m$^{-2}$.

The *climate sensitivity*, $\lambda$, is defined to be the coefficient of proportionality between the radiative forcing, $\Delta F$, and $\Delta T$, the associated change in equilibrium surface temperature that occurs over multicentury timescales, i.e., $\Delta T = \lambda \Delta F$. For the very simplest climate model, we found that the emitted radiation per unit area was $F = \sigma T_{\text{bb}}^4$, from which we inferred a value of $T_{\text{bb}} \approx 255$ K. Approximating $\Delta F$ by $\Delta F \approx \Delta T \, \mathrm{d}F/\mathrm{d}T$, the climate sensitivity in the absence of feedbacks is then given by $\lambda = (4\sigma T_{\text{bb}}^3)^{-1}$, or 0.26 K/(W m$^{-2}$). Using this to estimate the temperature increase at equilibrium due to a doubling of carbon dioxide (often called the *equilibrium climate sensitivity*) gives $\Delta T \approx 1$ K. In reality, feedbacks in the system (e.g., changes to the albedo and the water vapor content of the atmosphere) will influence the temperature change. Taking into account the feedbacks, the Intergovernmental Panel on Climate Change concluded in its Fifth Assessment Report that this equilibrium climate sensitivity probably lies in the range 1.5–4.5 K and that it is extremely unlikely to be less than 1 K and very unlikely to be greater than 6 K.[1]

### 3.2 Convection

In radiative equilibrium, the surface is warmer than the overlying atmosphere. This state is unstable to convective motions.

The first law of thermodynamics states that the increase in internal energy of a system $\delta U$ is equal to the heat supplied plus the work done on the system. This can be written as $\delta U = T\delta S - p\delta V$, where $T$ is the temperature, $V$ is the volume, and $S$ is the *entropy* of the system. If $Q$ is the amount of heat absorbed by the system, $\delta S = \delta Q/T$. The *specific heat capacity*, $c$, is the measure of the heat energy required to increase the temperature of a unit quantity of air by one unit. The specific heat of substances are typically measured under constant pressure ($c_{\text{p}}$). However, fluids may instead be measured at constant volume ($c_{\text{V}}$). Measurements under constant pressure produce greater values than those at constant volume because work must be performed in the former. For an ideal gas, $c_{\text{p}} = c_{\text{V}} + R$ (for dry air, $c_{\text{p}} = 1005$ J kg$^{-1}$ K$^{-1}$). Considering a unit mass of an ideal gas, for which $V = 1/\rho$, it can be shown that $U = c_{\text{V}}T$, and hence

$$\delta S = \frac{\delta Q}{T} = c_{\text{p}}\frac{\delta T}{T} - R\frac{\delta p}{p}. \tag{2}$$

An *adiabatic* process is one in which heat is neither lost nor gained, so $\delta S = 0$. In this case, (2) can be integrated to give $\theta = T(p_0/p)^{R/c_{\text{p}}}$, if $T = \theta$ when $p = p_0$. The quantity $\theta$, the *potential temperature*, is the temperature a portion of air would have if, starting from temperature $T$ and pressure $p$, it were compressed until its pressure equaled $p_0$.

1. Intergovernmental Panel on Climate Change reports are available from www.ipcc.ch.

From (2), the first law of thermodynamics can be written as

$$\frac{dT}{dt} = \frac{RT}{c_p p}\frac{dp}{dt} + \frac{1}{c_p}\frac{dQ}{dt}.$$

The first term on the right-hand side represents the rate of change of temperature due to adiabatic expansion or compression. In a typical weather system outside the tropics, air parcels in the middle troposphere undergo vertical displacements on the order of 100 hPa $\text{day}^{-1}$. Assuming that $T \approx 250$ K, the resulting adiabatic temperature change is about 15 K $\text{day}^{-1}$. The second term on the right-hand side is the diabatic heat sources and sinks, which include absorption of solar radiation, absorption and emittance of longwave radiation, and latent heat release. In general, in the troposphere the sum of diabatic terms is much smaller than the sum of adiabatic ones, with the net radiative contribution being small and the latent heat terms being comparable in magnitude with the adiabatic term only in small regions. For an adiabatically rising parcel of air, from (2) the change in temperature is given by

$$-\left(\frac{dT}{dz}\right)_{\text{parcel}} = \frac{RT}{c_p p}\left(\frac{dp}{dz}\right)_{\text{parcel}} = \frac{g}{c_p} \equiv \Gamma_a$$

if the atmosphere is in hydrostatic balance (see (1)). Here, $\Gamma_a \approx 10$ K $\text{km}^{-1}$ is known as the *adiabatic lapse rate.*

The observed decrease of temperature with height, the lapse rate $\Gamma = -dT/dz$, generally differs from the adiabatic lapse rate. If the observed background temperature falls more rapidly with height than the adiabatic lapse rate, i.e., $\Gamma > \Gamma_a$, then an adiabatically rising parcel will be warmer than its surroundings and will continue to rise. In this case, the atmosphere is unstable. On the other hand, if $\Gamma < \Gamma_a$ then the atmosphere is stable. In general, the atmosphere is stable to this dry convection, but it can be unstable in hot, arid regions such as deserts. Convection carries heat up and thus reduces the background lapse rate until the dry adiabatic lapse rate is reached.

Considering the buoyancy forces on a parcel that has been raised to a height $\delta z$ above its equilibrium position and applying Newton's second law gives

$$\rho\frac{d^2}{dt^2}\delta z = -g\delta\rho,$$

where $\delta\rho$ is the difference between the parcel density and that of the environment. The pressures inside and outside the parcel are the same, and so using the ideal gas law this can be rewritten after some manipulation as

$$\frac{d^2}{dt^2}\delta z + N^2\delta z = 0, \quad N^2 \equiv \frac{g}{T}(\Gamma_a - \Gamma).$$

For $N^2 < 0$, $\Gamma > \Gamma_a$, the solutions are exponential in time, and the atmosphere is unstable. For $N^2 > 0$, $\Gamma < \Gamma_a$, the motion is an oscillation with frequency $N$, and the atmosphere is stable. The quantity $N^2$ is a useful measure of atmospheric stratification and can be written in terms of the potential temperature:

$$N^2 = \frac{g}{\theta}\frac{d\theta}{dz}.$$

A region of the atmosphere is therefore statically stable if $\theta$ increases with height ($d\theta/dz > 0$), and it is statically unstable if it decreases with height.

Water vapor in the atmosphere plays an important role in the dynamics of the troposphere since latent heating and cooling can transfer heat from one location to another and because it influences convection. As a moist air parcel rises adiabatically, $p$ falls, so $T$ falls, the water vapor condenses, and latent heat is released, and hence the moist adiabatic lapse rate is less than for dry air (and thus is more easily exceeded). Convective processes in the atmosphere strongly influence the vertical temperature structure in the troposphere. Simple one-dimensional radiative equilibrium calculations predict that the temperature decreases sharply with height at the lower boundary, implying convective instability. Calculations including both radiative and convective effects—adjusting the temperature gradient to neutral stability where necessary, and taking into account the effects of moisture—predict a less sharp decline in temperature through the troposphere, in agreement with observations.

For descriptive purposes, the atmosphere can be divided into layers, defined by alternating negative and positive vertical temperature gradients. In the tropospheric layer from the ground up to about 10–15 km, the temperature decreases with height and the temperature structure is strongly influenced by convective processes. The temperature then increases with height to about 50 km in the *stratosphere*, where the temperature structure is determined predominantly by radiative processes. After this the temperature decreases again through the *mesosphere*.

## 4 Oceanic Properties

As noted above, the oceans are a key component of the coupled climate system. To understand their role, it is necessary to detail their properties and how they are forced.

The oceans are stratified by density, with the densest water near the seafloor and the least dense water near the surface. The density depends on temperature, salinity, and pressure in a complex and nonlinear way, but temperature typically influences density more than salinity in the parameter range of the open ocean. In discussions of ocean dynamics, the buoyancy $b = -g(\rho - \rho_*)/\rho$ is often used, where $\rho$ is the density of a parcel of water and $\rho_*$ is the density of the background. Thus, if $\rho < \rho_*$ the parcel will be positively buoyant and will rise. Since the density does not vary greatly in the ocean (by only a few percent), the buoyancy can be written as $b = -g(\sigma - \sigma_*)/\rho_0$, where $\sigma = \rho - \rho_0$ and $\rho_0 = 1000$ kg m$^{-3}$. The temperature and salinity, and hence density, vary little with depth over the surface layer of the ocean, typically 50–100 m, known as the *mixed layer*. Below this is a layer, called the *thermocline*, where the vertical gradients of temperature and density are greatest; this varies in depth from about 100 m to about 600 m. The waters of the thermocline are warmer and saltier than the deep ocean below, which is known as the *abyss*.

The forcing of the oceans is rather different to that of the atmosphere. A significant fraction of solar radiation passes through the atmosphere to heat the Earth's surface and drive atmospheric convection from below, whereas in the ocean, convection is driven by buoyancy loss from above as the ocean exchanges heat and freshwater at the surface (including through brine rejection in sea-ice formation). The heat flux at the ocean surface has four components: sensible heat flux (which depends on the wind speed and air/sea temperature difference), latent heat flux (from evaporation/precipitation), incoming shortwave radiation from the sun, and longwave radiation from the atmosphere and ocean. The net freshwater flux is given by evaporation minus precipitation, including the influences of river runoff and ice-formation processes. Winds blowing over the ocean surface exert a stress on it and directly drive ocean circulations, particularly in the upper kilometer or so. The wind stress is typically parametrized by $\tau_{\text{wind}} = \rho_a c_D u_{10}^2$, where $\rho_a$ is the density of air, $u_{10}$ is the wind speed at 10 m, and $c_D$ is a drag coefficient (a function of wind speed, atmospheric stability, and sea state). Below the surface, the winds, the flow over seafloor topography, the tides, and other processes indirectly influence the circulation.

## 5 Dynamics of the Atmosphere and the Oceans

The mathematics of fluid dynamics governs the motion of the atmosphere and the oceans. This framework can be used to understand key features of the Earth's weather and climate and to predict future change. Density stratification and the Earth's rotation provide strong constraints that organize the fluid flows.

### 5.1 Rotating Stratified Fluids

In studies of fluid dynamics it is useful to describe the evolution of a parcel of fluid as it follows the flow. This rate of change of a quantity is given by the Lagrangian derivative, $\mathrm{D}/\mathrm{D}t$, which is defined by

$$\frac{\mathrm{D}}{\mathrm{D}t} \equiv \frac{\partial}{\partial t} + \boldsymbol{u} \cdot \nabla,$$

where $\boldsymbol{u}$ is the velocity of the flow. When the wind blows or the ocean currents flow, they carry properties, such as heat and pollutants, with them. This is described by the term $\boldsymbol{u} \cdot \nabla$, which represents advection. There are five key variables relevant to the equations of motion for fluid flow: the velocity $\boldsymbol{u} = (u, v, w)$, the pressure $p$, and the temperature $T$. Correspondingly, there are five independent equations resulting from Newton's second law, conservation of mass, and the first law of thermodynamics.

Newton's second law states that in an inertial frame,

$$\frac{\mathrm{D}\boldsymbol{u}}{\mathrm{D}t} = -\frac{1}{\rho}\nabla p + \boldsymbol{g}^* + \mathcal{F},$$

where $-(1/\rho)\nabla p$ is the pressure gradient force of relevance for fluids, $\boldsymbol{g}^*$ is the gravitational force, and $\mathcal{F}$ is the sum of the frictional forces, all per unit mass. To represent the weather and climate, it is natural to describe the flow seen from the perspective of someone on the surface of the Earth, and thus we need to consider the motion in the rotating frame of the Earth. The angular rotation of the frame is given by the vector $\boldsymbol{\Omega}$ pointing in the direction of the axis of rotation, with magnitude equal to the Earth's angular rate of rotation $\Omega = 7.27 \times 10^{-5}$ s$^{-1}$ (one revolution per day). Newton's second law as described above holds in an inertial frame of reference. When it is translated into the rotating frame of reference, additional terms are introduced that are specific to that frame. The flow velocities in the rotating and inertial frames are related by $\boldsymbol{u}_{\text{inertial}} = \boldsymbol{u}_{\text{rotating}} + \boldsymbol{\Omega} \times \boldsymbol{r}$, and the Lagrangian derivative is given by

$$\left(\frac{\mathrm{D}\boldsymbol{u}_{\text{in}}}{\mathrm{D}t}\right)_{\text{in}} = \left(\frac{\mathrm{D}\boldsymbol{u}_{\text{rot}}}{\mathrm{D}t}\right)_{\text{rot}} + 2\boldsymbol{\Omega} \times \boldsymbol{u}_{\text{rot}} + \boldsymbol{\Omega} \times \boldsymbol{\Omega} \times \boldsymbol{r}.$$

The additional terms are therefore $2\boldsymbol{\Omega} \times \boldsymbol{u}_{\text{rot}}$, the *Coriolis acceleration*, and $\boldsymbol{\Omega} \times \boldsymbol{\Omega} \times \boldsymbol{r}$, the *centrifugal acceleration*. It is convenient to combine the centrifugal force with the gravitational force in one term $\boldsymbol{g} = -g\hat{\boldsymbol{z}} =$

$\boldsymbol{g}^* + \boldsymbol{\Omega} \times \boldsymbol{\Omega} \times \boldsymbol{r}$, where $\hat{\boldsymbol{z}}$ represents a unit vector parallel to the local vertical. The gravity, $g$, defined in this way is the gravity measured in the rotating frame, $g = 9.81 \text{ m s}^{-2}$.

The thinness of the atmosphere/ocean enables a local Cartesian coordinate system, which neglects the Earth's curvature, to be used for many problems. Taking the unit vectors $\hat{\boldsymbol{x}}$, $\hat{\boldsymbol{y}}$, and $\hat{\boldsymbol{z}}$ to be eastward (zonal), northward (meridional), and upward, respectively, the rotation vector can be written in this coordinate basis as $\boldsymbol{\Omega} = (0, \Omega\cos\phi, \Omega\sin\phi)$ for latitude $\phi$. The order of magnitude of zonal velocities in the atmosphere is $|u| \sim 10 \text{ m s}^{-1}$ (less in the ocean), so $\Omega u \ll g$. In addition, in both the atmosphere and the ocean, vertical velocities $w$, typically $\lesssim 10^{-1} \text{ m s}^{-1}$, are much smaller than horizontal velocities. Hence, $2\boldsymbol{\Omega} \times \boldsymbol{u}$ can be approximated by $f\hat{\boldsymbol{z}} \times \boldsymbol{u}$, where $f = 2\Omega\sin\phi$ is the Coriolis parameter. (In many instances it proves useful to approximate the Coriolis parameter by its constant value at a particular latitude $f = f_0 = 2\Omega\sin\phi_0$ or by its Taylor expansion for small latitudinal departures from a particular latitude $f = f_0 + \beta y$, where $\beta = \mathrm{d}f/\mathrm{d}y$; the latter is known as the *$\beta$-plane approximation*.)

With these various conventions and approximations, Newton's second law in a rotating frame becomes

$$\frac{\mathrm{D}\boldsymbol{u}}{\mathrm{D}t} + \frac{1}{\rho}\nabla p - \boldsymbol{g} + f\hat{\boldsymbol{z}} \times \boldsymbol{u} = \mathcal{F}. \tag{3}$$

For some purposes it is useful to write the gravity term as the gradient of a potential function $\Phi$, known as the *geopotential*, $\boldsymbol{\nabla}\Phi = -\boldsymbol{g}$. The geopotential $\Phi(z)$ at a height $z$ is the work required to raise a unit mass to height $z$ from mean sea level, $\Phi = \int_0^z g\,\mathrm{d}z$. The remaining two equations are the conservation of mass (the mass of a fixed volume can change only if $\rho$ changes, and this requires a mass flux into the volume),

$$\frac{\mathrm{D}\rho}{\mathrm{D}t} + \rho\nabla \cdot \boldsymbol{u} = 0, \tag{4}$$

and the first law of thermodynamics (see (2)),

$$\frac{\mathrm{D}Q}{\mathrm{D}t} = c_{\mathrm{p}}\frac{\mathrm{D}T}{\mathrm{D}t} - \frac{1}{\rho}\frac{\mathrm{D}p}{\mathrm{D}t}. \tag{5}$$

Here, $\mathrm{D}Q/\mathrm{D}t$ is the diabatic heating rate per unit mass, which in the atmosphere is mostly due to latent heating/cooling (condensation/evaporation) and radiative heating/cooling. In the ocean, analogous equations hold for temperature and salinity.

The equations of motion we have just derived are a simplified form of the equations that are at the heart of weather and climate models. Such models solve numerically discretized versions of the equations of motion, and computational constraints mean that there is a limit to the scale of motion that can be directly resolved. In the atmosphere, large-scale motion such as weather systems (~1000 km) are well captured, but smaller-scale motion such as convective storm systems (~1–100 km) generally need to be parametrized, i.e., represented approximately in terms of the larger-scale resolved variables. Similarly, in the ocean small-scale processes are parametrized (indeed, the scales of motion are typically ten times smaller in the ocean than in the atmosphere, making the problem even more challenging).

The forcing of an atmosphere-only model may include specified solar input, radiatively active gases, sea-surface temperature, and sea ice. What is and is not included in the model depends on whether the processes are important over the timescale for which the model is being used to project (hours to weeks for weather models, decades to centuries for climate models). For climate projections, COUPLED MODELS [IV.16 §5] are usually employed. In these, separate models of the atmosphere, ocean, ice, land, and some chemical cycles are linked together in such a way that changes in one may influence another.

### 5.2 Circulation of the Atmosphere and Ocean

Tropical regions receive more incoming solar radiation than polar regions because the solar beam is concentrated over a smaller area due to the spherical curvature of the Earth. If the Earth were not rotating, the atmospheric circulation would be driven by the pole-to-equator temperature difference, with warm air rising in the tropical regions and sinking in the polar regions. On the rotating Earth, as air moves away from the equator in the upper troposphere, it gains an eastward (westerly) velocity component from the Coriolis effect, as we will describe mathematically below. In the tropics the Coriolis parameter $f$ is small, but elsewhere it has a significant influence. The overturning circulation in the atmosphere is therefore confined to the tropics: if it extended all the way to the poles, the westerly component arising from the Coriolis effect would become infinite. Moist air rises near the equator and dry air descends in the subtropical desert regions at about 20–30°. This overturning circulation is known as the *Hadley circulation*. During the course of a year, the pattern of solar forcing migrates: north in northern summer, south in southern summer. The entire Hadley circulation shifts seasonally following the solar forcing.

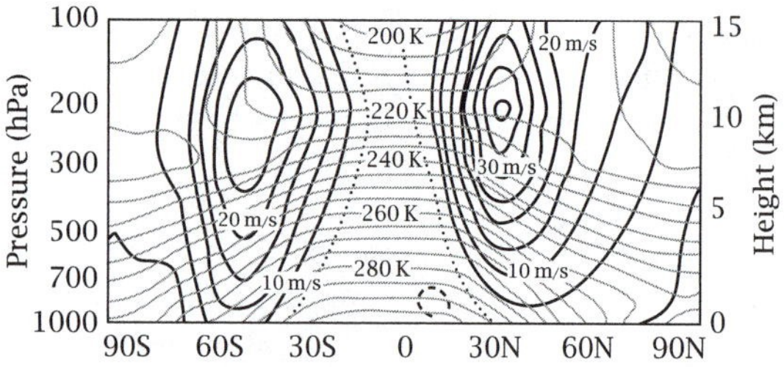


**Figure 2** Temperature (thin gray contours, interval 5 K) and westerly winds (thick black contours, interval 5 m $s^{-1}$; zero wind, dotted line; easterly winds, dashed line) in the troposphere. All quantities have been averaged in longitude. Values are typical of the Northern Hemisphere winter. Approximate altitude is also indicated.

In the upper troposphere, at the poleward extent of the Hadley circulation (about 30°), are the jet streams of strong westerly flow (see figure 2). They are strongest in winter, with average speeds of around 30 m $s^{-1}$. The equatorward return flow at the surface is weak. The influence of friction together with the Coriolis effect results in the northeasterly trade winds (i.e., winds originating from the northeast) in the Northern Hemisphere and the southeasterly trade winds in the Southern Hemisphere, as we will describe mathematically below.

The westerly flow in the jet streams is hydrodynamically unstable and can spontaneously break down into vortical structures known as *eddies*, which manifest themselves as traveling weather systems. Eddies play a vital role in transporting heat, moisture, and chemical species in the latitude/height plane. Observations indicate that in the annual mean the tropical regions emit less radiation back to space than they receive and that the polar regions emit more radiation than they receive. This implies that there must be a transport of energy from the equator to the pole that takes places in the atmosphere and/or the ocean. In the atmosphere, heat is transported poleward through the tropics by the Hadley circulation; at higher latitudes, eddies are mainly responsible for the heat transport.

In the ocean there is an overturning circulation that encompasses a system of surface and deep currents running through all basins. This circulation transports heat—and also salt, carbon, nutrients, and other substances—around the globe and connects the surface ocean and atmosphere with the huge reservoirs of the deep ocean. As such, it is of critical importance to the global climate system. We have discussed above the requirement for poleward heat transport in the atmosphere and/or ocean to explain the observed incoming/outgoing radiation profiles. Detailed calculations indicate that the atmosphere is responsible for the bulk of the transport in the middle and high latitudes, but the ocean makes up a considerable fraction, particularly in the tropics. Heat is transported poleward by the ocean in the overturning circulation if waters moving poleward are compensated by equatorward flow at colder temperatures. In the Atlantic heat transport is northward everywhere, but in the Pacific the heat transport is directed poleward in both hemispheres, while the Indian Ocean provides a poleward transport in the Southern Hemisphere. The net heat transport is poleward in each hemisphere.

The surface ocean currents are dominated by closed circulation patterns known as *gyres*. In the Northern Hemisphere there are gyres in the subtropics of the Pacific and Atlantic, with eastward flow at midlatitudes and westward flow at the equator. The current speed at the interior of these gyres is $\lesssim 10$ cm $s^{-1}$, but at the western edge there are strong poleward currents (the Kuroshio in the Pacific and the Gulf Stream in the Atlantic) with speeds $\gtrsim 100$ cm $s^{-1}$. Ocean surface waters are only dense enough to sink down to the deep abyss at a few key locations: particularly in the northern North Atlantic and around Antarctica. Deep ocean convection occurs only in these cold high-latitude regions, where the internal stratification is small and surface density can increase through direct cooling/evaporation or brine rejection in sea-ice formation. Dense water formed in the North Atlantic flows south as a deep western boundary current and eventually enters the Southern Ocean, where it mixes with other water masses. Ultimately, the deep water is brought up to the surface by vertical mixing (tides and winds are the primary sources of energy for this) and by the overturning circulation in the Southern Ocean. Hydrodynamic instabilities associated with the major currents can generate eddies, which are ubiquitous in the ocean. As in the atmosphere, these eddies are responsible for heat transport and they play a central role in driving the overturning circulation in the Southern Ocean.

## 6 Dynamical Processes

To understand the behavior of different dynamical processes in the atmosphere and the oceans, it is useful to consider the relative magnitudes of the different terms in the equation of motion (3). For this purpose we introduce the Rossby number *Ro*, which is the ratio of acceleration terms to Coriolis terms. If $U$ is a typical velocity scale and $L$ is a typical length scale, $Ro = U/fL$.

### 6.1 Ocean Surface Waves

For ocean waves crashing onshore, the influence of the Earth's rotation is small and so *Ro* is large and the Coriolis terms in (3) can be neglected. The wave motion results when the water surface is displaced above its equilibrium level and gravity acts to pull it downward. An oscillatory motion results as the water overshoots its equilibrium position in both vertical directions and is restored by pressure from the surrounding water mass in one direction and by gravity in the other.

For a fluid with uniform density $\rho_0$, a free surface at $z = 0$, and a bottom at $z = -H$, the equilibrium solution has zero velocity and the equilibrium pressure $p_*$ is given by integrating the hydrostatic balance equation (1): $p_*(z) = -g\rho z$, where $\rho$ takes the value $\rho_0$ in the fluid and zero above it. A perturbation to this system can be defined such that the perturbed position of the free surface is given by $z = \eta(x, y, t)$ and the perturbed pressure is given by $p = p_* + p'$. Neglecting the Coriolis and frictional terms, (3) gives

$$\frac{\mathrm{D}\boldsymbol{u}}{\mathrm{D}t} = -\frac{1}{\rho}\nabla p'.$$

Taking the $x$, $y$, and $z$ derivatives of the components of this equation and using the continuity equation (4) results in LAPLACE'S EQUATION [III.18] for $p'$, $\nabla^2 p' = 0$. The relevant boundary conditions are that

(i) there is no normal flow at the bottom ($w = 0$ at $z = -H$), from which it follows that $\partial p'/\partial z$ must vanish at $z = -H$ by considering the $z$ component of (3);
(ii) a particle in the free surface $z = \eta$ will remain in it, i.e., $\mathrm{D}(z - \eta)/\mathrm{D}t = 0$, which gives $w = \partial\eta/\partial t$ at $z = \eta$ for small perturbations; and
(iii) the pressure must vanish at the free surface, i.e., $p' = \rho g \eta$ at $z = \eta$.

The two conditions at $z = \eta$ can also be applied at $z = 0$ to good approximation.

A wavelike solution to this system can be sought that is assumed to correspond to a displacement $\eta$ of the same form, where $\hat{\eta}$ is the amplitude, $\boldsymbol{k} = (k, l)$ is the wave number, and $\omega$ is the frequency. This takes the form

$$p' = \mathrm{Re}\,\hat{\eta}\exp(\mathrm{i}(kx + ly - \omega t)).$$

From Laplace's equation, $\partial^2 p'/\partial z^2 - \kappa^2 p' = 0$, where $\kappa^2 = k^2 + l^2$. Applying the first and third boundary conditions, the solution is

$$p' = \frac{\rho g \hat{\eta}\cos(kx + ly - \omega t)\cosh\kappa(z + H)}{\cosh\kappa H}.$$

Using the $z$-component of (3) it can be shown that it is possible for only the second boundary condition to be met and the solution to be consistent with the assumed form for $\eta$ if

$$\omega^2 = g\kappa\tanh\kappa H.$$

This *dispersion relation* determines the frequency $\omega$ of waves of a given wave number and hence also the phase speed $c = \omega/\kappa$.

In the shallow-water or longwave limit when $\kappa H \ll 1$, $\tanh\kappa H \to \kappa H$ and hence $c = \sqrt{gH}$. This means that all long waves travel at the same speed. Earthquakes on the sea floor can excite TSUNAMIS [V.19]. These are long waves with wavelengths up to hundreds of kilometers on an ocean that is at most a few kilometers deep. A tsunami therefore propagates at speed $\sqrt{gH} \approx 200\ \mathrm{m\ s^{-1}}$ without dispersion, allowing its energy to be maintained as it crosses a vast expanse of ocean. The tsunami slows as it approaches the shore and the water depth shallows, and its wavelength $\lambda = 2\pi c/\omega$ decreases ($\omega$ is constant). For the same energy density, the amplitude increases until nonlinear effects become important and the wave breaks.

### 6.2 Midlatitude Weather Systems and Ocean Gyres

The midlatitudes are the regions between the tropics and the polar regions. In these regions the typical velocity scales and length scales of weather systems are $U \sim 10\ \mathrm{m\ s^{-1}}$ and $L \sim 10^6$ m, and $f \sim 10^{-4}\ \mathrm{s^{-1}}$; hence $Ro \sim 0.1$. In the ocean, in midlatitude gyres, the typical scales are $U \sim 0.1\ \mathrm{m\ s^{-1}}$ and $L \sim 10^6$ m, so $Ro \sim 10^{-3}$. Therefore, in both cases, because *Ro* is small we can neglect the acceleration terms in (3) in favor of the Coriolis terms. In addition, away from boundaries the friction is negligible, so in the horizontal we have

$$f\hat{\boldsymbol{z}} \times \boldsymbol{u} + \frac{1}{\rho}\nabla p = \boldsymbol{0}.$$

This approximation is known as *geostrophic balance*. It is a balance between the Coriolis force and the horizontal pressure gradient force and is used to define the geostrophic velocity $\boldsymbol{u}_{\mathrm{g}}$ given by

$$\boldsymbol{u}_{\mathrm{g}} = (u_{\mathrm{g}}, v_{\mathrm{g}}) \equiv \left(\frac{-1}{f\rho}\frac{\partial p}{\partial y}, \frac{1}{f\rho}\frac{\partial p}{\partial x}\right). \tag{6}$$

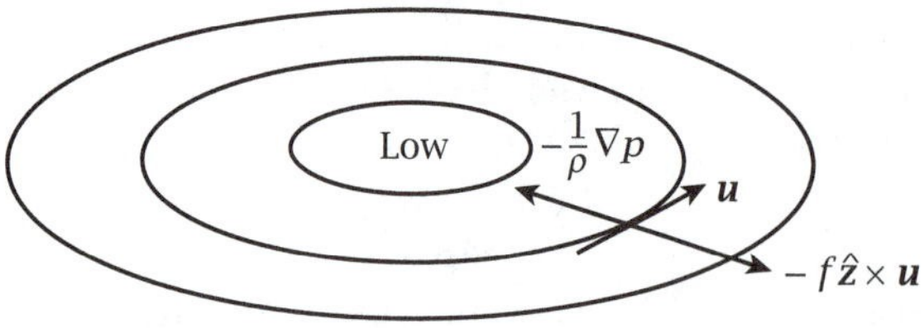


**Figure 3** Counterclockwise (cyclonic) geostrophic flow around a low-pressure center in the Northern Hemisphere. The effect of the Coriolis force deflecting the flow is balanced by the horizontal component of the pressure gradient force, directed from high to low pressure.

Considering the vertical direction in (3), if friction $\mathcal{F}_z$ and vertical acceleration $\mathrm{D}w/\mathrm{D}t$ are small (as is generally true for large-scale atmospheric and oceanic systems), then this reduces to the equation (1) for hydrostatic balance introduced earlier.

Geostrophically balanced flow is normal to the pressure gradient, i.e., along contours of constant pressure. In the Northern (Southern) Hemisphere, motion is therefore counterclockwise (clockwise) around the center of low-pressure systems (see figure 3). Note also that the speed depends on the magnitude of the pressure gradient: it is stronger when the isobars are closer. When the flow swirls counterclockwise in the Northern Hemisphere or clockwise in the Southern Hemisphere, it is called *cyclonic* flow (a hurricane is an example of this); the opposite direction is called *anticyclonic* flow.

Geostrophic balance can be used to explain many features that are observed in atmosphere and ocean flows. In the atmosphere, because of the pole-to-equator temperature gradient, there is a horizontal pressure gradient above the surface in the troposphere going from warm tropical latitudes to cold polar latitudes. This is in geostrophic balance with the Coriolis force associated with the westerly flow of the midlatitude jet streams (see figure 2). Instabilities of the jet streams form eddies consisting of cyclonic geostrophic flow around low-pressure systems (these are midlatitude weather systems). In the ocean the sea surface is higher in the warm subtropical gyre of the North Atlantic than it is further north in the cool subpolar gyre, resulting in a pressure gradient force directed northward to geostrophically balance the southward Coriolis force associated with the eastward-flowing Gulf Stream. Instabilities of the current generate eddies that are manifested as geostrophic flow around anomalies in sea-surface height.

Further analysis highlights the fact that rotation provides strong constraints on the flow. In the case where $\rho$ and $f$ are constant, by taking the horizontal derivatives of the geostrophic velocity it can be shown that it is horizontally nondivergent. Taking the vertical derivative and using the hydrostatic balance equation gives $(\partial u_g/\partial z, \partial v_g/\partial z) = 0$. The equation for the conservation of mass (4) can then be applied to show that $\partial w_g/\partial z = 0$. Hence, the geostrophic velocity does not vary in the vertical and is two dimensional. Under slightly more general conditions—namely for a slow, steady, frictionless flow of a *barotropic* (i.e., density depends only on pressure, so $\rho = \rho(p)$), incompressible fluid—it can be shown that the horizontal and vertical components of the velocity cannot vary in the direction of the rotation vector $\boldsymbol{\Omega}$, and hence the flow is two dimensional. This is known as the *Taylor–Proudman theorem*. It means that vertical columns of fluid remain vertical (they cannot be tilted or stretched).

In general, however, the density in the atmosphere and the oceans does vary on pressure surfaces, as it depends on temperature, for example. In this case the fluid is said to be *baroclinic*. If the density can be written as $\rho = \rho_0 + \sigma$, where $\rho_0$ is a constant reference density (usually taken to be 1000 kg m$^{-3}$ for the ocean) and $\sigma \ll \rho_0$ (this is generally the case in the ocean), then replacing $\rho$ by $\rho_0$ in the denominator of the geostrophic velocity (6), taking $\partial/\partial z$, and making use of the equation for hydrostatic balance (1) gives

$$\left(\frac{\partial u_g}{\partial z}, \frac{\partial v_g}{\partial z}\right) = \frac{g}{f\rho_0}\left(\frac{\partial \sigma}{\partial y}, -\frac{\partial \sigma}{\partial x}\right).$$

For the compressible atmosphere with larger density variations, it is often useful to consider the relevant equations with pressure as the vertical coordinate. This can be done through the introduction of a *log-pressure* coordinate defined as $\tilde{z} = -H\ln(p/p_0)$. Here, $p_0$ is a reference pressure (usually taken to be 1000 hPa) and the quantity $H = RT_0/g$, known as the *scale height*, is the height over which the pressure falls by a factor of e. If $T_0 = 250$ K (a typical value for the troposphere) then $H \approx 7.3$ km. From (3), the horizontal momentum equation in log-pressure in the case where friction can be neglected can be written as

$$\mathrm{D}\boldsymbol{u}/\mathrm{D}t + f\hat{\boldsymbol{z}} \times \boldsymbol{u} + \nabla\Phi = \mathbf{0},$$

where $\mathrm{D}/\mathrm{D}t = \partial/\partial t + \boldsymbol{u}\cdot\nabla_{\mathrm{h}} + \tilde{w}\partial/\partial\tilde{z}$ with $\nabla_{\mathrm{h}}$ representing the horizontal gradient components and where $\Phi$ is the geopotential. Additionally, the hydrostatic equation can be written as

$$\partial\Phi/\partial\tilde{z} = RT/H.$$

These equations can be used to show that

$$\left(\frac{\partial u_g}{\partial \tilde{z}}, \frac{\partial v_g}{\partial \tilde{z}}\right) = \frac{R}{fH}\left(-\frac{\partial T}{\partial y}, \frac{\partial T}{\partial x}\right).$$

In both the atmosphere and the ocean the vertical gradient in the geostrophic velocities is therefore related to the horizontal density gradient. This is known as the *thermal wind* relationship. As discussed before, there is a pole-to-equator temperature gradient in the atmosphere ($f^{-1}\partial T/\partial y < 0$ in both hemispheres), which implies $\partial u_g/\partial \tilde{z} > 0$ and, hence, that with increasing height the winds become more westerly in both hemispheres. Consistent with this, strong jet streams are observed in midlatitudes of both hemispheres in the upper troposphere (see figure 2).

### 6.3 Ekman Layers and Wind-Driven Ocean Circulation

Where frictional effects become important, such as close to boundaries, geostrophic balance no longer holds. For small *Ro* numbers, as found in the midlatitude atmosphere and oceans, the acceleration terms in (3) can again be neglected in favor of the Coriolis terms, and hence the horizontal velocity $\boldsymbol{u}$ is given by

$$f\hat{\boldsymbol{z}} \times \boldsymbol{u} + \frac{1}{\rho}\nabla p = \mathcal{F}. \tag{7}$$

The result is that friction introduces an *ageostrophic* component of flow (high to low pressure), $\boldsymbol{u} = \boldsymbol{u}_g + \boldsymbol{u}_{ag}$. This effect is important in the lower 1 km or so of the atmosphere and the upper 100 m or so of the ocean. The ageostrophic component explains, for example, the meridional (north–south) component of the trade winds.

The geostrophic flow is horizontally nondivergent (except on planetary scales), but the ageostrophic flow is not. Winds that deviate ageostrophically toward low-pressure systems near the surface are convergent, and through mass continuity there must be an associated vertical velocity away from the surface. This is known as *Ekman pumping*. In the atmosphere Ekman pumping produces ascent, cooling, clouds, and sometimes rain in low-pressure systems. In the ocean it is a key component of the circulation in gyres.

The wind stress at the ocean surface gives rise to a force per unit mass on a slab of ocean of

$$\mathcal{F}_{\text{wind}} = \frac{1}{\rho_0}\frac{\partial \tau_{\text{wind}}}{\partial z},$$

where $\rho_0$ is the density of the slab. This directly drives ocean circulations close to the surface in the *Ekman layer*. At the surface, $z = 0$, the stress is $\tau(0) = \tau_{\text{wind}}$, and this decays over the depth $\delta \sim$ 10–100 m of the Ekman layer, so $\tau(-\delta) = 0$. The ageostrophic component of motion, $\boldsymbol{u}_{ag}$, is obtained by substituting the force arising from the wind into (7), giving $f\hat{\boldsymbol{z}} \times \boldsymbol{u}_{ag} = (\rho_0)^{-1}\partial\tau/\partial z$. By integrating this equation over the depth of the Ekman layer, it can be shown that the lateral mass transport over the layer is given by

$$\boldsymbol{M}_{\text{Ek}} \equiv \int_{-\delta}^{0} \rho_0 \boldsymbol{u}_{ag}\, \mathrm{d}z = \frac{\tau_{\text{wind}} \times \hat{\boldsymbol{z}}}{f}. \tag{8}$$

The mass transport in the Ekman layer is therefore directed to the right (left) of the wind in the Northern (Southern) Hemisphere. Further analysis indicates that, in the Northern Hemisphere (directions are reversed in the Southern Hemisphere), (i) the horizontal currents at the surface are directed at 45° to the right of the surface wind and (ii) the currents spiral in an anticyclonic (clockwise) direction with depth through the ocean, decaying exponentially in magnitude away from the surface. Similar Ekman spirals exist at the bottom of the ocean and the atmosphere, but the direction of the flow is opposite. Winds at the surface are therefore 45° to the left of the winds in the lower troposphere above the planetary boundary layer, which means that the currents at the sea surface are nearly in the direction of the lower tropospheric winds.

In the anticyclonic subtropical gyres, Ekman transport results in the flow converging horizontally toward the center of the gyre. Mass conservation then implies downwelling through Ekman pumping (see figure 4). In the cyclonic subpolar gyres there is divergence and Ekman suction. In the incompressible ocean, the vertical velocity $w$ is given by $\nabla_h \cdot \boldsymbol{u}_{ag} + \partial w/\partial z = 0$, since the geostrophic flow is nondivergent. The vertical velocity at the surface is zero, and so integrating this equation over the Ekman layer gives a vertical velocity at the base of the Ekman layer $w_{\text{Ek}}$ of

$$w_{\text{Ek}} = \frac{1}{\rho_0}\nabla_h \cdot \boldsymbol{M}_{\text{Ek}} = \frac{1}{\rho_0}\hat{\boldsymbol{z}} \cdot \nabla \times \left(\frac{\tau_{\text{wind}}}{f}\right).$$

The Ekman pumping velocity $w_{\text{Ek}}$ therefore depends on the curl of $\tau_{\text{wind}}/f$, which is largely set by variations in the wind stress.

In the ocean interior, away from the surface or continental boundaries, friction is negligible, and to a good approximation, the flow is in geostrophic balance. However, the flow does respond to the pattern of vertical velocities imposed by the Ekman layer above. The horizontal divergence of geostrophic flow is associated with vertical stretching of water columns in the interior because $\nabla_h \cdot \boldsymbol{u}_g + \partial w/\partial z = 0$ for an incompressible flow.

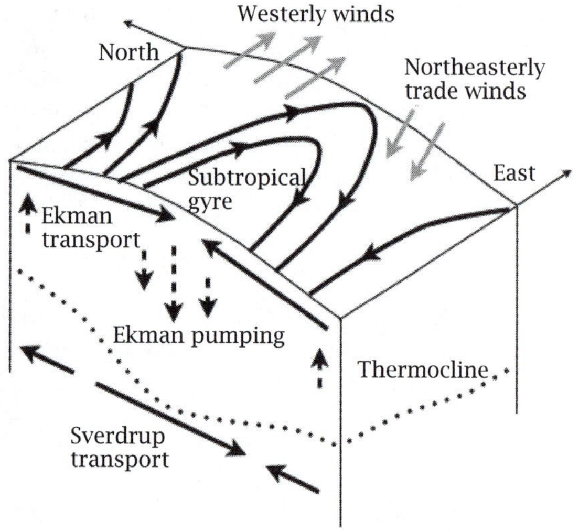


**Figure 4** Schematic indicating the Ekman and Sverdrup transports (solid lines) and Ekman pumping (dashed lines) associated with wind-driven ocean gyres.

From (6), the horizontal divergence of the geostrophic velocity is $\nabla_h \cdot \boldsymbol{u}_g = -(\beta/f)v_g$, where $\beta = df/dy$. Hence, the vertical and meridional (i.e., northward) currents are related by $\beta v_g = f\partial w/\partial z$. There is therefore expected to be an equatorward (poleward) component to the horizontal velocity where $w_{Ek} < 0$ ($w_{Ek} > 0$). This means that in the subtropical gyres the interior flow away from boundaries is equatorward (see figure 4). The interior flow must be consistent with the sense of the wind circulation, and so on the western side of the ocean basin, where frictional effects at the continental boundary mean geostrophy breaks down, there is a narrow return flow. In the subtropical gyres the wind puts anticyclonic vorticity into the ocean, which is removed by friction at the boundary as the flow returns poleward on the western side.

The full depth-integrated flow $V$ can be obtained by considering the generic equation for the meridional velocity $v$, obtained from the incompressibility condition together with (7):

$$\beta v = f\frac{\partial w}{\partial z} + \frac{1}{\rho_0}\frac{\partial}{\partial z}\left(\frac{\partial \tau_y}{\partial x} - \frac{\partial \tau_x}{\partial y}\right).$$

Integrating this from the bottom of the ocean ($z = -D$, $w = 0$, $\tau = 0$) to the surface ($z = 0$, $w = 0$, $\tau = \tau_{\text{wind}}$) gives

$$\beta V = \frac{1}{\rho_0}\hat{\boldsymbol{z}} \cdot \nabla \times \tau_{\text{wind}}.$$

Known as *Sverdrup balance*, this relates the depth of the integrated flow to the curl of the wind stress. It dictates the sense of the motion in the subpolar and subtropical gyres. In the Southern Ocean, which encircles Antarctica, at vertical levels where no topography exists to support east–west pressure gradients, there can be no mean meridional geostrophic flow and therefore the above Sverdrup approximation does not apply.

### 6.4 Quasigeostrophic Flow and Baroclinic Instability

For large-scale, low-frequency flows with small *Ro* number, the velocity can be split into a geostrophic part and an ageostrophic part, $\boldsymbol{u} = \boldsymbol{u}_g + \boldsymbol{u}_{ag}$, with $|\boldsymbol{u}_{ag}|/|\boldsymbol{u}_g| \sim O(Ro)$. In this case the momentum can be approximated by the geostrophic value, and the rate of change of momentum or temperature following horizontal motion can be approximated by the rate of change following the geostrophic flow. Thus, $\mathrm{D}\boldsymbol{u}/\mathrm{D}t \approx \mathrm{D_g}\boldsymbol{u}_g/\mathrm{D}t \equiv \partial \boldsymbol{u}_g/\partial t + \boldsymbol{u}_g \cdot \nabla \boldsymbol{u}_g$, and the remaining terms of the momentum equation in log-pressure coordinates give

$$\frac{\mathrm{D_g}\boldsymbol{u}_g}{\mathrm{D}t} = -f_0\hat{\boldsymbol{z}} \times \boldsymbol{u}_{ag} - \beta y \hat{\boldsymbol{z}} \times \boldsymbol{u}_g,$$

where the $\beta$-plane approximation has been used.

The geostrophic velocity is nondivergent, and so there exists a stream function $\psi$ such that $\boldsymbol{u}_g = \hat{\boldsymbol{z}} \times \nabla\psi$. Comparison with the geopotential indicates $\psi = (\Phi - \Phi_*)/f_0$, with $\Phi_*(\tilde{z})$ being a suitable reference geopotential profile. The continuity equation (4) in log-pressure coordinates becomes

$$\frac{\partial u_{ag}}{\partial x} + \frac{\partial v_{ag}}{\partial y} + \frac{1}{\rho_*}\frac{\partial(\rho_* \tilde{w})}{\partial \tilde{z}} = 0,$$

where $\rho_*(\tilde{z}) = \rho_0 e^{-\tilde{z}/H}$. The potential temperature $\theta$ is related to the temperature by $\theta = T\Xi$, where $\Xi = \exp(R\tilde{z}/Hc_p)$. Using the hydrostatic balance equation, it can be written as a reference profile, $\theta_*(\tilde{z}) = (H/R)\Xi\partial\Phi_*/\partial\tilde{z}$, plus a disturbance, $\theta' = (H/R)f_0\Xi\partial\psi/\partial\tilde{z}$. The thermodynamic equation can then be approximated by

$$\frac{\mathrm{D_g}\theta'}{\mathrm{D}t} + \tilde{w}\frac{\partial \theta_*}{\partial \tilde{z}} = \frac{1}{c_p}\frac{\mathrm{D}Q}{\mathrm{D}t}\exp(R\tilde{z}/Hc_p).$$

Together, these form the equations for *quasigeostrophic* flow. When friction and diabatic heating are neglected, these equations can be used to demonstrate that the quantity $q$, the quasigeostrophic *potential vorticity* given by

$$q = f_0 + \beta y + \nabla^2\psi + \frac{1}{\rho_*}\frac{\partial}{\partial \tilde{z}}\left(\rho_* \frac{f_0^2}{N^2}\frac{\partial \psi}{\partial \tilde{z}}\right),$$

is conserved following the flow: $\mathrm{D}_g q/\mathrm{D}t = 0$. Here, $N$ is a buoyancy frequency given by

$$N^2(\tilde{z}) = \frac{R}{H}\exp\left(-\frac{R\tilde{z}}{Hc_\mathrm{p}}\right)\frac{\partial\theta_*}{\partial\tilde{z}}.$$

Baroclinic instability is responsible for the development of eddies in the atmosphere and ocean. As noted earlier, such eddies play a central role in global heat transport. The instability is a common feature of flows in the atmosphere and the oceans because a rotating fluid adjusts to be in geostrophic balance, rather than rest, and in this configuration the fluid has potential energy that is available for conversion to other forms by a redistribution of mass. For example, the vertical shear of the westerly flow in the midlatitude jet streams is in thermal wind balance with a horizontal temperature gradient, and this provides available potential energy for baroclinic instability. However, having available potential energy in the fluid is not sufficient for instability since rotation tends to inhibit the release of this potential energy.

A highly simplified description of the instability process can be obtained by considering the evolution of a parcel of air in an idealized background state consistent with the typical mean state of the lower atmosphere, i.e., one with sloping surfaces of constant potential temperature in the height–latitude plane (see figure 5). If a parcel of air moves upward in a wedge between a sloping surface of constant potential temperature and the horizontal, and is replaced by a similar parcel moving downward (as indicated in figure 5), the warm air parcel (light gray) is then surrounded by denser air and will continue to rise, and vice versa for the cold air parcel (dark gray). The potential energy is reduced since the heavier parcel has moved lower and the lighter parcel has moved higher. The potential energy released can be converted to kinetic energy of eddying motions. This baroclinic instability process is associated with a poleward and upward transport of heat.

Analysis of idealized flows can provide an indication of the typical properties of baroclinic instability. The classic example, known as the *Eady problem*, considers the simplest possible model that satisfies the necessary conditions for instability. The density is treated as a constant except where it is coupled with gravity in the vertical momentum equation, and both the Coriolis parameter $f$ and the buoyancy frequency $N$ are assumed to be constant. This is known as the *Boussinesq approximation*. The model is set up with rigid lids at $\tilde{z} = 0$ and $\tilde{z} = H$. The flow $\boldsymbol{u}$ is considered to be

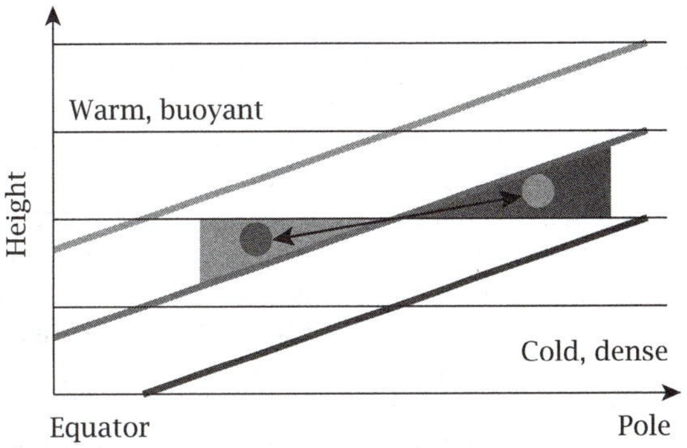


**Figure 5** Schematic indicating parcel trajectories relative to sloping surfaces of constant potential temperature (gray lines) within a wedge of instability (filled region) for a baroclinically unstable disturbance in a rotating frame.

a background mean field $\bar{\boldsymbol{u}} = (\bar{u}, 0, 0)$ with a constant vertical shear, $\partial\bar{u}/\partial\tilde{z} = \Lambda$, plus a disturbance, $\boldsymbol{u}'$.

With these conditions, $\partial\bar{q}/\partial y = 0$, which would imply that the flow was stable if it were not for the presence of the upper boundary. Linearizing the quasi-geostrophic potential vorticity equation gives

$$\left(\frac{\partial}{\partial t} + \bar{u}\frac{\partial}{\partial x}\right)q' = 0.$$

The linearized form of the thermodynamic equation gives

$$\left(\frac{\partial}{\partial t} + \bar{u}\frac{\partial}{\partial x}\right)\frac{\partial\psi'}{\partial\tilde{z}} - \frac{\partial\psi'}{\partial x}\frac{\partial\bar{u}}{\partial\tilde{z}} + \tilde{w}\frac{N^2}{f_0} = 0.$$

Vertical shear at the upper and lower boundaries implies a temperature gradient in the $y$-direction (i.e., in the cross-flow direction) through thermal wind balance. The advection of temperature at the upper and lower boundaries must therefore be taken into account. First considering the lower boundary only, with $\bar{u} = 0$ the equations become

$$\nabla^2\psi' + \frac{f_0^2}{N^2}\frac{\partial^2\psi'}{\partial\tilde{z}^2} = 0$$

and

$$\frac{\partial}{\partial t}\left(\frac{\partial\psi'}{\partial\tilde{z}}\right) - \Lambda\frac{\partial\psi'}{\partial x} = 0.$$

Looking for wavelike solutions of the form

$$\psi' = \mathrm{Re}\,\hat{\psi}(\tilde{z})\exp(\mathrm{i}(kx + ly + mz - \omega t))$$

leads to a dispersion relation $\omega = k\Pi/(l^2 + k^2)^{1/2}$, where $\Pi = \Lambda f_0/N$. This gives an eastward phase speed $c = \omega/k$. A poleward displacement of an air parcel on the lower boundary will induce a warm anomaly, which will be associated with a cyclonic circulation. A neighboring equatorward displacement will induce a cold anomaly and an anticyclonic circulation. Consideration of the induced circulation pattern shows that the temperature anomaly will propagate to the east. Considering the upper boundary only, $\tilde{u} = \Lambda H$ and this leads to a dispersion relation $\omega = k\Lambda H - k\Pi/(l^2 + k^2)^{1/2}$. This gives a westward phase speed relative to the flow. These edge waves by themselves do not transport heat and do not release any energy from the system. The process of baroclinic instability in the Eady model relies on the presence of the edge waves and, crucially, their interaction.

For the full problem, one can find solutions to the linearized quasigeostrophic potential vorticity equation of the same wavelike form, with $\hat{\psi}(\tilde{z}) = A\cosh\kappa\tilde{z} + B\sinh\kappa\tilde{z}$, where $\kappa^2 = (l^2+k^2)N^2/f_0^2$ and the boundary conditions define $A$ and $B$. In this case the dispersion relation is

$$\omega = k\Lambda H(\tfrac{1}{2} \pm \sqrt{\alpha}), \quad \alpha = \frac{1}{4} - \frac{\coth\kappa H}{\kappa H} + \frac{1}{\kappa^2 H^2}.$$

If $\alpha < 0$ the flow is unstable. Instability therefore requires $(l^2 + k^2) < k_c^2 \approx 5.76/L_R$, where $k_c$ is a critical wave number and $L_R = NH/f_0$ is the *Rossby radius*. So, there are only certain horizontal scales of waves that may grow exponentially, and the scale of these waves will depend on the rotation rate, the depth of the layer, and the static stability.

For a given zonal wave number $k$, the most unstable growing mode is that for which the meridional wave number $l$ is zero. The wave number for maximum growth is $k \approx 1.61/L_R$, corresponding to a wavelength of $L_{max} = 2\pi/k \approx 3.9L_R$ and a growth rate $\sigma \approx 0.31 f_0\Lambda/N$. Applying typical order-of-magnitude values for the atmosphere ($H \sim 10$ km, $U \sim 10$ m s$^{-1}$, and $N \sim 10^{-2}$ s$^{-1}$) gives $L_{max} \approx 4000$ km and $\omega \approx 0.26$ day$^{-1}$. For the ocean, $H \sim 1$ km, $U \sim 0.1$ m s$^{-1}$, and $N \sim 10^{-2}$ s$^{-1}$, giving $L_{max} \approx 400$ km and $\omega \approx 0.026$ day$^{-1}$. The atmospheric values are broadly consistent with the observed spatial and growth rates of midlatitude weather systems. In the ocean, the simple scenario on which these values are based is not quantitatively applicable, but the values give a qualitative sense of the scale and growth rate of instabilities in the ocean relative to the atmosphere.

### 6.5 Rossby and Kelvin Waves

Wave-like motions frequently occur in the atmosphere and oceans. One important class of waves is known as *Rossby waves.* The Rossby wave is a potential vorticity-conserving motion that owes its existence to an isentropic gradient of potential vorticity.

In midlatitudes, taking the $\beta$-plane approximation and considering a small-amplitude disturbance to a uniform zonal background flow $\boldsymbol{u} = (U, 0, 0)$, one finds a wavelike solution to the quasigeostrophic equations with a dispersion relation

$$\omega_{\text{Rossby}} = kU - \frac{\beta k}{k^2 + l^2 + f_0^2 m^2/N^2},$$

where $N$ has been assumed to be constant for simplicity. The zonal phase speed of the waves $c \equiv \omega/k$ always satisfies $U - c > 0$, i.e., the wave crests and troughs move westward with respect to the background flow.

The Coriolis parameter is much smaller in the tropics than in the extratropics, and consequently the *equatorial $\beta$-plane approximation*, in which $f \approx \beta y$ (where $\beta = 2\Omega$), $\sin\phi \approx y$, and $\cos\phi \approx 1$, is used to explore the dynamics. Eastward- and westward-propagating disturbances that are trapped about the equator (i.e., they decay away from the equatorial region) are possible solutions. Nondispersive waves that propagate eastward with phase speed $c_{\text{Kelvin}} = \sqrt{gH}$ (where $H$ is an equivalent depth) are known as equatorial *Kelvin waves.* Typical phase speeds in the atmosphere are $c_{\text{Kelvin}} \approx$ 20–80 m s$^{-1}$, and in the ocean $c_{\text{Kelvin}} \approx$ 0.5–3 m s$^{-1}$. Another class of possible solutions is equatorial Rossby waves, whose dispersion relation is

$$\omega_{\text{eqRossby}} = -\frac{\beta k}{(k^2 + (2n+1)\beta/c_{\text{Kelvin}})},$$

where $n$ is a positive integer. For very long waves (as the zonal wave number approaches zero), the nondispersive phase speed is approximately $c_{\text{eqRossby}} = -c_{\text{Kelvin}}/(2n+1)$. Hence, these equatorial Rossby waves move in the opposite direction to the Kelvin waves (i.e., they propagate westward) and at reduced speed. For $n = 1$ the speed is about a third that of a Kelvin wave, meaning it would take approximately six months to cross the Pacific Ocean basin.

## 7 Ocean–Atmosphere Coupling

The dynamics of the ocean and atmosphere in the tropics are highly coupled. On interannual timescales, the upper ocean responds to the past history of the wind stress, and the atmospheric circulation is largely determined by the distribution of sea-surface temperatures (SSTs).

The trade winds that converge on the equator supply water vapor to maintain convection. The convective heating produces large-scale midtropospheric temperature perturbations and associated surface and upper-level pressure perturbations, which maintain the low-level flow. The zonal mean of the vertical mass flux associated with this intertropical convergence zone constitutes the upward mass flux of the mean Hadley circulation. There are strong longitudinal variations associated with variations in the tropical SSTs mainly due to the effects of the wind-driven ocean currents. Overturning cells in the atmosphere along the equator are associated with diabatic heating over equatorial Africa, Central and South America, and Indonesia. The dominant cell is known as the *Walker circulation* and is associated with low surface pressure in the western Pacific and high surface pressure in the eastern Pacific, resulting in a pressure gradient that drives mean surface easterlies (the Coriolis force is negligible in this region). The easterlies provide a moisture source for the convection in the western Pacific in addition to that provided by the high evaporation rates caused by the warm SSTs there. The atmospheric circulation is closed by descent over the cooler water to the east.

Given a westward wind stress $\tau_x$ across the Pacific, assumed to be independent of $x$, equation (7) can be used to show that $\beta y v = -\rho_0^{-1}\partial\tau_x/\partial z$, assuming the response is also independent of longitude and using the equatorial $\beta$-plane approximation. A westward wind stress across the Pacific therefore gives rise to poleward flows either side of the equator in the oceanic Ekman layer, which by continuity drive upwelling near the equator. In addition, since the Pacific is bounded to the east and west, the westward wind stress results in the ocean thermocline being deeper in the west than the east. The cold deep water therefore upwells close to the surface in the east, cooling the SSTs there, whereas in the west the cold water does not reach the surface and the SSTs remain warm.

The upwelled region is associated with a geostrophic current in the direction of the winds, since in the limit $y \to 0$, (7) gives $\beta u = -\rho_0^{-1}\partial^2 p/\partial y$ by l'Hôpital's rule. The deepening of the thermocline causes the sea surface to be higher in the west, assuming that flow below the thermocline is weak. There is therefore an eastward pressure gradient along the equator in the ocean surface layers to a depth of a few hundred meters. Away from the equator, below the surface, this is balanced by an equatorward geostrophic flow. At the equator, where $f = 0$, there is a subsurface current directly down the pressure gradient (i.e., to the east): the *Equatorial Counter Current*.

The east–west pressure gradient across the Pacific undergoes irregular interannual variations with a period in the approximate range 2–7 years. This oscillation

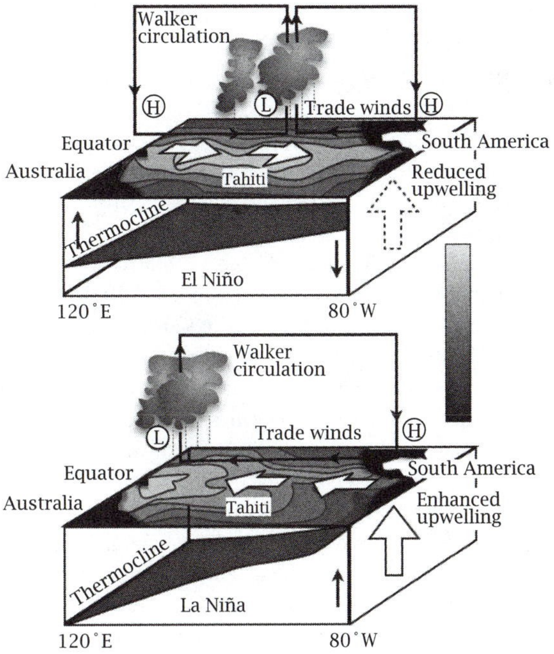


**Figure 6** Schematic of typical atmosphere/ocean conditions during (a) El Niño (negative SOI) and (b) La Niña (positive SOI) conditions (see text for details).

in pressure, and its associated patterns of wind, temperature, and precipitation, is called the *Southern Oscillation*, an index of which (the Southern Oscillation Index (SOI)) can be obtained by considering the pressure difference between Tahiti in the central Pacific and Darwin, Australia, in the western Pacific (see figure 6). The negative phase of the SOI represents below-normal sea level pressure (labeled "L" in the figure) at Tahiti and above-normal sea level pressure (labeled "H") at Darwin, and vice versa for the positive phase. SSTs in the eastern Pacific are negatively correlated with the SOI, i.e., the phase with warm SSTs (known as *El Niño*) coincides with a negative SOI, and vice versa for the phase with cold SSTs (known as *La Niña*). The entire coupled atmosphere–ocean response is known as the El Niño–Southern Oscillation (ENSO).

During an El Niño event, the region of warm SSTs is shifted eastward from the Indonesian region and with it the region of greatest convection and the associated atmospheric circulation pattern (see figure 6). The resulting adjustment of the Walker circulation leads to a weakening of the easterly trade winds, reinforcing the eastward shift of the warm SSTs. The sea-surface slope diminishes, raising sea levels in the east Pacific while lowering those in the west. Ekman-driven upwelling in the ocean reduces, allowing SSTs to increase. The ocean adjusts over the entire basin to a local anomaly in the forcing in the western Pacific through the excitation of internal waves in the upper ocean. We can explain the subsequent evolution using the properties of ocean waves derived in the previous section. An equatorial Kelvin wave propagates rapidly to the east, reinforcing the initial warm SST anomaly in a positive feedback, and equatorial Rossby waves propagate slowly (with group velocity about a third of the Kelvin wave) to the west. As the Kelvin wave propagates east it deepens the thermocline, relaxing the basin-wide slope. When it hits the coast on the eastern side (after about two months), its energy feeds westward Rossby waves and poleward coastal Kelvin waves. On the western side, when the Rossby waves hit the coast, some energy feeds an eastward-propagating Kelvin wave, which raises the thermocline back toward its original location, reducing the initial SST anomaly and providing a negative feedback. The propagation times of the waves mean that the negative feedback is lagged, resulting in a simple model of a delayed oscillator, which provides an explanation for the observed ENSO periodicity.

In the coupled system the ocean forces the atmospheric circulation (through the response to changed boundary conditions associated with the El Niño SST fluctuations) and the atmosphere forces the oceanic behavior (through the response to changed wind stress distribution associated with the Southern Oscillation).

## 8 Outlook

Mathematics has played, and continues to play, a central role in developing an understanding of the Earth system. It has provided insight into many of the fundamental processes that make up the weather and climate, and crucially it has also provided the necessary framework underpinning the numerical models used in future prediction. Some of the key areas of research in which mathematicians have a vital role include developing the tools that allow observational data to be assimilated into forecast models, applying concepts from dynamical systems theory to the evaluation of past climate and modeling of future variability, exploiting a range of mathematical techniques to develop novel model parameterizations for subgrid processes, and using advanced statistical techniques for data analysis.

### Further Reading

Andrews, D. G. 2010. *An Introduction to Atmospheric Physics.* Cambridge: Cambridge University Press.

Gill, A. E. 1982. *Atmosphere–Ocean Dynamics.* New York: Academic Press.

Holton, J. R. 2004. *An Introduction to Dynamic Meteorology.* Amsterdam: Elsevier.

Marshall, J., and R. A. Plumb. 2008. *Atmosphere, Ocean, and Climate Dynamics: An Introductory Text.* Amsterdam: Elsevier.

Richardson, L. F. 2007. *Weather Prediction by Numerical Process,* 2nd edn. Cambridge: Cambridge University Press. (First edition published in 1922.)

Vallis, G. K. 2006. *Atmosphere and Ocean Fluid Dynamics.* Cambridge: Cambridge University Press.

# IV.31 Effective Medium Theories

*Ross C. McPhedran*

## 1 Introduction

Effective medium theories arise in a variety of forms and in a wide variety of contexts in which one wants to model properties of structured media in a simplified way. The basic idea is to take into account the structure of the heterogeneous medium by calculating an equivalent homogeneous "effective medium" and to use the equivalent medium in further calculations. For example, one might be considering an optical material composed of two different components, the optical properties of each of which is known, and one might want to put the two components together in a structured material that behaves like a homogeneous material with properties differing from those of each component. Various of the effective medium theories that have been devised are of use in this particular example, which is one of the first technological examples of its use. (In fact, for several thousand years metals have been put into glass melts in ways designed to give particular optical coloration effects.)

A wide class of effective medium treatments have proved useful in the study of the transport properties of composite materials. The basic governing relation is Laplace's equation, and the solution is expressed in terms of a distribution of a field quantity, with its associated flux. Using the language of one such property, electrical conductivity, the field quantity is the electric field and its flux is the current. The materials constituting the composite are specified by their conductivity, and the effective medium equations give the effective conductivity of the composite. The basic equations are

$$\boldsymbol{j}(\boldsymbol{x}) = \boldsymbol{\sigma}(\boldsymbol{x})\boldsymbol{e}(\boldsymbol{x}), \quad \nabla \cdot \boldsymbol{j} = 0, \quad \nabla \times \boldsymbol{e} = \boldsymbol{0}. \qquad (1)$$

These vector fields may be real or complex, depending on the particular transport problem under consideration. If the electric field $\boldsymbol{e}$ is written as the gradient of a potential function, the last equation in (1) reduces to Laplace's equation.

Table 1 gives seven instances of this problem, with the equivalent terms for conductivity being given in the last column.

The seven examples of transport coefficients are all mathematically equivalent. However, different practical considerations result in particular effective medium approaches being better suited to some contexts than others. In particular, in optical applications of composite media involving metals mixed in with dielectrics, the electric permittivity of the metal is complex, as is the effective permittivity of the metal–dielectric composite. This may make it more difficult to establish an accurate effective permittivity formula in this case.

## 2 Conditions for Effective Medium Theories

Effective medium theories require a number of conditions concerning the characteristics of the structured medium to be modeled to be satisfied. The structured medium has to be divided into regions of the component materials that contain sufficient numbers of atoms or molecules for bulk properties to be used for those regions. It must also include a sufficiently large number of regions to permit construction of effective medium formulas that do not need to take into account finite-size effects. Typically, the construction of equivalent homogeneous regions will occur on a scale much smaller than the size of the sample of the structured medium but much larger than that of the regions of its component parts or of its separated parts. (An important class of composite media has a background continuous phase, sometimes called the matrix phase, into which separated components of other materials are inserted.)

The construction of appropriate effective medium formulas for composite structures is called *homogenization*. The process of homogenization has to take into account the geometry of the composite: whether it is two dimensional or three dimensional; whether it is isotropic or anisotropic; whether the different materials of which it is composed have similar spatial distributions, or whether there is a matrix phase into which the regions of other materials are inserted. The requirements on the effective medium formula in the problem under investigation also need to be understood. If a high-accuracy model is needed, then precise details of the geometry of the system will generally need to be incorporated into the theory. On the other hand, it will sometimes be enough to have upper and lower bounds on the transport coefficient, in which case much more general and simpler procedures will be adequate.

## 3 Effective Medium Formulas of Maxwell Garnett Type

The early history of formulas based on the effective electric permittivity of systems of particles characterized by the dipole moment they develop when placed in an external field is complicated. It spans the period from 1837 (when Faraday proposed a model for dielectric materials based on metallic spheres placed in an insulator) to 1904 (when J. C. Maxwell Garnett used formulas of this type in design studies on colored glass). Essentially similar formulas are given various names, depending on the first investigator to use them in a particular context, and the approximations inherent in them are sometimes glossed over. One example of the complexity of this history is that the formula put forward by Maxwell Garnett was essentially equivalent to one developed previously by his godfather, J. C. Maxwell, so the formula is often, justifiably, written Maxwell-Garnett.

What is generally called the Clausius–Mossotti formula gives the effective electric permittivity ($\varepsilon_{\mathrm{eff}}$) of a set of polarizable inclusions placed in a three-dimensional background material of electric permittivity $\varepsilon_{\mathrm{b}}$:

$$\varepsilon_{\mathrm{eff}} = \varepsilon_{\mathrm{b}}\left(1 + \frac{N\alpha}{1 - N\alpha/3}\right),$$

where $N$ is the number of inclusions per unit volume, and $\alpha$ is their polarizability. To arrive at the Maxwell Garnett formula, we replace $\alpha$ by the polarizability of a sphere of electric permittivity $\varepsilon_{\mathrm{p}}$ and radius $a$:

$$\alpha = 4\pi a^3\left(\frac{\varepsilon_{\mathrm{p}} - \varepsilon_{\mathrm{b}}}{\varepsilon_{\mathrm{p}} + 2\varepsilon_{\mathrm{b}}}\right).$$

Introducing the volume fraction $f = 4\pi N a^3/3$, the Maxwell Garnett result is

$$\varepsilon_{\mathrm{eff}} = \varepsilon_{\mathrm{b}}\left[1 + \frac{3f(\varepsilon_{\mathrm{p}} - \varepsilon_{\mathrm{b}})}{3\varepsilon_{\mathrm{b}} + (1-f)(\varepsilon_{\mathrm{p}} - \varepsilon_{\mathrm{b}})}\right]. \qquad (2)$$

The Maxwell Garnett formula is asymmetric in the variables $\varepsilon_{\mathrm{p}}$ and $\varepsilon_{\mathrm{b}}$, since the former corresponds to isolated particles and the latter to the continuous background phase. It is also a purely dipole formula, in that

**Table 1** Equivalent physical transport problems governed by Laplace's equation.

| Problem | $\boldsymbol{j}$ | $\boldsymbol{e}$ | $\boldsymbol{\sigma}$ |
|---|---|---|---|
| Electrical conduction | Electrical current $\boldsymbol{j}$ | Electric field $\boldsymbol{e}$ | Electric conductivity $\boldsymbol{\sigma}$ |
| Dielectrics | Displacement field $\boldsymbol{d}$ | Electric field $\boldsymbol{e}$ | Electric permittivity $\boldsymbol{\varepsilon}$ |
| Magnetism | Magnetic induction $\boldsymbol{b}$ | Magnetic field $\boldsymbol{h}$ | Magnetic permeability $\boldsymbol{\mu}$ |
| Thermal conduction | Heat current $\boldsymbol{q}$ | Temperature gradient $-\nabla T$ | Thermal conductivity $\boldsymbol{\kappa}$ |
| Diffusion | Particle current | Concentration gradient $-\nabla c$ | Diffusivity $\boldsymbol{D}$ |
| Flow in porous media | Weighted velocity $\eta_\mu \boldsymbol{v}$ | Pressure gradient $\nabla P$ | Fluid permeability $\boldsymbol{k}$ |
| Antiplane elasticity | Stress vector $(\tau_{13}, \tau_{23})$ | Vertical displacement gradient $\nabla u_3$ | Shear matrix $\boldsymbol{\mu}$ |

field expansions near the particles are limited to single dipole terms. As the volume fraction increases, the isolated particles come closer to each other, interact more strongly, and the dipole approximation increasingly becomes inaccurate. A consequence of it being a dipole formula is that the Maxwell Garnett equation has a single pole, located when the following equation for the permittivity ratio between particles and background material is satisfied:

$$\left(\frac{\varepsilon_p}{\varepsilon_b}\right)_\infty = -\left(\frac{2+f}{1-f}\right). \tag{3}$$

This equation specifies what is termed the *plasmon resonance* and gives the condition for the optical effects of the particles to be particularly strong; it is never satisfied exactly by physical systems, but it can be approximately satisfied. The other interesting case is when $\varepsilon_{\text{eff}} = 0$; this occurs when

$$\left(\frac{\varepsilon_p}{\varepsilon_b}\right)_0 = -\left(\frac{2-2f}{1+2f}\right).$$

In this case, the presence of the particles cannot be detected by measurements of $\varepsilon_{\text{eff}}$. The plasmon resonance occurs when the permittivity ratio is real, negative, and below $-2$, while the zero of $\varepsilon_{\text{eff}}$ for the Maxwell Garnett equation occurs when the permittivity ratio lies between $-2$ and 0.

As an example of the optical application of the Maxwell Garnett formula, in figure 1 we show the effective electric permittivity as a function of the wavelength of spherical silver particles in a silica matrix, with the silver occupying a small volume fraction (10%) of the composite. The strong permittivity resonance of the composite would occur for an ideal metal when $\varepsilon_p/\varepsilon_b = -2.33$ for $f = 0.10$; for the composite shown, the resonant condition (3) is most closely satisfied when the wavelength is around 0.40 $\mu$m, so the composite would have a strong reflectance in the violet spectral region. The closest approximation to $\varepsilon_{\text{eff}} = 0$ occurs just to the short-wavelength side of this peak.

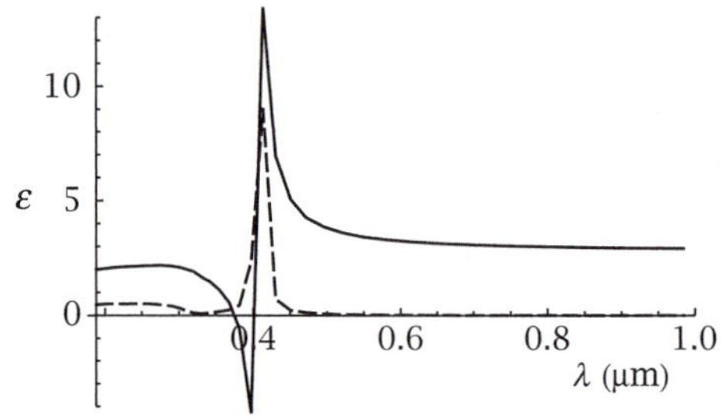


**Figure 1** The real (solid) and imaginary (dashed) parts of $\varepsilon_{\text{eff}}$, as a function of wavelength, for silver spheres occupying 10% by volume of a composite, in a background material of silica. The electric permittivity of silver is a strong function of the wavelength, and experimental data is used both for its permittivity and for that of silica.

Formulas can be developed that take into account the effect of higher-order multipoles, not just dipoles, assuming, for example, a specific regular arrangement of spherical particles. They generally use a method due to Lord Rayleigh published in 1892. Such formulas take into account the arrangement of the particles using lattice sums, one for each relevant spherical harmonic term. They also have one resonance for each multipole term, with the resonances tending to cluster around the permittivity ratio of $-2$.

The two-dimensional case of arrays of cylindrical particles placed in a background material is also of importance, particularly in the study of photonic crystals and metamaterials. The Maxwell Garnett formula based on circular cylinders that is equivalent to equation (2) is

$$\varepsilon_{\text{eff}}(\varepsilon_b, \varepsilon_p; f) = \varepsilon_b \left[\frac{\varepsilon_b + \varepsilon_p - f(\varepsilon_b - \varepsilon_p)}{\varepsilon_b + \varepsilon_p + f(\varepsilon_b - \varepsilon_p)}\right].$$

This exactly satisfies an important duality relationship due to J. B. Keller, true for cylinders of arbitrary cross section:

$$\varepsilon_{\text{eff}}(\varepsilon_b, \varepsilon_p; f)\varepsilon_{\text{eff}}(\varepsilon_p, \varepsilon_b; f) = \varepsilon_b \varepsilon_f.$$

The duality relationship does not have an equivalent in three dimensions. It has important consequences for the underlying structure of two-dimensional transport problems, since it provides a link between zeros and poles of the effective transport coefficient.

It should be noted that the Maxwell Garnett formula (2) is exact for a specific geometry, given by the Hashin–Shtrikman construction. The idea is to calculate the value ($\varepsilon_0$) of a background permittivity such that, if the electric field in this medium is uniform, a sphere of core permittivity $\varepsilon_p$ and shell permittivity $\varepsilon_b$ can be inserted into the medium of permittivity $\varepsilon_0$ without disturbing the uniform field there. The core and shell radii of the coated sphere must be such that the volume of the core divided by that of the shell is $f/(1-f)$. The value so calculated for $\varepsilon_0$ turns out to be exactly that given by (2). One can then continue to insert copies of such coated spheres repeatedly into the background, scaling them down when necessary so that all the volume corresponding to the material of permittivity $\varepsilon_0$ eventually disappears. All stages of the construction preserve the Maxwell Garnett (or Hashin–Shtrikman) effective electric permittivity.

## 4 Effective Medium Formulas of Bruggeman Type

As we have remarked, the Maxwell Garnett formula, and related expressions, take one material in a composite to extend infinitely while the second material takes the form of separated inclusions. In 1935 Bruggeman constructed an alternative form of theory that placed both materials on the same footing, i.e., both were assumed to have the same topology. The resultant formula is called the Bruggeman symmetric effective medium formula; it may also be called the coherent potential approximation, after a random alloy theory from solid-state physics.

We consider an aggregate structure composed of grains filling all space. The grains are of type 1 (volume fraction $f_1$, electric permittivity $\varepsilon_1$) or type 2 (volume fraction $f_2 = 1 - f_1$, electric permittivity $\varepsilon_2$). The effective permittivity of the composite will be denoted $\varepsilon_B$. To obtain a formula for it, we pick a representative sample of grains occupying a small volume fraction $\delta$ of the granular composite. The grains in the sample are assumed to be well separated from each other, and they must be chosen so that the sample has the correct volume fractions of each.

The derivation uses the self-consistency assumption that the effective permittivity of the composite remains equal to $\varepsilon_B$ to first order in $\delta$ when we replace the medium surrounding the representative grains by a homogeneous effective medium with electric permittivity $\varepsilon_B$. After this replacement has been made, we treat the representative grains as a dilute suspension of spherical grains embedded in the medium specified by $\varepsilon_B$. Correct to first order, the effective permittivity of the suspension is

$$\varepsilon_B\left[1 + \delta f_1 \frac{3(\varepsilon_1 - \varepsilon_B)}{\varepsilon_1 + 2\varepsilon_B} + \delta f_2 \frac{3(\varepsilon_2 - \varepsilon_B)}{\varepsilon_2 + 2\varepsilon_B}\right],$$

giving the equation

$$f_1 \frac{3(\varepsilon_1 - \varepsilon_B)}{\varepsilon_1 + 2\varepsilon_B} + f_2 \frac{3(\varepsilon_2 - \varepsilon_B)}{\varepsilon_2 + 2\varepsilon_B} = 0.$$

The result is a formula that is symmetric in the properties of the two types of grains:

$$\varepsilon_B = \tfrac{1}{4}[\gamma \pm (\gamma^2 + 8\varepsilon_1\varepsilon_2)^{1/2}]. \tag{4}$$

The choice of the plus or minus sign in (4) should be made so that the imaginary part of $\varepsilon_B$ has the same sign as the common sign of the imaginary parts of $\varepsilon_1$ and $\varepsilon_2$.

The Bruggeman expression (4) has several features that distinguish it from the Maxwell Garnett type of theory. As remarked, it has no preferred topology for the two media filling the composite. Rather than a resonant peak, it has a branch cut, arising when the switch is made between the two alternatives in (4). It also has a percolation threshold, which is manifest if we take $\varepsilon_2 = 0$ and $\varepsilon_1$ real. Imposing the criterion that $\varepsilon_B$ be nonnegative if $\varepsilon_1 > 0$, the result for this particular case is

$$\varepsilon_B = \begin{cases} (3f_1 - 1)\varepsilon_1/2 & \text{when } f_1 \geqslant \frac{1}{3}, \\ 0 & \text{when } f_1 < \frac{1}{3}. \end{cases}$$

This shows that in this case there is a percolation threshold at $f_1 = \frac{1}{3}$.

In figures 2 and 3 we show the results of the symmetric Bruggeman theory for the silver–silica composite of figure 1, but for two volume fractions: one below the percolation threshold and the other above it. The percolation threshold is not sharply defined in the case of complex electric permittivities, but there are clear changes that occur as the volume fraction of silver increases. For small metal volume fractions, the grains do not get close enough together to create the sort of current paths needed to form a strong imaginary part of the effective permittivity. Above the percolation threshold, the imaginary part gradually strengthens, dominating the real part above a volume fraction of 60%.

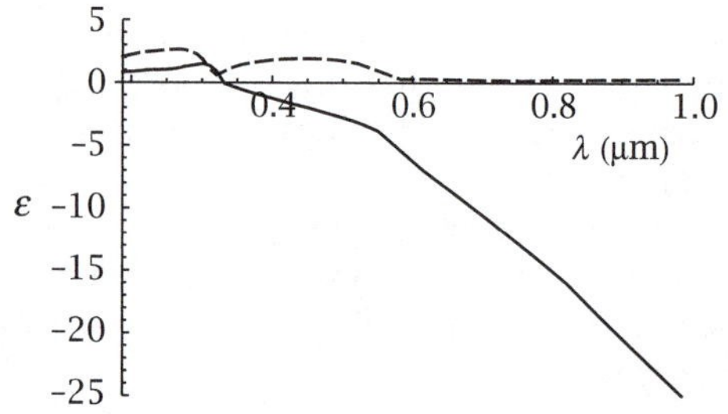


**Figure 2** The real (solid) and imaginary (dashed) parts of $\varepsilon_{\text{eff}}$, according to the Bruggeman formula, as a function of wavelength for silver occupying 30% by volume of a composite, with 70% of the composite volume being silica. Experimental data is used both for the permittivity of silver and for that of silica.

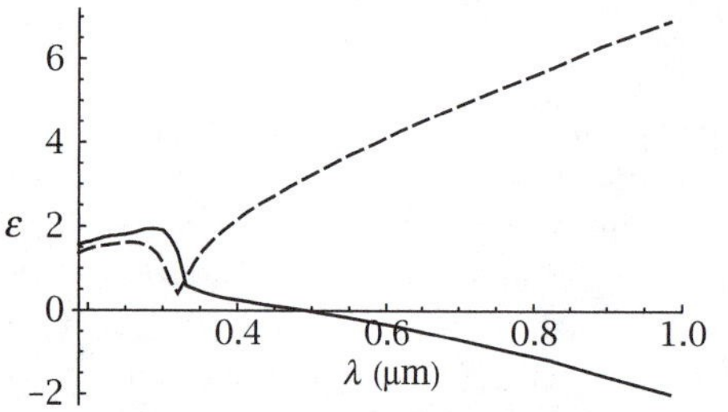


**Figure 3** As for figure 2, but now with a silver volume fraction of 60%, above the percolation threshold.

## 5 The Bergman–Milton Bounds

Bounds on the effective transport coefficients of composite materials are very useful in checking theoretical results and deducing information from experimental measurements. They can be particularly useful if they are *sharp*, i.e., there is at least one physical geometry for which the bounds are attained. In the following brief discussion of bounds we will use the notation $\sigma$ for the conductivity problem, with $\sigma_{\text{eff}}$ being the (real-valued) effective conductivity of a composite made of materials with two real conductivities $\sigma_1$ and $\sigma_2$ and volume fractions $f_1$ and $f_2$. We will use the notation $\varepsilon_{\text{eff}}$ for bounds appropriate to the case where $\varepsilon_1$ and $\varepsilon_2$ are allowed to be complex.

Taking $\sigma_1 > \sigma_2$, the Hashin–Shtrikman lower bound for three-dimensional composites is

$$\sigma_{\text{eff}} \geqslant \sigma_2\left[1 + \frac{3f_1(\sigma_1 - \sigma_2)}{3\sigma_2 + f_2(\sigma_1 - \sigma_2)}\right]. \tag{5}$$

This bound is sharp, being attained by an assemblage of coated spheres with phase 1 as the core and phase 2 as the coating. The corresponding upper bound is

$$\sigma_{\text{eff}} \leqslant \sigma_1\left[1 - \frac{3f_2(\sigma_1 - \sigma_2)}{3\sigma_1 - f_1(\sigma_1 - \sigma_2)}\right]. \tag{6}$$

This bound is attained by an assemblage of coated spheres with phase 2 as the core and phase 1 as the coating.

For the complex case, independent work by Bergman and Milton enabled the derivation of tight bounds on the effective permittivity $\varepsilon_{\text{eff}}$ of two-phase composites with known volume fractions. Instead of inequalities such as those in (5) and (6), the complex value of $\varepsilon_{\text{eff}}$ is constrained to lie in a specific region in the complex plane. The region is defined by straight lines and arcs of circles. If volume fractions are unknown, the region is defined to lie inside the area bounded by a straight line (the arithmetic mean of $\varepsilon_1$ and $\varepsilon_2$ as $f_1$ and $f_2 = 1 - f_1$ vary) and a circle (defined by the harmonic average, i.e., the reciprocal of the arithmetic mean of the quantities $1/\varepsilon_1$ and $1/\varepsilon_2$ as $f_1$ and $f_2 = 1 - f_1$ vary). Knowing the volume fractions $f_1$ and $f_2$, one can constrain $\varepsilon_{\text{eff}}$ to lie in a region between two arcs of circles defined by common endpoints and one extra point for each. The common endpoints correspond to the points on the outer region boundaries for the specified values of $f_1$ and $f_2$. The extra points correspond to the Hashin–Shtrikman coated-sphere assemblages with phase 1 and phase 2 as the coating. If it is known that the composite is isotropic, as well as having the specified volume fractions, then a still-smaller region in the complex plane can be defined.

This construction of bounds can be viewed as a recursive process. At each stage one has a region in the complex plane. As an extra piece of information about the composite is specified, two points on the boundary of the old region consistent with the new information are selected, and a smaller region is constructed with these new points on its boundary. It has been shown that as more and more information is specified, the bounds converge to a specific point: the exact complex permittivity for the (by now) uniquely specified composite material. The area in the complex plane between the bounds shrinks rapidly as more information is added if $\varepsilon_1$ and $\varepsilon_2$ are well away from the negative real axis. However, if their ratio is close to the negative real axis, the area shrinks much more slowly as information is added.

The Bergman–Milton bounds can also be used in the *inverse problem* for composites: given measured data on effective transport properties of a composite medium, what can be said about the medium's structure? In particular, one would like to be able to infer geometrical information, principally the volume fractions of the phases making up the composite. An important example of such an inverse problem is in geophysics and resource extraction, where it is of great value to be able to infer, say, the volume fraction of oil in a fluid-permeated sandstone structure using borehole electrical tomography. Another example in environmental science concerns the estimation of the ratio of brine to pure ice in SEA ICE [V.17]; this is of value when trying to deduce the melting–freezing history of sea ice over several seasons.

**Acknowledgments.** This article was written while the author held a Senior Marie Curie Fellowship at the University of Liverpool, with financial support from the European Community's Seventh Framework Programme under contract PIAPP-GA-284544-PARM-2.

### Further Reading

Bergman, D. J. 1978. The dielectric constant of a composite material: a problem in classical physics. *Physics Reports* 43(9):377–407.

Bonod, N., and S. Enoch, eds. 2012. *Plasmonics: From Basic Physics to Advanced Topics*. Berlin: Springer.

Hashin, Z. 1983. Analysis of composite materials: a survey. *Journal of Applied Mechanics* 50:481–505.

Landauer, R. 1978. Electrical conductivity in inhomogeneous media. In *Electrical Transport and Optical Properties of Inhomogeneous Media*, edited by J. C. Garland and D. B. Tanner, pp. 2–43. Woodbury, NY: American Institute of Physics.

Milton, G. W. 2002. *The Theory of Composites*. Cambridge: Cambridge University Press.

Torquato, S. 2001. *Random Heterogeneous Materials: Microstructure and Macroscopic Properties*. Berlin: Springer.

## IV.32 Mechanics of Solids

*L. B. Freund*

### 1 Introduction

The builders of ancient machines and structures must have been concerned with the strength and durability of their creations. Presumably, the standards applied had evolved over time through trial and error.

Early evidence of systematic scientific study of the strength of materials appears in the work of Leonardo da Vinci (1452–1519), who experimented with circular rods of various diameters in order to learn about the features that influence tensile breaking strength. A more systematic study of rods in tension, as well as beams in bending, was undertaken by Galileo (1564–1642). Among his conclusions were that the tensile breaking force of a uniform rod does not depend on its length but does depend proportionally on its cross-sectional area and that rupture of a cantilever beam loaded by a weight at its free end usually initiates on the top side of the beam near the support. The idea of the force applied to a solid object being related in a characteristic way to the deformation induced by that force was first stated for the linear elastic response of a watch spring by Hooke (1635–1703) in the form of a Latin anagram that, when unscrambled, reads "As the extension, so the force."

With the publication of the *Principia* by Newton (1642–1727), an approach emerged whereby the study of physical phenomena began with a statement of physical principles, with a view toward deducing the behavior of the physical world on the basis of those principles by mathematical means. Leonhard Euler (1707–83) and Daniel Bernoulli (1700–1782) developed the theory of elastic beams in bending that bears their names; Coulomb (1736–1806) put that theory into the form of a one-dimensional structural theory that is in common use today. Eventually, it was the great post-Renaissance engineer and mathematician Cauchy (1789–1857) who unified what was known at the time about stress, strain, and linear elastic material behavior to devise a full three-dimensional theory of the mechanics of a solid continuum. In the 250 years since, a long list of types of material behavior has evolved—theories for describing geometrically nonlinear deformation have been developed, material inertial effects have been incorporated into the quantitative description of the response of a solid to various types of applied loads, and special theories have been developed for solid configurations that are very thin in one or two dimensions, as for beams, plates, or shells. These contributions have been accompanied by the development of an array of mathematical and computational methodologies for solving boundary-value problems based on partial differential equations believed to describe these phenomena.

During the twentieth century, the mechanics of solids grew into an applied science that now underlies much of the practice of mechanical engineering and civil engineering. It is also widely applied within materials science, planetary geophysics, and biophysics, for example. The area includes a large and active experimental component, as well as a major computational component.

### 1.1 A Solid Material

The identification of matter filling some portion of space as a *solid* is based on the recognition of a constitutive property of that material; namely, any material that can resist imposed forces of a measurable magnitude that tend to shear the material, without evidence of ongoing deformation, is a solid. Otherwise, the material is a fluid. The separation of materials into these categories is not perfect, in that some materials behave as either fluid or solid depending on the rate of deformation and/or on the temperature.

A solid body is some solid matter filling a region of space. The constituent material has certain characteristics relevant to mechanical phenomena: mass per unit volume, resistance to deformation, and ultimate strength, for example. Furthermore, any piece of the solid that is removed from the whole exhibits these same material characteristics, and these properties remain unaltered as the solid is repeatedly divided into indefinitely smaller pieces. Although this *continuum* point of view overlooks the discrete small-scale structure of materials, it facilitates modeling and analysis at a scale insensitive to that structure. For example, suppose that a small volume of a material, say $v(p)$, bounded by a closed surface surrounding the particular point $p$ in the solid has total mass $m(v)$. Then the *mass density* at point $p$ is

$$\rho(p) = \lim_{v \to 0} \frac{m(v)}{v(p)}. \tag{1}$$

In general, this property of *indefinite divisibility* makes it possible to define both material properties and mechanical fields as continuous functions of position throughout a solid. The identification of the differential equations governing these fields, together with the solution of these equations subject to boundary conditions representing external influences, is the essence of solid mechanics.

### 1.2 A Conceptual Map

Before launching into the details of the subject, a conceptual map of solid mechanics is provided as an aid to understanding the relationships among various ideas that are central to the subject (see figure 1). Examination of the subject begins with the consideration of three fundamental ideas. The first of these is labeled *displacement*, a topic that encompasses rules for locating material particles in the Euclidean space in which we live and for representing changes in the positions of these particles. The next fundamental idea is that of *material strain*, which is focused on the way in which the deformation of a material is described quantitatively in terms of the change in the length of material line elements that join adjacent material particles or the change in angle between two material line elements emanating from the same material particle.

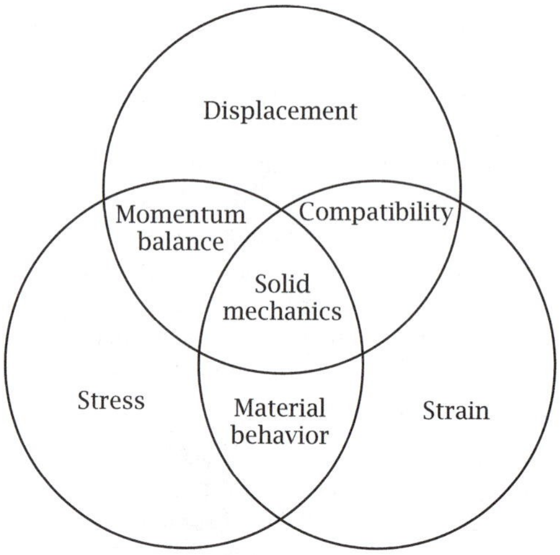


**Figure 1** A conceptual map of solid mechanics.

The third principal idea, *stress*, brings mechanical action into the picture. Generally, the concept of stress provides the basis for describing the mechanical force exerted by the material on one side of an interior surface on the material on the other side of that surface.

At the displacement–strain intersection, determining strain for any prescribed displacement distribution and the more challenging issue of determining displacement for a specified strain distribution are addressed. The stress–strain intersection is where characterization of the mechanical behavior of materials enters; many types of stress–strain relationships can be postulated, and experiment is always the final arbiter of ideas in this domain. Finally, the physical principle of conservation of momentum comes into play at the stress–displacement intersection. The concepts incorporated in these three intersection zones are the essential ingredients in the solution of any mathematical problem in the mechanics of solids, as suggested by their intersection at the center of the conceptual map. The next task is to describe the means by which these concepts are given useful mathematical forms.

## 2 Fundamental Concepts

As it deforms, a solid body occupies a continuous sequence of configurations. One of these configurations is adopted as a *reference configuration* for purposes of description; this is usually the configuration at the start of a deformation process of interest but the choice is arbitrary.

The positions of material points, changes in the positions of material points, and points in space are represented by vectors. For this purpose, a set of orthonormal basis vectors $\boldsymbol{e}_k$, $k = 1, 2, 3$, is introduced.

Any mathematical quantity represented by a symbol in boldface denotes a vector or tensor, with the precise meaning implied by context. As an aid in calculation, index notation is commonly adopted for components of vectors and tensors. The implied range of any index is 1–3, unless a note to the contrary is included; repeated indices in an expression indicate an inner product, with summation over their range implied. The rules governing the use of index notation are clearly summarized in Bower's *Applied Mechanics of Solids.*

The terms *material point*, *material line*, and *material surface* are used often in the sections that follow. It should be understood from the outset that these terms are to be interpreted literally. For example, a material line, once defined, always coincides with the same set of material points in the course of deformation.

### 2.1 Displacement

Every material point can be identified with the point in space with which it coincides in the reference configuration, and the position of the generic point is denoted by the vector

$$\boldsymbol{x} = x_1\boldsymbol{e}_1 + x_2\boldsymbol{e}_2 + x_3\boldsymbol{e}_3 \equiv x_k\boldsymbol{e}_k. \tag{2}$$

A set of basis vectors is shown in figure 2, which depicts a simple deformation. The reference configuration of the solid is a cube of edge length $\ell_0$, and the point at one of the vertices of the cube is identified by position $\boldsymbol{x}$. In (2), the repeated index is understood to represent summation over the implied range of that index; this summation convention is followed in all subsequent mathematical expressions.

The shaded object in figure 2 represents a possible deformed configuration of the solid body that had occupied the cubic region. The figure suggests that the

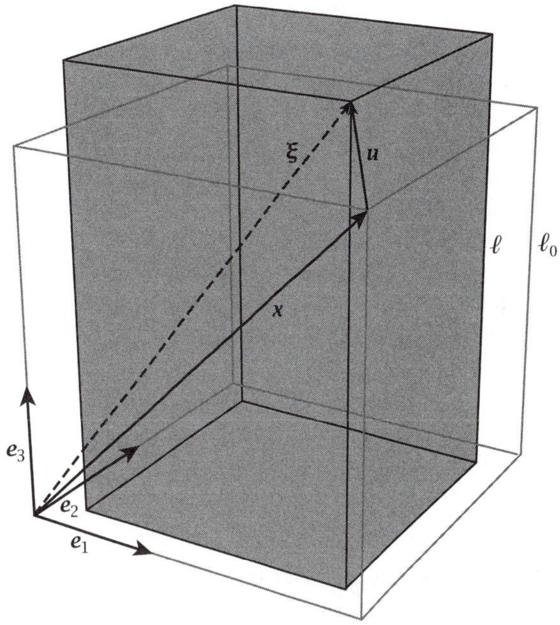


**Figure 2** A deformed configuration of a solid body that was a cube in its reference configuration.

solid has been stretched in the 3-direction and has been contracted in the 1- and 2-directions. Of particular note is the position of the corner point that was identified as $\boldsymbol{x}$ in the reference configuration; this particular point has displaced to the position $\boldsymbol{\xi} = \xi_i\boldsymbol{e}_i$ in the deformed configuration. The displacement of this point is the vector *difference* between these positions. If the displacement vector is denoted by $\boldsymbol{u}$, then the displacement of the corner point is

$$\boldsymbol{u} = \boldsymbol{\xi} - \boldsymbol{x} \quad \Leftrightarrow \quad u_i = \xi_i - x_i. \tag{3}$$

The displacement of any point in the reference configuration of the solid, uniquely identified by its coordinates $x_1$, $x_2$, $x_3$ with respect to the underlying reference frame, can be determined similarly. When defined in this way, the quantity $\boldsymbol{u}(x_1, x_2, x_3) \equiv \boldsymbol{u}(\boldsymbol{x})$ is seen to define a continuous *vector field* over the reference configuration of the solid that describes the complete displacement field associated with a deformation.

### 2.2 Strain

Commonly, the displacement field $\boldsymbol{u}(x_1, x_2, x_3)$ varies from point to point throughout the reference configuration of the solid body. Also, the length of a material line element connecting any two adjacent material points in the deformed configuration differs from the length of that line element in the reference configuration. The concept of strain is introduced to systematically quantify this difference.

Suppose that we focus on a generic material point within the solid that is identified in the reference configuration by its position $x_i$ (not necessarily the corner point illustrated in figure 2) and on a second material point an infinitesimal distance away at position $x_i + \mathrm{d}x_i$. The length and orientation of the material line joining these two points is defined by the vector $\mathrm{d}x_i$. After deformation, these points have moved to the positions $\xi_i$ and $\xi_i + \mathrm{d}\xi_i$, respectively. The mapping of each line element $\mathrm{d}x_i$ to $\mathrm{d}\xi_i$ characterizes the deformation at that material point.

Following deformation, the location of the endpoint of the infinitesimal material line of interest is

$$\xi_i + \mathrm{d}\xi_i = x_i + \mathrm{d}x_i + u_i(x_1 + \mathrm{d}x_1, \dots). \quad (4)$$

The displacement vector at this point is expanded in a Taylor series at the point $x_i$ and only the first-order terms in the increments $\mathrm{d}x_i$ are retained. Then

$$u_i(x_1 + \mathrm{d}x_1, \dots) \approx u_i(x_1, \dots) + \partial_j u_i(x_1, \dots)\,\mathrm{d}x_j,$$

where $\partial_j u_i$ denotes the matrix of components of the displacement gradient tensor in the reference configuration. In view of (4), the deformed material line element is

$$\mathrm{d}\xi_i = F_{ij}\mathrm{d}x_j, \quad F_{ij} = \delta_{ij} + \partial_j u_i, \quad (5)$$

where $\delta_{ij}$ is the identity matrix and $F_{ij}$ is the *deformation gradient*. Thus, the deformation gradient provides a complete description of the deformation of the neighborhood of a material point. Furthermore, the determinant of the deformation gradient at a point in a deformation field yields the ratio of the volume of a small material element at that point to the volume of that same material element in the reference configuration.

Suppose for the moment that the infinitesimal material line element represented by $\mathrm{d}x_i$ in the reference configuration has length $\mathrm{d}s_0$ and direction $m_i \boldsymbol{e}_i$, with $m_k m_k = 1$, and that $\mathrm{d}\xi_i\,\mathrm{d}\xi_i = \mathrm{d}s^2$; recall that a repeated index in an expression implies summation over its range. Then, forming the inner product of each side of $(5)_1$ with itself and dividing by $\mathrm{d}s_0^2$ yields

$$\frac{\mathrm{d}s^2 - \mathrm{d}s_0^2}{\mathrm{d}s_0^2} = 2E_{ij}m_i m_j, \quad (6)$$

$$E_{ij} = \tfrac{1}{2}(\partial_j u_i + \partial_i u_j + \partial_i u_k \partial_j u_k), \quad (7)$$

where $E_{ij}$ is the symmetric matrix of components of *Lagrange strain*.

To develop some sense of the geometrical character of $E_{ij}$, suppose that $\lambda_1 = \mathrm{d}s/\mathrm{d}s_0$: the *stretch ratio* of a material line element that has direction $\boldsymbol{m} = \boldsymbol{e}_1$ in the reference configuration. Then (6) implies that

$$\lambda_1^2 - 1 = 2E_{11}. \quad (8)$$

Similarly, for two infinitesimal material lines emanating from the same material point in the reference configuration, it is possible to express the angle between these two lines in the deformed configuration in terms of the Lagrange strain. For example, consider the line elements $\mathrm{d}x_i^a = \mathrm{d}s_0^a m_i^a$ and $\mathrm{d}x_i^b = \mathrm{d}s_0^b m_i^b$. For the case when $m_j^a m_j^b = 0$, the result is

$$\lambda_a \lambda_b \cos(\tfrac{1}{2}\pi - \gamma_{ab}) = 2E_{ij}m_i^a m_j^b, \quad (9)$$

where $\gamma_{ab}$ is the *reduction in angle* between the lines, called the *shear strain*, and $\lambda_a$ and $\lambda_b$ are the stretch ratios of the line elements. For the particular case when the $a$- and $b$-directions are the 1- and 2-directions and when the stretch ratios are equal to 1, this result reduces to $\sin(\gamma_{12}) = 2E_{12} = 2E_{21}$.

### 2.2.1 *Small Strain*

Up to this point in the discussion of strain, no assumption about the deformation has been invoked, other than continuity of the displacement field. In particular, the expressions for stretch and rotation are valid for deformations of arbitrary magnitude. These expressions simplify considerably in cases for which the deformation is "small" in some sense. Usually, the strain is understood to be small if $|\partial_i u_j| \ll 1$ for each choice of the indices $i$ and $j$. If the deformation of a solid material is small, then the strain (7) can be approximated by the small strain matrix

$$\varepsilon_{ij} = \tfrac{1}{2}(\partial_j u_i + \partial_i u_j). \quad (10)$$

The stretch ratio in the 1-direction is simply

$$\frac{\mathrm{d}s}{\mathrm{d}s_0} = 1 + \varepsilon_{11}, \quad (11)$$

and the expression for shear strain given in (9) becomes

$$\gamma_{ij} = 2\varepsilon_{ij}, \quad i \neq j. \quad (12)$$

Deformations falling outside the range of small strains are common, for example, when material line elements are stretched to several times their initial lengths in metal forming or when a thin compliant solid, such as a plastic ruler, is bent into a U-shape for which material line elements rotate by as much as $\pi/2$.

### 2.2.2 An Example of Deformation

Suppose that the solid body depicted as a cube of edge length $\ell_0$ in figure 2 deforms into a rectangular parallelepiped with edge length $\ell > \ell_0$ in the 3-direction and with every plane section $\xi_3 =$ const. being one and the same square shape. This is an example of a *homogeneous deformation*; that is, all initially cubic portions of the solid with edges aligned with the coordinate directions deform into the same shape. Two consequences of homogeneity are that the deformation gradient is constant throughout the reference configuration and that the displacement field is linear in $x_1$, $x_2$, and $x_3$. Finally, rigid body motions are ruled out by requiring that the material line initially along $x_1 = x_2 = \frac{1}{2}\ell_0$ remains along that line and that the surface $x_3 = 0$ remains in that plane. These features of the deformation constrain the displacement field $\boldsymbol{u}(\boldsymbol{x})$ in the reference configuration to be

$$\left.\begin{aligned} u_1(x_1,x_2,x_3) &= (\lambda - 1)(x_1 - \tfrac{1}{2}\ell_0), \\ u_2(x_1,x_2,x_3) &= (\lambda - 1)(x_2 - \tfrac{1}{2}\ell_0), \\ u_3(x_1,x_2,x_3) &= (\lambda_3 - 1)x_3, \end{aligned}\right\} \qquad (13)$$

where $\lambda_3 = \ell/\ell_0$ is the imposed stretch ratio in the 3-direction and $\lambda$ is the unknown stretch ratio in both the 1- and 2-directions.

What value of $\lambda$ ensures that the deformation is volume-preserving? This question can be addressed by noting that the local value of the determinant of the deformation gradient matrix is the ratio of the volume of a certain infinitesimal material element in the current configuration to the volume of that same material element in the reference configuration. In the present case, the deformation gradient is

$$F = \begin{bmatrix} \lambda & 0 & 0 \\ 0 & \lambda & 0 \\ 0 & 0 & \lambda_3 \end{bmatrix}. \qquad (14)$$

The requirement that $\det(F_{ij}) = 1$ leads to the conclusion that $\lambda$ has the value

$$\lambda = \sqrt{\ell_0/\ell}. \qquad (15)$$

## 2.3 Stress

The concept of stress provides the basis for quantifying the transmission of mechanical force across a material surface. Suppose that a solid body is in a state of equilibrium under the action of applied forces. Consider the resulting distributed force per unit area acting on a smooth material surface that divides the solid into two parts. Denote a unit vector normal to the surface at any point by $\boldsymbol{n}$.

Next, imagine that the portion of the solid into which $\boldsymbol{n}$ is directed is removed. The force per unit area acting on the exposed surface is a vector-valued function of position, say $\boldsymbol{t}(\boldsymbol{n})$, where the notation implies that the result depends not only on the location of that point on the surface but also on the orientation of the surface at that point; this vector quantity is commonly called the *traction* or the *stress vector*. By considering smooth surfaces for which $\boldsymbol{n} = \boldsymbol{e}_k$ for $k = 1, 2, 3$ successively, we are led to an array of nine quantities at any material point that can be expressed collectively as a matrix, say $\sigma_{ij}$. For example, traction on the surface with normal vector $\boldsymbol{e}_1$ is written in component form as

$$\boldsymbol{t}(\boldsymbol{e}_1) = \sigma_{11}\boldsymbol{e}_1 + \sigma_{12}\boldsymbol{e}_2 + \sigma_{13}\boldsymbol{e}_3. \qquad (16)$$

The matrix $\sigma_{ij}$ represents the components of a tensor $\boldsymbol{\sigma}$ at each point in the material called the *stress tensor*. It relates the local normal vector directed outward from an arbitrary material surface passing through that point to the corresponding force per unit area transmitted across that material surface according to

$$t_i(n) = \sigma_{ij}n_i. \qquad (17)$$

In order for the angular momentum of an infinitesimal material element to be conserved, the stress matrix must be symmetric; that is, $\sigma_{ij} = \sigma_{ji}$. The tensor character of stress is evident in (17), which identifies $\sigma_{ij}$ as a linear operator that, when applied to a direction $\boldsymbol{n}$ at a material point, yields the local traction transmitted across a surface passing through that point with local outward normal $\boldsymbol{n}$.

It was tacitly assumed in the foregoing discussion that stress is defined in a particular configuration of the solid body. If the configuration is the current deformed configuration, then stress is commonly called *true stress* or *Cauchy stress*. In the reference configuration, both the area and the orientation of any particular material surface element may be different, leading to a different but related description of stress called the *nominal stress*. This issue must be addressed directly in describing large deformation phenomena, but it may be overlooked without significant error when deformation is small.

The foregoing discussion has been based on transmission of force across a material surface interior to a solid body. The concept is also central to identifying boundary conditions on stress in the formulation of a boundary-value problem in solid mechanics. The

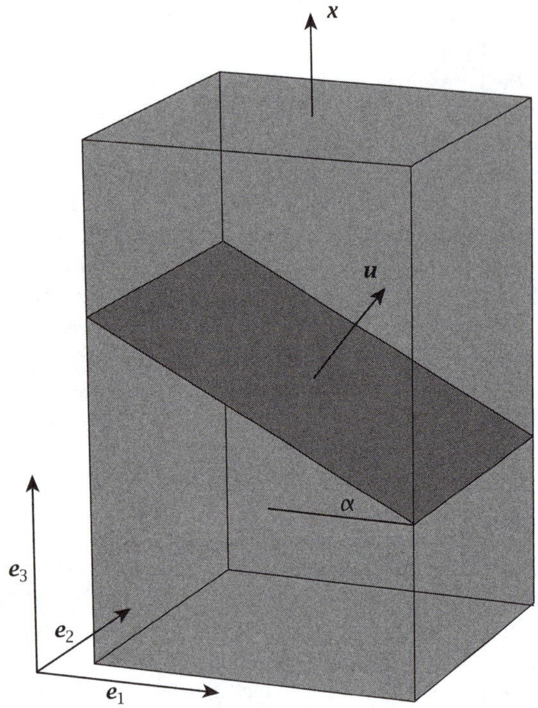


**Figure 3** Traction induced by applied tension on an interior plane with normal $\boldsymbol{n}$.

Cauchy relationship (17) can be used to infer boundary conditions for components of stress from the known imposed surface traction.

*2.3.1 An Example of Stress and Traction*

Recall that the solid block depicted in figure 2 has undergone stretching in the 3-direction and equibiaxial contraction in the 1- and 2-directions, so as to conserve the volume of the material. This deformation is induced by a traction distributed on each material plane perpendicular to the 3-direction with a resultant force, say $P$, as illustrated in figure 3. The stretch ratio $\lambda$ in the 1- and 2-directions is given in terms of the stretch ratio in the 3-direction, $\ell/\ell_0$, in (15).

The traction $\boldsymbol{t}(\boldsymbol{e}_3)$ is necessarily distributed uniformly over any plane perpendicular to the 3-direction. In particular, the traction $\boldsymbol{t}(e_3)$ has a single component in the 3-direction of magnitude $\sigma_{33} = P\ell/\ell_0^3$. This is the only nonzero component of true stress throughout the block.

The state of stress is uniform throughout the block, including everywhere on the interior plane with unit normal vector $\boldsymbol{n} = \sin\alpha\,\boldsymbol{e}_1 + \cos\alpha\,\boldsymbol{e}_3$. The traction acting on this material surface is provided by the Cauchy relation (17) as

$$\boldsymbol{t}(\boldsymbol{n}) = \sigma_{33}\cos\alpha\,\boldsymbol{e}_3. \tag{18}$$

The component normal to the inclined surface is the *normal stress* $\boldsymbol{t}\cdot\boldsymbol{n} = \sigma_{33}\cos^2\alpha$. The component tangent to the surface is the *shear stress* $\boldsymbol{t} - (\boldsymbol{t}\cdot\boldsymbol{n})\boldsymbol{n}$ with magnitude $\sigma_{33}\sin\alpha\cos\alpha$.

## 3 Governing Equations

With reference to the map in figure 2, the next step is to identify the equations comprising a boundary-value problem in the mechanics of solids. These equations provide a description of the response of the material relating stress and strain, the governing physical postulate relating stress and motion, and the compatibility equation relating strain and displacement. For a more formal introduction to these equations, see the article on CONTINUUM MECHANICS [IV.26].

### 3.1 Material Behavior

The variety of specific equations that have been adopted to describe the deformation of a material in response to applied stress is enormous and growing, driven by special applications, development of new materials, and increasingly stringent demands on precision in the use of traditional materials. Nonetheless, there are some basic requirements for any proposed description of response to be admissible. Briefly stated, the principal restrictions are that response must be consistent with the laws of thermodynamics and that the response must be independent of the frame of reference assumed by the observer describing it. The simple models for material behavior included here are consistent with these restrictions.

If all aspects of material response of a solid are identical from point to point within the body, then the material constituting that body is said to be *homogeneous*. If, on the other hand, the response of a solid at any material point due to an arbitrary but fixed state of stress is invariant under certain rotations of the material, with the state of stress unchanged, then the orthogonal transformations relating these particular orientations to the original orientation are collectively termed the *isotropy group*. If the isotropy group includes all possible rotations, then the material is said to be *isotropic* at that point; in common usage, a material is said to be *isotropic* if it is isotropic at each point. Both homogeneity and isotropy result in enormous simplification when dealing with boundary-value problems for solid bodies.

### 3.1.1 Linear Elastic Material

When used to describe material behavior, the term *elastic* usually implies that the deformation due to an applied stress cycle is reversible, repeatable, and independent of the rate of application of stress. Here, we also take it to imply that any material line will increase/decrease in length when its temperature is increased/decreased. Many elastic materials undergoing small strain exhibit a strain response that is linear in the applied stress or linear in the response to a temperature change $T$. The dependence of strain on stress at a material point in an isotropic linear elastic solid can be expressed compactly as

$$\varepsilon_{ij} = \frac{1+\nu}{E}\sigma_{ij} - \frac{\nu}{E}\sigma_{kk}\delta_{ij} + \alpha T\delta_{ij}. \qquad (19)$$

The constant $E > 0$ is called Young's modulus; it has physical dimensions of force per unit area, and it is the ratio of applied stress to induced strain in uniaxial tension of any stable elastic material. The dimensionless constant $\nu$ is called Poisson's ratio; it is the ratio of the contractive strain transverse to the tensile axis in uniaxial tension to the extensional strain along the tensile axis; it has values in the range $0 < \nu < \frac{1}{2}$ for homogeneous materials, but microstructures can be devised for which $-1 < \nu < \frac{1}{2}$. The constant $\alpha$ is called the coefficient of thermal expansion; it is the extensional strain of any material line element in the solid per degree increase in temperature.

Together, these three constants represent a complete description of the behavior of a homogeneous and isotropic elastic material. Any other material constant is necessarily representable in terms of all or some of these. For example, consider the ratio of the magnitude of an equitriaxial stress $\sigma_{11} = \sigma_{22} = \sigma_{33} = \sigma_v$ to the induced equitriaxial strain $\varepsilon_v$ at constant temperature. If the terms in (19) are each contracted over the indices $i$ and $j$ with $T = 0$, it follows that

$$3\varepsilon_v = \frac{1-2\nu}{E}3\sigma_v, \qquad (20)$$

where $3\varepsilon_v$ is the volume change to lowest order in strain. Consequently, the *bulk modulus* of an isotropic elastic material is expressible in terms of $E$ and $\nu$ as $E/(1-2\nu)$. It is evident that if $\nu = \frac{1}{2}$, then any triaxial state of stress induces no volume change whatsoever; that is, the material is incompressible if $\nu = \frac{1}{2}$.

Now suppose that the only nonzero component of stress is a shear component, say $\sigma_{12}$. Furthermore, suppose that the corresponding strain component $\varepsilon_{12}$ is expressed in terms of the actual shear strain $\gamma_{12}$ as observed in (12). The form of (19) for this case then shows that the elastic shear modulus is

$$\mu = \frac{\sigma_{12}}{\gamma_{12}} = \frac{E}{2(1+\nu)}. \qquad (21)$$

### 3.1.2 Elastic–Ideally Plastic Response

This terminology is applicable to the description of deformation of polycrystalline metals in metal forming or the plastic collapse of metal structures. The central idea underlying this type of behavior is that the range of stress values over which a material responds elastically is limited. For most metals, plastic deformation is insensitive to mean normal stress; it is a response to the total stress less the mean normal stress, called the *deviatoric stress*, which is defined as

$$s_{ij} = \sigma_{ij} - \tfrac{1}{3}\sigma_{kk}\delta_{ij}. \qquad (22)$$

The limit of elastic behavior is expressed in the form of a surface in stress space, usually called a *yield surface*. An example of such a surface is

$$\tfrac{3}{2}s_{ij}s_{ij} \begin{cases} < \sigma_y^2, & \text{elastic response,} \\ = \sigma_y^2, & \text{plastic flow,} \\ > \sigma_y^2, & \text{inaccessible,} \end{cases} \qquad (23)$$

where $\sigma_y > 0$ is the *yield stress* or *flow stress*, usually the magnitude of stress at which the elastic limit is reached in a uniform bar subjected to uniaxial tension or compression.

While the state of stress remains on the yield surface, plastic deformation can proceed without change in stress. This accounts for the phenomenon of *plastic collapse*, whereby metal structures appear to fail catastrophically at a constant level of load. Perhaps the simplest description of ongoing plastic flow assumes that the strain rate is proportional to the stress rate, so that the response is rate independent. In general, the current strain at a material point in a plastically deforming material depends on the entire history of strain at that point.

## 3.2 Stress Equilibrium

In general, the state of stress varies from point to point in a deforming solid, giving rise to local gradients in stress components. A gradient in stress across a small material element implies an imbalance in the force or moment on the element, so these gradients must be related in order to ensure equilibrium. To see how these gradients must be related, then, we begin by noting that, in the absence of any other applied forces or moments, the resultant of all surface tractions on the bounding surface $S$ of a solid occupying volume $V$ must be zero at equilibrium. When written as a surface integral, this requirement is ideally suited for application of the divergence theorem, so that

$$0 = \int_S \sigma_{ij} n_i \, dS = \int_V \partial_i \sigma_{ij} \, dV. \tag{24}$$

Next, we observe that not only is the entire solid subject to this requirement but so is every part of it. It follows that a spatially nonuniform distribution of Cauchy stress $\sigma_{ij}$ must satisfy the three conditions

$$\partial_i \sigma_{ij} = 0, \quad j = 1, 2, 3, \tag{25}$$

pointwise throughout the deformed configuration of the solid. This requirement is called the *stress equilibrium equation*. Once again, if the deformation is locally small, this condition can be imposed in the reference configuration without significant error.

If the solid being considered were subjected to a gravitational or electrostatic field, for example, then the left-hand side of the equilibrium equation (25) would include a distributed body force per unit material volume to represent the influence of such a field. Similarly, if the deformation of the material occurred at a rate sufficient to induce inertial resistance of the material to motion, the right-hand side of (25) would include a term in the form of the local rate of change of material momentum per unit volume.

### 3.3 Strain Compatibility

The notion of a strain compatibility requirement arises from the fact that, for a spatially nonuniform deformation, there are three independent displacement components at every point but six independent strain components. Given any distribution of the three displacement components, it is a straightforward matter to determine the corresponding strain distribution; see (7) or (10), for example. On the other hand, given an arbitrary distribution of each of six independent components of strain, it is not always possible to determine a displacement field from which that strain distribution can be deduced. The strain compatibility equations provide restrictions on the strain distribution, essentially in the form of integrability conditions, which ensure that three geometrically realizable components of displacement can be determined from the six prescribed components of strain.

The issue of strain compatibility is of central importance for the solution of boundary-value problems formulated in terms of stress. When confronted with problems such as these, the strain distribution that corresponds to the stress in question through a constitutive relationship must be a realizable deformation in a three-dimensional Euclidean space.

## 4 Global Formulations

In this section, methods that incorporate the local definitions of stress and deformation introduced above, but that begin from a global physical postulate governing behavior, are briefly introduced.

### 4.1 The Principle of Virtual Work

The principle of virtual work provides a gateway to understanding the deformation and failure of solid bodies from a broader perspective. Here, the subject will be briefly introduced in terms of how it applies to deformations under a small amount of strain, but its applicability extends to general deformation. Suppose that a solid body occupying the region $\mathcal{R}$ of space, with bounding surface $S$, is subjected to surface traction $t_i$ on part of the bounding surface $S_\sigma$ and to imposed $S_u = S - S_\sigma$. Let $\varepsilon_{ij}^*$ be an arbitrary distribution of compatible strain arbitrary distribution of compatible strain throughout $\mathcal{R}$ that is consistent with the condition that the corresponding displacement field satisfies $u_i^* = 0$ on $S_u$. The requirement

$$\int_{\mathcal{R}} \sigma_{ij} \varepsilon_{ij}^* \, d\mathcal{R} = \int_{S_\sigma} t_k u_k^* \, dS \tag{26}$$

for every admissible $\varepsilon_{ij}^*$ and $u_i^*$ then ensures that the stress distribution is an equilibrium distribution that is consistent with the assigned boundary values of traction. The aspect of this statement that makes it remarkable is that the stress field and the deformation field incorporated in the statement are completely uncoupled; the inference is indeed independent of material behavior.

For example, consideration of equilibrium of the material throughout the solid could begin with the requirement (26). Then, by exploiting the arbitrariness of the kinematic field, one is led to the conclusion that the stress field satisfies (25) pointwise throughout $\mathcal{R}$ and is consistent with the imposed traction through the Cauchy relation (17).

As a route to another result central to the study of elasticity boundary-value problems, suppose that the stress distribution appearing on the left-hand side in (26) is expressed in terms of the actual strain in $\mathcal{R}$ via Hooke's law (19) and that $t_k$ is the actual imposed boundary traction on $S_\sigma$. In addition, suppose that $\varepsilon^*_{ij} = \delta\varepsilon_{ij}$, a slight perturbation from the actual strain, and that $u^*_k = \delta u_k$, a slight perturbation from the actual displacement on $S_\sigma$. In this case, the principle of virtual work implies that a functional of deformation, the potential energy

$$\Phi[u_k] = \frac{1}{2}\int_{\mathcal{R}} \sigma_{ij}\varepsilon_{ij}\,\mathrm{d}\mathcal{R} - \int_S t_k u_k\,\mathrm{d}S, \qquad (27)$$

is stationary under arbitrary small variations in the deformation $\delta u_i$ that vanish on $S_u$.

### 4.2 The Finite-Element Method

Many computational techniques have been found to be useful in the field, but the one that has had the greatest impact is the finite-element method. The origin of the concepts underlying the method are thought to reside in early efforts to find approximate solutions of elliptic partial differential equations by enforcing these equations in the so-called weak sense, through variants of the principle of virtual work.

In the mechanics of solids, the basic idea of the finite-element method is to divide the domain of a continuous solid body into a finite number of subdomains, or elements. Within each element, particle displacement or some other mechanical field is assumed to be represented by its values at a number of discrete points, or nodes, on the boundary of the elements, and an interpolation scheme is adopted to extend the definition throughout each element. A model of material behavior is adopted, and the approximate deformation fields are required to abide by the principle of virtual work for arbitrary variations of the nodal displacements. This requirement generates as many equations governing the nodal values as there are nodes with unspecified values, and these are solved numerically.

The method is ideally suited for implementation on a digital computer. A great deal is understood about the accuracy and convergence of numerical methods applicable to broad problem classes, and computational mechanics has assumed its place, along with analytical mechanics and experimental mechanics, among the principal methodologies of the field.

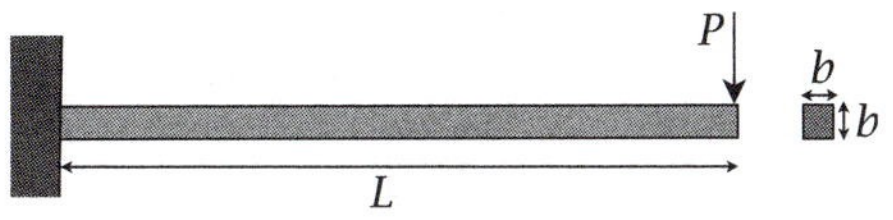


**Figure 4** An elastic–ideally plastic cantilever beam.

## 5 Selected Examples

This summary is concluded with brief descriptions of results obtained by analyzing particular phenomena. These provide some sense of its range and, at the same time, introduce relatively simple results with broad implications for understanding the mechanics of solids.

### 5.1 Plastic Limit Load

The beam in figure 4 has length $L$ and square $b \times b$ cross section; it is composed of an elastic–ideally plastic material with yield stress $\sigma_{\mathrm{Y}}$. Its left-hand end is rigidly constrained, and its right-hand end is subjected to a force of increasing magnitude $P$ in a direction transverse to the beam. The beam responds elastically until $P$ reaches a level sufficient to induce plastic deformation at the section closest to the cantilevered end, the section that bears the largest bending moment. Plastic flow begins at the outermost portions of the section, where the elastic stress is largest in magnitude. As $P$ is increased further, plastic yielding spreads inward from both the top and bottom of the section until the two regions coalesce at the beam center plane. The material can then "flow" with no further increase in load; the prevailing load is the *plastic limit load*

$$P_L = \sigma_{\mathrm{Y}} b^3/4L. \qquad (28)$$

It is noteworthy that the value of the limit load is independent of the details of the elastic deformation that leads to collapse. This feature makes it possible to calculate $P$ based only on the dimensions and the knowledge that the cross section is fully plastic; thus, it is possible to determine or estimate limit loads without reference to intervening elastic deformation.

### 5.2 Stress Concentration at a Hole

The large linearly elastic plate of uniform thickness $h$ in figure 5 contains a central hole of radius $a \gg h$ with a traction-free edge. The plate is subjected to a uniform remote traction of magnitude $\sigma_\infty$ along opposite edges. At points in the plate much farther than $a$ away from the hole, the state of stress is $\sigma_{11} = \sigma_\infty$, with other

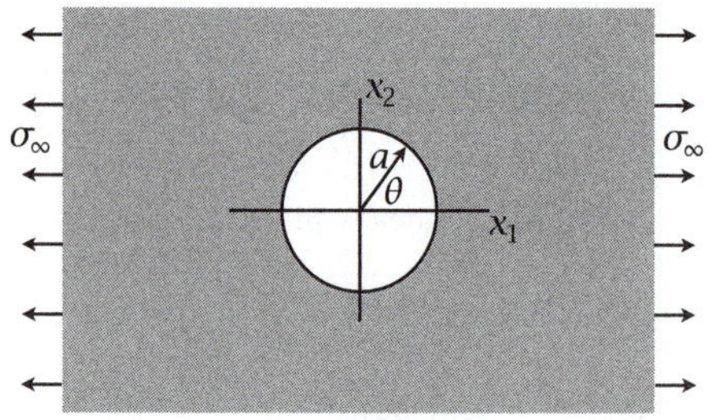


**Figure 5** Tension of an elastic plate with a hole.

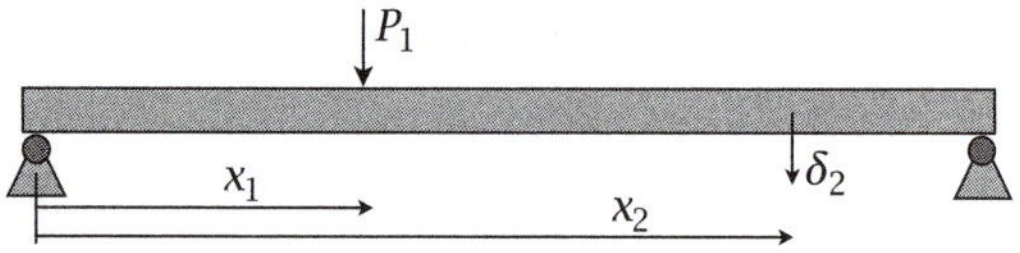


**Figure 6** A simply supported elastic beam.

stress components equal to zero. This stress distribution is not consistent with the condition of zero traction on the surface of the hole and, as a result, the stress field is perturbed in the vicinity of the hole. Analysis of this linear elastic boundary-value problem leads to the result that the stress adjacent to the edge of the hole at $\theta = \frac{1}{2}\pi$ or $\frac{3}{2}\pi$ is $\sigma_{11} = 3\sigma_\infty$ with all other components equal to zero, and at $\theta = 0$ or $\pi$ is $\sigma_{22} = -\sigma_\infty$ with all other components equal to zero. The ratio of the magnitude of the largest tensile stress component to the magnitude of the applied stress is called the *elastic stress concentration factor*, here equal to 3.

### 5.3 Elastic Reciprocity

Consider the simply supported beam illustrated in figure 6. Suppose a transverse force of magnitude $P_1$ at an arbitrary section at $x_1$ induces a transverse deflection $\delta_2$ at a second arbitrarily selected section at $x_2$. In addition, suppose that a transverse force of magnitude $P_2$ is applied at the section $x_2$ and that the resulting deflection at section $x_1$ is $\delta_1$. The elastic reciprocal theorem then states that

$$P_1\delta_2 = P_2\delta_1 \tag{29}$$

for any pair of section locations. An interesting corollary is that if any force $P$ applied at section $x_1$ induces the deflection $\delta$ at section $x_2$, then application of that force $P$ at section $x_2$ induces the same deflection $\delta$ at section $x_1$.

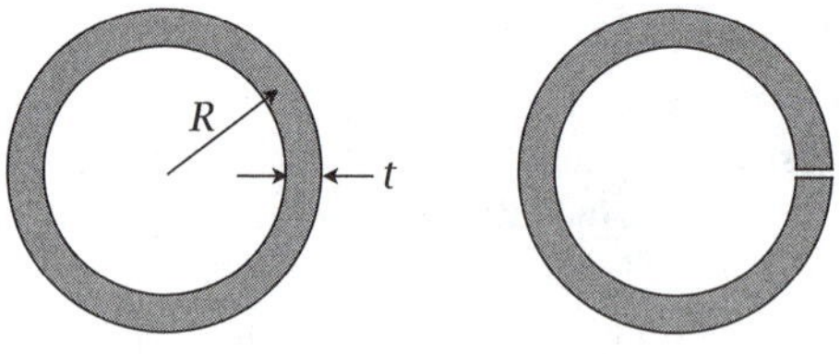


**Figure 7** Cross sections of tubes in torsion.

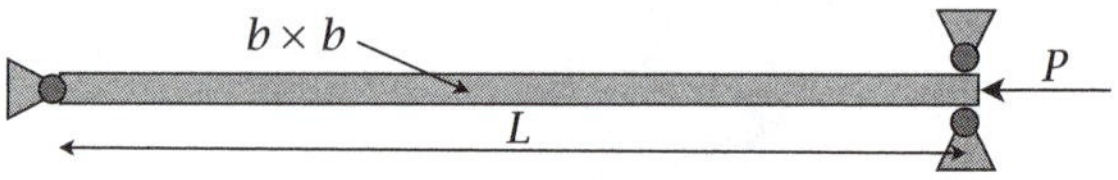


**Figure 8** A simply supported column under axial load.

### 5.4 Torsion of a Hollow Elastic Tube

A hollow tube of elastic material is a common configuration. It is used to transmit torque along the axis of the tube. The torsional stiffness of this configuration—the ratio of the torque transmitted to the relative rotation of cross sections a unit distance apart—depends strongly on details of the shape of the cross section. To illustrate this point, consider a tube like the one in figure 7 with a continuous circular cross section of mean radius $R$ and wall thickness $t \ll R$. The torsional stiffness is $2\pi\mu R^3 t$. Then, if the cross section is cut along the length of the tube but is otherwise unchanged, an estimate of the torsional stiffness leads to the result $\frac{2}{3}\pi\mu R t^3$. By conversion of the initially closed section to an open section of the same cross-sectional area, the torsional stiffness is reduced by a factor of $\frac{1}{3}(t/R)^2$. If $R/t \approx 10$, then the factor is approximately 0.0033, a dramatic reduction due only to a change from a "closed" section to an "open" section.

### 5.5 Euler Buckling

Suppose that the straight elastic column in figure 8, pinned at the left-hand end and constrained against deflection at the right-hand end, is subjected to an axial compressive load $P$. If the column is reasonably straight, it will support the load by undergoing uniform compression to generate the appropriate stress to resist the load. But what about the stability of the configuration? Several criteria are available to assess stability. For example, it might be assumed that the load is never perfectly axial but, instead, has a slight eccentricity. In this case, we seek to identify the lowest value of $P$ for which the transverse deflection will be indefinitely large, no matter how small the eccentricity. Alternatively, if the load is perfectly aligned, it might

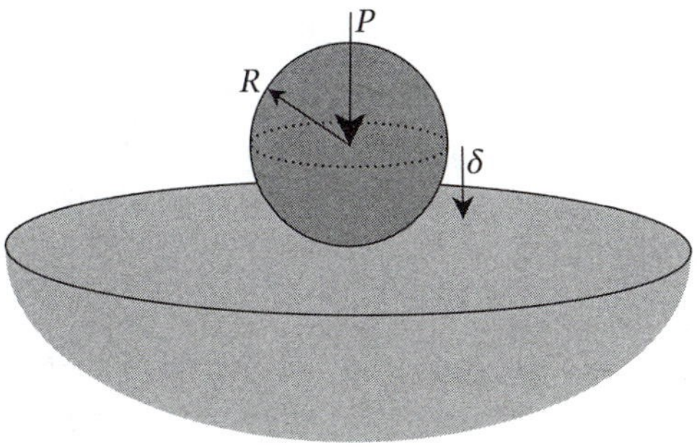


**Figure 9** A rigid sphere indenting an elastic solid.

be assumed that the column is given a slight transverse vibration. In this case, we identify the lowest value of $P$ at which the vibration amplitude becomes indefinitely large, no matter how small the initial magnitude. Yet another criterion of stability is based on energy comparisons, an idea drawn from thermodynamics. In this case, the lowest value of $P$ is sought for which the total energy of the system *decreases* when the straight column is given a slight perturbation in shape. Application of any of these criteria leads to the *Euler buckling load*

$$P_{\text{cr}} = \pi^2 Eb^4/12L^2 \tag{30}$$

with simply supported end constraints.

### 5.6 Hertzian Elastic Contact

Consider a smooth, nominally rigid sphere of radius $R$ being pressed with force $P$ into the plane surface of an isotropic elastic solid, as in figure 9. Initially, when $P = 0$, the sphere contacts the elastic solid at just one point on its surface. As the magnitude of $P$ is increased, an area of contact between the sphere and the elastic solid develops; this area is circular due to symmetry and has radius $a$, say. Even though the elastic solid is linear in its response, the radius $a$ does not increase in proportion to $P$. Instead, $a$ increases in proportion to $P^{1/3}$; the increase is sublinear because, as $P$ increases, the contact area also increases, resulting in an apparent stiffening of the response. The depth of penetration $\delta$, measured as displacement of the sphere from first contact, also increases nonlinearly with increasing values of $P$, with $\delta$ varying according to

$$\delta = \left[\frac{3}{4}\frac{(1-\nu^2)P}{E\sqrt{R}}\right]^{2/3}. \tag{31}$$

Again, the nonlinearity in response derives from the changing contact area, even though the material has a linear stress–strain response. Hertz contact theory has been remarkably useful in the field due to the combination of its relative simplicity and its broad range of applicability.

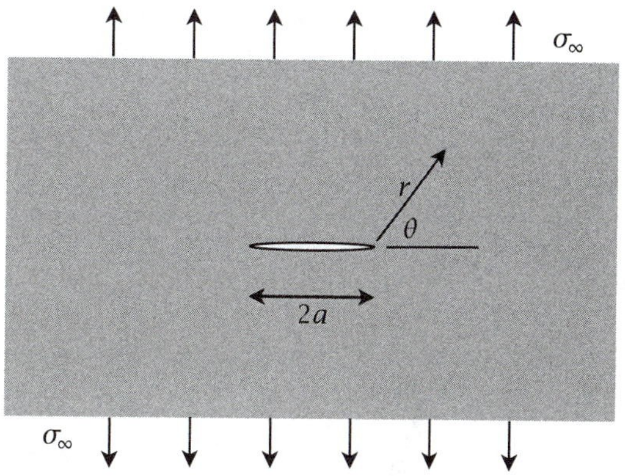


**Figure 10** Tensile loading of a cracked elastic plate.

### 5.7 Elastic Crack

An elastic plate is subjected to a uniform remotely applied tensile stress $\sigma_\infty$, as in figure 10. The plate is uniform except for an interior line that is unable to transmit traction from one side to the other, that is, a *crack*. Based on the results observed above for the case of a plate containing a hole within an otherwise uniform stress field, we might ask about the nature of the stress concentration in this case. Recognizing that the region near the edge of the crack is essentially a $2\pi$ wedge, the configuration lends itself to separation of variables in local polar coordinates. The dominant feature in such a solution shows that the stress components vary with position near the end of the crack according to

$$\sigma_{ij} \approx \frac{K}{\sqrt{2\pi r}} A_{ij}(\theta) \quad \text{as } r \to 0, \tag{32}$$

where $A_{ij}(-\theta) = A_{ij}(\theta)$ and $K$ is an amplitude called the *elastic stress intensity factor*. For the configuration considered here, $K = \sigma_\infty\sqrt{\pi a}$. The stress singularity should not be viewed literally. Instead, it is known that a stress field of this form surrounds the nonlinear crack edge region both in small laboratory samples and in large structural components. It is this feature that accounts for the wide use of elastic fracture mechanics for characterizing structural integrity.

### 5.8 Tensile Instability

Deformation of the block illustrated in figure 2 was discussed in section 2.2.2, and the state of stress driving that deformation was considered in section 2.3.1. Suppose now that the stretch ratio $\lambda_3$ in the 3-direction is related to the Cauchy stress $\sigma_{33}$ acting on each material plane perpendicular to the 3-direction according to $\sigma_{33} = A(\lambda_3 - 1)^{1/n}$, where $A > 0$ is a material constant and $n > 1$. Then, for any stretch $\lambda_3 > 1$, the total force acting on each cross section perpendicular to the 3-direction is $F(\lambda_3) = \sigma_{33}\ell_0^2\lambda^2$. Incompressibility implies that $\lambda^2 = \lambda_3^{-1}$. Under these conditions, the slope of the resulting force versus stretch relationship becomes zero at a stretch ratio of $\lambda_3 = n/(n-1)$, and it is negative for larger stretch ratios. The implication is that the load-carrying capacity of the tensile bar has been exhausted at that stretch ratio and, beyond that point, deformation can proceed with no further increase in load. This example illustrates the phenomenon of tensile instability, which is often associated with the onset of localized "necking" in a bar of ductile material under tension.

**Acknowledgments.** I thank Dr. Meredith Silberstein of Cornell University and Dr. Allan Bower of Brown University for useful input on the content of the article.

### Further Reading

Bower, A. F. 2010. *Applied Mechanics of Solids.* Boca Raton, FL: CRC Press.

Freund, L. B., and S. Suresh. 2003. *Thin Film Materials.* Cambridge: Cambridge University Press.

Fung, Y. C. 1965. *Foundations of Solid Mechanics.* Englewood Cliffs, NJ: Prentice Hall.

Gurtin, M. E. 1981. *An Introduction to Continuum Mechanics.* New York: Academic Press.

Johnson, K. L. 1987. *Contact Mechanics.* Cambridge: Cambridge University Press.

McClintock, F. A., and A. S. Argon. 1966. *Mechanical Behavior of Materials*, pp. 363-86. Reading, MA: Addison-Wesley.

Rice, J. R. 1968. Mathematical analysis in the mechanics of fracture. In *Fracture*, edited by H. Liebowitz, volume 2, pp. 191-311. New York: Academic Press

Strang, G., and G. Fix. 1988. *An Analysis of the Finite Element Method.* Wellesley, MA: Wellesley-Cambridge Press.

Timoshenko, S. P. 1983. *History of the Strength of Materials.* New York: Dover.

# IV.33 Soft Matter

*Randall D. Kamien*

## 1 Introduction

What makes matter soft? It is easily deformed but, consequently, it can easily self-assemble: sauce for the goose... Maybe it should be called "robust." Why is this article called "Soft Matter," then? Perhaps a better term would be "*la matière molle*," as used by Pierre-Gilles de Gennes. Soft matter typically refers to systems where entropy dominates energetic effects. In the case of hard materials, the energy scale is electronic. Let us note some energy scales: room temperature, around $T = 300$ K, corresponds to a thermal energy (the product of temperature $T$ and Boltzmann's constant $k_B$ ($1.381 \times 10^{-23}$ J K$^{-1}$)) of $\frac{1}{40}$ electron volts. Compared with binding energies in atoms and molecules, which are a few electron volts, the thermal energy is tiny. Of course, the precise division between soft and "hard" materials is more subtle, but this serves as a rule of thumb. In this article we will touch on some of the classic soft materials and discuss statistical mechanical effects. Quantities like pressure, forces, and energy are all to be considered Boltzmann-weighted statistical averages. We will in turn consider point-like particles (colloids), line-like objects (polymers), two-dimensional sheets (membranes), and finally the three-dimensional continuum theories of liquid crystals.

## 2 Colloids

Colloidal suspensions consist of microscopic particles, the "colloids," suspended in a fluid, often water. Idealized, the colloids are taken to be absolutely rigid, incompressible spheres that interact exclusively via *excluded volume*; that is, their only interaction is that they cannot overlap in space. For spheres of radius $R$, this could be modeled as a pair potential $U(r)$ taking the value 0 when $r \geqslant 2R$ and $\infty$ when $r < 2R$. At first glance, because the energy scale is infinite it might seem like this is not a soft system. We define the entropy of any collection of colloidal particles as $S = k_B \ln \Omega$, where $\Omega$ measures the volume of the configuration space of the colloids. Because the energy of any *allowed* finite-energy configuration remains 0, the free energy, defined as $F = E - TS$, is entirely entropic. Indeed, the problem of hard spheres reduces to a combinatorics problem: how many configurations of $N$ hard spheres of radius $R$ are possible at fixed volume $V$?

### 2.1 The Equation of State

The relation among pressure $p$, temperature $T$, and density $n = N/V$ is known as the equation of state. In the case of point particles, the equation of state is the famous ideal gas law, $p = nk_B T$, where $k_B$ is Boltzmann's constant defined above. The ideal gas law is accurate for point particles, but what happens for particles of finite radius? There is an expansion in powers of the volume fraction $\phi = nv_0$, where $v_0 = \frac{4}{3}\pi R^3$ is the volume per particle. Known as the virial expansion, its first few terms are

$$\frac{p}{nk_BT} = 1 + 4\phi + 10\phi^2 + 18.36\phi^3 + 28.22\phi^4 + 39.82\phi^5 + 53.34\phi^6 + \cdots .$$

A simple but brilliant approximation to this is the Carnahan–Starling formula:

$$\frac{p}{nk_BT} \approx \frac{1+\phi+\phi^2-\phi^3}{(1-\phi)^3} = 1 + 4\phi + 10\phi^2 + 18\phi^3 + 28\phi^4 + 40\phi^5 + 54\phi^6 + \cdots ,$$

which has *integer* coefficients that are remarkably close to the hard-won virial coefficients. This is a highly compact and highly accurate relation for low to modest $\phi$. Note, however, that the Carnahan–Starling formula suggests that the pressure diverges at volume fraction $\phi = 1$. This is certainly not the case, though, since the maximum packing fraction of hard spheres is $\phi = \pi\sqrt{2}/6 \approx 0.74$.

The virial expansion also explains a fluctuation-induced force known as the *depletion interaction*. Truncating at second order gives

$$p \approx k_BT(n + 4v_0n^2) \approx \frac{Nk_BT}{V - \frac{1}{2}(8Nv_0)}.$$

Comparing with the ideal gas law $p = Nk_BT/V$, we see that the total volume of the system has been reduced by $4Nv_0$, a measure of the *excluded volume* from the other spheres. Note that a hard sphere of radius $R$ excludes a volume of radius $2R$; a second sphere cannot have its center anywhere in that volume and so each isolated sphere occupies or *creates* an excluded volume of $8v_0$. Generalizing this, Asakura and Oosawa argued that two inclusions in a purely entropic fluid will *attract* in order to decrease the excluded volume or, equivalently, increase the *free volume*. Consider, for instance, two marked spheres with volume $v_0$ in the colloidal solution. Since each sphere excludes a volume of $8v_0$, the free volume available to the remaining spheres is reduced by $16v_0$ when the marked spheres are far apart. However, if the two spheres are brought together, their excluded volumes can *overlap* and it follows that the available volume will increase, reaching its maximum when the two spheres touch, at which point their excluded volume is $\frac{27}{2}v_0$. The change in the ideal gas law suggests a free energy of the form

$$F = -k_BTN\ln[(V - V_e)/N],$$

where $V_e$ is the excluded volume.

If large spheres of radius $R_L$ are placed in a colloidal solution of small spheres of radius $R_S$, then each excludes a volume of $\frac{4}{3}\pi(R_L + R_S)^3$; when the two large spheres touch, the extra free volume is $\frac{2}{3}\pi R_S^2(3R_L + 2R_S)$. For $r = R_L/R_S \gg 1$, this implies an entropic free energy gain of $\frac{3}{2}k_BTr\phi_S$ and so the depletion force is proportional to the volume fraction of small colloids, $\phi_S$.

## 2.2 Packing of Hard Spheres

At the other extreme, we have close-packed crystals, in which spheres touch neighboring spheres. In two dimensions, the triangular lattice, where each disk touches six adjacent neighbors, has an area fraction of $\pi/\sqrt{3}$. In three dimensions, the densest close packing consists of layers of two-dimensionally close-packed spheres stacked on top of each other, with the spheres on one layer fitting into the pockets on the next layer. Because there are two equivalent but different sets of pockets, there are an infinite number of degenerate close packings. It is standard to label the first layer A and then label the second layer B or C to indicate on which set of pockets the next layers sit. Thus, for instance, the face-centered cubic (FCC) lattice is ABCABCABC..., while the hexagonal close packed lattice is ABABAB.... Both of these configurations have volume fraction $\pi/\sqrt{18} \approx 0.74$, as does any other sequence of packing. A random sequence of As, Bs, and Cs *could* be called random close packed, but it *should not* (see the next subsection). Away from this close packing, for densities near $\phi_{FCC} = \pi/\sqrt{18}$, Kirkwood introduced "free volume theory" to calculate the equation of state. The volume fraction can be reduced by increasing the volume of the sample or, completely equivalently, by shrinking the spheres. Consider breaking space up into cells of equal volume, each of which contains one lattice site through the Voronoi tessellation. Because the spheres have been slightly shrunk, they all have room to jiggle about in their cells. By construction, if each sphere is rigorously kept inside its cell, then there are no overlaps and there is no interaction energy. The entropic contribution is again just the logarithm of the free volume, the volume available to the center of each sphere. For the FCC lattice, the pressure is

$$p = -\frac{\partial F}{\partial V}\bigg|_{N,T} = \frac{nk_{\mathrm{B}}T}{1-(\phi/\phi_{\mathrm{FCC}})^{1/3}}.$$

Note that this expression diverges at $\phi = \phi_{\mathrm{FCC}}$, as it must, since the lattice cannot be compressed any further. This is in contrast to the low-volume-fraction Carnahan–Starling formula that fails to capture close packing. This is not a surprise, since it is known that there is a discontinuous or *first-order* transition between the hard-sphere fluid and solid at a volume fraction around $\phi \approx \frac{1}{2}$. As a result, it is unreasonable to hope that the analytic behavior of the fluid phase should carry over to the solid phase.

### 2.3 Is There "Random Packing"?

Though it is ill-defined from a mathematical perspective, there is a notion of "random close packing." Unlike the random A, B, C stacking that one might consider for the densest sphere packings, this notion of random is associated with the densest fluid state, an arrangement of spheres with no particular translational symmetry but that cannot be packed any more densely. This is not precise, as local rearrangements can be made to increase the density without apparently generating long-range crystalline order. Despite this, numerous experiments and numerical simulations have given a value $\phi_{\mathrm{RCP}} \approx 0.64 \pm 0.02$ repeatedly and reliably; for example, this corresponds to pouring marbles into a bucket or packing sand together haphazardly. Why this number is apparently universal and whether it can be defined precisely remain open questions.

## 3 Polymers

Polymers are *everywhere*: from the plastics on which we sit, that cocoon us in our cars and airplanes, and that we eat as thickening agents, to proteins, ribonucleic acid, and the all-important informatic deoxyribonucleic acid (DNA) molecules. DNA, in particular, is a linear polymer, without the branches and junctions that are common and often uncontrollable in synthetic polymers. Biopolymers are, in general, much more uniform in length and structure owing to the magnificent molecular machinery of the cell. As a result, they are often the subject of study not for their biological properties but rather because of their purity.

### 3.1 Random Walks

How does one model a polymer? One can start with a microscopic model with chemical bonds connecting sections of molecule that are free or almost free to rotate about the bond axis (for single bonds only!). However, the elasticity of long rods has a universal behavior that allows us to avoid microscopic details. We consider a simple bending energy written in terms of the unit tangent vector $\hat{\boldsymbol{t}}$:

$$E = \tfrac{1}{2}k_{\mathrm{B}}TL_{\mathrm{p}}\int_{s_0}^{s_1}\left(\frac{\mathrm{d}\hat{\boldsymbol{t}}(s)}{\mathrm{d}s}\right)^2\mathrm{d}s,$$

where $k_{\mathrm{B}}TL_{\mathrm{p}}$ is the bending modulus written in terms of the temperature $T$ and a length known as the *persistence length*, $L_{\mathrm{p}}$, which may be temperature dependent. The probability of the tangent pointing along $\hat{\boldsymbol{t}}_1$ at $s_1$ given $\hat{\boldsymbol{t}} = \hat{\boldsymbol{t}}_0$ at $s_0$ is given by the functional integral (with measure $[\mathrm{d}\hat{\boldsymbol{t}}]$ over all functions satisfying the boundary conditions)

$$P(\hat{\boldsymbol{t}}_1, s_1; \hat{\boldsymbol{t}}_0, s_0) = \int_{\hat{\boldsymbol{t}}_0}^{\hat{\boldsymbol{t}}_1}\exp\{-E/(k_{\mathrm{B}}T)\}[\mathrm{d}\hat{\boldsymbol{t}}],$$

and it satisfies the diffusion equation on the sphere. This probability distribution implies that the autocorrelation function, or thermal average (denoted by $\langle\cdot\rangle$), is denoted by $\langle\hat{\boldsymbol{t}}(s')\cdot\hat{\boldsymbol{t}}(s)\rangle = \mathrm{e}^{-|s'-s|/L_{\mathrm{p}}}$, so that, at distances longer than the persistence length, the tangent vectors are decorrelated. It is therefore only a matter of length scale before a long linear object will behave as if it is composed of independent "freely jointed" elements of length on the order of $L_{\mathrm{p}}$. Indeed, integrating the tangent autocorrelation function, we find that the mean-squared end-to-end displacement for a chain of length $L$ is $\langle[\boldsymbol{R}(L)-\boldsymbol{R}(0)]^2\rangle = 2L_{\mathrm{p}}L$ for $L \gg L_{\mathrm{p}}$. $2L_{\mathrm{p}}$ is known as the *Kuhn length*, the length of each independent element.

### 3.2 Self-Avoiding Walks and Flory Theory

Polymers differ from random walks in one critical and consequential way: each element of the polymer chain, a monomer, cannot occupy an already occupied region of space. The polymer is a *self-avoiding walk*. There are sophisticated methods in statistical mechanics based on lattice models of polymers, where self-avoidance translates into no more than single occupancy of any site. However, a clever argument, due to Paul Flory, allows us to estimate how the mean-squared end-to-end displacement of the polymer chain scales with the chain length, $L$. A measure of the size of the polymer in three dimensions is the *radius of gyration* $R_{\mathrm{g}}$, defined through the average

$$R_{\mathrm{g}}^2 = \frac{1}{L}\int_0^L \boldsymbol{R}^2(\ell)\,\mathrm{d}\ell.$$

Alternatively, the mean-squared displacement is six times the radius of gyration squared,

$$\langle [\boldsymbol{R}(L) - \boldsymbol{R}(0)]^2 \rangle = 6R_g^2,$$

so both will scale in the same way. Both lengths govern the hydrodynamics and light scattering of the polymer chains.

For a chain of length $L$, the number of free "joints" or monomers is $N = L/(2L_p)$. The standard "phantom random walk" or "Gaussian chain" has a Gaussian probability distribution, $P(\boldsymbol{R})$, and so the free energy $F = -k_B T \ln Z \propto k_B T(R^2/(2N) - \alpha(d-1)\ln R)$, where $\alpha$ is a constant that depends on $L_p$ and $d$ is the dimensionality of space. The logarithmic term arises from the measure factor after integrating over the isotropic probability distribution. This free energy is interesting in its own right; in vector form,

$$F(\boldsymbol{R}) \propto k_B T |\boldsymbol{R}|^2/N,$$

so the ground state has $\boldsymbol{R} = \boldsymbol{0}$. Pulling the polymer away from its ground state gives a restoring force $-\nabla F \propto -T\boldsymbol{R}/N$, a Hookean spring with a stiffness that scales as $T/N$. The elasticity of rubber arises from entropy; a heated rubber band shrinks because $T$ grows.

To this free energy, Flory added a term to account for self-avoidance. In a similar vein to the notion of excluded volume and entropy loss, each time a monomer overlaps with another monomer, their mutual steric hindrance lowers the number of orientational conformations, lowering the entropy and raising the free energy. To model this entropic interaction, which is proportional to $k_B T$, we write the average monomer density as $\rho \propto N/R_g^d$. Neglecting monomer–monomer correlations, the number of near misses is $\rho N$, so the Flory free energy reads

$$F \propto k_B T \left[\frac{R_g^2}{2N} + v\frac{N^2}{R_g^d} - \alpha(d-1)\ln R_g\right],$$

where $v > 0$ is also some constant. Minimizing $F$ by varying $R_g$ for fixed $N$ gives

$$0 = \frac{R_g}{N} - vd\frac{N^2}{R_g^{d+1}} - \frac{\alpha(d-1)}{R_g}.$$

One might be concerned that the two arbitrary constants $v$ and $\alpha$ would make any prediction useless, but in the long-polymer limit $N \to \infty$, and it is only a matter of balancing two of the three terms with each other. One finds that $R_g \propto N^{3/(d+2)}$ for $d < 4$ and that $R_g \propto N^{1/2}$ for $d > 4$. In the dimension $d = 4$, known as the upper critical dimension, all terms contribute and the more precise statistical mechanical models predict logarithmic corrections to the scaling behavior. The Flory exponent $\nu_F = 3/(d+2)$ for $d < 4$ is known to be exact when $d = 2$ and differs by a small amount from the more precise prediction $\nu \approx 0.588$ compared with $\nu_F = 0.6$ when $d = 3$.

Above four dimensions, Flory's result agrees with the standard random walk. To see this, note that a standard random walk has a fractal or Hausdorff dimension of 2. In more than four dimensions, two two-dimensional sets do not generically intersect and so one would not expect any correction to the random-walk scaling. Finally, though it may be tempting to view the large-$N$ limit as a saddle point in some sort of steepest-descent calculation, there is not, at the current time, any known systematic expansion for $\nu_F$ in inverse powers of $N$.

Note that this discussion presumes that the polymer dissolves well in the surrounding solvent, known as a theta solvent when this condition is met. On the other hand, if the polymer self-attracts more strongly than it dissolves, $R_g \sim L^{1/3}$. The transition from the swollen state to the compact conformations is known as the *theta point*.

### 3.3 Polymer Melts: Where Self Is Lost and Self-Avoidance Vanishes

In solution, we may consider $N_p$ polymers at number density $n = N_p/V = c/N$, where $c$ is the monomer density. In the previous sections we considered single polymers in solution. This is applicable whenever the polymer volume fraction $\phi = n\frac{4}{3}\pi R_g^3 = c\frac{4}{3}\pi R_g^3/N < 1$, that is, when it is in the regime in which the polymers do not overlap. Because the monomers of different polymers can commingle, it is possible to consider concentrations $c > c^* = 3N/(4\pi R_g^3)$, the overlap concentration. It is essential to note that this concentration is *lower* than would be expected from Gaussian chains because of self-avoidance: $c^* \sim \phi^* \sim N^{1-3\nu}$, where $R_g \sim N^\nu$. Fortunately, the $c \gg c^*$ regime is amenable to scaling analysis.

The polymer melt regime corresponds to the completely incompressible limit where the volume fraction $\phi = 1$ everywhere. Note that this regime can be attained through increasing either $c$ or $R_g$; as a result, we can also consider the less dense, semidilute regime in which the volume fraction $\phi$ is both much smaller than 1 and much larger than $\phi^* = c^*(2L_p)^3$. Consider a specific chain in this regime and pick a monomer on it. In this regime we can define a correlation length scale or "mesh size" $\xi$ that is roughly the distance over which a single polymer segment does not interact with other segments. Note that $\xi$ cannot depend on the degree of polymerization $N$ for long polymers since the interaction with some other chain might as well be an interaction with a distant point of *the same chain.* By dimensional analysis, $\xi = R_g f(\phi^*/\phi)$ for some function $f(\cdot)$. However, for $\phi^*/\phi$ small, the $N$ dependence in $R_g$ must be canceled by the $N$ dependence in $\phi^*$, and so $f(x) = x^m$ must be a simple power law. As $R_g \sim N^\nu$ and $\phi^* \sim N^{1-3\nu}$, it follows that $\nu + m(1-3\nu) = 0$ or $m = -\nu/(1-3\nu)$.

Each of these correlation "blobs" has $g$ monomers of a single swollen polymer, and so $\xi = (2L_p)g^\nu$ and thus the number of such blobs is $b = N/g = N/[(\xi/2L_p)]^{1/\nu}$.

Since we have already accounted for chain–chain interaction, a polymer in the semidilute regime can be thought of as a string of $b$ uncorrelated blobs *without self-avoidance*, so that the radius of gyration in the semidilute regime is

$$R(\phi) = \xi b^{1/2} \sim \xi^{(2\nu-1)/2\nu} N^{1/2} \sim N^{1/2}\phi^{-(2\nu-1)/(3\nu-1)}.$$

We see that in this regime the end-to-end distance of the polymer scales as $N^{1/2}$, though there is a remnant volume fraction dependence. Physically, in a good solvent, $\nu$ must be at least $\frac{1}{2}$ since the phantom chain sets a lower bound, and, if the original polymers were phantom chains, $\nu = \frac{1}{2}$ and the $\phi$ dependence vanishes. Note also that in the melt regime, where $\phi = 1$, we recover the random walk result. Typically, $\nu$ is taken to be the Flory value of $\nu_F = \frac{3}{5}$, which gives $R(\phi)$ a $\phi^{-1/4}$ dependence. Again, this exponent is within a few percent of more precise calculations.

The dynamics of chains in the dilute, semidilute, and melt regimes continues to be an active area of research.

## 4 Membranes and Emulsions

So far we have discussed point-like objects (colloids) and one-dimensional objects (polymers). It is natural to move forward to the description of two-dimensional objects. These take two forms in soft matter. First, they exist as membranes such as the lipid bilayers in cell walls, soap films and bubbles, and freely floating polymerized sheets, ranging from highly ordered ones, such as graphene, to highly disordered ones, such as the spectrin networks of red blood cells. The second, more abstract, notion of a two-dimensional object is an *interface. Emulsions* are made by mixing two incompatible fluids together, along with a surfactant that allows them to mix on a supermolecular scale. The interface between the "oil" phase and the "water" phase, laden with surfactant, is also a surface. Unlike its bilayer cousin, however, the two sides of this membrane can be distinguished. This asymmetry can change the structure of the ground states, as we will discuss below.

### 4.1 The Young–Laplace Law

When we have lipid bilayers or surfactant monolayers in equilibrium with their single molecules in solution, there is a surface tension that acts as the chemical potential for area. In soap films, for instance, the soap molecules are in equilibrium between the surface and the fluid. There is another limit for these films when the individual molecules stretch, but we do not consider this high-tension, nonlinear limit here.

Consider a vesicle or bubble with total volume $V$ and a constant surface tension $\gamma$. The variation of the free energy at fixed volume is

$$dF = -A\,d\gamma - p\,dV.$$

Allowing the area to fluctuate gives the Gibbs free energy $G = F + \gamma A$, so $dG = \gamma\,dA - p\,dV$. In equilibrium, $dG = 0$ and we have $p = \gamma(dA/dV)$. How does the area vary with the volume? This is just $2H$, twice the mean curvature. We are led to the Young–Laplace law,

$$p = 2\gamma H = \gamma\left(\frac{1}{R_1} + \frac{1}{R_2}\right),$$

which relates the radii of curvature $R_1$ and $R_2$ to the pressure difference between the inside and the outside of the bubble. We have to be careful with signs, however. Recall that the relative signs of the radii of curvature are fixed by geometry, but the overall sign is *not.* To fix the sign we pick the outward normal to the surface and measure with respect to that. Fortunately, this also gives us a sense to measure the pressure difference between inside and outside. It follows that, if two bubbles are in contact with each other, the one with higher pressure bulges into the lower-pressure bubble, which fortunately agrees with intuition. For spherical bubbles this means that the smaller bubbles bulge into the larger bubbles.

### 4.2 Helfrich–Canham Free Energy

The free energy we have used to derive the shape of fluid membranes is not useful for studying local fluctuations of the surface. Helfrich and Canham proposed the following free energy functional, quadratic in the inverse radii of curvature:

$$F = \frac{1}{2}\int(\kappa H^2 + \bar{\kappa}K)\,\mathrm{d}A,$$

where $H$ is the mean curvature, $K$ is the Gaussian curvature, and $\kappa$ and $\bar{\kappa}$ are two independent bending moduli. (If applicable, this functional can be supplemented with a volume constraint.) In a one-component membrane, we would expect these moduli to be constants and, as a result, the second term, $\int \bar{\kappa}K\,\mathrm{d}A$, reduces to a topological invariant for the membrane.

There are two interesting embellishments of this energy. The first comes up whenever the two sides of the interface are different, in the case of an oil-water interface in an emulsion, for instance. Because the asymmetry identifies an "inside" and an "outside," the ambiguity of the surface normal is no longer an issue. As a result, the sign of $H$ is physical and terms linear in the mean curvature are allowed. Adding a constant to the free energy, we can write the free energy as

$$F' = \frac{1}{2}\int[\kappa(H - H_0)^2 + \bar{\kappa}K]\,\mathrm{d}A,$$

where $H_0$ is the preferred mean curvature. This leads to bent ground states and, in combination with boundary conditions, can induce a variety of complex morphologies and can be used to rationalize phases of diblock copolymers and lipid monolayers.

Finally, note that $\kappa$, $\bar{\kappa}$, and $H_0$ can vary over the membrane, either due to quenched-in or slow-moving impurities or because of composition variation in multicomponent systems. The latter—phase separation within a membrane and the subsequent localization of high or low curvature—plays an important role in biological processes such as budding and fusion.

### 4.3 Tethered Membranes

The membranes we have discussed are fluid, there is no order within the membrane and the molecules are free to flow. But a plastic sheet is *not* like this. The molecules have been glued in place, often through polymerization, and the connectivity and topology of the surface elements are no longer free to vary as they are in the liquid. As described in MECHANICS OF SOLIDS [IV.32 §3], elastic deformations are captured via the strain tensor $2u_{ij} = g_{ij} - \delta_{ij}$, where $g_{ij}$ is the induced metric of the otherwise flat surface. Recall the standard elastic energy in terms of $u_{ij}$, $\mu$, and $\lambda$, known as the *Lamé constants,*

$$F_{\mathrm{el}} = \frac{1}{2}\int\{\lambda(u_{ii})^2 + 2\mu u_{ij}u_{ij}\}\,\mathrm{d}A.$$

However, adding this to the curvature energy is not as straightforward as it may seem! Indeed, Gauss's *theorem egregium* relates the metric directly to the Gaussian curvature: $K = R_{1221}g$, where $R_{ijkl}$ is the Riemann curvature tensor and $g$ is the determinant of the metric. In other words, Gaussian curvature requires in-plane strain; bending and stretching are necessarily coupled. To lowest order in $u_{ij}$,

$$K = 2\partial_1\partial_2 u_{12} - \partial_1^2 u_{22} - \partial_2^2 u_{11},$$

and so the elastic and curvature deformations are coupled degrees of freedom.

## 5 Liquid Crystals

Is "liquid crystal" an oxymoron or simply an unfortunate name? Neither! Though there are many ways to characterize whether a material is liquid or solid, here we will use an approach based on broken symmetries. Recall that the gas and liquid phases join at their *critical point* in the phase diagram, so, from a symmetry perspective, they are both the same, and we will term them fluids. A fluid does not support static shear: if one exerts a step shear strain on one surface of a fluid (arbitrarily slowly), there will be no strain on the opposing surface. A crystal will support a step strain in all directions. Similarly, apply a finite rotation to the top surface of a fluid and the bottom surface will not rotate along with it, while crystals will support torques. Liquid crystals are in between these two cases. Some can transmit torques and not shear, some can transmit shear but only in a reduced set of directions; all of them are interesting.

### 5.1 Maier–Saupe Theory

Obviously, liquid crystals are of special interest because of their optical properties. They are typically rod-like molecules with differing dielectric constants $\varepsilon_{\parallel}$ along the long direction and $\varepsilon_{\perp}$ perpendicular to the long direction; since it is only a direction, its length and sign are arbitrary. We thus choose a unit vector $\hat{\boldsymbol{n}}$ *defined up to sign* to denote this direction. The dielectric tensor $\varepsilon_{ij}$ is written as the sum of an isotropic part and a traceless symmetric tensor $Q_{ij} = S(n_i n_j - \frac{1}{3}\delta_{ij})$ constructed from the directions $\hat{\boldsymbol{n}}$ with magnitude $S$:

$$\varepsilon_{ij} = \frac{\varepsilon_{\parallel} + 2\varepsilon_{\perp}}{3}\delta_{ij} + (\varepsilon_{\parallel} - \varepsilon_{\perp})Q_{ij}.$$

$S$, known as the Maier–Saupe order parameter, characterizes the amount of anisotropic order. When $S$ vanishes, the system is optically isotropic. Since $Q_{ij}$ is a thermodynamic average over the directions $\boldsymbol{v}_\alpha$ of the individual molecules labeled by $\alpha$, it follows that $Q_{ij} = \langle \boldsymbol{v}_i\boldsymbol{v}_j - \frac{1}{3}\delta_{ij}\rangle$. Note that the average is over molecules and fluctuations. Multiplying both sides by $n_k n_i$ and taking the trace yields $S = \frac{3}{2}\langle(\hat{\boldsymbol{n}}\cdot\boldsymbol{v})^2 - \frac{1}{3}\rangle = \langle P_2(\cos\theta)\rangle$, where $P_2(\cdot)$ is the second LEGENDRE POLYNOMIAL [II.29] and $\theta$ is the angle between the molecular direction $\boldsymbol{v}_\alpha$ and $\hat{\boldsymbol{n}}$. When $S = 1$ the long axis of every molecule is always along $\hat{\boldsymbol{n}}$; when $S = -\frac{1}{2}$, the molecules are all perpendicular to $\hat{\boldsymbol{n}}$, the so-called discotic phase.

The transition from the isotropic phase with $S = 0$ above $T = T_c$ to an ordered phase $S \neq 0$ below $T = T_c$ is modeled via Landau theory with free energy density

$$\begin{aligned} f &= f_0 + a(T - T_c)\operatorname{Tr}\boldsymbol{Q}^2 + b\operatorname{Tr}\boldsymbol{Q}^3 + c\operatorname{Tr}\boldsymbol{Q}^4 \\ &= f_0 + \tfrac{2}{3}a(T - T_c)S^2 + \tfrac{2}{9}bS^3 + \tfrac{2}{9}cS^4. \end{aligned}$$

Because in general $\operatorname{Tr} Q^3 \neq 0$, this indicates a discontinuous first-order phase transition between the nematic and isotropic states.

Note that a $3 \times 3$ traceless symmetric tensor has five independent components (three angles and two independent eigenvalues), whereas only three parameters appeared in the discussion above. In general,

$$Q_{ij} = S[n_i n_j - \tfrac{1}{3}\delta_{ij}] + W[m_i m_j - \tfrac{1}{2}(\delta_{ij} - n_i n_j)],$$

where $\hat{\boldsymbol{m}}$ is a unit vector perpendicular to $\hat{\boldsymbol{n}}$ and $W$ measures the *biaxial order*. There are, at present, only a few experimentally realized biaxial nematic phases, and we therefore focus on the uniaxial nematic ($W = 0$) below.

## 5.2 Frank Free Energy

The Frank free energy is the basis for studying elastic distortions of the nematic liquid crystal ground state. Starting with the unit vector $\hat{\boldsymbol{n}}$, the *director*, and recalling that it is defined up to sign, we require a free energy invariant under $\hat{\boldsymbol{n}} \to -\hat{\boldsymbol{n}}$. It should be noted that reducing the description to a vector is not always possible and requires that the director field be orientable. However, in orientable patches, we have the Frank free energy $F_{\text{nem}}[\hat{\boldsymbol{n}}] = \int f_{\text{nem}}\,\mathrm{d}^3x$, where $f_{\text{nem}}$ is the free energy density:

$$\begin{aligned} f_{\text{nem}} = \tfrac{1}{2}\{&K_1[\hat{\boldsymbol{n}}(\nabla\cdot\hat{\boldsymbol{n}})]^2 + K_2[\hat{\boldsymbol{n}}\cdot(\nabla\times\hat{\boldsymbol{n}})]^2 \\ &+ K_3[(\hat{\boldsymbol{n}}\cdot\nabla)\hat{\boldsymbol{n}}]^2 \\ &+ 2K_{24}\nabla\cdot[(\hat{\boldsymbol{n}}\cdot\nabla)\hat{\boldsymbol{n}} - \hat{\boldsymbol{n}}(\nabla\cdot\hat{\boldsymbol{n}})]\}. \end{aligned}$$

The elastic constants $K_1$, $K_2$, $K_3$, and $K_{24}$ are known, respectively, as the *splay*, *twist*, *bend*, and *saddle-splay* moduli. Note that each term has a precise geometric meaning: $\hat{\boldsymbol{n}}(\nabla\cdot\hat{\boldsymbol{n}})$ is twice the mean curvature vector of a surface with unit normal $\hat{\boldsymbol{n}}$; $\hat{\boldsymbol{n}}\cdot(\nabla\times\hat{\boldsymbol{n}})$ measures the deviation from the Frobenius integrability condition of the director field; $(\hat{\boldsymbol{n}}\cdot\nabla)\hat{\boldsymbol{n}}$ is the curvature of the integral lines of the director field; and finally, the saddle splay is twice the negative of the Gaussian curvature of a surface with unit normal $\hat{\boldsymbol{n}}$. The interpretations for the splay and saddle-splay break down, of course, whenever $\hat{\boldsymbol{n}}\cdot(\nabla\times\hat{\boldsymbol{n}}) \neq 0$, though they still measure elastic distortions. Note that the saddle-splay term is a total derivative and will therefore not contribute to the extremal equation for the director. However, it will contribute at boundaries, including at the locations of topological defects.

To study fluctuations around the ground state, for instance when the director is uniform along the $z$-axis $\hat{\boldsymbol{n}}_0 = \hat{z}$, it is usual to expand $\hat{\boldsymbol{n}} = \hat{z} + \delta\vec{n}$. Note that $\delta\vec{n}$ is a vector in the $xy$-plane, since deviations of the unit director are necessarily orthogonal to $\hat{z}$. In this case, the free energy in Fourier space

$$F_{\text{nem}} = \frac{1}{2}\int(2\pi)^{-3}\delta n_i(\boldsymbol{q})\delta n_j(-\boldsymbol{q})\Delta_{ij}(\boldsymbol{q})\,\mathrm{d}^3q,$$

where $\Delta_{ij}$ is

$$\Delta_{ij} = [K_1 q_\perp^2 + K_3 q_z^2]P_{ij}^{\text{L}} + [K_2 q_\perp^2 + K_3 q_z^2]P_{ij}^{\text{T}},$$

with $P_{ij}^{\text{L}} \equiv q_i q_j / q_\perp^2$ and $P_{ij}^{\text{L}} = \delta_{ij} - P_{ij}^{\text{L}}$ the longitudinal and transverse projection operators, and with $\vec{q}_\perp \equiv q_x\hat{x} + q_y\hat{y}$.

## 5.3 Smectics

The geometric interpretation of the Frank free energy becomes powerful in the smectic liquid crystal phase. This phase occurs at lower temperatures or higher densities than the nematic phase and, in addition to the director order already present in the nematic, the smectic has an additional one-dimensional, periodic, density modulation. To represent this, we construct a density field $\rho = \rho_0 + \rho_1\cos[2\pi\phi(\boldsymbol{x})/a_0]$, where $\rho_0$ is the background density, $\rho_1$ is the smectic order parameter, $a_0$ is the lattice constant, and $\phi(\boldsymbol{x})$ is the phase of the one-dimensional order. The ground state of the smectic has $\phi = \hat{\boldsymbol{k}}\cdot\boldsymbol{x}$, where $\hat{\boldsymbol{k}}$ is a unit vector along the periodic direction. When $\hat{\boldsymbol{k}}\|\hat{\boldsymbol{n}}$ this is known as the smectic A phase, and when $(\hat{\boldsymbol{k}}\cdot\hat{\boldsymbol{n}})(\hat{\boldsymbol{k}}\times\hat{\boldsymbol{n}}) \neq \boldsymbol{0}$ it is known as the smectic C phase. Many other embellishments of order and periodicity are possible.

We may interpret this system as being composed of layers, a distance $a_0$ apart, lying at the level sets of $\phi(\boldsymbol{x}) = ma_0$ with $m \in \mathbb{Z}$. Restricting our discussion to the smectic A phase, the layer normal $\hat{\boldsymbol{N}} = \hat{\boldsymbol{n}}$, and we therefore write $\hat{\boldsymbol{n}} = \nabla\phi/|\nabla\phi|$. In this case

$\hat{\boldsymbol{n}} \cdot (\nabla \times \hat{\boldsymbol{n}}) = 0$, and the terms in the Frank energy enjoy their full geometrical interpretation. The free energy vanishes identically for flat layers that are all parallel to each other. These layers, however, need not have a constant period; none of the terms in the Frank energy set the spacing. The smectic A free energy density $f_{\text{sm}} = f_{\text{c}} + f_{\text{nem}}$ has an additional term, the compression energy, that will vanish when $|\nabla\phi| = 1$; for instance, a term such as $f_{\text{sm}} = \frac{1}{2}B(|\nabla\phi| - 1)^2$ will favor equal spacing. Again, fluctuations around the ground state $\phi = z$ can be studied by writing $\phi = z - u$ and expanding to quadratic order in $u$ and lowest nontrivial order in derivatives:

$$F_{\text{sm}} = \frac{1}{2}\int \{B(\partial_z u)^2 + K_1(\nabla_\perp^2 u)^2\}\, \mathrm{d}^3x.$$

In this approximation, $\delta\boldsymbol{n} = -\nabla_\perp u$. It should be noted that the nonlinear elasticity is essential for studying large deformations and, more importantly, fluctuations and topological defects.

### 5.4 Cholesterics

In some sense, cholesterics represent the opposite situation in which the nematic director is *nowhere* integrable. To $f_{\text{nem}}$ we add $f_{\text{nem}}^* = K_2 q_0 \hat{\boldsymbol{n}} \cdot (\nabla \times \hat{\boldsymbol{n}})$. Note that under spatial inversion, $\nabla \to -\nabla$ and $f_{\text{nem}}^*$ changes sign. This term is therefore *chiral* and can only appear if the constituent molecules are not invariant under spatial inversion. The ground state of the cholesteric has a pitch axis, $\boldsymbol{P}$, perpendicular to all the molecules and along which the molecular orientation rotates; for example, $\hat{\boldsymbol{n}}_0 = (\cos q_0 z, \sin q_0 z, 0)$ when $\boldsymbol{P} = \hat{z}$. Since the cholesteric has one-dimensional periodic order, the fluctuations are described by a smectic-like energy functional ($F_{\text{sm}}$ above), with $B$ and $K_1$ replaced by combinations of the Frank constants and $q_0$. Writing a general deformation in terms of functions $\theta$ and $\phi$ as $\hat{\boldsymbol{n}} = [\sin\theta\cos(q_0 z + \phi), \sin\theta\sin(q_0 z + \phi), \cos\theta]$, $u$ in the expression for $F_{\text{sm}}$ is replaced by $\phi$.

### Further Reading


Chaikin, P. M., and T. C. Lubensky. 1995. *Principles of Condensed Matter Physics.* Cambridge: Cambridge University Press.

de Gennes, P.-G. 1970. *Scaling Concepts in Polymer Physics.* Ithaca, NY: Cornell University Press.

de Gennes, P.-G., and J. Prost. 1993. *The Physics of Liquid Crystals.* Oxford: Oxford University Press.

Doi, M., and S. F. Edwards. 1986. *The Theory of Polymer Dynamics.* Oxford: Oxford University Press.

Kamien, R. D. 2002. The geometry of soft materials: a primer. *Reviews of Modern Physics* 74:953–71.

Russel, W. B., D. A. Saville, and W. R. Schowalter. 1989. *Colloidal Dispersions.* Cambridge: Cambridge University Press.

Safran, S. A. 1994. *Statistical Thermodynamics of Surfaces, Interfaces, and Membranes.* New York: Addison-Wesley.


## IV.34 Control Theory


*Anders Rantzer and Karl Johan Åström*


### 1 Introduction

Feedback refers to a situation where the output of a dynamical system is connected to its input. For example, a feedback loop is created if room temperature in a building is measured and used to control the heating. Feedback is ubiquitous in nature as well as in engineering. Our body uses feedback to control body temperature, glucose levels, blood pressure, and countless other quantities. Similarly, feedback loops are crucial in all branches of engineering, such as the process industry, power networks, vehicles, and communication systems.

Control theory is a branch of applied mathematics devoted to analysis and synthesis of feedback systems. A wide range of mathematics is used, and the purpose of this article is to illustrate this. After some historical notes are given in section 2, we introduce some basic control engineering problems in sections 3 and 4. Section 5 illustrates how the theory of analytic functions can be used to derive fundamental limitations on achievable control performance. Multivariable control problems are discussed in section 6 using concepts from linear algebra and functional analysis. The important special case of linear quadratic control is presented in section 7, together with the idea of dynamic programming. Extensions to nonlinear systems are discussed in section 8. Finally, section 9 describes some current research challenges in the area of distributed control.

### 2 A Brief History

The use of feedback control in engineering dates back to at least the nineteenth century. A prime example is the centrifugal governor—an essential component in the Watt steam engine—the basic principles of which were discussed by J. C. Maxwell in his 1868 paper "On governors." From this early period until World War II, almost a hundred years later, control technology was developed independently in different branches of engineering, such as the process industry, the aerospace industry, and telecommunications.

It was not until the 1940s and 1950s, stimulated by the war effort, that the scientific subject of control theory was created and the main ideas were collected into a common mathematical framework, emphasizing a frequency-domain viewpoint of the subject. Several important new ideas, such as the Nyquist criterion and the Bode relations, came out of this process.

A second wave of progress, starting in the 1960s, was stimulated by the space race and the introduction of computer technology. Theories for optimal control were created, with fundamental contributions by L. P. Pontryagin and V. A. Yakubovich in Russia and R. Bellman and R. Kalman in the United States. Unlike earlier work, these new contributions to optimal control theory emphasized time-domain models in terms of differential and difference equations.

Between 1980 and 2010 two key words came to dominate the research arena: robustness and optimization. The need to quantify the effects of uncertainty and unmodeled dynamics led to new tools for robustness analysis and $H_\infty$-optimal control synthesis. Furthermore, efficient new algorithms for convex optimization are increasingly being used for the analysis, synthesis, and implementation of modern control systems.

## 3 A Simple Case of Proportional Feedback

Many of the basic ideas of control theory can be understood in the context of linear time-invariant systems. The analysis of such system is considerably simplified by the fact that their input-output relationship is completely characterized by the response to sinusoids. A general frequency-domain representation with input $u(t)$ and output $y(t)$ takes the form

$$Y(s) = P(s)U(s), \tag{1}$$

where

$$Y(s) = \int_0^\infty e^{-st} y(t)\,dt \quad \text{and} \quad U(s) = \int_0^\infty e^{-st} u(t)\,dt.$$

The function $P(s)$ is called the transfer function. In steady state, a sinusoidal input $u(t) = \sin \omega t$ gives the sinusoidal output

$$y(t) = |P(i\omega)| \sin(\omega t + \arg P(i\omega)).$$

Of particular interest are complex numbers $z$ and $p$ such that $P(z) = 0$ and $P(p) = \infty$. These are called *zeros* and *poles* of $P$, respectively.

Next, consider a simple feedback control law $u(t) = k[r(t) - y(t)]$, where $k$ is a constant and $r$ is a reference value for the output, such that the input $u$ is proportional to the deviation between the output $y$ and its reference value $r$. The control law in the frequency domain has the form

$$U(s) = k[R(s) - Y(s)]. \tag{2}$$

Eliminating $U$ from (1) and (2) gives $Y(s) = T(s)R(s)$, where

$$T(s) = \frac{kP(s)}{1 + kP(s)}.$$

In particular, we see that, if $kP(s)$ is significantly bigger than 1, then $T(s) \approx 1$ and $Y(s) \approx R(s)$. Hence, at first sight it may look like all that is needed to make the output $y(t)$ approximately follow the reference $r(t)$ is to apply the feedback law (2) with sufficiently large gain $k$. There are, however, several complicating issues.

A central issue is stability. The input-output relationship (1) is said to be stable if every bounded input gives a bounded output. It turns out that in the representation $U(s) = \int_0^\infty e^{-st} u(t)\,dt$, a bounded $u(t)$ corresponds to a $U(s)$ that is bounded and analytic in the right half-plane $\{s \mid \operatorname{Re} s > 0\}$. Similarly, the input-output relationship described by the transfer function $P(s)$ is stable if and only if $P(s)$ is bounded and analytic in the right half-plane.

To conclude, feedback interconnection of a stable process $Y(s) = P(s)U(s)$ with the controller $U(s) = k[R(s) - Y(s)]$ gives a stable closed-loop map from the reference signal $R$ to the output $Y$ if and only if $kP(s)/(1 + kP(s))$ is analytic and bounded in the right half-plane. However, as we will see in the next section, there are also degrees of stability and instability. This will be quantified by considering the effects of disturbances and measurement errors.

## 4 A More General Control Loop

We will now discuss a more general control structure like the one illustrated in figure 1. The process is still represented by a scalar transfer function $P(s)$.

The controller consists of two transfer functions, the feedback part $C(s)$ and the feedforward part $F(s)$. The control objective is to keep the process output $x$ close to the reference signal $r$, in spite of a load disturbance $d$. The measurement $y$ is corrupted by noise $n$.

Several types of specifications are relevant:

(I) reduce the effects of load disturbances,
(II) control the effects of measurement noise,
(III) reduce sensitivity to process variations, and
(IV) make the output follow command signals.

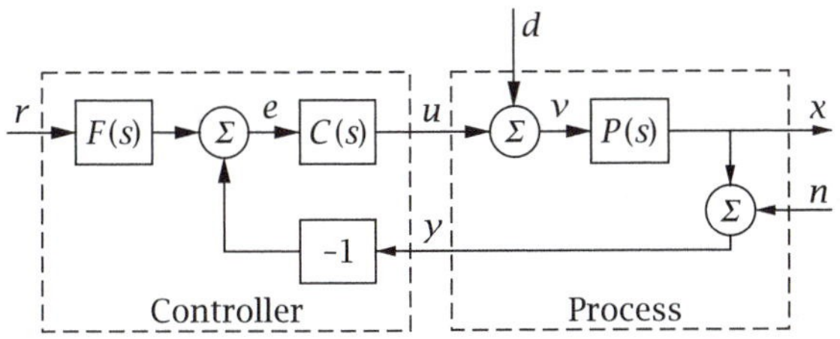


**Figure 1** A control system with three external signals: the reference value $r(t)$, an input disturbance $d(t)$, and a measurement error $n(t)$.

A useful synthesis approach is to first design $C(s)$ to meet specifications (I), (II), and (III), and then design $F(s)$ to deal with the response to reference changes, (IV). However, the two steps are not completely independent: a poor feedback design will also have a negative influence on the response to reference signals.

The following relations hold between the frequency-domain descriptions of the closed-loop signals:

$$X(s) = \frac{PCF}{1+PC}R(s) - \frac{PC}{1+PC}N(s) + \frac{P}{1+PC}D(s),$$

$$V(s) = \frac{CF}{1+PC}R(s) - \frac{C}{1+PC}N(s) + \frac{1}{1+PC}D(s),$$

$$Y(s) = \frac{PCF}{1+PC}R(s) + \frac{1}{1+PC}N(s) + \frac{P}{1+PC}D(s).$$

Note that the signals in the feedback loop are characterized by six transfer functions:

$$\frac{P(s)C(s)F(s)}{1+P(s)C(s)}, \quad \frac{P(s)C(s)}{1+P(s)C(s)}, \quad \frac{P(s)}{1+P(s)C(s)},$$

$$\frac{C(s)F(s)}{1+P(s)C(s)}, \quad \frac{C(s)}{1+P(s)C(s)}, \quad \frac{1}{1+P(s)C(s)}.$$

To fully understand the properties of the closed-loop system, it is necessary to look at all these transfer functions. (In particular, they all have to be stable.) It can be strongly misleading to show only properties of a few input–output maps, e.g., a step response from reference signal to process output.

The properties of the different transfer functions can be illustrated in several ways, by time responses or frequency responses. For a particular example, in figures 2 and 3 we show first the six frequency response amplitudes, and then the corresponding six step responses.

It is worthwhile to compare the frequency plots and the step responses and to relate their shapes to the specifications (I)–(IV).

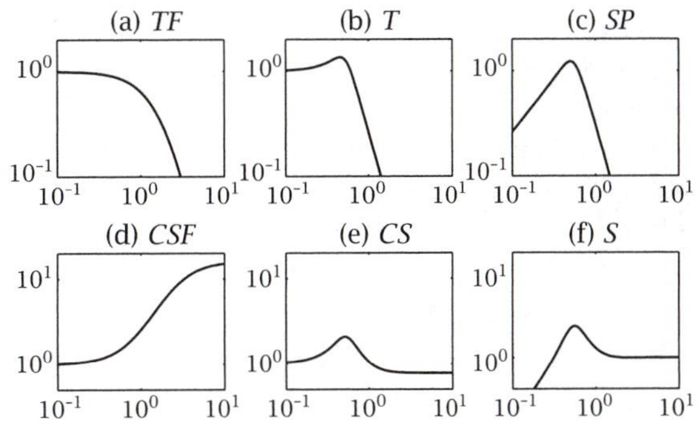


**Figure 2** Frequency response amplitudes for $P(s) = (s+1)^{-4}$ and $C(s) = 0.8(0.5s^{-2}+1)$ when $TF = (0.5s+1)^{-4}$. Here, the notation $S = (1+PC)^{-1}$ and $T = PC(1+PC)^{-1}$ is used.

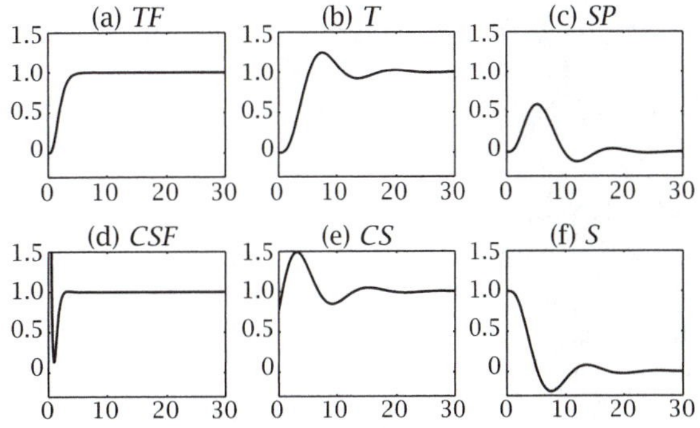


**Figure 3** Step responses for $P(s) = (s+1)^{-4}$ and $C(s) = 0.8(0.5s^{-2}+1)$ when $TF = (0.5s+1)^{-4}$.

*(I) Disturbance rejection*

Parts (c) and (f) of figure 2 show the effect of the disturbance $d$ in process output $x$ and input $v$, respectively. The resulting process error should not be too large and should settle to zero quickly enough. Corresponding step responses are shown in parts (c) and (f) of figure 3.

*(II) Suppression of measurement noise*

Figure 2(b) shows good attenuation of measurement noise above the "cutoff" frequency of 1 Hz (which in this example is mainly an effect of the process dynamics).

*(III) Robustness to process variations*

The robustness to process variations is determined by the *sensitivity function* $S = (1+PC)^{-1}$ and the *complementary sensitivity function* $T = PC(1+PC)^{-1}$. In fact, the closed-loop system remains stable as long as the relative error in the process model is less than $|T|^{-1}$. Most process models are inaccurate at high frequencies, so the complementary sensitivity function $T$ should be small for high frequencies.

*(IV) Command response*

Figure 3(a) shows how the process output $x$ responds to a step in $r$. Using the prefilter $F(s)$, it is possible to get a better step response here than in part (b). The price to pay is that the corresponding response in the control signal gets higher amplitude.

## 5 Fundamental Limitations

Clearly, the six transfer functions discussed in the previous section are not independent. In particular, the following identity holds trivially:

$$S(\mathrm{i}\omega) + T(\mathrm{i}\omega) = 1.$$

The first term describes the influence of load disturbances on process input. This should be small. The second term describes robustness to model errors and the influence of measurement noise. Ideally, this term should also be small, so controller design involves a trade-off between the two requirements. The essence of the conflict is that disturbances cannot be rejected unless measurements can be trusted. Usually, this is resolved by frequency separation: measurement noise is typically high-frequency dominant, so $T(\mathrm{i}\omega)$ should be small at high frequencies. This makes it impossible to remove the effects of fast load disturbances at high frequencies, but we can still cancel them on a slower timescale by making $S(\mathrm{i}\omega) \approx 0$ at low frequencies (see figure 4).

A popular approach to control synthesis, known as *loop shaping*, is to focus on the shape of the loop transfer function and to keep modifying the controller $C$ until the desired shape of $PC$ is obtained. However, a seriously complicating factor in loop shaping is the stability requirement, which mathematically means that all six of the transfer functions in section 4 need to be analytic in the right half-plane. This restricts the possibility to shape the transfer functions. In particular, *Bode's integral formula* shows that the effort required to make the sensitivity function $S(\mathrm{i}\omega)$ small is always a trade-off between different frequency regions: if $P(s)$, $C(s)$, and $S(s) = (1+P(s)C(s))^{-1}$ are stable and

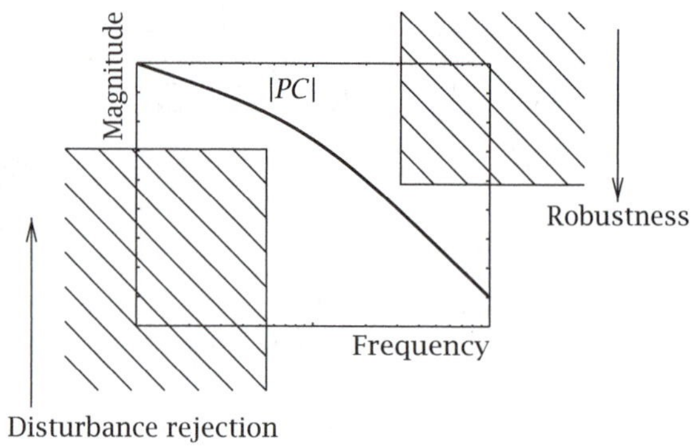


**Figure 4** Magnitude specifications on $T$ and $S$ can (approximately) be interpreted as specifications on the loop transfer function $P(\mathrm{i}\omega)C(\mathrm{i}\omega)$, which should have small norm at high frequencies and large norm at low frequencies.

$s^2P(s)C(s)$ is bounded, then it follows from CAUCHY'S INTEGRAL FORMULA [IV.1 §7] for analytic functions that

$$\int_0^\infty \log|S(\mathrm{i}\omega)|\,\mathrm{d}\omega = 0.$$

In the cases where there are unstable poles $p_i$ in $P(s)C(s)$, the integral formula changes into

$$\int_0^\infty \log|S(\mathrm{i}\omega)|\,\mathrm{d}\omega = \pi\sum_i \operatorname{Re} p_i$$

(see figure 5). Hence, unstable process poles make it harder to push down the sensitivity function! The faster the unstable modes, the harder it is. This can be used as an argument for why controllers with right half-plane poles should generally be avoided.

To further illustrate fundamental limitations imposed by plant dynamics, we will discuss the dynamics of a bicycle. A torque balance for the bicycle (figure 6) can be modeled as

$$J\frac{\mathrm{d}^2\theta}{\mathrm{d}t^2} = mg\ell\theta + \frac{mV_0\ell}{b}\left(V_0\beta + a\frac{\mathrm{d}\beta}{\mathrm{d}t}\right).$$

The transfer function from steering angle $\beta$ to tilt angle $\theta$ is

$$P(s) = \frac{mV_0\ell}{b}\frac{as+V_0}{Js^2 - mg\ell}.$$

This system has an unstable pole $p$ with time constant

$$p^{-1} = \sqrt{J/mg\ell} \quad (\approx 0.5 \text{ seconds}).$$

Moreover, the transfer function has a zero $z$ with

$$z^{-1} = -a/V_0 \quad (\approx 0.05 \text{ seconds}).$$

Riding the bicycle at normal speed, the zero is not really an obstacle for control. However, if one tries to ride the bicycle backward, $V_0$ gets a negative sign and the zero becomes unstable. Such zeros are sometimes

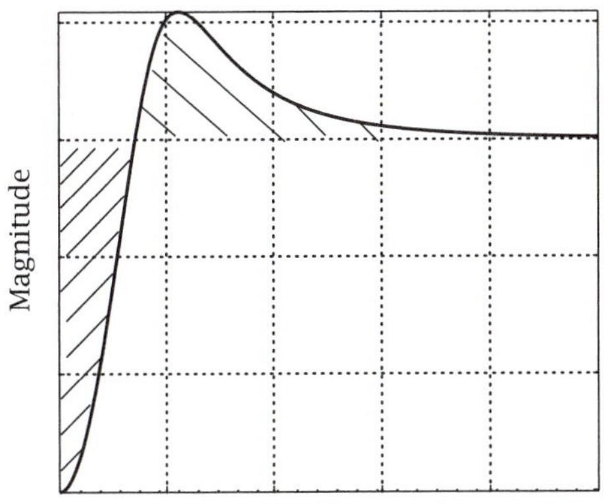


**Figure 5** The amplitude curve of the sensitivity function always encloses the same area below the level $|S| = 1$ as it does above it. This invariance is sometimes referred to as the "water bed" effect: if the designer tries to push the magnitude of the sensitivity function down at some point, it will inevitably pop up somewhere else!

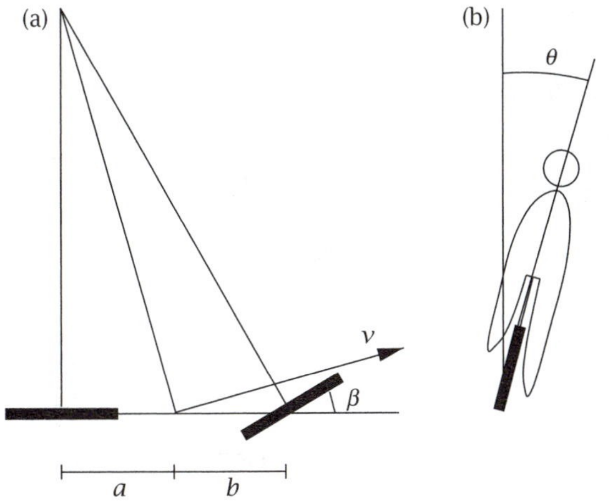


**Figure 6** A schematic of a bicycle. (a) The top view and (b) the rear view. The physical parameters have typical values as follows: mass $m = 70$ kg; rear-to-center distance $a = 0.3$ m; height over ground $\ell = 1.2$ m; center-to-front distance $b = 0.7$ m; moment of inertia $J = 120$ kg m$^2$; speed $V_0 = 5$ m s$^{-1}$; and acceleration of gravity $g = 9.81$ m s$^{-2}$.

called "minimum-phase zeros," and they further limit the achievable control performance. In particular, riding the bicycle backward at slow speed ($\approx 0.7$ m s$^{-1}$), there is an unstable pole-zero cancelation, and the bicycle becomes impossible to stabilize.

It is not hard to see that unstable open-loop dynamics puts a fundamental constraint on the necessary speed

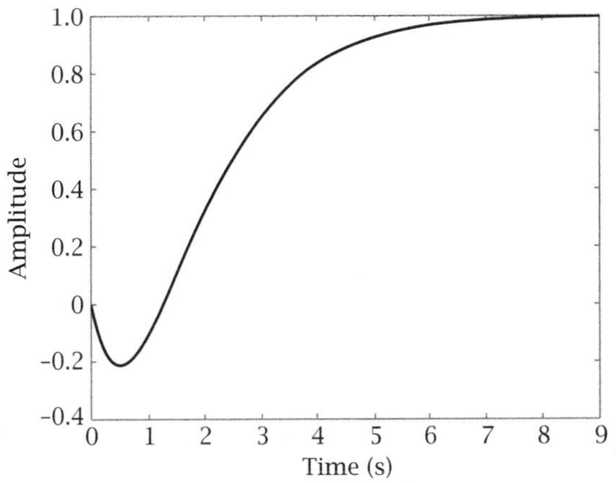


**Figure 7** The step response for a process with transfer function $(1-s)/(s+1)^2$. The response is initially negative, which is typical for processes with an unstable zero.

of control: *the feedback loop must be faster than the time constant of the fastest unstable pole.*

Unstable zeros also give rise to fundamental limitations. For a process with an unstable zero $z$, a step input generally yields an output that initially goes in the "wrong direction" (see figure 7). In fact, the definition of an unstable zero shows that the step response $y(t)$ must satisfy

$$0 = \int_0^\infty e^{-zt} y(t)\,dt,$$

so $y(t)$ cannot have the same sign for all $t$.

Think of the bicycle again. Riding it backward, we would operate with rear-wheel steering. Hence, when turning left, the center of mass will initially move to the right.

The delayed response due to an unstable zero gives a fundamental limitation on the possible speed of control: *the feedback loop cannot be faster than the time constant of the slowest unstable zero.* Similarly, the presence of time delays makes it impossible to achieve fast control: *the feedback loop cannot be faster than the time delay.*

Formal arguments about fundamental limitations due to unstable poles and zeros can be obtained using the theory of analytic functions. Recall that a controller is stabilizing if and only if the closed-loop transfer functions are analytic in the right half-plane.

If $p$ is an unstable pole of $P(s)$, then the complementary sensitivity function must satisfy $T(p) = 1$ *regardless of the choice of controller $C$*. The fact that $T$ has a hard constraint in the right half-plane also has

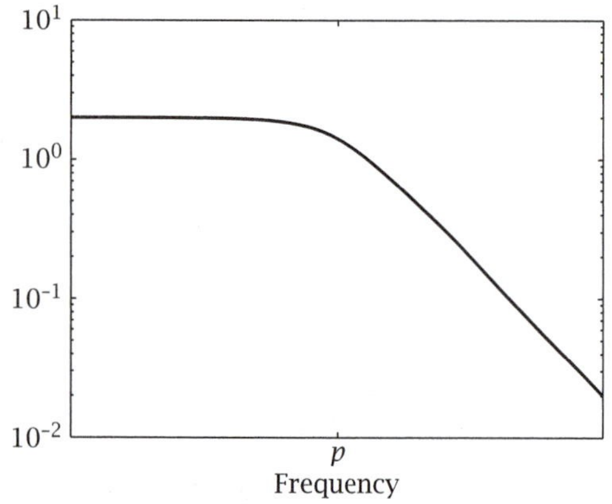


**Figure 8** Limitation from unstable pole. The complementary sensitivity function $T(\mathrm{i}\omega) = P(\mathrm{i}\omega)C(\mathrm{i}\omega)/(1 + P(\mathrm{i}\omega)C(\mathrm{i}\omega))$ should be small for high frequencies in order to tolerate measurement noise and model errors. However, if $P(s)$ has an unstable pole $p > 0$, then no stabilizing controller $C$ can push $|T(\mathrm{i}\omega)|$ entirely below the curve above. As a consequence, the unstable pole $p$ gives a lower limit on the bandwidth of the closed-loop system.

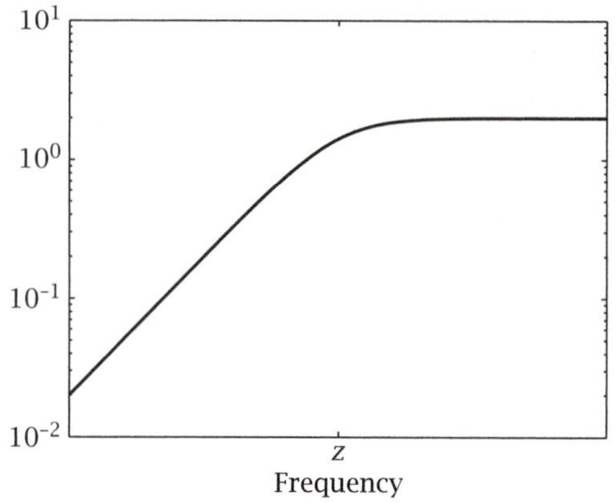


**Figure 9** Limitation from unstable zero. The sensitivity function $S(\mathrm{i}\omega) = (1 + P(\mathrm{i}\omega)C(\mathrm{i}\omega))^{-1}$ should be small for low frequencies in order to reject disturbances and follow reference signals at least on a slow timescale. However, if $P(z) = 0$, $z > 0$, then no stabilizing controller $C$ can push the sensitivity function $|S(\mathrm{i}\omega)|$ entirely below the curve above. As a consequence, the unstable zero $z$ puts an upper limit on the speed of disturbance rejection in the closed-loop system.

consequences on the imaginary axis. In particular, it follows from the maximum modulus theorem for analytic functions that the specification

$$|T(\mathrm{i}\omega)| < \frac{2}{\sqrt{1 + \omega^2/p^2}} \quad \text{for all } \omega$$

is impossible to satisfy (see figure 8). This is a rigorous mathematical statement of the heuristic idea that the feedback loop needs to be at least as fast as the fastest unstable pole.

Similarly, if $z$ is an unstable zero of $P(s)$, then the sensitivity function must satisfy $S(z) = 1$ for every stabilizing controller $C$. As a consequence, the specification

$$|S(\mathrm{i}\omega)| < \frac{2}{\sqrt{1 + z^2/\omega^2}} \quad \text{for all } \omega$$

is impossible to satisfy (see figure 9).

The situation becomes even worse if there is both an unstable pole $p$ and an unstable zero $z$, especially if they are close to each other. It follows from the maximum modulus theorem that

$$\max_{\omega \in \mathbb{R}} |S(\mathrm{i}\omega)| \geqslant \left| \frac{z + p}{z - p} \right|$$

for every stabilizing $C$. If $S$ is very large, then the same is true for $T$, since $S + T \equiv 1$. Hence, if $z \approx p$, then poor robustness to model errors and amplification of measurement noise make the system impossible to control.

## 6 Optimizing Multivariable Controllers

So far, we have only discussed systems with one input and one output. However, all the previous results have counterparts for multivariable systems, where $P(s)$ and $C(s)$ are matrices. The main difference with multivariable systems is that the number of relevant input–output maps is large, and there is therefore a stronger need to organize the variables and computations involved in control synthesis. It is common to use a framework with four categories of variables (see figure 10). The controller is a map from the measurement vector $y$ to the input vector $u$. Usually, the controller should be chosen such that the transfer matrix from disturbances $w$ to errors $z$ becomes "small" in some sense.

It turns out that all closed-loop transfer functions from $w$ to $z$ that are achievable using a stabilizing controller can be written in the *Youla parametrization* form:

$$P_{zw}(s) - P_{zu}(s)Q(s)P_{yw}(s), \tag{3}$$

where $P_{zw}(s)$, $P_{zu}(s)$, and $P_{yw}(s)$ are stable transfer matrices fixed by the process, and where every stable $Q(s)$ corresponds to a stabilizing controller.

The Youla parametrization is particularly simple to derive for stable processes; let the multivariable

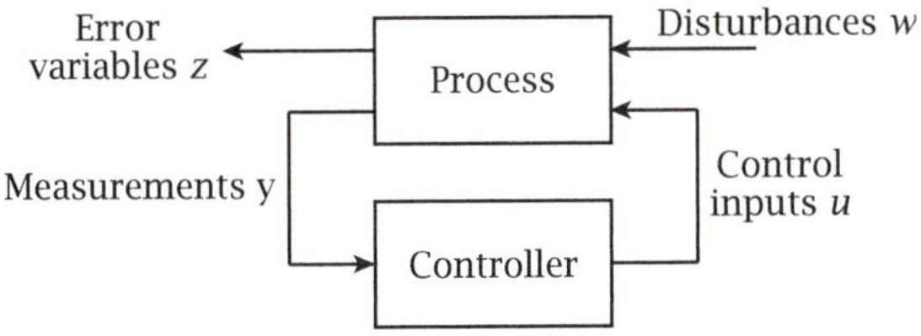


**Figure 10** A framework for optimization of multivariable controllers. The vector $w$ contains external signals, such as disturbances, measurement errors, and operator set points. The vector $z$ contains signals that should be kept small, usually deviations between actual values and desired values. The controller is chosen to minimize the effect of $w$ on $z$.

feedback system be given by the equations

$$\begin{bmatrix} z \\ y \end{bmatrix} = \begin{bmatrix} P_{zw}(s) & P_{zu}(s) \\ P_{yw}(s) & P_{yu}(s) \end{bmatrix} \begin{bmatrix} w \\ u \end{bmatrix}, \quad (4)$$

$$u = -C(s)y. \quad (5)$$

Eliminating $u$ and $y$ gives (3), with

$$Q(s) = C(s)[I + P_{yu}(s)C(s)]^{-1}.$$

Exploiting the stability of the process $P$, it is straightforward to verify that the closed-loop system is stable if and only if $Q(s)$ is stable. In the more general case of an unstable process, the Youla parametrization can be derived by first applying a stabilizing preliminary feedback and then proceeding as above. Given the Youla parametrization of all achievable transfer functions from $w$ to $z$, the control synthesis problem can be formulated as the problem of selecting a $Q(s)$ that gives the desired properties of $P_{zw} - P_{zu}QP_{yw}$. Recall that $w$ is a vector of external signals, and we know from section 4 that frequency separation of different signals is often essential. A natural design approach is therefore to select $Q(s)$ to minimize

$$\|W_z(P_{zw} - P_{zu}QP_{yw})W_w\|,$$

where $W_z$ and $W_w$ are frequency weights selected to emphasize relevant frequencies for the different signals. This approach connects control theory to the mathematical theory of functional analysis.

The choice of norm is important. The two most common norms are the $H_2$*-norm*

$$\|G\|_{H_2} = \sqrt{\frac{1}{2\pi}\int_{-\infty}^{\infty} \operatorname{tr}(G(\mathrm{i}\omega)^* G(\mathrm{i}\omega))\,\mathrm{d}\omega}$$

and the $H_\infty$*-norm*

$$\|G\|_{H_\infty} = \max_{\omega} \bar{\sigma}(G(\mathrm{i}\omega)),$$

where $\bar{\sigma}(\cdot)$ denotes the largest singular value. Analytic formulas can be given for the optimal $Q(s)$ and the corresponding controller $C(s)$ for both the $H_2$-norm and the $H_\infty$-norm. Some key ideas of the theory will be described in the next section.

## 7 Linear Quadratic Control

The frequency-domain viewpoint of control theory described in earlier sections dominated control theory before 1960 and has been of central importance ever since. However, between 1960 and 1980 there was a second wave of development, triggered by the introduction of computers for simulation and implementation. Process models in state-space form, either differential equations or difference equations, then started to play a central role. With reference to figure 10, a state-space model in continuous time can be expressed as follows:

$$\text{Process} \quad \begin{cases} \dfrac{\mathrm{d}}{\mathrm{d}t}x(t) = Ax(t) + Bu(t) + w_x(t), \\ y(t) = Cx(t) + w_y(t). \end{cases}$$

Here, the control objective is stated in terms of the map from $w = (w_x, w_y)$ to $z = (x, u)$. For example, in linear quadratic Gaussian optimal control, $w$ is modeled as Gaussian white noise and the control objective is to minimize the variance expression

$$\mathbb{E}(x^{\mathrm{T}}Qx + u^{\mathrm{T}}Ru),$$

where $Q$ and $R$ are symmetric positive-semidefinite matrices. This is equivalent to the $H_2$-norm minimization discussed previously (see also figure 11).

It turns out that the optimal controller can be written as a combination of two components:

$$\text{Controller} \quad \begin{cases} \dfrac{\mathrm{d}}{\mathrm{d}t}\hat{x}(t) = A\hat{x}(t) + Bu(t) \\ \qquad\qquad + K[y(t) - C\hat{x}(t)], \\ u(t) = -L\hat{x}(t). \end{cases}$$

The first is a state estimator (also called an observer or Kalman filter), which maintains a state estimate $\hat{x}(t)$ based on measurements of $y$ up to time $t$. The second is a state feedback law $u = -L\hat{x}$, which uses $\hat{x}(t)$ as if it were a measurement of the true state $x(t)$ and takes control action accordingly (see figure 12).

The main tuning parameters of the controller are the matrices $K$ and $L$. They can be determined independently by solving two different optimization problems. The state feedback gain $L$ is determined by solving the deterministic problem to compute

$$V^*(x_0) = \min_u \int_0^{\infty} (x^{\mathrm{T}}Qx + u^{\mathrm{T}}Ru)\,\mathrm{d}t$$

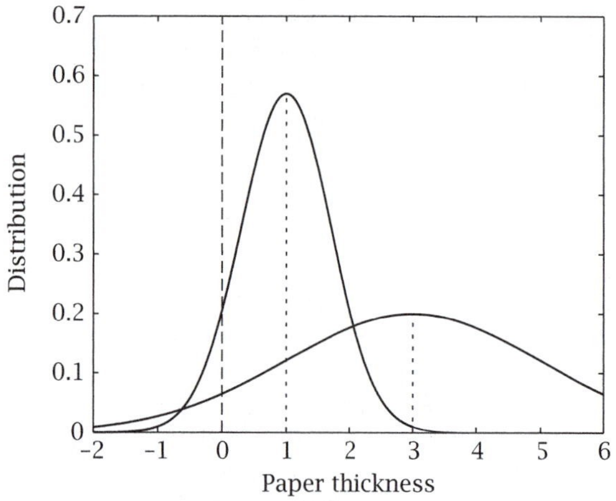


**Figure 11** Linear quadratic Gaussian control aims to minimize the output variance when the input has a given Gaussian distribution. This diagram illustrates two different probability distributions for the paper thickness in a paper machine. When the variance is small, a smaller mean value can be used without increasing the risk of violating quality requirements.

subject to $\dot{x} = Ax + Bu$, $x(0) = x_0$. This problem connects to classical CALCULUS OF VARIATIONS [IV.6] and can be conveniently solved using *dynamic programming*, implying that the optimal cost $V^*(x_0)$ must satisfy *Bellman's equation*:

$$0 = \min_u \Big( \underbrace{x_0^{\mathrm{T}} Q x_0 + u^{\mathrm{T}} R u}_{\text{current cost}} + \underbrace{\frac{\partial V}{\partial x}(A x_0 + B u)}_{\text{reduction of future cost}} \Big). \tag{6}$$

The solution is a quadratic function $V^*(x_0) = x_0^{\mathrm{T}} S x_0$. For linear dynamics and quadratic cost, Bellman's equation reduces to the *Riccati equation*, which is solved for $S$, and the accompanying optimal control law $u = -Lx_0$ gives the desired $L$.

The Kalman filter gain $K$ is determined to minimize the variance of the estimation error $\mathbb{E}|\hat{x} - x|^2$. It turns out that this problem is the dual of the state feedback problem and can also be solved by dynamic programming. In this way, the linear quadratic Gaussian optimal control problem was completely solved in the 1960s. With the terminology of section 6, this provided a solution to the $H_2$-norm optimization problem. Moreover, the structure of the observer-based controller is important regardless of optimality aspects, since it gives an interpretation to all the controller states. This is useful in, for example, diagnosis and fault detection.

In the 1980s it was proved that $H_\infty$-optimal controllers can be derived in a way that is similar to the way in which linear quadratic Gaussian controllers are derived. However, the previous minimization has to be replaced by a dynamic game:

$$V^*(x_0) = \min_u \max_w \int_0^\infty (x^{\mathrm{T}} Q x + u^{\mathrm{T}} R u - \gamma^2 w^{\mathrm{T}} w)\,\mathrm{d}t.$$

The input $u$ should therefore be selected to minimize the cost under the assumption that the disturbance $w$ acts in the worst possible way. If $V^*(x_0)$ is finite, it means that the resulting control law makes the $H_\infty$-norm of the map from $w$ to $z = (Q^{1/2}x, R^{1/2}u)$ smaller than $\gamma$. The optimum is found by iteration over gamma.

In many control applications it is necessary to put hard constraints on states and input variables. This results in nonlinear controllers, which cannot be represented by a transfer matrix. Instead, it is common to implement such controllers using *model-predictive control* (MPC), solving constrained optimization problems repeatedly in real time: at every sampling time, the state is measured and an optimal trajectory is computed for a fixed time horizon into the future. The computed trajectory determines the control action for the next sampling interval. After this, the state is measured again and a new trajectory is optimized. The method is therefore also called *receding-horizon control* (see figure 13).

MPC controllers have been used in the process industry since the 1970s, but the development of theory and supporting software has been most rapid during the past fifteen years.

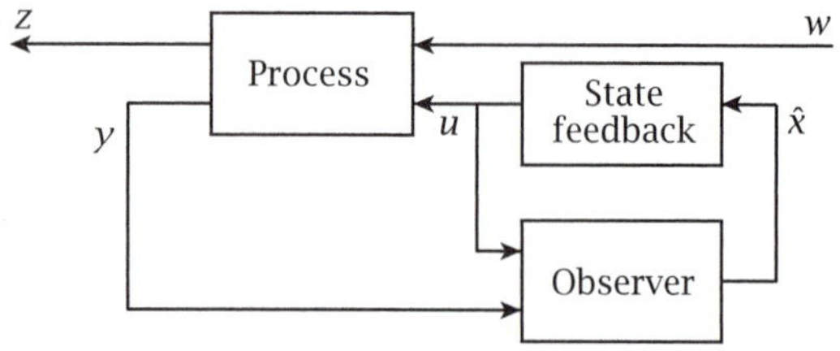


**Figure 12** The solution of the linear quadratic Gaussian optimal control problem has a very clean structure, where the controller has the same number of states as the process model. Every controller state $\hat{x}_k$ can be interpreted as an estimate of the corresponding state $x_k$ in the process. The optimal input is computed from the state estimates as if they were measurements of the true state.

## 8 Nonlinear Control

A remarkable feature of feedback is that it tends to reduce the effects of nonlinearities. In fact, one of the

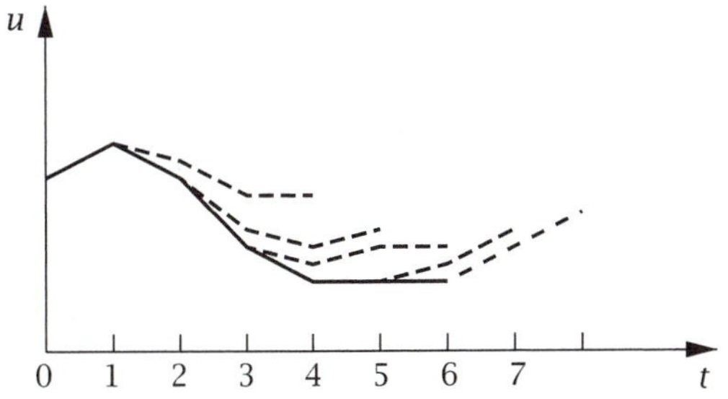


**Figure 13** In MPC a finite-horizon optimal control problem is solved at every time step. The state is measured and an optimal control sequence (dashed) is computed for a fixed time horizon into the future. The computed trajectory determines the control action (solid) for the next sampling interval. After this, the state is measured again and a new trajectory is optimized.

most important early success stories of feedback control was Black's invention of the feedback amplifier in 1927. The invention made it possible to build telecommunication lines over long distances by reducing the inevitable nonlinear distortion in the amplifiers.

For this reason, much of control theory has been developed based on linear models. Even though nonlinear effects are always present, they can often be ignored in the design of feedback controllers, since they tend to be attenuated by the feedback mechanism. Nevertheless, when high performance is required, nonlinear effects need to be considered.[1]

The most straightforward approach to dealing with nonlinearities is to ignore them at the design stage but still take them into account in the verification of the final solution. Usually, verification is done by extensive simulations using nonlinear models of high accuracy. However, it is also possible to do more formal analysis using a nonlinear process model. We will next describe some tools for this purpose.

The most well-known concept for the analysis of nonlinear systems is *Lyapunov functions*, i.e., nonnegative functions of the state $x$ that are decreasing along all the trajectories of the system. The existence of such a function proves stability of the nonlinear system. However, it is often desirable to also go beyond stability and prove bounds on the input–output gain from disturbance $w$ to error $z$. This can be done using the slightly more general concept of *storage functions*. For example, to prove that the output integral $\int |z|^p \,\mathrm{d}t$ is bounded by the input integral $\int |w|^p \,\mathrm{d}t$ along all trajectories starting and ending at an equilibrium, it is sufficient to find a storage function $V(x)$ such that

$$\frac{\mathrm{d}}{\mathrm{d}t}V(x(t)) \leqslant |w(t)|^p - |z(t)|^p$$

along all trajectories. Given $V$, the inequality can often be verified algebraically for each $t$ separately.

Lyapunov functions and storage functions can often be interpreted as measures of energy content in the system. A system is stable if the energy content decreases along all trajectories. Similarly, if the input integral measures the amount of energy that is injected into the system through the input, and the output integral measures how much energy is extracted through the output, then the input–output gain can be at most 1. Such interpretations are particularly useful when the states of the system have physical meaning.

A common way to derive Lyapunov functions and storage functions is via linearization. If a linear feedback law is designed based on a linearized process model, it often comes together with a quadratic Lyapunov function. For example, if it was obtained by optimization of a quadratic criterion, then the optimal cost defines a quadratic Lyapunov function. If an $H_\infty$-optimal control law was obtained by solving a min–max quadratic game, then the solution comes with a corresponding quadratic storage function. In both cases, the quadratic functions obtained in the linear design can be used for verification of a closed-loop system involving the nonlinear process. An alternative approach is to abandon the linear controllers and instead use the quadratic Lyapunov/storage functions as starting points for the design of nonlinear feedback laws.

Nonlinear controllers can also be designed using dynamic programming and Bellman's equation. A common problem formulation is to associate every control law $u = \mu(x)$ with a cost

$$V^\mu(x_0) = \int_0^\infty \ell(x(t), u(t))\,\mathrm{d}t,$$

where $\dot{x} = f(x, u)$, $u = \mu(x)$, and $x(0) = x_0$, and then to find the control law that gives the minimal cost $V^*(x_0)$. The corresponding Bellman equation is generally very difficult to solve, but it can still be used to derive approximatively optimal control laws. For example, given a function $V$ and a control law $\mu$, if the "Bellman inequalities"

$$\ell(x, \mu(x)) + \frac{\partial V}{\partial x} f(x, \mu(x)) \leqslant 0 \leqslant \alpha\ell(x, u) + \frac{\partial V}{\partial x} f(x, u)$$

1. Black's feedback amplifier traded power for reduced distortion. However, modern efforts to build energy-efficient electronics are pushing researchers in the opposite direction: not to give away power losses but instead to accept some level of nonlinear distortions.

hold for all $x$, $u$, and for some $\alpha \geqslant 1$, then $\mu$ approximates the optimal control law in the sense that $V^{\mu}(x) \leqslant \alpha V^{*}(x)$.

The statement is very general and widely applicable. If $\alpha = 1$, the inequalities reduce to the classical Bellman equation. Conversely, if $\alpha$ is big, then the inequalities are easier to satisfy, but the control law is likely to be further away from the optimum. The main difficulty is, in general, to come up with good candidates $\mu$ and $V$ for which the Bellman inequalities can be verified. Two main approaches have been discussed. One is to use linearization, which gives solutions that are also approximately optimal for the nonlinear problem. The second approach is to use MPC: the infinite-horizon cost can be approximated by a finite-horizon time-discrete cost, which can be optimized online. This means that $\mu$ is defined implicitly in terms of solutions to finite-dimensional optimization problems. The Bellman inequalities can then be used to form conclusions about the performance of the MPC controller.

## 9 Ongoing Research: Distributed Control

A new trend in control theory has emerged in the twenty-first century. The aim is to provide understanding and methodology for the control of large-scale systems, such as the power grid, the Internet, and living organisms. Classical control theory is insufficient for several reasons.

- Control action is taken in many different locations, but with only partial access to measurements and process dynamics (see figure 14).
- Even if optimal distributed controllers are computable, they are generally extremely complex to implement.
- The design and functionality of control and communication systems are increasingly intertwined with complex software engineering.

All of these issues have been recognized and discussed for a long time. However, encouraging progress has recently been made, leading to very stimulating developments in control theory.

Dual decomposition is an old idea for the solution of large-scale optimization problems. The idea is, for example, applicable to the minimization of an objective function with a large number of terms that are coupled by relatively few shared variables. The method means that the coupling between shared variables is removed and is replaced by a penalty for disagreement. The penalty is given by a price variable, also called a dual variable or Lagrange multiplier, which is then updated using a gradient algorithm.

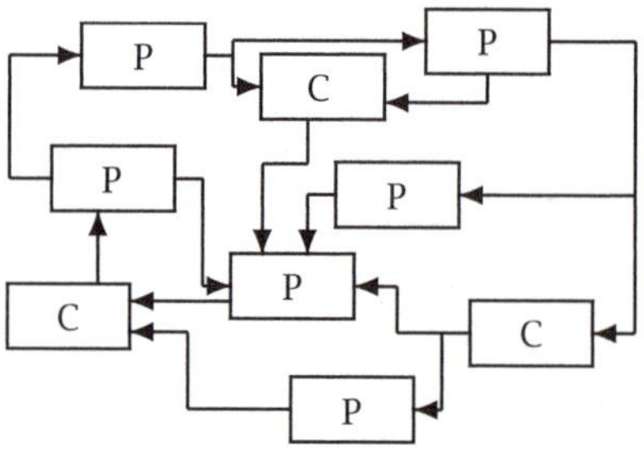


**Figure 14** The term "distributed control" refers to a situation in which action is taken in many different locations but with only partial access to measurements and process dynamics. In the diagram, "P" denotes fixed process dynamics, while "C" denotes controllers to be designed.

Dual decomposition has the important advantage that once the price variables are fixed, individual terms of the objective function can be optimized independently with access to only their own local variables. The need for communication in MPC applications is significant, since gradient updates of price variables requires comparison of shared variables along the entire optimization horizon. Nevertheless, the idea has shown great promise for use in the control of large-scale systems.

A second branch of research is devoted to problems with limited communication capabilities in the controller. It was already recognized by the 1960s that such problem formulations often lead to highly complex nonlinear controllers, even for very simple processes. The reason for this is that there is an incentive for the controller to use the process as a communication device. This incentive disappears as soon as the controller has access to communication links that are faster than the information transfer of the plant, an assumption which is often reasonable with modern technology. Hence, an interesting branch of research is now the investigating of how to derive optimal (linear) controllers when communication is limited but is faster than the process.

A third research field is devoted to a particular class of systems known as positive systems, or (in the nonlinear context) monotone systems. Positive systems arise, for example, in the study of transportation networks and vehicle formations. Unlike general linear systems, the positivity property makes it possible to do stability analysis and control synthesis using linear Lyapunov functions rather than quadratic ones. This makes a dramatic difference for large-scale systems, since the number of optimization parameters then grows linearly with the number of states and system components. Moreover, verification of the final solution can be done in a distributed way, where bounds on the input–output gain can be checked component by component.

Altogether, these different directions of progress (and others) show that control theory is still in a very active phase of development, and an article like this written ten years from now would probably include important sections that go far beyond what has been presented here.

**Acknowledgments.** The illustration in figure 11 is from Åström and Wittenmark (2011), and the bicycle example is from Åström et al. (2005).

### Further Reading

Åström, K. J., R. E. Klein, and A. Lennartsson. 2005. Bicycle dynamics and control. *IEEE Control Systems Magazine* 25(4):26–47.

Åström, K. J., and R. M. Murray. 2008. *Feedback Systems: An Introduction for Scientists and Engineers.* Princeton, NJ: Princeton University Press.

Åström, K. J., and B. Wittenmark. 2011. *Computer-Controlled Systems.* New York: Dover.

Bellman, R. E. 1957. *Dynamic Programming.* Princeton, NJ: Princeton University Press.

Doyle, J. C., B. A. Francis, and A. R. Tannenbaum. 1990. *Feedback Control Theory.* London: Macmillan.

Khalil, H. K. 1996. *Nonlinear Systems*, 2nd edn. Upper Saddle River, NJ: Prentice Hall.

Rawlings, J. B., and D. Q. Mayne. 2009. *Model Predictive Control: Theory and Design.* Madison, WI: Nob Hill Publishing.

# IV.35 Signal Processing

*John G. McWhirter and Ian Proudler*

## 1 Introduction

Since the advent of integrated circuit technology and the resulting exponential surge in the availability of affordable, high-performance, digital computers, digital signal processing (DSP) has become an important topic in applied mathematics. The purpose of this article is to provide a simplistic overview of this topic in order to illustrate where and why mathematics is required.

Since most naturally occurring signals, such as electromagnetic waves, seismic disturbances, or acoustic sounds, are continuous in nature, they must be sampled and digitized prior to digital signal processing. This is the role of an analog-to-digital converter, which takes the output from a sensor such as an antenna or a microphone and produces a sequence of uniformly sampled values $x(n)$ $(n = 0, 1, 2, \dots)$ representing the measured signal at the corresponding time instances $t_n$.

Low-frequency signals are usually digitized directly, resulting in a sequence of real numbers. High-frequency signals of a sinusoidal nature are often represented in terms of their phase and amplitude relative to a given high-frequency reference tone (pure sinusoid), resulting in a signal of much lower frequency, whose values are represented by complex numbers. This process, generally referred to as *down conversion*, is particularly important for electromagnetic signals in radio communications, for example.

The rate at which a signal is sampled depends on its spectral content and, in particular, on the highest-frequency component that it contains (possibly after down conversion to a lower-frequency band). Assuming that the highest frequency, measured in hertz (cycles per second), is $f$, the signal must be sampled every $\Delta t$ seconds, where $\Delta t \leqslant 1/(2f)$. This fundamental rule was established by Shannon, and the corresponding sampling frequency $f_s$ must satisfy $f_s \geqslant 2f$ (the *Nyquist rate*). Having decided on (or been restricted to) a sampling frequency $f_s$, it is important that any signal to be digitized at this rate does not contain any frequency components for which $f > f_s/2$, since these will be indistinguishable from their cyclic counterparts, i.e., those with the same frequency mod $f_s$. This cyclic wraparound in the frequency domain is referred to as *aliasing*, and it can cause severe signal distortion. To prevent this from occurring in practice, signals must be strictly band limited before analog-to-digital conversion.

In accordance with the discussion above, the input to a DSP device is a sequence of signal values $x(n)$, generally referred to as a time series, which may be processed on a sample-by-sample basis or in convenient blocks depending on the application.

## 2 Digital Finite Impulse Response Filtering

One of the most important tasks in DSP is the application of a particular linear time-invariant (LTI) operator $A$ that is defined in terms of its finite-length response to a unit impulse $[1, 0, \dots]$. The impulse

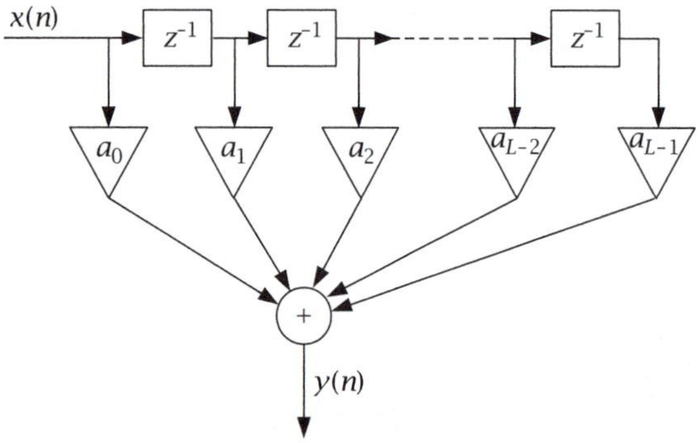


**Figure 1** A finite impulse response filter structure based on a tapped delay line, where $z^{-1}$ denotes a unit delay.

response may be expressed as $A \equiv [a_0, a_1, \ldots, a_{L-1}]$. In response to a general open-ended input time series $x(n)$, $n \in \mathbb{Z}$, it produces the output sequence $y(n) = \sum_{l=0}^{L-1} a_l x(n-l)$. In other words, it generates a convolution of the input sequence with its impulse response. In DSP, this type of operator is referred to as a *finite impulse response* (FIR) filter. Elements $a_l$ of the impulse response are therefore regarded as filter coefficients, the *order* of the filter being $L-1$. The reason for thinking of this LTI operator as a filter will become apparent shortly.

A digital FIR filter may be implemented quite naturally using a tapped delay line, as illustrated schematically in figure 1. The tapped delay line is assumed to be filled with zeros initially. Sample $x(0)$ arrives at time $t_0$ and is multiplied by coefficient $a_0$ to produce $y(0)$. The sample $x(0)$ is stored and moves one place along the delay chain on the next clock cycle, where it is multiplied by coefficient $a_1$. At the same time, the next sample $x(1)$ arrives. It is multiplied by $a_0$, and the product $a_0x(1)$ is added to $a_1x(0)$, thus generating $y(1)$. The process continues on every clock cycle, even when the tapped delay line is full, in which case one sample (the oldest) is forgotten every time. In effect, the stored data vector shifts one place to the right every sample time and it is this property that will lead to a TOEPLITZ DATA MATRIX [I.2 §18], as discussed in section 8.

Let the input to the filter $A$ take the form of a uniformly sampled sinusoid $x(n) = \mathrm{e}^{\mathrm{i}\omega n}$, where, for convenience, $\omega = 2\pi f$ is the angular frequency in radians per sample time. It follows that the output sequence $y(n) = \sum_{l=0}^{L-1} a_l \mathrm{e}^{\mathrm{i}\omega(n-l)}$ can be written as $y(n) = A(\omega)\mathrm{e}^{\mathrm{i}\omega n}$, where $A(\omega) = \sum_{l=0}^{L-1} a_l \mathrm{e}^{-\mathrm{i}\omega l}$ is referred to as the frequency response of the filter $A$. $A(\omega)$ is, of course, the Fourier transform of the finite discrete sequence of filter coefficients, and its inverse is given by

$$a_l = \frac{1}{2\pi}\int_{-\infty}^{\infty} A(\omega)\mathrm{e}^{\mathrm{i}\omega l}.$$

By taking the Fourier transform of the filter output sequence $y(n)$ as defined above, it is easy to show that

$$Y(\omega) = \sum_{l=0}^{L-1} a_l \sum_{n=-\infty}^{\infty} x(n-l)\mathrm{e}^{-\mathrm{i}\omega n}$$

and, hence, by substitution of indices, that $Y(\omega) = A(\omega)X(\omega)$. This is, of course, just a manifestation of the FOURIER CONVOLUTION THEOREM [II.19], which states that the Fourier transform of a convolution is simply the product of the Fourier transforms of the individual sequences. Clearly, then, the effect of convolving a digital sequence with a chosen set of filter coefficients is to modify the frequency content of the signal by multiplying its spectral components by those of the filter. For example, a low-pass filter is designed to multiply the high-frequency components by zero and the low-frequency components by one to a suitably high degree of approximation. With an FIR filter of the type outlined above, a large number of filter coefficients may be required to achieve the accuracy required for low-pass, high-pass, or band-pass filters, so digital filters often utilize infinite impulse response filters, as discussed in section 5.

## 3 Discrete Fourier Transform and Fast Fourier Transform

Section 2 introduced the Fourier transform of a finite discrete sequence $a_l$ as $A(\omega) = \sum_{l=0}^{L-1} a_l \mathrm{e}^{-\mathrm{i}\omega l}$, which describes, for example, the frequency response of the corresponding FIR filter. This is defined for any frequency $\omega$. However, due to the finite length of the sequence, not all values of $A(\omega)$ are independent. In fact, the entire response may be inferred from the response at the $L$ discrete frequency values $\omega_n = 2\pi n/L$ $(n = 0, 1, \ldots, L-1)$, i.e., in terms of the *discrete Fourier transform* (DFT) values

$$A_n \equiv A\left(\frac{2\pi n}{L}\right) = \sum_{l=0}^{L-1} a_l W_L^{-nl},$$

where $W_L \equiv \mathrm{e}^{2\pi\mathrm{i}/L}$ constitutes the $L$th root of unity in the complex number field. The original sequence $a_l$ can easily be recovered from these values (referred to as the DFT coefficients) by means of the inverse DFT, which is given by

$$a_l = \frac{1}{L}\sum_{n=0}^{L-1} A_n W_L^{nl} \quad (l = 0, 1, \ldots, L-1).$$

The inverse DFT can easily be proved by substituting for $A_n$ and noting that the geometric series sums to $L\delta_{lm}$, where $\delta_{lm}$ denotes the Kronecker delta, which takes the value unity when $l = m$ and zero otherwise.

In section 2 it was pointed out that the Fourier transform of the convolution of a discrete sequence with a set of filter coefficients is equivalent to the product of their individual Fourier transforms. A similar result applies in the case of fixed-length discrete sequences transformed using the DFT but the equivalence involves a cyclic convolution. Let $H_n$ and $X_n$ be the DFT coefficients corresponding to the finite-length sequences $h_k$ and $x_k$, and define $Y_n = H_nX_n$ $(n = 0, 1, \dots, L-1)$ so that

$$y_l = \frac{1}{L}\sum_{n=0}^{L-1} X_n W_L^{ln} \sum_{k=0}^{L-1} h_k W_L^{-kn}.$$

Swapping the order of the two summations leads to

$$y_l = \sum_{k=0}^{L-1} h_k \sum_{n=0}^{L-1} X_n W_L^{n\langle l-k\rangle_L},$$

where for any integer $r$, $\langle r\rangle_L$ denote $r \bmod (L-1)$. Consequently, $y_l = \sum_{k=0}^{L-1} h_k x_{\langle l-k\rangle_L}$, which takes the form of a cyclic convolution. This proves to be a very useful relationship, although, in order for it to apply to a general (noncyclic) convolution, it is necessary to use a technique such as zero padding, whereby the input sequence of length $M$, say, is extended to length $2M-1$ by adding new zero elements so that the output of the cyclic convolution contains, as a subsequence, the $M$ terms of the original noncyclic convolution. At first sight one might ask why one bothers doing a finite-length convolution in this roundabout manner. The answer, of course, lies in the ubiquitous FAST FOURIER TRANSFORM [II.10] (FFT), which, for a transform of length $N$, where $N = 2^n$, requires only $O(nN)$ arithmetic operations as opposed to $O(N^2)$ for the standard DFT. This reduction is so significant for large values of $N$ that in order to compute a convolution of length $M$, which would normally require $O(M^2)$ arithmetic operations, it is well worth the overhead of zero padding the sequence so that $M \to 2^m$, computing the FFT of both convolution sequences, multiplying the resulting FFT coefficients pairwise, and computing the inverse FFT. In this context it should be noted that the FFT is not just an efficient algorithm for transforming to the frequency domain in the classical sense but is also a very important "computing engine," finding application throughout the field of DSP and in the practical application of LTI operators more generally.

## 4 The $z$-Transform

A fundamental and widely used mathematical tool in DSP is the $z$-transform, which includes the Fourier transform as a special case but has much broader relevance. The $z$-transform of a time series $x(n)$ is denoted by $X(z)$ and is defined as $X(z) = \sum_{-\infty}^{\infty} x(n)z^{-n}$, where $z \in \mathbb{C}$. If $z$ is represented as $z = re^{i\theta}$, then it is evident that $X(z)$ is equivalent to the Fourier transform of the associated sequence $x(n)r^{-n}$, which converges provided $\sum_{-\infty}^{\infty} |x(n)r^{-n}| < \infty$. In effect, the sequence of values $x(n)$ constitutes the coefficients of a Laurent series expansion of $X(z)$ about the origin in the complex plane. Clearly, the $z$-transform of a sequence is meaningful only at values of $z$ for which the doubly infinite sum converges. This defines the *region of convergence* (ROC), which can be analyzed by splitting the transform into the sum of its *causal* component, involving only the coefficients for which $n \geqslant 0$, and its *anticausal* component, involving only coefficients for which $n < 0$. In general, the causal component converges at values of $z$ for which $|z|$ is large enough, i.e., $|z| > r_c$ for some value of $r_c$. Conversely, the anticausal component converges provided $|z|$ is small enough, i.e., $|z| < r_a$ for some value of $r_a$. Provided $r_c < r_a$, the ROC constitutes an annulus centered on the origin of the complex plane. If the unit circle lies within the ROC, then the $z$-transform evaluated on the unit circle is identical to the Fourier transform. It is worth noting that the inverse $z$-transform is given by

$$\frac{1}{2\pi i}\oint_C X(z)z^{n-1}\,dz,$$

where $C$ denotes a counterclockwise closed contour within the ROC that encloses the origin. Its validity can be demonstrated using the CAUCHY INTEGRAL THEOREM [IV.1 §7], which states that

$$\frac{1}{2\pi i}\oint_C z^{n-1}\,dz = \begin{cases} 1 & \text{if } n = 0, \\ 0 & \text{otherwise.} \end{cases}$$

Note that the $z$-transform is a linear operator. Note also that, if $y(n) = x(n-m)$ for a fixed integer $m$, i.e., the sequence $y(n)$ is simply the sequence $x(n)$ delayed by $m$ sample intervals, then $Y(z) = z^{-m}X(z)$. For this reason, $z^{-1}$ is generally referred to as the unit delay operator. The final property noted here is particularly useful and constitutes a generalization of the Fourier convolution theorem noted earlier. It states that, if two sequences $y(n)$ and $x(n)$ are related according to $y(n) = \sum_{l=0}^{L-1} a_l x(n-l)$, then $Y(z) = A(z)X(z)$. In other words, the $z$-transform of a discrete convolution

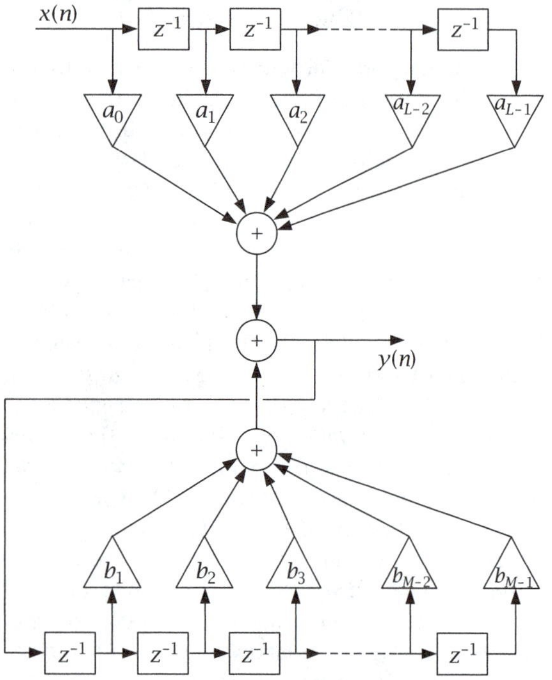


**Figure 2** A digital infinite impulse response filter composed of two tapped delay lines. The second one filters the *output* sequence and feeds the result back to be added to the output of the first.

is the product of the individual $z$-transforms. Although $L$ can be infinitely large, we have chosen to express the relationship here as it applies to an FIR filter with coefficients $a_l$ $(l = 0, 1, \dots, L-1)$. For this reason, an FIR filter is often represented for analytic purposes by its $z$-transform, which (for causal filters) takes the form of a polynomial in $z^{-1}$.

## 5 Digital Infinite Impulse Response Filtering

FIR filters are represented mathematically by a finite discrete convolution (LTI operator). Another important class of digital filters involves a discrete LTI operator with infinite impulse response. It has the form

$$y(n) = \sum_{l=0}^{L-1} a_l x(n-l) - \sum_{m=1}^{M-1} b_m y(n-m).$$

The key point to note here is that the filter output $y(n)$ consists of the filtered input samples, as in an FIR filter, as well as delayed values of the filter output $y(n)$, $n \in \mathbb{N}^+$, linearly combined using the coefficient vector $[b_1, b_2, \dots, b_{M-1}]$. Feeding the output of the filter back in this recursive manner, as illustrated in figure 2, leads to a filter whose response to a unit impulse is no longer finite in duration. As a result, the bounded-input bounded-output stability of these *infinite impulse response* (IIR) filters is no longer guaranteed and careful analysis is required. The IIR filter equation incorporates two distinct FIR filters, one applied to the input sequence and the other applied to the output sequence, before it is fed back. In terms of the associated $z$-transforms it may be expressed in the form $B(z)Y(z) = A(z)X(z)$ (where $b_0 = 1$), so $Y(z) = T(z)X(z)$, where the overall transfer function is given by the rational form $T(z) = A(z)/B(z)$. The zeros of the polynomial $A(z)$ are referred to as the zeros of the filter, while the zeros of $B(z)$ are referred to as the poles of the filter. The location of the poles in the complex plane is a key property for characterizing the ROC of $T(z)$. Taken together, the location of the poles and zeros plays an important role in determining the characteristics of an IIR filter.

The design of a digital filter, either FIR or IIR, intended to achieve specific objectives such as given pass-band frequencies, stop-band frequencies, and roll-off rates is now a well-established discipline. It is not possible to design a filter of finite order so that it has arbitrary properties. The approach therefore basically amounts to curve fitting. Techniques such as least-squares minimization, linear programming, minimax optimization, and Chebyshev approximation can all be used.

## 6 Correlation

So far in this article we have considered only deterministic signals, whereas, in practice, most measured signals are subject to random measurement noise and other statistical fluctuations. In the simplest case these can be modeled as an *additive white Gaussian noise* process with zero mean, so the measured sequence is described by $y'(n) = y(n) + v_n$, where $v_n$ represents a random noise sequence whose values are taken from a Gaussian distribution with $E[v_n] = 0$ and $E[v_i v_j^*] = \sigma^2 \delta_{ij}$. Throughout this article, $E[\cdot]$ denotes the statistical expectation operator.

Correlation is another important procedure that arises in DSP. The correlation between two open-ended random time series $x(k)$ and $y(k)$ is given by $C_{xy}(n) = E[x(k)y^*(n+k)]/\alpha_x\alpha_y$, where $\alpha_x = E[|x(k)|^2]^{1/2}$ and $\alpha_y = E[|y(k)|^2]^{1/2}$ are normalization factors here assumed to equal unity. The correlation is a function of the relative shift in sample time between the two sequences and, making the usual assumptions about ergodicity and wide-sense stationarity (stationarity over a suitably long time interval), it is estimated as $c_{xy}(n) = \sum_{k=-K}^{K} x(k)y^*(n+k)$, where the value of $K$ is set as large as necessary (or as large as possible in any practical situation). The specific case where the sequence $y(k)$ is identical to the sequence $x(k)$ yields the autocorrelation function $C_{xx}(n)$ with its corresponding estimate $c_{xx}(n) = \sum_{k=-K}^{K} x(k)x^*(n+k)$. The autocorrelation function is of particular interest since the Wiener–Kinchene theorem states that the Fourier transform of the autocorrelation function is equivalent to the power spectral density of the signal, i.e., $C_{xx}(\omega) = |X(\omega)|^2$, where $X(\omega)$ denotes the Fourier transform of $x(k)$. Assuming that the summation limits are infinite, this can easily be deduced as a simple corollary of the Fourier convolution theorem. The Wiener–Kinchene theorem is useful as there are numerous situations in DSP where it is simpler, or more convenient, to measure the autocorrelation function than the spectrum of the underlying signal. Note, however, that the power spectral density does not provide any phase information.

The case where an open-ended time series $y(k)$ is to be correlated with a finite sequence $x(k)$ leads to a correlation function estimate of the form $c_{xy}(n) = \sum_{k=0}^{L-1} x(k)y^*(n+k)$, where the number of sample values for each $n$ is simply the length $L$ of the finite sequence. Even if the value of $L$ is limited, this expression provides an unbiased estimator for the correlation function and so is very useful in practice. For a single instantiation of the sequence $y(k)$, it can be viewed as a measure of the similarity between the finite sequence $x(k)$ and a finite segment of the sequence $y(k)$ beginning at the $n$th sample. The value of $n$ for which this estimator takes its maximum value is of particular interest in the context of active radar and sonar. In such systems, an electromagnetic or acoustic pulse whose shape is described by the sequence $x(k)$ is transmitted by the antenna at sample time zero, say. After propagating to a reflective object in its path, some of the pulse energy returns to the antenna, delayed by its two-way time of flight and much weaker due to propagation losses. This much weaker pulse is also subject to electronic receiver noise and can be modeled as $y(n) = \beta x(n-m) + \nu_n$, where the round-trip delay is $m$ sample intervals and, as before, $\nu_n$ represents additive white Gaussian noise. In this situation,

$$c_{xy}(n) = \sum_{k=0}^{K-1} [x(k)x^*(n+k-m) + x(k)\nu_{n+k}]$$

and, for a suitably designed pulse, the expected value of $c_{xy}(n)$ will attain its maximum value for $n = m$. This provides an estimate of the time of flight of the pulse, and hence the distance to the corresponding reflector.

Computing the correlation function estimate for this purpose is generally referred to as matched filtering. The reason for this becomes clearer by rewriting the estimator in the form $c_{xy}(n) = \sum_{k=0}^{K-1} x(K-k-l)y(n+K-1-k)$ and introducing the order-reversed sequence $\bar{x}(k) = x(K+1-k)$ so that the estimator is given by $\sum_{k=0}^{K-1} \bar{x}(k)y(n+K-1-k)$. This takes the form of a discrete convolution corresponding to a digital FIR filter with coefficients $\bar{x}(k)$. In effect, the correlation function can be estimated by computing a convolution with the coefficients in reverse order.

## 7 Adaptive Filters

In the case of an adaptive filter, the coefficients are not specified in advance but must be computed as a function of the input data with a view to achieving a particular objective such as maximizing the output signal-to-noise ratio. This typically involves the minimization of a suitable cost function involving the filter coefficients (weight vector) and is sometimes implemented by feeding the filter output back to form a simple "closed-loop" control system.

Consider, for example, a signal estimation problem in which a desired signal $s(n)$ is subject to an unknown filtering operation $H(z)$ such that the signal received is given by $x(n) = \sum h_k s(n-k) + v(n)$, where $v(n)$ denotes an interference or noise signal uncorrelated with $s(n)$. With the aim of estimating the desired signal, the received signal is processed by an adaptive filter $W(z)$ whose output is $d(n) = \sum_{k=0}^{K-1} w_k x(n-k)$. Assume that a noisy version of the desired signal denoted by $y(n) = s(n) + \eta(n)$ is also available, where the noise $\eta(n)$ is again uncorrelated with $s(n)$. The coefficients of the adaptive filter are chosen to ensure that the *reference signal* $y(n)$ and the output of the adaptive filter $d(n)$ are as close as possible in terms of their mean squared error $J = E[|e(n)|^2]$, where $e(n) = y(n) - d(n)$. This approach works because, as we shall see below, the solution involves correlations between certain signals, and the two noise terms, being uncorrelated, have no effect on the result. An adaptive filter of this type is illustrated in figure 3.

Minimizing the cost function $J$ with respect to each of the adaptive filter coefficients (often referred to as

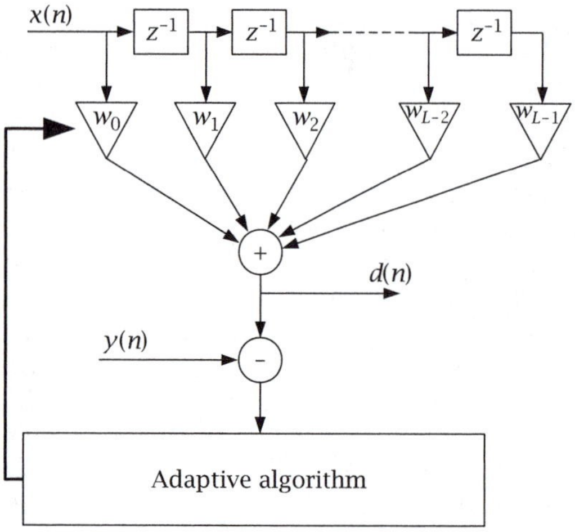


**Figure 3** Schematic of a basic adaptive filter. The difference between the output signal $d(n)$ and the reference signal $y(n)$ is used to compute an update to the weight vector using a suitable adaptive algorithm.

weights) leads directly to the *Wiener–Hopf* equations

$$r(k) = \sum_{j=0}^{K-1} w_j q(k-j) \quad (j = 0, 1, \dots, K-1),$$

where

$$r(k) = E[y(n)x^*(n-k)], \quad q(k) = E[x(n)x^*(n-k)]$$

denote the cross-correlation and auto-correlation functions, respectively.

Because these equations take the form of convolutions, the solution may be written formally in terms of the corresponding $z$-transforms as $W(z) = R(z)/Q(z)$. However, this expression is of little benefit when it comes to evaluating the filter weights in practice. Instead, the Wiener–Hopf equations may be written in the form $Mw = r$, where $w = [w_0, w_1, \dots, w_{K-1}]$ is the weight vector and $M \in \mathbb{C}^{K \times K}$ is termed the covariance matrix, with individual elements given by $m_{ij} = q(i-j)$, and $r \in \mathbb{C}^K$ is known as the cross-correlation vector with elements given by $r_i = r(i)$. This matrix equation may be solved using conventional linear algebra techniques, ideally exploiting the fact that the covariance matrix is of Toeplitz form.

Note that the Wiener–Hopf equations are formulated in terms of the ideal ensemble-averaged cross-correlation and auto-correlation components, which are not known and would have to be estimated from the data in any practical situation. It is generally more appropriate to estimate the filter coefficients directly from the data by defining a cost function ($f(w \mid D)$, say, where $D$ denotes the data). One can then solve $w = \arg\min f(w \mid D)$. Various different cost functions have been proposed. Some have the advantage that a reference signal is not required (*blind adaptive filtering*). For the specific scenario that led to the Wiener–Hopf equations, an appropriate cost function would be $f(w \mid D = [X, y]) = \|Xw - y\|_2$, where $\|x\|_2 = (x^*x)^{1/2}$, in which case $w$ is the solution to the linear system $Xw = y$.

However, note that, as in the signal estimation problem above, it is usually the case that $y$ is a noisy copy of the desired signal, i.e., $y = s + \eta$. Furthermore, as discussed below, it can be advantageous to take more measurements than the number of unknown filter coefficients, in which case there are more equations than unknowns and the matrix $X$ is rectangular. Thus, in general, there is no exact solution to the linear system $Xw = y$ and we have to turn to the LEAST-SQUARES SOLUTION [IV.10 §7.1]. There are many approaches to the least-squares solution of $Xw = y$, the appropriate choice depending on the properties of $X$. If $X^*X$ is of full rank, then $w$ is the unique solution to the linear equations $X^*Xw = X^*y$, which are known as the *normal equations*, and $w = (X^*X)^{-1}X^*y$ is known as the *Wiener solution*. Clearly, when $X^*X$ is not of full rank other methods have to be used to find the least-squares solution.

The adaptive filter problem therefore requires finding the solution to $X^*Xw = X^*y$, where, in the presence of noise, $y = s + \eta$. Here, $s$ is the noise-free signal. The Wiener solution can then be written $w = w_0 + (X^*X)^{-1}X^*\eta$, where $w_0$ is the noise-free least-squares solution. Note that the $i$th element of $X^*\eta$ is $\sum_{k=0}^{K} x^*(k-i+1)\eta(k)$, which is an ergodic estimate of the cross-correlation between $x(n-i+1)$ and $\eta(n)$. In the derivation of the Wiener–Hopf equations above, ensemble-averaged correlations were used, so $E[x^*(n-i+1)\eta(n)] = 0$ and $w = w_0$. However, with ergodic estimates this will not be the case as, in general, $\sum_{k=0}^{K} x^*(k-i+1)\eta(k) \neq 0$ for finite $K$. Clearly, the larger the value of $K$ in the ergodic estimate, the smaller the estimated cross-correlation will be. This is equivalent to taking many measurements of a noisy quantity and averaging to improve the accuracy of the resulting estimate. This is why adaptive filtering problems often have rectangular $X$ matrices.

Given that, in the expression for $w$, $X^*\eta \neq 0$ and it is multiplied by $(X^*X)^{-1}$, it is necessary to consider the CONDITION NUMBER [IV.10 §1] of the matrix $X$ as well as its rank. Inversion of an ill-conditioned matrix can amplify any noise that is present. Here, the noise ($X^*\eta$) comes from the signals and not from roundoff effects. Forming and solving the normal equations has the disadvantage that the 2-norm condition number of $X^*X$ is the square of that of $X$, which can cause the computed normal equations solution to be less accurate than the least-squares problem warrants in the presence of the noise. In this context, one particularly useful approach is to use the QR FACTORIZATION [IV.10 §2] $X = QR$ of the matrix $X$, which allows the calculation of $w$ without forming the covariance matrix $X^*X$ and thus squaring the condition number.

## 8 Implementation

A notable aspect of signal processing is the need to have low-complexity algorithms. This is because signal processing is often to be found on equipment that has limited computing power, such as battery-powered devices. There is therefore great interest in devising algorithms that require as little computation as possible.

The most notable method for reducing the computation in a signal-processing algorithm is to calculate the solution to one problem based on that of another, i.e., to use recursion. For adaptive filters the most widely used method is a recursion in time: calculate the solution at time $n$ from that at time $n-1$. This leads to the *recursive least-squares* (RLS) family of algorithms. The oldest member of this family is derived from the Wiener solution by means of the SHERMAN–MORRISON–WOODBURY FORMULA [IV.10 §3]. This lemma expresses the inverse of the matrix $X^*X$ in terms of the inverse of a smaller matrix. Specifically, the data matrix at time $n$ can be written $X(n) = [X^T(n-1) \quad x(n)]^T$ and it is possible to express $(X^*(n)X(n))^{-1}$ in terms of $(X^*(n-1)X(n-1))^{-1}$ and rank-one terms involving $x(n)$. In fact, when substituted into the Wiener solution $w(n) = (X^*(n)X(n))^{-1}X^*(n)y(n)$, the resulting formula can be simplified to an equation of the form $w(n) = w(n-1) + g(y(n) - x^*(n)w(n-1))$. The vector $g$ is known as the *Kalman gain vector* and is equal to $(X^*(n)X(n))^{-1}x(n)$. It specifies the direction of the update from $w(n-1)$ to $w(n)$. The advantage of this formula is that it requires only $O(N^2)$ operations rather than the $O(N^3)$ required to invert an $N \times N$ matrix. Note that the last term in the update formula can be seen as an a priori "fitting" error. This is a classic recursive update formula that appears in many iterative algorithms, such as the method of STEEPEST DESCENT [IV.11 §4.1] in optimization and the JACOBI ITERATION [IV.10 §9] for linear systems.

An important signal-processing algorithm in this class is the *least-mean-squares* (LMS) algorithm. The LMS algorithm can be derived in several ways. The original approach was statistical and was based on minimizing the mean square error. Another approach is to set the a posteriori fitting error $y(n) - x^*(n)w(n)$ to zero while minimizing the modulus of the change in the weight vector $|w(n) - w(n-1)|$. In any event, the LMS update formula is $w(n) = w(n-1) + x(n)(y(n) - x^*(n)w(n-1))$. Note that this is the same as the RLS update when the Kalman gain $g = x(n)$, that is, when $X^*(n)X(n) = I$. The LMS algorithm therefore behaves like the RLS algorithm when the input signal $x(n)$ is a white noise process. When the input signal does not have a white power spectral density, the LMS algorithm is slower to converge than the RLS algorithm. This can be shown to be related to the fact that $x(n)$ is now a poor estimate of the direction of steepest descent, unlike the Kalman gain vector. Nevertheless, the LMS algorithm is very simple to implement and has proved to be very robust to violations of the assumptions that were originally made in its derivation. In fact, the LMS algorithm can be derived using *minimax* optimization, where one minimizes the maximum possible error (for a given class of system). Broadly speaking, the LMS algorithm minimizes the energy of the a priori fitting errors given the worst-case input noise. This can be shown to be related to a concept known as the $H_\infty$-norm, which is the $L_\infty$-norm applied in the frequency domain: given a transfer function $T(z)$, $\|T(z)\|_{H_\infty} = \sup_\omega(|T(e^{i\omega})|)$. As such, the LMS algorithm is a very useful tool. It is worth noting that the QR factorization-based least-squares algorithm family also includes time-recursive versions, although these will not be discussed further in this article.

Another approach to reducing the computational load is to utilize any special structure in the problem. In adaptive filtering the rows of the data matrix $X$ are formed from the delayed signal values $x(n-l)$ that appear in the convolution that defines the filter, i.e., $y(n) = \sum_{l=0}^{L-1} a_l x(n-l)$. Each row of the data matrix corresponds to a different value of $n$. Hence the data matrix has a Toeplitz form. Thus, if $X_p(n)$ is the data matrix for a $p$th-order filter at time $n$, then

$$X_p(n) = [X_{p-1}(n), b_p(n)] = [f_p(n), X_{p-1}(n-1)],$$

where $b_p(n)$ and $f_p(n)$ depend on the data. In fact, $f_p(n)$ and $b_p(n)$ are related to specialized adaptive filtering problems known as *forward and backward linear prediction*, respectively. Here, the desired signal $y(n)$ is replaced by $f_p(n)$ or $b_p(n)$. This leads to a rank-two update equation for the covariance matrix. It is possible to incorporate time recursion with the order recursion resulting in an algorithm with $O(n)$ operations. Examples of this are the fast transversal filters algorithm and the RLS lattice algorithm.

One issue with all of the recursive algorithms is NUMERICAL STABILITY [I.2 §23]. The recursion formulas are valid only if all of the inputs are accurate. Inaccurate input values and numerical roundoff errors lead to errors in the output values that are then to be used as inputs in the next iteration. This can lead to explosive divergence of the algorithm. Because of the complexity introduced by recursion, the numerical stability of signal-processing algorithms is often explored through computer simulation. There has been relatively little theoretical work but it has led to some provably stable recursive algorithms as well as provably unstable ones.

## 9 Channel Equalization

Adaptive filters find application in many situations, such as echo cancelation, modern hearing aids, and seismology, but most notably they are used in modern digital communications systems. Broadly speaking there are two uses: system identification and system inversion. In system identification, given a system $G(z)$ we seek to find a filter $H(z)$ that minimizes $\|G(z)X(z) - H(z)X(z)\|$ for some fixed signal $X(z)$. That is, for the given $X(z)$, $H(z)$ behaves like $G(z)$. In system inversion, we seek to find a filter $H(z)$ that minimizes $\|H(z)G(z)X(z) - X(z)\|$. That is, $H(z)$ undoes the effect of $G(z)$. It is usual that the filter $H(z)$ is to be drawn from a given class of filters. Typically, the class of FIR filters is used as this leads to tractable algorithms, but see the discussion below on *decision feedback equalizers*.

System identification is used to investigate a real-world system by generating a model for it (the filter). The properties of the model (the position of poles and zeros, stability, etc.) can be used to infer the corresponding properties of the system. An example of this is system monitoring. If a set of models is generated over time, changes in the system can be detected. Ideally, deleterious changes can be detected before they cause a catastrophic failure in the system.

System inversion is more common and is used where the system $G(z)$ is a physical phenomenon that has a detrimental effect on the signal $X(z)$ and we want to mitigate this effect. The most common example of this is the transmission of radio signals. The radio waves can bounce off obstacles situated between the transmitter and the receiver, resulting in the reception of several signals each with different delays relative to one another. As these signals are coherent, the resultant summation can exhibit interference effects that adversely affect the ability of the receiver to recover the information in the radio signal. By incorporating an adaptive filter in the receiver, the effects of the propagation can be reduced. This is known as *channel equalization*. There is an obvious problem with system inversion: the system $G(z)$ may be ill-conditioned or, indeed, it may not have an inverse. In such cases various *regularization* techniques can be used; e.g., replace the least-squares problem $w = \arg\min\{\|Xw - y\|_2\}$ by a constrained one: $w = \arg\min\{\|Xw - y\|_2 \colon \|w\|_2 \leqslant \tau\}$ for some threshold $\tau$. In this case, however, the adaptive filter $H(z)$ cannot completely mitigate the effects of $G(z)$.

A more practical issue is the accuracy with which the system $G(z)$ can be inverted. In general, $G(z)$ can have poles as well as zeros. Its inverse therefore also has poles and zeros (see section 5). Note that even if $G(z)$ had only zeros, which is often the case in practice, its inverse will have poles. This causes an issue because most adaptive filter algorithms are based on an FIR filter; i.e., they have only zeros. Although some adaptive IIR filter algorithms have been developed, they have significant bounded-input bounded-output stability issues. Here we are referring not to numerical stability but to system stability. An IIR filter will be systematically unstable if it has poles outside of the unit circle. Most adaptive IIR filter algorithms are unable to ensure that this does not happen. Although this is mathematically well understood, it is very difficult to achieve in a signal-processing algorithm (often the solution of simultaneous nonlinear equations is required). Fortunately, it is possible to generate an FIR filter with a response that is arbitrarily close to that of an IIR filter, provided the FIR filter has sufficiently many coefficients. Thus, in practice, most system-inversion algorithms will use an adaptive FIR filter algorithm. A notable exception is, again, in the case of the transmission of radio signals. Recall that an IIR filter uses previous outputs $y(n-i)$ to calculate the current output $y(n)$ (see section 5). It is clear that trying to determine the filter coefficients $a_i$ and $b_i$ is going to be difficult since we cannot calculate the previous outputs $y(n-i)$ without already knowing the filter coefficients. However, in a modern radio the signals are digital; that is, they take only discrete values, for example $\pm 1$. Thus, if the filter coefficients $a_i$ and $b_i$ are correct, then $y(n) \in \{\pm 1\}$. When the coefficients $a_i$ and $b_i$ are incorrect but close to the correct values, the output of the filter $y(n)$ will in general not be $\pm 1$ but we will find that $y(n)$ is closer to one of the allowable values (i.e., $\pm 1$) than the other. If we denote the closest allowable value by $d(n)$, then we may assume that $y(n) = d(n)$ is the correct answer. Following this procedure we can replace the feedback of the previous outputs $y(n-i)$ by the "decisions" $d(n-i)$, i.e., $y(n) = \sum_{i=0}^{N} b_i x(n-i) + \sum_{i=1}^{M} a_i d(n-i)$. The calculation of the filter coefficients $a_i$ and $b_i$ is then more tractable. Furthermore, provided the filter coefficients can be initialized sufficiently close to their correct values, any errors caused by incorrect "decisions" do not seem to cause any problems. This type of adaptive IIR filter is known as a *decision-feedback equalizer*.

The decision-feedback equalizer makes "hard" decisions, in that $d(n) \in \{\pm 1\}$. An alternative approach is to make "soft" decisions. Here, $d(n) = Q(y(n))$, where the function $Q$ maps $y(n) \in \mathbb{R}$ to $d(n) \in [-1, 1]$ such that the distance between $d(n)$ and $+1$ or $-1$, as appropriate, is indicative of the "accuracy" of the decision. A soft-decision-feedback equalizer usually works better than a (hard-) decision-feedback equalizer. An obvious choice of the function $Q$ is the LIKELIHOOD FUNCTION [V.11]. Here, one would have to make assumptions about the statistics of the transmitted signal and the receiver noise. Let $y(n) = b(n) + \varepsilon(n)$, where $b(n) \in [-1, 1]$ is the correct value of the transmitted signal and $\varepsilon(n) \in \mathbb{R}$ represents noise and residual errors from incorrect decisions and filter coefficient values. An application of BAYES'S THEOREM [V.11] then gives $P(b(n) \mid y(n)) \propto P(y(n) \mid b(n))$. The term $P(y(n) \mid b(n))$ is easily calculated given the statistics of $\varepsilon(n)$ and is known as the likelihood function. The use of statistical estimation techniques leads to the field of *Bayesian signal processing* (see section 12). This approach has led to very powerful algorithms such as *turbo equalization*. Here, one attempts to estimate the coefficients of an equalizer but it is done iteratively using estimated probability density functions (pdfs) to capture the likely statistics of the parameters.

## 10 Adaptive Beamforming (ABF)

Another important signal-processing operation is the *beamformer*. Time-series filters (section 2) process a single time series and can vary the frequency content of the output signal. Whereas a filter performs a convolution, $y(n) = \sum_{i=0}^{N} w_i x(n-i)$, a beamformer processes separate signals, $y(n) = \sum_{i=0}^{N} w_i x_i(n)$. The input signals $x_i(n)$ for a beamformer come from physically separate sensors, as illustrated in figure 4. A beamformer can vary the content of its output signal according to the direction of arrival of the input signals. To see this, note that the signal picked up by each sensor from a given source is just a delayed version of the source signal. The delay will just be the time it takes for the radio frequency (RF) wave to travel from the source to the sensor. A modulated RF wave at time $n$ can be written $A(n)e^{-i\omega n}$, where $A(n)$ is the signal amplitude and $\omega$ is the angular frequency. The signal received at a sensor is therefore $A(n - d/c)e^{-i\omega(n-d/c)}$, where $d$ is the distance from the source to the sensor and $c$ is the velocity of light in meters per sample time. If, to a good approximation, $A(n - d/c) = A(n)$, the RF signal is called *narrowband* and the signal received by the sensor is just a phase-shifted version of the source signal: $A(n)e^{-i\omega n}e^{i\omega d/c}$. The case when $A(n - d/c) \neq A(n)$ is called *broadband* (see section 11).

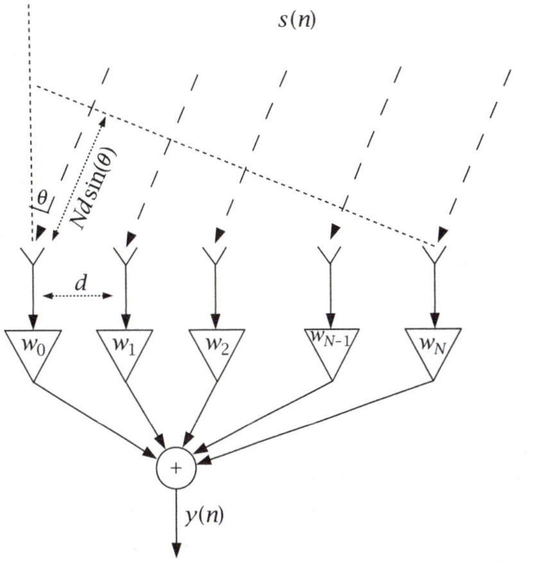


**Figure 4** A schematic of an antenna array beamformer where the elements are uniformly spaced along a straight line. Each of the Y-shaped symbols denotes an individual antenna. The received signals are weighted and combined to produce the array output signal $y(n)$.

In the narrowband case, the beamformer can apply phase shifts to the signal from each sensor, via the

$w_i$, before summing them coherently. So, for example, the beamformer could apply phase shifts that cancel out the phase shifts caused by the relative propagation delays expected for a source in a given direction, in which case any signal from that direction would be summed constructively, leading to a large-amplitude signal. Signals from other directions will tend to partially cancel in the summation and could even sum to zero. In fact, since the output of the beamformer is the sum of $N$ terms, it can be shown that there are $N - 1$ directions for which the beamformer response is exactly zero. These are known as the *nulls* of the beamformer. Thus a beamformer can "filter" signals on the basis of *direction of arrival.* Note that there is a requirement on the position of the sensors that is equivalent to the Nyquist sampling criterion: the sensors need to be close enough together to distinguish the highest spatial frequency present, otherwise one gets *spatial aliasing.*

From the above it is clear that, for a signal from a given direction, the vector of signals from the sensors $x(n)$ is proportional to a vector that is uniquely determined by the direction of arrival. This vector is known as the *steering vector* $a$. The proportionality constant is the instantaneous value of the transmitted signal $s(n)$, i.e., $x(n) = s(n)a$. A common architecture for a beamformer has the sensors equidistant from each other and in a straight line. This is known as a *uniform linear array,* and it is illustrated in figure 4. In this case, the steering vectors look like sampled complex exponentials: $\mathrm{e}^{ind\sin(\theta)/\lambda}$, where $n$ indexes the sensors, $d$ is the sensor separation, $\theta$ is the direction of arrival of the signal, and $\lambda$ is its wavelength. In such circumstances, the term $d\sin(\theta)/\lambda$ is known as the *spatial* frequency. Spatial frequency therefore has a sinusoidal dependency on the direction of arrival of the signal. In general, the steering vectors do not have this simple structure, but references to "spatial frequency" can still be found in the literature and, in general, beamformers are often referred to as *spatial filters.* There has been a lot of work on the design of fixed beamformers. As might be expected, there is much commonality with the design of fixed digital filters (see section 5). However, the emphasis here is mostly on controlling the response to the wanted signal (the *main lobe*) and on rejecting unwanted signals (the *side lobes*).

Like an adaptive filter, an adaptive beamformer calculates its coefficients (or "weights") $w_i$ based on the collected data. The mathematics is virtually identical to that of an adaptive filter (i.e., the least-squares solution to $Xw = y$). Here, too, techniques such as time recursion (see section 8) are used to reduce the computation required, but the data matrix $X$ is no longer Toeplitz, so there are no "fast" ABF algorithms. Another difference is the expected signal content. In an adaptive filter scenario one usually sees a significant (contiguous) range of frequencies present in the signal. In a beamforming application, one expects to see only a few discrete directions of arrival present. This leads to some algorithms that are specific to the beamforming application.

In some situations, the spectrum of the data covariance matrix $X^*X$ will break into two distinct sets: a set of large eigenvalues and a set of small ones. The large eigenvalues, or, more specifically, the associated eigenvectors, correspond to a subspace that contains information about the signals, while the small eigenvalues correspond to noise. This allows, for example, the noise to be rejected by means of an orthogonal projection of the data onto the "signal" subspace. This technique is in fact an example of PRINCIPAL-COMPONENT ANALYSIS [IV.17 §4] (PCA). In addition, by identifying those steering vectors that are orthogonal to the "noise" subspace, estimates of the direction of arrival of the signals can be obtained. This is the basis of the *MUSIC* direction-of-arrival estimation algorithm.

In the above example, the eigenvectors corresponding to the largest eigenvalues span a subspace that contains the signals. Note, however, that the eigenvectors do not correspond to steering vectors and further processing is required to separate the signals from one another. A simple approach is to set up a constrained least-squares problem: $w = \arg\min\{\|Xw - y\| : w^*c = 1\}$. Here, $c$ is chosen so that the response of the beamformer to a signal from a given direction is unity; i.e., for an input signal with steering vector $a$, $x = s(n)a$, we want $w^*x = s(n)$ so that $w^*a = 1$ and hence $c = a$. By trying all directions of arrival (or at least a finite set), the individual signals can be recovered. This technique is known as *minimum-variance distortionless-response* (MVDR) beamforming. Note that an MVDR beamformer is a mixture of an adaptive beamformer (the least-squares minimization) and a fixed one (the constraint). It is often advantageous to extend this concept and add further constraints to control other features of the beampattern, e.g., a derivative constraint to control the width of the main lobe in the desired "look direction." One issue with MVDR beamforming is inaccuracies in the steering vectors either due to practical issues (e.g., calibration measurements) or due to the finite search set. If the steering vector is not accurate, then the least-squares minimization can adjust the beampattern to remove the desired signal. Various approaches to mitigating this effect have been investigated, the gradient constraint mentioned above being one of them.

An alternative, non-search-based, technique for recovering the signals is based on *independent-component analysis* (ICA). This is an extension of PCA. If we write $X = SA$, where $S$ is a matrix of signal time series and $A$ is a matrix of steering vectors, it can be shown that PCA results in a set of orthonormal basis vectors $Y = SQ$, where $Q$ is a unitary matrix. If the signals $S$ are not Gaussian, then $S$ can be recovered from $Y$ by finding a unitary transformation $U$ that makes the columns of $YU$ statistically independent ($U = QP$, where $P$ is a permutation matrix). ICA is also known as *blind signal separation* since the matrix of steering vectors $A$ is not known a priori.

## 11 Broadband ABF and Multichannel Filtering

We have seen that temporal filtering and spatial filtering are useful practical operations. In some applications it is necessary to apply both forms of filtering simultaneously, i.e., *joint temporal and spatial filtering*. Examples of this are sonar, space-time adaptive processing (STAP) radar, and multiple-input multiple-output communications. The rationale for using both forms of filtering is the need to control the frequency content of the signal as well as its direction of arrival. This technique is called *broadband beamforming* by people familiar with conventional beamforming. This is because beamforming relies on applying time delays to the received signals. In conventional beamforming, the signals have narrow bandwidths and time delays are equivalent to phase shifts. When the signals are broadband, interpolation filters are required to implement time delays that are not integer multiples of the sample period (see section 1). Therefore, instead of the beamformer output being $y(n) = \sum_{i=0}^{N} w_i x_i(n)$, the products $w_i x_i(n)$ are replaced by filters: $y(n) = \sum_{i=0}^{N} \sum_{j=0}^{M} w_{ij} x_i(n-j)$. The oldest example of broadband ABF is found in *passive sonar*. Here, acoustic underwater signals are detected using arrays of hydrophones. Those familiar with time series filtering refer to joint temporal and spatial filtering as *multichannel filtering*. A multichannel filter is the same as the (scalar-valued) filter outlined in section 7 but with vector-valued time series instead.

Another example of broadband ABF is STAP radar, which is effectively a combination of a *phased-array* radar and a *pulse-Doppler* radar. As we have seen, the former radar is able to separate the radar echoes based on their direction of arrival. The latter can separate the radar echoes based on the velocity of the reflector (i.e., aircraft) by exploiting the Doppler effect. The radar echo from a moving aircraft will be Doppler shifted by virtue of its motion. This frequency shift can be detected by sending out multiple radar pulses and processing the resulting echoes as a time series. A frequency filter can then separate the echoes based on the velocity of the aircraft. One advantage of pulse-Doppler radar is that the echo from a low-flying aircraft can be separated from that due to reflectors on the ground since either the latter will be stationary or it will be moving very slowly with respect to the aircraft. By combining ABF and pulse-Doppler processing, STAP radar can separate the echoes based on both direction of arrival and velocity. In effect, it creates a two-dimensional filter in a space with direction of arrival on one axis and velocity on the other.

The standard approach to processing "space-time" data is to use the FFT to transform the data into a time-frequency space and then process each frequency slice separately using standard spatial algorithms. This has the advantage of computational efficiency as there is a "dimensionality curse" associated with moving from one dimension to two. The use of the FFT, which is computationally efficient, being $O(N \log N)$, transforms the two-dimensional problem to multiple one-dimensional ones. However, strictly speaking, the one-dimensional problems are not independent, and treating them as such incurs an approximation error. Fortunately, this error tends to reduce as the order of the FFT increases. An alternative approach to overcoming the dimensionality curse is to attempt to exploit the structure of the problem (see the discussion of fast adaptive filters in section 8). In the case of ABF, the data matrix is "block Toeplitz" rather than Toeplitz, but a similar computational reduction can nevertheless be obtained by exploiting related *multichannel* forward and backward linear prediction problems.

A more recent approach to this problem involves a generalization of matrix algorithms over the complex field to related algorithms over the ring of polynomials. In this context, a polynomial matrix (PM) $\underline{A}(z) \in \underline{\mathbb{C}}^{n \times m}$ is simply an $n \times m$ matrix whose elements are polynomials over $\mathbb{C}$. $\underline{A}(z)$ represents an $n \times m$ FIR filter and corresponds to its $z$-transform, except that it is treated as a polynomial in the indeterminate variable $z^{-1}$ rather than a function of $z$ to be evaluated at a point in the complex plane. The paraconjugate of a PM $\underset{\sim}{A}(z) \in \underset{\sim}{\mathbb{C}}^{n\times m}$ is denoted by $\tilde{\underset{\sim}{A}}(z) = \underset{\sim}{A}_*^{\mathrm{T}}(1/z)$, where the asterisk denotes complex conjugation of the polynomial coefficients. A parasymmetric PM $\underset{\sim}{M}(z) \in \underset{\sim}{\mathbb{C}}^{n\times n}$ is one that satisfies $\underset{\sim}{M}(z) = \tilde{\underset{\sim}{M}}(z)$, while a paraunitary PM $\underset{\sim}{H}(z) \in \underset{\sim}{\mathbb{C}}^{n\times n}$ (which corresponds to a multichannel allpass filter) satisfies $\underset{\sim}{H}(z)\tilde{\underset{\sim}{H}}(z) = \tilde{\underset{\sim}{H}}(z)\underset{\sim}{H}(z) = I$. As usual, $I$ denotes the unit matrix.

Algorithms have recently been developed for approximating the eigenvalue decomposition (EVD), the SINGULAR-VALUE DECOMPOSITION [II.32] (SVD), and the QR decomposition of polynomial matrices. For example, the EVD of a parasymmetric PM $\underset{\sim}{M}(z) \in \underset{\sim}{\mathbb{C}}^{n\times n}$ (this is a polynomial eigenvalue decomposition (PEVD)) can be computed as $\underset{\sim}{H}(z)\underset{\sim}{M}(z)\tilde{\underset{\sim}{H}}(z) \approxeq \underset{\sim}{D}(z)$, where $\underset{\sim}{H}(z) \in \underset{\sim}{\mathbb{C}}^{n\times n}$ is paraunitary and $\underset{\sim}{D}(z) \in \underset{\sim}{\mathbb{C}}^{n\times n}$ is diagonal. The exact decomposition does not exist for $\underset{\sim}{H}(z) \in \underset{\sim}{\mathbb{C}}^{n\times n}$, but the PM solution can lie arbitrarily close to a corresponding matrix of continuous functions for which it does exist. The PEVD defined above is clearly a generalization of the conventional matrix EVD to which it reduces for a PM of order zero. Just as the matrix EVD (or SVD) plays a fundamental role in the separation of signal and noise subspaces for narrowband ABF, so the PEVD (or polynomial SVD) may be used in broadband ABF. However, the range of potential applications for PM decomposition techniques in general is very much wider.

## 12 Bayesian Signal Processing/ Parameter Estimation

There are two main approaches to developing signal-processing algorithms: deterministic and statistical. The algorithms for adaptive filtering and ABF given above are deterministic, in the sense that we obtain a set of parameters (e.g., filter coefficients or beamformer weights) by solving a least-squares minimization problem. These algorithms can, however, be seen as special cases of more general *parameter-estimation* algorithms. The filter coefficients and beamformer weights can be seen as parameters that are to be estimated from the received data. Furthermore, when these estimated parameters are used in a filter or beamformer, one obtains an estimate of some signal of interest. For example, the MVDR beamformer recovers an estimate of the signal at a given location.

In general, parameter estimation is seen as a statistical estimation problem. The least-squares algorithms result when one has a linear problem with Gaussian noise. The most popular signal-processing approach to parameter estimation is based on BAYES'S THEOREM [V.11]. If a variable $x$ is dependent on a parameter $\theta$, then $P(\theta \mid x) \propto P(x \mid \theta)P(\theta)$. The pdf $P(x \mid \theta)$ is known, since it comes from the physics of the situation, and it encodes the dependency that $x$ has on $\theta$. Thus, on receipt of the measurement $x$, the a priori pdf $P(\theta)$ can be updated to the a posteriori pdf $P(\theta \mid x)$ merely by multiplying the former by $P(x \mid \theta)$ (known as the *likelihood function*). However, such a product will generally not have a closed-form expression. An exception to this rule is linear problems where the noise is Gaussian. In this case all variables are Gaussian and closed-form expressions for the means and variances are easy to find. In general, however, the pdfs have be represented in some computationally tractable form. One option is to model an arbitrary pdf as the sum of Gaussians (known as a *Gaussian mixture*). Another is to model the pdf by a discrete probability distribution. The latter approach is known as *particle filtering*. These modeling techniques are powerful as they allow nonlinear and non-Gaussian problems to be tackled. However, as with any modeling problem, there are issues to do with parameter choice, e.g., the number of Gaussians in the mixture, and the number of bins in the discrete-valued distribution. A particularly important issue is loss of resolution. Bayes's theorem required us to multiply the a priori pdf by the likelihood function. If the modeled a priori pdf has a zero value for some value of the variable, then the a posteriori pdf will also have a zero value for this value. Techniques are therefore needed to ensure that the modeled pdf has a zero value only when this is justified by the data. Note that these modeling approaches tend to require more computation than an algorithm based on closed-form expressions.

An important signal-processing area that is related to parameter estimation is *detection theory*. Here, we wish to decide if an event has taken place based on an observation. An example of this is the detection of aircraft using radar: the signal received by the radar could consist of the echo from the aircraft plus noise or it could just be noise. What is required is some optimal test that can confirm the presence of the aircraft. Typically, the radar signal is compared with a threshold. If the signal exceeds the threshold, then the aircraft is said to have been detected. Clearly, there could be false detection (false positives) and missed detection (false negatives). Detection theory uses statistical arguments to calculate the threshold with given properties such as a *constant false alarm rate*.

## 13 Tracking

As mentioned in the previous section, adaptive filters and beamformers can be seen as devices for estimating unknown parameters. In this case, however, the parameters are constants. If the unknown parameters are time varying, the problem is one of *tracking*.

Since the estimation of $N$ parameters requires at least $N$ pieces of data, it is not possible to estimate more than one arbitrary time-varying parameter from a single time series. It is therefore conventional to assume that the parameters evolve in a known manner; for example, $\theta(n) = F(\theta(n-1) \mid \Phi)$, where $\Phi$ are (known) parameters of the function $F$. Given this model for the time evolution of the parameter, it is then possible to formulate a parameter-estimation algorithm. As with adaptive filtering and beamforming, one can take a deterministic (i.e., least-squares) approach or a Bayesian approach. In the former case one ends up with the well-known *Kalman filter*, which is optimum for linear systems and Gaussian noise. In the latter case one ends up with a more powerful algorithm but with the computational issues mentioned above.

### Further Reading

Bernardo, J. M., and A. F. Smith. 2000. *Bayesian Theory*. New York: John Wiley.

Bunch, J. R., R. C. Le Borne, and I. K. Proudler. 2001. A conceptual framework for consistency, conditioning and stability issues in signal processing. *IEEE Transactions on Signal Processing* 49(9):1971–81.

Haykin, S. 2001. *Adaptive Filter Theory*, 4th edn. Englewood Cliffs, NJ: Prentice-Hall.

——. 2006. *Adaptive Radar Signal Processing*. New York: John Wiley.

McWhirter, J. G., P. D. Baxter, T. Cooper, S. Redif, and J. Foster. 2007. An EVD algorithm for para-Hermitian polynomial matrices. *IEEE Transactions on Signal Processing* 55(6):2158–69.

Proakis, J. G., C. Rader, F. Ling, and C. Nikias. 1992. *Advanced Digital Signal Processing*. London: Macmillan.

Proakis, J. G., and M. Salehi. 2007. *Digital Communications*, 5th edn. Columbus, OH: McGraw-Hill.

Rabiner, L. R., and B. Gold. 1975. *Theory and Application of Digital Signal Processing*. Englewood Cliffs, NJ: Prentice-Hall.

Shannon, C. E., and W. Weaver. 1949. *The Mathematical Theory of Communication*. Champaign, IL: University of Illinois Press.

Skolnik, M. I. 2002. *Introduction to Radar Systems*. Columbus, OH: McGraw-Hill.

# IV.36 Information Theory

*Sergio Verdú*

## 1 "A Mathematical Theory of Communication"

Rarely does a scientific discipline owe its existence to a single paper. Authored in 1948 by Claude Shannon (1916–2001), "A mathematical theory of communication" is the Magna Carta of the information age and information theory's big bang. Using the tools of probability theory, it formulates the central optimization problems in data compression and transmission, and finds the best achievable performance in terms of the statistical description of the information sources and communication channels by way of information measures such as entropy and mutual information. After a glimpse at the state of the art as it was in 1948, we elaborate on the scope of Shannon's masterpiece in the rest of this section.

### 1.1 Communication Theory before the Big Bang

Motivated by the improvement in telegraphy transmission rate that could be achieved by replacing the Morse code by an optimum code, both Nyquist (1924) and Hartley (1928) recognized the need for a measure of information devoid of "psychological factors" and put forward the logarithm of the number of choices as a plausible alternative. Küpfmüller (1924), Nyquist (1928), and Kotel'nikov (1933) studied the maximum telegraph signaling speed sustainable by band-limited linear systems at a time when Fourier analysis of signals was already a standard tool in communication engineering. Inspired by the telegraph studies, Hartley put forward the notion that the "capacity of a system to carry information" is proportional to the time-bandwidth product, a notion further elaborated by Gabor (1946). However, those authors failed to grapple with the random nature of both noise and the information-carrying signals. At the same time, the idea of using mathematics to design linear filters for combatting additive noise optimally had been put to use by Kolmogorov (1941) and Wiener (1942) for minimum mean-square error estimation and by North (1943) for the detection of radar pulses.

Communication systems such as FM and PCM in the 1930s and spread spectrum in the 1940s had opened up the practical possibility of using transmission bandwidth as a design parameter that could be traded off for reproduction fidelity and robustness against noise.

### 1.2 The Medium

In the title of Shannon's paper, "communication" refers to

- communication across space, namely, *information-transmission* systems like radio and television broadcasting, telephone wires, coaxial cables, optical fibers, microwave links, and wireless telephony; and
- communication across time, namely, *information-storage* systems, which typically employ magnetic (tape and disks), optical (CD, DVD, and BD), and semiconductor (volatile and flash) media.

Although, at some level, all transmission and storage media involve physical continuously variable analog quantities, it is useful to model certain media such as optical disks, computer memory, or the Internet as digital media that transmit or record digital signals (zeros/ones or data packets) with a certain reliability level.

### 1.3 The Message

The message to be stored or transmitted may be

- analog (such as sensor readings, audio, images, video, or, in general, any message intended for the human ear/eye) or
- digital (such as text, software, or data files).

An important difference between analog and digital messages is that, since noise is unavoidable in both sensing and transmission, it is impossible to reconstruct exactly the original analog message from the recorded or transmitted information. Lossy reproduction of analog messages is therefore inevitable. Even when, as is increasingly the case, sensors of analog signals output quantized information, it is often conceptually advantageous to treat those signals as analog messages.

### 1.4 The Coat of Arms

Shannon's theory is a paragon of *e pluribus unum*. Indeed, despite the myriad and diversity of communication systems encompassed by information theory, its key ideas and principles are all embracing and are applicable to any of them.

Reproduced from Shannon's paper, figure 1 encompasses most cases (see section 9) of communication

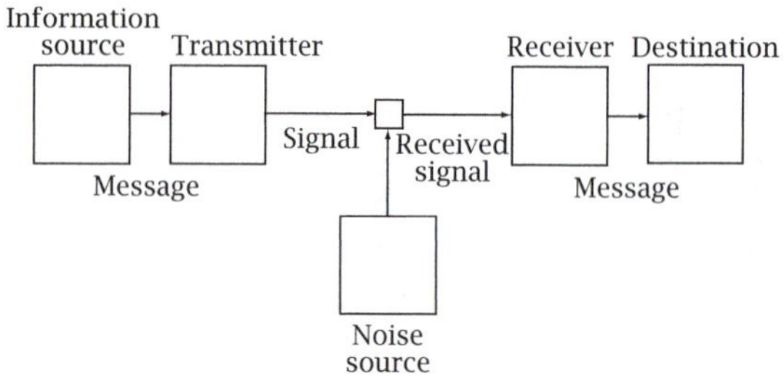


**Figure 1** A schematic of a general communication system (this is figure 1 in "A mathematical theory of communication").

across time or space between one sender and one destination.

The purpose of the *encoder* (or *transmitter*, in figure 1) is to translate the message into a signal suitable for the transmission or storage medium. Conversely, the *decoder* (or *receiver*, in figure 1) converts the received signal into an exact or approximate replica of the original message.

The communication medium that connects the transmitter to the receiver is referred to as the *channel*. Several notable examples, classified according to the various combinations of the nature of message and medium, are listed below.

**Analog message, analog medium.** Radio broadcasting and long-distance telephony were the primary applications of the first analog modulation systems, such as AM, SSB, and FM, developed in the early twentieth century. With messages intended for the ear/eye and the radio frequency spectrum as the medium, all current systems for radio and television (wireless) broadcasting are also examples of this case. However, in most modern systems (such as DAB and HDTV) the transmitter and receiver perform an internal intermediate conversion to digital, for reasons that are discussed in section 4.

**Analog message, digital medium.** This classification includes the audio compact disc, MP3, DVD, and Voice over Internet Protocol (VoIP). So "digital audio" or "digital video" refers to the medium rather than the message.

**Digital message, analog medium.** The earliest examples of optical and electrical systems for the transmission of digital information were the wired telegraph systems invented in the first half of the nineteenth century, while the second half of the century saw the advent of Marconi's wireless telegraph.

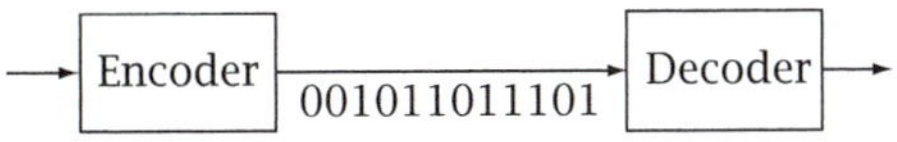


**Figure 2** A data-compression system.

Other examples developed prior to 1948 include teletype, fax, and spread spectrum. The last four decades of the twentieth century saw the development of increasingly fast general-purpose modems to transmit bit streams through analog media such as the voice-band telephone channel and radio frequency bands. Currently, modems that use optical, DSL, and CATV media to access the Internet are ubiquitous.

**Digital message, digital medium.** This classification includes data storage in an optical disk or flash memory.

Whether one is dealing with messages, channel inputs, or channel outputs, Shannon recognized that it is mathematically advantageous to view continuous-time analog signals as living in a finite-dimensional vector space. The simplest example is a real-valued signal of bandwidth $B$ and (approximate) duration $T$, which can be viewed as a point in the Euclidean space of dimension $2BT$. To that end, Shannon gave a particularly crisp version of the sampling theorem, precursors of which had been described by E. Whittaker (1915), J. Whittaker (1929), and Kotelnikov (1933), who discovered how to interpolate losslessly the sampled values of band-limited functions.

Three special cases of figure 1, dealt with in each of the next three sections, merit particular attention.

## 2 Lossless Compression

Although communication across time or space is always subject to errors or failures, it is useful to consider the idealized special case of figure 1 shown in figure 2, in which there is no channel and the input to the decoder is a digital sequence equal to the encoder output. This setup, also known as *source coding*, models the paradigm of compression in which the encoder acts as the compressor and the decoder acts as the decompressor. The task of the encoder is to remove *redundancy* from the message, which can be recovered exactly or approximately at the decoder from the compressed data itself.

Lossless, or reversible, conversion is possible only if the message is digital. Morse, Huffman, TIFF, and PDF are examples of lossless compression systems, where message redundancy (unequal likelihoods of the various choices) is exploited to compact the data by assigning shorter binary strings to more likely messages. As we discuss more precisely in section 6, the goal is to obtain a compression/decompression algorithm that generates, on average, the shortest encoded version of the message.

**Figure 3** A data-transmission system.

If the source is stationary, *universal* data compressors exploit its redundancy without prior knowledge of its probabilistic law. Found in every computer operating system (e.g., ZIP), the most widely used universal data compressors were developed by Lempel and Ziv between 1976 and 1978.

## 3 Lossy Compression

Depending on the nature of the message, we can distinguish two types of lossy compression.

**Analog-to-digital.** Early examples of analog-to-digital coding (such as the vocoder and pulse-code modulation (PCM)) were developed in the 1930s. The vocoder was the precursor to the speech encoders used in cellular telephony and in VoIP, while PCM remains in widespread use in telephony and in the audio compact disc. The conceptually simplest analog-to-digital compressor, used in PCM, is the scalar quantizer, which partitions the real line in $2^k$ segments, each of which is assigned a unique $k$-bit label. JPEG [VII.7 §5] and MPEG are contemporary examples of lossy compressors for images and audio/video, respectively. Even if the inputs to those algorithms are finite-precision numbers, their signal processing treats them as real numbers.

**Digital-to-digital.** Even in the case of digital messages, one may be willing to tolerate a certain loss of information for the sake of economy of transmission time or storage space (e.g., when emailing a digital image or when transmitting the analog-to-digitally compressed version of a sensor reading).

## 4 Data Transmission

Figure 3 depicts the paradigm, also known as *channel coding*, in which the message input to the encoder is incompressible or nonredundant, in the sense that it is chosen equiprobably from a finite set of alternatives (such as fair coin flips or "pure" bits, i.e., independent binary digits equally likely to be 0 or 1). The task of the encoder is to add redundancy to the message in order to protect it from channel noise and facilitate its recovery by the decoder from the noisy channel output. In general, this is done by assigning *codewords* to each possible message, which are different enough to be distinguishable at the decoder as long as the noise is not too severe. For example, in the case of a digital medium the encoder may use an error-correcting code that appends redundant bits to the binary message string. In the case of an analog medium such as a telephone channel, the codewords are continuous-time waveforms. Based on the statistical knowledge of the channel and the codebook (assignment of messages to codewords) used by the encoder, the decoder makes an intelligent guess about the transmitted message.

Remarkably, Shannon predicted the performance of the best possible codes at a time when very few error-correcting codes were known. Hamming, a coworker at Bell Laboratories, had just invented his namesake code (see APPLIED COMBINATORICS AND GRAPH THEORY [IV.37 §4]) that appends three parity-check bits to every block of four information bits in a way that makes all sixteen codewords differ from each other in at least three positions. Therefore, the decoder can correct any single error affecting every encoded block of seven bits.

## 5 Compression/Transmission

Figure 4 illustrates another special case of figure 1 in which the transmitter consists of the source encoder, or compressor, followed by the channel encoder, and the receiver consists of the channel decoder followed by the source decoder, or decompressor. This architecture capitalizes on the solutions found in the special cases in sections 2, 3, and 4. To that end, in the scheme shown in figure 4 the interfaces between source and channel encoders, and between channel and source decoders, are digital regardless of the message or medium. Inspired by the teachings of information theory, in which the bit emerges as the universal currency, the modular design in figure 4 is prevalent in most modern systems for the transmission of analog messages through either digital or analog media. It allows the source encoding/decoding system to be tailored particularly to the message, disregarding the nature of the channel. Analogously, it allows the

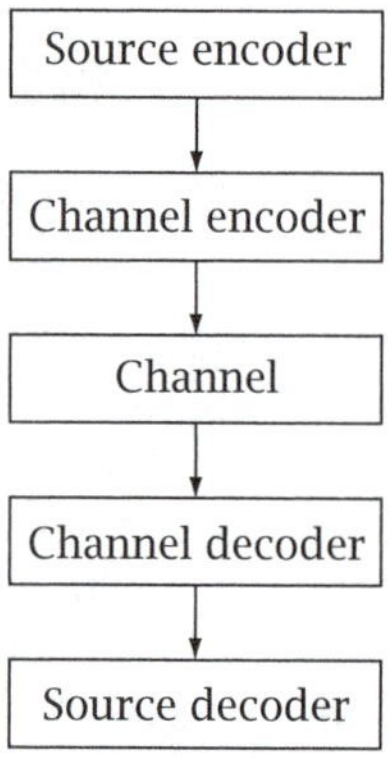


**Figure 4** A separate compression/transmission system.

channel encoding/decoding system to be focused on the reliable transmission of nonredundant bits by combatting the channel noise disregarding the nature of the original message. In this setup, the source encoder removes redundancy from the message in a way that is tuned to the information source, while the channel encoder adds redundancy in a way that is tuned to the channel. Under widely applicable sufficient conditions, such modular design is asymptotically optimal (in the sense of section 6) in the limit in which the length of the message goes to infinity and when both source and channel operate in the ergodic regime.

## 6 Performance Measures

The basic performance measures depend on the type of system under consideration.

**Lossless compression.** The compression *rate* (in bits per symbol) is the ratio of encoded bits to the number of symbols in the digital message.

**Lossy compression.** The quality of reproduction is measured by a distortion function of the original and reproduced signals, e.g., in the case of analog signals, the mean-square error (energy of the difference signal), and in the case of binary messages, the bit error rate. The rate (in bits per second, or per symbol) of a lossy compression system is the ratio of encoded bits to the duration of the message.

**Data transmission.** For a given channel and assuming that the message is incompressible, the performance of a data-transmission system is determined by the rate and the error probability. The rate (in bits per second, or per symbol) is the ratio of message duration to the time it takes to send it through the channel. Depending on the application, the reliability of the transmission is measured by the bit error rate or by the probability that the entire message is decoded correctly.

**Joint compression/transmission.** In the general case, the rate is measured as in the data-transmission case, with reliability measured either by a distortion measure or by the probability that the entire message is decoded correctly, depending on the nature of the message and the application.

## 7 Fundamental Limits

Instead of delving into the analysis and design of specific transmission systems or codes, the essence of Shannon's mathematical theory is to explore the best performance that an optimum encoder/decoder system (simply referred to as the *code*) can achieve. Information theory obtains *fundamental limits* without actually deriving the optimal codes, which are often unknown. For the three problems formulated by Shannon, the fundamental limits are as follows.

**Lossless compression:** the minimum achievable compression rate.

**Lossy compression:** the *rate-distortion function*, which is the minimum compression rate achievable as a function of the allowed average level of distortion.

**Data transmission:** the *channel capacity*, defined as the maximum transmission rate compatible with vanishing error probability. Capacity is often given in terms of channel parameters such as transmitted power. Before Shannon's paper, the common wisdom was that vanishing error probability would necessarily entail vanishing rate of information transmission.

The fundamental limits are very useful to the engineer because they offer a comparison of the performance of any given system with that ultimately achievable. Although, in Shannon's formulation, the growth of computational complexity as a function of the message size is not constrained in any way, decades of research on the constructive side of compression and transmission have yielded algorithms that can approach the Shannon limits with linear complexity. Often, information theory leads to valuable engineering conclusions that reveal that simple (or modular) solutions may perform at or near optimum levels. For example, as we mentioned, there is no loss in achievable performance if one follows the principle of separate compression/transmission depicted in figure 4. Fundamental limits can be, and often are, used to sidestep the need for cumbersome analysis in order to debunk performance claims made for a given system.

The fundamental limits turn out to depend crucially on the duration of the message. Since Shannon's 1948 paper, information theory has focused primarily, but not exclusively, on the fundamental limits in the regime of asymptotically long messages. By their very nature, the fundamental limits for a given source or channel are not technology dependent, and they do not become obsolete with improvements in hardware/software. On the contrary, technological advances pave the way for the design of coding systems that approach the ideal fundamental limits increasingly closely. Although the optimum compression and transmission systems are usually unknown, the methods of proof of the fundamental limits often suggest features that near-optimum practical communication systems ought to have, thereby offering design guidelines to approach the fundamental limits. Shannon's original proof of his channel coding theorem was one of the first nontrivial instances of the *probabilistic method*, now widely used in discrete mathematics; to show the existence of an object that satisfies a certain property it is enough to find a probability distribution on the set of all objects such that those satisfying the property have nonzero probability. In his proof, Shannon computed an upper bound to the error probability averaged with respect to an adequately chosen distribution on the set of all codes; at least one code must have error probability not exceeding the bound.

## 8 Information Measures

The fundamental performance limits turn out to be given in terms of so-called information measures, which have units such as bits. In this section we list the three most important information measures.

**Entropy:** a measure of the randomness of a discrete distribution $P_X$ defined on a finite or countably infinite alphabet $\mathcal{A}$, defined as

$$H(X) = \sum_{a \in \mathcal{A}} P_X(a) \log\left(\frac{1}{P_X(a)}\right).$$

In the limit as $n \to \infty$, a stationary ergodic random source $(X_1, \dots, X_n)$ can be losslessly encoded at its entropy rate

$$\lim_{n \to \infty} \frac{1}{n} H(X_1, \dots, X_n),$$

a limit that is easy to compute in the case of Markov chains. In the simplest case, asymptotically, $n$ flips of a coin with bias $p$ can be compressed losslessly at any rate exceeding $h(p)$ bits per coin flip with

$$h(p) = p \log \frac{1}{p} + (1-p) \log \left( \frac{1}{1-p} \right),$$

which is the entropy of the biased coin source. The ubiquitous linear-time Lempel–Ziv universal data-compression algorithms are able to achieve, asymptotically, the entropy rate of ergodic stationary sources. Therefore, at least in the long run, universality incurs no penalty.

**Relative entropy:** a measure of the dissimilarity between two distributions $P$ and $Q$ defined on the same measurable space $(\mathcal{A}, \mathcal{F})$, defined as

$$D(P \| Q) = \int \log \left( \frac{\mathrm{d}P}{\mathrm{d}Q} \right) \mathrm{d}P.$$

Relative entropy plays a central role not only in information theory but also in the analysis of the ability to discriminate between data models, and in particular in large-deviation results, which explore the exponential decrease (in the number of observations) of the probability of very unlikely events. Specifically, if $n$ independent data samples are generated with probability distribution $Q$, the probability that they will appear to be generated from a distribution in some class $\mathcal{P}$ behaves as

$$\exp\left( -n \inf_{P \in \mathcal{P}} D(P \| Q) \right).$$

Relative entropy was introduced by Kullback and Leibler in 1951 with the primary goal of extending Shannon's measure of information to nondiscrete cases.

**Mutual information:** a measure of the dependence between two (not necessarily discrete) random variables $X$ and $Y$ given by the relative entropy between the joint measure and the product of the marginal measures:

$$I(X;Y) = D(P_{XY} \| P_X \times P_Y).$$

Note that $I(X;X) = H(X)$ if $X$ is discrete.

For stationary channels that behave ergodically, the channel capacity is given by

$$C = \lim_{n \to \infty} \frac{1}{n} \max I(X_1, \dots, X_n; Y_1, \dots, Y_n),$$

where the maximum is over all joint distributions of $(X_1, \dots, X_n)$, and $(Y_1, \dots, Y_n)$ are the channel responses to $(X_1, \dots, X_n)$. If the channel is stationary memoryless, then the formula boils down to

$$C = \max I(X;Y).$$

The capacity of a channel that erases a fraction $\delta$ of the codeword symbols (drawn from an alphabet $\mathcal{A}$) is

$$C = (1 - \delta) \log |\mathcal{A}|,$$

as long as the location of the erased symbols is known to the decoder and the nonerased symbols are received error free. In the case of a binary channel that introduces errors independently with probability $\delta$, the capacity is given by

$$C = 1 - h(\delta),$$

while in the case of a continuous-time additive Gaussian noise channel with bandwidth $B$, transmission power $P$, and noise strength $N$, the capacity is

$$C = B \log \left( 1 + \frac{P}{BN} \right) \text{ bits per second,}$$

a formula that dispels the pre-1948 notion that the information-carrying capacity of a communication channel is proportional to its bandwidth and that is reminiscent of the fact that in a cellular phone the stronger the received signal the faster the download.

In lossy data compression of a stationary ergodic source $(X_1, X_2, \dots)$, the rate compatible with a given per-sample distortion level $d$ under a distortion measure $\mathsf{d}\colon \mathcal{A}^2 \to [0, \infty]$ is given by

$$R(d) = \lim_{n \to \infty} \frac{1}{n} \min I(X_1, \dots, X_n; Y_1, \dots, Y_n),$$

where the minimum is taken over the joint distribution of source $X^n$ and reproduction $Y^n$, with given $P_{X^n}$, and such that

$$\frac{1}{n} \sum_{i=1}^{n} \mathsf{d}(X_i, Y_i) \leqslant d.$$

For stationary memoryless sources, just as for capacity we obtain a "single-letter" expression $R(d) = \min I(X;Y)$.

It should be emphasized that the central concern of information theory is not the definition of information measures but the theorems that use them to describe the fundamental limits of compression and transmission. However, it is rewarding that entropy, mutual information, and relative information, as well as other related measures, have found applications in many fields beyond communication theory, including probability theory, statistical inference, ergodic theory, computer science, physics, economics, life sciences, and linguistics.

## 9 Beyond Figure 1

Work on the basic paradigm in figure 1 continues to this day, not only to tackle source and channel models inspired by new applications and technologies but in furthering the basic understanding of the capabilities of coding systems, particularly in the nonasymptotic regime. However, in order to analyze models of interest in practice, many different setups have been studied since 1948 that go beyond the original. We list a few of the ones that have received the most attention.

**Feedback.** A common feature of many communication links is the availability of another communication channel from receiver to transmitter. In what way can knowledge of the channel output aid the transmitter in a more efficient selection of codewords? In 1956 Shannon showed that, in the absence of channel memory, capacity does not increase even if the encoder knows the channel output instantaneously and noiselessly. Nevertheless, feedback can be quite useful to improve transmission rate in the nonasymptotic regime and in the presence of channel memory.

**Separate compression of dependent sources of information.** Suppose that there is one decompressor that receives the encoded versions of several sources produced by individual compressors. If, instead, a single compressor had access to all the sources, it could exploit the statistical dependence among them to encode at a rate equal to the overall entropy. Surprisingly, in 1973 Slepian and Wolf showed that even in the completely decentralized setup the sum of the encoded rates can be as low as in the centralized setting and still the decompressor is able to correctly decode with probability approaching 1. In the lossy setting the corresponding problem is not yet completely solved.

**Multiple-access channel.** If, as in the case of a cellular wireless telephony system, a single receiver obtains a signal with mutually interfering encoded streams produced by several transmitters, there is a trade-off among the achievable rates. The channel capacity is no longer a scalar but a *capacity region*.

**Interference channel.** As in the case of a wired telephone system subject to crosstalk, in this model there is a receiver for each transmitter, and the signal it receives not only contains the information transmitted by the desired user but is contaminated by the signals of all other users. It does not reduce to a special case of the multiple-access setup because each receiver is required to decode reliably only the message of its desired user.

**Broadcast channel.** A single transmitter sends a codeword, which is received by several geographically separated receivers. Each receiver is therefore connected to the transmitter by a different communication channel, but all those channels share the same input. If the broadcaster intends to send different messages to the various destinations, there is again a trade-off among the achievable rates.

**Relay channel.** The receiver obtains both a signal from the transmitter and a signal from a relay, which itself is allowed to process the signal it receives from the transmitter in any way it wants. In particular, the relay need not be able to fully understand the message sent by the transmitter.

Inspired by various information technologies, a number of information-theoretic problems have arisen that go beyond issues of eliminating redundancy (for compression) or adding redundancy (for transmission in the presence of noise). Some examples follow.

**Secrecy.** Simultaneously with communication theory, Shannon established the basic mathematical theory of cryptography and showed that iron-clad privacy requires that the length of the encryption key be as long as that of the message. Most modern cryptographic algorithms do not provide that level of security; they rely on the fact that certain computational problems, such as integer factorization, are believed to be inherently hard. A provable level of security is available using an information-theoretic approach pioneered by Wyner (1975), which guarantees that the eavesdropper obtains a negligible amount of information about the message.

**Random number generation for system simulation.** Random processes with prescribed distributions can be generated by a deterministic algorithm driven by a source of random bits. A key quantity that quantifies the "complexity" of the generated random process is the minimal rate of the source of random bits necessary to accomplish the task. The *resolvability* of a system is defined as the minimal randomness required to generate any desired input so that the output distributions are approximated with arbitrary accuracy. In 1993 Han and Verdú showed that the resolvability of a system is equal to its channel capacity.

**Minimum description length.** In the 1960s Kolmogorov and others took a nonprobabilistic approach to the compression of a message, which, like universal lossless compression, uses no prior knowledge: the algorithmic complexity of the message is the length of the shortest program that will output the message. Although this notion is useful only asymptotically, it has important links with information theory and has had an impact in statistical inference, primarily through the minimum description length statistical modeling principle put forward by Rissanen in 1978: the message is compressed according to a certain distribution, which is chosen from a predetermined model class and is also communicated to the decompressor. The distribution is chosen so that the sum of the lengths of its description and the compressed version of the message are minimized.

**Inequalities and convex analysis.** A principle satisfied by information measures is that processing cannot increase either the dependence between input and output as measured by mutual information or the relative entropy between any pair of distributions governing the input of the processor. Mathematically, the nonnegativity of relative entropy and those *data processing principles* are translated into convex inequalities, which have been used successfully in the rederivation of various inequalities, such as those of Hadamard and Brunn–Minkowski, and in the discovery of new inequalities.

**Portfolio theory.** One possible approach to PORTFOLIO SELECTION [V.10] (for a given number of stocks) is to choose the *log-optimal* portfolio, which maximizes the asymptotic appreciation growth rate. When their distribution is known, a simplistic model of independent identically distributed stock prices leads to limiting results with a strong information-theoretic flavor. Just as in data compression, under assumptions of stationarity and ergodicity, it is possible to deal with more realistic scenarios in which the distribution is not known a priori and the stock prices are interdependent.

**Identification.** Suppose that the transmitter sends the identity of an addressee to a multitude of possible users. Each user is interested only in finding out whether it is indeed the addressee or not. Allowing, as usual, a certain error probability, this setup can be captured as in figure 1, except that the decoder is free to declare a list of several messages (addresses) to be simultaneously "true." Each user simply checks whether its identity is in the list or not. How many messages can be transmitted while guaranteeing vanishing probability of erroneous information? The surprising answer found by Ahlswede and Dueck in 1989 is that the number of addresses grows doubly exponentially with the number of channel uses. Moreover, the second-order exponent is equal to the channel capacity.

Finally, we mention the discipline of *quantum information theory*, which deals with the counterparts of the fundamental limits discussed above for quantum mechanical models of sources and channels. Probability measures, conditional probabilities, and bits translate into density matrices, self-adjoint linear operators, and qubits. The quantum channel coding theorem was proved by Holevo in 1973, while the quantum source coding theorem was proved by Schumacher in 1995.

### Further Reading

Cover, T. M., and J. Thomas. 2006. *Elements of Information Theory*, 2nd edn. New York: Wiley Interscience.

Shannon, C. E. 1948. A mathematical theory of communication. *Bell System Technical Journal* 27:379–423, 623–56.

Verdú, S. 1998. Fifty years of Shannon theory. *IEEE Transactions on Information Theory* 44(6):2057–78.

# IV.37 Applied Combinatorics and Graph Theory

*Peter Winkler*

## 1 Introduction

Combinatorics and graph theory are the cornerstones of *discrete mathematics*, which has seen an explosion of activity since the middle of the twentieth century. The main reason for this explosion is the plethora of applications in a world where digital (as opposed to analog) computing has become the norm. Once considered more "recreational" than serious, combinatorics and graph theory now boast many fundamental and useful results, adding up to a cogent theory. Our objective here is to present the most elementary of these results in a format useful to those who may run into combinatorial problems in applications but have not studied combinatorics or graph theory.

Accordingly, we will begin each section with a (not necessarily serious, but representative) problem, introducing the basic techniques, algorithms, and theorems of combinatorics and graph theory in response.

We will assume basic familiarity with mathematics but none with computer science. Proofs, sometimes informal, are included when they are useful and short.

Algorithms will be discussed informally, with the term "efficient" used for those that can be executed in time bounded by a small-degree polynomial in the input length.

## 2 Counting Possibilities

Most counting problems can be phrased as, "How many ways are there to do $X$?" where $X$ is some task, either real or imaginary. If there are several alternative approaches, we may be able to make use of the following rule.

**The addition rule.** *If there are $k$ different approaches to the task, and the $i$th approach can be executed in $n_i$ ways, then the total number of different ways to do the task is the sum $\sum_{i=1}^{k} n_i$.*

If, on the other hand, we can break up the task into *stages*, we can use of the following alternative.

**The multiplication rule.** *If the task involves $k$ stages, the $i$th of which can be done $n_i$ ways regardless of choices made in previous stages, then the total number of different ways to do the task is the product $\prod_{i=1}^{k} n_i$.*

Sometimes it is easier to count the things that are *not* wanted, in which case we can make use of the subtraction rule.

**The subtraction rule.** *If a set $A$ consists of the elements of $C$ other than those of a subset $B$ of $C$, then $|A| = |C| - |B|$.*

Overcounting can be useful, and when everything is overcounted the same number of times, the following rule is the remedy.

**The division rule.** *If the number $n$ is obtained when every element of a set $S$ is counted $k$ times, then $|S| = n/k$.*

An arrangement, or *ordering*, of a set $A$ of distinguishable objects is called a *permutation* of $A$; from the multiplication rule, we see that, if $|A| = n$, then the number of permutations of $A$ is $n!$. If we want only to select and order $k$ of the elements of $A$, the number of ways of doing so is $n(n-1)(n-2)\cdots(n-k+1)$, which we can also write as $n!/(n-k)!$. (But note that *calculating* this number by computing $n!$ and then dividing by $(n-k)!$ might be a mistake, as it involves numbers that are unnecessarily large.)

Suppose we wish to select $k$ of the $n$ objects in $A$ but not to order them. We have then overcounted by a factor of $k!$, so the number of so-called *combinations* of $k$ objects out of $n$ is $n!/(k!(n-k)!)$, which we call "$n$ choose $k$" and denote by $\binom{n}{k}$. These expressions are called "binomial coefficients" on account of their appearance in the formula

$$(x+y)^n = \sum_{k=0}^{n} \binom{n}{k} x^k y^{n-k}.$$

The binomial coefficients possess an astonishing number of nice properties. You can easily verify, for example, that

$$\binom{n}{k} = \binom{n}{n-k} \quad \text{and} \quad \binom{n}{k} = \binom{n}{k-1} + \binom{n-1}{k-1}.$$

Note that logically

$$\binom{n}{0} = \binom{n}{n} = 1,$$

since there is just one way to select the empty set from $A$ and one way to pick the whole set $A$ from itself. This agrees with our formula if we stipulate that $0! = 1$.

## 3 Finding a Stable Matching

Some readers will recall mass public weddings performed by the Reverend Sun Myung Moon (1920–2012) under the banner of the Unification Church. Let us stretch our imaginations a bit and suppose that $n$ men and $n$ women wish to participate in such a ceremony but have not actually agreed upon their precise mates yet. Each submits a list of the $n$ members of the opposite sex in preference order; these $2n$ lists are submitted to the church elders, who somehow determine a *matching*, that is, a set of $n$ man–woman pairs to be married on the fateful day.

One might devise many reasonable criteria by which such a matching could be chosen; a particularly desirable one, from the church's point of view especially, is that the matching be *stable*. This means that there should *not* be any man–woman pair (say, Alice and Bill) who are not to be married to each other in the ceremony but who would rather be married to each other than to the persons they are supposed to marry. Such an Alice and Bill, whom we term an "unstable pair," would then be tempted to run away together and mess up the ceremony.

Remarkably, regardless of the preference lists, a stable matching is guaranteed to exist; moreover, there is an efficient algorithm to find one. Devised by David Gale and Lloyd Shapley in 1962, the algorithm is one of the most elegant in combinatorics and is widely used around the world, perhaps most famously in the annual matching of medical-school graduates with internships in the United States.

The algorithm proceeds as a series of mass proposals, from men to women, say. The game begins with every man proposing to the highest-ranked woman on his list; each woman accepts (perhaps temporarily) only the highest ranking of the men who propose to her, rejecting the others immediately.

In subsequent rounds, engaged men are idle, while each unengaged man proposes to the highest-ranked woman on his list who has not previously rejected him: namely, the woman just beneath the one who rejected him on the previous round. As before, each woman then rejects all proposals other than her highest-ranked proposing male. If she is unengaged or the latter is higher ranked (on *her* list, of course) than her current fiancé, she accepts the new proposal (again temporarily) and dumps the old beau. A woman might remain engaged to one man through many rounds but if, at any time, she gets a proposal from a man she likes better, she will unceremoniously dump her old beau and sign up with the best new one. Her old beau will then reenter the market at the next round, proposing to the next woman on his preference list.

The algorithm terminates when every male (and thus every female) is engaged; those final engagements constitute the output matching.

Suppose, for example, that the preferences are as shown in figure 1.

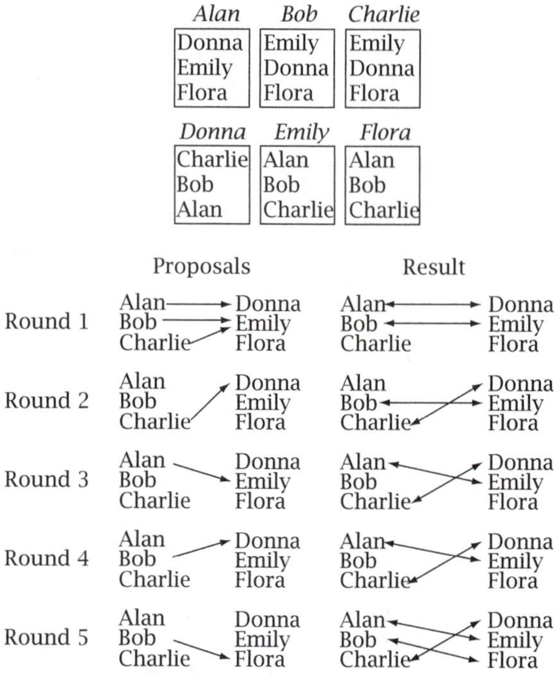


**Figure 1** Finding a stable matching.

In the first round Alan proposes to Donna while Bob and Charlie propose to Emily; Emily rejects Charlie, while the other proposals are provisionally accepted. In the next round Charlie proposes to his second choice, Donna; Donna accepts, putting Alan back into bachelorhood. In the third round Alan proposes to Emily, who accepts while ejecting Bob. In the fourth round Bob proposes to Donna, but she is happy with Charlie. In the fifth round Bob proposes to Flora and is accepted, ending the process with Alan matched to Emily, Bob to Flora, and Charlie to Donna.

As with any algorithm, you should be asking yourself: does it always terminate, and, if so, does it necessarily terminate in a solution to the problem?

Observe first that once a woman becomes engaged (which happens as soon as she gets her first proposal), she never becomes unengaged; she only "trades up." Men, on the other hand, are lowering their expectations as they accumulate rejections. But note that a man cannot run out of prospects because, if he were rejected by every woman, all the women would have to be engaged; an impossibility if he is not. During every round, either the number of engaged men goes up or at least one man is rejected and has his expectations lowered. Thus, the algorithm must end after at most $n^2$ rounds of proposals.

Is the result a stable matching? Yes. Suppose for the sake of reaching a contradiction that Alice and Bill are an unstable pair: Alice is matched to Clint (say) but prefers Bill, while Bill is matched to Dora and prefers Alice. But then, to get engaged to Dora, Bill must at some point have been rejected by Alice; at that time, Alice must have had a fiancé she preferred to Bill. Since she could only have traded up after that, she could not have ended up with Clint.

Usually, in practice (e.g., with the medical-school graduates and internships), no actual proposals are made; preference lists are submitted and a computer simulates the algorithm.

You might reasonably be asking: what happens if we do not have two sexes but just some even number of people who are to be paired up? In this version, often called the "stable roommates problem," each person has a preference list involving all the other people. Alas, for the roommates there may not be a stable matching;

nor (as far as we know) is there an efficient algorithm that is guaranteed to find a stable matching when one exists. Too bad, because there is a nice new application: organ transplants.

Getting a kidney transplant used to require finding one person who was willing to donate a kidney and who matched the kidney seeker in blood type (and possibly other characteristics as well). Nowadays, multiple transplants have made things much easier. Suppose Alice is willing to donate a kidney to Bob, but their blood types are incompatible. Meanwhile, elsewhere in the world, Cassie needs a kidney and has found a willing but poorly matched donor, Daniel. The point is that Alice and Daniel may be a good match, likewise Bob and Cassie. The four are brought together; by agreement, and simultaneously (so no one can chicken out between operations), Alice's kidney goes to Cassie while Daniel's goes to Bob.

Finding such pairs obviously requires some centralized database. Ideally, the database is used to rank, for each patient–donor pair such as Alice and Bob, all other patient–donor pairs according to how well their donor matches Bob. An algorithm that solves the stable roommate problem could be very useful.

In practice, heuristic algorithms seem to do a good job of finding exchangeable pairs. These days, larger cycles of pairs are sometimes organized; also useful are chains, sometimes quite long ones, catalyzed by a single generous individual who is willing to donate a kidney to the patient pool. For the kidney application and much more, Alvin Roth and Lloyd Shapley won the Nobel Memorial Prize in Economic Sciences in 2012.

There are many other consequences and generalizations of the stable marriage theorem; we mention just one curious fact. In many cases there is more than one stable solution, and of course the Gale–Shapley algorithm (as presented here) finds just one. You might think the algorithm is neutral between the sexes or even favors women, but in fact it is not hard to show that in it every man gets the *best* match that he can get in any stable marriage, and every woman gets the *worst* match that she could get in any stable marriage! In the annual matching of medical-school graduates to internships, it used to be the case that the internships did the proposing. Now it is the other way around.

## 4 Correcting Errors

Have you ever wondered why it is that, when you email a multi-megabyte file to a friend or colleague, it almost always comes through with not a single error? Can it be that the bits we send through air, wire, and fiber have error probability less than one in a trillion?

| Message (length $k = 4$) | Codeword (length $n = 7$) | Message (length $k = 4$) | Codeword (length $n = 7$) |
|---|---|---|---|
| 0000 | 0000000 | 1000 | 1000011 |
| 0001 | 0001111 | 1001 | 1001100 |
| 0010 | 0010110 | 1010 | 1010101 |
| 0011 | 0011001 | 1011 | 1011010 |
| 0100 | 0100101 | 1100 | 1100110 |
| 0101 | 0101010 | 1101 | 1101001 |
| 0110 | 0110011 | 1110 | 1110000 |
| 0111 | 0111100 | 1111 | 1111111 |

**Figure 2** A code that corrects one error.

The answer to that question is no: if we sent bits that carefully, the Internet would be a lot slower than it is. Errors do get made, but they are corrected. How?

We are going to consider only the alphabet $\{0, 1\}$, although much of what we say below can be generalized to larger alphabets. An *error-correcting code* is a set of sequences ("codewords") of some given length (say, $n$) that are used instead of sequences of a fixed *shorter* length (say, $k$), with the idea that a small number of errors in a codeword can be rectified. Figure 2 shows an example with $n = 7$ and $k = 4$.

The code pictured above is called a Hamming$(7, 4)$ code, and it has the following properties. First, there are $16 = 2^4$ codewords, one for each binary sequence of length 4, as we see. The codewords have length 7. To use the code, suppose the message we wish to send is 110010100110. We break that up into "blocks" of length 4: 1100 1010 0110. Each block is translated into its corresponding codeword: 1100011 1010101 0110110, and the concatenation 110001110101010110 110 is transmitted.

At the other end, the received bit sequence is broken up into substrings of length 7, each of which is then decoded to recover the original intended message. Decoding is easy, with this particular code, when there are no errors; the first four bits of each codeword identify the message sequence.

The disadvantages of the scheme are obvious. It takes work (well, computer work) to code and decode, and we end up sending nearly twice as many bits as we need to. The gain is that we can now correct errors: in particular, as long as there is no more than one flipped bit in each codeword, we can recover the exact original message.

To verify this, let us see how the code can be generated. First we label the seven codeword bit-positions with the numbers from one to seven in binary. It does

Labeling of positions

| | | | | | | |
|---|---|---|---|---|---|---|
| 011 | 101 | 110 | 111 | 100 | 010 | 001 |

Addition table for
three-digit nimbers

| | 000 | 001 | 010 | 011 | 100 | 101 | 110 | 111 |
|---|---|---|---|---|---|---|---|---|
| 000 | 000 | 001 | 010 | 011 | 100 | 101 | 110 | 111 |
| 001 | 001 | 000 | 011 | 010 | 101 | 100 | 111 | 110 |
| 010 | 010 | 011 | 000 | 001 | 110 | 111 | 100 | 101 |
| 011 | 011 | 010 | 001 | 000 | 111 | 110 | 101 | 100 |
| 100 | 100 | 101 | 110 | 111 | 000 | 001 | 010 | 011 |
| 101 | 101 | 100 | 111 | 110 | 001 | 000 | 011 | 010 |
| 110 | 110 | 111 | 100 | 101 | 010 | 011 | 000 | 001 |
| 111 | 111 | 110 | 101 | 100 | 011 | 010 | 001 | 000 |

**Figure 3** Labeling and nimber addition to create a Hamming(7, 4) code.

not actually matter how the labels are assigned, but it turns out to be convenient to let positions 5, 6, and 7 have labels 100, 010, and 001 (binary for 4, 2, and 1). We give the first four places the labels 011, 101, 110, and 111, respectively.

These labels will be treated not as binary numbers but as binary "nimbers," equivalently, as vectors in the three-dimensional vector space $\{0,1\}^3$ over a two-element field. That means that nimbers are added without carry; in other words, using the rules that $0+0=0$, $1+0=0+1=1$, and $1+1=0$, each column of numbers is added independently. For example, $011+101=110$. Addition is therefore the same as subtraction; in particular, adding any nimber to itself gives 000. Given any seven-digit word, we compute its *signature* by adding, as nimbers, the labels of the positions where a 1 is found. Thus, for example, the signature of 0100100 is $101+100=001$.

The position labeling and an addition table for three-digit nimbers are shown in figure 3.

The codewords are exactly the seven-digit words with signature 000. That makes the set of codewords a linear subspace of $\{0,1\}^7$, which goes a long way to making the code easy to deal with. Computing the code of a given four-digit message is a snap because the first four digits of the code are the same as those of the message and the other three function as "check bits." If the labels of the first four positions sum to 110, say, then the last three must be exactly 1, 1, 0.

The key property possessed by the set of codewords is that any two codewords differ in at least three of the seven positions. We know this because if two seven-digit words differ in only two places, with labels *abc* and *def*, then their signatures differ by *abc* − *def*, which cannot be 000. They cannot, therefore, both be codewords. (The argument is even easier if they differ in only one position.) Thus, if a bit in a codeword is flipped, it still differs in at least two places from all the *other* codewords, so there is only one codeword it could have come from. Determining that codeword is again easy: if the signature of the received word is *ghi* ≠ 000, then it must be the bit whose label is *ghi* that got flipped.

The Hamming codes are said to be "perfect one-error-correcting codes," meaning that every word of length $n$ is either a codeword or is one bit-flip away from a unique codeword. Another way to think of it is that (in the Hamming(7, 4) code) each codeword has seven neighbors, obtained by flipping one bit; together with the codeword they make a "ball" of size $2^3$, and these $2^4$ balls then partition the whole space of seven-digit words, whose size is of course $2^7$.

There is in fact an even easier one-error-correcting code than the Hamming(7, 4) code, with $n = 3$ and $k = 1$: just repeat each bit three times. If you receive "110" you will know that the third bit was flipped, and that the codeword should therefore have been "111" and that the intent had been to send the bit "1". Moreover, this code tolerates an error every three bits, while the previous could handle only one error every seven bits. But the simple code has a "rate" of 1/3, meaning that it sends only one bit of information per three bits transmitted; the Hamming(7, 4) code has the better rate 4/7.

In practice (e.g., on the Internet), codes with much greater block length, that can correct several errors, are used. Sophisticated error-correcting codes are also used in transmitting information (e.g., pictures) back to Earth from outer space. A simple kind of error-correcting code is used in most bank account numbers, to either detect errors or (as above) correct them.

Coding theory is a big subject with lots of linear algebra as well as combinatorics in it; moreover, it is a lively area of research, chockablock with theorems and applications.

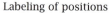

Note that the public often equates "codes" with "secret codes," but we are not talking about cryptography here. Error correcting is often done "on top of" cryptography in the following sense: a message is first encrypted using a secret code (this might not result in any lengthening), to create what is called cyphertext. The cyphertext is then coded with an error-correcting code before it is sent. At the other end, it is decoded to remove transmission errors and recover the correct cyphertext; then the cyphertext is decrypted to get the original message. Whew!

There is, nonetheless, an amusing direct application of error-correcting codes to espionage. Suppose that you are a spy in deep cover and your only means of communication with your headquarters is as follows. A local radio station transmits an unpredictable string of seven bits every day, and you can flip any one of those bits before they are broadcast. How many bits of information can you transmit to headquarters that way?

You can clearly transmit one bit by controlling the parity of the string, that is, whether the string has an even or an odd number of 1s. And you certainly cannot communicate more than three bits because there are only $8 = 2^3$ actions you can take (change any of the seven bits, or change none). But what good can these choices do if your control does not know what the original sequence was?

It is easy using our Hamming$(7,4)$ code. To communicate the three-bit message *abc*, just make the signature of the broadcast *abc*. How do you know you can do this? If the broadcast string was going to have signature *def*, compute the sum of *abc* and *def* as nimbers to get (say) *ghi*, and change the bit with label *ghi*. If, by chance, the broadcast string already had signature *abc*, do not change anything. Your control only has to compute the signature of what she hears on the radio in order to get your message.

Note that if your message is 000, the strings your control might get are exactly the Hamming$(7,4)$ codewords from figure 2.

## 5 Designing a Network

As the head of a new company, or perhaps the emperor of a new country, you may need to design a communications or transportation network to connect your offices or your cities. How can you do this as cheaply as possible?

This and many other questions can be formulated in terms of GRAPH THEORY [II.16], in which objects or places are abstracted as dots, any two of which may or may not bear a particular relation to each other. Graph theory is equally often thought of as a subfield and as a sister field of combinatorics. Like combinatorics, it was not highly regarded among "serious" mathematicians (between 1850 and 1950, roughly) until its importance in computer science emerged. The advent of the Internet, which begs to be modeled as a graph, helped quite a bit as well.

A *graph* $G = \langle V, E\rangle$ is a set (finite, for us) $V$ of *vertices* together with a collection $E$ of two-element subsets of $V$ called *edges*. The reason for this nomenclature is that the vertices and edges of a polyhedron constitute an archetypal graph. The *degree* of a vertex is the number of edges containing it; a graph is regular if all its vertices have the same degree. If $\{u,v\} \in E$, we write $u \sim v$ and say that $u$ is *adjacent* to $v$.

A *path* (of length $k$) in a graph is a sequence $v_0, v_1, \dots, v_k$ of vertices such that $v_i \sim v_{i+1}$ for each $i$, $0 \leqslant i < k$. We require that the $v_i$ are all distinct except that possibly $v_0 = v_k$, in which case the path is said to be *closed* and the sequence $v_0, v_1, \dots, v_{k-1}$ is said to constitute a *cycle*. A nonclosed path is said to *connect* its first and last vertices, and if any two vertices in $V$ can be so connected, $G$ itself is *connected*. The set of vertices adjacent to a given vertex $v$ is called the *neighborhood* of $v$ and is denoted $N(v)$.

A graph that has no more edges than it needs to be connected is called a *tree*. A tree has no cycles (if it had one, deleting one edge of the cycle could not possibly destroy connectivity). It is easily verified that all of the following are equivalent for a graph $G$ with $|V| = n$:

(1) $G$ is a tree;
(2) $G$ is connected and has at most $n-1$ edges;
(3) $G$ is cycle-free and has at least $n-1$ edges;
(4) between any two vertices of $G$ is a unique path;
(5) either $n = 1$ or $G$ has at least two vertices of degree 1 ("leaves"), the deletion of any of which, together with its incident edge, results in a tree.

A transportation network is a graph in which the vertices represent terminals and edges represent direct connections between pairs of nodes; communications networks are modeled similarly. We can already deduce that a network with $n$ vertices that is "efficient" in the sense that it has no more edges than it needs to be connected is a tree with $n-1$ edges. But which $n-1$ edges? Suppose that the network is to be set up from scratch, and for each pair $\{u,v\}$ of nodes there is an associated cost $C(u,v) = C(v,u)$ of building an edge between them. The cost of a tree is the sum of the cost of its edges; is there a good way to find the cheapest tree?

In fact, this could be one of the world's easiest problems; there is no way to go wrong. You can simply start

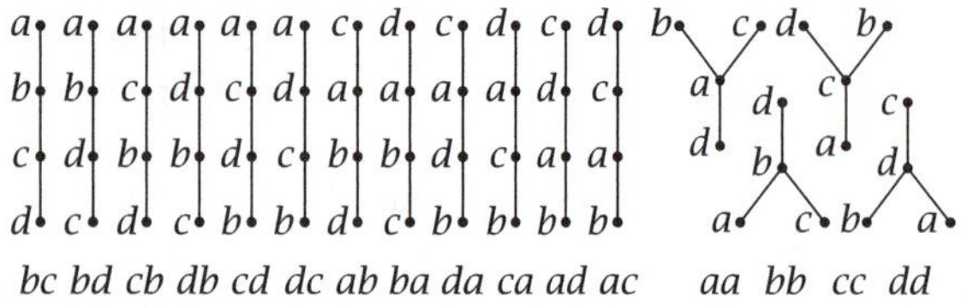


**Figure 4** The sixteen trees on vertex set $\{a, b, c, d\}$.

by taking the cheapest edge, then the next cheapest edge, etc., only making sure that no cycle is made; this is a classic example of what is called a *greedy algorithm.* Alternatively (although this takes longer), you could imagine that you begin with all the edges (the "complete graph" $K_n$), then eliminate edges, starting from the most expensive. Here you must ensure that you never disconnect the graph; if removing an edge would do so, that edge is kept (for good).

## 6 Enumerating Trees

Suppose you belong to a special-interest group, and you wish to set up a *notification tree.* The idea is this: as soon as anyone obtains some information of interest to the group (such as the time and place of the next meeting), she notifies her neighbors on the tree, each of whom notifies all of their neighbors other than the one from whom she got the information, and so on.

How many ways are there to choose such a tree? Here is an intuition test: if the group has, say, ten members, would you guess that the number of possible trees is less than a hundred million, or more?

Very importantly, the vertices of the trees we are counting here are *labeled,* in this case by people. Thus, for example, there are three possible trees if there are only three group members (any of the three could be the center vertex). A little more work will convince you that there are sixteen trees with four labeled vertices, shown in figure 4.

Beneath each tree you see a two-letter sequence called the *Prüfer code* of that tree. The multiplication rule tells us that there are $16 = 4^2$ two-letter sequences of symbols chosen from the set $\{a, b, c, d\}$, here, one for each tree.

How is the Prüfer code determined? We find the lowest-lettered leaf (that is, the leaf labeled by the earliest letter in the alphabet) of the tree and delete it, writing down not the label of the leaf but the label of the vertex it was adjacent to. We then repeat this process until only two vertices remain; thus, if the original tree had $n$ vertices, we will end up with a sequence of $n - 2$ elements, possibly with some repetitions, from a label set of size $n$.

The process is reversible. Suppose we have been given a sequence $x_1, x_2, \dots, x_{n-2}$ of $n - 2$ elements from an ordered set of size $n$; we claim there is exactly one way to reconstruct the labeled tree from which it arose. Note that the *missing* labels must correspond to the leaves of the tree; thus, the lowest missing label (say, $a$) is attached to $x_1$. We now cross out $x_1$ from the sequence and repeat; $a$ no longer counts as a missing label, since it is taken care of. But now $x_1$ will join the list of missing labels, unless it appears again later in the sequence.

When we reach the last entry of the sequence, we have put in $n - 1$ edges that connect all the labels to make our tree.

We have proved (albeit informally) that there is a one-to-one correspondence between trees with vertices labeled by a set of $n$ elements and sequences of length $n - 2$ of elements from that set. One consequence of this is Cayley's theorem (1854).

**Theorem (Cayley's theorem).** *The number of labeled trees on n vertices is* $n^{n-2}$.

This means that the number of possible notification trees for ten people is exactly a hundred million, so if you guessed that it was smaller or larger, you lose.

Prüfer's correspondence has other useful properties. For example, we have already seen that leaves do not appear in the code; more generally, the number of times a label appears in the code is one less than its degree. So, suppose that we want David ($d$), Eleanor ($e$), and Fred ($f$) to be leaves of our notification tree (because they are not so reliable), while George ($g$) has degree 3. Then our tree's code will have no $d$, $e$, or $f$ in it, but two $g$s. There are $\binom{8}{2} = 28$ ways to decide which places in the sequence to put the $g$s, and each of the other six entries can be chosen $10 - 4 = 6$ ways, so now there are only $28 \cdot 6^6 = 1\,306\,368$ ways to make the tree.

## 7 Maximizing a Flow

Suppose you are running an oil supply company whose pipelines constitute a graph $G$. Each edge $e$ of $G$ is a pipe with capacity $c(e)$ (measured, perhaps, in gallons per minute). You need to move a lot of oil from vertex $s$ (the "source") to vertex $t$ (the "tsink"); what's the maximum flow rate you can achieve in this task?

To define a "flow" abstractly we need to represent each edge $\{u, v\}$ by two *arcs,* $(u, v)$ (thought of as an edge going from $u$ to $v$) and $(v, u)$. A flow is a function $f$ from the arcs to the nonnegative real numbers that satisfies the following property.

*For any vertex $u$ other than $s$ or $t$, $\sum_{v \sim u} f(v,u) = \sum_{v \sim u} f(u,v)$.*

This condition says that the amount flowing into a vertex is equal to the amount flowing out, except at the source (where, normally, we only want stuff flowing out) and the tsink (where stuff only flows in). The amount leaving the source is then equal to the amount arriving at the tsink, and we call that amount the *magnitude* of the flow, denoted $|f|$.

A flow is *valid* if it respects capacities, that is, if $f(u,v) \leqslant c(u,v)$ for each arc $(u,v)$. (Initially, the capacity of each arc $(u,v)$ is the capacity of the edge $\{u,v\}$ that gave birth to it.) Our object is to maximize $|f|$ subject to $f$ being a valid flow.

The "max-flow problem" was formulated in the 1950s by T. E. Harris (modeling Soviet rail traffic) and solved by, among others, Lester Ford and Delbert Fulkerson in 1955. Their algorithm and theorem are presented below. Since that time, there have been many alternatives, variations, and improvements, e.g., by Dinitz, by Edmonds and Karp, and by Goldberg and Tarjan.

The Ford–Fulkerson algorithm proceeds by iteration and theoretically might not terminate when the capacities are not rational. If the capacities are integers, it yields an integer flow that is maximal among all flows. The idea of the algorithm is that given *some* flow $f_i$, we "subtract" it from the current graph $G_i$ to create what is called a *residual network* $G_{i+1}$, and then we look for a flow on this that can be added to $f_i$.

We can start with the zero flow and look for any path in $G_0 = G$ from $s$ to $t$. We can then send whatever is the minimum capacity (say, $c$) of the arcs on the path from $s$ to $t$.

If a step of the path is from $u$ to $v$, then in the residual graph $G_1$ we must reduce the capacity of the arc $(u,v)$ by $c$. Equally importantly, we must *increase* the capacity of the reverse arc $(v,u)$ by $c$ because we may later wish to send stuff from $v$ to $u$ and, in so doing, we get to cancel stuff sent from $u$ to $v$ as well as refill the pipe in the other direction. So, in the residual graphs the two arcs constituting a single edge will generally have different capacities, one of which may be zero. (We can have that in the original graph, too, if we want.)

At the next step we look for a path in $G_1$ from $s$ to $t$ involving arcs with positive capacity, and we repeat the procedure. If we can find no such path, we are done.

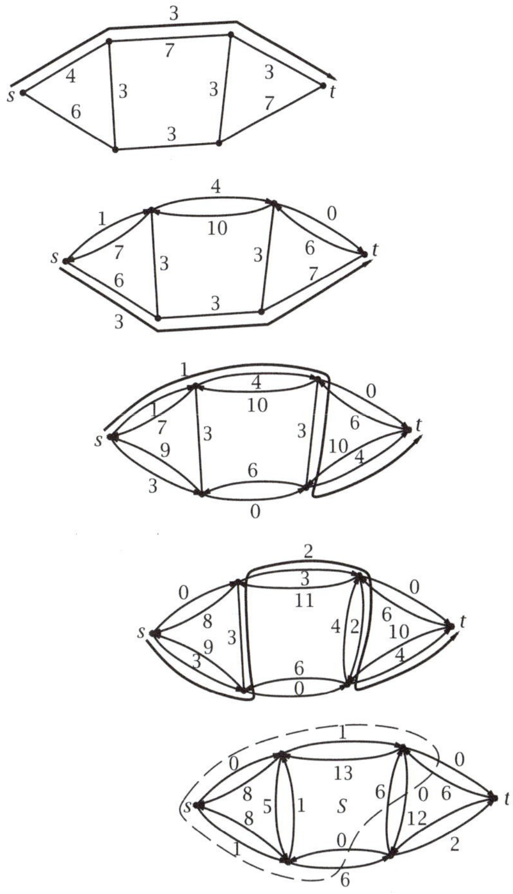


**Figure 5** A maximum flow and a minimum cut (value 9) are found.

How do we know there is no "positive" path? We can check each vertex, starting with the neighbors of $s$ and then moving to their neighbors, etc., to see whether flow can be sent there. Let $S$ be the set of all vertices, $s$ itself included, to which we can still send material. If $S$ does not contain $t$, it means that no flow can be sent out of $S$ in our current residual graph.

Figure 5 shows four Ford–Fulkerson steps leading to a total flow of value 9 and also to a "cut" (see below) of the same value. The heavy black line in each residual graph shows the chosen positive path.

When the algorithm terminates, after step $k$, say, the sum $f = f_1 + f_2 + \cdots + f_k$ of our flows is a maximum flow, and we can prove it. The reason we cannot escape the set $S$ of vertices that we can still ship to can only be that $f$ has used all the original arcs from $S$ to $V \setminus S$ to full capacity, and no valid flow on $G$ can do better than that. Any set of vertices containing $s$ and not $t$ is called a *cut*; the capacity of a cut is the sum of the capacities of the arcs leaving it. No flow's magnitude can exceed the capacity of *any* cut; succinctly, "max flow $\leqslant$ min cut." But when the Ford–Fulkerson algorithm terminates, we have correctly deduced that max flow = min cut. In fact, this is true whether the algorithm terminates or not, but we will not complete the proof here.

The point is that there are efficient algorithms (LINEAR PROGRAMMING [IV.11 §3] is one method) to determine both the maximum flow and the minimum cut in any $s$–$t$ network. The fact that these quantities are equal (which is itself a special case of linear programming duality) is elementary but powerful; we will see one of its myriad consequences in the next section.

## 8 Assigning Workers to Jobs

Suppose you are managing a business and have jobs to fill, together with a pool of workers from which to fill them. The difficulty is that each worker is qualified to do only certain jobs. Can the positions be filled?

The problem seems similar to finding a stable matching; we do want a matching, but here the criterion is simply that each job is matched to a worker who is qualified to do it. Clearly, such a matching might not be available; for example, there might be a job no one in the pool is qualified to do. How can we tell when all the positions can be filled?

The situation is modeled nicely by what we call a *bipartite* graph, one whose vertices can be split into two parts (here, workers and jobs) in such a way that all edges contain a vertex from each part. For this problem, let $X$ be the set of jobs and $Y$ the pool of workers; we draw an edge from $x \in X$ to $y \in Y$ if job $x$ can be performed by worker $y$. A *matching* is a disjoint set of edges. A *complete* matching (also called "a matching that covers $X$") is a disjoint set of edges that covers every vertex in $X$, i.e., fills every job.

To find a complete matching, we clearly need $|Y| \geqslant |X|$, and, as mentioned before, we need every $x \in X$ to have degree at least 1. In fact, we can generalize these two requirements as follows. For any set of jobs $A \subset X$, let $N(A)$ be the set of workers who are qualified to do at least one job in $A$. We then need $|N(A)| \geqslant |A|$ to have any chance of success. In words, then, in order for a complete matching to be possible, we require that for every set of $k$ jobs, we need to have at least $k$ workers

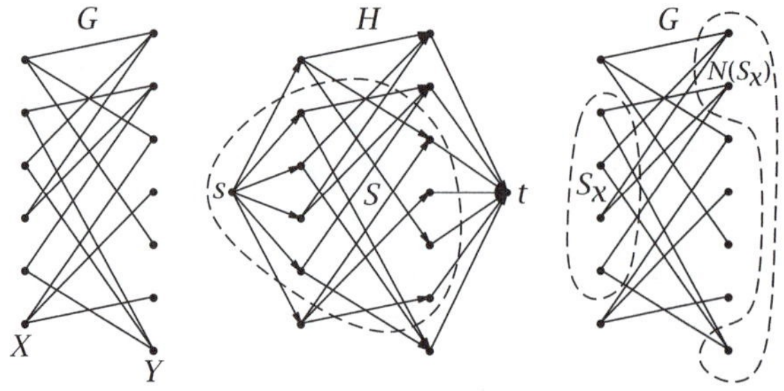


**Figure 6** A bipartite graph and its corresponding network.

who are qualified to do one or more of the jobs in the set.

That this set of criteria is sufficient is known as Hall's marriage theorem and is attributed to Philip Hall (1935).

**Theorem (Hall's marriage theorem).** *Let $G$ be a bipartite graph with parts $X$ and $Y$. A matching that covers $X$ is then possible if and only if for every subset $A \subset X$, the neighborhood $N(A)$ of $A$ in $Y$ has size at least the size of $A$.*

*Proof.* If there is a matching covering $X$, then any subset $A$ of $X$ is matched to a set in $Y$ of equal size, so $A$'s neighborhood in $Y$ must have been at least the size of $A$.

What remains is to show that, if $G$ does not have a matching that covers $X$, then there must be a subset $A$ of $X$ with $|N(A)| < |A|$. We do this by creating from $G$ an $s$–$t$ network $H$, as follows. We add a source $s$ adjacent to every vertex in $X$, and a tsink $t$ adjacent to every vertex in $Y$. All arcs from $s$ to $X$, $X$ to $Y$, and $Y$ to $t$ are given capacity 1, all other arcs 0 (see figure 6).

If there is a complete matching $M$, we can use the edges of $M$ and the new edges to get a flow on $H$ of magnitude $|X|$. Otherwise, by the max-flow min-cut theorem, there is a cut $S$ in $H$ of capacity less than $|X|$; let $S_X = S \cap X$, $S_Y = S \cap Y$. The capacity of $S$ is the number of arcs flowing out of it, which are of three types: edges from $s$ to $X \setminus S_X$, numbering $|X| - |S_X|$; edges from $S_X$ to $Y \setminus S_Y$, numbering, say, $b$; and edges from $S_Y$ to $t$, numbering $|S_Y|$. But $b + |S_Y|$ is at least the size of the neighborhood $N(S_X)$ of $S_X$ in $G$ because every such neighbor has an edge counted in $|S_Y|$ if it is in $S_Y$ and an edge counted in $b$ if it is not. Thus $|X| - |S_X| + |N(S_X)| < |X|$, and therefore $|N(S_X)| < |S_X|$, violating Hall's condition. □

## 9 Distributing Frequencies

Suppose you are setting up wireless phone towers in a developing country. Each tower communicates with the cell phones in its range using one of a small set of reserved frequencies. (Many cell phones can use the same frequency; in one method, called "time-division multiplexing," the tower divides each millisecond into parts and assigns one part to each active phone.) How can the frequencies be assigned so that no two towers that are close enough to cause interference use the same frequency?

We naturally model the towers by a graph $G$ whose vertices are the towers, two of which are adjacent if they have a potential interference problem. We then need to assign frequencies—think of them as colors—to the vertices in such a way that no two adjacent vertices get the same color. If we can get such a "proper" coloring with $k$ colors, $G$ is said to be "$k$-colorable." The smallest $k$ for which $G$ is $k$-colorable is the *chromatic number* of $G$, denoted $\chi(G)$.

Bipartite graphs are 2-colorable, since we can use one color for each part. But suppose $G$ is given to us without identifying parts; can we tell when it is 2-colorable?

The answer is yes; there is an efficient algorithm for this. We may assume that $G$ is connected (otherwise, we execute the algorithm on each connected piece of $G$). Pick any vertex $v$ and color it red; then color all of its neighbors blue. Now color all *their* neighbors red, and so on until all the vertices of $G$ are colored.

When will this fail? In order for two neighboring vertices, say $x$ and $y$, to get the same color, each must be reachable from $v$ by paths of the same parity (both even length, or both odd). Let $z$ be the closest point to $x$ and $y$ that shares this property with $v$; the paths from $z$ to $x$ and to $y$, together with the edge from $x$ to $y$, then form a cycle in $G$ of odd length.

But if there is an odd cycle in $G$, it was never possible to color $G$ with two colors. The algorithm will therefore work whenever $G$ is 2-colorable and will (quickly) come a cropper otherwise. Moreover, we have derived an equivalent condition for 2-colorability.

**Theorem.** *A graph $G$ has $\chi(G) \leqslant 2$ if and only if $G$ contains no odd cycle.*

Alas, the situation with larger numbers of colors is apparently quite different; unless P = NP, there is no efficient general algorithm for determining when $\chi(G) \leqslant k$ for fixed $k > 2$.

However, there *are* efficient algorithms for a related problem. Suppose the cell towers need to communicate *with each other*, using a special set $S$ of frequencies that cause interference only when two messages at the same frequency are both being transmitted, or both being received, by the same tower.

Then, if $|S| = k$ and some tower wants to transmit messages to more than $k$ other towers at the same time, or to receive from more than $k$ other towers at the same time, it is out of luck. So each tower limits its objectives accordingly, but does that mean that frequencies can be assigned to messages in such a way that no interference occurs?

Here, the graph we want (call it $H$) needs to have two vertices for every tower, one for its role as a transmitter, the other for its role as a receiver. We put an edge between a transmitter node $x$ and a receiver node $y$ if tower $x$ wants to send a message to tower $y$. $H$ is therefore a bipartite graph, and we could properly color its vertices with two colors if we wished.

However, it is not towers but messages that need frequencies here; in other words, we need to color the edges, not the vertices, of $H$, and we want no vertex to be in two edges of the same color. Equivalently, we want the set of edges that get any particular color to constitute a matching. If we have a vertex (transmitter or receiver) of degree $d$ and if $d$ exceeds the number $k$ of available colors, then as we have already noted, we are stuck. Surprisingly, and very usefully, we can, if $k$ is at least the maximum degree, always assign colors without a conflict.

**Theorem.** *Let $H$ be a bipartite graph all of whose vertices have degree at most $k$. The edges of $H$ can then be colored from a palette of $k$ colors in such a way that no vertex is contained in two edges of the same color.*

This theorem is often attributed to Dénes Kőnig (1931), although there are closely related results, both earlier and later, by others. To prove it, and indeed to actually find such a coloring, it suffices to find a matching that covers all the vertices of degree $k$; we can do that using Hall's marriage theorem or the max-flow min-cut theorem. We then paint all the edges in the matching with the $k$th color and then remove those edges to get a graph of maximum degree at most $k - 1$. Now we repeat the process until we are down to degree 0, with all edges colored.

## 10 Avoiding Crossings

Suppose you want to build a one-layer microchip with components connected as in a particular graph $G$. Can

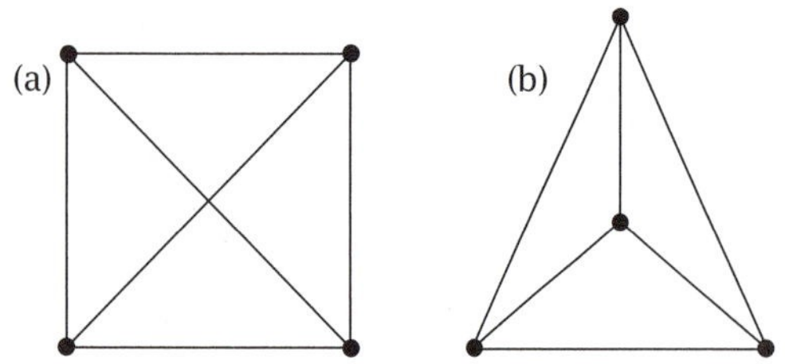


**Figure 7** Two drawings of the complete graph on four vertices.

you do it? More abstractly, given $G$, can you draw it in the plane so that no two edges cross? You might choose to draw the graph $K_4$ (four vertices, every pair constituting an edge) as pictured in figure 7(a), but with a little care you can redraw it as pictured in part (b), with no crossings.

A graph drawn on the plane without crossings is called a *plane graph*, and a graph that *can* be drawn that way is said to be *planar*. (The edges of a plane graph need not be *straight* line segments, although in fact they can always be made so.)

The famous "four-color map theorem," proved in 1976 by Ken Apel and Wolfgang Haken with computer help, says that every planar graph is 4-colorable. The connection between coloring the (contiguous) countries on a map and graph coloring is as follows. Given a map, we associate to every country a point in that country, connecting two points in adjacent countries by an edge that crosses their common border. Properly coloring the vertices of this "dual" graph is equivalent to coloring the countries in such a way that no two countries with a common border get the same color.

The four-color theorem may seem like a somewhat frivolous result—after all, most cartographers have plenty of colors and routinely use more than four—but, in fact, it is a fundamental and important theorem of graph theory. Planar graphs themselves constitute a crucial class of graphs, one that we would be singling out even if we did not live in space, had no use for planes, and had no aversion to crossings.

Since the graph $K_5$ requires five colors, it cannot be planar, if you believe Apel and Haken. But let us see this directly. Our primary tool is the Jordan curve theorem, which says that a simple closed curve in the plane divides the plane into two connected regions (the bounded region inside the curve, and the outside region). Assume that $K_5$ is drawn on the plane without crossings, with vertices labeled 1–5. Note that $K_5$ has $\binom{5}{2} = 10$ edges. The five edges constituting the cycle $1{\sim}2{\sim}3{\sim}4{\sim}5{\sim}1$ make a closed curve, and therefore of the five "noncycle" edges, either at least three must fall inside the curve or at least three must fall outside.

Suppose there are three inside (we can turn the argument inside out to do the other case). We may assume that one is the edge from 1 to 4 (say), but then only one other noncycle edge (from 1 to 3) fails to cross *something*, so we are stuck.

The complete bipartite graph $K_{3,3}$, in which (say) vertices 1, 2, and 3 are adjacent to 4, 5, and 6, is of course 2-colorable, but that does not make it planar. Indeed, assuming it were, consideration of the simple closed curve made by the edges of the cycle $1{\sim}4{\sim}2{\sim}5{\sim}3{\sim}6{\sim}1$ would again lead to a contradiction.

Adding new vertices *along* the edges of $K_5$ or $K_{3,3}$, thereby creating a "subdivision," makes no difference to the above proofs. The remarkable fact is that we have now reached sufficient criteria. The following result, usually known as Kuratowski's theorem, was proved by Kazimierz Kuratowski in 1930, as well as by others around that time.

**Theorem (Kuratowski's theorem).** *A graph $G$ is planar if and only if it does not contain, as a subgraph, any copy of $K_5$ or $K_{3,3}$, or any subdivision thereof.*

We will not give a proof here, but we do note that efficient algorithms exist to determine whether a given graph $G$ is planar and, if it is, to construct a drawing of $G$ with no crossings.

## 11 Delivering the Mail

Suppose you have signed up to be a postman and wish to devise a route that will traverse every street in your district exactly once. When is this possible? When it is, how can you find such a route?

In the abstract, you are given a graph $G$ and wish to devise a *walk* (a sequence of vertices, *not necessarily distinct*, any two consecutive vertices of which are adjacent) that traverses each edge exactly once. Perhaps you would also like to start at a particular vertex $s$ and end at a particular vertex $t$, possibly the same one. Such a walk is called an Eulerian tour or, if it starts and ends at the same point, an Eulerian circuit.

Euler famously observed, in connection with the problem of touring the Königsberg (now Kaliningrad) bridges, that such a walk is not possible if $G$ has more than two vertices of odd degree. The reason for this is simply that a vertex that is not the first or last must be exited the same number of times it is entered, thus using up its edges in pairs. What Euler did *not* do is prove that the possession of not more than two vertices of odd degree, together with connectedness, is sufficient as well as necessary for such a walk to exist.

**Theorem.** *If G is a connected graph with no vertices of odd degree, it has an Eulerian circuit.*

*Proof.* Let $P$ be a maximum-length walk traversing no edge twice. We claim that it is an Eulerian circuit.

Note first that if $P$ starts at $u$ it must end there as well, since $u$ is the only place it could get stuck. Why? To get stuck at a vertex $v \neq u$, you must have run out of edges containing $v$ after using $k$ of them to get to $v$ and $k-1$ to escape from $v$, for some $k$. But $v$, like all vertices of $G$, is supposed to have even degree.

If $P$ is not an Eulerian tour, then because $G$ is connected there is an unused edge of $G$ that connects a vertex of $P$, say $v$, with some vertex $w$ (that may or may not be on $P$). Make a new walk $Q$ starting at $w$, then traversing the edge $\{v, w\}$ to $v$, then following $P$ forward to $u$, then following the first part of $P$ from $u$ back to $v$. $Q$ is longer than $P$ (and is not even stuck yet), so our contradiction has been reached. □

If $G$ is connected and has precisely two vertices of odd degree, say $x$ and $y$, we can connect them by an edge to get a graph all of whose vertices have even degree, and then we can use the theorem to find an Eulerian circuit $P$. Tossing the new edge out of $P$ gives an Eulerian tour beginning at $x$ and ending at $y$. The new edge may duplicate an edge already present in $G$, but the theorem and proof above work fine for "multigraphs," which are allowed to have more than one edge containing the same two vertices.

In fact, the theorem and its proof also work fine for directed graphs, also called "digraphs," with arcs instead of edges that can be traversed only in one direction; some may even be loops from a vertex back to itself. The necessary, and again sufficient, condition for existence of an Eulerian circuit in a connected digraph $D$ is that the "indegree" (the number of arcs entering a vertex) must be equal to the outdegree for every vertex in $D$. Whether for graphs or digraphs, the proof above implicitly contains one of many efficient algorithms for actually generating Eulerian tours or circuits.

The Eulerian circuit theorem for digraphs has a nice application to *de Bruijn sequences.* Suppose you have written some software for a device that has $k$ buttons and you wish to ascertain that no sequence of $n$ button pushes will disable the device. You could separately enter all $k^n$ possible sequences, but it might make sense to save time by making use of overlap. For example, if there are just two buttons (labeled 0 and 1) and $n = 3$, you could enter the sequence 0001011100, which captures all eight length-three binary strings as substrings. Such a sequence, which contains exactly once each string of length $n$ over an alphabet of size $k$, is called a de Bruijn$(k, n)$ sequence, thus named on account of a 1946 paper by Nicolaas Govert de Bruijn. The sequences themselves date back to antiquity; one appears in Sanskrit prosody.

Typically, de Bruijn sequences are thought of as circular, so the above sequence might be shortened to 00010111 with the understanding that you are allowed to go "around the corner" to get 110 and 100. Since you cannot go around the corner in time, checking your device requires that you append a copy of the first $k-1$ characters of a circular sequence to the end.

To see that de Bruijn$(k, n)$ sequences exist for any positive integers $k$ and $n$, we create a digraph $D(k, n)$ as follows. The vertices of $D(k, n)$ are all $k^{n-1}$ sequences of length $k-1$ from our alphabet of size $n$. Arcs correspond to sequences $x_1, \dots, x_k$ of length $n$ and run from the vertex $x_1, \dots, x_{k-1}$ to the vertex $x_2, \dots, x_k$; note that this will be a loop when all the $x_i$ happen to be the same character.

The outdegree of every vertex in $D(k, n)$ will be $n$, since we can add a $k$th character in $n$ ways, and similarly the outdegrees will also all be $n$. To see that $D(k, n)$ is connected, let $x$ be one of the alphabetic characters, and observe that we can get from any vertex to $xxx \cdots x$ by repeatedly adding $x$ to the end and from $xxx \cdots x$ to any vertex by adding that vertex's characters one by one.

Applying the theorem gives an Eulerian circuit in $D$, exactly what is required for a de Bruijn$(k, n)$ sequence. In fact, it can be shown that there are $k!^{k^{n-1}}/k^n$ Eulerian circuits in $(k, n)$, and thus the same number of de Bruijn sequences, considered as cycles. Pictured in figure 8 is the digraph $D(4, 2)$, together with one of the 20 736 resulting de Bruijn$(4, 2)$ sequences.

It is a curiosity that, given any digraph, one can efficiently compute the precise number of Eulerian circuits, while for ordinary (undirected) graphs, no one knows an efficient algorithm that can even *estimate* that number. (In many cases, algorithms that work for digraphs work just as well for undirected graphs because you can replace an edge by an arc in each direction, as we did above in section 7. But doing that here would result in walks that traverse every edge *twice.*)

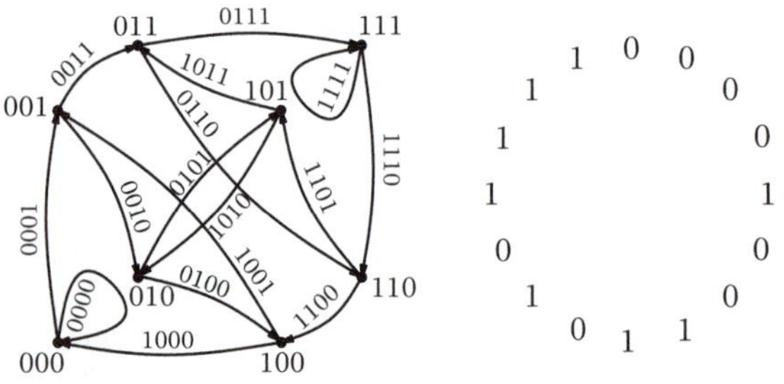


**Figure 8** The digraph $D(4,2)$ and the de Bruijn sequence arising from an Eulerian circuit.

We end this section, and indeed our whole discussion of applied combinatorics and graph theory, with one additional observation. Instead of the digraph $D(k,n)$, we could have created an undirected graph $G(k,n)$ whose vertices are all the sequences of length $k$, two being connected if they overlap in a string of length $k-1$. A de Bruijn$(k,n)$ sequence would then constitute a cycle in $G(k,n)$ that hits every *vertex* (instead of every edge) exactly once; such a cycle is called a *Hamilton circuit*.

But computability with respect to Hamilton circuits is startlingly different from the Eulerian case. Unless P = NP, there is no efficient way to determine whether an input graph has a Hamilton circuit or, if it does, to find one. It is hard to say why, apart from noting that in one case there is a theorem that provides an easy algorithm while in the other case no known theorem comes to the rescue. It seems that some combinatorial tasks are easy, and some are hard; that is just the way it is.

We have tried here to present some of the combinatorial problems that are easy, once you know how to do them. We hope that all your problems are similarly straightforward, but if they are not, some (of the many) books that will get you deeper into the above topics are listed below.

### Further Reading

Brualdi, R. 2010. *Introductory Combinatorics*, 5th edn. Upper Saddle River, NJ: Prentice Hall.

Gusfield, D., and R. W. Irving. 1989. *The Stable Marriage Problem: Structure and Algorithms*. Cambridge, MA: MIT Press.

Van Lint, J. H. 1999. *Introduction to Coding Theory*, 3rd edn. Berlin: Springer.

Stanley, R. P. 1986. *Enumerative Combinatorics*, volumes I and II. Monterey, CA: Wadsworth & Brooks/Cole.

West, D. B. 1996. *Introduction to Graph Theory*, 2nd edn. Upper Saddle River, NJ: Prentice Hall.

## IV.38 Combinatorial Optimization

*Jens Vygen*

Combinatorial optimization problems arise in numerous applications. In general, we look for an optimal element of a finite set. However, this set is too large to be enumerated; it is implicitly given by its combinatorial structure. The goal is to develop efficient algorithms by understanding and exploiting this structure.

### 1 Some Important Problems

First we give some classical examples. We refer to the article on GRAPH THEORY [II.16] for basic notation. In a directed graph, we denote by $\delta^+(X)$ and $\delta^-(X)$ the set of edges leaving and entering $X$, respectively; here, $X$ can be a vertex or a set of vertices. In an undirected graph, $\delta(X)$ denotes the set of edges with exactly one endpoint in $X$.

#### 1.1 Spanning Trees

In this problem we are given a finite connected undirected graph $(V,E)$ (so $V$ is the set of vertices and $E$ the set of edges) and weights on the edges, i.e., $c(e) \in \mathbb{R}$ for all $e \in E$. The task is to find a set $T \subseteq E$ such that $(V,T)$ is a (spanning) tree and $\sum_{e \in T} c(e)$ is minimized. (Recall that a *tree* is a connected graph without cycles.)

A set $V$ of eight points in the Euclidean plane is shown on the left of the figure below. Assuming that $(V,E)$ is the complete graph on these points (every pair of vertices is connected by an edge) and $c$ is the Euclidean distance, the right-hand side shows an optimal solution.

#### 1.2 Maximum Flows

Given a finite directed graph $(V,E)$, two vertices $s,t \in V$ (*source* and *sink*), and *capacities* $u(e) \in \mathbb{R}_{\geqslant 0}$ for all $e \in E$, we look for an $s$-$t$ *flow* $f\colon E \to \mathbb{R}_{\geqslant 0}$ with $f(e) \leqslant u(e)$ for all $e \in E$ and $f(\delta^-(v)) = f(\delta^+(v))$ for all $v \in V \setminus \{s,t\}$ (flow conservation: the total incoming flow equals the total outgoing flow at any vertex except $s$ and $t$). The goal is to maximize $f(\delta^-(t)) - f(\delta^+(t))$, i.e., the total amount of flow shipped from $s$ to $t$. This is called the *value* of $f$.

The figure below illustrates this: the left-hand side displays an instance, with the capacities shown next to the edges, and the right-hand side shows an $s$-$t$ flow of value 7. This is not optimal.

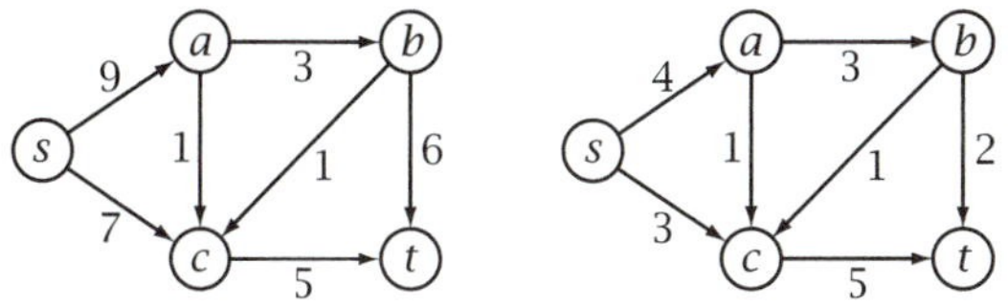


### 1.3 Matching

Given a finite undirected graph $(V, E)$, find a matching $M \subseteq E$ that is as large as possible. (A *matching* is a set of edges whose endpoints are all distinct.)

### 1.4 Knapsack

Given $n \in \mathbb{N}$, positive integers $a_i$, $b_i$ (the profit and weight of item $i$, for $i = 1, \dots, n$), and $B$ (the knapsack's capacity), find a subset $I \subseteq \{1, \dots, n\}$ with $\sum_{i \in I} b_i \leqslant B$ such that $\sum_{i \in I} a_i$ is as large as possible.

### 1.5 Traveling Salesman

Given a finite set $X$ with metric $d$, find a bijection $\pi \colon \{1, \dots, n\} \to X$ such that the length of the corresponding tour,

$$\sum_{i=1}^{n-1} d(\pi(i), \pi(i+1)) + d(\pi(n), \pi(1)),$$

is as small as possible.

### 1.6 Set Covering

Given a finite set $U$ and subsets $S_1, \dots, S_n$ of $U$, find the smallest collection of these subsets whose union is $U$, i.e., $I \subseteq \{1, \dots, n\}$ with $\bigcup_{i \in I} S_i = U$ and $|I|$ (the number of elements in the set $I$) minimized.

## 2 General Formulation and Goals

### 2.1 Instances and Solutions

These problems have many common features.

In each case there are infinitely many *instances*, each of which can be described (up to renaming) by a finite set of bits and in some cases by a finite set of real numbers.

For each instance, there is a set of *feasible solutions.* This set is finite in most cases. In the maximum-flow problem it is actually infinite, but even here one can, without loss of generality, restrict to a finite set of solutions (see below).

Given an instance and a feasible solution, we can easily compute its *value.* For example, in the matching problem, the instances are the finite undirected graphs; for each instance $G$, the set of feasible solutions are the matchings in $G$; and for each matching, its value is simply its cardinality.

Even if the number of feasible solutions is finite, it cannot be bounded by a polynomial in the *instance size* (the number of bits that is needed to describe the instance). For example, there are $n^{n-2}$ trees $(V, T)$ with $V = \{1, \dots, n\}$ (this is Cayley's formula). Similarly, the number of matchings on $n$ vertices, of subsets of an $n$-element set, and of permutations on $n$ elements grow exponentially in $n$. One cannot enumerate all of them in reasonable time except for very small $n$.

Whenever an instance contains real numbers, we assume that we can do elementary operations with them, or we assume them to be rationals with binary encoding.

### 2.2 Algorithms

The main goal of combinatorial optimization is to devise efficient algorithms for solving such problems.

*Efficient* usually means in *polynomial time,* that is, the number of elementary steps can be bounded by a polynomial in the instance size. Of course, the faster the better.

*Solving* a problem usually means always (i.e., for every given instance) computing a feasible solution with optimum value.

We give an example of an efficient algorithm solving the spanning tree problem in section 3.

However, for NP-HARD [I.4 §4.1] problems (like the last three examples in our list), an efficient algorithm that solves the problem does not exist unless P = NP, and consequently one is satisfied with less (see section 5).

### 2.3 Other Goals

Besides developing algorithms and proving their correctness and efficiency, combinatorial optimization (and related areas) also comprises other work:

- analyzing combinatorial structures, such as graphs, matroids, polyhedra, hypergraphs;
- establishing relations between different combinatorial optimization problems (reductions, equivalence, bounds, relaxations);

- proving properties of optimal (or near-optimal) solutions;
- studying the complexity of problems and establishing hardness results;
- implementing algorithms and analyzing their practical performance; and
- applying combinatorial optimization problems to real-world problems.

## 3 The Greedy Algorithm

The spanning tree problem has a very simple solution: the *greedy algorithm* does the job. We can start with the empty set and successively pick a cheapest edge that does not create a cycle until our subgraph is connected. Formally,

(1) sort $E = \{e_1, \dots, e_m\}$ so that $c(e_1) \leqslant \cdots \leqslant c(e_m)$,
(2) let $T$ be the empty set, and
(3) for $i = 1, \dots, m$ do
- if $(V, T \cup \{e_i\})$ contains no cycle,
- then add $e_i$ to $T$.

In our example, the first four steps would add the four shortest edges (shown on the left-hand side below). Then the dotted edge is examined, but it is not added as it would create a cycle. The right-hand side shows the final output of the algorithm.

This algorithm can be easily implemented so that it runs in $O(nm)$ time, where $n = |V|$ and $m = |E|$. With a little more care, a running time of $O(m \log n)$ can be obtained. This is therefore a polynomial-time algorithm.

This algorithm computes a maximal set $T$ such that $(V, T)$ contains no cycle. In other words, $(V, T)$ is a tree. It is not completely obvious that the output $(V, T)$ is always an optimal solution, i.e., a tree with minimum weight. Let us give a nice and instructive proof of this fact.

### 3.1 Proof of Correctness

Let $(V, T^*)$ be an optimal tree, and choose $T^*$ so that $|T^* \cap T|$ is as large as possible. Suppose $T^* \neq T$.

All spanning trees have exactly $|V| - 1$ edges, implying that $T^* \setminus T \neq \emptyset$. Let $j \in \{1, \dots, m\}$ be the smallest index with $e_j \in T^* \setminus T$.

Since the greedy algorithm did not add $e_j$ to $T$, there must be a cycle with edge set $C \subseteq \{e_j\} \cup (T \cap \{e_1, \dots, e_{j-1}\})$ and $e_j \in C$.

$(V, T^* \setminus \{e_j\})$ is not connected, so there is a set $X \subset V$ with $\delta(X) \cap T^* = \{e_j\}$. (Recall that $\delta(X)$ denotes the set of edges with exactly one endpoint in $X$.)

Now $|C \cap \delta(X)|$ is even (any cycle enters a set $X$ the same number of times that it leaves $X$), so it is at least two. Let $e_i \in (C \cap \delta(X)) \setminus \{e_j\}$. Note that $i < j$ and thus $c(e_i) \leqslant c(e_j)$.

Let $T^{**} := (T^* \setminus \{e_j\}) \cup \{e_i\}$. Then $(V, T^{**})$ is a tree with $c(T^{**}) = c(T^*) - c(e_j) + c(e_i) \leqslant c(T^*)$. So $T^{**}$ is also optimal. But $T^{**}$ has one more edge in common with $T$ (the edge $e_i$) than $T^*$, contradicting the choice of $T^*$.

### 3.2 Generalizations

In general (and for any of the other problems above), no simple greedy algorithm will always find an optimal solution.

The reason that the greedy approach works for spanning trees is that here the feasible solutions form the bases of a *matroid*. Matroids are a well-understood combinatorial structure that can in fact be characterized by the optimality of the greedy algorithm.

Generalizations such as optimization over the intersection of two matroids or minimization of submodular functions (given by an oracle) can also be solved in polynomial time, with more complicated combinatorial algorithms.

## 4 Duality and Min–Max Equations

The relationships between different problems can lead to many important insights and algorithms. We give some well-known examples.

### 4.1 The Max-Flow Min-Cut Theorem

We begin with the maximum-flow problem and its relation to $s$–$t$ cuts. An $s$–$t$ cut is the set of edges leaving $X$ (denoted by $\delta^+(X)$) for a set $X \subset V$ with $s \in X$ and $t \notin X$.

The total capacity of the edges in such an $s$–$t$ cut, denoted by $u(\delta^+(X))$, is an upper bound on the value of any $s$–$t$ flow $f$ in $(G, u)$. This is because this value is precisely $f(\delta^+(X)) - f(\delta^-(X))$ for every set $X$ containing $s$ but not $t$, and $0 \leqslant f(e) \leqslant u(e)$ for all $e \in E$.

The famous *max-flow min-cut theorem* says that the upper bound is tight: the maximum value of an $s$–$t$ flow equals the minimum capacity of an $s$-$t$ cut. In other words, if $f$ is any $s$-$t$ flow with maximum value, then there is a set $X \subset V$ with $s \in X$, $t \notin X$, $f(e) = u(e)$ for all $e \in \delta^+(X)$, and $f(e) = 0$ for all $e \in \delta^-(X)$.

Indeed, if no such set exists, we can find a directed path $P$ from $s$ to $t$ in which each edge $e = (v, w)$ is either an edge of $G$ with $f(e) < u(e)$ or the reverse: $e' := (w, v)$ is an edge of $G$ with $f(e') > 0$. (This follows from letting $X$ be the set of vertices that are reachable from $s$ along such paths.)

Such paths are called *augmenting paths* because along such a path we can augment the flow by increasing it on forward edges and decreasing it on backward edges. Some flow algorithms (but generally not the most efficient ones) start with the all-zero flow and successively find an augmenting path.

The figure below shows how to augment the flow shown in section 1.2 by one unit along the path $a$-$c$-$b$-$t$ (shown in bold on the left). The resulting flow, with value 8 (shown on the right), is optimal, as is proved by the $s$-$t$ cut $\delta^+(\{s, a, c\}) = \{(a, b), (c, t)\}$ of capacity 8.

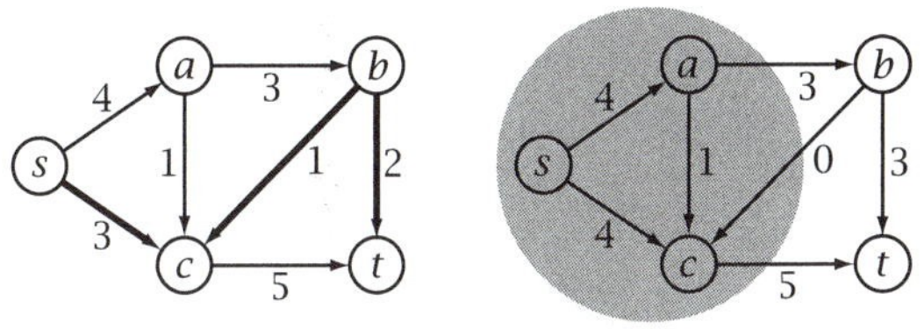


The above relation also shows that for finding an $s$-$t$ cut with minimum capacity, it suffices to solve the maximum-flow problem. This can also be used to compute a minimum cut in an undirected graph or to compute the connectivity of a given graph.

Any $s$-$t$ flow can be decomposed into flows on $s$-$t$ paths, and possibly on cycles (but cyclic flow is redundant as it does not contribute to the value). This decomposition can be done greedily, and the list of paths is then sufficient to recover the flow. This shows that one can restrict to a finite number of feasible solutions without loss of generality.

### 4.2 Disjoint Paths

If all capacities are integral (i.e., are integers), one can find a maximum flow by always augmenting by 1 along an augmenting path, until none exists anymore. This is not a polynomial-time algorithm (because the number of iterations can grow exponentially in the instance size), but it shows that in this case there is always an optimal flow that is integral. An integral flow can be decomposed into integral flows on paths (and possibly cycles).

Hence, in the special case of unit capacities an integral flow can be regarded as a set of pairwise edge-disjoint $s$-$t$ paths. Therefore, the max-flow min-cut theorem implies the following theorem, due to Karl Menger. Let $(V, E)$ be a directed graph and let $s, t \in V$. Then the maximum number of paths from $s$ to $t$ that are pairwise edge disjoint equals the minimum number of edges in an $s$-$t$ cut.

Other versions of Menger's theorem exist, for instance, for undirected graphs and for (internally) vertex-disjoint paths.

In general, finding disjoint paths with prescribed endpoints is difficult; for example, it is NP-COMPLETE [I.4 §4.1] to decide whether, in a given directed graph with vertices $s$ and $t$, there is a path $P$ from $s$ to $t$ and a path $Q$ from $t$ to $s$ such that $P$ and $Q$ are edge disjoint.

### 4.3 Linear Programming Duality

The maximum-flow problem (and also generalizations like minimum-cost flows and multicommodity flows) can be formulated as linear programs in a straightforward way.

Most other combinatorial optimization problems involve binary decisions and can be formulated naturally as (mixed-) integer linear programs. We give an example for the matching problem.

The matching problem can be written as the *integer linear program*

$$\max\{\mathbb{1}^{\mathrm{T}}x \colon Ax \leqslant \mathbb{1},\ x_e \in \{0,1\}\ \forall e \in E\},$$

where $A$ is the vertex-edge-incidence matrix of the given graph $G = (V, E)$, $\mathbb{1} = (1, 1, \dots, 1)^{\mathrm{T}}$ denotes an appropriate all-one vector (so $\mathbb{1}^{\mathrm{T}}x$ is just an abbreviation of $\sum_{e \in E} x_e$), and $\leqslant$ is meant componentwise. The feasible solutions to this integer linear program are exactly the incidence vectors of matchings in $G$.

Solving integer linear programs is NP-hard in general (see section 5.2), but linear programs (without integrality constraints) can be solved in polynomial time (see CONTINUOUS OPTIMIZATION [IV.11 §3]). This is one reason why it is often useful to consider the linear relaxation, which here is

$$\max\{\mathbb{1}^{\mathrm{T}}x \colon Ax \leqslant \mathbb{1},\ x \geqslant 0\},$$

where 0 and $\mathbb{1}$ denote appropriate all-zero and all-one vectors, respectively. The entries of $x$ can now be any real numbers between 0 and 1.

The dual linear program (LP) is

$$\min\{y^{\mathrm{T}}\mathbb{1}\colon y^{\mathrm{T}}A \geqslant \mathbb{1},\ y \geqslant 0\}.$$

By weak duality, every dual feasible vector $y$ yields an upper bound on the optimum. (Indeed, if $x$ is the incidence vector of a matching $M$ and $y \geqslant 0$ with $y^{\mathrm{T}}A \geqslant \mathbb{1}$, then $|M| = \mathbb{1}^{\mathrm{T}}x \leqslant y^{\mathrm{T}}Ax \leqslant y^{\mathrm{T}}\mathbb{1}$.)

If $G$ is bipartite, it turns out that these two LPs actually have integral optimal solutions. The minimal integral feasible solutions of the dual LP are exactly the incidence vectors of *vertex covers* (sets $X \subseteq V$ such that every edge has at least one endpoint in $X$).

In other words, in any bipartite graph $G$, the maximum size of a matching equals the minimum size of a vertex cover. This is a theorem of Dénes Kőnig. It can also be deduced from the max-flow min-cut theorem.

For general graphs, this is not the case, as, for example, the triangle (the complete graph on three vertices) shows. Nevertheless, the convex hull of incidence vectors of matchings in general graphs can also be described well; it is

$$\Big\{x\colon Ax \leqslant \mathbb{1},\ x \geqslant 0,\ \sum_{e\in E[A]} x_e \leqslant \Big\lfloor \frac{|A|}{2} \Big\rfloor\ \forall A \subseteq V\Big\},$$

where $E[A]$ denotes the set of edges whose endpoints both belong to $A$. This was shown by Jack Edmonds in 1965, who also found a polynomial-time algorithm for the matching problem. In contrast, the problem of finding a minimum vertex cover in a given graph is NP-hard.

## 5 Dealing with NP-Hard Problems

The other three problems mentioned in section 1 (knapsack, traveling salesman, and set covering) are NP-hard: they have a polynomial-time algorithm if and only if P = NP.

Since most researchers believe that P ≠ NP, they gave up looking for polynomial-time algorithms for NP-hard problems. Algorithms are sought with weaker properties, for example ones that

- solve interesting special cases in polynomial time;
- run in exponential time but faster than trivial enumeration;
- always compute a feasible solution whose value is at most $k$ times worse than the optimum (so-called *k-approximation algorithms* (see section 5.1));
- are efficient or compute good solutions for most instances, in some probabilistic model;
- are *randomized* (use random bits in their computation) and are expected to behave well; or
- run fast and produce good results in practice, although there is no formal proof (*heuristics*).

### 5.1 Approximation Algorithms

From a theoretical point of view, the notion of approximation algorithms has proved to be most fruitful. For example, for the knapsack problem (section 1.4) there is an algorithm that for any given instance and any given number $\varepsilon > 0$ computes a solution at most $1 + \varepsilon$ times worse than the optimum, and whose running time is proportional to $n^2/\varepsilon$. For the TRAVELING SALESMAN PROBLEM [VI.18] (see section 1.5), there is a $\frac{3}{2}$-approximation algorithm.

For set covering (section 1.6) there is no constant-factor approximation algorithm unless P = NP. But consider the special case where we ask for a minimum vertex cover in a given graph $G$; here, $U$ is the edge set of $G$ and $S_i = \delta(v_i)$ for $i = 1, \dots, n$, where $V = \{v_1, \dots, v_n\}$ is the vertex set of $G$. Here, we can use the above-mentioned fact that the size of any matching in $G$ is a lower bound. Indeed, if we take any (inclusion-wise) maximal matching $M$ (e.g., one found by the greedy algorithm), then the $2|M|$ endpoints of the edges in $M$ form a vertex cover. As $|M|$ is a lower bound on the optimum, this is a simple 2-approximation algorithm.

### 5.2 Integer Linear Optimization

Most classical combinatorial optimization problems can be formulated as integer linear programs

$$\min\{c^{\mathrm{T}}x\colon Ax \leqslant b,\ x \in \mathbb{Z}^n\}.$$

This includes all problems discussed in this chapter, except the maximum-flow problem, which is in fact a linear program. The variables are often restricted to 0 or 1. Sometimes, some variables are continuous, while others are discrete:

$$\min\{c^{\mathrm{T}}x\colon Ax + By \leqslant d,\ x \in \mathbb{R}^m,\ y \in \mathbb{Z}^n\}.$$

Such problems are called mixed-integer linear programs.

*Discrete optimization* comprises combinatorial optimization but also general (mixed-) integer optimization problems with no special combinatorial structure.

For general (mixed-) integer linear optimization all known algorithms have exponential worst-case running time. The most successful algorithms in practice use a combination of cutting planes and branch and bound (see sections 6.2 and 6.7). These are implemented in advanced commercial software. Since many practical problems (including almost all classical combinatorial optimization problems) can be described as (mixed-) integer linear programs, such software is routinely used in practice to solve small and medium-sized instances of such problems. However, combinatorial algorithms that exploit the specific structure of the given problem are normally superior and are often the only choice for very large instances.

## 6 Techniques

Since good algorithms have to exploit the structure of the problem, every problem requires different techniques. Some techniques are quite general and can be applied for a large variety of problems, but in many cases they will not work well. Nevertheless, we list the most important techniques that have been applied successfully to several combinatorial optimization problems.

### 6.1 Reductions

Reducing an unknown problem to a known (and solved) problem is of course the most important technique. To prove hardness, one proceeds the other way round: we reduce a problem that we know to be hard to a new problem (that must then also be hard). If reductions work in both ways, problems can actually be regarded to be equivalent.

### 6.2 Enumeration Techniques

Some problems can be solved by skillful enumeration. Dynamic programming is such a technique. It works if optimal solutions arise from optimal solutions to "smaller" problems by simple operations. DIJKSTRA'S SHORTEST-PATH ALGORITHM [VI.10] is a good example. Many algorithms on trees use dynamic programming.

*Branch and bound* is another well-known enumeration technique. Here, one enumerates only parts of a decision tree because lower and upper bounds tell us that the unvisited parts cannot contain a better solution. How well this works mainly depends on how good the available bounds are.

### 6.3 Reducing or Decomposing the Instance

Often, an instance can be preprocessed by removing irrelevant parts. In other cases one can compute a smaller instance or an instance with a certain structure whose solution implies a solution of the original instance.

Another well-known technique is DIVIDE AND CONQUER [I.4 §3]. In some problems, instances can be decomposed/partitioned into smaller instances whose solutions can then be combined in some way.

### 6.4 Combinatorial or Algebraic Structures

If the instances have a certain structure (like planarity or certain connectivity or sparsity properties of graphs, cross-free set families, matroid structures, submodular functions, etc.), this must usually be exploited.

Also, optimal solutions (of relaxations or the original problem) often have a useful structure. Sometimes (e.g., by sparsification or uncrossing techniques) such a structure can be obtained even if it is not there to begin with.

Many algorithms compute and use a combinatorial structure as a main tool. This is often a graph structure, but sometimes an algebraic view can reveal certain properties. For instance, the Laplacian matrix of a graph has many useful properties. Sometimes simple properties, like parity, can be extremely useful and elegant.

### 6.5 Primal–Dual Relations

We discussed linear programming duality, a key tool for many algorithms, above. Lagrangian duality can also be useful for nonlinear problems, and sometimes other kinds of duality, like planar duality or dual matroids, are very useful.

### 6.6 Improvement Techniques

It is natural to start with some solution and iteratively improve it. The greedy algorithm and finding augmenting paths can be considered as special cases. In general, some way of measuring progress is needed so that the algorithm will terminate.

The general principle of starting with any feasible solution and iteratively improving it by small local changes is called *local search*. Local-search heuristics are often quite successful in practice, but in many cases no reasonable performance guarantees can be given.

### 6.7 Relaxation and Rounding

Relaxations can arise combinatorially (by allowing solutions that do not have a certain property that was originally required for feasible solutions) or by omitting integrality constraints of a description as an optimization problem over variables in $\mathbb{R}^n$.

Linear programming formulations can imply polynomial-time algorithms even if they have exponentially many variables or constraints (by the equivalence of optimization and separation). Linear relaxations can be strengthened by adding further linear constraints, called *cutting planes*.

One can also consider nonlinear relaxations. In particular, *semidefinite relaxations* have been used for some approximation algorithms.

Of course, after solving a relaxation, the originally required property must be restored somehow. If a fractional solution is made integral, this is often called *rounding*. Note that rounding is used here in a general sense (deriving an integral solution from a fractional one), and not specifically meaning rounding to the nearest integer. Sophisticated rounding algorithms for various purposes have been developed.

### 6.8 Scaling and Rounding

Often, a problem becomes easier if the numbers in the instance are small integers. This can be achieved by scaling and rounding, of course at the cost of a loss of accuracy. The knapsack problem (see section 1.4) is a good example; the best algorithms use scaling and rounding and then solve the rounded instance by dynamic programming.

In some cases a solution of the rounded instance can be used in subsequent iterations to obtain more accurate, or even exact, solutions of the original instance more quickly.

### 6.9 Geometric Techniques

The role that geometric techniques play is also becoming more important. Describing (the convex hull of) feasible solutions by a polyhedron is a standard technique. Planar embeddings of graphs (if they exist) can often be exploited in algorithms. Approximating a certain metric space by a simpler one is an important technique in the design of approximation algorithms.

### 6.10 Probabilistic Techniques

Sometimes, a probabilistic view makes problems much easier. For example, a fractional solution can be viewed as a convex combination of extreme points, or as a probability distribution. Arguing over the expectation of some random variables can lead to simple algorithms and proofs. Many randomized algorithms can be derandomized, but this often complicates matters.

### Further Reading

Korte, B., and J. Vygen. 2012. *Combinatorial Optimization: Theory and Algorithms*, 5th edn. Berlin: Springer.

Schrijver, A. 2003. *Combinatorial Optimization: Polyhedra and Efficiency*. Berlin: Springer.

## IV.39 Algebraic Geometry

*Frank Sottile*

Physical objects and constraints may be modeled by polynomial equations and inequalities. For this reason, algebraic geometry, the study of solutions to systems of polynomial equations, is a tool for scientists and engineers. Moreover, relations between concepts arising in science and engineering are often described by polynomials. Whatever their source, once polynomials enter the picture, notions from algebraic geometry—its theoretical base, its trove of classical examples, and its modern computational tools—may all be brought to bear on the problem at hand.

As a part of applied mathematics, algebraic geometry has two faces. One is an expanding list of recurring techniques and examples that are common to many applications, and the other consists of topics from the applied sciences that involve polynomials. Linking these two aspects are algorithms and software for algebraic geometry.

### 1 Algebraic Geometry for Applications

We present here some concepts and objects that are common in applications of algebraic geometry.

#### 1.1 Varieties and Their Ideals

The fundamental object in algebraic geometry is a *variety* (or an affine variety), which is a set in the vector space $\mathbb{C}^n$ (perhaps restricted to $\mathbb{R}^n$ for an application) defined by polynomials,

$$V(S) := \{x \in \mathbb{C}^n \mid f(x) = 0 \ \forall f \in S\},$$

where $S \subset \mathbb{C}[x] = \mathbb{C}[x_1, \dots, x_n]$ is a set of polynomials. Common geometric figures—points, lines, planes, circles, conics, spheres, etc.—are all algebraic varieties. Questions about everyday objects may therefore be treated with algebraic geometry.

The real points of the line $x + y - 1 = 0$ are shown on the left-hand side of the figure below:

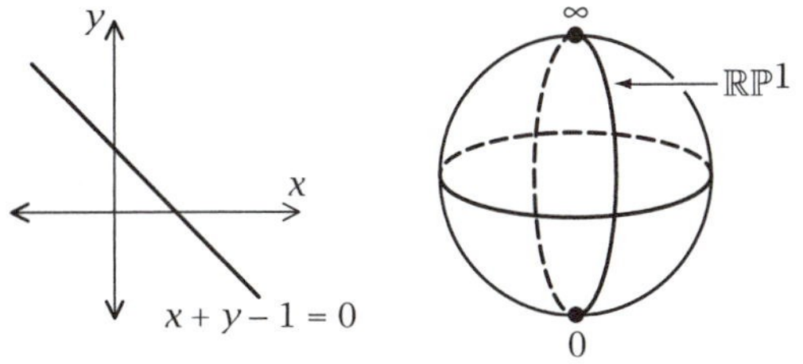


Its complex points are the Argand plane $\mathbb{C}$ embedded obliquely in $\mathbb{C}^2$.

We may compactify algebraic varieties by adding points at infinity. This is done in projective space $\mathbb{P}^n$, which is the set of lines through the origin in $\mathbb{C}^{n+1}$ (or $\mathbb{RP}^n$ for $\mathbb{R}^{n+1}$). This may be thought of as $\mathbb{C}^n$ with a $\mathbb{P}^{n-1}$ at infinity, giving directions of lines in $\mathbb{C}^n$. The projective line $\mathbb{P}^1$ is the Riemann sphere on the right-hand side of the above figure.

Points of $\mathbb{P}^n$ are represented by $(n+1)$-tuples of homogeneous coordinates, where $[x_0, \dots, x_n] = [\lambda x_0, \dots, \lambda x_n]$ if $\lambda \neq 0$ and at least one $x_i$ is nonzero. Projective varieties are subsets of $\mathbb{P}^n$ defined by homogeneous polynomials in $x_0, \dots, x_n$.

To a subset $Z$ of a vector space we associate the set of polynomials that vanish on $Z$:

$$I(Z) := \{f \in \mathbb{C}[x] \mid f(z) = 0 \ \forall z \in Z\}.$$

Let $f, g, h \in \mathbb{C}[x]$, with $f, g$ vanishing on $Z$. Both $f + g$ and $h \cdot f$ then vanish on $Z$, which implies that $I(Z)$ is an *ideal* of the polynomial ring $\mathbb{C}[x_1, \dots, x_n]$. Similarly, if $I$ is the ideal generated by a set $S$ of polynomials, then $V(S) = V(I)$.

Both $V$ and $I$ reverse inclusions with $S \subset I(V(S))$ and $Z \subset V(I(Z))$, with equality when $Z$ is a variety. Thus we have the correspondence

$$\{\text{ideals}\} \underset{I}{\overset{V}{\rightleftarrows}} \{\text{varieties}\}$$

linking algebra and geometry. By Hilbert's Nullstellensatz, this correspondence is bijective when restricted to radical ideals ($f^N \in I \Rightarrow f \in I$). This allows ideas and techniques to flow in both directions and is the source of the power and depth of algebraic geometry.

The fundamental theorem of algebra asserts that a nonconstant univariate polynomial has a complex root. The Nullstellensatz is a multivariate version, for it is equivalent to the statement that, if $I \subsetneq \mathbb{C}[x]$ is a proper ideal, then $V(I) \neq \varnothing$.

It is essentially for this reason that algebraic geometry works best over the complex numbers. Many applications require answers whose coordinates are real numbers, so results from algebraic geometry are often filtered through the lens of the real numbers when used in applications. While this restriction to $\mathbb{R}$ poses significant challenges for algebraic geometers, the generalization from $\mathbb{R}$ to $\mathbb{C}$ and then on to projective space often makes the problems easier to solve. The solution to this useful *algebraic relaxation* is often helpful in treating the original application.

### 1.2 Parametrization and Rationality

Varieties also occur as images of polynomial maps. For example, the map $t \mapsto (t^2 - 1, t^3 - t) = (x, y)$ has as its image the plane cubic $y^2 = x^3 + x^2$:

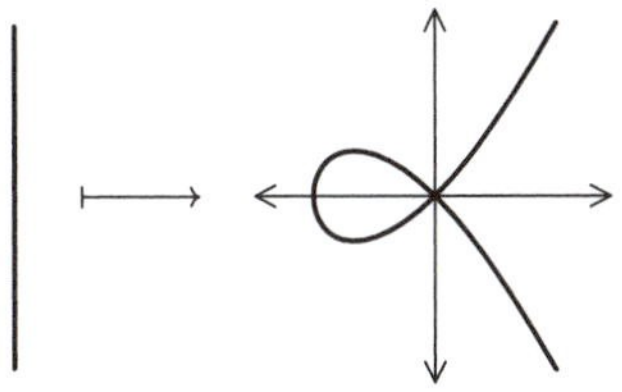

Given such a parametric representation of a variety (or any other explicit description), the *implicitization problem* asks for its ideal.

The converse problem is more subtle: can a given variety be parametrized? Euclid and Diophantus discovered the rational parametrization of the unit circle $x^2 + y^2 = 1$, $t \mapsto (x, y)$, where

$$x = \frac{2t}{1+t^2} \quad \text{and} \quad y = \frac{1-t^2}{1+t^2}. \tag{1}$$

This is the source of both Pythagorean triples and the rationalizing substitution $z = \tan(\frac{1}{2}\theta)$ of integral calculus. Homogenizing by setting $t = a/b$, (1) gives an isomorphism between $\mathbb{P}^1$ (with coordinates $[a, b]$) and the unit circle. Translating and scaling gives an isomorphism between $\mathbb{P}^1$ and any circle.

On the other hand, the cubic $y^2 = x^3 - x$ (on the left-hand side below) has no rational parametrization:

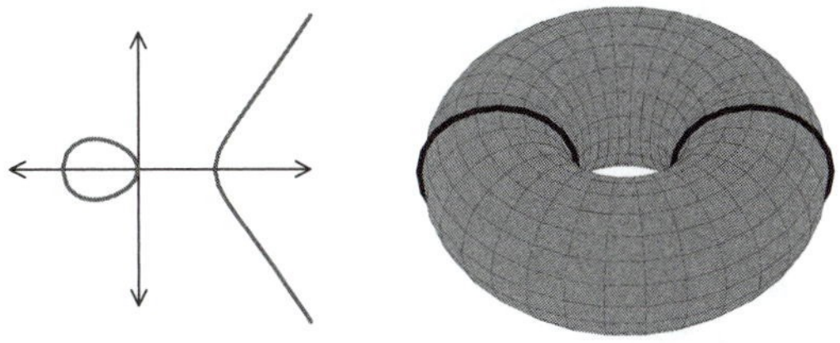

This is because the corresponding cubic in $\mathbb{P}^2$ is a curve of genus one (an elliptic curve), which is a torus (see the right-hand side above), and there is no nonconstant map from the Riemann sphere $\mathbb{P}^1$ to the torus. However, $(x, y) \mapsto x$ sends the cubic curve to $\mathbb{P}^1$ and is a two-to-one map except at the branch points $\{-1, 0, 1, \infty\}$. In fact, any curve with a two-to-one map to $\mathbb{P}^1$ having four branch points has genus one.

A smooth biquadratic curve also has genus one. The product $\mathbb{P}^1 \times \mathbb{P}^1$ is a compactification of $\mathbb{C}^2$ that is different from $\mathbb{P}^2$. Suppose that $C \subset \mathbb{P}^1 \times \mathbb{P}^1$ is defined by an equation that is separately quadratic in the two variables $s$ and $t$,

$$a_{00} + a_{10}s + a_{01}t + \cdots + a_{22}s^2t^2 = 0,$$

where $s$ and $t$ are coordinates for the $\mathbb{P}^1$ factors. Analyzing the projection onto one $\mathbb{P}^1$ factor, one can show that the map is two-to-one, except at four branch points, and so $C$ has genus one.

### 1.3 Toric Varieties

Varieties parametrized by monomials (toric varieties) often arise in applications, and they may be completely understood in terms of the geometry and combinatorics of the monomials.

Let $\mathbb{C}^*$ be the nonzero complex numbers. An integer vector $\alpha = (a_1, \dots, a_d) \in \mathbb{Z}^d$ is the exponent vector of a Laurent monomial $t^\alpha := t_1^{a_1} \cdots t_d^{a_d}$, where $t = (t_1, \dots, t_d) \in (\mathbb{C}^*)^d$ is a $d$-tuple of nonzero complex numbers. Let $\mathcal{A} = \{\alpha_0, \dots, \alpha_n\} \subset \mathbb{Z}^d$ be a finite set of integer vectors. The *toric variety* $X_\mathcal{A}$ is then the closure of the image of the map

$$\varphi_\mathcal{A} : (\mathbb{C}^*)^d \ni t \longmapsto [t^{\alpha_0}, t^{\alpha_1}, \dots, t^{\alpha_n}] \in \mathbb{P}^n.$$

The toric variety $X_\mathcal{A}$ has dimension equal to the dimension of the affine span of $\mathcal{A}$, and it has an action of $(\mathbb{C}^*)^d$ (via the map $\varphi_\mathcal{A}$) with a dense orbit (the image of $\varphi_\mathcal{A}$).

The implicitization problem for toric varieties is elegantly solved. Assume that $\mathcal{A}$ lies on an affine hyperplane, so that there is a vector $w \in \mathbb{R}^d$ with $w \cdot \alpha_i = w \cdot \alpha_j (\neq 0)$ for all $i, j$, where "$\cdot$" is the dot product. For $v \in \mathbb{R}^{n+1}$, write $\mathcal{A}v$ for $\sum_i \alpha_i v_i$.

**Theorem 1.** *The homogeneous ideal of $X_\mathcal{A}$ is spanned by binomials $x^u - x^v$, where $\mathcal{A}u = \mathcal{A}v$.*

The assumption that we have $w$ with $w \cdot \alpha_i = w \cdot \alpha_j$ for all $i, j$ is mild. Given any $\mathcal{A}$, if we append a new $(d+1)$th coordinate of 1 to each $\alpha_i$ and set $w = (0, \dots, 0, 1) \in \mathbb{R}^{d+1}$, then the assumption is satisfied and we obtain the same projective variety $X_\mathcal{A}$.

Applications also use the tight relation between $X_\mathcal{A}$ and the convex hull $\Delta_\mathcal{A}$ of $\mathcal{A}$, which is a polytope with integer vertices. The points of $X_\mathcal{A}$ with nonnegative coordinates form its *nonnegative part* $X^+_\mathcal{A}$. This is identified with $\Delta_\mathcal{A}$ through the *algebraic moment map*, $\pi_\mathcal{A} : \mathbb{P}^n \dashrightarrow \mathbb{P}^d$, which sends a point $x$ to $\mathcal{A}x$. (The broken arrow means that the map is not defined everywhere.) By Birch's theorem from statistics, $\pi_\mathcal{A}$ maps $X^+_\mathcal{A}$ homeomorphically to $\Delta_\mathcal{A}$.

There is a second homeomorphism $\beta_\mathcal{A} : \Delta_\mathcal{A} \xrightarrow{\sim} X^+_\mathcal{A}$ given by polynomials. The polytope $\Delta_\mathcal{A}$ is defined by linear inequalities,

$$\Delta_\mathcal{A} := \{x \in \mathbb{R}^d \mid \ell_F(x) \geqslant 0\},$$

where $F$ ranges over the codimension-one faces of $\Delta_\mathcal{A}$ and $\ell_F(F) \equiv 0$, with the coefficients of $\ell_F$ coprime integers. For each $\alpha \in \mathcal{A}$, set

$$\beta_\alpha(x) := \prod_F \ell_F(x)^{\ell_F(\alpha)}, \tag{2}$$

which is nonnegative on $\Delta_\mathcal{A}$. For $x \in \Delta_\mathcal{A}$, set

$$\beta_\mathcal{A}(x) := [\beta_{\alpha_0}(x), \dots, \beta_{\alpha_n}(x)] \in X^+_\mathcal{A}.$$

While $\pi_\mathcal{A}$ and $\beta_\mathcal{A}$ are homeomorphisms between the same spaces, they are typically not inverses.

A useful variant is to translate $X_\mathcal{A}$ by a nonzero weight, $\omega = (\omega_0, \dots, \omega_n) \in (\mathbb{C}^*)^{n+1}$,

$$X_{\mathcal{A},\omega} := \{[\omega_0 x_0, \dots, \omega_n x_n] \mid x \in X_\mathcal{A}\}.$$

This translated toric variety is spanned by binomials $\omega^v x^u - \omega^u x^v$ with $\mathcal{A}u = \mathcal{A}v$ as in Theorem 1, and it is parametrized by monomials via

$$\varphi_{\mathcal{A},\omega}(t) = (\omega_0 t^{\alpha_0}, \dots, \omega_n t^{\alpha_n}).$$

When the weights $\omega_i$ are positive real numbers, Birch's theorem holds, $\pi_\mathcal{A} : X^+_{\mathcal{A},\omega} \xrightarrow{\sim} \Delta_\mathcal{A}$, and we have the parametrization $\beta_{\mathcal{A},\omega} : \Delta_\mathcal{A} \to X^+_{\mathcal{A},\omega}$, where the components of $\beta_{\mathcal{A},\omega}$ are $\omega_i \beta_{\alpha_i}$.

**Example 2.** When $\mathcal{A}$ consists of the standard unit vectors $(1, 0, \dots, 0), \dots, (0, \dots, 0, 1)$ in $\mathbb{R}^{n+1}$, the toric variety is the projective space $\mathbb{P}^n$, and $\varphi_\mathcal{A}$ gives the usual homogeneous coordinates $[x_0, \dots, x_n]$ for $\mathbb{P}^n$. The nonnegative part of $\mathbb{P}^n$ is the convex hull of $\mathcal{A}$, which is the *standard n-simplex*, $\triangle^n$, and $\pi_\mathcal{A} = \beta_\mathcal{A}$ is the identity map.

**Example 3.** Let $\mathcal{A} = \{0, 1, \dots, n\}$ so that $\Delta_\mathcal{A} = [0, n]$, and choose weights $\omega_i = \binom{n}{i}$. Then $X_{\mathcal{A},\omega}$ is the closure of the image of the map

$$t \mapsto \left[1, nt, \binom{n}{2}t^2, \dots, nt^{n-1}, t^n\right] \in \mathbb{P}^n,$$

which is the (translated) moment curve. Its nonnegative part $X^+_{\mathcal{A},\omega}$ is the image of $[0,n]$ under the map $\beta_{\mathcal{A},\omega}$ whose components are

$$\beta_i(x) = \frac{1}{n^n}\binom{n}{i}x^i(n-x)^{n-i}.$$

Replacing $x$ by $ny$ gives the Bernstein polynomial

$$\beta_{i,n}(y) = \binom{n}{i}y^i(1-y)^{n-i}, \qquad (3)$$

and thus the moment curve is parametrized by the Bernstein polynomials. Because of this, we call the functions $\omega_i\beta_{\alpha_i}$ (2) *generalized Bernstein polynomials.*

The composition $\pi_{\mathcal{A}} \circ \beta_{\mathcal{A},\omega}(x)$ is

$$\frac{1}{n^n}\sum_{i=0}^{n} i\binom{n}{i}x^i(n-x)^{n-i}$$
$$= \frac{nx}{n^n}\sum_{i=1}^{n}\binom{n-1}{i-1}x^{i-1}(n-x)^{n-i} = x,$$

as the last sum is $(x+(n-x))^{n-1}$. Similarly, $(1/n)\pi_{\mathcal{A}} \circ \beta(y) = y$, where $\beta$ is the parametrization by the Bernstein polynomials. The weights $\omega_i = \binom{n}{i}$ are essentially the unique weights for which $\pi_{\mathcal{A}} \circ \beta_{\mathcal{A},\omega}(x) = x$.

**Example 4.** For positive integers $m$, $n$ consider the map $\varphi\colon \mathbb{C}^m \times \mathbb{C}^n \dashrightarrow \mathbb{P}(\mathbb{C}^{m\times n})$ defined by

$$(x,y) \mapsto [x_iy_j \mid i = 1,\dots,m,\ j = 1,\dots,n].$$

Its image is the *Segre variety*, which is a toric variety, as the map $\varphi$ is $\varphi_{\mathcal{A}}$, where $\mathcal{A}$ is

$$\{\boldsymbol{e}_i + \boldsymbol{f}_j \mid i = 1,\dots,m,\ j = 1,\dots,n\} \subset \mathbb{Z}^m \oplus \mathbb{Z}^n.$$

Here, $\{\boldsymbol{e}_i\}$ and $\{\boldsymbol{f}_j\}$ are the standard bases for $\mathbb{Z}^m$ and $\mathbb{Z}^n$, respectively.

If $z_{ij}$ are the coordinates of $\mathbb{C}^{m\times n}$, then the Segre variety is defined by the binomial equations

$$z_{ij}z_{kl} - z_{il}z_{kj} = \begin{vmatrix} z_{ij} & z_{il} \\ z_{kj} & z_{kl} \end{vmatrix}.$$

Identifying $\mathbb{C}^{m\times n}$ with $m \times n$ matrices shows that the Segre variety is the set of rank-one matrices.

Other common toric varieties include the *Veronese variety*, where $\mathcal{A}$ is $\mathcal{A}_{n,d} := n\triangle^d \cap \mathbb{Z}^{d+1}$, and the *Segre–Veronese variety*, where $\mathcal{A}$ is $\mathcal{A}_{m,d} \times \mathcal{A}_{n,e}$. When $d = e = 1$, $\mathcal{A}$ consists of the integer vectors in the $m\times n$ rectangle

$$\mathcal{A} = \{(i,j) \mid 0 \leqslant i \leqslant m,\ 0 \leqslant j \leqslant n\}.$$

## 2 Algorithms for Algebraic Geometry

Mediating between theory and examples and facilitating applications are algorithms developed to study, manipulate, and compute algebraic varieties. These come in two types: exact symbolic methods and approximate numerical methods.

### 2.1 Symbolic Algorithms

The words algebra and algorithm share an Arabic root, but they are connected by more than just their history. When we write a polynomial—as a sum of monomials, say, or as an expression such as a determinant of polynomials—that symbolic representation is an algorithm for evaluating the polynomial.

Expressions for polynomials lend themselves to algorithmic manipulation. While these representations and manipulations have their origin in antiquity, and methods such as Gröbner bases predate the computer age, the rise of computers has elevated symbolic computation to a key tool for algebraic geometry and its applications.

Euclid's algorithm, Gaussian elimination, and Sylvester's resultants are important symbolic algorithms that are supplemented by universal symbolic algorithms based on Gröbner bases. They begin with a *term order* $\prec$, which is a well-ordering of all monomials that is consistent with multiplication. For example, $\prec$ could be the lexicographic order in which $x^u \prec x^v$ if the first nonzero entry of the vector $v - u$ is positive. A term order organizes the algorithmic representation and manipulation of polynomials, and it is the basis for the termination of algorithms.

The initial term $\mathrm{in}_\prec f$ of a polynomial $f$ is its term $c_\alpha x^\alpha$ with the $\prec$-largest monomial in $f$. The initial ideal $\mathrm{in}_\prec I$ of an ideal $I$ is the ideal generated by initial terms of polynomials in $I$. This monomial ideal is a well-understood combinatorial object, and the passage to an initial ideal preserves much information about $I$ and its variety.

A *Gröbner basis* for $I$ is a finite set $G \subset I$ of polynomials whose initial terms generate $\mathrm{in}_\prec I$. This set $G$ generates $I$ and facilitates the transfer of information from $\mathrm{in}_\prec I$ back to $I$. This information may typically be extracted using linear algebra, so a Gröbner basis essentially contains all the information about $I$ and its variety.

Consequently, a bottleneck in this approach to symbolic computation is the computation of a Gröbner basis (which has high complexity due in part to its information content). Gröbner basis calculation also appears to be essentially serial; no efficient parallel algorithm is known.

The subject began in 1965 when Buchberger gave an algorithm to compute a Gröbner basis. Decades of development, including sophisticated heuristics and completely new algorithms, have led to reasonably efficient implementations of Gröbner basis computation. Many algorithms have been devised and implemented to use a Gröbner basis to study a variety. All of this is embedded in freely available software packages that are revolutionizing the practice of algebraic geometry and its applications.

### 2.2 Numerical Algebraic Geometry

While symbolic algorithms lie on the algebraic side of algebraic geometry, numerical algorithms, which compute and manipulate points on varieties, have a strongly geometric flavor.

These numerical algorithms rest upon Newton's method for refining an approximate solution to a system of polynomial equations. A system $F = (f_1, \dots, f_n)$ of polynomials in $n$ variables is a map $F\colon \mathbb{C}^n \to \mathbb{C}^n$ with solutions $F^{-1}(0)$. We focus on systems with finitely many solutions. A *Newton iteration* is the map $N_F\colon \mathbb{C}^n \to \mathbb{C}^n$, where

$$N_F(x) = x - DF_x^{-1}(F(x)),$$

with $DF_x$ the Jacobian matrix of partial derivatives of $F$ at $x$. If $\xi \in F^{-1}(0)$ is a solution to $F$ with $DF_\xi$ invertible, then when $x$ is sufficiently close to $\xi$, $N_F(x)$ is closer still, in that it has twice as many digits in common with $\xi$ as does $x$. Smale showed that "sufficiently close" may be decided algorithmically, which can allow the certification of output from numerical algorithms.

Newton iterations are used in numerical continuation. For a polynomial system $H_t$ depending on a parameter $t$, the solutions $H_t^{-1}(0)$ for $t \in [0,1]$ form a collection of arcs. Given a point $(x_t, t)$ of some arc and a step $\delta_t$, a predictor is called to give a point $(x', t+\delta_t)$ that is near to the same arc. Newton iterations are then used to refine this to a point $(x_{t+\delta_t}, t+\delta_t)$ on the arc. This numerical continuation algorithm can be used to trace arcs from $t = 0$ to $t = 1$.

We may use continuation to find all solutions to a system $F$ consisting of polynomials $f_i$ of degree $d$. Define a new system $H_t = (h_1, \dots, h_n)$ by

$$h_i := tf_i + (1-t)(x_i^d - 1).$$

At $t = 0$, this is $x_i^d - 1$, whose solutions are the $d$th roots of unity. When $F$ is general, $H_t^{-1}(0)$ consists of $d^n$ arcs connecting these known solutions at $t = 0$ to the solutions of $F^{-1}(0)$ at $t = 1$. These may be found by continuation.

While this *Bézout homotopy* illustrates the basic idea, it has exponential complexity and may not be efficient. In practice, other more elegant and efficient homotopy algorithms are used for numerically solving systems of polynomials.

These numerical methods underlie *numerical algebraic geometry*, which uses them to manipulate and study algebraic varieties on a computer. The subject began when Sommese, Verschelde, and Wampler introduced its fundamental data structure of a witness set, as well as algorithms to generate and manipulate witness sets.

Suppose we have a variety $V \subset \mathbb{C}^n$ of dimension $n-d$ that is a component of the zero set $F^{-1}(0)$ of $d$ polynomials $F = (f_1, \dots, f_d)$. A *witness set* for $V$ consists of a general affine subspace $L \subset \mathbb{C}^n$ of dimension $d$ (given by $d$ affine equations) and (approximations to) the points of $V \cap L$. The points of $V \cap L$ may be numerically continued as $L$ moves to sample points from $V$.

An advantage of numerical algebraic geometry is that path tracking is inherently parallelizable, as each of the arcs in $H_t^{-1}(0)$ may be tracked independently. This parallelism is one reason why numerical algebraic geometry does not face the complexity affecting symbolic methods. Another reason is that by computing approximate solutions to equations, complete information about a variety is never computed.

## 3 Algebraic Geometry in Applications

We illustrate some of the many ways in which algebraic geometry arises in applications.

### 3.1 Kinematics

Kinematics is concerned with motions of *linkages* (rigid bodies connected by movable joints). While its origins were in the simple machines of antiquity, its importance grew with the age of steam and today it is fundamental to ROBOTICS [VI.14]. As the positions of a linkage are solutions to a system of polynomial equations, kinematics has long been an area of application of algebraic geometry.

An early challenge important to the development of the steam engine was to find a linkage with a motion along a straight line. Watt discovered a linkage in 1784 that approximated straight line motion (tracing a curve near a flex), and in 1864 Peaucellier gave the first linkage with a true straight line motion (based on circle inversion):

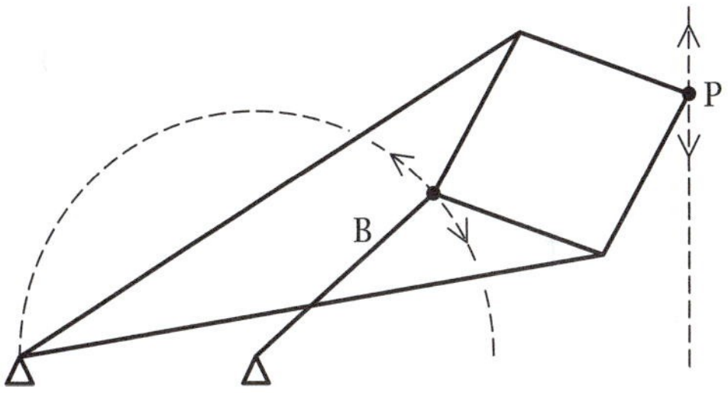


(When the bar B is rotated about its anchor point, the point P traces a straight line.)

Cayley, Chebyshev, Darboux, Roberts, and others made contributions to kinematics in the nineteenth century. The French Academy of Sciences recognized the importance of kinematics, posing the problem of determining the nontrivial mechanisms with a motion constrained to lie on a sphere for its 1904 Prix Vaillant, which was awarded to Borel and Bricard for their partial solutions.

The *four-bar linkage* consists of four bars in the plane connected by rotational joints, with one bar fixed. A triangle is erected on the *coupler bar* opposite the fixed bar, and we wish to describe the *coupler curve* traced by the apex of the triangle:

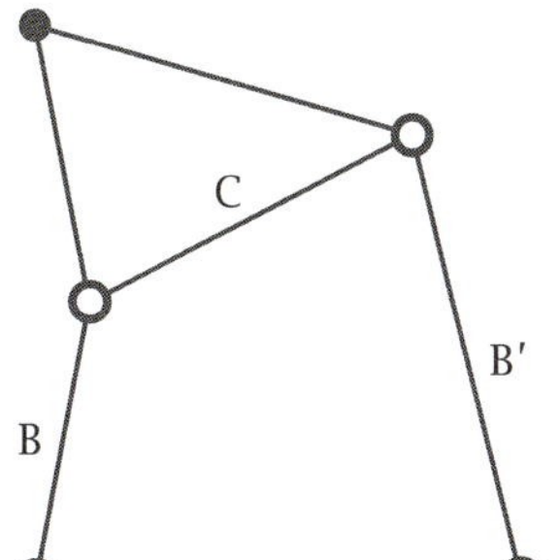


To understand the motion of this linkage, note that if we remove the coupler bar C, the bars B and B′ swing freely, tracing two circles, each of which we parameterize by $\mathbb{P}^1$ as in (1). The coupler bar constrains the endpoints of bars B and B′ to lie a fixed distance apart. In the parameters $s$, $t$ of the circles and if $b$, $b'$, $c$ are the lengths of the corresponding bars, the coupler constraint gives the equation

$$\begin{aligned} c^2 &= \left(b\frac{1-s^2}{1+s^2} - b'\frac{1-t^2}{1+t^2}\right)^2 + \left(b\frac{2s}{1+s^2} - b'\frac{2t}{1+t^2}\right)^2 \\ &= b^2 + b'^2 - 2bb'\frac{(1-s^2)(1-t^2)+4st}{(1+s^2)(1+t^2)}. \end{aligned}$$

Clearing denominators gives a biquadratic equation in the variety $\mathbb{P}^1 \times \mathbb{P}^1$ that parametrizes the rotations of bars B and B′. The coupler curve is therefore a genus-one curve and is irrational. The real points of a genus-one curve have either one or two components, which corresponds to the linkage having one or two assembly modes; to reach all points of a coupler curve with two components requires disassembly of the mechanism.

Roberts and Chebyshev discovered that there are three linkages (called Roberts cognates) with the same coupler curve, and they may be constructed from one another using straightedge and compass. The nine-point path synthesis problem asks for the four-bar linkages whose coupler curve contains nine given points. Morgan, Sommese, and Wampler used numerical continuation to solve the equations, finding 4326 distinct linkages in 1442 triplets of Roberts cognates. Here is one linkage that solves this problem for the indicated nine points:

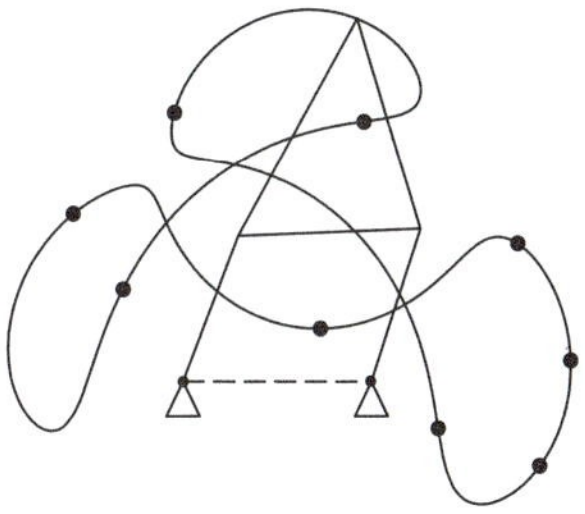

Such applications in kinematics drove the early development of numerical algebraic geometry.

### 3.2 Geometric Modeling

Geometric modeling uses curves and surfaces to represent objects on a computer for use in industrial design, manufacture, architecture, and entertainment. These applications of computer-aided geometric design and computer graphics are profoundly important to the world economy.

Geometric modeling began around 1960 in the work of de Casteljau at Citroën, who introduced what are now called Bézier curves (they were popularized by Bézier at Renault) for use in automobile manufacturing.

Bézier curves (along with their higher-dimensional analogues rectangular tensor-product and triangular

Bézier patches) are parametric curves (and surfaces) that have become widely used for many reasons, including ease of computation and the intuitive method to control shape by manipulating control points. They begin with Bernstein polynomials (3), which are nonnegative on $[0,1]$. Expanding $1^n = (t + (1-t))^n$ shows that

$$1 = \sum_{i=1}^{n} \beta_{i,n}(t).$$

Given control points $\boldsymbol{b}_0, \ldots, \boldsymbol{b}_n$ in $\mathbb{R}^2$ (or $\mathbb{R}^3$), we have the Bézier curve

$$[0,1] \ni t \longmapsto \sum_{i=0}^{n} \boldsymbol{b}_i \beta_{i,n}(t). \tag{4}$$

Here are two cubic ($n = 3$) Bézier curves in $\mathbb{R}^2$:

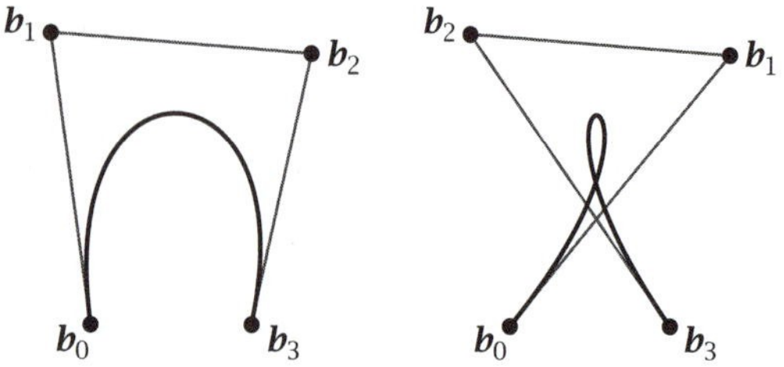


By (4), a Bézier curve is the image of the nonnegative part of the translated moment curve of example 3 under the map defined on projective space by

$$[x_0, \ldots, x_n] \mapsto \sum_{i=0}^{n} x_i \boldsymbol{b}_i.$$

On the standard simplex $\triangle^n$, this is the canonical map to the convex hull of the control points.

The tensor product patch of bidegree $(m,n)$ has basis functions

$$\beta_{i,m}(s)\beta_{j,m}(t)$$

for $i = 0, \ldots, m$ and $j = 0, \ldots, n$. These are functions on the unit square. Control points

$$\{\boldsymbol{b}_{i,j} \mid i = 0, \ldots, m,\ j = 0, \ldots, n\} \subset \mathbb{R}^3$$

determine the map

$$(s,t) \longmapsto \sum \boldsymbol{b}_{ij} \beta_{i,m}(s) \beta_{j,n}(t),$$

whose image is a rectangular patch.

Bézier triangular patches of degree $d$ have basis functions

$$\beta_{i,j;d}(s,t) = \frac{d!}{i!j!(d-i-j)!} s^i t^j (1-s-t)^{d-i-j}$$

for $0 \leqslant i, j$ with $i + j \leqslant d$. Again, control points give a map from the triangle with image a Bézier triangular patch. Here are two surface patches:

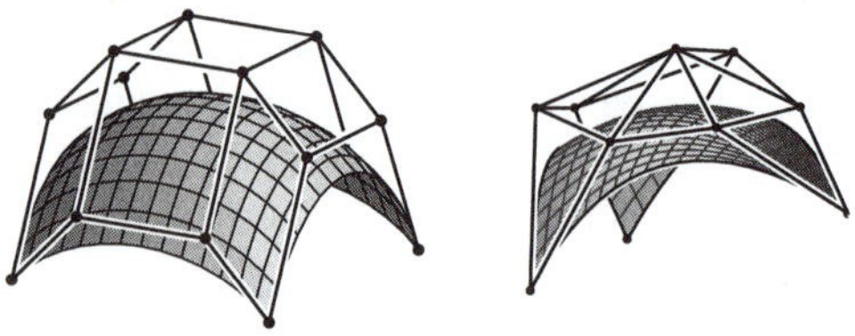

These patches correspond to toric varieties, with tensor product patches coming from Segre–Veronese surfaces and Bézier triangles from Veronese surfaces. The basis functions are the generalized Bernstein polynomials $\omega_i \beta_{\alpha_i}$ of section 1.3, and this explains their shape as they are images of $\Delta_{\mathcal{A}}$, which is a rectangle for the Segre–Veronese surfaces and a triangle for the Veronese surfaces.

An important question is to determine the intersection of two patches given parametrically as $F(x)$ and $G(x)$ for $x$ in some domain (a triangle or rectangle). This is used for trimming the patches or drawing the intersection curve. A common approach is to solve the implicitization problem for $G$, giving a polynomial $g$ which vanishes on the patch $G$. Then $g(F(x))$ defines the intersection in the domain of $F$. This application has led to theoretical and practical advances in algebra concerning resultants and syzygies.

### 3.3 Algebraic Statistics

Algebraic statistics applies tools from algebraic geometry to questions of statistical inference. This is possible because many statistical models are (part of) algebraic varieties, or they have significant algebraic or geometric structures.

Suppose that $X$ is a discrete random variable with $n+1$ possible states, $0, \ldots, n$ (e.g., the number of tails observed in $n$ coin flips). If $p_i$ is the probability that $X$ takes value $i$,

$$p_i := P(X = i),$$

then $p_0, \ldots, p_n$ are nonnegative and sum to 1. Thus $p$ lies in the standard $n$-simplex, $\triangle^n$. Here are two views of it when $n = 2$:

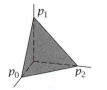

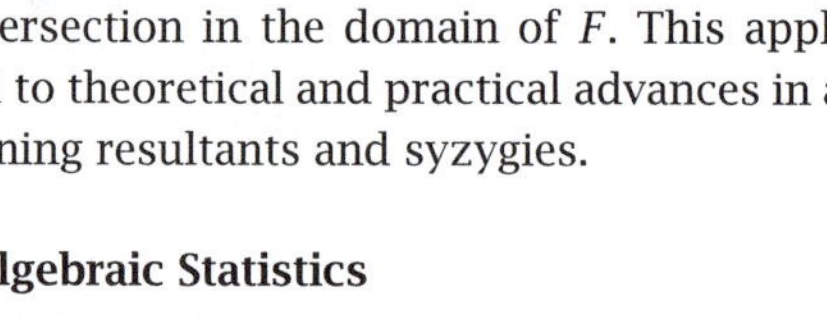


A *statistical model* $M$ is a subset of $\triangle^n$. If the point $(p_0, \ldots, p_n) \in M$, then we may think of $X$ as being explained by $M$.

**Example 5.** Let $X$ be a discrete random variable whose states are the number of tails in $n$ flips of a coin with a probability $t$ of landing on tails and $1 - t$ of heads. We may calculate that

$$P(X = i) = \binom{n}{i} t^i (1-t)^{n-i},$$

which is the Bernstein polynomial $\beta_{i,n}$ (3) evaluated at the parameter $t$. We call $X$ a *binomial random variable* or *binomial distribution*. The set of binomial distributions as $t$ varies gives the translated moment curve of example 3 parametrized by Bernstein polynomials. This curve is the model for binomial distributions. Here is a picture of this curve when $n = 2$:

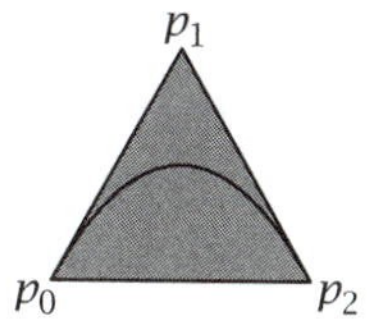


**Example 6.** Suppose that we have discrete random variables $X$ and $Y$ with $m$ and $n$ states, respectively. Their joint distribution has $mn$ states (cells in a table or a matrix) and lies in the simplex $\Delta^{mn-1}$. The *model of independence* consists of all distributions $p \in \Delta^{mn-1}$ such that

$$P(X = i, Y = j) = P(X = i)P(Y = j). \tag{5}$$

It is parametrized by $\Delta^{m-1} \times \Delta^{n-1}$ (probability simplices for $X$ and $Y$), and (5) shows that it is the nonnegative part of the Segre variety of example 4. The model of independence therefore consists of those joint probability distributions that are rank-one matrices.

Other common statistical models called discrete exponential families or *toric models* are also the nonnegative part $X^+_{\mathcal{A},\omega}$ of some toric variety. For these, the algebraic moment map $\pi_{\mathcal{A}} : \Delta^n \to \Delta_{\mathcal{A}}$ (or $u \mapsto \mathcal{A}u$) is a *sufficient statistic*. For the model of independence, $\mathcal{A}u$ is the vector of row and column sums of the table $u$.

Suppose that we have data from $N$ independent observations (or draws), each from the same distribution $p(t)$ from a model $M$, and we wish to estimate the parameter $t$ best explaining the data. One method is to maximize the *likelihood* (the probability of observing the data given a parameter $t$). Suppose that the data are represented by a vector $u$ of counts, where $u_i$ is how often state $i$ was observed in the $N$ trials. The likelihood function is

$$L(t|u) = \binom{N}{u} \prod_{i=0}^{n} p_i(t)^{u_i},$$

where $\binom{N}{u}$ is the multinomial coefficient.

Suppose that $M$ is the binomial distribution of example 5. It suffices to maximize the logarithm of $L(t \mid u)$, which is

$$C + \sum_{i=0}^{n} u_i(i \log t + (n-i)\log(1-t)),$$

where $C$ is a constant. By calculus, we have

$$0 = \frac{1}{t}\sum_{i=0}^{n} iu_i + \frac{1}{1-t}\sum_{i=0}^{n}(n-i)u_i.$$

Solving, we obtain that

$$t := \frac{1}{n}\sum_{i=0}^{n} i\frac{u_i}{N} \tag{6}$$

maximizes the likelihood. If $\hat{u} := (u/N) \in \Delta^n$ is the point corresponding to our data, then (6) is the normalized algebraic moment map $(1/n)\pi_{\mathcal{A}}$ of example 3 applied to $\hat{u}$. For a general toric model $X^+_{\mathcal{A},\omega} \subset \Delta^n$, and likelihood is maximized at the parameter $t$ satisfying $\pi_{\mathcal{A}} \circ \beta_{\mathcal{A},\omega}(t) = \pi_{\mathcal{A}}(\hat{u})$. An algebraic formula exists for the parameter that maximizes likelihood exactly when $\pi_{\mathcal{A}}$ and $\beta_{\mathcal{A},\omega}$ are inverses.

Suppose that we have data $u$ as a vector of counts as before and a model $M \subset \Delta^n$ and we wish to test the null hypothesis that the data $u$ come from a distribution in $M$. *Fisher's exact test* uses a score function $\Delta^n \to \mathbb{R}_{\geqslant}$ that is zero exactly on $M$ and computes how likely it is for data $v$ to have a higher score than $u$, when $v$ is generated from the same probability distribution as $u$. This requires that we sample from the probability distribution of such $v$.

For a toric model $X^+_{\mathcal{A},\omega}$, this is a probability distribution on the set of possible data with the same sufficient statistics:

$$\mathcal{F}_u := \{v \mid \mathcal{A}u = \mathcal{A}v\}.$$

For a parameter $t$, this distribution is

$$\begin{aligned} L(v \mid v \in \mathcal{F}_u, t) &= \frac{\binom{N}{v}\omega^v t^{\mathcal{A}v}}{\sum_{w\in\mathcal{F}_u}\binom{N}{w}\omega^w t^{\mathcal{A}w}} \\ &= \frac{\binom{N}{v}\omega^v}{\sum_{w\in\mathcal{F}_u}\binom{N}{w}\omega^w}, \end{aligned} \tag{7}$$

as $\mathcal{A}v = \mathcal{A}w$ for $v, w \in \mathcal{F}(u)$.

This sampling may be accomplished using a random walk on the fiber $\mathcal{F}_u$ with stationary distribution (7). This requires a connected graph on $\mathcal{F}_u$. Remarkably, any Gröbner basis for the ideal of the toric variety $X_{\mathcal{A}}$ gives such a graph.

### 3.4 Tensor Rank

The fundamental invariant of an $m \times n$ matrix is its rank. The set of all matrices of rank at most $r$ is defined by the vanishing of the determinants of all $(r+1) \times (r+1)$ submatrices. From this perspective, the simplest matrices are those of rank one, and the rank of a matrix $A$ is the minimal number of rank-one matrices that sum to $A$.

An $m \times n$ matrix $A$ is a linear map $V_2 = \mathbb{C}^n \to V_1 = \mathbb{C}^m$. If it has rank one, it is the composition

$$V_2 \xrightarrow{v_2} \mathbb{C} \xrightarrow{v_1} V_1$$

of a linear function $v_2$ on $V_2$ ($v_2 \in V_2^*$) and an inclusion given by $1 \mapsto v_1 \in V_1$. Thus, $A = v_1 \otimes v_2 \in V_1 \otimes V_2^*$, as this tensor space is naturally the space of linear maps $V_2 \to V_1$. A tensor of the form $v_1 \otimes v_2$ has rank one, and the set of rank-one tensors forms the Segre variety of example 4.

SINGULAR VALUE DECOMPOSITION [II.32] writes a matrix $A$ as a sum of rank-one matrices of the form

$$A = \sum_{i=1}^{\text{rank}(A)} \sigma_i v_{1,i} \otimes v_{2,i}, \tag{8}$$

where $\{v_{1,i}\}$ and $\{v_{2,i}\}$ are orthonormal and $\sigma_1 \geqslant \cdots \geqslant \sigma_{\text{rank}(A)} > 0$ are the singular values of $A$. Often, only the relatively few terms of (8) with largest singular values are significant, and the rest are noise. Letting $A_{\text{lr}}$ be the sum of terms with large singular values and $A_{\text{noise}}$ be the sum of the rest, then $A$ is the sum of the low-rank matrix $A_{\text{lr}}$ plus noise $A_{\text{noise}}$.

A *k-way tensor* ($k$-way table) is an element of the tensor space $V_1 \otimes \cdots \otimes V_k$, where each $V_i$ is a finite-dimensional vector space. A rank-one tensor has the form $v_1 \otimes \cdots \otimes v_k$, where $v_i \in V_i$. These form a toric variety, and the rank of a tensor $v$ is the minimal number of rank-one tensors that sum to $v$.

The (closure of) the set of rank-$r$ tensors is the *rth secant variety*. When $k = 2$ (matrices), the set of determinants of all $(r+1) \times (r+1)$ submatrices solves the implicitization problem for the $r$th secant variety. For $k > 2$ there is not yet a solution to the implicitization problem for the $r$th secant variety.

Tensors are more complicated than matrices. Some tensors of rank greater than $r$ lie in the $r$th secant variety, and these may be approximated by low-rank (rank-$r$) tensors. Algorithms for tensor decomposition generalize singular value decomposition. Their goal is often an expression of the form $v = v_{\text{lr}} + v_{\text{noise}}$ for a tensor $v$ as the sum of a low-rank tensor $v_{\text{lr}}$ plus noise $v_{\text{noise}}$.

Some mixture models in algebraic statistics are secant varieties. Consider an inhomogeneous population in which the fraction $\theta_i$ obeys a probability distribution $p^{(i)}$ from a model $M$. The distribution of data collected from this population then behaves in the same was as the convex combination

$$\theta_1 p^{(1)} + \theta_2 p^{(2)} + \cdots + \theta_r p^{(r)},$$

which is a point on the $r$th secant variety of $M$.

Theoretical and practical problems in complexity may be reduced to knowing the rank of specific tensors. Matrix multiplication gives a nice example of this.

Let $A = (a_{ij})$ and $B = (b_{ij})$ be $2 \times 2$ matrices. In the usual multiplication algorithm, $C = AB$ is

$$c_{ij} = a_{i1}b_{1j} + a_{i2}b_{2j}, \quad i, j = 1, 2. \tag{9}$$

This involves eight multiplications. For $n \times n$ matrices, the algorithm uses $n^3$ multiplications.

Strassen discovered an alternative that requires only seven multiplications (see the formulas in ALGORITHMS [I.4 §4]). If $A$ and $B$ are $2k \times 2k$ matrices with $k \times k$ blocks $a_{ij}$ and $b_{ij}$, then these formulas apply and enable the computation of $AB$ using $7k^3$ multiplications. Recursive application of this idea enables the multiplication of $n \times n$ matrices using only $n^{\log_2 7} \simeq n^{2.81}$ multiplications. This method is used in practice to multiply large matrices.

We interpret Strassen's algorithm in terms of tensor rank. The formula (9) for $C = AB$ is a tensor $\mu \in V \otimes V^* \otimes V^*$, where $V = M_{2\times 2}(\mathbb{C})$. Each multiplication is a rank-one tensor, and (9) exhibits $\mu$ as a sum of eight rank-one tensors, so $\mu$ has rank at most eight. Strassen's algorithm shows that $\mu$ has rank at most seven. We now know that the rank of any tensor in $V \otimes V^* \otimes V^*$ is at most seven, which shows how Strassen's algorithm could have been anticipated.

The fundamental open question about the complexity of multiplying $n \times n$ matrices is to determine the rank $r_n$ of the multiplication tensor. Currently, we have bounds only for $r_n$: we know that $r_n \geqslant o(n^2)$, as matrices have $n^2$ entries, and improvements to the idea behind Strassen's algorithm show that $r_n < O(n^{2.3728639})$.

### 3.5 The Hardy–Weinberg Equilibrium

We close with a simple application to Mendelian genetics.

Suppose that a gene exists in a population in two variants (alleles) $a$ and $b$. Individuals will have one of three *genotypes*, $aa$, $ab$, or $bb$, and their distribution $p = (p_{aa}, p_{ab}, p_{bb})$ is a point in the probability 2-simplex. A fundamental and originally controversial question in the early days of Mendelian genetics was the following: which distributions are possible in a population at equilibrium? (Assuming no evolutionary pressures, equidistribution of the alleles among the sexes, etc.)

The proportions $q_a$ and $q_b$ of alleles $a$ and $b$ in the population are

$$q_a = p_{aa} + \tfrac{1}{2}p_{ab}, \qquad q_b = \tfrac{1}{2}p_{ab} + p_{bb}, \qquad (10)$$

and the assumption of equilibrium is equivalent to

$$p_{aa} = q_a^2, \qquad p_{ab} = 2q_a q_b, \qquad p_{bb} = q_b^2. \qquad (11)$$

If

$$\mathcal{A} = \begin{pmatrix} 2 & 1 & 0 \\ 0 & 1 & 2 \end{pmatrix},$$

with

$$\begin{pmatrix} 2 \\ 0 \end{pmatrix} \leftrightarrow aa, \quad \begin{pmatrix} 1 \\ 1 \end{pmatrix} \leftrightarrow ab, \quad \text{and} \quad \begin{pmatrix} 0 \\ 2 \end{pmatrix} \leftrightarrow bb,$$

then (10) is $(q_a, q_b) = \frac{1}{2}\pi_{\mathcal{A}}(p_{aa}, p_{ab}, p_{bb})$, the normalized algebraic moment map of examples 3 and 5 applied to $(p_{aa}, p_{ab}, p_{bb})$. Similarly, the assignment $q \to p$ of (11) is the parametrization $\beta$ of the translated quadratic moment curve of example 3 given by the Bernstein polynomials.

Since $\frac{1}{2}\pi_{\mathcal{A}} \circ \beta(q) = q$, the population is at equilibrium if and only if the distribution $(p_{aa}, p_{ab}, p_{bb})$ of alleles lies on the translated quadratic moment curve, that is, if and only if it is a point in the binomial distribution, which we reproduce here:

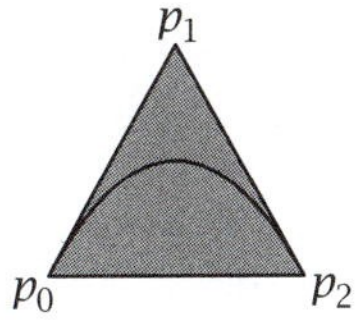


This is called the *Hardy–Weinberg equilibrium* after its two independent discoverers.

The Hardy in question is the great English mathematician G. H. Hardy, who was known for his disdain for applied mathematics, and this contribution came early in his career, in 1908. He was later famous for his work in number theory, a subject that he extolled for its purity and uselessness. As we all now know, Hardy was mistaken on this last point for number theory underlies our modern digital world, from the security of financial transactions via cryptography to using error-correcting codes to ensure the integrity of digitally transmitted documents, such as the one you have now finished reading.

### Further Reading

Cox, D., J. Little, and D. O'Shea. 2005. *Using Algebraic Geometry*, 2nd edn. Berlin: Springer.

——. 2007. *Ideals, Varieties, and Algorithms*, 3rd edn. Berlin: Springer.

Cox, D., J. Little, and H. Schenck. 2011. *Toric Varieties*. Providence, RI: American Mathematical Society.

Landsberg, J. M. 2012. *Tensors: Geometry and Applications*. Providence, RI: American Mathematical Society.

Sommese, A., and C. Wampler. 2005. *The Numerical Solution of Systems of Polynomials*. Singapore: World Scientific.

## IV.40 General Relativity and Cosmology

*George F. R. Ellis*

General relativity theory is currently the best classical theory of gravity. It introduced a major new theme into applied mathematics by treating geometry as a dynamic variable, determined by the EINSTEIN FIELD EQUATIONS [III.10] (EFEs). After outlining the basic concepts and structure of the theory, I will illustrate its nature by showing how the EFEs play out in two of its main applications: the nature of black holes, and the dynamical and observational properties of cosmology.

### 1 The Basic Structure: Physics in a Dynamic Space-Time

General relativity extended special relativity (SR) by introducing two related new concepts.

The first was that space-time could be dynamic. Not only was it not flat—so consideration of coordinate freedom became an essential part of the analysis—but it curved in response to the matter that it contained, via the EFEs (6). Consequently, as well as its dynamics, the boundary of space-time needs careful consideration, along with global causal relations and global topology.

Second, gravity is not a force like any other known force: it is inextricably entwined with inertia and can be transformed to zero by a change of reference frame. There is therefore no frame-independent gravitational force as such. Rather, its essential nature is encoded in space-time curvature, generating tidal forces and relative motions.

### 1.1 Riemannian Geometry and Physics

The geometry embodied in general relativity theory is four-dimensional Riemannian geometry, determined by a symmetric metric tensor $g_{ij}(x^k)$. The basic mathematical tool needed to investigate this geometry is tensor calculus. This generalizes vector calculus, where vector fields are quantities whose components have one free index, $X^i(x^m)$, to similar quantities whose components have an arbitrary number of indices, $T^{i\cdots j}{}_{k\cdots l}(x^m)$, some "up" and some "down," where in the case of general relativity theory the indices $i$, $j$, $k$ range over 0, 1, 2, 3 (there are four dimensions).

General coordinate transformations are allowed, so physical relations are best described through tensor equations relating tensors with the same kinds of indices, e.g., $T^{i\cdots j}{}_{k\cdots l}(x^m) = S^{i\cdots j}{}_{k\cdots l}(x^m)$. When one changes coordinates (denoted here by primed and unprimed indices), tensors transform in a linear way:

$$T^{a'b'}{}_{c'd'} = T^{ab}{}_{cd}\Lambda^{a'}{}_{a}\Lambda^{b'}{}_{b}\Lambda^{c}{}_{c'}\Lambda^{d}{}_{d'} \qquad (1)$$

for suitable inverse transformation matrices $\Lambda^{b'}{}_{b}$, $\Lambda^{c'}{}_{c}$, where we use the Einstein summation convention (i.e., repeated indices are summed over all index values). This leads to the key feature that *tensor equations that are true in one coordinate system will be true in any coordinate system.* One can form linear combinations, products, and contractions of tensors, all preserved under transformations (1). (More details of tensors and their manipulations are described in TENSORS AND MANIFOLDS [II.33].) One can thereby define symmetries in the indices: they can be symmetric (denoted by round brackets), antisymmetric (denoted by square brackets), or trace free. Symmetries are preserved when one changes coordinates, and they therefore define physically meaningful aspects of variables.

#### 1.1.1 *Special Relativity Applies Locally at Each Point*

The way in which SR applies near each point, with four dimensions (one time, three space), is determined by the metric tensor $g_{ab}(x^c) = g_{ba}(x^c)$. This determines distances along curves $x^a(\lambda)$ in space-time using the fundamental relation

$$L = \int \sqrt{|\mathrm{d}s^2|} = \int \sqrt{\left| g_{ab}\frac{\mathrm{d}x^a}{\mathrm{d}\lambda}\frac{\mathrm{d}x^b}{\mathrm{d}\lambda}\right|}\,\mathrm{d}v,$$

where the infinitesimal squared distance "$\mathrm{d}s^2$" can be positive or negative, as follows.

- $\mathrm{d}s^2 < 0$ means time-like curves (traced out by particles moving at less than the speed of light); in this case, $\mathrm{d}s^2 = -\mathrm{d}\tau^2 < 0$ determines *proper time*[1] $\tau$ along this curve, measured by perfect clocks.
- $\mathrm{d}s^2 = 0$ means null curves, representing motion at the speed of light.
- $\mathrm{d}s^2 > 0$ means this is not a possible particle path, as it implies motion at a speed greater than the speed of light; in this case, $\mathrm{d}s^2 = +\mathrm{d}l^2 > 0$ determines spatial distance $l$ along the curve.

To see how this metric geometry works, we choose Cartesian-like local coordinates

$$(x^0, x^1, x^2, x^3) = (t, x, y, z)$$

such that, at the point of interest,

$$g_{ab} = \mathrm{diag}(-1, 1, 1, 1),$$

with spatial distance given by $\mathrm{d}r^2 = \mathrm{d}x^2 + \mathrm{d}y^2 + \mathrm{d}z^2$. Along a time-like curve

$$x^a(\lambda) = (t(\lambda), x(\lambda), y(\lambda), z(\lambda))$$

moving at speed $v = \mathrm{d}r/\mathrm{d}t$ relative to these coordinates, we then have

$$\mathrm{d}\tau^2 = -\mathrm{d}s^2 = \mathrm{d}t^2(1 - v^2) \Leftrightarrow \frac{\mathrm{d}t}{\mathrm{d}\tau} = \frac{1}{\sqrt{1 - v^2}},$$

the standard time-dilation factor of SR. Setting the curve parameter $\lambda$ to be $\tau$, the 4-velocity is $u^a = \mathrm{d}x^a/\mathrm{d}\tau \Rightarrow g_{ab}u^a u^b = -1$. Changing the coordinates by a speed $v$ in the $x^1$-direction sets $\Lambda^{a'}{}_{a} = \cosh\beta\delta^{a'}_0\delta^0_a + \sinh\beta\delta^{a'}_1\delta^1_a$ and gives the standard Lorentz transformation results for spatial distances and time with $\sinh\beta = v$ (see INVARIANTS AND CONSERVATION LAWS [II.21] and TENSORS AND MANIFOLDS [II.33] for more on this area).

Space-time has one time dimension (represented by $g_{00} = -1$) and three spatial dimensions ($g_{11} = g_{22} = g_{33} = 1$). It unifies objects that are separately defined in Newtonian theory as four-dimensional entities, and it shows how they relate to each other when relative motion takes place. For example, the electric and magnetic fields are unified in a skew tensor $F_{ab} = F_{[ab]}$ such that $E_a = F_{ab}u^b$, $H_a = \frac{1}{2}\eta_{adef}u^f F^{de}$, where $\eta_{abcd}$ is the totally skew-symmetric volume tensor. The way in which these fields relate to each other when relative motion takes place follows from the tensor transformation law (1). The inverse metric $g^{mn}$: $g^{mn}g_{nk} = \delta^m_k$ is used to raise indices, $T^{ab} = g^{ac}g^{bd}T_{bd}$, while $g_{ab}$ lowers them.

1. We use units such that the speed of light $c = 1$.

### 1.1.2 Covariant Derivatives and Connections

A key question is: how does one define vectors to be mutually parallel at points P and Q that are distant from each other? In a curved space-time this question has no absolute meaning; it can be defined only in terms of parallel transport along a curve joining the points. This is defined via a covariant derivative operator (denoted by a semicolon) that acts on all tensor fields and gives the covariant derivative along any curve (denoted by $\delta$), so that if $\mathrm{d}x^a/\mathrm{d}\lambda = X^a$ is the tangent vector to the curve $x^a(\lambda)$, $\delta Y^a/\delta\lambda := Y^a{}_{;b}X^b$ is the covariant derivative of a vector field $Y^a$ along these curves, which vanishes if and only if $Y^a$ is *parallel transported* along $z^a(\lambda)$ (i.e., as $Y^a$ is translated along the curve, it is kept parallel to its previous direction at each infinitesimal step). It is determined by a geometric structure called a *connection* $\Gamma$, which specifies which vectors are parallel at neighboring points. Its components $\Gamma^a{}_{bc}$ represent the $a$-component of the covariant derivative of the $c$-basis vector in the direction of the $b$-basis vector. Then, for a vector field $Y^a$,

$$Y^a{}_{;b} = Y^a{}_{,b} + \Gamma^a{}_{bc}Y^c, \tag{2}$$

the first term $Y^a{}_{,b} := \partial Y^a/\partial x^b$ being the apparent derivative relative to the basis vectors, and the second term being due to the rate of change of the basis vectors along the curve.

In particular, $\delta X^a/\delta\lambda := X^a{}_{;b}X^b$ vanishes if and only if the curve direction is parallel transported along itself; its direction is unchanging, so it is a *geodesic* (the closest one can get to a straight line in a curved space-time). The associated curve parameter $\lambda$ is defined up to affine transformations $\lambda' = a\lambda + b$ ($a$, $b$ constants) and is therefore called an *affine parameter*. Geodesics represent the motion of matter moving subject only to the effects of gravity and inertia (time-like curves: $X^aX_a = -1$) and the paths of light rays (null curves: $X^aX_a = 0$).

The covariant derivative extends to arbitrary tensors by assuming that

(a) the covariant derivative of a function is just the partial derivative, $f_{;c} = f_{,c} := \partial f/\partial x^c$, and
(b) it is linear, obeys the Leibniz rule, and commutes with contractions.

All local physical equations should be written in terms of covariant derivatives, e.g., Maxwell's equations in a curved space-time take the form $F_{[ab;c]} = 0$, $F^{ab}{}_{;b} = J^a$, where $J^a$ is the 4-current.

### 1.1.3 Christoffel Relations

The connection $\Gamma$ is determined by assuming (i) that it is torsion free, $\Gamma^a{}_{bc} = \Gamma^a{}_{cb}$, and (ii) that the metric tensor is parallel propagated along arbitrary curves, $g_{ab;c} = 0$, which means that magnitudes are preserved under parallel propagation. Together these requirements determine the connection uniquely, linking parallelism to metric properties. The connection components are given by the *Christoffel relations*

$$\Gamma^m{}_{ab} = \tfrac{1}{2}g^{mn}(g_{an,b} + g_{bn,a} - g_{ab,n}). \tag{3}$$

The connection is therefore defined by derivatives of the metric.

This relation leads to the key result that *geodesics are extremal curves in space-time* (i.e., path length is maximized or minimized along them).

## 1.2 Space-Time Curvature and Field Equations

Covariant derivatives do not commute in general, leading to the concept of space-time curvature. For any vector field $X^a$, the Ricci identity

$$X^a{}_{;bc} - X^a{}_{;cb} = R^a{}_{dbc}X^d \tag{4}$$

determines this noncommutativity, where $R^a{}_{dbc}$ is the Riemann curvature tensor with symmetries

$$R_{abcd} = R_{[ab][cd]} = R_{cdab}, \qquad R_{a[bcd]} = 0.$$

This leads to *holonomy*: parallel transport of a vector around a closed loop from a point P back to P causes a change in the vector there (a rotation or Lorentz transformation because parallel transfer preserves magnitudes).

The Ricci tensor and the Ricci scalar are the first and second contractions of this tensor, defined by

$$R_{bf} = R_f{}^a{}_{ba} \Rightarrow R_{bf} = R_{fb}, \qquad R = g^{bc}R_{bf}.$$

By (4) and (2), the Ricci tensor is given in terms of the connection by

$$R_{bf} = \Gamma^a{}_{bf,a} - \Gamma^a{}_{af,b} + \Gamma^e{}_{bf}\Gamma^a{}_{ae} - \Gamma^e{}_{af}\Gamma^a{}_{be}.$$

On using the Christoffel relations (3), these are second-order differential expressions in terms of the metric tensor.

The curvature tensor obeys an important set of integrability conditions, namely, the *Bianchi identities*

$$R_{ab[cd;e]} = 0$$

(roughly, the curl of the curvature tensor vanishes), implying the key divergence relations

$$(R^a{}_b - \tfrac{1}{2}Rg^a_b)_{;b} = 0 \tag{5}$$

on contracting twice.

### 1.2.1 Field Equations

The matter present in space-time determines the space-time curvature through the EFEs

$$R_{ab} - \tfrac{1}{2}Rg_{ab} + \Lambda g_{ab} = \kappa T_{ab} \Rightarrow T^{ab}{}_{;b} = 0, \qquad (6)$$

where $T_{ab}$ is the energy–momentum–stress tensor for all matter present, $\kappa$ is the gravitational constant, and $\Lambda$ is the cosmological constant. The implication in (6) follows from the contracted Bianchi identities (5): if the EFEs are satisfied, energy and momentum are necessarily conserved.

Given suitable equations of state for the matter, these equations determine the dynamical evolution of the space-time, which through energy-momentum conservation relations determines the evolution of the matter in it. The commonly used model is $T_{ab} = (\rho+p)u_a u_b + p g_{ab}$, characterizing a "perfect fluid" with energy density $\rho$ and pressure $p$, which are related by an equation of state $p = p(\rho)$. If $p = 0$, we have the simplest case: pressure-free matter (e.g., "cold dark matter"). Whatever these relations may be, we usually require that various energy conditions hold (e.g., $\rho > 0$, $(\rho + p) > 0$) and additionally demand that the isentropic speed of sound $c_s = (p/\rho)_s$ obeys $0 \leqslant c_s \leqslant 1$, as required for local stability of matter (the lower bound) and causality (the upper bound).

The field equations are second-order hyperbolic equations for the metric tensor, with the matter tensor as the source. A generally used convenient form is given by the ADM formalism (as described by Anninos), as used, for example, in NUMERICAL RELATIVITY [V.15 §2.3] (the ADM formalism is named for its authors: Arnowitt, Deser, and Misner).

## 1.3 The Geodesic Deviation Equation and Tidal Forces

The geodesic deviation equation (GDE) determines the change in the deviation vectors (relative position vectors) linking a congruence of time-like geodesics. Consider the normalized tangent vector field $V^a := \mathrm{d}x^a(\tau)/\mathrm{d}\tau$ for such a congruence $x^a(\tau, w)$, with curves labeled by parameter $w$. Then $V_a V^a = -1$, $\delta V^a/\delta\tau = V^a{}_{;b}V^b = 0$. A deviation vector $\eta^a := \mathrm{d}x^a(w)/\mathrm{d}w$ can be thought of as linking pairs of neighboring geodesics in the congruence; it commutes with $V^a$, so $\delta\eta^a/\delta\tau = V^a{}_{;b}\eta^b$. Choosing the deviation vectors to be orthogonal to $v^a$: $\eta_a V^a = 0$, by (3) the GDE takes the form

$$\frac{\delta^2\eta^a}{\delta\tau^2} = -R^a{}_{bcd}V^b\eta^c V^d, \qquad (7)$$

showing how curvature causes relative acceleration of matter, i.e., tidal forces.

In this equation, the Ricci tensor is determined point by point through the EFEs, but this is only part of the curvature tensor. The rest of the curvature is given by the *Weyl tensor*, which is defined by

$$C_{ijkl} := R_{ijkl} + \tfrac{1}{2}(R_{ik}g_{jl} + R_{jl}g_{ik} - R_{il}g_{jk} - R_{jk}g_{il}) - \tfrac{1}{6}R(g_{ik}g_{jl} - g_{il}g_{jk}), \qquad (8)$$

implying that it shares the symmetries of the Riemann tensor but is also trace free: $C^a{}_{bad} = 0$. This tensor represents the free gravitational field; that is, it is the part of the curvature that is not determined pointwise, thus enabling nonlocal effects such as tidal forces and gravitational waves. As with the electromagnetic field, it can be decomposed into electric and magnetic parts, $E_{ac} = C_{abcd}u^b u^d$, $H_{ac} = \frac{1}{2}\eta_{abef}u^b C^{ef}{}_{cd}u^d$; through the GDE, these fields affect the relative motion of matter, causing tidal effects. Because of the Bianchi identities, they obey Maxwell-like equations, resulting in a wave equation underlying the existence of gravitational waves.

## 1.4 Types of Solutions and Generic Properties

Although the EFEs are complicated nonlinear equations for the metric tensor, numerous exact solutions are known (and are described well in Stephani et al.'s book *Exact Solutions of Einstein's Field Equations*). These are usually determined by imposing symmetries on the metric, often reducing hyperbolic equations to ordinary differential equations, and then solving for the free metric functions. Symmetries are generated by Killing vector fields $\xi^a(x^j)$; that is, vector fields that obey Killing's equation

$$\xi_{a;b} + \xi_{b;a} = 0$$

(the metric is dragged into itself invariantly along the integral curves of the vector fields). They obey the relation $\xi_{b;cd} = \xi^a R_{abcd}$, meaning that the initial data for them at some point are $\xi_a|_0$, $(\xi_{[a;b]})|_0$. They form a Lie algebra generating the symmetry group of a space-time; the isotropy group of a point P is generated by those Killing vectors for which $\xi_a|_0 = 0$ (they leave the point fixed).

The most important symmetries in practice are spherical symmetry and spatial homogeneity, which form the bases of the following two sections of this article, respectively. A space-time is stationary if it admits a time-like Killing vector field and is static if this field is additionally irrotational: $\xi_{[a;b}\xi_{c]} = 0$.

As well as finding exact and approximate solutions of the EFEs, one can find important generic relations such as the Raychaudhuri equation. This equation underlies generic singularity theorems. One can also prove generic existence and uniqueness theorems for vacuum- and matter-filled space-times. There has also been huge progress in finding numerical solutions of the EFEs, based on the ADM formalism.

## 2 The Schwarzschild Solution and Black Holes

To model the exterior field of the sun or of any spherical star, we look for a solution that is a

(1) vacuum solution (i.e., $T_{ab} = 0$), $R_{ab} = 0$, so the curvature is purely due to the Weyl tensor (this will be a good approximation outside some radius $r_S$ representing the surface of the central star in a solar system); and
(2) a spherically symmetric solution, as is true to high approximation in the solar system (here we ignore the rotation of the sun as well as the gravitational fields of the planets).

We choose coordinates for which the metric is

$$\mathrm{d}s^2 = -A(r,t)\,\mathrm{d}t^2 + B(r,t)\,\mathrm{d}r^2 + r^2\,\mathrm{d}\Omega^2,$$

where $\mathrm{d}\Omega^2 = \mathrm{d}\theta^2 + \sin^2\theta\,\mathrm{d}\phi^2$ is the metric of a unit 2-sphere. To work out the field equations, we must

(i) calculate the connection components determined by this metric,
(ii) determine, from these, the Ricci tensor components, and
(iii) set these components to zero, and solve for $A(r,t)$, $B(r,t)$.

Using the boundary conditions of asymptotic flatness, $\{r \to \infty\} \Rightarrow \{A(r,t) \to 1,\ B(r,t) \to 1\}$, and determining an integration constant by comparing with the Newtonian solution for a star of mass $m$ ($\equiv MG/c^2$ in physical units), one gets

$$\mathrm{d}s^2 = -\left(1-\frac{2m}{r}\right)\mathrm{d}t^2 + \left(1-\frac{2m}{r}\right)^{-1}\mathrm{d}r^2 + r^2\,\mathrm{d}\Omega^2. \quad (9)$$

This is the *Schwarzschild solution*. It is an *exact* solution of the EFEs, valid for $r > r_S$, where $r_S$ is the coordinate radius of the surface of the central massive object; we require that $r_S > 2m$, where $m$ is the *Schwarzschild radius* of the object. This is the mass in geometrical units. For the Earth $m \approx 8.8$ mm, and for the sun $m \approx 2.96$ km.

A remarkable result is hidden in the analysis above: $A$ and $B$ are in fact functions only of $r$. Consequently, a spherically symmetric vacuum exterior solution is necessarily static. This is *Birkhoff's theorem.*

Overall, this shows that the Schwarzschild solution is the valid exterior solution for every spherical object, no matter how it is evolving. It can be static, collapsing, expanding, pulsating; provided it is spherically symmetric, the exterior solution is always the Schwarzschild solution! This expresses the fact that general relativity does not allow monopole or dipole gravitational radiation, so spherical pulsations cannot radiate their mass away as energy.

The nature of particle orbits (time-like geodesics) follows from this metric, allowing circular orbits for any radius greater than $r = 3m$. The metric also determines the light ray paths: the radial null rays in the solution (9) have $\mathrm{d}s^2 = 0 = \mathrm{d}\theta^2 = \mathrm{d}\phi^2$, giving

$$\frac{\mathrm{d}t}{\mathrm{d}r} = \pm\frac{1}{1-2m/r} \Leftrightarrow t = \pm r^* + \text{const.}, \quad (10)$$

where

$$\mathrm{d}r^* = \frac{\mathrm{d}r}{1-2m/r} \Leftrightarrow r^* = r + 2m\ln\left(\frac{r}{2m}-1\right).$$

This is the equation of the local null cones, showing how light is bent by this gravitational field. In addition, outgoing light experiences redshifting depending on the radii $r_e$ and $r_0$, where the source and receiver are located. The redshift $z$ measured for an object emitting light of wavelength $\lambda_e$ that is observed with wavelength $\lambda_0$ is given by $1 + z := \lambda_0/\lambda_e$. Wavelength $\lambda$ is related to period $\Delta\tau$ by $\lambda = c\Delta\tau$, with proper time $\tau$ along a world line $\{r = \text{const.}\}$ determined by the metric (9). Because $\Delta t_0 = \Delta t_e$ follows from the null ray equation (10), the redshift $z$ is

$$1 + z := \frac{\lambda_0}{\lambda_e} = \frac{\Delta\tau_0}{\Delta\tau_e} = \left(\frac{1-2m/r_0}{1-2m/r_e}\right)^{1/2}. \quad (11)$$

This is gravitational redshift, caused by the change in potential energy as photons climb out of the potential well due to the central mass.

### 2.1 The "Singularity" at $r = 2m$

Something goes wrong with the metric at $r = 2m$. What happens there? The following all seem to be the case.

(1) **Singular metric.** The metric components are singular at $r = 2m$.
(2) **Radial infall.** Considering the radial infall of a test object (with nonzero rest mass), the proper time taken to fall from $r = r_0 > 2m$ to $r = 2m$ is finite.

However, the coordinate time is given by $dt/d\tau = (1-2m/r_0)^{1/2}/(1-2m/r)$, which is infinite as $r \to 2m$. This suggests that it never actually reaches that radial value, whereas the object measures a finite proper time for the trip.

**(3) Infinite redshift.** The redshift formula (11) shows that $z \to \infty$ as $r_e \to 2m$.

**(4) Unattainable surface.** On radial null geodesics (10), $dt/dr \to \pm\infty$ as $r \to 2m$. Accordingly, light rays from the outside region ($r > 2m$) cannot reach the surface $r = 2m$; rather, they become asymptotic to it as they approach it. The same will hold true for time-like geodesics.

**(5) Geodesic incompleteness.** These time-like and null geodesics are *incomplete*; that is, they cannot be extended to arbitrarily large values of their affine parameter in the region $2m < r < \infty$. In particular, for time-like geodesics the proper time along the geodesics from $r_0$ to $2m$ is finite. This shows that the geodesic, moving inward as $t$ increases, cannot be extended to infinite values of its affine parameter (namely, proper time).

**(6) Nonstatic interior.** Going to the interior part ($0 < r < 2m$), the solution changes from being static to being *spatially homogeneous* but *evolving with time*. This is because the essential metric dependence is still with the coordinate $r$, which has changed from being a coordinate measuring spatial distances to one measuring time changes. The nature of the space-time symmetry therefore changes completely for $r < 2m$.

**(7) A physical singularity?** To try to see if the singularity at $r = 2m$ is a physical singularity or a coordinate singularity, we look at scalars (because they are coordinate-independent quantities) constructed from mathematical objects that describe the curvature. Because the solution is a vacuum solution ($R_{ab} = 0$), both $R = R^a{}_a$ and $R^{ab}R_{ab}$ vanish. The simplest nonzero scalar is the *Kretschmann scalar*

$$R_{abcd}R^{abcd} = 48m^2/r^6.$$

This is finite at $r = 2m$ but diverges as $r \to 0$. This suggests (but does not prove) that $r = 2m$ is a coordinate singularity, that is, there is no problem with the space-time but rather the coordinates break down there, and so we can get rid of the singularity by choosing different coordinates. We prove that this supposition is correct by making extensions of the solution across the surface $r = 2m$. One can do this by attaching coordinates to either time-like or null geodesics that cross this surface.

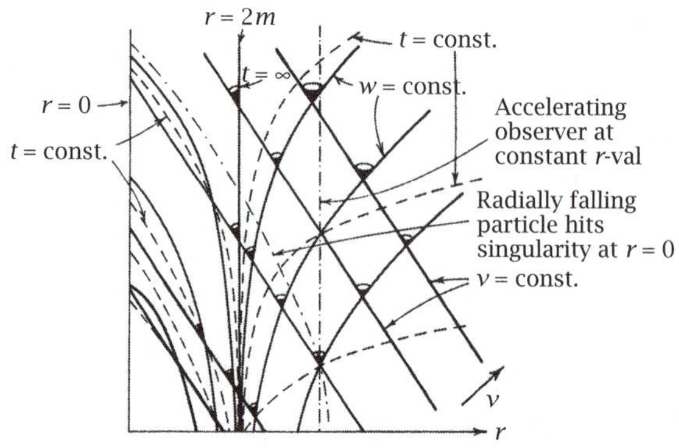


**Figure 1** Radial null geodesics and local light cones in the Schwarzschild solution in Eddington–Finkelstein coordinates (reproduced from Hawking and Ellis (1973), with permission).

## 2.2 Schwarzschild Null Coordinates

Defining the coordinates $v_+$, $v_-$ by $v_\pm = t \pm r^*$, then $dv_\pm = dt \pm dr/(1-2m/r)$; the outgoing null geodesics are $\{v_+ = \text{const.}\}$ and the ingoing ones are $\{v_- = \text{const.}\}$. We change to coordinates $(v, r, \theta, \phi)$, where $v_+ = v$. The metric is then

$$ds^2 = -\left(1 - \frac{2m}{r}\right) dv^2 + 2dv\,dr + r^2\,d\Omega^2. \qquad (12)$$

This is the *Eddington–Finkelstein form* of the metric. The transformation has succeeded in getting rid of the singularity at $r = 2m$. The coordinate transformation (which is singular at $r = 2m$) extends the original space-time region (denoted region I) defined by $2m < r < \infty$ to a new region (denoted region II) defined by $0 < r < 2m$. It is an analytic extension across the surface $r = 2m$ of the outside region I to the inside region II.

Plotting these local null cones and light rays in the space with coordinates $(t^*, r)$, we get the Eddington–Finkelstein diagram (figure 1). We can observe the following features in the figure.

(1) The surfaces of constant $r$ are vertical lines in this diagram. The surfaces of constant $t$ are nearly flat at large distances but bend down and never cross the surface $r = 2m$. In fact, $t$ diverges at this surface; and it is this bad behavior of the $t$-coordinate that is responsible for the coordinate singularity at $r = 2m$. This is why the coordinate time diverges for a freely falling particle that crosses this surface.

(2) The null cones tilt over, the inner ray always being at 45° inward, but the outer one (pointing outward for $r > 2m$) becomes vertical at $r = 2m$ and points inward for $r < 2m$. The surface $r = 2m$ is therefore a *null surface* (a light ray emitted outward at $r = 2m$ stays at that distance from the center forever). Because of this, it is a *trapping surface*: particles that have fallen in and crossed this surface from the outside region I to the inside II can never get out again. Once in region II their future is inevitably to be crushed by divergent tidal forces near the singularity at $r = 0$.

(3) Conversely, $r = 2m$ is an *event horizon*, hiding its interior from the view of outside observers. If we consider an observer who is static at $r = r_1 > 2m$, her world line is a vertical line in this diagram. Her past light cone never reaches inside $r = 2m$, so no signal from that region can reach her. This space-time may therefore reasonably be called a *black hole*, for no radiation emitted by the inside region II can reach the outside region I.

(4) If the outside observer drops a probe into the center, then it crosses the event horizon $r = 2m$ in a finite proper time but takes an infinite coordinate time $t$ to get there because $t$ diverges there. If it emits pulses at regular intervals (say, every second), these will be received by the outside observer at longer and longer time intervals. If the probe crosses the event horizon at 12:00 according to its internal clock and sends out a radio signal at that time, the signal will never reach the outside observer; it stays at $r = 2m$ forever. Every signal sent before then will (eventually) reach the outside observer and every signal sent afterward will fall into the central singularity. The infinite slowing down of the received signals as the probe approaches the event horizon will result in the redshift in received signals diverging as $r_e \to 2m$.

This analysis shows convincingly that $r = 2m$ is a null surface (the event horizon) at which the original coordinates go wrong; the space-time can be extended across this null surface by a change to new coordinates. However, there is a problem with what we have so far: namely, the original solution is time symmetric. The extension is not, as is obvious from the Eddington–Finkelstein diagram.

#### 2.2.1 The Time-Reversed Extension

We can make another similar Eddington–Finkelstein extension in which we choose the other direction of time for the extension by using the other null coordinate.

To be more precise, upon changing from $(t, r, \theta, \phi)$ to coordinates $(w, r, \theta, \phi)$, where $v_- = w$, the metric now takes the form

$$ds^2 = -\left(1 - \frac{2m}{r}\right) dw^2 - 2dw\, dr + r^2\, d\Omega^2,$$

which is the time-reverse of (12). This is also an Eddington–Finkelstein form of the metric, now extending the original space-time region I, defined by $2m < r < \infty$, to a further new region II′, defined by $0 < r < 2m$. This leads to a time-reversed Eddington–Finkelstein picture of the local light cones. This time it is the ingoing null geodesics that are badly behaved at $r = 2m$ (the outgoing ones cross this surface with no trouble). The surface $r = 2m$ is again a null cone, but this time it represents the surrounding horizon of a *white hole*: signals can come out of it but not go into it. The exterior observer can receive messages from region II′ but can never send signals there. The surface $r = 2m$ is now the same as $t = -\infty$. This divergence is the reason the original metric went wrong at this surface.

How do we know that region II is different from region II′? In the original metric, both the past and the future inward-pointing radial null geodesics through each event $q$ in region I were incomplete. The first extension completed the future-ingoing null geodesics but not the past-ingoing ones; the second extension completed the past-ingoing null geodesics but not the future-ingoing ones.

### 2.3 Kruskal–Szekeres Coordinates

The question now is whether we can make both extensions simultaneously. The time asymmetry of each extension is because we used only one null coordinate in each case. To obtain a time-symmetric extension, we must use both null coordinates. Indeed, if we change to coordinates $(v, w, \theta, \phi)$, we obtain the double null form of the metric:

$$ds^2 = -\left(1 - \frac{2m}{r}\right) dv\, dw + r^2\, d\Omega^2,$$

where $\frac{1}{2}(v - w) = r + 2m \ln((r/2m) - 1)$ defines $r(v, w)$ (both quantities being equal to $r^*$), and $t = \frac{1}{2}(v + w)$. This is a time-symmetric double null form but is singular at $r = 2m$. However, if we rescale the null coordinates, we can attain what we want. Defining $V = e^{v/4m}$, $W = -e^{-w/4m}$, we obtain a new set of double null coordinates for the Schwarzschild solution.

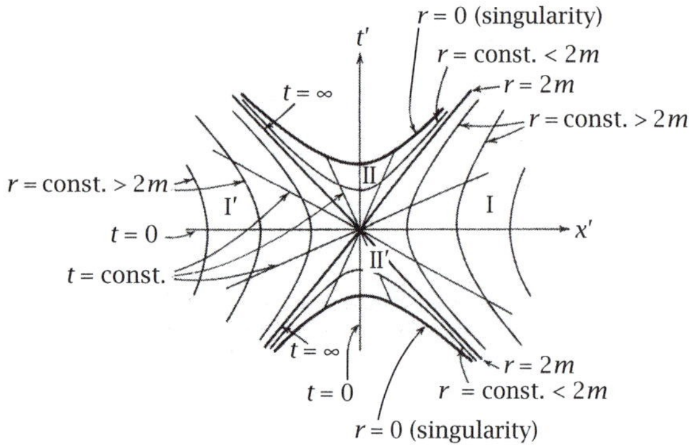


**Figure 2** The Kruskal diagram, representing the maximal extension of the Schwarzschild solution in conformally flat coordinates (reproduced from Hawking and Ellis (1973), with permission).

The metric becomes

$$ds^2 = -\frac{32m^3}{r}e^{-r/2m}\,dV\,dW + r^2\,d\Omega^2.$$

This metric form is regular at $r = 2m$ and gives us the extension we want. The light cones are given by $V = \text{const.}$, $W = \text{const.}$ It is convenient to define the associated time and radial coordinates $(t', x')$ by $x' = \frac{1}{2}(V - W)$, $t' = \frac{1}{2}(V + W)$. These are the *Kruskal–Szekeres coordinates*, giving the Kruskal diagram in figure 2, the complete time-symmetric extension of the Schwarzschild solution with light rays at $\pm 45°$.

It is important to remember that this is a cross section of the full space-time; in fact, each point represents a 2-sphere of area $4\pi r^2$ of the full space-time (we have suppressed the coordinates $(\theta, \phi)$ in order to draw this diagram). The whole solution is time symmetric, as desired. The most important new feature is that there is a new region I′ in addition to the three regions I, II, and II′ already identified. Let us see why this is so.

The region I in this diagram corresponds to the *same* region I in *both* the Eddington–Finkelstein diagrams. The region $t' > x'$ (bounded on the left by the line at $-45°$ through the origin) corresponds to the first extension to region II completing the future-ingoing null geodesics. This part of the Kruskal diagram corresponds point by point to the first Eddington–Finkelstein diagram; in particular, the vertical null geodesics at $r = 2m$ correspond to the null geodesic at $+45°$ through the origin in the Kruskal diagram. The region $t' < x'$ (bounded on the left by the line at $+45°$ through the origin) corresponds to the second extension to region II′, completing the past-ingoing null geodesics (moving in the opposite direction to the future-ingoing ones). This part of the Kruskal diagram corresponds point by point to a time-reversed Eddington–Finkelstein diagram.

Now consider a point in region II. The past-outgoing (i.e., moving to the right) null geodesics cross $r = 2m$ to the asymptotically flat region I. The past-ingoing (i.e., moving to the left) null geodesics are completely symmetric with them; they must cross $r = 2m$ to an asymptotically flat region I′ that is identical to region I. Similarly, for points in II′, the outgoing future-directed null geodesics cross $r = 2m$ to region I, and the ingoing future-directed null geodesics must cross $r = 2m$ to an identical region I″. What is perhaps not immediately obvious is that this is the *same* region as I′. The following features then arise.

(1) The surfaces of constant $r$ are given by $x'^2 - t'^2 = \text{const.}$ and therefore correspond to the hyperbola at constant distance from the origin in flat space-time. They are space-like for $0 < r < 2m$ and time-like for $r > 2m$ (with *two* surfaces occurring for each value of $r$). There are *two* singularities at $r = 0$: one in the past ($t' < 0$) and one in the future ($t' > 0$). The surfaces $r = 2m$ are the two intersecting null surfaces through the origin. The surfaces of constant $t$ are the straight lines through the origin. This coordinate diverges at both surfaces $r = 2m$.

(2) The null surface segments $\{r = 2m,\ t' > 0\}$ (obviously representing motion at the speed of light) are *trapping surfaces*: particles that have fallen in and crossed this surface from either outside region to the inside region II can never get out again; they inevitably fall into the future singularity at $\{r = 0,\ t' > 0\}$, where they are crushed by infinite tidal forces. Each of these segments is also an *event horizon*, hiding its interior from the view of outside observers. They bound the *black hole* region II of the space-time, from which no light or other radiation that is emitted can reach the outside. The details of dropping in a probe from the outside can be followed in this diagram, revealing again the same effects as discussed above. Thus we have (vacuum) black holes, bounded by event horizons—from which light cannot escape.

(3) There are two curvature singularities, each of which has a space-like nature, one in the past and one in the future. The complete solution has both a white hole and a black hole singularity (the former emitting particles and radiation into the space-time, the latter receiving them).

It is particularly important to note here that, unlike in the electromagnetic or Newtonian gravity cases, we do *not* find a singular time-like worldline at the center of the Schwarzschild solution, which would represent a particle generating the solution. Here, the strong gravitational field for $r < 2m$ profoundly alters the nature of the singularity from what we first expected.

(4) There is also an unexpected global topology, with two asymptotically flat spaces back to back, joined by a neck called a wormhole. To see this, consider the surface $\{t' = 0\}$; the area of the 4-spheres $\{r = \text{const.}\}$, which diverges as $x' \to \infty$, decreases to a minimum value $A_* = 4\pi(2m^2)$ at $x' = 0$ and then diverges again as $x' \to -\infty$. However, one cannot communicate between the two asymptotically flat regions through the wormhole because only space-like curves can pass through.

(5) The nature of the space-time symmetry changes at the surfaces $r = 2m$: here, there is a transition from a static to an evolving (Kantowski–Sachs) universe. The event horizon is therefore also a *Killing horizon*, generated by null Killing vectors that continue to be space-like on one side and time-like on the other.

(6) This solution is indeed the maximal extension of the initial Schwarzschild space-time: all geodesics in it are *either* complete (that is, they go to infinity) *or* they run into one of the singularities where $r = 0$. Hence, no further extensions are possible (all curves end up at a singularity or at infinity).

### 2.4 Gravitational Collapse to Black Holes

I have dealt with the maximal Schwarzschild solution at length because it is such a good example of general relativity. The subtleties of coordinates and their limits, the interpretation of time-like and null geodesics, and the examination of global topology are all involved in understanding all of this complex structure hidden in the seemingly innocuous Schwarzschild metric (9). But is it just a mathematical solution with no relevance to the real universe? What is its relation to astrophysics?

Black holes play an important role in high-energy astrophysics: they occur at the endpoint of the lifetime of massive stars; they are at the center of many galaxies, where they are surrounded by accretion disks; and they are the powerhouse for high-energy processes in quasistellar objects. The collapse of a spherical object to a black hole starts off in a regular space-time region with no horizons; these develop as matter coalesces and the gravitational field intensifies, causing future-directed light rays to be trapped. There is no singularity in the past, but one occurs in the future, hidden inside the event horizon. There is no wormhole through to another space-time region because the infalling matter cuts it off.

Diagrams of the resulting space-times are given in Hawking and Ellis (1973) and in Ellis and Williams (2000). Rotation will complicate matters considerably (we get a Kerr black hole instead of a Schwarzschild solution). The role of black holes in astrophysics is discussed in depth by Begelman and Rees (2009).

## 3 Cosmology

In the previous example the curvature was pure Weyl curvature (because $R_{ab} = 0$); space-time is unaffected by matter properties because it is a vacuum solution. In the next example there is, by contrast, pure Ricci curvature; there is no free gravitational field ($C_{abcd} = 0$). The evolution of space-time is determined by its matter content.

Gravity governs the universe in the large. On a large enough scale, the observed universe can be represented well by a Robertson–Walker metric, which is spatially homogeneous (no spatial point is preferred over any other) and isotropic (no direction in the sky is preferred over any other). Comoving coordinates can be chosen so that the metric takes the form

$$ds^2 = -dt^2 + S^2(t)\,d\sigma^2, \qquad u^a = \delta^a{}_0 \quad (a = 0, 1, 2, 3), \tag{13}$$

where $S(t)$ is the time-dependent scale factor, and the worldlines with tangent vector $u^a = dx^a/dt$ represent the histories of fundamental observers. The space sections $\{t = \text{const.}\}$ are surfaces of homogeneity and have maximal symmetry: they are 3-spaces of constant curvature $K = k/S^2(t)$, where $k$ is the sign of $K$. The metric $d\sigma^2$ is a 3-space of normalized constant curvature $k$; coordinates $(r, \theta, \phi)$ can be chosen such that

$$d\sigma^2 = dr^2 + f^2(r)(d\theta^2 + \sin^2\theta\, d\phi^2), \tag{14}$$

where $f(r) = \{\sin r, r, \sinh r\}$ if $k = \{+1, 0, -1\}$, respectively.

The metric is time-dependent, with distances between all fundamental observers scaling as $S(t)$ and the rate of expansion at time $t$ characterized by the *Hubble parameter* $H(t) = \dot{S}/S$. To determine the metric's evolution in time, one applies the EFEs to the metric (13), (14). Because of local isotropy, the matter tensor $T_{ab}$ necessarily takes a perfect fluid form relative to the preferred worldlines with tangent vector $u^a$. The integrability conditions (5) for the EFEs are the *energy-density conservation equations*, which reduce to

$$T^{ab}{}_{;b} = 0 \Leftrightarrow \dot{\rho} + (\rho + p)3\dot{S}/S = 0. \tag{15}$$

This becomes determinate when a suitable equation of state relates the pressure $p$ to the energy density $\rho$. Baryons have $p_{\mathrm{b}} = 0$ and radiation has $p_{\mathrm{r}} = \rho_{\mathrm{r}}/3$, $\rho_{\mathrm{r}} = aT_{\mathrm{r}}^4$, which by (15) implies that $\rho_{\mathrm{b}} \propto S^{-3}$, $\rho_{\mathrm{r}} \propto S^{-4}$, $T_{\mathrm{r}} \propto S^{-1}$. Radiation dominates at early times (the hot big bang era), matter dominates at late times; at very early times (the inflationary era), a scalar field may have been dynamically dominant.

The scale factor $S(t)$ generically obeys the *Raychaudhuri equation*:

$$3\ddot{S}/S = -\tfrac{1}{2}\kappa(\rho + 3p) + \Lambda, \tag{16}$$

where $\kappa$ is the gravitational constant and $\Lambda$ the cosmological constant. This shows that the active gravitational mass density of the matter and fields present is $\rho_{\mathrm{grav}} := \rho + 3p$, summed over all matter present. For ordinary matter this will be positive, so ordinary matter will tend to cause the universe to decelerate ($\ddot{S} < 0$). A positive cosmological constant on its own will cause an accelerating expansion ($\ddot{S} > 0$). When matter and a cosmological constant are both present, either result may occur depending on which effect is dominant. The first integral of equations (15), (16) when $\dot{S} \neq 0$ is the *Friedmann equation*:

$$\frac{\dot{S}^2}{S^2} = \frac{\kappa\rho}{3} + \frac{\Lambda}{3} - \frac{k}{S^2}. \tag{17}$$

Models with a Robertson–Walker geometry with metric (13), (14) and dynamics governed by equations (15), (16), (17) are called *Friedmann–Lemaître (FL) universes*.

The simplest solution is the *Einstein–de Sitter universe*, with flat spatial sections and baryonic matter content, and no cosmological constant. We then have

$$\{k = 0,\ p = 0,\ \Lambda = 0\} \Rightarrow S(t) = t^{2/3}.$$

The FL models are the standard models of modern cosmology, and they are surprisingly effective in view of their extreme geometrical simplicity. One of their great strengths is their explanatory role in terms of making explicit the way the local gravitational effect of matter and radiation determines the evolution of the universe as a whole, with this in turn forming the dynamic background for local physics (including determining the evolution of matter and radiation themselves).

### 3.1 An Initial Singularity?

The universe is presently expanding; following it back into the past, it was ever smaller. A key issue is whether it had a beginning or not. The Raychaudhuri equation (16) leads directly to the following result.

**The Friedmann–Lemaître universe singularity theorem.** *In an FL universe with $\Lambda \leqslant 0$ and $\rho + 3p > 0$ at all times, at any instant $t_0$ when $H_0 \equiv (\dot{S}/S)_0 > 0$, there is a finite time $t_*$ such that $S(t) \to 0$ as $t \to t_*$, where $t_0 - (1/H_0) < t_* < t_0$. If $\rho + p > 0$, the universe starts at a space-time singularity there: $\{t \to t_*\} \Rightarrow \{\rho \to \infty\}$.*

This is not merely a start to matter—it is a start to space, to time, to physics itself. It is the most dramatic event in the history of the universe; it is the start of the existence of everything. The underlying physical feature is the nonlinear nature of the EFEs: going back into the past, the more the universe contracts, the higher the active gravitational density, causing it to contract even more. The pressure $p$ that one might have hoped would help stave off the collapse makes it even worse because $p$ enters algebraically into the Raychaudhuri equation (16) with the same sign as the energy density $\rho$.

This conclusion can, in principle, be avoided by introducing a cosmological constant, but in practice, this cannot work because its value is too small. However, energy-violating matter components such as a scalar field can avoid this conclusion if they dominate at early enough times. In the case of a single scalar field $\phi$ with space-like surfaces of constant density, on choosing $u^a$ orthogonal to these surfaces, the stress tensor has a perfect fluid form with $\rho = \frac{1}{2}\dot{\phi}^2 + V(\phi)$, $p = \frac{1}{2}\dot{\phi}^2 - V(\phi)$, and so $\rho + 3p = 2\dot{\phi}^2 - 2V(\phi)$. The slow-rolling case is $\dot{\phi}^2 \ll V(\phi)$, leading to $\rho + p = 2\dot{\phi}^2 \approx 0 \Rightarrow \rho + 3p \simeq -2\mu < 0$. Quantum fields can in principle therefore avoid the conclusion that there was a start to the universe. The jury is still out as to whether or not this was the case.

### 3.2 Observational Relations

To be good models of the real universe, cosmological models must predict the right results for astronomical observations. To determine light propagation in a Robertson–Walker geometry, it suffices to consider only radial null geodesics: $\{\mathrm{d}s^2 = 0,\ \mathrm{d}\theta = 0 = \mathrm{d}\phi\}$ (by the symmetries of the model, these are equivalent to generic geodesics). From the Robertson–Walker metric (13), we then find that for light emitted at time $t_{\mathrm{e}}$ and received at time $t_0$, the comoving radial coordinate distance $u(t_0, t_e) := r_0 - r_e$ between comoving emitters and receivers is given by

$$u(t_0, t_e) = \int_{t_e}^{t_0} \frac{\mathrm{d}t}{S(t)} = \int_{S_e}^{S_0} \frac{\mathrm{d}S}{S\dot{S}}, \tag{18}$$

with $\dot{S}$ given by the Friedmann equation (17). The function $u(t_0, t_e)$ therefore encodes information on both spatial curvature and the space-time matter content.

The key quantities related to cosmological observations are redshift, area distance (or apparent size), and local volume, corresponding to some increment in distance (determining number counts). From (18), the cosmological redshift $z_c$ is given by

$$1 + z_c = \frac{S(t_0)}{S(t_e)}. \tag{19}$$

The same ratio of observed to emitted light holds for all wavelengths, a key identifying property of redshift.

The apparent angular size $\alpha$ of an object at redshift $z_c$ and of linear size $l$ is given by

$$l/\alpha = f(u)S(t_e) = f(u)(1+z)S(t_0). \tag{20}$$

Measures of apparent sizes as functions of redshift will therefore determine $f(u)$ if the source physical size is known. The flux of radiation $F$ measured from a point source of luminosity $L$ emitting radiation isotropically is given by the fraction of radiant energy emitted by the source in a unit of time that is received by the telescope:

$$F = \frac{L}{4\pi} \frac{1}{f^2(u)S^2(t_0)(1+z)^2} \tag{21}$$

(the two redshift factors account firstly for the time dilation between observer and source, and secondly for loss of energy due to redshifting of photons). Measures of apparent luminosity $m$ as a function of redshift will therefore also determine $f(u)$ if the source intrinsic luminosity is known.

The number of objects in a solid angle $\mathrm{d}\Omega$ for a distance increment $\mathrm{d}u$ (characterized by increments $\mathrm{d}z$, $\mathrm{d}m$ in the observable variables $z$ and $m$) is given by

$$\mathrm{d}N = n(t_e)S^3(t_e)f(u)\,\mathrm{d}u\,\mathrm{d}\Omega, \tag{22}$$

where $n(t_e)$ is the number density of objects at the time of emission. The observed total number $N$ of objects measured in a survey is given by integrating from the observer to the survey limit: in terms of the radial coordinate $r_e$ of the source (which can be related to redshifts or magnitudes), $N = \int_{r_0}^{r_e} \mathrm{d}N$. If the number of objects is conserved, $n(t_e) = n(t_0)(1+z)^3$, and we find from (22) that

$$N(t_e) = n(t_0)\,\mathrm{d}\Omega \int_{r_0}^{r_e} f(u)\,\mathrm{d}u. \tag{23}$$

A measure of numbers as a function of redshift or other distance indicators will therefore determine the number density of objects, which is then related to their mass and so to the energy density of these objects.

As $f(u)$ is determined by the Friedmann equation, the above equations enable us to determine observational relations between observable variables for any specific cosmological model and therefore to observationally test the model.

### 3.3 Causal and Visual Horizons

A fundamental feature affecting the formation of structure and our observational situation is the limits that arise because causal influences cannot propagate at speeds greater than the speed of light. The region that can causally influence us is therefore bounded by our past null cone. Combined with the finite age of the universe, this leads to the existence of particle horizons that limit the part of the universe with which we can have had causal connection.

A *particle horizon* is, by definition, composed of the limiting worldlines of the furthest matter that ever intersects our past null cone. This is the limit of matter that we can have had any kind of causal contact with since the start of the universe, which by (18) is characterized by the comoving radial coordinate value

$$u_{\mathrm{ph}} = \int_0^{t_0} \frac{\mathrm{d}t}{S(t)}. \tag{24}$$

The present physical distance to the matter constituting the horizon is

$$d_{\mathrm{ph}} = S(t_0)u_{\mathrm{ph}}.$$

The key question is whether the integral (24) converges or diverges as we go to the limit of the initial singularity where $S \to 0$. *Horizons will exist in standard FL cosmologies for all ordinary matter and radiation* because $u_{\mathrm{ph}}$ will be finite in those cases; for example, in the Einstein–de Sitter universe, $u_{\mathrm{ph}} = 3t_0^{1/3}$, $d_{\mathrm{ph}} = 3t_0$. We will then have seen only a fraction of what exists, unless we live in a universe with spatially compact sections so small that light has indeed had time to traverse the whole universe since its start; this will not be true for universes with the standard simply connected topology.

Penrose's powerful use of conformal methods gives a very clear geometrical picture of the nature of horizons. One switches to coordinates where the null cones are at ±45°; the initial singularity (the start of the universe) is then represented as a boundary to space-time. If this

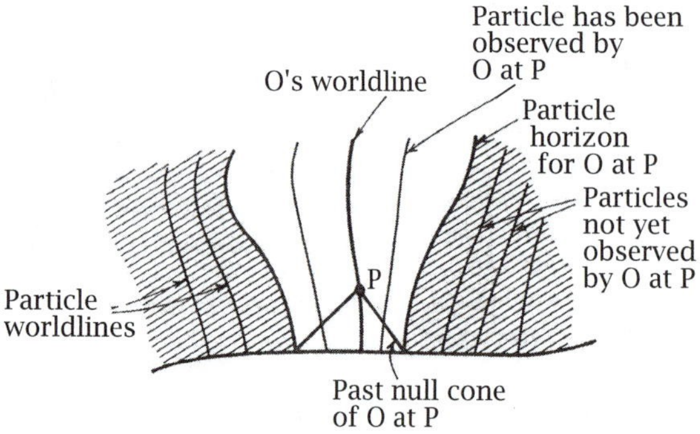


**Figure 3** Particle horizons in a world model (reproduced from Hawking and Ellis (1973), with permission). "O" is the observer (at the origin); the horizontal line at the bottom is the start of the universe. Here and now is the event "P."

is a space-like surface, then there exist galaxies beyond our causal horizon; we can receive no signals from them whatever (figure 3).

The importance of horizons is twofold: they underlie causal limitations that are relevant to the origin of structure and uniformity, and they represent absolute limits on what is testable in the universe. Many present-day speculations about the superhorizon structure of the universe (e.g., chaotic inflationary theory) are not observationally testable because one can obtain no definite information whatever about what lies beyond the visual horizon. This is one of the major limits to be taken into account in our attempts to test the veracity of cosmological models.

### 3.4 Structure Formation

A large amount of work in cosmology is concerned with structure formation: how galaxies come to exist by gravitational attraction acting on primordial perturbations. This raises a key technical issue, the choice of perturbation gauge, which is best solved by using a gauge-free perturbation formalism. When the perturbations become nonlinear, numerical simulations are needed to study the details of structure formation.

## 4 Issues

This article has emphasized the significance of geometry and topology in general relativity theory: a fascinating study of the interplay of geometry, analysis, and physics. Further topics I have not mentioned include how one derives the Newtonian limit of the general relativity equations; gravitational lensing; the emission and detection of gravitational radiation; the key issue of the nature of dark energy in cosmology, which has led to the acceleration of the universe in recent times; and the issue of dark matter—nonbaryonic matter that dominates structure formation and the gravitational dynamics of clusters of galaxies. A key requirement is to test the EFEs in every possible way, to try alternative theories of gravitation to see if any of them work better than Einstein's theory. So far it has withstood these tests; it is the best gravitational theory available.

### Further Reading

Anninos, P. 2001. Computational cosmology: from the early universe to the large scale structure. *Living Reviews in Relativity* 4:2.

Begelman, M., and M. Rees. 2009. *Gravity's Fatal Attraction: Black Holes in the Universe*. Cambridge: Cambridge University Press.

D'Inverno, R. 1992. *Introducing Einstein's Relativity*. Oxford: Clarendon Press.

Ehlers, J. 1993. Contributions to the relativistic mechanics of continuous media. *General Relativity and Gravitation* (Golden Oldie) 25:1225–66 (originally published in 1961).

Ellis, G. F. R. 2009. Relativistic cosmology. *General Relativity and Gravitation* (Golden Oldie) 41:581 (originally published in 1971).

Ellis, G. F. R., and R. M. Williams. 2000. *Flat and Curved Spacetimes*. Oxford: Oxford University Press.

Hawking, S. W., and G. F. R. Ellis. 1973. *The Large Scale Structure of Space-Time*. Cambridge: Cambridge University Press.

Maartens, R., and B. A. Bassett. 1998. Gravito-electromagnetism. *Classical and Quantum Gravity* 15:705.

Stephani, H., D. Kramer, M. MacCallum, and C. Hoenselaers. 2003. *Exact Solutions of Einstein's Field Equations*. Cambridge: Cambridge University Press.

# Part V
# Modeling

## V.1 The Mathematics of Adaptation (Or the Ten Avatars of Vishnu)

*David Krakauer and Daniel N. Rockmore*

### 1 Adaptation and Selection

In a celebrated letter to his sister and theologian Simone Weil, dated March 1940, André Weil (writing from a military prison in Paris) expounded on the powerful yet transient role that analogy plays in mathematics as a precursor to unification. He does this through another layer of analogy, that of the avatars of Vishnu, each of which reveal different facets of the deity. Weil had in mind various connections—yet to be supported by theorem and proof—that he was finding between problems in number theory and algebraic geometry, connections that today are encoded in one of the grandest current efforts in mathematics: the pursuit of the Langlands program, a Vishnu of mathematics. The value of analogy in pure mathematics in promoting identification of common foundations extends into its dealing with physical reality (even this aspect is evident in the Langlands program, realized through recent connections made with quantum field theory). This was already understood by Poincaré when he wrote that "the mathematical facts worthy of being studied are those which, by their analogy with other facts, are capable of leading us to the knowledge of a physical law."

The pursuit of a "biological law" (or even agreement on what would define one) is proving to be something of a challenge. But *adaptation* may be a context that supports such an achievement. Loosely construed, it is the collection of dynamics yielding evolved or learned traits that contribute positively to survival and propagation. Adaptation assumes a bewildering array of "avatars," spanning evolution by natural selection, reinforcement learning, Bayesian inference, and supervised machine learning, of which neural networks are perhaps the most explicitly adaptive. In all of these instances adaptation implies some element of (1) design or optimization, (2) differential selection or success, and (3) historical correlation between evolved trait and environment that grows stronger through time. Each of these elements has inspired mathematical descriptions, and it is only in the past decade that we have become aware that several fields concerned with the formal analysis of adaptation are in fact deeply analogous (an awareness that has been engendered by mathematics) and that these analogies suggest a form of unification that might be codified as a "biological law."

It is very difficult to discuss adaptation without considering *natural selection*, and in many treatments they are effectively synonymous. Stephen Jay Gould bemoaned the ambiguity whereby the word adaptation can be interpreted as both noun and verb. The noun adaptation refers to the constellation of properties that confer some locally optimal character on an agent, whereas the verb adaptation describes the process by which agents adapt through time to their environments. It is the latter active process that has attracted the greater mathematical effort and includes the often counterintuitive properties that are found in population dynamics. Adaptation at equilibrium (the noun form) simply requires that optimality be demonstrated, often through variational techniques derived from engineering. Examples include the optical properties of the lens, the excitability properties of neural membranes, and the kinematic properties of limbs and joints. They all fall into this equilibrium class of adaptive explanation and are as diverse in their mathematical treatments as material constraints demand. In this article, however, we focus on the genuinely novel dynamical features of the adaptive process, the extraordinary diversity of systems in which they continue to be realized, and the associated mathematical ideas. The context is generally one of populations evolving in distribution, and in this broad review we touch on the techniques of information geometry, evolutionary game theory, and mixtures of discrete dynamics and probability theory.

## 2 Selection Systems

### 2.1 Continuous Selection for Diploid Species

A historically accurate place to start exploring adaptation is with the continuous selection equation for diploid genomes from population genetics. In this setting we are interested in the time evolution of the distribution of genomic structures in the population as influenced by both endogenous and exogenous factors.

Consider a single coding site in a genome, which can assume multiple, discrete genetic forms or alleles, $A_i$. For a diploid organism (one whose genetic profile is encoded in two copies of each gene), each site will be represented by a pair of values $(A_i, A_j)$. The time evolution of a genome in terms of the frequency in the population of the various alleles is determined by the coefficients of its net growth rate in competition with a population of genomes. This growth rate is set by the so-called Malthusian fitness parameter.

Let us denote the frequency of allele $i$ by $x_i \in [0,1]$ and collect these frequencies in vector $\boldsymbol{x}$, which then has nonnegative elements that sum to 1.

Adaptive evolution proceeds according to

$$\dot{x}_i = x_i((M\boldsymbol{x})_i - \boldsymbol{x}^{\mathrm{T}} M\boldsymbol{x}), \tag{1}$$

where $M$ is the matrix of real-valued *Malthusian parameters* for the set of $n$ genomes. A positive derivative reflects an increasing frequency of an allele and is associated with a higher than average value for a given type in the $M$ matrix. Rewritten, with $m_i(x) = (Mx)_i$, this gives the so-called *replicator equation*

$$\dot{x}_i = x_i(m_i(\boldsymbol{x}) - \bar{m}(\boldsymbol{x})),$$

with $\bar{m}(\boldsymbol{x}) = \sum_i x_i m_i(\boldsymbol{x})$. This is a differential equation on the invariant simplex $S_n$ whose points represent all possible distribution of alleles, and it has deep correspondences across a range of fields. We shall come back to the importance of the replicator equation once we have described a few more properties of the more general selection equation.

The dynamics of (1) is determined by the properties of its stable equilibria. These are conventionally analyzed through the (weighted) incidence matrix of an undirected graph, which can serve as the matrix of Malthusian parameters defined by

$$m_{ij} = \begin{cases} 1 & \text{if } i \text{ and } j \text{ are joined } (i \neq j), \\ 0 & \text{if } i \text{ and } j \text{ are not joined}, \\ \frac{1}{2} & \text{if } i = j. \end{cases}$$

Consider the subsimplex $I \subseteq \{1, \dots, k\}$. The barycenter of the face $S_n(I)$ is defined as $\boldsymbol{p} = (1/k, \dots, 1/k, 0, \dots, 0)$ (the average of extremal coordinates). The mean fitness restricted to this face is

$$\boldsymbol{x}^{\mathrm{T}} M\boldsymbol{x} = 1 - \frac{1}{2}\sum_{i \in I} x_i^2.$$

Over the set of possible frequency vectors (having only positive entries with sum 1) this takes its maximal value $1 - (1/2k)$ at $\boldsymbol{p}$, which shows that $\boldsymbol{p}$ is asymptotically stable within $S_n(I)$. If $i \notin I$, then there is some $j \in I$ with $m_{ij} = 0$. Hence,

$$(M\boldsymbol{p})_j \leqslant \boldsymbol{p}^{\mathrm{T}} M\boldsymbol{p}.$$

Thus, $\boldsymbol{p}$ is not only saturated (maximal) but also asymptotically stable in $S_n$. Furthermore, for any system with $n - i$ alleles there are at most $2^i$ stable equilibria.

In this context, adaptation is a description of the changing frequencies of competing "species" or genotypes whose stable fixed points are governed by the values of a Malthusian matrix. This matrix is assumed to encode all of those factors that contribute to both mortality and increased viability or replicative potential.

A natural way to think about the process of adaptation is in terms of a flow on an appropriate manifold or the setting of *information geometry*. This framework was first introduced to population dynamics and the study of adaptation by Shahshahani. It is most easily treated for the symmetric case of the replicator equation. If we let $\dot{x}_i = m_i(\boldsymbol{x}) = \partial V/\partial x_i$ be a Euclidean gradient vector field on $\mathbb{R}^n$, then the replicator equation is a Shahshahani gradient on the interior of $S_n$ with the same potential function $V$. To be specific, the change in $V$ is equal to the variance of the values of $m_i(\boldsymbol{x})$ of the replicator equation:

$$\dot{V}(\boldsymbol{x}) = \sum_{i=1}^{n} x_i(m_i(\boldsymbol{x}) - \bar{m}(\boldsymbol{x}))^2.$$

This result is referred to as *Fisher's fundamental theorem of natural selection*, wherein the change in the potential—or, obversely, the increase in the mean population fitness—is proportional to the variance in fitness. Hence, populations will always increase their adaptedness assuming that there remains latent variability in a population.

### 2.2 Information and Adaptation: Game Dynamics

We have so far considered diploid population genetics, but this framework can as easily be thought of as describing haploid (single-genome, or "asexual," speciation) game dynamics. In this case the entries $m_{ij}$ of the Malthusian matrix $M$ describe payoffs associated with competition between an agent $A_i$ and an agent $A_j$ that meet with a probability $x_i x_j$. In the case of payoffs that are linear in the frequencies, the stable fixed point of the dynamics is given by a distribution $\hat{\boldsymbol{x}}$ in $S_n$. This is called an *evolutionarily stable state* (ESS) of the replicator equation if $\hat{\boldsymbol{x}}^{\mathrm{T}} M \hat{\boldsymbol{x}} > \boldsymbol{x}^{\mathrm{T}} M \boldsymbol{x}$ in the local neighborhood of $\hat{\boldsymbol{x}}$.

The ESS framework allows us to define an alternative potential function, or *Lyapunov function for adaptation*, through the use of information-theoretic concepts. We will assume that we are dealing with vector quantities at steady state, and for frugality we will drop the bold notation for vectors. One can think of the potential as "potential information" (a connection hinted at in the Shahshahani connection mentioned above) through use of divergence functions. In particular, let

$$V(x) = \sum_i \hat{x}_i \log \hat{x}_i - \sum_i \hat{x}_i \log x_i. \tag{2}$$

This is known as the RELATIVE ENTROPY [IV.36 §8] or *Kullback-Leibler divergence* between the two distributions and is also denoted as $D_{\mathrm{KL}}(\hat{x} \| x)$. Note that it is not symmetric in the arguments.

In a statistical setting it is thought of as the information that is lost when approximating $\hat{x}$ with $x$, and it quantifies in log units the additional information required to perfectly match the "optimal," "true," or desired distribution. Differentiating with respect to $x$ yields

$$\dot{V}(x) = -\sum_i \hat{x}_i \frac{\dot{x}_i}{x_i}.$$

Recall that the replicator equation has the form $\dot{x}_i = x_i(m_i(\boldsymbol{x}) - \bar{m}(\boldsymbol{x}))$, which upon substitution into the energy function gives us

$$\dot{V}(x) = -\sum_i \hat{x}_i m_i(x) + \sum_i \hat{x}_i \bar{m}(x).$$

Since the dynamics are restricted to the simplex, $\sum_i \hat{x}_i = 1$, and given the definition of the mean $\bar{m}(x)$, we obtain

$$\dot{V}(x) = -(\hat{x}^{\mathrm{T}} M x - x^{\mathrm{T}} M x) < 0$$

as long as the true distribution $\hat{x}$ is an ESS. For an ESS, the replicator equation will cause a system to adapt toward the fitness function as encoded in $\hat{x}$, and it will do so by extracting all information present in the "strategic environment" and storing this information in the population of agents.

We can also make the connection to information theory another way, through an identity with mutual information:

$$I(\hat{X}; X) = D_{\mathrm{KL}}(p(\hat{x}, x) \| p(\hat{x}) p(x)),$$

where $D_{\mathrm{KL}}$ is defined as in (2). As populations evolve they increase the information that they share with their environments, moving away from statistical independence.

One immediate value of an explicitly informational approach is that it can accommodate adaptive dynamics that do not reduce variance but tend to increase it. For example, the rock-paper-scissors game has a Nash equilibrium given by the uniform distribution over strategies. Fisher's fundamental theorem is violated because fitness is now determined by the population composition of strategies, and not by an invariant fitness landscape.

## 3 Evolution, Optimization, and Natural Gradients

In the previous discussion we have focused on adaptation in biological systems. The form of the equations of motion has a natural basis in the interaction between replication and competition. The adaptive process was thought of as a variational, i.e., energy-minimizing, dynamic that could be described in terms of divergence measures on statistical manifolds. We can, however, arrive at the same structures by considering nonlinear optimization. In this form, rather than emphasize biological properties we emphasize a form of hill-climbing algorithm suitable for dealing with nonconvex optimization or parameter estimation problems based on the so-called natural gradient of information geometry.

Standard gradient-based optimization assumes a locally differentiable function with well-behaved first and second derivatives and a search space that is isotropic in the sense that any departure from the optimum leads to an equivalent reduction in the gradient. Discrete-time gradient descent therefore updates solutions (which are still population frequencies in evolutionary systems) $\boldsymbol{X}(t)$ through a recursive algorithm:

$$\boldsymbol{X}(t+1) = \boldsymbol{X}(t) - k \nabla m(\boldsymbol{X}(t)).$$

These methods work reasonably well for single-peaked functions in Euclidean spaces. However, for many real-world problems, objective functions are many-peaked, nonisotropic, and live in non-Euclidean spaces. This broader setting means that gradient methods search space according to the Riemannian structure of probabilistic parameter spaces and employ the Fisher metric as the natural measure of distance. This provides information about the curvature of a statistical manifold and promotes more efficient adaptation to the optimum along an appropriate geodesic.

The updating rule for natural gradients includes additional geodesic information,

$$\boldsymbol{X}(t+1) = \boldsymbol{X}(t) - k\boldsymbol{F}_X^{-1}(t)\nabla m(\boldsymbol{X}(t)),$$

encoded in the *Fisher metric tensor* or *Fisher information matrix* $\boldsymbol{F}$. It is derived directly from the infinitesimal Kullback–Leibler divergence (shown above to be a suitable Lyapunov function for adaptation) between probability distributions $x$ and $y$:

$$D_{\mathrm{KL}}(x\|y) = \int x \ln\left(\frac{x}{y}\right) \mathrm{d}x.$$

This divergence is asymmetric. A true distance can be defined locally. When we expand $D_{\mathrm{KL}}$ to a second-order Taylor polynomial,

$$D_{\mathrm{KL}}(x\|y) \approx \int x \ln\left[1 + \left(\frac{y}{x} - 1\right)\right] + \frac{1}{2}\int x\left(\frac{y}{x} - 1\right)^2,$$

we see that the first-order term tends to zero for $y = x(1+\varepsilon)$. This leaves us with

$$D_{\mathrm{KL}}(x\|y) \approx \frac{1}{2}\int x\left(\frac{y}{x} - 1\right)^2.$$

Note that $\ln(y) - \ln(x) = \ln(1+\varepsilon) \approx \varepsilon$, and that $\ln(y) - \ln(x)$ is simply the derivative of $\ln(x)$ in the $\varepsilon$-direction, $\partial_\varepsilon \ln x$. Hence,

$$\begin{aligned} D_{\mathrm{KL}}(x\|y) &\approx \frac{1}{2}\int x(\ln(y) - \ln(x))^2\,\mathrm{d}x \\ &\approx \frac{1}{2}\int x(\partial_\varepsilon \ln x)(\partial_\varepsilon \ln x)\,\mathrm{d}x. \end{aligned}$$

We are interested in some finite distance between proximate distributions induced by some $\varepsilon$-directional perturbation, $y = x(1 + v^i\varepsilon_i)$:

$$F_{ij}(y) = \frac{1}{2}\int y \frac{\partial \ln y}{\partial \varepsilon_i}\frac{\partial \ln y}{\partial \varepsilon_j}\,\mathrm{d}y.$$

By convention, using the definition of partial information, $q = -\ln(y)$, the Fisher metric can be written as an expectation value of a quadratic form:

$$\begin{aligned} F_{ij}(y) &= \frac{1}{2}\int -\frac{\partial^2 q(y)}{\partial\varepsilon_i\partial\varepsilon_j} y\,\mathrm{d}y \\ &= \frac{1}{2}\left\langle -\frac{\partial^2 q(y)}{\partial\varepsilon_i\partial\varepsilon_j}\right\rangle, \end{aligned}$$

where the expectation value of $g(y)$ is

$$\langle g(y)\rangle = \int g(y) y\,\mathrm{d}y.$$

Technically, the $F_{ij}(y)$ values are the elements of a positive-definite matrix or Riemannian metric tensor. This tensor captures the curvature of a manifold in $N$-dimensional space. In Euclidean space this matrix reduces to the identity matrix. Hence, any nonlinear optimization that makes use of the true geometry of a "fitness" function will respect the dynamics of the replicator equation describing trajectories that minimize distances on statistical manifolds.

Adaptive dynamics are, therefore, precisely those trajectories on statistical manifolds that can be said to be maximizing the extraction of information from the environment.

## 4 Discrete and Stochastic Considerations

The logic of the continuous systems extends to the discrete case, but the delays intrinsic to maps introduce the possibility of periodic and chaotic behavior. The discrete replicator equation, or replicator mapping, has the form

$$x_i(t+1) = x_i(t) + x_i(t)((M\boldsymbol{x}(\boldsymbol{t}))_i - \boldsymbol{x}(\boldsymbol{t})^{\mathrm{T}}M\boldsymbol{x}(\boldsymbol{t})).$$

Even the low-dimensional, two-player discrete game exhibits a full range of complex dynamical behavior. If we consider the parametrized payoff matrix

$$M = \begin{bmatrix} 1 & S \\ T & 0 \end{bmatrix},$$

we can capture a range of so-called evolutionary dilemmas including the "hawk-dove game" ($T > 1$, $S > 0$), the "stag hunt game" ($T < 1$, $S > 0$), and the "prisoner's dilemma" ($T > 1$, $S < 0$). As a general rule, small or negative values of $S$ and $T$ induce periodic solutions, whereas large values of $S$ and $T$ produce chaotic solutions. In the constrained case where $T = S = A$, systematically increasing the value of $A$ leads to the period-doubling route to chaos.

In population genetics, the game selection model is typically written using the equivalent map

$$x_i(t) = x_i(t-1)\frac{\sum_j m_{ij}x_i(t-1)}{\sum_{pq} m_{pq}x_p(t-1)px(t-1)_q}.$$

Discreteness is important for adaptation. These results illustrate some of the challenges of evolving reliable strategies when fitness is frequency dependent. Deterministic chaos will severely limit the long-term stability of a lineage.

When models incorporate stochasticity, even greater instability ensues. The deterministic results apply most strongly in the limit of large populations overcoming the phenomenon of "neutral drift." If the difference in fitness values or payoff values between strategies is less than $O(1/N)$, then the dynamics are described by an $N$-dimensional random walk with the frequencies 0 and 1 acting as absorbing states. In small populations, sampling drift limits adaptation in the sense that we can no longer assume that the dynamics will minimize an appropriate energy function. The smaller the population, the greater the variance in the Malthusian parameters or payoff differences required in order to increase fitness. Stochasticity is treated formally in both the "neutral" and "near-neutral" theories of evolution.

### 4.1 Innovation Dynamics

The systems considered in the previous sections do not attempt to explain the "origin" of adaptations but instead focus on the efficacy of selection in promoting adaptation within an invariant population. In order to explore the origin of adaptive novelty, we need to introduce an intrinsic source of variability. In evolutionary theory this is accomplished by means of mutation and recombination operators. The natural extension of the replication equation to include mutation leads to the *quasispecies equation*

$$\dot{x}_i = \sum_j q_{ij} m_j x_j - x_i \bar{m}(x).$$

The operator $Q$ is a doubly stochastic matrix of transition probabilities encoding mutations from a strain $X_j \to X_i$. Geometrically, this operator ensures that the boundary of the simplex $S_n$ is no longer invariant: trajectories can flow from outside of the positive orthant into $S_n$ in such a way that new strains might emerge from existing strains that were not present in the initial population distribution.

In much the same way that neutral drift can eliminate adaptation, excessive mutation can abrogate hill climbing, replacing selection with diffusion over the simplex defined by the *mutation kernel* $Q$. This is known as the "error threshold." If each strain is encoded by a binary sequence of length $L$, then assuming a uniform error rate in transmission across each string, the transition probabilities can by written in binomial form as

$$q_{ij} = p^{d_{ij}}(1-p)^{L-d_{ij}},$$

where $d_{ij}$ is the Hamming distance between strains, and $p$ is the per bit per generation transition probability. For any choice of fitness function, the regime $p > 1/L$ will completely "flatten" the landscape, eliminating adaptation altogether. This result serves to illustrate costs and benefits of mutational transformation. If mutation is too low, there is a risk of extinction arising from the inability to adapt to a novel environment, whereas if mutation is too high, there is a risk of extinction arising from the complete loss of the adaptive capability through the dissipation of phylogenetic memory.

Recombination is formally more cumbersome but plays a role in modulating the rate of evolution within a population of recombining strains. For low rates of mutation, recombination tends to reduce population variability, leading to increased stability. For high rates of mutation that remain below the error threshold, recombination can tip the population over the error threshold and destabilize a population. Hence, rather like mutation, recombination is a double-edged adaptive sword.

## 5 Adaptation Is Algorithmic Information Acquisition

We have shown how adaptation represented through a symmetric form of the continuous selection equation (the replicator equation) can be thought of as a natural gradient that minimizes a potential representing information. Adaptive systems are precisely those that extract information from their environments (taken to include other agents) so as to minimize uncertainty. This is made explicit through the fact that the KL divergence between the joint distribution of a pair of random variables and the product distribution of these variables is equal to the mutual information between these variables. Adaptation is therefore measured by the degree of departure from independence of organisms and their environments, or from each other. In this framework, adaptation continues (driving up the mutual information) until there is no further functional information to be acquired and, thus, no latent dependencies remaining to be discovered. This is unlikely to occur in reality as environments are rarely static over long intervals of time. The very general nature of adaptation suggests that it might manifest in a variety of forms, and this is exactly what we find: adaptive dynamics possesses a number of "avatars," each of which is a framework for obtaining information through the iteration of a simple growth and culling process.

In this section we present a few of the mathematical avatars of replicator dynamics in order to demonstrate that adaptation is a ubiquitous nonlinear optimization procedure across a range of fields, many of which appear to be unrelated. This reduction to a canonical adaptive form constitutes evidence of functional equivalence with respect to information extraction in the service of increased prevalence, which should rightly be thought of as adaptive evolution.

### 5.1 Bayesian Inference

In 1763, Bayes's posthumous paper "An essay towards solving a problem in the doctrine of changes" was published in the *Philosophical Transactions of the Royal Society*. The paper presented a novel method for calculating probabilities with minimal or no knowledge of an event and indicated how repeated experiments can lead toward the confirmation of a conclusion.

BAYES'S RULE [V.11 §1] is the engine behind what we now call *Bayesian updating* and, when appropriately written, it is revealed to be a natural mathematical encoding of adaptation. In other words, it is an information-maximizing dynamic induced on the simplex. In dynamical terms it is a discrete particle system that through repeated iteration converges on an effective estimate of a true underlying probability distribution.

Whereas we think of adaptation in terms of the differential success of organisms, Bayesian updating treats adaptation as the differential success of hypotheses.

The basic updating equation is

$$P(X)_t = P(X)_{t-1}\frac{L(X)_{t-1}}{\langle L\rangle_{t-1}},$$

with the change in the concentration of the probability mass $P(X)$ delivered around the highest likelihood $L(X)$ values:

$$\begin{aligned}\Delta P(X)_t &= P(X)_{t-1}\frac{L(X)_{t-1}}{\langle L\rangle_{t-1}} - P(X)_{t-1}\\ &= P(X)_{t-1}\Big(\frac{L(X)_{t-1}}{\langle L\rangle_{t-1}} - 1\Big)\\ &= \frac{P(X)_{t-1}}{\langle L\rangle_{t-1}}(L(X_{t-1}) - \langle L\rangle_{t-1}).\end{aligned}$$

Note that the evolution of the probability distribution in continuous time reduces to the simplest form of the differential replicator equation:

$$\dot{x} = \frac{1}{\langle L\rangle}x(L_x - \langle L\rangle).$$

Bayes therefore constructed an adaptive avatar *avante le lettre* of Darwin's theory. A method for estimating the probability of a hypothesis through trial and error is the prequel to the differential replication of natural selection.

### 5.2 Imitation

In much the same way that Bayesian updating is an avatar of adaptation, learning through imitation similarly qualifies. Imitation is a cognitive strategy for arriving at a pattern of behavior based on sampling a population of model behaviors. After sufficient time has elapsed, the most prevalent behavior will be the one most frequently imitated.

Assume that each individual $i$ varies in its strategy choices $\boldsymbol{s}_i$. Think of this as a vector specifying a mixed strategy defining how individuals should interact. Interactions between strategies $\boldsymbol{s}_i$ and $\boldsymbol{s}_j$ give rise to pairwise payoffs or rewards $r_{ij}$. Reward information is used to decide which strategy to adopt in subsequent interactions. In particular, differences in payoffs lead to the adoption of more successful strategies. The rate of imitation leading strategy $j$ to imitate a strategy $i$, $\boldsymbol{s}_j \to \boldsymbol{s}_i$, can therefore be written in matrix form:

$$f_{ij} = [(R\boldsymbol{g})_i - (R\boldsymbol{g})_j]_+,$$

where $R$ is the matrix of rewards and $g$ the vector of genotypic or strategic frequencies. Notice that this function is only nonzero in the positive half-space, so lower payoff strategies are not imitated. The population of players will evolve in time according to an imitation learning dynamics that is easily seen to reduce to the game dynamical form of the replicator equation:

$$\begin{aligned}\dot{g}_i &= g_i\sum_j(f_{ij} - f_{ji})g_j\\ &= g_i\sum_j[(R\boldsymbol{g})_i - (R\boldsymbol{g})_j]g_j\\ &= g_i[(R\boldsymbol{g})_i - \boldsymbol{g}R\boldsymbol{g}].\end{aligned}$$

### 5.3 Reinforcement Learning

Skinner introduced the linear operator model of reinforcement to explain schedules of behavioral extinction and acquisition, and this framework has been enormously influential, giving rise to a very large class of learning rules. Any behavior that is correct is rewarded, while incorrect behavior is punished. Over time, reinforcement seeks to make correct behavior probable and incorrect behavior improbable. Consider a set of behaviors $X_i$ each associated with a probability $x_i$. For each behavior there is a reward $r_i$. The incremental change in the probability of any action, with a learning rate parameter $\alpha$, can be written as

$$\delta x_k = \alpha r_i(\delta_{ik} - x_k),$$

where $\delta_{ij}$ is the KRONECKER DELTA [I.2 §2, table 3]. This gives the updated probability of behavior assuming that a single action was performed. The average change of behavior takes into account the probability of an action $x_i$:

$$\Delta x_k = \sum_i x_i \delta x_k.$$

Combining these two equations we deduce that

$$\begin{aligned}\Delta x_k &= \sum_i x_i[\alpha r_i(\delta_{ik} - x_k)] \\ &= \alpha \sum_i x_i r_i \delta_{ik} - \alpha x_k \sum_i x_i r_i \\ &= \alpha x_k r_k - \alpha x_k \sum_i x_i r_i.\end{aligned}$$

The term $\sum_i x_i r_i$ is simply the mean reward, $\langle r \rangle$, and the change in behavior in continuous time is given by the simple replicator equation

$$\dot{x}_k = x_k(r_k - \langle r \rangle).$$

As with Bayesian updating and imitation learning, reinforcement learning is an algorithm for iteratively forming an accurate estimate of an underlying probability distribution imposed through a learning environment.

## 6 Final Remarks

In all that precedes we have focused on a very general context of adaptation and the related mathematics. This was the most expansive framing of adaptation as a form of optimization subject to the constraints of probability. When presented in this way we observe that adaptation is not unique to biology but is a ubiquitous phenomenon in nature. One might go so far as to assert that the property of adaptation is the defining signature of life in the universe and that the mathematics of adaptation represents a form of algorithmic biological law.

Many have explored the mathematics of complex adaptive systems, placing an emphasis on questions of "emergence." These are systems for which, as Nobel laureate Phil Anderson famously wrote "more is different" or "the whole is greater than the sum of the parts," a descriptor that suggests the variety of nonlinear phenomena they comprise. The formal relationship between emergence and adaptive dynamics remains somewhat unclear.

Familiar examples of such systems run the gamut from Ising models to economies. They are often characterized by hierarchical and multiscale structure as well as hysteresis. They have generally emerged over time and space, most through processes related to those that we have discussed, namely feedback and adaptation.

The mathematical framework presented above is a necessary component in the analysis of all such systems, but not all consider the process of evolution explicitly, or they prefer simulation approaches such as agent-based models and genetic or evolutionary programming. These involve the disciplines of programming and consideration of more sophisticated distributed algorithms.

### Further Reading

The bibliography below contains some excellent general resources on the ideas presented in this entry.

The exposition of section 2.1 follows closely the exposition presented in Hofbauer and Sigmund (1998). Section 5 follows the contents of the review by Krakauer (2011), which places the study of selection mechanics in the larger context of the study of complex elaborations of Maxwell's demon.


Amari, S.-I., and H. Nagaoka. 2007. *Methods of Information Geometry*. Providence, RI: American Mathematical Society.

Bell, G. 2008. *Selection: The Mechanism of Evolution*. Oxford: Oxford University Press.

Box, G. E. P., and G. C. Tiao. 1973. *Bayesian Inference in Statistical Analysis*. New York: John Wiley.

Gillespie, J. H. 2004. *Population Genetics: A Concise Guide*, 2nd edn. Baltimore, MD: Johns Hopkins University Press.

Haykin, S. 2008. *Neural Networks and Learning Machines*, 3rd edn. Englewood Cliffs, NJ: Prentice Hall.

Hofbauer, J., and K. Sigmund. 1998. *Evolutionary Games and Population Dynamics*. New York: Cambridge University Press.

Kimura, M. 1983. *The Neutral Theory of Molecular Evolution*. Cambridge: Cambridge University Press.

Krakauer, D. 2011. Darwinian demons, evolutionary complexity and information maximization. *Chaos* 21:037110-1-12.

Mitchell, M. 2009. *Complexity: A Guided Tour*. New York: Oxford University Press.

Sutton, R. S., and A. G. Barton. 1998. *Introduction to Reinforcement Learning*. Cambridge, MA: MIT Press.

## V.2 Sport

*Nicola Parolini and Alfio Quarteroni*

### 1 Introduction

The increasing importance of mathematical modeling is due to its improved ability to describe complex phenomena, predict the behavior of those phenomena, and, possibly, control their evolution. These improvements have been made possible by theoretical advances as well as advances in algorithms and computer hardware. In recent years mathematical models have received a lot of attention in the world of sport, where any (legal) practice that can improve performance is very welcome. By adopting appropriate mathematical models and efficient numerical methods for their solution, the level of accuracy required for optimizing the performance of an athlete or a team can be guaranteed.

Computational fluid dynamics (CFD) is the branch of computational mechanics that uses numerical methods to simulate problems that involve fluid flows. In the past two decades, CFD has become a key design tool for Formula 1 cars. However, F1 is not the only sport in which mathematical/numerical modeling has been applied. In this article we discuss the results of three research projects—one concerning sailing, one rowing, and one swimming—undertaken by the authors in recent years, with the objective of highlighting the role that mathematics and scientific computing can play in these fields.

### 2 Model Identification

In every physical process, phenomena of different types (mechanical, chemical, electrodynamical) act at the same time and interact with each other. However, in a practical modeling approach it is often possible to limit the analysis to a specific physical aspect. In the case of fluid dynamics, a mathematical model based on continuous mechanics principles, given by the *Navier-Stokes equations*, is usually adequate. Even in this framework, further model identification efforts are required. Indeed, specific flow regimes (laminar/turbulent, compressible/incompressible, single/multiphase) demand different specializations of the original mathematical model. Moreover, a suitable definition of the limited space/time computational domain and suitable choices for the initial and boundary conditions are required to obtain a well-posed mathematical problem.

The hydrodynamics of a boat (or a swimmer) is well described by an incompressible, turbulent, two-phase flow model. Denoting by $\Omega$ the three-dimensional computational domain around the boat (or the swimmer) that is occupied by either air or water, the NAVIER–STOKES EQUATIONS [III.23] read as follows. For all $\boldsymbol{x} \in \Omega$ and $0 < t < T$,

$$\frac{\partial \rho}{\partial t} + \nabla \cdot (\rho \boldsymbol{u}) = 0, \qquad (1)$$

$$\frac{\partial (\rho \boldsymbol{u})}{\partial t} + \nabla \cdot (\rho \boldsymbol{u} \otimes \boldsymbol{u}) - \nabla \cdot \boldsymbol{\tau}(\boldsymbol{u}, p) = \rho \boldsymbol{g}, \qquad (2)$$

$$\nabla \cdot \boldsymbol{u} = 0, \qquad (3)$$

where $\rho$ is the (variable) density, $\boldsymbol{u}$ the velocity, $p$ the pressure, $\boldsymbol{g} = (0, 0, g)^{\mathrm{T}}$ the gravitational acceleration, and $\boldsymbol{\tau}(\boldsymbol{u}, p) = \mu(\nabla \boldsymbol{u} + \nabla \boldsymbol{u}^{\mathrm{T}}) - p\boldsymbol{I}$ denotes the stress tensor with $\mu$ indicating the (variable) viscosity. Equation (1) prescribes mass conservation, equation (2) the balance of momentum, and equation (3) the incompressibility constraint. As mentioned above, equations (1)–(3) should be complemented with suitable initial and boundary conditions.

Different approaches are available to simulate the free-surface dynamics of the water–air interface. In *interface tracking* methods, the interface is explicitly reconstructed, while *interface capturing* techniques usually identify the interface as an (implicitly defined) isosurface of an auxiliary characteristic function. The flow solutions in air and water should satisfy the interface conditions

$$\boldsymbol{u}_{\mathrm{a}} = \boldsymbol{u}_{\mathrm{w}},$$

$$\boldsymbol{\tau}_{\mathrm{a}}(\boldsymbol{u}_{\mathrm{a}}, p_{\mathrm{a}}) \cdot \boldsymbol{n} = \boldsymbol{\tau}_{\mathrm{w}}(\boldsymbol{u}_{\mathrm{w}}, p_{\mathrm{w}}) \cdot \boldsymbol{n} + \kappa \sigma \boldsymbol{n},$$

which, respectively, enforce continuity of velocities and equilibrium of forces on the interface. The surface tension contribution $\kappa \sigma \boldsymbol{n}$, which depends on the coefficient $\sigma$, the local curvature $\kappa$, and the interface normal $\boldsymbol{n}$, is usually negligible in the applications considered in this work.

For the simulation of turbulent flows, which are commonly encountered in hydrodynamics applications, a turbulence model is usually adopted. These models are able to include the main effect of the turbulent nature of the flow without capturing all the space- and timescales that characterize the flow. They usually result in an additional turbulent viscosity $\mu_{\mathrm{T}}$ that is added to the physical viscosity $\mu$ in the stress tensor $\boldsymbol{\tau}$.

That is,

$$\tau = (\mu + \mu_T)(\nabla u + \nabla u^T) - pI.$$

The system of partial differential equations (1)–(3) can be discretized using different numerical schemes. For instance, the spatial discretization may be based on finite-difference, finite-volume, or finite-element methods. Here, the finite-volume method is considered. The domain is subdivided into small control volumes (forming a so-called *computational grid*) and the differential equations (in their integral form) are imposed locally on each control volume. In this way one can formulate a discrete version of the problem (1)–(3) that, at the end, leads to the solution of a (usually large) linear system. The dimension of the linear system is strictly related to the dimension of the computational grid. In turn, the size of the grid usually depends on the flow regime. In particular, with high-speed and multiphase flows with complex flow features (such as separation or laminar–turbulent transition), the size of the computational grid can easily exceed several million elements. The solution of the associated large linear system is carried out with state-of-the-art iterative methods (for example, multigrid methods) on parallel architectures.

Numerical models based on the discrete solution of the Navier–Stokes equations usually guarantee that the results come with a high level of accuracy, but this comes at a high computational cost. This is particularly true when viscous (and turbulent) effects play an important role in the simulated flows. In some cases—when one is mostly interested in reconstructing the shape of the free surface around a boat, for example—REDUCED-ORDER MODELS [II.26] based on the Euler equations or even on potential flow theory can be adopted. Reduced-order models are often used for preliminary analysis or multiquery problems (arising, for example, in genetic optimization). In other cases, the computational reduction strategy may be based on a geometrical multiscale approach, where the complete (and computationally expensive) three-dimensional model is only used in a very limited portion of the original domain and is coupled with spatial (two-dimensional/one-dimensional/zero-dimensional) reduced models that account for the rest. The correct identification of the most suitable model for a specific problem should be based on a compromise between the level of accuracy achieved by the model in estimating the objective function of the analysis at hand and its computational cost.

## 3 Numerical Models for Fluid Dynamics in Sport

### 3.1 The America's Cup

The America's Cup is a highly competitive sailing yacht race in which even the smallest details in the design of the boats' different components can make a big difference to the final result. To analyze and optimize a boat's performance, the aerodynamic and hydrodynamic flow around the whole boat should be simulated, taking into account the variability of wind and waves, and the presence of the opposing boat and its maneuvering, as well as the interaction between fluids (water and air) and the boat's structural components (its hull, appendages, sails, and mast).

The main objective for the designers is to select shapes that minimize water resistance on the hull and appendages and maximize the thrust produced by the sails. Numerical models allow different situations to be simulated, thus reducing the costs (in terms of both time and money) of running experiments in a towing tank or a wind tunnel.

To evaluate whether a new design idea is potentially advantageous, a velocity prediction program is often used to estimate the boat speed and attitude for a range of prescribed wind conditions and sailing angles. These numerical predictions are obtained by modeling the balance between the aerodynamic and hydrodynamic forces acting on the boat. The accuracy of speed and attitude predictions relies on the accuracy of the force estimation, which is usually computed by integrating experimental measurements in a towing tank, by analytical models, and by accurate CFD simulations.

In order to perform the latter, any new shape is reproduced in a computer-aided design model (usually defined by several hundred *nonuniform rational B-spline (NURBS) surfaces* for the whole boat). Based on this geometry, a computational grid can be generated. This step is usually crucial for obtaining accurate predictions. In particular, for free-surface hydrodynamic simulation it is often necessary to resort to block-structured grids in order to capture the strong gradients in the wall boundary layer (figure 1) and to reconstruct the free-surface geometry accurately.

A detailed simulation of the complex flow around the appendages (keel, bulb, and winglets) is displayed in figure 2, where the typical recirculation associated with the lift generated by the keel in upwind sailing can be observed.

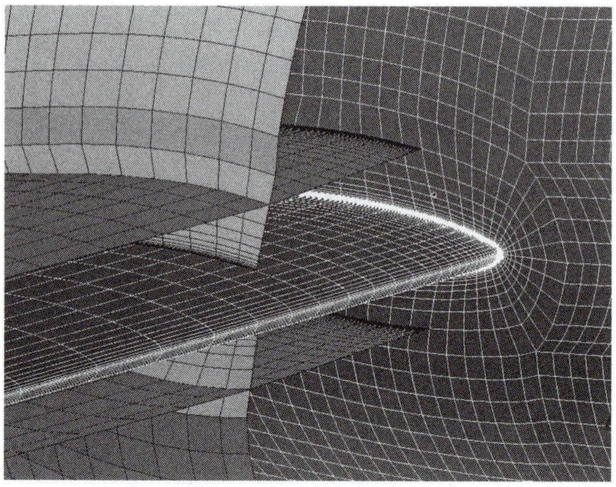

**Figure 1** Detail of the grid in the boundary layer.

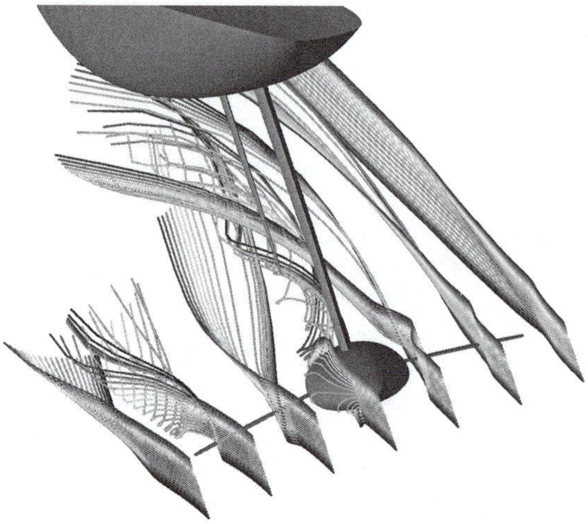

**Figure 2** Flow around the appendages in upwind sailing.

As mentioned, the air and water interact with the boat and can change its configuration. In this respect, two classes of problems can be considered:

- the rigid-body motion of the boat under the action of the fluid forces;
- the fluid–structure interaction between the flexible sails and the wind.

Different numerical schemes for the coupling between the flow solution and a structural model have been defined and are described in the following sections. Notice that in both cases, the Navier–Stokes equations have to be reformulated in the so-called *arbitrary Lagrangian–Eulerian* (ALE) formulation to deal with moving-domain problems.

### *3.1.1 Boat Dynamics*

The attitude of the boat advancing in calm water or on a wavy sea, as well as its dynamics, may strongly influence its performance. For this reason, a numerical tool for yacht design should be able to predict not only the performance of the boat in a steady configuration but also the boat's motion and its interaction with the flow field.

This can be achieved by coupling the flow solver with a six-degrees-of-freedom dynamical system describing the rigid-body motion of the boat. Consider an *inertial* reference system $(\mathbf{0}, X, Y, Z)$, moving forward with the mean boat speed, and a *body-fixed* reference system $(\boldsymbol{G}, x, y, z)$, centered in the boat center of mass $\boldsymbol{G}$, which translates and rotates with the boat. The boat dynamics is computed by integrating the equations of variation of linear and angular momentum in the inertial reference system, which are

$$m\ddot{\boldsymbol{X}}_G = \boldsymbol{F},$$
$$\boldsymbol{TIT}^{-1}\dot{\boldsymbol{\Omega}} + \boldsymbol{\Omega} \times \boldsymbol{TIT}^{-1}\boldsymbol{\Omega} = \boldsymbol{M}_G,$$

with $m$ denoting the boat mass; $\ddot{\boldsymbol{X}}_G$ the linear acceleration; $\boldsymbol{F}$ the force acting on the boat; $\dot{\boldsymbol{\Omega}}$ and $\boldsymbol{\Omega}$ the angular acceleration and velocity, respectively; and $\boldsymbol{M}_G$ the moment with respect to $\boldsymbol{G}$ acting on the boat. The tensor of inertia $\boldsymbol{I}$ is computed in the body-fixed reference system, and the rotation matrix $\boldsymbol{T}$ between the body-fixed and inertial reference systems is required to write the problem in the inertial reference system.

The forces and moments acting on the boat are in this case rather simple and include the flow force ($\boldsymbol{F}_{\text{Flow}}$) and moment ($\boldsymbol{M}_{\text{Flow}}$) and the gravitational force:

$$\boldsymbol{F} = \boldsymbol{F}_{\text{Flow}} + m\boldsymbol{g}, \qquad \boldsymbol{M}_G = \boldsymbol{M}_{\text{Flow}}. \tag{4}$$

Using this model it was possible to simulate the dynamics of the boat in different conditions, analyzing both calm water and wavy sea scenarios. In figure 3, an example of the free-surface distribution of a hull advancing in regular waves is shown.

### *3.1.2 Wind–Sails Interaction*

Sails are flexible structures that deform under the action of the wind. The pressure field acting on the sail changes its geometry and this, in turn, alters the flow field.

In mathematical terms, this problem can be defined as a coupled system that comprises a fluid problem $\mathcal{F}$ defined on the moving domain $\Omega_{\text{F}}(t)$ surrounding

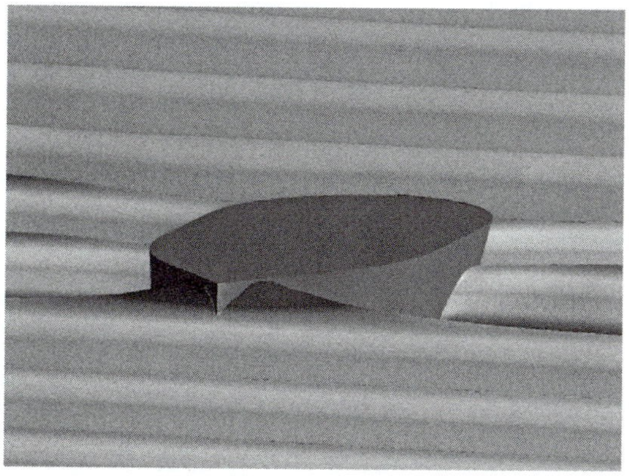

**Figure 3** Free-surface simulation in wavy conditions.

the sail, a structural problem $S$ defined on the moving domain $\Omega_S(t)$, and a mesh motion model $\mathcal{M}$. The structural model is usually based on a second-order elastodynamic equation depending on the sail deformation. Without entering into a detailed formulation of the structural and mesh motion models, we focus on the coupled nature of the problem.

In abstract form, the coupled problem can be formulated as

$$\left.\begin{aligned} \mathcal{F}(\boldsymbol{u},p,\boldsymbol{w}) &= 0 && \text{in } \Omega_F(t),\\ S(\boldsymbol{d}) &= 0 && \text{in } \Omega_S(t),\\ \mathcal{M}(\boldsymbol{\eta}) &= 0 && \text{in } \Omega_F^0,\\ \boldsymbol{u} &= \dot{\boldsymbol{d}} && \text{on } \Gamma(t),\\ \boldsymbol{\sigma}_F(\boldsymbol{u},p)\boldsymbol{n}_F &= \boldsymbol{\sigma}_S(\boldsymbol{d})\boldsymbol{n}_S && \text{on } \Gamma(t),\\ \boldsymbol{\eta} &= \boldsymbol{d} && \text{on } \Gamma^0, \end{aligned}\right\} \quad (5)$$

where the three (fluid, structure, and mesh motion) problems are coupled through three conditions over the interface $\Gamma(t)$ stating the continuity of velocity, the equilibrium of forces, and the geometric continuity, respectively. In (5), $\boldsymbol{d}$ is the structural displacement, $\boldsymbol{\eta}$ the mesh displacement, and $\Omega_F^0$ and $\Gamma^0$ denote a reference fluid domain and interface, respectively. The fluid mesh motion velocity $\boldsymbol{w}$, needed in the arbitrary Lagrangian–Eulerian formulation of the Navier–Stokes equations, depends on the mesh displacement $\boldsymbol{\eta}$ as $\boldsymbol{w} = \dot{\boldsymbol{\eta}}$.

For the solution of problem (5), different fluid–structure interaction schemes can be devised. In *monolithic* schemes a global system is assembled and solved for all the unknowns of the problem simultaneously. A different approach, usually preferred when one wants to exploit already existing fluid and structural solvers, is

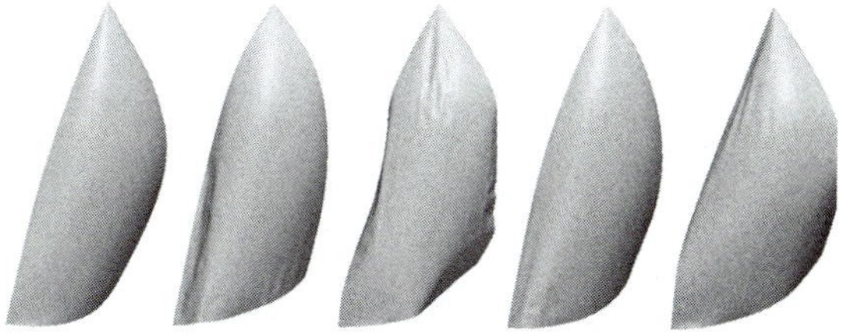

**Figure 4** An example of a transient fluid–structure interaction solution.

provided by *partitioned* schemes in which the fluid, structure, and mesh motion problem are solved iteratively. For the case at hand, a strongly coupled partitioned scheme has been devised and this scheme guarantees that at each time step equilibrium is reached between the different subproblems. To compute the solution at time $t^{n+1}$, a subiteration between the three subproblems is required. The structural solution is first computed by solving

$$\begin{gathered} S(\boldsymbol{d}_{k+1}^{n+1}) = 0 \quad \text{in } (\Omega_S)_k^{n+1},\\ \boldsymbol{\sigma}_S(\boldsymbol{d})_{k+1}^{n+1}(\boldsymbol{n}_S)_k^{n+1} = \boldsymbol{\sigma}_F(\boldsymbol{u},p)_k^{n+1}(\boldsymbol{n}_F)_k^{n+1} \quad \text{on } \Gamma, \end{gathered}$$

followed by solution of the mesh motion update

$$\begin{gathered} \mathcal{M}(\boldsymbol{\eta}_{k+1}^{n+1}) = 0 \quad \text{in } \Omega_F^0,\\ \boldsymbol{\eta}_{k+1}^{n+1} = \alpha\boldsymbol{d}_{k+1}^{n+1} + (1-\alpha)\boldsymbol{d}_k^{n+1} \quad \text{on } \Gamma^0, \end{gathered}$$

and finally undertaking the flow solution

$$\begin{gathered} \mathcal{F}(\boldsymbol{u}_{k+1}^{n+1}, p_{k+1}^{n+1}, \boldsymbol{w}_{k+1}^{n+1}) = 0 \quad \text{in } (\Omega_F)_{k+1}^{n+1},\\ \boldsymbol{u}_{k+1}^{n+1} = \alpha\dot{\boldsymbol{d}}_{k+1}^{n+1} + (1-\alpha)\dot{\boldsymbol{d}}_k^{n+1} \quad \text{on } \Gamma_{k+1}^{n+1}, \end{gathered}$$

where a relaxed value of the structure displacement at the current iteration $k+1$ and the previous subiteration $k$ is used. The iteration over the index $k$ is stopped when a suitable convergence criterion is fulfilled.

Several fluid–structure interaction simulations of sails have been carried out using this model, which has proved its ability in predicting both steady flying shapes and sail dynamics under different trimming conditions. An example of a transient fluid–structure interaction simulation can be seen in figure 4, where the shape of the sail at several time instants is displayed. The flow pattern around mainsail and gennaker in a downwind sail configuration is shown in figure 5.

### 3.2 Olympic Rowing

Can numerical models developed for the America's Cup be adopted in other hydrodynamics applications in sport? Consider, for instance, the prediction of a rowing boat's performance. In principle, the physical problem is very similar to one of those already described:

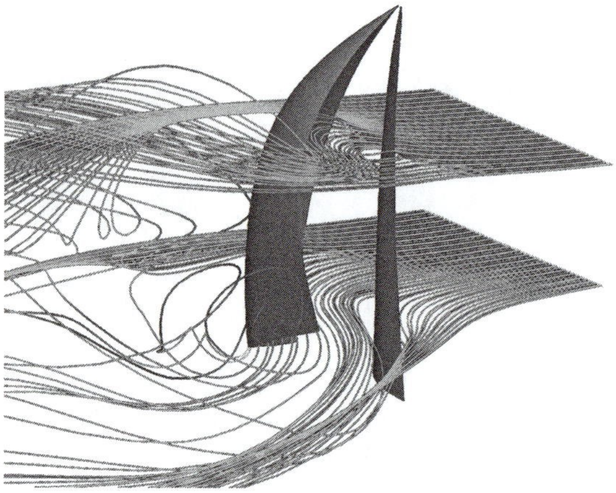

**Figure 5** Streamlines around mainsail and gennaker.

namely, a free-surface flow in a moving-domain framework. What characterizes the rowing action, however, is the motion of the rowers on board. Indeed, a men's quadruple scull weighs around 50 kg and has four 100 kg rowers on board who move with a cadence of around 40 strokes per minute. This motion has a strong effect on the global performance of the boat that should be captured by the simulation tool.

To account for the motion of the rowers, the forces and moments acting on the boat, which were simple in the rigid-body dynamics of the sailing boat (see (4)), become more involved in this case and should account for the additional forces that the rowers exert on the boat through the oars, the seats, and the footboards:

$$\left.\begin{aligned} \boldsymbol{F} &= \sum_{j=1}^{n} \boldsymbol{F}_{\mathrm{o}_j} + \sum_{j=1}^{n} \boldsymbol{F}_{\mathrm{s}_j} + \sum_{j=1}^{n} \boldsymbol{F}_{\mathrm{f}_j} + \boldsymbol{F}_{\mathrm{Flow}} + m\boldsymbol{g}, \\ \boldsymbol{M}_G &= \sum_{j=1}^{n} (\boldsymbol{X}_{\mathrm{o}_j} - \boldsymbol{G}) \times \boldsymbol{F}_{\mathrm{o}_j} + \sum_{j=1}^{n} (\boldsymbol{X}_{\mathrm{s}_j} - \boldsymbol{G}) \times \boldsymbol{F}_{\mathrm{s}_j} \\ &\qquad + \sum_{j=1}^{n} (\boldsymbol{X}_{\mathrm{f}_j} - \boldsymbol{G}) \times \boldsymbol{F}_{\mathrm{f}_j} + \boldsymbol{M}_{\mathrm{Flow}}, \end{aligned}\right\} \tag{6}$$

where $\boldsymbol{F}_{\mathrm{o}_j}$, $\boldsymbol{F}_{\mathrm{s}_j}$, and $\boldsymbol{F}_{\mathrm{f}_j}$ denote the forces exerted by the $j$th rower on his/her oarlock, seat, and footboard, respectively, and $\boldsymbol{X}_{\mathrm{o}_j}$, $\boldsymbol{X}_{\mathrm{s}_j}$, and $\boldsymbol{X}_{\mathrm{f}_j}$ are the corresponding application points. To compute these additional forcing terms it is necessary to reconstruct the kinematics of the rowers' action. A schematic of the boat/rowers/oars system is given in figure 6. An accurate prediction of these forces is crucial for the performance analysis of the system. These periodic forcing terms, in fact, induce relevant secondary motions of the boat that

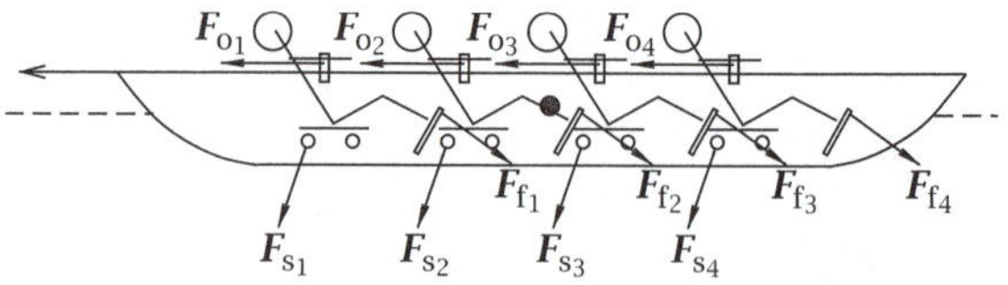


**Figure 6** Sketch of the boat/rowers/oars system.

**Figure 7** Free-surface flow around a rowing boat.

generate an additional wave radiation effect, increasing the global energy dissipation of the system. Indeed, the quality of a professional rower is evaluated not only by the force he/she is able to express during the active phase of the stroke but also on his/her ability to perform a well-balanced stroke cycle that minimizes the secondary motions.

The hydrodynamic forcing terms $F_{\mathrm{Flow}}$ and $M_{\mathrm{Flow}}$ in system (6) can be computed using either a complete Navier–Stokes flow model or a reduced hydrodynamic model based on a linear potential flow model. The former was used, for example, to investigate the performance of different boat designs, simulating the free-surface flow around them (see figure 7) and the rolling stability of the system (see figure 8); using the latter approach, full race simulations could be carried out in (almost) real time.

### 3.3 Swimming

In recent years, competition swimming has been unsettled by the introduction of high-tech swimsuits. These products resulted from scientific research on the development of low-drag fabrics and new techniques of fabric assembly. Thanks to the improved performance guaranteed by these new swimsuits, in a couple of years

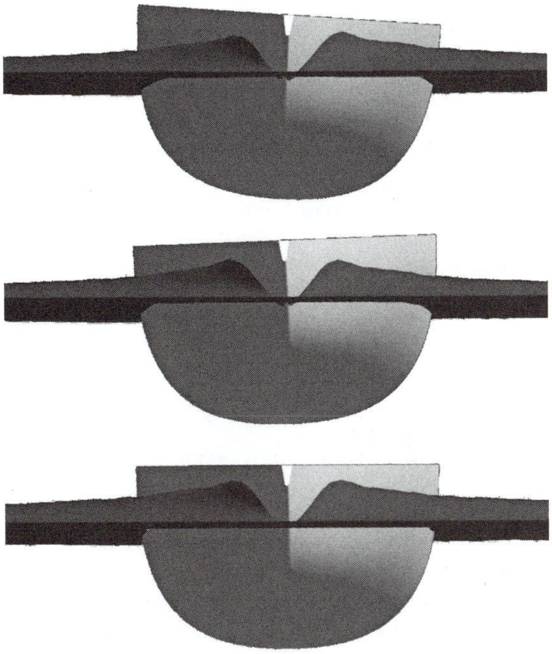

**Figure 8** Rolling motion of a rowing boat.

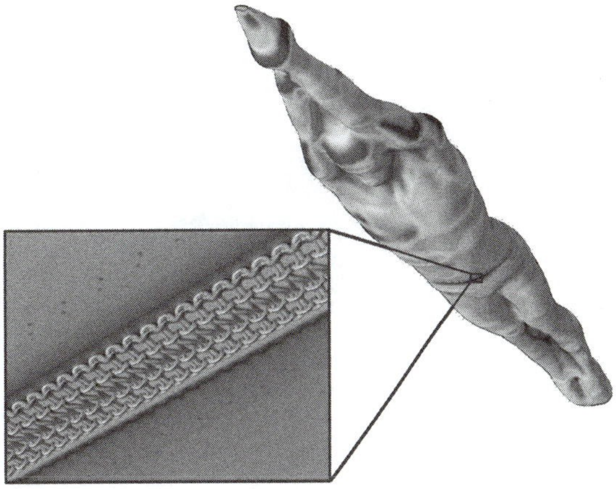

**Figure 9** Shear stress distribution over a swimmer at different spatial scales.

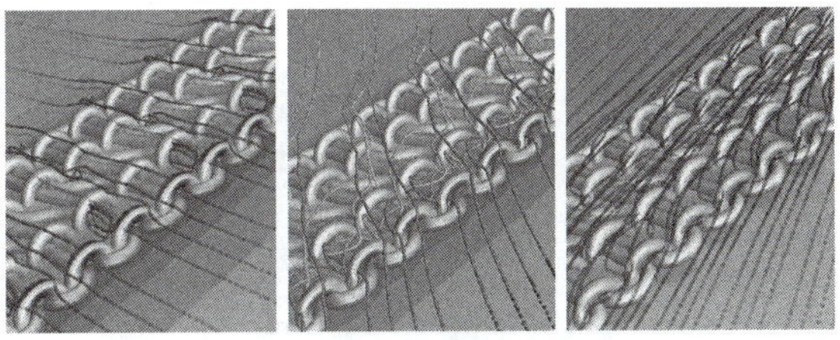

**Figure 10** Local flow around the sewing at different incidences.

(2008–9) more than 130 swimming world records were broken. In 2010 the International Swimming Federation (FINA) decided to ban the use of these products in official competitions.

In this context, the authors have (in collaboration with a leading swimsuit manufacturer) been involved in a research activity to estimate the potential gain associated with a new technique of fabric assembly based on a thermo-bonding system in which all the standard sewing was completely removed from the swimsuit surface. This new swimsuit was approved by FINA before the Beijing 2008 Olympic Games.

In order to quantify the advantages of removing the sewing, its impact on the flow around the swimmer needed to be analyzed. A complete simulation including the geometrical details of the swimmer's body as well as the sewing on the swimsuit is unaffordable, due to the different length scales involved (see figure 9). Model reduction was therefore mandatory. At first, the flow around the sewing was analyzed considering a local computational domain limited to a small patch of fabric surrounding the sewing.

The sewing geometry was reconstructed in detail to generate a computer-aided design model. Based on this geometry, an unstructured computational grid was generated and the flow around the sewing was simulated considering different flow velocities and orientations (see figure 10). The presence of the sewing strongly affects the flow in its boundary layer, and this results in an additional drag component.

Full-body simulations (see figure 11) were then carried out to estimate, for all the sewing distributed over the swimsuit, the local velocity and orientation, so that the results of the small-scale simulation could be integrated to obtain a global value of the drag component associated with the presence of the sewing.

To move from an estimate of the drag reduction to an estimate of the reduction in race time, a race model was developed, based on the Newton law:

$$m\frac{\mathrm{d}^2x(t)}{\mathrm{d}t^2} = P(t) - D(t), \quad x'(0) = 0,\ x(0) = 0,$$

where $m$ and $x$ are the mass and position of the swimmer, respectively. The resistance $D(t)$ can be estimated based on the results of the CFD simulation described above, while the propulsion $P(t)$ could also be computed through ad hoc simulations of the stroke action (see figure 12).

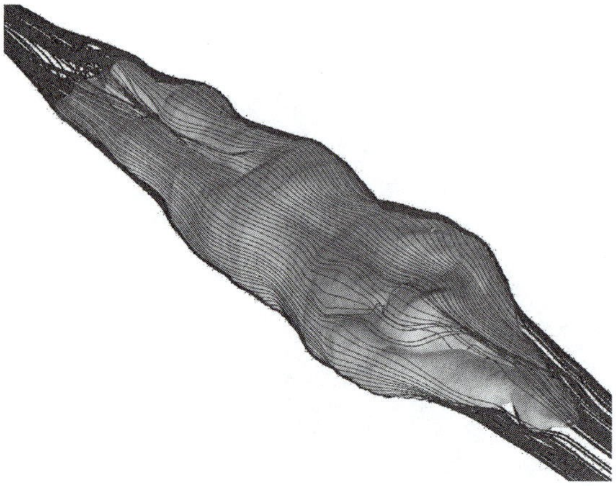

**Figure 11** Full-body simulation of the flow around a swimmer.

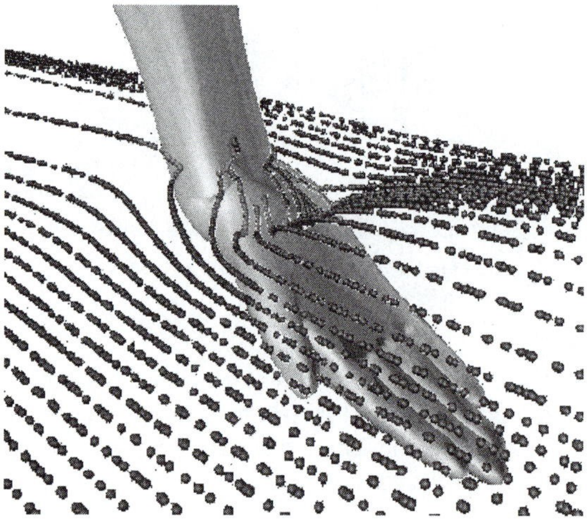

**Figure 12** Flow around the hand in a stroke simulation.

**Table 1** Time gains in seconds for different freestyle races.

| | Gliding | Length | Race |
|---|---|---|---|
| 50 m | 0.011 | 0.073 | 0.073 |
| 100 m | 0.011 | 0.077 | 0.154 |
| 200 m | 0.012 | 0.083 | 0.332 |
| 400 m | 0.013 | 0.087 | 0.696 |

Our analysis made it possible to quantify the potential gain in terms of race time associated with the new swimsuit design, which is the most useful metric of the performance improvement. The time gains for different race lengths are reported in table 1, where the gains in the gliding phase, over one length and over the whole race, are specified. These relevant race time improvements estimated by numerical simulations were confirmed by an experimental campaign carried out in the University of Liège's water tank by the biomechanics research group of the University of Reims (France) under the supervision of Redha Taiar.

### Further Reading

Detomi, D., N. Parolini, and A. Quarteroni. 2009. Numerical models and simulations in sailing yacht design. In *Computational Fluid Dynamics for Sport Simulation*, pp. 1–31. Lecture Notes in Computational Science and Engineering. New York: Springer.

Formaggia, L., E. Miglio, A. Mola, and N. Parolini. 2008. Fluid-structure interaction problems in free surface flows: application to boat dynamics. *International Journal for Numerical Methods in Fluids* 56(8):965–78.

Lombardi, M., M. Cremonesi, A. Giampieri, N. Parolini, and A. Quarteroni. 2012. A strongly coupled fluid–structure interaction model for sail simulations. In *Proceedings of the 4th High Performance Yacht Design Conference, 2012.* London: Royal Institution of Naval Architects.

Lombardi, M., N. Parolini, A. Quarteroni, and G. Rozza. 2012. Numerical simulation of sailing boats: dynamics, FSI, and shape optimization. In *Variational Analysis and Aerospace Engineering: Mathematical Challenges for Aerospace Design*, edited by G. Buttazzo and A. Frediani. New York: Springer.

Parolini, N., and A. Quarteroni. 2005. Mathematical models and numerical simulations for the America's Cup. *Computer Methods in Applied Mechanics and Engineering* 194: 1001–26.

——. 2007. Modelling and numerical simulation for yacht design. In *Proceedings of the 26th Symposium on Naval Hydrodynamics.* Arlington, VA: Strategic Analysis.

## V.3 Inerters

*Malcolm C. Smith*

### 1 Introduction

A new ideal mechanical modeling element, coined the "inerter," was introduced by the author in 2002 as a component that was needed for the solution of the following mathematical question (although it is not purely mathematical: it also touches on physics and engineering).

[P] What is the most general linear passive mechanical impedance function that can be realized physically?

An analogy with electrical circuits suggested that a solution ought to be straightforward. But this turned out not to be the case, and a new mechanical modeling element was needed. Embodiments of inerter devices were built and tested at the University of Cambridge's engineering department. The technology was subsequently developed by a Formula 1 team, and the inerter is now a standard component of many racing cars. The story of this development from mathematical theory to engineering practice is described below.

## 2 Network Analogies

There are two analogies between electrical and mechanical networks that are in common use. The oldest of these uses the correspondences force ↔ voltage and velocity ↔ current. An alternative analogy, usually attributed to Firestone (1933), makes use of the correspondences

force ↔ current,

velocity ↔ voltage,

mechanical ground ↔ electrical ground.

Although neither analogy can be said to be correct or incorrect, the force–current analogy has certain advantages. Firstly, from a topological point of view, network graphs are identical in the electrical and mechanical domains when the force–current analogy is used. This means that any electrical network has a mechanical analogue, and vice versa, whereas in the older force–voltage analogy only networks with planar graphs have an analogue in the other domain, since the dual graph is involved in finding the analogue. Secondly, in the force–current analogy, mechanical ground (namely, a fixed point in an inertial frame) corresponds to electrical ground (namely, a datum or reference voltage). Thirdly, and more fundamentally, the force–current analogy is underpinned by the notion of through and across variables. A *through variable* (such as force or current) involves a single measurement point and normally requires the system to be severed at that point to make the measurement. In contrast, an *across variable* (such as velocity or voltage) is the difference in an absolute quantity between two points and can in principle be measured without breaking into the system. The notion of through and across variables allows analogies to be developed with other domains (e.g., acoustic, thermal, fluid) in a systematic manner.

It is important to point out, since both force×velocity and current × voltage have the units of power, that both analogies are power preserving. Questions such as [P] that relate to passivity (meaning no internal power source) can therefore be correctly mapped from one domain to another.

## 3 The Inerter

The standard correspondences between circuit elements in the force–current analogy are as follows:

spring ↔ inductor,

damper ↔ resistor,

mass ↔ capacitor.

A well-known fact, which is only rarely emphasized, is that only five of the above six elements are genuine two-terminal devices. For mechanical devices, the terminals are the attachment points. Both the spring and the damper have two independently movable terminals. However, the mass element has only one attachment point: the center of mass. Its behavior is described by Newton's second law, which involves the acceleration of the mass relative to a fixed point in an inertial frame. This gives rise to the interpretation that one terminal of the mass is the ground and the other terminal is the position of the center of mass itself; effectively, that the mass is analogous to a *grounded* capacitor. This means that not every electrical circuit comprising inductors, capacitors, and resistors will have a spring–mass–damper analogue.

There is a further problem regarding the mass element in relation to [P]. Any mechanism that is constructed to realize a given impedance function is typically intended for connection between other masses, and must therefore have a small mass in relation to those. Any construction involving mass elements would need to ensure that the total mass employed could be kept as small as desired.

The above considerations were the motivation for the following question posed in Smith (2002): is it possible to construct a two-terminal mechanical device with the property that the equal and opposite force applied at the terminals is proportional to the relative acceleration between them? It was shown that devices of small mass could be constructed that approximate this behavior. Accordingly, the proposal was made to define an *ideal modeling element* as follows.

**Definition 1.** The (ideal) *inerter* is a mechanical two-terminal device with the property that the equal and opposite force applied at the terminals is proportional to the relative acceleration between them. That is,

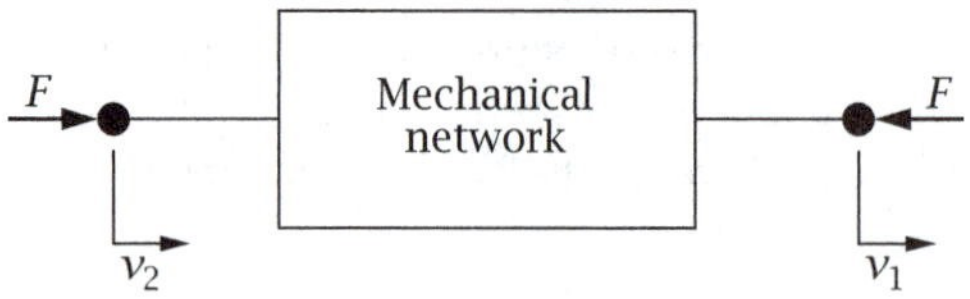


**Figure 1** A free-body diagram of a one-port (two-terminal) mechanical element or network with force–velocity pair $(F, v)$, where $v = v_2 - v_1$.

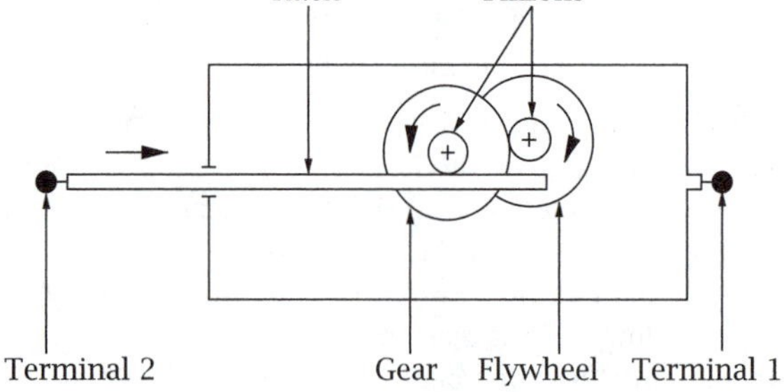


**Figure 2** Schematic of a mechanical model of an inerter employing a rack, pinion, gears, and flywheel.

$F = b(\dot{v}_2 - \dot{v}_1)$ in the notation of figure 1. The constant of proportionality $b$ is called the *inertance* and has units of kilograms.

As is the case for any modeling element, a distinction needs to be made between the ideal element (as defined above) and practical devices that approximate it. The manner in which practical realizations deviate from ideal behavior may determine whether a definition is sensible at all. In the case of the inerter, it has already been noted that realizations are needed with sufficiently small mass, independent of the inertance $b$. Another requirement is that the "available travel" (the relative displacement between the terminals) of the device can be specified independently of the inertance. One method of realization that can satisfy all these requirements makes use of a plunger sliding in a housing that drives a flywheel through a rack, pinion, and gears (see figure 2).

The circuit symbols of the six basic electrical and mechanical elements, with the inerter replacing the mass, is shown in table 1. The symbol chosen for the inerter represents a flywheel. The table also shows the defining (differential) equation for each element as well as the admittance function $Y(s) = Z(s)^{-1}$, where $Z(s)$ is the impedance function. Taking Laplace transforms of the defining equation, assuming zero initial

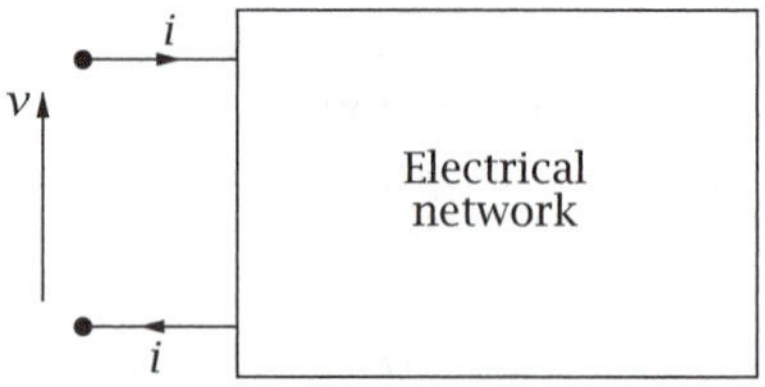


**Figure 3** A one-port electrical network.

conditions, the *impedance* for the electrical elements is defined by $Z(s) = \hat{\imath}(s)/\hat{v}(s)$, and similarly for the mechanical elements.

## 4 Passivity and Electrical Network Synthesis

Let us consider the problem in the electrical domain that is analogous to [P]. Figure 3 shows an electrical network with two external terminals (a "one-port") across which there is a voltage $v(t)$ and a corresponding current flow $i(t)$ through the network. The *driving-point impedance* of the network is defined by $Z(s) = \hat{\imath}(s)/\hat{v}(s)$. The network is defined to be *passive* if it contains no internal power source. More formally, we can define the network to be passive if, for all admissible $v$ and $i$ that are square integrable on $(-\infty, T]$,

$$\int_{-\infty}^{T} v(t)i(t)\,\mathrm{d}t \geqslant 0.$$

In his seminal 1931 paper, Brune introduced the notion of a positive-real function: for a rational function with real coefficients, $Z(s)$ is defined to be *positive-real* if $Z(s)$ is analytic and $\mathrm{Re}(Z(s)) \geqslant 0$ whenever $\mathrm{Re}(s) > 0$. Equivalently, $Z(s)$ is positive-real if and only if $Z(s)$ is analytic in $\mathrm{Re}(s) > 0$, $\mathrm{Re}(Z(\mathrm{j}\omega)) \geqslant 0$ for all $\omega$ at which $Z(\mathrm{j}\omega)$ is finite,[1] and any poles of $Z(s)$ on the imaginary axis or at infinity are simple and have a positive residue. Brune showed that a necessary condition for the network to be passive is that $Z(s)$ is positive-real.

Brune also showed the converse: for any (rational) positive-real function $Z(s)$ a network can be constructed (synthesized) comprising resistors, capacitors, inductors, and mutual inductances whose driving-point impedance is equal to $Z(s)$. In Brune's construction, mutual inductances were required with a coupling coefficient equal to 1, but that is difficult to achieve in practice. This led to the question of whether coupled coils (transformers) could be dispensed with altogether. This was settled in the affirmative in a famous 1949 paper of Bott and Duffin.

1. In the electrical engineering convention, $\mathrm{j} = \sqrt{-1}$.

**Table 1** Circuit symbols and correspondences with defining equations and admittance $Y(s)$.

| Mechanical | | Electrical | |
|---|---|---|---|
| 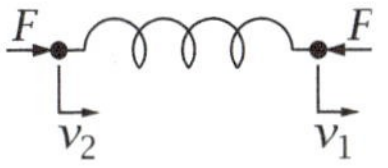  | $Y(s) = \frac{k}{s}$ | 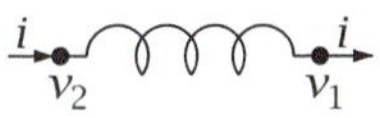  | $Y(s) = \frac{1}{Ls}$ |
| $\frac{\mathrm{d}F}{\mathrm{d}t} = k(v_2 - v_1)$ | Spring | $\frac{\mathrm{d}i}{\mathrm{d}t} = \frac{1}{L}(v_2 - v_1)$ | Inductor |
| 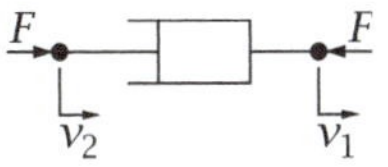  | $Y(s) = c$ |   | $Y(s) = \frac{1}{R}$ |
| $F = c(v_2 - v_1)$ | Damper | $i = \frac{1}{R}(v_2 - v_1)$ | Resistor |
| 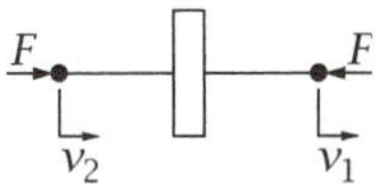  | $Y(s) = bs$ |   | $Y(s) = Cs$ |
| $F = b\frac{\mathrm{d}(v_2 - v_1)}{\mathrm{d}t}$ | Inerter | $i = C\frac{\mathrm{d}(v_2 - v_1)}{\mathrm{d}t}$ | Capacitor |

The Bott-Duffin construction certainly settled an important question in network synthesis. However, it also raised further questions since the construction appears, for some functions, to be rather lavish in the number of elements used. For example, the construction uses six energy storage elements for some biquadratic functions (ratios of two quadratics) when it might seem that two should suffice. The question of minimality remained unresolved when interest in passive circuit synthesis started to decline in the late 1960s due to the increasing prevalence of active circuits. Some of these issues are now being revived because of applications in the mechanical domain, where efficiency of realization is an important issue.

Let us now return to the problem [P] in the mechanical domain. All the ingredients are now in place to provide a solution. Defining the mechanical driving-point impedance by $Z(s) = \hat{F}(s)/\hat{v}(s)$, using the theorem of Bott and Duffin, the force-current analogy, and the ideal inerter, we see that $Z(s)$ can be realized as the impedance of a passive mechanical network if and only if $Z(s)$ is positive-real.

## 5 Vehicle Suspensions

A simple model for studying vehicle suspensions is the quarter-car vehicle model described by the equations

$$m_s \ddot{z}_s = u - F_s,$$
$$m_u \ddot{z}_u = -u + k_t(z_r - z_u),$$

where $m_s$ is the sprung mass, $m_u$ is the unsprung mass, $k_t$ is the tire (vertical) spring stiffness, $u$ is an active or passive control force, $F_s$ is a load disturbance input, $z_r$ is the vertical displacement of the road input, and $z_s$ and $z_u$ are the vertical displacements of the sprung and unsprung masses. A conventional passive suspension consists of a spring and a damper in parallel, which means that $u = k_s(z_u - z_s) + c_s(\dot{z}_u - \dot{z}_s)$, where $k_s$ and $c_s$ are the spring and damper constants, respectively. In the most general linear passive suspension, $\hat{u} = Q(s)s(\hat{z}_u - \hat{z}_s)$, where $Q(s)$ is a general positive-real admittance function, as depicted in figure 4. After a suitable positive-real admittance is found, circuit synthesis techniques come into play to find a network of springs, dampers, and inerters to realize $Q(s)$. Alternatively, the parameters of simple circuits can be optimized directly; for example, a spring, damper, and inerter in parallel.

When designing an active or passive suspension force law $u$, a number of different performance criteria are relevant. Some of these criteria are related to the response to road undulations and come under the category of "ride" performance. A common (simple) assumption is that $z_r$ is integrated white noise (Brownian motion). A second set of criteria relate to "handling" performance, which is the response to driver inputs, such as braking, accelerating, or cornering. In the quarter-car model these are sometimes approximated very crudely by the response to deterministic loads $F_s$. Variables of relevance include $\ddot{z}_s$ (comfort)

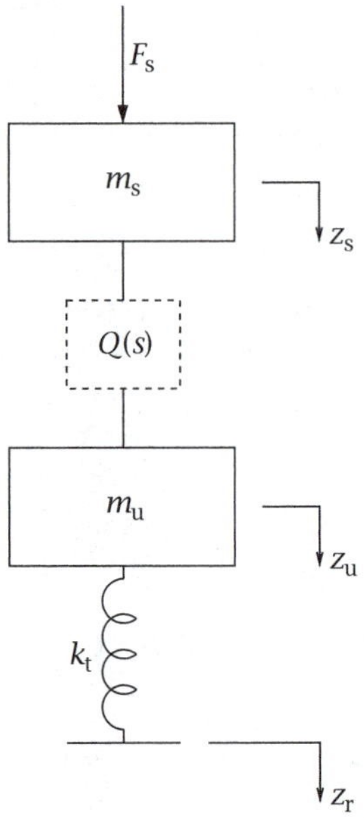


**Figure 4** The quarter-car vehicle model.

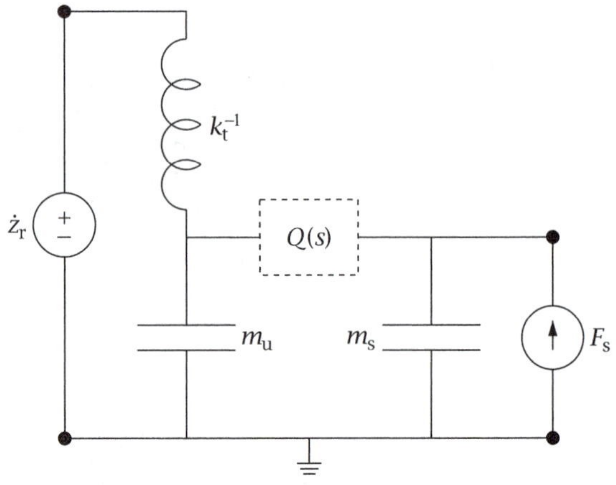


**Figure 5** The equivalent electrical circuit for the quarter-car model.

and $k_t(z_r - z_u)$ (tire grip). Performance measures can typically be improved by using a general linear passive suspension instead of a conventional spring and damper.

As an interesting aside—and an illustration of the force–current analogy—figure 5 shows the equivalent electrical circuit for the quarter-car model. It is always amusing to point out to vehicle dynamicists that the electrical circuit highlights the interpretation that the sprung mass and the unsprung mass are connected to the ground, but ... the tire is not: it is connected to the road (here modeled as a velocity source)!

## 6 The Inerter in Formula 1 Racing

For racing cars, a very important performance measure is "mechanical grip," which refers to the tire load fluctuations in response to unevenness of the road surface. Increased grip corresponds to decreased fluctuations. For stiffly sprung suspensions (invariably the case in racing cars), mechanical grip can be improved by the simple expedient of placing an inerter in parallel with the conventional spring and damper.

The story of the development and use of the inerter in Formula 1 racing has been recounted a number of times in the popular press and in magazines. The inerter was developed by McLaren Racing, under license from the University of Cambridge and a cloak of confidentiality. It was raced for the first time by Kimi Raikkonen at the 2005 Spanish Grand Prix, where he achieved McLaren's first victory of the season. During development, McLaren invented an internal decoy name for the inerter (the "J-damper") to make it difficult for personnel who might leave to join another team to make a connection with the technical literature on the inerter that was being published. This strategy succeeded in spectacular fashion during the 2007 Formula 1 "spy scandal," when a drawing of the McLaren J-damper came into the hands of the Renault engineering team. Renault made an attempt to get the device banned (unsuccessfully) on an erroneous interpretation of the device. The Fédération Internationale de l'Automobile World Motor Sport Council convened in Monaco on December 6, 2007, to investigate a spying charge brought by McLaren. The council found Renault to be in breach of the sporting code but issued no penalty. In Paragraph 8.7 of the World Motor Sport Council Decision, the council reasoned that

> the fact that Renault fundamentally misunderstood the operation of the system suggests that the "J-damper" drawing did not reveal to Renault enough about the system for the championship to have been affected.

During the hearing, neither the World Motor Sport Council nor McLaren made public what the J-damper was. Thereafter, speculation increased on Web sites and blogs about the function and purpose of the device. Finally, the truth was discovered by *Autosport* magazine, which revealed the Cambridge connection and that the J-damper was an inerter. Soon afterward, the University of Cambridge signed a licensing agreement

**Figure 6** A ballscrew inerter made in the University of Cambridge engineering department in 2003, with flywheel removed.

with Penske Racing Shocks to allow inerters to be supplied to other customers within Formula 1 and elsewhere. The use of inerters has now spread beyond the Formula 1 grid to IndyCars and several other motorsport formulas. A typical method of construction makes use of a ballscrew, as shown in figure 6.

### Further Reading

Bott, R., and R. J. Duffin. 1949. Impedance synthesis without use of transformers. *Journal of Applied Physics* 20:816.

Brune, O. 1931. Synthesis of a finite two-terminal network whose driving-point impedance is a prescribed function of frequency. *Journal of Mathematical Physics* 10: 191–236.

Jackson, A. 2001. Interview with Raoul Bott. *Notices of the American Mathematical Society* 48(4):374–82.

Shearer, J. L., A. T. Murphy, and H. H. Richardson. 1967. *Introduction to System Dynamics*. Reading, MA: Addison-Wesley.

Smith, M. C. 2002. Synthesis of mechanical networks: the inerter. *IEEE Transactions on Automatic Control* 47:1648–62.

## V.4 Mathematical Biomechanics

*Oliver E. Jensen*

### 1 Physical Forces in Biology

Biological organisms must extract resources from their environment in order to survive and reproduce. Animals need oxygen, water, and the chemical energy stored in food. Plants require a supply of carbon dioxide, water, and light. Different species have developed diverse mechanisms for harvesting these essential nutrients. For example, tiny single-celled microorganisms such as the bacterium *Escherichia coli* swim along gradients of nutrient, whereas large multicellular organisms such as humans have internal energy transport systems of great geometric complexity. Mathematics has an important role to play in understanding the relationship between the form and function of these remarkable natural systems, and in explaining how organisms have adapted to the physical constraints of the world they inhabit.

A nutrient in molecular form can be taken up by an organism by crossing a cell membrane, generally via molecular diffusion. While diffusion over such short distances is rapid, it is generally not an effective transport mechanism over long distances because diffusion time increases in proportion to the square of distance. Organs performing gas exchange (such as lungs, gills, or leaves) therefore exploit fluid flow (in which nutrients are carried by a moving gas or liquid) for long-distance transport, coupling this with an exchange surface that presents a large area for short-range diffusion. Similarly, most organisms that are larger than just a few cells generate internal flows to distribute nutrients around their bodies. The airways of our lungs, which have a surface area nearly as large as a tennis court but a volume of just a few liters, are therefore entwined with an elaborate network of blood vessels, which can deliver oxygen rapidly to other organs. In plants, light is harvested by maximizing the exposure of leaves to the rays of the sun, while nutrients generated through photosynthesis in leaves are carried to flowers and roots by an extensive internal vascular system. Naturally, many organisms will also move themselves to regions rich in nutrients using strategies (swimming, flying, running, etc.) that are appropriate to their environment.

Mathematical models of any of these systems allow one to encode and quantify physical laws and constraints. This may be done simply at the level of kinematics or transport, i.e., understanding motion in time and space or expressing conservation of mass; alternatively, models may involve analysis of physical forces through relationships such as Newton's laws. Often, mechanical models are coupled to descriptions of biological processes, such as cell signaling pathways, gene regulatory networks, or transport of hormones, allowing a quantitative "systems-level" description of a cell, organ, or organism to be developed. Such approaches can be essential to understand the full complexity of biological control mechanisms. A wealth of theoretical developments from other areas of science and engineering can be exploited in developing such models. However, biomechanical problems can have a flavor of their own, and they raise some particular mathematical challenges.

These become clear through a comparison with a "classical" field such as fluid mechanics. The governing equations describing the motion of most fluids were established in the first half of the nineteenth century and are firmly grounded on Boltzmann's kinetic theory. The NAVIER–STOKES EQUATIONS [III.23] are remarkable for many reasons: they can describe fluid motions over at least ten orders of magnitude (from submicron to planetary scales); they apply equally well to distinct phases of matter (liquids and gases); and in many situations it is sufficient to describe fluids with just a handful of physical parameters (a density, a viscosity, and possibly a bulk modulus), which can be reduced to an even smaller number of parameters (a Reynolds number, a Mach number) through NONDIMENSIONALIZATION [II.9]. Fluid flows arising in very different situations can therefore have essentially identical mathematical descriptions, often enabling experiments to be performed at comfortable bench-top scales involving nonintrusive measurements. While the field continues to raise profound mathematical questions (such as the origin and structure of TURBULENCE [V.21]), the topic as a whole is built on secure theoretical foundations.

It can therefore be disconcerting to face some of the awkward realities of the biological world, where features such as universality, scale independence, efficient parametrization, and nonintrusive experimentation can be illusory. Some specific challenges are as follows.

- Nature's remarkable biodiversity is reflected by a proliferation of different mathematical models, which often have to be adapted to describe a specific organ, organism, or process. A model might even be tailored to an individual, as is the case in the growing field of personalized medicine. However, models can sometimes be connected by unifying principles (such as the concept of natural selection driving optimization of certain structural features in a given environment).
- Biological materials are typically highly heterogeneous and have distinct levels of spatial and temporal organization spanning many orders of magnitude (from molecules and cells up to whole populations of organisms), requiring different mathematical descriptions at each level of the hierarchy. For example, the beating of our hearts over most of a century relies on the cooperative action of numerous molecular ion channels in heart muscle cells, flickering open and shut within milliseconds.
- Experimental measurements of key parameters can be exceptionally challenging, given that many biological materials function properly only in their natural living environment. Biological data gathered in vitro (for example, on tissue cultured in a petri dish) can misrepresent the true in vivo situation. Parameter uncertainty represents a profound difficulty in developing genuinely predictive computational (in silico) models.
- Biological processes do not generally operate in isolation: they are coupled to numerous others, monitoring and responding to their environment. For example, plant and animal cells have a variety of subcellular mechanisms of *mechanotransduction*, whereby stimuli such as stress or strain have direct biological consequences (such as protein production or gene expression).
- Unlike the relatively passive structures that dominate much of engineering science, biological materials generate forces (via osmotic pumps in plants, or via molecular motors, driven by chemical energy stored in adenosine triphosphate; arrays of motors can be organized to form powerful muscles). Organisms change their form through *growth* (the development of new tissues) or *remodeling* (the turnover of existing tissues in response to a changing environment). Examples of remodeling include the manner in which our arteries gradually harden as we age or the way a tree on a windswept hillside slowly adopts a bent shape.

This brief survey presents a few examples illustrating different applications of biomechanical modeling, all with a multiscale flavor. This is an endeavor that draws together research from numerous disciplines and should not be regarded as a purely mathematical exercise. However, mathematics has a key role to play in getting to grips with biological complexity.

## 2 Constitutive Modeling

Classical continuum mechanics is based on a number of simplifying assumptions. Under the so-called *continuum hypothesis*, one considers a representative volume element of material that is much larger than the underlying microstructure but much smaller than the size of the object that is undergoing deformation. The separation between the microscopic and macroscopic length scales allows the discrete nature of the microstructure to be approximated by smoothly varying scalar, vector, or tensor fields; physical relationships are then expressed by partial differential equations (PDEs) or integral equations at the macroscopic level. While this is a very effective strategy for homogeneous materials (such as water or steel), for which the granularity of the microstructure appears only at molecular length scales, it is less obviously appropriate for heterogeneous biological materials.

In describing a given material it is often necessary to make a *constitutive assumption*: namely, a choice over how to model the material properties, typically expressed as a relationship between the tensor fields describing stress, strain, and strain rate. This assumption has to be tested against observation and may require careful judgement that takes into account the question that the model seeks to address, the quality of available experimental data, and so on. Nevertheless, a number of canonical model choices are available. The material might be classified as elastic, viscous, viscoelastic, viscoplastic, etc., depending, for example, on whether deformations are reversible once any loading is removed (as in the elastic case). One might then consider whether a linear stress/strain/strain rate relationship is sufficient (this is appealing because of the small number of material parameters and the analytic and computational tractability) or whether a more complex nonlinear relationship is required.

The Chinese-American bioengineer Y. C. Fung has been prominent among those pioneering the translation of continuum mechanics to biological materials, particularly human tissues of interest in the biomedical area. As he and many others have shown, "off-the-shelf" constitutive models can be usefully deployed to describe materials such as tendon, bone, or soft tissue. By fitting such models to experimental data, the corresponding mechanical parameters (Young's modulus, viscosity, etc.) can be estimated, allowing predictive models to be developed. However, in many cases standard models turn out to be inappropriate; we discuss below the exotic properties of blood in small blood vessels, where a description as a Newtonian homogeneous fluid fails.

A particularly challenging area has been in characterizing the mechanical properties of individual cells. Animal cells have a phospholipid membrane that encloses a cytoplasm and a nucleus. The cytoplasm is a crowded soup of organelles and proteins, among which the components of the cytoskeleton (which includes fibers (actin), rods (microtubules), and molecular motors (myosin, dynein, etc.)) are of particular mechanical importance. Processes such as cell motility and division transcend any classical engineering description, and instead biomechanical models have been developed that take a "bottom-up" description of the evolving microstructure (accounting, for example, for dynamic actin polymerization and depolymerization). Among animal cells, the red blood cell is relatively simple in mechanical terms, because its properties are dominated by its membrane, which can be described as a passive structure that resists area changes, and which has measurable resistance to in-plane shear and bending. Plant cells, too, are typically dominated mechanically by the properties of their stiff cell walls rather than their cytoskeleton.

The properties of a multicellular tissue will be determined by the collective properties of a population of adherent cells. If the cells form a roughly periodic array, then the asymptotic technique of HOMOGENIZATION [II.17] can be used to derive a tissue-level description (e.g., in terms of PDEs) by careful averaging of properties at the cell level. This multiscale method works well for strictly periodic structures and can be applied (with some care) to disordered arrays. However, when there is sufficient heterogeneity at the microscale, the validity of a macroscale approximation (which assumes that properties vary over distances that are very long compared with individual cells) may become questionable. A variety of computational methods (cell-center models, vertex-based models, cellular Potts models, etc.) can instead be used to resolve the granular nature of multicellular tissues.

## 3 Blood Flow

Despite significant medical advances over recent decades, cardiovascular disease (heart attack, stroke, and related conditions) remains a leading cause of death in Europe and North America. There is therefore intense interest in the pump and plumbing of our blood-supply system and the mechanisms by which it fails. Mathematical modeling has contributed significantly to this effort.

Blood is a suspension of cells in plasma. The majority of cells are erythrocytes (red blood cells): small highly deformable capsules containing hemoglobin. These cells are efficient carriers of oxygen from lungs to tissues. While additional cells are present in blood, many of which have important roles in fighting infection, the rheology of blood is determined primarily by the collective properties of the red cells.

Red blood cells have a disk-like shape and a diameter of only 8 microns but are present in very high concentrations (typically 45% by volume). In large blood vessels, the suspension can therefore be well approximated as a homogeneous liquid, and a description as an incompressible "Newtonian" fluid (that is, one characterized by a constant viscosity $\mu$ and a constant density) is effective. Poiseuille's exact solution of the Navier–Stokes equations describes steady flow along a uniform pipe of radius $a$ and length $L$, via

$$u(r) = \frac{a^2 - r^2}{4\mu}\frac{\Delta p}{L}, \qquad Q = \frac{\pi a^4 \Delta p}{8\mu L}, \tag{1}$$

where $u(r)$ is the axial velocity component, $r$ is the radial coordinate, and $\Delta p$ is the pressure difference between the inlet and the outlet of the pipe. Equation (1) shows that the axial flow rate $Q$ (the volume of fluid passing a fixed location per unit time) driven by a fixed $\Delta p$ is proportional to $a^4$, implying that narrow vessels present a very high flow resistance. The *shear stress* (tangential force per unit area) exerted by the flow on the pipe wall is $-\mu u'(a) = 4\mu Q/\pi a^3$. While Poiseuille himself had a keen interest in understanding blood flow, few blood vessels are sufficiently long and straight, or have sufficiently steady flow within them, for his solution (1) to be directly relevant. For example, coronary arteries, feeding oxygenated blood to the muscle of the heart, have tortuous geometries and are either embedded in, or sit upon, muscle that undergoes vigorous periodic contractions. Vessel curvature, torsion, branching, nonuniformity, wall deformation, and spatial movement all lead to significant deviations from (1).

This observation's biomedical significance emerged in the 1980s when it was first appreciated how the endothelial cells that line blood vessels respond to stresses arising from blood flow (and in particular the shear stress, the component parallel to the vessel wall). Through mechanotransduction, cells align themselves with the flow direction; furthermore, regions of abnormally low and oscillatory wall shear stress are sites at which the arterial disease atherosclerosis tends to originate. In a coronary artery, this can be a precursor to a stenosis (a blockage), which can in turn cause angina and possibly an infarction (a heart attack). Since flow patterns are intimately associated with vessel shape, geometry becomes a direct risk factor for disease.

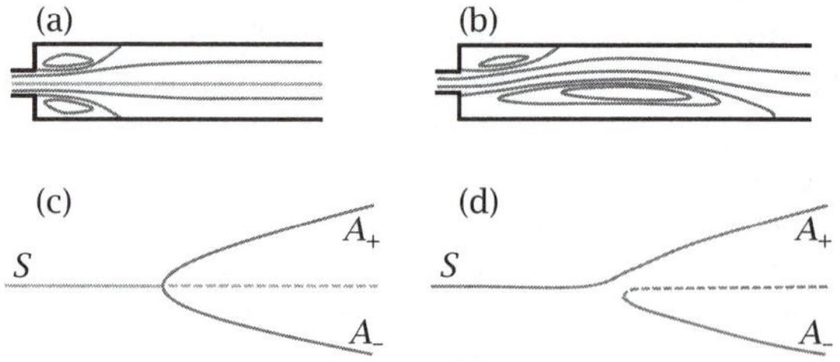


**Figure 1** (a), (b) Streamlines of steady flow in an expanding channel, mimicking flow past an obstruction in an artery; the flow is from left to right. (a) The symmetric state $S$. (b) One of two asymmetric states $A_\pm$, for which a large recirculation region exists on one or other wall. (c) $S$ loses stability to $A_\pm$ via a so-called PITCHFORK BIFURCATION [IV.21 §2] as the flow strength increases (the dashed line indicates an unstable solution). (d) A small geometric imperfection will bias the system, preferentially connecting $S$ to one of the asymmetric states.

Arterial flows are characterized by moderately high Reynolds numbers, implying that inertia dominates viscosity (friction) but without normally undergoing transition to full turbulence. The associated solutions of the Navier–Stokes equations nevertheless exhibit features characteristic of a nonlinear system, such as nonuniqueness and instability. Small changes in geometry (or other external factors) can therefore lead to large changes in the internal flow properties, as illustrated in figure 1. If modeling predictions are to help clinical decision making, it therefore becomes necessary to obtain detailed geometric information about an individual's arterial geometry, from X-ray tomography or magnetic resonance imaging, and use this as a geometric template on which to perform computational fluid dynamics. Even then, the robustness of predictions must be considered carefully given the natural variability of input conditions and geometric parameters, such as occurs between states of vigorous exercise and sleep.

In smaller blood vessels, of diameters below a few hundred microns, the discrete nature of the suspension becomes important and the description of blood as a Newtonian fluid breaks down. For practical purposes, a continuum description may still be useful but a number of unusual features emerge. First, it is necessary to consider the transport of red cells in addition to that of the suspension as a whole. This requires a distinction to be drawn between *tube hematocrit* $H_T$ (the volume fraction of red cells in a tube flow) and *discharge hematocrit* $H_D$ (the cross-sectionally averaged cell concentration weighted by the axial flow speed $u(r)$). These quantities differ because cells move away from the vessel wall into faster-moving parts of the flow through hydrodynamic interactions that are influenced by cell deformability. This has some important consequences beyond increasing $H_D$ relative to $H_T$: the peripheral cell-free layer lubricates the core and lowers the effective viscosity of the suspension (estimated experimentally by measuring $\pi a^4 \Delta p / 8LQ$, following (1)); and, at vessel bifurcations, plasma rather than cells may be preferentially drawn into side branches down which there is a weak flow (so-called *plasma skimming*).

The reduction in effective viscosity is of great importance in ensuring that blood can pass through the smallest capillaries without requiring enormous pressure drops to push them through and without generating huge shear stresses on endothelial cells. Red cells are sufficiently deformable that they can enter capillaries as small as 3 microns in diameter. Effects such as the dependence of effective viscosity on hematocrit and tube diameter have been carefully measured empirically, but for many years these resisted a full theoretical explanation. However, the problem of simulating the motion of multiple deformable cells within a tube has recently become computationally tractable, enabling "bottom-up" predictions to be made of the properties of whole blood starting from the properties of its components. This is an important precursor to understanding interactions with other cells, such as neutrophils that adhere transiently to the walls of blood vessels as part of the inflammatory response and platelets that form aggregates during the clotting process following a wound.

## 4 Tissue Growth

Growth of biological tissues takes a variety of forms. Soft tissues, such as those in a developing animal embryo, are formed of clumps or sheets of cells. The cells can expand (by increasing their volume), divide (duplicating genetic material in the cell nucleus and introducing new cell boundaries), and reorganize (by changing their neighbors). The process of proliferation (increase of cell number) can be accompanied by differentiation (in biological terminology, this describes the specification of the cell's type and that of its progeny). These processes are often exquisitely controlled in order that the embryo develops the right organs in the right place at the right time. Growth is therefore intrinsically coupled to a host of competing biological processes that "pattern" the developing organism. In addition to biochemical signals, cells have the capacity to respond to mechanical signals, enabling growth-induced stresses to feed back on further development. The growing organism is necessarily shaped by the physical forces it experiences, as was famously recognized a century ago by D'Arcy Thompson.

Even when fully grown, a biological organism operates in a state of homeostasis, with its tissues undergoing continual replacement and renewal. This is particularly striking in epithelia, the mucosal interface between the body and the external environment. For example, in the face of an onslaught of toxins, new cells in the lining of the gut are continually produced and removed. This rapid cell turnover predisposes such tissues to mutations that may lead to cancer. Tissues with structural properties, such as bone, muscle, or the walls of blood vessels, also undergo turnover, but at a slower rate and in a manner that responds to the biomechanical environment. Remodeling of bone has been described by Wolff's "law" of 1872, whereby the architecture of the trabecular matrix that constitutes bone aligns with the principal axes of the local stress tensor.

The structural rigidity of plant cells is provided primarily by a stiff cell wall. The primary cell wall is made of a pectin matrix, reinforced by fibers made of cellulose. When the fibers are uniformly aligned within the wall, it becomes mechanically anisotropic, being stiff in the direction parallel to the fibers but relatively soft in the perpendicular direction. A green plant is able to support itself against gravity and wind stresses by exploiting osmosis. By concentrating intracellular solutes, the cells draw water into intracellular vacuoles, pressurizing them (to levels comparable to a car tire) and generating tensions in the stiff cell walls. Then, by ensuring that cells adhere tightly to their neighbors, the rigidity of individual cells is conferred on the tissue as a whole. Even then, rapid growth is still possible. By regulated softening of cell walls, cells can elongate (in the direction orthogonal to the embedded fibers), driven by the high intracellular pressures. In order to grow to large heights, or to survive hostile environments, plants may develop a stiffer secondary cell wall, reinforced by lignin. Growth of a woody structure (such as thickening of a tree trunk or branch) occurs near its periphery in a thin layer beneath the bark known as the cambium. A horizontal branch can remodel itself to support its increasing weight by laying down new material known as *reaction wood*, either in tension within its upper surface or in compression on its lower surface.

These are manifestations of *residual stress*, a common feature of growing tissues. This is revealed when a piece of tissue is excised. A segment of artery, sliced axially along one wall, will spring open from an O-shaped to a C-shaped cross section. Similarly, a transverse slice across the stem of a green plant may cause the inner tissues to swell more than the periphery. One cause of such behavior is *differential growth*; for example, the inner part of the stem seeks to elongate more than the outer tissue. However, in the intact tissue, such differences are hidden by the requirement that the tissue remains continuous.

To model growth and residual stress at a continuum level, a popular formulation involves decomposition of the deformation gradient tensor $F$ into the product of two tensors as $F = AG$. (Here, the deformation is described via the map $\boldsymbol{X} \mapsto \boldsymbol{x}$, where $X$ labels points of the material in a reference configuration, while $\boldsymbol{x}$ denotes the location of such points in space; $F$ satisfies $F_{ij} = \partial x_i / \partial X_j$.) Growth is prescribed through the tensor $G$, which maps the reference configuration to an intermediate state. This state may violate compatibility conditions (e.g., two different parts of the body might map to the same location under $G$). However, a further deformation ($A$) is then required, mapping the intermediate state to the final state that accommodates both growth and the physical constraints placed on the body. This deformation might then be chosen to satisfy conditions of nonlinear elasticity. Growth that evolves in time can then be described by allowing $G$ to vary with time.

While residual stress may be hard to observe or measure without an invasive procedure, differential growth can often have a striking effect on the morphology of growing structures, particularly when they are thin in one or more directions. Elongated plant organs such as roots or shoots display a variety of *tropisms* (bending or twisting responses arising from signals such as light, gravity, or a nutrient). This can be achieved by ensuring that the cells on one side of the organ elongate slightly faster than those on the other, generating a bend. This enables a root penetrating a hard soil to navigate into the soft gaps between stones, as part of a thigmotropic (touch-sensitive) response, for example. While plant growth generally involves irreversible (viscous) deformations of plant cell walls, reversible deformations can also arise; in hygromorphic structures, adjacent layers of material elongate or contract differently in response to changes in humidity. This allows pine cones or pollen-bearing anthers to open or close within minutes.

Sheet-like structures such as leaves provide dramatic demonstrations of how nonuniform in-plane growth can generate exotic shapes. For example, the folds at the edge of a lettuce leaf can exhibit a cascade of wrinkles: just as in a pleated curtain, crinkly short-wavelength folds at the top give way to smoother longer-wavelength folds at the bottom. Thin objects are much easier to bend or twist than to stretch. Thus a leaf that undergoes growth primarily near its edge will tend to adopt configurations in which stretching is minimized, which is achieved through out-of-plane wrinkling. The separation of length scales in a thin sheet (with thickness much less than width) allows the sheet to be described using the Föppl–von Kármán equations of shell theory (an eighth-order nonlinear PDE system), modified to account for nonuniform growth. Instead of the tensor $G$, the growth pattern at the edge of a leaf can be prescribed through a non-Euclidean metric of the form

$$ds^2 = (1 + g(y))^2 \, dx^2 + dy^2,$$

where $x$ measures distance along the leaf edge in an initially undeformed planar configuration, and $y$ measures distance normal to the leaf edge, the leaf lying in $y \geqslant 0$. Here, $g(y)$ is assumed to be nonnegative and to approach zero as $y$ increases away from the leaf edge at $y = 0$. The function $g$ therefore defines the Gaussian curvature $K$ associated with the metric ($K \approx -g''(y) \leqslant 0$, for small $g$); recall that, because $K \neq 0$, both principal curvatures of the surface must be nonzero. However, from Gauss's *Theorema Egregium*, $K$ is invariant when the surface undergoes isometric deformations. The shape of a leaf, which may exhibit a self-similar cascade of wrinkles, can therefore be interpreted as an embedding of the surface in three dimensions in which there is no (or at least minimal) stretching. The out-of-plane deformation relieves the residual stress that would accumulate were the leaf confined to a plane.

## 5 Swimming Microorganisms

Single-celled microorganisms constitute a major proportion of global biomass. They occupy diverse habitats, from algae in deep oceans to acid-tolerant bacteria in our stomachs. For such organisms, locomotion can be essential in securing nutrients, given the limitations of diffusion as a transport mechanism. For organisms that are just a few microns in length, propulsion in a liquid environment is a battle against friction, inertia being negligible at very small length scales (the relevant REYNOLDS NUMBER [IV.28 §3.2] is very small). This demands specific swimming strategies. The cells are equipped with protruding appendages (cilia or flagellae). These may be driven in a wavelike fashion (such as synchronized waves in arrays of beating cilia on the alga *Volvox*), as a simple "breaststroke" (by a pair of flagellae on the alga *Chlamydomonas nivalis*), or as a helical propeller (*Escherichia coli* uses a bundle of flagellae that are driven by rotary motors embedded in its cell membrane). To generate a net forward motion, the appendages must be driven in nonreversible paths. This is a consequence of the linearity of the Stokes equations (the Navier–Stokes equations without the nonlinear inertial terms), from which it can be proved that reversible boundary motions will generate cyclic motion with no net drift.

The flagellar corkscrew of *Escherichia coli* satisfies this constraint admirably. However, the bacterium must also be able to direct itself toward a nutrient. It achieves this by interspersing "runs" with "tumbles," during which it reverses the rotation of one of the motors driving the flagellar bundle. These uncoil and the bacterium rotates to a new random orientation prior to the next run. The bacterium can bias its motion by prolonging the duration of runs in a favored direction. (Given the difficulty of measuring a spatial gradient over a short body length, the microswimmers are likely to measure concentration differences as they move from one region to another.) Thus *Escherichia coli* undertakes *chemotaxis* (seeking out a nutrient) via a form of biased RANDOM WALK [II.25]. Models of such processes must often be based on empirical descriptions of observed biological behavior rather than established physical principles.

Other species follow gradients in light (phototaxis) or oxygen (oxytaxis). In a dense suspension of the bacterium *Bacillus subtilis* beneath an air–liquid interface, accumulation of bacteria in the oxygen-rich region immediately beneath the interface will increase the density of the suspension relative to deeper regions of the liquid. The resulting heavy-over-light arrangement is unstable and the excess weight of the cell-rich liquid drives an instability in which dense plumes of cells will fall downward. This continual overturning of liquid is known as *bioconvection* and has many analogies with thermal or solutal convection in a liquid. It is a striking example of the generation of a large-scale pattern via the collective motion of individual microscopic swimmers. Mathematical models describing this process can involve coupled advection–diffusion PDEs for two scalar fields (the number of cells per unit volume $n(\boldsymbol{x},t)$ and the oxygen concentration $c(\boldsymbol{x},t)$) coupled to the Navier–Stokes equation for the average fluid velocity $\boldsymbol{u}(\boldsymbol{x},t)$, subject to the incompressibility constraint $\nabla \cdot \boldsymbol{u} = 0$. The cells are advected by the flow and move chemotactically toward the oxygen or light source; the buoyancy of the cells acts as a body force on the fluid, driving convective motions.

Algae such as *Chlamydomonas nivalis* exhibit *gyrotaxis*. These cells are bottom-heavy, and so experience a gravitational torque that aligns them head up. Thus, when they swim, propelled by their two flagellae, they move upward on average. However, in a region of shear (e.g., a vertical flow that varies horizontally), the cells experience a viscous torque that will rotate them away from the vertical. This leads to some striking phenomena, such as accumulation at the center of a pipe in a downward Poiseuille flow. Again, collective bioconvective phenomena can then emerge. In this case, continuum models track the mean swimming direction, represented by a unit vector $\boldsymbol{p}(\boldsymbol{x},t)$. A Fokker–Planck equation for the probability density $\Omega(\boldsymbol{x},\boldsymbol{p},t)$ (a distribution over position and orientation) can be used to account for reorientation of cells by hydrodynamic torques and rotational diffusion, which is coupled to transport equations for $n$ and $\boldsymbol{u}$.

*Escherichia coli* propel themselves from the rear, with the propulsive force being balanced by the resistive drag toward the front of the body. When viewed from afar, the generated flow can be approximated by that of a force dipole (point forces directed in opposite outward directions) in an arrangement known as a *pusher* (figure 2). The two forces, pushing fluid away from the body parallel to its axis, draw fluid inward from the sides. In contrast, *Chlamydomonas nivalis* is propelled by flagellae at the front of the body, with the drag arising behind. Viewed from afar, these are therefore *pullers*; the force dipole pushes fluid out sideways from the body. In a straining flow that tends to orient the swimmer in a particular direction, pushers will generate a flow that reinforces their orientation but pullers will do the opposite. A suspension of pushers is therefore more likely to exhibit large-scale collective motion known as *bacterial turbulence*, driven by the chemical energy that powers the swimming organisms. This has motivated the development of model equations that draw on earlier studies of flocking behavior, including extensions of the Navier–Stokes equations

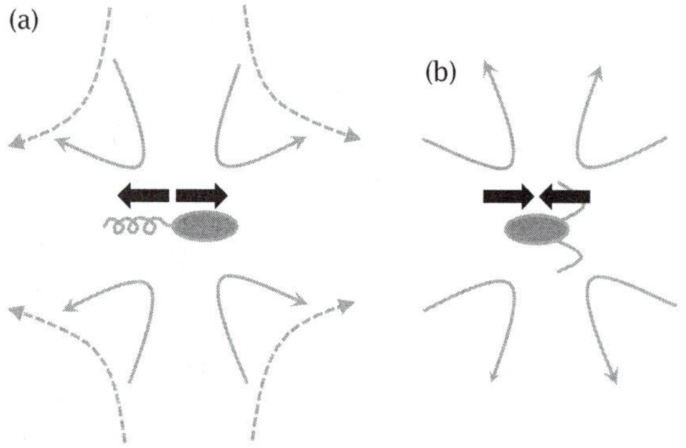


**Figure 2** Two classes of swimming microorganisms: (a) a pusher and (b) a puller. Short arrows show the force dipole, solid arrows show the flow induced by swimming, and dashed arrows show an external straining flow that orients the swimmer.

that incorporate additional nonlinear and higher-order terms that capture a variety of spatiotemporal patterns.

## 6 Coda

This brief summary of mathematical modeling in biomechanics has necessarily been selective. The increasingly vigorous engagement of mathematicians with biologists, bioengineers, and biophysicists is yielding significant advances in many underexplored areas, generating novel mathematical questions alongside new insights into biomechanical processes. Interactions between mathematics and computation have been necessary and effective, particularly in modeling the interaction of diverse processes spanning disparate length scales. However, major challenges remain in learning how to cope with sparse data, natural variability, and disorder—challenges that will require increasing use of statistical and probabilistic approaches.

### Further Reading

Audoly, B., and A. Boudaoud. 2003. Self-similar structures near boundaries in strained systems. *Physical Review Letters* 91:086105.

Fedosov, D. A., W. Pan, B. Caswell, G. Gompper, and G. E. Karniadakis. 2011. Predicting human blood viscosity in silico. *Proceedings of the National Academy of Sciences of the USA* 108:11772–77.

Fung, Y. C. 1993. *Biomechanics: Mechanical Properties of Living Tissues.* Berlin: Springer.

Koch, D., and G. Subramanian. 2011. Collective hydrodynamics of swimming microorganisms: living fluids. *Annual Review of Fluid Mechanics* 43:637–59.

Thompson, D. W. 1917. *On Growth and Form.* Cambridge: Cambridge University Press.

Wensink, H. H., J. Dunkel, S. Heidenreich, K. Drescher, R. E. Goldstein, H. Löwen, and J. M. Yeomans. 2012. Meso-scale turbulence in living fluids. *Proceedings of the National Academy of Sciences of the USA* 109:14308–13.

# V.5 Mathematical Physiology

*Anita T. Layton*

## 1 Introduction

Mathematics and physiology have enjoyed a long history of interaction. One aspect in which mathematics has proved useful is in the search for general principles in physiology, which involves organizing and describing the large amount of data available in more comprehensible ways. Mathematics is also helpful in the search for emergent properties, that is, in the identification of features of a collection of components that are not features of the individual components that make up the collection. This article considers a small selected set of mathematical models in physiology, showing how physiological problems can be formulated and studied mathematically and how such models give rise to interesting and challenging mathematical questions.

## 2 Cellular Physiology

The cell is the basic structural and functional unit of a living organism. Collectively, cells perform numerous functions to sustain life. Those functions are accomplished through the biochemical reactions that take place within the cell.

### 2.1 Biochemical Reactions

Chemical reactions are "governed" by the *law of mass action*, which describes the rate at which chemicals interact to form products. Suppose that two chemicals A and B react to form C:

$$\mathrm{A} + \mathrm{B} \xrightarrow{k} \mathrm{C}.$$

The rate of formation of the product C is proportional to the product of the concentrations of A and B, as well as the rate constant $k$, i.e.,

$$\frac{\mathrm{d}[\mathrm{C}]}{\mathrm{d}t} = k[\mathrm{A}][\mathrm{B}]. \tag{1}$$

For a reversible reaction

$$\mathrm{A} + \mathrm{B} \underset{k_-}{\overset{k_+}{\rightleftharpoons}} \mathrm{C},$$

where $k_+$ and $k_-$ denote the forward and reverse rate constants of reaction, the rate of change of [A] is given by

$$\frac{d[A]}{dt} = k_-[C] - k_+[A][B].$$

If $k_- = 0$, then the above reaction reduces to (1) with $k = k_+$ and $d[A]/dt = -d[C]/dt$ (from the reaction, the rate of consumption of A is equal to the rate of production of C).

Some important biochemical reactions are catalyzed by enzymes, many of which are proteins that help convert substrates into products but themselves emerge from the reactions unchanged. Consider the following simple reaction scheme, where an enzyme E converts a substrate A into the product B through a two-step process:

$$A + E \underset{k_{-1}}{\overset{k_1}{\rightleftharpoons}} C \overset{k_2}{\longrightarrow} B + E.$$

First, E combines with A to form a complex C, which then breaks down into the product B and releases E. This model is known as Michaelis-Menten kinetics.

We will determine the reaction rate, given by the rate at which the product is formed (i.e., $d[B]/dt$), by performing an equilibrium approximation. Let us assume that the substrate A is in instantaneous equilibrium with the complex C; thus,

$$k_1[A][E] = k_{-1}[C]. \quad (2)$$

Note that the total amount of enzyme is unchanged, i.e., $[E] + [C] = E_0$, where $E_0$ is a constant. Substituting this relation into (2) and rearranging yields

$$[C] = \frac{E_0[A]}{K_m + [A]},$$

where $K_m = k_{-1}/k_1$ is known as the Michaelis constant, which is the substrate concentration at which the reaction rate is half of its maximum. The overall reaction rate is given by

$$\frac{d[B]}{dt} = k_2[C] = \frac{V_{max}[A]}{K_m + [A]},$$

where $V_{max} = k_2 E_0$ is the maximum reaction rate. The above relation describes Michaelis-Menten kinetics, which is one of the simplest and best-known models of enzyme kinetics, named after biochemist Leonor Michaelis (1875-1949) and Canadian physician Maud Menten (1879-1960). A sample reaction curve is shown in figure 1.

## 2.2 Membrane Ion Channels

All animal cells are surrounded by a membrane composed of a lipid bilayer with proteins embedded in it.

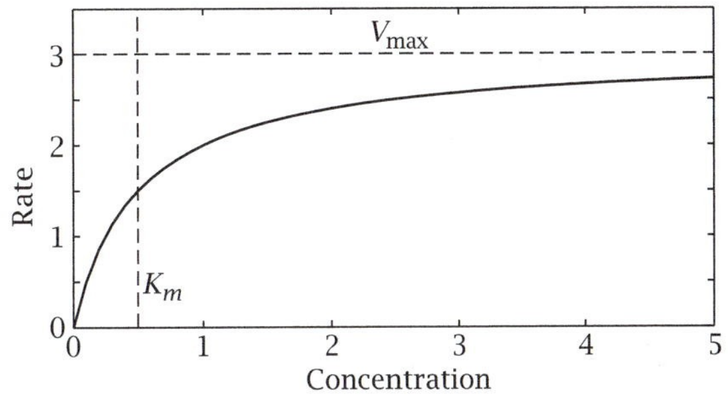


**Figure 1** An example Michaelis-Menten reaction rate curve, showing reaction rate as a function of substrate concentration, with parameters $V_{max} = 3.5$ and $K_m = 0.5$.

Found on most cellular membranes are ion channels, which are macromolecular pores that passively carry specific ions across the membrane. One of the driving forces for transport is ionic concentration gradient. Another is the transmembrane potential difference, which is the difference in electrical potential between the interior and exterior of the cell.

Ion channels are characterized by their current-voltage relationship, the parameters of which depend on the biophysical properties of the channel. The current $I$ across a population of $N$ channels can be written as

$$I = NP_o(V,t)i(V,t),$$

where $P_o$ is the fraction of open channels at time $t$ ($0 \leqslant P_o \leqslant 1$), and $i$ is the current across a single open channel. Note that the above expression assumes that both are functions of the transmembrane potential difference $V$. Below we describe common models for $i(V,t)$ and, subsequently, for $P_o(V,t)$.

The two most common models of current-voltage relationships are the modified form of Ohm's law and the Goldman-Hodgkin-Katz equation. The modified Ohm's law states that the flow of ions S across the membrane is a linear function of the difference between the membrane potential and the equilibrium (or Nernst) potential $E$ of the solute S. It can be written as

$$i_S = g_S(V - E),$$

where $g_S$ is the channel conductance, which is not necessarily constant. Indeed, it can vary with concentration, voltage, or other factors.

As an example, the conductance of inward rectifying potassium channels (Kir), which play an important role in maintaining the resting membrane potential, is a function of $V$. The activity of Kir channels is also modulated by the extracellular $K^+$ concentration ($C_K^e$).

Experimentally, the Kir conductance is often found to increase with the square root of $C_K^e$. Thus, the current across a Kir channel, denoted by $i_{Kir}$, can be expressed as

$$i_{Kir} = g_{Kir}(V - E_K),$$

$$g_{Kir} = \frac{g_{Kir}^{\circ}\sqrt{C_K^e}}{1 + \exp((V - V_{Kir})/k_{Kir})},$$

where $g_{Kir}^{\circ}$ is a constant, $V_{Kir}$ is the half-activation potential, and $k_{Kir}$ is a slope. The dependence of $g_{Kir}$ on the membrane potential stems from the presence of gating charges.

Other channels are better represented by the Goldman–Hodgkin–Katz current equation, which can be obtained by integrating the Nernst–Planck equation assuming a constant electric field across the membrane. The Nernst–Planck equation gives the flux of ion S when both concentration and electrical gradients are present:

$$J_S = -D\left(\nabla C_S + \frac{z_S F}{RT} C_S \nabla\psi\right).$$

The first term corresponds to Fick's law, a constitutive equation that describes diffusion driven by a concentration gradient; $D_S$ is the diffusivity of S. The second term corresponds to diffusion due to the electric field; the electric force exerted on a charged solute increases linearly with the electric field ($E = -\nabla\psi$). To derive the Goldman–Hodgkin–Katz equation for the simple one-dimensional case, we assume that the electrical field is constant across the membrane (i.e., $\mathrm{d}\psi/\mathrm{d}x = -V/L_m$, where $L_m$ denotes the length of the membrane) and obtain

$$J_S = -D_S\left(\frac{\mathrm{d}C_S}{\mathrm{d}x} - \frac{\xi}{L_m} C_S\right),$$

where $\xi = z_S F V_m/RT$. Since $J_S$ is constant, the above equation can be transformed into

$$\frac{\mathrm{d}C_S}{\mathrm{d}x} = \frac{\xi}{L_m} C_S - \frac{J_S}{D_S}$$

and then integrated to yield $C_S(x)$:

$$C_S(x) = a \exp\left(\frac{\xi x}{L_m}\right) + \frac{J_S L_m}{\xi D_S}.$$

To determine the integration constant $a$ and the unknown $J_S$, we use the two boundary conditions

$$C_S(0) = a + \frac{J_S L_m}{\xi D_S} = C_S^i,$$

$$C_S(L_m) = a \exp(\xi) + \frac{J_S L_m}{\xi D_S} = C_S^e,$$

where $C_S^i$ and $C_S^e$ denote the solute concentrations on the two sides of the membrane. These boundary conditions yield the Goldman–Hodgkin–Katz current equation

$$a = \frac{C_S^i - C_S^e}{1 - \exp(\xi)},$$

$$J_S = \frac{D_S}{L_m} \frac{z_S F V}{RT} \frac{C_S^e - C_S^i \exp(-z_S F V/RT)}{1 - \exp(-z_S F V/RT)},$$

where the definition of $\xi$ has been applied.

From the Goldman–Hodgkin–Katz equation one can obtain the Nernst potential, which yields zero flux when the transmembrane concentration gradient is nonzero, by setting $J_S$ to zero and solving for $V$:

$$V = -\frac{RT}{z_S F} \log\left(\frac{C_S^i}{C_S^e}\right). \tag{3}$$

As noted above, the current flowing across a population of channels is proportional to the number of open channels ($NP_o$). In excitable tissues such as smooth muscles, channels open in response to stimuli, such as adenosine-5′-triphosphate (a nucleoside triphosphate used in cells as a coenzyme, often called the "molecular unit of currency" of intracellular energy transfer), $Ca^{2+}$ concentration, or voltage. The simplest channel model assumes that the channel is either in the closed state C or in the open state O,

$$\mathrm{C} \overset{\alpha/\beta}{\longleftrightarrow} \mathrm{O},$$

where $\alpha$ and $\beta$ denote the rates of conversion from one state to the other. If $n$ is the fraction of channels in the open state, then $1 - n$ is the fraction of channels in the closed state and we have

$$\frac{\mathrm{d}n}{\mathrm{d}t} = \alpha(1 - n) - \beta n = \alpha - (\alpha + \beta)n.$$

The characteristic time of this equation is $\tau = 1/(\alpha+\beta)$, and the steady-state value of $n$ is $n_\infty = \alpha/(\alpha + \beta)$. The equation can be rewritten in terms of $\tau$:

$$\frac{\mathrm{d}n}{\mathrm{d}t} = \frac{n_\infty - n}{\tau}.$$

In general, $n_\infty$ and $\tau$ are voltage or concentration dependent, so the above equation cannot be solved analytically. Voltage-sensitive channels respond to electric potential variations; their voltage sensors consist of "gating" charges that move when $V$ is altered, thereby resulting in a conformational change. For these channels, $n_\infty$ can be described by a Boltzmann distribution:

$$n_\infty = \frac{1}{1 + k_o \exp(bFV/RT)},$$

where $b$ is a constant related to the number of gating charges on the channel and the distance over which they move during a conformational change.

## 3 Excitability

Membrane potential is used as a signal in cells such as muscle cells and neurons. Some of these cells are *excitable*. If a sufficiently strong current is applied, the membrane potential goes through a large excursion known as an *action potential* before eventually returning to rest. We will study one example of cellular excitability: the Morris–Lecar model.

The Morris–Lecar model is a two-dimensional "reduced" excitation model (reduced from the four-dimensional Hodgkin–Huxley model) that is applicable to systems having two noninactivating voltage-sensitive conductances. The model grew out of an experimental study of the excitability of the giant muscle fiber of the huge Pacific barnacle, *Balanus nubilus*. Synaptic depolarization of these muscle cells leads to the opening of $Ca^{2+}$ channels, allowing external $Ca^{2+}$ ions to enter deep into the cell interior to activate muscle contraction. The Morris–Lecar model describes this membrane with two conductances, $Ca^{2+}$ and $K^+$, the interplay of which yields qualitative phase portrait changes with small changes in experimental parameters, such as the relative densities of $Ca^{2+}$ and $K^+$ channels or the relative relaxation times of the conducting systems. This simple model is capable of simulating the entire panoply of (two-dimensional) oscillation phenomena that have been observed experimentally. Parameter maps in phase space can then be drawn to identify and classify the parametric regions having different types of stability.

The model equations describe the membrane potential ($V$) and the fraction of open potassium channels ($n$):

$$C\frac{\mathrm{d}V}{\mathrm{d}t} = -g_{\mathrm{L}}(V - V_{\mathrm{L}}) - g_{\mathrm{Ca}}m_{\infty}(V)(V - V_{\mathrm{Ca}}) - g_{\mathrm{K}}n(V - V_{\mathrm{K}}), \quad (4)$$

$$\frac{\mathrm{d}n}{\mathrm{d}t} = \phi_n \cosh\left(\frac{V - V_3}{2V_4}\right)(n_{\infty}(V) - n). \quad (5)$$

The first term on the right-hand side of (4) is the leak current, where $V_{\mathrm{L}}$ denotes the associated Nernst reversal potential. The second term is the calcium current, where $g_{\mathrm{Ca}}$ is the maximum whole-cell membrane conductance for the calcium current. This calcium current arises from the voltage-operated calcium channels, where $m_{\infty}$ represents the fraction of open channel states at equilibrium. Based on experimental data, $m_{\infty}$ is described as a function of membrane potential $v$ by

$$m_{\infty}(V) = 0.5\left(1 + \tanh\left(\frac{V - V_1}{V_2}\right)\right),$$

**Table 1** Model parameters used in the bifurcation study.

| | | | |
|---|---|---|---|
| $C$ | 20 pF | $V_1$ | −1.2 mV |
| $g_{\mathrm{L}}$ | 2 nS | $V_2$ | 18 mV |
| $g_{\mathrm{Ca}}$ | 4 nS | $V_{\mathrm{K}}$ | −85 mV |
| $g_{\mathrm{K}}$ | 8 nS | $v_3$ | 12 |
| $V_{\mathrm{L}}$ | −60 mV | $v_4$ | 17 |
| $V_{\mathrm{Ca}}$ | 120 mV | $\phi_n$ | 0.06666667 |

where $V_1$ is the voltage at which half of the channels are open and $V_2$ determines the spread of the distribution of open calcium channels at steady state. For very negative $V$, $\tanh((V - V_1)/V_2) \to -1$ and $m_{\infty} \to 0$, which implies that almost all calcium channels are closed. For large (more positive) $V$, $\tanh((V - V_1)/V_2) \to 1$ and $m_{\infty} \to 1$, which implies that most calcium channels are now open.

The third term in (4), $-g_{\mathrm{K}}n(V - V_{\mathrm{K}})$, represents the transmembrane potassium current induced by the opening of potassium channels. In (5), $n_{\infty}$ denotes the fraction of open $K^+$ channel at steady state; this fraction depends on the membrane potential $V$ through

$$n_{\infty}(V) = 0.5\left(1 + \tanh\left(\frac{V - V_3}{V_4}\right)\right),$$

which has a form similar to the equilibrium distribution of open $Ca^{2+}$ channel states ($m_{\infty}$). The potential $V_3$ determines the voltage at which half of the potassium channels are open; $V_4$ and $Ca_4$ are measures of the spread of the distributions of $n_{\infty}$ and $V_3$, respectively. Note also that both the potassium current (which we denote by $I_{\mathrm{K}} = -g_{\mathrm{K}}n(V - V_{\mathrm{K}})$) and the calcium current (denoted by $I_{\mathrm{Ca}} = -g_{\mathrm{Ca}}m_{\infty}(V)(V - V_{\mathrm{Ca}})$) depend on the membrane potential $V$, and both currents in turn change the membrane potential. Parameters for this model are given in table 1.

The model predicts drastically different behaviors depending on the initial conditions. When the membrane potential $v$ is initialized to −20 mV and $n$ is initialized to 0, the voltage and the currents $I_{\mathrm{K}}$ and $I_{\mathrm{Ca}}$ decay to rest (see figure 2). Qualitatively different behaviors are predicted with $v(0) = -10$ mV: the voltage rises substantially before decaying to rest (see figure 3(a)). To gain insight, one may consider the currents, which are shown in figure 3(b). The magnitude of both currents increases before decaying to rest. An interesting feature of this system is that the two currents go in opposite directions: $I_{\mathrm{K}}$ is outward-directed and thus hyperpolarizes the cell, whereas $I_{\mathrm{Ca}}$ is inward-directed and depolarizes the cell. Because the calcium channels are voltage gated, once the voltage crosses a

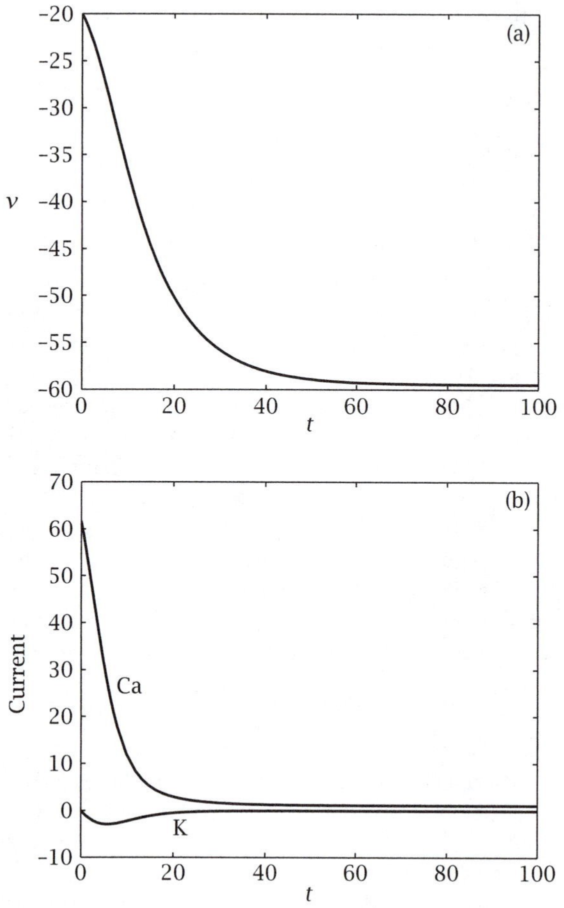


**Figure 2** Solution to the Morris–Lecar model with $v$ initialized to $-20$ mV: (a) voltage; (b) $K^+$ and $Ca^{2+}$ currents. The voltage and current both decay to zero.

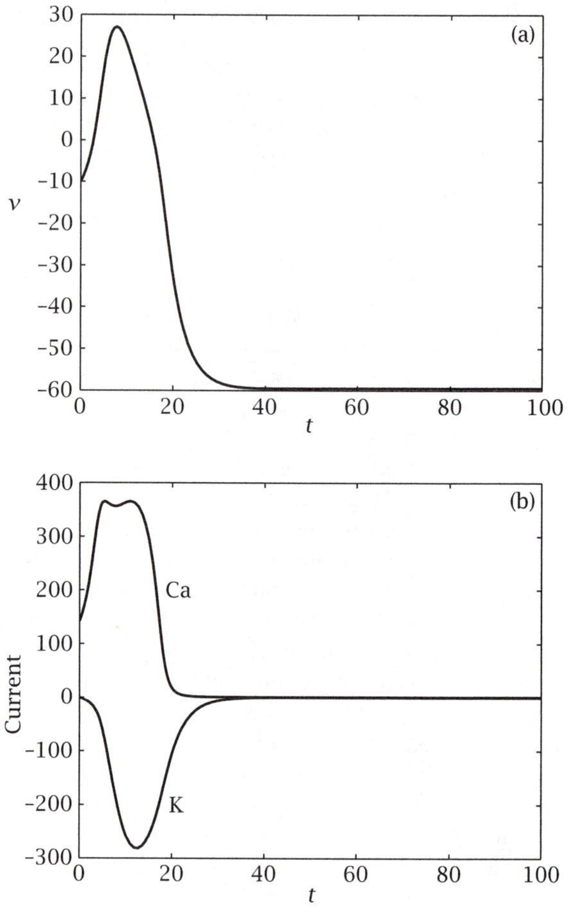


**Figure 3** Solution to the Morris–Lecar model with $v$ initialized to $-10$ mV: (a) voltage; (b) $K^+$ and $Ca^{2+}$ currents. The solution exhibits an initial bump (or crest) before decaying to rest.

threshold there is a large inward-directed calcium current. (This is therefore an example of an excitable cell.) The current causes the voltage to increase, and eventually the current slows down and reaches its peak. The large outward potassium current then comes into play, repolarizing the cell.

To understand this threshold behavior for excitable cells, one may study the nullclines of the system, which are curves where $dv/dt = 0$ and $dn/dt = 0$. The nullclines are depicted in figure 4. The first thing to note is that the two nullclines intercept at three equilibrium points. If the system is initialized at those $v$ and $n$ values, then the system will remain at equilibrium. The two nullclines divide the $v$-$n$ space into several regions, in which $v$ and $n$ are increasing or decreasing. The two previous simulations that produce qualitatively different behaviors correspond to initial $v$ and $n$ values that lie in different regions. With $v(0) = -20$ mV and $n(0) = 0$ (indicated by the asterisk in figure 4), the system lies in the region where $dv/dt < 0$ and $dn/dt > 0$, so the voltage decays to rest (see figure 2(a)), whereas the gating variable $n$, after an initial increase, also decays to 0. With $v(0) = -10$ mV (indicated by the open circle in figure 4), however, the system lies in the region where $dv/dt > 0$ and $dn/dt > 0$, so the voltage rises to a peak before decaying to rest (see figure 3(a)).

## 4 Kidneys

A group of cells that have similar function or structure may join together to form tissues. A group of tissues

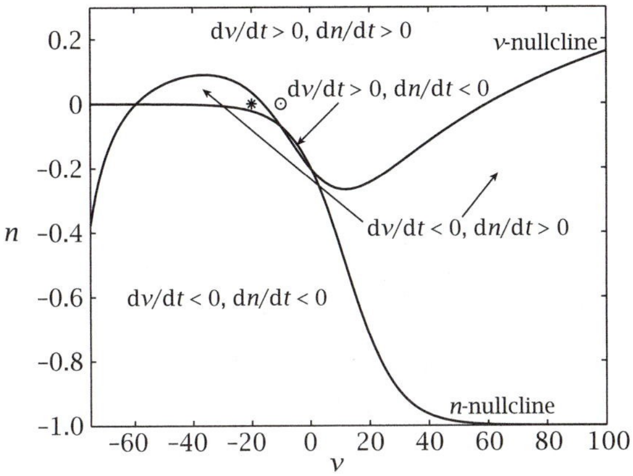


**Figure 4** The $v$- and $n$-nullclines divide the $v$–$n$ space into regions where $v$ and $n$ are increasing or decreasing.

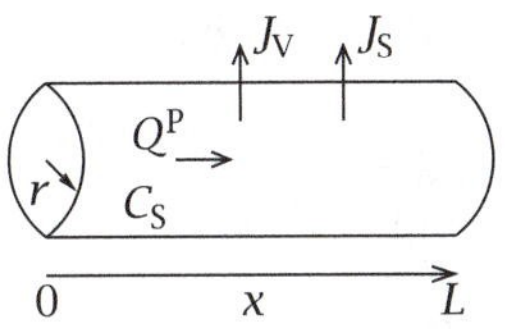


**Figure 5** Schematic of a model glomerular capillary.

may then collectively form an organ. A collection of organs then forms organ systems, an example of which is the excretory system.

The kidneys are part of the excretory system. They serve a number of essential regulatory roles. Besides their well-known function as filters, removing metabolic wastes and toxins from the blood and excreting them through urine, the kidneys also serve other essential functions. Through a number of regulatory mechanisms, the kidneys help maintain the body's water balance, electrolyte balance, and acid–base balance. Additionally, the kidneys produce or activate hormones that are involved in erythrogenesis, calcium metabolism, and the regulation of blood flow.

### 4.1 Glomerular Filtration

The first step in the formation of urine is the filtration of blood by glomerular capillaries. The filtrate is collected into a Bowman's capsule and subsequently flows through the tubular system, where it undergoes major changes in volume and composition.

To model the glomerular fluid filtration process, the glomerular capillaries are idealized as a network of identical, parallel capillaries with homogeneous properties along their entire length. The capillaries are represented as rigid cylinders of radius $r$ and length $L$ (figure 5). Let $P$ denote the plasma compartment and let $Q^P$ denote the volumetric rate of plasma flow in a capillary. At a given position $x$ along the capillary, a portion of the flow can be reabsorbed into (or secreted from) the surrounding medium via a transversal flux across the capillary wall. Assuming that the capillaries are rigid, there is no accumulation of fluid within the plasma compartment. Thus, for a single capillary, the conservation of fluid can be written as

$$\frac{dQ^P}{dx} = -2\pi r J_V,$$

where $J_V$ is the plasma volume flux, expressed as flow per unit of capillary area and taken to be positive for outward-directed flux. Fluid flow across a semipermeable membrane such as the capillary wall is driven by the hydrostatic, oncotic, and osmotic pressure differences:

$$J_V = L_p\Big(\Delta P - \Delta\Pi - RT\sum_S \sigma_S \Delta C_S\Big),$$

where $L_p$ is the hydraulic conductivity of the membrane; $\Delta P$ is the transmembrane difference in hydrostatic pressure; $\Delta\Pi$ is the oncotic pressure, due to the contribution of nonpenetrating solutes such as proteins; $R$ and $T$ are the gas constant and absolute temperature, respectively; $\sigma_S$ is the osmotic reflective coefficient of the membrane to solute S; and $\Delta C_S$ is the transmembrane concentration gradient (outside minus inside) of solute S.

To apply the above equation, which was derived for a single capillary, to one glomerulus, or to all the glomeruli of one or two kidneys, one writes

$$\frac{dQ^P}{dx} = -\frac{S^P}{L} J_V,$$

where $S^P$ denotes capillary surface area, $L$ is the capillary length, and the definition of plasma flow rate ($Q^P$) must be adjusted accordingly.

To determine water flux $J_V$, one needs to track the concentrations of small solutes and protein. To that end, consider the conservation of a solute S in plasma:

$$\frac{d(Q^P C_S^P)}{dx} = -2\pi r J_S,$$

where $J_S$ is the plasma flux of species S, expressed as a molar flow per unit of capillary area. Note that the product $Q^P C_S^P$ gives the plasma flow rate of solute S, in moles per unit time.

The transmembrane flux of an uncharged solute is given by the Kedem–Katchalsky equation

$$J_S = J_V(1 - \sigma_S)\bar{C}_S + P_S\Delta C_S,$$

where the first term represents the contribution from advection and the second term that from diffusion. $P_S$ is the permeability of the membrane to solute S, and $\bar{C}_S$ is an average membrane concentration, which in dilute solutions is given by

$$\bar{C}_S = \frac{\Delta C_S}{\Delta \ln C_S}.$$

Some proteins such as albumin are highly concentrated in plasma and thus exert a significant oncotic pressure, which opposes fluid filtration across the capillary wall. In healthy kidneys the fraction of plasma proteins that are filtered along the capillaries is negligible. Thus, if $C^{\mathrm{P}}_{\mathrm{pr}}$ denotes the plasma concentration of proteins, we have

$$\frac{\mathrm{d}(Q^{\mathrm{P}}C^{\mathrm{P}}_{\mathrm{pr}})}{\mathrm{d}x} = 0.$$

In summary, the set of equations that form the basis of models of glomerular filtration consist of equations that describe conservation of fluid and of solutes (small solutes and proteins). The boundary conditions are specified at the afferent end of the capillary:

$$Q^{\mathrm{P}}(x = 0) = Q^{\mathrm{A}}, \qquad C^{\mathrm{P}}_{\mathrm{pr}}(x = 0) = C^{\mathrm{A}}_{\mathrm{pr}},$$

where $Q^{\mathrm{A}}$ equals the afferent arteriolar plasma flow and $C^{\mathrm{A}}_{\mathrm{pr}}$ equals the total protein concentration in the afferent plasma. A similar boundary condition is imposed for solute S.

### 4.2 Urinary Concentration

During water deprivation, the kidney of a mammal can conserve water by producing urine that is more concentrated than blood plasma. This hypertonic urine is produced when water is reabsorbed, in excess of solutes, from the nephrons and into the renal vasculature, thereby concentrating the tubular fluid, which eventually emerges as urine. Here we will develop a model of the concentrating mechanism in an important segment of the renal tubular system, the loop of Henle. The loop of Henle is a hairpin-like tubule that lies mostly in the medulla and consists of a descending limb and an ascending limb (see figure 6).

The model represents a loop with a descending limb and an ascending limb. The two limbs are assumed to be in direct contact with each other. We make the simplifying assumption that the descending limb is water impermeable but infinitely permeable to solute. This results in the conservation equations

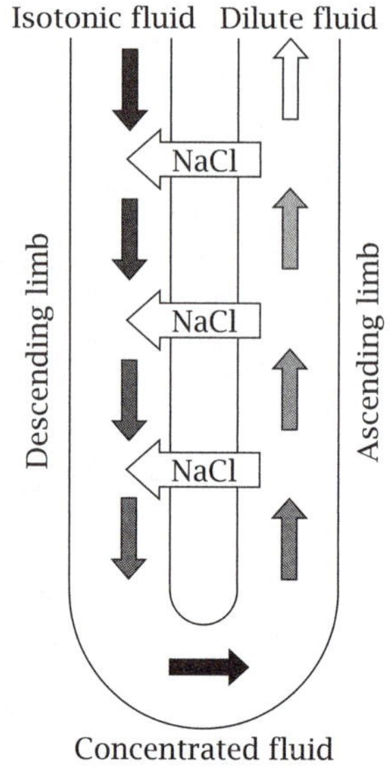


**Figure 6** Countercurrent multiplication by NaCl transfer from an ascending flow to a descending flow; the concentration of the descending flow is progressively concentrated by NaCl addition.

$$\frac{\partial}{\partial x}Q_{\mathrm{DL}}(x) = 0,$$

$$\frac{\partial}{\partial x}(Q_{\mathrm{DL}}(x)C_{\mathrm{DL}}(x)) = -2\pi r_{\mathrm{DL}}J_{\mathrm{DL},s}.$$

The notation is similar to the glomerular filtration model, with the subscript "DL" denoting the descending limb.

We assume that the ascending limb is water impermeable, and that the solute is pumped out of the ascending limb at a fixed rate $A$. (A more realistic description of this active transport would be the Michaelis–Menten kinetics discussed in section 2.1. Simplistic, fixed-rate active transport is assumed here to facilitate the analysis.) Additionally, we assume that all of that solute goes into the descending limb. Thus, $2\pi r_{\mathrm{DL}}J_{\mathrm{DL},s} = -A$ and $2\pi r_{\mathrm{AL}}J_{\mathrm{AL},s} = A$. The conservation equations for the ascending limb (denoted "AL") are

$$\frac{\partial}{\partial x}Q_{\mathrm{AL}}(x) = 0, \qquad \frac{\partial}{\partial x}(Q_{\mathrm{AL}}(x)C_{\mathrm{AL}}(x)) = -A.$$

Because the descending and ascending limbs are assumed to be contiguous, at the loop bend ($x = L$) we have

$$Q_{\mathrm{DL}}(L) = -Q_{\mathrm{AL}}(L), \qquad C_{\mathrm{DL}}(L) = C_{\mathrm{AL}}(L).$$

Finally, to complete the system, boundary conditions are imposed at the entrance to the descending limb:

$$Q_{\mathrm{DL}}(0) = Q_0, \qquad C_{\mathrm{DL}}(0) = C_0.$$

To determine $C_{\mathrm{DL}}(x)$ and $C_{\mathrm{AL}}(x)$, we note that, since the entire loop is water impermeable, $Q_{\mathrm{DL}}(x) = Q_{\mathrm{AL}}(x) = Q_0$. Thus,

$$Q_0 \frac{\partial}{\partial x} C_{\mathrm{DL}}(x) = A,$$

which can be integrated to yield

$$C_{\mathrm{DL}}(x) = C_0 + \left(\frac{A}{Q_0}\right)x.$$

To compute $C_{\mathrm{AL}}(x)$, we evaluate $C_{\mathrm{DL}}$ at $x = L$ and use that as the initial condition for the ordinary differential equation for $C_{\mathrm{AL}}(x)$ to get

$$C_{\mathrm{AL}}(x) = C_{\mathrm{DL}}(L) - \left(\frac{A}{Q_0}\right)(L - x) = C_{\mathrm{DL}}(x).$$

The simple model illustrates the principle of *countercurrent multiplication*, by which a transfer of solute from one tubule to another (called a "single" effect) augments ("multiplies," or reinforces) the axial osmolality gradient in the parallel flow. Thus, a small transverse osmolality gradient difference (a small single effect) is multiplied into a much larger osmolality difference along the axis of tubular flow. To summarize, the mode predicts that the concentrations along both limbs are the same at any given $x$, and that solute concentration increases linearly along $x$. Thus, the longer the loop, the higher the loop bend concentration.

### Further Reading

Due to space constraints, this article focuses on cellular transport and the kidneys. Mathematical models of other aspects of the kidney can be found in Layton and Edwards (2013). Mathematical models have been formulated for other organ systems, including the circulatory system, the digestive system, the endocrine system, the lymphatic system, the muscular system, the nervous system, the reproductive system, and the respiratory system. For these models, see Keener and Sneyd (2008).

Keener, J., and J. Sneyd. 2009. *Mathematical Physiology, Volume I: Cellular Physiology* and *Mathematical Physiology; Volume II: Systems Physiology*. New York: Springer.

Layton, A., and A. Edwards. 2013. *Mathematical Modeling of Renal Physiology*. New York: Springer.

## V.6 Cardiac Modeling

*Alexander V. Panfilov*

### 1 The Heart, Waves, and Arrhythmias

The main physiological function of the heart is mechanical: it pumps blood through the body. The pumping is controlled by electrical excitation waves, which propagate through the heart and initiate cardiac contraction. Anatomically, the human heart consists of four chambers. The two lower chambers, which are called ventricles, have thick (1–1.5 cm) walls, and it is contraction of the ventricles that pushes blood through the body. The upper two chambers, called the atria, have thin walls (about 0.3 cm thick). The atria collect the blood, and their contraction delivers it to the ventricles. Under normal conditions, excitation of the heart starts at the sinoatrial node located in the right atrium (figure 1). The cells in the sinoatrial node are oscillatory, and they periodically initiate excitation waves. Thereafter, the wave propagates through the two upper chambers of the heart (the atria), causing atrial contraction. After some delay at the atrioventricular (AV) node, the excitation enters the ventricles and initiates the main event: ventricular contraction.

Abnormal excitation of the heart results in cardiac arrhythmias, which have various manifestations. They might involve just one extra heartbeat initiated by a wave from another location, or they could manifest as an abnormally fast heart rate called tachycardia, which can occur in the atria or the ventricles. In some cases the excitation becomes spatially disorganized, which results in failure of contraction. If this occurs in the ventricles, cardiac arrest and sudden cardiac death result. Prediction and management of cardiac arrhythmias are therefore two of the greatest problems in modern cardiology. Sudden cardiac death due to arrhythmias is one of the largest causes of death in the industrialized world, accounting for approximately one in every ten deaths. Cardiac arrhythmias have important consequences in the pharmaceutical industry too: more than half of all drug withdrawals in recent years have been the result of the drugs in question potentiating the onset of arrhythmias or causing other cardiac side effects.

Mechanisms of the most dangerous cardiac arrhythmias are directly related to wave propagation. The properties of waves in the heart are quite different from those of most other types of nonlinear waves, the main

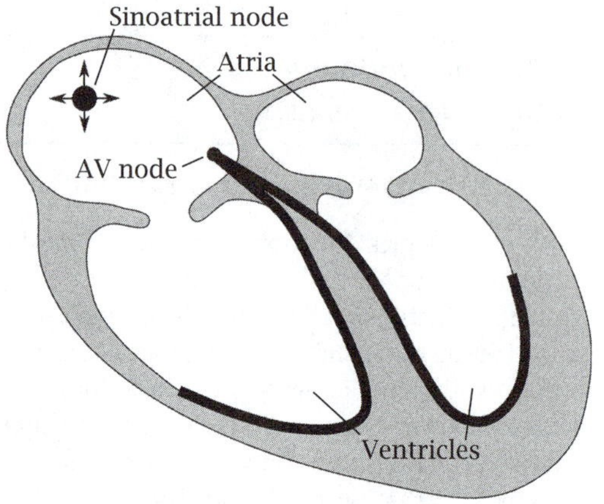


**Figure 1** Schematic of the heart's conduction pathway.

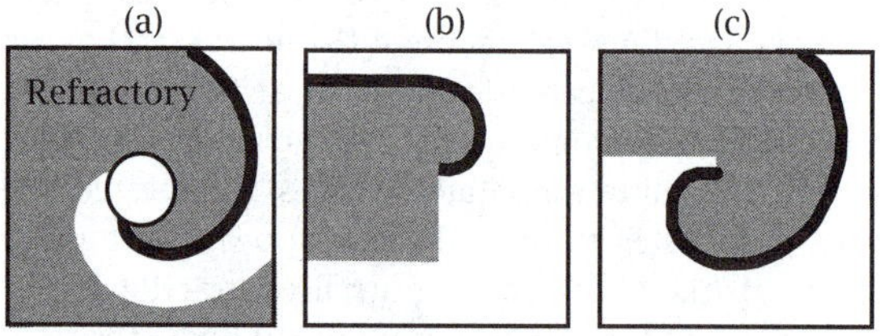


**Figure 2** Schematic of (a) a wave propagating around an obstacle and (b), (c) spiral wave formation.

difference being the presence of refractoriness. After excitation, a cardiac cell requires some time, called the refractory period, to recover its properties. During that refractory time it cannot be excited again, and vortices may appear in the heart as a result. Let us consider a wave propagating along an obstacle in cardiac tissue: around veins or arteries entering the heart, for example. If the travel time around the obstacle is longer than the refractory period, a sustained rotation (figure 2(a)) will take place. Such rotation produces periodic excitation of the heart with a frequency much faster than the frequency of the sinus node, and tachycardia occurs as a result.

It has also been shown that, because of refractoriness, such rotations can also occur without any anatomical obstacles. Refractory tissue is unexcitable, and the wave can propagate around it as it would do around anatomical obstacles (figure 2(b)). The only difference is that, as soon as the refractory period ends, the wave can enter this region (figure 2(c)). The wavefront therefore normally follows the refractory tail of the wave closely, and the period of rotation is very short. Such vortices are usually called spiral waves

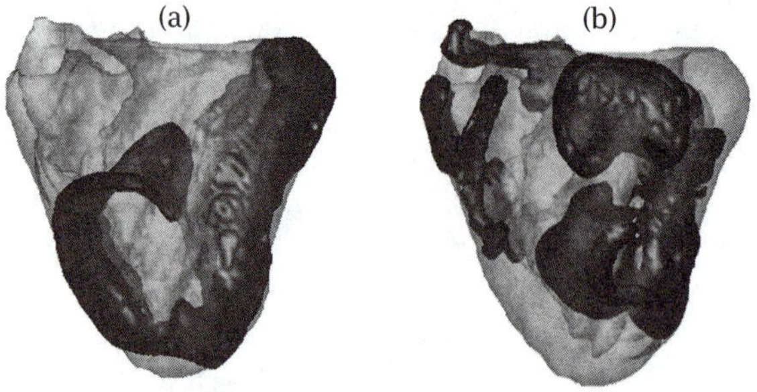


**Figure 3** (a) Spiral wave and (b) complex excitation pattern after the process of spiral breakup in an anatomical model of human ventricles. (Figure created by I. V. Kazbanov and A. V. Panfilov.)

because in large media they take the shape of rotating spirals (figure 3(a)). Other interesting types of dynamics include a breakup of spiral waves into complex spatiotemporal chaos (figure 3(b)). If this occurs in the heart, excitation becomes spatially disorganized, resulting in failure of organized cardiac contraction and sudden cardiac death.

Overall, we can say that the sources of abnormal excitation of the heart are not located in a single region but are the result of wave circulation involving millions of cardiac cells. Thus, if we want to understand the mechanisms of cardiac arrhythmias, we need to know how changes at the level of the individual building block of the heart (i.e., the single cell) will manifest themselves at the whole-organ level. It is in answering precisely these questions that mathematical modeling can be useful. Let us consider the main principles behind cardiac models.

### 1.1 Modeling Cardiac Cells

In the resting state, there is a difference in the potential between the two sides of the membrane of about −90 mV in cardiac cells. During excitation this voltage rapidly increases to about +10 mV before slowly returning to its resting value (figure 4(a)). A typical cardiac action potential is shown in figure 4. It has a sharp upstroke that lasts only 1–2 ms and then a repolarization phase of about 300 ms.

The voltage across the cardiac membrane changes as a result of the complex time dynamics of ionic currents through the membrane. The currents are conveyed by selective ionic channels that are permeable to various ions, the most important of which are $Na^+$, $K^+$, and $Ca^{2+}$. The process of excitation of the cardiac cell can

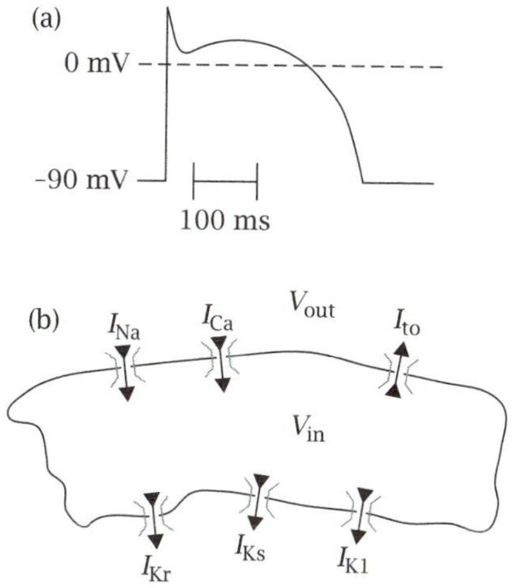


**Figure 4** (a) An action potential in the heart and (b) a schematic representation of a cardiac cell. $I_{Na}$, $I_{Ca}$, etc., denote different types of ionic channels; arrows indicate current direction.

be described by the following system:

$$C_m \frac{dV_m}{dt} = -I_{Na} - I_K - I_{Ca} + \cdots, \quad (1)$$

$$I_* = G_* g_*^{\alpha} g_\bullet^{\beta}(V_m - V_*), \quad (2)$$

$$\frac{dg_i}{dt} = \frac{g_i^{\infty}(V_m) - g_i}{\tau_i(V_m)}, \quad i = *, \bullet, \quad (3)$$

where the first equation describes the changes in transmembrane voltage $V_m = V_{in} - V_{out}$ as a result of the dynamics of various ionic currents. Each of the currents typically depends on $V_m$ and time. In most cases, time dependency is given by an exponential relaxation equation for gating variables $g_i$. For example, a hypothetical current $I_*$ that conveys ion "$*$" and has a maximal conductivity of $G_* =$ const. is represented by expression (2). The current is zero at $V_m = V_*$, where $V_*$ is the so-called Nernst potential for ion "$*$," which can be easily computed from the concentration of specific ions outside and inside the cardiac cell. The time dynamics of this current are governed by two gating variables $g_*$, $g_\bullet$ raised to the powers $\alpha$, $\beta$. The variables $g_*$, $g_\bullet$ approach their voltage-dependent steady state values $g_i^{\infty}(V_m)$ with characteristic time $\tau_i(V_m)$ (3). All the parameters and functions here are chosen to fit experimentally measured properties of the specific ionic current. Most of the ionic currents have one or two gating variables, with $\alpha = \beta = 1$.

The systems expressed by (1)–(3) are called detailed ionic models as they describe in detail the underlying biophysical mechanism of cardiac excitation and are based on direct measurement of the properties of cardiac cells. Such models may have from as few as 4 to more than 100 equations and contain hundreds of parameters. These models are mainly solved numerically and can be applied to detailed studies of drug actions, mutations, and other processes on cardiac cells. Another class of cardiac models is low-dimensional phenomenological models, which have just two or three differential equations. These are mainly used for generic studies of waves in the heart, which can be numerical as well as analytical.

### 1.2 Tissue and Whole-Organ Models

Wave propagation is a result of the successive excitation of cardiac cells, each of which is described by (1). In cardiac tissue, the excitable cells are connected to each other via resistors called gap junctions and equations can be obtained by representing tissue as a resistive network. One of the most widely used models involves the monodomain equations:

$$C_m \frac{\partial V_m}{\partial t} = \operatorname{div}(\boldsymbol{D}\nabla V_m) - I_{ion}, \quad (4)$$

where $\boldsymbol{D}$ is the diffusion tensor. The idea behind this equation is straightforward. The total current through the membrane has not only an ionic current $I_{ion}$ but also diffusive currents from the cell's neighbors ($\operatorname{div}(\boldsymbol{D}\nabla V_m)$). To understand why the diffusive current has such a form, consider a hypothetical cell that takes the form of a long cylinder. The current along the axis of the cylinder is proportional to the gradient of the voltage $\boldsymbol{D}\nabla V_m$ (Ohm's law). The divergence of this current is a membrane current, i.e., a current that goes through the surface of the cylinder, and it therefore has to be added to $I_{ion}$. As cardiac tissue consists of fibers, its resistance depends on the direction of these fibers, and this is accounted for by the diffusivity (conductivity) tensor $\boldsymbol{D}$. Recent studies have revealed that fibers in the heart are organized in myocardial sheets, giving resistivity in three main directions: along the fibers, across the fibers inside the sheets, and across the sheets. The measured velocities of these three directions have the approximate ratio $1{:}\frac{1}{2}{:}\frac{1}{4}$. The diffusive matrix can be directly calculated from the orientations of the fibers and the sheets. The main idea is that in a coordinate system that is locally aligned with the fibers and sheets, the matrix $\boldsymbol{D}$ is diagonal, with eigenvalues accounting for unequal resistivity along given directions. In a global coordinate system, $\boldsymbol{D}$ can be found by simple matrix transformation.

Modeling of excitation in the whole heart mostly involves finding a solution of (4) in the domain representing the heart shape. To solve this equation we will also need to specify the anisotropy of the tissue at each point (via the matrix $\boldsymbol{D}$) and the types of cardiac cells via models of cardiac cells ($I_{\text{ion}}$). In addition, boundary conditions also need to be imposed. Most of these data are easy to obtain, and the shape of the heart can be obtained from COMPUTERIZED TOMOGRAPHY [IV.4 §8] (CT) (see also [VII.19]) or MAGNETIC RESONANCE IMAGING [VII.10 §4.1] (MRI) scanning procedures, which are routinely used in clinical practice. However, obtaining $\boldsymbol{D}$ is much more difficult. Currently, it can be measured only on explanted hearts using direct histological measurements or by means of diffusion tensor MRI, which measures fiber directions by probing the diffusion rates of water in the tissue.

## 2 Analytical Methods in Cardiac Modeling

When it comes to the analytical study of cardiac models, the possibilities are very limited. Most of the results in this area deal with low-dimensional models, for which the main properties of solutions can be studied in the phase plane of the system. Another important direction of (semi)analytical research is the investigation of dynamical instabilities, which may result in the generation of spiral waves and their decay into complex turbulent patterns. One of the most interesting results was obtained in studies of the so-called alternans instability, which is simply an instability that occurs as a result of a period-doubling bifurcation for a one-dimensional discrete map describing a periodically forced cardiac cell. Such studies have predicted the necessary conditions for onset of the instability that leads to spiral breakup. These conditions were connected to measurable characteristics of cardiac tissue: the dependency of the duration of a cardiac pulse and the period of stimulation of cardiac cells. This hypothesis was tested not only in theoretical studies but also in experimental research, with the latter demonstrating that a decrease in the slope of the dependency can prevent the onset of electrical turbulence in the heart. Another recent development is a new way of describing anisotropy in the heart as a Riemannian manifold, whose metric is defined by the arrival of excitation at a given point. This viewpoint allows analytical equations to be obtained for wave velocity and spiral wave dynamics in two-dimensional and three-dimensional anisotropic tissue via properties of curvature tensors used in Riemannian geometry, such as the Ricci curvature tensor.

## 3 Numerical Approaches

Numerical approaches are the most important tool in cardiac modeling. The monodomain equations that describe cardiac excitation belong to the class of parabolic partial differential equations, and the numerical solution of these equations is straightforward. Most often, the simulations of (4) were made using explicit finite-difference methods, such as the Euler integration scheme. Equation (5) gives an example of this approach for the two-dimensional isotropic case, which updates the value of the voltage at each point $(i, j)$ on a two-dimensional grid at time $t + ht$ from the values of variables at time $t$:

$$V_{ij}^{t+ht} = V_{ij}^{t} - I_{ij}^{t} + \frac{htD}{C_{\text{m}}} \frac{V_{i+1,j}^{t} + V_{i,j+1}^{t} + V_{i-1,j}^{t} + V_{i,j+1}^{t} - 4V_{ij}^{t}}{hs^2}. \tag{5}$$

Here, $hs$ and $ht$ are the spatial and time integration steps, and $I_{ij}$ gives the value of the ionic current at a given point. Typical integration steps for cardiac models are $hs \approx 0.2$ mm and $ht \approx 0.01$ ms. Due to the big difference in timescales for the upstroke and depolarization phases (figure 4), many attempts have been made to develop algorithms that are adaptive in time and/or space. For whole-heart simulations, a big challenge is the proper representation of boundary conditions on domains of complex shape. To this end, as well as explicit finite-difference methods, more complex numerical methods (such as finite-volume or finite-element methods) have been used. Note that for most existing ionic models there is no longer any need to copy equations from the original literature: most of the models are present in Auckland University's Cell ML database (www.cellml.org).

## 4 Applications

Working with a whole-heart computer model is similar to experimental or clinical work, with the researcher having the same tools as experimentalists. For example, as in a real experiment we can put electrodes onto the virtual heart and initiate new waves just by adjusting the voltage at given points.

Modeling can be applied to practically important problems. For example, it turns out that most cardiac drugs are blockers of certain ionic channels. We can model the effects of these drugs by changing the conductance of the ionic channels $G_*$ in (1) and then studying how the changes affect the process of cardiac excitation. Similarly, many forms of inherited cardiac disease occur as a result of a mutation in the gene-coding ionic channels. For example, LQT1 syndrome is due to a mutation that modifies (decreases) the potassium current, while LQT3 syndrome is the result of a mutation that modifies (increases) the sodium current.

Thus, if one modifies the properties of ionic channels to reproduce the effect of a given mutation, the effect of this mutation on cell excitation and the onset of arrhythmias can be studied. Furthermore, it is possible to study the onset of arrhythmias due to various types of cardiac disease, such as ischemia or fibrosis, by changing the parameters of the model and by introducing new types of cells. Compared with real experimental research, modeling opens up many more possibilities for modification of the heart's properties, and it can also be used to study three-dimensional wave propagation in the heart.

There is a great deal of interest in applications that elucidate the mechanisms of the formation of spiral waves. Another important task is to understand and identify instabilities (bifurcations) that are responsible for the deterioration of a single spiral wave into spatiotemporal chaos.

Computational modeling also has potential applications in clinical interventions: cardiac resynchronization therapy and cardiac ablation, for example. In resynchronization therapy, several electrodes are placed on a patient's heart and a cardiologist then adjusts the delays of excitation of these electrodes in order to optimize the heart's pumping function. As this procedure is mainly used for patients who have suffered a heart attack, their hearts have abnormal properties and this procedure should thus be optimized on a patient-specific basis. Anatomically accurate modeling is one of the important tools needed for such optimization. Cardiac ablation is a procedure used by cardiologists to disrupt the pathological pathways along which the wave circulates in the heart during an arrhythmia. Here, too, modeling can help to identify such pathways and guide cardiologists during ablation. Although direct clinical applications of modeling are still under development, it is widely believed that they will become clinical tools during the next decade.

### Further Reading

Clayton, R. H., and A. V. Panfilov. 2008. A guide to modelling cardiac electrical activity in anatomically detailed ventricles. *Progress in Biophysics and Molecular Biology* 96:19–43.

Jalife, J., M. Delmar, J. Anumonwo, O. Berenfeld, and J. Kalifa. 2009. *Basic Cardiac Electrophysiology for the Clinician.* New York: Wiley-Blackwell.

Keener, J., and J. Sneyd. 2009. *Mathematical Physiology*, 2nd edn. New York: Springer.

Panfilov, A. V., and A. M. Pertsov. 2001. Ventricular fibrillation: evolution of the multiple-wavelet hypothesis. *Philosophical Transactions of the Royal Society of London* A 359:1315–26.

Rudy, Y., and J. R. Silva. 2006. Computational biology in the study of cardiac ion channels and cell electrophysiology. *Quarterly Reviews of Biophysics* 39:57–116.

## V.7 Chemical Reactions

*Martin Feinberg*

### 1 Introduction

Chemical reactions underlie a vast spectrum of natural phenomena and, as a result, play an indispensable role in many branches of science and engineering. An understanding of atmospheric chemistry, cell biology, or the efficient production of energy from fossil fuels requires, in one way or another, a framework for thinking systematically about how chemical reactions serve to convert certain molecules into others.

Chemical reactions can be studied at various conceptual levels. At a very fundamental level, for example, one is concerned with ways in which chemical bonds are broken and created within and between individual molecules during the occurrence of a single chemical reaction.

In this article our concerns will be different. In the biological cell and in the atmosphere there are a large number of distinct chemical species, and these are involved in a great variety of chemical reactions. A particular species might be consumed in several reactions and produced in several others. Thus, the reactions can be intricately coupled, and, as a result, the dynamics of the species population might be quite complex. It is this dynamics that will be our focus.

### 2 Some Primitives

It will be useful to lay out some primitive ideas upon which the mathematics of chemical reaction systems can be built.

## 2.1 Species

We begin by supposing that a chemical mixture we might wish to study comprises a fixed set of chemical *species*, which we will denote by $A_1, A_2, \ldots, A_N$. Thus, $A_1$ might be carbon monoxide, $A_2$ might be oxygen, and so on. In applications, there is some judgment required when selecting the species set, for one might choose to ignore molecular entities that are deemed inconsequential to the task at hand. At each place in a mixture and at each instant, we associate with every species $A_L$ a molar concentration $c_L$. In rough terms, $c_L$ is the local number of molecules of $A_L$ per unit volume, divided by Avogadro's number, approximately $6 \times 10^{23}$. In this way, with each place in the mixture and with each instant we can associate a local *composition vector* $c = [c_1, c_2, \ldots, c_N]$ in the standard vector space of $N$-tuples, $\mathbb{R}^N$. The mathematics of chemical reactions is, for the most part, aimed at a description of how the composition vector varies with time and spatial position.

## 2.2 Reactions

For the mixture under study we suppose that the various species interact through the occurrence of a fixed set of chemical reactions that constitute a *reaction network*. The network is often displayed in a reaction diagram such as the one shown below:

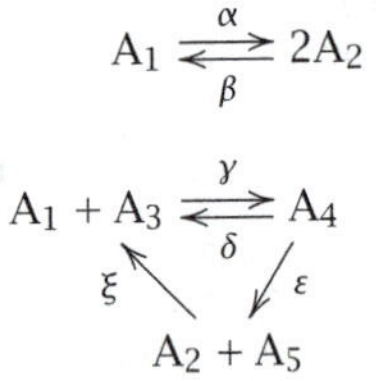

$$A_1 \underset{\beta}{\overset{\alpha}{\rightleftharpoons}} 2A_2$$

$$A_1 + A_3 \underset{\delta}{\overset{\gamma}{\rightleftharpoons}} A_4 \tag{1}$$
$$\xi \nwarrow \quad \swarrow \varepsilon$$
$$A_2 + A_5$$

The diagram is meant to suggest that a molecule of $A_1$ can decompose to form two molecules of $A_2$, that two molecules of $A_2$ can combine to form one molecule of $A_1$, that a molecule of $A_1$ can combine with a molecule of $A_3$ to form a molecule of $A_4$, and so on. (The Greek letters alongside the reaction arrows will play a role in the next section.)

Here again, there is a need for some judgment about which reactions to include. Certain reactions might be deemed to occur so slowly as to be safely ignored for the purposes of the analysis.

## 2.3 Reaction Rate Functions

The local occurrence rates of individual reactions are presumed to be given by *reaction rate functions*—one

for each reaction in the network at hand—that indicate how the individual reaction rates depend on the local mixture composition vector and the local temperature. Thus, in our example the function $\mathscr{K}_{A_1 \to 2A_2}(\cdot, \cdot)$ tells us how the local occurrence rate per unit volume of reaction $A_1 \to 2A_2$ depends on local mixture conditions; in particular, $\mathscr{K}_{A_1 \to 2A_2}(c, T)$ is the local molar occurrence rate per unit volume when the local composition vector is $c$ and the local temperature is $T$. (Roughly speaking, the local molar occurrence rate per unit volume is the number of times the reaction occurs per unit time per unit volume divided by Avogadro's number.) Thus, reaction rate functions take nonnegative values.

These functions are presumed, in general, to be smooth and to have certain natural relationships with the particular reactions they describe. For example, it is often supposed—and we shall suppose here—that, for a reaction such as $A_1 + A_3 \to A_4$, $\mathscr{K}_{A_1 + A_3 \to A_4}(c, T)$ takes a strictly positive value at composition $c$ if and only if both $A_1$ and $A_3$ are actually present—that is, if and only if the composition vector $c$ is such that both $c_1$ and $c_3$ are positive.

A specification of a reaction rate function for each of the reactions in a network is called a *kinetics* for the network. By a *kinetic system* we will mean a reaction network together with a kinetics.

### 2.3.1 Mass Action Kinetics

By a *mass action kinetics* for a reaction network we mean a kinetics in which the individual reaction rate functions have a special, and very natural, form.

For a reaction such as $A_1 \to 2A_2$ it is generally supposed that the local occurrence rate is simply proportional to $c_1$, the local concentration of $A_1$; after all, the more molecules of $A_1$ there are locally, the more occurrences of the reactions there will be. Thus, it is presumed that $\mathscr{K}_{A_1 \to 2A_2}(c, T) \equiv \alpha(T)c_1$, where $\alpha(T)$ is called the (positive) *rate constant* for the reaction $A_1 \to 2A_2$.

For the reaction $2A_2 \to A_1$ the situation is different. In this case, two molecules of $A_2$ must meet if the reaction is to proceed at all, and the probability of such a meeting is taken to be proportional to $(c_2)^2$. With this as motivation, the reaction rate function is presumed to have the form $\mathscr{K}_{2A_2 \to A_1}(c, T) \equiv \beta(T)(c_2)^2$, where, again, $\beta(T)$ is the rate "constant" for the reaction $2A_2 \to A_1$. Similarly, we would have $\mathscr{K}_{A_1 + A_3 \to A_4}(c, T) \equiv \gamma(T)c_1 c_3$, and so on.

When the kinetics is mass action, it is common, as in (1), to decorate the reaction diagram with symbols for the various rate constants alongside the corresponding reaction arrows.

### 2.3.2 Other Kinetics

Reaction networks deemed to describe chemistry at the fine level of so-called elementary reactions are usually presumed to be governed by mass action kinetics. Sometimes, however, coarser modeling is called into play, and the "reactions" considered are actually intelligent but nevertheless approximate descriptions of what is actually happening. A model might speak in terms of an "overall" reaction that describes, in effect, the result of a sequence of more elementary chemical events.

For example, a model might invoke an overall reaction such as $A_1 \to A_2$ that is, in reality, mediated by an enzyme, present at very low concentration, that acts through an elementary reaction sequence such as

$$A_1 + E \rightleftarrows A_1E \to A_2E \to A_2 + E.$$

The idea here is that $A_1$ binds reversibly to the enzyme E to form $A_1E$, whereupon the bound $A_1$ is rapidly transformed by the enzyme into $A_2$. The newly formed $A_2$ then unbinds, leaving behind naked enzyme, which is then free to rework its magic.

This sequence of elementary steps is presumed to be governed by mass action kinetics, which, through a suitable approximation procedure, can be made to yield a pseudokinetics for the pseudoreaction $A_1 \to A_2$. Such a kinetics might take a Michaelis–Menten form:

$$\mathscr{K}_{A_1 \to A_2}(c, T) \equiv \frac{k(T)c_1}{k'(T) + c_1}.$$

Although the (roughly constant) concentration of the enzyme might influence, in a latent way, values of the parameters (in particular $k$), the approximate rate function for the pseudoreaction $A_1 \to A_2$ is written explicitly only in terms of the concentration of $A_1$.

## 3 The Species-Formation-Rate Function for a Kinetic System

Given a reaction network endowed with a kinetics, we are in a position to calculate for each species its local net molar production rate per unit volume when the local composition vector is $c$ and the local temperature is $T$.

For our sample reaction network, let us focus for the moment on species $A_1$. Whenever the reaction $A_1 \to 2A_2$ occurs, we *lose* one molecule of $A_1$, and the local occurrence rate per unit volume of that reaction is $\mathscr{K}_{A_1 \to 2A_2}(c, T)$. Whenever the reaction $2A_2 \to A_1$ occurs we *gain* a molecule of $A_1$, and the local occurrence rate per unit volume of that reaction is $\mathscr{K}_{2A_2 \to A_1}(c, T)$. After also taking into account the contributions of the reactions $A_1 + A_3 \to A_4$, $A_4 \to A_1 + A_3$, and $A_2 + A_5 \to A_1 + A_3$ to the production of species $A_1$, we can calculate the *net molar production rate per unit volume of* $A_1$, denoted $r_1(c, T)$, as follows:

$$\begin{aligned} r_1(c, T) \equiv & -\mathscr{K}_{A_1 \to 2A_2}(c, T) + \mathscr{K}_{2A_2 \to A_1}(c, T) \\ & - \mathscr{K}_{A_1 + A_3 \to A_4}(c, T) + \mathscr{K}_{A_4 \to A_1 + A_3}(c, T) \\ & + \mathscr{K}_{A_2 + A_5 \to A_1 + A_3}(c, T). \end{aligned}$$

For species $A_2$, we can proceed in the same way, but we need to take cognizance of the fact that with each occurrence of the reaction $A_1 \to 2A_2$ there is a gain of *two* molecules of $A_2$, and with each occurrence of the reverse reaction there is a loss of *two* molecules of $A_2$. Thus we write

$$\begin{aligned} r_2(c, T) \equiv & \; 2\mathscr{K}_{A_1 \to 2A_2}(c, T) - 2\mathscr{K}_{2A_2 \to A_1}(c, T) \\ & + \mathscr{K}_{A_4 \to A_2 + A_5}(c, T) - \mathscr{K}_{A_2 + A_5 \to A_1 + A_3}(c, T). \end{aligned}$$

In this way we can formulate, for a kinetic system with $N$ species, the $N$ functions $r_L(\cdot, \cdot)$, $L = 1, 2, \dots, N$. For each $L$, $r_L(\cdot, \cdot)$ is just the sum, over all reactions, of the individual reaction rate functions, each multiplied by the net number of molecules of species $A_L$ produced with each occurrence of the corresponding reaction.

For use later on, we shall find it convenient to form the vectorial *species-formation-rate function* $r(\cdot, \cdot)$ for a kinetic system, which is defined by

$$r(c, T) := [r_1(c, T), r_2(c, T), \dots, r_N(c, T)] \in \mathbb{R}^N.$$

## 4 How Kinetic Systems Give Rise to Differential Equations

There are various physicochemical settings in which a kinetic system might exert itself, and there are corresponding differences in how the governing differential equations are formulated. In each setting, however, the presence of chemical reactions is manifested through the species-formation-rate function for the kinetic system at hand.

Our focus here will be almost entirely on what chemical engineers call the spatially homogeneous (well-stirred) fixed-volume *batch reactor*, for that is the setting in which chemical reactions exert themselves in the most pristine way, uncomplicated by composition changes resulting from bulk or diffusive transport of matter.

### 4.1 The Homogeneous Batch Reactor

Imagine that a mixture fills a vessel of fixed volume $V$ and that the mixture is stirred so effectively that its composition and temperature are always independent of spatial position. Imagine also that heat can be added to and removed from the vessel by an operator in any desired way and that, in particular, the temperature variation with time can be controlled to an exquisite degree. One possibility is that the temperature can be maintained at a fixed value for all time, in which case the operation is *isothermal.*

We suppose that the chemistry is well-modeled, as before, by a kinetic system with $N$ species, $A_1, A_2, \dots, A_N$. In this case, the spatially independent composition vector at time $t$ is $c(t) = [c_1(t), c_2(t), \dots, c_N(t)] \in \mathbb{R}^N$. In fact, our interest is in the temporal variation of the mixture composition.

If we denote by $n_L(t)$ the number of moles of species $A_L$ in the reactor at time $t$—that is, the number of molecules of $A_L$ divided by Avogadro's number—and if we recall that $r_L(c(t), T(t))$ is the net rate of production *per unit volume* of $A_L$ due to the occurrence of all reactions in the underlying reaction network, then we can write $\dot{n}_L(t) = V r_L(c(t), T(t))$. (An overdot always indicates time differentiation.) Dividing by $V$, we get the system of $N$ scalar differential equations

$$\dot{c}_L = r_L(c(t), T(t)), \quad L = 1, 2, \dots, N, \tag{2}$$

or, equivalently, the single vector equation

$$\dot{c}(t) = r(c(t), T(t)). \tag{3}$$

A possibility that we have not considered is that the reactor is *adiabatic*, which is to say that the vessel is perfectly insulated. In this case, energetic considerations must be brought into play through an additional differential equation for the temperature. In particular, one must take account of the fact that the occurrence of chemical reactions might, by itself, serve to transiently raise or lower the mixture temperature.

It is instructive to write out the isothermal batch reactor differential equations (2) for reaction network (1) when the kinetics is mass action:

$$\left.\begin{aligned}
\dot{c}_1 &= -\alpha c_1 + \beta(c_2)^2 - \gamma c_1 c_3 + \delta c_4 + \xi c_2 c_5,\\
\dot{c}_2 &= 2\alpha c_1 - 2\beta(c_2)^2 + \varepsilon c_4 - \xi c_2 c_5,\\
\dot{c}_3 &= -\gamma c_1 c_3 + \delta c_4 + \xi c_2 c_4,\\
\dot{c}_4 &= \gamma c_1 c_3 - (\delta + \varepsilon) c_4,\\
\dot{c}_5 &= \varepsilon c_4 - \xi c_2 c_5.
\end{aligned}\right\} \tag{4}$$

Note that this is a system of coupled polynomial differential equations in five dependent variables and in which six parameters (rate constants) appear.

Polynomial differential equations are notoriously difficult to study. In fact, one of David Hilbert's famous problems (the sixteenth) posed in 1900 at the International Congress of Mathematicians in Paris was about polynomial differential equations in just two variables. It remains unsolved.

### 4.2 A Few Words about Other Physicochemical Settings

In the batch reactor, composition changes result solely from the occurrence of chemical reactions. In other settings, composition variations in time or in spatial position might result from a coupling of reactions with bulk or diffusive transport of matter.

A *continuous-flow stirred-tank reactor* is very much like a batch reactor, but fresh mixture is added to the reactor continuously at a fixed flow rate, while mixture is simultaneously removed from the reactor at that same flow rate. When the mixture density can be presumed to be independent of composition and temperature, the governing vector differential equation takes the form

$$\dot{c}(t) = \frac{1}{\theta}(c^F - c(t)) + r(c(t), T(t)).$$

Here $c^F \in \mathbb{R}^N$ is the feed composition, and $\theta$ is the reactor *residence time*, which is the reactor volume divided by the volumetric flow rate.

In still other settings, there might be continuous variation in the composition not only in time but also in spatial position. In such instances, mixtures are usually modeled by systems of *reaction–diffusion equations*, about which there is a large literature.

Finally, we note that certain systems open to transport of matter are sometimes modeled *as if* they were batch reactors by adding pseudoreactions of the form $0 \to A_L$ or $A_L \to 0$ to account for the supply and removal of species $A_L$.

## 5 Seeing Things Geometrically

We have already pointed out that, even for a relatively simple reaction network such as (1), the corresponding differential equations can be difficult to study. What we have not yet exploited, however, is the fact that these equations bear an intimate structural relationship to the reaction network from which they derive, a relationship that imposes severe geometric constraints on the way that composition trajectories can travel through $\mathbb{R}^N$. An understanding of that geometry can help considerably in studying the dynamics to which the equations give rise and in posing the right questions at the outset.

To begin, we will need to make more mathematically transparent the way in which the species-formation-rate function $r(\cdot,\cdot)$ relates structurally to the underlying reaction network.

### 5.1 Complexes and Complex Vectors

By the *complexes* of a reaction network we mean the objects that sit at the heads and tails of the reaction arrows. Thus, for network (1) the complexes are $A_1$, $2A_2$, $A_1 + A_3$, $A_4$, and $A_2 + A_5$.

With each complex in an $N$-species network we can associate a vector of $\mathbb{R}^N$ in a natural way. This is best explained in terms of our sample network (1), for which $N = 5$: with the complex $A_1$ we associate the *complex vector* $y_1 := [1,0,0,0,0] \in \mathbb{R}^5$; with $2A_2$ we associate the vector $y_2 := [0,2,0,0,0]$; with $A_1 + A_3$ we associate the vector $y_3 := [1,0,1,0,0]$; with $A_4$ we associate the vector $y_4 := [0,0,0,1,0]$; and with $A_2 + A_5$ we associate the vector $y_5 := [0,1,0,0,1]$.

### 5.2 Reactions and Reaction Vectors

Note that we have numbered the complexes in an arbitrary way. Hereafter, we write $y_i \to y_j$ (or sometimes just $i \to j$) to indicate the reaction whereby the $i$th complex reacts to the $j$th complex.

In fact, we will associate with each reaction $y_i \to y_j$ in a network the corresponding *reaction vector* $y_j - y_i \in \mathbb{R}^N$, with $N$ again denoting the number of species. In our example, the reaction vector corresponding to the reaction $A_1 \to 2A_2$ is $[-1,2,0,0,0]$. The reaction vector corresponding to $A_2 + A_5 \to A_1 + A_3$ is $[1,-1,1,0,-1]$. *Note that the $L$th component of the reaction vector for a particular reaction is just the net number of molecules of species $A_L$ produced with each occurrence of that reaction.*

### 5.3 The Species-Formation-Rate Function Revisited

Consider a kinetic system, for which we denote by $\mathscr{R}$ the set of reactions and by $\{\mathscr{K}_{i\to j}(\cdot,\cdot)\}_{i\to j\in\mathscr{R}}$ the corresponding set of reaction rate functions. We claim that the (vector) species-formation-rate function can be written in terms of the reaction vectors $\{y_j - y_i\}_{i\to j\in\mathscr{R}}$ as

$$r(c,T) = \sum_{i\to j\in\mathscr{R}} \mathscr{K}_{i\to j}(c,T)(y_j - y_i) \qquad (5)$$

for all $c$ and $T$. To see that this is so, it helps to inspect the component of (5) corresponding to species $A_L$:

$$r_L(c,T) = \sum_{i\to j\in\mathscr{R}} \mathscr{K}_{i\to j}(c,T)(y_{jL} - y_{iL}). \qquad (6)$$

Recall how the species formation rate is calculated for species $A_L$: by summing, over all reactions, the individual reaction rate functions, each multiplied by the net number of molecules of $A_L$ produced when the corresponding reaction occurs once. This is precisely what is done in (6).

### 5.4 What the Vector Formulation Tells Us

A glance at (5) indicates that, no matter what the local mixture composition and temperature are, *the species-formation-rate vector is invariably a linear combination (in fact, a nonnegative linear combination) of the reaction vectors for the operative chemical reaction network.* This tells us that the species-formation-rate vector is constrained to point only in certain directions in $\mathbb{R}^N$; it can point only along the linear subspace of $\mathbb{R}^N$ spanned by the reaction vectors.

This is called the *stoichiometric subspace* for the reaction network, and we denote it by the symbol $S$:

$$S := \operatorname{span}\{y_j - y_i \in \mathbb{R}^N : y_i \to y_j \in \mathscr{R}\}.$$

(Stoichiometry is the part of elementary chemistry that takes account of conservation of various entities (e.g., mass, charge, atoms of various kinds) that are conserved even in the presence of change caused by chemical reactions.)

We will have interest in the dimension of the stoichiometric subspace. This is identical to the rank of the network's reaction vector set—that is, the number of vectors in the largest linearly independent set that can be formed from the network's reaction vectors. This number, denoted $s$, is called the *rank* of the network:

$$s := \dim S = \operatorname{rank}\{y_j - y_i \in \mathbb{R}^N : y_i \to y_j \in \mathscr{R}\}.$$

In language we now have available, we can say that, for any composition and temperature, $r(c,T)$ is constrained to point only along the stoichiometric subspace. Provided that $S$ is indeed smaller than $\mathbb{R}^N$ (i.e., $s < N$), this has important dynamical implications. We will examine these implications soon, but first we should see that, for networks that respect conservation of mass, $S$ *must* be smaller than $\mathbb{R}^N$.

### 5.5 An Elementary Property of Mass-Conserving Reaction Networks

Suppose that, for our sample network (1), $M_1, M_2, \ldots, M_5$ are the molecular weights of the five species $A_1, A_2, \ldots, A_5$. Although a chemical transformation takes place when the reaction $A_1 \to 2A_2$ occurs, we nevertheless expect the reaction to conserve mass. That is, we expect that the total molecular weight of molecules on the reactant side of the arrow will be the same as the total molecular weight of molecules on the product side. More precisely, if the reaction does indeed respect mass conservation, we expect that $M_1 = 2M_2$. Similarly, if the reaction $A_1 + A_3 \to A_4$ is consistent with mass conservation, then we should have $M_1 + M_3 = M_4$.

More generally, if $y_i \to y_j$ is a reaction in a network with $N$ species, then the reaction will be consistent with mass conservation only if

$$\sum_{L=1}^{N} y_{iL} M_L = \sum_{L=1}^{N} y_{jL} M_L,$$

where, as in our example, $M_1, M_2, \ldots, M_N$ are the molecular weights of the species. Letting $M = [M_1, M_2, \ldots, M_N]$ be the vector of molecular weights, we can express this condition in vector terms through the standard scalar product in $\mathbb{R}^N$:

$$y_i \cdot M = y_j \cdot M.$$

If all reactions of the network respect mass conservation, then we have

$$(y_j - y_i) \cdot M = 0, \quad \forall y_i \to y_j \in \mathscr{R}.$$

This is to say that $M$ is orthogonal to each of the reaction vectors and, therefore, to their span, whereupon $S$ is indeed smaller than $\mathbb{R}^N$.

### 5.6 The Batch Reactor Viewed Geometrically

It is in the batch reactor setting that we can see most clearly the implications of the fact that the species-formation-rate function takes values in the stoichiometric subspace for the underlying reaction network.

Recall that the governing vector differential equation is $\dot{c}(t) = r(c(t), T(t))$. Here $r(\cdot,\cdot)$ is the species-formation-rate function for the operative kinetic system. For that system we will suppose that the stoichiometric subspace is smaller than $\mathbb{R}^N$. To be concrete, we will also suppose that the temperature variation with time is controlled by an external agent. (The temperature might be maintained constant, but that is not essential to what we will say.)

Our interest will be in understanding how the composition vector moves around the nonnegative orthant of $\mathbb{R}^N$ if an initial composition $c(0) = c^0 \in \mathbb{R}^N$ is specified. By means of the following analogy we can see why the motion is very highly constrained by the reaction network itself, no matter what the kinetics might be.

#### *5.6.1 A Bug Analogy*

Imagine that a friendly bug is sitting at the very top of the doorknob in your bedroom. At time $t = 0$ the bug begins flying around, and we are interested in predicting where in your bedroom the bug will be a minute later. (It is winter. The doors and windows are closed. We are assuming that the bug suffers no violence.) Without further information, it is hard to say anything meaningful in advance about the bug's position, even qualitatively.

But suppose that, for some deep entomological reason, the bug chooses to have its velocity vector *always* point along your bedroom floor. Then it is intuitively clear that, for all $t \geqslant 0$, the bug *must* be on the plane parallel to the floor that passes through the top of the doorknob. True, we do not know precisely where the bug will be on that plane, but we do know *a lot* about where he or she cannot be.

#### *5.6.2 Stoichiometric Compatibility*

For stoichiometric rather than entomological reasons, the batch reactor behaves in essentially the same way. Because $r(\cdot,\cdot)$ invariably takes values in $S$, the composition "velocity" $\dot{c}$ must always point along $S$. It is at least intuitively clear, then, that for all $t \geqslant 0$, $c(t)$ must lie in the translate of $S$ containing $c^0$.

In figure 1 we depict schematically the situation for a kinetic system based on a three-species network with reactions $A_1 \rightleftarrows A_2$ and $A_1 + A_2 \rightleftarrows A_3$. The stoichiometric subspace is two dimensional: it is a plane through the origin that sits behind the nonnegative orthant, $\bar{\mathbb{R}}^3_+$. We denote by $c^0 + S$ the translate of the stoichiometric subspace containing the initial composition $c^0$.

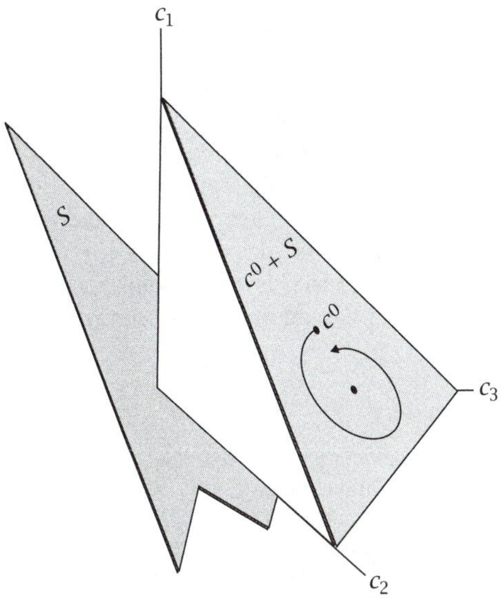


**Figure 1** A stoichiometric compatibility class.

Thus, we expect the resulting composition trajectory to reside entirely within the intersection of $c^0 + S$ with $\bar{\mathbb{R}}^3_+$. That intersection appears as a shaded triangle in figure 1.

The set of compositions within the triangle is an example of what we shall call a *stoichiometric compatibility class.* A composition outside the triangle is inaccessible from a composition within the triangle; the two compositions are not stoichiometrically compatible.

More generally, for a reaction network with stoichiometric subspace $S \subset \mathbb{R}^N$ we say that two compositions $c$ and $c'$ are *stoichiometrically compatible* if $c' - c$ is a member of $S$. Thus, the set $\bar{\mathbb{R}}^N_+$ of all possible compositions can be partitioned into *stoichiometric compatibility classes*: that is, subsets of compositions that are mutually stoichiometrically compatible. As with our example, each of these is a convex set of the form $(c^0 + S) \cap \bar{\mathbb{R}}^N_+$ for some vector $c^0 \in \bar{\mathbb{R}}^N_+$. A batch reactor composition trajectory that begins within a given stoichiometric compatibility class will never leave it.

## 6 A Little Chemical Reaction Network Theory

Although we have discovered, at least for the batch reactor, that composition trajectories are severely constrained geometrically, the deepest questions are about what might happen *within* the various stoichiometric compatibility classes.

Indeed, when questions are posed about the capacity of a given kinetic systems to admit, for example, multiple equilibria, one is generally asking about the possibility of two or more equilibria that are stoichiometrically compatible with each other: that is, that reside within the same stoichiometric compatibility class. Similarly, when one asks about the stability of a given equilibrium, one is generally interested in stability relative to initial conditions that are stoichiometrically compatible with it.

If we think back to the mass action differential equations (4) that derived from our relatively simple network example, it is easy to see that these questions can be extremely difficult, and answers might depend on parameter values (e.g., rate constants). Nevertheless, in the early 1970s a body of theory began to emerge that, especially in the mass action case, draws firm connections between qualitative properties of the differential equations for a kinetic system and the structure of the underlying network of chemical reactions. We need just a little more vocabulary to state, as an important example, an early reaction network theory result: the *deficiency zero theorem,* due to F. Horn, R. Jackson, and M. Feinberg.

### 6.1 Some More Vocabulary

It should be noted that, in our display of reaction network (1), each complex was written just once, and then arrows were drawn to indicate how the complexes are connected by reactions. Formulated this way, the so-called *standard reaction diagram* becomes a directed graph, with the complexes serving as vertices and the reaction arrows serving as directed edges. Note that the diagram in (1) has two connected components: one containing the complexes $A_1$ and $2A_2$, and the other containing the complexes $A_1 + A_3$, $A_4$, and $A_2 + A_5$. In the language of chemical reaction network theory the (not necessarily strong) components of the standard reaction diagram are called the *linkage classes* of the network.

By a *reversible* reaction network a chemist usually means one in which each reaction is accompanied by its reverse. By a *weakly reversible* reaction network we mean one for which, in the standard reaction diagram, each reaction arrow is contained in a directed cycle. Every reversible network is weakly reversible. Network (1) is weakly reversible but not reversible.

The *deficiency* of a reaction network is a nonnegative integer index with which reaction networks can be classified. It is defined by

deficiency := #complexes − #linkage classes − rank.

The deficiency is not a measure of a network's size. A deficiency zero network can have hundreds of species and hundreds of reactions.

### 6.2 The Deficiency Zero Theorem

In what follows, it will be understood that we are considering the *isothermal* version of (3), $\dot{c} = r(c)$. When we speak of a *positive stoichiometric compatibility class* we mean a (nonempty) intersection of a translate of the stoichiometric subspace with the strictly positive orthant of $\mathbb{R}^N$ (i.e., the interior of $\mathbb{R}^N$).

It is well known that mass action systems can admit, within a positive stoichiometric compatibility class, unstable equilibria, multiple equilibria, and cyclic composition trajectories. The following theorem tells us that *we should not expect such phenomena when the underlying reaction network has a deficiency of zero, no matter how complicated it might be.*

**Theorem 1.** *For any reaction network of deficiency zero the following statements hold true.*

(i) *If the network is not weakly reversible, then for any kinetics, not necessarily mass action, the corresponding differential equations cannot admit an equilibrium in which all species concentrations are positive, nor can they admit a cyclic composition trajectory that passes through a composition in which all species concentrations are positive.*

(ii) *If the network is weakly reversible, then, when the kinetics is mass action (but regardless of the positive values the rate constants take), each positive stoichiometric compatibility class contains precisely one equilibrium, and it is asymptotically stable. Moreover, there is no nontrivial cyclic composition trajectory along which all species concentrations are positive.*

Proof of the theorem is a little complicated. Its second part involves a Lyapunov function suggested by classical thermodynamics.

Recall the system of isothermal mass action differential equations (4) for our sample network (1). Without the help of an overarching theory it would be difficult to determine, even for *specified* values of the rate constants, the presence, for example, of multiple stoichiometrically compatible positive equilibria or a positive cyclic composition trajectory.

Note, however, that the network has five complexes, two linkage classes, and rank three. Thus, its deficiency is zero. Moreover, the network is weakly reversible. The theorem tells us immediately that, *regardless of the rate constant values*, the dynamics must be of the relatively dull, stable kind that the theorem describes.

For an introduction to more recent and very different reaction network theory results, see the article by Craciun et al., which is more graph-theoretical in spirit and which has an emphasis on enzyme-driven chemistry.

### Further Reading

Aris, R. 1989. *Elementary Chemical Reactor Analysis*. Mineola, NY: Dover Publications.

Craciun, G., Y. Tang, and M. Feinberg. 2006. Understanding bistability in complex enzyme-driven reaction networks. *Proceedings of the National Academy of Sciences of the USA* 103:8697–702.

Feinberg, M. 1979. Lectures on chemical reaction networks. Mathematics Research Center, University of Wisconsin. (Available at www.crnt.osu.edu/LecturesOnReactionNetworks.)

Feinberg, M. 1987. Chemical reaction network structure and the stability of complex isothermal reactors. I. The deficiency zero and deficiency one theorems. *Chemical Engineering Science* 42:2229–69.

Rawlings, J. B., and J. G. Ekerdt. 2002. *Chemical Reactor Analysis and Design Fundamentals*. Madison, WI: Nob Hill Publishing.

## V.8 Divergent Series: Taming the Tails

*Michael V. Berry and Christopher J. Howls*

### 1 Introduction

By the seventeenth century, in the field that became the theory of convergent series, it was beginning to be understood how a sum of infinitely many terms could be finite; this is now a fully developed and largely standard element of every mathematician's education. Contrasting with those series, we have the theory of series that do not converge, especially those in which the terms first get smaller but then increase factorially: this is the class of "asymptotic series," encountered frequently in applications, with which this article is mainly concerned. Although now a vibrant area of research, the development of the theory of divergent series has been tortuous and has often been accompanied by controversy. As a pedagogical device to explain the subtle concepts involved, we will focus on the contributions of individuals and describe how the ideas developed during several (overlapping) historical epochs, often driven by applications (ranging from wave physics to number theory). This article complements PERTURBATION THEORY AND ASYMPTOTICS [IV.5] elsewhere in this volume.

## 2 The Classical Period

In 1747 the Reverend Thomas Bayes (better known for his theorem in probability theory) sent a letter to Mr. John Canton (a Fellow of the Royal Society); it was published posthumously in 1763. Bayes demonstrated that the series now known as Stirling's expansion for $\log(z!)$, "asserted by some eminent mathematicians," does not converge. Arguing from the recurrence relation relating successive terms of the series, he showed that the coefficients "increase at a greater rate than what can be compensated by an increase of the powers of $z$, though $z$ represent a number ever so large." We would now say that this expansion of the factorial function is a factorially divergent asymptotic series. The explicit form of the series, written formally as an equality, is

$$\log(z!) = (z + \tfrac{1}{2}) \log z + \log \sqrt{2\pi} - z + \frac{1}{2\pi^2 z} \sum_{r=0}^{\infty} (-1)^r \frac{a_r}{(2\pi z)^{2r}},$$

where

$$a_r = (2r)! \sum_{n=1}^{\infty} \frac{1}{n^{2r+2}}.$$

Bayes claimed that Stirling's series "can never properly express any quantity at all" and the methods used to obtain it "are not to be depended upon."

Leonhard Euler, in extensive investigations of a wide variety of divergent series beginning several years after Bayes sent his letter, took the opposite view. He argued that such series have a precise meaning, to be decoded by suitable resummation techniques (several of which he invented): "if we employ [the] definition ... that ... the sum of a series is that quantity which generates the series, all doubts with respect to divergent series vanish and no further controversy remains."

With the development of rigorous analysis in the nineteenth century, Euler's view, which as we will see is the modern one, was sidelined and even derided. As Niels Henrik Abel wrote in 1828: "Divergent series are the invention of the devil, and it is shameful to base on them any demonstration whatsoever." Nevertheless, divergent series, especially factorially divergent ones, repeatedly arose in application. Toward the end of the century they were embraced by Oliver Heaviside, who used them in pioneering studies of radio-wave propagation. He obtained reliable results using undisciplined semiempirical arguments that were criticized by mathematicians, much to his disappointment: "It is not easy to get up any enthusiasm after it has been artificially cooled by the wet blanket of rigorists."

## 3 The Neoclassical Period

In 1886 Henri Poincaré published a definition of asymptotic power series, involving a large parameter $z$, that was both a culmination of previous work by analysts and the foundation of much of the rigorous mathematics that followed. A series of the form $\sum_{n=0}^{\infty} a_n/z^n$ is defined as asymptotic by Poincaré if the error resulting from truncation at the term $n = N$ vanishes as $K/z^{N+1}$ ($K > 0$) as $|z| \to \infty$ in a certain sector of the complex $z$-plane. In retrospect, Poincaré's definition seems a retrograde step because, although it encompasses convergent as well as divergent series in one theory, it fails to address the distinctive features of divergent series that ultimately lead to the correct interpretation that can also cure their divergence.

It was George Stokes, in research inspired by physics nearly four decades before Poincaré, who laid the foundations of modern asymptotics. He tackled the problem of approximating an integral devised by George Airy to describe waves near caustics, the most familiar example being the rainbow. This is what we now call the Airy function $\mathrm{Ai}(z)$, defined by the oscillatory integral

$$\mathrm{Ai}(z) \equiv \frac{1}{2\pi} \int_{-\infty}^{\infty} \exp\left(\frac{\mathrm{i}}{3}t^3 + \mathrm{i}zt\right) \mathrm{d}t,$$

the rainbow-crossing variable being $\mathrm{Re}\, z$ (figure 1) and the light intensity being $\mathrm{Ai}^2(z)$. Stokes derived the asymptotic expansion representing the Airy function for $z > 0$ and showed that it is factorially divergent. His innovation was to truncate this series not at a fixed order $N$ but at its smallest term (*optimal truncation*), corresponding to an order $N(z)$ that increases with $z$. By studying the remainder left after optimal truncation, he showed that it is possible to achieve exponential accuracy (figure 1), far beyond the power-law accuracy envisaged in Poincaré's definition. We will call such optimal truncation *superasymptotics*.

Superasymptotics enabled Stokes to understand a much deeper phenomenon—one that is fundamental to the understanding of divergent series. In $\mathrm{Ai}(z)$, $z > 0$

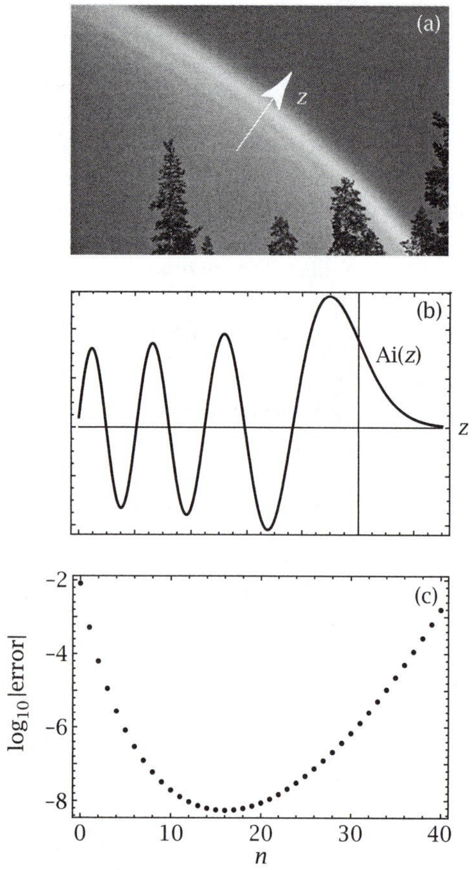


**Figure 1** (a) The rainbow-crossing variable, along any line transverse to the rainbow curve. (b) The Airy function. (c) The error from truncating asymptotic series for $\mathrm{Ai}(z)$ at the term $n - 1$, for $z = 5.24148$; optimal truncation occurs at the nearest integer to $F = 4z^{3/2}/3$, i.e., at $n = 16$.

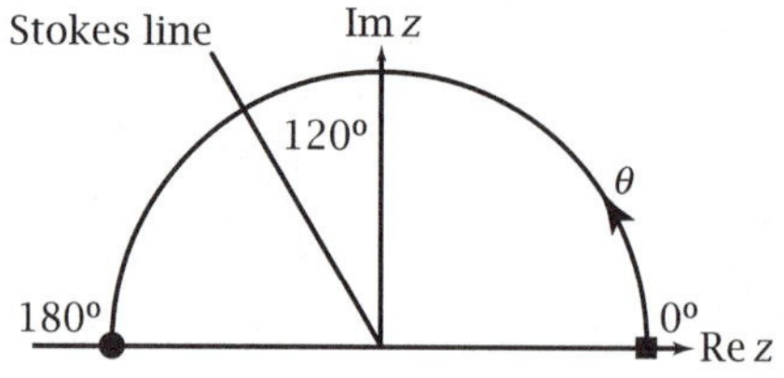


**Figure 2** The complex plane of argument $z$ (whose real part is the $z$ of figure 1) of $\mathrm{Ai}(z)$, showing the Stokes line at $\arg z = 120°$.

corresponds to the dark side of the rainbow, where the function decays exponentially; physically, this represents an evanescent wave. On the bright side $z < 0$, the function oscillates trigonometrically, that is, as the sum of two complex exponential contributions, each representing a wave; the interference of these waves generates the "supernumerary rainbows" (whose observation was one of the phenomena earlier adduced by Thomas Young in support of his view that light is a wave phenomenon). One of these complex exponentials is the continuation across $z = 0$ of the evanescent wave on the dark side. But where does the other originate?

Stokes's great discovery was that this second exponential appears during continuation of $\mathrm{Ai}(z)$ in the complex plane from positive to negative $z$, across what is now called a "Stokes line," where the dark-side exponential reaches its maximum size. Alternatively stated, the small (subdominant) exponential appears when maximally hidden behind the large (dominant) one. For $\mathrm{Ai}(z)$ the Stokes line is $\arg z = 120°$ (figure 2).

Stokes thought that the least term in the asymptotic series, representing the large exponential, constituted an irreducible vagueness in the description of $\mathrm{Ai}(z)$ in his superasymptotic scheme. By quantitative analysis of the size of this least term, Stokes concluded that only at maximal dominance could this obscure the small exponential, which could then appear without inconsistency. As we will explain later, Stokes was wrong to claim that superasymptotics—optimal truncation—represents the best approximation that can be achieved within asymptotics. But his identification of the Stokes line as the place where the small exponential is born (figure 3) was correct. Moreover, he also appreciated that the concept was not restricted to $\mathrm{Ai}(z)$ but applies to a wide variety of functions arising from integrals, solutions of differential equations and recurrence equations, etc., for which the associated asymptotic series are factorially divergent.

This *Stokes phenomenon*, connecting different exponentials representing the same function, is central to our current understanding of such divergent series, and it is the feature that distinguishes them most sharply from convergent ones. In view of this seminal contribution, it is ironic that George ("G. H.") Hardy makes no mention of the Stokes phenomenon in his textbook *Divergent Series*. Nor does he exempt Stokes from his devastating assessment of nineteenth-century English mathematics: "there [has been] no first-rate subject, except music, in which England has occupied so consistently humiliating a position. And what have been the peculiar characteristics of such English mathematics?... For the most part, amateurism, ignorance, incompetence, and triviality."

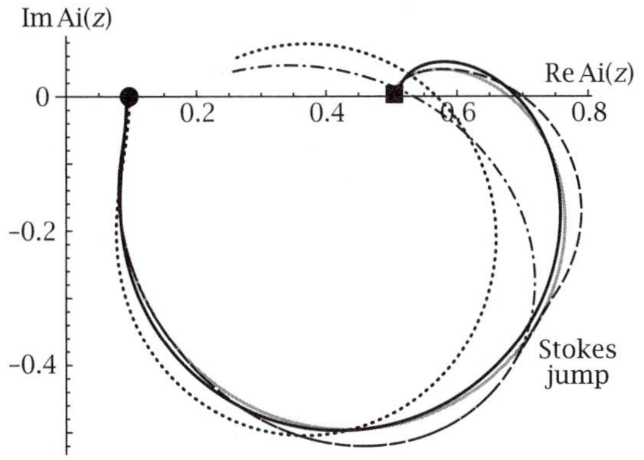


**Figure 3** Approximations to $\mathrm{Ai}(ze^{i\theta})$ in the Argand plane, (Re(Ai), Im(Ai)) for $z = 1.31\dots$, i.e., $F = 4z^{3/2}/3 = 2$, plotted parametrically from $\theta = 0°$ (black circle) to $\theta = 180°$ (black square). The curves are exact Ai (solid black line), lowest-order asymptotics (no correction terms; dotted black line), optimal truncation without the Stokes jump (dot-dashed black line), optimal truncation including the Stokes jump (dashed black line), and optimal truncation including the smoothed Stokes jump (solid gray line). For this value of $F$, the optimally truncated sum contains only two terms; on the scale shown, the Stokes jump would be invisible for larger $F$. Note that without the Stokes jump (solid and dotted black lines), the asymptotics must deviate from the exact function beyond the Stokes line at $\theta = 120°$.

## 4 The Modern Period

Late in the nineteenth century, Jean-Gaston Darboux showed that for a wide class of functions the high derivatives diverge factorially. This would become an important ingredient in later research, for the following reason. Asymptotic expansions (particularly those encountered in physics and applied mathematics) are often based on local approximations: the steepest-descent method for approximating integrals is based on local expansion about a saddle point; the phase-integral method for solving differential equations (e.g., the Wentzel–Kramers–Brillouin (WKB) approximation to Schrödinger's equation in quantum mechanics) is based on local expansions of the coefficients; and so on. Therefore, successive orders of approximation involve successive derivatives, and the high orders, responsible for the divergence of the series, involve high derivatives.

Another major late-nineteenth-century ingredient of our modern understanding was Émile Borel's development of a powerful summation method in which the factorials causing the high orders to diverge are tamed by replacing them with their integral representations. Often this enables the series to be summed "under the integral sign." Underlying the method is the formal equality

$$\sum_{r=0}^{\infty} \frac{a_r}{z^r} = \sum_{r=0}^{\infty} \frac{a_r r!}{z^r r!} = \int_0^{\infty} \mathrm{d}t\, \mathrm{e}^{-t} \sum_{r=0}^{\infty} \frac{a_r}{r!}\left(\frac{t}{z}\right)^r.$$

Reading this from right to left is instructive. Interchanging summation and integration shows why the series on the left diverges if the $a_r$ increase factorially (as in the cases we are considering): the integral is over a semi-infinite range, yet the sum in the integrand converges only for $|t/z| < 1$. Borel summation effectively repairs an analytical transgression that may have caused the divergence of the series. The power of Borel summation is that, as was fully appreciated only later, it can be analytically continued across Stokes lines, where some other summation techniques (e.g., Padé approximants) fail.

Now we come to the central development in modern asymptotics. In a seminal and visionary advance, motivated initially by mathematical difficulties in evaluating some integrals occurring in solid-state physics and developed in a series of papers culminating in a book published in 1973, Robert Dingle synthesized earlier ideas into a comprehensive theory of factorially divergent asymptotic series.

Dingle's starting point was Euler's insight that divergent series are obtained by a sequence of precisely specified mathematical operations on the integral or differential equation defining the function being approximated, so the resulting series must represent the function exactly, albeit in coded form, which it is the task of asymptotics to decode. Next was the realization that Darboux's discovery that high derivatives diverge factorially implies that the high orders of a wide class of asymptotic series also diverge factorially. This in turn means that the terms beyond Stokes's optimal truncation—representing the tails of such series beyond superasymptotics—can all be Borel-summed in the same way.

The next insight was Dingle's most original contribution. Consider a function represented by several different formal asymptotic series (e.g., those corresponding to the two exponentials in $\mathrm{Ai}(z)$), each representing the function differently in sectors of the complex plane separated by Stokes lines. Since each series is a formally complete representation of the function, each must contain, coded into its high orders, information about all the other series. Darboux's factorials are therefore simply the first terms of asymptotic expansions of each of the late terms of the original series. Dingle appreciated that the natural variables implied by Darboux's theory are the differences between the various exponents; these are usually proportional to the large asymptotic parameter. In the simplest case, where there are only two exponentials, there is one such variable, which Dingle called the *singulant*, denoted by $F$. For the Airy function $\mathrm{Ai}(z)$, $F = 4z^{3/2}/3$.

We display Dingle's expression for the high orders for an integral with two saddle points $a$ and $b$, corresponding to exponentials $\mathrm{e}^{-F_a}$ and $\mathrm{e}^{-F_b}$ with $F_{ab} = F_b - F_a$ and series with terms $T_r^{(a)}$ and $T_r^{(b)}$; for $r \gg 1$, the terms of the $a$ series are related to those of the $b$ series by

$$T_r^{(a)} = K\frac{(r-1)!}{F_{ab}^r}\left(T_0^{(b)} + \frac{F_{ab}}{r-1}T_1^{(b)} + \frac{F_{ab}^2}{(r-1)(r-2)}T_2^{(b)} + \cdots\right),$$

in which $K$ is a constant. This shows that, although the early terms $T_0^{(a)}, T_1^{(a)}, T_2^{(a)}, \dots$ of an asymptotic series can rapidly get extremely complicated, the high orders display a miraculous functional simplicity.

With Borel's as the chosen summation method, Dingle's late-terms formula enabled the divergent tails of series to be summed in terms of certain *terminant* integrals and then reexpanded to generate new asymptotic series that were exponentially small compared with the starting series. He envisaged that "these terminant expansions can themselves be closed with new terminants; and so on, stage after stage." Such resummations, beyond superasymptotics, were later called *hyperasymptotics.*

Dingle therefore envisaged a universal technique for repeated resummation of factorially divergent series, to obtain successively more accurate exponential improvements far beyond that achievable by Stokes's optimal truncation of the original series. The meaning of universality is that although the early terms—the ones that get successively smaller—can be very different for different functions, the summation method for the tails is always the same, involving terminant integrals that are the same for a wide variety of functions. The method automatically incorporates the Stokes phenomenon. Although Dingle clearly envisaged the hyperasymptotic resummation scheme as described above, he applied it only to the first stage; this was sufficient to illustrate the high improvement in numerical accuracy compared with optimal truncation.

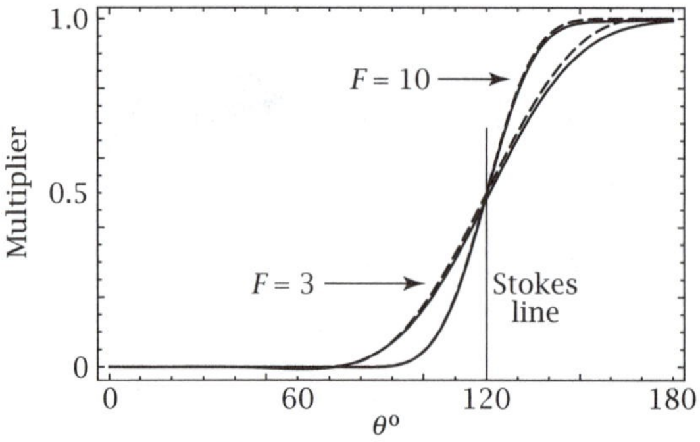


**Figure 4** The Stokes multiplier (solid lines) and error-function smoothing (dashed lines) for $\mathrm{Ai}(z\mathrm{e}^{\mathrm{i}\theta})$, for $z = 1.717\dots$ (i.e., $F = 4z^{3/2}/3 = 3$) and $z = 3.831\dots$ (i.e., $F = 4z^{3/2}/3 = 10$).

Like Stokes before him, Dingle presented his new ideas not in the "lemma, theorem, proof" style familiar to mathematicians but in the discursive manner of a theoretical physicist, perhaps explaining why it has taken several decades for the originality of his approach to be widely appreciated and accepted. Meanwhile, his explicit relation, connecting the early and late terms of different asymptotic series representing the same function, was rediscovered independently by several people. In particular, Jean Écalle coined the term *resurgence* for the phenomenon, and in a sophisticated and comprehensive framework applied it to a very wide class of functions.

## 5 The Postmodern Period

One of the first steps beyond superasymptotics into hyperasymptotics was an application of Dingle's ideas to give a detailed description of the Stokes phenomenon. In 1988 one of us (Michael Berry) resummed the divergent tail of the dominant series of the expansion, near a Stokes line, of a wide class of functions, including $\mathrm{Ai}(z)$. The analysis demonstrated that the subdominant exponential did not change suddenly, as had been thought; rather, it changed smoothly and, moreover, in a universal manner. In terms of Dingle's singulant $F$, now defined as the difference between the exponents of the dominant and subdominant exponentials, the Stokes line corresponds to the positive real axis in the complex $F$-plane; asymptotics corresponds to $\mathrm{Re}\, F \gg 1$; and the Stokes phenomenon corresponds to crossing the Stokes line, that is $\mathrm{Im}\, F$ passing through zero.

The result of the resummation is that the change in the coefficient of the small exponential—the Stokes multiplier—is universal for all factorially divergent series and it is proportional to

$$\frac{1}{2}\left(1 + \text{Erf}\left(\frac{\text{Im}\,F}{\sqrt{2\,\text{Re}\,F}}\right)\right).$$

In the limit $\text{Re}\,F \to \infty$ this becomes the unit step. For $\text{Re}\,F$ large but finite, the formula describes the smooth change in the multiplier (figure 4) and makes precise the description given by Stokes in 1902 after thinking about divergent series for more than half a century:

> The inferior term enters as it were into a mist, is hidden for a little from view, and comes out with its coefficient changed. The range during which the inferior term remains in a mist decreases indefinitely as the [large parameter] increases indefinitely.

The smoothing shows that the "range" referred to by Stokes (that is, the effective thickness of the Stokes line) is of order $\sqrt{\text{Re}\,F}$.

Implementation of the full hyperasymptotic repeated resummation scheme envisaged by Dingle has been carried out in several ways. We (the present authors) investigated one-dimensional integrals with several saddle points, each associated with an exponential and its corresponding asymptotic series. With each "hyperseries" truncated at its least term, this incorporated all subdominant exponentials and all associated Stokes phenomena; and the accuracy obtained far exceeded superasymptotics (figure 5) but was nevertheless limited.

It was clear from the start that in many cases unlimited accuracy could, in principle, be achieved with hyperasymptotics by truncating the hyperseries not at the smallest term but beyond it (although this introduces numerical stability issues associated with the cancelation of larger terms). This version of the hyperasymptotic program was carried out by Adri Olde Daalhuis, who reworked the whole theory, introducing mathematical rigor and effective algorithms for computing Dingle's terminant integrals and their multidimensional generalizations, and applied the theory to differential equations with arbitrary finite numbers of transition points.

There has been an explosion of further developments. Écalle's rigorous formal theory of resurgence has been developed in several ways, based on the Borel (effectively inverse-Laplace) transform. This converts the factorially divergent series into a convergent one,

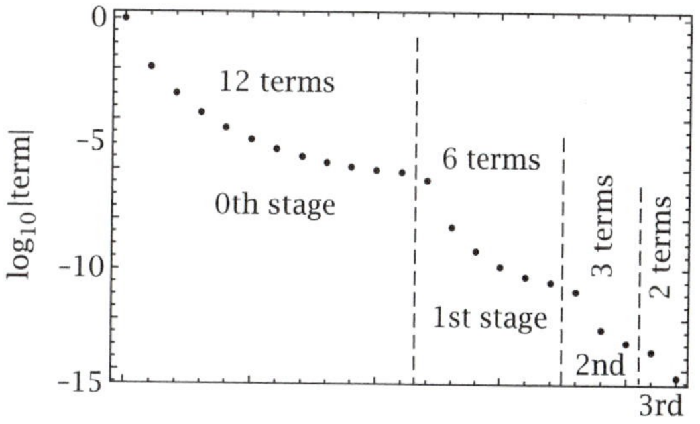


**Figure 5** The magnitude of terms in the first four stages in the hyperasymptotic approximation to $\text{Ai}(4.326\ldots) = 4.496\ldots \times 10^{-4}$, i.e., $F = 4z^{3/2}/3 = 12$, normalized so that the lowest approximation is unity. For the lowest approximation, i.e., no correction terms, the fractional error is $\varepsilon \approx 0.01$; after stage 0 of hyperasymptotics, i.e., optimal truncation of the series (superasymptotics), $\varepsilon \approx 3.6 \times 10^{-7}$; after stage 1, $\varepsilon \approx 1.3 \times 10^{-11}$; after stage 2, $\varepsilon \approx 4.4 \times 10^{-14}$; after stage 3, $\varepsilon \approx 6.1 \times 10^{-15}$. At each stage, the error is of the same order as the first neglected term.

with the radius of convergence determined by singularities on a Riemann sheet. These singularities are responsible for the divergence of the original series, and for integrals discussed above, they correspond to the adjacent saddle points. In the Borel plane, complex and microlocal analysis allows the resurgence linkages between asymptotic contributions to be uncovered and exact remainder terms to be established. Notable results include exponentially accurate representations of quantum eigenvalues (R. Balian and C. Bloch, A. Voros, F. Pham, E. Delabaere); this inspired the work of the current authors on quantum eigenvalue counting functions, linking the divergence of the series expansion of smoothed spectral functions to oscillatory corrections involving the classical periodic orbits.

T. Kawai and Y. Takei have extended "formally exact," exponentially accurate WKB analysis to several areas, most notably to Painlevé equations. They have also developed a theory of "virtual turning points" and "new Stokes curves." In the familiar WKB situation, with only two wavelike asymptotic contributions, several Stokes lines emerge from classical turning points, and (if they are nondegenerate) they never cross. With three or more asymptotic contributions, Stokes lines can cross in the complex plane at points where the WKB solutions are not singular. Local analysis shows that an extra, active, "new Stokes line" sprouts from one side only of this regular point; this can be shown to emerge from a distant virtual turning point, where, unexpectedly, the WKB solutions are not singular. This discovery has been explained by C. J. Howls, P. Langman, and A. B. Olde Daalhuis in terms of the Riemann sheet structure of the Borel plane and linked hyperasymptotic expansions, and independently by S. J. Chapman and D. B. Mortimer in terms of matched asymptotics.

Groups led by S. J. Chapman and J. R. King have developed and applied the work of M. Kruskal and H. Segur to a variety of nonlinear and partial differential equation problems. This involves a local matched-asymptotic analysis near the distant Borel singularities that generate the factorially divergent terms in the expansion to identify the form of late terms, thereby allowing for an optimal truncation and exponentially accurate approach. Applications include selection problems in viscous fluids, gravity–capillary solitary waves, oscillating shock solutions in Kuramoto–Sivashinsky equations, elastic buckling, nonlinear instabilities in pattern formation, ship wave modeling, and the seeking of reflectionless hull profiles. Using a similar approach, O. Costin and S. Tanveer have identified and quantified the effect of "daughter singularities" that are not present in the initial data of partial differential equation problems but that are generated at infinitesimally short times. In so doing, they have also found a formally exact Borel representation for small-time solutions of three-dimensional Navier–Stokes equations, offering a promising tool to explore the global existence problem.

Additional applications include quantum transitions, quantum spectra, the Riemann zeta function high on the critical line, nonperturbative quantum field theory and string theory, and even the philosophy of representing physical theories describing phenomena at different scales by singular relations.

### Further Reading

Batterman, R. W. 1989. Uniform asymptotic smoothing of Stokes's discontinuities. *Proceedings of the Royal Society of London* A 422:7–21.

——. 2002. *The Devil in the Details: Asymptotic Reasoning in Explanation, Reduction and Emergence.* Oxford: Oxford University Press.

Berry, M. V., and C. J. Howls. 1991. Hyperasymptotics for integrals with saddles. *Proceedings of the Royal Society of London* A 434:657–75.

Delabaere, E., and F. Pham. 1999. Resurgent methods in semi-classical mechanics. *Annales de l'Institut Henri Poincaré* 71:1–94.

Digital Library of Mathematical Functions. 2010. *NIST Handbook of Mathematical Functions*, chapters 9 and 36. Cambridge: Cambridge University Press.

Dingle, R. B. 1973. *Asymptotic Expansions: Their Derivation and Interpretation.* New York: Academic Press.

Howls, C. J. 1997. Hyperasymptotics for multidimensional integrals, exact remainder terms and the global connection problem. *Proceedings of the Royal Society of London* A 453:2271–94.

Kruskal, M. D., and H. Segur. 1991. Asymptotics beyond all orders in a model of crystal growth. *Studies in Applied Mathematics* 85:129–81.

Olde Daalhuis, A. B. 1998. Hyperasymptotic solutions of higher order linear differential equations with a singularity of rank one. *Proceedings of the Royal Society of London* A 454:1–29.

Stokes, G. G. 1864. On the discontinuity of arbitrary constants which appear in divergent developments. *Transactions of the Cambridge Philosophical Society* 10:106–28.

## V.9 Financial Mathematics

*René Carmona and Ronnie Sircar*

The complexity, unpredictability, and evolving nature of financial markets continue to provide an enormous challenge to mathematicians, engineers, and economists in identifying, analyzing, and quantifying the issues and risks they pose. This has led to problems in stochastic analysis, simulation, differential equations, statistics and big data, and stochastic control and optimization (including dynamic game theory), all of which are reflected in the core of financial mathematics. Problems range from modeling a single risky stock and the risks of derivative contracts written on it, to understanding how intricate interactions between financial institutions may bring down the whole financial edifice and in turn the global economy: the problem of systemic risk. At the same time, hitherto specialized markets, such as those in commodities (metals, agriculturals, and energy), have become more *financialized*, which has led to our need to understand how financial reduced-form models combine with supply and demand mechanisms.

In its early days, financial mathematics used to rest on two pillars that could be characterized roughly as *derivatives pricing* and PORTFOLIO SELECTION [V.10]. In this article we outline its broadening into newer topics including, among others, energy and commodities markets, systemic risk, dynamic game theory and equilibrium, and understanding the impact of algorithmic and high-frequency trading. We also touch on the

2008 financial crisis (and others) and the extent to which such increasingly frequent tremors call for more mathematics, not less, in understanding and regulating financial markets and products.

## 1 The Pricing and Development of Derivative Markets

Central to the development of quantitative finance, as distinguished from classical economics, was the modeling of uncertainty about future price fluctuations as being phenomenological, rather than something that could be accurately captured by models of fundamentals, or demand and supply. The introduction of randomness into models of (initially) stock prices took off in the 1950s and 1960s, particularly in the work of the economist Paul Samuelson at MIT, who adopted the continuous-time Brownian motion based tools of stochastic calculus that had been developed by physicists and mathematicians such as Einstein in 1905; Wiener in the 1920s; Lévy, Ornstein, and Uhlenbeck in the 1930s; and Chandrasekhar in the 1940s. Samuelson, however, came to this mathematical machinery not through physics but via a little-known doctoral dissertation by Bachelier from 1900, which had formulated Brownian motion, for the purpose of modeling the Paris stock market, five years before Einstein's landmark paper in physics ("On the motion of small particles suspended in a stationary liquid, as required by the molecular kinetic theory of heat"). Combined with the stochastic calculus developed by Itô in the early 1940s, these kinds of models became, and still remain, central to the analysis of a wide range of financial markets.

The Black–Scholes paradigm for equity derivatives was originally introduced in the context of Samuelson's model, in which stocks evolve according to geometric Brownian motions. If we consider a single stock for the sake of exposition, one assumes that the value at time $t$ of one share is $S_t$, and $S_t$ evolves according to the stochastic differential equation

$$dS_t = \mu S_t \, dt + \sigma S_t \, dW_t, \tag{1}$$

with $\mu$ representing the expected growth rate, and $\sigma > 0$ the volatility. Here, $(W_t)_{t \geqslant 0}$ is a standard Brownian motion. A contingent claim (or derivative security) on this stock is defined by its payoff at a later date $T$, called its maturity, and modeled as a random variable $\xi$ whose uncertain value will be revealed at time $T$. The typical example is given by a European call option with maturity $T$ and strike $K$, in which case the payoff to the buyer of the option would be $\xi = (S_T - K)^+$, where we use the notation $x^+ = \max\{x, 0\}$ for the positive part of the real number $x$.

A desirable property of a model is to exclude arbitrage opportunities, namely, the possibility of making money with no risk (often called a "free lunch"). If this property holds, then the value of owning the claim should be equal to the value of an investment in the stock and a safe bank account that could be managed by dynamic rebalancing into the same value as $\xi$ at maturity $T$. Such a replicating portfolio would represent a perfect hedge, mitigating the uncertainty in the outcome of the option payoff $\xi$. The remarkable discovery in the early 1970s of Fischer Black and Myron Scholes, and independently by Robert Merton, was that it was possible to identify such a perfect replicating portfolio and compute the initial investment needed to set it up by solving a linear partial differential equation (PDE) of parabolic type. They also provided formulas for the no-arbitrage prices of European call and put options that are now known as Black–Scholes formulas. Later, in two papers that both appeared in 1979, the result was given its modern formulation, first by Cox, Ross, and Rubinstein in a simpler discrete model and then by Harrison and Kreps in more general settings. Accordingly, if there are no arbitrage opportunities, the prices of all traded securities (stocks, futures, options) are given by computing the expectation of the present value (i.e., the discounted value) of their future payoffs but with respect to a "risk-neutral" probability measure under which the growth rate $\mu$ in (1) is replaced by the riskless interest rate of the bank account.

While pricing by expectation is one of the important consequences of the original works of Black, Scholes, and Merton, it would be misleading and unfair to reduce their contribution to this aspect, even if it is at the origin of a wave (in retrospect, it was clearly a tsunami) of interest in derivatives and the explosion of new and vibrant markets. Even though the original rationale was based on a hedging argument aimed at mitigating the uncertainties in the future outcomes of the value of the stock underlying the option, pricing, and especially pricing by expectation (whether the expectations were computed by Monte Carlo simulations or by PDEs provided by the Feynman–Kac formula), became the main motivation for many research programs.

### 1.1 Stochastic Volatility Models

It has long been recognized (even by Black and Scholes, and many others, in the 1970s) that the lognormal distribution inherent in the geometric Brownian motion model (1) is not reflected by historical stock price data and that volatility is not constant over time. Market participants were pricing options, even on a given day $t$, as if the volatility parameter $\sigma$ depended upon the strike $K$ and the time to maturity $T - t$ of the option.

To this day, two extensions of the model (1) have been used with great success to account for these stylized facts observed in empirical market data. They both involve replacing the volatility parameter by a stochastic process, so they can be viewed as stochastic volatility models. The first extension is to replace (1) by

$$\mathrm{d}S_t = \mu S_t \,\mathrm{d}t + \sigma_t S_t \,\mathrm{d}W_t, \tag{2}$$

where the volatility parameter $\sigma$ is replaced by the value at time $t$ of a stochastic process $(\sigma_t)_{t\geqslant 0}$ whose time evolution could be (for example) of the form

$$\mathrm{d}\sigma_t = \lambda(\bar{\sigma} - \sigma_t)\,\mathrm{d}t + \gamma\sqrt{\sigma_t}\,\mathrm{d}\tilde{W}_t \tag{3}$$

for some constants $\gamma$ (known as volvol), $\bar{\sigma}$ (the mean reversion level), and $\lambda$ (the rate of mean reversion) and where $(\tilde{W}_t)_{t\geqslant 0}$ is another Brownian motion that is typically negatively correlated with $(W_t)_{t\geqslant 0}$ to capture the fact that, when volatility rises, prices usually decline. Stochastic volatility models of this type have been (and are still) very popular.

Having two sources of random shocks (whether or not they are independent) creates some headaches for the quants, as the no-arbitrage prices are now plentiful, and in the face of the nonuniqueness of derivative prices, tricky *calibration* issues have to be resolved. So rather than dealing with the incompleteness of these stochastic volatility models, a more minimalist approach was proposed to capture the empirical properties of the option prices while at the same time keeping only one single source of shocks and, hence, retaining the completeness of the model. These models go under the name of *local volatility models*. They are based on dynamics given by Markovian stochastic differential equations of the form

$$\mathrm{d}S_t = \mu S_t \,\mathrm{d}t + \sigma(t, S_t)\,\mathrm{d}W_t,$$

where $(t, s) \hookrightarrow \sigma(t, s)$ is a deterministic function, which can be computed from option prices using what is known as Dupire's formula in lieu of the geometric Brownian motion equation (1). The other stochastic volatility models are also the subject of active research.

According to the Black–Scholes theory, prices of contingent claims appear as risk-adjusted expectations of the discounted cash flows triggered by the settlement of the claims. In the case of contingent claims with European exercises, the random variable giving the payoff is often given by a function of an underlying Markov process at the time of maturity of the claim. Using the Feynman–Kac formula, these types of expectations appear as solutions of PDEs of parabolic type, showing that prices can be computed by solving PDEs. The classical machinery of the numerical analysis of linear PDEs is the cornerstone of most of the pricers of contingent claims in low dimensions. However, the increasing size of the *baskets* of instruments underlying derivative contracts and the complexity of the exercise contingencies have limited the efficacy of the PDE solvers because of the high dimensionality. This is one of the main reasons for the increasing popularity of Monte Carlo methods. Combined with regression ideas, they provide robust algorithms capable of pricing options (especially options with American exercises) on large portfolios while avoiding the CURSE OF DIMENSIONALITY [I.3 §2] that plagues traditional PDE methods. Moreover, the ease with which one can often generate Monte Carlo scenarios for the sole purpose of backtesting has added to the popularity of these random simulation methods.

### 1.2 Bond Pricing and Fixed-Income Markets

The early and mid-1990s saw growth in the bond markets (a sudden increase in the traded volumes in Treasury, municipal, sovereign, corporate, and other bonds), and the fixed-income desks of many investment banks and other financial institutions became major sources of profit. The academic financial mathematics community took notice, and a burst of research into mathematical models for fixed-income instruments followed.

Parametric and nonparametric models for the term structure of interest rates (yield curves describing the evolution of forward interest rates as a function of the maturity tenors of the bonds) were successfully developed from classical data-analysis procedures. While spline smoothing was often used, principal component analysis of the yield data clearly points to a small number of easily identified factors, and least-squares regressions can be used to identify the term structure in parameterized families of curves successfully. Despite the fact that infinite-dimensional functional analysis and stochastic PDEs were brought to bear in order to describe the data, model calibration ended up being easier than in the case of the equity markets, where the nonlinear correspondence between option prices and the implied volatility surface creates challenges that, to this day, still remain mostly unsolved.

The development of fixed-income markets was very rapid, and the complexity of interest rate derivatives (swaps, swaptions, floortions, captions, etc.) motivated the scaling-up of mathematical models from the mere analysis of one-dimensional stochastic differential equations to the study of infinite-dimensional stochastic systems and stochastic PDEs. Current research in financial mathematics is geared toward the inclusion of jumps in these models and the understanding of the impact of these jumps on calibration, pricing, and hedging procedures.

### 1.3 Default Models and Credit Derivative Markets

Buying a bond is just making a loan; issuing a bond is nothing but borrowing money. While the debt of the U.S. government is still regarded by *most* as default free, sovereign bonds and most corporate bonds carry a significant risk associated with the nonnegligible probability that the issuer may not be able to make good on his debt and may default by the time the principal of the loan is to be returned. Unsurprisingly, models of default were first successfully included in the pricing of corporate bonds.

*Structural models* of default based on the fundamentals of a firm and the competing roles of its assets and liabilities were first introduced by Merton in 1974. Their popularity was due to a rationale that was solidly grounded on fundamental financial principles and data reporting. However, murky data and a lack of transparency have plagued the use of these models for the purpose of pricing corporate bonds and their derivatives.

On the other hand, *reduced-form models* based on stochastic models for the intensity of arrival of the time of default have gained in popularity because of the versatility of the intensity-based models and the simplicity and robustness of the calibration from credit default swap (CDS) data. Indeed, one of the many reasons for the success of reduced-form models is readily available data. Quotes of the spreads on CDSs for most corporations are easy to get, and the fact that they are plentiful contrasts with the scarcity of corporate bond quotes, which are few and far between. In fact, because the CDS market exploded in the early to mid-2000s, gauging the creditworthiness of a company is easily read off the CDS spreads instead of the bond spread (the difference between the interest paid on a riskless government bond and a corporate bond).

A CDS is a form of insurance against the default of a corporation, say X. Such a CDS contract involves two counterparties, say Y and Z, the latter receiving a regular premium payment from Y as long as X does not default and paying a lump sum to Y in the case of default of X before the maturity of the CDS contract. The existence of a CDS contract between Y and Z therefore seems natural if the financial health of one of these two counterparties depends upon the survival of X and the other counterparty is willing to take the opposite side of the transaction. However, neither of the counterparties Y or Z who are *gambling* on the possibility of a serious credit event concerning X (both having different views on the likelihood of default) need to have any financial interest, direct or indirect, in X. In other words, two agents can enter into a deal involving a third entity just as a pure bet on the creditworthiness of this third entity. While originally designed as a credit insurance, CDSs ended up increasing the overall risk in the system through the multiplication of *private* bets. As they spread like uncontrolled brush fires, they created an intricate network of complex dependencies between institutions, making it practically impossible to trace the sources of the risks.

### 1.4 Securitization

Investment is a risky business, and the risks of large portfolios of defaultable instruments (corporate bonds, loans, mortgages, CDSs) were clearly a major source of fear, at least until the spectacular growth of *securitization* in recent years. A financial institution bundles together a large number of such defaultable instruments, *slices* the portfolio according to the different levels of default risk, forming a small number of tranches (say five), keeps the riskiest one (called the "equity tranche"), and sells the remaining tranches (known as mezzanine or senior tranches) to trusting investors as an investment far safer than the original portfolio itself. These instruments are called collateralized debt obligations (CDOs). Pooling risks and tranching them as reinsurance contracts to pass on to investors with differing risk profiles is natural, and it has been the basis of insurance markets for dozens of years at least. But, unlike insurance products, CDOs were not regulated. They enjoyed tremendous success for most of the 2000s; credit desks multiplied, and academics and mathematicians tried to understand how practitioners were pricing them. Needless to say, no effort was made at hedging the risk exposure as not only was it not understood but there was no reason to worry since it seemed that we could only *make* money in this game! Attempts at understanding their risk structure intensified after the serious warnings of 2004 and 2005, but they did not come early enough to seriously impact the onset of the financial crisis, which was most certainly triggered, or at least exacerbated by, the housing freeze and the ensuing collapse of the huge market in mortgage-backed securities (MBSs), which are complex CDOs on pools of mortgages. Parts II and III of the report of the Financial Crisis Inquiry Commission are an enlightening read for the connections between CDOs and the credit crisis.

The overuse of credit derivatives, particularly in the mortgage arena, contributed massively to the financial crisis. While the May 2005 ripple in the corporate CDO market served as an early warning of worse things to come, its lessons were largely ignored as the risks abated. The attraction of unfunded returns on default protection proved too great, and the culture of unbounded bonus-based compensation for traders led to excessive risk taking that jeopardized numerous long-standing and once fiscally conservative financial institutions.

In the midst of the crisis, some questioned the role that quantitative models played in motivating or justifying these trades. A *Wall Street Journal* article from 2005 highlights the practice of playing CDO tranches off against each other to form a dubious hedge. In some sense this was "quantified" using the Gaussian copula (and similar) models (particularly implied correlations and their successor, so-called base correlations). These models were developed in-house by quants hired by and answering to traders. The raison d'etre for these models was twofold: simplicity and speed. Unfortunately, the short cut used to achieve these goals was rather drastic, capturing a complex dependence structure among hundreds of correlated risks using a single correlation parameter. Many (surviving) banks have since restructured their quant/modeling teams so that they now report directly to management instead of to traders.

Numerous media outlets and commentators have blamed the crisis on financial models, and called for *less* quantification of risk. This spirit is distilled in Warren Buffet's "beware of geeks bearing formulas" comment. Indeed, the caveat about the use of models to *justify a posteriori* "foolproof" hedges is warranted. But this crisis, like past crises, only highlights the need for more mathematics and for quantitatively trained people at the highest level, not least at ratings and regulatory agencies.

The real damage was done in the highly *unquantified* market for MBSs. Here the notions of independence and diversification through tranching were taken to ludicrous extremes. The MBS desks at major banks were relatively free of quants and quantitative analysis compared with the corporate CDO desks, even though the MBS books were many times larger. The excuse given was that these products were AAA and therefore, like U.S. government bonds, did not need any risk analysis. As it turned out, this was a mass delusion willingly played into by banks, hedge funds, and the like. The quantitative analysis performed by the ratings agencies tested outcomes on only a handful of scenarios, in which, typically, the worst-case scenario was that U.S. house prices would appreciate at the rate of 0.0%. The rest is history.

## 2 Portfolio Selection and Investment Theory

A second central foundational pillar underlying modern financial mathematics research is the problem of optimal investment in uncertain market conditions. Typically, optimality is with respect to the expected utility of portfolio value, where utility is measured by a concave increasing *utility function*, as introduced by von Neumann and Morgenstern in the 1940s. A major breakthrough in applying continuous-time stochastic models to this problem came with the work of Robert Merton that was published in 1969 and 1971. In these works Merton derived optimal strategies when stock prices have constant expected returns and volatilities and when the utility function has a specific convenient form. This remains among the few examples of explicit solutions to fully nonlinear Hamilton–Jacobi–Bellman (HJB) PDEs motivated by stochastic control applications.

To explain the basic analysis, suppose an investor has a choice between investing his capital in a single risky stock (or a market index such as the S&P 500) or in a riskless bank account. The stock price $S$ is uncertain and is modeled as a geometric Brownian motion (1). The choice (or control) for the investor is therefore over $\pi_t$, the dollar amount to hold in the stock at time $t$, with his remaining wealth deposited in the bank, earning interest at the constant rate $r$. If $X_t$ denotes the portfolio value at time $t$, and if we assume that the portfolio is *self-financing* in the sense that no other monies flow in or out, then

$$dX_t = \pi_t \frac{dS_t}{S_t} + r(X_t - \pi_t)\,dt$$

is the equation governing his portfolio time evolution. We will take $r = 0$ for simplicity of exposition, and so, from (1),

$$dX_t = \mu \pi_t\,dt + \sigma \pi_t\,dW_t.$$

Increasing the stock holding $\pi$ increases the growth rate of $X$ while also increasing its volatility. The investor is assumed to have a smooth terminal utility function $U(x)$ on $\mathbb{R}^+$ that satisfies the "usual conditions" (the Inada conditions and asymptotic elasticity)

$$U'(0^+) = \infty, \qquad U'(\infty) = 0, \qquad \lim_{x\to\infty} x\frac{U'(x)}{U(x)} < 1,$$

and he wants to maximize $\mathbb{E}\{U(X_T)\}$, his expected utility of wealth at a fixed time horizon $T$. To apply dynamic programming principles, it is standard to define the value function as

$$V(t,x) = \sup_{\pi} \mathbb{E}\{U(X_T) \mid X_t = x\},$$

where the supremum is taken over admissible strategies that satisfy $\mathbb{E}\{\int_0^T \pi_t^2\,dt\} < \infty$. Then $V(t,x)$ is the solution of the HJB PDE problem

$$V_t - \tfrac{1}{2}\lambda^2 \frac{V_x^2}{V_{xx}} = 0, \qquad V(T,x) = U(x),$$

where $\lambda := \mu/\sigma$ is known as the *Sharpe ratio*. Given the value function $V$, the optimal stock holding is given by

$$\pi_t^* = -\frac{\mu}{\sigma^2}\frac{V_x}{V_{xx}}(t, X_t).$$

Remarkably, Merton discovered an explicit solution when the utility is a power function:

$$U(x) = \frac{x^{1-\gamma}}{1-\gamma}, \quad \gamma > 0,\ \gamma \neq 1.$$

Here, $\gamma$ measures the concavity of the utility function and is known as the constant of relative risk aversion. With this choice,

$$V(t,x) = \frac{x^{1-\gamma}}{1-\gamma}\exp\Big(\tfrac{1}{2}\lambda^2\Big(\frac{1-\gamma}{\gamma}\Big)(T-t)\Big),$$

and, more importantly,

$$\pi_t^* = \frac{\mu}{\sigma^2\gamma}X_t.$$

That is, the optimal strategy is to hold the fixed fraction $\mu/(\sigma^2\gamma)$ of current wealth in the stock and the rest in the bank. As the stock price rises, this strategy says to sell some stock so that the fraction of the portfolio comprised of the risky asset remains the same. This fixed-mix result generalizes to multiple securities as long as they are also assumed to be (correlated) geometric Brownian motions.

Since Merton's work, the basic problem has been generalized in many directions. In particular, developments in duality theory (or the martingale method) led to a revolution in thinking as to how these problems should be studied in abstract settings, culminating in very general results in the context of semimartingale models of incomplete markets. One of the most challenging problems was to extend the theory in the presence of transaction costs, as this required the fine analysis of singular stochastic control problems.

Optimal investment still generates challenging research problems as models evolve to try to incorporate realistic features such as uncertain volatility or random jumps in prices using modern technology such as forward–backward stochastic differential equations, asymptotic approximations for fully nonlinear HJB PDEs, and related numerical methods.

Early portfolio optimization was achieved through the well-known Markowitz linear quadratic optimization problems based on the analysis of the mean vector and the covariance matrix. The use of increasingly large portfolios and the introduction of exchange-traded fund (ETF) tracking indexes (S&P 500, Russell 2000, etc.) created the need for more efficient estimation methods for large covariance matrices, typically using sparsity and robustness arguments. This problem has been the major impetus behind a significant proportion of the *big data* research in statistics.

## 3 Growing Research Areas

### 3.1 Systemic Risk

The mathematical theory of risk measures, like value at risk, expected shortfall, or maximum drawdown, was introduced to help policy makers and portfolio managers quantify risk and define unambiguously a form of capital requirement to preserve solvency. It enjoyed immediate success among the mathematicians working on financial applications. However, their dynamical analogues did not enjoy the same popular support, mostly because of their complexity and the challenges in attempting to aggregate firm-level risks into system behavior.

Research on systemic risk began in earnest after the September 11 attacks in 2001. The financial crisis of 2008 subsequently brought counterparty risk and the propagation of defaults to the forefront. The analysis of large complex systems in which all the entities behave rationally at the individual level but which produce calamities at the aggregate level is a very exciting challenge for mathematicians, who are now developing models for what some economists have called *rational irrationality*.

### 3.2 Energy and the New Commodity Markets

As for the models of defaults used in fixed-income markets, and subsequently in credit markets, models for commodities can, roughly speaking, be divided into two distinct categories: reduced-form models and structural models.

Commodities are physical in nature and are overwhelmingly traded on a forward basis, namely, for future delivery. So, as with fixed-income markets, a snapshot of the state of a market is best provided by a term structure of forwards of varying maturities. But unlike the forward rates, forward prices are actually prices of traded commodities, and as a consequence they should be modeled as martingales (pure fluctuation processes with no trend or bias). Despite this fundamental difference, the first models for commodities were borrowed (at least in spirit) from the models developed by sophisticated researchers in fixed-income markets. However, the shortcomings of these ad hoc transplants are now recognized, and fundamental research in this area now focuses on structural models that are more in line with equilibrium arguments and the economic rationale of supply and demand. A case in point is the pricing of electricity contracts: figure 1 shows the time evolution of electric power spot prices in a recently deregulated market. Clearly, none of the mathematical models used for equities, currencies, or interest rates can be calibrated to be consistent with these data, and structural models involving the factors that drive demand (like weather) and supply (like means of production) need to be used to give a reasonable account of the spikes.

Electricity production is one of the major sources of greenhouse gas emissions, and various forms of market mechanisms have been touted to control these harmful externalities. Most notable is the implementation of the Kyoto protocol in the form of mandatory cap-and-trade for $CO_2$ allowance certificates by the European Union. While policy issues are still muddying the final form that the control of $CO_2$ emissions will take in the United States, cap-and-trade schemes already exist in

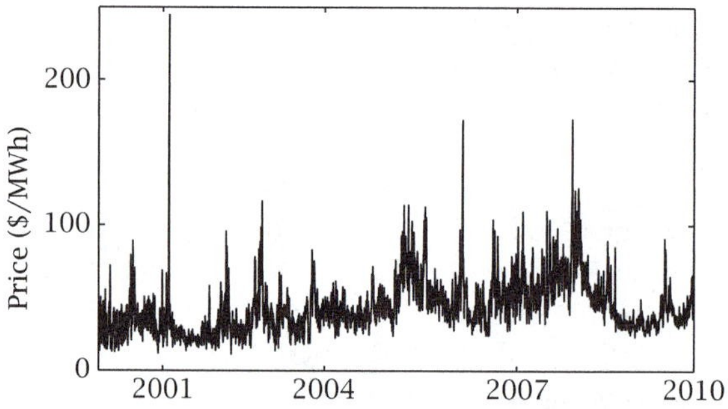


**Figure 1** Historical daily prices of electricity from the PJM market in the northeastern United States.

the northeast of the country (the Regional Greenhouse Gas Initiative) and, more recently, in California, giving a new impetus to theoretical research into these markets.

The proliferation of commodity indexes and the dramatic increase in investors gaining commodity exposure through ETFs that track indexes have changed the landscape of the commodity markets and increased the correlations between commodities and equity, and the correlations among commodities included in the same indexes. Figure 2 illustrates this striking change in correlations. These changes are difficult to explain if one relies solely on the fundamentals of these markets. They seem to be part of a phenomenon, known as *financialization* of the commodity markets, that has been taking place over the last ten years and that is now being investigated by a growing number of economists, econometricians, and mathematicians. Also of great interest to theoretical research is the fact that financial institutions, including hedge funds, endowment funds, and pension funds, have realized that the road to success in the commodity markets was more often than not to rely on managing portfolios that included both physical and financial assets. Once more, this combination raises new issues that are not addressed by traditional financial mathematics or financial engineering. Finally, the physical nature of energy production led to the introduction of new financial markets in areas such as weather, freight, and emissions, the design, regulation, and investigation of which pose new mathematical challenges.

### 3.3 High-Frequency Trading

In the last few years, the notion of a single publicly known price at which transactions can happen in arbitrary sizes has been challenged. The existence and the importance of liquidity frictions and price impact

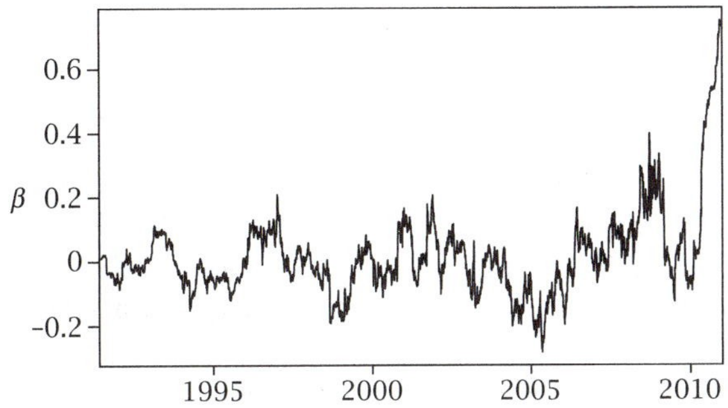


**Figure 2** Instantaneous dependence ($\beta$) of the Goldman Sachs Total Return Commodity Index on S&P 500 returns.

due to the size and frequency of trades are recognized as the source of many of the financial industry's most spectacular failures (Long-Term Capital Management, Amaranth, Lehman Brothers, etc.), prompting new research into applications of stochastic optimization to optimal execution and predatory trading, for example.

Another important driver in the change of direction in quantitative finance research is the growing role of algorithmic and high-frequency trading. Indeed, it is commonly accepted that between 60% and 70% of trading is electronic nowadays. Market makers and brokers are now mostly electronic, and while they are claimed to be liquidity providers (the jury is still out on that one), occurrences like the flash crash of May 6, 2010, and computer glitches like those which took down Knight Capital, have raised serious concerns. Research into the development of limit order book models and their impact on trading is clearly one of the emerging topics in quantitative finance research.

### 3.4 Back to Basics: Stochastic Equilibrium and Stochastic Games

More recently, these tools have been adapted to problems involving multiple "agents" optimizing for themselves but interacting with each other through a market. These problems involve analyzing and computing an equilibrium, which may come from a market clearing condition, for example, or in enforcing the strong competition of a Nash equilibrium. There has been much recent progress in stochastic differential games. We mention, for example, recent works of Lasry and Lions, who consider mean-field games in which there are a large number of players and competition is felt only through an average of one's competitors, with each player's impact on the average being negligible.

Problems arising from price impact, stability, liquidity, and the formation of bubbles remain of vital interest and have produced very interesting mathematics along the way. Not long after the 1987 crash there was much concern about the extent to which the large drop was caused or exacerbated by program traders whose computers were automatically hedging options positions, causing them to sell mechanically when prices went down, pushing them down further. In a continuous-time framework, this type of feedback model of price impact was initiated in the late 1990s, and since then there have been many influential studies of situations in which an investor is not simply a "price taker" and where large stock positions are sold off in pieces to avoid large price-depressing trades: the epitome of an optimal execution algorithm.

### Further Reading

The books by Fouque et al. and Gatheral describe recent research taking place with stochastic volatility models. A textbook reference for Monte Carlo methods for financial problems is the book by Glasserman. For more on infinite-dimensional interest rate models, see, for example, the recent textbooks by Carmona and Tehranchi, and by Filipovic. Merton's classic works on portfolio optimization are reprinted in his 1992 collection. The recent handbook edited by Fouque and Langsam provides many viewpoints on the analysis of systemic risk. For further reading on energy and commodities markets, see the recent survey by Carmona and Coulon and the book by Swindle. The recent book by Lehalle and Laruelle discusses market microstructure in the context of high-frequency trading. A recent survey of work on asset price bubbles (and subsequent crashes) can be found in the survey article by Protter.

Carmona, R., and M. Coulon. 2013. A survey of commodity markets and structural approaches to modeling electricity. In *Energy Markets: Proceedings of the WPI Special Year*, edited by F. Benth, pp. 1–42. New York: Springer.

Carmona, R., and M. Tehranchi. 2006. *Interest Rate Models: An Infinite Dimensional Stochastic Analysis Perspective.* New York: Springer.

Filipovic, D. 2009. *Term Structure Models: A Graduate Course.* New York: Springer.

Fouque, J.-P., and J. Langsam, eds. 2012. *Handbook of Risk.* Cambridge: Cambridge University Press.

Fouque, J.-P., G. Papanicolaou, R. Sircar, and K. Sølna. 2011. *Multiscale Stochastic Volatility for Equity, Interest-Rate and Credit Derivatives.* Cambridge: Cambridge University Press.

Gatheral, J. 2006. *The Volatility Surface: A Practitioner's Guide.* New York: John Wiley.
Glasserman, P. 2003. *Monte Carlo Methods in Financial Engineering.* New York: Springer.
Lehalle, C., and S. Laruelle. 2013. *Market Microstructure in Practice.* Singapore: World Scientific.
Merton, R. 1992. *Continuous-Time Finance.* Oxford: Blackwell.
Protter, P. 2013. A mathematical theory of financial bubbles. In *Paris-Princeton Lectures on Mathematical Finance 2013,* edited by V. Henderson and R. Sircar. New York: Springer.
Swindle, G. 2014. *Valuation and Risk Management in Energy Markets.* Cambridge: Cambridge University Press.

# V.10 Portfolio Theory

*Thomas J. Brennan, Andrew W. Lo, and Tri-Dung Nguyen*

## 1 Basic Mean–Variance Analysis

Pioneered by the Nobel Prize-winning economist Harry Markowitz over half a century ago, portfolio theory is one of the oldest branches of modern financial economics. It addresses the fundamental question faced by an investor: how should money best be allocated across a number of possible investment choices? That is, what collection or portfolio of financial assets should be chosen? In this article, we describe the fundamentals of portfolio theory and methods for its practical implementation. We focus on a fixed time horizon for investment, which we generally take to be a year, but the period may be as short as days or as long as several years. We summarize many important innovations over the past several decades, including techniques for better understanding how financial prices behave, robust methods for estimating input parameters, Bayesian methods, and resampling techniques.

A *portfolio* is a collection of financial securities, often called *assets,* and the *return* to an asset or a portfolio is the uncertain incremental percentage financial gain or loss that results from holding the asset or portfolio over a particular time horizon. If the price of an asset $i$ at date $t$ is denoted $p_{it}$, then its return $r_{i,t+1}$ between $t$ and $t+1$ is defined as

$$r_{i,t+1} \equiv \frac{p_{i,t+1} + d_{i,t+1}}{p_{it}} - 1, \tag{1}$$

where $d_{i,t+1}$ denotes any cash payouts made by asset $i$ between $t$ and $t+1$, such as a dividend payment. The return on a portfolio of assets is defined in a similar manner.

Markowitz's seminal idea was to choose an optimal portfolio using two key features of the distribution of a portfolio's return: its mean and variance. For any target level of variance, an allocation yielding the greatest mean (or expected) return should be chosen. Similarly, for any target level of expected return, an allocation yielding the lowest possible level of variance should be chosen. So far, this does not allow a unique portfolio to be selected since either the target variance or the target mean needs to be specified in advance. However, this does already allow us to sketch a curve in mean–variance space corresponding to the characteristics of the portfolios that are *efficient* inasmuch as they satisfy Markowitz's requirements. This curve is known as the *mean–variance efficient frontier* (see figure 1 for an example).

Variance is often replaced by its square root, standard deviation, resulting in an equivalent curve of efficient portfolios in the space defined by mean and standard deviation. We refer to this latter curve as the *efficient frontier.*

Consider a collection of $n$ financial assets that are available as investment choices and let them be indexed by $i = 1, \dots, n$, with $r_i$ denoting the return of asset $i$ over the applicable investment horizon (we suppress the time subscript $t$ to simplify notation). An initial investment of one dollar in $i$ thus yields $1 + r_i$ dollars at the end of the period of the investment. The expected return of $i$ is $\mu_i = \sum_s r_i(s) p_s$, and the covariance of the returns of $i$ and $j$ is $\sigma_{i,j} = \sum_s (r_i(s) - \mu_i)(r_j(s) - \mu_j) p_s$. Here we assume that the returns have finite distributions and that all the different possible sets of returns $\boldsymbol{r} = (r_1, \dots, r_n)$ are indexed by the parameter $s$, with $\boldsymbol{r}_s = (r_1(s), \dots, r_n(s))$. The probability of state $s$ is denoted by $p_s$. We write $\boldsymbol{\mu} = [\mu_i]_i$ and $\boldsymbol{\Sigma} = [\sigma_{i,j}]_{i,j}$. All vectors denote column vectors unless stated otherwise.

An allocation of funds among the assets can be thought of as a vector $\boldsymbol{\omega} = [\omega_i]_i$ of weights, with $\omega_i$ representing the proportion of available funds invested in asset $i$. The weights are subject to the constraint $\boldsymbol{e}^t \boldsymbol{\omega} = 1$, where $\boldsymbol{e}$ is a vector of all 1s. This is necessary so that exactly 100% of the available funds are invested. In the general case, weights are permitted to be negative, with the interpretation that these correspond to *short sales* of those assets. A short sale is a specific financial transaction in which an investor can sell a security that he does not own by borrowing it from a third party, such as a broker, with the promise to return it at a later date. Short sales allow investors who expect

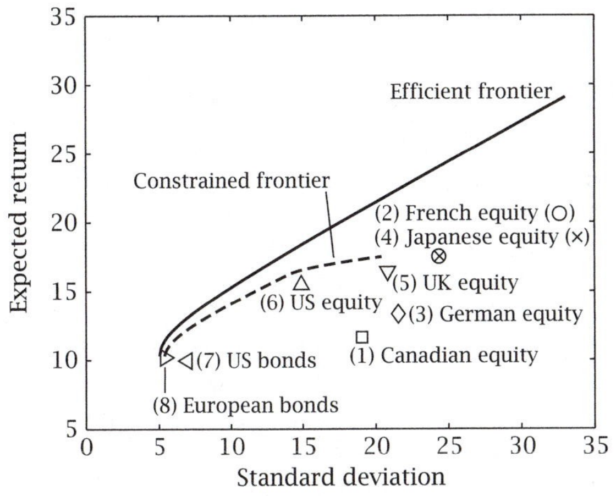


**Figure 1** Efficient frontier for the collection of eight assets with characteristics described in table 1. The dashed line is the constrained frontier that does not allow negative asset weights in portfolios. For each individual asset the standard deviation and expected return are indicated on the graph.

an asset's price to decline to benefit from this expectation. When less generality is desired, more constraints can be added, such as the restriction that $\omega_i \geqslant 0$ for all $i$, implying that no short sales are allowed.

A portfolio formed with the weight vector $\boldsymbol{\omega}$ has expected return $\boldsymbol{\mu}^{\mathrm{T}}\boldsymbol{\omega}$ and variance $\boldsymbol{\omega}^{\mathrm{T}}\boldsymbol{\Sigma}\boldsymbol{\omega}$. Thus, to determine the efficient frontier, we need to solve the minimization problem

$$\min \boldsymbol{\omega}^{\mathrm{T}}\boldsymbol{\Sigma}\boldsymbol{\omega} \quad \text{such that } \boldsymbol{\mu}^{\mathrm{T}}\boldsymbol{\omega} = \mu_{\mathrm{p}}\boldsymbol{e}^{\mathrm{T}}\boldsymbol{\omega} = 1 \qquad (2)$$

for each value of the portfolio expected return $\mu_{\mathrm{p}}$. The solution weight vector $\boldsymbol{\omega}^*$ will be the optimal portfolio corresponding to the target expected return $\mu_{\mathrm{p}}$.

Figure 1 illustrates the efficient frontier for the set of eight assets with expected returns, standard deviations, and correlations listed in table 1. In the figure, as well as in what follows, we generally restrict our attention to the "upper branch" of the efficient frontier, meaning that we do not include portfolios on the frontier that have returns lower than the return of the global minimum-variance portfolio. For any such portfolios, a higher level of return is possible for the same amount of risk.

### 1.1 Analytical Solutions

We can find an exact analytic solution to the problem (2). The method of LAGRANGE MULTIPLIERS [I.3 §10] is applicable, and its use in this context was pioneered by

**Table 1** Expected returns, standard deviations, and correlations for a collection of eight assets. The asset numbers correspond to the named assets in figure 1. The values are annualized versions of the statistics appearing in Michaud (1998) and reflect historical data from 1978 through 1995. (ER, expected return; SD, standard deviation.)

| | ER (%) | SD (%) | Correlations 1 | 2 | 3 | 4 | 5 | 6 | 7 | 8 |
|---|---|---|---|---|---|---|---|---|---|---|
| 1 | 11.64 | 19.05 | 1.00 | 0.41 | 0.30 | 0.25 | 0.58 | 0.71 | 0.26 | 0.33 |
| 2 | 17.52 | 24.35 | | 1.00 | 0.62 | 0.42 | 0.54 | 0.44 | 0.22 | 0.26 |
| 3 | 13.32 | 21.55 | | | 1.00 | 0.35 | 0.48 | 0.34 | 0.27 | 0.28 |
| 4 | 17.52 | 24.39 | | | | 1.00 | 0.40 | 0.22 | 0.14 | 0.16 |
| 5 | 16.44 | 20.82 | | | | | 1.00 | 0.56 | 0.25 | 0.29 |
| 6 | 15.48 | 14.90 | | | | | | 1.00 | 0.36 | 0.42 |
| 7 | 9.96 | 6.96 | | | | | | | 1.00 | 0.92 |
| 8 | 10.20 | 5.40 | | | | | | | | 1.00 |

Merton. The appropriate Lagrangian is

$$\mathcal{L} = \boldsymbol{\omega}^{\mathrm{T}}\boldsymbol{\Sigma}\boldsymbol{\omega} + \lambda_1(\boldsymbol{\mu}^{\mathrm{T}}\boldsymbol{\omega} - \mu_{\mathrm{p}}) + \lambda_2(\boldsymbol{e}^{\mathrm{T}}\boldsymbol{\omega} - 1). \qquad (3)$$

A calculation setting the gradient of $\mathcal{L}$ equal to zero shows that the solution to the minimization problem is

$$\boldsymbol{\omega}^* = \frac{\mu_{\mathrm{p}}C - B}{D}\boldsymbol{\Sigma}^{-1}\boldsymbol{\mu} + \frac{A - \mu_{\mathrm{p}}B}{D}\boldsymbol{\Sigma}^{-1}\boldsymbol{e}, \qquad (4)$$

where $A = \boldsymbol{\mu}^{\mathrm{T}}\boldsymbol{\Sigma}^{-1}\boldsymbol{\mu}$, $B = \boldsymbol{\mu}^{\mathrm{T}}\boldsymbol{\Sigma}^{-1}\boldsymbol{e}$, $C = \boldsymbol{e}^{\mathrm{T}}\boldsymbol{\Sigma}^{-1}\boldsymbol{e}$, and $D = AC - B^2$.

The optimal weight of (4) combined with the formula for the variance of the portfolio, namely $\boldsymbol{\omega}^{\mathrm{T}}\boldsymbol{\Sigma}\boldsymbol{\omega}$, allows us to calculate the minimum variance possible for a given level of expected return $\mu_{\mathrm{p}}$. Specifically, we have

$$(\boldsymbol{\omega}^*)^{\mathrm{T}}\boldsymbol{\Sigma}\boldsymbol{\omega}^* = \frac{C\mu_p^2 - 2B\mu_{\mathrm{p}} + A}{D}. \qquad (5)$$

The minimum variance is thus simply a quadratic function of the level of expected return. Moreover, we can use (5) to see that the global minimum variance is $1/C$ and occurs at the return level $\mu_{\mathrm{p}} = B/C$. The efficient frontier can therefore be generated as the set of points

$$\begin{aligned}\mathcal{F} &= \left\{(\sigma, \mu) : \sigma = \frac{\sqrt{C\mu^2 - 2B\mu + A}}{\sqrt{D}} \mu \geqslant \frac{B}{C}\right\} \\ &= \left\{(\sigma, \mu) : \mu = \frac{B}{C} + \frac{\sqrt{D(C\sigma^2 - 1)}}{C}\sigma \geqslant \frac{1}{\sqrt{C}}\right\}. \qquad (6)\end{aligned}$$

### 1.2 Inequality Constraints

It is often desirable to add further constraints to the optimization problem presented in (2). For example, it may be required that the weight of a particular asset be exactly 10%, that all elements of $\boldsymbol{\omega}$ be positive, or that no one element of $\boldsymbol{\omega}$ be greater than 50%. These constrained portfolio selection problems do not generally have simple closed-form solutions. Instead, numerical methods must be employed to compute optimal portfolios.

To the extent that more constraints are stated in the form of equalities, the Lagrange multiplier formula in (3) can be extended to have additional appropriate terms. If some of the constraints are inequalities, then the method of Lagrange multipliers can be extended using the KARUSH–KUHN–TUCKER CONDITIONS [IV.11 §2].

In practice, computer software packages are used to calculate frontiers with constraints beyond the basic requirement that $\boldsymbol{e}^{\mathrm{T}}\boldsymbol{\omega} = 1$. In figure 1, we illustrate a portion of the efficient frontier, as well as an example of a constrained frontier, for the collection of assets with characteristics described in table 1. We computed the points on the efficient frontier analytically, but those on the constrained frontier were computed using quadratic programming software.

### 1.3 Finding the "Best" Portfolio

We have not yet introduced a way to choose one "best" portfolio for an investor from among all those on the efficient frontier. To do so, we need to know something about the investor's *value function*, i.e., how he ranks different combinations of risk and return. In this section, we describe several commonly used methods for an investor to rank different risk–return profiles. We will refer to these ranking methodologies in subsequent sections.

*The Minimum-Variance Portfolio*

In an extreme case, for an investor who cares only about risk and not at all about expected return, the portfolio with the lowest level of risk should be chosen. As we saw in section 1.1, this minimum value corresponds to a variance of $1/C$ or a standard deviation of $1/\sqrt{C}$ and is displayed in figure 2. The corresponding set of weights can also be easily derived by solving the Lagrange multiplier problem corresponding to the Lagrangian $\mathcal{L} = \boldsymbol{\omega}^{\mathrm{T}}\boldsymbol{\Sigma}\boldsymbol{\omega} + \lambda_1(\boldsymbol{e}^{\mathrm{T}}\boldsymbol{\omega} - 1)$. The resulting portfolio weight is seen to be $\boldsymbol{\omega}_{\min} = (1/C)\boldsymbol{\Sigma}^{-1}\boldsymbol{e}$.

*Standard Value Functions*

More realistically, an investor may assign higher value to portfolios with higher expected return and lower value to those with higher risk. A commonly used simple value function along these lines is

$$V_\gamma(\sigma_{\mathrm{p}}, \mu_{\mathrm{p}}) = \mu_{\mathrm{p}} - \tfrac{1}{2}\gamma\sigma_{\mathrm{p}}^2. \qquad (7)$$

In this case, the value of a portfolio with parameters $\mu_{\mathrm{p}}$ and $\sigma_{\mathrm{p}}$ is a linear function of expected return ($\mu_{\mathrm{p}}$) and variance ($\sigma_{\mathrm{p}}^2$), and $\gamma$ is an investor-specific parameter indicating the investor's tolerance for risk. This value function is most appropriate when $\mu_{\mathrm{p}}$ and $\sigma_{\mathrm{p}}$ are relatively small, as may happen when the investment time horizon is short. For longer time periods, more complex functions are generally needed. These value functions are often derived from more primitive assumptions about an investor's *utility* for wealth, $U(W)$. By setting $W = W_0\boldsymbol{\omega}^{\mathrm{T}}\boldsymbol{r}$, where $W_0$ is initial wealth, the value function $V$ is given by $\mathbb{E}[U(W_0\boldsymbol{\omega}^{\mathrm{T}}\boldsymbol{r})]$ according to the axioms of Von Neumann and Morgenstern's *expected utility theory*.

For the special case of the value function defined in (7), we can find the unconstrained optimal portfolio weights $\boldsymbol{\omega}_V$ and determine the corresponding highest achievable value using an appropriate Lagrangian. We find that

$$\boldsymbol{\omega}_V = \frac{1}{\gamma}\boldsymbol{\Sigma}^{-1}\boldsymbol{\mu} + \frac{\gamma - B}{\gamma C}\boldsymbol{\Sigma}^{-1}\boldsymbol{e}.$$

The expected return and risk at this point of optimal value are

$$\mu_{\mathrm{p}} = \frac{B}{C} + \frac{D}{\gamma C} \quad \text{and} \quad \sigma_{\mathrm{p}} = \frac{1}{\gamma}\sqrt{\frac{D + \gamma^2}{C}}.$$

*Sharpe Ratio*

An investor may also look at the reward-to-variability ratio represented by a portfolio, i.e., the expected excess return he receives from the portfolio divided by the risk the portfolio represents. This ratio is referred to as the *Sharpe ratio* and can be written as

$$S(\sigma_{\mathrm{p}}, \mu_{\mathrm{p}}) = \frac{\mu_{\mathrm{p}} - r_{\mathrm{f}}}{\sigma_{\mathrm{p}}},$$

where $\sigma_{\mathrm{p}}$ and $\mu_{\mathrm{p}}$ are the risk and expected return of the portfolio, respectively, and where $r_{\mathrm{f}}$ is the risk-free rate that the investor would receive if he did not invest in the portfolio. By subtracting the risk-free rate, we are measuring the incremental reward in excess of the risk-free rate that the investor receives by investing in the portfolio.

Using the characterization of points on the efficient frontier given in (6), we can calculate the optimal Sharpe ratio and corresponding portfolio weights. Specifically, we can express $\sigma_{\mathrm{p}}$, and hence $S$, as a function of $\mu_{\mathrm{p}}$,

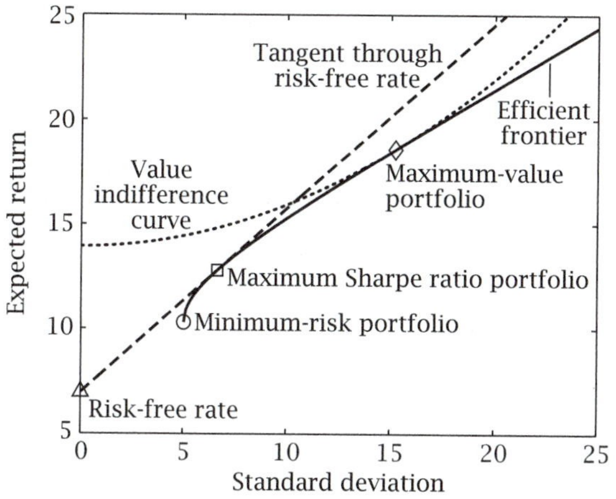


**Figure 2** Optimal points on the efficient frontier for the collection of eight assets with characteristics described in table 1. The maximum Sharpe ratio is computed assuming a risk-free rate of $r_f = 6.96\%$, consistent with Michaud (1998), and the maximum value is calculated using a tolerance for risk of $\gamma = 4$.

and we then maximize $S$ as a univariate function of $\mu_p$ and find that

$$S_{\max} = \sqrt{Cr_f^2 - 2Br_f + A}$$

for values of $r_f$ less than the return of the minimum-variance portfolio (see figure 2). The portfolio weights are given by (4), with

$$\mu_p = \frac{A - Br_f}{B - Cr_f},$$

namely

$$\boldsymbol{\omega}_{\text{Sharpe}} = \frac{1}{B - Cr_f}\boldsymbol{\Sigma}^{-1}(\boldsymbol{\mu} - r_f\boldsymbol{e}).$$

### 1.4 Connections to the Capital Asset Pricing Model (CAPM)

The mean–variance framework developed by Markowitz was fundamental to the development of the CAPM. In the 1950s and 1960s, Tobin, Sharpe, and Lintner derived the equilibrium implications under the assumption that all investors held efficient portfolios, and this led to the *capital market line*, the line connecting the risk-free rate on the expected-return axis with the tangency portfolio on the efficient frontier in mean–standard deviation space, which is the same as the optimal Sharpe ratio portfolio derived above. Under the assumptions of the model, all investors hold a combination of positions in this tangency portfolio and the risk-free asset.

In the CAPM, aggregate positions in the risk-free asset net to zero, and the tangency portfolio represents the aggregate position in risky assets across all investors. Accordingly, this portfolio is referred to as the *market portfolio*. In the CAPM world, the expected return, $\mu_i$, for a particular asset $i$ can be expressed in terms of the expected return and risk of the market portfolio, as well as the covariance of the asset with the market portfolio. The precise relationship is

$$\mu_i = r_f + \beta_i(\mu_{\text{mkt}} - r_f),$$

where $\mu_{\text{mkt}}$ is the expected return of the market portfolio; $\beta_i = \text{cov}(r_i, r_{\text{mkt}})/\sigma_{\text{mkt}}^2$; $r_i$ and $r_{\text{mkt}}$ represent the returns on asset $i$ and the market portfolio, respectively; and $\sigma_{\text{mkt}}$ represents the standard deviation of the market portfolio. The return $r_i$ can be written as

$$r_i = r_f + \beta_i(r_{\text{mkt}} - r_f) + \epsilon_i,$$

where $\epsilon_i$ is a stochastic variable that is uncorrelated with $r_{\text{mkt}}$ and has zero mean. The risk represented by the $\epsilon_i$ is known as *idiosyncratic risk*. In matrix notation, the expected return vector and the covariance matrix of asset returns under the CAPM assumptions are thus

$$\left.\begin{aligned} \boldsymbol{\mu}_{\text{CAPM}} &= r_f\boldsymbol{e} + \boldsymbol{\beta}(\mu_{\text{mkt}} - r_f), \\ \boldsymbol{\Sigma}_{\text{CAPM}} &= \sigma_{\text{mkt}}^2\boldsymbol{\beta}\boldsymbol{\beta}^{\text{T}} + \boldsymbol{\Omega}_\epsilon, \end{aligned}\right\} \qquad (8)$$

where $\boldsymbol{\beta}$ is the vector of $\beta$ values for the assets, and where $\boldsymbol{\Omega}_\epsilon$ is the covariance matrix for the idiosyncratic risk.

#### *Efficiency of the Market Portfolio*

The importance of the mean–variance efficiency of the market portfolio was recognized early on by many authors and led to a series of debates on the testable implications of the CAPM. Markowitz has argued that empirical deviations from the CAPM are not surprising in light of the counterfactual assumptions on which the CAPM is based. In particular, he observes that: "When one clearly unrealistic assumption of the CAPM is replaced by a real-world version, some of the dramatic CAPM conclusions no longer follow." An example is the fact that unlimited borrowing and lending at identical interest rates is not possible in practice, and this limitation implies that the market portfolio need not be mean–variance efficient in equilibrium.

#### *Impossible Frontiers*

If an efficient frontier contains no portfolio with all positive weights, it is incompatible with the CAPM and is defined to be *impossible* because, by definition, the market portfolio must have a positive weight for each asset, where the weight is proportional to the market capitalization of that asset. Brennan and Lo have demonstrated that, as the number of assets grows large, all efficient frontiers are almost surely impossible for a randomly drawn (with respect to Haar measure) covariance matrix. This result explains the near-universal disdain with which professional portfolio managers regard standard mean–variance optimization techniques: the vast majority of them are constrained to hold long-only portfolios, and hence an impossible frontier is, in fact, literally impossible for them to implement.

## 2 Techniques for Practical Implementation

In practice, we generally do not know the exact nature of the distributions of returns for the assets we can invest in. In fact, we do not even know the exact values of the inputs required for mean–variance optimization, i.e., $\boldsymbol{\mu}$ and $\boldsymbol{\Sigma}$. Instead, we must find ways to estimate these quantities and then use the estimates when carrying out optimizations.

In this section we discuss several methods for the practical determination of optimal portfolios. In section 2.1 we describe the simple approach of using historical data to compute unbiased statistics under the assumption that asset returns follow a stable distribution over time. In section 2.2 we detail several methods for reducing noise, with a focus on estimating the covariance matrix, $\boldsymbol{\Sigma}$, including methods that incorporate theoretical predictions for the structure of $\boldsymbol{\Sigma}$. In section 2.3 we describe Bayesian methods that allow for better estimation of $\boldsymbol{\mu}$, as well as $\boldsymbol{\Sigma}$, and that also allow for incorporation of investor beliefs and theoretical models for the nature of asset returns. In section 2.4 we survey improved and more robust methods for selecting optimal portfolios.

### 2.1 Unbiased Estimators Using Historical Data

If we assume that each period's data represent an independent draw from a stable process governing asset returns, we can treat the observed historical returns as a sample from which we can estimate the desired statistics $\boldsymbol{\mu}$ and $\boldsymbol{\Sigma}$. For $T$ observations of historical returns, unbiased estimators of $\boldsymbol{\mu}$ and $\boldsymbol{\Sigma}$ are

$$\hat{\mu}_i = \frac{1}{T}\sum_{t=1}^{T} r_{it}, \qquad \hat{\boldsymbol{\Sigma}} = \frac{1}{T-1}\boldsymbol{H}^{\mathrm{T}}\boldsymbol{H}, \tag{9}$$

where $r_{it}$ is the observation of the return on $i$ in period $t$, and $\boldsymbol{H} = [(r_{it} - \hat{\mu}_i)]_{it}$ is a $T \times n$ matrix. An unbiased estimator of $\boldsymbol{\Sigma}^{-1}$ is $\tilde{\boldsymbol{\Sigma}}^{-1}$, with

$$\tilde{\boldsymbol{\Sigma}} = \frac{T-1}{T-n-2}\hat{\boldsymbol{\Sigma}}.$$

The estimator $\tilde{\boldsymbol{\Sigma}}$ may be more appropriate than $\hat{\boldsymbol{\Sigma}}$ when the object of interest is $\boldsymbol{\Sigma}^{-1}$ instead of $\boldsymbol{\Sigma}$, as in the formulas for the values $A$, $B$, and $C$ in (4).

### 2.2 Covariance Matrix Estimation

Unbiased estimates suffer from estimation error; we describe several techniques designed to provide better estimates, with a particular focus on the covariance matrix, $\boldsymbol{\Sigma}$. These include methods for the reduction of noise as well as for the incorporation of theoretical results regarding the structure of $\boldsymbol{\Sigma}$.

*Factor Analysis*

The returns of the $n$ available assets will generally not be independent. In fact, the returns may be driven in large part by a small number of common factors. If this is the case, the covariance matrix, $\boldsymbol{\Sigma}$, is less complex and estimators taking this into account may be expected to contain less noise. It may also be assumed that the common factors determine the risk-free vector, $\boldsymbol{\mu}$, but for expositional purposes we allow $\boldsymbol{\mu}$ to remain fully general and focus solely on reducing the complexity of $\boldsymbol{\Sigma}$.

The return vector for the $n$ assets takes the form

$$\boldsymbol{r} = \boldsymbol{\mu} + \boldsymbol{V}^{\mathrm{T}}\boldsymbol{\Lambda} + \boldsymbol{\epsilon},$$

where $\boldsymbol{\mu}$ is a vector of constants; $\boldsymbol{\Lambda}$ is a stochastic vector of $n_{\mathrm{f}}$ factors; $\boldsymbol{V}$ is a constant $n_{\mathrm{f}} \times n$ matrix of factor loadings; and $\boldsymbol{\epsilon}$ is a stochastic vector of residual returns for the $n$ assets, each having zero mean. The factors represented by $\boldsymbol{\Lambda}$ are generally thought of as historically observable aggregate market variables, and various economic models give rise to factor structures for asset returns. For example, the CAPM yields a factor model with $n_{\mathrm{f}} = 1$ in which the single factor is the market-value-weighted average of all asset returns.

Because we allow $\boldsymbol{\mu}$ to remain completely general, the factors represented by the elements of $\boldsymbol{\Lambda}$ may each be assumed to have zero mean. This assumption involves no loss of generality because an adjustment to $\boldsymbol{\mu}$ can be made, if necessary, to ensure that the expected returns of the factors are zero. In addition to having zero mean, it is often assumed that $\boldsymbol{\Lambda}$ follows a multivariate normal distribution with covariance matrix $\boldsymbol{\Sigma}_{\boldsymbol{\Lambda}}$. The vector of residual returns $\boldsymbol{\epsilon}$ is also assumed to follow a multivariate normal distribution, with covariance matrix $\boldsymbol{\Sigma}_{\epsilon}$, and to be independent of $\boldsymbol{\Lambda}$. These normality assumptions do result in loss of generality of the distribution of returns, but they greatly simplify the analysis to follow. Indeed, under these assumptions $\boldsymbol{r}$ follows a multivariate normal distribution, with

$$\boldsymbol{r} = \boldsymbol{\mu} + \boldsymbol{V}^{\mathrm{T}}\boldsymbol{\Lambda} + \boldsymbol{\epsilon} \sim \mathcal{N}(\boldsymbol{\mu}, \boldsymbol{V}^{\mathrm{T}}\boldsymbol{\Sigma}_{\Lambda}\boldsymbol{V} + \boldsymbol{\Sigma}_{\epsilon}).$$

We use the notation $\mathcal{N}(\cdot,\cdot)$ to denote a multivariate normal distribution with mean specified by the first argument and covariance matrix specified by the second argument.

An estimator $\hat{\boldsymbol{\Sigma}}_{\Lambda}$ of the covariance matrix of the factors can be found using historical data for factor returns with a methodology similar to that described in section 2.1. Estimators for the matrix of factor loadings $\hat{\boldsymbol{V}}$ and the covariance matrix of the error terms $\hat{\boldsymbol{\Sigma}}_{\epsilon}$ can be obtained using historical data and asset-by-asset linear regressions. The regression for the $i$th asset yields estimators both for the $n_{\mathrm{f}}$ elements in the $i$th column of $\boldsymbol{V}$ and for the variance of the $i$th element of $\boldsymbol{\epsilon}$, as well as for the $i$th element of $\boldsymbol{\mu}$. An estimator for the full covariance matrix is, therefore,

$$\hat{\boldsymbol{\Sigma}}_{\text{factor}} = \hat{\boldsymbol{V}}^{\mathrm{T}}\hat{\boldsymbol{\Sigma}}_{\Lambda}\hat{\boldsymbol{V}} + \hat{\boldsymbol{\Sigma}}_{\epsilon}.$$

### *Covariance Shrinkage*

To reduce the estimation error of the covariance matrix, $\hat{\boldsymbol{\Sigma}}$, one can take a weighted average of $\hat{\boldsymbol{\Sigma}}$ and a known covariance matrix $\boldsymbol{F}$, a process known as *shrinking* the covariance matrix $\hat{\boldsymbol{\Sigma}}$ toward the target matrix $\boldsymbol{F}$. The resulting revised estimator for the covariance matrix is

$$\hat{\boldsymbol{\Sigma}}_{\mathrm{LW}} = \alpha\boldsymbol{F} + (1-\alpha)\hat{\boldsymbol{\Sigma}}.$$

This is a special case of *Bayesian shrinkage estimators* in which $\boldsymbol{F}$ plays the role of a *prior* and the *posterior* is given by $\hat{\boldsymbol{\Sigma}}_{\mathrm{LW}}$ (see section 2.3). There are various possibilities for the choice of $\boldsymbol{F}$, but the basic motivation is to select an $\boldsymbol{F}$ that has a known structure that is a plausible alternative to $\hat{\boldsymbol{\Sigma}}$. In this way, the shrinkage process effectively reduces estimation error while still keeping a portion of the characteristics of the unbiased estimator $\hat{\boldsymbol{\Sigma}}$. One possible choice for $\boldsymbol{F}$ is the CAPM covariance matrix, $\boldsymbol{\Sigma}_{\mathrm{CAPM}}$, described in (8), with the assumption that the covariance matrix $\boldsymbol{\Omega}_{\epsilon}$ for the idiosyncratic return components is diagonal.

To determine the appropriate $\alpha$ for the shrinkage procedure, Ledoit and Wolf, who first applied shrinkage estimation to covariance-matrix estimation, derive a consistent estimator for $\alpha$ that minimizes the norm:

$$\|(\alpha\boldsymbol{F} + (1-\alpha)\hat{\boldsymbol{\Sigma}}) - \boldsymbol{\Sigma}\|^2,$$

where $\|\boldsymbol{M}\|^2$ is defined to be the sum of the squares of all the entries of a matrix $\boldsymbol{M}$.

### *Random-Matrix Theory*

Simple estimators for $\boldsymbol{\Sigma}$ based upon historical data become less reliable as the ratio of observation periods to assets ($q = T/n$) decreases. In the extreme case in which $q < 1$, the estimator in (9) is degenerate because the rank of $\boldsymbol{H}$ is less than $n$. Laloux, Cizeau, Potters, and Bouchaud have addressed the problem of noisy estimation when $q$ is not much larger than 1 by arguing that $\hat{\boldsymbol{\Sigma}}$ should behave like a random matrix in such cases. To the extent that it exhibits behavior other than that of a random matrix, there should be actual information present, and this insight leads to a procedure for purging the matrix of its random noise.

Let $\boldsymbol{H}_{\mathrm{r}}$ be a $T \times n$ matrix with elements independently drawn from a normal random distribution with mean zero and standard deviation $\sigma_{\mathrm{r}}$. The matrix $\boldsymbol{M}_{\mathrm{r}} = \boldsymbol{H}_{\mathrm{r}}^{\mathrm{T}}\boldsymbol{H}_{\mathrm{r}}$ follows a Wishart distribution, and as $T \to \infty$, $n \to \infty$, and the ratio $q = T/n$ remains constant, the eigenvalues of $\boldsymbol{M}_{\mathrm{r}}$ asymptotically all lie within an interval $[\lambda_-, \lambda_+]$, where

$$\lambda_{\pm} = \sigma_{\mathrm{r}}^2(1 + 1/q \pm 2\sqrt{1/q}).$$

If a matrix has, with only a few exceptions, eigenvalues that lie in this range, then it may be argued that the outliers correspond to the information content of the matrix while the other eigenvalues correspond to random noise.

Instead of focusing on the estimated covariance matrix $\hat{\boldsymbol{\Sigma}}$ directly, it is more convenient to consider the corresponding correlation matrix $\hat{\boldsymbol{C}}$ because RANDOM-MATRIX THEORY [IV.24] is most easily applied to cases in which all variances are equal. $\hat{\boldsymbol{C}}$ is defined by its entries $\hat{C}_{i,j} = \hat{\Sigma}_{i,j}/\sqrt{\hat{\Sigma}_{i,i}\hat{\Sigma}_{j,j}}$ and is itself a covariance matrix for the returns of $n$ assets but with each return rescaled so it has unit variance.

Let $\hat{\boldsymbol{C}} = \hat{\boldsymbol{Q}}\hat{\boldsymbol{\Lambda}}\hat{\boldsymbol{Q}}^{\mathrm{T}}$ be an eigendecomposition of $\hat{\boldsymbol{C}}$ with the eigenvalues on the diagonal of $\hat{\boldsymbol{\Lambda}}$ and the corresponding eigenvectors in the columns of $\hat{\boldsymbol{Q}}$. We compare the distribution of the eigenvalues $\hat{\lambda}_i$ to the theoretical distribution predicted for a random matrix $\boldsymbol{M}_{\mathrm{r}} = \boldsymbol{H}_{\mathrm{r}}^{\mathrm{T}}\boldsymbol{H}_{\mathrm{r}}$, where the elements of $\boldsymbol{H}_{\mathrm{r}}$ are drawn independently from a normal distribution with unit variance. The eigenvalues may be separated into two groups by using a suitable method, e.g., by specifying a threshold, $c_{\mathrm{RM}} > 1$, such that any eigenvalue larger than $c_{\mathrm{RM}}\lambda_+$ is deemed to convey information, with the rest deemed to be random noise.

The correlation matrix can be *cleaned* by replacing eigenvalues corresponding to noise with the average of such eigenvalues and leaving the other eigenvalues as they are. This set of cleaned eigenvalues can be used to create a cleaned covariance matrix in two steps. First, define $\tilde{\boldsymbol{C}}$ to be $\hat{\boldsymbol{Q}}\tilde{\boldsymbol{\Lambda}}\hat{\boldsymbol{Q}}^{\mathrm{T}}$, where $\tilde{\boldsymbol{\Lambda}}$ is the diagonal matrix with diagonal entries equal to the cleaned eigenvalues. Second, define the cleaned covariance matrix by

$$\hat{\boldsymbol{\Sigma}}_{\mathrm{RM}} = [\tilde{\boldsymbol{C}}_{ij}(\hat{\Sigma}_{ii}\hat{\Sigma}_{jj})^{1/2}]_{i,j}.$$

This covariance matrix can now be used along with $\hat{\boldsymbol{\mu}}$ as a basis for portfolio optimization.

It is important to underscore that the approach for cleaning a covariance matrix using the theory of random matrices is most appropriate when the value of $q = T/n$ is not much larger than 1.

*Nearest Correlation Matrix*

A problem that is related to the estimation of a covariance matrix is that of computing a correlation matrix that is closest in some metric to a given matrix. This problem arises routinely in applications of portfolio optimization in which financial analysts wish to impose their priors by altering various entries of the correlation matrix to conform to their beliefs, e.g., that the correlation between stock A and stock B is 0. Arbitrary changes to elements of a bona fide correlation matrix can easily violate positive-semidefiniteness, implying negative variances for certain portfolios. Nonpositive-semidefiniteness can also arise if elements of the correlation matrix are estimated individually rather than via a matrix estimator.

Under the Frobenius norm, a unique solution to the nearest correlation matrix (NCM) problem exists, and Higham has shown that it can be computed via an alternating projections algorithm that projects onto the space of matrices with unit diagonal and the cone of symmetric positive-semidefinite matrices. Although this algorithm is guaranteed to converge, it does so at a linear rate, which can be slow for large matrices. Using the dual of the NCM problem, it is possible to achieve global quadratic convergence by applying Newton's method, as shown by Qi and Sun. The NCM problem has also been extended to the case where the true correlation matrix is assumed to have a $k$-factor structure, where $k$ is much less than the dimension of the correlation matrix.

*Finding a Possible Frontier*

It is possible to modify a covariance matrix so that the corresponding efficient frontier is possible, rather than impossible, in the sense described in section 1.4. The tangency portfolio of the CAPM should satisfy $\boldsymbol{\omega}_{\mathrm{mkt}} = (1/(B - Cr_{\mathrm{f}}))\boldsymbol{\Sigma}^{-1}(\boldsymbol{\mu} - r_{\mathrm{f}}\boldsymbol{e})$, up to scaling by a positive constant. If this equality does not hold, then an adjustment to $\boldsymbol{\Sigma}$ can be made to restore the equality. This approach was introduced by Brennan and Lo and is related to the Black–Litterman method of asset allocation with prior information, which is described further in section 2.3.

Brennan and Lo's covariance matrix, $\boldsymbol{\Sigma}_{\mathrm{poss}}$, has the desired property that $\boldsymbol{\omega}_{\mathrm{mkt}} = (1/(B - Cr_{\mathrm{f}}))\boldsymbol{\Sigma}_{\mathrm{poss}}^{-1}(\boldsymbol{\mu} - r_{\mathrm{f}}\boldsymbol{e})$, and it is constructed to be the matrix requiring the least amount of change to the original covariance matrix $\boldsymbol{\Sigma}$. Specifically, the nature of their proposed change alters $\boldsymbol{\Sigma}^{-1}$ only with respect to the one-dimensional vector space spanned by $\boldsymbol{\mu} - r_{\mathrm{f}}\boldsymbol{e}$, and the product of $\boldsymbol{\Sigma}^{-1}$ with any vector orthogonal to $\boldsymbol{\mu} - r_{\mathrm{f}}\boldsymbol{e}$ is unaffected. Brennan and Lo argue that this covariance matrix should be used by those whose best estimate of the covariance matrix is otherwise $\boldsymbol{\Sigma}$ but who also have a strong conviction that the CAPM must hold and that $\boldsymbol{\mu}$ and $\boldsymbol{\omega}_{\mathrm{mkt}}$ are, in fact, the correct expected returns and market weights.

## 2.3 Bayesian Methods

Estimators for the risk-free vector, $\boldsymbol{\mu}$, are often considered particularly problematic because of their large estimation errors. To improve the estimation of $\boldsymbol{\mu}$, as well as $\boldsymbol{\Sigma}$, techniques applying BAYES'S THEOREM [V.11 §1] have been proposed. The basic idea is that an investor specifies certain *prior information* about the nature of asset returns, updates this prior with additional information and observations, and finally obtains a *posterior* distribution for $\boldsymbol{r}$. Improved estimates for $\boldsymbol{\mu}$ and $\boldsymbol{\Sigma}$ are then recovered from this posterior distribution. In general, there is no restriction on the nature of the possible prior or additional information used in Bayesian methods. For the purposes of our discussion, we will restrict attention to the simple assumption that asset returns follow a normal distribution, so that $\boldsymbol{r} \sim \mathcal{N}(\boldsymbol{\mu}, \boldsymbol{\Sigma})$. The values of $\boldsymbol{\mu}$ and $\boldsymbol{\Sigma}$ are unknown a priori, but the additional information and Bayesian updating procedure will allow estimates for these parameters to be made.

The additional information used throughout our discussion consists of a list of $m$ separate pieces of information, each specifying a probability distribution for the value of a linear transformation of $\boldsymbol{\mu}$. For each $j = 1, \dots, m$, the applicable transformation is specified by a $k_j \times n$ matrix $\boldsymbol{P}_j$, and the $j$th piece of information is the assumption that $\boldsymbol{P}_j\boldsymbol{\mu} \sim \mathcal{N}(\boldsymbol{Q}_j, \boldsymbol{S}_j)$ for some $k_j$-dimensional vector $\boldsymbol{Q}_j$ and some $k_j \times k_j$ covariance matrix $\boldsymbol{S}_j$. This information specifies that the transformation of $\boldsymbol{\mu}$ given by $\boldsymbol{P}_j\boldsymbol{\mu}$ is normally distributed with mean $\boldsymbol{Q}_j$ and covariance matrix $\boldsymbol{S}_j$. The value of $k_j$ may vary with $j$.

With the prior and additional information in hand, we can apply Bayes's theorem for continuous distributions to obtain a posterior distribution for asset returns. The procedure is to evaluate the integral with respect to $\boldsymbol{\mu}$ of the probability density for $\boldsymbol{r}$, with unknown values of $\boldsymbol{\mu}$ and $\boldsymbol{\Sigma}$, multiplied by the product of all the probability densities corresponding to the additional information. The result is a posterior distribution for $\boldsymbol{r}$ that is normally distributed with mean and covariance given by

$$\boldsymbol{\mu}_{\text{post}} = \left(\sum_{j=1}^{m} \boldsymbol{P}_j^{\mathrm{T}}\boldsymbol{S}_j^{-1}\boldsymbol{P}_j\right)^{-1} \sum_{j=1}^{m} \boldsymbol{P}_j^{\mathrm{T}}\boldsymbol{S}_j^{-1}\boldsymbol{Q}_j \qquad (10)$$

and

$$\boldsymbol{\Sigma}_{\text{post}} = \boldsymbol{\Sigma}\left(\boldsymbol{I}_n - \left(\boldsymbol{I}_n + \sum_{j=1}^{m} \boldsymbol{P}_j^{\mathrm{T}}\boldsymbol{S}_j^{-1}\boldsymbol{P}_j\boldsymbol{\Sigma}\right)^{-1}\right)^{-1},$$

where $\boldsymbol{I}_n$ denotes the $n \times n$ identity matrix. Note that the expression for $\boldsymbol{\Sigma}_{\text{post}}$ still involves the unknown covariance matrix $\boldsymbol{\Sigma}$. Additional information can be assumed about $\boldsymbol{\Sigma}$, and a further Bayesian posterior distribution can be calculated to eliminate the dependence on $\boldsymbol{\Sigma}$. However, for our purposes we will simply replace $\boldsymbol{\Sigma}$ with the unbiased estimator $\hat{\boldsymbol{\Sigma}}$ derived from historical returns, as discussed in section 2.1. A shortcut of this sort is typical in many applications of Bayesian analysis because elimination of $\boldsymbol{\Sigma}$ generally involves more complicated integrals.

There is an alternate derivation for the expression of $\boldsymbol{\mu}_{\text{post}}$ from (10): $\boldsymbol{\mu}_{\text{post}}$ is the value of $\boldsymbol{\mu}$ that minimizes the function

$$\mathcal{F}(\boldsymbol{\mu}) = \sum_{j=1}^{m} \|\boldsymbol{P}_j\boldsymbol{\mu} - \boldsymbol{Q}_j\|^2_{\boldsymbol{S}_j^{-1}},$$

where we have used the notation $\|\boldsymbol{v}\|^2_W = \boldsymbol{v}^{\mathrm{T}}\boldsymbol{W}\boldsymbol{v}$. That is, $\boldsymbol{\mu}_{\text{post}}$ is the point most "compatible" with the constraints $\boldsymbol{P}_j\boldsymbol{\mu} = \boldsymbol{Q}_j$, with the uncertainty around each constraint specified by $\boldsymbol{S}_j$. To see that $\boldsymbol{\mu}_{\text{post}}$ indeed minimizes $\mathcal{F}$, compute the gradient of $\mathcal{F}$ and solve for the value of $\boldsymbol{\mu}$ that makes the gradient equal to $\boldsymbol{0}$. The resulting formula is the same as the formula for $\boldsymbol{\mu}_{\text{post}}$ in (10).

### *The Grand Mean*

Bayesian techniques can be used to determine a posterior mean and covariance matrix based on a combination of historical information and information about a *grand mean*, an $n$-dimensional vector with all components equal to some real number $\eta$. The assumption is that asset returns will ultimately tend toward a common average value and, by incorporating this tendency into a Bayesian analysis, it may be possible to obtain better values for $\boldsymbol{\mu}_{\text{post}}$ and $\boldsymbol{\Sigma}_{\text{post}}$.

Jorion uses a total of $m = T + 1$ pieces of additional information to update the prior that $\boldsymbol{r}$ is normally distributed, where $T$ is the sample size of the available return data. The first $T$ items are based on the historical data observations and are of the form $\boldsymbol{\mu} = \boldsymbol{r}_j$ for $1 \leqslant j \leqslant T$, where $\boldsymbol{r}_j$ is the $j$th observed return, with uncertainty $\boldsymbol{\Sigma}$, the historical covariance matrix. The final item has the form $\boldsymbol{\mu} = \eta\boldsymbol{e}$, the vector with all elements equal to the common grand mean, with uncertainty specified by $(1/\lambda)\boldsymbol{\Sigma}$ for a suitable $\lambda > 0$.

The appropriate value of $\lambda$ is estimated as

$$\hat{\lambda} = \frac{n+2}{(\hat{\boldsymbol{\mu}} - \eta_0\boldsymbol{e})^{\mathrm{T}}\tilde{\boldsymbol{\Sigma}}^{-1}(\hat{\boldsymbol{\mu}} - \eta_0\boldsymbol{e})},$$

where $\eta_0$ is an estimate for the expected return of the minimum-risk portfolio. After performing the necessary Bayesian updating, Jorion finds estimators for the expected return and covariance that depend only on historical data, namely

$$\hat{\boldsymbol{\mu}}_J = \frac{T}{T+\lambda}\hat{\boldsymbol{\mu}} + \frac{\lambda}{T+\lambda}\eta_0\boldsymbol{e},$$

$$\hat{\boldsymbol{\Sigma}}_J = \left(1 + \frac{1}{T+\lambda}\right)\tilde{\boldsymbol{\Sigma}} + \frac{\lambda}{T(T+1+\lambda)}\left(\frac{\boldsymbol{e}\boldsymbol{e}^{\mathrm{T}}}{\boldsymbol{e}^{\mathrm{T}}\tilde{\boldsymbol{\Sigma}}^{-1}\boldsymbol{e}}\right).$$

The values $\hat{\boldsymbol{\mu}}_J$ and $\hat{\boldsymbol{\Sigma}}_J$ tend toward $\hat{\boldsymbol{\mu}}$ and $\tilde{\boldsymbol{\Sigma}}$ as $T \to \infty$, but for smaller values of $T$ there are "correction" terms that take on a greater weight.

### *Market Equilibrium and Investor Beliefs*

One can approach portfolio optimization by starting from a neutral market-implied set of expected returns and allowing investors to overlay their views to inform and modify these values. This method avoids the well-known difficulty of correctly predicting future returns from historical data and also provides a mechanism through which investors can express views different from those implied by history or the market.

A Bayesian technique along these lines was developed by Black and Litterman, who started with the usual prior that the distribution of $\boldsymbol{r}$ is normally distributed but assumed that the mean and covariance matrix are unknown. This prior is then updated with two types of additional information. The first is that the value of $\boldsymbol{\mu}$ has a distribution centered at $\boldsymbol{\mu}_{\text{mkt}}$, the expected return vector implied by the market, discussed further below. The second is information provided by the investor regarding his beliefs about the distribution of $\boldsymbol{\mu}$. Using all this additional information, a posterior normal distribution for $\boldsymbol{r}$ is computed.

To determine the expected return vector implied by the market, assume that the CAPM holds and that the tangency portfolio is given by $\boldsymbol{\omega}_{\text{mkt}} = (1/\gamma)\boldsymbol{\Sigma}^{-1}(\boldsymbol{\mu} - r_{\text{f}}\boldsymbol{e})$, where $\gamma$ is a constant that can be interpreted as the level of investor risk tolerance. Because the portfolio of market weights is easily observed, it is useful to rewrite this formula as an expression for $\boldsymbol{\mu}$ and to define this to be the value of $\boldsymbol{\mu}$ implied by the market, namely $\boldsymbol{\mu}_{\text{mkt}} = r_{\text{f}}\boldsymbol{e} + \gamma\boldsymbol{\Sigma}\boldsymbol{\omega}_{\text{mkt}}$. The value $\gamma$ can be estimated using historical data as the level of risk tolerance that is compatible with the risk in the market portfolio and the excess return to the market portfolio over the risk-free rate. Black and Litterman assume that the uncertainty around $\boldsymbol{\mu}_{\text{mkt}}$ is $\tau\boldsymbol{\Sigma}$, where $\tau$ is a small positive number. The uncertainty is therefore assumed to be proportional to the covariance matrix for returns, but small. The exact value of $\tau$ should be chosen in such a way as to reflect the uncertainty in the mean estimator $\boldsymbol{\mu}_{\text{mkt}}$. The value of $\tau$ should generally be close to zero to reflect the idea that the uncertainty in the market-implied mean return vector should not be very large.

The second step of Black and Litterman's method is the incorporation of specified investor beliefs. An investor is allowed to express a number of views of the form

$$p_{k,1}\mu_1 + \cdots + p_{k,n}\mu_n = q_k + \epsilon_k,$$

where $k = 1, \dots, K$ runs through the list of views expressed. The quantity $\epsilon_k$ represents a normally distributed random variable with zero mean, reflecting the uncertainty in the $k$th investor belief. The $p_{k,i}$ and $q_k$ are real numbers, and $\mu_i$ represents the expected return for the $i$th asset. This set of beliefs can be written compactly in matrix form as

$$\boldsymbol{P}_{\text{inv}}\boldsymbol{\mu} = \boldsymbol{Q}_{\text{inv}} + \boldsymbol{\epsilon}_{\text{inv}}, \tag{11}$$

where $\boldsymbol{P}_{\text{inv}} = [p_{k,i}]_{k,i}$ is a $K \times n$ matrix, $\boldsymbol{Q}_{\text{inv}} = [q_k]_k$ is a $K$-dimensional vector, and $\boldsymbol{\epsilon}_{\text{inv}}$ is a multivariate normal random variable with zero mean and diagonal covariance matrix $\boldsymbol{\Omega}_{\text{inv}}$.

To combine the investor beliefs of (11) with the market-implied returns and determine a posterior distribution for $\boldsymbol{r}$, one can follow the methodology described at the start of section 2.3 with $m = 2$ pieces of information. To incorporate market-implied returns, they set $\boldsymbol{P}_1 = \boldsymbol{I}_n$, $\boldsymbol{Q}_1 = \boldsymbol{\mu}_{\text{mkt}}$, and $\boldsymbol{S}_1 = \tau\boldsymbol{\Sigma}$. To incorporate investor beliefs, they set $\boldsymbol{P}_2 = \boldsymbol{P}_{\text{inv}}$, $\boldsymbol{Q}_2 = \boldsymbol{Q}_{\text{inv}}$, and $\boldsymbol{S}_2 = \boldsymbol{\Omega}_{\text{inv}}$. They then find that the posterior mean is

$$\begin{aligned}\boldsymbol{\mu}_{BL} = {} & ((\tau\boldsymbol{\Sigma})^{-1} + \boldsymbol{P}_{\text{inv}}^{\text{T}}\boldsymbol{\Omega}_{\text{inv}}^{-1}\boldsymbol{P}_{\text{inv}})^{-1} \\ & \times ((\tau\boldsymbol{\Sigma})^{-1}\boldsymbol{\mu}_{\text{mkt}} + \boldsymbol{P}_{\text{inv}}^{\text{T}}\boldsymbol{\Omega}_{\text{inv}}^{-1}\boldsymbol{Q}_{\text{inv}}).\end{aligned}$$

This is the expected value of $\boldsymbol{r}$ that is most consistent (subject to the specified levels of uncertainty) with both the expected returns implied by the market and investor beliefs.

### 2.4 Other Approaches and Metrics

It is possible to extend the basic framework of mean–variance optimization by using measures of risk and reward other than the mean vector, $\boldsymbol{\mu}$, and the covariance matrix, $\boldsymbol{\Sigma}$. It is also possible to find additional ways to implement the basic mean–variance framework beyond the methods already described. In the remainder of this section we describe a few of these approaches.

*One-Sided Risk Measures*

Some of the most common alternative measures of portfolio risk are "one sided," in that they focus on downside risk rather than symmetric risk around an expected return. Examples include *value at risk* (VaR) and *shortfall risk*. To measure VaR, an investor must specify a threshold $0 < \theta < 1$. The VaR is an amount such that the probability that portfolio losses will equal or exceed such an amount is exactly equal to $\theta$. Thus, an investor knows that with probability $1 - \theta$ losses will not exceed the VaR level. To measure shortfall risk, an investor must specify a benchmark, $b$, relative to which performance can be measured. The shortfall risk is the probability that the portfolio return will fall below $b$ multiplied by the average amount by which the portfolio return falls below $b$ conditional on being below $b$. The shortfall risk thus provides an investor with an expected value of downside exposure.

By replacing variance with a one-sided measure in choosing optimal portfolios, an investor is able to control his worst-case scenarios to within a certain confidence threshold (for VaR) or his average loss below a benchmark (for shortfall risk). A difficulty with using these measures, however, is that they do not yield closed-form expressions and they are less intuitive. Also, from a computational perspective, nonlinear optimization will generally be necessary to find optimal portfolios with respect to these measures, which is far less efficient than the linear and quadratic programming algorithms applicable to standard mean–variance optimization problems. Nonetheless, despite the additional computational complexity, it is possible to optimize relative to these alternative measures of risk, and doing so may be desirable for investors with preferences or asset-return dynamics that are especially asymmetric.

*Resampling the Efficient Frontier*

A technique known as *resampling* (which is closely related to the *bootstrap technique* in statistics) can be used to determine portfolio weights. The idea is to smooth out errors arising from uncertainty in estimators for $\boldsymbol{\mu}$ and $\boldsymbol{\Sigma}$ by generating a large number of alternative possibilities for these values from a single data set, constructing a resampled efficient frontier in each alternative case, and then averaging all of the alternatives to find an average resampled frontier. Optimal portfolios are then selected from among the points on the average resampled frontier.

To construct a resampled frontier, Michaud generates a hypothetical alternative history of realized asset returns. These are chosen from a multivariate normal distribution with mean and covariance equal to the unbiased estimates $\hat{\boldsymbol{\mu}}$ and $\hat{\boldsymbol{\Sigma}}$, determined based on the true history. The alternative history is then used to calculate portfolios on a resampled efficient frontier. This process is repeated a large number of times, with the alternative history being chosen independently each time. The number of portfolios computed on the frontier is fixed across all resamplings. In addition, an upper bound for possible returns on the resampled frontiers can be specified in order to limit the returns on all portfolios considered to a finite range. This limit may be taken to be the largest element in $\hat{\boldsymbol{\mu}}$, for example.

The individual portfolios on the resampled frontiers are averaged together to form an average resampled frontier. All these frontiers are discrete, rather than continuous, but if the number of portfolios computed is large, a close approximation to a continuous frontier is obtained. We calculate expected returns and standard deviations for each portfolio on the average resampled frontier using estimates for mean and return such as $\hat{\boldsymbol{\mu}}$ and $\hat{\boldsymbol{\Sigma}}$. These values allow us to determine which portfolio is optimal under a specified metric, such as minimum risk, maximum utility, or maximum Sharpe ratio. Alternatively, we could compute the optimal portfolio along each resampled frontier and average the results. The answer in this situation is generally not the same as the optimal portfolio on the average resampled frontier, but it is generally computationally much easier because it avoids the need to compute all points on the frontier for each resampling.

*Ordering of Returns*

Mean–variance portfolio optimization can be extended by allowing an investor to specify less information about asset returns than is encompassed by complete knowledge of the vector $\boldsymbol{\mu}$ of expected returns. For example, an investor may specify a list of inequalities and interrelationships that will hold for elements of the return vector $\boldsymbol{r}$. This type of information leads to a much larger set of "efficient" portfolios than mean–variance optimization, and it is thus more complicated to select a single optimal portfolio from among all the efficient ones. Almgren and Chriss resolve this difficulty by introducing a methodology for ranking portfolios in light of the information specified by the investor. They also cast their methodology in a manner that makes it computationally feasible to determine an optimal portfolio.

**Acknowledgments.** We thank Nick Higham for many helpful comments and suggestions and Jayna Cummings for excellent editorial assistance. Research support from the MIT Laboratory for Financial Engineering and the Northwestern University School of Law Faculty Research Program is gratefully acknowledged.

### Further Reading

Almgren, R., and N. Chriss. 2006. Optimal portfolios from ordering information. *Journal of Risk* 9:1–47.

Black, F., and R. Litterman. 1992. Global portfolio optimization. *Financial Analysts Journal* 48(5):28–43.

Brennan, T., and A. Lo. 2010. Impossible frontiers. *Management Science* 56:905–23.

Higham, N. J. 2002. Computing the nearest correlation matrix—a problem from finance. *IMA Journal of Numerical Analysis* 22:329–43.

Jorion, P. 1986. Bayes–Stein estimation for portfolio analysis. *Journal of Financial and Quantitative Analysis* 21: 279–92.

Laloux, L., P. Cizeau, M. Potters, and J.-P. Bouchaud. 2000. Random matrix theory and financial correlations. *International Journal of Theoretical and Applied Finance* 3(3): 391–97.

Markowitz, H. 1987. *Mean-Variance Analysis in Portfolio Choice and Capital Markets.* Oxford: Basil Blackwell.

Merton, R. 1972. An analytical derivation of the efficient portfolio frontier. *Journal of Financial and Quantitative Analysis* 7(4):1851–72.

Michaud, R. 1998. *Efficient Asset Management.* Boston, MA: Harvard Business School Press.

# V.11 Bayesian Inference in Applied Mathematics

*Desmond J. Higham*

## 1 Use of Bayes's Theorem

Deterministic models can be used to represent a huge variety of physical systems. Applied mathematicians are traditionally schooled in a framework where a given mathematical model is presented, or developed, with known values for all input data, such as problem domain, initial conditions, boundary conditions, and model parameters. Our task is then to analyze properties of the solution by whatever means we have at our disposal, for example, finding the exact solution, developing approximations under asymptotic limits, or applying computational methods. However, even when a deterministic model is appropriate, there are many realistic scenarios where uncertainty arises and it becomes beneficial to employ tools from statistics. Here are four illustrative examples.

(1) A partial differential equation describes the spread of a pollutant after an environmental disaster. However, the initial location and quantity of the pollutant can only be estimated. What is our uncertainty in the pollutant level after one week, given the uncertainty in the initial data? This is a *problem-sensitivity* or *conditioning* question.

(2) A pair of dice are rolled on a table. It is not practical to measure the initial location and velocity of the dice and then solve their equations of motion until they come to a halt. Instead, we can model each die independently as a random variable that is equally likely to take the value $1, 2, \dots, 6$. Here, we are introducing randomness as a convenient *modeling approximation.*

(3) A large chemical reaction network is represented by a system of ordinary differential equations. However, the value of one reaction rate constant is not currently known and is too tricky to measure directly from a laboratory experiment. In this case, we would like to *infer* a value for the unknown rate constant using whatever data is available about the full system. This is a *parameter-fitting*, or *model-calibration*, problem, where the key question is what parameter value causes the model to best fit the data or, from a statistical inference perspective: for each possible choice of the parameter value, what is my degree of belief that this choice is correct?

(4) Biologists have two competing, and incompatible, theories for the mechanism by which signals are passed through a transduction network. These lead to two different deterministic mathematical models, each having one or more unknown modeling parameters. Given some experimental data, representing outputs from the network, which of the two models is most likely to be correct? This is a *model-comparison* problem.

Questions of this type, where models based on mechanistic laws of motion meet real data, lie at the intersection between applied mathematics and applied statistics. In recent years the term UNCERTAINTY QUANTIFICATION [II.34] has been coined to describe this field, although the phrase also has many other connotations.

A powerful tool for statistical inference, and hence for uncertainty quantification, is Bayes's theorem (also commonly referred to as Bayes' theorem), which allows us to update our beliefs according to data. The theorem is named after the Reverend Thomas Bayes (1701–61), a British mathematician and Presbyterian minister.

Considering item (3) from the list above, let $y$ denote the unknown problem parameter in our model and let $Y$ denote observational data consisting of a time series of chemical species levels. Bayes's theorem may then be written

$$P(y \mid Y) = \frac{P(Y \mid y)P(y)}{P(Y)}.$$

Here, $P(y \mid Y)$, known as the *posterior distribution*, answers our question. It quantifies the probability of the model parameter $y$ given the data $Y$. On the right-hand side we have $P(Y \mid y)$, known as the *likelihood*, quantifying the probability of the data $Y$ arising given the model parameter $y$. This value is available to us, since we have access to the mathematical model. Also appearing on the right-hand side is the *prior distribution*, $P(\gamma)$. This factor quantifies our original degree of belief in the parameter value $\gamma$, before we see any data.

The denominator $P(Y)$ does not depend on the value of the model parameter $\gamma$, so we have the proportionality relationship

$$P(\gamma \mid Y) \propto P(Y \mid \gamma)P(\gamma).$$

In words, the posterior is proportional to the product of the likelihood and the prior.

## 2 Example

To illustrate these ideas we consider the very simple production and decay system

$$\varnothing \xrightarrow{1} X,$$
$$X \xrightarrow{\gamma} \varnothing.$$

The first reaction indicates that a species $X$ is created at a constant, unit rate. Conversely, the second reaction tells us that $X$ also degrades at a rate proportional to its current level, with rate constant $\gamma$. The corresponding mass action ordinary differential equation for the abundance of $X$ at time $t$ is given by

$$\frac{\mathrm{d}x(t)}{\mathrm{d}t} = 1 - \gamma x(t),$$

with solution

$$x(t) = \frac{1}{\gamma} + \left(\frac{\gamma x(0) - 1}{\gamma}\right)\mathrm{e}^{-\gamma t}.$$

Suppose that our data takes the form $Y = \{t_j, y_j\}_{j=1}^{D}$, where $y_j$ denotes the observed level of $x(t)$ at time $t_j$. Our aim is to use this data to infer the unknown rate constant, $\gamma$. To emphasize that $\gamma$ is not known, we will also write the solution as $x_\gamma(t)$.

To define a likelihood, we must impose some assumptions. The simplest choice is to suppose that the data contains experimental measurement errors that are independent across time points and normally distributed about the "exact" value, with a common standard deviation, $\sigma$. This leads to a likelihood function of the form

$$P(Y \mid \gamma) = \prod_{j=1}^{D} \frac{1}{\sqrt{2\pi\sigma^2}} \exp\left(-\frac{(y_j - x_\gamma(t_j))^2}{2\sigma^2}\right).$$

In words, the right-hand side quantifies the probability of the data $Y$ arising given that the rate constant is $\gamma$.

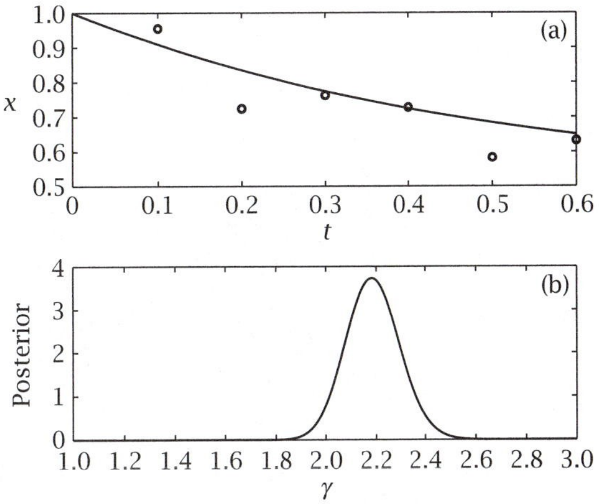


**Figure 1** (a) The solid curve shows the underlying exact solution, and the circles indicate the synthetic data generated by adding Gaussian noise. (b) The posterior distribution for the model parameter, $\gamma$, whose "correct" value is known by construction to be $\gamma = 2$.

Figure 1(a) uses a solid line to plot the solution when $x(0) = 1$ and $\gamma = 2$. To generate some synthetic data, we take the solution $x(t_j)$ at the time points $t_1 = 0.1$, $t_2 = 0.2, \dots, t_6 = 0.6$ and add independent normally distributed noise with mean zero and standard deviation $\sigma = 0.05$. These data points are shown as circles. Figure 1(b) plots the resulting posterior distribution for the rate constant $\gamma$. Here, we have used the known values $x(0) = 1$ and $\sigma = 0.05$ and taken the prior to be uniform over $[1, 3]$; that is, $P(\gamma) = \frac{1}{2}$ for $1 \leqslant \gamma \leqslant 3$ and $P(\gamma) = 0$ otherwise. Because we generated the data ourselves, we are able to judge this result. We see that the inference procedure has assigned nontrivial weight to the "correct" value of $\gamma = 2$ but it assigns more weight to slightly higher values of $\gamma$, with a peak at around 2.18. Intuitively, this mismatch has arisen because, by chance, the two noisiest data points, at $t = 0.2$ and $t = 0.5$, are both below the true solution, encouraging the decay rate to be overestimated. Of course, the inference would become more accurate if we were to increase the number of data points (or reduce the noise level).

It is slightly more realistic to consider the case where the standard deviation, $\sigma$, for the experimental noise is not known. Conceptually, this presents no added difficulty, since Bayes's theorem holds when there are multiple variables to be inferred. Figure 2 shows the resulting posterior when we regard $\sigma$ as a second unknown parameter, with a uniform prior over $0.01 \leqslant \sigma \leqslant 0.1$.

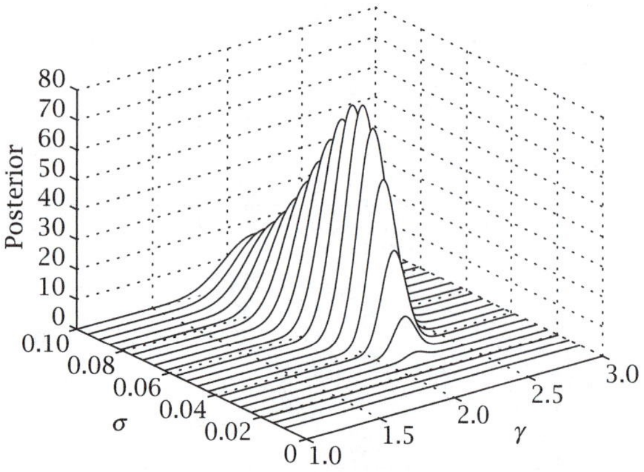


**Figure 2** The posterior distribution for the experiment in figure 1 when both the model parameter, $\gamma$, and the experimental noise level, $\sigma$, are regarded as unknowns. Here, the "correct" values are known by construction to be $\gamma = 2$ and $\sigma = 0.05$.

In this very simple setting, with an assumption of independent Gaussian noise, the posterior distribution is, after taking logarithms, effectively a least-squares measure of the mismatch between model and data. With a uniform prior, optimizing this least-squares objective function corresponds to computing a maximum-likelihood estimate: a point where the posterior is largest. More generally, using a nonuniform prior corresponds to adding a penalty term to the least-squares objective function, which is a standard approach in nonlinear optimization for constraining the solution to lie in a predetermined domain. However, a key philosophical difference between a least-squares best fit and the full computation of a Bayesian posterior is that in the latter case we are concerned with all possible parameter values and wish to know which regions of parameter space support likely values, even if they are not globally optimal. Moreover, the posterior gives access to higher levels of inference. For example, we may sample parameter values from the posterior and run the model forward in order to build up a distribution for the output. Figure 3 illustrates this idea: in part (a) we show solution curves arising from five independent samples of $\gamma$ from the posterior in figure 1, so values where the posterior is larger are more likely to be chosen; in part (b) we show a histogram of the $X$ level at $t = 1$ based on 10 000 samples of $\gamma$ from the posterior. Knowledge of the full posterior distribution also allows us to compute integrals over parameter space, with more weight attributed to the more likely values. With this approach, given competing models we may use so-called *Bayes factors* in order to judge systematically which one best describes the data.

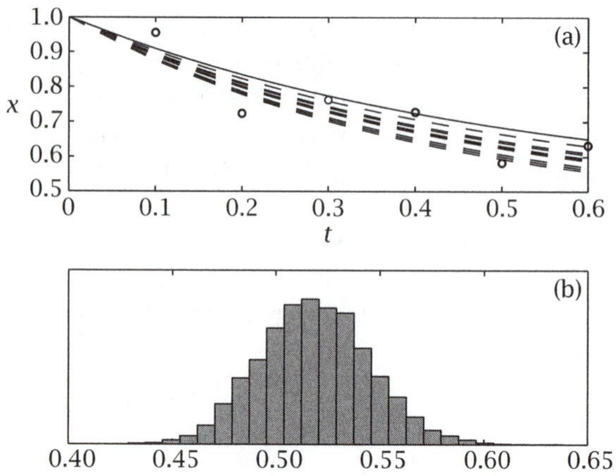


**Figure 3** (a) As in figure 1, the solid curve shows the underlying exact solution and the circles indicate the synthetic data generated by adding Gaussian noise. Five output solutions are shown as dashed curves, where the rate constant, $\gamma$, has been sampled from the posterior shown in figure 1(b). (b) A histogram for the predicted level of species $X$ at the future time $t = 1$, based on 10 000 samples of $\gamma$ from the posterior. The "correct" value is known to be $x(1) = 0.568$.

## 3 Challenges

The growing popularity of Bayesian inference can be attributed, at least in part, to the availability of increasing computing power and better-quality/more abundant data sets. Generic challenges in the field include the following.

**Priors.** A fundamental tenet of Bayesian inference is that we must quantify our inherent beliefs at the outset. This allows expert opinions to be incorporated into the process but, of course, also places a burden on the researcher and introduces a level of subjectivity. Even the use of uniform priors, as an indication of no inherent preferences, is problematic; for example, being indifferent about $\gamma$, and hence imposing a uniform prior, is not equivalent to being indifferent about $\gamma^2$.

**Identifiability.** The task of inferring model parameters may be inherently ill-defined. For example, there may be insufficient data or, more fundamentally, model parameters may be very highly correlated. In a dynamical system, for instance, increasing a decay rate may have a very similar effect to decreasing a production rate, making it difficult to uncover a precise value for each, that is, causing the posterior to have a flat, elongated peak in that direction.

**High dimension.** Many models in applied mathematics involve a large number of parameters. Moving from the simple two-parameter example in figure 2 into very high dimensions poses significant challenges in terms of both exploring the parameter space and visualizing the results. Progress in this area is largely based on the use of *Markov chain Monte Carlo* algorithms, where a MARKOV CHAIN [II.25] is constructed whose equilibrium distribution matches the desired posterior, allowing approximate samples to be computed via long-time path simulation. Here, the applied mathematics/applied statistics interface comes back into focus, with many Markov chain Monte Carlo notions having counterparts in the fields of optimization and numerical time stepping.

In the applied mathematics setting, a further issue of fundamental importance is the construction of a suitable likelihood function. As we saw in our simple example, a deterministic model does not sit well with the notion of likelihood, that is, the probability of seeing this data, given the model parameters. With a deterministic model, we either recover the data exactly or we do not recover it at all. To get around the issue we took the commonly used step of placing all the blame on the data—assuming that the mismatch is caused entirely by measurement error. This step clearly ignores the fundamental truth that mathematical models are based on idealizations and can never reflect all aspects of a physical system.

At first sight, this issue seems to disappear if we begin with a stochastic model; for example, the simple production and decay system used in our tests could be cast as a discrete-state birth and death process, or a continuous-state stochastic differential equation. In either case, we are not forced to assume that the data-model mismatch is driven solely by experimental error. Given the model parameters, by construction a stochastic model assigns a probability to any observed data, and hence we have an automatic likelihood. However, although stochastic modeling offers a seamless transition to a likelihood function, it clearly does not overcome objections that the model itself is a source of error. Overall, the systematic incorporation of uncertainties arising from modeling, discretization, and experimental observation remains a high-profile challenge.

### Further Reading

Sivia, D. S., and J. Skilling. 2006. *Data Analysis: A Bayesian Tutorial*, 2nd edn. Oxford: Oxford University Press.

Smith, R. C. 2013. *Uncertainty Quantification*. Philadelphia, PA: SIAM.

# V.12 A Symmetric Framework with Many Applications

*Gilbert Strang*

## 1 Introduction

This article describes one of the structures on which applied mathematics is built. It has a discrete form, expressed by matrix equations, and a continuous form, expressed by differential equations. The matrices are symmetric and the differential equations are self-adjoint. These properties appear naturally when there is an underlying minimum principle.

The important fact about this structure is that it is truly useful (and extremely widespread). Problems from mechanics, physics, economics, and more fit into this framework. The temptation to overwhelm the reader with a list of applications is almost irresistible. Instead, we go directly to the *KKT matrix*, which shows the form that all of these applications share:

$$M = \begin{bmatrix} C^{-1} & A \\ A^{\mathrm{T}} & 0 \end{bmatrix} \begin{matrix} m \text{ rows} \\ n \text{ rows} \end{matrix}$$

$C$ is square, symmetric, and positive-definite (and so is $C^{-1}$). $A$ is rectangular with $n$ independent columns ($n \leqslant m$). The block matrix $M$ is then symmetric and invertible—but certainly not positive-definite. This is a *saddle-point matrix*.

The first $m$ pivots are positive, coming from $C^{-1}$. The elimination steps multiply the first block row by $A^{\mathrm{T}}C$ and subtract from the second block row to produce zeros below $C^{-1}$:

$$\begin{bmatrix} I & 0 \\ -A^{\mathrm{T}}C & I \end{bmatrix} \begin{bmatrix} C^{-1} & A \\ A^{\mathrm{T}} & 0 \end{bmatrix} = \begin{bmatrix} C^{-1} & A \\ 0 & -A^{\mathrm{T}}CA \end{bmatrix}.$$

Elimination has produced the $A^{\mathrm{T}}CA$ *matrix* that is fundamental to so many problems in computational science. This $n \times n$ matrix appears in equation (2), and it is the focus of section 3.

The *Schur complement* $-A^{\mathrm{T}}CA$ in the last block is negative-definite, so the last $n$ pivots of $M$ will all be negative. Then $M$ must have $m$ positive eigenvalues and $n$ negative eigenvalues (pivots and eigenvalues have the same signs for symmetric matrices). A graph of the quadratic,

$$Q(w,u) = \tfrac{1}{2}w^{\mathrm{T}}C^{-1}w + u^{\mathrm{T}}A^{\mathrm{T}}w$$
$$= \tfrac{1}{2}\begin{bmatrix} w_1 \cdots w_m \; u_1 \cdots u_n \end{bmatrix}\begin{bmatrix} C^{-1} & A \\ A^{\mathrm{T}} & 0 \end{bmatrix}\begin{bmatrix} w \\ u \end{bmatrix},$$

would go upward in $m$ eigenvector directions and downward in the (orthogonal) $n$ eigenvector directions. The saddle point is at $(w,u) = (0,0)$.

### 1.1 Constrained Least Squares

Let us see how these matrices appear in a broad class of optimization problems. The minimum principle is quadratic with linear constraints:

$$\text{minimize } F(w) = \tfrac{1}{2}w^{\mathrm{T}}C^{-1}w - w^{\mathrm{T}}b$$
$$\text{subject to } A^{\mathrm{T}}w = f.$$

The $n$ constraints $A^{\mathrm{T}}w = f$ lead to $n$ Lagrange multipliers $u_1, \dots, u_n$. Lagrange's beautiful idea was to build the constraints (using the multipliers) into the function $L$:

$$L(w,u) = \tfrac{1}{2}w^{\mathrm{T}}C^{-1}w - w^{\mathrm{T}}b + u^{\mathrm{T}}(A^{\mathrm{T}}w - f).$$

The optimal $w$ and $u$ are found at a stationary point (not a minimum!) of $L$. The coefficient matrix is $M$:

$$\left.\begin{aligned} \frac{\partial L}{\partial w} &= 0 \to C^{-1}w + Au = b, \\ \frac{\partial L}{\partial u} &= 0 \to A^{\mathrm{T}}w = f. \end{aligned}\right\} \qquad (1)$$

Examples show convincingly that when the variables $w_1, \dots, w_m$ have physical meaning (that is, when they are currents in circuit theory, stresses in mechanics, velocity in fluids, momentum in physics), the dual variables $u_1, \dots, u_n$ have meaning too. These Lagrange multipliers represent the forces needed to impose the constraints $A^{\mathrm{T}}w = f$ (they are voltages in circuit theory, displacements in mechanics, pressure in fluids, position in physics).

The $w$s and $u$s are *dual unknowns*. The fundamental theorem of optimization is the *minimax* or *duality* theorem. A minimization over $w$ (the primal problem) connects to a maximization over $u$ (the dual problem). The first is a minimax of $L(w,u)$, the second is the maximin—and they are equal.

## 2 Numerical Optimization/Finite Elements

These two big subjects are not usually combined. The FINITE-ELEMENT METHOD [II.12] is an approach (a successful one) to solving differential equations. OPTIMIZATION [IV.11] is generally seen in a different world. But both subjects present us with the same choices, precisely because equation (1) is central to both. Here are three options for computing $u$ and $w$.

(i) The *mixed method* for finite elements and the *primal-dual method* for optimization solve for $u$ and $w$ together. We work with $M$.
(ii) The *stress method* (or *nullspace method*) finds the best $w$ among all candidates that satisfy the constraints $A^{\mathrm{T}}w = f$. If $w_p$ is one particular solution, we add any solution of $A^{\mathrm{T}}w = 0$ (Kirchhoff's current law). We therefore need a basis for the nullspace of $A^{\mathrm{T}}$.
(iii) The *displacement method* eliminates $w = C(b - Au)$ from equation (1) and solves for $u$. Multiply the first row by $A^{\mathrm{T}}C$ and subtract from the second row, and then reverse all signs to work with $A^{\mathrm{T}}CA$:

$$A^{\mathrm{T}}CAu = A^{\mathrm{T}}Cb - f. \qquad (2)$$

Equation (2) is a central problem of scientific computing. The matrix $A^{\mathrm{T}}CA$ is symmetric positive-definite (sign reversal produced a minimization). $K = A^{\mathrm{T}}CA$ is the *stiffness matrix* in finite elements. It is the *graph Laplacian matrix* that appears everywhere in applied mathematics. In statistics and linear regression, we have the *weighted normal equations*.

For solid mechanics, this $A^{\mathrm{T}}CA$ approach is the nearly universal choice. For fluid mechanics, with pressures and velocities, the mixed method is often preferred. This mathematical contest is reflected in battles between software packages: variations of SPICE for electronic circuits and of finite-element codes like NASTRAN, FEMLAB, and ABAQUS.

An excellent reference for the numerical solution of saddle-point problems is the survey by Benzi et al. (2005). For fluid problems and mixed methods (with preconditioning), Elman et al. (2014) is outstanding.

## 3 The Framework

Here is the key sentence from Strang (1988): "I believe that this $A^{\mathrm{T}}CA$ pattern is the central framework for problems of equilibrium ... *continuous and discrete, linear and nonlinear, equations and variational principles.*" $A^{\mathrm{T}}CA$ is the *system matrix* or *stiffness matrix* that contains the geometry in $A$, the material properties in $C$, and the physical laws in $A^{\mathrm{T}}$. The problems that we need to solve are

$$A^{\mathrm{T}}CAu = f, \qquad A^{\mathrm{T}}CAx = \lambda x,$$

$$\frac{\mathrm{d}u}{\mathrm{d}t} = -A^{\mathrm{T}}CAu, \qquad \frac{\mathrm{d}^2u}{\mathrm{d}t^2} = -A^{\mathrm{T}}CAu.$$

The mathematical model produces $A^{\mathrm{T}}CA$. The computational problem then deals with that matrix. We build up the equations in three steps ($A$ and $C$ and $A^{\mathrm{T}}$). Then we attack the equations numerically and analytically.

The continuous model has functions $F$ instead of vectors $f$, and derivatives $A = \mathrm{d}/\mathrm{d}x$ and $A =$ gradient instead of matrices. But $A^{\mathrm{T}}CA$ is still present:

**Continuous linear ODE:**

$$-\frac{\mathrm{d}}{\mathrm{d}x}\left(c(x)\frac{\mathrm{d}u}{\mathrm{d}x}\right) = F(x), \quad A = \frac{\mathrm{d}}{\mathrm{d}x}.$$

**Continuous linear partial differential equation:**

$$-\operatorname{div}(c \operatorname{grad} u) = F(x, y).$$

**Continuous nonlinear partial differential equation:**

$$\operatorname{div}\left(\frac{\operatorname{grad} u}{\sqrt{1 + |\operatorname{grad} u|^2}}\right) = 0.$$

When $A$ is $\mathrm{d}/\mathrm{d}x$, its transpose is $-\mathrm{d}/\mathrm{d}x$. The diagonal matrix $C$, which multiplies every component of a vector $Au$, is replaced by $c(x)$, which multiplies $\mathrm{d}u/\mathrm{d}x$ at every point.

The partial differential equations with $A =$ grad and $A^{\mathrm{T}} = -$ div involve partial derivatives:

$$A = \text{gradient} = \begin{bmatrix} \partial/\partial x \\ \partial/\partial y \end{bmatrix},$$

$$A^{\mathrm{T}} = -\text{divergence} = \begin{bmatrix} -\dfrac{\partial}{\partial x} & -\dfrac{\partial}{\partial y} \end{bmatrix}.$$

Between $A^{\mathrm{T}}$ and $A$ comes $C$, often linear but not always. We will write $e$ for $Au$:

the three-step framework,

$$e = Au, \qquad w = Ce, \qquad F = A^{\mathrm{T}}w,$$

$$\text{potential } u(x,y) \xrightarrow{A} \text{gradient } e(x,y) \xrightarrow{C} \text{flow } w(x,y) \xrightarrow{A^{\mathrm{T}}} \text{source } F(x,y).$$

For flow in a network, $A$ and $A^{\mathrm{T}}$ express Kirchhoff's voltage and current laws and $w = Ce$ is Ohm's law:

voltages $u \to$ voltage drops $e$ $\to$ currents $w \to$ sources $F$.

$A$, $C$, and $A^{\mathrm{T}}$ are matrices or differential operators. They combine into a positive-definite stiffness matrix $K = A^{\mathrm{T}}CA$. Positive-definite matrices are associated with minimum principles for the *total energy* in the system:

$$E(u) = \tfrac{1}{2}u^{\mathrm{T}}Ku - u^{\mathrm{T}}F = \tfrac{1}{2}(Au)^{\mathrm{T}}C(Au) - u^{\mathrm{T}}F.$$

The minimum of $E(u)$ occurs where $Ku = F$. We have recovered $A^{\mathrm{T}}CAu = F$. When $C$ is a matrix multiplication, the energy $E(u)$ is quadratic and its derivative is linear. When $C$ is nonlinear, $CAu$ is really $C(Au)$. In this case, complicated by nonlinearity, minimum principles are often most natural.

## 4 Nonlinear Equations and Minimum Principles

When the material properties are *nonlinear*, $w = Ce$ becomes $w = C(e)$. At any point on that curve, the tangent gives the local linearization $\Delta w \approx C'(e)\Delta e$. A linear to nonlinear example is the step from Newton's law,

$$F = ma = \frac{\mathrm{d}}{\mathrm{d}t}\left(m\frac{\mathrm{d}u}{\mathrm{d}t}\right),$$

to Einstein's law. For relativity the mass $m$ is not constant. The momentum is not a linear function:

$$\text{Newton's momentum} \quad p = mv,$$

$$\text{Einstein's momentum} \quad p(v) = \frac{m_0 v}{\sqrt{1 - v^2/c^2}}.$$

Returning from $p$, $m$, $v$ to $w$, $c$, $e$, we want the equation $w = C(e)$ to appear in the derivative of the energy. The energy is, therefore, an integral:

$$\text{energy} = \int C(e)\,\mathrm{d}e,$$

$$\text{linear case} \int ce\,\mathrm{d}e = \tfrac{1}{2}ce^2.$$

The reverse direction uses the variable $w$. In the linear case we simply divide: $e = w/c$. In the nonlinear case $w = C(e)$ is monotonic, and $e = w/c$ changes to the inverse function $e = C^{-1}(w)$. When we want the energy as a function of $w$, we integrate as before:

$$\text{energy} = \int C^{-1}(w)\,\mathrm{d}w,$$

$$\text{linear case} \int \frac{w}{c}\,\mathrm{d}w = \frac{1}{2}\frac{w^2}{c}.$$

These two functions $F^*(e)$ and $F(w)$, the integrals of inverse functions $C(e)$ and $C^{-1}(w)$, are related by the *Legendre–Fenchel transform*.

## 5 The Graph Laplacian Matrices: $A^TA$ and $A^TCA$

If we plan to offer one example in greater detail, the graph Laplacian must be our choice. *Graphs are the dominant model in discrete applied mathematics.*

Start with $n$ nodes. Connect them with $m$ undirected edges (a complete graph would have all $\frac{1}{2}n(n-1)$ edges; a spanning tree would have $n-1$ edges and no loops). The figure below shows $n = 4$ and $m = 5$. We may imagine that each edge leaves from its lower numbered node, indicated by $-1$ in the *edge-node incidence matrix* $A$:

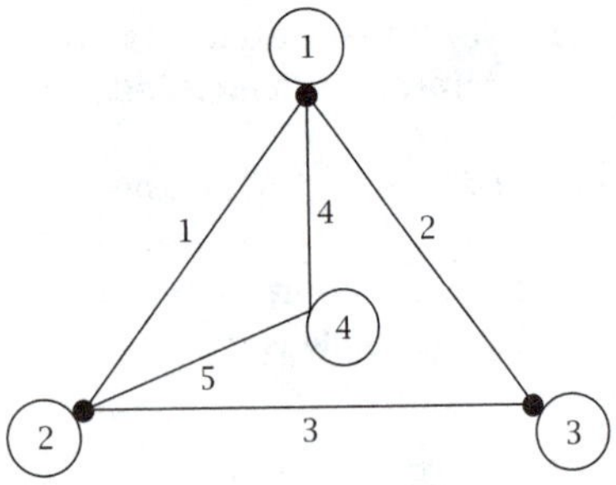


$$A = \begin{bmatrix} -1 & 1 & 0 & 0 \\ -1 & 0 & 1 & 0 \\ 0 & -1 & 1 & 0 \\ -1 & 0 & 0 & 1 \\ 0 & -1 & 0 & 1 \end{bmatrix} \quad \begin{matrix} \text{edge } 1 \\ 2 \\ 3 \\ 4 \\ 5 \end{matrix}$$

The $n$ columns are not independent. $A$ is a "difference matrix." All differences between pairs of $u$s are zero when the voltages $u_1$, $u_2$, $u_3$, $u_4$ are equal:

$$A\begin{bmatrix} u_1 \\ u_2 \\ u_3 \\ u_4 \end{bmatrix} = \begin{bmatrix} u_2 - u_1 \\ u_3 - u_1 \\ u_3 - u_2 \\ u_4 - u_1 \\ u_4 - u_2 \end{bmatrix},$$

$$Au = 0 \quad \text{for } u = (c, c, c, c).$$

With the vector $(1, 1, 1, 1)$ in the nullspace, the sum of the four columns is the zero column. The rank of $A$ is 3. For every connected graph the rank is $n-1$, and the one-dimensional nullspace contains the constant vectors $u_c = (c, \dots, c)$.

Since these vectors have $Au_c = 0$, they also have $A^TCAu_c = 0$. The Laplacian $A^TA$ and the weighted Laplacian matrix $A^TCA$ will be *positive semidefinite* but not positive-definite. Each row will sum to zero. When we fix one voltage at zero and remove it from the set of $n$ unknown $u$s, this removes a column of $A$ and row of $A^T$ to leave *reduced Laplacians* of size $n-1$. Then $A^TA$ and $A^TCA$ are positive-definite.

For networks, $C$ is an $m \times m$ diagonal matrix that comes from Ohm's law $w_i = c_ie_i$. The numbers $c_1, \dots, c_m$ on the diagonal are positive (so $C$ is positive-definite, which is the property we need).

Matrix multiplication produces $A^TA$ (and then we identify the pattern):

$$A^TA = \begin{bmatrix} 3 & -1 & -1 & -1 \\ -1 & 3 & -1 & -1 \\ -1 & -1 & 2 & 0 \\ -1 & -1 & 0 & 2 \end{bmatrix}.$$

The diagonal entries are the *degrees* of the nodes: the number of adjacent edges. The off-diagonal entry in the $j, k$ position is $-1$ if an edge connects nodes $j$ and $k$. The zero entries in the 3, 4 and 4, 3 positions reflect the nonexistence of an edge between those nodes.

It is useful to separate the unweighted Laplacian $A^TA$ into $D - W$. The diagonal matrix $D = \text{diag}(3, 3, 3, 2)$ shows the degrees of the nodes. The off-diagonal matrix $W$ is the *adjacency matrix*, with $W_{jk} = 1$ when nodes $j$ and $k$ are connected.

Finally, we compute $A^TCA$. Why not multiply matrices as columns times rows instead of always rows times columns? Column 1 of $A^T$ multiplies row 1 of $CA$ (which is just $c_1$ times row 1 of $A$). So the first piece of $A^TCA$ *comes entirely from edge* 1 *in the network*, connecting nodes 1 and 2:

$$\begin{bmatrix} -1 & 1 & 0 & 0 \end{bmatrix}^T c_1 \begin{bmatrix} -1 & 1 & 0 & 0 \end{bmatrix} = \begin{bmatrix} c_1 & -c_1 & 0 & 0 \\ -c_1 & c_1 & 0 & 0 \\ 0 & 0 & 0 & 0 \\ 0 & 0 & 0 & 0 \end{bmatrix}.$$

This is the *element stiffness matrix* for edge 1. Each edge $i$ will produce such a matrix, full size but with only four nonzeros. For the edge connecting nodes $j$ and $k$, the only nonzeros will be in those rows and columns.

The product $A^TCA$ then assembles (adds) these five element matrices containing $c_1, \dots, c_5$. They overlap only on the main diagonal, to produce $D - W$:

$$D = \text{diag}\left(\begin{bmatrix} c_1 + c_2 + c_4 \\ c_1 + c_3 + c_5 \\ c_2 + c_3 \\ c_4 + c_5 \end{bmatrix}\right),$$

$$W = \begin{bmatrix} 0 & c_1 & c_2 & c_4 \\ c_1 & 0 & c_3 & c_5 \\ c_2 & c_3 & 0 & 0 \\ c_4 & c_5 & 0 & 0 \end{bmatrix}.$$

The weighted adjacency matrix $W$ contains the $c_i$ (previously all 1s). The weighted degree matrix $D$ shows the edges touching each node. More simply, the diagonal $D$ ensures that every row of $A^TCA$ adds to zero. The smallest eigenvalue is then $\lambda_1 = 0$.

The next eigenvalue, $\lambda_2$, plays an important role in applications. Its eigenvector $x_2$ will be orthogonal (since $A^TCA$ is symmetric) to the first eigenvector $x_1 = (1,1,1,1)$. Separating positive and negative components of this *Fiedler eigenvector* $x_2$ is a useful way to *cluster the nodes*. More heavily weighted edges lie within the two clusters than between them. When a graph comes to us with no clear structure or interpretation, clustering is an important step toward understanding.

So we end with modern applications of the same graph Laplacian that is at the heart of network theory and mass-spring systems. The fundamental constructions of electronics and mechanics—the laws of Ohm and Kirchhoff and Hooke—produce the saddle-point matrix $M$ in equation (1) before elimination removes the $w$s (currents and stresses). From that elimination the graph Laplacian matrix appears!

$A^TCA$ expresses a structure that is fundamental for all graphs and networks: in engineering, in statistics (where $C^{-1}$ is the covariance matrix), in science, and in mathematics.

### Further Reading

Benzi, M., G. H. Golub, and J. Liesen. 2005. Numerical solution of saddle point problems. *Acta Numerica* 14: 1-137.

Ekeland, I., and R. Temam. 1976. *Convex Analysis and Variational Problems*. Amsterdam: North-Holland.

Elman, H. C., D. J. Silvester, and A. J. Wathen. 2014. *Finite Elements and Fast Iterative Solvers: With Applications in Incompressible Fluid Dynamics*, 2nd edn. Oxford: Oxford University Press.

Strang, G. 1988. A framework for equilibrium equations. *SIAM Review* 30:283-97. (Reprinted with new commentary, 2012, in *Essays in Linear Algebra*. Wellesley, MA: Wellesley-Cambridge Press.)

Strang, G. 2007. *Computational Science and Engineering*. Wellesley, MA: Wellesley-Cambridge Press.

Tonti, E. 1975. *On the Formal Structure of Physical Theories*. Milan: Istituto di Politecnica di Milano.

## V.13 Granular Flows

*Joe D. Goddard*

### 1 Introduction

> Granular materials represent a major object of human activities: as measured in tons, the first material manipulated on Earth is water; the second is granular matter.... This may show up in very different forms: rice, corn, powders for construction.... In our supposedly modern age, we are extraordinarily clumsy with granular systems.
>
> P. G. de Gennes, "From Rice to Snow," Nishina Memorial Lecture 2008

This quote by a French Nobel laureate is familiar to many in the field of granular mechanics and reflects a long-standing scientific fascination and practical interest. This state of affairs is acknowledged by many others and summarized in the articles listed in the list of further reading at the end of the article.

As a general definition, we understand by *granular medium* a particle assembly dominated by pairwise nearest-neighbor interactions and usually limited to particles larger than 1 micrometer in diameter, for which the direct mechanical effects of van der Waals and ordinary thermal ("Brownian") forces are negligible. This includes a large class of materials, such as cereal grains, pharmaceutical tablets and capsules, geomaterials such as sand, and the masses of rock and ice in planetary rings.

Most of this article is concerned either with dry granular materials in which there are negligible effects of air or other gases in the interstitial space, or with granular materials that are completely saturated by an interstitial liquid. In this case, liquid surface tension at "capillary necks" between grains, as in the wet sand of sand castles, or related forms of cohesion are largely negligible.

There are several important and interrelated aspects of granular mechanics:

(i) experiment and industrial or geotechnical applications;
(ii) analytical and computational micromechanics (dynamics at the grain level);
(iii) homogenization ("upscaling" or "coarse graining") to obtain smoothed continuum models from the Newtonian mechanics of discrete grains;

(iv) mathematical classification and solution of the continuum field equations for granular flow; and
(v) development of continuum models ("constitutive equations") for stress-deformation behavior.

This article focuses on item (v), since the study of most of the preceding items either depends on it or is strongly motivated by it. The following discussion of (iv) sets the stage for the subsequent coverage of (v).

## 2 Field Equations and Constitutive Equations

By *field equations* we mean the partial differential equations (PDEs) (with time $t$ and spatial position $\boldsymbol{x} \hat{=} [x_i]$ as independent variables) that represent the continuum-level mass and linear momentum balances governing the density $\rho(\boldsymbol{x}, t)$, the velocity $\boldsymbol{v}(\boldsymbol{x}, t) \hat{=} [v_i]$, and the symmetric stress tensor $\boldsymbol{T}(\boldsymbol{x}, t) \hat{=} [\tau_{ij}]$ *for any material*:

$$\dot{\rho} = -\rho \nabla \cdot \boldsymbol{v} \quad \text{and} \quad \rho \dot{\boldsymbol{v}} = \nabla \cdot \boldsymbol{T}. \tag{1}$$

The indices refer to components in Cartesian coordinates, and the notation "$\hat{=} [\cdot]$" indicates components of vectors and tensors in those coordinates. Written out, (1) becomes $\partial_t \rho + \partial_j(\rho v_j) = 0$ and $\partial_t(\rho v_i) = \partial_j(\tau_{ij} - \rho v_i v_j)$ for $i = 1, 2, 3$, where $\dot{w} = \partial_t w + v_j \partial_j w$ for any quantity $w$, sums from $j = 1$ to 3 are taken over terms with repeated indices, $\partial_i w = \partial w / \partial x_i$, $\partial_t w = \partial w / \partial t$, and the product rule has been used. For more information on (1), see CONTINUUM MECHANICS [IV.26].

There are ten dependent variables ($\rho$, $v_i$, and $\tau_{ij} = \tau_{ji}$) but only four equations in (1). In order to "close" them (to make them soluble), we need six more equations. For example, the closure representing the constitutive equations for an incompressible Newtonian fluid (see NAVIER–STOKES EQUATIONS [III.23]) is

$$\boldsymbol{T}' = 2\eta \boldsymbol{D}' \quad \text{and} \quad \rho = \text{const.}, \tag{2}$$

where $\eta$ is a coefficient of shear viscosity and $\boldsymbol{D}$ denotes the *rate of deformation*, also a symmetric second-rank tensor, $\boldsymbol{D} = \operatorname{sym} \nabla \boldsymbol{v} \hat{=} \frac{1}{2}[\partial_i v_j + \partial_j v_i]$, where "sym" denotes the symmetric part of a second-rank tensor. The prime denotes the *deviator* ("traceless" or "shearing" part) $\boldsymbol{X}' = \boldsymbol{X} - \frac{1}{3}(\operatorname{tr} \boldsymbol{X})\boldsymbol{I}$ or $X'_{ij} = X_{ij} - \frac{1}{3} X_{kk} \delta_{ij}$, where $\delta_{ij}$ are the components of the identity $\boldsymbol{I}$.

For rheologically more complicated ("complex") fluids, such as *viscoelastic liquids*, the stress at a material point may depend on the entire past history of the velocity gradient $\nabla \boldsymbol{v}$ at that point. Viscoelasticity is exemplified by the simplest form of the *Maxwell fluid*, giving the rate of change of $\boldsymbol{T}'$ as a linear function of $\boldsymbol{D}'$ and $\boldsymbol{T}'$:

$$\mathring{\boldsymbol{T}}' = 2\mu \boldsymbol{D}' - \lambda \boldsymbol{T}' = \lambda(2\eta \boldsymbol{D}' - \boldsymbol{T}'), \tag{3}$$

where $\mathring{\boldsymbol{X}} = \dot{\boldsymbol{X}} + \operatorname{sym}(\boldsymbol{X}\boldsymbol{W})$, $\boldsymbol{X}\boldsymbol{W} \hat{=} [X_{ik} W_{kj}]$, $\boldsymbol{W}$ is the skew part of the velocity gradient, $\boldsymbol{W} \hat{=} \frac{1}{2}[\partial_i v_j - \partial_j v_i]$, $\lambda$ is the *relaxation rate* or inverse of the *relaxation time*, and $\mu$ is the elastic modulus. In this model, $\mu$ can be identified with the elastic shear modulus $G$, which is discussed below.

According to (3), which was proposed in a slightly different form by James Clerk Maxwell in 1867, the coefficient of viscosity and the elastic modulus are connected by $\eta = \mu/\lambda$, and a material responds elastically on a timescale $\lambda^{-1}$ with viscosity and viscous dissipation arising from relaxing elastic stress.

One obtains from (3) elastic-solid or viscous-fluid behavior in the respective limits $\lambda \to 0$ or $\lambda \to \infty$. This is illustrated by the quintessential viscoelastic material "silly putty," which bounces elastically but undergoes viscous flow in slow deformations. Rheologists employ a *Deborah number* of the form $\dot{\gamma}/\lambda$ to distinguish between rapid solid-like deformations and slow fluid-like deformations.

The superposed "$\circ$" in (3) denotes a *Jaumann derivative*, which gives the time rate of change in a frame translating with a material point and rotating with the local material spin. The Jaumann derivative is the simplest *objective* time derivative that embodies the *principle of material frame indifference*. This principle, which is not a fundamental law of mechanics, is tantamount to the assumption that arbitrary rigid-body rotations superimposed on a given motion of a material merely rotate stresses in the same way. Stated more generally, accelerations of a body relative to an inertial ("Newtonian") frame do not affect the stress-deformation behavior. The principle is already reflected in the simpler constitutive equation (2), where only the symmetric part of $\nabla \boldsymbol{v}$ is allowed.

While frame-indifference is a safe assumption for most molecular materials, which are dominated at the molecular level by random thermal motion, it could conceivably break down for granular flows, where random granular motions are comparable to those associated with macroscopic shearing. Nevertheless, the assumption will be adopted in the constitutive models considered here.

Note that, for a fixed material particle $\boldsymbol{x}^\circ$, (3) represents a set of five simultaneous ordinary differential equations, which we call *Lagrangian ordinary differential equations* (LODEs). In the *Eulerian* or spatial description, they must be combined with (1) to provide nine equations in the nine dependent variables, $p = -\frac{1}{3}\operatorname{tr}\boldsymbol{T}$, $\boldsymbol{v}$, and $\boldsymbol{T}'$, governing the velocity field of the Maxwell fluid. Solving the problem, subject to various initial and boundary conditions, is no easy task and usually requires advanced numerical methods.

We identify below a broad class of plausible constitutive equations, which basically amount to generalizations of (3). They include the so-called *hypoplastic* models for granular plasticity. It turns out that all of these can be obtained from elastic and inelastic potentials, which are familiar in classical theories of elastoplasticity. The above models can be enlarged to include models of visco-elastoplasticity that are broadly applicable to all the prominent regimes of granular flow. Following the brief review in the following section of phenomenology and flow regimes, a systematic outline of the genesis of such models will be presented.

This article does not deal with the rather large body of literature on numerical methods, neither the direct simulation of micromechanics by the distinct element method (DEM) nor solution of the continuum field equations based on finite-element methods (FEMs) or related techniques.

## 3 Phenomenological Aspects

We consider here the important physical parameters and dimensionless groups that characterize the various regimes of granular mechanics and flows. The review article of Forterre and Pouliquen that appears in the further reading section at the end of this article provides more quantitative comparisons for fairly simple shearing flows.

### 3.1 Key Parameters and Dimensionless Groups

Apart from various dimensionless parameters describing grain shape, the most prominent physical parameters for noncohesive granular media are grain elastic (shear) modulus $G_s$, intrinsic grain density $\rho_s$, representative grain diameter $d$, intergranular (Coulomb) contact-friction coefficient $\mu_s$ or macroscopic counterpart $\mu_C$, and confining pressure $p_s$. These parameters define the key dimensionless groups that serve to delineate various regimes of granular flow: namely, an *elasticity number*, an *inertia number*, and a *viscosity number*, given, respectively, by

$$\mathbb{E} = G_s/p_s, \qquad \mathbb{I} = \dot{\gamma} d\sqrt{\rho_s/p_s}, \qquad \mathbb{H} = \eta_s\dot{\gamma}/p_s,$$

and involving $\dot{\gamma}$, a representative value of the shear rate $|\boldsymbol{D}'|$. We will also refer briefly below to a Knudsen number based on the ratio of microscopic to macroscopic length scales.

Note that $\mathbb{I}$ is the analogue of the Deborah number mentioned above, but it now involves a relaxation rate that represents the competition between grain inertia and Coulomb friction, $\mu_C p_s$. The quantity $\mathbb{I}^2$ represents the ratio of representative granular kinetic energy $\rho_s d^2\dot{\gamma}^2$ to frictional confinement.

The transition from a fluid-saturated granular medium to a dense fluid–particle suspension occurs when the viscosity number $\mathbb{H} \approx 1$, where viscous and frictional contact forces are comparable.

In the case of fluid–particle suspensions, it is customary to identify $\mathbb{I}^2/\mathbb{H}$ as the "Stokes number" or the "Bagnold number" (after a pioneer in granular flow), representing the magnitude of grain-inertial forces to viscous forces.

In the following sections we focus our attention on dry granular media, including only a very brief mention of fluid–particle systems.

### 3.2 Granular Flow Regimes

Although granular materials are devoid of intrinsic thermal motion at the grain level, they nevertheless exhibit states that resemble the solid, liquid, and gaseous states of molecular systems. All these granular states may coexist in the same flow field as the analogs of "multiphase flow." There are several open questions as to the proper matching of solid-like immobile states with the rapidly sheared states, but there is not enough space in this brief article to address them here.

With $\tau$ denoting a representative shear stress and $\gamma$ a representative shear strain relative to a rest state (in which $\tau/p_s = 0$ and $\mathbb{I} = 0$), the various flow regimes are shown in the qualitative and highly simplified sketch in figure 1 and in table 1. In the figure, the dimensionless ratio $\tau/p_s$ is represented as a function of a single dimensionless variable representing an interpolating form:

$$X \sim \mu_C\mathbb{E}\gamma/(\mathbb{E}\gamma + \mu_C) + \mathbb{I}^2. \tag{4}$$

(A compound representation, closer to the constitutive models considered below and illustrated in figure 5, is $\tau/p_s = \mu_C + \mathbb{I}^2$, with $\gamma_E = (\tau/p_s)\mathbb{E}$, of which the first represents a Coulomb–Bagnold interpolation found in certain constitutive models and where $\gamma_E$ is elastic deformation at any stress state.)

Figure 1 fails to capture the strong nonlinearity and history dependence of granular plasticity in regime Ib:

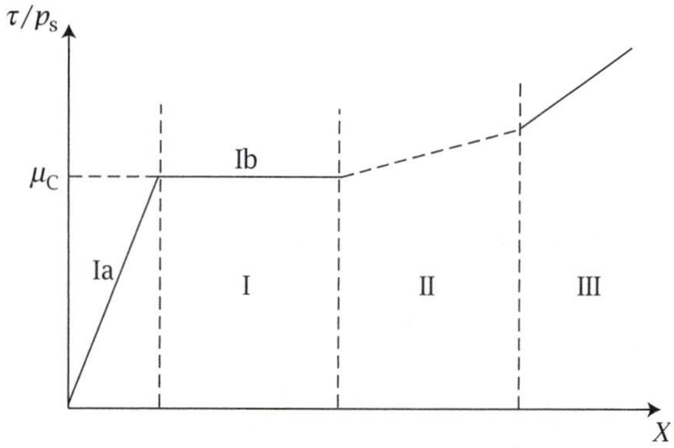


**Figure 1** Schematic of granular-flow regimes.

**Table 1** The granular-flow regimes shown in figure 1 (the last column gives the scaling of the stress $\tau$).

| | | |
|---|---|---|
| I | Quasi-static: (Hertz–Coulomb) elastoplastic "solid" | |
| Ia | (Hertz) elastic | $G_s\gamma$ |
| Ib | (Coulomb) elastoplastic | $\mu_C$ |
| II | Dense-rapid: viscoplastic "liquid" | $p_s f(\mathbb{I})$ |
| III | Rarified-rapid: (Bagnold) viscous "granular gas" | $\rho_s d^2 \dot{\gamma}^2$ |

a matter that is addressed by the constitutive models discussed below.

Like the liquid states of molecular systems, dense rapid flow, represented as a general dimensionless function $f(\mathbb{I})$ in table 1, may be the most poorly understood regime of granular mechanics. It involves important phenomena such as the fascinating granular size segregation. There is some evidence that this regime may involve an additional dependence on $\mathbb{E}$ for soft granular materials.

### 3.2.1 *The Elastic Regime*

The geometry of contact between quasispherical linear (Hookean) elastic particles should lead to nonlinear elasticity of a granular mass at low confining pressures $p_s$ and to an interesting scaling of elastic moduli and elastic wave speeds with pressure.

Based upon Hertzian contact mechanics, a rough "mean-field" estimate of the continuum shear modulus $G$ in terms of grain shear modulus $G_s$ is given by $G/G_s \sim \mathbb{E}^{-1/3}$, which indicates a $\frac{1}{3}$-power dependence on pressure. For example, with $p_s \sim 100$ kPa and $G_s \sim 100$ GPa (a rather stiff geomaterial), one finds $\mathbb{E} \sim 10^6$ and $G \sim 10^{-2}G_s$, amounting to a huge reduction of global stiffness due to relatively soft Hertzian contact.

In principle, we should replace $G_s$ by $G$ in table 1 and (4), and $\mathbb{E} = G_s/p_s$ by $G/p_s = \mathbb{E}^{-2/3}$. In so doing, we obtain an estimate of the limiting elastic strain for the onset of Coulomb slip (with $\mu_C \sim 1$) to be $\gamma_E \sim \mathbb{E}^{-2/3} \sim 10^{-4}$, given the above numerical value. Although crude, this provides a reasonable estimate of the small elastic range of stiff geomaterials such as sand.

### 3.2.2 *The Elastoplastic Regime*

Because of its venerable history, dating back to the classical works of Coulomb, Rankine, and others in the eighteenth and nineteenth centuries, and its enduring relevance to geomechanics, the field of elastoplasticity is the most thoroughly studied area of granular mechanics. Here we touch on a few salient phenomena, and the theoretical issues surrounding them, by way of background for the discussion of constitutive modeling that follows.

Figure 2 shows the results of Wolfgang Ehlers's two-dimensional DEM simulations of a quasistatic hopper discharge and a biaxial compression test, both of which illustrate a well-known localization of deformation into shear bands. A hallmark of granular plasticity, this localized slip (or "failure") may be implicated in dynamic "arching," with large transient stresses on bounding surfaces such as hopper walls or structural retaining walls. Similar phenomena are implicated in large-scale landslides.

Figure 3 shows the development of shear bands in a standard experimental quasistatic compression test on a sand column surrounded by a thin elastic membrane. The literature abounds with many interesting experimental observations and numerical simulations (see, for example, the book by Tejchman that appears in the further reading below).

The occurrence of shear bands can be viewed mathematically as material *bifurcation* and *instability* arising from loss of convexity in the underlying constitutive equations, accompanied by a *change of type* in the PDEs involved in the field equations. We recall that similar changes of type, e.g., from elliptic PDEs to hyperbolic PDEs, are associated with phenomena such as the gas dynamic transition from subsonic to supersonic flows with formation of thin "shocks."

To help understand the elastoplastic instability, figure 4 presents a qualitative sketch of the typical stress-strain/dilatancy behavior in the axial compression of

(a)

(b)

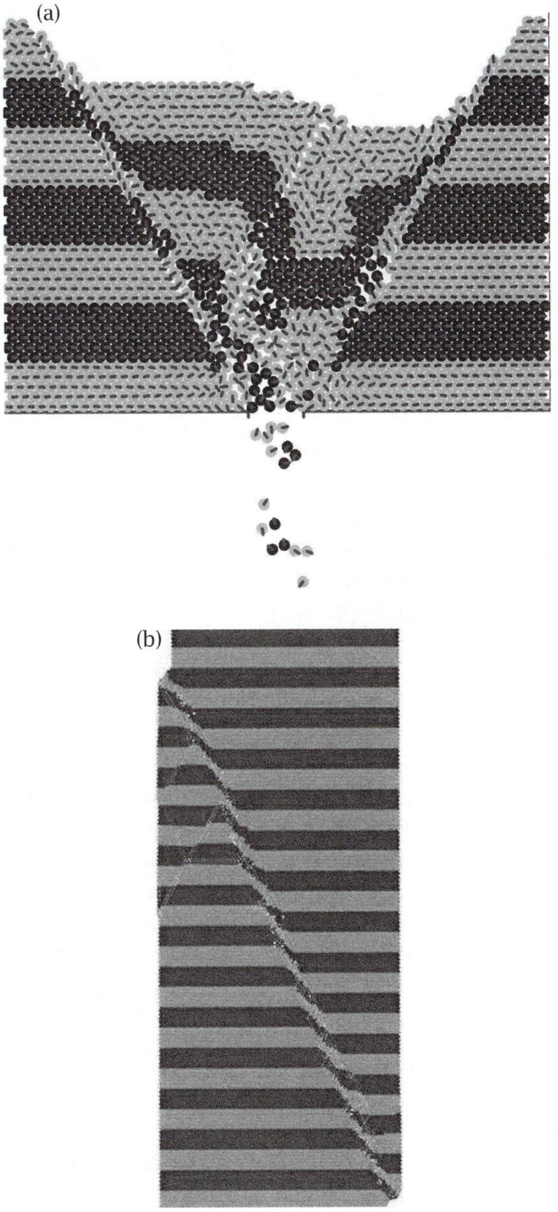

**Figure 2** DEM simulations: (a) shear bands in slow hopper flow and (b) shear bands in axial compression. (Courtesy of Wolfgang Ehlers, University of Stuttgart.)

(a)

(b)

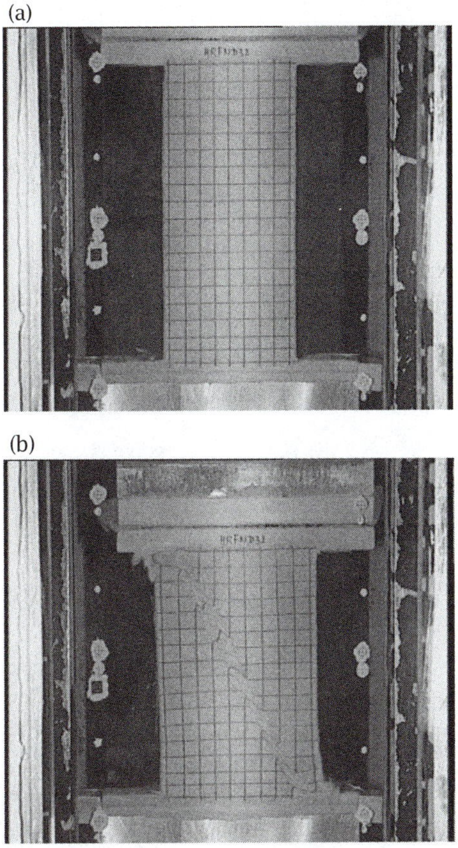

**Figure 3** Experimental axial compression of a dry Hostun sand specimen: (a) before compression and (b) after compression. (Courtesy of Wolfgang Ehlers.)

dense and loose sands, with $\sigma$ denoting compressive stress, $\varepsilon$ compressive strain, and $\varepsilon_{\mathrm{V}} = \log(V/V_0)$ volumetric strain (where a volume $V_0$ has been deformed into $V$). While no numerical scales are shown on the axes, the peak stress and the change of $\varepsilon_{\mathrm{V}}$ from negative (contraction) to positive (dilation) typically occur at strains of a few percent ($\varepsilon \sim 0.01$–$0.05$) for dense sands.

Although it is tempting to regard the initial growth of $\sigma$ with $\varepsilon$ as elastic in nature, the elastic regime is represented by much smaller strains (corresponding to the estimates of order $10^{-4}$ given above), corresponding to a nearly vertical unloading from any point on the $\sigma$–$\varepsilon$ curve. It is therefore much more plausible that the initial stress growth represents an *almost completely dissipative* plastic "hardening" associated with compaction ($\varepsilon_{\mathrm{V}} < 0$) accompanied by growth of contact number density $n_{\mathrm{c}}$ and contact anisotropy, whereas the maximum in stress can be attributed to the subsequent decrease in $n_{\mathrm{c}}$ that accompanies dilation ($\varepsilon_{\mathrm{V}} > 0$).

As will be discussed below, one can rationalize the formation of shear bands as the result of "strain softening" or unstable "wrong-way" behavior (a decrease of incremental force with incremental displacement)

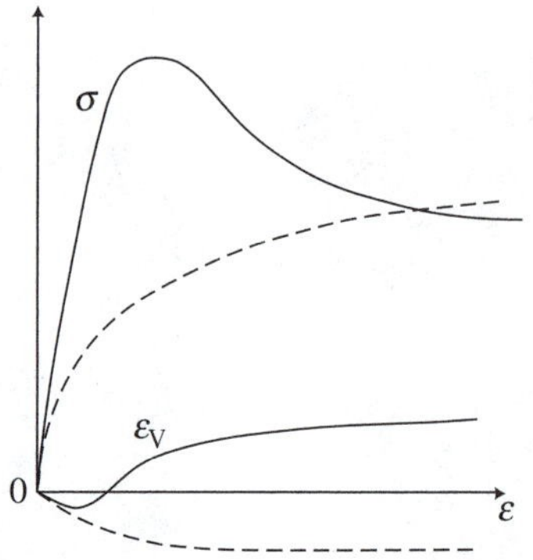


**Figure 4** Schematic of triaxial stress/dilatation–strain curves for initially dense (solid curves) and loose (dashed curves) sands.

following the peak stress. According to certain analyses, this can arise as a purely dissipative process associated with the decrease of plastic stress, whereas others hypothesize that it may be the result of a peculiar quasielastic response.

Whatever the precise nature of the material instability, it generally calls for *multipolar* or "higher-gradient" constitutive models involving an intrinsic material length scale. Hence, to our list of important dimensionless parameters we must now add a *Knudsen number* $\mathbb{K}$, as discussed below. Here, it suffices to say that a material length scale $\ell$ is necessary to *regularize* field equations in order to avoid sharp discontinuities in strain rate by assigning finite thickness to zones of strain localization. Such a scale, clearly evident in the DEM simulations and experiments of figures 2 and 3, cannot be predicted with the *nonpolar* models that constitute the main subject of this article.

### 3.2.3 Springs and Slide Blocks

Setting aside length-scale effects, the simple mechanical model in figure 5 provides a useful intuitive view of the continuum model considered below. There, the applied force $\boldsymbol{T}'$ is the analogue of continuum-mechanical shear stress, and the rate of extension of the device represents deviatoric deformation rate $\boldsymbol{D}'$.

The serrated slide block at the top of figure 5 is a modification of the standard flat plastic slide block, where sliding stress $\tau$ represents a pressure-independent yield stress, or a frictional slide block with $\tau/p_{\mathrm{s}} = \mu_{\mathrm{C}}$ representing the (Amontons–Coulomb) coefficient of sliding friction. The serrated version represents the granular "interlocking" model of Taylor

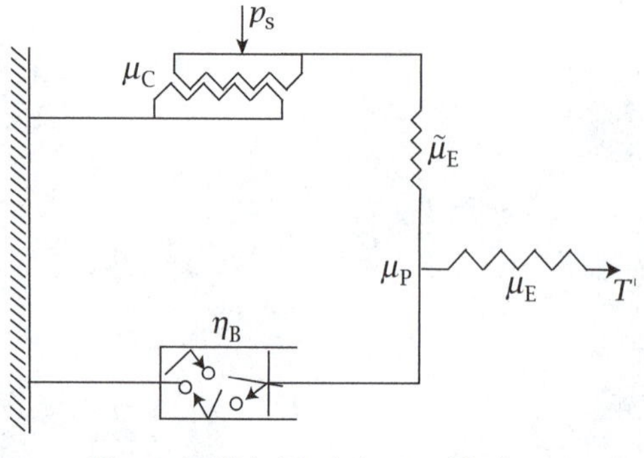


**Figure 5** Slide block/spring/dashpot analogue of visco-elastoplasticity.

(1948) or the "sawtooth" model of Rowe (1962) based on the concept of granular *dilatancy* introduced in 1885 by Osborne Reynolds (who is also renowned for his work in various fields of fluid mechanics). According to Reynolds, the shearing resistance of a granular medium is partly due to volumetric expansion against the confining pressure. Thus, if the sawtooth angle vanishes, then one obtains a standard model of plasticity, with resistance due solely to yield stress or pressure-dependent friction.

At the point of sliding instability in the model, where the maximum volume expansion occurs, part of the stored volumetric energy must be dissipated by subsequent collapse and collisional impact. This may be assumed to occur on extremely short timescales, giving rise to an apparent sliding friction coefficient $\mu_{\mathrm{C}}$ even if there is no sliding friction between grains ($\mu_{\mathrm{s}} = 0$). This rapid energy dissipation is emblematic of tribological and plastic-flow processes, where, owing to topological roughness and instability *in the small*, stored energy is thermalized on negligibly small timescales giving rise to rate-independent forces *in the large*.

As a second special feature of the model, the spring with constant $\tilde{\mu}_{\mathrm{E}}$ converts plastic deformation into "frozen elastic energy" and also provides an elementary model of plastic work hardening. This elastic element is intended to illustrate the fact that some of the stored elastic energy can never be entirely recovered solely by the mechanical action of $\boldsymbol{T}'$, which leads to history effects and associated complications in the thermodynamic theories of plasticity.

### 3.2.4 Viscoplasticity

For rapid granular flow, the viscous dashpot with (Bagnold) viscosity $\eta_{\mathrm{B}}$ (see figure 5) adds a rate-dependent force associated with granular kinetic energy and collisional dissipation. This leads to a form of *Bingham plasticity* that is discussed below and is described by certain rheological models, providing a rough interpolation between regimes I and III in figure 1. The viscous dashpot can also represent the effects of interstitial fluids. Note that removal of the plastic slide block gives the Maxwell model of viscoelasticity (3) whenever the spring and dashpot are linear.

Apart from presenting constitutive models that encompass those currently employed, we do not deal directly with the numerous issues and challenges in modeling the dense-rapid flow regime.

## 4 Constitutive Models

Inevitably, constitutive models for granular flow are complicated because of the wide range of phenomenology observed. We use a class of models that relate Cauchy stress $\boldsymbol{T}$ to deformation rate $\boldsymbol{D}$, often referred to as the *Eulerian description*. In particular, we focus on a class of generalized *hypoplastic* models, based on a stress-space description that allows for nonlinear functions of stress. Although the approximation of linear elasticity is suitable for many granular materials, particularly stiff geomaterials, the nonlinear theory may find application in the field of soft granular materials such as pharmaceutical capsules and clayey soils.

### 4.1 Hypoplasticity

Let us start with an isotropic nonlinearly elastic material. This is one for which the stress $\boldsymbol{T}$ and the finite (Eulerian) strain tensor $\boldsymbol{E} = \frac{1}{2}(\boldsymbol{F}\boldsymbol{F}^{\mathrm{T}} - \boldsymbol{I})$ are connected by isotropic relations of the form

$$\boldsymbol{T} = \mathrm{t}(\boldsymbol{E}) \quad \text{or} \quad \boldsymbol{E} = \mathrm{t}^{-1}(\boldsymbol{T}). \tag{5}$$

Here, $\boldsymbol{F} = \partial \boldsymbol{x}/\partial \boldsymbol{x}^{\circ}$, where $\boldsymbol{x}^{\circ}$ and $\boldsymbol{x}(\boldsymbol{x}^{\circ})$ denote the reference position and current placement, respectively, of material points. *Hyperelasticity* (or "Green elasticity") is based on the thermodynamic consistency condition that the functions in (5) be derivable from elastic potentials. Truesdell's *hypoelasticity*—a generalization of elasticity that allows for a more general rate-independent but path-dependent relation between stress and strain—is represented by the LODE

$$\mathring{\boldsymbol{T}} = \boldsymbol{\mu}_{\mathrm{E}}(\boldsymbol{T}) : \boldsymbol{D}, \tag{6}$$

where $\boldsymbol{\mu}_{\mathrm{E}}$ is the fourth-rank hypoelastic modulus. With suitable integrability conditions on this modulus, it is possible to show that (6) implies an elastic relation of the form (5).

To obtain *visco-elastoplasticity* from (6), we adapt the *first fundamental postulate* of incremental plasticity: that the deformation rate can be decomposed into elastic and inelastic (or "plastico-viscous") parts, $\boldsymbol{D} = \boldsymbol{D}_{\mathrm{E}} + \boldsymbol{D}_{\mathrm{P}}$. Then replace $\boldsymbol{D}$ in the relations above by $\boldsymbol{D}_{\mathrm{E}} = \boldsymbol{D} - \boldsymbol{D}_{\mathrm{P}}$ and provide a constitutive equation for $\boldsymbol{D}_{\mathrm{P}}$.

As the *second fundamental postulate*, we assume that the elastic stress $\boldsymbol{T}$ conjugate to $\boldsymbol{D}_{\mathrm{E}}$ is identical to the inelastic stress conjugate to $\boldsymbol{D}_{\mathrm{P}}$, which follows from an assumption of internal equilibrium of the type suggested by the simple model in figure 5.

Finally, as the *third fundamental postulate*, we assume that the system is strongly dissipative such that $\boldsymbol{T}$ and $\boldsymbol{D}_{\mathrm{P}}$ are related by dual inelastic potentials. *Viscoelasticity* follows and yields a nonlinear form of the Maxwell fluid (3).

*Plasticity*, as defined by rate-independent stress, represents a singular exception to the above: the magnitude $|\boldsymbol{D}_{\mathrm{P}}|$ becomes indeterminate. The assumption of overall independence of rate and history implies a relation of the form

$$|\boldsymbol{D}_{\mathrm{P}}| = |\boldsymbol{D}|\vartheta(\boldsymbol{T}, \hat{\boldsymbol{D}}),$$

where $\hat{\boldsymbol{D}} = \boldsymbol{D}/|\boldsymbol{D}|$ and $|\boldsymbol{D}| = \sqrt{D_{ij}D_{ij}}$, and, hence, a generalized form of *isotropic hypoplasticity*,

$$\mathring{\boldsymbol{T}} = \boldsymbol{\mu}_{\mathrm{H}}(\boldsymbol{T}, \hat{\boldsymbol{D}}) : \boldsymbol{D}, \tag{7}$$

with a formula for $\boldsymbol{\mu}_{\mathrm{H}}$ in terms of the "inelastic clock" function $\vartheta$. In the standard theory of hypoplasticity, $\vartheta(\boldsymbol{T}, \hat{\boldsymbol{D}})$ is independent of $\hat{\boldsymbol{D}}$, there is no distinction between elastic and plastic deformation rates, and the constitutive equation reduces to (7) with hypoplastic modulus taking the form

$$\boldsymbol{\mu}_{\mathrm{H}} = \boldsymbol{\mu}_{\mathrm{E}}(\boldsymbol{T}) - \boldsymbol{K}(\boldsymbol{T}) \otimes \hat{\boldsymbol{D}}, \tag{8}$$

where $\boldsymbol{A} \otimes \boldsymbol{B} \mathrel{\hat{=}} [A_{ij}B_{kl}]$. Such models are popular with many in the granular mechanics community (particularly those interested in geomechanics), mainly because they do not rely directly on the concept of a yield surface or inelastic potentials (whence the qualifier "hypo"), although they do involve limit states where Jaumann stress rate vanishes asymptotically under the action of a constant deformation rate $\boldsymbol{D}$. This asymptotic state may be identified with the so-called critical state of soil mechanics, where granular dilatancy vanishes and the granular material essentially flows like an incompressible liquid.

The problem with the general form (8) is that it is exceedingly difficult to establish mathematical restrictions that will guarantee its thermodynamic admissibility, e.g., such that steady periodic cycles of deformation do not give positive work output. By contrast, models built up from elastic and dissipative potentials are much more likely to be satisfactory in that regard.

The treatise by Kolymbas that appears in the further reading provides an excellent overview and history of hypoplastic modeling.

### 4.2 Anisotropy, Internal Variables, and Parametric Hypoplasticity

Initially isotropic granular masses exhibit flow-induced anisotropy, particularly in flow that is quasistatic elastoplastic, since the grain kinetic energy or "temperature" is insufficient to randomize granular microstructure. This anisotropy is often modeled by an assumed dependence of elastic and inelastic potentials, and associated moduli, on a symmetric second-rank *fabric tensor*, $\boldsymbol{A} \mathrel{\hat{=}} [A_{ij}]$, which is subject to a rate-independent evolution equation of the form

$$\mathring{\boldsymbol{A}} = \boldsymbol{\alpha}(\boldsymbol{T}, \hat{\boldsymbol{D}}, \boldsymbol{A}) : \boldsymbol{D}.$$

This represent a special case of the "isotropic extension" of anisotropic constitutive relations, where the anisotropic dependence on stress $\boldsymbol{T}$ or strain $\boldsymbol{E}$ is achieved by the introduction of a set of "structural tensors," consisting in the simplest case of a set of second-rank tensors and vectors. The structural tensors represent in turn a special case of a set $\boldsymbol{X}$ of evolutionary internal variables consisting of scalars, vectors, and second-order tensors.

At present, the origins of the evolution equations are unclear, although there may be a possibility of obtaining them from a generalized balance of *internal forces* derived on elastic and inelastic potentials that depend on the internal variables or their rate of change.

Whatever the origin of their evolution equations, it is easy to see that one can enlarge the set of dependent variables from $\boldsymbol{T}$ to $\{\boldsymbol{T}, \boldsymbol{X}\}$ with the evolution equations leading to *parametric hypoplasticity* defined by a generalization of (7). This gives a set of LODEs that describe the effects of initial anisotropy, density, or volume fraction, etc., represented as initial conditions, and their subsequent evolution under flow. This formulation includes most of the current nonpolar models of the evolutionary plasticity of granular media.

### 4.3 Multipolar Effects

In a system that is strongly inhomogeneous, such as the shear field associated with shear bands, we expect to encounter departures from the response of a classical simple (nonpolar) material having no intrinsic length scale. The situation is generally characterized by nonnegligible magnitudes of a Knudsen number $\mathbb{K} = \ell/L$, where $\ell$ is a characteristic microscale and $L$ is a characteristic macroscale.

In the case of elastoplasticity, various empirical models suggest taking $\ell$ to be about five to ten times the median grain diameter: a scale that is most plausibly associated with the length of the ubiquitous *force chains* in static granular assemblies. These are chain-like structures in which the contact forces are larger than the mean force, and it is now generally accepted that they represent the microscopic force network that supports granular shear stress and rapidly reorganize under a change of directional loading. The micromechanics determining the length of force chains is still poorly understood.

Whatever the physical origins, a microscopic length scale serves as a parameter in various *enhanced continuum* models, including various *micropolar* and *higher-gradient* models, all of which represent a form of *weak nonlocality* referred to by the blanket term *multipolar*. Perhaps the simplest is the Cosserat model, often referred to as *micropolar*. This model owes its origins to a highly influential treatise on structured continua by the Cosserat brothers, Eugène and François, which was celebrated internationally in 2009 on the occasion of the centenary of its original publication. Tejchman's book provides a comprehensive summary of a fairly general form of Cosserat hypoplasticity.

#### 4.3.1 *Particle Migration and Size Segregation*

Another fascinating aspect of granular mechanics is the shear-driven separation or "unmixing" of large particles from an initially uniform mixture of large and small particles. Various models of particle migration in fluid–particle suspensions or size segregation in granular media involve diffusion-like terms that suggest multipolar effects. While some models involve gravitationally driven sedimentation opposed by diffusional remixing, other models involve a direct effect of gradients in shearing akin to those found in fluid–particle suspensions.

It may be significant that many granular size-segregation effects are associated with dense flow in thin layers, which again suggests that Knudsen-number or multipolar effects are likely. Whatever the origins of particle migration, it can probably be treated as a strictly dissipative process, implying that it can be represented as a generalized velocity in a dissipation potential, thus suggesting a convenient way of formulating properly invariant constitutive relations.

*4.3.2 Relevance to Material Instability*

There is a bewildering variety of instabilities in granular flow. These range from the shear-banding instabilities in quasistatic flow discussed above to gravitational layering in moderately dense rapid flow and clustering instabilities in granular gases.

The author advocates a distinction between material or constitutive instability, representing the instability of homogeneous states in the absence of boundary influences, and the dynamical or geometric instability that occurs in materially stable media, such as elastic buckling and inertial instability of viscous flows. With this distinction, it is easier to assess the importance of multipolar and other effects.

Past studies reveal multipolar effects on elastoplastic instability, not only on post-bifurcation features such as the width of shear bands but also on the material instability itself. This represents an interesting and challenging area for further research based on the parametric viscoelastic or hypoplastic models of the type discussed above. The general question is whether and how the length scales that lend dimensions to subsequent patterned states enter into the initial instability leading to those patterns.

**Further Reading**

Forterre, Y., and O. Pouliquen. 2008. Flows of dense granular media. *Annual Review of Fluid Mechanics* 40:1–24.

Goddard, J. 2014. Continuum modeling of granular media. *Applied Mechanics Reviews* 66(5):050801.

Kolymbas, D. 2000. *Introduction to Hypoplasticity.* Rotterdam: A. A. Balkema.

Lubarda, V. A. 2002. *Elastoplasticity Theory.* Boca Raton, FL: CRC Press.

Ottino, J. M., and D. V. Khakhar. 2000. Mixing and segregation of granular materials. *Annual Review of Fluid Mechanics* 32:55–91.

Rao, K., and P. Nott. 2008. *An Introduction to Granular Flow.* Cambridge: Cambridge University Press.

Tejchman, J. 2008. *Shear Localization in Granular Bodies with Micro-Polar Hypoplasticity.* Berlin: Springer.

# V.14 Modern Optics

*Miguel A. Alonso*

## 1 Introduction

Since ancient times, optics has been inextricably linked to geometry and other branches of mathematics. Euclid himself wrote the earliest known treatise on optics, in which he postulated the laws of perspective. Similarly, Hero of Alexandria formulated perhaps the earliest variational principle when he observed that a light ray traveling between two points with an intermediate reflection by a mirror corresponds to the shortest such path. The parallel development of optics and geometry continued in the Arabic world with the likes of Ibn-Sahl and Ibn al-Haytham (also known as Alhazen). More recently, modern vector calculus was inspired by the studies of Gibbs and Heaviside in electromagnetic theory.

Modern optics is ubiquitous in contemporary science and technology. It provides direct tests for fundamental physics and is the basis of applications including telecommunications, data storage, integrated circuit manufacture, astronomy, environmental monitoring, and medical diagnosis and therapy. Optics and photonics incorporate many mathematical methods covered by typical undergraduate curricula in mathematics, physics, and engineering. Topics such as linear algebra, Fourier theory, separation of variables, and the theory of analytic functions acquire meanings that are palpably physical and, by the very nature of the topic, visual. This article is an overview of the mathematical techniques involved in the study of light, emphasizing free-space propagation and some simple photonic devices.

## 2 Free-Space Propagation

Consider first the simple case of light propagating through free space. It is well known that MAXWELL'S EQUATIONS [III.22] for the electromagnetic field can be combined into a vector wave equation whose solutions include waves that travel at a speed given by a combination of the free-space electric permittivity and the magnetic permeability. Maxwell, aware of the numerical similarity of this combination with recent measurements of the speed of light, corrected the equations of electricity and magnetism in such a way that they could be combined into an equation that also described the phenomenon of light.

As well as explaining the speed of light, Maxwell's equations also explain the fact that light is a transverse wave: for an extended wave with fairly uniform intensity, the electric field (and magnetic field, which we do not consider further) is a vector whose direction is perpendicular to the wave's direction of propagation. For simplicity, however, we ignore the vector nature of the electric field and consider a scalar wave that satisfies a similar wave equation (this may be a component of the vector field). We make the further simplifying assumption that at every spatial point, the dependence of the field on time $t$ is a sinusoidal wave with fixed angular frequency $\omega$. Since our eyes translate temporal frequency into color, such fields are referred to as *monochromatic*. A real monochromatic field $E(\boldsymbol{r},t)$ can be defined as

$$E(\boldsymbol{r},t) = A(\boldsymbol{r})\cos[\Phi(\boldsymbol{r}) - \omega t],$$

where $A$ is called the *amplitude* and $\Phi$ the *phase*, both being real functions of position $\boldsymbol{r} = (x,y,z)$. The *intensity* of light at each point is given by $I \equiv A^2$. The optical wave can therefore be written as

$$E(\boldsymbol{r},t) = \mathrm{Re}\{U(\boldsymbol{r})\mathrm{e}^{-\mathrm{i}\omega t}\},$$

where Re denotes the real part and $U(\boldsymbol{r}) = A(\boldsymbol{r})\mathrm{e}^{\mathrm{i}\Phi(\boldsymbol{r})}$ is a complex function of $\boldsymbol{r}$, independent of $t$. The wave equation then reduces to the Helmholtz equation

$$\nabla^2 U(\boldsymbol{r}) + k^2 U(\boldsymbol{r}) = 0, \tag{1}$$

with wave number $k \equiv \omega/c = 2\pi/\lambda$, with $c$ the speed of light in vacuum and $\lambda$ the wavelength.

### 2.1 Modes of Free Space

In many practical problems in optics one seeks to model the field's propagation away from a given initial plane where it is known. For this purpose, one of the Cartesian axes, by convention the $z$-axis, is singled out as a propagation parameter. Suppose that we are interested in modeling only the field for $z \geqslant 0$ and that the field's source is located within the half-space $z < 0$, so within our region of interest light travels with increasing $z$. Planes of constant $z$ are referred to as *transverse planes*. This allows us to introduce the concept of an optical *mode*. In this context, a mode is a field whose dependence on $z$ is separable. The simplest mode is a plane wave, for which the amplitude is completely uniform:

$$U^{\mathrm{PW}}(\boldsymbol{r};\boldsymbol{u}) = \mathrm{e}^{\mathrm{i}k\boldsymbol{u}\cdot\boldsymbol{r}},$$

where $\boldsymbol{u} = (u_x, u_y, u_z)$ is a unit vector that specifies the wave's propagation direction. Since we assume forward propagation in $z$, plane waves for which $u_z < 0$ are not allowed, so $u_z$ can be expressed as a function of $u_x$ and $u_y$, i.e.,

$$u_z(u_x,u_y) = \sqrt{1 - u_x^2 - u_y^2}. \tag{2}$$

This in fact allows solutions where $u_x^2 + u_y^2 > 1$, for which $u_z$ is purely imaginary. In this case, one must choose the branch of the square root that is positive imaginary, so that the field remains bounded for $z > 0$ and $U^{\mathrm{PW}}$ decays exponentially under propagation in $z$. Such waves are called *evanescent waves*, and they play an important role in fields close to interfaces.

Forward-propagating plane-wave modes, together with forward-decaying evanescent modes, form a complete set over the initial transverse plane. Fields propagating toward larger values of $z$ can therefore be expressed as a linear superposition of these waves:

$$U(\boldsymbol{r}) = \iint \tilde{U}(u_x,u_y)\mathrm{e}^{\mathrm{i}k\boldsymbol{u}\cdot\boldsymbol{r}}\,\mathrm{d}u_x\,\mathrm{d}u_y, \tag{3}$$

where we assume that integrals are over all real numbers unless otherwise specified. $\tilde{U}$ is a function known as the *angular spectrum* that determines the amplitude of each plane wave. Note that the angular spectrum is easily determined from the known field at $z = 0$, since at this initial plane (3) takes the form of a Fourier superposition, which can be easily inverted:

$$\tilde{U}(u_x,u_y) = \left(\frac{k}{2\pi}\right)^2 \iint U(x,y,0)\mathrm{e}^{-\mathrm{i}k(xu_x+yu_y)}\,\mathrm{d}x\,\mathrm{d}y.$$

This gives the key result that the two-dimensional spatial Fourier transform of a known monochromatic field at some initial plane gives access to the amplitudes of the plane waves that compose such a field.

This description of optical propagation can be summarized in terms of Fourier transforms, i.e.,

$$U(x,y,z) = \hat{F}^{-1}[\mathrm{e}^{\mathrm{i}kzu_z(u_x,u_y)}\hat{F}U(x,y,0)],$$

where

$$\hat{F}g = \left(\frac{k}{2\pi}\right)^2 \iint g(x,y)\mathrm{e}^{-\mathrm{i}k(xu_x+yu_y)}\,\mathrm{d}x\,\mathrm{d}y,$$

$$\hat{F}^{-1}\tilde{g} = \iint \tilde{g}(u_x,u_y)\mathrm{e}^{\mathrm{i}k(xu_x+yu_y)}\,\mathrm{d}u_x\,\mathrm{d}u_y.$$

Free propagation can thus be modeled by suitable Fourier transformations of the known two-dimensional initial field via the propagation *transfer function* $\mathrm{e}^{\mathrm{i}kzu_z}$. This is not surprising, as it is well known that systems that are both linear and translation invariant in $x$ and $y$ can be modeled by a transfer function. Free space is also translation invariant in $z$, and hence propagations over consecutive intervals must equal the single propagation over the total distance. In such systems, the transfer function must be an exponential whose exponent is proportional to $z$, so the product of the transfer functions for various distances equals the transfer function for their sum.

The transfer function depends on the directional parameters $u_x$ and $u_y$ only through the combination $u_x^2 + u_y^2 \equiv u_\perp^2$. Therefore, linear superpositions of plane waves for which $u_\perp^2$ is fixed preserve their transverse functional form and are therefore also modes that could be more useful in particular situations. A natural choice for cylindrical light beams is the mode set labeled by polar coordinates $u_\perp$, $\varphi$ in the Fourier plane, so $u_x = u_\perp \cos\varphi$ and $u_y = u_\perp \sin\varphi$, for which the transfer function is $e^{ikz(1-u_\perp^2)^{1/2}}$. Alternative complete families of modes result from combining mode families with different dependence on the azimuth $\varphi$. For instance, the angular spectrum may be expanded as a Fourier series in this angle,

$$\tilde{U}(u_\perp \cos\varphi, u_\perp \sin\varphi) = \sum_m \frac{\bar{U}_m(u_\perp)}{2\pi \mathrm{i}^m} e^{\mathrm{i}m\varphi},$$

where the sum is over all integers. The plane-wave superposition is then replaced by

$$U(\boldsymbol{r}) = \sum_m \int_0^\infty \bar{U}_m(u_\perp) J_m(ku_\perp \rho) e^{\mathrm{i}(m\phi + ku_z z)}\, du_\perp,$$

where $J_m(t)$ are BESSEL FUNCTIONS [III.2] of the first kind, and $\rho = \sqrt{x^2+y^2}$ and $\phi = \arctan(x,y)$ are polar coordinates in the spatial transverse plane. The functions $J_m(ku_\perp\rho)e^{\mathrm{i}(m\phi+ku_z z)}$ are a new set of complete orthogonal modes, known as *Bessel beams*. (When $u_\perp > 1$, $u_z = (1-u_\perp^2)^{1/2}$ is purely imaginary, so the corresponding Bessel beam is evanescent.) Unlike plane waves, Bessel beams have a nonuniform intensity distribution over transverse planes, taking maximum values near a ring of radius $|m|/ku_\perp$. Furthermore, for $m \neq 0$, these modes have an $m$th-order phase vortex on the $z$-axis. That is, on the line $x = y = 0$ the amplitude is zero and the phase varies by $2\pi m$ as $\phi$ increases from 0 to $2\pi$. Optical phase vortices have been the focus of significant attention recently; among other reasons, this is because they endow beams with angular momentum. For example, a beam with a phase vortex along its axis is capable of making a small particle spin. Modeling such transfer of angular momentum, however, requires knowledge of the theory of light–matter interaction, which is not discussed here.

Plane waves and Bessel beams are only two of an infinite range of possible modes, those whose transverse profiles are separable in Cartesian and polar coordinates, respectively. Other separable families include so-called Mathieu and parabolic beams (separable in elliptic and parabolic cylindrical coordinates, respectively), and there are also numerous nonseparable families. Notice that for all of these modes, the integral of the intensity over the transverse plane diverges, implying that they are not physically realizable fields in a strict sense, as they would require an infinite amount of power. However, appropriate continuous superpositions of them *are* square integrable over transverse planes and they are therefore physical. Furthermore, finite-power approximations to individual modes are achievable and useful for a range of applications.

### 2.2 Paraxial Approximation

Light beams often have strong directionality, in the sense that they can be expressed as superpositions of plane waves that all travel within a narrow angular range with respect to a central direction, conveniently the $z$-axis of the angular spectrum decomposition. In this case $\tilde{U}$ differs significantly from zero only for $|u_x|, |u_y| \ll 1$. Therefore, from (2), $u_z \approx 1 - (u_x^2 + u_y^2)/2$, so that the plane-wave expansion becomes

$$U \approx \iint \tilde{U} \exp[\mathrm{i}k[z + u_x x + u_y y - \tfrac{1}{2}z(u_x^2 + u_y^2)]]\, du_x\, du_y. \quad (4)$$

Of course, since an approximation was performed, the superposition in this expression is no longer a solution of the Helmholtz equation. In fact, $U^{\mathrm{p}} \equiv Ue^{-\mathrm{i}kz}$ is a general solution to the so-called *paraxial wave equation*, written here as

$$\frac{\mathrm{i}}{k}\frac{\partial U^{\mathrm{p}}}{\partial z} = -\frac{1}{2k^2}\nabla_\perp^2 U^{\mathrm{p}}, \quad (5)$$

where $\nabla_\perp^2 \equiv \partial^2/\partial x^2 + \partial^2/\partial y^2$. This equation has the same mathematical form as the Schrödinger equation that determines the time evolution of a wave function in QUANTUM MECHANICS [IV.23], with zero potential and $z$ as the time propagation parameter. The similarity between the paraxial and Schrödinger equations suggests that the latter is also an approximation of a more precise equation under some limit. Solutions of the Schrödinger equation are approximations to solutions of relativistic quantum mechanical equations, such as the DIRAC EQUATION [III.9], when all velocities involved are small compared with $c$. In other words, nonrelativistic wave functions are "paraxial" around the time axis, since they are composed of plane-wave components forming small angles with respect to the time axis (scaled by $c$ to have spatial units) in space-time.

Substituting $\tilde{U}$ as a Fourier transform into (4) leads to an alternative expression for paraxial field propagation in the form of a convolution, known as the *Fresnel diffraction formula*,

$$U(\boldsymbol{r}) = \iint U(x', y', 0) K(x - x', y - y', z)\, dx'\, dy',$$

where the *Fresnel propagator* $K$ is proportional to the paraxial approximation to a spherical wave:

$$K(x,y,z) = \frac{1}{\mathrm{i}\lambda z}\exp\left(\mathrm{i}k\left(z + \frac{x^2+y^2}{2z}\right)\right).$$

This description is related to Huygens's interpretation of wave propagation: each point along a transverse plane can be regarded as a secondary wave whose amplitude and phase of emission are determined by the field at that point. This interpretation can be extended to quantum mechanics via the Schrödinger equation, where it is the basis of Feynman's path integral formulation.

The paraxial approximation allows the closed-form calculation of several types of fields. For example, let the initial field be a Gaussian function of $x$ and $y$ with height $U_0$ and width $w_0$, i.e., $U^{\mathrm{G}}(x,y,0) = U_0\mathrm{e}^{-(x^2+y^2)/2w_0^2}$. The propagated paraxial field is

$$U^{\mathrm{G}}(\boldsymbol{r}) = \frac{U_0}{1+\mathrm{i}z/z_{\mathrm{R}}}\exp\left(\mathrm{i}k\left[z - \frac{x^2+y^2}{2w_0^2(1+\mathrm{i}z/z_{\mathrm{R}})}\right]\right),$$

where $z_{\mathrm{R}} = kw_0^2$ is known as the *Rayleigh range*. The beam's intensity remains Gaussian under propagation, broadening away from the *waist plane* $z = 0$:

$$|U^{\mathrm{G}}(\boldsymbol{r})|^2 = |U_0|^2\frac{w_0^2}{w^2(z)}\exp\left(-\frac{x^2+y^2}{w^2(z)}\right),$$

where $w(z) = w_0(1 + z^2/z_{\mathrm{R}}^2)^{1/2}$. $U^{\mathrm{G}}$ is a good approximation to the beams generated by standard continuous-wave lasers, such as laser pointers. This is because, within the paraxial approximation, they are modes of three-dimensional resonant laser cavities with slightly concave, partially reflecting surfaces resembling a segment of a sphere. The light is transmitted out of the cavity through one of these surfaces after many reflections while keeping the same transverse profile.

There is an interesting relation between Gaussian beams and the Fresnel propagator: in the limit of small $w_0$, $U^{\mathrm{G}}$ becomes this propagator, i.e., the paraxial counterpart to a diverging spherical wave:

$$K(x,y,z) = \lim_{w_0\to 0}\frac{U^{\mathrm{G}}(\boldsymbol{r})}{2\pi w_0^2 U_0}.$$

Conversely, a Gaussian beam of arbitrary width $w_0$ can be expressed as proportional to a spherical wave that has been displaced in $z$ by an imaginary distance $z_{\mathrm{R}} = kw_0^2$:

$$U^{\mathrm{G}}(\boldsymbol{r}) = U_0\lambda z_{\mathrm{R}}\mathrm{e}^{-kz_{\mathrm{R}}}K(x,y,z-\mathrm{i}z_{\mathrm{R}}).$$

There are families of fields that, like Gaussian beams, are modes of cavities with curved mirrors. While these are not strictly speaking modes of free space, their intensity profile is preserved upon propagation, with width proportional to $w(z)$ defined above, and a corresponding longitudinal attenuation proportional to the inverse square of this factor. One such family is the so-called *Hermite–Gaussian beams*, where the initial field is the product of a Gaussian and appropriately scaled HERMITE POLYNOMIALS [II.29] $H_n$ of different orders in $x$ and $y$:

$$U_{m,n}^{\mathrm{HG}}(\boldsymbol{r}) = \left[\frac{w_0}{w(z)}\left(1+\mathrm{i}\frac{z}{z_{\mathrm{R}}}\right)\right]^{m+n} \times H_m\left[\frac{x}{w(z)}\right]H_n\left[\frac{y}{w(z)}\right]U^{\mathrm{G}}(\boldsymbol{r})$$

for $m,n = 0,1,\dots$. Similarly, *Laguerre–Gaussian beams* are defined as

$$U_{\ell,p}^{\mathrm{LG}}(\boldsymbol{r}) = \left[\frac{w_0}{w(z)}\left(1+\mathrm{i}\frac{z}{z_{\mathrm{R}}}\right)\right]^{2p+|\ell|} \times \rho^{|\ell|}\mathrm{e}^{\mathrm{i}\ell\phi}L_p^{(|\ell|)}\left[\frac{\rho^2}{w^2(z)}\right]U^{\mathrm{G}}(\boldsymbol{r})$$

for integer $\ell$ and $p = 0,1,\dots$, and where $L_p^{(\ell)}$ is an associated LAGUERRE POLYNOMIAL [II.29]. Like Bessel beams, Laguerre–Gaussian beams with $\ell \neq 0$ have an $\ell$th-order phase vortex at the $z$-axis and therefore carry orbital angular momentum. For fixed $z$, the Hermite–Gaussian beams are separable in two-dimensional Cartesian coordinates, while the Laguerre–Gaussian beams are separable in polar coordinates (and Gaussian beams are separable in each). A third family, known as the Ince–Gaussian beams, is separable in elliptical coordinates. With appropriate constant prefactors, each of these families is orthonormal in any plane of constant $z$. Therefore, any paraxial field can be expressed as a discrete linear superposition of Hermite–Gaussian or Laguerre–Gaussian beams as

$$U(\boldsymbol{r}) = \sum_{m,n=0}^{\infty} a_{m,n}U_{m,n}^{\mathrm{HG}} = \sum_{p=0}^{\infty}\sum_{\ell=-\infty}^{\infty} b_{m,n}U_{\ell,p}^{\mathrm{LG}},$$

where $a_{m,n}$ and $b_{\ell,p}$ are appropriate coefficients.

### 2.3 Connection with Ray Optics

Maxwell's equations give a description of light as an electromagnetic wave. However, the older and simpler ray model is often closer to our experience-based intuition and is sufficient for many practical purposes including the design of simple imaging systems. To properly understand the connection between the ray and wave models, one must delve into the theory of ASYMPTOTICS [II.1]. In this section, however, we give a simplified version of the connection for basic optical systems in the paraxial approximation.

In the ray model, optical power travels along mutually independent lines called *rays*, which are straight for propagation through homogeneous media such as free space. A ray crossing a plane of constant $z$ is labeled by its intersection $\boldsymbol{r}_\perp = (x, y)$, together with its direction, specified by the transverse part of its direction vector, $\boldsymbol{u}_\perp = (u_x, u_y)$. These define the ray *state vector* $\boldsymbol{v} \equiv (\boldsymbol{r}_\perp, \boldsymbol{u}_\perp)$. In the paraxial approximation and for simple systems, $\boldsymbol{v}$ evolves under propagation according to simple rules. Under free propagation between two planes of constant $z$ separated by a distance $d$, the ray direction is constant while the transverse position changes by approximately $d\boldsymbol{u}_\perp$. In this case, $\boldsymbol{v} \to \boldsymbol{F}(d) \cdot \boldsymbol{v}$ under free propagation, determined by the $4 \times 4$ matrix

$$\boldsymbol{F}(d) = \begin{pmatrix} \mathbf{1} & d\mathbf{1} \\ \boldsymbol{O} & \mathbf{1} \end{pmatrix},$$

with $\mathbf{1}$ and $\boldsymbol{O}$ being the $2 \times 2$ identity and zero matrices, respectively. Similarly, propagation across a thin lens with focal distance $f$ does not affect $\boldsymbol{r}_\perp$ but does change the ray's direction in proportion to $\boldsymbol{r}_\perp$, so that the effect of the lens on the ray follows the linear relation $\boldsymbol{v} \to \boldsymbol{L}(f) \cdot \boldsymbol{v}$, with

$$\boldsymbol{L}(f) = \begin{pmatrix} \mathbf{1} & \boldsymbol{O} \\ -\mathbf{1}/f & \mathbf{1} \end{pmatrix}.$$

Other basic optical elements are also described by matrices that multiply the state vector. When a ray propagates along a series of elements, the state vector must be multiplied by the corresponding succession of matrices, so the complete system is characterized by a single matrix $\boldsymbol{S}$ given by the product of the matrices for the corresponding elements. This matrix is generally written in terms of its four $2 \times 2$ submatrices as

$$\boldsymbol{S} = \begin{pmatrix} \boldsymbol{A} & \boldsymbol{B} \\ \boldsymbol{C} & \boldsymbol{D} \end{pmatrix},$$

and in the literature it is somewhat clumsily referred to as an *ABCD matrix*.

A related description of propagation exists in the wave domain. It can be shown that the propagation of the wave field is given by the so-called *Collins formula*:

$$U(\boldsymbol{r}) = \iint U(x', y', 0) K^{\mathrm{C}}(x', y'; x, y)\, \mathrm{d}x'\, \mathrm{d}y',$$

where the Collins propagator is

$$K^{\mathrm{C}} = \frac{\mathrm{e}^{\frac{1}{2}\mathrm{i}k(\boldsymbol{r}'_\perp \cdot \boldsymbol{B}^{-1}\boldsymbol{A} \cdot \boldsymbol{r}'_\perp - 2\boldsymbol{r}'_\perp \cdot \boldsymbol{B}^{-1} \cdot \boldsymbol{r}_\perp + \boldsymbol{r}_\perp \cdot \boldsymbol{D}\boldsymbol{B}^{-1} \cdot \boldsymbol{r}_\perp)}}{\mathrm{i}\lambda\sqrt{\det(\boldsymbol{B})}}.$$

That is, for simple paraxial systems, the same matrices that describe the propagation of rays also describe the propagation of waves. For free propagation, $\boldsymbol{S} = \boldsymbol{F}(z)$, and the Collins propagator reduces to the Fresnel propagator.

### 2.4 Phase Retrieval

When measuring light, CCD arrays and photographic film detect only a field's intensity $I = |U|^2$, so all information about the phase is lost. However, many applications require accurate knowledge of the phase, which contains most of the information about the source that generated the field and the medium it traveled through. For example, in astronomical and ophthalmic measurements it is important to determine phase fluctuations caused by the atmosphere or the eye's imperfections for their real-time correction through adaptive elements.

Several methods have been developed to retrieve the phase based only on the intensity. For example, knowing the intensity at two nearby planes of constant $z$ is nearly equivalent to knowing the intensity and its derivative in $z$. By writing $U^{\mathrm{p}} = \sqrt{I}\mathrm{e}^{\mathrm{i}\Phi}$ and taking the real part of (5) times $-\mathrm{i}\mathrm{e}^{-\mathrm{i}\Phi}$, one finds the so-called *transport-of-intensity equation*:

$$\frac{\partial I}{\partial z} + \nabla_\perp \cdot (I \nabla_\perp \Phi) = 0.$$

Therefore, knowledge of $I$ and its $z$ derivative over a plane of constant $z$ allows, in principle, the solution for $\Phi(x, y)$ at that plane, provided appropriate boundary conditions are used.

In other situations, the intensity is known at two distant planes, between which there might be some optical elements. (In some cases, the intensity is known at one plane and only certain restrictions are known at the second, e.g., the support of the field.) Several strategies exist for estimating the phase in these cases, the best-known one being the *Gerchberg–Saxton iterative algorithm*. The form of the propagator between the planes (such as the Fresnel or Collins propagator) is assumed, then an initial field estimate at plane 1 is proposed whose amplitude is the square root of the known intensity and whose phase is an ansatz. This field estimate is then propagated numerically to plane 2, where the resulting intensity typically does not match the known one. A corrected estimate is then proposed by replacing the amplitude with the square root of the known intensity (or enforcing the known restrictions) while leaving the phase unchanged. This corrected estimate is then propagated back to plane 1, where the same amplitude replacement procedure is applied. Usually, after several such iterations, the estimate converges to the correct field, giving access to the phase. However, there can be issues with convergence as well as with uniqueness of the solution, depending on the system's propagator. For example, this approach would not be able to resolve the sign of $\ell$ of the propagating Laguerre–Gauss field introduced above.

## 3 Guided Waves

The great success of photonic technology stems in large part from the convenience of using light for transmitting and/or processing information. The basis of such technology is the ability to *guide* light through specially designed channels whose transverse dimensions can be of the order of the light's wavelength.

### 3.1 Waveguides and Optical Fibers

The propagation of monochromatic light through a medium is determined approximately by a modified version of the Helmholtz equation (1):

$$\nabla^2 U(\boldsymbol{r}) + k^2 n^2(\boldsymbol{r}) U(\boldsymbol{r}) = 0,$$

where $n(\boldsymbol{r})$ is known as the *refractive index*, which can vary with position. The real part of the refractive index, $n_{\mathrm{r}}$, determines the speed at which the field's wavefronts travel at that point in the medium, according to $c/n_{\mathrm{r}}$. On the other hand, the imaginary part, $n_{\mathrm{i}}$, determines the rate of absorption of light by the material: after propagation by a distance $d$ in a homogeneous medium, the amplitude is damped by a factor $\mathrm{e}^{-kn_{\mathrm{i}}d}$.

In order to understand light guiding, we consider a medium in which $n$ depends only on the transverse variables $x$, $y$. By using the paraxial approximation, one finds a modified version of the paraxial wave equation (5),

$$\frac{\mathrm{i}}{kn_0}\frac{\partial U^{\mathrm{p}}}{\partial z} = -\frac{1}{2k^2 n_0^2}\nabla_{\perp}^2 U^{\mathrm{p}} + \frac{n_0^2 - n^2(x,y)}{2n_0} U^{\mathrm{p}},$$

where $U^{\mathrm{p}} \equiv U\mathrm{e}^{-\mathrm{i}kn_0 z}$ with $n_0 = n(0,0)$. We simplify further by setting $n$ as purely real and by assuming that it takes its largest value along the line $x = y = 0$, away from which it decreases monotonically as transverse distance increases. In this case, the paraxial wave equation is analogous to Schrödinger's equation for a quantum particle in a two-dimensional attractive potential, and therefore it accepts solutions that are concentrated within a region surrounding the $z$-axis. To see this, consider the case of a slab of transparent material with refractive index $n_0$ restricted to $|x| \leqslant a$, surrounded by a transparent medium of refractive index $n_1 (< n_0)$. Note that this structure is independent of $y$. The paraxial wave equation then accepts separable solutions $U^{\mathrm{p}} = f(x)\mathrm{e}^{-\mathrm{i}kn_0\delta z}$, where $\delta$ is a real constant to be determined. Let $\Delta \equiv n_0 - n_1 \ll 1$, so that $n_0^2 - n_1^2 \approx 2n_0\Delta$. Symmetric solutions for $f$ can be found to approximately have the form

$$f(x) = \begin{cases} f_0 \cos(kn_0\sqrt{2\delta}x), & |x| \leqslant a, \\ f_1 \exp[-kn_0\sqrt{2(\Delta-\delta)}|x|], & |x| > a, \end{cases}$$

where $\delta$ and the constant amplitudes $f_0$ and $f_1$ must be chosen so that $f(x)$ and its derivative are continuous. It is easy to find that this condition leads to a transcendental equation for $\delta$:

$$\tan(kn_0\sqrt{2\delta}a) = \sqrt{\frac{\Delta-\delta}{\delta}}. \tag{6}$$

This equation has a finite, discrete set of solutions for $\delta$, restricted to the range $(0, \Delta)$. Therefore, only a finite set of even solutions for $f$ exist where the field is confined exponentially to the central slab. Odd solutions also exist where, for $|x| \leqslant a$, $f$ is proportional to a sine rather than a cosine and the tangent in (6) is replaced by minus a cotangent. These even and odd solutions correspond to the modes of the waveguide, each of which preserves its transverse profile while accumulating a phase $\mathrm{e}^{\mathrm{i}kn_0(1-\delta)z}$ under propagation in $z$. (In more physical models, $n_0$ and $n_1$ have small imaginary parts that cause a small amount of absorption.) Different modes accumulate different phases under propagation (given their different $\delta$ values) since they propagate at different speeds. This effect, known as *modal dispersion*, can pose a problem for applications in data transmission, since multiple replicas of a signal arrive at different times at the end of the transmission line, scrambling the transmitted message. This problem can be avoided by using waveguides with sufficiently small width $a$ and/or index mismatch $\Delta$ for only one mode to exist.

Now consider an optical fiber, where the refractive index depends only on $\rho$, the radial distance from the $z$-axis. So-called *step-index fibers* are composed of a cylindrical *core* of radius $a$ with refractive index $n_0$ surrounded by a *cladding* with refractive index $n_1 < n_0$. Solutions of the relevant paraxial wave equation are similar to those for the slab. Using solutions (modes) separated in cylindrical coordinates $U^{\mathrm{p}} = f(\rho)\mathrm{e}^{\mathrm{i}m\phi}\mathrm{e}^{-\mathrm{i}kn_0\delta z}$, one finds that, for $\rho \leqslant a$, the solutions for $f$ are proportional to Bessel functions $J_m(kn_0\sqrt{2\delta}\rho)$ while for $\rho > a$ they are modified Bessel functions $I_m[kn_0\sqrt{2(\Delta-\delta)}\rho]$, which decay rapidly as

$\rho$ increases. The conditions of continuity and smoothness of $f$ at $\rho = a$ lead once more to a transcendental equation (now involving Bessel functions) that restricts the allowed values of $\delta$. Again, for small enough $a$ and $\Delta$, only one mode is guided, and the structure is referred to as a *single-mode fiber*, free of modal dispersion. For some applications, though, it is convenient to use *multimode fibers* accepting many modes. Note that the modes for which $m \neq 0$ carry orbital angular momentum.

### 3.2 Surface Plasmons

Guided modes can also be localized around a single interface between two media provided that one of them is a metal. The refractive index of a metal, due to the presence of free electrons, is dominated by its imaginary part, explaining why these materials are opaque. To understand the resulting modes, known as *surface plasmons*, we need to consider the vector character of the field. These vector mode solutions are exponentially localized near the interface between a metal and a transparent material, chosen to be the $x = 0$ plane,

$$\boldsymbol{E}(\boldsymbol{r}) = \begin{cases} \boldsymbol{a}\mathrm{e}^{-\gamma_0 x}\mathrm{e}^{\mathrm{i}\beta z}, & x > 0, \\ \boldsymbol{b}\mathrm{e}^{\gamma_1 x}\mathrm{e}^{\mathrm{i}\beta z}, & x < 0, \end{cases}$$

where the constant vectors $\boldsymbol{a}$ and $\boldsymbol{b}$ have zero $y$ components for reasons that will soon become apparent. Let the refractive index of the transparent material ($x > 0$) and the metal ($x < 0$) be $n_0$ and $n_1 = \mathrm{i}\nu$, respectively. In both media, $\boldsymbol{E}$ satisfies the Helmholtz equation $(\nabla^2 + k^2 n^2)\boldsymbol{E} = \boldsymbol{0}$ and the divergence condition $\nabla \cdot \boldsymbol{E} = 0$, leading to the constraints $\gamma_0^2 = \beta^2 - k^2 n_0^2$, $\gamma_1^2 = \beta^2 + k^2\nu^2$, $a_x\gamma_0 = \mathrm{i}a_z\beta$, and $b_x\gamma_1 = -\mathrm{i}b_z\beta$.

Different boundary conditions apply to field components that are tangent and normal to the interface: the tangent component must be continuous, as must be the longitudinal one times the square of the refractive index (for media with negligible magnetic response). This results in the relations $a_z = b_z$ and $n_0^2 a_x = -\nu^2 b_x$, which in combination with the four constraints found earlier give

$$\frac{\gamma_0}{n_0^2} = \frac{\gamma_1}{\nu^2} = \frac{\beta}{n_0\nu} = \frac{k}{\sqrt{\nu^2 - n_0^2}}.$$

That is, for surface plasmons to exist, the imaginary part of the metal's refractive index must be larger than the real part of the transparent medium's refractive index, and hence the exponential localization within the metal is faster.

Surface plasmons are also supported by curved interfaces such as the surfaces of wires or particles. They are the basis for many modern photonic devices like sensors (given the strong dependence of their behavior on shape, frequency, and material properties) and field enhancement techniques used in microscopy and solar cells.

## 4 Time Dependence and Causality

The ideas described so far rely on the assumption that the field is monochromatic, so that the temporal dependence is accounted for simply through a factor $\mathrm{e}^{-\mathrm{i}\omega t}$. While this assumption is useful when modeling light emitted by highly coherent continuous-wave lasers, it is an idealization as it implies that the field presents this behavior for all times. Fields with more general time dependence require a superposition of temporal frequencies:

$$E(\boldsymbol{r};t) = \int_0^\infty U(\boldsymbol{r};\omega)\mathrm{e}^{-\mathrm{i}\omega t}\,\mathrm{d}\omega.$$

Note that this choice of Fourier superposition extends only over positive $\omega$, which implies that $E$ is a complex function. While the physical field represented by $E$ corresponds to its real part, it is convenient to keep the imaginary part for several reasons. One is that the effective spectral support (the range in $\omega$ where $U$ takes on significant values) is reduced by at least half if negative frequencies are excluded, which is advantageous for, say, sampling purposes. The second and more important reason is that this complex representation allows us to preserve the concept of *phase*, which is very useful as it is often connected to physical parameters in the problem. This complex field representation, known as the *analytic signal representation*, can be obtained from the real field by Fourier transforming, removing the negative-frequency components and doubling the positive-frequency ones, and then inverse Fourier transforming. Equivalently, one can simply add to the real field its temporal Hilbert transform times the imaginary unit i.

The optical properties of a material or system depend on the field's frequency, and can also be represented by complex functions of $\omega$. These properties include the refractive index $n(\omega)$ or the propagation transfer functions $\mathrm{e}^{\mathrm{i}\beta(\omega)z}$ for guided modes in waveguides (with $\beta = \omega n_0(1-\delta)/c$) or plasmons. Such properties appear in *products* with the field, so that in the time domain they appear in *convolutions* with the field:

$$\hat{F}^{-1}[\tilde{g}(\omega)U(\omega)] = \int E(t')g(t-t')\,\mathrm{d}t',$$

where $\tilde{g}$ denotes any such quantity. The integral in $t'$ can be interpreted as the superposition of medium responses $g$ (that depend on the time delay $t - t'$) at the "present" $t$ due to external field stimuli $E$ at all "past times" $t'$. However, from physical considerations the medium cannot predict the future, so the integral should only extend over $t' \in (-\infty, t]$, which can be enforced if $g(\tau) = 0$ for all $\tau < 0$. This implies that $\tilde{g}(\omega)$ must be analytic over the upper half complex $\omega$-plane, all singularities being at complex frequencies with negative imaginary part. The locations of these singularities are directly related to physical properties of the medium; the real part usually corresponds to a resonant frequency, while the imaginary part is related to the resonance's width. Given these properties, it can easily be shown through standard residue theory that one can calculate (via Hilbert transformation) the imaginary part of $\tilde{g}$ given the knowledge of the corresponding real part for all real frequencies, and vice versa. Consider the case of the refractive index of a material. We can measure $n_{\mathrm{r}}(\omega)$ for a sufficiently wide range in $\omega$ by measuring refraction angles at an interface and using Snell's law. Similarly, we can measure $n_{\mathrm{i}}(\omega)$ by measuring light absorption for the same range of frequencies. The analytic relation reveals that these two physical aspects of wave propagation that appear to be completely separate are in fact intimately linked, to the point that in theory only one of the measurements is needed to predict the results of the other.

## 5 Concluding Remarks

The brief overview of mathematical aspects of optics and photonics given here is by no means comprehensive. Optics is an extremely broad discipline encompassing pure science and engineering, and as such the range of mathematics used in its study is immense. It is the author's hope that this brief overview will motivate the reader to explore the vast literature in this field.

### Further Reading

Fienup, J. 2013. Phase retrieval algorithms: a personal tour. *Applied Optics* 52:45–56.

Gbur, G. J. 2011. *Mathematical Methods for Optical Physics and Engineering*. Cambridge: Cambridge University Press.

Hecht, E. 2002. *Optics*, 4th edn. Reading, MA: Addison-Wesley.

Lakshminarayanan, V., M. L. Calvo, and T. Alieva, eds. 2013. *Mathematical Optics*. Boca Raton, FL: CRC Press.

# V.15 Numerical Relativity

*Ian Hawke*

## 1 The Size of the Problem

There is a beautiful simplicity in EINSTEIN'S FIELD EQUATIONS [III.10] (EFEs),

$$G_{ab} = 8\pi G T_{ab}, \tag{1}$$

relating the curvature of space-time, encoded in the Einstein tensor $G_{ab}$, to the matter content, encoded in the stress-energy tensor $T_{ab}$. When expanded as equations determining the space-time metric $g_{ab}$, the EFEs are a complex system of coupled partial differential equations, which are analytically tractable only in situations with a high degree of symmetry or if one solves them perturbatively. For many mathematically and physically interesting cases—such as the evolution of the very early universe or the formation, evolution, and merger of black holes—the fully nonlinear EFEs must be tackled; this requires a numerical approach. The aim of this approach is to construct, from given initial data, the space-time metric $g_{ab}$ and the matter variables that form the stress–energy tensor $T_{ab}$ over as much of the space-time as possible.

At first glance, we could take the partial differential equations for $g_{ab}$ and solve them using standard numerical methods such as those based on FINITE DIFFERENCES [IV.13 §3]. However, numerical calculations require specific coordinate systems and cannot cope with the infinities associated with singularities, whether real or due to purely coordinate effects. The fundamental steps of numerical relativity are, therefore, to first transform the EFEs (1) into a form suitable for numerical calculation, then to choose a suitable coordinate system that covers as much of the space-time as possible while avoiding singularities, and finally to interpret the results.

## 2 Formulating the Equations

Our first task is to break the coordinate-free nature of the EFEs and introduce "time" and "space" coordinates. This will allow us to write the EFEs in a form reminiscent of the WAVE EQUATION [III.31] and therefore suitable for numerical evolution. Even before discussing explicit choices of coordinates, significant mathematical questions arise; while the motivation of this section is to prepare the ground for numerical work, nothing numerical will occur until we explicitly choose coordinates in section 3.

### 2.1 The 3 + 1 Decomposition

There are many ways of splitting the space-time that are suitable for numerical evolution. The standard approach starts by introducing a coordinate time $t$, a scalar field on space-time. The four-dimensional space-time is then *foliated* into three-dimensional "slices," where on each slice $t$ is constant. We must restrict $t$ so that each slice is space-like. The line element thus takes the form

$$\begin{aligned} ds^2 &= g_{ab}dx^a\,dx^b \\ &= (-\alpha^2 + \beta_i\beta^i)\,dt^2 + 2\beta_i\,dt\,dx^i + \gamma_{ij}\,dx^i\,dx^j. \end{aligned}$$

The three-dimensional metric $\gamma_{ij}$ measures proper distances *within* the slice of constant $t \equiv dx^0$, and here the spatial indices $i, j, \ldots = 1, 2, 3$, while the space-time indices $a, b, \ldots = 0, 1, 2, 3$. Relativistic index notation, as used in GENERAL RELATIVITY AND COSMOLOGY [IV.40] and TENSORS AND MANIFOLDS [II.33], is used throughout. The *lapse* function $\alpha$ measures the proper time $\tau$ between neighboring slices: $d\tau = \alpha\,dt$. The shift vector $\beta^i$ measures the relative velocity between observers moving perpendicular (usually called *normal*) to the slices and the lines of constant spatial coordinates, $x^i_{t+dt} = x^i_t - \beta^i\,dt$. All of the new functions that we have introduced, $\{\alpha, \beta^i, \gamma_{ij}\}$, are themselves functions of the coordinates $(t, x^i)$.

The coordinate freedom is contained within the lapse and shift. The full metric is determined from the field equations (1), which must be rewritten to determine the spatial metric $\gamma_{ij}$.

### 2.2 A Slice Embedded in Space-Time

The *intrinsic* curvature of the slice can be found from the spatial metric $\gamma_{ij}$ in the same way as the full space-time curvature is found from the space-time metric. The remaining information about the space-time curvature, necessary to describe the full space-time at a point, is contained in the *extrinsic* curvature that describes how the three-dimensional slice is embedded in the four-dimensional space-time.

We require two key vectors. The first is the vector normal to the slice, $n_a = (-\alpha, 0)$. The second is the time vector $t^a = \alpha n^a + \beta^a$, which is tangent to the lines of constant spatial coordinates. The projection operator $P^a_b = \delta^a_b + n^a n_b$ acts on any tensor to project it into the slice. For example, $P^a_b t^b = \beta^a$, as the piece normal to the slice is "projected away." The extrinsic curvature

$$K_{ab} = -P^c_a \nabla_c n_b = -(\nabla_a n_b + n_a n^c \nabla_c n_b)$$

measures how $n_a$ changes within the slice, where $\nabla_a$ is the covariant derivative.

By defining the spatial (three-dimensional) covariant derivative $D_i = P^a_i \nabla_a$, the definition of the extrinsic curvature can be used to show that

$$\partial_t \gamma_{ij} = -2\alpha K_{ij} + D_i\beta_j + D_j\beta_i. \qquad (2)$$

So if the extrinsic curvature is known, we can solve for the spatial metric.

### 2.3 The ADM Equations

The field equations (1) can be projected in two ways: either normal to the slice using $n_a$ or into the slice using $P^a_i$. Projections including $n_a$ give

$$^{(3)}R + K^2 - K_{ij}K^{ij} = 16\pi n^a n^b T_{ab} := 16\pi\rho, \qquad (3)$$

$$D_j(K^{ij} - \gamma^{ij}K) = -8\pi P^{ia} n^b T_{ab} := 8\pi j^i. \qquad (4)$$

The 3-Ricci scalar $^{(3)}R$ is determined from the Riemann curvature tensor of the spatial metric $\gamma_{ij}$, and $K = K^i_i$ is the trace of the extrinsic curvature. These *constraint equations* do not involve time derivatives and are independent of the coordinate gauge functions $\alpha$, $\beta^i$. They constrain the admissible solutions inside a slice based on its geometry and the matter sources.

Projecting the field equations purely into the slice gives an evolution equation for the extrinsic curvature, which can be written as

$$\begin{aligned} \partial_t K_{ij} &= \beta^k \partial_k K_{ij} + K_{ki}\partial_j\beta^k + K_{kj}\partial_i\beta^k - D_i D_j \alpha \\ &\quad + \alpha[^3R_{ij} + KK_{ij} - 2K_{ik}K^k_j] \\ &\quad + 4\pi\alpha[\gamma_{ij}(S-\rho) - 2S_{ij}]. \qquad (5) \end{aligned}$$

Here we have defined the projected stress term $S_{ab} := P^c_a P^d_b T_{cd}$. The equations (2) and (5) (which, when combined with a choice of coordinate gauge, determine the evolution of the space-time) are typically referred to as the *ADM equations* (after Arnowitt, Deser, and Misner) in the numerical relativity literature. However, this system is mathematically *not* equivalent to the original system derived by Arnowitt, Deser, and Misner. The difference is the addition of a term proportional to the Hamiltonian constraint (3), so the solutions should be physically equivalent.

This highlights a theme of research in numerical relativity over the past twenty years: the search for "better" formulations of the equations, either by changing variables or by adding multiples of the constraint equations (which should, for physical solutions, vanish). One mathematical problem is finding criteria for the system of equations that ensure accurate and reliable numerical evolutions.

### 2.4 Hyperbolicity

Two key criteria have been considered in numerical relativity, both borrowed from general numerical analysis. First, the system of equations should be *well-posed*: solutions to the system should depend continuously on the initial data. Formally, we define this to mean that solutions $\boldsymbol{u}(t,x)$ must in some norm $\|\cdot\|$ satisfy

$$\|\boldsymbol{u}(t,x)\| \leqslant C\mathrm{e}^{kt}\|\boldsymbol{u}(0,x)\|, \tag{6}$$

where the constants $C$, $k$ must be independent of the initial data. The state vector $\boldsymbol{u}$, which for the ADM equations is $\boldsymbol{u} = (\alpha, \beta_j, \gamma_{ij}, K_{ij})^{\mathrm{T}}$, should not be confused with a space-time vector. An example system that is not well-posed is

$$\partial_t \begin{pmatrix} u_1 \\ u_2 \end{pmatrix} = \begin{pmatrix} 1 & 1 \\ 0 & 1 \end{pmatrix} \partial_x \begin{pmatrix} u_1 \\ u_2 \end{pmatrix}$$
$$\Longrightarrow \begin{pmatrix} u_1 \\ u_2 \end{pmatrix} = \begin{pmatrix} \mathrm{i}kAt + B \\ A \end{pmatrix} \mathrm{e}^{\mathrm{i}k(t+x)},$$

where $A$, $B$ are constants and $\mathrm{i} = \sqrt{-1}$. As the solution for $u_1$ grows as $t\mathrm{e}^{kt}$, the bound (6) does not hold. Systems of this form are particularly relevant for the ADM equations.

A closely related concept is that of *hyperbolicity*. For systems of the form

$$\partial_t \boldsymbol{u} + M^i \partial_i \boldsymbol{u} = \boldsymbol{s}(\boldsymbol{u}),$$

we can analyze the system solely from the set of matrices $M^i$. Specifically, choosing an arbitrary direction specified by a unit vector $n_i$, the *principal symbol* $P(n_i) := M^i n_i$ is a matrix determining the hyperbolicity of the system. If $P$ has real eigenvalues and a complete set of eigenvectors, the system is *strongly hyperbolic*; if its eigenvectors are not complete, the system is *weakly hyperbolic*. If the eigenvalues are not real, the system is not hyperbolic.

The central result is that only strongly hyperbolic systems are well-posed. The original ADM equations are not hyperbolic. The standard ADM equations in section 2.3 are weakly hyperbolic in general (in low-dimensional cases such as spherical symmetry, the system may become strongly hyperbolic). Therefore, neither system is useful for large simulations without symmetries.

### 2.5 Current Formulations

As the standard ADM equations are unsuitable for numerical evolution, considerable effort has been put into finding better formulations. There have been two approaches: the first, more experimental, approach started from the ADM equations and attempted to find the minimal set of modifications that lead to an acceptable formulation; the second approach starts from the mathematical requirements of well-posedness and hyperbolicity and systematically constructs acceptable formulations. The two most widely used formulations are as follows.

#### 2.5.1 *BSSNOK*

The BSSNOK formulation originates from a modification of the ADM equations introduced by Nakamura, Oohara, and Kojima in 1987, which was then modified by Shibata and Nakamura and systematically studied by Baumgarte and Shapiro. It introduces a number of new auxiliary variables with constraints resulting from their definition. It is written in terms of the conformal metric $\tilde{\gamma}_{ij}$, which is the rescaling $\tilde{\gamma}_{ij} := \mathrm{e}^{-4\phi}\gamma_{ij}$ such that $\tilde{\gamma}_{ij}$ has unit determinant. The evolution equation for the conformal metric now depends on the conformal traceless extrinsic curvature $\tilde{A}_{ij} = \mathrm{e}^{-4\phi}[K_{ij} - \frac{1}{3}\gamma_{ij}K]$. Finally, the contracted Christoffel symbols of the conformal metric $\tilde{\Gamma}^i := \tilde{\gamma}^{jk}\tilde{\Gamma}^i_{jk} = -\partial_j\tilde{\gamma}^{ij}$ are used, leading to a formulation containing the seventeen equations

$$\frac{\mathrm{d}}{\mathrm{d}t}\begin{pmatrix} \tilde{\gamma}_{ij} \\ \phi \\ \tilde{A}_{ij} \\ K \\ \tilde{\Gamma}^i \end{pmatrix} = \begin{pmatrix} -2\alpha\tilde{A}_{ij} \\ -\frac{1}{6}\alpha K \\ \mathrm{e}^{-4\phi}H_{ij} + \alpha(K\tilde{A}_{ij} - 2\tilde{A}_{ik}\tilde{A}^k_j) \\ -\mathrm{D}_i\mathrm{D}^i\alpha + \alpha(\tilde{A}_{ij}\tilde{A}^{ij} + \frac{1}{3}K^2) \\ \tilde{\gamma}^{jk}\partial_j\partial_k\beta^i + \frac{1}{3}\tilde{\gamma}^{ij}\partial_j\partial_k\beta^k \end{pmatrix} + \text{matter terms}, \tag{7}$$

where $H_{ij}$ is the trace-free part of $-\mathrm{D}_i\mathrm{D}_j\alpha + \alpha R_{ij}$. Note that the constraints (3), (4) have been used to eliminate certain terms, and $\mathrm{d}/\mathrm{d}t = \partial_t - \mathcal{L}_{\boldsymbol{\beta}}$, where $\mathcal{L}_{\boldsymbol{v}}$ is the Lie derivative giving the change of the quantity along integral curves of $\boldsymbol{v}$, which obeys, for example,

$$\begin{aligned} \mathcal{L}_{\boldsymbol{v}}\phi &= v^c\nabla_c\phi, \\ \mathcal{L}_{\boldsymbol{v}}w^a &= v^c\nabla_c w^a - w^c\nabla_c v^a, \\ \mathcal{L}_{\boldsymbol{v}}w_a &= v^c\nabla_c w_a + w_c\nabla_a v^c, \\ \mathcal{L}_{\boldsymbol{v}}T_{ab} &= v^c\nabla_c T_{ab} + T_{cb}\nabla_a v^c + T_{ac}\nabla_b v^c. \end{aligned}$$

The motivations for introducing these new variables include

- better control over coordinate conditions, as discussed in section 3, and
- the fact that all second derivative terms in (7) appear as simple scalar Laplace operators.

Empirically, the key step was using the momentum constraint (4) in the reformulation of the evolution equation for the $\tilde{\Gamma}^i$. It was only considerably after its introduction that the BSSNOK system was shown to be strongly hyperbolic.

#### 2.5.2 The Generalized Harmonic Formalism

The generalized harmonic formalism starts from the original proof, by Choquet-Bruhat, that the EFEs are well-posed. That used a specific coordinate system (*harmonic coordinates* defined by d'Alembert's equation $0 = \square x^c =: \nabla_b \nabla^b x^c$) to show that the EFEs (1) could be rewritten such that the principal part resembled the simple wave equation. Mathematical requirements such as well-posedness are therefore straightforward to prove.

For numerical evolution, relying on a specific choice of coordinates is frequently a bad idea, as will be seen later. Instead, the generalized harmonic formulation relies on a set of *arbitrary* functions $H^c = \square x^c$, which allow the EFEs (1) to be written as

$$\begin{aligned}\tfrac{1}{2} g^{cd} g_{ab,cd} &+ g^{cd}_{(,a} g_{b)d,c} \\ &+ H_{(a,b)} - H_d \Gamma^d_{ab} + \Gamma^c_{bd} \Gamma^d_{ac} = -8\pi (T_{ab} - \tfrac{1}{2} g_{ab} T).\end{aligned}$$

Again the principal part of the system can be related to the wave equation and, provided suitable equations are given to evolve the arbitrary functions $H^c$, the system is manifestly hyperbolic. The arbitrary functions $H^c$ do not play exactly the same role as the gauge functions $(\alpha, \beta^i)$ do in standard formulations such as BSSNOK but instead determine their evolution, as expressing the definition of the $H^c$ functions in 3 + 1 terms gives

$$(\partial_t - \beta^k \partial_k) \begin{pmatrix} \alpha \\ \beta^i \end{pmatrix} = \begin{pmatrix} -\alpha(H_t - \beta^k H_k + \alpha K) \\ \alpha g^{ij} [\alpha(H_j + g^{kl} \Gamma_{jkl}) - \partial_j \alpha] \end{pmatrix}.$$

It is expected that for every coordinate choice there will be a suitable choice of functions $H^c$, and vice versa, so the standard approach within the field is to consider the gauge functions $(\alpha, \beta^i)$.

## 3 The Choice of Coordinates

In the 3 + 1 picture we are free to choose the gauge $(\alpha, \beta^i)$ as we wish. However, to be suitable for numerical evolution, the gauge must satisfy some additional criteria. In particular, the choice of gauge must avoid the formation of coordinate singularities (as illustrated by standard Schwarzschild coordinates) and avoid reaching physical singularities, as the representation of the infinities involved in these singularities is problematic in numerical calculations. In addition, the conditions must be well well-behaved mathematically, retaining the well-posedness of the system. Ideally, the conditions should also easy to implement numerically, respect underlying symmetries of the space-time, and not complicate the analysis of the results. Clearly this is a difficult list of criteria to meet!

The slicing condition, specifying the lapse function $\alpha$, has been the focus of most work in this area. As the lapse describes the relation between one slice and the next, while the shift $\beta^i$ "merely" describes how the 3-coordinates are arranged on the slice, it is clear that the lapse determines whether the slice intersects with the physical singularity. To discuss the effect of the gauge it is useful to borrow the idea of an Eulerian observer from fluid dynamics: an idealized observer who stays at a fixed spatial coordinate location. A useful quantity in analyzing slicing conditions is the acceleration $a^c$ of the Eulerian observers $n^c$ given by $a^c := n^b \nabla_b n^c$. Expressing this in terms of the gauge, we obtain

$$a_t = \beta^k \partial_k \log(\alpha), \quad a_i = \partial_i \log(\alpha),$$

where $a_t = a_0$ is the "time" component of the acceleration, illustrating the problems that would arise from coordinate singularities. We also note that the definition of the extrinsic curvature implies

$$\nabla_c n^c = -K,$$

illustrating the potentially catastrophic growth (or collapse) of the volume elements (given by $n^c$) should the extrinsic curvature diverge.

### 3.1 Geodesic Slicing

The simplest slicing condition is to set $\alpha \equiv 1$; each slice is equally spaced in coordinate time. This *geodesic slicing* condition means that the acceleration of the Eulerian observers vanishes; they are in free fall. This is catastrophically bad for two reasons.

First, the standard calculation for a particle freely falling into a (Schwarzschild) black hole shows that it reaches the singularity within a time $t = \pi M$, where $M$

is the black hole mass. The geodesic slicing condition will not therefore avoid physical singularities.

Second, considering the evolution equations for the extrinsic curvature (and simplifying such that $\beta^i \equiv 0$) we find, for geodesic slicing, that

$$\partial_t K = K_{ij}K^{ij} + 4\pi(\rho + S).$$

Assuming the strong energy condition, the right-hand side is strictly positive and hence $K$ increases without bound, meaning the volume elements $n^c$ collapse to zero. This indicates that a coordinate singularity is inevitable when the extrinsic curvature is nontrivial.

### 3.2 Maximal Slicing

To avoid the "focusing" problem suffered by geodesic slicing, we could insist that the volume elements remain constant, i.e., $K = 0 = \partial_t K$. This condition, known as *maximal slicing*, satisfies

$$\mathrm{D}^2\alpha = \alpha[K_{ij}K^{ij} + 4\pi(\rho + S)].$$

The "maximal" part of the name comes from the proof that, when $K = 0$, the volume of the slice is maximal with respect to small variations of the hypersurface itself.

Maximal slicing leads to an elliptic equation, with the key advantages that the slice will be smooth and will avoid physical singularities. To illustrate the behavior of the slice, Smarr and York introduced a model in which the space-time is spherically symmetric, the spatial metric $\gamma_{ij}$ is flat, the scalar curvature $R$ is a constant $R_0$ within a sphere of volume $r_0$ and zero outside, and there is no matter. This is inconsistent (as the scalar curvature clearly depends on the spatial metric) but sufficiently accurate to be useful when interpreted in a perturbative sense. Using the Hamiltonian constraint (3), the maximal slicing equation in vacuum can be rewritten as $\mathrm{D}^2\alpha = \alpha R$. We thus find that for this model the maximal slicing equation becomes

$$\frac{1}{r^2}\frac{\mathrm{d}}{\mathrm{d}r}\left(r^2\frac{\mathrm{d}}{\mathrm{d}r}\alpha\right) = \begin{cases} \alpha R_0 & r < r_0, \\ 0 & \text{otherwise,} \end{cases}$$

with solution

$$\alpha(r) = \frac{1}{r\sqrt{R_0}}\frac{\sinh(r\sqrt{R_0})}{\cosh(r_0\sqrt{R_0})}, \quad r < r_0.$$

The minimal value of the lapse occurs at the origin and is $\alpha_{\min} = \alpha(0) = 1/\cosh(r_0\sqrt{R_0}) \sim \mathrm{e}^{-r_0}$. This exponential *collapse of the lapse* is a standard feature of singularity-avoiding slicings.

While maximal slicing avoids both physical and coordinate singularities, it has a number of problems that restrict its use in numerical simulations. First, as an elliptic condition it can be difficult to solve efficiently and accurately on grids adapted for solving hyperbolic equations. Elliptic equations also depend strongly on the boundary conditions, which for astrophysical problems should be enforced at infinity, where the space-time is expected to be flat. This is not possible on a standard finite numerical grid, leading to inherent approximations. Most important, however, is the phenomenon of *slice stretching*. In the vicinity of a black hole the slice is *sucked* toward the singularity (as observers infall) and also *wraps* around it (as the slice avoids the singularity). This leads to a peak in the metric components; for a Schwarzschild black hole the peak is in the radial components and grows as $\tau^{4/3}$, where $\tau$ is proper time at infinity.

### 3.3 Hyperbolic Slicings

While the key problem with maximal slicing is the stretching of the slice (a 3-coordinate effect that requires a shift condition to rectify), considerable effort has been invested in hyperbolic slicing conditions. The two main reasons for this are that the computational effort required is considerably reduced when using simple numerical techniques and that the analysis of the well-posedness of the *full* system, coupling any formulation in section 2 to the gauge, is considerably simplified.

The simplest hyperbolic slicing condition is the harmonic gauge condition $\Box x^c = 0$ introduced in section 2.5.2. Written in terms of the gauge variables, this becomes

$$\frac{\mathrm{d}}{\mathrm{d}t}\alpha = -\alpha^2 K \quad \Rightarrow \quad \frac{\mathrm{d}}{\mathrm{d}t}\left(\frac{\alpha}{\sqrt{\gamma}}\right) = 0,$$

where $\gamma$ is the determinant of the 3-metric $\gamma_{ij}$. Harmonic slicing thus directly relates the lapse to the volume elements of Eulerian observers. However, numerical experiments rapidly showed that harmonic slicing is not singularity avoiding.

The *Bona–Massó* family of slicings is an ad hoc generalization satisfying the condition

$$\frac{\mathrm{d}}{\mathrm{d}t}\alpha = -\alpha^2 f(\alpha)K,$$

where $f(\alpha)$ is an arbitrary, positive function. Note that, if $f(\alpha) = 2/\alpha$, this can be directly integrated to give $\alpha = 1 + \log(\gamma)$, the "1 + log slicing" condition, which closely mimics maximal slicing. Harmonic slicing is clearly given by $f(\alpha) = 1$.

To study the singularity-avoiding properties of the slicing, note that as $\partial_t \gamma^{1/2} = -\gamma^{1/2}\alpha K$ (when the shift vanishes) we have

$$\mathrm{d}\log(\gamma^{1/2}) = \frac{\mathrm{d}\alpha}{\alpha f(\alpha)}.$$

Thus, $\gamma^{1/2} \propto \exp[\int(\alpha f(\alpha))^{-1}\,\mathrm{d}\alpha]$, and it follows that, as $f$ is positive, the lapse must collapse as the volume elements go to zero in the approach to the singularity. For the specific case of $1 + \log$ slicing we have $\gamma^{1/2} \propto \mathrm{e}^{\alpha/2}$, which is finite as the lapse collapses, implying singularity avoidance. For the case of harmonic slicing we have $\gamma^{1/2} \propto \alpha$, so as the lapse collapses the volume elements do as well. This *marginal* singularity avoidance is insufficient for numerical simulations.

### 3.4 Shift Conditions

As with the development of hyperbolic slicing conditions, the standard hyperbolic shift conditions are motivated by elliptic conditions. The *distortion tensor* $\Sigma_{ij} = \frac{1}{2}\gamma^{1/3}\mathcal{L}_t\tilde{\gamma}_{ij}$ is essentially the velocity of the conformal metric. The minimal distortion condition, which minimizes the integral of $\Sigma_{ij}\Sigma^{ij}$ over the slice, is

$$\mathrm{D}_j\Sigma^{ij} = 0.$$

When rewritten in terms of variables introduced for the BSSNOK formulation, this becomes the *Gamma freezing* condition

$$\partial_t\tilde{\Gamma}^i = 0. \qquad (8)$$

*Driver conditions*—hyperbolic equations that aim to asymptotically satisfy elliptic conditions such as (8)—have become the method of choice in the field. The Gamma driver condition

$$\partial_t\beta^i = \tfrac{3}{4}B^i, \qquad \partial_t B^i = \partial_t\tilde{\Gamma}^i - \eta B^i$$

is intended to resemble a wave equation. With it, we assume there exists a (slowly varying) stationary state for the shift satisfying (8), to which the "current" shift is a small perturbation. The driver condition then propagates and damps (using the hand-tuned coefficient $\eta$) these perturbations, driving the final result to a stationary Gamma freezing state. While this condition has been used successfully for evolving single black holes, once a black hole moves on the grid (as it must in the example of binary black holes), it is necessary to modify the gauge to the *moving puncture* condition, where $\partial_t \to \partial_t - \beta^k\partial_k$.

The combination of $1 + \log$ slicing with moving puncture gauges produces the "trumpet" slice, which has been shown to cover the exterior and part of the interior of the space-time of a Schwarzschild black hole, without coordinate singularities, while settling down to a steady state. The slice does not cover the entire space-time; in the wormhole picture, the slice just penetrates the "neck" or "throat," as shown in figure 1.

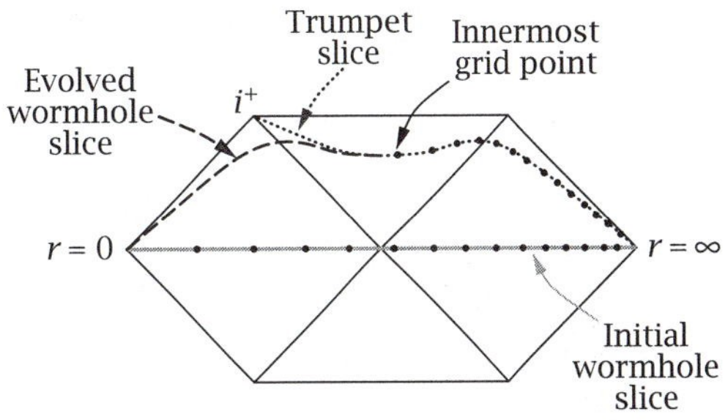


**Figure 1** A Penrose diagram showing the relationship between the initial and evolved wormhole slices and a trumpet slice. The heavy dots represent the distribution of numerical grid points. (Reprinted figure with permission from David J. Brown, *Physical Review* D 80:084042 (2009). Copyright (2009) by the American Physical Society.)

## 4 Covering the Space-Time

The simplest example of a numerical space-time considers 1 + 1 dimensions with Cartesian coordinates and no matter. Unfortunately, there can be no interesting dynamics: the Riemann tensor has only one independent component, which must vanish, so the space-time is Minkowski. However, this case can be used to illustrate the techniques and, in particular, the gauge dynamics. In this restricted case we will use the notation $g = g_{xx}$ and $K = K^x_x$ for the only nontrivial components of the metric and the extrinsic curvature.

The ADM equations together with $1 + \log$ slicing can be written, in 1 + 1 dimensions, in the following form:

$$\partial_t \begin{pmatrix} \alpha \\ g \\ D_\alpha \\ D_g \\ g^{1/2}K \end{pmatrix} + \partial_x \begin{pmatrix} 0 \\ 0 \\ 2K \\ 2\alpha K \\ \alpha D_\alpha/g^{1/2} \end{pmatrix} = \begin{pmatrix} -2\alpha K \\ -2\alpha g K \\ 0 \\ 0 \\ 0 \end{pmatrix},$$

where the notation $D_\alpha := \partial_x \log(\alpha)$, $D_g := \partial_x \log(g)$ has been used. Only the last three components contribute to the principal symbol

$$P(n_i = (1,0,0)) = \begin{pmatrix} 0 & 0 & 2/g^{1/2} \\ 0 & 0 & 2\alpha/g^{1/2} \\ \alpha/g^{1/2} & 0 & 0 \end{pmatrix},$$

which has eigenvalues and eigenvectors

$$\lambda_0 = 0, \qquad \lambda_\pm = \pm\sqrt{\frac{2\alpha}{g}},$$

$$\boldsymbol{e}_0 = \begin{pmatrix} 0 \\ 1 \\ 0 \end{pmatrix}, \qquad \boldsymbol{e}_\pm = \begin{pmatrix} 2/\alpha \\ 2 \\ \pm\sqrt{2/\alpha} \end{pmatrix}.$$

These are real and distinct when the lapse is positive, so in this special case the system is strongly hyperbolic.

By diagonalizing the principal symbol, we see that the eigenfunctions $\omega_\pm = g^{1/2}K \pm D_\alpha(\alpha/2)^{1/2}$ satisfy the advection equations

$$\partial_t \omega_\pm + \partial_x(\lambda_\pm \omega_\pm) = 0,$$

while the eigenfunction $\omega_0 = \alpha D_\alpha/2 - D_g/2$ does not evolve. However, the characteristic speeds $\lambda_\pm$ are themselves evolving. It is straightforward to check that

$$\partial_t \lambda_\pm + \lambda_\pm \partial_x \lambda_\pm = \frac{\alpha\lambda_\pm}{g^{1/2}}\left[\left(1 + \frac{1}{\alpha}\right)\omega_\mp \pm \sqrt{\frac{2}{\alpha}}\omega_0\right].$$

If we start from a region with, for example, $\omega_- = 0 = \omega_0$, then this equation for $\lambda_+$ is equivalent to the BURGERS EQUATION [III.4], for which the solutions generically become discontinuous in finite time. A discontinuity in $\lambda_+$ must correspond to a discontinuity in $g$ or $\alpha$, a gauge pathology that is completely unphysical!

It should be noted that these gauge pathologies rarely occur in evolutions of physically interesting space-times. Firstly, the restriction to a flat space-time significantly simplifies the analysis but drops the curvature terms, which are important in most cases. Secondly, the use of a shift condition, such as the moving puncture gauge outlined in section 3.4, adds a number of terms to the principal symbol, changing the nonlinear dynamics. The results in this section do illustrate, however, the essential importance of ensuring a well-behaved coordinate system in numerical evolutions.

## 5 Evolving Black Holes

To evolve fully general space-times without symmetries, one of the complete formulations outlined in section 2.5 must be used. In special circumstances such as spherical symmetry, it is possible to use reduced systems; it is even possible to use variants of the simple ADM equations. This allows detailed understanding of the numerical behavior of the evolution to be built up in simpler situations: for example, the only spherically symmetric black hole space-time is Schwarzschild. As detailed in GENERAL RELATIVITY AND COSMOLOGY [IV.40], there are a wide range of coordinate representations of the Schwarzschild metric, but (as explained in section 4) not all are suitable for numerical evolution. The additional regularity problems at the origin in spherical symmetry mean that some care is needed that is not necessary in the general case.

One key question that can be studied in spherical symmetry is whether the formulation used for the numerical evolution respects the physics being modeled. For a black hole space-time, the main point is that no information should leave the horizon. As the eigenvalues of the principal symbol correspond to the "speed" with which information (specified by the associated eigenvalues) is propagating, the mathematical question to check is whether the radially "outgoing" eigenvalues are negative with respect to the horizon.

This question depends on the choice of formulation and gauge. For a specific variant of the BSSNOK formulation with the standard gauge, the eigenvalues are

$$\pm 1, \ \pm\sqrt{2/\alpha}, \ 0, \ \hat{\beta}^r, \ \pm(\alpha\chi)^{-1/2},$$

where $\chi = e^{-4\phi}$ is related to the metric determinant, and $\hat{\beta}^r = \sqrt{g^{rr}/\chi}\,\beta^r/\alpha$ is the proper shift per unit length per unit time. Note that some of the information can propagate faster than light, particularly where the lapse collapses. However, by checking the corresponding eigenvectors it can be shown that no information about the physics propagates outside a region within the horizon.

There are additional numerical reasons for performing such a mathematical analysis for black hole space-times. While the use of moving puncture gauges has been remarkably successful for evolving black hole space-times, the physics suggests another solution for avoiding the singularity: excising it from the numerical grid completely. This requires placing a boundary *inside* the numerical domain, within the horizon, so that the infinities associated with the singularity have no effect. Of course, this means that boundary conditions must be imposed on this surface and the inherent errors introduced from these conditions must not be allowed to propagate outside of the horizon. As seen for the horizon above, the sign of the eigenvalues of the principal symbol will determine which information is "coming into" the numerical domain and hence must be set by the boundary conditions. This technique has been particularly successful when combined with the generalized harmonic formalism described in section 2.5.2.

## 6 Extensions and Further Reading

The extension of the methods outlined above to cases without symmetries (which is necessary for studying cutting-edge cosmological models or the astrophysical problem of binary black hole merger) is well covered in a number of references. This article has relied heavily on Alcubierre's excellent book, and a complementary viewpoint is given by Baumgarte and Shapiro. For the inclusion of relativistic matter, necessary for both modeling the early universe and for astrophysical applications such as supernovas and neutron stars, see the book by Rezzolla and Zanotti.

In addition, relativity in general is extremely well served by *Living Reviews*, a series of online review articles covering the field that can be found at www.livingreviews.org. Within numerical relativity there is a range of articles on important issues not touched on above. One such issue is the construction of initial data for the numerical evolution that satisfies any constraint equations while also being physically meaningful. A second issue concerns black holes in a dynamical space-time: what *types* of horizons are meaningful and useful, and how can they be found in numerical data?

Finally, we should note the key topic that has driven large parts of modern numerical relativity: the quest to detect gravitational waves. These "ripples in spacetime" carry energy and information from violent astrophysical events, such as binary black hole mergers. There are a number of currently running detectors that, when they successfully measure gravitational waves, will give unprecedented insight into the physics of strongly gravitating systems. The development of the mathematical theory of gravitational waves can be traced from the classic texts of Misner, Thorne, and Wheeler, and d'Inverno through to the recent monographs and review articles mentioned above. With recent advances in computational power, numerical methods, and mathematical techniques, numerical relativity can simulate the gravitational waves from binary black hole mergers with sufficient accuracy for current detectors, and research into relativistic matter, such as binary neutron star simulations, continues.

Alcubierre, M. 2008. *Introduction to 3 + 1 Numerical Relativity*. Oxford: Oxford University Press.

Arnowitt, R., S. Deser, and C. W. Misner. 1962. The dynamics of general relativity. In *Gravitation: An Introduction to Current Research*, edited by L. Witten, pp. 227–65. New York: John Wiley.

Baumgarte, T. W., and S. L. Shapiro. 2010. *Numerical Relativity: Solving Einstein's Equations on the Computer*. Cambridge: Cambridge University Press.

Brown, J. D. 2009. Probing the puncture for black hole simulations. *Physical Review* D 80:084042.

d'Inverno, R. 1992. *Introducing Einstein's Relativity*. Oxford: Clarendon.

Misner, C. W., K. S. Thorne, and J. A. Wheeler. 1973. *Gravitation*. London: Macmillan.

Rezzolla, L., and O. Zanotti. 2013. *Relativistic Hydrodynamics*. Oxford: Oxford University Press.

Smarr, L., and J. W. York Jr. 1978. Kinematical conditions in the construction of spacetime. *Physical Review* D 17: 2529–51.

# V.16 The Spread of Infectious Diseases

*Fred Brauer and P. van den Driessche*

## 1 Introduction

The first recorded example of a mathematical model for an infectious disease is the study by Daniel Bernoulli in 1760 of the effect of vaccination against smallpox on life expectancy. This study illustrates the use of mathematical modeling to try to predict the outcome of a control strategy. An underlying goal of much mathematical modeling in epidemiology is to estimate the effect of a control strategy on the spread of disease or to compare the effects of different control strategies. Another striking example is the work of Ronald Ross on malaria. He received the second Nobel Prize in Physiology or Medicine for his demonstration of the mechanism of the transmission of malaria between humans and mosquitoes. However, his conclusion that malaria could be controlled by controlling mosquitoes was originally dismissed on the grounds that it would be impossible to eradicate mosquitoes from a region. Subsequently, Ross formulated a mathematical model predicting that it would suffice to reduce the mosquito population below a critical level, and this conclusion was supported by field trials.

The idea that there is a critical level of transmissibility for a disease is a fundamental one in epidemiology, and it developed from models rather than from experimental data. Some of the predictions of infectious disease models may be counterintuitive. While it is considered obvious that treatment of a disease should decrease the prevalence of the disease (i.e., the proportion of people infected at a given time), there are situations in which drug treatment may incite the development of a drug-resistant strain of the disease and an increase in the treatment level may actually increase the prevalence of disease, with the histories of tuberculosis and HIV/AIDS being good examples of this phenomenon. Modeling is essential to identify the possibility of such counterintuitive effects.

While the foundations of mathematical epidemiology were laid by public health physicians, there have been many theoretical elaborations. An elaborate mathematical theory has developed, and there has been a divergence of interests between mathematicians and public health professionals. One result of this has been that there are both strategic models, which concentrate on general classes of models and theoretical understanding, and tactical models, which attempt to incorporate as much detail as possible into a model in order to be able to make good quantitative predictions. In recent years there have been serious attempts to encourage interaction between mathematical modelers and public health professionals to improve understanding of different views.

We concentrate here on relatively simple models and qualitative understanding, but we recognize that the models used to make quantitative predictions must have much more detail and incorporate data. They must therefore usually be solved numerically. The development of high-speed computers has made it possible to simulate very complicated models quickly, and this has had a great influence on disease-outbreak management.

There are many different types of infectious disease transmission models. We consider only deterministic models suitable for human disease, although there is an extensive parallel theory of stochastic models. Also, we confine our attention to continuous models, though it can be argued that since disease transmission data are obtained at discrete times it would be reasonable to use discrete models.

For modeling an infectious disease, as for modeling other natural phenomena, it is essential to understand the biology, then to translate the biological problem into a mathematical framework for the important features, and finally to translate results back to the biology, bearing in mind the assumptions made. An important distinction is made between short-term models, as for disease outbreaks and epidemics, in which demographic effects such as births and deaths can be ignored, and long-term models, as for endemic situations, in which it is necessary to include demographic effects. In fact, an essential difference between these two categories of models is the absence or presence of a flow of new susceptible people into the population, and such a flow may come from demographic effects or from recovery of infectious individuals without immunity against reinfection.

We mainly concentrate on *compartmental* models, originally introduced by Kermack and McKendrick, in which a given population is separated into compartments identified by the disease status of the individuals in those compartments. For example, there are SIR (susceptible–infectious–removed) models in which individuals are susceptible to infection; are infectious; or are removed from infection by recovery from infection with immunity against reinfection, by inoculation against being infected, or by death from the disease. The model is described by assumptions about the rates of passage between compartments. In an SIR model it is assumed that recovery provides complete immunity against reinfection. There are also SIS (susceptible–infectious–susceptible) models, describing a situation in which recovered individuals are immediately susceptible to reinfection rather than being removed. These two types of model describe diseases with fundamentally different properties. Typically, diseases caused by a virus (e.g., influenza) are of SIR type, while diseases caused by a bacterium (e.g., gonorrhea) are of SIS type. More elaborate compartmental structures are possible, such as SLIR (susceptible–latent–infectious–removed) models, in which there is a latent period between becoming infected and becoming infectious (the compartment L represents individuals who are infected but not yet infectious); models in which some individuals are selected for treatment; or models in which some individuals are asymptomatic, i.e., infectious but without disease symptoms. Also, there are diseases of SIRS type, for which recovery provides immunity but only temporarily. For example, influenza strains mutate rapidly, and recovery from one strain provides immunity against reinfection only until the strain mutates enough to act as a different strain. We consider SIR and SIS as the basic disease types and describe their analysis, but we also indicate how more complicated compartmental structures may be described.

Our aim is to give a taste of this exciting area of applied mathematics, and to encourage further exploration, rather than to present a complete portrait of mathematical epidemiology.

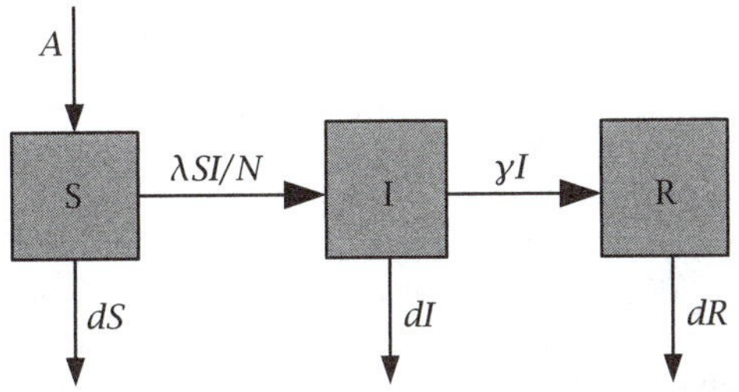


**Figure 1** An SIR model flow chart with demographics.

## 2 SIR Models: Measles and Influenza

Measles and influenza are viral diseases and, in most cases, an individual who has been infected and has subsequently recovered has lifelong immunity to the disease; thus, an SIR model is appropriate. Measles (without vaccination) often stays endemic in a population. By contrast, seasonal influenza is a relatively short-lived epidemic, with the number of infectious individuals peaking and then going to zero. We use these two diseases as examples of general SIR models.

### 2.1 Models with Demographics: Measles

We want to know what happens if a small number of people infected with measles are introduced into a susceptible population of $N$ people. To investigate this, we formulate an ordinary differential equation (ODE) system that determines the change in the numbers of susceptible, infectious, and recovered people with time, denoted by $S, I$, and $R$, respectively, and includes demographics but assumes that there are no deaths due to measles. A flow chart for the model is given in figure 1. Let $A > 0$ be the number of newborns per unit time in the population, let $d > 0$ be the natural death rate, let $1/\gamma$ be the average duration of measles infection (this is about five days), and let $\lambda$ be the disease transmission parameter. The rate of change of $S$ is then given as the input rate ($A$) minus the output rate ($\lambda SI/N + dS$), and similarly for the rates of change of $I$ and $R$. Thus, measles evolves in time according to the equations

$$\frac{\mathrm{d}S}{\mathrm{d}t} = A - \frac{\lambda SI}{N} - dS,$$
$$\frac{\mathrm{d}I}{\mathrm{d}t} = \frac{\lambda SI}{N} - (d+\gamma)I,$$
$$\frac{\mathrm{d}R}{\mathrm{d}t} = \gamma I - dR.$$

Here $N = S + I + R$, and it is assumed that an average infectious person makes $\lambda$ contacts in unit time, of which a fraction $S/N$ are with susceptible people and thus transmit infection, giving $\lambda SI/N$ new infections in unit time. This is called *standard incidence*. The model above assumes that all newborns enter the susceptible class, thus ignoring passive immunity from maternal antibodies.

The $R$ equation is not in fact needed as $R$ does not enter into the other equations: it can therefore be determined from $S$ and $I$. The first two equations always have one equilibrium: that is, a constant solution with $\mathrm{d}S/\mathrm{d}t = \mathrm{d}I/\mathrm{d}t = 0$. This is given by $(S,I) = (A/d, 0)$ and is called the *disease-free equilibrium* (DFE). The equations may have another equilibrium $(S^*, I^*)$ with $I^* > 0$, and this is called an *endemic equilibrium*.

Assume that a small number of people in a susceptible population are infected with measles. Will measles die out or become endemic? Analysis of the equilibria show that this depends on the product of the disease transmission parameter and the average death-adjusted infectious duration. This product, denoted by $\mathcal{R}_0$, with $\mathcal{R}_0 = \lambda/(d+\gamma)$, is called the *basic reproduction number* and is the average number of new infections caused by an infectious person introduced into a susceptible population. If $\mathcal{R}_0 < 1$, then the solution tends to the DFE and measles dies out in the population. If $\mathcal{R}_0 > 1$, then measles tends to an endemic level with the number of infectious people:

$$I^* = \frac{dN}{d+\gamma}\left(1 - \frac{1}{\mathcal{R}_0}\right),$$

with $N = A/d$. Thus $\mathcal{R}_0 = 1$ acts as a critical level, or *threshold*, at which the model exhibits a *forward bifurcation*. There is an exchange of stability between the disease-free and endemic equilibria. Stochastic models usually exhibit a similar threshold, but if $\mathcal{R}_0 > 1$, then there is a finite probability of disease extinction.

In order to prevent measles from becoming endemic, a fraction $p > 1 - 1/\mathcal{R}_0$ of newborns needs to be vaccinated to reduce $\mathcal{R}_0$ below 1 and give the population *herd immunity*. This was achieved worldwide for smallpox (eradicated in 1977), which had $\mathcal{R}_0 \approx 5$, so that theoretically 80% vaccination provided herd immunity. However, measles has a higher $\mathcal{R}_0$ value, up to 18 in some populations, making it unrealistic to achieve herd immunity. Data on measles in the pre-vaccine era show biennial oscillations about an endemic level, so more complicated models including seasonal forcing are needed for accurate predictions.

### 2.2 Models Ignoring Demographics: Influenza

Interest in epidemic models, which had been largely ignored for many years, was reignited by the severe

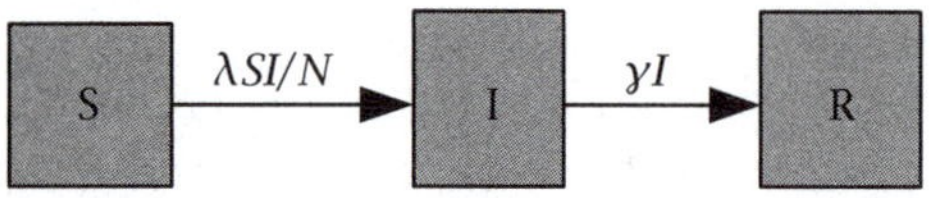


**Figure 2** An SIR model flow chart with no demographics.

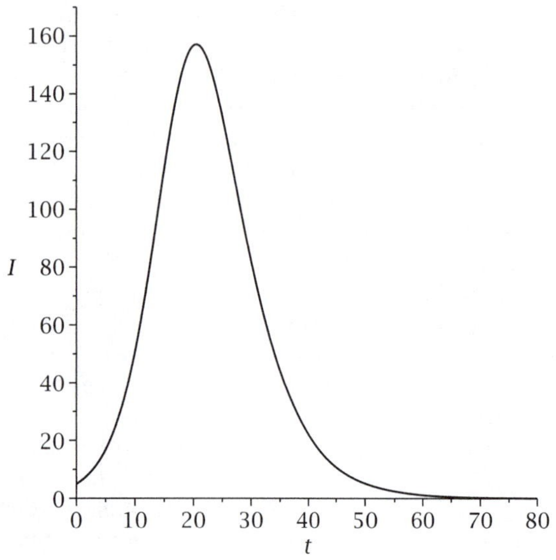


**Figure 3** Simulation of the influenza SIR model showing the number of infectious people against time, with $\lambda = 0.5$, $\gamma = 0.25$, $N = 1000$, $I(0) = 5$.

acute respiratory syndrome (SARS) outbreak of 2002–3, and this interest carried over into concerns about the possibility of an influenza pandemic.

For seasonal influenza, birth and death can be ignored, since this is a single outbreak of short duration and the demographic timescale is much slower than the epidemiological timescale (see figure 2 for a flow chart). Setting $A = d = 0$ in the measles model gives a model with no endemic equilibrium since $I = 0$ is the only equilibrium condition. With an initial number of infectious people $I(0) > 0$, the parameter $\mathcal{R}_0 = \lambda/\gamma$ still acts as a threshold. Influenza first increases to a peak and then decreases to zero (an influenza epidemic) if $\mathcal{R}_0 > 1$, but it simply decays to zero (no epidemic) if $\mathcal{R}_0 < 1$. Figure 3 shows a numerical solution of the model equations (done with Maple) with parameters that give $\mathcal{R}_0 = 2$ for a population of 1000 starting with 5 infectious people. With $\mathcal{R}_0 > 1$, the model shows that not all susceptible people are infected before the epidemic dies out. In fact, the number of susceptible people remaining uninfected, $S(\infty)$, is given implicitly as the positive root of the *final size equation*:

$$\log \frac{S(0)}{S(\infty)} = \mathcal{R}_0\left(1 - \frac{S(\infty)}{N}\right),$$

where $S(0) \approx N$ is the initial number of susceptible people. The total number infected by the disease is $I(0) + S(0) - S(\infty)$. The *attack rate* (the fraction of the population infected by the disease) is $1 - S(\infty)/N$. The epidemic first grows approximately exponentially, since for small time,

$$I(t) \approx I(0)\exp\{\gamma(\mathcal{R}_0 - 1)t\}.$$

This initial growth rate can be determined from data, and together with a value of $\gamma$ from data it can be used to estimate $\mathcal{R}_0$. For seasonal influenza, $\mathcal{R}_0$ is usually found to be in the range 1.4–2.4. The above formula for herd immunity also applies for estimating the fraction of the population that would need to be vaccinated to theoretically eradicate influenza. However, at the beginning of a disease outbreak, stochastic effects are important.

For more realistic models, other factors such as a latent period, deaths due to influenza, asymptomatic cases (people with no symptoms of disease but able to transmit infection), and age structure need to be included. These can all be put in a more general framework in which the essential properties continue to hold. However, the quantitative analysis of such complicated models requires numerical simulations. Antiviral treatment is also available for influenza, and model results can guide public health policy on vaccination and antiviral schedules. In fact, model predictions helped to guide such policies during the 2009 H1N1 influenza pandemic. Treatment with antiviral drugs can cause the emergence of resistant strains. Some recent models for optimal treatment strategies during an influenza pandemic suggest that it is best to do nothing initially and then quickly ramp up treatment. However, policy makers must weigh modeling predictions (and their inherent assumptions) with ethical, economic, and political concerns.

As noted in our introduction, recovery from influenza may provide temporary immunity, indicating an SIRS model, which has a basic reproduction number $\mathcal{R}_0$ such that $\mathcal{R}_0 = 1$ acts as a sharp threshold (between the DFE and the endemic equilibrium), as it does in the SIR model with demographics. Models of this type that take into account that vaccination is not totally effective and also wanes with time can give rise to a *backward bifurcation*, in which $\mathcal{R}_0$ must be reduced below

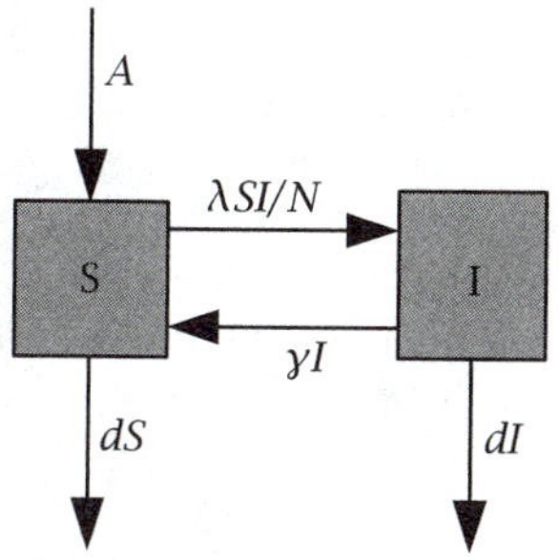


**Figure 4** An SIS model flow chart.

a value $\mathcal{R}_c$ that is less than one to eradicate the disease. In models that exhibit backward bifurcation, for $\mathcal{R}_c < \mathcal{R} < 1$ there are two endemic equilibria, with the larger $I$ value locally stable and the smaller $I$ value unstable, in addition to the locally stable DFE. Thus, in this range of parameter values the disease may die out or tend to the larger endemic equilibrium depending on the initial conditions.

If we assume that the period of temporary immunity is a constant, then a delay differential equation SIRS model can be formulated, and it predicts that there are periodic solutions arising through a HOPF BIFURCATION [IV.21 §2] about an endemic equilibrium for some values of this period of temporary immunity. Data on disease dynamics often exhibit periodicity; it is therefore an interesting challenge to determine what factors contribute to this oscillatory behavior. Delay differential equations and systems that assume more general distributions pose challenging mathematical problems.

## 3 SIS Models: Gonorrhea

Bacterial diseases do not usually give immunity on recovery, so an SIS model is appropriate as a simple model for such diseases (see figure 4 for a flow chart). Ignoring death due to the disease, taking the total population at its equilibrium value $N = A/d$, and using $S = N - I$, the model can be described by the equation for the rate of change of the number of infectious people with time:

$$\frac{dI}{dt} = \lambda I \frac{N - I}{N} - (d + \gamma)I.$$

Given an initial number of infectious people, this logistic equation can be solved by separation of variables to give the number of infectious people at any later time. Alternatively, an analysis of the equilibria can be used to show that the basic reproduction number $\mathcal{R}_0 = \lambda/(d + \gamma)$ acts as a threshold, with the disease dying out if $\mathcal{R}_0 < 1$ and going to an endemic level $I^* = N(1 - 1/\mathcal{R}_0)$ if $\mathcal{R}_0 > 1$. Even if demographics are ignored, these conclusions still hold, unlike the epidemic found for an SIR model.

Now consider a model for the heterosexual transmission of gonorrhea: a sexually transmitted bacterial disease. It is necessary to divide the population into females and males and to model transmission of the disease between them. Assume that there are no deaths due to disease, so that the populations of females $N_F$ and of males $N_M$ are constants. Set $\mu_F = d + \gamma_F$ and $\mu_M = d + \gamma_M$, which are the natural death-corrected transition rates. The equations governing the number of infectious females $I_F$ and infectious males $I_M$ are

$$\frac{dI_F}{dt} = \lambda_{MF} \frac{N_F - I_F}{N_F} I_M - \mu_F I_F,$$
$$\frac{dI_M}{dt} = \lambda_{FM} \frac{N_M - I_M}{N_M} I_F - \mu_M I_M,$$

where $\lambda_{MF}$ and $\lambda_{FM}$ are the transmission coefficients from male to female and female to male, and $1/\gamma_F$ and $1/\gamma_M$ are the mean infectious periods for females and males. Suppose that there is initially a small number of males or females infected with gonorrhea. Will the outbreak persist or will it die out? The equations have a DFE with $(I_F, I_M) = (0, 0)$, and the behavior close to this can be determined by linearizing the system (i.e., by neglecting terms of higher order than linear). Linear stability of the DFE is determined by setting $I_F = C_1 e^{zt}$ and $I_M = C_2 e^{zt}$ in this linearized system and determining the sign of the real part of $z$ for nonzero constants $C_1, C_2$. This is equivalent to finding the eigenvalues of the Jacobian matrix $J$ at the DFE, where

$$J = \begin{pmatrix} -\mu_F & \lambda_{MF} \\ \lambda_{FM} & -\mu_M \end{pmatrix}.$$

The eigenvalues of $J$ are given by the roots of the characteristic equation

$$z^2 + z(\mu_F + \mu_M) + \mu_F \mu_M - \lambda_{MF} \lambda_{FM} = 0.$$

Both roots of this quadratic equation have negative real parts if and only if the constant term is positive. With this condition, small perturbations from the DFE decay, and the DFE is linearly stable. Drawing on experience from the previous models and taking into account biological considerations, $\mathcal{R}_0$ can be determined from this condition (see the formula for $\mathcal{R}_0$ given below). Alternatively, there is a systematic way to find $\mathcal{R}_0$. Write $J$ as $J = F - V$, where $F$ is the matrix associated with new infections and $V$ is the matrix associated with transitions. Then $\mathcal{R}_0 = \rho(FV^{-1})$, where $\rho$ denotes the spectral radius (i.e., the eigenvalue with the largest absolute value) and $FV^{-1}$ is the *next-generation matrix*. For the above model,

$$F = \begin{pmatrix} 0 & \lambda_{MF} \\ \lambda_{FM} & 0 \end{pmatrix}, \qquad V = \begin{pmatrix} \mu_F & 0 \\ 0 & \mu_M \end{pmatrix},$$

giving

$$\mathcal{R}_0 = \sqrt{\frac{\lambda_{MF}\lambda_{FM}}{\mu_M\mu_F}}.$$

This expression contains a square root because the basic reproduction number is the geometric mean of the reproduction number for each sex, namely $\lambda_{MF}/\mu_F$ and $\lambda_{FM}/\mu_M$. Linear stability is governed by $\mathcal{R}_0$, but in fact global results also hold. If $\mathcal{R}_0 < 1$, then the disease dies out in both sexes; whereas if $\mathcal{R}_0 > 1$, the disease tends to an endemic state $(I_F^*, I_M^*)$ and persists in both sexes. This endemic state can be found by solving for the positive equilibrium, giving

$$I_M^* = \frac{(\lambda_{MF}\lambda_{FM} - \mu_F\mu_M)N_F N_M}{\mu_M\lambda_{MF}N_M + \lambda_{MF}\lambda_{FM}N_F},$$

with a corresponding formula for $I_F^*$. The global stability of the endemic state can be proved by using Lyapunov functions, but as far as we know it is still an open problem to prove this if deaths due to the disease are incorporated into the model.

## 4 Models for HIV/AIDS

### 4.1 Population Models

The acquired immune deficiency syndrome (AIDS) epidemic that began in the early 1980s spread worldwide and has prompted much research on this and other sexually transmitted diseases. AIDS is caused by the human immunodeficiency virus (HIV) and once infected, individuals never recover. Data show that after initial infection with the virus an individual has high infectivity, but infectivity then drops to a low level for a period of up to ten years before rising sharply due to the onset of AIDS. Infectivity therefore has a "bathtub" shape. Compartmental ODE models with several stages of infectivity subdividing the infectious state can be formulated to approximate this shape. Alternatively, this can be handled by partial differential equation models using age of infection as a variable. Currently, HIV can be treated with highly active antiretroviral therapy drugs, which both reduce the symptoms and prolong the period of low infectivity.

Models for HIV/AIDS often focus on one population: a male homosexual community, for example, or a group of female sex workers and their male clients. Such models need to take account of many factors, including level of sexual activity, drug use, condom use, and sexual contact network, and this results in large-scale systems with many parameters that need to be estimated from data. The homogeneous models described previously therefore need to be extended.

To illustrate this with a simple model with one infectious stage, consider a male homosexual community in which there are $k$ risk groups that are delineated by the numbers of partners an individual has each month, thus introducing heterogeneity. Letting $S_i$, $I_i$, and $A_i$ be the proportion of males with $i$ partners each month who are susceptible, infectious, and have developed AIDS, respectively, the ODEs for the system are

$$\frac{dS_i}{dt} = dn_i - \sum_{j=1}^{k} \beta_{ij}S_iI_j - dS_i,$$

$$\frac{dI_i}{dt} = \sum_{j=1}^{k} \beta_{ij}S_iI_j - dI_i - \gamma I_i,$$

$$\frac{dA_i}{dt} = \gamma I_i - dA_i - mA_i,$$

for $i = 1, \dots, k$, where $n_i$ is the proportion of males with $i$ contacts per month, $dn_i$ is the recruitment into the male sexually active population with $i$ partners per month, $m$ is mortality due to AIDS, and $\beta_{ij}$ contains the contact and transmission rates between a susceptible male with $i$ partners per month and an infectious male with $j$ partners per month. This formulation assumes that all infectious males proceed to AIDS with rate $\gamma$ and at that stage they withdraw from sexual activity and do not continue to contribute to disease spread. The contact matrix among the males in the population $(\beta_{ij})$ can be determined by surveying their distribution and the number of sexual partners they have. This matrix is sometimes termed a WAIFW (who acquires infection from whom) matrix. If it is assumed that partnerships are formed at random but in proportion to their expected number of partners, then

$$\beta_{ij} = \beta \frac{ij}{\sum_{\ell=1}^{k} \ell n_\ell}.$$

Making this assumption and taking $\mu_i = \mu$ and $\gamma_i = \gamma$, the next-generation matrix method has $F$ matrix with rank 1, and $V$ a scalar matrix. Thus, the nonzero eigenvalue of $FV^{-1}$ is equal to its trace, giving the basic reproduction number as

$$\mathcal{R}_0 = \frac{\beta}{d+\gamma} \frac{\sum_{i=1}^{k} i^2 n_i}{\sum_{i=1}^{k} i n_i} = \frac{\beta}{d+\gamma}\left(M + \frac{V}{M}\right),$$

where $M$ and $V$ are the mean and variance of the number of sexual partners per month. If all males have the same number of partners, then $\mathcal{R}_0 = \beta M/(d+\gamma)$, and the above formula therefore shows that $\mathcal{R}_0$ increases with the variance in the number of partners. Other assumptions about the formation of partnerships can be incorporated into this model structure. For example, males could be more likely to form partnerships with others having the same or a similar number of contacts per month (assortative mixing).

To extend the above model to study the heterosexual spread of HIV/AIDS, the population must be divided into two sexes as well as being split according to activity level. Contact rates together with parameters related to the disease need to be determined to parametrize the model. It is generally observed that the probability of transmission per sexual act from an infectious male to a susceptible female is greater than that from an infectious female to a susceptible male, so there is asymmetry in the model.

Control strategies such as condoms, highly active antiretroviral therapy treatment, and education can be incorporated into the models for HIV/AIDS to guide planning so that control can be optimized. However, HIV develops resistance against the drugs that are currently used for its treatment, and this, as well as a lack of treatment compliance, needs to be built into models. In addition, individuals who are HIV positive have a higher than average risk of developing other diseases, such as tuberculosis and pneumonia. Rather than developing such complicated models further, we turn instead to a brief discussion of intra-host models.

### 4.2 Virus Dynamics

When HIV virus enters an individual's body it attacks the cells susceptible to infection, called target cells, and produces infected cells that in turn produce new virus. A basic model of intra-host virus infection consists of the following three ODEs for the number of susceptible target cells $T$, actively infected target cells $I$, and free mature virus particles (virions) $V$:

$$\frac{\mathrm{d}T}{\mathrm{d}t} = R - d_{\mathrm{T}}T - kVT,$$
$$\frac{\mathrm{d}I}{\mathrm{d}t} = kVT - d_{\mathrm{I}}I,$$
$$\frac{\mathrm{d}V}{\mathrm{d}t} = pI - d_{\mathrm{V}}V.$$

Here, $d_{\mathrm{T}}$, $d_{\mathrm{I}}$, and $d_{\mathrm{V}}$ are the death rates of target cells, infected cells, and virus particles, respectively; target cells are infected by virus with rate constant $k$ assuming *mass action* incidence; $R$ is the production rate of new target cells; and $p$ is the production rate of new virus per infected cell. A flow chart for this model is given in figure 5.

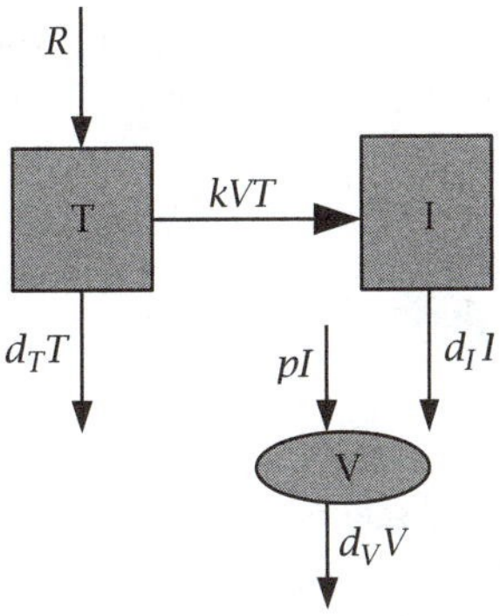


**Figure 5** Virus dynamics flow chart.

If a person is initially uninfected and then a small amount of HIV virus is introduced, will the virus be successful in establishing a persistent HIV infection? This question can be addressed by a linear stability analysis, as developed in previous sections. The DFE is given by $(T,V,I) = (R/d_{\mathrm{T}}, 0, 0)$, and the endemic equilibrium (if it exists) is given by

$$(T,V,I) = \left(\frac{d_{\mathrm{I}}d_{\mathrm{V}}}{kp}, \frac{R}{d_{\mathrm{I}}} - \frac{d_{\mathrm{T}}d_{\mathrm{V}}}{kp}, \frac{Rp}{d_{\mathrm{I}}d_{\mathrm{V}}} - \frac{d_{\mathrm{T}}}{k}\right).$$

Assuming that the production of new virus (i.e., the $pI$ term in the equation for $V$) is not considered to be a new infection, the next-generation matrix method gives

$$\mathcal{R}_0 = \frac{kpR}{d_{\mathrm{T}}d_{\mathrm{I}}d_{\mathrm{V}}}.$$

However, if this production is assumed to be a new infection, then this term must go in the $F$ matrix and gives

$$\mathcal{R}_0 = \sqrt{\frac{kpR}{d_{\mathrm{T}}d_{\mathrm{I}}d_{\mathrm{V}}}}.$$

The endemic equilibrium exists and persistent HIV infection occurs if these numbers are greater than 1, which can be proved by using a Lyapunov function. In fact, this differential equation system for the early stage of HIV infection is equivalent to the population-level SLIR system with a constant host population size. Virus dynamic models therefore have features in common with disease transmission models even though the biological background is different.

Notice that these formulas for $\mathcal{R}_0$ agree at the 1-threshold value. This example illustrates the fact that the formula for $\mathcal{R}_0$ depends on the biological assumptions of the model, and several formulas can give the same expression at the 1-threshold value. However, the sensitivity of $\mathcal{R}_0$ to perturbations in parameters depends on the exact formula. Drug treatment aims to reduce the parameter $k$ so that $\mathcal{R}_0$ is below 1. The parameters and effects of drugs can be estimated from data on individuals who are HIV positive.

## 5 Model Extensions

Some infectious diseases have features that may necessitate models with compartments additional to those described above. For example, some diseases are transmitted to humans by vectors, as is the case for malaria and West Nile virus, which are transmitted by mosquitos. For such diseases, vector compartments are needed. Waterborne diseases such as cholera can be transmitted directly from person to person or indirectly via contaminated water; thus some cholera models include a pathogen compartment. Other infectious diseases, for example HIV/AIDS and hepatitis B, can be transmitted vertically from mother to offspring. In the case of hepatitis B, the mother may even be asymptomatic. This alternative route of transmission can be modeled by adding an input term into the infectious class that represents infectious newborns, and this modifies the basic reproduction number. For Ebola, recently dead bodies are an important source of infection that needs to be included in a model. From the recent outbreak in West Africa, it appears that health-care workers have an increased risk of infection, and there may be a significant number of asymptomatic cases; these compartments should therefore be explicitly included in an Ebola model.

The simplest compartmental models assume homogeneous mixing of individuals, but it is possible to extend the structure to include heterogeneity of mixing, as briefly described above in the model for HIV/AIDS in a male heterosexual population. Network models take this still further by concentrating on the frequencies of contacts between individuals. Agent-based models separate the population into individuals, leading to very large systems that can be analyzed only by numerical simulations. A particularly important heterogeneity in disease transmission is age structure, since in many diseases, especially childhood diseases such as measles, most transmission of infection occurs between individuals of similar ages. Spatial heterogeneity appears in two quite different forms: namely, local motion such as diffusion and long-distance travel such as by airlines between distant locations. The former is usually modeled by partial differential equations, whereas the latter is usually modeled by a large system of ODEs, giving a metapopulation model. With the availability of good travel data, metapopulation models are especially important for public health planning for mass gatherings such as the Olympic Games. New or newly emerging infectious diseases often call for new modeling ideas; for example, metapopulation models were further developed for SARS, and coinfection models have been developed for HIV and tuberculosis. In addition, social behavior and the way that people change their behavior during an epidemic are factors that should be integrated into models, especially those designed for planning vaccination and other control strategies.

For interested readers the literature listed in the further reading section below, as well as current journal articles and online resources, will provide more information about these and other models.

### Further Reading

Anderson, R. M., and R. M. May. 1991. *Infectious Diseases of Humans.* Oxford: Oxford Science Publications.

Brauer, F., and C. Castillo-Chavez. 2012. *Mathematical Models in Population Biology and Epidemiology.* New York: Springer.

Brauer, F., P. van den Driessche, and J. Wu, eds. 2008. *Mathematical Epidemiology.* Lecture Notes in Mathematics, Mathematical Biosciences Subseries 1945. New York: Springer.

Daley, D. J., and J. Gani. 1999. *Epidemic Modelling: An Introduction.* Cambridge: Cambridge University Press.

Diekmann, O., and J. A. P. Heesterbeek. 2000. *Mathematical Epidemiology of Infectious Diseases: Model Building, Analysis and Interpretation.* New York: John Wiley.

Greenwood, P. E., and L. F. Gordillo. 2009. Stochastic epidemic modeling. In *Mathematical and Statistical Estimation Approaches in Epidemiology*, edited by G. Chowell, J. M. Hyman, L. M. A. Bettencourt, and C. Castillo-Chavez, pp. 31–52. New York: Springer.

Hethcote, H. W. 2000. The mathematics of infectious diseases, *SIAM Review* 42:599–653.

Nowak, M. A., and R. M. May. 2000. *Virus Dynamics: Mathematical Principles of Immunology and Virology.* Oxford: Oxford University Press.

# V.17 The Mathematics of Sea Ice

*Kenneth M. Golden*

## 1 Introduction

Among the large-scale transformations of the Earth's surface that are apparently due to global warming, the sharp decline of the summer Arctic sea ice pack is probably the most dramatic. For example, the area of the Arctic Ocean covered by sea ice in September of 2012 was less than half of its average over the 1980s and 1990s. While global climate models generally predict declines in the polar sea ice packs over the twenty-first century, these precipitous losses have significantly outpaced most projections. Here we will show how mathematics is being used to better understand the role of sea ice in the climate system and improve projections of climate change. In particular, we will focus on how mathematical models of composite materials and statistical physics are being used to study key sea ice structures and processes, as well as represent sea ice more rigorously in global climate models. Also, we will briefly discuss these climate models as systems of partial differential equations (PDEs) solved using computer programs with millions of lines of code, on some of the world's most powerful computers, with particular focus on their sea ice components.

### 1.1 Sea Ice and the Climate System

Sea ice is frozen ocean water, which freezes at a temperature of about −1.8 °C, or 28.8 °F. As a material, sea ice is quite different from the glacial ice in the world's great ice sheets covering Antarctica and Greenland. When salt water freezes, the result is a composite of pure ice with inclusions of liquid brine, air pockets, and solid salts. As the temperature of sea ice increases, the porosity or volume fraction of brine increases. The brine inclusions in sea ice host extensive algal and bacterial communities that are essential for supporting life in the polar oceans. For example, krill feed on the algae, and in turn they support fishes, penguins, seals, and Minke whales, and on up the food chain to the top predators: killer whales, leopard seals, and polar bears. The brine microstructure also facilitates the flow of salt water through sea ice, which mediates a broad range of processes, such as the growth and decay of seasonal ice, the evolution of ice pack reflectance, and biomass buildup.

As the boundary between the ocean and the atmosphere in the polar regions of the Earth, sea ice plays a critical role as both a leading indicator of climate change and as a key player in the global climate system. Roughly speaking, most of the solar radiation that is incident on snow-covered sea ice is reflected, while most of the solar radiation that is incident on darker sea water is absorbed. The sea ice packs serve as part of Earth's polar refrigerator, cooling it and protecting it from absorbing too much heat from sunlight. The ratio of reflected sunlight to incident sunlight is called *albedo*. While the albedo of snow-covered ice is usually larger than 0.7, the albedo of sea water is an order of magnitude smaller, around 0.06.

#### *1.1.1 Ice–Albedo Feedback*

As warming temperatures melt more sea ice over time, fewer bright surfaces are available to reflect sunlight, more heat escapes from the ocean to warm the atmosphere, and the ice melts further. As more ice is melted, the albedo of the polar oceans is lowered, leading to more solar absorption and warming, which in turn leads to more melting, creating a positive feedback loop. It is believed that this so-called *ice–albedo feedback* has played an important role in the recent dramatic declines in summer Arctic sea ice extent.

Thus even a small increase in temperature can lead to greater warming over time, making the polar regions the most sensitive areas to climate change on Earth. Global warming is *amplified* in the polar regions. Indeed, global climate models consistently show amplified warming in the high-latitude Arctic, although the magnitude varies considerably across different models. For example, the average surface air temperature at the North Pole by the end of the twenty-first century is predicted to rise by a factor of about 1.5 to 4 times the predicted increase in global average surface air temperature.

While global climate models generally predict declines in sea ice area and thickness, they have significantly underestimated the recent losses observed in summer Arctic sea ice. Improving projections of the fate of Earth's sea ice cover and its ecosystems depends on a better understanding of important processes and feedback mechanisms. For example, during the melt season the Arctic sea ice cover becomes a complex, evolving mosaic of ice, melt ponds, and open water. The albedo of sea ice floes is determined by melt pond evolution. Drainage of the ponds, with a resulting increase in albedo, is largely controlled by the fluid permeability of the porous sea ice underlying the ponds. As ponds develop, ice–albedo feedback enhances the melting process. Moreover, this feedback loop is the driving mechanism in mathematical models developed to address the question of whether we have passed a so-called *tipping point* or *critical threshold* in the decline of summer Arctic sea ice. Such studies often focus on the existence of saddle-node bifurcations in

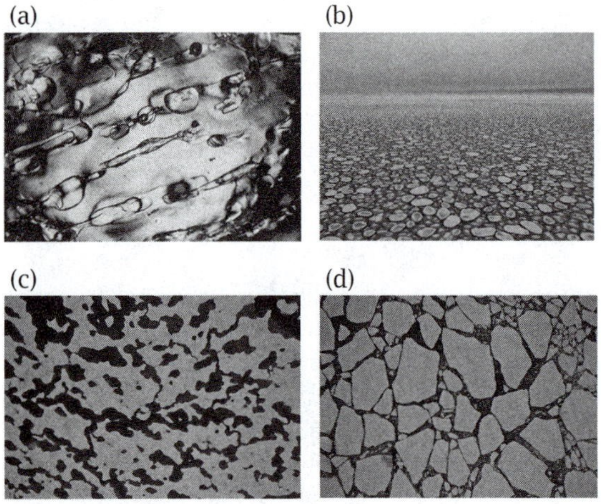


**Figure 1** Sea ice exhibits composite structure on length scales over many orders of magnitude: (a) the submillimeter scale brine inclusions in sea ice (credit: CRREL (U.S. Army Cold Regions Research and Engineering Lab) report); (b) pancake ice in the Southern Ocean, with microstructural scale on the order of tens of centimeters; (c) melt ponds on the surface of Arctic sea ice with meter-scale microstructure (courtesy of Donald Perovich); and (d) ice floes in the Arctic Ocean on the kilometer scale (courtesy of Donald Perovich).

dynamical system models of sea ice coverage of the Arctic Ocean. In general, sea ice albedo represents a significant source of uncertainty in climate projections and a fundamental problem in climate modeling.

### *1.1.2 Multiscale Structure of Sea Ice*

One of the fascinating, yet challenging, aspects of modeling sea ice and its role in global climate is the sheer range of relevant length scales of structure, over ten orders of magnitude, from the submillimeter scale to hundreds of kilometers. In figure 1 we show examples of sea ice structure illustrating such a range of scales. Modeling sea ice on a large scale depends on some understanding of the physical properties of sea ice at the scale of individual floes, and even on the submillimeter scale since the brine phase in sea ice is such a key *determinant* of its bulk physical properties. Today's climate models challenge the most powerful supercomputers to their fullest capacity. However, even the largest computers still limit the horizontal resolution to tens of kilometers and require clever approximations and parametrizations to model the basic physics of sea ice. One of the central themes of this article is how to use information on smaller scales to predict behavior on larger scales. We observe that this central problem of climate science shares commonality with, for example, the key challenges in theoretical computations of the effective properties of composites.

Here we will explore some of the mathematics used in studying sea ice and its role in the climate system, particularly through the lens of sea ice albedo and processes related to its evolution.

## 2 Global Climate Models and Sea Ice

Global climate models, also known as general circulation models, are systems of PDEs derived from the basic laws of physics, chemistry, and fluid motion. They describe the state of the ocean, ice, atmosphere, and land, as well as the interactions between them. The equations are solved on very powerful computers using three-dimensional grids of the air–ice–ocean–land system, with horizontal grid sizes on the order of tens of kilometers. Consideration of general climate models will take us too far off course, but here we will briefly consider the sea ice components of these large-scale models.

The polar sea ice packs consist primarily of open water, thin first-year ice, thicker multiyear ice, and pressure ridges created by ice floes colliding with each other. The dynamic and thermodynamic characteristics of the ice pack depend largely on how much ice is in each thickness range. One of the most basic problems in sea ice modeling is thus to describe the evolution of the ice thickness distribution in space and time. The ice thickness distribution $g(x,t,h)\,\mathrm{d}h$ is defined (informally) as the fractional area covered by ice in the thickness range $(h, h + \mathrm{d}h)$ at a given time $t$ and location $x$. The fundamental equation controlling the evolution of the ice thickness distribution, which is solved numerically in sea ice models, is

$$\frac{\partial g}{\partial t} = -\nabla \cdot (g\boldsymbol{u}) - \frac{\partial}{\partial h}(\beta g) + \Psi,$$

where $\boldsymbol{u}$ is the horizontal ice velocity, $\beta$ is the rate of thermodynamic ice growth, and $\Psi$ is a ridging redistribution function that accounts for changes in ice thickness due to ridging and mechanical processes, as illustrated in figure 2.

The momentum equation, or Newton's second law for sea ice, can be deduced by considering the forces on a single floe, including interactions with other floes:

$$m\frac{\mathrm{D}\boldsymbol{u}}{\mathrm{D}t} = \nabla \cdot \boldsymbol{\sigma} + \boldsymbol{\tau}_{\mathrm{a}} + \boldsymbol{\tau}_{\mathrm{w}} - m\alpha\boldsymbol{n} \times \boldsymbol{u} - mg\nabla H,$$

where each term has units of force per unit area of the sea ice cover, $m$ is the combined mass of ice and snow

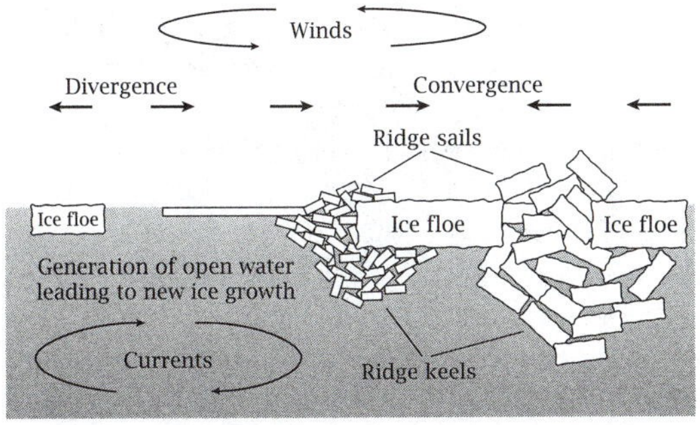


**Figure 2** Different factors contributing to the evolution of the ice thickness distribution $g(x,t,h)$. (Adapted, courtesy of Christian Haas.)

per unit area, $\boldsymbol{\tau}_a$ and $\boldsymbol{\tau}_w$ are wind and ocean stresses, and $D/Dt = (\partial/\partial t) + \boldsymbol{u}\cdot\nabla$ is the material or convective derivative. This is a two-dimensional equation obtained by integrating the three-dimensional equation through the thickness of the ice in the vertical direction.

The strength of the ice is represented by the internal stress tensor $\sigma_{ij}$. The other two terms on the right-hand side are, in order, stresses due to Coriolis effects and the sea surface slope, where $\boldsymbol{n}$ is a unit normal vector in the vertical direction, $\alpha$ is the Coriolis parameter, $H$ describes the sea surface, and in this equation $g$ is the acceleration due to gravity.

The temperature field $T(x,t)$ inside the sea ice (and snow layer), which couples to the ocean below and the atmosphere above through appropriate boundary conditions, satisfies an advection–diffusion equation

$$\frac{\partial T}{\partial t} = \nabla\cdot(D(T)\nabla T) - \boldsymbol{v}\cdot\nabla T,$$

where $D = K/\rho C$ is the thermal diffusivity of sea ice, $K$ is its thermal conductivity, $\rho$ is its bulk density, $C$ is the specific heat, and $\boldsymbol{v}$ is an averaged brine velocity field in the sea ice.

The bulk properties of low-Reynolds-number flow of brine of viscosity $\eta$ through sea ice can be related to the geometrical properties of the porous brine microstructure using HOMOGENIZATION THEORY [II.17]. The volume fractions of brine and ice are $\phi$ and $1-\phi$. The local velocity and pressure fields in the brine satisfy the Stokes equations for incompressible fluids, where the length scale of the microstructure (e.g., the period in periodic media) is $\varepsilon$. Under appropriate assumptions, in the homogenization limit as $\varepsilon \to 0$, the averaged velocity $\boldsymbol{v}(x)$ and pressure $p(x)$ satisfy Darcy's law and the incompressibility condition

$$\boldsymbol{v} = -\frac{1}{\eta}\boldsymbol{k}\nabla p, \qquad \nabla\cdot\boldsymbol{v} = 0. \tag{1}$$

Here, $\boldsymbol{k}$ is the permeability tensor, with vertical component $k_{zz} = k$ in units of $\mathrm{m}^2$. The permeability $\boldsymbol{k}$ is an example of an *effective* or *homogenized parameter*. The existence of the homogenized limits $\boldsymbol{v}$, $\boldsymbol{k}$, and $p$ in (1) can be proven under broad assumptions, such as for media with inhomogeneities that are periodic or have translation-invariant statistics.

Obtaining quantitative information on $\boldsymbol{k}$ or other effective transport coefficients—such as electrical or thermal conductivity and how they depend on, say, the statistical properties of the microstructure—is a central problem in the theory of composites. A broad range of techniques have been developed to obtain rigorous bounds, approximate formulas, and general theories of effective properties of composite and inhomogeneous media in terms of partial information about the microstructure. This problem is, of course, quite similar in nature to the fundamental questions of calculating bulk properties of matter from information on molecular interactions, which is central to statistical mechanics.

We note that it is also the case that one of the fundamental challenges of climate modeling is how to rigorously account for *sub-grid scale* processes and structures. That is, how do we incorporate important effects into climate models when the scale of the relevant phenomena being incorporated is far smaller than the grid size of the numerical model, which may be tens of kilometers. For example, it is obviously unrealistic to account for every detail of the submillimeter-scale brine microstructure in sea ice in a general circulation model! However, the volume fraction and connectedness properties of the brine phase control whether or not fluid can flow through the ice. The *on–off switch* for fluid flow in sea ice, known as the *rule of fives* (see below), in turn controls such critical processes as melt pond drainage, snow-ice formation (where sea water percolates upward, floods the snow layer on the sea ice surface, and subsequently freezes), the evolution of salinity profiles, and nutrient replenishment. It is the homogenized transport coefficient (the effective fluid permeability) that is incorporated into sea ice and climate models to account for these and related physical and biogeochemical processes. This effective coefficient is a well-defined parameter (under appropriate assumptions about the microstructure) that captures the relevant microstructural transitions and determines how a number of sea ice processes evolve. In this example we will see that rigorous mathematical methods can be employed to analyze effective sea ice behavior on length scales much greater than the submillimeter scale of the brine inclusions.

## 3 Mathematics of Composites

Here we give a brief overview of some of the mathematical models and techniques that are used in studying the effective properties of sea ice.

### 3.1 Percolation Theory

Percolation theory was initiated in 1957 with the introduction of a simple lattice model to study the flow of air through permeable sandstones used in miner's gas masks. In subsequent decades this theory has been used to successfully model a broad array of disordered materials and processes, including flow in porous media like rocks and soils; doped semiconductors; and various types of disordered conductors like piezoresistors, thermistors, radar-absorbing composites, carbon nanotube composites, and polar firn. The original percolation model and its generalizations have been the subject of intensive theoretical investigations, particularly in the physics and mathematics communities. One reason for the broad interest in the percolation model is that it is perhaps the simplest purely probabilistic model that exhibits a type of phase transition.

The simplest form of the lattice percolation model is defined as follows. Consider the $d$-dimensional integer lattice $\mathbb{Z}^d$, and the square or cubic network of bonds joining nearest-neighbor lattice sites. We assign to each bond a conductivity $\sigma_0 > 0$ (not to be confused with the stress tensor above) with probability $p$, meaning it is open, and a conductivity 0 with probability $1-p$, meaning it is closed. Two examples of lattice configurations are shown in figure 3, with $p = \frac{1}{3}$ in (a) and $p = \frac{2}{3}$ in (b). Groups of connected open bonds are called *open clusters*. In this model there is a critical probability $p_c$, $0 < p_c < 1$, called the *percolation threshold*, at which the average cluster size diverges and an infinite cluster appears. For the two-dimensional bond lattice, $p_c = \frac{1}{2}$. For $p < p_c$, the density of the infinite cluster $P_\infty(p)$ is 0, while for $p > p_c$, $P_\infty(p) > 0$ and near the threshold,

$$P_\infty(p) \sim (p - p_c)^\beta, \quad p \to p_c^+,$$

where $\beta$ is a universal critical exponent, that is, it depends only on dimension and not on the details of the lattice. Let $x, y \in \mathbb{Z}^d$ and let $\tau(x, y)$ be the probability that $x$ and $y$ belong to the same open cluster. The correlation length $\xi(p)$ is the mean distance between points on an open cluster, and it is a measure of the linear size of finite clusters. For $p < p_c$, $\tau(x, y) \sim e^{-|x-y|/\xi(p)}$, and $\xi(p) \sim (p_c - p)^{-\nu}$ diverges with a universal critical exponent $\nu$ as $p \to p_c^-$, as shown in figure 3(c).

The effective conductivity $\sigma^*(p)$ of the lattice, now viewed as a random resistor (or conductor) network, defined via Kirchhoff's laws, vanishes for $p < p_c$ as does $P_\infty(p)$ since there are no infinite pathways, as shown in figure 3(e). For $p > p_c$, $\sigma^*(p) > 0$, and near $p_c$,

$$\sigma^*(p) \sim \sigma_0(p - p_c)^t, \quad p \to p_c^+,$$

where $t$ is the conductivity critical exponent, with $1 \leqslant t \leqslant 2$ if $d = 2, 3$ (for an idealized model), and numerical values $t \approx 1.3$ if $d = 2$ and $t \approx 2.0$ if $d = 3$. Consider a random pipe network with effective fluid permeability $k^*(p)$ exhibiting similar behavior $k^*(p) \sim k_0(p - p_c)^e$, where $e$ is the permeability critical exponent, with $e = t$ for lattices. Both $t$ and $e$ are believed to be universal; that is, they depend only on dimension and not on the type of lattice. Continuum models, like the so-called *Swiss cheese model*, can exhibit nonuniversal behavior with exponents different from the lattice case and $e \neq t$.

### 3.2 Analytic Continuation and Spectral Measures

*Homogenization* is where one seeks to find a homogeneous medium that behaves the same macroscopically as a given inhomogeneous medium. The methods are focused on finding the effective properties of inhomogeneous media such as composites. We will see that the *spectral measure* in a Stieltjes integral representation for the effective parameter provides a powerful tool for upscaling geometrical information about a composite into calculations of effective properties.

We now briefly describe the *analytic continuation method* for studying the effective transport properties of composite materials. This method has been used to obtain rigorous bounds on effective transport coefficients of two-component and multicomponent composite materials. The bounds follow from the special analytic structure of the representations for the effective parameters and from partial knowledge of the microstructure, such as the relative volume fractions of the phases in the case of composite media. The analytic continuation method was later adapted to treating the effective diffusivity of passive tracers in incompressible fluid velocity fields.

We consider the effective complex permittivity tensor $\boldsymbol{\varepsilon}^*$ of a two-phase random medium, although the

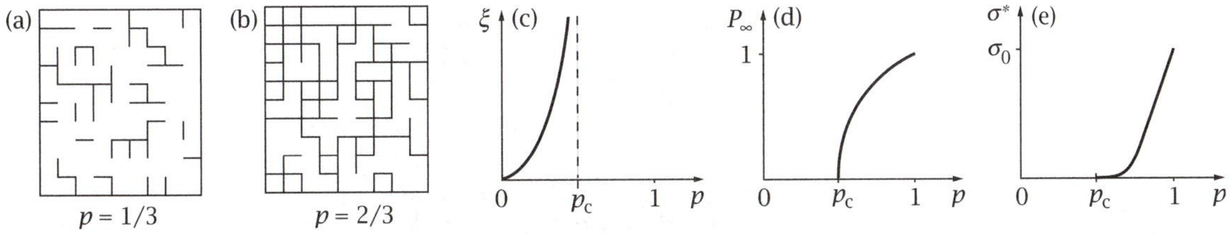


**Figure 3** The two-dimensional square lattice percolation model (a) below its percolation threshold of $p_c = \frac{1}{2}$ and (b) above it (courtesy of Salvatore Torquato). (c) Divergence of the correlation length as $p$ approaches $p_c$. (d) The infinite cluster density of the percolation model, and (e) the effective conductivity.

method applies to any classical transport coefficient. Here, $\varepsilon(\boldsymbol{x},\omega)$ is a (spatially) stationary random field in $\boldsymbol{x} \in \mathbb{R}^d$ and $\omega \in \Omega$, where $\Omega$ is the set of all geometric realizations of the medium, which is indexed by the parameter $\omega$ representing one particular realization, and the underlying probability measure $P$ is compatible with stationarity.

As in sea ice, we assume we are dealing with a two-phase locally isotropic medium, so that the components $\varepsilon_{jk}$ of the local permittivity tensor $\boldsymbol{\varepsilon}$ satisfy $\varepsilon_{jk}(\boldsymbol{x},\omega) = \varepsilon(\boldsymbol{x},\omega)\delta_{jk}$, where $\delta_{jk}$ is the Kronecker delta and

$$\varepsilon(\boldsymbol{x},\omega) = \varepsilon_1\chi_1(\boldsymbol{x},\omega) + \varepsilon_2\chi_2(\boldsymbol{x},\omega). \tag{2}$$

Here, $\varepsilon_j$ is the *complex* permittivity for medium $j = 1, 2$, and $\chi_j(\boldsymbol{x},\omega)$ is its characteristic function, equaling 1 for $\omega \in \Omega$ with medium $j$ at $\boldsymbol{x}$, and 0 otherwise, with $\chi_2 = 1 - \chi_1$.

When the wavelength is much larger than the scale of the composite microstructure, the propagation properties of an electromagnetic wave in a given composite medium are determined by the quasistatic limit of Maxwell's equations:

$$\nabla \times \boldsymbol{E} = 0, \qquad \nabla \cdot \boldsymbol{D} = 0, \tag{3}$$

where $\boldsymbol{E}(\boldsymbol{x},\omega)$ and $\boldsymbol{D}(\boldsymbol{x},\omega)$ are stationary electric and displacement fields, respectively, related by the local constitutive equation $\boldsymbol{D}(\boldsymbol{x}) = \varepsilon(\boldsymbol{x})\boldsymbol{E}(\boldsymbol{x})$, and $\boldsymbol{e}_k$ is a standard basis vector in the $k$th direction. The electric field is assumed to have unit strength on average, with $\langle \boldsymbol{E} \rangle = \boldsymbol{e}_k$, where $\langle\cdot\rangle$ denotes ensemble averaging over $\Omega$ or spatial averaging over all of $\mathbb{R}^d$. The effective complex permittivity tensor $\boldsymbol{\varepsilon}^*$ is defined by

$$\langle \boldsymbol{D} \rangle = \boldsymbol{\varepsilon}^* \langle \boldsymbol{E} \rangle, \tag{4}$$

which is a homogenized version of the local constitutive relation $\boldsymbol{D} = \varepsilon \boldsymbol{E}$.

For simplicity, we focus on one diagonal coefficient $\varepsilon^* = \varepsilon^*_{kk}$, with $\varepsilon^* = \langle \varepsilon \boldsymbol{E} \cdot \boldsymbol{e}_k \rangle$. By the homogeneity of $\varepsilon(\boldsymbol{x},\omega)$ in (2), $\varepsilon^*$ depends on the contrast parameter $h = \varepsilon_1/\varepsilon_2$, and we define $m(h) = \varepsilon^*/\varepsilon_2$, which is a Herglotz function that maps the upper half $h$-plane to the upper half-plane and is analytic in the entire complex $h$-plane except for the negative real axis $(-\infty, 0]$.

The key step in the method is obtaining a Stieltjes integral representation for $\varepsilon^*$. This integral representation arises from a resolvent representation of the electric field $\boldsymbol{E} = s(sI - \Gamma\chi_1)^{-1}\boldsymbol{e}_k$, where $\Gamma = \nabla(\Delta^{-1})\nabla\cdot$ acts as a projection from $L^2(\Omega, P)$ onto the Hilbert space of curl-free random fields, and $\Delta^{-1}$ is based on convolution with the free-space Green function for the Laplacian $\Delta = \nabla^2$. Consider the function $F(s) = 1 - m(h)$, $s = 1/(1-h)$, which is analytic off $[0,1]$ in the $s$-plane. Then, writing $F(s) = \langle \chi_1[(sI - \Gamma\chi_1)^{-1}\boldsymbol{e}_k] \cdot \boldsymbol{e}_k \rangle$ yields

$$F(s) = \int_0^1 \frac{\mathrm{d}\mu(\lambda)}{s - \lambda}, \tag{5}$$

where $\mu(\mathrm{d}\lambda) = \langle \chi_1 Q(\mathrm{d}\lambda)\boldsymbol{e}_k \cdot \boldsymbol{e}_k \rangle$ is a positive *spectral measure* on $[0,1]$, and $Q(\mathrm{d}\lambda)$ is the (unique) projection valued measure associated with the bounded, self-adjoint operator $\Gamma\chi_1$.

Equation (5) is based on the spectral theorem for the resolvent of the operator $\Gamma\chi_1$. It provides a Stieltjes integral representation for the effective complex permittivity $\varepsilon^*$ that separates the component parameters in $s$ from the complicated geometrical information contained in the measure $\mu$. (Extensions of (5) to multicomponent media with $\varepsilon = \varepsilon_1\chi_1 + \varepsilon_2\chi_2 + \varepsilon_3\chi_3 + \cdots + \varepsilon_n\chi_n$ involve several complex variables.) Information about the geometry enters through the moments

$$\mu_n = \int_0^1 \lambda^n \,\mathrm{d}\mu(\lambda) = \langle \chi_1[(\Gamma\chi_1)^n \boldsymbol{e}_k] \cdot \boldsymbol{e}_k \rangle, \qquad n = 0, 1, 2, \dots.$$

For example, the mass $\mu_0$ is given by $\mu_0 = \langle \chi_1 \boldsymbol{e}_k \cdot \boldsymbol{e}_k \rangle = \langle \chi_1 \rangle = \phi$, where $\phi$ is the volume or area fraction of material of phase 1, and $\mu_1 = \phi(1-\phi)/d$ if the material

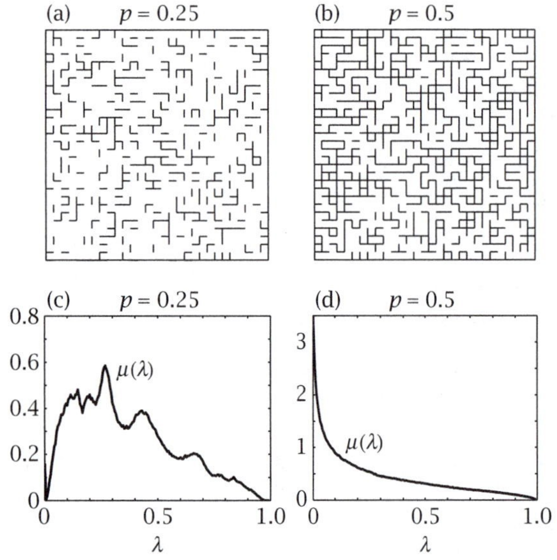


**Figure 4** Realizations of the two-dimensional lattice percolation model are shown in (a) and (b), and the corresponding spectral functions (averaged over 5000 random realizations) are shown in (c) and (d). In (c), there is a spectral gap around $\lambda = 1$, indicating the lack of long-range order or connectedness. The gap collapses in (d) when the percolation threshold of $p = p_c = 0.5$ has been reached, and the system exhibits long-range connectedness. Note the difference in vertical scale in the graphs in (c) and (d).

is statistically isotropic. In general, $\mu_n$ depends on the $(n+1)$-point correlation function of the medium.

Computing the spectral measure $\mu$ for a given composite microstructure involves first discretizing a two-phase image of the composite into a square lattice filled with 1s and 0s corresponding to the two phases. The key operator $\Gamma\chi_1$, which depends on the geometry via $\chi_1$, then becomes a self-adjoint matrix. The spectral measure may be calculated directly from the eigenvalues and eigenvectors of this matrix. Examples of these spectral measures for the percolation model on the two-dimensional square lattice are shown in figure 4.

## 4 Applications to Sea Ice

### 4.1 Percolation Theory

Given a sample of sea ice at temperature $T$ in degrees Celsius and bulk salinity $S$ in parts per thousand (ppt), the brine volume fraction $\phi$ is given (approximately) by the equation of Frankenstein and Garner:

$$\phi = \frac{S}{1000}\left(\frac{49.185}{|T|} + 0.532\right). \tag{6}$$

As temperature increases for fixed salinity, the volume fraction $\phi$ of liquid brine in the ice also increases. The inclusions become larger and more connected, as illustrated in parts (a)–(c) of plate 6, which show images of the brine phase in sea ice (in gold) obtained from X-ray tomography scans of sea ice single crystals.

As the connectedness of the brine phase increases with rising temperature, the ease with which fluid can flow through sea ice—its fluid permeability—should increase as well. In fact, sea ice exhibits a *percolation threshold*, or critical brine volume fraction $\phi_c$, or critical temperature $T_c$, below which columnar sea ice is effectively impermeable to vertical fluid flow and above which the ice is permeable, and increasingly so as temperature rises. This critical behavior of fluid transport in sea ice is illustrated in plate 6(d). The data on the vertical fluid permeability $k(\phi)$ display a rapid rise just above a threshold value of about $\phi_c \approx 0.05$ or 5%, similar to the conductivity (or permeability) in figure 3(e). This type of behavior is also displayed by data on brine drainage, with the effects of drainage shutting down for brine volume fractions below about 5%. Roughly speaking, we can refer to this phenomenon as the *on–off switch* for fluid flow in sea ice. Through the Frankenstein–Garner relation in (6), the critical brine volume fraction $\phi_c \approx 0.05$ corresponds to a critical temperature $T_c \approx -5$ °C, for a typical salinity of 5 ppt. This important threshold behavior has therefore become known as the *rule of fives*.

In view of this type of critical behavior, it is reasonable to try to find a theoretical percolation explanation. However, with $p_c \approx 0.25$ for the $d = 3$ cubic bond lattice, it was apparent that key features of the geometry of the brine microstructure in sea ice were being missed by lattices. The threshold $\phi_c \approx 0.05$ was identified with the critical probability in a continuum percolation model for compressed powders that exhibit microstructural characteristics similar to sea ice. The identification explained the rule of fives, as well as data on algal growth and snow-ice production. The compressed powders shown in figure 5 were used in the development of so-called *stealthy* or radar-absorbing composites.

When we applied the compressed powder model to sea ice, we had no *direct* evidence that the brine microstructure undergoes a transition in connectedness at a critical brine volume fraction. This lack of

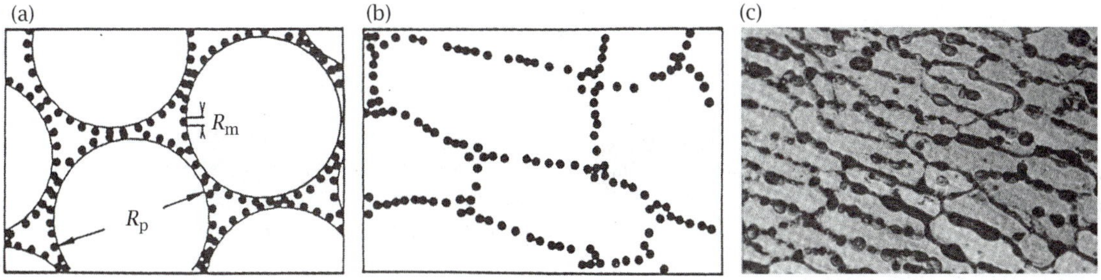


**Figure 5** (a) A powder of large polymer spheres mixed with smaller metal spheres. (b) When the powder is compressed, its microstructure is similar to that of the sea ice in (c). (Parts (a) and (b) are adapted from Golden, K. M., S. F. Ackley, and V. I. Lytle. *Science* 18 December 1998:282 (5397), 2238–2241. Part (c) is adapted from CRREL report 87-20 (October 1987).)

evidence was partly due to the difficulty of imaging and quantitatively characterizing the brine inclusions in three dimensions, particularly the thermal evolution of their connectivity. Through X-ray computed tomography and pore structure analysis we have now analyzed the critical behavior of the thermal evolution of brine connectedness in sea ice single crystals over a temperature range from −18 °C to −3 °C. We have mapped three-dimensional images of the pores and throats in the brine phase onto graphs of nodes and edges, and analyzed their connectivities as functions of temperature and sample size. Realistic network models of brine inclusions can be derived from porous media analysis of three-dimensional microtomography images. Using finite-size scaling techniques largely confirms the rule of fives, as well as confirming that sea ice is a natural material that exhibits a strong anisotropy in percolation thresholds.

Now we consider the application of percolation theory to understanding the fluid permeability of sea ice. In the continuum, the permeability and conductivity exponents $e$ and $t$ can take nonuniversal values and need not be equal, as in the case of the three-dimensional Swiss cheese model. Continuum models have been studied by mapping to a lattice with a probability density $\psi(g)$ of bond conductances $g$. Nonuniversal behavior can be obtained when $\psi(g)$ is singular as $g \to 0^+$. However, for a lognormal conductance distribution arising from intersections of lognormally distributed inclusions, as in sea ice, the behavior is *universal*. Thus $e \approx 2$ for sea ice.

The permeability scaling factor $k_0$ for sea ice is estimated using *critical path analysis*. For media with $g$ in a wide range, the overall behavior is dominated by a critical *bottleneck* conductance $g_c$, the smallest conductance such that the critical path $\{g\colon g \geqslant g_c\}$ spans the sample. With most brine channel diameters between 1.0 mm and 1.0 cm, spanning fluid paths have a smallest characteristic radius $r_c \approx 0.5$ mm, and we estimate $k_0$ by the pipe-flow result $r_c^2/8$. Thus,

$$k(\phi) \sim 3(\phi - \phi_c)^2 \times 10^{-8}\ \mathrm{m}^2, \quad \phi \to \phi_c^+. \tag{7}$$

In plate 6(f), field data with $\phi$ in [0.055, 0.15], just above $\phi_c \approx 0.05$, are compared with (7) and show close agreement. The striking result that, for sea ice, $e \approx 2$, the universal *lattice* value in three dimensions, is due to the general lognormal structure of the brine inclusion distribution function. The general nature of our results suggests that similar types of porous media, such as saline ice on extraterrestrial bodies, may also exhibit universal critical behavior.

### 4.2 Analytic Continuation

#### 4.2.1 *Bounds on the Effective Complex Permittivity*

Bounds on $\varepsilon^*$, or $F(s)$, are obtained by fixing $s$ in (5), varying over admissible measures $\mu$ (or admissible geometries), such as those that satisfy only

$$\mu_0 = \phi, \tag{8}$$

and finding the corresponding range of values of $F(s)$ in the complex plane. Two types of bounds on $\varepsilon^*$ are obtained. The first bound $R_1$ assumes only that the relative volume fractions $p_1 = \phi$ and $p_2 = 1 - p_1$ of the brine and ice are known, so that (8) is satisfied. In this case, the admissible set of measures forms a compact, convex set. Since (5) is a linear functional of $\mu$, the extreme values of $F$ are attained by extreme points of the set of admissible measures, which are the Dirac point measures of the form $p_1\delta_z$. The values of $F$ must lie inside the circle $p_1/(s - z)$, $-\infty \leqslant z \leqslant \infty$, and the region $R_1$ is bounded by circular arcs, one of which is parametrized in the $F$-plane by

$$C_1(z) = \frac{p_1}{s - z}, \quad 0 \leqslant z \leqslant p_2. \tag{9}$$

To display the other arc, it is convenient to use the auxiliary function

$$E(s) = 1 - \frac{\varepsilon_1}{\varepsilon^*} = \frac{1 - sF(s)}{s(1 - F(s))}, \tag{10}$$

which is a Herglotz function like $F(s)$, analytic off $[0,1]$. Then, in the $E$-plane, we can parametrize the other circular boundary of $R_1$ by

$$\hat{C}_1(z) = \frac{p_2}{s - z}, \quad 0 \leqslant z \leqslant p_1. \tag{11}$$

In the $\varepsilon^*$-plane, $R_1$ has vertices $V_1 = \varepsilon_1/(1 - \hat{C}_1(0)) = (p_1/\varepsilon_1 + p_2/\varepsilon_2)^{-1}$ and $W_1 = \varepsilon_2(1 - C_1(0)) = p_1\varepsilon_1 + p_2\varepsilon_2$, and collapses to the interval

$$(p_1/\varepsilon_1 + p_2/\varepsilon_2)^{-1} \leqslant \varepsilon^* \leqslant p_1\varepsilon_1 + p_2\varepsilon_2 \tag{12}$$

when $\varepsilon_1$ and $\varepsilon_2$ are real, which are the classical arithmetic (upper) and harmonic (lower) mean bounds, also called the elementary bounds. The complex elementary bounds (9) and (11) are optimal and can be attained by a composite of uniformly aligned spheroids of material 1 in all sizes coated with confocal shells of material 2, and vice versa. These arcs are traced out as the aspect ratio varies.

If the material is further assumed to be statistically isotropic, i.e., if $\varepsilon^*_{ik} = \varepsilon^* \delta_{ik}$, then $\mu_1 = \phi(1-\phi)/d$ must be satisfied as well. A convenient way of including this information is to use the transformation

$$F_1(s) = \frac{1}{p_1} - \frac{1}{sF(s)}. \tag{13}$$

The function $F_1(s)$ is, again, a Herglotz function, which has the representation

$$F_1(s) = \int_0^1 \frac{\mathrm{d}\mu^1(z)}{s - z}.$$

The constraint $\mu_1 = \phi(1 - \phi)/d$ on $F(s)$ is then transformed to a restriction of only the mass, or zeroth moment $\mu_0^1$, of $\mu^1$, with

$$\mu_0^1 = p_2/p_1 d.$$

Applying the same procedure as for $R_1$ yields a region $R_2$ whose boundaries are again circular arcs. When $\varepsilon_1$ and $\varepsilon_2$ are real with $\varepsilon_1 \geqslant \varepsilon_2$, the region collapses to a real interval, whose endpoints are known as the Hashin–Shtrikman bounds. We remark that higher-order correlation information can be conveniently incorporated by iterating (13).

### 4.2.2 *Inverse Homogenization*

It has been shown that the spectral measure $\mu$, which contains all geometrical information about a composite, can be uniquely reconstructed if measurements of the effective permittivity $\varepsilon^*$ are available on an arc in the complex $s$-plane. If the component parameters depend on frequency $\omega$ (not to be confused with realizations of the random medium above), variation of $\omega$ in an interval $(\omega_1, \omega_2)$ gives the required data. The reconstruction of $\mu$ can be reduced to an inverse potential problem. Indeed, $F(s)$ admits a representation through a logarithmic potential $\Phi$ of the measure $\mu$,

$$F(s) = \frac{\partial \Phi}{\partial s}, \qquad \Phi(s) = \int_0^1 \ln|s - z| \,\mathrm{d}\mu(z), \tag{14}$$

where $\partial/\partial s = \partial/\partial x - \mathrm{i}\partial/\partial y$. The potential $\Phi$ satisfies the Poisson equation $\Delta\Phi = -\rho$, where $\rho(z)$ is a density on $[0,1]$. A solution to the forward problem is given by the Newtonian potential with $\mu(\mathrm{d}z) = \rho(z)\,\mathrm{d}z$. The inverse problem is to find $\rho(z)$ (or $\mu$) given values of $\Phi$, $\partial\Phi/\partial n$, or $\nabla\Phi$. The inverse problem is ill-posed and requires REGULARIZATION [IV.4 §7] to develop a stable numerical algorithm.

When frequency $\omega$ varies across $(\omega_1, \omega_2)$, the complex parameter $s$ traces an arc $C$ in the $s$-plane. Let $A$ be the integral operator in (14),

$$A\mu = \frac{\partial}{\partial s} \int_0^1 \ln|s - \lambda| \,\mathrm{d}\mu(\lambda),$$

mapping the set of measures $\mathcal{M}[0,1]$ on the unit interval onto the set of derivatives of complex potentials defined on a curve $C$. To construct the solution we consider the problem of minimizing $\|A\mu - F\|^2$ over $\mu \in \mathcal{M}$, where $\|\cdot\|$ is the $L^2(C)$-norm, $F(s)$ is the measured data, and $s \in C$. The solution does not depend continuously on the data, and regularization based on constrained minimization is needed. Instead of $\|A\mu - F\|^2$ being minimized over all functions in $\mathcal{M}$, it is minimized over a convex subset satisfying $J(\mu) \leqslant \beta$ for a stabilizing functional $J(\mu)$ and some $\beta > 0$. The advantage of using quadratic $J(\mu) = \|L\mu\|^2$ is the linearity of the corresponding Euler equation, resulting in efficiency of the numerical schemes. However, the reconstructed solution necessarily possesses a certain smoothness. Nonquadratic stabilization imposes constraints on the variation of the solution. The total variation penalization, as well as a nonnegativity constraint, does not imply smoothness, permitting more general recovery, including the important Dirac measures.

We have also solved a reduced inverse spectral problem exactly by bounding the volume fraction of the constituents, an inclusion separation parameter $q$, and the spectral gap of $\Gamma\chi_1$. We developed an algorithm based on the Möbius transformation structure of the forward bounds whose output is a set of algebraic curves in parameter space bounding regions of admissible parameter values. These results advance the development of techniques for characterizing the microstructure of composite materials, and they have been applied to sea ice to demonstrate electromagnetically that the brine inclusion separations vanish as the percolation threshold is approached.

## 5 Geometry of the Marginal Ice Zone

Dense pack ice transitions to open ocean over a region of broken ice termed the marginal ice zone (MIZ), a highly dynamic region in which the ice cover lies close to an open ocean boundary and where intense atmosphere–ice–ocean interactions take place. The width of the MIZ is a fundamental length scale for polar dynamics, in part because it represents the distance over which ocean waves and swell penetrate into the sea ice cover. Wave penetration can break a smooth ice layer into floes, meaning that the MIZ acts as a buffer zone that protects the stable morphology of the inner ice. Waves also promote the formation of pancake ice, as shown in plate 7. Moreover, the width of the MIZ is an important spatial dimension of the marine polar habitat and impacts human accessibility to high latitudes. Using a conformal mapping method to quantify MIZ width (see below), a dramatic 39% widening of the summer Arctic MIZ, based on three decades of satellite-derived data (1979–2012), has been reported.

Challenges associated with objective measurement of the MIZ width include the MIZ's shape, which is in general not geodesically convex, as illustrated by the shaded example in plate 8(a). Sea ice concentration ($c$) is used here to define the MIZ as a body of marginal ice ($0.15 \leqslant c \leqslant 0.80$) adjoining both pack ice ($c > 0.80$) and sparse ice ($c < 0.15$). To define an objective MIZ width applicable to such shapes, an idealized sea ice concentration field $\psi(x, y)$ satisfying LAPLACE'S EQUATION [III.18] within the MIZ,

$$\nabla^2 \psi = 0, \tag{15}$$

was introduced. We use $(x, y)$ to denote a point in two-dimensional space, and it is understood that we are working on the spherical Earth. Boundary conditions for (15) are $\psi = 0.15$ where MIZ borders a sparse ice region and $\psi = 0.80$ where the MIZ borders a pack ice region. The solutions to (15) for the examples in parts (a) and (b) of plate 8 are illustrated by colored shading. Any curve $\gamma$ orthogonal to the level curves of $\psi$ and connecting two points on the MIZ perimeter (a black field line through the gradient field $\nabla\psi$, as in plate 8(b)) is contained in the MIZ, and its length provides an objective measure of MIZ width ($\ell$). Defined in this way, $\ell$ is a function of distance along the MIZ perimeter ($s$) from an arbitrary starting point, and this dependence is denoted by $\ell = \ell(s)$. Analogous applications of Laplace's equation have been introduced in medical imaging to measure the width or thickness of human organs.

Derivatives in (15) were numerically approximated using second-order finite differences, and solutions were obtained in the data's native stereographic projection since solutions of Laplace's equation are invariant under conformal mapping. For a given day and MIZ, a summary measure of MIZ width ($w$) can be defined by averaging $\ell$ with respect to distance along the MIZ perimeter:

$$w = \frac{1}{L_M} \int_M \ell(s)\, \mathrm{d}s, \tag{16}$$

where $M$ is the closed curve defining the MIZ perimeter and $L_M$ is the length of $M$. Averaging $w$ over July–September of each available year reveals the dramatic widening of the summer MIZ, as illustrated in plate 8(c).

## 6 Geometry of Arctic Melt Ponds

From the first appearance of visible pools of water, often in early June, the area fraction $\phi$ of sea ice covered by melt ponds can increase rapidly to over 70% in just a few days. Moreover, the accumulation of water at the surface dramatically lowers the albedo where the ponds form. There is a corresponding critical drop-off in average albedo. The resulting increase in solar absorption in the ice and upper ocean accelerates melting, possibly triggering *ice–albedo feedback*. Similarly, an increase in open-water fraction lowers albedo, thus increasing solar absorption and subsequent melting. The spatial coverage and distribution of melt ponds on the surface of ice floes and the open water between the floes thus exerts primary control of ice pack albedo and the partitioning of solar energy in the ice–ocean system. Given the critical role of ice–albedo feedback in the recent losses of Arctic sea ice, ice pack albedo and the formation and evolution of melt ponds are of significant interest in climate modeling.

While melt ponds form a key component of the Arctic marine environment, comprehensive observations or theories of their formation, coverage, and evolution remain relatively sparse. Available observations of melt ponds show that their areal coverage is highly variable, particularly for first-year ice early in the melt season, with rates of change as high as 35% per day. Such

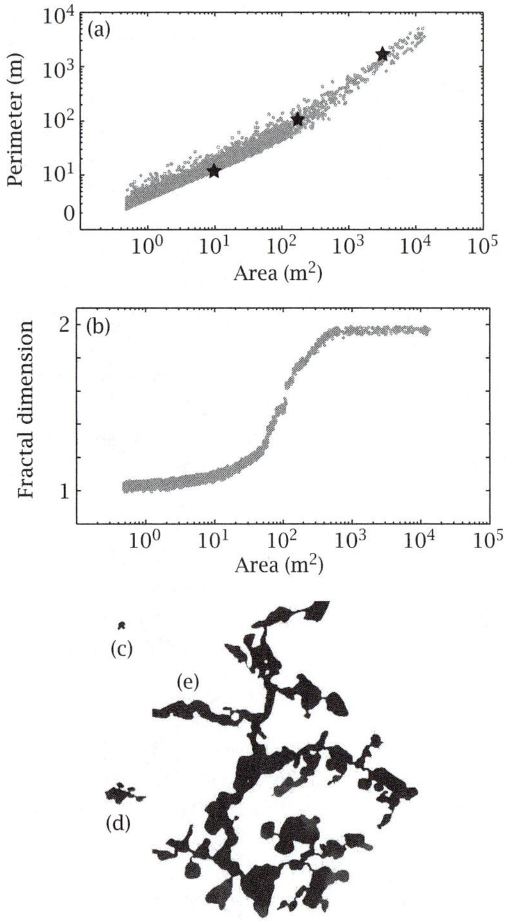


**Figure 6** (a) Area-perimeter data for 5269 Arctic melt ponds, plotted on logarithmic scales. (b) Melt pond fractal dimension $D$ as a function of area $A$, computed from the data in (a). Ponds corresponding to the three black stars in (a), from left to right, are denoted by (c), (d), and (e), respectively, in the bottom diagram. The transitional pond in (d) has a horizontal scale of about 30 m. (Adapted from Hohenegger, C., B. Alali, K. R. Steffen, D. K. Perovich, and K. M. Golden. 2012. Transition in the fractal geometry of Arctic melt ponds. *The Cryosphere* 6:1157--62 (doi:10.5194/tc-6-1157-2012).)

variability, as well as the influence of many competing factors controlling the evolution of melt ponds and ice floes, makes the incorporation of realistic treatments of albedo into climate models quite challenging. Small- and medium-scale models of melt ponds that include some of these mechanisms have been developed, and melt pond parametrizations are being incorporated into global climate models.

The surface of an ice floe is viewed here as a two-phase composite of dark melt ponds and white snow or ice. The onset of ponding and the rapid increase in coverage beyond the initial threshold is similar to critical phenomena in statistical physics and composite materials. It is natural, therefore, to ask if the evolution of melt pond geometry exhibits universal characteristics that do not necessarily depend on the details of the driving mechanisms in numerical melt pond models. Fundamentally, the melting of Arctic sea ice is a phase-transition phenomenon, where a solid turns to liquid, albeit on large regional scales and over a period of time that depends on environmental forcing and other factors. We thus look for features of melt pond evolution that are mathematically analogous to related phenomena in the theories of phase transitions and composite materials. As a first step in this direction, we consider the evolution of the geometric complexity of Arctic melt ponds.

By analyzing area-perimeter data from hundreds of thousands of melt ponds, we have discovered an unexpected separation of scales, where the pond fractal dimension $D$ exhibits a transition from 1 to 2 around a critical length scale of 100 $m^2$ in area, as shown in figure 6. Small ponds with simple boundaries coalesce or percolate to form larger connected regions. Pond complexity increases rapidly through the transition region and reaches a maximum for ponds larger than 1000 $m^2$, whose boundaries resemble space-filling curves with $D \approx 2$. These configurations affect the complex radiation fields under melting sea ice, the heat balance of sea ice and the upper ocean, under-ice phytoplankton blooms, biological productivity, and biogeochemical processes.

Melt pond evolution also appears to exhibit a *percolation threshold*, where one phase in a composite becomes connected on macroscopic scales as some parameter exceeds a critical value. An important example of this phenomenon in the microphysics of sea ice (discussed above), which is fundamental to the process of melt pond drainage, is the percolation transition exhibited by the brine phase in sea ice, or the *rule of fives* discussed on page 697. When the brine volume fraction of columnar sea ice exceeds approximately 5%, the brine phase becomes macroscopically connected so that fluid pathways allow flow through the porous microstructure of the ice. Similarly, even casual inspection of the aerial photos in plate 9 shows that the melt pond phase of sea ice undergoes a percolation transition where disconnected ponds evolve into much larger-scale connected structures with complex boundaries. Connectivity of melt ponds promotes further melting and the breakup of floes, as well as horizontal transport of meltwater and drainage through cracks, leads, and seal holes.

**Acknowledgments.** I gratefully acknowledge support from the Division of Mathematical Sciences and the Division of Polar Programs at the U.S. National Science Foundation (NSF) through grants DMS-1009704, ARC-0934721, DMS-0940249, and DMS-1413454. I am also grateful for support from the Arctic and Global Prediction Program and the Applied Computational Analysis Program at the Office of Naval Research through grants N00014-13-10291 and N00014-12-10861. Finally, I would like to thank the NSF Math Climate Research Network for their support of this work, as well as many colleagues and students who contributed so much to the research represented here, especially Steve Ackley, Hajo Eicken, Don Perovich, Tony Worby, Court Strong, Elena Cherkaev, Jingyi Zhu, Adam Gully, Ben Murphy, and Christian Sampson.

### Further Reading

Avellaneda, M., and A. Majda. 1989. Stieltjes integral representation and effective diffusivity bounds for turbulent transport. *Physical Review Letters* 62:753–55.

Bergman, D. J., and D. Stroud. 1992. Physical properties of macroscopically inhomogeneous media. *Solid State Physics* 46:147–269.

Cherkaev, E. 2001. Inverse homogenization for evaluation of effective properties of a mixture. *Inverse Problems* 17: 1203–18.

Eisenman, I., and J. S. Wettlaufer. 2009. Nonlinear threshold behavior during the loss of Arctic sea ice. *Proceedings of the National Academy of Sciences of the USA* 106(1):28–32.

Feltham, D. L. 2008. Sea ice rheology. *Annual Review of Fluid Mechanics* 40:91–112.

Flocco, D., D. L. Feltham, and A. K. Turner. 2010. Incorporation of a physically based melt pond scheme into the sea ice component of a climate model. *Journal of Geophysical Research* 115:C08012.

Golden, K. M. 2009. Climate change and the mathematics of transport in sea ice. *Notices of the American Mathematical Society* 56(5):562–84 (and issue cover).

Hohenegger, C., B. Alali, K. R. Steffen, D. K. Perovich, and K. M. Golden. 2012. Transition in the fractal geometry of Arctic melt ponds. *The Cryosphere* 6:1157–62.

Hunke, E. C., and W. H. Lipscomb. 2010. CICE: the Los Alamos sea ice model. Documentation and Software User's Manual, version 4.1. LA-CC-06-012, t-3 Fluid Dynamics Group, Los Alamos National Laboratory.

Milton, G. W. 2002. *Theory of Composites*. Cambridge: Cambridge University Press.

Orum, C., E. Cherkaev, and K. M. Golden. 2012. Recovery of inclusion separations in strongly heterogeneous composites from effective property measurements. *Proceedings of the Royal Society of London* A 468:784–809.

Perovich, D. K., J. A. Richter-Menge, K. F. Jones, and B. Light. 2008. Sunlight, water, and ice: extreme Arctic sea ice melt during the summer of 2007. *Geophysical Research Letters* 35:L11501.

Stauffer, D., and A. Aharony. 1992. *Introduction to Percolation Theory*, 2nd edn. London: Taylor & Francis.

Stroeve, J., M. M. Holland, W. Meier, T. Scambos, and M. Serreze. 2007. Arctic sea ice decline: faster than forecast. *Geophysical Research Letters* 34:L09501.

Strong, C., and I. G. Rigor. 2013. Arctic marginal ice zone trending wider in summer and narrower in winter. *Geophysical Research Letters* 40(18):4864–68.

Thomas, D. N., and G. S. Dieckmann, editors. 2009. *Sea Ice*, 2nd edn. Oxford: Wiley-Blackwell.

Thompson, C. J. 1988. *Classical Equilibrium Statistical Mechanics*. Oxford: Oxford University Press.

Torquato, S. 2002. *Random Heterogeneous Materials: Microstructure and Macroscopic Properties*. New York: Springer.

Untersteiner, N. 1986. *The Geophysics of Sea Ice*. New York: Plenum.

Washington, W., and C. L. Parkinson. 2005. *An Introduction to Three-Dimensional Climate Modeling*, 2nd edn. Herndon, VA: University Science Books.

## V.18 Numerical Weather Prediction

*Peter Lynch*

### 1 Introduction

The development of computer models for numerical simulation and prediction of the atmosphere and oceans is one of the great scientific triumphs of the past fifty years. Today, numerical weather prediction (NWP) plays a central and essential role in operational weather forecasting, with forecasts now having accuracy at ranges beyond a week. There are several reasons for this: enhancements in model resolution, better numerical schemes, more realistic parametrizations of physical processes, new observational data from satellites, and more sophisticated methods of determining the initial conditions. In this article we focus on the fundamental equations, the formulation of the numerical algorithms, and the variational approach to data assimilation. We present the mathematical principles of NWP and illustrate the process by considering some specific models and their application to practical forecasting.

### 2 The Basic Equations

The atmosphere is governed by the fundamental laws of physics, expressed in terms of mathematical equations. They form a system of coupled nonlinear partial differential equations (PDEs). These equations can be used to predict the evolution of the atmosphere and to simulate its long-term behavior.

The primary variables are the fluid velocity $\boldsymbol{V}$ (with three components, $u$ eastward, $v$ northward, and $w$ upward), pressure $p$, density $\rho$, temperature $T$, and humidity $q$. Using Newton's laws of motion and the principles of conservation of energy and mass, we can obtain a system whose solution is well determined by the initial conditions.

The central components of the system, governing fluid motion, are the NAVIER–STOKES EQUATIONS [III.23]. We write them in vector form:

$$\frac{\partial \boldsymbol{V}}{\partial t} + \boldsymbol{V} \cdot \nabla \boldsymbol{V} + 2\boldsymbol{\Omega} \times \boldsymbol{V} + \frac{1}{\rho}\nabla p = \boldsymbol{F} + \boldsymbol{g}.$$

The equations are relative to the rotating Earth and $\boldsymbol{\Omega}$ is the Earth's angular velocity. In order, the terms of this equation represent local acceleration, nonlinear advection, Coriolis term, pressure gradient, friction, and gravity. The friction term $\boldsymbol{F}$ is small in the free atmosphere but is crucially important in the boundary layer (roughly, the first 1 km above the Earth's surface). The apparent gravity $\boldsymbol{g}$ includes the centrifugal force, which depends only on position.

The temperature, pressure, and density are linked through the equation of state

$$p = R\rho T,$$

where $R$ is the gas constant for dry air. In practice, a slight elaboration of this is used that takes account of moisture in the atmosphere.

Energy conservation is embodied in the first law of thermodynamics,

$$c_{\mathrm{v}}\frac{\mathrm{d}T}{\mathrm{d}t} + RT\nabla \cdot \boldsymbol{V} = Q,$$

where $c_{\mathrm{v}}$ is the specific heat at constant volume and $Q$ is the diabatic heating rate. Conservation of mass is expressed in terms of the continuity equation:

$$\frac{\mathrm{d}\rho}{\mathrm{d}t} + \rho\nabla \cdot \boldsymbol{V} = 0.$$

Finally, conservation of water substance is expressed by the equation

$$\frac{\mathrm{d}q}{\mathrm{d}t} = S,$$

where $q$ is the specific humidity and $S$ represents all sources and sinks of water vapor.

Once initial conditions, appropriate boundary conditions, and external forcings, sources, and sinks are given, the above system of seven (scalar) equations provides a complete description of the evolution of the seven variables $\{u, v, w, p, \rho, T, q\}$.

For large-scale motions the vertical component of velocity is very much smaller than the horizontal components, and we can replace the vertical equation by a balance between the vertical pressure gradient and gravity. This yields the hydrostatic equation

$$\frac{\partial p}{\partial z} + g\rho = 0.$$

Hydrostatic models were used for the first fifty years of NWP but nonhydrostatic models are now coming into widespread use.

## 3 The Emergence of NWP

The idea of calculating the changes in the weather by numerical methods emerged around the turn of the twentieth century. Cleveland Abbe, an American meteorologist, viewed weather forecasting as an application of hydrodynamics and thermodynamics to the atmosphere. He also identified a system of mathematical equations, essentially those presented in section 2 above, that govern the evolution of the atmosphere. This idea was developed in greater detail by the Norwegian Vilhelm Bjerknes, whose stated goal was to make meteorology an exact science: a true physics of the atmosphere.

### 3.1 Richardson's Forecast

During World War I, Lewis Fry Richardson, an English Quaker mathematician, calculated the changes in the weather variables directly from the fundamental equations and presented his results in a book, *Weather Prediction by Numerical Process*, in 1922. His prediction of pressure changes was utterly unrealistic, being two orders of magnitude too large. The primary cause of this failure was the inaccuracy and imbalance of the initial conditions. Despite the outlandish results, Richardson's methodology was unimpeachable, and is essentially the approach we use today to integrate the equations.

Richardson was several decades ahead of his time. For computational weather forecasting to become a practical reality, advances on a number of fronts were required. First, an observing system for the troposphere, the lowest layer of the atmosphere, extending to about 12 km, was established to serve the needs of aviation; this also provided the initial data for weather forecasting. Second, advances in numerical analysis led to the design of stable and accurate algorithms for solving the PDEs. Third, progress in meteorological theory, especially the development of the quasigeostrophic equations and improved understanding of atmospheric balance, provided a means to eliminate the spurious high-frequency oscillations that had spoiled Richardson's forecast. Finally, the invention of high-speed digital computers enabled the enormous computational task of solving the equations to be undertaken.

### 3.2 The ENIAC Integrations

The first forecasts made using an automatic computer were completed in 1950 on the ENIAC (Electronic Numerical Integrator and Computer), the first programmable general-purpose computer. The forecasts used a highly simplified model, representing the atmosphere as a single layer and assuming conservation of absolute vorticity expressed by the barotropic vorticity equation,

$$\frac{\mathrm{d}}{\mathrm{d}t}(\zeta + f) = 0,$$

where $\zeta$ is the vorticity of the flow and $f = 2\Omega \sin\phi$ is the Coriolis parameter, with $\Omega$ the angular velocity of the Earth and $\phi$ the latitude. The Lagrangian time derivative

$$\frac{\mathrm{d}}{\mathrm{d}t} = \frac{\partial}{\partial t} + \boldsymbol{V} \cdot \boldsymbol{\nabla}$$

includes the nonlinear advection by the flow. The equation was approximated by finite differences in space and time with a grid size of 736 km (at the North Pole) and a time step of three hours. The resulting forecasts, while far from perfect, were realistic and provided a powerful stimulus for further work.

Baroclinic, or multilevel, models that enabled realistic representation of the vertical structure of the atmosphere were soon developed. Moreover, the simplified equations were replaced by more accurate primitive equations, that is, the equations presented in section 2 but with the hydrostatic approximation. As these equations simulate high-frequency gravity waves in addition to the motions that are important for weather, the initial conditions must be carefully balanced. Techniques for ensuring this were developed. Most notable among these was the normal-mode initialization method: the flow is resolved into normal modes and modified to ensure that the tendencies, or rates of change, of the gravity wave components vanish. This suppresses spurious oscillations.

## 4 Solving the Equations

Analytical solution of the equations is impossible, so approximate methods must be employed. We consider methods of discretizing the spatial domain to reduce the PDEs to an algebraic system and of advancing the solution in time.

### 4.1 Time-Stepping Schemes

Let $Q$ denote a typical dependent variable, governed by an equation of the form

$$\frac{\mathrm{d}Q}{\mathrm{d}t} = F(Q).$$

We replace the continuous-time domain $t$ by a sequence of discrete times $\{0, \Delta t, 2\Delta t, \dots, n\Delta t, \dots\}$, with the solution at these times denoted by $Q^n = Q(n\Delta t)$. If this solution is known up to time $t = n\Delta t$, the right-hand term $F^n = F(Q^n)$ can be computed. The time derivative is now approximated by a centered difference

$$\frac{Q^{n+1} - Q^{n-1}}{2\Delta t} = F^n,$$

so the "forecast" value $Q^{n+1}$ may be computed from the old value $Q^{n-1}$ and the tendency $F^n$:

$$Q^{n+1} = Q^{n-1} + 2\Delta t F^n.$$

This is called the *leapfrog scheme*. The process of stepping forward from moment to moment is repeated a large number of times, until the desired forecast range is reached.

The leapfrog scheme is limited by a stability criterion that restricts the size of the time step $\Delta t$. One way of circumventing this is to use an implicit scheme such as

$$\frac{Q^{n+1} - Q^{n-1}}{2\Delta t} = \frac{F^{n-1} + F^{n+1}}{2}.$$

The time step is now unconstrained by stability, but the scheme requires the solution of the equation

$$Q^{n+1} - \Delta t F^{n+1} = Q^{n-1} + \Delta t F^{n-1},$$

which is prohibitive unless $F(Q)$ is a linear function. Normally, implicit schemes are used only for particular (linear) terms of the equations.

### 4.2 Spatial Finite Differencing

For the PDEs that govern atmospheric dynamics we must replace continuous variations in space by discrete variables. The primary way to do this is to substitute finite-difference approximations for the spatial derivatives. It then transpires that the stability depends on the *relative* sizes of the space and time steps. A realistic solution is not guaranteed by reducing their sizes independently.

We consider the simple one-dimensional wave equation

$$\frac{\partial Q}{\partial t} + c\frac{\partial Q}{\partial x} = 0,$$

where $Q(x,t)$ depends on both $x$ and $t$, and where the advection speed $c$ is constant. We consider the sinusoidal solution $Q = Q^0 \mathrm{e}^{\mathrm{i}k(x-ct)}$ of wavelength $L = 2\pi/k$. We use centered difference approximations in both space and time:

$$\frac{Q_m^{n+1} - Q_m^{n-1}}{2\Delta t} + c\left(\frac{Q_{m+1}^n - Q_{m-1}^n}{2\Delta x}\right) = 0,$$

where $Q_m^n = Q(m\Delta x, n\Delta t)$. We seek a solution of the form $Q_m^n = Q^0 \mathrm{e}^{\mathrm{i}k(m\Delta x - Cn\Delta t)}$. For real $C$, this is a wavelike solution. However, if $C$ is complex, this solution will behave exponentially, quite unlike the solution of the continuous equation. Substituting $Q_m^n$ into the finite-difference equation, we find that

$$C = \frac{1}{k\Delta t}\sin^{-1}\left[\left(\frac{c\Delta t}{\Delta x}\right)\sin k\Delta x\right].$$

If the argument of the inverse sine is less than unity, $C$ is real. Otherwise, $C$ is complex, and the solution will grow with time. Thus, the condition for stability of the solution is

$$\left|\frac{c\Delta t}{\Delta x}\right| \leqslant 1.$$

This is the Courant–Friedrichs–Lewy criterion, discovered in 1928. It imposes a strong constraint on the relative sizes of the space and time grids. The limitation on stability can be circumvented by means of implicit finite differencing. Then

$$C = \frac{2}{k\Delta t}\tan^{-1}\left[\left(\frac{c\Delta t}{2\Delta x}\right)\sin k\Delta x\right].$$

The numerical phase speed $C$ is always real, so the implicit scheme is unconditionally stable, but the cost is that a linear system must be solved at each time step.

### 4.3 Spectral Method

In the spectral method, each field is expanded in a series of spherical harmonics:

$$Q(\lambda,\phi,t) = \sum_{n=0}^{\infty}\sum_{m=-n}^{n} Q_n^m(t)Y_n^m(\lambda,\phi),$$

where the coefficients $Q_n^m(t)$ depend only on time, and where $Y_n^m(\lambda,\phi)$ are the spherical harmonics

$$Y_n^m(\lambda,\phi) = \mathrm{e}^{\mathrm{i}m\lambda}P_n^m(\phi)$$

for longitude $\lambda$ and latitude $\phi$. The coefficients $Q_n^m$ of the harmonics provide an alternative to specifying the field values $Q(\lambda,\phi)$ in the spatial domain. When the model equations are transformed to spectral space they become a coupled set of equations (ordinary differential equations) for the spectral coefficients $Q_n^m$. These are used to advance the coefficients in time, after which the new physical fields may be computed.

In practice, the series expansion must be truncated at some point:

$$Q(\lambda_i,\phi_j,t) = \sum_{n=0}^{N}\sum_{m=-n}^{n} Q_n^m(t)Y_n^m(\lambda_i,\phi_j).$$

This is called *triangular truncation*, and the value of $N$ indicates the resolution of the model. There is a computational grid, called the Gaussian grid, corresponding to the spectral truncation.

## 5 Initial Conditions

Numerical weather prediction is an initial-value problem; to integrate the equations of motion we must specify the values of the dependent variables at an initial time. The numerical process then generates the values of these variables at later times. The initial data are ultimately derived from direct observations of the atmosphere.

The *optimal interpolation* analysis method was, for several decades, the most popular method of automatic analysis for NWP. This method optimizes the combination of information in the background (forecast) field and in the observations, using the statistical properties of the forecast and observation errors to produce an analysis that, in a precise statistical sense, is the best possible analysis.

An alternative approach to data assimilation is to find the analysis field that minimizes a *cost function.* This is called *variational assimilation* and it is equivalent to the statistical technique known as the maximum-likelihood estimate, subject to the assumption of Gaussian errors. When applied at a specific time, the method is called three-dimensional variational assimilation, or 3D-Var for short. When the time dimension is also taken into account, we have 4D-Var.

### 5.1 Variational Assimilation

The cost function for 3D-Var may be defined as the sum of two components:

$$J = J_{\mathrm{B}} + J_{\mathrm{O}}.$$

We represent the model state by a high-dimensional vector $\boldsymbol{X}$. The term

$$J_{\mathrm{B}} = \tfrac{1}{2}(\boldsymbol{X} - \boldsymbol{X}_{\mathrm{B}})^{\mathrm{T}}\boldsymbol{B}^{-1}(\boldsymbol{X} - \boldsymbol{X}_{\mathrm{B}})$$

represents the distance between the model state $X$ and the background field $X_B$ weighted by the background error covariance matrix $B$. The term

$$J_O = \tfrac{1}{2}(Y - HX)^T R^{-1}(Y - HX)$$

represents the distance between the analysis and the observed values $Y$ weighted by the observation error covariance matrix $R$. The observation operator $H$ is a rectangular matrix that converts the background field into first-guess values of the observations. More generally, the observation operator is nonlinear but, for ease of description, we assume here that it is linear.

The minimum of $J$ is attained at $X = X_A$, where

$$\nabla_X J = 0,$$

that is, where the gradient of $J$ with respect to each of the analyzed values is zero. Computing this gradient, we get

$$\nabla_X J = B^{-1}(X - X_B) + H^T R^{-1}(Y - HX).$$

Setting this to zero we can deduce the expression

$$X = X_B + K(Y - HX_B).$$

Thus, the analysis is obtained by adding to the background field a weighted sum of the difference between observed and background values. The matrix $K$, the *gain matrix*, is given by

$$K = BH^T(R + HBH^T)^{-1}.$$

The analysis error covariance is then given by

$$A = (I - KH)B.$$

The minimum of the cost function is found using a descent algorithm such as the CONJUGATE GRADIENT METHOD [IV.11 §4.1]; 3D-Var solves the minimization problem directly, avoiding computation of the gain matrix.

The 3D-Var method has enabled the direct assimilation of satellite radiance measurements. The error-prone inversion process, whereby temperatures are deduced from the radiances before assimilation, is thus eliminated. Quality control of these data is also easier and more reliable. As a consequence, the accuracy of forecasts has improved markedly since the introduction of variational assimilation. The accuracy of medium-range forecasts is now about equal for the two hemispheres (see figure 1). This is due to better satellite data assimilation. Satellite data are essential for the Southern Hemisphere as conventional data are in such short supply. The extraction of useful information from satellite soundings has been one of the great research triumphs of NWP over the past forty years.

### 5.2 Inclusion of the Time Dimension

Whereas conventional meteorological observations are made at the main synoptic hours, satellite data are distributed continuously in time. To assimilate these data, it is necessary to perform the analysis over a time interval rather than for a single moment. This is also more appropriate for observations that are distributed inhomogeneously in space. Four-dimensional variational assimilation, or 4D-Var for short, uses all the observations within an interval $t_0 \leqslant t \leqslant t_N$. The cost function has a term $J_B$ measuring the distance to the background field $X_B$ at the initial time $t_0$, just as in 3D-Var. It also contains a summation of terms measuring the distance to observations at each time step $t_n$ in the interval $[t_0, t_N]$:

$$J = J_B + \sum_{n=0}^{N} J_O(t_n),$$

where $J_B$ is defined as for 3D-Var and $J_O(t_n)$ is given by

$$J_O(t_n) = (Y_n - H_n X_n)^T R_n^{-1}(Y_n - H_n X_n).$$

The state vector $X_n$ at time $t_n$ is generated by integration of the forecast model from time $t_0$ to $t_n$, written $X_n = \mathcal{M}_n(X_0)$. The vector $Y_n$ contains the observations valid at time $t_n$.

Just as the observation operator had to be linearized to obtain a quadratic cost function, we linearize the model operator $\mathcal{M}_n$ about the trajectory from the background field, obtaining what is called the *tangent linear model* operator $M_n$. Then we find that 4D-Var is formally similar to 3D-Var with the observation operator $H$ replaced by $H_n M_n$. Just as the minimization of $J$ in 3D-Var involved the transpose of $H$, the minimization in 4D-Var involves the transpose of $H_n M_n$, which is $M_n^T H_n^T$. The operator $M_n^T$, the transpose of the tangent linear model, is called the *adjoint model*. The *control variable* for the minimization of the cost function is $X_0$, the model state at time $t_0$, and the sequence of analyses $X_n$ satisfies the model equations, that is, the model is used as a strong constraint.

The 4D-Var method finds initial conditions $X_0$ such that the forecast best fits the observations within the assimilation interval. This removes an inherent disadvantage of optimal interpolation and 3D-Var, where all observations within a fixed time window (typically of six hours) are assumed to be valid at the analysis time. The introduction of 4D-Var at the European Centre for Medium-Range Weather Forecasts (ECMWF) led to a significant improvement in the quality of operational medium-range forecasts.

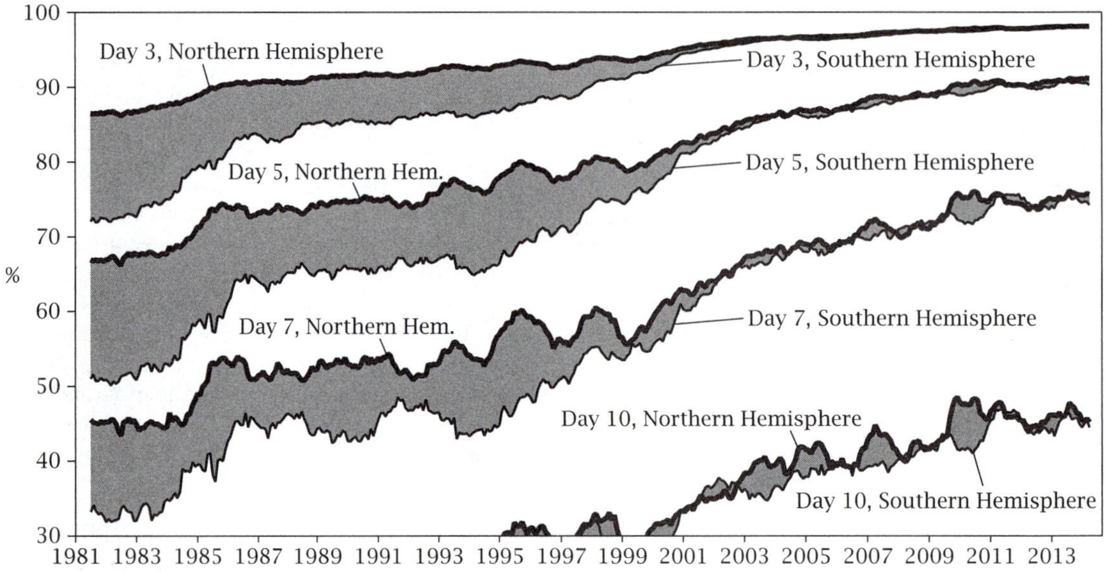


**Figure 1** Anomaly correlation (%) of 500 hPa geopotential height: twelve-month running mean (©ECMWF).

## 6 Forecasting Models

Operational forecasting today is based on output from a suite of computer models. Global models are used for predictions of several days ahead, while shorter-range forecasts are based on regional or limited-area models.

### 6.1 The ECMWF Global Model

As an example of a global model we consider the integrated forecast system (IFS) of the ECMWF (which is based in Reading, in the United Kingdom). The ECMWF produces a wide range of global atmospheric and marine forecasts and disseminates them on a regular schedule to its thirty-four member and cooperating states. The primary products are deterministic forecasts for the atmosphere out to ten days ahead, based on a high-resolution model, and probabilistic forecasts, extending to a month, made using a reduced resolution and an ensemble of fifty-one model runs.

The basis of the NWP operations at the ECMWF is the IFS. It uses a *spectral representation* of the meteorological fields. The IFS system underwent major resolution upgrades in 2006 and in 2010. Table 1 compares the spatial resolutions of the three model cycles, indicating the substantial improvements in model resolution in recent years. The truncation of the deterministic model is now T1279; that is, the spectral expansion is terminated at total wave number 1279. This is equivalent to a spatial resolution of 16 km. The number of model levels in the vertical has recently been increased to 137. The new Gaussian grid for the IFS has about $2 \times 10^6$ points. With 137 levels and five primary prognostic variables at each point, about $1.2 \times 10^9$ numbers are required to specify the atmospheric state at a given time. That is, the model has about a billion degrees of freedom. The computational task of making forecasts with such high resolution is truly formidable. The ECMWF carries out its operational program using a powerful and complex computer system. At the heart of this system is a Cray XC30 high-performance computer, comprising some 160 000 processors, with a sustained performance of over 200 teraflops ($2 \times 10^{14}$ floating-point operations per second).

**Table 1** Upgrades to the ECMWF IFS in 2006 and 2010. The spectral resolution is indicated by the triangular truncation number, and the effective resolution of the associated Gaussian grid is indicated. The number of model levels, or layers used to represent the vertical structure of the atmosphere, is also given.

| | Before 2006 | 2006–9 | After 2009 |
|---|---|---|---|
| Spectral truncation | T511 | T799 | T1279 |
| Effective resolution | 39 km | 25 km | 16 km |
| Model levels | 60 | 91 | 137 |

### 6.2 Mesoscale Modeling

Short-range forecasting requires detailed guidance that is updated frequently. Many national meteorological services run limited-area models with high resolution to provide such forecast guidance. These models permit a free choice of geographical area and spatial resolution, and forecasts can be run as frequently as required. Limited-area models make available a comprehensive range of outputs, with a high time resolution. Nested grids with successively higher resolution can be used to provide greater local detail.

The Weather Research and Forecasting Model is a next-generation mesoscale NWP system developed in a partnership involving American national agencies (the National Centers for Environmental Prediction and the National Center for Atmospheric Research) and universities. It is designed to serve the needs of both operational forecasting and atmospheric research. The Weather Research and Forecasting Model is suitable for a broad range of applications, from meters to thousands of kilometers, and it is currently in operational use at several national meteorological services.[1]

### 6.3 Ensemble Prediction

The chaotic nature of atmospheric flow is now well understood. It imposes a limit on predictability, as unavoidable errors in the initial state grow rapidly and render the forecast useless after some days. As a result of our increased understanding of the inherent difficulties of making precise predictions, there has been a paradigm shift in recent years from deterministic to probabilistic prediction. A forecast is now considered incomplete without an accompanying error bar, or quantitative indication of confidence.

The most successful way of producing a probabilistic prediction is to run a series, or ensemble, of forecasts, each starting from a slightly different initial state and each randomly perturbed during the forecast to simulate model errors. The ensemble of forecasts is used to deduce probabilistic information about future changes in the atmosphere. Since the early 1990s this systematic method of providing an a priori measure of forecast accuracy has been operational at both the ECMWF and at the National Centers for Environmental Prediction in Washington. In the ECMWF's ensemble prediction system, an ensemble of fifty-one forecasts is performed, each having a resolution half that of the deterministic forecast. Probability forecasts for a wide range of weather events are generated and disseminated for use in the operational centers, and they have become the key tools for medium-range prediction.

## 7 Verification of ECMWF Forecasts

Forecast accuracy has improved dramatically in recent decades. This can be measured by the *anomaly correlation*. The anomaly is the difference between a forecast value and the corresponding climate value, and the agreement between the forecast anomaly and the observed anomaly is expressed as the anomaly correlation. The higher this score the better; by general agreement, values in excess of 60% imply skill in the forecast. In figure 1, the twelve-month running mean anomaly correlations (in percentages) of the three-, five-, seven-, and ten-day 500 hPa height forecasts are shown for the extratropical Northern Hemisphere and Southern Hemisphere. The lines above each shaded region are for the Northern Hemisphere and the lines below are for the Southern Hemisphere, with the shading showing the difference in scores between the two.

The plots in figure 1 show a continuing improvement in forecast accuracy, especially for the Southern Hemisphere. By the turn of the millennium, the accuracy was comparable for the two hemispheres. Predictive ability has improved steadily over the past thirty years, and there is now accuracy out to eight days ahead. This record is confirmed by a wealth of other data. Predictive skill has been increasing by about one day per decade, and there are reasons to hope that this trend will continue for several more decades.

## 8 Applications of NWP

NWP models are used for a wide range of applications. Perhaps the most important purpose is to provide timely warnings about weather extremes. Great financial losses can be caused by gales, floods, and other anomalous weather events. The warnings that result from NWP guidance can greatly diminish losses of both life and property. Transportation, energy consumption, construction, tourism, and agriculture are all sensitive to weather conditions. There are expectations from all these sectors of increasing accuracy and detail in short-range forecasts, as decisions with heavy financial implications must continually be made.

NWP models are used to generate special guidance for the marine community. Predicted winds are used to

1. Full details of the system are available at www.wrf-model.org.

drive wave models, which predict sea and swell heights and periods. Prediction of road ice is performed by specially designed models that use forecasts of temperature, humidity, precipitation, cloudiness, and other parameters to estimate the conditions on the road surface. Trajectories are easily derived from limited-area models. These are vital for modeling pollution drift, for nuclear fallout, smoke from forest fires, and so on. Aviation benefits significantly from NWP guidance, which provides warnings of hazards such as lightning, icing, and clear-air turbulence.

### 9 The Future

Progress in NWP over the past sixty years can be accurately described as revolutionary. Weather forecasts are now consistently accurate and readily available. Nevertheless, some formidable challenges remain. Sudden weather changes and extremes cause much human hardship and damage to property. These rapid developments often involve intricate interactions between dynamical and physical processes, both of which vary on a range of timescales. The effective computational coupling between the dynamical processes and the physical parametrizations is a significant challenge. *Nowcasting* is the process of predicting changes over periods of a few hours. Guidance provided by current numerical models occasionally falls short of what is required to take effective action and avert disasters. Greatest value is obtained by a systematic combination of NWP products with conventional observations, radar imagery, satellite imagery, and other data. But much remains to be done to develop optimal nowcasting systems, and we may be optimistic that future developments will lead to great improvements in this area.

At the opposite end of the timescale, the chaotic nature of the atmosphere limits the validity of deterministic forecasts. Interaction between the atmosphere and the ocean becomes a dominant factor at longer forecast ranges, as does coupling to SEA ICE [V.17]. Also, a more accurate description of aerosols and trace gases should improve long-range forecasts. Although good progress in seasonal forecasting for the tropics has been made, the production of useful long-range forecasts for temperate regions remains to be tackled by future modelers. Another great challenge is the modeling and prediction of climate change, a matter of increasing importance and concern.

### Further Reading

Lynch, P. 2006. *The Emergence of Numerical Weather Prediction: Richardson's Dream*. Cambridge: Cambridge University Press.

## V.19 Tsunami Modeling

*Randall J. LeVeque*

### 1 Introduction

The general public's appreciation of the danger of tsunamis has soared since the Indian Ocean tsunami of December 26, 2004, killed more than 200 000 people. Several other large tsunamis have occurred since then, including the devastating March 11, 2011, Great Tohoku tsunami generated off the coast of Japan. The international community of tsunami scientists has also grown considerably since 2004, and an increasing number of applied mathematicians have contributed to the development of better models and computational tools for the study of tsunamis. In addition to its importance in scientific studies and public safety, tsunami modeling also provides an excellent case study to illustrate a variety of techniques from applied and computational mathematics. This article combines a brief overview of tsunami science and hazard mitigation with descriptions of some of these mathematical techniques, including an indication of some challenging problems of ongoing research.

The term "tsunami" is generally used to refer to any large-scale anomalous motion of water that propagates as a wave in a sizable body of water. Tsunamis differ from familiar surface waves in several ways. Typically, the fluid motion is not confined to a thin layer of water near the surface, as it is in wind-generated waves. Also the wavelength of the waves is much longer than normal: sometimes hundreds of kilometers. This is orders of magnitude larger than the depth of the ocean (which is about 4000 m on average), and tsunamis are therefore also sometimes referred to as "long waves" in the scientific literature. In the past, tsunamis were often called "tidal waves" in English because they share some characteristics with tides, which are the visible effect of very long waves propagating around the Earth. However, tsunamis have nothing to do with the gravitational (tidal) forcing that drives the tides, and so this term is misleading and is no longer used. The Japanese word "tsunami" means "harbor wave," apparently because sailors would sometimes return home to find their harbor destroyed by mysterious waves they did not observe while at sea. Strong currents and vortices in harbors often cause extensive damage to ships and infrastructure even when there is no onshore inundation. Although the worst effects of a tsunami are often observed in harbors, the effects can be devastating in any coastal region. Because tsunamis have such a long wavelength, they frequently appear onshore as a flood that can continue flowing inward for tens of minutes or even hours before flowing back out. The flow velocities can also be quite large, with the consequence that even a tsunami wave with an amplitude of less than a meter can sweep people off their feet and do considerable damage to structures. Tsunamis arising from large earthquakes often result in flow depths greater than a meter, particularly along the coast closest to the earthquake, where run-up can reach tens of meters.

Tsunamis are generated whenever a large mass of water is rapidly displaced, either by the motion of the seafloor due to an earthquake or submarine landslide, or when a solid mass enters the water from a landslide, volcanic flow, or asteroid impact. The largest tsunamis in recent history, such as the 2004 and 2011 events mentioned above, were all generated by megathrust subduction zone earthquakes at the boundary of oceanic and continental plates. Offshore from many continents there is a subduction zone where plates are converging. The denser material in the oceanic plate subducts beneath the lighter continental crust. Rather than sliding smoothly, stress builds up at the interface and is periodically released when one plate suddenly slips several meters past the other, causing an earthquake during which the seafloor is lifted up in some regions and depressed in others. All of the water above the seafloor is lifted or falls along with it, creating a disturbance on the sea surface that propagates away in the form of waves. See figure 1 for an illustration of tsunami generation and figure 2 for a numerical simulation of waves generated by the 2011 Tohoku earthquake off the coast of Japan. This article primarily concerns tsunamis caused by subduction zone earthquakes since they are a major concern in risk management and have been widely studied.

## 2 Mathematical Models and Equations of Motion

Tsunamis are modeled by solving systems of partial differential equations (PDEs) arising from the theory of fluid dynamics. The motion of water can be very

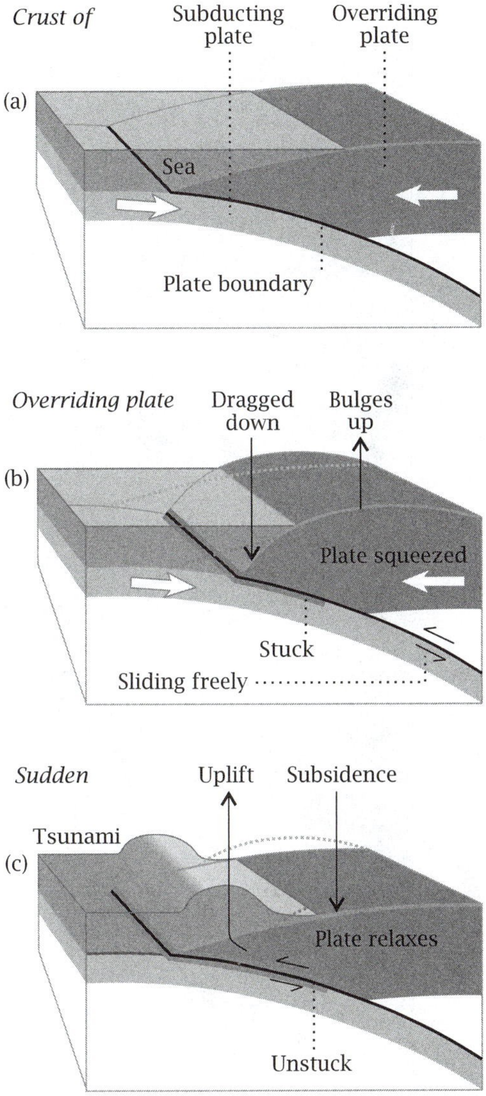


**Figure 1** Illustration of the generation of a tsunami by a subduction zone earthquake. (a) Overall, a tectonic plate descends, or "subducts," beneath an adjoining plate, but it does so in a stick–slip fashion. (b) Between earthquakes the plates slide freely at great depth, where hot and ductile, but at shallow depth, where cool and brittle, they stick together; slowly squeezed, the overriding plate thickens. (c) During an earthquake the leading edge of the overriding plate breaks free, springing seaward and upward; behind, the plate stretches and its surface falls; the vertical displacements set off a tsunami. (Image courtesy of Brian Atwater and taken from Atwater, B., et al. 2005. *The Orphan Tsunami of 1700—Japanese Clues to a Parent Earthquake in North America.* Washington, DC: University of Washington Press.)

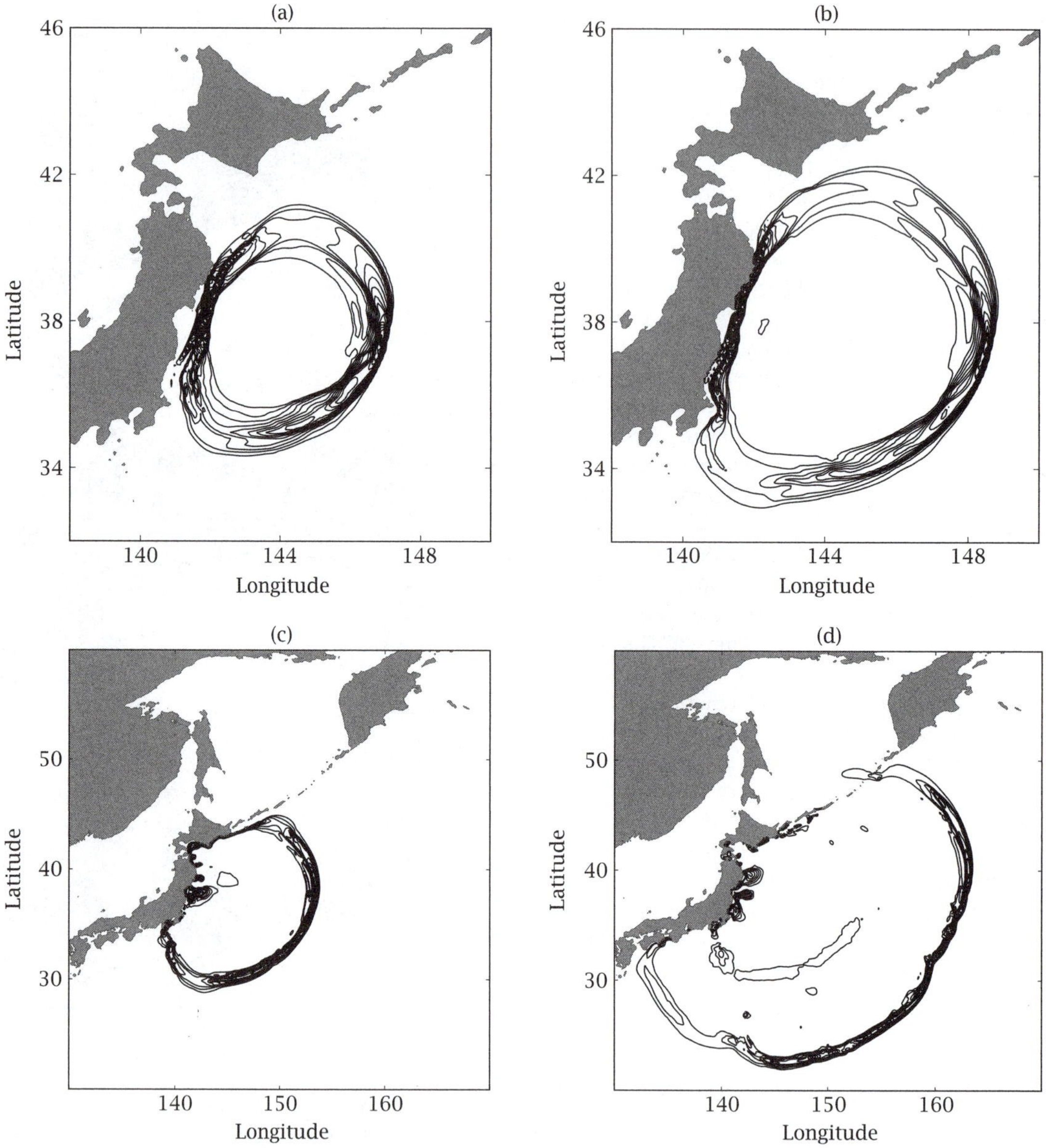


**Figure 2** Propagation of the tsunami arising from the March 11, 2011, Tohoku earthquake off the coast of Japan at four different times: (a) 20 minutes, (b) 30 minutes, (c) 1 hour, and (d) 2 hours after the earthquake. Waves propagate away from the source region with a velocity that varies with the local depth of the ocean. Contour lines show sea surface elevation above sea level, in increments of 0.2 m (in (a) and (b): at early times) and 0.1 m (in (c) and (d): at later times). There is a wave trough behind the leading wave peak shown here, but for clarity the contours of elevation below sea level are not shown.

well modeled by the NAVIER–STOKES EQUATIONS [III.23] for an incompressible viscous fluid. However, these are rarely used directly in tsunami modeling since they would have to be solved in a time-varying three-dimensional domain, bounded by a free surface at the top and by moving boundaries at the edges of the ocean as the wave inundates or retreats at the shoreline. Fortunately, for most tsunamis it is possible to use "depth-averaged" systems of PDEs, obtained by integrating in the (vertical) $z$-direction to obtain equations in two space dimensions (plus time). In these formulations, the depth of the fluid at each point is modeled by a function $h(x, y, t)$ that varies with location and time. The velocity of the fluid is described by two functions $u(x, y, t)$ and $v(x, y, t)$ that represent depth-averaged values of the velocity in the $x$- and $y$-directions, respectively. In addition to a reduction from three to two space dimensions, this eliminates the free surface boundary in $z$; the location of the sea surface is now determined directly from the depth $h(x, y, t)$. These equations are solved in a time-varying two-dimensional $xy$-domain since the moving boundaries at the shoreline must still be dealt with.

A variety of depth-averaged equations can be derived, depending on the assumptions made about the flow. For large-scale tsunamis, the so-called SHALLOW-WATER EQUATIONS [III.27] (also called the Saint Venant or long-wave equations) are frequently used and have been shown to be very accurate. The assumption with these equations is that the fluid is sufficiently shallow relative to the wavelength of the wave being studied. This is generally true for tsunamis generated by earthquakes, where the wavelength is typically 10–100 times greater than the ocean depth.

The two-dimensional shallow-water equations have the form

$$\left.\begin{aligned} h_t + (hu)_x + (hv)_y &= 0, \\ (hu)_t + (hu^2 + \tfrac{1}{2}gh^2)_x + (huv)_y &= -ghB_x, \\ (hv)_t + (huv)_x + (hv^2 + \tfrac{1}{2}gh^2)_y &= -ghB_y, \end{aligned}\right\} \quad (1)$$

where subscripts denote partial derivatives, e.g., $h_t = \partial h/\partial t$. In addition to the variables $h$, $u$, $v$ already introduced, which are each functions of $(x, y, t)$, these equations involve the gravitational force $g$ and the topography or seafloor bathymetry (underwater topography), which is denoted by $B(x, y)$. Typically, $B = 0$ represents sea level, while $B > 0$ is onshore topography and $B < 0$ represents seafloor bathymetry. Water is present wherever $h > 0$, and $\eta(x, y, t) = h(x, y, t) + B(x, y)$ is the elevation of the water surface. See figure 3 for a diagram in one space dimension. During an earthquake, $B$ should also be a function of $t$ in the region where the seafloor is deforming. In practice it is often sufficient to include this deformation in $B(x, y)$, while the initial conditions for the depth $h(x, y, 0)$ are based on the undeformed topography. The seafloor deformation then appears instantaneously in the initial surface $\eta(x, y, 0)$, which initializes the tsunami. In the remainder of this article, the term *topography* will be used for both $B > 0$ and $B < 0$ for simplicity.

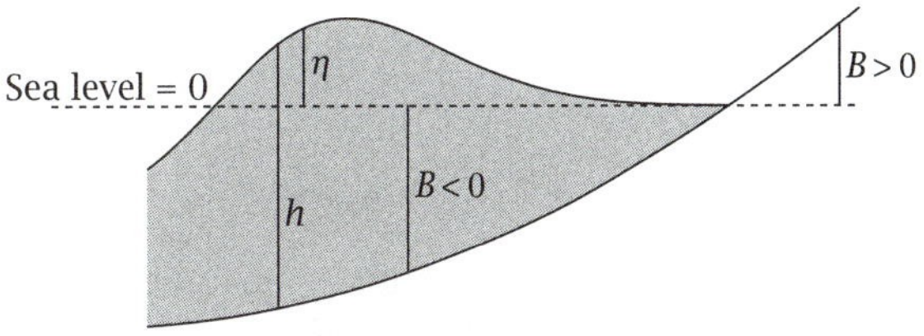


**Figure 3** Illustration showing the notation used in the shallow-water equations (1).

If $B(x, y) < 0$ is constant (a flat bottom), then the "source terms" on the right-hand side of these equations drop out and the equations model the conservation of mass ($h$) and momentum ($hu, hv$). Over a varying bottom, mass is still conserved but momentum is affected by the terrain, as seen in the reflection of waves at a shoreline, for example, and in partial reflection when a wave interacts with underwater features. The term $\frac{1}{2}gh^2$ appearing in the momentum equations is the depth-averaged "hydrostatic pressure" in a column of water of depth $h$. (This and all other terms in (1) should in fact also involve the fluid density $\rho$, but this cancels out everywhere if the density is assumed to be constant.)

The equations (1) are a nonlinear system of equations of hyperbolic type. Hyperbolic PDEs frequently arise when waves are modeled mathematically; the prototype is the WAVE EQUATION [III.31] itself. The amplitude of a tsunami in the deep ocean is generally very small relative to the water depth, typically less than a meter even for a large megathrust tsunami. Away from the coast these equations could be approximated by linearized equations with variable coefficients arising from the varying topography. Near the shore, however, the amplitude of the wave is large relative to the depth of the fluid and the full nonlinear equations must be used to accurately model the interaction of a tsunami with the nearshore topography and the onshore inundation that occurs. Solutions to nonlinear hyperbolic PDEs can become discontinuous if a SHOCK [II.30] develops. In the case of the shallow-water equations, a shock is also called a "hydraulic jump" and is a mathematical idealization of a thin zone in which the depth and velocity both undergo rapid transitions from one value to another. Such regions frequently appear as a turbulent wave front (sometimes called a turbulent bore) once the tsunami moves into sufficiently shallow water. The shallow-water equations do not model this turbulent zone directly, but they are frequently adequate to capture important quantities such as the depth and fluid velocities behind the bore and its propagation speed.

## 3 Uses of Tsunami Modeling

The PDEs describing a tsunami cannot be solved exactly, in general, and numerical methods must therefore be used to simulate the propagation and inundation of a tsunami. A brief description of how this might be done and some of the challenges that arise is given in section 4, but first we motivate the need for numerical models by describing some common uses of such models.

### 3.1 Real-Time Warning Systems

One natural use of a numerical model is to assist in issuing warnings in real time as a tsunami propagates across the ocean and to determine which coastal regions should be evacuated. There are many challenges to doing this quickly and accurately. Accurate assessment is critical not only to ensure that people in areas that are at risk are properly warned but also to avoid triggering evacuation in areas where it is not necessary, which can itself cause loss of life, have a serious financial impact, and decrease the likelihood that attention will be paid to future warnings. For a subduction zone megathrust earthquake it is often impossible to issue tsunami warnings quickly enough for areas along the nearby coastline. The tsunami may arrive in less than an hour, and it is critical that residents understand the need to move to high ground when a major earthquake occurs. On the other hand, across the ocean the earthquake itself is not felt and so provides no direct indication of an impending tsunami, but several hours are available in which to perform simulations and issue warnings.

### 3.2 Tsunami Source Inversion

To perform tsunami simulations it is necessary to estimate the source, i.e., the deformation of the sea floor that generates the tsunami, since this determines the initial conditions that are used to numerically solve the PDEs modeling tsunami propagation and inundation. There is generally no way to measure this directly, and so some form of INVERSE PROBLEM [IV.15] must be solved to obtain an estimate of the deformation based on measurements that can be made, such as seismometer recordings of the earthquake or measurements of the tsunami itself. Initial estimates of the location and magnitude of an earthquake generally come from analyzing recordings of seismic waves, which are compression and shear waves that travel through the Earth with much higher velocity than tsunamis and that are routinely recorded at hundreds of seismometers widely scattered around the world. From the measured waveforms at many locations it is possible to construct an estimate of how the Earth must have moved to produce this set of data. This relies ultimately on solving an inverse problem for the PDEs modeling wave motion in elastic materials. Seismic inversions generally estimate the slip of the Earth along the earthquake fault, which may be tens of kilometers below the sea floor. Converting this slip on the fault plane to deformation of the sea floor requires solving another elasticity problem, whose solution is often approximated by the *Okada model*. This is based on the Green function for the deformation of the boundary of an elastic half-space caused by a delta function dislocation. Integrating this over a finite-sized patch of a fault plane gives an estimate of the resulting seafloor displacement.

While the results of seismic inversions are invaluable in modeling tsunamis, performing an accurate inversion requires collecting and processing a large amount of data and this may not be feasible in real time. In order to gather better information about tsunamis as they propagate, a number of pressure gauges have recently been deployed on the seafloor that are able to measure water pressure extremely accurately. From the hydrostatic pressure it is possible to estimate the depth of the water at these locations with enough precision to capture variations due to a tsunami passing by. Direct measurement of a tsunami at one or more of these gauges can then be combined with seismic models identifying the approximate source location and geophysical knowledge of the faults that are most likely to produce tsunamis. This information, together with accurate tsunami propagation models, allows an inverse problem to be solved that in turn enables us to estimate the seafloor deformation that caused the tsunami more accurately and quickly than is possible using seismic information alone.

### 3.3 Hazard Modeling and Mitigation

Real-time simulations of tsunamis are used to issue warnings, but tsunami modeling has many ongoing uses beyond this. Protecting communities requires adequate planning long before a tsunami takes place, and tsunami models are used to simulate the effect of tsunamis arising from hypothetical earthquake events. The results of such models can be used to determine what regions of a community are most at risk and what regions can be designated as safe zones for evacuation. Modeling the arrival time and pattern of the waves can be used in connection with traffic-flow models of evacuation. Some communities in tsunami-prone regions build sea walls or gates that can be closed for protection against tsunamis, or build "vertical evacuation structures" in regions where there is no easily accessible high ground for large-scale evacuation. These structures may take the form of multiuse buildings built to withstand tsunamis and tall enough that the upper floors are safe havens, or they may consist of large berms that form artificial high ground. Designing such structures requires modeling the flow depth and often also the fluid velocities of hypothetical tsunamis.

Of course it is impossible to know exactly what the seafloor deformation will be for future earthquakes, but quite a bit is known about the major subduction zones and the likely locations and magnitudes of large earthquakes based on geology and past history. There is always a question of how large a tsunami one should design for. Sometimes an estimate of the "credible worst case" tsunami for that location is used, but this may correspond to an event with very low probability of occurrence that would require enormous expenditure to protect against—money that might be better spent protecting against more likely events at additional locations. To better understand these trade-offs, there has recently been increased interest in *probabilistic tsunami hazard assessment*, in which a set of possible events are assigned probabilities or an entire spectrum of possible events is assigned some probability density function, typically over a very high-dimensional stochastic space. The goal is then to obtain from this a probabilistic description of the resulting inundation patterns, flow depths, velocities, etc. This is a form of UNCERTAINTY QUANTIFICATION [II.34], a rapidly growing field of importance in many fields of computational science where simulations are based on many uncertain inputs and the goal is a probabilistic description of the resulting outputs rather than a single simulation result. Applied mathematicians and statisticians have a large role to play in the development of new techniques to efficiently solve these problems.

### 3.4 The Study of Past Tsunamis and Earthquakes

Another major use of tsunami modeling is the study of past tsunamis. A wealth of data has been collected following recent tsunami events by "tsunami survey teams" that measure inundation and runup along affected coasts. There are also data available from seafloor pressure gauges, tide gauges along the coast, and other data-collection facilities. Models of the seafloor deformation produced by solving the source inversion problem can then be used as initial data for tsunami models and the computed results compared with measurements. Such studies are important in verifying that a tsunami model gives a sufficiently accurate approximation to a real tsunami that it can be used with confidence for warning or hazard mitigation purposes. Validated models are also used in performing tsunami source inversion to estimate the seafloor deformation, and this can give additional insight into the earthquake mechanism that is useful to seismologists. Tsunami models can also help explain unusual features of past events by providing a laboratory for exploring the fluid dynamics taking place during the event.

Tsunami models can also help reconstruct events that happened in the more distant past, for which there are no pressure gauge or tide gauge data and perhaps only limited historical records of the regions inundated, or no human records in the case of prehistoric events or those that occurred on uninhabited coastlines. Luckily, for many events a geological record of the tsunami inundation is recorded in the form of *tsunami deposits*. As a tsunami approaches shore it typically becomes turbulent and picks up sediment from the seafloor, such as sand and marine microorganisms. This material is carried inland during the flooding stage and typically settles out of the flow as the flow decelerates and reverses, leaving behind a layer of deposits, often far inland. In tsunami-prone areas there are often many layers of tsunami deposits that have been built up over thousands of years, separated by layers of soil that slowly build up between tsunamis. Core samples or trenches can reveal many past events that can often be dated using radiocarbon dating of organic matter or interspersed tephra layers from known volcanic eruptions. The study of tsunami deposits is a major source of information about the magnitude, location, and recurrence times of past earthquakes. This information is critical in developing probabilistic models of tsunami or earthquake hazards, as well as to obtaining a better scientific understanding of earthquake processes. Numerical tsunami models can be used to help identify the location and magnitude of seafloor deformation that would lead to the patterns of inundation recorded by tsunami deposits. Models that include sediment erosion, transport, and deposition are also being used to better understand the fluid dynamics of the creation of tsunami deposits, and this will ultimately lead to more accurate descriptions of the tsunamis that caused observed deposits.

## 4 Numerical Modeling

Systems of nonlinear PDEs such as the nonlinear shallow-water equations (1) typically cannot be solved exactly except for very simple cases: a one-dimensional wave on a linear beach for example. Realistic tsunami modeling always relies on numerical solution of the PDEs. This requires discretizing the equations in some manner: replacing the differential equations describing the continuum solution (defined for all $(x, y, t)$ in some domain) by a finite set of discrete algebraic equations whose solution can be computed in finite time on a computer. There are many ways to do this, and general discussions of numerical solution of PDEs are given in NUMERICAL SOLUTION OF PARTIAL DIFFERENTIAL EQUATIONS [IV.13].

*Finite-difference methods* are often used, in which a discrete grid is introduced consisting of a finite number of grid points $(x_i, y_j)$ covering the domain, and the solution is approximated only at these points at a discrete set of times $t_0, t_1, t_2, \dots$. Derivatives in the PDE are replaced by finite-difference approximations based on the approximate solution at neighboring grid points, giving a discrete set of algebraic equations that can be solved on a computer. Another popular approach is to use a *finite-volume method*, in which the domain is subdivided into a finite number of grid cells and the approximate solution consists of average values of the solution over each grid cell. Integrating the PDEs over a grid cell gives an expression for the time derivative of the cell average that can be used to update the cell averages from one time $t_n$ to the next time $t_{n+1}$. To obtain better accuracy, methods in which the solution on each grid cell is approximated by a polynomial rather than by only the cell average (which can be interpreted as a constant function, or polynomial of degree 0, over each cell) are sometimes used. In this case, the higher-order coefficients of each polynomial must be updated from one time step to the next. A method of this type that has recently become popular for tsunami modeling is the *discontinuous Galerkin method*, in which the piecewise polynomial function obtained from the polynomials defined on each cell is not assumed to be continuous at the interface between one cell and its neighbor. The term "Galerkin" refers to a *finite-element* approach to deriving equations for evolving the polynomial coefficients in time.

### 4.1 Nonlinearity and Shock Formation

A prominent feature of nonlinear hyperbolic PDEs is that *shocks* can form in the solutions: shocks are discontinuities in the depth and velocity that can arise even from smooth initial conditions. As mentioned in section 2, these correspond to hydraulic jumps or bores that are seen in tsunamis as they approach the shore. Sharp discontinuities are only an approximation of the true behavior, but they often give a good approximation of the flow. Incorporating more accurate fluid dynamics models would lead to systems of PDEs that are much more computationally expensive to solve.

The presence of discontinuities in the solution can lead to difficulties in solving the PDEs numerically, since derivatives are infinite at a point of discontinuity, and finite-difference approximations to derivatives generally diverge. This has led to the increased popularity of both finite-volume and discontinuous Galerkin methods, which are better able to robustly capture discontinuities in the solution. Methods designed to do this well are often called *shock-capturing methods*.

### 4.2 Inundation and the Moving Shoreline

Another computational challenge in modeling tsunamis, or any other geophysical flow over topography, is the need to handle the moving boundary of the flow at the shoreline. Many early tsunami models did not capture this moving boundary at all. Instead, the equations were solved over a fixed domain defined by the original shoreline with some boundary conditions imposed at this fixed boundary, such as an impermeable wall. While this approach could not be used to model inundation directly, it could still give some indication of the tsunami runup based on recording the depths and velocities along this wall boundary. Other mathematical or physical models were then used to estimate inundation from these values.

Most recently developed tsunami models attempt to model inundation directly. For simple problems it may be possible to use a grid that moves with time so that one edge of the grid is always along the shoreline. For realistic problems this is generally infeasible, since the shoreline can be very complex and can break into pieces as islands or isolated pools of water form. Most tsunami models instead use a fixed grid and implement some form of *wetting and drying algorithm* to keep track of which grid points or cells are dry ($h = 0$) and which are wet ($h > 0$). Standard approaches to approximating the PDEs typically break down near the shoreline, and it is a major challenge in developing tsunami models to deal with this phenomenon robustly and accurately, particularly since this is often the region of primary interest in terms of the model results.

### 4.3 Mesh Refinement

Another challenge arises from the vast differences in spatial scale between the ocean over which a tsunami propagates and a section of coastline such as a harbor where the solution is of interest. In the shoreline region it may be necessary to have a fine grid, with perhaps 10 m or less between grid points, in order to resolve the flow at a scale that is useful. It is clearly impractical, and luckily also unnecessary, to resolve the entire ocean to this resolution. The wavelength of a tsunami is typically more than 100 km, so the grid point spacing in the ocean can be more like 1–10 km. Moreover, we need even lower resolution over most of the ocean, particularly before the tsunami arrives.

To deal with the variation in spatial scales, virtually all tsunami codes use unequally spaced grids, often by starting with a coarse grid over the ocean and then refining portions of the grid to higher resolution where needed. Some models only use static refinement, in which the grid does not change with time but has finer grids in regions of interest along the coast. Other computer codes use *adaptive mesh refinement*, in which the regions of refinement change with time to adapt to the evolving solution. For example, areas of refinement might be used to follow the propagating wave with a finer grid than is used over the rest of the ocean, and additional levels of refinement added near the coastal region of interest only when the tsunami is approaching shore.

A related issue is the choice of time steps for advancing the solution. Stability conditions generally require that the time step multiplied by the maximum wave speed should be no greater than the width of a grid cell. This is because the explicit methods that are typically used for solving hyperbolic PDEs, such as the shallow-water equations, update the solution in each grid cell based only on data from the neighboring cells in each time step. If a wave can propagate more than one grid cell in one time step, then the method becomes unstable. This necessary condition for stability is called the CFL CONDITION [V.18 §4.2], after fundamental work on the convergence of numerical methods by Courant, Friedrichs, and Lewy in the 1920s. For the shallow-water equations the wave speed is $\sqrt{gh}$, which varies dramatically from the shoreline, where $h \approx 0$, to the deepest parts of the ocean, where $h$ can reach 10 000 m. Additional difficulties arise in implementing an adaptive mesh refinement algorithm: if the grid is refined in part of the domain by a factor of ten, say, in each spatial dimension, then typically the time step must also be decreased by the same factor. Hence, for every time step on the coarse grid it is necessary to take ten time steps on the finer grid, and information must be exchanged between the grids to maintain an accurate and stable solution near the grid interfaces.

### 4.4 Dispersive Terms

In some situations, tsunamis are generated with short wavelengths that are not sufficiently long relative to the fluid depth for the shallow-water equations to be valid. This most frequently happens with smaller localized sources such as a submarine landslide rather than with large-scale earthquakes. In this case it is often still possible to use depth-averaged two-dimensional equations, but the equations obtained typically include additional terms involving higher-order derivatives. These are generally dispersive terms that can better model the observed effect that waves with different wavelengths propagate at different speeds.

The introduction of higher-order derivatives typically requires the use of *implicit methods* to efficiently solve the equations, since the stability constraint for an explicit method generally requires a time step that is much smaller than is desirable. Implicit methods result in an algebraic system of equations that need to be solved at each time step, coupling the solution at all grid points. This is typically much more time-consuming than an explicit method.

## Further Reading

See Bourgeois (2009) for a recent survey of tsunami sedimentology and Geist et al. (2009) for a general introduction to probabilistic modeling of tsunamis. Some detailed descriptions of numerical methods for tsunami simulation can be found, for example, in Giraldo and Warburton (2008), Grilli et al. (2007), Kowalik et al. (2005), LeVeque et al. (2011), Lynett et al. (2002), and Titov and Synolakis (1998). The use of dispersive equations for modeling submarine landslides is discussed, for example, in Watts et al. (2003).

Bourgeois, J. 2009. Geologic effects and records of tsunamis. In *The Sea*, edited by E. N. Bernard and A. R. Robinson, volume 15, pp. 55–92. Cambridge, MA: Harvard University Press.

Geist, E. L., T. Parsons, U. S. ten Brink, and H. J. Lee. 2009. Tsunami probability. In *The Sea*, edited by E. N. Bernard and A. R. Robinson, volume 15, pp. 201–35. Cambridge, MA: Harvard University Press.

Giraldo, F. X., and T. Warburton. 2008. A high-order triangular discontinuous Galerkin oceanic shallow water model. *International Journal for Numerical Methods in Fluids* 56(7):899–925.

Grilli, S. T., M. Ioualalen, J. Asavanant, F. Shi, J. T. Kirby, and P Watts. 2007. Source constraints and model simulation of the December 26, 2004, Indian Ocean tsunami. *Journal of Waterway, Port, Coastal, and Ocean Engineering* 133: 414–28.

Kowalik, Z., W. Knight, T. Logan, and P. Whitmore. 2005. Modeling of the global tsunami: Indonesian tsunami of 26 December 2004. *Science of Tsunami Hazards* 23(1):40–56.

LeVeque, R. J., D. L. George, and M. J. Berger. 2011. Tsunami modeling with adaptively refined finite volume methods. *Acta Numerica* 20:211–89.

Lynett, P. J., T. R. Wu, and P. L. F. Liu. 2002. Modeling wave runup with depth-integrated equations. *Coastal Engineering* 46(2):89–107.

Titov, V. V., and C. E. Synolakis. 1998. Numerical modeling of tidal wave runup. *Journal of Waterway, Port, Coastal, and Ocean Engineering* 124:157–71.

Watts, P., S. Grilli, J. Kirby, G. J. Fryer, and D. R. Tappin. 2003. Landslide tsunami case studies using a Boussinesq model and a fully nonlinear tsunami generation model. *Natural Hazards and Earth System Sciences* 3:391–402.

# V.20 Shock Waves

*C. J. Chapman*

## 1 Introduction

A shock wave is a sudden or even violent change in pressure occurring in a thin layer-like region of a continuous medium such as air. Other properties of the medium also change suddenly: notably the velocity, temperature, and entropy. In a strong shock, produced for example by a spacecraft while still in the Earth's atmosphere, the increase in temperature is great enough to break up the molecules of the air and produce ionization. Shock waves that are "weak" in the mathematical sense (i.e., those that produce a pressure change that is only a minute fraction of atmospheric pressure) are, unfortunately, extremely strong in the subjective sense because of the exquisite sensitivity of our hearing. This fact is of great importance for civil aviation in limiting the possible routes of supersonic aircraft, which produce shock waves with focusing properties and locations that depend sensitively on atmospheric conditions.

In the modeling of a shock wave, the thin region of rapid variation may nearly always be replaced by a surface across which the properties of the medium are regarded as discontinuous. The governing equations of motion, representing conservation of mass, momentum, and energy, still apply if they are expressed in integrated form, and their application to a region containing the surface of discontinuity leads to simultaneous algebraic equations relating the limiting values of physical quantities on opposite sides of this surface. In addition, the entropy of the medium must increase on passage through the shock. The resulting algebraic relations, in conjunction with the partial differential equations of motion applied everywhere except on the shock surface, suffice for the solution of many important practical problems. For example, they determine the location and speed of propagation of the shock wave, which are usually not given in advance but have to be determined mathematically as part of the process of solving a problem.

The theory of shock waves has important military applications, especially to the design of high-speed missiles and the properties of blast waves produced by explosions. On occasion, this has stimulated world-class mathematicians (and even philosophers) to make contributions to the subject that have turned out to be enduringly practical. Ernst Mach, who with Peter Salcher in the 1880s was the first person to photograph a shock wave, explained the properties of the high-speed bullets that had been used in the Franco-Prussian war (1870–71), and in World War II John von Neumann analyzed the effect of blast waves on tanks and buildings, obtaining the surprising result that a shock wave striking obliquely exerts a greater pressure than if it strikes head-on. Richard von Mises and Lev Landau also worked on shock waves in this period, as well as fluid and solid mechanics "full-timers" such as Geoffrey I. Taylor and Theodore von Karman.

## 2 Mathematical Theory

### 2.1 Mass, Momentum, and Energy

For definiteness, consider a shock wave in air. First we present the equations that apply to a *normal shock*, through which the air flows at right angles (see also SHOCKS [II.30]). Later we consider an *oblique shock*, for which arbitrary orientations (within wide limits) are possible between the incoming and outgoing air, and the shock itself.

Three jump conditions across a normal shock are

$$\rho_1 u_1 = \rho_2 u_2,$$
$$p_1 + \rho_1 u_1^2 = p_2 + \rho_2 u_2^2,$$
$$e_1 + \frac{p_1}{\rho_1} + \tfrac{1}{2}u_1^2 = e_2 + \frac{p_2}{\rho_2} + \tfrac{1}{2}u_2^2.$$

The subscripts 1 and 2 here refer to opposite sides of the shock, and the frame of reference is that in which the shock is at rest. The variables are the densities $\rho_1$, $\rho_2$; the velocity components $u_1$, $u_2$; the pressures $p_1$, $p_2$; and the internal energies per unit mass $e_1$, $e_2$. All quantities may be taken to be positive, with the air flowing in from side 1 and out from side 2.

The first of the above equations represents conservation of mass, since the left-hand side is the mass of air per unit area per unit time flowing *into* the shock and the right-hand side is the corresponding quantity flowing *out*. The second equation accounts for the change in the momentum of air per unit volume from $\rho_1 u_1$ to $\rho_2 u_2$. This change occurs at a rate proportional to the velocity, giving the terms $\rho_1 u_1^2$, $\rho_2 u_2^2$, and is driven by the net force arising from the jump in pressure, giving the terms $p_1$, $p_2$. The third equation accounts for the change in the energy of the air per unit mass from $e_1 + \frac{1}{2}u_1^2$ to $e_2 + \frac{1}{2}u_2^2$, comprising internal energy and kinetic energy. The forces that produce this change in energy come from the pressure and give the terms $p_1/\rho_1$, $p_2/\rho_2$. It is convenient to combine each internal energy and pressure term into a single quantity: the enthalpy. Then, in terms of the enthalpies $h_1$, $h_2$, defined by $h_1 = e_1 + p_1/\rho_1$ and $h_2 = e_2 + p_2/\rho_2$, the energy equation is

$$h_1 + \tfrac{1}{2}u_1^2 = h_2 + \tfrac{1}{2}u_2^2.$$

## 2.2 Entropy

A fourth jump condition across a shock concerns the entropy, which is a measure of the energy in the disordered molecular motions that is not available to perform work at the macroscopic scale. The condition is that the entropy of the material passing through the shock must increase. Although the entropy condition is "only" an inequality, it is of fundamental importance in determining the type of shocks that can or cannot occur. For example, the condition shows that, in almost all materials, a rarefaction shock (at which the pressure suddenly falls) is impossible. A shock in air therefore has the property that on passing through it the air undergoes an increase in pressure, density, temperature, internal energy, and enthalpy and a decrease in velocity.

For air in conditions that are not too extreme, the entropy per unit mass per degree on the absolute temperature scale is proportional to $\ln(p/\rho^{\gamma})$, where $\gamma$ is the ratio of specific heats, which is approximately a constant. Thus, with the sign convention that $u_2$ and $u_1$ in the jump conditions are positive, the entropies $s_2$ and $s_1$ on opposite sides of the shock satisfy the condition $s_2 > s_1$, so that

$$p_2/\rho_2^{\gamma} > p_1/\rho_1^{\gamma}.$$

## 2.3 The Rankine–Hugoniot Relation

Simple algebraic manipulation of the jump conditions gives

$$h_2 - h_1 = \frac{1}{2}\left(\frac{1}{\rho_1} + \frac{1}{\rho_2}\right)(p_2 - p_1).$$

This is an especially useful relation because $h_2$ and $h_1$ are functions of $(p_2, \rho_2)$ and $(p_1, \rho_1)$, respectively. Thus, if $(p_1, \rho_1)$ are fixed, representing known upstream conditions, the Rankine–Hugoniot relation gives the *shock adiabatic*, relating $p_2$ to $\rho_2$, i.e., the pressure and density downstream.

## 2.4 The Mach Number

A basic property of a normal shock, deducible from the four jump conditions, is that the flow into it is supersonic, i.e., faster than the speed of sound, and the flow out of it is subsonic, i.e., slower than the speed of sound. This provides a strong hint that we should use variables based on the incoming and outgoing *Mach numbers*, $M_1$ and $M_2$, defined as the ratio of the flow speeds to the local speed of sound, denoted by $c_1$ and $c_2$. Thus

$$M_1 = \frac{u_1}{c_1} > 1, \qquad M_2 = \frac{u_2}{c_2} < 1.$$

We also have available the equation of state for air, $p = \rho RT$, and the innumerable formulas of thermodynamics, from which we select merely

$$c^2 = \gamma RT = \gamma p/\rho = (\gamma - 1)h.$$

Here $R$ is the gas constant, $T$ is absolute temperature, and the formulas apply with subscript 1 or subscript 2 (but the same values of $\gamma$ and $R$) on each side of the shock. We now have simultaneous equations in abundance, and a little algebra starting from the jump conditions gives

$$M_2^2 = \frac{1 + \frac{1}{2}(\gamma - 1)M_1^2}{\gamma M_1^2 - \frac{1}{2}(\gamma - 1)}.$$

### 2.5 Pressure and Temperature Rise

With the aid of the relation between $M_2^2$ and $M_1^2$, the change in any quantity at a normal shock may readily be expressed in terms of $M_1^2$ and $\gamma$ alone. Such a representation is highly practical in calculations. For example, an important aspect of a shock wave is that it can produce considerable increases in pressure and temperature. But how do these increases depend on the Mach number of the incoming flow? The answer for a shock wave in air is

$$\frac{p_2}{p_1} = 1 + \frac{2\gamma}{\gamma + 1}(M_1^2 - 1)$$

and

$$\frac{T_2}{T_1} = \frac{(1 + \frac{1}{2}(\gamma - 1)M_1^2)(\gamma M_1^2 - \frac{1}{2}(\gamma - 1))}{\frac{1}{4}(\gamma + 1)^2 M_1^2}.$$

The functional form of this last equation is important for modeling purposes. Because the specific-heat ratio $\gamma$ is of "order one" (it is approximately 1.4 for atmospheric air), the equation shows that at high Mach numbers the absolute temperature ratio $T_2/T_1$ scales with $M_1^2$. Thus, during the acceleration phase of the flight of a hypersonic vehicle, the simple equation of state we have been using, $p = \rho RT$, ceases to apply long before operating speed is attained. Physical processes beyond the reach of this "ideal gas equation" must be included, and the subject of shock waves rapidly starts to overlap with the chemistry of the breakup of the molecules and with the physics of ionized gases. Applied mathematicians such as Sir James Lighthill and John F. Clarke have played an important role in developing the required theory of such strong shock waves.

## 3 Oblique Shock Waves

### 3.1 Usefulness in Modeling

Oblique shock waves provide a versatile tool for modeling the flow around a supersonically moving body of any shape. The reason for this is that the local change in flow direction produced by the surface of the body always produces an oblique shock somewhere in the flow. A concave corner, for example, produces an oblique shock attached to the corner, and the apex of a forward-facing wedge produces oblique shocks on each side of the wedge, attached to the apex.

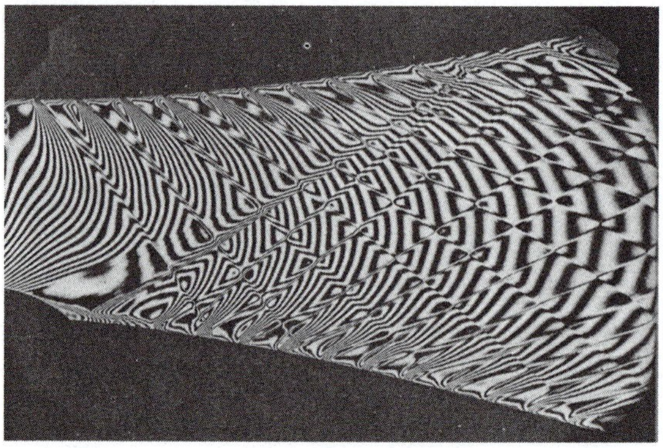

**Figure 1** Weak shock waves (Mach surfaces) in a supersonic nozzle.

Moreover, a smooth concave part of the body surface produces a flow field in which disturbances propagate on surfaces whose orientation is that of oblique shocks of vanishingly small strength. These *Mach surfaces* come to a focus a short distance away from a concave boundary to produce a full-strength oblique shock. In general, a Mach surface is a surface on which a weak disturbance in pressure and associated quantities can propagate in accordance with the equations of acoustics as a *Mach wave*. The theory of weak shock waves is subsumed under acoustic theory in such a way that a weak shock corresponds to a characteristic surface of the acoustic wave equation. Shocks and characteristic surfaces do not in general coincide, but they do in the limit of zero shock strength.

Oblique shock waves are fundamental to the modeling of shock wave intersections and reflections. In a fluid, shock intersections produce vorticity, localized in surfaces called *slip surfaces*, across which there is a jump in the tangential component of velocity. Shock reflections can be of different types, depending on the orientation and strength of the incoming shock. A common type is the *Mach reflection*, in which a nearly straight shock wave, known as the stem, extends from the body surface to a triple intersection of shocks, from which there also emerges a slip surface.

Many excellent photographs of oblique shock wave patterns may be found in Van Dyke (1982). Figures 1 and 2 are examples of science as art.

### 3.2 Flow Deflection

The most basic question one can ask about an oblique shock is the following: if the incoming flow is at Mach number $M_1$ and the flow is deflected by an angle $\theta$ from

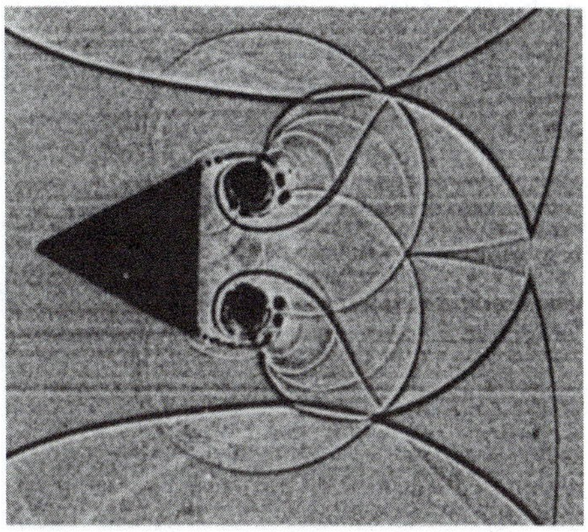

**Figure 2** Reflection of a shock wave by a wedge, and rolling up of the vortex sheets (slip surfaces) produced.

its original direction, what are the possible oblique shock angles $\phi$ that can bring this deflection about? The angle $\phi$ is that between the shock surface and the incoming flow, so that a normal shock corresponds to $\theta = 0$ and $\phi = \pi/2$. The question is readily answered by starting with the formulas for a normal shock and superposing a uniform flow parallel to the shock surface. A little algebra and trigonometry give

$$\tan\theta = \frac{(M_1^2 \sin^2\phi - 1)\cot\phi}{1 + (\frac{1}{2}(\gamma + 1) - \sin^2\phi)M_1^2}.$$

At a single value of $M_1$, the graph of $\theta$ against $\phi$ rises from $\theta = 0$ at $\phi = \sin^{-1}(1/M_1)$ to a maximum $\theta = \theta_{\max}(M_1)$ at a higher value of $\phi$, before falling back to $\theta = 0$ at $\phi = \pi/2$. Thus the shock angle and flow deflection are limited to definite intervals. The angle $\sin^{-1}(1/M_1)$ is the *Mach angle* and gives the Mach wave described above.

As $M_1$ is varied, there is a maximum possible value of $\theta_{\max}(M_1)$, attained in the limit of large $M_1$, and there is a corresponding value of $\phi$. For air in conditions that are not too extreme, with a ratio of specific heats close to $\gamma = 1.4$, we obtain the useful result that the greatest possible flow deflection at an oblique shock is 46°, and the corresponding shock angle is 68°.

For given $M_1$, and a deflection angle in the allowable range, there are two possible shock angles $\phi$. The smaller angle gives the weaker shock, and, except in a narrow range of deflection angles just below $\theta_{\max}(M_1)$, a supersonic outflow; the outflow from the stronger shock, at the larger angle, is always subsonic. Thus, although the inflow must be supersonic, the outflow may be subsonic or supersonic, depending on the magnitude of the equivalent superposed uniform flow on a normal shock referred to above. The ultimate explanation of such matters is the entropy condition. Whether or not the outflow is supersonic is important in practice because a supersonic outflow may contain further shock waves. Which shock pattern actually occurs may depend on the downstream conditions.

## 4 Dramatic Examples of Shock Waves

### 4.1 Asteroid Impact

The Earth has a long memory: the shock wave produced by the asteroid that extinguished the dinosaurs (and hence led to human life, via the rise of mammals) left a signature that is still visible in the Earth's crust. The asteroid struck Earth 66 million years ago near Chicxulub in the Yucátan Peninsula in Mexico, producing a crater 180 km in diameter. Besides generating a TSUNAMI [V.19] in the oceans, which mathematically speaking evolves into a type of shock wave near the shore, it generated shock waves in the rocks of the Earth's crust. These shock waves propagated around the Earth, triggering an abundance of violent events, such as earthquakes and the eruption of volcanoes.

Geologists have determined which metamorphic features of rock can be produced only by the intense pressure in the shock waves from a meteorite or asteroid, one example being the patterns of closely spaced parallel planes that are visible under a microscope in grains of shocked quartz. The mathematical theory of shock waves in solids, such as metal and rock, is highly developed.

### 4.2 The Trinity Atomic Bomb Test

The world's first nuclear explosion took place in July 1945, in the desert of New Mexico in the United States, when the U.S. army tested an atomic bomb with the code name Trinity. The explosion was filmed, and the film, which showed the hemispherical shock wave rising into the atmosphere, was released not long afterward.

For all the complexity of the physical processes and fluid dynamics in the explosion, it is remarkable that a simple scaling law describes to high accuracy the radius $r$ of the shock wave as a function of time $t$ after detonation, in which the only parameters are the explosion energy $E$ and the initial air density $\rho$. The scaling law is

$$r = C\left(\frac{Et^2}{\rho}\right)^{1/5},$$

where $C$ is a constant close in value to 1. With the aid of this scaling law, and a logarithmic plot, G. I. Taylor was able to determine the energy of the explosion rather accurately using only the film as data.

### 4.3 Shock Wave Lithotripsy

A medical treatment for a kidney stone is to break up the stone by applying a pulsed sound wave of high intensity that is brought to a focus at the stone. The focusing causes the pulses to become shock waves, in which the sudden rise in pressure provides the required disintegrative force on the stone. The machine used is called a lithotripter; it is designed so that the frequency and intensity of the shock waves can be controlled and varied as the treatment progresses. A water bath is applied to the patient's back, so that the pulse propagates through water and then tissue.

### Further Reading

Chapman, C. J. 2000. *High Speed Flow*. Cambridge: Cambridge University Press.

Ockendon, H., and J. R. Ockendon. 2004. *Waves and Compressible Flow*. New York: Springer.

Van Dyke, M. 1982. *An Album of Fluid Motion*. Stanford, CA: Parabolic Press.

(a)

(b)

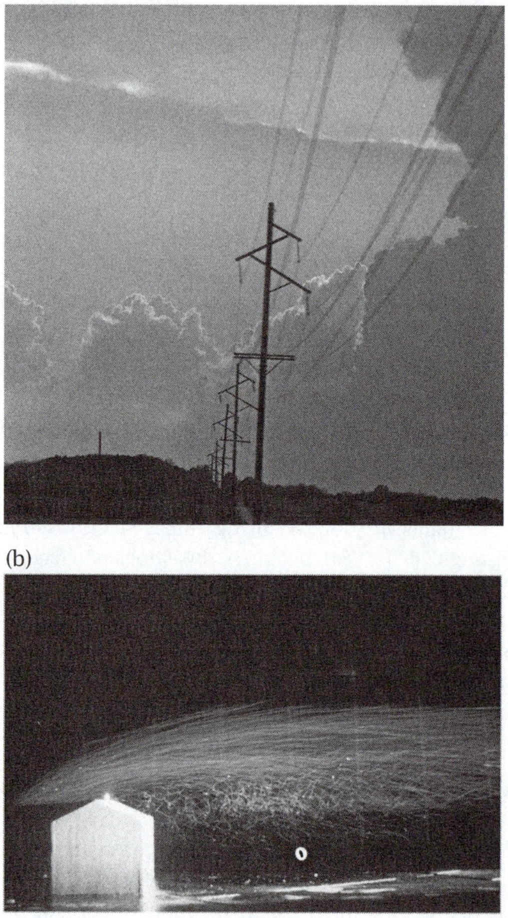

**Figure 1** (a) Random edges of turbulent clouds. (b) Turbulence in the wake of a model building in a wind tunnel, showing variations in random motions of fluid particles. (Courtesy of D. Hall.)

# V.21 Turbulence

*Julian C. R. Hunt*

## 1 An Introduction to the Physical and Mathematical Aspects of Turbulence

Turbulence in artificially and naturally generated fluid flows consists of a wide range of random unsteady eddy motions that differ over varying sizes and frequencies, examples being those observed in the wake of a model building in a wind tunnel and in clouds (figure 1). In the eddies with the greatest energy, the velocity fluctuations are of order $u_0$ and their typical size is of order $L_0$ (figure 2).

Turbulent flows differ greatly from LAMINAR FLOWS [IV.28], which are stable and predictable even if they are unsteady. The two kinds of flow depend differently on the kinematic viscosity $\nu$, which is a property of fluids that is independent of the velocity and is determined by very small molecular motions on scales unrelated to the scale of the flow. The study of turbulence begins with defining, mathematically and conceptually, first the statistical framework for the kinematics and dynamics of random flow fields and then the mechanisms for transition processes between laminar and turbulent flows, which occur when the dimensionless *Reynolds number* $Re = u_0L_0/\nu$ is large enough for the fluctuating viscous stresses to be small. The goal of turbulence research is not, realistically, to find an overall theory but rather to describe and explain characteristics, patterns, and statistical properties found in different types of turbulent flows or particular regions of turbulent flows. Remarkably, the smallest-scale eddy motions have an approximately universal statistical structure.

An equally significant but more practical result is the Prandtl–Karman theory, which states that the mean velocity near all kinds of resistive surfaces has a

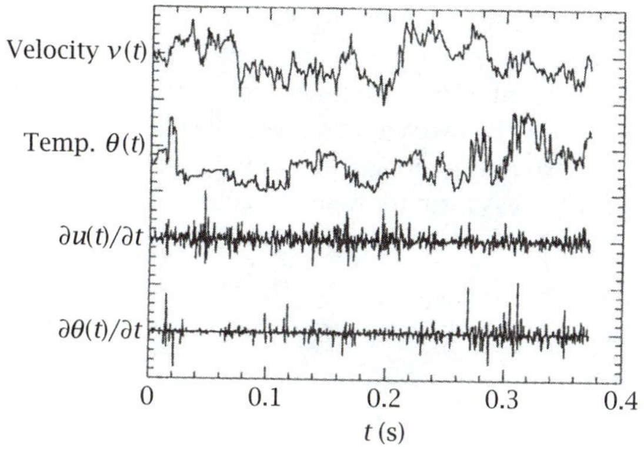


**Figure 2** Measurements of turbulent velocities in a wind tunnel at $Re \sim 10^5$ (courtesy of Z. Warhaft: taken from "Passive scalars in turbulent flows" (*Annual Review of Fluid Mechanics* 32:203–40, 2000)).

logarithmic profile. Turbulence concepts apply in engineering and environmental flows, as well as in oceans and planetary atmospheres, where the eddy sizes range over factors of hundreds to thousands, while in the interior of the Earth, on other planets, in stars, and even in the far constellations, the range can extend to factors of millions or more. Partially turbulent motions are observed in the flows along air passages and in the larger blood vessels of humans and large mammals, where they greatly influence the vital transport of liquids, gases, and particles.

The fluid media in which turbulence occurs may consist of liquids, gases, or multiphase mixtures with droplets, bubbles, or particles. Some complex fluid media behave like fluids only under certain conditions and for limited periods of time, examples being low-density ionized gases in space or mixtures of solids and fluids in mud and volcanoes.

In the early twentieth century, leading scientists and mathematicians including Ludwig Prandtl, G. I. Taylor, Theodore von Kármán, and Lewis Fry Richardson established the systematic study of turbulence based on the principles of fluid dynamics and statistical analysis that had been developed during this period. The combination of these mathematical approaches, together with limited experimental results and hypotheses based on the ideas of statistical physics, led to Andrey Kolmogorov and Alexander Obukhov's general statistical theory in 1941. New measurement technologies and greater computational speed and capacity have shown how eddy structure changes at very high Reynolds number ($Re \geqslant 10^5$), and this is leading to revision and clarification of physical concepts and statistical models.

There are differences between turbulence and other kinds of random motions in fluids, such as waves on liquid surfaces or the chaotic motions of colliding solid particles. In turbulence, localized flow patterns adjust faster to local conditions than typical wave motions.

The aim of this article is to show how mathematically based research into turbulent flows, with the aid of experiments and computations, has led to methods for exploring the qualitative and quantitative aspects of turbulence. Key physical and statistical results are explained.

## 2 Randomness and Structure

The mathematical concepts and terminology that are used to measure, describe, and analyze the random fields that occur in turbulent flows are the same as those used for other random processes. The turbulent velocity field, $\boldsymbol{v}^*$, is broadly like other continuous three-dimensional random processes. It and other variables at given points in space ($\boldsymbol{x}$) and time ($t$) cannot be exactly predicted when the Reynolds number *Re* exceeds a critical value for the region of flow where turbulence exists, $\mathcal{D}_{\mathrm{T}}$ (this may be a subregion of the whole flow).

A large number of experimental or computational trials are needed to calculate for a given flow the probability, defined as $\mathrm{pr}(v^*)\,\mathrm{d}v^*$, of a single component $v^*$ of the variable lying between two values $v^*$ and $v^* + \mathrm{d}v^*$. The *mean value* of $v^*$ is

$$\langle v^* \rangle = \int_{-\infty}^{\infty} v^* \,\mathrm{pr}(v^*)\,\mathrm{d}v^*.$$

The fluctuation of $v^*$ relative to its mean value is

$$v = v^* - \langle v^* \rangle.$$

These "ensemble" statistical properties are mainly expressed in terms of the moments, correlations, and spectra of the fluctuation $v$, but particular features of the flow can be defined by calculating conditional statistics when variables lie within certain ranges or have particular properties.

The $m$th moment of the fluctuation is defined as

$$M^{(m)} = \langle v^m \rangle = \langle (v^* - \langle v^* \rangle)^m \rangle.$$

Thus $M^{(1)} = \langle v \rangle = 0$ and $M^{(2)} = \langle v^2 \rangle \equiv v'^2$, the variance of $v$ ($v'$ is the standard deviation), which defines the width of the distribution and, broadly, the magnitude of fluctuations. The *skewness* Sk, which is defined by $\mathrm{Sk} = M^{(3)}/v'^3$, indicates whether large fluctuations are positive or negative (see figure 2, where Sk = 0).

In many types of turbulent flows, such as in a wind tunnel, the statistics do not change with time and are spatially homogeneous (in one or more directions). Time averages taken over periods between $t_0$ and $t_0 + \mathcal{T}$ are then used to calculate mean or higher $m$th moments, denoted by overbars. For example,

$$\overline{(v^*)^m_{\mathcal{T}}}(\boldsymbol{x}) = \frac{1}{\mathcal{T}} \int_{t_0}^{t_0+\mathcal{T}} [v^*(\boldsymbol{x}, t)]^m \, \mathrm{d}t.$$

Similarly, spatial averages can be taken over distances between two points, $x_{10}$ and $x_{10} + X_1$:

$$\overline{(v^*)^m_{X_1}}(x_2, x_3, t) = \frac{1}{X_1} \int_{x_{10}}^{x_{10}+X_1} [v^*(x_1, x_2, x_3, t)]^m \, \mathrm{d}x_1.$$

In "ergodic" flows the time and/or space averages are well defined and are the same for all experiments (e.g., by turning on and off the flow). The space and time means ($m = 1$) are equal to the ensemble mean $\langle v^* \rangle$.

However, there are also well-defined "non-ergodic" turbulent flows, where time means in one experiment differ significantly from those in other experiments in the same flow. In these types of flows, a number of different, but persistent, large-scale flow patterns occur, given the same initial or boundary conditions. Such nonuniqueness is found in flows in containers with one or more axes of symmetry, for example, in the decay of swirling flows in ellipsoidal containers. In such cases, the ensemble mean does not correspond to the space/time average in any given flow.

In most turbulent flows, $\mathrm{pr}(v)$ has a single maximum (or mode) where $v = \hat{v}$, which is the most frequent value of the fluctuations. Where $\mathrm{Sk} > 0$, as in the vertical velocity in cumulus clouds (figure 1), there is a greater probability of large values of positive $v$ than of large negative values. But since the latter are the most frequent values, $\hat{v} < 0$.

When studying probability distributions it is usual to start by comparing them with $\mathrm{pr}(v)$ for the most general and least repeatable type of random variable that can be conceived: namely, the sum of an infinite number of independent random variables. This is the Gaussian or normal distribution:

$$p_{\mathrm{G}}(v) = \frac{1}{\sqrt{2\pi v'}} \exp(-(v - \langle v \rangle)^2/2v'^2).$$

For many of the random variables observed in turbulent flows this is a useful approximation. But if $\mathrm{pr}(v)$ slightly differs from $p_{\mathrm{G}}(v)$, this is physically significant because it indicates repeatability and structure.

The average sizes of turbulent eddies with different levels of energy can be estimated from statistical measurements of space-time two-point correlations, defined as the average of the product of the velocities at two points along a line in the $x_1$-direction. The dimensionless form for homogeneous turbulence is

$$C(r_1, x_1, t) = \frac{\langle v(x_1, t) v(x_1 + r_1, t) \rangle}{\langle v^2 \rangle (x_1, t)}.$$

Integrals of $C$ define integral timescales and length scales,

$$T = \int_{-\infty}^{\infty} C \, \mathrm{d}t, \qquad L = \int_{-\infty}^{\infty} C \, \mathrm{d}r_1.$$

Turbulence structure and statistics are also described in terms of characteristic random modes. Consider a variable $v(x_1)$, expressed as a sum of the products of space filling, nonrandom modes $\phi_n(x)$ (e.g., sinusoidal waves with wave numbers $k_n = 2\pi n/X_1$) with random coefficients $a_n$:

$$v(x) = \sum_{n=1}^{N} a_n \phi_n(x).$$

The modes are assumed to be orthogonal to each other over the space $X_{\mathrm{L}}$ to $X_{\mathrm{U}}$, i.e.,

$$\int_{X_{\mathrm{L}}}^{X_{\mathrm{U}}} \phi_m(x) \phi_n(x) \, \mathrm{d}x = \delta_{mn},$$

where $\delta_{mn}$ is the KRONECKER DELTA [I.2 §2, table 3]. The *spectrum* $E^{(v)}(k_n)$ of $v(x)$ is defined as the mean square of the amplitude coefficients:

$$\langle a_n^2 \rangle = E^{(v)}(k_n).$$

The sum of the energy of the modes is equal to the variance, $\sum_n E^{(v)}(k_n) = v'^2$. The Karhunen–Loève and Wiener–Khintchine theorems show how the correlation $C$ and the spectrum are linked. For example, fluctuations correlated over small distances $r_1$ ($\ll L$) also contribute most of the energy over this length scale, i.e., $E(k_n)$, where $k_n \sim L/r_1$.

In an infinite space, $E(k_n)$ tends to a continuous spectrum $E(k)$ as $N \to \infty$. Note that the spectra of velocity gradients $\partial v/\partial x \equiv v_x$, i.e., $E^{(v_x)}(k)$, are equal to $k^2 E(k)$, so that $E^{(v_x)}(k)$ is greatest where $k$ is large and eddies are small.

Sometimes there is a "spike" in the spectrum where energy is concentrated at a particular wave number $k_n$ or frequency $\omega_n$, such as vortex shedding downwind of an aircraft wing.

In turbulence, the fluctuating velocity is a three-dimensional vector $\boldsymbol{v}$ in three-dimensional space with three components, $\boldsymbol{v} = (v_1, v_2, v_3)(\boldsymbol{x}, t)$, the total variance being $v'^2 = \langle v_j v_j \rangle$ (summing over $j$ where indices are repeated).

The anisotropy of turbulence variables can be expressed independently of the frame of reference in terms of the variances of the components and of the correlations between the components in different directions, e.g., $\langle v_i v_j \rangle / v'^2$.

## 3 Dynamics in a Statistical Framework

Based on Newton's second law, the rate of change of momentum of a fluid element (which is much larger than any molecules, but much smaller than the scale of fluid motions), with density $\rho^*(\boldsymbol{x}^*, t^*)$ and moving with a velocity $v_i^*(\boldsymbol{x}^*, t^*)$, is equal to the molecular forces acting on it caused by the gradients of fluid pressure $p^*(\boldsymbol{x}^*, t^*)$, viscous stresses $\tau_{ij}^*$, and a specified body force $F_i^*(\boldsymbol{x}^*, t^*)$.

The momentum and continuity equations can be expressed in nondimensional forms by expressing distances in relation to a fixed overall scale $L^*$, velocity vectors in terms of a reference velocity $v_0^*$, and the time variation in terms of $L^*/v_0^*$. From the NAVIER–STOKES EQUATIONS [III.23],

$$\frac{\mathrm{D}u_i}{\mathrm{D}t} = \frac{\partial u_i}{\partial t} + u_j \frac{\partial u_i}{\partial x_j} = -\frac{\partial}{\partial x_i}(\bar{p} + p) + \frac{1}{Re}\nabla^2 u_i + \bar{f}_i + f_i, \qquad (1)$$

where $x_i = x_i^*/L^*$, $t = t^* v_0^*/L^*$, $u_i = v_i^*/v_0^*$, and the mean and fluctuating values for $p^*$, $F_i^*$ are

$$\bar{p} + p = \frac{p^*}{\rho_0 (v_0^*)^2}, \qquad \bar{f}_i + f_i = \frac{F_i^*}{\rho_0 (v_0^*)^2 / L^*}.$$

Since the densities of fluid elements change little in most natural and engineering turbulent flows, the fluctuating velocity field is not divergent, i.e., $\nabla \cdot \boldsymbol{u} = \partial u_j / \partial x_j = 0$.

Note that the equations now contain the single nondimensional Reynolds number $Re = v_0^* L^* / \nu$, so that the effect of varying the velocity scale, the length scale, or the kinematic viscosity (by the same amount) is the same as varying the single parameter $Re$. For turbulent flows, $v_0^*$ may be taken as the (dimensional) standard deviation $v'^*$.

To study the flow field's statistical properties, the nondimensional velocity and pressure are expressed in terms of their mean and fluctuating values, $U_i(\boldsymbol{x}, t)$ and $u_i(\boldsymbol{x}, t)$ (see section 2).

Taking the divergence of (1) gives

$$\nabla^2(\bar{p} + p) = -(\Sigma^2 - \tfrac{1}{2}\omega^2) + \frac{\partial(\bar{f}_i + f_i)}{\partial x_i},$$

where $\boldsymbol{\omega} = \nabla \times \boldsymbol{u}$ is the vorticity vector, $\omega = |\boldsymbol{\omega}|$, and

$$\Sigma^2 = \Sigma_{ij}\Sigma_{ij}, \quad \Sigma_{ij} = \frac{1}{2}\left(\frac{\partial u_i}{\partial x_j} + \frac{\partial u_j}{\partial x_i}\right).$$

Thus, the pressure, which is determined by the divergence of the inertial and body force terms, has maximum or minimum values where the strain $\Sigma$ is greater than or less than $\omega/\sqrt{2}$.

For incompressible fluids the viscous stresses, $\tau_{ij}$, are proportional to $\Sigma_{ij}$, $\tau_{ij} = (2/Re)\Sigma_{ij}$. Note that $\tau_{ij}$ is zero both in pure rotational flows (i.e., $\omega \neq 0$ and $\Sigma = 0$), such as at the center of a vortex, and in inviscid flows, where $Re \to \infty$. But the local viscous force at a point in the flow, which is equal to the stress gradient, $\partial \tau_{ij}/\partial x_j = -Re^{-1}(\nabla \times \boldsymbol{\omega})_i$, is zero in pure straining flows (i.e., $\omega = 0$) and where the vorticity is uniform.

The dissipation of energy is given by $\varepsilon = \Sigma_{ij}\tau_{ij}$. It is proportional to $Re^{-1}$ and is also zero locally for pure rotational flows, but its total integral in an isolated flow region $\mathcal{D}_{\mathrm{T}}$ is proportional to $\omega^2$, i.e., $\varepsilon = (2/Re)\int \omega^2 \,\mathrm{d}V$.

Taking the curl of (1) eliminates the pressure gradient and leads to an equation for the vorticity $\boldsymbol{\omega}(\boldsymbol{x}, t)$. In dimensionless vector terms

$$\frac{\mathrm{D}\boldsymbol{\omega}}{\mathrm{D}t} = (\boldsymbol{\omega} \cdot \nabla)\boldsymbol{u} + \frac{1}{Re}\nabla^2 \boldsymbol{\omega} + \nabla \times (\bar{\boldsymbol{f}} + \boldsymbol{f}), \qquad (2)$$

which shows how the inertial straining term leads to high and low variations in $\boldsymbol{\omega}$. The main effect of viscosity in (2) is to "diffuse" vorticity from regions of high vorticity to regions of low vorticity, especially from/to boundaries. If $Re \ll 1$, vorticity is diffused and there are no sharp peaks in vorticity.

The body force affects only the vorticity and, from there, the velocity field if it is rotational; for example, if the force is determined by a potential $\Phi$, $\bar{\boldsymbol{f}} + \boldsymbol{f} = \nabla\Phi$, it affects only the pressure.

The momentum equation (1) defines how the mean and fluctuating components of the turbulent flow field, denoted by $U_i = \langle u_i \rangle$ and $u_i$, are connected dynamically. From there, the ensemble average of the rate of change of the turbulent kinetic energy of the fluctuations $K = \frac{1}{2}\langle u_i u_i \rangle$ is

$$\frac{\partial K}{\partial t} + U_i \frac{\partial K}{\partial x_i} = -\langle u_i u_j \rangle \frac{\partial U_i}{\partial x_j} + \frac{1}{Re}\langle u_i \nabla^2 u_i \rangle + \frac{\partial}{\partial x_j}(\langle u_j p \rangle + \tfrac{1}{2}\langle u_k u_k u_j \rangle) + \langle u_i f_i \rangle, \qquad (3)$$

or, schematically, $\{A\} = \{P\} - \langle\varepsilon\rangle + \{ET\} + \{F\}$. Here, $\{A\}$ is the time variation and advection of $K$; $\{P\}$ is the production of $K$ by interactions between the fluctuations and gradients of the mean velocity; $\langle\varepsilon\rangle$ is the viscous dissipation by the smallest eddies, determined by the correlation between $u_i$ and viscous stresses; $\{ET\}$ is the transfer of energy by the fluctuations, i.e., by gradients of the cross correlations of $u_j$ and $p$, and by gradients of third-order moments; and $\{F\}$ is the rate of production of $K$ by the forcing term, which is equal to the correlation between $u_i$ and $f_i$.

For homogeneous flows without mean gradients or fluctuating body forces, (3) simply shows that $K$ decays in proportion to the dissipation:

$$\frac{\mathrm{D}K}{\mathrm{D}t} = -\langle\varepsilon\rangle. \tag{4}$$

This process in a wake is depicted in figure 1(b).

## 4 Transition to Fully Developed Turbulence

The transition from laminar flow to well-developed turbulence may take place in part of a flow, such as over an aircraft wing, or in the whole flow, such as in a pipe. The turbulence may develop for a certain time and then decay back into laminar flow.

There are different stages in the transition to turbulence, which evolves from growing small-amplitude fluctuations, depending on the boundary conditions and the Reynolds number, $Re$. One class of initially transitional fluctuations are "modal," when all the components in different directions of the flow grow at the same rate within a defined flow domain $\mathcal{D}_{\mathrm{T}}$. The most unstable modes have a defined wavelength and frequency, so that the initial energy spectrum may be in a "spike." By contrast, in "nonmodal" fluctuations, different components have different rates of growth despite interacting with each other (see FLUID MECHANICS [IV.28 §8]).

At the beginning of the transition process (in space or time, or as $Re$ increases), the dynamics are linear. In shear layers with velocity profile $U(x_3)$ with an inflection point where $\mathrm{d}^2U/\mathrm{d}x_3^2 = 0$, the undulating "Kelvin–Helmholtz" modal fluctuations $\tilde{u}$ are exponentially growing at a rate defined by $\sigma$, where $\tilde{u} \propto \exp(\sigma t)$. In other parts of the flow, fluctuations may not exist. In swirling flow between circular cylinders, the axisymmetric vortex-like modal fluctuations of G. I. Taylor extend across the whole flow. These modal fluctuations can grow only when $Re$ exceeds the critical value $Re_{\mathrm{crit}} = \Delta U_0^* L_0^*/\nu$, where $\Delta U_0^*$ is the change in mean velocity across the flow and $L_0^*$ is the length scale. $Re_{\mathrm{crit}}$ is independent of the magnitude of $\tilde{u}$ provided it is small. Different kinds of nonlinear development occur as these fluctuations grow or change into other modal forms.

Some basic laminar flows $U_i(\boldsymbol{x}, t)$, such as shear flow between planes or in the center of a vortex, are stable to small modal disturbances, but low-amplitude nonmodal disturbances can grow in these flows if $Re$ exceeds a critical value, $Re'_{\mathrm{crit}}$, defined by root mean square fluctuations. In a shear flow, arbitrary initial disturbances are transformed into well-defined eddy structures such as the elongated streaks of high and low streamwise fluctuations. The spectra change from a narrow to a broad band shape as modal and nonmodal fluctuations develop and become nonlinear.

The transition process affects the structure of the fluctuations and their spectral energy distribution, ranging from large scales to the smallest viscously affected eddy motions. Typically, the changes occur on a timescale of the most energetic eddies, so that interactions take place between all the scales down to the smallest. In general, whatever the initial transition process, and whatever the boundary conditions, all types of turbulence retain some of their initial characteristic features over extended periods, as is discussed in the next section.

However, most types of "well-developed turbulence" also have a number of common features so long as the Reynolds number exceeds its critical value for transition, $Re_{\mathrm{trans}}$. There may be multiple values of $Re_{\mathrm{trans}}$ above which further changes in the structure can occur.

## 5 Homogeneous Turbulence

The basic mechanisms and statistical structure of turbulence are best studied where the velocity fluctuations are homogeneous and where they are not affected by boundaries or the eddy structures generated by large-scale gradients of the mean and fluctuating velocities. Nevertheless, even homogeneous turbulent flows depend on the initial or continuing forcing.

Turbulent flows consist of random fluctuations and eddies, i.e., local regions of vortical flow confined within moving envelopes. At high Reynolds number, the boundaries of flow regions like clouds tend to be sharply defined, with high local gradients of velocity. Nonlinear fluctuations and eddies evolve internally and interact with other eddies.

In a fixed frame of reference (or one moving with the mean velocity), the one-dimensional spatial spectrum $E_{11}(k_1)$ for wave number $k_1$ in the $x_1$-direction is always finite at $k_1 = 0$, i.e., $E_{11}(0) = \langle u_1^2 \rangle L_x/\pi$. The integral scale $L_x$ defines the size of the eddies containing most of the energy.

Also, $E_{11}(k_1)$ tends monotonically to zero as $k_1 L_x \to \infty$, which implies that the smallest eddies on average have small velocities. Over the range $0 < k_1 L_x \ll 1$, the spectrum remains of the order of $E_{11}(0)$. For most well-developed flows, $E_{11}(k_1)$ has a single maximum where $k_1 L_x \sim 1$.

The "Eulerian" frequency spectrum for the $x_1$ component is denoted by $E_{11}^E(\omega)$. The integral timescale $\mathcal{T}_\mathrm{E} = (2\pi u_1')^{-1} E_{11}^E(0)$, where $u_1' = \langle u_1^2 \rangle^{1/2}$, is broadly similar to the spatial spectrum. In shear flows with "coherent structures," $\mathcal{T}_\mathrm{E}$ can be significantly larger.

At high Reynolds number, over smaller scales and higher frequencies, the frequency and spatial spectra are proportional to each other,

$$E_{11}^E(\omega) \sim (1/u_1')E_{11}(k_1),$$

where $k_1 = \omega u_1'$. These results demonstrate how large eddies advect small eddies randomly at speeds approximately equal to the root mean square velocity $u_1'$, which explains why $\mathcal{T}_\mathrm{E} \sim L_x/u_1'$.

The time dependence of the velocity or displacement $X_i(t; X_i(t_0))$ of a fluid particle released at $X_i(t_0)$ over numerous experiments defines the "Lagrangian" time-dependent statistics of its velocity $u_i(t; X_i(t_0)) \equiv u_i(t)$. The auto-correlation (for stationary, homogeneous turbulence) of the velocity over a time interval $\tau$, defined as $C_{ii}^L(\tau) = \langle u_i(t) u_i(t+\tau) \rangle$, where $C_{ii}^L(0) = \langle u_i^2 \rangle$, also defines the Lagrangian timescale $\mathcal{T}_\mathrm{L} = (u')^{-2} \int_0^\infty C_{ii}^L(\tau)\,\mathrm{d}\tau$. $\mathcal{T}_\mathrm{L}$, which is the timescale for the velocity of a fluid particle as it is carried around a large eddy, is of the same order as $\mathcal{T}_\mathrm{E}$ and also the time for pairs of particles to separate. Thus $\mathcal{T}_\mathrm{L}$ is also the "interaction" time, $\mathcal{T}_\mathrm{i}$, for the large eddies to interact with each other.

The decay of homogeneous turbulence with time depends on the mean loss of energy by viscous dissipation, $\langle \varepsilon \rangle$, as shown by (4).

Since $\varepsilon$ is proportional to the square of the velocity gradients, which are greatest for the smallest scales, the mean value is related to the energy spectrum by

$$\langle \varepsilon \rangle = \frac{1}{Re} \int_0^\infty k^2 E(k)\,\mathrm{d}k.$$

At high *Re*, these dissipative eddies are quasilaminar thin shear flows and vortices, with typical velocities $u_0$ and length scale $L_0$, and characteristic thickness $\ell$ of order $L_0 Re^{-1/2}$. Consequently, the dissipation rate within the layer, $\varepsilon$, is of order $u_0^3/L_0$, which is independent of the value of *Re*. Outside these layers, $\varepsilon$ is very small. Even smaller-scale scattered vortices contribute significantly to $\langle \varepsilon \rangle$ (see below). High-resolution computer simulations demonstrate that

$$\langle \varepsilon \rangle = C_\varepsilon K^{3/2}/L_x, \tag{5}$$

where $C_\varepsilon$ decreases as *Re* increases. When *Re* is above about $10^4$, $C_\varepsilon$ reaches a constant value of order unity, depending slightly on the initial form of the turbulence.

The turbulence is fully developed at time $t_0$ and then decays with a self-similar structure for $t > t_0$. The integral timescales increase in proportion to the development time $t - t_0$, i.e., $\mathcal{T}_\mathrm{E} \sim \mathcal{T}_\mathrm{L} \sim L_x/K^{1/2} \sim (t - t_0)$.

Combining the above equations leads to the *decay law of turbulence*,

$$\frac{\mathrm{d}K}{\mathrm{d}t} = -\frac{A_\mathrm{d} K}{t - t_0},$$

where $A_\mathrm{d}$ has a constant value for each type of turbulence as it decays, usually in the range 1.1–1.3. The turbulence energy $K$ and the length scale $L_x$ then have related power-law variations in terms of their values, $K_\mathrm{d}$ and $L_{x\mathrm{d}}$, at time $t_\mathrm{d} > t_0$:

$$\frac{K}{K_\mathrm{d}} = \left(\frac{t - t_0}{t_\mathrm{d} - t_0}\right)^{-A_\mathrm{d}}, \qquad \frac{L_x}{L_{x\mathrm{d}}} = \left(\frac{t - t_0}{t_\mathrm{d} - t_0}\right)^{1 - A_\mathrm{d}/2}.$$

Since $A_\mathrm{d} > 1$, the Reynolds number of the turbulence decreases from its initial value $Re_0$ at $t = t_0$ in proportion to $K^{1/2} L_x \propto (t - t_0)^{1 - A_\mathrm{d}}$. Eventually, the local value of *Re* becomes so small that the eddy motions are smeared out by viscous stresses and the energy decays rapidly in proportion to $\exp(-k^2(t - t_0)/Re_0)$.

The rate of change of the energy spectrum $E(k)$ for eddies of scale $k^{-1}$ is equal to the net inertial transfer of energy to eddies on this scale, denoted by $-\mathrm{d}\Pi(k)/\mathrm{d}k$, minus the spectrum of dissipation of energy $\tilde{\varepsilon}(k)$ at wave number $k$:

$$\frac{\mathrm{d}E(k)}{\mathrm{d}t} = -\frac{\mathrm{d}\Pi(k)}{\mathrm{d}k} - \tilde{\varepsilon}(k).$$

The growth of the integral scale in decaying turbulence results from the net upscale transfer of energy to eddies larger than $L_x$, where $\tilde{\varepsilon}(k)$ is negligible. For eddy scales less than $L_x$, downscale transfer exceeds upscale transfer, as a result of negative straining (i.e., $\partial u/\partial x < 0$) in eddies larger than $k^{-1}$ amplifying gradients $(\partial u/\partial x)^2$ associated with scales less than $k^{-1}$. Since there is a greater probability that large-velocity gradients are negative, the normalized skewness of $\partial u/\partial x$,

$$\mathrm{Sk}_{\partial u/\partial x} = \frac{\langle(\partial u/\partial x)^3\rangle}{\langle(\partial u/\partial x)^2\rangle^{3/2}},$$

has a finite negative value, typically $-0.5 \pm 0.2$.

For the smallest-scale motions with typical velocity $u_{\mathrm{visc}}$ and large velocity gradients (where $k \sim k_{\mathrm{visc}} \ll 1/L_x$), the transfer of energy is balanced by the viscous dissipation of energy, i.e., $\mathrm{d}\Pi/\mathrm{d}k \sim \tilde{\varepsilon}(k) \sim Re^{-1}k_{\mathrm{visc}}u_{\mathrm{visc}}^2$, so that $\mathrm{d}E(k)/\mathrm{d}t$ is small. These eddy motions must also be energetic and large enough for inertio-viscous interactions with larger scales to sustain the fluctuations, i.e., their local Reynolds number must be large enough that $Re\,u_{\mathrm{visc}}/k_{\mathrm{visc}} \sim 1$.

In the inertial range, where $k \ll k_{\mathrm{visc}}$, since the dissipation and energy decay terms are small, the inertial transfer gradient term is very small, i.e., $\mathrm{d}\Pi/\mathrm{d}k \ll u'^3$, from which $\Pi(k) \sim$ const. $= \Pi_{\mathrm{i}}$. By integrating the spectrum equation over the inertio-viscous range between $k_{\mathrm{visc}}$ and $\infty$, it follows that the mean dissipation over all wave numbers is equal to $\Pi_{\mathrm{i}}$, i.e., $\int_0^\infty \tilde{\varepsilon}(k)\,\mathrm{d}k = \langle\varepsilon\rangle = \Pi_{\mathrm{i}}$.

These equations define the viscous or micro-length scales in terms of $Re$, since $\langle\varepsilon\rangle \sim 1$. Thus, $\ell_{\mathrm{visc}} = 1/k_{\mathrm{visc}} \sim Re^{-3/4}$, which may be 0.1% of the integral scales $L_x$ for large-scale or high-speed flows. From this, we infer that the typical eddy velocities in the viscous range are $u_{\mathrm{visc}} \sim (\langle\varepsilon\rangle/Re)^{1/4} \sim Re^{-1/4}u'$.

At very high Reynolds number, highly dissipative fluctuations ($\varepsilon \gg \langle\varepsilon\rangle$) and large velocities of order $u'$ are generated intermittently on the length scale $\ell_{\mathrm{visc}}$ within thin shear layers of thickness $\ell$. But they contribute little to the overall values of $C_{11}$ and $E(k)$.

Kolmogorov and Obukhov used this statistical-physics analysis (inspired by the poetic description of eddy motions and downscale energy transfer by Richardson) to hypothesize that the cross-correlation $C_{11}(r)$ of fluctuating velocities $u_1$ over small distances in the inertial range where $\ell_{\mathrm{visc}} \ll r \ll L_x$ is determined by $\Pi_{\mathrm{i}} = \langle\varepsilon\rangle$. From this, using dimensional analysis (or local scaling) we obtain $\hat{C}_{11}(r) = \langle u_1^2\rangle - C_{11}(r) = \alpha\langle\varepsilon\rangle^{2/3}r^{2/3}$.

Experiments show that the coefficient $\alpha \sim 10$. This similarity extends to correlations of third moments of velocity differences. Consequently, for these scales the energy spectrum $E(k)$ in the inertial range also has a self-similar form, $E(k) = \alpha_k\langle\varepsilon\rangle^{2/3}k^{-5/3}$, where the coefficient $\alpha_k = \alpha/3$.

The statistics of the random displacements of fluid particles (released at time $t_0$ from $X_1 = 0$) in the $x_1$-direction determine how particles are dispersed, and the distance between them increases with time $t - t_0$. Since $\mathrm{d}X_1/\mathrm{d}t = u_1(t; X_1(t_0) = 0)$, the mean square displacement is

$$\frac{\mathrm{d}\langle X_1^2\rangle}{\mathrm{d}t} = 2\left\langle X_1\frac{\mathrm{d}X_1}{\mathrm{d}t}\right\rangle = 2\int_0^{t-t_0} C_{11}^L(\tau)\,\mathrm{d}\tau.$$

Properties of the auto-correlation show that the initial dispersion at small time depends on the whole spectrum, i.e., for $(t - t_0) \ll T_{\mathrm{L}}$, $\langle X_1^2\rangle \sim u'^2(t-t_0)^2$. Later, large scales determine the dispersion (as in any diffusion process):

$$\langle X_1^2\rangle \sim 2u'^2T_{\mathrm{L}}(t - t_0). \tag{6}$$

The rate of spreading of pairs of particles released at $t = t_0$ and located at $\boldsymbol{X}^{(a)}(t)$ and $\boldsymbol{X}^{(b)}(t)$ is $\mathrm{d}(\Delta X)/\mathrm{d}t$, where $\Delta X = |\boldsymbol{X}^{(a)} - \boldsymbol{X}^{(b)}|$. This rate is mostly determined by the eddy motions on the scale of the separation distance $\Delta X$. Thus, in the inertial range scales, i.e., $\ell_{\mathrm{visc}} < \Delta X < L_x$, $\mathrm{d}(\Delta X)/\mathrm{d}t \sim \langle\varepsilon\rangle^{1/3}(\Delta X)^{1/3}$, from which $\langle(\Delta X)^2\rangle \sim \langle\varepsilon\rangle(t-t_0)^3$. Over longer time intervals and larger separations the particle motions become decorrelated. Then $\langle(\Delta X)^2\rangle$ is given by twice the single particle dispersion in (6). In this random eddying, some pairs of particles stay close together over considerable distances; this explains why odors can often be detected over large distances.

## 6 Physical and Computational Aspects of Inhomogeneous Turbulence

Most types of inhomogeneous turbulent flows in engineering and geophysics, as well as in laboratory experiments, are influenced by rigid or flexible boundaries, or by interfaces between different regions of turbulent flow, such as those at the edges of wakes, jets, and shear layers. The overall widths of each of these flows is denoted by $\Lambda$. There may or may not be a gradient of the mean velocity, e.g., where $\partial U_1/\partial x_3 \neq 0$.

The eddy motions and dynamics in these flows can be represented schematically on a "state-space" map of relative timescales and length scales, as a guide to models, statistical analyses, and computational methods.

The $x$-axis of the map, $\tilde{T}$, is the ratio of the Lagrangian timescale $\mathcal{T}_{\mathrm{L}}$ to the imposed distortion time $T_{\mathrm{dis}}$ from when turbulence is generated or distorted. The $y$-axis is $\tilde{L}$, the ratio of the spatial integral scale $L_x$ to the overall width $\Lambda$.

When $\tilde{T}$ and $\tilde{L}$ are significantly less than unity, the turbulence is in local equilibrium and is determined by local dynamics, and by slow, large-scale random forcing such as that in engines.

In free shear flows (jets, wakes, and shear layers) the turbulence lies between these limits in the quasi-equilibrium zone, where $\tilde{T}$ and $\tilde{L}$ are both of order one, even though the energy and the eddy structure may be changing significantly.

Models of the mean flow $U_i(x_j, t)$ and variances or eddy shear stress derived from $\langle u_i u_j \rangle$ of such turbulent flows are based on the momentum equation (1) and on approximate relations between $U_i$ and $\langle u_i u_j \rangle$. For quasi-equilibrium turbulence, $\langle u_i u_j \rangle$ is proportional to the gradients of the mean velocity,

$$-\langle u_i u_j \rangle = \nu_e \left( \frac{\partial U_i}{\partial x_j} + \frac{\partial U_j}{\partial x_i} \right),$$

where $\nu_e$ is the eddy viscosity, whose order of magnitude is $L_x \sqrt{K}$, where the kinetic energy and length scale are determined by physical arguments, (3) and (5).

In some flows, $L_x$ is prescribed. The mean flow and $K$ are then derived from the coupled equations for $U_i$ and $K$ and from initial and boundary conditions. Generally, however, $L_x$ is calculated from (5) using the approximate equation for $\langle \varepsilon \rangle$ in terms of $U$ and $K$. Then $U_i$, $K$, and $\langle \varepsilon \rangle$ are calculated using up to twelve coupled partial differential equations. This is the widely used $K$–$\varepsilon$ computational method.

In nonequilibrium turbulent flows, or in inhomogeneous regions of turbulent flows (such as near obstacles) when $\tilde{T} > 1$ and/or $\tilde{L} > 1$, quasilinear rapid-distortion analysis or time-dependent numerical solutions demonstrate the sensitive effects of turbulence structure.

Now consider the basic types of inhomogeneous turbulence. In a uniform shear flow when $U(x_3, t) = U_0 + Sx_3$, the energy equation (3) reduces to $\{A\} = \{P\} - \langle \varepsilon \rangle$, because the turbulence is quasihomogeneous, i.e., $\tilde{L} \ll 1$. When $S \gg u'/L_x$, the turbulent energy $u'^2$ and the length scale $L_x$ both increase with time. Generally, the effect of dissipation is to reduce the growth rate.

The mean shear distorts the eddies into elongated accelerating and decelerating "streaks," with length comparable to $L_x$. The shear also limits the length scale $L_x^{(3)}$ of the vertical fluctuations and determines the momentum transport across the shear. Thus, following Prandtl's concept, the eddy viscosity $\nu_e \sim u' L_x^{(3)} \sim u'^2/S$. The spectra for these anisotropic flows are distorted but broadly similar to the spectra for homogeneous turbulence.

Shear free boundary layers occur where velocity fluctuations, with root mean square velocity $u'$ and length scale $L_x^{(3)}$, vary in the $x_3$-direction and the mean velocity gradient, $dU_1/dx_3$, is zero. These layers occur in stirred or heated flows near rigid surfaces or near interfaces between gas and liquid flows. Here, eddy transport and forcing balance finite dissipation, i.e., $0 = \{ET\} + \{F\} - \langle \varepsilon \rangle$. The normal component of the turbulence is blocked by the interface at $x_3 = 0$, i.e., $L_x^{(3)}$ is reduced and $u'_3 \sim \langle \varepsilon \rangle^{1/3} x_3^{1/3}$. Very close to resistive interfaces, thin viscous or roughness sublayers form with thickness $\ell_s$, so that the length scale $L_x^{(3)} \sim (x_3 - \ell_s)$.

At the edges of clouds, jets, etc., randomly moving thin interfacial shear layers with thickness $\ell_i$ form, centered, for example, at $x_3 = z_i(x_1, x_2, t)$. The local energy balance, $0 = \{P\} + \{ET\} - \langle \varepsilon \rangle$, shows that the local gradients of turbulence amplify the tangential components of vorticity while keeping the thickness $\ell_i$ of the layer much less than the integral scale $L_x^{(3)}$.

Turbulent shear boundary layers with thickness $\Lambda$ flow over resistive surfaces at $x_3 = 0$, where $U_1 = u_i = 0$. Outside the boundary layers, there are uniform flows with velocity $U_0$ in the $x_1$-direction. In the surface layer, where $x_3 \ll \Lambda$, the eddy structure is similar to that in uniform shear flow. But the energy equation reduces to a balance between production and dissipation, $0 = \{P\} - \langle \varepsilon \rangle$, because the energy here is not increasing ($\{A\} = 0$). As in shear free boundary layers, the blocking effect of the viscous/rough surface determines the length scale $L_x^{(3)} \sim (x_3 - \ell_s)$. Using this length scale and the estimated relations between characteristic velocity fluctuations $u^*$ near the surface (which are of order $u' \sim \sqrt{K}$), the gradient of the mean velocity is derived, namely, $dU_1/dx_3 = u^*/(\kappa x_3)$, where Karman's constant coefficient, $\kappa \approx 0.4$, depends on the general relation between $u^*$, $x_3$, and $\langle \varepsilon \rangle (x_3)$. Here, $u^{*2}$ is equal to the mean shear stress at the surface.

Integrating from the sublayer, where the boundary condition at the resistive surface is $U_1 \to 0$ as $x_3 \to 0$, leads to the Karman–Prandtl profile outside the viscous sublayer:

$$U_1(x_3) \sim \frac{u^*}{\kappa} \log \left( \frac{x_3 - \ell_s}{\ell_s} \right) \quad \text{for } x_3 > \ell_s,$$

where $\ell_s \sim (Re^{-1})/u^*$.

Close to a smooth surface, where $x_3 \ll \ell_s$, the turbulence is damped and the profile is determined by viscous stresses only. The profile at the bottom of the viscous sublayer is

$$U_1(x_3) = x_3 Re\, u^{*2}, \quad \text{where } \ell_s > x_3 > 0.$$

In the upper part of the turbulent shear boundary layer the eddies are affected by the mean shear flow and by interactions with the thin shear layer of the undulating interface. The mean "jump" velocity across the interface, $\Delta U$, is found to be of the order of $u'$, as occurs in interfacial shear layers. When averaged, the mean velocity profile is a smooth adjustment from the surface layer to the free stream velocity, where $x_3 \sim \Lambda$.

Evidently, different types of thin layers affect the structure of most turbulent flows and, therefore, their sensitivity to factors such as buoyancy, or changes to the external mean flow.

## Further Reading

Davidson, P. A. 2004. *Turbulence: An Introduction for Scientists and Engineers.* Oxford: Oxford University Press.

Tennekes, H., and J. L. Lumley. 1972. *A First Course in Turbulence.* Cambridge, MA: MIT Press.

# Part VI
# Example Problems

## VI.1 Cloaking
*Kurt Bryan and Tanya Leise*

### 1 Imaging

An object is *cloaked* if its presence cannot be detected by an observer using electromagnetic or other forms of imaging. In fact, the observer should not notice that cloaking is even occurring. Cloaking has a long history in science fiction, but recent developments have put the idea on a firmer mathematical and physical basis.

Suppose that an observer seeks to image some bounded region $\Omega$ in space. This is to be accomplished by injecting energy into $\Omega$ from the outside, then observing the response: that is, the energy that comes out of $\Omega$. An example of this process is radar: the injected energy consists of electromagnetic waves and the observed response consists of the waves reflected by objects in the region. Many other types of energy can be used to form images, for example, acoustic (sonar), mechanical, electrical, or thermal. In each case energy is injected into $\Omega$, response data is collected, and from this information an image may be formed by solving an inverse problem. To cloak an object the relevant physical properties of $\Omega$ must be altered so that the energy "flows around" the object, as if the object were not there. The challenge is to do this in a way that is mathematically rigorous and physically implementable. One successful approach to cloaking is based on the idea of *transformation optics*, which we will now discuss in the context of *impedance imaging*.

In impedance imaging, the bounded region $\Omega \subset \mathbb{R}^n$ to be imaged consists of an electrically conductive medium, with $n = 2$ or $n = 3$. Let the vector $x$ represent position in a Cartesian coordinate system, let $u = u(x)$ represent the electrical potential inside $\Omega$, and let $J = J(x)$ represent the electric current flux in $\Omega$. We assume a linear relation $J = -\sigma\nabla u$ (a form of Ohm's law), where for each $x$ the quantity $\sigma = \sigma(x)$ is a symmetric positive-definite $n \times n$ matrix. The matrix $\sigma$ is called the *conductivity* of $\Omega$ and dictates how variations in the potential induce current flow. If $\sigma = \gamma I$ for some scalar function $\gamma(x) > 0$ ($I$ is the $n \times n$ identity matrix), then the conductivity is said to be *isotropic*: there is no preferred direction for current flow. Otherwise, $\sigma$ is said to be *anisotropic*. In impedance imaging the goal is to recover the internal conductivity of $\Omega$ using external measurements.

Specifically, to image a region $\Omega$, the observer applies an electric current flux density $g$ with $\int_{\partial\Omega} g\,\mathrm{d}s = 0$ to the boundary $\partial\Omega$. If charge is conserved inside $\Omega$, then $\nabla \cdot J = 0$ in $\Omega$ and the potential $u$ satisfies the boundary-value problem

$$\nabla \cdot \sigma\nabla u = 0 \quad \text{in } \Omega, \tag{1}$$

$$(\sigma\nabla u) \cdot n = g \quad \text{on } \partial\Omega. \tag{2}$$

Here $n$ denotes an outward unit normal vector field on $\partial\Omega$, and we assume that each component of the matrix $\sigma$ is suitably smooth. The boundary-value problem (1), (2) has a solution $u$ that is uniquely defined up to an arbitrary additive constant. This solution depends on $g$, of course, but also on the conductivity $\sigma(x)$. By applying different current inputs $g$ and measuring the resulting potential $f = u|_{\partial\Omega}$, an observer builds up information about the conductivity $\sigma$. If $\sigma = \gamma I$ is isotropic, then knowledge of the response $f$ for every input $g$ uniquely determines $\gamma(x)$; that is, the observer can "image" an isotropic conductivity with this type of input current/measured voltage data. However, a general anisotropic conductivity cannot be uniquely determined from this type of data, opening the way for cloaking.

### 2 Cloaking: First Ideas

Suppose a region $\Omega$ has isotropic conductivity $\sigma = I$, so $\gamma \equiv 1$. We wish to hide an object with different conductivity inside $\Omega$ in such a way that the object is invisible to an observer using impedance imaging. One approach is to remove the conductive material from some subregion $D \subset \Omega$, where $D \cap \partial\Omega = \varnothing$, thereby creating a nonconductive hole for hiding an object. For example, take $D$ to be a ball $B_\rho(p)$ of radius $\rho$ centered at $p \in \Omega$. In this case the boundary-value problem (1), (2) must be amended with the boundary condition $\nabla u \cdot n = 0$ on $\partial D$, since current cannot flow over $\partial D$. Unfortunately, this means that the hole $D$ is likely to be visible to the observer, for the region $\Omega \setminus D$ typically yields a different potential response on $\partial\Omega$ than does the region $\Omega$. The difference in input-response mappings for $\Omega$ versus $\Omega \setminus B_\rho(p)$ grows like $O(\rho^n)$ with the radius $\rho$ of the hole (measured via an operator norm). The visibility of the hole is therefore proportional to its area or volume. To hide something nontrivial, we need $\rho$ to be large, but the observer can then easily detect it.

## 3 Cloaking via a Transformation

Let us consider the special case in which $\Omega$ is the unit disk in $\mathbb{R}^2$. We will show how to make a large nonconductive hole in $\Omega$, say a hole $B_{1/2}(0)$ of radius $\frac{1}{2}$ centered at the origin, essentially undetectable to an observer. We do this by "wrapping" the hole with a carefully designed layer of anisotropic conductor. Let $\Omega_\rho$ denote $\Omega \setminus B_\rho(0)$ and let $r$ denote distance from the origin. Choose $\rho \in (0, \frac{1}{2})$ and let $\phi$ be a smooth invertible mapping from $\overline{\Omega_\rho}$ to $\overline{\Omega_{1/2}}$ with smooth inverse, with the properties that $\phi$ maps $r = \rho$ to $r = \frac{1}{2}$ and $\phi$ fixes a neighborhood $\frac{1}{2} < \rho_0 < r \leqslant 1$ of the outer boundary $r = 1$. Such mappings are easily constructed. Let $y = \phi(x)$ and define a function $v(y) = u(\phi^{-1}(y))$ on $\Omega_{1/2}$. The function $v$ satisfies the boundary-value problem

$$\begin{aligned} \nabla \cdot \sigma \nabla v &= 0 \quad \text{in } \Omega_{1/2}, \\ \nabla v \cdot n &= g \quad \text{on } \partial\Omega, \\ (\sigma \nabla v) \cdot n &= 0 \quad \text{on } r = \tfrac{1}{2}, \end{aligned}$$

where $\sigma$ is the symmetric positive-definite matrix

$$\sigma(y) = \frac{\mathrm{D}\phi(x)(\mathrm{D}\phi(x))^{\mathrm{T}}}{|\det(\mathrm{D}\phi(x))|}$$

and $\mathrm{D}\phi$ is the Jacobian of $\phi$. Also, because $\phi$ fixes $r = 1$, we have $v \equiv u$ on the outer boundary.

The quantity $\sigma$ may be interpreted as a conductivity consisting of an anisotropic shell around the hole, but with $\sigma \equiv I$ near $\partial\Omega$. For any input current $g$, the potential $v$ on the region $\Omega_{1/2}$ with conductivity $\sigma$ has the same value on $r = 1$ as the potential $u_\rho$ on $\Omega_\rho$. If $\rho \approx 0$, then $u_\rho - u_0 = O(\rho^2)$, where $u_0$ is the potential on the region $\Omega$ with no hole. The anisotropic conductivity $\sigma$ thus has the effect of making the hole $B_{1/2}(0)$ appear to

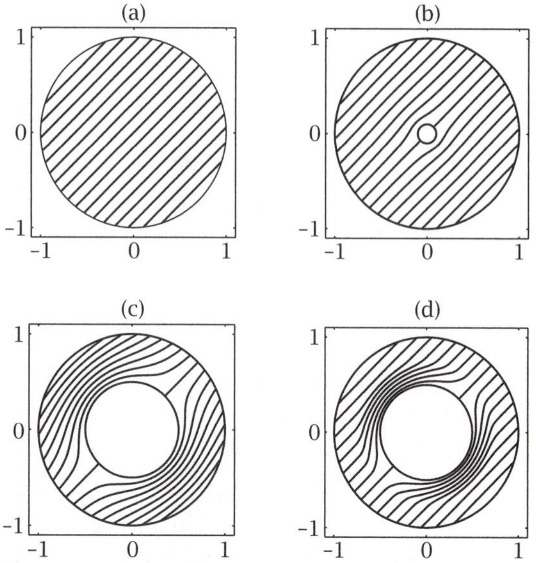


**Figure 1** Comparison of flow lines of the current $J = -\sigma\nabla u$ on annuli $\Omega \subset \mathbb{R}^2$, with input flux $g(\theta) = (\sigma\nabla u) \cdot n|_{r=1} = \cos\theta + \sin\theta$ (in polar coordinates). (a) Empty disk. (b) Uncloaked $\Omega_{1/10}$. (c) Uncloaked $\Omega_{1/2}$. (d) Cloaked $\Omega_{1/2}$. The regions in (a)–(c) have isotropic conductivity 1, while the near-cloaked annulus in (d) has conductivity $\sigma$ given by the transformation described in section 3. For (a), (b), and (c), the resulting boundary potential is $f(\theta) = (\cos\theta + \sin\theta)(1+\rho^2)/(1-\rho^2)$, with $\rho = 0$, $\rho = \frac{1}{10}$, and $\rho = \frac{1}{2}$, respectively, so that these regions will be distinguishable to an observer. In (d), the near-cloaked annulus has anisotropic conductivity $\sigma$ corresponding to $\rho = \frac{1}{10}$, for which the mapping between $f$ and $g$ is identical to that in (b).

be a hole of radius $\rho$, effectively cloaking the larger hole if $\rho \approx 0$ (see figure 1). In the limit $\rho \to 0^+$, the resulting cloaking conductivity is singular but cloaks perfectly.

## 4 Generalizations

The transformation optics approach to cloaking is not limited to impedance imaging. Many forms of imaging with physics governed by a suitable partial differential equation may be amenable to cloaking. The key is to define a suitable mapping $\phi$ from the region containing something to be hidden to a region that looks "empty." Under the appropriate change of variable, one obtains a new partial differential equation that describes how to cloak the region, and the coefficients in this new partial differential equation may possess a reasonable physical interpretation (e.g., an anisotropic conductivity). The transformation thus prescribes the required physical properties the medium must possess in order to direct the flow of energy around an obstacle in order to cloak it. The construction of materials with the required properties is an active and very challenging area of research.

### Further Reading

Bryan, K., and T. Leise. 2010. Impedance imaging, inverse problems, and Harry Potter's cloak. *SIAM Review* 52:359–77.

Greenleaf, A., Y. Kurylev, M. Lassas, and G. Uhlmann. 2009. Cloaking devices, electromagnetic wormholes, and transformation optics. *SIAM Review* 51:3–33.

Kohn, R. V., H. Shen, M. S. Vogelius, and M. I. Weinstein. 2008. Cloaking via change of variables in electric impedance tomography. *Inverse Problems* 24:1–21.

## VI.2 Bubbles

*Andrea Prosperetti*

Ordinarily, the term "bubble" designates a mass of gas and/or vapor enclosed in a different medium, most often a liquid. Soap bubbles and the bubbles in a boiling pot are very familiar examples, but bubbles occur in many, and very diverse, situations of great importance in science and technology: air entrainment in breaking waves, with implications for the acidification of the oceans; water oxygenation, with consequences for, for example, marine life and water purification; volcanic eruptions; vapor generation, e.g., for heat transfer, power generation, and distillation; cavitation and cavitation damage, e.g., in hydraulic machinery and on ship propellers; medicine, e.g., in decompression sickness (the "bends"), kidney stone fragmentation (lithotripsy), plaque removal in dentistry, blood flow visualization, and cancer treatment; beverage carbonation; curing of concrete; bread making; and many others. There is therefore a very extensive literature on this subject across a wide variety of fields.

It is often the case that liquid masses in an immiscible liquid are also referred to as bubbles, rather than, more properly, "drops." This is more than a semantic issue as the most characteristic—and, indeed, defining—feature of bubbles is their large compressibility. When a bubble contains predominantly vapor (i.e., a gas below its critical point), condensation and evaporation are strong contributors to volume changes, which can be so extreme as to lead to the complete disappearance of the bubble or, conversely, to its explosive growth. Bubbles containing predominantly an incondensible gas are usually less compressible but still far more so than the surrounding medium. Some unexpectedly large effects are associated with this compressibility because, in a sense, a bubble represents a singularity for the host liquid.

To appreciate this feature one can look at the simplest mathematical model, in which the bubble is spherical with a time-dependent radius $R(t)$ and is surrounded by an infinite expanse of incompressible liquid. The dynamics of the bubble volume is governed by the so-called Rayleigh–Plesset equation, which, after neglecting surface tension and viscous effects, takes the form

$$R(t)\frac{\mathrm{d}U}{\mathrm{d}t} + \tfrac{3}{2}U^2 = \frac{1}{\rho}(p_i - p_\infty). \tag{1}$$

Here, $U = \mathrm{d}R/\mathrm{d}t$, $\rho$ is the liquid density, and $p_i$, $p_\infty$ are the bubble's internal pressure and the ambient pressure (i.e., the pressure far from the bubble). If these two pressures can be approximated as constants, this equation has an energy first integral, which, for a vanishing initial velocity, is given by

$$U(t) = \pm\sqrt{\frac{2(p_i - p_\infty)}{3\rho}\left[1 - \frac{R^3(0)}{R^3(t)}\right]}, \tag{2}$$

with the upper sign for growth ($p_i > p_\infty$, $R(0) \leqslant R(t)$) and the lower sign for collapse ($p_i < p_\infty$, $R(0) \geqslant R(t)$). In this latter case, by the time that $R(t)$ has become much smaller than $R(0)$, this expression would predict $U \propto -R^{-3/2}$, which diverges as $R(t) \to 0$. Of course, many physical effects that are neglected in this simple model (particularly the ultimate increase of $p_i$ but also liquid compressibility, loss of sphericity, viscosity, and others) prevent an actual divergence from occurring, but, nevertheless, this feature is qualitatively robust and responsible for the unexpected violence of many bubble phenomena.

The approximation $p_i \simeq$ const. is reasonable for much of the lifetime of a bubble that contains mostly a low-density vapor. A frequently used model for an incondensible gas bubble assumes a polytropic pressure–volume relation of the form $p_i \times (\text{volume})^\kappa =$ const., with $\kappa$ a number between 1 (for an isothermal bubble) and the ratio of the gas specific heats (for an adiabatic bubble). The internal pressure $p_i$ in the simple model (1) is then replaced by $p_i = p_{i0}(R_0/R)^{3\kappa}$ for some reference values $R_0$ and $p_{i0}$.

The interaction of bubbles with pressure disturbances, such as sound, is often of interest, e.g., in underwater sound propagation, medical ultrasonics, ultrasonic cleaning, and other fields. Provided the wavelength is much larger than $R$ (which is usually the case in the vast majority of situations of interest), the effect of a sound field with frequency $f$ can be accounted for in the model (1) by setting $p_\infty = p_0 + p_A \cos(2\pi f t)$, where $p_0$ is the static pressure and $p_A$ the sound pressure amplitude at the bubble location. With this adaptation and $p_i$ given by the polytropic model, (1) describes strongly nonlinear oscillations and gives an explanation of the richness of the acoustic spectrum scattered by bubbles. In the limit of small-amplitude oscillations, it is easy to deduce from (1) an expression for the resonance frequency $f_0$ of the bubble, namely, $2\pi f_0 R_0 = \sqrt{3\kappa p_0/\rho}$, which, for an air-water system at normal pressure, gives approximately $f_0 R_0 \simeq 3$ kHz $\times$ mm.

The gas in an oscillating gas bubble alternately cools and heats up. The phase mismatch between the resulting heat exchanges with the liquid and the driving sound pressure is responsible for major energy losses, which, for bubbles larger than a few microns, dominate over other losses (e.g., those due to a moderate viscosity or sound reradiation). This energy loss is the cause of the strikingly dull sound that a glass full of carbonated beverage emits when struck with a solid object. If the sound intensity (e.g., that produced by a transducer in a resonant system) is sufficiently strong, then the increase in the temperature of the gas can be so large as to give rise to exotic chemical reactions (sonochemistry) and even to the generation of a plasma, which emits brief flashes of light synchronous with the periodic collapses of the bubble (sonoluminescence).

Bubbles containing predominantly vapor can form due to a temperature increase, as in boiling, but also as a result of a lowering of local pressure, as in cavitation. Such bubbles are much more unstable than bubbles that contain a significant amount of incondensible gas. They may keep growing and be ultimately removed by buoyancy, as in fully developed boiling, or they might violently collapse as soon as they encounter cooler liquid (hence the hissing noise of a pot that is just about to boil) or the pressure recovers.

While the spherical model is useful to understand many aspects of bubble phenomena, in many cases it is an oversimplification. For example, the spherical shape can be unstable when a bubble compresses because surface perturbations grow in amplitude as they are confined into a smaller and smaller surface or as the bubble contents begins to push out to limit the collapse velocity, thus giving rise to a Rayleigh–Taylor unstable situation. Spherical instability is strongly favored in the neighborhood of a solid; the collapsing bubble develops an involuted shape with a liquid jet that traverses the bubble and is directed against the solid surface. The velocity of the jets thus formed can reach several tens of meters per second and can contribute strongly to the effects of cavitation, both undesirable (metal fatigue and failure, vibration, noise) and desirable (dental plaque removal, kidney stone comminution, cleaning of jewelry and small electronic components).

Surface tension is the physical process that tends to keep bubbles spherical against the action of other agents, such as the instabilities just mentioned, but also gravity and translational motion, as in the case of the buoyant rise of bubbles in a liquid. Deformation due to gravity is often quantified by the Eötvös number $Eo = \rho d^2 g/\sigma$ (where $d$ is the diameter of a sphere of equal volume; $g$ is the acceleration due to gravity; and $\sigma$ is the surface tension coefficient), which expresses the balance between gravity and surface tension. The effect of translation is quantified by the Weber number $We = \rho u^2 d/\sigma$ (where $u$ is the translational velocity with respect to the liquid), which expresses the balance between inertia and surface tension. It is also a common observation that bubbles often fail to ascend along a rectilinear path: the flattening of the bubble into a roughly ellipsoidal shape causes a loss of stability of the rectilinear trajectory, a phenomenon known as Leonardo's paradox.

In a boiling pot or in a champagne glass, bubbles are often seen to originate from preferred spots rather than randomly over the entire surface of the container or in the bulk of the liquid. The reason for this phenomenon is that intermolecular forces (the origin of the macroscopic surface tension) make it very difficult to nucleate (i.e., expand) a bubble starting from a "molecular hole." In the vast majority of cases bubbles originate from preexisting micron-size nuclei, mostly consisting of gas pockets stabilized somehow, either on floating particles or, more frequently, on the surface of the container. The additional pressure necessary to expand these nuclei to visible size against surface tension causes, for example, the boiling incipience of water at atmospheric pressure to occur at a temperature a few degrees higher than 100 °C or a carbonated beverage not to rush in a foamy mess out of a newly opened can (provided it has not been shaken beforehand, an action that injects a multitude of nuclei into the liquid).

When a bubble detaches from a nozzle, or the jet in a bubble collapsing near a wall strikes the other side of the bubble surface, or a bubble is entrapped in a liquid due to a surface disturbance (e.g., an impacting drop, a breaking wave), the liquid surface undergoes a topological change that, in a macroscopic sense, amounts to a mathematical singularity. Processes of this type represent a significant numerical challenge in computational modeling.

In many cases bubbles occur in groups, or clouds, which, for certain purposes, can be considered as turning the liquid into a homogeneous medium with effective properties different from those of the pure liquid. Examples are the bubble clouds produced by a breaking wave, which endow the liquid mass in which they reside with a compressibility much larger than that of the surrounding bubble-free liquid. As a consequence, these clouds can execute volume pulsations that result in a significant low-frequency (from about 50 Hz to over 1000 Hz) ambient noise in the ocean. Clouds of bubbles deliberately injected at the bottom of ponds or in some industrial processes (e.g., glass making, chemical industry) are used to destratify and mix the liquid, or to greatly increase the gas–liquid contact area to facilitate the occurrence of chemical reactions. There is a significant body of literature on the application of homogenization and various statistical methods to the derivation of effective properties for such systems.

Some types of flow cavitation are characterized by the formation of large vapor cavities that are periodically shed and break up into bubble clouds when they are transported into regions of higher pressure. The collapse of these clouds proceeds by a cascade process that greatly enhances the destructive action of cavitation phenomena.

The massive formation of vapor bubble clouds that can occur, for example, when a pressurized liquid-filled container is opened, when liquefied natural gas comes into contact with water ("cold explosion"), or when a liquid that is highly supersaturated with a gas is exposed to lower pressure (e.g., the $CO_2$ eruption of Lake Nyos, Cameroon, in 1986) can be a violent and devastating phenomenon. The mathematical modeling of these processes offers numerous challenges, many of which are still far from having been satisfactorily met.

### Further Reading

Brenner, M. P., S. Hilgenfeldt, and D. Lohse. 2002. Single-bubble sonoluminescence. *Reviews of Modern Physics* 74: 425–84.

Leighton, T. G. 1994. *The Acoustic Bubble*. New York: Academic Press.

Plesset, M. S., and A. Prosperetti. 1977. Bubble dynamics and cavitation. *Annual Review of Fluid Mechanics* 9:145–85.

Prosperetti, A. 2004. Bubbles. *Physics of Fluids* 16:1852–65.

## VI.3 Foams

*Denis Weaire and Stefan Hutzler*

### 1 Introduction

Ever since J. A. F. Plateau laid down the foundations of much of the theory of liquid foam structures in the middle of the nineteenth century, we have understood the problem in terms of the minimization of the total areas of soap films, under appropriate constraints. The subject therefore relates to the theory of minimal surfaces, which poses important pure mathematical problems, one of which is an existence theorem for a single film, named in honor of Plateau himself. More practical problems deal with the complex disordered arrangements of bubbles, as in figure 1. Theory is often confined to static equilibrium or slowly varying (quasistatic) cases.

Figure 1 is an idealized representation of a *dry* foam, that is, one of very low liquid content, so that it consists entirely of thin films, represented by surfaces that meet symmetrically at 120°, three at a time, in lines (Plateau borders), as required for stable equilibrium. Furthermore, the lines themselves meet symmetrically, four at a time, at tetrahedral angles of 109.47°, another necessary condition for equilibrium and stability.

A foam may also be *wet* and have a finite liquid fraction $\phi$. Generalizing the idealized model, all the liquid is contained in the Plateau borders, whose cross section swells as $\phi$ increases. At a maximum value of $\phi_c \simeq 0.36$, corresponding to the porosity of a random packing of spheres, the bubbles become spherical (the wet limit), as in figure 2.

The main complication in determining or describing these structures in detail is the awkward geometrical form of their constituent surfaces. Each has constant total curvature (by the law of Laplace and Young, $\Delta P = 2\gamma(C_1 + C_2)$, where $\Delta P$ is the pressure difference between neighboring bubbles, and $\gamma$ denotes the surface tension), but the two principal curvatures $C_1$ and $C_2$ can vary. Only in very special cases can the form of such a surface be captured by explicit mathematical expressions. Computer simulation therefore plays a strong role in the subject.

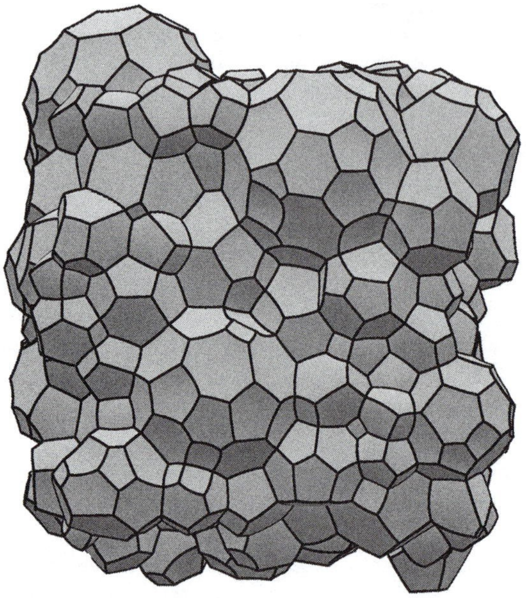

**Figure 1** A computer simulation of a dry foam using Ken Brakke's *Surface Evolver* program. (This simulation was carried out by Andy Kraynik and is reproduced with his kind permission.)

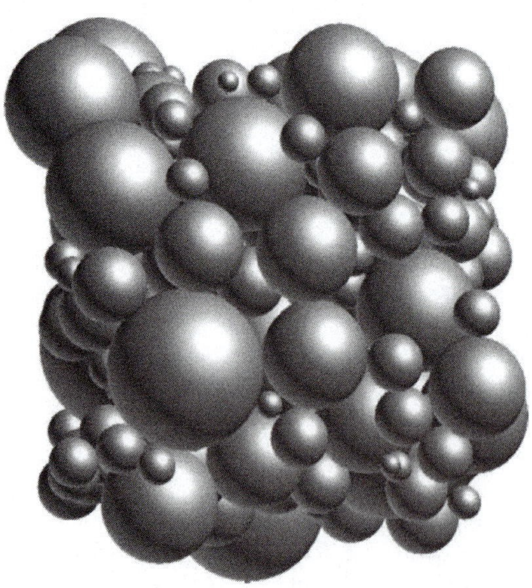

**Figure 2** Wet foams resemble sphere packings. (The example shown was computed by Andy Kraynik (Kraynik 2006) and is reproduced with kind permission of Wiley-VCH Verlag GmbH & Co. KGaA.)

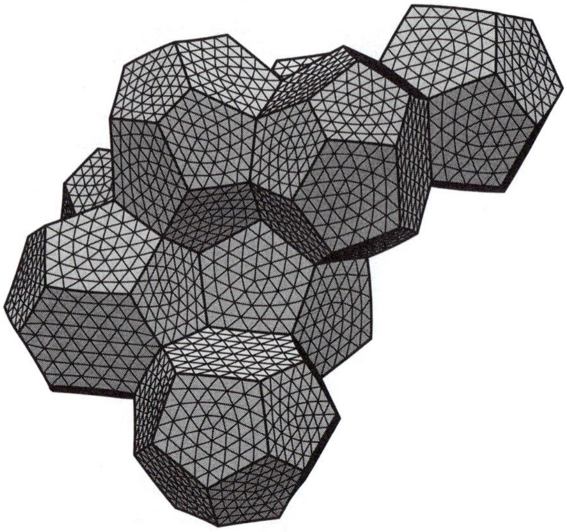

**Figure 3** *Surface Evolver* represents the surfaces of a bubble as a triangular mesh that can be refined to a high degree. The example shown is the Weaire–Phelan structure for dry foam: a space-filling arrangement of equal-volume bubbles and minimal surface area.

## 2 Simulation

For more than two decades, Ken Brakke's *Surface Evolver* has been the favorite method of simulation.[1] It represents the surfaces as tessellations (figure 3) and can deal with a variety of constraints and refinements. Samples of thousands of bubbles have been simulated (figure 1), and many physical properties have been determined. These include the statistical properties of the structure, elastic moduli, yield stress, and coarsening (the evolution that results from diffusion of gas between bubbles).

## 3 Heuristic Models

Simpler, more approximate models are often invoked to avoid the heavy computational demands of the accurate *Surface Evolver* simulations. Bubbles are often represented by overlapping spheres (or circles, in two dimensions), with a repulsive force associated with the overlap. Figure 4 shows the variation of shear modulus against liquid fraction for a typical simulation. Similar models are used in the theory of granular matter, and the two subjects find common ground when foams are studied close to the wet limit, when the bubbles

1. *Surface Evolver* can be download for free at www.susqu.edu/brakke/evolver/evolver.html.

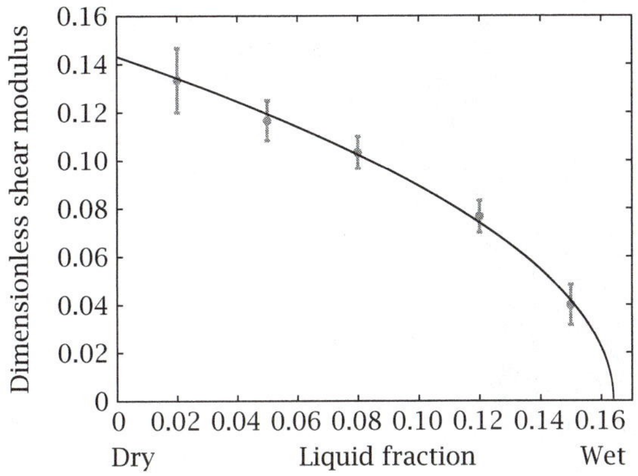


**Figure 4** The variation of shear modulus with liquid fraction for two-dimensional random foam. The data points are results from bubble model simulations for a number of samples; the solid line is a least-squares fit. As the shear modulus approaches zero (at a value of $\phi_c \simeq 0.16$ in two dimensions), the foam loses its rigidity and is better described as a bubbly liquid.

may (paradoxically) be described as hard spheres. A complex of fascinating questions arises in that limit, generally gathered under the heading of "jamming."

## 4 Continuum Models

In the context of engineering applications, foams are often described by semi-empirical continuum representations, which may be justified to some extent by appeal to the detailed microscopic models that we have described.

In rheology, the foam may be taken to be a continuous medium with the characteristics of the Bingham or Herschel–Bulkley models, which describe the foam as an elastic solid, for stress $S$ below a threshold (the yield stress $S_y$). When subjected to stress above this threshold, the foam flows according to

$$S = S_y + c\dot{\gamma}^a, \tag{1}$$

where $\dot{\gamma}$ is the strain rate, $c$ is called the consistency, and the exponent $a$ may be determined from simulations or experiments. In recent years such models of non-Newtonian fluids have been the basis of debates on *shear localization*. In a nonlinear system such as this, the response to an imposed shear stress may be flowing within a shear band.

For foam drainage (the passage of liquid through a foam under gravity and pressure gradients), a partial differential equation may be developed for the evolution of liquid fraction $\phi(x,t)$ as a function of vertical

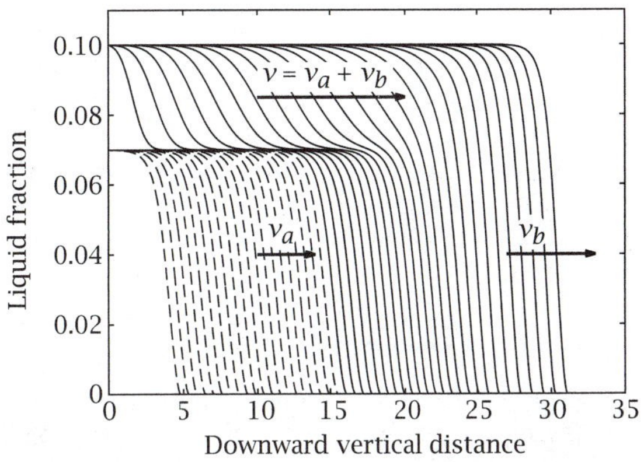


**Figure 5** The merging of two solitary waves with respective velocities $v_a$ and $v_b$ (shown here as numerical solution of the foam drainage equation) has also been seen in foam drainage experiments.

position and time. In its most elementary form, it has been called the *foam drainage equation*, and it is given by

$$\frac{\partial\phi}{\partial t} = \frac{\partial}{\partial x}\left(c_2\sqrt{\phi}\frac{\partial\phi}{\partial x} - c_1\phi^2\right), \tag{2}$$

where $c_1$ and $c_2$ are parameters containing values for the viscosity, density, and surface tension of the liquid, the mean bubble diameter, and geometrical constants related to foam structure.

Various elaborations have been advanced, e.g., taking into account different boundary conditions at the gas–liquid interfaces. In simple circumstances (one-dimensional flow), the foam drainage equation has interesting analytic solutions, such as a solitary wave propagating with velocity $v$:

$$\phi(x,t) = \begin{cases} \dfrac{v}{c_1}\tanh^2\left[\dfrac{\sqrt{c_1 v}}{c_2}(x - vt)\right] & \text{if } x \leqslant vt, \\ 0 & \text{if } x > vt. \end{cases}$$

The merging of two such waves is shown in figure 5.

## 5 More General Computational Models

More recently, attention has turned to phenomena that require models and methods that go beyond the quasistatic regime.

In 2013 Saye and Sethian developed a formalism to accurately describe local fluid motion within bubbles, soap films, and their junctions, and they applied this to the evolution of a bubble cluster. Three phases of evolution were identified and separated for the purposes of simulation. These are the approach to equilibrium, involving rearrangements of bubbles, followed by liquid drainage through the films and Plateau borders, and finally film rupture caused by thinning. This last event throws the system far out of equilibrium so that we may return to the first phase, and so on.

Further progress in both theory and computation is required to model the types of foams that feature in many applications and that are often wet and laden with particles.

## 6 The Kelvin Problem

The *Kelvin problem* asks the following question. For a dry foam of equal-sized bubbles, what arrangement has lowest energy (i.e., total surface area)? As a problem in discrete geometry, this seems rather intractable. Kelvin himself offered an inspired conjecture in 1887, in which the bubbles (or cells) were arranged in the body-centered cubic crystal structure. This remained the best candidate until 1994, when Weaire and Phelan identified a structure of lower energy, using *Surface Evolver* (figure 3).

In terms of rigorous proof, the Kelvin problem remains open, but few doubt that the Weaire–Phelan structure will prevail. It comes closest to satisfying some criteria for average numbers of cell faces (and other properties) that may be loosely argued as follows. One may conceive an ideal polyhedron for area minimization that has flat faces and the tetrahedral angle at the vertices, but this cannot be realized since it is easily shown that it has noninteger numbers of faces and edges. Average values for the Weaire–Phelan structure come close to this ideal.

## 7 Two Dimensions

Many of the basic questions of foam physics may be pursued in two dimensions, which may be realized experimentally as a sandwich of bubbles between two plates (figure 6).

In two dimensions much of the mathematics is radically simplified: ideally, the bubbles occupy polygonal cells with circular sides. Much more can be adduced by way of exact results; for example, the implication of Euler's theorem that the average number of sides of a cell is *six*.

Von Neumann pointed out that if gas diffusion between cells is proportional to pressure difference, each cell grows (or shrinks) at a rate determined only by its number of sides $n$, apart from a constant of proportionality:

$$\frac{\mathrm{d}A_n}{\mathrm{d}t} \propto (n-6), \tag{3}$$

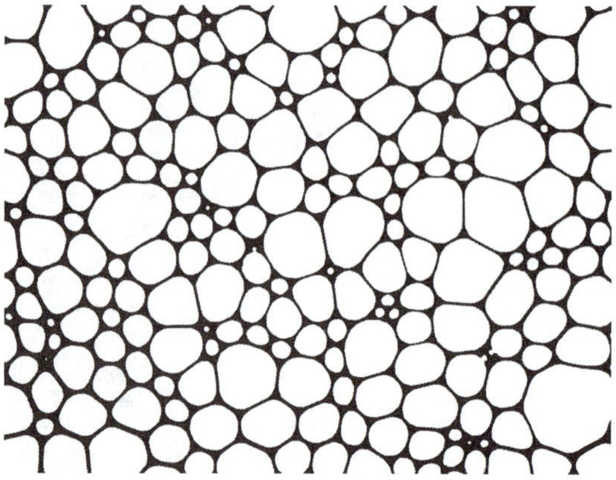

**Figure 6** The simplified geometry of a two-dimensional foam, realized as bubbles squeezed between two plates, as in this photograph, makes the computation of its properties more tractable. Many of the results are also relevant for three dimensions.

where $A_n$ is the area of an $n$-sided cell. The effect is a gradual *coarsening* of the structure, as cells progressively vanish.

Another example of the tractability of problems in two dimensions is the counterpart of the Kelvin problem: Thomas Hales has produced a fairly elementary proof that the honeycomb pattern has the minimum line length.

Empirical correlations are also found, such as that between the number of sides $n$ of a cell and $m$, the average number of sides of its neighbors (this is the Aboav–Weaire law):

$$m = 6 - a + \frac{6a + \mu_2}{n}. \tag{4}$$

Here, $\mu_2$ is the second moment of the cell side distribution about the mean, and $a \simeq 1.2$ is an empirical parameter.

Such relations all have their counterparts in a three-dimensional foam, but only as debatable approximations.

### Further Reading

Cantat, I., S. Cohen-Addad, F. Elias, F. Graner, R. Höhler, O. Pitois, F. Rouyer, and A. Saint-Jalmes. 2013. *Foams: Structure and Dynamics*. Oxford: Oxford University Press.

Kraynik, A. M. 2006. The structure of random foam. *Advanced Engineering Materials* 8:900–906.

Liu, A. J., and S. R. Nagel, eds. 2001. *Jamming and Rheology: Constrained Dynamics on Microscopic and Macroscopic Scale*. Boca Raton, FL: CRC Press.

Oprea, J. 2000. *The Mathematics of Soap Films: Explorations with Maple.* Providence, RI: American Mathematical Society.

Plateau, J. A. F. 1873. *Statique Expérimentale et Théorique des Liquides Soumis aux Seules Forces Moléculaires.* Paris: Gauthier-Villars. (The original text plus a translation into English may be found at www.susqu.edu/brakke/.)

Saye, R. I., and J. A. Sethian. 2013. Multiscale modeling of membrane rearrangement, drainage, and rupture in evolving foams. *Science* 340:720–24.

Weaire, D., ed. 1996. *The Kelvin Problem.* London: Taylor & Francis.

Weaire, D., and S. Hutzler. 1999. *The Physics of Foams.* Oxford: Clarendon Press.

## VI.4 Inverted Pendulums

*David Acheson*

In 1908 a mathematician at Manchester University called Andrew Stephenson discovered that a rigid pendulum can be stabilized *upside down* if its pivot is vibrated up and down at high frequency.

Suppose, then, that we let $a$ denote the amplitude of the pivot motion and $\omega$ the pivot frequency, so that the height of the pivot is $a \sin \omega t$ at time $t$.

The simplest case is for a single light rod of length $l$ with a point mass at one end. Stephenson assumed in his mathematical analysis that $a \ll l$ and showed that the inverted state will then be stable if $a\omega > \sqrt{2gl}$.

He confirmed his results experimentally, and with $l = 10$ cm and $a = 1$ cm, say, the critical value of $\omega/2\pi$ turns out to be about 22 Hz.

This is all very different, incidentally, from balancing an upturned pole on the palm of one hand, for there is no "feedback" in Stephenson's experiment, and the pivot vibrations are completely regular and strictly up and down.

The phenomenon was rediscovered and brought to wider attention by P. L. Kapitza in the 1950s, and it is now a well-known curiosity of classical mechanics.

### 1 An Inverted Pendulums Theorem (1993)

It is less well known, perhaps, that the same "trick" can be performed *with any finite number of linked pendulums, all balanced on top of one another.*

The theorem in question works by relating the stability of the inverted state to just two simple properties of the pendulum system in its downward-hanging, unvibrated state.

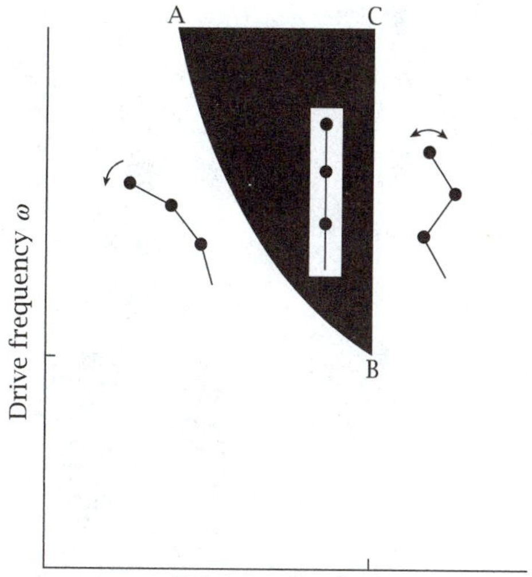


**Figure 1** The inverted pendulums theorem.

Suppose, then, that we have $N$ pendulums hanging down, one from another, with the uppermost one attached to a (fixed) pivot. There will be $N$ modes of small oscillation of this system about the downward state, each with its own natural frequency. In the lowest-frequency mode, for example, all the pendulums swing in the same direction at any given moment, while in the highest-frequency mode adjacent pendulums swing in opposite directions.

Let $\omega_{\min}$ and $\omega_{\max}$ denote the lowest and highest natural frequencies, and suppose, too, that $\omega_{\max}^2 \gg \omega_{\min}^2$, which is usually the case when $N \geqslant 2$.

The whole system can then be stabilized in its upside-down state by vibrating the pivot up and down with amplitude $a$ and frequency $\omega$ such that

$$a < \frac{0.450g}{\omega_{\max}^2} \tag{1}$$

and

$$a\omega > \frac{\sqrt{2}\,g}{\omega_{\min}}. \tag{2}$$

The stable region in the $a\omega$-plane therefore has the characteristic shape indicated in figure 1, where the straight line BC corresponds to (1) and the curve AB to (2).

The sketches indicate what happens if we gradually leave the stable region. If the frequency $\omega$ is reduced far enough that (2) is violated, the pendulums collapse by slowly wobbling down on one side of the vertical. If, instead, the drive amplitude $a$ is increased enough for (1) to be violated, the pendulums first lose their

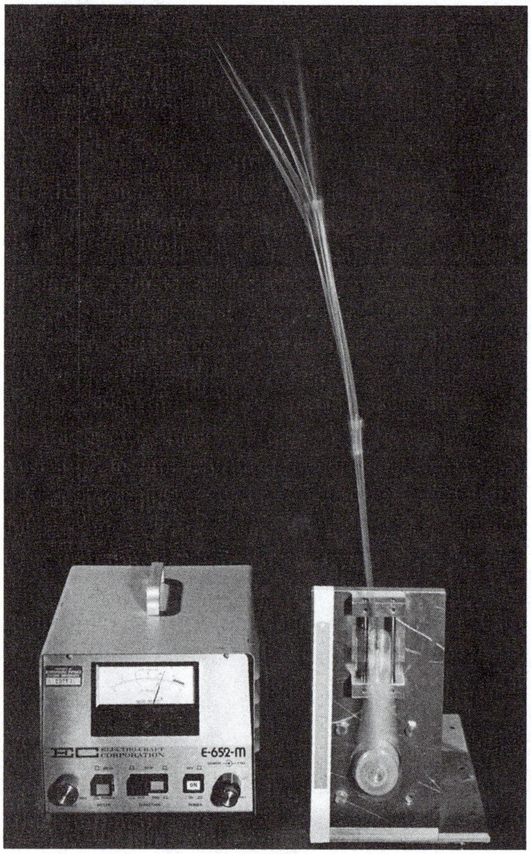


**Figure 2** An inverted triple pendulum.

stability to an upside-down buckling oscillation at frequency $\frac{1}{2}\omega$. Further increases in $a$ then cause a rather more dramatic collapse.

The numerical values of $\omega_{\min}$ and $\omega_{\max}$ depend, of course, on the number of pendulums involved, and their various shapes, sizes, and mass distributions. If many pendulums are involved, $\omega_{\max}$ is typically quite large, so the pivot amplitude $a$ has to be small to satisfy (1). This in turn means that the pivot frequency $\omega$ needs to be very large in order to satisfy (2), and in this way it becomes comparatively difficult to stabilize a long chain of pendulums in its inverted state, as one would expect.

## 2 Experiments

These strange theoretical predictions were verified experimentally—for two-, three-, and four-pendulum systems—by Tom Mullin in the early 1990s.

Figure 2 shows a three-pendulum system, stabilized upside down, recovering from a substantial disturbance and gradually wobbling back to the upward vertical.

In this particular case the rods all have length $l$ = 19 cm, and they are joined by two low-friction bearings, each of which weighs nearly twice as much as one of the rods. As a result,

$$\omega_{\min} = 0.729\left(\frac{g}{l}\right)^{1/2} \quad \text{and} \quad \omega_{\max} = 2.174\left(\frac{g}{l}\right)^{1/2},$$

so the theorem predicts that this particular system will be stable in the inverted state if

$$\frac{a}{l} < 0.095 \quad \text{and} \quad \frac{a}{l}\frac{\omega}{(g/l)^{1/2}} > 1.94.$$

These predictions were confirmed by the experimental data, and in figure 2 $a$ is 1.4 cm and $\omega/2\pi$ = 35 Hz.

## 3 Nonlinear Dynamics

We were particularly surprised, in fact, by just how stable the inverted multiple-pendulum system could be in the actual experiment. The theorem itself is based on linear stability theory and therefore guarantees stability only with respect to infinitesimally small disturbances. Yet in the experiments we found that we could gently push the pendulum column over by as much as 45° or so and, provided we kept it reasonably straight in the process, it would then recover and gradually settle again on the upward vertical.

This kind of robust behavior had been evident, however, in our computer simulations of one-, two- and three-pendulum systems based on the full, nonlinear differential equations of motion (including a small amount of friction).

And these numerical simulations revealed another surprising feature: in some regions of parameter space for which the inverted state is stable, there is a second, entirely different, way in which the pendulums can avoid falling over. For a certain range of initial conditions they will settle instead into curious "multiple-nodding" oscillations about the upward vertical.

The simplest of these is a double-nodding mode, at frequency $\frac{1}{4}\omega$, with adjacent pendulums on opposite sides of the upward vertical at any given moment, and each pendulum nodding twice on one side before flipping over to the other. For reasonably large values of $\omega$ this mode exists when $a$ is between about 70% and 90% of the maximum value allowed by (1).

There is also a triple-nodding mode, at frequency $\frac{1}{6}\omega$, with each pendulum nodding three times in succession on each side, for $a$ between about 60% and 70% of its maximum value. Again, the system will only gradually settle into such an oscillation if given a sufficient nudge toward it; if the inverted, upright state is disturbed sufficiently slightly, the pendulums just gradually settle back again on the upward vertical.

In principle, there exist even more exotic oscillations of this general kind. With an inverted double pendulum, for instance, it is possible to get an asymmetric oscillation of double-nodding type in which both upside-down pendulums stay on one side of the upward vertical throughout. This was discovered by first setting up a standard, symmetrical double-nodding oscillation and then moving in the general direction of the point B in the stability diagram by gradually changing both $a$ and $\omega$. At present, however, all of these strange nonlinear oscillations are just theoretical predictions, and they have yet to receive proper experimental study.

## 4 Not Quite the Indian Rope Trick

The principal results of the theorem above have had a fair amount of media attention over the years, largely because of a (loose!) connection with the legendary Indian rope trick.

Yet one clear prediction from the theory is that our particular gravity-defying "trick" cannot be done with a length of rope that is perfectly flexible, with no bending stiffness. This is because if we model such a rope as $N$ freely linked pendulums of fixed *total* length, and then let $N \to \infty$, we find that $\omega_{\min}$ tends to a finite limit but $\omega_{\max} \to \infty$. The theorem then requires that $a \to 0$ in order to satisfy (1) and, therefore, that $\omega \to \infty$ in order to satisfy (2).

Even without this difficulty, however, we would be a long way from the genuine Indian rope trick, as described down the generations, for this would involve making a small boy climb up the vibrating apparatus before disappearing at the top.

### Further Reading


Acheson, D. J. 1993. A pendulum theorem. *Proceedings of the Royal Society of London* A 443:239–45.

Acheson, D. J., and T. Mullin. 1993. Upside-down pendulums. *Nature* 366:215–16.

Kapitza, P. L. 1965. Dynamic stability of a pendulum with an oscillating point of suspension. In *Collected Papers of P. L. Kapitza*, edited by D. ter Haar, volume 2, pp. 714–26. London: Pergamon Press.

Stephenson, A. 1908. On a new type of dynamical stability. *Memoirs and Proceedings of the Manchester Literary and Philosophical Society* 52:1–10.


# VI.5 Insect Flight

*Z. Jane Wang*

## 1 How Do Insects Fly?

All things fall. Apples and leaves fall to the ground, the moon falls toward the Earth, and the Earth toward the sun. We would fall too if we did not will ourselves to stand upright. To walk, we lift one leg and let ourselves fall again. Why any organism should decide to fight against the inevitable fate of falling is a mystery of evolution. But there they are, insects and birds, flying in the face of gravity.

What insects and birds have discovered are ways to push the air around them; and the air, in turn, pushes the Earth away. How do insects flap their wings so as to create the necessary aerodynamic forces to hover? How do they adjust their wing motion to dart forward or to turn? Have they found efficient wing strokes? Can we emulate them?

Insects, millions of species in all, are small creatures, but they span a wide range of sizes. One of the largest, a hawkmoth, has a wing span of 5 cm and weighs 1.5 g. One of the smallest, a chalcid wasp, measures less than a millimeter and weighs 0.02 mg. The smaller they are, the faster they flap their wings. A hawkmoth flaps its wings at about 20 Hz and a chalcid wasp at about 400 Hz. This inverse scaling of the wing beat frequency with the wing length implies a relatively constant wing tip speed, about 1 m $s^{-1}$, which holds over three orders of magnitude in length scale across all the different species. Curiously, 1 m $s^{-1}$ is a common speed for natural locomotion: we walk and swim at a similar pace.

Each flapping wing creates a whirlwind around itself, governed by the Navier–Stokes equations. Insects have stumbled upon families of solutions to the Navier–Stokes equations that provide their wings with the necessary thrust (plate 10). Given the wing size and the wing speed, the flow generated by each wing is characterized by a Reynolds number (*Re*) in the range 10–10 000, which is neither small enough to be in the Stokesian regime, where the viscous force dominates, nor large enough to be in the inviscid regime, where the viscous force is negligible. The interplay of the viscous and inertial effects, especially near the wing, often leads to unexpected behavior and eludes simple theories. Another important aspect of the flow is that it is intrinsically unsteady, and this implies that the timing is critical. Part of an insect's skill is to tame the unruly flow and to coordinate the timing of its own movement with the timing of the flow.

For instance, dragonflies can adjust the timing of their forewings and hindwings during different maneuvers. When hovering, their forewings and hindwings tend to beat out of phase to gain stability and save power. When taking off, the beats of the two sets of wings will be closer to being in phase, and the interaction of the flow leads to a higher thrust. To stay aloft, each wing flaps up and down along an inclined stroke plane as if rowing in air. The downstroke has an angle of attack of about 60°, while the upstroke has a smaller angle of attack, close to 10°. Such an asymmetry results in an upward aerodynamic drag that supports much of its weight. Fruit flies, on the other hand, use only two wings to fly. Their halteres, the much-reduced hindwings, have evolved into a gyroscopic sensor that measures the rotational velocity of the body. The wings of fruit flies flap back and forth with an angle of attack of about 40°, and they support their weight with aerodynamic lift, much like a helicopter. The angles of attack used by insects, much greater than the 10–15° used by an airfoil in steady flight, are determined by the weight balance, given the limited wing speed that the insects can generate. Associated with the large angle of attack is the flow separation. The wings must then take advantage of dynamic stall, which provides a high transient force during the wing beat. In such a flow regime, the two flight strategies, asymmetric rowing and symmetric flapping, can be similarly effective.

Can a flapping wing be more efficient than a steady translating wing? A generic flapping wing motion is almost always less efficient than the optimal steady flight. This is not to say that all flapping wing motions are less efficient. Computational optimization of Navier–Stokes solutions finds some solutions, rare as they are, that are more efficient than the optimal steady flight at insect scales. One trick that works to the advantage of the flapping wing is that it can catch its own wake as it reverses, allowing the wing to gain an added lift with almost no energy cost.

## 2 Solving Navier–Stokes Equations Coupled to Flapping Wings

To understand the nature of unsteady flows and to mathematically quantify the aerodynamic forces, flight efficiency, and the timing of the wing strokes, it is necessary to solve the governing Navier–Stokes equations coupled to the dynamics of a flapping wing.

The wing drives the flow, and the flow modulates the wing motion. The dynamics of the wing is governed by Newton's equation. The fluid velocity $\boldsymbol{u}(\boldsymbol{x},t)$ and pressure $p(\boldsymbol{x},t)$ are governed by the conservation of momentum and mass of the fluid, the NAVIER–STOKES EQUATIONS [III.23]:

$$\frac{\partial \boldsymbol{u}}{\partial t} + (\boldsymbol{u}\cdot\nabla)\boldsymbol{u} = -\frac{\nabla p}{\rho} + \nu\nabla^2\boldsymbol{u},$$

$$\nabla\cdot\boldsymbol{u} = 0,$$

where $\rho$ is the density of the fluid and $\nu$ is the kinematic viscosity. By choosing a length scale, $L$, and a velocity scale, $U$, the equation can be expressed in a nondimensional form containing the Reynolds number, $Re = UL/\nu$. The flow velocity is far smaller than the speed of sound, and the flow is therefore nearly incompressible. The coupling between the wing and the fluid lies in the no-slip boundary condition at the wing surface, $\boldsymbol{u}_{\text{bd}} = \boldsymbol{u}_{\text{s}}$, which states that the flow velocity at the wing surface is the same as the wing velocity.

What kinds of solutions do we expect from these coupled partial differential equations? Imagine that a wing is detached from the insect and falls with a weight attached to it. The dynamics of this passive falling wing is governed by the same set of equations described above, and the way it falls would reveal a specific solution to the governing equations. To view some of these solutions, we can drop a piece of paper. It may tumble and flutter erratically. The falling style of a piece of paper depends on its geometry and density. If we drop a business card, we would find it tumbling about its span axis while drifting away. That is, a card driven by its own weight while interacting with air can lead to a periodic motion (figure 1). To reason this backward, we deduce that the observed periodic movement of the paper generates a thrust that on average balances the weight of the paper. If we tilt the falling trajectories so that they move along in a horizontal direction, they begin to resemble a forward-flapping flight. These periodic motions, though not identical to insect wing motions, are in essence similar to the ones that nature has found.

To quantify the flow dynamics and to tease out the key elements that are responsible for the thrust, we turn to computers and experiments. Much of the interesting behavior of the flow originates near the sharp tips of the wing, yet computational schemes often

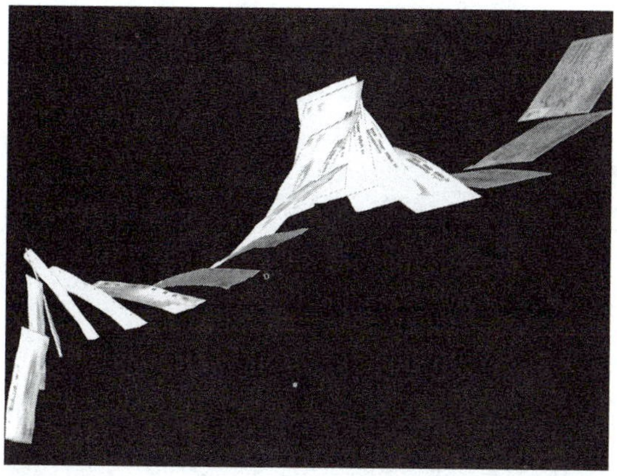

**Figure 1** Periodic motion of free falling paper. From U. Pesavento and Z. J. Wang (2004), Falling paper: Navier–Stokes solutions, model of fluid forces, and center of mass elevation, *Physical Review Letters* 93(14):144501.

encounter great difficulty in resolving moving sharp interfaces. This is a known difficulty in nearly all fluid-structure simulations. To resolve the flows, there is no one-size-fits-all scheme. If we are interested in a single rigid wing flapping in a two-dimensional fluid, we can take advantage of the two-dimensional aspect of the problem. We can use a conformal mapping technique to generate a naturally adaptive grid around the wing such that the grids are exponentially refined at the tip. We can also solve the equation in a frame that is comoving with the wing to avoid grid regeneration. The movement of the wing then gets translated into far-field boundary conditions on the flow field. The conformal map allows us to simulate the flow in a large computational domain, where the vorticity is small at the far field. The solution near the far field can be approximated analytically, and it can be implemented in the far-field boundary conditions. These treatments lead to high-order numerical computations to resolve the flow efficiently.

A three-dimensional flexible wing needs a different treatment. Computational methods may also need to handle multiple moving objects. One class of techniques is based on the idea of the immersed interface method and its predecessor, the immersed boundary method. In these Cartesian-grid methods, the interfaces cut through the grids. The problem of treating the moving interfaces therefore becomes the problem of handling singular forces along the interfaces or the discontinuities in the fluid variables across each interface. One of the challenges with these methods is to go beyond first-order accuracy at the interfaces where singularities reside. Recent work has shown that with careful treatment of the discontinuities across moving interfaces, it is possible to obtain second-order accuracy. This makes a difference when resolving the critical part of the flow.

## 3 How Do Insects Turn?

Unsteady aerodynamics is one of the many puzzles when it comes to understanding insect flight. Flapping flight, like fixed-wing flight, is intrinsically unstable. Without proper circuitry for sensing and control, an insect would fall. The same control circuitry that insects use for stabilization may also be used for acrobatic maneuvers. So how do insects modulate their wings to turn?

We end the article with the example of how a fruit fly makes a sharp yaw turn, or a saccade (plate 11). A fruit fly beats its wings about 250 times per second, and it can make a saccade in about 20 wing beats, or about 80 ms. The wing beat frequency is too fast for direct beat-to-beat control by neurons. Instead, the insects have learned to shift their "gear" and wait a few wing beats before shifting the gear back. We now have some idea of how the "gear shift" for the yaw turn works. The wing hinge acts as if it is a torsional spring. To adjust its wing motion, the wing hinge shifts the equilibrium position of the effective torsional spring, and this leads to a slight shift of the angle of attack of that wing. The asymmetry of the left and right wings creates a drag imbalance that causes the insect to turn. To turn 120° degrees, the asymmetry in the wing angle of attack is only about 5°.

The torques acting at the wing base have been computed using aerodynamic models applied to real-time tracking of the three-dimensional wing and body kinematics during free flight, which were filmed with multiple high-speed cameras. The wing and body orientations were extracted using computer vision algorithms. In the case of fruit flies, this amounts to reconstructing three rigid bodies based on the three sets of silhouettes recorded by the cameras. For larger insects, one can also add markers to help with tracking. Recent progress in tracking algorithms allows for semiautomatic processing of vast amounts of data. Although this is still time-consuming, it is a welcome departure from the earlier days of working with miles of cine films, a technique invented in the late nineteenth century.

### 4 From Flight Dynamics to Control Algorithms

In a natural environment, insects are constantly being knocked about by wind or visual and mechanical perturbations. And yet they appear to be unperturbed and are able to correct their course with ease. The halteres, mentioned earlier, provide a fast gyroscopic sensor that enables a fruit fly to keep track of its angular rotational rate. Recent work has found that when a fruit fly's body orientation is perturbed with a torque impulse, it automatically adjusts its wing motion to create a corrective torque. If the perturbation is small, the correction is almost perfect.

Exactly how their brains orchestrate this is a question for neural science as well as for mathematical modeling of the whole organism. By examining how insects turn and respond to external perturbations, we can begin to learn about their thoughts.

#### Further Reading

Chang, S., and Z. J. Wang. 2014. Predicting fruit fly's sensing rate with insect flight simulations. *Proceedings of the National Academy of Sciences of the USA* 111(31): 11246–51.

Ristroph, L., A. Bergou, G. Ristroph, G. Berman, J. Guckenheimer, Z. J. Wang, and I. Cohen. 2012. Dynamics, control, and stabilization of turning flight in fruit flies. In *Natural Locomotion in Fluids and on Surfaces*, edited by S. Childress, A. Hosoi, W. W. Schultz, and Z. J. Wang, pp. 83–99. New York: Springer.

Wang, Z. J. 2000. Two dimensional mechanism for insect hovering. *Physical Review Letters* 85(10):2035.

——. 2005. Dissecting insect flight. *Annual Review of Fluid Mechanics* 37:183–210.

Xu, S., and Z. J. Wang. 2006. An immersed interface method for simulating the interaction of a fluid with moving boundaries. *Journal of Computational Physics* 201:454–93.

## VI.6 The Flight of a Golf Ball

*Douglas N. Arnold*

A skilled golfer hitting a drive can accelerate his club head from zero to 120 miles per hour in the quarter of a second before making contact with the ball. As a result, the ball leaves the tee with a typical speed of 175 miles per hour and at an angle of 11° to the ground. From that moment the golfer no longer exercises control. The trajectory of the ball is determined by the laws of physics.

**Figure 1** The actual trajectory of a golf ball is far from parabolic.

In elementary calculus we learn to model the trajectory of an object under the influence of gravity. The horizontal component of its velocity is constant, while it experiences a vertical acceleration down toward Earth at 32.2 feet per second per second. This results in a parabolic trajectory that can be described exactly. Over a flat course, a ball traveling with the initial speed and launch angle mentioned above would return to Earth at a point 256 yards from the tee. In fact, observation of golf ball trajectories reveals that their shape is far from parabolic, as illustrated in figure 1, and that golfers often drive the ball significantly higher and farther than the simple formulas from calculus predict, even on a windless day. The discrepancy can be attributed to the fact that these formulas assume that gravity is the only force acting on the ball during its flight. They neglect the forces that the atmosphere exerts on the ball passing through it. Surprisingly, this air resistance can help to *increase* the range of the ball.

### 1 Drag and Lift

Instead of decomposing the air resistance force vector into its horizontal and vertical components, it is more convenient to make a different choice of coordinate directions: namely, the direction opposite to the motion of the ball, and the direction orthogonal to that and directed skyward (see figure 2). The corresponding components of the force of air resistance are then called the *drag* and the *lift*, respectively. Drag is the same force you feel pushing on your arm if you stick it

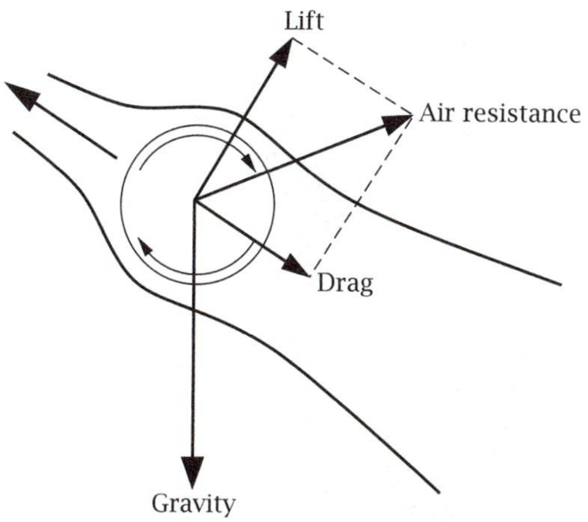


**Figure 2** Air resistance is decomposed into drag and lift.

out of the window of a moving car. Golfers want to minimize it, so their ball will travel farther. Lift is largely a consequence of the back spin of the ball, which speeds the air passing over the top of the ball and slows the air passing under it. By Bernoulli's principle, the result is lower pressure above and therefore an upward force on the ball. Lift is advantageous to golfers, since it keeps the ball aloft far longer than would otherwise be the case, allowing it to achieve more distance.

Drag and lift are very much affected by how the air interacts with the surface of the ball. In the middle of the nineteenth century, when rubber golf balls were introduced, golfers noticed that old scuffed golf balls traveled farther than new smooth balls, although no one could explain this unintuitive behavior. This eventually gave rise to the modern dimpled golf ball. Along the way a great deal was learned about aerodynamics and its mathematical modeling. Hundreds of different dimple patterns have been devised, marketed, and patented. However, even today the optimal dimple pattern lies beyond our reach, and its discovery remains a tough challenge for applied mathematics and computational science.

## 2 Reynolds Number

Drag and lift—which are also essential to the design of aircraft and ships, the swimming of fish and the flight of birds, the circulation of blood cells, and many other systems—are not easy to model mathematically. In this article, we shall concentrate on drag. It is caused by two main sources: the friction between the ball's surface and the air, and the difference in pressure ahead of and behind the ball. The size and relative importance of these contributions depends greatly on the flow regime. In the second half of the nineteenth century, George Stokes and Osborne Reynolds realized that a single number could be assigned to a flow that captured a great deal about its qualitative behavior. Low *Reynolds number* flows are slow, orderly, and laminar. Flows with high Reynolds number are fast, turbulent, and mixing.

The Reynolds number has a simple formula in terms of four fundamental characteristics of the flow: (1) the diameter of the key features (e.g., of the golf ball), (2) the flow speed, (3) the fluid density, and (4) the fluid viscosity. The formula is simple: the Reynolds number is simply the product of the first three of these divided by the fourth. This results in a dimensionless quantity: it does not matter what units you use to compute the four fundamental characteristics as long they are used consistently. The *viscosity*, which enters the Reynolds number, measures how thick the fluid is: water, for example, is a moderately thin fluid and has viscosity $5 \times 10^{-4}$ lb/ft s, while honey, which is much thicker, has a viscosity of 5 in the same units, and pitch, which is practically solid, has a viscosity of about 200 000 000.

Using the diameter of a golf ball (0.14 feet), its speed (257 feet per second), and the density (0.74 pounds per cubic foot) and viscosity (0.000012 lb/ft s) of air, we compute the Reynolds number for a professionally hit golf ball in flight as about 220 000, much more than a butterfly flying (4000) or a minnow swimming (1), but much less than a Boeing 747 (2 000 000 000).

## 3 The Mysterious Drag Crisis

At the very beginning of the twentieth century, as the Wright brothers made the first successful airplane flight, aerodynamics was a subject of intense interest. The French engineer Alexandre Gustave Eiffel, renowned for his famous tower, dedicated his later life to the study of aerodynamics. He built a laboratory in the Eiffel tower and a wind tunnel on its grounds and measured the drag on various objects at various Reynolds numbers. In 1912 Eiffel made a shocking discovery: the *drag crisis*. Although one would expect that drag increases with increasing speed, Eiffel found that for flow around a smooth sphere, there is a paradoxical *drop* in drag as the flow speed increases past Reynolds number 200 000. This is illustrated in figure 3. Of great importance in some aerodynamical regimes, the drag crisis begged for an explanation.

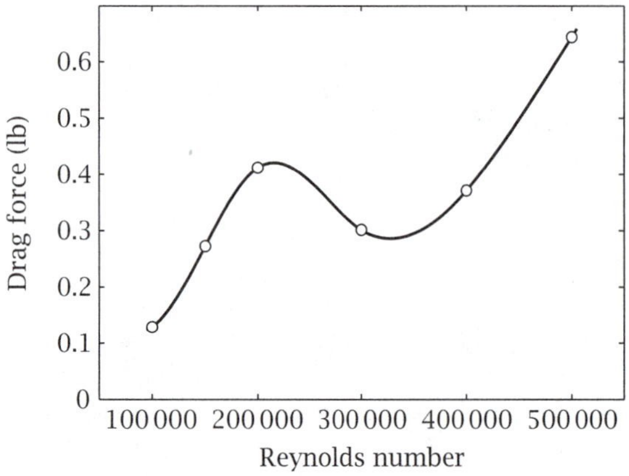


**Figure 3** A smooth sphere moving through a fluid exhibits the drag crisis: between Reynolds numbers of approximately 200 000 and 300 000, the drag decreases as the speed increases.

## 4 The Drag Crisis Resolved

The person who was eventually to explain the drag crisis was Ludwig Prandtl. Eight years before Eiffel discovered the crisis, Prandtl had presented one of the most important papers in the field of fluid dynamics at the International Congress of Mathematicians. In his paper he showed how to mathematically model flow in the *boundary layer*. As a ball flies through the air, a very accurate mathematical model of the flow is given by the system of partial differential equations known as the NAVIER–STOKES EQUATIONS [III.23]. If we could solve these equations, we could compute the drag and thereby elucidate the drag crisis. But the solution of the Navier–Stokes equations is too difficult. Prandtl showed how parts of the equations could be safely ignored in certain parts of the flow: namely, in the extremely thin layer where the air comes into contact with the ball. His equations demonstrated how the air speed increased rapidly from zero (relative to the ball) at the surface of the ball to the ball speed outside a thin layer around the ball surface. Prandtl also described very accurately the phenomenon of *boundary-layer separation*, by which higher pressure behind the ball (the pressure being lower on the top and bottom of the ball, by Bernoulli's principle) forces the boundary layer off the ball and leads to a low-pressure trailing wake behind the ball, much like the wake left behind by a ship. This low-pressure trailing wake is a major source of drag.

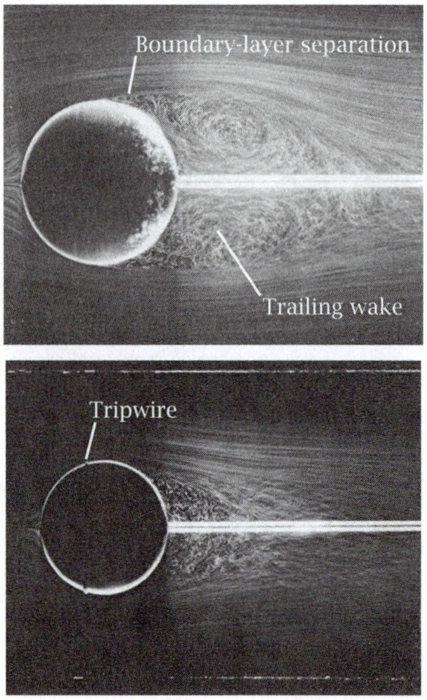


**Figure 4** Flow past a smooth sphere, clearly exhibiting boundary-layer separation and the resulting trailing wake. A tripwire has been added to the lower sphere. The resulting turbulence in the boundary layer delays separation and so leads to a smaller trailing wake. (Photos from *An Album of Fluid Motion*, Milton Van Dyke.)

In 1914 Prandtl used these tools to give the following explanation of the drag crisis.

(1) At high speed, the boundary layer become turbulent. For a smooth sphere, this happens at a Reynolds number of about 250 000.
(2) The turbulence mixes fast-moving air outside the boundary layer into the slow air of the boundary layer, thereby speeding it up.
(3) The air in the boundary layer can therefore resist the high-pressure air from behind the ball for longer, and boundary-layer separation occurs farther downwind.
(4) The low-pressure trailing wake is therefore narrower, reducing drag.

Prandtl validated this subtle line of reasoning experimentally by measuring the drag on a sphere in an air stream and then adding a small tripwire to the sphere to induce turbulence. As you can see in a reproduction of this experiment shown in figure 4, the result is indeed a much smaller trailing wake.

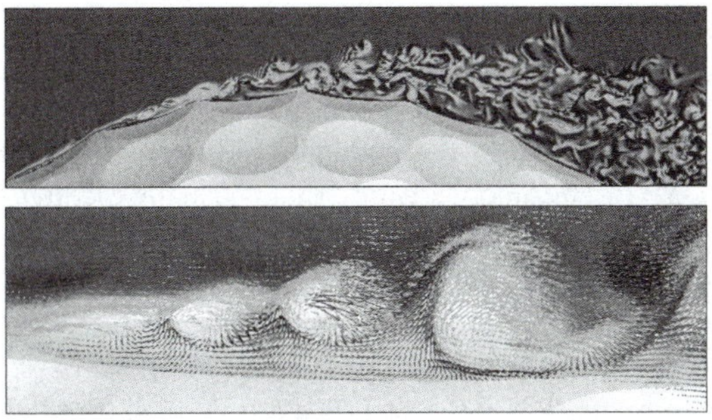

**Figure 5** Simulation of flow over dimples.

## 5 The Role of Dimples

The drag crisis means that when a smooth sphere reaches a Reynolds number of 250 000 or so, it experiences a large decrease in drag and can travel farther. This would be a great boon to golfers were it not for one fact: a golf ball–size sphere would need to travel over 200 miles per hour to achieve that Reynolds number, a speed that is not attained in golf. So why is the drag crisis relevant to golfers? The answer lies in the dimples. Just as a tripwire can be added to a smooth sphere to induce turbulence and precipitate the drag crisis, so can other perturbations of the surface. By suitably roughening the surface of a golf ball, e.g., by adding dimples, the Reynolds number at which the drag crisis occurs can be lowered to about 50 000, well within the range of any golfer. The resulting drag reduction doubles the distance flown by the ball over what can be achieved with a smooth ball.

## 6 Stalking the Optimal Golf Ball

As we have seen, dimples dramatically affect the flight of a golf ball, so a natural question is how to design an optimally dimpled ball. How many dimples should there be and in what pattern should they be arranged? What shape of dimple is best: round, hexagonal, triangular, ..., some combination? What size should they be? How deep and with what profile? There are countless possibilities, and the thousands of dimple patterns that have been tested, patented, and marketed encompass only a small portion of the relevant design space. Modern computational science offers the promise that this space can be explored in depth with computational simulation, and indeed great progress has been made. For example, in 2010 a detailed simulation of flow over a golf ball with about 300 spherical dimples at a Reynolds number of 110 000 was carried out by Smith et al. (2012). The computation was based on a finite-difference discretization of the Navier–Stokes equations using about a billion unknowns and it required hundreds of hours on a massive computing cluster to solve. It furnished fascinating insights into the role of the dimples in boundary-layer detachment and reattachment, hinted at in figure 5. But even such an impressive computation neglects some important and difficult aspects, such as the spin of the golf ball, and once those issues have been addressed the coupling of the simulation to effective optimization procedures will be no small task. The understanding of the flight of a golf ball has challenged applied mathematicians for over a century, and the end is not yet in sight.

### Further Reading

Smith, C. E., N. Beratlis, E. Balaras, K. Squires, and M. Tsunoda. 2012. Numerical investigation of the flow over a golf ball in the subcritical and supercritical regimes. *International Journal of Heat and Fluid Flow* 31:262–73.

# VI.7 Automatic Differentiation

*Andreas Griewank*

## 1 From Analysis to Algebra

In school, many people have suffered the pain of having to find derivatives of algebraic formulas. As in some other domains of human endeavor, everything begins with just a few simple rules:

$$(u + cv)' = u' + cv', \qquad (uv)' = u'v + uv'. \quad (1)$$

With a constant factor $c$, the first identity means that differentiation is a *linear process*; the second identity is known as the *product rule*. Here we have assumed that $u$ and $v$ are smooth functions of some variable $x$, and differentiation with respect to $x$ is denoted by a prime. Alternatively, one writes $u' = u'(x) = \mathrm{d}u/\mathrm{d}x$ and also calls the derivative a differential quotient. To differentiate composite functions, suppose the independent variable $x$ is first mapped into an intermediate variable $z = f(x)$ by the function $f$, and then $z$ is mapped by some function $g$ into the dependent variable $y$. One then obtains, for the composite function $y = h(x) \equiv g(f(x))$,

$$h'(x) = g'(f(x))f'(x) = \frac{\mathrm{d}y}{\mathrm{d}x} = \frac{\mathrm{d}y}{\mathrm{d}z}\frac{\mathrm{d}z}{\mathrm{d}x}. \quad (2)$$

This expression for $h'(x)$ as the product of the derivatives $g'$ and $f'$ evaluated at $z = f(x)$ and $x$, respectively, is known as the *chain rule*. One also needs to know the derivatives $\varphi'(x)$ of all *elemental functions* $\varphi(x)$, such as $\sin'(x) = \cos(x)$.

Once these elemental derivatives are known, we can combine them using the three rules mentioned above to differentiate any expression built up from elemental functions, however complicated. In other words, we are then doing *algebra* on a finite set of symbols representing elemental functions in one variable and additions or multiplications combining two variables. In principle this is simple, but as one knows from school, the size of the resulting algebraic expressions can quickly become unmanageable for humans. The painstaking execution of this *differentiation by algebra* task is therefore best left to computers.

## 2 From Formulas to Programs

Even the set of *undifferentiated* formulas describing moderately complicated systems like a robot arm, a network of pipes, or a grinding process for mechanical components may run over several pages. People therefore stop thinking of them as formulas and instead call them programs, coded in languages such as Fortran and C, or in systems such as Mathematica and Maple.

This is particularly true when we have many, say $n \gg 1$, independent (input) variables $x_j$ and $m \gg 1$ dependent (output) variables $y_i$. For the sake of notational simplicity, we collect them into vectors $\boldsymbol{x} = (x_j) \in \mathbb{R}^n$ and $\boldsymbol{y} = (y_i) \in \mathbb{R}^m$. The computer program will typically generate a large number of quantities $v_k$, $k = 1, \dots, \ell$, which we may interpret and thus differentiate as functions of the input vector $\boldsymbol{x}$. We can include all the quantities of interest by defining $v_k \equiv x_{k+n}$ for $k = 1 - n, \dots, 0$ and $v_{\ell-m+k} \equiv y_k$ for $k = 1, \dots, m$.

Assuming that there are no branches, we find that the $v_i$ with $i > 0$ are computed through a fixed, finite sequence of $\ell$ assignments:

$$v_i = v_j \circ v_k \quad \text{or} \quad v_i = \varphi_i(v_j), \quad i = 1, \dots, \ell. \tag{3}$$

Each operation $\circ$ is an addition, $+$, or a multiplication, $*$, and

$$\varphi_i \in \{c, \mathrm{rec}, \mathrm{sqrt}, \sin, \cos, \exp, \log, \dots\} \tag{4}$$

is an elemental function from a given library, including a constant setting $v = c$, the reciprocal $v = \mathrm{rec}(u) = 1/u$, and the square root $v = \mathrm{sqrt}(u) = \sqrt{u}$. Note that a subtraction $u - w = u + (-1) * w$ can be performed as an addition after multiplication of the second argument by $-1$, and a division $u/w = u * \mathrm{rec}(w)$ can be performed as a multiplication after computing the reciprocal of the second argument.

Assuming that there are no cyclic dependencies, we may order the variables such that it always holds that $j < i$ and $k < i$ in (3). Hence, with respect to data dependency the $v_i$ form the nodes of a DIRECTED ACYCLIC GRAPH [II.16], an observation that is usually credited to the Russian mathematician Kantorovich. All $v_i = v_i(\boldsymbol{x})$ for $i = 1, \dots, \ell$ will therefore be uniquely defined as functions of the independent variable vector $\boldsymbol{x}$. This applies in particular to the dependent variables $y_i = v_{\ell-m+i}$, so that the program loop actually evaluates some vector function $\boldsymbol{y} = \boldsymbol{F}(\boldsymbol{x})$.

For example, we may first specify a vector function $\boldsymbol{y} = F(\boldsymbol{x})\colon \mathbb{R}^3 \mapsto \mathbb{R}^2$ by a formula

$$F(x) = \left[ x_1^2 \frac{x_2 + x_3}{\sin(x_1)}, \frac{\cos(x_3)}{x_2 + x_3} \right].$$

Then we may decompose the formula into the following sequence of elemental operations:

| $v_{-2} = x_1,\ v_{-1} = x_2,\ v_0 = x_3$ | |
|---|---|
| $v_1 = v_{-2} * v_{-2}$ | $\varphi_1 = *$ |
| $v_2 = \sin(v_{-2})$ | $\varphi_2 = \sin$ |
| $v_3 = v_{-1} + v_0$ | $\varphi_3 = +$ |
| $v_4 = v_1 * v_3$ | $\varphi_4 = *$ |
| $v_5 = \cos(v_{-2})$ | $\varphi_5 = \cos$ |
| $v_6 = 1/v_2$ | $\varphi_6 = \mathrm{rec}$ |
| $v_7 = v_4 * v_6$ | $\varphi_7 = *$ |
| $v_8 = v_5 * v_6$ | $\varphi_8 = *$ |
| $y_1 = v_7,\ y_2 = v_8$ | |

Of course, there may be several such evaluation programs yielding the same mathematical mapping $\boldsymbol{F}$, and they may differ significantly with respect to efficiency and accuracy. While computer algebra packages may attempt to simplify or optimize the code, in the *automatic differentiation* approach described below, one generally unquestioningly accepts the given evaluation program as the proper problem specification.

## 3 Propagating Partial Derivatives

So far we have tacitly assumed that one wishes to compute the derivatives of all dependents $y_i$ with respect to all $x_j$. These form a rectangular $m \times n$ matrix, the so-called Jacobian $\boldsymbol{F}'(\boldsymbol{x})$, which can be quite costly to evaluate. Each column of the Jacobian is the partial derivative of $\boldsymbol{F}(\boldsymbol{x})$ with just one component $x_j$ of $\boldsymbol{x}$ considered to be variable. Similarly, each row of the Jacobian is the gradient of just one component $y_i = F_i$ with respect to the complete vector $\boldsymbol{x}$. Therefore, we will consider the univariate case $n = 1$ first, and in the next section we begin with the scalar-valued case $m = 1$.

In celestial mechanics or for a complicated system like a mechanical clock, the only really independent variable may be the time $x_1 = t$, so that $n = 1$ arises naturally. Then the derivatives $y_i'$ of the dependent variables $y_i$, and also those $z_k'$ of the intermediates $z_k$, are in fact velocities. In the general case $n = 1$, we can interpret them as rates of change.

So how can we compute them? Very simply, by applying the differentiation rules (1), (2) for addition, multiplication, and elemental functions, we obtain for $i = 1, \dots, \ell$ one of the three instructions

$$v_i' = v_j' + v_k', \quad v_i' = v_j' v_k + v_j v_k', \quad v_i' = \varphi_i'(v_j) v_j'. \quad (5)$$

As is the case for $\varphi(v_j) = \sin(v_j)$, evaluating the elemental derivatives $\varphi'(v_j)$ usually comes at about the same cost as the elementals $\varphi(v_j)$ themselves. Therefore, compared with (3) the propagation of the partial derivatives $v_i'$ costs just a few additional arithmetic operations, so that overall

$$\mathrm{OPS}\{\boldsymbol{y}'\} \leqslant 3\,\mathrm{OPS}\{\boldsymbol{F}(\boldsymbol{x})\}. \quad (6)$$

Here, $\boldsymbol{y}' = \boldsymbol{F}'(x)$ and OPS counts arithmetic operations and memory accesses. When $x$ is in fact a vector $\boldsymbol{x}$, we have to propagate partial derivatives with respect to each one of $n$ components to obtain the full Jacobian, at a cost of

$$\mathrm{OPS}\{\boldsymbol{F}'(\boldsymbol{x})\} \leqslant 3n\,\mathrm{OPS}\{\boldsymbol{F}(\boldsymbol{x})\}. \quad (7)$$

## 4 Comparison with Other Approaches

A very similar operation count to the bound (7) is obtained if one approximates Jacobians by one-sided or central FINITE DIFFERENCES [II.11]. However, this classical technique cannot deliver fully accurate approximations, that is, with an accuracy equal to the precision of the underlying floating-point arithmetic. In contrast, applying the differentiation rules in the way described here imposes no inherent limitation on the accuracy.

Sometimes, automatic differentiation is portrayed as a halfway house between numerical differentiation by differencing and the kind of symbolic differentiation performed by computer algebra systems like Maple and Mathematica. In fact, though, automatic differentiation is much closer to the latter. The main distinction is that our derivatives $v_i'$ are propagated as floating-point numbers at the current point $\boldsymbol{x} \in \mathbb{R}^n$, whereas fully symbolic differentiation propagates algebraic expressions in terms of the independent variables $x_j$ for $j = 1, \dots, n$.

By first transforming the original code (3) to (5) according to the differentiation rules, we actually perform symbolic differentiation in some way. At the software level, this transition is usually implemented by operator overloading or source transformation. Both the transformation and the subsequent evaluation of derivatives in floating-point arithmetic have a complexity that is proportional to the complexity of the code for evaluating $F$ itself. This is also true for the following *reverse mode* of automatic differentiation.

## 5 Propagating Partial Adjoints

Data fitting and other large-scale computational tasks require the unconstrained optimization of a scalar function $y = F(\boldsymbol{x})$, where $m = 1$ and $n \gg 1$. One then needs the gradient vector $\nabla F(\boldsymbol{x}) = F'(\boldsymbol{x}) \in \mathbb{R}^n$ to guarantee iterative decent. According to (7), its cost relative to that of $F(\boldsymbol{x})$ would grow like the domain dimension $n$.

Fortunately, just as one can propagate forward the scalar derivatives $v_i'$ of intermediates $v_i$ with respect to a single independent $x$, it is possible to propagate backward the so-called adjoint derivatives $\bar{v}_i$ of a single dependent $y$ with respect to the intermediates $v_i$ through the evaluation loop. More specifically, starting from $\bar{y} = \bar{v}_\ell = 1$ and $\bar{v}_i = 0$ for $1 - n \leqslant i < \ell$ initially, we need to execute for $i = \ell, \dots, 1$ the incremental operations

$$\begin{aligned} \bar{v}_j &:= \bar{v}_j + \bar{v}_i, & \bar{v}_k &:= \bar{v}_k + \bar{v}_i, \\ \bar{v}_j &:= \bar{v}_j + \bar{v}_i v_k, & \bar{v}_k &:= \bar{v}_k + \bar{v}_i v_k, \end{aligned}$$
$$\bar{v}_j := \bar{v}_j + \bar{v}_i \varphi'(v_j)$$

in the case of additions, multiplications, and elemental functions, respectively. In all three cases the adjoints of the arguments $v_j$, and possibly $v_k$, are updated by a contribution proportional to the adjoint $\bar{v}_i$ of the value $v_i$. Provided these composite operations are executed in reverse order to the original function evaluation, at the end one obtains the desired gradient:

$$\nabla F(\boldsymbol{x}) = \bar{\boldsymbol{x}} = (\bar{v}_{j-n})_{j=1,\dots,n}.$$

Moreover, just like in the forward mode one has to do just a few additional operations, so that

$$\mathrm{OPS}\{\bar{\boldsymbol{x}}\} \leqslant 4\,\mathrm{OPS}\{\boldsymbol{F}(\boldsymbol{x})\}.$$

This remarkable result is called the *cheap gradient principle*. The same scheme can be applied componentwise to vector-valued $\boldsymbol{F}$ with $m > 1$. The resulting Jacobian cost is

$$\mathrm{OPS}\{\boldsymbol{F}'(\boldsymbol{x})\} \leqslant 4m\,\mathrm{OPS}\{\boldsymbol{F}(\boldsymbol{x})\},$$

which is more advantageous than (7) if there are significantly fewer dependents than independents.

### 6 Generalizations and Extensions

We have considered here only the propagation of first derivatives in the forward or reverse mode. Naturally, there are hybrid procedures combining elements of forward and backward propagation of derivatives. Some of them exploit sparsity in Jacobians and also Hessians, i.e., symmetric matrices of second-order partial derivatives. Automatic differentiation techniques can be used to evaluate univariate Taylor polynomials of any degree $d$, with the effort being quadratic in $d$. These Taylor coefficients can then be used to determine derivative tensors of any degree and any order.

#### Further Reading

The basic theory of automatic differentiation is laid out by Griewank and Walther (2008), and a more computer science-oriented introduction is given by Naumann (2012). For references to current activities in automatic differentiation, check www.autodiff.org. Diverse topics including software tool development and specific applications are covered in a series of proceedings volumes of which Forth et al. (2012) is the most recent.

Forth, S., P. Hovland, E. Phipps, J. Utke, and A. Walther. 2012. *Recent Advances in Automatic Differentiation*. Lecture Notes in Computational Science and Engineering, volume 67. Berlin: Springer.

Griewank, A., and A. Walther. 2008. *Evaluating Derivatives: Principles and Techniques of Algorithmic Differentiation*, 2nd edn. Philadelphia, PA: SIAM.

Naumann, U. 2012. *The Art of Differentiating Computer Programs: An Introduction to Algorithmic Differentiation*. Philadelphia, PA: SIAM.

## VI.8 Knotting and Linking of Macromolecules

*Dorothy Buck*

Knot theory has its roots in the natural sciences: Lord Kelvin conjectured that elements were composed of knotted vortices in the ether. This motivated Peter Guthrie Tait to begin classifying knots. Since then, knot theory has grown into a rich and fundamental area of mathematics with surprising connections to algebra, geometry, and mathematical physics.

Since the 1980s, when knotted and linked deoxyribonucleic acid (DNA) molecules were discovered, knot theorists have played a central role in exploring the biological ramifications of the topological aspects of DNA. This has been a successful symbiotic dialogue: problems in biology have suggested novel, interesting questions to knot theorists. In turn, new topological ideas arose that allowed wider biological problems to be considered.

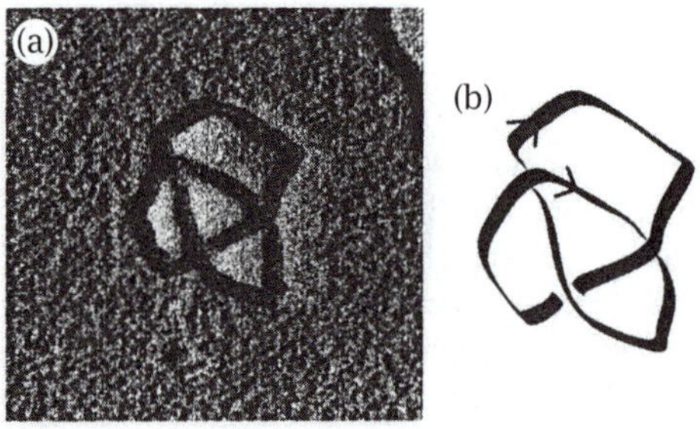


**Figure 1** (a) An electron microscopy image of a knotted DNA trefoil. (b) A trefoil knot. (Images courtesy of Shailja Pathania.)

### 1 Knots and Links

A *knot* is a curve in 3-space whose ends meet and that does not self-intersect, i.e., a simple closed curve. A planar circle is an example of the trivial knot; see figure 1(b) for an example of the simplest nontrivial knot, the trefoil, also known as the $(2,3)$-torus knot. Several, say $n$, knots can be intertwined together (again without self-intersections) to form an $n$-*component link*. Two knots or links are *equivalent* if it is possible to smoothly deform, without cutting and resealing, one into the other (more technically, if there is an ambient isotopy of 3-space that takes one to the other).

The fundamental question in knot theory is: when are two different-looking knots equivalent? In particular, when is a complicated-looking knot actually unknotted? And when knotted, what is the most efficient way to unknot it? Much of 3-manifold topology has developed in order to provide ways of telling knots apart.

### 2 DNA Molecules Can Knot and Link

Anyone who has untangled their headphone cords after retrieving them from their backpack or handbag intuitively understands how a long flexible object confined to a small volume can become entangled. DNA molecules in the cell are the molecular analogue of this situation.

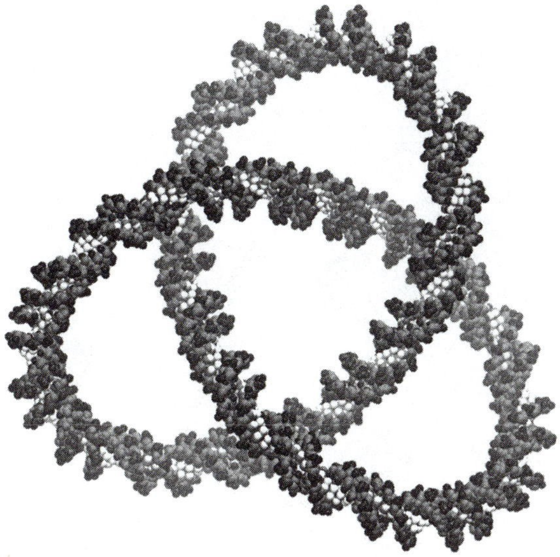

**Figure 2** A DNA trefoil knot.
(Figure courtesy of Massa Shoura.)

DNA is composed of two intertwined strands, each composed of many repeated units (*nucleotides*), each in turn composed of a phosphate group, a sugar, and one of four bases (adenine, cytosine, thymine, and guanine). These two intertwined strands, like a spiral staircase, form the famous double helix of DNA, wrapping around an imaginary central axis. Unlike a staircase, though, DNA molecules are flexible, and in living cells the central axis bends and contorts. In some cases—such as with bacterial DNA, chloroplast DNA, and our own mitochondrial DNA—the top and bottom of this twisted ladder are bonded together, so that the central axis is actually circular or even knotted (figures 1(a) and 2).

When this circular DNA is copied before cell division, the resulting two DNA molecules are linked as a $(2, n)$-torus two-component link, where $n$ is proportional to the number of nucleotides of the original DNA molecule. These knotted and linked molecules can be simple or complex: for example, mitochondrial DNA from the parasite that causes sleeping sickness forms an intricate network of thousands of small linked circles that resemble chain mail.

In the cell, this (circular or linear) DNA is confined to a roughly spherical volume whose diameter is many magnitudes smaller than the length of the molecule itself. (For example, in humans, a meter of DNA resides in a nucleus whose diameter is $10^{-6}$ meters.) This extreme confinement means that the DNA axis can writhe (forming *supercoils*). As discussed in more detail below, when certain enzymes act on circular DNA, they can knot or link—or indeed unknot or unlink—DNA molecules, converting these supercoils into knot or link nodes (or vice versa).

Linear DNA, such as human chromosomal DNA, is often "stapled down" at regular intervals via a protein scaffolding. These bulky proteins ensure that if a DNA segment in between becomes knotted, it cannot be untied. Even linear DNA can therefore exhibit nontrivial knotting.

### 2.1 DNA Forms Only Certain Knots and Links

Experimentally, DNA knots and links can be resolved in two ways: via electron microscopy or via electrophoretic migration. DNA molecules have been visualized by electron microscopy (figure 1(a)). However, this process can be laborious and difficult: even deciphering the sign of crossings on an electron micrograph is nontrivial. A much more widespread technique to partially separate DNA knot and link types is that of agarose gel electrophoresis. Gel electrophoresis is straightforward and requires relatively small amounts of DNA. The distance a DNA knot or link migrates through the gel is proportional to the minimal crossing number (MCN), the fewest number of crossings with which a knot can be drawn. Under standard conditions, knots of greater MCN migrate more rapidly than those with lesser MCN. However, there are 1 701 936 knots with MCN $\leqslant 16$, so both mathematicians and experimentalists have been actively developing new methods for determining (or predicting) the precise DNA knot or link type. They have demonstrated that a very small subfamily of knots and links (including the torus knots and links above) appear in DNA.

## 3 DNA Knotting and Linking Is Biologically Important

The helical nature of DNA leads to a fundamental topological problem: the two strands of DNA are wrapped around each other once for every 10.5 base pairs, or 0.6 billion times in every human cell, and they must be unlinked so that the DNA can be copied at every cell division. Additionally, in the cell, DNA knots can inhibit important cellular functions (such as transcription) and are therefore lethal.

Unsurprisingly, then, there are a host of proteins, in every organism, that carefully regulate this knotting and linking.

## 4 Unknotting and Unlinking DNA

In organisms as varied as bacteria, mice, humans, and plants, there are proteins (*type II topoisomerases*) whose sole function is to unlink and unknot entangled DNA. These act by performing crossing changes to convert nontrivial knots and links into planar circles. There has been significant interdisciplinary research to understand how these proteins that act locally at DNA crossings perform a global reaction (unknotting/unlinking).

These topoisomerases are major antibiotic and chemotherapeutic drug targets. With topoisomerases inactivated (or stranded in an intermediate state of the reaction), the resulting trapped linked DNA molecules prevent proliferation of the bacteria (or cancerous cells) as well as being lethal to the original cell.

## 5 Knotting and Linking DNA

There are also other important proteins, *site-specific recombinases*, that affect the knot/link type of DNA. These site-specific recombinases are standard tools for genetically modifying organisms and in synthetic biology. The proteins alter the order of the sequence of the DNA base pairs by deletion, insertion, or inversion of a DNA segment. While changing DNA knotting or linking is not the primary function of these site-specific recombinases, it is a by-product of the reaction when the original circular DNA is supercoiled.

Using tools from knot theory, mathematicians have been able to help biologists better understand the ways in which these proteins interact with DNA. For example, mathematicians have developed models of how the recombinase proteins reshuffle the DNA sequence. These models can then predict various new features of these interactions, such as the particular geometric configuration the DNA takes when the protein is attached or the biochemical pathway the reaction proceeds through.

## 6 Mathematical Methods of Study

As discussed above, topology has been used to understand topoisomerases and recombinases and their interactions with DNA. The rough character of these arguments is to begin by modeling the circular DNA molecule in terms of tangles (a tangle is a 3-ball with two properly embedded arcs (figure 3)). The action of the protein on the DNA, which converts it from one knot type to another, is then represented as tangle surgery: replacing one tangle by another. If the tangles involved are rational (a tangle obtained from two vertical arcs by alternately twisting the two arcs horizontally then vertically around one another, such as in figure 3), then this tangle surgery corresponds to so-called Dehn surgery in the corresponding double branch cover. Questions about which tangles are involved then turn into questions about Dehn surgery, often yielding lens spaces. These latter questions are central to 3-manifold topology, and a rich array of methods have been developed to answer them, including, most recently, knot homologies.

**Figure 3** Two segments (lighter and darker) of DNA in a rational tangle. (Figure courtesy of Kenneth L. Baker.)

## 7 Knotted Proteins

We conclude by discussing one other family of macromolecules that can become entangled. While almost all proteins are linear biopolymers, the rigid geometry of a folded protein means that entanglement can become trapped. These entanglements can be knots, slipknots, or ravels. Originally thought to be experimental artifacts, many "knotted" proteins have recently been discovered in the Protein DataBase of the National Center for Biotechnology Information. The function of these knots is still being fully explored, but they may contribute to thermal and/or mechanical stability.

If the ends of the linear entangled protein chain are at the surface of the folded protein, then joining them together to characterize the underlying knot is fairly straightforward. However, if these ends are buried, then depending on the closure one can achieve a variety of knots. Characterizing this entanglement has thus become a new focus for interdisciplinary researchers, including knot theorists.

### Further Reading


Buck, D. 2009. DNA topology. In *Applications of Knot Theory*, edited by D. Buck and E. Flapan, pp. 47–79. Providence, RI: American Mathematical Society.

Buck, D., and K. Valencia. 2011. Characterization of knots and links arising from site-specific recombination on twist knots. *Journal of Physics A* 44:045002.

Hoste, J., M. Thistlethwaite, and J. Weeks. 1998. The first 1,701,936 knots. *Mathematical Intelligencer* 20:33–48.

Maxwell, A., and A. Bates. 2005. *DNA Topology*. Oxford: Oxford University Press.

Millett, K., E. Rawdon, A. Stasiak, and J. Sulkowska. 2013. Identifying knots in proteins. *Biochemical Society Transactions* 41:533–37.


## VI.9 Ranking Web Pages

*David F. Gleich and Paul G. Constantine*

Google's search engine enables Internet users around the world to find web pages that are relevant to their queries. Early on, Google distinguished its search algorithm from competing methods by combining a measure of a page's textual relevance with a measure of the page's global importance; this latter measure was dubbed *PageRank*. Google's PageRank scores help distinguish important pages like Purdue University's home page, www.purdue.edu, from its array of subpages.

If we view the web as a huge directed graph, then PageRank scores are the stationary distribution of a particular MARKOV CHAIN [II.25] on the graph. In the *web graph*, each web page is a node and there is a directed edge from node $i$ to node $j$ if web page $i$ has a hypertext reference, or link, to web page $j$. A small sample of the web graph from Wikipedia pages is shown in figure 1. The ADJACENCY MATRIX [II.16] for the graph from the figure is

$$A = \begin{bmatrix} 0&1&1&1&1&0&1&0&0&0\\ 1&0&0&0&0&0&0&0&0&0\\ 0&0&0&0&1&1&1&0&0&0\\ 1&1&0&0&1&1&0&1&0&0\\ 1&1&1&1&0&0&0&0&1&0\\ 0&0&1&0&0&0&1&0&1&1\\ 0&0&0&0&0&1&0&0&0&1\\ 0&0&0&0&0&0&0&0&1&0\\ 0&0&0&0&0&0&0&1&0&0\\ 0&0&0&0&0&0&0&0&0&0 \end{bmatrix} \begin{array}{c} \text{PageRank}\\ \text{Google}\\ \text{Adjacency matrix}\\ \text{Markov chain}\\ \text{Eigenvector}\\ \text{Directed graph}\\ \text{Graph}\\ \text{Linear system}\\ \text{Vector space}\\ \text{Multiset} \end{array}$$

Google's founders Sergey Brin and Larry Page imagined an idealized web surfer with the following behavior. At a given page, the surfer flips a coin with probability $\alpha$ of heads and probability $1 - \alpha$ of tails. On heads, the surfer clicks a link chosen uniformly at random from all the links on the page. If the page has no links, then we call it a *dangling node*. On tails, and at dangling nodes as well, the surfer jumps to a page chosen uniformly at random from the whole graph. (There are alternative models for handling dangling nodes.) This simple model of a *random surfer* creates a Markov chain on the web graph; transitions depend only on the current page and not the web surfer's browsing history. The PageRank vector is the stationary distribution of this Markov chain, which depends on the value of $\alpha$. For the graph in figure 1, the transition matrix $P$ for the PageRank Markov chain with $\alpha = 0.85$ is (to two decimal places)

$$P = \begin{bmatrix} 0.02&0.19&0.19&0.19&0.19&0.02&0.19&0.02&0.02&0.02\\ 0.86&0.02&0.02&0.02&0.02&0.02&0.02&0.02&0.02&0.02\\ 0.02&0.02&0.02&0.02&0.30&0.30&0.30&0.02&0.02&0.02\\ 0.19&0.19&0.02&0.02&0.19&0.19&0.02&0.19&0.02&0.02\\ 0.19&0.19&0.19&0.19&0.02&0.02&0.02&0.02&0.19&0.02\\ 0.02&0.02&0.23&0.02&0.02&0.02&0.23&0.02&0.23&0.23\\ 0.02&0.02&0.02&0.02&0.02&0.44&0.02&0.02&0.02&0.44\\ 0.02&0.02&0.02&0.02&0.02&0.02&0.02&0.02&0.86&0.02\\ 0.02&0.02&0.02&0.02&0.02&0.02&0.02&0.86&0.02&0.02\\ 0.10&0.10&0.10&0.10&0.10&0.10&0.10&0.10&0.10&0.10 \end{bmatrix} \begin{array}{c} \text{PR}\\ \text{Go}\\ \text{AM}\\ \text{MC}\\ \text{Ei}\\ \text{DG}\\ \text{Gr}\\ \text{LS}\\ \text{VS}\\ \text{MS} \end{array}$$

The PageRank vector $x$ is

$$x = \begin{bmatrix} 0.08\\ 0.05\\ 0.06\\ 0.04\\ 0.06\\ 0.07\\ 0.07\\ 0.24\\ 0.25\\ 0.06 \end{bmatrix} \begin{array}{c} \text{PR}\\ \text{Go}\\ \text{AM}\\ \text{MC}\\ \text{Ei}\\ \text{DG}\\ \text{Gr}\\ \text{LS}\\ \text{VS}\\ \text{MS} \end{array}$$

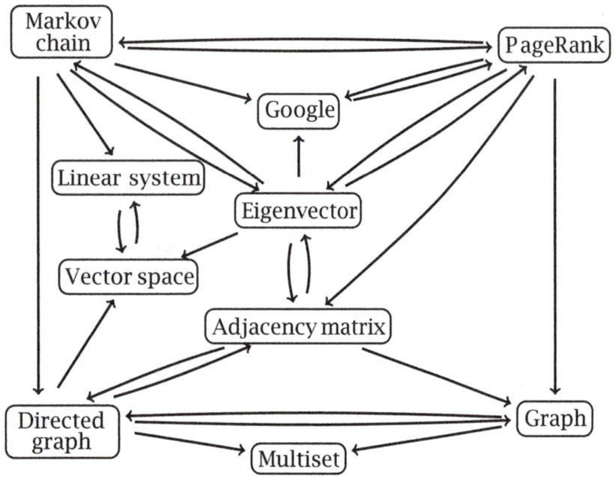


**Figure 1** A subgraph of the web from Wikipedia.

To define the PageRank Markov chain generally:

- let $G = (V, E)$ be the web graph;
- let $A$ be the adjacency matrix;
- let $n = |V|$ be the number of nodes;
- let $D$ be a diagonal matrix where $D_{ii}$ is 1 divided by the number of outlinks for page $i$, or 0 if page $i$ has no outlinks;
- let $d$ be an $n$-vector where $d_i = 1$ if page $i$ is dangling with no outlinks, and 0 otherwise; and
- let $e$ be the $n$-vector of all ones.

Then

$$P = \alpha DA + \alpha/n \cdot de^{\mathrm{T}} + (1-\alpha)/n \cdot ee^{\mathrm{T}}$$

is the PageRank Markov chain's transition matrix. The PageRank vector $x$ is the eigenvector of $P^{\mathrm{T}}$ with eigenvalue 1:

$$P^{\mathrm{T}}x = x.$$

This eigenvector always exists, is nonnegative, and is unique up to scaling because the matrix $P$ is irreducible. By convention, the vector $x$ is normalized to be a probability distribution, $e^{\mathrm{T}}x = 1$. Therefore, $x$ is unique for a given web graph and $0 < \alpha < 1$. The vector $x$ also satisfies the nonsingular linear system

$$(I - \alpha(A^{\mathrm{T}}D + 1/n \cdot ed^{\mathrm{T}}))x = (1-\alpha)/n \cdot e.$$

Thus, PageRank is both an eigenvector of the Markov chain transition matrix and the solution of a nonsingular linear system. This duality gives rise to a variety of efficient algorithms to compute $x$ for a graph as large as the web. As the value of $\alpha$ tends to 1, the matrix $I - \alpha(A^{\mathrm{T}}D + 1/n \cdot ed^{\mathrm{T}})$ becomes more ill-conditioned, and computing PageRank becomes more difficult. However, when $\alpha$ is too close to 1, the quality of the ranking degrades. For the graph in figure 1, as $\alpha$ tends to 1, the PageRank vector concentrates all its mass on the pair "Linear system" and "vector space." This happens because this pair is a terminal strong component of the graph. The same behavior occurs in the web graph; consequently, the PageRank scores of important pages such as www.purdue.edu are extinguished as $\alpha \to 1$. We recommend $0.5 \leqslant \alpha \leqslant 0.99$ and note that $\alpha = 0.85$ is a standard choice.

The canonical algorithm for computing PageRank scores is the POWER METHOD [IV.10 §5.5] applied to the eigenvector equation $P^{\mathrm{T}}x = x$ with the normalization $e^{\mathrm{T}}x = 1$. If we start with $x^{(0)} = e/n$ and iterate,

$$x^{(k+1)} = \alpha A^{\mathrm{T}}Dx^{(k)} + \alpha(d^{\mathrm{T}}x^{(k)})/n \cdot e + (1-\alpha)/n \cdot e,$$

then after $k$ steps of this method, $\|x^{(k)} - x\|_1 \leqslant 2\alpha^k$. This iteration converges quickly when $\alpha \leqslant 0.99$. With $\alpha = 0.85$, it gives a useful approximation with 10–15 iterations.

PageRank was not the first web ranking to use the structure of the web graph. Shortly before Brin and Page proposed PageRank, Jon Kleinberg proposed HYPERLINK-INDUCED TOPIC SEARCH [I.1 §3.1] (HITS) scores to estimate the importance of pages in a query-dependent subset of the web. HITS scores are computed only on a subgraph of the web graph, with the top 1000 textually relevant pages and all inlink and outlink neighbors within distance 2. The left and right dominant SINGULAR VECTORS [II.32] of the adjacency matrix are *hub* and *authority* scores for each page, respectively. An authority is a page with many hubs pointing to it, and a hub is a page that points to many authorities. The now-defunct search engine Teoma used scores related to HITS.

Modern search engines use complex algorithms to produce a ranked list of web pages in response to a query; scores such as PageRank and HITS may be one component of much larger systems. To judge the quality of a ranking, search engine architects must compare algorithmic results with human judgments of the relevance of pages to a particular query. There are a few common measures for such comparisons. *Precision* is the percentage of the search engine's results that are relevant to the human. *Recall* is the percentage of all relevant results identified by the search engine. Recall is not often used for web search as there are often many more relevant results than would fit on a top 10, or even a top 1000, list. *Normalized discounted cumulative gain* is a weighted score that rewards a search engine for placing more highly relevant documents before less relevant documents. Architects study these measures over a variety of queries to optimize a search engine and choose components of their final ranking procedure. Two active research areas on ranking algorithms include new types of regression problems to automatically optimize ranked lists of results and multiarmed bandit problems to generate personalized rankings for both web search engines and content recommendation services like Netflix.

### Further Reading

Langville and Meyer's book contains a complete treatment of mathematical web ranking metrics that were publicly available in 2005. Manning et al. give a modern treatment of search algorithms including web search. While PageRank's influence in Google's ranking may have decreased over time, its importance as a tool for finding important nodes in a graph has grown tremendously. PageRank vectors for graphs from biology (Singh et al.), chemistry (Mooney et al.), and ecology (Allesina and Pascual) have given domain scientists important new insights.

Allesina, S., and M. Pascual. 2009. Googling food webs: can an eigenvector measure species' importance for coextinctions? *PLoS Computational Biology* 5:e1000494.

Langville, A. N., and C. D. Meyer. 2006. *Google's PageRank and Beyond.* Princeton, NJ: Princeton University Press.

Manning, C. D., P. Raghavan, and H. Schütze. 2008. *An Introduction to Information Retrieval.* Cambridge: Cambridge University Press.

Mooney, B. L., L. R. Corrales, and A. E. Clark. 2012. MoleculaRnetworks: an integrated graph theoretic and data mining tool to explore solvent organization in molecular simulation. *Journal of Computational Chemistry* 33: 853–60.

Singh, R., J. Xu, and B. Berger. 2008. Global alignment of multiple protein interaction networks with application to functional orthology detection. *Proceedings of the National Academy of Sciences of the USA* 105:12763–68.

## VI.10 Searching a Graph

*Timothy A. Davis*

When you ask your smartphone or GPS to find the best route from Seattle to New York, how does it find a good route? If you ask it to help you drive from Seattle to Hawaii, how does it know you cannot do that? If you post something on a social network for only your friends and friends of friends, how many people can see it? These questions can all be posed in terms of searching a graph (also called *graph traversal*).

For an unweighted graph $G = (V, E)$ of $n$ nodes, storing a binary adjacency matrix $A$ in a conventional array using $O(n^2)$ memory works well in a graph traversal algorithm if the graph is small or if there are edges between many of the nodes. In this matrix, $a_{ij} = 1$ if $(i, j)$ is an edge, and 0 otherwise. With edge weights, the value of $a_{ij}$ in a real matrix can represent the weight of the edge $(i, j)$, although this assumes that in the problem at hand an edge with zero weight is the same as no edge at all.

A road network can be represented as a weighted directed graph, with each node an intersection and each edge a road between them. The graph is *directed* because some roads are one-way, and it is *unconnected* because you cannot drive from Seattle to Hawaii. The edge weight $a_{ij}$ represents the distance along the road from $i$ to $j$; all edge weights must therefore be greater than zero.

An adjacency matrix works well for a single small town, but it takes too much memory for the millions of intersections in the North American road network. Fortunately, only a handful of roads intersect at any one node, and thus the adjacency matrix is mostly zero, or *sparse*. A matrix is sparse if it pays to exploit its many zero entries. This kind of matrix or graph is best represented with adjacency lists, where each node $i$ has a list of nodes $j$ that it is adjacent to, along with the weights of the corresponding edges.

Searching a graph from a single *source* node, $s$, discovers all nodes reachable from $s$ via path(s) in the graph. Nodes are marked when visited, so to search the entire graph, a single-source algorithm can be repeated by starting a new search from each node in the graph, ignoring nodes that have already been visited. The graph traversal then performs actions on each node visited, and also records the order in which the graph was traversed. Often, only a small part of the graph needs to be searched. This can greatly speed up the solution of problems such as route planning.

*Breadth-first search* (BFS) is the simplest way to search a graph. It ignores edge weights, so it is suited only for unweighted graphs. It starts by placing the source node $s$ at distance $d(s) = 0$; the distance of all other nodes starts as $d(i) = \infty$. At the $k$th step (starting at $k = 0$), all nodes $i$ at distance $d(i) = k$ are examined, and any neighbors $j$ with $d(j) = \infty$ have their distance $d(j)$ set to $k+1$. The process halts when step $k$ finds no such neighbors; $d(j)$ is then the length of the shortest path from $s$ to $j$, or $d(j) = \infty$ if there is no such path. In a social network, your friends are at level one and your friends of friends are at level two in a BFS starting at your node. The BFS algorithm can also be expressed in terms of the binary adjacency matrix: $(A^{\mathrm{T}} + I)^k_{is}$ is nonzero if and only if there is a path of length $k$ or less from $s$ to $i$.

*Depth-first search* (DFS) visits the same nodes as BFS but in a different order. If it sees an unvisited node $j$ while examining node $i$, it fully discovers all unvisited nodes reachable from $j$ and then backtracks to node $i$ to consider the remainder of the nodes adjacent to $i$. It is best described recursively, as below. All nodes start out unvisited.

**DFS($i$):**
  mark $i$ as visited
  **for all** nodes $j$ adjacent to $i$ **do:**
    **if** node $j$ is not visited
      DFS($j$)

Both BFS and DFS describe a tree; $i$ is the parent of $j$ if the unvisited node $j$ is discovered while examining node $i$. The DFS tree has a rich set of mathematical properties. For example, if "($i$" is printed at the start of DFS($i$) and "$i$)" when it finishes (after traversing all its neighbors $j$), then the result is an expression with properly nested and matching parentheses. The parentheses of two nodes $i$ and $j$ are either nested one within the other, or they are disjoint. Many problems beyond the scope of this article rely on properties of the DFS tree such as this.

If the graph is stored in adjacency list form, both BFS and DFS take an amount of time that is linear in the size of the graph: $O(|V| + |E|)$, where $|V|$ and $|E|$ are the number of nodes and edges, respectively.

Examples of a BFS traversal and a DFS traversal of an undirected graph are shown in figure 1. The search starts at node 1. The levels of the BFS search and the parentheses of the DFS are shown. The DFS of figure 1 assumes that neighbors are searched in ascending order of node number (the order makes no difference in the BFS). Both the BFS and DFS trees are shown via arrows, which point from the parent to the child in the tree; they do not denote edge directions since the graph is undirected.

*Dijkstra's algorithm* finds the shortest path from the source node $s$ to all other nodes in a weighted graph. It is roughly similar to BFS, except that it keeps track of a distance $d(j)$ (the shortest path known so far to each node) for each node $j$. Instead of examining all nodes in the next level, it prioritizes them by the distance $d$ and picks just one unvisited node $i$ with the smallest $d(i)$, whose distance $d(i)$ is now final, and updates tentative distance $d$ for all its neighbors.

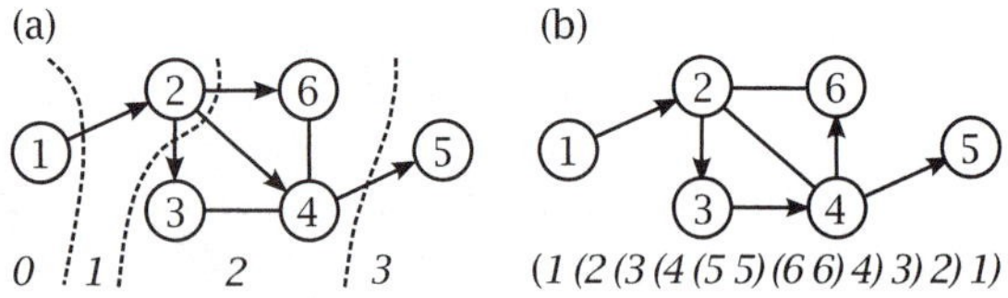


**Figure 1** (a) The BFS and (b) the DFS of an undirected graph. Arrows denote tree edges, not edge directions. Part (a) is labeled with the BFS level of each node, starting at level zero, where the BFS started, and ending at the last node discovered (node 5) at level 4. In (b), if "($i$" is printed when the DFS starts at node $i$, and "$i$)" is printed when node $i$ finishes, then the result when starting a DFS at node 1 is (1 (2 (3 (4 (5 5) (6 6) 4) 3) 2) 1), which describes how the DFS traversed the graph.

The algorithm uses a *heap* to keep track of its unvisited nodes $j$, each with a metric $d(j)$. Removing the item with smallest metric takes $O(\log n)$ time if the heap contains $n$ items. If an item's metric changes but it remains in the heap, it takes $O(\log n)$ time to adjust its position in the heap. Initializing a heap of $n$ items takes $O(n)$ time.

Nodes no longer in the heap have been visited, and their $d(j)$ is the shortest path from $s$ to $j$. The shortest path can be traced backward from $j$ to $s$ by walking the tree from $j$ to its parent $p(j)$, then to $p(p(j))$, and so on until reaching $s$.

Nodes in the heap have not been visited, and their $d(j)$ is tentative. They split into two kinds of nodes: those with finite $d$ (in the *frontier*), and those with infinite $d$. Each node in the frontier is incident on at least one edge from the visited nodes. The node in the frontier with the smallest $d(j)$ has a very useful property on which the algorithm relies: its $d(j)$ is the true shortest distance of the path from $s$ to $j$. The algorithm selects this node and then updates its neighbors as $j$ moves from the frontier into the set of visited nodes.

The algorithm finds the shortest path from $s$ to all other nodes in $O((|V| + |E|) \log |V|)$ time; this asymptotic time can be reduced with a *Fibonacci heap*, but in practice a conventional heap is faster. More importantly, the search can halt early if a particular target node $t$ is sought, although in the worst case the entire graph must still be searched.

```
Dijkstra(G, s):
  set d(j) = ∞ and p(j) = 0 for all nodes
  d(s) = 0
  initialize a heap for all nodes with metric d
  while the heap is not empty do:
    remove i from heap with smallest d(i)
    for all nodes j adjacent to i do:
      if d(j) > d(i) + a_ij
        d(j) = d(i) + a_ij
        adjust j in the heap
        p(j) = i
```

For a route from Seattle to New York, Dijkstra's algorithm acts like BFS, searching for New York in ever-widening circles centered at Seattle. It considers a route through Alaska before finding New York, which is clearly nonoptimal. $A^*$ search (pronounced A-star) is one method that can exploit the known locations of the nodes of a road network to reduce the search. It modifies the heap metric via a heuristic that uses a lower bound on the shortest distance from $s$ to $j$. To find the shortest path from the source node $s$ to all other nodes, both algorithms find the same result in the same time, but if they halt early when finding the target, A* is typically faster than Dijkstra's algorithm. Your smartphone uses additional methods to speed up the search, but they are based on the ideas described here.

### Further Reading


Kepner, J., and J. Gilbert. 2011. *Graph Algorithms in the Language of Linear Algebra.* Philadelphia, PA: SIAM.

Rosen, K. H. 2012. *Discrete Mathematics and Its Applications,* 7th edn. Columbus, OH: McGraw-Hill.


## VI.11 Evaluating Elementary Functions

*Florent de Dinechin and Jean-Michel Muller*

### 1 Introduction

Elementary (transcendental) functions, such as exponentials, logarithms, sines, etc., play a central role in computing. It is therefore important to evaluate them quickly and accurately.

The basic operations that are easily and efficiently implemented on integrated circuits are addition, subtraction, and comparison (at the same cost), and multiplication (at a larger cost). The only functions of one variable that can be implemented using these

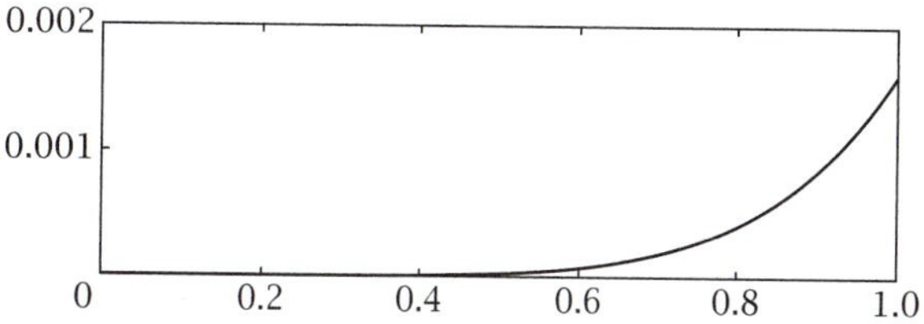


**Figure 1** Taylor approximation is local: it becomes very inaccurate for values distant from 0 (solid line, Taylor approximation error).

primitive operations are piecewise *polynomials.* Section 2 therefore studies the approximation of a function by a polynomial. As such, approximations are better on smaller intervals, and section 3 therefore introduces techniques that reduce the interval size. Section 4 discusses the evaluation of a polynomial by a machine.

If we also consider division to be a primitive operation, we may compute *rational functions,* to which most of these techniques can be generalized. Finally, the last important parameter is memory, which may be used to store tables of precomputed values to speed up the evaluation.

For manipulating *multiple-precision numbers* (i.e., numbers with thousands of digits), special algorithms (based on arithmetic–geometric iteration, for instance) are used.

### 2 Polynomial Approximation

Consider as an example the approximation of the exponential function $f(x) = \mathrm{e}^x$ on the interval $[0, 1]$ by a polynomial of degree 5. We might first try using a Taylor approximation. The Taylor formula at 0 gives

$$f(x) \approx p(x) = 1 + x + \frac{x^2}{2} + \frac{x^3}{6} + \frac{x^4}{24} + \frac{x^5}{120}.$$

Figure 1 plots the approximation error, i.e., the difference $f(x) - p(x)$, on $[0, 1]$.

Taylor approximation is focused on a point. The REMEZ ALGORITHM [IV.9 §3.5] converges to a polynomial that minimizes the maximum error over the whole interval. As figure 2 shows, this is achieved by ensuring that the error $f - p$ oscillates in the interval, which means that the error achieves its maximal absolute value at several points on the interval, with the sign of the error oscillating from one point to the next. The Remez approximation is much more accurate than the Taylor approximation. Unfortunately, the Remez algorithm computes real-valued coefficients, and rounding

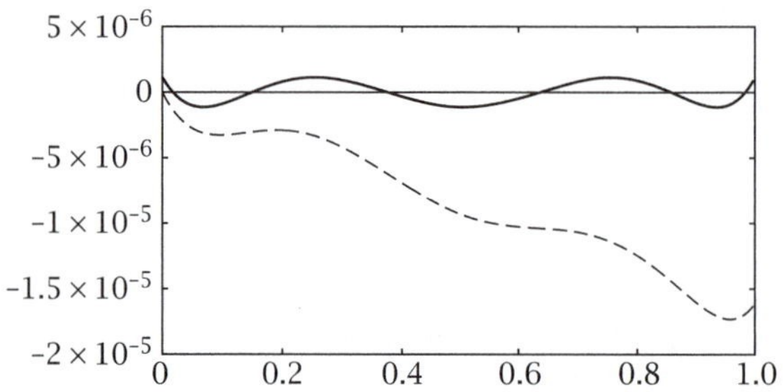


**Figure 2** The Remez error oscillates around zero (solid line), but this is lost when its coefficients are rounded to 16-bit numbers (dashed line).

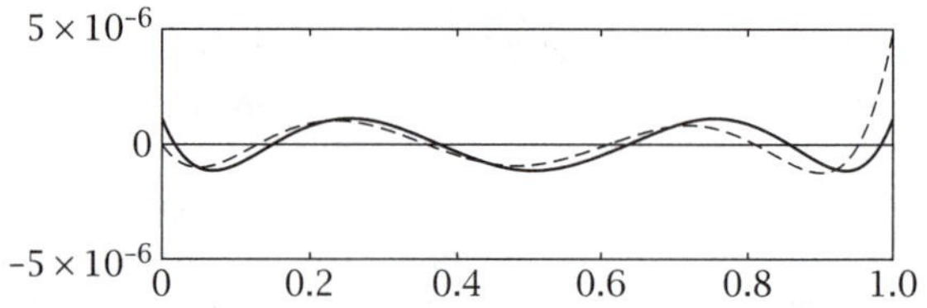


**Figure 3** The best polynomial with 16-bit coefficients (solid line, Remez approximation error; dashed line, error of best 16-bit approximation).

**Table 1** $\|e^x - p(x)\|_\infty$ versus degree and interval size.

| | Degree | | | |
|---|---|---|---|---|
| | 2 | 3 | 4 | 5 |
| $[0,1]$ | $1.41 \times 10^{-2}$ | $8.76 \times 10^{-4}$ | $5.07 \times 10^{-5}$ | $4.86 \times 10^{-6}$ |
| $[0,\frac{1}{2}]$ | $1.08 \times 10^{-3}$ | $3.40 \times 10^{-5}$ | $8.33 \times 10^{-7}$ | $5.27 \times 10^{-8}$ |
| $[0,\frac{1}{4}]$ | $1.07 \times 10^{-4}$ | $1.90 \times 10^{-6}$ | $4.57 \times 10^{-8}$ | $1.94 \times 10^{-9}$ |

these coefficients to machine-representable numbers degrades the approximation quality. (This is illustrated by figure 2 in the case where the coefficients are rounded to 16-bit numbers, corresponding to about five significant decimal digits.) A recent variation on the Remez algorithm that is implemented in the open-source Sollya tool computes the best approximation with 16-bit coefficients (see figure 3).

Whatever the approximation scheme, the approximation error decreases with the degree and increases with the interval size, as illustrated by table 1. Evaluating polynomials of high degree is time-consuming. Also, the evaluation error may increase with the degree. This suggests that we investigate *range reduction* techniques to reduce the interval size.

## 3 Range Reduction

Range reduction exploits identities that are satisfied by the function to reduce the evaluation on an interval to an evaluation (of $f$, or of another function) on a smaller interval. Techniques depend on the function itself but also on the number format used for inputs. The literature therefore provides many range reduction techniques for each function.

Let us take the example of the exponential on $[0,1]$, for which we will use the identity $e^{a+b} = e^a \times e^b$. An input $x$ with $n$-bit binary representation $x = \sum_{i=1}^{n} x_i 2^{-i}$ can be decomposed into $x = a + b$ with $a = \sum_{i=1}^{k} x_i 2^{-i}$ and $b = \sum_{i=k+1}^{n} x_i 2^{-i}$ for some $k$. This decomposition is free in terms of hardware and costs very little—just a few shifts and logical operations—in terms of software. We may now tabulate $e^a$. The table will have $2^k$ entries, and $k$ will be chosen such that the table size is acceptable: typically, $k \approx 8$. Furthermore, we observe that $b \in [0, 2^{-k}]$, and therefore $e^b$ can be evaluated using a polynomial of much lower degree than would be required in the full range $[0,1]$. The reconstruction $e^x = e^a \times e^b$ will cost only one multiplication.

With the same input decomposition, another more general technique is to tabulate one polynomial $p_a$ for each value of $a$, with all the $p_a$ having the same degree. In this case the reconstruction costs us nothing, as $f(x) \approx p_a(b)$. However, each table entry is larger.

Since cos and sin are periodic functions, range reduction for these functions consists of deducing, from the input variable $x$, an integer $n$ and a number $y$ such that

$$y \approx x - n\pi. \qquad (1)$$

This range reduction step should not be overlooked: a naive implementation of (1) may lead to very inaccurate results for very large arguments. In such cases, a sophisticated range reduction technique has been suggested by Payne and Hanek.

## 4 Evaluation

Once a suitable approximation polynomial has been found, there are several ways of evaluating it. For instance, a degree-5 polynomial may be evaluated in Horner form:

$$p(x) = a_0 + x(a_1 + x(a_2 + x(a_3 + x(a_4 + xa_5)))).$$

This form minimizes the number of multiplications in general, but it is sequential. On modern machines, one may improve the performance by exploiting parallelism. We can write

$$p(x) = a_0 + a_2x^2 + a_4x^4 + x(a_1 + a_3x^2 + a_5x^4)$$
$$\equiv p_e(x^2) + xp_o(x^2).$$

This costs one extra multiplication, but $p_e(x^2)$ and $p_o(x^2)$ can be evaluated in parallel. This idea can be exploited recursively to express more parallelism if needed. Function-specific techniques can be used here, too: exploiting the fact that the Taylor formula for sine has only odd coefficients, for example.

### Further Reading


Brent, R. P., and P. Zimmermann. 2010. *Modern Computer Arithmetic.* Cambridge: Cambridge University Press.

Ercegovac, M. D., and T. Lang. 2004. *Digital Arithmetic.* San Francisco, CA: Morgan Kaufmann.

Higham, N. J. 2002. *Accuracy and Stability of Numerical Algorithms*, 2nd edn. Philadelphia, PA: SIAM.

Knuth, D. E. 1998. *The Art of Computer Programming*, 3rd edn, volume 2. Reading, MA: Addison-Wesley.

Muller, J.-M. 2006. *Elementary Functions, Algorithms and Implementation*, 2nd edn. Boston, MA: Birkhäuser.

Payne, M., and R. Hanek. 1983. Radian reduction for trigonometric functions. *SIGNUM Newsletter* 18:19–24.


## VI.12 Random Number Generation

*Harald Niederreiter*

The notion of randomness is frequently encountered in everyday life. For instance, we view the outcome of a fair coin toss as random and unpredictable, and the same holds for a throw of dice, a perfect shuffle of a card deck, and the drawing of numbers in a lottery. Such examples may lead to an intuitive understanding of random objects (heads and tails, integers from 1 to 6, permutations of a finite set, bits 0 and 1, or real numbers). Grasping randomness in scientific terms is much harder and may even prove elusive, depending on viewpoints that verge on the philosophical.

Some theoretical framework for randomness is needed, however, since there are important methods in computational and applied mathematics that rely on random objects. A prime example is offered by *Monte Carlo methods*, which can be described as numerical methods based on random sampling. The random samples usually consist of real numbers or points in a Euclidean space. In its standard form, a Monte Carlo method approximates the expected value of a random variable (or, in other words, the integral of an integrable function) by a sample mean (or, in other words, the average value of the integrand taken over a random sample). Random objects are also required as inputs in probabilistic algorithms that are used in mathematics and computer science. Further applications appear in computational statistics, operations research, and computer games, among many other areas. Random bits are employed specifically in cryptography in the context of fast and powerful schemes for data encryption.

The problem of generating random objects (be they bits, integers from a given interval, or points in a Euclidean space) can, in most cases, be reduced to that of generating random (real) numbers. The task of random number generation presents itself in the following form: given a target distribution function $F$ on the real line $\mathbb{R}$, generate a sequence of real numbers that simulates a sequence of independent and identically distributed random variables with distribution function $F$. This task can be simplified by concentrating on an easy standardized distribution function: the uniform distribution function $U$ on $\mathbb{R}$ given by $U(t) = 0$ for $t < 0$, $U(t) = t$ for $0 \leqslant t \leqslant 1$, and $U(t) = 1$ for $t > 1$. Random numbers for which the target distribution function is $U$ are called *uniform random numbers*. Since the probability measure corresponding to $U$ is supported on the interval $[0, 1]$, uniform random numbers can be taken from the same interval and they satisfy the property that the probability of a uniform random number from $[0, 1]$ falling into the subinterval $[0, t]$ is equal to $t$.

Random numbers for which the target distribution function $F$ is different from $U$ are obtained from uniform random numbers by transformation methods. For instance, if $F$ is strictly increasing and continuous on $\mathbb{R}$ and $x_0, x_1, \dots$ is a sequence of uniform random numbers from the open interval $(0, 1)$, then $F^{-1}(x_0), F^{-1}(x_1), \dots$ can be taken as a sequence of random numbers with target distribution function $F$. For obvious reasons, this transformation method is called the *inversion method*. Many other transformation methods have been developed, such as the rejection method, the composition method, and the ratio-of-uniforms method.

From a practical point of view, truly random numbers obtained, for example, by physical means (such as tossing coins, throwing dice, or spinning a roulette wheel) are too cumbersome because of the need for storage and extensive statistical testing. Therefore, users have resorted to *pseudorandom numbers* that are generated in a computer by deterministic algorithms with relatively few input parameters. In this way, problems of storage and reproducibility of the numbers do not arise. Furthermore, well-chosen algorithms for pseudorandom number generation can be subjected to rigorous theoretical analysis that may reduce the need for extensive statistical testing of randomness properties.

In principle, it should be clear that a deterministic sequence of numbers cannot pass all possible tests for randomness. When using pseudorandom numbers in a numerical method such as a Monte Carlo method, you should therefore be aware of the specific desirable statistical properties of the random numbers in the computational problem at hand and choose pseudorandom numbers that are known to pass the corresponding statistical tests. For instance, if all that is needed is the statistical independence of pairs of two successive uniform random numbers, then goodness-of-fit tests relative to a two-dimensional distribution (in this case, the two-dimensional uniform distribution supported on the unit square) are quite sufficient for this particular purpose.

For the same reasons as above, we can focus on *uniform pseudorandom numbers*, that is, pseudorandom numbers with the target distribution function $U$, when facing the problem of generating pseudorandom numbers. The history of uniform pseudorandom number generation is nearly as old as that of the Monte Carlo method and goes back to the 1940s. One of the earliest recorded methods is the *middle-square method* of John von Neumann. The algorithm here works with a fixed finite precision and an iterative procedure: it takes the square of the previous uniform pseudorandom number (which can be taken to belong to the interval $[0,1]$) and extracts the middle digits to form a new uniform pseudorandom number in $[0,1]$. Theoretical analysis of this method has shown, however, that the algorithm is actually quite deficient, since the generated numbers tend to run into a cycle with a short period length. If a finite precision is fixed (which is standard practice when working with a computer), then periodic patterns seem unavoidable when using an iterative algorithm for uniform pseudorandom number generation. It is a primary requirement of a good algorithm for uniform pseudorandom number generation that the period of its generated sequences is very long.

Another early method for generating uniform pseudorandom numbers that performs much better with regard to periodicity properties is the *linear congruential method*, which was introduced by the number theorist Derrick Henry Lehmer. Here we choose a large integer $M$ (usually a prime or a power of 2), an integer $g$ relatively prime to $M$ and of high multiplicative order modulo $M$, and, as an initial value, an integer $y_0$ with $1 \leqslant y_0 \leqslant M-1$ that is relatively prime to $M$. A sequence $y_0, y_1, \dots$ of integers from the set $\{1, 2, \dots, M-1\}$ is then generated recursively by $y_{n+1} \equiv gy_n \pmod{M}$ for $n = 0, 1, \dots$. The numbers $x_n = y_n/M$ for $n = 0, 1, \dots$ belong to the interval $[0,1]$ and are called *linear congruential pseudorandom numbers*. The sequence $x_0, x_1, \dots$ is periodic with period length equal to the multiplicative order of $g$ modulo $M$. This periodicity property suggests that $M$ should be of size at least $2^{30}$ in practice. Given the modulus $M$, it is important that the parameter $g$ be chosen judiciously. Bad choices of $g$ lead to sequences of linear congruential pseudorandom numbers that fail simple statistical tests for randomness, and there are in fact infamous cases of published generators that used bad values of $g$. The linear congruential method is still popular and many recommendations for good choices of parameters are available in the literature. For instance, the GNU Scientific Library recommends the CRAY pseudorandom number generator, which uses the linear congruential method with parameters $M = 2^{48}$ and $g = 44\,485\,709\,377\,909$.

Most of the currently employed methods for uniform pseudorandom number generation use number-theoretic or algebraic techniques. A simple extension of the linear congruential method, the *multiple-recursive method*, replaces the first-order linear recursion in the linear congruential method by linear recursions of higher order. This leads to larger period lengths for the same value of the modulus $M$. Another family of methods is formed by *shift-register methods*, where uniform pseudorandom numbers are generated by means of linear recurring sequences modulo 2. A related family of methods uses vector recursions modulo 2 of higher order, and this family includes the very popular *Mersenne twister* MT19937. The MT19937 produces sequences of uniform pseudorandom numbers with period length $2^{19937} - 1$ that possess 623-dimensional equidistribution up to 32 bits accuracy. Furthermore, the generated sequences pass numerous statistical tests for randomness. There are also methods based on nonlinear recursions, but these methods are harder to analyze.

### Further Reading

Devroye, L. 1986 *Non-Uniform Random Variate Generation.* New York: Springer.

Fishman, G. S. 1996 *Monte Carlo: Concepts, Algorithms, and Applications.* New York: Springer.

Gentle, J. E. 2003 *Random Number Generation and Monte Carlo Methods*, 2nd edn. New York: Springer.

# VI.13 Optimal Sensor Location in the Control of Energy-Efficient Buildings

*Jeffrey T. Borggaard, John A. Burns, and Eugene M. Cliff*

## 1 Motivation and Introduction

Buildings are responsible for more than a third of all global greenhouse gas emissions and consume approximately 40% of global energy. To fully appreciate the scale of this problem, consider the fact that a 10% reduction in buildings' energy usage is equivalent to all renewable energy generated in the United States each year. Moreover, a 70% reduction would be the equivalent of eliminating the energy consumption of the entire U.S. transportation sector. Building systems are highly complex and uncertain dynamical systems. Optimization and control of these systems offer unique modeling, mathematical, and computational challenges if we are to develop new tools for the design, construction, and operation of energy-efficient buildings. In particular, research into new mathematical methods and computational algorithms is needed to control, estimate, and optimize systems governed by partial differential equations (PDEs); research into the model-reduction techniques that are required to efficiently simulate fully coupled dynamic building phenomena is also needed.

These PDE control and estimation problems must be approximated by a "control-appropriate" numerical scheme. It is well known that unless these approximate models preserve certain system properties, they may not be suitable for control or optimization.

In this article we consider a single problem from this large research area: an optimal sensor location problem. We show how rather abstract mathematical theory can be employed to solve the problem and illustrate where approximation theory plays an important role in developing numerical methods. We present an example motivated by the design and control of hospitals. A U.S. Energy Information Administration report has shown that large hospitals accounted for less than 1% of all commercial buildings in the United States in 2007 but consumed 5.5% of the total delivered energy used by the commercial sector. Hospitals alone therefore represent a significant consumer of energy, and the design, optimization, and control of energy use at the scale of individual rooms has been the subject of several recent studies.

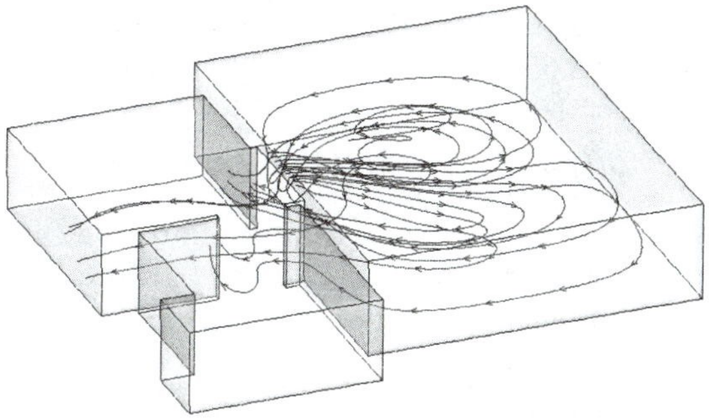

**Figure 1** A hospital suite.

## 2 Mathematical Formulation

Consider the problem of locating a sensor somewhere on a wall in the main zone of the hospital suite in figure 1 such that the sensor provides the best estimate of the temperature field in the entire suite. We assume that the room is configured so that the control inlet diffusers and outlet vents are fixed and the flow field is in steady state. The goal is to find a single thermostat location that provides a sensed output that can be used by the Kalman filter to optimally estimate the state of the thermal conditions inside the room.

Let $\Omega \subset \mathbb{R}^3$ be an open bounded domain with boundary $\partial\Omega$ of Lipschitz class. Consider an advection-diffusion process in the region $\Omega$ with boundary $\partial\Omega$ described by the PDE

$$T_t(t,x)+[v(x)\cdot\nabla T(t,x)] = \kappa\nabla^2 T(t,x)+w(t,x), \quad (1)$$

with boundary conditions

$$T(t,x)|_{\Gamma_{\mathrm{I}}} = b_T(x)u, \qquad \boldsymbol{\eta}(x)\cdot[\kappa\nabla T(t,x)]|_{\Gamma_{\mathrm{O}}} = 0.$$

Here, $T$ is the state, $u$ is a fixed constant temperature at the inflow, $b_T(\cdot)$ is a function describing the inflow shape profile, $\boldsymbol{\eta}(x)$ is the unit outer normal, $\Gamma_{\mathrm{I}}$ is the part of the boundary where the inflow vents are located, and $\Gamma_{\mathrm{O}} = \partial\Omega - \Gamma_{\mathrm{I}}$ is the remaining part of the boundary. The function $v(x)$ is a given velocity field, and $w(t,x)$ is a spatial disturbance. We assume that the disturbance $w(t,x)$ is a spatially averaged random field such that

$$w(t,x) = \iiint_{\Omega} g(x,y)\eta(t,y)\,\mathrm{d}y,$$

where $\eta(t,\cdot) \in L^2(\Omega)$ for all $t \geqslant 0$, and $g(\cdot,\cdot) \in L^2(\Omega\times\Omega)$ is given; this is a very reasonable assumption for problems of this type.

We assume that there is a single sensor (a thermostat) located at a point $q \in \partial\Omega$ that produces a local spatial average of the state $T(t,x)$. In particular,

$$y(t) = \iiint_{B_\delta(q)\cap\Omega} h(x)T(t,x)\,\mathrm{d}x + \omega(t), \qquad (2)$$

where $h(\cdot) \in L^2(\Omega)$ is a given weighting function, $B_\delta(q)$ is a ball of radius $\delta$ about $q$, and $\omega(t)$ is the sensor noise. Define the output map $C(q)\colon L^2(\Omega) \longrightarrow \mathbb{R}^1$ by

$$C(q)\varphi(\cdot) = \iiint_{B_\delta(q)\cap\Omega} h(x)\varphi(x)\,\mathrm{d}x,$$

in terms of which the measured output defined by (2) has the abstract form

$$y(t) = C(q)T(t,\cdot) + \omega(t). \qquad (3)$$

The standard formulation of the distributed parameter model for the convection–diffusion system (1) with output (3) leads to an infinite-dimensional system in the Hilbert space $\mathcal{H} = L^2(\Omega)$ given by

$$\dot{z}(t) = Az(t) + G\eta(t) \in \mathcal{H},$$

with output

$$y(t) = C(q)z(t) + \omega(t),$$

where the state of the distributed parameter system is $z(t)(\cdot) = T(t,\cdot) \in \mathcal{H} = L^2(\Omega)$, $A$ is the usual convection diffusion operator, and $G\colon L^2(\Omega) \to L^2(\Omega)$ is defined by

$$[G\eta(\cdot)](x) = \iiint_\Omega g(x,y)\eta(y)\,\mathrm{d}y.$$

Typically, the measurement $y(t)$ produced by the thermostat on the wall is the only information available to feed into a controller (the heating, ventilation, and air conditioning system) to adjust the room temperature. However, this information can be used to estimate the temperature in the entire room by using a mathematical model. The most well-known method for constructing this approximation is the so-called Kalman filter. In this PDE setting, the optimal Kalman filter produces an estimate $z_e(t)$ of $z(t)$ by solving the system

$$\dot{z}_e(t) = Az_e(t) + F(q)[y(t) - C(q)z_e(t)],$$

where the observer gain operator $F(q)$ is given by

$$F(q) = \Sigma[C(q)]^*$$

and the operator $\Sigma = \Sigma(q)$ satisfies the infinite-dimensional RICCATI OPERATOR EQUATION [III.25]

$$A\Sigma + \Sigma A^* - \Sigma[C(q)]^*C(q)\Sigma + GG^* = 0. \qquad (4)$$

The solution $\Sigma(q)$ is the state estimation covariance operator, and the estimation error is given by

$$\mathbb{E}\Big(\int_0^{+\infty} \|z_e(s,q) - z(s)\|^2_{L^2(\Omega)}\,\mathrm{d}s\Big) = \mathrm{trace}(\Sigma(q)), \quad (5)$$

where $\mathbb{E}(y)$ denotes the expected value of the random variable $y$ and $\mathrm{trace}(\Lambda)$ denotes the trace of the operator $\Lambda$.

We note that in finite dimensions the trace $\mathrm{trace}(\Sigma)$ of an operator $\Sigma$ is the usual trace of the matrix. However, in infinite dimensions not all operators have finite trace, and hence one needs to establish that $\Sigma(q)$ is of trace class (i.e., has finite trace) in order for (5) to be finite. One can show that for the problem here, $\Sigma = \Sigma(q)$ is of trace class.

The optimal sensor management problem becomes a distributed parameter optimization problem with the "state" $\Sigma$ defined by the Riccati system (4) and the cost functional defined in terms of the trace of $\Sigma$ by

$$J(\Sigma,q) = \mathrm{trace}(\Sigma) + R(q),$$

where $\Sigma$ satisfies (4) and the function $R(q)$ could be selected to impose penalties on "bad" regions to be avoided. This cost function is selected because $\mathrm{trace}(\Sigma(q))$ is the estimation error when the Kalman filter is used as a state estimator. We assume that for all $q \in \partial\Omega$ the Riccati equation (4) has a unique positive-definite solution $\Sigma = \Sigma(q)$. The optimal sensor placement problem can be stated as the following optimal control problem, where we note that the assumption that for each $q \in \partial\Omega$ the Riccati equation (4) produces a unique $\Sigma = \Sigma(q)$ allows us to introduce the *reduced functional* $\tilde{J}(q) = J(\Sigma(q),q)$.

**The optimal sensor location problem.** *Find* $q^{\mathrm{opt}}$ *such that*

$$\tilde{J}(q) = J(\Sigma(q),q) = \mathrm{trace}(\Sigma(q)) + R(q) \qquad (6)$$

*is minimized, where* $\Sigma = \Sigma(q)$ *is a solution of the system (4).*

As observed above, one can show that $\Sigma(q)$ is of trace class and that the optimization problem is well defined. This formulation of the problem allows us to directly apply standard optimization algorithms.

## 3 Approximation and Convergence

One of the main issues that needs to be addressed in order to develop practical numerical schemes for solving the optimal sensor location problem is the approximation of the Riccati operator equation (4), which is required to compute $\Sigma = \Sigma(q)$. Here, $C(\cdot)$ varies continuously with respect to the Hilbert–Schmidt norm on the set of trace class operators. The practical implication of this fact is that $\mathrm{trace}(\Sigma(\cdot))$ also varies continuously with respect to $q$. Thus, we have the basic foundations that allow the application of numerical optimization methods. In particular, we make use of known results on existence and approximations dealing with convergence in the space of Hilbert–Schmidt operators on $\mathcal{H}$.

The key observation is that numerical approaches to solve the Riccati equation (4) that produce approximations $\Sigma^N(q)$ of $\Sigma(q)$ must have specific properties if one expects to obtain convergence of $\mathrm{trace}(\Sigma^N(q))$ to $\mathrm{trace}(\Sigma(q))$.

Writing the Riccati equation (4) as

$$A\Sigma + \Sigma A^* - \Sigma D(q)\Sigma + F = 0,$$

where $D(q) = C(q)^*C(q)$ and $F = GG^*$, one clearly sees that one must construct convergent approximations $A^N$, $C^N(q)$, and $G^N$ of the operators $A$, $C(q)$, and $G$, respectively. What is sometimes overlooked is that the dual operators also need to be approximated. Thus, one also needs to construct convergent approximations $[A^*]^N$, $[C^*(q)]^N$, and $[G^*]^N$ of the operators $A^*$, $C^*(q)$, and $G^*$, respectively. It is important to note that, in general, $[A^N]^*$, $[C^N(q)]^*$, and $[G^N]^*$ may not converge to $A^*$, $C^*(q)$, and $G^*$ in a suitable topology, and hence convergence of $\mathrm{trace}(\Sigma^N(q))$ to $\mathrm{trace}(\Sigma(q))$ can fail. In addition to dual convergence, the numerical scheme must preserve some stabilizability and detectability conditions. Again, it is important to note that even standard numerical schemes may not preserve these important control-system properties.

Consider a sequence of approximating problems defined by $(\mathcal{H}^N, A^N, C^N(q), G^N)$, where $\mathcal{H}^N \subset \mathcal{H}$ is a sequence of finite-dimensional subspaces of $\mathcal{H}$, and $A^N \in \mathcal{L}(\mathcal{H}^N, \mathcal{H}^N)$, $C^N(q) \in \mathcal{L}(\mathcal{H}^N, \mathbb{R}^1)$, and $G^N \in \mathcal{L}(\mathcal{H}^N, \mathcal{H}^N)$ are bounded linear operators. Here, $\mathcal{L}(\mathcal{X}, \mathcal{Y})$ denotes the usual space of bounded linear operators from $\mathcal{X}$ to $\mathcal{Y}$. In this article we use standard finite-element methods so that $\mathcal{H}^N$ is a finite-element space of piecewise polynomial functions. Let $P^N\colon \mathcal{H} \to \mathcal{H}^N$ denote the orthogonal projection of $\mathcal{H}$ onto $\mathcal{H}^N$ satisfying $\|P^N\| \leqslant 1$ and such that $\|P^N z - z\| \to 0$ as $N \to \infty$ for all $z \in \mathcal{H}$. For each $N = 1, 2, 3, \dots$ consider the finite-dimensional approximations of (4) given by

$$A^N\Sigma^N + \Sigma^N[A^*]^N - \Sigma^N D^N(q)\Sigma^N + F^N = 0, \quad (7)$$

and assume that (7) has solutions $\Sigma^N$. The approximate optimal sensor location problem is now to minimize

$$\tilde{J}^N(q) = \mathrm{trace}(\Sigma^N(q)) + R(q),$$

subject to the constraint (7).

In order to discuss convergence of the finite-dimensional approximating Riccati operators, we need to assume that the numerical scheme preserves the basic stabilizability and detectability conditions needed to guarantee that the Riccati equation (7) has a unique positive-definite solution.

The required assumptions, which we will not state here, break into four distinct conditions concerning the convergence of the operators, convergence of the adjoint operators, preservation of uniform stabilizability/detectability under the approximation, and compactness requirements on the input operator. Under these assumptions one can show that the Riccati equation (7) has a unique nonnegative solution $\Sigma^N(q)$ for all sufficiently large $N$ and that

$$\lim_{N\to+\infty} \mathrm{trace}(\Sigma^N(q)) = \mathrm{trace}(\Sigma(q)).$$

One can also prove that

$$\lim_{N\to+\infty} \mathrm{trace}(\Sigma^N(q)P^N - P^N\Sigma(q)) = 0,$$

which states that the operators $\Sigma^N(q)$ converge to $\Sigma(q)$ in the trace norm. For the specific case treated here, one can establish that these assumptions hold for standard finite-element schemes.

## 4 Numerical Results for the Hospital Suite

For simplicity, we consider the two-dimensional version of the hospital suite shown in figure 2. The "bed area" is the large zone on the right, and the remaining zones are bath and dressing areas. There are four inlet vents, and the outflow is into a hall through the door on the left. The optimal sensor location problem is not very sensitive to the outflow location since the flow is dominated by the outflow from the bed area to the other two zones.

For the numerical runs, we assumed a nearly uniform disturbance, such that

$$w(t,x) = [G_\varepsilon \eta(t,\cdot)] = \iiint_\Omega \delta_\varepsilon(x-y)\eta(t,y)\,\mathrm{d}y,$$

where $\delta_\varepsilon(y)$ is a smooth approximation of the delta function. We also set $R(q) = 0$.

In figure 2 we show the inflow, the outflow, and the flow through the suite. The asterisks on the north and south walls are the points $q_1$ and $q_2$, where the cost function $\tilde{J}(q) = J(\Sigma(q)) = \mathrm{trace}(\Sigma(q))$ has local minima. Figure 3 provides a plot of $\tilde{J}(q) = \mathrm{trace}(\Sigma(q))$ as $q$ moves from the northwest corner along the north wall, down the east wall, across the south wall, and then up the west side of the bed area. Observe that the optimal sensor location problem has two solutions: one on the north wall between the two inlets and one on the south

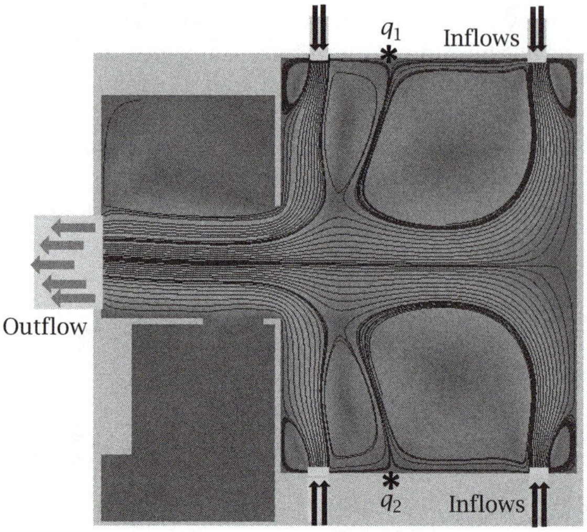


**Figure 2** A two-dimensional hospital suite with four inlets.

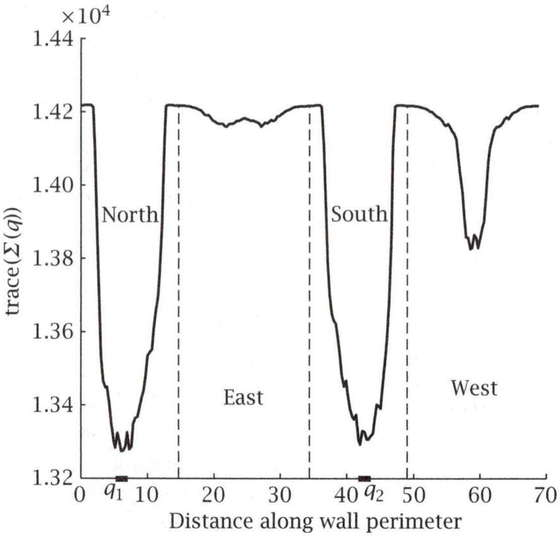


**Figure 3** The value of trace($\Sigma(q)$) along the walls.

wall between the two inlets there. There is also a local minimum located in the opening from the bed area to the bath area where there is no wall. Although there is also a local minimum on the east wall, this minimum is nearly the same as the maximum value.

This example illustrates that the optimal sensor placement problem is reasonably well behaved. Even though there are multiple local minima, most numerical optimization algorithms can handle this type of problem. The "roughness" near $q_1$ and $q_2$ is caused primarily by the coarse grid used in the flow solver. However, even with the coarse grid used for these calculations, it is clear that the model is accurate enough to roughly identify the neighborhoods where the global minimizers are located. This trend has been noticed in several other problems of this type and suggests that a MULTIGRID-LIKE [IV.13 §3] approach to the optimization problem might improve the overall algorithm. In particular, we first solve the optimization problem on a coarse grid to identify a rough neighborhood of the minimizer and then conduct a fine-grid optimization starting in this neighborhood. This idea has been successfully applied to similar problems.

## 5 Closing Comments

One goal of this article was to illustrate how a rather abstract mathematical framework can be used to formulate and solve some very practical (optimal) design problems that arise in the energy sector. A second takeaway is that one must be careful when introducing approximations to numerically solve the resulting optimization problem. In particular, additional appropriate assumptions are needed to guarantee convergence of the operators.

Another issue concerns the finite-dimensional optimization problem itself. Although we do not have the space to fully address this issue, it is clear that applying modern gradient-based optimization to the approximate optimal sensor location problem requires that the mapping $q \to [\tilde{J}^N(q)] = \mathrm{trace}(\Sigma^N(q))$ be at least $C^1$. This smoothness requirement also places additional restrictions on the types of approximations that can be used to discretize the infinite-dimensional system. There are several approaches one might consider to approximate the gradient $\nabla_q[\tilde{J}^N(q)]$. These range from direct finite-difference methods to more advanced continuous sensitivity and adjoint methods. In theory there is no difference between these methods as long as they produce consistent gradients. However, in practice the choice can greatly influence the speed and accuracy of the optimization algorithm. We are therefore reminded of the following quote attributed to Manfred Eigen:

> In theory, there is no difference between theory and practice. But, in practice, there is.

**Acknowledgments.** This research was supported by the Air Force Office of Scientific Research under grant FA9550-10-1-0201, and by the Department of Energy contract DE-EE0004261 under subcontract #4345-VT-DOE-4261 from Penn State University.

**Further Reading**

Borggaard, J., J. Burns, E. Cliff, and L. Zietsman. 2012. An optimal control approach to sensor/actuator placement for optimal control of high performance buildings. In *Proceedings of the 2nd International High Performance Buildings Conference, Purdue University*, pp. 34661–67.

Brown, D. L., J. A. Burns, S. Collis, J. Grosh, C. A. Jacobson, H. Johansen, I. Mezic, S. Narayanan, and M. Wetter. 2010. Applied and computational mathematics challenges for the design and control of dynamic energy systems. Technical Report, Lawrence Livermore National Laboratory, Arlington, VA.

Mendez, C., J. F. San Jose, J. M. Villafruel, and F. Castro. 2008. Optimization of a hospital room by means of CFD for more efficient ventilation. *Energy and Buildings* 40: 849–54.

Narayanan, S. 2012. A systems approach to high performance buildings: a computational systems engineering R&D program to increase DoD energy efficiency. Final Report for Project EW-1709, Strategic Environmental Research and Development Program.

U.S. Energy Information Administration (EIA). 2007. Energy characteristics and energy consumed in large hospital buildings in the United States in 2007. Commercial Buildings Energy Consumption Survey Report. Available at www.eia.gov/consumption/commercial/reports/2007/large-hospital.cfm.

Yao, R. 2013. *Design and Management of Sustainable Built Environments*. New York: Springer.


# VI.14 Robotics

*Charles W. Wampler*

## 1 Robot Kinematics

Robotics is an interdisciplinary field involving mechanical and electrical engineering, computer science, and applied mathematics that also draws inspirations from biology, psychology, and cognitive science. From robots in a factory to interplanetary rovers, one of the fundamental capabilities a robot must have is knowledge of where it is and how to get to where it needs to go. This includes knowing where its appendages are and how to move them effectively. While in technical jargon these appendages may be called "manipulators" with "end-effectors" that directly interact with workpieces, they are more commonly referred to as robot "arms" and "hands," especially when the robot links are connected end to end in serial fashion. Locating a hand involves finding both its position and its orientation. For example, to pick an object out of an open jar, it is important for the hand to not only reach the position of the object but also to be oriented through the opening of the jar. Orientation is fully specified by a $3 \times 3$ orthogonal matrix, sometimes called a rotation matrix or a direction cosine matrix.

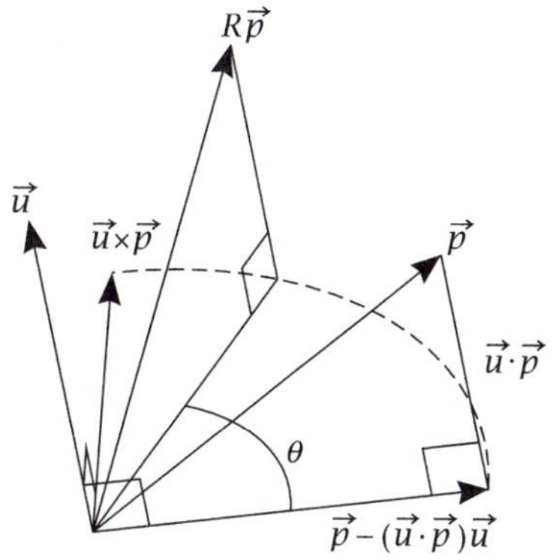


**Figure 1** Rotation of $\vec{p}$ around axis $\vec{u}$.

It turns out that the geometric motion characteristics, that is, the *kinematics*, of most robots (and most mechanical systems in general) are well modeled by systems of polynomial equations. In particular, this can be seen in the most common element used in mechanism work: the rotational joint. When a series of links are connected end-to-end by rotational joints, each link turns circles with respect to its neighbors, and circles can be described by polynomial equations. The polynomial nature of the models allows one to apply algebraic geometry to solving kinematics problems.

Consider a serial-link robot arm with rotational joints canted at various angles so that the hand maneuvers in three-dimensional space, not just in one plane. To be more precise, consider first just a single joint and let $\vec{u} \in \mathbb{R}^3$ be a unit vector ($\vec{u} \cdot \vec{u} = 1$) along its joint axis, assumed to be passing through the origin. As illustrated in figure 1, the rotation of an arbitrary vector $\vec{p} \in \mathbb{R}^3$ through an angle of $\theta$ around unit vector $\vec{u}$ is

$$R(\vec{u},\theta)\vec{p} = (\vec{u}\cdot\vec{p})\vec{u} + [\vec{p} - (\vec{u}\cdot\vec{p})\vec{u}]\cos\theta + \vec{u}\times\vec{p}\,\sin\theta,$$

where $R(\vec{u},\theta)$ is a $3 \times 3$ matrix expression formed from the matrix interpretation of the vectorial operations on the right-hand side.

The trigonometric expression for the rotation can be converted to an algebraic one by replacing $(\cos\theta, \sin\theta)$ by $(c, s)$ subject to the unit-circle condition $c^2 + s^2 = 1$. Abusing notation, we call the reformulated rotation matrix $R(\vec{u}, c, s)$, which in matrix form becomes

$$R(\vec{u}, c, s) = \vec{u}\vec{u}^{\mathrm{T}} + (I - \vec{u}\vec{u}^{\mathrm{T}})c + \Lambda(\vec{u})s,$$

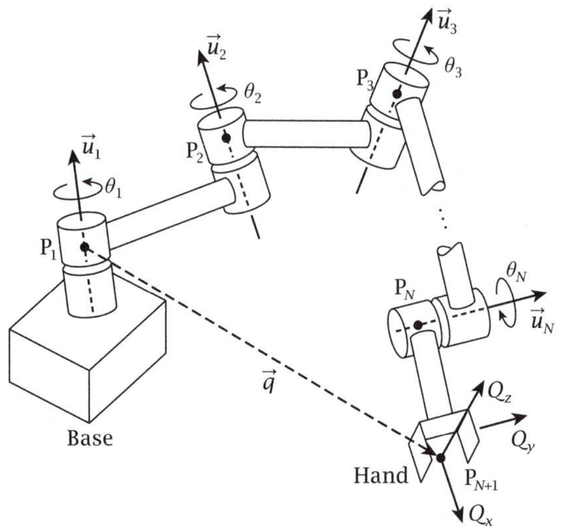


**Figure 2** Serial-link robot schematic.

where

$$\Lambda\left(\begin{bmatrix} x \\ y \\ z \end{bmatrix}\right) = \begin{bmatrix} 0 & -z & y \\ z & 0 & -x \\ -y & x & 0 \end{bmatrix}.$$

For example, if $\vec{u} = [1\ \ 0\ \ 0]^{\mathrm{T}}$, then $R(\vec{u}, c, s)$ is a rotation in the $yz$-plane. We note that $R(\vec{u}, c, s)$ is linear in $(c, s)$. As the joint turns, the vector $\vec{u}$ defining it stays constant while $c$ and $s$ vary, but $R(\vec{u}, c, s)$ is always orthogonal, and its determinant is 1.

This generalizes to a multilink arm. Consider links numbered 0 to $N$ from the robot's base to its hand, connected in series by rotational joints. To formulate the position and orientation of the hand with respect to the base, we need geometric information about the links and their joints. As illustrated in figure 2, mark a point $P_1$ on the joint axis between links 0 and 1, and similarly mark points $P_2, \dots, P_N$ on the succeeding axes. Also, mark a reference point in the hand as $P_{N+1}$. Next, freeze the arm in some initial pose, and at this configuration let $\vec{u}_i$ be a unit vector along the axis of the joint between link $i-1$ and link $i$. Finally, in the initial pose, let the vector from $P_i$ to $P_{i+1}$ be $\vec{p}_i$ and let the initial orientation of the hand be $Q_0 \in \mathrm{SO}(3)$. With these definitions and the shorthand $R_i := R(\vec{u}_i, c_i, s_i)$, we may write the orientation, $Q = [Q_x\ \ Q_y\ \ Q_z] \in \mathrm{SO}(3)$, of the hand with respect to the base and the position vector, $\vec{q}$, from $P_1$ to $P_N$ as

$$Q = R_1 R_2 \cdots R_{N-1} R_N Q_0, \tag{1}$$

$$\vec{q} = R_1(\vec{p}_1 + R_2(\vec{p}_2 + \cdots + R_N(\vec{p}_N) \cdots)). \tag{2}$$

Given the joint rotations $(c_i, s_i)$, $i = 1, \dots, N$, one may evaluate these expressions to obtain $Q$ and $\vec{q}$. This solves the *forward kinematics problem* for serial-link arms.

## 2 Inverse Kinematics

Evaluation of the forward kinematics formulas tells us where a serial-link robot's hand is relative to its base. More challenging is to reverse this by answering the *inverse kinematics problem*: what joint rotations $(c_i, s_i)$, $i = 1, \dots, N$, will cause the hand to attain a desired location $(Q, \vec{q})$? The space of all rigid-body motions is SE(3), a six-dimensional space parameterizable by three rotations and three translations. Thus, we need $N \geqslant 6$ joints to place the hand in any arbitrary position and orientation within the working volume of the arm.

Let us consider the important case of $N = 6$, where we expect to have a finite number of solutions to the equations (1), (2) along with

$$c_i^2 + s_i^2 = 1, \quad i = 1, \dots, 6. \tag{3}$$

Although (1) has nine entries, only three are independent since $Q^{\mathrm{T}}Q = I$. The isolated solutions of the system are preserved if one takes three random linear combinations of these nine, which with the rest of the equations makes a system of twelve polynomials in twelve unknowns. By Bézout's theorem, the number of isolated solutions of a system of $N$ polynomials in $N$ unknowns cannot exceed the total degree, defined as the product of the degrees of the equations. For the system at hand, this comes to $6^6 2^6 = 2\,985\,984$. It turns out that this upper bound is rather loose.

The root count can be reduced by algebraically manipulating the equations. First, using the fact that $R_i^{-1} = R_i^{\mathrm{T}}$, one may rewrite (1), (2) as

$$R_3^{\mathrm{T}} R_2^{\mathrm{T}} R_1^{\mathrm{T}} Q = R_4 R_5 R_6 Q_0, \tag{4}$$

$$R_3^{\mathrm{T}}(R_2^{\mathrm{T}}(R_1^{\mathrm{T}}\vec{q} - \vec{p}_1) - \vec{p}_2) = \vec{p}_3 + R_4(\vec{p}_4 + R_5(\vec{p}_5 + R_6\vec{p}_6)). \tag{5}$$

These equations are now cubic, reducing the total degree to $3^6 2^6 = 46\,656$. But the equations are far from being general cubics because the $(c_i, s_i)$ pairs each appear linearly. Grouping the unknowns into three groups as $\{c_1, s_1, c_4, s_4\}$, $\{c_2, s_2, c_5, s_5\}$, and $\{c_3, s_3, c_6, s_6\}$, equations (4), (5) are recognized as being trilinear. Based on this grouping of the variables, a variant of Bézout's theorem known as the "multihomogeneous Bézout number" bounds the maximum possible number of isolated roots by the coefficient of $\alpha^4\beta^4\gamma^4$ in $(\alpha+\beta+\gamma)^6(2\alpha)^2(2\beta)^2(2\gamma)^2$, i.e., 5760.

This is just the beginning of the algebraic manipulations that one can perform on the way to showing that the "six-revolute" (or 6R) inverse kinematic problem has at most sixteen isolated roots.

From an early statement of the problem by Pieper in 1968 to a numerical solution by Tsai and Morgan in 1985 using continuation through to the first algebraic derivation of an eliminant equation of degree 16 by Lee and Liang in 1988, this problem was one of the top conundrums for kinematicians for twenty years. Now, though, powerful computer algorithms can be applied to solve the problem in minutes with either symbolic computer algebra, based on variants of Buchberger's algorithm, or numerical algebraic geometry, based on continuation methods. In the latter approach, no further manipulation of the equations is required, as one can set up a homotopy that continuously deforms an appropriate start system into a target 6R example. Using the variable groups mentioned above, this homotopy has 5760 paths, which can be tracked in parallel on multiple processors. The endpoints of these paths include the sixteen isolated solutions (real and complex) of the example 6R problem. Once solved, the general target example can serve as the start system for a sixteen-path parameter homotopy to solve any other 6R inverse kinematic problem.

## 3 Generalizations

In addition to serial-link arms, robots and similar mechanisms can have a variety of topologies, composed of serial chains connected together to form closed-chain loops. Both forward and inverse kinematics problems become challenging, but the kinematics remain algebraic and modern algorithms derived from algebraic geometry apply. These mathematical methods for kinematic chains also find application in biomechanical models of humans and animals and in studies of protein folding.

### Further Reading

Raghavan, M., and B. Roth. 1995. Solving polynomial systems for the kinematic analysis and synthesis of mechanisms and robot manipulators. *Journal of Mechanical Design* 117(B):71–79.

Sommese, A. J., and C. W. Wampler. 2005. *The Numerical Solution of Systems of Polynomials Arising in Engineering and Science.* Singapore: World Scientific.

Wampler, C. W., and A. J. Sommese. 2011. Numerical algebraic geometry and algebraic kinematics. *Acta Numerica* 20:469–567.

# VI.15 Slipping, Sliding, Rattling, and Impact: Nonsmooth Dynamics and Its Applications

*Chris Budd*

## 1 Overview

We all live in a nonsmooth world. Events stop and start, by accident, design, or control. Once started, mechanical parts in bearings, bells, and other machines can come into contact and impact with each other. Once in contact, such parts can slide against each other, intermittently sticking and slipping. If these parts are rocks, then the result is an earthquake, a system that is notoriously hard to predict. While on a small scale these systems might be smooth, on a macroscopic scale they lose smoothness in one or more derivatives, and their motion is best approximated, and analyzed, by assuming that they are nonsmooth. To study such problems—which may be as simple as a ball bouncing on a table or as complex as a collapsing building, or even the climate—we need to extend classical dynamical systems theory to allow for nonsmooth effects. Typically, the problems considered, including systems with impact, sticking, and sliding, are *piecewise smooth* (PWS) and comprise smooth trajectories interrupted by instantaneous nonsmooth events. Such dynamical systems are tractable to analysis, which reveals a rich variety of behavior, including chaos, new routes to chaos, and bifurcations that are not observed in smooth dynamical systems.

## 2 What Is a Piecewise-Smooth Dynamical System?

Piecewise-smooth systems comprise flows, maps, and a hybrid mixture of the two. To define them, let the whole system be given by $\boldsymbol{x}(t) \in \Omega \subset \mathbb{R}^n$, with $\Omega$ divided into subsets $S_i$, $i = 1, \dots, N$, with the boundary between $S_i$ and $S_j$ being the surface $\Sigma_{ij}$ of codimension $p \geqslant 1$. A PWS *flow*, often called a Filippov system, is given by

$$\frac{\mathrm{d}\boldsymbol{x}}{\mathrm{d}t} = \boldsymbol{F}_i(\boldsymbol{x}) \quad \text{if } \boldsymbol{x} \in S_i, \tag{1}$$

where $\boldsymbol{F}_i$ is smooth in the set $\Omega_i$. As $\boldsymbol{x}$ crosses the set $\Sigma_{ij}$, the right-hand side of (1) will typically lose smoothness. An example of a Filippov system is a room heater controlled by a thermostat that switches a heating element on when the temperature of the room $T(t)$ falls below specified temperature $T_0$. In this case we have

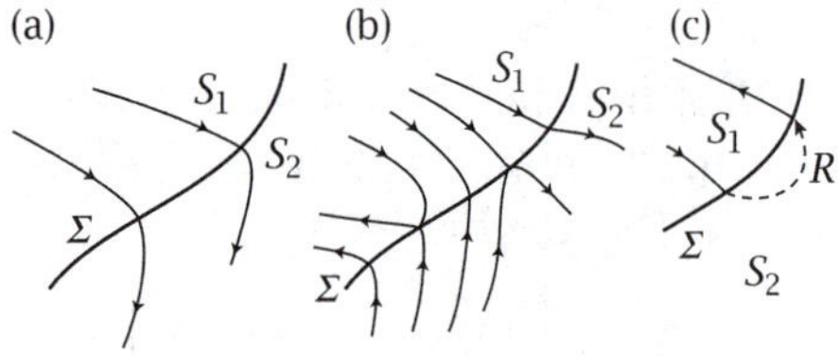


**Figure 1** (a) Piecewise-smooth flow, (b) sliding flow, and (c) hybrid flow.

two regions in which the operation of the heating system is smooth, and the set $\Sigma_{12}$ is the surface $T - T_0 = 0$. The trajectories of a Filippov system are themselves PWS, losing regularity when they intersect $\Sigma_{ij}$.

Particularly interesting behavior (illustrated in figure 1) occurs on those subsets of $\Sigma_{ij}$ for which both vector fields $\boldsymbol{F}_i$ and $\boldsymbol{F}_j$ point toward $\Sigma_{ij}$. In this case we observe *sliding motion* in which the dynamical system has solutions that *slide* along $\Sigma_{ij}$. On the sliding surface the system obeys a different set of equations obtained by certain averages of the vector field to give a regularized *sliding vector field*. Sliding problems arise naturally in studying friction and in many types of control problems and they are typically studied using the theory of *differential inclusions*, which studies differential equations with set-valued right-hand sides.

A PWS *map* is defined by

$$\boldsymbol{x}_{n+1} = \boldsymbol{F}_i(\boldsymbol{x}_n) \quad \text{if } \boldsymbol{x}_n \in S_i, \tag{2}$$

with a *piecewise-linear map* given by the special case of taking

$$\boldsymbol{F}_i(\boldsymbol{x}_n) = A_i \boldsymbol{x}_n + \boldsymbol{b}_i, \qquad \Sigma = \{\boldsymbol{x} : \boldsymbol{c} \cdot \boldsymbol{x} = 0\}, \tag{3}$$

where the $A_i$ are matrices and the $\boldsymbol{b}_i$ are vectors. Maps such as (3), often called *Feigin maps* if they are continuous or *maps with a gap* otherwise, arise naturally in their own right, in control problems and in circle maps, but they also appear in neuroscience, where the condition $\boldsymbol{c} \cdot \boldsymbol{x}_n \geqslant 0$ is a threshold for neurons firing. They also occur in the context of Poincaré maps for systems of the form (1). Despite their apparent simplicity, the dynamics of systems of the form (3) is especially rich, particularly if the $A_i$ have complex eigenvalues. Another such map is the remarkable *Nordmark map*, which arises very often in the context of the *grazing bifurcations* described in the next section. This takes the form

$$\boldsymbol{x}_{n+1} = \begin{cases} A_1 \boldsymbol{x}_n + \boldsymbol{b}_1 \sqrt{\boldsymbol{c} \cdot \boldsymbol{x}_n}, & \boldsymbol{c} \cdot \boldsymbol{x}_n \geqslant 0, \\ A_2 \boldsymbol{x}_n + \boldsymbol{b}_2, & \text{otherwise.} \end{cases} \tag{4}$$

The square-root term means that close to the line $\Sigma = \{\boldsymbol{x} : \boldsymbol{c} \cdot \boldsymbol{x} = 0\}$ the map has unbounded derivatives and can introduce infinite stretching in the phase plane. It is therefore no surprise that such naturally occurring maps lead to chaotic behavior over a wide range of parameter values.

Finally, we have HYBRID SYSTEMS [II.18], which are a combination of PWS flows in the regions $S_i$ and maps $R: \Sigma \to \Sigma$ that act when the solution trajectories intersect the set $\Sigma$ (see figure 1). Hybrid systems arise naturally in control problems, and are also a natural description of the many types of phenomena that involve instantaneous impacts, such as bouncing balls, rattling gears, or constrained mechanical motion such as that of bearings. The dynamics of hybrid systems can also be described by using *measure differential inclusions*.

## 3 What Behavior Is Observed?

Many of the phenomena associated with smooth dynamical systems can be found in PWS ones. For example, the systems (1) and (2) can have states such as fixed points, periodic trajectories, and chaotic solutions. If the fixed points are distant from the discontinuity surfaces $\Sigma_{ij}$ or if any trajectories intersect such surfaces *transversally*, then we can apply the implicit function theorem to study them, and they have bifurcations (saddle–node, period-doubling, Hopf) in the same manner as smooth systems. The main differences arise when, as a parameter varies, these states intersect $\Sigma_{ij}$ for the first time, so that a fixed point evolves to lie on $\Sigma_{ij}$, or a trajectory evolves so that it has a tangential, or *grazing*, intersection with $\Sigma_{ij}$. These intersections typically lead to dramatic changes in the behavior of the solution in a *discontinuity-induced bifurcation*. Such bifurcations lead to new routes to chaos and many new phenomena. As an example consider the system (2) in the simplest case when $x_n$ is a scalar and for some $0 < \lambda < 1$ we have

$$x_{n+1} = \begin{cases} \lambda x_n + \mu - 1, & x_n \geqslant 0, \\ \lambda x_n + \mu, & x_n < 0. \end{cases}$$

If $\mu > 1$ then the map defined above has a fixed point at $x^* = (\mu - 1)/(1 - \lambda)$. Similarly, if $\mu < 0$ the map has a fixed point at $x^* = \mu/(1 - \lambda)$. At $\mu = 1$ and at $\mu = 0$, respectively, one of these two fixed points lies on the discontinuity set $\Sigma = \{x : x = 0\}$. If $0 < \mu < 1$ we see complex behavior, and a bifurcation diagram for the case of $\lambda = 0.7$ is given in figure 2. This has the remarkable structure of a *period-adding cascade* in which we see a period-$(n + m)$ orbit existing for values of the parameter $\mu$ that lie between the values of $\mu$

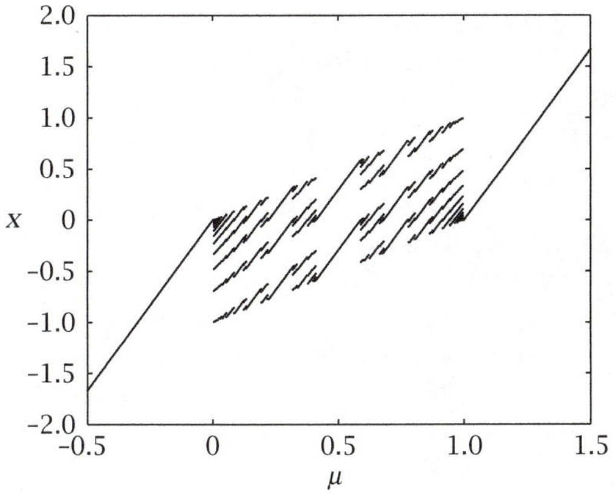


**Figure 2** A discontinuity-induced bifurcation leading to a period-adding cascade in a piecewise-linear map.

for which we see period-$n$ and period-$m$ orbits. Period adding (and the closely related phenomenon of period incrementing) is a characteristic feature of the bifurcations observed in PWS systems. More generally, very similar phenomena are observed when periodic orbits of PWS flows graze with $\Sigma_{ij}$, leading to *grazing bifurcations*, which can be studied using the Nordmark map (4) and observed experimentally. Even more complex behavior is observed in the bifurcations associated with the onset of sliding motion.

Overall, the study of bifurcations in PWS systems is still in a relatively early stage of development. Much needs to be done both to classify them and understand them mathematically, and to study their rich behavior in the many applications in which they arise.

### Further Reading

di Bernardo, M., C. Budd, A. Champneys, and P. Kowalczyk. 2008. *Piecewise-Linear Dynamical Systems: Theory and Applications*. New York: Springer.

Guckenheimer, J., and P. Holmes. 1983. *Bifurcations and Chaos*. New York: Springer.

Kuznetsov, Y. A. 2004. *Elements of Applied Bifurcation Theory*, 3rd edn. New York: Springer.

## VI.16 From the *N*-Body Problem to Astronomy and Dark Matter

*Donald G. Saari*

### 1 A Dark Mystery from Astronomy

Puzzles coming from astronomy have proved to be an intellectual temptation for mathematicians, which is why for centuries these two areas have enjoyed a symbiotic relationship. The challenge of describing the motion of the planets, for instance, led to calculus and Newton's equations of motion, which in turn revolutionized astronomy and physics. But so many mathematical results have been found about the Newtonian $N$-body problem that only a flavor can be offered here. To unify the description, topics are selected to indicate how they shed light on the intriguing astronomical mystery of *dark matter*.

The dark matter enigma can be seen, for example, in the fact that the predicted mass level needed to keep galaxies from dissipating vastly exceeds what is known to exist. This huge difference between predicted and known mass amounts is believed to consist of unobserved mass called dark matter. While no current evidence exists about the nature of this missing matter, whatever it may be, it appears to dominate the mass of galaxies and the universe.

What is not appreciated is that mathematics is a major player in this mystery story. It must be; the predicted mass is a result of a mathematical analysis of the Newtonian $N$-body problem. A review of the mathematics is accompanied with descriptions of other $N$-body puzzles.

#### 1.1 The Mass of Our Sun

Start with something simpler; how can the mass of our sun be determined? The answer involves Newton's laws and planetary rotational velocities. Let $\boldsymbol{r}(t)$ be the vector position of a planet (with mass $m$) relative to the sun (with mass $M$). If $r = |\boldsymbol{r}|$, and $G$ is the gravitational constant, Newton's inverse-square force law requires the acceleration $\boldsymbol{r}''$ to satisfy

$$m\boldsymbol{r}'' = -\frac{GMm}{r^2}\left(\frac{\boldsymbol{r}}{r}\right) = -\frac{GMm\boldsymbol{r}}{r^3}. \qquad (1)$$

If $r$ is the scalar length of $\boldsymbol{r}$, then its value is given by the dot product $r^2 = \boldsymbol{r}\cdot\boldsymbol{r} = (\boldsymbol{r})^2$. By differentiating this expression twice and using $[\boldsymbol{r}\cdot\boldsymbol{v}]^2 + [\boldsymbol{r}\times\boldsymbol{v}]^2 = \boldsymbol{r}^2\boldsymbol{v}^2$, where $\boldsymbol{v} = \boldsymbol{r}'$ is the velocity, it follows by substitution into (1) that the scalar acceleration satisfies

$$mr'' = -\frac{GMm}{r^2} + m\frac{[\boldsymbol{r}\times\boldsymbol{v}]^2}{r^3} = -\frac{GMm}{r^2} + m\frac{v_{\text{rot}}^2}{r}, \qquad (2)$$

where $v_{\text{rot}}$ is the planet's rotational velocity (the $\boldsymbol{v}$ component that is orthogonal to $\boldsymbol{r}$).

Our Earth has an essentially circular orbit, so $r'' \approx 0$; from (2) we can then deduce that

$$M \approx \frac{rv_{\text{rot}}^2}{G}. \qquad (3)$$

Using Earth's distance from the sun (the $r$ value) and the time it takes to orbit the sun (a year), the rotational velocity and $M$, the sun's mass, can be computed.

Conversely, (3) asserts that the sun's specified mass $M$ is needed to sustain Earth's rotational velocity with its nearly circular orbit. According to (2), a $v_{\text{rot}}$ larger than

$$v_{\text{rot}} \approx [GM/r]^{1/2} \tag{4}$$

would force a larger $r'' > 0$ and introduce the possibility that Earth might escape from the solar system. Luckily for us, this does not happen.

But there is a potential problem: the planet masses are negligible compared with that of the sun, so what would it mean if information about a farther out planet—Neptune, say—predicted a significantly larger mass for the sun? Three natural options are as follows.

(1) Newton's law is wrong.
(2) The larger mass prediction means that Neptune's velocity is too large for the planet to be kept in the solar system ((2), (4)). Our system may therefore dissipate.
(3) The predicted mass is there. Rather than reflecting the sun's mass, it is unobserved and hiding between the orbits of Earth and Neptune. This is a "solar system dark matter" concern.

Fortunately, we escape the onerous task of selecting among these undesired alternatives because Kepler's third law requires the numerator in (3), $rv_{\text{rot}}^2$, to essentially be a constant, i.e., the rotational velocities of planets decrease to zero as $r$ increases according to $v_{\text{rot}}^2 = C/r$. Kepler's law, then, offers us reassurance that Newton's law is correct: we will not lose a planet (at least in the near future), and there is no solar system dark matter concern.

The intent of this hypothetical scenario is to demonstrate the close relationship that connects Newton's law of attraction, mass values, and limits on the rotational velocities needed to sustain a stable system. But while Kepler's third law ensures that problems do not arise in our solar system, what about galaxies? Could the rotational velocities of stars be too large to be sustained by the amount of known mass? This is discussed next.

### 1.2 The Mass of a Galaxy

The minuscule planetary mass values allow our solar system to be modeled as a collection of two-body problems; this is what allows us to use (2) to determine the sun's mass. But "two-body" approximations are not realistic for a galaxy, with its billions of stars. To determine mass values from galactic dynamics would require a deep understanding of the behavior of $N$-body systems in which $N$ is in the billions. Unfortunately, a complete mathematical solution is known only for the two-body problem, so something else is needed.

Astrophysicists address this concern in several clever ways. Pictures of galaxies, for instance, convey a sense of a thick "star soup." This appearance suggests approximating a system of $N$ discrete bodies with a continuum model in which the force on a particle is determined by Newton's first and second laws.

As developed in calculus classes, if a body in a continuum symmetric setting is inside a spherical shell, it experiences no net gravitational force from the shell. If the body is outside a spherical ball, the symmetry causes the gravitational force to behave as though the ball's mass is concentrated at its center. With these assumptions, an equation can be derived to determine $M(r)$, which is a galaxy's total mass up to distance $r$ from the center of mass. This equation, which closely resembles (3), is

$$M(r) \approx \frac{rv_{\text{rot}}^2}{G}. \tag{5}$$

There are other ways to predict galactic masses, but (5) is often used to justify the amount of mass that is needed to keep stars in circular orbits.

Herein lies the problem: rather than behaving in a way that is consistent with Kepler's third law, where $v_{\text{rot}}$ values decrease as $r$ increases, observations prove that the $v_{\text{rot}}$ values start with a sharp, almost linear increase before tapering off to an essentially constant or increasing value! According to (5), a constant $v_{\text{rot}}$ means that the predicted mass must grow linearly with the distance from the center of the galaxy. Observations, however, cannot account for anywhere near this much mass! As with the solar system dark matter story, this vast difference between predictions and observations puts forth uncomfortable options.

(1) Newton's laws are incorrect, at least at the large light-year distances in a galaxy.
(2) The velocities are too large for the mass values; expect the galaxy to fly apart.
(3) The difference between the predicted $M(r) \approx Cr$ and the known mass is there but cannot be seen; it is due to unobserved "dark matter."

Most, but not all, astronomers and astrophysicists find the first two options to be unpalatable, so they concentrate on the third choice.

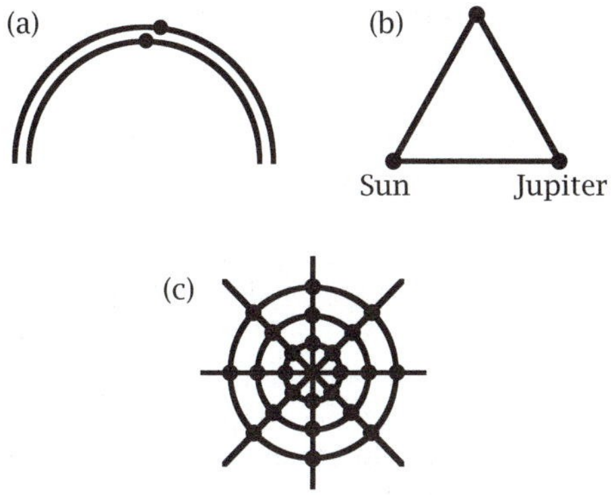


**Figure 1** Discrete-body interactions: (a) interactions, (b) central configuration, and (c) spider web.

## 2 A Mathematical Analysis

While astrophysicists attack this concern with observations and imaginative experiments, an applied mathematician would examine the mathematics. Is (5) being used properly? Namely, could it be that dark matter is a mathematical error in the predicted mass values? The following outline of how to analyze these questions is intended to encourage others to explore these issues.

As (5) relies on continuum models to approximate Newtonian systems with $N$ discrete bodies, it is natural to examine this assumption. Clearly, the behavior of "star soup" continuum models need not agree with that of systems of $N$ discrete bodies. After all, if two discrete bodies on circular orbits come close to each other (figure 1(a)), their mutual attraction can dwarf the effects of other particles, so neither acceleration is directed toward the center of mass. Further violating what continuum models require, the faster body can drag along the slower one, as suggested by both figure 1(a) and pictures of arms of galaxies with the appearance of stars being pulled along by others.

For discrete systems, then, actual rotational velocities involve combinations of the total mass $M(r)$ and a tugging, rather than just $M(r)$ as required by (5). This added pulling requires the stars to have velocities larger than those strictly permitted by $M(r)$: larger velocities force exaggerated predicted mass values. The mathematical issue is whether this reality of exaggerated mass values invalidates the use of (5). A way to analyze this concern is to use analytic $N$-body solutions in which the precisely known mass value is compared with (5) prediction. First, though, other properties of $N$-body systems are described.

### 2.1 Central Configurations

Interestingly, certain $N$-body solutions can permanently retain the same geometric shape. These shapes are *central configurations*; they occur when each body's vector position (relative to the center of mass), $\boldsymbol{r}_j$, lines up with its acceleration (or force) so that with the same negative constant $\lambda$,

$$\lambda \boldsymbol{r}_j = \boldsymbol{r}_j'', \quad j = 1, \dots, N. \tag{6}$$

This balancing scenario (6) requires carefully positioned bodies. For example, in 1769 Euler proved that there is a unique three-body positioning on a line that defines such a configuration. This separation of particles depends on the mass values. For intuition, if the middle one is too close to one of the end ones, it would be pulled toward that end. As this is also true for each of the end bodies, the intermediate-value theorem suggests that there is a balancing position for the middle body where it defines a central configuration. Then, in 1772, Lagrange discovered that the only noncollinear three-body central configuration is an equilateral triangle (figure 1(b)); this is true independent of mass values!

The mutual tugging (6) among the bodies allows each body to behave as if it were in a two-body system ((1)). Indeed, with appropriate initial conditions, a planar central configuration will rotate forever in a circular or elliptic manner, while retaining its shape. This happens in our solar system; in 1906 the astronomer Max Wolf discovered that the sun, Jupiter, and some asteroids, which he named the Trojans, form the rotating equilateral triangle configuration seen in figure 1(b)!

### 2.2 Rings of Saturn and Spiderwebs

In the nineteenth century, the mathematician James Clerk Maxwell used central configurations to study properties of the rings of Saturn. To indicate what he did, take a circle and any number, $k$, of evenly spaced lines passing through the circle's center, i.e., each adjacent pair is separated by the same angle. Place a body with mass $m$ at every point where a line passes through the circle; these $2k$ bodies model the particles in a ring. By symmetry, the force acting on each body is the same, so the system satisfies the central configuration condition (6). To include Saturn, place a body with any mass value, $m^*$, at the circle's center $\boldsymbol{r} = \boldsymbol{0}$. By symmetry, the forces of the $2k$ bodies cancel, so $\boldsymbol{r}'' = \boldsymbol{0}$. As $\lambda\boldsymbol{0} = \boldsymbol{0}$, this $(N = 2k + 1)$-body system satisfies (6). By placing this system in a circular motion, Maxwell created a dynamical model for a ring of Saturn.

As represented in figure 1(c), Maxwell's approach can be extended by using any number of circles, say $n$ of them. Wherever a line passes through the $j$th circle, place a mass $m_j$, $j = 1, \dots, k$. As each line passes through a circle twice, this creates the $(N = 2nk)$-body problem. Whatever the mass choices, by using an argument similar to the one used for the collinear three-body configuration, which involves adjusting the radii of the circles, it follows that a balancing point can be found where (6) is satisfied. This spiderweb-style configuration can be placed in a circular orbit.

### 2.3 Unsolved Problems

Beyond the central configurations described above and many others that are known, it remains interesting to discover new classes and kinds of central configurations. This means that we need to know when all of them have been found; could there be an infinite number of them? Not for three- and four-body problems; the Euler and Lagrange solutions are the only three-body choices, and Hampton and Moeckel have proved that there are a finite number of four-body central configurations. A fundamental unsolved problem for $N \geqslant 5$ is whether, for given mass values, there are only a finite number of central configurations.

## 3 Mass Values

The main objective is to determine whether (5) can be trusted to always provide an accurate, or at least reasonably accurate, mass prediction when applied to systems of $N$ discrete bodies. For instance, it may be acceptable for the predicted value to be twice that of the actual value. A way to analyze this question is to select mass values for the spiderweb configuration. Remember, no matter what mass values are selected, adjusting the distances between circles ensures that it defines a central configuration. In fact, because a rotation or scale change of a central configuration is again that same central configuration, let the minimum spacing between circles be one unit.

As closely as it is possible to do with $N$ discrete bodies, this configuration resembles the symmetric star soup continuum setting. By choosing the number of lines to be sufficiently large, say $k = 10\,000\,000$, the mass distribution is very symmetric. By selecting a large $n$ value, say $n = 10\,000$, this $(N = 2nk)$-body problem involves billions of bodies.

Whatever the choice of mass values $m_j$, the spiderweb rotates like a rigid body. This means that there is a constant $D > 0$ such that the common rotational velocity of the bodies at distance $r$ is $v_{\text{rot}} = Dr$. The (5) mass prediction for constant $E > 0$ is, therefore,

$$M(r) \approx Er^3. \tag{7}$$

Equation (7) holds independent of the choice of the masses, which allows an infinite number of examples to be created. For instance, if the masses on the $j$th circle are $m_j = 1/2kj$, $j = 1, \dots, n$, then the ring's total mass is $1/j$. A computation shows that the precise mass value out to the $s$th ring is

$$\sum_{j=1}^{s} \frac{1}{j} \approx \ln(s).$$

Remember, the minimum spacing between circles is one unit, so to reach the $s$th ring, it must be that $r \geqslant s$. Thus, to have the $M(s) = 20$ mass value, it must be that $\ln(r) \geqslant \ln(s) = 20$, or $r \geqslant e^{20}$. This forces the (7) predicted value to exceed $[Ee^{20}]^3$, which means that the (5) prediction exponentially exaggerates the precise value by a multiple in the trillions. Clearly, this is unacceptable. Indeed, any consistent multiple larger than 100 would seriously undermine claims about the existence and amount of dark matter.

As this simple analysis proves that (5) cannot be reliably used in all situations, the wonderful mathematical challenge is to determine when, where, and whether it does provide reasonably valid predictions. Of course, it follows from Newtonian dynamics, which requires a dragging effect, that (5) must exaggerate predicted mass values. So the issue is to determine by how much in general. An even more interesting mathematical challenge is to discover equations that will predict mass values for discrete systems that can be trusted.

### Further Reading

Binney, J., and S. Tremaine. 2008. *Galactic Dynamics*, 2nd edn. Princeton, NJ: Princeton University Press.

Hampton, M., and R. Moeckel. 2006. Finiteness of relative equilibria of the four-body problem. *Inventiones Mathematicae* 163:289–312.

Pollard, H. 1976. *Celestial Mechanics*. Washington, DC: Mathematical Association of America.

Rubin, V. C. 1993. Galaxy dynamics and the mass density of the universe. *Proceedings of the National Academy of Sciences of the USA* 90:4814–21.

Saari, D. G. 2005. *Collisions, Rings, and Other Newtonian N-Body Problems*. Providence, RI: American Mathematical Society.

# VI.17 The $N$-Body Problem and the Fast Multipole Method

*Lexing Ying*

## 1 Introduction

The $N$-body problem in astrophysics studies the motion of a large group of celestial objects that interact with each other. An essential step in the simulation of an $N$-body problem is the following evaluation of the gravitational potentials that involve pairwise interaction among these objects: given a set $P \subset \mathbb{R}^3$ of $N$ stars and the masses $\{f(y) : y \in P\}$ of the stars, the goal is to compute for each star $x \in P$ the gravitational potential $u(x)$ defined by

$$u(x) = \sum_{y \in P, y \neq x} G(x,y)f(y),$$

where $G(x,y) = 1/|x - y|$ is the Newtonian gravitational kernel. Similar computational tasks appear in many other areas of physics. For example, in electromagnetics, the evaluation of the electrostatic potentials takes exactly the same form, with the kernel $G(x,y) = 1/|x - y|$ in three dimensions or $G(x,y) = \ln(1/|x - y|)$ in two dimensions; evaluation of magnetic induction with the Biot–Savart law also takes a similar form. In acoustic scattering, a similar computation shows up with more oscillatory kernel functions $G(x,y)$. Finally, problems related to heat diffusion often require this computation with Gaussian-type kernel $G(x,y)$.

A direct computation of $u(x)$ for all $x \in P$ clearly takes $O(N^2)$ steps. For many of these applications, the number of objects $N$ can be in the millions, if not more, so a computation of order $O(N^2)$ can be very time-consuming. The fast multipole method, developed by Greengard and Rokhlin, computes an accurate approximate solution in about $O(N)$ steps. The word *multipole* refers to a series expansion that is used frequently in electromagnetics for describing a field in a region that is well separated from the source charges. Each term of this multipole expansion is a product of an inverse power in the radial variable and a spherical harmonic function in the angular variables. Since the multipole expansions often converge rapidly, they can be truncated after just a few terms—a fact that plays a key role in the efficiency of the fast multipole method.

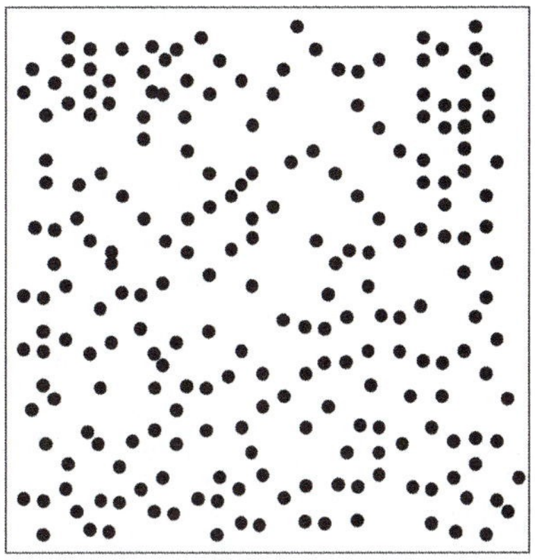

**Figure 1** $N$ points quasiuniformly distributed in a unit box $[0,1]^2$.

## 2 Algorithm Description

To illustrate the algorithm, we consider the two-dimensional case and assume that the point set $P$ is distributed *quasiuniformly* inside the unit box $\Omega = [0,1]^2$ (see figure 1). The algorithms for three-dimensional and nonuniform distributions are more involved, but the main ideas remain the same.

We start with a slightly simpler problem, where $B$ and $A$ are two disjoint cubes of the same size, each containing $O(n)$ points. Consider the evaluation at points in $A$ of the potentials induced by the points in $B$, i.e., for each $x \in A \cap P$, compute

$$u(x) = \sum_{y \in B \cap P} G(x,y)f(y).$$

Though direct computation of $u(x)$ takes $O(n^2)$ steps, there is a much faster way to approximate the calculation if $A$ and $B$ are well separated. Let us imagine $A$ and $B$ as two galaxies. When $A$ and $B$ are far away from each other, instead of considering all pairwise interactions, one can sum up the mass in $B$ to obtain $f_B = \sum_{y \in B \cap P} f(y)$ and place it at the center $c_B$ of $B$, evaluate the potential $u_A = G(c_A, c_B)f_B$ at the center $c_A$ of $A$ as if all the mass is located at $c_B$, and finally use $u_A$ as the approximation of the potential at each point $x$ in $A$. A graphical description of this three-step procedure is given in figure 2 and it takes only $O(n)$ steps instead of $O(n^2)$ steps. The procedure works well when $A$ and $B$ are sufficiently far away from each other, but it gives poor approximation when $A$ and $B$ are close. For the time being, however, let us assume that the procedure provides a valid approximation whenever the

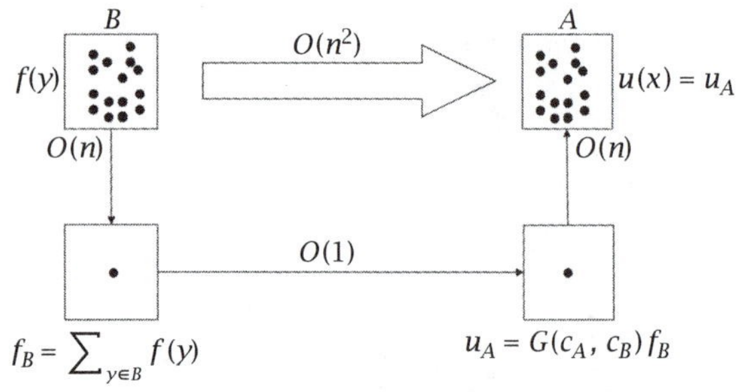


**Figure 2** A three-step procedure that efficiently approximates the potential in $A$ induced by the sources in $B$. The computational cost is reduced from $O(n^2)$ to $O(n)$.

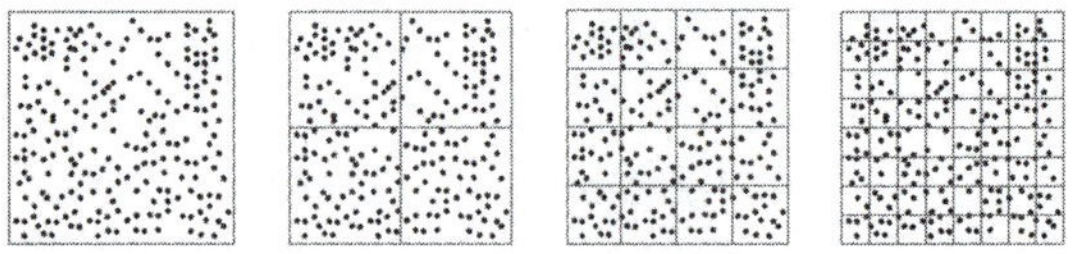

**Figure 3** The domain is partitioned with an octree structure until the number of points in each leaf node is bounded by a small constant.

distance between $A$ and $B$ is greater than or equal to the width of $A$ and $B$.

From an algebraic point of view, the three-step procedure is a rank-1 approximation of the interaction between $A$ and $B$:

$$G(A, B) \approx \begin{bmatrix} 1 \\ \vdots \\ 1 \end{bmatrix} G(c_A, c_B) \begin{bmatrix} 1 & \cdots & 1 \end{bmatrix},$$

where $G(A, B)$ is the matrix with entries given by $(G(x, y))_{x\in A\cap P, y\in B\cap P}$. $f_A$ and $u_B$ are now the intermediate results of applying this rank-1 approximation.

This simple problem considers the potential in $A$ only from points in $B$. However, we are interested in the interaction between all points in $P$. To get around this, we partition the domain hierarchically with an octree structure until the number of points in each leaf box is less than a prescribed $O(1)$ constant (see figure 3). The whole octree then has $O(\log N)$ levels, and we define the top level to be level 0. At level $\ell$, there are $4^\ell$ cubes and each cube has $O(N/4^\ell)$ points due to the quasiuniform point distribution.

The algorithm starts from level 2. Let $B$ be one of the boxes on level 2 (see figure 4(a)). The *near field* $N(B)$ of $B$ is the union of $B$ and its neighboring boxes, and the *far field* $F(B)$ of $B$ is the complement of the near field $N(B)$. For a box $A$ in $B$'s far field, the computation

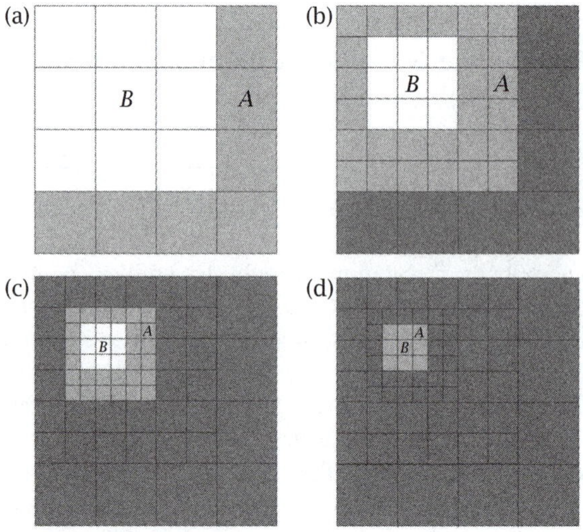


**Figure 4** The algorithm at different levels. $B$ stands for a source box. $A$ is a target box for which the interaction with $B$ is processed at the current level. Dark gray denotes boxes for which the interaction has already been considered by the previous level. Light gray denotes boxes for which the interaction is being considered at the current level. For the first three plots, a three-step procedure is used to accelerate the interaction between well-separated boxes. For the last plot, the nearby interaction is handled directly at the leaf level.

of the potentials at points in $A$ induced by the points in $B$ can be accelerated with the three-step procedure. There are $4^2$ possibilities for $B$, and for each $B$ there are $O(1)$ choices for $A$ (see figure 4(a)). Since both $A$ and $B$ contain $O(N/4^2)$ points due to the quasiuniformity assumption, the cost for all the interaction that can be taken care of on this level is

$$4^2 \cdot O(1) \cdot O(N/4^2) = O(N).$$

We cannot process the interaction between $B$ and its near field on this level, so we go down one level in the octree.

We again use $B$ to denote a cube at level 3 (see figure 4(b)). We do not need to consider $B$'s interaction with the far field of $B$'s parent since it has already been taken care of at the previous level. Only the interaction between $B$ and its parent's near field needs to be considered. At this level, there are at most $6^2$ boxes in its parent's near field. Out of these boxes, 27 of them are, typically, well separated from $B$ and the interaction between $B$ and these boxes can be accelerated using the three-step procedure. The set of these boxes is called the *interaction list* of $B$. Since each box on this level contains $O(N/4^3)$ points and there are $4^3$ possibilities for $B$, the total cost for the three-step procedures performed on this level is again

$$4^3 \cdot O(1) \cdot O(N/4^3) = O(N).$$

However, for the interaction of $B$ and the boxes in its near field, we need to go down again.

For a general level $\ell$, there are $4^\ell$ choices for $B$ (see figure 4(c)). For each $B$, there are at most 27 possibilities for $A$. Since each box on this level has $O(N/4^\ell)$ points, the cost of far-field computation is

$$4^\ell \cdot O(1) \cdot O(N/4^\ell) = O(N).$$

Once we reach the leaf level, there is still the interaction between a leaf box and its neighbors to consider (see figure 4(d)). For that, we just use direct computation. Since there are $O(N)$ leaf boxes, each containing $O(1)$ points and having $O(1)$ neighbors, the total cost of direct computation is

$$O(N) \cdot O(1) \cdot O(1) = O(N).$$

For the three-step procedure between a pair of well-separated boxes $A$ and $B$, it is clear that the first step depends only on $B$ and the last step depends only on $A$. Therefore, there is an opportunity for reusing computation. Taking this observation into consideration, we can write the algorithm as follows.

(1) For each level $\ell$ and each box $A$ on level $\ell$, set $u_A$ to be zero.
(2) For each level $\ell$ and each box $B$ on level $\ell$, compute $f_B = \sum_{y \in B \cap P} f(y)$.
(3) For each level $\ell$ and each box $B$ on level $\ell$, and for each box $A$ in $B$'s interaction list, update $u_A = u_A + G(c_A, c_B) f_B$.
(4) For each level $\ell$ and each box $A$ on this level, update $u(x) = u(x) + u_A$ for each $x \in A \cap P$.
(5) For each box $A$ at the leaf level, update $u(x) = u(x) + \sum_{y \in N(A) \cap P} G(x, y) f(y)$ for each $x \in A \cap P$.

Since the cost at each level is $O(N)$ and there are $O(\log N)$ levels, the whole cost of the algorithm is $O(N \log N)$.

The question now is whether we can do it in fewer steps. The answer is that we can, and it is based on the following simple observation.

Let $B_1, \dots, B_4$ be $B$'s children. Since $B = B_1 \cup \cdots \cup B_4$ and all the $B_i$ are disjoint, we can conclude that (see figure 5(a))

$$f_B = f_{B_1} + f_{B_2} + f_{B_3} + f_{B_4}.$$

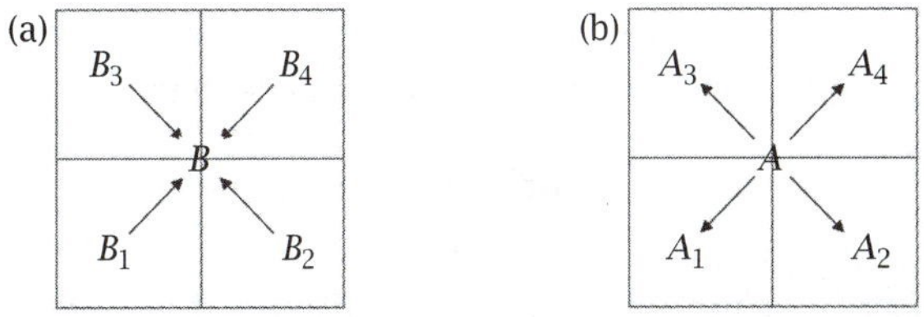


**Figure 5** A basic observation that speeds up the computation of $f_B$ and $u_A$. (a) $f_B$ can be computed directly from $f_{B_i}$, where the $B_i$ are $B$'s children. (b) Instead of adding $u_A$ directly to all the points in $A$, we only add to $u_{A_i}$, where the $A_i$ are $A$'s children.

Therefore, assuming that the $f_{B_i}$ are ready, using the previous line to compute $f_B$ is much more efficient than summing over all $f(y)$ in $B$ for large $B$.

Similarly, for each $A$, we update $u(x) := u(x) + u_A$ for each $x \in A \cap P$. Assume that $A_1, \dots, A_4$ are the children boxes of $A$. Then, since we perform the same step for each $A_i$ and each $x$ belongs to one such $A_i$, we can simply update $u_{A_i} := u_{A_i} + u_A$ instead, which is much more efficient (see figure 5(b)).

Notice that in order to carry out these two improvements, we make the assumption that for $f_B$ we visit the parent after the children, while for $u_A$ we visit the children after the parent. This requires us to traverse the octree with different orders at different stages of the algorithm. Putting the pieces together, we reach the following algorithm.

(1) For each level $\ell$ and each box $A$ on level $\ell$, set $u_A$ to zero.
(2) Traverse the tree from level $L - 1$ up to level 0, and for each box $B$, if $B$ is a leaf box, set $f_B = \sum_{y \in B \cap P} f(y)$. If $B$ is not a leaf box, set $f_B = f_{B_1} + \cdots + f_{B_4}$.
(3) For each level $\ell$, for each $B$, and for each $A$ in $B$'s interaction list, update $u_A = u_A + G(c_A, c_B) f_B$.
(4) Traverse the tree from level 0 down to level $L - 1$, and for each box $A$, if $A$ is not a leaf box, update $u_{A_i} = u_{A_i} + u_A$ for each child $A_i$ of $A$. If $A$ is a leaf box, update $u(x) = u(x) + u_A$.
(5) For each box $A$ at the leaf level, update $u(x) = u(x) + \sum_{y \in N(A) \cap P} G(x, y) f(y)$ for each $x \in A \cap P$.

Steps (1), (3), and (5) are the same as in the previous algorithm, and their cost is $O(N)$ each. For steps (2) and (4), since there are at most $O(N)$ boxes in the tree and the algorithm spends $O(1)$ steps per box, the cost is again $O(N)$. As a result, the total cost is, as promised,

$O(N)$. This is essentially the fast multipole method algorithm proposed by Greengard and Rokhlin.

## 3 Algorithmic Details

We have so far been ignoring the issue of accuracy. In fact, if we use only $f_B$ and $u_A$ as described, the accuracy is low, since $A$ and $B$ can be only one box away from each other. Recall that the three-step procedure that we have used so far is a poor rank-1 approximation of the interaction between $A$ and $B$. The low-rank approximation used in the fast multipole method by Greengard and Rokhlin is based on the *truncated* multipole and local expansions. The resulting $f_B$ and $u_A$ are the coefficients of the truncated multipole and local expansions, respectively. In addition, there are natural multipole-to-multipole operators $T_{B,B_i}$ that transform $f_{B_i}$ to $f_B$ ($f_B = \sum_i T_{B,B_i} f_{B_i}$), local-to-local operators $T_{A_i,A}$ that take $u_A$ to $u_{A_i}$ ($u_{A_i} = u_{A_i} + T_{A_i,A} u_A$), and multipole-to-local operators $T_{A,B}$ that take $f_B$ to $u_A$ ($u_A = u_A + T_{A,B} f_B$). We will not go into the details of these representation here, except for two essential points.

- For any fixed accuracy, both $f_B$ from the multipole expansions and $u_A$ from the local expansions contain only $O(1)$ numbers.
- The translation operators are maps from $O(1)$ numbers to $O(1)$ numbers, and applying them takes $O(1)$ steps. A lot of effort has been devoted to further optimizing the implementation of these operators.

From these two points it is clear that the overall $O(N)$ complexity of the fast multipole method remains the same when more accurate low-rank approximations are used. There are many other ways to implement the low-rank approximations and the translation operators between them. Some examples include the $H^2$ matrices of Hackbusch et al., the fast multipole method without multipole expansion of Anderson, and the kernel-independent fast multipole method.

**Acknowledgments.** Part of the text and all of the figures are reproduced from Ying (2012) with the permission of the publisher.

### Further Reading

Greengard, L. 1988. *The Rapid Evaluation of Potential Fields in Particle Systems*. Cambridge, MA: MIT Press.

Greengard, L., and V. Rokhlin. 1987. A fast algorithm for particle simulations. *Journal of Computational Physics* 73(2):325–48.

Hackbusch, W., B. Khoromskij, and S. A. Sauter. 2000. On $H^2$-matrices. In *Lectures on Applied Mathematics*, edited by H.-J. Bungartz, R. H. W. Hoppe, and C. Zenger, pp. 9–29. Berlin: Springer.

Ying, L. 2012. A pedestrian introduction to fast multipole methods. *Science China Mathematics* 55(5):1043–51.

# VI.18 The Traveling Salesman Problem

*William Cook*

The *traveling salesman problem*, or TSP for short, is a prominent model in DISCRETE OPTIMIZATION [IV.38]. In its general form, we are given a set of cities and the cost to travel between each pair of them. The problem is to find the cheapest route to visit each city and to return to the starting point. The TSP owes its fame to its success as a benchmark for developing and testing algorithms for computationally difficult problems in discrete mathematics, operations research, and computer science. The decision version of the problem is a member of the NP-HARD [I.4 §4.1] complexity class. Applications of the TSP arise in logistics, machine scheduling, computer-chip design, genome mapping, data analysis, guidance of telescopes, circuit-board drilling, machine-translation systems, and in many other areas.

The origin of the TSP's catchy name is somewhat of a mystery. It first appears in print in a 1949 RAND Corporation research report by Julia Robinson, but she uses the term in an offhand way, suggesting it was a familiar concept at the time. The origin of the problem itself goes back to the early 1930s, when it was proposed by Karl Menger at a mathematics colloquium in Vienna and by Hassler Whitney in a seminar at Princeton University. The problem also has roots in graph theory and in the study of Hamiltonian circuits.

## 1 Exact Algorithms

A route that visits all cities and returns to the starting point is called a *tour*; finding an optimal tour (that is, one that is cheapest) is the goal of the TSP. This can be accomplished by simply enumerating all permutations of the cities and evaluating the costs of the corresponding tours. For an $n$-city TSP, this approach requires time proportional to $n$ factorial. In 1962, Bellman and the team of Held and Karp each showed that any instance of the problem can solved in time proportional to $n^2 2^n$. This is significantly faster than brute-force enumeration, but the algorithm takes both exponential time and exponential space.

The heart of the Bellman–Held–Karp method can be described by a simple recursive equation. To set this up, denote the cities by the labels $1, 2, \dots, n$ and, for any pair of cities $(i, j)$, let $c_{ij}$ denote the cost of traveling from city $i$ to city $j$. We may set city 1 as the fixed starting point for the tours; let $N = \{2, \dots, n\}$ denote the remaining cities. For any $S \subseteq N$ and for any $j \in S$, let $\text{opt}(S, j)$ denote the minimum cost of a path starting at 1, running through all points in $S$, and ending at $j$. We have

$$\text{opt}(S, j) = \min(\text{opt}(S \setminus \{j\}, i) + c_{ij} \colon i \in S \setminus \{j\}). \quad (1)$$

Moreover, the optimal value of the TSP is

$$v^* = \min(\text{opt}(N, j) + c_{j1} \colon j \in N). \quad (2)$$

Observe that for all $j \in N$ we have $\text{opt}(\{j\}, j) = c_{1j}$. Starting with these values, the recursive equation (1) is used to build the values $\text{opt}(S, j)$ for all $S \subseteq N$ and $j \in S$, working our way through sets with two elements, then sets with three elements, and step by step up to the full set $N$. Once we have the values $\text{opt}(N, j)$ for all $j \in N$, we use (2) to find $v^*$. Now, in a second pass, the optimal tour is computed by first identifying a city $v_{n-1}$ such that $\text{opt}(N, v_{n-1}) + c_{v_{n-1}1} = v^*$, then identifying a city $v_{n-2} \in N \setminus \{v_{n-1}\}$ such that $\text{opt}(N \setminus \{v_{n-1}\}, v_{n-2}) + c_{v_{n-1}v_{n-2}} = \text{opt}(N, v_{n-1})$, and so on until we have $v_1$. The optimal tour is $(1, v_1, \dots, v_{n-1})$. This second pass is to permit the algorithm to store only the values $\text{opt}(S, j)$ and not the actual paths that determine these values.

The running-time bound arises from the fact that in an $n$-city problem there are $2^{n-1}$ subsets $S$ that do not contain the starting point. For each of these we consider at most $n$ choices for the end city $j$, and the computation of the $\text{opt}(S, j)$ value involves fewer than $n$ additions and $n$ comparisons. Multiplying $2^{n-1}$ by $n$ by $2n$, we have that the total number of steps is no more than $n^2 2^n$.

Despite the attention that the TSP receives, no algorithm with a better worst-case running time has been discovered for general instances of the problem in the fifty years plus since the work of Bellman, Held, and Karp. It is a major open challenge in TSP research to improve upon the $n^2 2^n$ bound.

## 2 Approximation Algorithms

The problem of determining whether or not a given graph has a Hamiltonian circuit, that is, a circuit that visits every vertex of the graph, is also a member of the NP-hard complexity class. This problem can be encoded as a TSP by letting the cities correspond to the vertices of the graph, assigning cost 0 to all pairs of cities corresponding to edges in the graph and assigning cost 1 to all other pairs of cities. A Hamiltonian circuit has cost 0, and any nonoptimal tour has cost at least 1. A method that returns a solution within any fixed percentage of optimality will therefore provide an answer to the yes/no question. Thus, unless NP = P, there can be no $\alpha$-approximation algorithm for the TSP that runs in polynomial time, that is, no algorithm that is guaranteed to produce a tour of cost no more than $\alpha$ times the cost of an optimal tour.

An important case of the TSP is when the travel costs are *symmetric*, that is, when the cost of traveling from city A to city B is the same as the cost of traveling from B to A. A further natural restriction, known as the *triangle inequality*, states that for any three cities A, B, and C, the cost of traveling from A to B plus the cost of traveling from B to C must not be less than the cost of traveling from A directly to C. Under these two restrictions, there is a polynomial-time algorithm due to Christofides that finds a tour the cost of which is guaranteed to be no more than $\frac{3}{2}$ times the cost of an optimal tour.

A symmetric instance of the TSP can be described by a complete graph $G = (V, E)$ having vertex set $V = \{1, \dots, n\}$ and edge set $E$ consisting of all pairs of vertices, together with travel costs $(c_e \colon e \in E)$. Christofides's algorithm begins by finding a minimum-cost spanning tree $T$ in $G$; the cost of $T$ is no more than the cost of an optimal TSP tour. The algorithm then computes a minimum-cost perfect matching $M$ in the subgraph of $G$ spanned by the vertices that meet an odd number of edges in the tree $T$; the cost of $M$ is no more than $\frac{1}{2}$ times the cost of an optimal TSP tour since any tour is the union of two perfect matchings. The union of the edges in $T$ and the edges in $M$ form a graph $H$ on the vertices $V$ such that every vertex meets an even number of edges. Thus $H$ has an Eulerian cycle, that is, a route that starts and ends at the same vertex and visits every edge of $H$. The Eulerian cycle can be transformed into a TSP tour by traversing the cycle and skipping over any vertices that have already been visited. The cost of the resulting tour is no more than the cost of the cycle due to the triangle inequality. Thus, the cost of the tour is no more than $\frac{3}{2}$ times the cost of the optimal tour.

Christofides proved his $\frac{3}{2}$-approximation result in 1976. Again, despite considerable attention from the research community, the factor of $\frac{3}{2}$ has not been improved upon for the case of general symmetric costs satisfying the triangle inequality. This is in contrast to the Euclidean TSP, where cities correspond to points in the plane and the travel cost between two cities is the Euclidean distance between the points. For the Euclidean TSP, it is known that for any $\alpha$ greater than 1 there exists a polynomial-time $\alpha$-approximation algorithm. Such a result is not possible in the general triangle-inequality case unless NP = P.

The $\frac{3}{2}$ result of Christofides applies only to symmetric instances of the TSP. For asymmetric instances satisfying the triangle inequality, there exists a randomized polynomial-time algorithm that with high probability produces a tour of cost no more than a factor proportional to $\log n / \log\log n$ times the cost of an optimal tour.

## 3 Exact Computations

In her 1949 paper Robinson formulates the TSP as finding "the shortest route for a salesman starting from Washington, visiting all the state capitals and then returning to Washington." The fact that the TSP is in the NP-hard class does not imply that exactly solving a particular instance of the problem, such as Robinson's example, is an insurmountable task. Indeed, five years after Robinson's paper, Dantzig, Fulkerson, and Johnson initiated a computational approach to the problem by building a tour through 49 cities in the United States, together with a proof that their solution was the shortest possible. This line of work has continued over the years, including a 120-city solution though Germany in 1977, a 532-city U.S. tour in 1987, and a 24 978-city tour through Sweden in 1998. Each of these studies works with symmetric integer-valued travel costs that approximate either the road distances or the Euclidean distances between city locations.

The dominant approach to the exact solution of large-scale symmetric instances of the TSP is the LINEAR-PROGRAMMING [IV.11 §3.1] (LP) technique developed by Dantzig et al. in their original 1954 paper. For a TSP specified by a complete graph $G = (V, E)$, an LP relaxation of the TSP instance can be formulated using variables $(x_e : e \in E)$. A tour corresponds to the LP solution obtained by setting $x_e = 1$ if the edge $e$ is included in the tour, and setting $x_e = 0$ otherwise. The Dantzig et al. *subtour relaxation* of the symmetric TSP is the LP model

$$\text{minimize} \sum (c_e x_e : e \in E), \tag{3}$$

$$\text{subject to} \sum (x_e : v \text{ is an end of } e) = 2 \quad \forall v \in V, \tag{4}$$

$$\sum (x_e : e \text{ has exactly one end in } S) \geqslant 2 \quad \forall \varnothing \neq S \subsetneq V, \tag{5}$$

$$0 \leqslant x_e \leqslant 1 \quad \forall e \in E. \tag{6}$$

The equations (4) are called the *degree equations* and the inequalities (5) are called the *subtour-elimination constraints*. For any tour, the corresponding 0/1-vector $x$ satisfies all the degree equations and subtour-elimination constraints. Thus, the optimal LP value is a lower bound on the cost of an optimal solution to the TSP. If the travel costs satisfy the triangle inequality, then the optimal TSP value is at most $\frac{3}{2}$ times the optimal subtour value. The $\frac{4}{3}$ *conjecture* states that this $\frac{3}{2}$ factor can be improved to $\frac{4}{3}$; its study is a focal point of efforts to improve Christofides's approximation algorithm.

In computational studies, the subtour relaxation is solved by an iterative process called the *cutting-plane method*. To begin, an optimal LP solution vector $x^\star$ is computed for the much smaller model having only the degree equations as constraints. The method iteratively improves the relaxation by adding to the model individual subtour-elimination constraints that are violated by $x^\star$ and then computing the optimal LP solution again. When $x^\star$ satisfies all subtour-elimination constraints, then it is an optimal solution to the subtour relaxation. The success of the method relies in practice on the fact that typically only a very small number of the exponentially many subtour-elimination constraints need to be represented explicitly in the LP relaxation.

To improve the lower bound provided by the subtour relaxation, the exact-solution process continues by considering further classes of linear inequalities that are satisfied by all tours, again adding individual inequalities to the LP model in an iterative fashion. If the lower bound provided by the LP model is sufficiently strong for a test instance, then the TSP can be solved via a branch-and-bound search using the LP model as the bounding mechanism. The overall process is called the *branch-and-cut method*; it was developed for the TSP, and it has been successfully applied to many other problem classes in discrete optimization.

## 4 Heuristic Solution Methods

In many applied settings, a near-optimal solution to the TSP is satisfactory, that is, a tour that has cost close to that of an optimal tour but is not necessarily the cheapest possible route. The search for heuristic methods that perform well for such applications is the most actively studied topic in TSP research, and this activity has helped to create and improve many of the best-known schemes for general heuristic search, such as simulated annealing, genetic algorithms, and local-improvement methods. This line of research differs from that of approximation algorithms in that the methods do not come with any strong worst-case guarantee on the cost of the discovered tour, but the performance of typical examples can be much better than results produced by the known $\alpha$-approximation techniques.

Heuristic methods range from extremely fast algorithms based on space-filling curves through sophisticated combinations of local-improvement and genetic-algorithm techniques. The choice of method in a practical computation involves a trade-off between hoped-for tour quality and anticipated running time. The fastest techniques construct a tour step by step, typically in a greedy fashion, such as the nearest-neighbor method, where at each step the next city to be added to the route is the one that can be reached by the least travel cost from the current city. The higher-quality heuristic algorithms create a sequence of tours, seeking modifications that can lower an existing tour's total cost. The simplest of these tour-improvement methods, for symmetric costs, is the 2-*opt algorithm*, which repeatedly searches for a pair of links in the tour that can be replaced by a cheaper pair reconnecting the resulting paths. The 2-opt algorithm serves as a building block for many of the best-performing heuristic methods, such as the Lin–Kernighan $k$-opt algorithm.

In a genetic algorithm for the TSP, an initial collection of tours is generated, say, by repeatedly applying the nearest-neighbor algorithm with random starting cities. In each iteration, some pairs of members of the collection are chosen to *mate* and produce *child* tours. A new collection of tours is then selected from the old population and the children. The process is repeated a large number of times and the best tour in the collection is chosen as the winner. In the mating step, subpaths from the parent tours are combined in some fashion to produce the child tours. In the best-performing variants, tour-improvement methods are used to lower the costs of the child tours that are produced by the mating process.

### Further Reading

Applegate, D., R. Bixby, V. Chvátal, and W. Cook. 2006. *The Traveling Salesman Problem: A Computational Study.* Princeton, NJ: Princeton University Press.

Cook, W. 2012. *In Pursuit of the Traveling Salesman: Mathematics at the Limits of Computation.* Princeton, NJ: Princeton University Press.

# Part VII
# Application Areas

## VII.1 Aircraft Noise

*C. J. Chapman*

### 1 Introduction

Aircraft noise is a subject of intense public interest, and it is rarely out of the news for long. It became especially important early in the 1950s with the growth in the use of jet engines and again in the 1960s with the development of the Anglo-French supersonic aircraft Concorde (now retired). Aircraft noise is currently a major part of the public debate about where to build new airports and whether they should be built at all.

The general public might be surprised to know that our understanding of the principles of aircraft noise, and of the main methods that are available to control it, have come largely from mathematicians. In particular, Sir James Lighthill, one of the great mathematical scientists of the twentieth century, created a new scientific discipline, called aerodynamic sound generation, or aeroacoustics, when he published the first account of how a jet generates sound in 1952. Prior to this theory, which he created by mathematical analysis of the equations of fluid motion, no means were available to estimate the acoustic power from a jet even to within a factor of a million.

Moreover, Lighthill's theory had an immediate practical consequence. The resulting eighth-power scaling law for sound generation as a function of Mach number (flow speed divided by sound speed) meant that the amount of thrust taken from the jet would have to be strictly controlled; this was achieved by the development of high bypass-ratio turbofan aeroengines, in which a high proportion of the thrust is generated by the large fan at the front of the engine. Such a fan produces a relatively small increase in the velocity of a large volume of air, i.e., in the bypass air flow surrounding the jet, in contrast to the jet itself, which produces a large increase in the velocity of a small volume of air. The engineering consequence of a mathematical theory is therefore plain to see whenever a passenger ascends the steps to board a large aircraft and admires the elegant multibladed fan so clearly on display.

### 2 Lighthill's Acoustic Analogy

The idea behind Lighthill's theory is that the equations of fluid dynamics may be written exactly like the wave equation of acoustics, with certain terms (invariably written on the right-hand side) regarded as acoustic sources. Solution of this equation gives the sound field generated by the fluid flow. A key feature of the method is that the source terms may be estimated using the simpler aerodynamic theory of flow in which acoustic waves have been filtered out. The simpler theory, which includes the theory of incompressible flow as a limiting case, had become very highly developed during the course of the two world wars and in the period thereafter and was therefore available for immediate use in Lighthill's theory of sound generation.

This innocuous-sounding description of the method hides many subtleties, which are in no way alleviated by the fact that Lighthill's equation is exact. But let us first give the equation in its standard form and define its terms. The equation is

$$\left(\frac{\partial^2}{\partial t^2} - c_0^2\nabla^2\right)\rho' = \frac{\partial^2 T_{ij}}{\partial x_i \partial x_j}.$$

Here, $t$ is time and $\nabla^2 = \partial^2/\partial x_1^2 + \partial^2/\partial x_2^2 + \partial^2/\partial x_3^2$ is the Laplacian operator, where $(x_1, x_2, x_3)$ is the position vector in Cartesian coordinates. Away from the source region, the fluid (taken to be air) is assumed to have uniform properties, notably sound speed $c_0$ and density $\rho_0$. At arbitrary position and time, the fluid density is $\rho$, and the deviation from the uniform surrounding value is $\rho' = \rho - \rho_0$, the density perturbation. The left-hand side of Lighthill's equation is therefore simply the wave operator, for a uniform sound speed $c_0$, acting on the density perturbation produced throughout the fluid by a localized jet.

The right-hand side of the equation employs the summation convention for indices (here over $i, j = 1, 2, 3$) and so is the sum of nine terms in the second derivatives of $T_{11}, T_{12}, \dots, T_{33}$. Collectively, these nine quantities $T_{ij}$ form the Lighthill stress tensor, defined by

$$T_{ij} = \rho u_i u_j + (p' - c_0^2 \rho')\delta_{ij} - \tau_{ij}.$$

Here, $(u_1, u_2, u_3)$ is the velocity vector, $p' = p - p_0$ is the pressure perturbation, defined analogously to $\rho'$, and $\tau_{ij}$ is the viscous stress tensor, representing the forces due to fluid friction. Because of the symmetry of $T_{ij}$ with respect to $i$ and $j$, there are only six independent components, of which the three with $i = j$ are longitudinal and the three with $i \neq j$ are lateral.

The solution of Lighthill's equation, together with the acoustic relation $p' = c_0^2 \rho'$ that applies in the radiated sound field (though not in the source region), gives the outgoing sound field $p'(\boldsymbol{x}, t)$ in the form

$$p' = \frac{1}{4\pi} \frac{\partial^2}{\partial x_i \partial x_j} \int_V \frac{T_{ij}(\boldsymbol{y}, t - |\boldsymbol{x} - \boldsymbol{y}|/c_0)}{|\boldsymbol{x} - \boldsymbol{y}|} \, \mathrm{d}V.$$

Here, $\boldsymbol{x}$ is the observer position, $\boldsymbol{y} = (y_1, y_2, y_3)$ is an arbitrary source position, and $|\boldsymbol{x} - \boldsymbol{y}|$ is the distance between them. The integration is over the source region $V$, and the volume element is $\mathrm{d}V = \mathrm{d}y_1 \, \mathrm{d}y_2 \, \mathrm{d}y_3$. The retarded time $t - |\boldsymbol{x} - \boldsymbol{y}|/c_0$ allows for the time it takes a signal traveling at speed $c_0$ to traverse the distance from $\boldsymbol{y}$ to $\boldsymbol{x}$.

Lighthill's equation is referred to as an acoustic analogy because the uniform "acoustic fluid," in which acoustic perturbations are assumed to be propagating, is entirely fictional in the source region! For example, the speed of sound there is not uniform but varies strongly with position in the steep temperature gradients of an aircraft jet; and the fluid velocity itself, which contributes to the propagation velocity of any physical quantity such as a density perturbation, is conspicuously absent from a wave operator that contains only $c_0$. Nevertheless, Lighthill's power as a mathematical modeler enabled him to see that the acoustic analogy, with $T_{ij}$ estimated from nonacoustic aerodynamic theory, would provide accurate predictions of jet noise in many important operating conditions, e.g., in subsonic jets at Mach numbers that are not too high.

The increasing difficulty in applying Lighthill's theory at higher Mach numbers derives from the fact that an aircraft jet is TURBULENT [V.21] and contains swirling eddies of all sizes that move irregularly and unpredictably throughout the flow. The higher the Mach number, the more that must be known about the individual eddies or their statistical properties in order to take account of the delicate cancelation of the sound produced by neighboring eddies. As the Mach number increases, small differences in retarded times become increasingly important in determining the precise amount of this cancelation. Lighthill's equation therefore provides an impetus to develop ever more sophisticated theories of fluid turbulence for modeling the stress tensor $T_{ij}$ to greater accuracy.

### 2.1 Quadrupoles

A noteworthy feature of Lighthill's equation is that the source terms occur as second spatial derivatives of $T_{ij}$. This implies that the sound field produced by an individual source has a four-lobed cloverleaf pattern, referred to as a quadrupole directivity. A crucial aspect of Lighthill's theory of aerodynamic sound generation is, therefore, that the turbulent fluid motion creating the sound field is regarded as a continuous superposition of quadrupole sources of sound. Physically, this is the correct point of view because away from boundaries the fluid has "nothing to push against except itself"; that is, since the pressure inside a fluid produces equal and opposite forces on neighboring elements of fluid, the total dipole strength, arising from the sum of the acoustic effects of these internal forces, must be identically zero.

Mathematically, the four-lobed directivity pattern arises from the expansion of $p'$ in powers of $1/|\boldsymbol{x}|$ for large $|\boldsymbol{x}|$. The dominant term, proportional to $1/|\boldsymbol{x}|$, gives the radiated sound field; its coefficient contains angular factors $x_i x_j / |\boldsymbol{x}|^2$ corresponding to the derivatives $\partial^2 / \partial x_i \partial x_j$ acting on the integral containing $T_{ij}$. For example, in spherical polar coordinates $(r, \theta, \phi)$ with $r = |\boldsymbol{x}|$ and

$$(x_1, x_2, x_3) = (r \sin\theta \cos\phi, r \sin\theta \sin\phi, r \cos\theta),$$

the lateral $(x_1, x_2)$ quadrupole has the directivity pattern

$$\frac{x_1 x_2}{|\boldsymbol{x}|^2} = \tfrac{1}{2} \sin^2\theta \sin 2\phi.$$

This pattern has lobes centered on the meridional half-planes $\phi = \pi/4$, $3\pi/4$, $5\pi/4$, and $7\pi/4$, separated by the half-planes $\phi = 0, \pi/2, \pi$, and $3\pi/2$, in which there is no sound radiation.

### 2.2 The Eighth-Power Law

Lighthill's theory predicts that the power generated by a subsonic jet is proportional to the eighth power of the Mach number. This follows from a beautifully simple scaling argument based on only the equations given above, and we will now present this argument in full, at the same time giving the scaling laws for the important physical quantities in the subsonic jet noise problem.

Assume that a jet emerges with speed $U$ from a nozzle of diameter $a$ into a fluid in which the ambient density is $\rho_0$ and the speed of sound is $c_0$. The Mach number of the jet is therefore $M = U/c_0$, and the timescale of turbulent fluctuations in the jet is $a/U$. The frequency therefore scales with $U/a$, and the wavelength $\lambda$ of the radiated sound scales with $c_0 a/U$, i.e., $a/M$. We are considering the subsonic regime $M < 1$, so $a < \lambda$, and the source region is compact.

The Lighthill stress tensor $T_{ij}$ is dominated by the terms $\rho u_i u_j$, which scale with $\rho U^2$, and the integration over the source region gives a factor $a^3$. In the far field, the term $|\boldsymbol{x} - \boldsymbol{y}|$ appearing in the denominator of the integrand may be approximated by $|\boldsymbol{x}|$, and the spatial derivatives applied to the integral introduce a factor $1/\lambda^2$. The far-field acoustic quantities are the perturbation pressure, density, and velocity, i.e., $p'$, $\rho'$, and $u'$, which satisfy the acoustic relations $p' = c_0^2 \rho' = \rho_0 c_0 u'$. Putting all this together gives the basic far-field scaling law

$$\frac{p'}{\rho_0 c_0^2} = \frac{\rho'}{\rho_0} = \frac{u'}{c_0} \sim M^4 \frac{a}{|\boldsymbol{x}|}.$$

In a sound wave, the rate of energy flow, measured per unit area per unit time, is proportional to $p'u'$. Since a sphere of radius $|\boldsymbol{x}|$ has area proportional to $|\boldsymbol{x}|^2$, it follows that the acoustic power $W'$ of the jet, i.e., the total acoustic energy that it radiates per unit time in all directions, obeys the scaling law

$$\frac{W'}{\rho_0 c_0^3 a^2} \sim M^8.$$

This is the famous eighth-power law for subsonic jet noise. It would be hard to find a more striking example of the elegance and usefulness of the best applied mathematics.

The significance of the high exponent in the scaling law is that at very low Mach numbers a jet is inefficient at producing sound; but this inefficiency does not last when the Mach number is increased. This is why jet noise became such a problem in the early 1950s, as aircraft engines for passenger flight became more powerful. Unfortunately, there is no way to evade a fundamental scaling law imposed by the laws of mechanics, and it is impossible to make a very-high-speed jet quiet. This was the original impetus to develop high-bypass-ratio turbofan aeroengines to replace jet engines.

## 3 Further Acoustic Analogies

The acoustic analogy has been extended in many ways. For example, the Ffowcs Williams–Hawkings equation accounts for boundary effects by means of a vector $J_i$ and a tensor $L_{ij}$ to represent mass and momentum flux through a surface, which may be moving. These terms are given by

$$J_i = \rho(u_i - v_i) + \rho_0 v_i$$

and

$$L_{ij} = \rho u_i(u_j - v_j) + (p - p_0)\delta_{ij} - \tau_{ij},$$

where $(v_1, v_2, v_3)$ is the velocity of the surface and $\tau_{ij}$ is the viscous stress tensor. Derivatives of $J_i$ and $L_{ij}$, with suitable delta functions included to localize them on the surface, are then used as source terms for the acoustic wave equation. The Ffowcs Williams–Hawkings equation has been of enduring importance in the study of aircraft noise and is widely used in computational aeroacoustics.

Variations of Lighthill's equation can be obtained by redistributing terms between the right-hand side, where they are regarded as sources, and the left-hand side, where they are regarded as contributing to the wave propagation operator; different choices of the variable (or the combination of variables) on which the wave operator acts are also possible. Although the equations obtained in this way are always exact, the question of how useful they are, and whether the physical interpretations they embody are "real," has at times been contentious and led to interminable discussion! Among the widely accepted equations are those of Lilley and Goldstein, which place the convective effect of the jet mean flow in the wave operator and modify the source terms on the right-hand side accordingly. Morfey's acoustic analogy is also useful; this provides source terms for sound generation by unsteady dissipation and nonuniformities in density, such as occur in the turbulent mixing of hot and cold fluids.

## 4 Howe's Vortex Sound Equation

Since the vorticity of a flow field is so important in generating sound, both in a turbulent jet and in other flows, it is desirable to have available an equation that contains it explicitly as a source, rather than implicitly, as in the Lighthill stress tensor. The vorticity $\boldsymbol{\omega}$, defined as the curl of the velocity field, is a measure of the local rotation rate of a flow. In vector notation, $\boldsymbol{\omega} = \nabla \wedge \boldsymbol{u}$, where $\wedge$ indicates the cross product; in components,

$$(\omega_1, \omega_2, \omega_3) = \left(\frac{\partial u_3}{\partial x_2} - \frac{\partial u_2}{\partial x_3}, \frac{\partial u_1}{\partial x_3} - \frac{\partial u_3}{\partial x_1}, \frac{\partial u_2}{\partial x_1} - \frac{\partial u_1}{\partial x_2}\right).$$

An extremely useful equation with vorticity as a source is Howe's vortex sound equation, in which the independent acoustic variable is the total enthalpy defined by

$$B = \int \frac{\mathrm{d}p}{\rho} + \tfrac{1}{2}|\boldsymbol{u}|^2.$$

The symbol $B$ is used to stand for Bernoulli because this variable appears in Bernoulli's equation. Howe's vortex sound equation is a far-reaching generalization of Bernoulli's equation, and is

$$\left(\frac{\mathrm{D}}{\mathrm{D}t}\left(\frac{1}{c^2}\frac{\mathrm{D}}{\mathrm{D}t}\right) - \frac{1}{\rho}\nabla \cdot (\rho\nabla)\right)B = \frac{1}{\rho}\nabla \cdot (\rho\boldsymbol{\omega} \wedge \boldsymbol{u}),$$

where $c$ is the speed of sound and $\mathrm{D}/\mathrm{D}t = \partial/\partial t + \boldsymbol{u} \cdot \nabla$ is the convective derivative. In the limit of low Mach number, this simplifies to

$$\left(\frac{1}{c_0^2}\frac{\partial^2}{\partial t^2} - \nabla^2\right)B = \nabla \cdot (\boldsymbol{\omega} \wedge \boldsymbol{u}),$$

and the pressure in the radiated sound field is $p = \rho_0 B$.

Howe's equation is used to calculate the sound produced by the interaction of vorticity with any type of surface. Examples include aircraft fan blades interacting with turbulent inflows or with the vortices shed by any structure upstream of the blades. An advantage of Howe's equation is that it provides analytical solutions to a number of realistic problems, and hence gives the dependence of the sound field on the design parameters of an aircraft.

## 5 Current Research in Aircraft Noise

Aircraft noise is such an important factor in limiting the future growth of aviation that research is carried out worldwide to limit its generation. Jet noise is dominant at takeoff, fan and engine noise are dominant in flight, and airframe noise is dominant at landing, so there is plenty for researchers to do. Sources of engine noise are combustion and turbomachinery, and sources of airframe noise are flaps, slats, wing tips, nose gear, and main landing gear. The propagation of sound once it has been generated is also an important area of aircraft noise research, especially the propagation of sonic boom.

A new subject, computational aeroacoustics, was created thirty years ago, and this is now the dominant tool in aircraft noise research. The reader might wonder why its elder sibling, computational fluid dynamics, is not adequate for the task of predicting aircraft noise simply by "adding a little bit of compressibility" to an established code. The answer lies in one of the most pervasive ideas in mathematics, that of an invariant—here, the constant in Bernoulli's theorem. In an incompressible flow, a consequence of Bernoulli's theorem is that pressure fluctuations are entirely balanced by corresponding changes in velocity. Specialized computational techniques must therefore be developed to account for the minute changes in the Bernoulli "constant" produced by fluid compressibility and satisfying the wave equation. Computational aeroacoustics takes full account of such matters, which are responsible for the fact that only a minute proportion of near-field energy (the province of computational fluid dynamics) propagates as sound.

Research on aircraft noise takes place in the world's universities, companies, and government research establishments, and all of it relies heavily on mathematics. It is remarkable how many of the most practical contributions to the subject are made by individuals who are mathematicians; no fewer than six of the authors in the further reading section below took their first degree in mathematics (many with consummate distinction) and have used mathematics throughout their careers.

### Further Reading


Dowling, A. P., and J. E. Ffowcs Williams. 1983. *Sound and Sources of Sound.* Chichester: Ellis Horwood.

Goldstein, M. E. 1976. *Aeroacoustics.* New York: McGraw-Hill.

Howe, M. S. 2003. *Theory of Vortex Sound.* Cambridge: Cambridge University Press.

Lighthill, M. J. 1952. On sound generated aerodynamically. I. General theory. *Proceedings of the Royal Society* A 211: 564–87.

McAlpine, A., and R. J. Astley. 2012. Aeroacoustics research in Europe: the CEAS-ASC report on 2011 highlights. *Journal of Sound and Vibration* 331:4609–28.

Peake, N., and A. B. Parry. 2012. Modern challenges facing turbomachinery aeroacoustics. *Annual Review of Fluid Mechanics* 44:227–48.

Smith, M. J. T. 1989. *Aircraft Noise.* Cambridge: Cambridge University Press.

# VII.2 A Hybrid Symbolic–Numeric Approach to Geometry Processing and Modeling

*Thomas A. Grandine*

## 1 Background

Numerical computations for scientific and engineering purposes were the very first applications of the electronic computer. Originally built to calculate artillery firing tables, the first use of ENIAC (Electronic Numerical Integrator and Computer) was performing calculations for the hydrogen bomb being developed at Los Alamos. Indeed, all of the applications of early computers were numerical applications. After the art of computer programming had matured considerably, Macsyma at MIT introduced the notion of symbolic computing to the world. Unlike numerical computing, symbolic computing solves problems by producing analytic expressions in closed form.

More recently, tools have been developed to combine these two approaches in powerful and intriguing ways, Chebfun and Sage being prime examples. At Boeing, my colleagues and I have also been pursuing this idea for some time, and we continue to discover new ways of leveraging this technology to solve problems.

Geometric models are required throughout Boeing for many reasons: digital preassembly, engineering analysis, direct manufacturing, supply chain management, cost estimation, and myriad other applications. Depending on the application, models may be very simple (see figure 1) or very complex (see figure 2). Regardless of complexity, many computations need to be performed with these models. Properly conveying the algebraic formulation of these geometry problems so that the computations can be performed is not always straightforward.

Boeing has software that takes advantage of the combined strength of numeric and symbolic methods. It is named Geoduck (pronounced GOO-ee-duck) after a large clam indigenous to the Pacific Northwest in the United States. Like Sage, it is based on PYTHON [VII.11] programming. The tool enables construction of geometry models at arbitrary levels of detail and rapid query and processing of those models by most downstream applications.

Like most commercially available geometric modeling tools, Geoduck represents geometric entities as

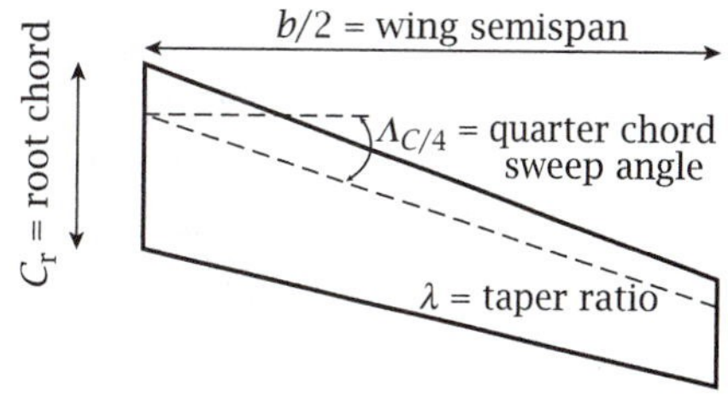


**Figure 1** A simple two-dimensional airplane wing planform model.

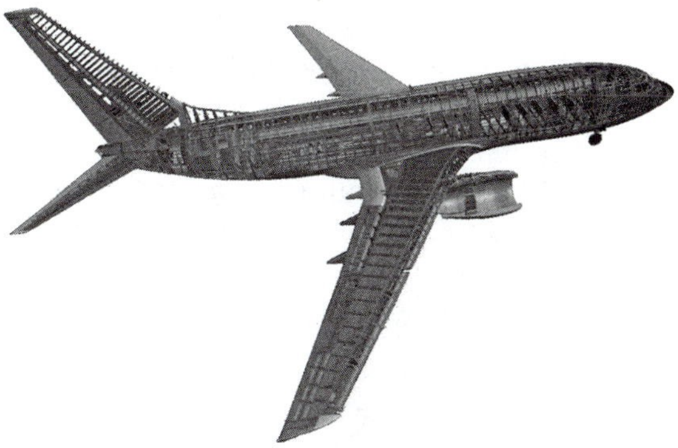

**Figure 2** A detailed model of the Boeing 777.

mathematical maps from a rectangular domain (parameter space) to a range (model space). For example, the cylinder shown in figure 3 depicts the range of the map

$$S(u,v) = \begin{pmatrix} \cos 2\pi u \\ \sin 2\pi u \\ 4v \end{pmatrix}.$$

Geoduck uses tensor product SPLINES [IV.9 §2.3] as the primary functional form. The most important reasons for this are that they have a convenient basis for computation (B-splines), they live in nestable linear function spaces, and they are made up of polynomial pieces that can be integrated and differentiated symbolically.

## 2 Some Simple Examples

Given a planar curve represented in Geoduck as a parametric spline map $c$ from $[0,1]$ to two-dimensional model space, one frequent problem is finding the closest and farthest points from a given point $p$, specifically to determine the extreme points of

$$\tfrac{1}{2}(c(u)-p)\cdot(c(u)-p).$$

Like Chebfun and Sage, Geoduck permits functions to be assigned to variables, and this makes calculations

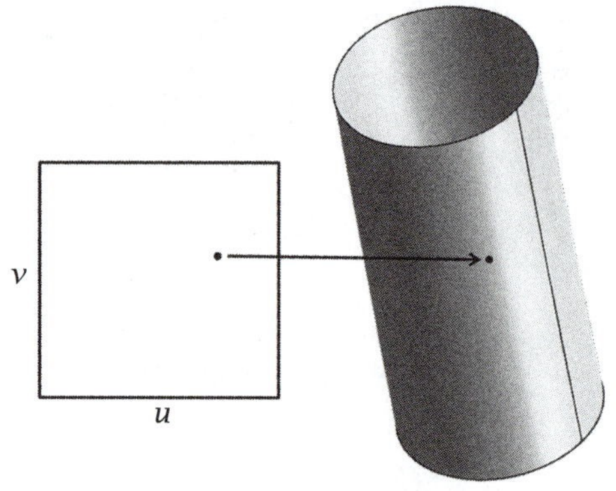


**Figure 3** A simple parametric surface map.

with spline functions easy to code and understand in Geoduck. After accounting for the endpoints, the extreme points are found by using the following Geoduck code to differentiate the expression and set the resulting expression to zero:

```
cder = c.Differentiate ()
der = Dot (cder, c - p)
expts = der.Zeros ()
```

Arithmetic operators have been overloaded in Geoduck to work on splines, and this is done in the assignment to `der`. Since splines are piecewise polynomial functions, any linear combination of them is also a spline and can be represented as a B-spline series, and that is how Geoduck implements overloaded operators.

Since `c` is a spline function, so is its derivative, and the `Differentiate` method performs the expected function. Similarly, the `Dot` function forms the spline function that is the dot product of `cder` and `c - p`. Note that the product of two splines is also a spline.

All of the operations up to this point have been carried out analytically. The `Zeros` method, however, is a numerical method because it is designed to work on piecewise polynomial functions of arbitrary degree. In Geoduck, the method is a straightforward implementation of a 1989 algorithm of mine, and it finds all of the real zeros of the spline function. The results of this code are depicted in figure 4. Here, the curve `c` is shown along with line segments drawn from the point `p` to the calculated points on the curve `c`. The segments corresponding to local maxima are shown as dashed lines, while the local minima are shown as dotted lines.

Of course, the basic recipe of performing a derivation for a geometric calculation of interest then encoding that derivation into a Geoduck script that combines

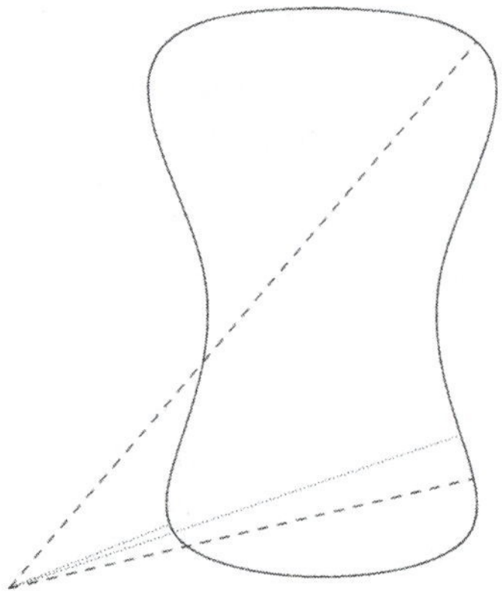

**Figure 4** The local extrema of the distance from a point to a curve.

symbolic and numeric calculations can be followed over and over again to solve a very rich collection of interesting problems.

As a second example, consider the problem of finding common tangents to a pair of curves $c$ and $d$. As before, both curves will be represented as vector-valued spline functions in two dimensions. If points on $c$ are given by $c(u)$ and points on $d$ are given by $d(v)$, then a common tangent will be the line segment between points $c(u)$ and $d(v)$ that has the property that the vector $c(u) - d(v)$ is parallel to the tangents of both curves, given by $c'(u)$ and $d'(v)$. The $(u, v)$ pairs that satisfy the following system of equations correspond to the endpoints of the desired tangent lines:

$$c'(u) \times (c(u) - d(v)) = 0,$$
$$d'(v) \times (c(u) - d(v)) = 0,$$

where $\times$ is the scalar-valued cross product in $\mathbb{R}^2$, i.e., the determinant of the $2 \times 2$ matrix whose columns are the operands.

Consider the following Geoduck code:

```
cminusd = TensorProduct (c, '-', d)
cder = cminusd.Differentiate (0)
dder = cminusd.Differentiate (1)
f1 = Cross (cder, cminusd)
f2 = Cross (dder, cminusd)
tanpts = Zeros ([f1, f2])
```

This example introduces the `TensorProduct` function, which takes a pair of spline functions as inputs and returns a new spline function defined over the tensor product of the domains of the original functions. In this

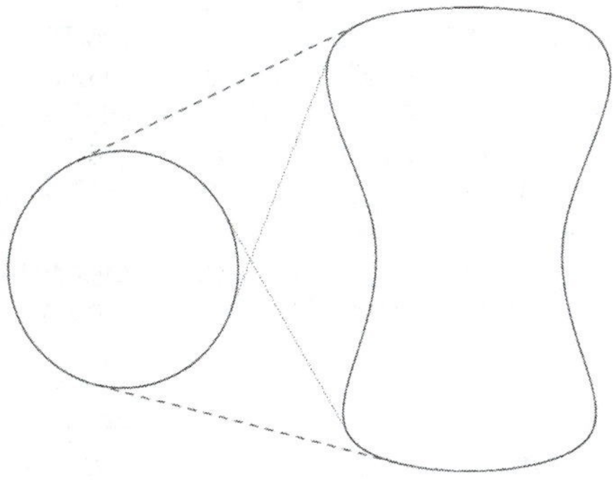

**Figure 5** Calculated common tangents to a pair of curves.

case, both c and d are defined over the interval $[0,1]$, so the new spline function `cminusd` is defined over $[0,1]^2$.

The `Differentiate` method is similar to before, except now it is applied to functions of more than one variable, so it will return the partial derivative with respect to the variable indicated by the index passed in. The `Cross` function here is similar to `Dot` in the first example, and it also requires the two input splines to be defined over the same domain, which in this case is $[0,1]^2$.

Finally, the `Zeros` function solves the given spline system of equations. The curves c and d are shown in figure 5 along with the four computed tangent lines.

## 3 Industrial Examples

Although Geoduck excels at solving simple geometric-analysis problems as just illustrated, it also solves real-life geometric-modeling and geometry-processing problems faced by Boeing. Consider the wing depicted in figure 6. Following standard practice, a surface $w$ is constructed that models the wing as a tensor product spline. Usually, the points $w(u,v)$ on the wing are arranged so that the variable $v$ increases with wingspan and the functions $w(\cdot,v)$ for each fixed value of $v$ describe individual airfoils. Each such airfoil has the property that $w(0,v) = w(1,v)$ is the point on the trailing edge, and as $u$ increases, the point on the airfoil traverses first the lower part of the airfoil and then the upper part.

One practical problem that needs to be solved as a preprocessing step for many aerodynamic analysis simulations is to determine the leading edge of the wing. For each airfoil section, the point on the leading edge is characterized by being the point that is farthest from the trailing edge. Thus, for each $v$ it can be determined by minimizing $(w(u,v)-w(0,v))\cdot(w(u,v)-w(0,v))$. Since $v$ is (temporarily) fixed, this problem can be solved by differentiating with respect to $u$ and setting the resulting expression equal to 0:

$$w_u(u,v)\cdot(w(u,v)-w(0,v)) = 0.$$

**Figure 6** A wing model with leading edge curve shown.

Since this equation can be solved for $u$ for every value of $v$, that leads to an entire one-parameter family of solutions of the form $w(u(v),v)$ for the implicit function $u$ given by the equation. The resulting leading edge curve is shown in figure 6, and it is generated with the following Geoduck code:

```
w0 = w.Trim ([[0.0, 0.0], [0.0, 1.0]])
zero = Line ([0.0], [0.0])
w0uv = TensorProduct (zero, '+', w0)
wu = w.Differentiate (0)
f = Dot (wu, w - w0uv)
le = f.Intersect ([1.0, 0.0])
```

The spline function `zero` is just the function defined over $[0,1]$ whose value is zero everywhere. It is needed here so that `w0uv` is a function of two variables.

This is very similar to the previous examples, with only two new wrinkles. The first is the use of the `Trim` method to construct a new spline identical to the original except that its domain of definition is restricted to a subset of the original. Since the value of $u$ is restricted in this case to be 0, the function `w0` is a function of only one variable. Making `w0` a function of two variables necessitates the use of `TensorProduct` so that `Dot` is once again provided with input functions with identical domains. The last wrinkle is the use of the `Intersect` method. In this case, it finds the zero set of an underdetermined system in the domain of $f$. The resulting locus of points in that domain is the preimage of the leading edge curve we are seeking.

As a final example, consider the problem of locating a spar of maximum depth along the span of a wing, something that is very desirable for structural reasons. This time, imagine the wing split into two pieces $w_1$

**Figure 7** The maximum-depth spar for the given wing. Only the lower part of the wing is shown here so that the spar is visible.

and $w_2$. Suppose that $w_1$ is the upper part of the wing while $w_2$ is the lower part. As before, the second parameter increases with the span of the wing. Imagine that both surfaces have the same parametrization, so that the $y$ component of the two pieces satisfies $y_1(\cdot, v_1) = y_2(\cdot, v_2)$ whenever $v_1 = v_2$. Thus, for each value of $v_1$ the vertical distance between the upper and lower surfaces will be maximized whenever the corresponding tangent vectors are parallel, i.e., when

$$\frac{\partial}{\partial u_1}\begin{pmatrix} x_1(u_1, v_1) \\ z_1(u_1, v_1) \end{pmatrix} \parallel \frac{\partial}{\partial u_2}\begin{pmatrix} x_2(u_2, v_2) \\ z_2(u_2, v_2) \end{pmatrix}.$$

With this in mind, new surfaces $\hat{s}_1$ and $\hat{s}_2$ can be defined by

$$\hat{s}_1(u_1, v_1) = \begin{pmatrix} x_1(u_1, v_1) \\ y_1(u_1, v_1) \\ \dfrac{(\partial/\partial u_1) z_1(u_1, v_1)}{(\partial/\partial u_1) x_1(u_1, v_1)} \end{pmatrix},$$

$$\hat{s}_2(u_2, v_2) = \begin{pmatrix} x_2(u_2, v_2) \\ y_2(u_2, v_2) \\ \dfrac{(\partial/\partial u_2) z_2(u_2, v_2)}{(\partial/\partial u_2) x_2(u_2, v_2)} \end{pmatrix}.$$

Note that the third component of each of these new surfaces is the slope of the curves that are the sections of the surfaces. The tangent vectors are parallel when these slopes are the same. Thus, curves along each of the two original surfaces $s_1$ and $s_2$ can be calculated as the intersection of the new surfaces $\hat{s}_1$ and $\hat{s}_2$. The resulting spar is the ruled surface between these two curves, which is shown in figure 7.

### Further Reading

de Boor, C. 2000. *A Practical Guide to Splines*. New York: Springer.

Eröcal, B., and W. Stein. 2010. The Sage project: unifying free mathematical software to create a viable alternative to Magma, Maple, Mathematica and MATLAB. In *Mathematical Software—ICMS 2010*, edited by K. Fukuda, J. Hoeven, M. Joswig, and N. Takayama, pp. 12–27. Lecture Notes in Computer Science, volume 6327. Berlin: Springer.

Goldstine, H. H. 1972. *The Computer from Pascal to von Neumann*. Princeton, NJ: Princeton University Press.

Grandine, T. A. 1989. Computing zeros of spline functions. *Computer Aided Geometric Design* 6:129–36.

Moses, J. 2012. Macsyma: a personal history. *Journal of Symbolic Computation* 47:123–30.

Trefethen, L. N. 2007. Computing numerically with functions instead of numbers. *Mathematics in Computer Science* 1:9–19.

# VII.3 Computer-Aided Proofs via Interval Analysis

*Warwick Tucker*

## 1 Introduction

The aim of this article is to give a brief introduction to the field of computer-aided proofs. We will do so by focusing on the problem of solving nonlinear equations and inclusions using techniques from interval analysis. Interval analysis is a framework designed to make numerical computations mathematically rigorous. Instead of computing *approximations* to sought quantities, the aim is to compute *enclosures* of the same. This requires taking both rounding and discretization errors into account.

### 1.1 What Is a Computer-Aided Proof?

Computer-aided proofs come in several flavors. The types of proofs that we will address here are those that involve the continuum of the real line. Thus, the problems we are trying to solve are usually taken from analysis, rather than from, say, combinatorics.

Rounding errors can be handled by performing all calculations with a set-valued arithmetic, such as interval arithmetic with outward rounding. The discretization errors, however, require some very careful analysis, and this usually constitutes the genuinely hard, analytical, part of the preparation of the proof. Quite often, the problem must be reformulated as a fixed-point equation in some suitable function space, and this can require some delicate arguments from functional analysis. Moreover, the discretization bounds must be made completely rigorous and explicit; in fact, they must be computable in finite time.

### 1.2 Examples from Mathematics

The field of INTERVAL ANALYSIS [II.20] has reached a high level of maturity, and its techniques have been used in solving many notoriously hard problems in mathematics.

As a testament to the advances made in recent years, we mention a few results that have been obtained using computer-aided proofs:

- Hass and Schlafly (2000) solved the double bubble conjecture (a problem stemming from the theory of minimal surfaces);
- Gabai et al. (2003) showed that homotopy hyperbolic 3-manifolds have noncoalescable insulator families;
- the long-standing Kepler conjecture (concerning how to densely pack spheres) was recently settled by Hales (2005); and
- the present author solved Smale's fourteenth problem about the existence of the Lorenz attractor in 1999.

## 2 Solving Nonlinear Equations

In this section we describe some computer-aided techniques for solving nonlinear equations. As we shall see later, this is a basic (hard) ingredient in many problems from analysis. Some of the presented techniques can be extended to the infinite-dimensional setting (e.g., fixed-point equations in function spaces), but in order to keep the exposition simple and short we will cover only the finite-dimensional case here.

To be precise, given a function $f$, together with a search domain $\mathcal{X}$, our task is to establish the existence (and uniqueness) of all zeros of $f$ residing inside $\mathcal{X}$. Due to the nonlinearity of $f$, and the global nature of $\mathcal{X}$, this is indeed not a simple task.

A basic principle of interval analysis is to extend our original problem to a set-valued version that satisfies the *inclusion principle*. In the setting we are interested in (solving $f(x) = 0$), this means that we must find an interval extension of $f$: that is, an interval-valued function $F$ satisfying

$$\operatorname{range}(f;\boldsymbol{x}) = \{f(x) : x \in \boldsymbol{x}\} \subseteq F(\boldsymbol{x}). \qquad (1)$$

Here, $\boldsymbol{x}$ denotes an interval and $x$ denotes a real number. When $f$ is an elementary function, $F$ is obtained by substituting all appearing arithmetic operators and standard functions with their interval-valued counterparts. In higher dimensions, we consider the components of $f$ separately.

We will now briefly describe two techniques that we use in our search for the zeros of $f$.

### 2.1 Interval Bisection

Interval analysis provides simple criteria for excluding regions in the search space where no solutions to $f(x) = 0$ can reside. The entire search region $\mathcal{X}$ is adaptively subdivided by set-valued bisection into subrectangles $\boldsymbol{x}_i$, each of which must withstand the test $0 \in F(\boldsymbol{x}_i)$. Failing to do so results in the subrectangle being discarded from further search; by (1), the set $\boldsymbol{x}_i$ cannot contain any solutions. The bisection phase ends when all remaining subrectangles have reached a sufficiently small size. What we are left with is a collection of rectangles whose union is guaranteed to contain all zeros of $f$ (if there are any) within the domain $\mathcal{X}$.

As a simple example, consider the function $f(x) = \sin x(x - \cos x)$ on the domain $\mathcal{X} = [-10, 10]$ (see figure 1(a)). This function clearly has eight zeros in $\mathcal{X}$: $\{\pm 3\pi, \pm 2\pi, \pm\pi, 0, x^*\}$, where $x^*$ is the unique (positive) zero of $x - \cos x = 0$. Applying the interval bisection method with the stopping tolerance 0.001 produces the nine intervals listed below. Note, however, that intervals 4 and 5 are adjacent. This always happens when a zero is located exactly at a bisection point of the domain.

```
Domain         : [-10,10]
Tolerance      : 0.001
Function calls: 227
Solution list :
1: [-9.42505,-9.42444]   6: [+0.73853,+0.73914]
2: [-6.28357,-6.28296]   7: [+3.14148,+3.14209]
3: [-3.14209,-3.14148]   8: [+6.28296,+6.28357]
4: [-0.00061,+0.00000]   9: [+9.42444,+9.42505]
5: [+0.00000,+0.00061]
```

The mathematical content of this computation is that all zeros of $f$, restricted to $[-10, 10]$, are contained in the union of the nine output intervals (see figure 1(b)). No claims can be made about existence at this point.

### 2.2 The Interval Newton Method

Another (equally important) tool of interval analysis is a checkable criterion for proving the existence and (local) uniqueness of solutions to nonlinear equations. The underlying tool is based on a set-valued extension of Newton's method, which incorporates the Kantorovich condition, ensuring that the basin of attraction has been reached. Let $\boldsymbol{x}$ denote a subrectangle that survived the bisection process, and let $\check{x}$ be a point in the interior of $\boldsymbol{x}$ (e.g., its midpoint). We now form the *Newton image* of $\boldsymbol{x}$:

$$N(\boldsymbol{x}) = \check{x} - [\mathrm{D}F(\boldsymbol{x})]^{-1} f(\check{x}).$$

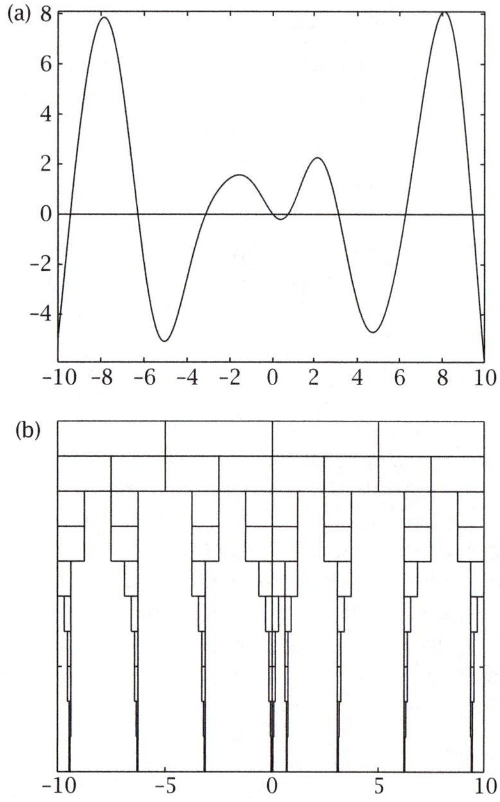


**Figure 1** (a) The function $f(x) = \sin x(x - \cos x)$ on the domain $X = [-10, 10]$. (b) Increasingly tight enclosures of the zeros.

This is a set-valued version of the standard Newton method: note that the correction term involves solving a linear system with interval entries. If we cannot solve this linear system, we apply some more bisection steps to the set $\boldsymbol{x}$. The Newton image carries some very powerful information. If $N(\boldsymbol{x}) \cap \boldsymbol{x} = \varnothing$, then $\boldsymbol{x}$ cannot contain any solution to $f(x) = 0$ and is therefore discarded. On the other hand, if $N(\boldsymbol{x}) \subseteq \boldsymbol{x}$, then $\boldsymbol{x}$ contains a unique zero of $f$. In the remaining case, we can shrink $\boldsymbol{x}$ into $N(\boldsymbol{x}) \cap \boldsymbol{x}$ and redo the Newton test. If successful, this stage will give us an exact count of the number of zeros of $f(x)$ within the domain $X$.

In figure 2 we illustrate the geometric construction of the Newton image.

## 3 An Application to the Restricted *n*-Body Problem

The Newtonian $n$-body problem addresses the dynamics of $n$ bodies (which can be assumed to be point

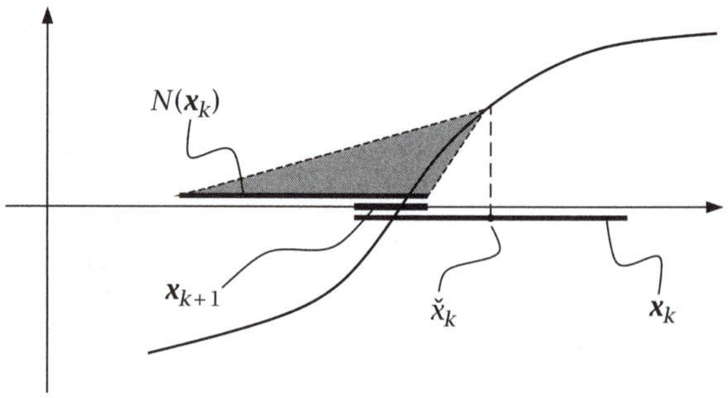


**Figure 2** One iteration of the interval Newton method.

particles) with masses $m_i > 0$, moving according to Newton's laws of motion. A *relative equilibrium* is a *planar* solution to Newton's equations that performs a rotation of uniform angular velocity about the system's center of mass. Thus, in a rotating coordinate frame, the constellation of bodies is fixed.

A long-standing question, raised by Aurel Wintner in 1941, concerns the finiteness of the number of (equivalence classes of) relative equilibria. In 1998, Fields medalist Steven Smale listed a number of challenging problems for the twenty-first century. Problem number 6 reads:

> Is the number of relative equilibria finite, in the $n$-body problem of celestial mechanics, for any choice of positive real numbers $m_1, \dots, m_n$ as the masses?

At the time of writing, this problem remains open for $n \geqslant 5$.

Recently, Kulevich et al. (2009) established that the number of equilibria in the planar circular restricted four-body problem (PCR4BP) is finite for any choice of masses. The PCR4BP under consideration consists of three large bodies (primaries) with arbitrary masses at the vertices of an equilateral triangle rotating on circular orbits about their common center of mass. A fourth infinitesimal mass, subject to the gravitational attraction of the primaries, is inserted into their plane of motion and is assumed to have no effect on their circular orbits. In an appropriately selected rotating frame, the primaries are fixed (see figure 3).

Since we are concerned only with equivalence classes of solutions, we may normalize the system's total mass such that $\sum_i m_i = 1$. Without loss of generality, we may also assume that the equilateral triangle of primaries has unit sides. This makes the problem compact and thus amenable to a global search.

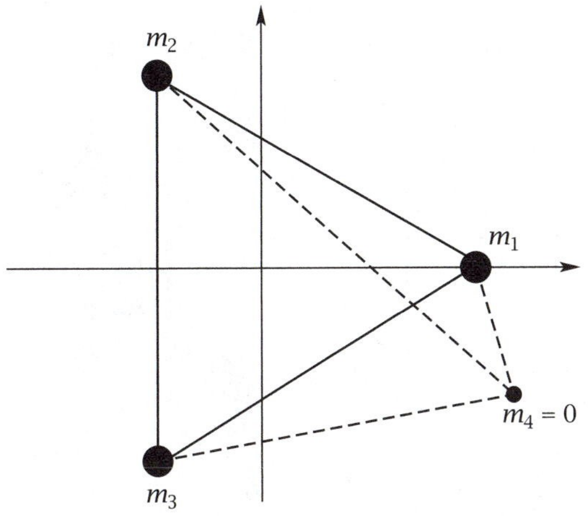


**Figure 3** The basic configuration with Lagrange's equilateral triangle for the PCR4BP.

Let $x_1, x_2, x_3, c \in \mathbb{R}^2$ denote the (fixed) positions of the three primary bodies and their center of mass, respectively. Let $z = (z_1, z_2)$ be the position of the fourth (infinitesimal mass) body. Then its motion is governed by the system of differential equations

$$\ddot{z}_1 = \frac{\partial V}{\partial z_1} + 2\dot{z}_2,$$

$$\ddot{z}_2 = \frac{\partial V}{\partial z_2} - 2\dot{z}_1,$$

where the potential $V$ is given by

$$V(z) = \tfrac{1}{2}\|z - c\|_2^2 + \sum_{k=1}^{3} \frac{m_i}{\|x_i - z\|_2}. \qquad (2)$$

Thus, the relative equilibria are given by the critical points of $V$. In essence, we have reduced the problem to that of solving nonlinear equations.

Kulevich et al. showed not only that the number of relative equilibria in the PCR4BP is finite, but that there are at most 196 equilibrium points. This upper bound is believed to be a large overestimation. Numerical explorations by Simó (1978) indicate that there are eight, nine, or ten equilibria, depending on the values of the masses.

The proof of Kulevich et al. is based on techniques from Bernstein–Khovanskii–Kushnirenko (BKK) theory. This provides checkable conditions determining if a system of polynomial equations has a finite number of solutions for which all variables are nonzero. The techniques utilized stem from algebraic geometry, such as the computation of Newton polytopes, and are more general than Gröbner basis methods.

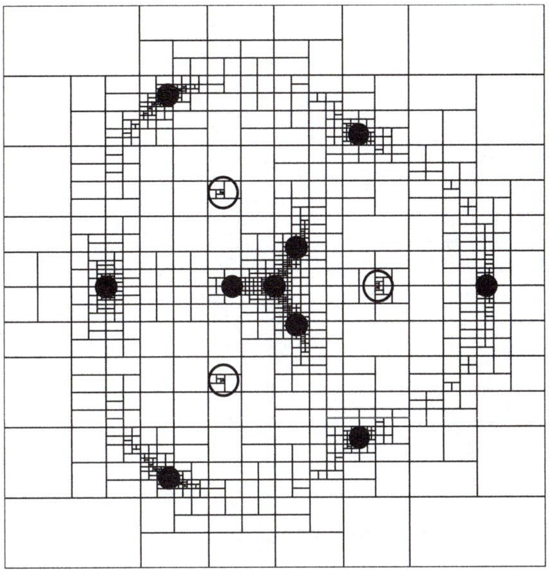

**Figure 4** The PCR4BP in the equal-mass case $m_1 = m_2 = m_3 = \frac{1}{3}$. The three primaries are marked by circles; the ten equilibria are marked by disks. The rectangular grid stems from the subdivision in the adaptive elimination process.

An interesting special case is taking all masses equal: $m_1 = m_2 = m_3 = \frac{1}{3}$. The PCR4BP then has exactly ten solutions, as was first established by Lindow in 1922. We will illustrate how rigorous zero-finding techniques can readily deal with this scenario.

By the normalization of the primaries and masses, we can restrict our search to the square $\mathcal{X} = [-\frac{3}{2}, +\frac{3}{2}]^2$. Let $f(z) = \nabla V(z)$ denote the gradient of $V$, and extend it to a set-valued function satisfying the inclusion principle (1). The problem is now reduced to finding (or simply counting) the zeros of $f$ restricted to $\mathcal{X}$. Using a combination of the set-valued bisection and the interval Newton operator described above, it is straightforward to establish that the PCR4BP in the equal-mass case has exactly ten relative equilibria. Using a stopping tolerance of $10^{-2}$ in the bisection stage, we arrive at 1546 discarded subrectangles, with 163 remaining for the next stage. The Newton stage discards another 150 subrectangles and produces ten isolating neighborhoods for the relative equilibria (and three for the primaries) (see figure 4).

As far as our approach is concerned, there is nothing special about choosing the masses to be equal in (2). Taking, for example, $m = (\frac{1}{10}, \frac{2}{10}, \frac{7}{10})$, we can repeat the computations and find that there are exactly eight relative equilibria (see figure 5).

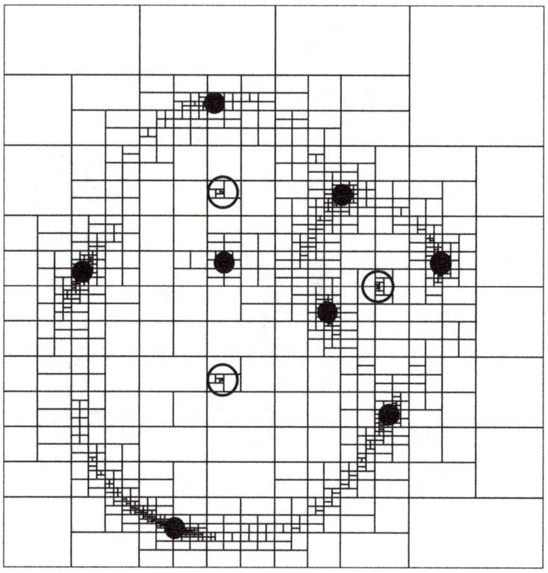

**Figure 5** The PCR4BP in the case $m = (\frac{1}{10}, \frac{2}{10}, \frac{7}{10})$.

The computations involved take a fraction of a second to perform on a standard laptop.

## 4 Solving Nonlinear Inclusions

In the set-valued framework it makes a lot of sense to extend the concept of solving equations of the form $f(x) = y$ to include solving for inclusions $f(x) \in \boldsymbol{y}$, where $\boldsymbol{y}$ is a given *set*. This is a very natural problem formulation in the presence of uncertainties; think of $\boldsymbol{y}$ as representing noisy data with error bounds.

If $\boldsymbol{y}$ is a set with nonempty interior, we should expect the solution set $S$ to share this property. Therefore, we should be able to approximate the solution set both from the inside and from the outside. In other words, we would like to compute two sets $\underline{S}$ and $\overline{S}$ that satisfy

$$\underline{S} \subseteq S \subseteq \overline{S}.$$

$\underline{S}$ and $\overline{S}$ are called the *inner* and *outer* approximations of $S$, respectively. By measuring the size of their difference, $\overline{S} \setminus \underline{S}$, we can obtain reliable information about how close we are to the solution set $S$.

Given a partition $\mathcal{P}(X)$ of the domain $X$, the outer approximation $\overline{S}$ is computed precisely as in section 2.1: it contains all partition elements whose interval images have nonempty intersection with the range $\boldsymbol{y}$. The inner approximation contains all partition elements whose interval images are contained in the range $\boldsymbol{y}$. In other words, we have

$$\underline{S} = \{\boldsymbol{x} \in \mathcal{P}(X) \colon F(\boldsymbol{x}) \subseteq \boldsymbol{y}\},$$
$$\overline{S} = \{\boldsymbol{x} \in \mathcal{P}(X) \colon F(\boldsymbol{x}) \cap \boldsymbol{y} \neq \varnothing\}.$$

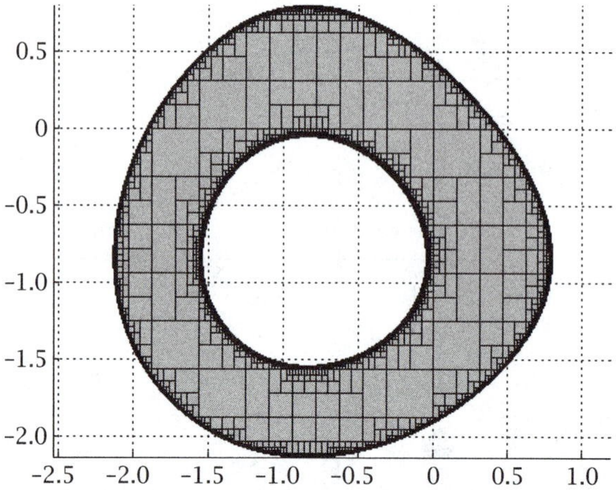

**Figure 6** An outer approximation $\overline{S}$ of the solution set $S$ (shaded). All rectangles of width greater than $10^{-2}$ belong to the inner approximation $\underline{S}$.

As an example, consider the nonlinear function

$$f(x) = \sin x_1 + \sin x_2 + \tfrac{2}{5}(x_1^2 + x_2^2)$$

and suppose we want to find the set

$$S = \{x \in [-5, +5]^2 \colon f(x) \in [-0.5, 0.5]\}.$$

By a simple bisection procedure, we adaptively partition the domain $X = [-5, 5]^2$ into subrectangles and discard rectangles $\boldsymbol{x}$ such that $F(\boldsymbol{x}) \cap [-0.5, 0.5] = \varnothing$. The remaining rectangles are classified according to whether they belong to $\underline{S}$ and/or $\overline{S}$. Note that, as soon as a subrectangle is determined to belong to $\underline{S}$, it undergoes no further bisection. Only subrectangles whose interval images intersect $\partial S$ are subdivided. This is illustrated in figure 6, in which a stopping tolerance of $10^{-2}$ was used.

The ability to solve nonlinear inclusions is of great importance in parameter estimation. Given a finitely parametrized model function $f(x; p) = y$ together with a set of uncertain data $(x_1, \boldsymbol{y}_1), \dots, (x_n, \boldsymbol{y}_n)$ and a search space $\mathcal{P}$, the task is to solve for the set

$$S = \{p \in \mathcal{P} \colon f(x_i; p) \in \boldsymbol{y}_i,\ i = 1, \dots, n\}.$$

This can be done by computing inner and outer approximations of $S$, exactly as described above. Today there exist very efficient techniques (e.g., constraint propagation) for solving precisely such problems.

## 5 Recent Developments

In this article we have focused exclusively on nonlinear equation solving, but computer-aided proofs are also used in many other areas. Global optimization is a field that is well suited to rigorous techniques, and there are many software suites that rely on interval analysis and set-valued computations. A great deal of effort has been expended developing methods for rigorously solving ordinary differential equations, and several mature software packages that produce validated results at a reasonable computational cost now exist. There have also been many successful endeavors in the realm of partial differential equations, but here each problem requires its own set of tools and there is no natural one-size-fits-all approach to this vast area. Another area of application is parameter estimation, where set-valued techniques are used to model the uncertainties in the estimated states. With the recent interest in uncertainty quantification, rigorous computations have a bright future.

### Further Reading


Gabai, D., R. G. Meyerhoff, and N. Thurston. 2003. Homotopy hyperbolic 3-manifolds are hyperbolic. *Annals of Mathematics* 157(2):335–431.

Hales, T. C. 2005. A proof of the Kepler conjecture. *Annals of Mathematics* 162:1065–185.

Hass, J., and R. Schlafly. 2000. Double bubbles minimize. *Annals of Mathematics* 151(2):459–515.

Jaulin, L., M. Kieffer, O. Didrit, and E. Walter. 2001. *Applied Interval Analysis: With Examples in Parameter and State Estimation, Robust Control and Robotics.* Berlin: Springer.

Kulevich, J., G. Roberts, and C. Smith. 2009. Finiteness in the planar restricted four-body problem. *Qualitative Theory of Dynamical Systems* 8(2):357–70.

Moore, R. E., R. B. Kearfott, and M. J. Cloud. 2009. *Introduction to Interval Analysis.* Philadelphia, PA: SIAM.

Neumaier, A. 1990. *Interval Methods for Systems of Equations.* Cambridge: Cambridge University Press.

Simó, C. 1978. Relative equilibrium solutions in the four body problem. *Celestial Mechanics* 18:165–84.

Tucker, W. 1999. The Lorenz attractor exists. *Comptes Rendus de l'Académie des Sciences Paris—Série 1: Mathématique* 328(12):1197–202.

Tucker, W. 2011. *Validated Numerics: A Short Introduction to Rigorous Computations.* Princeton, NJ: Princeton University Press.


# VII.4 Applications of Max-Plus Algebra

*David S. Broomhead*

## 1 Basics

Max-plus algebra is a term loosely applied to algebraic manipulations using the operations max and $+$. More precisely, consider the set of real numbers $\mathbb{R}$ and augment this with a smallest (with respect to the total ordering on $\mathbb{R}$) element $\varepsilon = -\infty$. Let us say $\bar{\mathbb{R}} = \mathbb{R} \cup \{\varepsilon\}$. A rich arithmetic on $\bar{\mathbb{R}}$ is then obtained by defining the binary operations $\oplus$ and $\otimes$:

$$a \oplus b = \max\{a, b\} \quad \forall a, b \in \bar{\mathbb{R}},$$
$$a \otimes b = a + b \quad \forall a, b \in \bar{\mathbb{R}}.$$

In this algebra, $\varepsilon$ acts as the "zero" element in the sense that $a \oplus \varepsilon = a$ and $a \otimes \varepsilon = \varepsilon$ for any $a \in \bar{\mathbb{R}}$, while 0 acts as the unit element since $a \otimes 0 = a$ for any $a \in \bar{\mathbb{R}}$. Both $\oplus$ and $\otimes$ are commutative and associative operations, and $\otimes$ distributes over $\oplus$ because the identity

$$a + \max\{b, c\} = \max\{a + b, a + c\}$$

translates directly to the distributive law

$$a \otimes (b \oplus c) = (a \otimes b) \oplus (a \otimes c).$$

Max-plus powers of any $a \in \bar{\mathbb{R}}$ can be defined recursively by $a^{\otimes(k+1)} = a \otimes a^{\otimes k}$, with $a^{\otimes 0} = 0$. Note that $a^{\otimes k} = k \times a$. This can be extended to any real-valued power, $\alpha$, by defining $a^{\otimes \alpha} = \alpha \times a$.

The arithmetic that is taught in childhood differs from this max-plus system in important ways. In particular, $a \oplus a = a$ for all $a \in \bar{\mathbb{R}}$, so $\oplus$ is idempotent. A consequence of this is that there is no additive inverse, i.e., there is not a nice analogue of the negative of a number. (The negative numbers in $\bar{\mathbb{R}}$ actually correspond to the *multiplicative* inverse since $x \otimes (-x) = 0$). Mathematically, $(\bar{\mathbb{R}}, \oplus, \otimes)$ is a *semi*field, albeit one that is commutative and idempotent.

## 2 Tropical Mathematics

Tropical mathematics is a recent, broader name for algebra and geometry based on the max-plus semifield and related semifields. In many applications, it can be more convenient to use the min-plus semifield, which is defined as above but using the min operation rather than max. In this case, the binary operations are defined over $\mathbb{R} \cup \{\infty\}$. The max-plus and min-plus semifields are isomorphic by the map $x \mapsto -x$. A third commutative idempotent semifield is called max-times. It is based on the nonnegative reals with operations max and times and is isomorphic to max-plus by the map $x \mapsto \log x$.

## 3 An Overview of Applications

Tropical mathematics has been applied to such a wide range of problems that it is impossible to cover them all here. It has been used to analyze the timing of asynchronous events, such as timetabling for transport networks, the scheduling of complex tasks, the control of discrete-event systems, and the design of asynchronous circuits in microelectronics. There are deep connections with algebraic geometry, which have led to applications in the analysis of deoxyribonucleic acid (DNA) sequences and their relation to phylogenetic trees. There is also work on optimization that is connected with semiclassical limits of quantum mechanics, Hamilton–Jacobi theory, and the asymptotics of zero-temperature statistical mechanics. In all these cases the mathematical structures, built on the semifield structure described above, lead to simple, often linear, formulations of potentially complicated nonlinear problems.

Consider a simple example of scheduling for asynchronous processes. At time $x_1$ an engineer begins to make a component that takes $a_1$ hours to complete. In the max-plus notation, the time in which the task is completed is $a_1 \otimes x_1$. A second engineer begins to make a different component at time $x_2$ that takes $a_2$ hours. If both components are to be combined to make the finished product, the product cannot be completed before $(a_1 \otimes x_1) \oplus (a_2 \otimes x_2)$, i.e., before the last component is finished. The two operations $\otimes$ and $\oplus$ occur naturally here, and there is also a hint of linear algebra, since the earliest time for completion appears to be the scalar product, in a max-plus sense, of a vector of starting times $(x_1, x_2)$ and the vector of durations of each task $(a_1, a_2)$.

## 4 Timing on Networks

The timetable for a rail network has to coordinate the movements of many independent trains in order to provide a safe and predictable service. A range of issues have to be addressed: railway stations have limited numbers of platforms; parts of the network may have single-track lines; passengers need to make connections; etc. Max-plus algebra provides useful tools to do this.

As a basic example consider a railway, from A to B, a section of which is single track. For safety, there is a token that must be given by a signalman to the driver about to enter the single-track section. There is only one token, and when the driver leaves the section he returns it to a second signalman, who can give it to the driver of a train traveling in the opposite direction.

Let the timetable be such that trains from either direction arrive at the single-track section every $T$ hours, and let $x_A(k)$ be the time at which the $k$th train from A to B enters the section, and similarly with $x_B(k)$. The time taken for a train to traverse the section is $\tau$. Then, for $1 < k \in \mathbb{N}$,

$$x_A(k) = \tau \otimes x_B(k-1) \oplus T^{\otimes k},$$
$$x_B(k) = \tau \otimes x_A(k) \oplus T^{\otimes k},$$

with the initial condition $x_A(1) = T$. The lack of symmetry between these expressions arises because the first train is assumed to be traveling from A to B. In words, these formulas mean that a train cannot enter the section before it arrives there or before the previous train traveling in the opposite direction has left. By substitution, the following linear nonhomogeneous equation for $x_A(k)$ is found:

$$x_A(k) = \tau^{\otimes 2} \otimes x_A(k-1) \oplus \tau \otimes T^{\otimes k-1} \oplus T^{\otimes k}.$$

This equation can be solved by introducing $x_A(k) = \tau^{\otimes 2k} \otimes y(k)$ with $y(1) = \tau^{\otimes -2} \otimes T$. Assuming that $T > \tau$ gives

$$y(k) = y(k-1) \oplus \lambda^{\otimes k} \Rightarrow y(k) = \bigoplus_{j=1}^{k} \lambda^{\otimes j},$$

where $\lambda = \tau^{\otimes -2} \otimes T = T - 2\tau$.

There are two cases that are distinguished by the sign of the parameter $\lambda$. If $\lambda > 0$, the sequence of $\lambda^{\otimes j}$ is increasing, so that $y(k) = \lambda^{\otimes k}$. In this case, therefore, $x_A(k) = T^{\otimes k}$: trains enter the single-track section from A every $T$ hours, i.e., at the same rate that they arrive from A. If $\lambda < 0$, the sequence of $\lambda^{\otimes j}$ is now decreasing, so that $y(k) = \lambda$. Hence $x_A(k) = T \otimes \tau^{\otimes 2(k-1)}$: trains enter the single-track section from A every $2\tau$ hours, and, since this is less than the rate at which they arrive from A, a queue develops.

Calculations such as this can be used to assess the stability of timetables. Here, the parameter $\lambda$ is the difference between the timetabled interval between trains arriving at the single-track section and the time it takes a token to return to the first signalman. A timetable that sets $T$ to be such that $\lambda$ has a small positive value is vulnerable to unexpected delays in the single-track section. Large-scale timetabling of rail networks has been carried out using linear systems of max-plus equations.

## 5 Linear Algebra

Matrices and vectors based on the semifield $(\bar{\mathbb{R}}, \oplus, \otimes)$ can be defined in the obvious way as $A = (a_{ij}) \in \bar{\mathbb{R}}^{m\times n}$. In particular, when $n = 1$, this is an element of $(\bar{\mathbb{R}}^m, \oplus, \otimes)$ that, although it is not quite a vector space (the negative of a vector is not defined), is a semimodule defined over $(\bar{\mathbb{R}}, \oplus, \otimes)$.

Matrices can be combined using a natural generalization of the usual rules of matrix addition and multiplication:

$$(A \oplus B)_{ij} = a_{ij} \oplus b_{ij},$$

with $A, B \in \bar{\mathbb{R}}^{m\times n}$, and

$$(A \otimes B)_{ij} = a_{i1} \otimes b_{1j} \oplus \cdots \oplus a_{il} \otimes b_{lj} = \bigoplus_{k=1}^{l} a_{ik} \otimes b_{kj},$$

where $A \in \bar{\mathbb{R}}^{m\times l}$ and $B \in \bar{\mathbb{R}}^{l\times n}$.

A diagonal matrix, say $D = \operatorname{diag}(d_1, \dots, d_n)$, has all its off-diagonal elements equal to $\varepsilon$. In particular, the $n \times n$ identity matrix is

$$\mathbb{I} = \operatorname{diag}(0, \dots, 0).$$

The only max-plus matrices that are invertible are the diagonal matrices and those nondiagonal matrices that can be obtained by permutation of the rows and columns of a diagonal matrix. The inverse of the diagonal matrix $D$ is $D^{\otimes -1} = \operatorname{diag}(-d_1, \dots, -d_n)$.

The paucity of invertible max-plus matrices is due to the lack of subtraction, or additive inverses, as noted in section 1, but there are tools for solving systems of max-plus linear equations. Define conjugate operations on $\bar{\mathbb{R}} \cup \infty$ as

$$a \oplus' b = \min\{a, b\},$$
$$a \otimes' b = a + b$$

(with the convention that $\varepsilon \otimes' \infty = \infty$ and $\varepsilon \otimes \infty = \varepsilon$), and extend these definitions to matrices and vectors as before. Consider the one-sided linear equation

$$A \otimes x = d,$$

where $x \in \bar{\mathbb{R}}^n$ is an unknown vector and $d \in \bar{\mathbb{R}}^m$, $A \in \bar{\mathbb{R}}^{m\times n}$ are known. The product $A \otimes x$ is a max-plus linear combination of the columns of $A$, and so the existence of a solution corresponds to $d$ being an element of the span of the columns of $A$. If a solution exists in $\bar{\mathbb{R}} \cup \infty$, it is $x = A^- \otimes' d$, where $A^- = -A^{\mathrm{T}}$ is the conjugate of $A$. Even if no solution exists, $A^- \otimes' d$ is the greatest solution (with respect to the product order on $\bar{\mathbb{R}}^n$) of the system of inequalities $A \otimes x \leqslant d$.

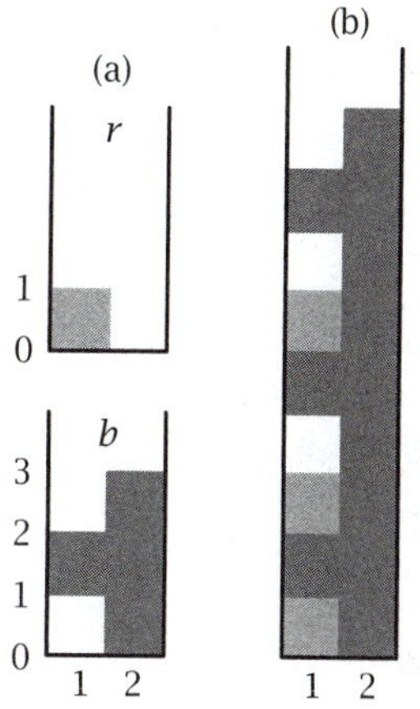


**Figure 1** A heaps of pieces example. The two pieces are shown separately in (a), and the heap following the sequence of pieces $rbrbrb$ is shown in (b).

## 6 An Example Application

Imagine a set of resources $\mathcal{R} = \{1, \dots, n\}$ and a set of basic tasks $\mathcal{A} = \{\omega_1, \dots, \omega_m\}$. For each task it is known which resources will be required, in what order, and for how long. This problem can be studied using a *heaps of pieces* model, which resembles the game Tetris. Figure 1 illustrates this for a simple example based on two tasks $\omega_1 = r$ and $\omega_2 = b$ and two resources labeled 1 and 2. Part (a) shows how the pieces are employed to represent the use of resources on the basic tasks: for $r$, resource 1 is required for one time unit and resource 2 is not required at all; for $b$, the task takes three time units, initially resource 2 is required, then resources 1 and 2 work in parallel, and finally resource 2 completes the task. Complex scheduling of these tasks is represented by piling the pieces into heaps such that each piece touches but does not overlap the pieces below it (unlike in Tetris, no rotation or horizontal movement of pieces is allowed). Part (b) shows the heap created when the pieces are introduced alternately, starting with $r$.

Each piece is characterized by a pair of functions $l, u\colon \mathcal{R} \to \bar{\mathbb{R}}$ that give, respectively, the lower and upper contours of the piece, i.e., the earliest and latest times that each resource is in use. Since $l$ and $u$ are functions defined on a discrete set, $\mathcal{R}$, they will be treated as vectors with dimension equal to the cardinality of $\mathcal{R}$. If a piece does not use a given resource, the corresponding components of $l$ and $u$ are set to $\varepsilon$, so in the example, $l(r) = (0, \varepsilon)^{\mathrm{T}}$, $u(r) = (1, \varepsilon)^{\mathrm{T}}$, $l(b) = (1, 0)^{\mathrm{T}}$, and $u(b) = (2, 3)^{\mathrm{T}}$. Associated with each piece there is a subset $S \subset \mathcal{R}$ of resources that the corresponding task requires; in this example, $S(r) = \{1\}$ and $S(b) = \{1,2\}$. Generally, for each piece, $\omega \in \mathcal{A}$, $u(\omega) \geqslant l(\omega)$, and $l(\omega)$ is defined such that $\min_{k \in S(\omega)} l_k(\omega) = 0$.

These models can be used to study the efficiency of different scenarios, for example, introducing the tasks in various periodic patterns or randomly. The rate at which the maximum height of the heap grows indicates how long the total job takes as the number of repetitions of the tasks is increased. A comparison between the rates of growth of the maximum and minimum heights is a measure of how uniformly the work is distributed over the various resources available.

A heap model can be expressed as a linear max-plus system. Consider the upper contour of a heap containing $p$ pieces, $x(p)\colon \mathcal{R} \to \bar{\mathbb{R}}$. What is the new contour if a new piece is added? Let the new piece be called $\omega_{j_{p+1}} \in \mathcal{A}$. If this piece is placed at some height $h$, this can be represented by adding $h$ to its upper and lower contour functions or using the max-plus notation: $h \otimes u(\omega_{j_{p+1}})$ and $h \otimes l(\omega_{j_{p+1}})$. The smallest possible value of $h$ such that $h \otimes l(\omega_{j_{p+1}}) \geqslant x(p)$ must now be chosen. This is achieved when $\max_{i \in S(\omega_{j_{p+1}})}\{x_i(p) - h \otimes l_i(\omega_{j_{p+1}})\} = 0$ or, in max-plus notation,

$$h = \bigoplus_{i \in S(\omega_{j_{p+1}})} l_i^{-1}(\omega_{j_{p+1}}) \otimes x_i(p) = l^-(\omega_{j_{p+1}}) \otimes x(p),$$

where $l^-(\omega_{j_{p+1}})$ is a row vector whose components are $-l_i$ if $l_i$ is finite and $\varepsilon$ otherwise. The upper contour of the heap is now given by

$$x(p+1) = M(\omega_{j_{p+1}}) \otimes x(p), \tag{1}$$

where $M(\omega)$ is an $n \times n$ matrix

$$M(\omega) = \mathbb{I} \oplus u(\omega) \otimes l^-(\omega).$$

This is a linear equation that relates successive upper contours of the heap.

Returning to the example, use of this formula gives

$$M(r) = \begin{pmatrix} 1 & \varepsilon \\ \varepsilon & 0 \end{pmatrix}, \qquad M(b) = \begin{pmatrix} 1 & 2 \\ 2 & 3 \end{pmatrix}.$$

The iteration process given by equation (1) should begin with no pieces present, i.e., $x(0) = (0, \dots, 0)^{\mathrm{T}}$, and requires a specification of the matrix at each step. The sequence of matrices corresponds to the sequence of pieces introduced. In figure 1, the example shows a periodic sequence alternating pieces $r$ and $b$. This begins

$$x(1) = \begin{pmatrix} 1 & \varepsilon \\ \varepsilon & 0 \end{pmatrix} \otimes \begin{pmatrix} 0 \\ 0 \end{pmatrix} = \begin{pmatrix} 1 \\ 0 \end{pmatrix},$$

$$x(2) = \begin{pmatrix} 1 & 2 \\ 2 & 3 \end{pmatrix} \otimes \begin{pmatrix} 1 \\ 0 \end{pmatrix} = \begin{pmatrix} 2 \\ 3 \end{pmatrix},$$

and so on.

The regularity of this sequence means that the matrix describing every second iterate is given by the product $M(br) = M(b) \otimes M(r)$:

$$M(br) = \begin{pmatrix} 1 & 2 \\ 2 & 3 \end{pmatrix} \otimes \begin{pmatrix} 1 & \varepsilon \\ \varepsilon & 0 \end{pmatrix} = \begin{pmatrix} 2 & 2 \\ 3 & 3 \end{pmatrix}.$$

It is interesting to note that the vector $x(2)$ is an eigenvector of $M(br)$:

$$M(br) \otimes \begin{pmatrix} 2 \\ 3 \end{pmatrix} = 3 \otimes \begin{pmatrix} 2 \\ 3 \end{pmatrix},$$

where the eigenvalue is 3. This means that, if pieces continue to be added alternately, the heap will grow in height linearly at a rate of 3 units every cycle. More generally, if the scheduling of tasks is periodic with period $p$ and the sequence of tasks is given by $w \in \mathcal{A}^p$, then the growth rate of the completion time might be obtained by finding the eigenvalue of $M(w)$. This leaves two questions: do iterates of the initial condition converge to an eigenvector of $M(w)$, and does $M(w)$ generally possess an eigenvalue (and, if so, how is it computed)?

## 7 Eigenvalues and Eigenvectors

In general, the eigenvalue problem for square matrices is defined as one would expect. Given an $n \times n$ matrix $A$, look for a $\lambda \in \bar{\mathbb{R}}$ and a nontrivial vector $x \in \bar{\mathbb{R}}^n$ such that

$$A \otimes x = \lambda \otimes x.$$

A graphical interpretation of $A$ is helpful. Associate with $A$ a weighted directed graph $G_A$ with $n$ vertices, such that, for each $a_{ij} \neq \varepsilon$ there is an edge from the vertex $j$ to the vertex $i$. Each edge is assigned a weight given by the corresponding matrix element $a_{ij}$. Conventionally, these are known as *communication graphs*. The following theorem links the existence of a (unique) eigenvalue of a matrix to the structure of the corresponding communication graph.

**Theorem 1.** *If $G_A$ is strongly connected, then $A$ has a unique, nonzero, i.e., not equal to $\varepsilon$, eigenvalue equal to the maximal average weight of circuits in $G_A$.*

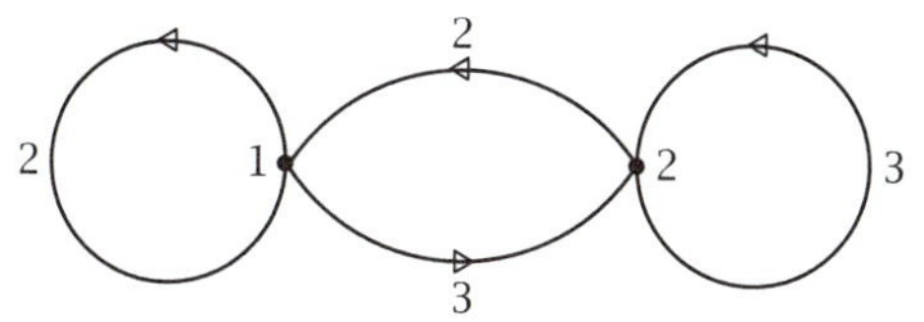


**Figure 2** The communication graph of the matrix $M(br)$.

A circuit is a path in $G_A$ that leads from a vertex back to itself. There are two obvious notions of the length of a circuit: the number of vertices or edges (generally known as the circuit length) and the sum of the weights on the edges (known as the circuit weight). The average weight of a circuit is found by dividing the latter by the former.

Figure 2 shows the communication graph for the matrix $M(br)$ in the heaps of pieces example. There are three circuits, two of which are loops containing one edge and one vertex each. The remaining circuit has two edges, and these edges connect both vertices of the graph. The maximum average weight of these circuits is 3, which is the eigenvalue found above by substituting an eigenvector into the eigenvalue equation.

## 8 Cyclicity and Convergence

An *elementary circuit* of the communication graph $G_A$ is a circuit that does not intersect itself. The cyclicity of a strongly connected graph is the greatest common divisor of the lengths of all its elementary circuits. If the graph consists of more than one strong component (i.e., more than one maximal strongly connected subgraph), its cyclicity is the least common multiple of the cyclicities of its strong components.

Cyclicity is a topological property as it depends only on the circuits and their lengths. The matrix $A$ contains more information than this since it associates a weight with each edge of $G_A$. This information can be captured by considering the *critical graph* associated with $G_A$. The critical circuits of $G_A$ are those elementary circuits with maximum mean weight (i.e., the total weight divided by the length). The critical graph associated with $G_A$ is the subgraph consisting of vertices and directed edges found in the critical circuits. The *cyclicity of the matrix* $A$, denoted $\sigma(A)$, is the cyclicity of the critical graph associated with $G_A$.

As an example consider $M(br)$ and its corresponding communication graph shown in figure 2. The graph has two kinds of elementary circuit: the self-loops (weights 2 and 3; length 1) and the cycle involving both vertices (weight 5; length 2). There is one critical circuit: the self-loop with weight 3. The critical graph therefore consists of the vertex labeled "2" in figure 2 together with the self-loop. The cyclicity of this, and therefore of $M(br)$, is unity.

Integer max-plus powers of an $n \times n$ matrix, $A$, can be defined recursively as in the scalar case. The following theorem is about the asymptotics of max-plus powers of a matrix and therefore provides information about the dynamics if a max-plus square matrix is applied repeatedly to a general vector.

**Theorem 2.** *Let $G_A$ be strongly connected and let the matrix $A$ have eigenvalue $\lambda$ and cyclicity $\sigma(A)$. Then there exists a positive integer $K$ such that*

$$A^{\otimes(k+\sigma(A))} = \lambda^{\otimes\sigma(A)} \otimes A^{\otimes k}$$

*for all $k \geqslant K$.*

The theorem guarantees that after sufficiently many iterations the system in question converges to a regular behavior dominated by the eigenvalue of the matrix. In the example and in terms of the upper contour of the heap, there is a $K$ such that

$$M((br)^{q+K}) \otimes x(0) = 3^{\otimes q} \otimes x(K)$$

for all positive integers $q$. The rate at which the maximum height of the heap grows asymptotically is then

$$\lim_{q\to\infty} \frac{\|M((br)^{q+K}) \otimes x(0)\|_{\max}}{2q} = \frac{3}{2},$$

where $\|x\|_{\max}$ is the maximum component of $x$. The same calculation using the minimum component shows that it also grows at a rate $\frac{3}{2}$. Note that theorem 2 implies that these results are independent of the initial condition $x(0)$. Together, theorems 1 and 2 provide the means to calculate these measures of efficiency for any periodic sequence of tasks.

## 9 Stochastic Models

If the heap model is of a system responding to external random influences, the iteration process given by equation (1) will involve randomly chosen sequences of matrices rather than the periodic sequences discussed. What can now be said of the asymptotic behavior of the heap? Rather than eigenvalues, the behavior of long random sequences of matrices is determined by the Lyapunov exponents of the system. Assume that the matrices in the sequence are independent and identically distributed and that they are regular with probability one (each row contains at least one element that is not $\varepsilon$). Then the Lyapunov exponent $\lambda_{\max}$ defined as

$$\lim_{q\to\infty} \frac{\|x(q)\|_{\max}}{q} = \lambda_{\max}$$

(and analogously for $\lambda_{\min}$) exists for almost all sequences and is independent of the initial (finite) choice of $x(0)$. This result shows that, in principle, it is possible to calculate the efficiency measures discussed above. The main difference is that, although the Lyapunov exponents exist, there is no nice prescription for their calculation. The development of efficient numerical algorithms is an open problem that the interested reader might, perhaps, wish to consider!

### Further Reading

Butkovič, P. 2010. *Max-linear Systems: Theory and Algorithms.* London: Springer.

Gaubert, S., and M. Plus. 1997. Methods and applications of (max,+) linear algebra. In *STACS '97: 14th Annual Symposium on Theoretical Aspects of Computer Science.* Lecture Notes in Computer Science, volume 1200, pp. 261–82. Berlin: Springer.

Heidergott, B., G. J. Olsder, and J. van der Woude. 2006. *Max Plus at Work: Modeling and Analysis of Synchronized Systems: A Course on Max-Plus Algebra and Its Applications.* Princeton, NJ: Princeton University Press.

Litvinov, G. L. 2007. Maslov dequantization, idempotent and tropical mathematics: a brief introduction, *Journal of Mathematical Sciences* 140:426–44.

Pachter, L., and B. Sturmfels. 2005. *Algebraic Statistics for Computational Biology.* Cambridge: Cambridge University Press.

## VII.5 Evolving Social Networks, Attitudes, and Beliefs—and Counterterrorism

*Peter Grindrod*

### 1 Introduction

The central objects of interest here are

(i) an evolving digital network of peer-to-peer communication,
(ii) the dynamics of information, ideas, and beliefs that can propagate through that digital network, and
(iii) how such networks become important in matters of national security and defense.

The applications of this theory spread far beyond these topics though. For example, more than a quarter of all marketing and advertising spending in the United Kingdom and the United States is now spent online: so digital media marketing ("buzz" marketing), though in its infancy, requires a deeper understanding of the nature of social communication networks. This is an important area of complexity theory, since there is no closed theory (analogous to conservation laws for molecular dynamics or chemical reactions) available at the microscopic "unit" level. Instead, here we must consider irrational, inconsistent, and ever-changing people. Moreover, while the passage of ideas is mediated by the networking behavior, the very existence of such ideas may cause communication to take place: systems can therefore be fully coupled.

In observing peer-to-peer communication in mobile phone networks, messaging, email, and online chats, the size of communities is a substantial challenge. Equally, from a conceptual modeling perspective, it is clear that being able to simulate, anticipate, and infer behavior in real time, or on short timescales, may be critical in designing interventions or spotting sudden aberrations. This field therefore requires and has inspired new ideas in both applied mathematical models and methods.

### 2 Evolving Networks in Continuous and Discrete Time

Consider a population of $N$ individuals (agents/actors) connected through a dynamically evolving undirected network representing pairwise voice calls or online chats. Let $A(t)$ denote the $N \times N$ binary adjacency matrix for this network at time $t$, having a zero diagonal. At future times, $A(t)$ is a stochastic object defined by a probability distribution over the set of all possible adjacency matrices. Each edge within this network will be assumed to evolve independently over time, though it is conditionally dependent upon the current network (so any edges conditional on related current substructures may well be highly correlated over time). Rather than model a full probability distribution for future network evolution, conditional on its current structure, say $\mathcal{P}_{\delta t}(A(t+\delta t) \mid A(t))$, it is enough to specify its expected value $E(A(t+\delta t) \mid A(t))$ (a matrix containing all edge probabilities, from which edges may be generated independently). Their equivalence is trivial, since

$$E(A(t+\delta t) \mid A(t)) = \sum_B B\mathcal{P}_{\delta t}(B \mid A(t)),$$

and

$$\mathcal{P}_{\delta t}(B \mid A(t)) = \prod_{i=1,j=i+1}^{N-1,N} W_{i,j}^{B_{i,j}} (1 - W_{i,j})^{1-B_{i,j}},$$

where $W = E(A(t + \delta t) \mid A(t))$. Hence we shall specify our model for the stochastic network evolution via

$$E(A(t + \delta t) \mid A(t)) = A(t) + \delta t \mathcal{F}(A(t)), \qquad (1)$$

valid as $\delta t \to 0$. Here the real matrix-valued function $\mathcal{F}$ is symmetric, it has a zero diagonal, and all elements within the right-hand side will be in $[0, 1]$. We write

$$\mathcal{F}(A(t)) = -A(t) \circ \omega(A(t)) + (C - A(t)) \circ \alpha(A(t)).$$

Here $C$ denotes the adjacency matrix for the *clique* where all $\frac{1}{2}N(N-1)$ edges are present (all elements are 1s except for 0s on the diagonal), so $C - A(t)$ denotes the adjacency matrix for the graph complement of $A(t)$; $\omega(A(t))$ and $\alpha(A(t))$ are both real nonnegative symmetric matrix functions containing conditional edge death rates and conditional edge birth rates, respectively; and $\circ$ denotes the Hadamard, or element-wise, matrix product.

In many cases we can usefully consider a discrete-time version of the above evolution. Let $\{A_k\}_{k=1}^{K}$ denote an ordered sequence of adjacency matrices (binary, symmetric with zero diagonals) representing a discrete-time evolving network with value $A_k$ at time step $t_k$. We shall then assume that edges evolve independently from time step to time step with each new network conditionally dependent on the previous one. A first-order model is given by a Markov process

$$E(A_{k+1} \mid A_k) = A_k \circ (C - \tilde{\omega}(A_k)) + (C - A_k) \circ \tilde{\alpha}(A_k). \quad (2)$$

Here $\tilde{\omega}(A_k)$ is a real nonnegative symmetric matrix function containing conditional death probabilities, each in $[0, 1]$, and $\tilde{\alpha}(A_k)$ is a real nonnegative symmetric matrix function containing conditional edge birth probabilities, each in $[0, 1]$.

As before, the edge independence assumption implies that $P(A_{k+1} \mid A_k)$ can be reconstructed from $E(A_{k+1} \mid A_k)$.

A generalization of KATZ [IV.18 §3.4] centrality for such discrete-time evolving networks can be obtained. In particular, if $0 < \mu < 1/\max\{\rho(A_k)\}$, then the communicability matrix

$$\mathcal{Q} = (I - \mu A_1)^{-1}(I - \mu A_2)^{-1} \cdots (I - \mu A_K)^{-1}$$

provides a weighted count of all possible dynamic paths between all pairs of vertices. It is nonsymmetric (due to time's arrow) and its row sums represent the abilities of the corresponding people to send messages to others, while its column sums represent the abilities of the corresponding people to receive messages from others. Such performance measures are useful in identifying influential people within evolving networks. This idea has recently been extended so as to successively discount the older networks in order to produce better inferences.

## 3 Nonlinear Effects: Seen and Unseen

In the sociology literature the simplest form of nonlinearity occurs when people introduce their friends to each other. So, in (2), if two nonadjacent people are connected to a common friend at step $k$, then it is more likely that those two people will be directly connected at step $k+1$. To model this *triad closure* dynamic we may use $\tilde{\omega}(A_k) = \gamma C$, so all edges have the same step-to-step death probability, $\gamma \in [0, 1]$, and

$$\tilde{\alpha}(A_k) = \delta C + \varepsilon A_k^2.$$

Here $\delta$ and $\varepsilon$ are positive and such that $\delta + \varepsilon(N-2) < 1$. The element $(A^2)_{i,j}$ counts the number of mutual connections that person $i$ and person $j$ have at step $k$. This equation is ergodic and yet it is destined to spend most of its time close to states where the density of edges means that there is a balance between edge births and deaths. A mean-field approach can be applied, approximating $A_k$ with its expectation, which may be assumed to be of the form $p_k C$ (an ERDŐS–RÉNYI RANDOM GRAPH [IV.18 §4.1] with edge density $p_k$). In the mean-field dynamic one obtains

$$p_{k+1} = p_k(1 - \gamma) + (1 - p_k)(\delta + (N - 2)\varepsilon p_k^2). \qquad (3)$$

If $\delta$ is small and $\omega < \frac{1}{4}\varepsilon(N - 2)$, then this nonlinear iteration has three fixed points: two stable ones, at $\delta/\gamma + O(\delta^2)$ and $\frac{1}{2} + (\frac{1}{4} - \gamma/(\varepsilon(N - 2)))^{1/2} + O(\delta)$, and one unstable one in the middle. Thus the extracted mean-field behavior is bistable. In practice, one might observe the edge density of such a network approaching one or other stable mean-field equilibrium and jiggling around it for a very long time, without any awareness that another type of orbit or pseudostable edge density could exist. Direct comparisons of transient orbits from (2), incorporating triad closure, with their mean-field approximations in (3) are very good over short to medium timescales. Yet though we have captured the nonlinear effects well in (3), the stochastic nature of (2) must *eventually* cause orbits to diverge from the deterministic stability seen in (3).

The phenomenon seen here explains the events of a new undergraduate's first week at university are so important in forming high-density connected social networks among student year groups. If we do not perturb them with a mix of opportunities to meet, they may be condemned to remain close to the low-density (few-friends) state for a very long time.

## 4 Fully Coupled Systems

There is a large literature within psychology that is based on individuals' attitudes and behaviors being in a tensioned equilibrium between excitory (activating) processes and inhibiting processes. Typically, the state of an individual is represented by a set of state variables, some measuring activating elements and some measuring the inhibiting elements.

Activator–inhibitor systems have had an impact within mathematical models where a uniformity equilibrium across a population of individual systems becomes destabilized by the very act of simple "passive" coupling between them. Such Turing instabilities can sometimes seem counterintuitive.

Homophily is a term that describes how associations are more likely to occur between people who have similar attitudes and views. Here we show how individuals' activator–inhibitor dynamics coupled through a homophilic evolving network produce systems that have pseudoperiodic consensus and fractionation.

Consider a population of $N$ identical individuals, each described by a set of $m$ state variables that are continuous functions of time $t$. Let $x_i(t) \in \mathbb{R}^m$ denote the $i$th individual's attitudinal state. Let $A(t)$ denote the adjacency matrix for the communication network, as it does in (1). Then consider

$$\dot{x}_i = f(x_i) + D \sum_{j=1}^{N} A_{ij}(x_j - x_i), \quad i = 1, \dots, N. \quad (4)$$

Here $f$ is a given smooth field over $\mathbb{R}^m$, drawn from a class of activator–inhibitor systems, and is such that $f(x^*) = 0$ for some $x^*$, and the Jacobian there, $\mathrm{d}f(x^*)$, is a stability matrix (that is, all its eigenvalues have negative real parts). $D$ is a real diagonal nonnegative matrix containing the maximal transmission coefficients (diffusion rates) for the corresponding attitudinal variables between adjacent neighbors. Let $X(t)$ denote the $m \times N$ matrix with $i$th column given by $x_i(t)$, and let $F(X)$ be the $m \times N$ matrix with $i$th column given by $f(x_i(t))$. Then (4) may be written as

$$\dot{X} = F(X) - DX\Delta. \quad (5)$$

Here $\Delta(t)$ denotes the graph Laplacian for $A(t)$, given by $\Delta(t) = \Gamma(t) - A(t)$, where $\Gamma(t)$ is the diagonal matrix containing the degrees of the vertices. This system has an equilibrium at $X = X^*$, say, where the $i$th column of $X^*$ is given by $x^*$ for all $i = 1, \dots, N$.

Now consider an evolution equation for $A(t)$, in the form of (1), coupled to the states $X$:

$$\begin{aligned} E(A(t + \delta t) \mid A(t)) &= A(t) + \delta t(-A(t) \circ (C - \Phi(X(t)))\gamma \\ &\quad + (C - A(t)) \circ \Phi(X(t))\delta). \quad (6) \end{aligned}$$

Here $\delta$ and $\gamma$ are positive constants representing the maximum birth rate and the maximum death rate, respectively; and the homophily effects are governed by the *pairwise similarity* matrix, $\Phi(X(t))$, such that each term $\Phi(X(t))_{i,j} \in [0, 1]$ is a monotonically decreasing function of a suitable seminorm $\|x_j(t) - x_i(t)\|$. We shall assume that $\Phi(X(t))_{i,j} \sim 1$ for $\|x_j(t) - x_i(t)\| < \varepsilon$, and $\Phi(X(t))_{i,j} = 0$ otherwise, for some suitably chosen $\varepsilon > 0$.

There are equilibria at $X = X^*$ with either $A = 0$ or $A = C$ (the full clique). To understand their stability, let us assume that $\delta$ and $\gamma \to 0$. Then $A(t)$ evolves very slowly via (6). Let $0 = \lambda_1 \leqslant \lambda_2 \leqslant \cdots \leqslant \lambda_N$ be the eigenvalues of $\Delta$. Then it can be shown that $X^*$ is asymptotically stable only if all $N$ matrices, $\mathrm{d}f(x^*) - D\lambda_i$, are simultaneously stability matrices; and conversely, it is unstable in the $i$th mode of $\Delta$ if $\mathrm{d}f(x^*) - D\lambda_i$ has an eigenvalue with positive real part.

Now one can see the possible tension between homophily and the attitude dynamics.

Consider the spectrum of $\mathrm{d}f(x^*) - D\lambda$ as a function of $\lambda$. If $\lambda$ is small then this is dominated by the stability of the uncoupled system, $\mathrm{d}f(x^*)$. If $\lambda$ is large, then this is again a stability matrix, since $D$ is positive-definite. The situation, dependent on some collusion between choices of $D$ and $\mathrm{d}f(x^*)$, where there is a *window of instability* for an intermediate range of $\lambda$, is known as a Turing instability. Note that, as $A(t) \to C$, we have $\lambda_i \to N$, for $i > 1$. So if $N$ lies within the window of instability, we are assured that the systems can never reach a stable consensual fully connected equilibrium. Instead, Turing instabilities can drive the breakup (weakening) of the network into relatively well-connected subnetworks. These in turn may restabilize the equilibrium dynamics (as the eigenvalues leave the window of instability), and then the whole process can begin again as homophily causes any absent edges to reappear. Thus we expect a pseudocyclic emergence and diminution of patterns, representing transient variations in attitudes. In simulations, by projecting the network $A(t)$ onto two dimensions using the Frobenius matrix inner product, one may observe directly the cyclic nature of consensus and division.

Even if the stochastic dynamics in (6) are replaced by deterministic dynamics for a weighted communication adjacency matrix, one obtains a system that exhibits aperiodic, wandering, and also sensitive dependence. In such cases the orbits are chaotic: we know that they will oscillate, but we cannot predict whether any specific individuals will become relatively inhibited or relatively activated within future cycles. This phenomenon even occurs when $N = 2$.

These models show that, when individuals, who are each in a dynamic equilibrium between their activational and inhibitory tendencies, are coupled in a homophilic way, we should expect a relative lack of global social convergence to be the norm. Radical and conservative behaviors can coexist across a population and are in a constant state of flux. While the macroscopic situation is predictable, the journeys for individuals are not, within both deterministic and stochastic versions of the model. There are some commentators in socioeconomic fields who assert that divergent attitudes, beliefs, and social norms require leaders and are imposed on populations; or else they are driven by partial experiences and events. But here we can see that the transient existence of locally clustered subgroups, holding diverse views, can be an emergent behavior within fully coupled systems. This can be the normal state of affairs within societies, even without externalities and forcing terms.

Sociology studies have in the past focused on rather small groups of subjects under experimental conditions. Digital platforms and modern applied mathematics will transform this situation: computation and social science can use vast data sets from very large numbers of users of online platforms (Twitter, Facebook, blogs, group discussions, multiplayer online games) to analyze how norms, opinions, emotions, and collective action emerge and spread through local interactions.

## 5 Networks on Security and Defense

"It takes a network to defeat a network" is the mantra expressed by the most senior U.S. command in Afghanistan and Iraq. This might equally be said of the threats posed by terrorists, or in post-conflict peacekeeping (theaters of asymmetric warfare), and even by the recent summer riots and looting within U.K. cities. But what type of networks must be defeated, and what type of network thinking will be required?

So far we have discussed peer-to-peer networks in general terms. But we are faced with some specific challenges that stress the importance of social and communications networks in enabling terrorist threats:

- the analysis of very large communications networks, in real time;
- the identification of influential individuals;
- inferring how such networks *should* evolve in the future (and thus spotting aberrant behavior); and
- recognizing that fully coupled systems may naturally lead to diverse views, and pattern formation.

All of these things become ever more essential. Population-wide data from digital platforms requires efficient and effective applicable mathematics.

Modern adversaries may be most likely to be

- organized through an *actor network* of transient affiliations appropriate for (i) time-limited opportunities and *trophy* or *inspired* goals; (ii) procurement, intelligence, reconnaissance and planning; and (iii) empowering individuals and encouraging both innovation and replication through competition;
- employing an operational digital *communication network* that enables and empowers action while maximizing agility (self-adaptation and reducing the time to act) through the flow of information, ideas, and innovations; and
- reliant upon a third-party *dissemination network* within the public and media space (social media, broadcast media, and so forth) so as to maximize the impact of their actions.

There are thus at least three independent networks operating on the side of those who would threaten security.

Here we have set out a framework for analyzing the form and dynamics of large evolving peer-to-peer communications networks. It seems likely that the challenge of modeling their behavior may lead us to develop new models and methods in the future.

### Further Reading

Bohannon, J. 2009. Counterterrorism's new tool: "metanetwork" analysis. *Science* July 24:409–11.

Grindrod, P. 1995. *Patterns and Waves*, 2nd edn. Oxford: Oxford University Press.

Grindrod, P., and D. J. Higham. 2013. A matrix iteration for dynamic network summaries. *SIAM Review* 55(1):118–28.

Grindrod, P., D. J. Higham, and M. C. Parsons. 2012. Bistability through triadic closure. *Internet Mathematics* 8(4): 402–23.

Grindrod, P., D. J. Higham, M. C. Parsons, and E. Estrada. 2011. Communicability across evolving networks. *Physical Review* E 83:046120.

Grindrod, P., and M. C. Parsons. 2012. Complex dynamics in a model of social norms. Technical Report, University of Reading.

McCrystal, S. A. 2011. It takes a network: the new front line of modern warfare. *Foreign Policy* (March/April).

Ward, J. A., and P. Grindrod. 2012. Complex dynamics in a deterministic model of cultural dissemination. Technical Report, University of Reading.

# VII.6 Chip Design

*Stephan Held, Stefan Hougardy, and Jens Vygen*

## 1 Introduction

An *integrated circuit* or *chip* contains a collection of electronic circuits—composed of *transistors*—that are connected by wires to fulfill some desired functionality. The first integrated circuit was built in 1958 by Jack Kilby. It contained a single transistor. As predicted by Gordon Moore in 1965, the number of transistors per chip doubles roughly every two years. The process of creating chips soon became known as *very-large-scale integration* (VLSI). In 2014 the most complex chips contain billions of transistors on a few square centimeters.

In this article we concentrate on the design of digital logic chips. Analog integrated circuits have many fewer transistors and more complex design rules and are therefore still largely designed manually. In a memory chip, the transistors are packed in a very regular structure, which makes their design rather easy. In contrast, the design of VLSI digital logic chips is impossible without advanced mathematics.

New technological challenges, exponentially increasing transistor counts, and shifting objectives like decreased power consumption or increased yield constantly create new and challenging mathematical problems. This has made chip design one of the most interesting application areas for mathematics during the last forty years, and we expect this to continue to be the case for at least the next two decades, although technology scaling might slow down at some point.

### 1.1 Hierarchical Chip Design

Due to its enormous complexity, the design of VLSI chips is usually done hierarchically. A hierarchical design makes it possible to distribute the design task to different teams. Moreover, it can reduce the overall effort, and it makes the design process more predictable and more manageable.

For hierarchical design, a chip is subdivided into logical units, each of which may be subdivided into several levels of smaller units. An obvious advantage of hierarchical design is that components that are used multiple times need to be designed only once. In particular, almost all chips are designed based on a *library* of so-called *books*, predesigned integrated circuits that realize simple logical functions such as AND or NOT or a simple memory element. A chip often contains many instances of the same book; these instances are often called *circuits*.

The books are composed of relatively few transistors and are predesigned at an early stage. For their design one needs to work at the *transistor level* and hence follow more complicated rules. Once a book (or any hierarchical unit) is designed, the properties it has that are relevant for the design of the next higher level (e.g., minimum-distance constraints, timing behavior, power consumption) are computed and stored. Most books are designed so that they have a rectangular shape and the same height, making it easier to place them in rows or columns.

### 1.2 The Chip-Design Process

The first step in chip design is the specification of the desired functionality and the technology that will be used. In *logic design*, this functionality is made precise using some hardware description language. This hardware description is converted into a *netlist* that specifies which circuits have to be used and how they have to be connected to achieve the required functionality.

The *physical-design* step takes this netlist as input and outputs the physical location of each circuit and each wire on the chip. It will also change the netlist (in a logically equivalent way) in order to meet timing constraints.

Before fabricating the chip (or fixing a hierarchical unit for later use on the next level up), one conducts *physical verification* to confirm that the physical layout meets all constraints and implements the desired functionality, and *timing analysis* checks that all signals arrive in time. Further testing will be done with the hardware once a chip is manufactured.

From a mathematical point of view, physical design is the most interesting part of chip design as it requires the solution of several different challenging mathematical problems. We will therefore describe this in more detail below.

### 1.3 Physical Design

Two inputs of the physical-design stage are a netlist and a *chip area*. The *netlist* contains a set of *circuits*. Each circuit is an instance of a book and has some *pins* that must be connected to some other pins. Moreover, the netlist includes pins on the chip area that are called input/output-ports and connect the chip to the outside. The set of all pins is partitioned into *nets*. All pins that belong to the same net have to be connected to each other by wires.

The task of the physical-design step is to assign a location to each circuit on the chip area (*placement*) and to specify locations for all the wires that are needed to realize the netlist (*routing*). Placement and routing are also called *layout* (see figure 1).

A layout has to satisfy many constraints. For example, *design rules* specify the minimum width of a wire, the minimum distance between two different wires, or legal positions for the circuits.

Moreover, a chip works correctly (at the desired clock frequency) only if all signals arrive in time (neither too early nor too late). This is described by *timing constraints*. It is usually impossible to meet all timing constraints without changing the netlist. This is called *timing optimization*. Of course, any changes must ensure that the netlist remains logically equivalent.

Due to the complexity of the physical-design problem, placement, routing, and timing optimization are treated largely as independent subproblems, but they are of course not independent. Placement must ensure that a feasible routing can be found and that timing constraints can be met. Changes in timing optimization must be reflected by placement and routing. Finally, routing must also consider timing constraints.

We describe some of the mathematical aspects of these three subproblems in what follows.

(a)

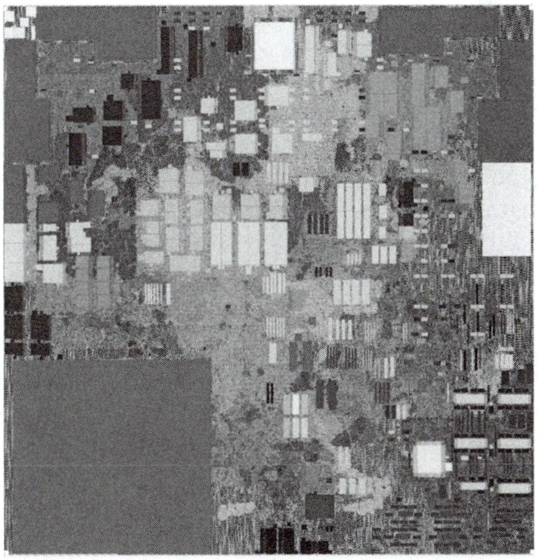

(b)

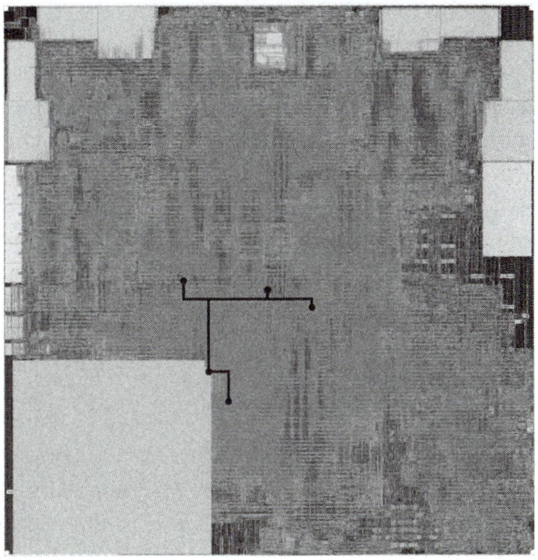

**Figure 1** (a) Placement and (b) routing of a chip with 4 496 492 circuits and 5 091 819 nets, with 762 meters of wires. Large rectangles are predesigned units, e.g., memory arrays or microprocessors. A Steiner tree connecting the five pins of one net is highlighted in (b).

## 2 Placement

All circuits must be placed within the chip area so that no two circuits overlap. This is called a feasible placement. In addition we want to minimize a given objective function. Normally, the chip area and all circuits have a rectangular shape. Thus, finding a feasible placement is equivalent to placing a set of small rectangles disjointly within some larger rectangle. This is known as the *rectangle packing problem*.

No efficient algorithm is known that is guaranteed to solve the rectangle packing problem for all possible instances. However, finding an arbitrary feasible placement is usually easy in practice.

### 2.1 Netlength Minimization

The location of the circuits on the chip area is primarily responsible for the total wire length that is needed for wiring all the nets in the netlist. If the total wire length is too long, a chip will not be routable. Moreover, the length of the wires greatly impacts the signal delays and the power consumption of a chip. Thus, a reasonable objective function of the placement problem is to minimize the total wire length.

As this cannot be computed efficiently, estimates are used. Most notably, the *bounding box length* of a net is obtained by taking half the perimeter of a smallest axis-parallel rectangle that contains all its pins. A commonly used *quadratic netlength* estimate is obtained by summing up the squared Euclidean distances between each pair of pins in the net and dividing this value by one less than the number of pins.

No efficient algorithms are known for finding a placement that minimizes the total netlength with respect to any such estimate, even if we ask only for a solution that is worse than an optimum solution by an arbitrarily large constant factor. Under additional assumptions, such as that all circuits must have exactly the same size, one can find a placement in polynomial time whose total netlength is $O(\log n)$ worse than an optimum placement, where $n$ denotes the number of circuits.

### 2.2 Placement in Practice

As mentioned above, finding an arbitrary feasible placement is usually easy. Moreover, one can define local changes to a feasible placement that results in another feasible placement. General local search-based heuristics (such as *simulated annealing*) can therefore be applied. However, such methods are prohibitively slow for today's instances, with several million circuits.

Another paradigm, motivated by some theoretical work, is called *min-cut*. Here, the netlist is partitioned into two parts, each with roughly half of the circuits, such that as few nets as possible cross the cut. The two parts will be placed on the left and right parts of the chip area and then partitioned further recursively. Unfortunately, the bipartitioning problem cannot be solved easily, and the overall paradigm lacks stability properties and the results are inferior.

A third paradigm, *analytical placement*, is the one that is predominantly used in practice today. It begins by ignoring the constraint that circuits must not overlap; minimizing netlength (bounding box or quadratic) is then relatively easy. For several reasons (it is faster to solve; it is more stable; it gives better spreading), quadratic netlength is minimized in practice. This is equivalent to solving a system of linear equations with a sparse positive-definite matrix.

The placement that minimizes quadratic netlength typically has many overlapping circuits. Two strategies for working toward a feasible placement exist: either the objective function is modified in order to pull circuits away from overloaded regions or a geometric partitioning is done. For geometric partitioning one can assign the circuits efficiently to four quadrants (or more than four regions) such that no region contains more circuits than fit into it and such that the total (linear or quadratic) movement is minimized. The assignment to the regions can then be translated into a modified quadratic optimization problem.

Both strategies (as well as min-cut placement) are iterated until the placement is close to legal. This roughly means until there exists a legal placement in which all circuits are placed nearby. This ends the *global-placement* phase.

After global placement (whether analytic or min-cut), the solution must be *legalized*. Here, given an illegal placement as input, we ask for a legal placement that differs from the input as little as possible. The common measure is the sum of the squared distances. Unfortunately, only special cases of this problem can be solved optimally in polynomial time, even when all circuits have the same height and are to be arranged in rows.

## 3 Routing

In routing, we must connect the set of each net's pins by wires. The positions of the pins are determined by the placement. Wires can run on different wiring planes (sometimes more than ten), which are separated by insulating material. Wires of adjacent planes can be connected by so-called *vias*. In almost all current technologies, all wire segments run horizontally or vertically. For efficient packing, every plane is used predominantly in one direction; horizontal and vertical planes alternate.

Wires can have different widths, complicated spacing requirements, and other rules to obey. Although important, such rules do not change the overall nature of the problem.

Before all nets are routed, some areas are already used by power supply or clock grids. These too must be designed, but this task is still largely a manual one.

### 3.1 Steiner Trees

The minimal connections for a net can be modeled as Steiner trees. A *Steiner tree* for a given set of terminals (pins) is a minimal connected graph containing these terminals and possibly other vertices (see figure 1). If wiring is restricted to predefined *routing tracks*, the space available for routing a single net can be modeled as an undirected graph. Finding a shortest Steiner tree for a given set of terminals in a graph is NP-hard, and the same holds even for shortest rectilinear Steiner trees in the plane. Moreover, shortest is not always best when it comes to meeting timing constraints, and the routing graph is huge (it can have more than $10^{11}$ vertices). Therefore, routing algorithms mostly use fast variants of Dijkstra's algorithm in order to find a shortest path between two components and then compose the Steiner trees of such paths. If done carefully, this is at most a factor $2(t-1)/t$ worse than optimal, where $t$ is the number of terminals (pins in the net).

### 3.2 Packing Steiner Trees

Since finding just one shortest Steiner tree is hard, it is not surprising that finding vertex-disjoint Steiner trees in a given graph is even harder. In fact, it is NP-hard even if every net has only two pins and the graph is a planar grid. Nevertheless, it would be possible to solve such problems if the instances were not too large.

Current detailed routing algorithms route the nets essentially sequentially, revising earlier decisions as necessary (*rip-up and reroute*). To speed up the sequential routing approach and to improve the quality of results, a *global routing* step is performed at the beginning. Here, the routing space is modeled by a coarser graph, whose vertices normally correspond to rectangular areas (induced by a grid) on a certain plane. Two vertices are connected if they correspond to the same area on adjacent planes or to horizontally or vertically (depending on the routing direction of the plane) adjacent areas on the same plane. Edges have capacities, depending on how many wires we can pack between the corresponding areas.

Global routing then asks us to find a Steiner tree for each net such that the number (more generally, the total width) of Steiner trees using an edge does not exceed its capacity. This problem is still NP-hard, and the global routing graphs can still be large (they often have more than $10^7$ vertices). Nevertheless, global routing can be solved quite well in both theory and practice; the best approach with a theoretical guarantee is based on first approximately solving a fractional relaxation (called min-max resource sharing, a generalization of multicommodity flows), then applying randomized rounding to obtain an integral solution, and finally correcting local violations (induced by rounding).

Global routing is also done at earlier stages of the design flow, e.g., during placement, in order to estimate routability and exhibit areas with possible routing congestion.

## 4 Timing Optimization

A chip performs its computations in cycles. In each cycle electrical signals start from registers or chip inputs, traverse some circuits and nets, and finally enter registers or chip outputs.

Timing optimization has to ensure that all signals arrive within a given cycle time. Under this constraint, the power consumption shall be minimized. However, achieving the cycle time is a difficult problem on its own.

### 4.1 Logic Synthesis

The structure of a Boolean circuit has a big impact on the performance and power consumption of a chip. On the one hand, the *depth*, i.e., the maximum number of logic circuits on a combinatorial path, should be small so that the cycle time is met. On the other hand, the total number of circuits to realize a function should not be too big.

Almost all Boolean functions have a minimum representation size that is exponential in the number of input variables. Hence, functions that are realized in hardware are quite special.

Some very special functions—such as adders, certain symmetric functions, or paths consisting alternately of AND and OR circuits—can be implemented optimally or near-optimally by divide-and-conquer or dynamic-programming algorithms, but general logic synthesis is done by (mostly local) heuristics today.

### 4.2 Repeater Trees

Another central task is to distribute a signal from a source to a set of sinks. As the delay along a wire grows almost quadratically with its length, repeaters, i.e., circuits implementing the identity function or inversion, have to be inserted to strengthen the signal and linearize the growth.

For a given Steiner tree, repeaters can be inserted arbitrarily close to optimally in polynomial time using dynamic programming.

A more difficult problem asks for the structure of the Steiner tree (into which repeaters can be inserted). A minimum-length Steiner tree can have very long source–sink paths. In addition, every bifurcation from a path adds capacitance and delay. Trees should therefore not only be short; they should also consist of short paths with few bifurcations.

Combining approximation algorithms for minimum Steiner trees with Huffman coding, bicriteria algorithms can be derived, trading off total length and path delays.

### 4.3 Circuit Sizing

In *circuit sizing*, the channel widths of the underlying transistors are optimized. A wider channel charges the capacitance of the output net faster but increases the input capacitances and, thus, the delays of the predecessors. Assuming continuously scalable circuits and a simplified delay model, the problem of finding optimum sizes for all circuits can be transformed into a geometric program. This can be solved by interior-point methods or by the subgradient method and Lagrangian relaxation.

However, rounding such a continuous solution to discrete circuit sizes can corrupt the result. Theoretical models for discrete timing optimization, such as the *discrete time-cost trade-off problem*, are not yet well understood. Local search is therefore used extensively for post-optimization.

### 4.4 Clock-Tree Construction

One of the few problems that can be solved efficiently in theory and in practice is *clock skew scheduling*. Each register triggers its stored bit once per cycle. The times at which the signals are released can be optimized such that the cycle time is minimized. To this end a register graph is constructed, with each register represented by a vertex. There is an arc if there is a signal path between the corresponding registers. An arc is weighted by the maximum delay of a path between the two registers. The minimum possible cycle time is now given by the maximum mean arc weight of a cycle in the graph. This reduces to the well-studied *minimum mean cycle problem*.

The challenging problem is then to distribute a clock signal such that the optimal trigger times are met. Here, *facility location* algorithms for bottom-up tree construction are combined with dynamic programming for repeater insertion.

### Further Reading

Alpert, C. J., D. P. Mehta, and S. S. Sapatnekar, eds. 2009. *Handbook of Algorithms for Physical Design Automation*. Boca Raton, FL: Taylor & Francis.

Held, S., B. Korte, D. Rautenbach, and J. Vygen. 2011. Combinatorial optimization in VLSI design. In *Combinatorial Optimization: Methods and Applications*, edited by V. Chvátal, pp. 33–96. Amsterdam: IOS Press.

International Technology Roadmap for Semiconductors. 2013. *SEMATECH, Austin, TX, 2013*. Available (and annually updated) at www.itrs.net.

## VII.7 Color Spaces and Digital Imaging

*Nicholas J. Higham*

### 1 Vector Space Model of Color

The human retina contains photoreceptors called cones and rods that act as sensors for the human imaging system. The cones come in three types, with responses that peak at wavelengths corresponding to red, green, and blue light, respectively (see figure 1). Rods are of a single type and produce only monochromatic vision; they are used mainly for night vision. Because there are three types of cones, color theory is replete with terms having the prefix "tri." In particular, *trichromacy*, developed by Young, Grassmann, Maxwell, and Helmholtz, is the theory that shows how to match any color with an appropriate mixture of just three suitably chosen primary colors.

We can model the responses of the three types of cones to light by the integrals

$$c_i(f) = \int_{\lambda_{\min}}^{\lambda_{\max}} s_i(\lambda) f(\lambda)\,\mathrm{d}\lambda, \quad i = 1:3, \tag{1}$$

where $f$ describes the spectral distribution of the light hitting the retina, $s_i$ describes the sensitivity of the $i$th cone to different wavelengths, and $[\lambda_{\min}, \lambda_{\max}] \approx [400\ \text{nm}, 700\ \text{nm}]$ is the interval of wavelengths of the visible spectrum. Note that this model is linear ($c_i(f+g) = c_i(f) + c_i(g)$) and it projects the spectrum onto the space spanned by the $s_i(\lambda)$—the "human visual subspace."

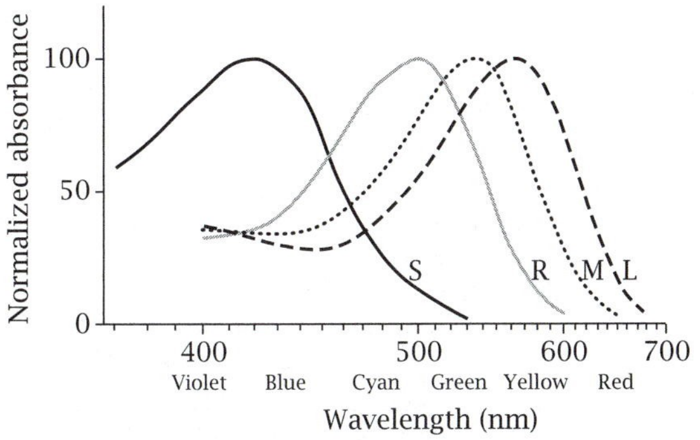


**Figure 1** Response curves for the cones and the rods (solid gray line). S, M, and L denote the cones most sensitive to short (blue, solid black line), medium (green, dotted black line), and long (red, dashed black line) wavelengths, respectively. (After Bowmaker, J. K., and H. J. A. Dartnall, 1980, Visual pigments of rods and cones in a human retina, *Journal of Physiology* 298:501–11.)

For computational purposes a grid of $n$ equally spaced points $\lambda_i$ on the interval $[\lambda_{\min}, \lambda_{\max}]$ is introduced, and the repeated rectangle (or midpoint) quadrature rule is applied to (1), yielding

$$c = S^{\mathrm{T}} f, \quad c \in \mathbb{R}^3,\ S \in \mathbb{R}^{n\times 3},\ f \in \mathbb{R}^n,$$

where the $i$th column of the matrix $S$ has samples of $s_i$ at the grid points, the vector $f$ contains the values of $f(\lambda)$ at the grid points, and the vector $c$ absorbs constants from the numerical integration. In practice, a value of $n$ around 40 is typically used.

Let the columns of $P = [p_1\ p_2\ p_3] \in \mathbb{R}^{n\times 3}$ represent *color primaries*, defined by the property that the $3 \times 3$ matrix $S^{\mathrm{T}}P$ is nonsingular. For example, $p_1$, $p_2$, and $p_3$ could represent red, blue, and green, respectively. We can write

$$S^{\mathrm{T}} f = S^{\mathrm{T}} P (S^{\mathrm{T}} P)^{-1} S^{\mathrm{T}} f \equiv S^{\mathrm{T}} P a(f), \qquad (2)$$

where $a(f) = (S^{\mathrm{T}}P)^{-1} S^{\mathrm{T}} f \in \mathbb{R}^3$. This equation shows that the color of any spectrum $f$ (or more precisely the response of the cones to that spectrum) can be matched by a linear combination, $Pa(f)$, of the primaries. A complication is that we need all the components of $a$ to be nonnegative for this argument, as negative intensities of primaries cannot be produced. A way around this problem is to write $a(f) = a_1 - a_2$, where $a_1$ contains the nonnegative components of $a(f)$ and $a_2$ has positive components, and rewrite (2) as

$$S^{\mathrm{T}}(f + Pa_2) = S^{\mathrm{T}} P a_1.$$

This equation says that $Pa_1$ matches $f$ with appropriate amounts of some of the primaries added. This rearrangement is a standard trick in *colorimetry*, which is the science of color measurement and description.

To summarize, the color of a visible spectrum $f$ can be matched by *tristimulus values* $a(f) = A^{\mathrm{T}} f$, where $A^{\mathrm{T}} = (S^{\mathrm{T}}P)^{-1} S^{\mathrm{T}}$, because $S^{\mathrm{T}} f = S^{\mathrm{T}} P a(f)$. The columns of $A \in \mathbb{R}^{3\times n}$ are called (samples of) color-matching functions for the given primaries.

To determine $A$, a human observer is asked to match light of single wavelengths $\lambda_i$ by twiddling knobs to mix light sources constituting the three primaries until a match is obtained. Light of a single wavelength corresponds to a vector $f = e_i$, where $e_i$ has a 1 in the $i$th position and zeros everywhere else, and the vector $a(f) = A^{\mathrm{T}} f$ that gives the match is therefore the $i$th column of $A^{\mathrm{T}}$. In this way we can determine the color-matching matrix $A$ corresponding to the given primaries.

This vector space model of color is powerful. For example, since the $3 \times n$ matrix $S^{\mathrm{T}}$ has a nontrivial null space, it tells us that there exist spectra $f$ and $g$ with $f \neq g$ such that $S^{\mathrm{T}} f = S^{\mathrm{T}} g$. Hence two colors can look the same to a human observer but have a different spectral decomposition, which is the phenomenon of *metamerism*. This is a good thing in the sense that color output systems (such as computer monitors) exploit metamerism to reproduce color. There is another form of metamerism that is not so welcome: when two colors appear to match under one light source but do not match under a different light source. An example of this is when you put on socks in the bedroom with the room lights on and they appear black, but when you view them in daylight one sock turns out to be blue.

The use of linear algebra in understanding color was taken further by Jozef Cohen (1921–95), whose work is summarized in the posthumous book *Visual Color and Color Mixture: The Fundamental Color Space* (2001). Cohen stresses the importance of what he calls the "matrix $R$," defined by

$$R = S(S^{\mathrm{T}}S)^{-1}S^{\mathrm{T}} = SS^{+},$$

where $S^{+}$ denotes the MOORE–PENROSE PSEUDOINVERSE [IV.10 §7.3] of $S$. Mathematically, $R$ is the orthogonal projector onto range($S$). Cohen noted that $R$ is independent of the choice of primaries used for color matching, that is, $R$ is unchanged under transformations $S \leftarrow SZ$ for nonsingular $Z \in \mathbb{R}^{3\times 3}$ and so is an invariant. He also showed how in the factorization $S = QL$, where $Q \in \mathbb{R}^{n\times 3}$ has orthonormal columns and $L \in \mathbb{R}^{3\times 3}$ is lower triangular, the factor $Q$ (which he called $F$) plays an important role in color theory through the use of "tricolor coordinates" $Q^{\mathrm{T}}f$.

We do not all see color in the same way: about 8% of males and 0.5% of females are affected by color blindness. The first investigation into color vision deficiencies was by Manchester chemist John Dalton (1766–1844), who described his own color blindness in a lecture to the Manchester Literary and Philosophical Society. He thought that his vitreous humor was tinted blue and instructed that his eyes be dissected after his death. No blue coloring was found but his eyes were preserved. A deoxyribonucleic acid (DNA) analysis in 1985 concluded that Dalton was a deuteranope, meaning that he lacked cones sensitive to the medium wavelengths (green). The color model and analogues of figure 1 for different cone deficiencies help us to understand color blindness.

Given the emphasis in this section on trichromacy, one might wonder why printing is usually done with a four-color CMYK model when three colors should be enough. CMYK stands for cyan-magenta-yellow-black, and cyan, magenta, and yellow are complementary colors to red, green, and blue, respectively. Trichromatic theory says that a CMY system is entirely adequate for color matching, so the K component is redundant. The reason for using K is pragmatic. Producing black in a printing process by overlaying C, M, and Y color plates uses a lot of ink, makes the paper very wet, and does not produce a true, deep black due to imperfections in the inks. In CMYK printing, *gray component replacement* is used to replace proportions of the CMY components that produce gray with corresponding amounts of K. (A naive algorithm to convert from CMY to CMYK is $K = \min(C, M, Y)$, $C \leftarrow C - K$, $M \leftarrow M - K$, $Y \leftarrow Y - K$, though in practice slightly different amounts of C, M, and Y are required to produce black.)

## 2 Standardization

The Commission Internationale de l'Éclairage (CIE) is responsible for standardization of color metrics and terminology. Figure 2 shows the standard RGB color-matching functions produced by the CIE in 1931 and 1964. They are based on color-matching experiments and correspond to primaries at 700 nm (red), 546.1 nm (green), and 435.8 nm (blue). The red curve takes negative values as shown in the figure, but nonnegative functions were preferred for calculations in the precomputer era as they avoided the need for subtractions. So

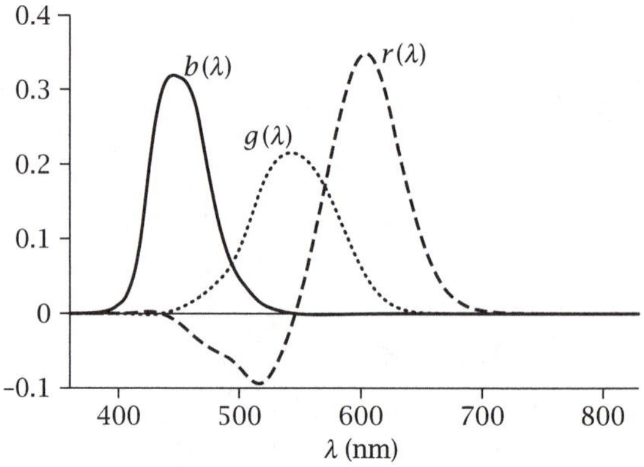


**Figure 2** CIE RGB color-matching functions from the 1931 standard. (File adapted from an original on Wikimedia Commons.)

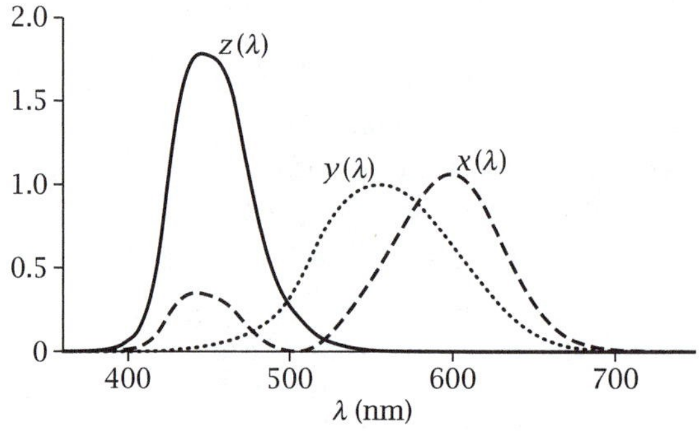


**Figure 3** CIE XYZ color-matching functions from the 1931 standard. (File adapted from an original on Wikimedia Commons.)

a CIE XYZ space was defined that has nonnegative color-matching functions (see figure 3) and that is obtained via the linear mapping[1]

$$\begin{bmatrix} X \\ Y \\ Z \end{bmatrix} = \begin{bmatrix} 0.49 & 0.31 & 0.20 \\ 0.17697 & 0.81240 & 0.01063 \\ 0 & 0.01 & 0.99 \end{bmatrix} \begin{bmatrix} R \\ G \\ B \end{bmatrix}.$$

Two of the choices made by the CIE that led to this transformation are that the $Y$ component approximates the perceived brightness, called the *luminance*, and that $R = G = B = 1$ corresponds to $X = Y = Z = 1$, which requires that the rows of the matrix sum to 1.

Because the XYZ space is three dimensional it is not easy to visualize the subset of it corresponding to

1. The coefficients of the matrix are written here in a way that indicates their known precision. Thus, for example, the (1, 1) element is known to two significant digits but the (2, 1) element is known to five significant digits.

the visible spectrum. It is common practice to use a projective transformation

$$x = \frac{X}{X+Y+Z}, \quad y = \frac{Y}{X+Y+Z} \quad (z = 1 - x - y)$$

to produce a *chromaticity diagram* in terms of the $(x, y)$ coordinates (see plate 12). The visible spectrum forms a convex set in the shape of a horseshoe. The curved boundary of the horseshoe is generated by light of a single wavelength (pure color) as it varies across the visible spectrum, while at the bottom the "purple line" is generated by combinations of red and blue light. The diagram represents color and not luminance, which is why there is no brown (a dark yellow). White is at $(\frac{1}{3}, \frac{1}{3})$, and pure colors lying at opposite ends of a line passing through the white point are complementary: a combination of them produces white. Any point outside this region represents an "imaginary color," a distribution of light that is not visible to us.

A common use of the chromaticity diagram is in reviews of cameras, scanners, displays, and printers, where the gamut of the device (the range of producible colors) is overlaid on the diagram. Generally, the closer the gamut is to the visible spectrum the better, but since images are passed along a chain of devices starting with a camera or scanner, a key question is how the gamuts of the devices compare and whether colors are faithfully translated from one device to another. Color management deals with these issues, through the use of International Color Consortium (ICC) profiles that describe the color attributes of each device by defining a mapping between the device space and the CIE XYZ reference space. Calibrating a device involves solving nonlinear equations, which is typically done by NEWTON'S METHOD [II.28].

## 3 Nonlinearities

So far, basic linear algebra and a projective transformation have been all that we need to develop color theory, and one might hope that by using more sophisticated techniques from matrix analysis one can go further. To some extent, this is possible; for example, the Binet–Cauchy theorem on determinants finds application in several problems in colorimetry. But nonlinearities cannot be avoided for long because human eyes respond to light nonlinearly, in contrast to a digital camera's sensor, which has a linear response. The relative difference in brightness that we see between a dark cellar and bright sunlight is far smaller than the relative difference in the respective number of photons reaching our eyes, and this needs to be incorporated into the model. One way of doing this is described in the next section.

## 4 LAB Space

A problem with the CIE XYZ and RGB spaces is that they are far from being perceptually uniform, which means that there is not a linear relation between distances in the tristimulus space and perceptual differences. This led the CIE to search for nonlinear transformations that give more uniform color spaces, and in 1976 they came up with two standardized systems, L*u*v* and L*a*b* (or LAB, pronounced "ell-A-B"). In the LAB space the L coordinate represents lightness, the A coordinate is on a green–magenta axis, and the B coordinate is on a blue–yellow axis. For a precise definition in terms of the XYZ space, if $X_n$, $Y_n$, and $Z_n$ are the tristimuli of white then

$$\begin{aligned} L &= 116 f(Y/Y_n) - 16, \\ A &= 500[f(X/X_n) - f(Y/Y_n)], \\ B &= 200[f(Y/Y_n) - f(Z/Z_n)], \end{aligned}$$

where

$$f(x) = \begin{cases} x^{1/3}, & x \geqslant 0.008856, \\ 7.787x + \frac{16}{116}, & x \leqslant 0.008856. \end{cases}$$

The cube root term tries to capture the nonlinear perceptual response of human vision to brightness. The two cases in the formula for $f$ bring in a different formula for low tristimulus values, i.e., low light. The lightness coordinate L ranges from 0 to 100. The A and B coordinates are typically in the range $-128$ to 128 (though not explicitly constrained), and $A = B = 0$ denotes lack of color, i.e., a shade of gray from black ($L = 0$) to white ($L = 100$). In colorimetry, color differences are expressed as Euclidean distances between LAB coordinates and are denoted by $\Delta E$.

An interesting application of LAB space is to the construction of color maps, which are used to map numbers to colors when plotting data. The most commonly used color map is the rainbow color map, which starts at dark blue and progresses through cyan, green, yellow, orange, and red, through colors of increasing wavelength. In recent years the rainbow color map has been heavily criticized for a number of reasons, which include

- it is not perceptually uniform, in that the colors appear to change at different rates in different regions (faster in the yellow, slower in the green);

- it is confusing, because people do not always remember the ordering of the colors, making interpretation of an image harder;
- it loses information when printed on a monochrome printer, since high and low values map to similar shades of gray.

These particular criticisms can be addressed by using a color map constructed in LAB space with colors having monotonically increasing L values and linearly spaced A and B values. Color maps based on such ideas have supplanted the once-ubiquitous rainbow color map as the default in MATLAB and in some visualization software.

For image manipulation there are some obvious advantages to working in LAB space, as luminosity and color can easily be independently adjusted, which is not the case in RGB space. However, LAB space has some strange properties. For example, $(L, A, B) = (0, 128, -128)$ represents a brilliant magenta as black as a cellar! LAB space contains many such imaginary colors that cannot exist and are not representable in RGB. For many years LAB was regarded as a rather esoteric color space of use only for intermediate representations in color management and the like, though it is supported in high-end software such as Adobe Photoshop and the MATLAB Image Processing Toolbox. However, in recent years this view has changed, as photographers and retouchers have realized that LAB space, when used correctly, is a very powerful tool for manipulating digital images. The book *Photoshop LAB Color* (2006) by Dan Margulis describes the relevant techniques, which include ways to reduce noise (blur the A and B channels), massively increase color contrast (stretch the A and B channels), and change the color of colorful objects in a scene while leaving the less colorful objects apparently unchanged (linear transformations of the A and B channels). As an example of the latter technique, plate 13(a) shows an RGB image of a building at the University of Manchester, while plate 13(b) shows the result of converting the image to LAB, flipping the sign of the A channel, then converting back to RGB. The effect is to change the turquoise paint to pink without, apparently, significantly changing any other color in the image including the blue sky. In truth, all the colors have changed, but mostly by such a small amount that the changes are not visible, due to the colors having small A components in LAB coordinates.

## 5 JPEG

JPEG is a compression scheme for RGB images that can greatly reduce file size, though it is lossy (throws information away). The JPEG process first converts from RGB to the $\mathrm{YC_bC_r}$ color space, where Y represents luminance and $\mathrm{C_b}$ and $\mathrm{C_r}$ represent blue and red chrominances, respectively, using the linear transformation

$$\begin{bmatrix} Y \\ C_b \\ C_r \end{bmatrix} = \begin{bmatrix} 0.299 & 0.587 & 0.114 \\ -0.1687 & -0.3313 & 0.5 \\ 0.5 & -0.4187 & -0.0813 \end{bmatrix} \begin{bmatrix} R \\ G \\ B \end{bmatrix}.$$

The motivation for this transformation is that human vision has a poor response to spatial detail in colored areas of the same luminance, so the $\mathrm{C_b}$ and $\mathrm{C_r}$ components can take greater compression than the Y component. The image is then broken up into $8 \times 8$ blocks and for each block a two-dimensional discrete cosine transform is applied to each of the components, after which the coefficients are rounded, more aggressively for the $\mathrm{C_b}$ and $\mathrm{C_r}$ components. Of course, it is crucial that the $3 \times 3$ matrix in the above transformation is nonsingular, as the transformation needs to be inverted in order to decode a JPEG file.

The later JPEG2000 standard replaces the discrete cosine transform with a WAVELET TRANSFORM [I.3 §3.3] and uses larger blocks. Despite the more sophisticated mathematics underlying it, JPEG2000 has not caught on as a general-purpose image format, but it is appropriate in special applications such as storing fingerprints, where it is much better than JPEG at reproducing edges.

## 6 Nonlinear RGB

The CIE LAB and XYZ color spaces are device independent: they are absolute color spaces defined with reference to a "standard human observer." The RGB images one comes across in practice, such as JPEG images from a digital camera, are in device-*dependent* RGB spaces. These spaces are obtained from a linear transformation of CIE XYZ space followed by a nonlinear transformation of each coordinate that modifies the gamma (brightness), analogously to the definition of the L channel in LAB space. They also have specifications in $(x, y)$ chromaticity coordinates of the primaries red, green, and blue, and the white point (as there is no unique definition of white). The most common nonlinear RGB space is sRGB, defined by Hewlett-Packard and Microsoft in the late 1990s for use by consumer digital devices and now the default space for images on the Web. The sRGB gamut is the convex hull of the red, green, and blue points, and it is shown on the chromaticity diagram in plate 12.

## 7 Digital Image Manipulation

Digital images are typically stored as 8-bit RGB images, that is, as arrays $A \in \mathbb{R}^{m \times n \times 3}$, where $a_{ijk}$, $k = 1{:}3$, contain the RGB values of the $(i, j)$ pixel. In practice, each element $a_{ijk}$ is an integer in the range 0 to 255, but for notational simplicity we will assume the range is instead $[0, 1]$. Image manipulations correspond to transforming the array $A$ to another array $B$, where

$$b_{ijk} = f_{ijk}(a_{ijk})$$

for some functions $f_{ijk}$. In practice, the $3mn$ functions $f_{ijk}$ will be highly correlated. The simplest case is where $f_{ijk} \equiv f$ is independent of $i$, $j$, and $k$, and an example is $f_{ijk}(a_{ijk}) = \min(a_{ijk} + 0.2, 1)$, which brightens an image by increasing the RGB values of every pixel by 0.2. Another example is

$$f_{ijk}(a_{ijk}) = \begin{cases} 2a_{ijk}^2 & a_{ijk} \leqslant 0.5, \\ 1 - 2(1 - a_{ijk})^2 & a_{ijk} \geqslant 0.5. \end{cases} \quad (3)$$

This transformation increases contrast by making RGB components less than 0.5 smaller (darkening the pixel) and those greater than 0.5 larger (lightening the pixel). These kinds of global manipulations are offered by all programs for editing digital images, but the results they produce are usually crude and unprofessional. For realistic high-quality results, the transformations need to be local and, from the photographic point of view, *proportional*. For example, the brightening transformation above will change any RGB triplet $(r, g, b)$ with $\min(r, g, b) \geqslant 0.8$ to $(1, 1, 1)$, which is pure white, almost certainly producing an artificial result. The transformation (3) that increases contrast will change the colors, since it modifies the RGB values at each pixel independently.

Advanced digital image manipulation avoids these problems by various techniques, a very powerful one being to apply a global transformation through a mask that selectively reduces the effect of the transformation in certain parts of the image. The power of this technique lies in the fact that the image itself can be used to construct the mask.

However, there are situations where elementary calculations on images are useful. A security camera might detect movement by computing the variance of several images taken a short time apart. A less trivial example is where one wishes to photograph a busy scene such as an iconic building without the many people and vehicles that are moving through the scene at any one time. Here, a solution is to put the camera on a tripod and take a few tens of images, each a few seconds apart, and then take the median of the images (each pixel gets the median R, G, and B values of all the pixels in that location in the image). With luck, and assuming the lighting conditions remained constant through the procedure, the median pixel will in every case be the unobscured one that was wanted!

### Further Reading

Austin, D. 2008. What is ... JPEG? *Notices of the American Mathematical Society* 55(2):226–29.

Hunt, D. M., K. S. Dulai, J. K. Bowmaker, and J. D. Mollon. 1995. The chemistry of John Dalton's color blindness. *Science* 267:984–88.

Ings, S. 2007. *The Eye: A Natural History*. London: Bloomsbury.

Johnson, C. S. 2010. *Science for the Curious Photographer: An Introduction to the Science of Photography*. Natick, MA: A. K. Peters.

Margulis, D. 2006. *Photoshop LAB Color: The Canyon Conundrum and Other Adventures in the Most Powerful Colorspace*. Berkeley, CA: Peachpit Press.

Moreland, K. 2009. Diverging color maps for scientific visualization. In *Advances in Visual Computing*, edited by G. Bebis et al, pp. 92–103. Lecture Notes in Computer Science, volume 5876. Berlin: Springer.

Reinhard, E., E. A. Khan, A. O. Akyüz, and G. M. Johnson. 2008. *Color Imaging: Fundamentals and Applications*. Natick, MA: A. K. Peters.

Sharma, G., and H. J. Trussell. 1997. Digital color imaging. *IEEE Transactions on Image Processing* 6(7):901–32.

# VII.8 Mathematical Image Processing

*Guillermo Sapiro*

## 1 What Is Image Processing?

In order to illustrate what image processing is, let us use examples from different applications and from some superb contributions to image (and video) processing research from the last few decades.

Without doubt, the most important contribution to the field is JPEG, the standard for image compression. Together with bar code readers, JPEG is the most widely used algorithm in the area. Virtually all the images on our cell phones and on popular image-sharing Web sites such as Facebook and Flickr are compressed with JPEG. The basic idea behind JPEG is to first transform every $8 \times 8$ image patch via a discrete cosine transform (DCT; the real-valued component of the fast Fourier transform), attempting to approximate the decorrelation that could be optimally achieved only via the data-dependent Karhunen–Loeve transform. The DCT coefficients are then quantized, and while a simple uniform quantization is used per coefficient position, interesting mathematical theory lies behind this in the form of the optimal Max–Lloyd quantizer (and vector quantization in general). Finally, the quantized coefficients are encoded via Huffman coding, putting to work one of the most successful (and, along with the Lempel–Ziv universal compression framework, most widely used) information theory algorithms. Part of the beauty of JPEG is its simplicity: the basic algorithm can be implemented in a morning by any undergraduate student, and of course it can be executed very efficiently for large images using simple hardware. And yet despite its simplicity, a lot of theory stands behind JPEG.

Image compression does not stop with JPEG, of course, and we also have lossless compression techniques like JPEG-LS. This is another technique based on beautiful mathematics in the context modeling and Golomb coding areas, and it has been used by NASA's Jet Propulsion Laboratory for both Mars rovers expeditions; the Mars rovers also incorporate a WAVELETS-BASED [I.3 §3.3] lossy compression algorithm (see figure 1, which was acquired on Mars and transmitted to Earth with these compression techniques). The image in figure 1 is composed of multiple high-resolution images covering different regions of the scene (with some overlapping), aligned together to form a larger image. Such alignment is often obtained via nonlinear optimization techniques that penalize disagreements between the overlapping regions. Finally, videos are compressed via MPEG, in which, as in the JPEG-LS technique, predictive coding plays a fundamental role. The idea behind predictive coding is to use past data (e.g., past frames in a video) to predict future ones, encoding only the difference.

Consumers are familiar with image processing beyond compression. We all go to the movies, and some often notice mistakes in the films they see: a cameraman who is accidentally included in a shot, say. In order to repair such mistakes, the areas of the corresponding objects have to be inpainted, or filled in, with information from the background or other sources (see plate 14). The mathematics behind image inpainting is borrowed from the calculus of variations,

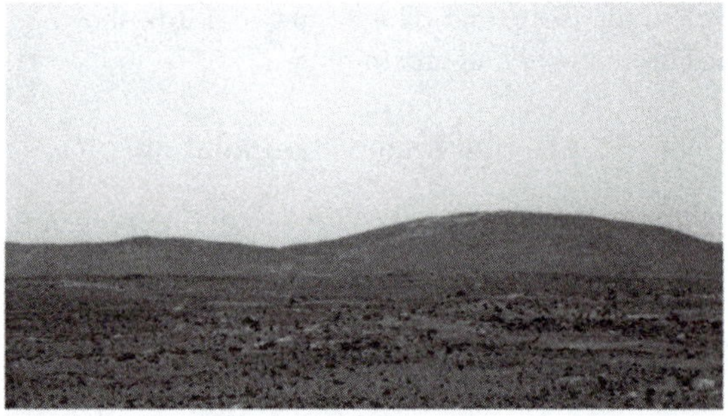

**Figure 1** A beautiful image from Mars that we can enjoy thanks to image compression. (Image courtesy of NASA/Jet Propulsion Laboratory.)

partial differential equations, transport problems, and the Navier–Stokes equations; think in terms of colors being transported rather than fluids. Some more recent techniques are based on exploiting the redundancy and self-similarity in images and automatically performing a cut-and-paste approach. This borrows ideas that go back to Shannon's thoughts on the English language: by looking at the past in a given text, and constructing letter and word distributions, we can predict the future, or at least construct words and sentences that look like correct grammar. These techniques, which can also be formalized with tools from the calculus of variations, are based on finding the best possible information, in the form of image regions, edges, or *patches*, to "copy" from, and then "pasting" such information into the zone to be inpainted.

Image inpainting can be considered an extreme case of image denoising, where the "noise" is the region to be inpainted. Noisy images (in the more standard sense, where the noise is spread around the whole image (e.g., additive Gaussian or Poisson noise)) have also motivated considerable research in image processing, with one of the most famous mathematical approaches being total variation (TV). In TV we optimize for the absolute norm of the image gradient, this being a critical regularizer for inverse problems, which are intrinsically ill-posed. TV also appears in COMPRESSED SENSING [VII.10] and in image-segmentation formulations like that of Mumford and Shah, where the idea is to segment an image by fitting it with, for example, piecewise-constant functions. TV appears both because of its piecewise requirement (total variation zero for each piece) and the fact that the length of a curve can be measured via the TV of the corresponding indicator function. And yet TV is also part of the famous

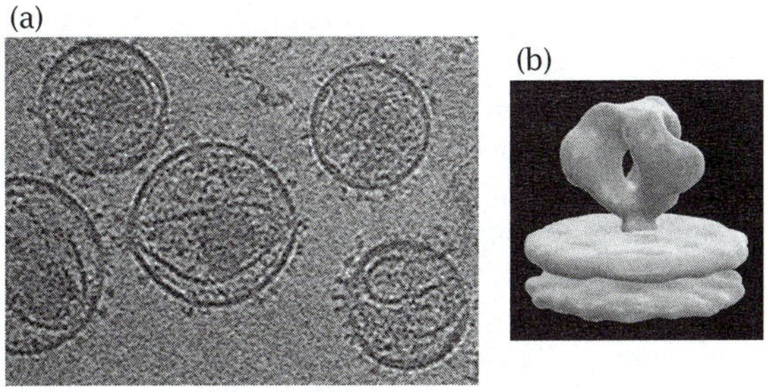


**Figure 2** (a) Cryotomography of HIV viruses (one slice of the actual three-dimensional volume) and (b) a reconstruction from such data of the HIV env protein via mathematical image processing. The very noisy HIV particles, as they appear in the raw data illustrated in (a), are classified and aligned/registered to obtain the three-dimensional reconstructed HIV env protein. (Courtesy of Liu, Bartesaghi, Borgnia, Subramaniam, and Sapiro.)

active contour model, where a curve is geometrically deformed toward object boundaries, an effect that can be achieved by computing geodesics (plate 15).

Special effects are very useful in images and movies. We can play with them ourselves, at an amateur level, on our computers, but at the professional level they are included in virtually all current movies (virtually all movies are now "touched up" by mathematical imaging tools before they are released). At the core of this is the idea of image segmentation, where we isolate objects and then paste them into new backgrounds (see plate 16).

While image processing is widely used in commercial applications such as those just mentioned, medical and biological imaging is another key area in which image processing has made an important contribution. In figure 2 we see an example of human immunodeficiency virus (HIV) research.

Finally, the analysis of images and videos is also key, e.g., identifying the objects in an image or the activity in a video (see plate 17).

To recap, with image processing we go to Hollywood, Mars, and the hospital, and mathematics is everywhere. Problems range from compression to reconstruction and recognition, and as we will discuss further below, challenging problems emerge for a number of branches of applied mathematics.

### 1.1 Who Uses Image Processing?

The answer to this question is "everybody," as we have seen above. Consumers use image processing every time they take a digital picture, and they experience its benefits every time they go to the movies. Doctors increasingly use image processing for improving radiology pictures, as well as for performing automatic analysis. Surveillance applications are abundant. Industrial automatization is now a ubiquitous client of image processing. Imaging-related companies and, therefore, image processing itself are simply everywhere.

## 2 Who Works in Image Processing?

One of the beauties of this area is that regardless of one's interest in applied mathematics, there is always an important and challenging problem and application in the area of image and video processing. Let us present a few examples.

**Harmonic analysis.** Wavelets come to mind immediately, of course; they are the basic component behind the JPEG2000 image compression standard and they are behind numerous other image reconstruction and enhancement techniques. Moreover, harmonic analysis is the precursor of *compressed sensing* and *sparse modeling*, two of the most successful and interesting ideas in image analysis, leading to the design of new cameras and state-of-the-art algorithms in applications ranging from image denoising to object and activity recognition.

**The calculus of variations and partial differential equations.** This is the mathematics behind image inpainting and leading image-segmentation techniques. An integral part of this is, of course, numerical analysis, from classical techniques to more modern ones such as LEVEL SET METHODS. [II.24]

**Optimization.** Image processing has been greatly influenced by modern optimization techniques, whether it be in the search for patches for image inpainting or in segmentation techniques based on graph partitions. Images are large, and efficient computational techniques are therefore critical to the success of image processing.

**Statistics and probability.** Bayesian and non-Bayesian formulations appear all the time, in numerous problems and to give different interpretations to other formulations, such as the standard optimization one for sparse modeling.

**Topology.** Although not obviously related to image processing, topology, and in particular persistent homology, has made significant contributions to image processing.

But as well as benefiting from tremendous mathematical advances, image processing has also opened up new questions that have led to the development of new mathematical results, ranging from the viscosity solutions of partial differential equations to fundamental theorems in harmonic analysis and optimization. There is therefore a perfect synergy between image processing and applied mathematics, each benefiting and feeding the other.

## 3 Concluding Thoughts and Perspectives

Digital images are more and more becoming part of our daily lives, and the more images we have, the more challenges we face. Applied mathematics has been critical to image processing, and all the best image-processing algorithms have fundamental mathematics behind them. Image processing also asks new and challenging questions of mathematics, thereby opening the door to new and exciting results. As one famous mathematician once told me, "If it works, there should be a math explanation behind it."

### Further Reading

Bronstein, A., M. Bronstein, and R. Kimmel. 2003. *The Numerical Geometry of Non-rigid Shapes.* New York: Springer.

Elad, M. 2010. *Sparse and Redundant Representations: From Theory to Applications in Signal and Image Processing.* New York: Springer.

Desolneux, A., L. Moisan, and J.-M. Morel. 2010. *From Gestalt Theory to Image Analysis: A Probabilistic Approach.* New York: Springer.

Mallat, S. 2008. *A Wavelet Tour of Signal Processing: The Sparse Way*, 3rd edn. New York: Academic Press.

Morel, J.-M., and S. Solimini. 2012. *Variational Methods in Image Segmentation: With Seven Image Processing Experiments.* Boston, MA: Birkhäuser.

Mumford, D., and A. Desolneux. 2010. *Pattern Theory: The Stochastic Analysis of Real-World Signals.* Natick, MA: A. K. Peters.

Osher, S., and R. Fedkiw. 2002. *Level Set Methods and Dynamic Implicit Surfaces.* New York: Springer.

Sapiro, G. 2006. *Geometric Partial Differential Equations and Image Analysis.* Cambridge: Cambridge University Press.

# VII.9 Medical Imaging

*Charles L. Epstein*

## 1 Introduction

Over the past fifty years the processes and techniques of medical imaging have undergone a veritable explosion, calling into service increasingly sophisticated mathematical tools. Mathematics provides a language to describe the measurement processes that lead, eventually, to algorithms for turning the raw data into high-quality images. There are four principal modalities in wide application today: X-ray computed tomography (X-ray CT), ultrasound, magnetic resonance imaging (MRI), and emission tomography (positron emission tomography (PET) and single-photon emission computed tomography (SPECT)). Each modality uses a different physical process to produce image contrast: X-ray CT produces a map of the X-ray attenuation coefficient, which is strongly correlated with density; ultrasound images are produced by mapping absorption and reflection of acoustic waves; in their simplest form, magnetic resonance images show the density of water protons, but the subtlety of the underlying physics provides many avenues for producing clinically meaningful contrasts in this modality; PET and SPECT give spatial maps of the chemical activity of metabolites, which are bound to radioactive elements. It has recently been found useful to merge different modalities. For example, a fused MRI/PET image shows metabolic activity produced by PET, at a fairly low spatial resolution, against the background of a detailed anatomic image produced by MRI. Plate 18 shows a PET image, a PET image fused with a CT image, and the CT image as well.

In this article we consider mathematical aspects of PET, whose underlying physics we briefly explain. Positron emission is a mode of radioactive decay stemming from the reaction

$$\text{proton} \to \text{neutron} + \text{positron} + \text{neutrino} + \text{energy}. \quad (1)$$

Two isotopes, of clinical importance, that undergo this type of decay are $F^{18}$ and $C^{11}$. The positron, which is the positively charged antiparticle of the electron, is typically very short-lived as it is annihilated, along with the first electron it encounters, producing a pair of 0.511 MeV photons. This usually happens within a millimeter or two of the site of the radioactive decay. Due to conservation of momentum, these two photons travel in nearly opposite directions along a straight line (see figure 1). The phenomenon of pair annihilation underlies the operation of a PET scanner.

A short-lived isotope that undergoes the reaction in (1) is incorporated into a metabolite, e.g., fluorodeoxyglucose, which is then injected into the patient. This metabolite is taken up differentially by various structures in the body. For example, many types of cancerous tumors have a very rapid metabolism and quickly take up available fluorodeoxyglucose. The detector in a PET scanner is a ring of scintillation crystals that surrounds some portion of the patient. The high-energy photon interacts with the crystal to produce a flash of light. These flashes are fed into photomultiplier tubes with electronics that localize, to some extent, where the flash of light occurred and measure the energy of the photon that produced it. Finally, different arrival times are compared to determine which events are likely to be "coincidences," caused by a single pair annihilation. Two photons detected within a time window of about 10 nanoseconds are assumed to be the result of a single annihilation event. The measured locations of a pair of coincident photons then determines a line. If the photons simply exited the patient's body without further interactions, then the annihilation event must have occurred somewhere along this line (see figure 1). It is not difficult to imagine that sufficiently many such measurements could be used to reconstruct an approximation for the distribution of sources, a goal that is facilitated by a more quantitative model.

## 2 A Quantitative Model

Radioactive decay is usually modeled as a Poisson random process. Recall that $Y$ is a Poisson random variable of intensity $\lambda$ if

$$\operatorname{Prob}(Y = k) = \frac{\lambda^k e^{-\lambda}}{k!}. \tag{2}$$

A simple calculation shows that $E[Y] = \lambda$ and $\operatorname{Var}[Y] = \lambda$ as well. Let $H$ denote the region within the scanner that is occupied by the patient, and, for $p \in H$, let $\rho(p)$ denote the concentration of radioactive metabolite as a function of position. If $\rho$ is measured in the correct units, then the probability of $k$ decay events originating from a small volume $dV$ centered at $p$, in a time interval of unit length, is

$$\operatorname{Prob}(k; p) = \frac{\rho(p)^k e^{-\rho(p)}}{k!}\, dV. \tag{3}$$

Decays originating at different spatial locations are regarded as independent events.

Assume, for the moment, that

(i) there are many decay events, so that we are justified in replacing this probabilistic law by its mean, $\rho(p)$,
(ii) the high-energy photons simply exit the patient without interaction, and
(iii) we are equally likely to detect a given decay event on any line passing through the source point.

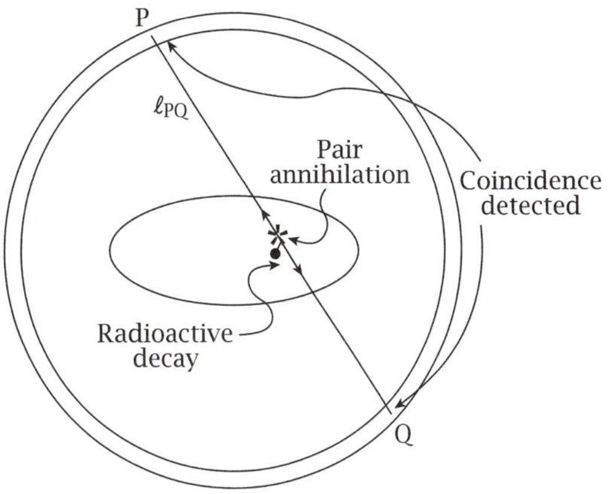


**Figure 1** A radioactive decay leading to a positron–electron annihilation, exiting along $\ell_{PQ}$, which is detected as a coincidence event at P and Q in the detector ring.

Let $\ell_{PQ}$ be the line joining the two detector positions P and Q where photons are simultaneously detected. With these assumptions we see that by counting up the coincidences observed at P and Q we are finding an approximation to the line integral

$$X\rho(\ell_{PQ}) = \int_{\ell_{PQ}} \rho(p)\, dl,$$

where $dl$ is the arc length along the line $\ell_{PQ}$. This is nothing other than a sample of the three-dimensional X-ray transform of $\rho$, which, if it could be approximately measured with sufficient accuracy, for a sufficiently dense set of lines, could then be inverted to produce a good approximate value for $\rho$. This is essentially what is done in X-RAY CT [VII.19].

For the moment, we restrict our attention to lines that lie in a plane $\pi_0$ intersecting the patient and choose coordinates $(x, y, z)$ so that $\pi_0 = \{z = z_0\}$. The lines in this plane are parametrized by an angle $\theta \in [0, \pi]$ and a real number $s$, with

$$\ell_{\theta,s} = \{(s\cos\theta, s\sin\theta, z_0) + t(-\sin\theta, \cos\theta, 0) : t \in (-\infty, \infty)\}$$

(see figure 2). In terms of $s$ and $\theta$, the two-dimensional X-ray transform is given by the integral

$$X\rho(s,\theta,z_0) = \int_{-\infty}^{\infty} \rho((s\cos\theta, s\sin\theta, z_0) + t(-\sin\theta, \cos\theta, 0))\, dt.$$

The inverse of this transform is usually represented as a composition of two operations: a filter acting on $X\rho(s,\theta,z_0)$ in the $s$ variable, followed by the back-projection operator. If $g(s,\theta)$ is a function on the space

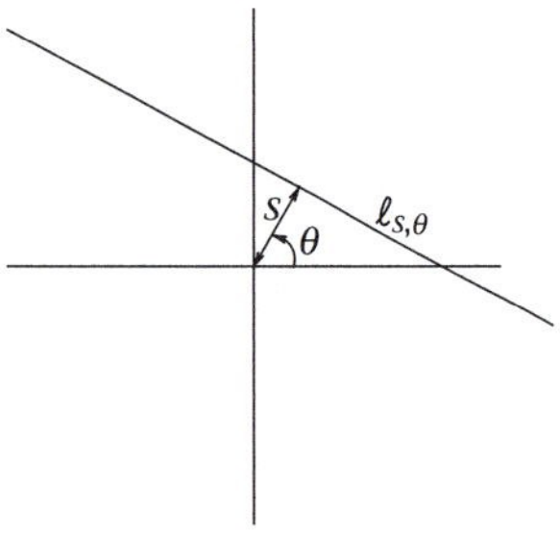


**Figure 2** The lines in the plane $\pi_0$ are labeled by $\theta$, the angle the normal makes with the $x$-axis, and $s$, the distance from the line to the origin.

of lines in a plane, then the filter operation can be represented by $\mathcal{F}g(s,\theta) = \partial_s \mathcal{H}g(s,\theta)$, where $\mathcal{H}$ is a constant multiple of the Hilbert transform acting in the $s$ variable. The back-projection operator $X^*g$ defines a function of $(x,y) \in \pi_0$ that is the average of $g$ over all lines passing through $(x,y)$:

$$X^*g(x,y) = \frac{1}{\pi}\int_0^{\pi} g(x\cos\theta + y\sin\theta, \theta)\,\mathrm{d}\theta.$$

Putting together the pieces, we get the filtered back-projection (FBP) operator, which inverts the two-dimensional X-ray transform: $\rho(x,y,z_0) = [X^* \circ \mathcal{F}] \cdot X\rho$. By using this approach for a collection of parallel planes, the function $\rho$ could be reconstructed in a volume. This provides a possible method for reconstruction of PET images, and indeed the discrete implementations of this method have been extensively studied. In the early days of PET imaging this approach was widely used, and it remains in use today. Note, however, that using only data from lines lying in a set of parallel planes is very wasteful and leads to images with low signal-to-noise ratio.

Assumption (i) implies that our measurement is a good approximation to the X-ray transform of $\rho$, $X\rho(s,\theta,z_0)$. Because of the very high energies involved in positron emission radioactivity, only very small amounts of short-lived isotopes can be used. The measured count rates are therefore low, which leads to measurements dominated by Poisson noise that are not a good approximation to the mean. Because the FBP algorithm involves a derivative in $s$, the data must be significantly smoothed before this approach to image reconstruction can be applied. This produces low-resolution images that contain a variety of artifacts due to systematic measurement errors, which we describe below.

At this point it is useful to have a more accurate description of the scanner and the measured data. We

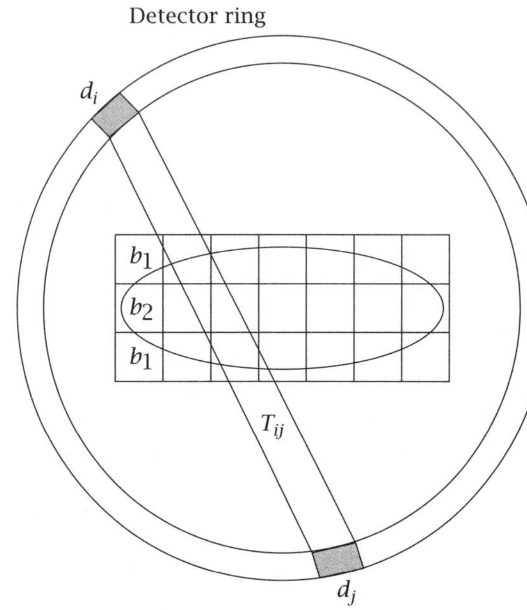


**Figure 3** The detector ring is divided into finitely many detectors of finite size. Each pair $(d_i, d_j)$ defines a tube $T_{ij}$ in the region occupied by the patient. This region is divided into boxes $\{b_k\}$.

model the detector as a cylindrical ring surrounding the patient, which is partitioned into a finite set of regions $\{d_1, \dots, d_n\}$. The scanner can localize a scintillation event as having occurred in one of these regions, which we heretofore refer to as detectors. This instrument design suggests that we divide the volume inside the detector ring into a collection of tubes, $\{T_{ij}\}$, with each tube defined as the union of lines joining points in $d_i$ to points in $d_j$ (see figure 3). A measurement $n_{ij}$ is the number of coincidence events observed by the pair of detectors $(d_i, d_j)$. The simplest interpretation of $n_{ij}$ is as a sample of a Poisson random variable with mean proportional to

$$\int_{\ell_{\mathrm{PQ}} \subset T_{ij}} X\rho(\ell_{\mathrm{PQ}}). \qquad (4)$$

Below we will see that this interpretation requires several adjustments.

Assumption (ii) fails as the photons tend to interact quite a lot with the bulk of the patient's body. Large fractions of the photons are absorbed, or scattered, with each member of an annihilation pair meeting its fate independently of the other. This leads to three distinct types of measurement errors.

**Randoms.** These are coincidences that are observed by a pair of detectors but that do not correspond to a single annihilation event. These can account for 10–30% of the observed events (see figure 4(a)).

**Scatter.** If one or both photons is/are scattered and then both are detected, this may register as a coincidence at a pair of detectors $(d_i, d_j)$, but the annihilation event did not occur at a point lying near $T_{ij}$ (see figure 4(b)).

**Attenuation.** Most photon pairs (often 95%) are simply absorbed, leading to a substantial underestimate of the number of events occurring along a given line.

Below we discuss how the effects of these sorts of measurement errors can be incorporated into the model and the reconstruction algorithm. To get quantitatively meaningful, artifact-free images, these errors must be corrected before application of any image-reconstruction method.

Assumption (iii) is false in that the detector array, which is usually a ring of scintillation counters, encloses only part of the patient. Many lines through the patient will therefore be disjoint from the detector only intersect it or at one end. This problem can, to some extent, be mitigated by using only observations coming from lines that lie in planes that intersect the detector in a closed curve. If the detector is a section of a cylinder, then each point $p$ lies in a collection of such planes $\{\pi_{\psi,\phi}\}$ whose normal vectors $\{\nu_{\psi,\phi}\}$ fill a disk $D_p$ lying on the unit sphere. If $\rho^{\psi,\phi}(p)$ denotes the approximate value for $\rho(p)$ determined using the FBP algorithm in the plane $\pi_{\psi,\phi}$, then an approximate value with improved signal-to-noise ratio is obtained as the average:

$$\bar{\rho}(p) = \frac{1}{|D_p|} \int_{D_p} \rho^{\psi,\phi}(p)\, \mathrm{d}S(\psi, \phi),$$

where $\mathrm{d}S(\psi, \phi)$ is the spherical areal measure. A particular implementation of this idea that is often used in PET scanners goes under the name of the "Colsher filter." Other methods use a collection of parallel two-dimensional planes to reconstruct an approximate image from which the missing data for the three-dimensional X-ray transform can then be approximately computed.

In addition to these inherent physical limitations on the measurement process, there are a wide range of instrumentation problems connected to the detection and spatial localization of high-energy photons, as well as the discrimination of coincidence events. Effective solutions to these problems are central to the success of a PET scanner, but they are beyond the scope of this article.

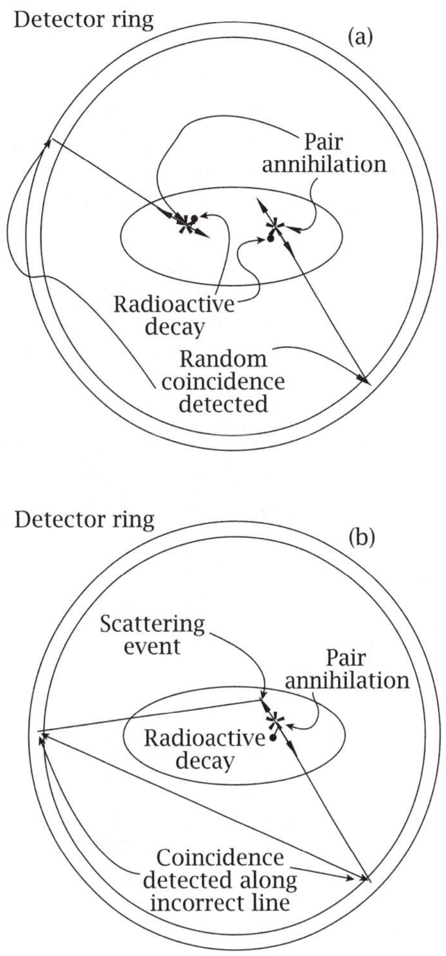


**Figure 4** The measurement process in PET scanners is subject to a variety of systematic errors. (a) Randoms are detected coincidences that do not result from a single decay event. (b) Scatter is the result of one or both of the photons scattering off an object before being detected as a coincidence event but along the wrong line.

## 3 Correcting Measurement Errors

To reconstruct images that are quantitatively meaningful and reasonably free of artifacts, the measured data $\{n_{ij}\}$ must *first* be corrected for randoms, scatter, and attenuation (see figure 4). This requires both additional measurements and models for the processes that lead to these errors.

### 3.1 Randoms

We first discuss how to correct for randoms. Let $R_{ij}$ denote the number of coincidences detected on the pair $(d_i, d_j)$ that are not caused by a decay event in $T_{ij}$. In practice, coincidences are considered to be two events that are observed within a certain time window $\tau$ (usually about 10 nanoseconds). In addition to coincidences between two detectors, the numbers of single counts, $\{n_i\}$, observed at $\{d_i\}$ are recorded. In fact, the number of "singles" is usually one or two *orders of magnitude* larger than the number of coincidences. From the measured number of singles observed over a known period of time we can infer rates of singles events $\{r_i\}$ for each detector. Assuming that each of these singles processes is independent, a reasonable estimate for the number of coincidences observed on the detector pair $(d_i, d_j)$ over the course of $T$ units of time that are actually randoms is $R_{ij} \simeq \tau T r_i r_j$. A somewhat more accurate estimate is obtained if one accounts for the decay of the radioactive source.

There are other measurement techniques for estimating $R_{ij}$, though these estimates tend to be rather noisy. Simply subtracting $R_{ij}$ from $n_{ij}$ can increase the noise in the measurements and also change their basic statistical properties. There is a useful technique for replacing $R_{ij}$ with a lower-variance estimate. Let $A$ be a collection of contiguous detectors including $d_i$ that are joined to $d_j$, and let $B$ be a similar collection, including $d_j$, that are joined to $d_i$. Suppose that $\tilde{R}_{mn}$ are estimates for the randoms detected in the pairs $\{(d_n, d_m): n \in B,\ m \in A\}$. The expression

$$\hat{R}_{ij} = \frac{[\sum_{m\in A} \tilde{R}_{im}][\sum_{n\in B} \tilde{R}_{nj}]}{\sum_{m\in A,\, n\in B} \tilde{R}_{mn}}$$

provides an estimate for $R_{ij}$ with reduced noise variance.

### 3.2 Scatter

The next source of error we consider is scatter, which results from one or both photons in the annihilation pair scattering off some matter in the patient before being recorded as a coincidence at a pair of detectors $(d_i, d_j)$. If the scattering angle is not small, then the annihilation event will not have occurred in expected tube $T_{ij}$. Some part of $n_{ij}$, which we denote $S_{ij}$, therefore corresponds to radioactive decays that did not occur in $T_{ij}$. Scattered photons tend to lose energy, so many approaches to estimating the amount of scatter are connected to measurement of the energies of detected photons. Depending on the design of the scanner, scatter can account for 15–40% of the observed coincidences. There are many methods for estimating the contribution of scatter but most of them are related to the specific design of the PET scanner and are therefore beyond the purview of this article.

### 3.3 Attenuation

Once contributions from randoms and scatter have been removed to obtain corrected observations $\tilde{n}_{ij} = n_{ij} - R_{ij} - S_{ij}$, we still need to account for the fact that many photon pairs are simply absorbed. This process is described by Beer's law, which is the basis for X-ray CT. Suppose that an annihilation event takes place at a point $p_0$ within the patient and that the photons travel along the rays $\ell_+$ and $\ell_-$ originating at $p_0$. The attenuation coefficient is a function $\mu(p)$ defined throughout the patient's body such that the probabilities of detecting photons traveling along $\ell_\pm$ are

$$P_\pm = \exp\left(-\int_{\ell_\pm} \mu\, \mathrm{d}l\right).$$

As we "count" only coincidences, and the two photons are independent, the probability of observing the coincidence is simply the product

$$P_+P_- = \exp\left(-\int_{\ell} \mu\, \mathrm{d}l\right),$$

where $\ell = \ell_+ \cup \ell_-$. In other words, the attenuation of coincidence counts due to photons traveling along $\ell$ does not depend on the location of the annihilation event along this line! This factor can therefore be measured by observing the fraction of photons, of the given energy, emitted *outside* the patient's body that pass through the patient along $\ell$ and are detected on the opposite side.

For each pair $(d_i, d_j)$ we can therefore determine an attenuation coefficient $q_{ij}$. The extent of the intersection of $T_{ij}$ with the patient's body has a marked effect on the size of $q_{ij}$, which can range from approximately 1 for tubes lying close to the skin to approximately 0.1 for those passing through a significant part of the body. The corrected data, which is passed to a reconstruction algorithm, is therefore

$$N_{ij} = \frac{\tilde{n}_{ij}}{q_{ij}} = \frac{n_{ij} - R_{ij} - S_{ij}}{q_{ij}}.$$

In addition to the corrections described above, there are a variety of adjustments that are needed to account for measurement errors attributable to the details of the behavior of the detector and the operation of the electronics. Applying an FBP algorithm to the corrected data, we can obtain a discrete approximation $\rho^{\text{FBP}}(p)$ to $\rho(p)$. In the next section we describe iterative algorithms for PET image reconstruction. While the FBP algorithm is linear, and efficient, iterative algorithms allow for incorporation of more information about the measurement process and are better suited to low signal-to-noise ratio data.

## 4 Iterative Reconstruction Algorithms

While filtered back projection provides a good starting point for image reconstruction in PET, the varying statistical properties of different measurements cannot be easily incorporated into this algorithm. A variety of approaches have been developed that allow for the exploitation of such information. To describe these algorithms we need to provide a discrete measurement model that is somewhat different from that discussed above. The underlying idea is that we are developing a statistical estimator for the strengths of the Poisson processes that produce the observed measurements. Note that these measurements must still be corrected as described in the previous section.

In the previous discussion we described the region, $H$, occupied by the patient as a continuum, with $\rho(p)$ the strength of the radioactive decay processes, a continuous function of a continuous variable $p \in H$. We now divide the measurement volume into a finite collection of boxes, $\{b_1, \dots, b_B\}$. The radioactive decay of the tracer in box $b_k$ is modeled as a Poisson process of strength $\lambda_k$. For each point $p \in H$ and each detector pair $(d_i, d_j)$, we let $c(p; i, j)$ denote the probability that a decay event at $p$ is detected as a coincidence in this detector pair. The patient's body will produce scatter and attenuation that will in turn alter the values of $c(p; i, j)$ from what they would be in its absence, i.e., the area fraction of a small sphere centered at $p$ intercepted by lines joining points in $d_i$ to points in $d_j$.

In the simplest case, the measurements $\{n_{ij}\}$ would be interpreted as samples of Poisson random variables with means

$$\sum_{k=1}^{B} p(k; i, j)\lambda_k,$$

where

$$p(k; i, j) = \frac{1}{V(b_k)} \int_{b_k} c(p; i, j)\, \mathrm{d}p$$

is the probability that a decay event in $b_k$ is detected in the pair $(d_i, d_j)$. Here, $V(b_k)$ is the volume of $b_k$. Assuming that the attenuation coefficient does not vary rapidly within the tube $T_{ij}$, we can incorporate attenuation into this model, as above, by replacing $p(k; i, j)$ with $p(k; i, j) \to q_{ij} p(k; i, j) = p^a(k; i, j)$. Ignoring scatter and randoms, the expected value of $n_{ij}$ would then satisfy

$$E[n_{ij}] = \sum_{k=1}^{B} p^a(k; i, j)\lambda_k = \bar{n}_{ij}.$$

In this model, scatter and randoms are regarded as independent Poisson processes, with means $\lambda^{\text{s}}(i, j)$ and $\lambda^{\text{r}}(i, j)$, respectively. Including these effects, we see that the measurement $n_{ij}$ is then a sample of a Poisson random variable with mean $\bar{n}_{ij} + \lambda^{\text{s}}(i, j) + \lambda^{\text{r}}(i, j)$. The reconstruction problem is then to infer estimates for the intensities of the sources $\{\lambda_k\}$ from the observations $\{n_{ij}\}$. There are a variety of approaches to solving this problem.

First we consider the reconstruction problem ignoring the contributions of scatter and randoms. The measurement model suggests that we look for a solution, $(\lambda_1^*, \dots, \lambda_B^*)$, to the system of equations

$$n_{ij} = \sum_{k=1}^{B} p^a(k; i, j)\lambda_k^*.$$

If the array of detectors is three dimensional, there are likely to be many more detector pairs than boxes in the volume. The number of such pairs is quadratic in the number of detectors. This system of equations is therefore highly overdetermined, so a least-squares solution is a reasonable choice. That is, $\lambda^*$ could be defined as

$$\lambda^* = \underset{\{y:\, 0 \leqslant y_k\}}{\arg\min} \sum_{i,j} \Bigl(n_{ij} - \sum_{k=1}^{B} p^a(k; i, j) y_k\Bigr)^2.$$

Note that we constrain the variables $\{y_k\}$ to be nonnegative, as this is certainly true of the actual intensities. The least-squares solution can be interpreted as a maximum-likelihood (ML) estimate for $\lambda$, when the likelihood of observing $n$ given the intensities $y$ is given by the product of Gaussians:

$$L_g(y) = \prod_{i,j} \exp\Bigl[-\Bigl(n_{ij} - \sum_{k=1}^{B} p^a(k; i, j) y_k\Bigr)^2\Bigr].$$

It is assumed that the various observations are samples of independent processes. If we have estimates for the variances $\{\sigma_{ij}\}$ of these measurements, then we could instead consider a weighted least-squares solution and look for

$$\lambda_{g,\sigma}^* = \underset{\{y:\, 0 \leqslant y_k\}}{\arg\min} \sum_{i,j} \frac{1}{\sigma_{ij}} \Bigl(n_{ij} - \sum_{k=1}^{B} p^a(k; i, j) y_k\Bigr)^2.$$

Because the data tends to be very noisy, in addition to the "data term" many algorithms include a regularization term, such as

$$R(y) = \sum_{k=1}^{B} \sum_{k \in N(k)} |y_k - y_{k'}|^2,$$

where for each $k$, the $N(k)$ are the indices of the boxes contiguous to $b_k$. The $\beta$-regularized solution is then defined as

$$\lambda^*_{g,\sigma,\beta} = \underset{\{y:\,0 \leqslant y_k\}}{\arg\min} \left[ \sum_{i,j} \frac{1}{\sigma_{ij}} \left( n_{ij} - \sum_{k=1}^{B} p^a(k; i,j) y_k \right)^2 + \beta R(y) \right].$$

As noted above, these tend to be very large systems of equations and are therefore usually solved via iterative methods. Indeed, a great deal of the research effort in PET is connected with finding data structures and algorithms to enable solution of such optimization problems in a way that is fast enough and stable enough for real-time imaging applications.

Given the nature of radioactive decay, it is perhaps more reasonable to consider an expression for the likelihood in terms of Poisson processes. With

$$\mu(i,j)(y) = \sum_{k}^{B} p^a(k; i,j) y_k,$$

we get the Poisson likelihood function

$$L_p(y) = \prod_{i,j} \frac{e^{-\mu(i,j)(y)} [\mu(i,j)(y)]^{n_{ij}}}{n_{ij}!}.$$

The expectation–maximization (EM) algorithm provides a means to iteratively find the nonnegative vector that maximizes $\log L_p(y)$. After choosing a nonnegative starting vector $y^{(0)}$, the map from $y^{(m)}$ to $y^{(m+1)}$ is given by the formula

$$y_k^{(m+1)} = y_k^{(m)} \frac{1}{\sum_{i,j} p^a(k; i,j)} \sum_{i,j} \left[ \frac{p^a(k; i,j) n_{ij}}{\sum_k p^a(k; i,j) y_k^{(m)}} \right]. \tag{5}$$

This algorithm has several desirable features. Firstly, if the initial guess $y^{(0)}$ is positive, then the positivity condition on the components of $y^{(m)}$ is automatic. Secondly, before convergence, the likelihood is monotonically increasing; that is, $L(y^{(m)}) < L(y^{(m+1)})$. Note that if $n_{ij} = \sum_k p^a(k; i,j) y_k^{(m)}$ for all $(i,j)$, then $y^{(m+1)} = y^{(m)}$. The algorithm defined in (5) converges too slowly to be practical in clinical applications. There are several methods to accelerate its convergence, which also include regularization terms.

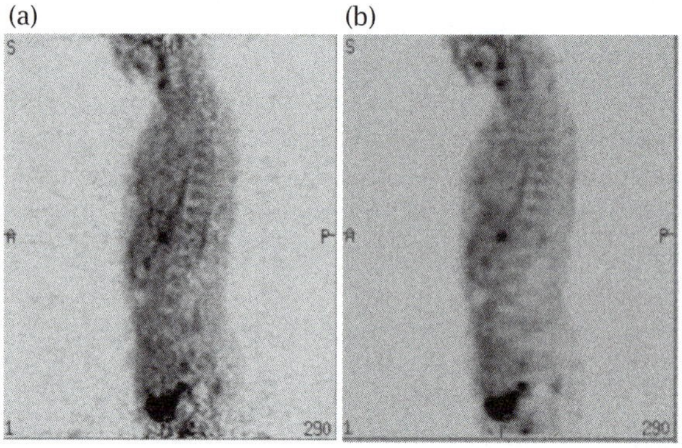


**Figure 5** Two reconstructions from the same PET data illustrating the superior noise and artifact suppression attainable using iterative algorithms: (a) image reconstructed using the FBP algorithm and (b) image reconstructed using an iterative ML algorithm. Images courtesy of Dr. Joel Karp, Hospital of the University of Pennsylvania.

We conclude this discussion by explaining how to include estimates for the contributions of scatter and randoms to $n_{ij}$ in an ML reconstruction algorithm. We suppose that $n_{ij}$ can be decomposed as a sum of three terms:

$$n_{ij} = \sum_{k=1}^{B} p^a(k; i,j) \lambda_k + R_{ij} + S_{ij}. \tag{6}$$

With this decomposition, it is clear how to modify the update rule in (5):

$$y_k^{(m+1)} = y_k^{(m)} \frac{1}{\sum_{i,j} p^a(k; i,j)} \times \sum_{i,j} \left[ \frac{p^a(k; i,j) n_{ij}}{\sum_k p^a(k; i,j) y_k^{(m)} + R_{ij} + S_{ij}} \right].$$

In addition to the ML-based algorithms, there are many other iterative approaches to solving these optimization problems that go under the general rubric of "algebraic reconstruction techniques," or ART. Figure 5 shows two reconstructions of a PET image: part (a) is the result of using the FBP algorithm, while part (b) shows the output of an iterative ML–EM algorithm.

## 5 Outlook

PET imaging provides a method for directly visualizing and spatially localizing metabolic processes. As is clear from our discussion, the physics involved in interpreting the measurements and designing detectors is rather complicated. In this article we have touched on only some of the basic ideas used to model the measurements and develop reconstruction algorithms. The

FBP algorithm gives the most direct method for reconstructing images, but the images tend to have low resolution, streaking artifacts, and noise. One can easily incorporate much more of the physics into iterative techniques based on probabilistic models, and this should lead to much better images. Because of the large number of detector pairs for three-dimensional volumes, naive implementations of iterative algorithms require vast computational resources. At the time of writing, both reconstruction techniques and the modality as a whole are rapidly evolving in response to the development of better detectors and faster computers, and because of increased storage capabilities.

### Further Reading

A comprehensive overview of PET is given in Bailey et al. (2005). A review article on instrumentation and clinical applications is Muehllehner and Karp (2006). An early article on three-dimensional reconstruction is Colsher (1980). An early paper on the use of ML algorithms in PET is Vardi et al. (1985). Review articles on reconstruction methods from projections and reconstruction methods in PET are Lewitt and Matej (2003) and Reader and Zaidi (2007), respectively.

Bailey, D. L., D. W. Townsend, P. E. Valk, and M. N. Maisey, eds. 2005. *Positron Emission Tomography*. London: Springer.

Colsher, J. G. 1980. Fully three-dimensional positron emission tomography. *Physics in Medicine and Biology* 25: 103–15.

Lewitt, R. M., and S. Matej. 2003. Overview of methods for image reconstruction from projections in emission computed tomography. *Proceedings of the IEEE* 91:1588–611.

Muehllehner, G., and J. Karp. 2006. Positron emission tomography. *Physics in Medicine and Biology* 51:R117–37.

Reader, A. J., and H. Zaidi. 2007. Advances in PET image reconstruction. *PET Clinics* 2:173–90.

Vardi, Y., L. A. Shepp, and L. Kaufman. 1985. A statistical model for positron emission tomography. *Journal of the American Statistical Association* 80:8–20.

# VII.10 Compressed Sensing

*Yonina C. Eldar*

## 1 Introduction

Compressed sensing (CS) is an exciting, rapidly growing field that has attracted considerable attention in electrical engineering, applied mathematics, statistics, and computer science. CS offers a framework for simultaneous sensing and compression of finite-dimensional vectors that relies on linear dimensionality reduction. Quite surprisingly, it predicts that sparse high-dimensional signals can be recovered from highly incomplete measurements by using efficient algorithms.

To be more specific, let $x$ be an $n$-vector. In CS we do not measure $x$ directly but instead acquire $m < n$ linear measurements of the form $y = Ax$ using an $m \times n$ CS matrix $A$. Ideally, the matrix is designed to reduce the number of measurements as much as possible while allowing for recovery of a wide class of signals from their measurement vectors $y$. Thus, we would like to choose $m \ll n$.

Since $A$ has fewer rows than columns, it has a nonempty null space. This implies that for any particular signal $x_0$, an infinite number of signals $x$ yield the same measurements $y = Ax = Ax_0$. To enable recovery, we must therefore limit ourselves to a special class of input signals $x$.

Sparsity is the most prevalent signal structure used in CS. In its simplest form, sparsity implies that $x$ has only a small number of nonzero values but we do not know which entries are nonzero. Mathematically, we express this condition as $\|x\|_0 \leqslant k$, where $\|x\|_0$ denotes the $\ell_0$-"norm" of $x$, which counts the number of nonzeros in $x$ (note that $\|\cdot\|_0$ is not a true norm, since in general $\|\alpha x\|_0 \neq |\alpha|\,\|x\|_0$ for $\alpha \in \mathbb{R}$). More generally, CS ideas can be applied when a suitable representation of $x$ is sparse. A signal $x$ is $k$-sparse in a basis $\Psi$ if there exists a vector $\theta \in \mathbb{R}^n$ with only $k \ll n$ nonzero entries such that $x = \Psi\theta$. As an example, the success of many compression algorithms, such as JPEG 2000 [VII.7 §5], is tied to the fact that natural images are often sparse in an appropriate wavelet transform.

Finding a sparse vector $x$ that satisfies the measurement equation $y = Ax$ can be performed by an exhaustive search over all possible sets of size $k$. In general, however, this is impractical; in fact, the task of finding such an $x$ is known to be NP-HARD [I.4 §4.1]. The surprising result at the heart of CS is that, if $x$ (or a suitable representation of $x$) is $k$-sparse, then it can be recovered from $y = Ax$ using a number of measurements $m$ that is on the order of $k \log n$, under certain conditions on the matrix $A$. Furthermore, recovery is possible using polynomial-time algorithms that are robust to noise and mismodeling of $x$. In particular, the essential results hold when $x$ is *compressible*, namely, when it is well approximated by its best $k$-term representation

$\min_{\|v\|_0 \leqslant k} \|x - v\|$, where the norm in the objective is arbitrary.

CS has led to a fundamentally new approach to signal processing, analog-to-digital converter (ADC) design, image recovery, and compression algorithms. Consumer electronics, civilian and military surveillance, medical imaging, radar, and many other applications rely on efficient sampling. Reducing the sampling rate in these applications by making efficient use of the available degrees of freedom can improve the user experience; increase data transfer; improve imaging quality; and reduce power, cost, and exposure time.

## 2 The Design of Measurement Matrices

The ability to recover $x$ from a small number of measurements $y = Ax$ depends on the properties of the CS matrix $A$. In particular, $A$ should be designed so as to enable unique identification of a $k$-sparse signal $x$. Let the support $S$ of $x$ be the set of indices over which $x$ is nonzero, and denote by $x_S$ the vector $x$ restricted to its support. We similarly denote by $A_S$ the columns of $A$ corresponding to this support, so that $y = Ax = A_S x_S$. When the support is known, we can recover $x$ from $y$ via $x_S = (A_S^{\mathrm{T}} A_S)^{-1} A_S^{\mathrm{T}} y$, assuming that $A_S$ has full column rank. The difficulty in CS arises from the fact that the support of $x$ is not known in advance. Therefore, determining conditions on $A$ that ensure recovery is more involved.

As a first step, we would like to choose $A$ such that every two distinct signals $x, x'$ that are $k$-sparse lead to different measurement vectors $Ax \neq Ax'$. This can be ensured if the *spark* of $A$ satisfies $\operatorname{spark}(A) \geqslant 2k + 1$, where $\operatorname{spark}(A)$ is the smallest number of columns of $A$ that are linearly dependent. Since $\operatorname{spark}(A) \in [2, m + 1]$, this yields the requirement that $m \geqslant 2k$.

Unfortunately, computing the spark of a general matrix $A$ has combinatorial computational complexity, since one must verify that all sets of columns of a certain size are linearly independent. Instead, one can provide (suboptimal) recovery guarantees using the *coherence* $\mu(A)$, which is easily computable and is defined as

$$\mu(A) = \max_{1 \leqslant i \neq j \leqslant n} \frac{|a_i^{\mathrm{T}} a_j|}{\|a_i\|_2 \|a_j\|_2},$$

where $a_i$ is the $i$th column of $A$. For any $A$,

$$\operatorname{spark}(A) \geqslant 1 + \frac{1}{\mu(A)}.$$

Therefore, if

$$k < \frac{1}{2}\left(1 + \frac{1}{\mu(A)}\right), \tag{1}$$

then for any $y \in \mathbb{R}^m$ there exists at most one $k$-sparse signal $x \in \mathbb{R}^n$ such that $y = Ax$.

In order to ensure stable recovery from noisy measurements $y = Ax + w$, where $w$ represents noise, more stringent requirements on $A$ are needed. One such condition is the *restricted isometry property* (RIP). A matrix $A$ has the $(k, \delta)$-RIP for $\delta \in (0, 1)$ if, for all $k$-sparse vectors $x$,

$$(1 - \delta)\|x\|_2^2 \leqslant \|Ax\|_2^2 \leqslant (1 + \delta)\|x\|_2^2.$$

This means that all submatrices of $A$ of size $m \times k$ are close to an isometry and are therefore distance preserving. Clearly, if $A$ has the $(2k, \delta)$-RIP with $0 < \delta < 1$, then $\operatorname{spark}(A) \geqslant 2k + 1$.

The RIP enables recovery guarantees that are much stronger than those based on spark and coherence. Another property used to characterize $A$ is the null-space condition. This requirement ensures that the null space of $A$ does not contain vectors that are concentrated on a small subset of indices. If a matrix satisfies the RIP, then it also has the null-space property. However, checking whether $A$ satisfies either of these conditions has combinatorial computational complexity.

Random matrices $A$ of size $m \times n$ with $m < n$, whose entries are independent and identically distributed with continuous distributions, have $\operatorname{spark}(A) = m + 1$ with high probability. When the distribution has zero mean and finite variance, then in the asymptotic regime (as $m$ and $n$ grow) the coherence converges to $\mu(A) = 2\sqrt{\log n / m}$. Random matrices from Gaussian, Rademacher, or (more generally) a sub-Gaussian distribution have the $(k, \delta)$-RIP with high probability if $m = \mathcal{O}(k \log(n/k)/\delta^2)$. Similarly, it can be shown that a partial Fourier matrix with $m = \mathcal{O}(k \log^4 n / \delta^2)$ rows, namely a matrix formed from the $n \times n$ Fourier matrix by taking $m$ of its rows uniformly at random, satisfies the RIP of order $k$ with high probability. A similar result holds for random submatrices of orthogonal matrices.

There are also deterministic matrices that satisfy the spark and RIP conditions. For example, an $m \times n$ Vandermonde matrix constructed from $n$ distinct scalars has spark equal to $m + 1$. Unfortunately, these matrices are poorly conditioned for large values of $n$, rendering the recovery problem numerically unstable. It is also possible to construct deterministic CS matrices of size $m \times n$ that have the $(k, \delta)$-RIP for $k = \mathcal{O}(\sqrt{m} \log m / \log(n/m))$.

## 3 Recovery Algorithms

Many algorithms have been proposed to recover a sparse vector $x$ from measurements $y = Ax$. When the measurements are noise free, and when $A$ satisfies the spark requirement, the unique sparse vector $x$ can be found by solving the optimization problem

$$\hat{x} = \arg\min_{x \in \mathbb{R}^n} \|x\|_0 \quad \text{subject to } y = Ax. \tag{2}$$

Solving (2) relies on an exhaustive search, so a variety of computationally feasible alternatives have been developed.

One popular approach to obtain a tractable problem is to replace the $\ell_0$-norm by the $\ell_1$-norm, which is convex. The resulting adaptation of (2), known as basis pursuit, is defined by

$$\hat{x} = \arg\min_{x \in \mathbb{R}^n} \|x\|_1 \quad \text{subject to } y = Ax.$$

This algorithm can be implemented as a LINEAR PROGRAM [IV.11 §3], making its computational complexity polynomial in $n$. Basis pursuit is easily modified to allow for noisy measurements by changing the constraint to $\|y - Ax\|_2^2 \leqslant \varepsilon$, where $\varepsilon$ is an appropriately chosen bound on the noise magnitude. The Lagrangian relaxation of the resulting problem is given by

$$\hat{x} = \arg\min_{x \in \mathbb{R}^n} \|x\|_1 + \lambda \|y - Ax\|_2^2,$$

and it is known as basis pursuit denoising (BPDN). Many fast methods have been developed in order to find BPDN solutions.

An alternative to optimization-based techniques are greedy algorithms for sparse signal recovery. These methods are iterative in nature and select columns of $A$ according to their correlation with the measurements $y$. Several greedy methods can be shown to have performance guarantees that match those obtained for BPDN.

For example, the matching-pursuit and orthogonal matching pursuit algorithms proceed by finding the column $a_j$ of $A$ most correlated to the signal residual, where

$$j = \arg\max_i \frac{|a_i^{\mathrm{T}} r|^2}{\|a_i\|_2^2}.$$

The residual $r$ is obtained by subtracting the contribution of a partial estimate of the signal from $y$: $r = y - A_S x_S$, where $S$ is the current guess of the support set. The convergence criterion used to find sparse representations consists of checking whether $y = Ax$ exactly or approximately. The difference between the two techniques is in the coefficient update stage. While in orthogonal matching pursuit in each stage all nonzero elements are chosen so as to minimize the residual error $\|y - A_S x_S\|_2^2$, in matching pursuit only the component associated with the currently selected column is updated to $a_j^{\mathrm{T}} r / \|a_j\|_2^2$.

Another popular approach is known as iterative hard thresholding: starting from an initial estimate $\hat{x}_0 = 0$, the algorithm iterates a gradient descent step followed by hard thresholding, i.e.,

$$\hat{x}_i = H_k(\hat{x}_{i-1} + A^{\mathrm{T}}(y - A\hat{x}_{i-1})).$$

Here, $H_k(v)$ returns the $k$ entries of $v$ that are largest in absolute value.

Many of the CS algorithms above come with guarantees on their performance. For example, basis pursuit and orthogonal matching pursuit recover a $k$-sparse vector from noiseless measurements when the matrix $A$ satisfies (1). There also exist coherence-based guarantees designed for measurements corrupted with arbitrary noise. In general, though, results based on coherence typically suffer from the so-called square-root bottleneck: they require $m = O(k^2)$ measurements to ensure good recovery.

Stronger guarantees are available based on the RIP, which motivates the popularity of random CS matrices. In particular, orthogonal matching pursuit recovers a $k$-sparse vector from exact measurements if $A$ has the $(k+1, \delta)$-RIP with a small enough value of $\delta$. More generally, a sparse vector $x$ can be recovered with small error from noisy measurements using iterative hard thresholding and BPDN when $A$ has the $(ck, \delta)$-RIP, with appropriate values of $c$ and $\delta$. These results also hold when $x$ is not exactly sparse but only compressible. The recovery error in this case is proportional to that of the best $k$-sparse approximation of $x$ and to the norm of the noise. Since random matrices satisfying the RIP can be constructed as long as $m = O(k \log(n/k))$, it follows that with high probability on the order of $k \log(n/k)$ measurements suffice to guarantee recovery of sparse vectors in the noise-free setting and to ensure recovery with small error in the noisy case.

## 4 Applications

### 4.1 Imaging

One of the first applications of CS was the single-pixel camera. This camera uses a single photon detector (the single pixel) to measure $m$ inner products of a desired image, represented by a vector $x$ in $\mathbb{R}^n$, and a set of test vectors. Each vector represents the pattern of a digital micromirror device, which consists of $n$ tiny mirrors that are individually oriented in a pseudorandom fashion either toward the photodiode (representing a 1) or away from it (representing a 0). The incident light field is reflected off the digital micromirror device, collected by a lens, and then focused onto the photodiode, which computes the inner product between $x$ and the random digital micromirror device pattern. This process is repeated $m$ times with different patterns. Good recovery of the underlying image has been obtained using about 60% fewer random measurements than the number of reconstructed pixels, assuming sparsity of the image in the wavelet domain.

Another application in which the measurements are performed in the transform domain is magnetic resonance imaging (MRI). In MRI the measurements correspond to samples of the image's two-dimensional continuous Fourier transform. By exploiting the principles of CS, one can recover an MRI image from fewer Fourier-domain measurements assuming sparsity of the image in an appropriate transform domain. For example, magnetic resonance angiography images are typically sparse in the pixel domain. The sparsity of these images can be increased by considering spatial finite differences. Brain MRIs are known to be sparse in the wavelet domain, and dynamic MRI is sparse in the temporal domain.

MRI scanning time strictly depends on the number of samples taken during acquisition. Therefore, applications of CS to MRI offer significant improvement in image acquisition speed. As performing an MRI scan currently takes at least thirty minutes, rapid MRI will reduce patient discomfort and image distortion due to patient movement during acquisition. An important factor affecting the performance of CS-based MRI recovery is the sampling trajectory chosen in the frequency domain. Pure random sampling is impractical, due to hardware and physiological constraints. This directly impacts the RIP and the coherence of the resulting measurement matrix. Different applications of MRI impose varying constraints on the possible trajectories, which must be taken into account when designing a CS-based MRI system.

### 4.2 Analog-to-Digital Conversion

To date, essentially all ADCs follow the celebrated Shannon–Nyquist theorem, which states that, in order to avoid information loss when converting an analog signal to a digital one, the sampling rate must be at least twice the signal bandwidth. Ongoing demand for data, as well as advances in radio frequency technology and the desire to improve resolution, have promoted the use of high-bandwidth signals. The resulting rates dictated by the Shannon–Nyquist theorem impose severe challenges both on the acquisition hardware and on subsequent storage and processing.

(a) (b)

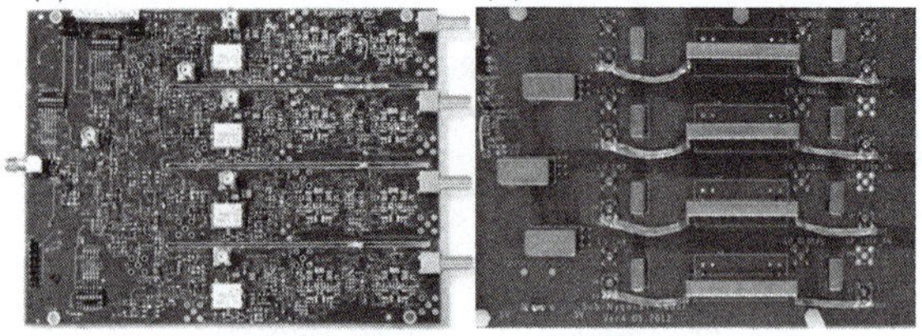

**Figure 1** Sub-Nyquist hardware prototypes for (a) cognitive radio and (b) radar.

Combining ideas from sampling theory with the principles of CS, several new paradigms have been developed that allow the sampling and processing of a wide class of analog signals at sub-Nyquist rates using practical hardware architectures. One such framework is referred to as Xampling, and it has led to sub-Nyquist prototypes for a variety of problems including cognitive radio, radar, ultrasound imaging, ultra-wideband communication, and more. Two of the hardware boards developed for cognitive radio and radar are presented in figure 1.

In a cognitive radio setting, the signal $x(t)$ is modeled as a multiband input with sparse spectra, such that its continuous-time Fourier transform is supported on $N$ frequency intervals with individual widths not exceeding $B$ Hz. Each interval is centered around an unknown carrier frequency $f_i$ that is no larger than a maximum frequency $f_{\max}$. Using the Xampling paradigm, a sub-Nyquist prototype referred to as the modulated wideband converter has been developed that can sample and process such signals at rates as low as $2NB$, despite the fact that the signal may be spread over a very wide frequency range. This rate is much lower than the Nyquist rate, corresponding to $f_{\max}$.

The modulated wideband converter modulates the incoming signal with a pseudorandom periodic sequence, applies a lowpass filter to the result, and then samples the output at a low rate. The mixing operation aliases the spectrum to baseband with different weights for each frequency interval. The signal is recovered using CS techniques that account for the signal structure. The board in figure 1 samples signals with

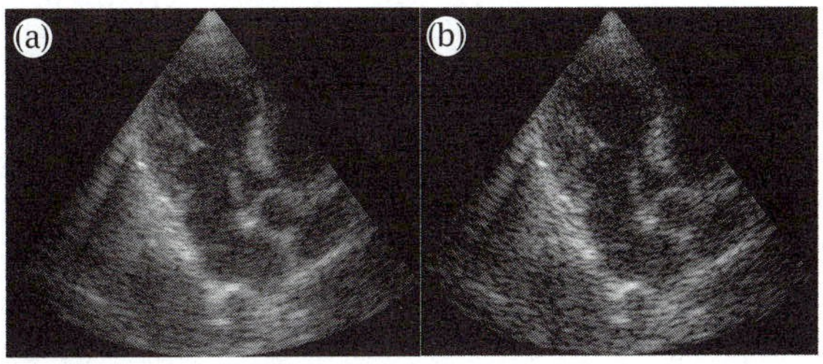


**Figure 2** Sub-Nyquist ultrasound imaging. (a) A standard image and (b) an image formed at a twenty-eighth of the Nyquist rate.

a Nyquist rate of 2.4 GHz and a spectral occupancy of 120 MHz at a rate of 280 MHz.

Another signal class that can be sampled at sub-Nyquist rates are streams of pulses:

$$x(t) = \sum_{\ell=1}^{L} a_\ell h(t - t_\ell), \quad t \in [0, \tau],$$

where the time delays $t_\ell$ and amplitudes $a_\ell$ are unknown. Such signals arise, for example, in communication channels that introduce multipath fading, ultrasound imaging, and radar. Here, again, the Xampling paradigm can be used to sample and process such signals at rates as low as $2L/\tau$ irrespective of the signal's bandwidth.

The board in figure 1 allows, for example, detection of radar signals at a thirtieth of the signal's Nyquist rate. Figure 2 demonstrates fast ultrasound imaging using Xampling. Part (a) shows an ultrasound frame obtained by standard imaging techniques, while the image in part (b) is formed from samples at a twenty-eighth of the Nyquist rate. All the processing is performed at this low rate as well.

## 5 Extensions

In recent years, the area of CS has branched out to many new fronts and has worked its way into several application areas. This, in turn, necessitates a fresh look at many of the basics of CS.

A significant part of recent work on CS can be classified into three major areas. The first of these consists of theory and applications related to CS matrices that are not completely random, or entirely deterministic, and that often exhibit considerable structure. This largely follows from efforts to model the way in which samples are acquired in practice, which leads to sensing matrices that inherit their structure from the real world.

The second area includes signal representations that exhibit structure beyond sparsity as well as broader classes of signals, such as low-rank matrices and matrix completion, exploiting the distribution of the nonzero coefficients or other structured knowledge about the nonzero entries of $x$, and continuous-time signals with finite- or infinite-dimensional representations. In the context of analog signals, large amounts of effort are being devoted to the development of efficient ADC prototypes that achieve sub-Nyquist sampling in practice.

Finally, a very recent trend in CS is to move away from the linear measurement model and consider various types of nonlinear measurements. One particular example of this is phase-retrieval problems in which the measurements have the form $y_i = |a_i^{\mathrm{T}} x|^2$ for a set of vectors $a_i$. Note that only the magnitude of $a_i^{\mathrm{T}} x$ is measured here, and not the phase. Phase-retrieval problems arise in many areas of optics, where the detector can measure only the magnitude of the received optical wave. Several important applications of phase retrieval include X-ray crystallography, transmission electron microscopy, and coherent diffractive imaging. Exploiting sparsity and ideas related to low-rank matrix representations results in efficient algorithms for phase retrieval with provable recovery guarantees. Another example of nonlinear measurements are quantized measurements.

### Further Reading

Bruckstein, A. M., D. L. Donoho, and M. Elad. 2009. From sparse solutions of systems of equations to sparse modeling of signals and images. *SIAM Review* 51(1):34–81.

Duarte, M. F., M. A. Davenport, D. Takhar, J. N. Laska, T. Sun, K. F. Kelly, and R. G. Baraniuk. 2008. Single pixel imaging via compressive sampling. *IEEE Signal Processing Magazine* 25(2):83–91.

Eldar, Y. C., and G. Kutyniok, eds. 2012. *Compressed Sensing: Theory and Applications.* Cambridge: Cambridge University Press.

Foucart, S., and H. Rauhut. 2013. *A Mathematical Introduction to Compressive Sensing.* Boston, MA: Birkhäuser.

Lustig, M., D. Donoho, and J. M. Pauly. 2007. Sparse MRI: the application of compressed sensing for rapid MR imaging. *Magnetic Resonance in Medicine* 58(6):1182–95.

Mishali, M., and Y. C. Eldar. 2011. Sub-Nyquist sampling. *IEEE Signal Processing Magazine* 28(6):98–124.

Tropp, J., and S. Wright. 2010. Computational methods for sparse solution of linear inverse problems. *Proceedings of the IEEE* 98(6):948–58.

Wagner, N., Y. C. Eldar, and Z. Friedman. 2012. Compressed beamforming in ultrasound imaging. *IEEE Transactions on Signal Processing* 60(9):4643–57.

# VII.11 Programming Languages: An Applied Mathematics View

*Nicholas J. Higham*

The purpose of this article is to give an overview of computer programming languages from the point of view of applied mathematics. The historical development is emphasized because modern languages have been strongly influenced by those that came before, and indeed the oldest language of all, Fortran, is still widely used. Figure 1 shows the major relationships between the languages discussed in this article.

## 1 The Early Days

The first digital stored-program computers were programmed by directly entering the low-level instructions, represented by binary numbers, that the central processing unit understood (see figure 2). This was tedious and error prone, so assembly languages were developed that allowed the instructions to be entered as mnemonics, which were then translated by software (the assembler) into the corresponding binary instructions. To add 5 to a number stored in a memory location, one might write a sequence of assembly language instructions such as `LDA P` (load the contents of memory location `P` into the accumulator), `ADC #5` (add 5 to the accumulator), `STA Q` (store the contents of the accumulator in memory location `Q`). Assembly language requires the programmer to work at the level of individual machine instructions and is far removed from mathematical notation.

It was a major step forward when John Backus and his colleagues at IBM designed the language *Fortran* and in 1957 distributed a Fortran compiler for the IBM 704 computer. A *compiler* translates a program written in a high-level language into a sequence of machine instructions that can then be directly executed. Standing for "formula translation," Fortran allowed mathematical expressions to be expressed in a natural algebraic notation, such as `Q = P + 5` for the example above. It also included many of the features we take for granted in programming languages today, such as loops, conditional tests, arrays, and elementary functions. Fortran was a huge success, and it became the first programming language to be standardized, as American National Standards Institute (ANSI) standard Fortran 66 (where the digits denote the year of adoption of the standard). Standardization was important for expanding the number of compiler implementations of the language and aiding the portability of programs from one system to another. Fortran 66 included subroutines and functions, and also supported three floating-point data types: real, double precision, and complex.

Subroutines and functions are examples of *subprograms*, sequences of code forming essentially separate programs that can be called with different input arguments from a main program or other subprogram. They are essential in mathematical computation for encapsulating basic operations such as adding two vectors or finding a norm of a vector, as well as for higher-level tasks such as finding the roots of a polynomial or solving a differential equation. Arguments to a subprogram can be passed in at least two ways. In *call by value* the argument is evaluated at the time of the subprogram call and its value is copied into the formal parameter inside the subprogram. In *call by reference* the address of the parameter is passed, so that the actual and formal parameters effectively share the same memory locations. An important difference between call by value and call by reference is that in the latter case any change made to the argument within the subprogram also changes the actual argument. In Fortran all parameters are passed by reference.

The first commercial Fortran textbook was *A Guide to Fortran Programming* by Daniel McCracken, published in 1961. The author has stated that only a couple of programs in the book had been tested because machine time cost $46 per hour!

*Lisp*, invented by John McCarthy in 1958, is the second oldest language still in wide use. The name stands for "list processor" and, as the name suggests, Lisp is based on list data structures. Lisp is well suited to *functional programming*, in which programs are entirely expressed in terms of mathematical functions (in particular, function application is the only control structure) and functions have no "side effects"; that is, they do not do anything except return a value. Lisp programs look completely different to those written in an *imperative language* such as Fortran, not least due to the profusion of parentheses and the use of prefix notation (the sum `1 + 2 + 3` is expressed as `(+ 1 2 3)`; see section 5.4). While Lisp is rarely used for floating-point computation, it is well suited to symbolic computation, and it is the language in which the popular Emacs editor is mostly written. Lisp also has intrinsic mathematical interest due to its close relation to Alonzo Church's lambda calculus. The "if–then–else" construct

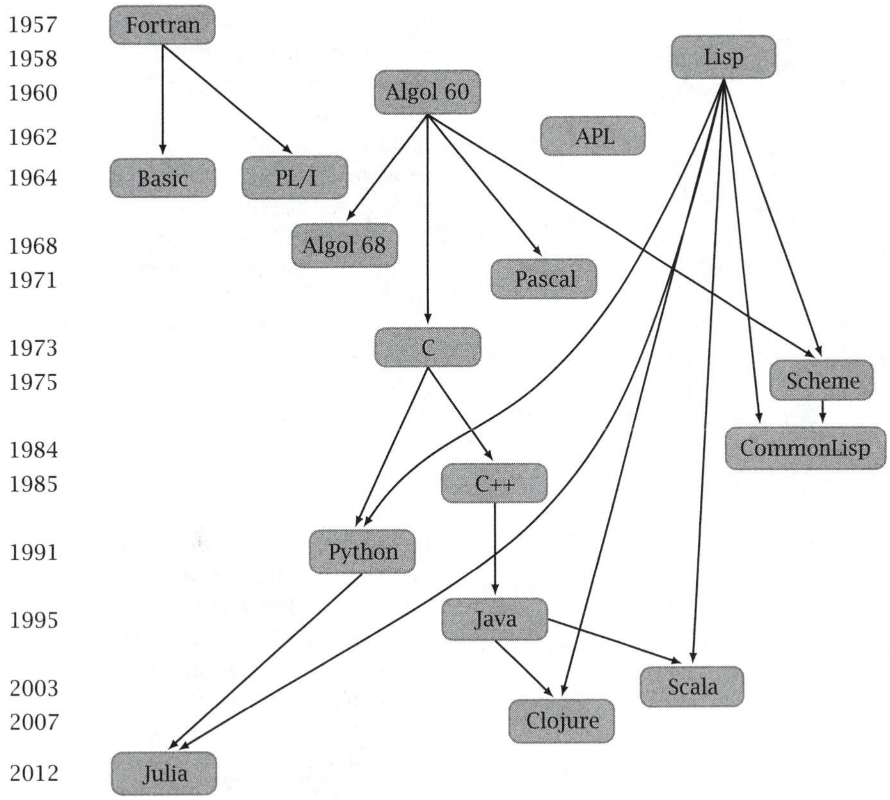


**Figure 1** Time line of selected programming languages, with major influences denoted by arrows.

first appeared in Lisp. Lisp has various dialects, including Scheme (used particularly for teaching) and Common Lisp.

Fortran had been developed in an ad hoc manner. A different language called *Algol 60* was produced in 1960 through the efforts of an international committee. The language was described in a formal notation, later called Backus–Naur form. Algol 60 was based on nested blocks delimited by `begin` and `end` statements, with the scope of a variable (the region of the program in which it is valid) restricted to the enclosing block, and it allowed for dynamic arrays, whose size is determined during execution of the program. It became the "official" language for publishing mathematical software in the 1960s (notably for the first six years of the journal *Communications of the ACM*, which began in 1960) and a strong competitor to Fortran for practical use. However, the language ultimately did not succeed for a variety of reasons, including the fact that it did not define any input–output facilities (making it impossible to write a portable "Hello, world!" program). Nevertheless, some influential early mathematical software was published in Algol 60, notably linear algebra software in the journal *Numerische Mathematik*, later collected into a 1971 volume of the *Handbook for Automatic Computation* series. In late 2014 it was reported that a language called JOVIAL based on a 1958 version of Algol was still in use in the UK air traffic control system.

Algol 60 greatly influenced future languages, such as *Algol 68* (1968), a more rigorously defined language designed by a working group of the International Federation for Information Processing, which was mainly of interest to computer science researchers, and *Pascal*, published in 1971 by Niklaus Wirth, which is a much smaller and simpler language than Algol 68. Pascal was widely taught in universities through the 1980s, as it promoted the notion of structured programming (see section 5.17) and so provided a way to avoid the hard-to-read "spaghetti code" that could easily be produced in Fortran 66. It also achieved wide use in industry, thanks to the availability of compilers on early

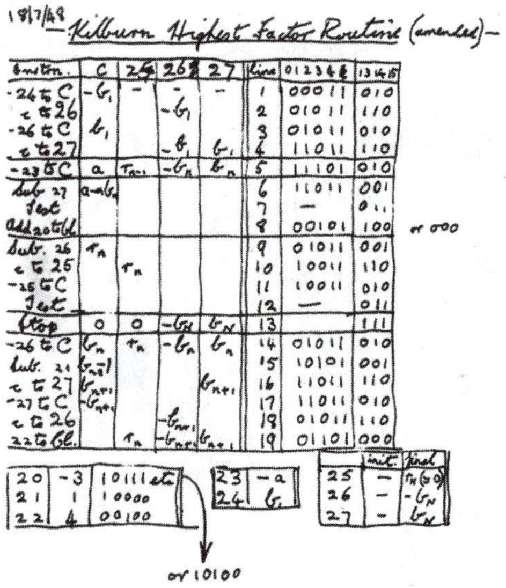

**Figure 2** The first program for a stored-program computer: a version of Tom Kilburn's highest factor routine that first ran on the Manchester "Baby" on June 21, 1948. Taken from G. C. Tootill's notebook. Copyright of The University of Manchester.

PCs and Macintosh computers. However, Pascal was not well suited to numerical programming, not least due to its support for only one type of floating-point variable (real) and the absence of an exponentiation operator.

An influential early textbook was George Forsythe and Cleve Moler's *Computer Solution of Algebraic Equations*, published in 1967. It contained listings of programs written in Algol 60, Fortran, and PL/I for solving a square linear system of equations $Ax = b$. (*PL/I* (1964) was a large language that was not widely adopted for scientific computing; Edsger Dijkstra said that "Using PL/I must be like flying a plane with 7,000 buttons, switches, and handles to manipulate in the cockpit.") These codes were part of a long sequence that led to the Fortran linear system solver in the LINPACK (1979) library.

*Basic* was invented at Dartmouth College in 1964 by John Kemeny and Thomas Kurtz in order to teach programming to students who did not necessarily have a science background. At Dartmouth, Basic was used on a time-sharing system, which allowed the programmer to interact with the computer via a terminal, as opposed to the usual batch processing of the time, in which jobs were prepared on punched cards and handled by computer operators. The original Basic was in some respects a simplified Fortran, with only one data type (double precision), free-form input, the key word LET required before every assignment, numbered lines, and a GOTO command whose destination was a line number. Many early personal computers, including the IBM PC, provided versions of Basic (usually based on Microsoft Basic), typically built into the firmware of the machine. *Visual Basic*, introduced by Microsoft in 1991, included features to aid in the development of graphical user interface (GUI) applications, and it continues to exist as part of the .NET framework. Although a language often associated with writing games (such as the classic Star Trek game originating on 1970s minicomputers), Basic was a capable language for numerical computations, and its accessibility on microcomputers led to it being widely used in mathematical research and teaching, including by this author. Basic was often implemented with an *interpreter*, which translates and executes each statement in the source code before going on to the next statement.

Another 1960s development was the language *APL*, implemented at IBM in 1965. It takes its name, and much of its notation, from the 1962 book *A Programming Language* by Kenneth Iverson. It is unusual in using non-ASCII characters to represent operators and functions, which make possible very concise programs that are often criticized as being cryptic. The notation $\lfloor\cdot\rfloor$ and $\lceil\cdot\rceil$ for the floor and ceiling functions originates in Iverson's book and is used (as functions $\lfloor$ and $\lceil$) in APL. Indeed, APL has been described as an "executable notation." APL has powerful array processing and is normally interpreted. It was never widely adopted but has been influential and is still in use today.

## 2 The Modern Era

The language *C* (1973), by Dennis Ritchie, is a compact language in which the Unix operating system was mainly implemented. C has float and long floating-point data types, corresponding to single and double precision, respectively. Arguments to C functions are passed by value, but a pointer can be passed in order to achieve call by reference. The syntax is terse and powerful. C has been remarkably successful for several reasons. First, its operations and types map directly to the hardware, making it easy to write programs that carry out low-level system tasks and making it possible for compilers to produce very efficient code. Second, an ANSI/ISO standard was produced in 1989 (and revised in 1999 and 2011), aiding portability of programs. Third, C has remained more free from proprietary extensions than other languages.

The language *C++* by Bjarne Stroustrup (1985) is a descendant of C that is a superset, but for minor details, and adds better type checking, flexible data abstraction mechanisms, and support for object-oriented programming. *Data abstraction* allows the programmer to specify user-defined types, called *classes* in C++, and isolates how they are represented from how they are used; in other words, it hides the implementation details within the implementation of the types. *Object-oriented programming* is a methodology based on a hierarchy of classes and *objects*, which are specific instances of the classes with their own characteristics. One of the most popular uses of object-oriented programming is in developing GUIs. C++ also supports "generic programming," through the use of templates, whereby code can be written with parametrized types.

Throughout the history of computing, new programming languages have regularly been designed, with various goals, including providing a better general-purpose language or providing a language tailored to specific purposes, such as system programming tasks.

*Java*, developed by James Gosling at Sun Microsystems in 1995, is a widely used object-oriented language with a syntax similar to that of C++. It compiles to a machine-independent bytecode that runs in the Java virtual machine (JVM), and a JVM is provided for each machine on which Java is to be used. The initial version of Java required bitwise reproducibility of floating-point arithmetic across different machines. While superficially an attractive feature, it inhibited common compiler optimizations as well as the use of extended precision registers. These over-restrictive floating-point semantics were relaxed in later versions of Java, but other aspects such as the lack of complex arithmetic continue to hinder its use for numerical computation. JVMs exploit *just-in-time compilation*, in which Java bytecode is compiled into native machine code at run time. The JVM has importance beyond Java: some more recent languages such as *Scala* (2003) and *Clojure* (a dialect of Lisp created in 2007) compile to JVM bytecode.

Of the many languages introduced since C++, the most important from the computational mathematics point of view is *Python* (1991), designed by Guido van Rossum. Python is a *dynamic language*, which means that it lies somewhere between an interpreted language and a compiled language, with many features of the latter. It supports several programming paradigms, including object orientation and functional programming. Its success in scientific computing stems to a large extent from its libraries, which provide core computational and graphics capabilities (NumPy, SciPy, and matplotlib), and from its ability to integrate components written in other languages, such as C and Fortran. It has been said that "one doesn't need to switch to Python, only to know where to use it." Python was designed to be a readable language, and its expression syntax is similar to that of C.

The newest language discussed here is *Julia* (2012), designed specifically for high-performance scientific computing. Julia is a dynamic language that achieves speed approaching that of compiled C code, in part due to just-in-time compilation using the LLVM compiler infrastructure. A distinctive feature of Julia is its exploitation of *multiple dispatch*, which allows a function to exist in several forms operating on different data types, with the appropriate version being called at run time based on the actual arguments supplied. An interesting feature of Julia is that it allows the user to view the underlying assembly language that the language generates. Viewing these low-level operations can provide much insight into how the language works and its efficiency (see figure 3).

The Fortran standard has undergone regular revisions, known as "Fortran xy," where xy is 77, 90, 95, 2003, or 2008 and is related to the year of publication of the standard. Fortran 77 introduced an if–then–else construct, improved input/output, and a character data type. Fortran 90 incorporated dynamic array allocation, operations on arrays, modules (a mechanism for packaging data, derived types, subprograms, and interface blocks), recursive subprograms, numeric inquiry functions, and parametrized intrinsic types. Later revisions have introduced support for object-oriented programming and for handling exceptions in IEEE floating-point arithmetic, and interoperability with C.

## 3 Parallelism

Most of the languages mentioned above do not include facilities for managing execution of codes in parallel, that is, for specifying how a computation is to be broken up and executed by different processors simultaneously. Various extensions of existing languages have been proposed for parallel computing, but generally they have not achieved widespread or long-term

```
In [1]: f(x,y) = x*y
Out[1]: f (generic function with 1 method)
In [2]: @code_native f(3,5)
.text
Filename: In[1]
Source line: 1
        push RBP
        mov  RBP, RSP
Source line: 1
        imul RDI, RSI
        mov  RAX, RDI
        pop  RBP
        ret
In [3]: @code_native f(3.0,5.0)
.text
Filename: In[1]
Source line: 1
        push RBP
        mov  RBP, RSP
Source line: 1
        mulsd XMM0, XMM1
        pop   RBP
        ret
```

**Figure 3** A short Julia session, run from within a Jupyter notebook. The text that follows "`In [·]:`" on a line is user input. The definition of the function `f` does not specify the types of the arguments. Julia generates different x86 assembler code depending on whether the actual arguments are integers (as in `In [2]`) or floating-point numbers (as in `In [3]`).

use. Two widely used systems for parallel computing are the *Message Passing Interface* (MPI) for distributed memory systems and Open MP for shared-memory systems. Both are implemented as application programming interfaces (APIs) that can be invoked from languages such as C, C++, and Fortran. For expressing parallelism on specialist devices such as graphics processing units (GPUs), specialist languages are available, such as *CUDA* for NVIDIA GPUs and *Open Computing Language* (OpenCL) for GPUs and heterogeneous platforms in general.

## 4 Problem-Solving Environments

Nowadays, a large part of scientific computing is done within environments that provide a programming language, an interactive command window with the display of graphics, and the ability to export graphics and more generally publish documents to HTML, PDF, TeX, and so on. They usually also have the ability to mix numerical and symbolic computing and by default display the result of assignments in the command window. The term *problem-solving environments* (PSEs) is used for such systems, of which there are many.

PSEs have dynamic languages that, combined with the interactive interface, avoid the edit-compile-run cycle of languages such as C and Fortran. They allow quick coding without the need to define the types of variables before use. Moreover, a PSE language typically includes high-level constructs that would correspond in a traditional language to many lines of code, such as a command to find the indices of the largest element(s) of an array or to compute the eigensystem of a matrix. Since it is generally accepted that a programmer's productivity, measured in the number lines of code written, is independent of the language, it follows that using a higher-level language should allow the programmer to achieve more in a given time. On the other hand, PSEs usually do not execute code as fast as a compiled language.

The oldest PSE is *MATLAB*, originally written in Fortran in 1978 by Cleve Moler as a means of providing students with easy access to the EISPACK and LINPACK linear algebra program libraries. Rewritten in C, MATLAB was released as a commercial product in 1984 by The MathWorks. The fundamental data type in MATLAB is a matrix, and MATLAB fully supports complex arithmetic.

An interesting feature of MATLAB is that much of it is written in MATLAB, in the form of M-files containing MATLAB commands. Certain key functions are written in C or call vendor-supplied BASIC LINEAR ALGEBRA SUBPROGRAMS (BLAS) [IV.10 §13] or LAPACK [IV.10 §13] codes. MATLAB programs tend to be much shorter than their equivalents in compiled languages, and yet, depending on the nature of the code, they can run at similar speed. Because of the ease and economy of coding, and the interactive interface that aids debugging, MATLAB is often used as a *prototyping tool*, an environment for developing and testing ideas before implementing them in a language such as C or Fortran.

*GNU Octave* is free software with many of the features of MATLAB and a largely compatible syntax, so that carefully coded programs can run in both MATLAB and Octave. *Scilab* is another open-source alternative to MATLAB, but it is less compatible with MATLAB than Octave.

*Maple* started out as a computer algebra system developed at the University of Waterloo in 1980. It is now a commercial product sold by Waterloo Maple and has all the usual features of a PSE.

*Mathematica*, by Wolfram Research Software, had a notebook interface from its first release in 1988, showing program code, output with typeset mathematics, and graphics in a single window. It supports procedural, functional, rule-based, and pattern-based programming paradigms. It is particularly popular in the physics community.

*R* is a freely available PSE targeted at statistical computing and data analysis. Many contributed R packages are available on the Comprehensive R Archive Network (CRAN).

*Sage* is an open-source, Python-based PSE that builds on many other open-source packages. It has a browser-based notebook.

*Project Jupyter* (formerly known as IPython) is an open-source project that includes a network protocol for interactive computing in any programming language, a browser-based notebook interface, and tools for sharing and converting these notebooks into multiple output formats, including HTML and PDF. This makes the Jupyter Notebook a full-fledged PSE for Julia, Python, R, and other languages.

## 5 Programming Miscellany

We now focus on a variety of different aspects of programming that have a particular relevance to applied mathematics.

### 5.1 Pseudocode

In the early days of computing it was common to include a complete program listing in an article, as can be seen in the 1950s issues of the journal *Mathematical Tables and Aids to Computation*. This practice is now uncommon, not least because of the ease of distributing code over the Web. It is now usual to describe in print the underlying algorithm in terms of a *pseudocode* that the author bases informally on the control structures and other syntax of a particular programming language (MATLAB being a common example). A good pseudocode combines precision, brevity, and readability. For examples of pseudocode see the article on ALGORITHMS [I.4].

### 5.2 Abstraction

The mathematical concept of abstraction has proved to be important in programming, where it refers to separating concepts from implementation details. Subprograms take input arguments, carry out a computation, and then return output. How they do it need not be known to the programmer who invokes them, so a subprogram is an abstraction of the computation it carries out. Abstraction applies to both procedures and data, and is used to the full in object-oriented programming.

### 5.3 Influence on Mathematics

While mathematics has had a strong influence on programming language design, programming languages have also influenced mathematics. We already noted that APL introduced the ceiling and floor notation. The array subscripting (or slicing) notation `A(i:j,p:q)`—used in Algol 68, MATLAB, and other languages to denote the subarray comprising the intersection of rows `i` to `j` and columns `p` to `q` of the two-dimensional array `A`—is now widely used in numerical linear algebra, especially in pseudocode.

In a 1928 paper, Kurt Hensel suggested the notation $A\backslash B$ for $A^{-1}B$ and $A/B$ for $AB^{-1}$, but it did not catch on. Cleve Moler independently introduced the notation in MATLAB, and the term "backslash" is now commonly used to mean solving a matrix equation.

### 5.4 Notation for Expressions

In mathematics we normally write expressions in the conventional *infix notation* illustrated by `a + b(c-d)`, using parentheses and the usual precedence rules to specify the order of operations. In Lisp and related languages, the expression above is written

```
+ a * (b (- c d))        (1)
```

in which each arithmetic operator is followed by its two arguments. This *prefix notation* (also called Polish notation) is easier for computers to parse. The evaluation proceeds left to right, with the arguments of each operator evaluated recursively (in practice, using a stack), and no knowledge of the precedence of the operators is necessary.

The parentheses in (1) are not strictly necessary, but they are required in Lisp because operators can take multiple arguments: `+ 1 2 3` evaluates to `6`.

In *reverse Polish notation* (RPN) the operator follows rather than precedes the operands (as in the expression $n!$ for a factorial). The expression (1) is written

```
a b c d - * +
```

which is again evaluated left to right, with the variables `a` and `b` set aside until it is time to use them. An alternative way to write the expression that mingles the data and the operands is

```
c d - b * a +
```

RPN is used in the languages *Forth* and *PostScript*, and on HP pocket calculators.

### 5.5 Syntactic Peculiarities and Pitfalls

While there is much commonality between different languages, certain differences can catch the unwary programmer out. In most languages a single equals sign denotes an assignment: `x = 1`. A test for equality is written with a double equals in C and MATLAB: `if (x == y)`. If the test is written `if (x = y)` then in C this results in `y` being assigned to `x` and the `if` test being passed if `x` is nonzero because `(x = y)` evaluates as a true Boolean expression. Algol and Pascal use `:=` for assignment, but this is not common in modern languages. R has two assignment operators, `<-` and `=`, of which only the former can be used anywhere in a program. The test for "not equal" is even more varied: `~=` in MATLAB, `!=` in C, R, and Python, `.NE.` in Fortran 77, `/=` in Fortran 90, and `<>` in Basic and Pascal.

A common operation is to increment a variable, which is typically done using a statement such as `x = x + 1`. Some languages provide a shorthand notation for this operation: in Python it is `x += 1` and in C, C++, and Java it is `x++`. A subtlety is illustrated by the C code

```
i = 1; j = 1; a = i++; b = ++j;
```

which results in `a = 1` and `i = j = b = 2` because the assignment is done before the incrementation with `i++` and after for `++j`.

Another aspect of syntax that varies among languages is operator precedence in expressions. An expression `a*b + c` is interpreted as $ab + c$ in most languages, but as $a(b + c)$ in APL, which does not have any operator precedence and always evaluates right to left. However, it is for relational and logical operators that differences are most common. An expression of the form `x or y and z` (with symbols such as `|` and `&` replacing `or` and `and` in many languages) can mean `x or (y and z)` or `(x or y) and z` depending on the language. In Lisp, expressions must be fully parenthesized, so they always have an unambiguous mathematical meaning.

### 5.6 Booleans

A Boolean, or logical, data type contains two possible values: true and false. Many languages denote these values `true` and `false`. Exceptions include Lisp (`t` and `nil`) and Fortran (`.true.` and `.false.`).

C does not have a Boolean data type and instead regards any nonzero numerical value as representing true and zero as representing false. In MATLAB, logical values are converted to 0 (for false) or 1 (for true) in numerical expressions, and this can be useful in a one-line expression such as

```
(exp(x) - 1 + (x == 0)) /( x + (x == 0))
```

which evaluates to $(e^x - 1)/x$ when $x \neq 0$ and to $1 = \lim_{x \to 0}(e^x - 1)/x$ when $x = 0$, avoiding what would otherwise be a division by zero.

### 5.7 Array Storage and Array Indexing

Fortran stores arrays in column major order, meaning that a two-dimensional array is stored sequentially in memory, with the elements of the first column being followed by those of the second, and so on. C and many other languages store arrays in row major order. This difference is inconvenient when calling Fortran codes from other languages. Knowledge of the storage format is crucial because for efficiency it is important to access elements of arrays in the order in which they are stored.

Programming languages differ as to the starting index for arrays. In Fortran and MATLAB, for example, arrays start at index 1 (`a(1), a(2), ...`), whereas in C and Python the first index is 0 (`a[0], a[1], ...`). Note that the type of brackets used for array indices, round or square, also varies, as illustrated. Mathematical descriptions of an algorithm may use 0 or 1 as the starting index, depending on the notation in effect.

The syntax for array slices also differs between languages. While in Fortran and MATLAB `a(i:j)` extracts `a(i), a(i+1), ..., a(j)`, in Python `a[i:j]` extracts `a[i], a[i+1], ..., a[j-1]`, so `a[j]` is omitted. These differences can be a cause of confusion and bugs. One needs to be aware of them and program with care.

### 5.8 Complex Arithmetic

Computations with complex numbers are ubiquitous in applied mathematics. From its earliest versions Fortran has had a complex data type that can be used in expressions such as `a + b*c`, just like the real and double-precision data types. In some other languages, functions implementing complex arithmetic can be written but then expressions must be converted to a sequence of function calls, such as `cadd(a,cmult(b,c))`. The PSEs mentioned above all support complex arithmetic, as do C (introduced in the 1999 standard), C++, Julia (which uses `im` rather than `i` for the imaginary unit), and Python (which uses `j` for the imaginary unit).

It cannot necessarily be assumed that the compiler or interpreter implements complex arithmetic in the most accurate and robust way. For example, if the modulus of a complex number is computed as $|a + ib| = (a^2 + b^2)^{1/2}$, then the intermediate sum of squares can overflow even when $|a + ib|$ is representable as a finite floating-point number. The possibility of overflow is easily avoided by evaluating $|a|(1 + (b/a)^2)^{1/2}$ when $|a| \geqslant |b|$ and an analogous expression when $|b| > |a|$. Operations such as complex division and evaluation of complex elementary functions are more difficult to implement reliably.

### 5.9 Variable Names

In mathematics, variable names are usually one letter, Greek or Roman, in lowercase or uppercase. Since Fortran introduced the possibility of variable names having more than one letter (albeit limited to six letters in Fortran 77 and earlier versions), multiletter names have been common. Due to the use of long variable names comprising several words joined together, several naming conventions have been introduced, illustrated by `endOfFile` or `EndOfFile` (camel case), `end-of-file`, and `end_of_file` (pothole case). Of course, which characters are allowed in variable names depends on the language. The use of long variable names is facilitated by text editors that allow autocompletion.

### 5.10 Floating-Point Semantics

Many mathematical relations fail to hold in FLOATING-POINT ARITHMETIC [II.13] because of the effects of rounding errors. For example, `(a+b)+c` and `a+(b+c)` will in general evaluate to results differing at the roundoff level. Unfortunately, much more subtle issues can cause mathematical relations to break down. Intel x86 chips have 80-bit registers whose precision exceeds that of 64-bit double-precision variables. After the assignment `x = 1.0/3.0` to a double-precision variable `x`, a test `if x == 1.0/3.0` can return false with some optimizing compilers if `1.0/3.0` is temporarily stored in an extended precision register.

Some processors offer a *fused multiply-add* (FMA) instruction that evaluates an expression `x*y + z` with just a single rounding error; that is, the result is the exact value of `x*y + z` rounded to the target precision. The behavior of a program can then depend on the compiler in subtle ways. For example, the discriminant $b^2 - 4ac$ of a quadratic equation can evaluate as negative when $b^2 \geqslant 4ac$ if an FMA is used. These kinds of behavior make it very difficult to prove rigorous correctness results for computer programs executed in floating-point arithmetic.

### 5.11 Floating-Point Parameters

Programs that perform floating-point computation often need to use parameters of the floating-point arithmetic, such as the UNIT ROUNDOFF [II.13] (typically in a convergence test) or the overflow level. Some languages, such as Fortran, provide direct access to these parameters via intrinsic functions. For those that do not, there are ways to compute them at run time, though these may not be entirely reliable when used with optimizing compilers.

### 5.12 High-Precision Computations

The IEEE floating-point arithmetic standard defines single- and double-precision formats corresponding to about eight and sixteen significant decimal digits, respectively. Most programming languages support two floating-point data types that map onto these formats. A 2008 revision of the IEEE standard added a 128-bit quadruple-precision format, which corresponds to about thirty-two significant decimal digits. Quadruple precision is not yet available in hardware, so arithmetic of precision higher than double must currently be provided in software.

In Fortran 90 and later versions of Fortran the availability of different precisions can be queried, through the `selected_real_kind` function. This allows access to quadruple precision if it is supported by the compiler.

A number of open-source libraries are available that implement arbitrary precision floating-point arithmetic. The GNU MPFR library is a C library that provides correctly rounded arithmetic and mathematical functions, and it is used by Julia's BigFloat data type. The GNU MPC library builds on MPFR to handle complex arithmetic. Mpmath is a Python library for arbitrary precision floating-point arithmetic.

High-precision arithmetic has many uses, including in EXPERIMENTAL APPLIED MATHEMATICS [VIII.6] and for obtaining accurate solutions to ill-conditioned problems. For a researcher developing or testing a numerical algorithm, high precision provides a way to compute reference solutions that allow the accuracy of the algorithm to be tested.

### 5.13 Types

A number of subtle issues in programming languages revolve around the data type of a variable or expression: integer, floating point, logical, string, and so on. Some languages, such as C and Java, require the type of a variable to be explicitly declared before an assignment is made to that variable. For example, in C the statement `double x = 1.1` both declares `x` to be a double-precision variable and gives it the value 1.1. Some languages make specifying the type of a variable optional or not possible at all. Fortran uses implicit typing: if the type is not specified then a default type is assigned based on the first letter of the variable (integer for `i` to `p` and real otherwise). However, it is regarded as good practice to turn off this implicit typing with the statement `implicit none`. PSEs tend to determine the type at the point of assignment.

The type of a variable or expression might be fixed or it might be able to change during the execution of a program. For example, some languages allow a string to be added to a number and define the result to be either a string or a number. The terms *weakly typed* and *strongly typed* are often used in this context to characterize a language's type system, but these terms have no commonly agreed definition.

Type systems have an important influence on programs in at least two main ways. First, many programming errors are caused by variables (or constants) having an incorrect type. An apocryphal story tells of the loss of a 1960s NASA rocket due to the Fortran 66 software controlling the rocket having a line of the form `DO 10 k = 1.3` instead of the intended `DO 10 k = 1,3`, which starts a loop. The mistyping of a period for a comma in the former statement causes the Fortran compiler to interpret it as the assignment of 1.3 to the variable `DO10k`, since spaces are unimportant in Fortran 66 source code, and the implicit typing of Fortran causes the variable `DO10k` to be created with real type.

The second influence of a type system is on efficiency, since the speed at which a code executes will depend on how much the compiler or interpreter knows about the types of the variables. The computation of `x*y` will run much slower than it might if at run time the types of `x` and `y` must be checked to decide whether to issue an integer multiplication or a floating-point multiplication instruction. Figure 3 illustrates the point, but in this instance the decision is made at compile time, with no loss of efficiency.

### 5.14 Complexity Analysis

Several measures of the complexity of a code have been proposed. They can be used to estimate the probability of bugs, the difficulty of testing, and the cost of maintenance of the code. The metrics apply to individual components such as functions, subroutines, and procedures, and a large complexity measure can be reduced by breaking the component into smaller pieces.

The simplest metric is the number of executable lines of source code. The *cyclomatic complexity*, or *McCabe complexity*, of a code is defined in terms of the DIRECTED GRAPH [II.16] that has nodes given by blocks of code containing no decisions or branches and edges corresponding to branches between nodes. The cyclomatic complexity is given by the formula edges $-$ nodes $+$ 2, and it turns out to be equal to one plus the number of predicates (logical tests). The *Npath metric* is the number of possible execution paths through the code, which can be much larger than the cyclomatic complexity.

Tools are needed to compute these metrics. In MATLAB the function `checkcode` computes the cyclomatic complexity.

### 5.15 Formatting of Source Code

Mathematicians are used to having complete freedom in how they lay out their written mathematics on the page. Programming languages vary in their prescriptiveness of the layout of the source code. Most impose few restrictions and allow one to collapse a program block onto a single very long line provided comments are removed and (if necessary) statement separators are added. When computers had small memories, such a transformation would sometimes be done in order to save having to store the carriage return and line feed characters. Sometimes further code obfuscation is done in order to conceal the purpose of a code, for security reasons.

Fortran 77 requires code to lie between columns 7 and 72, with columns 1–5 reserved for statement numbers and column 6 for indicating a continuation line. These restrictions stem from the punched cards used to enter programs into early computers and were removed in Fortran 90.

Some text editors provide automatic indentation tailored to the language being edited, and various pretty printing tools are available to format code for readability or to impose a particular house style. The use of such tools can aid debugging and make it simpler to compare different versions of a program with a diff command. Python is unusual in that it uses indentation to define if statements, for loops, while loops, and so on, whereas most languages use braces, brackets, or key words to delimit code blocks.

### 5.16 Readability

There are often several ways to write a piece of code. A balance needs to be struck between length of code, efficiency, and understandability. In C++, for an integer variable `n` one can compute the expression `2*n+1` as `n << 1 | 1`, where << is the bit-shift left operator and | is the bitwise or. The latter version is, however, rather inscrutable and may not be any faster than the former under a good compiler.

Sometimes one needs to make a variable cycle between several values. If the values are 0 and 1 then the assignment `n = 1 - n` flips between them and the purpose of the assignment is reasonably clear. Suppose, though, that we wish to make $n$ take on the values 1, 2, 3, repeatedly. If we can find a polynomial $p$ such that $p(1) = 2$, $p(2) = 3$, and $p(3) = 1$, then the assignment `n = p(n)` will do the trick. Such a $p$ is a POLYNOMIAL INTERPOLANT [I.3 §3.1] to the given data, and the $p$ of lowest degree is the quadratic $p(x) = -\frac{3}{2}x^2 + \frac{11}{2}x - 2$. However, the purpose of the assignment with $p$ is not obvious, and its correctness is not trivial to check. An `if` statement of the form

```
if n == 1
   n = 2
elseif n == 2
   n = 3
else
   n = 1
end
```

does the job in a more transparent fashion. Alternatively, an assignment replacing `n` by $(\text{n} \bmod 3)+1$ could be used, supplemented by a comment explaining its purpose.

### 5.17 Structured Programming

In Fortran a `go to` statement causes a jump to a labeled statement anywhere in the program. In 1968 Edsger Dijkstra wrote a letter to the editor of the journal *Communications of the ACM* in which he claimed that the use of `go to` statements, which were very common in Fortran 66 programs, represented poor programming practice. The letter was published with the title "Go to statement considered harmful." The notion of *structured programming* subsequently became popular. Structured programming enforces a logical structure on the program that makes it easier to understand and modify through the use of certain canonical control structures together with modular composition of programs. A long 1974 paper by Donald Knuth titled "Structured programming with `go to` statements" presents a balanced analysis of the pros and cons of `go to` statements.

### 5.18 Literate Programming

In the 1980s Knuth championed the idea of *literate programming*, in which a document contains a combination of source code and documentation for the code (in TeX format), and both the code and the documentation can be generated from it. He used this approach to great effect in writing TeX and associated programs using his WEB system (which has no connection with the World Wide Web, which it predates). Nowadays, literate programming is mainly used in two forms. In the first, documentation is embedded in comment lines of a program's source code and documentation generation tools are used to extract it to HTML, PDF, etc. In the second form, a document contains code that carries out the computational experiments needed for a paper, and a separate "preprocessor" executes the code and inserts its output (numeric or graphical) back into the source document. This approach facilitates REPRODUCIBLE RESEARCH [VIII.5] and is typically done with "weave" tools available for R and Python or in Emacs Org mode.

### 5.19 Interoperability

Interoperability refers to the ability to call a program written in one language from a program written in a different language. Historically, the degree of interoperability that is available has depended on which operating system and compiler is in use, as well as on the languages themselves. Even when cross-language calls are possible there are pitfalls to watch out for, such as the potentially different ways in which multidimensional arrays are stored in different languages (see section 5.7). There is a strong trend to mixed-language programming, encouraged by languages such as C++, Julia, and Python that have been designed with interoperability in mind, by the support provided in PSEs for calling or being called by another language, and by languages that are built on the same virtual machine (such as Java, Scala, and Clojure).

### 5.20 Domain-Specific Languages

A domain-specific language (DSL) is a language focused on a particular problem domain, examples being HTML for Web pages, SQL for databases, and TeX FOR MATHEMATICAL TYPESETTING [VIII.4 §1]. An important benefit of a DSL is that it can allow programming at a high level of abstraction that fully exploits knowledge of the problem domain and thereby reduces the total time to deliver a solution to a problem.

Applied mathematics has a variety of DSLs, and these often involve symbolic manipulation as part of the code-generation process. The *General Algebraic Modeling System* (GAMS) is a high-level modeling system for mathematical optimization. It includes a DSL in which optimization problems of several different types can be specified. A number of DSLs are associated with software for solving partial differential equations. For example, the *Unified Form Language* in the FEniCS project is a DSL for finite-element discretizations, implemented as a Python module.

DSLs for plotting graphics are plentiful, even being built on top of other DSLs (e.g., the various graphics packages for LaTeX).

### 5.21 Translation between Languages

In mathematics we are used to translating between different notations and moving from one space or basis to another. It is natural to ask whether a program can be transformed from one language to another without any change in its behavior. One reason for wanting to do so is to convert programs that were written many years ago but are still used today (legacy codes) into a more modern language. Such translation tools are available, but they are used out of necessity rather than as a standard tool. A tool called f2c written at Bell Labs in the 1990s could convert Fortran 77 codes to C, though the resulting code was not meant to be readable by humans. There are more recent tools for converting Fortran to C++ that produce more readable code.

**Table 1** Extract from the TIOBE Programming Community Index for February 2015. Clojure, Forth, Mathematica, and OpenCL are all ranked in the range 51–100.

| Language | Rank |
|---|---|
| C | 1 |
| Java | 2 |
| C++ | 3 |
| Python | 8 |
| Visual Basic | 9 |
| MATLAB | 17 |
| R | 18 |
| Pascal | 19 |
| PostScript | 24 |
| Fortran | 31 |
| Lisp | 32 |
| Scheme | 38 |
| Scala | 41 |
| PL/I | 45 |

### 5.22 Popularity of Languages

An interesting question is which are the most popular programming languages. This question is both hard to define precisely and hard to answer. One attempt is provided by the TIOBE Programming Community Index (www.tiobe.com), which is produced once a month based on "the number of skilled engineers world-wide, courses and third party vendors," as found via popular search engines. Table 1 shows a ranking of most of the languages mentioned in this article. These rankings are quite volatile and should not be taken too seriously, but an interesting implication is that old languages such as Fortran and Lisp continue to compete with their younger counterparts.

### 5.23 Language of the Future

An old joke goes, "I don't know what language we'll be using in fifty years time, but it will be called Fortran." Fortran has been under attack since the 1960s but shows no signs of dying, as noted in the previous subsection. The frequent revisions to the Fortran standard have kept the language up to date, while the huge amount of legacy code means that in many applications it is difficult or impossible to switch to alternative languages. The improved interoperability of languages and compilers enables binaries of compiled Fortran libraries such as LAPACK and the commercial NAG Library to be readily called from other languages and even Excel spreadsheets. Perhaps the future is inherently multilingual, with programs being written in a modern language such as C++ or Python and calling kernels written in C, Fortran, or assembly language tuned for particular processors by the manufacturer.

One thing we can be sure of is that new languages will continue to be developed, each trying to combine the best features of existing languages with new ideas that resonate with developments in hardware and software. However, it is important that language designers remember the lessons of the past and contemplate the comment of Tony Hoare about Algol 60: "Here is a language so far ahead of its time, that it was not only an improvement on its predecessors, but also on nearly all its successors."

### Further Reading

The following list is very selective and merely provides a starting point for further exploration.

Abelson and Sussman (1996) is a classic introduction to programming based on Scheme that emphasizes ideas such as abstraction and recursion over syntax. It has many interesting mathematical examples, including symbolic differentiation.

The longevity of Fortran, with its multiple revisions, is such that its history, as told by Metcalf (2011), provides a prism into the history of programming languages.

The Turing Award of the Association for Computing Machinery (ACM) is an annual award that is to computer science what the Fields Medal is to mathematics. The book of lectures from the first twenty years of awards is full of insights into programming languages. It includes lectures by, among those mentioned in this article, Backus, Dijkstra, Iverson, Knuth, McCarthy, Ritchie, and Wirth.

An excellent source for the history of programming languages (and computing) is the journal *IEEE Annals of the History of Computing*.

Abelson, H., and G. J. Sussman. 1996. *Structure and Interpretation of Computer Programs*, 2nd edn. Cambridge, MA: MIT Press.

ACM. 1987. *ACM Turing Award Lectures: The First Twenty Years, 1966–1985*. Reading, MA: Addison-Wesley.

Bentley, J. L. 1988. *More Programming Pearls: Confessions of a Coder*. Reading, MA: Addison-Wesley.

Kernighan, B. W., and P. J. Plauger. 1978. *The Elements of Programming Style*, 2nd edn. New York: McGraw-Hill.

Kernighan, B. W., and D. M. Ritchie. 1988. *The C Programming Language*, 2nd edn. Englewood Cliffs, NJ: Prentice-Hall.

Knuth, D. E. 2003. *Selected Paper on Computer Languages*. Stanford, CA: Center for the Study of Language and Information.

Metcalf, M. 2011. The seven ages of Fortran. *Journal of Computer Science and Technology* 11(1):1–8.

Stroustrup, B. 2013. *The C++ Programming Language*, 4th edn. Upper Saddle River, NJ: Addison-Wesley.

## VII.12 High-Performance Computing

*Jack Dongarra*

### 1 Historical Overview

Looking back on the last four decades, high-performance computing (HPC) has been characterized by rapid change in vendors, architectures, technologies, algorithms, software, and system usage. Despite all these changes, performance, as measured in terms of the number of flops[1] per second, has evolved steadily. Often cited in this context is Moore's law, which states that the number of transistors on integrated circuits doubles approximately every two years. Figure 1 plots the peak performance of various computers over the last six decades, all supercomputers of their time, and demonstrates how well Moore's law holds for performance for nearly the entire lifespan of modern computing.

The initial success of vector computers in the 1970s, which could carry out operations on whole vectors at a time, was driven by raw performance. The introduction of this type of computer system started the modern supercomputing era. In the 1980s the availability of standard development environments and application software packages became more important. In addition to performance, these criteria determined the success of multiprocessor vector systems, especially with industrial customers.

Massively parallel computers, which share the work among a large number of processors, became successful in the early 1990s due to their better price-performance ratios, which were made possible by the improving performance of "off the shelf" microprocessors. At the lower end of the market and for mid-priced systems, massively parallel processing computers were replaced by microprocessor-based symmetric multiprocessing systems (systems in which identical processors share the same memory) in the middle of

1. A flop is an elementary floating-point operation: addition, subtraction, multiplication, or division.

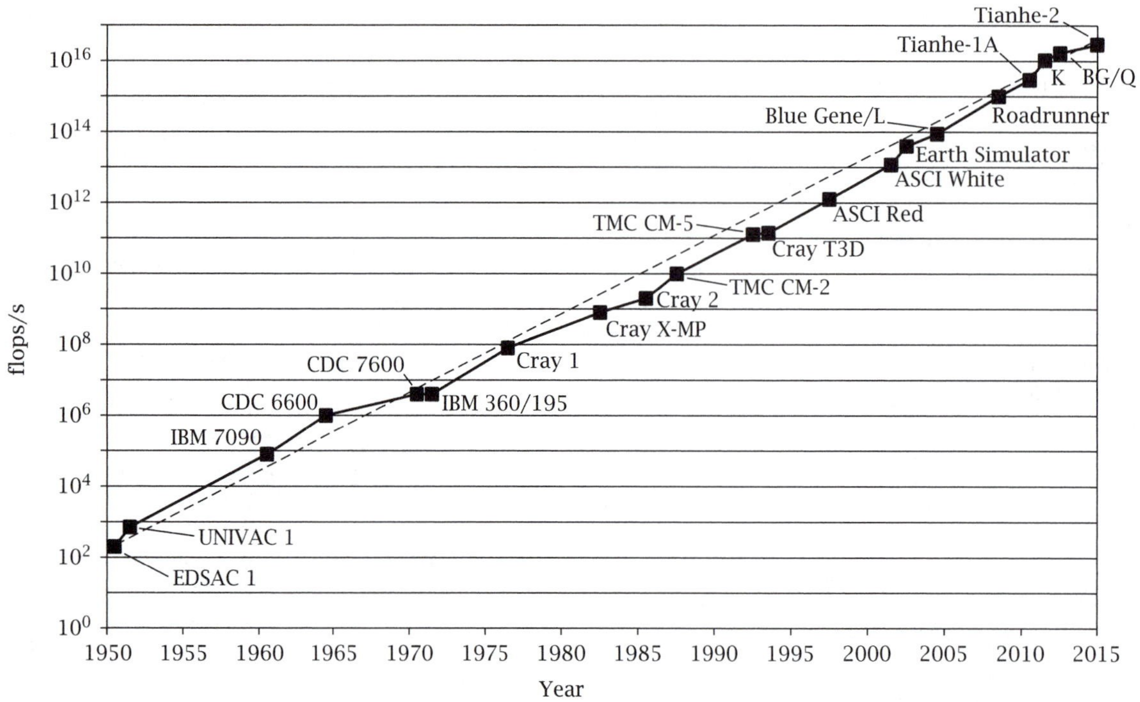


**Figure 1** Peak performance of the fastest computer systems over the last six decades.

the 1990s. The success of microprocessor-based symmetric multiprocessing systems, even for very high-end systems, was the basis for the emergence of cluster concepts in the early 2000s. During the first half of the decade clusters of personal computers and workstations became the prevalent architecture for many application areas. However, the Japanese Earth Simulator vector system (2002) demonstrated that many scientific applications could benefit greatly from a different computer architecture, creating renewed interest in new architectures and new programming paradigms within the scientific HPC community.

The IBM Roadrunner system at Los Alamos National Laboratory, which employs a hybrid design built from commodity parts, broke the petaflops ($10^{15}$ flops per second) threshold in June 2008. The next major target is exascale computing ($10^{18}$ flops per second), a thousandfold increase over petascale, which is not expected to be achieved before 2020.

## 2 Challenges

Science priorities lead to scientific models, and models are implemented in the form of algorithms. Algorithm selection is based on various criteria, such as accuracy, verification, convergence, performance, parallelism, and scalability. Models and associated algorithms are not selected in isolation but must be evaluated in the context of the existing computer hardware environment. Algorithms that perform well on one type of computer hardware may become obsolete on newer hardware, so selections must be made carefully and may change over time. Moving forward to exascale computing will put heavier demands on algorithms in at least two areas: the need for increasing amounts of data locality in order to perform computations efficiently and the need to obtain much higher factors of fine-grained parallelism as high-end systems support increasing numbers of compute threads. As a consequence, parallel algorithms must adapt to this environment, and new algorithms and implementations must be developed to exploit the computational capabilities of the new hardware. The transition from current sub-petascale and petascale computing to exascale computing will be at least as disruptive as the transition from vector to parallel computing was in the 1990s.

We now describe some of the particular challenges that lie ahead in the use of HPCs.

### 2.1 New Algorithms for Multicore Architectures

Multicore processors, in which a single chip contains two or more independent processing units called cores, are now ubiquitous from the desktop through to HPC systems. Scalable multicore systems will increase the cost of communication relative to computation. Within a node (a single multicore processor) data transfer between cores is relatively inexpensive, but across nodes the cost of data transfer is becoming very large. This trend is addressed by new approaches such as communication-avoiding algorithms (see section 2.4), algorithms that support simultaneous computation and communication, and algorithms that vectorize well and have a large volume of functional parallelism.

### 2.2 Adaptive Response to Load Imbalance

Adaptive multiscale algorithms are an important part of many applications because they apply computational power precisely where it is needed. However, they introduce dynamically changing computation that results in processor workload imbalances because the distribution of tasks is static. As we move toward systems with billions of processors, even naturally load-balanced algorithms on homogeneous hardware will present many of the same daunting problems with adaptive load balancing that are observed in today's adaptive codes. For example, software-based recovery mechanisms for fault tolerance or energy management will create substantial load imbalances as tasks are delayed by rollback to a previous state or correction of detected errors. Scheduling based on a directed acyclic graph also requires new approaches to optimize resource utilization without compromising spatial locality. These challenges require development and deployment of sophisticated software approaches to rebalance computation dynamically in response to changing workloads and conditions of the operating environment.

### 2.3 Multiple-Precision Algorithms and Software

One instance of the increasingly adaptive nature of libraries is the capability to recognize and exploit the presence of mixed-precision arithmetic. Motivation comes from the fact that, on modern architectures, 32-bit (single-precision) floating-point operations can execute at least twice as fast as 64-bit (double-precision) operations. The performance of algorithms for solving linear systems or computing eigenvalues or singular values can be significantly enhanced by applying a given method in single precision and then using a few steps of ITERATIVE REFINEMENT [IV.10 §2] in double precision to elevate the accuracy of the result from single to double precision. This technique can be applied not only to conventional processors but also to other technologies such as graphics processing units, and it can therefore utilize heterogeneous hardware more effectively. The use of mixed precision exploits not only the greater speed of single-precision arithmetic but also the fact that there is a reduction in the amount of storage needed and in the amount of information moved from memory for 32-bit or single-precision data when compared with 64-bit or double-precision arrays.

### 2.4 Communication-Avoiding Algorithms

Algorithmic complexity is usually expressed in terms of the number of operations performed rather than the quantity of data movement within memory. However, in modern systems memory movement is increasingly expensive compared with the cost of computation. It is therefore necessary to develop algorithms that reduce communication to a minimum while not unduly increasing the amount of computation. A general approach is to derive bandwidth and latency lower bounds for various dense and sparse linear algebra algorithms on parallel and sequential machines, e.g., by extending the well-known lower bounds for the usual $O(n^3)$ matrix multiplication algorithm, and then to seek new algorithms that (nearly) attain these lower bounds. The study of communication-avoiding algorithms is in its infancy, but it is already leading to new algorithmic ideas and approaches.

### 2.5 Auto-tuning

Numerical libraries need to be able to adapt to the possibly heterogeneous environment in which they have to operate in order to achieve good performance, energy efficiency, load balancing, and so on. The objective is to provide a consistent library interface that remains the same for users independent of scale and processor heterogeneity but that achieves good performance and efficiency by binding to different underlying code, depending on the configuration. In addition, the auto-tuning has to be extended to frameworks that go beyond library limitations and are able to optimize data layout (such as blocking strategies for sparse matrix kernels), stencil auto-tuners (since stencil kernels, which update array elements according to a fixed pattern, are diverse and not amenable to library calls), and even tuning of the optimization strategy for multigrid solvers (optimizing the transition between the multigrid coarsening cycle and the coarse grid solver to minimize run time). Adding heuristic search techniques and combining them with traditional compiler techniques will enhance the ability to address generic problems.

### 2.6 Fault Tolerance and Robustness for Large-Scale Systems

Modern personal computers may run for weeks without rebooting and most data servers are expected to run for years. However, because of their scale and complexity, today's supercomputers run for only a few days before a reboot is needed. The major challenge in fault tolerance is that faults in extreme-scale systems, with their millions of processors, will be continuous rather than exceptional events. This requires a major shift from today's software infrastructure. On today's supercomputers every failure kills the application running on the affected resources. These applications have to be restarted from the beginning or from their last checkpoint. The checkpoint/restart technique will not scale to highly parallel systems because a new fault will occur before the application can be restarted, causing the application to become stuck in a state of constant restarts. New fault-tolerant paradigms need to be developed and integrated into both the system software and user applications.

### 2.7 Building Energy Efficiency into Algorithm Foundations

Energy consumption is becoming a major issue in HPC, with energy costs for some of the largest machines already exceeding a million dollars per year. Minimizing power consumption must now be added to the traditional goals of algorithm design: namely, correctness and performance. The emerging metric of merit is performance per watt. Energy reduction depends on software as well as hardware, so it is essential to build power and energy awareness, control, and efficiency into the foundations of numerical libraries.

### 2.8 Sensitivity Analysis

As the high-fidelity solution of models becomes possible, the next challenge is to study the sensitivity of the model to parameter variability and uncertainty and to seek an optimal solution over a range of parameter values. The most basic form of analysis—the forward method for either local or global sensitivity analysis—simultaneously runs many instances of the model or its linearization, leading to an embarrassingly parallel execution model. Such high-throughput computing tasks are well suited to using spare cycles on pools of personal computers, e.g., running at night or over weekends.

### 2.9 Numerical Pitfalls

Problems that warrant the use of the fastest computers are necessarily among the largest problems ever to be solved, according to any appropriate measure of problem dimension. Various mathematical or numerical difficulties can potentially arise as dimensions grow ever larger, including slower convergence of an iterative method that has performed well for smaller problems, computed results having lower accuracy due to an increased number of rounding errors, and overflow of intermediate results. A good example of what can go wrong concerns the use of RANDOM NUMBER GENERATORS [VI.12] to construct the matrix $A$ and vector $b$ for the linear system $Ax = b$ to be solved by Gaussian elimination with partial pivoting for benchmarking purposes. The obvious approach is to fill the columns of the matrix $A$, one by one, with the output from a pseudorandom number generator. A few years ago, after a computation of this form lasting 20 hours, the computed result was found to be incorrect. The cause was eventually identified as a singular matrix $A$: the number of matrix elements exceeded the period of the random number generator, with the result that columns repeated and the matrix was singular. By itself, singularity should not affect the computation, since rounding errors usually ensure that the matrix is numerically nonsingular. However, the presence of exactly repeated columns eventually leads to "zero pivots," which cause algorithm failure. The moral of the story is that code that has worked perfectly up to a certain problem size can fail in subtle ways for larger problems.

One desirable numerical property of extreme-scale computing is bitwise reproducibility of results for any fixed processor count. But current computing frameworks and libraries do not guarantee reproducibility. The nonreproducibility is usually caused by a parallel reduction operation. While the corresponding operation is mathematically associative, associativity may not hold in floating-point arithmetic. For example, the natural way to evaluate the sum $a + b + c + d$ is from left to right, but alternatives are $(a + b) + (c + d)$ and $(a + c) + (b + d)$, which are trivial examples of a parallel reduction operation, and these three expressions will usually produce different results in floating-point arithmetic. In general, one cannot make assumptions about the order in which reduction operations are carried out in parallel, so the values computed in floating-point arithmetic may depend on the number of threads of execution. This makes it much harder to debug programs. At extreme scale it may be possible to construct faster algorithms if the order of evaluation is not prespecified: through the use of dynamic task scheduling, for example. Thus, there may trade-offs between speed and reproducibility. Furthermore, it may be possible to more cheaply ensure a bound on the variability between different runs than to guarantee strict reproducibility, by using extra precision in selected parts of an algorithm, for example. Many users may prefer non-reproducible results produced very quickly along with a bound on the variability.

## 3 Outlook

The move to extreme-scale computing will require collaboration between hardware architects, systems software experts, designers of programming models, and implementers of the science applications that provide the rationale for these systems. The various issues discussed in this article will need to be considered from a whole-system perspective, and the different tools will need to interoperate. As new ideas and approaches are identified and pursued, some will fail. As with past experience, there may be breakthroughs in hardware technologies that result in different micro- and macro-architectures becoming feasible and desirable, and these will require rethinking of algorithms and system software.

### Further Reading

Ballard, G., J. Demmel, O. Holtz, and O. Schwartz. 2011. Minimizing communication in numerical linear algebra. *SIAM Journal on Matrix Analysis and Applications* 32(3): 866–901.

Dongarra, J. J., et al. 2011. International exascale software project roadmap. *Computing Applications* 25(1):3–60.

Dongarra, J. J., I. S. Duff, D. C. Sorensen, and H. A. Van der Vorst. 1998. *Numerical Linear Algebra for High-Performance Computers*. Philadelphia, PA: SIAM.

Dongarra, J. J., and J. Langou. 2009. The problem with the Linpack benchmark 1.0 matrix generator. *Performance Computing Applications* 23(1):5–13.

Dongarra, J. J., and A. J. van der Steen. 2012. High-performance computing systems: status and outlook. *Acta Numerica* 21:379–474.

# VII.13 Visualization

*Kristin Potter and Chris R. Johnson*

## 1 Introduction

Schroeder, Martin, and Lorensen have offered the following useful definition of visualization:

> Scientific visualization is the formal name given to the field in computer science that encompasses user interface, data representation and processing algorithms, visual representations, and other sensory presentation such as sound or touch. The term data visualization is another phrase to describe visualization. Data visualization is generally interpreted to be more general than scientific visualization, since it implies treatment of data sources beyond the sciences and engineering.... Another recently emerging term is information visualization. This field endeavors to visualize abstract information such as hyper-text documents on the World Wide Web, directory/file structures on a computer, or abstract data structures.

The field of visualization is focused on creating images that convey salient information about underlying data and processes. In recent decades, there has been unprecedented growth in computational and acquisition technologies, and this has resulted in an increased ability to sense the physical world in precise detail and to model and simulate complex physical phenomena. As such, visualization plays a crucial role in our ability to comprehend such large amounts of complex data and to convey insight into diverse scientific applications.

Shown in figure 1, the "visualization pipeline" is one way to describe the process of visualization.[1] The *filtering* step involves processing raw data and includes operations such as resampling, compression, and other image-processing algorithms such as feature-preserving noise suppression. In what can be considered the core of the visualization process, the *mapping* stage transforms the preprocessed filtered data into geometric primitives along with additional visual attributes, such as color or opacity, determining the

1. The figures in this article are all reproduced with the permission of the SCI Institute, apart from plate 21, which is reproduced with the permission of Miriah Meyer. Citations in figure captions refer to sources of further information on the techniques and methods for creating the visualization.

visual representation of the data. *Rendering* utilizes computer graphics techniques to generate the final image using the geometric primitives from the mapping process.

While the range of visualization applications is vast, it can be useful to classify techniques based on the type of data to be presented. These types can be summarized as whether they are

- scalar fields (temperature, voltage, density, vector magnitudes, most image data),
- vector fields (pressure, velocity, electric field, magnetic field), or
- tensor fields (diffusion, electrical and thermal conductivity, stress, strain).

## 2 Scalar Field Visualization

Scalar fields are among the most common data sets in scientific visualization and have therefore received the most research attention.

### 2.1 Direct Volume Rendering

Direct volume rendering is a method of displaying three-dimensional volumetric scalar data as two-dimensional images and is probably one of the simplest ways to visualize volume data. As shown in plate 19, the individual values in the data set are made visible by an assignment to a *transfer function* of optical properties, such as color and opacity, which are then projected and composited to form an image. As a tool for scientific visualization, the appeal of direct volume rendering is that no intermediate geometric information need be calculated, so the process maps from the data set "directly" to an image. This is in contrast to other rendering techniques such as isosurfacing or segmentation, in which one must first extract elements from the data before rendering them. To create an effective visualization with direct volume rendering the researcher must find the right transfer function to highlight regions and features of interest.

### 2.2 Isosurface Extraction

Isosurface extraction is a powerful tool for investigating volumetric scalar fields. The isosurface in a scalar volume is the surface on which the data value is constant, separating regions of higher and lower value. Given the physical or biological significance of the data value, the position of an isosurface, as well as its relation to other neighboring isosurfaces, can provide clues to the underlying structure of the scalar field. A dynamic use of isosurfaces can provide better visualization of complex space- or time-dependent behaviors in many scientific applications. Another powerful technique for analyzing and computing isosurfaces and moving fronts is level set methods.

## 3 Vector Field Visualization

Visualizing vector field data is challenging because no existing natural representation can visually convey large amounts of three-dimensional directional information. Vector field methods must balance the conflicting goals of displaying large amounts of directional information while maintaining an informative and uncluttered display. Researchers have developed a number of vector field visualization techniques using iconic representations, particle tracing methods, and stream constructions. These methods are useful for showing certain field characteristics, but they inherently result in visual clutter when applied globally.

In physical fluid flow experiments, external materials such as dye, hydrogen bubbles, or heat energy are injected into the flow. As these external materials are carried through the flow, scientists can track them visually, using them to examine the underlying flow structure. For example, to understand the flow patterns of river currents, scientists might release dye into the river to expose currents, eddies, and turbulence. Similarly, to understand the air flow around an aircraft wing, scientists may release smoke into wind tunnel experiments to provide visual information about flow patterns, turbulent regions, and vortex formation. Analogs to these experimental techniques have been adopted by scientific visualization researchers, particularly in the computational fluid dynamics field. These researchers have used numerical methods and three-dimensional computer graphics techniques to produce graphical icons such as arrows, motion particles, and other representations that highlight different aspects of the flow. Advection methods numerically integrate along paths defined by the vector field. For example, the researcher can create streamlines by advecting points through an instantaneous, static vector field and tracing their path. They create pathlines, on the other hand, by advecting points through a dynamic, time-varying vector field.

## 4 Tensor Field Visualization

In physical and biological systems, representing intrinsic material properties is an essential part of accurate

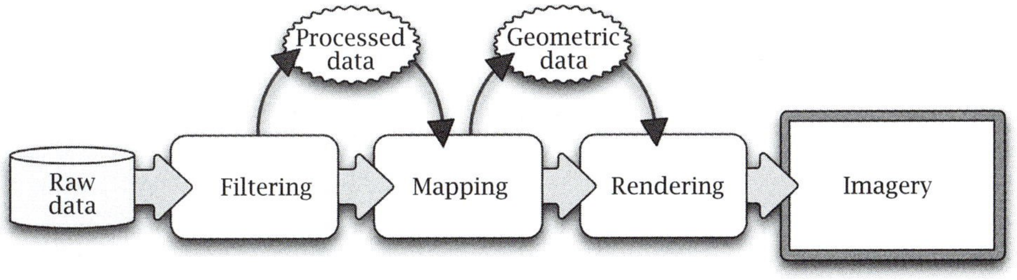


**Figure 1** The visualization pipeline.

modeling and imaging. Electrical conductivity and molecular diffusivity are examples of material properties that describe the ability of particles (such as electrons and water molecules) to pass through a given material. In the simplest situation, the property is constant in all directions, a condition called *isotropy*. On the other hand, *anisotropic* materials are those that exhibit directional sensitivity in the rate of transport: for example, water diffuses through paper faster in one direction than another. In real-world problems, material properties are often *inhomogeneous*: varying as a function of position within the material. Thus, proper modeling of conductivity and diffusivity requires a field of tensor values sampled in three dimensions. Gaining insight into the structure of three-dimensional tensor fields is a significant and ongoing problem in tensor visualization.

Creating meaningful images or models from diffusion tensor data is challenging because each sample point has six independent degrees of freedom. As with vector visualization, simple attempts at encoding all the tensor variables at all sample locations rapidly produce unintelligible visual clutter. In some application instances, such as diffusion tensor magnetic resonance imaging of nerve tissue, the degree of anisotropy has a biological significance relating to the white matter structure, as shown in plate 20(b). An effective way to avoid clutter is, therefore, to display only those tensors that exhibit anisotropy of a certain degree or greater, often using graphical icons such as three-dimensional quadrics to indicate the six degrees of freedom through shape and size. Another popular technique of feature extraction is fiber tractography, which seeks to create pathways to illustrate directional tissue structure. Standard visualization methods for this type of data use hyper-streamlines to illustrate the directional path of a fiber bundle, as shown in plate 20(a). Such methods are often augmented with edge bundling to reduce clutter, or blurring to indicate areas of noise or uncertainty.

## 5 Applications

The most important aspect of visualization is its role in analysis, exploration, and discovery in the scientific process. State-of-the-art technologies are combined in creative ways to facilitate data understanding. It is the role of the visualization researcher to understand how to combine appropriate visualization techniques with hypotheses about the data to reveal answers to scientific questions. Here, we briefly discuss three real-world applications of visualization.

### 5.1 Genomics

MizBee is a multiscale *synteny* browser for exploring conservation relationships in comparative genomics data (see plate 21). Synteny refers to the presence of two or more genes on the same chromosome and can be used to answer questions about evolution and genomic function by comparing the genomes of different species to find regions of shared sequences. Using side-by-side linked views, MizBee enables efficient data browsing across a range of scales, from the genome to the gene. To present pairwise comparisons of similar regions between different chromosomes, concentric circular layouts are used to show two genes and lines are drawn between regions to show pairs. On the right-hand side of the figure, the chromosome view presents a detailed look at user-selected blocks of interest, along with statistical information and layered annotations. The block view (rightmost) is the most detailed view, providing information about the conservation relationships of features within the selected block related to proximity/location, size, orientation, and similarity. Each view is optimized for viewing the different types of relevant information; the views can communicate relevant information and degree of similarity for each region. The design of MizBee is grounded in perceptual principles, and it includes several techniques, such as edge bundling and layering, to enhance visual cues about relationships and the consistent use of an 8-channel color map to effectively distinguish regions.

### 5.2 Climate and Weather

The EnsembleVis framework was developed to explore ensembles of weather forecast models. This work uses summary statistics to manage the display of the large, time-varying data set. Because of its richness, it was important to tease out a number of summaries across various dimensions of the data. Rather then deciding upon a single visualization technique to show all aspects of the data, the system links multiple displays to allow the user to simultaneously explore a variety of data characteristics. Thus, the EnsembleVis framework, as shown in plate 22, combines a number of aggregation windows, each designed to answer a specific question. The main window presents a summary of the data across the spatial domain. Alongside the main window, a number of smaller windows show summaries across other dimensions, such as a filmstrip view of individual time steps, graphs of the individual model responses for a data subset, and a query dialog to filter results according to specific parameters. This collection of visualizations reveals insight into general trends of the model ensemble and highlights outlier model runs. This system provides weather scientists with the ability to explore the outputs of the simulation to understand the consensus of the ensemble, the probability of the consensus outcome, and model configurations that lead to outlying results and to identify biased models that may hint at errors in the model construction or missing atmospheric phenomena.

### 5.3 Bioelectric Fields

Bioelectric fields are produced by the living cells responsible for the action of muscles and the transmission of information in nerves. Many different imaging modalities have been developed to assess this activity, including electrocardiography of the heart and electrocardiography and magnetoencephalography of the brain. In addition to their clinical applications, these imaging techniques provide the basis for a computational reconstruction of the physiological activity at the origin of the electric signal. This allows researchers to gain insight into the mechanisms and consequences of conditions such as myocardial ischemia: a shortfall in blood supply that can, in extreme cases, lead to a complete blockage of blood, resulting in a heart attack. To explore the electric fields and electric current densities associated with the cardiovascular and cerebral activity in humans, advanced vector visualization techniques for visual analysis have been developed.

To understand the bioelectric field in the direct vicinity of the epicardium of the heart, stream surfaces are used to capture the geometry of the current induced by the cardiac source (see plate 23(a)). The surfaces also provide an effective representation of the interconnections that exist between different regions on the heart's surface. A rainbow color map is used along each curve to visualize the stretching of the return current as it propagates through the torso.

The three-dimensional electric current within the brain can be visualized using texturing techniques to show the asymmetry of the electric patterns (see plate 23(b)). Textures, computed on a clipping plane, reveal the dipolar source of electric current and its interaction with the surrounding tissue. The electric current is clearly diverted by the presence of white matter tracts that lie close to the source. The field also changes direction very rapidly as it approaches the skull just beneath the surface of the head.

Focusing efforts on vector-field visualizations of electric current, rather than scalar representations of potentials, offers a more meaningful global depiction of the continuous flow. This permits a deeper understanding of the three-dimensional shape of the bioelectric sources and their fields. Such an approach provides new insight into the impact of tissue characteristics, such as directional dependence, on the resulting bioelectric fields.

## 6 Outlook for Visualization

The greatest successes in visualization come when researchers are able to explore much more information than previously possible. Often, tools designed for specific applications turn out to address general problems and can be applied to other domains and applications. With the many software packages, prototype examples, and tool suites that are available, visualization is becoming accessible at a variety of levels, and we expect to see it become an integral part of the scientific work flow.

### Further Reading

Hansen, C., and C. Johnson, eds. 2005. *The Visualization Handbook*. Amsterdam: Elsevier.

Kindlmann, G. L., D. M. Weinstein, A. D. Lee, A. W. Toga, and P. M. Thompson. 2004. Visualization of anatomic covariance tensor fields. In *26th Annual International Conference of the IEEE Engineering in Medicine and Biology Society*, volume 1, pp. 1842–45. Piscataway, NJ: IEEE.

Kniss, J., P. McCormick, A. McPherson, J. Ahrens, J. Painter, A. Keahey, and C. Hansen. 2001. Interactive texture-based volume rendering for large datasets. *Computer Graphics and Applications* 21(4):52–61.

Meyer, M., T. Munzner, and H.-P. Pfister. 2009. Mizbee: a multiscale synteny browser. *IEEE Transactions on Visualization and Computer Graphics* 15(6):897–904.

Potter, K., A. Wilson, P.-T. Bremer, D. Williams, C. Doutriaux, V. Pascucci, and C. R. Johnson. 2009. Ensemble-vis: a framework for the statistical visualization of ensemble data. In *IEEE Workshop on Knowledge Discovery from Climate Data: Prediction, Extremes*, pp. 233–40. Piscataway, NJ: IEEE.

Schott, M., T. Martin, A. V. P. Grosset, C. Brownlee, T. Hollt, B. P. Brown, S. T. Smith, and C. D. Hansen. 2012. Combined surface and volumetric occlusion shading. In *Pacific Visualization Symposium (PacificVis)*, pp. 169–76. Piscataway, NJ: IEEE.

Schroeder, W., K. Martin, and W. Lorensen. 1997. *The Visualization Toolkit: An Object-Oriented Approach to 3D Graphics*. Englewood Cliffs, NJ: Prentice Hall.

Tricoche, X., and G. Scheuermann. 2003. Topology simplification of symmetric, second order 2D tensor fields. In *Geometric Modeling Methods in Scientific Visualization*, pp. 171–84. New York: Springer.


# VII.14 Electronic Structure Calculations (Solid State Physics)

*Eric Cancès*

## 1 Introduction

The study of the electronic properties of solids has led to major scientific discoveries (superconductivity, the quantum Hall effect, giant magnetoresistance), as well as to a large number of applications that have revolutionized our daily lives: computers and electronic devices being prime examples.

Besides its importance in terms of applications, the modeling of the electronic structure of solids is an inexhaustible source of fascinating problems in mathematical physics, the analysis of partial differential equations, numerical analysis, and scientific computing.

In solid state physics, the fundamental components of matter are atomic nuclei and electrons in Coulomb interaction. Starting with the chemical composition of a material—that is, the number of atoms of each chemical element that are present—it is possible to write down the $n$-body Schrödinger equation encoding most of its physical, and all of its chemical, properties. Unfortunately, this equation is much too complicated to be solved: indeed, it reads as a $3n$-dimensional partial differential equation, where $n$ is the total number of particles (atomic nuclei and electrons) in the material under consideration. For a macroscopic solid, $n$ is of the order of $10^{20}$ or larger, so the numerical solution of the $n$-body Schrödinger equation is way out of reach of even the most powerful computers conceivable at the present time. Several approximations are therefore adopted. The first, called the Born–Oppenheimer approximation, is based on the fact that nuclei are thousands of times heavier than electrons, and it can be mathematically justified by means of adiabatic limits, letting the mass ratio between electrons and nuclei go to zero. According to the Born–Oppenheimer approximation, it is possible to compute the electronic structure of the system at each time $t$ by solving a time-independent Schrödinger equation, parametrized by the (time-dependent) positions of the nuclei. The second step consists of replacing the electronic time-independent Schrödinger equation by simpler models that are amenable to numerical simulation. Electronic structure calculation is concerned with the design and simulation of such models.

The purpose of this article is to introduce the reader to

- independent-particle models, which enable us to qualitatively understand the electronic structure of crystals, and
- the Kohn–Sham model, based on density functional theory.

The latter model allows us to run numerical simulations that are in quantitative agreement with experimental data. It can be used to predict the properties of new molecules, materials, and nanostructures and therefore has an extremely broad range of applications. For instance, in the field of energy technology, it can be used to design new materials for nuclear power plants, fuel cells, or solar cells.

## 2 Electronic States

Any quantum system is characterized by a Hamiltonian: that is, a self-adjoint operator $H$ that acts on some

HILBERT SPACE [I.2 §19.4] $\mathcal{H}$. For independent-electron models, as well as for mean-field models derived from density functional theory, $H$ is a Schrödinger operator of the form

$$H = -\tfrac{1}{2}\Delta + V(\boldsymbol{r})$$

that acts on the Hilbert space $L^2(\mathbb{R}^3)$ of complex-valued square-integrable functions on $\mathbb{R}^3$, where

$$\Delta = \frac{\partial^2}{\partial x^2} + \frac{\partial^2}{\partial y^2} + \frac{\partial^2}{\partial z^2}$$

is the Laplace operator and where $V$ is a real-valued function on $\mathbb{R}^3$. We adopt the system of atomic units here, in which $\hbar = 1$, $e = 1$, $m_e = 1$, and $4\pi\varepsilon_0 = 1$, where $\hbar$ is the reduced Planck constant, $e$ the elementary charge, $m_e$ the mass of the electron, and $\varepsilon_0$ the dielectric permittivity of the vacuum.

The SPECTRAL THEORY OF SELF-ADJOINT OPERATORS [IV.8] is the key mathematical tool in electronic structure calculation. In some sense, a self-adjoint operator on $L^2(\mathbb{R}^3)$ can be seen as an infinite-dimensional Hermitian matrix. A finite-dimensional Hermitian matrix $A \in \mathbb{C}^{n\times n}$ has exactly $n$ real eigenvalues (taking multiplicities into account), and can be diagonalized in an orthonormal basis set. By definition, the spectrum of $A$ is the set $\sigma(A)$ of the complex numbers $\lambda$ such that $\lambda I - A$ is noninvertible. A finite-dimensional square matrix is noninvertible if and only if it is noninjective. Therefore, $\sigma(A)$ is also the set of the eigenvalues of $A$: namely, the set of the complex numbers $\lambda$ for which there exists $x \in \mathbb{C}^n \setminus \{0\}$ such that $Ax = \lambda x$. On the other hand, as $L^2(\mathbb{R}^3)$ is an infinite-dimensional Hilbert space, the set of the eigenvalues of $H$ (the point spectrum of $H$) is, in general, a strict subset of the spectrum $\sigma(H)$ of $H$. For instance, the point spectrum $\sigma_p(H_0)$ of the kinetic energy operator $H_0 = -\frac{1}{2}\Delta$, acting on $L^2(\mathbb{R}^3)$, is empty (there is no nonzero function $\phi \in L^2(\mathbb{R}^3)$ such that $H_0\phi = E\phi$ for some $E \in \mathbb{C}$), while its spectrum $\sigma(H_0)$ is equal to $[0, +\infty)$. The elements of $\sigma(H_0)$ can be interpreted as *generalized* eigenvalues since, for each $\boldsymbol{k} \in \mathbb{R}^3$,

$$H_0 e^{i\boldsymbol{k}\cdot\boldsymbol{r}} = \frac{|\boldsymbol{k}|^2}{2} e^{i\boldsymbol{k}\cdot\boldsymbol{r}}.$$

The plane wave $\boldsymbol{r} \mapsto e_{\boldsymbol{k}}(\boldsymbol{r}) = e^{i\boldsymbol{k}\cdot\boldsymbol{r}}$ satisfies the equation $H_0 e_{\boldsymbol{k}} = E e_{\boldsymbol{k}}$ with $E = \frac{1}{2}|\boldsymbol{k}|^2$, but it is not a *true* eigenfunction, since it does not belong to the space $L^2(\mathbb{R}^3)$. It is called a *generalized* eigenfunction. In physics terminology, true eigenmodes are called *bound states*, while generalized eigenmodes are called *scattering states*. The nature and the location of the spectrum of a Schrödinger operator $H = -\frac{1}{2}\Delta + V$ depends on the potential $V$. For $V(\boldsymbol{r}) = \frac{1}{2}\omega^2|\boldsymbol{r}|^2$ (a three-dimensional harmonic oscillator), the spectrum of $H$ is pure point: that is, it is composed of only the true eigenvalues $E_n = (n + \frac{3}{2})\omega$, $n \in N$. For $V(\boldsymbol{r}) = -Z/|\boldsymbol{r}|$, $Z \in \mathbb{N}^*$ (a hydrogen-like ion), the spectrum of $H$ consists of an infinite sequence of true eigenvalues $E_n = -Z^2/(2n^2)$, $n \in \mathbb{N}^*$, and of a continuum $[0, +\infty)$ of generalized eigenvalues.

The purpose of electronic structure calculation is to identify the electronic ground state, and possibly the lowest-energy excited states, of a molecular system. In the framework of independent-electron models, these states are easily obtained from the spectral decomposition of the Hamiltonian $H$. Assume for simplicity that the bottom of the spectrum of $H$ consists of an increasing sequence of eigenvalues $\varepsilon_1 \leqslant \varepsilon_2 \leqslant \cdots \leqslant \varepsilon_p$ (taking multiplicities into account) and that the molecular system of interest contains an even number $N = 2p$ of electrons. The ground state energy and the density are then given by

$$E_0 = 2\sum_{i=1}^{p} \varepsilon_i \quad \text{and} \quad \rho_0(\boldsymbol{r}) = 2\sum_{i=1}^{p} |\phi_i(\boldsymbol{r})|^2,$$

respectively, where $(\phi_i)_{1\leqslant i\leqslant p}$ is an orthonormal set of eigenfunctions of $H$ associated with the eigenvalues $\varepsilon_1, \dots, \varepsilon_p$:

$$H\phi_i = \varepsilon_i\phi_i \quad \text{and} \quad \int_{\mathbb{R}^3} \phi_i^*\phi_j = \delta_{ij}.$$

In physics, the eigenvalues $\varepsilon_i$ are called energy levels. The ground state is therefore obtained by putting the electrons in the lowest energy levels, under the constraint that there are at most two electrons per energy level. This constraint, called the Pauli principle, is related to the fact that electrons are Fermions of spin $\frac{1}{2}$. We will not elaborate further on the concept of spin here and simply mention that, from a mathematical perspective, the spin labels the projective representations of the rotation group SO(3). Excited states correspond to other distributions of the $N$ electrons in the energy levels. For instance, if $\varepsilon_{p+1} > \varepsilon_p$, the ground state is nondegenerate and the energy of the first excited state is $E_1 = 2\sum_{i=1}^{p-1}\varepsilon_i + \varepsilon_p + \varepsilon_{p+1}$ (see figure 1).

## 3 Noninteracting Electrons in Crystals

As in solid state physics textbooks, we will first focus on the case of perfect crystals, which are periodic arrangements of atoms. We will show in particular that a simple model of noninteracting electrons in a perfect crystal allows us to qualitatively understand why

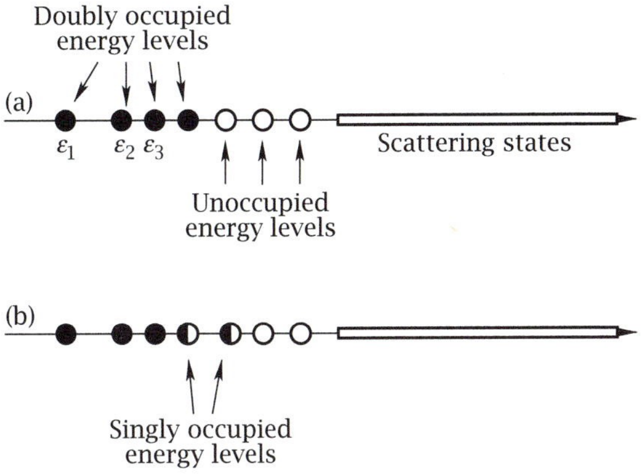


**Figure 1** (a) The ground state and (b) the first excited state for $N = 8$ electrons ($p = 4$ electron pairs).

some crystalline materials are insulators while others are conductors. The mathematical framework for investigating the electronic structure of perfect crystals is Bloch's theory, the basics of which are sketched below. We will then see how to model crystals with defects and disordered materials such as doped semiconductors or alloys.

In noninteracting electron models, a perfect crystal is characterized by a lattice $\mathcal{L}$ (a discrete subgroup of $\mathbb{R}^3$) and an $\mathcal{L}$-periodic potential $V_{\text{per}}\colon \mathbb{R}^3 \to \mathbb{R}$. For the sake of simplicity we will deal with the case when $\mathcal{L} = \mathbb{Z}^3$, so that

$$\forall \boldsymbol{R} \in \mathbb{Z}^3,\ \forall \boldsymbol{r} \in \mathbb{R}^3, \quad V_{\text{per}}(\boldsymbol{r}+\boldsymbol{R}) = V_{\text{per}}(\boldsymbol{r}),$$

and we denote the unit cell by $\Gamma = [-\frac{1}{2}, \frac{1}{2})^3$. Introducing the translation operators $\tau_{\boldsymbol{R}}$ defined by $(\tau_{\boldsymbol{R}}\phi)(\boldsymbol{r}) = \phi(\boldsymbol{r}-\boldsymbol{R})$, the above periodicity condition can be formulated as $\tau_{\boldsymbol{R}} V_{\text{per}} = V_{\text{per}}$ for all $\boldsymbol{R} \in \mathbb{Z}^3$. The quantum Hamiltonian

$$H_{\text{per}} = -\tfrac{1}{2}\Delta + V_{\text{per}}$$

is called a periodic Schrödinger operator. Under some local integrability assumptions on $V_{\text{per}}$, it is self-adjoint on $L^2(\mathbb{R}^3)$. It has no true eigenvalues ($\sigma_{\text{p}}(H_{\text{per}}) = \varnothing$), and its spectrum is bounded below and composed of a countable number of possibly overlapping intervals of the real line, called *bands*. To establish this fundamental result, which is key to explaining insulating, semiconducting, and conducting behaviors, we need Bloch's theory.

Denoting by $\Gamma^* = [-\pi, \pi)^3$ the Brillouin zone associated with the lattice $\mathbb{Z}^3$, any function $u \in L^2(\mathbb{R}^3)$ can be decomposed as

$$u(\boldsymbol{r}) = \frac{1}{(2\pi)^3}\int_{\Gamma^*} u_{\boldsymbol{q}}(\boldsymbol{r})\mathrm{e}^{\mathrm{i}\boldsymbol{q}\cdot\boldsymbol{r}}\,\mathrm{d}\boldsymbol{q},$$

where

$$u_{\boldsymbol{q}}(\boldsymbol{r}) = \sum_{\boldsymbol{R}\in\mathbb{Z}^3} u(\boldsymbol{r}+\boldsymbol{R})\mathrm{e}^{-\mathrm{i}\boldsymbol{q}(\boldsymbol{r}+\boldsymbol{R})}.$$

For each $\boldsymbol{q} \in \Gamma^*$, the function $u_{\boldsymbol{q}}$ is in $L^2_{\text{per}}$, the space of complex-valued, $\mathbb{Z}^3$-periodic, locally square-integrable functions. This decomposition, which is reminiscent of the Fourier transform, is called the *Bloch–Floquet transform*. Bloch's theorem states that, if an operator $A$ on $L^2(\mathbb{R}^3)$ commutes with the translations of the lattice $\mathbb{Z}^3$ (i.e., if it satisfies $A\tau_{\boldsymbol{R}} = \tau_{\boldsymbol{R}}A$ for all $\boldsymbol{R} \in \mathbb{Z}^3$), then there exists a family of operators $(A_{\boldsymbol{q}})_{\boldsymbol{q}\in\Gamma^*}$ on $L^2_{\text{per}}$ such that

$$(Au)(\boldsymbol{r}) = \frac{1}{(2\pi)^3}\int_{\Gamma^*}(A_{\boldsymbol{q}}u_{\boldsymbol{q}})(\boldsymbol{r})\mathrm{e}^{\mathrm{i}\boldsymbol{q}\cdot\boldsymbol{r}}\,\mathrm{d}\boldsymbol{q}.$$

The above formula means that the operator $A$ is block diagonalized by the Bloch–Floquet transform. The operators $(H_{\text{per}})_{\boldsymbol{q}}$ have simple expressions:

$$(H_{\text{per}})_{\boldsymbol{q}} = -\tfrac{1}{2}\Delta - \mathrm{i}\boldsymbol{q}\cdot\nabla + \tfrac{1}{2}|\boldsymbol{q}|^2 + V_{\text{per}}.$$

For each $\boldsymbol{q} \in \Gamma^*$, the operator $(H_{\text{per}})_{\boldsymbol{q}}$ can be diagonalized in an orthonormal basis set of $L^2_{\text{per}}$,

$$(H_{\text{per}})_{\boldsymbol{q}}\phi_{n,\boldsymbol{q}} = \varepsilon_{n,\boldsymbol{q}}\phi_{n,\boldsymbol{q}}, \qquad \int_\Gamma \phi^*_{m,\boldsymbol{q}}\phi_{n,\boldsymbol{q}} = \delta_{mn},$$

and the sequence $(\varepsilon_{n,\boldsymbol{q}})_{n\geqslant 1}$ of its eigenvalues converges to $+\infty$. Besides, as the mapping $\boldsymbol{q} \mapsto (H_{\text{per}})_{\boldsymbol{q}}$ is analytic, the eigenvalues $\varepsilon_{n,\boldsymbol{q}}$ can be indexed in such a way that $\varepsilon_{1,0} \leqslant \varepsilon_{2,0} \leqslant \cdots$ and the mapping $\boldsymbol{q} \mapsto \varepsilon_{n,\boldsymbol{q}}$ is even and continuous (and, in fact, analytic in each direction). As a consequence (see figure 2),

$$\sigma((H_0)_{\text{per}}) = \bigcup_{\boldsymbol{q}\in\Gamma^*}\bigcup_{n\geqslant 1}\{\varepsilon_{n,\boldsymbol{q}}\} = \bigcup_{n\geqslant 1}[\Sigma_n^-, \Sigma_n^+],$$

with

$$\Sigma_n^- = \min_{\boldsymbol{q}\in\Gamma^*}\varepsilon_{n,\boldsymbol{q}} \quad \text{and} \quad \Sigma_n^+ = \max_{\boldsymbol{q}\in\Gamma^*}\varepsilon_{n,\boldsymbol{q}}.$$

The interval $[\Sigma_n^-, \Sigma_n^+]$ is called the $n$th band of the spectrum of $H_{\text{per}}$. If $\Sigma_{n+1}^- > \Sigma_n^+$, the interval $(\Sigma_n^+, \Sigma_{n+1}^-)$ is called a spectral gap. It is easily checked that the functions $\psi_{n,\boldsymbol{q}}(\boldsymbol{r}) = \phi_{n,\boldsymbol{q}}(\boldsymbol{r})\mathrm{e}^{\mathrm{i}\boldsymbol{q}\cdot\boldsymbol{r}}$ are generalized eigenfunctions of $H_{\text{per}}$:

$$H_{\text{per}}\psi_{n,\boldsymbol{q}} = \varepsilon_{n,\boldsymbol{q}}\psi_{n,\boldsymbol{q}}.$$

The functions $\psi_{n,\boldsymbol{q}}$ are not periodic but quasiperiodic, in the sense that for each $\boldsymbol{R} \in \mathbb{Z}^3$, $(\tau_{\boldsymbol{R}}\psi_{n,\boldsymbol{q}})(\boldsymbol{r}) = \mathrm{e}^{\mathrm{i}\boldsymbol{q}\cdot\boldsymbol{R}}\psi_{n,\boldsymbol{q}}(\boldsymbol{r})$. They are called Bloch waves.

The electronic ground state is then obtained by filling the lowest energy levels. In this particular setting, this

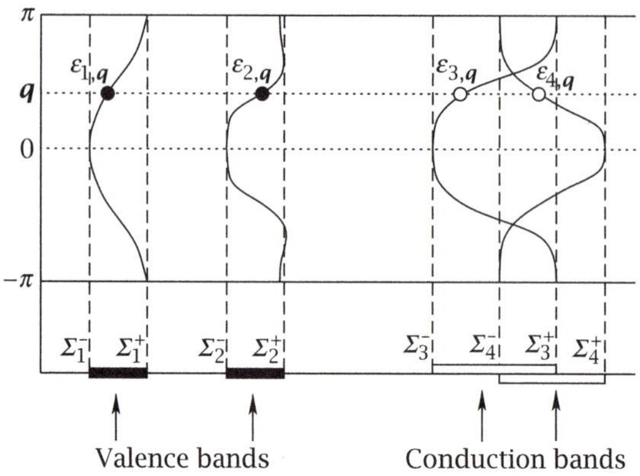


**Figure 2** The electronic structure of an insulating crystal with $N = 4$ electrons per unit cell.

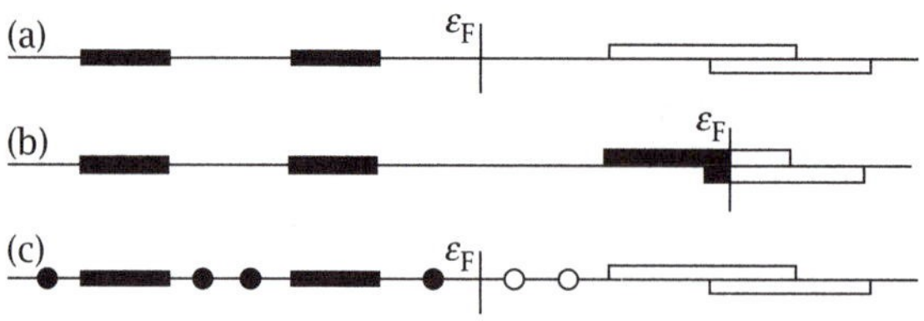


**Figure 3** Spectra of (a) an insulating perfect crystal with $N = 4$ electrons per unit cell, (b) a conducting perfect crystal with $N = 6$ electrons per unit cell, and (c) an insulating crystal with a local defect.

means that, if the crystal contains $N$ electrons per unit cell, the ground state energy per unit volume and the ground state density are, respectively, given by

$$E_0 = 2\sum_{n=1}^{+\infty}\int_{\Gamma^*}\mathcal{H}(\varepsilon_F - \varepsilon_{n,\boldsymbol{q}})\varepsilon_{n,\boldsymbol{q}}\,\mathrm{d}\boldsymbol{q},$$

$$\rho_0(\boldsymbol{r}) = 2\sum_{n=1}^{+\infty}\int_{\Gamma^*}\mathcal{H}(\varepsilon_F - \varepsilon_{n,\boldsymbol{q}})|\psi_{n,\boldsymbol{q}}(\boldsymbol{r})|^2\,\mathrm{d}\boldsymbol{q},$$

where $\mathcal{H}$ is the Heaviside function ($\mathcal{H}(x) = 0$ if $x < 0$; $\mathcal{H}(x) = 1$ if $x \geqslant 0$). In the above formulas, $\varepsilon_F$ is a real number, called the *Fermi level*, such that $2\mathcal{N}(\varepsilon_F) = N$, where the function

$$\mathcal{N}(E) = \sum_{n=1}^{+\infty}\int_{\Gamma^*}\mathcal{H}(E - \varepsilon_{n,\boldsymbol{q}})\,\mathrm{d}\boldsymbol{q}$$

denotes the so-called integrated density of states.

Filling one band therefore corresponds to putting two electrons into each unit cell. One of two situations then arises (see figure 3). The first corresponds to the case when $N = 2p$ is even and there is a gap $g := \Sigma_{p+1}^- - \Sigma_p^+ > 0$ between the $p$th and the $(p+1)$st band. In this case, the lowest $p$ bands (the valence bands) are completely filled, while the other bands (the conducting bands) are empty. The minimum energy required to excite an electron from the valence bands to the conducting bands, in which electrons are free to travel, is then equal to $g$. In the second situation ($N = 2p - 1$, or $N = 2p$ and the $p$th and $(p+1)$st bands overlap), the $p$th band is not completely filled and an infinitesimal amount of energy is sufficient to create an electronic excitation. In the former case the crystal behaves as an insulator (large gap) or a semiconductor (small gap), while in the latter case it behaves as a metal.

Perfect crystals do not exist in nature. Besides the fact that real crystals are of course finite, the arrangement of atoms in such materials is not perfectly periodic. Indeed, real crystals contain both local defects (vacancies, interstitial atoms, impurities, dislocation loops) and extended defects (dislocation lines, grain boundaries). For the sake of brevity we focus on local defects in this article.

A single local defect (or a finite number of them) is modeled by a perturbed periodic Schrödinger operator $H = -\frac{1}{2}\Delta + V_{\text{per}} + W$, where $W$ is a potential that vanishes at infinity. The spectrum of $H$ contains the spectrum of $H_{\text{per}}$ but may also contain discrete eigenvalues located in the spectral gaps of $H_{\text{per}}$ (see figure 3). These eigenvalues correspond to bound states localized in the vicinity of the defects, and they play an important role in physics.

Doped semiconductors are key materials in electronics. Their properties are due to the presence of impurities randomly distributed throughout the crystal. A few impurities per million atoms can dramatically modify the electronic properties of the crystal. Doped semiconductors can be modeled by random Schrödinger operators of the form $H_\omega = -\frac{1}{2}\Delta + V_\omega$, with

$$V_\omega(\boldsymbol{r}) = \sum_{\boldsymbol{R}\in\mathcal{R}}[(1-\omega_{\boldsymbol{R}})v(\boldsymbol{r}-\boldsymbol{R}) + \omega_{\boldsymbol{R}}w(\boldsymbol{r}-\boldsymbol{R})],$$

where $v$ and $w$ are compactly supported functions and the $\omega_{\boldsymbol{R}}$ are independent, identically distributed random variables. Consider, for instance, the case when $\omega_{\boldsymbol{R}}$ is a Bernoulli random variable, meaning that $\omega_{\boldsymbol{R}} = 1$ with probability $p$ and $\omega_{\boldsymbol{R}} = 0$ with probability $1-p$. If $p = 0$, the potential is periodic (equal to $\sum_{\boldsymbol{R}\in\mathcal{R}} v(\boldsymbol{r}-\boldsymbol{R})$). On the other hand, the operator $H_\omega$ depends on the realization of the random variables $\omega_{\boldsymbol{R}}$. Similar stochastic models can be used to describe alloys. The study of the spectral properties of random Schrödinger operators is an active field of research. It is known that the spectrum of $H_\omega$ is almost surely independent of the realization $\omega = (\omega_{\boldsymbol{R}})_{\boldsymbol{R}\in\mathbb{Z}^3}$. This result is in fact a consequence of the Birkhoff ergodic theorem of probability theory. A fundamental result is that disorder tends to localize the electrons. While there are no bound states for $p = 0$, true eigenvalues that correspond to bound states appear in the vicinity of the edges of the bands as soon as $p > 0$. This phenomenon, called *Anderson localization*, is central to physics. Many mathematical questions about the spectral properties of random Schrödinger operators remain open.

## 4 Density Functional Theory

Let us finally turn to the modeling of *interacting* electrons in crystals. We limit ourselves to the case of perfect crystals. Extending these models to the cases of crystals with defects and alloys is at the edge of current research.

The introduction of the Kohn–Sham model in the mid-1960s revolutionized the modeling and simulation of the electronic structure of solids. Indeed, this model allows us to obtain simulation results that are in quantitative agreement with experimental data.

The Kohn–Sham model is a mean-field model that allows computation of the ground state density. For a perfect crystal, the Kohn–Sham Hamiltonian reads as follows:

$$H_{\text{per}}^{\rho_0} = -\tfrac{1}{2}\Delta + V_{\text{per}}^{\text{C},\rho_0} + V_{\text{per}}^{\text{xc},\rho_0},$$

where $V_{\text{per}}^{\text{C},\rho_0}$ and $V_{\text{per}}^{\text{xc},\rho_0}$ are, respectively, the Coulomb and exchange-correlation potentials. The former is the electrostatic potential generated by the total (nuclear and electronic) charge distribution. It is obtained by solving the Poisson equation

$$-\Delta V_{\text{per}}^{\text{C},\rho_0} = 4\pi(\rho_0 - \rho^{\text{nuc}}), \quad V_{\text{per}}^{\text{C},\rho_0}\ \mathcal{R}\text{-periodic},$$

where $\rho^{\text{nuc}}$ is the periodic nuclear charge distribution and $\rho_0$ is the periodic electronic ground state density. Several expressions for $V_{\text{per}}^{\text{xc},\rho_0}$ have been formulated by physicists and chemists, based on theoretical physics arguments, and parametrized by quantum Monte Carlo simulations of the homogeneous (interacting) electron gas. The simplest model for $V_{\text{per}}^{\text{xc},\rho_0}$ is the $X\alpha$ potential introduced by Slater, for which

$$V_{\text{per}}^{\text{xc},\rho_0}(\boldsymbol{r}) = -C\rho_0(\boldsymbol{r})^{1/3},$$

where $C$ is a given positive constant. Much more elaborate forms for $V_{\text{per}}^{\text{xc},\rho_0}$ are used in practice. As in the noninteracting case, the electronic ground state density is obtained by filling up the lowest bands of the periodic Schrödinger operator $H_{\text{per}}^{\rho_0}$. But the crucial difference is that, this time, the Hamiltonian $H_{\text{per}}^{\rho_0}$ depends on the ground state density. The Kohn–Sham model is therefore a *nonlinear* eigenvalue problem. Note that the Kohn–Sham equations are the Euler equations (the first-order optimality conditions) of a constrained optimization problem consisting of minimizing the Kohn–Sham energy functional on the set of admissible electronic states.

From a numerical point of view, the ground state can be obtained either by minimizing the Kohn–Sham energy functional or by solving the Kohn–Sham equations by an iterative procedure called a *self-consistent field algorithm.* The Kohn–Sham model for perfect crystals is usually discretized as follows. First, the Brillouin zone $\Gamma^*$ is meshed using a regular grid $\mathcal{G}_L$ of step length $\Delta q = 2\pi/L$, $L \in \mathbb{N}^*$. Then, for each $\boldsymbol{q} \in \mathcal{G}_L = \Delta q\mathbb{Z}^3 \cap \Gamma^*$, the lowest-energy eigenvalues and eigenfunctions of the operator $(H_{\text{per}})_{\boldsymbol{q}}$ are computed numerically by a Rayleigh–Ritz approximation in the space spanned by the Fourier modes $(e_{\boldsymbol{k}})_{\boldsymbol{k}\in 2\pi\mathbb{Z}^3, |\boldsymbol{k}|\leqslant\sqrt{2E_{\text{c}}}}$, where $E_{\text{c}}$ is an energy cutoff. This amounts to diagonalizing the Hermitian matrix $(\langle e_{\boldsymbol{k}}|(H_{\text{per}})_{\boldsymbol{q}}|e_{\boldsymbol{k}'}\rangle)_{\boldsymbol{k},\boldsymbol{k}'\in 2\pi\mathbb{Z}^3, |\boldsymbol{k}|\leqslant\sqrt{2E_{\text{c}}}, |\boldsymbol{k}'|\leqslant\sqrt{2E_{\text{c}}}}$, where

$$\begin{aligned}\langle e_{\boldsymbol{k}}|(H_{\text{per}})_{\boldsymbol{q}}|e_{\boldsymbol{k}'}\rangle &= \int_\Gamma e_{\boldsymbol{k}}(\boldsymbol{r})^*((H_{\text{per}})_{\boldsymbol{q}}e_{\boldsymbol{k}'})(\boldsymbol{r})\,\mathrm{d}\boldsymbol{r} \\ &= \frac{|\boldsymbol{k}+\boldsymbol{q}|^2}{2}\delta_{\boldsymbol{k}\boldsymbol{k}'} + \int_\Gamma V_{\text{per}}(\boldsymbol{r})\mathrm{e}^{\mathrm{i}(\boldsymbol{k}'-\boldsymbol{k})\cdot\boldsymbol{r}}\,\mathrm{d}\boldsymbol{r}.\end{aligned}$$

When the two numerical parameters of the simulation, $L$ and $E_{\text{c}}$, go to infinity, the numerical results (the ground state energy per unit volume and the ground state density) converge to the exact solution of the Kohn–Sham model.

### Further Reading

Kohn, W. 1999. Nobel Lecture: Electronic structure of matter—wave functions and density functionals. *Reviews of Modern Physics* 71:1253–66.

Martin, R. M. 2004. *Electronic Structure: Basic Theory and Practical Methods.* Cambridge: Cambridge University Press.

Reed, M., and B. Simon. 1978–1981. *Methods of Modern Mathematical Physics*, volumes I–IV. New York: Academic Press.

# VII.15 Flame Propagation

*Moshe Matalon*

## 1 Introduction

A fundamental problem in combustion theory is the determination of the propagation speed of a *premixed* flame through a gaseous combustible mixture. In a premixed system the fuel and oxidizer are already mixed, so following ignition a chemical reaction takes place and spreads into the remaining unburned mixture until one of the reactants is completely depleted. The propagation speed depends on whether the combustible mixture is quiescent or in a state of motion and on whether the underlying flow is laminar or turbulent. Its practical importance is evident, allowing, for example, the determination of the mean fuel consumption rate in a combustor or an estimation of the spread of, and damage caused by, an explosion in a combustible system.

It is seldom the case that the original reactants in a combustion system interact among themselves in a single step and produce the final product. In most cases there is a large number of steps with intermediate species involved before the final formation of products. It is convenient, however, to use a global representation

$$\text{fuel} + \nu\,\text{oxidizer} \to (1+\nu)\,\text{products} + \{Q\}$$

for the chemical description. Accordingly, a mass $\nu$ of oxidizer is consumed for each unit mass of fuel, producing a mass $(1+\nu)$ of products and releasing a thermal energy $Q$. The differential equation describing the mass balance of fuel is

$$\frac{\mathrm{D}Y_{\mathrm{F}}}{\mathrm{D}t} - \mathcal{D}_{\mathrm{F}}\nabla^2 Y_{\mathrm{F}} = -\frac{\omega}{\rho}, \qquad (1)$$

where the operator $\mathrm{D}/\mathrm{D}t \equiv \partial/\partial t + \boldsymbol{v}\cdot\nabla$ is the convective derivative, i.e., $\mathrm{D}\phi/\mathrm{D}t$ is the rate of change of the property $\phi$ of a material element taken while following the fluid motion, with $\boldsymbol{v}$ the velocity vector and $\nabla$ the gradient operator taken with respect to the three spatial coordinates (the dot signifies the inner product). Here, $\rho$ is the density of the entire mixture, the fuel mass fraction $Y_{\mathrm{F}}$ is the ratio of the mass of fuel to the total mass of the mixture, and the coefficient $\mathcal{D}_{\mathrm{F}}$ is the diffusivity of fuel relative to the bulk (nitrogen for combustion in air). In lean mixtures, where the reaction consumes only a very small amount of oxidizer, the oxidizer mass fraction $Y_{\mathrm{O}}$ can be treated as constant; otherwise, it is an additional variable, and a similar equation to (1) must be written for its consumption. The differential equation for the energy balance expressed in terms of the temperature $T$ is

$$\frac{\mathrm{D}T}{\mathrm{D}t} - \alpha\nabla^2 T = \frac{Q}{\rho c_{\mathrm{p}}}\omega, \qquad (2)$$

where $\alpha$ is the thermal diffusivity and $c_{\mathrm{p}}$ is the specific heat (at constant pressure) of the mixture. Equations (1) and (2) state that fuel is consumed and energy released (fuel and energy are the two main ingredients of any combustion system) at a rate $\omega$. The reaction rate typically obeys an Arrhenius law of the form

$$\omega = \mathcal{B}\rho Y_{\mathrm{F}}\mathrm{e}^{-E/\mathcal{R}T},$$

with $E$ the activation energy (the minimum energy required for the reaction to be possible) and $\mathcal{R}$ the universal gas constant.

The large temperature variations within the combustion field produce large density variations, which, in turn, modify the flow field. Since flames propagate very slowly compared with sound waves, the process is nearly isobaric and the density is inversely proportional to the temperature. The velocity $\boldsymbol{v}$ obeys the Navier–Stokes equations

$$\frac{\mathrm{D}\rho}{\mathrm{D}t} + \rho\nabla\cdot\boldsymbol{v} = 0, \qquad (3)$$

$$\rho\frac{\mathrm{D}\boldsymbol{v}}{\mathrm{D}t} = -\nabla p + \mu\nabla\cdot\boldsymbol{\Sigma} \qquad (4)$$

describing conservation of mass and momentum. Here, $p$ is the dynamic pressure (the small deviations from the ambient pressure), $\boldsymbol{\Sigma}$ is the viscous stress tensor, and $\mu$ is the viscosity of the mixture.

Despite the numerous simplifications introduced in the above formulation, its mathematical complexity is apparent, involving coupled partial differential equations that, in addition to the quadratic nonlinearity of ordinary fluid flow problems, contain the highly nonlinear exponential term. Recent advances in computational capabilities permit such problems to be addressed. However, the computations are quite intensive and are usually carried out for a particular set of parameters. Moreover, when dealing with multidimensional flows and turbulence, they are often unable to provide the required spatial and temporal resolutions. Fundamental understanding has been achieved primarily by analyzing simplified mathematical models that elucidate the physical interactions taking place between the various mechanisms involved in a given process. Because of the disparities between the spatial scales and timescales involved in combustion problems, the techniques of perturbation and asymptotic methods have become the primary tools of mathematical analysis.

## 2 The Planar Adiabatic Flame

One of the simplest problems in combustion is a planar flame propagating into a quiescent mixture of temperature $T_u$, under adiabatic conditions. The solution, of the form $\phi = \phi(x + S_L t)$, corresponds to a combustion wave propagating from right to left along the $x$-axis at a speed $S_L$ that is uniquely determined by the thermodynamics and the transport and chemical kinetic properties of the combustible mixture. The determination of $S_L$, known as *the laminar flame speed*, has been the subject of a large number of theoretical, numerical, and experimental investigations. An analytical solution is not available, but an asymptotic approximation can be obtained when $E/\mathcal{R}T_u \gg 1$, i.e., when the activation energy of the chemical reaction is much larger than the energy of the fresh mixture, a condition that characterizes most combustion systems. In this limit, the reaction is confined to a thin region like that illustrated in figure 1. Elsewhere, the chemical reaction rate $\omega$ is negligible either because the exponential factor is vanishingly small (the preheat zone) or because the fuel has been entirely consumed (the post-flame zone). The asymptotic solution is then constructed by obtaining solutions of the simplified equations in each of the separate regions. Asymptotic matching then yields an expression for the flame speed that, for a lean mixture under the adopted simplifications, is of the form

$$S_L = \sqrt{\frac{\rho_b}{\rho_u} \frac{2\alpha^2 \mathcal{B}}{\mathcal{D}_F} \frac{\mathcal{R} T_a^2}{E(T_a - T_u)}} \mathrm{e}^{-E/2\mathcal{R}T_a}, \qquad (5)$$

where

$$T_a = T_u + QY_{F_u}/c_p \qquad (6)$$

is the (adiabatic) flame temperature, a direct consequence of energy conservation. The subscripts "u" and "b" identify conditions in the unburned and burned states, respectively. The flame speed and temperature given by (5), (6) are two of the most important properties that characterize a premixed flame. The flame temperature increases with increasing heat release $Q$, and its appearance in the highly temperature-dependent exponential in (5) implies that it exerts the strongest influence on the flame speed, meaning that reactions with larger values of heat release propagate faster. The flame speed is also affected by preheating the mixture, i.e., increasing $T_u$, or diluting the gas with an inert substance whose properties can change the thermal and/or mass diffusivities remarkably.

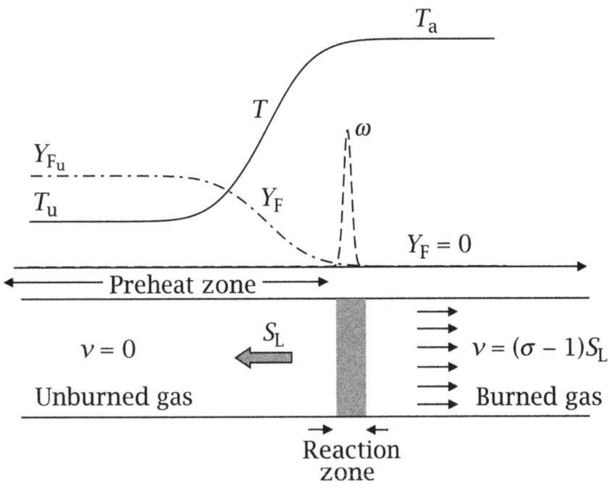


**Figure 1** The structure of a planar premixed flame.

The mechanism of propagation is associated with the heat conducted back from the reaction zone, precipitating a rise in temperature in the adjacent gas layer that triggers the chemical reaction. Since, in a one-dimensional flow, the mass flux relative to the wave is constant, the gas traveling through the flame expands and speeds up. The hot burned gas moves away in the direction opposite to flame propagation at a velocity larger than $S_L$ by a factor $(\sigma - 1)$, where $\sigma = \rho_u/\rho_b$ is the thermal expansion parameter.

The realization nearly fifty years ago that the planar propagation problem could be solved by means of asymptotic techniques paved the way for much of the theoretical development that has taken place in recent years, particularly the mathematical description of unsteady multidimensional laminar and turbulent flames, which is discussed next.

## 3 Multidimensional Laminar Flames

Although planar flames can be observed in the laboratory if appropriate measures are taken, real flames are seldom flat. A Bunsen flame, for example, maintains a conical shape when the gas velocity at the exit of the burner, $V$, is greater than $S_L$. The geometry then determines the cone opening angle $\theta = 2\sin^{-1}(S_L/V)$. The shape of the flame stabilized in the laboratory around a porous sphere is affected by natural convection and resembles a teardrop that lacks spherical symmetry. Buoyancy plays a significant role when the representative Froude number $Fr = V^2L/g$ is small; here, $V$ is a characteristic flow velocity, $L$ is a measure of the flame height, and $g$ is gravitational acceleration. A more fundamental reason why flames are

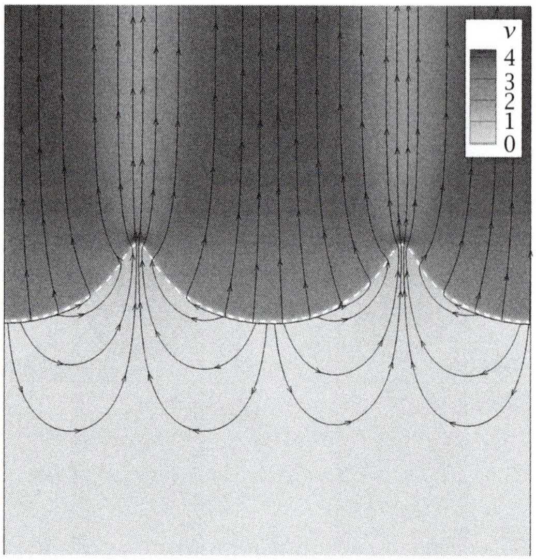


**Figure 2** A steadily propagating corrugated flame.

often corrugated and propagate in an unsteady manner is associated with instabilities, the most prominent of which, in premixed flames, is the hydrodynamic, or Darrieus–Landau (DL), instability. The gas expansion induces hydrodynamic disturbances that tend to enhance perturbations of the flame front. Diffusion effects that often have stabilizing influences act only on the short-wavelength disturbances, so large flames acquire cusp-like conformations with elongated intrusions pointing toward the burned gas region, as shown in figure 2. These structures are *stable* and, because of their larger surface area, propagate at a speed $U_L$ that is substantially larger than the laminar flame speed $S_L$. The dashed curve in the figure is the flame front, the solid curves are selected streamlines, and the various shades of gray correspond to regions of increased velocity. The flow pattern demonstrates the deflection of streamlines upon crossing the flame front, which is a consequence of gas expansion, and the induced vortical motion in the unburned gas that is otherwise at rest, which is responsible for sustaining the cellular structure by "pushing" the crests upward.

The mathematical description of multidimensional flames exploits the disparity among the length scales associated with the fluid dynamic field, the diffusion processes, and the highly temperature-sensitive reaction rate. On the largest scale, the entire flame, consisting of the preheat and reaction zones, may be viewed as a surface of discontinuity separating burned from unburned gases, described by $\psi(\boldsymbol{x}, t) = 0$. Equations (3)–(4), with densities $\rho_u$, $\rho_b$, must then be solved on either side of the flame front with appropriate jump conditions for the pressure and velocities across $\psi = 0$. The internal flame structure facilitated by the large-activation-energy assumption is resolved on the smaller diffusion length scale, and through asymptotic matching it provides the aforementioned jump relations in the form of generalized RANKINE–HUGONIOT RELATIONS [V.20 §2.3], as well as giving us an expression for the flame speed. It is customary in combustion to define the*flame speed* $S_f$ relative to the unburned gas, namely, $S_f \equiv v_n^* - V_f$, where $v_n = \boldsymbol{v} \cdot \boldsymbol{n}$ denotes the normal component of the gas velocity and $V_f = -\psi_t / |\nabla\psi|$ is the propagation speed with respect to a fixed coordinate system; here, the unit normal $\boldsymbol{n}$ is taken positive when pointing toward the burned gas region, the asterisk superscript indicates that the velocity is to be evaluated at the flame front on its unburned side, and the subscript $t$ stands for time differentiation. The expression for the flame speed takes the form

$$S_f = S_L - \mathcal{L}K, \tag{7}$$

where $K$ is the *stretch rate*, a measure of the flame front deformation that results from the normal propagating motion and the nonuniform underlying flow field. A spherically expanding flame is stretched because of the increase in its surface area that occurs at a rate proportional to the instantaneous curvature. A planar flame stabilized in a stagnation point flow is stretched as a result of the diverging flow at a rate proportional to the hydrodynamic strain. For a general surface, the local stretch rate can be obtained from kinematic considerations and is found to depend on the (mean) curvature of the flame surface $\kappa$ and the underlying hydrodynamic strain, namely,

$$K = \kappa S_L - \boldsymbol{n} \cdot \boldsymbol{E} \cdot \boldsymbol{n},$$

where $\boldsymbol{E}$ is the strain rate tensor. The coefficient $\mathcal{L}$, known as the Markstein length, is of the order of the flame thickness and incorporates the effects of diffusion and chemical reaction. It can take positive or negative values depending on the mixture composition. In an experimental setting, changes in $\mathcal{L}$ could be accommodated by varying the fuel type and mixture composition or the system's pressure. The mathematical formulation is thus composed of a nonlinear free-boundary problem for the pressure and velocity fields with the free surface (the flame front) determined by solving

$$\psi_t + \boldsymbol{v}^* \cdot \nabla\psi = S_f |\nabla\psi|, \tag{8}$$

with $S_f$ given by (7). This formulation, known as the *hydrodynamic theory*, has been a useful framework for understanding intricate flame–flow interactions and, in particular, the development of flame instabilities. The flame shown in figure 2, for example, was obtained using a variable-density Navier–Stokes solver in conjunction with a front-capturing technique.

## 4 Turbulent Flames

Analogous to the definition of the laminar flame speed, the *turbulent flame speed* $S_T$ can be defined as the mean propagation speed of a premixed flame within an isotropic homogeneous turbulent field of zero mean velocity. If the incident fluid velocity is decomposed into a mean (denoted by an overline) and a fluctuating component, and the flame is held statistically stationary by adjusting the mean longitudinal flow velocity $\overline{u_1}$, then the turbulent flame speed is $S_T = \overline{u_1}$, as illustrated in figure 3. The mean mass flow rate through the entire flame shown in the figure is given by $\dot{m} = \rho_u A S_T$. Since all the reactants pass through the wrinkled flame of area $A_f$, it can also be calculated from the total contributions of mass flowing through the differential segments comprising the wrinkled flame, assuming each segment propagates normal to itself at a speed $S_L$. We then have $\dot{m} = \rho_u \overline{A_f} S_L$, implying that $S_T/S_L = \overline{A_f}/A$; i.e., the increase in the speed of the turbulent flame is due to the increase in the surface area of the flame front. This relation was first noted by Damköhler, who resorted to geometrical arguments with analogy to a Bunsen flame to further deduce an explicit dependence on the turbulence intensity $v_c'$, defined as the root mean square of the velocity fluctuations. His result and those of numerous other phenomenological studies conform to expressions of the form

$$S_T/S_L = 1 + C(v_c'/S_L)^n, \tag{9}$$

with various constants $C$ and adjustable exponents $n$. Although the experimental record exhibits a wide scatter due to the variable accuracy of the methods and the varied operating conditions, the data appears to confirm that $S_T$ increases with increasing intensity of turbulence for low to moderate values of $v_c'$. At higher turbulence intensities, however, $S_T$ increases only slightly and then levels off, an observation known as the *bending effect*.

The hydrodynamic model provides a more rigorous approach to the determination of the turbulent flame speed. The thin-flame assumption implies that the internal structure of the flame is not disturbed by the turbulence and retains its laminar structure. The results of this model are, strictly speaking, applicable to only the wrinkled and corrugated flamelet regimes of turbulent combustion, which encompass many combustion applications and most laboratory experiments. It excludes, in particular, the distributed-reaction regime, where the basic structure of a flame no longer exists and the notion of turbulent flame speed becomes ambiguous. Since in the hydrodynamic description every segment of the wrinkled flame propagates at a speed $S_f$ that depends on the local mixture composition and flow conditions, as given by (7), the mean mass flowing through the entire flame is now $\dot{m} = \rho_u \overline{S_f A_f}$. This yields

$$S_T = \overline{S_f |\nabla \psi|}, \tag{10}$$

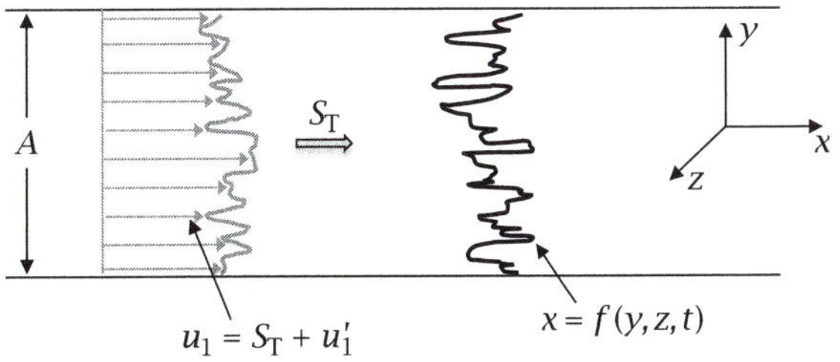


**Figure 3** A schematic of a statistically stationary turbulent premixed flame.

which must be determined through calculations similar to the one used to generate figure 2 after replacing the quiescent setting with a turbulent flow field. A pregenerated homogeneous isotropic turbulent flow, characterized by intensity $v_c'$ and integral scale $\ell$, is thus fed as an inflow at the bottom of the integration domain, as shown in figure 4(a). The flame is retained at a prescribed location, on average, by controlling the mean inflow velocity. Figure 4 illustrates such calculations. The turbulent nature of the flow is elucidated by the clockwise/counterclockwise (dashed/solid) vorticity contours, and the flame front is represented by the solid dark curve. One notes the vorticity generated downstream near the “cusp” of the flame front by baroclinic torque and a significant decrease in the vorticity elsewhere; this is the result of volumetric expansion.

If the local flame speed is assumed to be constant and equal to the laminar flame speed, (10) yields $S_T/S_L = \overline{|\nabla \psi|}$, implying that the increase in the speed of the turbulent flame is due to the increase in the flame surface area, as in Damköhler's proposition. Equation (7) in conjunction with (10) demonstrates that in turbulent propagation an important role is played by the

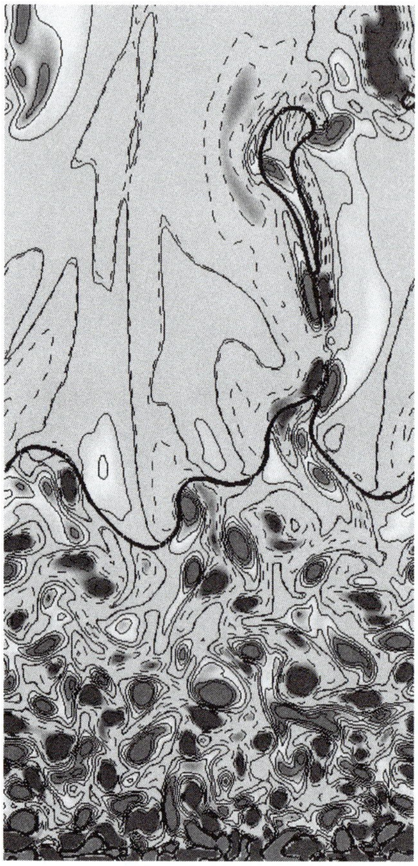


**Figure 4** Flame propagation in turbulent flow; the instantaneous portrait shows a pocket of unburned gas that pinched off from a folded segment of the flame surface.

mean stretching of the flame $\bar{K}$, which influences the mean local flame speed $\overline{S_f}$ and, consequently, $S_T$. Scaling laws accounting for stretch effects were recently obtained using numerical simulations within the context of the hydrodynamic theory, for "two-dimensional" turbulence and for $\mathcal{L} > 0$. Although two-dimensional flow lacks some features of real turbulence, the results extend the current understanding of turbulent flame propagation and yield expressions for $S_T$ that are free of any turbulence modeling assumptions and/or ad hoc adjustment parameters. The condition $\mathcal{L} > 0$ excludes the development of the thermodiffusive instabilities that are observed in rich hydrocarbon–air or lean hydrogen–air mixtures; instabilities that may further contaminate the flame surface with small structures increase its overall surface and its propagation speed. Three regimes have been identified depending on the mixture composition, the thermal expansion coefficient, and the turbulence intensity:

(i) a regime in which, on average, the flame brush remains planar (i.e., has zero mean curvature) and is unaffected by the DL instability;
(ii) a regime in which the DL effects, which are responsible for frequent intrusions of the flame front into the burned gas region, have a marked influence on the flame front, which remains partially resilient to turbulence; and
(iii) a highly turbulent regime in which the influences of the DL instability play a limited role and the flame propagation is totally controlled by the turbulence.

The third regime is affected by frequent folding of the flame front and formation of pockets of unburned gas that detach from the main flame surface and are rapidly consumed, causing a significant reduction in the turbulent flame speed. Expressions for the turbulent flame speed of the form

$$\frac{S_T}{S_L} = \left[1 - \frac{\mathcal{L}\bar{K}}{S_L}\right]\left[a + \frac{b(\ell, \mathcal{L})c(\sigma)}{(\mathcal{L}/L)^m}\left(\frac{v_c'}{S_L}\right)^n\right]$$

have been obtained, with coefficients that depend on the functional parameters in the various operating regimes. The constant $a$ depends on the nature of the stable laminar flame when $v_c' = 0$; it is equal to 1 when the stable flame is planar and equal to $U_L/S_L$ when the stable flame is the cusp-like conformation (shown in figure 2) resulting from the DL instability. The coefficient $c$ increases with increasing $\sigma$ and plateaus as the thermal expansion reaches sufficiently high values. The coefficient $b$ is of relatively small magnitude, reaching a maximum at an intermediate scale that disturbs the flame most effectively. Variations in the mixture composition and ambient conditions, exhibited through the Markstein length $\mathcal{L}$, appear as a power law with exponent $m$ less than 1. Of greatest significance is the dependence of $S_T$ on turbulence intensity; at low $v_c'$ the dependence of $S_T$ on turbulence intensity is quadratic ($n = 2$), in accordance with Damköhler's heuristic result, but at higher turbulence levels the dependence is sublinear ($n < 1$), which explains the bending effect that is observed experimentally.

## Further Reading

Matalon, M. 2009. Flame dynamics. *Proceedings of the Combustion Institute* 32:57–82.

Matalon, M., and J. Bechtold. 2009. A multi-scale approach to the propagation of non-adiabatic premixed flames. *Journal of Engineering Mathematics* 63:309–26.

Matalon, M., and F. Creta. 2012. The "turbulent flame speed" of wrinkled premixed flames. *Comptes Rendus Mécanique* 340:845–58.
Matalon, M., C. Cui, and J. Bechtold. 2003. Hydrodynamic theory of premixed flames: effects of stoichiometry, variable transport coefficients and reaction orders. *Journal of Fluid Mechanics* 487:179–210.

# VII.16 Imaging the Earth Using Green's Theorem

*Roel Snieder*

## 1 Introduction

The Earth is a big place: its radius is about 6400 km. In comparison, the deepest boreholes so far drilled are about 10 km deep. We therefore have little opportunity to take direct measurements or samples inside the Earth: it is mostly inaccessible. And even for the upper 10 km that we can sample, the cost of drilling deep boreholes is very high. This means that inferences about the Earth's interior are largely based on physical and chemical measurements taken at the Earth's surface, or even from space. Investigating the inside of the Earth therefore resembles that classical black-box problem: determine the contents of a closed box when you can do anything except open the box.

If one could measure physical fields, such as the gravitational field or the elastic wave field, inside the Earth, one could infer the local properties of the Earth by inserting the measured field into the equation that governs that field and then extracting the physical parameters, such as the mass density, from the field equation. However, the fields are measured at, or sometimes even above, the Earth's surface. One therefore needs a recipe for propagating the measured field from its surface of observation into the Earth's interior. This is a problem where mathematics comes to the rescue in the form of Green's theorem. This relates measurements taken at a surface bounding a volume to the fields inside that volume: a principle called *downward continuation.*

In what follows we apply Green's theorem to a large class of physical systems and show that the theorem only relates measurements at a surface to measurements in the interior when the equations are the same regardless of whether one moves toward the future or toward the past. Such equations are said to be invariant for time reversal. We focus in particular on seismic imaging because this is the technique that provides the highest spatial resolution.

## 2 Green's Theorem for General Systems

Consider physical systems that satisfy the following partial differential equation for a field $u(\boldsymbol{r},t)$ that is excited by sources $q(\boldsymbol{r},t)$:

$$\sum_{n=0}^{N} a_n(\boldsymbol{r})\frac{\partial^n u}{\partial t^n} = \nabla\cdot(B(\boldsymbol{r})\nabla u(\boldsymbol{r},t)) + q(\boldsymbol{r},t). \quad (1)$$

This equation captures many specific equations. An example is the wave equation:

$$\frac{1}{\kappa(\boldsymbol{r})}\frac{\partial^2 u}{\partial t^2} + \gamma(\boldsymbol{r})\frac{\partial u}{\partial t} = \nabla\cdot\left(\frac{1}{\rho(\boldsymbol{r})}\nabla u\right) + q(\boldsymbol{r},t), \quad (2)$$

where $\kappa$ is the bulk modulus, $\gamma$ is a damping parameter, and $\rho$ is the density. Another example of equation (1) is the diffusion equation:

$$\frac{\partial u(\boldsymbol{r},t)}{\partial t} = \nabla\cdot(D(\boldsymbol{r})\nabla u(\boldsymbol{r},t)) + q(\boldsymbol{r},t), \quad (3)$$

with $D(\boldsymbol{r})$ the diffusion constant. This equation is used to describe flow in porous media such as aquifers and hydrocarbon reservoirs. It also accounts for heat conduction and for diffusive spreading of pollutants. A variant of (3) is the Schrödinger equation, which accounts for the dynamics of microscopic particles:

$$\mathrm{i}\hbar\frac{\partial\psi(\boldsymbol{r},t)}{\partial t} - V(\boldsymbol{r})\psi(\boldsymbol{r},t) = -\frac{\hbar^2}{2m}\nabla^2\psi(\boldsymbol{r},t). \quad (4)$$

Here $\hbar$ is Planck's constant divided by $2\pi$, $m$ is the mass of the particle, and $V(\boldsymbol{r})$ is the real potential in which the particle moves. Yet another example is the gravitational potential, which plays an important role in geophysics because it helps constrain the mass density inside the Earth. The gravitational field satisfies Poisson's equation:

$$0 = \nabla^2 u(\boldsymbol{r}) - 4\pi G\rho(\boldsymbol{r}), \quad (5)$$

where $G$ is the gravitational constant. This equation does not depend on time.

Note that the applications (2)–(5) are special forms of the general equation (1). In these applications $B(\boldsymbol{r})$ is real, hence we use $B = B^*$ in the following, with the asterisk denoting complex conjugation. We also use the Fourier convention: $f(t) = \int f(\omega)\mathrm{e}^{-\mathrm{i}\omega t}\,\mathrm{d}\omega$. With this, the general equation (1) reduces to

$$\sum_{n=0}^{N}(-\mathrm{i}\omega)^n a_n(\boldsymbol{r})u(\boldsymbol{r},\omega) = \nabla\cdot(B(\boldsymbol{r})\nabla u(\boldsymbol{r},\omega)) + q(\boldsymbol{r},\omega). \quad (6)$$

Each time derivative is replaced by a multiplication by $-\mathrm{i}\omega$. The treatment that follows is valid in the frequency domain. For brevity we omit the frequency dependence of variables.

Let us consider two different field states, $u_P(\boldsymbol{r})$ and $u_Q(\boldsymbol{r})$, excited by sources $q_P(\boldsymbol{r})$ and $q_Q(\boldsymbol{r})$, respectively. Take equation (6) for state $P$, multiply by $u_Q^*$, and integrate over volume. Next take the complex conjugate of (6) for state $Q$, multiply by $u_P$, and integrate over volume. Subtracting these two volume integrals and applying Green's theorem to the terms containing $B(\boldsymbol{r})$ gives

$$\sum_{n=0}^{N}(-\mathrm{i}\omega)^n \int \mathcal{A}_n u_P u_Q^* \,\mathrm{d}V$$
$$= \oint B\left(\frac{\partial u_P}{\partial n}u_Q^* - u_P\frac{\partial u_Q^*}{\partial n}\right)\mathrm{d}S + \int (u_Q^* q_P - u_P q_Q^*)\,\mathrm{d}V, \tag{7}$$

where $\partial/\partial n$ denotes the outward normal derivative to the surface $S$ that bounds the volume over which we integrate, and

$$\mathcal{A}_n(\boldsymbol{r}) = a_n(\boldsymbol{r}) - (-1)^n a_n^*(\boldsymbol{r}). \tag{8}$$

The Green function $G(\boldsymbol{r},\boldsymbol{r}_0)$, defined as the solution of (6) to a point excitation, $q(\boldsymbol{r}) = \delta(\boldsymbol{r}-\boldsymbol{r}_0)$, plays a key role in what follows. An important property of $G$ is reciprocity: $G(\boldsymbol{r},\boldsymbol{r}') = G(\boldsymbol{r}',\boldsymbol{r})$. Under suitable boundary conditions, this property is valid for all applications that follow.

Consider the case where $u_P$ is source free ($q_P = 0$) and $u_Q$ is excited by a point source, $q_Q(\boldsymbol{r}) = \delta(\boldsymbol{r}-\boldsymbol{r}_Q)$. Then $u_Q(\boldsymbol{r}) = G(\boldsymbol{r},\boldsymbol{r}_Q) = G(\boldsymbol{r}_Q,\boldsymbol{r})$. Using this in (7), denoting $u_P$ by $u$, and replacing $\boldsymbol{r}$ with $\boldsymbol{r}'$ and $\boldsymbol{r}_Q$ with $\boldsymbol{r}$ gives

$$u(\boldsymbol{r}) = -\sum_{n=0}^{N}(-\mathrm{i}\omega)^n \int \mathcal{A}_n(\boldsymbol{r}')G^*(\boldsymbol{r},\boldsymbol{r}')u(\boldsymbol{r}')\,\mathrm{d}V'$$
$$+ \oint B(\boldsymbol{r}')\left(G^*(\boldsymbol{r},\boldsymbol{r}')\frac{\partial u}{\partial n'} - \frac{\partial G^*(\boldsymbol{r},\boldsymbol{r}')}{\partial n'}u\right)\mathrm{d}S'. \tag{9}$$

## 3 Moving the Field into the Interior

Equation (9) is a powerful tool for propagating measurements taken at the boundary of a system into the interior of that system. This is of particular importance in earth science. First we illustrate this principle for the acoustic wave equation (2), which is a prototype of the equations that govern seismic imaging. In the notation of (1), the wave equation (2) has $N = 2$, $a_2 = 1/\kappa$, $a_1 = \gamma$, $a_0 = 0$, and $B = 1/\rho$. The coefficients $a_n$ enter (9) in a volume integral through the term $A_n$ defined in equation (8), which equals

$$\mathcal{A}_n = \begin{cases} 2\mathrm{i}\,\mathrm{Im}(a_n) & \text{for } n \text{ even,} \\ 2\,\mathrm{Re}(a_n) & \text{for } n \text{ odd,} \end{cases} \tag{10}$$

where Re and Im denote the real and imaginary parts, respectively. According to (10), $a_2$ does not contribute because $\kappa$ is real. Consider first the case when there is no attenuation, so that $a_1 = \gamma = 0$ and (9) reduces to the *representation theorem*:

$$u(\boldsymbol{r}) = \oint \frac{1}{\rho}\left(G^*(\boldsymbol{r},\boldsymbol{r}')\frac{\partial u}{\partial n'} - \frac{\partial G^*(\boldsymbol{r},\boldsymbol{r}')}{\partial n'}u\right)\mathrm{d}S'. \tag{11}$$

This expression relates measurements at the surface in the integral on the right-hand side to the wave field in the interior on the left-hand side.

What happens if there is attenuation? In that case, $a_1 = \gamma > 0$, and according to (9) and (10), equation (11) must be extended with a volume term $2\mathrm{i}\omega \times \int \gamma(\boldsymbol{r}')G^*(\boldsymbol{r},\boldsymbol{r}')u(\boldsymbol{r}')\,\mathrm{d}V'$ on the right-hand side. This term contains the wave field in the interior that we seek to determine, so that this field does not follow from measurements at the surface only. In principle, attenuation makes seismic imaging impossible, but in practice, attenuation in the Earth is weak and the offending volume integral can therefore be ignored.

Can diffuse fields be imaged? For the diffusion equation (3) the only nonzero terms in (1) are $a_1 = 1$ and $B = D$. Inserting these into (9) gives

$$u(\boldsymbol{r}) = 2\mathrm{i}\omega \int G^*(\boldsymbol{r},\boldsymbol{r}')u(\boldsymbol{r}')\,\mathrm{d}V'$$
$$+ \oint D(\boldsymbol{r}')\left(G^*(\boldsymbol{r},\boldsymbol{r}')\frac{\partial u}{\partial n'} - \frac{\partial G^*(\boldsymbol{r},\boldsymbol{r}')}{\partial n'}u\right)\mathrm{d}S'.$$

Just as for attenuating acoustic waves, the right-hand side contains the unknown field in the interior. This means that measurements of diffusive fields taken at the surface cannot be used for imaging using Green's theorem.

The Schrödinger equation (4) is first order in time, and for this reason one might think that as for the diffusion equation one cannot infer the field values within a volume from measurements taken at the boundary. For this equation, $N = 1$, $a_1 = \mathrm{i}\hbar$, $a_0 = -V$, and $B = -\hbar^2/(2m)$. According to (9) and (10), and assuming that the potential $V$ is real, the volume integral depends on $\mathrm{Im}(a_0) = \mathrm{Im}(-V) = 0$ for $n = 0$ and on $\mathrm{Re}(a_1) = \mathrm{Re}(\mathrm{i}\hbar) = 0$ for $n = 1$. The volume integral therefore vanishes and field values in the interior can be determined from field values measured at the boundary.

For the gravitational potential, field equation (5), all $a_n = 0$ and $B = 1$, so that in a source-free region the field satisfies

$$u(\boldsymbol{r}) = \oint \left(G(\boldsymbol{r},\boldsymbol{r}')\frac{\partial u}{\partial n'} - \frac{\partial G(\boldsymbol{r},\boldsymbol{r}')}{\partial n'}u(\boldsymbol{r}')\right)\mathrm{d}S'. \tag{12}$$

The potential field does not depend on time, and as a result both the field $u$ and the Green function $G$ are real functions; there are therefore no complex conjugates in (12). This expression makes it possible to infer the gravitational field above the Earth when the field is known at the Earth's surface. Expression (12) can be used for *upward continuation*, where one infers the gravitational field at higher elevations from measurements taken at the Earth's surface. This can be used, for example, to compute the trajectories of satellites. Similarly, one can use this expression for *downward continuation*, where one computes the gravitational field at lower elevations from measurements taken higher up. An application of downward continuation is to infer the gravitational field at the Earth's surface from measurements taken from satellites or aircraft. It is, however, not possible to use (12) to compute the gravitational field inside the Earth. In the interior, the mass density $\rho(\boldsymbol{r})$ is nonzero, and according to (5), the source $q(\boldsymbol{r})$ is nonzero. This violates the assumption $q_P = 0$ used in the derivation of (9). For this reason, Green's theorem cannot be used to infer the mass density in the Earth from measurements taken at the surface.

In general, the property that the field in the interior follows from field measurements taken at the boundary is valid for systems that are *invariant for time reversal.* These are systems that obey equations that are invariant when time is reversed and $t$ is replaced by $-t$. This is true for the wave equation in the absence of attenuation, but attenuation breaks the symmetry between past and future. The diffusion equation is not invariant under time reversal; heat diffuses away when moving forward in time. Like the diffusion equation, Schrödinger's equation is first order in time, and one might think it is not invariant for time reversal. One can show, however, that, when $\psi(\boldsymbol{r},t)$ is a solution, then so is $\psi^*(\boldsymbol{r},-t)$. According to the principles of quantum mechanics, one cannot make a distinction between the wave function and its complex conjugate, and therefore the equation is effectively invariant for time reversal and, as we have seen, measurements at the surface suffice to determine the field in the interior.

## 4 Seismic Imaging

In this section we discuss the application of the representation theorem (11) to seismic imaging. A typical marine seismic experiment is shown in figure 1. In the figure a ship tows a streamer (shown by the dashed line), which is a long tube with hydrophones (pressure sensors) and/or geophones (motion sensors) that act as recording devices. An air gun (a device delivering an impulsive bubble of air) acts as a seismic source just behind the ship. The waves reflected by layers in the Earth are recorded by sensors in the streamer.

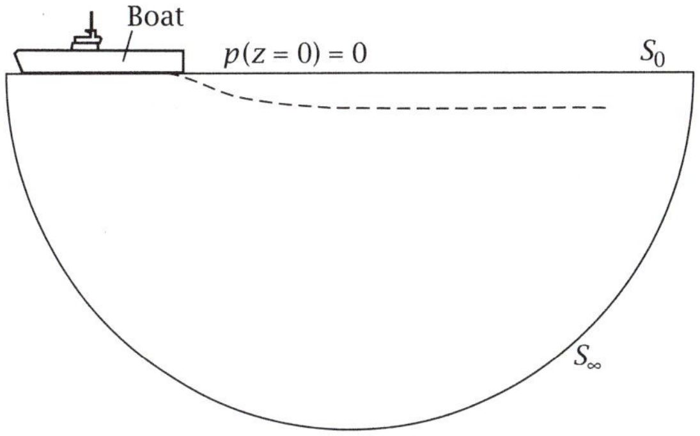


**Figure 1** The geometry of a marine seismic survey.

The water surface is a free surface, hence the pressure $p$ vanishes there: $p(z=0)=0$. However, the particle motion does not vanish. According to Newton's law, the acceleration $\boldsymbol{a}$ is related to the pressure by $\rho\boldsymbol{a} = -\nabla p$. The vertical component of this expression is given by

$$\rho a_z = -\frac{\partial p}{\partial z}. \qquad (13)$$

We use this relation in the representation theorem (11) for the pressure $p$. For the boundary we take the combination of the sea surface $S_0$ and a hemisphere $S_\infty$ with radius $R$ (figure 1). In the presence of a tiny amount of attenuation, the pressure $p$ and Green function $G$ decay as $\mathrm{e}^{-\alpha R}$, with $\alpha$ an attenuation coefficient, and the contribution of $S_\infty$ vanishes as $R \to \infty$. The closed surface integral thus reduces to the contribution of the free surface $S_0$. Since that surface is horizontal, the normal derivative is just the derivative in the $-z$-direction. (We have chosen a coordinate system with positive $z$ pointing down.) As the pressure vanishes at the free surface, expression (11) reduces to

$$p(\boldsymbol{r}) = -\int_{S_0} \rho^{-1}(\boldsymbol{r}')G^*(\boldsymbol{r},\boldsymbol{r}')\left(\frac{\partial p(\boldsymbol{r}')}{\partial z'}\right)\mathrm{d}S'.$$

Eliminating $\partial p/\partial z'$ using (13) gives

$$p(\boldsymbol{r},\omega) = \int_{S_0} G^*(\boldsymbol{r},\boldsymbol{r}',\omega)a_z(\boldsymbol{r}',\omega)\,\mathrm{d}S', \qquad (14)$$

having restored the frequency dependence. This formula relates the pressure in the subsurface to the motion recorded at the sea surface.

The reader may have wondered why the complex conjugate of $G$ was used, since most of the expressions also hold when the complex conjugation is not applied. The time-domain Green function is related to the frequency-domain Green function by $G(\boldsymbol{r},\boldsymbol{r}',t) = \int G(\boldsymbol{r},\boldsymbol{r}',\omega)\mathrm{e}^{-\mathrm{i}\omega t}\,\mathrm{d}\omega$, and hence the time-reversed Green function satisfies $G(\boldsymbol{r},\boldsymbol{r}',-t) = \int G^*(\boldsymbol{r},\boldsymbol{r}',\omega)\mathrm{e}^{-\mathrm{i}\omega t}\,\mathrm{d}\omega$. This means that $G^*(\boldsymbol{r},\boldsymbol{r}',\omega)$ corresponds in the time domain to the time-reversed Green function $G(\boldsymbol{r},\boldsymbol{r}',-t)$. As a consequence, equation (14) corresponds in the time domain to

$$p(\boldsymbol{r},t) = \int_{S_0} G(\boldsymbol{r},\boldsymbol{r}',-t) \star a_z(\boldsymbol{r}',t)\,\mathrm{d}S', \qquad (15)$$

where the star ($\star$) denotes convolution. The Green function $G(\boldsymbol{r},\boldsymbol{r}',t)$ is causal, meaning that it is only nonzero after the point source acts at $t = 0$. It then moves the waves forward in time away from the point of excitation. Consequently, the time-reversed Green function $G(\boldsymbol{r},\boldsymbol{r}',-t)$ is nonzero only for $t < 0$, and it propagates the wave backward in time. In (15), the time-reversed Green function $G(\boldsymbol{r},\boldsymbol{r}',-t)$ is convolved with the recorded acceleration. This means that it takes the waves that are recorded at the streamer and propagates them backward in time. This is a desirable property: in order to find the reflectors in the Earth, one needs to know the wave field at the moment when it was reflected off the reflectors. The recorded waves thus need to be propagated back in time so that we know them at earlier times when they were reflecting inside the Earth. This is the reason why the time-reversed Green function is used, and ultimately this is the reason why the theory presented here used the complex conjugate $G^*(\boldsymbol{r},\boldsymbol{r}',\omega)$ instead of $G(\boldsymbol{r},\boldsymbol{r}',\omega)$. If we had used $G(\boldsymbol{r},\boldsymbol{r}',t)$ instead of $G(\boldsymbol{r},\boldsymbol{r}',-t)$, equation (15) would have given the pressure field inside the Earth *after* it has been recorded at the receivers. This field does not give information about the interaction of waves with reflectors *before* the waves propagated to the surface where they are recorded. For this reason the theory in section 3 is based on $G^*$ rather than $G$.

## 5 A Chicken and Egg Problem

As shown here, Green's theorem makes it possible to infer the value of a physical field in the interior of the Earth from measurements taken at, or above, the Earth's surface. There is, however, a catch. In order to downward continue fields measured at the Earth's surface, one must know the Green function: see, for example, (14). For the wave equation (2), the space and time derivative fields are multiplied by the mass density and bulk modulus of the Earth, respectively. The Green function needed for downward continuation of seismic waves thus depends on the properties of the Earth, but it is these properties that one seeks to determine. We therefore need the properties of the Earth to determine the properties of the Earth!

Fortunately, there is a way out of this conundrum. It turns out that for seismic imaging it suffices to have an estimate of the Green function that positions the wavefronts at more or less the correct location. Such an estimated Green function is computed from a smooth velocity model. The velocity used is obtained from a procedure called *velocity estimation*, where one determines a smooth velocity model from measured arrival times from reflected seismic waves. The success of the seismic method in the hydrocarbon industry shows that this procedure works in practice.

### Further Reading


Blakeley, R. J. 1995. *Potential Theory in Gravity and Magnetics*. Cambridge: Cambridge University Press.

Bleistein, N., J. K. Cohen, and J. W. Stockwell Jr. 2001. *Mathematics of Multidimensional Seismic Imaging, Migration, and Inversion*. New York: Springer.

Coveney, P., and R. Highfield. 1991. *The Arrow of Time*. London: HarperCollins.

Schneider, W. A. 1978. Integral formulation for migration in two and three dimensions. *Geophysics* 43:49–76.

Snieder, R. 2002. Time-reversal invariance and the relation between wave chaos and classical chaos. In *Imaging of Complex Media with Acoustic and Seismic Waves*, edited by M. Fink, W. A. Kuperman, J. P. Montagner, and A. Tourin, pp. 1–15. Berlin: Springer.

——. 2004. *A Guided Tour of Mathematical Methods for the Physical Sciences*, 2nd edn. Cambridge: Cambridge University Press.

Yilmaz, O. 1987. *Seismic Data Processing*. Tulsa, OK: Society of Exploration Geophysicists.


# VII.17 Radar Imaging

*Margaret Cheney and Brett Borden*

## 1 Background

"Radar" is an acronym for *ra*dio *d*etection *a*nd *r*anging. Radar was originally developed as a technique for detecting objects and determining their positions by means of *echolocation*, and this remains the principal function of modern radar systems. Radar can provide very accurate distance (range) measurements, and can also measure the rate at which this range is changing. However, radar systems have evolved over more than seven decades to perform an additional variety of very complex functions; one such function is imaging.

Radar imaging has much in common with optical imaging: both processes involve the use of electromagnetic waves to form images. The main difference between the two is that the wavelengths of radar are much longer than those of optics. Because the resolving ability of an imaging system depends on the ratio of the wavelength to the size of the aperture, radar imaging systems require an aperture many thousands of times larger than optical systems in order to achieve comparable resolution. Since kilometer-sized antennas are not practicable, fine-resolution radar imaging has come to rely on so-called *synthetic apertures*, in which a small antenna is used to sequentially sample a much larger measurement region.

There are many advantages to using radar for remote sensing. Unlike many optical systems, radar systems can be used day or night. Because the long radar wavelengths pass through clouds, smoke, etc., radar systems can be used in all weather conditions. Moreover, some radar systems can penetrate foliage, buildings, dry soil, and other materials.

Radar waves scatter mainly from objects and features whose size is on the same order as the wavelength. This means that radar is sensitive to objects whose length scales range from centimeters to meters, and many objects of interest are in this range.

Radar has many applications, both military and civilian. Radar systems are widely used in aviation and transportation, for navigation, for collision avoidance, and for low-altitude flight. Most of us are familiar with police radar for monitoring vehicle speed. Radar is also used to monitor weather, including Doppler measurements of precipitation and wind velocity. Imaging radar is used for land-use monitoring, for agricultural monitoring, and for environmental monitoring. Radar-based techniques are used to map the Earth's surface topography and dynamic evolution. Medical microwave tomography is currently under development.

## 2 Mathematical Modeling

Mathematical modeling is based on MAXWELL'S EQUATIONS [III.22], or, more commonly in the case of propagation through dry air, the scalar approximation:

$$\left(\nabla^2 - \frac{1}{c^2}\frac{\partial^2}{\partial t^2}\right)\mathcal{E}(t,\boldsymbol{x}) = s(t,\boldsymbol{x}), \tag{1}$$

where $\mathcal{E}(t,\boldsymbol{x})$ denotes the (electric) field transmitted and measured by the radar, $\boldsymbol{x}$ and $t$ are position and time variables, and $c$ denotes the speed of light in vacuum. In constant-wave-velocity radar problems, the source $s$ is a sum of two terms, $s = s^{\text{in}} + s^{\text{sc}}$, where $s^{\text{in}}$ models the source due to the transmitting antenna, and $s^{\text{sc}}$ models the effects of target scattering. The solution $\mathcal{E}$ to equation (1), which is written as $\mathcal{E}^{\text{tot}}$, therefore splits into two parts: $\mathcal{E}^{\text{tot}} = \mathcal{E}^{\text{in}} + \mathcal{E}^{\text{sc}}$. The first term, $\mathcal{E}^{\text{in}}$, satisfies the wave equation for the known, prescribed source $s^{\text{in}}$, usually corresponding to the current density on an antenna. This part we call the *incident* field; it is the field in the absence of scatterers. The second part of $\mathcal{E}^{\text{tot}}$ is due to the presence of scattering targets, and this part is called the *scattered* field.

One approach to finding the scattered field is to simply solve (1) directly using, for example, numerical time-domain techniques. For many purposes, however, it is convenient to reformulate the scattering problem in terms of an integral equation.

In scattering problems the source term $s^{\text{sc}}$ represents the target's *response* to an incident field. This part of the source function will generally depend on the geometric and material properties of the target and on the form and strength of the incident field. Consequently, $s^{\text{sc}}$ can be quite complicated to describe analytically.

Usually, one makes the *Born* or *single-scattering* approximation, namely

$$s^{\text{sc}}(t,\boldsymbol{x}) = \int V(\boldsymbol{x})\mathcal{E}^{\text{in}}(t',\boldsymbol{x})\,\mathrm{d}t', \tag{2}$$

where $V(\boldsymbol{x})$ is called the *reflectivity function* and depends on target orientation. This results in a linear formula for $\mathcal{E}^{\text{sc}}$ in terms of $V$:

$$\begin{aligned}\mathcal{E}^{\text{sc}}(t,\boldsymbol{x}) &\approx \mathcal{E}_B(t,\boldsymbol{x})\\ &\equiv \iint g(t-\tau,\boldsymbol{x}-\boldsymbol{z})V(\boldsymbol{z})\mathcal{E}^{\text{in}}(\tau,\boldsymbol{z})\,\mathrm{d}\tau\,\mathrm{d}\boldsymbol{z},\end{aligned} \tag{3}$$

where $g$ is the *outgoing fundamental solution* or *(outgoing) Green function*:

$$g(t,\boldsymbol{x}) = \frac{\delta(t-|\boldsymbol{x}|/c)}{4\pi|\boldsymbol{x}|} = \int \frac{\mathrm{e}^{-\mathrm{i}\omega(t-|\boldsymbol{x}|/c)}}{8\pi^2|\boldsymbol{x}|}\,\mathrm{d}\omega. \tag{4}$$

Here, $|\boldsymbol{x}| = \sqrt{\boldsymbol{x}\cdot\boldsymbol{x}}$.

The Born approximation is very useful because it makes the scattering problem linear. It is not, however, always a good approximation.

The incident field in (3) is typically of the form of an antenna beam pattern multiplied by the convolution $g * p$, where $g$ denotes the Green function (4) and $p$ denotes the waveform fed to the antenna.

Consequently, we obtain the following model for the scattered field:

$$\mathcal{E}_B^{\text{sc}}(t, \boldsymbol{x}^0) = \int \frac{p(t - 2|\boldsymbol{x}^0 - \boldsymbol{z}|/c)}{(4\pi|\boldsymbol{x}^0 - \boldsymbol{z}|)^2} V(\boldsymbol{z})\,\mathrm{d}\boldsymbol{z}, \qquad (5)$$

where we have neglected the antenna beam pattern. Under the Born approximation, the scattered field can be viewed as a superposition of scattered fields from targets that are point-like (i.e., $V(\boldsymbol{z}') \propto \delta(\boldsymbol{z} - \boldsymbol{z}')$) in the sense that they scatter isotropically. No shadowing, obscuration, or multiple scattering effects are included.

Equation (5) is an expression for the field, which is proportional to the voltage measured on a receiving antenna. Note that the received power, which is proportional to the square of the voltage, is proportional to $1/R^4$, where $R = |\boldsymbol{x}^0 - \boldsymbol{z}|$ is the distance between the antenna at position $\boldsymbol{x}^0$ and the scatterer at position $\boldsymbol{z}$. This $1/R^4$ dependence is the reason that radar signals are typically extremely weak and are often swamped by thermal noise in the receiving equipment. Radar receivers typically correlate the incoming signal with the transmitted pulse, a process called *pulse compression* or *matched filtering*, in order to separate the signal from the noise.

Radar data do not normally consist simply of the backscattered field. Radar systems typically demodulate the scattered field measurements to remove the rapidly oscillating carrier signal and convert the remaining real-valued voltages to in-phase (I) and quadrature (Q) components, which become the real and imaginary parts of a complex-valued *analytic signal* (for which the signal phase is well defined). For the purposes of this article, however, we ignore the effects of this processing and work simply with the scattered field.

## 3 A Survey of Radar Imaging Methods

Synthetic-aperture radar (SAR) imaging relies on a number of very specific simplifying assumptions about radar scattering phenomenology and data-collection scenarios.

- Most imaging radar systems make use of the *start–stop approximation*, in which both the radar sensor and the scattering object are assumed to be stationary during the time interval in which the pulse interacts with the target.
- The target or scene is assumed to behave as a rigid body.
- SAR imaging methods assume a linear relationship between the data and the scene.

Different geometrical configurations for the sensor and target are associated with different terminology.

### 3.1 Inverse Synthetic-Aperture Radar

A fixed radar system staring at a rotating target is equivalent (by change of reference frame) to a stationary target viewed by a radar moving (from pulse to pulse) on a circular arc. This circular arc will define, over time, a synthetic aperture, and sequential radar pulses can be used to sample those data that would be collected by a much larger radar antenna. Radar imaging based on such a data-collection configuration is known as *inverse synthetic-aperture radar* (ISAR) imaging. This imaging scheme is typically used for imaging airplanes, spacecraft, and ships. In these cases, the target is relatively small and is usually isolated.

The *small-scene* approximation, namely

$$|\boldsymbol{x} - \boldsymbol{y}| = |\boldsymbol{x}| - \hat{\boldsymbol{x}} \cdot \boldsymbol{y} + O\left(\frac{|\boldsymbol{y}|^2}{|\boldsymbol{x}|}\right), \qquad (6)$$

where $\hat{\boldsymbol{x}}$ denotes a unit vector in the direction $\boldsymbol{x}$, is often applied to situations in which the scene to be imaged is small in comparison with its average distance from the radar. This approximation is valid for $|\boldsymbol{x}| \gg |\boldsymbol{y}|$.

Using (6) in (5) and shifting the time origin show that, when the small-scene approximation is valid, the radar data are approximately the Radon transform of the reflectivity function.

### 3.2 Synthetic-Aperture Radar

SAR involves a moving antenna, and usually the antenna is pointed toward the Earth. For an antenna viewing the Earth, we need to include a model for the antenna beam pattern, which describes the directivity of the antenna. For highly directive antennas, we often simply refer to the antenna "footprint," which is the illuminated area on the ground.

Most SAR systems use a single antenna for both transmitting and receiving. For a pulsed system, we assume that pulses are transmitted at times $t_n$, and we denote the antenna position at time $t_n$ by $\boldsymbol{\gamma}_n$. The data can then be written in the form

$$\mathcal{E}_B^{\text{sc}}(t, n) = \int \mathrm{e}^{-\mathrm{i}\omega[t - 2|\boldsymbol{\gamma}_n - \boldsymbol{y}|/c]} A(\omega, n, \boldsymbol{y})\,\mathrm{d}\omega V(\boldsymbol{y})\,\mathrm{d}\boldsymbol{y}, \qquad (7)$$

where $A$ incorporates the geometrical spreading factors $|\boldsymbol{x}^0 - \boldsymbol{y}|^{-2}$, the transmitted waveform, and the antenna beam pattern. (More details can be found in Cheney and Borden (2009).)

Because the timescale on which the antenna moves is much slower than the timescale on which the electromagnetic waves propagate, the timescales have been separated into a *slow time*, which corresponds to the $n$ of $t_n$, and a *fast time* $t$.

The goal of SAR is to determine $V$ from radar data that are obtained from the scattered field $E^{\text{sc}}$ by I/Q demodulation (mentioned at the end of section 2) and matched filtering. Again, for the purposes of this article, we neglect the processing done by the radar system and work simply with the scattered field.

Assuming that $\boldsymbol{y}$ and $A$ are known, the scattered field (7) depends on two variables, so we expect to form a two-dimensional image. For typical radar frequencies, most of the scattering takes place in a thin layer at the surface. We therefore assume that the ground reflectivity function $V$ is supported on a known surface. For simplicity we take this surface to be a flat plane, so that $V(\boldsymbol{x}) = V(\boldsymbol{x})\delta(x_3)$, where $\boldsymbol{x} = (x_1, x_2)$.

SAR imaging comes in two basic varieties: *spotlight* SAR and *stripmap* SAR.

### 3.2.1 Spotlight SAR

Spotlight SAR is illustrated in figure 1. Here, the moving radar system stares at a specific location (usually on the ground) so that at each point in the flight path the same target is illuminated from a different direction. When the ground is assumed to be a horizontal plane, the constant-range curves are large circles whose centers are directly below the antenna at $\boldsymbol{y}_n$. If the radar antenna is highly directional and the antenna footprint is sufficiently far away, then the circular arcs within the footprint can be approximated as lines. Consequently, the imaging method is mathematically the same as that used in ISAR.

As in the ISAR case, the time-domain formulation of spotlight SAR leads to a problem of inverting the Radon transform.

### 3.2.2 Stripmap SAR

Stripmap SAR is illustrated in figure 2. Just as the time-domain formulations of ISAR and spotlight SAR reduce to inversion of the Radon transform, which is a tomographic inversion of an object from its integrals over lines or planes, stripmap SAR also reduces to a tomographic inversion of an object from its integrals over circles or spheres.

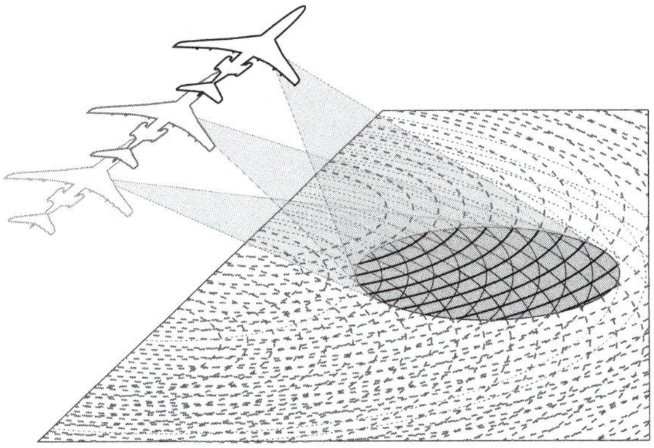

**Figure 1** In spotlight SAR the radar is trained on a particular location as the radar moves. In this figure the equirange circles (dotted lines) are formed from the intersection of the radiated spherical wavefront and the surface of a (flat) Earth.

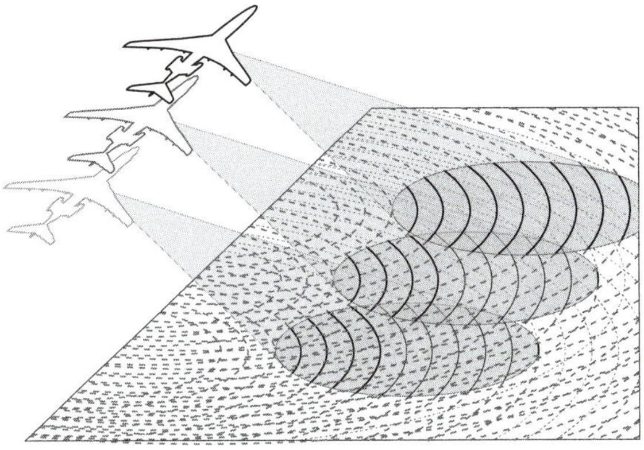

**Figure 2** Stripmap SAR acquires data without staring. The radar typically has fixed orientation with respect to the flight direction and the data are acquired as the beam footprint sweeps over the ground.

### 3.2.3 Interferometric SAR

Interferometric SAR is a sort of binocular radar imaging system that can provide height information. These systems use two antennas that create separate SAR images. These images are complex, and height information is encoded in the phase difference between the two images.

## 4 Future Directions for Research

In the decades since the invention of SAR imaging, there has been much progress, but many open problems still remain. In particular, as outlined at the beginning of section 2, SAR imaging is based on specific assumptions that may not be satisfied in practice. When they are not satisfied, artifacts appear in the image. Consequently, a large number of the outstanding problems can be grouped into two major areas.

**Problems related to unmodeled motion.** Both SAR and ISAR are based on known relative motion between target and sensor, e.g., including the assumption that the target behaves as a rigid body. When this is not the case, the images are blurred or uninterpretable.

**Problems related to unmodeled scattering physics.** The Born approximation leaves out many physical effects, including not only multiple scattering and creeping waves but also shadowing, obscuration, and polarization changes. Neglecting these effects can lead to image artifacts. But without the Born approximation (or the Kirchhoff approximation, which is similar), the imaging problem is nonlinear.

### Further Reading

Buderi, R. 1996. *The Invention That Changed the World.* New York: Simon & Schuster.

Carrara, W. C., R. G. Goodman, and R. M. Majewski. 1996. *Spotlight Synthetic Aperture Radar: Signal Processing Algorithms.* Boston, MA: Artech House.

Cheney, M., and B. Borden. 2009. *Fundamentals of Radar Imaging.* Philadelphia, PA: SIAM.

Cook, C. E., and M. Bernfeld. 1967. *Radar Signals.* New York: Academic Press.

Franceschetti, G., and R. Lanari. 1999. *Synthetic Aperture Radar Processing.* New York: CRC Press.

Jakowatz, C. V., D. E. Wahl, P. H. Eichel, D. C. Ghiglia, and P. A. Thompson. 1996. *Spotlight-Mode Synthetic Aperture Radar: A Signal Processing Approach.* Boston, MA: Kluwer.

Natterer, F. 2001. *The Mathematics of Computerized Tomography.* Philadelphia, PA: SIAM.

Newton, R. G. 2002. *Scattering Theory of Waves and Particles.* Mineola, NY: Dover.

Soumekh, M. 1999. *Synthetic Aperture Radar Signal Processing with MATLAB Algorithms.* New York: John Wiley.

Sullivan, R. J. 2004. *Radar Foundations for Imaging and Advanced Concepts.* Raleigh, NC: SciTech.

## VII.18 Modeling a Pregnancy Testing Kit

*Sean McKee*

### 1 Device Description

In this article we describe a general medical diagnostic tool, based on antibody/antigen technology, whose principal, and certainly most lucrative, application is as a pregnancy testing kit. The fluorescence capillary-fill device (FCFD) consists of two plates of glass separated by a narrow gap. The lower plate is coated with an immobilized layer of specific antibody and acts like an optical waveguide. The upper plate has an attached reagent layer of antigen (or hapten[1]) labeled with a fluorescent dye. When the sample is presented at one end of the FCFD, it is drawn into the gap by capillary action and dissolves the reagent. The fluorescently labeled antigen in the reagent now competes with the sample antigen for the limited number of antibody sites on the lower glass plate (see figure 1). The FCFD plate structure may be regarded as a composite waveguide: the intensities of the distinct optical paths, depending on whether they originate from a fluorescent molecule that is free in the solution or from a molecule bound close to the surface of the plate, are picked up by a photodetector.

We shall not be concerned with the optical aspects of this device but rather with the competitive reaction between the antigen and the fluorescent antigen for those antibody sites, and we shall focus on its use as a pregnancy testing kit. When a woman is pregnant the presence of a "foreign body" causes the production of antigen molecules (X), which are then countered by specific antibodies (Y). Denoting the fluorescently labeled antigen by $X_F$, figure 2 (representing the "blow-ups" in figure 1) displays the situation when a woman is, and is not, pregnant both at time $t = 0$ (when the sample is first presented to the device) and at $t = t_f$ (when the reaction is finished). Upon completion, a light is shone down the lower plate, and, if the woman is pregnant, the beam of greater light intensity will be detected at a larger angle to the plate (as shown in figure 1).

### 2 Mathematical Model

When a sample (urine in this case) is presented to the FCFD, it dissolves the labeled antigen but not the antibody that is fixed on the lower wall. The device is then left in a stationary position to allow the antigen and the labeled antigen to diffuse across the gap and compete for the antibody sites. Thus the mathematical model consists of two one-dimensional diffusion equations coupled through nonlinear and nonlocal boundary conditions, and a number of conservation

1. Haptens are low-molecular-weight molecules that contain an antigenic determinant; these have substantially larger diffusion coefficients than the antigen molecules.

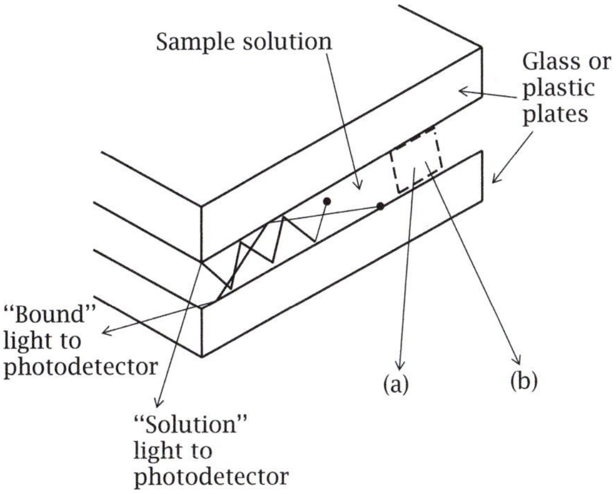


**Figure 1** A schematic of the fluorescent capillary-fill device. More details can be found in Badley et al. (1987).

relationships. Since the equations themselves are linear, Laplace transforms may be employed to recharacterize the system as two coupled nonlinear Volterra integro-differential equations: in nondimensional form they are

$$\frac{dw_1}{dt}(t) = \gamma_1\delta\Big[(1 - w_1(t) - w_2(t)) \times \Big(\mu - m\int_0^t \frac{dw_1}{d\tau}(\tau)K(\delta(t-\tau))\,d\tau\Big) - L_1 w_1(t)\Big], \quad (1)$$

$$\frac{dw_2}{dt}(t) = \gamma_2\Big[(1 - w_1(t) - w_2(t)) \times \Big(g(t) - m\int_0^t \frac{dw_2}{d\tau}(\tau)K(t-\tau)\,d\tau\Big) - L_2 w_2(t)\Big]. \quad (2)$$

Here, $w_1(t)$ and $w_2(t)$ denote the concentrations of the complexes XY and $X_FY$, respectively, while $\delta$, $\mu$, $m$, $\gamma_1$, and $\gamma_2$ are nondimensional constants. The functions $g(t)$ and $K(t)$ are given by

$$g(t) = 1 - e^{-\lambda t} + 2\lambda\sum_{n=1}^{\infty}\frac{(-1)^n}{(\lambda - n^2\pi^2)}(e^{-n^2\pi^2 t} - e^{-\lambda t}),$$

$$K(t) = \frac{1}{\sqrt{\pi t}}\Big(1 + 2\sum_{n=1}^{\infty} e^{-n^2/t}\Big),$$

where $\lambda$ is a further nondimensional constant representing the dissolution rate of the bound $X_F$.

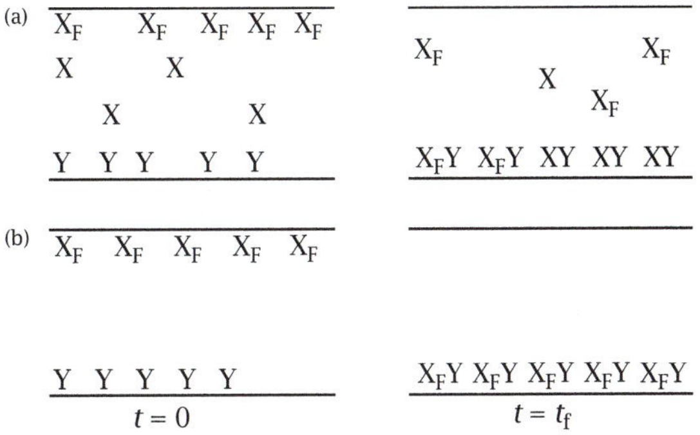


**Figure 2** (a) Woman is pregnant. (b) Woman is not pregnant. Here, X denotes the sample antigen, $X_F$ the antigen or hapten with a fluorescent label, and Y the specific antibody for the antigen or hapten.

Equations (1) and (2) admit a regular perturbation solution for small $m$.

Extending results of Jumarhon and McKee, a further recharacterization may be obtained in the form of a system of four coupled, nonlinear, (weakly) singular Volterra integral equations of the second kind: the four dependent variables in this case, in nondimensional form, are

$$[X](1,t), \qquad [X_F](1,t), \qquad \int_0^1 [X](x,t)\,dx,$$

and

$$\int_0^1 [X_F](x,t)\,dx,$$

where [X] denotes the concentration of the antigen X, etc. It is from this system that one is able to deduce (global) existence and uniqueness of a solution of the original diffusion problem, though the proofs are not trivial.

It is also from these characterizations that one is able to obtain small- and large-$t$ asymptotic results; interestingly, the large-$t$ results require the explicit solution of a quartic.

## 3 Design Considerations

The objective of this work was to provide a quantitative design tool for the bioscientists. Indeed, this model—or, more precisely, an earlier considerably simplified model—was ultimately employed in the development of Clearblue, the well-known pregnancy testing kit. The model (and its associated code) allowed bioscientists to see that the device could be made small (and, consequently, be produced very cheaply) and in large batches, suitable for hospital use. Furthermore, it provided an indication of both the plate separation distance and how much antibody and labeled antigen were required to be affixed to the plate surfaces. In short, the tool obviated a great deal of experimentation, thus saving time and allowing Unilever Research, through a company that was then called Unimed, to bring the product to market early.

### Further Reading

Badley, R. A., R. A. L. Drake, I. A. Shanks, A. M. Smith, and P. R. Stephenson. 1987. Optical biosensors for immunoassays: the fluorescence capillary-fill device. *Philosophical Transactions of the Royal Society of London* B 316:143–60.

Jumarhon, B., and S. McKee. 1995. On the heat equation with nonlinear and nonlocal boundary conditions. *Journal of Mathematical Analysis and Applications* 190:806–20.

Rebelo, M., T. Diogo, and S. McKee. 2012. A mathematical treatment of the fluorescence capillary-fill device. *SIAM Journal of Applied Mathematics* 72(4):1081–112.

## VII.19 Airport Baggage Screening with X-Ray Tomography

*W. R. B. Lionheart*

### 1 The Security Screening Problem

In airport security, baggage carried in an aircraft's hold needs to be screened for explosive devices. Traditionally, an X-ray machine is used that gives a single two-dimensional projection of the X-ray attenuation of the contents of the bag. In some systems, several views are used to reveal threats that may be obscured by large dense objects. An extension of this idea is to use a large number of projections to reconstruct a three-dimensional volume image of the luggage. This is the X-ray computed tomography (CT) technology that is familiar from medical applications (see MEDICAL IMAGING [VII.9]). The image can then be viewed by an operator from any desired angle and threat-detection software can be used that, for example, segments the volume image, identifying objects that have a similar X-ray attenuation to explosives. In some airports, a two-stage system is used in which bags that cannot be cleared by an automatic system analyzing a two-dimensional projection are then passed to a much slower X-ray tomography system.

Airport baggage handling systems operate using conveyor belts traveling at around 0.5 m/s. Medical CT machines use a gantry supporting the X-ray source and an array of detectors that rotates in a horizontal plane while the patient is translated in the direction of the rotation axis. Relative to the patient, the source describes a helical trajectory. By contrast, small laboratory CT machines rotate and translate the sample while the source and detector remain fixed. Neither of these is practical for scanning luggage at the desired speed: the mass of the gantry is too great to rotate fast enough and rotating the bag would displace the contents.

### 2 Real-Time Tomography

The company Rapiscan Systems has developed a system called real-time tomography (RTT) that uses multiple X-ray sources fixed in a circular configuration that can be switched electronically, removing the need for a rotating gantry. A cylindrical array of detectors is used, and this is coaxial with the sources, but the X-rays cannot penetrate the detectors, so the detectors are offset relative to the sources (see figure 1).

A reasonable mathematical model of X-ray tomography is that the line integrals of the linear attenuation are measured for all lines joining sources to detectors. For a helical source trajectory and detectors covering a set called the *Tam–Danielson window*, there is an exact reconstruction algorithm due to Katsevich expressed in terms of derivatives and integral operators applied to the data. Most medical and industrial CT machines use an approximation of this using overdetermined data.

The RTT presents several mathematical challenges.

- The data is incomplete, resulting in an ill-posed INVERSE PROBLEM [IV.15] to solve.
- The reconstruction must be completed quickly to ensure the desired throughput of luggage.
- The sources can be fired in almost any order. In fact, sequential firing that approximates a rotating gantry and a single-threaded helix trajectory is the most difficult due to heat dissipation issues in clusters of sources. What is the optimal firing order?

### 3 Sampling Data and Sufficiency

As the RTT was originally conceived as a fast helical scan machine, it is natural to think of the sources firing sequentially at equal time intervals as a discrete approximation to a curve. A firing sequence in which a fixed number of sources is skipped at each time step approximates a multithreaded helix. Figure 2 illustrates

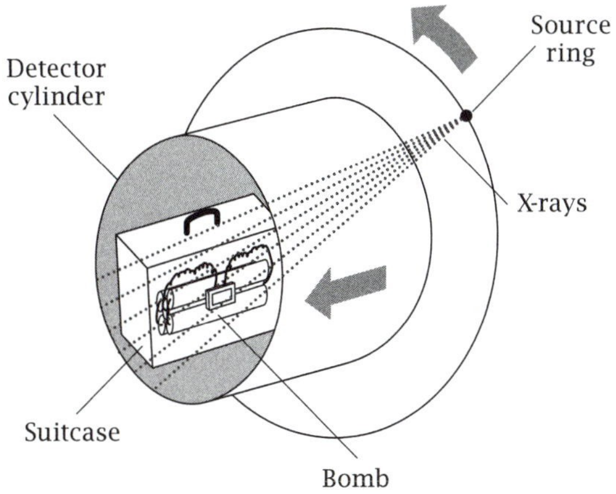


**Figure 1** A cartoon of an RTT scanner showing the source circle and detector cylinder.

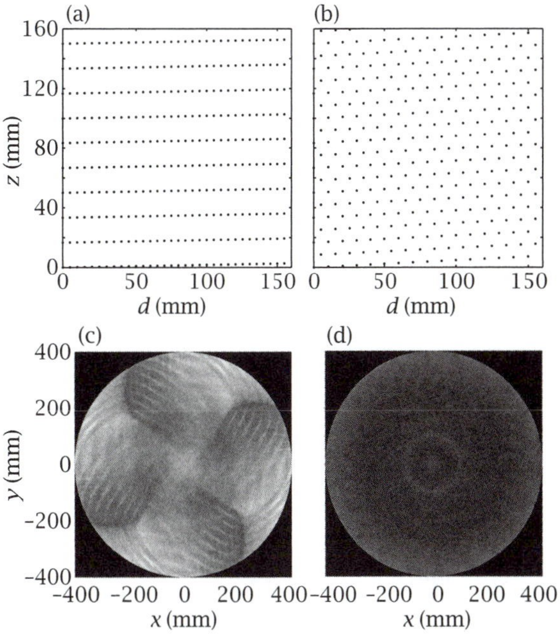


**Figure 2** (a), (b) Two multithreaded helix source firing sequences; (b) a lattice on the cylinder of source positions. (c), (d) The number of rays intersecting each voxel for the firing orders above. The second sequence ((b), (d)) has more uniform coverage of the cylinder and produces a more uniform distribution of rays.

two possible firing orders, the first approximating a multithreaded helix. The second could be interpreted as a multithreaded helix (with two different pitches), but it is more natural to view it as a triangular lattice that samples the two-dimensional surface of a cylinder rather than a curve.

The manifold of lines in three-dimensional space is four dimensional and data from the X-ray transform of a function of three variables satisfies a consistency condition called John's ultrahyperbolic equation. In conventional helical CT, the lines through the helix in the Tam–Danielson window are sufficient data to solve the reconstruction problem (and consequently solve the Dirichlet problem for John's equation). We can interpret the RTT data in figure 1 intersected with the detector array as a discrete sampling scheme for an open subset of the four-dimensional space of lines. Using the Fourier slice theorem for the Radon plane transform along with the Payley–Wiener theorem, we can see that for continuum data the inverse problem has a unique solution, but an inversion using this method would be highly unstable. In a practical problem with a discrete sampling scheme and noisy data, we would expect to need some regularization for a numerically stable solution.

## 4 Inversion

The inversion of the RTT data can be considered as the solution of a sparse linear system of equations relating the attenuation coefficients in voxels within the region of interest to the measured data. As such we can employ the standard techniques of numerical linear algebra and linear inverse problems, including iterative solution methods and regularization. The matrix of the system we solve is generally too large for direct solution, but one method that works well for this problem is conjugate gradient least squares applied to the generalized Tikhonov regularized system.

This approach allows us to apply a systematic choice of regularization penalty, equivalent to an assumption about the covariance of the prior distribution in Bayesian terms. The matrix of the system we solve is generally too large for direct solution with the computing hardware available, but looking at scaled-down systems we have found that the condition number of the (unregularized) matrix is smallest for firing orders that uniformly sample space. Another way to assess the merits of a firing order is to look at the distribution of directions of rays intersecting each voxel, and this criterion leads to the same conclusion as the studies of condition number (see figure 2). The resulting reconstruction results also demonstrate an improvement over the lattice source firing.

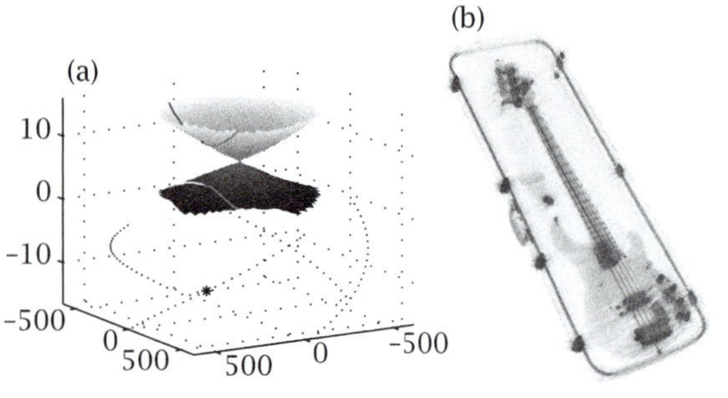


**Figure 3** (a) Optimal two-sheet surface with an exaggerated axial scale. (b) Reconstruction of a bass guitar from data collected using a prototype RTT80 baggage scanner.

Even using contemporary graphics processing units, iterative methods are too slow to solve the reconstruction problem in real time. Rebinning methods interpolate data from a (multithreaded) helix to approximate data taken on a plane. Radon transform inversion on the planes can be performed very quickly using filtered back projection. A variation of this method is surface rebinning. For each source location $\lambda$ the image values on a surface $z = \zeta_\lambda(x, y)$ (where $z$ is the axial direction) are approximated using two-dimensional Radon transform inversion of the data corresponding to rays close to that surface (see figure 3). For conventional helical CT, the optimal surface to use is close to a plane, but for the RTT geometry a good solution is to find a surface with two sheets and reconstruct the sum of the attenuation coefficients at points on those sheets with the same $(x, y)$ coordinates. There is then a separate very sparse linear system to solve to find the values of the attenuation coefficient at a voxel, which will depend on values on the upper and lower sheet for different source positions. The optimal surface for a given geometry is conveniently found as a fixed point of a contraction mapping.

### Further Reading

Betcke, M. M., and W. R. B. Lionheart. 2013. Multi-sheet surface rebinning methods for reconstruction from asymmetrically truncated cone beam projections. I. Approximation and optimality. *Inverse Problems* 29:115003.

Ramm, A. G., and A. I. Katsevich. 1996. *The Radon Transform and Local Tomography*. Boca Raton, FL: CRC Press.

Thompson, W. M., W. R. B. Lionheart, and D. Öberg. 2013. Reduction of periodic artefacts for a switched-source X-ray CT machine by optimising the source firing pattern. In *Proceedings of the 12th International Meeting on Fully Three-Dimensional Image Reconstruction in Radiology and Nuclear Medicine, Lake Tahoe, CA*.

## VII.20 Mathematical Economics

*Ivar Ekeland*

Economic theory is traditionally divided into microeconomics and macroeconomics. Microeconomics deal with individuals, macroeconomics with society. At the present time, microeconomics is seen as foundational, meaning that society is understood as no more than a contract (implicit or explicit) between individuals, so that macroeconomics should be derived from the behavior of individuals, just as all the laws of physics should be derived from the behavior of atoms. Let me hasten to add that we are no closer to this grand unification in economics than we are in physics. Macroeconomics and microeconomics are largely separate fields, with the latter being much more conceptually mature, whereas the basic principles of the former are still under discussion.

### 1 Individuals

Individuals are seen as utility maximizers. Each of us lives in an environment in which decisions are to be made: choosing a point $x \in A$, say, where $A$ is a closed subset of Euclidean space. My preferences are characterized by a continuous function $U: A \to R$: I prefer $x$ to $y$ if $U(x) > U(y)$, and I am indifferent if $U(x) = U(y)$. Note that the preference relation thus defined is transitive: if I prefer $x$ to $y$ and $y$ to $z$, then I prefer $x$ to $z$. As we shall see later on, this transitivity holds only for individuals, not for groups. The utility function characterizes the individual: I have mine; you have yours; they are different. Note that it is taken as given that some of us are selfish, some are not, some of us are drug addicts, some of us prefer guns—it will all show up in our utility. At this point, economic theory is *positive*, meaning that it does not tell people what they should aim for; it tells them how best to reach their goals.

It is assumed that each individual chooses the point that maximizes his or her utility function $U$ over the set of possible choices $A$. Of course, we need this maximizer to be unique; otherwise we would have to choose between all possible maximizers, and the decision problem would not be solved. In order to achieve uniqueness, the set $A$ is assumed to be convex and the function $U$ concave, one of them strictly so. The admissible set $A$ is usually defined by budgetary constraints. For instance, we may consider an economy with $D$ goods, with $x = (x^1, \dots, x^D)$ being a goods bundle (meaning that $x^d$ is the quantity of good $d$ in the bundle). If the price of good $d$ is $p_d$, the price of the bundle $x$ is given by

$$px = \sum_{d=1}^{D} p_d x^d.$$

Consumers are then characterized by their utility function $U$ and their wealth $w$, and they will choose the bundle $x$ that they prefer among all those they can afford. This translates into the convex optimization problem

$$\max_x U(x), \; px \leqslant w, \; x_d \geqslant 0 \; \forall d. \tag{1}$$

Strange as it may seem, this model has testable consequences, and the data supports it. It is worth going into this in a little more detail. One first needs to distinguish what is *observable* from what is not. The utility function $U$, for instance, cannot be observed: if I am asked what my utility function is, I do not know how to answer. However, the map $p \to x(p)$ (the demand function) that associates with each price system the corresponding minimum in problem (1) is observable: it is conceivable that my consumption pattern could be observed under price changes. The demand function is known to satisfy the so-called Slutsky relations. These are a system of first-order partial differential equations (PDEs) that can be expressed compactly by stating that the matrix $S(x)$ defined by

$$S(x)^{m,n} = \frac{\partial x^m}{\partial p_n} - \sum_k x^m \frac{\partial x^n}{\partial p_k} p_k$$

is symmetric and negative-definite. Browning and Chiappori have tested the Slutsky relations on Canadian data and found them to be satisfied, so the consumer model stands. Unfortunately, the model is too restricted: it is not enough to weigh one decision against another when the consequences are immediate and certain. Most of the time, the consequences will occur more or less far into the future, and they are affected by various degrees of uncertainty.

The standard way to take into account uncertainty goes back to von Neumann and Morgenstern (VNM). Assume that uncertainty is modeled by a space $\Omega$ of events, with a $\sigma$-algebra $\mathcal{A}$, and that the individual (who already has a utility function $U$) puts on $(\Omega, \mathcal{A})$ a probability $P$. The utility of a random variable $\tilde{X}$ will then be its expected utility, namely, $\mathbb{E}[U(\tilde{X})]$. In this framework, the concavity of $U$ can be interpreted as risk aversion.

The standard way to account for late consequences is to discount future utilities. More precisely, it is assumed that there is some $\delta > 0$ (the psychological rate of time preference) such that if $x$ is consumed at time $t \geqslant 0$, the resulting utility at time $t = 0$ is $e^{-\delta t} U(x)$. Combining both VNM utility and discounting, we can give the present value of an uncertain consumption flow (a stochastic process) $\tilde{X}_t$, $t \geqslant 0$, as

$$\mathbb{E}\left[\int_0^\infty e^{-\delta t} U(\tilde{X}_t)\, dt\right]. \tag{2}$$

One more equation is needed to describe how the consumption flow $\tilde{X}_t$ is generated or paid for. With this model one can then describe how people allocate consumption and investment over time. It is the workhorse of economic theory (both micro and macro) and finance. The mathematical tools are optimal control, both stochastic and deterministic, both from the PDE point of view (leading to the Hamilton–Jacobi–Bellman equation) and from the point of view of the Pontryagin maximum principle (leading, in the stochastic case, to a backward stochastic differential equation). All economics textbooks are replete with examples.

This model is now facing criticism due to accumulated psychological evidence.

On the one hand, people do not seem to maximize expected utility when facing uncertainty, e.g., they give undue importance to events with a small probability, and they are more sensitive to losses than to gains. Allais (very early on) and (later) Ellsberg have pointed out paradoxes, that is, experiments in which the actual choices could not be explained by VNM utilities. Various alternative models have been suggested to take these developments into account, the most popular of which is the prospect theory of Kahneman and Tversky. This is a modification of the VNM model in which probabilities are distorted before the expectation is computed.

On the other hand, people do discount future utilities but not at a constant rate: the present value of receiving $x$ at a time $t$ is $h(t)U(x)$, where $h(t)$ is decreasing from 1 to 0, but there is no reason it should be an exponential, $h(t) = e^{-\delta t}$. Psychological evidence suggests that it looks more like $h(t) = (1 + kt)^{-1}$ (hyperbolic discounting). This means that (2) should be replaced by

$$\mathbb{E}\left[\int_0^\infty h(t) U(\tilde{X}_t)\, dt\right].$$

Unfortunately, this implies that the preferences of the decision maker now depend on time! More precisely, suppose that different flows ($X_t$ and $Y_t$, say) are given for $t \geqslant T$ (assume they are deterministic; uncertainty has nothing to do with it), and for $0 \leqslant s \leqslant T$ set

$$I_s(X) := \int_T^{\infty} h(t-s)U(X_t)\,dt.$$

If $h(t)$ is not an exponential, it may well be the case that, for $s_1 < s_2 < T$, we have $I_{s_1}(X) < I_{s_1}(Y)$ but $I_{s_2}(X) > I_{s_2}(Y)$. This phenomenon is called time inconsistency. It implies that seeking an optimal solution is useless, since the decision maker will change his or her idea of optimality as time goes by.

Non-VNM utilities and nonexponential discounts are both the subject of active research because they seem to stick closer to the actual behavior of economic agents, notably investors, than do VNM utilities. This creates major mathematical challenges, since it introduces nonconvexities in optimization and control problems (prospect theory), and challenges the very concept of optimality (time inconsistency), which then has to be replaced by a Nash equilibrium of the game between successive decision makers.

## 2 Groups

The main thing to understand about groups is that they are not individuals; in particular, they do not have a utility function. If there are $N$ members in a group, each of them with a utility function $U_n$, $1 \leqslant n \leqslant N$, and a decision has to be made, then one cannot maximize all the $U_n$ at the same time, except in very particular cases. Each member has his of her own preferred choice $x_n$, and the collective choice will be the result of a decision process. There is a large literature on collective choice, which is mainly axiomatic: one specifies certain properties that such a process should satisfy, and then one seeks to identify the solution, if there is one. The main result in this direction is the Arrow impossibility theorem, which states the following. Suppose there is a procedure that transforms any set of individual rankings into a collective ranking. Suppose this procedure satisfies two very mild conditions, namely:

- if every voter prefers alternative X over alternative Y, then the group prefers X over Y; and
- if another alternative Z is introduced and if every voter's preference between X and Y remains unchanged, then the group's preference between X and Y will also remain unchanged (even if voters' preferences between other pairs, like X and Z, Y and Z, or Z and W, change).

The procedure then consists of conforming in every circumstance to the ranking of a fixed member of the group (the dictator). The importance of Arrow's theorem consists of emphasizing that in any situation where a collective decision is to be made, the result will be as much a function of the procedure chosen as of the individual preferences of the members. In other words, if the premises of microeconomic theory are to be accepted, there is no such thing as the common good. In any society there are only individual interests, and as soon as they do not agree, there is no overriding concern that would resolve the conflict; there are only procedures, and one is as good as another. A rich literature on collective choice has sprung from this well. It is axiomatic in nature, i.e., one seeks procedures that will satisfy certain axioms posited a priori. For instance, one might want to satisfy some criterion of fairness. Unfortunately, there is not currently a formal model of fairness or a clear understanding of distributive justice.

However, there is a generally accepted notion of *efficiency*. Choose coefficients $\lambda_n \geqslant 0$, with $\sum \lambda_n = 1$, and maximize $\sum \lambda_n U_n(x)$. The point $x$ obtained in this way will have the property that one cannot find another point $y$ such that $U_n(y) > U_n(x)$ for some $n$, and $U_n(y) \geqslant U_n(x)$ for all $n$. It is called a *Pareto optimum*, and of course it depends on the choice of the coefficients $\lambda_n$. If $x$ is not a Pareto optimum, then resources are being wasted, since one could give more to one person without hurting anyone else. Note that if $x$ is a Pareto optimum, the choice $x$ is efficient, since there is no waste, but it need not be fair: taking $\lambda_1 = 1$ and $\lambda_n = 0$ for $n \geqslant 0$ gives a Pareto optimum.

Groups are ubiquitous in economic theory: the smallest economic unit is not the individual but the household, two or more people living together and sharing resources. How the resources are shared depends on cultural, social, and psychological constraints, and these would be extremely hard to model. However, one can get a long way with the analysis by making the simple assumption that the sharing of resources within the group is efficient, i.e., Pareto optimal. For an efficient household, composed of $H$ members, each one with his/her own utility function $U_h(x_h, X)$, the consumption problem (1) becomes

$$\max\left\{ \sum_h \lambda_h(p)U_h(x_h) \;\middle|\; px \leqslant w \right\},$$

where $x_h$ is the consumption of each member, $p$ is the price system, and the $\lambda_h$ are the Pareto coefficients that reflect power within the group (and depend on the price system). The econometrician observes the sum $\sum x_h(p)$, that is, he or she observes the aggregate consumption of the group, not the individual consumption of each member. It is possible to extend the Slutsky relations to this situation and identify the individual preferences of each member. This is done by using exterior differential calculus, notably the Darboux and Cartan–Kähler theorems.

## 3 Markets

Markets are groups that have chosen a particular way to allocate resources. The standard model for competitive markets is due to Arrow and Debreu. There are global resources, namely a bundle of goods $\bar{x} \in R^D$, that should either be shared outright (an exchange economy) or used to produce more goods, which will then be shared (a production economy) among the $N$ members of the group, each of whom has his or her own utility function $U_n$. There is also a production technology, consisting of $K$ firms, each of which is characterized by a production set $Y_k$: if $(x, y) \in Y_k$, then firm $k$ can produce the bundle $x$ by consuming the bundle $y$. Time runs from $t = 0$ to infinity, and there is some uncertainty about the future, modeled by a set $\Omega$ of possible states of the world.

Each good is characterized by its date of delivery $t \geqslant 0$ and is contingent on some state of the world $\omega$: I pay now, and I get the good delivered at time $t$ if $\omega$ occurs (so, in effect, one can insure oneself against any possible event). We start from an initial allocation (sharing) of the global resources: each individual $n$ in the economy starts with a bundle $x_n \in R^D$, so that $\sum_n x_n = \bar{x}$. However, this initial allocation may not be Pareto optimal, hence the need for trading. The main result of Arrow and Debreu states that, provided utilities are concave and production functions are convex, there is an equilibrium, namely, a price system such that all markets are clear (demand equals supply for each good). The main mathematical tool used to prove the Arrow–Debreu result is the Brouwer fixed-point theorem, usually in its multivalued variant, the Kakutani fixed-point theorem. Of course, the equilibrium that is reached depends on the initial allocation.

Note that the theory is not entirely satisfactory, even within its own set of assumptions. On the one hand, there may be several equilibria, and the theory has nothing to say about which one is chosen. On the other, the theory is purely static: it does not say how an equilibrium is reached from a nonequilibrium situation. Mathematicians have expended much effort on finding dynamics that would always converge to an equilibrium, and economists have tried to find procedures by which market participants would infer the equilibrium prices, but progress has so far been limited.

## 4 How Markets Fail

The idea that markets are always right has permeated all economic policy since the 1970s, and it inspired the wave of deregulation that has swept through the U.S. and U.K. economies since the Reagan era. Economic theory, though, does not support this view. The strength of the market mechanism lies in the first and second welfare theorems, which state that every equilibrium is an efficient allocation of resources and that, conversely, every efficient allocation of resources can be achieved as a no-trade equilibrium (if it is chosen as the initial allocation, it is also an equilibrium allocation for an appropriate price system). However, the conditions required for these theorems to hold, and indeed for all of Arrow–Debreu theory, are very rarely achieved in practice; in fact, I cannot think of a single real-world market in which they are. A huge part of economic theory is devoted to studying the various ways in which markets fail and to how they can be fixed.

First of all, markets may exist and not be competitive. All major markets (weapons, oil, minerals, cars, pharmaceutical firms, operating systems, search engines) are either monopolistic or oligopolistic, meaning that producers charge noncompetitive prices, so the resulting allocation of resources is not efficient. In addition, Arrow–Debreu theory eliminates uncertainty by assuming that one can insure oneself against any future event, and that is not the case. Note also that classical economic theory assumes that consumption goods are homogeneous (a Coke is a Coke is a Coke), whereas many actually differ in subtle ways (Romanée-Conti is not just another wine, and the 2000 vintage is not the same as the 2001 vintage). Prices then have to take quality into account, and consumers tend to behave differently: if you become richer, you do not necessarily buy more wine, but you certainly buy better wine. Such markets are called hedonic, and interesting connections have been found with the mathematical theory of optimal transportation.

Second of all, markets do not price externalities. An externality occurs when the fact that I consume something affects you. For instance, the competitive market price for cigarettes would take into account the demand of smokers and the supply of tobacco manufacturers but not the inconvenience to nonsmokers nor the damage to their health, resulting in an inefficient allocation of resources. To restore efficiency, one would have to charge the smokers for smoking and transfer the money to nonsmokers, a nonmarket mechanism. Markets do not price public goods either. A public good is, according to an old saying, "like a mother's love: everyone has a share, and everyone enjoys it fully." More precisely, it is a good that you can consume even if you did not pay for it or contribute to producing it. Sunshine, clean air, and a temperate climate are public goods: even the worst polluter can enjoy them. Other goods, like roads, education, or health, can be either public or private: if they are free, they are public goods; if you put a toll on a road, if you charge tuition fees, or if you have to pay for medical treatment, they become private goods. The problem with making them private is that they have large positive externalities (the road supports economic activity, having an educated workforce is good for business, living with healthy individuals reduces one's chances of falling ill) that the market will not price, and the resulting situation will therefore be inefficient.

## 5 Asymmetry of Information

Another imperfection of the Arrow-Debreu model is that there is full information on the goods that are traded: a loaf of bread is a loaf of bread; there can hardly be a surprise. However, there are many situations in which information is asymmetric: typically, the seller, who already owns the good, or who has produced it, knows more about it than the buyer. Is that important? For a long time it was thought that only marginal adjustments to the Arrow-Debreu model would be required to take care of asymmetry of information, but then a seminal 1970 paper by Akerlof showed that, in fact, this asymmetry is so important that the Arrow-Debreu model breaks down entirely.

The Akerlof example is so important that I will sketch it here. Consider a population of $2N$ people. Half of them have a used car and want to sell it; the other half do not have a car and want to buy one. So there are $N$ cars for sale. The quality of each car is modeled by a number $x$. A car of quality $x$ is worth $x$ to the seller, and $\frac{3}{2}x$ to the buyer.

The Arrow-Debreu situation is the case when buyer and seller both know the value of $x$. Cars of different quality will then be considered as different goods and traded at different prices: quality $x$ will trade at any price $p \in [x, \frac{3}{2}x]$. The market then functions and all cars are sold. However, this is not realistic: as we all know, the seller, who has used the car for some time, knows more about the quality than the buyer, who will only get to drive it around the block. Suppose, then, that the seller knows $x$ but all the buyer knows is its probability law, say that $x$ is uniformly distributed between 0 and 1. Since buyers cannot distinguish between them, all cars must have the same price $p$. If the buyer buys at that price, the expected quality he gets is $x = \frac{1}{2}p$, which to him is worth $\frac{3}{4}p < p$, which is less than he paid. So no one will buy, at any price, and the market simply does not function: there are people willing to buy, people willing to sell, both would be better off trading, and yet trade does not occur.

The lesson from the Akerlof example is that competitive markets cannot function when there is asymmetry of information. Since that paper, economists have sought alternative procedures to allocate resources, but at the time of writing, after many years of effort, dealing with groups of three or more agents at once under asymmetry of information is still beyond reach. The standard model therefore deals with only two: the principal and the agent. The agent has information that the principal does not have, but the principal knows that the agent has that information. Trading proceeds as follows: the principal makes an offer (the contract) to the agent, and the agent takes it or leaves it (no bargaining).

Contract theory is divided into two main branches: adverse selection and moral hazard. *Adverse selection* occurs when the agent has a characteristic that he or she knows but can do nothing about and that the principal would dearly like to know but does not. This is typically the case in insurance, but it occurs in other situations as well. If I am looking for health insurance, insurers would like to know as much as possible about my medical condition and my genetic map, which may be precisely what I do not want them to know. Airlines provide basically the same service to all passengers (leaving from point A at time $t$ and arriving at point B at time $T$), and they would like to know how much each passenger is willing to pay. Since they cannot access that information, they separate economy class from business class in the hope that customers will sort themselves out. *Moral hazard* occurs when the principal contracts the agent to perform an action that he or she, the principal, cannot monitor. The contract then has to be devised in such a way that the agent finds it in his or her own interest to comply, even if noncompliance cannot be observed and therefore cannot be punished. This is typically the case in the finance industry. If I entrust my wealth to a money manager, how am I to know that he or she is competent, or even that he or she works hard? The track record of the financial industry is not encouraging in that respect. The solution, if you can afford it (this is what happens in hedge funds, for instance), is to give the money manager such a large part of your earnings that your interests become aligned with theirs. It would be even better to have him or her share the losses, but this is prohibited by legislation (limited liability rules).

Informational asymmetry has spurred many mathematical developments. Adverse selection translates into problems in the calculus of variations with convexity constraints. A typical example is the following, due to Rochet and Choné. Given a square $\Omega = [a,b]^2$, with $0 < a < b$, and a number $c > 0$, solve the problem

$$\min \int (\tfrac{1}{2}c\nabla u^2 - x\nabla u + u)\,dx,$$
$$u \in H^1,\ u \geqslant 0,\ u \text{ convex}.$$

If it were not for the convexity constraint, this would be a straightforward obstacle problem. As it is, we have existence, uniqueness, and $C^1$-regularity, but we do not yet have usable optimality conditions.

For a long time, there was no comparable progress on moral hazard problems, but in 2008 there was a breakthrough by Sannikov, who showed how to solve them in a dynamic setting. His basic model is quite simple. Consider a stream of revenue

$$dX_t = a_t\,dt + \sigma\,dB_t,$$

where $\sigma$ is a constant and $B_t$ is Brownian motion. The revenue accrues to the principal, but $a_t$ depends on the agent. More precisely, at each time $t$ the agent decides to perform an effort $a_t \geqslant 0$, at a personal cost $h(a_t)$, with $h(0) = 0$ and $h$ increasing with $a$. The effort is unobservable by the principal; all he or she can observe is $dX_t$, so that the agent's effort is hidden in the randomness. The problem is then for the principal to devise a compensation scheme for the agent, depending only on the past history of $X_t$, that will maximize the principal's profit. Sannikov solves it by a clever use of the martingale representation theorem, thereby opening up a new avenue for research.

## 6 Conclusion

I hope this brief survey has shown the breath of mathematics used in modern economic theory: optimal control, both deterministic and stochastic; diffusion theory and Malliavin calculus; exterior differential calculus; optimal transportation; and the calculus of variations. Economic theory is still trying to assimilate all the consequences of informational asymmetry. This revolution is far from over; the theory will probably be very different in 2050 from what it is today. These are exciting times.

### Further Reading

Akerlof, G. 1970. The market for lemons: quality uncertainty and the market mechanism. *Quarterly Journal of Economics* 84:488–500.

Carlier, G., and T. Lachand-Robert. 2001. Regularity of solutions for some variational problems subject to a convexity constraint. *Communications on Pure and Applied Mathematics* 54:583–94.

Chernozhukov, V., P. A. Chiappori, and M. Henry. 2010. Symposium on transportation methods. *Economic Theory* 42: 271–459.

Chiappori, P. A., and I. Ekeland. 2009. *The Economics and Mathematics of Aggregation: Formal Models of Efficient Group Behavior.* Foundations and Trends in Microeconomics, volume 5. Hanover, MA: Now.

Rochet, J. C., and P. Choné. 1998. Ironing, sweeping and multidimensional screening. *Econometrica* 66:783–826.

Sannikov, Y. 2008. A continuous-time version of the principal-agent problem. *Review of Economic Studies* 75:957–84.

# VII.21 Mathematical Neuroscience

*Bard Ermentrout*

## 1 Introduction

Mathematical neuroscience is a branch of applied mathematics and, more specifically, an area of nonlinear dynamics concerned with the analysis of equations that arise from models of the nervous system. The basic units of the nervous system are neurons, which are cells that reside in the brain, spinal cord, and throughout many other regions of the bodies of animals. Indeed, the presence of neurons is what separates animals from other living organisms such as plants and fungi. Neurons and their interactions are what allow organisms to respond to a rapidly changing and unpredictable environment. The number of neurons in an organism ranges from just a few hundred in the nematode (a small worm) to around 100 billion in the human brain. Most neurons are complex cells that consist (roughly speaking) of a cell body, an axon, and dendrites. A typical neuron will generate transient changes in its membrane potential on the order of 100 millivolts (called action potentials) that communicate electrochemically with other neurons through *synapses*. Neurons are typically threshold devices, so that if their membrane potential exceeds a particular value, they will "fire" an action potential. The synapses from one neuron can be either excitatory or inhibitory: either they promote action potentials in the receiving neuron or they prevent them. There are about 100 trillion synapses in the human brain. The properties of individual neurons are reasonably well characterized as are the dynamics of individual synapses. However, the behavior of even a single neuron can be very complex and so, of course, is the behavior of even a small piece of brain.

The cell membrane of a neuron is what separates the neuron from the surrounding medium, and it is through this membrane that the complex firing patterns of neurons are mediated. Various ions such as sodium, potassium, calcium, and chloride are contained within the neuron and in the space between the neurons. The membrane is dotted with thousands of proteins called *channels* that selectively allow one or more ions to pass into and out of the cell. By controlling the instantaneous permeability of specific channels, the cell is able to create large transient fluctuations in its potential (the above-mentioned action potential). While perhaps not at the level of Newton's laws or the Navier–Stokes equations for fluids, there are standard models for the neuronal membrane, for the propagation of action potentials down axons, for the dynamics of synapses, and for how all this information is integrated into the cell body. Mathematical neuroscience analyzes the dynamics of these models using various tools from applied mathematics.

## 2 Single-Cell Dynamics

The modeling of a single neuron ranges from a simple scalar ordinary differential equation (ODE) to a system of many coupled partial differential equations (PDEs). For simulation purposes, the latter model is simplified to a large number of coupled ODEs representing the different parts of the neuron, which are called *compartments*. For example, the soma or cell body (see figure 1) might be represented by one compartment, while the dendrites and possibly the axon might be broken into dozens of compartments. A neuron represented by a single compartment is called a *point neuron* and that is where we will focus our attention. A point neuron is modeled by ODEs representing the permeabilities of one or more channels whose properties have been experimentally measured. The easiest way to think of a single compartment is through an *equivalent circuit* (see figure 1) defined by the transmembrane voltage, $V(t)$, the membrane capacitance, $C$, and the conductances of the various channels, $g_j(t)$. The first neuronal membrane to be mathematically modeled was that of the squid giant axon, for which Hodgkin and Huxley received a Nobel Prize. The differential equations represented by the figure have the form

$$\left.\begin{aligned} C\frac{\mathrm{d}V}{\mathrm{d}t} &= -g_{\mathrm{K}}n^4(V-E_{\mathrm{K}}) - g_{\mathrm{L}}(V-E_{\mathrm{L}}) \\ &\qquad - g_{\mathrm{Na}}m^3h(V-E_{\mathrm{Na}}) + I(t), \\ \frac{\mathrm{d}n}{\mathrm{d}t} &= \frac{n_\infty(V)-n}{\tau_n(V)}, \\ \frac{\mathrm{d}m}{\mathrm{d}t} &= \frac{m_\infty(V)-m}{\tau_m(V)}, \\ \frac{\mathrm{d}h}{\mathrm{d}t} &= \frac{h_\infty(V)-h}{\tau_h(V)}. \end{aligned}\right\} \quad (1)$$

Here, $I(t)$ represents either the electrical current that the experimenter can inject into the neuron or the current from synapses from other neurons (see below). The functions $n_\infty(V)$, etc., were measured through very clever experiments and then fitted to particular functions. The constants $g_{\mathrm{L}}$, $E_{\mathrm{L}}$, etc., were also experimentally determined. While Hodgkin and Huxley (HH) did this work nearly sixty years ago, the basic formulation has not changed, and the thousands of papers devoted to the modeling of channels and neurons follow this general format: for each compartment in the representation of the neuron, there is a voltage, $V(t)$, and auxiliary variables such as $n(t)$, $m(t)$, $h(t)$ whose time evolutions depends on $V$. One goal in mathematical neuroscience is to characterize the dynamics of these nonlinear ODEs as some of the parameters vary. In particular, a typical experiment is to inject a constant current, $I$, into the neuron and to then look at its dynamics. It is also possible to pharmacologically block or reduce the magnitude of the channels, so that the parameters $g_{\mathrm{K}}$, $g_{\mathrm{Na}}$, etc., may also be manipulated. The most common way to analyze the dynamics of equations like (1) is to compute their *fixed points* and then study their stability by *linearizing* about a steady state. For systems like the HH equations, the equilibria are easy to find by solving for the variables $m, h, n$ as functions of voltage, obtaining a single equation of the form $I = F(V)$, where $I$ is the constant current. Plotting $V$ against $I$ allows one

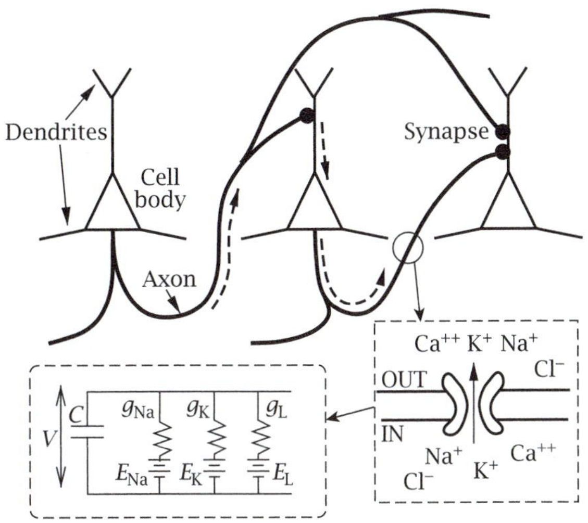


**Figure 1** Parts of neurons. Three neurons are shown, with the input (dendrites), cell body, and output (axon) labeled on one neuron. Dashed arrows show the direction of information flow. The two insets show a magnified schematic of the cell membrane and its equivalent electrical circuit.

to identify all the equilibria. Whether or not the equilibria are stable is determined by the eigenvalues of the $4 \times 4$ Jacobian matrix. This procedure allows one to find BIFURCATIONS [IV.21] to limit cycles through the *Andronov–Hopf bifurcation*. In general, numerical methods are used to find the solutions and their stability. Figure 2 shows the diagram for the HH equations as the applied current varies. There seems to be only a single equilibrium point, which loses stability via an Andronov–Hopf bifurcation and spawns a family of periodic orbits. The stable periodic solutions represent repetitive firing of the neuron. Note that there is a range of currents (shown by the dashed lines) for which a stable equilibrium and a stable periodic orbit coexist. This *bistability* was experimentally confirmed in the 1970s.

The other common approach to the analysis of equations like (1) is to exploit the differences in timescales. For example, the dynamics of $V$, $m$ are very fast compared with the dynamics of $n$, $h$, so it becomes possible to use *singular-perturbation* methods to analyze the dynamics. Because of this difference in timescales, even the simple HH model can show complex dynamics. For example, figure 2 shows that reducing the rate of change of $h$ by half results in a completely different type of dynamics (compare the top and bottom plots of the voltage, $V$). By treating this as a singularly perturbed system, it is possible to explain the complex spiking patterns seen in the HH model.

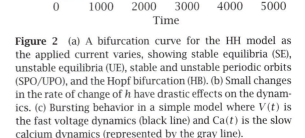

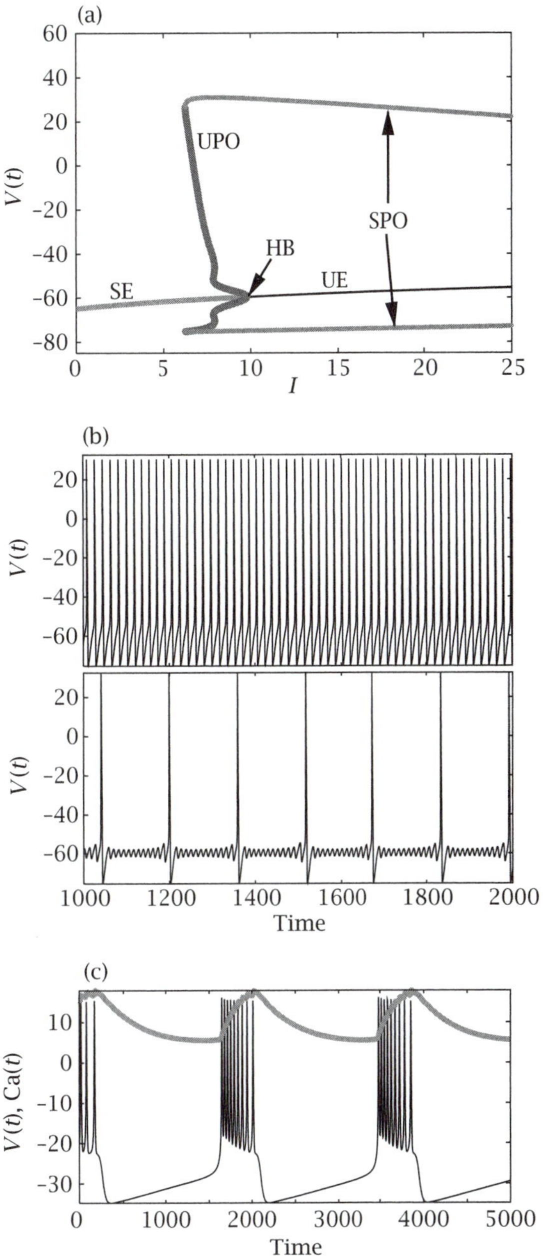


**Figure 2** (a) A bifurcation curve for the HH model as the applied current varies, showing stable equilibria (SE), unstable equilibria (UE), stable and unstable periodic orbits (SPO/UPO), and the Hopf bifurcation (HB). (b) Small changes in the rate of change of $h$ have drastic effects on the dynamics. (c) Bursting behavior in a simple model where $V(t)$ is the fast voltage dynamics (black line) and $\mathrm{Ca}(t)$ is the slow calcium dynamics (represented by the gray line).

Singular perturbation also plays an important role in the study of so-called bursting dynamics, where the voltage shows periods of high-frequency spiking followed by periods of quiescence; an example of this can be seen in part (c) of the figure. To understand the dynamics of, say, the bursting, we exploit the differences in timescales to write the system as

$$\frac{dx}{dt} = f(x, y), \qquad \frac{dy}{dt} = \varepsilon g(x, y),$$

where $x$ represents the fast variables (typically, voltage and possibly other variables) and $y$ represents the slow variable(s) (typically, a slow current). One then looks at the dynamics of $x$ treating $y$ as a constant parameter. This leads to parametrized dynamics, $x(t; y)$, which is then substituted into the $y$ equation before *averaging* is performed (if $x$ is time-periodic) to resolve the dynamics of $y$. If the $x$ dynamics is bistable, say, between a limit cycle and a fixed point, then, as $y$ slowly changes (at the timescale $\varepsilon t$), the $x$ dynamics can vary between equilibrium behavior and oscillations, thus producing the burst. Figure 2(c) shows the slow dynamics $y$ (the thicker gray line, representing the slow change in calcium concentration of the model) and the fast dynamics $x$ (the thinner black line, representing voltage) for a simple bursting model.

## 3 Networks

The complexity of the dynamics of even a simple cell dictates that some type of simplification is necessary in order to study large networks of neurons. There are two commonly taken paths: either one uses simpler models for individual neurons or one takes "mean-field" approaches, where neurons are approximated by an abstract variable called the *firing rate*, or activity. In the former case, a classic model for simplifying the dynamics of a single neuron is the *leaky integrate-and-fire model*, in which $\tau V' = -V + V_0$, along with the condition that, if $V(t^-) = V_1$, then $V(t^+) = V_2$, where $V_1 > V_2$. Each reset event results in a simulated action potential or spike of the neuron. (While the dynamics of this class of models is quite simple, their discontinuous nature presents many mathematical challenges.) With this model, a general network has the form

$$\tau_i \frac{dV_i}{dt} = -V_i + U_i(t) - \sum_j w_{ij}(t)(V_i - E_j), \qquad (2)$$

where $E_i$, $\tau_i$ are constants, $U_i(t)$ are inputs (often consisting of broadband random signals), and $w_{ij}(t)$ is some prescribed function of the time, $t$, since neuron $j$ fired that decays to zero for large positive $t$ and is zero for $t < 0$. For example, $w_{ij}(t) = W_{ik}(t - t_j)^+ \exp(-(t - t_j)^+)$, where $x^+$ is the positive part of $x$ and $W_{ij}$ is a constant. If $E_j < 0$ ($E_j > 0$), then the synapse is called *inhibitory* (*excitatory*). If $w_{ij}(t)$ depends only on $i - j$, then we can regard the network as being spatially distributed. In this case, various patterns such as traveling waves and spatially structured dynamics are possible. The analysis of these networks is difficult, but methods from statistical physics have been applied when the number of neurons becomes large and there is no structure in the connectivity. With certain assumptions about the connectivity patterns and strengths, the firing of individual neurons in (2) is chaotic and the correlations between the firing times of neurons are nearly zero. This state is called the *asynchronous state*, and in this case it makes sense to define a population firing rate, $r(t)$. As the number of neurons in a given population (say, excitatory cells) tends to infinity, the population rate, $r(t)\Delta t$, is defined as the fraction of the population that has fired an action potential between $t$ and $t + \Delta t$. One of the important questions in theoretical neuroscience is to relate $r(t)$ to the dynamics of the network of complex spiking neurons and, thus, derive the dynamics of $r(t)$. One can use various approaches to reduce scalar models such as (2) to a PDE for the distribution of the voltages that can have discontinuities and has the form

$$\frac{\partial P(V,t)}{\partial t} = -\frac{\partial}{\partial V}\left(f(V,t)P(V,t) - D\frac{\partial g(V,t)P(V,t)}{\partial V}\right).$$

The firing rate is proportional to the flux through the firing threshold, $V_1$. This PDE can be very difficult to solve; if there are several different populations, then the same number of coupled PDEs must be solved. Thus, there have been many attempts at arriving at simplified models for the rate equations. Here, we offer a heuristic description of a rate model for several interacting populations of neurons. Let $I_p(t)$ be the current entering a representative neuron from population $p$. If $I_p$ is large enough, the neuron will fire at a rate $u_p(t) = F_p(I_p(t))$ (see, for example, figure 2, which gives the firing rate of the HH neuron as the current changes). Each time a neuron from population $p$ fires, it will produce current in population $q$, so that the new current into $q$ from $p$ is

$$J_{qp}(t) = \int_0^\infty \eta_{qp}(t')u_p(t - t')\,dt'.$$

$\eta_{qp}$ is positive for excitatory currents and negative for inhibitory ones. *Dale's principle* states that the sign of $\eta_{qp}$ depends only on $p$. Summing up the currents from all populations gives us a closed system:

$$I_q(t) = \sum_q \int_0^\infty \eta_{qp}(t')F_p(I_p(t-t'))\,dt'. \qquad (3)$$

Often, an equation for the rates is desired, so that

$$u_q(t) = F_q\Big(\sum_p \int_0^\infty \eta_{qp}(t')u_p(t-t')\,dt'\Big).$$

In either case, the result is a set of coupled integro-differential equations. If the functions $\eta_{qp}(t)$ are sums of exponentials, then (3) can be inverted to create an ODE. For example, if $\eta_{qp}(t) = W_{qp}\exp(-t/\tau_q)/\tau_q$, then (3) becomes

$$\tau_q \frac{dI_q}{dt} = -I_q + \sum_p W_{qp}F_p(I_p). \qquad (4)$$

In the case where $W_{qp}$ is symmetric, it is easy to construct a Lyapunov function for this system and prove that all solutions converge to fixed points. Networks such that the sign of $W_{qp}$ depends only on $p$ are called *excitatory-inhibitory* networks, and they also obey Dale's principle. If $W_{qp} > 0$ ($W_{qp} < 0$), then population $p$ is called excitatory (inhibitory). Networks with both excitatory and inhibitory populations have been shown to exhibit complex dynamics such as limit cycle oscillations and chaos.

## 4 Spatially Extended Networks

Equations of the form (4) have an obvious extension to spatially distributed populations:

$$\tau_e \frac{\partial I_e(x,t)}{\partial t} = W_{ee}(x) * F_e(I_e(x,t)) - W_{ei}(x) * F_i(I_i(x,t)) - I_e(x,t), \qquad (5)$$

$$\tau_i \frac{\partial I_i(x,t)}{\partial t} = W_{ie}(x)(x) * F_e(I_e(x,t)) - W_{ii}(x) * F_i(I_i(x,t)) - I_i(x,t), \qquad (6)$$

where $W(x) * U(x)$ is a convolution over the spatial domain of the network (often the infinite line or plane; or on a circle or torus), and $I_e$, $I_i$ are the respective excitatory and inhibitory currents for the neuron at position $x$. The functions $W_{qp}(x)$ generally depend only on $|x|$ and decay rapidly for large arguments. These models have been used to explain spatiotemporal patterns of activity found in experiments. While the equations represent a great simplification, they have been quite successful in explaining various experimental findings.

### 4.1 Wavefronts

When inhibitory circuits in the brain are damaged or pharmacologically blocked, activity can pathologically spread across your cortex. Experiments have shown that this activity takes the form of a traveling wave. A natural starting point for the study of equations of the form (5), (6) is, therefore, to consider models with no inhibitory population. In this case, it is possible to prove the existence of traveling fronts under certain circumstances. In particular, let $w_{ee} = \int_R W_{ee}(x)\,dx$ and

(i) suppose that $g(u) = -u + w_{ee}F_e(u)$ has three positive roots $u_1 < u_2 < u_3$,
(ii) let $W_{ee}(r)$ be a decreasing nonnegative function for $r \geqslant 0$, and
(iii) assume that $g'(u_{1,3}) < 0$ and $g'(u_2) > 0$.

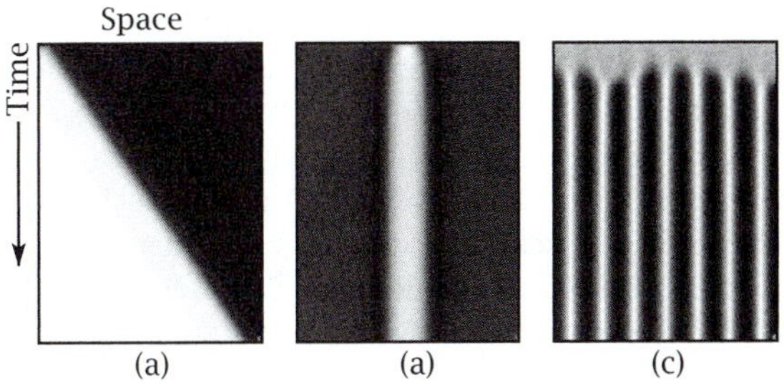


**Figure 3** Three examples of spatial activity in a population model: (a) a traveling wavefront, (b) a stationary pulse, and (c) spatially periodic activity.

There is then a unique constant-speed wavefront joining $u_1$ with $u_3$. This wave is analogous to the wave seen in the scalar bistable reaction–diffusion equation. Figure 3(a) shows an example of a traveling wave when $W_{ee}$ is an exponentially decaying function of space.

### 4.2 Stationary Pulses

Suppose that you are asked to remember the location of a flash of light for a short period of time, and after that time you must move your eyes to the place where you recalled the light. This form of memory is called short-term or working memory. By making recordings of the activity in certain brain regions of monkeys while they perform such a task, experimentalists have shown that the neural correlate of working memory consists of persistent neural activity in a spatially localized region. In the context of the model equations (5), (6), working memory is represented by a stationary spatially localized region of activity. Solutions to (5) can be constructed with the assumptions that $W_{ei}(x) = 0$, $F_e(I)$ is a step function, and $W_{ee}(x)$ has the *lateral inhibition* property. That is, let $q(x) = \int_0^x W_{ee}(y)\,dy$. The lateral inhibition property states that $q(x)$ is increasing up to $x = a > 0$ and then decreases, approaching a finite value $q_\infty$ as $x \to \infty$. While it is possible to relax the assumptions on the shape of $F$ to some extent, there is still not a general existence theorem for stationary pulses for this class of equations. Figure 3(b) shows a simulation of (5) when $W_{\text{ee}}(x)$ is the difference between two exponential functions. One can formally construct stationary pulse solutions to the full excitatory–inhibitory network, (5), (6), using singular-perturbation theory.

### 4.3 Regular Patterns

Numerous authors have suggested that the neural analogue of simple geometric visual hallucinations is spontaneous spatially periodic activity in the visual cortex. The connectivity of the visual system is such that under certain circumstances, the spatially uniform solution loses stability due to spatially periodic perturbations in a restricted range of spatial frequencies. This phenomenon is generally known as a *symmetry-breaking bifurcation* and, in biological contexts, a *Turing bifurcation*. Figure 3(c) shows an example of such a bifurcation in equations (5), (6) in one spatial dimension. In the late 1970s bifurcation methods were applied to prove the existence of stable spatial patterns in two spatial dimensions that are the analogue of simple geometric visual hallucinations.

## 5 Plasticity

One of the hallmarks of the nervous system is that it is almost infinitely reconfigurable. If part of a region is lost due to injury, other regions will rewire their connections to compensate. The connection strengths between neurons are believed to be the physiological correlates of learning and memory. This ability to change over time and alter connection strengths is called *synaptic plasticity*; it can take several forms, over timescales that range from fractions of a second to decades. For simplicity, we can divide plasticity into short-term and long-term plasticity. To make things concrete, we consider two neurons A and B, with A sending the signal (the *presynaptic neuron*) and B receiving the signal (the *postsynaptic neuron*). Short-term plasticity typically involves the weakening or strengthening of connections in a usage-dependent manner and depends only on the presynaptic activity. In contrast, long-term plasticity, presumably responsible for learning and lifetime memory, depends on the activity of both presynaptic and postsynaptic activity.

### 5.1 Short-Term Plasticity

In short-term plasticity, only the activity of the presynaptic neuron matters, and two types of phenomena occur: depression and facilitation. With depression (respectively, facilitation), the strength of the synapse weakens (respectively, strengthens) with each successive spike of neuron A and then recovers back to baseline if there are no subsequent spikes. Let $t^*$ denote the time of a presynaptic spike and define two variables $f(t)$ and $q(t)$ corresponding to facilitation and depression, respectively. Let $\tau_f$ and $\tau_d$ be the respective time constants, $f_0$ and $d_0$ the starting values, and $a_f$ and $a_d$ the degree of facilitation and depression. We then have

$$\frac{\mathrm{d}f}{\mathrm{d}t} = \frac{f_0 - f}{\tau_f} + a_f\delta(t - t^*)(1 - f),$$

$$\frac{\mathrm{d}q}{\mathrm{d}t} = \frac{d_0 - q}{\tau_d} - a_d\delta(t - t^*)q.$$

Typically, $f_0 = 0$, $d_0 = 1$, with $\tau_{d,f}$ of the order of hundreds of milliseconds. The strength of the synapse is then multiplied by $q(t)f(t)$. For example, if there is no facilitation, then $f_0 = 1$, $a_f = 0$, and if there is no depression, $d_0 = 1$, $a_d = 0$.

### 5.2 Long-Term Plasticity

Long-term plasticity depends on the activity of both the presynaptic and postsynaptic neurons and is based on an idea first proposed by Donald Hebb that "neurons that fire together, wire together." This type of *Hebbian plasticity* is believed to be responsible for both wiring up the nervous system as it matures and for the creation of new long-term memories (in contrast to the working memory described above). If $w_{\text{BA}}$ is the weight or strength of the connection of neuron A to neuron B and $r_{\text{A,B}}$ are the firing rates of the two neurons over some period, then, typically,

$$\frac{\mathrm{d}w_{\text{BA}}}{\mathrm{d}t} = \varepsilon_1(r_{\text{A}} - \theta_{\text{A}})(r_{\text{B}} - \theta_{\text{B}}) + \varepsilon_2(W_{\text{BA}} - w_{\text{BA}}), \quad (7)$$

with the additional proviso that, if the terms multiplied by $\varepsilon_1$ are both negative, the term is set to 0. This model says that the weight will increase if both A and B are firing at sufficiently high rates. If A is firing at a high rate and B is firing at a low rate (or vice versa), then the weight will decrease. Learning equations of the form (7) are used to adjust the connectivities in rate models such as equation (4) and are capable of encoding many "memories" in the strength of the weights. The term with $\varepsilon_2$ represents a tendency to decay to some baseline weight.

Another form of Hebbian plasticity incorporates causality in the sense that the weight from A to B will be increased (decreased) if A fires slightly before (after) B. A typical equation for the synaptic weight takes the form

$$\frac{\mathrm{d}w_{\mathrm{BA}}}{\mathrm{d}t} = F(t_{\mathrm{B}} - t_{\mathrm{A}})g(w_{\mathrm{BA}}) + \varepsilon_2(W_{\mathrm{BA}} - w_{\mathrm{BA}}),$$

where $t_{\mathrm{A,B}}$ are the times of the spikes of neurons A and B. $F(t)$ is often an exponentially decaying function of $|t|$ and is negative for $t > 0$ (B fires before A) and positive for $t < 0$ (A fires before B). The function $g(w)$ can be constant, linear in $w$, or of a more complicated form. This form of plasticity is called *spike-time-dependent plasticity* and allows a group of neurons to connect up sequentially and thus learn sequential tasks.

## 6 Summary

Mathematics plays a prominent role in the understanding of the behavior of the nervous system. Methods from nonlinear dynamics have been used to explain both normal and pathological behavior (such as epilepsy, Parkinson's disease, and schizophrenia). Conversely, models that arise in neuroscience provide a wide range of problems that are awaiting rigorous mathematical treatment.

### Further Reading

Amari, S. I. 1977. Dynamics of pattern formation in lateral-inhibition type neural fields. *Biological Cybernetics* 27(2): 77–87.

Ermentrout, B. 1998. Neural networks as spatio-temporal pattern-forming systems. *Reports on Progress in Physics* 61(4):353–430.

Ermentrout, G. B., and D. H. Terman. 2010. *Mathematical Foundations of Neuroscience*, volume 35. New York: Springer.

Rinzel, J. 1987. A formal classification of bursting mechanisms in excitable systems. In *Mathematical Topics in Population Biology, Morphogenesis and Neurosciences*, pp. 267–81. Berlin: Springer.

# VII.22 Systems Biology

*Qing Nie*

## 1 Introduction

Biological organisms are complex interconnected systems comprising an enormous number of components that interact in highly choreographed ways. These systems, which are constantly evolving, have been designed to carry out numerous tasks under diverse environmental conditions. Because of limitations in terms of the available experimental tools and technology, biologists have for many years focused primarily on how the individual components in these systems work. More recently, advancements in genomics, proteomics, metabolomics, and live tissue imaging have led to an explosion of data at increasingly fine spatial and temporal scales. In order to describe and understand our increasingly complex and interconnected view of the biological world, which is far more intricate than any physical system, an integrated systems approach is needed.

Systems biology, which initially gained momentum in the 1990s, focuses on studying interactions between components that give rise to emergent behavior. In this view, a system is more than just the sum of its parts. To better understand such systems, an interdisciplinary approach that brings together experimental tools and theoretical methodologies from diverse disciplines must be developed. In 2009 the US National Research Council published *A New Biology for the 21st Century*, which recommends a "new biology" approach and asks for "greater integration within biology, and closer collaboration with physical, computational, and earth scientists, mathematicians and engineers." In many ways, this new biology is systems biology.

In many cases, the goals and questions of interests in systems biology are different from those in classical biology. First, in addition to asking "how" (e.g., "How does a transcriptional factor lead to a feedback loop between two regulators?"), one often asks "why" (e.g., "Why is such a loop useful and what is its benefit?"). Second, a primary goal of systems biology is to uncover common principles governing many different molecular mechanisms inferred from building blocks in diverse organisms. Toward this goal, biological components are scrutinized at different spatial or temporal scales to explore novel and irreducible emergent properties, the biological equivalent of "first principles," arising from their interactions.

Mathematics through modeling plays a critical role in systems biology (figure 1). Useful models span the range from simple to highly complex. In some cases, simple models of two components may be sufficient to capture the key behavior of a network composed of thousands of components, whereas in other cases a twenty-component model might be required

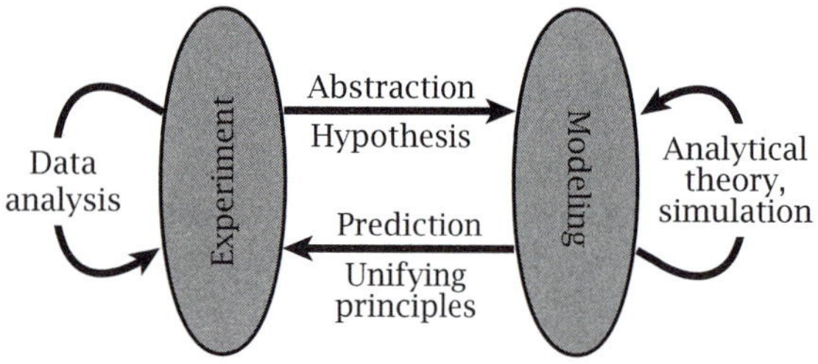


**Figure 1** Mathematics in systems biology. All arrows require various mathematical and computational tools that are critical in systems biology.

to unravel complex dynamics and account for critical features such as variability between gene products. Furthermore, these pieces can either be conceptual or they can be identified with specific biological components. These complexities introduce many challenges for mathematics. Moving from a descriptive to a predictive model requires incorporation of massive experimental data to constrain parameters. Furthermore, elucidating complex interactions to obtain laws and emerging properties often requires exploration of these models. The biological questions of interest dictate the choice of models, which in turn demand different mathematical and computational tools. Here, we illustrate this systems biology approach and the mathematics associated with it at three fundamental biological scales: gene, cell, and tissue. We also discuss its role in elucidating the function of an often-overlooked but important element that permeates all scales: noise.

## 2 The Information Flows from and to the Gene

The genetic information coded in deoxyribonucleic acid (DNA) needs to be processed into ribonucleic acid (RNA) and proteins in order to function in a cell. This process, in which DNA makes RNA and RNA makes protein, is referred to in biology as the central dogma. Through "transcription," genetic information contained in parts of DNA is transferred to messenger RNA (mRNA). mRNA then interacts with a ribosome, a complex molecular machine that builds proteins, to "translate" RNA into protein. Transcription and translation are two major steps for passing genetic codes to cellular functions. Transcription factors—a special group of proteins that facilitate the transcription process—are critical in linking one genetic code to another, in many cases creating loops of gene-to-gene regulation. Different genes make different proteins that may regulate various transcriptions to control production of other proteins.

Modeling transcription and translation requires a mechanistic description of gene action and regulation from the sequences in regulatory regions of DNA. Statistical physics provides a natural approach to examine how a gene is activated and repressed through transcription factors. Stochastic and probabilistic methods (e.g., STOCHASTIC DIFFERENTIAL EQUATIONS [IV.14] and MARKOV PROCESS MODELS [II.25]) are the foundations for this approach. In a typical transcription model, all possible states of an enhancer with a statistical weight are first catalogued. The action of a gene, estimated through a probability, depends on whether a regulatory region (e.g., binding site) is bound or unbound, and this probability is estimated as a fraction of the states bound to activators. Transcriptional regulations such as cooperativity, competitive interactions among transcription factors, inhibition of activators by repressors, and other epigenetic regulations (e.g., chromatin structure and modification, or DNA methylation) can be incorporated into this type of probabilistic model. This approach has been used to identify numerous regulatory motifs, to elucidate the function of regulatory regions of DNA, and to predict novel targets in the genome.

## 3 Signal Transduction into and out of the Cell

The cell, a "building block" of life, is the smallest information-processing unit that is able to respond to environmental changes and make decisions on its own. Proteins outside of a cell, which often serve as a stimulus, can bind receptors on the cell membrane or be trafficked into the cell to initiate cascades of BIOCHEMICAL REACTIONS [V.7]. Through these cascades, a specific stimulus activates one or a group of genes that contribute to a specific program or task (e.g., cell division for replication or changing cell fates). This process of turning an external signal into an internal response is often called signal transduction. One such example is the MAPK/ERK pathway that transmits growth factor stimuli through cell surface receptors to induce cell growth and division. Beyond understanding the process of signal transduction, the critical factors that lead to the pathway's malfunction are also of interest since, in many cases, this leads to cancer and other disorders.

Modeling signal transductions requires an accurate description of biochemical reactions for different types of biomolecules. When the number of molecules is relatively small (e.g., smaller than ten), one can use discrete, probabilistic approaches based on *chemical master equations*. When a relatively large number of molecules are present in a "well-stirred system," chemical concentrations provide a good approximation of the numbers of molecules, and a continuum model can describe the system. Common, fundamental concepts in this approach include rate equations, Michaelis-Menten kinetics, and the law of mass action. These models lead to systems of ordinary differential equations that are often stiff when drastically different reaction rates are present. Numerical simulations of such models require specialized algorithms of good absolute stability properties.

Regulations often exist in these signaling pathways. *Feedback* represents interactions in which downstream signals influence upstream signaling elements, whereas *feedforward* regulation refers to connections that pass upstream signals directly to the downstream signaling components. Mathematically, feedback and feedforward regulations can be described by Hill functions, which encode saturating tunable response curves. Similar to a thermostat that adjusts the influx of heat to stabilize the temperature around a prescribed one, negative feedback is often used to control homeostasis (i.e., stable steady state) in biology. In such a controller, two distinct timescales for a fast "detector" and a slow "reactor" are critical to maintaining stable homeostasis. When these two timescales become close, oscillatory dynamics arise, leading to periodic solutions. Positive feedbacks (e.g., excitation–contraction coupling in the heart, cellular differentiation, interaction between cytokines and immune cells) can amplify signals leading to bistable or ultrasensitive responses. In a bistable system one input can lead to two distinct outputs depending on the state of the system, commonly referred to as hysteresis. BIFURCATION ANALYSIS [IV.21] (e.g., through pseudo-arc-length continuation methods) is a major tool for studying bistability. In feedforward regulations, a *coherent* loop has two parallel paths of the same regulation type (both positive or both negative), whereas an *incoherent* loop has different types of regulations in the two pathways. Positive coherent loops can induce delay in response to an ON signal without putting any delay effect on the OFF signal. On the other hand, incoherent loops can speed up the response to ON signals as well as produce adaptation in which a response can disappear even though the signal is still present; this is a common feature in sensory systems.

Many pathways consist of hundreds of elements and connections that often share components. While this "cross-talk" between pathways allows the limited number of components in a cell to perform multiple tasks, insulating mechanisms must exist to enable specificity and fidelity of signals and avoid paradoxical situations in which an input specific to one pathway activates another pathway's output or responds to another pathway's input more strongly than its own. To understand how a cell directs information flows from diverse extracellular stimuli to multiple gene responses, new mathematics on modularity, coarse-graining, and sensitivity are needed to map the dynamics of systems consisting of many components, drastic temporal scales, and stochastic effects.

## 4 Communication and Organization

How cells effectively and correctly recognize and respond to their microenvironment is critical to development, repair, and homeostasis of tissue, and to the immunity of the animal body. Different tissues—such as nerves, muscle, and epithelium—are derived from different parts of embryos. Errors made during communication between cells cause abnormal development or lead to disease. Typically, cells communicate with other cells, either through releasing diffusive molecules into the extracellular space for signaling or by direct cell-to-cell contact. In the first mechanism, the molecules interact with the target cells through their receptors on the cell membrane, which in turn activate pathways leading to different cellular functions. The action range of diffusive molecules can be short (e.g., within a distance of a few cells) or long (e.g., a distance of tens of cells) depending on the relative relationship between diffusion and reactions. These spatial dynamics are naturally described by reaction–diffusion equations when tissues are modeled as continuum media. Analysis of such equations is possible through linear or weakly linear stability when the systems have fewer than three components, but numerical investigations of their dynamics are required when more components or higher spatial dimensions are involved. To create complex spatial patterns in tissue, such as stripes and spots, systems often need to utilize multiple types of signaling molecules. One primary strategy is to use two diffusive molecules of drastically different diffusion speeds, e.g., a slow activator and a fast inhibitor. In this short-range activation/long-range inhibition mechanism, small initial disturbances can spontaneously create complex patterns that describe processes such as pigmentation, branching morphogenesis, or the skeletal shape in the limb.

Cell-to-cell contact through interactions between membrane-bound proteins in different cells is another major mechanism for patterning. A common example of such signaling is lateral inhibition, which often results in clusters of cells of one fate surrounded by cells of different fates. This mechanism, which often involves the notch signaling pathway, can increase the contrast and sharpness of responses (e.g., in the visual system) and is important in the central nervous system, in angiogenesis, and in endocrine development. Active motility driven by, for example, a chemical gradient (chemotaxis), a light gradient (phototaxis), or a stiffness gradient (durotaxis) is another patterning mechanism that is often employed in wound healing, early development (e.g., sperm swimming toward an egg), and migration of neurons.

Hybrid models that combine ordinary differential equations for the temporal dynamics of signaling and discrete and probabilistic descriptions of cells are important in capturing these patterning mechanisms when the role of individual cells is important. Subcellular element methods and cellular Potts models are effective for simulating collective dynamics and emerging properties, both of which arise from these kinds of direct cell-to-cell communications. The multiscale features of these systems—which involve molecules in extracellular spaces, signal pathways with feedbacks inside cells, mechanics in the surrounding tissues, and their physiological consequences—present great challenges for modeling and computation, requiring innovation and transformative mathematical developments.

## 5 Noise: Detrimental or Beneficial

Noise and randomness exist at all scales of living organisms. The small numbers of binding sites on DNA, the fluctuations in biochemical reactions, the complex physical structures of intracellular and extracellular spaces, and the noise associated with environmental inputs all introduce uncertainty. Information that flows from one level to the next (e.g., from gene to cell) may be distorted if the noise propagates or is amplified.

What are the general principles that give rise to noise attenuation? For a switching system consisting of ON and OFF states (e.g., calcium signaling and p53 regulation) stimulated by a temporal pulse input, a critical intrinsic quantity, termed the signed activation time (SAT), succinctly captures the system's ability to maintain a robust ON state under noise disturbances. When the value of the SAT, defined as the difference between the deactivation and activation times, is small, the system is susceptible to noise in the ON state, whereas systems with a large SAT buffer noise in the ON state. This theory, which was developed based on fluctuation dissipation theory and multi-timescale analysis, can also be used to scrutinize the noise in the OFF state through the concept of iSAT (input-dependent SAT).

While noise is typically thought to impose a threat that cells must carefully eliminate, noise is also found to be beneficial in biology. For example, noise in gene expression can induce switching of cell fates, enabling the sharpening of boundaries between gene expression domains. Without such gene noise, there would be an undesirable salt and pepper patterning region caused by noisy morphogen signals. Other examples of the benefits of noise include that noise can induce bimodal responses in cases of positive transcriptional feedback loops without bistability, noise can synchronize oscillations in cell-to-cell signaling, rapid signal fluctuations can lead to stochastic focusing, and signaling noise can enhance the chemotactic drift of cells. Apparently, life seems to find ways to deal with noise: control and attenuate it whenever possible, but exploit it otherwise. Systematic exploration of noise in complex interconnected biological systems requires new tools in stochastic analysis and computation.

## 6 Conclusion

The rapid increase that has been seen in the amount of available biological data provides a tremendous opportunity for systems biology. Integration of new and old data on multiple scales, abstracting connections among key components, considering randomness and noise, and deriving principles from commonality will soon be at the center of scientific discovery. Systems biology, which has stimulated the development of many new mathematical and computational methods, is becoming an increasingly important approach in the era of big data. As systems biology holds the key to greater understanding of life, what systems biology will do for mathematics in the twenty-first century will be similar to what physics and mechanics did in the nineteenth century: it will give mathematics a new life.

## Further Reading

Alon, U. 2006. *An Introduction to Systems Biology: Design Principles of Biological Circuits*. Boca Raton, FL: CRC Press.

Christley, S., B. Lee, X. Dai, and Q. Nie. 2010. Integrative multicellular biological modeling: a case study of 3D epidermal development using GPU algorithms. *BMC Systems Biology* 4:107.

National Research Council. 2009. *A New Biology for the 21st Century*. Washington, DC: The National Academies Press.

Wang, L., J. Xin, and Q. Nie. 2010. A critical quantity for noise attenuation in feedback systems. *PLoS Computational Biology* 6(4):e1000764.

Zhang, L., K. Radtke, L. Zheng, A. Q. Cai, T. F. Schilling, and Q. Nie. 2012. Noise drives sharpening of gene expression boundaries in zebrafish hindbrain. *Molecular Systems Biology* 8:613.

# VII.23 Communication Networks

*Walter Willinger*

## 1 Introduction

A communication network is generally defined as a collection of objects (e.g., devices, people, cells) that use communication media such as optical fiber cables, sound waves, or signaling pathways to exchange information with one another using a common set of protocols (i.e., rules and conventions that specify the minute details of the actual information exchange). A particularly well-known, and arguably the largest, man-made communication network is the Internet, and it will feature prominently in this article. This is not to say that other examples, especially some of biology's communication networks, are not interesting or that they are less important than the Internet. But studying the Internet has unique and appealing advantages when contrasted with biology. For one, we have full access to the Internet designers' original intentions and to an essentially complete record of the entire evolutionary process. We also know in detail how the network's individual components should work and how the components interconnect to create system-level behavior. Moreover, the Internet offers unique opportunities for measurement and for collecting massive and often detailed data sets that can be used for an in-depth study of the network's properties and features.

As such, it is the design, operation, and evolution of this large-scale, highly engineered system that we focus on in this article. We use it as a concrete example for illustrating the types of mathematical concepts, approaches, and theories that are being developed in support of a rigorous first-principles treatment of large-scale communication networks.

## 2 Internet Hourglass Architecture

The early reasoning behind the design philosophy that shaped the Internet's architecture consists of two main arguments. First, the primary *objective* was internetworking, that is, the development of an effective technique for multiplexed utilization of already existing interconnected (but typically separately administered) networks. Second, the primary *requirement* was robustness; that is, the network needed to be resilient to uncertainty in its components (e.g., router failures) and usage patterns (e.g., traffic-demand variations), and (on even longer timescales) to unanticipated changes in networking technologies and offered services.

To achieve the stated key objective of internetworking, the fundamental structure of the original architecture was the result of a combination of known technologies, conscientious choices, and visionary thinking. It led to a packet-switched network in which a number of separate networks are connected together using packet communications processors called gateways. Communication is based on a single universal logical addressing scheme with a simple (net, host) hierarchy, and the gateways or routers implement a store-and-forward packet-forwarding algorithm.

Similarly, to satisfy the crucial requirement for robustness, the early developers of the Internet relied heavily on two proven guidelines for system design, namely, the *layering principle* and the *end-to-end argument*. To illustrate this, in the context of packet-switched networks the layering principle argues against implementing a complex task (e.g., a file transfer between two end hosts) as a single module. Instead, it favors breaking the task up into subtasks, each of which is relatively simple and can be implemented separately. The modules corresponding to the different subtasks can then be thought of as being arranged in a vertical stack, where each layer in the stack is responsible for performing a well-defined set of functionalities. The role of the end-to-end argument is then to help with the specifications of these functionalities and guide their placement among the different layers. It achieves these tasks by expressing a clear bias against low-layer function implementation and arguing for bottom layers that are kept as general and simple as possible.

This design process was largely responsible for defining the main suite of protocols used in today's

Internet, and it is known as the five-layer *transmission control protocol/Internet protocol* (TCP/IP) *protocol stack*. This "vertical" decomposition encompasses (from the bottom up) the physical, link, internetwork, transport, and application layers. The same design process was also instrumental in creating the closely related "hourglass" metaphor for the architecture of the Internet: an arrangement of the multilayer suite of protocols in which each protocol interfaces only with those in the layers immediately above and below it, and where IP, the protocol for the internetworking layer of TCP/IP, occupies a separate layer at the hourglass's waist, where it provides a generic packet-delivery service.

This abstract bit-level network service at the hourglass's waist consists of a minimal set of widely agreed-upon features. These features have to be implemented according to an Internet-wide standard, they must be supported by all the routers in the network, and they are key to enabling communication across the global Internet. In this context, global robustness and scalability of routing, a key objective for IP, is achieved and implemented in a fully decentralized and asynchronous manner in TCP/IP ("horizontal" decomposition). The layers below the waist deal with the wide variety of existing transmission and link technologies and provide the protocols for running IP over whatever bit-carrying network infrastructure is in place ("IP over everything"). Above the waist is where enhancements to IP (e.g., reliable packet delivery) are provided that simplify the process of writing application-level protocols (e.g., the Hypertext Transfer Protocol (HTTP) for the World Wide Web) through which users ultimately interact with the Internet ("everything over IP").

From today's perspective, "IP over everything and everything over IP" is nothing but an ingenious heuristic solution to an extremely challenging engineering problem that the early Internet designers faced. This problem manifested itself in the form of great uncertainty—uncertainty about the ways technological advances challenge or do away with conventional wisdom, about the manner in which the network will be used in the future, and about the ways a network comprised in part of unreliable components can fail. Indeed, during its transition from a small-scale research network some fifty years ago to a critical component of today's world economy and a social phenomenon, the Internet has experienced drastic changes with respect to practically all imaginable aspects of networking. At the same time, its ability to scale (i.e., adapt to explosive growth), to support innovation (i.e., foster technological progress below and above the hourglass waist), and to ensure flexibility (i.e., follow a design that can evolve over time in response to changing conditions) has been remarkable. In short, the past fifty years have been testimony to the ingenuity of the early designers of the Internet architecture. The fact that, by and large, their original design has been maintained and has guided the development of the network through a "sea of change" to what we call today's Internet is an astounding engineering achievement.

## 3 Mathematics Meets Engineering

Despite the Internet's enormous success, numerous things have happened since its inception that have questioned some of the fundamental architectural design choices that were made by its early designers. For one, the trust model assumed as part of the original design turned out to be the very opposite of what is required today (i.e., "trust no one" instead of "trust everyone"). In fact, this part of the design is the main culprit behind many of the security problems that plague today's Internet. Moreover, some of the problems that have been encountered have a tendency to create a patchwork of technical solutions that, while addressing particular short-term needs, may severely restrict the future use of the network and force it down an evolutionary dead end. To rule out such undesirable "solutions" and prevent the problems from occurring in the first place, a mathematical language for reasoning about the design aspects and evolutionary paths of Internet-like systems is needed. The following examples illustrate recent progress toward formulating such a language and providing the foundations for a rigorous theory of Internet-like systems.

### 3.1 Internet (Big) Data

The Internet is, by and large, a collection of interconnected computers, and computers excel in performing measurements. This inherent ability for measurement has turned the Internet into an early source for what is nowadays referred to as "big data." Readily accessible public data sets and carefully guarded proprietary data have transformed the scientific studies of many Internet-related problems into prime examples of measurement-driven research activities. In turn, they have provided unprecedented opportunities to test the soundness of proposed Internet theories or check the validity of highly publicized "emergent phenomena": measurement-driven discoveries that come as a complete surprise, baffle the experts (at first), cannot be explained nor predicted within the framework of well-established mathematical theories, and typically invite significant subsequent scientific scrutiny.

One such discovery was made in the early 1990s when the first data sets of measured Internet traffic—in the form of high-quality (i.e., high-time-resolution time stamps), complete (i.e., no missing data), and high-volume (i.e., over hours and days) packet traces—became available for analysis. The picture that emerged from mining this data challenged conventional wisdom because the observed traffic was the precise opposite of what was assumed. Rather then being smooth, with either no correlations or ones that decayed exponentially quickly, measured traffic rates on network links (i.e., the total number of packets per time unit) were "bursty" over a wide range of timescales. Importantly, the observed burstiness was fully consistent with *(asymptotic second-order) self-similarity*, or, equivalently, *long-range dependence*, that is, autocorrelations that decay algebraically or like a power law.

A second example of an emergent phenomenon was reported in the late 1990s when different data sets for studying the various types of connectivity structures that are enabled by the layered architecture of the Internet became available. While mathematicians focused primarily on more "virtual," or higher-layer, connectivity structures, such as the Web graph (i.e., the link structure between virtual entities, in the form of Web pages), networking researchers were mainly interested in more "physical," or lower-layer, structures, such as the Internet's router topology, that is, the Internet's actual physical link structure that connects its physical components (i.e., routers and switches). Using the available data to infer this physical router-level Internet led to the surprising discovery that the observed node degree distributions had a completely unexpected characteristic. Instead of exhibiting the expected exponentially fast decaying right tail behavior that essentially rules out the occurrence of high-degree nodes in the Internet, the right tail of the measured node degree distribution decayed *algebraically, like a power law*. As a result, while most nodes have small degrees, high-degree nodes are bound to exist and have orders of magnitude more neighbors than a "typical" node in the Internet.

### 3.2 A Tale of Two Discoveries

#### *Self-Similarity and Heavy Tails*

Given that there was no explicit mechanism in the original design of the Internet that predicted the observed self-similarity property of actual Internet traffic, its discovery fueled the development of new mathematical models that could explain this phenomenon. Of particular interest were generative models that could explain how the observed scaling behavior can arise naturally within the confines of the Internet's hourglass architecture. Some of the simplest such models were inspired by the pioneering works of Mandelbrot on renewal reward processes and Cox on birth–immigration processes. When applied to modeling Internet traffic, they describe the aggregate traffic rate on a network link (i.e., the total number of packets per time unit) as a superposition of many individual *on–off processes*. To a first approximation, each on–off process can be thought of as describing the activity of a single user as seen at the IP layer, and the user's activity is assumed to alternate between sending packets at regular intervals when active (or "on") and not sending any packets when inactive (or "off").

An important distinguishing feature of these models is that under suitable conditions on the individual on–off processes, and when properly rescaled, the aggregate traffic rate processes converge to a limiting process as the length of the time unit and the number of users tend to infinity. Provided that, for example, the length of a "typical" on period is described by a *heavy-tailed distribution with infinite variance*, this limiting process can be shown to be *fractional Gaussian noise*, the unique stationary Gaussian process that is long-range dependent or, equivalently, *exactly (second-order) self-similar*. The appeal of these constructive models is that they identify a basic characteristic of individual user behavior as the main reason for aggregate Internet traffic being self-similar. Intuitively, this basic characteristic implies that the activity of individual users is highly variable: while most on periods are very short and see only a few packets, there are a few very long on periods during which the bulk of an individual user's traffic is sent.

These generative traffic models have been instrumental in adding long-range dependence and heavy-tailed distributions to the mathematical modeler's toolkit. They have also sparked important subsequent research efforts that have significantly advanced our understanding of these two mathematical concepts, which were viewed as esoteric and of no real practical value just two decades ago. Moreover, they have raised the bar with respect to model validation by identifying new types of measurements at the different layers within the TCP/IP protocol stack that can be used to test for the presence of heavy-tailed behavior in the packet traces generated by individual users. Not only has the initial discovery of self-similarity in Internet traffic withstood the test of time, but some twenty years later, the ubiquitous nature of the heavy-tailed characteristic of individual traffic components is now viewed as an invariant of Internet traffic. Similarly, the existence of long-range dependence in measured traffic, which reveals itself in terms of self-similar scaling behavior, is considered a quintessential property of modern-day Internet traffic.

*Engineered versus Random*

In contrast to the discovery of self-similarity in Internet traffic, the reported power-law claim for the Internet's router topology did not withstand subsequent scrutiny and collapsed when tested with alternate measurements or examined by domain experts. This collapse ruled out the use of the popular SCALE-FREE [IV.18 §3.1] graphs as realistic models of the Internet's router topology. In contrast to the traditional ERDŐS–RÉNYI RANDOM GRAPHS [IV.18 §4.1], which cannot be used to obtain power-law node degrees, scale-free graphs can easily achieve this objective by following a simple stochastic rule whereby a new node connects with higher probability to an already highly connected node in the exiting graph (i.e., *preferential attachment*). For the last decade, the study of scale-free graphs and variants thereof has fueled the emergence and popularity of *network science*, a new scientific discipline dedicated to the study of large-scale man-made or naturally occurring networked systems.

The failure of scale-free graphs, in particular, and network science, in general, to cope with man-made physical structures such as the Internet's router topology motivated a new approach to modeling highly engineered systems. Radically different from the inherently random connectivity that results from constructs such as Erdős–Rènyi or scale-free graphs, this new approach is motivated by practical engineering considerations. It posits that the physical connectivity of the Internet is not the result of a series of (biased) coin tosses but is in fact designed; that is, it is based on decisions that are driven by objectives and reflect hard trade-offs between what is technologically feasible and what is economically sensible. Importantly, randomness does not enter in the form of (biased) coin tosses but in the form of the uncertainty that exists about the "environment" (i.e., the traffic demands that the network is expected to carry), and "good" designs are expected to be robust with respect to changes in this environment.

The mathematical modeling language that naturally reflects such a decision-making process under uncertainty is *constrained optimization*. As a result, this approach is typically not concerned with network designs that are "optimal" in a strictly mathematical sense and are also likely to be NP-HARD [I.4 §4.1]. Instead, it aims at solutions that are "heuristically optimal," that is, solutions that achieve "good" performance subject to the hard constraints that technology imposes on the network's physical entities (i.e., routers and links) and the economic considerations that influence network design (e.g., budget limits). Such models have been discussed in the context of *highly organized/optimized tolerances/trade-offs (HOT)*, and they show that the contrast with random structures such as scale-free graph-based networks could not be more dramatic. In particular, they highlight the fact that substituting randomness for any architecture- and protocol-specific design choices leaves nothing but an abstract graph structure that is incapable of providing any Internet-relevant insight and is, in addition, inconsistent with even the most basic types of available router topology-related measurements.

### 3.3 Networks as Optimizers

The ability to identify the root cause of self-similarity in Internet traffic and to explain the Internet's router-level structure from first principles are two examples of an ongoing research effort aimed at solving a "grand" challenge for communication networks. This challenge consists of developing a relevant mathematical language for systematically reasoning about network architecture and protocol design in large-scale communication networks, including the Internet.

At this point, the most promising candidate for providing such a common language is *layering as optimization decomposition*. This approach is based on two key ideas. First, it treats network protocol stacks holistically and views them as distributed solutions of some global optimization problems. The latter are typically formulated as some generalized network utility-maximization problems ("networks as optimizers") and are often the result of successful *reverse engineering*, that is, the challenging task of starting from an observed network design that is largely based on engineering heuristics and discovering and formulating the mathematical problems being solved by the implemented design. Second, by invoking the mathematical language of decomposition theory for constrained optimization, the approach enables the principled study of how optimal solutions of a given global optimization problem can be attained in a modularized (i.e., layered) and distributed way ("layering as decomposition"). In the process, it facilitates the creative task of *forward engineering*, that is, the systematic comparison of alternate solutions and the informed selection of new designs based on their provably superior performance, efficiency, or robustness.

Note that the HOT approach complements recent successful attempts at reverse engineering existing protocols within the TCP/IP stack (e.g., TCP and its variants, the Border Gateway Protocol for routing, and various contention-based medium-access control protocols). It also illustrates that the basic idea of "networks as optimizers" extends beyond protocols and is directly applicable to problems concerned with network design. On the other hand, since different optimal solutions typically correspond to different layering architectures, "layering as decomposition" allows for a principled treatment of "how and how not to layer." Importantly, it also provides a rigorous framework for comparing different protocol designs in terms of optimality, efficiency, or robustness with respect to perturbations of the original global optimization problem. The uncertainty due to heavy-tailed user activity—the root cause of the self-similar nature of Internet traffic—is one such perturbation with respect to which the designed protocols ought to be robust.

## 4 Outlook

As our understanding of Internet-related communication networks deepens, it becomes more and more apparent that in terms of architecture and protocol design, technological networks are strikingly similar to highly evolved systems that arise, for example, in genomics or molecular biology, despite having completely different material substrates, evolution, and assembly. A constructive scientific discourse about architectural designs and protocols arising in the context of highly engineered or highly evolved systems therefore looms as a promising future research objective. The examples and recent developments discussed in this article offer hope that such a discussion can and will be based on a rich and relevant mathematical theory that succeeds in making everything "as simple as possible but not simpler."

### Further Reading

Chiang, M., S. H. Low, A. R. Calderbank, and J. C. Doyle. 2007. Layering as optimization decomposition: a mathematical theory of network architectures. *Proceedings of the IEEE* 95(1):255–312.

Clark, D. D. 1988. The design philosophy of the DARPA Internet protocols. *Computer Communication Reviews* 18(4):106–14.

Doyle, J. C., and M. Csete. 2011. Architecture, constraints, and behavior. *Proceedings of the National Academy of Sciences of the USA* 108:15624–30.

Leland, W. E., M. S. Taqqu, W. Willinger, and D. V. Wilson. 1993. On the self-similar nature of ethernet traffic. *Computer Communication Review* 23(4):183–93.

Willinger, W., D. Alderson, and J. C. Doyle. 2009. Mathematics and the Internet: a source of enormous confusion and great potential. *Notices of the American Mathematical Society* 56(5):586–99.

# VII.24 Text Mining

*Dian I. Martin and Michael W. Berry*

## 1 Introduction

Text mining is the automated analysis of natural language text data to derive items of meaning contained in that data and find information of interest. Automation is required due to the volume of text involved. Further complicating matters is the fact that text collections are typically unstructured, or very minimally structured. Text mining activities involve searching the data collection for specific information, classifying items within the collection to derive useful characteristics, and analyzing the content of a collection to arrive at an overall understanding of the body of information as a whole. The ability to partition the data into understandable, meaningful units for a user is a challenge.

While some methods for text mining attempt to utilize linguistic properties and grammars, most of the methods employed in text mining are based on mathematics. These methods include statistical and probabilistic approaches, simple vector space models, latent semantic indexing, and nonnegative matrix factorization.

## 2 Statistical and Probabilistic Evaluation of Text

Statistical methods for evaluating text collections have been in use since the 1950s. These methods look at term frequency and probability of term usage to assign significance to particular words or phrases. These methods may be further informed by the use of dictionaries, grammars, stop lists, or other language-specific information. A *stop list*, for example, is a list of unimportant or unwanted words that are discarded during document parsing and are not used as referents for any document. The end result of this analysis is to produce a significance value for particular terms, sentences, or passages within a collection. These approaches can be susceptible to changing language use and are difficult to generalize across widely separated collections of information and different languages. However, these techniques are still in use today in the areas of search engine optimization and the generation of tag clouds.

## 3 The Vector Space Model

The *vector space model* (VSM) was developed to handle text retrieval from a large information database with heterogeneous text and varied vocabulary. The underlying formal mathematical model of the VSM defines unique vectors for each term and document by representing terms and documents in a large sparse matrix. The columns are considered the document vectors and the rows are considered the term vectors. The matrix is populated with the term frequency of each term in each document. A weighting can then be applied to each entry of the matrix in order to increase or decrease the importance of terms within documents and across the entire document collection. Similarity or distance measurements can then be calculated within the context of the vector space. Two such measures, cosine and Euclidean distance, are easily calculated between any two vectors in the space and are frequently used.

One of the first systems to use a traditional VSM was the system for the mechanical analysis and retrieval of text (SMART), developed by Gerard Salton at Cornell University in the 1960s. Among the notable characteristics of the VSM used by SMART is the premise that the meaning of a document can be derived from its components. The VSM is useful for lexical matching and exploiting term co-occurrence among documents. Searching the VSM for items similar to a given document, or query, from outside of the collection is possible by constructing a query vector within the VSM. Given the nature of the term-by-document matrix, a query vector is formed in the same way a document is constructed in the matrix: by giving those rows corresponding to the terms from the query a frequency number, followed by a weighting, and finding documents that are relevant to the query vector based on a similarity measure. Similarities between queries and documents are then based on concepts or similar semantic content. Exploiting the mathematical foundation of a VSM for a document collection by first creating the term-by-document matrix and then calculating similarities between queries and documents as just described is beneficial for searching through a large amount of information efficiently.

## 4 Latent Semantic Indexing

Expanding on the term-by-document matrix of the VSM, *latent semantic indexing* (LSI) uses a matrix factorization to simultaneously map the contents of a document collection on a set of orthogonal axes, scale those mapping vectors across the collection according to the singular values, and reduce the dimensionality of the representation to obtain the *latent* structure of a document collection. The terms and documents are organized into a single large matrix, just as in the vector space model, and diagonal row scaling is used to effect a particular term-weighting scheme. Prior to scaling, each matrix cell is simply the nonzero frequency of a term within a document. This matrix is then processed using the SINGULAR VALUE DECOMPOSITION [II.32] (SVD).

The SVD produces a dense multidimensional hyperspace representation of the information collection (an LSI space) containing vectors corresponding to the terms and documents of the collection. Within this *semantic space*, the meaning of a term is represented as the average effect that it has on the meaning of documents in which it occurs. Similarly, the meaning of a document is represented as the sum of the effects of all the terms it contains. The position of terms and documents in the vector space represents the semantic relationships between those terms and documents. Terms that are close to one another in the LSI space are considered to have similar meaning regardless of whether or not they appear in the same document. Likewise, documents are identified as similar to each other if they have close proximity in the LSI space regardless of the specific terms they contain.

### 4.1 Dimensionality

The SVD allows for the adjustment of the representation of terms and documents in the vector space by choosing the number of dimensions: that is, the number of singular values that are kept (the remainder being discarded or, equivalently, set to zero). This controls the number of parameters by which a word or document is described. The number of dimensions is usually much smaller than either the number of terms or documents, but it is still considered a large number of dimensions (typically 200–500). This reduction of dimensionality is the key to sufficiently capturing the underlying semantic structure of a document collection. It reduces the noise associated with the variability in word usage and causes minor differences in terminology to be ignored.

Selecting the optimal dimensionality is an important factor in the performance of LSI. The conceptual space for a large document collection needs more than a few underlying independent concepts to define it. Using a low number of dimensions is undesirable as it does not produce enough differentiation between terms and documents, whereas full dimensionality provides little semantic grouping of terms and documents and in effect treats every term and document as being unique in meaning.

Figure 1 depicts a simple three-dimensional representation of the LSI space composed of both term vectors and document vectors. This illustration is an extremely simplified representation using only three dimensions. In practice, the LSI hyperspace will typically have anywhere from 300 to 500 dimensions or more.

When this processing is completed, information items are left clustered together based on the latent semantic relationships between them. The result of this clustering is that terms that are similar in meaning are clustered close to each other in the space and dissimilar terms are distant from each other. In many ways, this is how a human brain organizes the information that an individual accumulates over a lifetime.

### 4.2 The Difference between the Vector Space Model and Latent Semantic Indexing

The difference between the traditional VSM and the reduced-dimensional VSM used by LSI is that terms form the dimensions or the axes of the vector space in the traditional VSM and documents are represented as

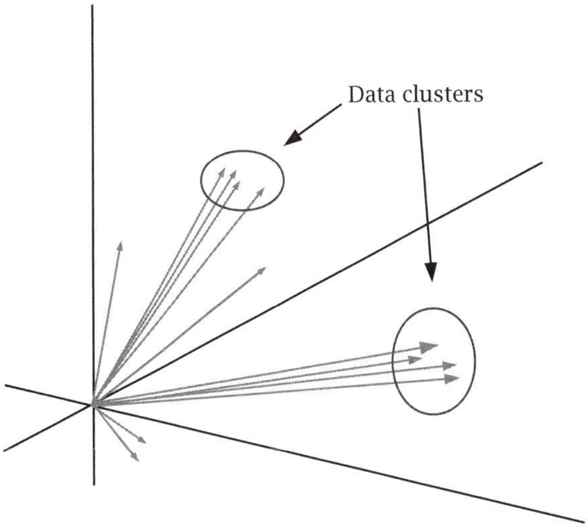


**Figure 1** Three-dimensional representation of the LSI space.

vectors in the term space, whereas the axes or dimensions in LSI are derived from the SVD. Therefore, since the terms are the axes of the vector space in the traditional VSM, they are orthogonal to each other, causing documents that do not contain a term in a user's query to have a similarity of zero with the query. The result is that terms have their own unique independent meanings. With LSI, the derived dimensions from the SVD are orthogonal, but terms (as well as documents) are vectors in the reduced-dimensional space, not the axes. Terms no longer have unique meaning on their own, nor are they independent. Terms get their meanings from their mappings in the semantic space.

### 4.3 Application

LSI can be used to search, compare, evaluate, and understand the information in a collection in an automated, efficient way. In fact, LSI provides a computational model that can be used to perform many of the cognitive tasks that humans do with information essentially as well as humans do them. The effectiveness and power of LSI lies in the mathematical calculation of the needed part of the SVD: that is, the partial SVD of reduced dimension. Given a large term-by-document matrix $A$, where terms are in the rows and documents are in the columns, the SVD computation becomes a problem of finding the $k$ largest eigenvalues and eigenvectors of the matrix $B = A^{\mathrm{T}}A$.

Finding the eigenvectors and eigenvalues of $B$ produces the document vectors and the singular values (the nonnegative square roots of the eigenvalues of $B$). The term vectors are then produced by back multiplying. Thus, the SVD computation is based on solving a large, sparse symmetric eigenproblem. The approach most effectively used to compute the SVD for this application is based on the Lanczos algorithm. The Lanczos algorithm, which is an iterative method, is proven to be accurate and efficient in solving large, sparse symmetric eigenproblems where only a modest number of the largest or smallest eigenvalues of a matrix are desired. While the computation of the reduced-dimensional vector space for a term-by-document matrix is a nontrivial calculation, advanced implementations of LSI have been shown to be scalable to address large problem sizes.

## 5 Nonnegative Matrix Factorization

LSI can be quite robust in identifying which documents are related. While it produces a set of orthogonal axes that forms a mapping analogous to the cognitive representation of meaning, it does not produce a set of conveniently labeled features that can be examined intuitively. *Nonnegative matrix factorization* (NMF) is another approach that produces decompositions that can be readily interpreted. Lee and Seung (1999) introduced the NMF in the context of text retrieval. They demonstrated the application of NMF in both text mining and image analysis. In our context, NMF decomposes and preserves the nonnegativity of the original term-by-document matrix whereby the resulting nonnegative matrix factors produce interpretable features of text that tend to represent usage patterns of words that are common across the given document corpus.

### 5.1 Constrained Optimization

To approximate the original (and possibly term-weighted) term-by-document matrix $A$, NMF derives two reduced-rank nonnegative matrix factors $W$ and $H$ such that $A \approx WH$. The sparse matrix $W$ is commonly referred to as the *feature matrix* containing feature (column) vectors representing usage patterns of prominent weighted terms, while $H$ is referred to as the *coefficient matrix* because its columns describe how each document spans each feature and to what degree (see figure 2).

In general, the NMF problem can be stated as follows: given a nonnegative real-valued $m \times n$ matrix $A$ and an integer $k$ such that $0 \leqslant k \leqslant \min(m,n)$, find

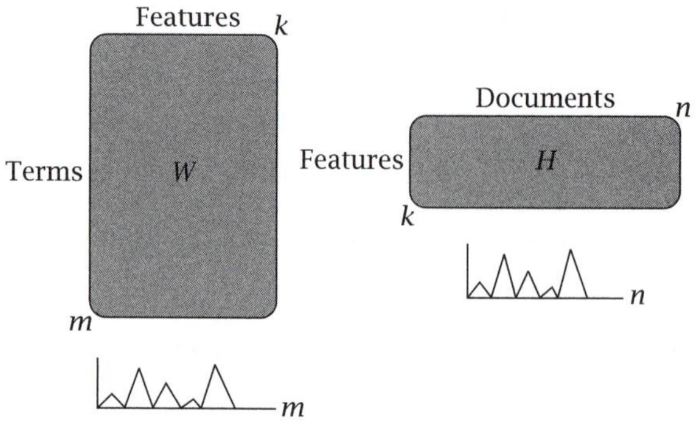


**Figure 2** Nonnegative feature and coefficient matrices from NMF; peaks or large components in individual feature (weight) vectors are dominant terms (documents). The subgraphs under each matrix factor ($W$ and $H$) reflect the values of the nonnegative components of each column.

two nonnegative matrices $W$ $(m \times k)$ and $H$ $(k \times n)$ that minimize the cost function

$$f(W,H) = \|A - WH\|_F^2,$$

where the norm is the FROBENIUS NORM [I.2 §20].

### 5.2 Convergence Issues

The minimization of $f(W,H)$ can be challenging due to the existence of local minima owing to the fact that $f(W,H)$ is nonconvex in both $W$ and $H$. Due to the underlying iterative process, NMF-based methods do not necessarily converge to a unique solution, so the resulting matrix factors $(W,H)$ depend on the initial conditions. One approach to remedy the nonunique solution problem is to avoid the use of randomization for the initial $W$ and $H$ factors. One common approach is to use the positive components of the truncated SVD factors of the original term-by-document matrix. The LSI factors described above can therefore be used to seed the iterative NMF process with a fixed starting point (initial $W$ and $H$) that will converge to the same minima (final $W$ and $H$) for repeated factorizations. Having multiple NMF solutions does not necessarily mean that any of the solutions must be erroneous. However, having a consistent ordering of interpretable features is definitely advantageous for knowledge-discovery applications.

## 6 Automation and Scalability

All of the methods discussed above leverage mathematical decompositions to analyze the meaning of natural language text where the structure is not explicit. These methods provide a means of processing text in an automated way that is capable of handling large volumes of information. With the continued expansion of the amount of text to be analyzed, techniques such as LSI and NMF can be used for scalable yet robust document clustering and classification.

### Further Reading


Berry, M. W., and M. Browne. 2005. *Understanding Search Engines: Mathematical Modeling and Text Retrieval*, 2nd edn. PA: SIAM.

Berry, M. W., M. Browne, A. Langville, V. P. Pauca, and R. Plemmons. 2006. Algorithms and applications for approximate nonnegative matrix factorization. *Computational Statistics & Data Analysis* 52:155–73.

Berry, M. W., and D. Martin. 2005. Principal component analysis for information retrieval. In *Handbook of Parallel Computing and Statistics*, edited by E. J. Kontoghiorghes. Boca Raton, FL: Chapman & Hall/CRC.

Lee, D., and H. Seung. 1999. Learning the parts of objects by non-negative matrix factorization. *Nature* 401:788–91.

Martin, D., and M. W. Berry. 2007. Mathematical foundations behind latent semantic analysis. In *Handbook of Latent Semantic Analysis*, edited by T. Landauer et al. Hillsdale, NJ: Lawrence Erlbaum.


## VII.25 Voting Systems
*Donald G. Saari*

### 1 Paradoxical Outcomes

As only addition is needed to tally ballots, I once mistakenly believed that "mathematical voting theory" was an oxymoron. To explore this thought, consider a simple setting in which 19 voters are selecting a social club president from among Ann, Barb, and Connie. Suppose the profile (i.e., a list of how many voters prefer each ranking) is as in table 1.

**Table 1** Social club president voting profile.

| Number | Ranking | Number | Ranking |
|---|---|---|---|
| 2 | $A \succ B \succ C$ | 4 | $C \succ B \succ A$ |
| 6 | $A \succ C \succ B$ | 4 | $B \succ C \succ A$ |
| 0 | $C \succ A \succ B$ | 3 | $B \succ A \succ C$ |

It just takes counting to prove that Ann wins; the plurality (vote-for-one) outcome is $A \succ B \succ C$ with tally 8:7:4. Counting even identifies the "vote-for-two" winner as ... *Connie.* This is a surprise because Connie is the plurality loser! Not only does this voting rule change the "winner," but its election ranking of $C \succ B \succ A$ (with tally 14:13:11) *reverses* the plurality ranking. To add to the confusion, the *Borda count* winner (ascertained from assigning 2, 1, and 0 points, respectively, to a ballot's top-, second-, and third-positioned candidate) is *Barb.* So Ann wins with one rule, Barb with another, and Connie with a third. Perhaps clarity comes from the majority-vote paired comparisons because Barb beats Ann (11:8) and Ann beats Connie (11:8), indicating the $B \succ A \succ C$ outcome. But no, Connie beats Barb (10:9), thus creating a cycle, so this rule's outcome suggests that no candidate is favored!

This example captures a concern that should bother everyone: *election outcomes can more accurately reflect the choice of a voting method rather than the intent of the voters.* A crucial role played by mathematics is to explain why this can happen and to identify which rules have outcomes that most accurately reflect the voters' views. More generally, because "voting" serves as a prototype for general aggregation methods, such as those used in statistics or in approaches that are central to the social sciences, expect the kinds of problems that arise with voting rules to identify difficulties that can arise in these other areas.

To tackle these issues, mathematical structures must be found. To do so, assign each alternative to a vertex of an equilateral triangle, as in figure 1(a). The ranking assigned to a point in the triangle is determined by its proximity to each vertex: for example, points on the vertical perpendicular bisector represent indifference, or an $A$–$B$ tie, denoted by $A \sim B$. The perpendicular bisectors divide figure 1(a) into six regions, with each small triangle representing all of the points with a particular ranking. For example, region 1 points are closest to $A$, and next closest to $B$, so they have the $A \succ B \succ C$ ranking; region 5 represents the $B \succ C \succ A$ ranking.

To exploit this geometry, place the introductory example's profile entries in the associated region; this is illustrated in figure 1(b). Two voters have the region 1 ranking, for instance, so place "2" in this triangle; three voters have the region 6 ranking, so place "3" in that triangle. An advantage gained by this geometry is that it separates the entries in a manner that simplifies the tallying process. With the $\{A, B\}$ vote, for example, all voters preferring $A$ to $B$ are in the shaded region to the left of the vertical line in figure 1(b). Thus, to compute this majority-vote tally, just add the numbers on each side of this line. Doing the same with each of figure 1(b)'s perpendicular bisectors leads to the tallies listed by each edge.

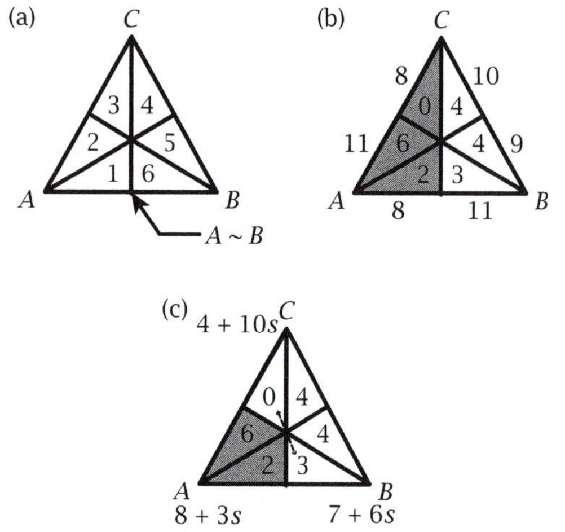


**Figure 1** Profile representations, and tallying ballots: (a) ranking regions; (b) paired comparisons; and (c) positional outcomes.

The plurality tally is equally simple. Regions sharing a vertex share the same top-ranked candidate (e.g., both shaded regions in figure 1(c) have $A$ as the top-ranked candidate), so add these values. This summation defines the values that are listed by the vertices after selecting the $s$ value (which is introduced next) to be $s = 0$, or the $A \succ B \succ C$ tally of 8:7:4.

"Positional rules" use specific weights $(w_1, w_2, w_3)$, $w_j \geqslant w_{j+1}$, where $w_3 = 0$ and $w_1 > 0$. To tally a ballot, assign $w_j$ points to the $j$th-positioned candidate. The plurality vote, then, is $(1, 0, 0)$, the vote-for-two rule gives $(1, 1, 0)$, and the Borda count gives $(2, 1, 0)$. To simplify the process, normalize the values by dividing by $w_1$ to obtain $\boldsymbol{w}_s = (1, s, 0)$, where $s = w_2/w_1$ is the "second place" value; for example, the normalized Borda count is $\boldsymbol{w}_{1/2} = (1, \frac{1}{2}, 0)$.

With this normalization, a candidate's $\boldsymbol{w}_s$ tally becomes "her plurality tally plus {$s$ times her second place votes}." For instance, as $A$ is second ranked in the two regions with arrows in figure 1(c), her $\boldsymbol{w}_s$ tally is $8 + 3s$; the $B$ and $C$ tallies are similarly computed and listed by their vertices. To illustrate, the "vote-for-two" $C \succ B \succ A$ tally of 14:13:11 in the introductory paragraph is recovered by using $s = 1$; the normalized Borda ($s = \frac{1}{2}$) $B \succ A \succ C$ tally is $10{:}9\frac{1}{2}{:}9$, so the standard $(2, 1, 0)$ Borda tally is double that: 20:19:18.

This tallying approach allows questions to be answered with simple algebra; for example, to find which $\boldsymbol{w}_s$ rule causes an $A \sim B$ tie, the equation $8 + 3s = 7 + 6s$ proves it is $s = \frac{1}{3}$, which is equivalent to a $(3, 1, 0)$ rule.

Similar algebraic computations prove that the profile in figure 1(b) admits *seven different positional outcomes*, four of which are strict rankings, that is, rankings without any ties.

In general and by using the geometric structures of higher-dimensional simplexes, we now know that, for $n$ candidates, profiles exist that allow precisely $k$ different strict election rankings, where $k$ is any integer satisfying $1 \leqslant k \leqslant (n-1)[(n-1)!]$; for example, with $n = 10$ candidates, a profile can be created that has 3 265 920 *different* positional election rankings. Among these rankings, it is possible for each candidate to be first, then second, then third, then ..., then last ranked just by using different positional voting rules!

As is typically true in applied mathematics, results are often discovered via experimentation. So, let me invite the reader to use the above methods to solve the following four problems. Answers (given later) motivate the mathematical structures of voting rules.

(i) Not all voters want $A$ to be the figure 1(c) plurality winner. Identify all voters who could vote "strategically" to force a personally preferred outcome.

(ii) Create a profile with the $A \succ B \succ C$ plurality outcome and a 10:9:8 tally in which the paired comparison (majority-vote) outcome is an $A \succ B$, $B \succ C$, $C \succ A$ cycle.

(iii) Create a different profile with the same plurality tallies but where $B$ now beats both $A$ and $C$ in majority votes; by beating everyone, Barb is the *Condorcet winner.*

(iv) Create a third profile with these plurality tallies but the vote-for-two outcome is whatever you want: $B \succ C \succ A$, say.

## 2 Two Central Results

Solving these questions makes it clear that the information used in paired comparisons differs from that used with plurality or any $\boldsymbol{w}_s$ method. This assertion is supported by the shaded regions in parts (b) and (c) of figure 1: $A$'s plurality tally uses information from two regions, while $A$'s majority vote uses three. Differences become more pronounced with $\boldsymbol{w}_s$ methods where $A$'s outcome involves information from four regions! But if rules use different information, then different outcomes must be expected. This raises an interesting mathematical challenge: to invent a voting rule that is free from paradoxical behaviors.

In examining this issue, Kenneth Arrow put forth the following ground rules.

(i) Voters have complete, transitive preferences, i.e., they rank each pair of candidates, and if a voter prefers $X \succ Y$, and $Y \succ Z$, then the voter prefers $X \succ Z$.

(ii) To ensure reasonable outcomes, the societal ranking is also required to be complete and transitive. (The "societal ranking" is the outcome of the group decision rule; if the rule is an election, then this is the "election ranking.")

(iii) The rule satisfies a unanimity condition (called the Pareto condition) where for any pair $\{X, Y\}$, if *everyone* prefers $X \succ Y$, then $X \succ Y$ is the societal outcome.

(iv) Why just unanimity? When determining the {Ann, Barb} societal ranking, what voters think about Connie should not matter. Arrow's *independence of irrelevant alternatives* (IIA) condition requires each pair's societal ranking to depend only on how each voter ranks that particular pair; other information is irrelevant.

While these conditions appear to be innocuous, the surprising fact is that only one rule satisfies them: a *dictator* (i.e., the rule is a function of one variable)! Namely, the rule can be treated as selecting a particular voter, say Mikko. Then, for all elections the rule's outcome merely reports Mikko's preferences as the societal outcome. From a mathematical perspective, *Arrow's theorem* proves that the information used by paired comparisons (figure 1(b)) is incapable of determining transitive rankings for three or more candidates. What kind of information is appropriate? (For a different interpretation of Arrow's result, see the last listed reference in the further reading section below.)

Another central result addresses whether strategic voting can be avoided. A "strategic vote" is one where a voter votes in a crafty manner to obtain a personally preferred outcome. This option is not available with two candidates, say Ann and Barb, where Ann will win a majority vote. Someone supporting Barb has precisely two choices, but neither is strategic. This is because voting for Barb is sincere, while voting for Ann is counterproductive, i.e., a "two-candidate" setting does not provide enough wiggle room to be strategic.

Three candidates admit more possibilities. With figure 1(c), a nonsincere vote is counterproductive for supporters of $A$ or $B$. What remains are those $C$ voters who prefer $B$ to $A$; by strategically voting for $B$, rather than $C$, they achieve a personally preferred outcome ($B$ over $A$); for example, strategic voting moves the four votes by the $C$ vertex toward $B$. A similar analysis holds for any positional rule; for example, the figure 1(c) vote-for-two winner is $C$, while $B$ is the Borda winner. In each case, there are ways for certain voters to be strategic.

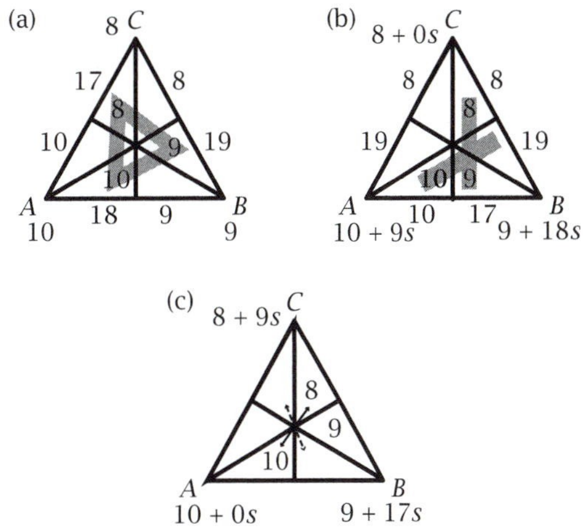


**Figure 2** Creating "paradoxes": (a) cycle; (b) Condorcet $B$; and (c) vote-for-two $B \succ C \succ A$.

For essentially these geometric reasons, the *Gibbard–Satterthwaite theorem* asserts that, for any group decision or election rule involving three or more candidates, there exist situations where some voter can be strategic. Technical conditions are added to ensure that, say, it is possible for each candidate to win with some profile (e.g., the rule is not a de facto comparison of one pair). While the proofs are combinatoric, the mathematical reasons are essentially as above; this theorem is a directional derivative result where three or more alternatives are needed to provide enough "directions."

## 3 The Mathematical Structure of Positional Rules

Answers to questions (ii)–(iv) (figure 2) identify the basic mathematical structure of voting rules. The plurality tally constraint requires placing numbers in the triangle that will have the specified sum by each vertex, e.g., the two numbers by the $A$ vertex must sum to 10. Combinatorics proves that there are $11 \times 10 \times 9 = 990$ ways to do so. To identify choices preferring $A$ over $B$ in a majority vote, insert a shaded bar in the $A \succ B$ region to the left of the vertical line (figure 2(a)). Similarly, to highlight $B$ over $C$, place a shaded bar on $B$'s side of the perpendicular bisection, with a similar bar for $C$ over $A$. Next, select values to stress the shaded triangle; emphasizing its vertices creates the profile in figure 2(a), which will have a heavy cyclic vote. To answer (iii), place shaded bars (figure 2(b)) to identify where $B$ dominates each of the other candidates.

Indeed, it is possible to select *any* ranking for each pair, and profiles can be generated where the plurality tallies and paired outcomes are as specified. As the rules use different information, the mathematical assertion is that consistent election outcomes for these rules cannot be expected, nor are they even very likely. This negative conclusion holds for any number of candidates and most positional rules.

To solve (iv) with the figure 1(c) $\boldsymbol{w}_s$-tallying method, emphasize "second-place votes." To keep plurality winner $A$ from receiving a larger tally, avoid placing numbers in her two figure 2(c) "second-place" regions, which are indicated by the dashed double-headed arrow. To assist $B$, place values in her second-place regions indicated by the solid double-headed arrow. An extreme (figure 2(c)) profile has the plurality winner $A$ but a vote-for-two ranking $B \succ C \succ A$ with tally 26:17:10.

With these tools, the reader can create examples to illustrate almost all of the possible three-candidate paradoxical behaviors. Of more value is to use these tools to extract the mathematical structures that will explain these millennia-old puzzles. To do so, notice how paired comparison differences (figure 2(b)) involve the vertices of the shaded equilateral triangles; they create combinations of profiles of the form

$$A \succ B \succ C, \qquad B \succ C \succ A, \qquad C \succ A \succ B, \tag{1}$$

with regions 120° apart. Positional differences are created by

$$A \succ B \succ C, \qquad C \succ B \succ A, \tag{2}$$

expressions with regions diametrically apart.

In particular, the symmetry structures exposed by my triangle approach identify the source of voting complexities. To capture these symmetries, express them as orbits of symmetry groups. The simplest group is $Z_2 = \{I, R\}$ consisting of the identity map $I$ and a reversal $R$, where $R^2 = R \circ R = I$; applying this group to any ranking $X \succ Y \succ Z$ yields the structure in (2); $I$ keeps the ranking and mapping $R$ reverses it.

It is easy to show that differences in paired comparisons are immune to these $Z_2$ structures, which cause differences in positional outcomes, i.e., this orbit is in the kernel of differences between paired comparison tallies. This is the mathematical property that permits outcomes for positional methods to differ as radically as desired from paired comparison rankings. The unique exception is the Borda count, which also has this $Z_2$ orbit structure in its kernel. Thus only Borda rankings must be related to paired comparison rankings. More precisely, for any other positional method, profiles exist for which the positional outcome can be whatever is wished, say $A \succ B \succ C$, but for which the paired comparison is $C \succ B \succ A$; this kind of behavior can never happen with Borda. Assertions of this form extend to any number of candidates.

The next-simplest permutation group is defined by $P = (1, 3, 2)$, where the first-placed entry of $X \succ Y \succ Z$ is moved to third place, the third-placed entry is moved to second, and the second-placed entry is moved to first place, creating $Y \succ Z \succ X$; orbits of this $\{I, P, P^2\}$ group, where $P^3 = I$ (so, permutation $P$ is applied three times to a ranking), generate the behavior in (1), which causes all possible paired comparison differences. As this $Z_3$ structure has each candidate in each position precisely once, the $Z_3$ symmetry affects paired comparisons, but it never affects differences in tallies for any positional rule. Thus, this symmetry structure in a profile creates differences between paired and positional outcomes; again, Borda is immune to its effects.

Surprisingly, for three candidates, these structures completely explain all paradoxical outcomes, and they answer all classical questions concerning differences in positional and paired comparisons (details can be found in the last two listed references in the further reading section). Indeed, to create the introductory example, I started with one voter preferring $B \succ C \succ A$, added appropriate $Z_2$ reversal structures to create desired positional outcomes (they never affect Borda or paired rankings), and then added a sufficiently strong $Z_3$ paired comparison component (they never affect positional or Borda rankings) to create a cyclic paired comparison outcome.

Note that only the Borda count is immune to $Z_2$ and $Z_3$ symmetry structures that create paradoxical conclusions. This is what makes it easy to construct arguments showing why the Borda count is the unique rule that most accurately represents voter interests.

Elections involving more candidates require finding appropriate symmetry group structures that affect one kind of subset of alternatives but not others. As an example, orbits of the Klein four-group (capturing symmetries of squares) do not affect rankings for paired comparisons or four-candidate positional rankings, but they change positional rankings for triplets. Again, only the Borda count (assign $n - j$ points to a $j$th-positioned candidate), which is a linear function, places all of these orbit structures in its kernel. As an immediate consequences and for any number of candidates, only Borda is not affected by the symmetries that cause paradoxical outcomes. That is, the Borda count is the unique rule that minimizes the kinds and likelihoods of paradoxical outcomes.

Another class of voting theory explores what can happen by changing a profile. An obvious example is the above mentioned Gibbard–Satterthwaite strategic voting result, where a sincere profile is converted into a strategic one. Because this theorem proves that all rules are susceptible to strategic efforts, the next natural question is to find a positional rule that is *least affected* by this behavior, that is, find the rule that is most unlikely to allow successful strategic actions. (The answer is the Borda count.) Other issues include understanding how a winning candidate can lose an election while receiving *more support*, or how, by not voting, a voter can be rewarded with a personally preferred outcome.

These results involve the geometric structure of sets of profiles in profile space. The Borda assertion about strategic voting, for instance, reflects its symmetry structure, where the differences between successive $(2, 1, 0)$ weights agree. To indicate how this symmetry plays a role, recall how geometric symmetries reduce boundary sizes, e.g., of all rectangles with area one, the rectangle with the smallest boundary (perimeter) is a square. Similarly, for the set of profiles defining a given election ranking, the positional rule with the smallest boundary is the Borda count; this boundary consists of tie votes. But for a strategic voter to successfully change an election outcome, the profile's election outcome must be nearly a tie, i.e., the profile must be near a boundary. So, by having the smallest boundary, the Borda count admits the smallest number of strategic opportunities. In contrast, the plurality vote, with its larger boundary, offers the greatest number of strategic opportunities, which means, as one would expect, that it is highly susceptible to successful strategic actions. Notice the troubling conclusion: the voting rule that is most commonly used to make group decisions—the plurality vote—is the most likely to have questionable outcomes, and it is the most susceptible to strategic actions.

Other results—such as the situation in which receiving more support hurts a candidate or the situation in which not voting helps a voter—involve the geometry of regions of profile space combined with directional derivative sorts of arguments. As described above, for instance, the two rules used in a "runoff election" (a positional rule for the first election, a majority vote for the runoff) involve different profile structures; this difference forces the set of all profiles in which a particular candidate is the "winner" to have dents in its structure (it is nonconvex). The lack of convexity admits settings in which a winning candidate loses by receiving more votes; namely, the straight line created by adding supporting voters moves the new profile outside of the "winning region." (This could happen, for example, if the added support for the previously winning candidate changed her runoff opponent.)

### Further Reading


Arrow, K. J. 1963. *Social Choice and Individual Values*. New York: John Wiley. (First edition: 1951.)

Arrow, K. J., A. Sen, and K. Suzumura. 2010. *Handbook of Social Choice and Welfare*, volume 2. San Diego, CA: Elsevier.

Nurmi, H. 2002. *Voting Procedures under Uncertainty*. New York: Springer.

Saari, D. G. 1999. Explaining all three-alternative voting outcomes. *Journal of Economic Theory* 87:313–55.

——. 2001. *Chaotic Elections! A Mathematician Looks at Voting*. Providence, RI: American Mathematical Society.

——. 2008. *Disposing Dictators, Demystifying Voting Paradoxes*. Cambridge: Cambridge University Press.

# Part VIII
# Final Perspectives

## VIII.1 Mathematical Writing
*Timothy Gowers*

### 1 Introduction

The purpose of this article is not to offer advice about how to write mathematics well. Such advice can be found in many places. However, I do have three pieces of very general advice, which inform the rest of the article. The first is to be clear about your intended readership; for example, if you want what you write to be understood by an undergraduate, then do not assume knowledge of any terminology that is not standardly taught in undergraduate mathematics courses. The second is to aim for this readership to be as wide as possible. If, with a small amount of extra explanation, you can make what you write comprehensible to an expert in another field of mathematics, then put in that extra effort. Whatever you do, do not worry that experts will not need the explanation; if that is true (which it often is not), then they can easily skip it. The third, which is related to the second, is to set the scene before you start. Some people who read what you have written will do so because they want to understand all the technical details and use them in their own work, but the vast majority will not. Most readers, including people who need to make quick judgments that will profoundly affect your career, will want to read the introduction quickly to see what you have done and assess how important it is. However much you might wish everyone to read what you have written in complete detail, you should be realistic and cater for readers who just want to skim it.

For the rest of this article I shall discuss various choices that one must make when writing a mathematical document. I will not advocate choosing one way rather than another, since the choices you should make depend on what you want to achieve; my final piece of general advice is merely that you should make the choices consciously rather than by accident.

### 2 Formality versus Informality

There are (at least) two goals that one might have when writing a mathematical document. One is to establish a mathematical result by whatever means are appropriate to the field; in pure mathematics the usual requirement is unambiguous definitions and rigorous proofs, whereas in applied mathematics other forms of evidence, such as heuristic arguments and experimental backing, may be acceptable. The other is to convey mathematical ideas to the reader.

These two goals are often in tension. If a pure mathematician discovers a complicated proof of a theorem, then that proof will be hard to understand. However, sometimes the apparent complication of a proof is misleading; what is really going on is that the author had one or two key ideas, and the complication of the argument is the natural working out of the details of those ideas. For the expert reader, an informal explanation of the ideas that drive the proof may well be more valuable than the proof itself.

Nobody would advocate writing papers with *just* informal explanations of ideas, since plausible-looking ideas often turn out not to work. However, there is still a choice to make, since it *is* considered acceptable to display proofs and not explain the underlying ideas. There may sometimes be circumstances where this is appropriate; for example, perhaps the proof is short, and explaining the ideas that generate it will double the length of what you are writing and put off readers. But usually, the advice I gave earlier—to broaden your readership if it not too difficult to do so—would dictate that technical arguments should be accompanied by informal explanations.

### 3 Giving Full Detail versus Leaving Details to the Reader

When you are writing you need to decide how much detail to give. If you give too little, then the reader you are aiming at will not be able to understand what you have written. But you also want to avoid making too many points that the reader will find completely obvious.

Of these two potential problems, the first is undoubtedly more serious. It is much easier for readers to skip details that they find too obvious to be worth saying than it is for them to fill in details that they do not find obvious at all.

The question of how much detail to give is related to the question of how formal to be, but it is not the same question. It is true that there is a tendency in informal mathematical writing to leave out details, but with even the most formal writing a decision has to be made about how much detail to give; it is just that in formal writing one probably wants to signal more carefully when details have been left out. This can be done in various ways. One can use expressions such as "It is an easy exercise to check that...," or "The second case is similar," which basically say to the reader, "I have decided not to spell out this part of the argument." One can also give small hints, such as "By compactness," or "An obvious inductive argument now shows that...," or "Interchanging the order of summation and simplifying, we obtain...."

If you do decide to leave out detail, it is a good idea to signal to the reader how difficult it would be to put that detail in. A mistake that some writers make is to give references to other papers for arguments that can easily be worked out by the reader, without saying that the particular result that is needed is easy. This is straightforwardly misleading; it suggests that the best thing to do is to go and look up the other paper when in fact the best thing to do is to work out the argument for oneself.

## 4 Letters versus Words

The following is problem 10 of book 1 of an English translation of Diophantus's *Arithmetica.*

> Given two numbers, to add to the lesser and to subtract from the greater the same (required) number so as to make the sum in the first case have to the difference in the second case a given ratio.

A modern writer would express the same problem more like this.

> Given two numbers $a$ and $b$ and a ratio $\rho$, find $x$ such that $a+x=\rho(b-x)$.

The main difference between these two ways of describing the problem is that in the second formulation the numbers under discussion have been given *names.* These names take the form of letters, which allow us to replace wordy expressions such as "the second number" or "the given ratio" by letters such as "$b$" and "$\rho$."

The advantage of modern notation is that it is much more concise. This is not just a matter of saving paper; the extra length of "to make the sum in the first case have to the difference in the second case a given ratio" over "such that $a+x=\rho(b-x)$" makes it significantly harder to understand because it is difficult to take in the entire phrase at once.

However, the concision that comes from naming mathematical objects comes at a cost: one has to learn the names. In the example above, that is very easy and the cost is negligible. However, sometimes it is far from negligible. The following proposition comes from a paper in Banach space theory.

**Proposition.** *Let* $0 \leqslant \alpha \leqslant \frac{1}{2}$ *and* $1/p = \frac{1}{2} - \alpha$. *Then*

$$\mathfrak{P}_2(E,F) \subset \mathfrak{L}^{(a)}_{p,\infty}(E,F)$$

*for all Banach spaces F if and only if* $E \subset \boldsymbol{\Gamma}_\alpha$.

Just before the proposition, the reader has been told that $\mathfrak{P}_2$ is the ideal of 2-summing operators from $E$ to $F$, which is a standard definition in the area. As for $\mathfrak{L}^{(a)}_{p,\infty}(E,F)$, this has been defined early in the paper as follows. (It is not necessary to understand these definitions to understand the point I am making.)

> Given an operator $T$, the *approximation number* $a_n(T)$ is defined to be $\inf\{\|T-L\|: \operatorname{rank}(L) < n\}$. Then $\mathfrak{L}^{(a)}_{s,w}(E,F)$ is the set of operators $T$ such that the sequence $(a_n(T))_{n=1}^{\infty}$ belongs to the Marcinkiewicz space $\ell_{s,w}$.

The definition of the Marcinkiewicz space is again standard in the area. Finally, the set $\Gamma_\alpha$ is defined to be the set of all Banach spaces of weak Hilbert type $\alpha$. That is not a standard definition, but it is given earlier on in the paper.

Thus, another way of stating the proposition is as follows.

**Proposition.** *Let* $0 \leqslant \alpha \leqslant \frac{1}{2}$, *let* $1/p = \frac{1}{2} - \alpha$, *and let E be a Banach space. Then the following two statements are equivalent.*

(1) *For every Banach space F and every 2-summing operator* $T\colon E \to F$, *the sequence* $(a_n(T))_{n=1}^{\infty}$ *of approximation numbers belongs to the Marcinkiewicz space* $\ell_{p,\infty}$.
(2) *E is a space of weak Hilbert type* $\alpha$.

This time, it is the *wordier* definition that places a smaller burden on the reader's memory. If you know what 2-summing operators, approximation numbers, the Marcinkiewicz space, and weak Hilbert type are, most of which are standard definitions in the area, then you can understand without much effort what the proposition is claiming. With the first formulation, there is an extra step you have to perform to unpack the notation into those standard definitions. Another advantage of the second formulation is that a sufficiently expert reader who is skimming the paper will be able to understand it without having to look back in the paper to find out what everything means. The first formulation does not leave that option open.

Thus, in more complicated mathematical writing, there is another source of tension. If you use too little notation, your sentences will become hopelessly clumsy and repetitive, but if you use too much, you are placing excessive demands on the memory of your readers.

This may be a delicate balance to strike, but there is one principle that applies universally: if you do decide to use some nonstandard notation, then make sure that the reader can easily find where it is defined. This can be done by means of a section devoted to preliminary definitions, though it will often be kinder to give definitions just before they are used. If that is not possible, one can give reminders of definitions, or at the very least pointers to where they can be found.

## 5 Single Long Arguments versus Arguments Broken Up into Modules

If you are trying to justify a mathematical statement and the justification is long and complicated, then what you write may well be hard to understand unless you can somehow break the argument up into smaller "modules" that fit together to give you what you want. In pure mathematics, these modules usually take the form of *lemmas*. If you are proving a theorem and you do not want the proof to become unwieldy, then you try to identify parts of the argument that can be extracted and proved separately. One can then simply quote these results in the main argument. Lemmas play a role in proofs that is similar to the role of subroutines in computer programs.

For breaking up an argument to be a good idea, it greatly helps if the part of the argument you want to extract is not too context dependent. If the statement of a lemma requires a long piece of scene setting, then it is probably better to leave it in the main body of the argument, where the scene has already been set. However, if it can be stated without reference to the particular context, which usually means that it is more general than the particular application needed of it in the main argument, then it is more appropriate to extract it. Again, this is a matter of judgment.

A disadvantage of more modular arguments is that extracting lemmas, or more general modules, forces you to put them somewhere where they do not arise naturally. If you put them before the main argument, so that they will be available when needed, then the reader is presented with statements of no obvious use and is expected to remember them. If they are particularly memorable, then that is not a problem, but often they are not; for instance, they may depend on two or three slightly odd conditions that just happen to be satisfied in the later argument. If you put them after the main argument, then the reader keeps being told, "We will prove this claim later" and reaches the end of the argument with the uneasy feeling that the proof is incomplete. A third possibility is to state and prove lemmas *within* an argument, but nested statements of this kind can be fairly ugly.

With some complicated arguments, there may be no truly satisfactory solution to these problems. In that case, the best thing to do may well be to choose an unsatisfactory solution and mitigate the problems somehow. The default option is probably to state lemmas before they are used. If you choose that option and the lemmas are somewhat complicated and hard to remember, then you can always add a few words of explanation about the role that the lemma will play. If even that is hard to do, then another option is to advise the reader to read the main argument first and return to the lemma only when the need for it has become clear. (An experienced reader may well do that anyway, but it is still helpful to be told by the author that it is a good approach to understanding the argument.)

## 6 Logical Order versus Order of Discovery

Suppose you wish to present the fact that a sequence of continuous functions that converges pointwise does not have to converge uniformly. Here is one way that you might do it.

**Theorem.** *There exists a sequence of continuous functions* $f_n\colon [0,1] \to [0,1]$ *that converges pointwise but not uniformly.*

*Proof.* For each positive integer $n$ and each $x \in [0,1]$, let $f_n(x) = nx\mathrm{e}^{-nx}$. Then, for $x > 0$ we have $\mathrm{e}^{-x} < 1$, so $n\mathrm{e}^{-nx} = n(\mathrm{e}^{-x})^n \to 0$ as $n \to \infty$. It follows that $f_n(x) \to 0$ as $n \to \infty$. Also, when $x = 0$ we have $f_n(x) = 0$ for every $n$, so again $f_n(x) \to 0$ as $n \to \infty$. Therefore, $f_n(x) \to 0$ pointwise.

However, the convergence is not uniform. To see this, observe that $f_n(n^{-1}) = \mathrm{e}^{-1}$ for every $n$. Thus, for every $n$ there exists $x$ such that $|f_n(x) - 0| \geqslant \mathrm{e}^{-1}$. □

This proof has a feature that is common in mathematics: it is easier to follow the steps than it is to see where the steps came from. If you are told to try the functions $f_n(x) = nx\mathrm{e}^{-nx}$, then checking that they satisfy the conditions is a straightforward exercise, but what made anybody think of that particular sequence of functions?

Here is what we might write if we wanted to make the answer to that last question clearer.

*Proof.* If $f_n \to f$ pointwise but not uniformly, then $f_n - f \to 0$ pointwise but not uniformly, so we may as well look for functions that converge to zero. In order to ensure that they do not converge uniformly to zero, we need a positive number $\theta$ such that for infinitely many $n$ there exists $x \in [0,1]$ with $|f_n(x)| \geqslant \theta$. Since infinitely many of these $f_n(x)$ will have the same sign, and since we can multiply all functions by $\theta^{-1}$, we may as well look for a sequence of functions $f_n$ that converges pointwise to 0 such that for every $n$ there exists $x_n$ with $f_n(x_n) \geqslant 1$.

Now, if $f_n(x_n) \geqslant 1$ and $f_n$ is continuous, then there exists an open interval $I_n = (x_n - \delta_n, x_n + \delta_n)$ around $x_n$ such that $f_n(y) \geqslant \frac{1}{2}$ for every $y \in I_n$. We are going to have to make sure that we do not have infinitely many of these intervals overlapping in some point $u$, since then we would have $f_n(u) \geqslant \frac{1}{2}$ for infinitely many $n$, which would imply that $f_n(u)$ does not tend to zero.

How can we find infinitely many open intervals without infinitely many of them overlapping? The simplest way of doing it is to take intervals of the form $(a, b_n)$ for a sequence $(b_n)$ that converges to $a$. So, for example, we could take $I_n$ to be the interval $(0, 1/n)$.

This suggests that we should let $f_n$ be a continuous function that takes the value 1 somewhere inside the interval $(0, 1/n)$ and is small outside that interval. One way of defining a function that reaches 1 for a small value of $x$ and then quickly drops back down again is to take a function that grows rapidly to 1, such as $g_n(x) = \lambda x$, and multiply it by a function that is roughly 1 for a little while and then decays rapidly, such as $\mathrm{e}^{-\mu x}$. The rapid decay of $\mathrm{e}^{-\mu x}$ starts when $x$ is around $1/\mu$, which suggests that we should take $\mu$ to be around $n$. Since we want $g_n(x)$ to reach 1 in the interval $(0, 1/n)$, we should probably take $\lambda$ to be around $n$ as well.

It is now easy to check that the functions $f_n(x) = nx\mathrm{e}^{-nx}$ converge pointwise to zero but not uniformly. □

Of course, one might well give a detailed proof that the functions $nx\mathrm{e}^{-nx}$ do the job.

As with the other choices, there are advantages and disadvantages that need to be weighed up when deciding how much to explain the origin (or at least a possible origin) of the ideas one presents. If one's main concern is *verification* of a result—that is, convincing the reader of its truth—then it may not matter too much where the ideas come from as long as they work. But if the aim is to *teach the reader* how to solve problems of a certain kind, then presenting solutions that appear out of nowhere as if by magic is not helpful. What is more, demonstrating where the ideas come from gives the reader a much clearer idea of which features are essential and which merely incidental. For example, in the argument above it is clear from the second presentation that there is nothing special about the functions $f_n(x) = nx\mathrm{e}^{-nx}$: for $f_n(x)$ one could take any non-negative function such that $f_n(0) = 0$, $f_n(1/n) \geqslant c$ (for some fixed constant $c$), and $f_n(x)$ is small for every $x \geqslant 2/n$. For instance, one could take a "witch's hat" that equals $nx$ when $0 \leqslant x \leqslant 1/n$, $2 - nx$ when $1/n \leqslant x \leqslant 2/n$, and 0 when $2/n \leqslant x \leqslant 1$.

That is not to say that a diligent reader cannot look at a presentation of the first kind and work out for him/herself where the idea might have come from. In this case, if one sketches the graph of $f_n(x)$, one sees that it grows and shrinks rapidly in a small interval near 0 and is small thereafter, and then it becomes clear why these functions are suitable. However, one needs experience to be able to do this with an argument. So the extent to which you should explain where your arguments come from depends largely on the level of experience of your intended reader—both generally and in the specific area you are writing about.

## 7 Definitions First versus Examples First

Suppose that one wanted to write an explanation of what a topological manifold is. An obvious approach would be to start by giving the definition. That could be done as follows.

**Definition.** A *d-dimensional topological manifold* is a topological space $X$ such that every point $x$ in $X$ has a neighborhood that is homeomorphic to a connected open subset of $\mathbb{R}^d$.

Having done that, one would give a few examples of topological manifolds, such as spheres and tori, to illuminate the definition.

An alternative approach is to start with a brief discussion of the examples. One could point out, for instance, that it is easy to come up with a satisfactory coordinate system for any small region of the world but that it is not possible to find a good coordinate system for the world in its entirety; there will always be annoying problems such as the poles not having well-defined longitudes. A discussion of that kind will give the reader the informal concept of a space that is "locally like $\mathbb{R}^d$" and after that the formal definition is motivated: it is the formal expression of an informal idea that the reader already has.

The advantage of the second approach is that an abstract definition is often much easier to understand if one has a good idea of what it is abstracting. One will read the definition with strong expectations of what it will look like, and all one will have to commit to memory is the ways in which the definition does not quite fit those expectations. If the definition is presented first, then one will be expected to hold the whole thing in one's head, rather than what one might think of as the difference between the definition and one's prior expectation of it.

Whether or not this advantage makes it worth presenting examples before giving a definition depends on how difficult you expect it to be for your reader to grasp the definition. To give an example where it might not be worth giving examples first, suppose that you want to introduce the notion of a commutative ring for a reader who is already familiar with groups and fields. A natural way of doing it would be to list the axioms for a commutative ring and make the remark that what you have listed is very similar to the list of axioms for a field but you no longer assume that elements have multiplicative inverses, and sometimes you do not even assume that your rings have multiplicative identities.

Once you have said that, it will still be a very good idea to give some important examples, such as the ring $\mathbb{Z}$ of all integers, the ring $\mathbb{Z}[x]$ of all polynomials with integer coefficients, and the ring $\mathbb{Z}[\sqrt{2}]$ of all numbers of the form $a + b\sqrt{2}$ where $a$ and $b$ are integers. However, the argument for presenting these examples *first* is weaker than it was for topological manifolds, for two reasons.

The first reason is that the definition is easy to grasp: rings are like fields but without multiplicative inverses. Therefore, giving the definition straight away does not place a burden on the reader's memory. Of course, the reader will want reassurance that there are interesting examples, but that can be given immediately after the definition.

The second reason is that the necessity for this particular abstraction is less obvious than it is for manifolds. Given examples such as spheres and tori, it is natural to think that they are all examples of the same basic "thing" and then to try to work out what that "thing" is. But the benefits of thinking of the integers and the polynomials with integer coefficients as examples of the same underlying algebraic structure are not clear in advance; they become clear only after one has developed a considerable amount of theory. So it is more natural in this case to think of the abstraction as primary, at least in the first instance.

As ever, the decision about how to present a new mathematical concept involves a judgment that is sometimes quite delicate. Broadly speaking, the harder a definition is to grasp, the more helpful it will be to the reader to have some examples in mind when reading it. But that depends both on the reader and on the intrinsic complexity of the definition. However, one general piece of advice is still possible here, which is at least to *consider* the possibility of starting with examples. It may not always be appropriate to do so, but many mathematical writers like to start with definitions under all circumstances, and the result is that many expositions are harder to understand than they need to be.

Let me close this section by pointing out that the examples-first device is quite a general one. Indeed, I have used it in a number of places in this article; see the openings of sections 4 and 6 and of this very section.

## 8 Traditional Methods of Dissemination versus New Methods

A person who wishes to produce mathematical writing today faces a choice that did not exist twenty years ago. Until recently, almost all mathematical writing took the form of books or journal articles. But now the Internet has given us new methods of dissemination, which have already had an impact and are likely to have a much bigger impact in the future.

In particular, the existence of the Internet affects every single one of the considerations discussed in this article. Let me take them in turn.

### 8.1 Level of Formality

The main task of each generation of mathematicians is to add to the body of mathematical knowledge. However, there is a second task that is almost as important as the first, and not entirely separate from it, which is to digest this new knowledge and present it in a form that subsequent generations will find as easy as possible to grasp. This process of digestion can of course happen many times to the same piece of mathematics.

Sometimes, digesting a piece of mathematics is itself a significant advance in mathematical knowledge. For example, a theory may be developed that yields quite easily a number of already existing and seemingly disparate results. The traditional publication system is well suited to this situation; one can just write an article about the theory and get it published in the normal way.

Sometimes, however, digesting a piece of mathematics does not constitute a mathematical advance. It can be something more minor, such as thinking of a way of looking at an argument that makes it clearer where the ideas have come from or drawing an informal analogy between one piece of mathematics and another that is simpler or better known. Insights of this kind can be hard earned and extremely valuable to other mathematicians, but they do not lead to publishable papers.

With the Internet, there are many ways that more informal mathematical thoughts can be shared. An obvious one is to write a conventional mathematical text and make it available on one's home page. Another option, which an increasing number of mathematicians have adopted, is to have a blog. The advantage of this is that one obtains feedback from one's readers, and experience has shown that the quality of much of this feedback is very high.

There are other forms of mathematical literature that would not be conventionally publishable but that could be extremely valuable. For example, an article about a serious but failed attempt to solve a problem would not be accepted by a journal, and the result is a great deal of duplication of work; if the problem is important and the attempt looks plausible to begin with, then many people will try it. A database of failed proof attempts would be very useful, and in principle the Internet makes it easy to set up, though so far nobody has done so.

In general, the Internet allows us much greater freedom in choosing the level of formality at which we wish to write and allows us to publish documents that do not fit the mould of a standard journal article.

### 8.2 Level of Detail

Suppose that you use a mathematical result or definition that will be familiar to some readers but not to others. In a print document you have to decide whether to explain it and, if so, how elaborate an explanation to give.

In a hyperlinked document on the web, one is no longer forced to make this choice. One can write a version for experts, but with certain key words and phrases underlined, so that readers who need these words and phrases explained further can click on them and read explanations. This kind of writing has become very common on Wikipedia and other wikis.

It also introduces a new balance that needs to be struck. Sometimes wiki articles are hard to read because the writers use the existence of links to other wiki pages as a license not to explain terms that they might otherwise have explained. The result is that unless one is familiar with most of the definitions in the original article, one can get lost in a complicated graph of linked wiki pages as one finds that the page that explains an unfamiliar concept itself requires one to click through to several other pages. So if you are going to exploit hyperlinks, you need to think carefully about what the experience of following those links will be like for your intended readers.

Another inconvenience of hyperlinks is that they require you to visit an entirely new page, which makes it easy to forget where you were before (especially if you backtrack and then follow some other sequence of links). However, there is plenty of software that gets round this problem. For example, on some sites one can incorporate "sliders," pieces of text that insert themselves into what you are reading when you click on an appropriate box and disappear when you click on it again. So if, for example, one wrote, "by the second isomorphism theorem," one could have a box with the words "What does that say?" on it, so that readers who needed it could click on the box and have a short paragraph about the second isomorphism theorem inserted into the text. One can have sliders within sliders, so perhaps within that slider one could have the option of bringing up a proof of the theorem as well.

The main point is that the Internet has made it possible to write new kinds of documents where one is no longer forced to make choices such as of how much detail to give. One can leave that decision to the reader. Such documents have a huge potential to improve the way mathematics is presented, and this potential will only increase as technology improves.

### 8.3 Letters versus Words

I will not say much about this, since most of what I have to say is very similar to what I have already said about the level of detail in which a document is written. With the kinds of electronic documents that are now possible, one can save the reader the trouble of searching through a paper to find out what a letter stands for by incorporating a reminder that appears when you click on the letter. Perhaps better still, it could appear in a little box when you hover over the letter. One could also have condensed statements involving lots of letters with the option of converting them into equivalent wordier statements. Again, the point is that there are many more options now.

### 8.4 Modularity

The kinds of electronic documents I have been discussing make possible a form of top-down mathematical writing that would be far less convenient in a print document. One could write a high-level account of some piece of mathematics, giving the reader the option of expanding any part of that high-level account into a lower-level account that justifies it in more detail. And there could be many levels of this, so that if you clicked on everything you would end up with a presentation of the entire argument in full gory detail.

A less ambitious possibility is one that solves the problem discussed earlier about where to place a lemma. The difficulty was that in a print document you will either put it before the proof where it is used, in which case it is not adequately motivated, or during the proof, in which case it looks ugly, or after the proof, in which case the proof itself leaves you with awkward promises to fill in gaps later. But with an electronic document, putting a lemma exactly where it is needed is no longer ugly. During the proof, one can say, "We are now going to make use of the following statement," and give the reader a button to click on that will bring up a proof of that statement.

### 8.5 Order of Presentation

If you do not want to decide whether to give an abstract definition first or start with motivating examples, then you can give the reader the choice. Just start with a page of headings and invite the reader to decide whether to click on "Motivating examples" first or "The formal definition" first.

To some extent, the same goes for the decision about whether to present arguments in their logical order or in a way that brings out how they were discovered. If at some point the logical order requires you to draw a rabbit out of a hat, you could at the very least introduce a slider that explains where that rabbit actually came from.

# VIII.2 How to Read and Understand a Paper

*Nicholas J. Higham*

Whether you are a mathematician or work in another discipline and need to use mathematical results, you will need to read mathematics papers—perhaps lots of them. The purpose of this article is to give advice on how to go about reading mathematics papers and gaining understanding from them.

The advice is particularly aimed at inexperienced readers. A professional mathematician may read from tens to hundreds of papers every year, including published papers, manuscripts sent for refereeing by journals, and draft papers written by students and colleagues. To a large extent the suggestions I make here are ones that you naturally adopt after reading sufficiently many papers.

Mathematics papers fall into two main types: primary research papers and review papers. Review papers give an overview of an area and usually contain a substantial amount of background material. By design they tend to be easier to read than papers presenting new research, although they are often longer. The suggestions in this article apply to both types of papers.

## 1 The Anatomy of a Paper

Mathematics papers are fairly rigid in format, having some or all of the following components.

**Title.** The title should indicate what the paper is about and give a hint about the paper's contributions.

**Abstract.** The abstract describes the problem being tackled and summarizes the contributions of the paper. The length and the amount of detail both vary greatly. The abstract is meant to be able to stand alone. Often it is visible to everyone on a journal's Web site, while the paper is visible only to subscribers.

**Introduction.** The first section of the paper, almost always called "Introduction," sets out the context and problem being addressed in more detail than the abstract. Depending on how the paper has been written, the introduction may or may not describe the results and conclusions. Some papers lend themselves to a question being posed in the introduction but fully answered only in a conclusions section.

**Conclusions.** Many, but not all, papers contain a final section with a title such as "Conclusion" or "Concluding Remarks" that summarizes the main conclusions of the paper. Omission of such a section indicates that the conclusions have been stated in the introduction or perhaps at the end of a section describing experiments, or that no explicit summary has been provided. This section is often used to identify open questions and describe areas for future research, and such suggestions can be very useful if you are looking for problems to work on.

**Appendix.** Some papers contain one or more appendices, which contain material deemed best separated from the main paper, perhaps because it would otherwise clutter up the development or because it contains tedious details.

**References.** The references section contains a list of publications that are referred to in the text and that the reader might want to consult.

**Supplementary materials.** A relatively new concept in mathematics is the notion of additional materials that are available on the publisher's Web site along with the paper but are not actually part of the paper. These might include figures, computer programs, data, and other further material and might not have been refereed even if the paper itself has. It is not always easy to tell if a paper has supplementary materials, as different journals have different conventions for referring to them. They might be mentioned at the end of the paper or in a footnote on the first page, and they may be referred to with "see the supplementary materials" or via an item in the reference list.

## 2 Deciding Whether to Read a Paper

A common scenario is that you come across a paper that, based on the title, you think you might need to read. For example, you may be signed up to receive alerts from a journal or search engine and become aware of a new paper on a topic related to your interests. How do you decide whether to read the paper? The abstract should contain enough information about the context of the work and the paper's results for you to make a decision. However, abstracts are sometimes very short and are not always well written, so it may be necessary to skim through the introduction and conclusions sections of the paper.

The reference list is worth perusing. If few of the references are familiar, this may mean that the paper presents a rather different view on the topic than you expected, perhaps because the authors are from a different field. If papers that you know are relevant are missing, this is a warning that the authors may not be fully aware of past work on the problem.

If the main results of the paper are theorems, read those to see whether it is worth spending further time on the paper. Consider also the reputation of the journal and the authors, and, unless the paper is very recent, check how often (and how, and by whom) it has been cited in order to get a feel for what other people think about it. (Citations can be checked using online tools, such as Google Scholar or one of several other services, most of which require a subscription.)

## 3 Getting an Overview

A paper does not have to be read linearly. You may want to make multiple passes, beginning by reading the abstract, introduction, and conclusions, as well as looking at the tables, figures, and references.

Many authors end the introduction with a paragraph that gives an overview of what appears in each part of the paper. Sometimes, though, a glance at the paper's section headings provides a more easily assimilated summary of the content and organization.

Another way in which you might get an overview of the paper is by reading the main results first: the lemmas, theorems, algorithms, and associated definitions, omitting proofs. The usefulness of this approach will depend on the topic and your familiarity with it.

## 4 Understanding

It is often hard to understand what you are reading. After all, research papers are meant to contain original ideas, and ideas that you have not seen before can be hard to grasp. You may want to stop and ponder an argument, perhaps playing with examples.

I strongly recommend making notes, to help you understand the text and avoid having to retrace your steps in grasping a tricky point if you come back to the paper in the future. It is also a good idea to write a summary of your overall thoughts on the paper; when you go back to the paper a few months or years later, your summary will be the first thing to look at. I recommend dating your notes and summary, as in the future it can be useful to know when they were written. Indeed, I have papers that I have read several times, and the notes show how my understanding changed on each reading. (There exist papers for which multiple readings are needed to appreciate fully the contents, perhaps because the paper is deep, because it is badly written, or both!)

As well as writing notes, it is a good idea to mark key sentences, theorems, and so on. I do this either by putting a vertical line in the margin that delineates the area of interest or by marking the relevant text with a highlighter pen.

I write my notes on a hard copy of the paper. Many programs are available that will allow you to annotate PDF files on-screen, though using mathematical notation may be problematic; one solution is to handwrite notes and then scan them in and append them to the PDF file.

A good exercise, especially if you are inexperienced at writing papers, is to write your own abstract for the paper (100–200 words, say).

Writing while you read turns you from a passive reader into an active one, and being an active reader helps you to understand and remember the contents. One useful technique is to try out special cases of results. If a theorem is stated for analytic functions, see what it says for polynomials or for the exponential. If a theorem is stated for $n \times n$ matrices, check it for $n = 1, 2, 3$. Another approach is to ask yourself what would happen if one of the conditions in a theorem were to be removed: where would the proof break down?

When you reach a point that you do not understand, it may be best to jump to the end of the argument and go back over the details later to avoid getting bogged down. Keep in mind that some ideas and techniques are so well known to researchers in the relevant field that they might not be spelled out. If you are new to the field you may at first need a bit of help from a more experienced colleague to fill in what appear to be gaps in arguments.

It is important to keep in mind that what you are reading may be badly explained or just wrong. Typographical errors are quite common, especially in preprints and in papers that have not been copy edited. Mathematical errors also occur, and even the best journals occasionally have to print corrections ("errata") to previously published articles.

In mathematical writing certain standard phrases are used that have particular meanings. "It follows that" or "it is easy to see that" mean that the next statement can be proved without using any new ideas and that giving the details would clutter the text. The detail may, however, be tedious. The shorter "hence," "therefore," or "so" imply a more straightforward conclusion. "It can be shown that" again implies that details are not felt to be worth including but is noncommittal about the difficulty of the proof.

## 5 Documenting Your Reading

I advise keeping a record of which papers you have read, even if you have read them only partially. If you are a beginning Ph.D. student this may seem unnecessary, as at first you will be able to keep the papers in your mind. But at some point you will forget which papers you have read and having this information readily available will be very useful.

A few decades ago papers existed only as hard copies, and one would file them by author or subject. Today, most papers are obtained as PDF downloads that can be stored on our computers. Various computer programs are available for managing collections of papers. One of those, or a BibTeX database, can serve to record what you have read and provide links to the PDF files.

## 6 Screen or Print?

Should you read papers on a computer screen or in print form? This is a personal choice. People brought up in the digital publishing era may be happy reading on-screen, but others, such as me, may feel that they can properly read a paper only in hard copy form. There is no doubt that hard copy allows easier viewing of multiple pages at the same time, while a PDF file makes it easier to search for a particular term and can be zoomed to whatever size is most comfortable to read. It is important to try both and use whatever combination of screen and print works best for you.

If you do read on-screen, keep in mind that most PDF readers allow you to customize the colors. White or yellow text on a black background may be less strain on the eyes than the default black on white. In Adobe Acrobat the colors can be changed with the menu option Preferences-Accessibility-Document Colors Options.

### 7 Reading for Writing

One of the reasons to read is to become a better writer. When you read an article that you think is particularly well written, analyze it to see what techniques, words, and phrases seemed to work so well. Reading also expands your knowledge and experience, and can improve your ability to do research. Donald Knuth put it well when he said:

> In general when I'm reading a technical paper... I'm trying to get into the author's mind, trying to figure out what the concept is. The more you learn to read other people's stuff, the more able you are to invent your own in the future.

### 8 What Next?

Having read the paper you should ask yourself not only what the authors have achieved but also what questions remain. Can you identify open questions that you could answer? Can you see how to combine ideas from this paper with other ideas in a new way? Can you obtain stronger or more general results?

## VIII.3 How to Write a General Interest Mathematics Book

*Ian Stewart*

> *I've always wanted to write a book.*
> *Then why don't you?*
> —Common party conversation

Popular science is a well-developed genre in its own right, and popular mathematics is an established subgenre. Several hundred popular mathematics books now appear every year, ranging from elementary introductions through school-level topics to substantial volumes about research breakthroughs. Writing about mathematics for the general public can be a rewarding experience for anyone who enjoys and values communication. Established authors include journalists, teachers, and research mathematicians; subjects are limited only by the imagination of authors and publishers' assessments of what booksellers are willing to stock. Even that is changing as the growth of e-books opens the way for less orthodox offerings. The style may be serious or lighthearted, preferably avoiding extremes of solemnity or frivolity. On the whole, most academic institutions no longer look down on "outreach" activities of this kind, and many place great value on them, both as publicity exercises and for their educational aspects. So do government funding bodies.

### 1 What Is Popular Mathematics?

For many people the phrase is an oxymoron. To them, mathematics is *not* popular. Never mind: popularization is the art of making things popular when they were not originally. It is also the art of presenting advanced material to people who are genuinely interested but do not have the technical background required to read professional journals. Generally speaking, most popular mathematics books address this second audience. It would be wonderful to write a book that would open up the beauty, power, and utility of mathematics to people who swore off the subject when they were five, hate it, and never want to see it or hear about it again—but, by definition, very few of them would read such a book, so you would be wasting your time.

Already we see a creative tension between the wishes of the author and the practicalities of publishing. As e-books start to take off, the whole publication model is changing. One beneficial aspect is that new kinds of book start to become publishable. If an e-book fails commercially, the main thing wasted is the author's time and energy. That may or may not be an issue—an author with a track record can use his/her time to better effect by avoiding things that are likely to fail—but it will not bankrupt the publisher.

Popular mathematics is a genre, a specific class of books with common features, attractive to fans and often repellent to everybody else. In this respect it is on a par with science fiction, detective novels, romantic fantasy, and bodice rippers. Genres have their own rules, and although these rules may not be explicit, fans notice if you break them. If you want to write a popular mathematics book, it is good preparation to read a few of them first. Many writers started that way; they began as fans and ended up as authors, motivated by the books they enjoyed reading.

Most popular mathematics books fall into a relatively small number of types. Many fall into several simultaneously. The rest are as diverse as human imagination can make them. The main classifiable types are:

(1) Children's books.
   (a) Basic school topics.
   (b) More exciting things.
(2) History.
(3) Biography.
(4) Fun and games.
(5) Big problems.
(6) Major areas.
   (a) Classical.
   (b) Modern.
(7) Applications.
(8) Cultural links.
(9) Philosophy.

Children's books are a special case. They involve many considerations that are irrelevant to books for adults, such as deciding which words are simple enough to include—I will say no more about them, lacking experience. Histories and biographies have a natural advantage over more technical types of books: there is generally plenty of human interest. (If not, you chose the wrong topic or the wrong person.) Books about fun and games are lighthearted, even if they have a more serious side. Martin Gardner was the great exponent of this form of writing. Authors have written in depth about individual games, while others have compiled miscellanies of "fun" material.

Big problems, and major areas of mathematical research, are the core of popular mathematics. Examples are Fermat's last theorem, the Poincaré conjecture, chaos, and fractals. To write about such a topic you need to understand it in more depth than you will reveal to your readers. That will give you confidence, help you find illuminating analogies, and generally grease the expository wheels. You should choose a topic that is timely, has not been exhausted by others, and stands some chance of being explained to a nonexpert. Within both of these subgenres you can present significant topics from the past—Fourier analysis, say—or you can go for the latest hot research area—wavelets, maybe. It is possible to combine both if there is a strong historical thread from past to present.

It is always easier to explain a mathematical idea if it has concrete applications. People can relate to the applications when the mathematics alone starts to become impenetrable. The same goes for cultural links, such as perspective in Renaissance art and the construction of musical scales.

Finally, there are deep conceptual issues, "philosophical" aspects of mathematics: infinity, many dimensions, chance, proof, undecidability, computability. Even simple ideas like zero or the empty set could form the basis of a really fascinating book, and have done so. The main thing is to have something to say that is worth saying. That is true for all books, but it is especially vital for philosophical ones, which can otherwise seem woolly and vague.

## 2 Why Write a Popular Mathematics Book?

Authors write for many reasons. When Frederik Pohl, a leading science fiction writer, was being inducted into the U.S. military he was asked his profession and replied "writer." This was received with a degree of concern; writers are often impractical idealists who criticize everything and cause trouble. So Pohl was asked *why* he wrote. "To make money," he replied. This was received with relief as an entirely sensible and comprehensible reason. Another leading American writer, Isaac Asimov, produced more than 300 books of science fiction, popular science, and other genres. When asked why he wrote so many books, he replied that he found it impossible to stop. He also went out of his way not to recommend his prolific approach to anyone else. Some authors write one book and are satisfied, even if it becomes a best seller; others keep writing whether or not they receive commercial success. Some write one book and vow never to repeat the experience. Some want to put a message across that strikes them as being of vital importance—a new area of mathematics, a social revolution, a political innovation. Some just like writing. There are no clear archetypes, no hard and fast rules; everything is diverse and fluid.

The best reason for writing a popular mathematics book, in my opinion, is that you desperately want to tell the world about something you find inspiring and interesting. Books work better when the author is excited and enthusiastic about the topic. The excitement and enthusiasm will shine through of their own accord, and it is best not to be too explicit about them. Far too many television presenters seem to imagine that if they keep telling the viewers how excited they are, viewers will also become excited. This is a mistake. Do not *tell* them you are excited; *show* them you are. It is the same with a book. Tell the story, bring out its inherent interest, and you are well on the way. Popularization is not about *making* mathematics fun (interesting, useful, beautiful, ...); it is about showing people that it already *is* fun (interesting, useful, beautiful, ...).

In the past I have called mathematics the Cinderella science. It does all the hard work but never gets to go to the ball. Our subject definitely suffers, compared with many, because of a negative image among a large sector of the public. One reason for writing mathematics books for general readers is to combat this image. It is mostly undeserved, but as a profession, mathematicians do not always help their own cause. Even nowadays, when some area of mathematics attracts attention from the media, some mathematicians immediately go into demolition mode and complain loudly about "hype," exaggeration, and lack of precision.

When James Gleick's *Chaos* became a best seller, and the U.S. government noticed and began considering increasing the funding for nonlinear dynamics, a few distinguished mathematicians went out of their way to inform the government that it was all nonsense and the subject should be ignored. This might have been a good idea if they had been right—not about the popular image of "chaos theory," which was at best a vague approximation to the reality, but about the reality itself—but they were wrong. Nonlinear dynamics, of which chaos is one key component, is one of the great success stories of the late twentieth century, and it is powering ahead into the twenty-first. I remember one letter to a leading mathematics journal claiming that chaos and fractals had no applications whatsoever at a time when you could not open the pages of *Science* or *Nature* without finding papers that made excellent scientific use of these topics. Conclusion: many mathematicians have no idea what is going on in the rest of the scientific world, perhaps because they do not read *Science* or *Nature*.

To ensure that our subject is valued and supported, we mathematicians need to explain to ordinary people that mathematics is vital to their society, to their economic and social welfare, to their health, and to their children's future. No one else is going to do it for us. But we will not succeed if no one is allowed to mention a manifold without explaining that it has to be Hausdorff and paracompact as well as locally Euclidean. We have to grab our audience's attention with things they can understand; that necessarily implies using imprecise language, making broad-brush claims, and selecting areas that can be explained simply rather than others of equal or greater academic merit that cannot.

I am not suggesting that we should mislead the public about the importance of mathematics. But whenever a scientific topic attracts public attention, its media image is seldom a true reflection of the technical reality. Provided the technical reality is useful and important, a bit of overexcitement does no serious harm. By all means try to calm it down but not at the expense of ruining the entire enterprise. Grabbing public attention and then leveraging that (as the bankers would say) into a more informed understanding is fine. Grabbing public attention and then self-destructing because a few fine points have not quite been understood is silly.

## 3 Choosing a Topic

An article on popular writing is especially appropriate in a companion to applied mathematics because most people understand mathematics better if they can see what it is good for. This is one reason why chaos and fractals have grabbed public attention but algebraic $K$-theory has not. This is not a value judgment; algebraic $K$-theory is core mathematics, hugely important—it may even be more important than nonlinear dynamics. But there is little point in arguing their relative merits because each enriches mathematics. We are not obliged to choose one and reject the other. As research mathematicians, we probably do not want to work in both, but it is not terribly sensible to insist that your own area of mathematics is the only one that matters. Think how much competition there would be if everyone moved into your area. This happens quite a lot in physics, and at times it turns the subject into a fashion parade.

The key to most popular mathematics books is simple: *tell a story*.

In refined literary circles, the role of narrative is often downplayed. Whatever you think of *Finnegans Wake*, few would consider it a rip-roaring yarn. But popular science, like all genre writing, does not move in refined literary circles. What readers want, what authors must supply, is a story. Well, usually: all rules in this area have exceptions. In genre writing, humans are not *Homo sapiens*, wise men. They are *Pan narrans*, storytelling apes. Look at the runaway success of *The Da Vinci Code*—all story and little real sense.

A story has structure. It has a beginning, a middle, and an end. It often involves a conflict and its eventual resolution. If there are people in it, that is a plus; it is what made Simon Singh's *Fermat's Last Theorem* a best seller. But the protagonist of your book could well be the monster simple group or the four-color theorem. Human interest helps, in some subgenres, but it is not essential.

## 4 How to Write for the General Public

I wish I knew.

The standard advice to would-be authors is to ask, Who am I writing for? In principle, that is sound advice, but in practice there is a snag: it is often impossible to know. You may be convinced you are writing for conscientious parents who want to help their teenage children pass mathematics exams. The buying public may decide, by voting with their wallets, that the correct audience for your book is retired lawyers and bank managers who always regretted not knowing a bit more mathematics and have spotted an opportunity to bone up on the subject now that they have got the time.

Some books are aimed at specific age groups—young children, teenagers, adults—and of course you need to bear the age of your readers in mind if you are writing that kind of book. But for the majority of popular mathematics books, the audience turns out to be very broad, not concentrated in any very obvious demographic, and difficult to characterize. "The sort of people who buy popular math books" is about as close as you can get, so I do not think you should worry too much about your audience.

The things that matter most are audience independent. Write at a consistent level. If the first chapter of your book assumes the reader does not know what a fraction is and chapter two is about $p$-adic cohomology, you may be in trouble. The traditional advice to colloquium lecturers—start with something easy so that everyone can follow; then plough into the technicalities for the experts—was always bad advice even for colloquia because it lost most of the audience after five minutes of trivia. It is a complete disaster for a popular mathematics book.

It is common for the level of difficulty to ramp up *gradually* as the book progresses. After all, your reader is gaining insight into the topic as your limpid prose passes through their eyeballs to their brain. Chapter 10 *ought* to be a bit more challenging than chapter 1; if it is not, you are not doing your job properly.

One useful technique—when writing anything, be it for the public or for the editors and readers of the *Annals of Mathematics*—is self-editing. You need to develop an editor's instincts and apply them to your own work. You can do it as you go, rejecting poor sentences before your fingers touch the keyboard, but I find that slows me down and can easily lead to "writer's block." I hardly ever suffer from that affliction because I leave the Maoist self-criticism sessions for later. The great mathematical expositor Paul Halmos always said that the key to writing a book was to write it—however scruffily, however badly organized. When you have got most of it down on paper, or in the computer, you can go through the text systematically and decide what is good, what is bad, what is in the wrong place, what is missing, or what is superfluous. It is much easier to sort out these structural issues if you have something concrete to look at.

Word processors have made this process much easier. I generally write 10–15% more words than the book needs and then throw the excess away. I find it is quicker if I do that than if I agonize over each sentence as I type it. As George Bernard Shaw wrote: "I'm sorry this letter is so long, I didn't have time to make it shorter." When editing your work, here are a few things to watch out for.

- If you are using a term that you have not explained already, and it seems likely to puzzle readers, find a place to set it up. It might be a few chapters back; it might be just before you use it. Whatever you do, do not put it in the middle of the thing you are using it for; that can be distracting: "Poincaré conjectured that if a three-dimensional manifold (that is, a space ... [several sentences] ... three coordinates, that is, numbers that ... [several sentences] ... a kind of generalized surface) such that every closed loop can be shrunk to a point..." is unreadable.
- If some side issue starts to expand too much, as you add layer upon layer of explanation, ask whether you really need it. David Tall and I spent ten years struggling to explain the basics of homology in a complex analysis book for undergraduates without getting into algebraic topology as such. Eventually we realized we could omit that chapter altogether, whereupon we finished the book in two weeks.
- If you are really proud of the classy writing in some section, worry that it might be overwritten and distracting. Cut it out (saving it in case you decide to put it back later) and reread the result. Did you need it? "Fine writing" can be the enemy of effective communication.
- Above all, try to keep everything simple. Put yourself in the shoes of your reader. What would *they* want you to explain? Do you need the level of detail you have supplied? Would something less specific do the same job? Do not start telling them about the domain and range of a function if all they need

to know is that a function is a rule for turning one number into another. They are not going to take an exam.

- Do not forget that many ideas that are bread and butter to you (increment cliché count) are things most people have never heard of. *You* know what a factorial is; they may not. *You* know what an equivalence relation is; most of your readers do not. They can grasp how to compose two loops, or even homotopy classes, if you tell them to run along one and then the other but not if you give them the formula.
- It is said that the publishers of Stephen Hawking's *A Brief History of Time* told him to avoid equations, every one of which would allegedly "halve his sales." To some extent this advice was based on the publisher's hang-ups rather than on what readers could handle. Look at Roger Penrose's *The Road to Reality*, a huge commercial success with equations all over most pages. However, Hawking's publishers had a point: do not use an equation if you can say the same thing in words or pictures.

Your main aims, to which any budding author of popular science should aspire, are to keep your readers interested, entertain them, inform them, and—the ultimate—make them feel like geniuses. Do that, and they will think that *you* are a genius.

## 5 Technique

Every author develops his or her own characteristic style. The style may be different for different kinds of books, but there is not some kind of universal house style that is perfect for a book of a given kind. Some people write in formal prose, some are more conversational in style (my own preference), some go for excitement, some like to keep the story smooth and calm. On the whole, it is best to write in a natural style—one that you feel comfortable with—otherwise you spend most of the time forcing words into what for you are unnatural patterns when you should be concentrating on telling the story.

There are, however, some useful guidelines. You do not need to follow them slavishly; the main point here is to illustrate the kinds of issues that an author should be aware of.

Use correct grammar. If in doubt, consult a standard reference such as *Fowler's Modern English Usage*. Bear in mind that some aspects of English usage have changed since his day. Be aware that informal language is often grammatically impure, but even when writing informally, be a little conservative in that respect. For example, it is impossible nowadays to escape phrases like "the team are playing well." Technically, "team" is singular, and the correct phrase is "the team is playing well." Sometimes the technically correct usage sounds so pedantic and awkward that it might be better not to use it, but on the whole it is better to be correct. At all costs avoid being inconsistent: "the team is playing well and they have won nine of its last ten games."

I have some pet hates. "Hopefully" is one. It can be used correctly, meaning "with hope," but much more often it is used to mean "I hope that," which is wrong. "The fact that" is another: it is almost always a sign of sloppy sentence construction, and most of the time it is verbose and unnecessary. Replace "in view of the fact that" and similar phrases by the simple English word "because." Try deleting "the fact" and leaving just "that." If that fails to work, you can usually see an easy way to fix things up. Another unnecessarily convoluted phrase is "the way in which." Plain "how" usually does the same job, better.

Avoid Latin abbreviations: e.g., i.e., etc. They are obsolete even in technical mathematical writing—Latin is no longer the language of science—and they certainly have no place in popular writing. Replace with plain English. The writing will be easier to understand and less clumsy. Avoid clichés (Wikipedia has links to lists), but bear in mind that it is impossible to avoid them altogether. Attentive readers will find some in this article: "bear in mind" for example! Stay away from crass ones like "run it up the flagpole," and keep your cliché quotient small.

Metaphors and analogies are great ... provided they work. "DNA is a double helix, like two spiral staircases winding around each other" conveys a vivid image—although it is perilously close to being a cliché. Do not mix metaphors: I once wrote "the Galois group is a vital weapon in the mathematician's toolkit" and a helpful editor explained that weapons belong in an armory and "vital" they are not. The human mind is a metaphor machine; it grasps analogies intuitively, and its demand for understanding can often be satisfied by finding an analogy that goes to the heart of the matter. A well-chosen analogy can make an entire book. A poor one can break it.

Those of us with an academic background need to work very hard to avoid standard academic reflexes. "First tell them what you are going to tell them, then tell them, then tell them what you have told them" is fine for teaching—and a gesture in that direction can help readers make sense of a chapter, or even an entire book—but the reflex can easily become formulaic. Worse, it can destroy the suspense. Imagine if *Romeo and Juliet* opened with "Behold! Here comes Romeo! He will take a sleeping potion and fair Juliet will think him dead and kill herself."

Some sort of road map often helps readers understand what you are doing. My recent *Mathematics of Life* opens with this:

> Biology used to be about plants, animals, and insects, but five great revolutions have changed the way scientists think about life.
>
> A sixth is on its way.
>
> The first five revolutions were the invention of the microscope, the systematic classification of the planet's living creatures, evolution, the discovery of the gene, and the structure of DNA. Let's take them in turn, before moving on to my sixth, more contentious, revolution.

The reader, thus informed, understands why the next few chapters, in a book ostensibly about mathematics, are about historical high points of biology. Notice that I did not tell them what revolution six *is*. It is mathematics, of course, and if they have read the blurb on the back of the book they will know that, but you do not need to rub it in.

Writing a popular science book is not like writing a textbook; it is closer to fiction. You are telling a story, not teaching a course. You need to think about pacing the story and about what to reveal up front and what to keep up your sleeve. Your reader may need to know that a manifold is a multidimensional analogue of a surface, and you may have to simplify that to "curved many-dimensional space." Do not start talking about charts and atlases and $C^\infty$ overlap maps; avoid all mention of "paracompact" and "Hausdorff." To some extent, be willing to tell lies: lies of omission, white lies that slide past technical considerations that would get in the way if they were mentioned. Jack Cohen and I call this technique "lies to children." It is an educational necessity: what experts need to know is different from what the public needs to know.

Above all, remember that you are not trying to teach a class. You are trying to give an intelligent but uninformed person some idea of what is going on.

## 6 How Do I Get My Book Published?

Experienced writers know how to do this. If you are a new writer, it is probably better at present to work with a recognized publisher. However, this advice could well become obsolete as e-books grow in popularity. A publisher will bear the cost of printing the book, organize publicity and distribution, and deal with the publication process. In return it will keep most of the income, passing on about 10% to the author in the form of royalties. An alternative is to publish the book yourself through Web sites that print small quantities of books at competitive prices. The stigma that used to be attached to "vanity publishing" is fast disappearing as more and more authors cut out the middle man. You will have to handle the marketing, probably using a Web site, but distribution is no longer a great problem thanks to the Internet. An even simpler method is to publish your work as an e-book. Amazon offers a simple publishing service; the author need do little more than register, upload a Word file, proofread the result, and click the "publish" button. At the moment the author gets 70% of all revenue. Many authors' societies are starting to recommend this method of publication.

Assuming that you follow the traditional route, you will need to make contact with one or more publishers. An agent will know which publishers to approach and will generally make this process quicker and more effective. The agency fee (between 10% and 20%) is usually outweighed by improved royalty rates or advances, where the publisher pays an agreed sum that is set against subsequent royalties. The advance is not refundable provided the book appears in print, but you will not receive further payment until the book "earns out." Advice about finding an agent (and much else) can be obtained from authors' societies. In the absence of an agent, find out which publishers have recently produced similar types of books. Send an outline and perhaps a sample chapter, with a short covering letter. If the publisher is interested, an editor will reply, typically asking for further information. This may take a while; if so, be patient, but not if it is taking months.

If a publisher accepts your book, it will send a contract. It always looks official, but you should not hesitate to tear it to pieces and scribble all over it. By all means negotiate with the publisher about these changes, but if you do not like it, do not sign it. Read the contract carefully, even if your agent is supposed to do this for you. Delete any clauses that tie your hands on future books, especially "this will be the author's next book." A few publishers routinely demand an option on your next work: authors should equally routinely cross out that clause. Try to keep as many subsidiary rights (translations, electronic, serializations, and so on) as you can, but be aware that you may have to grant them all to the publisher before it agrees to accept your book. Some compromises are necessary.

### 7 Organization

Different authors write in different ways. Some set themselves a specific target of, say, 2000 words per day. Some write when they are in the mood and keep going until the feeling disappears. Some start at the front and work their way through the book page by page. Some jump in at random, writing whichever section appeals to them at the time, filling in the gaps later.

A plan, even if it is a page of headings, is almost indispensable. Planning a book in outline helps you decide what it is about and what should go in it. Popular science books usually tell a story, so it is a good idea to sort out what the main points of the story are going to be. However, most books evolve as they are being written, so you should think of your plan as a loose guide, not as a rigid constraint. Be willing to redesign the plan as you proceed. Your book will talk to you; listen to what it says.

To avoid writer's block—a common malady in which the same page is rewritten over and over again, getting worse every time, until the author grids to a halt, depressed—avoid rewriting material until you have completed a rough draft of the entire book. You will have a far better idea of how to rewrite chapter 1 when you have finished chapter 20. When you are about halfway through, the rest is basically downhill and the writing often gets easier. Do not put off that feeling by being needlessly finicky early on. Once you know you have a book, tidying it up and improving it becomes a pleasure. Terry Pratchett called this "scattering fairydust."

### 8 What Else You Will Have to Do

Your work is not finished when you submit the final manuscript (typescript, Word file, LaTeX file, whatever). It will be read by an editor, who may suggest broad changes: "move chapter 4 after chapter 6," "cut chapter 14 in half," "add two pages about the origin of topology," and so on. There will also be a copy editor, whose main job is to prepare the manuscript for the printer and to correct typographical, factual, and grammatical errors. Some of them may suggest minor changes: "Why not put in a paragraph telling readers what a Möbius band is?" Many will change your punctuation, paragraphing, and choice of words. If they seem to be overdoing this—this is happening when you feel they are imposing their own style, instead of preparing yours for the printer—ask them to stop and change everything back the way it was. It is your book.

After several months proofs will arrive, to be read and corrected. Often the time allowed will be short. The publisher will issue guidelines about this process, such as which symbols to use and which color of ink. (Many still work with paper copy. Others will send a PDF file, or a Word file with "track changes" enabled, and explain how they want you to respond.) Avoid making further changes to the text, unless absolutely necessary, even if you have just thought of a far better way to explain what cohomology is. You may be charged for excessive alterations that are not typesetting errors—see your contract, about 25 sections in.

With the proofs corrected, you might think that you have done your bit and can relax: you would be mistaken. Nowadays, books are released alongside a barrage of publicity material: podcasts, webcasts, blogs, tweets, and articles for printed media like *New Scientist* or newspapers. Your publisher's publicity department will try to get you on radio or television, and it will expect you to turn up for interviews if anyone bites. You may be asked to give public lectures, especially at literary festivals and science festivals. Your contract may oblige you to take part in such activities unless you have good reason not to or the demands become excessive. It is part of the job, so do not be surprised. If it seems to be getting out of hand, talk to your publisher. It won't be in the publisher's best interest to wear you out or take away too much time from writing another book it can publish.

Always wanted to write a book? Then do so. Just be aware of what you are letting yourself in for.

## VIII.4 Workflow

*Nicholas J. Higham*

Workflow refers to everything involved in producing a mathematical paper other than the actual research. It is about the practicalities of how to do things, including how best to use different kinds of software for different tasks. The characteristics of a good workflow are that it allows the end result to be achieved efficiently, repeatably, and in a way that allows easy recovery from mistakes.

## 1 Typesetting: TeX and LaTeX

In the days before personal computers, articles would be handwritten, then typed on a typewriter by a secretary, and ultimately typeset by a publisher. Nowadays, almost every author prepares the article herself or himself on a computer, and the publisher works from the author's files. In many areas of academia it is the custom to use Microsoft Word, or an open-source equivalent. In mathematics, computer science, and physics LaTeX has become the de facto standard.

TeX is a typesetting system invented by Donald Knuth in the late 1970s that has a particular strength in handling mathematics. LaTeX is a macro package, written originally by Leslie Lamport, that sits on top of TeX. A TeX or LaTeX file is an ASCII (plain text) file that contains commands that specify how the output is to be formatted, and it must be compiled to produce the final output (nowadays usually a Portable Document Format (PDF) file). This contrasts with a WYSIWYG ("what you see is what you get") word processor, such as Microsoft Word, that displays on the screen a representation of what the output will look like. TeX allows finer control than word processors (the latter are sometimes described as "what you see is all you get"), and the ability of LaTeX to use style files that set various typesetting parameters makes it very easy to adjust the format of an article to match a particular journal. LaTeX is also well suited to large projects such as books. Indeed, this volume is typeset in LaTeX, and the editors and production editor find it hard to imagine having produced the volume in any other way.

Figure 1 shows some LaTeX source code. Although how the code is formatted makes no difference to the output, it is good practice to make the source as readable as possible, with liberal use of spaces. I like to start new sentences on new lines, which makes it easier to cut and paste them during editing.

```
Polynomials are one of the simplest and most
familiar classes of functions and they find
wide use in applied mathematics.
A degree $n$ \py\
$$
    p_n(x) = a_0 + a_1 x + \cdots + a_n x^n
$$
is defined by its $n+1$ coefficients
$a_0,\dots,a_n \in \C$ (with $a_n \ne 0$).
```

**Figure 1** LaTeX source for part of THE LANGUAGE OF APPLIED MATHEMATICS [I.2 §14]. `\py` and `\C` are user-defined macros. `\cdots` and `\dots` are built-in TeX macros. Dollar signs delimit mathematics mode.

TeX and LaTeX are open-source software and are available in various distributions. In particular, the TeX Live distribution is available for Windows, Linux, and Mac systems (and as the augmented MacTeX for the latter).

How does one go about using LaTeX? There are two approaches. The first is to edit the LaTeX source in a general-purpose text editor such as Emacs, Vim, or a system-specific text editor. Ideally, the editor is customized so that its syntax highlights the LaTeX source, can directly compile the document, can pinpoint the location of compilation errors in the source, and can invoke a preview of the compiled document with two-way synchronization between the location of the cursor in the source and the page of the preview. I use Emacs together with the AUCTeX and RefTeX packages, which provides an extremely powerful LaTeX environment; indeed I use Emacs for all my editing tasks, ranging from programming to writing emails. A popular alternative is to use a program designed specifically for editing LaTeX documents, which typically comes with an integrated previewer. Such programs tend to be system specific.

TeX compiles to its own DVI (device independent) file format, which can then be translated into PostScript, a file format commonly used for printing. The standard format for distributing documents is now PDF. While PostScript can be converted to PDF, versions of TeX and LaTeX that compile directly to PDF are available (typically invoked as `pdftex` and `pdflatex`). Whether one is using the DVI-based or PDF-based versions of LaTeX affects how one generates graphics files for inclusion in figures. For DVI, included figures are typically in encapsulated PostScript format, whereas for PDF graphics files are typically in PDF or JPEG format and PostScript files are not allowed. I use a PDF workflow, though the *Companion* itself uses DVI and PostScript, because many of the figures in the book needed fine-tuning and this is more easily done in PostScript than in PDF. It is important to note that the Adobe Acrobat program is not suitable for use as a PDF previewer in the edit–compile–preview cycle as it does not refresh the view when a PDF file is updated on disk. Various open-source alternatives are available that do not have this limitation.

## 2 Preparing a Bibliography

A potentially time-consuming and error-prone part of writing a paper is preparing the bibliography, which

```
@book{knut86,
  author = "Donald E. Knuth",
  title = "The {\TeX\ book}",
  publisher = "Addison-Wesley",
  address = "Reading, MA, USA",
  year = 1986,
  pages = "ix+483",
  isbn = "0-201-13448-9"}
```

**Figure 2** An example of a BIBTEX bibliography entry.

contains the bibliographic details of the articles that are cited. In a LATEX document the bibliography entries are cited with a command of the form `\cite{smit65}`, where `smit65` is a key that uniquely specifies the entry in a `bibliography` environment that contains the item being cited. LATEX has a companion program called BIBTEX that extracts bibliography entries from a database contained in a `bib` file (an ASCII file of a special structure with a `.bib` extension) and automatically creates the `bibliography` environment. Figure 2 shows an example of a `bib` file entry. Using BIBTEX is a great time-saver and ensures accurate bibliographies, assuming that the `bib` file is accurate and kept up to date. Most journal Web sites allow BIBTEX entries for papers to be downloaded, so it is easy to build up a personal `bib` file. There exist open-source BIBTEX reference managers (such as JabRef) that facilitate creating and maintaining `bib` files.

A digital object identifier (DOI) is a character string that uniquely identifies an electronic document. It can be resolved into a uniform resource locator (URL) by preceding it with the string http://dx.doi.org/. Nowadays most papers (and many books) have DOIs and many older papers have been assigned DOIs. A DOI remains valid even if the location of the document changes, provided that the publisher updates the metadata. It is recommended to record DOIs in BIBTEX databases, and it is then possible with the use of a suitable BIBTEX style file and the LATEX `hyperref` package to include clickable links in a paper's bibliography (for example, from a paper's title).

## 3 Graphics

Mathematics papers often contain figures that plot functions, depict physical setups, or graph experimental results. These can be produced in many different ways. In a LATEX workflow one can generate a graphic outside LATEX and then include it as an external JPEG, PostScript, or PDF file or generate it from within LATEX. The most popular LATEX packages for graphics are TikZ and PGFPlots, which are built on top of the low-level primitives provided by the PGF (portable graphics format) package. Most of the figures in part I of this volume were generated using these packages. A major benefit of them is that they can incorporate LATEX commands and fonts, thus providing consistency with the main text. These powerful packages are not easy to use, but one can usually find an example online that provides a starting point for modification.

## 4 Version Control and Backups

Every good workflow contains procedures for making regular backups of files and recording a history of different versions of the files. Backups store one or more copies of the current version of key files on a separate disk or machine, so that a hard disk failure or the loss of a complete machine does not result in loss of files. Version control serves a different purpose, which is to record in a repository intermediate states of files so that authors can revert to an earlier version of a file or reinstate part of one. The use of the plural in "authors" refers to the fact that version control systems allow more than one user to contribute to a repository and allow any user to check out the latest versions of files. Although version control originated in software development, it is equally useful for documents. As long as the repository is kept on a different disk or machine, version control also provides a form of backup.

Of course a simple version control system is to regularly copy a file to another directory, renaming it with a version number (paper1.tex, paper2.tex, ...). However, this is tedious and error prone. A proper version control system keeps files in a database and stores only the lines that have changed between one version and the next. Popular version control systems include Git and Subversion (SVN). Although these are command based and can be difficult to learn, graphical user interfaces (GUIs) are available that simplify their usage.

Microsoft Word's "track changes" feature provides annotations of who made what changes to a document (and is a primitive and widely used form of version control). In LATEX a similar effect can be achieved by using the `latexdiff` command-line program provided with some LATEX distributions, which takes as input two different version of a LATEX file and produces a third LATEX file that marks up the differences between them; see figure 3 for an example.

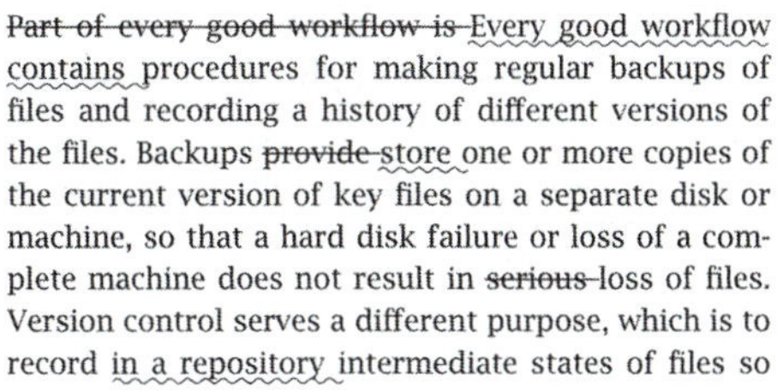
~~Part of every good workflow is~~ Every good workflow contains procedures for making regular backups of files and recording a history of different versions of the files. Backups ~~provide~~ store one or more copies of the current version of key files on a separate disk or machine, so that a hard disk failure or loss of a complete machine does not result in ~~serious~~ loss of files. Version control serves a different purpose, which is to record in a repository intermediate states of files so

**Figure 3** Example of output from `latexdiff`.

## 5 Computational Experiments

In papers that involve computational experiments, one needs to include figures or tables summarizing the results. It is typical that as a paper is developed the experiments are refined and repeated. One therefore needs an efficient way to regenerate tables and figures. Cutting and pasting the output of a program into the paper is not a good approach. It is much better to make the program output the results in a form that can be directly included in the paper (e.g., via an `\input` command in LaTeX). LITERATE PROGRAMMING [VII.11] techniques allow program code to be included within the source code for a paper, and they automate the running of the programs and the insertion of the results back into the paper source code.

## 6 Putting It All Together

Here are the things I do when I start to write a paper. I create a directory (folder) with a name that denotes the project in question. In that directory I copy a file `paper.tex` from a recent paper that I have written and use it as a template. I delete most of the content of `paper.tex` but keep the macros and some of the basic structural commands. I set up a repository in my version control system and commit `paper.tex` to it. I create a subdirectory for the computer programs I will write and a subdirectory named `figs` into which the PDF figures will be placed.

## 7 Presentations

As well as writing a paper about a piece of research, one may want to give a presentation about it in a seminar or at a conference. This will normally involve preparing slides or a poster, although it is still sometimes possible to give a blackboard talk. LaTeX has excellent tools for preparing slides and posters.

The Beamer class is the most widely used way to prepare slides in LaTeX. It can create overlays, allowing a slide to change dynamically (perhaps as an equation is built up, a piece at a time). Slide color and background, and elements such as a header (which may contain a mini-table of contents) and a footer, are all readily customized. Figure 4 shows an example slide.

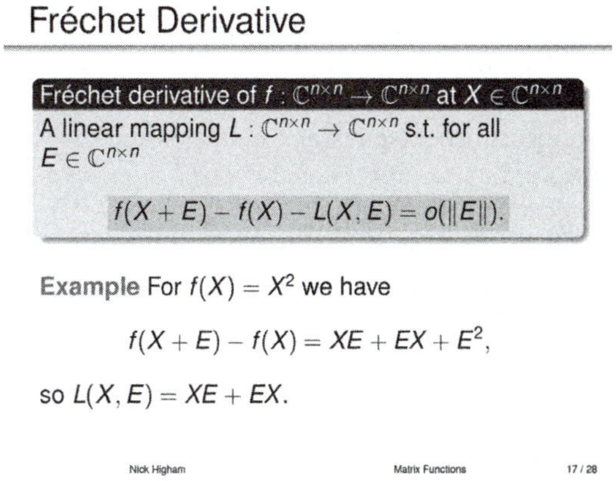


**Figure 4** A Beamer slide.

Various LaTeX packages are available for producing posters, at up to A0 paper size. A popular one is the beamerposter package built on Beamer.

## 8 Collaboration

In the early days of the Internet the most common way for authors to collaborate was to email documents back and forth. A regularly encountered problem was that Unix mailers would insert a greater than sign in front of any word "from" that appeared at the start of a line of a plain text message, so LaTeX files would often have stray > characters. For many people, email still serves as a useful mechanism for collaborative writing, but more sophisticated approaches are available. A file-hosting service such as Dropbox enables a group of users to share and synchronize a folder on their disks via the cloud. Version control based on a shared repository hosted on the Internet is the most powerful approach; it is widely used by programmers (e.g., on sites such as GitHub and SourceForge) and is increasingly popular with authors of papers.

## 9 Workflow for This Book

I wrote my articles using Emacs and TeX Live, with all files under version control with Git. I edited some of my figures in Adobe Photoshop. The production editor/typesetter, Sam Clark, used WinEdt and TeX Live with a

PostScript-based workflow, editing PostScript figures in Adobe Illustrator.

I produced a draft index for the articles that I authored using LaTeX indexing commands and the MakeIndex program. A professional indexer then expanded the index to cover the whole book.

The font used for this book is Lucida Bright, which has a full set of mathematical symbols that work well in TeX. It is from the same family as the Lucida Grande sans serif font that was used throughout the Mac OS X user interface up until version 10.9.

### Further Reading

Of the many good references on LaTeX I recommend Griffiths and D. J. Higham (1997) for a brief introduction and Kopka and Daly (2004) for a more comprehensive treatment. Knuth (1986) continues to be worth reading, even for those who use only LaTeX. Various aspects of workflow are covered in Higham (1998). Version control is best explored with the many freely available Web resources.

A good place to start looking for information about TeX and LaTeX is the Web site of the TeX Users Group, http://tug.org. A large collection of LaTeX packages is available at the Comprehensive TeX Archive Network (CTAN), http://www.ctan.org.

Griffiths, D. F., and D. J. Higham. 1997. *Learning LaTeX*. Philadelphia, PA: SIAM.

Higham, N. J. 1998. *Handbook of Writing for the Mathematical Sciences*, 2nd edn. Philadelphia, PA: SIAM.

Knuth, D. E. 1986. *The TeXbook*. Reading, MA: Addison-Wesley.

Kopka, H., and P. W. Daly. 2004. *Guide to LaTeX*, 4th edn. Boston, MA: Addison-Wesley.

## VIII.5 Reproducible Research in the Mathematical Sciences

*David L. Donoho and Victoria Stodden*

### 1 Introduction

Traditionally, mathematical research was conducted via mental abstraction and manual symbolic manipulation. Mathematical journals published theorems and completed proofs, while other sorts of evidence were gathered privately and remained in the shadows. For example, long after Riemann had passed away, historians discovered that he had developed advanced techniques for calculating the Riemann zeta function and that his formulation of the Riemann hypothesis—often depicted as a triumph of pure thought—was actually based on painstaking numerical work. In fact, Riemann's computational methods remained far ahead of what was available to others for decades after his death. This example shows that mathematical researchers have been "covering their (computational) tracks" for a long time.

Times have been changing. On the one hand, mathematics has grown into the so-called mathematical sciences, and in this larger endeavor, proposing new computational methods has taken center stage and documenting the behavior of proposed methods in test cases has become an important part of research activity (witness current publications throughout the mathematical sciences, including statistics, optimization, and computer science). On the other hand, even pure mathematics has been affected by the trend toward computational evidence; Tom Hales's brilliant article "Mathematics in the age of the Turing machine" points to several examples of important mathematical regularities that were discovered empirically and have driven much subsequent mathematical research, the Birch and Swinnerton-Dyer conjecture being his lead example. This conjecture posits deep relationships between the zeta function of elliptic curves and the rank of elliptic curves, and it was discovered by counting the number of rational points on individual elliptic curves in the early 1960s.

We can expect that, over time, an ever-increasing fraction of what we know about mathematical structures will be based on computational experiments, either because our work (in applied areas) is explicitly about the behavior of computations or because (in pure mathematics) the leading questions of the day concern empirical regularities uncovered computationally.

Indeed, with the advent of cluster computing, cloud computing, graphics processing unit boards, and other computing innovations, it is now possible for a researcher to direct overwhelming amounts of computational power at specific problems. With mathematical programming environments like Mathematica, MATLAB, and Sage, it is possible to easily prototype algorithms that can then be quickly scaled up using the cloud. Such direct access to computational power is an irresistible force. Reflect for a moment on the fact that the Birch and Swinnerton-Dyer conjecture was discovered using the rudimentary computational resources of the early 1960s. Research in the mathematical sciences can now be dramatically more ambitious in scale and

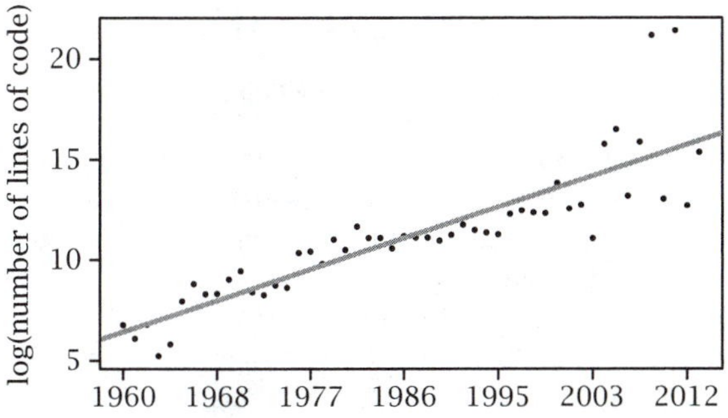


**Figure 1** The number of lines of code published in *ACM Transactions on Mathematical Software*, 1960–2012, on a log scale. The proportion of articles that published code remained roughly constant at about a third, with standard error of about 0.12, and the journal consistently published around thirty-five articles each year.

scope. This opens up very exciting possibilities for discovery and exploration, as explained in EXPERIMENTAL APPLIED MATHEMATICS [VIII.6].

The expected scaling up of experimental and computational mathematics is, at the same time, problematic. Much of the knowledge currently being generated using computers is not of the same quality as traditional mathematical knowledge. Mathematicians are very strict and demanding when it comes to understanding the basis of a theorem, the assumptions used, the prior theorems on which it depends, and the chain of inference that establishes the theorem. As it stands, the way in which evidence based on computations is typically published leaves "a great deal to the imagination," and computational evidence therefore simply does not have the same epistemological status as a rigorously proved theorem.

Algorithms are becoming ever more complicated. Figure 1 shows the number of lines of code published in the journal *ACM Transactions on Mathematical Software* from 1960 to 2012. The number of lines has increased exponentially, from 875 in 1960 to nearly 5 million in 2012, including libraries. The number of articles in the journal that contain code has been roughly constant; individual algorithms are requiring ever more code, even though modern languages are ever more expressive.

Algorithms are also being combined in increasingly complicated processing pipelines. Individual algorithms of the kind that have traditionally been documented in journal articles increasingly represent only a small fraction of the code making up a computational science project. Scaling up projects to fully exploit the potential of modern computing resources requires complex workflows to pipeline together numerous algorithms, with problems broken into pieces and farmed out to be run on numerous processors and the results harvested and combined in project-specific ways. As a result, a given computational project may involve much infrastructure not explicitly described in journal articles. In that environment, journal articles become simply *advertisements*: pointers to a complex body of software development, experimental outcomes, and analyses, in which there is really no hope that "outsiders" can understand the full meaning of those summaries.

The computational era seems to be thrusting the mathematical sciences into a situation in which mathematical knowledge in the wide sense, also including solidly based empirical discoveries, is broader and more penetrating but far less transparent and far less "common property" than ever. Individual researchers report that over time they are becoming increasingly uncertain about what other researchers have done and about the strength of evidence underlying the results those other researchers have published.

The phrase *mathematical sciences* contains a key to improving the situation. The traditional laboratory sciences evolved, over hundreds of years, a set of procedures for enabling the *reproducibility of findings* in one laboratory by other laboratories. As the mathematical sciences evolve toward ever-heavier reliance on computation, they should likewise develop a discipline for documenting and sharing algorithms and empirical mathematical findings. Such a disciplined approach to scholarly communication in the mathematical sciences offers two advantages: it promotes scientific progress, and it resolves uncertainties and controversies that spread a "fog of uncertainty."

## 2 Reproducible Research

We fully expect that in two decades there will be widely accepted standards for communication of findings in computational mathematics. Such standards are needed so that computational mathematics research can be used and believed by others.

The raw ingredients that could enable such standards seem to already be in place today. *Problem solving environments* (PSEs) like MATLAB, R, IPython, Sage, and Mathematica, as well as open-source operating systems and software, now enable researchers to share their code and data with others. While such sharing is not nearly as common as it should be, we expect that it soon will be.

In a 2006 lecture, Randall J. LeVeque described well the moment we are living through. On the one hand, many computational mathematicians and computational scientists do not work reproducibly:

> Even brilliant and well-intentioned computational scientists often do a poor job of presenting their work in a reproducible manner. The methods are often very vaguely defined, and even if they are carefully defined they would normally have to be implemented from scratch by the reader in order to test them. Most modern algorithms are so complicated that there is little hope of doing this properly.

On the other hand, LeVeque continues, the ingredients exist:

> The idea of "reproducible research" in scientific computing is to archive and make publicly available all of the codes used to create the figures or tables in a paper in such a way that the reader can download the codes and run them to reproduce the results. The program can then be examined to see exactly what has been done. The development of very high level programming languages has made it easier to share codes and generate reproducible research.... These days many algorithms can be written in languages such as MATLAB in a way that is both easy for the reader to comprehend and also executable, with all details intact.

While the technology needed for reproducible research exists today, mathematical scientists do not yet agree on exactly how to use this technology in a disciplined way. At the time of writing, there is a great deal of activity to define and promote standards for reproducible research in computational mathematics.

A number of publications address reproducibility and verification in computational mathematics; topics covered include computational scale and proof checking, probabilistic model checking, verification of numerical solutions, standard methods in uncertainty quantification, and reproducibility in computational research. This is not an exhaustive account of the literature in these areas, of course, merely a starting point for further investigation.

In this article we review some of the available tools that can enable reproducible research and conclude with a series of "best-practice" recommendations based on modern examples and research methods.

## 3 Script Sharing Based on PSEs

### 3.1 PSEs Offer Power and Simplicity

A key precondition for reproducible computational research is the ability for researchers to run the code that generated results in some published paper of interest. Traditionally, this has been problematic. Researchers were often unprepared or unwilling to share code, and even if they did share it, the impact was minimal as the code depended on a specific computational environment (hardware, operating system, compiler, etc.) that others could not access.

PSEs like R, Mathematica, and MATLAB have, over the last decade, dramatically simplified and uniformized much computational science.

Each PSE offers a high-level language for describing computations, often a language that is very compatible with standard mathematical notation. PSEs also offer graphics capabilities that make it easy to produce often quite sophisticated figures for inclusion in research papers. The researcher is gaining extreme ease of access to fundamental capabilities like matrix algebra, symbolic integration and optimization, and statistical model fitting; in many cases, a whole research project, involving a complex series of variations on some basic computation, can be encoded in a few compact command scripts.

The popularity of this approach to computing is impressive. Figure 2 shows that the PSEs with the most impact on research (by number of citations) are the commercial closed-source packages Mathematica and MATLAB, which revolutionized technical computing in the 1980s and 1990s. However, these systems are no longer rapidly growing in impact, while the recent growth in popularity of R and Python is dramatic.

### 3.2 PSEs Facilitate Reproducibility

As LeVeque pointed out in the quote above, a side effect of the power and compactness of coding in PSEs is that reproducible research becomes particularly straightforward, as the original researcher can supply some simple command scripts to interested researchers, who can then rerun the experiment or variations of it privately in their own local instances of the relevant PSE.

In some fields, authors of research papers are already heavily committed to a standard of reproducing results in published papers by sharing PSE scripts. In statistics, for example, papers often seek to introduce new tools

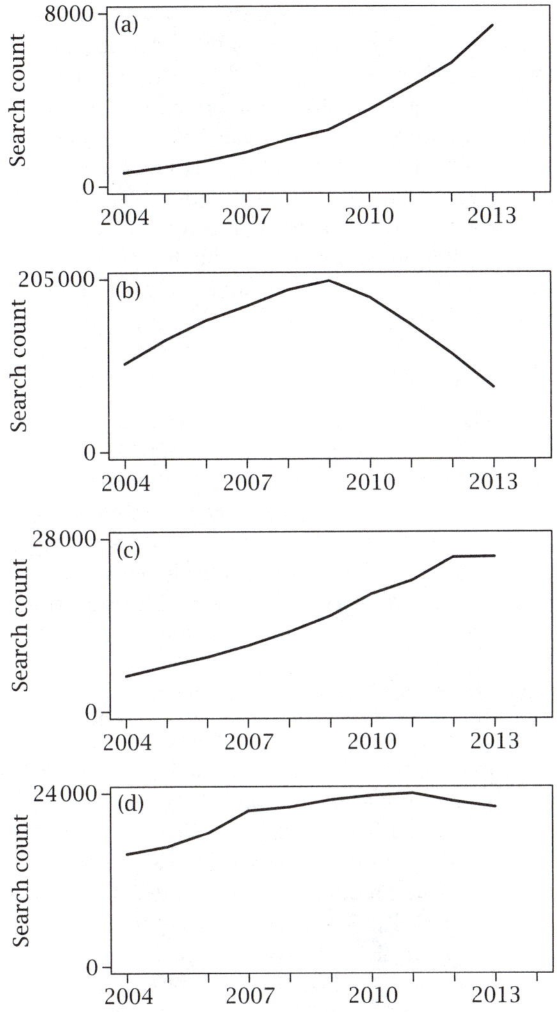


**Figure 2** The total number of hits on Google Scholar for each of the four search terms: (a) "R software", (b) MATLAB, (c) Python, and (d) Mathematica. The search was carried out for each year in the decade 2004–13. Note that the $y$-axes are on different scales to show the increase or decrease in software use over time. R and Python are open source, whereas MATLAB and Mathematica are not.

that scientists can apply to their data. Many authors would like to increase the visibility and impact of such methodological papers and are persuaded that a good way to do this is to make it as easy as possible for users to try the newly proposed tools. Traditional theoretical statistics journal papers might be able to expect citations in the single or low double digits; there are numerous recent examples of articles that were supplemented by easy access to code and that obtained hundreds of readers and citations. It became very standard for authors in statistics to offer access to code using *packages* in one specific PSE, R. To build such a package, authors document their work in a standard LaTeX format and bundle up the R code and documentation in a defined package structure. They post their package on CRAN, the Comprehensive R Archive Network. All R users can access the code from within R by simple invocations (`require("package_name")`) that direct R to locate, download, and install the package from CRAN. This process takes only seconds. Consequently, all that a user needs to know today to begin applying a new methodology is the name of the package. CRAN offered 5519 packages as of May 8, 2014. A side effect of authors making their methodology available in order to attract readers is, of course, that results in their original articles may become easily reproducible.[1]

### 3.3 Notebooks for Sharing Results

A notebook interface to a PSE stores computer instructions alongside accompanying narrative, which can include mathematical expressions, and allows the user to execute the code and store the output, including figures, all in one document. Because all the steps leading to the results are saved in a single file, notebooks can be shared online, which provides a way to communicate reproducible computational results.

The Jupyter Notebook (formerly known as the IPython Notebook), provides an interface to back-end computations, for example in Python or R, that displays code and output, including figures, with LaTeX used to typeset mathematical notation (see figure 3). A Jupyter Notebook permits the researcher to track and document the computational steps that generate results and can be shared with others online using `nbviewer` (see http://nbviewer.ipython.org).

## 4 Open-Source Software: A Key Enabler

PSEs and notebook interfaces are having a very substantial effect in promoting reproducibility, but they have their limits. They make many research computations

1. In fields like statistics, code alone is not sufficient to reproduce published results. Computations are performed on data sets from specific scientific projects; the data may result from experiments, surveys, or costly measurements. Increasingly, data repositories are being used by researchers to share such data across the Internet. Since 2010, arXiv has partnered with Data Conservancy to facilitate external hosting of data associated with publications uploaded to arXiv (see, for example, http://arxiv.org/abs/1110.3649v1, where the data files are accessible from the paper's arXiv page). Such practices are not yet widespread, but they are occurring with increasing frequency.

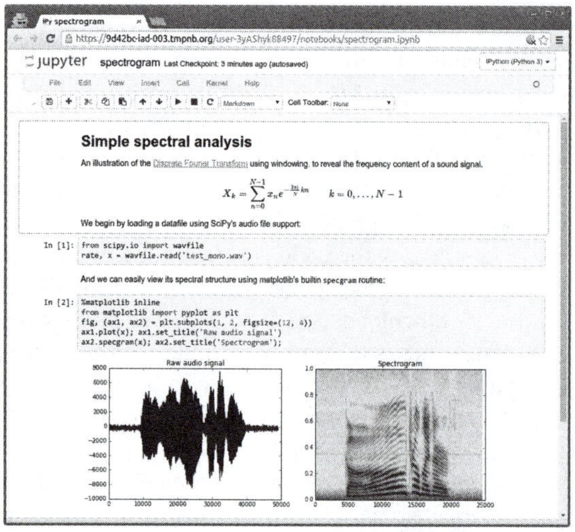


**Figure 3** A snapshot of the interactive Jupyter Notebook.

convenient and easy to share with others, but ambitious computations often demand more capability than they can offer. Historically, this would have meant that ambitious projects had to be idiosyncratically coded and difficult to export to new computing environments.

The open-source revolution has largely changed this. Today, it is often possible to develop all of an ambitious computational project using code that is freely available to others. Moreover, this code can be hosted on an open-source operating system (Linux) and run within a standard virtual machine that hides hardware details. The open-source "spirit" also makes researchers more open to sharing code; attribution-only open-source licenses may also allow them to do this while retaining some assurance that the shared code will not be misappropriated.

Several broad classes of software are now being shared in ways that we describe in this section. These various classes of software are becoming, or have already become, part of the standard approaches to reproducible research.

## 4.1 Fundamental Algorithms and Packages

In table 1 we consider some of the fundamental problems that underly modern computational mathematics, such as FAST FOURIER TRANSFORMS [II.10], LINEAR EQUATIONS [IV.10], and NONLINEAR OPTIMIZATION [IV.11], and we give examples of some of the many families of open-source codes that have become available for enabling high-quality mathematical computation. The table includes the packages' inception dates, their current release numbers, and the total numbers of citations that the packages have garnered since inception.[2] The different packages within each section of the table may offer very different approaches to the same underlying problem. As the reader can see, a staggering amount of basic functionality is being developed worldwide by many teams and authors in particular subdomains, and it is being made available for broad use. The citation figures in the table testify to the significant impact these enablers are having on published research.

## 4.2 Specialized Systems

The packages tabulated in table 1 are broadly useful in computational mathematics; it is perhaps not surprising that developers would rise to the challenge of creating such broadly useful tools. We have been surprised to see the rise of systems that attack very specific problem areas and offer extremely powerful environments to formulate and solve problems in those narrow domains. We give three examples.

### *4.2.1 Hyperbolic Partial Differential Equations (PDEs)*

Clawpack is an open-source software package designed to compute numerical solutions to hyperbolic PDEs using a wave propagation approach. According to the system's lead author, Randall J. LeVeque, "the development and use of the Clawpack software implementing [high-resolution finite-volume methods for solving hyperbolic PDEs] serves as a case study for a more general discussion of mathematical aspects of software development and the need for more reproducibility in computational research."

The package has been used in the creation of reproducible mathematical research. For example, the figures for LeVeque's book *Finite Volume Methods for Hyperbolic Problems* were generated using Clawpack; instructions are provided for recreating those figures.

Clawpack is now a framework that offers numerous extensions including PyClaw (with a Python interface to a number of advanced capabilities) and GeoClaw (developed for TSUNAMI MODELING [V.19] and the modeling of other geophysical flows). Open-source software practices have apparently enabled not only reproducibility but also code extension and expansion into new areas.

2. The data for citation counts was collected via Google Scholar in August 2013. Note that widely used packages such as LAPACK, FFTW, ARPACK, and Suitesparse are built into other software (e.g., MATLAB), which do not generate citations for them directly.

**Table 1** Software for some fundamental problems underlying modern computational mathematics.

| Package | Year of inception | Current release | Citations |
|---|---|---|---|
| *Dense linear algebra* | | | |
| LAPACK | 1992 | 3.4.2 | 7600 |
| JAMA | 1998 | 1.0.3 | 129 |
| IT++ | 2006 | 4.2 | 14 |
| Armadillo | 2010 | 3.900.7 | 105 |
| EJML | 2010 | 0.23 | 22 |
| Elemental | 2010 | 0.81 | 51 |
| *Sparse-direct solvers* | | | |
| SuperLU | 1997 | 4.3 | 317 |
| MUMPS | 1999 | 4.10.0 | 2029 |
| Amesos | 2004 | 11.4 | 104 |
| PaStiX | 2006 | 5.2.1 | 114 |
| Clique | 2010 | 0.81 | 12 |
| *Krylov-subspace eigensolvers* | | | |
| ARPACK | 1998 | 3.1.3 | 2624 |
| SLEPc | 2002 | 3.4.1 | 293 |
| Anasazi | 2004 | 11.4 | 2422 |
| PRIMME | 2006 | 1.1 | 61 |
| *Fourier-like transforms* | | | |
| FFTW | 1997 | 3.3.3 | 1478 |
| P3DFFT | 2007 | 2.6.1 | 14 |
| DIGPUFFT | 2011 | 2.4 | 17 |
| DistButterfly | 2013 | | 27 |
| PNFFT | 2013 | | 215 |
| *Fast multipole methods* | | | |
| KIFMM3d | 2003 | | 1780 |
| Puma-EM | 2007 | 0.5.7 | 32 |
| PetFMM | 2009 | | 29 |
| GemsFMM | 2010 | | 16 |
| ExaFMM | 2011 | | 28 |
| *PDE frameworks* | | | |
| PETSc | 1997 | 3.4 | 2695 |
| Cactus | 1998 | 4.2.0 | 669 |
| deal.II | 1999 | 8.0 | 576 |
| Clawpack | 2001 | 4.6.3 | 131 |
| Hypre | 2001 | 2.9.0 | 384 |
| libMesh | 2003 | 0.9.2.1 | 260 |
| Trilinos | 2003 | 11.4 | 3483 |
| Feel++ | 2005 | 0.93.0 | 405 |
| Lis | 2005 | 1.4.11 | 29 |
| *Finite-element analysis* | | | |
| Code Aster | | 11.4.03 | 48 |
| CalculiX | 1998 | 2.6 | 69 |
| deal.II | 1999 | 8.0 | 576 |
| DUNE | 2002 | 2.3 | 325 |
| Elmer | 2005 | 6.2 | 97 |
| FEniCS Project | 2009 | 1.2.0 | 418 |
| FEBio | 2010 | 1.6.0 | 32 |

**Table 1** (*Continued.*)

| Package | Year of inception | Current release | Citations |
|---|---|---|---|
| *Optimization* | | | |
| MINUIT/MINUIT2 | 2001 | 94.1 | 2336 |
| CUTEr | 2002 | r152 | 1368 |
| IPOPT | 2002 | 3.11.2 | 1517 |
| CONDOR | 2005 | 1.11 | 1019 |
| OpenOpt | 2007 | 0.50.0 | 24 |
| ADMB | 2009 | 11.1 | 175 |
| *Graph partitioning* | | | |
| Scotch | 1992 | 6.0.0 | 435 |
| ParMeTIS | 1997 | 4.0.3 | 4349 |
| kMeTIS | 1998 | 1.5.3 | 3449 |
| Zoltan-HG | 2008 | r362 | 125 |
| KaHIP | 2011 | 0.52 | 71 |
| *Adaptive mesh refinement* | | | |
| AMRClaw | 1994 | 4.6.3 | 4800 |
| PARAMESH | 1999 | 4.1 | 409 |
| SAMRAI | 1998 | | 185 |
| Carpet | 2001 | 4 | 579 |
| BoxLib | 2000 | | 155 |
| Chombo | 2000 | 3.1 | 198 |
| AMROC | 2003 | 1.1 | 342 |
| p4est | 2007 | 0.3.4.1 | 227 |

### 4.2.2 *Parabolic and Elliptic PDEs: DUNE*

The Distributed and Unified Numerics Environment (DUNE) is an open-source modular software toolbox for solving PDEs using grid-based methods. It was developed by Mario Ohlberger and other contributors and supports the implementation of methods such as finite elements, finite volumes, finite differences, and discontinuous Galerkin methods.

DUNE was envisioned to permit the integrated use of both legacy libraries and new ones. The software uses modern C++ programming techniques to enable very different implementations of the same concepts (i.e., grids, solvers, linear algebra, etc.) using a common interface with low overhead, meaning that DUNE prioritizes efficiency in scientific computations and supports high-performance computing applications. DUNE has a variety of downloadable modules including various grid implementations, linear algebra solvers, quadrature formulas, shape functions, and discretization modules.

DUNE is based on several main principles: the separation of data structures and algorithms by abstract interfaces, the efficient implementation of these interfaces using generic programming techniques, and reuse of existing finite-element packages with a large body of functionality. The finite-element codes UG, ALBERTA, and ALUGrid have been adapted to the DUNE framework, showing the value of open-source development not only for reproducibility but for acceleration of discovery through code reuse.[3]

### 4.2.3 Computer-Aided Theorem Proving

COMPUTER-AIDED THEOREM PROVING [VII.3] has made extremely impressive strides in the last decade. This progress ultimately rests on the underlying computational tools that are openly available and that a whole community of researchers is contributing to and using. Indeed, one can only have justified belief in a computationally enabled proof with transparent access to the underlying technology and broad discussion.

There are, broadly speaking, two approaches to computer-aided theorem-proving tools in experimental mathematics. The first type encompasses machine-human collaborative proof assistants and interactive theorem-proving systems to verify mathematics and computation, while the second type includes automatic proof checking, which occurs when the machine verifies previously completed human proofs or conjectures.

Interactive theorem-proving systems include coq, Mizar, HOL4, HOL Light, Isabelle, LEGO, ACL2, Veritas, NuPRL, and PVS. Such systems have been used to verify the four-color theorem and to reprove important classical mathematical results. Thomas Hales's Flyspeck project is currently producing a formal proof of the Kepler conjecture, using HOL Light and Isabelle. The software produces machine-readable code that can be reused and repurposed into other proof efforts. Examples of open-source software for automatic theorem proving include E and Prover9/Mace 4.

## 5 Scientific Workflows

Highly ambitious computations today often go beyond single algorithms to combine different pieces of software in complex pipelines. Moreover, modern research often considers a whole pipeline as a single object of study and makes experiments varying the pipeline itself. Experiments involving many moving parts that must be combined to produce a complete result are often called *workflows.*

Kepler is an open-source project structured around *scientific workflows*: "an executable representation of the steps required to generate results," or the capture of experimental details that permit others to reproduce computational findings. Kepler provides a graphical interface that allows users to create and share these workflows. An example of a Kepler workflow is given in figure 4, solving a model of two coupled differential equations and plotting the output. Kepler maintains a component repository where workflows can be uploaded, downloaded, searched, and shared with the community or designated users, and it contains a searchable library with more than 350 processing components. Kepler operates on data stored in a variety of formats, locally and over the Internet, and can merge software from different sources such as R scripts and compiled C code by linking in their inputs and outputs to perform the desired overall task.

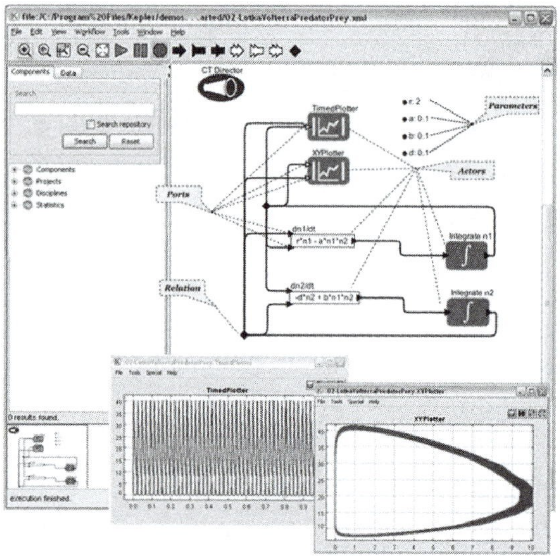


**Figure 4** An example of the Kepler interface, showing a workflow solving the classic Lotka–Volterra predator–prey dynamics model.

## 6 Dissemination Platforms

Dissemination platforms are Web sites that serve specialized content to interested visitors. They offer an interesting method for facilitating reproducibility; we describe here the Image Processing OnLine (IPOL) project and ResearchCompendia.org.

IPOL is an open-source journal infrastructure developed in Python that publishes relevant image-processing and image-analysis algorithms. The journal peer reviews article contributions, including code, and

3. See also FEniCS (http://fenicsproject.org) for another example of an open-source finite-element package.

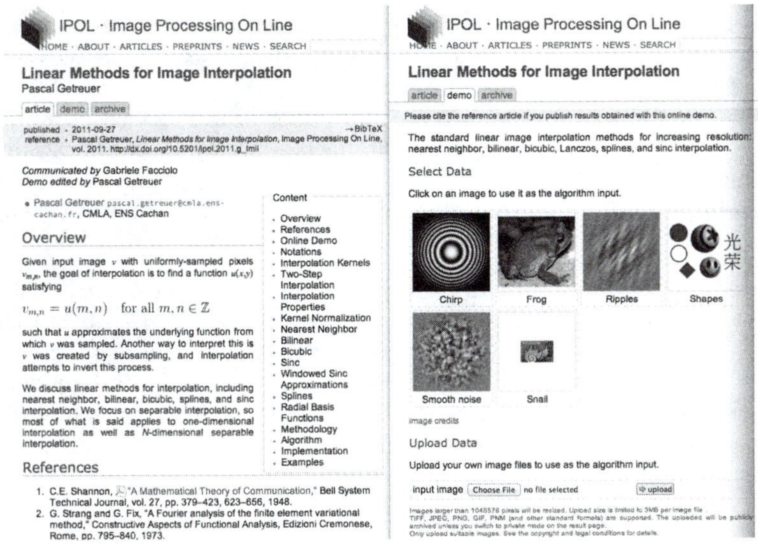

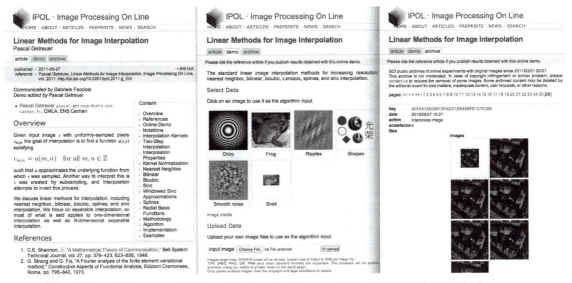

**Figure 5** An example IPOL publication. The three panels from left to right include the manuscript, the cloud-executable demo, and the archive of all previous executions.

publishes accepted papers in a standardized format that includes

- a manuscript containing a detailed description of the algorithm, its bibliography, and documented examples;
- a downloadable software implementation of the algorithm;
- an online demo, where the algorithm can be tested on data sets, for example images, uploaded by the users; and
- an archive containing a history of the online experiments.

Figure 5 displays these components for a sample IPOL publication.

ResearchCompendia, which one of the authors is developing, is an open-source platform designed to link the published article with the code and data that generated the results. The idea is based on the notion of a "research compendium": a bundle including the article and the code and data needed to recreate the findings. For a published paper, a Web page is created that links to the article and provides access to code and data as well as metadata, descriptions, and documentation, and code and data citation suggestions. Figure 6 shows an example compendium page.

ResearchCompendia assigns a Digital Object Identifier (DOI) to all citable objects (code, data, compendium page) in such a way as to enable bidirectional

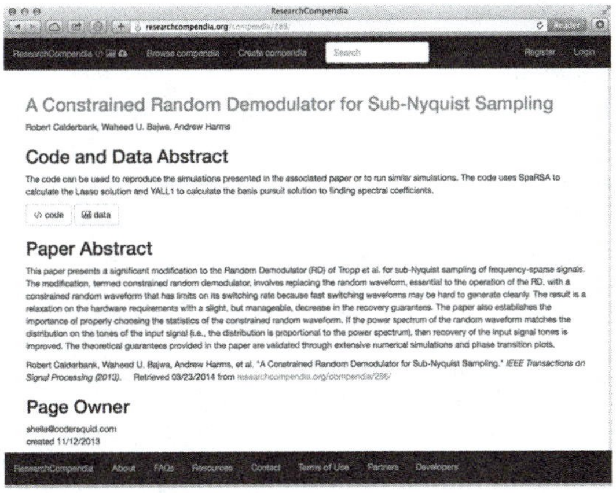


**Figure 6** An example compendium page on ResearchCompendia.org. The page links to a published article and provides access to the code and data that generated the published results.

linking between related digital scholarly objects, such as the publication and the data and code that generated its results (see www.stm-assoc.org/2012_06_14_STM_DataCite_Joint_Statement.pdf). DOIs are well-established and widely used unique persistent identifiers for digital scholarly objects. There are other PSE-independent methods of sharing such as via GitHub (which can now assign DOIs to code: https://guides.github.com/activities/citable-code) and via supplementary materials on journal Web sites. A DOI is affixed to a certain version of software or data that generates a certain set of results. For this reason, among others, VERSION CONTROL [VIII.4 §4] for scientific codes and data is important for reproducibility.[4]

## 7 Best Practices for Reproducible Computational Mathematics

Best practices for communicating computational mathematics have not yet become standardized. The workshop "Reproducibility in Computational and Experimental Mathematics", held at the Institute for Computational and Experimental Research in Mathematics (ICERM) at Brown University in 2012, recommended the following for every paper in computational mathematics.

- A precise statement of assertions made in the paper.
- A statement of the computational approach and why it constitutes a rigorous test of the hypothesized assertions.
- Complete statements of, or references to, every algorithm employed.
- Salient details of auxiliary software (both research and commercial software) used in the computation.
- Salient details of the test environment, including hardware, system software, and the number of processors utilized.
- Salient details of data-reduction and statistical-analysis methods.
- Discussion of the adequacy of parameters such as precision level and grid resolution.
- A full statement (or at least a valid summary) of experimental results.
- Verification and validation tests performed by the author(s).
- Availability of computer code, input data, and output data, with some reasonable level of documentation.
- Curation. Where are code and data available? With what expected persistence and longevity? Is there a site for future updates, e.g., a version control repository of the code base?
- Instructions for repeating computational experiments described in the paper.
- Terms of use and licensing. Ideally code and data "default to open," i.e., a permissive reuse license, if nothing opposes it.
- Avenues of exploration examined throughout development, including information about negative findings.
- Proper citation of all code and data used, including that generated by the authors.

These guidelines can, and should, be adapted to different research contexts, but the goal is to provide readers with the information (such as metadata including parameter settings and workflow documentation), data, and code they require to independently verify computational findings.

## 8 The Outlook

The recommendations of the ICERM workshop listed in the previous section are the least we would hope for today. They commendably propose that authors give enough information for readers to understand at some high level what was done.

They do not actually require sharing of all code and data in a form that allows precise reexecution and reproduction of results, and as such, the recommendations are very far from where we hope to be in twenty years.

One can envision a day when every published research document will be truly reproducible in a deep sense, where others can repeat published computations utterly mechanically. The reader of such a reproducible research article would be able to deeply study any specific figure, for example, viewing the source code and data that underlie a figure, recreating the original figure from scratch, examining input parameters that define this particular figure, and even changing their settings in order to study the effect on the resulting figure.

Reproducibility at this ambitious level would enable more than just individual understanding; it would enable metaresearch. Consider the "dream applications" mentioned in Gavish and Donoho (2012), where robots automatically crawl through, reproduce, and vary research results. Reproducible work can be automatically extended and generalized; it can be optimized, differentiated, extrapolated, and interpolated. A reproducible data analysis can be statistically bootstrapped to automatically place confidence statements on the whole analysis.

Coming back down to earth, what is likely to happen in the near future? We confidently predict increasing

4. Other reasons include good coding practices enabling reuse, assigning explicit credit for bug fixing and code extensions or applications, efficiency in code organization and development, and the ability to join collaborative coding communities such as GitHub.

computational transparency and increasing computational reproducibility in coming years. We imagine that PSEs will continue to be very popular and that authors will increasingly share their scripts and data, if only to attract readership. Specialized platforms like Clawpack and DUNE will come to be seen as standard platforms for whole research communities, who will naturally then be able to reproduce work in those areas. We expect that as the use of cloud computing grows and workflows become more complex, researchers will increasingly document and share the workflows that produce their most ambitious results. We expect that code will be developed on common platforms and will be stored in the cloud, enabling the code to run for many years after publication.

We expect that over the next two decades such practices will become standard and will be based on tools of the kind discussed in this article. The direction of increasing transparency and increasing sharing seem clear, but it is still unclear which combinations of tools and approaches will come to be standard.

**Acknowledgments.** We would like to thank Jennifer Seiler for outstanding research assistance.

### Further Reading

Bailey, D. H., J. Borwein, and V. Stodden. 2013. Set the default to "open". *Notices of the American Mathematical Society* 60(6):679–80.

Buckheit, J., and D. Donoho. 1995. Wavelab and reproducible research. In *Wavelets and Statistics*, edited by A. Antoniadis, pp. 55–81. New York: Springer.

Donoho, D., A. Maleki, M. Shahram, I. Ur Rahman, and V. Stodden. 2009. Reproducible research in computational harmonic analysis. *IEEE Computing in Science & Engineering* 11(1):8–18.

Gavish, M., and D. Donoho. 2012. Three dream applications of verifiable computational results. *Computing in Science and Engineering* 14(4):26–31

Gentleman, R., and D. Temple Lang. 2007. Statistical analyses and reproducible research. *Journal of Computational and Graphical Statistics* 16:1–23.

Hales, T. C. 2014. Mathematics in the age of the Turing machine. In *Turing's Legacy: Developments from Turing's Ideas in Logic*, edited by R. Downey. Cambridge: Cambridge University Press/Association for Symbolic Logic.

LeVeque, R. J. 2006. Wave propagation software, computational science, and reproducible research. In *Proceedings of the International Congress of Mathematicians, Madrid, 2006*, volume 3. Zurich: European Mathematical Society.

Stodden, V. 2009a. Enabling reproducible research: licensing for scientific innovation. *International Journal of Communications Law and Policy* 13:1–25.

——. 2009b. The legal framework for reproducible scientific research: licensing and copyright. *IEEE Computing in Science & Engineering* 11(1):35–40.

Stodden, V., P. Guo, and Z. Ma. 2013. Toward reproducible computational research: an empirical analysis of data and code policy adoption by journals. *PLoS ONE* 8(6):e67111.

## VIII.6 Experimental Applied Mathematics

*David H. Bailey and Jonathan M. Borwein*

### 1 Introduction

"Experimental applied mathematics" is the name given to the use of modern computer technology as an active agent of research. It is used for gaining insight and intuition, for discovering new patterns and relationships, for testing conjectures, and for confirming analytically derived results, in much the same spirit that laboratory experimentation is employed in the physical sciences. It is closely related to what is known as "experimental mathematics" in pure mathematics, as has been described elsewhere, including in *The Princeton Companion to Mathematics*.

In one sense, most applied mathematicians have for decades aggressively integrated computer technology into their research. What is meant here is computationally assisted applied mathematical research that features one or more of the following characteristics:

(i) computation for exploration and discovery;
(ii) symbolic computing;
(iii) high-precision arithmetic;
(iv) integer relation algorithms;
(v) graphics and visualization;
(vi) connections with nontraditional mathematics.

Depending on the context, the role of rigorous proof in experimental applied mathematics may be either much reduced or unchanged from that of its pure sister. There are many complex applied problems for which there is little point in proving the validity of a minor component rather than finding strong evidence for the appropriateness of the general method.

#### *High-Precision Arithmetic*

Most work in scientific or engineering computing relies on either 32-bit IEEE FLOATING-POINT ARITHMETIC [II.13] (roughly 7-decimal-digit precision) or 64-bit IEEE floating-point arithmetic (roughly 16-decimal-digit precision). But for an increasing body of applied mathematical studies, even 16-digit arithmetic is not sufficient. The most common form of high-precision arithmetic is "double-double" or "quad" precision, which is equivalent to roughly 31-digit precision. Other studies require hundreds or thousands of digits.

Algorithms for performing arithmetic and evaluating common transcendental functions with high-precision data structures have been known for some time, although challenges remain. Mathematical software packages such as Maple and Mathematica typically include facilities for arbitrarily high precision, but for some applications researchers rely on Internet-available software, such as the GNU multiprecision package.

*Integer Relation Detection*

Given a vector of real or complex numbers $x_i$, an *integer relation algorithm* attempts to find a nontrivial set of integers $a_i$ such that $a_1x_1 + a_2x_2 + \cdots + a_nx_n = 0$. One common application of such an algorithm is to find new identities involving computed numeric constants.

For example, suppose one suspects that an integral (or any other numerical value) $x_1$ might be a linear sum of a list of terms $x_2, x_3, \dots, x_n$. One can compute the integral and all the terms to high precision (typically several hundred digits) and then provide the vector $(x_1, x_2, \dots, x_n)$ to an integer relation algorithm. It will either determine that there is an integer-linear relation among these values, or it will provide a lower bound on the Euclidean norm of any integer relation vector $(a_i)$ that the input vector might satisfy. If the algorithm does produce a relation, then solving it for $x_1$ produces an experimental identity for the original integral. The most commonly employed integer relation algorithm is the "PSLQ" algorithm of mathematician-sculptor Helaman Ferguson, although the Lenstra–Lenstra–Lovasz algorithm can also be adapted for this purpose.

## 2 Historical Examples

The best way to clarify what is meant by experimental applied mathematics is to show some examples of the paradigm in action.

*Gravitational Boosting*

One interesting space-age example is the unexpected discovery of *gravitational boosting* by Michael Minovitch at NASA's Jet Propulsion Laboratory in 1961. Minovitch described how he discovered that *Hohmann transfer ellipses* were not, as was then believed, the minimum-energy way to reach the outer planets. Instead, he discovered computationally that spacecraft orbits that pass close to other planets could gain a substantial boost in speed (compensated by an extremely small change in the orbital velocity of the planet) on their way to a distant location via a "slingshot effect." Until this demonstration, "most planetary mission designers considered the gravity field of a target planet to be somewhat of a nuisance, to be cancelled out, usually by onboard Rocket thrust."

Without such a boost from Jupiter, Saturn, and Uranus, the Voyager mission would have taken more than 30 years to reach Neptune; instead, Voyager reached Neptune in only 10 years. Indeed, without gravitational boosting we would still be waiting! We would have to wait much longer still for Voyager to leave the solar system, as it now appears to be doing.

*Fractals and Chaos*

One prime example of twentieth-century applied experimental mathematics is the development of *fractal theory*, as exemplified by the works of Benoit Mandelbrot. Mandelbrot studied numerous examples of fractal sets, many of them with direct connections to nature. Applications include analyses of the shapes of coastlines, mountains, biological structures, blood vessels, galaxies, even music, art, and the stock market. For example, Mandelbrot found that the coast of Australia, the west coast of Britain, and the land frontier of Portugal all satisfy shapes given by a fractal dimension of approximately 1.75.

In the 1960s and early 1970s, applied mathematicians began to computationally explore features of chaotic iterations that had previously been studied by analytic methods. May, Lorenz, Mandelbrot, Feigenbaum, Ruelle, York, and others led the way in utilizing computers and graphics to explore this realm, as chronicled for example in Gleick's *Chaos: Making a New Science*.

*The Uncertainty Principle*

We finish this section with a principle that, while discovered early in the twentieth century by conventional formal reasoning, could have been discovered much more easily with computational tools.

Most readers have heard of the UNCERTAINTY PRINCIPLE [IV.23] from quantum mechanics, which is often expressed as the fact that the position and momentum of a subatomic particle cannot simultaneously be prescribed or measured to arbitrary accuracy. Others may be familiar with the uncertainty principle from signal processing theory, which is often expressed as the fact that a signal cannot simultaneously be both "time-limited" and "frequency-limited." Remarkably, the precise mathematical formulations of these two principles are identical.

Consider a real, continuously differentiable, $L^2$ function $f(t)$, which satisfies $|t|^{3/2+\varepsilon} f(t) \to 0$ as $|t| \to \infty$ for some $\varepsilon > 0$. (This ensures convergence of the integrals below.) For convenience, we assume $f(-t) = f(t)$, so the Fourier transform $\hat{f}(x)$ of $f(t)$ is real, although this is not necessary. Define

$$\left.\begin{aligned} E(f) &= \int_{-\infty}^{\infty} f^2(t)\,\mathrm{d}t, & V(f) &= \int_{-\infty}^{\infty} t^2 f^2(t)\,\mathrm{d}t, \\ \hat{f}(x) &= \int_{-\infty}^{\infty} f(t)\mathrm{e}^{-\mathrm{i}tx}\,\mathrm{d}t, & Q(f) &= \frac{V(f)}{E(f)}\,\frac{V(\hat{f})}{E(\hat{f})}. \end{aligned}\right\} \tag{1}$$

Then the uncertainty principle is the assertion that $Q(f) \geqslant \frac{1}{4}$, with equality if and only if $f(t) = a\mathrm{e}^{-(bt)^2/2}$ for real constants $a$ and $b$. The proof of this fact is not terribly difficult but is hardly enlightening (see, for example, Borwein and Bailey 2008, pp. 183–88).

Let us approach this problem as an experimental mathematician might. It is natural when studying Fourier transforms (particularly in the context of signal processing) to consider the "dispersion" of a function and to compare this with the dispersion of its Fourier transform. Noting what appears to be an inverse relationship between these two quantities, we are led to consider $Q(f)$ in (1). With the assistance of Maple or Mathematica, one can explore examples, as shown in table 1. Note that each of the entries in the last column is in the range $(\frac{1}{4}, \frac{1}{2})$. Can one get any lower?

To further study this problem experimentally, note that the Fourier transform $\hat{f}$ of $f(t)$ can be closely approximated with a *fast Fourier transform*, after suitable discretization. The integrals $V$ and $E$ can also be evaluated numerically.

One can then adopt a search strategy to minimize $Q(f)$, starting, say, with a "tent function," then perturbing it up or down by some $\varepsilon$ on a regular grid with spacing $\delta$, thus creating a continuous, piecewise-linear function. *When, for a given $\delta$, a minimizing function $f(t)$ has been found, reduce $\varepsilon$ and $\delta$ and repeat. Terminate when $\delta$ is sufficiently small: say, $10^{-6}$ or so.* (For details, see Borwein and Bailey (2008).)

**Table 1** $Q$ values for various functions.

| $f(t)$ | Interval | $\hat{f}(x)$ | $Q(f)$ |
|---|---|---|---|
| $1 - t\operatorname{sgn} t$ | $[-1, 1]$ | $\dfrac{2(1-\cos x)}{x^2}$ | $\frac{3}{10}$ |
| $1 - t^2$ | $[-1, 1]$ | $\dfrac{4(\sin x - x\cos x)}{x^3}$ | $\frac{5}{14}$ |
| $\dfrac{1}{1+t^2}$ | $[-\infty, \infty]$ | $\pi\exp(-x\operatorname{sgn} x)$ | $\frac{1}{2}$ |
| $\mathrm{e}^{-\|t\|}$ | $[-\infty, \infty]$ | $\dfrac{2}{1+x^2}$ | $\frac{1}{2}$ |
| $1 + \cos t$ | $[-\pi, \pi]$ | $\dfrac{2\sin(\pi x)}{x - x^3}$ | $\frac{1}{9}(\pi^2 - \frac{15}{2})$ |

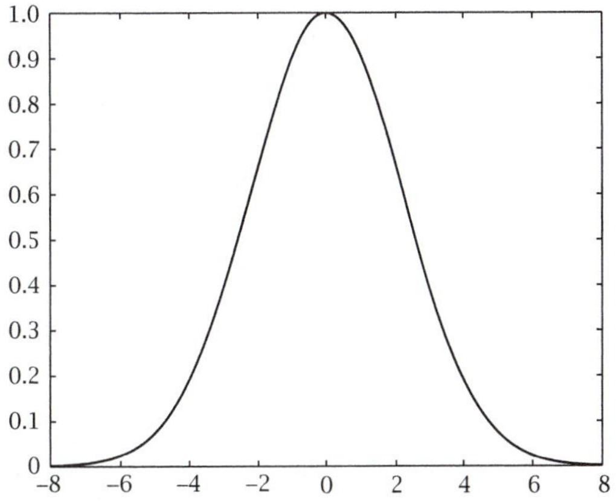


**Figure 1** $Q$-minimizer and matching Gaussian.

The resulting function $f(t)$ is shown in figure 1. Needless to say, its shape strongly suggests a *Gaussian* probability curve. Figure 1 shows both $f(t)$ and the function $\mathrm{e}^{-(bt)^2/2}$, where $b = 0.45446177$; to the precision of the plot, they are identical!

In short, it is a relatively simple matter, using twenty-first-century computational tools, to numerically "discover" the uncertainty principle. Doubtless the same is true of many other historical principles of physics, chemistry, and other fields.

## 3 Twenty-First-Century Studies

It is fair to say that the computational–experimental approach in applied mathematics has greatly accelerated in the twenty-first century. We present a few specific illustrative examples here. These include several by the present authors because we are familiar with them. There are doubtless many others that we are not aware of that are similarly exemplary of the experimental paradigm.

### 3.1 Chimera States in Oscillator Arrays

One interesting example of experimental applied mathematics was the 2002 discovery by Kuramoto, Battogtokh, and Sima of "chimera" states, which arise in arrays of identical oscillators, where individual oscillators are correlated with oscillators some distance away in the array. These systems can arise in a wide range of physical systems, including Josephson junction arrays, oscillating chemical systems, epidemiological models, neural networks underlying snail shell patterns, and "ocular dominance stripes" observed in the visual cortex of cats and monkeys. In chimera states—named for the mythological beast that incongruously combines features of lions, goats, and serpents—the oscillator array bifurcates into two relatively stable groups: the first composed of coherent, phased-locked oscillators, and the second composed of incoherent, drifting oscillators.

According to Abrams and Strogatz, who subsequently studied these states in detail, most arrays of oscillators quickly converge into one of four typical patterns:

(i) synchrony, with all oscillators moving in unison;
(ii) solitary waves in one dimension or spiral waves in two dimensions, with all oscillators locked in frequency;
(iii) incoherence, where phases of the oscillators vary quasiperiodically, with no global spatial structure; and
(iv) more complex patterns, such as spatiotemporal chaos and intermittency.

In chimera states, however, phase locking and incoherence are simultaneously present in the same system.

The simplest governing equation for a continuous one-dimensional chimera array is

$$\frac{\partial \phi}{\partial t} = \omega - \int_0^1 G(x-x') \sin[\phi(x,t) - \phi(x',t) + \alpha]\,\mathrm{d}x', \tag{2}$$

where $\phi(x,t)$ specifies the phase of the oscillator given by $x \in [0,1)$ at time $t$, and $G(x-x')$ specifies the degree of nonlocal coupling between the oscillators $x$ and $x'$. A discrete, computable version of (2) can be obtained by replacing the integral with a sum

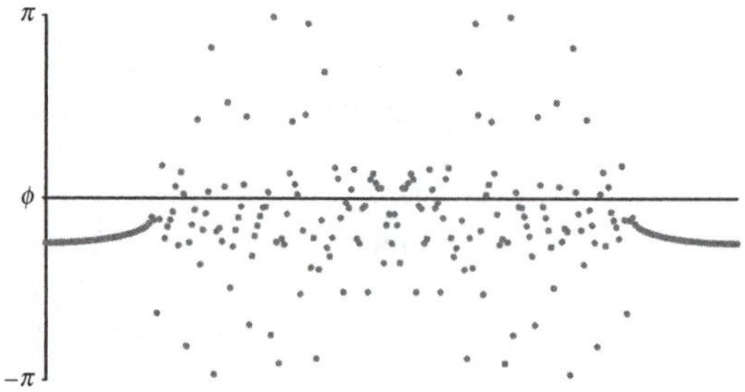


**Figure 2** Phase of oscillations for a chimera system. The $x$-axis runs from 0 to 1 with periodic boundaries.

over a one-dimensional array ($x_k$, $0 \leqslant k < N$), where $x_k = k/N$. Kuramoto and Battogtokh took $G(x-x') = C\exp(-\kappa|x-x'|)$ for constant $C$ and parameter $\kappa$.

Specifying $\kappa = 4$, $\alpha = 1.457$, array size $N = 256$, and time step size $\Delta t = 0.025$, and starting from $\phi(x) = 6\exp[-30(x-1/2)^2]r(x)$, where $r$ is a uniform random variable on $[-\frac{1}{2}, \frac{1}{2})$, gives rise to the phase patterns shown in figure 2. Note that the oscillators near $x = 0$ and $x = 1$ appear to be phase-locked, moving in near-perfect synchrony with their neighbors, but those oscillators in the center drift wildly in phase, with respect to both their neighbors and the locked oscillators.

Numerous researchers have studied this phenomenon since its initial numerical discovery. Abrams and Strogatz studied the coupling function given by $G(x) = (1 + A\cos x)/(2\pi)$, where $0 \leqslant A \leqslant 1$, for which they were able to solve the system analytically, and then extended their methods to more general systems. They found that chimera systems have a characteristic life cycle: a uniform phase-locked state, followed by a spatially uniform drift state, then a modulated drift state, then the birth of a chimera state, followed by a period of stable chimera, then a saddle-node bifurcation, and finally an unstable chimera state.

### 3.2 Winfree Oscillators

One development closely related to chimera states is the resolution of the Quinn-Rand-Strogatz (QRS) constant. Quinn, Rand, and Strogatz had studied the *Winfree model* of coupled nonlinear oscillators, namely

$$\dot{\theta}_i = \omega_i + \frac{\kappa}{N}\sum_{j=1}^{N} -(1+\cos\theta_j)\sin\theta_i \tag{3}$$

for $1 \leqslant i \leqslant N$, where $\theta_i(t)$ is the phase of oscillator $i$ at time $t$, the parameter $\kappa$ is the coupling strength, and the frequencies $\omega_i$ are drawn from a symmetric unimodal density $g(w)$. In their analyses, they were led to the formula

$$0 = \sum_{i=1}^{N} \left( 2\sqrt{1 - s^2(1 - 2(i-1)/(N-1))^2} - \frac{1}{\sqrt{1 - s^2(1 - 2(i-1)/(N-1))^2}} \right),$$

implicitly defining a phase offset angle $\phi = \sin^{-1} s$ due to bifurcation. The authors conjectured, on the basis of numerical evidence, the asymptotic behavior of the $N$-dependent solution $s$ to be

$$1 - s_N \sim \frac{c_1}{N} + \frac{c_2}{N^2} + \frac{c_3}{N^3} + \cdots,$$

where $c_1 = 0.60544365\ldots$ is now known as the QRS constant.

In 2008, the present authors together with Richard Crandall computed the numerical value of this constant to 42 decimal digits, obtaining

$$c_1 \approx 0.60544365719673274947892284244\ldots.$$

With this numerical value in hand, it was possible to demonstrate that $c_1$ is the unique zero of the *Hurwitz zeta function* $\zeta(\frac{1}{2}, \frac{1}{2}z)$ on the interval $0 \leqslant z \leqslant 2$. What is more, they found that $c_2 = -0.104685459\ldots$ is given analytically by

$$c_2 = c_1 - c_1^2 - 30\frac{\zeta(-\frac{1}{2}, \frac{1}{2}c_1)}{\zeta(\frac{3}{2}, \frac{1}{2}c_1)}.$$

### 3.3 High-Precision Dynamics

Periodic orbits form the "skeleton" of a dynamical system and provide much useful information, but when the orbits are unstable, high-precision numerical integrators are often required to obtain numerically meaningful results.

For instance, figure 3 shows computed symmetric periodic orbits for the $(7+2)$-ring problem using double and quadruple precision. The $(n+2)$*-body ring problem* describes the motion of an infinitesimal particle attracted by the gravitational field of $n+1$ primary bodies, $n$ of which are in the vertices of a regular polygon rotating in its own plane about its center with constant angular velocity. Each point corresponds to the initial conditions of one symmetric periodic orbit, and the gray areas correspond to regions of forbidden motion (delimited by the limit curve). To avoid "false" initial conditions it is useful to check if the initial conditions generate a periodic orbit up to a given tolerance level; but for highly unstable periodic orbits, double precision is not enough, resulting in gaps in the

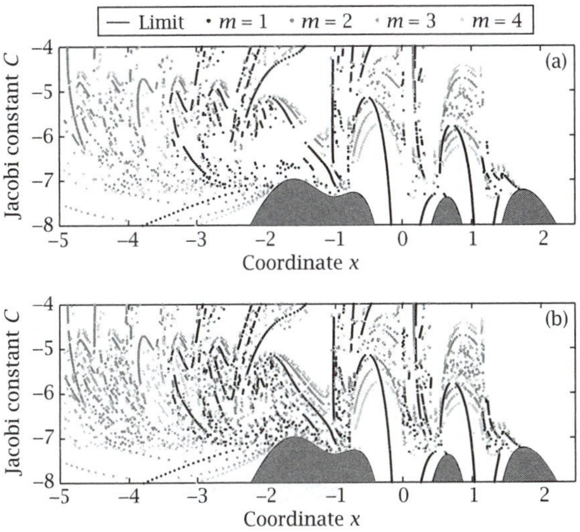


**Figure 3** Symmetric periodic orbits ($m$ denotes the multiplicity of the periodic orbit) in the most chaotic zone of the $(7+2)$-ring problem using (a) double and (b) quadruple precision. Note the "gaps" in the double precision plot. (Reproduced by permission of Roberto Barrio.)

figure that are not present in the more accurate quad precision run.

Hundred-digit precision arithmetic plays a fundamental role in a 2010 study of the fractal properties of the LORENZ ATTRACTOR [III.20]; see figure 4. The first plot in the figure shows the intersection of an arbitrary trajectory on the Lorenz attractor with the section $z = 27$, in a rectangle in the $xy$-plane. All subsequent plots zoom in on a tiny region (too small to be seen by the unaided eye) at the center of the red rectangle of the preceding plot to show that what appears to be a line is in fact many lines.

The Lindstedt–Poincaré method for computing periodic orbits is based on Lindstedt–Poincaré perturbation theory, Newton's method for solving nonlinear systems, and Fourier interpolation. Viswanath has used this in combination with high-precision libraries to obtain periodic orbits for the Lorenz model at the classical Saltzman parameter values. This procedure permits one to compute, to high accuracy, highly unstable periodic orbits more efficiently than with conventional schemes, in spite of the additional cost of high-precision arithmetic. For these reasons, high-precision arithmetic plays a fundamental role in the study of the fractal properties of the Lorenz attractor (see figures 4 and 5) and in the consistent formal development of complex singularities of the Lorenz system using

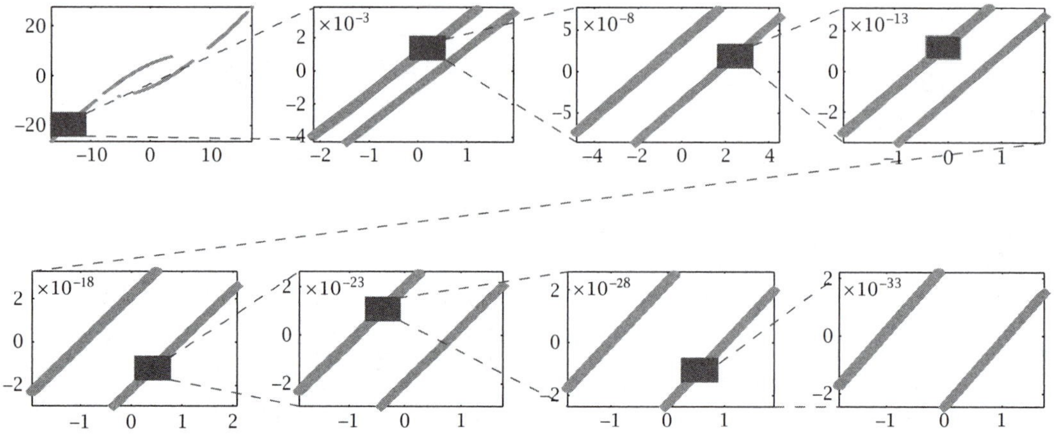


**Figure 4** The fractal property of the Lorenz attractor. (Reproduced by permission of Divakar Viswanath.)

infinite series. (For additional details and references, see Bailey et al. (2012).)

### 3.4 Ising Integrals

The previously mentioned study employed 100-digit arithmetic. Much higher precision has proven essential in studies with Richard Crandall (see Borwein and Bailey 2008; Bailey and Borwein 2011) of the following integrals that arise in the Ising theory of mathematical physics and in quantum field theory:

$$C_n = \frac{4}{n!}\int_0^\infty \cdots \int_0^\infty \frac{1}{(\sum_{j=1}^n (u_j + 1/u_j))^2}\,\mathrm{d}U,$$

$$D_n = \frac{4}{n!}\int_0^\infty \cdots \int_0^\infty \frac{\prod_{i<j}((u_i - u_j)/(u_i + u_j))^2}{(\sum_{j=1}^n (u_j + 1/u_j))^2}\,\mathrm{d}U,$$

$$E_n = 2\int_0^1 \cdots \int_0^1 \biggl(\prod_{1\leqslant j<k\leqslant n} \frac{u_k - u_j}{u_k + u_j}\biggr)^2 \mathrm{d}T,$$

where

$$\mathrm{d}U = \frac{\mathrm{d}u_1}{u_1}\cdots\frac{\mathrm{d}u_n}{u_n}, \quad \mathrm{d}T = \mathrm{d}t_2\cdots\mathrm{d}t_n, \quad u_k = \prod_{i=1}^k t_i.$$

Note that $E_n \leqslant D_n \leqslant C_n$.

Direct computation of these integrals from their defining formulas is very difficult, but for $C_n$ it can be shown that

$$C_n = \frac{2^n}{n!}\int_0^\infty pK_0^n(p)\,\mathrm{d}p,$$

where $K_0$ is the MODIFIED BESSEL FUNCTION [IV.7 §9]. Thousand-digit numerical values so computed were used with the PSLQ algorithm to deduce results such as $C_4 = \frac{7}{12}\zeta(3)$ and furthermore to discover that

$$\lim_{n\to\infty} C_n = 0.63047350\cdots = 2\mathrm{e}^{-2\gamma},$$

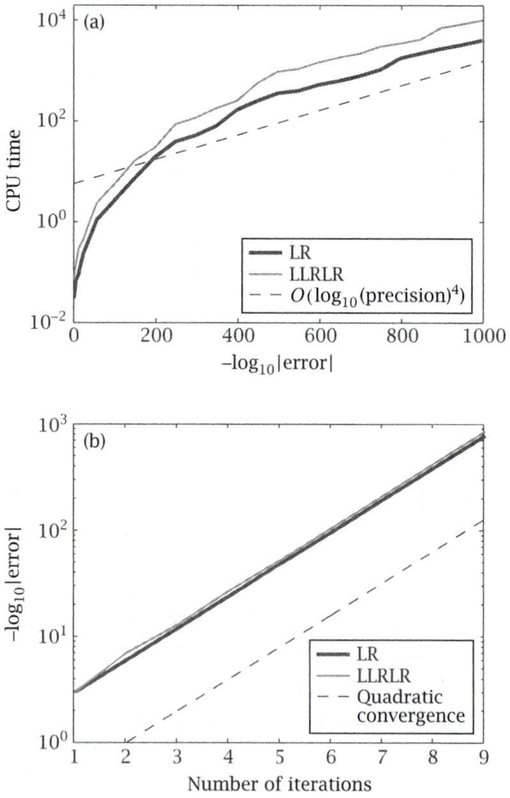


**Figure 5** Computational relative error versus (a) CPU time and (b) number of iterations in a 1000-digit computation of the periodic orbits LR and LLRLR of the Lorenz model. (Reproduced by permission of Roberto Barrio.)

with additional higher-order terms in an asymptotic expansion. One intriguing experimental result (that has not yet been proven) is the following:

$$\begin{aligned}E_5 \overset{?}{=} 42 - 1984\,\mathrm{Li}_4(\tfrac{1}{2}) + \tfrac{189}{10}\pi^4 - 74\zeta(3) \\ - 1272\zeta(3)\log 2 + 40\pi^2\log^2 2 - \tfrac{62}{3}\pi^2 \\ + \tfrac{40}{3}\pi^2\log 2 + 88\log^4 2 + 464\log^2 2 - 40\log 2.\end{aligned}$$

This was found by a multi-hour computation on a highly parallel computer system and confirmed to 250-digit precision. Here, $\mathrm{Li}_4(z) = \sum_{k\geqslant 1} z^k/k^4$ is the standard order-4 polylogarithm.

### 3.5 Ramble Integrals and Short Walks

Consider, for complex $s$, the $n$-dimensional *ramble integrals* (Bailey and Borwein 2012)

$$W_n(s) = \int_{[0,1]^n} \left|\sum_{k=1}^{n} \mathrm{e}^{2\pi x_k \mathrm{i}}\right|^s \mathrm{d}x, \tag{4}$$

which occur in the theory of uniform random walk integrals in the plane, where at each step a unit step is taken in a random direction, as first studied by Pearson, Rayleigh, and others 100 years ago. Integrals such as (4) are the $s$th moment of the distance to the origin after $n$ steps. As is well known, various types of random walks arise in fields as diverse as aviation, ecology, economics, psychology, computer science, physics, chemistry, and biology.

*Walks and Measures*

In work from 2010 by Borwein, Straub, Wan, and Zudilin, using a combination of analysis and high-precision numerical computation, results such as

$$W_n'(0) = -n\int_0^\infty \log(x)J_0^{n-1}(x)J_1(x)\,\mathrm{d}x$$

were obtained, where $J_n(x)$ denotes the Bessel function of the first kind and $\gamma$ denotes Euler's constant. These results, in turn, lead to various closed forms and have been used to confirm, to 600-digit precision, the following *Mahler measure* conjecture adapted from Villegas:

$$\begin{aligned}W_5'(0) \overset{?}{=} \left(\frac{15}{4\pi^2}\right)^{5/2}\int_0^\infty \{\eta^3(e^{-3t})\eta^3(\mathrm{e}^{-5t}) \\ + \eta^3(\mathrm{e}^{-t})\eta^3(\mathrm{e}^{-15t})\}t^3\,\mathrm{d}t,\end{aligned}$$

where the *Dedekind eta-function* can be computed from

$$\begin{aligned}\eta(q) &= q^{1/24}\prod_{n\geqslant 1}(1-q^n) \\ &= q^{1/24}\sum_{n=-\infty}^{\infty}(-1)^n q^{n(3n+1)/2}.\end{aligned}$$

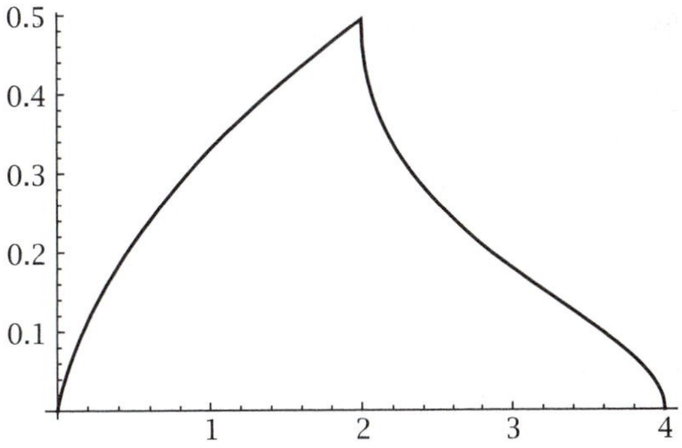


**Figure 6** The "shark-fin" density of a four-step walk.

There are remarkable connections between diverse parts of pure, applied, and computational mathematics lying behind these results. As is often the case, there is a fine interplay between developing better computational tools—especially for special functions and polylogarithms—and discovering new structure.

*Densities of Short Walks*

One of the deepest related discoveries is the following closed form for the *radial density* of a four-step uniform random walk in the plane: for $2 \leqslant \alpha \leqslant 4$ one has the real hypergeometric form

$$p_4(\alpha) = \frac{2}{\pi^2}\frac{\sqrt{16-\alpha^2}}{\alpha}\,{}_3F_2\left(\begin{matrix}\frac{1}{2},\frac{1}{2},\frac{1}{2}\\ \frac{5}{6},\frac{7}{6}\end{matrix}\;\middle|\;\frac{(16-\alpha^2)^3}{108\alpha^4}\right).$$

Remarkably, the real part of the right-hand side of this identity is valid everywhere on $[0,4]$, as plotted in figure 6. This was an entirely experimental discovery—involving at least one fortunate error—but is now fully proven.

### 3.6 Moments of Elliptic Integrals

The study on ramble integrals that was discussed in the previous subsection also led to a comprehensive analysis of moments of elliptic integral functions of the form

$$\int_0^1 x^{n_0}K^{n_1}(x)K'^{n_2}(x)E^{n_3}(x)E'^{n_4}(x)\,\mathrm{d}x,$$

where the elliptic functions $K$ and $E$ and their complementary versions are given by

$$\begin{aligned}K(x) &= \int_0^1 \frac{\mathrm{d}t}{\sqrt{(1-t^2)(1-x^2t^2)}}, \\ E(x) &= \int_0^1 \frac{\sqrt{1-x^2t^2}}{\sqrt{1-t^2}}\,\mathrm{d}t, \\ K'(x) &= K(\sqrt{1-x^2}), \\ E'(x) &= E(\sqrt{1-x^2}).\end{aligned}$$

Computations of these integrals to 3200-digit precision, combined with searches for relations using the PSLQ algorithm, yielded thousands of unexpected relations among these integrals (see Bailey and Borwein 2012). The scale of the computation was required due to the number of integrals under investigation.

### 3.7 Snow Crystals

Computational experimentation has even been useful in the study of snowflakes. In a 2007 study, Janko Gravner and David Griffeath used a sophisticated computer-based simulator to study the process of formation of these structures, known in the literature as snow crystals and informally as *snowfakes*. Their model simulated each of the key steps, including diffusion, freezing, and attachment, and thus enabled researchers to study dependence on melting parameters. Snow crystals produced by their simulator vary from simple stars, to six-sided crystals with plate-ends, to crystals with dendritic ends, and they look remarkably similar to natural snow crystals. Among the findings uncovered by their simulator is the fact that these crystals exhibit remarkable overall symmetry, even in the process of dynamically changing parameters. Their simulator is publicly available at http://psoup.math.wisc.edu/Snowfakes.htm.

## 4 Limits of Computation

Developments such as the above have led to reexamination of the role of computation in formal mathematical work. To begin with, a legitimate question is whether one can truly trust—in the mathematical sense—the result of a computation, since there are many possible sources of errors: unreliable numerical algorithms; bug-ridden computer programs implementing these algorithms; system software or compiler errors; hardware errors, either in processing or storage; insufficient numerical precision; and obscure errors of hardware, software, or programming that surface only in particularly large or difficult computations.

As a single example of the sorts of difficulties that can arise, the present authors found that neither Maple nor Mathematica was able to numerically evaluate constants of the form

$$\frac{1}{2\pi}\int_0^{2\pi} f(\mathrm{e}^{\mathrm{i}\theta})\,\mathrm{d}\theta,$$

where

$$f(\theta) = \mathrm{Li}_1(\theta)^m\,\mathrm{Li}_1^{(1)}(\theta)^p\,\mathrm{Li}_1(\theta+\pi)^n\,\mathrm{Li}_1^{(1)}(\theta-\pi)^q$$

(for $m, n, p, q \geqslant 0$ integers), to high precision in reasonable run time. In part this was because of the challenge of computing polylogs and polylog derivatives (with respect to order) for complex arguments. The version of Mathematica that we were using was able to numerically compute $\partial\,\mathrm{Li}_s(z)/\partial s$ to high precision, which is required here, but such evaluations were not only many times slower than computation of $\mathrm{Li}_s(z)$ itself but in some cases did not even return a tenth of the requested number of digits correctly.

For such reasons, experienced programmers of mathematical or scientific computations routinely insert validity checks into their code. Typically, such checks take advantage of known high-level mathematical facts, such as the fact that the product of two matrices used in the calculation should always give the identity or that the results of a convolution of integer data, done using a fast Fourier transform, should all be very close to integers.

For instance, Kanada's 2002 computation of $\pi$ to 1.3 trillion decimal digits involved first computing slightly over one trillion hexadecimal (base-16) digits. He found that the 20 hex digits of $\pi$ beginning at position $10^{12}+1$ are

B4466E8D21 5388C4E014.

Kanada then calculated these hex digits using the "Bailey–Borwein–Plouffe" algorithm. The result was

B4466E8D21 5388C4E014,

dramatically confirming that both results are almost certainly correct. While one cannot rigorously assign a "probability" to this event, the chances that two random strings of 20 hex digits perfectly agree is one in $16^{20} \approx 1.2089 \times 10^{24}$.

Even so, researchers are well advised to be cautious with experimentation. Consider

$$\int_0^\infty \cos(2x)\prod_{n=1}^{\infty}\cos(x/n)\,\mathrm{d}x = 0.392699081698724154807830422909937860524645434187231595926\dots. \quad (5)$$

(In the printed value the digits 5434187231595926 are underlined.)

At first glance, this appears to be $\pi/8$, but upon comparison with the numerical value,

$$\pi/8 = 0.392699081698724154807830422909937860524646174921888227621\dots,$$

(In the printed value the digits 6174921888227621 are underlined.)

the two values disagree after the 42nd digit! Richard Crandall later explained this mystery via a physically motivated analysis of *running out of fuel* random walks.

He found the following very rapidly convergent series expansion, of which formula (5) is the first term:

$$\frac{\pi}{8} = \sum_{m=0}^{\infty} \int_0^{\infty} \cos[2(2m+1)x] \prod_{n=1}^{\infty} \cos(x/n)\,\mathrm{d}x.$$

Two series terms suffice for 500-digit agreement.

As a final sobering example, consider

$$\sigma_p = \sum_{n=-\infty}^{\infty} \operatorname{sinc}(n/2)\operatorname{sinc}(n/3)\cdots\operatorname{sinc}(n/p)\,\mathrm{d}x$$
$$\stackrel{?}{=} \int_{-\infty}^{\infty} \operatorname{sinc}(x/2)\operatorname{sinc}(x/3)\cdots\operatorname{sinc}(x/p)\,\mathrm{d}x,$$

where in each line the divisors range over *all* primes up to $p$. Provably, the following is true. The "sum equals integral" identity for $\sigma_p$ remains valid at least for $p$ among roughly the first $10^{176}$ primes; but it stops holding after some larger prime, and thereafter the "sum less the integral" is strictly positive, but *they always differ by much less than one part in a googolplex* $= 10^{10^{100}}$. An even stronger estimate is possible assuming the generalized Riemann hypothesis.

**Acknowledgments.** This work was supported in part by the director, Office of Computational and Technology Research, Division of Mathematical, Information, and Computational Sciences of the U.S. Department of Energy, under contract number DE-AC02-05CH11231.

### Further Reading

Abrams, D. M., and S. H. Strogatz. 2006. Chimera states in a ring of nonlocally coupled oscillators. *International Journal of Bifurcations and Chaos* 16:21–37.

Bailey, D. H., R. Barrio, and J. M. Borwein. 2012. High-precision computation: mathematical physics and dynamics. *Applied Mathematics and Computation* 218:10106–21.

Bailey, D. H., and J. M. Borwein. 2011. Exploratory experimentation and computation. *Notices of the American Mathematical Society* 58:1410–19.

——. 2012. Hand-to-hand combat with thousand-digit integrals. *Journal of Computational Science* 3:77–86.

Borwein, J. M., and D. H. Bailey. 2008. *Mathematics by Experiment: Plausible Reasoning in the 21st Century.* Natick, MA: A. K. Peters.

Borwein, J. M., and K. Devlin. 2008. *The Computer as Crucible: An Introduction to Experimental Mathematics.* Natick, MA: A. K. Peters.

Gleick, J. 1987. *Chaos: The Making of a New Science.* New York: Viking Penguin.

Gravner, J., and D. Griffeath. 2009. Modeling snow crystal growth: a three-dimensional mesoscopic approach. *Physical Review* E 79:011601.

Mandelbrot, B. B. 1982. *The Fractal Geometry of Nature.* New York: W. H. Freeman.

Minovitch, M. 1961. A method for determining interplanetary free-fall reconnaissance trajectories. Jet Propulsion Laboratory Technical Memo TM-312-130, pp. 38–44 (August 23).

Wilf, H. S. 2008. Mathematics: an experimental science. In *The Princeton Companion to Mathematics*, edited by T. Gowers. Princeton, NJ: Princeton University Press.

Wolfram, S. 2002. *A New Kind of Science.* Champaign, IL: Wolfram Media.

## VIII.7 Teaching Applied Mathematics

How can we enthuse the next generation of students about applied mathematics? The four contributors to this group of articles, who have all thought deeply about this question, were asked to give their personal views. The resulting articles provide a variety of perspectives and will be of interest to anyone who wishes to inspire their students to pursue the subject.

### I. David Acheson: What's the Big Picture?

Let A and B be two teachers of applied mathematics (at any level) and suppose that, generally speaking, A is a much better teacher than B.

*Why* is A's teaching so much better? Even without any further information, can we at least hazard a guess?

I wonder, for instance, if you might be prepared to bet that A is more trained in "communication skills"? Or perhaps A knows more mathematics than B or is nearer to the cutting edge of research? Then again, maybe A just has a more lively personality?

All these things can be advantageous, of course, but I would not actually bet on any of them.

In fact, in the absence of any further information, there is only one thing that I would be prepared to bet good money on. I would be prepared to bet that A's teaching is so much better—so inspirational, at best—mainly because A *wants* it to be that way, for reasons that we will probably never learn and that A may not even know.

This is only an opinion, of course, but it comes from thinking back to my own inspirational teachers when I was young. Some were notable for their scholarship, some for their eccentricity, but—so far as I can see—they only really had one thing in common: they had a great story to tell, *and they really wanted to tell it.*

## "Removing Some of the Rubbish..."

It is simple common sense with applied mathematics teaching—and possibly with mathematics teaching of any kind—to start with the basics and work up. In other words, "Don't try to run before you can walk."

But I believe it is a terrible mistake not to also bear in mind a very different piece of advice: namely, "If you have no idea where you are going, do not be too surprised if you never get there."

This is, I suspect, what the author John Ward meant a long time ago, in his *Plain and Easie Introduction to the Mathematicks* (1729), when he wrote:

> Tis Honour enough for me to be accounted as one of the under Labourers in Clearing the Ground a little, and Removing some of the Rubbish that lies in the way to Knowledge.

In any event, I believe that a major difficulty with mathematics, at all levels, is that people can easily get bogged down in things of little consequence instead of engaging with things that really matter.

And what could help them, more than anything else, is some kind of "big picture."

My own big picture of mathematics starts with *wonderful theorems*, by which I mean major results, usually with considerable generality and often an element of surprise. Secondly, *beautiful proofs*, i.e., concise deductive arguments, possibly containing a truly "lightbulb" moment when all suddenly becomes clear. And finally, *great applications*, particularly to physics, and hence to our understanding of how the world really works.

I would argue, in fact, that mathematics is at its very best when you get *all three things at once*. That, in my view, is when you should really open the champagne.

More controversially, perhaps, I believe that we can, and should, offer some such big picture to virtually anyone, including very young children and the rest of the general public.

Nonetheless, the majority of my teaching experience has been with university students, and that is where I would like to turn next.

## Lectures and Classes

One way of bringing a student lecture to life is through a picture or video, but best of all, perhaps, is a live experiment, and my own subject, fluid dynamics, lends itself particularly well to this.

As an example, take two glass plates and put a blob of dishwashing liquid on one of them. (I dye the blob red, with food coloring, for dramatic effect.) Now press down with the other plate, so that the narrowing gap causes the blob to spread out in a nice, symmetric fashion, with a more or less circular boundary.

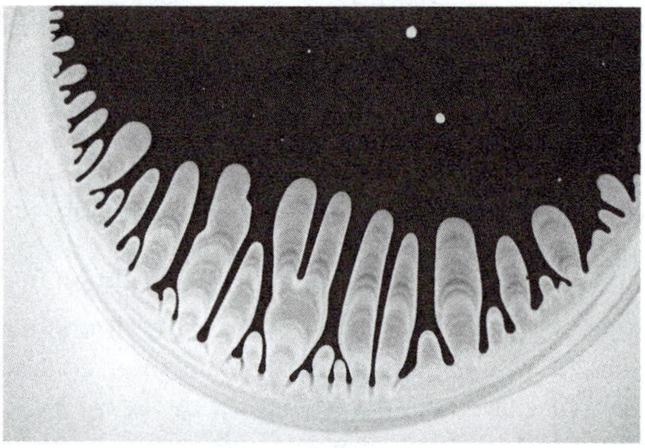

**Figure 1** Viscous fingering.

But if we now gradually pull the plates apart again, the *reverse* motion is hopelessly unstable; tiny ripples appear in the boundary, for no apparent reason, and grow rapidly into long viscous fingers (see figure 1).

If performed on an overhead projector or visualizer, this experiment can often make an audience gasp with astonishment.

But my real point here is a little more subtle. For why do a demonstration like this only in an advanced course on fluid mechanics, along with all the associated theory? Why not stimulate interest by first showing it much earlier, perhaps even in an elementary course on particle dynamics, as soon as the whole idea of stability and instability first arises?

Another way of helping people see the "big picture" is through the history of the subject, provided that the history in question has some real scholarship and depth to it.

In a first course on particle dynamics, for example, my experience is that students find it genuinely interesting to actually see, with their own eyes, that their textbook treatment of planetary motion is spectacularly different from the one in Newton's *Principia* and that it was not until about sixty years later, in the subsequent works of Euler and others, that dynamics came to be done in more or less the way it is done today.

To take a more lightweight example, my own research in fluid mechanics once gave a new twist to a hundred-year-old problem in vortex motion, first studied by Augustus Love in 1894. And whenever I present this (as a short diversion) in student lectures, I am convinced that it is enlivened by snippets from Love's original paper, to say nothing of an early photograph of Love and his striking Victorian moustache (which was apparently much admired at the time).

But there is another, possibly more unusual, way in which it is possible to bring student lectures to life.

Imagine, if you will, that you have just arrived at what you perceive to be the high point of the lecture, where the next line in the mathematical argument is very clever or inventive in some way. (I would even include here taking the curl of the momentum equation in fluid mechanics, to eliminate the pressure.)

For many years now, whenever this happens I tend to ask the audience whether they have any idea what the next, clever step might be.

Now, conventional wisdom is, I think, that if you can do this sort of thing at all, you can do it only with very small audiences. In my experience, however, even with audiences of 200 or more, once they realize that no one can possibly be *expected* to know the answer and that they are being invited—just for a moment—to more or less put themselves in the shoes of some genius like Newton or Euler, the suggestions will start coming *if you hold your nerve.*

Like so many things of this kind, it all depends on just how much you want to do it.

### Books

A well-known publisher once said:

> Everybody has a book inside them. And it should usually stay there.

However true that may be, it could be argued that the sheer impact and reach of a sufficiently original book can completely dwarf what its author might ever hope to achieve through direct, face-to-face teaching, and I am a great optimist about the future of books in the teaching of applied mathematics.

And while, as far as I can tell, it takes considerable imagination and skill to write either an outstanding textbook or a successful popular mathematics book, I have long wondered if a real breakthrough in the future may instead come from some thoroughly original approach that combines the best elements of both.

### Public Engagement

One of the most striking developments in recent years has been the rapid increase in popular mathematics lectures for either school students or the wider public.

I count myself fortunate to have been involved in several mathematical shows of this kind, mainly for teenagers. They are often held in mainstream city center theaters, with all the paraphernalia of stage lighting, sound technicians, etc., and the pressure to be entertaining as well as informative is therefore intense. So, to illustrate applied mathematics, I often use the formula for the frequency of a vibrating string, thereby smuggling in a practical demonstration of harmonics (and a self-composed tune!) on my electric guitar.

But so-called community lectures (which are usually held in the evening) can be even more rewarding because the age range at them can be enormous: from grandparents to very young children indeed. All you can really assume on these occasions is that each family group includes at least one person who is good at sums.

It was at one of these events, at a school in North London, that I was midway through a "proof by pizza" (for the sum of an infinite series) when I happened to notice a particular little boy, age about ten, in the audience. A split-second after delivering the punch line of my proof—at the moment when a deep idea suddenly becomes almost obvious—I practically saw the "light-bulb" go on in his head, and he got so excited that he fell off his chair.

And, in a sense, that fleeting moment says it all.

For mathematics at its best, at any level, lifts the human spirit, by showing us that the world—whether the world of the mind or the actual physical world in which we live—is an even more weird and wonderful place than we thought.

## II. Peter R. Turner: Computation, Modeling, and Projects

### Introduction

This article presents a personal philosophy for teaching applied, and particularly computational, mathematics at the undergraduate level. It is largely drawn from my own experience over more than 40 years, mostly at three institutions in the United Kingdom and the United States.

That experience has been enhanced by my activities on the Society for Industrial and Applied Mathematics (SIAM) Education Committee, including four years as vice president for education. During this time, I have gained awareness of broader aspects of the role of applied mathematics education at the undergraduate level. It is important here to note that I am not a mathematics education specialist but a mathematician who is interested in education.

Important among these broader aspects was the February 2012 report from the President's Council of Advisors on Science and Technology entitled *Engage to Excel: Producing One Million Additional College Graduates with Degrees in Science, Technology, Engineering, and Mathematics*, which emphasized the role of a good and relevant applied mathematics education within the framework of STEM (science, technology, engineering, and mathematics) education. The national emphasis in the United States on STEM has been a hallmark of recent education policy development.

One of the key points raised was the "math gap." This is a term used to highlight the difficulty in transition from high school to undergraduate study in STEM disciplines—a problem that is exacerbated by what the colleges perceive as a lack of mathematical preparation in high school. There is a gap between colleges' expectations and reality. The fundamental thesis is that this can be addressed through a stronger founding in mathematical modeling of real-world situations, and the solution, analysis, and validation of these models using computational and theoretical applied mathematics.

The issue of college mathematics, or broader STEM, readiness has been an area of interest for many people—some at a local and highly detailed level and others at a broader big-picture level, such as the studies carried out by the Mathematical Association of America. The rest of this article addresses a few ideas about how applied and computational mathematics might improve the situation. My basic thesis is that the use of projects that in turn require some modeling and (computational) problem solving enhances almost all (not just applied) mathematics teaching *in all mathematics classes*. For example, calculus can be taught with applied projects replacing endless drills once basic skills are acquired.

## Use of Projects in Numerical Methods Classes

Well over a decade ago, the basic structure of my undergraduate numerical methods/scientific computing courses changed to being entirely based on projects. The topical syllabus remained essentially unaltered, covering the fundamentals of nonlinear equations, linear systems, polynomial and spline interpolation, quadrature and numerical solution of ordinary differential equations, with each major theme being approached through an extended project.

The scheme was modified to incorporate more homework assignments to avoid the issue of procrastination. The homework assignments included some of the theoretical background and some of the preliminary steps in addressing the projects. The motivation for the changes was an (oversimplistic but illustrative) model that can explain why good students often found their first computational course difficult. To a faculty member, the class had the beauty of bringing together much of the students' prior experience in calculus, linear algebra, differential equations, and perhaps modeling, too—together with drawing on their programming skills, or even learning some algorithmic programming for the first time.

This same set of properties was the primary source of difficulty for the students. Suddenly, and for the first time, students were required to synthesize methods and solutions from multiple courses. Furthermore, their mathematical and programming experience had typically been totally disjoint up to that point. Thus we had an audience who may have been "good B students" in both their mathematical and programming ability, but simplistically multiplying these independent 0.85 probabilities (the middle of the common B grade range), we had a success rate of only 61%. In other words, these good students were struggling to get a D in numerical methods.

More importantly, the effect on students' attitudes to the course was impacted by a failure to see the wood/forest for the trees. The perception was of a required course they had to endure rather than an exciting culmination of all that had gone before. The use of projects was broadly successful in countering this.

The particular choice of projects is one that is important to the success of the course but also one that can be tailored to the individual instructor and audience. The initial set of topics I chose (and which have been modified successively over the years, both by me and by others who have taken over the immediate teaching responsibility) is described briefly here and illustrates the linkage to the main syllabus topics.

### The Length of a Telephone Cable

A cable above even ground, and with physical parameters in the model simplified to a specified sag, introduces iterative solution of a single nonlinear equation. The full project referred to a multi-loop cable above undulating ground with a profile determined by geographic data, connected with a simple cubic spline. The problem for each loop is then a nonlinear system of two equations, and the solutions over different pieces have to be matched to ensure continuity of the cable.

### Rats in a Maze

Based on simple psychology experiments on rats' learning abilities, this was an open-ended project that introduced iterative solution of linear systems. Even a simple rectangular, say $6 \times 5$, maze results in a $30 \times 30$ linear system. This is an eye-opener for many students, who rarely see systems much larger than $3 \times 3$ in introductory linear algebra. The basic problem is to compute the probability of a rat successfully finding food at some set of exits of the maze from an arbitrary starting point. These results provide the baseline against which to measure the rats' success at learning the maze by comparing actual performance with the simulated random decisions.

Students are then required to modify the maze in ways of their choosing: adding diagonal passages, removing certain links, adding a second level, and modifying the decision model from purely random to some bias (perhaps to go straight ahead) are all variations that students came up with and solved. One even tried to apply some artificial intelligence to simulate the rat learning.

### Reproduce a Picture

Although the concept of splines had been mentioned in the telephone cable problem, this project was the real introduction to interpolation. The objective was simply to reproduce a chosen picture or line drawing using interpolation. Polynomial interpolation was explored and usually quickly discarded for all but the simplest of shapes. Splines and other functions were introduced. A more modern treatment would probably extend this to using subdivision surfaces.

### The Gamma Function

When I started this project-based course, most students were concurrently enrolled in an applied statistics course—hence the choice of computing the gamma function as the quadrature-based project. One benefit is that this necessarily requires modification of conventional quadrature routines to handle both singularities and an infinite range of integration. The basic idea was simply to find appropriate bounds for the infinite tail and a region close to the singularity, and then to compute the major contribution from the resulting bounded integral. Using the recurrence to reduce the need to compute for all values of the argument $\alpha$ improved computational efficiency but necessitated introducing careful, though fairly simple, error analysis to control the required accuracy.

### Human Cannonball

A shooting problem for a projectile with nonlinear air resistance was the vehicle for the introduction of numerical solution of differential equations. The setting was finding the appropriate launch angle in order to hit a specified target (described as an escape window to escape from the course).

The particular list of projects above is certainly not intended to be prescriptive. Many improvements have been made, while other changes have, of course, proved less successful! The list here is only intended to illustrate the feasibility of such an approach and advance my thesis that project-based learning can significantly enhance the success of introductory scientific computing courses.

## Modeling across the Curriculum

The emphases on projects, computation, and modeling have combined more recently in a general "modeling across the curriculum" philosophy. This started to take shape in the work of a SIAM Education Committee working group, which led to a 2011 *SIAM Review* paper on undergraduate computational science and engineering (CSE) programs. The report emphasized that curriculum design needs to fit local conditions, but it also stressed student experiences such as internships. Other key points in the paper concerned the role that undergraduate CSE education plays in regard to both industrial expectations for graduates and feeding the educational pipeline.

Undergraduate CSE programs can take many forms. A few are full-fledged undergraduate majors in Computational X, where X could be physics, biology, or finance, for example. More common is some form of minor that accompanies an undergraduate major in either (applied) mathematics or some field of science or engineering. The latter model seems better suited to ensuring some depth in a core discipline while maintaining the breadth that such a minor introduces to the program.

Undergraduate education in applied and computational mathematics feeds the K-12 (preschool through completion of high school) education system, industrial appointments, and, of course, graduate schools in all areas of science and engineering as well as in applied mathematics itself.

One of the main obstacles here is that teacher education in mathematics in the United States is often very light in applied content (including statistics), and good programs of continuing education and professional development for teachers are therefore a necessary precursor to any real change. With colleagues, I have been involved in a very successful design-based summer activity for middle and high school students, including some professional development for their teachers. Students, and their teachers, are introduced to the mathematics and physics of designing a physical roller coaster and to modeling software to simulate their design. This has been successful in bringing appropriately adapted real-world applications and relevance to mathematics education. An increasing proportion of these students, mostly from economically disadvantaged backgrounds or other under-represented groups, have subsequently entered STEM college programs, demonstrating the benefit of exposure to such applied content. The benefit is realized not just in their mathematics but also in the science that accompanies it.

### Final Thoughts

The main point I am making is that students learn better when they perceive their studies as being relevant to their lives and future careers. In the case of mathematics, this provides a strong motivation to increase the applied and computational content at all stages of a student's development.

Early emphasis on problem solving leads to more advanced projects and full-scale modeling experiences as the students' abilities and background knowledge develop. This essay addresses some of those issues at a "big-picture" level rather than in detail because the details have to be right for the combination of institutional philosophy, instructor, and students where they are to be applied.

Understanding applied mathematics is inherently difficult because of the combined demands of the theoretical basis, the modeling and understanding of the application field, and the computational abilities that are needed to solve the problems. Determining that a "solution" really addresses the original issue, and if necessary refining it and solving again, are important aspects that only add to the inherent difficulty of the subject.

In my opinion, these difficulties oblige the educational community to address them throughout the curriculum. For example, I believe that the call for more emphasis on modeling and applications in the K-12 curriculum needs to be heard and that this move should be implemented soon. This must continue into the undergraduate program to help to address the "math gap" identified in the President's Council of Advisors on Science and Technology report. In summary, "early and often" is perhaps not sufficient to achieve the desired improvements. I advocate, instead, "early and always" for modeling and applications throughout the STEM curriculum.

#### Further Reading

President's Council of Advisors on Science and Technology. 2012. *Engage to Excel: Producing One Million Additional College Graduates with Degrees in Science, Technology, Engineering, and Mathematics.* Office of Science and Technology Policy Report (February). Available at www.whitehouse.gov/sites/default/files/microsites/ostp/pcast-engage-to-excel-final_2-25-12.pdf.

Schalk, P. D., D. P. Wick, P. R. Turner, and M. W. Ramsdell. 2009. IMPACT: integrated mathematics and physics assessment for college transition. In *Frontiers in Education Conference, San Antonio, TX, October 2009.* IEEE Press.

——. 2011. Predictive assessment of student performance for early strategic guidance. In *Frontiers in Education Conference, Rapid City, SD, October 2011.* IEEE Press.

SIAM Working Group on CSE Undergraduate Education. 2011. Undergraduate computational science and engineering education. *SIAM Review* 53:561–74.

Turner, P. R. 2001. Teaching scientific computing through projects. *Journal of Engineering Education* 90:79–83.

——. 2008. A predictor-corrector process with refinement for first-year calculus transition support. *Primus* 18:370–93.

Turner, P. R., and K. R. Fowler. 2010. A holistic approach to applied mathematics education for middle and high schools. In *Proceedings of the Education Interface between Mathematics and Industry Conference, EIMI 2010, Lisbon, Portugal*, pp. 521–31. International Commission on Mathematical Instruction/International Council of Industrial and Applied Mathematics.

## III. Gilbert Strang: What to Teach and How?

In expressing these brief but strongly held thoughts about teaching, I would like to distinguish between two separate questions.

- What should we teach?
- How should we teach it?

## What to Teach?

In my experience, a good decision on what to teach tells students that you care about them: you are thinking about them and their needs. This *sincerity of effort*—what you are contributing and what you want for them—is the most important message you could send to students. We all respond favorably to sincerity.

The teacher may think that he or she has no freedom. In calculus that may be partly true (and partly untrue). In applied mathematics, the syllabus is seldom so rigid. Our subject is extremely large! There is no hope of presenting or comprehending all the important directions, so there is an opening for independent thought. And a teacher who thinks independently will pass on an all-important message to students: that they can begin to work by themselves and think for themselves.

I will compare and contrast my thoughts about linear algebra and about computational science. Linear algebra in 1970 was in a very abstract and unsatisfactory state as a basic undergraduate course. Its enormous importance in practice was quite unappreciated by many of the standard textbooks. *Finite-Dimensional Vector Spaces* by Paul Halmos, for example, was written carefully and concisely, but it was written for mathematics students—stronger ones or weaker ones—and not for the much larger number who needed to *use* the subject.

Outside the classroom, every year brought new applications of linear algebra. Matrices became part of the core language of science and engineering and economics. To solve differential equations or to study large networks, linearity is the first step. At the same time, a flood of data began to arrive from much better sensors, and it often came in matrix form. The challenge was (and still is) to interpret that data and extract what is useful.

Along with these big external changes came, inevitably, a new set of options for our teaching. Examples became more important. Instead of inventing subspaces to study, the subspaces came from the matrices themselves: the row space and the column space of $A$ and $A^{\mathrm{T}}$. The ideas of basis and dimension and orthogonality apply to those concrete subspaces. The abstraction of a linear transformation need not come first!

Instead of starting with the abstract case, understanding emerged from the examples themselves. This is how most mathematicians think and learn. Why should the minds of our students not work in the same way?

There is certainly a danger that the new approach could also become too rigid. There seems to be general acceptance of an overall syllabus by textbook authors and textbook choosers. Algebraic ideas like independence, basis, and dimension combine with algorithms like $LU$ and Gram–Schmidt. Eigenvalues are introduced for a purpose: to decouple the variables and make the matrix diagonal. Computing $A$ and $e^{At}$ then becomes a one-dimensional problem. Factorizations of $A$ express the essential facts: $A = LU$, $A = QR$, $A = S\Lambda S^{-1}$, $A = U\Sigma V^{\mathrm{T}}$. Matrices that are important in practice (reflections, rotations, differences) provide genuine examples.

This subject is still partly driven by what happens outside the classroom. Further changes in the syllabus will come. In all of this renaissance I want to emphasize that the beautiful ideas of this subject are not suppressed! The very opposite, in fact; they become better understood and appreciated.

Now for the comparison and contrast with applied mathematics and computational science as a whole. This is a vast subject and our classroom time is so limited. I do not see rapid convergence to one fully developed core curriculum. My own course did converge over a twenty-five-year period to focus on key ideas, and those ideas went into a textbook. But more examples keep emerging, and new codes. Courses on applied and engineering mathematics are still in a (healthy) state of flux.

## How to Teach?

The heading here is a question and not a statement! Teaching is far too difficult, and success is too uncertain and ill-conditioned, to give an algorithm for success. I will give suggestions below, not rules.

A key point: *the subject is to be uncovered, not covered.*

It is natural to prepare for a class by deciding on a plan. Start with a question that it is important to answer. What is the inverse of a matrix and which matrices have inverses?

The requirements $A^{-1}A = I$ and $AA^{-1} = I$ are straightforward. But those are only letters! Examples are needed right away. Write down

$$\begin{bmatrix} 0 & 1 \\ 1 & 0 \end{bmatrix}, \quad \begin{bmatrix} 1 & 1 \\ 1 & 1 \end{bmatrix}, \quad \begin{bmatrix} 1 & 1 \\ 0 & 1 \end{bmatrix}, \quad \begin{bmatrix} a & b \\ c & d \end{bmatrix}.$$

Invert the first and third of these and describe $A^{-1}$ in words. Show that the second matrix is not invertible. (The best way to do this is to see a specific vector $x$ in the nullspace.) The fourth case above can reduce to a test for parallel rows or parallel columns. Determinants can be mentioned, as can nonzero pivots. Multiple ways of recognizing invertibility or its opposite are highly valuable.

The important point is that in working with simple examples you are giving the students a chance. The key is to build their confidence as active users of mathematics. The teacher has to be saying, in many ways but not in so many words, *you can do it.*

In an applied mathematics course, a correspondingly simple question might be the following.

What are the solutions to $\mathrm{d}^2y/\mathrm{d}x^2 = \lambda y(x)$?

Here we are looking for eigenfunctions. The matrix in $Ay = \lambda y$ is replaced by the second derivative. No boundary conditions have been imposed so far.

Now add the conditions $y(0) = 0$ and $y(1) = 0$. That should leave only the eigenfunctions $y = \sin k\pi x$ with their eigenvalues $\lambda = -k^2\pi^2$. The goal is to make the idea of an "eigenfunction" familiar. The answer was already known, and it illuminates the question.

I would like to emphasize the importance of *the teacher's voice.* All of us watch for verbal signals from a speaker: "This is exciting." "This is very ordinary." "Pay attention to this!" Boredom or enthusiasm come through so clearly. We are virtually announcing low expectations or raised expectations. If we are not interested ourselves, that message overrides our words. And fortunately, if our own curiosity about where a particular example leads is aroused, students understand what applied mathematics mostly is: *following an example to the end.*

Instead of consulting published references, I recommend a severe critique of lecture videos. You can find the author's own courses on the OpenCourseWare site (ocw.mit.edu): 18.06 (linear algebra) and 18.085 (computational science). The version 18.06SC of the former involves brief lectures on problem solving by six teaching assistants. What is it that makes each of them succeed or fail?

If only we knew more about teaching, we could define success.

## IV. Rachel Levy: Industrial Mathematics Inspires Mathematical Modeling Tasks

### Motivation

A common refrain from high school mathematics students goes something like this: "We have to learn this stuff for twelve years! Nobody ever even uses it!" As a mathematics professor and former middle and high school mathematics teacher, I am sadly unsurprised. Much of the mathematics written in textbooks, even when it is framed as practical, can strike students as overly academic and divorced from reality. Real problems require skills that most students never have the opportunity to practice. In this article I outline a set of skills that we can incorporate into assignments—initially one at a time, and eventually all together—to prepare students for the types of problems they will face in the real world, especially in science, technology, engineering, and mathematics fields. While I have compiled the list below with undergraduate students in mind, students would ideally have opportunities to practice simple versions of these skills throughout their mathematics education.

I used to think modeling meant story problems. And why not? After all, story problems come from real situations. But they do not ring true because they are too neat. Traditional story problems provide the task, the methodology, and the exact information needed to complete the task. Real problems are messy. Someone actually cares whether we solve them or not. We do not usually know how to solve them a priori. To prepare students to solve ill-defined, messy problems, we need to increase the cognitive demand we place on students by incorporating genuine and engaging modeling problems into our curricula. By cognitive demand I mean the complexity of the tasks and the degree of decision making required.

What types of mathematical modeling tasks can we provide? Of course, as modelers we might start with a model of modeling itself. Modeling can be viewed as an iterative process in which assumptions are made and then questioned, models are proposed and then refined, and data from real situations help to test the validity of the model through multiple versions of the solution. Models generally balance simplicity with realism, just as computations must often balance efficiency with accuracy. Every step of the modeling process may not be necessary and the process may not always proceed in the same order, but the spirit is one of iteration and balance.

With this iterative process in mind, we can begin to design tasks that enable students to practice modeling in meaningful ways. My ideas about mathematical modeling stem from my experience in modeling camps, British-style industrial mathematics study groups, and the Harvey Mudd College Mathematics Clinic. In these intensive experiences, groups of faculty and students gather to solve problems posed by companies, governmental organizations, and sometimes individuals. Rather than create a hierarchical taxonomy, I will suggest elements that can be incorporated into modeling tasks to increase their cognitive demand and make problems richer and more realistic.

## The Modeling Tasks

The order of the tasks below roughly follows one iteration of the modeling process. However, individual elements can be incorporated into assignments, to provide practice with various challenges from real-world mathematics.

**Tackle realistic problems.** Problems that can be described as "real," of course, are ones that have not yet been solved. The task of identifying interesting problems with an appropriate level of challenge that are still tractable could be an interesting assignment for students, or it could fall to the faculty. Many businesses have problems that they need to solve or processes they would like to better understand.

**Interact with a vested party.** Ideally, the modeling problem will have been posed by a party who cares about the outcome. When students interact with the problem sponsors (e.g., from a company) and end users of a project, they can practice professional communication and negotiation. If the students communicate with the sponsor on a regular basis, the solution will be more likely to satisfy the sponsor. The sponsor can provide realistic constraints—such as time, money, hardware, and software limitations—as well as information about company/sponsor culture and policy. Communication with the sponsor can motivate the students, since the problem they must solve matters to someone outside the classroom.

**Define the problem statement.** When a problem is first proposed in writing, the ideas put down on paper may not reflect exactly what the sponsor wants. Sometimes the sponsor will have omitted critical information or constraints that necessitate a specific approach. The suggested method of solution may not be the best approach for the proposed problem. After researching background on the problem, students can propose reasonable goals and their preferred approach. This often involves negotiation among the students about how to proceed as well as with the sponsor about the approach and final deliverables.

**Disregard extraneous information.** Modeling projects often begin with a literature search, in which students seek relevant approaches, data, or parameter values. The variety and scope of digitally available information can both facilitate and overwhelm problem-solving efforts. Sophisticated search strategies, such as those taught by librarians, can help students navigate the information overload. Students need to be given an opportunity to decide which of the available information about a problem should be used to develop their solution. Few textbook problems contain extraneous information, whereas with a real problem any available information (data, mathematical techniques, approaches) can be brought to bear. Students must determine which ideas and information are most salient to the problem at hand.

**Cope with messy data.** Textbooks (as well as many models) assume that data are normally distributed or that they follow some other distinguishable distribution, pattern, or trend. Real data often do not fall into a nice category and furthermore contain outliers, faulty entries, and missing values that need to be identified. In addition, the governing principles underlying the data may not be known.

**Define and justify assumptions.** Models simplify reality via assumptions. Even when students are presented with a model, they can discern and question the model's assumptions. They can also predict what might happen if the various assumptions are relaxed. When students do have the opportunity to make modeling assumptions, they can learn to identify ways to simplify problems and to balance generality with specificity.

**Choose an approach.** When we present mathematical models in lectures, we often have a particular approach in mind and actively steer the discussion in that direction. For example, when I introduce the mass–spring system in a differential equations class, I know which equation I want the students to use to model the system. Ideally, students will tackle modeling problems for which a variety of approaches could succeed. When that is not possible, students who debate the motivations for a particular approach can begin to see modeling as a set of choices based on underlying principles rather than an application of absolute and obvious laws.

**Combine mathematical, statistical, and computational skills.** In academic classes, students generally learn to apply a specific set of skills designated by the current text/topic, chapter, and problem set. In many texts, the title of the section will indicate which technique to use. Real problems leave more to the imagination. In addition, for most industrial problems, mathematics students must bring their programming skills to bear, including collaborative coding, version control, commenting, testing, validating, and debugging. Projects sometimes also require proficiency in and use of a particular piece of software, even if the students have a different preferred platform or solution.

**Validate and test results.** Once students have run their models, how can they conduct a sanity check on their results? If they have used data to create a model or algorithm, did they reserve some data for testing purposes? Under what circumstances should the results be most accurate? What metrics best define a good solution? Can the model run in a reasonable amount of time? Does the solution meet the requirements of the sponsor? Students must report the accuracy and reliability of their data, assumptions, and model as well as estimate the error introduced by the parameter values and data and methods they have used.

**Iterate to refine the solution.** Iteration can begin at many stages, as rethinking becomes necessary. Preliminary results may uncover unnecessary assumptions or new model requirements. While iteration is always a potentially useful step in the process, it is less likely to be feasible for short-term projects.

**Draw conclusions.** Once students have designed, refined, and run their model, they must decide what conclusions they can draw from the results. The conclusions will of course depend on the assumptions and choices that were made in the previous parts of the modeling process, and on the results.

**Communicate results to both general and technical audiences.** Deliverables can provide opportunities for students to practice mathematical writing, software documentation, peer review, and public speaking, as well as visual communication of key ideas and relationships. People interested in the results of their model will likely have various levels of understanding of the problem and its solution. Students will need to be able to communicate effectively with managers and officers as well as with technical teams. Regular communication with a sponsor, mid-project oral updates, and periodic written reports can all provide opportunities to practice these skills before deliverables are due.

## Practical Matters

Typical textbook story problems rarely require any of the skills discussed above, even though many of the tasks may be required in jobs that involve mathematical modeling. Modeling camps, study groups, and industrial mathematics workshops provide opportunities to practice combinations of the skills, but not every student will have those opportunities or be prepared to fully participate. As a way to provide practice, one or more of the skills can be incorporated into a mathematics course through class activities, projects, and homework problems. In this way students can benefit from practicing each skill in isolation before attempting to combine them.

A simple place to start might be to incorporate some extraneous information into a problem set. Another possibility would be to give students a raw data set and ask them how they would approach the information to draw a conclusion. Governmental organizations, such as the U.S. Environmental Protection Agency, make data and models available online. A third option could require students to analyze a problem and suggest a course of action. The Mathematical Contest in Modeling provides 10 years of online example problems and winning papers.

Long-term projects, such as the year-long Mathematics Clinic at Harvey Mudd College, require students to practice all of the skills above. With long-term projects students have the opportunity to work in teams and practice project management. With some guidance, students can learn to prepare reasonable time lines, plan for failure and other contingencies, and assign tasks so that each team member contributes to the solution.

As we design modeling activities for students, we should consider how much iteration is feasible. Ideally, we would discuss the rationale for iteration with students even when they do not have time to implement the iteration themselves. As we teach mathematical modeling and provide mentoring, we also need to decide how much scaffolding we will provide to lead their modeling efforts in a desirable direction. Cognitive research in mathematics education supports the idea that in the early stages of learning, worked examples can promote skill acquisition, but problem solving is superior during later stages. While we may want to provide students with a framework for approaching particular problems, scaffolding can unintentionally reduce the cognitive demand of the task. Students need to struggle in order to develop strategies for real, messy problems. Therefore, as students advance, we must gradually take away the scaffolding and make sure that the problems are tackled by the students rather than being demonstrated to them.

As we teach mathematical modeling, the tasks outlined above can provide students with opportunities to develop individual skills they can apply to real problems. When they can combine the skills, work well individually or in teams, and produce relevant results for real unsolved problems, they will be ready to make valuable contributions to modeling problems in science, technology, engineering, and mathematics fields. We can provide these challenges through problems with high cognitive demand, inspired by industrial mathematics.

### Further Reading

Borrelli, R. 2010. The doctor is in. *Notices of the American Mathematical Society* 57(9):1127–29. (This article describes the Harvey Mudd College Mathematics Clinic.)

Renkl, A., and R. Atkinson. 2003. Structuring the transition from example study to problem solving in cognitive skill acquisition: a cognitive load perspective. *Educational Psychologist* 38:15–22.

Soon, W., L. T. Lioe, and B. McInnes. 2011. Understanding the difficulties faced by engineering undergraduates in learning mathematical modelling. *International Journal of Mathematical Education in Science and Technology* 42: 1023–39.

Stein, M. K., M. S. Smith, M. A. Henningsen, and E. A. Silver. 2000. *Implementing Standards-Based Mathematics Instruction: A Casebook for Professional Development*. New York: Teachers College Press.

Williams, J. A. S., and R. C. Reid. 2010. Developing problem solving and communication skills through memo assignments in a management science course. *Journal of Education for Business* 85:323–29.

## VIII.8 Mediated Mathematics: Representations of Mathematics in Popular Culture and Why These Matter

*Heather Mendick*

### 1 Introduction

As geek becomes increasingly chic, media representations of mathematics and mathematicians proliferate. Writing this from England, I can currently tune into regular television episodes of U.S. mathematics-solves-crime drama *Numb3rs* and U.S. sitcom *The Big Bang Theory*, which features three physicists and an engineer among its five central characters. In 2012 U.K. comedy channel Dave launched a new show called *Dara Ó Briain: School of Hard Sums*, in which host and mathematical physics graduate Ó Briain competes against fellow comedians at tasks set by professor of mathematics (and occasional television presenter) Marcus du Sautoy, with two mathematics undergraduates in the studio as backup problem solvers. In this article I will be exploring such popular cultural texts, analyzing the ways in which they portray mathematics and mathematicians.

First, something about me. It feels strange to be writing for the *Princeton Companion to Applied Mathematics*. Although my first degree was in mathematics and I taught it for more than seven years, as an academic I identify and work as a sociologist of education. The editors kindly agreed that I be excused from the requirement to use LaTeX, a program of which I was previously ignorant. My move from mathematics to sociology has been accompanied by many changes. For example, my hair is no longer waist length and unstyled but short and dyed a vivid shade of red. My epistemological view has also shifted. In their book *The Mathematical Experience*, Philip Davis and Reuben Hersh joke that "the typical working mathematician is a Platonist on weekdays and a formalist on Sundays." Indeed, when I studied and taught mathematics I took a Platonic position: I felt that I was experiencing and relating to absolute, objective knowledge. Now I see mathematics, like all knowledge, as a social construct arising from human action in the world rather than being external to it.

So, following this approach, mathematics comes into being through the constellations of meanings that circulate about it, the stories that we (choose to) tell about it, and the ways in which people take up positions within those stories and are positioned by them. Looking at it this way, the idea that mathematical knowledge is absolute is a story about mathematics, albeit one that is very powerful and that I used to make sense of what I was doing for about the first thirty years of my life. This story undoubtedly had the social function of binding me to an "imagined community" of mathematicians, who had the ability to access this unique form of knowledge. This story about mathematics is thus also one about who I am (and who I am not); it is part of the way that I constructed my*self* (and the story that I am telling here, of moving away from this, is part of how I construct myself now). Looking back, I wonder how this impacted on my views of other people and other ways of knowing. Perhaps this story had a psychic function:

what role did this conception of certain knowledge, and the promise of control that it offers, play for me?

Thus, my intellectual project now is to identify the stories we tell about mathematics—the meanings that we give to mathematics—and to understand their patterning and their effects. We can, of course, find these meanings in workplaces, school classrooms, and university lecture halls. But, in contemporary society, the media are a primary site for the circulation of meanings, so this too is a good place to look for mathematics. My purpose in identifying media stories is to open up different ways of thinking about, engaging with, and teaching mathematics. This article is split into four main parts. The first two focus on two types of mathematics (everyday and esoteric) playing out in media representations and the identities associated with these. And in the next two I focus on tracking popular representations of mathematics via key emotions: pleasure and excitement, fear and boredom. I show that, while all are held responsible for learning mathematics, not all are perceived to have mathematical ability. I end by offering some brief reflections on using popular culture within university mathematics education.

## 2 The Paradox of Relevance—or Do We All Use Mathematics Every Day?

In a reflective mood, Abraham Lincoln, played by Daniel Day Lewis in *Lincoln*, Hollywood's 2013 film account of the abolition of slavery in the United States, recalls reading Euclid's *Elements*:

> Euclid's first common notion is this: "things which are equal to the same thing are equal to each other." That's a rule of mathematical reasoning. It's true because it works. Has done and always will do. In this book, Euclid says this is "self evident." You see, there it is, even in that 2000-year-old book of mechanical law. It is a self-evident truth that things which are equal to the same thing are equal to each other.

Here, Lincoln is presented as using Euclid to support the obviousness of race equality. While there is no evidence that he actually did this, what is interesting is that a Hollywood movie associates mathematics with empathy and justice. This is a common association in popular mathematics. My favorite example comes from the classic teen comedy *Mean Girls* from 2004. In an interschool competition for teams of "mathletes," the film's heroine, Cady Heron (played by Lindsay Lohan), is in the sudden-death round of the final. Standing at the podium she has a revelation. Her internal monologue during this runs:

> Calling somebody else fat won't make you any skinnier. Calling someone stupid doesn't make you any smarter. And ruining [frenemy] Regina George's life definitely didn't make me any happier. All you can do in life is try to solve the problem in front of you.

This insight is part of the reasoning process through which Cady secures the right answer to the limit problem in front of her, thereby securing victory for her team.

In these examples, mathematics is constructed as part of our common humanity, intrinsic to our everyday being and reasoning. There are many popular examples that do this in other related ways. Game shows (such as *Countdown*, *Deal or No Deal*, and *Who Wants to Be a Millionaire*) present mathematics as part of "general knowledge" rather than being the province of only a lucky (or unlucky) few. The success of puzzles like sudokus and kakuros, and of computer games like Tetris and Dr Kawashima's Brain Training, bring mathematics into people's lives, as does interacting with mathematics in magazine quizzes and through sporting events (try working out finishes in a game of darts, for example). Via the media, we are continually shown or told that mathematics is useful or even necessary for familiar everyday practices such as calculating profits (in children's fantasy *Matilda*) and counting calories (in *Mean Girls*) or for unfamiliar but important ones such as winning wars (in World War II films *Enigma* and *Dambusters* and Cold War movie *A Beautiful Mind*), criminal activity (in glamorous film *Ocean's 11*), and crime fighting (in television series *Numb3rs*).

Given the ubiquity and utility of mathematics in popular culture, we may expect this to have resolved what Ole Skovsmose refers to as the paradox of relevance: that "on the one hand, mathematics has a pervasive social influence and, on the other hand, students and children are unable to recognise this relevance." However, recent research shows that this paradox persists. I think this is because of the many contradictions in how popular mathematics is woven into our "life-worlds." Some of these are evident when newspapers reassure people that sudokus are *not* mathematics or when the normally erudite presenter Melvyn Bragg describes himself as "blinking in the face of this [mathematical] assault" and "completely intrigued by this out of space thought that you mathematicians go in for." Mathematics is *both* everyday activity *and* "out

**Figure 1** The Eppes brothers, Charlie and Don, show off their contrasting approaches to fighting crime.

of space thinking"; it is *both* for all *and* the specialized province of "you mathematicians." Now I want to explore these contradictions by looking in detail at the television series *Numb3rs*.

*Numb3rs* (2005–10) ran for six seasons (118 episodes in all) in the United States on the CBS network and continues to be shown in the United Kingdom on the station 5USA. The narrative centers on two brothers. Older brother Don Eppes (Rob Morrow) is the lead Federal Bureau of Investigation (FBI) agent at the Los Angeles violent crime squad, while his younger brother Charlie (David Krumholtz) works as a professor of mathematics at the local university, the California Institute of Science (CalSci). Each episode shows Don and Charlie joining forces to solve a new crime. They work alongside Alan (their self-proclaimed "FBI dad"), Don's FBI team, and two of Charlie's CalSci colleagues: Amita Ramanujan, initially his doctoral student and later his colleague and girlfriend, and quirky physicist Larry Fleinhardt. *Numb3rs* contains perhaps the most blatant and pedagogic instance of popular culture presenting mathematics as an everyday activity undertaken by all: the voice-over during the credits sequence for the first two seasons (after this, the show followed the increasing trend of eliminating the opening credits). The season one voice-over insists:

> We all use math every day: to forecast weather; to handle money. We also use math to analyze crime: reveal patterns; predict behaviors. Using numbers we can solve the biggest mysteries we know.

This plays out in the show, as mathematical "genius" Charlie seems to spend as much time in the FBI offices as at his university, appearing largely unencumbered by the seminars and meetings that fill my working life as an academic. He is therefore always on hand to provide an application of mathematics. His uses of mathematics go beyond analyzing, and inevitably solving, crime after crime. In a notable example, in season four, Charlie writes a self-help book on the mathematics of friendship (*The Attraction Equation*) that quickly becomes a best-seller. Later, in season five, he, Larry, and Amita apply mathematics and physics to basketball in an attempt to halt CalSci's entrenched losing streak.

Throughout the seasons we see Charlie—and to a lesser extent, Amita, Larry, and other characters—talking about mathematics. But each episode also contains some distinct "math-bits." These focus on the use of analogies to convey a mathematical idea. For example, in an episode where a supercomputer is suspected of murder, a math-bit is used to explain the Turing test via an analogy between computers and roses. Roses may be real or artificial or they may be genetically modified so as to be artificial but indistinguishable from real, hence passing the Turing test for roses (if one existed). While the rest of the series has a common realistic visual style looking much like any mainstream U.S. crime series, these math-bits look very different. They feature selective use of vivid colors against a black and white three-dimensional grid-like background; images of objects and formulas move around rapidly as if choreographed in time, with a voice-over from Charlie (or occasionally, in later seasons, another character). In an interesting move, the transitions into and out of the ads, and some other scenes, use the three-dimensional grid-like graphic from these math-bits. This has the effect of encouraging us to frame the evolving events as manipulable mathematically, which can be understood as an instance of what Skovsmose calls the formatting power of mathematics. This names the way "that mathematics produces new inventions in reality, not only in the sense that new insights may change interpretations, but also in the sense that mathematics colonises part of reality and reorders it."

Watching the math-bits, even when I follow the analogies, while they help me understand what mathematics can do, they give me no access to the mathematics itself. Charlie is repeatedly referred to as a genius, and these math-bits seem to be intended to provide us with insights into mathematical minds like his rather than to indicate something that all minds could do. So rendering the everyday mathematical requires both a process of transformation and a person to conduct that transformation. In this way even the idea of mathematics as within the everyday relies on a construction of mathematics as an esoteric skill carried out by an elite, something I now discuss.

## 3 The Mathematical Mystery Tour

In the last section I showed that a tension runs through popular culture and, more widely, between stories of mathematics as everyday and accessible to all and as esoteric and accessible only to an elite. By looking at *Numb3rs* I showed that the same text can tell both stories simultaneously by suggesting that mathematics is useful to all but inaccessible to most. In *Numb3rs*, while Charlie and the others who do mathematics may occasionally express uncertainty and need help, the mathematics itself has no similar vulnerabilities; it is hard, absolute, certain. This idea of mathematics has become common sense. It is perhaps most explicitly stated by G. H. Hardy in his book *A Mathematician's Apology*:

> A chair or a star is not the least like what it seems to be; the more we think of it, the fuzzier its outlines become in the haze of sensation which surrounds it; but "2" or "317" has nothing to do with sensation, and its properties stand out the more closely we scrutinize it ... 317 is a prime, not because we think so, or because our minds are shaped in one way rather than another, but *because it is so* [original emphasis], because mathematical reality is built in that way.

As I said above, this is not a position I share. But I am not concerned here with arguing against it; instead, I want to look at how this story (of mathematics as absolute and objective) is told within popular culture and at what effects it has. In particular, I want to track its relationship with mathematical elitism.

In popular culture, the absoluteness of mathematics is often supported by associations with spirituality, mystery, or magic. For example, in Dan Brown's 2003 novel *The Da Vinci Code*, cryptographer Sophie Neveu (played in the film version by Audrey Tatou) is revealed to be a direct descendant of Jesus and Mary Magdalene, an embodiment of the "sacred feminine," something that must remain in balance with the masculine. In the Darren Aronofsky film *Pi* (from 1998), a Jewish follower of Kabbalah believes that the lost name of God can be found within the decimal expansion of $\pi$. While (as I discuss further below) in Michael Crichton's 1991 novel (and Crichton and Steven Spielberg's 1993 film) *Jurassic Park*, the predictive power of the mathematics of chaos takes on a magical quality.

This leads to a representation of those who can access the secrets of mathematics as gifted and special. For example, Charlie Eppes (of *Numb3rs*) was five years ahead of his age at school, entered Princeton at thirteen, and got his first journal article published at the tender age of fourteen. Indeed, the mathematicians we see on big and small screens are nearly always "geniuses," from the Nobel laureate John Nash (played by Russell Crowe) in the 2001 biopic *A Beautiful Mind* to *Pi*'s Max Cohen (played by Sean Gullette), who searches for patterns in $\pi$. This specialness is reinforced by their presentation as unstable, with Nash suffering from lifelong schizophrenia and Max pursuing his work beyond obsession. These mental-health problems are often directly linked to their mathematical "gifts." Scenes showing people doing mathematics usually involve frenetic scribbling on any available surface, including mirrors and windows when the more usual whiteboards and blackboards are unavailable. For me, the most striking example of a direct link between madness and mathematics is a fantasy sequence in which Max is depicted applying an electric drill to his own head, metaphorically excising the mathematical "gift" from his brain before he can go on to a happier and more relational future, but one in which he has lost both the desire and the ability to calculate. For these and the other big- and small-screen geniuses, mathematics is viewed as defining their whole personality and infusing every aspect of their lives. In an extreme but not atypical example, the married pedophile mathematician in Kate Atkinson's 2005 novel *Case Histories*:

> didn't really feel the need for another person in his life, in fact he found the concept of "sharing" a life bizarre. He had mathematics, which filled up his time almost completely, so he wasn't entirely sure what he wanted with a wife. Women seemed to him to be in possession of all kinds of undesirable properties, chiefly madness, but also a multiplicity of physical drawbacks—blood, sex, children—which were unsettling and *other* [original emphasis].

This also makes explicit how, despite mathematics being represented as absolute and therefore outside of society (or perhaps even because of it), the default mathematician is a (white, middle-class, heterosexual) man.

This gendering of mathematics is apparent in the only scene in *A Beautiful Mind* where we see some actual mathematics: Nash's work on game theory. Here is a brief description.

> The scene is a bar with upbeat music playing. The camera focuses on a tall blonde (among a group of women)

and then on a group of male mathematics graduate students staring at her. The exception is John Nash, who is working, surrounded by papers and books, piled haphazardly and with a pint of beer. His fellow students draw his attention to "the blonde." They look at the group of young women, who then look back at them. Nash looks uncomfortable. One student, Martin, makes reference to Adam Smith's theory that "in competition, individual ambition serves the common good." "The blonde" looks at Nash. His fellow students joke about Nash's lack of success with women. There is a change in Nash's posture and a change from upbeat jazzy music to softer piano music. He smiles and says "Adam Smith needs revision." He explains that if they all go for "the blonde" they will block each other and upset the other women; however, if they cooperate, and none of them go for "the blonde," they can all be successful. During this exposition the images become surreal and blur slightly, as if the characters are puppets illustrating Nash's conjectures. We get an aerial view where, in a geometrical pattern, we see all the men going for "the blonde," then going for the other women. The camera pans from a close up of Nash to his mathematical "visions." This happens alongside changes in Nash's tone of voice, from nervous to authoritative, and the loss of his bodily twitches. This sequence ends with Nash saying, "that's the only way we win, that's the only way we all get laid," as the music returns to jazzy. Hastily, Nash gathers his papers and leaves. He pauses by "the blonde," says "thank you," and rushes out. She looks puzzled.

There are close links between this and the math-bits in *Numb3rs*, as we leave the realistic space of the drama and enter the figurative space of the mathematical model. Paying attention to how women are positioned within this narrative raises questions about who is represented as being able to do mathematics. "The blonde" acts as the silent muse for the creativity of the great male genius. This positioning of women as handmaidens to, and inspiration for, creativity, but not as creative agents in their own right, is common in texts; it speaks to the ways in which our very notions of creative (mathematical) thought are gendered, a theme that is explored in detail in the work of Valerie Walkerdine.

There are women doing mathematics in popular drama but they mostly exist, like Amita in *Numb3rs*, as daughters, students, and/or love interests of more central/established male mathematicians. Even Amita's surname, Ramanujan, suggests her heritage from a male mathematician. I have already briefly mentioned Cady Heron from *Mean Girls*, one of an emerging generation of screen "smart girls." Yet she spends most of the film pretending to be bad at mathematics in order to attract the attention of Aaron Samuels, the best-looking boy in her calculus class. In the mathlete-sudden-death scene discussed earlier, Cady has to literally and metaphorically see past Aaron, the object of her desire, in order to get her question right. He is a distraction from mathematics rather than an inspiration and support for it; he also gets both a name and a voice, unlike "the blonde" in *A Beautiful Mind*. *Numb3rs* does briefly introduce a senior woman, Mildred Finch, but she operates largely as a manager rather than a mathematician and is a love interest for the more established male character of Alan Eppes. After nine episodes she disappears from the series without explanation.

Turning to social class, the 1997 film *Good Will Hunting* contains a rare example of a mathematical genius, Will Hunting (played by Matt Damon), from a working-class Irish–American background. Will, entirely self-taught, works as a janitor at the Massachusetts Institute of Technology (MIT), and the film tells the consequences of his "discovery" by MIT mathematics professor and Fields medallist Gerald Lambeau (played by Stellan Skarsgård). Will is depicted differently from the middle-class mathematicians. As Marie-Pierre Moreau, Debbie Epstein, and I noted in 2009:

> There is a scene when he loses control reacting with severe violence when he meets a man who abused him as a child. The element of physical violence in the way Will expresses his emotions contrasts with [*Numb3rs*] Charlie's (more middle-class) ways. This incident happens prior to Will entering the mathematical community, embarking on a course of therapy, and falling in love with a wealthy Harvard student, thus suggesting that the story of Will is also one of redemption through incorporation of middle-class practices.

So *Good Will Hunting* is very much the story of Will's "middle-classification," as, in becoming a mathematician, he is required to embrace the values of the middle class and to leave his (working-class) neighborhood, friends, and job behind. Will—like Charlie Eppes, Max Cohen, and John Nash—reinforce the association of mathematics with whiteness. We know that this white male middle-class image is easily called up when young people are asked to imagine a mathematician. In the next two sections I want to continue tracking who is positioned as "un/able" to do mathematics in popular culture by turning to the dominant emotions evoked in relation to mathematics in that space. I have crudely divided them into the positive emotions of pleasure and excitement and the negative emotions of boredom and fear, although, as I hope to show, things are more mixed up than that taxonomy suggests.

## 4 Pleasure, Excitement, and the "Ability" to Do Mathematics

If you are reading this, you are probably more likely than most to find pleasure and excitement in mathematics. However, such feelings are often laced with fear. In my own research into advanced mathematics in England, I vividly remember one young woman, who had chosen to continue with the subject and was doing well at it, saying "sometimes I dread going into [mathematics]"; while Robert Early (1992, p. 15), in his research into students' feelings about mathematical challenge, quotes one powerful and terrifying account: "I felt as though I was jumping rope on a razor blade, and with each jump blood trickled onto the blank paper below me." In some ways, the greater the personal investment in the subject, the greater the fear, for there is more to be gained, or lost, by success or failure. I look at how these and other fears are mobilized in popular culture in the next section. Here, I focus on representations of mathematical pleasure and excitement.

As illustrated above, while lots of people *do* mathematics in popular culture, there is very little actual mathematics. Sound and vision indicate mathematics as beautiful and pattern oriented but avoid details; formulas abound but without any indication of how to read them. Indeed, I have written elsewhere about the problems that people have in seeing mathematics as enjoyable, showing how they usually construct it as the opposite of music and other forms of popular entertainment. This polarization is evident in how those who do enjoy mathematics are positioned as being a particular type of person. As I have shown, they are usually geniuses, and they are also usually nerds or geeks.

These terms are difficult to define precisely and to distinguish from one another; they intersect with genius. Essentially, they capture a figure who is male but often physically weak and/or overweight, heterosexual but awkward with women, white (or occasionally East or South Asian), and academically intelligent but socially incompetent. The Urban Dictionary Web site offers a range of definitions, the pithiest and most highly rated of which is: "The people you pick on in high school and wind up working for as an adult." This captures the tension between awe and derision within the cultural gaze on geeks, a tension that reaches its apotheosis in "geek chic." Here are three definitions of "geek chic," also taken from Urban Dictionary:

> An obvious oxymoron, "geek chic" emerges from oxygen deprived hallucination in which geeks evolve into actual existence as a sort of technocracy radiating the holy aura of cool. You will find no more crushing an argument against geek chic than Bill Gates, who despite being the richest man in the world sports an apparent $5 coiffure and birth control glasses.
>
> Clothing or accessories that are very geeky/nerdy and yet, at the same time, says [sic] "I'm cool because I'm proud of the fact that I'm a nerd, and am not afraid to dress the part."
>
> Geek Chic within it's self [sic] is an oxymoron and ironic twist of events by people who are generally not geeks, attractive men and women who want to act like geeks but don't actually have the passion for geeky hobbies or the intellect. Real geeks got picked on, and were proud to call themselves geeks because they could construct their own PCs or were very good at math or something, not because they spend massive abouts [sic] of times [sic] on Xbox play [sic] *Call Of Duty* or in [sic] some consumer electronic [sic] like an iPod or iPhone. Real geeks consider iPods and iPhones basic shit.

In the first definition we see the intensity of the hatred that can be directed at geeks via a strong objection to the claim that geeks could ever be chic. Contrasting with this, the next two definitions are written from geek positions' they both locate value in geeks for being proud of who they are despite others teasing and bullying. The first and third definitions assert that "geek chic" is oxymoronic but for different reasons: one rejects geek chic, asserting geeks' inbuilt inferiority; the other constructs an authentic geekness, asserting geeks' superiority over those who have the desire to be "real geeks" but lack the passion and intelligence. All three definitions support the construction of mathematics and those who do this as "special." They differ on whether it is good to be a geek or not, but they agree that you either are one or you are not. There is no sense of engaging with and enjoying mathematics or technology as one might other school subjects, like history or Spanish, as an interested, informed amateur.

The most successful series to capitalize on the rise of geek chic is the sitcom *The Big Bang Theory* (2007–). The program centers on four male geeks and their unlikely friendship with the blonde, stylish, sociable, would-be actress Penny. Although they are physicists and engineers, they are often shown doing mathematics. Three are white (one of whom is Jewish) and one

**Figure 2** Howard, Sheldon, Leonard, Penny, and Raj (left to right) pictured in the apartment that Sheldon and Leonard share.

is Indian. Leonard, the most conventional and sexually successful of the four, is the most ashamed of his geekiness. For example, he conceals the fact that he plays word games in the science-fiction language Klingon. He suffers from a range of health problems and nervous disorders including lactose intolerance, sleep apnoea, migraines, carsickness, nose bleeds, and asthma. He tilts his head when he talks and has a tendency to whine; he regularly applies excessive hair gel and wears mismatched clothing. At the other extreme is Sheldon, who is unashamedly geeky. He interprets everything literally (needing to be taught to recognize irony), feels compelled to order his surroundings (even when this involves breaking into his new neighbor's flat), and worries that massaging his own shoulders involves excessive physical contact. In between these two on the geekiness scale are Howard, who lives with his overbearing mother, and Raj, who cannot talk to women while sober. Women have a strange position in the show, as Moreau and I noted in *Debates in Mathematics Education*:

> Most of the female characters, like Penny, exist to contrast with the male geeks, and implicitly emphasise their heterosexuality. The two female geeks are peripheral characters introduced as potential love interests for the men. They stand out like the one female geek and one gay geek included in a total of 42 geeks featuring across five seasons of US Reality TV show *Beauty and the Geek*.

While mathematics may be depicted as neutral knowledge, geek stories like those of geniuses reinforce the idea that the "ability" to do mathematics belongs in particular bodies. In these and other shows, popular culture makes jokes with and about geeks but also values them for being intelligent and unafraid of being different. In *The Big Bang Theory*, Penny clearly enjoys hanging out with the geek gang and she dates Leonard and has a one-night stand with Raj. Their ability to find fun in mathematics, science, and technology seems to extend into other areas.

From Howard's dependence on his mother to Leonard's preponderance of childhood illnesses, there is a boyishness to the way geeks are depicted. All of *The Big Bang Theory*'s characters own a large number of comic, fantasy, and science-fiction themed toys and memorabilia and are regularly shown playing games. However, their skills are also depicted as being immensely valuable, securing significant government funding; this is exemplified when Howard's technical expertise wins him a place on a space mission despite his physical inadequacies when compared with the other astronauts. Within popular culture, this power of mathematical knowledge is inseparable from excitement.

I end this section with one last example of power and excitement: *Jurassic Park*, which also contains perhaps the only screen example of a "cool" mathematician. In both the book and film versions, genetic engineering has been used to manufacture live dinosaurs in order to create a prehistoric amusement park. The narrative takes place on a weekend inspection of this project prior to opening by two paleontologists, a lawyer, and a mathematician, Ian Malcolm, accompanied by the park's driving force, John Hammond, his two grandchildren, and other park employees. Malcolm, played by Jeff Goldblum in the film, wears black throughout (including shades and a leather jacket), flirts outrageously, and "suffers from a deplorable excess of personality, especially for a mathematician." He is clearly able to use chaos theory to predict the ultimate failure of Jurassic Park and its end in disaster and death. Even when he is told that all the dinosaurs are female, he correctly predicts that they will sexually reproduce: "life will find a way." This predictive power is most apparent in the book, where each section is introduced by a quotation from Malcolm and by a growing image of a fractal:

> At the earliest drawings of the fractal curve, few clues to the underlying mathematical structure will be seen.
>
> With subsequent drawings of the fractal curve, sudden changes may appear.
>
> Details emerge more clearly as the fractal curve is redrawn.

> Inevitably, underlying instabilities begin to appear.
>
> Flaws in the system will now become severe.
>
> System recovery may prove impossible.
>
> Increasingly, the mathematics will demand the courage to face its implications.

In the book, and to a lesser extent in the film, Malcolm articulates a critique of scientific progress. He compares the actions of Hammond at Jurassic Park to those of a child with his father's gun: since he and his scientists did not work for the knowledge they are using, they exercise no responsibility in relation to it. He questions the value of discovery, calling it a violent and penetrative act and comparing it to rape. But even as "progress" is radically challenged, this is undermined by the role that mathematics plays in understanding and ultimately controlling events. The unknowing audience look on in wonder and relief at both the dinosaurs and at the power of mathematics and of the "mathemagician" who wields it. In the final section of this article I consider the way in which popular culture portrays those of us in the audience looking on.

## 5 Boredom, Fear, and the Responsibility to Do Mathematics

In the "Brotherhood" episode of television drama series *Six Feet Under*, teenager Claire Fisher is shown in a high school algebra class, unable to explain the formula on the board. Her teacher reprimands her, "well maybe if you paid attention in class instead of reading." Claire responds, "well maybe if you talked about something that was actually gonna be useful to me I would," and returns to reading her book. Her teacher persists: "Oh algebra is useless. Hmm, know a lot of physicists who'd beg to differ." Claire, sullenly looking up, replies, "well I don't want to be a physicist." Her teacher, arms unfolded, speaks lyrically: "algebra forces your mind to solve problems logically, it's one of the only perfect sciences." Claire interrupts: "Do you think the world runs on logic? Open your eyes." Her teacher's final sally is "ok, I'll see you after class Miss Fisher." At this point Claire stares very intensely at her teacher from whose head steam is beginning to rise; her head then explodes, and Claire laughs.

Here we see a familiar trope in popular culture: teenage alienation. Mathematics stands for all that is most boring and pointless about education. The scene offers a vivid dramatization of the failure of mathematics to resolve the paradox of relevance—the juxtaposition of a personal feeling of the futility of mathematics with a generalized belief in its utility. Here and elsewhere mathematics figures as a counterpoint to imagination and creativity. When asked to draw a teacher, the most common response of primary-school children is to draw someone smartly dressed, standing behind a desk and/or in front of a blackboard, and teaching formal mathematics. In another example taken from the "Bart the Genius" episode of *The Simpsons*, ten-year-old Bart, faced with a mathematics test, spends the whole of the allocated time engaged in a daydream provoked by the first test question: a word problem about two trains headed in opposite directions. Both these examples suggest that the boredom response is not simply about the dullness of mathematics but also indicates a fear of engaging with the subject. Perhaps boredom partly serves as a defense against failure in mathematics. To understand why such a defense is so important to so many people, we need to look at the status conferred upon mathematics.

In the example from *Six Feet Under*, we can also see the obligation placed upon Claire, by her teacher and by society more widely, to learn mathematics. Policy and public discourses commonly present mathematics as a key to both national progress (economic and scientific/technological) and individual progress (empowerment, employment, and success). For example, in 2004 Adrian Smith was commissioned by the U.K. government to undertake a major enquiry into mathematics education after the age of fourteen. Near the start of his report (published under the title "Making mathematics count") he states:

> It has been widely recognised that mathematics occupies a rather special position. It is a major intellectual discipline in its own right, as well as providing the underpinning language for the rest of science and engineering and, increasingly, for other disciplines in the social and medical sciences. It underpins major sectors of modern business and industry, in particular, financial services and ICT. It also provides the individual citizen with empowering skills for the conduct of private and social life and with key skills required at virtually all levels of employment.

Here we can see how individual fears associated with mathematics (of individual failure, social exclusion, being judged) are brought together with national fears (of national failure, economic exclusion, being uncompetitive). Mathematics can bring together these fears because it serves as a signifier of intelligence and technological progress. The books *Do You Panic about Mathematics?* (1981) and *Overcoming Math Anxiety*

(1978) by Laurie Buxton and Sheila Tobias, respectively, tracked how mathematics anxiety and panic derive from people's feelings of being judged via mathematics and found wanting. Research has shown that the introduction of national testing into primary schools in England has meant that ever-younger children are exposed to high-stakes assessment and to grouping by "ability" in mathematics, thereby increasing the anxiety and fear attached to the subject.

Government advertising to promote take-up of adult mathematics provision has played on these fears. In these advertising campaigns, we find people who are unable to do mathematics and who are urged to take up the opportunity to "get rid of your mathematics gremlin" and "become mathematics confident." Thus I end this overview of popular culture with a discussion of advertisements from the two most prominent recent U.K. adult numeracy campaigns. The first is called "Bad Dad" (it can be viewed online at www.youtube.com/watch?v=CzQvD9oBx-0) and comes from a series of ads that use gremlins as a metaphor for people's gaps in mathematical and English skills. It is filmed with very little color, perhaps suggesting the darkness of ignorance, in which its central character resides, the eponymous "bad dad." It opens with the camera on him: white, aged 35–40, overweight, casually dressed, sitting on a drab sofa, watching television. This figure draws on clichés of the "feckless working class," whose unhealthy bodies are imagined to be permanently stuck in front of the television.

> The camera turns to his daughter, about ten years old, with neatly combed hair, wearing a school uniform, who calls from the next room: "Can you help me with my maths?" At this her dad's face shows panic and we see his ugly grey gremlin lounging on the sofa with pointy ears and nose. The gremlin speaks with obvious disdain: 'You. Maths. That's a good one." The camera focuses on dad's worried face as he suggests: "Ask your mother." When his daughter tells him "She's out," the gremlin taunts, "Oooooh! She's gone out. What are you going to do now?" Dad, becoming more panicked, seems out of breath (again suggesting his unhealthy body), asks: "Why don't you use your calculator?" Daughter: "That's cheating!" The gremlin scolds: "Bad dad. Very bad dad!" The daughter enters the room and looks at the television screen with confusion: "Dad. I need to do it now!" Her dad looks ashamed.

The contrast between the hard-working, neat girl and her lazy, untidy dad is striking. It directly invokes fears of being a bad parent, who cannot support their child's education and fails to inspire them to "high" aspirations. The solution offered is to take advantage of the available adult education classes and so transform oneself.

**Figure 3** Beryl as a school girl "too scared" to put up her hand and ask about mathematics.

The second ad centers on a woman called Beryl, a name associated in England with elderly working-class women (this can be viewed online at www.youtube.com/watch?v=-xrQoU6qZLU). Although her name is not revealed in the ad itself, her age and social class are indicated visually via her hair curlers, cup of tea, and other signs of "ordinariness." All the characters are created using hand puppetry and a small number of props.

> Beryl describes her difficult and troubling experiences with mathematics when she was at school: "Well you know looking back I know the exact time and place where I lost my confidence with mathematics. It was back in class 4C and I lost my ways with times tables." A male teacher, carrying glasses and wearing a mortarboard and bow tie (both symbols of middle-class teacherly authority), appears in order to chant times tables: "Seven eights are fifty six, eight eights are sixty four, ...." Beryl goes on to explain how taking the adult mathematics course has helped to improve her confidence: "I was too scared to put my hand up and say it's all gobbledygook to me. But I've just taken this free adult mathematics course at my local college. They start you at the point where you got lost and now my fear of mathematics has just gone away."

The contrasts between the teacher and Beryl emphasize her social class and gender.

In both these ads an obligation is placed on the individual to repair the damage done by school or by their own earlier irresponsibility, otherwise they are failing: failing their child, failing themselves, failing their nation. Wider fears of mathematics are deliberately invoked to motivate action. The gender and class positions of the characters in these ads are almost opposites of those of the geniuses and geeks we met earlier. In this way some people are presented as always and already failures in relation to mathematics, in need of remediation, but also as having a personal responsibility to take up the offered support. Others are presented as self-taught, ahead of their age: our hope for the future.

## 6 Conclusion

In this article I have surveyed some of the many ways in which mathematics and people doing mathematics are represented in contemporary popular culture. I have tried to draw out the contradictions, notably those between, on the one hand, mathematics as accessible, open, useful, and exciting and, on the other hand, mathematics as hard, closed, useless, and boring. In 2004 Sarah Greenwald and Andrew Nestler offered some helpful examples of how to use the popular when teaching mathematics within universities, showing how this can provide more points of engagement and identification for students as they interact mathematically with characters and storylines. In the intervening decade there has been an increase in the possibilities for such an approach, with the proliferation of mathematics within the popular, including much material on YouTube (from Web series "Maths Warriors," through quirky educational channel "Numberphile," to another channel (from missionastar) that attempts to teach quadratics via a Bollywood spoof). However, popular culture can include some and exclude others. For, as I have shown, while society confers on all a responsibility to become mathematically literate, it suggests that only a special few possess mathematical "ability." It overwhelmingly depicts this ability as belonging in white, male, middle-class, heterosexual bodies. Thus, its use requires care and thought. I return to *Numb3rs* briefly to suggest what such care and thought entail.

Texas Instruments have produced materials for high school teachers to use to explore the mathematics in *Numb3rs*, and the Wolfram Web site has produced something similar for teachers of undergraduate mathematics. But ethically, can we talk about the mathematics and not about the politics of how mathematicians are represented in the series? Can we use these materials without endorsing the excessive and glamorized violence on the program? Or without taking into account the difficult topics it tackles (sexual violence, war, terrorism, stalking), which students or members of their families may have experienced? In one episode, Charlie persuades the FBI to buy up all supplies of a new drug and so manipulate the market in such a way as to reduce demand to zero. Charlie intends this as a way to cut off crime at its source rather than simply dealing with the symptoms of crime. At the end of the episode he briefly reflects that even this success is meaningless since another drug will always be waiting in the wings to fill the gap. However, this discussion quickly leads to a play fight between Charlie and Don, and there's no space to think more broadly about the underlying causes of drug crime (racism, poverty, inequality) or to discuss alternative approaches to tackling drug addiction, such as decriminalization. What is lost, and gained, by extracting and abstracting the mathematics used in the show without addressing the wider questions raised by its use? Whether and in what ways these questions are part of mathematics goes to the heart of why we teach mathematics and what we want people to learn as a result.

**Acknowledgments.** This article draws on work carried out in collaboration with Debbie Epstein, Marie-Pierre Moreau, and Sumi Hollingworth. It owes a great deal to their thinking. I am grateful to them for their support and to the Economic and Social Research Council (RES-000-23-1454) and the U.K. Resource Centre for Women in Science, Engineering and Technology for funding this initial research.

### Further Reading

Davis, P. J., and R. Hersh. 1982. *The Mathematical Experience*. Brighton, UK: Harvester.

Early, R. E. 1992. The alchemy of mathematical experience: a psychoanalysis of student writings. *For the Learning of Mathematics* 12(1):15–20.

Greenwald, S. J., and A. Nestler. 2004. Using popular culture in the mathematics and mathematics education classroom. *PRIMUS—Problems, Resources, and Issues in Mathematics Undergraduate Studies* 14(1):1–4.

Hardy, G. H. 1969. *A Mathematician's Apology*. Cambridge: Cambridge University Press.

Mendick, H. 2006. *Masculinities in Mathematics*. Maidenhead, UK: Open University Press.

Mendick, H., and M. Moreau. 2013. From *Good Will Hunting* to *Deal or No Deal*: using popular culture in the mathematics classroom. In *Debates in Mathematics Education*, edited by D. Leslie and H. Mendick. London: Routledge.

Moreau, M.-P., H. Mendick, and D. Epstein. 2009. Constructions of mathematical masculinities in popular culture. In *Pimps, Wimps, Studs, Thugs and Gentlemen: Essays on Media Images of Masculinity*, edited by E. Watson, pp. 141–56. Jefferson, NC: McFarland.

Skovsmose, O. 1994. *Towards a Philosophy of Critical Mathematics Education*. Dordrecht: Kluwer Academic.

Walkerdine, V. 1990. *Schoolgirl Fictions*. London: Verso.

## VIII.9 Mathematics and Policy

The continued success of mathematics as a subject of study and research depends on its importance being appreciated by those who control the funding that supports it, as well as those who are in a position to make use of mathematics-based advice in formulating policy. Mathematicians arguably find it more difficult to make the case for their subject than colleagues in other "more practical" disciplines, such as biology, chemistry, and engineering, yet the case for mathematics is continually being made, and with much success. The four contributors to this group of articles, from four different countries, were asked to give their perspective on how to influence government as a mathematician. Their anecdotes and advice should be of interest to all, and in particular to those who wish to be involved in promoting mathematics to politicians.

### I. Ya-xiang Yuan: How Chinese Mathematicians Influence Government

The longest-running television program in the United Kingdom, *The Sky at Night* (first broadcast in 1957), is on astronomy. However, the public's appetite for astronomy was not so great prior to the second half of the twentieth century. In the preface to the third volume of his classic *Science and Civilisation in China* published in 1959, Josef Needham (1912–96) included the following quote from Franz Kohnert (Vienna, 1888):

> Probably another reason why many Europeans consider the Chinese such barbarians is on account of the support they give to their Astronomers—people regarded by our cultivated Western mortals as completely useless. Yet there they rank with Heads of Departments and Secretaries of State. What frightful barbarism!

In ancient times almost all astronomers were also mathematicians. China therefore has a long tradition of mathematicians being well respected, and famous mathematicians are often put into high-ranking positions in government. For example, Hua Loo Geng (1910–85) was a vice chairman of the Chinese People's Political Consultative Conference (China's top advisory body), and Ding Shi Sun (1927–), former president of Peking University, was a vice president of the National People's Congress of China (the Chinese parliament). An interesting phenomenon is that many university presidents in China are mathematicians, but this is unusual in the West. Another example of the popularity of mathematicians in China is that two of their number, Hua Loo Geng and Chen Jing Run (1933–96), were chosen by China Central Television (the official Chinese government television station) when it ran a competition to select the 100 most deserving winners of the "Touching China" award between 1949 and 2009.

During the great Cultural Revolution in the 1960s, Hua Loo Geng was able to convince Chairman Mao Zedong (1893–1976) that mathematics could help the country to modernize. Hua traveled all over China teaching the golden section search method—a method for maximizing a unimodal function in an interval by successively reducing the length of the interval by the golden ratio ($\frac{1}{2}(\sqrt{5}-1)$)—and other operational research techniques in factories, coal mines, and oil fields. The Chinese Academy of Sciences sent many research teams to the countryside to solve practical problems in the areas of transportation and production planning. A few years earlier, during the Great Leap Forward (1958–60), Chinese scientists were encouraged to solve real-world problems to help Chairman Mao's ambitious campaign to rapidly transform the country from an agrarian economy into a modern communist society. In 1960, twenty-six-year-old Mei-Ko Kwan, a young lecturer at Shangdong Normal University, published the famous "Chinese postman problem" in graph theory.

In 2002 the Chinese Mathematical Society hosted the International Congress of Mathematicians (ICM) in Beijing. As the general secretary of that congress, along with other mathematicians in China I had the chance to lobby government officials to support the meeting. We emphasized that sciences are vitally important for China's economic development and that mathematics is the foundation of all other sciences. The president of the People's Republic of China at that time, Jiang Zemin, attended the opening ceremony and handed the medals to the Fields Medal winners Laurent Lafforgue and Vladimir Voevodsky. The success of ICM 2002 promoted mathematics research, increased public awareness of mathematics, and attracted more young talented people to study mathematics in China.

Chinese mathematicians are very influential in science and technology policy making in the country. For example, in the 1980s the late Shiing S. Chern (1911–2004) was able to persuade the Chinese government to increase its support of mathematics, and this resulted in the establishment of the Tian Yuan

Foundation within the National Natural Science Foundation of China (NSFC). Following recommendations from many famous Chinese mathematicians, such as Shiing S. Chern, Shing Tung Yau, Lo Yang, and Gang Tian, the Chinese government has spent an enormous amount of money in the past twenty years setting up a number of mathematical centers in China. These include the Shiing S. Chern Institute of Mathematics at Nankai University, the Morningside Center of Mathematics at the Chinese Academy of Sciences, the Beijing International Center for Mathematical Research at Peking University, and the Shanghai Mathematics Center at Fudan University.

The main source of funding for mathematicians in China is the NSFC, which is very similar to the National Science Foundation in the United States. The NSFC splits its programs into three categories: research promotion, talent fostering, and infrastructure construction for basic research. The category of research promotion is further subdivided into programs called the General Program, the Key Program, the Major Program, the Major Research Plan, and the Joint Funds to International Joint Research Program. General Program grants are open to any application, while Key Program, Major Program, and Major Research Plan grants are for applications relating to specific topics. Mathematicians have to lobby the NSFC if they want a particular topic to be listed as part of the Key Program or the Major Program or if they want to set up a Major Research Plan.

Since my return to China from the University of Cambridge in 1988 I have been involved in events that have given me the honor of witnessing firsthand the ways in which mathematicians are able to influence the Chinese government. In the late 1980s and the early 1990s, the late Feng Kang (1920–93) was able to convince the government to develop scientific computing. Due to his initiative and the proposals he put forward, the Ministry of Science and Technology of China founded the State Key Laboratory of Scientific and Engineering Computing in 1991 and, in the same year, launched the National Key Research Project "Large Scale Scientific and Engineering Computing." The latter project eventually developed into the huge National Basic Research Project of China (also known as the 973 Program) and was approved by Chinese leader Deng Xiaoping in March 1997.

Mathematicians also played an important role in stimulating China's success in developing high-performance computing (HPC) hardware. In December 2002 China's supercomputer DeepComp 1800 ranked forty-third in the worldwide TOP500 list (www.top500.org). This breakthrough marked the beginning of China's ascent in the field of HPC hardware development, and in 2012 China built what was for a time the world's most powerful supercomputer: Tianhe-1A. The rapid development of HPC hardware in China in turn stimulated HPC research in the country, and in 2012 the NSFC launched a four-year Major Research Plan called "Algorithms for high performance scientific computing and computable modeling" with a budget of 180 million Chinese yuans (about US$30 million).

In recent years the economic systems of developing countries such as China have been rapidly and drastically reformed in response to economic globalization, and new and challenging problems have therefore been raised in energy, transportation, telecommunication, financial engineering, urban planning, health care, environmental pollution, natural resource consumption, and transnational logistics. I am sure that mathematics will play an increasingly important role in modeling and solving these practical problems.

## II. Maria Esteban: A Personal Experience in France and Europe of How to Influence Government as a Mathematician

I am a Basque–French mathematician currently working in France, the president-elect of the International Council for Industrial and Applied Mathematics, past president of the Société de Mathématiques Appliquées et Industrielles, and past chair of the Applied Mathematics Committee of the European Mathematical Society. In recent years I have been involved with a European Science Foundation Forward Look project on "Mathematics and Industry." These three activities have helped to shape my views about how mathematicians can (or cannot) influence government science policy.

In this article I will not only relate my personal experience but will also describe how European colleagues have fared in their interactions with officials from their own governments and from European institutions. To avoid overgeneralization, it is important to bear in mind that the opportunities that a scientist has to influence government policies and initiatives depend very much on which country they operate in: its size; its history; and how knowledgeable, on average, its politicians are about science.

In France, scientists have a long history of close relationships with politicians. It is usually straightforward for us to gain access to members of the government or to members of parliament and regional politicians. This is not the case for every mathematician, of course, but it is reasonably easy for those in leadership positions who invest the necessary time and energy in building contacts. Prominent scientists (Fields Medalists and recipients of Nobel Prizes, for example) may be contacted directly by politicians. Ministers of research or higher education and other top government officials may ask these scientists for meetings to exchange ideas or to receive advice on important decisions concerning science. Additionally, there are mathematicians who have been appointed to important positions: as managers of institutes or big projects, and as chairs or members of committees created to advise the government on particular issues of scientific or technological importance, such as high-performance computing, the treatment of nuclear waste, food safety, and so on. Mathematicians who have been invited to participate in the crafting of advisory documents have influence on the government through these documents but also through personal meetings and discussions with government officials.

The presidents of the mathematical societies, together with mathematicians who are involved with the management of institutions relevant to the mathematics community at a high or national level, can gain access to policy makers if they need to pass on information or intervene in an important issue. Contact does not usually start at the highest level; one often starts by talking to ministerial advisors. On the other hand, if the matter is important, it is relatively easy to arrange a meeting with someone at a higher level. It needs to be said, though, that talking to politicians does not necessarily translate into success in convincing them of what (in our opinions) is good for mathematics or for science; but it is a start.

All this may make it sound as if France is a paradise for mathematicians, especially when it comes to their relationships with politicians and decision makers. While it is true that the situation in France is much better than in many other European countries, it is not always easy to convey our achievements or what our community needs or is looking for. By talking to policy makers, however, we build relationships that make further contact and exchanges of views easier.

A recent example of how mathematicians in France have managed to change something of importance to them is their contribution to the call, initiated three years ago, for the creation of centers of excellence. When the project was about to be launched, a senior official in the Ministry of Higher Education and Research learned about two projects dear to mathematicians that would not be covered by the original wording of the initiative. One of the projects pertained to the French Institutes of Mathematics: the Institut Henri Poincaré, which hosts thematic three-month programs; the Institut des Hautes Études Scientifiques, a high-level research institution south of Paris; the Centre International de Rencontres Mathématiques, an international center near Marseille that hosts mathematics conferences; and the Centre International de Mathématiques Pures et Appliquées, which promotes mathematical research in developing countries. The other project was concerned with the creation of an agency designed to facilitate relationships between mathematicians and industry. When the official learned about these two projects he thought them interesting enough to ask for a slight change in the wording of the policy document. The result of this was that these two projects were covered in the initial deliberations and were indeed both selected and funded. This would not have happened without the intervention of the mathematicians. Of course, for this to have happened, information had to be available and (some) mathematicians needed to keep up to date with news about official projects. This is a big task and requires a lot of time and effort. Fortunately, in France most mathematicians in positions of power are willing to play the collaborative game, sharing information and jointly developing strategies. Collaboration is very important for achieving mathematicians' desired outcomes.

I do not have firsthand knowledge about other European countries, only impressions gleaned from conversations with colleagues and observations made during joint projects when colleagues from various countries tried to reach their politicians for input or advice. The situation that mathematicians face clearly varies by country; as with many other issues in Europe, there is unfortunately no unified approach to scientists' interactions with their governments. There are countries like France where mathematicians have managed to influence high-level politicians and decision makers, and where scientists are often consulted when new decisions concerning science and scientific programs need to be made. On the other hand, there are other countries where mathematicians do not seem to have any access to politicians and decision makers, at least at an institutional level.

Whether mathematicians can influence the European decision makers is a very different story. I have had several negative experiences in this regard, and other mathematicians I know have echoed this sentiment. Mathematics is almost entirely absent from European scientific policy decision making. In the last European framework program, mathematics was invisible. When asked about this, European officials would say, "But what are you saying? Mathematics is everywhere; mathematics is linked to so many scientific fields that are explicitly part of the European programs!" But because mathematics is "everywhere," when it comes to funding it is nowhere. Mathematics does not seem to be a priority for the European Union, which chooses not to fund general scientific fields, funding only applied, concrete fields, like life sciences or nanotechnology. The only exception to this is the European Research Council, which funds excellence through various programs such as "Starting Grants" and "Advanced Grants." Mathematics is clearly present there, as are other scientific fields.

The European Mathematical Society has tried to open a dialogue with Brussels for many years, and it has tried more intensely still in the last six years or so. Several mathematicians have had meetings with high-level officials in Brussels and in the European Parliament and have tried to explain our community's situation and its needs. Sometimes it has seemed like the officials were really listening to them, but the final outcome has never been positive. For instance, it is clear that in order to build important infrastructure at the European level (concerning digital mathematics, publishing, mathematics and industry, etc.), European funding is very important, but up to this point very little (almost nothing) has been forthcoming, despite the best efforts of the mathematicians and many promises from Brussels.

Up to now I have discussed only direct interactions between mathematicians and politicians, but there could be, indeed must be, other ways for mathematicians to influence government. For instance, we should not forget the important role that the media can play in educating the public about the contributions of mathematics to science and to society, about important mathematical results and their implications and possible applications. A case in point is the broad attention French media have devoted in recent years to French mathematicians when they receive important international prizes. Politicians pay close attention to the media and the ideas it promotes. The presence of mathematics in the media can therefore help mathematicians reach beyond their own community to the spheres of power.

Mathematicians may also influence public opinion and policy makers by taking action when news coverage misrepresents mathematics and its applications. For example, statistics is often used to justify decisions concerning health and drug design, but the statistical methods that are used are often incorrect and unreasonable. I know of several recent cases where statisticians have strongly attacked such misuse of statistics. This is an excellent way of defending the quality label that mathematics can provide and to refuse to let the subject be misused. In drug and food safety, the intervention of statisticians has been widely acknowledged, and statisticians have consequently been appointed to supervisory committees that give advice to decision makers.

Of course, you need to be good at what you do to have influence. Only a well-organized mathematics community with high scientific standards will have a societal impact. The commitment of this community to scientific excellence also gives mathematicians the authority to exert their influence at the educational level. The mathematics community has to participate in education programs, and it needs to shape the training of mathematics teachers as well as the way mathematics is being taught in schools. Education and research are the pillars on which the mathematics community is built, and mathematicians should therefore be involved with all committees and at all levels where decisions on these subjects are made.

In closing, let me stress that all these activities can be time-consuming for the individuals involved. Building networks, being well informed, and keeping information channels open take time and energy. But if we mathematicians want to be able to influence governments in terms of both the projects they promote that could affect the organization and funding of our community and their perception of the ways in which mathematics can advance society, there is no way around the fact that some of us will have to be ready to spend the necessary time trying to promote our goals.

## III. James M. Crowley: SIAM and Science Policy in the United States

The Society for Industrial and Applied Mathematics (SIAM), founded in 1952, is an international organization based in Philadelphia, Pennsylvania. Since the mid-1990s, SIAM has actively engaged in science policy and advocacy on behalf of the mathematical sciences and computing from the perspective of applied mathematics and computational science.

## Science Policy and Funding in the United States

Perhaps more so than in Europe and other parts of the world, in the United States science policy and funding is distributed across many agencies and involves many players. This leads to a system that is robust but that is sometimes difficult to understand in all its facets.

Many different agencies affect science policy and funding in the United States in the areas of the mathematical and computational sciences. For traditional areas of mathematics, the National Science Foundation (NSF) certainly plays the lead role, but in applied and computational mathematics, the Department of Energy's Office of Science, the various Department of Defense funding agencies (the Air Force Office of Scientific Research, the Office of Naval Research, the Army Research Office, and the Defense Advanced Research Projects Agency), and parts of the National Institutes of Health are also important players. More recently, privately funded foundations, like the Simons Foundation, have also played an increasing role.

"Mission agencies" like the Department of Energy and the Department of Defense generally place a greater emphasis on research that contributes to solving specific application problems of interest to that agency; therefore, they make grants to the areas within the mathematical sciences that they deem to be most relevant. NSF grants, on the other hand, tend to cover the whole spectrum of the mathematical sciences, from research deep within a core area of the discipline to multidisciplinary research taking in other areas of science or engineering. However, in both cases there is a spectrum of grants in each agency's portfolio: from those that include very basic research within a given field to those that are research motivated and driven by application goals. In the case of the NSF, for example, multidisciplinary research can team applied mathematicians with scientists in other disciplines to simultaneously advance mathematical/computational methods as well as applications.

The funding agencies are staffed by program managers: a mix of permanent employees and "rotators"—people from the research community who come to the agency for a two- or three-year period. Normally, both groups (permanent staff and rotators) are experts in the fields they manage (typically they have a related doctorate). The permanent staff provide stability and memory; rotators provide new ideas and immediate contact with the research community from which they came. In mathematical sciences at the NSF, for example, the mix of permanent staff to rotators is roughly 50:50.

Program managers in mission agencies may have more discretion over funding decisions because they are able to factor in the relevance of the proposal to the agency's mission and also to consider the likelihood of application. The NSF relies most heavily on external peer review from the scientific community, but most funding agencies rely on peer advice to some degree.

To get a sense of the scale of funding by the various agencies, consider the fiscal year 2011 (the fiscal year for the federal government in the United States runs from October 1 to September 30). The budget for the NSF Division of Mathematical Sciences that year was $240 million, of which approximately a third supports applied and computational mathematics. The Department of Energy contributed another $99 million to the mathematical sciences, much of this in applied and computational mathematics. The various agencies of the Department of Defense also supported $118 million in basic research in the mathematical sciences. Estimates were not available for the National Institutes of Health.

What distinguishes funding in the United States from that in Canada and in many parts of Europe is the importance of individual grants to single researchers or to small groups, as opposed to block grants to departments or universities. Such grants can provide funding for a researcher's summer salary (typically one month, but possibly two) and/or for graduate student support. Such grants also provide funds for travel. Research grants to individuals and small groups account for the majority of funding in the mathematical sciences. Individual/small-group grants are also a major part of the applied mathematics program at the Department of Energy's Office of Science, although recent policy seems to indicate a shift toward more large-scale grants. Block grants to academic departments or to institutes account for only a small portion of the NSF Division of Mathematical Sciences portfolio, and they play little or no role in the portfolios of other agencies.

Within the mathematics portfolio at the NSF, institutes play a significant role. While accounting for only 10% or so of the mathematical sciences budget at the NSF, the eight NSF Mathematical Sciences Institutes involve a large number of postdocs and visitors from within the community in their programs each year (6000 in fiscal year 2011).

### The Budget Process

In the United States, Congress has the power to appropriate funds while the administration implements the spending of the funds through individual agencies. The federal budget is determined on an annual cycle through presidential and congressional action.

Each year, the president issues a budget proposal to fund all the agencies of the federal government for the following year. The administration arrives at this budget request after a lengthy process during which each agency is charged with developing its budget with guidance from the Office of Management and Budget, which reviews each agency's request and assembles all of the agency plans into the president's budget request.

Congress then reviews the president's budget (or the "request") and develops legislation to fund the agencies called appropriations bills. These bills sometimes provide money at the agency or subagency level and at other times determine the specific amount available to individual programs or spell out other directives about the budget. Once appropriations bills are passed, agencies can then spend the funding to support research. In many years Congress fails to pass appropriations bills and instead passes a "continuing resolution" that enables an agency to continue to spend a designated amount of funds but generally prevents the beginning of new programs or initiatives.

### Advocacy

Due to the very distributed system of budget development and appropriations, there are many points of access for scientific societies to interact with the process. Societies, including SIAM, play an important role in providing information, advice, and feedback from the communities they represent to those with responsibility for the budget and policy in the administration, in Congress, and across agencies.

SIAM has a Committee on Science Policy (CSP) that regularly meets with key leaders to both gather information for our community and to provide feedback from the community to officials. Key points of contact include leaders within the funding agencies, congressional staffers and members of Congress, and representatives of the White House Office of Science Technology Policy and Office of Management and Budget. SIAM presidents and the executive director also regularly provide testimony and meet with staff from relevant congressional committees, such as the House Appropriations Subcommittee on Commerce, Justice, and Science.

CSP discussions with government officials often focus on the budget in applied and computational mathematics but may from time to time also take in specific policy issues. The House Science, Space, and Technology Committee, for example, has in the past asked the CSP to discuss issues related to high-performance computing and the role of computational science in developing new models and computational methods for solving grand challenge problems on emerging computer architectures.

Another example of policy discussions was SIAM participation in 2012 debates on undergraduate education, which led to the President's Council of Advisors on Science and Technology issuing the report "Engage to Excel: Producing One Million Additional College Graduates with Degrees in Science, Technology, Engineering, and Mathematics," which called for a national initiative to promote science, technology, engineering, and mathematics (STEM) education in the first two years of college. A key finding of the report is that mathematics education is a critical component of all undergraduate STEM degrees and that it is the current deficiencies in mathematics learning that are partly to blame for the loss of STEM majors in the early college years. However, many SIAM members expressed reservations about a suggestion in the report that nonmathematicians should be engaged in the teaching of undergraduate mathematics for nonmajors. In response to the report, and to defend the role of mathematicians in mathematics education, the SIAM Education Committee in partnership with the CSP prepared a formal white paper. The white paper highlights the importance of collaboration for effective mathematics education and makes recommendations on ways to strengthen K-16 (kindergarten to four-year degree) mathematics education. The white paper was sent to the Office of Science and Technology Policy and the NSF.

### SIAM's History in Advocacy

In the late 1970s, the major mathematics societies in the United States (the American Mathematical Society (AMS), the Mathematical Association of America (MAA), and SIAM) jointly supported a congressional fellow, an individual from the community who served in a congressional office (or related position) for a year in order to obtain experience in science policy and funding issues.

However, it was not until the publication of the David Report in May 1984 that mathematics in general, and SIAM in particular, became seriously involved in the arena of science policy. The report, titled "Renewing U.S. Mathematics: Critical Resource for the Future," proved pivotal in energizing funding support for the mathematical sciences and for propelling mathematics societies into engaging in science policy discussions. The report was produced by the Ad Hoc Committee on Resources for the Mathematical Sciences, chaired by Edward David of Exxon Research and Engineering Company and with Kenneth Hoffman as executive director of the committee; it was the result of a project commissioned by the National Research Council and supported by SIAM and the AMS, along with six companies and five government funding agencies.

Following completion of the David Report in the summer of 1983, the Ad Hoc Committee on Resources for the Mathematical Sciences was terminated, but Kenneth Hoffman agreed to continue activities in Washington and submitted a budget to the three major societies (the AMS, the MAA, and SIAM) to support those activities. The SIAM board approved their share of this budget at their June 1983 meeting.

Hoffman became the Executive Secretary for National Affairs under the aegis of the Joint Policy Board for Mathematics (JPBM) in 1984. In March 1985 the JPBM contracted Kathleen Holmay and Associates to do public relations for the public understanding of science (specifically mathematical sciences). This led to a decade or more of discussion among the members of the SIAM board about the role of SIAM in science policy.

Seeking more oversight of the activities of the JPBM that it helped fund, the SIAM board asked the SIAM Committee on Relations with the Federal Government to play this role and changed the name of the committee to the SIAM Committee on Science Policy in October 1985.

In the late 1980s the name of the office within the JPBM was changed to the Office of Governmental and Public Affairs, and by 1989 this had grown to become an office employing a full-time director, two administrative assistants, a legislative liaison, and a public relations consultant. SIAM's degree of support wavered at times during the late 1980s and early 1990s over various issues, but the JPBM remained SIAM's sole presence in Washington.

SIAM used the JPBM as its forum for discussions with funding agencies and legislative staffers on issues related to science policy and funding in areas of interest to SIAM.

When the AMS created its own Washington office, support for the JPBM waned, and in 1999 it was restructured to become a coordinating body among member societies (eliminating its paid staff), a role that it continues to play today.

With the elimination of the JPBM's legislative liaison (which had been a full-time staff position) and its public affairs consultant, SIAM sought to grow its own voice in science policy through its CSP. The CSP meets twice a year in Washington and coordinates the writing of white papers and other policy activities throughout the year. To enable the CSP to carry out its role, in 2001 SIAM contracted Lewis-Burke Associates LLC, a government relations firm that specializes in science and technology policy, to provide support for its policy activities.

SIAM, through the CSP, now regularly engages in discussions with various federal agencies on issues related to applied mathematics and computational science.

## IV. Alistair D. Fitt: Making the Case for U.K. Mathematics Research in a Rapidly Changing Environment

### The Need to Make the Case (and Not Just for Funding)

First, we have to be clear, in the U.K. system there is no alternative to "making the case for mathematics." We all need to show in as many ways as possible that mathematics research is a key contributor to the nation's overall success. As always, perhaps the most important issue is research funding. It is a long time since public money was handed out to universities and they were simply trusted to do a good job with it (a system that still remains in surprisingly many nations), and there seems no chance of those days (or the minute academic salaries that went with them) ever returning. Public funding for science (and universities in general) increased significantly during the long tenure of the 1997–2010 Labour government, but it has been more uncertain since. This period of uncertainty has necessitated much more involvement from the community to ensure that funding streams are protected. In addition, 2011 saw the biggest change for decades in the way in which the U.K. higher-education sector is funded, and by many financial definitions, universities in the United Kingdom are now close to being private, rather than public, ventures.

Virtually all that is left of the public funding that used to allow us to charge relatively low student fees is the research money, currently about £5 billion per annum. Though by all measures our "dual research funding system" (where about two-thirds of the money is given out by the funding councils and the other third is allotted as a result of a research assessment exercise of some sort) has served us well since its introduction in 1986, funding for mathematics research has certainly not increased. In fact, there is a case (depending on who you believe) for saying that the share of the research funding that finds its way to mathematics has shrunk considerably. This makes it all the more crucial to be sure that our subject is seen as being worth supporting.

Furthermore, a new overriding political mission has evolved in the United Kingdom: transparency. The public should be able to hold the spending of public money to account and readily observe that their taxes have been well spent. This is a potential problem for our discipline, for mathematics is one of the very few subjects where many members of the public take proud delight in proclaiming the depths of their ignorance. Even those whose school experience was more positive are unlikely to have much real understanding of what mathematics research consists of. This means that it is absolutely crucial for us to have a portfolio of stories, arguments, citations, and reasons why we contribute so much to what is so horribly called U.K. PLC. Though it will prove hard to make the case to the public, we can influence politicians and governments, and these are the people who make the important decisions.

## A Single Point of Contact

Now that we have agreed that "making the case" is unavoidable if mathematics is to thrive, we have to consider the best way of influencing government (by which I mean politicians, civil servants, funding agencies, and anybody else who might be in a position of power). Undoubtedly we need individuals who are skilled influencers, but this alone is not enough; we also need the people who matter in government to want to hear us and to actively seek our views. One of the hardest problems for ministers to deal with is knowing who to turn to when they want information or wish to solicit views. Experience has shown that politicians and civil servants like to have a small number of trusted contacts who they regard as their key "go-to" people. Professional and learned societies are important, but having too many of them to canvass is distinctly unhelpful. We currently have four learned societies in the mathematical sciences in England (the Institute of Mathematics and its Applications, the London Mathematical Society, the Royal Statistical Society, and the Operational Research Society), and there is also the Edinburgh Mathematical Society and possibly others too. The Council for Mathematical Sciences does a valiant job in trying to synthesize consensus from its member societies to present a single face to influencers in the outside world. Though the Council for Mathematical Sciences has undoubtedly had successes, a structure that involves so many different societies is hard for outsiders to understand. The result of the fragmentation of mathematical learned societies (and their unwillingness to merge and become a single voice) is that mathematics as a discipline is unable to exert the traction that, for example, the Royal Society of Chemistry or the Institute of Physics can, as single and well-known voices in their disciplines.

Others will no doubt disagree that we need a single society, but there is ample evidence to suggest that, where the ability to influence government is key, a single point of contact is extremely important.

## What Works Best

The rules for how to influence government have not changed: they are timeless and constant. Nevertheless, they do not tend to sit easily with the nature of mathematics as a discipline. Bluntly put, the most important detailed points are as follows.

**Limit your number of messages.** Ministers are deluged with information. You can probably afford to get only one or two points across, so do not overcomplicate and risk confusing the minister.

**Be clear and concise in your points.** Mathematicians tend to worry about detail, and detail rarely matters when it comes to influence. Do not obfuscate, and do not dilute the main points that you are making.

**Use a few "killer" statistics and stories again and again.** Ministers like simple one-line facts to remember, and they will have to be told them again and again. Prepare your ammunition well in advance, and decide carefully what the minister needs to know and will want to use again.

**Offer help.** There is no future in complaining or being grudging. Government officials spend much of their lives under attack, so offer them assistance rather than criticism.

**Concentrate on what is good first; then point out the threats.** Even if you are vehemently against every one of the government's policies, you need to begin with some positives. You can mention the possible difficulties and threats later.

**Understand the political difficulties.** The government may want to help you but be unable to for political reasons. You need to understand these before you can have a good idea of what might reasonably be achieved.

**Speak for mathematics, not your own university or department.** The government wants a broad view. If you are seen to be acting from any kind of self-interest, you will rapidly be discarded.

**Use the media to make your case.** Politicians see the news media as a vital barometer of public opinion and are much more likely to be influenced by something that attracts headlines.

**Get included on missions abroad.** Traveling with government officials on delegations abroad can provide a considerable opportunity to talk in detail to the people that matter.

**Hire people.** Ministers and "their people" are very used to dealing with publicists and professional lobbyists. There is no reason why we cannot spend hard cash hiring professional influencers who know how to deal with the government and how to get results.

## Fact: We Do Not Do It Very Well

This short article gives plenty of suggestions and solutions for how to influence government, so how successful have we been in this area in the United Kingdom? Unfortunately, the answer in my view is simple: not very. If we are brutally honest, we have to admit that, as a profession, too often we do not take a strategic view, we fight among ourselves, and we give the overall appearance of a bunch of whingers who do not understand the wider agenda. Is this lack of success confined to the United Kingdom? Other articles in this volume will allow the reader to judge whether mathematicians in the rest of the world are better influencers than we are, but notable successes seem to me to be well hidden.

Of course, influencing the government is very hard work. Politics is a transitory business, and it is frustrating to build good connections only to see them vanish as a new election sweeps new faces into power. However, though politicians change, the civil servants that are so important to them normally remain. We *can* influence these people, and we should be thinking about this aspect of our subject and doing a much better job.

# Index

A page number followed by "f" indicates a figure, a page number followed by "t" refers to a table, and a footnote reference is indicated by "n."